General, Organic, and Biological Chemistry

STRUCTURES OF LIFE

General, Organic, and Biological Chemistry

STRUCTURES OF LIFE

Sixth Edition

Karen Timberlake

Contributions by

MaryKay Orgill, Ph.D.
University of Nevada, Las Vegas

Pearson

330 Hudson Street, NY NY 10013

Courseware Portfolio Manager: Scott Dustan
Director, Courseware Portfolio Management: Jeanne Zalesky
Content Producer: Melanie Field
Managing Producer: Kristen Flathman
Courseware Analyst: Coleen Morrison
Courseware Director, Content Development: Jennifer Hart
Courseware Editorial Assistant: Fran Falk and Leslie Lee
Rich Media Content Producer: Paula Iborra
Full-Service Vendor: SPi Global
Full-Service Project Manager: Christian Arsenault
Copyeditor: Karen Slaght
Design Manager: Maria Guglielmo Walsh
Cover and Interior Designer: Tamara Newnam
Courseware Portfolio Analyst, Content Development, Art: Jay McElroy
Photo and Illustration Project Manager: Stephanie Marquez, Imagineering Art
Rights and Permissions Project Manager: Matt Perry
Rights and Permissions Management: Ben Ferrini
Photo Researcher: Clare Maxwell
Manufacturing Buyer: Stacey Weinberger
Product Marketing Manager: Elizabeth Ellsworth Bell
Cover Image Credit: Zoonar GmbH/Alamy Stock Photo

Library of Congress Cataloging-in-Publication Data

Names: Timberlake, Karen C., author. | Orgill, MaryKay, 1974-
Title: General, organic, and biological chemistry : structures of life / Karen Timberlake ; contributions by MaryKay Orgill, Ph.D., professor of chemistry, University of Nevada, Las Vegas.
Description: Sixth edition. | San Francisco, CA : Pearson Education, Inc., c2019.
Identifiers: LCCN 2017041702| ISBN 9780134730684 (hardcover) | ISBN 0134730682
Subjects: LCSH: Chemistry--Textbooks.
Classification: LCC QD33.2 .T56 2019 | DDC 540--dc23
LC record available at https://lccn.loc.gov/2017041702

1 18

ISBN-10: 0-134-73068-2
ISBN-13: 978-0-134-73068-4

Pearson

www.pearsonhighered.com

Brief Contents

Contents

1 Chemistry in Our Lives 1

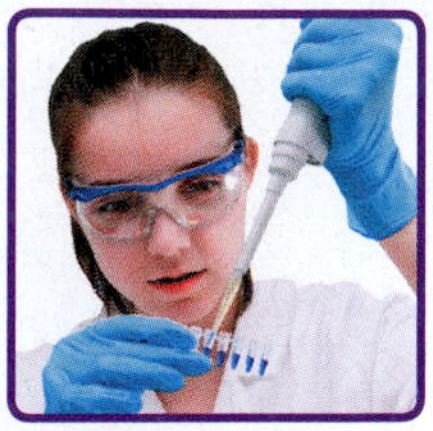

2 Chemistry and Measurements 26

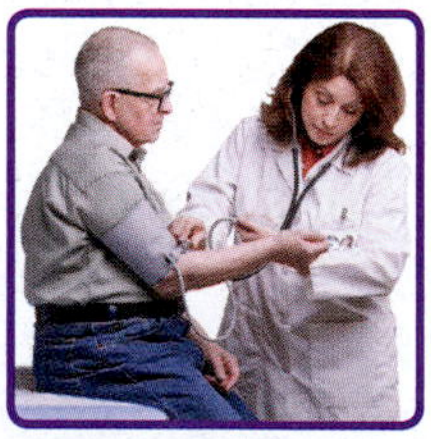

3 Matter and Energy 63

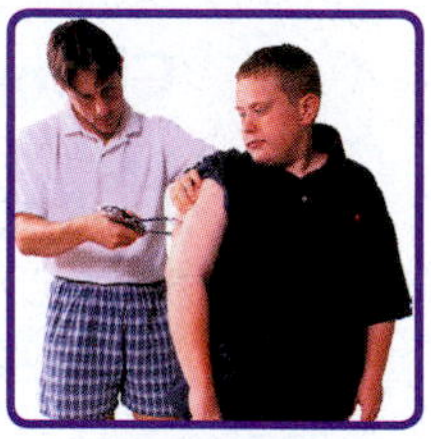

4 Atoms and Elements 99

5 Nuclear Chemistry 145

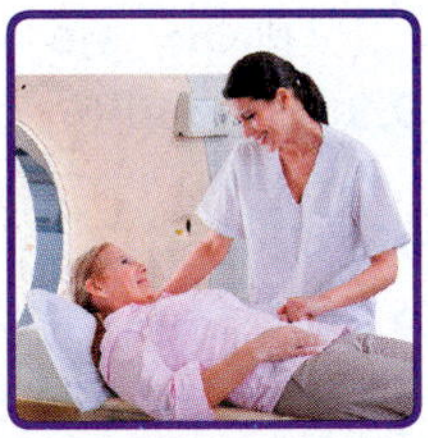

6 Ionic and Molecular Compounds 174

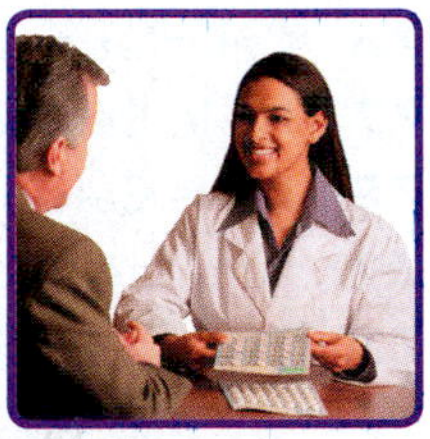

7 Chemical Reactions and Quantities 223

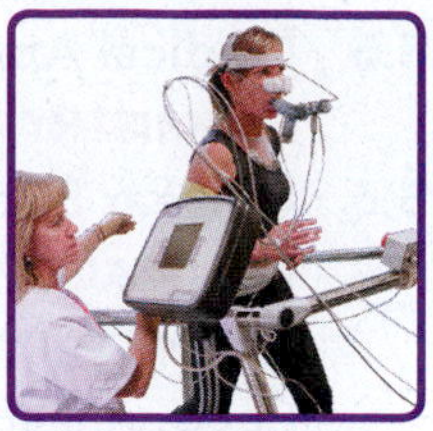

8 Gases 275

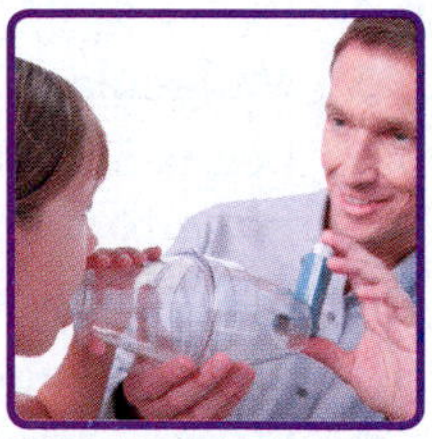

9 Solutions 310

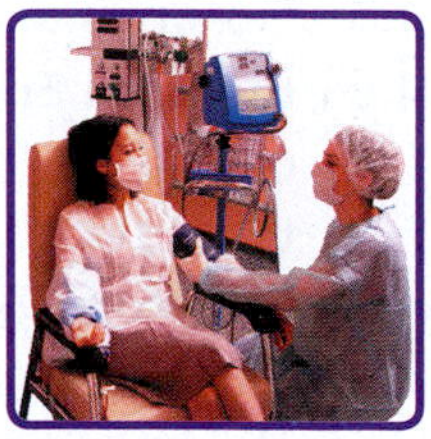

10 Reaction Rates and Chemical Equilibrium 355

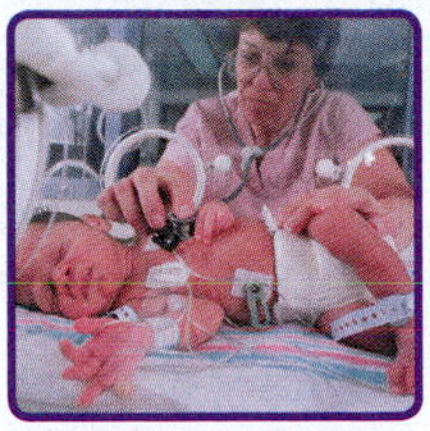

11 Acids and Bases 382

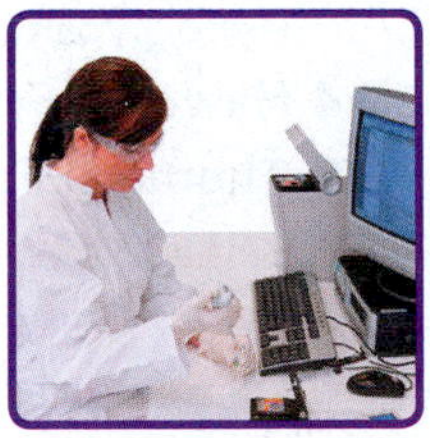

12 Introduction to Organic Chemistry: Hydrocarbons 426

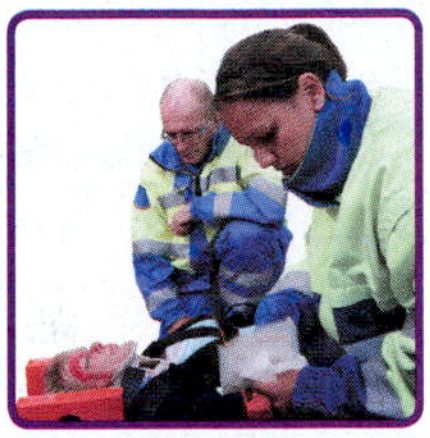

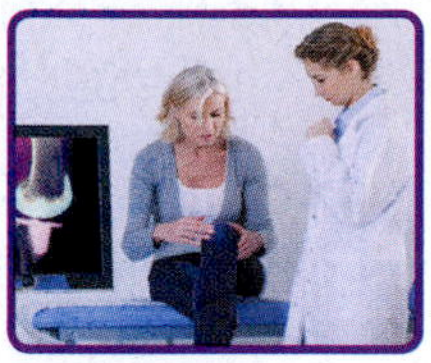

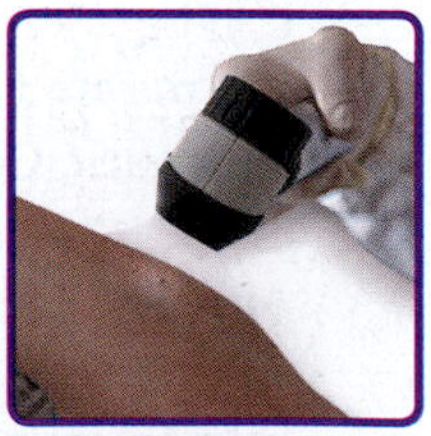

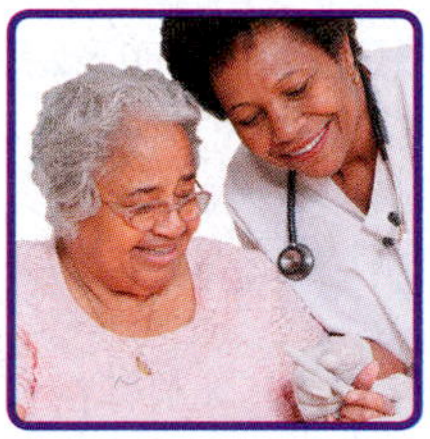

16 Carboxylic Acids and Esters 560

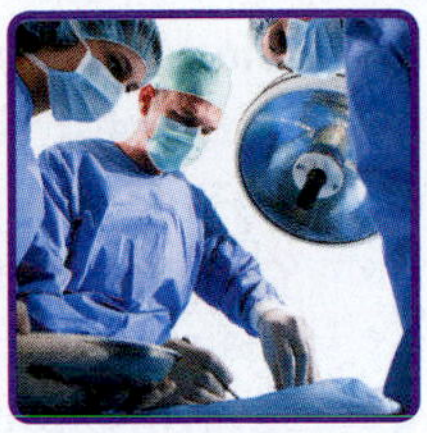

17 Lipids 586

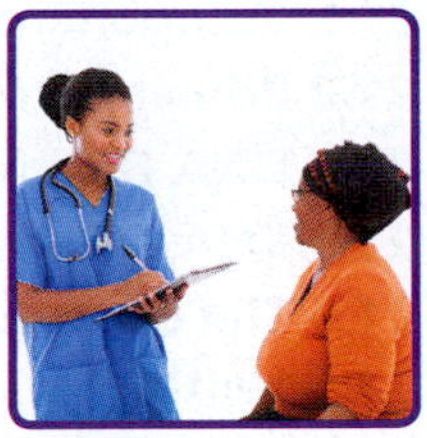

18 Amines and Amides 624

19 Amino Acids and Proteins 660

20 Enzymes and Vitamins 688

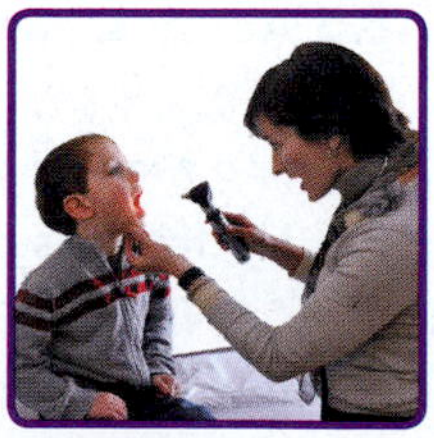

21 Nucleic Acids and Protein Synthesis 721

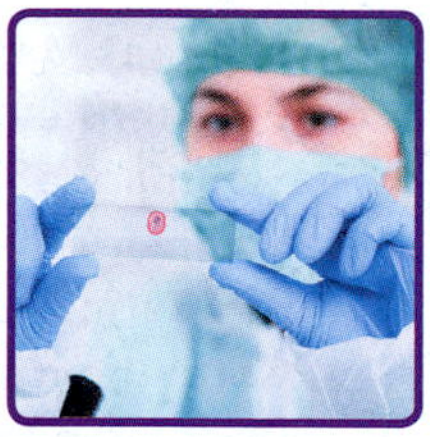

22 Metabolic Pathways for Carbohydrates 764

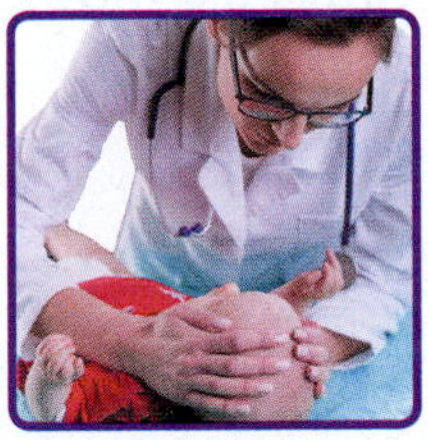

23 Metabolism and Energy Production 802

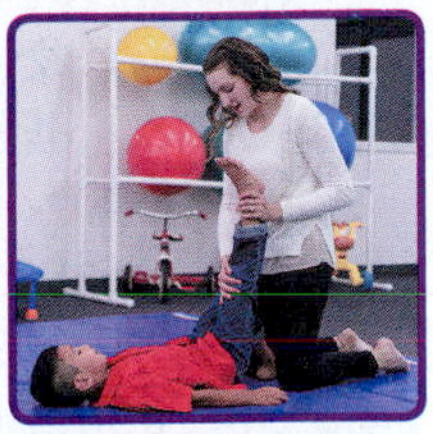

24 Metabolic Pathways for Lipids and Amino Acids 825

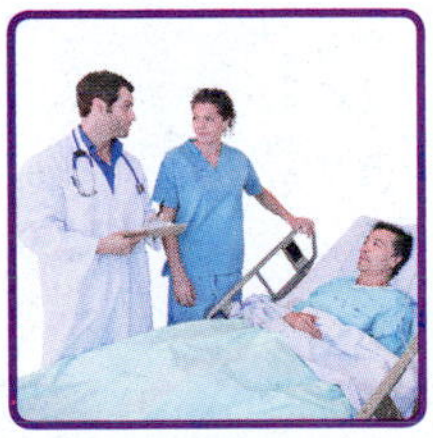

Applications and Activities

KEY MATH SKILLS

CORE CHEMISTRY SKILLS

Interactive Videos

About the Author

KAREN TIMBERLAKE is Professor Emerita of Chemistry at Los Angeles Valley College, where she taught chemistry for allied health and preparatory chemistry for 36 years. She received her bachelor's degree in chemistry from the University of Washington and her master's degree in biochemistry from the University of California at Los Angeles.

Professor Timberlake has been writing chemistry textbooks for 40 years. During that time, her name has become associated with the strategic use of pedagogical tools that promote student success in chemistry and the application of chemistry to real-life situations. More than one million students have learned chemistry using texts, laboratory manuals, and study guides written by Karen Timberlake. In addition to *General, Organic and Biological Chemistry*, sixth edition, she is also the author of *An Introduction to General, Organic, and Biological Chemistry*, thirteenth edition, with the accompanying *Study Guide and Selected Solutions Manual*, *Laboratory Manual*, and *Essential Laboratory Manual for General, Organic, and Biological Chemistry*, and *Basic Chemistry*, fifth edition, with the accompanying *Study Guide and Selected Solutions Manual*.

Professor Timberlake belongs to numerous scientific and educational organizations including the American Chemical Society (ACS) and the National Science Teachers Association (NSTA). She has been the Western Regional Winner of the Excellence in College Chemistry Teaching Award given by the Chemical Manufacturers Association. She received the McGuffey Award in Physical Sciences from the Textbook Authors Association for her textbook *Chemistry: An Introduction to General, Organic, and Biological Chemistry*, eighth edition, which has demonstrated her excellence over time. She received the "Texty" Textbook Excellence Award from the Textbook Authors Association for the first edition of *Basic Chemistry*. She has participated in education grants for science teaching including the Los Angeles Collaborative for Teaching Excellence (LACTE) and a Title III grant at her college. She attends and speaks at chemistry conferences and educational meetings on the teaching methods in chemistry that promote the learning success of students.

When Professor Timberlake is not writing textbooks, she and her husband relax by playing tennis, ballroom dancing, traveling, trying new restaurants, and cooking.

DEDICATION

I dedicate this book to

- My husband, Bill, for his patience, loving support, and preparation of late meals
- My son, John, daughter-in-law, Cindy, grandson, Daniel, and granddaughter, Emily, for the precious things in life
- The wonderful students over many years whose hard work and commitment always motivated me and put purpose in my writing

FAVORITE QUOTES

The whole art of teaching is only the art of awakening the natural curiosity of young minds.

—Anatole France

One must learn by doing the thing; though you think you know it, you have no certainty until you try.

—Sophocles

Discovery consists of seeing what everybody has seen and thinking what nobody has thought.

—Albert Szent-Györgyi

I never teach my pupils; I only attempt to provide the conditions in which they can learn.

—Albert Einstein

Preface

Welcome to the sixth edition of *General, Organic, and Biological Chemistry, Structures of Life*. This chemistry text was written and designed to help you prepare for a career in a health-related profession, such as nursing, dietetics, respiratory therapy, and environmental and agricultural science. This text assumes no prior knowledge of chemistry. My main objective in writing this text is to make the study of chemistry an engaging and positive experience for you by relating the structure and behavior of matter to its role in health and the environment. This new edition introduces more problem-solving strategies, more problem-solving guides, new Analyze the Problem with Connect features, new Try It First and Engage features, conceptual and challenge problems, and new sets of combined problems.

It is my goal to help you become a critical thinker by understanding scientific concepts that will form a basis for making important decisions about issues concerning health and the environment. Thus, I have utilized materials that

- help you to learn and enjoy chemistry
- relate chemistry to careers that may interest you
- develop problem-solving skills that lead to your success in chemistry
- promote learning and success in chemistry

New for the Sixth Edition

New and updated features have been added throughout this sixth edition, including the following:

- **NEW AND UPDATED! Chapter Openers** provide engaging clinical stories in the health profession and introduce the chemical concepts in each chapter.
- **NEW! Clinical Updates** added at the end of each chapter continue the story of the Chapter Opener and describe the follow-up treatment.
- **NEW! Engage** feature in the margin asks students to think about the paragraph they are reading and to test their understanding by answering the Engage question.
- **NEW! Try It First** precedes the Solution section of each Sample Problem to encourage the student to work on the problem before reading the given Solution.
- **NEW! Connect** feature added to **Analyze the Problem** boxes indicates the relationships between *Given* and *Need*.
- **NEW! Clinical Applications** added to Practice Problems show the relevance between the chemistry content and medicine and health.
- **NEW! Strategies for Learning Chemistry** are added that describe successful ways to study and learn chemistry.
- **NEW! Expanded Study Checks in Sample Problems** now contain multiple questions to give students additional self-testing practice.
- **NEW!** The names and symbols for the newest elements 113, Nihonium, Nh, 115, Moscovium, Mc, 117, Tennessine, Ts, and 118, Oganesson, Og.
- **NEW!** The **Steps in the Sample Problems** include a worked-out Solution plan for solving the problem.
- **NEW! Table Design** now has cells that highlight and organize related data.
- **NEW! Test** feature added in the margin encourages students to solve related Practice Problems to practice retrieval of content for exams.
- **NEW! Interactive Videos** give students the experience of step-by-step problem solving for problems from the text.
- **NEW! Review** topics are now placed in the margin at the beginning of a Section, listing the Key Math Skills and Core Chemistry Skills from the previous chapters, which provide the foundation for learning new chemistry principles in the current chapter.
- **UPDATED! Key Math Skills** review basic math relevant to the chemistry the students are learning throughout the text. A **Key Math Skill Review** at the end of each chapter summarizes and gives additional examples.
- **UPDATED! Core Chemistry Skills** identify the key chemical principles in each chapter that are required for successfully learning chemistry. A **Core Chemistry Skill Review** at the end of each chapter helps reinforce the material and gives additional examples.
- **UPDATED! Analyze the Problem** features included in the Solutions of the Sample Problems strengthen critical-thinking skills and illustrate the breakdown of a word problem into the components required to solve it.
- **UPDATED! Practice Problems**, **Sample Problems**, and **Art** demonstrate the connection between the chemistry being discussed and how these skills will be needed in professional experience.
- **UPDATED! Combining Ideas** features offer sets of integrated problems that test students' understanding and develop critical thinking by integrating topics from two or more previous chapters.
- **UPDATED!** New zoom design highlights macro-to-micro art and captions are now on a gray screen to emphasize the art and text content.
- **UPDATED! Concept Maps** are updated with new design that shows a clearer path linking concept to concept.
- **UPDATED!** Biochemistry chapters 15, 17, and 19 to 24 have been rewritten to strengthen connections between sections, and include new Study Checks and new Chemistry Links to Health.

Chapter Organization of the Sixth Edition

In each textbook I write, I consider it essential to relate every chemical concept to real-life issues. Because a chemistry course may be taught in different time frames, it may be difficult to cover all the chapters in this text. However, each chapter is a complete package, which allows some chapters to be skipped or the order of presentation to be changed.

Chapter 1, Chemistry in Our Lives, discusses the Scientific Method in everyday terms, guides students in developing a study plan for learning chemistry, with a section of Key Math Skills that reviews the basic math, including scientific notation, needed in chemistry calculations.

- The Chapter Opener tells the story of two murders and features the work and career of forensic scientists.
- A new Clinical Update feature describes the forensic evidence that helps to solve the murders and includes Clinical Applications.
- Scientific Method: Thinking Like a Scientist is expanded to include *law* and *theory*.
- An updated Section 1.3 Studying and Learning Chemistry expands the discussion of strategies that improve learning and understanding of content.
- New Section 1.5 Writing Numbers in Scientific Notation is added.
- Key Math Skills are: Identifying Place Values, Using Positive and Negative Numbers in Calculations, Calculating Percentages, Solving Equations, Interpreting Graphs, and Writing Numbers in Scientific Notation.

Chapter 2, Chemistry and Measurements, looks at measurement and emphasizes the need to understand numerical relationships of the metric system. Significant figures are discussed in the determination of final answers. Prefixes from the metric system are used to write equalities and conversion factors for problem-solving strategies. Density is discussed and used as a conversion factor.

- The Chapter Opener tells the story of a patient with high blood pressure and features the work and career of a registered nurse.
- The Clinical Update describes the patient's status and follow-up visit with his doctor.
- Sample Problems relate problem solving to health-related topics such as the measurements of blood volume, omega-3 fatty acids, radiological imaging, body fat, cholesterol, and medication orders.
- Clinical Applications feature questions about measurements, daily values for minerals and vitamins, and equalities and conversion factors for medications.
- The Key Math Skill is: Rounding Off.
- Core Chemistry Skills are: Counting Significant Figures, Using Significant Figures in Calculations, Using Prefixes, Writing Conversion Factors from Equalities, Using Conversion Factors, and Using Density as a Conversion Factor.

Chapter 3, Matter and Energy, classifies matter and states of matter, describes temperature measurement, and discusses energy, specific heat, energy in nutrition, and changes of state. Physical and chemical properties and physical and chemical changes are discussed.

- The Chapter Opener describes diet and exercise for an overweight adolescent at risk for type 2 diabetes and features the work and career of a dietitian.
- The Clinical Update describes the diet prepared with a dietitian for weight loss.
- Practice Problems and Sample Problems include high temperatures used in cancer treatment, the energy produced by a high-energy shock output of a defibrillator, body temperature lowering using a cooling cap, ice bag therapy for muscle injury, dental implants, and energy values for food.
- Core Chemistry Skills are: Identifying Physical and Chemical Changes, Converting Between Temperature Scales, Using Energy Units, Using the Heat Equation, and Calculating Heat for Change of State.
- The interchapter problem set, Combining Ideas from Chapters 1 to 3, completes the chapter.

Chapter 4, Atoms and Elements, introduces elements and atoms and the periodic table. The names and symbols for the newest elements 113, Nihonium, Nh, 115, Moscovium, Mc, 117, Tennessine, Ts, and 118, Oganesson, Og, are added to the periodic table. Electron configurations are written for atoms and the trends in periodic properties are described. Atomic numbers and mass numbers are determined for isotopes. The most abundant isotope of an element is determined by its atomic mass. Atomic mass is calculated using the masses of the naturally occurring isotopes and their abundances. Electron arrangements are written using orbital diagrams, electron configurations, and abbreviated electron configurations.

- The Chapter Opener and Clinical Update feature the improvement in crop production by a farmer.
- Atomic number and mass number are used to calculate the number of protons and neutrons in an atom.
- The number of protons and neutrons are used to calculate the mass number and to write the atomic symbol for an isotope.
- The trends in periodic properties are described for valence electrons, atomic size, ionization energy, and metallic character.
- Core Chemistry Skills are: Counting Protons and Neutrons, Writing Atomic Symbols for Isotopes, Writing Electron Configurations, Using the Periodic Table to Write Electron Configurations, Identifying Trends in Periodic Properties, and Drawing Lewis Symbols.

Chapter 5, Nuclear Chemistry, looks at the types of radiation emitted from the nuclei of radioactive atoms. Nuclear equations are written and balanced for both naturally occurring radioactivity and artificially produced radioactivity. The half-lives of radioisotopes are discussed, and the amount of time for a sample to decay is calculated. Radioisotopes important in the field of nuclear medicine are described. Fission and fusion and their role in energy production are discussed.

- The Chapter Opener describes a patient with possible coronary heart disease who undergoes a nuclear stress test and features the work and career of a radiation technologist.
- The Clinical Update discusses the results of cardiac imaging using the radioisotope Tl-201.
- Sample Problems and Practice Problems use nursing and medical examples, including phosphorus-32 for the treatment of leukemia, titanium seeds containing a radioactive isotope implanted in the body to treat cancer, yttrium-90 injections for arthritis pain, and millicuries in a dose of phosphorus-32.
- New art includes the illustration of the organs of the body where medical radioisotopes are used for diagnosis and treatment.
- Core Chemistry Skills are: Writing Nuclear Equations and Using Half-Lives.

Chapter 6, Ionic and Molecular Compounds, describes the formation of ionic and covalent bonds. Chemical formulas are written, and ionic compounds—including those with polyatomic ions—and molecular compounds are named.

- The Chapter Opener describes the chemistry of aspirin and features the work and career of a pharmacy technician.
- The Clinical Update describes several types of compounds at a pharmacy and includes Clinical Applications.
- Section 6.6 is now titled Lewis Structures for Molecules and Polyatomic Ions, and 6.9 is now titled Intermolecular Forces in Compounds.
- New material on polyatomic ions compares the names of *ate* ions and *ite* ions, the charge of sulfate and sulfite, phosphate and phosphite, carbonate and hydrogen carbonate, and the formulas and charges of halogen polyatomic ions with oxygen.
- Core Chemistry Skills are: Writing Positive and Negative Ions, Writing Ionic Formulas, Naming Ionic Compounds, Writing the Names and Formulas for Molecular Compounds, Drawing Lewis Structures, Using Electronegativity, Predicting Shape, Identifying Polarity of Molecules, and Identifying Intermolecular Forces.
- The interchapter problem set, Combining Ideas from Chapters 4 to 6, completes the chapter.

Chapter 7, Chemical Reactions and Quantities, shows students how to balance chemical equations and to recognize the types of chemical reactions: combination, decomposition, single replacement, double replacement , and combustion. Students are introduced to moles and molar masses of compounds, which are used in calculations to determine the mass or number of particles in a given quantity as well as limiting reactants and percent yield. The chapter concludes with a discussion of energy in reactions.

- The Chapter Opener describes the symptoms of heart and pulmonary disease and discusses the career of an exercise physiologist.
- A new Clinical Update, Improving Natalie's Overall Fitness, discusses her test results and suggests exercise to improve oxygen intake.
- A new order of topics begins with Section 7.5 Molar Mass, 7.6 Calculations Using Molar Mass, 7.7 Mole Relationships in Chemical Equations, and 7.8 Mass Calculations for Chemical Reactions, Section 7.9 Limiting Reactants and Percent Yield, and 7.10 Energy in Chemical Reactions.
- New Sample Problems are: Oxidation and Reduction, and Exothermic and Endothermic Reactions.
- New expanded art shows visible evidence of several types of chemical reactions.
- Core Chemistry Skills are: Balancing a Chemical Equation, Classifying Types of Chemical Reactions, Identifying Oxidized and Reduced Substances, Converting Particles to Moles, Calculating Molar Mass, Using Molar Mass as a Conversion Factor, Using Mole–Mole Factors, Converting Grams to Grams, Calculating Quantity of Product from a Limiting Reactant, Calculating Percent Yield, and Using the Heat of Reaction.

Chapter 8, Gases, discusses the properties of gases and calculates changes in gases using the gas laws: Boyle's, Charles's, Gay-Lussac's, Avogadro's, Dalton's, and the Ideal Gas Law. Problem-solving strategies enhance the discussion and calculations with the ideal gas laws.

- The Chapter Opener features the work and career of a respiratory therapist who uses oxygen to treat a child with asthma.
- The Clinical Update describes exercise to manage exercise-induced asthma. Clinical Applications are related to lung volume and gas laws.
- Sample Problems and Challenge Problems use nursing and medical examples, including, calculating the volume of oxygen gas delivered through a face mask during oxygen therapy, preparing a heliox breathing mixture for a scuba diver, and home oxygen tanks.
- Core Chemistry Skills are: Using the Gas Laws, Using the Ideal Gas Law, Calculating Mass or Volume of a Gas in a Chemical Reaction, and Calculating Partial Pressure.
- The interchapter problem set, Combining Ideas from Chapters 7 and 8, completes the chapter.

Chapter 9, Solutions, describes solutions, electrolytes, saturation and solubility, insoluble salts, concentrations, and osmosis. The concentrations of solutions are used to determine volume or mass of solute. The volumes and molarities of solutions are used in calculations of dilutions and titrations. Properties of solutions, freezing and boiling points, osmosis in the body, and dialysis are discussed.

- The Chapter Opener describes a patient with kidney failure and dialysis treatment and features the work and career of a dialysis nurse.
- The Clinical Update explains dialysis treatment and electrolyte levels in dialysate fluid.

- A new example of suspensions used to purify water in treatment plants is added.
- New art illustrates the freezing point decrease and boiling point increase for aqueous solutions with increasing number of moles of solute in one kilogram of water.
- Core Chemistry Skills are: Using Solubility Rules, Calculating Concentration, Using Concentration as a Conversion Factor, Calculating the Quantity of a Reactant or Product for a Chemical Reaction in Solution, and Calculating the Boiling Point/Freezing Point of a Solution.

Chapter 10, Reaction Rates and Chemical Equilibrium, looks at the rates of reactions and the equilibrium condition when forward and reverse rates for a reaction become equal. Equilibrium expressions for reactions are written and equilibrium constants are calculated. Le Châtelier's principle is used to evaluate the impact on concentrations when stress is placed on the system.

- The Chapter Opener describes the symptoms of infant respiratory distress syndrome (IRDS) and discusses the career of a neonatal nurse.
- The Clinical Update describes a child with anemia, hemoglobin–oxygen equilibrium, and a diet that is high in iron-containing foods.
- Core Chemistry Skills are: Writing the Equilibrium Expression, Calculating an Equilibrium Constant, Calculating Equilibrium Concentrations, and Using Le Châtelier's Principle.

Chapter 11, Acids and Bases, discusses acids and bases and their strengths, and conjugate acid–base pairs. The dissociation of strong and weak acids and bases is related to their strengths as acids or bases. The dissociation of water leads to the water dissociation expression, K_w, the pH scale, and the calculation of pH. Chemical equations for acids in reactions are balanced and titration of an acid is illustrated. Buffers are discussed along with their role in the blood. The pH of a buffer is calculated.

- The Chapter Opener describes a blood sample for an emergency room patient sent to the clinical laboratory for analysis of blood pH and CO_2 gas and features the work and career of a clinical laboratory technician.
- The Clinical Update describes the symptoms and treatment for acid reflux disease (GERD).
- Key Math Skills are: Calculating pH from $[H_3O^+]$ and Calculating $[H_3O^+]$ from pH.
- Core Chemistry Skills are: Identifying Conjugate Acid–Base Pairs, Calculating $[H_3O^+]$ and $[OH^-]$ in Solutions, Writing Equations for Reactions of Acids and Bases, Calculating Molarity or Volume of an Acid or Base in a Titration, and Calculating the pH of a Buffer.
- The interchapter problem set, Combining Ideas from Chapters 9 to 11, completes the chapter.

Chapter 12, Introduction to Organic Chemistry: Hydrocarbons, compares inorganic and organic compounds, and describes the structures and naming of alkanes, alkenes including cis–trans isomers, alkynes, and aromatic compounds.

- The Chapter Opener describes a fire victim and the search for traces of accelerants and fuel at the arson scene and features the work and career of a firefighter/emergency medical technician.
- The Clinical Update describes the treatment of burns in the hospital and the types of fuels identified in the fire.
- Subsections in 12.4 Solubility and Density and 12.5 Identifying Alkenes and Alkynes are revised for clarity.
- More line-angle formulas for organic structures in Practice Problems have been added.
- Core Chemistry Skills are: Naming and Drawing Alkanes and Writing Equations for Hydrogenation, Hydration, and Polymerization.

Chapter 13, Alcohols, Phenols, Thiols, and Ethers, describes the functional groups and names of alcohols, phenols, thiols, and ethers.

- The new Chapter Opener describes local anesthetics for surgery to repair a torn anterior cruciate ligament (ACL) and features the work and career of a nurse anesthetist.
- The Clinical Update describes some foods added to a diet plan including a comparison of their functional groups.
- New art includes new career photo of a nurse anesthetist, ball-and-stick models added to primary, secondary, and tertiary alcohol structures in Section 13.3 to visualize the classification of alcohols, anesthesia apparatus for delivery of isoflurane, exhausted athlete, and perming hair.
- Chemistry Link to Health "Hand Sanitizers" is revised and "Methanol Poisoning" is moved into "Oxidation of Alcohol in the Body" at the end of Section 13.4.
- Core Chemistry Skills are: Identifying Alcohols, Phenols, and Thiols, Naming Alcohols and Phenols, Writing Equations for the Dehydration of Alcohols, and Writing Equations for the Oxidation of Alcohols.

Chapter 14, Aldehydes and Ketones, discusses the nomenclature, structures, and oxidation and reduction of aldehydes and ketones. The chapter discusses the formation of hemiacetals and acetals.

- The Chapter Opener describes the risk factors for melanoma and discusses the career of a dermatology nurse.
- The Clinical Update discusses melanoma, skin protection, and functional groups of sunscreens.
- New art using line-angle formulas is drawn for separate equations of hemiacetal and acetal formation.
- Sections 14.3 Oxidation and Reduction of Aldehydes and Ketones and 14.4 Addition of Alcohols: Hemiacetals and Acetals are revised for clarity.
- A summary of the Tollens' and Benedict's tests is added to section 14.3.
- Core Chemistry Skills are: Naming Aldehydes and Ketones, and Forming Hemiacetals and Acetals.
- New structures of pamplemousse acetal in grapefruit and rose acetal in perfume are added.
- The interchapter problem set, Combining Ideas from Chapters 12 to 14, completes the chapter.

Chapter 15, Carbohydrates, describes the carbohydrate molecules monosaccharides, disaccharides, and polysaccharides and their formation by photosynthesis. Monosaccharides are classified as aldo or keto pentoses or hexoses. Chiral molecules are discussed along with Fischer projections and D and L notations. The formation of glycosidic bonds in disaccharides and polysaccharides is described.

- The Chapter Opener describes a diabetes patient and her diet and features the work and career of a diabetes nurse.
- The Clinical Update describes a diet and exercise program to lower blood glucose.
- New art accompanies content on tooth decay and use of xylitol, the structures of amino sugars and uronic acids, and hyaluronic acid used as facial fillers.
- New Chemistry Links to Health are: Dental Cavities and Xylitol Gum, and Varied Biological Roles of Carbohydrate Polymers: The Case of Glycosaminoglycans.
- New Study Checks include penicillamine to treat rheumatoid arthritis, and ethambutol to treat tuberculosis.
- Section on Chirality is moved to Chapter 15.
- Core Chemistry Skills are: Identifying Chiral Molecules, Identifying D and L Fischer Projections for Carbohydrates, and Drawing Haworth Structures.

Chapter 16, Carboxylic Acids and Esters, discusses the functional groups and naming of carboxylic acids and esters. Chemical reactions include esterification and acid and base hydrolysis of esters.

- The Chapter Opener describes heart surgery and discusses the work and career of a surgical technician.
- The Clinical Update describes the chemistry and use of liquid bandages.
- More line-angle structures for carboxylic acids and esters have been added.
- New art of ester-containing fruit has been added.
- Core Chemistry Skills are: Naming Carboxylic Acids and Hydrolyzing Esters.

Chapter 17, Lipids, discusses fatty acids and the formation of ester bonds in triacylglycerols and glycerophospholipids. Chemical properties of fatty acids and their melting points along with the hydrogenation of unsaturated triacylglycerols are discussed. Steroids, such as cholesterol and bile salts, are described. The role of phospholipids in the lipid bilayer of cell membranes is discussed as well as the lipids that function as steroid hormones.

- The updated Chapter Opener describes a patient with symptoms of familial hypercholesterolemia and features the work and career of a clinical lipid specialist.
- The Clinical Update describes medications a program to and a diet to lower cholesterol.
- New art diagrams include glaucoma and its treatment with a prostaglandin, healthy and nonhealthy livers, and the steroid structure of spironolactone.
- Chemistry Links to Health are: Omega-3 Fatty Acids in Fish Oils and Infant Respiratory Distress Syndrome (IRDS).
- New Chemistry Links to Health are: A Prostaglandin-like Medication for Glaucoma That Also Thickens Eyelashes, and A Steroid Receptor Antagonist That Prevents the Development of Male Sexual Characteristics.
- Core Chemistry Skills are: Identifying Fatty Acids, Drawing Structures for Triacylglycerols, Drawing the Products for the Hydrogenation, Hydrolysis, and Saponification of a Triacylglycerol, and Identifying the Steroid Nucleus.

Chapter 18, Amines and Amides, emphasizes the nitrogen atom in their functional groups and their names. Properties of amines including classification, boiling point, solubility in water, and use as neurotransmitters are included. Alkaloids are discussed as the naturally occurring amines in plants. Chemical reactions include dissociation and neutralization of amines, amidation, and acid and base hydrolysis of amides.

- The Chapter Opener describes pesticides and pharmaceuticals used on a ranch and discusses the career of an environmental health practitioner.
- The Clinical Update describes the collection of soil and water samples for testing of insecticides and antibiotics.
- New line-angle formulas are drawn for amines, alkaloids, heterocyclic amines, and neurotransmitters.
- Introduction to Section 18.5, Amides is revised.
- Chemistry Link to Health Synthesizing Drugs and Opioids is revised.
- Clinical Applications include novocaine, lidocaine, ritalin, niacin, serotonin, histamine, acetylcholine, dose calculations of pesticides and antibiotics, enrofloxacin, and voltaren.
- Core Chemistry Skills are: Forming Amides and Hydrolyzing Amides.
- The interchapter problem set, Combining Ideas from Chapters 15 to 18, completes the chapter.

Chapter 19, Amino Acids and Proteins, discusses amino acids, formation of peptide bonds and the primary, secondary, tertiary, and quaternary structural levels of proteins. The ionized structures of amino acids are drawn at physiological pH.

- A new Chapter Opener discusses the symptoms of sickle-cell anemia in a child, the mutation in amino acids that causes the crescent shape of abnormal red blood cells, and the career of a hematology nurse.
- A new Clinical Update discusses the diagnosis of sickle-cell anemia using electrophoresis and its treatment.
- The protein structure sections are reorganized as: 19.2 Proteins: Primary Structure; 19.3 Proteins: Secondary Structure; and 19.4 Proteins: Tertiary and Quaternary Structures.
- Chemistry Links to Health are: Essential Amino Acids and Complete Proteins, Protein Secondary Structures and Alzheimer's Disease, and Sickle-Cell Anemia.
- New Chemistry Links to Health are: Cystinuria, and Keratoconus.
- New art includes normal cornea, cornea with keratoconus, collagen fibers in keratoconus, and insoluble fiber formation in sickle-cell anemia.

- New Sample Problems are: 19.3 Identifying a Tripeptide and 19.4 Drawing a Peptide.
- Core Chemistry Skills are: Drawing the Structure for an Amino Acid at Physiological pH and Identifying the Primary, Secondary, Tertiary, and Quaternary Structures of Proteins.

Chapter 20, Enzymes and Vitamins, relates the importance of the three-dimensional shape of proteins to their function as enzymes. The shape of an enzyme and its substrate are factors in enzyme regulation. End products of an enzyme-catalyzed sequence can increase or decrease the rate of an enzyme-catalyzed reaction. Other regulatory processes include allosteric enzymes, covalent modification and phosphorylation, and zymogens. Proteins change shape and lose function when subjected to pH changes and high temperatures. The important role of water-soluble vitamins as coenzymes is related to enzyme function.

- The Chapter Opener discusses the symptoms of lactose intolerance and describes the career of a physician assistant.
- The Clinical Update describes the hydrogen breath test to confirm lactose intolerance and a diet that is free of lactose and use of Lactaid.
- Chemistry Link to Health is: Isoenzymes as Diagnostic Tools.
- New Chemistry Links to Health are: Fabry Disease and Taking Advantage of Enzyme Inhibition to Treat Cancer: Imatinib.
- New art includes the structure of galactosidase A and enzyme inhibition of imatinib used to treat myeloid leukemia.
- Core Chemistry Skills are: Describing Enzyme Action, Classifying Enzymes, Identifying Factors Affecting Enzyme Activity, and Describing the Role of Cofactors.

Chapter 21, Nucleic Acids and Protein Synthesis, describes the nucleic acids and their importance as biomolecules that store and direct information for the synthesis of cellular components. The role of complementary base pairing is discussed in both DNA replication and the formation of mRNA during protein synthesis. The role of RNA is discussed in the relationship of the genetic code to the sequence of amino acids in a protein. Mutations describe ways in which the nucleotide sequences are altered in genetic diseases.

- The Chapter Opener describes a patient's diagnosis and treatment of breast cancer and discusses the work and career of a histology technician.
- A Clinical Update describes estrogen-positive tumors, the impact of the altered genes BRCA1 and BRCA2 on the estrogen receptor, and medications to suppress tumor growth.
- A new Section discusses recombinant DNA, polymerase chain reaction, and DNA fingerprinting.
- The Chemistry Link to Health Protein Sequencing was moved from Chapter 19 to Chapter 21.
- New Chemistry Links to Health are: Cataracts and Ehlers–Danlos Syndrome.
- Core Chemical Skills are: Writing the Complementary DNA Strand, Writing the mRNA Segment for a DNA Template, and Writing the Amino Acid for an mRNA Codon.
- The interchapter problem set, Combining Ideas from Chapters 19 to 21, completes the chapter.

Chapter 22, Metabolic Pathways for Carbohydrates, describes the stages of metabolism and the digestion of carbohydrates, our most important fuel. The breakdown of glucose to pyruvate is described using glycolysis, which is followed under aerobic conditions by the decarboxylation of pyruvate to acetyl CoA. The synthesis of glycogen and the synthesis of glucose from noncarbohydrate sources are discussed.

- The Chapter Opener describes the symptoms of a glycogen storage disease and discusses the career of a hepatology nurse.
- The Clinical Update describes medical treatment of frequent feedings of glucose for von Gierke's disease, in which a child has a defective glucose-6-phosphatase and cannot break down glucose-6-phosphate to glucose.
- Chemistry Link to Health is: Glycogen Storage Diseases (GSDs).
- New Chemistry Links to Health are: Galactosemia and Glucocorticoids, and Steroid-Induced Diabetes.
- Sections 22.4 "Glycolysis: Oxidation of Glucose", 22.6 "Glycogen Synthesis and Degradation", and 22.7 "Gluconeogenesis: Glucose Synthesis" are revised for clarity.
- New art includes diagrams of normal lactose oxidation compared to galactosemia, and the impact of glucocorticoids on glucose metabolism.
- Core Chemistry Skills are: Identifying Important Coenzymes in Metabolism, Identifying the Compounds in Glycolysis, and Identifying the Compounds and Enzymes in Glycogenesis and Glycogenolysis.

Chapter 23, Metabolism and Energy Production, looks at the entry of acetyl CoA into the citric acid cycle and the production of reduced coenzymes for electron transport, oxidative phosphorylation, and the synthesis of ATP. The malate–aspartate shuttle describes the transport of NADH from the cytosol into the mitochondrial matrix.

- The new Chapter Opener discusses a child with mitochondrial myopathy and discusses the work and career of a physical therapist.
- A new Clinical Update discusses treatment that helps increase a child's functional capacity.
- New Clinical Applications include problems about diseases associated with enzyme deficiencies.
- New material discusses diseases of enzymes in the citric acid cycle such as fumarase deficiency that causes neurological impairment, developmental delay, and seizures.
- Feedback Control, Covalent Modification, and Enzyme Inhibition subsections are expanded to enhance student understanding.

- A new subsection Diseases of the Citric Acid Cycle is added to Section 23.1.
- Section 23.2 Electron Transport and ATP is revised for clarity.
- Chemistry Links to Health are: Toxins: Inhibitors of Electron Transport, Uncouplers of ATP Synthase, and Efficiency of ATP Production.
- Core Chemistry Skills are: Describing the Reactions in the Citric Acid Cycle and Calculating the ATP Produced from Glucose.

Chapter 24, Metabolic Pathways for Lipids and Amino Acids, discusses the digestion of lipids and proteins and the metabolic pathways that convert fatty acids and amino acids into energy. Discussions include the conversion of excess carbohydrates to triacylglycerols in adipose tissue and how the intermediates of the citric acid cycle are converted to nonessential amino acids.

- The Chapter Opener describes a liver profile with elevated levels of liver enzymes for a patient with chronic hepatitis C infection and discusses the career of a public health nurse.
- The Clinical Update describes interferon and ribavirin therapy for hepatitis C.
- New material discusses the digestion of triacylglycerols and dietary fats, lipase deficiency, eruptive xanthomas, calculating ATP from beta oxidation of an unsaturated fatty acid, and ketoacidosis.
- Sections 24.1 Digestion of Triacylglycerols, 24.2 Oxidation of Fatty Acids, and 24.3 ATP and Fatty Acid Oxidation are revised for clarity.
- New art includes xanthomas, ackee fruit, and injection of interferon.
- Chemistry Links to Health are: Diabetes and Ketone Bodies and Phenylketonuria (PKU).
- A new Chemistry Link to Health discusses Jamaican vomiting sickness.
- Clinical Applications include new problems about Jamaican vomiting sickness caused by an inhibitor of acyl CoA dehydrogenase, and inhibitors of beta oxidation.
- Core Chemistry Skills are: Calculating the ATP from Fatty Acid Oxidation (β Oxidation), Describing How Ketone Bodies are Formed, and Distinguishing Anabolic and Catabolic Pathways.
- The interchapter problem set, Combining Ideas from Chapters 22 to 24, completes the chapter.

Acknowledgments

The preparation of a new text is a continuous effort of many people. I am thankful for the support, encouragement, and dedication of many people who put in hours of tireless effort to produce a high-quality book that provides an outstanding learning package. I am thankful for the outstanding contributions of Professor MaryKay Orgill whose updates and clarifications enhanced the content of the biochemistry chapters 15, 17, and 19 to 24. The editorial team at Pearson has done an exceptional job. I want to thank Jeanne Zalesky, Director, Courseware Portfolio Management, and Scott Dustan, Courseware Portfolio Manager, who supported our vision of this sixth edition.

I appreciate all the wonderful work of Melanie Field, Content Producer, who skillfully brought together files, art, web site materials, and all the things it takes to prepare a book for production. I appreciate the work of Christian Arsenault at SPi Global, who brilliantly coordinated all phases of the manuscript to the final pages of a beautiful book. Thanks to Mark Quirie, manuscript and accuracy reviewer, and Karen Williams, who precisely analyzed and edited the manuscripts and pages to make sure the words and problems were correct to help students learn chemistry. Their keen eyes and thoughtful comments were extremely helpful in the development of this text.

Thanks to Kristen Flathman, Managing Producer, Coleen Morrison, Courseware Analyst, and Jennifer Hart, Courseware Director for their excellent review of pages and helpful suggestions.

I am especially proud of the art program in this text, which lends beauty and understanding to chemistry. I would like to thank Jay McElroy, Art Courseware Analyst and Stephanie Marquez, Photo and Illustration Project Manager; Maria Guglielmo Walsh, Design Manager, and Tamara Newnam, Cover and Interior Designer, whose creative ideas provided the outstanding design for the cover and pages of the book. I appreciate the tireless efforts of Clare Maxwell, Photo Researcher, and Matt Perry, Rights and Permissions Project Manager in researching and selecting vivid photos for the text so that students can see the beauty of chemistry. Thanks also to *Bio-Rad Laboratories* for their courtesy and use of *KnowItAll ChemWindows*, drawing software that helped us produce chemical structures for the manuscript. The macro-to-micro illustrations designed by Jay McElroy and Imagineering Art give students visual impressions of the atomic and molecular organization of everyday things and are a fantastic learning tool. I also appreciate all the hard work in the field put in by the marketing team and Elizabeth Ellsworth Bell, Marketing Manager.

I am extremely grateful to an incredible group of peers for their careful assessment of all the new ideas for the text; for their suggested additions, corrections, changes, and deletions; and for providing an incredible amount of feedback about improvements for the book. I admire and appreciate every one of you.

If you would like to share your experience with chemistry, or have questions and comments about this text, I would appreciate hearing from you.

Karen Timberlake
Email: khemist@aol.com

Career Focus Engages Students

Best-selling author Karen Timberlake, joined by new contributing author MaryKay Orgill, connects chemistry to real-world and career applications like no one else. The 6th edition of *General, Organic, and Biological Chemistry: Structures of Life* engages students by helping them see the connections between chemistry, the world around them, and future careers.

11

Acids and Bases

Larry, a 30-year-old man, is brought to the emergency room after an automobile accident where he is unresponsive. One of the emergency room nurses takes a blood sample, which is then sent to Brianna, a clinical laboratory technician, who begins the process of analyzing the pH, the partial pressures of O_2 and CO_2, and the concentrations of glucose and electrolytes.

Brianna determines that Larry's blood pH is 7.30 and the partial pressure of CO_2 gas is above the desired level. Blood pH is typically in the range of 7.35 to 7.45, and a value less than 7.35 indicates a state of acidosis. Respiratory acidosis occurs because an increase in the partial pressure of CO_2 gas in the bloodstream prevents the biochemical buffers in blood from making a change in the pH.

Brianna recognizes these signs and immediately contacts the emergency room to inform them that Larry's airway may be blocked. In the emergency room, they provide Larry with an IV containing bicarbonate to increase the blood pH and begin the process of unblocking his airway. Shortly afterward, Larry's airway is cleared, and his blood pH and partial pressure of CO_2 gas return to normal.

CAREER

Clinical Laboratory Technician

Clinical laboratory technicians, also known as medical laboratory technicians, perform a wide variety of tests on body fluids and cells that help in the diagnosis and treatment of patients. These tests range from determining blood concentrations of glucose and cholesterol to determining drug levels in the blood for transplant patients or a patient undergoing treatment. Clinical laboratory technicians also prepare specimens in the detection of cancerous tumors and type blood samples for transfusions. Clinical laboratory technicians must also interpret and analyze the test results, which are then passed on to the physician.

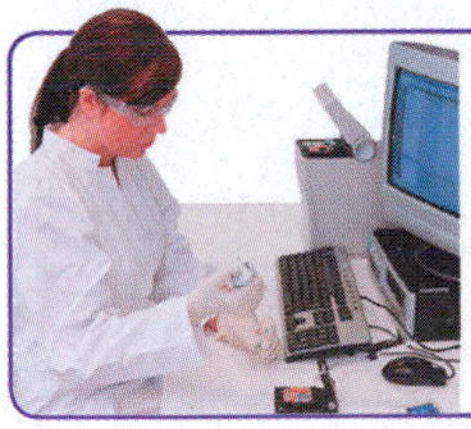

CLINICAL UPDATE

Acid Reflux Disease

After Larry was discharged from the hospital, he complained of a sore throat and dry cough, which his doctor diagnosed as acid reflux. You can view the symptoms of acid reflux disease (GERD) in the **CLINICAL UPDATE** **Acid Reflux Disease**, pages 414–415, and learn about the pH changes in the stomach and how the condition is treated.

382

Chapter Openers emphasize **clinical connections** by showing students relevant, engaging, and topical examples of how health professionals use chemistry everyday in their careers.

Clinical Updates added at the end of each chapter continue the story of the chapter opener and describe the follow-up treatment, helping students see the connections to the chemistry learned in the chapter.

Chemistry Links to Health and **Chemistry Links to the Environment** apply chemical concepts to health and medical topics as well as topics in the environment, such as bone density, weight loss and weight gain, alcohol abuse, kidney dialysis, dental cavities and xylitol gum, hyperglycemia and hypoglycemia, Alzheimer's disease, sickle-cell anemia, cancer, cataracts, galactosemia, and steroid-induced diabetes, illustrating the importance of understanding chemistry in real-life situations.

Chemistry Link to Health

Stomach Acid, HCl

Gastric acid, which contains HCl, is produced by parietal cells that line the stomach. When the stomach expands with the intake of food, the gastric glands begin to secrete a strongly acidic solution of HCl. In a single day, a person may secrete 2000 mL of gastric juice, which contains hydrochloric acid, mucins, and the enzymes pepsin and lipase.

The HCl in the gastric juice activates a digestive enzyme from the chief cells called *pepsinogen* to form *pepsin*, which breaks down proteins in food entering the stomach. The secretion of HCl continues until the stomach has a pH of about 2, which is the optimum for activating the digestive enzymes without ulcerating the stomach lining. In addition, the low pH destroys bacteria that reach the stomach. Normally, large quantities of viscous mucus are secreted within the stomach to protect its lining from acid and enzyme damage. Gastric acid may also form under conditions of stress when the nervous system activates the production of HCl. As the contents of the stomach move into the small intestine, cells produce bicarbonate that neutralizes the gastric acid until the pH is about 5.

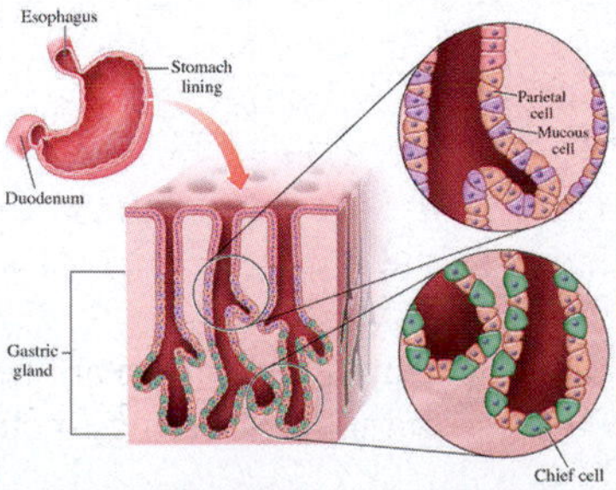

Esophagus
Stomach lining
Duodenum
Gastric gland
Parietal cell
Mucous cell
Chief cell

Parietal cells in the lining of the stomach secrete gastric acid HCl.

Builds Students' Critical Thinking

One of Karen Timberlake's goals is to help students become critical thinkers. Color-coded tips found throughout each chapter are designed to provide guidance and encourage students to really think about what they are reading and help develop important critical-thinking skills.

in $[H_3O^+]$ and a decrease in $[OH^-]$, which makes an acidic solution. If base is added, $[OH^-]$ increases and $[H_3O^+]$ decreases, which gives a basic solution. However, for any aqueous solution, whether it is neutral, acidic, or basic, the product $[H_3O^+][OH^-]$ is equal to K_w (1.0×10^{-14} at 25 °C) (see **TABLE 11.6**).

TABLE 11.6 Examples of $[H_3O^+]$ and $[OH^-]$ in Neutral, Acidic, and Basic Solutions

Type of Solution	$[H_3O^+]$	$[OH^-]$	K_w (25 °C)
Neutral	1.0×10^{-7} M	1.0×10^{-7} M	1.0×10^{-14}
Acidic	1.0×10^{-2} M	1.0×10^{-12} M	1.0×10^{-14}
Acidic	2.5×10^{-5} M	4.0×10^{-10} M	1.0×10^{-14}
Basic	1.0×10^{-8} M	1.0×10^{-6} M	1.0×10^{-14}
Basic	5.0×10^{-11} M	2.0×10^{-4} M	1.0×10^{-14}

TEST
Try Practice Problems 11.35 and 11.36

NEW! Test feature found in the margin throughout each chapter encourages students to solve related Practice Problems to practice retrieval of content for exams.

Using the K_w to Calculate $[H_3O^+]$ and $[OH^-]$ in a Solution

If we know the $[H_3O^+]$ of a solution, we can use the K_w to calculate $[OH^-]$. If we know the $[OH^-]$ of a solution, we can calculate $[H_3O^+]$ from their relationship in the K_w, as shown in Sample Problem 11.6.

ENGAGE
If you know the $[H_3O^+]$ of a solution, how do you use the K_w to calculate the $[OH^-]$?

$$K_w = [H_3O^+][OH^-]$$

$$[OH^-] = \frac{K_w}{[H_3O^+]} \qquad [H_3O^+] = \frac{K_w}{[OH^-]}$$

▸SAMPLE PROBLEM 11.6 Calculating the $[H_3O^+]$ of a Solution

CORE CHEMISTRY SKILL
Calculating $[H_3O^+]$ and $[OH^-]$ in Solutions

UPDATED! Core Chemistry Skills found throughout each chapter identify the fundamental chemistry concepts that students need to understand in the current chapter.

TRY IT FIRST

A vinegar solution has a $[OH^-] = 5.0 \times 10^{-12}$ M at 25 °C. What is the $[H_3O^+]$ of the vinegar solution? Is the solution acidic, basic, or neutral?

SOLUTION

STEP 1 State the given and needed quantities.

ANALYZE THE PROBLEM	Given	Need	Connect
	$[OH^-] = 5.0 \times 10^{-12}$ M	$[H_3O^+]$	$K_w = [H_3O^+][OH^-]$

STEP 2 Write the K_w for water and solve for the unknown $[H_3O^+]$.

$$K_w = [H_3O^+][OH^-] = 1.0 \times 10^{-14}$$

Solve for $[H_3O^+]$ by dividing both sides by $[OH^-]$.

$$\frac{K_w}{[OH^-]} = \frac{[H_3O^+][OH^-]}{[OH^-]}$$

$$[H_3O^+] = \frac{1.0 \times 10^{-14}}{[OH^-]}$$

STEP 3 Substitute the known $[OH^-]$ into the equation and calculate.

$$[H_3O^+] = \frac{1.0 \times 10^{-14}}{[5.0 \times 10^{-12}]} = 2.0 \times 10^{-3}\ M$$

Because the $[H_3O^+]$ of 2.0×10^{-3} M is larger than the $[OH^-]$ of 5.0×10^{-12} M, the solution is acidic.

ENGAGE
Why does the $[H_3O^+]$ of an aqueous solution increase if the $[OH^-]$ decreases?

NEW! Engage feature asks students to think about what they are reading and immediately assess their understanding by answering the Engage question, which is related to the topic. With regular self-assessment, students connect new concepts to prior knowledge to help them retrieve that content during exams.

and Problem-Solving Skills

New problem-solving features enhance Karen Timberlake's unmatched problem-solving strategies and help students deepen their understanding of content while improving their problem-solving skills.

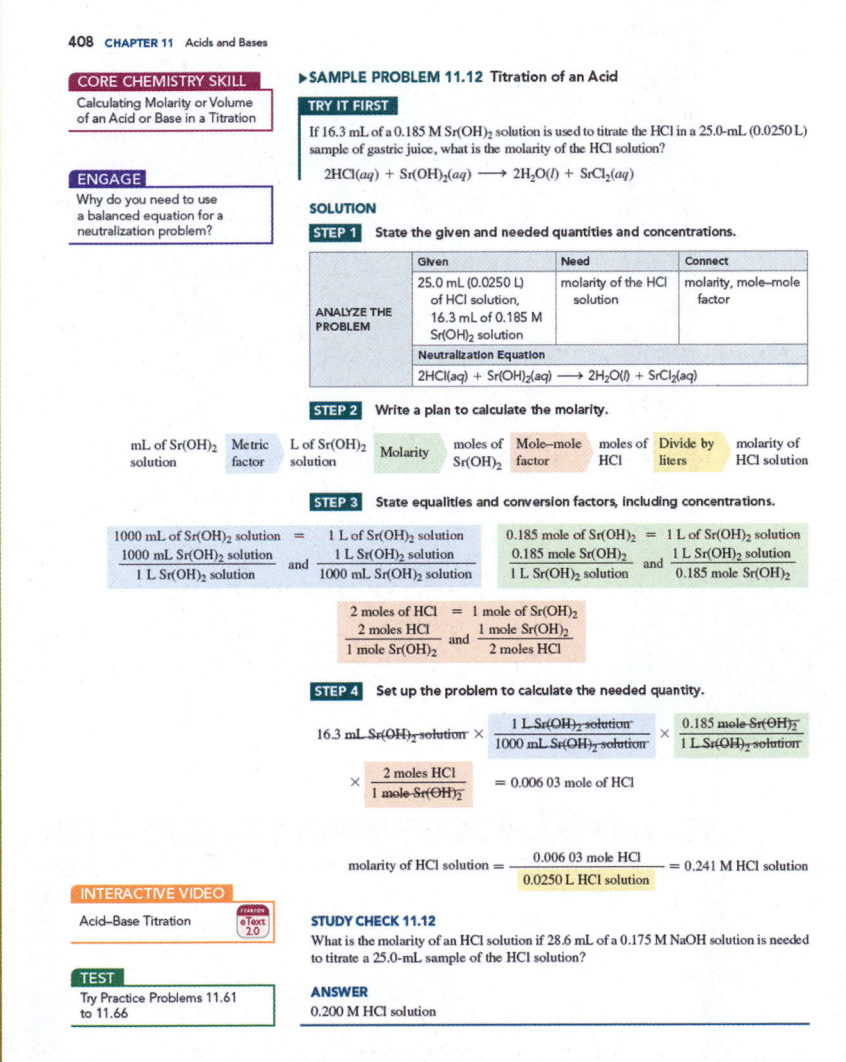

408 CHAPTER 11 Acids and Bases

CORE CHEMISTRY SKILL
Calculating Molarity or Volume of an Acid or Base in a Titration

ENGAGE
Why do you need to use a balanced equation for a neutralization problem?

▶SAMPLE PROBLEM 11.12 Titration of an Acid

TRY IT FIRST

If 16.3 mL of a 0.185 M $Sr(OH)_2$ solution is used to titrate the HCl in a 25.0-mL (0.0250 L) sample of gastric juice, what is the molarity of the HCl solution?

$$2HCl(aq) + Sr(OH)_2(aq) \longrightarrow 2H_2O(l) + SrCl_2(aq)$$

SOLUTION

STEP 1 State the given and needed quantities and concentrations.

ANALYZE THE PROBLEM	Given	Need	Connect
	25.0 mL (0.0250 L) of HCl solution, 16.3 mL of 0.185 M $Sr(OH)_2$ solution	molarity of the HCl solution	molarity, mole–mole factor
	Neutralization Equation		
	$2HCl(aq) + Sr(OH)_2(aq) \longrightarrow 2H_2O(l) + SrCl_2(aq)$		

STEP 2 Write a plan to calculate the molarity.

mL of $Sr(OH)_2$ solution → Metric factor → L of $Sr(OH)_2$ solution → Molarity → moles of $Sr(OH)_2$ → Mole–mole factor → moles of HCl → Divide by liters → molarity of HCl solution

STEP 3 State equalities and conversion factors, including concentrations.

1000 mL of $Sr(OH)_2$ solution = 1 L of $Sr(OH)_2$ solution

$$\frac{1000 \text{ mL } Sr(OH)_2 \text{ solution}}{1 \text{ L } Sr(OH)_2 \text{ solution}} \text{ and } \frac{1 \text{ L } Sr(OH)_2 \text{ solution}}{1000 \text{ mL } Sr(OH)_2 \text{ solution}}$$

0.185 mole of $Sr(OH)_2$ = 1 L of $Sr(OH)_2$ solution

$$\frac{0.185 \text{ mole } Sr(OH)_2}{1 \text{ L } Sr(OH)_2 \text{ solution}} \text{ and } \frac{1 \text{ L } Sr(OH)_2 \text{ solution}}{0.185 \text{ mole } Sr(OH)_2}$$

2 moles of HCl = 1 mole of $Sr(OH)_2$

$$\frac{2 \text{ moles HCl}}{1 \text{ mole } Sr(OH)_2} \text{ and } \frac{1 \text{ mole } Sr(OH)_2}{2 \text{ moles HCl}}$$

STEP 4 Set up the problem to calculate the needed quantity.

$$16.3 \text{ mL } Sr(OH)_2 \text{ solution} \times \frac{1 \text{ L } Sr(OH)_2 \text{ solution}}{1000 \text{ mL } Sr(OH)_2 \text{ solution}} \times \frac{0.185 \text{ mole } Sr(OH)_2}{1 \text{ L } Sr(OH)_2 \text{ solution}} \times \frac{2 \text{ moles HCl}}{1 \text{ mole } Sr(OH)_2} = 0.006\,03 \text{ mole of HCl}$$

$$\text{molarity of HCl solution} = \frac{0.006\,03 \text{ mole HCl}}{0.0250 \text{ L HCl solution}} = 0.241 \text{ M HCl solution}$$

INTERACTIVE VIDEO
Acid–Base Titration

STUDY CHECK 11.12

What is the molarity of an HCl solution if 28.6 mL of a 0.175 M NaOH solution is needed to titrate a 25.0-mL sample of the HCl solution?

ANSWER

0.200 M HCl solution

TEST
Try Practice Problems 11.61 to 11.66

NEW! Try It First precedes the Solution section of each Sample Problem to encourage the student to work on the problem before reading the given Solution.

NEW! Connect feature added to **Analyze the Problem** boxes indicates the relationships between **Given** and **Need**.

NEW! Solution provides steps for successful problem solving within the Sample Problem.

Continuous Learning
Before, During, and After Class

BEFORE CLASS

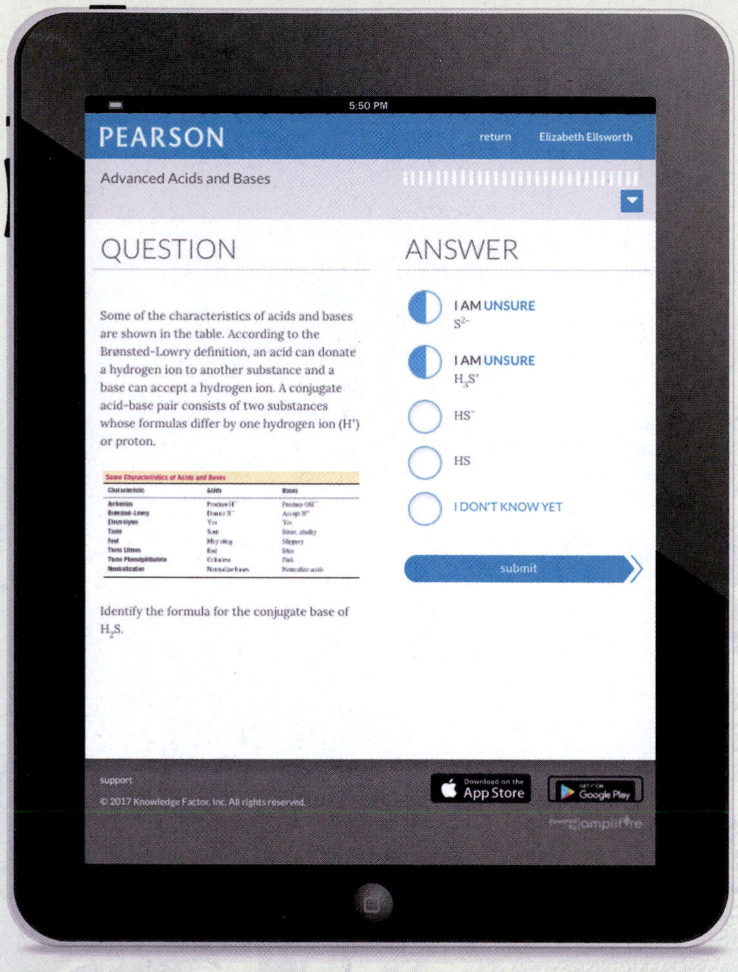
5:50 PM
PEARSON
return
Elizabeth Ellsworth
Advanced Acids and Bases
QUESTION
ANSWER
Some of the characteristics of acids and bases are shown in the table. According to the Brønsted-Lowry definition, an acid can donate a hydrogen ion to another substance and a base can accept a hydrogen ion. A conjugate acid-base pair consists of two substances whose formulas differ by one hydrogen ion (H⁺) or proton.
Identify the formula for the conjugate base of H₂S.
I AM UNSURE
I AM UNSURE
HS⁻
HS
I DON'T KNOW YET
submit
support
© 2017 Knowledge Factor, Inc. All rights reserved.
Download on the App Store
Google Play

NEW! 66 Dynamic Study Modules, specific to General, Organic, and Biological Chemistry, help students study effectively on their own by continuously assessing their activity and performance in real time.

Students complete a set of questions with a unique answer format that also asks them to indicate their confidence level. Questions repeat until the student can answer them all correctly and confidently.

Once completed, Dynamic Study Modules explain the concept. These are available as graded assignments prior to class, and accessible on smartphones, tablets, and computers.

NEW! Mastering Chemistry Primer tutorials are focused on remediating students taking their first college chemistry course.

Topics include math in the context of chemistry, chemical skills and literacy, as well as some basics of balancing chemical equations, mole–mole factors, and mass–mass calculations—all of which were chosen based on extensive surveys of chemistry professors across the country.

The primer is offered as a prebuilt assignment that is automatically generated with all chemistry courses.

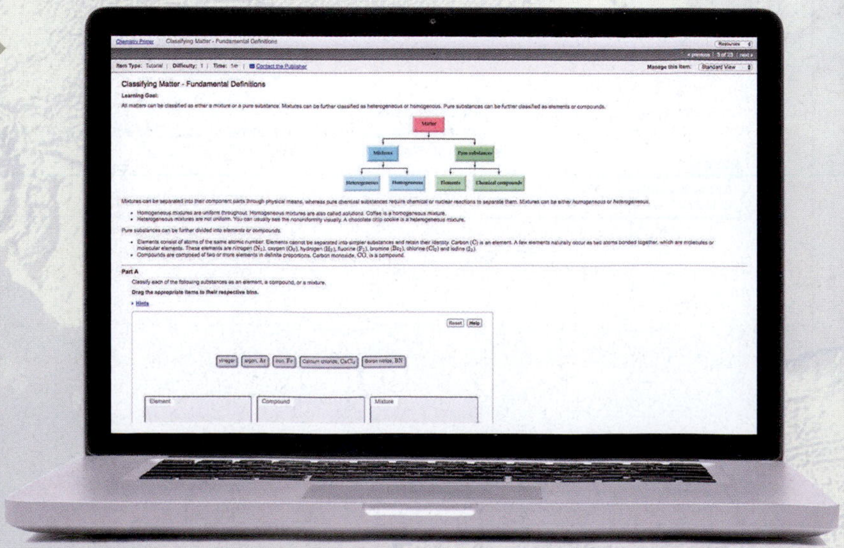
Classifying Matter - Fundamental Definitions

with Mastering Chemistry

DURING CLASS

Learning Catalytics generates class discussion, guides your lecture, and promotes peer-to-peer learning with real-time analytics.

MasteringChemistry™ with eText now provides Learning Catalytics—an interactive student response tool that uses students' smartphones, tablets, or laptops to engage them in more sophisticated tasks and thinking. Instructors can:

- Upload a full PowerPoint® deck for easy creation of slide questions.
- Help students develop critical-thinking skills.
- Monitor responses to find out where students are struggling.
- Rely on real-time data to adjust teaching strategies.
- Automatically group students for discussion, teamwork, and peer-to-peer learning.

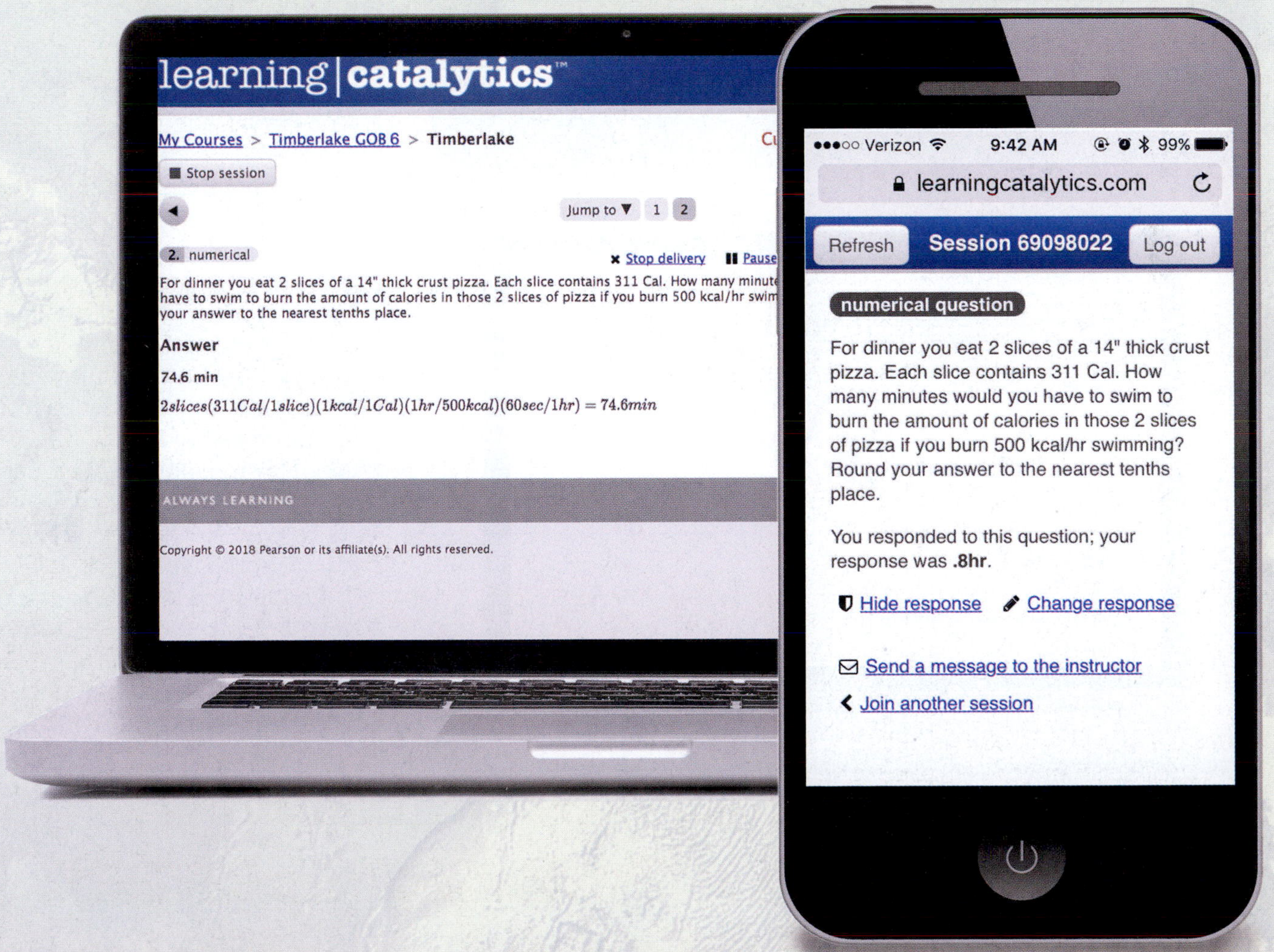
learning | catalytics™
My Courses > Timberlake GOB 6 > Timberlake
Stop session
Jump to ▼ 1 2
2. numerical
Stop delivery
Answer
74.6 min
2slices(311Cal/1slice)(1kcal/1Cal)(1hr/500kcal)(60sec/1hr) = 74.6min
ALWAYS LEARNING
Copyright © 2018 Pearson or its affiliate(s). All rights reserved.
Verizon 9:42 AM 99%
learningcatalytics.com
Refresh
Session 69098022
Log out
numerical question
For dinner you eat 2 slices of a 14" thick crust pizza. Each slice contains 311 Cal. How many minutes would you have to swim to burn the amount of calories in those 2 slices of pizza if you burn 500 kcal/hr swimming? Round your answer to the nearest tenths place.
You responded to this question; your response was .8hr.
Hide response
Change response
Send a message to the instructor
Join another session

Mastering Chemistry

AFTER CLASS

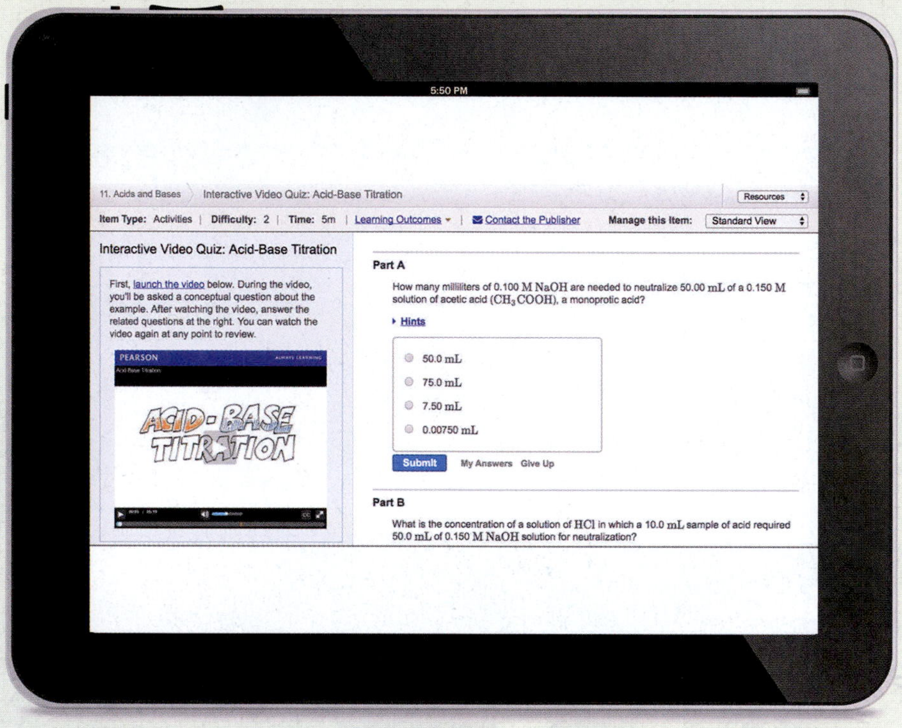
5:50 PM
11. Acids and Bases
Interactive Video Quiz: Acid-Base Titration
Resources
Item Type: Activities | Difficulty: 2 | Time: 5m | Learning Outcomes | Contact the Publisher
Manage this Item: Standard View
Interactive Video Quiz: Acid-Base Titration
First, launch the video below. During the video, you'll be asked a conceptual question about the example. After watching the video, answer the related questions at the right. You can watch the video again at any point to review.
PEARSON
ACID-BASE TITRATION
Part A
How many milliliters of 0.100 M NaOH are needed to neutralize 50.00 mL of a 0.150 M solution of acetic acid (CH3COOH), a monoprotic acid?
Hints
50.0 mL
75.0 mL
7.50 mL
0.00750 mL
Submit
My Answers
Give Up
Part B
What is the concentration of a solution of HCl in which a 10.0 mL sample of acid required 50.0 mL of 0.150 M NaOH solution for neutralization?

TEN NEW! Interactive Videos indicated by icons in the margins, have been created. These videos give students an opportunity to connect what they just learned by showing how chemistry works in real life and introducing a bit of humor into chemical problem solving and demonstrations. **Sample Calculations** walk students through the most challenging chemistry problems and provide a successful strategy on how to approach problem solving. Topics include: Using Conversion Factors, Mass Calculations for Reactions, Concentration of Solutions, and Acid–Base Titration.

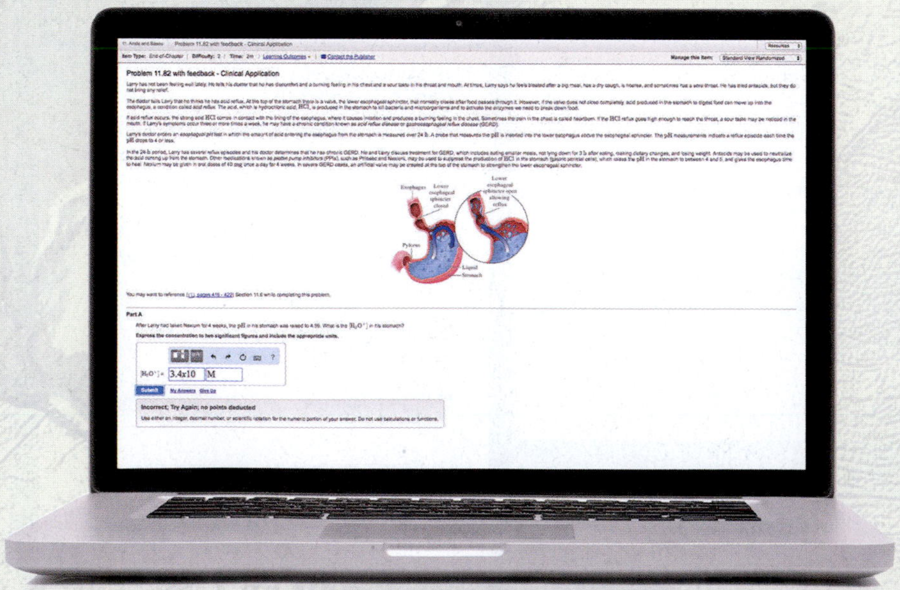
Problem 11.82 with feedback - Clinical Application

MasteringChemistry™ offers a wide variety of problems, ranging from multi-step tutorials with extensive hints and feedback to multiple-choice **End-of-Chapter Problems** and **Test Bank** questions. To provide additional scaffolding for students moving from **Tutorial Problems** to End-of-Chapter Problems, we created New! **Enhanced End-of-Chapter Problems** that now contain specific wrong-answer feedback.

Pearson eText

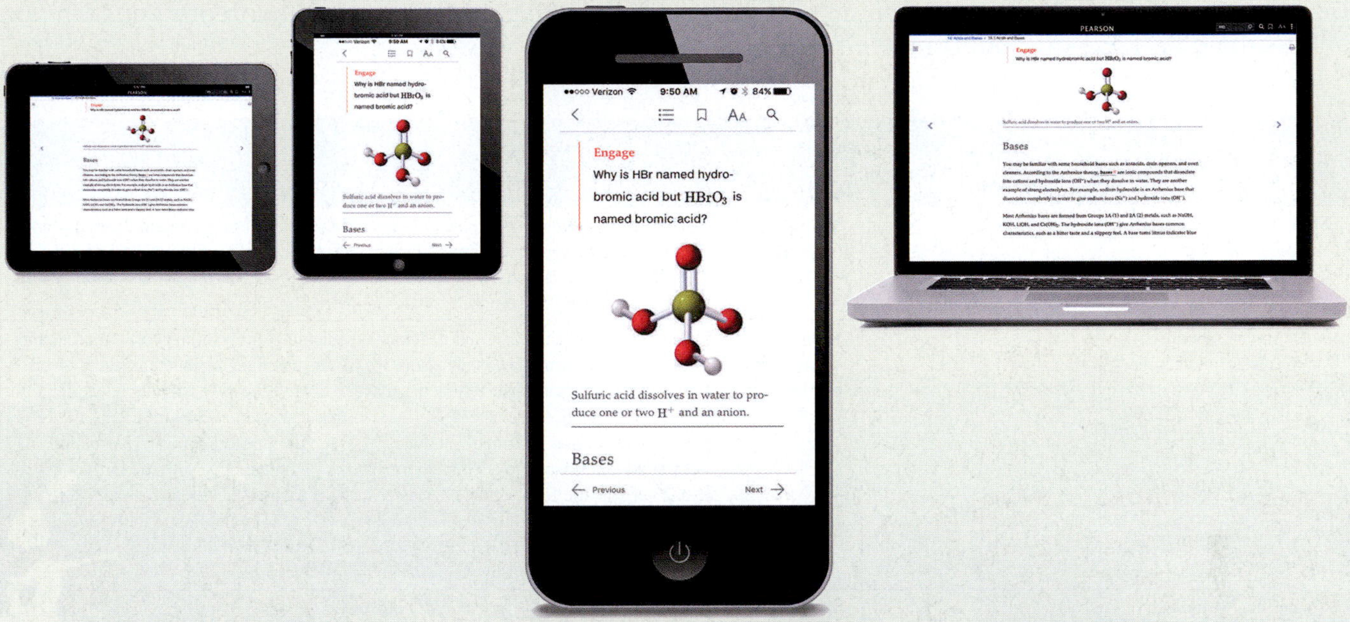

Verizon
9:50 AM
84%
Engage
Why is HBr named hydro-bromic acid but HBrO3 is named bromic acid?
Sulfuric acid dissolves in water to pro-duce one or two H+ and an anion.
Bases
Previous
Next
PEARSON

NEW! Pearson eText 2.0, optimized for mobile, seamlessly integrates videos and other rich media with the text and gives students access to their textbook anytime, anywhere. Pearson eText 2.0 is available with MasteringChemistry™ when packaged with new books, or as an upgrade students can purchase online. The Pearson eText 2.0 mobile app offers:

- Offline access on most iOS and Android phones/tablets
- Accessibility (screen-reader ready)
- Configurable reading settings, including resizable type and night-reading mode
- Instructor and student note-taking, highlighting, bookmarking, and search tools
- Embedded videos for a more interactive learning experience

Instructor and Student Supplements

General, Organic, and Biological Chemistry: Structures of Life sixth edition, provides an integrated teaching and learning package of support material for both students and professors.

Name of Supplement	Available in Print	Available Online	Instructor or Student Supplement	Description
Study Guide and Selected Solutions Manual (9780134814735)	✓		Supplement for Students	The *Study Guide and Selected Solutions Manual*, by Karen Timberlake and Mark Quirie, promotes active learning through a variety of exercises with answers as well as practice tests that are connected directly to the learning goals of the textbook. Complete solutions to odd-numbered problems are included.
Mastering™ Chemistry (www.masteringchemistry.com) (9780134787312)		✓	Supplement for Students and Instructors	Mastering™ Chemistry from Pearson is the leading online homework, tutorial, and assessment system, designed to improve results by engaging students with powerful content. Instructors ensure students arrive ready to learn by assigning educationally effective content and encourage critical thinking and retention with in-class resources such as Learning Catalytics™. Students can further master concepts through traditional and adaptive homework assignments that provide hints and answer specific feedback. The Mastering™ gradebook records scores for all automatically-graded assignments in one place, while diagnostic tools give instructors access to rich data to assess student understanding and misconceptions. http://www.masteringchemistry.com.
Mastering™ Chemistry with Pearson eText (9780134813011)		✓	Supplement for Students	The sixth edition of *General, Organic, and Biological Chemistry: Structures of Life* features a Pearson eText enhanced with media within Mastering™ Chemistry. In conjunction with Mastering™ assessment capabilities, new **Interactive Videos** will improve student engagement and knowledge retention. Each chapter contains a balance of interactive animations, videos, sample calculations, and self-assessments/quizzes embedded directly in the eText. Additionally, the Pearson eText offers students the power to create notes, highlight text in different colors, create bookmarks, zoom, and view single or multiple pages.
Laboratory Manual by Karen Timberlake (9780321811851)	✓		Supplement for Students	This best-selling lab manual coordinates 35 experiments with the topics in *General, Organic, and Biological Chemistry: Structures of Life*, sixth edition, uses laboratory investigations to explore chemical concepts, develop skills of manipulating equipment, reporting data, solving problems, making calculations, and drawing conclusions.
Instructor's Solutions Manual-Download Only (9780134814773)		✓	Supplement for Instructors	Prepared by Mark Quirie, the Instructor's Solutions Manual highlights chapter topics, and includes answers and solutions for all Practice Problems in the text.
Instructor Resource Materials–Download Only (9780134814780)		✓	Supplement for Instructors	Includes all the art, photos, and tables from the book in JPEG format for use in classroom projection or when creating study materials and tests. In addition, the instructors can access modifiable PowerPoint™ lecture outlines. Also available are downloadable files of the Instructor's Solutions Manual. Visit the Pearson Education catalog page for Timberlake's *General, Organic, Biological Chemistry: Structures of Life*, sixth edition, at www.pearsonhighered.com to download available instructor supplements.
TestGen Test Bank–Download Only (9780134814766)		✓	Supplement for Instructors	Prepared by William Timberlake, this resource includes more than 1600 questions in multiple-choice, matching, true/false, and short-answer format.
Online Instructor Manual for Laboratory Manual (9780321812858)		✓	Supplement for Instructors	This manual contains answers to report sheet pages for the *Laboratory Manual* and a list of the materials needed for each experiment with amounts given for 20 students working in pairs, available for download at www.pearsonhighered.com.

1

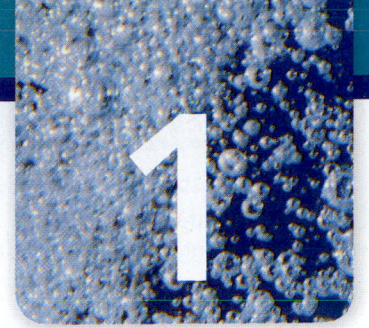

Chemistry in Our Lives

A call came in to 911 from a man who arrived home from work to find his wife Gloria lying on the floor of their living room. When the police arrived, they pronounced the woman dead. There was no blood at the scene, but the police did find a glass on the side table that contained a small amount of liquid. In an adjacent laundry room, the police found a half-empty bottle of antifreeze, which contains the toxic compound ethylene glycol. The bottle, glass, and liquid were bagged and sent to the forensic laboratory. At the morgue, the victim's height was measured as 1.573 m, and her mass was 40.5 kg.

In another 911 call, a man was found lying on the grass outside his home. Blood was present on his body, and some bullet casings were found on the grass. Inside the victim's home, a weapon was recovered. The bullet casings and the weapon were bagged and sent to the forensic laboratory.

Sarah, a forensic scientist, uses scientific procedures and chemical tests to examine the evidence from law enforcement agencies. She analyzes blood, stomach contents, and the unknown liquid from the first victim's home, as well as the fingerprints on the glass. She also looks for the presence of drugs, poisons, and alcohol. She will also match the characteristics of the bullet casings to the weapon that was found at the second crime scene.

CAREER

Forensic Scientist

Most forensic scientists work in crime laboratories that are part of city or county legal systems where they analyze bodily fluids and tissue samples collected by crime scene investigators. In analyzing these samples, forensic scientists identify the presence or absence of specific chemicals within the body to help solve the criminal case. Some of the chemicals they look for include alcohol, illegal or prescription drugs, poisons, arson debris, metals, and various gases such as carbon monoxide. To identify these substances, a variety of instruments and highly specific methodologies are used. Forensic scientists analyze samples from criminal suspects, athletes, and potential employees. They also work on cases involving environmental contamination and animal samples for wildlife crimes. Forensic scientists usually have a bachelor's degree that includes courses in math, chemistry, and biology.

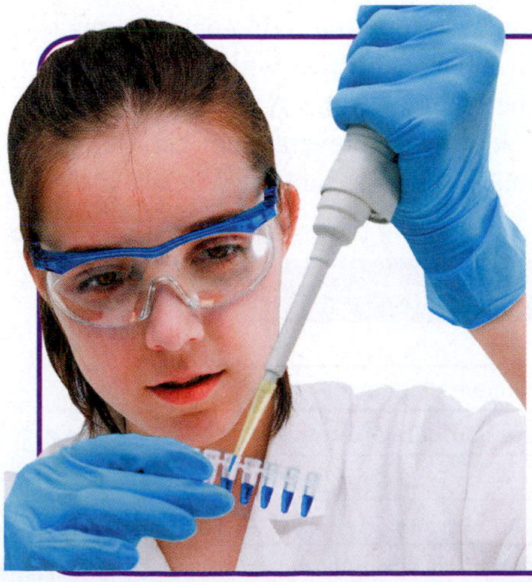

CLINICAL UPDATE

Forensic Evidence Helps Solve the Crime

In the forensic laboratory, Sarah analyzes the victim's stomach contents and blood for toxic compounds. You can view the results of the tests on the forensic evidence in the **CLINICAL UPDATE Forensic Evidence Helps Solve the Crime**, page 20, and determine if the victim ingested a toxic level of ethylene glycol (antifreeze).

LOOKING AHEAD

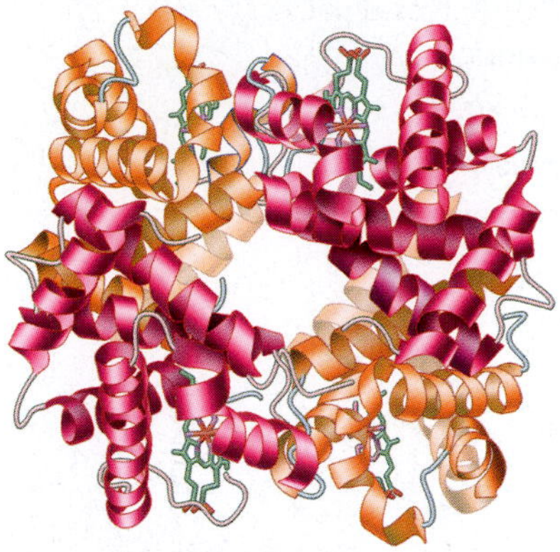

In the blood, hemoglobin transports oxygen to the tissues and carbon dioxide to the lungs.

Antacid tablets undergo a chemical reaction when dropped into water.

ENGAGE

Why is water a chemical?

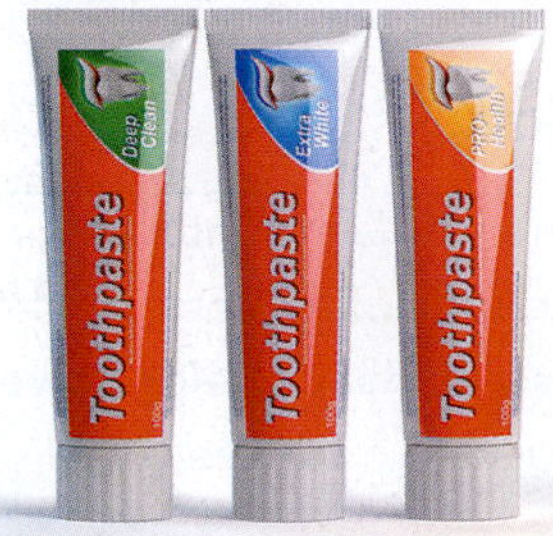
Toothpaste
Toothpaste
Toothpaste

Toothpaste is a combination of many chemicals.

TEST

Try Practice Problems 1.1 to 1.6

1.1 Chemistry and Chemicals

LEARNING GOAL Define the term chemistry, and identify chemicals.

Now that you are in a chemistry class, you may be wondering what you will be learning. What questions in science have you been curious about? Perhaps you are interested in what hemoglobin does in the blood or how aspirin relieves a headache. Just like you, chemists are curious about the world we live in.

What does hemoglobin do in the body? Hemoglobin consists of four polypeptide chains, each containing a heme group with an iron atom that binds to oxygen (O_2) in the lungs. From the lungs, hemoglobin transports oxygen to the tissues of the body, where it is used to provide energy. Once the oxygen is released, hemoglobin binds to carbon dioxide (CO_2) for transport to the lungs where it is released.

Why does aspirin relieve a headache? When a part of the body is injured, substances called prostaglandins are produced, which cause inflammation and pain. Aspirin acts to block the production of prostaglandins, reducing inflammation and pain. Chemists in the medical field develop new treatments for diabetes, genetic defects, cancer, AIDS, and other diseases. For the forensic scientist, the nurse, the dietitian, the chemical engineer, or the agricultural scientist, chemistry plays a central role in understanding problems and assessing possible solutions.

Chemistry

Chemistry is the study of the composition, structure, properties, and reactions of matter. *Matter* is another word for all the substances that make up our world. Perhaps you imagine that chemistry takes place only in a laboratory where a chemist is working in a white coat and goggles. Actually, chemistry happens all around you every day and has an impact on everything you use and do. You are doing chemistry when you cook food, add bleach to your laundry, or start your car. A chemical reaction has taken place when silver tarnishes or an antacid tablet fizzes when dropped into water. Plants grow because chemical reactions convert carbon dioxide, water, and energy to carbohydrates. Chemical reactions take place when you digest food and break it down into substances that you need for energy and health.

Chemicals

A **chemical** is a substance that always has the same composition and properties wherever it is found. All the things you see around you are composed of one or more chemicals. Often the terms *chemical* and *substance* are used interchangeably to describe a specific type of matter.

Every day, you use products containing substances that were developed and prepared by chemists. Soaps and shampoos contain chemicals that remove oils on your skin and scalp. In cosmetics and lotions, chemicals are used to moisturize, prevent deterioration of the product, fight bacteria, and thicken the product. Perhaps you wear a ring or watch made of gold, silver, or platinum. Your breakfast cereal is probably fortified with iron, calcium, and phosphorus, whereas the milk you drink is enriched with vitamins A and D. When you brush your teeth, the substances in toothpaste clean your teeth, prevent plaque formation, and stop tooth decay. Some of the chemicals used to make toothpaste are listed in **TABLE 1.1**.

TABLE 1.1 Chemicals Commonly Used in Toothpaste

Chemical	Function
Calcium carbonate	Used as an abrasive to remove plaque
Sorbitol	Prevents loss of water and hardening of toothpaste
Sodium lauryl sulfate	Used to loosen plaque
Titanium dioxide	Makes toothpaste white and opaque
Sodium fluorophosphate	Prevents formation of cavities by strengthening tooth enamel
Methyl salicylate	Gives toothpaste a pleasant wintergreen flavor

PRACTICE PROBLEMS

1.1 Chemistry and Chemicals

In every chapter, odd-numbered exercises in the *Practice Problems* are paired with even-numbered exercises. The answers for the magenta, odd-numbered *Practice Problems* are given at the end of each chapter. The complete solutions to the odd-numbered *Practice Problems* are in the *Study Guide and Student Solutions Manual*.

1.1 Write a one-sentence definition for each of the following:
a. chemistry **b.** chemical

1.2 Ask two of your friends (not in this class) to define the terms in problem 1.1. Do their answers agree with the definitions you provided?

Clinical Applications

1.3 Obtain a bottle of multivitamins, and read the list of ingredients. What are four chemicals from the list?

1.4 Obtain a box of breakfast cereal, and read the list of ingredients. What are four chemicals from the list?

1.5 Read the labels on some items found in your medicine cabinet. What are the names of some chemicals contained in those items?

1.6 Read the labels on products used to wash your dishes. What are the names of some chemicals contained in those products?

1.2 Scientific Method: Thinking Like a Scientist

LEARNING GOAL Describe the activities that are part of the scientific method.

When you were very young, you explored the things around you by touching and tasting. As you grew, you asked questions about the world in which you live. What is lightning? Where does a rainbow come from? Why is the sky blue? As an adult, you may have wondered how antibiotics work or why vitamins are important to your health. Every day, you ask questions and seek answers to organize and make sense of the world around you.

When the late Nobel Laureate Linus Pauling described his student life in Oregon, he recalled that he read many books on chemistry, mineralogy, and physics. "I mulled over the properties of materials: why are some substances colored and others not, why are some minerals or inorganic compounds hard and others soft?" He said, "I was building up this tremendous background of empirical knowledge and at the same time asking a great number of questions." Linus Pauling won two Nobel Prizes: the first, in 1954, was in chemistry for his work on the nature of chemical bonds and the determination of the structures of complex substances; the second, in 1962, was the Peace Prize.

Linus Pauling won the Nobel Prize in Chemistry in 1954.

The Scientific Method

The process of trying to understand nature is unique to each scientist. However, the **scientific method** is a process that scientists use to make observations in nature, gather data, and explain natural phenomena.

1. **Observations** The first step in the scientific method is to make observations about nature and ask questions about what you observe. When an observation always seems to be true, it may be stated as a *law* that predicts that behavior and is often measurable. However, a law does not explain that observation. For example, we can use the *Law of Gravity* to predict that if we drop our chemistry book, it would fall on the floor, but this law does not explain why our book falls.
2. **Hypothesis** A scientist forms a hypothesis, which gives a possible explanation of an observation or a law. The hypothesis must be stated in such a way that it can be tested by experiments.
3. **Experiments** To determine if a hypothesis is *true* or *false*, experiments are done to find a relationship between the hypothesis and the observations. The results of the experiments may confirm the hypothesis. However, if the experiments do not confirm the hypothesis, it is modified or discarded. Then new experiments will be designed to test the hypothesis.
4. **Conclusion/Theory** When the results of the experiments are analyzed, a conclusion is made about whether the hypothesis is *true* or *false*. When experiments give consistent results, the hypothesis may be stated to be true. Even then, the hypothesis continues to be tested and, based on new experimental results, may need to be modified or replaced. If many additional experiments by a group of scientists continue to support the hypothesis, it may become a *scientific theory*, which gives an explanation for the initial observations.

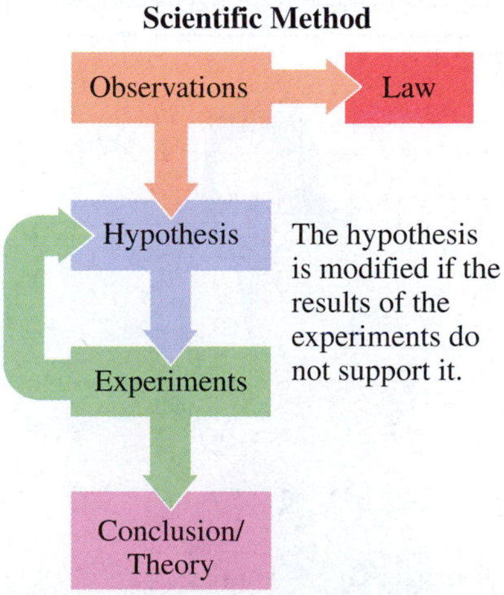
Scientific Method
Observations
Law
Hypothesis
The hypothesis is modified if the results of the experiments do not support it.
Experiments
Conclusion/ Theory

The scientific method develops a conclusion or theory about nature using observations, hypotheses, and experiments.

Chemistry Link to Health

Early Chemist: Paracelsus

For many centuries, chemistry has been the study of changes in matter. From the time of the ancient Greeks to the sixteenth century, alchemists described matter in terms of four components of nature: earth, air, fire, and water. By the eighth century, alchemists believed that they could change metals such as copper and lead into gold and silver. Although these efforts failed, the alchemists provided information on the chemical reactions involved in the extraction of metals from ores. The alchemists also designed some of the first laboratory equipment and developed early laboratory procedures. These early efforts were some of the first observations and experiments using the scientific method.

Paracelsus (1493–1541) was a physician and an alchemist who thought that alchemy should be about preparing new medicines. Using observation and experimentation, he proposed that a healthy body was regulated by a series of chemical processes that could be unbalanced by certain chemical compounds and rebalanced by using minerals and medicines. For example, he determined that inhaled dust caused lung disease in miners. He also thought that goiter was a problem caused by contaminated water, and he treated syphilis with compounds of mercury. His opinion of medicines was that the right dose makes the difference between a poison and a cure. Paracelsus changed alchemy in ways that helped establish modern medicine and chemistry.

Swiss physician and alchemist Paracelsus (1493–1541) believed that chemicals and minerals could be used as medicines.

Through observation you may think that you are allergic to cats.

ENGAGE

Why would the following statement "Today I placed two tomato seedlings in the garden, and two more in a closet. I will give all the plants the same amount of water and fertilizer." be considered an experiment?

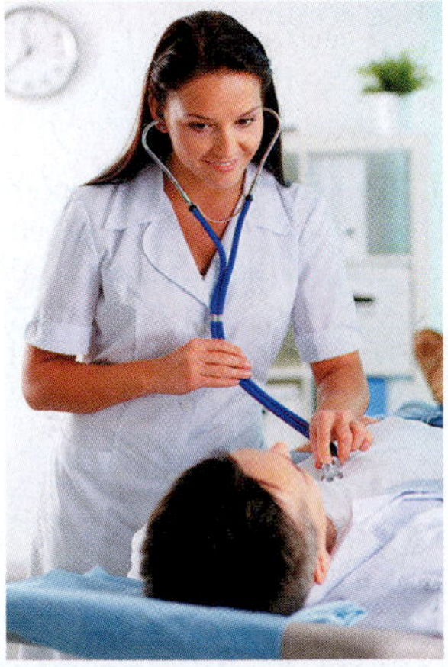

Nurses make observations in the hospital.

Using the Scientific Method in Everyday Life

You may be surprised to realize that you use the scientific method in your everyday life. Suppose you visit a friend in her home. Soon after you arrive, your eyes start to itch and you begin to sneeze. Then you observe that your friend has a new cat. Perhaps you form the hypothesis that you are allergic to cats. To test your hypothesis, you leave your friend's home. If the sneezing stops, perhaps your hypothesis is correct. You test your hypothesis further by visiting another friend who also has a cat. If you start to sneeze again, your experimental results support your hypothesis, and you come to the conclusion that you are allergic to cats. However, if you continue sneezing after you leave your friend's home, your hypothesis is not supported. Now you need to form a new hypothesis, which could be that you have a cold.

▶ SAMPLE PROBLEM 1.1 Scientific Method

TRY IT FIRST

Identify each of the following as an observation, a hypothesis, an experiment, or a conclusion:

a. During an assessment in the emergency room, a nurse writes that the patient has a resting pulse of 30 beats/min.
b. Repeated studies show that lowering sodium in the diet leads to a decrease in blood pressure.
c. A nurse thinks that an incision from a recent surgery that is red and swollen is infected.

SOLUTION

a. observation **b.** conclusion **c.** hypothesis

STUDY CHECK 1.1

Identify each of the following as an observation, a hypothesis, an experiment, or a conclusion:

a. Drinking coffee at night keeps me awake.
b. I will try drinking coffee only in the morning.
c. If I stop drinking coffee in the afternoon, I will be able to sleep at night.

d. When I drink decaffeinated coffee, I sleep better at night.
e. I am going to drink only decaffeinated coffee.
f. I sleep better at night because I stopped drinking caffeinated drinks.

ANSWER

a. observation **b.** experiment **c.** hypothesis
d. observation **e.** experiment **f.** conclusion

TEST
Try Practice Problems 1.7 to 1.10

PRACTICE PROBLEMS

1.2 Scientific Method: Thinking Like a Scientist

1.7 Identify each activity, **a** to **f**, as an observation, a hypothesis, an experiment, or a conclusion.

At a popular restaurant, where Chang is the head chef, the following occurred:

Customers rated the sesame seed dressing as the best.

a. Chang determined that sales of the house salad had dropped.
b. Chang decided that the house salad needed a new dressing.
c. In a taste test, Chang prepared four bowls of sliced cucumber, each with a new dressing: sesame seed, olive oil and balsamic vinegar, creamy Italian, and blue cheese.
d. Tasters rated the sesame seed salad dressing as the favorite.
e. After two weeks, Chang noted that the orders for the house salad with the new sesame seed dressing had doubled.
f. Chang decided that the sesame seed dressing improved the sales of the house salad because the sesame seed dressing enhanced the taste.

1.8 Identify each activity, **a** to **f**, as an observation, a hypothesis, an experiment, or a conclusion.

Lucia wants to develop a process for dyeing shirts so that the color will not fade when the shirt is washed. She proceeds with the following activities:

a. Lucia notices that the dye in a design fades when the shirt is washed.
b. Lucia decides that the dye needs something to help it combine with the fabric.
c. She places a spot of dye on each of four shirts and then places each one separately in water, salt water, vinegar, and baking soda and water.
d. After one hour, all the shirts are removed and washed with a detergent.
e. Lucia notices that the dye has faded on the shirts in water, salt water, and baking soda, whereas the dye did not fade on the shirt soaked in vinegar.
f. Lucia thinks that the vinegar binds with the dye so it does not fade when the shirt is washed.

Clinical Applications

1.9 Identify each of the following as an observation, a hypothesis, an experiment, or a conclusion:
a. One hour after drinking a glass of regular milk, Jim experienced stomach cramps.
b. Jim thinks he may be lactose intolerant.
c. Jim drinks a glass of lactose-free milk and does not have any stomach cramps.
d. Jim drinks a glass of regular milk to which he has added lactase, an enzyme that breaks down lactose, and has no stomach cramps.

1.10 Identify each of the following as an observation, a hypothesis, an experiment, or a conclusion:
a. Sally thinks she may be allergic to shrimp.
b. Yesterday, one hour after Sally ate a shrimp salad, she broke out in hives.
c. Today, Sally had some soup that contained shrimp, but she did not break out in hives.
d. Sally realizes that she does not have an allergy to shrimp.

1.3 Studying and Learning Chemistry

LEARNING GOAL Identify strategies that are effective for learning. Develop a study plan for learning chemistry.

Here you are taking chemistry, perhaps for the first time. Whatever your reasons for choosing to study chemistry, you can look forward to learning many new and exciting ideas.

Strategies to Improve Learning and Understanding

Success in chemistry utilizes good study habits, connecting new information with your knowledge base, rechecking what you have learned and what you have forgotten, and retrieving what you have learned for an exam. Let's take a look at ways that can help you study and learn chemistry. Suppose you were asked to indicate if you think each of the following common study habits is helpful or not helpful:

	Helpful	**Not helpful**
Highlighting		
Underlining		
Reading the chapter many times		
Memorizing the key words		
Testing practice		
Cramming		
Studying different ideas at the same time		
Retesting a few days later		

Learning chemistry requires us to place new information in our long-term memory, which allows us to remember those ideas for an exam, a process called retrieval. Thus, our study habits need to help us to recall knowledge. The study habits that are not very helpful in retrieval include highlighting, underlining, reading the chapter many times, memorizing key words, and cramming. If we want to recall new information, we need to connect it with prior knowledge. This can be accomplished by doing a lot of practice testing that requires us to retrieve new information. We can determine how much we have learned by going back a few days later and retesting. Another useful learning strategy is to study different ideas at the same time, which allows us to connect those ideas and to differentiate between them. Although these study habits may take more time and seem more difficult, they help us find the gaps in our knowledge and connect new information with what we already know.

Tips for Using New Study Habits for Successful Learning

1. **Do not keep rereading text or notes.** Reading the same material over and over will make that material seem familiar but does not mean that you have learned it. You need to test yourself to find out what you do and do not know.
2. **Ask yourself questions as you read.** Asking yourself questions as you read requires you to interact continually with new material. For example, you might ask yourself how the new material is related to previous material, which helps you make connections. By linking new material with long-term knowledge, you make pathways for retrieving new material.
3. **Self-test by giving yourself quizzes.** Using problems in the text or sample exams, practice taking tests frequently.
4. **Study at a regular pace rather than cramming.** Once you have tested yourself, go back in a few days and practice testing and retrieving information again. We do not recall all the information when we first read it. By frequent quizzing and retesting, we identify what we still need to learn. Sleep is also important for strengthening the associations between newly learned information. Lack of sleep may interfere with retrieval of information as well. So staying up all night to cram for your chemistry exam is not a good idea. Success in chemistry is a combined effort to learn new information and then to retrieve that information when you need it for an exam.
5. **Study different topics in a chapter, and relate the new concepts to concepts you know.** We learn material more efficiently by relating it to information we already know. By increasing connections between concepts, we can retrieve information when we need it.

Helpful	**Not helpful**
Testing practice	Highlighting
Studying different ideas at the same time	Underlining
Retesting a few days later	Reading the chapter many times
	Memorizing the key words
	Cramming

ENGAGE

Why is self-testing helpful for learning new concepts?

▶ SAMPLE PROBLEM 1.2 Strategies for Learning Chemistry

TRY IT FIRST

Predict which student will obtain the best exam score.

a. Bill, who reads the chapter four times.
b. Jennifer, who reads the chapter two times and works all the problems at the end of each Section.
c. Mark, who reads the chapter the night before the exam.

SOLUTION

b. Jennifer, who reads the chapter two times and works all the problems at the end of each Section has interacted with the content in the chapter using self-testing to make connections between concepts and practicing retrieving information learned previously.

STUDY CHECK 1.2

What are two more ways that Jennifer could improve her retrieval of information?

ANSWER

Jennifer could wait two or three days and practice working the problems in each Section again to determine how much she has learned. Retesting strengthens connections between new and previously learned information for longer lasting memory and more efficient retrieval. She could also ask questions as she reads and try to study at a regular pace to avoid cramming.

Features in This Text That Help You Study and Learn Chemistry

This text has been designed with study features to complement your individual learning style. On the inside of the front cover is a periodic table of the elements. On the inside of the back cover are tables that summarize useful information needed throughout your study of chemistry. Each chapter begins with *Looking Ahead*, which outlines the topics in the chapter. At the beginning of each Section, a *Learning Goal* describes the topics to learn. *Review* icons in the margins refer to Key Math Skills or Core Chemistry Skills from previous chapters that relate to new material in the chapter. *Key Terms* are bolded when they first appear in the text and are summarized at the end of each chapter. They are also listed and defined in the comprehensive *Glossary and Index*, which appears at the end of the text. *Key Math Skills* and *Core Chemistry Skills* that are critical to learning chemistry are indicated by icons in the margin, and summarized at the end of each chapter.

Before you begin reading, obtain an overview of a chapter by reviewing the topics in Looking Ahead. As you prepare to read a Section of the chapter, look at the Section title, and turn it into a question. Asking yourself questions about new topics builds new connections to material you have already learned. For example, for Section 1.1, "Chemistry and Chemicals," you could ask, "What is chemistry?" or "What are chemicals?" At the beginning of each Section, a *Learning Goal* states what you need to understand. As you read the text, you will see *Engage* questions in the margin, which remind you to pause your reading and test yourself with a question related to the material.

Several *Sample Problems* are included in each chapter. The *Try It First* feature reminds you to work the problem before you look at the Solution. It is helpful to try to work a problem first because it helps you link what you know to what you need to learn. The *Analyze the Problem* feature includes *Given*, the information you have; *Need*, what you have to accomplish; and *Connect*, how you proceed. Sample Problems include a *Solution* that shows the steps you can use for problem solving. Work the associated *Study Check*, and compare your answer to the one provided.

At the end of each chapter Section, you will find a set of *Practice Problems* that allows you to apply problem solving immediately to the new concepts. Throughout each Section, *Test* icons remind you to solve the indicated Practice Problems as you study.

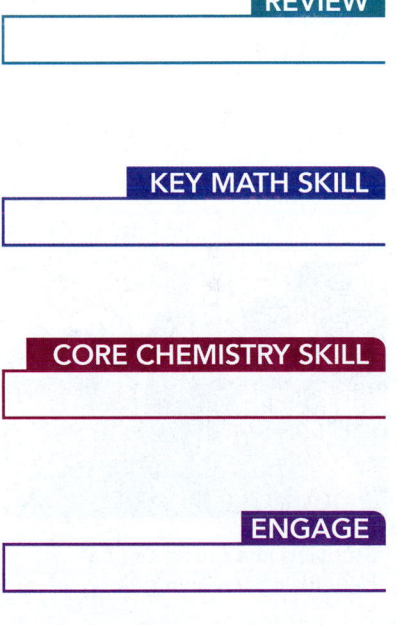

REVIEW

KEY MATH SKILL

CORE CHEMISTRY SKILL

ENGAGE

TRY IT FIRST

ANALYZE THE PROBLEM	Given	Need	Connect

TEST

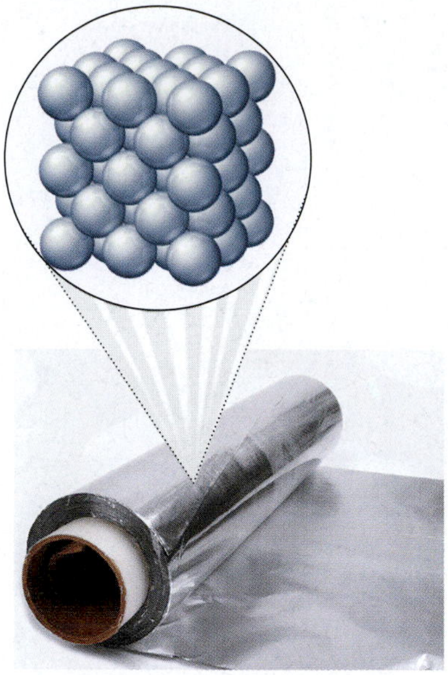

Illustrating the atoms of aluminum in aluminum foil is an example of macro-to-micro art.

INTERACTIVE VIDEO

The *Clinical Applications* in the Practice Problems relate the content to health and medicine. The problems are paired, which means that each of the odd-numbered problems is matched to the following even-numbered problem. At the end of each chapter, the answers to all the odd-numbered problems are provided. If the answers match yours, you most likely understand the topic; if not, you need to study the Section again.

Throughout each chapter, boxes titled *Chemistry Link to Health* and *Chemistry Link to the Environment* help you relate the chemical concepts you are learning to real-life situations. Many of the figures and diagrams use macro-to-micro illustrations to depict the atomic level of organization of ordinary objects, such as the atoms in aluminum foil. These visual models illustrate the concepts described in the text and allow you to "see" the world in a microscopic way. *Interactive Video* suggestions illustrate content as well as problem solving.

At the end of each chapter, you will find several study aids that complete the chapter. *Chapter Reviews* provide a summary in easy-to-read bullet points, and *Concept Maps* visually show the connections between important topics. *Understanding the Concepts* are problems that use art and models to help you visualize concepts and connect them to your background knowledge. *Additional Practice Problems* and *Challenge Problems* provide additional exercises to test your understanding of the topics in the chapter. *Answers* to all of the odd-numbered problems complete the chapter, allowing you to compare your answers to the ones provided.

After some chapters, problem sets called *Combining Ideas* test your ability to solve problems containing material from more than one chapter.

Many students find that studying with a group can be beneficial to learning. In a group, students motivate each other to study, fill in gaps, and correct misunderstandings by teaching and learning together. Studying alone does not allow the process of peer correction. In a group, you can cover the ideas more thoroughly as you discuss the reading and problem solve with other students.

Making a Study Plan

As you embark on your journey into the world of chemistry, think about your approach to studying and learning chemistry. You might consider some of the ideas in the following list. Check those ideas that will help you successfully learn chemistry. Commit to them now. *Your* success depends on *you*.

Studying in a group can be beneficial to learning.

My study plan for learning chemistry will include the following:

________ reading the chapter before class
________ going to class
________ reviewing the Learning Goals
________ keeping a problem notebook
________ reading the text
________ working the Test problems as I read each Section
________ answering the Engage questions
________ trying to work the Sample Problem before looking at the Solution
________ working the Practice Problems and checking answers
________ studying different topics at the same time
________ organizing a study group
________ seeing the professor during office hours
________ reviewing Key Math Skills and Core Chemistry Skills
________ attending review sessions
________ studying as often as I can

▶ SAMPLE PROBLEM 1.3 A Study Plan for Learning Chemistry

TRY IT FIRST

Which of the following activities should you include in your study plan for learning chemistry successfully?

a. reading the chapter over and over until you think you understand it
b. going to the professor's office hours
c. self-testing during and after reading each Section
d. waiting to study until the night before the exam
e. trying to work the Sample Problem before looking at the Solution
f. retesting on new information a few days later

SOLUTION

Your success in chemistry can be improved by:

b. going to the professor's office hours
c. self-testing during and after reading each Section
e. trying to work the Sample Problem before looking at the Solution
f. retesting on new information a few days later

STUDY CHECK 1.3

Which of the following will help you learn chemistry?

a. skipping review sessions
b. working problems as you read a Section
c. staying up all night before an exam
d. reading the assignment before class
e. highlighting the key ideas in the text

ANSWER

b and **d**

TEST

Try Practice Problems 1.11 to 1.14

PRACTICE PROBLEMS

1.3 Studying and Learning Chemistry

1.11 What are four things you can do to help yourself to succeed in chemistry?

1.12 What are four things that would make it difficult for you to learn chemistry?

1.13 A student in your class asks you for advice on learning chemistry. Which of the following might you suggest?
a. forming a study group
b. skipping class
c. asking yourself questions while reading the text
d. waiting until the night before an exam to study
e. answering the Engage questions

1.14 A student in your class asks you for advice on learning chemistry. Which of the following might you suggest?
a. studying different topics at the same time
b. not reading the text; it's never on the test
c. attending review sessions
d. working the problems again after a few days
e. keeping a problem notebook

1.4 Key Math Skills for Chemistry

LEARNING GOAL Review math concepts used in chemistry: place values, positive and negative numbers, percentages, solving equations, and interpreting graphs.

During your study of chemistry, you will work many problems that involve numbers. You will need various math skills and operations. We will review some of the key math skills that are particularly important for chemistry. As we move through the chapters, we will also reference the key math skills as they apply.

KEY MATH SKILL

Identifying Place Values

Identifying Place Values

For any number, we can identify the *place value* for each of the digits in that number. These place values have names such as the ones place (first place to the left of the decimal point) or the tens place (second place to the left of the decimal point). A premature baby has a mass of 2518 g. We can indicate the place values for the number 2518 as follows:

Digit	Place Value
2	thousands
5	hundreds
1	tens
8	ones

ENGAGE

In the number 8.034, how do you know the 0 is in the tenths place?

We also identify place values such as the tenths place (first place to the right of the decimal point) and the hundredths place (second place to the right of the decimal point). A silver coin has a mass of 6.407 g. We can indicate the place values for the number 6.407 as follows:

Digit	Place Value
6	ones
4	ten**ths**
0	hundred**ths**
7	thousand**ths**

Note that place values ending with the suffix *ths* refer to the decimal places to the right of the decimal point.

▶SAMPLE PROBLEM 1.4 Identifying Place Values

TRY IT FIRST

A bullet found at a crime scene has a mass of 15.24 g. What are the place values for each of the digits in the mass of the bullet?

SOLUTION

Digit	Place Value
1	tens
5	ones
2	tenths
4	hundredths

STUDY CHECK 1.4

Identify the place values for each of the following:

a. the victim's height of 1.573 m
b. the victim's mass of 40.5 kg

ANSWER

a.

Digit	Place Value
1	ones
5	tenths
7	hundredths
3	thousandths

b.

Digit	Place Value
4	tens
0	ones
5	tenths

TEST

Try Practice Problems 1.15 and 1.16

Using Positive and Negative Numbers in Calculations

KEY MATH SKILL

Using Positive and Negative Numbers in Calculations

A *positive number* is any number that is greater than zero and has a positive sign (+). Often the positive sign is understood and not written in front of the number. For example, the number +8 is usually written as 8. A *negative number* is any number that is less than zero and is written with a negative sign (−). For example, a negative eight is written as −8.

Multiplication and Division of Positive and Negative Numbers

When two positive numbers or two negative numbers are multiplied, the answer is positive (+).

$2 \times 3 = 6$ The + sign (+6) is understood.

$(-2) \times (-3) = 6$

When a positive number and a negative number are multiplied, the answer is negative (−).

$2 \times (-3) = -6$

$(-2) \times 3 = -6$

The rules for the division of positive and negative numbers are the same as the rules for multiplication. When two positive numbers or two negative numbers are divided, the answer is positive (+).

$$\frac{6}{3} = 2 \qquad \frac{-6}{-3} = 2$$

When a positive number and a negative number are divided, the answer is negative (−).

$$\frac{-6}{3} = -2 \qquad \frac{6}{-3} = -2$$

Addition of Positive and Negative Numbers

When positive numbers are added, the sign of the answer is positive.

$3 + 4 = 7$ The + sign (+7) is understood.

When negative numbers are added, the sign of the answer is negative.

$-3 + (-4) = -7$

When a positive number and a negative number are added, the smaller number is subtracted from the larger number, and the result has the same sign as the larger number.

$12 + (-15) = -3$

ENGAGE

Why does $-5 + 4 = -1$, whereas $-5 + (-4) = -9$?

Subtraction of Positive and Negative Numbers

When two numbers are subtracted, change the sign of the number to be subtracted and follow the rules for addition shown above.

$12 - (\mathbf{+5}) = 12 \mathbf{- 5} = 7$

$-12 - (\mathbf{-5}) = -12 \mathbf{+ 5} = -7$

TEST

Try Practice Problems 1.17 and 1.18

Calculator Operations

On your calculator, there are four keys that are used for basic mathematical operations. The change sign [+/−] key is used to change the sign of a number.

To practice these basic calculations on the calculator, work through the problem going from the left to the right, doing the operations in the order they occur. If your calculator has a change sign [+/−] key, a negative number is entered by pressing the number and then pressing the change sign [+/−] key. At the end, press the equals [=] key or ANS or ENTER.

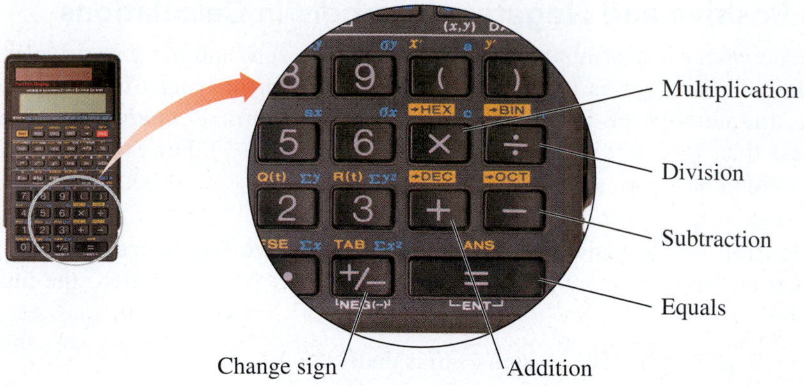
Multiplication
Division
Subtraction
Equals
Change sign
Addition

Addition and Subtraction		Multiplication and Division	
Example 1:	$15 - 8 + 2 =$	**Example 3:**	$2 \times (-3) =$
Solution:	15 [−] 8 [+] 2 [=] 9	**Solution:**	2 [×] 3 [+/−] [=] −6
Example 2:	$4 + (-10) - 5 =$	**Example 4:**	$\frac{8 \times 3}{4} =$
Solution:	4 [+] 10 [+/−] [−] 5 [=] −11	**Solution:**	8 [×] 3 [÷] 4 [=] 6

KEY MATH SKILL

Calculating Percentages

Calculating Percentages

To determine a percentage, divide the parts by the total (whole) and multiply by 100%. For example, if an aspirin tablet contains 325 mg of aspirin (active ingredient) and the tablet has a mass of 545 mg, what is the percentage of aspirin in the tablet?

$$\frac{325 \text{ mg aspirin}}{545 \text{ mg tablet}} \times 100\% = 59.6\% \text{ aspirin}$$

ENGAGE

Why is the value of 100% used in the calculation of a percentage?

When a value is described as a percentage (%), it represents the number of parts of an item in 100 of those items. If the percentage of red balls is 5, it means there are 5 red balls in every 100 balls. If the percentage of green balls is 50, there are 50 green balls in every 100 balls.

$$5\% \text{ red balls} = \frac{5 \text{ red balls}}{100 \text{ balls}} \qquad 50\% \text{ green balls} = \frac{50 \text{ green balls}}{100 \text{ balls}}$$

▶SAMPLE PROBLEM 1.5 Calculating a Percentage

TRY IT FIRST

A bullet found at a crime scene may be used as evidence in a trial if the percentage of metals is a match to the composition of metals in a bullet from the suspect's ammunition. Sarah's analysis of the bullet showed that it contains 13.9 g of lead, 0.3 g of tin, and 0.9 g of antimony. What is the percentage of each metal in the bullet? Express your answers to the ones place.

1

A bullet casing at a crime scene is marked as evidence.

SOLUTION

Total mass = 13.9 g + 0.3 g + 0.9 g = 15.1 g

Percentage of lead

$$\frac{13.9 \text{ g}}{15.1 \text{ g}} \times 100\% = 92\% \text{ lead}$$

Percentage of tin

$$\frac{0.3 \text{ g}}{15.1 \text{ g}} \times 100\% = 2\% \text{ tin}$$

Percentage of antimony

$$\frac{0.9 \text{ g}}{15.1 \text{ g}} \times 100\% = 6\% \text{ antimony}$$

STUDY CHECK 1.5

A bullet seized from the suspect's ammunition has a composition of lead 11.6 g, tin 0.5 g, and antimony 0.4 g.

a. What is the percentage of each metal in the bullet? Express your answers to the ones place.

b. Could the bullet removed from the suspect's ammunition be considered as evidence that the suspect was at the crime scene mentioned in Sample Problem 1.5?

ANSWER

a. The bullet from the suspect's ammunition is lead 93%, tin 4%, and antimony 3%.

b. The composition of this bullet does not match the bullet from the crime scene and cannot be used as supporting evidence.

TEST

Try Practice Problems 1.19 and 1.20

Solving Equations

KEY MATH SKILL

Solving Equations

In chemistry, we use equations that express the relationship between certain variables. Let's look at how we would solve for x in the following equation:

$$2x + 8 = 14$$

Our overall goal is to rearrange the items in the equation to obtain x on one side.

ENGAGE

Why is the number 8 subtracted from both sides of this equation?

1. *Place all like terms on one side*. The numbers 8 and 14 are like terms. To remove the 8 from the left side of the equation, we subtract 8. To keep a balance, we need to subtract 8 from the 14 on the other side.

$$2x + \cancel{8} - \cancel{8} = 14 - 8$$
$$2x = 6$$

2. *Isolate the variable you need to solve for*. In this problem, we obtain x by dividing both sides of the equation by 2. The value of x is the result when 6 is divided by 2.

$$\frac{\cancel{2}x}{\cancel{2}} = \frac{6}{2}$$
$$x = 3$$

3. *Check your answer*. Check your answer by substituting your value for x back into the original equation.

$$2(3) + 8 = 14$$
$$6 + 8 = 14$$
$$14 = 14 \quad \text{Your answer } x = 3 \text{ is correct.}$$

Summary: To solve an equation for a particular variable, be sure you perform the same mathematical operations on *both* sides of the equation.

If you eliminate a symbol or number by subtracting, you need to subtract that same symbol or number from both sides.

If you eliminate a symbol or number by adding, you need to add that same symbol or number to both sides.

If you cancel a symbol or number by dividing, you need to divide both sides by that same symbol or number.

If you cancel a symbol or number by multiplying, you need to multiply both sides by that same symbol or number.

When we work with temperature, we may need to convert between degrees Celsius and degrees Fahrenheit using the following equation:

$$T_F = 1.8(T_C) + 32$$

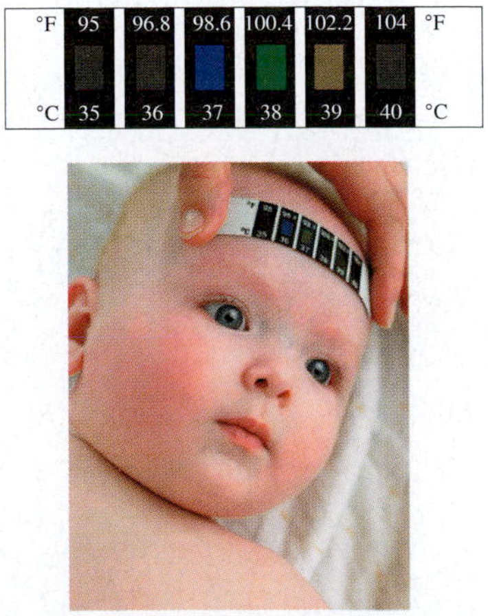
°F 95 96.8 98.6 100.4 102.2 104 °F
°C 35 36 37 38 39 40 °C

A plastic strip thermometer changes color to indicate body temperature.

To obtain the equation for converting degrees Fahrenheit to degrees Celsius, we subtract 32 from both sides.

$$T_F = 1.8(T_C) + 32$$
$$T_F - 32 = 1.8(T_C) + \cancel{32} - \cancel{32}$$
$$T_F - 32 = 1.8(T_C)$$

To obtain T_C by itself, we divide both sides by 1.8.

$$\frac{T_F - 32}{1.8} = \frac{\cancel{1.8}(T_C)}{\cancel{1.8}} = T_C$$

▶ SAMPLE PROBLEM 1.6 Solving Equations

TRY IT FIRST

Solve the following equation for V_2:

$P_1V_1 = P_2V_2$

INTERACTIVE VIDEO

Solving Equations

PEARSON eText 2.0

SOLUTION

$P_1V_1 = P_2V_2$

To solve for V_2, divide both sides by the symbol P_2.

$$\frac{P_1V_1}{P_2} = \frac{\cancel{P_2}V_2}{\cancel{P_2}}$$
$$V_2 = \frac{P_1V_1}{P_2}$$

ENGAGE

Why is the numerator divided by P_2 on both sides of the equation?

STUDY CHECK 1.6

Solve each of the following equations for m:

a. $\text{heat} = m \times \Delta T \times SH$ **b.** $D = \frac{m}{V}$

ANSWER

a. $m = \frac{\text{heat}}{\Delta T \times SH}$ **b.** $m = D \times V$

TEST

Try Practice Problems 1.21 and 1.22

KEY MATH SKILL

Interpreting Graphs

Interpreting Graphs

A graph is a diagram that represents the relationship between two variables. These quantities are plotted along two perpendicular axes, which are the *x* axis (horizontal) and *y* axis (vertical).

Example

In the graph Volume of a Balloon versus Temperature, the volume of a gas in a balloon is plotted against its temperature.

Title

Look at the title. What does it tell us about the graph? The title indicates that the volume of a balloon was measured at different temperatures.

Vertical Axis

Look at the label and the numbers on the vertical (*y*) axis. The label indicates that the volume of the balloon was measured in liters (L). The numbers, which are chosen to include the low and high measurements of the volume of the gas, are evenly spaced from 22.0 L to 30.0 L.

Horizontal Axis

The label on the horizontal (*x*) axis indicates that the temperatures of the balloon, measured in degrees Celsius (°C), are evenly spaced from 0 °C to 100 °C.

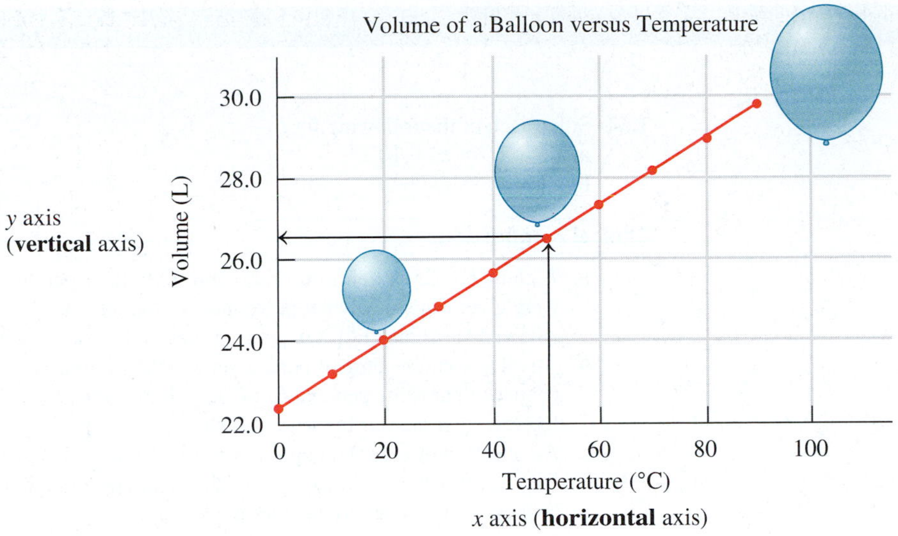

Volume of a Balloon versus Temperature
y axis
(vertical axis)
Volume (L)
30.0
28.0
26.0
24.0
22.0
0
20
40
60
80
100
Temperature (°C)
x axis (horizontal axis)

ENGAGE

Why are the numbers on the vertical and horizontal axes placed at regular intervals?

Points on the Graph

Each point on the graph represents a volume in liters that was measured at a specific temperature. When these points are connected, a line is obtained.

Interpreting the Graph

From the graph, we see that the volume of the gas increases as the temperature of the gas increases. This is called a *direct relationship*. Now we use the graph to determine the volume at various temperatures. For example, suppose we want to know the volume of the gas at 50 °C. We would start by finding 50 °C on the *x* axis and then drawing a line up to the plotted line. From there, we would draw a horizontal line that intersects the *y* axis and read the volume value where the line crosses the *y* axis as shown on the graph above.

▸SAMPLE PROBLEM 1.7 Interpreting a Graph

TRY IT FIRST

A nurse administers Tylenol to lower a child's fever. The graph shows the body temperature of the child plotted against time.

a. What is measured on the vertical axis?
b. What is the range of values on the vertical axis?
c. What is measured on the horizontal axis?
d. What is the range of values on the horizontal axis?

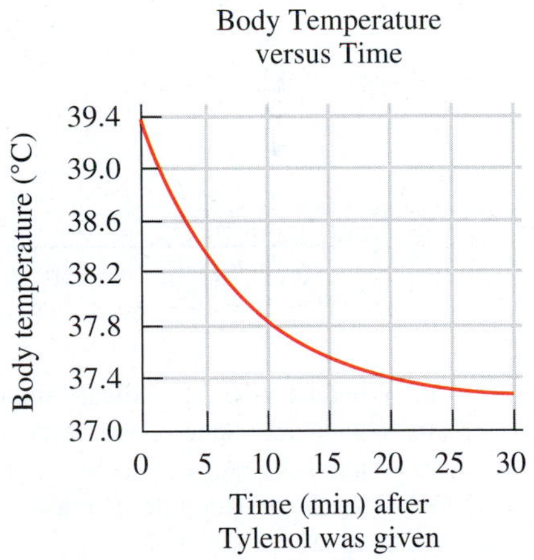

Body Temperature
versus Time
Body temperature (°C)
39.4
39.0
38.6
38.2
37.8
37.4
37.0
0 5 10 15 20 25 30
Time (min) after
Tylenol was given

SOLUTION

a. body temperature, in degrees Celsius
b. 37.0 °C to 39.4 °C
c. time, in minutes, after Tylenol was given
d. 0 min to 30 min

STUDY CHECK 1.7

a. Using the graph in Sample Problem 1.7, what was the child's temperature 15 min after Tylenol was given?
b. How many minutes elapsed for the temperature to decrease from 39.4 °C to 38.0 °C?
c. What was the decrease, in degrees Celsius, between 5 min and 20 min?

ANSWER

a. 37.6 °C **b.** 8 min **c.** 0.9 °C

TEST

Try Practice Problems 1.23 to 1.26

PRACTICE PROBLEMS

1.4 Key Math Skills for Chemistry

1.15 What is the place value for the bold digit?
a. 7.32**8**8
b. 1**6**.1234
c. 4**6**75.99

1.16 What is the place value for the bold digit?
a. 97.**5**689
b. 375.88
c. 46.1**0**00

1.17 Evaluate each of the following:
a. $15 - (-8) =$ ____
b. $-8 + (-22) =$ ____
c. $4 \times (-2) + 6 =$ ____

1.18 Evaluate each of the following:
a. $-11 - (-9) =$ _____
b. $34 + (-55) =$ _____
c. $\frac{-56}{8} =$ _____

Use the following graph for problems 1.19 and 1.20:

Temperature of Tea versus Time for Cooling

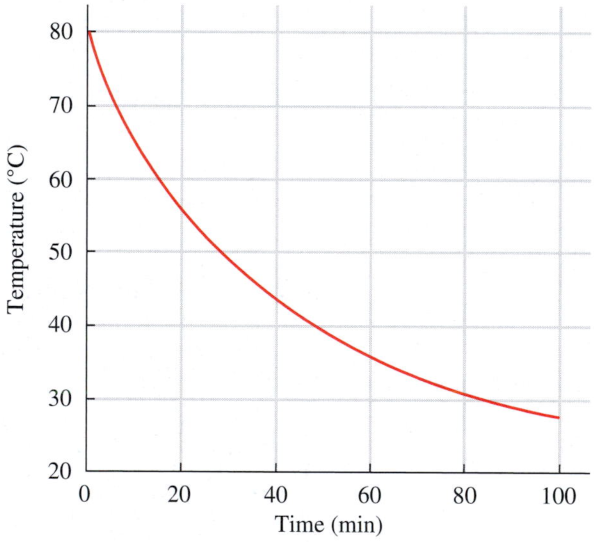
80
70
60
50
40
30
20
Temperature (°C)
0
20
40
60
80
100
Time (min)

1.19 **a.** What does the title indicate about the graph?
b. What is measured on the vertical axis?
c. What is the range of values on the vertical axis?
d. Does the temperature increase or decrease with an increase in time?

1.20 **a.** What is measured on the horizontal axis?
b. What is the range of values on the horizontal axis?
c. What is the temperature of the tea after 20 min?
d. How many minutes were needed to reach a temperature of 45 °C?

1.21 Solve each of the following for a:
a. $4a + 4 = 40$
b. $\frac{a}{6} = 7$

1.22 Solve each of the following for b:
a. $2b + 7 = b + 10$
b. $3b - 4 = 24 - b$

Clinical Applications

1.23 **a.** A clinic had 25 patients on Friday morning. If 21 patients were given flu shots, what percentage of the patients received flu shots? Express your answer to the ones place.
b. An alloy contains 56 g of pure silver and 22 g of pure copper. What is the percentage of silver in the alloy? Express your answer to the ones place.
c. A collection of coins contains 11 nickels, 5 quarters, and 7 dimes. What is the percentage of dimes in the collection? Express your answer to the ones place.

1.24 **a.** At a local hospital, 35 babies were born in May. If 22 were boys, what percentage of the newborns were boys? Express your answer to the ones place.
b. An alloy contains 67 g of pure gold and 35 g of pure zinc. What is the percentage of zinc in the alloy? Express your answer to the ones place.
c. A collection of coins contains 15 pennies, 14 dimes, and 6 quarters. What is the percentage of pennies in the collection? Express your answer to the ones place.

Use the following graph for problems 1.25 and 1.26:

Post-Mortem Body Temperature versus Time

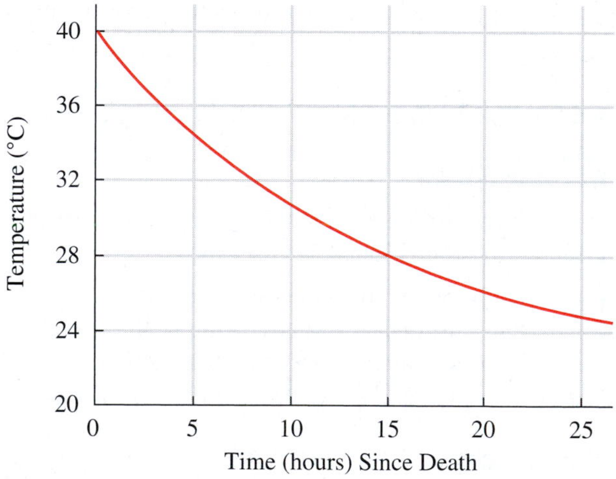
40
36
32
28
24
20
Temperature (°C)
0
5
10
15
20
25
Time (hours) Since Death

1.25 **a.** What does the title indicate about the graph?
b. What is measured on the vertical axis?
c. What is the range of values on the vertical axis?
d. Does the temperature increase or decrease with an increase in time?

1.26 **a.** What is measured on the horizontal axis?
b. What is the range of values on the horizontal axis?
c. How many hours were needed to reach a temperature of 28 °C?
d. The coroner measured Gloria's body temperature at 9 P.M. as 34 °C. What was the time of her death?

1.5 Writing Numbers in Scientific Notation

LEARNING GOAL Write a number in scientific notation.

In chemistry, we often work with numbers that are very large and very small. We might measure something as tiny as the width of a human hair, which is about 0.000 008 m. Or perhaps we want to count the number of hairs on the average human scalp, which is about 100 000 hairs. In this text, we add spaces between sets of three digits when it helps make the places easier to count. However, we will see that it is more convenient to write large and small numbers in *scientific notation*.

1×10^5 hairs
8×10^{-6} m

Humans have an average of 1×10^5 hairs on their scalps. Each hair is about 8×10^{-6} m wide.

Standard Number	Scientific Notation
0.000 008 m	8×10^{-6} m
100 000 hairs	1×10^5 hairs

A number written in **scientific notation** has two parts: a coefficient and a power of 10. For example, the number 2400 is written in scientific notation as 2.4×10^3. The coefficient, 2.4, is obtained by moving the decimal point to the left to give a number that is at least 1 but less than 10. Because we moved the decimal point three places to the left, the power of 10 is a positive 3, which is written as 10^3. When a number greater than 1 is converted to scientific notation, the power of 10 is positive.

KEY MATH SKILL
Writing Numbers in Scientific Notation

Standard Number — **Scientific Notation**

$$2400. = 2.4 \times 10^3$$

← 3 places; Coefficient; Power of 10

ENGAGE
Why is 530 000 written as 5.3×10^5 in scientific notation?

In another example, 0.000 86 is written in scientific notation as 8.6×10^{-4}. The coefficient, 8.6, is obtained by moving the decimal point to the right. Because the decimal point is moved four places to the right, the power of 10 is a negative 4, written as 10^{-4}. When a number less than 1 is written in scientific notation, the power of 10 is negative.

Standard Number — **Scientific Notation**

$$0.00086 = 8.6 \times 10^{-4}$$

4 places →; Coefficient; Power of 10

ENGAGE
Why is 0.000 053 written as 5.3×10^{-5} in scientific notation?

TABLE 1.2 gives some examples of numbers written as positive and negative powers of 10. The powers of 10 are a way of keeping track of the decimal point in the number. **TABLE 1.3** gives several examples of writing measurements in scientific notation.

TABLE 1.2 Some Powers of 10

Standard Number	Multiples of 10	Scientific Notation	
10 000	$10 \times 10 \times 10 \times 10$	1×10^4	Some positive powers of 10
1000	$10 \times 10 \times 10$	1×10^3	
100	10×10	1×10^2	
10	10	1×10^1	
1	0	1×10^0	
0.1	$\frac{1}{10}$	1×10^{-1}	Some negative powers of 10
0.01	$\frac{1}{10} \times \frac{1}{10} = \frac{1}{100}$	1×10^{-2}	
0.001	$\frac{1}{10} \times \frac{1}{10} \times \frac{1}{10} = \frac{1}{1000}$	1×10^{-3}	
0.0001	$\frac{1}{10} \times \frac{1}{10} \times \frac{1}{10} \times \frac{1}{10} = \frac{1}{10\,000}$	1×10^{-4}	

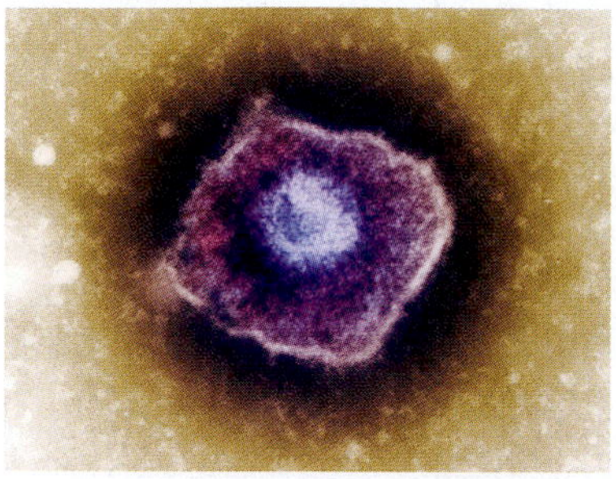

A chickenpox virus has a diameter of 3×10^{-7} m.

TABLE 1.3 Some Measurements Written as Standard Numbers and in Scientific Notation

Measured Quantity	Standard Number	Scientific Notation
Volume of gasoline used in the United States each year	550 000 000 000 L	5.5×10^{11} L
Diameter of Earth	12 800 000 m	1.28×10^7 m
Average volume of blood pumped in 1 day	8500 L	8.5×10^3 L
Time for light to travel from the Sun to Earth	500 s	5×10^2 s
Mass of a typical human	68 kg	6.8×10^1 kg
Mass of stirrup bone in ear	0.003 g	3×10^{-3} g
Diameter of a chickenpox (*Varicella zoster*) virus	0.000 000 3 m	3×10^{-7} m
Mass of bacterium (mycoplasma)	0.000 000 000 000 000 000 1 kg	1×10^{-19} kg

▶ SAMPLE PROBLEM 1.8 Writing a Number in Scientific Notation

TRY IT FIRST

Write each of the following in scientific notation:

a. 3500 **b.** 0.000 016

SOLUTION

ANALYZE THE PROBLEM	Given	Need	Connect
	standard number	scientific notation	coefficient is at least 1 but less than 10

a. 3500

STEP 1 **Move the decimal point to obtain a coefficient that is at least 1 but less than 10.** For a number greater than 1, the decimal point is moved to the left three places to give a coefficient of 3.5.

STEP 2 **Express the number of places moved as a power of 10.** Moving the decimal point three places to the left gives a power of 3, written as 10^3.

STEP 3 **Write the product of the coefficient multiplied by the power of 10.**

3.5×10^3

b. 0.000 016

STEP 1 **Move the decimal point to obtain a coefficient that is at least 1 but less than 10.** For a number less than 1, the decimal point is moved to the right five places to give a coefficient of 1.6.

STEP 2 **Express the number of places moved as a power of 10.** Moving the decimal point five places to the right gives a power of negative 5, written as 10^{-5}.

STEP 3 **Write the product of the coefficient multiplied by the power of 10.**
1.6×10^{-5}

STUDY CHECK 1.8

Write each of the following in scientific notation:

a. 425 000 **b.** 0.000 000 86 **c.** 0.007 30 **d.** 978×10^5

ANSWER

a. 4.25×10^5 **b.** 8.6×10^{-7} **c.** 7.30×10^{-3} **d.** 9.78×10^7

TEST

Try Practice Problems 1.27 and 1.28

Scientific Notation and Calculators

You can enter a number in scientific notation on many calculators using the [EE or EXP] key. After you enter the coefficient, press the [EE or EXP] key and enter the power 10. To enter a negative power of 10, press the [+/−] key or the [−] key, depending on your calculator.

Number to Enter	Procedure	Calculator Display
4×10^6	4 [EE or EXP] 6	4 06 or 4^{06} or 4E06
2.5×10^{-4}	2.5 [EE or EXP] [+/−] 4	2.5−04 or 2.5^{-04} or 2.5E−04

When a calculator answer appears in scientific notation, the coefficient is shown as a number that is at least 1 but less than 10, followed by a space or E and the power of 10. To express this display in scientific notation, write the coefficient value, write $\times$ 10, and use the power of 10 as an exponent.

ENGAGE

Describe how you enter a number in scientific notation on your calculator.

Calculator Display	Expressed in Scientific Notation
7.52 04 or 7.52^{04} or 7.52E04	7.52×10^4
5.8−02 or 5.8^{-02} or 5.8E−02	5.8×10^{-2}

On many calculators, a number is converted into scientific notation using the appropriate keys. For example, the number 0.000 52 is entered, followed by pressing the 2nd or 3rd function key (2nd F) and the SCI key. The scientific notation appears in the calculator display as a coefficient and the power of 10.

0.000 52 [2nd F] [SCI] = 5.2−04 or 5.2^{-04} or 5.2E−04 = 5.2×10^{-4}

Calculator display

PRACTICE PROBLEMS

1.5 Writing Numbers in Scientific Notation

1.27 Write each of the following in scientific notation:
a. 55 000 **b.** 480 **c.** 0.000 005
d. 0.000 14 **e.** 0.0072 **f.** 670 000

1.28 Write each of the following in scientific notation:
a. 180 000 000 **b.** 0.000 06 **c.** 750
d. 0.15 **e.** 0.024 **f.** 1500

1.29 Which number in each of the following pairs is larger?
a. 7.2×10^3 or 8.2×10^2 **b.** 4.5×10^{-4} or 3.2×10^{-2}
c. 1×10^4 or 1×10^{-4} **d.** 0.000 52 or 6.8×10^{-2}

1.30 Which number in each of the following pairs is smaller?
a. 4.9×10^{-3} or 5.5×10^{-9} **b.** 1250 or 3.4×10^2
c. 0.000 000 4 or 5.0×10^2 **d.** 2.50×10^2 or 4×10^5

CLINICAL UPDATE Forensic Evidence Helps Solve the Crime

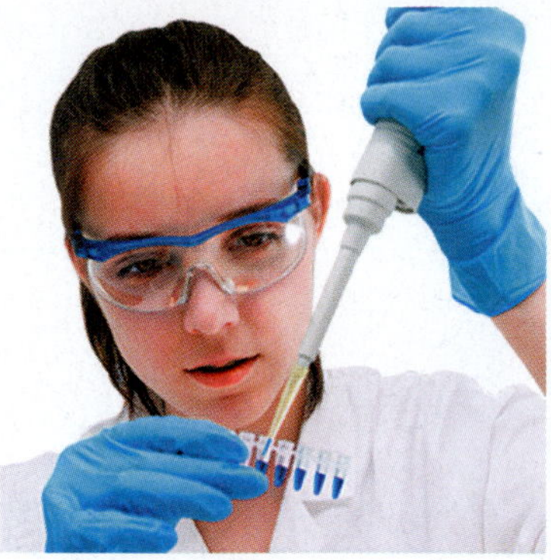

Using a variety of laboratory tests, Sarah finds ethylene glycol in the victim's blood. The quantitative tests indicate that the victim had ingested 125 g of ethylene glycol. Sarah determines that the liquid in a glass found at the crime scene was ethylene glycol that had been added to an alcoholic beverage. Ethylene glycol is a clear, sweet-tasting, thick liquid that is odorless and mixes with water. It is easy to obtain since it is used as antifreeze in automobiles and in brake fluid. Because the initial symptoms of ethylene glycol poisoning are similar to being intoxicated, the victim is often unaware of its presence.

If ingestion of ethylene glycol occurs, it can cause depression of the central nervous system, cardiovascular damage, and kidney failure. If discovered quickly, hemodialysis may be used to remove ethylene glycol from the blood. A toxic amount of ethylene glycol is 1.5 g of ethylene glycol/kg of body mass. Thus, 75 g could be fatal for a 50-kg (110-lb) person.

Sarah determines that fingerprints on the glass containing the ethylene glycol were those of the victim's husband. This evidence along with the container of antifreeze found in the home led to the arrest and conviction of the husband for poisoning his wife.

Clinical Applications

1.31 Identify each of the following comments in the police report as an observation, a hypothesis, an experiment, or a conclusion:

a. Gloria may have had a heart attack.
b. Test results indicate that Gloria was poisoned.
c. The liquid in the glass was analyzed.
d. The antifreeze in the pantry was the same color as the liquid in the glass.

1.32 Identify each of the following comments in the police report as an observation, a hypothesis, an experiment, or a conclusion:

a. Gloria may have committed suicide.
b. Sarah ran blood tests to identify any toxic substances.
c. The temperature of Gloria's body was 34 °C.
d. The fingerprints found on the glass were determined to be her husband's.

1.33 A container was found in the home of the victim that contained 120 g of ethylene glycol in 450 g of liquid. What was the percentage of ethylene glycol? Express your answer to the ones place.

1.34 If the toxic quantity is 1.5 g of ethylene glycol per 1000 g of body mass, what percentage of ethylene glycol is fatal?

CONCEPT MAP

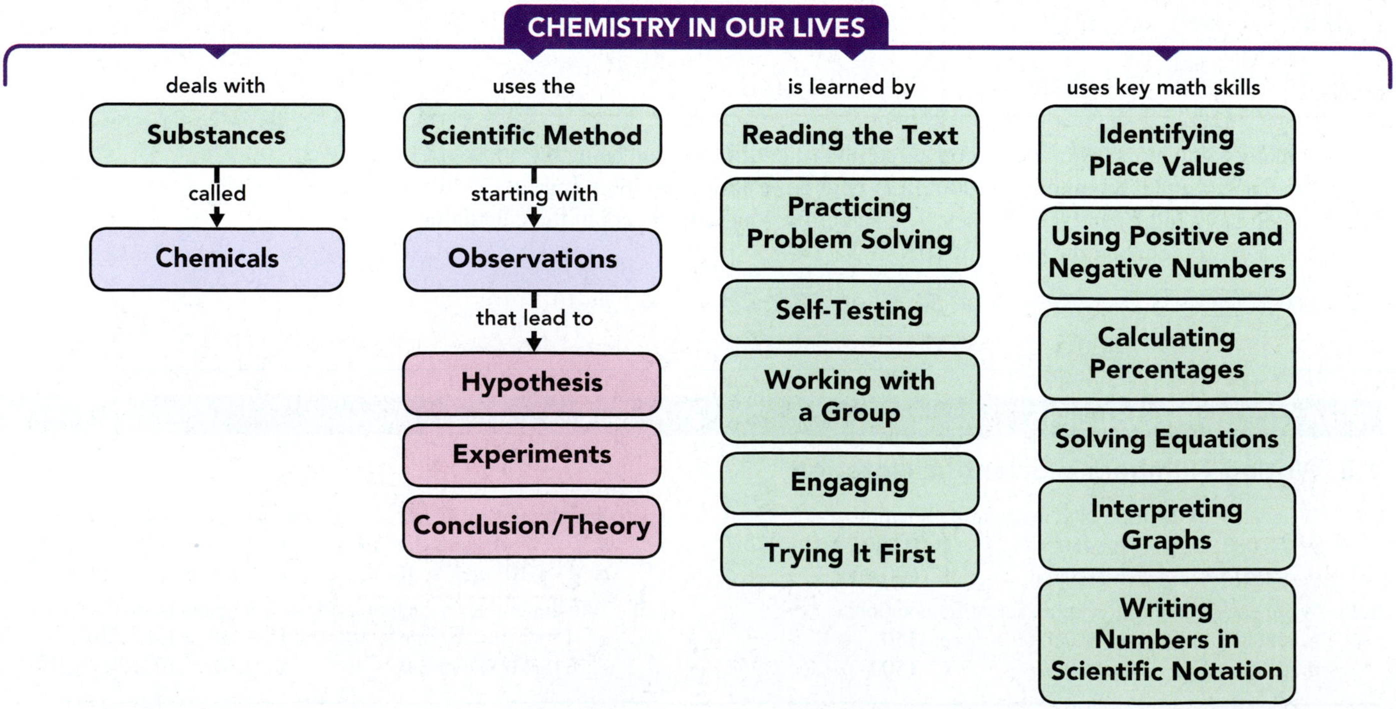

CHEMISTRY IN OUR LIVES
deals with
Substances
called
Chemicals
uses the
Scientific Method
starting with
Observations
that lead to
Hypothesis
Experiments
Conclusion/Theory
is learned by
Reading the Text
Practicing Problem Solving
Self-Testing
Working with a Group
Engaging
Trying It First
uses key math skills
Identifying Place Values
Using Positive and Negative Numbers
Calculating Percentages
Solving Equations
Interpreting Graphs
Writing Numbers in Scientific Notation

CHAPTER REVIEW

1.1 Chemistry and Chemicals

LEARNING GOAL Define the term chemistry, and identify chemicals.

- Chemistry is the study of the composition, structure, properties, and reactions of matter.
- A chemical always has the same composition and properties wherever it is found.

1.2 Scientific Method: Thinking Like a Scientist

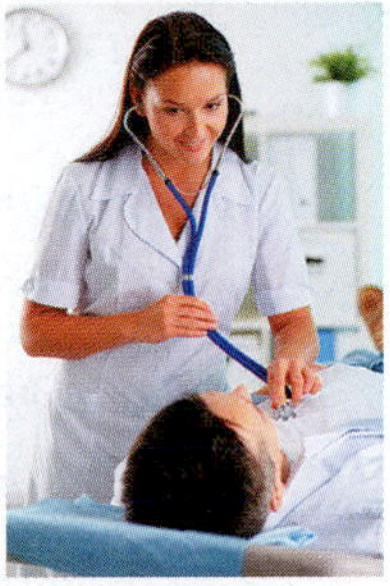

LEARNING GOAL Describe the activities that are part of the scientific method.

- The scientific method is a process of explaining natural phenomena beginning with making observations, forming a hypothesis, and performing experiments.
- After repeated successful experiments, a hypothesis may become a theory.

1.3 Studying and Learning Chemistry

LEARNING GOAL Identify strategies that are effective for learning. Develop a study plan for learning chemistry.

- A plan for learning chemistry utilizes the features in the text that help develop a successful approach to learning chemistry.
- By using the *Learning Goals*, *Reviews*, *Analyze the Problems*, *Try It First* in the chapter, and working the *Sample Problems*, *Study Checks*, and the *Practice Problems* at the end of each Section, you can successfully learn the concepts of chemistry.

1.4 Key Math Skills for Chemistry

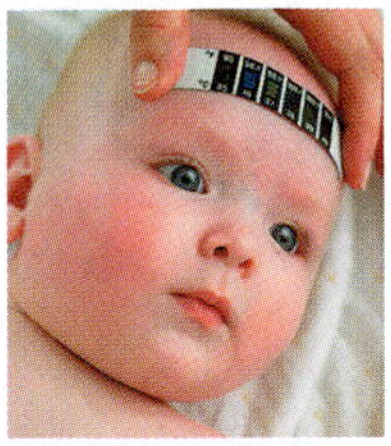

LEARNING GOAL Review math concepts used in chemistry: place values, positive and negative numbers, percentages, solving equations, and interpreting graphs.

- Solving chemistry problems involves a number of math skills: identifying place values, using positive and negative numbers, calculating percentages, solving equations, and interpreting graphs.

1.5 Writing Numbers in Scientific Notation

LEARNING GOAL Write a number in scientific notation.

1×10^5 hairs
8×10^{-6} m

- A number written in scientific notation has two parts, a coefficient and a power of 10.
- When a number greater than 1 is converted to scientific notation, the power of 10 is positive.
- When a number less than 1 is written in scientific notation, the power of 10 is negative.

KEY TERMS

chemical A substance that has the same composition and properties wherever it is found.

chemistry The study of the composition, structure, properties, and reactions of matter.

conclusion An explanation of an observation that has been validated by repeated experiments that support a hypothesis.

experiment A procedure that tests the validity of a hypothesis.

hypothesis An unverified explanation of a natural phenomenon.

observation Information determined by noting and recording a natural phenomenon.

scientific method The process of making observations, proposing a hypothesis, and testing the hypothesis; after repeated experiments validate the hypothesis, it may become a theory.

scientific notation A form of writing large and small numbers using a coefficient that is at least 1 but less than 10, followed by a power of 10.

theory An explanation for an observation supported by additional experiments that confirm the hypothesis.

KEY MATH SKILLS

The chapter Section containing each Key Math Skill is shown in parentheses at the end of each heading.

Identifying Place Values (1.4)

- The place value identifies the numerical value of each digit in a number.

Example: Identify the place value for each of the digits in the number 456.78.

Answer:

Digit	Place Value
4	hundreds
5	tens
6	ones
7	tenths
8	hundredths

Using Positive and Negative Numbers in Calculations (1.4)

- A *positive number* is any number that is greater than zero and has a positive sign (+). A *negative number* is any number that is less than zero and is written with a negative sign (−).
- When two positive numbers are added, multiplied, or divided, the answer is positive.
- When two negative numbers are multiplied or divided, the answer is positive. When two negative numbers are added, the answer is negative.
- When a positive and a negative number are multiplied or divided, the answer is negative.
- When a positive and a negative number are added, the smaller number is subtracted from the larger number and the result has the same sign as the larger number.
- When two numbers are subtracted, change the sign of the number to be subtracted, then follow the rules for addition.

Example: Evaluate each of the following:

a. $-8 - 14 =$ _____ **b.** $6 \times (-3) =$ _____

Answer: **a.** -22 **b.** -18

Calculating Percentages (1.4)

- A percentage is the part divided by the total (whole) multiplied by 100%.

Example: A drawer contains 6 white socks and 18 black socks. What is the percentage of white socks?

Answer: $\frac{6 \text{ white socks}}{24 \text{ total socks}} \times 100\% = 25\%$ white socks

Solving Equations (1.4)

An equation in chemistry often contains an unknown. To rearrange an equation to obtain the unknown factor by itself, you keep it balanced by performing matching mathematical operations on both sides of the equation.

- If you eliminate a number or symbol by subtracting, subtract that same number or symbol from both sides.
- If you eliminate a number or symbol by adding, add that same number or symbol to both sides.
- If you cancel a number or symbol by dividing, divide both sides by that same number or symbol.
- If you cancel a number or symbol by multiplying, multiply both sides by that same number or symbol.

Example: Solve the equation for a: $3a - 8 = 28$

Answer: *Add 8 to both sides* $3a - \cancel{8} + \cancel{8} = 28 + 8$

$3a = 36$

Divide both sides by 3 $\frac{\cancel{3}a}{\cancel{3}} = \frac{36}{3}$

$a = 12$

Check: $3(12) - 8 = 28$

$36 - 8 = 28$

$28 = 28$

Your answer $a = 12$ is correct.

Interpreting Graphs (1.4)

- A graph represents the relationship between two variables.
- The quantities are plotted along two perpendicular axes, which are the x axis (horizontal) and y axis (vertical).
- The title indicates the components of the x and y axes.
- Numbers on the x and y axes show the range of values of the variables.
- The graph shows the relationship between the component on the y axis and that on the x axis.

Example:

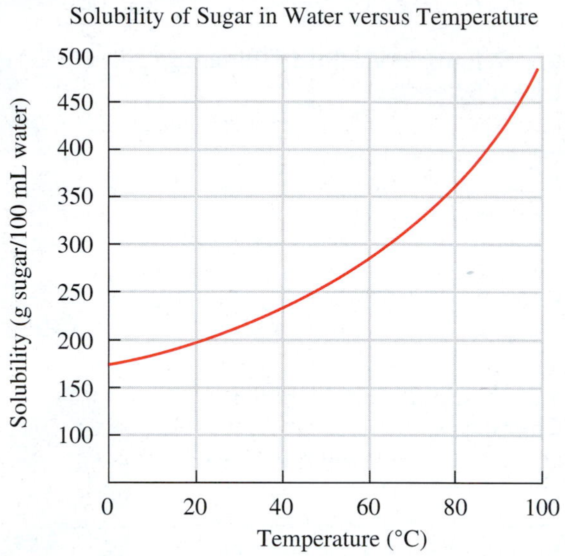

Solubility of Sugar in Water versus Temperature
500
450
400
350
300
250
200
150
100
Solubility (g sugar/100 mL water)
0
20
40
60
80
100
Temperature (°C)

a. Does the amount of sugar that dissolves in 100 mL of water increase or decrease when the temperature increases?

b. How many grams of sugar dissolve in 100 mL of water at 70 °C?

c. At what temperature (°C) will 275 g of sugar dissolve in 100 mL of water?

Answer: **a.** increase
b. 320 g
c. 55 °C

Writing Numbers in Scientific Notation (1.5)

- A number written in scientific notation consists of a coefficient and a power of 10.

A number is written in scientific notation by:

- Moving the decimal point to obtain a coefficient that is at least 1 but less than 10.
- Expressing the number of places moved as a power of 10. The power of 10 is positive if the decimal point is moved to the left, negative if the decimal point is moved to the right.

Example: Write the number 28 000 in scientific notation.

Answer: Moving the decimal point four places to the left gives a coefficient of 2.8 and a positive power of 10, 10^4. The number 28 000 written in scientific notation is 2.8×10^4.

UNDERSTANDING THE CONCEPTS

The chapter Sections to review are shown in parentheses at the end of each problem.

1.35 A "chemical-free" shampoo includes the following ingredients: water, cocamide, glycerin, and citric acid. Is the shampoo truly "chemical-free"? (1.1)

1.36 A "chemical-free" sunscreen includes the following ingredients: titanium dioxide, vitamin E, and vitamin C. Is the sunscreen truly "chemical-free"? (1.1)

1.37 According to Sherlock Holmes, "One must follow the rules of scientific inquiry, gathering, observing, and testing data, then formulating, modifying, and rejecting hypotheses, until only one remains." Did Holmes use the scientific method? Why or why not? (1.2)

1.38 In *A Scandal in Bohemia*, Sherlock Holmes receives a mysterious note. He states, "I have no data yet. It is a capital mistake to theorize before one has data. Insensibly one begins to twist facts to suit theories, instead of theories to suit facts." What do you think Holmes meant? (1.2)

1.39 For each of the following, indicate if the answer has a positive or negative sign: (1.4)
- **a.** Two negative numbers are added.
- **b.** A positive and negative number are multiplied.

1.40 For each of the following, indicate if the answer has a positive or negative sign: (1.4)
- **a.** A negative number is subtracted from a positive number.
- **b.** Two negative numbers are divided.

Clinical Applications

1.41 Classify each of the following statements as an observation or a hypothesis: (1.2)
- **a.** A patient breaks out in hives after receiving penicillin.
- **b.** Dinosaurs became extinct when a large meteorite struck Earth and caused a huge dust cloud that severely decreased the amount of light reaching Earth.
- **c.** A patient's blood pressure was 110/75.

1.42 Classify each of the following statements as an observation or a hypothesis: (1.2)
- **a.** Analysis of 10 ceramic dishes showed that four dishes contained lead levels that exceeded federal safety standards.
- **b.** Marble statues undergo corrosion in acid rain.
- **c.** A child with a high fever and a rash may have chickenpox.

ADDITIONAL PRACTICE PROBLEMS

1.43 Select the correct phrase(s) to complete the following statement: If experimental results do not support your hypothesis, you should (1.2)
- **a.** pretend that the experimental results support your hypothesis
- **b.** modify your hypothesis
- **c.** do more experiments

1.44 Select the correct phrase(s) to complete the following statement: A hypothesis is confirmed when (1.2)
- **a.** one experiment proves the hypothesis
- **b.** many experiments validate the hypothesis
- **c.** you think your hypothesis is correct

1.45 Which of the following will help you develop a successful study plan? (1.3)
- **a.** skipping class and just reading the text
- **b.** working the Sample Problems as you go through a chapter
- **c.** self-testing
- **d.** reading through the chapter, but working the problems later

1.46 Which of the following will help you develop a successful study plan? (1.3)
- **a.** studying all night before the exam
- **b.** forming a study group and discussing the problems together
- **c.** working problems in a notebook for easy reference
- **d.** highlighting important ideas in the text

1.47 Evaluate each of the following: (1.4)
- **a.** $4 \times (-8) =$ _____
- **b.** $-12 - 48 =$ _____
- **c.** $\frac{-168}{-4} =$ _____

1.48 Evaluate each of the following: (1.4)
- **a.** $-95 - (-11) =$ _____
- **b.** $\frac{152}{-19} =$ _____
- **c.** $4 - 56 =$ _____

1.49 A bag of gumdrops contains 16 orange gumdrops, 8 yellow gumdrops, and 16 black gumdrops. (1.4)
- **a.** What is the percentage of yellow gumdrops? Express your answer to the ones place.
- **b.** What is the percentage of black gumdrops? Express your answer to the ones place.

1.50 On the first chemistry test, 12 students got As, 18 students got Bs, and 20 students got Cs. (1.4)
- **a.** What is the percentage of students who received Bs? Express your answer to the ones place.
- **b.** What is the percentage of students who received Cs? Express your answer to the ones place.

1.51 Write each of the following in scientific notation: (1.5)
- **a.** 120 000
- **b.** 0.000 000 34
- **c.** 0.066
- **d.** 2700

1.52 Write each of the following in scientific notation: (1.5)
- **a.** 0.0042
- **b.** 310
- **c.** 890 000 000
- **d.** 0.000 000 056

Clinical Applications

1.53 Identify each of the following as an observation, a hypothesis, an experiment, or a conclusion: (1.2)
- **a.** A patient has a high fever and a rash on her back.
- **b.** A nurse tells a patient that her baby who gets sick after drinking milk may be lactose intolerant.
- **c.** Numerous studies have shown that omega-3 fatty acids lower triglyceride levels.

1.54 Identify each of the following as an observation, a hypothesis, an experiment, or a conclusion: (1.2)
- **a.** Every spring, you have congestion and a runny nose.
- **b.** An overweight patient decides to exercise more to lose weight.
- **c.** Many research studies have linked obesity to heart disease.

1.55 How will each of the following increase your chance of success in chemistry? (1.3)
a. self-testing
b. forming a study group
c. reading an assignment before class

1.56 How will each of the following decrease your chance of success in chemistry? (1.3)
a. studying only the night before an exam
b. not going to class
c. not practicing the problems in the text

CHALLENGE PROBLEMS

The following problems are related to the topics in this chapter. However, they do not all follow the chapter order, and they require you to combine concepts and skills from several Sections. These problems will help you increase your critical thinking skills and prepare for your next exam.

1.57 Classify each of the following as an observation, a hypothesis, an experiment, or a conclusion: (1.2)
a. The bicycle tire is flat.
b. If I add air to the bicycle tire, it will expand to the proper size.
c. When I added air to the bicycle tire, it was still flat.
d. The bicycle tire has a leak in it.

1.58 Classify each of the following as an observation, a hypothesis, an experiment, or a conclusion: (1.2)
a. A big log in the fire does not burn well.
b. If I chop the log into smaller wood pieces, it will burn better.
c. The small wood pieces burn brighter and make a hotter fire.
d. The small wood pieces are used up faster than burning the big log.

1.59 Solve each of the following for x: (1.4)
a. $2x + 5 = 41$ **b.** $\frac{5x}{3} = 40$

1.60 Solve each of the following for z: (1.4)
a. $3z - (-6) = 12$ **b.** $\frac{4z}{-12} = -8$

Use the following graph for problems 1.61 and 1.62:

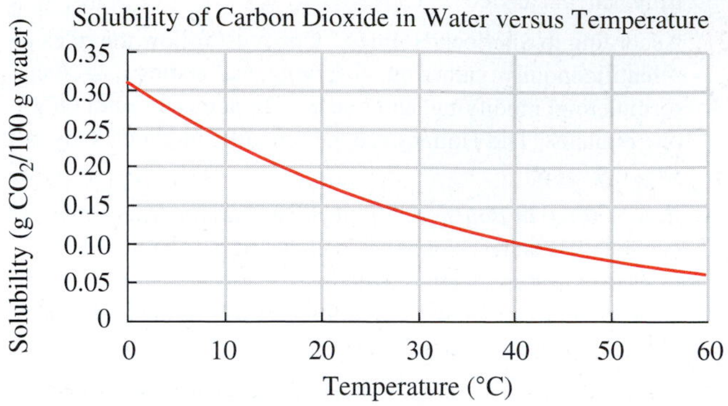
Solubility of Carbon Dioxide in Water versus Temperature
Solubility (g CO_2/100 g water)
0.35
0.30
0.25
0.20
0.15
0.10
0.05
0
0
10
20
30
40
50
60
Temperature (°C)

1.61 **a.** What does the title indicate about the graph? (1.4)
b. What is measured on the vertical axis?
c. What is the range of values on the vertical axis?
d. Does the solubility of carbon dioxide increase or decrease with an increase in temperature?

1.62 **a.** What is measured on the horizontal axis? (1.4)
b. What is the range of values on the horizontal axis?
c. What is the solubility of carbon dioxide in water at 25 °C?
d. At what temperature does carbon dioxide have a solubility of 0.20 g/100 g water?

ANSWERS

1.1 **a.** Chemistry is the study of the composition, structure, properties, and reactions of matter.
b. A chemical is a substance that has the same composition and properties wherever it is found.

1.3 Many chemicals are listed on a vitamin bottle such as vitamin A, vitamin B_3, vitamin B_{12}, vitamin C, and folic acid.

1.5 Typical items found in a medicine cabinet and some of the chemicals they contain are as follows:

Antacid tablets: calcium carbonate, cellulose, starch, stearic acid, silicon dioxide

Mouthwash: water, alcohol, thymol, glycerol, sodium benzoate, benzoic acid

Cough suppressant: menthol, beta-carotene, sucrose, glucose

1.7 **a.** observation **b.** hypothesis **c.** experiment
d. observation **e.** observation **f.** conclusion

1.9 **a.** observation **b.** hypothesis
c. experiment **d.** experiment

1.11 There are several things you can do that will help you successfully learn chemistry: forming a study group, retesting, doing Try It First before reading the Solution, checking Review, working Sample Problems and Study Checks, working Practice Problems and checking Answers, reading the assignment ahead of class, and keeping a problem notebook.

1.13 **a**, **c**, and **e**

1.15 **a.** thousandths **b.** ones **c.** hundreds

1.17 **a.** 23 **b.** −30 **c.** −2

1.19 **a.** 84% **b.** 72% **c.** 30%

1.21 **a.** 9 **b.** 42

1.23 **a.** The graph shows the relationship between the temperature of a cup of tea and time.
b. temperature, in °C
c. 20 °C to 80 °C
d. decrease

1.25 **a.** This graph shows the relationship between body temperature and time since death.
b. temperature, in °C
c. 20 °C to 40 °C
d. decrease

1.27 **a.** 5.5×10^4 **b.** 4.8×10^2 **c.** 5×10^{-6}
d. 1.4×10^{-4} **e.** 7.2×10^{-3} **f.** 6.7×10^5

1.29 **a.** 7.2×10^3 **b.** 3.2×10^{-2}
c. 1×10^4 **d.** 6.8×10^{-2}

1.31 **a.** hypothesis **b.** conclusion
c. experiment **d.** observation

1.33 27% ethylene glycol

1.35 No. All of the ingredients are chemicals.

1.37 Yes. Sherlock's investigation includes making observations (gathering data), formulating a hypothesis, testing the hypothesis, and modifying it until one of the hypotheses is validated.

1.39 **a.** negative **b.** negative

1.41 **a.** observation **b.** hypothesis **c.** observation

1.43 **b** and **c**

1.45 **b** and **c**

1.47 **a.** -32 **b.** -60 **c.** 42

1.49 **a.** 20% **b.** 40%

1.51 **a.** 1.2×10^5 **b.** 3.4×10^{-7}
c. 6.6×10^{-2} **d.** 2.7×10^3

1.53 **a.** observation **b.** hypothesis **c.** conclusion

1.55 **a.** Self-testing allows you to check on what you understand.
b. Forming a study group can motivate you to study, fill in gaps, and correct misunderstandings by teaching and learning together.
c. Reading the assignment before class prepares you to learn new material.

1.57 **a.** observation **b.** hypothesis
c. experiment **d.** conclusion

1.59 **a.** 18 **b.** 24

1.61 **a.** The graph shows the relationship between the solubility of carbon dioxide in water and temperature.
b. solubility of carbon dioxide (g CO_2/100 g water)
c. 0 to 0.35 g of CO_2/100 g of water
d. decrease

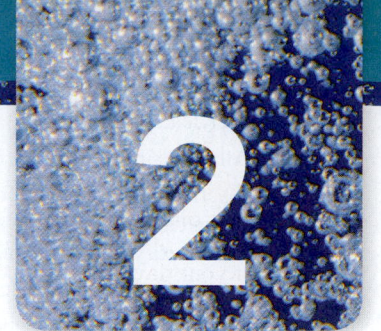

Chemistry and Measurements

2

During the past few months, Greg has had an increased number of headaches, dizzy spells, and nausea. He goes to his doctor's office, where Sandra, the registered nurse, completes the initial part of his exam by recording several measurements: weight 164 lb, height 5 ft 8 in., temperature 37.2 °C, and blood pressure 155/95. Normal blood pressure is 120/80 or lower.

When Greg sees his doctor, he is diagnosed with high blood pressure, or hypertension. The doctor prescribes 80. mg of Inderal (propranolol), which is available in 40.-mg tablets. Inderal is a beta blocker, which relaxes the muscles of the heart. It is used to treat hypertension, angina (chest pain), arrhythmia, and migraine headaches.

Two weeks later, Greg visits his doctor again, who determines that Greg's blood pressure is now 152/90. The doctor increases the dosage of Inderal to 160. mg. Sandra informs Greg that he needs to increase his daily dosage from two tablets to four tablets.

CAREER

Registered Nurse

In addition to assisting physicians, registered nurses work to promote patient health and prevent and treat disease. They provide patient care and help patients cope with illness. They take measurements such as a patient's weight, height, temperature, and blood pressure; make conversions; and calculate drug dosage rates. Registered nurses also maintain detailed medical records of patient symptoms and prescribed medications.

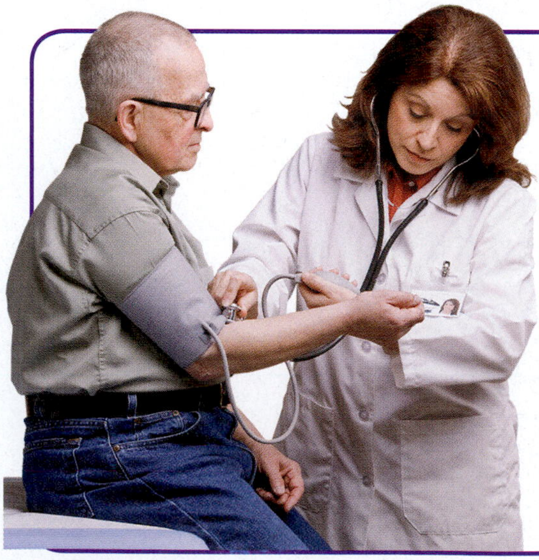

CLINICAL UPDATE

Greg's Visit with His Doctor

A few weeks later, Greg complains to his doctor that he is feeling tired. He has a blood test to determine if his iron level is low. You can see the results of Greg's blood serum iron level in the **CLINICAL UPDATE Greg's Visit with His Doctor**, page 54, and determine if Greg should be given an iron supplement.

2.1 Units of Measurement

LEARNING GOAL Write the names and abbreviations for the metric and SI units used in measurements of volume, length, mass, temperature, and time.

Think about your day. You probably took some measurements. Perhaps you checked your weight by stepping on a bathroom scale. If you made rice for dinner, you added two cups of water to one cup of rice. If you did not feel well, you may have taken your temperature. Whenever you take a measurement, you use a measuring device such as a scale, a measuring cup, or a thermometer.

Scientists and health professionals throughout the world use the **metric system** of measurement. It is also the common measuring system in all but a few countries in the world. The **International System of Units (SI)**, or Système International, is the official system of measurement throughout the world except for the United States. In chemistry, we use metric units and SI units for volume, length, mass, temperature, and time, as listed in **TABLE 2.1**.

LOOKING AHEAD

TABLE 2.1 Units of Measurement and Their Abbreviations

Measurement	Metric	SI
Volume	liter (L)	cubic meter (m^3)
Length	meter (m)	meter (m)
Mass	gram (g)	kilogram (kg)
Temperature	degree Celsius (°C)	kelvin (K)
Time	second (s)	second (s)

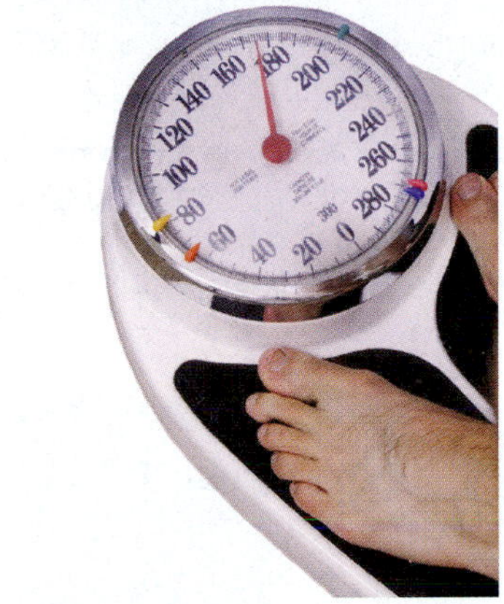

120
140
160
180
200
220
240
260
280
100
80
60
40
20
0

Your weight on a bathroom scale is a measurement.

Suppose you walked 1.3 mi to campus today, carrying a backpack that weighs 26 lb. The temperature was 72 °F. Perhaps you weigh 128 lb and your height is 65 in. These measurements and units may seem familiar to you because they are stated in the U.S. system of measurement. However, in chemistry, we use the *metric system* in making our measurements. Using the metric system, you walked 2.1 km to campus, carrying a backpack that has a mass of 12 kg, when the temperature was 22 °C. You have a mass of 58.0 kg and a height of 1.7 m.

1.7 m (65 in.)
22 °C (72 °F)
58.0 kg (128 lb)
12 kg (26 lb)
2.1 km (1.3 mi)

There are many measurements in everyday life.

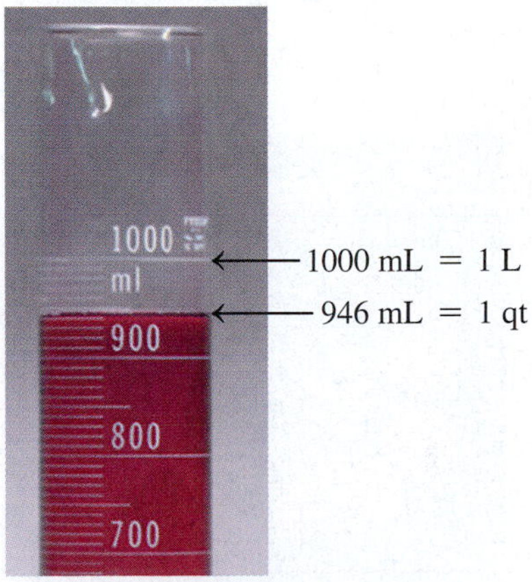

1000 ml
900
800
700
1000 mL = 1 L
946 mL = 1 qt

FIGURE 2.1 ▶ In the metric system, volume is based on the liter.

Q How many milliliters are in 1 quart?

Volume

Volume (*V*) is the amount of space a substance occupies. The metric unit for volume is the **liter (L)**, which is slightly larger than a quart (qt). In a laboratory or a hospital, chemists work with metric units of volume that are smaller and more convenient, such as the **milliliter (mL)**. There are 1000 mL in 1 L (see **FIGURE 2.1**). Some relationships between units for volume are

1 L = 1000 mL   1 L = 1.06 qt   946 mL = 1 qt

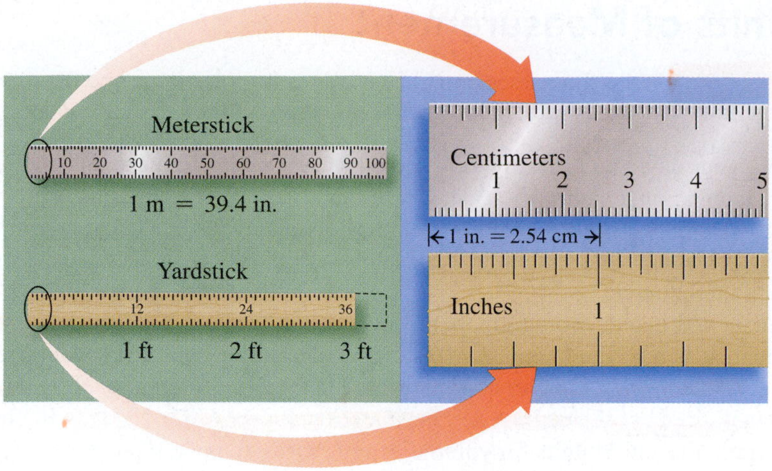
Meterstick
10 20 30 40 50 60 70 80 90 100
1 m = 39.4 in.
Yardstick
12 24 36
1 ft 2 ft 3 ft
Centimeters
1 2 3 4 5
1 in. = 2.54 cm
Inches
1

FIGURE 2.2 ▶ Length in the metric system (SI) is based on the meter, which is slightly longer than a yard.

Q How many centimeters are in a length of 1 inch?

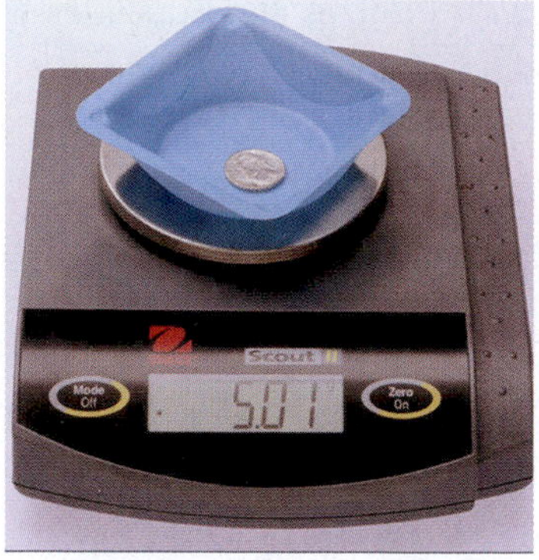
5.01

FIGURE 2.3 ▶ On an electronic balance, the digital readout gives the mass of a nickel, which is 5.01 g.

Q What is the mass of 10 nickels?

FIGURE 2.4 ▶ A thermometer is used to determine temperature.

Q What kinds of temperature readings have you made today?

A stopwatch is used to measure the time of a race.

Length

The metric and SI unit of length is the **meter (m)**. The **centimeter (cm)**, a smaller unit of length, is commonly used in chemistry and is about equal to the width of your little finger (see **FIGURE 2.2**). Some relationships between units for length are

1 m = 100 cm 1 m = 39.4 in. 1 m = 1.09 yd 2.54 cm = 1 in.

Mass

The **mass** of an object is a measure of the quantity of material it contains. The SI unit of mass, the **kilogram (kg)**, is used for larger masses such as body mass. In the metric system, the unit for mass is the **gram (g)**, which is used for smaller masses. There are 1000 g in 1 kg. One pound (lb) is equal to 454 g. Some relationships between units for mass are

1 kg = 1000 g 1 kg = 2.20 lb 454 g = 1 lb

You may be more familiar with the term *weight* than with mass. Weight is a measure of the gravitational pull on an object. On Earth, an astronaut with a mass of 75.0 kg has a weight of 165 lb. On the Moon, where the gravitational pull is one-sixth that of Earth, the astronaut has a weight of 27.5 lb. However, the mass of the astronaut is the same as on Earth, 75.0 kg. Scientists measure mass rather than weight because mass does not depend on gravity. In a chemistry laboratory, an electronic balance is used to measure the mass in grams of a substance (see **FIGURE 2.3**).

Temperature

Temperature tells us how hot something is, tells us how cold it is outside, or helps us determine if we have a fever (see **FIGURE 2.4**). In the metric system, temperature is measured using Celsius temperature. On the **Celsius (°C) temperature scale**, water freezes at 0 °C and boils at 100 °C, whereas on the Fahrenheit (°F) scale, water freezes at 32 °F and boils at 212 °F. In the SI system, temperature is measured using the **Kelvin (K) temperature scale** on which the lowest possible temperature is 0 K. A unit on the Kelvin scale is called a kelvin (K) and is not written with a degree sign.

Time

We typically measure time in units such as years (yr), days, hours (h), minutes (min), or seconds (s). Of these, the SI and metric unit of time is the **second (s)**. The standard now used to determine a second is an atomic clock. Some relationships between units for time are

1 day = 24 h 1 h = 60 min 1 min = 60 s

▶ SAMPLE PROBLEM 2.1 Units of Measurement

TRY IT FIRST

On a typical day, a nurse encounters several situations involving measurement. State the name and type of measurement indicated by the units in each of the following:

a. A patient has a temperature of 38.5 °C.
b. A physician orders 1.5 g of cefuroxime for injection.
c. A physician orders 1 L of a sodium chloride solution to be given intravenously.
d. A medication is to be given to a patient every 4 h.

SOLUTION

a. A degree Celsius is a unit of temperature.
b. A gram is a unit of mass.
c. A liter is a unit of volume.
d. An hour is a unit of time.

STUDY CHECK 2.1

State the names and types of measurements for an infant that is 54.6 cm long with a mass of 5.2 kg.

ANSWER

A centimeter (cm) is a unit of length. A kilogram (kg) is a unit of mass.

TEST

Try Practice Problems 2.1 to 2.8

PRACTICE PROBLEMS

2.1 Units of Measurement

2.1 Write the abbreviation for each of the following:
a. gram **b.** degree Celsius
c. liter **d.** pound
e. second

2.2 Write the abbreviation for each of the following:
a. kilogram **b.** kelvin
c. quart **d.** meter
e. centimeter

2.3 State the type of measurement in each of the following statements:
a. I put 12 L of gasoline in my gas tank.
b. My friend is 170 cm tall.
c. Earth is 385 000 km away from the Moon.
d. The horse won the race by 1.2 s.

2.4 State the type of measurement in each of the following statements:
a. I rode my bicycle 15 km today.
b. My dog weighs 12 kg.
c. It is hot today. It is 30 °C.
d. I added 2 L of water to my fish tank.

2.5 State the name of the unit and the type of measurement indicated for each of the following quantities:
a. 4.8 m **b.** 325 g **c.** 1.5 mL
d. 4.8×10^2 s **e.** 28 °C

2.6 State the name of the unit and the type of measurement indicated for each of the following quantities:
a. 0.8 L **b.** 3.6 cm **c.** 4 kg
d. 3.5 h **e.** 373 K

Clinical Applications

2.7 On a typical day, medical personnel may encounter several situations involving measurement. State the name and type of measurement indicated by the units in each of the following:
a. The clotting time for a blood sample is 12 s.
b. A premature baby weighs 2.0 kg.
c. An antacid tablet contains 1.0 g of calcium carbonate.
d. An infant has a temperature of 39.2 °C.

2.8 On a typical day, medical personnel may encounter several situations involving measurement. State the name and type of measurement indicated by the units in each of the following:
a. During open-heart surgery, the temperature of a patient is lowered to 29 °C.
b. The circulation time of a red blood cell through the body is 20 s.
c. A patient with a persistent cough is given 10. mL of cough syrup.
d. The amount of iron in the red blood cells of the body is 2.5 g.

REVIEW

Writing Numbers in Scientific Notation (1.5)

2.2 Measured Numbers and Significant Figures

LEARNING GOAL Identify a number as measured or exact; determine the number of significant figures in a measured number.

When you make a measurement, you use some type of measuring device. For example, you may use a meterstick to measure your height, a scale to check your weight, or a thermometer to take your temperature.

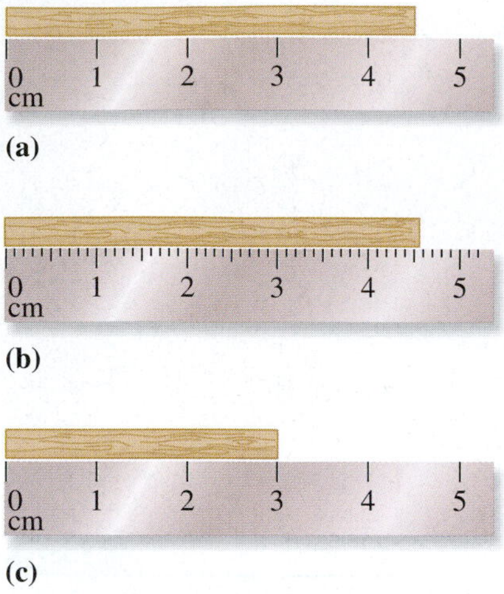

0 cm
1
2
3
4
5
(a)
(b)
(c)

FIGURE 2.5 ▶ The lengths of the rectangular objects are measured as **(a)** 4.5 cm and **(b)** 4.55 cm.

Q Why is the length of the object in (c) reported as 3.0 cm not 3 cm?

Measured Numbers

Measured numbers are the numbers you obtain when you measure a quantity such as your height, weight, or temperature. Suppose you are going to measure the lengths of the objects in **FIGURE 2.5**. To report the length of the object, you observe the numerical values of the marked lines at the end of the object. Then you can *estimate* by visually dividing the space between the marked lines. This estimated value is the final digit in a measured number.

For example, in Figure 2.5a, the end of the object is between the marks of 4 cm and 5 cm, which means that the length is more than 4 cm but less than 5 cm. If you estimate that the end of the object is halfway between 4 cm and 5 cm, you would report its length as 4.5 cm. Another student might report the length of the same object as 4.4 cm because people do not estimate in the same way.

The metric ruler shown in Figure 2.5b is marked at every 0.1 cm. Now you can determine that the end of the object is between 4.5 cm and 4.6 cm. Perhaps you report its length as 4.55 cm, whereas another student reports its length as 4.56 cm. Both results are acceptable.

In Figure 2.5c, the end of the object appears to line up with the 3-cm mark. Because the end of the object is on the 3-cm mark, the estimated digit is 0, which means the measurement is reported as 3.0 cm.

CORE CHEMISTRY SKILL

Counting Significant Figures

Significant Figures

In a measured number, the **significant figures (SFs)** are all the digits including the estimated digit. Nonzero numbers are always counted as significant figures. However, a zero may or may not be a significant figure depending on its position in a number. **TABLE 2.2** gives the rules and examples of counting significant figures.

TABLE 2.2 Significant Figures in Measured Numbers

Rule	Measured Number	Number of Significant Figures
1. A number is a *significant figure* if it is		
a. not a zero	4.5 g 122.35 m	2 5
b. a zero between nonzero digits	205 °C 5.008 kg	3 4
c. a zero at the end of a decimal number	50. L 16.00 mL	2 4
d. in the coefficient of a number written in scientific notation	4.8×10^5 m 5.70×10^{-3} g	2 3
2. A zero is *not significant* if it is		
a. at the beginning of a decimal number	0.0004 s 0.075 cm	1 2
b. used as a placeholder in a large number without a decimal point	850 000 m 1 250 000 g	2 3

TEST

Try Practice Problems 2.9 to 2.12

Significant Zeros and Scientific Notation

In this text, we will place a decimal point after a significant zero at the end of a number. For example, if a measurement is written as 500. g, the decimal point after the second zero indicates that *both zeros* are significant. To show this more clearly, we can write it as 5.00×10^2 g. When the first zero in the measurement 300 m is a significant zero, but the second zero is not, the measurement is written as 3.0×10^2 m. We will assume that all zeros at the end of large standard numbers without a decimal point are not significant. Therefore, we write 400 000 g as 4×10^5 g, which has only one significant figure.

ENGAGE

Why is the zero in the coefficient of 3.20×10^4 cm a significant figure?

TEST

Try Practice Problems 2.13 to 2.16

Exact Numbers

Exact numbers *are those numbers obtained by counting items or using a definition that compares two units in the same measuring system.* Suppose a friend asks you how many classes you are taking. You would answer by counting the number of classes in your schedule. Suppose you want to state the number of seconds in one minute. Without using any measuring device, you would give the definition: There are 60 s in 1 min. *Exact numbers are not measured, do not have a limited number of significant figures, and do not affect the number of significant figures in a calculated answer.* For more examples of exact numbers, see **TABLE 2.3**.

TABLE 2.3 Examples of Some Exact Numbers

Counted Numbers	Defined Equalities	
Items	Metric System	U.S. System
8 doughnuts	1 L = 1000 mL	1 ft = 12 in.
2 baseballs	1 m = 100 cm	1 qt = 4 cups
5 capsules	1 kg = 1000 g	1 lb = 16 oz

The number of baseballs is counted, which means 2 is an exact number.

For example, a mass of 42.2 g and a length of 5.0×10^{-3} cm are measured numbers because they are obtained using measuring tools. There are three SFs in 42.2 g because all nonzero digits are always significant. There are two SFs in 5.0×10^{-3} cm because all the digits in the coefficient of a number written in scientific notation are significant. However, a quantity of three eggs is an exact number that is obtained by counting. In the equality 1 kg = 1000 g, the masses of 1 kg and 1000 g are both exact numbers because this is a definition in the metric system.

▶SAMPLE PROBLEM 2.2 Measured and Exact Numbers

TRY IT FIRST

Identify each of the following numbers as measured or exact, and give the number of significant figures (SFs) in each of the measured numbers:

a. 0.170 L
b. 4 knives
c. 6.3×10^{-6} s
d. 1 m = 100 cm

SOLUTION

a. measured; three SFs
b. exact
c. measured; two SFs
d. exact

STUDY CHECK 2.2

Identify each of the following numbers as measured or exact and give the number of significant figures (SFs) in each of the measured numbers:

a. 0.020 80 kg
b. 5.06×10^4 h
c. 4 chemistry books
d. 85 600 s

ANSWER

a. measured; four SFs
b. measured; three SFs
c. exact
d. measured; three SFs

TEST

Try Practice Problems 2.17 to 2.22

PRACTICE PROBLEMS

2.2 Measured Numbers and Significant Figures

2.9 How many significant figures are in each of the following?
a. 11.005 g **b.** 0.000 32 m
c. 36 000 000 km **d.** 1.80×10^4 kg
e. 0.8250 L **f.** 30.0 °C

2.10 How many significant figures are in each of the following?
a. 20.60 mL **b.** 1036.48 kg
c. 4.00 m **d.** 20.8 °C
e. 60 800 000 g **f.** 5.0×10^{-3} L

2.11 In which of the following pairs do both numbers contain the same number of significant figures?
a. 11.0 m and 11.00 m
b. 0.0250 m and 0.205 m
c. 0.000 12 s and 12 000 s
d. 250.0 L and 2.5×10^{-2} L

2.12 In which of the following pairs do both numbers contain the same number of significant figures?
a. 0.005 75 g and 5.75×10^{-3} g
b. 405 K and 405.0 K
c. 150 000 s and 1.50×10^4 s
d. 3.8×10^{-2} L and 3.80×10^5 L

2.13 Indicate if the zeros are significant in each of the following measurements:
a. 0.0038 m **b.** 5.04 cm
c. 800. L **d.** 3.0×10^{-3} kg
e. 85 000 g

2.14 Indicate if the zeros are significant in each of the following measurements:
a. 20.05 °C **b.** 5.00 m
c. 0.000 02 g **d.** 120 000 yr
e. 8.05×10^2 L

2.15 Write each of the following in scientific notation with two significant figures:
a. 5000 L **b.** 30 000 g
c. 100 000 m **d.** 0.000 25 cm

2.16 Write each of the following in scientific notation with two significant figures:
a. 5 100 000 g **b.** 26 000 s
c. 40 000 m **d.** 0.000 820 kg

2.17 Identify the numbers in each of the following statements as measured or exact:
a. A patient has a mass of 67.5 kg.
b. A patient is given 2 tablets of medication.
c. In the metric system, 1 L is equal to 1000 mL.
d. The distance from Denver, Colorado, to Houston, Texas, is 1720 km.

2.18 Identify the numbers in each of the following statements as measured or exact:
a. There are 31 students in the laboratory.
b. The oldest known flower lived 1.20×10^8 yr ago.
c. The largest gem ever found, an aquamarine, has a mass of 104 kg.
d. A laboratory test shows a blood cholesterol level of 184 mg/dL.

2.19 Identify the measured number(s), if any, in each of the following pairs of numbers:
a. 3 hamburgers and 6 oz of hamburger
b. 1 table and 4 chairs
c. 0.75 lb of grapes and 350 g of butter
d. 60 s = 1 min

2.20 Identify the exact number(s), if any, in each of the following pairs of numbers:
a. 5 pizzas and 50.0 g of cheese
b. 6 nickels and 16 g of nickel
c. 3 onions and 3 lb of onions
d. 5 miles and 5 cars

Clinical Applications

2.21 Identify each of the following as measured or exact, and give the number of significant figures (SFs) in each measured number:
a. The mass of a neonate is 1.607 kg.
b. The Daily Value (DV) for iodine for an infant is 130 mcg.
c. There are 4.02×10^6 red blood cells in a blood sample.
d. In November, 23 babies were born in a hospital.

2.22 Identify each of the following as measured or exact, and give the number of significant figures (SFs) in each measured number:
a. An adult with the flu has a temperature of 103.5 °F.
b. A blister (push-through) pack of prednisone contains 21 tablets.
c. The time for a nerve impulse to travel from the feet to the brain is 0.46 s.
d. A brain contains 1.20×10^{10} neurons.

REVIEW
Identifying Place Values (1.4)
Using Positive and Negative Numbers in Calculations (1.4)

2.3 Significant Figures in Calculations

LEARNING GOAL Give the correct number of significant figures for a calculated answer.

In the sciences, we measure many things: the length of a bacterium, the volume of a gas sample, the temperature of a reaction mixture, or the mass of iron in a sample. The number of significant figures in measured numbers determines the number of significant figures in the calculated answer.

Using a calculator will help you perform calculations faster. However, calculators cannot think for you. It is up to you to enter the numbers correctly, press the correct function keys, and give the answer with the correct number of significant figures.

Rounding Off

KEY MATH SKILL
Rounding Off

Suppose you decide to buy carpeting for a room that has a length of 5.52 m and a width of 3.58 m. To determine how much carpeting you need, you would calculate the area of the room by multiplying 5.52 times 3.58 on your calculator. The calculator shows the number 19.7616 in its display. Because each of the original measurements has only three significant figures, the calculator display (19.7616) is *rounded off* to three significant figures, 19.8.

5.52 m	×	3.58 m	=	19.7616	=	19.8 m^2
Three SFs		Three SFs		Calculator display		Final answer, rounded off to three SFs

Therefore, you can order carpeting that will cover an area of 19.8 m^2.

Each time you use a calculator, it is important to look at the original measurements and determine the number of significant figures that can be used for the answer. You can use the following rules to round off the numbers shown in a calculator display.

A technician uses a calculator in the laboratory.

Rules for Rounding Off

1. If the first digit to be dropped is *4 or less*, then it and all following digits are simply dropped from the number.
2. If the first digit to be dropped is *5 or greater*, then the last retained digit of the number is increased by 1.

Number to Round Off	Three Significant Figures	Two Significant Figures
8.4234	8.42 (drop 34)	8.4 (drop 234)
14.780	14.8 (drop 80, increase the last retained digit by 1)	15 (drop 780, increase the last retained digit by 1)
3256	3260* (drop 6, increase the last retained digit by 1, add 0) (3.26×10^3)	3300* (drop 56, increase the last retained digit by 1, add 00) (3.3×10^3)

*The value of a large number is retained by using placeholder zeros to replace dropped digits.

ENGAGE
Why is 10.072 rounded off to three significant figures equal to 10.1?

▶SAMPLE PROBLEM 2.3 Rounding Off

TRY IT FIRST

Round off or add zeros to each of the following measurements to give three significant figures:

a. 35.7823 m **b.** 0.002 621 7 L **c.** 3.8268×10^3 g **d.** 8 s

SOLUTION

a. 35.8 m **b.** 0.002 62 L **c.** 3.83×10^3 g **d.** 8.00 s

STUDY CHECK 2.3

Round off each of the measurements in Sample Problem 2.3 to two significant figures.

ANSWER

a. 36 m **b.** 0.0026 L **c.** 3.8×10^3 g **d.** 8.0 s

TEST
Try Practice Problems 2.23 to 2.26

Multiplication and Division with Measured Numbers

CORE CHEMISTRY SKILL
Using Significant Figures in Calculations

In multiplication or division, the final answer is written so that it has the same number of significant figures (SFs) as the measurement with the fewest SFs. An example of rounding off a calculator display follows:

Perform the following operations with measured numbers:

$$\frac{2.8 \times 67.40}{34.8} =$$

A calculator is helpful in working problems and doing calculations faster.

When the problem has multiple steps, the numbers in the numerator are multiplied and then divided by each of the numbers in the denominator.

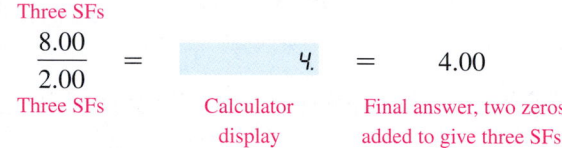

2.8	×	67.40	÷	34.8	=	5.422988506	=	5.4
Two SFs		Four SFs		Three SFs		Calculator display		Answer, rounded off to two SFs

Because the calculator display has more digits than the significant figures in the measured numbers allow, we need to round it off. Using the measured number that has the smallest number (two) of significant figures, 2.8, we round off the calculator display to an answer with two SFs.

Adding Significant Zeros

Sometimes, a calculator display gives a small whole number. For example, suppose the calculator display is 4, but you used measurements that have three significant numbers. Then two significant zeros are *added* to give 4.00 as the correct answer.

$$\frac{8.00}{2.00} = 4. = 4.00$$

Three SFs (8.00); Three SFs (2.00); Calculator display (4.); Final answer, two zeros added to give three SFs (4.00)

ENGAGE

Why is the answer for the multiplication of 0.3 × 52.6 written with one significant figure?

▶ SAMPLE PROBLEM 2.4 Significant Figures in Multiplication and Division

TRY IT FIRST

Perform the following calculations with measured numbers. Write each answer with the correct number of significant figures.

a. 56.8 × 0.37 **b.** $\frac{(2.075)(0.585)}{(8.42)(0.0245)}$ **c.** $\frac{25.0}{5.00}$

SOLUTION

ANALYZE THE PROBLEM	Given	Need	Connect
	multiplication and division	answer with SFs	rules for rounding off, adding zeros

STEP 1 Determine the number of significant figures in each measured number.

a. 56.8 (Three SFs) × 0.37 (Two SFs)

b. $\frac{(2.075)(0.585)}{(8.42)(0.0245)}$ — Four SFs, Three SFs (numerator); Three SFs, Three SFs (denominator)

c. $\frac{25.0}{5.00}$ — Three SFs (numerator); Three SFs (denominator)

STEP 2 Perform the indicated calculation.

a. 21.016 (Calculator display) **b.** 5.884313345 (Calculator display) **c.** 5. (Calculator display)

STEP 3 Round off (or add zeros) to give the same number of significant figures as the measurement having the fewest significant figures.

a. 21 **b.** 5.88 **c.** 5.00

STUDY CHECK 2.4

Perform the following calculations with measured numbers, and give the answers with the correct number of significant figures:

a. 45.26 × 0.01088 **b.** 2.6 ÷ 324 **c.** $\frac{4.0 \times 8.00}{16}$

ANSWER

a. 0.4924 **b.** 0.0080 or 8.0×10^{-3} **c.** 2.0

TEST

Try Practice Problems 2.27 and 2.28

Addition and Subtraction with Measured Numbers

In addition or subtraction, the final answer is written so that it has the same number of decimal places as the measurement having the fewest decimal places.

2.045	Thousandths place
+ 34.1	Tenths place
36.145	Calculator display
36.1	Answer, rounded off to the tenths place

When numbers are added or subtracted to give an answer ending in zero, the zero does not appear after the decimal point in the calculator display. For example, 14.5 g − 2.5 g = 12.0 g. However, if you do the subtraction on your calculator, the display shows 12. To write the correct answer, a significant zero is written after the decimal point.

ENGAGE

Why is the answer for the addition of 55.2 and 2.506 written with one decimal place?

▶ SAMPLE PROBLEM 2.5 Decimal Places in Addition and Subtraction

TRY IT FIRST

Perform the following calculations with measured numbers and give each answer with the correct number of decimal places:

a. 104 + 7.8 + 40 **b.** 153.247 − 14.82

SOLUTION

ANALYZE THE PROBLEM	Given	Need	Connect
	addition and subtraction	correct number of decimal places	rules for rounding off, adding zeros

STEP 1 Determine the number of decimal places in each measured number.

a.

104	Ones place
7.8	Tenths place
+ 40	Tens place

b.

153.247	Thousandths place
− 14.82	Hundredths place

STEP 2 Perform the indicated calculation.

a. 151.8 Calculator display **b.** 138.427 Calculator display

STEP 3 Round off (or add zeros) to give the same number of decimal places as the measured number having the fewest decimal places.

a. 150 Rounded off to the tens place
b. 138.43 Rounded off to the hundredths place

STUDY CHECK 2.5

Perform the following calculations with measured numbers, and give each answer with the correct number of decimal places:

a. 82.45 + 1.245 + 0.000 56
b. 4.259 − 3.8

ANSWER

a. 83.70 **b.** 0.5

TEST

Try Practice Problems 2.29 and 2.30

PRACTICE PROBLEMS

2.3 Significant Figures in Calculations

2.23 Round off each of the following calculator answers to three significant figures:
a. 1.854 kg **b.** 88.2038 L
c. 0.004 738 265 cm **d.** 8807 m
e. 1.832×10^5 s

2.24 Round off each of the calculator answers in problem 2.23 to two significant figures.

2.25 Round off or add zeros to each of the following to three significant figures:
a. 56.855 m **b.** 0.002 282 g
c. 11 527 s **d.** 8.1 L

2.26 Round off or add zeros to each of the following to two significant figures:
a. 3.2805 m **b.** 1.855×10^2 g
c. 0.002 341 mL **d.** 2 L

2.27 Perform each of the following calculations, and give an answer with the correct number of significant figures:
a. 45.7×0.034 **b.** $0.002\,78 \times 5$
c. $\dfrac{34.56}{1.25}$ **d.** $\dfrac{(0.2465)(25)}{1.78}$
e. $(2.8 \times 10^4)(5.05 \times 10^{-6})$ **f.** $\dfrac{(3.45 \times 10^{-2})(1.8 \times 10^5)}{(8 \times 10^3)}$

2.28 Perform each of the following calculations, and give an answer with the correct number of significant figures:
a. 400×185
b. $\dfrac{2.40}{(4)(125)}$
c. $0.825 \times 3.6 \times 5.1$
d. $\dfrac{(3.5)(0.261)}{(8.24)(20.0)}$
e. $\dfrac{(5 \times 10^{-5})(1.05 \times 10^4)}{(8.24 \times 10^{-8})}$
f. $\dfrac{(4.25 \times 10^2)(2.56 \times 10^{-3})}{(2.245 \times 10^{-3})(56.5)}$

2.29 Perform each of the following calculations, and give an answer with the correct number of decimal places:
a. $45.48 + 8.057$
b. $23.45 + 104.1 + 0.025$
c. $145.675 - 24.2$
d. $1.08 - 0.585$
e. $2300 + 196.11$
f. $145.111 - 22.9 + 34.49$

2.30 Perform each of the following calculations, and give an answer with the correct number of decimal places:
a. $5.08 + 25.1$
b. $85.66 + 104.10 + 0.025$
c. $24.568 - 14.25$
d. $0.2654 - 0.2585$
e. $66.77 + 17 - 0.33$
f. $460 - 33.77$

2.4 Prefixes and Equalities

LEARNING GOAL Use the numerical values of prefixes to write a metric equality.

The special feature of the metric system is that a **prefix** can be placed in front of any unit to increase or decrease its size by some factor of 10. For example, the prefixes *milli* and *micro* are used to make the smaller units, milligram (mg) and microgram (μg).

The U.S. Food and Drug Administration (FDA) has determined the Daily Values (DV) for nutrients for adults and children age 4 or older. Examples of these recommended Daily Values, some of which use prefixes, are listed in **TABLE 2.4**.

TABLE 2.4 Daily Values (DV) for Selected Nutrients

Nutrient	Amount Recommended
Calcium	1.0 g
Copper	2 mg
Iodine	150 μg (150 mcg)
Iron	18 mg
Magnesium	400 mg
Niacin	20 mg
Phosphorus	800 mg
Potassium	3.5 g
Selenium	70. μg (70. mcg)
Sodium	2.4 g
Zinc	15 mg

The prefix *centi* is like cents in a dollar. One cent would be a “centidollar” or 0.01 of a dollar. That also means that one dollar is the same as 100 cents. The prefix *deci* is like dimes in a dollar. One dime would be a “decidollar” or 0.1 of a dollar. That also means that one dollar is the same as 10 dimes. **TABLE 2.5** lists some of the metric prefixes, their symbols, and their numerical values.

The relationship of a prefix to a unit can be expressed by replacing the prefix with its numerical value. For example, when the prefix *kilo* in kilometer is replaced with its value of 1000, we find that a kilometer is equal to 1000 m. Other examples follow:

1 **kilo**meter (1 km) = **1000** meters ($1000 \text{ m} = 10^3 \text{ m}$)

1 **kilo**liter (1 kL) = **1000** liters ($1000 \text{ L} = 10^3 \text{ L}$)

1 **kilo**gram (1 kg) = **1000** grams ($1000 \text{ g} = 10^3 \text{ g}$)

TABLE 2.5 Metric and SI Prefixes

Prefix	Symbol	Numerical Value	Scientific Notation	Equality
Prefixes That Increase the Size of the Unit				
peta	P	1 000 000 000 000 000	10^{15}	$1\ Pg = 1 \times 10^{15}\ g$ $1\ g = 1 \times 10^{-15}\ Pg$
tera	T	1 000 000 000 000	10^{12}	$1\ Ts = 1 \times 10^{12}\ s$ $1\ s = 1 \times 10^{-12}\ Ts$
giga	G	1 000 000 000	10^{9}	$1\ Gm = 1 \times 10^{9}\ m$ $1\ m = 1 \times 10^{-9}\ Gm$
mega	M	1 000 000	10^{6}	$1\ Mg = 1 \times 10^{6}\ g$ $1\ g = 1 \times 10^{-6}\ Mg$
kilo	k	1 000	10^{3}	$1\ km = 1 \times 10^{3}\ m$ $1\ m = 1 \times 10^{-3}\ km$
Prefixes That Decrease the Size of the Unit				
deci	d	0.1	10^{-1}	$1\ dL = 1 \times 10^{-1}\ L$ $1\ L = 10\ dL$
centi	c	0.01	10^{-2}	$1\ cm = 1 \times 10^{-2}\ m$ $1\ m = 100\ cm$
milli	m	0.001	10^{-3}	$1\ ms = 1 \times 10^{-3}\ s$ $1\ s = 1 \times 10^{3}\ ms$
micro	μ*	0.000 001	10^{-6}	$1\ \mu g = 1 \times 10^{-6}\ g$ $1\ g = 1 \times 10^{6}\ \mu g$
nano	n	0.000 000 001	10^{-9}	$1\ nm = 1 \times 10^{-9}\ m$ $1\ m = 1 \times 10^{9}\ nm$
pico	p	0.000 000 000 001	10^{-12}	$1\ ps = 1 \times 10^{-12}\ s$ $1\ s = 1 \times 10^{12}\ ps$
femto	f	0.000 000 000 000 001	10^{-15}	$1\ fs = 1 \times 10^{-15}\ s$ $1\ s = 1 \times 10^{15}\ fs$

*In medicine, the abbreviation *mc* for the prefix *micro* is used because the symbol μ may be misread, which could result in a medication error. Thus, 1 μg would be written as 1 mcg.

CORE CHEMISTRY SKILL
Using Prefixes

TEST
Try Practice Problems 2.31 to 2.38

ENGAGE
Why is 60. mg of vitamin C the same as 0.060 g of vitamin C?

▸SAMPLE PROBLEM 2.6 Prefixes and Equalities

TRY IT FIRST K h d m d c m

An endoscopic camera has a width of 1 mm. Complete each of the following equalities involving millimeters:

a. 1 m = 1000 mm **b.** 1 cm = 10 mm

SOLUTION

a. 1 m = 1000 mm **b.** 1 cm = 10 mm

STUDY CHECK 2.6

a. What is the relationship between millimeters and micrometers?
b. What is the relationship between liters and centiliters?

ANSWER

a. 1 mm = 1000 μm (mcm) **b.** 1 L = 100 cL

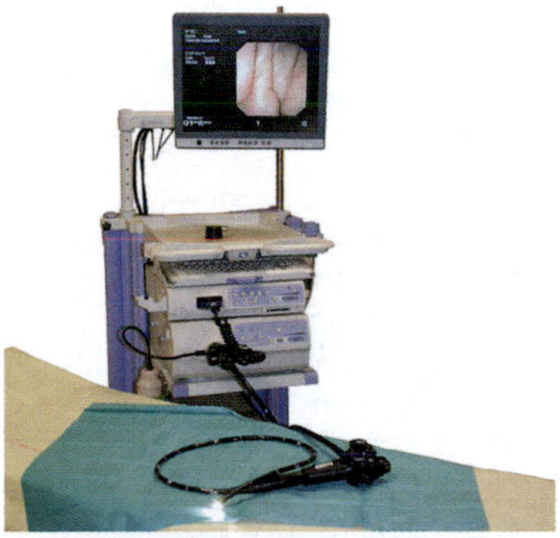

An endoscope has a video camera with a width of 1 mm attached to the end of a thin cable.

Measuring Length

An ophthalmologist may measure the diameter of the retina of an eye in centimeters (cm), whereas a surgeon may need to know the length of a nerve in millimeters (mm). When the prefix *centi* is used with the unit meter, it becomes *centimeter*, a length that is one-hundredth

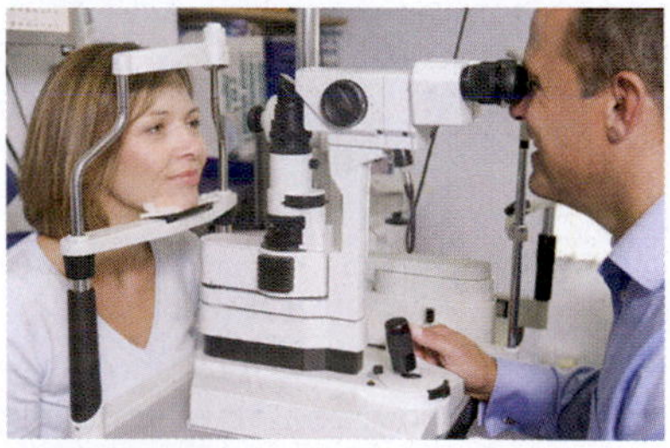

Using a retinal camera, an ophthalmologist photographs the retina of an eye.

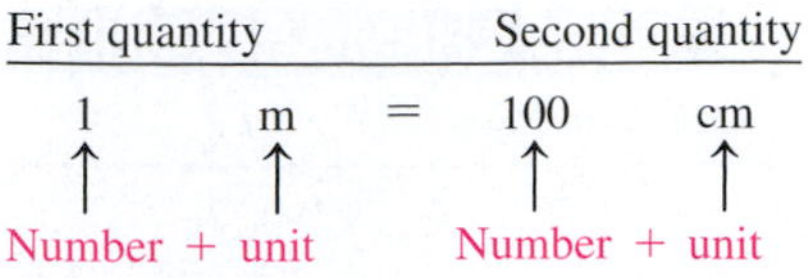
First quantity
Second quantity
1 m = 100 cm
Number + unit
Number + unit

This example of an equality shows the relationship between meters and centimeters.

of a meter (0.01 m). When the prefix *milli* is used with the unit meter, it becomes *millimeter*, a length that is one-thousandth of a meter (0.001 m). There are 100 cm and 1000 mm in a meter.

If we compare the lengths of a millimeter and a centimeter, we find that 1 mm is 0.1 cm; there are 10 mm in 1 cm. These comparisons are examples of **equalities**, which show the relationship between two units that measure the same quantity. Examples of equalities between different metric units of length follow:

$$1\text{ m} = 100\text{ cm} = 1 \times 10^2\text{ cm}$$
$$1\text{ m} = 1000\text{ mm} = 1 \times 10^3\text{ mm}$$
$$1\text{ cm} = 10\text{ mm} = 1 \times 10^1\text{ mm}$$

Some metric units for length are compared in **FIGURE 2.6**.

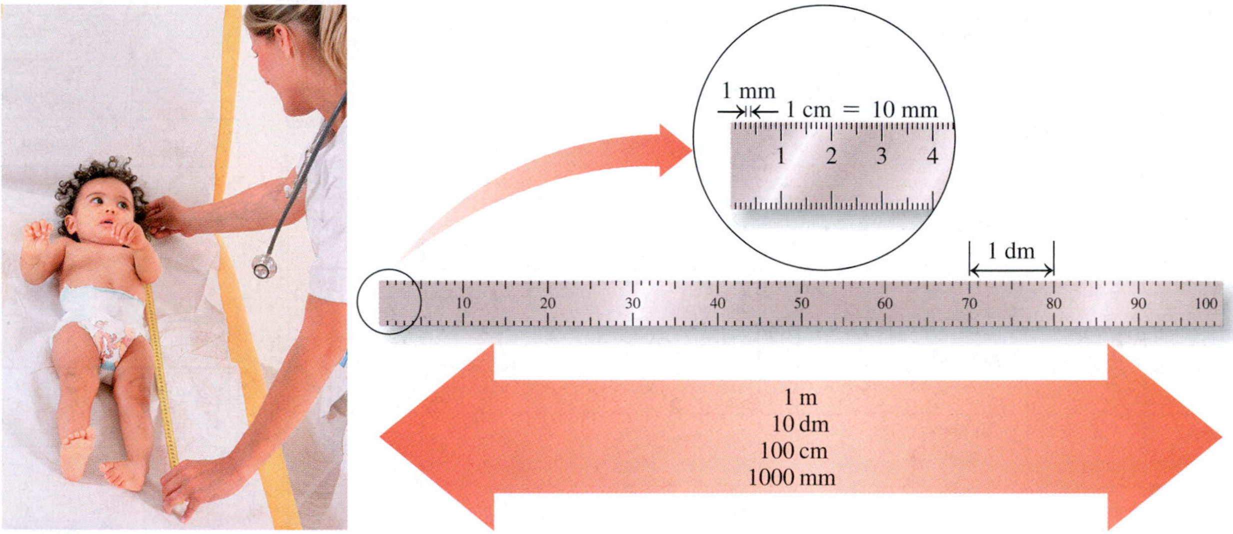
1 mm
1 cm = 10 mm
1 2 3 4
1 dm
10 20 30 40 50 60 70 80 90 100
1 m
10 dm
100 cm
1000 mm

FIGURE 2.6 ▶ The metric length of 1 m is the same length as 10 dm, 100 cm, or 1000 mm.

Q How many millimeters (mm) are in 1 centimeter (cm)?

ENGAGE

Why can the relationship of centimeters and meters be written as 1 m = 100 cm or 0.01 m = 1 cm?

Measuring Volume

Volumes of 1 L or smaller are common in the health sciences. When a liter is divided into 10 equal portions, each portion is a deciliter (dL). There are 10 dL in 1 L. Laboratory results for bloodwork are often reported in mass per deciliter. **TABLE 2.6** lists normal laboratory test values for some substances in the blood.

TABLE 2.6 Some Normal Laboratory Test Values

Substance in Blood	Normal Range
Albumin	3.5–5.4 g/dL
Ammonia	20–70 μg/dL (mcg/dL)
Calcium	8.5–10.5 mg/dL
Cholesterol	105–250 mg/dL
Iron (male)	80–160 μg/dL (mcg/dL)
Protein (total)	6.0–8.5 g/dL

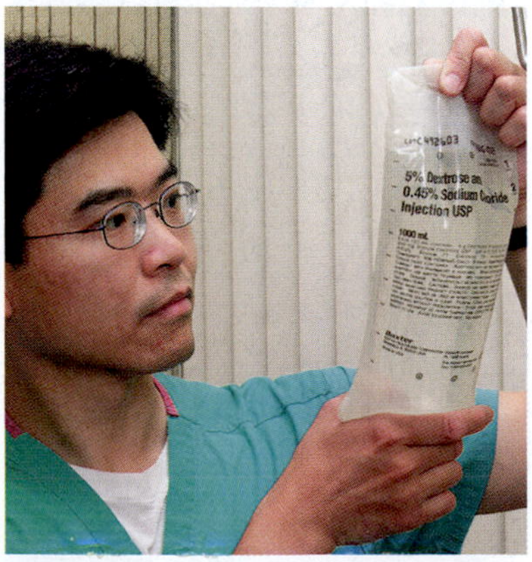
5% Dextrose and
0.45% Sodium Chloride
Injection USP

FIGURE 2.7 ▶ A plastic intravenous fluid container contains 1000 mL.

Q How many liters of solution are in the intravenous fluid container?

When a liter is divided into a thousand parts, each of the smaller volumes is a milliliter (mL). In a 1-L container of physiological saline, there are 1000 mL of solution (see **FIGURE 2.7**). Examples of equalities between different metric units of volume follow:

$$1\text{ L} = 10\text{ dL} = 1 \times 10^1\text{ dL}$$
$$1\text{ L} = 1000\text{ mL} = 1 \times 10^3\text{ mL}$$
$$1\text{ dL} = 100\text{ mL} = 1 \times 10^2\text{ mL}$$
$$1\text{ mL} = 1000\ \mu\text{L (mcL)} = 1 \times 10^3\ \mu\text{L (mcL)}$$

The **cubic centimeter** (abbreviated as **cm^3** or **cc**) is the volume of a cube whose dimensions are 1 cm on each side. A cubic centimeter has the same volume as a milliliter, and the units are often used interchangeably.

$$1\ cm^3 = 1\ cc = 1\ mL$$

When you see *1 cm*, you are reading about length; when you see *1 cm^3* or *1 cc* or *1 mL*, you are reading about volume. A comparison of units of volume is illustrated in **FIGURE 2.8**.

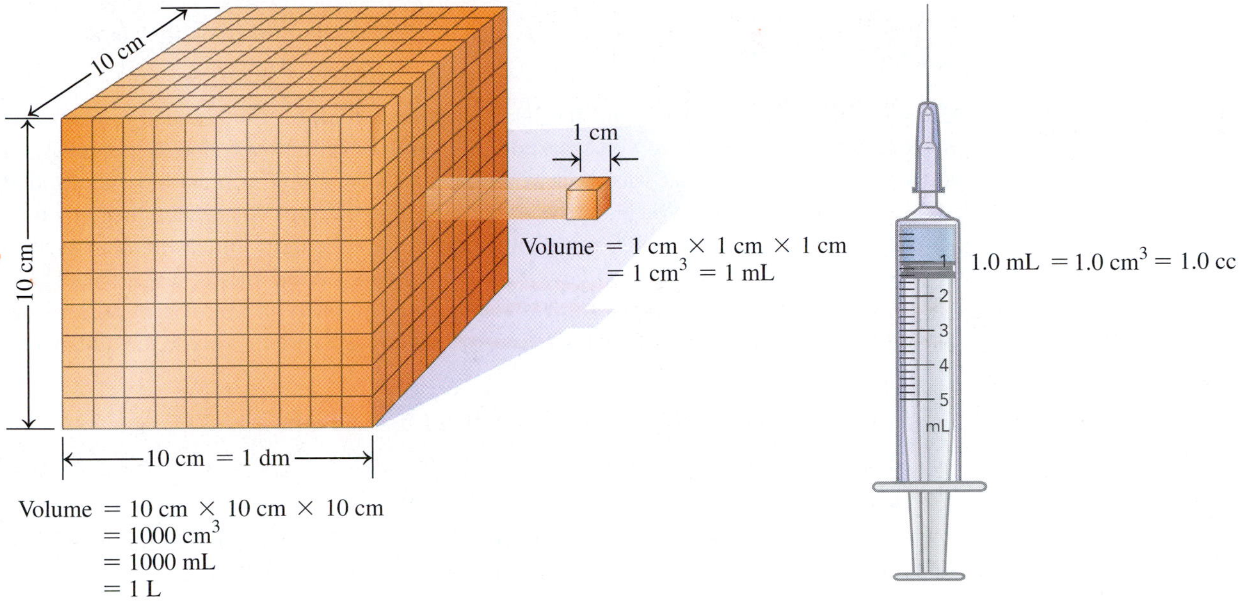

10 cm
10 cm
10 cm = 1 dm
1 cm
Volume = 1 cm × 1 cm × 1 cm
= 1 cm³ = 1 mL
Volume = 10 cm × 10 cm × 10 cm
= 1000 cm³
= 1000 mL
= 1 L
1
2
3
4
5
mL
1.0 mL = 1.0 cm³ = 1.0 cc

FIGURE 2.8 ▶ A cube measuring 10 cm on each side has a volume of 1000 cm^3 or 1 L; a cube measuring 1 cm on each side has a volume of 1 cm^3 (cc) or 1 mL.

Q What is the relationship between a milliliter (mL) and a cubic centimeter (cm^3)?

Measuring Mass

When you go to the doctor for a physical examination, your mass is recorded in kilograms, whereas the results of your laboratory tests are reported in grams, milligrams (mg), or micrograms (μg or mcg). A kilogram is equal to 1000 g. One gram represents the same mass as 1000 mg, and one mg equals 1000 μg (or 1000 mcg). Examples of equalities between different metric units of mass follow:

$$1\ kg = 1000\ g = 1 \times 10^3\ g$$
$$1\ g = 1000\ mg = 1 \times 10^3\ mg$$
$$1\ mg = 1000\ \mu g\ (mcg) = 1 \times 10^3\ \mu g\ (mcg)$$

TEST

Try Practice Problems 2.39 to 2.42

PRACTICE PROBLEMS

2.4 Prefixes and Equalities

2.31 Write the abbreviation for each of the following units:
a. milligram **b.** deciliter
c. kilometer **d.** picogram

2.32 Write the abbreviation for each of the following units:
a. gigagram **b.** megameter
c. microliter **d.** femtosecond

2.33 Write the complete name for each of the following units:
a. cL **b.** kg **c.** ms **d.** Pm

2.34 Write the complete name for each of the following units:
a. dL **b.** Ts **c.** mcg **d.** pm

2.35 Write the numerical value for each of the following prefixes:
a. centi **b.** tera **c.** milli **d.** deci

2.36 Write the numerical value for each of the following prefixes:
a. giga **b.** micro **c.** mega **d.** nano

2.37 Use a prefix to write the name for each of the following:
a. 0.1 g **b.** 10^{-6} g **c.** 1000 g **d.** 0.01 g

2.38 Use a prefix to write the name for each of the following:
a. 10^9 m **b.** 10^6 m **c.** 0.001 m **d.** 10^{-12} m

2.39 Complete each of the following metric relationships:
a. 1 m = ________ cm **b.** 1 m = ________ nm
c. 1 mm = ________ m **d.** 1 L = ________ mL

2.40 Complete each of the following metric relationships:
a. 1 Mg = ________ g **b.** 1 mL = ________ μL
c. 1 g = ________ kg **d.** 1 g = ________ mg

2.41 For each of the following pairs, which is the larger unit?
a. milligram or kilogram **b.** milliliter or microliter
c. m or km **d.** kL or dL
e. nanometer or picometer

2.42 For each of the following pairs, which is the smaller unit?
a. mg or g **b.** centimeter or nanometer
c. millimeter or micrometer **d.** mL or dL
e. centigram or megagram

REVIEW
Calculating Percentages (1.4)

CORE CHEMISTRY SKILL
Writing Conversion Factors from Equalities

TEST
Try Practice Problems 2.43 and 2.44

2.5 Writing Conversion Factors

LEARNING GOAL Write a conversion factor for two units that describe the same quantity.

Many problems in chemistry and the health sciences require you to change from one unit to another unit. Suppose you worked 2.0 h on your homework, and someone asked you how many minutes that was. You would answer 120 min. You must have multiplied 2.0 h × 60 min/h because you knew the equality (1 h = 60 min) that related the two units. When you expressed 2.0 h as 120 min, you changed only the unit of measurement used to express the time. *Any equality can be written as fractions called* **conversion factors** *with one of the quantities in the numerator and the other quantity in the denominator.* Two conversion factors are always possible from any equality. Be sure to include the units when you write the conversion factors.

Two Conversion Factors for the Equality: 1 h = 60 min

$$\frac{\text{Numerator}}{\text{Denominator}} \begin{matrix}\longrightarrow \\ \longrightarrow\end{matrix} \quad \frac{60\text{ min}}{1\text{ h}} \quad \text{and} \quad \frac{1\text{ h}}{60\text{ min}}$$

These factors are read as "60 minutes per 1 hour" and "1 hour per 60 minutes." The term *per* means "divide." Some common relationships are given in **TABLE 2.7**.

TABLE 2.7 Some Common Equalities

Quantity	Metric (SI)	U.S.	Metric–U.S.
Length	1 km = 1000 m	1 ft = 12 in.	2.54 cm = 1 in. (exact)
	1 m = 1000 mm	1 yd = 3 ft	1 m = 39.4 in.
	1 cm = 10 mm	1 mi = 5280 ft	1 km = 0.621 mi
Volume	1 L = 1000 mL	1 qt = 4 cups	946 mL = 1 qt
	1 dL = 100 mL	1 qt = 2 pt	1 L = 1.06 qt
	1 mL = 1 cm^3	1 gal = 4 qt	473 mL = 1 pt
	1 mL = 1 cc*		5 mL = 1 t (tsp)*
			15 mL = 1 T (tbsp)*
Mass	1 kg = 1000 g	1 lb = 16 oz	1 kg = 2.20 lb
	1 g = 1000 mg		454 g = 1 lb
	1 mg = 1000 mcg*		
Time	1 h = 60 min	1 h = 60 min	
	1 min = 60 s	1 min = 60 s	

*Used in medicine.

The numbers in any definition between two metric units or between two U.S. system units are exact. Because numbers in a definition are exact, they are not used to determine significant figures. For example, the equality of 1 g = 1000 mg is a definition, which means that both of the numbers 1 and 1000 are exact.

When an equality consists of a metric unit and a U.S. unit, one of the numbers in the equality is obtained by measurement and counts toward the significant figures in the answer. For example, the equality of 1 lb = 454 g is obtained by measuring the grams in exactly 1 lb. In this equality, the measured quantity 454 g has three significant figures, whereas the 1 is exact. An exception is the relationship of 1 in. = 2.54 cm, which has been defined as exact.

Metric Conversion Factors

We can write two metric conversion factors for any of the metric relationships. For example, from the equality for meters and centimeters, we can write the following factors:

Metric Equality	Conversion Factors
1 m = 100 cm	$\frac{100\text{ cm}}{1\text{ m}}$ and $\frac{1\text{ m}}{100\text{ cm}}$

Both are proper conversion factors for the relationship; one is just the inverse of the other. *The usefulness of conversion factors is enhanced by the fact that we can turn a conversion factor over and use its inverse.* The numbers 100 and 1 in this equality, and its conversion factors are both *exact* numbers.

ENGAGE

Why does the equality 1 day = 24 h have two conversion factors?

Metric–U.S. System Conversion Factors

Suppose you need to convert from pounds, a unit in the U.S. system, to kilograms in the metric system. A relationship you could use is

1 kg = 2.20 lb

The corresponding conversion factors would be

$$\frac{2.20\text{ lb}}{1\text{ kg}} \quad \text{and} \quad \frac{1\text{ kg}}{2.20\text{ lb}}$$

FIGURE 2.9 illustrates the contents of some packaged foods in both U.S. and metric units.

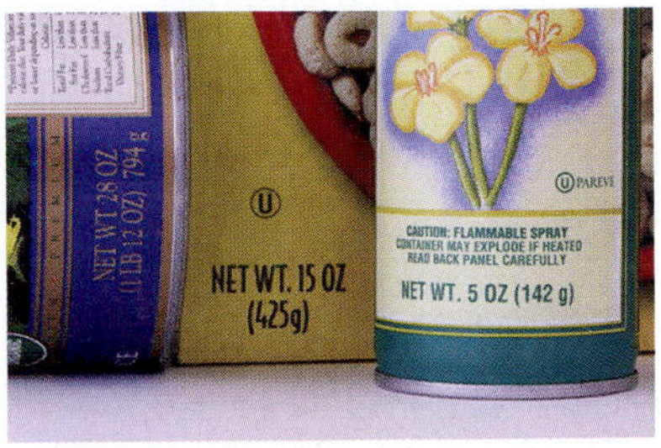
NET WT. 15 OZ (425g)
CAUTION: FLAMMABLE SPRAY
NET WT. 5 OZ (142 g)

FIGURE 2.9 ▶ In the United States, the contents of many packaged foods are listed in both U.S. and metric units.

Q What are some advantages of using the metric system?

Equalities and Conversion Factors Stated Within a Problem

An equality may also be stated within a problem that applies only to that problem. For example, the speed of a car in kilometers per hour or the price of onions in dollars per pound would be specific relationships for that problem only. From each of the following statements, we can write an equality and two conversion factors, and identify each number as exact or give the number of significant figures.

The car was traveling at a speed of 85 km/h.

Equality	Conversion Factors	Significant Figures or Exact
1 h = 85 km	$\frac{85\text{ km}}{1\text{ h}}$ and $\frac{1\text{ h}}{85\text{ km}}$	The 85 km is measured: It has two significant figures. The 1 h is exact.

The price of onions is \$1.24 per pound.

Equality	Conversion Factors	Significant Figures or Exact
1 lb = \$1.24	$\frac{\$1.24}{1\text{ lb}}$ and $\frac{1\text{ lb}}{\$1.24}$	The \$1.24 is measured: It has three significant figures. The 1 lb is exact.

TEST

Try Practice Problems 2.45 and 2.46

Conversion Factors from Dosage Problems

Equalities stated within dosage problems for medications can also be written as conversion factors. Keflex (cephalexin), an antibiotic used for respiratory and ear infections, is available in 250-mg capsules. Vitamin C, an antioxidant, is available in 500-mg tablets. These dosage relationships can be used to write equalities from which two conversion factors can be derived.

One capsule contains 250 mg of Keflex.

Equality	Conversion Factors	Significant Figures or Exact
1 capsule = 250 mg of Keflex	$\frac{250\text{ mg Keflex}}{1\text{ capsule}}$ and $\frac{1\text{ capsule}}{250\text{ mg Keflex}}$	The 250 mg is measured: It has two significant figures. The 1 capsule is exact.

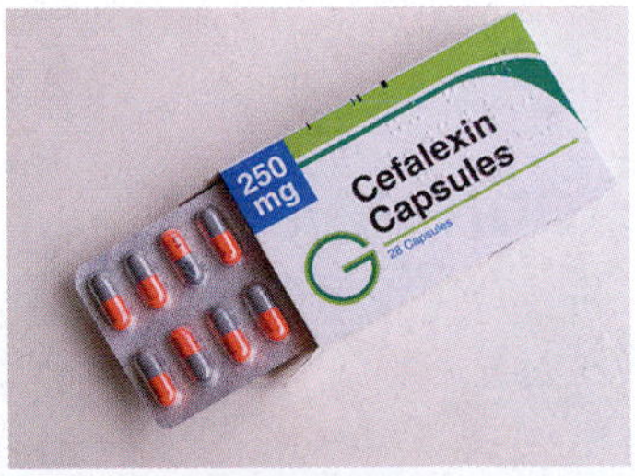
250 mg
Cefalexin Capsules

Keflex (cephalexin), used to treat respiratory infections, is available in 250-mg capsules.

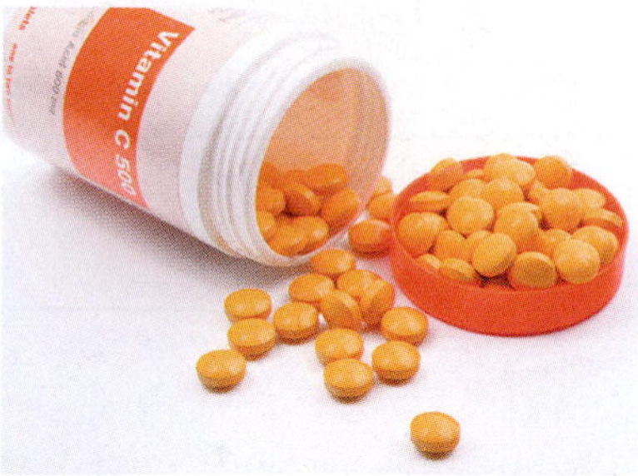
Vitamin C 500

Vitamin C is an antioxidant needed by the body.

One tablet contains 500 mg of vitamin C.

Equality	Conversion Factors	Significant Figures or Exact
1 tablet = 500 mg of vitamin C	$\frac{500 \text{ mg vitamin C}}{1 \text{ tablet}}$ and $\frac{1 \text{ tablet}}{500 \text{ mg vitamin C}}$	The 500 mg is measured: It has one significant figure. The 1 tablet is exact.

Conversion Factors from a Percentage, ppm, and ppb

A percentage (%) is written as a conversion factor by choosing a unit and expressing the numerical relationship of the parts of this unit to 100 parts of the whole. For example, a person might have 18% body fat by mass. The percentage quantity can be written as 18 mass units of body fat in every 100 mass units of body mass. Different mass units such as grams (g), kilograms (kg), or pounds (lb) can be used, but both units in the factor must be the same.

ENGAGE

How is a percentage used to write an equality and two conversion factors?

Equality	Conversion Factors	Significant Figures or Exact
100 kg of body mass = 18 kg of body fat	$\frac{18 \text{ kg body fat}}{100 \text{ kg body mass}}$ and $\frac{100 \text{ kg body mass}}{18 \text{ kg body fat}}$	The 18 kg is measured: It has two significant figures. The 100 kg is exact.

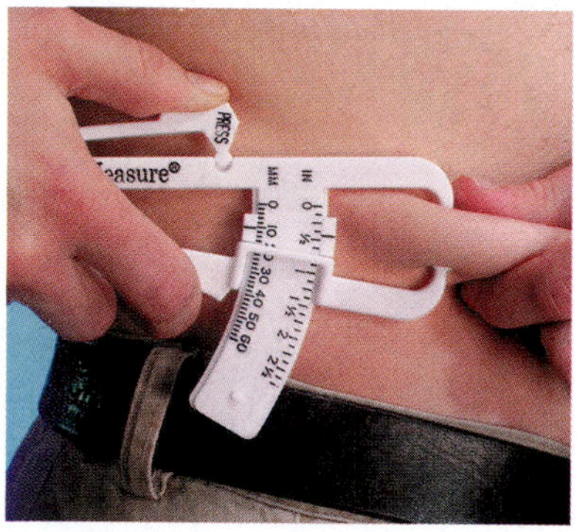

The thickness of the skin fold at the abdomen is used to determine the percentage of body fat.

When scientists want to indicate very small ratios, they use numerical relationships called *parts per million* (ppm) or *parts per billion* (ppb). The ratio of parts per million is the same as the milligrams of a substance per kilogram (mg/kg). The ratio of parts per billion equals the micrograms per kilogram (μg/kg, mcg/kg).

Ratio	Units
parts per million (ppm)	milligrams per kilogram (mg/kg)
parts per billion (ppb)	micrograms per kilogram (μg/kg, mcg/kg)

For example, the maximum amount of lead that is allowed by the FDA in glazed pottery bowls is 2 ppm, which is 2 mg/kg.

Equality	Conversion Factors	Significant Figures or Exact
1 kg of glaze = 2 mg of lead	$\frac{2 \text{ mg lead}}{1 \text{ kg glaze}}$ and $\frac{1 \text{ kg glaze}}{2 \text{ mg lead}}$	The 2 mg is measured: It has one significant figure. The 1 kg is exact.

▶SAMPLE PROBLEM 2.7 Equalities and Conversion Factors in a Problem

TRY IT FIRST

Write the equality and two conversion factors, and identify each number as exact or give the number of significant figures for each of the following:

a. The medication that Greg takes for his high blood pressure contains 40. mg of propranolol in 1 tablet.

b. Cold-water fish such as salmon contains 1.9% omega-3 fatty acids by mass.

c. The U.S. Environmental Protection Agency (EPA) has set the maximum level for mercury in tuna at 0.5 ppm.

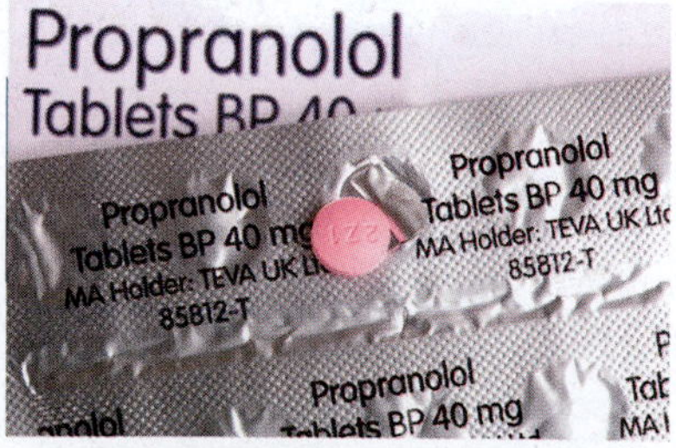
Propranolol
Tablets BP 40
Propranolol
Tablets BP 40 mg
MA Holder: TEVA UK Ltd
85812-T
Propranolol
Tablets BP 40 mg

Propranolol is used to lower high blood pressure.

SOLUTION

a.

Equality	Conversion Factors	Significant Figures or Exact
1 tablet = 40. mg of propranolol	$\frac{40. \text{ mg propranolol}}{1 \text{ tablet}}$ and $\frac{1 \text{ tablet}}{40. \text{ mg propranolol}}$	The 40. mg is measured: It has two significant figures. The 1 tablet is exact.

b.

Equality	Conversion Factors	Significant Figures or Exact
100 g of salmon = 1.9 g of omega-3 fatty acids	$\frac{1.9\text{ g omega-3 fatty acids}}{100\text{ g salmon}}$ and $\frac{100\text{ g salmon}}{1.9\text{ g omega-3 fatty acids}}$	The 1.9 g is measured: It has two significant figures. The 100 g is exact.

Salmon contains high levels of omega-3 fatty acids.

c.

Equality	Conversion Factors	Significant Figures or Exact
1 kg of tuna = 0.5 mg of mercury	$\frac{0.5\text{ mg mercury}}{1\text{ kg tuna}}$ and $\frac{1\text{ kg tuna}}{0.5\text{ mg mercury}}$	The 0.5 mg is measured: It has one significant figure. The 1 kg is exact.

The maximum amount of mercury allowed in tuna is 0.5 ppm.

STUDY CHECK 2.7

Write the equality and its corresponding conversion factors, and identify each number as exact or give the number of significant figures for each of the following:

a. Levsin (hyoscyamine), used to treat stomach and bladder problems, is available as drops with 0.125 mg of Levsin per 1 mL of solution.

b. The EPA has set the maximum level of cadmium in rice as 0.4 ppm.

ANSWER

a. 1 mL of solution = 0.125 mg of Levsin

$$\frac{0.125\text{ mg Levsin}}{1\text{ mL solution}} \quad \text{and} \quad \frac{1\text{ mL solution}}{0.125\text{ mg Levsin}}$$

The 0.125 mg is measured: It has three significant figures. The 1 mL is exact.

b. 1 kg of rice = 0.4 mg of cadmium

$$\frac{0.4\text{ mg cadmium}}{1\text{ kg rice}} \quad \text{and} \quad \frac{1\text{ kg rice}}{0.4\text{ mg cadmium}}$$

The 0.4 mg is measured: It has one significant figure. The 1 kg is exact.

TEST

Try Practice Problems 2.47 to 2.54

PRACTICE PROBLEMS

2.5 Writing Conversion Factors

2.43 Why can two conversion factors be written for an equality such as 1 m = 100 cm?

2.44 How can you check that you have written the correct conversion factors for an equality?

2.45 Write the equality and two conversion factors for each of the following pairs of units:
- **a.** centimeters and meters
- **b.** nanograms and grams
- **c.** liters and kiloliters
- **d.** seconds and milliseconds
- **e.** millimeters and decimeters

2.46 Write the equality and two conversion factors for each of the following pairs of units:
- **a.** centimeters and inches
- **b.** kilometers and miles
- **c.** pounds and grams
- **d.** liters and deciliters
- **e.** grams and picograms

2.47 Write the equality and two conversion factors, and identify the numbers as exact or give the number of significant figures for each of the following:
- **a.** One yard is 3 ft.
- **b.** One kilogram is 2.20 lb.
- **c.** A car goes 27 mi on 1 gal of gas.
- **d.** Sterling silver is 93% silver by mass.
- **e.** One minute is 60 s.

2.48 Write the equality and two conversion factors, and identify the numbers as exact or give the number of significant figures for each of the following:
- **a.** One liter is 1.06 qt.
- **b.** At the store, oranges are $1.29 per lb.
- **c.** There are 7 days in 1 week.
- **d.** One deciliter contains 100 mL.
- **e.** An 18-carat gold ring contains 75% gold by mass.

2.49 Write the equality and two conversion factors, and identify the numbers as exact or give the number of significant figures for each of the following:

a. A bee flies at an average speed of 3.5 m per second.
b. The Daily Value (DV) for potassium is 3.5 g.
c. An automobile traveled 26.0 km on 1 L of gasoline.
d. The pesticide level in plums was 29 ppb.
e. Silicon makes up 28.2% by mass of Earth's crust.

2.50 Write the equality and two conversion factors, and identify the numbers as exact or give the number of significant figures for each of the following:

a. The Daily Value (DV) for iodine is 150 mcg.
b. The nitrate level in well water was 32 ppm.
c. Gold jewelry contains 58% gold by mass.
d. The price of a liter of milk is $1.65.
e. A metric ton is 1000 kg.

Clinical Applications

2.51 Write the equality and two conversion factors, and identify the numbers as exact or give the number of significant figures for each of the following:

a. A calcium supplement contains 630 mg of calcium per tablet.
b. The Daily Value (DV) for vitamin C is 60 mg.
c. The label on a bottle reads 50 mg of atenolol per tablet.
d. A low-dose aspirin contains 81 mg of aspirin per tablet.

2.52 Write the equality and two conversion factors, and identify the numbers as exact or give the number of significant figures for each of the following:

a. The label on a bottle reads 10 mg of furosemide per 1 mL.
b. The Daily Value (DV) for selenium is 70. mcg.
c. An IV of normal saline solution has a flow rate of 85 mL per hour.
d. One capsule of fish oil contains 360 mg of omega-3 fatty acids.

2.53 Write an equality and two conversion factors for each of the following medications:

a. 10 mg of Atarax per 5 mL of Atarax syrup
b. 0.25 g of Lanoxin per 1 tablet of Lanoxin
c. 300 mg of Motrin per 1 tablet of Motrin

2.54 Write an equality and two conversion factors for each of the following medications:

a. 2.5 mg of Coumadin per 1 tablet of Coumadin
b. 100 mg of Clozapine per 1 tablet of Clozapine
c. 1.5 g of Cefuroxime per 1 mL of Cefuroxime

2.6 Problem Solving Using Unit Conversion

LEARNING GOAL Use conversion factors to change from one unit to another.

The process of problem solving in chemistry often requires one or more conversion factors to change a given unit to the needed unit. For the problem, the unit of the given and the unit of the needed are identified. From there, the problem is set up with one or more conversion factors used to convert the given unit to the needed unit as seen in Sample Problem 2.8.

Given unit × one or more conversion factors = needed unit

▶SAMPLE PROBLEM 2.8 Using Conversion Factors

TRY IT FIRST

Greg's doctor has ordered a PET scan of his heart. In radiological imaging, dosages of pharmaceuticals are based on body mass. If Greg weighs 164 lb, what is his body mass in kilograms?

SOLUTION

STEP 1 State the given and needed quantities.

ANALYZE THE PROBLEM	Given	Need	Connect
	164 lb	kilograms	U.S.–metric conversion factor

STEP 2 Write a plan to convert the given unit to the needed unit.

pounds → U.S.–Metric factor → kilograms

STEP 3 **State the equalities and conversion factors.**

$$1 \text{ kg} = 2.20 \text{ lb}$$
$$\frac{2.20 \text{ lb}}{1 \text{ kg}} \quad \text{and} \quad \frac{1 \text{ kg}}{2.20 \text{ lb}}$$

STEP 4 **Set up the problem to cancel units and calculate the answer.** Write the given, 164 lb, and multiply by the conversion factor that has lb in the denominator (bottom number) to cancel lb in the given.

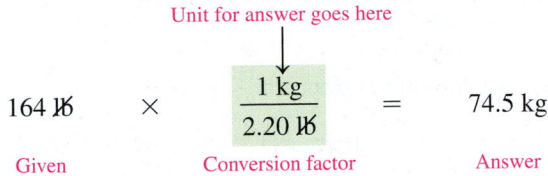

Unit for answer goes here

$$164 \cancel{\text{lb}} \times \frac{1 \text{ kg}}{2.20 \cancel{\text{lb}}} = 74.5 \text{ kg}$$

Given — Conversion factor — Answer

The given unit lb cancels out and the needed unit kg is in the numerator. *The unit you want in the final answer is the one that remains after all the other units have canceled out.* This is a helpful way to check that you set up a problem properly.

$$\cancel{\text{lb}} \times \frac{\text{kg}}{\cancel{\text{lb}}} = \text{kg}$$ Unit needed for answer

The calculator display gives the numerical answer, which is adjusted to give a final answer with the proper number of significant figures (SFs). The value of 74.5 combined with the unit, kg, gives the final answer of 74.5 kg.

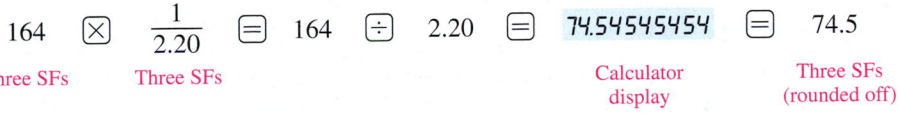

164	×	$\frac{1}{2.20}$	=	164	÷	2.20	=	74.54545454	=	74.5
Three SFs		Exact / Three SFs						Calculator display		Three SFs (rounded off)

STUDY CHECK 2.8

a. A total of 2500 mL of a boric acid antiseptic solution is prepared from boric acid concentrate. How many quarts of boric acid have been prepared?

b. A vial contains 65 mg of phenobarbital per 1 mL of solution. How many milliliters are needed for an order of 40. mg of phenobarbital?

ANSWER

a. 2.6 qt **b.** 0.62 mL

ENGAGE
When you convert one unit to another, how do you know which unit of the conversion factor to place in the denominator?

INTERACTIVE VIDEO
PEARSON eText 2.0 Conversion Factors

TEST
Try Practice Problems 2.55 and 2.56

Using Two or More Conversion Factors

In problem solving, two or more conversion factors are often needed to complete the change of units. In setting up these problems, one factor follows the other. Each factor is arranged to cancel the preceding unit until the needed unit is obtained. Once the problem is set up to cancel units properly, the calculations can be done without writing intermediate results. In this text, when two or more conversion factors are required, the final answer will be based on obtaining a final calculator display and rounding off (or adding zeros) to give the correct number of significant figures as shown in Sample Problem 2.9.

CORE CHEMISTRY SKILL
Using Conversion Factors

▶SAMPLE PROBLEM 2.9 Using Two Conversion Factors

TRY IT FIRST

Greg has been diagnosed with diminished thyroid function. His doctor prescribes a dosage of 0.150 mg of Synthroid to be taken once a day. If tablets in stock contain 75 mcg of Synthroid, how many tablets are required to provide the prescribed medication?

ENGAGE
How can two conversion factors be utilized in a problem setup?

SOLUTION

STEP 1 **State the given and needed quantities.**

ANALYZE THE PROBLEM	Given	Need	Connect
	0.150 mg of Synthroid	number of tablets	metric conversion factor, clinical conversion factor

STEP 2 **Write a plan to convert the given unit to the needed unit.**

milligrams → Metric factor → micrograms → Clinical factor → number of tablets

STEP 3 **State the equalities and conversion factors.**

$$1\ \text{mg} = 1000\ \text{mcg}$$
$$\frac{1000\ \text{mcg}}{1\ \text{mg}} \quad \text{and} \quad \frac{1\ \text{mg}}{1000\ \text{mcg}}$$

$$1\ \text{tablet} = 75\ \text{mcg of Synthroid}$$
$$\frac{75\ \text{mcg Synthroid}}{1\ \text{tablet}} \quad \text{and} \quad \frac{1\ \text{tablet}}{75\ \text{mcg Synthroid}}$$

STEP 4 **Set up the problem to cancel units and calculate the answer.** The problem can be set up using the metric factor to cancel milligrams, and then the clinical factor to obtain the number of tablets as the final unit.

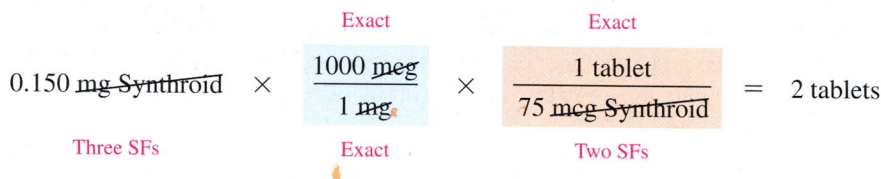

$$0.150\ \cancel{\text{mg Synthroid}} \times \frac{1000\ \cancel{\text{mcg}}}{1\ \cancel{\text{mg}}} \times \frac{1\ \text{tablet}}{75\ \cancel{\text{mcg Synthroid}}} = 2\ \text{tablets}$$

Three SFs; Exact (1000 mcg / 1 mg, both exact); Exact (1 tablet), Two SFs (75 mcg Synthroid)

One teaspoon of cough syrup is measured for a patient.

STUDY CHECK 2.9

a. A bottle contains 120 mL of cough syrup. If one teaspoon (5 mL) is given four times a day, how many days will elapse before a refill is needed?

b. A patient is given a solution containing 0.625 g of calcium carbonate. If the calcium carbonate solution contains 1250 mg per 5 mL, how many milliliters of the solution were given to the patient?

TEST

Try Practice Problems 2.57 to 2.60

ANSWER

a. 6 days **b.** 2.5 mL

▶ SAMPLE PROBLEM 2.10 Using a Percentage as a Conversion Factor

TRY IT FIRST

A person who exercises regularly has 16% body fat by mass. If this person weighs 155 lb, what is the mass, in kilograms, of body fat?

Exercising regularly helps reduce body fat.

SOLUTION

STEP 1 **State the given and needed quantities.**

ANALYZE THE PROBLEM	Given	Need	Connect
	155 lb body weight	kilograms of body fat	U.S.–metric conversion factor, percentage conversion factor

STEP 2 **Write a plan to convert the given unit to the needed unit.**

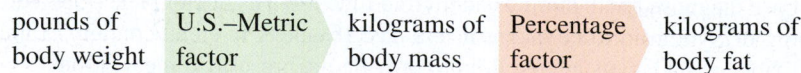

pounds of body weight → U.S.–Metric factor → kilograms of body mass → Percentage factor → kilograms of body fat

STEP 3 State the equalities and conversion factors.

1 kg of body mass = 2.20 lb of body weight

$$\frac{2.20 \text{ lb body weight}}{1 \text{ kg body mass}} \text{ and } \frac{1 \text{ kg body mass}}{2.20 \text{ lb body weight}}$$

100 kg of body mass = 16 kg of body fat

$$\frac{16 \text{ kg body fat}}{100 \text{ kg body mass}} \text{ and } \frac{100 \text{ kg body mass}}{16 \text{ kg body fat}}$$

STEP 4 Set up the problem to cancel units and calculate the answer.

$$\underset{\text{Three SFs}}{155 \cancel{\text{lb body weight}}} \times \overset{\text{Exact}}{\underset{\text{Three SFs}}{\frac{1 \cancel{\text{kg body mass}}}{2.20 \cancel{\text{lb body weight}}}}} \times \overset{\text{Two SFs}}{\underset{\text{Exact}}{\frac{16 \text{ kg body fat}}{100 \cancel{\text{kg body mass}}}}} = \underset{\text{Two SFs}}{11 \text{ kg of body fat}}$$

STUDY CHECK 2.10

a. A package contains 1.33 lb of ground round. If it contains 15% fat, how many grams of fat are in the ground round?

b. A cream contains 4.0% (by mass) lidocaine to relieve back pain. If a patient uses 12 grams of cream, how many milligrams of lidocaine are used?

ANSWER

a. 180 g of fat **b.** 480 mg

TEST

Try Practice Problems 2.61 to 2.66

Chemistry Link to Health

Toxicology and Risk–Benefit Assessment

Each day, we make choices about what we do or what we eat, often without thinking about the risks associated with these choices. We are aware of the risks of cancer from smoking or the risks of lead poisoning, and we know there is a greater risk of having an accident if we cross a street where there is no light or crosswalk.

The LD_{50} of caffeine is 192 mg/kg.

A basic concept of toxicology is the statement of Paracelsus that the dose is the difference between a poison and a cure. To evaluate the level of danger from various substances, natural or synthetic, a risk assessment is made by exposing laboratory animals to the substances and monitoring the health effects. Often, doses very much greater than humans might ordinarily encounter are given to the test animals.

Many hazardous chemicals or substances have been identified by these tests. One measure of toxicity is the LD_{50}, or lethal dose, which is the concentration of the substance that causes death in 50% of the test animals. A dosage is typically measured in milligrams per kilogram (mg/kg) of body mass or micrograms per kilogram (mcg/kg) of body mass.

Other evaluations need to be made, but it is easy to compare LD_{50} values. Parathion, a pesticide, with an LD_{50} of 3 mg/kg, would be highly toxic. This means that 3 mg of parathion per kg of body mass would be fatal to half the test animals. Table salt (sodium chloride) with an LD_{50} of 3300 mg/kg would have a much lower toxicity. You would need to ingest a huge amount of salt before any toxic effect would be observed. Although the risk to animals can be evaluated in the laboratory, it is more difficult to determine the impact in the environment since there is also a difference between continued exposure and a single, large dose of the substance.

TABLE 2.8 lists some LD_{50} values and compares substances in order of increasing toxicity.

TABLE 2.8 Some LD_{50} Values for Substances Tested in Rats

Substance	LD_{50} (mg/kg)
Table sugar	29 700
Boric acid	5140
Baking soda	4220
Table salt	3300
Ethanol	2080
Aspirin	1100
Codeine	800
Oxycodone	480
Caffeine	192
DDT	113
Cocaine (injected)	95
Dichlorvos (pesticide strips)	56
Ricin	30
Sodium cyanide	6
Parathion	3

PRACTICE PROBLEMS

2.6 Problem Solving Using Unit Conversion

2.55 Perform each of the following conversions using metric conversion factors:

a. 44.2 mL to liters
b. 8.65 m to nanometers
c. 5.2×10^8 g to megagrams
d. 0.72 ks to milliseconds

2.56 Perform each of the following conversions using metric conversion factors:

a. 4.82×10^{-5} L to picoliters
b. 575.2 dm to kilometers
c. 5×10^{-4} kg to micrograms
d. 6.4×10^{10} ps to seconds

2.57 Perform each of the following conversions using metric and U.S. conversion factors:

a. 3.428 lb to kilograms
b. 1.6 m to inches
c. 4.2 L to quarts
d. 0.672 ft to millimeters

2.58 Perform each of the following conversions using metric and U.S. conversion factors:

a. 0.21 lb to grams
b. 11.6 in. to centimeters
c. 0.15 qt to milliliters
d. 35.41 kg to pounds

2.59 Use metric conversion factors to solve each of the following problems:

a. If a student is 175 cm tall, how tall is the student in meters?
b. A cooler has a volume of 5000 mL. What is the capacity of the cooler in liters?
c. A hummingbird has a mass of 0.0055 kg. What is the mass, in grams, of the hummingbird?
d. A balloon has a volume of 3500 cm^3. What is the volume in liters?

2.60 Use metric conversion factors to solve each of the following problems:

a. The Daily Value (DV) for phosphorus is 800 mg. How many grams of phosphorus are recommended?
b. A glass of orange juice contains 3.2 dL of juice. How many milliliters of orange juice are in the glass?
c. A package of chocolate instant pudding contains 2840 mg of sodium. How many grams of sodium are in the pudding?
d. A jar contains 0.29 kg of olives. How many grams of olives are in the jar?

2.61 Solve each of the following problems using one or more conversion factors:

a. A container holds 0.500 qt of liquid. How many milliliters of lemonade will it hold?
b. What is the mass, in kilograms, of a person who weighs 175 lb?
c. An athlete has 15% body fat by mass. What is the weight of fat, in pounds, of a 74-kg athlete?
d. A plant fertilizer contains 15% nitrogen (N) by mass. In a container of soluble plant food, there are 10.0 oz of fertilizer. How many grams of nitrogen are in the container?

Agricultural fertilizers applied to a field provide nitrogen for plant growth.

2.62 Solve each of the following problems using one or more conversion factors:

a. Wine is 12% alcohol by volume. How many milliliters of alcohol are in a 0.750-L bottle of wine?
b. Blueberry high-fiber muffins contain 51% dietary fiber by mass. If a package with a net weight of 12 oz contains six muffins, how many grams of fiber are in each muffin?
c. A jar of crunchy peanut butter contains 1.43 kg of peanut butter. If you use 8.0% of the peanut butter for a sandwich, how many ounces of peanut butter did you take out of the container?
d. In a candy factory, the nutty chocolate bars contain 22.0% pecans by mass. If 5.0 kg of pecans were used for candy last Tuesday, how many pounds of nutty chocolate bars were made?

Clinical Applications

2.63 Using conversion factors, solve each of the following clinical problems:

a. You have used 250 L of distilled water for a dialysis patient. How many gallons of water is that?
b. A patient needs 0.024 g of a sulfa drug. There are 8-mg tablets in stock. How many tablets should be given?
c. The daily dose of ampicillin for the treatment of an ear infection is 115 mg/kg of body weight. What is the daily dose for a 34-lb child?
d. You need 4.0 oz of a steroid ointment. How many grams of ointment does the pharmacist need to prepare?

2.64 Using conversion factors, solve each of the following clinical problems:

a. The physician has ordered 1.0 g of tetracycline to be given every six hours to a patient. If your stock on hand is 500-mg tablets, how many will you need for one day's treatment?
b. An intramuscular medication is given at 5.00 mg/kg of body weight. What is the dose for a 180-lb patient?
c. A physician has ordered 0.50 mg of atropine, intramuscularly. If atropine were available as 0.10 mg/mL of solution, how many milliliters would you need to give?
d. During surgery, a patient receives 5.0 pt of plasma. How many milliliters of plasma were given?

2.65 Using conversion factors, solve each of the following clinical problems:

a. A nurse practitioner prepares 500. mL of an IV of normal saline solution to be delivered at a rate of 80. mL/h. What is the infusion time, in hours, to deliver 500. mL?
b. A nurse practitioner orders Medrol to be given 1.5 mg/kg of body weight. Medrol is an anti-inflammatory administered as an intramuscular injection. If a child weighs 72.6 lb and the available stock of Medrol is 20. mg/mL, how many milliliters does the nurse administer to the child?

2.66 Using conversion factors, solve each of the following clinical problems:

a. A nurse practitioner prepares an injection of promethazine, an antihistamine used to treat allergic rhinitis. If the stock bottle is labeled 25 mg/mL and the order is a dose of 12.5 mg, how many milliliters will the nurse draw up in the syringe?
b. You are to give ampicillin 25 mg/kg to a child with a mass of 67 lb. If stock on hand is 250 mg/capsule, how many capsules should be given?

2.7 Density

LEARNING GOAL Calculate the density of a substance; use the density to calculate the mass or volume of a substance.

The mass and volume of any object can be measured. If we compare the mass of the object to its volume, we obtain a relationship called **density**.

$$\text{Density} = \frac{\text{mass of substance}}{\text{volume of substance}}$$

Every substance has a unique density, which distinguishes it from other substances. For example, lead has a density of 11.3 g/mL, whereas cork has a density of 0.26 g/mL. From these densities, we can predict if these substances will sink or float in water. *If an object is less dense than a liquid, the object floats when placed in the liquid.* If a substance, such as cork, is less dense than water, it will float. However, a lead object sinks because its density is greater than that of water (see **FIGURE 2.10**).

ENGAGE

If a piece of iron sinks in water, how does its density compare to that of water?

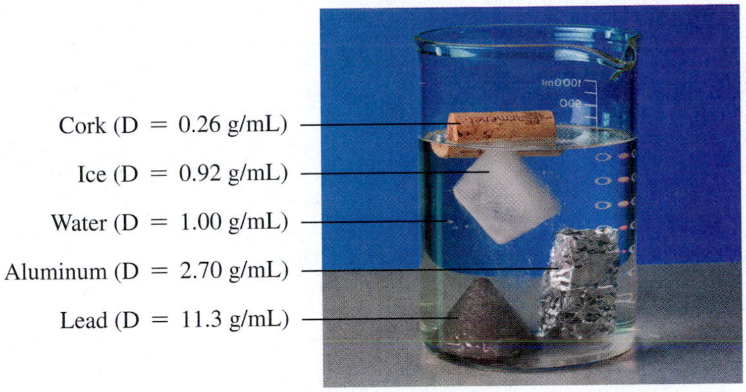
Cork (D = 0.26 g/mL)
Ice (D = 0.92 g/mL)
Water (D = 1.00 g/mL)
Aluminum (D = 2.70 g/mL)
Lead (D = 11.3 g/mL)

FIGURE 2.10 ▶ Objects that sink in water are more dense than water; objects that float are less dense.

Q Why does an ice cube float and a piece of aluminum sink?

Density is used in chemistry in many ways. If we calculate the density of a pure metal as 10.5 g/mL, then we could identify it as silver, but not gold or aluminum. Metals such as gold and silver have higher densities, whereas gases have low densities. In the metric system, the densities of solids and liquids are usually expressed as grams per cubic centimeter (g/cm^3) or grams per milliliter (g/mL). The densities of gases are usually stated as grams per liter (g/L). **TABLE 2.9** gives the densities of some common substances.

TABLE 2.9 Densities of Some Common Substances

Solids (at 25 °C)	Density (g/mL)	Liquids (at 25 °C)	Density (g/mL)	Gases (at 0 °C)	Density (g/L)
Cork	0.26	Gasoline	0.74	Hydrogen	0.090
Body fat	0.909	Ethanol	0.79	Helium	0.179
Ice (at 0 °C)	0.92	Olive oil	0.92	Methane	0.714
Muscle	1.06	Water (at 4 °C)	1.00	Neon	0.902
Sugar	1.59	Urine	1.003–1.030	Nitrogen	1.25
Bone	1.80	Plasma (blood)	1.03	Air (dry)	1.29
Salt (NaCl)	2.16	Milk	1.04	Oxygen	1.43
Aluminum	2.70	Blood	1.06	Carbon dioxide	1.96
Iron	7.86	Mercury	13.6		
Copper	8.92				
Silver	10.5				
Lead	11.3				
Gold	19.3				

Calculating Density

We can calculate the density of a substance from its mass and volume as shown in Sample Problem 2.11.

▶ SAMPLE PROBLEM 2.11 Calculating Density

TRY IT FIRST

High-density lipoprotein (HDL) is a type of cholesterol, sometimes called "good cholesterol," that is measured in a routine blood test. If a 0.258-g sample of HDL has a volume of 0.215 mL, what is the density, in grams per milliliter, of the HDL sample?

SOLUTION

STEP 1 State the given and needed quantities.

ANALYZE THE PROBLEM	Given	Need	Connect
	0.258 g of HDL, 0.215 mL	density (g/mL) of HDL	density expression

STEP 2 Write the density expression.

$$\text{Density} = \frac{\text{mass of substance}}{\text{volume of substance}}$$

STEP 3 Express mass in grams and volume in milliliters.

Mass of HDL sample = 0.258 g
Volume of HDL sample = 0.215 mL

STEP 4 Substitute mass and volume into the density expression and calculate the density.

$$\text{Density} = \frac{\overset{\text{Three SFs}}{0.258\ \text{g}}}{\underset{\text{Three SFs}}{0.215\ \text{mL}}} = \frac{1.20\ \text{g}}{1\ \text{mL}} = \underset{\text{Three SFs}}{1.20\ \text{g/mL}}$$

STUDY CHECK 2.11

a. Low-density lipoprotein (LDL), sometimes called "bad cholesterol," is also measured in a routine blood test. If a 0.380-g sample of LDL has a volume of 0.362 mL, what is the density, in grams per milliliter, of the LDL sample?

b. Osteoporosis is a condition in which bone deteriorates to cause a decreased bone mass. If a bone sample has a mass of 2.15 g and a volume of 1.40 cm^3, what is its density in grams per cubic centimeter?

ANSWER

a. 1.05 g/mL **b.** 1.54 g/cm^3

TEST

Try Practice Problems 2.67 to 2.70

Density of Solids Using Volume Displacement

The volume of a solid can be determined by volume displacement. When a solid is completely submerged in water, it displaces a volume that is equal to the volume of the solid. In **FIGURE 2.11**, the water level rises from 35.5 mL to 45.0 mL after the zinc object is added. This means that 9.5 mL of water is displaced and that the volume of the object is 9.5 mL.

The density of the zinc is calculated using volume displacement as follows:

$$\text{Density} = \frac{\overset{\text{Four SFs}}{68.60 \text{ g Zn}}}{\underset{\text{Two SFs}}{9.5 \text{ mL}}} = \underset{\text{Two SFs}}{7.2 \text{ g/mL}}$$

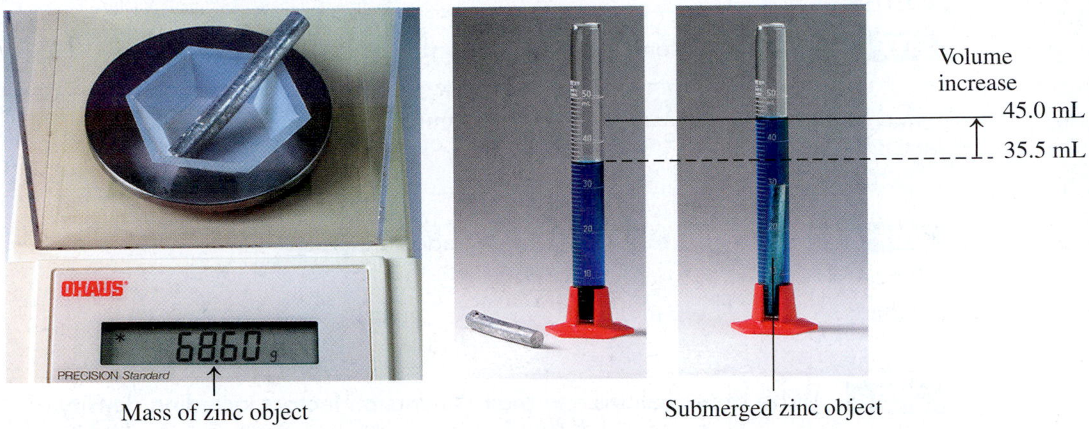

OHAUS
68.60 g
PRECISION Standard
Mass of zinc object
Volume increase
45.0 mL
35.5 mL
Submerged zinc object

FIGURE 2.11 ▶ The density of a solid can be determined by volume displacement because a submerged object displaces a volume of water equal to its own volume.

Q How is the volume of the zinc object determined?

Problem Solving Using Density

CORE CHEMISTRY SKILL
Using Density as a Conversion Factor

Density can be used as a conversion factor. For example, if the volume and the density of a sample are known, the mass in grams of the sample can be calculated as shown in Sample Problem 2.12.

Chemistry Link to Health

Bone Density

Our bones' density is a measure of their health and strength. Our bones are constantly gaining and losing calcium, magnesium, and phosphate. In childhood, bones form at a faster rate than they break down. As we age, bone breakdown occurs more rapidly than new bone forms. As bone loss increases, bones begin to thin, causing a decrease in mass and density. Thinner bones lack strength, which increases the risk of fracture. Hormonal changes, disease, and certain medications can also contribute to the bone thinning. Eventually, a condition of severe bone thinning known as *osteoporosis*, may occur. *Scanning electron micrographs* (SEMs) show **(a)** normal bone and **(b)** bone with osteoporosis due to loss of bone minerals.

Bone density is often determined by passing low-dose X-rays through the narrow part at the top of the femur (hip) and the spine **(c)**. These locations are where fractures are more likely to occur, especially as we age. Bones with high density will block more of the X-rays compared to bones that are less dense. The results of a bone density test are compared to a healthy young adult as well as to other people of the same age.

Recommendations to improve bone strength include calcium and vitamin D supplements. Weight-bearing exercise such as walking and lifting weights can also improve muscle strength, which in turn increases bone strength.

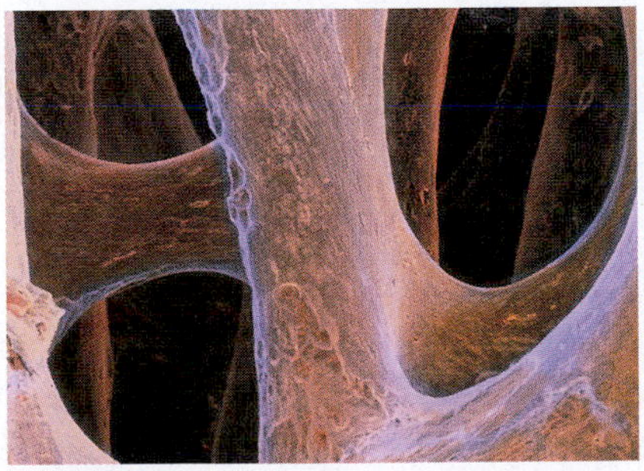

(a) Normal bone

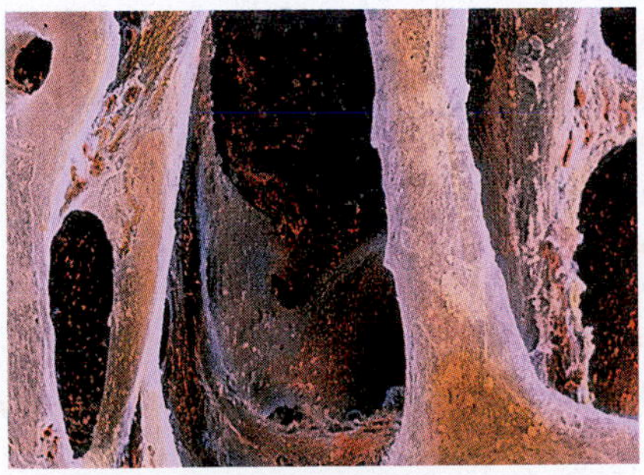

(b) Bone with osteoporosis

(c) Viewing a low-dose X-ray of the spine

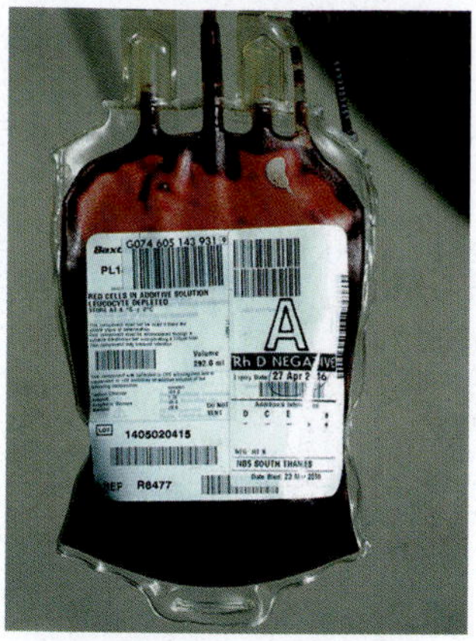

1 pt of blood contains 473 mL.

▶SAMPLE PROBLEM 2.12 Problem Solving Using Density

TRY IT FIRST

Greg has a blood volume of 5.9 qt. If the density of blood is 1.06 g/mL, what is the mass, in grams, of Greg's blood?

SOLUTION

STEP 1 State the given and needed quantities.

	Given	Need	Connect
ANALYZE THE PROBLEM	5.9 qt of blood	grams of blood	U.S.–metric conversion factor, density conversion factor

STEP 2 Write a plan to calculate the needed quantity.

quarts → U.S.–Metric factor → milliliters → Density factor → grams

STEP 3 Write the equalities and their conversion factors including density.

$$1\text{ qt} = 946\text{ mL} \qquad \frac{946\text{ mL}}{1\text{ qt}} \text{ and } \frac{1\text{ qt}}{946\text{ mL}}$$

$$1\text{ mL of blood} = 1.06\text{ g of blood} \qquad \frac{1.06\text{ g blood}}{1\text{ mL blood}} \text{ and } \frac{1\text{ mL blood}}{1.06\text{ g blood}}$$

STEP 4 Set up the problem to calculate the needed quantity.

$$5.9\ \cancel{\text{qt blood}} \times \frac{946\ \cancel{\text{mL}}}{1\ \cancel{\text{qt}}} \times \frac{1.06\text{ g blood}}{1\ \cancel{\text{mL blood}}} = 5900\text{ g of blood}$$

Two SFs (5.9 qt blood); Three SFs / Exact (946 mL / 1 qt); Three SFs / Exact (1.06 g blood / 1 mL blood); Two SFs (5900 g of blood)

STUDY CHECK 2.12

a. During surgery, a patient receives 3.0 pt of blood. How many kilograms of blood (density = 1.06 g/mL) were needed for the transfusion?

b. A woman receives 1280 g of type A blood. If the blood has a density of 1.06 g/mL, how many liters of blood did she receive?

ANSWER

a. 1.5 kg **b.** 1.21 L

TEST

Try Practice Problems 2.71 to 2.76

Specific Gravity

Specific gravity (sp gr) is a relationship between the density of a substance and the density of water. Specific gravity is calculated by dividing the density of a sample by the density of water, which is 1.00 g/mL at 4 °C. A substance with a specific gravity of 1.00 has the same density as water (1.00 g/mL).

$$\text{Specific gravity} = \frac{\text{density of sample}}{\text{density of water}}$$

TEST

Try Practice Problems 2.77 and 2.78

Specific gravity is one of the few unitless values you will encounter in chemistry. The specific gravity of urine helps evaluate the water balance in the body and the substances in the urine. In **FIGURE 2.12**, a hydrometer is used to measure the specific gravity of urine. The normal range of specific gravity for urine is 1.003 to 1.030. The specific gravity can decrease with *type 2 diabetes* and kidney disease. Increased specific gravity may occur with dehydration, kidney infection, and liver disease. In a clinic or hospital, a dipstick containing chemical pads is used to evaluate specific gravity.

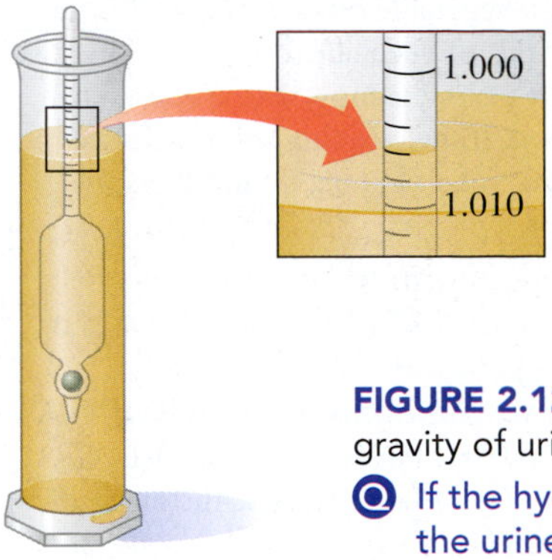
1.000
1.010

FIGURE 2.12 ▶ A hydrometer is used to measure the specific gravity of urine, which, for adults, is 1.003 to 1.030.

Q If the hydrometer reading is 1.006, what is the density of the urine?

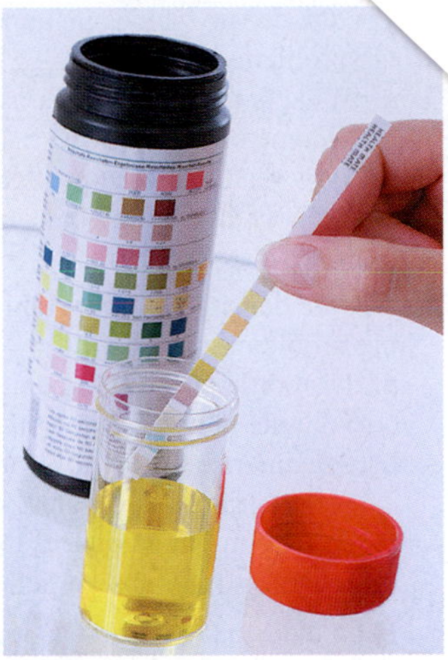

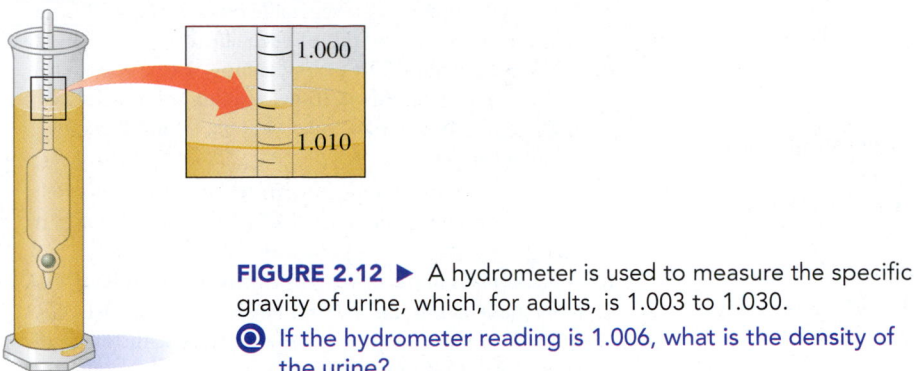

A dipstick is used to measure the specific gravity of a urine sample.

PRACTICE PROBLEMS

2.7 Density

2.67 Determine the density (g/mL) for each of the following:

a. A 20.0-mL sample of a salt solution has a mass of 24.0 g.

b. A cube of butter weighs 0.250 lb and has a volume of 130.3 mL.

c. A gem has a mass of 4.50 g. When the gem is placed in a graduated cylinder containing 12.00 mL of water, the water level rises to 13.45 mL.

d. A 3.00-mL sample of a medication has a mass of 3.85 g.

2.68 Determine the density (g/mL) for each of the following:

a. The fluid in a car battery has a volume of 125 mL and a mass of 155 g.

b. A plastic material weighs 2.68 lb and has a volume of 3.5 L.

c. A 4.000-mL urine sample from a person suffering from diabetes mellitus has a mass of 4.004 g.

d. A solid object has a mass of 1.65 lb and a volume of 170 mL.

2.69 What is the density (g/mL) of each of the following samples?

a. A lightweight head on a golf club is made of titanium. The volume of a sample of titanium is 114 cm^3, and the mass is 514.1 g.

Lightweight heads on golf clubs are made of titanium.

b. A syrup is added to an empty container with a mass of 115.25 g. When 0.100 pt of syrup is added, the total mass of the container and syrup is 182.48 g.

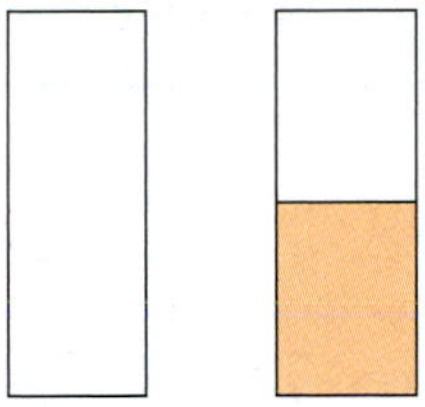

115.25 g **182.48 g**

c. A block of aluminum metal has a volume of 3.15 L and a mass of 8.51 kg.

2.70 What is the density (g/mL) of each of the following samples?

a. An ebony carving has a mass of 275 g and a volume of 207 cm^3.

b. A 14.3-cm^3 sample of tin has a mass of 0.104 kg.

c. A bottle of acetone (fingernail polish remover) contains 55.0 mL of acetone with a mass of 43.5 g.

2.71 Use the density values in Table 2.9 to solve each of the following problems:

a. How many liters of ethanol contain 1.50 kg of ethanol?

b. How many grams of mercury are present in a barometer that holds 6.5 mL of mercury?

c. A sculptor has prepared a mold for casting a silver figure. The figure has a volume of 225 cm^3. How many ounces of silver are needed in the preparation of the silver figure?

2.72 Use the density values in Table 2.9 to solve each of the following problems:

a. A graduated cylinder contains 18.0 mL of water. What is the new water level, in milliliters, after 35.6 g of silver metal is submerged in the water?

b. A thermometer containing 8.3 g of mercury has broken. What volume, in milliliters, of mercury spilled?

c. A fish tank holds 35 gal of water. How many kilograms of water are in the fish tank?

2.73 Use the density values in Table 2.9 to solve each of the following problems:

a. What is the mass, in grams, of a cube of copper that has a volume of 74.1 cm^3?
b. How many kilograms of gasoline fill a 12.0-gal gas tank?
c. What is the volume, in cubic centimeters, of an ice cube that has a mass of 27 g?

2.74 Use the density values in Table 2.9 to solve each of the following problems:

a. If a bottle of olive oil contains 1.2 kg of olive oil, what is the volume, in milliliters, of the olive oil?
b. A cannon ball made of iron has a volume of 115 cm^3. What is the mass, in kilograms, of the cannon ball?
c. A balloon filled with helium has a volume of 7.3 L. What is the mass, in grams, of helium in the balloon?

2.75 In an old trunk, you find a piece of metal that you think may be aluminum, silver, or lead. You take it to a lab, where you find it has a mass of 217 g and a volume of 19.2 cm^3. Using Table 2.9, what is the metal you found?

2.76 Suppose you have two 100-mL graduated cylinders. In each cylinder, there is 40.0 mL of water. You also have two cubes: one is lead, and the other is aluminum. Each cube measures 2.0 cm on each side. After you carefully lower each cube into the water of its own cylinder, what will the new water level be in each of the cylinders? Use Table 2.9 for density values.

Clinical Applications

2.77 Solve each of the following problems:

a. A urine sample has a density of 1.030 g/mL. What is the specific gravity of the sample?
b. A 20.0-mL sample of a glucose IV solution has a mass of 20.6 g. What is the density of the glucose solution?
c. The specific gravity of a vegetable oil is 0.92. What is the mass, in grams, of 750 mL of vegetable oil?
d. A bottle containing 325 g of cleaning solution is used to clean hospital equipment. If the cleaning solution has a specific gravity of 0.850, what volume, in milliliters, of solution was used?

2.78 Solve each of the following problems:

a. A glucose solution has a density of 1.02 g/mL. What is its specific gravity?
b. A 0.200-mL sample of very-low-density lipoprotein (VLDL) has a mass of 190 mg. What is the density of the VLDL?
c. Butter has a specific gravity of 0.86. What is the mass, in grams, of 2.15 L of butter?
d. A 5.000-mL urine sample has a mass of 5.025 g. If the normal range for the specific gravity of urine is 1.003 to 1.030, would the specific gravity of this urine sample indicate that the patient could have type 2 diabetes?

CLINICAL UPDATE Greg's Visit with His Doctor

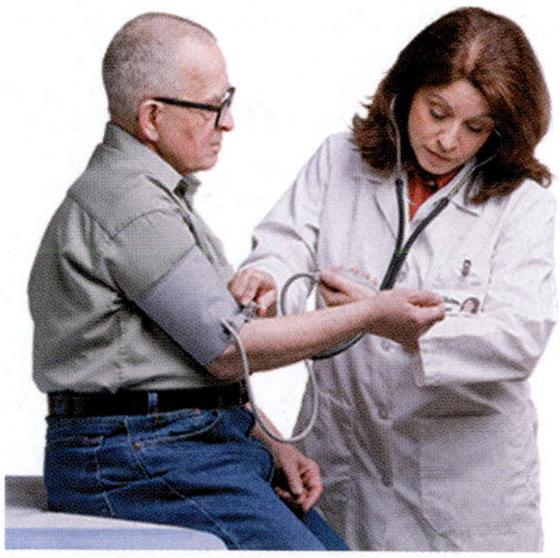

On Greg's visit to his doctor, he complains of feeling tired. Sandra, the registered nurse, withdraws 8.0 mL of blood, which is sent to the lab and tested for iron. When the iron level is low, a person may have fatigue and decreased immunity.

The normal range for serum iron in men is 80 to 160 mcg/dL. Greg's iron test shows a blood serum iron level of 42 mcg/dL, which indicates that Greg has *iron-deficiency anemia*. His doctor orders an iron supplement. One tablet of the iron supplement contains 50 mg of iron.

Clinical Applications

2.79 **a.** Write an equality and two conversion factors for Greg's serum iron level.
b. How many micrograms of iron were in the 8.0-mL sample of Greg's blood?

2.80 **a.** Write an equality and two conversion factors for one tablet of the iron supplement.
b. How many grams of iron will Greg consume in one week, if he takes two tablets each day?

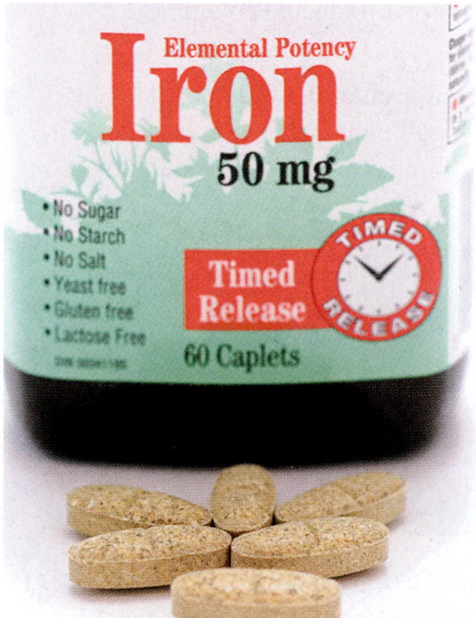
Elemental Potency
Iron
50 mg
•No Sugar
•No Starch
•No Salt
•Yeast free
•Gluten free
•Lactose Free
Timed Release
TIMED RELEASE
60 Caplets

Each tablet contains 50 mg of iron, which is given for iron supplementation.

CONCEPT MAP

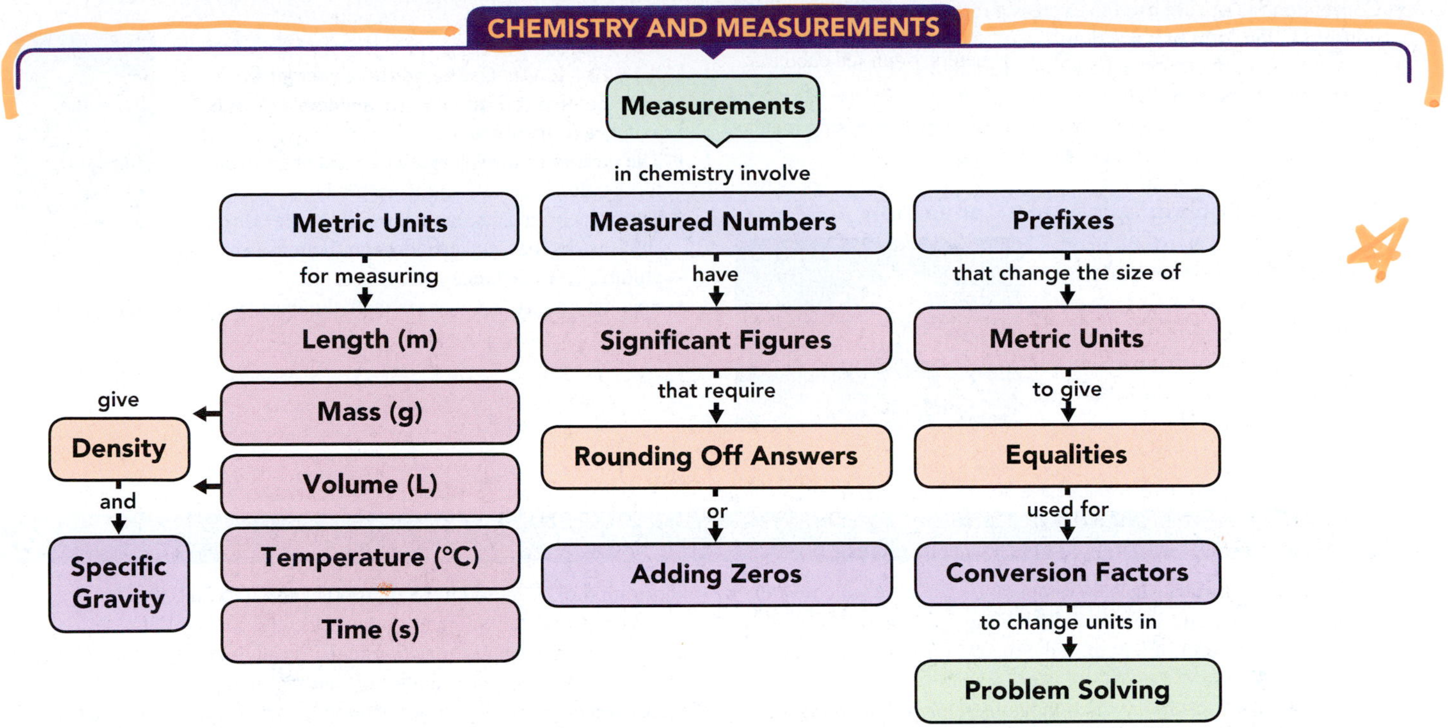
CHEMISTRY AND MEASUREMENTS
Measurements
in chemistry involve
Metric Units
for measuring
Length (m)
Mass (g)
Volume (L)
Temperature (°C)
Time (s)
give
Density
and
Specific Gravity
Measured Numbers
have
Significant Figures
that require
Rounding Off Answers
or
Adding Zeros
Prefixes
that change the size of
Metric Units
to give
Equalities
used for
Conversion Factors
to change units in
Problem Solving

CHAPTER REVIEW

2.1 Units of Measurement

LEARNING GOAL Write the names and abbreviations for the metric and SI units used in measurements of volume, length, mass, temperature, and time.

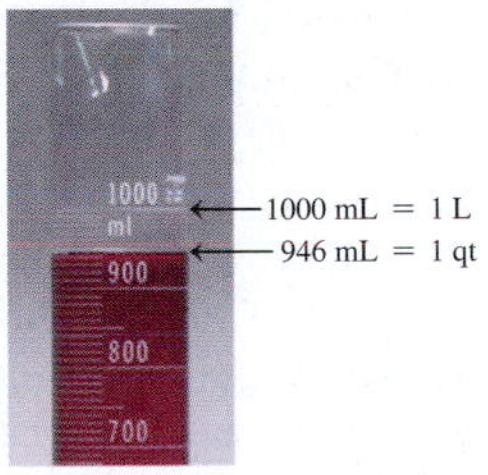
1000 mL = 1 L
946 mL = 1 qt

- In science, physical quantities are described in units of the metric or International System of Units (SI).
- Some important units are liter (L) for volume, meter (m) for length, gram (g) and kilogram (kg) for mass, degree Celsius (°C) and kelvin (K) for temperature, and second (s) for time.

2.2 Measured Numbers and Significant Figures

LEARNING GOAL Identify a number as measured or exact; determine the number of significant figures in a measured number.

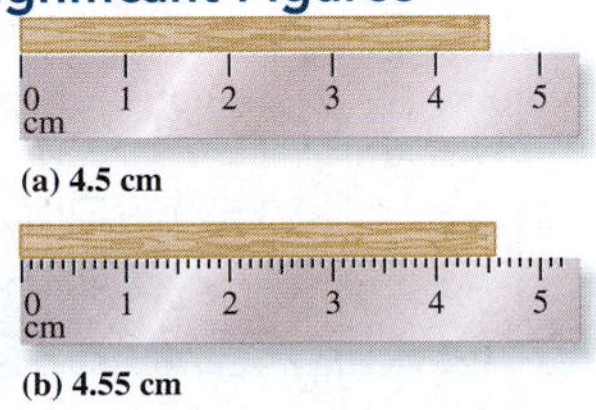
0 1 2 3 4 5
cm
(a) 4.5 cm
0 1 2 3 4 5
cm
(b) 4.55 cm

- A measured number is any number obtained by using a measuring device.
- An exact number is obtained by counting items or from a definition; no measuring device is needed.
- Significant figures are the numbers reported in a measurement including the estimated digit.
- Zeros in front of a decimal number or at the end of a nondecimal number are not significant.

2.3 Significant Figures in Calculations

LEARNING GOAL Give the correct number of significant figures for a calculated answer.

- In multiplication and division, the final answer is written so that it has the same number of significant figures as the measurement with the fewest significant figures.
- In addition and subtraction, the final answer is written so that it has the same number of decimal places as the measurement with the fewest decimal places.

2.4 Prefixes and Equalities

LEARNING GOAL Use the numerical values of prefixes to write a metric equality.

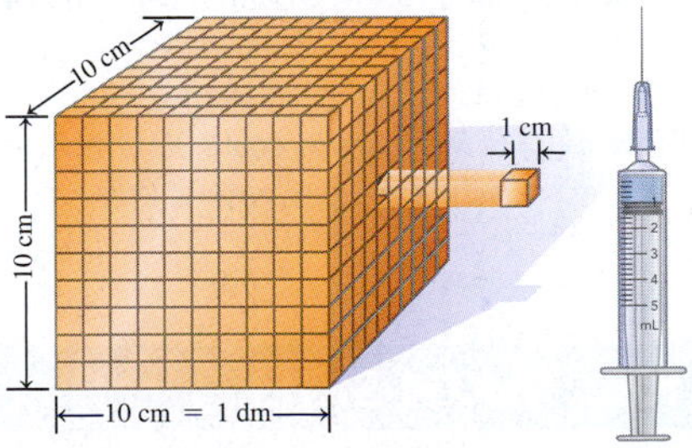
10 cm
10 cm
1 cm
10 cm = 1 dm

- A prefix placed in front of a metric or SI unit changes the size of the unit by factors of 10.
- Prefixes such as *centi*, *milli*, and *micro* provide smaller units; prefixes such as *kilo*, *mega*, and *tera* provide larger units.
- An equality shows the relationship between two units that measure the same quantity of volume, length, mass, or time.
- Examples of metric equalities are 1 L = 1000 mL, 1 m = 100 cm, 1 kg = 1000 g, and 1 min = 60 s.

2.5 Writing Conversion Factors

NET WT. 15 OZ (425g)
NET WT. 5 OZ (142 g)

LEARNING GOAL Write a conversion factor for two units that describe the same quantity.

- Conversion factors are used to express a relationship in the form of a fraction.
- Two conversion factors can be written for any relationship in the metric or U.S. system.
- A percentage is written as a conversion factor by expressing matching units as the parts in 100 parts of the whole.

2.6 Problem Solving Using Unit Conversion

LEARNING GOAL Use conversion factors to change from one unit to another.

- Conversion factors are useful when changing a quantity expressed in one unit to a quantity expressed in another unit.
- In the problem-solving process, a given unit is multiplied by one or more conversion factors that cancel units until the needed answer is obtained.

2.7 Density

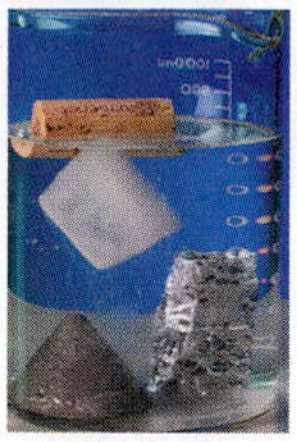

LEARNING GOAL Calculate the density of a substance; use the density to calculate the mass or volume of a substance.

- The density of a substance is a ratio of its mass to its volume, usually g/mL or g/cm^3.
- The units of density can be used to write conversion factors that convert between the mass and volume of a substance.
- Specific gravity (sp gr) compares the density of a substance to the density of water, 1.00 g/mL.

KEY TERMS

Celsius (°C) temperature scale A temperature scale on which water has a freezing point of 0 °C and a boiling point of 100 °C.

centimeter (cm) A unit of length in the metric system; there are 2.54 cm in 1 in.

conversion factor A ratio in which the numerator and denominator are quantities from an equality or given relationship. For example, the two conversion factors for the equality 1 kg = 2.20 lb are written as

$$\frac{2.20\text{ lb}}{1\text{ kg}} \quad \text{and} \quad \frac{1\text{ kg}}{2.20\text{ lb}}$$

cubic centimeter (cm^3, cc) The volume of a cube that has 1-cm sides; 1 cm^3 is equal to 1 mL.

density The relationship of the mass of an object to its volume expressed as grams per cubic centimeter (g/cm^3), grams per milliliter (g/mL), or grams per liter (g/L).

equality A relationship between two units that measure the same quantity.

exact number A number obtained by counting or by definition.

gram (g) The metric unit used in measurements of mass.

International System of Units (SI) The official system of measurement throughout the world, except for the United States, that modifies the metric system.

Kelvin (K) temperature scale A temperature scale on which the lowest possible temperature is 0 K.

kilogram (kg) A metric mass of 1000 g, equal to 2.20 lb. The kilogram is the SI standard unit of mass.

liter (L) The metric unit for volume that is slightly larger than a quart.

mass A measure of the quantity of material in an object.

measured number A number obtained when a quantity is determined by using a measuring device.

meter (m) The metric unit for length that is slightly longer than a yard. The meter is the SI standard unit of length.

metric system A system of measurement used by scientists and in most countries of the world.

milliliter (mL) A metric unit of volume equal to one-thousandth of a liter (0.001 L).

prefix The part of the name of a metric unit that precedes the base unit and specifies the size of the measurement. All prefixes are related on a decimal scale.

second (s) A unit of time used in both the SI and metric systems.

SI See International System of Units (SI).

significant figures (SFs) The numbers recorded in a measurement.

specific gravity (sp gr) A relationship between the density of a substance and the density of water:

$$\text{sp gr} = \frac{\text{density of sample}}{\text{density of water}}$$

temperature An indicator of the hotness or coldness of an object.

volume (*V*) The amount of space occupied by a substance.

KEY MATH SKILL

The chapter Section containing each Key Math Skill is shown in parentheses at the end of each heading.

Rounding Off (2.3)

Calculator displays are rounded off to give the correct number of significant figures.

- If the first digit to be dropped is *4 or less*, then it and all following digits are simply dropped from the number.
- If the first digit to be dropped is *5 or greater*, then the last retained digit of the number is increased by 1.

One or more significant zeros are added when the calculator display has fewer digits than the needed number of significant figures.

Example: Round off each of the following to three significant figures:

a. 3.608 92 L
b. 0.003 870 298 m
c. 6 g

Answer: **a.** 3.61 L **b.** 0.003 87 m **c.** 6.00 g

CORE CHEMISTRY SKILLS

The chapter Section containing each Core Chemistry Skill is shown in parentheses at the end of each heading.

Counting Significant Figures (2.2)

The significant figures (SFs) are all the measured numbers including the last, estimated digit.

- All nonzero digits
- Zeros between nonzero digits
- Zeros within a decimal number
- All digits in a coefficient of a number written in scientific notation

An *exact* number is obtained from counting or a definition and has no effect on the number of significant figures in the final answer.

Example: State the number of significant figures in each of the following:

a. 0.003 045 mm
b. 15 000 m
c. 45.067 kg
d. 5.30×10^3 g
e. 2 cans of soda

Answer:
a. four SFs
b. two SFs
c. five SFs
d. three SFs
e. exact

Using Significant Figures in Calculations (2.3)

- In multiplication or division, the final answer is written so that it has the same number of significant figures as the measurement with the fewest SFs.
- In addition or subtraction, the final answer is written so that it has the same number of decimal places as the measurement having the fewest decimal places.

Example: Perform the following calculations using measured numbers, and give answers with the correct number of SFs or decimal places:

a. $4.05 \text{ m} \times 0.6078 \text{ m}$
b. $\dfrac{4.50 \text{ g}}{3.27 \text{ mL}}$
c. $0.758 \text{ g} + 3.10 \text{ g}$
d. $13.538 \text{ km} - 8.6 \text{ km}$

Answer: **a.** 2.46 m^2 **b.** 1.38 g/mL **c.** 3.86 g **d.** 4.9 km

Using Prefixes (2.4)

In the metric and SI systems of units, a prefix attached to any unit increases or decreases its size by some factor of 10.

- When the prefix *centi* is used with the unit meter, it becomes centimeter, a length that is one-hundredth of a meter (0.01 m).
- When the prefix *milli* is used with the unit meter, it becomes millimeter, a length that is one-thousandth of a meter (0.001 m).

Example: Complete each of the following metric relationships:

a. 1000 m = 1 ____ m **b.** 0.01 g = 1 ____ g

Answer: **a.** 1000 m = 1 km **b.** 0.01 g = 1 cg

Writing Conversion Factors from Equalities (2.5)

- A conversion factor allows you to change from one unit to another.
- Two conversion factors can be written for any equality in the metric, U.S., or metric–U.S. systems of measurement.
- Two conversion factors can be written for a relationship stated within a problem.

Example: Write two conversion factors for the equality: 1 L = 1000 mL.

Answer: $\dfrac{1000 \text{ mL}}{1 \text{ L}}$ and $\dfrac{1 \text{ L}}{1000 \text{ mL}}$

Using Conversion Factors (2.6)

In problem solving, conversion factors are used to cancel the given unit and to provide the needed unit for the answer.

- State the given and needed quantities.
- Write a plan to convert the given unit to the needed unit.
- State the equalities and conversion factors.
- Set up the problem to cancel units and calculate the answer.

Example: A computer chip has a width of 0.75 in. What is the width in millimeters?

Answer: $0.75 \cancel{\text{in.}} \times \dfrac{2.54 \cancel{\text{cm}}}{1 \cancel{\text{in.}}} \times \dfrac{10 \text{ mm}}{1 \cancel{\text{cm}}} = 19 \text{ mm}$

Using Density as a Conversion Factor (2.7)

Density is an equality of mass and volume for a substance, which is written as the *density expression.*

$$\text{Density} = \frac{\text{mass of substance}}{\text{volume of substance}}$$

Density is useful as a conversion factor to convert between mass and volume.

Example: The element tungsten used in light bulb filaments has a density of 19.3 g/cm^3. What is the volume, in cubic centimeters, of 250 g of tungsten?

Answer: $250 \cancel{\text{g}} \times \dfrac{1 \text{ cm}^3}{19.3 \cancel{\text{g}}} = 13 \text{ cm}^3$

UNDERSTANDING THE CONCEPTS

The chapter Sections to review are shown in parentheses at the end of each problem.

2.81 In which of the following pairs do both numbers contain the same number of significant figures? (2.2)
a. 2.0500 m and 0.0205 m
b. 600.0 K and 60 K
c. 0.000 75 s and 75 000 s
d. 6.240 L and 6.240×10^{-2} L

2.82 In which of the following pairs do both numbers contain the same number of significant figures? (2.2)
a. 3.44×10^{-3} g and 0.0344 g
b. 0.0098 s and 9.8×10^4 s
c. 6.8×10^3 m and 68 000 m
d. 258.000 g and 2.58×10^{-2} g

2.83 Indicate if each of the following is answered with an exact number or a measured number: (2.2)

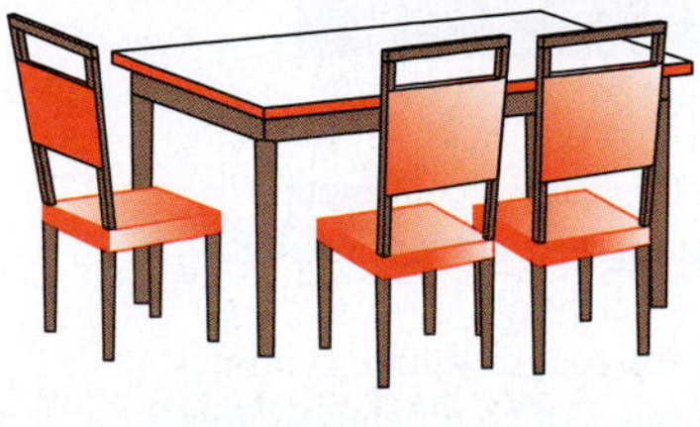

a. number of legs
b. height of table
c. number of chairs at the table
d. area of tabletop

2.84 Measure the length of each of the objects in diagrams **(a)**, **(b)**, and **(c)** using the metric ruler in the figure. Indicate the number of significant figures for each and the estimated digit for each. (2.2)

0
cm
1
2
3
4
5

(a)

0
cm
1
2
3
4
5

(b)

0
cm
1
2
3
4
5

(c)

2.85 State the temperature on the Celsius thermometer to the correct number of significant figures: (2.3)

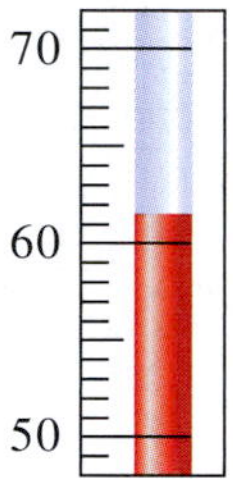
70
60
50

2.86 State the temperature on the Celsius thermometer to the correct number of significant figures: (2.3)

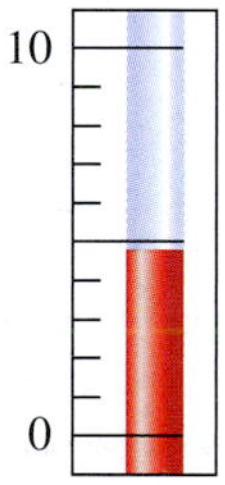
10
0

2.87 The length of this rug is 38.4 in. and the width is 24.2 in. (2.3, 2.6)

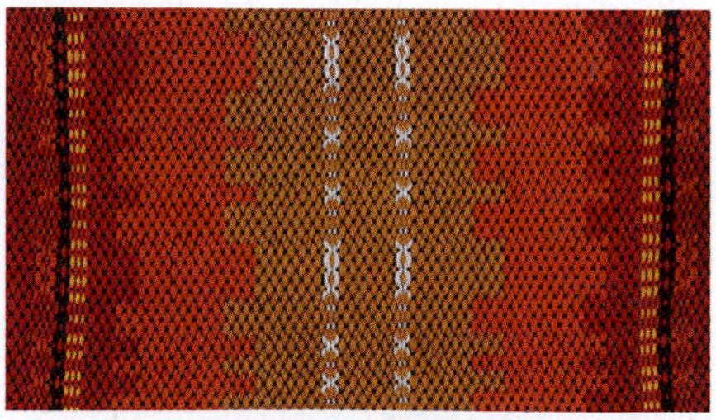

a. What is the length of this rug, in centimeters?
b. What is the width of this rug, in centimeters?
c. How many significant figures are in the length measurement?
d. Calculate the area of the rug, in square centimeters, to the correct number of significant figures. (Area = Length × Width)

2.88 A shipping box has a length of 7.00 in., a width of 6.00 in., and a height of 4.00 in. (2.3, 2.6)

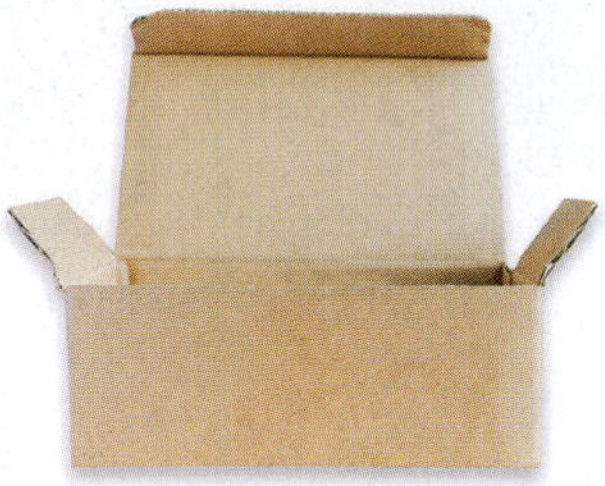

a. What is the length of the box, in centimeters?
b. What is the width of the box, in centimeters?
c. How many significant figures are in the width measurement?
d. Calculate the volume of the box, in cubic centimeters, to the correct number of significant figures. (Volume = Length × Width × Height)

2.89 Each of the following diagrams represents a container of water and a cube. Some cubes float while others sink. Match diagrams **1**, **2**, **3**, or **4** with one of the following descriptions and explain your choices: (2.7)

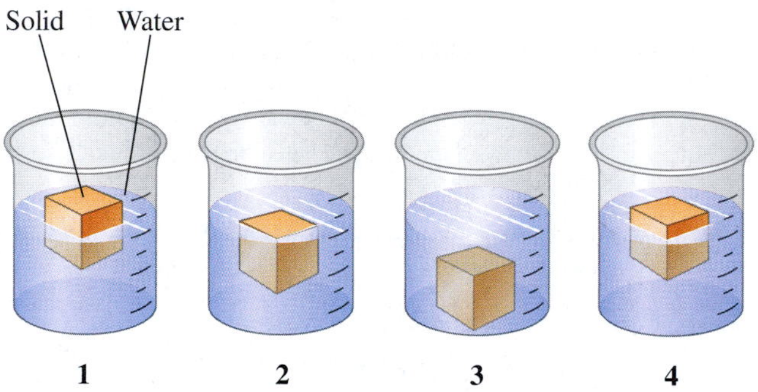
Solid
Water
1
2
3
4

a. The cube has a greater density than water.
b. The cube has a density that is 0.80 g/mL.
c. The cube has a density that is one-half the density of water.
d. The cube has the same density as water.

2.90 What is the density of the solid object that is weighed and submerged in water? (2.7)

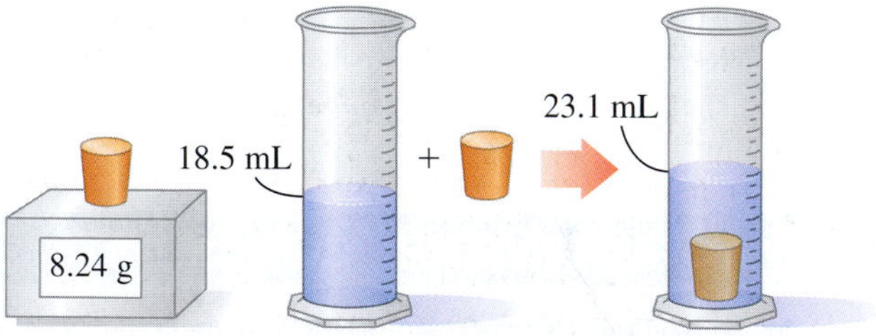
8.24 g
18.5 mL
+
23.1 mL

2.91 Consider the following solids. The solids **A**, **B**, and **C** represent aluminum (D = 2.70 g/mL), gold (D = 19.3 g/mL), and silver (D = 10.5 g/mL). If each has a mass of 10.0 g, what is the identity of each solid? (2.7)

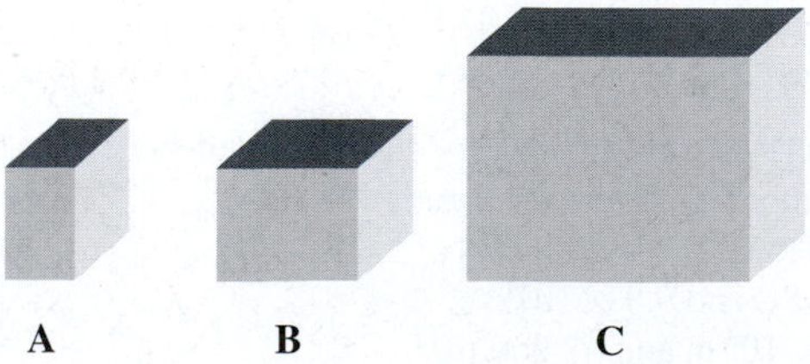
A
B
C

2.92 A graduated cylinder contains three liquids **A**, **B**, and **C**, which have different densities and do not mix: mercury (D = 13.6 g/mL), vegetable oil (D = 0.92 g/mL), and water (D = 1.00 g/mL). Identify the liquids **A**, **B**, and **C** in the cylinder. (2.7)

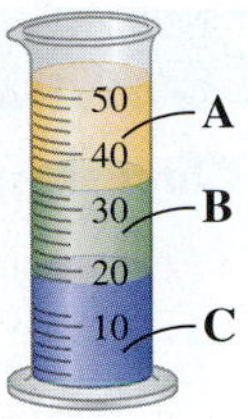
50
40
30
20
10
A
B
C

2.93 The gray cube has a density of 4.5 g/cm^3. Is the density of the green cube the same, lower than, or higher than that of the gray cube? (2.7)

2.94 The gray cube has a density of 4.5 g/cm^3. Is the density of the green cube the same, lower than, or higher than that of the gray cube? (2.7)

ADDITIONAL PRACTICE PROBLEMS

2.95 Round off or add zeros to the following calculated answers to give a final answer with three significant figures: (2.2)
a. 0.000 012 58 L **b.** 3.528 × 10^2 kg
c. 125 111 m **d.** 34.9673 s

2.96 Round off or add zeros to the following calculated answers to give a final answer with three significant figures: (2.2)
a. 58.703 mL **b.** 3 × 10^{-3} s
c. 0.010 826 g **d.** 1.7484 × 10^3 ms

2.97 A dessert contains 137.25 g of vanilla ice cream, 84 g of fudge sauce, and 43.7 g of nuts. (2.3, 2.6)
a. What is the total mass, in grams, of the dessert?
b. What is the total weight, in pounds, of the dessert?

2.98 A fish company delivers 22 kg of salmon, 5.5 kg of crab, and 3.48 kg of oysters to your seafood restaurant. (2.3, 2.6)
a. What is the total mass, in kilograms, of the seafood?
b. What is the total number of pounds?

2.99 In France, grapes are 1.95 euros per kilogram. What is the cost of grapes, in dollars per pound, if the exchange rate is 1.14 dollars/euro? (2.6)

2.100 In Mexico, avocados are 48 pesos per kilogram. What is the cost, in cents, of an avocado that weighs 0.45 lb if the exchange rate is 18 pesos to the dollar? (2.6)

2.101 Bill's recipe for onion soup calls for 4.0 lb of thinly sliced onions. If an onion has an average mass of 115 g, how many onions does Bill need? (2.6)

2.102 The price of 1 lb of potatoes is $1.75. If all the potatoes sold today at the store bring in $1420, how many kilograms of potatoes did grocery shoppers buy? (2.6)

2.103 During a workout at the gym, you set the treadmill at a pace of 55.0 m/min. How many minutes will you walk if you cover a distance of 7500 ft? (2.6)

2.104 The distance between two cities is 1700 km. How long will it take, in hours, to drive from one city to the other if your average speed is 63 mi/h? (2.6)

2.105 The water level in a graduated cylinder initially at 215 mL rises to 285 mL after a piece of lead is submerged. What is the mass, in grams, of the lead (see Table 2.9)? (2.7)

2.106 A graduated cylinder contains 155 mL of water. A 15.0-g piece of iron and a 20.0-g piece of lead are added. What is the new water level, in milliliters, in the cylinder (see Table 2.9)? (2.7)

2.107 How many milliliters of gasoline have a mass of 1.2 kg (see Table 2.9)? (2.7)

2.108 What is the volume, in quarts, of 3.40 kg of ethanol (see Table 2.9)? (2.7)

Clinical Applications

2.109 The following nutrition information is listed on a box of crackers: (2.6)

Serving size 0.50 oz (6 crackers)

Fat 4 g per serving; Sodium 140 mg per serving

a. If the box has a net weight (contents only) of 8.0 oz, about how many crackers are in the box?
b. If you ate 10 crackers, how many ounces of fat did you consume?
c. How many servings of crackers in part **a** would it take to obtain the Daily Value (DV) for sodium, which is 2.4 g?

2.110 A dialysis unit requires 75 000 mL of distilled water. How many gallons of water are needed? (2.6)

2.111 To treat a bacterial infection, a doctor orders 4 tablets of amoxicillin per day for 10 days. If each tablet contains 250 mg of amoxicillin, how many ounces of the medication are given in 10 days? (2.6)

2.112 Celeste's diet restricts her intake of protein to 24 g per day. If she eats 1.2 oz of protein, has she exceeded her protein limit for the day? (2.6)

2.113 A doctor orders 5.0 mL of phenobarbital elixir. If the phenobarbital elixir is available as 30. mg per 7.5 mL, how many milligrams is given to the patient? (2.6)

2.114 A doctor orders 2.0 mg of morphine. The vial of morphine on hand is 10. mg/mL. How many milliliters of morphine should you administer to the patient? (2.6)

CHALLENGE PROBLEMS

The following problems are related to the topics in this chapter. However, they do not all follow the chapter order, and may require you to combine concepts and skills from several Sections. These problems will help you increase your critical thinking skills and prepare for your next exam.

2.115 A balance measures mass to 0.001 g. If you determine the mass of an object that weighs about 31 g, would you record the mass as 31 g, 31.1 g, 31.08 g, 31.075 g, or 31.0750? Explain your choice by writing two to three complete sentences that describe your thinking. (2.3)

2.116 When three students use the same meterstick to measure the length of a paper clip, they obtain results of 5.8 cm, 5.75 cm, and 5.76 cm. If the meterstick has millimeter markings, what are some reasons for the different values? (2.3)

2.117 A car travels at 55 mi/h and gets 11 km/L of gasoline. How many gallons of gasoline are needed for a 3.0-h trip? (2.6)

2.118 A sunscreen preparation contains 2.50% benzyl salicylate by mass. If a tube contains 4.0 oz of sunscreen, how many kilograms of benzyl salicylate are needed to manufacture 325 tubes of sunscreen? (2.6)

2.119 How many milliliters of olive oil have the same mass as 1.50 L of gasoline (see Table 2.9)? (2.7)

2.120 A 50.0-g silver object and a 50.0-g gold object are both added to 75.5 mL of water contained in a graduated cylinder. What is the new water level, in milliliters, in the cylinder (see Table 2.9)? (2.7)

Clinical Applications

2.121 **a.** An athlete with a body mass of 65 kg has 3.0% body fat. How many pounds of body fat does that person have? (2.6)
b. In liposuction, a doctor removes fat deposits from a person's body. If body fat has a density of 0.909 g/mL and 3.0 L of fat is removed, how many pounds of fat were removed from the patient?

2.122 A mouthwash is 21.6% ethanol by mass. If each bottle contains 1.06 pt of mouthwash with a density of 0.876 g/mL, how many kilograms of ethanol are in 180 bottles of the mouthwash? (2.6, 2.7)

ANTISEPTIC
mouth
rinse

A mouthwash may contain over 20% ethanol.

ANSWERS

2.1 **a.** g **b.** °C **c.** L **d.** lb **e.** s

2.3 **a.** volume **b.** length **c.** length **d.** time

2.5 **a.** meter, length **b.** gram, mass
c. milliliter, volume **d.** second, time
e. degree Celsius, temperature

2.7 **a.** second, time **b.** kilogram, mass
c. gram, mass **d.** degree Celsius, temperature

2.9 **a.** five SFs **b.** two SFs **c.** two SFs
d. three SFs **e.** four SFs **f.** three SFs

2.11 b and c

2.13 **a.** not significant **b.** significant
c. significant **d.** significant
e. not significant

2.15 **a.** 5.0×10^3 L **b.** 3.0×10^4 g
c. 1.0×10^5 m **d.** 2.5×10^{-4} cm

2.17 **a.** measured **b.** exact
c. exact **d.** measured

2.19 **a.** 6 oz **b.** none
c. 0.75 lb, 350 g **d.** none (definitions are exact)

2.21 **a.** measured, four SFs **b.** measured, two SFs
c. measured, three SFs **d.** exact

2.23 **a.** 1.85 kg **b.** 88.2 L
c. 0.004 74 cm **d.** 8810 m
e. 1.83×10^5 s

2.25 **a.** 56.9 m **b.** 0.002 28 g
c. 11 500 s (1.15×10^4 s) **d.** 8.10 L

2.27 **a.** 1.6 **b.** 0.01
c. 27.6 **d.** 3.5
e. 0.14 (1.4×10^{-1}) **f.** 0.8 (8×10^{-1})

2.29 **a.** 53.54 **b.** 127.6
c. 121.5 **d.** 0.50
e. 2500 **f.** 156.7

2.31 **a.** mg **b.** dL **c.** km **d.** pg

2.33 **a.** centiliter **b.** kilogram
c. millisecond **d.** petameter

2.35 **a.** 0.01 **b.** 10^{12} **c.** 0.001 **d.** 0.1

2.37 **a.** decigram **b.** microgram
c. kilogram **d.** centigram

2.39 **a.** 100 cm **b.** 1×10^9 nm
c. 0.001 m **d.** 1000 mL

2.41 **a.** kilogram **b.** milliliter
c. km **d.** kL
e. nanometer

2.43 A conversion factor can be inverted to give a second conversion factor.

2.45 **a.** 1 m = 100 cm; $\frac{100\text{ cm}}{1\text{ m}}$ and $\frac{1\text{ m}}{100\text{ cm}}$

b. 1 g = 1×10^9 ng; $\frac{1 \times 10^9\text{ ng}}{1\text{ g}}$ and $\frac{1\text{ g}}{1 \times 10^9\text{ ng}}$

c. 1 kL = 1000 L; $\frac{1000\text{ L}}{1\text{ kL}}$ and $\frac{1\text{ kL}}{1000\text{ L}}$

d. 1 s = 1000 ms; $\frac{1000\text{ ms}}{1\text{ s}}$ and $\frac{1\text{ s}}{1000\text{ ms}}$

e. 1 dm = 100 mm; $\frac{100\text{ mm}}{1\text{ dm}}$ and $\frac{1\text{ dm}}{100\text{ mm}}$

2.47 **a.** 1 yd = 3 ft; $\frac{3\text{ ft}}{1\text{ yd}}$ and $\frac{1\text{ yd}}{3\text{ ft}}$

The 1 yd and 3 ft are both exact.

b. 1 kg = 2.20 lb; $\frac{2.20\text{ lb}}{1\text{ kg}}$ and $\frac{1\text{ kg}}{2.20\text{ lb}}$

The 2.20 lb is measured: It has three SFs. The 1 kg is exact.

c. 1 gal = 27 mi; $\frac{27\text{ mi}}{1\text{ gal}}$ and $\frac{1\text{ gal}}{27\text{ mi}}$

The 27 mi is measured: It has two SFs. The 1 gal is exact.

d. 100 g of sterling = 93 g of silver;

$\frac{93\text{ g silver}}{100\text{ g sterling}}$ and $\frac{100\text{ g sterling}}{93\text{ g silver}}$

The 93 g is measured: It has two SFs. The 100 g is exact.

e. 1 min = 60 s; $\frac{60\text{ s}}{1\text{ min}}$ and $\frac{1\text{ min}}{60\text{ s}}$

The 1 min and 60 s are both exact.

2.49 **a.** 1 s = 3.5 m; $\frac{3.5\text{ m}}{1\text{ s}}$ and $\frac{1\text{ s}}{3.5\text{ m}}$

The 3.5 m is measured: It has two SFs. The 1 s is exact.

b. 1 day = 3.5 g of potassium;

$\frac{3.5\text{ g potassium}}{1\text{ day}}$ and $\frac{1\text{ day}}{3.5\text{ g potassium}}$

The 3.5 g is measured: It has two SFs. The 1 day is exact.

c. 1 L = 26.0 km; $\frac{26.0\text{ km}}{1\text{ L}}$ and $\frac{1\text{ L}}{26.0\text{ km}}$

The 26.0 km is measured: It has three SFs. The 1 L is exact.

d. 1 kg of plums = 29 mcg of pesticide;

$\frac{29\text{ mcg pesticide}}{1\text{ kg plums}}$ and $\frac{1\text{ kg plums}}{29\text{ mcg pesticide}}$

The 29 mcg is measured: It has two SFs. The 1 kg is exact.

e. 100 g of crust = 28.2 g of silicon;

$\frac{28.2\text{ g silicon}}{100\text{ g crust}}$ and $\frac{100\text{ g crust}}{28.2\text{ g silicon}}$

The 28.2 g is measured: It has three SFs. The 100 g is exact.

2.51 **a.** 1 tablet = 630 mg of calcium;

$\frac{630\text{ mg calcium}}{1\text{ tablet}}$ and $\frac{1\text{ tablet}}{630\text{ mg calcium}}$

The 630 mg is measured: It has two SFs. The 1 tablet is exact.

b. 1 day = 60 mg of vitamin C;

$\frac{60\text{ mg vitamin C}}{1\text{ day}}$ and $\frac{1\text{ day}}{60\text{ mg vitamin C}}$

The 60 mg is measured: It has one SF. The 1 day is exact.

c. 1 tablet = 50 mg of atenolol;

$\frac{50\text{ mg atenolol}}{1\text{ tablet}}$ and $\frac{1\text{ tablet}}{50\text{ mg atenolol}}$

The 50 mg is measured: It has one SF. The 1 tablet is exact.

d. 1 tablet = 81 mg of aspirin;

$\frac{81\text{ mg aspirin}}{1\text{ tablet}}$ and $\frac{1\text{ tablet}}{81\text{ mg aspirin}}$

The 81 mg is measured: It has two SFs. The 1 tablet is exact.

2.53 **a.** 5 mL of syrup = 10 mg of Atarax;

$\frac{10\text{ mg Atarax}}{5\text{ mL syrup}}$ and $\frac{5\text{ mL syrup}}{10\text{ mg Atarax}}$

b. 1 tablet = 0.25 g of Lanoxin;

$\frac{0.25\text{ g Lanoxin}}{1\text{ tablet}}$ and $\frac{1\text{ tablet}}{0.25\text{ g Lanoxin}}$

c. 1 tablet = 300 mg of Motrin;

$\frac{300\text{ mg Motrin}}{1\text{ tablet}}$ and $\frac{1\text{ tablet}}{300\text{ mg Motrin}}$

2.55 **a.** 0.0442 L **b.** 8.65×10^9 nm **c.** 5.2×10^2 Mg **d.** 7.2×10^5 ms

2.57 **a.** 1.56 kg **b.** 63 in. **c.** 4.5 qt **d.** 205 mm

2.59 **a.** 1.75 m **b.** 5 L **c.** 5.5 g **d.** 3.5 L

2.61 **a.** 473 mL **b.** 79.5 kg **c.** 24 lb **d.** 43 g

2.63 **a.** 66 gal **b.** 3 tablets **c.** 1800 mg **d.** 110 g

2.65 **a.** 6.3 h **b.** 2.5 mL

2.67 **a.** 1.20 g/mL **b.** 0.871 g/mL **c.** 3.10 g/mL **d.** 1.28 g/mL

2.69 **a.** 4.51 g/mL **b.** 1.42 g/mL **c.** 2.70 g/mL

2.71 **a.** 1.9 L of ethanol **b.** 88 g of mercury **c.** 83.3 oz of silver

2.73 **a.** 661 g **b.** 34 kg **c.** 29 cm^3

2.75 Because we calculate the density to be 11.3 g/cm^3, we identify the metal as lead.

2.77 **a.** 1.03 **b.** 1.03 g/mL **c.** 690 g **d.** 382 mL

2.79 **a.** 1 dL of blood = 42 mcg of iron;

$\frac{42\text{ mcg iron}}{1\text{ dL blood}}$ and $\frac{1\text{ dL blood}}{42\text{ mcg iron}}$

b. 3.4 mcg of iron

2.81 c and **d**

2.83 **a.** exact **b.** measured **c.** exact **d.** measured

2.85 61.5 °C

2.87 **a.** 97.5 cm **b.** 61.5 cm **c.** three SFs **d.** 6.00×10^3 cm^2

2.89 **a.** Diagram 3; a cube that has a greater density than the water will sink to the bottom.

b. Diagram 4; a cube with a density of 0.80 g/mL will be about four-fifths submerged in the water.

c. Diagram 1; a cube with a density that is one-half the density of water will be one-half submerged in the water.
d. Diagram 2; a cube with the same density as water will float just at the surface of the water.

2.91 **A** would be gold; it has the highest density (19.3 g/mL) and the smallest volume.
B would be silver; its density is intermediate (10.5 g/mL) and the volume is intermediate.
C would be aluminum; it has the lowest density (2.70 g/mL) and the largest volume.

2.93 The green cube has the same volume as the gray cube. However, the green cube has a larger mass on the scale, which means that its mass/volume ratio is larger. Thus, the density of the green cube is higher than the density of the gray cube.

2.95 **a.** 0.000 012 6 L (1.26×10^{-5} L)
b. 353 kg (3.53×10^{2} kg)
c. 125 000 m (1.25×10^{5} m)
d. 35.0 s

2.97 **a.** 265 g **b.** 0.584 lb

2.99 $1.01 per lb

2.101 16 onions

2.103 42 min

2.105 790 g

2.107 1600 mL (1.6×10^{3} mL)

2.109 **a.** 96 crackers **b.** 0.2 oz of fat
c. 17 servings

2.111 0.35 oz

2.113 20. mg

2.115 You would record the mass as 31.075 g. Because the balance will weigh to the nearest 0.001 g, the mass value would be reported to 0.001 g.

2.117 6.4 gal

2.119 1200 mL

2.121 **a.** 4.3 lb of body fat **b.** 6.0 lb

3

Matter and Energy

Charles is 13 years old and overweight. His doctor is worried that Charles is at risk for type 2 diabetes and advises his mother to make an appointment with a dietitian. Daniel, a dietitian, explains to them that choosing the appropriate foods is important to living a healthy lifestyle, losing weight, and preventing or managing diabetes.

Daniel also explains that food contains potential or stored energy, and different foods contain different amounts of potential energy. For instance, carbohydrates contain 4 kcal/g (17 kJ/g), whereas fats contain 9 kcal/g (38 kJ/g). He then explains that diets high in fat require more exercise to burn the fats, as they contain more energy. When Daniel looks at Charles's typical daily diet, he calculates that Charles obtains 2500 kcal in one day. The American Heart Association recommends 1800 kcal for boys 9 to 13 years of age. Daniel encourages Charles and his mother to include whole grains, fruits, and vegetables in their diet instead of foods high in fat. They also discuss food labels and the fact that smaller serving sizes of healthy foods are necessary to lose weight. Daniel also recommends that Charles exercises at least 60 minutes every day. Before leaving, Charles and his mother make an appointment for the following week to look at a weight loss plan.

CAREER

Dietitian

Dietitians specialize in helping individuals learn about good nutrition and the need for a balanced diet. This requires them to understand biochemical processes, the importance of vitamins and food labels, as well as the differences between carbohydrates, fats, and proteins in terms of their energy value and how they are metabolized. Dietitians work in a variety of environments, including hospitals, nursing homes, school cafeterias, and public health clinics. In these roles, they create specialized diets for individuals diagnosed with a specific disease or create meal plans for those in a nursing home.

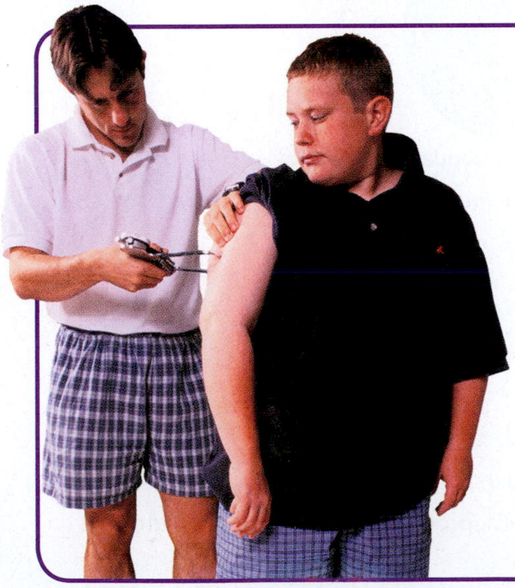

CLINICAL UPDATE

A Diet and Exercise Program

When Daniel sees Charles and his mother, they discuss a menu for weight loss. Charles is going to record his food intake and return to discuss his diet with Daniel. You can view the results in the **CLINICAL UPDATE A Diet and Exercise Program**, page 88, and calculate the kilocalories that Charles consumes in one day and also the weight that Charles has lost.

LOOKING AHEAD

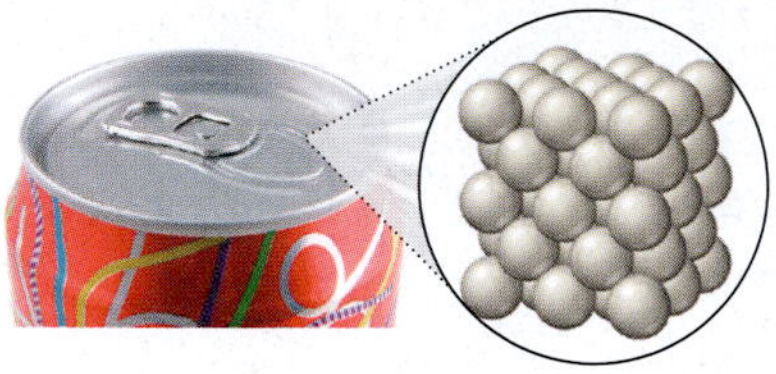

An aluminum can consists of many atoms of aluminum.

ENGAGE

Why are elements and compounds both pure substances?

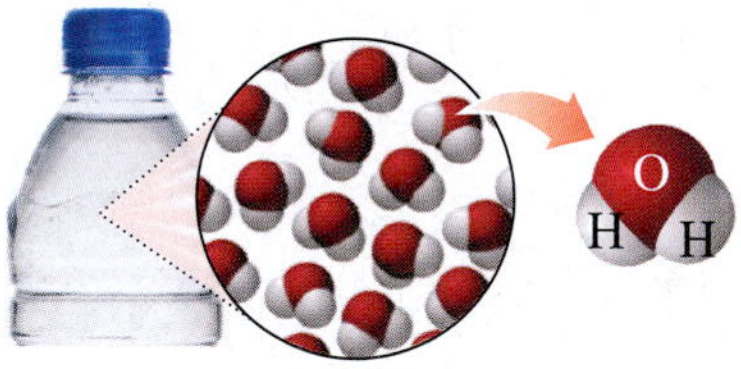
O
H
H

Water, H_2O, consists of two atoms of hydrogen (white) for one atom of oxygen (red).

TEST

Try Practice Problems 3.1 and 3.2

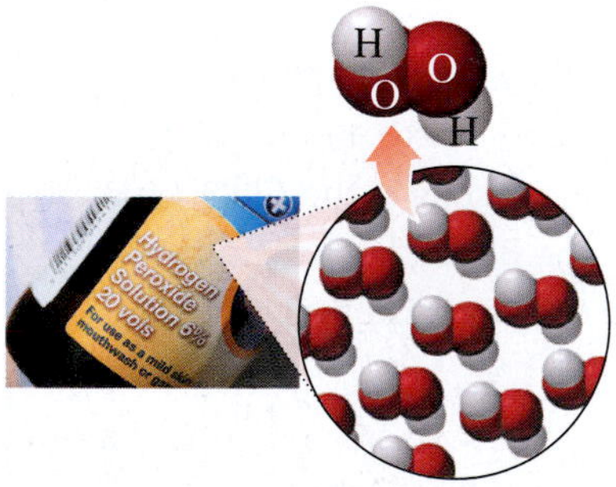
H
O
O
H
Hydrogen Peroxide Solution 6%
20 vols

Hydrogen peroxide, H_2O_2, consists of two atoms of hydrogen (white) for every two atoms of oxygen (red).

3.1 Classification of Matter

LEARNING GOAL Classify examples of matter as pure substances or mixtures.

Matter is anything that has mass and occupies space. Matter is everywhere around us: the orange juice we had for breakfast, the water we put in the coffee maker, the plastic bag we put our sandwich in, our toothbrush and toothpaste, the oxygen we inhale, and the carbon dioxide we exhale. To a scientist, all of this material is matter. The different types of matter are classified by their composition.

Pure Substances: Elements and Compounds

A **pure substance** is matter that has a fixed or definite composition. There are two kinds of pure substances: *elements* and *compounds*. An **element**, the simplest type of a pure substance, is composed of only one type of material such as silver, iron, or aluminum. Every element is composed of *atoms*, which are extremely tiny particles that make up each type of matter. Silver is composed of silver atoms, iron of iron atoms, and aluminum of aluminum atoms. A full list of the elements is found on the inside front cover of this text.

A **compound** is also a pure substance, but it consists of atoms of two or more elements always chemically combined in the same proportion. For example, in the compound water, there are two hydrogen atoms for every one oxygen atom, which is represented by the formula H_2O. This means that water always has the same composition of H_2O. Another compound that consists of a chemical combination of hydrogen and oxygen is hydrogen peroxide. It has two hydrogen atoms for every two oxygen atoms and is represented by the formula H_2O_2. Thus, water (H_2O) and hydrogen peroxide (H_2O_2) are different compounds that have different properties even though they contain the same elements, hydrogen and oxygen.

Pure substances that are compounds can be broken down by chemical processes into their elements. They cannot be broken down through physical methods such as boiling or sifting. For example, ordinary table salt consists of the compound NaCl, which can be separated by chemical processes into sodium metal and chlorine gas, as seen in **FIGURE 3.1**. Elements cannot be broken down further.

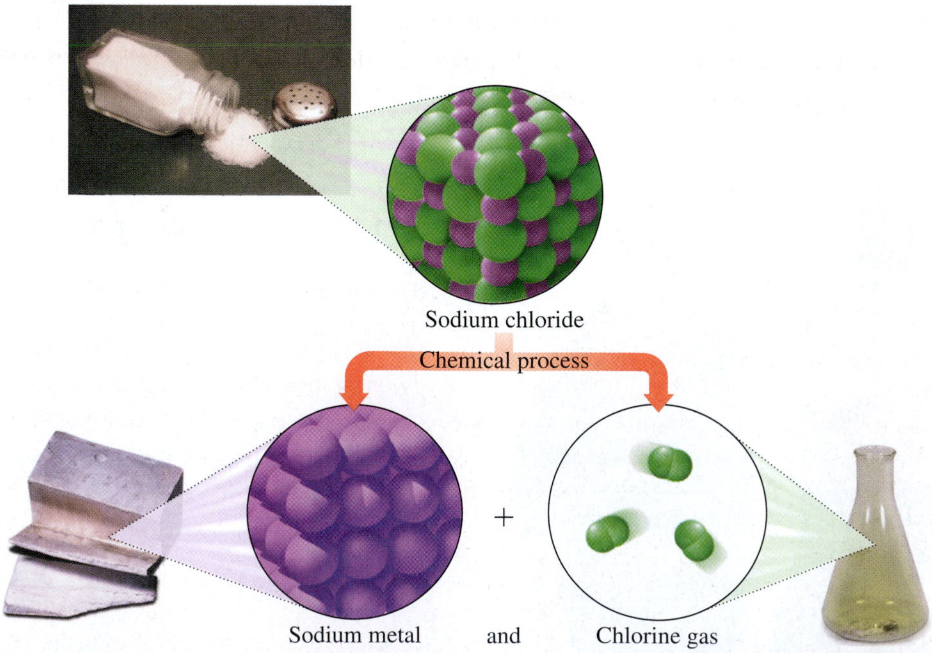
Sodium chloride
Chemical process
+
Sodium metal
and
Chlorine gas

FIGURE 3.1 ▶ The decomposition of salt, NaCl, produces the elements sodium and chlorine.

Q How do elements and compounds differ?

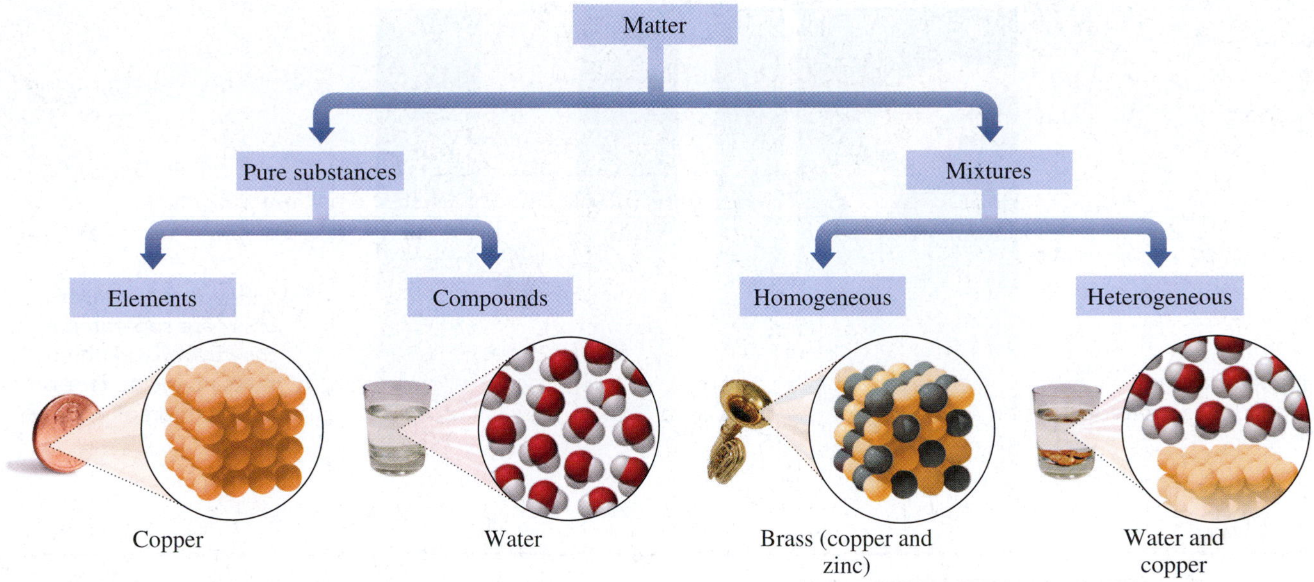

Matter
Pure substances
Mixtures
Elements
Compounds
Homogeneous
Heterogeneous
Copper
Water
Brass (copper and zinc)
Water and copper

FIGURE 3.2 ▶ Matter is organized by its components: elements, compounds, and mixtures.

Ⓠ Why are copper and water pure substances, but brass is a mixture?

Mixtures

In a **mixture**, two or more different substances are physically mixed, but not chemically combined. Much of the matter in our everyday lives consists of mixtures. The air we breathe is a mixture of mostly oxygen and nitrogen gases. The steel in buildings and railroad tracks is a mixture of iron, nickel, carbon, and chromium. The brass in doorknobs and musical instruments is a mixture of copper and zinc (see **FIGURE 3.2**). Tea, coffee, and ocean water are mixtures too. Unlike compounds, the proportions of substances in a mixture are not consistent but can vary. For example, two sugar–water mixtures may look the same, but the one with the higher ratio of sugar to water would taste sweeter.

Physical processes can be used to separate mixtures because there are no chemical interactions between the components. For example, different coins, such as nickels, dimes, and quarters, can be separated by size; iron particles mixed with sand can be picked up with a magnet; and water is separated from cooked spaghetti by using a strainer (see **FIGURE 3.3**).

FIGURE 3.3 ▶ A mixture of spaghetti and water is separated using a strainer, a physical method of separation.

Ⓠ Why can physical methods be used to separate mixtures but not compounds?

Types of Mixtures

Mixtures are classified further as homogeneous or heterogeneous. In a *homogeneous mixture*, also called a *solution*, the composition is uniform throughout the sample. We cannot see the individual components, which appear as one state. Familiar examples of homogeneous mixtures are air, which contains oxygen and nitrogen gases, and seawater, a solution of salt and water.

In a *heterogeneous mixture*, the components do not have a uniform composition throughout the sample. The components appear as two separate regions. For example, a mixture of oil and water is heterogeneous because the oil floats on the surface of the water. Other examples of heterogeneous mixtures are a cookie with raisins and orange juice with pulp.

In the chemistry laboratory, mixtures are separated by various methods. Solids are separated from liquids by *filtration*, which involves pouring a mixture through a filter paper, set in a funnel. The solid (residue) remains in the filter paper, and the filtered liquid (filtrate) moves through. In *chromatography*, different components of a liquid mixture separate as they move at different rates up the surface of a piece of chromatography paper.

ENGAGE

Why is a pizza heterogeneous, whereas vinegar is a homogeneous mixture?

(a) A mixture of a liquid and a solid is separated by filtration.

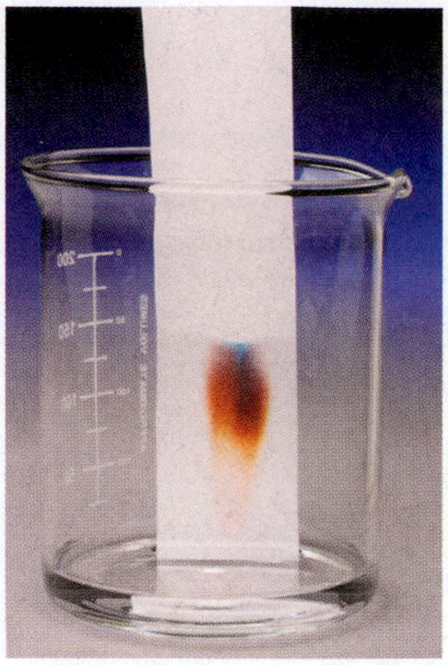

(b) Different substances in ink are separated as they travel at different rates up the surface of chromatography paper.

Chemistry Link to Health

Breathing Mixtures

The air we breathe is composed mostly of the gases oxygen (21%) and nitrogen (79%). The homogeneous breathing mixtures used by scuba divers differ from the air we breathe depending on the depth of the dive. Nitrox is a mixture of oxygen and nitrogen, but with more oxygen gas (up to 32%) and less nitrogen gas (68%) than air. A breathing mixture with less nitrogen gas decreases the risk of *nitrogen narcosis* associated with breathing regular air while diving. Heliox contains oxygen and helium, which is typically used for diving to more than 200 ft. By replacing nitrogen with helium, nitrogen narcosis does not occur. However, at dive depths over 300 ft, helium is associated with severe shaking and a drop in body temperature.

A breathing mixture used for dives over 400 ft is trimix, which contains oxygen, helium, and some nitrogen. The addition of some nitrogen lessens the problem of shaking that comes with breathing high levels of helium. Heliox and trimix are used only by professional, military, or other highly trained divers.

In hospitals, heliox may be used as a treatment for respiratory disorders and lung constriction in adults and premature infants. Heliox is less dense than air, which reduces the effort of breathing and helps distribute the oxygen gas to the tissues.

A nitrox mixture is used to fill scuba tanks.

▶SAMPLE PROBLEM 3.1 Classifying Mixtures

TRY IT FIRST

Classify each of the following as a pure substance (element or compound) or a mixture (homogeneous or heterogeneous):

a. copper in wire
b. a chocolate-chip cookie
c. nitrox, a combination of oxygen and nitrogen used to fill scuba tanks

SOLUTION

a. Copper is an element, which is a pure substance.
b. A chocolate-chip cookie does not have a uniform composition, which makes it a heterogeneous mixture.
c. The gases oxygen and nitrogen have a uniform composition in nitrox, which makes it a homogeneous mixture.

STUDY CHECK 3.1

a. A salad dressing is prepared with oil, vinegar, and chunks of blue cheese. Is this a homogeneous or heterogeneous mixture?

b. A mouthwash used to reduce plaque and clean gums and teeth contains several ingredients, such as menthol, alcohol, hydrogen peroxide, and a flavoring. Is this a homogeneous or heterogeneous mixture?

ANSWER

a. heterogeneous mixture **b.** homogeneous mixture

TEST
Try Practice Problems 3.3 to 3.6

PRACTICE PROBLEMS

3.1 Classification of Matter

3.1 Classify each of the following pure substances as an element or a compound:
a. a silicon (Si) chip **b.** hydrogen peroxide (H_2O_2)
c. oxygen gas (O_2) **d.** rust (Fe_2O_3)
e. methane (CH_4) in natural gas

3.2 Classify each of the following pure substances as an element or a compound:
a. helium gas (He) **b.** sulfur (S)
c. sugar ($C_{12}H_{22}O_{11}$) **d.** mercury (Hg) in a thermometer
e. lye (NaOH)

3.3 Classify each of the following as a pure substance or a mixture:
a. baking soda ($NaHCO_3$) **b.** a blueberry muffin
c. ice (H_2O) **d.** zinc (Zn)
e. trimix (oxygen, nitrogen, and helium) in a scuba tank

3.4 Classify each of the following as a pure substance or a mixture:
a. a soft drink **b.** propane (C_3H_8)
c. a cheese sandwich **d.** an iron (Fe) nail
e. salt substitute (KCl)

Clinical Applications

3.5 A dietitian includes one of the following mixtures in the lunch menu. Classify each of the following as homogeneous or heterogeneous:
a. vegetable soup
b. tea
c. fruit salad
d. tea with ice and lemon slices

3.6 A dietitian includes one of the following mixtures in the lunch menu. Classify each of the following as homogeneous or heterogeneous:
a. nonfat milk
b. chocolate-chip ice cream
c. peanut butter sandwich
d. cranberry juice

3.2 States and Properties of Matter

LEARNING GOAL Identify the states and the physical and chemical properties of matter.

On Earth, matter exists in one of three *physical forms* called the **states of matter**: *solids*, *liquids*, and *gases* (see **TABLE 3.1**). Water is a familiar substance that we routinely observe in all three states. In the solid state, water can be an ice cube or a snowflake. It is a liquid when it comes out of a faucet or fills a pool. Water forms a gas, or vapor, when it evaporates from wet clothes or boils in a pan. A **solid**, such as a pebble or a baseball, has a definite shape and volume. You can probably recognize several solids within your reach right now such as books, pencils, or a computer mouse. In a solid, strong attractive forces hold the particles close together. The particles in a solid are arranged in such a rigid pattern, their only movement is to vibrate slowly in fixed positions. For many solids, this rigid structure produces a crystal such as that seen in amethyst. **TABLE 3.1** compares the three states of matter.

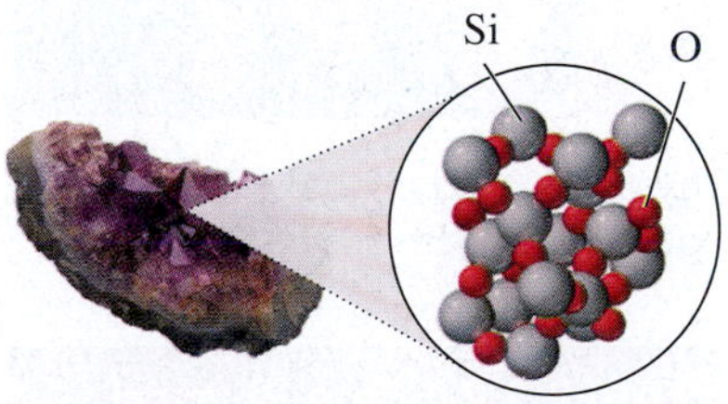
Si
O

Amethyst, a solid, is a purple form of quartz that contains atoms of Si and O.

TABLE 3.1 A Comparison of Solids, Liquids, and Gases

Characteristic	Solid	Liquid	Gas
Shape	Has a definite shape	Takes the shape of the container	Takes the shape of the container
Volume	Has a definite volume	Has a definite volume	Fills the volume of the container
Arrangement of Particles	Fixed, very close	Random, close	Random, far apart
Interaction between Particles	Very strong	Strong	Essentially none
Movement of Particles	Very slow	Moderate	Very fast
Examples	Ice, salt, iron	Water, oil, vinegar	Water vapor, helium, air

A **liquid** has a definite volume, but not a definite shape. In a liquid, the particles move in random directions but are sufficiently attracted to each other to maintain a definite volume, although not a rigid structure. Thus, when water, oil, or vinegar is poured from one container to another, the liquid maintains its own volume but takes the shape of the new container.

A **gas** does not have a definite shape or volume. In a gas, the particles are far apart, have little attraction to each other, and move at high speeds, taking the shape and volume of their container. When you inflate a bicycle tire, the air, which is a gas, fills the entire volume of the tire. The propane gas in a tank fills the entire volume of the tank.

ENGAGE

Why does a gas take both the shape and volume of its container?

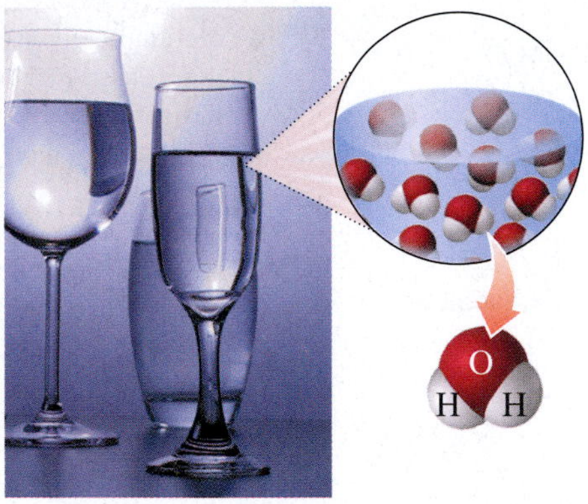
O
H
H

Water as a liquid takes the shape of its container.

He

A gas takes the shape and volume of its container.

TEST

Try Practice Problems 3.7 and 3.8

Physical Properties and Physical Changes

One way to describe matter is to observe its properties. For example, if you were asked to describe yourself, you might list characteristics such as your height and weight, the color of your eyes and skin, or the length, color, and texture of your hair.

Physical properties are those characteristics that can be observed or measured without affecting the identity of a substance. In chemistry, typical physical properties include the shape, color, melting point, boiling point, and physical state of a substance. For example, some of the physical properties of a penny include its round shape, orange-red color (from copper), solid state, and shiny luster. **TABLE 3.2** gives more examples of physical properties of copper, which is found in pennies, electrical wiring, and copper pans.

Water is a substance that is commonly found in all three states: solid, liquid, and gas. When matter undergoes a **physical change**, its state, size, or appearance will change, but its composition remains the same. The solid state of water, snow or ice, has a different appearance than its liquid or gaseous state, but all three states are water. In a physical change of state, no new substances are produced.

Copper, used in cookware, is a good conductor of heat.

TABLE 3.2 Some Physical Properties of Copper

State at 25 °C	Solid
Color	Orange-red
Odor	Odorless
Melting Point	1083 °C
Boiling Point	2567 °C
Luster	Shiny
Conduction of Electricity	Excellent
Conduction of Heat	Excellent

Chemical Properties and Chemical Changes

Chemical properties describe the ability of a substance to change into a new substance. When a **chemical change** takes place, the original substance is converted into one or more new substances, which have new physical and chemical properties. For example, the rusting or corrosion of a metal, such as iron, is a chemical property. In the rain, an iron (Fe) nail undergoes a chemical change when it reacts with oxygen (O_2) to form rust (Fe_2O_3). A chemical change has taken place: Rust is a new substance with new physical and chemical properties. **TABLE 3.3** gives examples of some physical and chemical changes. **TABLE 3.4** summarizes physical and chemical properties and changes.

TABLE 3.3 Examples of Some Physical and Chemical Changes

Physical Changes	Chemical Changes
Water boils to form water vapor.	Shiny, silver metal reacts in air to give a black, grainy coating.
Copper is drawn into thin copper wires.	A piece of wood burns with a bright flame and produces heat, ashes, carbon dioxide, and water vapor.
Sugar dissolves in water to form a solution.	Heating sugar forms a smooth, caramel-colored substance.
Paper is cut into tiny pieces of confetti.	Iron, which is gray and shiny, combines with oxygen to form orange-red rust.

A chemical change occurs when sugar is heated, forming a caramelized topping for flan.

TABLE 3.4 Summary of Physical and Chemical Properties and Changes

	Physical	Chemical
Property	A characteristic of a substance: color, shape, odor, luster, size, melting point, or density.	A characteristic that indicates the ability of a substance to form another substance: paper can burn, iron can rust, silver can tarnish.
Change	A change in a physical property that retains the identity of the substance: a change of state, a change in size, or a change in shape.	A change in which the original substance is converted to one or more new substances: paper burns, iron rusts, silver tarnishes.

CORE CHEMISTRY SKILL

Identifying Physical and Chemical Changes

▸ SAMPLE PROBLEM 3.2 Physical and Chemical Changes

TRY IT FIRST

Classify each of the following as a physical or chemical change:

a. A gold ingot is hammered to form gold leaf.
b. Gasoline burns in air.
c. Garlic is chopped into small pieces.
d. Milk left in a warm room turns sour.

FEINGOLD 999.9
100g

A gold ingot is hammered to form gold leaf.

SOLUTION

a. A physical change occurs when the gold ingot changes shape.
b. A chemical change occurs when gasoline burns and forms different substances with new properties.
c. A physical change occurs when the size of the garlic pieces changes.
d. A chemical change occurs when milk turns sour and forms new substances.

STUDY CHECK 3.2

Classify each of the following as a physical or chemical change:

a. Water freezes on a pond.
b. Gas bubbles form when baking powder is placed in vinegar.
c. A log is cut for firewood.
d. Butter melts in a warm room.

ANSWER

a. physical change
b. chemical change
c. physical change
d. physical change

INTERACTIVE VIDEO

PEARSON eText 2.0

Chemical vs. Physical Changes

TEST

Try Practice Problems 3.9 to 3.14

PRACTICE PROBLEMS

3.2 States and Properties of Matter

3.7 Indicate whether each of the following describes a gas, a liquid, or a solid:
- **a.** The breathing mixture in a scuba tank has no definite volume or shape.
- **b.** The neon atoms in a lighting display do not interact with each other.
- **c.** The particles in an ice cube are held in a rigid structure.

3.8 Indicate whether each of the following describes a gas, a liquid, or a solid:
- **a.** Lemonade has a definite volume but takes the shape of its container.
- **b.** The particles in a tank of oxygen are very far apart.
- **c.** Helium occupies the entire volume of a balloon.

3.9 Describe each of the following as a physical or chemical property:
- **a.** Chromium is a steel-gray solid.
- **b.** Hydrogen reacts readily with oxygen.
- **c.** A patient has a temperature of 40.2 °C.
- **d.** Ammonia will corrode iron.
- **e.** Butane gas in an igniter burns in oxygen.

3.10 Describe each of the following as a physical or chemical property:
- **a.** Neon is a colorless gas at room temperature.
- **b.** Apple slices turn brown when they are exposed to air.
- **c.** Phosphorus will ignite when exposed to air.
- **d.** At room temperature, mercury is a liquid.
- **e.** Propane gas is compressed to a liquid for placement in a small cylinder.

3.11 What type of change, physical or chemical, takes place in each of the following?
- **a.** Water vapor condenses to form rain.
- **b.** Cesium metal reacts explosively with water.
- **c.** Gold melts at 1064 °C.
- **d.** A puzzle is cut into 1000 pieces.
- **e.** Cheese is grated.

3.12 What type of change, physical or chemical, takes place in each of the following?
- **a.** Pie dough is rolled into thin pieces for a crust.
- **b.** A silver pin tarnishes in the air.
- **c.** A tree is cut into boards at a saw mill.
- **d.** Food is digested.
- **e.** A chocolate bar melts.

3.13 Describe each of the following properties for the element fluorine as physical or chemical:
- **a.** is highly reactive
- **b.** is a gas at room temperature
- **c.** has a pale, yellow color
- **d.** will explode in the presence of hydrogen
- **e.** has a melting point of −220 °C

3.14 Describe each of the following properties for the element zirconium as physical or chemical:
- **a.** melts at 1852 °C
- **b.** is resistant to corrosion
- **c.** has a grayish white color
- **d.** ignites spontaneously in air when finely divided
- **e.** is a shiny metal

REVIEW

Using Positive and Negative Numbers in Calculations (1.4)
Solving Equations (1.4)
Counting Significant Figures (2.2)

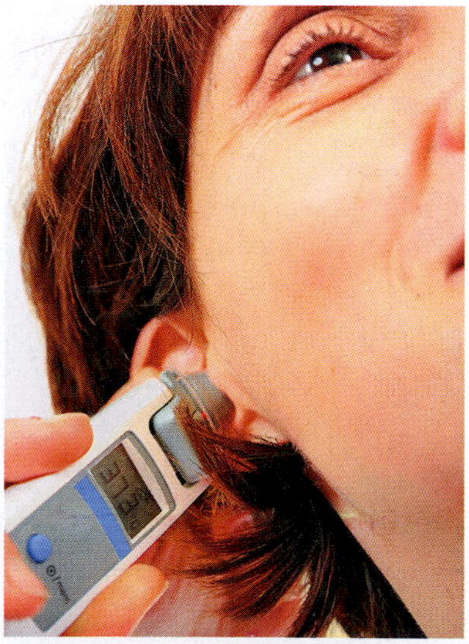

A digital ear thermometer is used to measure body temperature.

3.3 Temperature

LEARNING GOAL Given a temperature, calculate the corresponding temperature on another scale.

Temperatures in science are measured and reported in ***Celsius*** (°C) units. On the Celsius scale, the reference points are the freezing point of water, defined as 0 °C, and the boiling point, 100 °C, which means they are exact. In the United States, everyday temperatures are commonly reported in ***Fahrenheit*** (°F) units. On the Fahrenheit scale, water freezes at 32 °F and boils at 212 °F. A typical room temperature of 22 °C would be the same as 72 °F. Normal human body temperature is 37.0 °C, which is the same temperature as 98.6 °F.

On the Celsius and Fahrenheit temperature scales, the temperature difference between freezing and boiling is divided into smaller units called *degrees*. On the Celsius scale, there are 100 degrees Celsius between the freezing and boiling points of water, whereas the Fahrenheit scale has 180 degrees Fahrenheit between the freezing and boiling points of water. That makes a degree Celsius almost twice the size of a degree Fahrenheit: 1 °C = 1.8 °F (see **FIGURE 3.4**).

$$180 \text{ degrees Fahrenheit} = 100 \text{ degrees Celsius}$$

$$\frac{180 \text{ degrees Fahrenheit}}{100 \text{ degrees Celsius}} = \frac{1.8\ °\text{F}}{1\ °\text{C}}$$

We can write a temperature equation that relates a Fahrenheit temperature and its corresponding Celsius temperature.

$$T_F = 1.8(T_C) + 32$$ Temperature equation to obtain degrees Fahrenheit

(1.8: Changes °C to °F; 32: Adjusts freezing point)

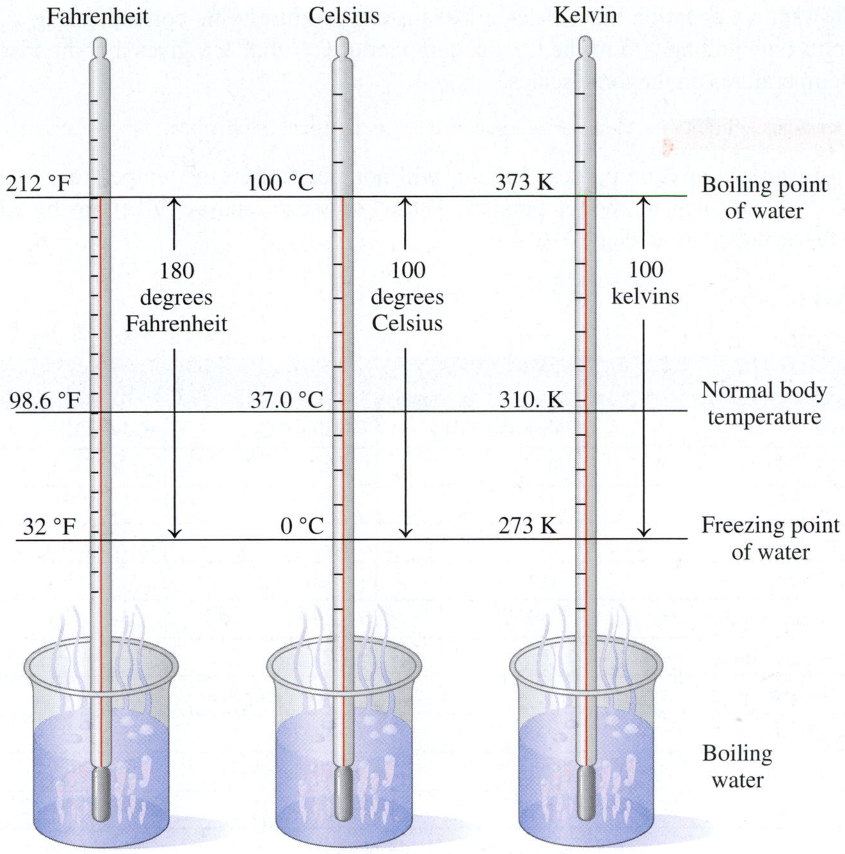
Fahrenheit
Celsius
Kelvin
212 °F
100 °C
373 K
Boiling point of water
180 degrees Fahrenheit
100 degrees Celsius
100 kelvins
98.6 °F
37.0 °C
310. K
Normal body temperature
32 °F
0 °C
273 K
Freezing point of water
Boiling water

FIGURE 3.4 ▶ A comparison of the Fahrenheit, Celsius, and Kelvin temperature scales between the freezing and boiling points of water.

Q What is the difference in the freezing points of water on the Celsius and Fahrenheit temperature scales?

ENGAGE

Why is a degree Celsius a larger unit of temperature than a degree Fahrenheit?

In the equation, the Celsius temperature is multiplied by 1.8 to change °C to °F; then 32 is added to adjust the freezing point from 0 °C to the Fahrenheit freezing point, 32 °F. The values, 1.8 and 32, used in the temperature equation are exact numbers and are not used to determine significant figures in the answer.

To convert from degrees Fahrenheit to degrees Celsius, the temperature equation is rearranged to solve for T_C. First, we subtract 32 from both sides because we must apply the same operation to both sides of the equation.

CORE CHEMISTRY SKILL

Converting between Temperature Scales

$$T_F - 32 = 1.8(T_C) + \cancel{32} - \cancel{32}$$

$$T_F - 32 = 1.8(T_C)$$

Second, we solve the equation for T_C by dividing both sides by 1.8.

$$\frac{T_F - 32}{1.8} = \frac{\cancel{1.8}(T_C)}{\cancel{1.8}}$$

$$\frac{T_F - 32}{1.8} = T_C \quad \text{Temperature equation to obtain degrees Celsius}$$

Scientists have learned that the coldest temperature possible is −273 °C (more precisely, −273.15 °C). On the *Kelvin* scale, this temperature, called *absolute zero*, has the value of 0 K. Units on the Kelvin scale are called kelvins (K); *no degree symbol is used.* Because there are no lower temperatures, the Kelvin scale has no negative temperature values. Between the freezing point of water, 273 K, and the boiling point, 373 K, there are 100 kelvins, which makes a kelvin equal in size to a degree Celsius.

$$1\ K = 1\ °C$$

We can write an equation that relates a Celsius temperature to its corresponding Kelvin temperature by adding 273 to the Celsius temperature. **TABLE 3.5** gives a comparison of some temperatures on the three scales.

$$T_K = T_C + 273$$ Temperature equation to obtain kelvins

An antifreeze mixture in a car radiator will not freeze until the temperature drops to −37 °C. We can calculate the temperature of the antifreeze mixture in kelvins by adding 273 to the temperature in degrees Celsius.

$$T_K = -37\ °C + 273 = 236\ K$$

ENGAGE

Show that −40. °C is the same temperature as −40. °F.

TABLE 3.5 A Comparison of Temperatures

Example	Fahrenheit (°F)	Celsius (°C)	Kelvin (K)
Sun	9937	5503	5776
A hot oven	450	232	505
Water boils	212	100	373
A high fever	104	40	313
Normal body temperature	98.6	37.0	310
Room temperature	70	21	294
Water freezes	32	0	273
A northern winter	−66	−54	219
Nitrogen liquefies	−346	−210	63
Absolute zero	−459	−273	0

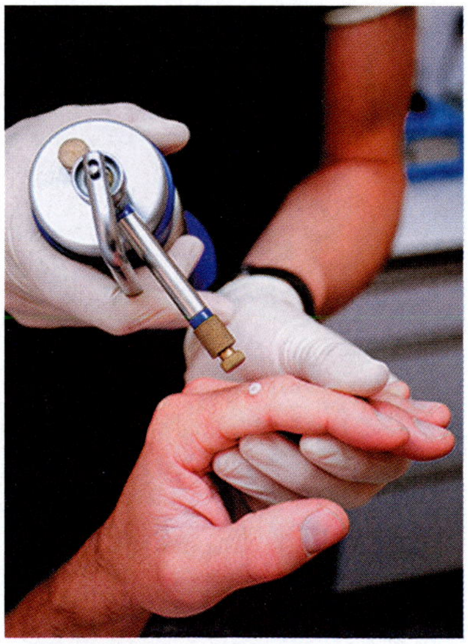

The low temperature of cryogenic nitrogen is used to destroy skin lesions.

▶SAMPLE PROBLEM 3.3 Calculating Temperature

TRY IT FIRST

A dermatologist uses cryogenic nitrogen at −196 °C to remove skin lesions and some skin cancers. What is the temperature, in degrees Fahrenheit, of the nitrogen?

SOLUTION

STEP 1 State the given and needed quantities.

ANALYZE THE PROBLEM	Given	Need	Connect
	−196 °C	*T* in degrees Fahrenheit	temperature equation

STEP 2 Write a temperature equation.

$$T_F = 1.8(T_C) + 32$$

STEP 3 Substitute in the known values and calculate the new temperature.

$$T_F = 1.8(-196) + 32$$ 1.8 is exact; 32 is exact

$$= -353 + 32 = -321\ °F$$ Answer to the ones place

STUDY CHECK 3.3

a. In the process of making ice cream, rock salt is added to crushed ice to chill the ice cream mixture. If the temperature drops to −11 °C, what is it in degrees Fahrenheit?

b. A hot tub reaches a temperature of 40.6 °C. What is that temperature in degrees Fahrenheit?

ANSWER

a. 12 °F **b.** 105 °F

TEST

Try Practice Problems 3.15 to 3.18

▶ SAMPLE PROBLEM 3.4 Calculating Degrees Celsius and Kelvins

TRY IT FIRST

In a type of cancer treatment called *thermotherapy*, temperatures as high as 113 °F are used to destroy cancer cells or make them more sensitive to radiation. What is that temperature in degrees Celsius? In kelvins?

SOLUTION

STEP 1 State the given and needed quantities.

ANALYZE THE PROBLEM	Given	Need	Connect
	113 °F	*T* in degrees Celsius, kelvins	temperature equations

STEP 2 Write a temperature equation.

$$T_C = \frac{T_F - 32}{1.8} \qquad T_K = T_C + 273$$

STEP 3 Substitute in the known values and calculate the new temperature.

$$T_C = \frac{(113 - 32)}{1.8}$$ 32 is exact; 1.8 is exact

$$= \frac{81}{1.8} = 45\ °C$$

(81: Two SFs; 1.8: Exact; 45 °C: Two SFs)

Using the equation that converts degrees Celsius to kelvins, we substitute in degrees Celsius.

$$T_K = 45 + 273 = 318\ K$$

(45: Ones place; 273: Ones place; 318 K: Ones place)

STUDY CHECK 3.4

a. A child has a temperature of 103.6 °F. What is this temperature on a Celsius thermometer?

b. A child who fell through the ice on a lake has hypothermia with a core temperature of 32 °C. What is this temperature in degrees Fahrenheit?

ANSWER

a. 39.8 °C **b.** 90. °F

TEST

Try Practice Problems 3.19 and 3.20

PRACTICE PROBLEMS

3.3 Temperature

3.15 Your friend who is visiting from Canada just took her temperature. When she reads 99.8 °F, she becomes concerned that she is quite ill. How would you explain this temperature to your friend?

3.16 You have a friend who is using a recipe for flan from a Mexican cookbook. You notice that he set your oven temperature at 175 °F. What would you advise him to do?

3.17 Calculate the unknown temperature in each of the following:

a. 37.0 °C = _______ °F

b. 65.3 °F = _______ °C

c. −27 °C = _______ K

d. 62 °C = _______ K

e. 114 °F = _______ °C

3.18 Calculate the unknown temperature in each of the following:

a. 25 °C = _______ °F

b. 155 °C = _______ °F

c. −25 °F = _______ °C

d. 224 K = _______ °C

e. 145 °C = _______ K

Clinical Applications

3.19 **a.** A patient with hyperthermia has a temperature of 106 °F. What does this read on a Celsius thermometer?

b. Because high fevers can cause convulsions in children, the doctor needs to be called if the child's temperature goes over 40.0 °C. Should the doctor be called if a child has a temperature of 103 °F?

3.20 **a.** Water is heated to 145 °F. What is the temperature of the hot water in degrees Celsius?

b. During extreme hypothermia, a child's temperature dropped to 20.6 °C. What was his temperature in degrees Fahrenheit?

Chemistry Link to Health

Variation in Body Temperature

Normal body temperature is considered to be 37.0 °C, although it varies throughout the day and from person to person. Oral temperatures of 36.1 °C are common in the morning and climb to a high of 37.2 °C between 6 P.M. and 10 P.M. Individuals who are involved in prolonged exercise may also experience elevated temperatures. Body temperatures of marathon runners can range from 39 °C to 41 °C as heat production during exercise exceeds the body's ability to lose heat. Temperatures above 37.2 °C for a person at rest are usually an indication of illness.

Hyperthermia occurs at body temperatures above 41 °C. Sweat production stops, and the skin becomes hot and dry. The pulse rate is elevated, and respiration becomes weak and rapid. The person can become lethargic and lapse into a coma. High body temperatures can lead to convulsions, particularly in children, which may cause permanent brain damage. Damage to internal organs is a major concern, and treatment, which must be immediate, may include immersing the person in an ice-water bath.

At the low temperature extreme of *hypothermia*, body temperature can drop as low as 28.5 °C. The person may appear cold and pale and have an irregular heartbeat. Unconsciousness can occur if the body temperature drops below 26.7 °C. Respiration becomes slow and shallow, and oxygenation of the tissues decreases. Treatment involves providing oxygen and increasing blood volume with glucose and saline fluids. Injecting warm fluids (37.0 °C) into the peritoneal cavity may restore the internal temperature.

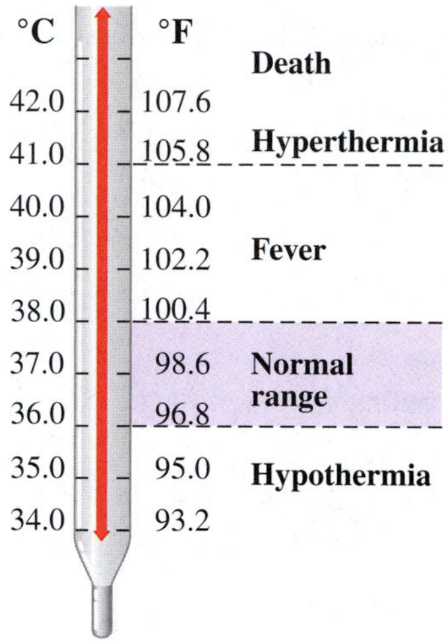

°C
°F
Death
42.0
107.6
41.0
105.8
Hyperthermia
40.0
104.0
39.0
102.2
Fever
38.0
100.4
37.0
98.6
Normal range
36.0
96.8
35.0
95.0
Hypothermia
34.0
93.2

REVIEW

Rounding Off (2.3)

Using Significant Figures in Calculations (2.3)

Writing Conversion Factors from Equalities (2.5)

Using Conversion Factors (2.6)

3.4 Energy

LEARNING GOAL Identify energy as potential or kinetic; convert between units of energy.

Almost everything you do involves energy. When you are running, walking, dancing, or thinking, you are using energy to do *work*, any activity that requires energy. In fact, **energy** is defined as the ability to do work. Suppose you are climbing a steep hill and you become too tired to go on. At that moment, you do not have the energy to do any more work. Now suppose you sit down and have lunch. In a while, you will have obtained some energy from the food, and you will be able to do more work and complete the climb.

ENGAGE

Why does a book have more potential energy when it is on the top of a high table than when it is on the floor?

Kinetic and Potential Energy

Energy can be classified as either kinetic energy or potential energy. **Kinetic energy** is the energy of motion. Any object that is moving has kinetic energy. **Potential energy** is determined by the position of an object or by the chemical composition of a substance. A boulder resting on top of a mountain has potential energy because of its location. If the boulder rolls down the mountain, the potential energy becomes kinetic energy. Water stored in a reservoir has potential energy. When the water goes over the dam and falls to the stream below, its potential energy is converted to kinetic energy. Foods and fossil fuels have potential energy. When you digest food or burn gasoline in your car, potential energy is converted to kinetic energy to do work.

TEST

Try Practice Problems 3.21 to 3.24

Heat and Energy

Heat is the energy associated with the motion of particles. An ice cube feels cold because heat flows from your hand into the ice cube. The faster the particles move, the greater the heat or thermal energy of the substance. In the ice cube, the particles are moving very slowly. As heat is added, the motion of the particles in the ice cube increases. Eventually, the particles have enough energy to make the ice cube melt as it changes from a solid to a liquid.

Water at the top of the dam stores potential energy. When the water flows over the dam, potential energy is converted to hydroelectric power.

Units of Energy

The SI unit of energy and work is the **joule (J)** (pronounced "jewel"). The joule is a small amount of energy, so scientists often use the kilojoule (kJ), 1000 joules. To heat water for one cup of tea, you need about 75 000 J or 75 kJ of heat. **TABLE 3.6** shows a comparison of energy in joules for several energy sources or uses.

CORE CHEMISTRY SKILL

Using Energy Units

You may be more familiar with the unit **calorie (cal)**, from the Latin *caloric*, meaning "heat." The calorie was originally defined as the amount of energy (heat) needed to raise the temperature of 1 g of water by 1 °C. Now, one calorie is defined as exactly 4.184 J. This equality can be written as two conversion factors:

$$1 \text{ cal} = 4.184 \text{ J (exact)} \qquad \frac{4.184 \text{ J}}{1 \text{ cal}} \text{ and } \frac{1 \text{ cal}}{4.184 \text{ J}}$$

One *kilocalorie* (kcal) is equal to 1000 calories, and one *kilojoule* (kJ) is equal to 1000 joules. The equalities and conversion factors follow:

$$1 \text{ kcal} = 1000 \text{ cal} \qquad \frac{1000 \text{ cal}}{1 \text{ kcal}} \text{ and } \frac{1 \text{ kcal}}{1000 \text{ cal}}$$

$$1 \text{ kJ} = 1000 \text{ J} \qquad \frac{1000 \text{ J}}{1 \text{ kJ}} \text{ and } \frac{1 \text{ kJ}}{1000 \text{ J}}$$

TABLE 3.6 A Comparison of Energy for Various Resources and Uses

Energy in Joules	
10^{27}	Energy radiated by the Sun in 1 s (10^{26})
10^{24}	World reserves of fossil fuel (10^{23})
10^{21}	Energy consumption for 1 yr in the United States (10^{20})
10^{18}	Solar energy reaching Earth in 1 s (10^{17})
10^{15}	
10^{12}	Energy use per person in 1 yr in the United States (10^{11})
10^{9}	Energy from 1 gal of gasoline (10^{8})
10^{6}	Energy from one serving of pasta, a doughnut, or needed to bicycle for 1 h (10^{6})
10^{3}	Energy used to sleep for 1 h (10^{5})
10^{0}	

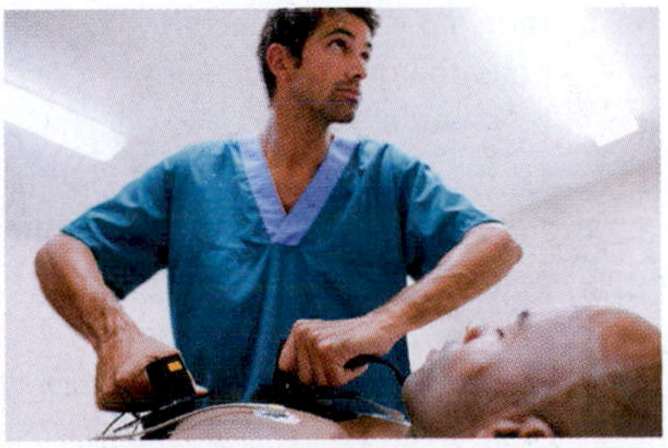

A defibrillator provides electrical energy to heart muscle to re-establish normal rhythm.

▶SAMPLE PROBLEM 3.5 Energy Units

TRY IT FIRST

A defibrillator gives a high-energy shock of 360 J. What is this quantity of energy in calories?

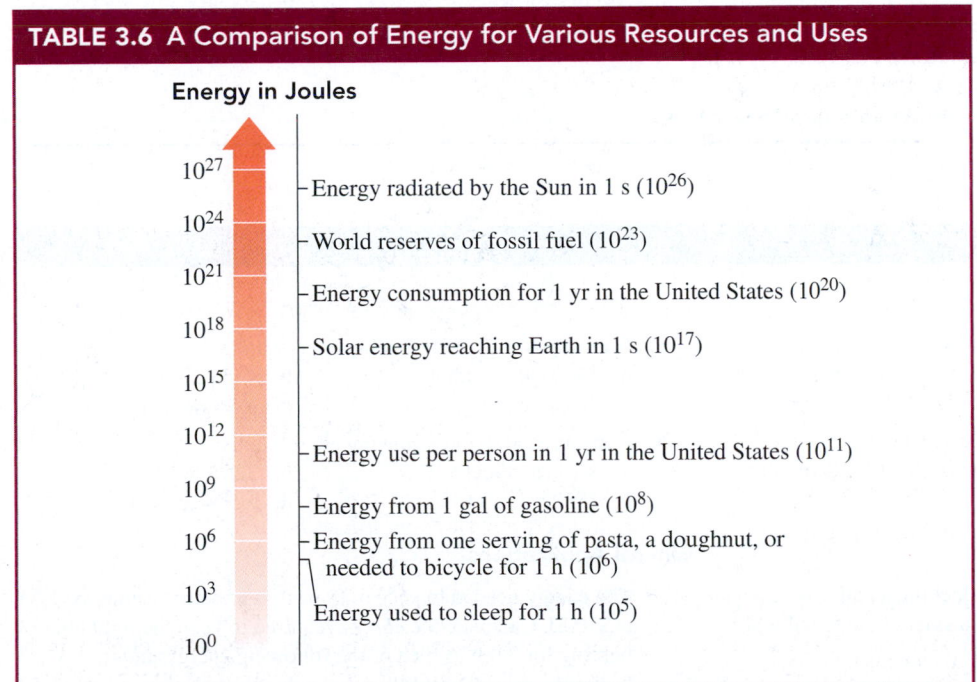

SOLUTION

STEP 1 State the given and needed quantities.

ANALYZE THE PROBLEM	Given	Need	Connect
	360 J	calories	energy factor

STEP 2 Write a plan to convert the given unit to the needed unit.

joules → Energy factor → calories

STEP 3 State the equalities and conversion factors.

$$1\text{ cal} = 4.184\text{ J}$$

$$\frac{4.184\text{ J}}{1\text{ cal}} \quad \text{and} \quad \frac{1\text{ cal}}{4.184\text{ J}}$$

STEP 4 Set up the problem to calculate the needed quantity.

$$360\,\cancel{\text{J}} \times \frac{1\text{ cal}}{4.184\,\cancel{\text{J}}} = 86\text{ cal}$$

Two SFs (360 J); Exact (1 cal / 4.184 J); Two SFs (86 cal)

STUDY CHECK 3.5

a. When 1.0 g of glucose is metabolized in the body, it produces 3900 cal. How many joules are produced?

b. A swimmer expends 855 kcal during practice. How many kilojoules did the swimmer expend?

ANSWER

a. 16 000 J

b. 3580 kJ or 3.58×10^3 kJ

TEST

Try Practice Problems 3.25 to 3.28

PRACTICE PROBLEMS

3.4 Energy

3.21 Discuss the changes in the potential and kinetic energy of a roller-coaster ride as the roller-coaster car climbs to the top and goes down the other side.

3.22 Discuss the changes in the potential and kinetic energy of a ski jumper taking the elevator to the top of the jump and going down the ramp.

3.23 Indicate whether each of the following statements describes potential or kinetic energy:
a. water at the top of a waterfall **b.** kicking a ball
c. the energy in a lump of coal **d.** a skier at the top of a hill

3.24 Indicate whether each of the following statements describes potential or kinetic energy:
a. the energy in your food **b.** a tightly wound spring
c. a car speeding down the freeway **d.** an earthquake

3.25 Convert each of the following energy units:
a. 3500 cal to kcal **b.** 415 J to cal
c. 28 cal to J **d.** 4.5 kJ to cal

3.26 Convert each of the following energy units:
a. 8.1 kcal to cal **b.** 325 J to kJ
c. 2550 cal to kJ **d.** 2.50 kcal to J

Clinical Applications

3.27 The energy needed to keep a 75-watt light bulb burning for 1.0 h is 270 kJ. Calculate the energy required to keep the light bulb burning for 3.0 h in each of the following energy units:
a. joules **b.** kilocalories

3.28 A person uses 750 kcal on a long walk. Calculate the energy used for the walk in each of the following energy units:
a. joules **b.** kilojoules

3.5 Energy and Nutrition

LEARNING GOAL Use the energy values to calculate the kilocalories (kcal) or kilojoules (kJ) for a food.

The food we eat provides energy to do work in the body, which includes the growth and repair of cells. Carbohydrates are the primary fuel for the body, but if the carbohydrate reserves are exhausted, fats and then proteins are used for energy.

For many years in the field of nutrition, the energy from food was measured as Calories or kilocalories. The nutritional unit *Calorie*, *Cal* (with an uppercase C), is the same as 1000 cal, or 1 kcal. The international unit, kilojoule (kJ), is becoming more prevalent. For example, a baked potato has an energy content of 100 Calories, which is 100 kcal or 440 kJ. A typical diet that provides 2100 Cal (kcal) is the same as an 8800 kJ diet.

TEST

Try Practice Problems 3.29 and 3.30

$$1 \text{ Cal} = 1 \text{ kcal} = 1000 \text{ cal}$$

$$1 \text{ Cal} = 4.184 \text{ kJ} = 4184 \text{ J}$$

In the nutrition laboratory, foods are burned in a steel container called a *calorimeter* to determine their *energy value* (kcal/g or kJ/g) (see **FIGURE 3.5**). A measured amount of water is added to fill the area surrounding the combustion chamber. The food sample is burned, releasing heat that increases the temperature of the water. From the known mass of the food and water, as well as the measured temperature increase, the energy value for the food is calculated. We assume that the energy absorbed by the calorimeter is negligible.

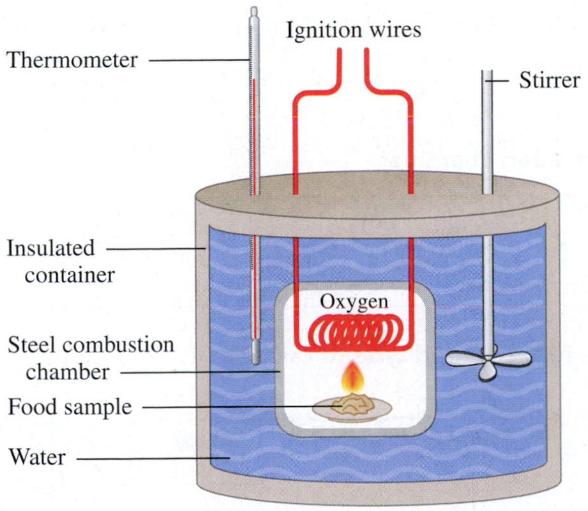
Ignition wires
Thermometer
Stirrer
Insulated container
Oxygen
Steel combustion chamber
Food sample
Water

FIGURE 3.5 ▶ Heat released from burning a food sample in a calorimeter is used to determine the energy value for the food.

Q What happens to the temperature of water in a calorimeter during the combustion of a food sample?

Energy Values for Foods

The **energy values** for food are the kilocalories or kilojoules obtained from burning 1 g of carbohydrate, fat, or protein, which are listed in **TABLE 3.7**. Using these energy values, we can calculate the total energy for a food if the mass of each food type is known.

$$\text{kilocalories} = \cancel{g} \times \frac{\text{kcal}}{\cancel{g}} \qquad \text{kilojoules} = \cancel{g} \times \frac{\text{kJ}}{\cancel{g}}$$

TABLE 3.7 Typical Energy Values for the Three Food Types

Food Type	kcal/g	kJ/g
Carbohydrate	4	17
Fat	9	38
Protein	4	17

On packaged food, the energy content is listed on the Nutrition Facts label, usually in terms of the number of Calories or kilojoules for one serving. The composition and energy content for some foods are given in **TABLE 3.8**. The total energy in kilocalories for each food type was calculated using energy values in kilocalories. Total energy in kilojoules was calculated using energy values in kilojoules. The energy for each food type was rounded off to the tens place.

ENGAGE

What type of food provides the most energy per gram?

TABLE 3.8 Composition and Energy Content for Some Foods

Food	Carbohydrate (g)	Fat (g)	Protein (g)	Energy
Apple, 1 medium	15	0	0	60 kcal (260 kJ)
Banana, 1 medium	26	0	1	110 kcal (460 kJ)
Beef, ground, 3 oz	0	14	22	220 kcal (900 kJ)
Broccoli, 3 oz	4	0	3	30 kcal (120 kJ)
Carrots, 1 cup	11	0	2	50 kcal (220 kJ)
Chicken, no skin, 3 oz	0	3	20	110 kcal (450 kJ)
Egg, 1 large	0	6	6	70 kcal (330 kJ)
Milk, nonfat, 1 cup	12	0	9	90 kcal (350 kJ)
Potato, baked	23	0	3	100 kcal (440 kJ)
Salmon, 3 oz	0	5	16	110 kcal (460 kJ)
Steak, 3 oz	0	27	19	320 kcal (1350 kJ)

Snack Crackers
Nutrition Facts
Serving Size 14 crackers (31 g)
Servings Per Container About 7
Amount Per Serving
Calories 130 Calories from Fat 40
% Daily Value*
Total Fat 4 g 6%
Saturated Fat 0.5 g 3%
Trans Fat 0 g
Polyunsaturated Fat 0.5%
Monounsaturated Fat 1.5 g
Cholesterol 0 mg 0%
Sodium 310 mg 13%
Total Carbohydrate 19 g 6%
Dietary Fiber Less than 1 g 4%
Sugars 2 g
Proteins 2 g

The Nutrition Facts include the total Calories and the grams of carbohydrate, fat, and protein per serving.

▸SAMPLE PROBLEM 3.6 Calculating the Energy from a Food

TRY IT FIRST

While working on his diet log, Charles observed that the Nutrition Facts label for Snack Crackers states that one serving contains 19 g of carbohydrate, 4 g of fat, and 2 g of protein. If Charles eats one serving of Snack Crackers, what is the energy, in kilocalories, from each food type and the total kilocalories? Round off the kilocalories for each food type to the tens place.

SOLUTION

STEP 1 State the given and needed quantities.

ANALYZE THE PROBLEM	Given	Need	Connect
	19 g of carbohydrate, 4 g of fat, 2 g of protein	total number of kilocalories	energy values

STEP 2 Use the energy value for each food type to calculate the kilocalories, rounded off to the tens place.

Food Type	Mass		Energy Value		Energy
Carbohydrate	19 g	$\times$	$\frac{4\ \text{kcal}}{1\ \text{g}}$	=	80 kcal
Fat	4 g	$\times$	$\frac{9\ \text{kcal}}{1\ \text{g}}$	=	40 kcal
Protein	2 g	$\times$	$\frac{4\ \text{kcal}}{1\ \text{g}}$	=	10 kcal

STEP 3 Add the energy for each food type to give the total energy from the food.

Total energy = 80 kcal + 40 kcal + 10 kcal = 130 kcal

STUDY CHECK 3.6

a. Using the mass of carbohydrate listed in the Nutrition Facts, calculate the energy, in kilojoules, for the carbohydrate in one serving of Snack Crackers. Round off the kilojoules to the tens place.

b. Using the energy calculations from Sample Problem 3.6, calculate the percentage of energy obtained from fat in one serving of Snack Crackers.

ANSWER

a. 320 kJ **b.** 31%

TEST

Try Practice Problems 3.31 to 3.36

Chemistry Link to Health

Losing and Gaining Weight

The number of kilocalories or kilojoules needed in the daily diet of an adult depends on gender, age, and level of physical activity. Some typical levels of energy needs are given in **TABLE 3.9**.

A person gains weight when food intake exceeds energy output. The amount of food a person eats is regulated by the hunger center in the hypothalamus, which is located in the brain. Food intake is normally proportional to the nutrient stores in the body. If these nutrient stores are low, you feel hungry; if they are high, you do not feel like eating.

A person loses weight when food intake is less than energy output. Many diet products contain cellulose, which has no nutritive value but provides bulk and makes you feel full. Some diet drugs depress the hunger center and must be used with caution because they excite the nervous system and can elevate blood pressure. Because muscular exercise is an important way to expend energy, an increase in daily exercise aids weight loss. **TABLE 3.10** lists some activities and the amount of energy they require.

One hour of swimming uses 2100 kJ of energy.

TABLE 3.9 Typical Energy Requirements for Adults

Gender	Age	Moderately Active kcal (kJ)	Highly Active kcal (kJ)
Female	19–30	2100 (8800)	2400 (10 000)
	31–50	2000 (8400)	2200 (9200)
Male	19–30	2700 (11 300)	3000 (12 600)
	31–50	2500 (10 500)	2900 (12 100)

TABLE 3.10 Energy Expended by a 70.0-kg (154-lb) Adult

Activity	Energy (kcal/h)	Energy (kJ/h)
Sleeping	60	250
Sitting	100	420
Walking	200	840
Swimming	500	2100
Running	750	3100

PRACTICE PROBLEMS

3.5 Energy and Nutrition

3.29 Calculate the kilocalories for each of the following:
- **a.** one stalk of celery that produces 125 kJ when burned in a calorimeter
- **b.** a waffle that produces 870. kJ when burned in a calorimeter

3.30 Calculate the kilocalories for each of the following:
- **a.** one cup of popcorn that produces 131 kJ when burned in a calorimeter
- **b.** a sample of butter that produces 23.4 kJ when burned in a calorimeter

3.31 Using the energy values for foods (see Table 3.7), determine each of the following (round off the answer for each food type to the tens place):
- **a.** the total kilojoules for one cup of orange juice that contains 26 g of carbohydrate, no fat, and 2 g of protein
- **b.** the grams of carbohydrate in one apple if the apple has no fat and no protein and provides 72 kcal of energy
- **c.** the kilocalories in one tablespoon of vegetable oil, which contains 14 g of fat and no carbohydrate or protein
- **d.** the grams of fat in one avocado that has 410 kcal, 13 g of carbohydrate, and 5 g of protein

3.32 Using the energy values for foods (see Table 3.7), determine each of the following (round off the answer for each food type to the tens place):
- **a.** the total kilojoules in two tablespoons of crunchy peanut butter that contains 6 g of carbohydrate, 16 g of fat, and 7 g of protein
- **b.** the grams of protein in one cup of soup that has 110 kcal with 9 g of carbohydrate and 6 g of fat
- **c.** the kilocalories in one can of cola if it has 40. g of carbohydrate and no fat or protein
- **d.** the total kilocalories for a diet that consists of 68 g of carbohydrate, 9 g of fat, and 150 g of protein

Clinical Applications

3.33 For dinner, Charles had one cup of clam chowder, which contains 16 g of carbohydrate, 12 g of fat, and 9 g of protein. How much energy, in kilocalories and kilojoules, is in the clam chowder? (Round off the answer for each food type to the tens place.)

3.34 For lunch, Charles consumed 3 oz of skinless chicken, 3 oz of broccoli, 1 medium apple, and 1 cup of nonfat milk (see Table 3.8). How many kilocalories did Charles obtain from the lunch?

3.35 A patient receives 3.2 L of intravenous (IV) glucose solution. If 100. mL of the solution contains 5.0 g of glucose (carbohydrate), how many kilocalories did the patient obtain from the glucose solution?

3.36 A high-protein diet contains 70.0 g of carbohydrate, 5.0 g of fat, and 150 g of protein. How much energy, in kilocalories and kilojoules, does this diet provide? (Round off the answer for each food type to the tens place.)

3.6 Specific Heat

LEARNING GOAL Use specific heat to calculate heat loss or gain.

Every substance has its own characteristic ability to absorb heat. When you bake a potato, you place it in a hot oven. If you are cooking pasta, you add the pasta to boiling water. You already know that adding heat to water increases its temperature until it boils. Certain substances must absorb more heat than others to reach a certain temperature.

TABLE 3.11 Specific Heats for Some Substances

Substance	cal/g °C	J/g °C
Elements		
Aluminum, Al(*s*)	0.214	0.897
Copper, Cu(*s*)	0.0920	0.385
Gold, Au(*s*)	0.0308	0.129
Iron, Fe(*s*)	0.108	0.452
Silver, Ag(*s*)	0.0562	0.235
Titanium, Ti(*s*)	0.125	0.523
Compounds		
Ammonia, $NH_3(g)$	0.488	2.04
Ethanol, $C_2H_6O(l)$	0.588	2.46
Sodium chloride, NaCl(*s*)	0.207	0.864
Water, $H_2O(l)$	1.00	4.184
Water, $H_2O(s)$	0.485	2.03

The energy requirements for different substances are described in terms of a physical property called *specific heat* (see **TABLE 3.11**). The **specific heat (*SH*)** for a substance is defined as the amount of heat needed to raise the temperature of exactly 1 g of a substance by exactly 1 °C. This temperature change is written as ΔT (*delta T*), where the delta symbol means "change in."

$$\text{Specific heat } (SH) = \frac{\text{heat}}{\text{mass } \Delta T} = \frac{\text{cal (or J)}}{\text{g } ^\circ\text{C}}$$

The specific heat for water is written using our definition of the calorie and joule.

$$SH \text{ for } H_2O(l) = \frac{1.00 \text{ cal}}{\text{g } ^\circ\text{C}} = \frac{4.184 \text{ J}}{\text{g } ^\circ\text{C}}$$

TEST

Try Practice Problems 3.37 and 3.38

The high specific heat of water keeps temperatures more moderate in summer and winter.

The specific heat of water is about five times larger than the specific heat of aluminum. Aluminum has a specific heat that is about twice that of copper. Adding the same amount of heat (1.00 cal or 4.184 J) will raise the temperature of 1 g of aluminum by about 5 °C and 1 g of copper by about 10 °C. The low specific heats of aluminum and copper mean they transfer heat efficiently, which makes them useful in cookware.

The high specific heat of water has a major impact on the temperatures in a coastal city compared to an inland city. A large mass of water near a coastal city can absorb or release five times the energy absorbed or released by the same mass of rock near an inland city. This means that in the summer a body of water absorbs large quantities of heat, which cools a coastal city, and then in the winter that same body of water releases large quantities of heat, which provides warmer temperatures. A similar effect happens with our bodies, which contain 70% water by mass. Water in the body absorbs or releases large quantities of heat to maintain body temperature.

Calculations Using Specific Heat

When we know the specific heat of a substance, we can calculate the heat lost or gained by measuring the mass of the substance and the initial and final temperature. We can substitute

CORE CHEMIST

Using the Heat Equa

these measurements into the specific heat expression that is rearranged to solve for heat, which we call the *heat equation*.

Heat	=	mass	×	temperature change	×	specific heat
Heat	=	m	×	ΔT	×	SH
cal	=	$\cancel{g}$	×	$\cancel{°C}$	×	$\frac{\text{cal}}{\cancel{g}\,\cancel{°C}}$
J	=	$\cancel{g}$	×	$\cancel{°C}$	×	$\frac{\text{J}}{\cancel{g}\,\cancel{°C}}$

▶ SAMPLE PROBLEM 3.7 Using Specific Heat

TRY IT FIRST

During surgery or when a patient has suffered a cardiac arrest or stroke, lowering the body temperature will reduce the amount of oxygen needed by the body. Some methods used to lower body temperature include cooled saline solution, cool water blankets, or cooling caps worn on the head. How many kilojoules are lost when the body temperature of a surgery patient with a blood volume of 5500 mL is cooled from 38.5 °C to 33.2 °C? (Assume that the specific heat and density of blood are the same as for water.)

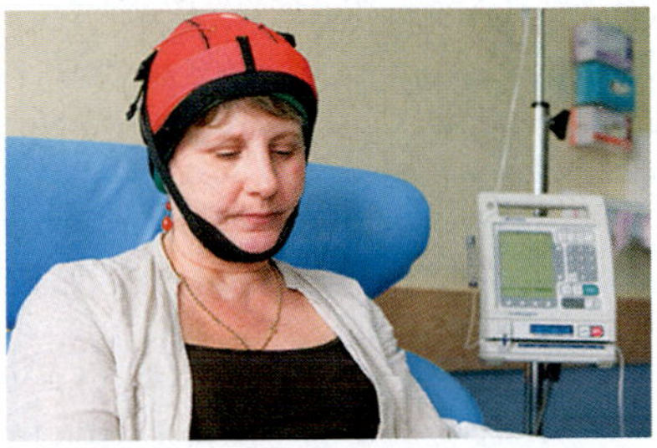

A cooling cap lowers the body temperature to reduce the oxygen required by the tissues.

SOLUTION

STEP 1 State the given and needed quantities.

	Given	Need	Connect
ANALYZE THE PROBLEM	5500 mL of blood = 5500 g of blood, cooled from 38.5 °C to 33.2 °C	kilojoules removed	heat equation, specific heat of water

STEP 2 Calculate the temperature change (ΔT).

$$\Delta T = 38.5\ °\text{C} - 33.2\ °\text{C} = 5.3\ °\text{C}$$

STEP 3 Write the heat equation and needed conversion factors.

$$\text{Heat} = m \times \Delta T \times SH$$

$$SH_{\text{water}} = \frac{4.184\ \text{J}}{\text{g}\ °\text{C}}$$

$$\frac{4.184\ \text{J}}{\text{g}\ °\text{C}} \text{ and } \frac{\text{g}\ °\text{C}}{4.184\ \text{J}}$$

$$1\ \text{kJ} = 1000\ \text{J}$$

$$\frac{1000\ \text{J}}{1\ \text{kJ}} \text{ and } \frac{1\ \text{kJ}}{1000\ \text{J}}$$

STEP 4 Substitute in the given values and calculate the heat, making sure units cancel.

$$\text{Heat} = \underset{\text{Two SFs}}{5500\ \cancel{g}} \times \underset{\text{Two SFs}}{5.3\ \cancel{°C}} \times \underset{\text{Exact}}{\overset{\text{Four SFs}}{\frac{4.184\ \cancel{J}}{\cancel{g}\ \cancel{°C}}}} \times \underset{\text{Exact}}{\overset{\text{Exact}}{\frac{1\ \text{kJ}}{1000\ \cancel{J}}}} = \underset{\text{Two SFs}}{120\ \text{kJ}}$$

STUDY CHECK 3.7

a. Dental implants use titanium because it is strong, is nontoxic, and can fuse to bone. How many calories are needed to raise the temperature of 16 g of titanium from 36.0 °C to 41.3 °C (see Table 3.11)?

b. Some cooking pans have a layer of copper on the bottom. How many kilojoules are needed to raise the temperature of 125 g of copper from 22 °C to 325 °C (see Table 3.11)?

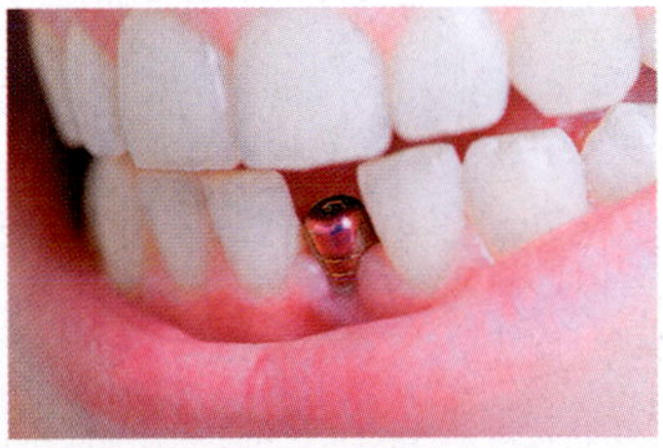

A dental implant contains the metal titanium that is attached to a crown.

ANSWER

a. 11 cal **b.** 14.6 kJ

TEST

Try Practice Problems 3.39 to 3.42

PRACTICE PROBLEMS

3.6 Specific Heat

3.37 If the same amount of heat is supplied to samples of 10.0 g each of aluminum, iron, and copper, all at 15.0 °C, which sample would reach the highest temperature (see Table 3.11)?

3.38 Substances A and B are the same mass and at the same initial temperature. When they are heated, the final temperature of A is 55 °C higher than the temperature of B. What does this tell you about the specific heats of A and B?

3.39 Use the heat equation to calculate the energy for each of the following (see Table 3.11):
- **a.** calories to heat 8.5 g of water from 15 °C to 36 °C
- **b.** joules lost when 25 g of water cools from 86 °C to 61 °C
- **c.** kilocalories to heat 150 g of water from 15 °C to 77 °C
- **d.** kilojoules to heat 175 g of copper from 28 °C to 188 °C

3.40 Use the heat equation to calculate the energy for each of the following (see Table 3.11):
- **a.** calories lost when 85 g of water cools from 45 °C to 25 °C
- **b.** joules to heat 75 g of water from 22 °C to 66 °C
- **c.** kilocalories to heat 5.0 kg of water from 22 °C to 28 °C
- **d.** kilojoules to heat 224 g of gold from 18 °C to 185 °C

3.41 Use the heat equation to calculate the energy, in joules and calories, for each of the following (see Table 3.11):
- **a.** to heat 25.0 g of water from 12.5 °C to 25.7 °C
- **b.** to heat 38.0 g of copper from 122 °C to 246 °C
- **c.** lost when 15.0 g of ethanol, C_2H_6O, cools from 60.5 °C to −42.0 °C
- **d.** lost when 125 g of iron cools from 118 °C to 55 °C

3.42 Use the heat equation to calculate the energy, in joules and calories, for each of the following (see Table 3.11):
- **a.** to heat 5.25 g of water from 5.5 °C to 64.8 °C
- **b.** lost when 75.0 g of water cools from 86.4 °C to 2.1 °C
- **c.** to heat 10.0 g of silver from 112 °C to 275 °C
- **d.** lost when 18.0 g of gold cools from 224 °C to 118 °C

REVIEW
Interpreting Graphs (1.4)

3.7 Changes of State

LEARNING GOAL Describe the changes of state between solids, liquids, and gases; calculate the energy released or absorbed.

Matter undergoes a **change of state** when it is converted from one state to another (see **FIGURE 3.6**).

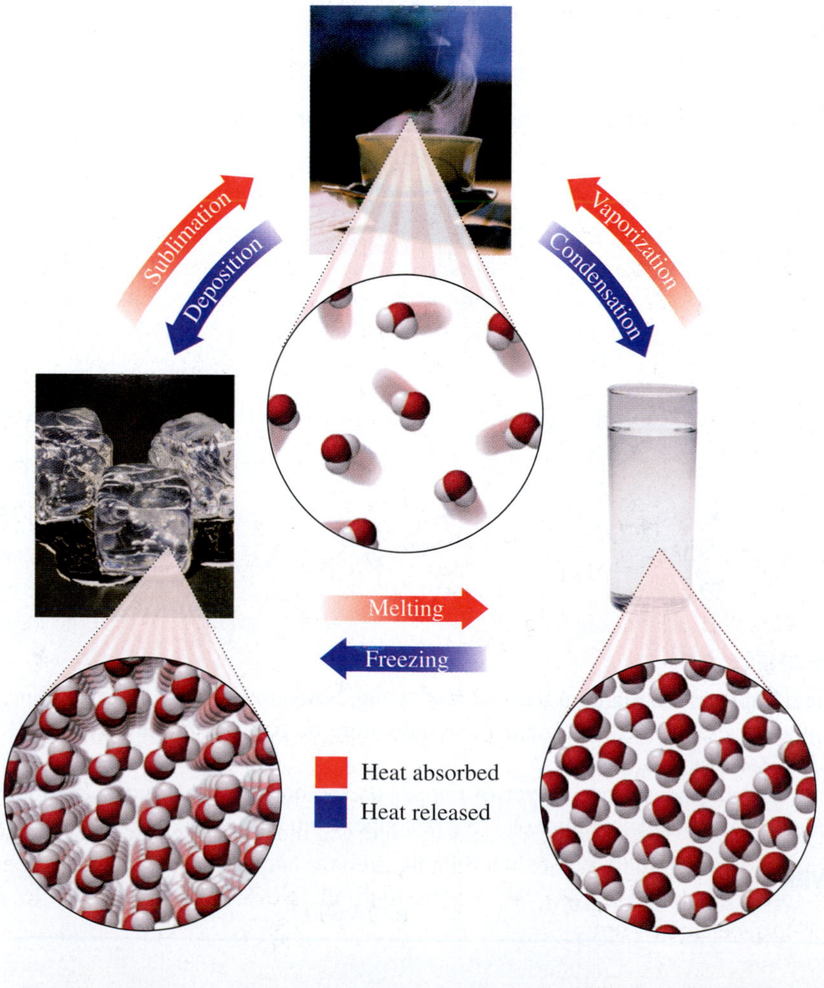
Sublimation
Deposition
Vaporization
Condensation
Melting
Freezing
Heat absorbed
Heat released

FIGURE 3.6 ▶ Changes of state include melting and freezing, vaporization and condensation, sublimation and deposition.

Q Is heat added or released when liquid water freezes?

Melting and Freezing

When heat is added to a solid, the particles move faster. At a temperature called the **melting point (mp)**, the particles of a solid gain sufficient energy to overcome the attractive forces that hold them together. The particles in the solid separate and move about in random patterns. The substance is **melting**, changing from a solid to a liquid.

If the temperature of a liquid is lowered, the reverse process takes place. Kinetic energy is lost, the particles slow down, and attractive forces pull the particles close together. The substance is **freezing**. A liquid changes to a solid at the **freezing point (fp)**, which is the same temperature as its melting point. Every substance has its own freezing (melting) point: Solid water (ice) melts at 0 °C when heat is added, and liquid water freezes at 0 °C when heat is removed.

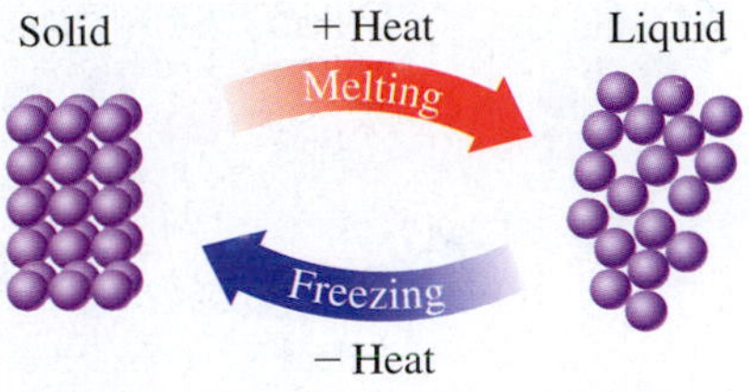
Solid
+ Heat
Liquid
Melting
Freezing
− Heat

Melting and freezing are reversible processes.

Heat of Fusion

During melting, the **heat of fusion** is the energy that must be added to convert exactly 1 g of solid to liquid at the melting point. For example, 80. cal (334 J) of heat is needed to melt exactly 1 g of ice at its melting point (0 °C).

$$H_2O(s) + 80.\ \text{cal/g (or 334 J/g)} \longrightarrow H_2O(l)$$

$$\frac{80.\ \text{cal}}{1\ \text{g}\ H_2O} \text{ and } \frac{1\ \text{g}\ H_2O}{80.\ \text{cal}} \qquad \frac{334\ \text{J}}{1\ \text{g}\ H_2O} \text{ and } \frac{1\ \text{g}\ H_2O}{334\ \text{J}}$$

The heat of fusion (80. cal/g or 334 J/g) is also the quantity of heat that must be removed to freeze exactly 1 g of water at its freezing point (0 °C).

$$H_2O(l) \longrightarrow H_2O(s) + 80.\ \text{cal/g (or 334 J/g)}$$

The heat of fusion can be used as a conversion factor in calculations. For example, to determine the heat needed to melt a sample of ice, the mass of ice, in grams, is multiplied by the heat of fusion.

ENGAGE

Why is the freezing point of water the same as its melting point?

ENGAGE

What units are used for the heat of fusion?

Calculating Heat to Melt (or Freeze) Water

$$\text{Heat} = \text{mass} \times \text{heat of fusion}$$

$$\text{cal} = \cancel{\text{g}} \times \frac{80.\ \text{cal}}{\cancel{\text{g}}} \qquad \text{J} = \cancel{\text{g}} \times \frac{334\ \text{J}}{\cancel{\text{g}}}$$

CORE CHEMISTRY SKILL

Calculating Heat for Change of State

Sublimation and Deposition

In a process called **sublimation**, the particles on the surface of a solid change directly to a gas without going through the liquid state. In the reverse process called **deposition**, gas particles change directly to a solid. For example, dry ice, which is solid carbon dioxide, sublimes at −78 °C. It is called "dry" because it does not form a liquid as it warms. In extremely cold areas, snow does not melt but sublimes directly to water vapor.

When frozen foods are left in the freezer for a long time, so much water sublimes that foods, especially meats, become dry and shrunken, a condition called *freezer burn*. Deposition occurs in a freezer when water vapor forms ice crystals on the surface of freezer bags and frozen food.

Freeze-dried foods prepared by sublimation are convenient for long-term storage and for camping and hiking. A frozen food is placed in a vacuum chamber where it dries as the ice sublimes. The dried food retains all of its nutritional value and needs only water to be edible. Freeze-dried foods do not need refrigeration because bacteria cannot grow without moisture.

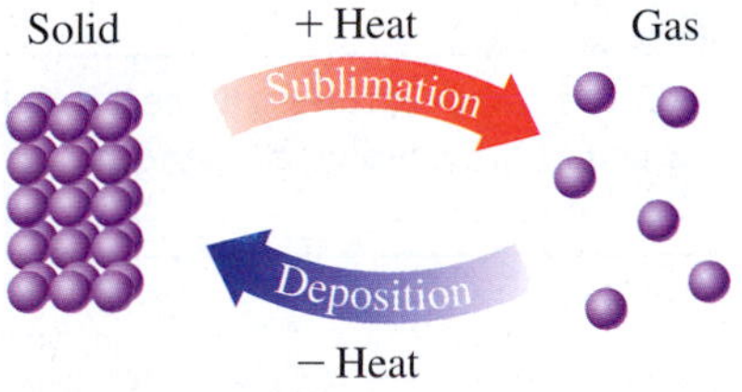
Solid
+ Heat
Gas
Sublimation
Deposition
− Heat

Sublimation and deposition are reversible processes.

TEST

Try Practice Problems 3.43 to 3.46

Evaporation, Boiling, and Condensation

Water in a mud puddle disappears, unwrapped food dries out, and clothes hung on a line dry. **Evaporation** is taking place as water particles with sufficient energy escape from the liquid surface and enter the gas phase. The loss of the "hot" water particles removes heat, which cools the liquid water. As heat is added, more and more water particles evaporate. At the **boiling point (bp)**, the particles within a liquid have enough energy to overcome their attractive forces and become a gas. We observe the **boiling** of a liquid as gas bubbles form, rise to the surface, and escape.

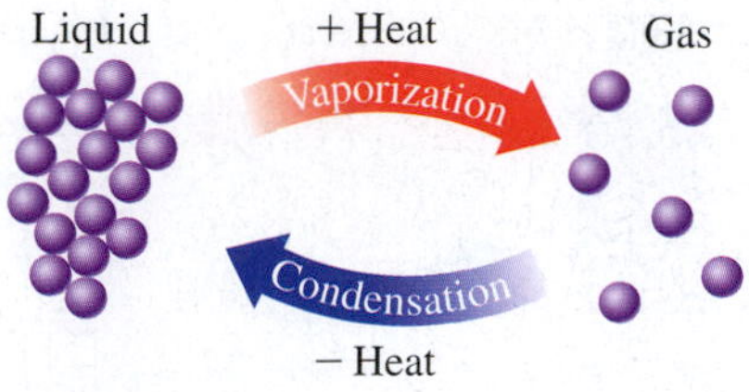
Liquid
+ Heat
Gas
Vaporization
Condensation
− Heat

Vaporization and condensation are reversible processes.

When heat is removed, a reverse process takes place. In **condensation**, water vapor is converted back to liquid as the water particles lose kinetic energy and slow down. Condensation occurs at the same temperature as boiling. You may have noticed that condensation occurs when you take a hot shower and the water vapor forms water droplets on a mirror.

Heat of Vaporization and Condensation

The **heat of vaporization** is the energy that must be added to convert exactly 1 g of liquid to gas at its boiling point. For water, 540 cal (2260 J) is needed to convert 1 g of water to vapor at 100 °C.

$$H_2O(l) + 540\ \text{cal/g (or 2260 J/g)} \longrightarrow H_2O(g)$$

This same amount of heat is released when 1 g of water vapor (gas) changes to liquid at 100 °C.

$$H_2O(g) \longrightarrow H_2O(l) + 540\ \text{cal/g (or 2260 J/g)}$$

Therefore, 540 cal/g or 2260 J/g is also the *heat of condensation* of water.

The heat of vaporization and condensation can each be written as two conversion factors using calories or joules.

$$\frac{540\ \text{cal}}{1\ \text{g}\ H_2O} \text{ and } \frac{1\ \text{g}\ H_2O}{540\ \text{cal}} \qquad \frac{2260\ \text{J}}{1\ \text{g}\ H_2O} \text{ and } \frac{1\ \text{g}\ H_2O}{2260\ \text{J}}$$

TEST

Try Practice Problems 3.47 and 3.48

To determine the heat needed to boil a sample of water, the mass, in grams, is multiplied by the heat of vaporization. Because the temperature remains constant during a change of state (melting, freezing, boiling, or condensing), there is no temperature change used in the calculation, as shown in Sample Problem 3.8.

Calculating Heat to Condense (or Boil) Water

$$\text{Heat} = \text{mass} \times \text{heat of vaporization}$$

$$\text{cal} = \cancel{g} \times \frac{540\ \text{cal}}{\cancel{g}} \qquad \text{J} = \cancel{g} \times \frac{2260\ \text{J}}{\cancel{g}}$$

TEST

Try Practice Problems 3.49 and 3.50

▶ SAMPLE PROBLEM 3.8 Calculating Heat for Changes of State

TRY IT FIRST

In a sauna, 122 g of water is converted to steam at 100 °C. How many kilojoules of heat are needed?

ENGAGE

Why is temperature not used in calculations involving changes of state?

SOLUTION

STEP 1 State the given and needed quantities.

ANALYZE THE PROBLEM	Given	Need	Connect
	122 g of water at 100 °C	kilojoules to convert to steam at 100 °C	heat of vaporization, 2260 J/g

STEP 2 Write a plan to convert the given quantity to the needed quantity.

grams of water → Heat of vaporization → joules → Metric factor → kilojoules

STEP 3 Write the heat conversion factor and any metric factor.

$$1\ \text{g of}\ H_2O(l \longrightarrow g) = 2260\ \text{J}$$

$$\frac{2260\ \text{J}}{1\ \text{g}\ H_2O} \text{ and } \frac{1\ \text{g}\ H_2O}{2260\ \text{J}}$$

$$1\ \text{kJ} = 1000\ \text{J}$$

$$\frac{1000\ \text{J}}{1\ \text{kJ}} \text{ and } \frac{1\ \text{kJ}}{1000\ \text{J}}$$

STEP 4 Set up the problem and calculate the needed quantity.

$$\text{Heat} = 122\ \cancel{\text{g}\ H_2O} \times \frac{2260\ \cancel{\text{J}}}{1\ \cancel{\text{g}\ H_2O}} \times \frac{1\ \text{kJ}}{1000\ \cancel{\text{J}}} = 276\ \text{kJ}$$

Three SFs (122 g H₂O) · Three SFs / Exact (2260 J / 1 g H₂O) · Exact / Exact (1 kJ / 1000 J) · Three SFs (276 kJ)

STUDY CHECK 3.8

a. When steam from a pan of boiling water reaches a cool window, it condenses. How much heat, in kilojoules, is released when 25.0 g of steam condenses at 100 °C?

b. Ice bags are used by sports trainers to treat muscle injuries. If 260. g of ice is placed in an ice bag, how much heat, in kilojoules, will be released when all the ice melts at 0 °C?

ANSWER

a. 56.5 kJ **b.** 86.8 kJ

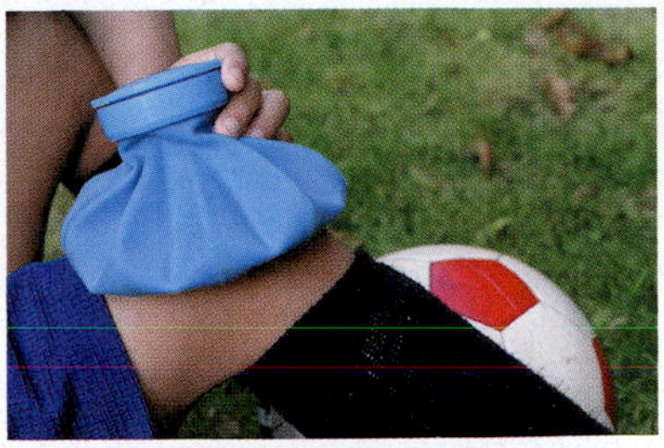

An ice bag is used to treat a sports injury.

Heating and Cooling Curves

All the changes of state that take place during the heating or cooling of a substance can be illustrated visually. On a **heating curve** or **cooling curve**, the temperature is shown on the vertical axis and the loss or gain of heat is shown on the horizontal axis.

Steps on a Heating Curve

The first diagonal line indicates the warming of a solid as heat is added. When the melting temperature is reached, a horizontal line, or plateau, indicates that the solid is melting. As melting takes place, the solid is changing to liquid without any change in temperature (see **FIGURE 3.7**).

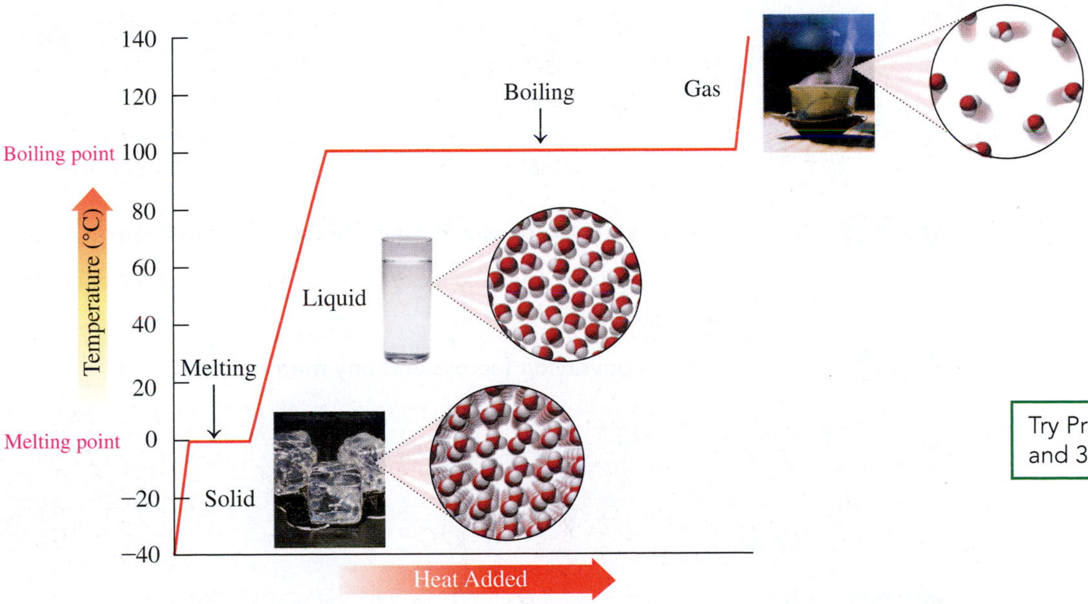
140
120
Boiling point 100
80
60
40
20
Melting point 0
−20
−40
Temperature (°C)
Boiling
Gas
Liquid
Melting
Solid
Heat Added

FIGURE 3.7 ▶ A heating curve diagrams the temperature increases and changes of state as heat is added.

Q What does the plateau at 100 °C represent on the heating curve for water?

TEST

Try Practice Problems 3.51 and 3.52

When all the particles are in the liquid state, adding more heat will increase the temperature of the liquid, which is shown as a diagonal line. Once the liquid reaches its boiling point, a horizontal line indicates that the liquid changes to gas at constant temperature. Because the heat of vaporization is greater than the heat of fusion, the horizontal line at the boiling point is longer than the horizontal line at the melting point. When all the liquid becomes gas, adding more heat increases the temperature of the gas.

Steps on a Cooling Curve

A cooling curve is a diagram in which the temperature decreases as heat is removed. Initially, a diagonal line is drawn to show that heat is removed from a gas until it begins to condense. At the condensation point, a horizontal line indicates a change of state as gas condenses to form a liquid. When all the gas has changed to liquid, further cooling lowers the

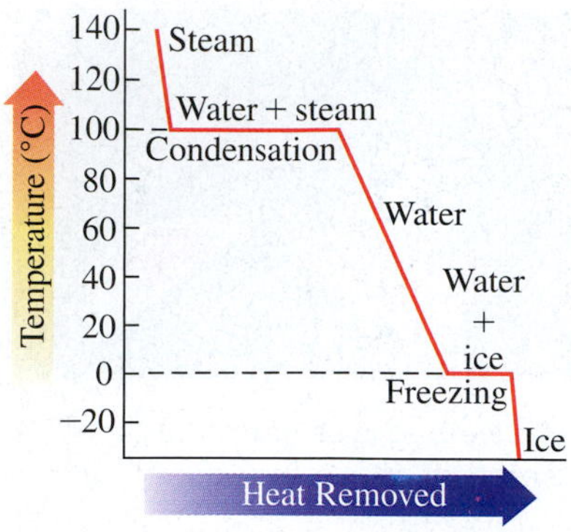
Steam
Water + steam
Condensation
Water
Water + ice
Freezing
Ice
Temperature (°C)
140
120
100
80
60
40
20
0
−20
Heat Removed

A cooling curve for water illustrates the change in temperature and changes of state as heat is removed.

temperature. The decrease in temperature is shown as a diagonal line from the condensation point to the freezing point, where another horizontal line indicates that liquid is changing to solid. When all the substance is frozen, the removal of more heat decreases the temperature of the solid below its freezing point, which is shown as a diagonal line.

Combining Energy Calculations

Up to now, we have calculated one step in a heating or cooling curve. However, some problems require a combination of steps that include a temperature change as well as a change of state. The heat is calculated for each step separately and then added together to find the total energy, as seen in Sample Problem 3.9.

▶SAMPLE PROBLEM 3.9 Combining Heat Calculations

TRY IT FIRST

Charles has increased his activity by doing more exercise. After a session of using weights, he has a sore arm. An ice bag is filled with 125 g of ice at 0 °C. How much heat, in kilojoules, is absorbed to melt the ice and raise the temperature of the water to body temperature, 37.0 °C?

SOLUTION

STEP 1 State the given and needed quantities.

	Given	Need	Connect
ANALYZE THE PROBLEM	125 g of ice at 0 °C	total kilojoules to melt ice at 0 °C and to raise temperature of water to 37.0 °C	combine heat from change of state (heat of fusion) and temperature change (specific heat of water)

STEP 2 Write a plan to convert the given quantity to the needed quantity.

Total heat = kilojoules needed to melt the ice at 0 °C and heat the water from 0 °C (freezing point) to 37.0 °C

STEP 3 Write the heat conversion factors and any metric factor.

$1 \text{ g of } H_2O\ (s \rightarrow l) = 334 \text{ J}$

$$\frac{334 \text{ J}}{1 \text{ g } H_2O} \text{ and } \frac{1 \text{ g } H_2O}{334 \text{ J}}$$

$SH_{\text{water}} = \dfrac{4.184 \text{ J}}{\text{g °C}}$

$$\frac{4.184 \text{ J}}{\text{g °C}} \text{ and } \frac{\text{g °C}}{4.184 \text{ J}}$$

$1 \text{ kJ} = 1000 \text{ J}$

$$\frac{1000 \text{ J}}{1 \text{ kJ}} \text{ and } \frac{1 \text{ kJ}}{1000 \text{ J}}$$

STEP 4 Set up the problem and calculate the needed quantity.

$\Delta T = 37.0 \text{ °C} - 0 \text{ °C} = 37.0 \text{ °C}$

Heat needed to change ice (solid) to water (liquid) at 0 °C:

$$\text{Heat} = 125 \text{ g ice} \times \frac{334 \text{ J}}{1 \text{ g ice}} \times \frac{1 \text{ kJ}}{1000 \text{ J}} = 41.8 \text{ kJ}$$

Three SFs (125 g ice); Three SFs (334 J); Exact (1 g ice); Exact (1 kJ); Exact (1000 J); Three SFs (41.8 kJ)

Heat needed to warm water (liquid) from 0 °C to water (liquid) at 37.0 °C:

$$\text{Heat} = 125 \text{ g} \times 37.0 \text{ °C} \times \frac{4.184 \text{ J}}{\text{g °C}} \times \frac{1 \text{ kJ}}{1000 \text{ J}} = 19.4 \text{ kJ}$$

Three SFs (125 g); Three SFs (37.0 °C); Exact (4.184 J/g °C); Exact (1 kJ/1000 J); Three SFs (19.4 kJ)

Calculate the total heat:

Melting ice at 0 °C	41.8 kJ
Heating water (0 °C to 37.0 °C)	19.4 kJ
Total heat needed	61.2 kJ

STUDY CHECK 3.9

a. How many kilojoules are released when 75.0 g of steam at 100 °C condenses, cools to 0 °C, and freezes at 0 °C? (*Hint*: The solution will require three energy calculations.)

b. How many kilocalories are needed to convert 18 g of ice at 0 °C to steam at 100 °C? (*Hint*: The solution will require three energy calculations.)

ANSWER

a. 226 kJ **b.** 13 kcal

TEST

Try Practice Problems 3.53 to 3.56

Chemistry Link to Health

Steam Burns

Hot water at 100 °C will cause burns and damage to the skin. However, getting steam on the skin is even more dangerous. If 25 g of hot water at 100 °C falls on a person's skin, the temperature of the water will drop to body temperature, 37 °C. The heat released during cooling can cause severe burns. The amount of heat can be calculated from the mass, the temperature change, 100 °C − 37 °C = 63 °C, and the specific heat of water, 4.184 J/g °C.

$$25\ \cancel{g} \times 63\ \cancel{°C} \times \frac{4.184\ \text{J}}{\cancel{g}\ \cancel{°C}} = 6600\ \text{J of heat released when water cools}$$

The condensation of the same quantity of steam to liquid at 100 °C releases much more heat. The heat released when steam condenses can be calculated using the heat of vaporization, which is 2260 J/g for water at 100 °C.

$$25\ \cancel{g} \times \frac{2260\ \text{J}}{1\ \cancel{g}} = 57\,000\ \text{J released when water (gas) condenses to water (liquid) at 100 °C}$$

The total heat released is calculated by combining the heat from the condensation at 100 °C and the heat from cooling of the steam from 100 °C to 37 °C (body temperature). We can see that most of the heat is from the condensation of steam.

Condensation (100 °C)	= 57 000 J
Cooling (100 °C to 37 °C)	= 6 600 J
Heat released	= 64 000 J (rounded off)

The amount of heat released from steam is almost ten times greater than the heat from the same amount of hot water. This large amount of heat released on the skin is what causes the extensive damage from steam burns.

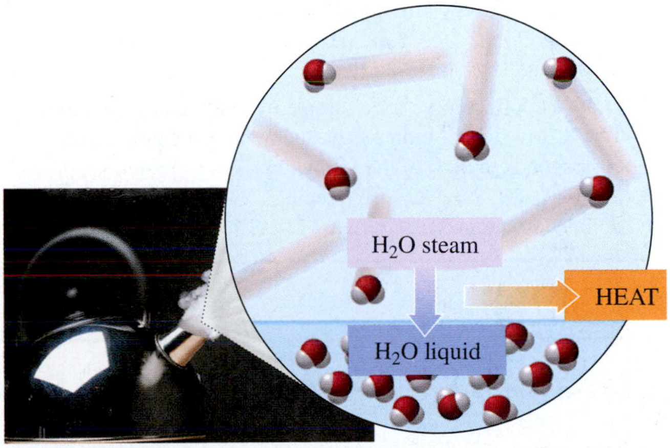
H2O steam
HEAT
H2O liquid

When 1 g of steam condenses, 2260 J or 540 cal is released.

PRACTICE PROBLEMS

3.7 Changes of State

3.43 Identify each of the following changes of state as melting, freezing, sublimation, or deposition:

a. The solid structure of a substance breaks down as liquid forms.
b. Coffee is freeze-dried.
c. Water on the street turns to ice during a cold wintry night.
d. Ice crystals form on a package of frozen corn.

3.44 Identify each of the following changes of state as melting, freezing, sublimation, or deposition:

a. Dry ice in an ice-cream cart disappears.
b. Snow on the ground turns to liquid water.
c. Heat is removed from 125 g of liquid water at 0 °C.
d. Frost (ice) forms on the walls of a freezer unit of a refrigerator.

3.45 Calculate the heat change at 0 °C for each of the following, and indicate whether heat was absorbed/released:

a. calories to melt 65 g of ice
b. joules to melt 17.0 g of ice
c. kilocalories to freeze 225 g of water
d. kilojoules to freeze 50.0 g of water

3.46 Calculate the heat change at 0 °C for each of the following, and indicate whether heat was absorbed/released:

a. calories to freeze 35 g of water
b. joules to freeze 275 g of water
c. kilocalories to melt 140 g of ice
d. kilojoules to melt 5.00 g of ice

3.47 Identify each of the following changes of state as evaporation, boiling, or condensation:
- **a.** The water vapor in the clouds changes to rain.
- **b.** Wet clothes dry on a clothesline.
- **c.** Lava flows into the ocean and steam forms.
- **d.** After a hot shower, your bathroom mirror is covered with water.

3.48 Identify each of the following changes of state as evaporation, boiling, or condensation:
- **a.** At 100 °C, the water in a pan changes to steam.
- **b.** On a cool morning, the windows in your car fog up.
- **c.** A shallow pond dries up in the summer.
- **d.** Your teakettle whistles when the water is ready for tea.

3.49 Calculate the heat change at 100 °C for each of the following, and indicate whether heat was absorbed/released:
- **a.** calories to vaporize 10.0 g of water
- **b.** joules to vaporize 5.00 g of water
- **c.** kilocalories to condense 8.0 kg of steam
- **d.** kilojoules to condense 175 g of steam

3.50 Calculate the heat change at 100 °C for each of the following, and indicate whether heat was absorbed/released:
- **a.** calories to condense 10.0 g of steam
- **b.** joules to condense 7.60 g of steam
- **c.** kilocalories to vaporize 44 g of water
- **d.** kilojoules to vaporize 5.00 kg of water

3.51 Draw a heating curve for a sample of ice that is heated from −20 °C to 150 °C. Indicate the segment of the graph that corresponds to each of the following:
a. solid **b.** melting **c.** liquid
d. boiling **e.** gas

3.52 Draw a cooling curve for a sample of steam that cools from 110 °C to −10 °C. Indicate the segment of the graph that corresponds to each of the following:
a. solid **b.** freezing **c.** liquid
d. condensation **e.** gas

3.53 Using the values for the heat of fusion, specific heat of water, and/or heat of vaporization, calculate the amount of heat energy in each of the following:
- **a.** joules needed to melt 50.0 g of ice at 0 °C and to warm the liquid to 65.0 °C
- **b.** kilocalories released when 15.0 g of steam condenses at 100 °C and the liquid cools to 0 °C
- **c.** kilojoules needed to melt 24.0 g of ice at 0 °C, warm the liquid to 100 °C, and change it to steam at 100 °C

3.54 Using the values for the heat of fusion, specific heat of water, and/or heat of vaporization, calculate the amount of heat energy in each of the following:
- **a.** joules released when 125 g of steam at 100 °C condenses and cools to liquid at 15.0 °C
- **b.** kilocalories needed to melt a 525-g ice sculpture at 0 °C and to warm the liquid to 15.0 °C
- **c.** kilojoules released when 85.0 g of steam condenses at 100 °C, cools, and freezes at 0 °C

Clinical Applications

3.55 A patient arrives in the emergency room with a burn caused by steam. Calculate the heat, in kilocalories, that is released when 18.0 g of steam at 100 °C hits the skin, condenses, and cools to body temperature of 37.0 °C.

3.56 A sports trainer applies an ice bag to the back of an injured athlete. Calculate the heat, in kilocalories, that is absorbed if 145 g of ice at 0 °C is placed in an ice bag, melts, and warms to body temperature of 37.0 °C.

CLINICAL UPDATE A Diet and Exercise Program

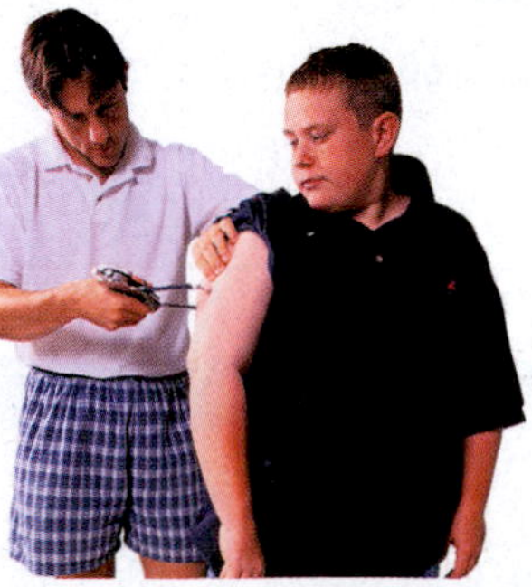

After Charles has been on his diet and exercise plan for a month, he and his mother meet again with the dietitian. Daniel looks at the diary of food intake and exercise, and weighs Charles to evaluate how well the diet is working. The following is what Charles ate in one day:

Breakfast

1 banana, 1 cup of nonfat milk, 1 egg

Lunch

1 cup of carrots, 3 oz of ground beef, 1 apple, 1 cup of nonfat milk

Dinner

6 oz of skinless chicken, 1 baked potato, 3 oz of broccoli, 1 cup of nonfat milk

Clinical Applications

3.57 Using energy contents from Table 3.8, determine each of the following:
- **a.** the total kilocalories for each meal
- **b.** the total kilocalories for one day
- **c.** If Charles consumes 1800 kcal per day, he will maintain his weight. Would he lose weight on his new diet?
- **d.** If expending 3500 kcal is equal to a loss of 1.0 lb, how many days will it take Charles to lose 5.0 lb?

3.58
- **a.** During one week, Charles swam for a total of 2.5 h and walked for a total of 8.0 h. If Charles expends 340 kcal/h swimming and 160 kcal/h walking, how many total kilocalories did he expend for one week?
- **b.** For the amount of exercise that Charles did for one week in part **a**, if expending 3500 kcal is equal to a loss of 1.0 lb, how many pounds did he lose?
- **c.** How many hours would Charles have to walk to lose 1.0 lb?
- **d.** How many hours would Charles have to swim to lose 1.0 lb?

CONCEPT MAP

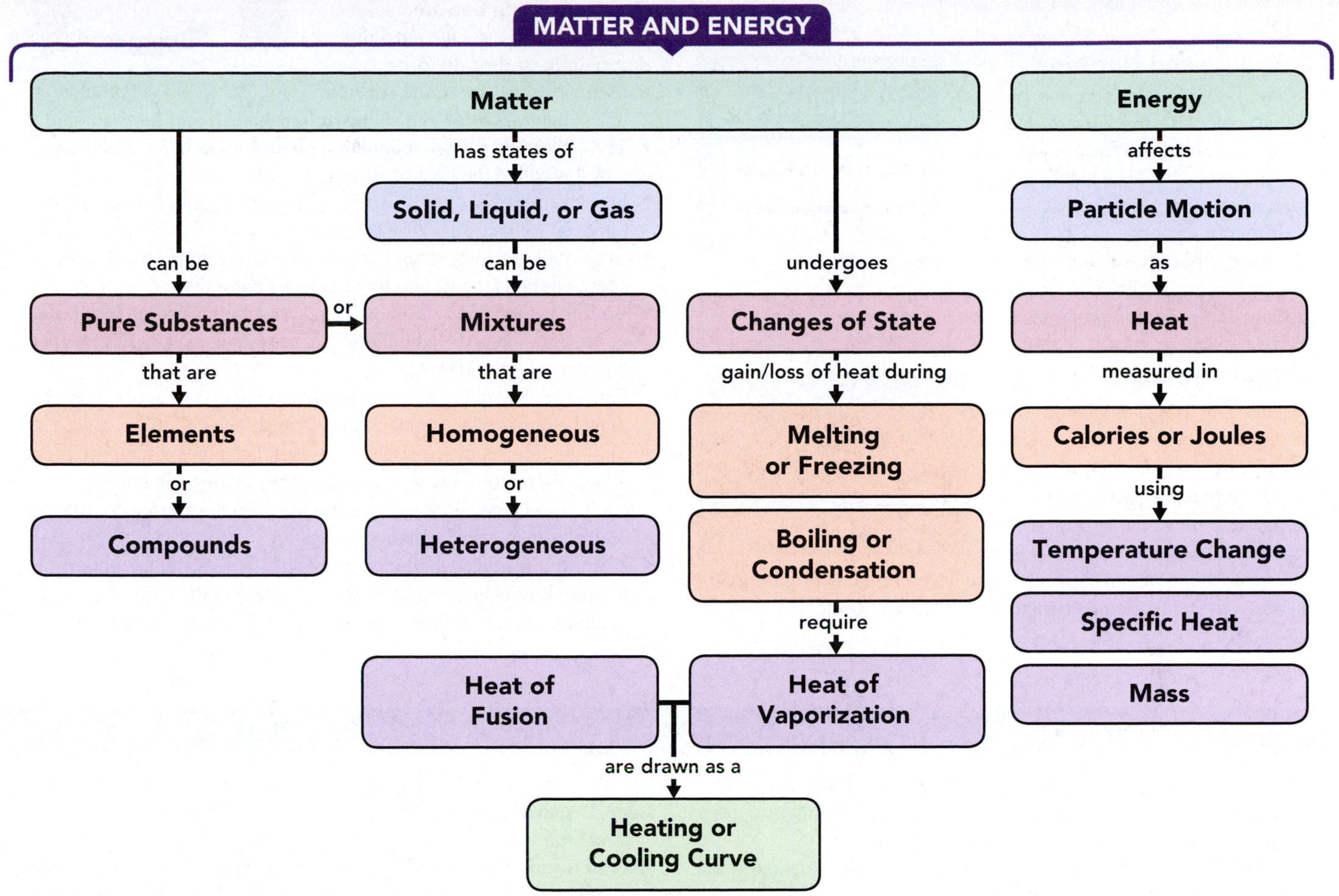
MATTER AND ENERGY
Matter
has states of
Solid, Liquid, or Gas
can be
Pure Substances
or
Mixtures
that are
Elements
or
Compounds
can be
that are
Homogeneous
or
Heterogeneous
undergoes
Changes of State
gain/loss of heat during
Melting or Freezing
Boiling or Condensation
require
Heat of Fusion
Heat of Vaporization
are drawn as a
Heating or Cooling Curve
Energy
affects
Particle Motion
as
Heat
measured in
Calories or Joules
using
Temperature Change
Specific Heat
Mass

CHAPTER REVIEW

3.1 Classification of Matter

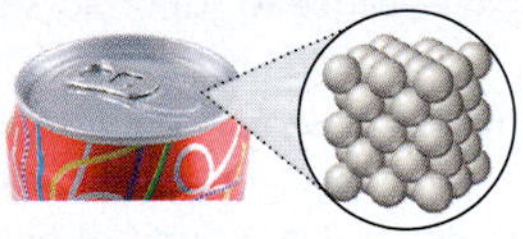

LEARNING GOAL Classify examples of matter as pure substances or mixtures.

- Matter is anything that has mass and occupies space.
- Matter is classified as pure substances or mixtures.
- Pure substances, which are elements or compounds, have fixed compositions, and mixtures have variable compositions.
- The substances in mixtures can be separated using physical methods.

3.2 States and Properties of Matter

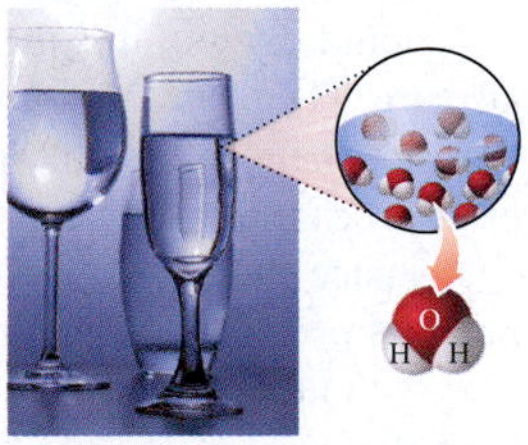
O
H
H

LEARNING GOAL Identify the states and the physical and chemical properties of matter.

- The three states of matter are solid, liquid, and gas.
- A physical property is a characteristic of a substance that can be observed or measured without affecting the identity of the substance.
- A physical change occurs when physical properties change, but not the composition of the substance.
- A chemical property indicates the ability of a substance to change into another substance.
- A chemical change occurs when one or more substances react to form a new substance with new physical and chemical properties.

3.3 Temperature

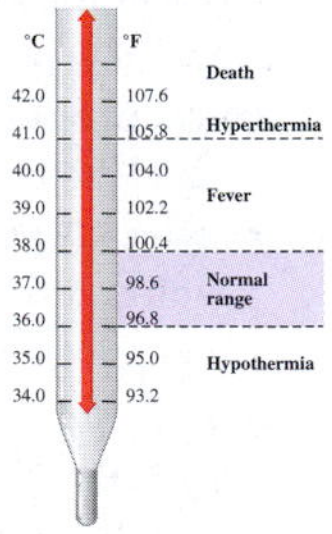
°C
°F
Death
42.0
107.6
41.0
105.8
Hyperthermia
40.0
104.0
39.0
102.2
Fever
38.0
100.4
37.0
98.6
Normal range
36.0
96.8
35.0
95.0
Hypothermia
34.0
93.2

LEARNING GOAL Given a temperature, calculate the corresponding temperature on another scale.

- In science, temperature is measured in degrees Celsius (°C) or kelvins (K).
- On the Celsius scale, there are 100 units between the freezing point (0 °C) and the boiling point (100 °C) of water.
- On the Fahrenheit scale, there are 180 units between the freezing point (32 °F) and the boiling point (212 °F) of water. A Fahrenheit temperature is related to its Celsius temperature by the equation $T_F = 1.8(T_C) + 32$.
- The SI unit, kelvin, is related to the Celsius temperature by the equation $T_K = T_C + 273$.

3.4 Energy

LEARNING GOAL Identify energy as potential or kinetic; convert between units of energy.

- Energy is the ability to do work.
- Potential energy is stored energy; kinetic energy is the energy of motion.

- Common units of energy are the calorie (cal), kilocalorie (kcal), joule (J), and kilojoule (kJ).
- One calorie is equal to 4.184 J.

3.5 Energy and Nutrition

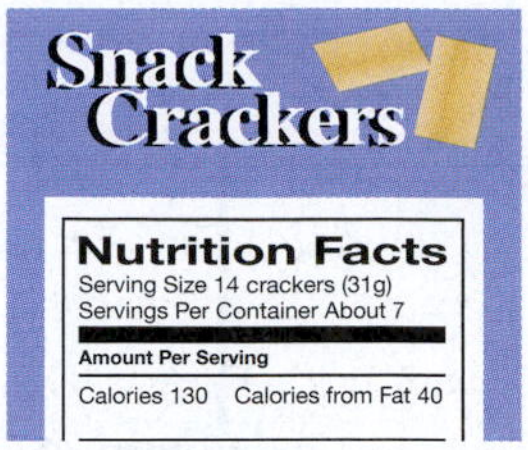
Snack Crackers
Nutrition Facts
Serving Size 14 crackers (31g)
Servings Per Container About 7
Amount Per Serving
Calories 130 Calories from Fat 40

LEARNING GOAL Use the energy values to calculate the kilocalories (kcal) or kilojoules (kJ) for a food.

- The nutritional Calorie is the same amount of energy as 1 kcal or 1000 calories.
- The energy of a food is the sum of kilocalories or kilojoules from carbohydrate, fat, and protein.

3.6 Specific Heat

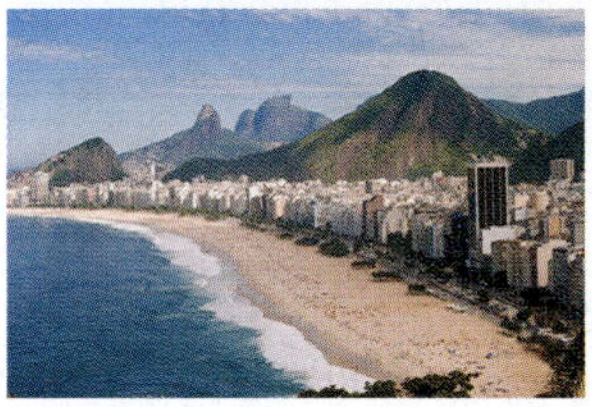

LEARNING GOAL Use specific heat to calculate heat loss or gain.

- Specific heat is the amount of energy required to raise the temperature of exactly 1 g of a substance by exactly 1 °C.
- The heat lost or gained by a substance is determined by multiplying its mass, the temperature change, and its specific heat.

3.7 Changes of State

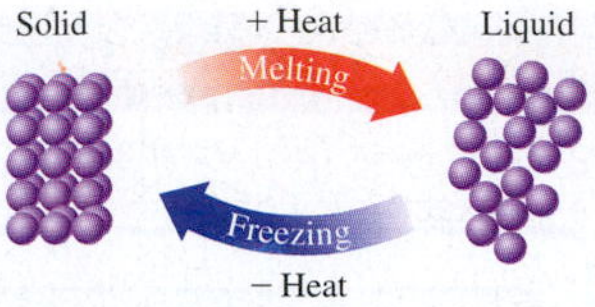
Solid
+ Heat
Liquid
Melting
Freezing
− Heat

LEARNING GOAL Describe the changes of state between solids, liquids, and gases; calculate the energy released or absorbed.

- Melting occurs when the particles in a solid absorb enough energy to break apart and form a liquid.
- The amount of energy required to convert exactly 1 g of solid to liquid is called the heat of fusion.
- For water, 80. cal (334 J) is needed to melt 1 g of ice or must be removed to freeze 1 g of water at 0 °C.
- Evaporation occurs when particles in a liquid state absorb enough energy to break apart and form gaseous particles.
- Boiling is the vaporization of liquid at its boiling point. The heat of vaporization is the amount of heat needed to convert exactly 1 g of liquid to vapor at 100 °C.
- For water, 540 cal (2260 J) is needed to vaporize 1 g of water or must be removed to condense 1 g of steam at 100 °C.
- Sublimation is a process whereby a solid changes directly to a gas.
- A heating or cooling curve illustrates the changes in temperature and state as heat is added to or removed from a substance. Plateaus on the graph indicate changes of state.
- The total heat absorbed or removed from a substance undergoing temperature changes and changes of state is the sum of energy calculations for change(s) of state and change(s) in temperature.

KEY TERMS

boiling The formation of bubbles of gas throughout a liquid.

boiling point (bp) The temperature at which a liquid changes to gas (boils) and gas changes to liquid (condenses).

calorie (cal) The amount of heat energy that raises the temperature of exactly 1 g of water by exactly 1 °C.

change of state The transformation of one state of matter to another; for example, solid to liquid, liquid to solid, liquid to gas.

chemical change A change during which the original substance is converted into a new substance that has a different composition and new physical and chemical properties.

chemical properties The properties that indicate the ability of a substance to change into a new substance.

compound A pure substance consisting of two or more elements, with a definite composition, that can be broken down into simpler substances only by chemical methods.

condensation The change of state from a gas to a liquid.

cooling curve A diagram that illustrates temperature changes and changes of state for a substance as heat is removed.

deposition The change of a gas directly into a solid; the reverse of sublimation.

element A pure substance containing only one type of matter, which cannot be broken down by chemical methods.

energy The ability to do work.

energy value The kilocalories (or kilojoules) obtained per gram of the food types: carbohydrate, fat, and protein.

evaporation The formation of a gas (vapor) by the escape of high-energy particles from the surface of a liquid.

freezing The change of state from liquid to solid.

freezing point (fp) The temperature at which a liquid changes to a solid (freezes) and a solid changes to a liquid (melts).

gas A state of matter that does not have a definite shape or volume.

heat The energy associated with the motion of particles in a substance.

heat of fusion The energy required to melt exactly 1 g of a substance at its melting point. For water, 80. cal (334 J) is needed to melt 1 g of ice; 80. cal (334 J) is released when 1 g of water freezes at 0 °C.

heat of vaporization The energy required to vaporize exactly 1 g of a substance at its boiling point. For water, 540 cal (2260 J) is needed to vaporize 1 g of liquid; 1 g of steam gives off 540 cal (2260 J) when it condenses at 100 °C.

heating curve A diagram that illustrates the temperature changes and changes of state of a substance as it is heated.

joule (J) The SI unit of heat energy; 4.184 J = 1 cal.

kinetic energy The energy of moving particles.

liquid A state of matter that takes the shape of its container but has a definite volume.

matter The material that makes up a substance and has mass and occupies space.

melting The change of state from a solid to a liquid.

melting point (mp) The temperature at which a solid becomes a liquid (melts). It is the same temperature as the freezing point.

mixture The physical combination of two or more substances that are not chemically combined.

physical change A change in which the physical properties of a substance change but its identity stays the same.

physical properties The properties that can be observed or measured without affecting the identity of a substance.

potential energy A type of energy related to position or composition of a substance.

pure substance A type of matter that has a definite composition.
solid A state of matter that has its own shape and volume.
specific heat (*SH*) A quantity of heat that changes the temperature of exactly 1 g of a substance by exactly 1 °C.
states of matter Three forms of matter: solid, liquid, and gas.
sublimation The change of state in which a solid is transformed directly to a gas without forming a liquid first.

CORE CHEMISTRY SKILLS

The chapter Section containing each Core Chemistry Skill is shown in parentheses at the end of each heading.

Identifying Physical and Chemical Changes (3.2)

- Physical properties can be observed or measured without changing the identity of a substance.
- Chemical properties describe the ability of a substance to change into a new substance.
- When matter undergoes a physical change, its state or its appearance changes, but its composition remains the same.
- When a chemical change takes place, the original substance is converted into a new substance, which has different physical and chemical properties.

Example: Classify each of the following as a physical or chemical property:

a. Helium in a balloon is a gas.
b. Methane, in natural gas, burns.
c. Hydrogen sulfide smells like rotten eggs.

Answer: **a.** A gas is a state of matter, which makes it a physical property.
b. When methane burns, it changes to different substances with new properties, which is a chemical property.
c. The odor of hydrogen sulfide is a physical property.

Converting between Temperature Scales (3.3)

- The temperature equation $T_F = 1.8(T_C) + 32$ is used to convert from Celsius to Fahrenheit and can be rearranged to convert from Fahrenheit to Celsius.
- The temperature equation $T_K = T_C + 273$ is used to convert from Celsius to Kelvin and can be rearranged to convert from Kelvin to Celsius.

Example: Convert 75.0 °C to degrees Fahrenheit.

Answer:
$$T_F = 1.8(T_C) + 32$$
$$T_F = 1.8(75.0) + 32 = 135 + 32$$
$$= 167\ °F$$

Example: Convert 355 K to degrees Celsius.

Answer: $T_K = T_C + 273$

To solve the equation for T_C, subtract 273 from both sides.

$$T_K - 273 = T_C + \cancel{273} - \cancel{273}$$
$$T_C = T_K - 273$$
$$T_C = 355 - 273$$
$$= 82\ °C$$

Using Energy Units (3.4)

- Equalities for energy units include 1 cal = 4.184 J, 1 kcal = 1000 cal, and 1 kJ = 1000 J.
- Each equality for energy units can be written as two conversion factors:

$$\frac{4.184\ J}{1\ cal} \text{ and } \frac{1\ cal}{4.184\ J} \quad \frac{1000\ cal}{1\ kcal} \text{ and } \frac{1\ kcal}{1000\ cal} \quad \frac{1000\ J}{1\ kJ} \text{ and } \frac{1\ kJ}{1000\ J}$$

- The energy unit conversion factors are used to cancel given units of energy and to obtain the needed unit of energy.

Example: Convert 45 000 J to kilocalories.

Answer: Using the conversion factors above, we start with the given 45 000 J and convert it to kilocalories.

$$45000\ \cancel{J} \times \frac{1\ \cancel{cal}}{4.184\ \cancel{J}} \times \frac{1\ kcal}{1000\ \cancel{cal}} = 11\ kcal$$

Using the Heat Equation (3.6)

- The quantity of heat absorbed or lost by a substance is calculated using the heat equation.

$$\text{Heat} = m \times \Delta T \times SH$$

- Heat, in calories, is obtained when the specific heat of a substance in cal/g °C is used.
- Heat, in joules, is obtained when the specific heat of a substance in J/g °C is used.
- To cancel, the unit grams is used for mass, and the unit °C is used for temperature change.

Example: How many joules are required to heat 5.25 g of titanium from 85.5 °C to 132.5 °C?

Answer: $m = 5.25$ g, $\Delta T = 132.5\ °C - 85.5\ °C = 47.0\ °C$
SH for titanium = 0.523 J/g °C

The known values are substituted into the heat equation making sure units cancel.

$$\text{Heat} = m \times \Delta T \times SH = 5.25\ \cancel{g} \times 47.0\ \cancel{°C} \times \frac{0.523\ J}{\cancel{g}\ \cancel{°C}}$$
$$= 129\ J$$

Calculating Heat for Change of State (3.7)

- At the melting/freezing point, the heat of fusion is absorbed/released to convert 1 g of a solid to a liquid or 1 g of liquid to a solid.
- For example, 334 J of heat is needed to melt (freeze) exactly 1 g of ice at its melting (freezing) point, 0 °C.
- At the boiling/condensation point, the heat of vaporization is absorbed/released to convert exactly 1 g of liquid to gas or 1 g of gas to liquid.
- For example, 2260 J of heat is needed to boil (condense) exactly 1 g of water/steam at its boiling (condensation) point, 100 °C.

Example: What is the quantity of heat, in kilojoules, released when 45.8 g of steam (water) condenses at its boiling (condensation) point?

Answer:
$$45.8\ \cancel{g\ steam} \times \frac{2260\ \cancel{J}}{1\ \cancel{g\ steam}} \times \frac{1\ kJ}{1000\ \cancel{J}} = 104\ kJ$$

UNDERSTANDING THE CONCEPTS

The chapter Sections to review are shown in parentheses at the end of each problem.

3.59 Identify each of the following as an element, a compound, or a mixture. Explain your choice. (3.1)

a.

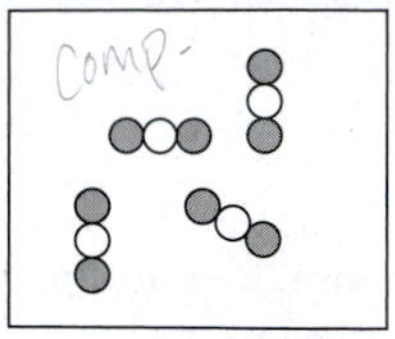

b.

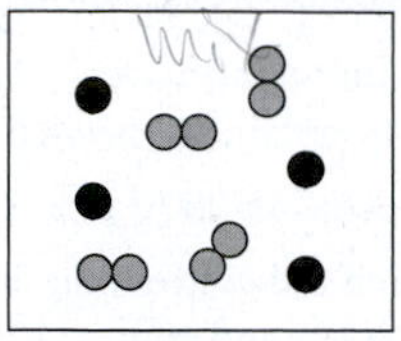

c.

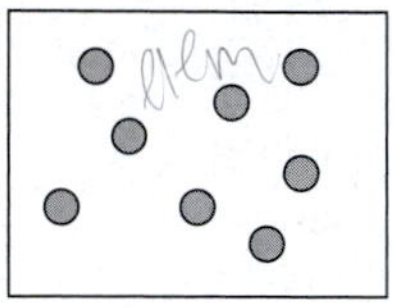

3.60 Identify each of the following as a homogeneous or heterogeneous mixture. Explain your choice. (3.1)

a.

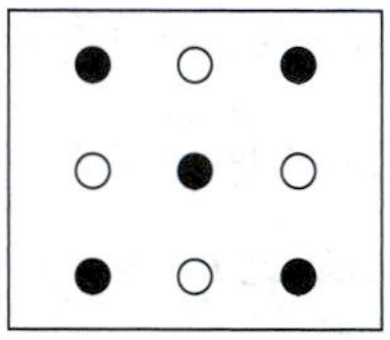

b.

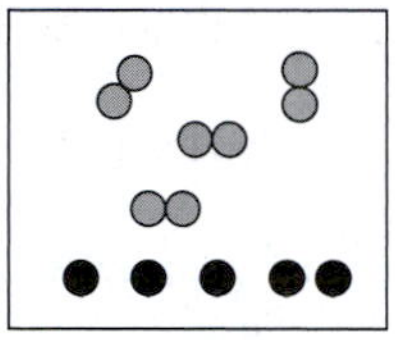

c.

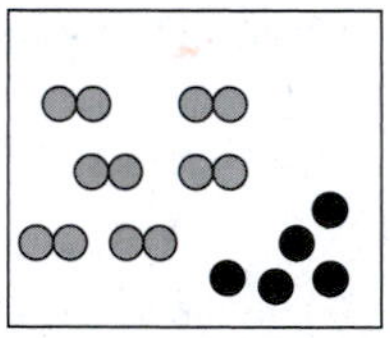

3.61 Classify each of the following as a homogeneous or heterogeneous mixture: (3.1)

a. lemon-flavored water

b. stuffed mushrooms

c. eye drops

3.62 Classify each of the following as a homogeneous or heterogeneous mixture: (3.1)

a. ketchup **b.** hard-boiled egg **c.** tortilla soup

3.63 State the temperature on the Celsius thermometer and convert to Fahrenheit. (3.3)

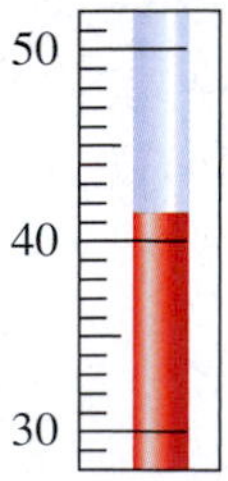
50
40
30

3.64 State the temperature on the Celsius thermometer and convert to Fahrenheit. (3.3)

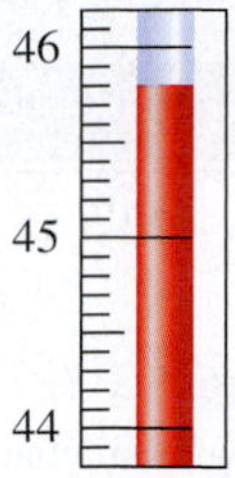
46
45
44

3.65 Compost can be made at home from grass clippings, kitchen scraps, and dry leaves. As microbes break down organic matter, heat is generated, and the compost can reach a temperature of 155 °F, which kills most pathogens. What is this temperature in degrees Celsius? In kelvins? (3.3)

Compost produced from decayed plant material is used to enrich the soil.

3.66 After a week, biochemical reactions in compost slow, and the temperature drops to 45 °C. The dark brown organic-rich mixture is ready for use in the garden. What is this temperature in degrees Fahrenheit? In kelvins? (3.3)

3.67 Calculate the energy to heat two cubes (gold and aluminum) each with a volume of 10.0 cm^3 from 15 °C to 25 °C. Refer to Tables 2.9 and 3.11. (3.6)

3.68 Calculate the energy to heat two cubes (silver and copper), each with a volume of 10.0 cm^3 from 15 °C to 25 °C. Refer to Tables 2.9 and 3.11. (3.6)

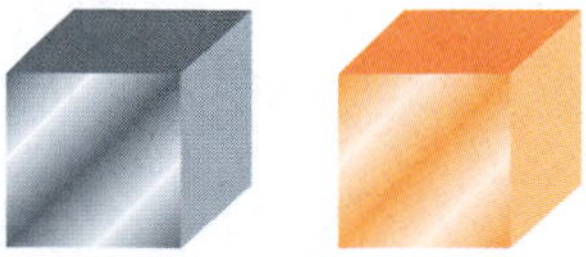

Clinical Applications

3.69 A 70.0-kg person had a quarter-pound cheeseburger, french fries, and a chocolate shake. (3.5)

Item	Carbohydrate (g)	Fat (g)	Protein (g)
Cheeseburger	46	40.	47
French fries	47	16	4
Chocolate shake	76	10.	10.

a. Using Table 3.7, calculate the total kilocalories for each food type in this meal (round off the kilocalories to the tens place).
b. Determine the total kilocalories for the meal.
c. Using Table 3.10, determine the number of hours of sleep needed to burn off the kilocalories in this meal.
d. Using Table 3.10, determine the number of hours of running needed to burn off the kilocalories in this meal.

3.70 Your friend, who has a mass of 70.0 kg, has a slice of pizza, a cola soft drink, and ice cream. (3.5)

Item	Carbohydrate (g)	Fat (g)	Protein (g)
Pizza	29	10.	13
Cola	51	0	0
Ice cream	44	28	8

a. Using Table 3.7, calculate the total kilocalories for each food type in this meal (round off the kilocalories to the tens place).
b. Determine the total kilocalories for the meal.
c. Using Table 3.10, determine the number of hours of sitting needed to burn off the kilocalories in this meal.
d. Using Table 3.10, determine the number of hours of swimming needed to burn off the kilocalories in this meal.

ADDITIONAL PRACTICE PROBLEMS

3.71 Classify each of the following as an element, a compound, or a mixture: (3.1)
a. carbon in pencils element
b. carbon monoxide (CO) in automobile exhaust comp.
c. orange juice mixture

3.72 Classify each of the following as an element, a compound, or a mixture: (3.1)
a. neon gas in lights element
b. a salad dressing of oil and vinegar mix.
c. sodium hypochlorite (NaClO) in bleach comp.

3.73 Classify each of the following mixtures as homogeneous or heterogeneous: (3.1)
a. hot fudge sundae hetero.
b. herbal tea homo
c. vegetable oil homo

3.74 Classify each of the following mixtures as homogeneous or heterogeneous: (3.1)
a. water and sand Het
b. mustard Homo
c. blue ink Homo

3.75 Identify each of the following as solid, liquid, or gas: (3.2)
a. vitamin tablets in a bottle Solid
b. helium in a balloon gas
c. milk in a bottle liquid
d. the air you breathe gas
e. charcoal briquettes on a barbecue solid

3.76 Identify each of the following as solid, liquid, or gas: (3.2)
a. popcorn in a bag S
b. water in a garden hose L
c. a computer mouse S
d. air in a tire g
e. hot tea in a teacup L

3.77 Identify each of the following as a physical or chemical property: (3.2)
a. Gold is shiny. physical
b. Gold melts at 1064 °C. phys
c. Gold is a good conductor of electricity. phys
d. When gold reacts with sulfur, a black sulfide compound forms. chem

3.78 Identify each of the following as a physical or chemical property: (3.2)
a. A candle is 10 cm high and 2 cm in diameter. P
b. A candle burns. C
c. The wax of a candle softens on a hot day. P
d. A candle is blue. P

3.79 Identify each of the following as a physical or chemical change: (3.2)
a. A plant grows a new leaf. chem.
b. Chocolate is melted for a dessert. physi.
c. Wood is chopped for the fireplace. phys
d. Wood burns in a woodstove. chem.

3.80 Identify each of the following as a physical or chemical change: (3.2)
a. Aspirin tablets are broken in half.
b. Carrots are grated for use in a salad.
c. Malt undergoes fermentation to make beer.
d. A copper pipe reacts with air and turns green.

3.81 Calculate each of the following temperatures in degrees Celsius and kelvins: (3.3)
a. The highest recorded temperature in the continental United States was 134 °F in Death Valley, California, on July 10, 1913.
b. The lowest recorded temperature in the continental United States was −69.7 °F in Rodgers Pass, Montana, on January 20, 1954.

3.82 Calculate each of the following temperatures in kelvins and degrees Fahrenheit: (3.3)

a. The highest recorded temperature in the world was 58.0 °C in El Azizia, Libya, on September 13, 1922.

b. The lowest recorded temperature in the world was −89.2 °C in Vostok, Antarctica, on July 21, 1983.

3.83 What is −15 °F in degrees Celsius and in kelvins? (3.3)

3.84 What is 56 °F in degrees Celsius and in kelvins? (3.3)

3.85 A 0.50-g sample of vegetable oil is placed in a calorimeter. When the sample is burned, 18.9 kJ is given off. What is the energy value (kcal/g) for the oil? (3.5)

3.86 A 1.3-g sample of rice is placed in a calorimeter. When the sample is burned, 22 kJ is given off. What is the energy value (kcal/g) for the rice? (3.5)

3.87 On a hot day, the beach sand gets hot but the water stays cool. Would you predict that the specific heat of sand is higher or lower than that of water? Explain. (3.6)

On a sunny day, the sand gets hot but the water stays cool.

3.88 On a hot sunny day, you get out of the swimming pool and sit in a metal chair, which is very hot. Would you predict that the specific heat of the metal is higher or lower than that of water? Explain. (3.6)

3.89 The following graph is a heating curve for chloroform, a solvent for fats, oils, and waxes: (3.7)

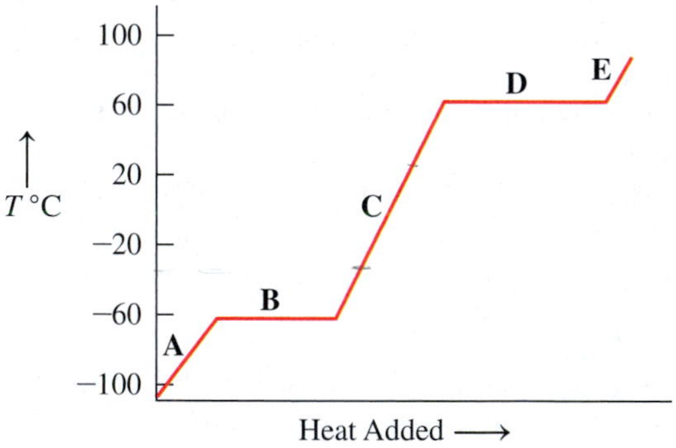
100
60
20
−20
−60
−100
T °C
A
B
C
D
E
Heat Added ⟶

a. What is the approximate melting point of chloroform?

b. What is the approximate boiling point of chloroform?

c. On the heating curve, identify the segments **A**, **B**, **C**, **D**, and **E** as solid, liquid, gas, melting, or boiling.

d. At each of the following temperatures, is chloroform a solid, liquid, or gas?

−80 °C; −40 °C; 25 °C; 80 °C

3.90 Associate the contents of the beakers (**1** to **5**) with segments (**A** to **E**) on the following heating curve for water: (3.7)

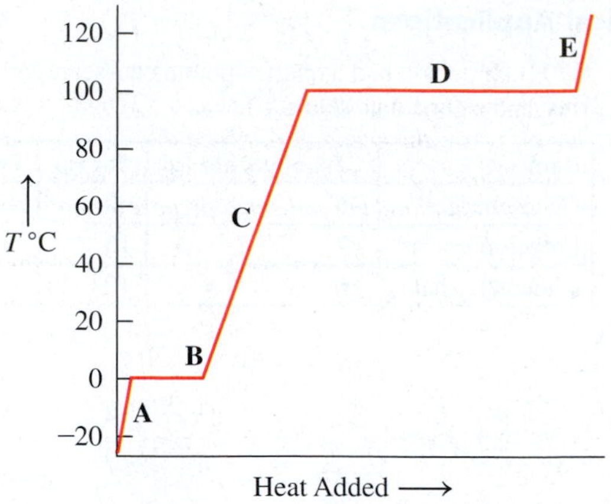
120
100
80
60
40
20
0
−20
T °C
A
B
C
D
E
Heat Added ⟶

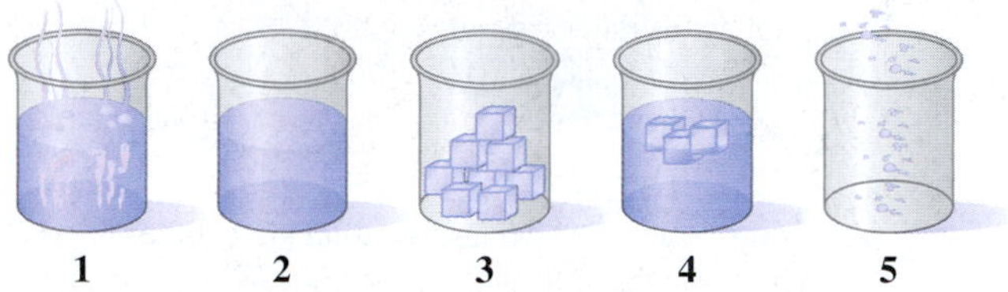
1
2
3
4
5

3.91 The melting point of dibromomethane is −53 °C and its boiling point is 97 °C. Sketch a heating curve for dibromomethane from −100 °C to 120 °C. (3.7)

a. What is the state of dibromomethane at −75 °C?

b. What happens on the curve at −53 °C?

c. What is the state of dibromomethane at −18 °C?

d. What is the state of dibromomethane at 110 °C?

e. At what temperature will both solid and liquid be present?

3.92 The melting point of benzene is 5.5 °C and its boiling point is 80.1 °C. Sketch a heating curve for benzene from 0 °C to 100 °C. (3.7)

a. What is the state of benzene at 15 °C?

b. What happens on the curve at 5.5 °C?

c. What is the state of benzene at 63 °C?

d. What is the state of benzene at 98 °C?

e. At what temperature will both liquid and gas be present?

Clinical Applications

3.93 If you want to lose 1 lb of "body fat," which is 15% water, how many kilocalories do you need to expend? (3.5)

3.94 A young patient drinks whole milk as part of her diet. Calculate the total kilocalories if the glass of milk contains 12 g of carbohydrate, 9 g of fat, and 9 g of protein. (Round off the answer for each food type to the tens place.) (3.5)

3.95 A hot-water bottle for a patient contains 725 g of water at 65 °C. If the water cools to body temperature (37 °C), how many kilojoules of heat could be transferred to sore muscles? (3.6)

3.96 The highest recorded body temperature that a person has survived is 46.5 °C. Calculate that temperature in degrees Fahrenheit and in kelvins. (3.3)

CHALLENGE PROBLEMS

The following problems are related to the topics in this chapter. However, they do not all follow the chapter order, and they require you to combine concepts and skills from several Sections. These problems will help you increase your critical thinking skills and prepare for your next exam.

3.97 When a 0.66-g sample of olive oil is burned in a calorimeter, the heat released increases the temperature of 370 g of water from 22.7 °C to 38.8 °C. What is the energy value for the olive oil in kcal/g? (3.5, 3.6)

3.98 A 45-g piece of ice at 0 °C is added to a sample of water at 8.0 °C. All of the ice melts and the temperature of the water decreases to 0 °C. How many grams of water were in the sample? (3.6, 3.7)

3.99 In a large building, oil is used in a steam boiler heating system. The combustion of 1.0 lb of oil provides 2.4×10^7 J. (3.4, 3.6)

a. How many kilograms of oil are needed to heat 150 kg of water from 22 °C to 100 °C?

b. How many kilograms of oil are needed to change 150 kg of water to steam at 100 °C?

3.100 When 1.0 g of gasoline burns, it releases 11 kcal. The density of gasoline is 0.74 g/mL. (3.4, 3.6)

a. How many megajoules are released when 1.0 gal of gasoline burns?

b. If a television requires 150 kJ/h to run, how many hours can the television run on the energy provided by 1.0 gal of gasoline?

3.101 An ice bag containing 275 g of ice at 0 °C was used to treat sore muscles. When the bag was removed, the ice had melted and the liquid water had a temperature of 24.0 °C. How many kilojoules of heat were absorbed? (3.6, 3.7)

3.102 A 115-g sample of steam at 100 °C is emitted from a volcano. It condenses, cools, and falls as snow at 0 °C. How many kilojoules were released? (3.6, 3.7)

3.103 A 70.0-g piece of copper metal at 54.0 °C is placed in 50.0 g of water at 26.0 °C. If the final temperature of the water and metal is 29.2 °C, what is the specific heat, in J/g °C, of copper? (3.6)

3.104 A 125-g piece of metal is heated to 288 °C and dropped into 85.0 g of water at 12.0 °C. The metal and water come to the same temperature of 24.0 °C. What is the specific heat, in J/g °C, of the metal? (3.6)

3.105 A metal is thought to be titanium or aluminum. When 4.7 g of the metal absorbs 11 J, its temperature rises by 4.5 °C. (3.6)

a. What is the specific heat, in J/g °C, of the metal?

b. Would you identify the metal as titanium or aluminum (see Table 3.11)?

3.106 A metal is thought to be copper or gold. When 18 g of the metal absorbs 58 cal, its temperature rises by 35 °C. (3.6)

a. What is the specific heat, in cal/g °C, of the metal?

b. Would you identify the metal as copper or gold (see Table 3.11)?

ANSWERS

3.1 **a.** element **b.** compound **c.** element **d.** compound **e.** compound

3.3 **a.** pure substance **b.** mixture **c.** pure substance **d.** pure substance **e.** mixture

3.5 **a.** heterogeneous **b.** homogeneous **c.** heterogeneous **d.** heterogeneous

3.7 **a.** gas **b.** gas **c.** solid

3.9 **a.** physical **b.** chemical **c.** physical **d.** chemical **e.** chemical

3.11 **a.** physical **b.** chemical **c.** physical **d.** physical **e.** physical

3.13 **a.** chemical **b.** physical **c.** physical **d.** chemical **e.** physical

3.15 In the United States, we still use the Fahrenheit temperature scale. In °F, normal body temperature is 98.6. On the Celsius scale, her temperature would be 37.7 °C, a mild fever.

3.17 **a.** 98.6 °F **b.** 18.5 °C **c.** 246 K **d.** 335 K **e.** 46 °C

3.19 **a.** 41 °C
b. No. The temperature is equivalent to 39 °C.

3.21 When the roller-coaster car is at the top of the ramp, it has its maximum potential energy. As it descends, potential energy changes to kinetic energy. At the bottom, all the energy is kinetic.

3.23 **a.** potential **b.** kinetic **c.** potential **d.** potential

3.25 **a.** 3.5 kcal **b.** 99.2 cal **c.** 120 J **d.** 1100 cal

3.27 **a.** 8.1×10^5 J **b.** 190 kcal

3.29 **a.** 29.9 kcal **b.** 208 kcal

3.31 **a.** 470 kJ **b.** 18 g **c.** 130 kcal **d.** 38 g

3.33 210 kcal, 880 kJ

3.35 640 kcal

3.37 Copper has the lowest specific heat of the samples and will reach the highest temperature.

3.39 **a.** 180 cal **b.** 2600 J **c.** 9.3 kcal **d.** 10.8 kJ

3.41 **a.** 1380 J, 330. cal **b.** 1810 J, 434 cal **c.** 3780 J, 904 cal **d.** 3600 J, 850 cal

3.43 **a.** melting **b.** sublimation **c.** freezing **d.** deposition

3.45 **a.** 5200 cal absorbed **b.** 5680 J absorbed **c.** 18 kcal released **d.** 16.7 kJ released

3.47 **a.** condensation **b.** evaporation **c.** boiling **d.** condensation

3.49 **a.** 5400 cal absorbed **b.** 11 300 J absorbed **c.** 4300 kcal released **d.** 396 kJ released

3.51

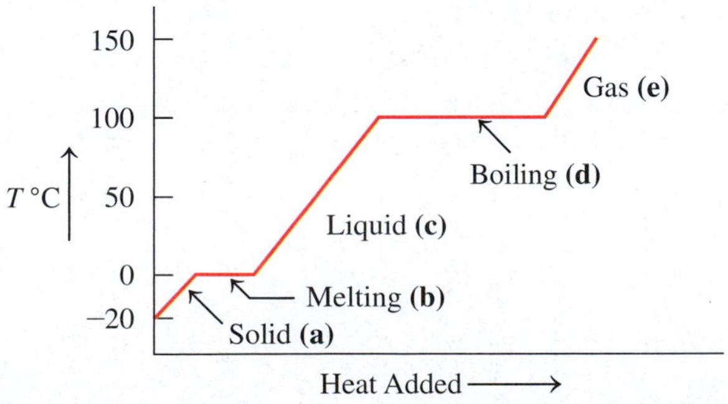

150
100
50
0
−20
T °C
Gas (e)
Boiling (d)
Liquid (c)
Melting (b)
Solid (a)
Heat Added

3.53 **a.** 30 300 J **b.** 9.6 kcal **c.** 72.2 kJ

3.55 10.8 kcal

3.57 **a.** Breakfast, 270 kcal; Lunch, 420 kcal; Dinner, 440 kcal
b. 1130 kcal total
c. Yes. Charles should be losing weight.
d. 26 days

3.59 **a.** compound, the particles have a 2:1 ratio of atoms
b. mixture, has two different kinds of particles
c. element, has a single kind of atom

3.61 **a.** homogeneous **b.** heterogeneous
c. homogeneous

3.63 41.5 °C, 107 °F

3.65 68.3 °C, 341 K

3.67 gold, 250 J or 60. cal; aluminum, 240 J or 58 cal

3.69 **a.** carbohydrate, 680 kcal; fat, 590 kcal; protein, 240 kcal
b. 1510 kcal
c. 25 h
d. 2.0 h

3.71 **a.** element **b.** compound **c.** mixture

3.73 **a.** heterogeneous **b.** homogeneous **c.** homogeneous

3.75 **a.** solid **b.** gas **c.** liquid **d.** gas **e.** solid

3.77 **a.** physical **b.** physical **c.** physical **d.** chemical

3.79 **a.** chemical **b.** physical **c.** physical **d.** chemical

3.81 **a.** 56.7 °C, 330. K **b.** −56.5 °C, 217 K

3.83 −26 °C, 247 K

3.85 9.0 kcal/g

3.87 The same amount of heat causes a greater temperature change in the sand than in the water; thus the sand must have a lower specific heat than that of water.

3.89 **a.** about −60 °C
b. about 60 °C
c. The diagonal line A represents the solid state as temperature increases. The horizontal line B represents the change from solid to liquid or melting of the substance. The diagonal line C represents the liquid state as temperature increases. The horizontal line D represents the change from liquid to gas or boiling of the liquid. The diagonal line E represents the gas state as temperature increases.
d. At −80 °C, solid; at −40 °C, liquid; at 25 °C, liquid; at 80 °C, gas

3.91

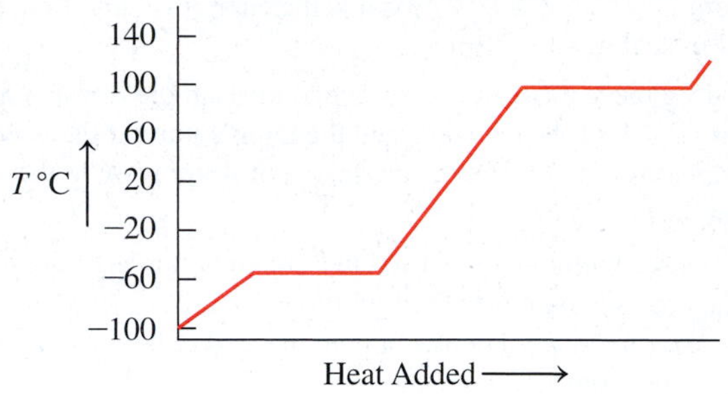
140
100
60
20
−20
−60
−100
T °C
Heat Added

a. solid **b.** solid dibromomethane melts
c. liquid **d.** gas
e. −53 °C

3.93 3500 kcal

3.95 85 kJ

3.97 9.0 kcal/g

3.99 **a.** 0.93 kg **b.** 6.4 kg

3.101 119.5 kJ

3.103 Specific heat = 0.385 J/g °C

3.105 **a.** 0.52 J/g °C **b.** titanium

COMBINING IDEAS from Chapters 1 to 3

CI.1 Gold, one of the most sought-after metals in the world, has a density of 19.3 g/cm^3, a melting point of 1064 °C, and a specific heat of 0.129 J/g °C. A gold nugget found in Alaska in 1998 weighed 20.17 lb. (2.4, 2.6, 2.7, 3.3, 3.5)

Gold nuggets, also called native gold, can be found in streams and mines.

a. How many significant figures are in the measurement of the weight of the nugget?
b. Which is the mass of the nugget in kilograms?
c. If the nugget were pure gold, what would its volume be in cubic centimeters?
d. What is the melting point of gold in degrees Fahrenheit and kelvins?
e. How many kilocalories are required to raise the temperature of the nugget from 500. °C to 1064 °C?
f. If the price of gold is \$42.06 per gram, what is the nugget worth in dollars?

CI.2 The mileage for a motorcycle with a fuel-tank capacity of 22 L is 35 mi/gal. (2.5, 2.6, 2.7, 3.4)

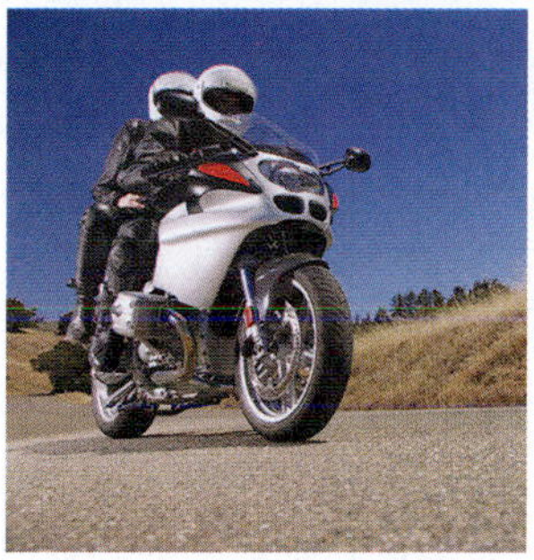

When 1.00 g of gasoline burns, 47 kJ of energy is released.

a. How long a trip, in kilometers, can be made on one full tank of gasoline?
b. If the price of gasoline is \$2.82 per gallon, what would be the cost of fuel for the trip?
c. If the average speed during the trip is 44 mi/h, how many hours will it take to reach the destination?
d. If the density of gasoline is 0.74 g/mL, what is the mass, in grams, of the fuel in the tank?
e. When 1.00 g of gasoline burns, 47 kJ of energy is released. How many kilojoules are produced when the fuel in one full tank is burned?

CI.3 Answer the following for the water samples **A** and **B** shown in the diagrams: (3.1, 3.2, 3.3, 3.5)

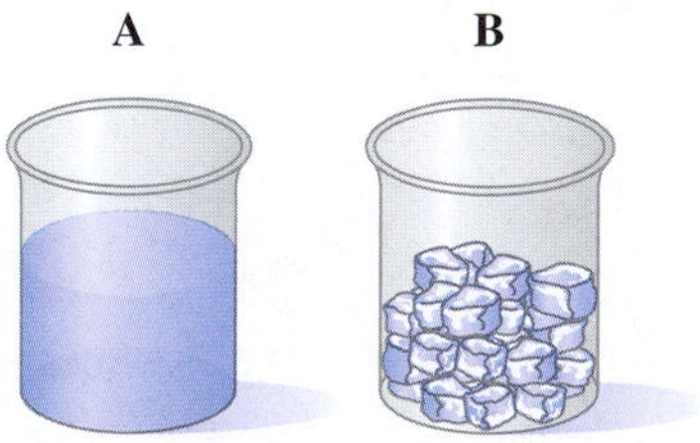
A
B

a. In which sample (**A** or **B**) does the water have its own shape?
b. Which diagram (**1** or **2** or **3**) represents the arrangement of particles in water sample **A**?
c. Which diagram (**1** or **2** or **3**) represents the arrangement of particles in water sample **B**?

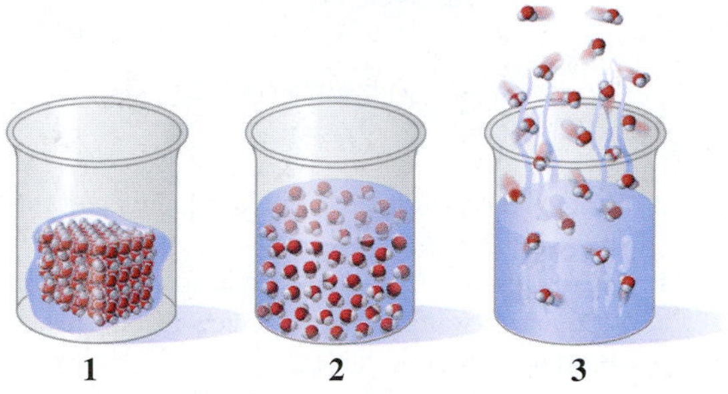
1
2
3

Answer the following for diagrams **1**, **2**, and **3**:
d. The state of matter indicated in diagram **1** is a ____; in diagram **2**, it is a ____; and in diagram **3**, it is a ____.
e. The motion of the particles is slowest in diagram ____.
f. The arrangement of particles is farthest apart in diagram ____.
g. The particles fill the volume of the container in diagram ____.
h. If the water in diagram **2** has a mass of 19 g and a temperature of 45 °C, how much heat, in kilojoules, is removed to cool the liquid to 0 °C?

CI.4 The label of an energy bar with a mass of 68 g lists the nutrition facts as 39 g of carbohydrate, 5 g of fat, and 10. g of protein. (2.5, 2.6, 3.4, 3.6)

An energy bar contains carbohydrate, fat, and protein.

a. Using the energy values for carbohydrates, fats, and proteins (see Table 3.7), what are the total kilocalories listed for the energy bar? (Round off the answer for each food type to the tens place.)
b. What are the kilojoules for the energy bar? (Round off the answer for each food type to the tens place.)
c. If you obtain 160 kJ, how many grams of the energy bar did you eat?
d. If you are walking and using energy at a rate of 840 kJ/h, how many minutes will you need to walk to expend the energy from two energy bars?

CI.5 In one box of nails, there are 75 iron nails weighing 0.250 lb. The density of iron is 7.86 g/cm^3. The specific heat of iron is 0.452 J/g °C. The melting point of iron is 1535 °C. (2.5, 2.6, 2.7, 3.4, 3.5)

Nails made of iron have a density of 7.86 g/cm^3.

a. What is the volume, in cubic centimeters, of the iron nails in the box?
b. If 30 nails are added to a graduated cylinder containing 17.6 mL of water, what is the new level of water, in milliliters, in the cylinder?
c. How much heat, in joules, must be added to the nails in the box to raise their temperature from 16 °C to 125 °C?
d. How much heat, in joules, is required to heat one nail from 25 °C to its melting point?

CI.6 A hot tub is filled with 450 gal of water. (2.5, 2.6, 2.7, 3.3, 3.4, 3.5)

A hot tub filled with water is heated to 105 °F.

a. What is the volume of water, in liters, in the tub?
b. What is the mass, in kilograms, of water in the tub?
c. How many kilocalories are needed to heat the water from 62 °F to 105 °F?
d. If the hot-tub heater provides 5900 kJ/min, how long, in minutes, will it take to heat the water in the hot tub from 62 °F to 105 °F?

ANSWERS

CI.1 **a.** four SFs
b. 9.17 kg
c. 475 cm^3
d. 1947 °F; 1337 K
e. 159 kcal
f. $386 000

CI.3 **a.** **B**
b. **A** is represented by diagram **2**.
c. **B** is represented by diagram **1**.
d. solid; liquid; gas
e. diagram **1**
f. diagram **3**
g. diagram **3**
h. 3.6 kJ

CI.5 **a.** 14.4 cm^3
b. 23.4 mL
c. 5590 J
d. 1030 J

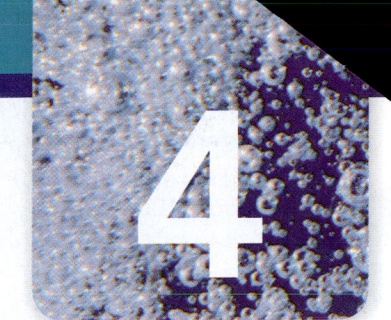

4

Atoms and Elements

John prepares for the next growing season by deciding how much of each crop to plant and where on his farm. The quality of the soil, including the pH, the amount of moisture, and the nutrient content in the soil helps John make his decision. After performing a few chemical tests on the soil, John determines that several of his fields need additional fertilizer before the crops can be planted. John considers several different types of fertilizers, as each supplies different nutrients to the soil. Plants need three elements for growth: potassium, nitrogen, and phosphorus. Potassium (K on the periodic table) is a metal, whereas nitrogen (N) and phosphorus (P) are nonmetals. Fertilizers may also contain several other elements including calcium (Ca), magnesium (Mg), and sulfur (S). John applies a fertilizer containing a mixture of all of these elements to his soil and plans to recheck the soil nutrient content in a few days.

CAREER

Farmer

Farming involves much more than growing crops and raising animals. Farmers must understand how to perform chemical tests and how to apply fertilizer to soil and pesticides or herbicides to crops. Pesticides are chemicals used to kill insects that could destroy the crop, whereas herbicides are chemicals used to kill weeds that would compete with the crops for water and nutrients. Farmers must understand how to use these chemicals safely and effectively. In using this information, farmers are able to grow crops that produce a higher yield, greater nutritional value, and better taste.

CLINICAL UPDATE

Improving Crop Production

Last year, John noticed that the potatoes from one of his fields had brown spots on the leaves and were undersized. Last week, he obtained soil samples from that field to check the nutrient levels. You can see how John improved the soil and crop production in the **CLINICAL UPDATE Improving Crop Production**, page 135, and learn what kind of fertilizer John used and the quantity applied.

KING AHEAD

ENGAGE

What are the names and symbols of some elements that you encounter every day?

Aluminum

Carbon

Gold

4.1 Elements and Symbols

LEARNING GOAL Given the name of an element, write its correct symbol; from the symbol, write the correct name.

All matter is composed of *elements*, of which there are 118 different kinds. Of these, 88 elements occur naturally and make up all the substances in our world. Many elements are already familiar to you. Perhaps you use aluminum in the form of foil or drink soft drinks from aluminum cans. You may have a ring or necklace made of gold, silver, or perhaps platinum. If you play tennis or golf, then you may have noticed that your racket or clubs may be made from the elements titanium or carbon. In our bodies, calcium and phosphorus form the structure of bones and teeth, iron and copper are needed in the formation of red blood cells, and iodine is required for thyroid function.

Elements are pure substances from which all other things are built. Elements cannot be broken down into simpler substances. Over the centuries, elements have been named for planets, mythological figures, colors, minerals, geographic locations, and famous people. Some sources of names of elements are listed in **TABLE 4.1**. A complete list of all the elements and their symbols are found on the inside front cover of this text.

TABLE 4.1 Some Elements, Symbols, and Source of Names

Element	Symbol	Source of Name
Uranium	U	The planet Uranus
Titanium	Ti	Titans (mythology)
Chlorine	Cl	*Chloros*: "greenish yellow" (Greek)
Iodine	I	*Ioeides*: "violet" (Greek)
Magnesium	Mg	Magnesia, a mineral
Tennessine	Ts	Tennessee
Curium	Cm	Marie and Pierre Curie
Copernicium	Cn	Nicolaus Copernicus

Chemical Symbols

Chemical symbols are one- or two-letter abbreviations for the names of the elements. Only the first letter of an element's symbol is capitalized. If the symbol has a second letter, it is lowercase so that we know when a different element is indicated. If two letters are

TABLE 4.2 Names and Symbols of Some Common Elements

Name*	Symbol	Name*	Symbol	Name*	Symbol
Aluminum	Al	Gallium	Ga	Oxygen	O
Argon	Ar	Gold (*aurum*)	Au	Phosphorus	P
Arsenic	As	Helium	He	Platinum	Pt
Barium	Ba	Hydrogen	H	Potassium (*kalium*)	K
Boron	B	Iodine	I	Radium	Ra
Bromine	Br	Iron (*ferrum*)	Fe	Silicon	Si
Cadmium	Cd	Lead (*plumbum*)	Pb	Silver (*argentum*)	Ag
Calcium	Ca	Lithium	Li	Sodium (*natrium*)	Na
Carbon	C	Magnesium	Mg	Strontium	Sr
Chlorine	Cl	Manganese	Mn	Sulfur	S
Chromium	Cr	Mercury (*hydrargyrum*)	Hg	Tin (*stannum*)	Sn
Cobalt	Co	Neon	Ne	Titanium	Ti
Copper (*cuprum*)	Cu	Nickel	Ni	Uranium	U
Fluorine	F	Nitrogen	N	Zinc	Zn

*Names given in parentheses are ancient Latin or Greek words from which the symbols are derived.

capitalized, they represent the symbols of two different elements. For example, the element cobalt has the symbol Co. However, the two capital letters CO specify two elements, carbon (C) and oxygen (O).

Although most of the symbols use letters from the current names, some are derived from their ancient names. For example, Na, the symbol for sodium, comes from the Latin word *natrium*. The symbol for iron, Fe, is derived from the Latin name *ferrum*. **TABLE 4.2** lists the names and symbols of some common elements. Learning their names and symbols will greatly help your learning of chemistry.

Silver

Sulfur

▶ SAMPLE PROBLEM 4.1 Names and Symbols of Chemical Elements

TRY IT FIRST

Complete the following table with the correct name or symbol for each element:

Name	Symbol
nickel	
nitrogen	
	Zn
	K

SOLUTION

Name	Symbol
nickel	Ni
nitrogen	N
zinc	Zn
potassium	K

STUDY CHECK 4.1

Write the chemical symbols for the elements silicon, sulfur, silver, and seaborgium.

ANSWER

Si, S, Ag, and Sg

TEST

Try Practice Problems 4.1 to 4.6

PRACTICE PROBLEMS

4.1 Elements and Symbols

4.1 Write the symbols for the following elements:
a. copper **b.** platinum **c.** calcium **d.** manganese
e. iron **f.** barium **g.** lead **h.** strontium

4.2 Write the symbols for the following elements:
a. oxygen **b.** lithium **c.** uranium **d.** titanium
e. hydrogen **f.** chromium **g.** tin **h.** gold

Clinical Applications

4.3 Write the name for the symbol of each of the following elements found in the body:
a. C **b.** Cl **c.** I **d.** Se
e. N **f.** S **g.** Zn **h.** Co

4.4 Write the name for the symbol of each of the following elements found in the body:
a. V **b.** P **c.** Na **d.** As
e. Ca **f.** Mo **g.** Mg **h.** Si

4.5 Write the names for the elements in each of the following formulas of compounds used in medicine:
a. table salt, $NaCl$
b. plaster casts, $CaSO_4$
c. Demerol, $C_{15}H_{22}ClNO_2$
d. treatment of bipolar disorder, Li_2CO_3

4.6 Write the names for the elements in each of the following formulas of compounds used in medicine:
a. salt substitute, KCl
b. dental cement, $Zn_3(PO_4)_2$
c. antacid, $Mg(OH)_2$
d. contrast agent for X-ray, $BaSO_4$

4.2 The Periodic Table

LEARNING GOAL Use the periodic table to identify the group and the period of an element; identify an element as a metal, a nonmetal, or a metalloid.

By the late 1800s, scientists recognized that certain elements looked alike and behaved the same way. In 1869, a Russian chemist, Dmitri Mendeleev, arranged the 60 elements known at that time into groups with similar properties and placed them in order of increasing atomic masses. Today, this arrangement of 118 elements is known as the **periodic table** (see **FIGURE 4.1**).

Periodic Table of Elements

Representative elements: Groups 1A–2A (Alkali metals, Alkaline earth metals) and Groups 3A–8A (Halogens, Noble gases). Transition elements: Groups 3B–2B (3–12).

Period numbers	1 Group 1A (Alkali metals)	2 Group 2A (Alkaline earth metals)	3 3B	4 4B	5 5B	6 6B	7 7B	8 8B	9 8B	10 8B	11 1B	12 2B	13 Group 3A	14 Group 4A	15 Group 5A	16 Group 6A	17 Group 7A (Halogens)	18 Group 8A (Noble gases)
1	1 H																	2 He
2	3 Li	4 Be											5 B	6 C	7 N	8 O	9 F	10 Ne
3	11 Na	12 Mg											13 Al	14 Si	15 P	16 S	17 Cl	18 Ar
4	19 K	20 Ca	21 Sc	22 Ti	23 V	24 Cr	25 Mn	26 Fe	27 Co	28 Ni	29 Cu	30 Zn	31 Ga	32 Ge	33 As	34 Se	35 Br	36 Kr
5	37 Rb	38 Sr	39 Y	40 Zr	41 Nb	42 Mo	43 Tc	44 Ru	45 Rh	46 Pd	47 Ag	48 Cd	49 In	50 Sn	51 Sb	52 Te	53 I	54 Xe
6	55 Cs	56 Ba	57* La	72 Hf	73 Ta	74 W	75 Re	76 Os	77 Ir	78 Pt	79 Au	80 Hg	81 Tl	82 Pb	83 Bi	84 Po	85 At	86 Rn
7	87 Fr	88 Ra	89† Ac	104 Rf	105 Db	106 Sg	107 Bh	108 Hs	109 Mt	110 Ds	111 Rg	112 Cn	113 Nh	114 Fl	115 Mc	116 Lv	117 Ts	118 Og

*Lanthanides	58 Ce	59 Pr	60 Nd	61 Pm	62 Sm	63 Eu	64 Gd	65 Tb	66 Dy	67 Ho	68 Er	69 Tm	70 Yb	71 Lu
†Actinides	90 Th	91 Pa	92 U	93 Np	94 Pu	95 Am	96 Cm	97 Bk	98 Cf	99 Es	100 Fm	101 Md	102 No	103 Lr

Metals Metalloids Nonmetals

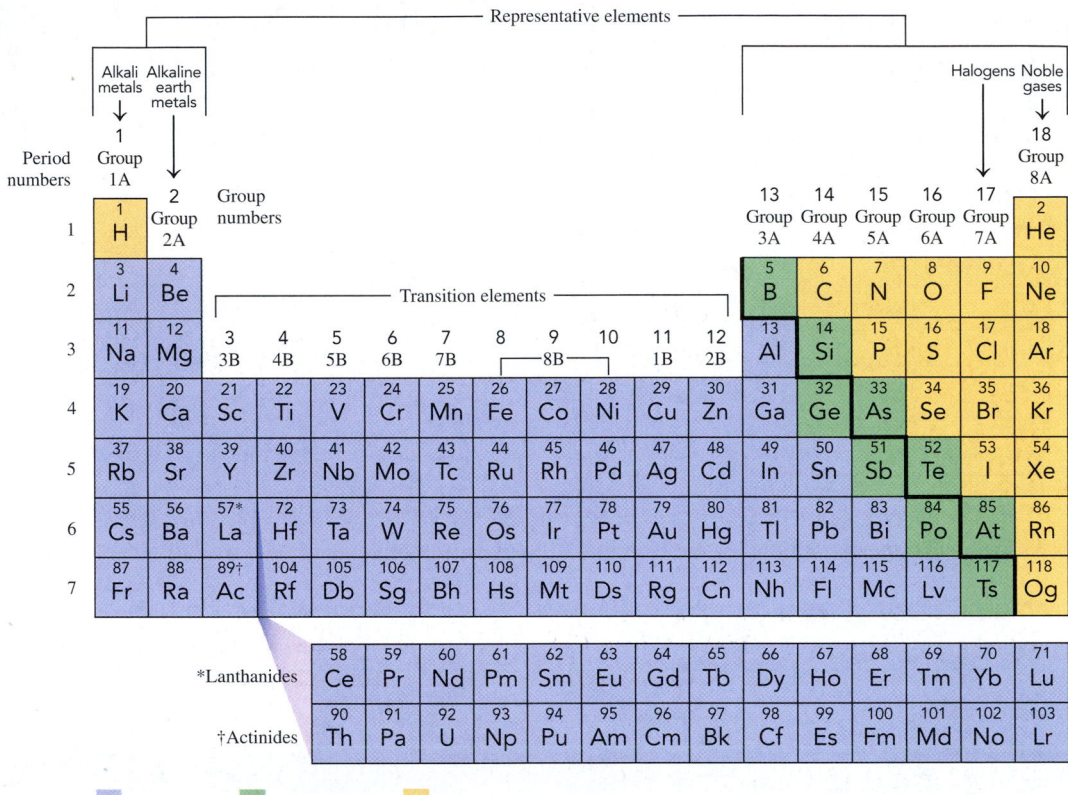

FIGURE 4.1 ▶ On the periodic table, groups are the elements arranged as vertical columns, and periods are the elements in each horizontal row.

Q What is the symbol of the alkali metal in Period 3?

Groups and Periods

Each vertical column on the periodic table contains a **group** (or family) of elements that have similar properties. A **group number** is written at the top of each vertical column (group) in the periodic table. For many years, the **representative elements** have had group numbers 1A to 8A. In the center of the periodic table is a block of elements known as the **transition elements**, which have numbers followed by the letter "B." A newer system assigns numbers 1 to 18 to the groups going left to right across the periodic table. Because both systems are in use, they are shown on the periodic table and are included in our discussions of elements and group numbers. The two rows of 14 elements called the *lanthanides* and *actinides* (or the inner transition elements) are placed at the bottom of the periodic table to allow them to fit on a page.

Each horizontal row in the periodic table is a **period**. The periods are counted down from the top of the table as Periods 1 to 7. The first period contains two elements: hydrogen (H) and helium (He). The second period contains eight elements: lithium (Li), beryllium (Be), boron (B), carbon (C), nitrogen (N), oxygen (O), fluorine (F), and neon (Ne). The third period also contains eight elements beginning with sodium (Na) and ending with argon

(Ar). The fourth period, which begins with potassium (K), and the fifth period, which begins with rubidium (Rb), have 18 elements each. The sixth period, which begins with cesium (Cs), has 32 elements. The seventh period contains 32 elements, for a total of 118 elements.

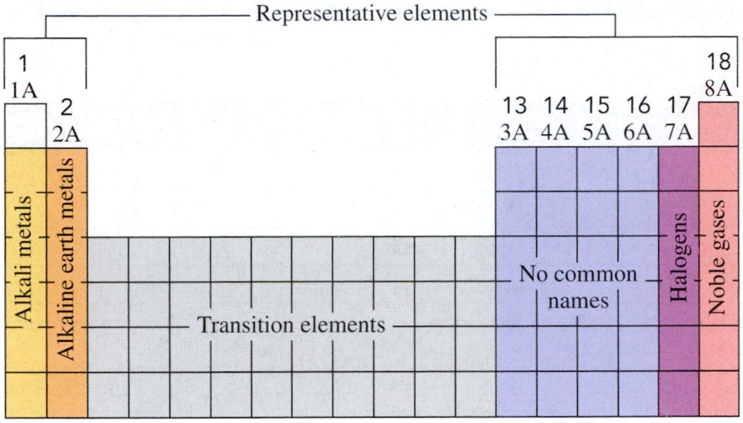
Representative elements
1
1A
2
2A
13 14 15 16 17
3A 4A 5A 6A 7A
18
8A
Alkali metals
Alkaline earth metals
Transition elements
No common names
Halogens
Noble gases

FIGURE 4.2 ▶ Certain groups on the periodic table have common names.

Q What is the common name for the group of elements that includes helium and argon?

Names of Groups

Several groups in the periodic table have special names (see **FIGURE 4.2**). Group 1A (1) elements—lithium (Li), sodium (Na), potassium (K), rubidium (Rb), cesium (Cs), and francium (Fr)—are a family of elements known as the **alkali metals** (see **FIGURE 4.3**). The elements within this group are soft, shiny metals that are good conductors of heat and electricity and have relatively low melting points. Alkali metals react vigorously with water and form white products when they combine with oxygen.

Although hydrogen (H) is at the top of Group 1A (1), it is not an alkali metal and has very different properties than the rest of the elements in this group. Thus, hydrogen is not included in the alkali metals.

The **alkaline earth metals** are found in Group 2A (2). They include the elements beryllium (Be), magnesium (Mg), calcium (Ca), strontium (Sr), barium (Ba), and radium (Ra). The alkaline earth metals are shiny metals like those in Group 1A (1), but they are not as reactive.

The **halogens** are found on the right side of the periodic table in Group 7A (17). They include the elements fluorine (F), chlorine (Cl), bromine (Br), iodine (I), astatine (At), and tennessine (Ts) (see **FIGURE 4.4**). The halogens, especially fluorine and chlorine, are highly reactive and form compounds with most of the elements.

The **noble gases** are found in Group 8A (18). They include helium (He), neon (Ne), argon (Ar), krypton (Kr), xenon (Xe), radon (Rn), and oganesson (Og). They are quite unreactive and are seldom found in combination with other elements.

Metals, Nonmetals, and Metalloids

Another feature of the periodic table is the heavy zigzag line that separates the elements into the *metals* and the *nonmetals*. *Except for hydrogen*, the metals are to the left of the line with the nonmetals to the right.

In general, most **metals** are shiny solids, such as copper (Cu), gold (Au), and silver (Ag). Metals can be shaped into wires (ductile) or hammered into a flat sheet (malleable). Metals are good conductors of heat and electricity. They usually melt at higher temperatures than nonmetals. All the metals are solids at room temperature, except for mercury (Hg), which is a liquid.

Nonmetals are not especially shiny, ductile, or malleable, and they are often poor conductors of heat and electricity. They typically have low melting points and low densities. Some examples of nonmetals are hydrogen (H), carbon (C), nitrogen (N), oxygen (O), chlorine (Cl), and sulfur (S).

Except for aluminum and oganesson, the elements located along the heavy line are **metalloids**: B, Si, Ge, As, Sb, Te, Po, At, and Ts. Metalloids are elements that exhibit some

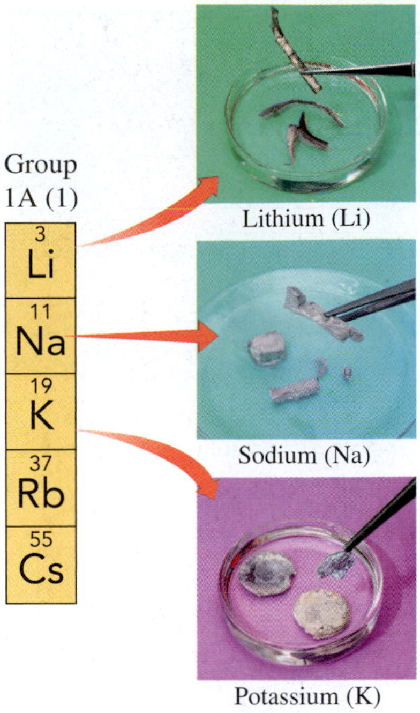
Group
1A (1)
3
Li
11
Na
19
K
37
Rb
55
Cs
Lithium (Li)
Sodium (Na)
Potassium (K)

FIGURE 4.3 ▶ Lithium (Li), sodium (Na), and potassium (K) are alkali metals from Group 1A (1).

Q What other properties do these alkali metals have in common?

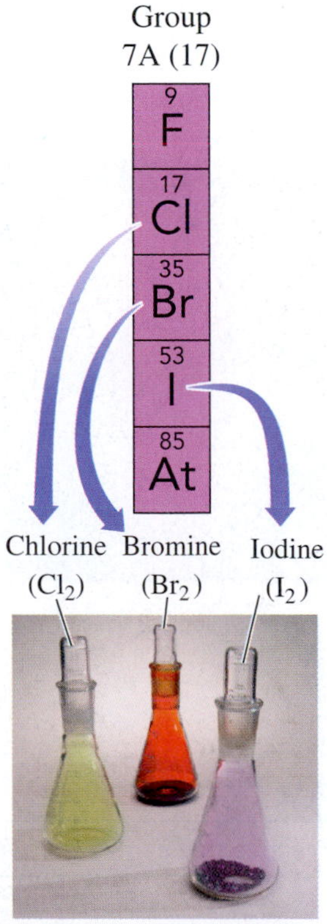
Group
7A (17)
9
F
17
Cl
35
Br
53
I
85
At
Chlorine
(Cl_2)
Bromine
(Br_2)
Iodine
(I_2)

FIGURE 4.4 ▶ Chlorine (Cl_2), bromine (Br_2), and iodine (I_2) are halogens from Group 7A (17).

Q What other elements are in the halogen group?

Silver
Sulfur
Antimony

Silver is a metal, antimony is a metalloid, and sulfur is a nonmetal.

properties that are typical of the metals and other properties that are characteristic of the nonmetals. For example, they are better conductors of heat and electricity than the nonmetals, but not as good as the metals. The metalloids are *semiconductors* because they can be modified to function as conductors or insulators. **TABLE 4.3** compares some characteristics of silver, a metal, with those of antimony, a metalloid, and sulfur, a nonmetal.

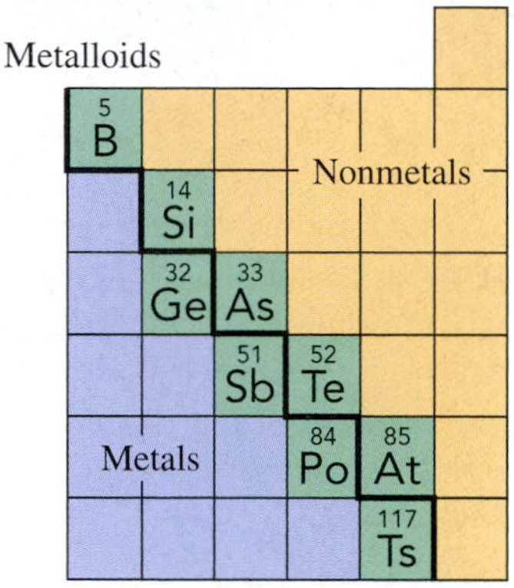
Metalloids
5 B
14 Si
32 Ge
33 As
51 Sb
52 Te
84 Po
85 At
117 Ts
Nonmetals
Metals

The metalloids on the zigzag line exhibit characteristics of both metals and nonmetals.

TABLE 4.3 Some Characteristics of a Metal, a Metalloid, and a Nonmetal

Silver (Ag)	Antimony (Sb)	Sulfur (S)
Metal	Metalloid	Nonmetal
Shiny	Blue-gray, shiny	Dull, yellow
Extremely ductile	Brittle	Brittle
Can be hammered into sheets (malleable)	Shatters when hammered	Shatters when hammered
Good conductor of heat and electricity	Poor conductor of heat and electricity	Poor conductor of heat and electricity, good insulator
Used in coins, jewelry, tableware	Used to harden lead, color glass and plastics	Used in gunpowder, rubber, fungicides
Density 10.5 g/mL	Density 6.7 g/mL	Density 2.1 g/mL
Melting point 962 °C	Melting point 630 °C	Melting point 113 °C

▶SAMPLE PROBLEM 4.2 Metals, Nonmetals, and Metalloids

TRY IT FIRST

Use the periodic table to classify each of the following elements by its group and period, group name (if any), and as a metal, a nonmetal, or a metalloid:

a. Na, important in nerve impulses, regulates blood pressure metal
b. I, needed to produce thyroid hormones NON
c. Si, needed for tendons and ligaments metalloid

SOLUTION

a. Na (sodium), Group 1A (1), Period 3, is an alkali metal.
b. I (iodine), Group 7A (17), Period 5, halogen, is a nonmetal.
c. Si (silicon), Group 4A (14), Period 3, is a metalloid.

STUDY CHECK 4.2

Strontium is an element that gives a brilliant red color to fireworks.

a. In what group is strontium found? 2A
b. What is the name of this chemical family? Alkaline
c. In what period is strontium found? 5
d. Is strontium a metal, a nonmetal, or a metalloid? metal

Strontium provides the red color in fireworks.

ANSWER

a. Group 2A (2)
b. alkaline earth metals
c. Period 5
d. a metal

TEST

Try Practice Problems 4.7 to 4.16

Chemistry Link to Health

Elements Essential to Health

Of all the elements, only about 20 are essential for the well-being and survival of the human body. Of those, four elements—oxygen, carbon, hydrogen, and nitrogen—which are representative elements in Period 1 and Period 2 on the periodic table, make up 96% of our body mass. Most of the food in our daily diet provides these elements to maintain a healthy body. These elements are found in carbohydrates, fats, and proteins. Most of the hydrogen and oxygen is found in water, which makes up 55 to 60% of our body mass.

The *macrominerals*—Ca, P, K, Cl, S, Na, and Mg—are located in Period 3 and Period 4 of the periodic table. They are involved in the formation of bones and teeth, maintenance of heart and blood vessels, muscle contraction, nerve impulses, acid–base balance of body fluids, and regulation of cellular metabolism. The macrominerals are present in lower amounts than the major elements, so that smaller amounts are required in our daily diets.

The other essential elements, called *microminerals* or *trace elements*, are mostly transition elements in Period 4 along with Si in Period 3 and Mo and I in Period 5. They are present in the human body in very small amounts, some less than 100 mg. In recent years, the detection of such small amounts has improved so that researchers can more easily identify the roles of trace elements. Some trace elements such as arsenic, chromium, and selenium are toxic at high levels in the body but are still required by the body. Other elements, such as tin and nickel, are thought to be essential, but their metabolic role has not yet been determined. Some examples and the amounts present in a 60.-kg person are listed in **TABLE 4.4**.

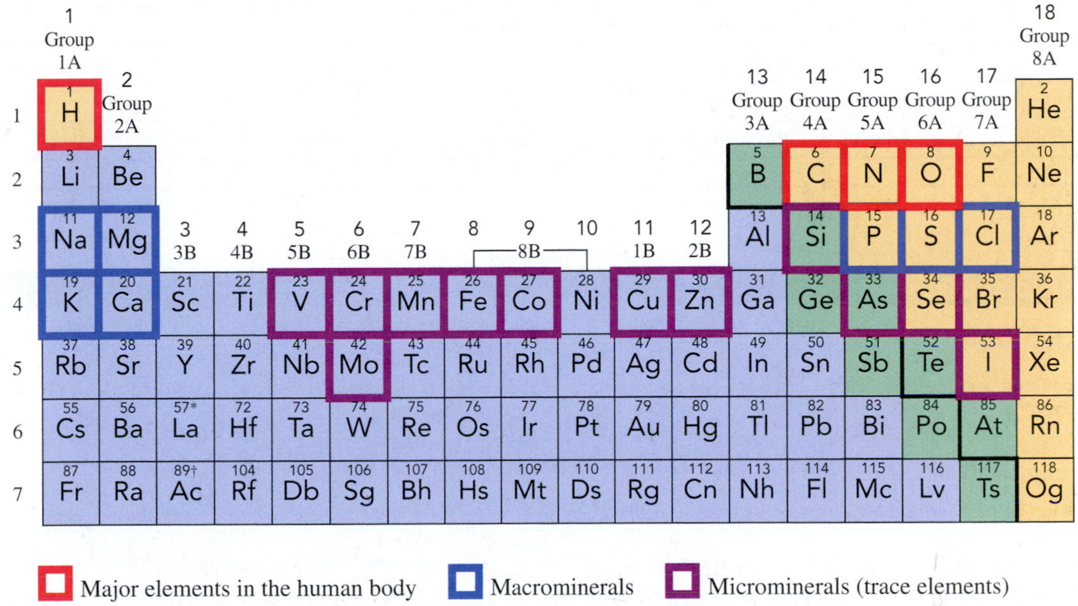
1 Group 1A
2 Group 2A
13 Group 3A
14 Group 4A
15 Group 5A
16 Group 6A
17 Group 7A
18 Group 8A
3 3B
4 4B
5 5B
6 6B
7 7B
8
9 8B
10
11 1B
12 2B
1 H
2 He
3 Li
4 Be
5 B
6 C
7 N
8 O
9 F
10 Ne
11 Na
12 Mg
13 Al
14 Si
15 P
16 S
17 Cl
18 Ar
19 K
20 Ca
21 Sc
22 Ti
23 V
24 Cr
25 Mn
26 Fe
27 Co
28 Ni
29 Cu
30 Zn
31 Ga
32 Ge
33 As
34 Se
35 Br
36 Kr
37 Rb
38 Sr
39 Y
40 Zr
41 Nb
42 Mo
43 Tc
44 Ru
45 Rh
46 Pd
47 Ag
48 Cd
49 In
50 Sn
51 Sb
52 Te
53 I
54 Xe
55 Cs
56 Ba
57* La
72 Hf
73 Ta
74 W
75 Re
76 Os
77 Ir
78 Pt
79 Au
80 Hg
81 Tl
82 Pb
83 Bi
84 Po
85 At
86 Rn
87 Fr
88 Ra
89† Ac
104 Rf
105 Db
106 Sg
107 Bh
108 Hs
109 Mt
110 Ds
111 Rg
112 Cn
113 Nh
114 Fl
115 Mc
116 Lv
117 Ts
118 Og
Major elements in the human body
Macrominerals
Microminerals (trace elements)

Elements essential to health include the major elements H, C, N, O (red), the macrominerals (blue), and the microminerals (purple).

TABLE 4.4 Typical Amounts of Essential Elements in a 60.-kg Adult

Element	Quantity	Function
Major Elements		
Oxygen (O)	39 kg	Building block of biomolecules and water (H_2O)
Carbon (C)	11 kg	Building block of organic and biomolecules
Hydrogen (H)	6 kg	Component of biomolecules, water (H_2O), regulates pH of body fluids, stomach acid (HCl)
Nitrogen (N)	2 kg	Component of proteins and nucleic acids
Macrominerals		
Calcium (Ca)	1000 g	Needed for bones and teeth, muscle contraction, nerve impulses
Phosphorus (P)	600 g	Needed for bones and teeth, nucleic acids
Potassium (K)	120 g	Most common positive ion (K^+) in cells, muscle contraction, nerve impulses
Chlorine (Cl)	100 g	Most common negative ion (Cl^-) in fluids outside cells, stomach acid (HCl)
Sulfur (S)	86 g	Component of proteins, vitamin B_1, insulin
Sodium (Na)	60 g	Most common positive ion (Na^+) in fluids outside cells, water balance, muscle contraction, nerve impulses
Magnesium (Mg)	36 g	Component of bones, required for metabolic reactions

(continued)

Chemistry Link to Health (*continued*)

TABLE 4.4 Typical Amounts of Essential Elements in a 60.-kg Adult (*Continued*)

Element	Quantity	Function
Microminerals (Trace Elements)		
Iron (Fe)	3600 mg	Component of oxygen carrier hemoglobin
Silicon (Si)	3000 mg	Needed for growth and maintenance of bones and teeth, tendons and ligaments, hair and skin
Zinc (Zn)	2000 mg	Needed for metabolic reactions in cells, DNA synthesis, growth of bones, teeth, connective tissue, immune system
Copper (Cu)	240 mg	Needed for blood vessels, blood pressure, immune system
Manganese (Mn)	60 mg	Needed for growth of bones, blood clotting, metabolic reactions
Iodine (I)	20 mg	Needed for proper thyroid function
Molybdenum (Mo)	12 mg	Needed to process Fe and N from food
Arsenic (As)	3 mg	Needed for growth and reproduction
Chromium (Cr)	3 mg	Needed for maintenance of blood sugar levels, synthesis of biomolecules
Cobalt (Co)	3 mg	Component of vitamin B_{12}, red blood cells
Selenium (Se)	2 mg	Used in the immune system, health of heart and pancreas
Vanadium (V)	2 mg	Needed in the formation of bones and teeth, energy from food

PRACTICE PROBLEMS

4.2 The Periodic Table

4.7 Identify the group or period number described by each of the following: 92
- **a.** contains C, N, and O
- **b.** begins with helium 1
- **c.** contains the alkali metals 1
- **d.** ends with neon 2

4.8 Identify the group or period number described by each of the following:
- **a.** contains Na, K, and Rb 1
- **b.** begins with Be 2A
- **c.** contains the noble gases 8
- **d.** contains B, N, and F 2

4.9 Give the symbol of the element described by each of the following:
- **a.** Group 4A (14), Period 2 Si
- **b.** the noble gas in Period 1 He
- **c.** the alkali metal in Period 3 mg
- **d.** Group 2A (2), Period 4 Ca
- **e.** Group 3A (13), Period 3 Ga

4.10 Give the symbol of the element described by each of the following:
- **a.** the alkaline earth metal in Period 2 Be
- **b.** Group 5A (15), Period 3 As
- **c.** the noble gas in Period 4 Kr
- **d.** the halogen in Period 5 I
- **e.** Group 4A (14), Period 4 Sn

4.11 Identify each of the following elements as a metal, a nonmetal, or a metalloid:
- **a.** calcium m
- **b.** sulfur NM
- **c.** a shiny element m
- **d.** an element that is a gas at room temperature NM
- **e.** located in Group 8A (18) NM
- **f.** bromine NM
- **g.** boron ML
- **h.** silver m

4.12 Identify each of the following elements as a metal, a nonmetal, or a metalloid:
- **a.** located in Group 2A (2) m
- **b.** a good conductor of electricity m
- **c.** chlorine NM
- **d.** arsenic ML
- **e.** an element that is not shiny NM
- **f.** oxygen NM
- **g.** nitrogen NM
- **h.** tin ML

Clinical Applications

4.13 Using Table 4.4, identify the function of each of the following in the body and classify each as an alkali metal, an alkaline earth metal, a transition element, or a halogen:
a. Ca **b.** Fe **c.** K **d.** Cl

4.14 Using Table 4.4, identify the function of each of the following in the body, and classify each as an alkali metal, an alkaline earth metal, a transition element, or a halogen:
a. Mg **b.** Cu **c.** I **d.** Na

4.15 Using the Chemistry Link to Health: Elements Essential to Health, answer each of the following:
- **a.** What is a macromineral?
- **b.** What is the role of sulfur in the human body?
- **c.** How many grams of sulfur would be a typical amount in a 60.-kg adult?

4.16 Using the Chemistry Link to Health: Elements Essential to Health, answer each of the following:
- **a.** What is a micromineral?
- **b.** What is the role of iodine in the human body?
- **c.** How many milligrams of iodine would be a typical amount in a 60.-kg adult?

4.3 The Atom

LEARNING GOAL Describe the electrical charge of a proton, neutron, and electron and identify their location within an atom.

All the elements listed on the periodic table are made up of atoms. An **atom** is the smallest particle of an element that retains the characteristics of that element. Imagine that you are tearing a piece of aluminum foil into smaller and smaller pieces. Now imagine that you have a microscopic piece so small that it cannot be divided any further. Then you would have a single atom of aluminum.

The concept of the atom is relatively recent. Although the Greek philosophers in 500 B.C.E. reasoned that everything must contain minute particles they called *atomos*, the idea of atoms did not become a scientific theory until 1808. Then, John Dalton (1766–1844) developed an atomic theory that proposed that atoms were responsible for the combinations of elements found in compounds.

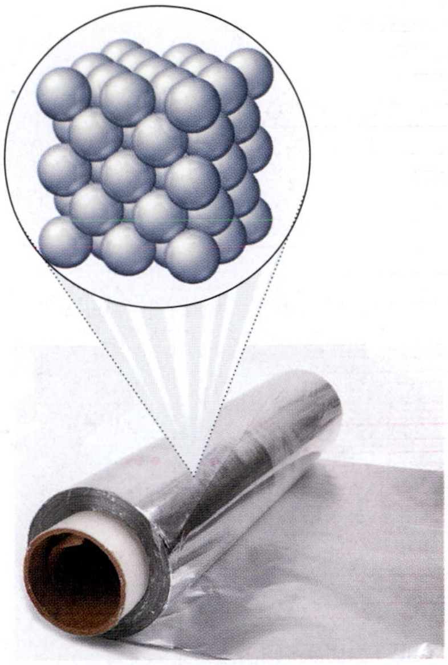

Aluminum foil consists of atoms of aluminum.

Dalton's Atomic Theory

1. All matter is made up of tiny particles called atoms.
2. All atoms of a given element are the same and different from atoms of other elements.
3. Atoms of two or more different elements combine to form compounds. A particular compound is always made up of the same kinds of atoms and always has the same number of each kind of atom.
4. A chemical reaction involves the rearrangement, separation, or combination of atoms. Atoms are neither created nor destroyed during a chemical reaction.

ENGAGE

What does Dalton's atomic theory say about the atoms in a compound?

Dalton's atomic theory formed the basis of current atomic theory, although we have modified some of Dalton's statements. We now know that atoms of the same element are not completely identical to each other and consist of even smaller particles. However, an atom is still the smallest particle that retains the properties of an element.

Although atoms are the building blocks of everything we see around us, we cannot see an atom or even a billion atoms with the naked eye. However, when billions and billions of atoms are packed together, the characteristics of each atom are added to those of the next until we can see the characteristics we associate with the element. For example, a small piece of the element gold consists of many, many gold atoms. A special kind of microscope called a *scanning tunneling microscope* (STM) produces images of individual atoms (see **FIGURE 4.5**).

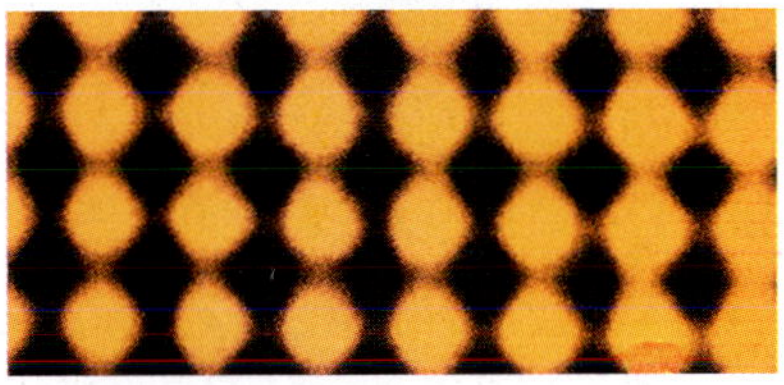

FIGURE 4.5 ▶ Images of gold atoms magnified 16 million times by a scanning tunneling microscope.

Q Why is a microscope with extremely high magnification needed to see these atoms?

Electrical Charges in an Atom

By the end of the 1800s, experiments with electricity showed that atoms were not solid spheres but were composed of smaller bits of matter called **subatomic particles**, three of which are the *proton*, *neutron*, and *electron*. Two of these subatomic particles were discovered because they have electrical charges.

An electrical charge can be positive or negative. Experiments show that like charges repel or push away from each other. When you brush your hair on a dry day, electrical charges that are alike build up on the brush and in your hair. As a result, your hair flies away from the brush. However, opposite or unlike charges attract. The crackle of clothes taken from the clothes dryer indicates the presence of electrical charges. The clinginess of the clothing results from the attraction of opposite, unlike, charges (see **FIGURE 4.6**).

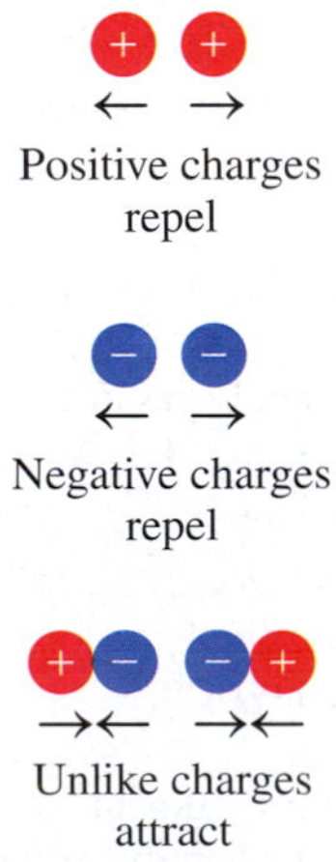
Positive charges repel
Negative charges repel
Unlike charges attract

FIGURE 4.6 ▶ Like charges repel and unlike charges attract.

Q Why are the electrons attracted to the protons in the nucleus of an atom?

Structure of the Atom

In 1897, J. J. Thomson, an English physicist, applied electricity to electrodes sealed in a glass tube, which produced streams of small particles called *cathode rays*. Because these rays were attracted to a positively charged electrode, Thomson realized that the particles in the rays must be negatively charged. In further experiments, these particles called **electrons** were found to be much smaller than the atom and to have extremely small masses. Because atoms are neutral, scientists soon discovered that atoms contained positively charged particles called **protons** that were much heavier than the electrons.

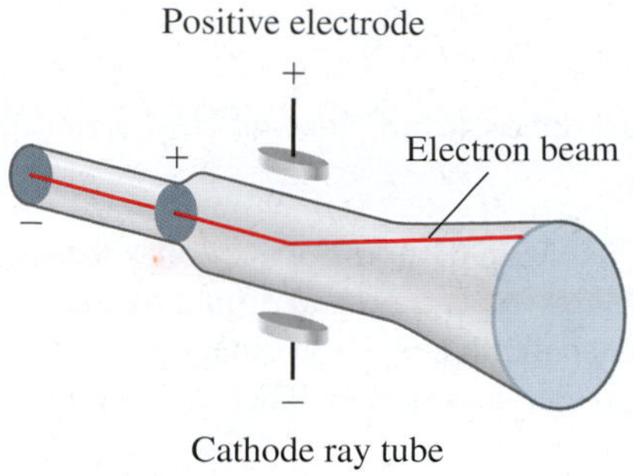
Positive electrode
+
+
Electron beam
−
−
Cathode ray tube

Negatively charged cathode rays (electrons) are attracted to the positive electrode.

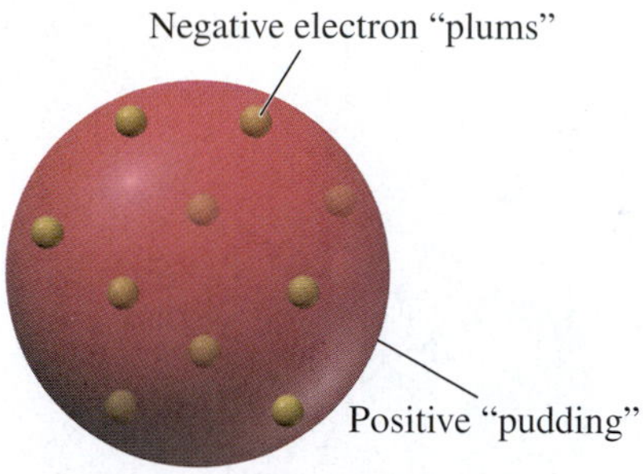
Negative electron "plums"
Positive "pudding"

Thomson's "plum-pudding" model had protons and electrons scattered throughout the atom.

Thomson proposed a "plum-pudding" model for the atom in which the electrons and protons were randomly distributed in a positively charged cloud like plums in a pudding. In 1911, Ernest Rutherford worked with Thomson to test this model. In Rutherford's experiment, positively charged particles were aimed at a thin sheet of gold foil (see **FIGURE 4.7**). If the Thomson model were correct, the particles would travel in straight paths through the gold foil. Rutherford was greatly surprised to find that some of the particles were deflected as they passed through the gold foil, and a few particles were deflected so much that they went back in the opposite direction. According to Rutherford, it was as though he had shot a cannonball at a piece of tissue paper, and it bounced back at him.

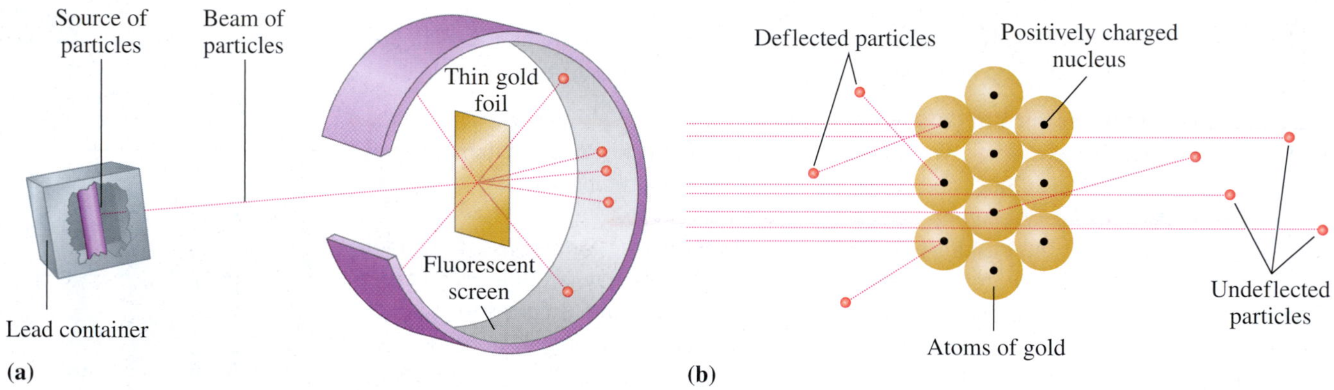
Source of particles
Beam of particles
Thin gold foil
Fluorescent screen
Lead container
(a)
Deflected particles
Positively charged nucleus
Undeflected particles
Atoms of gold
(b)

FIGURE 4.7 ▶ **(a)** Positive particles are aimed at a piece of gold foil. **(b)** Particles that come close to the atomic nuclei are deflected from their straight path.

Q Why are some particles deflected, whereas most pass through the gold foil undeflected?

INTERACTIVE VIDEO
Rutherford's Gold-Foil Experiment
PEARSON eText 2.0

From his gold-foil experiments, Rutherford realized that the protons must be contained in a small, positively charged region at the center of the atom, which he called the **nucleus**. He proposed that the electrons in the atom occupy the space surrounding the nucleus through which most of the particles traveled undisturbed. Only the particles that came near this dense, positive center were deflected. If an atom were the size of a football stadium, the nucleus would be about the size of a golf ball placed in the center of the field.

Scientists knew that the nucleus was heavier than the mass of the protons, so they looked for another subatomic particle. Eventually, they discovered that the nucleus also contained a particle that is neutral, which they called a **neutron**. Thus, the masses of the protons and neutrons in the nucleus determine the mass of an atom (see **FIGURE 4.8**).

Mass of the Atom

All the subatomic particles are extremely small compared with the things you see around you. One proton has a mass of 1.67×10^{-24} g, and the neutron is about the same. However, the electron has a mass 9.11×10^{-28} g, which is much less than the mass of either a proton

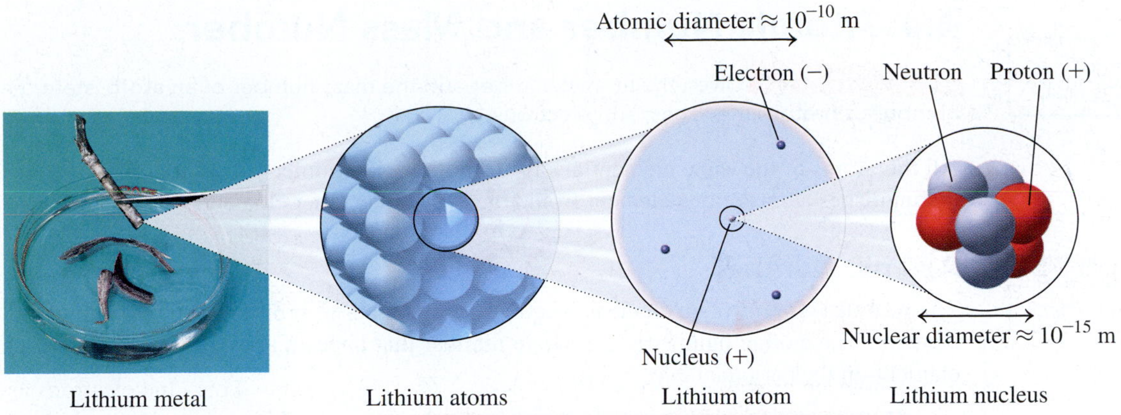

Atomic diameter ≈ 10^{-10} m
Electron (−)
Neutron
Proton (+)
Nucleus (+)
Nuclear diameter ≈ 10^{-15} m
Lithium metal
Lithium atoms
Lithium atom
Lithium nucleus

FIGURE 4.8 ▶ In an atom, the protons and neutrons that make up almost all the mass are packed into the tiny volume of the nucleus. The rapidly moving electrons (negative charge) surround the nucleus and account for the large volume of the atom.

Q Why can we say that the atom is mostly empty space?

or neutron. Because the masses of subatomic particles are so small, chemists use a very small unit of mass called an **atomic mass unit (amu)**. An amu is defined as one-twelfth of the mass of a carbon atom, which has a nucleus containing six protons and six neutrons. In biology, the atomic mass unit is called a *Dalton* (Da) in honor of John Dalton. On the amu scale, the proton and neutron each have a mass of about 1 amu. Because the electron mass is so small, it is usually ignored in atomic mass calculations. **TABLE 4.5** summarizes some information about the subatomic particles in an atom.

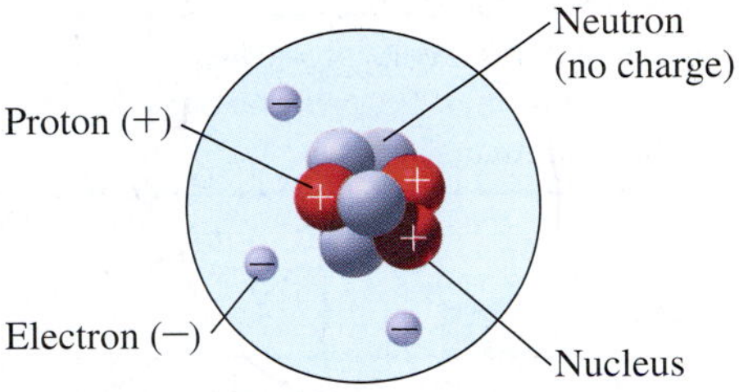

Neutron (no charge)
Proton (+)
Electron (−)
Nucleus

The nucleus of a typical lithium atom contains three protons and four neutrons.

TABLE 4.5 Subatomic Particles in the Atom

Particle	Symbol	Charge	Mass (amu)	Location in Atom
Proton	p or p^+	1+	1.007	Nucleus
Neutron	n or n^0	0	1.008	Nucleus
Electron	e^-	1−	0.000 55	Outside nucleus

TEST

Try Practice Problems 4.17 to 4.24

PRACTICE PROBLEMS

4.3 The Atom

4.17 Identify each of the following as describing either a proton, a neutron, or an electron:
a. has the smallest mass
b. has a 1+ charge
c. is found outside the nucleus
d. is electrically neutral

4.18 Identify each of the following as describing either a proton, a neutron, or an electron:
a. has a mass about the same as a proton
b. is found in the nucleus
c. is attracted to the protons
d. has a 1− charge

4.19 What did Rutherford determine about the structure of the atom from his gold-foil experiment?

4.20 How did Thomson determine that the electrons have a negative charge?

4.21 Is each of the following statements *true* or *false*?
a. A proton and an electron have opposite charges.
b. The nucleus contains most of the mass of an atom.
c. Electrons repel each other.
d. A proton is attracted to a neutron.

4.22 Is each of the following statements *true* or *false*?
a. A proton is attracted to an electron.
b. A neutron has twice the mass of a proton.
c. Neutrons repel each other.
d. Electrons and neutrons have opposite charges.

4.23 On a dry day, your hair flies apart when you brush it. How would you explain this?

4.24 Sometimes clothes cling together when removed from a dryer. What kinds of charges are on the clothes?

REVIEW

Using Positive and Negative Numbers in Calculations (1.4)

4.4 Atomic Number and Mass Number

LEARNING GOAL Given the atomic number and the mass number of an atom, state the number of protons, neutrons, and electrons.

All the atoms of the same element always have the same number of protons. This feature distinguishes atoms of one element from atoms of all the other elements.

CORE CHEMISTRY SKILL

Counting Protons and Neutrons

Atomic Number

The **atomic number** of an element is equal to the number of protons in every atom of that element. The atomic number is the whole number that appears above the symbol of each element on the periodic table.

Atomic number = number of protons in an atom

The periodic table on the inside front cover of this text shows the elements in order of atomic number from 1 to 118. We can use an atomic number to identify the number of protons in an atom of any element. For example, a lithium atom, with atomic number 3, has 3 protons. Any atom with 3 protons is always a lithium atom. In the same way, we determine that a carbon atom, with atomic number 6, has 6 protons. Any atom with 6 protons is carbon.

An atom is electrically neutral. That means that the number of protons in an atom is equal to the number of electrons, which gives every atom an overall charge of zero. Thus, the atomic number also gives the number of electrons.

ENGAGE

Why does every atom of barium have 56 protons and 56 electrons?

Mass Number

We now know that the protons and neutrons determine the mass of the nucleus. Thus, for a single atom, we assign a **mass number**, which is the total number of protons and neutrons in its nucleus. However, the mass number does not appear on the periodic table because it applies to single atoms only.

ENGAGE

Which subatomic particles in an atom determine its mass number?

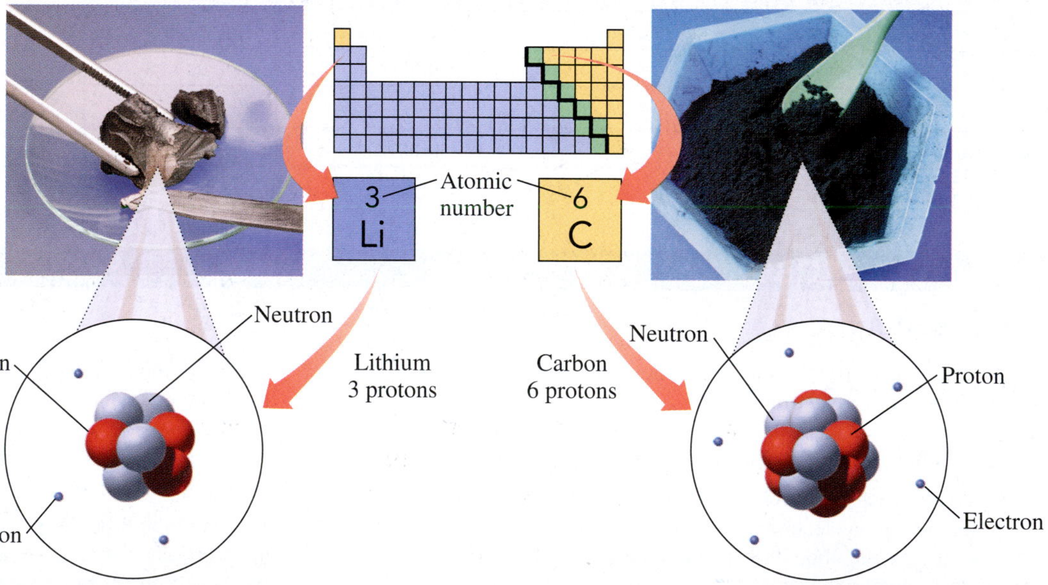
3
Li
Atomic
number
6
C
Neutron
Proton
Lithium
3 protons
Carbon
6 protons
Neutron
Proton
Electron
Electron

All atoms of lithium contain three protons and three electrons.

All atoms of carbon contain six protons and six electrons.

Mass number = number of protons + number of neutrons

For example, the nucleus of an oxygen atom that contains 8 protons and 8 neutrons has a mass number of 16. An atom of iron that contains 26 protons and 32 neutrons has a mass number of 58.

If we are given the mass number of an atom and its atomic number, we can calculate the number of neutrons in its nucleus.

Number of neutrons in a nucleus = mass number − number of protons

For example, if we are given a mass number of 37 for an atom of chlorine (atomic number 17), we can calculate the number of neutrons in its nucleus.

Number of neutrons = 37 (mass number) − 17 (protons) = 20 neutrons

TABLE 4.6 illustrates the relationships between atomic number, mass number, and the number of protons, neutrons, and electrons in examples of single atoms for different elements.

ENGAGE

How many neutrons are in an atom of tin that has a mass number of 102?

TABLE 4.6 Composition of Some Atoms of Different Elements

Element	Symbol	Atomic Number	Mass Number	Number of Protons	Number of Neutrons	Number of Electrons
Hydrogen	H	1	1	1	0	1
Nitrogen	N	7	14	7	7	7
Oxygen	O	8	16	8	8	8
Chlorine	Cl	17	37	17	20	17
Iron	Fe	26	58	26	32	26
Gold	Au	79	197	79	118	79

TEST

Try Practice Problems 4.25 to 4.28

▶SAMPLE PROBLEM 4.3 Calculating Numbers of Protons, Neutrons, and Electrons

TRY IT FIRST

Zinc, a micromineral, is needed for metabolic reactions in cells, DNA synthesis, the growth of bone, teeth, and connective tissue, and the proper functioning of the immune system. For an atom of zinc that has a mass number of 68, determine the number of:

a. protons **b.** neutrons **c.** electrons

SOLUTION

ANALYZE THE PROBLEM	Given	Need	Connect
	zinc (Zn), mass number 68	number of protons, neutrons, electrons	periodic table, atomic number

a. Zinc (Zn), with an atomic number of 30, has 30 protons.

b. The number of neutrons in this atom is found by subtracting the number of protons (atomic number) from the mass number.

Mass number	−	atomic number	=	number of neutrons
68	−	30	=	38

c. Because a zinc atom is neutral, the number of electrons is equal to the number of protons. A zinc atom has 30 electrons.

STUDY CHECK 4.3

a. How many neutrons are in the nucleus of a bromine atom that has a mass number of 80?

b. What is the mass number of a cesium atom that has 71 neutrons?

ANSWER

a. 45 **b.** 126

TEST

Try Practice Problems 4.29 to 4.32

Chemistry Link to the Environment

Many Forms of Carbon

Carbon has the symbol C. However, its atoms can be arranged in different ways to give several different substances. Two forms of carbon—diamond and graphite—have been known since prehistoric times. A diamond is transparent and harder than any other substance, whereas graphite is black and soft. In diamond, carbon atoms are arranged in a rigid structure. In graphite, carbon atoms are arranged in flat sheets that slide over each other. Graphite is used as pencil lead and as a lubricant.

Two other forms of carbon have been discovered more recently. In the form called *Buckminsterfullerene* or *buckyball* (named after R. Buckminster Fuller, who popularized the geodesic dome), 60

(continued)

Chemistry Link to the Environment (*continued*)

carbon atoms are arranged as rings of five and six atoms to give a spherical, cage-like structure. When a fullerene structure is stretched out, it produces a cylinder with a diameter of only a few nanometers called a *nanotube*. Practical uses for buckyballs and nanotubes are not yet developed, but they are expected to find use in lightweight structural materials, heat conductors, computer parts, and medicine. Recent research has shown that carbon nanotubes (CNT) can carry many drug molecules that can be released once the CNT enter the targeted cells.

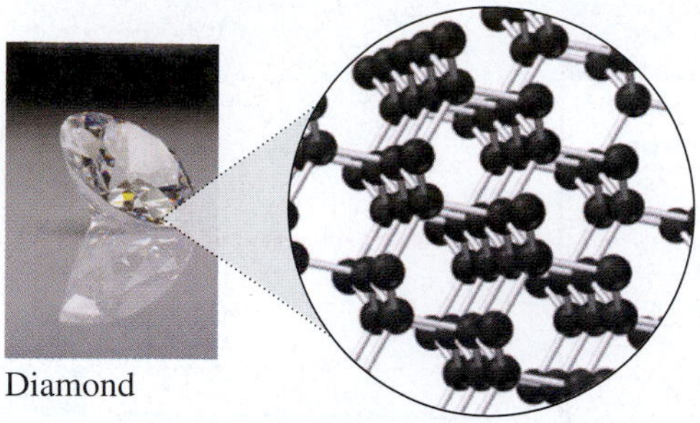

Diamond

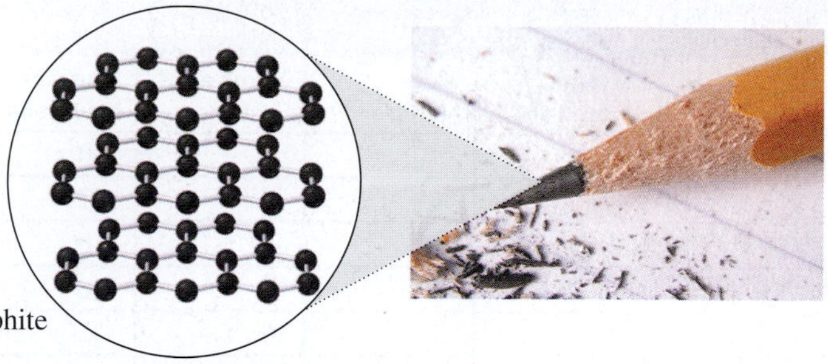

Graphite

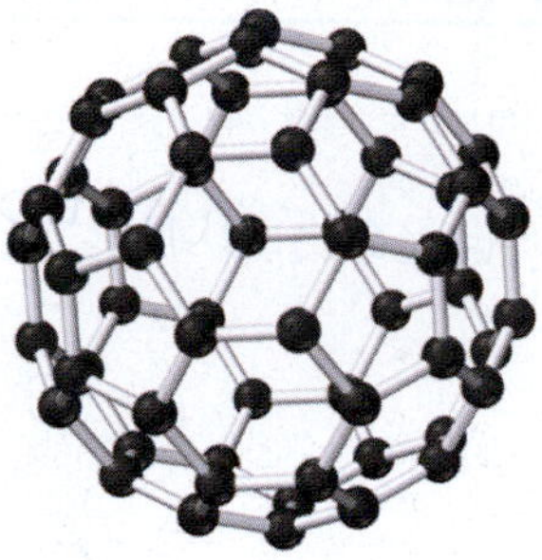

Buckminsterfullerene

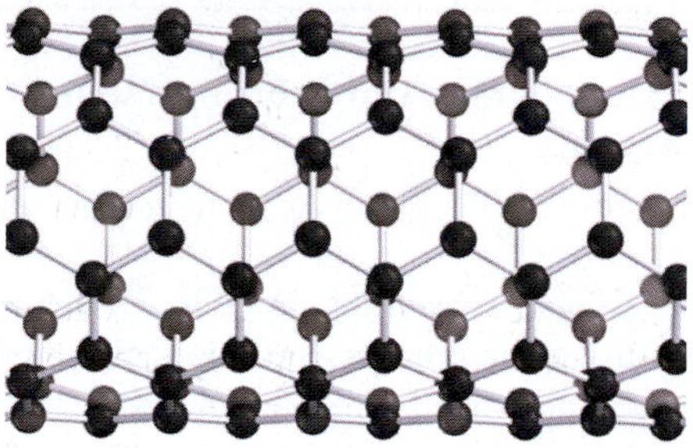

Nanotubes

Carbon atoms can form different types of structures.

PRACTICE PROBLEMS

4.4 Atomic Number and Mass Number

4.25 Would you use the atomic number, mass number, or both to determine each of the following?
- **a.** number of protons in an atom Atomic
- **b.** number of neutrons in an atom Both
- **c.** number of particles in the nucleus ~~mass~~ Both
- **d.** number of electrons in a neutral atom Atomic

4.26 Identify the type of subatomic particles described by each of the following:
- **a.** atomic number P
- **b.** mass number N
- **c.** mass number − atomic number N
- **d.** mass number + atomic number m

4.27 Write the names and symbols for the elements with the following atomic numbers:
a. 3 Li **b.** 9 F **c.** 20 Ca **d.** 30 Zn
e. 10 Ne **f.** 14 Si **g.** 53 I **h.** 8 O

4.28 Write the names and symbols for the elements with the following atomic numbers:
a. 1 H **b.** 11 Na **c.** 19 K **d.** 82 Pb
e. 35 Br **f.** 47 Ag **g.** 15 P **h.** 2 He

4.29 How many protons and electrons are there in a neutral atom of each of the following elements?
a. argon **b.** manganese **c.** iodine **d.** cadmium

4.30 How many protons and electrons are there in a neutral atom of each of the following elements?
a. carbon **b.** fluorine **c.** tin **d.** nickel

Clinical Applications

4.31 Complete the following table for atoms of essential elements in the body:

Name of the Element	Symbol	Atomic Number	Mass Number	Number of Protons	Number of Neutrons	Number of Electrons
	Zn		66			
		12			12	
Potassium					20	
				16	15	
			56			26

4.32 Complete the following table for atoms of essential elements in the body:

Name of the Element	Symbol	Atomic Number	Mass Number	Number of Protons	Number of Neutrons	Number of Electrons
	N		15			
Calcium			42			
				53	72	
		14			16	
		29	65			

4.5 Isotopes and Atomic Mass

LEARNING GOAL Determine the number of protons, neutrons, and electrons in one or more of the isotopes of an element; calculate the atomic mass of an element using the percent abundance and mass of its naturally occurring isotopes.

REVIEW
Calculating Percentages (1.4)

We have seen that all atoms of the same element have the same number of protons and electrons. However, the atoms of any one element are not entirely identical because the atoms of most elements have different numbers of neutrons. When a sample of an element consists of two or more atoms with differing numbers of neutrons, those atoms are called *isotopes*.

Atoms and Isotopes

CORE CHEMISTRY SKILL
Writing Atomic Symbols for Isotopes

Isotopes are atoms of the same element that have the same atomic number but different numbers of neutrons. For example, all atoms of the element magnesium (Mg) have an atomic number of 12. Thus every magnesium atom always has 12 protons. However, some naturally occurring magnesium atoms have 12 neutrons, others have 13 neutrons, and still others have 14 neutrons. The different numbers of neutrons give the magnesium atoms different mass numbers but do not change their chemical behavior. The three isotopes of magnesium have the same atomic number but different mass numbers.

To distinguish between the different isotopes of an element, we write an **atomic symbol** for a particular isotope that indicates the mass number in the upper left corner and the atomic number in the lower left corner.

An isotope may be referred to by its name or symbol, followed by its mass number, such as magnesium-24 or Mg-24. Magnesium has three naturally occurring isotopes, as

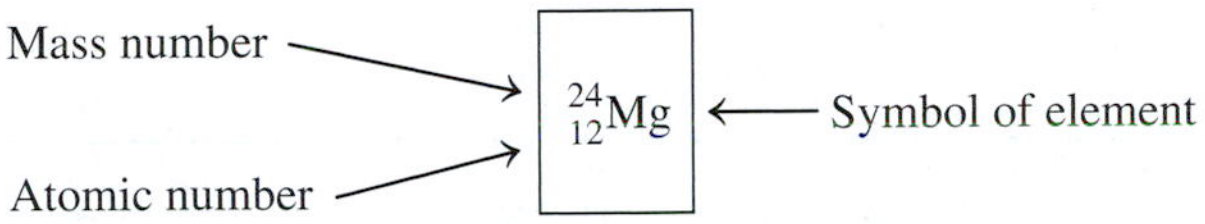
Mass number
$^{24}_{12}Mg$
Symbol of element
Atomic number

Atomic symbol for an isotope of magnesium, Mg-24.

shown in **TABLE 4.7**. In a large sample of naturally occurring magnesium atoms, each type of isotope can be present as a low percentage or a high percentage. For example, the Mg-24 isotope makes up almost 80% of the total sample, whereas Mg-25 and Mg-26 each make up only about 10% of the total number of magnesium atoms.

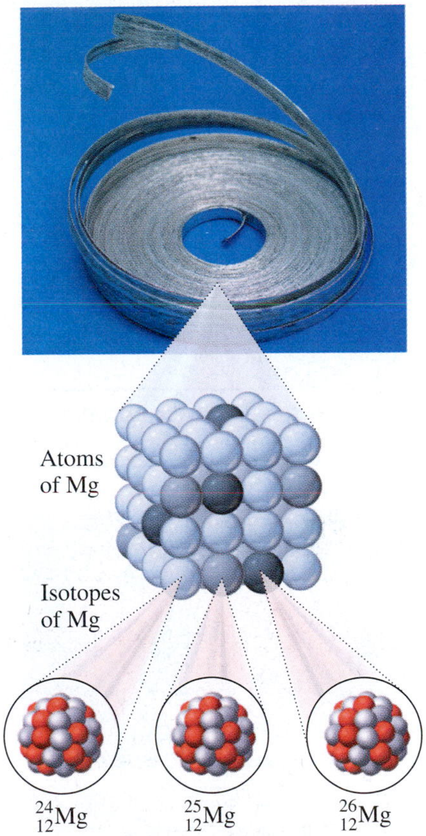
Atoms of Mg
Isotopes of Mg
$^{24}_{12}Mg$
$^{25}_{12}Mg$
$^{26}_{12}Mg$

The nuclei of three naturally occurring magnesium isotopes have the same number of protons but different numbers of neutrons.

TABLE 4.7 Isotopes of Magnesium

Atomic Symbol	$^{24}_{12}Mg$	$^{25}_{12}Mg$	$^{26}_{12}Mg$
Name	Mg-24	Mg-25	Mg-26
Number of Protons	12	12	12
Number of Electrons	12	12	12
Mass Number	24	25	26
Number of Neutrons	12	13	14
Mass of Isotope (amu)	23.99	24.99	25.98
Percent Abundance	78.70	10.13	11.17

▸SAMPLE PROBLEM 4.4 Identifying Protons and Neutrons in Isotopes

TRY IT FIRST

Chromium, needed for maintenance of blood sugar levels, has four naturally occurring isotopes. Calculate the number of protons and number of neutrons in each of the following isotopes:

a. $^{50}_{24}Cr$ **b.** $^{52}_{24}Cr$ **c.** $^{53}_{24}Cr$ **d.** $^{54}_{24}Cr$

SOLUTION

	Given	Need	Connect
ANALYZE THE PROBLEM	atomic symbols for Cr isotopes	number of protons, number of neutrons	periodic table, atomic number

In the atomic symbol, the mass number is shown in the upper left corner of the symbol, and the atomic number is shown in the lower left corner of the symbol. Thus, each isotope of Cr, atomic number 24, has 24 protons. The number of neutrons is found by subtracting the number of protons (24) from the mass number of each isotope.

Atomic Symbol	Atomic Number	Mass Number	Number of Protons	Number of Neutrons
a. ${}^{50}_{24}\text{Cr}$	24	50	24	26 (50 − 24)
b. ${}^{52}_{24}\text{Cr}$	24	52	24	28 (52 − 24)
c. ${}^{53}_{24}\text{Cr}$	24	53	24	29 (53 − 24)
d. ${}^{54}_{24}\text{Cr}$	24	53	24	30 (54 − 24)

ENGAGE

What is different and what is the same for an atom of Sn-105 and an atom of Sn-132?

STUDY CHECK 4.4

a. Vanadium is a micromineral needed in the formation of bones and teeth. Write the atomic symbol for the isotope of vanadium that has 27 neutrons.

b. Molybdenum, a micromineral needed to process Fe and N from food, has seven naturally occurring isotopes. Write the atomic symbols for two isotopes, Mo-97 and Mo-98.

ANSWER

a. ${}^{50}_{23}\text{V}$ **b.** ${}^{97}_{42}\text{Mo}$ ${}^{98}_{42}\text{Mo}$

TEST

Try Practice Problems 4.33 to 4.36

Atomic Mass

In laboratory work, a chemist generally uses samples with many atoms that contain all the different atoms or isotopes of an element. Because each isotope has a different mass, chemists have calculated an **atomic mass** for an "average atom," which is a *weighted average* of the masses of all the naturally occurring isotopes of that element. On the periodic table, the atomic mass is the number including decimal places that is given below the symbol of each element. Most elements consist of two or more isotopes, which is one reason that the atomic masses on the periodic table are seldom whole numbers.

ENGAGE

What is the difference between mass number and atomic mass?

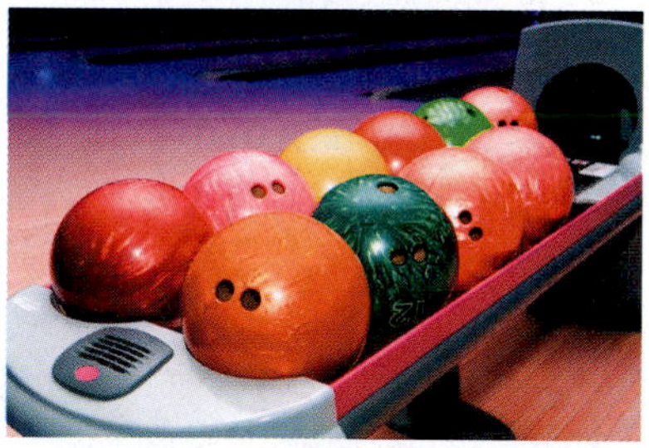

The weighted average of 8-lb and 14-lb bowling balls is calculated using their percent abundance.

Weighted Average Analogy

To understand how the atomic mass as a weighted average for a group of isotopes is calculated, we will use an analogy of bowling balls with different weights. Suppose that a bowling alley has ordered 5 bowling balls that weigh 8 lb each and 20 bowling balls that weigh 14 lb each. In this group of bowling balls, there are more 14-lb balls than 8-lb balls. The abundance of the 14-lb balls is 80.% (20/25), and the abundance of the 8-lb balls is 20.% (5/25). Now we can calculate a weighted average for the "average" bowling ball using the weight and percent abundance of the two types of bowling balls.

Item	Weight (lb)		Percent Abundance		Weight from Each Type
14-lb bowling ball	14	×	$\frac{80.}{100}$	=	11.2 lb
8-lb bowling ball	8	×	$\frac{20.}{100}$	=	1.6 lb
Weighted average of a bowling ball				=	12.8 lb
"Atomic mass" of a bowling ball				=	12.8 lb

Calculating Atomic Mass

To calculate the atomic mass of an element, we need to know the percentage abundance and the mass of each isotope, which must be determined experimentally. For example, a large sample of naturally occurring chlorine atoms consists of 75.76% of ${}^{35}_{17}\text{Cl}$ atoms and 24.24% of ${}^{37}_{17}\text{Cl}$ atoms. The ${}^{35}_{17}\text{Cl}$ isotope has a mass of 34.97 amu, and the ${}^{37}_{17}\text{Cl}$ isotope has a mass of 36.97 amu.

$$\text{Atomic mass of Cl} = \underbrace{\text{mass of } {}^{35}_{17}\text{Cl} \times \frac{{}^{35}_{17}\text{Cl}\%}{100\%}}_{\text{amu from } {}^{35}_{17}\text{Cl}} + \underbrace{\text{mass of } {}^{37}_{17}\text{Cl} \times \frac{{}^{37}_{17}\text{Cl}\%}{100\%}}_{\text{amu from } {}^{37}_{17}\text{Cl}}$$

Atomic Symbol	Mass (amu)	Percent Abundance	Contribution to the Atomic Mass
${}^{35}_{17}\text{Cl}$	34.97 ×	$\frac{75.76}{100}$ =	26.49 amu
${}^{37}_{17}\text{Cl}$	36.97 ×	$\frac{24.24}{100}$ =	8.962 amu
		Atomic mass of Cl =	35.45 amu (weighted average mass)

The atomic mass of 35.45 amu is the weighted average mass of a sample of Cl atoms, although no individual Cl atom actually has this mass. An atomic mass of 35.45, which is closer to the mass number of Cl-35, indicates there is a higher percentage of ${}^{35}_{17}\text{Cl}$ atoms in the chlorine sample. In fact, there are about three atoms of ${}^{35}_{17}\text{Cl}$ for every one atom of ${}^{37}_{17}\text{Cl}$ in a sample of chlorine atoms.

TABLE 4.8 lists the naturally occurring isotopes of some selected elements and their atomic masses along with their most abundant isotopes.

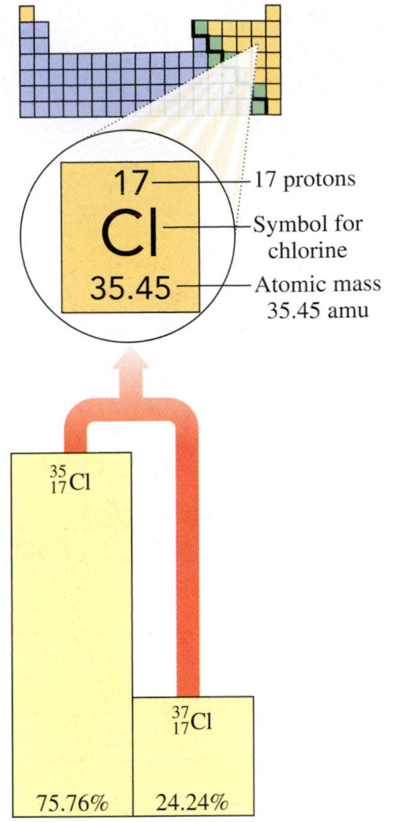
17
Cl
35.45
17 protons
Symbol for chlorine
Atomic mass 35.45 amu
${}^{35}_{17}\text{Cl}$
${}^{37}_{17}\text{Cl}$
75.76%
24.24%

Chlorine, with two naturally occurring isotopes, has an atomic mass of 35.45 amu.

TABLE 4.8 The Atomic Mass of Some Elements

Element	Atomic Symbols	Atomic Mass (weighted average)	Most Abundant Isotope
Lithium	${}^{6}_{3}\text{Li}$, ${}^{7}_{3}\text{Li}$	6.941 amu	${}^{7}_{3}\text{Li}$
Carbon	${}^{12}_{6}\text{C}$, ${}^{13}_{6}\text{C}$, ${}^{14}_{6}\text{C}$	12.01 amu	${}^{12}_{6}\text{C}$
Oxygen	${}^{16}_{8}\text{O}$, ${}^{17}_{8}\text{O}$, ${}^{18}_{8}\text{O}$	16.00 amu	${}^{16}_{8}\text{O}$
Fluorine	${}^{19}_{9}\text{F}$	19.00 amu	${}^{19}_{9}\text{F}$
Sulfur	${}^{32}_{16}\text{S}$, ${}^{33}_{16}\text{S}$, ${}^{34}_{16}\text{S}$, ${}^{36}_{16}\text{S}$	32.07 amu	${}^{32}_{16}\text{S}$
Potassium	${}^{39}_{19}\text{K}$, ${}^{40}_{19}\text{K}$, ${}^{41}_{19}\text{K}$	39.10 amu	${}^{39}_{19}\text{K}$
Copper	${}^{63}_{29}\text{Cu}$, ${}^{65}_{29}\text{Cu}$	63.55 amu	${}^{63}_{29}\text{Cu}$

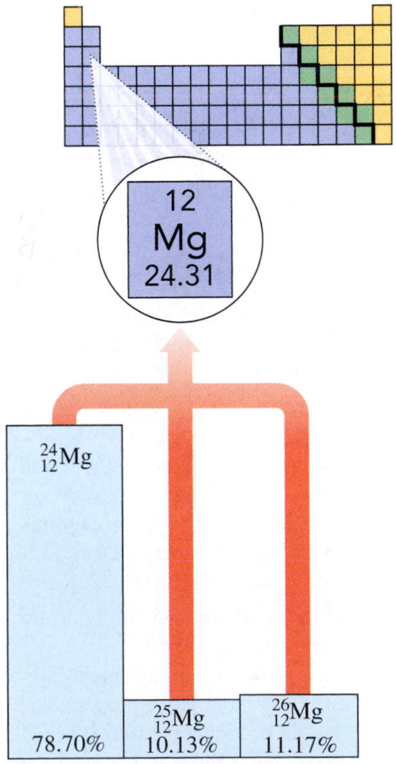
12
Mg
24.31
${}^{24}_{12}\text{Mg}$
78.70%
${}^{25}_{12}\text{Mg}$
10.13%
${}^{26}_{12}\text{Mg}$
11.17%

Magnesium, with three naturally occurring isotopes, has an atomic mass of 24.31 amu.

▶SAMPLE PROBLEM 4.5 Calculating Atomic Mass

TRY IT FIRST

Magnesium is a macromineral needed in the contraction of muscles and metabolic reactions. Using Table 4.7, calculate the atomic mass for magnesium using the weighted average mass method.

SOLUTION

ANALYZE THE PROBLEM	Given	Need	Connect
	percent abundance, atomic mass	atomic mass of Mg	weighted average mass

STEP 1 Multiply the mass of each isotope by its percent abundance divided by 100.

Atomic Symbol	Mass (amu)		Percent Abundance		Contribution to the Atomic Mass
${}^{24}_{12}Mg$	23.99	×	$\frac{78.70}{100}$	=	18.88 amu
${}^{25}_{12}Mg$	24.99	×	$\frac{10.13}{100}$	=	2.531 amu
${}^{26}_{12}Mg$	25.98	×	$\frac{11.17}{100}$	=	2.902 amu

STEP 2 Add the contribution of each isotope to obtain the atomic mass.

Atomic mass of Mg = 18.88 amu + 2.531 amu + 2.902 amu
= 24.31 amu (weighted average mass)

INTERACTIVE VIDEO

Isotopes and Atomic Mass

PEARSON eText 2.0

STUDY CHECK 4.5

There are two naturally occurring isotopes of boron. The isotope ${}^{10}_{5}B$ has a mass of 10.01 amu with an abundance of 19.80%, and the isotope ${}^{11}_{5}B$ has a mass of 11.01 amu with an abundance of 80.20%. Calculate the atomic mass for boron using the weighted average mass method.

ANSWER

10.81 amu

TEST

Try Practice Problems 4.37 to 4.44

PRACTICE PROBLEMS

4.5 Isotopes and Atomic Mass

4.33 What are the number of protons, neutrons, and electrons in the following isotopes?
a. ${}^{89}_{38}Sr$ **b.** ${}^{52}_{24}Cr$ **c.** ${}^{34}_{16}S$ **d.** ${}^{81}_{35}Br$

4.34 What are the number of protons, neutrons, and electrons in the following isotopes?
a. ${}^{2}_{1}H$ **b.** ${}^{14}_{7}N$ **c.** ${}^{26}_{14}Si$ **d.** ${}^{70}_{30}Zn$

4.35 Write the atomic symbol for the isotope with each of the following characteristics:
a. 15 protons and 16 neutrons
b. 35 protons and 45 neutrons
c. 50 electrons and 72 neutrons
d. a chlorine atom with 18 neutrons
e. a mercury atom with 122 neutrons

4.36 Write the atomic symbol for the isotope with each of the following characteristics:
a. an oxygen atom with 10 neutrons
b. 4 protons and 5 neutrons
c. 25 electrons and 28 neutrons
d. a mass number of 24 and 13 neutrons
e. a nickel atom with 32 neutrons

4.37 Argon has three naturally occurring isotopes, with mass numbers 36, 38, and 40.
a. Write the atomic symbol for each of these atoms.
b. How are these isotopes alike?
c. How are they different?
d. Why is the atomic mass of argon listed on the periodic table not a whole number?
e. Which isotope is the most abundant in a sample of argon?

4.38 Strontium has four naturally occurring isotopes, with mass numbers 84, 86, 87, and 88.
a. Write the atomic symbol for each of these atoms.
b. How are these isotopes alike?
c. How are they different?
d. Why is the atomic mass of strontium listed on the periodic table not a whole number?
e. Which isotope is the most abundant a sample of strontium?

4.39 Copper consists of two isotopes, ${}^{63}_{29}Cu$ and ${}^{65}_{29}Cu$. If the atomic mass for copper on the periodic table is 63.55, are there more atoms of ${}^{63}_{29}Cu$ or ${}^{65}_{29}Cu$ in a sample of copper?

4.40 A fluorine sample consists of only one type of atom, ${}^{19}_{9}F$, which has a mass of 19.00 amu. How would the mass of a ${}^{19}_{9}F$ atom compare to the atomic mass listed on the periodic table?

4.41 There are two naturally occurring isotopes of thallium: ${}^{203}_{81}Tl$ and ${}^{205}_{81}Tl$. Use the atomic mass of thallium listed on the periodic table to identify the more abundant isotope.

4.42 Zinc consists of five naturally occurring isotopes: ${}^{64}_{30}Zn$, ${}^{66}_{30}Zn$, ${}^{67}_{30}Zn$, ${}^{68}_{30}Zn$, and ${}^{70}_{30}Zn$. None of these isotopes has the atomic mass of 65.41 listed for zinc on the periodic table. Explain.

4.43 Two isotopes of gallium are naturally occurring, with ${}^{69}_{31}Ga$ at 60.11% (68.93 amu) and ${}^{71}_{31}Ga$ at 39.89% (70.92 amu). Calculate the atomic mass for gallium using the weighted average mass method.

4.44 Two isotopes of rubidium occur naturally, with ${}^{85}_{37}Rb$ at 72.17% (84.91 amu) and ${}^{87}_{37}Rb$ at 27.83% (86.91 amu). Calculate the atomic mass for rubidium using the weighted average mass method.

4.6 Electron Energy Levels

LEARNING GOAL Describe the energy levels, sublevels, and orbitals for the electrons in an atom.

When we listen to a radio, use a microwave oven, turn on a light, see the colors of a rainbow, or have an X-ray taken, we are experiencing various forms of *electromagnetic radiation*. Light and other electromagnetic radiation consist of energy particles that move as waves of energy. In an electromagnetic wave, just like the waves in an ocean, the distance between the peaks is called the *wavelength*. All forms of electromagnetic radiation travel in space at the speed of light, 3.0×10^8 m/s, but differ in energy and wavelength. High-energy radiation has short wavelengths compared to low-energy radiation, which has longer wavelengths. The *electromagnetic spectrum* shows the arrangement of different types of electromagnetic radiation in order of increasing energy (see **FIGURE 4.9**).

ENGAGE

Which type of electromagnetic radiation has lower energy, ultraviolet or infrared?

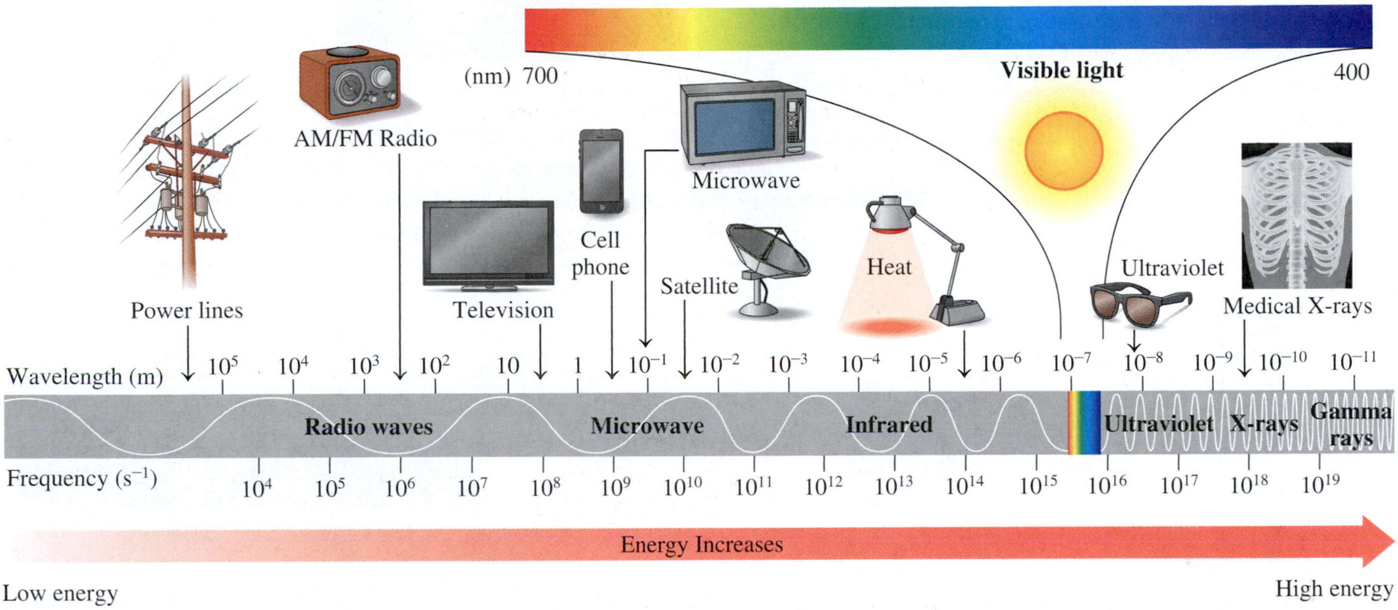

(nm) 700
Visible light
400
AM/FM Radio
Microwave
Cell phone
Power lines
Television
Satellite
Heat
Ultraviolet
Medical X-rays
Wavelength (m)
10^5 10^4 10^3 10^2 10 1 10^-1 10^-2 10^-3 10^-4 10^-5 10^-6 10^-7 10^-8 10^-9 10^-10 10^-11
Radio waves
Microwave
Infrared
Ultraviolet
X-rays
Gamma rays
Frequency (s^-1)
10^4 10^5 10^6 10^7 10^8 10^9 10^10 10^11 10^12 10^13 10^14 10^15 10^16 10^17 10^18 10^19
Energy Increases
Low energy
High energy

FIGURE 4.9 ▶ The electromagnetic spectrum shows the arrangement of wavelengths of electromagnetic radiation. The visible portion consists of wavelengths from 700 nm to 400 nm.

Q How does the wavelength of ultraviolet light compare to that of a microwave?

Chemistry Link to Health

Biological Reactions to UV Light

Our everyday life depends on sunlight, but exposure to sunlight can have damaging effects on living cells, and too much exposure can even cause their death. The light energy, especially ultraviolet (UV), excites electrons and may lead to unwanted chemical reactions. The list of damaging effects of sunlight includes sunburn; wrinkling; premature aging of the skin; changes in the DNA of the cells, which can lead to skin cancers; inflammation of the eyes; and perhaps cataracts. Some drugs, like the acne medications Accutane and Retin-A, as well as antibiotics, diuretics, sulfonamides, and estrogen, make the skin extremely sensitive to light.

Phototherapy uses light to treat certain skin conditions, including psoriasis, eczema, and dermatitis. In the treatment of psoriasis, for example, oral drugs are given to make the skin more photosensitive; then exposure to UV radiation follows. Low-energy radiation (blue light) with wavelengths from 390 to 470 nm is used to treat babies with neonatal jaundice, which converts high levels of bilirubin to water-soluble compounds that can be excreted from the body. Sunlight is also a factor in stimulating the immune system.

(continued)

Chemistry Link to Health (*continued*)

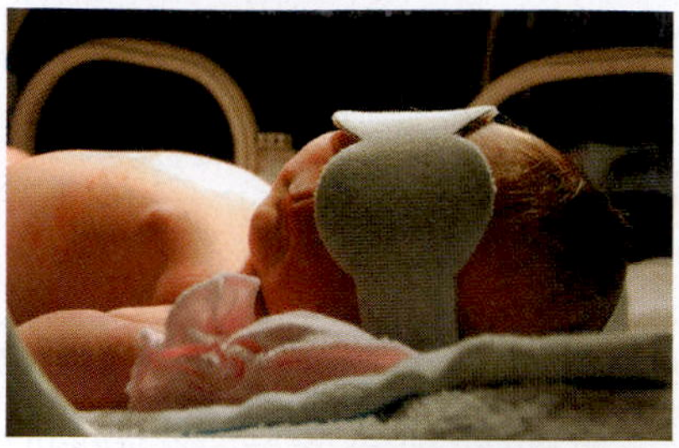

Phototherapy is used to treat babies with neonatal jaundice.

In a disorder called *seasonal affective disorder*, or SAD, people experience mood swings and depression during the winter. Some research suggests that SAD is the result of a decrease in serotonin, or an increase in melatonin, when there are fewer hours of sunlight. One treatment for SAD is therapy using bright light provided by a lamp called a light box. A daily exposure to blue light (460 nm) for 30 to 60 min seems to reduce symptoms of SAD.

A light box is used to provide light, which reduces symptoms of SAD.

When the light from the Sun passes through a prism, the light separates into a continuous color spectrum, which consists of the colors we see in a rainbow. In contrast, when light from a heated element passes through a prism, it separates into distinct lines of color separated by dark areas called an *atomic spectrum*. Each element has its own unique atomic spectrum.

Light passes through a slit
Prism
Film
Ba
Light
Barium light spectrum

In an atomic spectrum, light from a heated element separates into distinct lines.

Electron Energy Levels

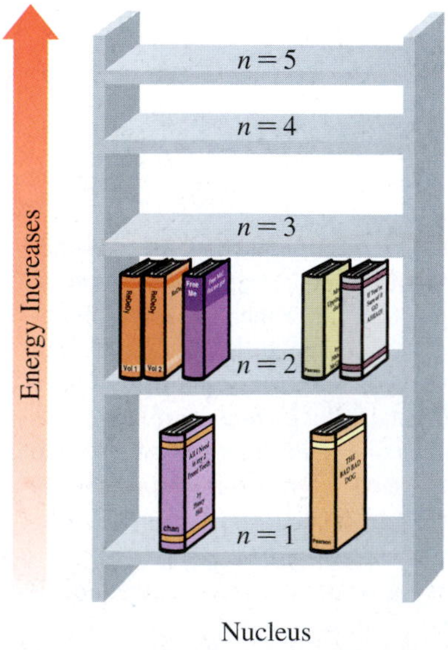
Energy Increases
$n = 5$
$n = 4$
$n = 3$
$n = 2$
$n = 1$
Nucleus

An electron can have only the energy of one of the energy levels in an atom.

Scientists have now determined that the lines in the atomic spectra of elements are associated with changes in the energies of the electrons. In an atom, each electron has a specific energy known as its **energy level**, which is assigned values called *principal quantum numbers* (n), ($n = 1, n = 2, \ldots$). Generally, electrons in the lower energy levels are closer to the nucleus, whereas electrons in the higher energy levels are farther away. The energy of an electron is *quantized*, which means that the energy of an electron can only have specific energy values, but cannot have values between them.

Principal Quantum Number (n)

$$1 < 2 < 3 < 4 < 5 < 6 < 7$$

Lowest energy $\longrightarrow$ Highest energy

All the electrons with the same energy are grouped in the same energy level. As an analogy, we can think of the energy levels of an atom as similar to the shelves in a bookcase. The first shelf is the lowest energy level; the second shelf is the second energy level, and so on. If we are arranging books on the shelves, it would take less energy to fill the bottom shelf first, and then the second shelf, and so on. However, we could never get any book to stay in the space between any of the shelves. Similarly, the energy of an electron must be at specific energy levels, and not between.

Unlike standard bookcases, however, there is a large difference between the energy of the first and second levels, but then the higher energy levels are closer together. Another difference is that the lower electron energy levels hold fewer electrons than the higher energy levels.

Changes in Electron Energy Level

An electron can change from one energy level to a higher level only if it absorbs the energy equal to the difference in energy levels. When an electron changes to a lower energy level, it emits energy equal to the difference between the two levels (see **FIGURE 4.10**). If the energy emitted is in the visible range, we see one of the colors of visible light. The yellow color of sodium streetlights and the red color of neon lights are examples of electrons emitting energy in the visible color range.

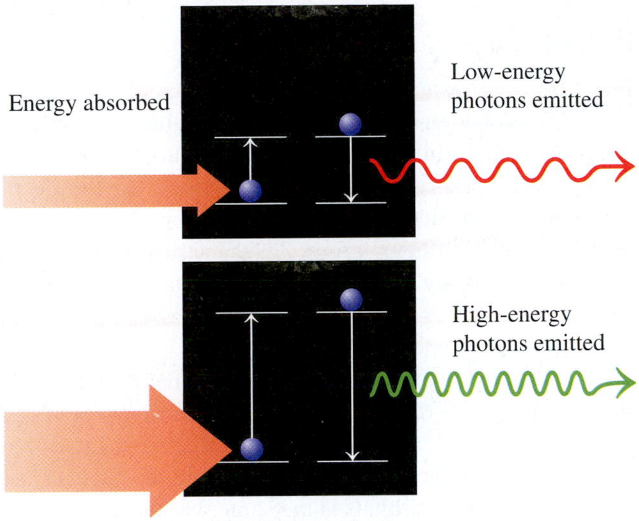
Energy absorbed
Low-energy
photons emitted
High-energy
photons emitted

FIGURE 4.10 ▶ Electrons absorb a specific amount of energy to move to a higher energy level. When electrons lose energy, a specific quantity of energy is emitted.

Q What causes electrons to move to higher energy levels?

Sublevels

We have seen that the protons and neutrons are contained in the small, dense nucleus of an atom. However, it is the electrons within the atoms that determine the physical and chemical properties of the elements. Therefore, we will look at the arrangement of electrons within the large volume of space surrounding the nucleus.

Each of the energy levels consists of one or more **sublevels**, in which electrons with identical energy are found. The sublevels are identified by the letters *s*, *p*, *d*, and *f*. The number of sublevels within an energy level is equal to the principal quantum number, *n*. For example, the first energy level ($n = 1$) has only one sublevel, 1*s*. The second energy level ($n = 2$) has two sublevels, 2*s* and 2*p*. The third energy level ($n = 3$) has three sublevels, 3*s*, 3*p*, and 3*d*. The fourth energy level ($n = 4$) has four sublevels, 4*s*, 4*p*, 4*d*, and 4*f*. Energy levels $n = 5$, $n = 6$, and $n = 7$ also have as many sublevels as the value of *n*, but only *s*, *p*, *d*, and *f* sublevels are needed to hold the electrons in atoms of the 118 known elements (see **FIGURE 4.11**).

Within each energy level, the *s* sublevel has the lowest energy. If there are additional sublevels, the *p* sublevel has the next lowest energy, then the *d* sublevel, and finally the *f* sublevel.

Order of Increasing Energy of Sublevels in an Energy Level

$s < p < d < f$

Lowest energy ⟶ Highest energy

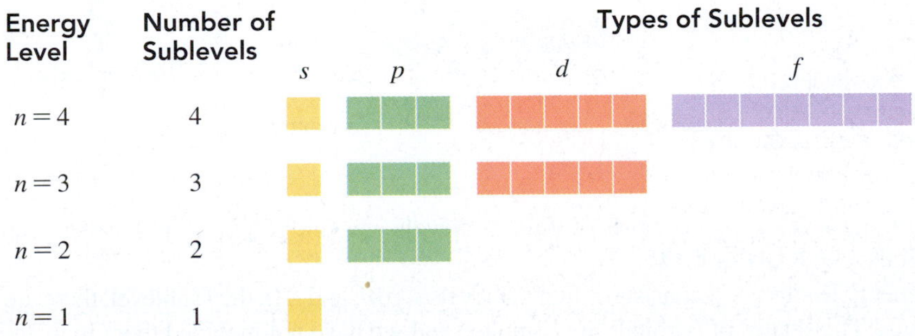
Energy Level
Number of Sublevels
Types of Sublevels
s
p
d
f
n = 4
4
n = 3
3
n = 2
2
n = 1
1

FIGURE 4.11 ▶ The number of sublevels in an energy level is the same as the principal quantum number, *n*.

Q How many sublevels are in the $n = 5$ energy level?

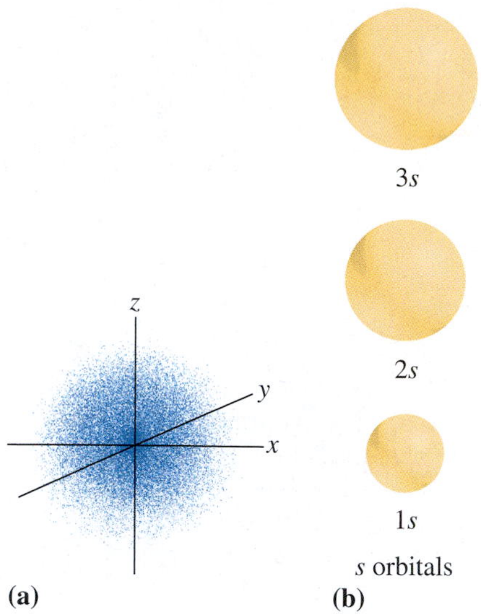
z
y
x
(a)
3s
2s
1s
s orbitals
(b)

FIGURE 4.12 ▶ **(a)** The electron cloud of an *s* orbital represents the highest probability of finding an *s* electron. **(b)** The *s* orbitals are shown as spheres. The sizes of the *s* orbitals increase because they contain electrons at higher energy levels.

Q Is the probability high or low of finding an *s* electron outside an *s* orbital?

Orbitals

There is no way to know the exact location of an electron in an atom. Instead, scientists describe the location of an electron in terms of probability. The **orbital** is the three-dimensional volume in which electrons have the highest probability of being found.

As an analogy, imagine that you draw a circle with a 100-m radius around your chemistry classroom. There is a high probability of finding you within that circle when your chemistry class is in session. But once in a while, you may be outside that circle because you were sick or your car did not start.

Shapes of Orbitals

Each type of orbital has a unique three-dimensional shape. Electrons in an *s* orbital are most likely found in a region with a spherical shape. Imagine that you take a picture of the location of an electron in an *s* orbital every second for an hour. When all these pictures are overlaid, the result, called a *probability density*, would look like the electron cloud shown in **FIGURE 4.12a**. For convenience, we draw this electron cloud as a sphere called an *s* orbital. There is one *s* orbital for every energy level starting with $n = 1$. For example, in the first, second, and third energy levels, there are *s* orbitals designated as 1*s*, 2*s*, and 3*s*. As the principal quantum number increases, there is an increase in the size of the *s* orbitals, although the shape is the same (see **FIGURE 4.12b**).

The orbitals occupied by *p*, *d*, and *f* electrons have three-dimensional shapes different from those of the *s* electrons. There are three *p* orbitals, starting with $n = 2$. Each *p* orbital has two lobes like a balloon tied in the middle. The three *p* orbitals are arranged in three perpendicular directions, along the *x*, *y*, and *z* axes around the nucleus (see **FIGURE 4.13**). As with *s* orbitals, the shape of *p* orbitals is the same, but the volume increases at higher energy levels.

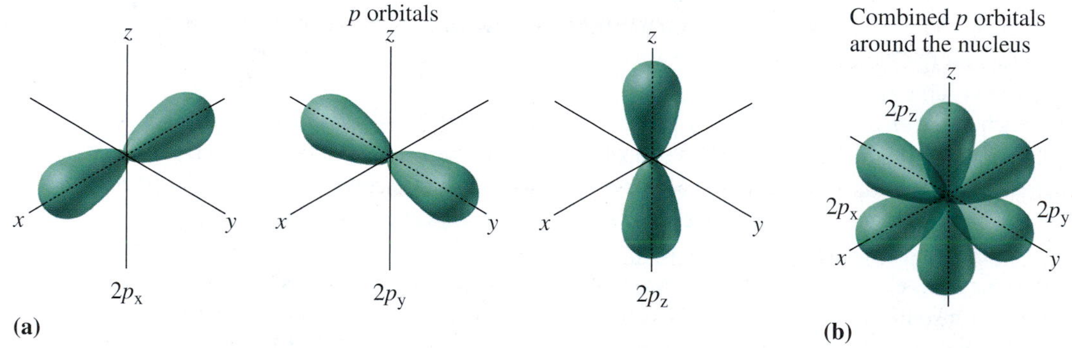
p orbitals
z
x
y
$2p_x$
$2p_y$
$2p_z$
(a)
Combined p orbitals around the nucleus
$2p_z$
$2p_x$
$2p_y$
(b)

FIGURE 4.13 ▶ A *p* orbital has two regions of high probability, which gives a "dumb-bell" shape. **(a)** Each *p* orbital is aligned along a different axis from the other *p* orbitals. **(b)** All three *p* orbitals are shown around the nucleus.

Q What are some similarities and differences of the *p* orbitals in the $n = 3$ energy level?

In summary, the $n = 2$ energy level, which has 2*s* and 2*p* sublevels, consists of one *s* orbital and three *p* orbitals.

Energy level $n = 2$ consists of one 2*s* orbital and three 2*p* orbitals.

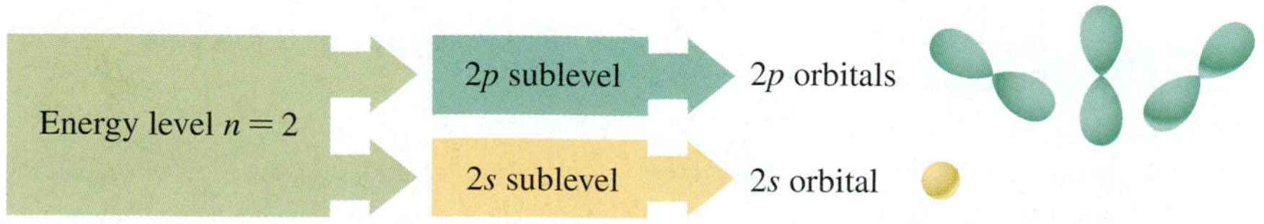
Energy level $n = 2$
2p sublevel
2p orbitals
2s sublevel
2s orbital

Energy level $n = 3$ consists of three sublevels *s*, *p*, and *d*. The *d* sublevels contain five *d* orbitals (see **FIGURE 4.14**).

Energy level $n = 4$ consists of four sublevels *s*, *p*, *d*, and *f*. In the *f* sublevel, there are seven *f* orbitals. The shapes of *f* orbitals are complex, and we have not included them in this text.

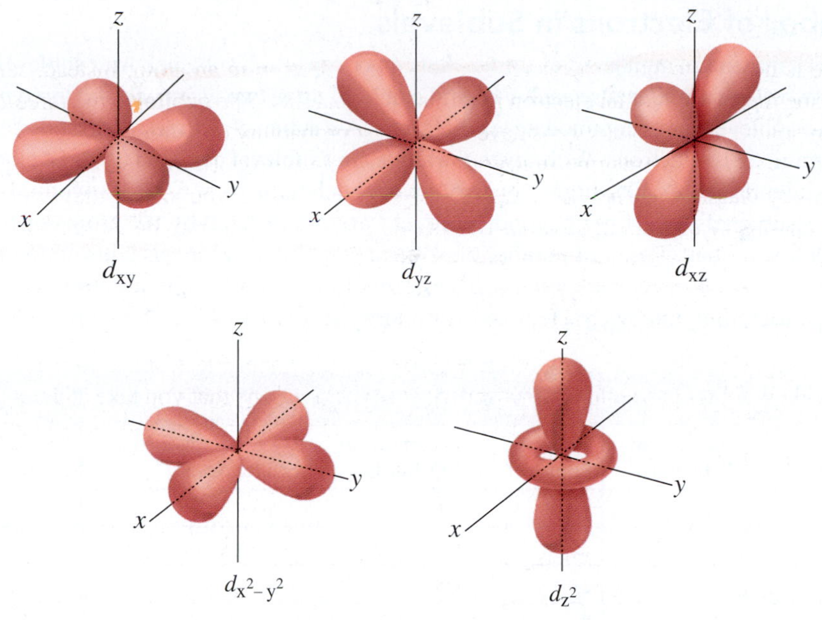
d_{xy}
d_{yz}
d_{xz}
$d_{x^2-y^2}$
d_{z^2}
x
y
z

FIGURE 4.14 ▶ Four of the five *d* orbitals consist of four lobes that are aligned along or between different axes. One *d* orbital consists of two lobes and a doughnut-shaped ring around its center.

Q How many orbitals are there in the 5*d* sublevel?

▶ SAMPLE PROBLEM 4.6 Energy Levels, Sublevels, and Orbitals

TRY IT FIRST

Indicate the type and number of orbitals in each of the following energy levels or sublevels:

a. 3*p* sublevel **b.** $n = 2$ **c.** $n = 3$ **d.** 4*d* sublevel

SOLUTION

	Given	Need	Connect
ANALYZE THE PROBLEM	energy level, sublevel	orbitals	*n*, orbitals in *s*, *p*, *d*, *f*

a. The 3*p* sublevel contains three 3*p* orbitals.
b. The $n = 2$ energy level consists of one 2*s* and three 2*p* orbitals.
c. The $n = 3$ energy level consists of one 3*s*, three 3*p*, and five 3*d* orbitals.
d. The 4*d* sublevel contains five 4*d* orbitals.

STUDY CHECK 4.6

a. What is similar and what is different for 1*s*, 2*s*, and 3*s* orbitals?
b. How many orbitals would be in the 4*f* sublevel?

ANSWER

a. The 1*s*, 2*s*, and 3*s* orbitals are all spherical, but they increase in volume because the electron is most likely to be found farther from the nucleus for higher energy levels.
b. The 4*f* sublevel has seven orbitals.

TEST

Try Practice Problems 4.45 to 4.50

Orbital Capacity and Electron Spin

The *Pauli exclusion principle* states that each orbital can hold a maximum of two electrons. According to a useful model for electron behavior, an electron is seen as spinning on its axis, which generates a magnetic field. When two electrons are in the same orbital, they will repel each other unless their magnetic fields cancel. This happens only when the two electrons spin in opposite directions. We can represent the spins of the electrons in the same orbital with one arrow pointing up and the other pointing down.

Electron spinning counterclockwise | Electron spinning clockwise

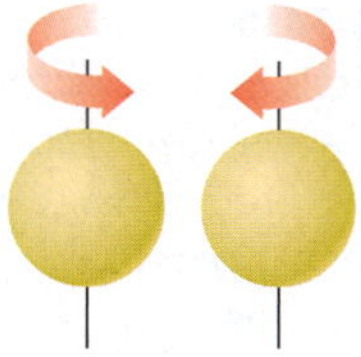

Opposite spins of electrons in an orbital

An orbital can hold up to two electrons with opposite spins.

TEST

Try Practice Problems 4.51 and 4.52

Number of Electrons in Sublevels

There is a maximum number of electrons that can occupy each sublevel. An *s* sublevel holds one or two electrons. Because each *p* orbital can hold up to two electrons, the three *p* orbitals in a *p* sublevel can accommodate six electrons. A *d* sublevel with five *d* orbitals can hold a maximum of 10 electrons. With seven *f* orbitals, an *f* sublevel can hold up to 14 electrons.

As mentioned earlier, higher energy levels such as $n = 5, 6,$ and 7 would have 5, 6, and 7 sublevels, but those beyond sublevel *f* are not utilized by the atoms of the elements known today. The total number of electrons in all the sublevels adds up to give the electrons allowed in an energy level. The number of sublevels, the number of orbitals, and the maximum number of electrons for energy levels 1 to 4 are shown in **TABLE 4.9**.

TABLE 4.9 Electron Capacity in Sublevels for Energy Levels 1 to 4

Energy Level (*n*)	Number of Sublevels	Type of Sublevel	Number of Orbitals	Maximum Number of Electrons	Total Electrons
1	1	1*s*	1	2	2
2	2	2*s*	1	2	
		2*p*	3	6	8
3	3	3*s*	1	2	
		3*p*	3	6	
		3*d*	5	10	18
4	4	4*s*	1	2	
		4*p*	3	6	
		4*d*	5	10	
		4*f*	7	14	32

PRACTICE PROBLEMS

4.6 Electron Energy Levels

4.45 Describe the shape of each of the following orbitals:
a. 1*s* **b.** 2*p* **c.** 5*s*

4.46 Describe the shape of each of the following orbitals:
a. 3*p* **b.** 6*s* **c.** 4*p*

4.47 Match statements **1** to **3** with **a** to **d**.
1. They have the same shape.
2. The maximum number of electrons is the same.
3. They are in the same energy level.

a. 1*s* and 2*s* orbitals **b.** 3*s* and 3*p* sublevels
c. 3*p* and 4*p* sublevels **d.** three 3*p* orbitals

4.48 Match statements **1** to **3** with **a** to **d**.
1. They have the same shape.
2. The maximum number of electrons is the same.
3. They are in the same energy level.

a. 5*s* and 6*s* orbitals **b.** 3*p* and 4*p* orbitals
c. 3*s* and 4*s* sublevels **d.** 2*s* and 2*p* orbitals

4.49 Indicate the number of each in the following:
a. orbitals in the 3*d* sublevel
b. sublevels in the $n = 1$ energy level
c. orbitals in the 6*s* sublevel
d. orbitals in the $n = 3$ energy level

4.50 Indicate the number of each in the following:
a. orbitals in the $n = 2$ energy level
b. sublevels in the $n = 4$ energy level
c. orbitals in the 5*f* sublevel
d. orbitals in the 6*p* sublevel

4.51 Indicate the maximum number of electrons in the following:
a. 2*p* orbital **b.** 3*p* sublevel
c. $n = 4$ energy level **d.** 5*d* sublevel

4.52 Indicate the maximum number of electrons in the following:
a. 3*s* sublevel **b.** 4*p* orbital
c. $n = 3$ energy level **d.** 4*f* sublevel

4.7 Electron Configurations

LEARNING GOAL Draw the orbital diagram and write the electron configuration for an element.

We can now look at how electrons are arranged in the orbitals within an atom. An **orbital diagram** shows the placement of the electrons in the orbitals in order of increasing energy (see **FIGURE 4.15**). In this energy diagram, we see that the electrons in the 1*s* orbital have the lowest energy level. The energy level is higher for the 2*s* orbital and is even higher for the 2*p* orbitals.

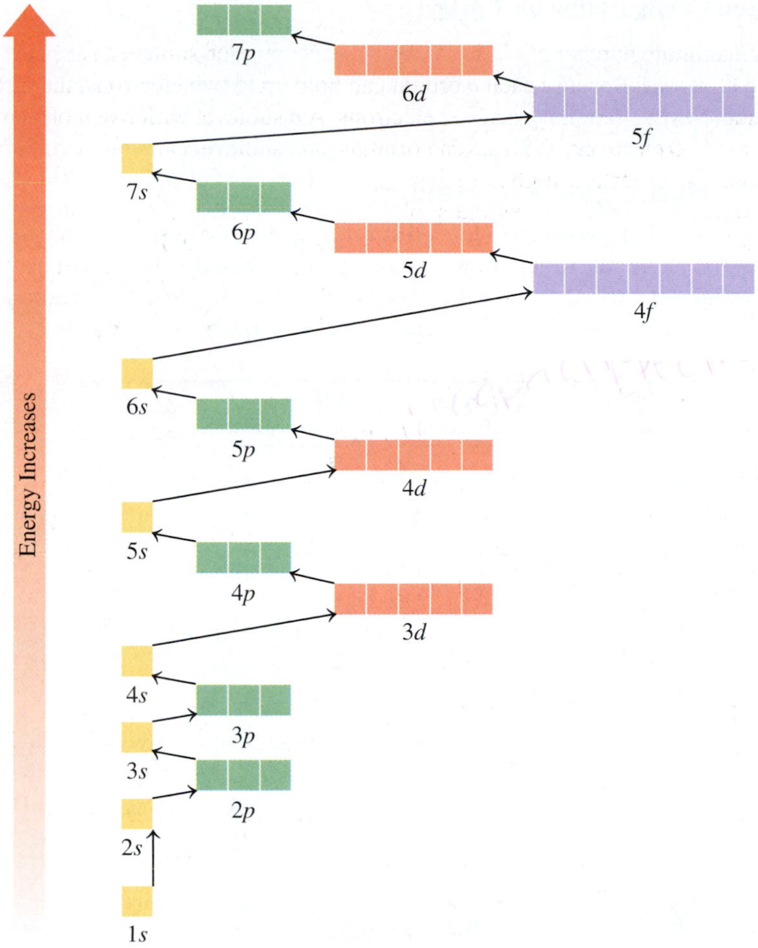
Energy Increases
7p
6d
5f
7s
6p
5d
4f
6s
5p
4d
5s
4p
3d
4s
3p
3s
2p
2s
1s

FIGURE 4.15 ▶ The orbitals in an atom fill in order of increasing energy beginning with 1*s*.

Q Why do 3*d* orbitals fill after the 4*s* orbital?

We begin our discussion of electron configuration by using orbital diagrams in which boxes represent the orbitals. Any orbital can have a maximum of two electrons.

To draw an orbital diagram, the lowest energy orbitals are filled first. For example, we can draw the orbital diagram for carbon. The atomic number of carbon is 6, which means that a carbon atom has six electrons. The first two electrons go into the 1*s* orbital; the next two electrons go into the 2*s* orbital. In the orbital diagram, the two electrons in the 1*s* and 2*s* orbitals are shown with opposite spins; the first arrow is up, and the second is down. The last two electrons in carbon begin to fill the 2*p* sublevel, which has the next lowest energy. However, there are three 2*p* orbitals of equal energy. Because the negatively charged electrons repel each other, they are placed in separate 2*p* orbitals. With few exceptions (which will be noted later in this chapter) lower energy sublevels are filled first, and then the "building" of electrons continues to the next lowest energy sublevel that is available until all the electrons are placed.

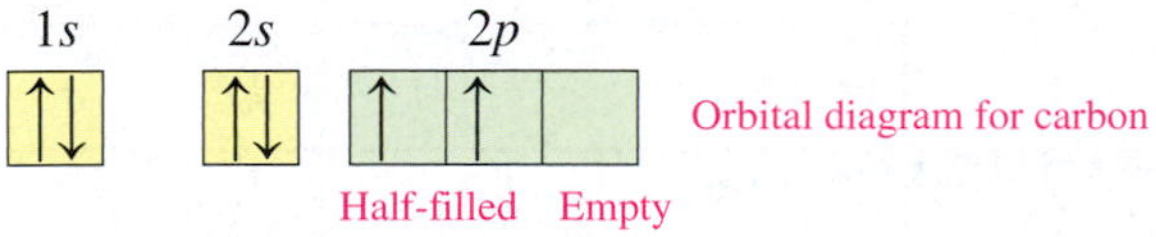
1s
2s
2p
Orbital diagram for carbon
Half-filled
Empty

TEST

Try Practice Problems 4.53 to 4.56

Electron Configurations

CORE CHEMISTRY SKILL

Writing Electron Configurations

Chemists use a notation called the **electron configuration** to indicate the placement of the electrons of an atom in order of increasing energy. The electron configuration is written with the lowest energy sublevel first, followed by the next lowest energy sublevel. The number of electrons in each sublevel is shown as a superscript.

Electron Configuration for Carbon

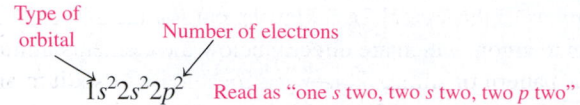

Type of orbital

Number of electrons

$1s^22s^22p^2$ Read as "one s two, two s two, two p two"

Period 1: Hydrogen and Helium

We will draw orbital diagrams and write electron configurations for the elements H and He in Period 1. The 1*s* orbital (which is the 1*s* sublevel) is written first because it has the lowest energy. Hydrogen has one electron in the 1*s* sublevel; helium has two. In the orbital diagram, the electrons for helium are shown as arrows pointing in opposite directions.

Element	Atomic Number	Orbital Diagram	Electron Configuration
1s Block			
		1*s*	
H	1	[↑]	$1s^1$
He	2	[↑↓]	$1s^2$

Period 2: Lithium to Neon

Period 2 begins with lithium, which has three electrons. The first two electrons fill the 1*s* orbital, whereas the third electron goes into the 2*s* orbital, the sublevel with the next lowest energy. In beryllium, another electron is added to complete the 2*s* orbital. The next six electrons are used to fill the 2*p* orbitals. The electrons are added one at a time from boron to nitrogen, which gives three half-filled 2*p* orbitals.

From oxygen to neon, the remaining three electrons are paired up using opposite spins to complete the 2*p* sublevel. In writing the electron configurations for the elements in Period 2, begin with the 1*s* orbital followed by the 2*s* and then the 2*p* orbitals.

An electron configuration can also be written in an *abbreviated configuration*. The electron configuration of the preceding noble gas is replaced by writing its element symbol inside square brackets. For example, the electron configuration for lithium, $1s^22s^1$, can be abbreviated as [He]$2s^1$ where [He] replaces $1s^2$.

Element	Atomic Number	Orbital Diagram	Electron Configuration	Abbreviated Electron Configuration
2s Block				
		1*s* 2*s*		
Li	3	[↑↓] [↑]	$1s^22s^1$	[He]$2s^1$
Be	4	[↑↓] [↑↓]	$1s^22s^2$	[He]$2s^2$
2p Block				
		2*p*		
B	5	[↑↓] [↑↓] [↑][ ][ ]	$1s^22s^22p^1$	[He]$2s^22p^1$
C	6	[↑↓] [↑↓] [↑][↑][ ]	$1s^22s^22p^2$	[He]$2s^22p^2$
		Unpaired electrons		
N	7	[↑↓] [↑↓] [↑][↑][↑]	$1s^22s^22p^3$	[He]$2s^22p^3$
O	8	[↑↓] [↑↓] [↑↓][↑][↑]	$1s^22s^22p^4$	[He]$2s^22p^4$
F	9	[↑↓] [↑↓] [↑↓][↑↓][↑]	$1s^22s^22p^5$	[He]$2s^22p^5$
Ne	10	[↑↓] [↑↓] [↑↓][↑↓][↑↓]	$1s^22s^22p^6$	[He]$2s^22p^6$

Period 3: Sodium to Argon

In Period 3, electrons enter the orbitals of the 3*s* and 3*p* sublevels, but not the 3*d* sublevel. We notice that the elements sodium to argon, which are directly below the elements lithium to neon in Period 2, have a similar pattern of filling their *s* and *p* orbitals. In sodium and magnesium, one and two electrons go into the 3*s* orbital. The electrons for aluminum, silicon, and phosphorus go into separate 3*p* orbitals. The remaining electrons in sulfur, chlorine, and argon are paired up (opposite spins) with the electrons already in the 3*p* orbitals. For the abbreviated electron configurations of Period 3, the symbol [Ne] replaces $1s^22s^22p^6$.

Element	Atomic Number	Orbital Diagram (3*s* and 3*p* orbitals only)	Electron Configuration	Abbreviated Electron Configuration
3*s* Block				
Na	11	[Ne] 3*s* [↑] 3*p* [ ][ ][ ]	$1s^22s^22p^63s^1$	$[Ne]3s^1$
Mg	12	[Ne] [↑↓] [ ][ ][ ]	$1s^22s^22p^63s^2$	$[Ne]3s^2$
3*p* Block				
Al	13	[Ne] [↑↓] [↑][ ][ ]	$1s^22s^22p^63s^23p^1$	$[Ne]3s^23p^1$
Si	14	[Ne] [↑↓] [↑][↑][ ]	$1s^22s^22p^63s^23p^2$	$[Ne]3s^23p^2$
P	15	[Ne] [↑↓] [↑][↑][↑]	$1s^22s^22p^63s^23p^3$	$[Ne]3s^23p^3$
S	16	[Ne] [↑↓] [↑↓][↑][↑]	$1s^22s^22p^63s^23p^4$	$[Ne]3s^23p^4$
Cl	17	[Ne] [↑↓] [↑↓][↑↓][↑]	$1s^22s^22p^63s^23p^5$	$[Ne]3s^23p^5$
Ar	18	[Ne] [↑↓] [↑↓][↑↓][↑↓]	$1s^22s^22p^63s^23p^6$	$[Ne]3s^23p^6$

▶SAMPLE PROBLEM 4.7 Orbital Diagrams and Electron Configurations

TRY IT FIRST

Silicon is a micromineral needed for the growth of bones, tendons, and ligaments. Draw or write each of the following for silicon:

a. orbital diagram
b. abbreviated orbital diagram
c. electron configuration
d. abbreviated electron configuration

SOLUTION

ANALYZE THE PROBLEM	Given	Need	Connect
	silicon (Si)	orbital diagram, abbreviated orbital diagram, electron configuration, abbreviated electron configuration	periodic table, atomic number, order of filling orbitals

a. From the periodic table, we see that Si, with atomic number 14, has 14 electrons. Place two electrons in each orbital, and single electrons in the highest energy level.

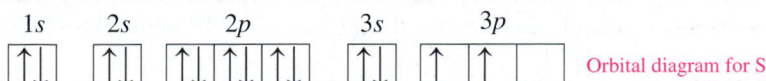

1*s* [↑↓] 2*s* [↑↓] 2*p* [↑↓][↑↓][↑↓] 3*s* [↑↓] 3*p* [↑][↑][ ] Orbital diagram for Si

b. Write the symbol for the preceding noble gas followed by the orbitals for the remaining electrons.

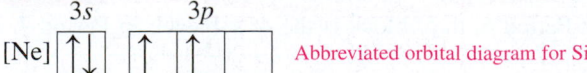

[Ne] 3*s* [↑↓] 3*p* [↑][↑][ ] Abbreviated orbital diagram for Si

c. List the sublevels in order of filling and add the number of electrons.
$1s^22s^22p^63s^23p^2$ Electron configuration for Si

d. Write the symbol for the preceding noble gas followed by the configuration for the remaining electrons.
$[Ne]3s^23p^2$ Abbreviated electron configuration for Si

STUDY CHECK 4.7

a. Write the complete and abbreviated electron configurations for sulfur, a macromineral in proteins, vitamin B_1, and insulin.

b. Write the complete and abbreviated electron configurations for phosphorus, needed for bones and teeth.

ANSWER

a. $1s^22s^22p^63s^23p^4$, $[Ne]3s^23p^4$ **b.** $1s^22s^22p^63s^23p^3$, $[Ne]3s^23p^3$

TEST

Try Practice Problems 4.57 to 4.64

Electron Configurations and the Periodic Table

The electron configurations of the elements are also related to their position on the periodic table. Different sections or blocks within the periodic table correspond to the *s*, *p*, *d*, and *f* sublevels (see **FIGURE 4.16**).

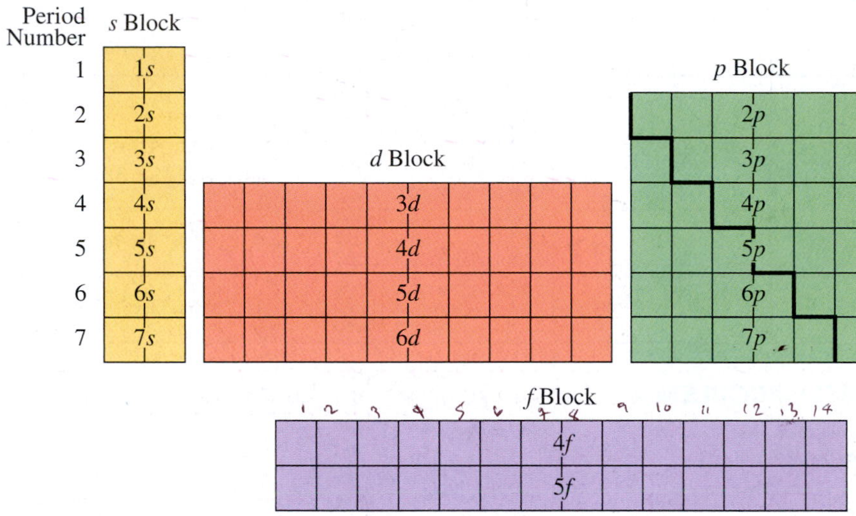
Period Number
s Block
1
2
3
4
5
6
7
1s
2s
3s
4s
5s
6s
7s
d Block
3d
4d
5d
6d
p Block
2p
3p
4p
5p
6p
7p
f Block
4f
5f

FIGURE 4.16 ▶ Electron configurations follow the order of occupied sublevels on the periodic table.

Q If neon is in Group 8A (18), Period 2, how many electrons are in the 1*s*, 2*s*, and 2*p* sublevels of neon?

CORE CHEMISTRY SKILL

Using the Periodic Table to Write Electron Configurations

Blocks on the Periodic Table

1. The ***s* block** includes hydrogen and helium as well as the elements in Group 1A (1) and Group 2A (2). This means that the final one or two electrons in the elements of the *s* block are located in an *s* orbital. The period number indicates the particular *s* orbital that is filling: 1*s*, 2*s*, and so on.
2. The ***p* block** consists of the elements in Group 3A (13) to Group 8A (18). There are six *p* block elements in each period because three *p* orbitals can each hold up to six electrons. The period number indicates the particular *p* sublevel that is filling: 2*p*, 3*p*, and so on.
3. The ***d* block**, containing the transition elements, first appears after calcium (atomic number 20). There are 10 elements in each period of the *d* block because five *d* orbitals can hold up to 10 electrons. The particular *d* sublevel is one less ($n - 1$) than the period number. For example, in Period 4, the *d* block is the 3*d* sublevel. In Period 5, the *d* block is the 4*d* sublevel.
4. The ***f* block**, the inner transition elements, are the two rows at the bottom of the periodic table. There are 14 elements in each *f* block because seven *f* orbitals can hold up to 14 electrons. Elements that have atomic numbers higher than 57 (La) have electrons in the 4*f* block. The particular *f* sublevel is two less ($n - 2$) than the period number. For example, in Period 6, the *f* block is the 4*f* sublevel. In Period 7, the *f* block is the 5*f* sublevel.

Writing Electron Configurations Using Sublevel Blocks

Now we can write electron configurations using the sublevel blocks on the periodic table. As before, each configuration begins at H. But now we move across the table from left to right, writing down each sublevel block we come to until we reach the element for which we are writing an electron configuration.

Electron Configurations for Period 4 and Above

Up to Period 4, the filling of the sublevels has progressed in order. However, if we look at the sublevel blocks in Period 4, we see that the 4*s* sublevel fills before the 3*d* sublevel. This occurs because the electrons in the 4*s* sublevel have slightly lower energy than the electrons in the 3*d* sublevel. This order occurs again in Period 5 when the 5*s* sublevel fills before the 4*d* sublevel, in Period 6 when the 6*s* fills before the 5*d*, and in Period 7 when the 7*s* fills before the 6*d*.

At the beginning of Period 4, the electrons in potassium (19) and calcium (20) go into the 4*s* sublevel. In scandium, the next electron added goes into the 3*d* block, which continues to fill until it is complete with 10 electrons at zinc (30). Once the 3*d* block is complete, the next six electrons, gallium to krypton, go into the 4*p* block.

Element	Atomic Number	Electron Configuration	Abbreviated Electron Configuration	
4s Block				
K	19	$1s^22s^22p^63s^23p^64s^1$	$[Ar]4s^1$	
Ca	20	$1s^22s^22p^63s^23p^64s^2$	$[Ar]4s^2$	
3d Block				
Sc	21	$1s^22s^22p^63s^23p^64s^23d^1$	$[Ar]4s^23d^1$	
Ti	22	$1s^22s^22p^63s^23p^64s^23d^2$	$[Ar]4s^23d^2$	
V	23	$1s^22s^22p^63s^23p^64s^23d^3$	$[Ar]4s^23d^3$	
Cr*	24	$1s^22s^22p^63s^23p^64s^13d^5$	$[Ar]4s^13d^5$	half-filled *d* sublevel is stable
Mn	25	$1s^22s^22p^63s^23p^64s^23d^5$	$[Ar]4s^23d^5$	
Fe	26	$1s^22s^22p^63s^23p^64s^23d^6$	$[Ar]4s^23d^6$	
Co	27	$1s^22s^22p^63s^23p^64s^23d^7$	$[Ar]4s^23d^7$	
Ni	28	$1s^22s^22p^63s^23p^64s^23d^8$	$[Ar]4s^23d^8$	
Cu*	29	$1s^22s^22p^63s^23p^64s^13d^{10}$	$[Ar]4s^13d^{10}$	filled *d* sublevel is stable
Zn	30	$1s^22s^22p^63s^23p^64s^23d^{10}$	$[Ar]4s^23d^{10}$	
4p Block				
Ga	31	$1s^22s^22p^63s^23p^64s^23d^{10}4p^1$	$[Ar]4s^23d^{10}4p^1$	
Ge	32	$1s^22s^22p^63s^23p^64s^23d^{10}4p^2$	$[Ar]4s^23d^{10}4p^2$	
As	33	$1s^22s^22p^63s^23p^64s^23d^{10}4p^3$	$[Ar]4s^23d^{10}4p^3$	
Se	34	$1s^22s^22p^63s^23p^64s^23d^{10}4p^4$	$[Ar]4s^23d^{10}4p^4$	
Br	35	$1s^22s^22p^63s^23p^64s^23d^{10}4p^5$	$[Ar]4s^23d^{10}4p^5$	
Kr	36	$1s^22s^22p^63s^23p^64s^23d^{10}4p^6$	$[Ar]4s^23d^{10}4p^6$	

*Exceptions to the order of filling

We show how to write the electron configuration for selenium (atomic number 34) using the sublevel blocks on the periodic table in Sample Problem 4.8.

▸SAMPLE PROBLEM 4.8 Using Sublevel Blocks to Write Electron Configurations

TRY IT FIRST

Selenium is a micromineral used in the immune system and cardiac health. Use the sublevel blocks on the periodic table to write the electron configuration for selenium.

SOLUTION

STEP 1 **Locate the element on the periodic table.** Selenium is in Period 4 and Group 6A (16).

STEP 2 **Write the filled sublevels in order, going across each period.**

Period	Sublevel Block Filling	Sublevel Block Notation (filled)
1	1*s* (H → He)	$1s^2$
2	2*s* (Li → Be) and 2*p* (B → Ne)	$2s^22p^6$
3	3*s* (Na → Mg) and 3*p* (Al → Ar)	$3s^23p^6$
4	4*s* (K → Ca) and 3*d* (Sc → Zn)	$4s^23d^{10}$

STEP 3 **Complete the configuration by counting the electrons in the last occupied sublevel block.** Because selenium is the fourth element in the 4*p* block, there are four electrons to place in the 4*p* sublevel.

Period	Sublevel Block Filling	Sublevel Block Notation
4	4*p* (Ga → Se)	$4p^4$

The electron configuration for selenium (Se) is: $1s^22s^22p^63s^23p^64s^23d^{10}4p^4$.

STUDY CHECK 4.8

Use the sublevel blocks on the periodic table to write the electron configuration for each of the following:

a. cobalt, a micromineral needed for vitamin B_{12}
b. iodine, a micromineral needed for thyroid function

ANSWER

a. $1s^22s^22p^63s^23p^64s^23d^7$
b. $1s^22s^22p^63s^23p^64s^23d^{10}4p^65s^24d^{10}5p^5$

TEST

Try Practice Problems 4.65 to 4.70

Some Exceptions in Sublevel Block Order

Within the filling of the 3*d* sublevel, exceptions occur for chromium and copper. In Cr and Cu, the 3*d* sublevel is close to being a half-filled or filled sublevel, which is particularly stable. Thus, the electron configuration for chromium has only one electron in the 4*s* and five electrons in the 3*d* sublevel to give the added stability of a half-filled *d* sublevel. This is shown in the abbreviated orbital diagram for chromium:

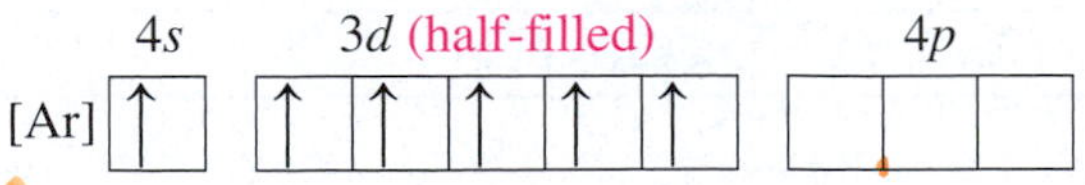
4s
3d (half-filled)
4p
[Ar]

Abbreviated orbital diagram for chromium

A similar exception occurs when copper achieves a stable, filled 3*d* sublevel with 10 electrons and only one electron in the 4*s* orbital. This is shown in the abbreviated orbital diagram for copper:

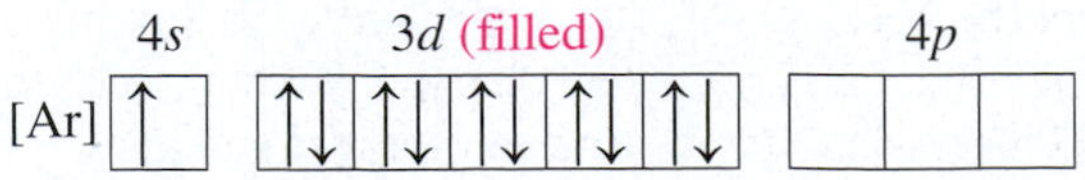
4s
3d (filled)
4p
[Ar]

Abbreviated orbital diagram for copper

PRACTICE PROBLEMS

4.7 Electron Configurations

4.53 Compare the terms electron configuration and abbreviated electron configuration.

4.54 Compare the terms orbital diagram and electron configuration.

4.55 Draw the orbital diagram for each of the following:
a. boron **b.** aluminum
c. phosphorus **d.** argon

4.56 Draw the orbital diagram for each of the following:
a. fluorine **b.** sulfur
c. magnesium **d.** beryllium

4.57 Write the complete electron configuration for each of the following:
a. boron **b.** sodium
c. lithium **d.** magnesium

4.58 Write the complete electron configuration for each of the following:
a. nitrogen **b.** chlorine
c. oxygen **d.** neon

4.59 Write the abbreviated electron configuration for each of the following:
a. carbon **b.** zirconium
c. antimony **d.** fluorine

4.60 Write the abbreviated electron configuration for each of the following:
a. magnesium **b.** oxygen
c. sulfur **d.** nitrogen

4.61 Give the symbol of the element with each of the following electron or abbreviated electron configurations:
a. $1s^22s^1$ **b.** $1s^22s^22p^63s^23p^4$
c. $[Ne]3s^23p^2$ **d.** $[He]2s^22p^5$

4.62 Give the symbol of the element with each of the following electron or abbreviated electron configurations:
a. $1s^22s^22p^4$ **b.** $[Ne]3s^2$
c. $1s^22s^22p^63s^23p^6$ **d.** $[Ne]3s^23p^1$

4.63 Give the symbol of the element that meets the following conditions:
a. has three electrons in the $n = 3$ energy level
b. has two $2p$ electrons
c. completes the $3p$ sublevel
d. completes the $2s$ sublevel

4.64 Give the symbol of the element that meets the following conditions:
a. has five electrons in the $3p$ sublevel
b. has four $2p$ electrons
c. completes the $3s$ sublevel
d. has one electron in the $3s$ sublevel

4.65 Use the sublevel blocks on the periodic table to write the complete electron configuration and abbreviated electron configuration for an atom of each of the following:
a. arsenic **b.** iron
c. indium **d.** bromine
e. titanium

4.66 Use the sublevel blocks on the periodic table to write the complete electron configuration and abbreviated electron configuration for an atom of each of the following:
a. calcium **b.** nickel
c. gallium **d.** cadmium
e. tin

4.67 Use the periodic table to give the symbol of the element that meets the following conditions:
a. has three electrons in the $n = 4$ energy level
b. has three $2p$ electrons
c. completes the $5p$ sublevel
d. has two electrons in the $4d$ sublevel

4.68 Use the periodic table to give the symbol of the element that meets the following conditions:
a. has five electrons in the $n = 3$ energy level
b. has one electron in the $6p$ sublevel
c. completes the $7s$ sublevel
d. has four $5p$ electrons

4.69 Use the periodic table to give the number of electrons in the indicated sublevels for the following:
a. $3d$ in zinc **b.** $2p$ in sodium
c. $4p$ in arsenic **d.** $5s$ in rubidium

4.70 Use the periodic table to give the number of electrons in the indicated sublevels for the following:
a. $3d$ in manganese **b.** $5p$ in antimony
c. $6p$ in lead **d.** $3s$ in magnesium

4.8 Trends in Periodic Properties

LEARNING GOAL Use the electron configurations of elements to explain the trends in periodic properties.

SPRING
SUMMER
AUTUMN
WINTER

The change in temperature with the seasons is a periodic property.

The electron configurations of atoms are an important factor in the physical and chemical properties of the elements and in the properties of the compounds that they form. In this section, we will look at the *valence electrons* in atoms, the trends in *atomic size, ionization energy,* and *metallic character*. Going across a period, there is a pattern of regular change in these properties from one group to the next. Known as *periodic properties*, each property increases or decreases across a period, and then the trend is repeated in each successive period. We can use the seasonal changes in temperatures as an analogy for periodic properties. In the winter, temperatures are cold and become warmer in the spring. By summer, the temperatures are hot but begin to cool in the fall. By winter, we expect cold temperatures again as the pattern of decreasing and increasing temperatures repeats for another year.

CORE CHEMISTRY SKILL
Identifying Trends in Periodic Properties

Group Number and Valence Electrons

The chemical properties of representative elements are mostly due to the **valence electrons**, which are the electrons in the outermost energy level. These valence electrons occupy the *s* and *p* sublevels with the highest principal quantum number *n*. The group numbers indicate the number of valence (outer) electrons for the elements in each vertical column. For example, the elements in Group 1A (1), such as lithium, sodium, and potassium, all have one electron in an *s* orbital. Looking at the sublevel block, we can represent the valence electron in the alkali metals of Group 1A (1) as ns^1. All the elements in Group 2A (2), the alkaline earth metals, have two valence electrons, ns^2. The halogens in Group 7A (17) have seven valence electrons, ns^2np^5.

We can see the repetition of the outermost *s* and *p* electrons for the representative elements for Periods 1 to 4 in **TABLE 4.10**. Helium is included in Group 8A (18) because it is a noble gas, but it has only two electrons in its complete energy level.

ENGAGE
What do the group numbers indicate about the electron configurations of the elements in Groups 1A (1) to 8A (18)?

ENGAGE
What is similar and what is different about the valence electrons of the elements in a group?

TABLE 4.10 Valence Electron Configurations for Representative Elements in Periods 1 to 4

1A (1)	2A (2)	3A (13)	4A (14)	5A (15)	6A (16)	7A (17)	8A (18)
1 H $1s^1$							2 He $1s^2$
3 Li $2s^1$	4 Be $2s^2$	5 B $2s^22p^1$	6 C $2s^22p^2$	7 N $2s^22p^3$	8 O $2s^22p^4$	9 F $2s^22p^5$	10 Ne $2s^22p^6$
11 Na $3s^1$	12 Mg $3s^2$	13 Al $3s^23p^1$	14 Si $3s^23p^2$	15 P $3s^23p^3$	16 S $3s^23p^4$	17 Cl $3s^23p^5$	18 Ar $3s^23p^6$
19 K $4s^1$	20 Ca $4s^2$	31 Ga $4s^24p^1$	32 Ge $4s^24p^2$	33 As $4s^24p^3$	34 Se $4s^24p^4$	35 Br $4s^24p^5$	36 Kr $4s^24p^6$

▶SAMPLE PROBLEM 4.9 Using Group Numbers

TRY IT FIRST

Using the periodic table, write the group number, the period, the number of valence electrons, and the valence electron configuration for each of the following:

a. calcium **b.** iodine **c.** lead

SOLUTION

The valence electrons are the outermost *s* and *p* electrons. Although there may be electrons in the *d* or *f* sublevels, they are not valence electrons.

a. Calcium is in Group 2A (2), Period 4. It has two valence electrons, and a valence electron configuration of $4s^2$.

b. Iodine is in Group 7A (17), Period 5. It has seven valence electrons, and a valence electron configuration of $5s^25p^5$.

c. Lead is in Group 4A (14), Period 6. It has four valence electrons, and a valence electron configuration of $6s^26p^2$.

STUDY CHECK 4.9

What are the group numbers, the periods, the number of valence electrons, and the valence electron configurations for sulfur and strontium?

ANSWER

Sulfur is in Group 6A (16), Period 3, has six valence electrons, and a $3s^23p^4$ valence electron configuration. Strontium is in Group 2A (2), Period 5, has two valence electrons, and a $5s^2$ valence electron configuration.

Lewis Symbols

CORE CHEMISTRY SKILL
Drawing Lewis Symbols

A **Lewis symbol** is a convenient way to represent the valence electrons, which are shown as dots placed on the sides, top, or bottom of the symbol for the element. One to four valence electrons are arranged as single dots. When there are five to eight electrons, one or more electrons are paired. Any of the following would be acceptable Lewis symbol for magnesium, which has two valence electrons:

Lewis Symbols for Magnesium

Ṁg• Ṁg •Ṁg •Mg• Mg• •Mg

Lewis symbols for selected elements are given in **TABLE 4.11**.

TABLE 4.11 Lewis Symbols for Selected Elements in Periods 1 to 4

Group Number Number of Valence Electrons	1A (1) 1	2A (2) 2	3A (13) 3	4A (14) 4	5A (15) 5	6A (16) 6	7A (17) 7	8A (18) 8
	Number of Valence Electrons Increases							
Lewis Symbol	H•							He:*
	Li•	Ḃe•	•Ḃ•	•Ċ•	•N̈•	•Ö:	•F̈:	:N̈e:
	Na•	Ṁg•	•Ȧl•	•Ṡi•	•P̈•	•S̈:	•C̈l:	:Är:
	K•	Ċa•	•Ġa•	•Ġe•	•Äs•	•S̈e:	•B̈r:	:K̈r:

*Helium (He) is stable with two valence electrons.

▶ SAMPLE PROBLEM 4.10 Drawing Lewis Symbols

TRY IT FIRST

Draw the Lewis symbol for each of the following:

a. bromine **b.** aluminum

SOLUTION

a. The Lewis symbol for bromine, which is in Group 7A (17), has seven valence electrons. Thus, three pairs of dots and one single dot are drawn on the sides of the Br symbol.

•B̈r:

b. The Lewis symbol for aluminum, which is in Group 3A (13), has three valence electrons drawn as single dots on the sides of the Al symbol.

•Ȧl•

STUDY CHECK 4.10

Draw the Lewis symbol for phosphorus, a macromineral needed for bones and teeth.

ANSWER

•P̈•

TEST
Try Practice Problems 4.71 to 4.78

Atomic Size

The **atomic size** of an atom is determined by the distance of the valence electrons from the nucleus. For each group of representative elements, the atomic size *increases* going from the top to the bottom because the outermost electrons in each energy level are farther from the nucleus. For example, in Group 1A (1), Li has a valence electron in energy level 2; Na has a valence electron in energy level 3; and K has a valence electron in energy level 4. This means that a K atom is larger than a Na atom and a Na atom is larger than a Li atom (see **FIGURE 4.17**).

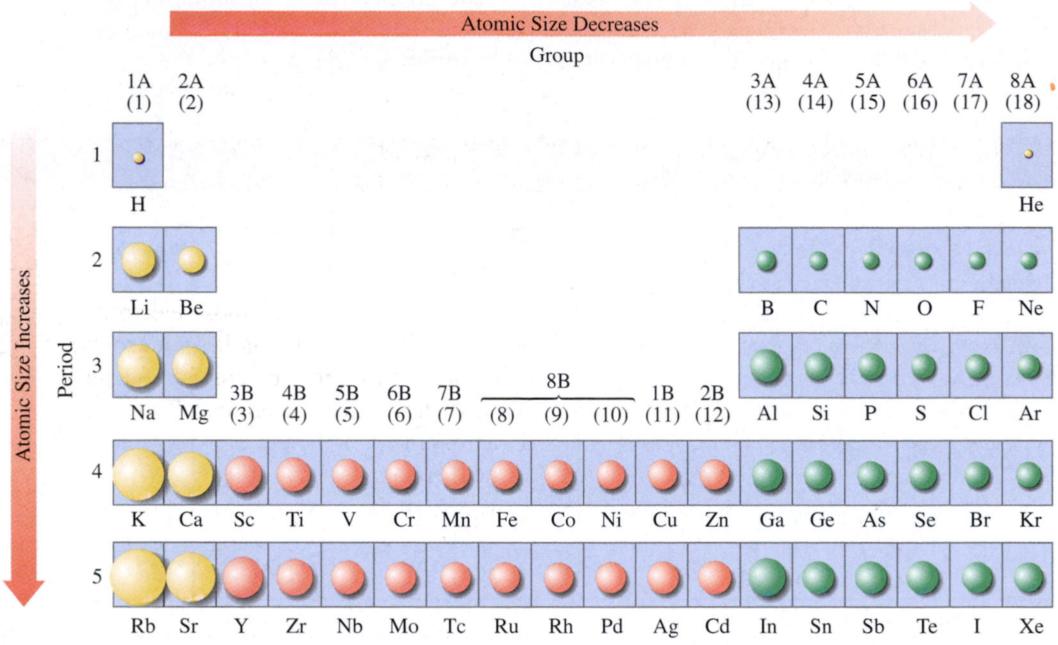
Atomic Size Decreases
Group
Atomic Size Increases
Period
1A (1)
2A (2)
3A (13)
4A (14)
5A (15)
6A (16)
7A (17)
8A (18)
3B (3)
4B (4)
5B (5)
6B (6)
7B (7)
8B
(8)
(9)
(10)
1B (11)
2B (12)
1 H He
2 Li Be B C N O F Ne
3 Na Mg Al Si P S Cl Ar
4 K Ca Sc Ti V Cr Mn Fe Co Ni Cu Zn Ga Ge As Se Br Kr
5 Rb Sr Y Zr Nb Mo Tc Ru Rh Pd Ag Cd In Sn Sb Te I Xe

FIGURE 4.17 ▶ The atomic size increases going down a group but decreases going from left to right across a period.

Q Why does the atomic size increase going down a group?

ENGAGE

Why is a phosphorus atom larger than a nitrogen atom but smaller than a silicon atom?

The atomic size of representative elements is affected by the attractive forces of the protons in the nucleus on the electrons in the outermost level. For the elements going across a period, the increase in the number of protons in the nucleus increases the positive charge of the nucleus. As a result, the electrons are pulled closer to the nucleus, which means that the atomic size of representative elements decreases going from left to right across a period.

The size of atoms of transition elements within the same period changes only slightly because electrons are filling *d* orbitals rather than the outermost energy level. Because the increase in nuclear charge is canceled by an increase in *d* electrons, the attraction of the valence electrons by the nucleus remains about the same. Because there is little change in the nuclear attraction for the valence electrons, the atomic size remains relatively constant for the transition elements.

▶ SAMPLE PROBLEM 4.11 Size of Atoms

TRY IT FIRST

Identify the smaller atom in each of the following pairs:

a. N or F **b.** K or Kr **c.** Ca and Sr

SOLUTION

a. The F atom has a greater positive charge on the nucleus, which pulls electrons closer, and makes the F atom smaller than the N atom. Atomic size decreases going from left to right across a period.

b. The Kr atom has a greater positive charge on the nucleus, which pulls electrons closer, and makes the Kr atom smaller than the K atom. Atomic size decreases going from left to right across a period.

c. The outer electrons in the Ca atom are closer to the nucleus than in the Sr atom, which makes the Ca atom smaller than the Sr atom. Atomic size increases going down a group.

STUDY CHECK 4.11

Which atom has the largest atomic size, P, As, or Se?

ANSWER

As

Ionization Energy

In an atom, negatively charged electrons are attracted to the positive charge of the protons in the nucleus. Therefore, energy is required to remove an electron from an atom. The **ionization energy** is the energy needed to remove one electron from an atom in the gaseous (*g*) state. When an electron is removed from a neutral atom, a cation with a 1+ charge is formed.

$$Na(g) + \text{energy (ionization)} \longrightarrow Na^+(g) + e^-$$

1s	2s	2p			3s		1s	2s	2p			3s	
↑↓	↑↓	↑↓	↑↓	↑↓	↑		↑↓	↑↓	↑↓	↑↓	↑↓		+ ↑

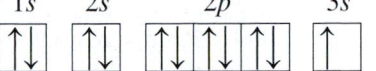

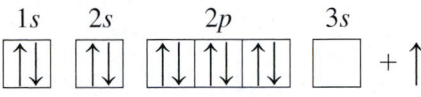

The attraction of a nucleus for the outermost electrons decreases as those electrons are farther from the nucleus. Thus the ionization energy decreases going down a group (see **FIGURE 4.18**). However, going across a period from left to right, the positive charge of the nucleus increases because there is an increase in the number of protons. Thus the ionization energy increases going from left to right across the periodic table.

In summary, the ionization energy is low for the metals and high for the nonmetals. The high ionization energies of the noble gases indicate that their electron configurations are especially stable.

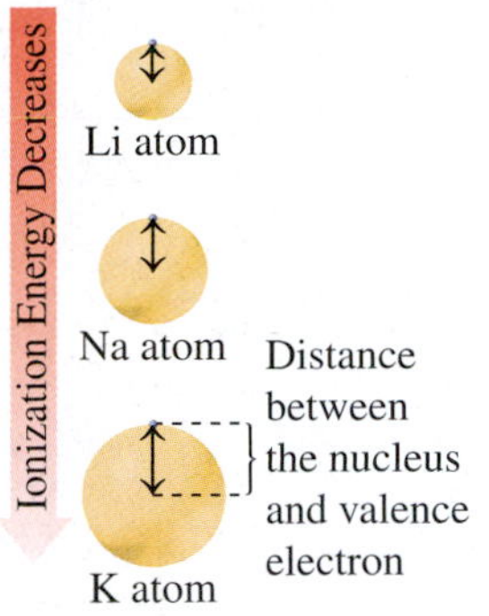
Ionization Energy Decreases
Li atom
Na atom
K atom
Distance between the nucleus and valence electron

FIGURE 4.18 ▶ As the distance from the nucleus to the valence electron in Li, Na, and K atoms increases in Group 1A (1), the ionization energy decreases, and less energy is required to remove the valence electron.

Q Why would Cs have a lower ionization energy than K?

ENGAGE

When a magnesium atom ionizes to form a magnesium ion Mg^{2+}, which electrons are lost?

▶ SAMPLE PROBLEM 4.12 Ionization Energy

TRY IT FIRST

Indicate the element in each set that has the higher ionization energy and explain your choice.

a. K or Na **b.** Mg or Cl **c.** F, N, or C

SOLUTION

a. Na. In Na, an electron is removed from a sublevel closer to the nucleus, which requires a higher ionization energy for Na compared with K.

b. Cl. The increased nuclear charge of Cl increases the attraction for the valence electrons, which requires a higher ionization energy for Cl compared to Mg.

c. F. The increased nuclear charge of F increases the attraction for the valence electrons, which requires a higher ionization energy for F compared to C or N.

STUDY CHECK 4.12

a. Arrange Sr, I, and Sn in order of increasing ionization energy.

b. Arrange Ba, Mg, and Ca in order of decreasing ionization energy.

ANSWER

a. Ionization energy increases going from left to right across a period: Sr, Sn, I.

b. Ionization energy decreases going down a group: Mg, Ca, Ba.

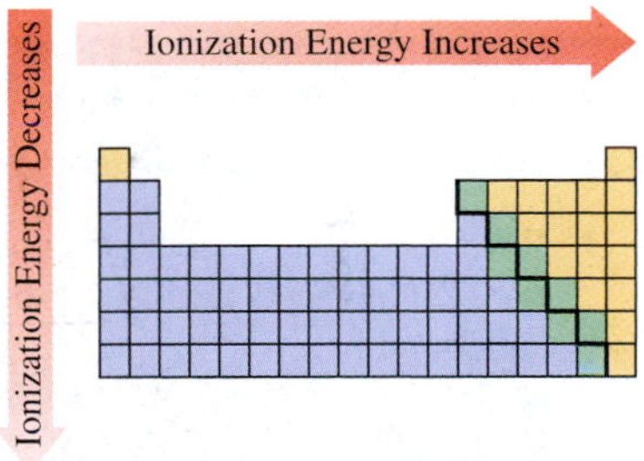
Ionization Energy Increases
Ionization Energy Decreases

Ionization energy decreases going down a group and increases going from left to right across a period.

Metallic Character

An element that has **metallic character** is an element that loses valence electrons easily. Metallic character is more prevalent in the elements on the left side of the periodic table (metals) and decreases going from left to right across a period. The elements on the right side of the periodic table (nonmetals) do not easily lose electrons, which means they are less metallic. Most of the metalloids between the metals and nonmetals tend to lose electrons,

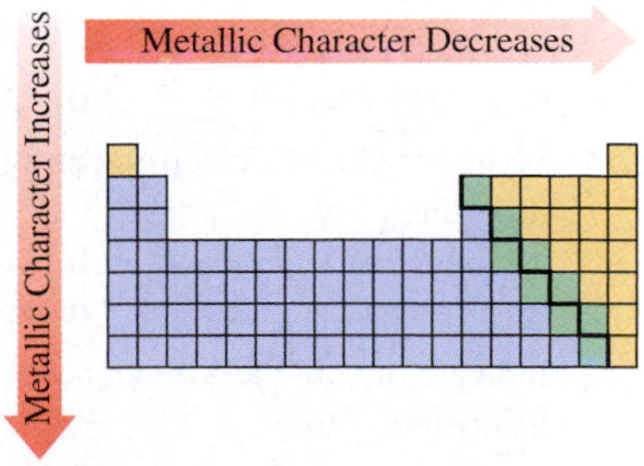
Metallic Character Decreases
Metallic Character Increases

Metallic character increases going down a group and decreases going from left to right across a period.

ENGAGE

Why is magnesium more metallic than aluminum?

but not as easily as the metals. Thus, in Period 3, sodium, which loses electrons most easily, would be the most metallic. Going across from left to right in Period 3, metallic character decreases to argon, which has the least metallic character.

For elements in the same group of representative elements, metallic character increases going from top to bottom. Atoms at the bottom of any group have more electron levels, which makes it easier to lose electrons. Thus, the elements at the bottom of a group on the periodic table have lower ionization energy and are more metallic compared to the elements at the top.

A summary of the trends in periodic properties we have discussed is given in **TABLE 4.12**.

TABLE 4.12 Summary of Trends in Periodic Properties of Representative Elements

Periodic Property	Top to Bottom within a Group	Left to Right across a Period
Valence Electrons	Remains the same	Increases
Atomic Size	Increases because there is an increase in the number of energy levels	Decreases as the number of protons increases, which strengthens the attraction of the nucleus for the valence electrons, and pulls them closer to the nucleus
Ionization Energy	Decreases because the valence electrons are easier to remove when they are farther from the nucleus	Increases as the number of protons increases, which strengthens the attraction of the nucleus for the valence electrons, and more energy is needed to remove a valence electron
Metallic Character	Increases because the valence electrons are easier to remove when they are farther from the nucleus	Decreases as the number of protons increases, which strengthens the attraction of the nucleus for the valence electrons, and makes it more difficult to remove a valence electron

TEST

Try Practice Problems 4.79 to 4.94

PRACTICE PROBLEMS

4.8 Trends in Periodic Properties

4.71 Write the group number using both A/B and 1 to 18 notations for elements that have the following outer electron configuration:
a. $2s^2$ **b.** $3s^23p^3$
c. $4s^23d^5$ **d.** $5s^24d^{10}5p^4$

4.72 Write the group number using both A/B and 1 to 18 notations for elements that have the following outer electron configuration:
a. $4s^23d^{10}4p^5$ **b.** $4s^1$
c. $4s^23d^8$ **d.** $5s^24d^{10}5p^2$

4.73 Write the valence electron configuration for each of the following:
a. alkali metals **b.** Group 4A (14)
c. Group 7A (17) **d.** Group 5A (15)

4.74 Write the valence electron configuration for each of the following:
a. halogens **b.** Group 6A (16)
c. Group 13 **d.** alkaline earth metals

4.75 Indicate the number of valence electrons in each of the following:
a. aluminum **b.** Group 5A (15)
c. barium **d.** F, Cl, Br, and I

4.76 Indicate the number of valence electrons in each of the following:
a. Li, Na, K, Rb, and Cs **b.** Se
c. C, Si, Ge, Sn, and Pb **d.** Group 8A (18)

4.77 Write the group number and draw the Lewis symbol for each of the following elements:
a. sulfur **b.** nitrogen
c. calcium **d.** sodium
e. gallium

4.78 Write the group number and draw the Lewis symbol for each of the following elements:
a. carbon **b.** oxygen
c. argon **d.** lithium
e. chlorine

4.79 Select the larger atom in each pair.
a. Na or Cl **b.** Na or Rb
c. Na or Mg **d.** Rb or I

4.80 Select the larger atom in each pair.
a. S or Ar **b.** S or O
c. S or K **d.** S or Mg

4.81 Place the elements in each set in order of decreasing atomic size.
a. Al, Si, Mg **b.** Cl, I, Br
c. Sb, Sr, I **d.** P, Si, Na

4.82 Place the elements in each set in order of decreasing atomic size.
a. Cl, S, P **b.** Ge, Si, C
c. Ba, Ca, Sr **d.** S, O, Se

4.83 Select the element in each pair with the higher ionization energy.
a. Br or I **b.** Mg or Sr
c. Si or P **d.** I or Xe

4.84 Select the element in each pair with the higher ionization energy.
a. O or Ne **b.** K or Br
c. Ca or Ba **d.** N or Ne

4.85 Arrange each set of elements in order of increasing ionization energy.
a. F, Cl, Br **b.** Na, Cl, Al
c. Na, K, Cs **d.** As, Ca, Br

4.86 Arrange each set of elements in order of increasing ionization energy.
a. O, N, C **b.** S, P, Cl
c. P, As, N **d.** Al, Si, P

4.87 Place the following in order of decreasing metallic character:
Br, Ge, Ca, Ga

4.88 Place the following in order of increasing metallic character:
Na, P, Al, Ar

4.89 Fill in each of the following blanks using *higher* or *lower*, *more* or *less*: Sr has a ______________ ionization energy and is ______________ metallic than Sb.

4.90 Fill in each of the following blanks using *higher* or *lower*, *more* or *less*: N has a ________________ ionization energy and is ________________ metallic than As.

4.91 Complete each of the following statements **a** to **d** using **1**, **2**, or **3**:
1. decreases **2.** increases
3. remains the same

Going down Group 6A (16),
a. the ionization energy __________
b. the atomic size __________
c. the metallic character __________
d. the number of valence electrons __________

4.92 Complete each of the following statements **a** to **d** using **1**, **2**, or **3**:
1. decreases **2.** increases
3. remains the same

Going from left to right across Period 3,
a. the ionization energy __________
b. the atomic size __________
c. the metallic character __________
d. the number of valence electrons __________

4.93 Which statements completed with **a** to **e** will be *True* and which will be *False*?

An atom of N compared to an atom of Li has a larger (greater)
a. atomic size **b.** ionization energy
c. number of protons **d.** metallic character
e. number of valence electrons

4.94 Which statements completed with **a** to **e** will be *True* and which will be *False*?

An atom of C compared to an atom of Sn has a larger (greater)
a. atomic size **b.** ionization energy
c. number of protons **d.** metallic character
e. number of valence electrons

CLINICAL UPDATE Improving Crop Production

Plants need potassium (K) for many metabolic processes, including the regulation of growth, protein synthesis, photosynthesis, and ionic balance. Potassium-deficient potato plants show purple or brown spots and reduced plant, root, and seed growth John notices that the leaves of his last crop of potatoes have brown spots, the potatoes are undersized, and the crop yield is low.

Tests on soil samples show that the potassium levels are below 100 ppm, indicating that the potato plants need supplemental potassium. John applies a fertilizer containing potassium chloride (KCl). To obtain the correct amount of potassium, John calculates that he needs to apply 170 kg of fertilizer per hectare.

Clinical Applications

4.95 **a.** What is the group number and name of the group that contains potassium?
b. Is potassium a metal, a nonmetal, or a metalloid?

A mineral deficiency of potassium causes brown spots on potato leaves.

c. How many protons are in an atom of potassium?
d. Potassium has three naturally occurring isotopes. They are K-39, K-40, and K-41. Using the atomic mass of potassium, determine which isotope of potassium is the most abundant.

4.96 **a.** How many neutrons are in K-41?
b. Write the electron configuration for potassium.
c. Which is the larger atom, K or Cs?
d. Which is the smallest atom, K, As, or Br?
e. If John's potato field has an area of 34.5 hectares, how many pounds of fertilizer does John need to use?

CONCEPT MAP

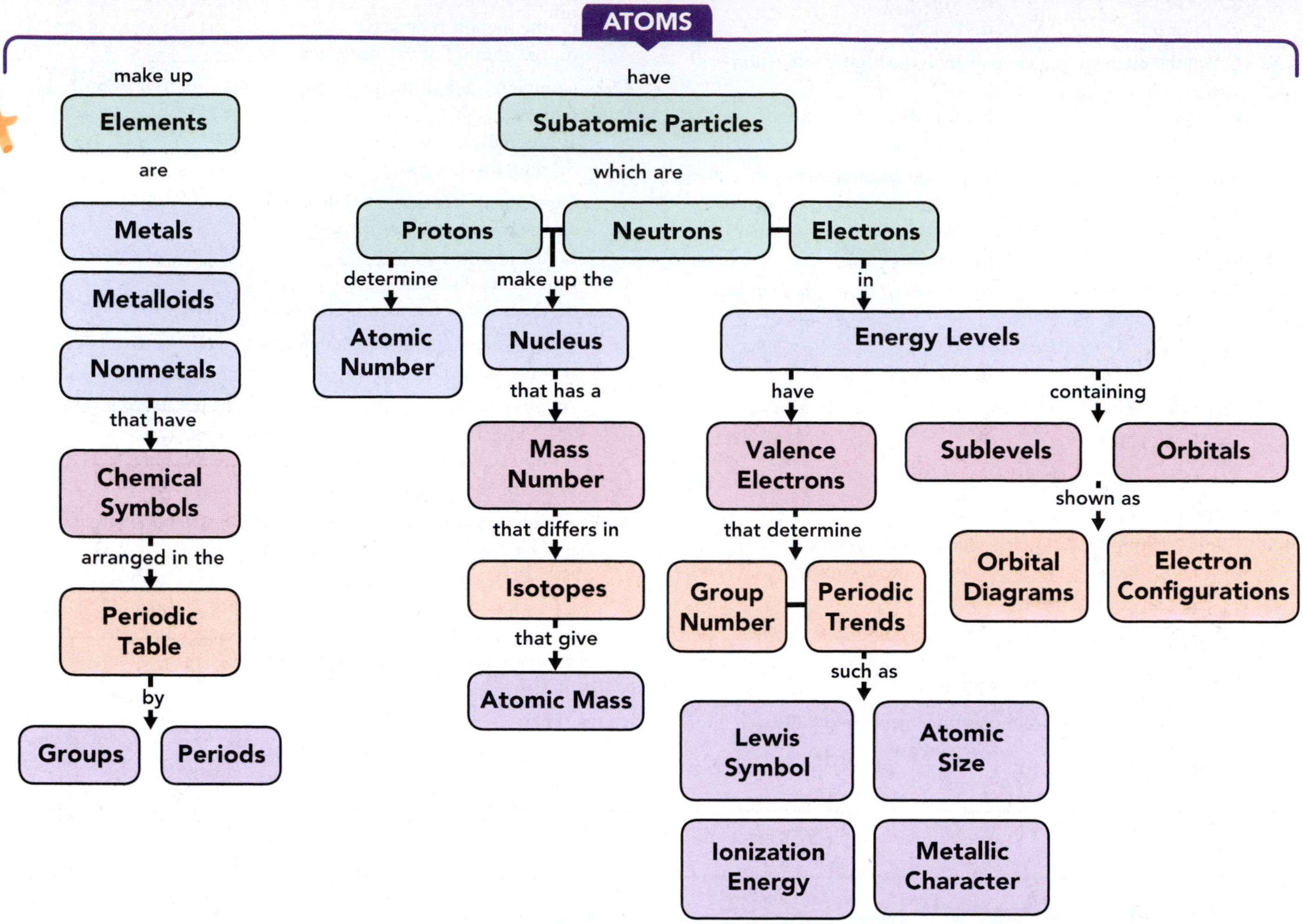
ATOMS
make up
Elements
are
Metals
Metalloids
Nonmetals
that have
Chemical Symbols
arranged in the
Periodic Table
by
Groups
Periods
have
Subatomic Particles
which are
Protons
Neutrons
Electrons
determine
Atomic Number
make up the
Nucleus
that has a
Mass Number
that differs in
Isotopes
that give
Atomic Mass
in
Energy Levels
have
Valence Electrons
that determine
Group Number
Periodic Trends
such as
Lewis Symbol
Atomic Size
Ionization Energy
Metallic Character
containing
Sublevels
Orbitals
shown as
Orbital Diagrams
Electron Configurations

CHAPTER REVIEW

4.1 Elements and Symbols

LEARNING GOAL Given the name of an element, write its correct symbol; from the symbol, write the correct name.

- Elements are the primary substances of matter.
- Chemical symbols are one- or two-letter abbreviations of the names of the elements.

4.2 The Periodic Table

LEARNING GOAL Use the periodic table to identify the group and the period of an element; identify an element as a metal, a nonmetal, or a metalloid.

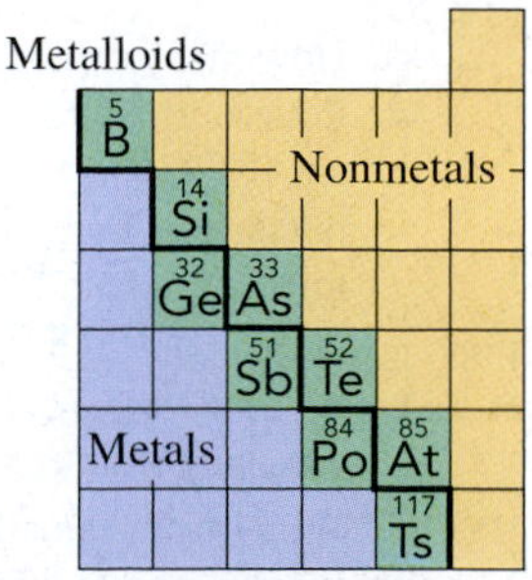
Metalloids
5 B
14 Si
Nonmetals
32 Ge
33 As
51 Sb
52 Te
Metals
84 Po
85 At
117 Ts

- The periodic table is an arrangement of the elements by increasing atomic number.
- A horizontal row is called a *period*.
- A vertical column on the periodic table containing elements with similar properties is called a *group*.
- Elements in Group 1A (1) are called the *alkali metals*; Group 2A (2), the *alkaline earth metals*; Group 7A (17), the *halogens*; and Group 8A (18), the *noble gases*.
- On the periodic table, *metals* are located on the left of the heavy zigzag line, and *nonmetals* are to the right of the heavy zigzag line.
- Except for aluminum and tennessine, elements located along the heavy zigzag line are called *metalloids*.

4.3 The Atom

LEARNING GOAL Describe the electrical charge of a proton, neutron, and electron and identify their location within an atom.

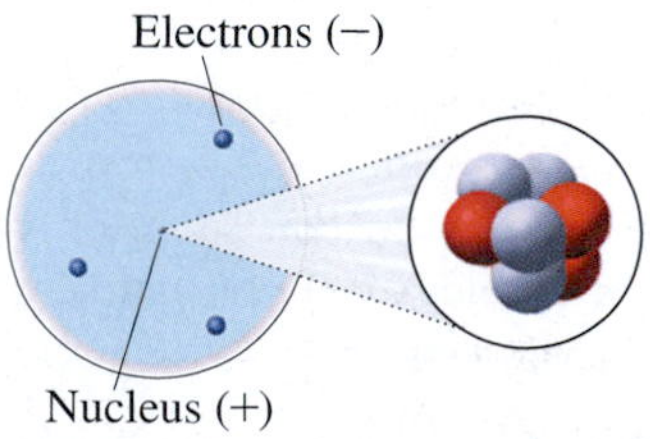
Electrons (−)
Nucleus (+)

- An atom is the smallest particle that retains the characteristics of an element.
- Atoms are composed of three types of subatomic particles.
- Protons have a positive charge (+), electrons carry a negative charge (−), and neutrons are electrically neutral.
- The protons and neutrons are found in the tiny, dense nucleus; electrons are located outside the nucleus.

4.4 Atomic Number and Mass Number

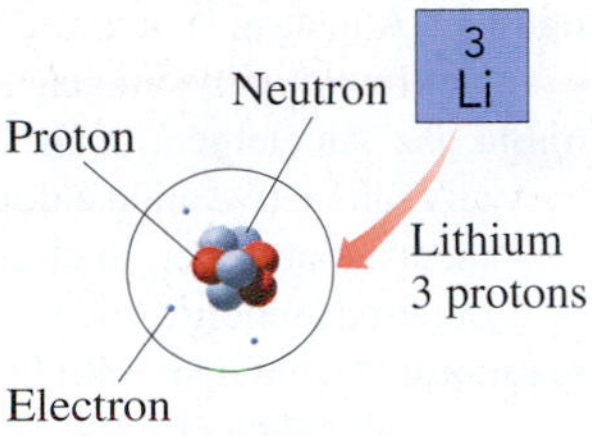
3
Li
Neutron
Proton
Lithium
3 protons
Electron

LEARNING GOAL Given the atomic number and the mass number of an atom, state the number of protons, neutrons, and electrons.

- The atomic number gives the number of protons in all the atoms of the same element.
- In a neutral atom, the number of protons and electrons is equal.
- The mass number is the total number of protons and neutrons in an atom.

4.5 Isotopes and Atomic Mass

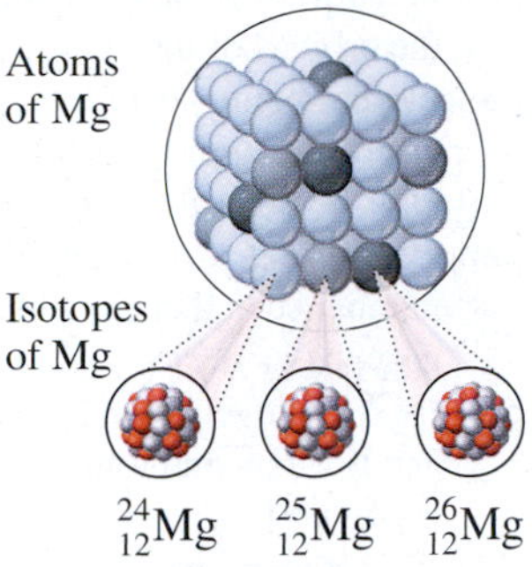
Atoms
of Mg
Isotopes
of Mg
$^{24}_{12}Mg$
$^{25}_{12}Mg$
$^{26}_{12}Mg$

LEARNING GOAL Determine the number of protons, neutrons, and electrons in one or more of the isotopes of an element; calculate the atomic mass of an element using the percent abundance and mass of its naturally occurring isotopes.

- Atoms that have the same number of protons but different numbers of neutrons are called *isotopes*.
- The atomic mass of an element is the weighted average mass of all the isotopes in a naturally occurring sample of that element.

4.6 Electron Energy Levels

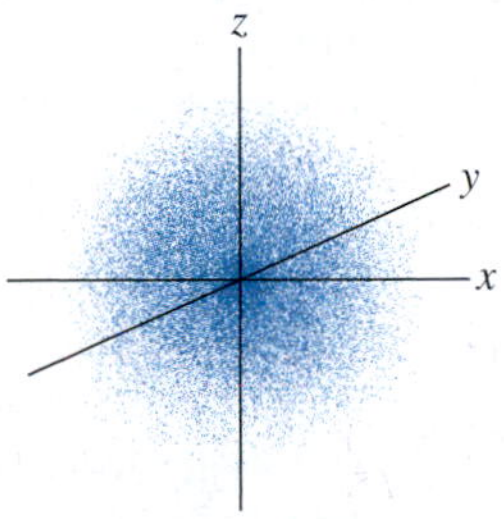
z
y
x

LEARNING GOAL Describe the energy levels, sublevels, and orbitals for the electrons in an atom.

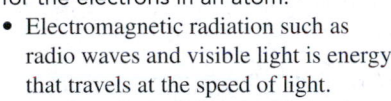

- Electromagnetic radiation such as radio waves and visible light is energy that travels at the speed of light.
- Each particular type of radiation has a specific wavelength and frequency.
- Long-wavelength radiation has low frequencies, whereas short-wavelength radiation has high frequencies.
- Radiation with a high frequency has high energy.
- The atomic spectra of elements are related to the specific energy levels occupied by electrons.
- When an electron absorbs energy, it attains a higher energy level. When an electron drops to a lower energy level, energy is emitted.
- An orbital is a region around the nucleus where an electron with a specific energy is most likely to be found.
- Each orbital holds a maximum of two electrons, which must have opposite spins.
- In each energy level (n), electrons occupy orbitals of identical energy within sublevels.
- An s sublevel contains one s orbital, a p sublevel contains three p orbitals, a d sublevel contains five d orbitals, and an f sublevel contains seven f orbitals.

4.7 Electron Configurations

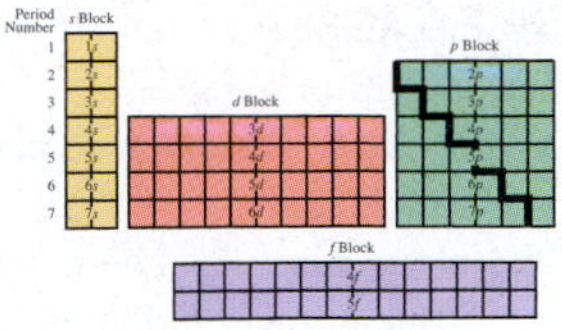
Period Number
s Block
p Block
d Block
f Block
1
2
3
4
5
6
7

LEARNING GOAL Draw the orbital diagram, and write the electron configuration for an element.

- Within a sublevel, electrons enter orbitals in the same energy level one at a time until all the orbitals are half-filled.
- Additional electrons enter with opposite spins until the orbitals in that sublevel are filled with two electrons each.
- The orbital diagram for an element such as silicon shows the orbitals that are occupied by paired and unpaired electrons:

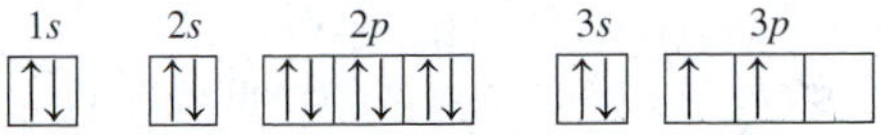
1s
2s
2p
3s
3p

- The electron configuration for an element such as silicon shows the number of electrons in each sublevel: $1s^22s^22p^63s^23p^2$.
- An abbreviated electron configuration for an element such as silicon places the symbol of a noble gas in brackets to represent the filled sublevels: $[Ne]3s^23p^2$.
- The periodic table consists of s, p, d, and f sublevel blocks.
- An electron configuration can be written following the order of the sublevel blocks on the periodic table.
- Beginning with $1s$, an electron configuration is obtained by writing the sublevel blocks in order going across each period on the periodic table until the element is reached.

4.8 Trends in Periodic Properties

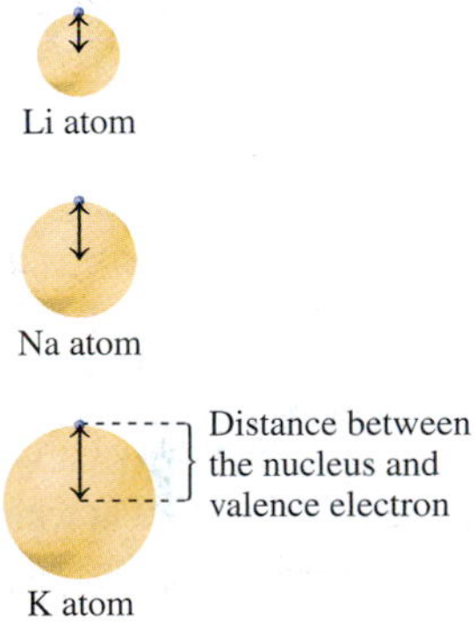
Li atom
Na atom
K atom
Distance between the nucleus and valence electron

LEARNING GOAL Use the electron configurations of elements to explain the trends in periodic properties.

- The properties of elements are related to the valence electrons of the atoms.
- With only a few exceptions, each group of elements has the same arrangement of valence electrons differing only in the energy level.
- The Lewis symbol of an atom shows the valence electrons, represented as dots, around the symbol of the element.
- The size of an atom increases going down a group and decreases going from left to right across a period.
- The energy required to remove a valence electron is the ionization energy, which decreases going down a group, and generally increases going from left to right across a period.
- The metallic character of an element increases going down a group and decreases going from left to right across a period.

KEY TERMS

alkali metal An element in Group 1A (1), except hydrogen, that is a soft, shiny metal with one electron in its outermost energy level.

alkaline earth metal An element in Group 2A (2) that has two electrons in its outermost energy level.

atom The smallest particle of an element that retains the characteristics of the element.

atomic mass The weighted average mass of all the naturally occurring isotopes of an element.

atomic mass unit (amu) A small mass unit used to describe the mass of extremely small particles such as atoms and subatomic particles; 1 amu is equal to one-twelfth the mass of a $^{12}_{6}C$ atom.

atomic number A number that is equal to the number of protons in an atom.

atomic size The distance between the outermost electrons and the nucleus.

atomic symbol An abbreviation used to indicate the mass number and atomic number of an isotope.

chemical symbol An abbreviation that represents the name of an element.

***d* block** The block of ten elements from Groups 3B (3) to 2B (12) in which electrons fill the five *d* orbitals in the *d* sublevels.

electron A negatively charged subatomic particle having a minute mass that is usually ignored in mass calculations; its symbol is e^-.

electron configuration A list of the number of electrons in each sublevel within an atom, arranged by increasing energy.

energy level A group of electrons with similar energy.

***f* block** The block of 14 elements in the rows at the bottom of the periodic table in which electrons fill the seven *f* orbitals in the *f* sublevels.

group A vertical column in the periodic table that contains elements having similar physical and chemical properties.

group number A number that appears at the top of each vertical column (group) in the periodic table and indicates the number of electrons in the outermost energy level.

halogen An element in Group 7A (17)—fluorine, chlorine, bromine, iodine, astatine, and tennessine—that has seven electrons in its outermost energy level.

ionization energy The energy needed to remove the least tightly bound electron from the outermost energy level of an atom.

isotope An atom that differs only in mass number from another atom of the same element. Isotopes have the same atomic number (number of protons), but different numbers of neutrons.

Lewis symbol The representation of an atom that shows valence electrons as dots around the symbol of the element.

mass number The total number of protons and neutrons in the nucleus of an atom.

metal An element that is shiny, malleable, ductile, and a good conductor of heat and electricity. The metals are located to the left of the heavy zigzag line on the periodic table.

metallic character A measure of how easily an element loses a valence electron.

metalloid An element with properties of both metals and nonmetals located along the heavy zigzag line on the periodic table.

neutron A neutral subatomic particle having a mass of about 1 amu and found in the nucleus of an atom; its symbol is n or n^0.

noble gas An element in Group 8A (18) of the periodic table, generally unreactive and seldom found in combination with other elements, that has eight electrons (helium has two electrons) in its outermost energy level.

nonmetal An element with little or no luster that is a poor conductor of heat and electricity. The nonmetals are located to the right of the heavy zigzag line on the periodic table.

nucleus The compact, extremely dense center of an atom, containing the protons and neutrons of the atom.

orbital The region around the nucleus where electrons of a certain energy are more likely to be found. The *s* orbitals are spherical; the *p* orbitals have two lobes.

orbital diagram A diagram that shows the distribution of electrons in the orbitals of the energy levels.

***p* block** The elements in Groups 3A (13) to 8A (18) in which electrons fill the *p* orbitals in the *p* sublevels.

period A horizontal row of elements in the periodic table.

periodic table An arrangement of elements by increasing atomic number such that elements having similar chemical behavior are grouped in vertical columns.

proton A positively charged subatomic particle having a mass of about 1 amu and found in the nucleus of an atom; its symbol is p or p^+.

representative element An element in the first two columns on the left of the periodic table and the last six columns on the right that has a group number of 1A through 8A or 1, 2, and 13 through 18.

***s* block** The elements in Groups 1A (1) and 2A (2) in which electrons fill the *s* orbitals.

subatomic particle A particle within an atom; protons, neutrons, and electrons are subatomic particles.

sublevel A group of orbitals of equal energy within an energy level. The number of sublevels in each energy level is the same as the principal quantum number (n).

transition element An element in the center of the periodic table that is designated with the letter "B" or the group number of 3 through 12.

valence electrons Electrons in the outermost energy level of an atom.

CORE CHEMISTRY SKILLS

The chapter Section containing each Core Chemistry Skill is shown in parentheses at the end of each heading.

Counting Protons and Neutrons (4.4)

- The atomic number of an element is equal to the number of protons in every atom of that element. The atomic number is the whole number that appears above the symbol of each element on the periodic table.

  Atomic number = number of protons in an atom

- Because atoms are neutral, the number of electrons is equal to the number of protons. Thus, the atomic number gives the number of electrons.
- The mass number is the total number of protons and neutrons in the nucleus of an atom.

  Mass number = number of protons + number of neutrons

- The number of neutrons is calculated from the mass number and atomic number.

  Number of neutrons = mass number − number of protons

Example: Calculate the number of protons, neutrons, and electrons in a krypton atom with a mass number of 80.

Answer:

Element	Atomic Number	Mass Number	Number of Protons	Number of Neutrons	Number of Electrons
Kr	36	80	equal to atomic number 36	equal to mass number − number of protons 80 − 36 = 44	equal to number of protons 36

Writing Atomic Symbols for Isotopes (4.5)

- Isotopes are atoms of the same element that have the same atomic number but different numbers of neutrons.
- An atomic symbol is written for a particular isotope, with its mass number (protons and neutrons) shown in the upper left corner and its atomic number (protons) shown in the lower left corner.

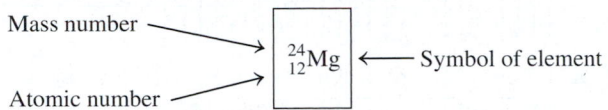

Mass number → $^{24}_{12}Mg$ ← Symbol of element

Atomic number →

Example: Calculate the number of protons and neutrons in the cadmium isotope $^{112}_{48}Cd$.

Answer:

Atomic Symbol	Atomic Number	Mass Number	Number of Protons	Number of Neutrons
$^{112}_{48}Cd$	number in lower left corner 48	number in upper left corner 112	equal to atomic number 48	equal to mass number − number of protons 112 − 48 = 64

Writing Electron Configurations (4.7)

- The electron configuration for an atom specifies the energy levels and sublevels occupied by the electrons of an atom.
- An electron configuration is written starting with the lowest energy sublevel, followed by the next lowest energy sublevel.
- The number of electrons in each sublevel is shown as a superscript.

Example: Write the electron configuration for palladium.

Answer: Palladium has atomic number 46, which means it has 46 protons and 46 electrons.

$1s^22s^22p^63s^23p^64s^23d^{10}4p^65s^24d^8$

Using the Periodic Table to Write Electron Configurations (4.7)

- An electron configuration corresponds to the location of an element on the periodic table, where different blocks within the periodic table are identified as the *s*, *p*, *d*, and *f* sublevels.
- The number of valence electrons remains the same going down a group and increases going from left to right across a period.

Example: Use the periodic table to write the electron configuration for sulfur.

Answer: Sulfur (atomic number 16) is in Group 6A (16) and Period 3.

Period	Sublevel Block Filling	Sublevel Block Notation Filled
1	$1s$ (H → He)	$1s^2$
2	$2s$ (Li → Be) and $2p$ (B → Ne)	$2s^22p^6$
3	$3s$ (Na → Mg)	$3s^2$
	$3p$ (Al → S)	$3p^4$

The electron configuration for sulfur (S) is: $1s^22s^22p^63s^23p^4$.

Identifying Trends in Periodic Properties (4.8)

- The size of an atom increases going down a group and decreases going from left to right across a period.
- The ionization energy decreases going down a group and increases going from left to right across a period.
- The metallic character of an element increases going down a group and decreases going from left to right across a period.

Example: For Mg, P, and Cl, identify which has the

a. largest atomic size
b. highest ionization energy
c. most metallic character

Answer: **a.** Mg **b.** Cl **c.** Mg

Drawing Lewis Symbols (4.8)

- The valence electrons are the electrons in the outermost energy level.
- The number of valence electrons is the same as the group number for the representative elements.
- A Lewis symbol represents the number of valence electrons shown as dots placed around the symbol for the element.

Example: Give the group number and number of valence electrons, and draw the Lewis symbol for each of the following:

a. Rb **b.** Se **c.** Xe

Answer: **a.** Group 1A (1), one valence electron, Rb·
b. Group 6A (16), six valence electrons, ·Se: (with dots)
c. Group 8A (18), eight valence electrons, :Xe: (with dots)

UNDERSTANDING THE CONCEPTS

The chapter Sections to review are shown in parentheses at the end of each problem.

4.97 According to Dalton's atomic theory, which of the following are *True* or *False*? If *False*, correct the statement to make it *True*. (4.3)
a. Atoms of an element are identical to atoms of other elements.
b. Every element is made of atoms.
c. Atoms of different elements combine to form compounds.
d. In a chemical reaction, some atoms disappear and new atoms appear.

4.98 Use Rutherford's gold-foil experiment to answer each of the following: (4.3)
a. What did Rutherford expect to happen when he aimed particles at the gold foil?
b. How did the results differ from what he expected?
c. How did he use the results to propose a model of the atom?

4.99 Match the subatomic particles (**1** to **3**) to each of the descriptions below. (4.4)
1. protons **2.** neutrons **3.** electrons
a. atomic mass
b. atomic number
c. positive charge
d. negative charge
e. mass number − atomic number

4.100 Match the subatomic particles (**1** to **3**) to each of the descriptions below. (4.4)
1. protons **2.** neutrons **3.** electrons
a. mass number
b. surround the nucleus
c. in the nucleus
d. charge of 0
e. equal to number of electrons

4.101 Consider the following atoms in which X represents the chemical symbol of the element: (4.4, 4.5)

$^{16}_{8}X$ $^{16}_{9}X$ $^{18}_{10}X$ $^{17}_{8}X$ $^{18}_{8}X$

a. What atoms have the same number of protons?
b. Which atoms are isotopes? Of what element?
c. Which atoms have the same mass number?

4.102 Consider the following atoms in which X represents the chemical symbol of the element: (4.4, 4.5)

$^{124}_{47}X$ $^{116}_{49}X$ $^{116}_{50}X$ $^{124}_{50}X$ $^{116}_{48}X$

a. What atoms have the same number of protons?
b. Which atoms are isotopes? Of what element?
c. Which atoms have the same mass number?

4.103 Complete the following table for three of the naturally occurring isotopes of germanium, which is a metalloid used in semiconductors: (4.4, 4.5)

	Atomic Symbol		
	$^{70}_{32}Ge$	$^{73}_{32}Ge$	$^{76}_{32}Ge$
Atomic Number			
Mass Number			
Number of Protons			
Number of Neutrons			
Number of Electrons			

4.104 Complete the following table for the three naturally occurring isotopes of silicon, the major component in computer chips: (4.4, 4.5)

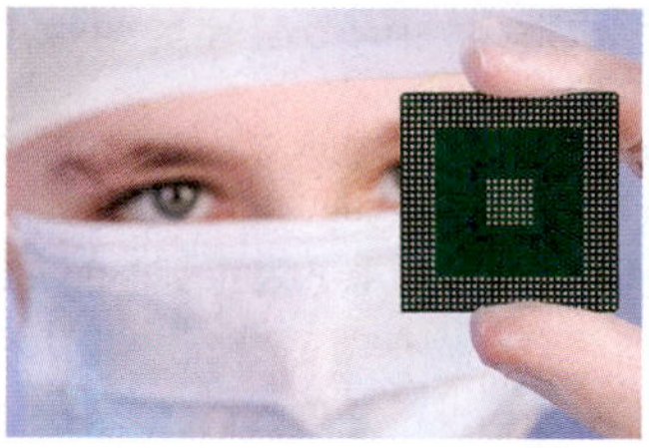

Computer chips consist primarily of the element silicon.

	Atomic Symbol		
	$^{28}_{14}Si$	$^{29}_{14}Si$	$^{30}_{14}Si$
Atomic Number			
Mass Number			
Number of Protons			
Number of Neutrons			
Number of Electrons			

4.105 For each representation of a nucleus **A** through **E**, write the atomic symbol, and identify which are isotopes. (4.4, 4.5)

Proton ●
Neutron ●

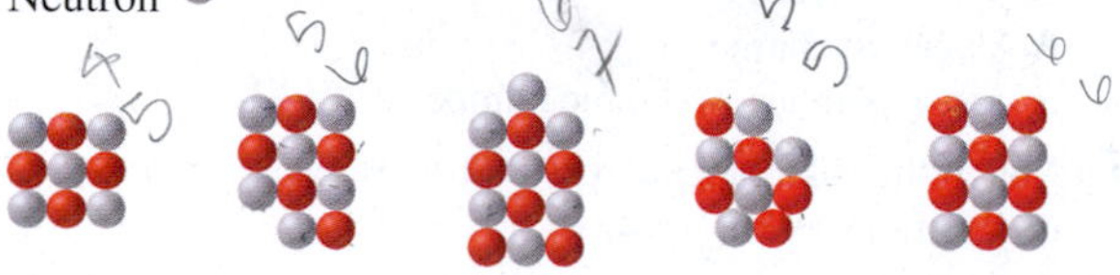

A B C D E

4.106 Identify the element represented by each nucleus **A** through **E** in problem 4.105 as a metal, a nonmetal, or a metalloid. (4.3)

4.107 Indicate whether or not the following orbital diagrams are possible and explain. When possible, indicate the element it represents. (4.7)

a.
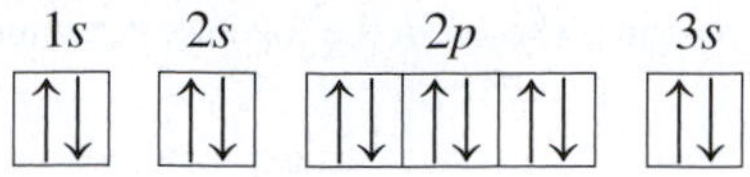
1s
2s
2p
3s

b.
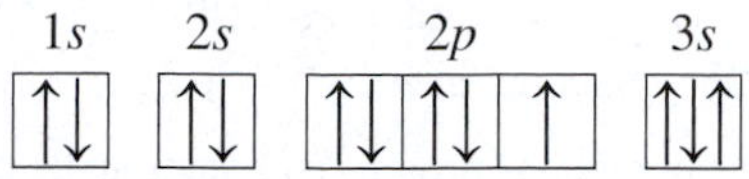
1s
2s
2p
3s

4.108 Indicate whether or not the following abbreviated orbital diagrams are possible and explain. When possible, indicate the element it represents. (4.7)

a.
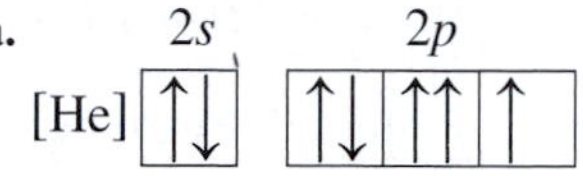
2s
2p
[He]

b.
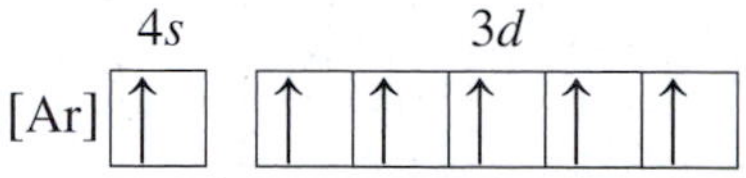
4s
3d
[Ar]

4.109 Match the spheres **A** through **D** with atoms of Li, Na, K, and Rb. (4.8)

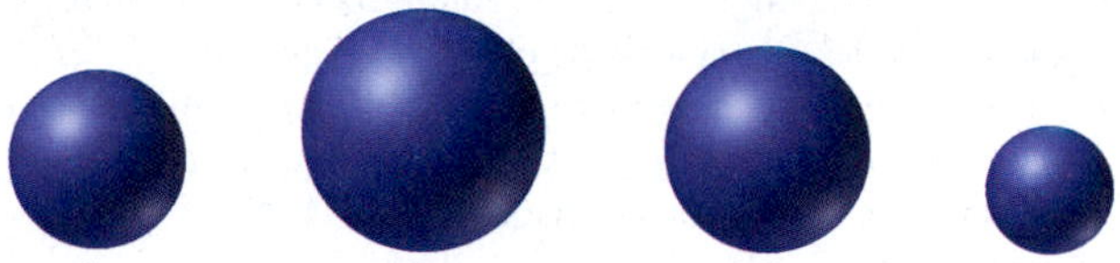

A B C D

4.110 Match the spheres **A** through **D** with atoms of K, Ge, Ca, and Kr. (4.8)

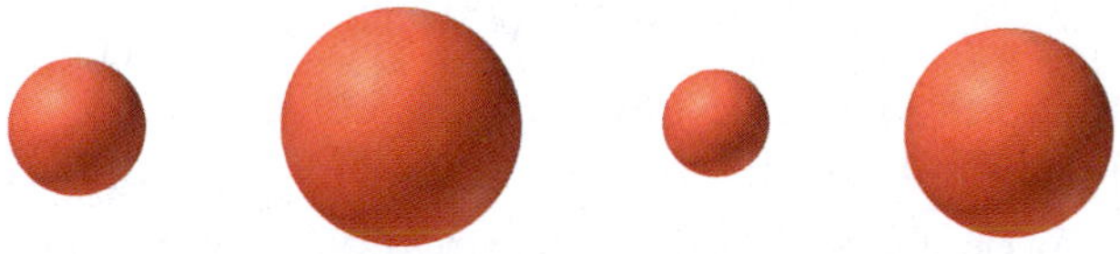

A B C D

4.111 Of the elements Na, Mg, Si, S, Cl, and Ar, identify one that fits each of the following: (4.2, 4.6, 4.7, 4.8)

a. largest atomic size
b. a halogen
c. electron configuration $1s^22s^22p^63s^23p^2$
d. highest ionization energy
e. in Group 6A (16)
f. most metallic character
g. two valence electrons

4.112 Of the elements Sn, Xe, Te, Sr, I, and Rb, identify one that fits each of the following: (4.2, 4.6, 4.7, 4.8)

a. smallest atomic size
b. an alkaline earth metal
c. a metalloid
d. lowest ionization energy
e. in Group 4A (14)
f. least metallic character
g. seven valence electrons

ADDITIONAL PRACTICE PROBLEMS

4.113 Give the group and period number for each of the following elements: (4.2)
a. bromine **b.** argon **c.** lithium **d.** radium

4.114 Give the group and period number for each of the following elements: (4.2)
a. radon **b.** tin **c.** carbon **d.** magnesium

4.115 The following trace elements have been found to be crucial to the functions of the body. Indicate each as a metal, a nonmetal, or a metalloid. (4.2)
a. zinc **b.** cobalt **c.** manganese **d.** iodine

4.116 The following trace elements have been found to be crucial to the functions of the body. Indicate each as a metal, a nonmetal, or a metalloid. (4.2)
a. copper **b.** selenium **c.** arsenic **d.** chromium

4.117 Indicate if each of the following statements is *true* or *false*: (4.3)
a. The proton is a negatively charged particle.
b. The neutron is 2000 times as heavy as a proton.
c. The atomic mass unit is based on a carbon atom with six protons and six neutrons.
d. The nucleus is the largest part of the atom.
e. The electrons are located outside the nucleus.

4.118 Indicate if each of the following statements is *true* or *false*: (4.3)
a. The neutron is electrically neutral.
b. Most of the mass of an atom is because of the protons and neutrons.
c. The charge of an electron is equal, but opposite, to the charge of a neutron.
d. The proton and the electron have about the same mass.
e. The mass number is the number of protons.

4.119 For the following atoms, give the number of protons, neutrons, and electrons: (4.3)
a. ${}^{114}_{48}Cd$ **b.** ${}^{98}_{43}Tc$ **c.** ${}^{199}_{79}Au$
d. ${}^{222}_{86}Rn$ **e.** ${}^{136}_{54}Xe$

4.120 For the following atoms, give the number of protons, neutrons, and electrons: (4.3)
a. ${}^{202}_{80}Hg$ **b.** ${}^{127}_{53}I$ **c.** ${}^{75}_{35}Br$
d. ${}^{133}_{55}Cs$ **e.** ${}^{195}_{78}Pt$

4.121 Complete the following table: (4.3)

Name of the Element	Atomic Symbol	Number of Protons	Number of Neutrons	Number of Electrons
	${}^{80}_{34}Se$			
		28	34	
Magnesium			14	
	${}^{228}_{88}Ra$			

4.122 Complete the following table: (4.3)

Name of the Element	Atomic Symbol	Number of Protons	Number of Neutrons	Number of Electrons
Potassium			22	
	${}^{51}_{23}V$			
		48	64	
Barium			82	

4.123 **a.** What electron sublevel starts to fill after completion of the 3*s* sublevel? (4.7)
b. What electron sublevel starts to fill after completion of the 4*p* sublevel?
c. What electron sublevel starts to fill after completion of the 3*d* sublevel?
d. What electron sublevel starts to fill after completion of the 3*p* sublevel?

4.124 **a.** What electron sublevel starts to fill after completion of the 5*s* sublevel? (4.7)
b. What electron sublevel starts to fill after completion of the 4*d* sublevel?
c. What electron sublevel starts to fill after completion of the 4*f* sublevel?
d. What electron sublevel starts to fill after completion of the 5*p* sublevel?

4.125 **a.** How many 3*d* electrons are in Fe? (4.7)
b. How many 5*p* electrons are in Ba?
c. How many 4*d* electrons are in I?
d. How many 7*s* electrons are in Ra?

4.126 **a.** How many 4*d* electrons are in Cd? (4.7)
b. How many 4*p* electrons are in Br?
c. How many 6*p* electrons are in Bi?
d. How many 5*s* electrons are in Zn?

4.127 Name the element that corresponds to each of the following: (4.7, 4.8)
a. $1s^22s^22p^63s^23p^3$
b. alkali metal with the smallest atomic size
c. $[Kr]5s^24d^{10}$
d. Group 5A (15) element with the highest ionization energy
e. Period 3 element with the largest atomic size

4.128 Name the element that corresponds to each of the following: (4.7, 4.8)
a. $1s^22s^22p^63s^23p^64s^13d^5$
b. $[Xe]6s^24f^{14}5d^{10}6p^5$
c. halogen with the highest ionization energy
d. Group 2A (2) element with the smallest ionization energy
e. Period 4 element with the smallest atomic size

4.129 Of the elements Na, P, Cl, and F, which (4.2, 4.7, 4.8)
a. is a metal?
b. is in Group 5A (15)?
c. has the highest ionization energy?
d. loses an electron most easily?
e. is found in Group 7A (17), Period 3?

4.130 Of the elements K, Ca, Br, and Kr, which (4.2, 4.7, 4.8)
a. is a halogen?
b. has the smallest atomic size?
c. has the lowest ionization energy?
d. requires the most energy to remove an electron?
e. is found in Group 2A (2), Period 4?

CHALLENGE PROBLEMS

The following problems are related to the topics in this chapter. However, they do not all follow the chapter order, and they require you to combine concepts and skills from several Sections. These problems will help you increase your critical thinking skills and prepare for your next exam.

4.131 The most abundant isotope of lead is $^{208}_{82}Pb$. (4.4)
- **a.** How many protons, neutrons, and electrons are in $^{208}_{82}Pb$?
- **b.** What is the atomic symbol of another isotope of lead with 132 neutrons?
- **c.** What is the name and symbol of an atom with the same mass number as in part **b** and 131 neutrons?

4.132 The most abundant isotope of silver is $^{107}_{47}Ag$. (4.4)
- **a.** How many protons, neutrons, and electrons are in $^{107}_{47}Ag$?
- **b.** What is the symbol of another isotope of silver with 62 neutrons?
- **c.** What is the name and symbol of an atom with the same mass number as in part **b** and 61 neutrons?

4.133 Give the symbol of the element that has the (4.8)
- **a.** smallest atomic size in Group 6A (16)
- **b.** smallest atomic size in Period 3
- **c.** highest ionization energy in Group 4A (14)
- **d.** lowest ionization energy in Period 3
- **e.** most metallic character in Group 2A (2)

4.134 Give the symbol of the element that has the (4.8)
- **a.** largest atomic size in Group 1A (1)
- **b.** largest atomic size in Period 4
- **c.** highest ionization energy in Group 2A (2)
- **d.** lowest ionization energy in Group 7A (17)
- **e.** least metallic character in Group 4A (14)

4.135 Silicon has three naturally occurring isotopes: Si-28 that has a percent abundance of 92.23% and a mass of 27.977 amu, Si-29 that has a 4.68% abundance and a mass of 28.976 amu, and Si-30 that has a percent abundance of 3.09% and a mass of 29.974 amu. Calculate the atomic mass for silicon using the weighted average mass method. (4.5)

4.136 Antimony (Sb) has two naturally occurring isotopes: Sb-121 that has a percent abundance of 57.21% and a mass of 120.90 amu, and Sb-123 that has a percent abundance of 42.79% and a mass of 122.90 amu. Calculate the atomic mass for antimony using the weighted average mass method. (4.5)

4.137 Consider three elements with the following abbreviated electron configurations: (4.2, 4.7, 4.8)

$X = [Ar]4s^2$ $\quad$ $Y = [Ne]3s^23p^4$

$Z = [Ar]4s^23d^{10}4p^4$

- **a.** Identify each element as a metal, a nonmetal, or a metalloid.
- **b.** Which element has the largest atomic size?
- **c.** Which elements have similar properties?
- **d.** Which element has the highest ionization energy?
- **e.** Which element has the smallest atomic size?

4.138 Consider three elements with the following abbreviated electron configurations: (4.2, 4.7, 4.8)

$X = [Ar]4s^23d^5$ $\quad$ $Y = [Ar]4s^23d^{10}4p^1$

$Z = [Ne]3s^23p^2$

- **a.** Identify each element as a metal, a nonmetal, or a metalloid.
- **b.** Which element has the smallest atomic size?
- **c.** Which elements have similar properties?
- **d.** Which element has the highest ionization energy?
- **e.** Which element has a half-filled sublevel?

ANSWERS

4.1 **a.** Cu **b.** Pt **c.** Ca **d.** Mn **e.** Fe **f.** Ba **g.** Pb **h.** Sr

4.3 **a.** carbon **b.** chlorine **c.** iodine **d.** selenium **e.** nitrogen **f.** sulfur **g.** zinc **h.** cobalt

4.5 **a.** sodium, chlorine
b. calcium, sulfur, oxygen
c. carbon, hydrogen, chlorine, nitrogen, oxygen
d. lithium, carbon, oxygen

4.7 **a.** Period 2 **b.** Group 8A (18) **c.** Group 1A (1) **d.** Period 2

4.9 **a.** C **b.** He **c.** Na **d.** Ca **e.** Al

4.11 **a.** metal **b.** nonmetal **c.** metal **d.** nonmetal **e.** nonmetal **f.** nonmetal **g.** metalloid **h.** metal

4.13 **a.** needed for bones and teeth, muscle contraction, nerve impulses; alkaline earth metal
b. component of hemoglobin; transition element
c. muscle contraction, nerve impulses; alkali metal
d. found in fluids outside cells; halogen

4.15 **a.** A macromineral is an element essential to health, which is present in the body in amounts from 5 to 1000 g.
b. Sulfur is a component of proteins, vitamin B_1, and insulin.
c. 86 g

4.17 **a.** electron **b.** proton **c.** electron **d.** neutron

4.19 Rutherford determined that an atom contains a small, compact nucleus that is positively charged.

4.21 **a.** true **b.** true **c.** true **d.** false A proton is attracted to an electron.

4.23 In the process of brushing hair, strands of hair become charged with like charges that repel each other.

4.25 **a.** atomic number **b.** both **c.** mass number **d.** atomic number

4.27 **a.** lithium, Li **b.** fluorine, F **c.** calcium, Ca **d.** zinc, Zn **e.** neon, Ne **f.** silicon, Si **g.** iodine, I **h.** oxygen, O

4.29 **a.** 18 protons and 18 electrons
b. 25 protons and 25 electrons
c. 53 protons and 53 electrons
d. 48 protons and 48 electrons

4.31

Name of the Element	Symbol	Atomic Number	Mass Number	Number of Protons	Number of Neutrons	Number of Electrons
Zinc	Zn	30	66	30	36	30
Magnesium	Mg	12	24	12	12	12
Potassium	K	19	39	19	20	19
Sulfur	S	16	31	16	15	16
Iron	Fe	26	56	26	30	26

4.33 **a.** 38 protons, 51 neutrons, 38 electrons
b. 24 protons, 28 neutrons, 24 electrons
c. 16 protons, 18 neutrons, 16 electrons
d. 35 protons, 46 neutrons, 35 electrons

4.35 **a.** $^{31}_{15}P$ **b.** $^{80}_{35}Br$ **c.** $^{122}_{50}Sn$
d. $^{35}_{17}Cl$ **e.** $^{202}_{80}Hg$

4.37 **a.** $^{36}_{18}Ar$ $^{38}_{18}Ar$ $^{40}_{18}Ar$
b. They all have the same number of protons and electrons.
c. They have different numbers of neutrons, which gives them different mass numbers.
d. The atomic mass of Ar listed on the periodic table is the weighted average atomic mass of all the naturally occurring isotopes.
e. The isotope Ar-40 is most abundant because its mass is closest to the atomic mass of Ar on the periodic table.

4.39 Since the atomic mass of copper is closer to 63 amu, there are more atoms of $^{63}_{29}Cu$.

4.41 Since the atomic mass of thallium is 204.4 amu, the most abundant isotope is $^{205}_{81}Tl$.

4.43 69.72 amu

4.45 **a.** spherical **b.** two lobes **c.** spherical

4.47 **a.** **1** and **2** **b.** **3**
c. **1** and **2** **d.** **1**, **2**, and **3**

4.49 **a.** There are five orbitals in the 3*d* sublevel.
b. There is one sublevel in the $n = 1$ energy level.
c. There is one orbital in the 6*s* sublevel.
d. There are nine orbitals in the $n = 3$ energy level.

4.51 **a.** There is a maximum of two electrons in a 2*p* orbital.
b. There is a maximum of six electrons in the 3*p* sublevel.
c. There is a maximum of 32 electrons in the $n = 4$ energy level.
d. There is a maximum of 10 electrons in the 5*d* sublevel.

4.53 The electron configuration shows the number of electrons in each sublevel of an atom. The abbreviated electron configuration uses the symbol of the preceding noble gas to show completed sublevels.

4.55 **a.**
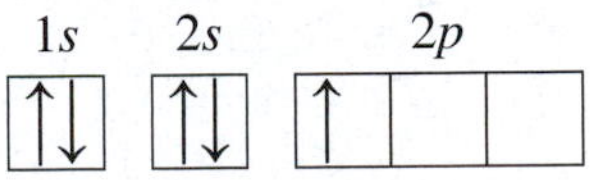
1s
2s
2p

b.
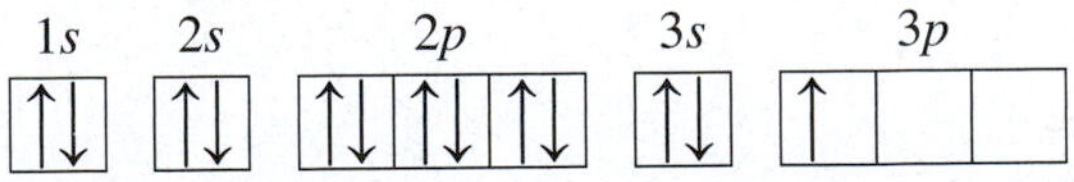
1s
2s
2p
3s
3p

c.
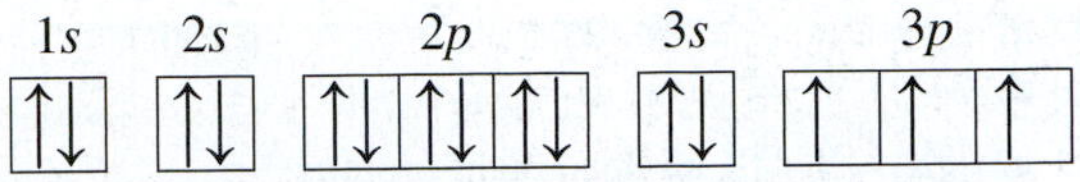
1s
2s
2p
3s
3p

d.
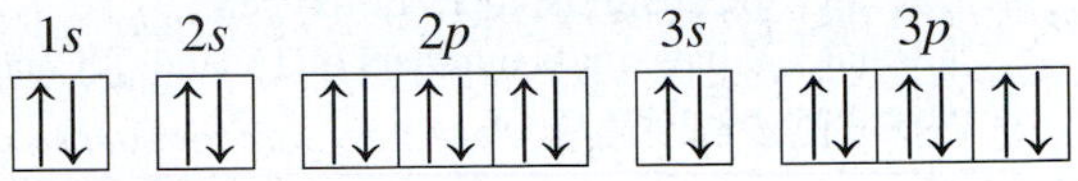
1s
2s
2p
3s
3p

4.57 **a.** $1s^22s^22p^1$
b. $1s^22s^22p^63s^1$
c. $1s^22s^1$
d. $1s^22s^22p^63s^2$

4.59 **a.** $[He]2s^22p^2$
b. $[Kr]5s^24d^2$
c. $[Kr]5s^24d^{10}5p^3$
d. $[He]2s^22p^5$

4.61 **a.** Li **b.** S **c.** Si **d.** F

4.63 **a.** Al **b.** C **c.** Ar **d.** Be

4.65 **a.** $1s^22s^22p^63s^23p^64s^23d^{10}4p^3$, $[Ar]4s^23d^{10}4p^3$
b. $1s^22s^22p^63s^23p^64s^23d^6$, $[Ar]4s^23d^6$
c. $1s^22s^22p^63s^23p^64s^23d^{10}4p^65s^24d^{10}5p^1$, $[Kr]5s^24d^{10}5p^1$
d. $1s^22s^22p^63s^23p^64s^23d^{10}4p^5$, $[Ar]4s^23d^{10}4p^5$
e. $1s^22s^22p^63s^23p^64s^23d^2$, $[Ar]4s^23d^2$

4.67 **a.** Ga **b.** N **c.** Xe **d.** Zr

4.69 **a.** 10 **b.** 6 **c.** 3 **d.** 1

4.71 **a.** 2A (2) **b.** 5A (15) **c.** 7B (7) **d.** 6A (16)

4.73 **a.** ns^1 **b.** ns^2np^2 **c.** ns^2np^5 **d.** ns^2np^3

4.75 **a.** 3 **b.** 5 **c.** 2 **d.** 7

4.77 **a.** Sulfur is in Group 6A (16); $\cdot\ddot{\text{S}}\colon$
b. Nitrogen is in Group 5A (15); $\cdot\ddot{\text{N}}\cdot$
c. Calcium is in Group 2A (2); $\dot{\text{Ca}}\cdot$
d. Sodium is in Group 1A (1); Na•
e. Gallium is in Group 3A (13); $\cdot\dot{\text{Ga}}\cdot$

4.79 **a.** Na **b.** Rb **c.** Na **d.** Rb

4.81 **a.** Mg, Al, Si **b.** I, Br, Cl
c. Sr, Sb, I **d.** Na, Si, P

4.83 **a.** Br **b.** Mg **c.** P **d.** Xe

4.85 **a.** Br, Cl, F **b.** Na, Al, Cl
c. Cs, K, Na **d.** Ca, As, Br

4.87 Ca, Ga, Ge, Br

4.89 lower, more

4.91 **a.** **1.** decreases
b. **2.** increases
c. **2.** increases
d. **3.** remains the same

4.93 **a.** false **b.** true **c.** true **d.** false **e.** true

4.95 **a.** Group 1A (1), alkali metals **b.** metal
c. 19 protons **d.** K-39
e. 39.10 amu

4.97 **a.** false All atoms of a given element are different from atoms of other elements.
b. true
c. true
d. false In a chemical reaction, atoms are neither created nor destroyed.

4.99 **a.** 1 + 2 **b.** 1 **c.** 1 **d.** 3 **e.** 2

4.101 **a.** $^{16}_{8}X$, $^{17}_{8}X$, $^{18}_{8}X$ all have eight protons.
b. $^{16}_{8}X$, $^{17}_{8}X$, $^{18}_{8}X$ are all isotopes of oxygen.
c. $^{16}_{8}X$ and $^{16}_{9}X$ have mass numbers of 16, and $^{18}_{10}X$ and $^{18}_{8}X$ have mass numbers of 18.

4.103

	Atomic Symbol		
	$^{70}_{32}Ge$	$^{73}_{32}Ge$	$^{76}_{32}Ge$
Atomic Number	32	32	32
Mass Number	70	73	76
Number of Protons	32	32	32
Number of Neutrons	38	41	44
Number of Electrons	32	32	32

4.105 **a.** $^{9}_{4}Be$ **b.** $^{11}_{5}B$ **c.** $^{13}_{6}C$ **d.** $^{10}_{5}B$ **e.** $^{12}_{6}C$
B and **D** are isotopes of boron; **C** and **E** are isotopes of carbon.

4.107 **a.** This is possible. This element is magnesium.
b. Not possible. The 2*p* sublevel would fill before the 3*s*, and only two electrons are allowed in an *s* orbital.

4.109 Li is **D**, Na is **A**, K is **C**, and Rb is **B**.

4.111 **a.** Na **b.** Cl **c.** Si **d.** Ar **e.** S **f.** Na **g.** Mg

4.113 **a.** Group 7A (17), Period 4 **b.** Group 8A (18), Period 3 **c.** Group 1A (1), Period 2 **d.** Group 2A (2), Period 7

4.115 **a.** metal **b.** metal **c.** metal **d.** nonmetal

4.117 **a.** false **b.** false **c.** true **d.** false **e.** true

4.119 **a.** 48 protons, 66 neutrons, 48 electrons
b. 43 protons, 55 neutrons, 43 electrons
c. 79 protons, 120 neutrons, 79 electrons
d. 86 protons, 136 neutrons, 86 electrons
e. 54 protons, 82 neutrons, 54 electrons

4.121

Name of the Element	Atomic Symbol	Number of Protons	Number of Neutrons	Number of Electrons
Selenium	$^{80}_{34}Se$	34	46	34
Nickel	$^{62}_{28}Ni$	28	34	28
Magnesium	$^{26}_{12}Mg$	12	14	12
Radium	$^{228}_{88}Ra$	88	140	88

4.123 **a.** 3*p* **b.** 5*s* **c.** 4*p* **d.** 4*s*

4.125 **a.** 6 **b.** 6 **c.** 10 **d.** 2

4.127 **a.** phosphorus **b.** lithium (H is a nonmetal.) **c.** cadmium **d.** nitrogen **e.** sodium

4.129 **a.** Na **b.** P **c.** F **d.** Na **e.** Cl

4.131 **a.** 82 protons, 126 neutrons, 82 electrons
b. $^{214}_{82}Pb$
c. $^{214}_{83}Bi$

4.133 **a.** O **b.** Ar **c.** C **d.** Na **e.** Ra

4.135 28.09 amu

4.137 **a.** X is a metal; Y and Z are nonmetals.
b. X has the largest atomic size.
c. Y and Z have six valence electrons and are in Group 6A (16).
d. Y has the highest ionization energy.
e. Y has the smallest atomic size.

5

Nuclear Chemistry

Simone's doctor is concerned about her elevated cholesterol, which could lead to coronary heart disease and a heart attack. He sends her to a nuclear medicine center to undergo a cardiac stress test. Pauline, the radiation technologist, explains to Simone that a stress test measures the blood flow to her heart muscle at rest and then during stress. The test is performed in a similar way as a routine exercise stress test, but images are produced that show areas of blood flow through the heart. It involves taking two sets of images of her heart, one when the heart is at rest, and one when she is walking on a treadmill.

Pauline tells Simone that she will inject thallium-201 into her bloodstream. She explains that Tl-201 is a radioactive isotope that has a half-life of 3.0 days. Simone is curious about the term "half-life." Pauline explains that a half-life is the amount of time it takes for one-half of a radioactive sample to break down. She assures Simone that after four half-lives, the radiation emitted will be almost zero. Pauline tells Simone that Tl-201 decays to Hg-201 and emits energy similar to X-rays. When the Tl-201 reaches any areas within her heart with restricted blood supply, smaller amounts of the radioisotope will accumulate.

CAREER

Radiation Technologist

A radiation technologist works in a hospital or imaging center where nuclear medicine is used to diagnose and treat a variety of medical conditions. In a diagnostic test, a radiation technologist uses a scanner, which converts radiation into images. The images are evaluated to determine any abnormalities in the body. A radiation technologist performs tests such as computed tomography (CT), magnetic resonance imaging (MRI), and positron emission tomography (PET) and operates the instrumentation and computers associated with these tests. In addition, they must physically and mentally prepare patients for imaging. A radiation technologist must know how to handle radioisotopes safely, use the necessary type of shielding, and give radioactive isotopes to patients. A patient may be given radioactive tracers such as technetium-99m, iodine-131, gallium-67, and thallium-201 that emit gamma radiation, which is detected and used to develop an image of the kidneys or thyroid or to follow the blood flow in the heart muscle.

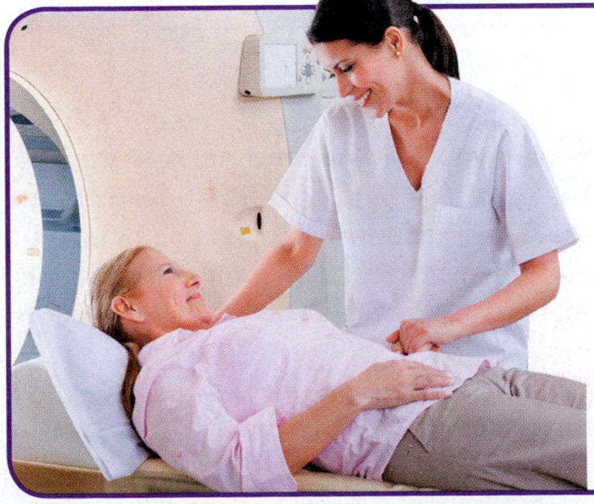

CLINICAL UPDATE

Cardiac Imaging Using a Radioisotope

When Simone arrives at the clinic for her nuclear stress test, a radioactive dye is injected that will show images of her heart muscle. You can read about Simone's scans and the results of her nuclear stress test in the **CLINICAL UPDATE** **Cardiac Imaging Using a Radioisotope**, page 168.

LOOKING AHEAD

REVIEW

Writing Atomic Symbols for Isotopes (4.5)

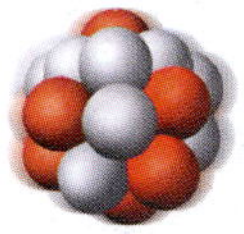

$^{14}_{6}C$ carbon-14 C-14

6 protons (red)

8 neutrons (white)

5.1 Natural Radioactivity

LEARNING GOAL Describe alpha, beta, positron, and gamma radiation.

Most naturally occurring isotopes of elements up to atomic number 19 have stable nuclei. Elements with atomic numbers 20 and higher usually have one or more isotopes that have unstable nuclei in which the nuclear forces cannot offset the repulsions between the protons. An unstable nucleus is *radioactive*, which means that it spontaneously emits small particles of energy called **radiation** to become more stable. Radiation may take the form of alpha (α) and beta (β) particles, positrons (β^+), or pure energy such as gamma (γ) rays. An isotope of an element that emits radiation is called a **radioisotope**. For most types of radiation, there is a change in the number of protons in the nucleus, which means that an atom is converted into an atom of a different element. Elements with atomic numbers of 93 and higher are produced artificially in nuclear laboratories and consist only of radioactive isotopes.

Symbols for Radioisotopes

The atomic symbols for the different isotopes are written with the mass number in the upper left corner and the atomic number in the lower left corner. The mass number is the sum of the number of protons and neutrons in the nucleus, and the atomic number is equal to the number of protons. For example, a radioactive isotope of carbon used for archaeological dating has a mass number of 14 and an atomic number of 6.

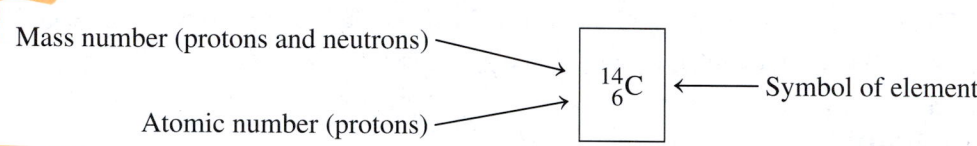

Mass number (protons and neutrons) → $^{14}_{6}C$ ← Symbol of element

Atomic number (protons) → $^{14}_{6}C$

Radioactive isotopes are identified by writing the mass number after the element's name or symbol. Thus, in this example, the isotope is called carbon-14 or C-14. **TABLE 5.1** compares some stable, nonradioactive isotopes with some radioactive isotopes.

TABLE 5.1 Stable and Radioactive Isotopes of Some Elements

Magnesium	Iodine	Uranium
Stable Isotopes		
$^{24}_{12}Mg$	$^{127}_{53}I$	None
Magnesium-24	Iodine-127	
Radioactive Isotopes		
$^{23}_{12}Mg$	$^{125}_{53}I$	$^{235}_{92}U$
Magnesium-23	Iodine-125	Uranium-235
$^{27}_{12}Mg$	$^{131}_{53}I$	$^{238}_{92}U$
Magnesium-27	Iodine-131	Uranium-238

Types of Radiation

Alpha particle $^{4}_{2}He$ or α

Beta particle $^{0}_{-1}e$ or β

By emitting radiation, an unstable nucleus forms a more stable, lower-energy nucleus. One type of radiation consists of *alpha particles*. An **alpha particle** is identical to a helium (He) nucleus, which has two protons and two neutrons. An alpha particle has a mass number of 4, an atomic number of 2, and a charge of 2+. The symbol for an alpha particle is the Greek letter alpha (α) or the symbol of a helium nucleus except that the 2+ charge is omitted.

Another type of radiation is a **beta particle**, which is a high-energy electron, with a charge of 1− and a mass number of 0. It is represented by the Greek letter beta (β) or by the symbol for the electron including the mass number and the charge ($^{0}_{-1}e$).

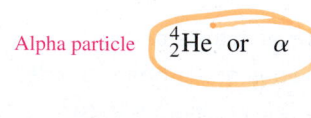

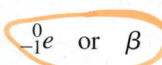

A **positron**, similar to a beta particle, has a positive (1+) charge with a mass number of 0. It is represented by the Greek letter beta with a 1+ charge, β^+, or by the symbol, ${}^{0}_{+1}e$. A positron is an example of *antimatter*, a term physicists use to describe a particle that is the opposite of another particle, in this case, an electron.

Positron ${}^{0}_{+1}e$ or β^+

Gamma rays are high-energy radiation, released when an unstable nucleus undergoes a rearrangement of its particles to give a more stable, lower-energy nucleus. Gamma rays are often emitted along with other types of radiation. A gamma ray is written as the Greek letter gamma, γ. Because gamma rays are energy only, zeros are used to show that a gamma ray has no mass or charge (${}^{0}_{0}\gamma$).

Gamma ray ${}^{0}_{0}\gamma$ or γ

TABLE 5.2 summarizes the types of radiation we use in nuclear equations.

TABLE 5.2 Some Forms of Radiation

Type of Radiation	Symbol		Mass Number	Charge
Alpha Particle	${}^{4}_{2}He$	α	4	2+
Beta Particle	${}^{0}_{-1}e$	β	0	1−
Positron	${}^{0}_{+1}e$	β^+	0	1+
Gamma Ray	${}^{0}_{0}\gamma$	γ	0	0
Proton	${}^{1}_{1}H$	p	1	1+
Neutron	${}^{1}_{0}n$	n	1	0

ENGAGE

What is the charge and the mass number of an alpha particle emitted by a radioactive atom?

▶ SAMPLE PROBLEM 5.1 Radiation Particles

TRY IT FIRST

Identify and write the symbol for each of the following types of radiation:

a. contains two protons and two neutrons
b. has a mass number of 0 and a 1− charge

SOLUTION

a. An alpha particle, ${}^{4}_{2}He$ or α, has two protons and two neutrons.
b. A beta particle, ${}^{0}_{-1}e$ or β, has a mass number of 0 and a 1− charge.

TEST

Try Practice Problems 5.1 to 5.10

STUDY CHECK 5.1

a. Identify and write the symbol for the type of radiation that has a mass number of zero and a 1+ charge.
b. Identify and write the symbol for the type of radiation that has a mass number of 0 and a 0 charge.

ANSWER

a. A positron, ${}^{0}_{+1}e$, has a mass number of 0 and a 1+ charge.
b. Gamma radiation, ${}^{0}_{0}\gamma$, has a mass number of 0 and a 0 charge.

Biological Effects of Radiation

When radiation strikes molecules in its path, electrons may be knocked away, forming unstable ions. If this *ionizing radiation* passes through the human body, it may interact with water molecules, removing electrons and producing H_2O^+, which can cause undesirable chemical reactions.

The cells most sensitive to radiation are the ones undergoing rapid division—those of the bone marrow, skin, reproductive organs, and intestinal lining, as well as all cells of growing children. Damaged cells may lose their ability to produce necessary materials. For example, if radiation damages cells of the bone marrow, red blood cells may no longer be produced. If sperm cells, ova, or the cells of a fetus are damaged, birth defects may result.

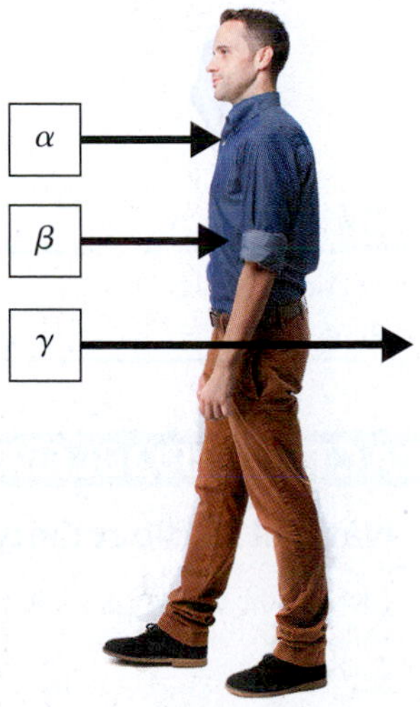
α
β
γ

Different types of radiation penetrate the body to different depths.

In contrast, cells of the nerves, muscles, liver, and adult bones are much less sensitive to radiation because they undergo little or no cellular division.

Cancer cells are another example of rapidly dividing cells. Because cancer cells are highly sensitive to radiation, large doses of radiation are used to destroy them. The normal tissue that surrounds cancer cells divides at a slower rate and suffers less damage from radiation. However, radiation may cause malignant tumors, leukemia, anemia, and genetic mutations.

Radiation Protection

Radiation technologists, chemists, doctors, and nurses who work with radioactive isotopes must use proper radiation protection. Proper *shielding* is necessary to prevent exposure. Alpha particles, which have the largest mass and charge of the radiation particles, travel only a few centimeters in the air before they collide with air molecules, acquire electrons, and become helium atoms. A piece of paper, clothing, and our skin are protection against alpha particles. Lab coats and gloves will also provide sufficient shielding. However, if alpha emitters are ingested or inhaled, the alpha particles they give off can cause serious internal damage.

Beta particles have a very small mass and move much faster and farther than alpha particles, traveling as much as several meters through air. They can pass through paper and penetrate as far as 4 to 5 mm into body tissue. External exposure to beta particles can burn the surface of the skin, but they do not travel far enough to reach the internal organs. Heavy clothing such as lab coats and gloves are needed to protect the skin from beta particles.

Gamma rays travel great distances through the air and pass through many materials, including body tissues. Because gamma rays penetrate so deeply, exposure to gamma rays can be extremely hazardous. Only very dense shielding, such as lead or concrete, will stop them. Syringes used for injections of radioactive materials use shielding made of lead or heavyweight materials such as tungsten and plastic composites.

When working with radioactive materials, medical personnel wear protective clothing and gloves and stand behind a shield (see **FIGURE 5.1**). Long tongs may be used to pick up vials of radioactive material, keeping them away from the hands and body. **TABLE 5.3** summarizes the shielding materials required for the various types of radiation.

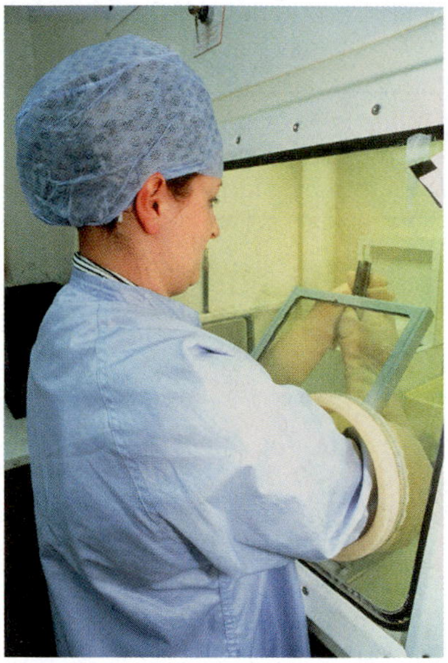

FIGURE 5.1 ▶ In a nuclear pharmacy, a person working with radioisotopes wears protective clothing and gloves and uses a lead glass shield on a syringe.

Q What types of radiation does a lead shield block?

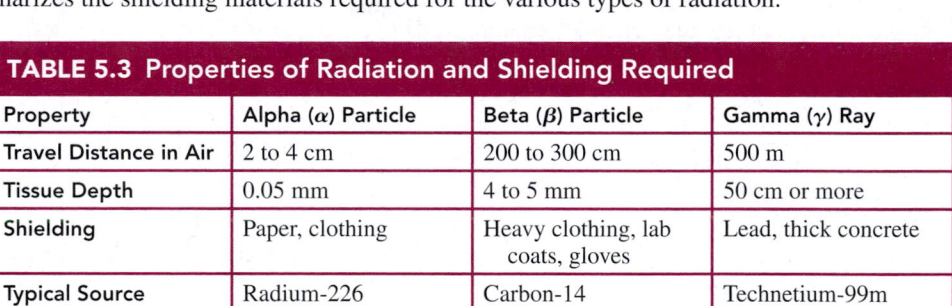

TABLE 5.3 Properties of Radiation and Shielding Required

Property	Alpha (α) Particle	Beta (β) Particle	Gamma (γ) Ray
Travel Distance in Air	2 to 4 cm	200 to 300 cm	500 m
Tissue Depth	0.05 mm	4 to 5 mm	50 cm or more
Shielding	Paper, clothing	Heavy clothing, lab coats, gloves	Lead, thick concrete
Typical Source	Radium-226	Carbon-14	Technetium-99m

If you work in an environment where radioactive materials are present, such as a nuclear medicine facility, try to keep the time you spend in a radioactive area to a minimum. Remaining in a radioactive area twice as long exposes you to twice as much radiation.

Keep your distance! The greater the distance from the radioactive source, the lower the intensity of radiation received. By doubling your distance from the radiation source, the intensity of radiation drops to $\left(\frac{1}{2}\right)^2$, or one-fourth of its previous value.

TEST

Try Practice Problems 5.11 and 5.12

PRACTICE PROBLEMS

5.1 Natural Radioactivity

5.1 Identify the type of particle or radiation for each of the following:

a. $^{4}_{2}He$ **b.** $^{0}_{+1}e$ **c.** $^{0}_{0}\gamma$

5.2 Identify the type of particle or radiation for each of the following:

a. $^{0}_{-1}e$ **b.** $^{1}_{1}H$ **c.** $^{1}_{0}n$

5.3 Naturally occurring potassium consists of three isotopes: potassium-39, potassium-40, and potassium-41.

a. Write the atomic symbol for each isotope.

b. In what ways are the isotopes similar, and in what ways do they differ?

5.4 Naturally occurring iodine is iodine-127. Medically, radioactive isotopes of iodine-125 and iodine-131 are used.

a. Write the atomic symbol for each isotope.

b. In what ways are the isotopes similar, and in what ways do they differ?

5.5 Identify each of the following:

a. ${}^{0}_{-1}X$ **b.** ${}^{4}_{2}X$ **c.** ${}^{1}_{0}X$

d. ${}^{38}_{18}X$ **e.** ${}^{14}_{6}X$

5.6 Identify each of the following:

a. ${}^{1}_{1}X$ **b.** ${}^{81}_{35}X$ **c.** ${}^{0}_{0}X$

d. ${}^{59}_{26}X$ **e.** ${}^{0}_{+1}X$

Clinical Applications

5.7 Write the atomic symbol for each of the following isotopes used in nuclear medicine:

a. copper-64 **b.** selenium-75

c. sodium-24 **d.** nitrogen-15

5.8 Write the atomic symbol for each of the following isotopes used in nuclear medicine:

a. indium-111 **b.** palladium-103

c. barium-131 **d.** rubidium-82

5.9 Supply the missing information in the following table:

Medical Use	Atomic Symbol	Mass Number	Number of Protons	Number of Neutrons
Heart imaging	${}^{201}_{81}Tl$			
Radiation therapy		60	27	
Abdominal scan			31	36
Hyperthyroidism	${}^{131}_{53}I$			
Leukemia treatment		32		17

5.10 Supply the missing information in the following table:

Medical Use	Atomic Symbol	Mass Number	Number of Protons	Number of Neutrons
Cancer treatment	${}^{131}_{55}Cs$			
Brain scan			43	56
Blood flow		141	58	
Bone scan		85		47
Lung function	${}^{133}_{54}Xe$			

5.11 Match the type of radiation (**1** to **3**) with each of the following statements:

1. alpha particle
2. beta particle
3. gamma radiation

a. does not penetrate skin
b. shielding protection includes lead or thick concrete
c. can be very harmful if ingested

5.12 Match the type of radiation (**1** to **3**) with each of the following statements:

1. alpha particle
2. beta particle
3. gamma radiation

a. penetrates farthest into skin and body tissues
b. shielding protection includes lab coats and gloves
c. travels only a short distance in air

5.2 Nuclear Reactions

LEARNING GOAL Write a balanced nuclear equation for radioactive decay, showing mass numbers and atomic numbers.

REVIEW

Using Positive and Negative Numbers in Calculations (1.4)

Solving Equations (1.4)

Counting Protons and Neutrons (4.4)

In a process called **radioactive decay**, a nucleus spontaneously breaks down by emitting radiation. This process is shown by writing a *nuclear equation* with the atomic symbols of the original radioactive nucleus on the left, an arrow, and the new nucleus and the type of radiation emitted on the right.

$$\text{Radioactive nucleus} \longrightarrow \text{new nucleus} + \text{radiation } (\alpha, \beta, \beta^+, \gamma)$$

In a nuclear equation, the sum of the mass numbers and the sum of the atomic numbers on one side of the arrow must equal the sum of the mass numbers and the sum of the atomic numbers on the other side.

CORE CHEMISTRY SKILL

Writing Nuclear Equations

Alpha Decay

An unstable nucleus may emit an alpha particle, which consists of two protons and two neutrons. Thus, the mass number of the radioactive nucleus decreases by 4, and its atomic number decreases by 2. For example, when uranium-238 emits an alpha particle, the new nucleus that forms has a mass number of 234 and an atomic number of 90. Uranium is transformed into a different element, thorium, an example of *transmutation*.

ENGAGE

What happens to the U-238 nucleus when an alpha particle is emitted?

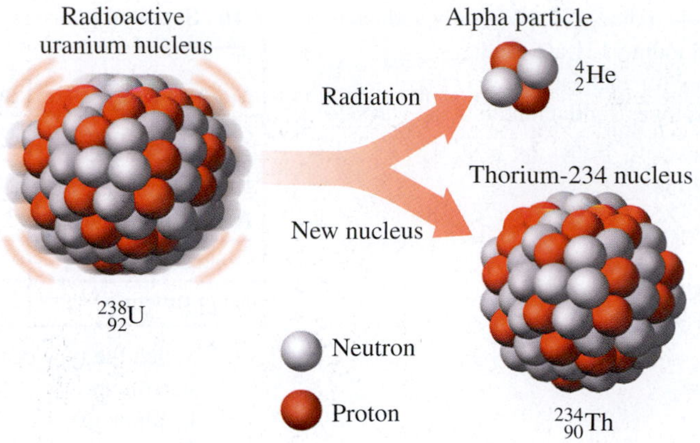

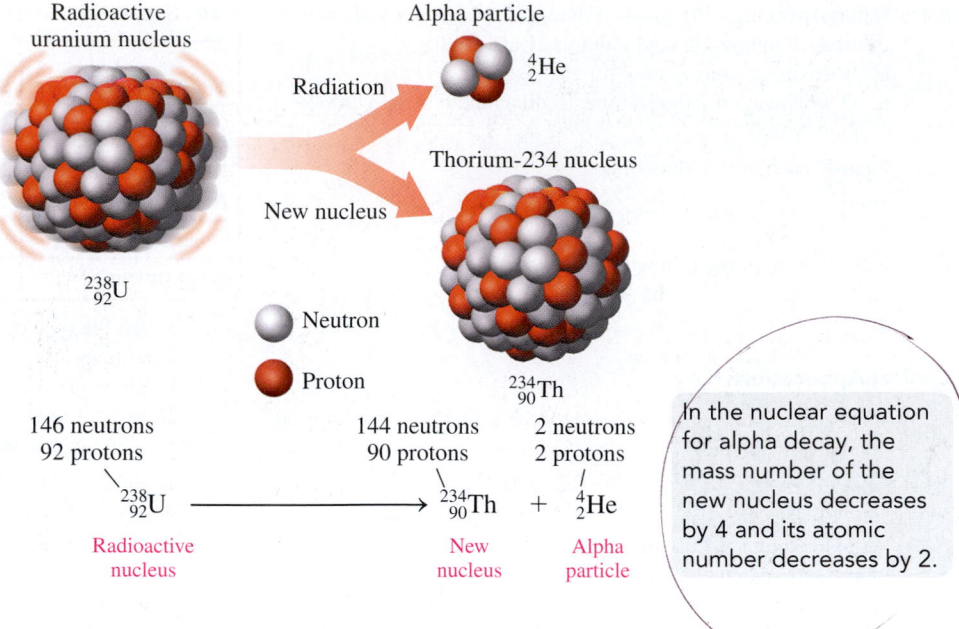

Radioactive uranium nucleus
Alpha particle
$^{4}_{2}He$
Radiation
Thorium-234 nucleus
New nucleus
$^{238}_{92}U$
Neutron
Proton
$^{234}_{90}Th$

146 neutrons 92 protons		144 neutrons 90 protons		2 neutrons 2 protons

$$^{238}_{92}U \longrightarrow \, ^{234}_{90}Th + \, ^{4}_{2}He$$

Radioactive nucleus — New nucleus — Alpha particle

In the nuclear equation for alpha decay, the mass number of the new nucleus decreases by 4 and its atomic number decreases by 2.

We can look at writing a balanced nuclear equation for americium-241, which undergoes alpha decay as shown in Sample Problem 5.2.

A smoke detector contains the radioactive isotope americium-241, which undergoes alpha decay.

▶ SAMPLE PROBLEM 5.2 Writing a Nuclear Equation for Alpha Decay

TRY IT FIRST

Smoke detectors that are used in homes contain americium-241, which undergoes alpha decay. When alpha particles collide with air molecules, charged particles are produced that generate an electrical current. If smoke particles enter the detector, they interfere with the formation of charged particles in the air, and the electrical current is interrupted. This causes the alarm to sound and warns the occupants of the danger of fire. Write the balanced nuclear equation for the alpha decay of americium-241.

SOLUTION

ANALYZE THE PROBLEM	Given	Need	Connect
	Am-241, alpha decay	balanced nuclear equation	mass number, atomic number of new nucleus

STEP 1 **Write the incomplete nuclear equation.**

$$^{241}_{95}Am \longrightarrow ? + \, ^{4}_{2}He$$

STEP 2 **Determine the missing mass number.** In the equation, the mass number, 241, is equal to the sum of the mass numbers of the new nucleus and the alpha particle.

$$241 = ? + 4$$
$$? = 241 - 4 = 237 \text{ (mass number of new nucleus)}$$

STEP 3 **Determine the missing atomic number.** The atomic number, 95, must equal the sum of the atomic numbers of the new nucleus and the alpha particle.

$$95 = ? + 2$$
$$? = 95 - 2 = 93 \text{ (atomic number of new nucleus)}$$

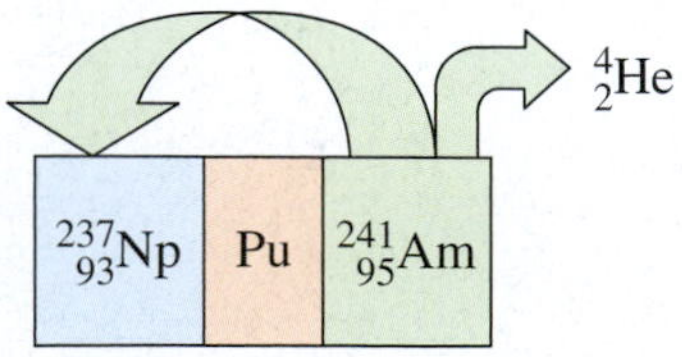

$^{4}_{2}He$
$^{237}_{93}Np$
Pu
$^{241}_{95}Am$

STEP 4 **Determine the symbol of the new nucleus and complete the nuclear equation.** On the periodic table, the element that has atomic number 93 is neptunium, Np. The atomic symbol for this isotope of Np is written $^{237}_{93}Np$.

$$^{241}_{95}Am \longrightarrow \, ^{237}_{93}Np + \, ^{4}_{2}He$$

STUDY CHECK 5.2

a. Write the balanced nuclear equation for the alpha decay of Po-214.

b. Bismuth-213, used to treat leukemia, emits alpha particles that destroy cancer cells. Write the balanced nuclear equation for the alpha decay of Bi-213.

ANSWER

a. $^{214}_{84}\text{Po} \longrightarrow {}^{210}_{82}\text{Pb} + {}^{4}_{2}\text{He}$

b. $^{213}_{83}\text{Bi} \longrightarrow {}^{209}_{81}\text{Tl} + {}^{4}_{2}\text{He}$

TEST

Try Practice Problems 5.13 and 5.14

Chemistry Link to Health

Radon in Our Homes

The presence of radon gas has become an environmental and health issue because of the radiation danger it poses. Radioactive isotopes such as radium-226 are naturally present in many types of rocks and soils. Radium-226 emits an alpha particle and is converted into radon gas, which diffuses out of the rocks and soil.

$$^{226}_{88}\text{Ra} \longrightarrow {}^{222}_{86}\text{Rn} + {}^{4}_{2}\text{He}$$

Outdoors, radon gas poses little danger because it disperses in the air. However, if the radioactive source is under a house or building, the radon gas can enter the house through cracks in the foundation or other openings. Those who live or work there may inhale the radon. In the lungs, radon-222 emits alpha particles to form polonium-218, which is known to cause lung cancer.

$$^{222}_{86}\text{Rn} \longrightarrow {}^{218}_{84}\text{Po} + {}^{4}_{2}\text{He}$$

The U.S. Environmental Protection Agency (EPA) estimates that radon causes about 20 000 lung cancer deaths in one year. The EPA recommends that the maximum level of radon not exceed 4 picocuries (pCi) per liter of air in a home. One picocurie (pCi) is equal to 10^{-12} curies (Ci); curies are described in Section 5.3. The EPA estimates that more than 6 million homes have radon levels that exceed this maximum.

SHORT-TERM
RADON
TEST KIT
FREE SHIPPING
& ANALYSIS!
FIRST CLASS MAIL
AccuStar

A radon test kit is used to determine radon levels in buildings.

Beta Decay

A beta particle forms as the result of the breakdown of a neutron into a proton and an electron (beta particle). Because the proton remains in the nucleus, the number of protons increases by one, whereas the number of neutrons decreases by one. Thus, in a nuclear equation for beta decay, the mass number of the radioactive nucleus and the mass number of the new nucleus are the same. However, the atomic number of the new nucleus increases by one. For example, the beta decay of a carbon-14 nucleus produces a nitrogen-14 nucleus.

ENGAGE

What happens to a C-14 nucleus when a beta particle is emitted?

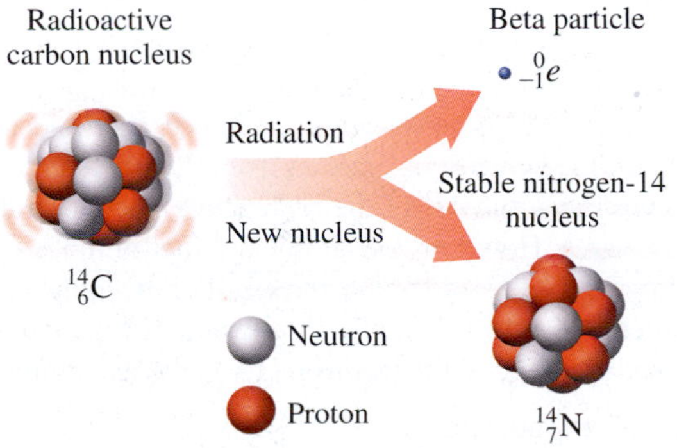
Radioactive
carbon nucleus
Beta particle
$^{0}_{-1}e$
Radiation
Stable nitrogen-14
nucleus
New nucleus
$^{14}_{6}\text{C}$
Neutron
Proton
$^{14}_{7}\text{N}$

8 neutrons
6 protons

7 neutrons
7 protons

0 neutrons
0 protons
1− charge

$$^{14}_{6}\text{C} \longrightarrow {}^{14}_{7}\text{N} + {}^{0}_{-1}e$$

Radioactive nucleus — New nucleus — Beta particle

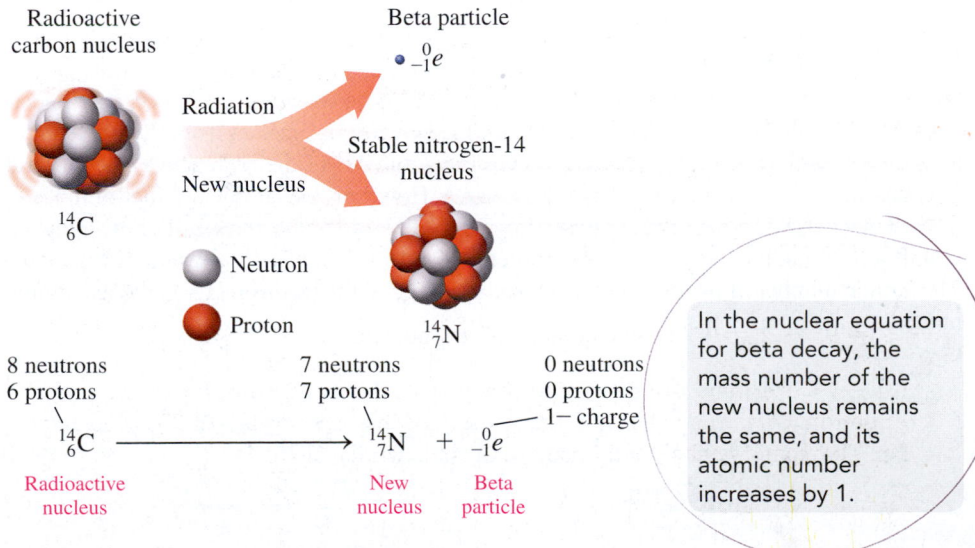

In the nuclear equation for beta decay, the mass number of the new nucleus remains the same, and its atomic number increases by 1.

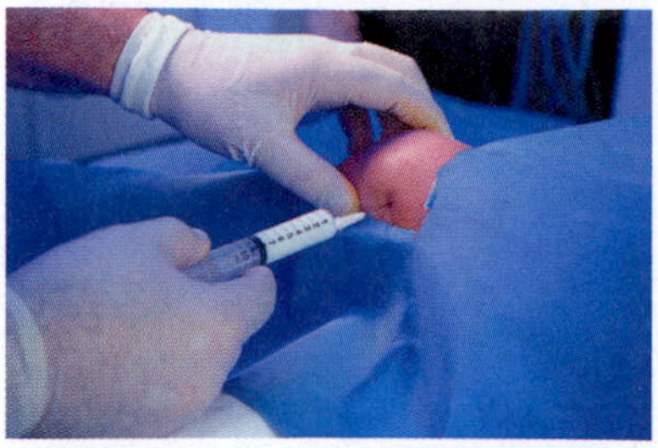

A radioisotope is injected into the joint to relieve the pain caused by arthritis.

▶ SAMPLE PROBLEM 5.3 Writing a Nuclear Equation for Beta Decay

TRY IT FIRST

The radioactive isotope yttrium-90, a beta emitter, is used in cancer treatment and as a colloidal injection into joints to relieve arthritis pain. Write the balanced nuclear equation for the beta decay of yttrium-90.

SOLUTION

ANALYZE THE PROBLEM	Given	Need	Connect
	Y-90, beta decay	balanced nuclear equation	mass number, atomic number of new nucleus

STEP 1 **Write the incomplete nuclear equation.**

$$^{90}_{39}\mathrm{Y} \longrightarrow ? + ^{\;\,0}_{-1}e$$

STEP 2 **Determine the missing mass number.** In the equation, the mass number, 90, is equal to the sum of the mass numbers of the new nucleus and the beta particle.

$90 = ? + 0$

$? = 90 - 0 = 90$ (mass number of new nucleus)

STEP 3 **Determine the missing atomic number.** The atomic number, 39, must equal the sum of the atomic numbers of the new nucleus and the beta particle.

$39 = ? - 1$

$? = 39 + 1 = 40$ (atomic number of new nucleus)

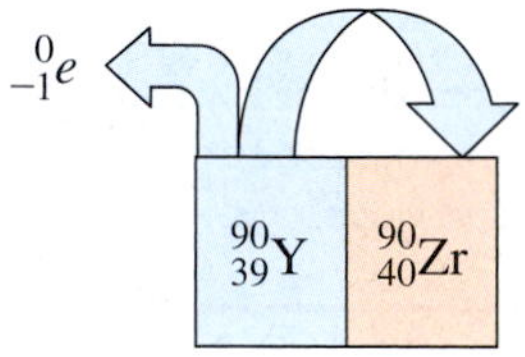

$^{\;\,0}_{-1}e$
$^{90}_{39}\mathrm{Y}$
$^{90}_{40}\mathrm{Zr}$

STEP 4 **Determine the symbol of the new nucleus and complete the nuclear equation.** On the periodic table, the element that has atomic number 40 is zirconium, Zr. The atomic symbol for this isotope of Zr is written $^{90}_{40}\mathrm{Zr}$.

$$^{90}_{39}\mathrm{Y} \longrightarrow ^{90}_{40}\mathrm{Zr} + ^{\;\,0}_{-1}e$$

STUDY CHECK 5.3

a. Write the balanced nuclear equation for the beta decay of chromium-51.

b. Write the balanced nuclear equation for the beta decay of Cu-64, used in cancer therapy.

ANSWER

a. $^{51}_{24}\mathrm{Cr} \longrightarrow ^{51}_{25}\mathrm{Mn} + ^{\;\,0}_{-1}e$

b. $^{64}_{29}\mathrm{Cu} \longrightarrow ^{64}_{30}\mathrm{Zn} + ^{\;\,0}_{-1}e$

TEST

Try Practice Problems 5.15 and 5.16

Positron Emission

+
Proton in the nucleus
New neutron remains in the nucleus
Positron emitted

In positron emission, a proton in an unstable nucleus is converted to a neutron and a positron. The neutron remains in the nucleus, but the positron is emitted from the nucleus. In a nuclear equation for positron emission, the mass number of the radioactive nucleus and the mass number of the new nucleus are the same. However, the atomic number of the new nucleus decreases by one, indicating a change of one element into another. For example, an aluminum-24 nucleus undergoes positron emission to produce a magnesium-24 nucleus. The atomic number of magnesium (12) and the charge of the positron (1+) give the atomic number of aluminum (13).

$$^{24}_{13}\mathrm{Al} \longrightarrow ^{24}_{12}\mathrm{Mg} + \underset{\text{Positron}}{^{\;\,0}_{+1}e}$$

▶ SAMPLE PROBLEM 5.4 Writing a Nuclear Equation for Positron Emission

TRY IT FIRST

Write the balanced nuclear equation for the positron emission of manganese-49.

ENGAGE

Why does the atomic number but not the mass number change in positron emission?

SOLUTION

	Given	Need	Connect
ANALYZE THE PROBLEM	Mn-49, positron emission	balanced nuclear equation	mass number, atomic number of new nucleus

STEP 1 **Write the incomplete nuclear equation.**

$$^{49}_{25}\text{Mn} \longrightarrow ? + ^{0}_{+1}e$$

STEP 2 **Determine the missing mass number.** In the equation, the mass number, 49, is equal to the sum of the mass numbers of the new nucleus and the positron.

$49 = ? + 0$

$? = 49 - 0 = 49$ (mass number of new nucleus)

STEP 3 **Determine the missing atomic number.** The atomic number, 25, must equal the sum of the atomic numbers of the new nucleus and the positron.

$25 = ? + 1$

$? = 25 - 1 = 24$ (atomic number of new nucleus)

STEP 4 **Determine the symbol of the new nucleus and complete the nuclear equation.** On the periodic table, the element that has atomic number 24 is chromium, Cr. The atomic symbol for this isotope of Cr is written $^{49}_{24}\text{Cr}$.

$$^{49}_{25}\text{Mn} \longrightarrow ^{49}_{24}\text{Cr} + ^{0}_{+1}e$$

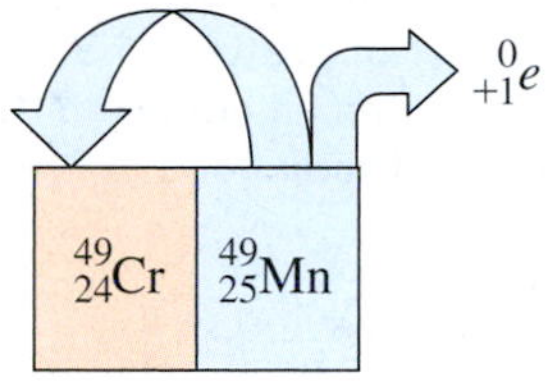

$^{0}_{+1}e$
$^{49}_{24}\text{Cr}$
$^{49}_{25}\text{Mn}$

STUDY CHECK 5.4

a. Write the balanced nuclear equation for the positron emission of xenon-118.

b. Write the balanced nuclear equation for the positron emission of arsenic-74, used in cancer imaging.

ANSWER

a. $^{118}_{54}\text{Xe} \longrightarrow ^{118}_{53}\text{I} + ^{0}_{+1}e$

b. $^{74}_{33}\text{As} \longrightarrow ^{74}_{32}\text{Ge} + ^{0}_{+1}e$

TEST

Try Practice Problems 5.17 and 5.18

Gamma Emission

Pure gamma emitters are rare, although gamma radiation accompanies most alpha and beta radiation. In radiology, one of the most commonly used gamma emitters is technetium (Tc). The unstable isotope of technetium is written as the *metastable* (symbol m) isotope technetium-99m, Tc-99m, or $^{99m}_{43}\text{Tc}$. By emitting energy in the form of gamma rays, the nucleus becomes more stable.

$$^{99m}_{43}\text{Tc} \longrightarrow ^{99}_{43}\text{Tc} + ^{0}_{0}\gamma$$

FIGURE 5.2 summarizes the changes in the nucleus for alpha, beta, positron, and gamma radiation.

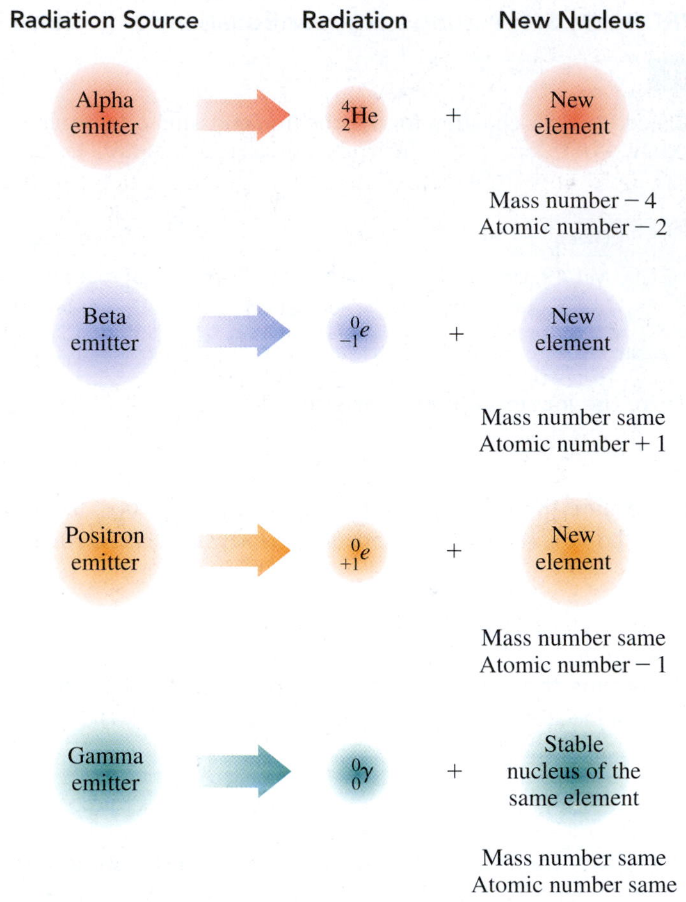

Radiation Source
Radiation
New Nucleus
Alpha emitter
$^{4}_{2}He$
+
New element
Mass number − 4
Atomic number − 2
Beta emitter
$^{0}_{-1}e$
+
New element
Mass number same
Atomic number + 1
Positron emitter
$^{0}_{+1}e$
+
New element
Mass number same
Atomic number − 1
Gamma emitter
$^{0}_{0}\gamma$
+
Stable nucleus of the same element
Mass number same
Atomic number same

FIGURE 5.2 ▶ When the nuclei of alpha, beta, positron, and gamma emitters emit radiation, new, more stable nuclei are produced.

Q What changes occur in the number of protons and neutrons when an alpha emitter gives off radiation?

TEST

Try Practice Problems 5.19 and 5.20

Producing Radioactive Isotopes

Today, many radioisotopes are produced in small amounts by bombarding stable, nonradioactive isotopes with high-speed particles such as alpha particles, protons, neutrons, and small nuclei. When one of these particles is absorbed, the stable nucleus is converted to a radioactive isotope and usually some type of radiation particle.

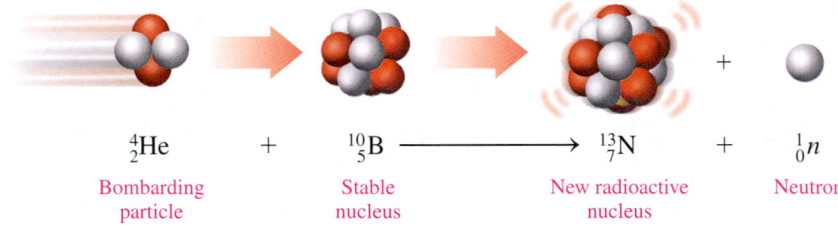

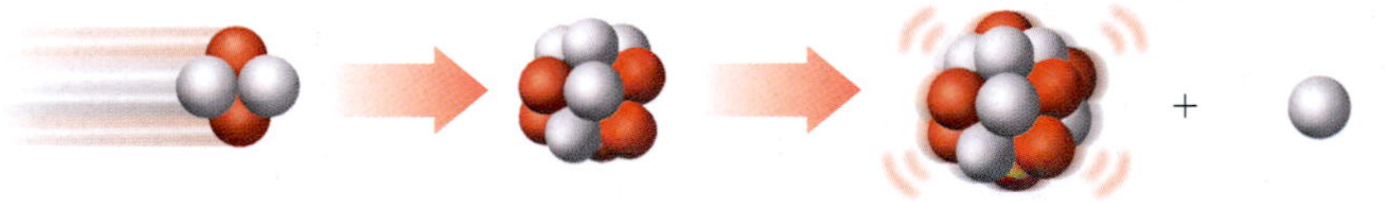

$$^{4}_{2}He + ^{10}_{5}B \longrightarrow ^{13}_{7}N + ^{1}_{0}n$$

Bombarding particle — Stable nucleus — New radioactive nucleus — Neutron

When nonradioactive B-10 is bombarded by an alpha particle, the products are radioactive N-13 and a neutron.

All elements that have an atomic number greater than 92 have been produced by bombardment. Most have been produced in only small amounts and exist for only a short time, making it difficult to study their properties. For example, when californium-249 is bombarded with nitrogen-15, the radioactive element dubnium-260 and four neutrons are produced.

$$^{15}_{7}N + ^{249}_{98}Cf \longrightarrow ^{260}_{105}Db + 4^{1}_{0}n$$

Technetium-99m is a radioisotope used in nuclear medicine for several diagnostic procedures, including brain tumor detection and liver and spleen examinations. The source of technetium-99m is molybdenum-99, which is produced in a nuclear reactor by neutron bombardment of molybdenum-98.

$$^{1}_{0}n + ^{98}_{42}Mo \longrightarrow ^{99}_{42}Mo$$

INTERACTIVE VIDEO

Writing Equations for an Isotope Produced by Bombardment

PEARSON eText 2.0

Many radiology laboratories have small generators containing molybdenum-99, which decays to the technetium-99m radioisotope.

$$^{99}_{42}\text{Mo} \longrightarrow {}^{99m}_{43}\text{Tc} + {}^{0}_{-1}e$$

The technetium-99m radioisotope decays by emitting gamma rays. Gamma emission is desirable for diagnostic work because the gamma rays pass through the body to the detection equipment.

$$^{99m}_{43}\text{Tc} \longrightarrow {}^{99}_{43}\text{Tc} + {}^{0}_{0}\gamma$$

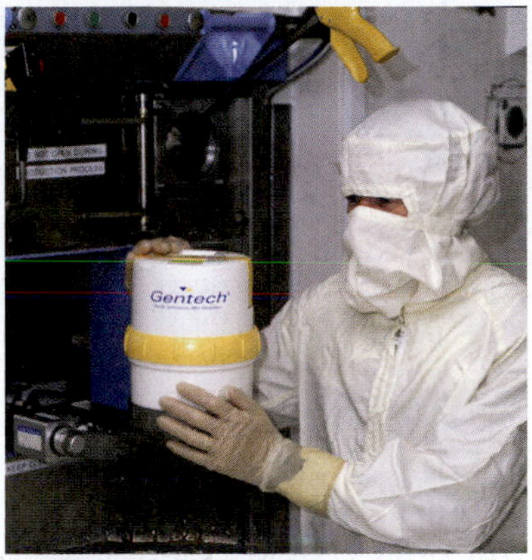
Gentech

A generator is used to prepare technetium-99m.

▶SAMPLE PROBLEM 5.5 Writing a Nuclear Equation for an Isotope Produced by Bombardment

TRY IT FIRST

Write the balanced nuclear equation for the bombardment of nickel-58 by a proton, $^{1}_{1}\text{H}$, which produces a radioactive isotope and an alpha particle.

SOLUTION

	Given	Need	Connect
ANALYZE THE PROBLEM	Ni-58, proton bombardment	balanced nuclear equation	mass number, atomic number of new nucleus

STEP 1 **Write the incomplete nuclear equation.**

$$^{1}_{1}\text{H} + {}^{58}_{28}\text{Ni} \longrightarrow ? + {}^{4}_{2}\text{He}$$

STEP 2 **Determine the missing mass number.** In the equation, the sum of the mass numbers of the proton, 1, and the nickel, 58, must equal the sum of the mass numbers of the new nucleus and the alpha particle.

$$1 + 58 = ? + 4$$
$$? = 59 - 4 = 55 \text{ (mass number of new nucleus)}$$

STEP 3 **Determine the missing atomic number.** The sum of the atomic numbers of the proton, 1, and nickel, 28, must equal the sum of the atomic numbers of the new nucleus and the alpha particle.

$$1 + 28 = ? + 2$$
$$? = 29 - 2 = 27 \text{ (atomic number of new nucleus)}$$

STEP 4 **Determine the symbol of the new nucleus and complete the nuclear equation.** On the periodic table, the element that has atomic number 27 is cobalt, Co. The atomic symbol for this isotope of Co is written $^{55}_{27}\text{Co}$.

$$^{1}_{1}\text{H} + {}^{58}_{28}\text{Ni} \longrightarrow {}^{55}_{27}\text{Co} + {}^{4}_{2}\text{He}$$

ENGAGE

If the bombardment of N-14 by an alpha particle produces a proton, how do you know the other product is O-17?

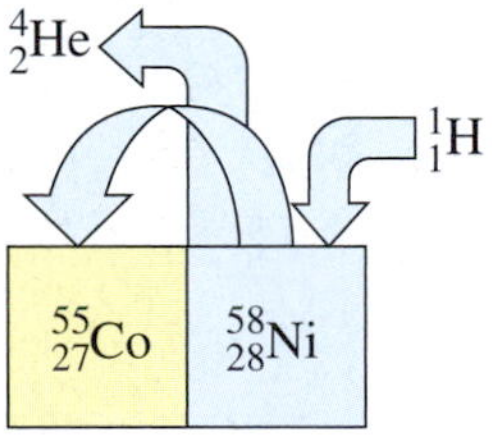
$^{4}_{2}$He
$^{1}_{1}$H
$^{55}_{27}$Co
$^{58}_{28}$Ni

STUDY CHECK 5.5

a. The first radioactive isotope was produced in 1934 by the bombardment of aluminum-27 by an alpha particle to produce a radioactive isotope and one neutron. Write the balanced nuclear equation for this bombardment.

b. When iron-54 is bombarded by an alpha particle, the products are a new isotope and two protons. Write the balanced nuclear equation for this bombardment.

ANSWER

a. $^{4}_{2}\text{He} + {}^{27}_{13}\text{Al} \longrightarrow {}^{30}_{15}\text{P} + {}^{1}_{0}n$

b. $^{4}_{2}\text{He} + {}^{54}_{26}\text{Fe} \longrightarrow {}^{56}_{26}\text{Fe} + 2{}^{1}_{1}\text{H}$

TEST

Try Practice Problems 5.21 and 5.22

PRACTICE PROBLEMS

5.2 Nuclear Reactions

5.13 Write a balanced nuclear equation for the alpha decay of each of the following radioactive isotopes:
a. $^{208}_{84}Po$ **b.** $^{232}_{90}Th$ **c.** $^{251}_{102}No$ **d.** radon-220

5.14 Write a balanced nuclear equation for the alpha decay of each of the following radioactive isotopes:
a. curium-243 **b.** $^{252}_{99}Es$ **c.** $^{251}_{98}Cf$ **d.** $^{261}_{107}Bh$

5.15 Write a balanced nuclear equation for the beta decay of each of the following radioactive isotopes:
a. $^{25}_{11}Na$ **b.** $^{20}_{8}O$ **c.** strontium-92 **d.** iron-60

5.16 Write a balanced nuclear equation for the beta decay of each of the following radioactive isotopes:
a. $^{44}_{19}K$ **b.** iron-59 **c.** potassium-42 **d.** $^{141}_{56}Ba$

5.17 Write a balanced nuclear equation for the positron emission of each of the following radioactive isotopes:
a. silicon-26 **b.** cobalt-54 **c.** $^{77}_{37}Rb$ **d.** $^{93}_{45}Rh$

5.18 Write a balanced nuclear equation for the positron emission of each of the following radioactive isotopes:
a. boron-8 **b.** $^{15}_{8}O$ **c.** $^{40}_{19}K$ **d.** nitrogen-13

5.19 Complete each of the following nuclear equations, and describe the type of radiation:
a. $^{28}_{13}Al \longrightarrow ? + {}^{0}_{-1}e$ **b.** $^{180m}_{73}Ta \longrightarrow {}^{180}_{73}Ta + ?$
c. $^{66}_{29}Cu \longrightarrow {}^{66}_{30}Zn + ?$ **d.** $? \longrightarrow {}^{234}_{90}Th + {}^{4}_{2}He$
e. $^{188}_{80}Hg \longrightarrow ? + {}^{0}_{+1}e$

5.20 Complete each of the following nuclear equations, and describe the type of radiation:
a. $^{11}_{6}C \longrightarrow {}^{11}_{5}B + ?$ **b.** $^{35}_{16}S \longrightarrow ? + {}^{0}_{-1}e$
c. $? \longrightarrow {}^{90}_{39}Y + {}^{0}_{-1}e$ **d.** $^{210}_{83}Bi \longrightarrow ? + {}^{4}_{2}He$
e. $? \longrightarrow {}^{89}_{39}Y + {}^{0}_{+1}e$

5.21 Complete each of the following bombardment reactions:
a. ${}^{1}_{0}n + {}^{9}_{4}Be \longrightarrow ?$ **b.** ${}^{1}_{0}n + {}^{131}_{52}Te \longrightarrow ? + {}^{0}_{-1}e$
c. ${}^{1}_{0}n + ? \longrightarrow {}^{24}_{11}Na + {}^{4}_{2}He$ **d.** ${}^{4}_{2}He + {}^{14}_{7}N \longrightarrow ? + {}^{1}_{1}H$

5.22 Complete each of the following bombardment reactions:
a. $? + {}^{40}_{18}Ar \longrightarrow {}^{43}_{19}K + {}^{1}_{1}H$ **b.** ${}^{1}_{0}n + {}^{238}_{92}U \longrightarrow ?$
c. ${}^{1}_{0}n + ? \longrightarrow {}^{14}_{6}C + {}^{1}_{1}H$ **d.** $? + {}^{64}_{28}Ni \longrightarrow {}^{272}_{111}Rg + {}^{1}_{0}n$

5.3 Radiation Measurement

LEARNING GOAL Describe how radiation is detected and measured.

One of the most common instruments for detecting beta and gamma radiation is the Geiger counter. It consists of a metal tube filled with a gas such as argon. When radiation enters a window on the end of the tube, it forms charged particles in the gas, which produce an electrical current. Each burst of current is amplified to give a click and a reading on a meter.

$$Ar + radiation \longrightarrow Ar^{+} + e^{-}$$

Measuring Radiation

Radiation is measured in several different ways. When a radiology laboratory obtains a radioisotope, the *activity* of the sample is measured in terms of the number of nuclear disintegrations per second. The **curie (Ci)**, the original unit of activity, was defined as the number of disintegrations that occur in 1 s for 1 g of radium, which is equal to 3.7×10^{10} disintegrations/s. The unit was named for Polish scientist Marie Curie, who along with her husband, Pierre, discovered the radioactive elements radium and polonium. The SI unit of radiation activity is the **becquerel (Bq)**, which is 1 disintegration/s.

Radiation levels in workers at the Fukushima Daiichi nuclear power plant are measured.

Meter
Electrons
High-voltage power supply
Radiation

A Geiger counter detects alpha particles, beta particles, and gamma rays using the ionization effect produced in a Geiger–Müller tube.

The **rad (radiation absorbed dose)** is a unit that measures the amount of radiation absorbed by a gram of material such as body tissue. The SI unit for absorbed dose is the **gray (Gy)**, which is defined as the joules of energy absorbed by 1 kg of body tissue. The gray is equal to 100 rad.

The **rem (radiation equivalent in humans)** is a unit that measures the biological effects of different kinds of radiation. Although alpha particles do not penetrate the skin, if they should enter the body by some other route, they can cause extensive damage within a short distance in tissue. High-energy radiation, such as beta particles, high-energy protons, and neutrons that travel into tissue, causes more damage. Gamma rays are damaging because they travel a long way through body tissue.

To determine the **equivalent dose** or rem dose, the absorbed dose (rad) is multiplied by a factor that adjusts for biological damage caused by a particular form of radiation. For beta and gamma radiation the factor is 1, so the biological damage in rems is the same as the absorbed radiation (rad). For high-energy protons and neutrons, the factor is about 10, and for alpha particles it is 20.

$$\text{Biological damage (rem)} = \text{Absorbed dose (rad)} \times \text{Factor}$$

Often, the measurement for an equivalent dose will be in units of millirems (mrem). One rem is equal to 1000 mrem. The SI unit is the **sievert (Sv)**. One sievert is equal to 100 rem. **TABLE 5.4** summarizes the units used to measure radiation.

ENGAGE

What is the difference between becquerels and rems?

TABLE 5.4 Units of Radiation Measurement

Measurement	Common Unit	SI Unit	Relationship
Activity	curie (Ci) $1\text{ Ci} = 3.7 \times 10^{10}$ disintegrations/s	becquerel (Bq) $1\text{ Bq} = 1$ disintegration/s	$1\text{ Ci} = 3.7 \times 10^{10}\text{ Bq}$
Absorbed Dose	rad	gray (Gy) $1\text{ Gy} = 1\text{ J/kg}$ of tissue	$1\text{ Gy} = 100\text{ rad}$
Biological Damage	rem	sievert (Sv)	$1\text{ Sv} = 100\text{ rem}$

Chemistry Link to Health

Radiation and Food

Foodborne illnesses caused by pathogenic bacteria such as *Salmonella*, *Listeria*, and *Escherichia coli* (*E. coli*) have become a major health concern in the United States. *E. coli* has been responsible for outbreaks of illness from contaminated ground beef, fruit juices, lettuce, and alfalfa sprouts.

The U.S. Food and Drug Administration (FDA) has approved the use of 0.3 to 1 kGy of radiation produced by cobalt-60 or cesium-137 for the treatment of foods. The irradiation technology is much like that used to sterilize medical supplies. Radioactive cobalt pellets are placed in stainless steel tubes, which are arranged in racks. When food moves through the series of racks, the gamma rays pass through the food and kill the bacteria.

It is important for consumers to understand that when food is irradiated, it never comes in contact with the radioactive source. The gamma rays pass through the food to kill bacteria, but that does not make the food radioactive. The radiation kills bacteria because it stops their ability to divide and grow. We cook or heat food thoroughly for the same purpose. Radiation, as well as heat, has little effect on the food itself because its cells are no longer dividing or growing. Thus irradiated food is not harmed, although a small amount of vitamins B_1 and C may be lost.

Currently, tomatoes, blueberries, strawberries, and mushrooms are being irradiated to allow them to be harvested when completely ripe and extend their shelf life (see **FIGURE 5.3**). The FDA has also approved the irradiation of pork, poultry, and beef to decrease potential infections and to extend shelf life. Currently, irradiated vegetable and meat products are available in retail markets in more than 40 countries. In the United States, irradiated foods such as tropical fruits, spinach, and ground meats are found in some stores. *Apollo 17* astronauts ate irradiated foods on the Moon, and some U.S. hospitals and nursing homes now use irradiated poultry to reduce the possibility of salmonella infections among residents. The extended shelf life of irradiated food also makes it useful for campers and military personnel. Soon, consumers concerned about food safety will have a choice of irradiated meats, fruits, and vegetables at the market.

(a) (b)

FIGURE 5.3 ▶ **(a)** The FDA requires this symbol to appear on irradiated retail foods. **(b)** After two weeks, the irradiated strawberries on the right show no spoilage. Mold is growing on the nonirradiated ones on the left.

Q Why are irradiated foods used on spaceships and in nursing homes?

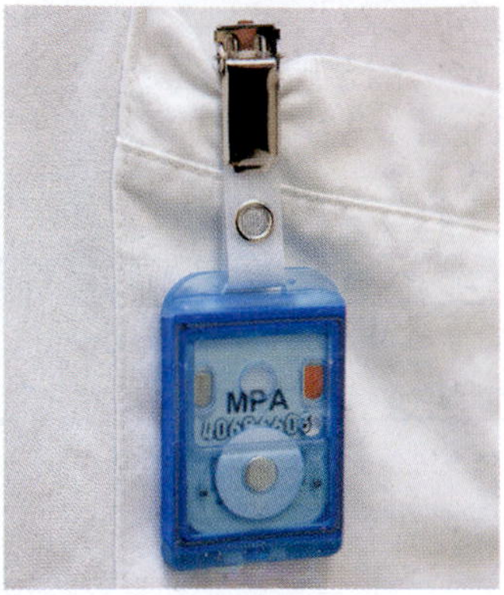
MPA

A dosimeter measures radiation exposure.

TEST

Try Practice Problems 5.23 to 5.28

TABLE 5.5 Average Annual Radiation Received by a Person in the United States

Source	Dose (mSv)
Natural	
Ground	0.2
Air, water, food	0.3
Cosmic rays	0.4
Wood, concrete, brick	0.5
Medical	
Chest X-ray	0.2
Dental X-ray	0.2
Mammogram	0.4
Hip X-ray	0.6
Lumbar spine X-ray	0.7
Upper gastrointestinal tract X-ray	2
Other	
Nuclear power plants	0.001
Television	0.2
Air travel	0.1
Radon	2*

*Varies widely.

TABLE 5.6 Lethal Doses of Radiation for Some Life Forms

Life Form	LD_{50} (Sv)
Insect	1000
Bacterium	500
Rat	8
Human	5
Dog	3

People who work in radiology laboratories wear dosimeters attached to their clothing to determine any exposure to radiation such as X-rays, gamma rays, or beta particles. A dosimeter can be thermoluminescent (TLD), optically stimulated luminescence (OSL), or electronic personal (EPD). Dosimeters provide real-time radiation levels measured by monitors in the work area.

▶ SAMPLE PROBLEM 5.6 Radiation Measurement

TRY IT FIRST

One treatment for bone pain involves intravenous administration of the radioisotope phosphorus-32, which is incorporated into bone. A typical dose of 7.0 mCi can produce up to 450 rad in the bone. What is the difference between the units of mCi and rad?

SOLUTION

The millicuries (mCi) indicate the activity of the P-32 in terms of the number of nuclei that break down in 1 s. The radiation absorbed dose (rad) is a measure of the amount of radiation absorbed by the bone.

STUDY CHECK 5.6

a. What is the absorbed dose of 450 rad in grays (Gy)?
b. What is the activity of 7.0 mCi in becquerels (Bq)?

ANSWER

a. 4.5 Gy
b. 2.6×10^8 Bq

Exposure to Radiation

Every day, we are exposed to low levels of radiation from naturally occurring radioactive isotopes in the buildings where we live and work, in our food and water, and in the air we breathe. For example, potassium-40, a naturally occurring radioactive isotope, is present in any potassium-containing food. Other naturally occurring radioisotopes in air and food are carbon-14, radon-222, strontium-90, and iodine-131. The average person in the United States is exposed to about 3.6 mSv of radiation annually. Medical sources of radiation, including dental, hip, spine, and chest X-rays and mammograms, add to our radiation exposure. **TABLE 5.5** lists some common sources of radiation.

Another source of background radiation is cosmic radiation produced in space by the Sun. People who live at high altitudes or travel by airplane receive a greater amount of cosmic radiation because there are fewer molecules in the atmosphere to absorb the radiation. For example, a person living in Denver receives about twice the cosmic radiation as a person living in Los Angeles. A person living close to a nuclear power plant normally does not receive much additional radiation, perhaps 0.001 mSv in 1 yr. However, in the accident at the Chernobyl nuclear power plant in 1986 in Ukraine, it is estimated that people in a nearby town received as much as 10 mSv/h.

Radiation Sickness

The larger the dose of radiation received at one time, the greater the effect on the body. Exposure to radiation of less than 0.25 Sv usually cannot be detected. Whole-body exposure of 1 Sv produces a temporary decrease in the number of white blood cells. If the exposure to radiation is greater than 1 Sv, a person may suffer the symptoms of radiation sickness: nausea, vomiting, fatigue, and a reduction in white-cell count. A whole-body dosage greater than 3 Sv can decrease the white-cell count to zero. The person suffers diarrhea, hair loss, and infection. Exposure to radiation of 5 Sv is expected to cause death in 50% of the people receiving that dose. This amount of radiation to the whole body is called the *lethal dose for one-half the population*, or the LD_{50}. The LD_{50} varies for different life forms, as **TABLE 5.6** shows. Whole-body radiation of 6 Sv or greater would be fatal to all humans within a few weeks.

PRACTICE PROBLEMS

5.3 Radiation Measurement

5.23 Match each property (**1** to **3**) with its unit of measurement.
1. activity
2. absorbed dose
3. biological damage

a. rad **b.** mrem
c. mCi **d.** Gy

5.24 Match each property (**1** to **3**) with its unit of measurement.
1. activity
2. absorbed dose
3. biological damage

a. mrad **b.** gray
c. becquerel **d.** Sv

Clinical Applications

5.25 Two technicians in a nuclear laboratory were accidentally exposed to radiation. If one was exposed to 8 mGy and the other to 5 rad, which technician received more radiation?

5.26 Two samples of a radioisotope were spilled in a nuclear laboratory. The activity of one sample was 8 kBq and the other 15 mCi. Which sample produced the higher amount of radiation?

5.27 **a.** The recommended dosage of iodine-131 is 4.20 μCi/kg of body mass. How many microcuries of iodine-131 are needed for a 70.0-kg person with hyperthyroidism?
b. A person receives 50 rad of gamma radiation. What is that amount in grays?

5.28 **a.** The dosage of technetium-99m for a lung scan is 20. μCi/kg of body mass. How many millicuries of technetium-99m should be given to a 50.0-kg person (1 mCi $=$ 1000 μCi)?
b. Suppose a person absorbed 50 mrad of alpha radiation. What would be the equivalent dose in millisieverts?

5.4 Half-Life of a Radioisotope

REVIEW
Interpreting Graphs (1.4)
Using Conversion Factors (2.6)

LEARNING GOAL Given the half-life of a radioisotope, calculate the amount of radioisotope remaining after one or more half-lives.

The **half-life** of a radioisotope is the amount of time it takes for one-half of a sample to decay. For example, $^{131}_{53}I$ has a half-life of 8.0 days. As $^{131}_{53}I$ decays, it produces the nonradioactive isotope $^{131}_{54}Xe$ and a beta particle.

$$^{131}_{53}I \longrightarrow ^{131}_{54}Xe + ^{0}_{-1}e$$

Suppose we have a sample that initially contains 20. mg of $^{131}_{53}I$. In 8.0 days, one-half (10. mg) of all the I-131 nuclei in the sample will decay, which leaves 10. mg of I-131. After 16 days (two half-lives), 5.0 mg of the remaining I-131 decays, which leaves 5.0 mg of I-131. After 24 days (three half-lives), 2.5 mg of the remaining I-131 decays, which leaves 2.5 mg of I-131 nuclei still capable of producing radiation.

ENGAGE
If a 24-mg sample of Tc-99m has a half-life of 6.0 h, why are only 3 mg of Tc-99m radioactive after 18 h?

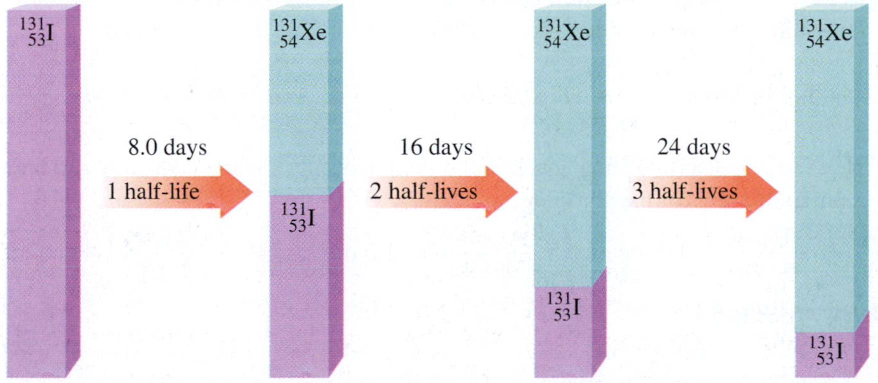

$^{131}_{53}I$
8.0 days
1 half-life
$^{131}_{54}Xe$
$^{131}_{53}I$
16 days
2 half-lives
$^{131}_{54}Xe$
$^{131}_{53}I$
24 days
3 half-lives
$^{131}_{54}Xe$
$^{131}_{53}I$

$$20.\text{ mg of } ^{131}_{53}I \xrightarrow[8.0\text{ days}]{1\text{ half-life}} 10.\text{ mg of } ^{131}_{53}I \xrightarrow[16\text{ days}]{2\text{ half-lives}} 5.0\text{ mg of } ^{131}_{53}I \xrightarrow[24\text{ days}]{3\text{ half-lives}} 2.5\text{ mg of } ^{131}_{53}I$$

In one half-life, the activity of an isotope decreases by half.

A **decay curve** is a diagram of the decay of a radioactive isotope. **FIGURE 5.4** shows such a curve for the $^{131}_{53}I$ we have discussed.

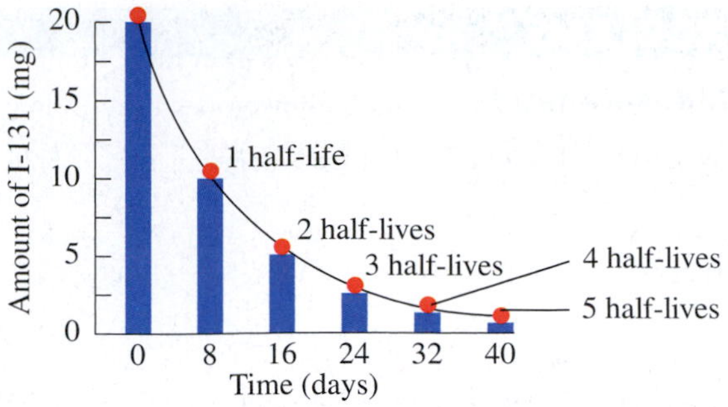

Amount of I-131 (mg)
20
15
10
5
0
1 half-life
2 half-lives
3 half-lives
4 half-lives
5 half-lives
0
8
16
24
32
40
Time (days)

FIGURE 5.4 ▶ The decay curve for iodine-131 shows that one-half of the radioactive sample decays and one-half remains radioactive after each half-life of 8.0 days.

Q How many milligrams of the 20.-mg sample remain radioactive after two half-lives?

CORE CHEMISTRY SKILL

Using Half-Lives

▶ SAMPLE PROBLEM 5.7 Using Half-Lives of a Radioisotope in Medicine

TRY IT FIRST

Phosphorus-32, a radioisotope used in the treatment of leukemia, has a half-life of 14.3 days. If a sample contains 8.0 mg of phosphorus-32, how many milligrams of phosphorus-32 remain after 42.9 days?

SOLUTION

STEP 1 **State the given and needed quantities.**

	Given	Need	Connect
ANALYZE THE PROBLEM	8.0 mg of P-32, 42.9 days elapsed, half-life = 14.3 days	milligrams of P-32 remaining	number of half-lives

STEP 2 **Write a plan to calculate the unknown quantity.**

days → Half-life → number of half-lives

milligrams of $^{32}_{15}P$ → Number of half-lives → milligrams of $^{32}_{15}P$ remaining

STEP 3 **Write the half-life equality and conversion factors.**

1 half-life = 14.3 days

$$\frac{14.3 \text{ days}}{1 \text{ half-life}} \quad \text{and} \quad \frac{1 \text{ half-life}}{14.3 \text{ days}}$$

STEP 4 **Set up the problem to calculate the needed quantity.** First, we determine the number of half-lives in the amount of time that has elapsed.

$$\text{number of half-lives} = 42.9 \cancel{\text{days}} \times \frac{1 \text{ half-life}}{14.3 \cancel{\text{days}}} = 3.00 \text{ half-lives}$$

Now we can determine how much of the sample decays in three half-lives and how many milligrams of the phosphorus remain.

$$8.0 \text{ mg of } ^{32}_{15}P \xrightarrow[14.3 \text{ days}]{1 \text{ half-life}} 4.0 \text{ mg of } ^{32}_{15}P \xrightarrow[28.6 \text{ days}]{2 \text{ half-lives}} 2.0 \text{ mg of } ^{32}_{15}P \xrightarrow[42.9 \text{ days}]{3 \text{ half-lives}} 1.0 \text{ mg of } ^{32}_{15}P$$

STUDY CHECK 5.7

a. Iron-59 has a half-life of 44 days. If a nuclear laboratory receives a sample of 8.0 μg of iron-59, how many micrograms of iron-59 are still active after 176 days?

b. Rubidium-82, used to diagnose myocardial disease, has a half-life of 1.27 min. If the initial dose is 1480 MBq, how many half-lives have elapsed, and what is the activity of Rb-82 after 5.08 min?

ANSWER

a. 0.50 μg of iron-59 **b.** 4 half-lives; 92.5 MBq

INTERACTIVE VIDEO

Half-Lives

PEARSON eText 2.0

Naturally occurring isotopes of the elements usually have long half-lives, as shown in **TABLE 5.7**. They disintegrate slowly and produce radiation over a long period of time, even hundreds or millions of years. In contrast, the radioisotopes used in nuclear medicine have much shorter half-lives. They disintegrate rapidly and produce almost all their radiation in a short period of time. For example, technetium-99m emits half of its radiation in the first 6 h. This means that a small amount of the radioisotope given to a patient is essentially gone within two days. The decay products of technetium-99m are totally eliminated by the body.

ENGAGE

Why do radioisotopes used in nuclear medicine have shorter half-lives than naturally occurring radioisotopes?

TABLE 5.7 Half-Lives of Some Radioisotopes

Element	Radioisotope	Half-Life	Type of Radiation
Naturally Occurring Radioisotopes			
Carbon-14	${}^{14}_{6}C$	5730 yr	Beta
Potassium-40	${}^{40}_{19}K$	1.3×10^9 yr	Beta, gamma
Radium-226	${}^{226}_{88}Ra$	1600 yr	Alpha
Strontium-90	${}^{90}_{38}Sr$	38.1 yr	Alpha
Uranium-238	${}^{238}_{92}U$	4.5×10^9 yr	Alpha
Some Medical Radioisotopes			
Carbon-11	${}^{11}_{6}C$	20. min	Positron
Chromium-51	${}^{51}_{24}Cr$	28 days	Gamma
Iodine-131	${}^{131}_{53}I$	8.0 days	Gamma
Oxygen-15	${}^{15}_{8}O$	2.0 min	Positron
Iron-59	${}^{59}_{26}Fe$	44 days	Beta, gamma
Radon-222	${}^{222}_{86}Rn$	3.8 days	Alpha
Technetium-99m	${}^{99m}_{43}Tc$	6.0 h	Gamma

TEST

Try Practice Problems 5.29 to 5.34

Chemistry Link to the Environment

Dating Ancient Objects

Radiological dating is a technique used by geologists, archaeologists, and historians to determine the age of ancient objects. The age of an object derived from plants or animals (such as wood, fiber, natural pigments, bone, and cotton or woolen clothing) is determined by measuring the amount of carbon-14, a naturally occurring radioactive form of carbon. In 1960, Willard Libby received the Nobel Prize for the work he did developing carbon-14 dating techniques during the 1940s. Carbon-14 is produced in the upper atmosphere by the bombardment of ${}^{14}_{7}N$ by high-energy neutrons from cosmic rays.

$$\underset{\text{Neutron from cosmic rays}}{{}^{1}_{0}n} + \underset{\text{Nitrogen in atmosphere}}{{}^{14}_{7}N} \longrightarrow \underset{\text{Radioactive carbon-14}}{{}^{14}_{6}C} + \underset{\text{Proton}}{{}^{1}_{1}H}$$

The carbon-14 reacts with oxygen to form radioactive carbon dioxide, ${}^{14}_{6}CO_2$. Living plants continuously absorb carbon dioxide, which incorporates carbon-14 into the plant material. The uptake of carbon-14 stops when the plant dies. As the carbon-14 undergoes beta decay, the amount of radioactive carbon-14 in the plant material steadily decreases.

$${}^{14}_{6}C \longrightarrow {}^{14}_{7}N + {}^{0}_{-1}e$$

In a process called *carbon dating*, scientists use the half-life of carbon-14 (5730 yr) to calculate the length of time since the plant died. For example, a wooden beam found in an ancient dwelling might have one-half of the carbon-14 found in living plants today. Because one half-life of carbon-14 is 5730 yr, the dwelling was constructed about 5730 yr ago. Carbon-14 dating was used to determine that the Dead Sea Scrolls are about 2000 yr old.

A radiological dating method used for determining the age of much older items is based on the radioisotope uranium-238, which decays through a series of reactions to lead-206. The uranium-238 isotope has an incredibly long half-life, about 4×10^9 yr. Measurements of the amounts of uranium-238 and lead-206 enable geologists to determine the age of rock samples. The older rocks will have a higher percentage of lead-206 because more of the uranium-238 has decayed. The age of rocks brought back from the Moon by the *Apollo* missions, for example, was determined using uranium-238. They were found to be about 4×10^9 yr old, approximately the same age calculated for Earth.

The age of the Dead Sea Scrolls was determined using carbon-14.

The age of a bone sample from a skeleton can be determined by carbon dating.

▶ SAMPLE PROBLEM 5.8 Dating Using Half-Lives

TRY IT FIRST

The bones of humans and animals assimilate carbon until death. Using radiocarbon dating, the number of half-lives of carbon-14 from a bone sample determines the age of the bone. Suppose a sample is obtained from a prehistoric animal and used for radiocarbon dating. We can calculate the age of the bone or the years elapsed since the animal died by using the half-life of carbon-14, which is 5730 yr. A bone sample from the skeleton of a prehistoric animal has 25% of the activity of C-14 found in a living animal. How many years ago did the prehistoric animal die?

SOLUTION

STEP 1 State the given and needed quantities.

ANALYZE THE PROBLEM	Given	Need	Connect
	half-life = 5730 yr, 25% of initial C-14 activity	years elapsed	number of half-lives

STEP 2 Write a plan to calculate the unknown quantity.

$$\text{Activity:}\quad 100\%\ (\text{initial}) \xrightarrow{1.0\ \text{half-life}} 50\% \xrightarrow{2.0\ \text{half-lives}} 25\%$$

STEP 3 Write the half-life equality and conversion factors.

$$1\ \text{half-life} = 5730\ \text{yr}$$
$$\frac{5730\ \text{yr}}{1\ \text{half-life}} \quad \text{and} \quad \frac{1\ \text{half-life}}{5730\ \text{yr}}$$

STEP 4 Set up the problem to calculate the needed quantity.

$$\text{Years elapsed} = 2.0\ \cancel{\text{half-lives}} \times \frac{5730\ \text{yr}}{1\ \cancel{\text{half-life}}} = 11\,000\ \text{yr}$$

We would estimate that the animal died 11 000 yr ago.

STUDY CHECK 5.8

Suppose that a piece of wood found in a cave had one-eighth of its original carbon-14 activity. About how many years ago was the wood part of a living tree?

ANSWER

17 000 yr

PRACTICE PROBLEMS

5.4 Half-Life of a Radioisotope

5.29 For each of the following, indicate if the number of half-lives elapsed is:
1. one half-life
2. two half-lives
3. three half-lives

a. a sample of Pd-103 with a half-life of 17 days after 34 days
b. a sample of C-11 with a half-life of 20 min after 20 min
c. a sample of At-211 with a half-life of 7 h after 21 h

5.30 For each of the following, indicate if the number of half-lives elapsed is:
1. one half-life
2. two half-lives
3. three half-lives

a. a sample of Ce-141 with a half-life of 32.5 days after 32.5 days
b. a sample of F-18 with a half-life of 110 min after 330 min
c. a sample of Au-198 with a half-life of 2.7 days after 5.4 days

Clinical Applications

5.31 Technetium-99m is an ideal radioisotope for scanning organs because it has a half-life of 6.0 h and is a pure gamma emitter. Suppose that 80.0 mg were prepared in the technetium generator this morning. How many milligrams of technetium-99m would remain after each of the following intervals?
a. one half-life **b.** two half-lives
c. 18 h **d.** 1.5 days

5.32 A sample of sodium-24 with an activity of 12 mCi is used to study the rate of blood flow in the circulatory system. If sodium-24 has a half-life of 15 h, what is the activity after each of the following intervals?
a. one half-life **b.** 30 h
c. three half-lives **d.** 2.5 days

5.33 Strontium-85, used for bone scans, has a half-life of 65 days.
a. How long will it take for the radiation level of strontium-85 to drop to one-fourth of its original level?
b. How long will it take for the radiation level of strontium-85 to drop to one-eighth of its original level?

5.34 Fluorine-18, which has a half-life of 110 min, is used in PET scans.
a. If 100. mg of fluorine-18 is shipped at 8:00 A.M., how many milligrams of the radioisotope are still active after 110 min?
b. If 100. mg of fluorine-18 is shipped at 8:00 A.M., how many milligrams of the radioisotope are still active when the sample arrives at the radiology laboratory at 1:30 P.M.?

5.5 Medical Applications Using Radioactivity

LEARNING GOAL Describe the use of radioisotopes in medicine.

The first radioactive isotope was used to treat a person with leukemia at the University of California at Berkeley. In 1946, radioactive iodine was successfully used to diagnose thyroid function and to treat hyperthyroidism and thyroid cancer. Radioactive isotopes are now used to produce images of organs including the liver, spleen, thyroid, kidneys, brain, and heart.

To determine the condition of an organ in the body, a radiation technologist may use a radioisotope that concentrates in that organ. The cells in the body do not differentiate between a nonradioactive atom and a radioactive one, so these radioisotopes are easily incorporated. Then the radioactive atoms are detected because they emit radiation. Some radioisotopes used in nuclear medicine are listed in **TABLE 5.8**.

TABLE 5.8 Medical Applications of Radioisotopes

Isotope	Half-Life	Radiation	Medical Application
Au-198	2.7 days	Beta	Liver imaging; treatment of abdominal carcinoma
Bi-213	46 min	Alpha	Treatment of leukemia
Ce-141	32.5 days	Beta	Gastrointestinal tract diagnosis; measuring blood flow to the heart
Cs-131	9.7 days	Gamma	Prostate brachytherapy
F-18	110 min	Positron	Positron emission tomography (PET)
Ga-67	78 h	Gamma	Abdominal imaging; tumor detection
Ga-68	68 min	Positron	Detection of pancreatic cancer
I-123	13.2 h	Gamma	Treatment of thyroid, brain, and prostate cancer
I-131	8.0 days	Beta	Treatment of Graves' disease, goiter, hyperthyroidism, thyroid and prostate cancer
Ir-192	74 days	Gamma	Treatment of breast and prostate cancer
P-32	14.3 days	Beta	Treatment of leukemia, excess red blood cells, and pancreatic cancer
Pd-103	17 days	Gamma	Prostate brachytherapy
Sm-153	46 h	Beta	Treatment of bone cancer
Sr-85	65 days	Gamma	Detection of bone lesions; brain scans
Tc-99m	6.0 h	Gamma	Imaging of skeleton and heart muscle, brain, liver, heart, lungs, bone, spleen, kidney, and thyroid; most widely used radioisotope in nuclear medicine
Xe-133	5.2 days	Beta	Pulmonary function diagnosis
Y-90	2.7 days	Beta	Treatment of liver cancer

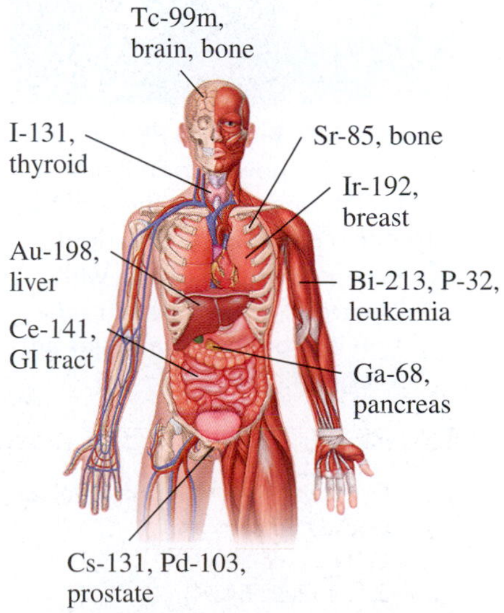
Tc-99m, brain, bone
I-131, thyroid
Sr-85, bone
Ir-192, breast
Au-198, liver
Bi-213, P-32, leukemia
Ce-141, GI tract
Ga-68, pancreas
Cs-131, Pd-103, prostate

Different radioisotopes are used to diagnose and treat a number of diseases.

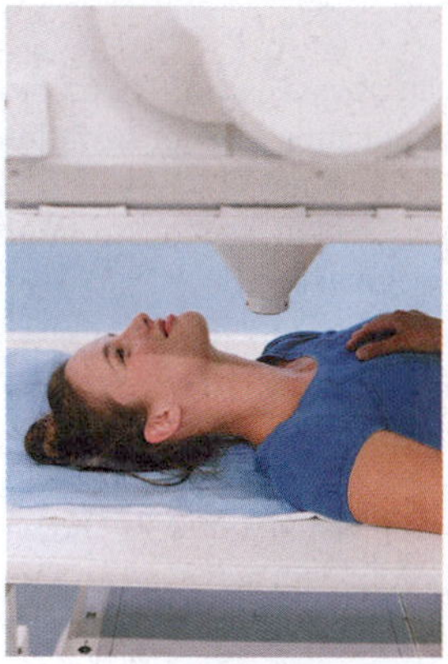

(a)

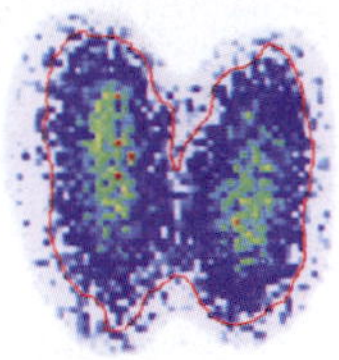

(b)

FIGURE 5.5 ▶ **(a)** A scanner detects radiation from a radioisotope in an organ. **(b)** A scan shows radioactive iodine-131 in the thyroid.

Q What type of radiation would move through body tissues to create a scan?

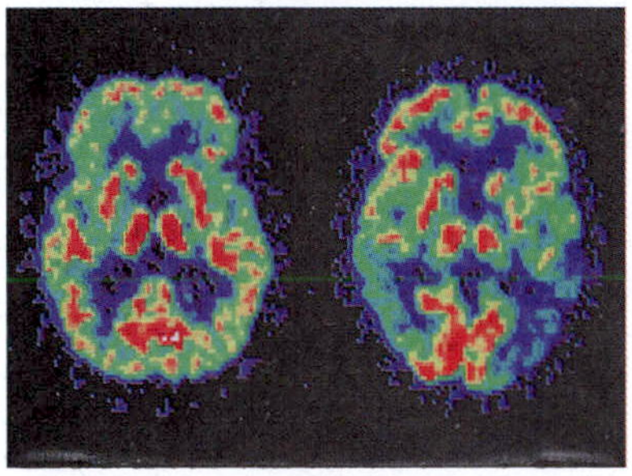

FIGURE 5.6 ▶ These PET scans of the brain show a normal brain on the left and a brain affected by Alzheimer's disease on the right.

Q When positrons collide with electrons, what type of radiation is produced that gives an image of an organ?

Scans with Radioisotopes

After a person receives a radioisotope, the radiation technologist determines the level and location of radioactivity emitted by the radioisotope. An apparatus called a *scanner* is used to produce an image of the organ. The scanner moves slowly across the body above the region where the organ containing the radioisotope is located. The gamma rays emitted from the radioisotope in the organ can be used to expose a photographic plate, producing a *scan* of the organ. On a scan, an area of decreased or increased radiation can indicate conditions such as a disease of the organ, a tumor, a blood clot, or edema.

A common method of determining thyroid function is the use of *radioactive iodine uptake*. Taken orally, the radioisotope iodine-131 mixes with the iodine already present in the thyroid. Twenty-four hours later, the amount of iodine taken up by the thyroid is determined. A detection tube held up to the area of the thyroid gland detects the radiation coming from the iodine-131 that has located there (see **FIGURE 5.5**).

A person with a hyperactive thyroid will have a higher than normal level of radioactive iodine, whereas a person with a hypoactive thyroid will have lower values. If a person has hyperthyroidism, treatment is begun to lower the activity of the thyroid. One treatment involves giving a therapeutic dosage of radioactive iodine, which has a higher radiation level than the diagnostic dose. The radioactive iodine goes to the thyroid where its radiation destroys some of the thyroid cells. The thyroid produces less thyroid hormone, bringing the hyperthyroid condition under control.

Positron Emission Tomography

Positron emitters with short half-lives such as carbon-11, oxygen-15, nitrogen-13, and fluorine-18 are used in an imaging method called *positron emission tomography* (PET). A positron-emitting isotope such as fluorine-18 combined with substances in the body such as glucose is used to study brain function, metabolism, and blood flow.

$$^{18}_{9}\text{F} \longrightarrow {}^{18}_{8}\text{O} + {}^{0}_{+1}e$$

As positrons are emitted, they combine with electrons to produce gamma rays that are detected by computerized equipment to create a three-dimensional image of the organ (see **FIGURE 5.6**).

Nonradioactive Imaging

Computed Tomography

Another imaging method used to scan organs such as the brain, lungs, and heart is *computed tomography* (CT). A computer monitors the absorption of 30 000 X-ray beams directed at successive layers of the target organ. Based on the densities of the tissues and fluids in the organ, the differences in absorption of the X-rays provide a series of images of the organ. This technique is successful in the identification of hemorrhages, tumors, and atrophy.

Magnetic Resonance Imaging

Magnetic resonance imaging (MRI) is a powerful imaging technique that does not involve nuclear radiation. It is the least invasive imaging method available. MRI is based on the absorption of energy when the protons in hydrogen atoms are excited by a strong magnetic field. The difference in energy between the two states is released, which produces the

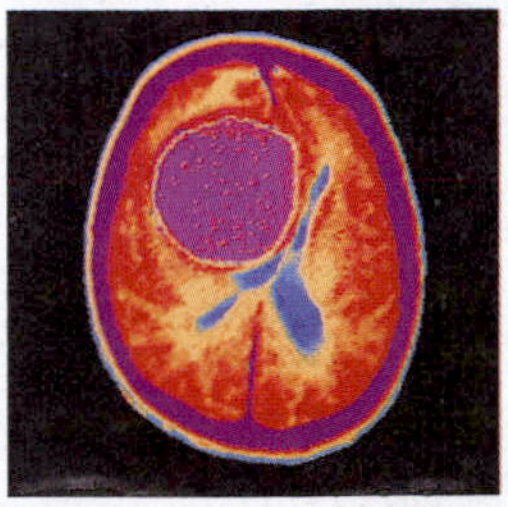

A CT scan shows a tumor (purple) in the brain.

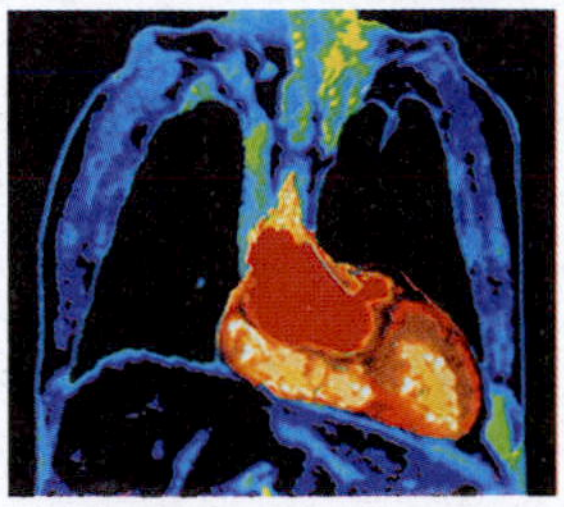

An MRI scan provides images of the heart and lungs.

electromagnetic signal that the scanner detects. These signals are sent to a computer system, where a color image of the body is generated. MRI is particularly useful in obtaining images of soft tissues, which contain large amounts of hydrogen atoms.

TEST

Try Practice Problems 5.35 to 5.40

Chemistry Link to Health

Brachytherapy

The process called *brachytherapy*, or seed implantation, is an internal form of radiation therapy. The prefix *brachy* is from the Greek word for short distance. With internal radiation, a high dose of radiation is delivered to a cancerous area, whereas normal tissue sustains minimal damage. Because higher doses are used, fewer treatments of shorter duration are needed. Conventional external treatment delivers a lower dose per treatment, but requires six to eight weeks of treatment.

Permanent Brachytherapy

One of the most common forms of cancer in males is prostate cancer. In addition to surgery and chemotherapy, one treatment option is to place 40 or more titanium capsules, or "seeds," in the malignant area. Each seed, which is the size of a grain of rice, contains radioactive iodine-125, palladium-103, or cesium-131, which decay by gamma emission. The radiation from the seeds destroys the cancer by interfering with the reproduction of cancer cells with minimal damage to adjacent normal tissues. Ninety percent (90%) of the radioisotopes decay within a few months because they have short half-lives.

Isotope	I-125	Pd-103	Cs-131
Radiation	Gamma	Gamma	Gamma
Half-Life	60 days	17 days	10 days
Time Required to Deliver 90% of Radiation	7 months	2 months	1 month

Almost no radiation passes out of the patient's body. The amount of radiation received by a family member is no greater than that received on a long plane flight. Because the radioisotopes decay to products that are not radioactive, the inert titanium capsules can be left in the body.

Temporary Brachytherapy

In another type of treatment for prostate cancer, long needles containing iridium-192 are placed in the tumor. However, the needles are removed after 5 to 10 min, depending on the activity of the iridium isotope. Compared to permanent brachytherapy, temporary brachytherapy can deliver a higher dose of radiation over a shorter time. The procedure may be repeated in a few days.

Brachytherapy is also used following breast cancer lumpectomy. An iridium-192 isotope is inserted into the catheter implanted in the space left by the removal of the tumor. Radiation is delivered primarily to the tissue surrounding the cavity that contained the tumor and where the cancer is most likely to recur. The procedure is repeated twice a day for five days to give an absorbed dose of 34 Gy (3400 rad). The catheter is removed, and no radioactive material remains in the body.

In conventional external beam therapy for breast cancer, a patient is given 2 Gy once a day for six to seven weeks, which gives a total absorbed dose of about 80 Gy or 8000 rad. The external beam therapy irradiates the entire breast, including the tumor cavity.

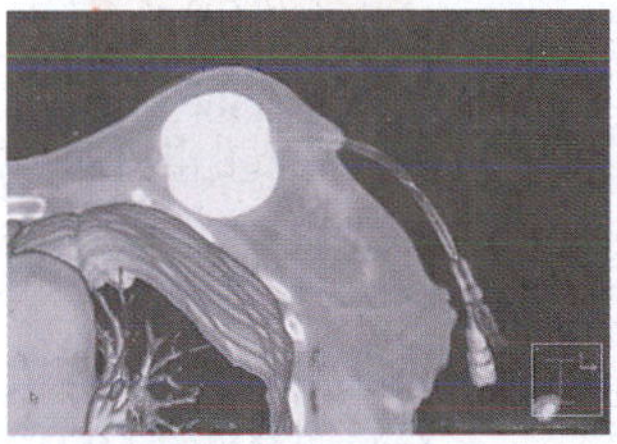

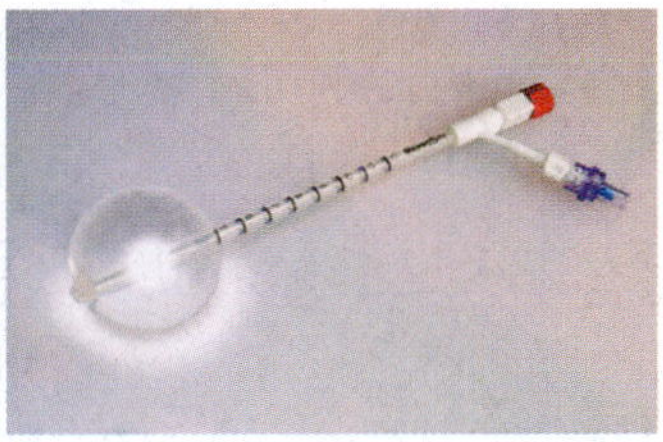

A catheter placed temporarily in the breast for radiation from Ir-192.

PRACTICE PROBLEMS

5.5 Medical Applications Using Radioactivity

Clinical Applications

5.35 Bone and bony structures contain calcium and phosphorus.
- **a.** Why would the radioisotopes calcium-47 and phosphorus-32 be used in the diagnosis and treatment of bone diseases?
- **b.** During nuclear tests, scientists were concerned that strontium-85, a radioactive product, would be harmful to the growth of bone in children. Explain.

5.36
- **a.** Technetium-99m emits only gamma radiation. Why would this type of radiation be used in diagnostic imaging rather than an isotope that also emits beta or alpha radiation?
- **b.** A person with *polycythemia vera* (excess production of red blood cells) receives radioactive phosphorus-32. Why would this treatment reduce the production of red blood cells in the bone marrow of the patient?

5.37 In a diagnostic test for leukemia, a person receives 4.0 mL of a solution containing selenium-75. If the activity of the selenium-75 is 45 μCi/mL, what dose, in microcuries, does the patient receive?

5.38 A vial contains radioactive iodine-131 with an activity of 2.0 mCi/mL. If a thyroid test requires 3.0 mCi in an "atomic cocktail," how many milliliters are used to prepare the iodine-131 solution?

5.39 Gallium-68 is taken up by tumors; the emission of positrons allows the tumors to be located.
- **a.** Write an equation for the positron emission of Ga-68.
- **b.** If the half-life is 68 min, how much of a 64-mcg sample is active after 136 min?

5.40 Xenon-133 is used to test lung function; it decays by emitting a beta particle.
- **a.** Write an equation for the beta decay of Xe-133.
- **b.** If the half-life of Xe-133 is 5.2 h, how much of a 20.-mCi sample is still active after 15.6 h?

5.6 Nuclear Fission and Fusion

LEARNING GOAL Describe the processes of nuclear fission and fusion.

During the 1930s, scientists bombarding uranium-235 with neutrons discovered that the U-235 nucleus splits into two smaller nuclei and produces a great amount of energy. This was the discovery of **nuclear fission**. The energy generated by splitting the atom was called *atomic energy*. A typical equation for nuclear fission is

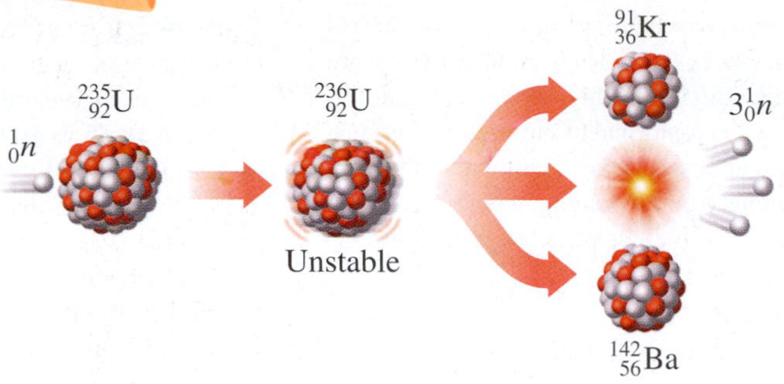

$^{91}_{36}Kr$
$^{235}_{92}U$
$^{236}_{92}U$
$^{1}_{0}n$
$3^{1}_{0}n$
Unstable
$^{142}_{56}Ba$

$$^{1}_{0}n + ^{235}_{92}U \longrightarrow ^{236}_{92}U \longrightarrow ^{91}_{36}Kr + ^{142}_{56}Ba + 3^{1}_{0}n + \text{energy}$$

If we could determine the mass of the products krypton, barium, and three neutrons with great accuracy, we would find that their total mass is slightly less than the mass of the starting materials. The missing mass has been converted into an enormous amount of energy, consistent with the famous equation derived by Albert Einstein:

$$E = mc^2$$

where E is the energy released, m is the mass lost, and c is the speed of light, 3×10^8 m/s. Even though the mass loss is very small, when it is multiplied by the speed of light squared, the result is a large value for the energy released. The fission of 1 g of uranium-235 produces about as much energy as the burning of 3 tons of coal.

Chain Reaction

ENGAGE

Why is a critical mass of uranium-235 necessary to sustain a nuclear chain reaction?

Fission begins when a neutron collides with the nucleus of a uranium atom. The resulting nucleus is unstable and splits into smaller nuclei. This fission process also releases several neutrons and large amounts of gamma radiation and energy. The neutrons emitted have high energies and bombard other uranium-235 nuclei. In a **chain reaction**, there is a rapid increase in the number of high-energy neutrons available to react with more uranium. To sustain a nuclear chain reaction, sufficient quantities of uranium-235 must be brought together to provide a *critical mass* in which almost all the neutrons immediately collide with more uranium-235 nuclei. So much heat and energy build up that an atomic explosion can occur (see **FIGURE 5.7**).

TEST

Try Practice Problems 5.41 to 5.44

Nuclear Fusion

In **fusion**, two small nuclei combine to form a larger nucleus. Mass is lost, and a tremendous amount of energy is released, even more than the energy released from nuclear fission. However, a fusion reaction requires a temperature of 100 000 000 °C to overcome the repulsion of the hydrogen nuclei and cause them to undergo fusion. Fusion reactions occur continuously in the Sun and other stars, providing us with heat and light. The huge amounts of energy produced by our sun come from the fusion of 6×10^{11} kg of hydrogen every second. In a fusion reaction, isotopes of hydrogen combine to form helium and large amounts of energy.

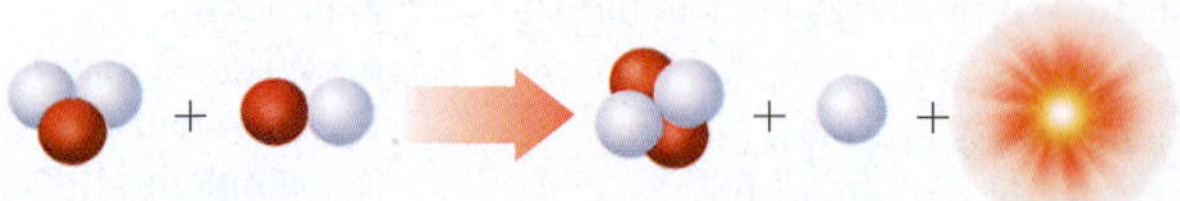

$$^{3}_{1}H + ^{2}_{1}H \longrightarrow ^{4}_{2}He + ^{1}_{0}n + \text{energy}$$

Hydrogen isotopes combine in a fusion reaction to produce helium, a neutron, and energy.

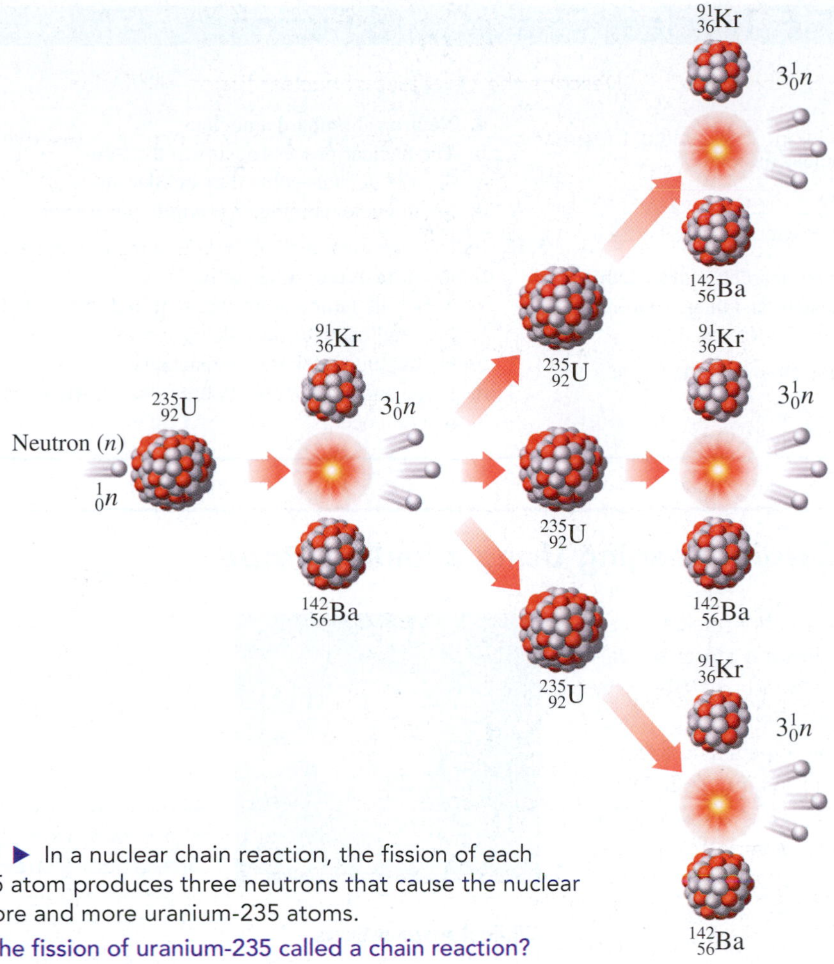

$^{91}_{36}Kr$
3^1_0n
$^{142}_{56}Ba$
$^{235}_{92}U$
Neutron (n)
^1_0n
$^{235}_{92}U$
$^{91}_{36}Kr$
3^1_0n
$^{142}_{56}Ba$
$^{235}_{92}U$
$^{235}_{92}U$
$^{91}_{36}Kr$
3^1_0n
$^{142}_{56}Ba$
$^{91}_{36}Kr$
3^1_0n
$^{142}_{56}Ba$

FIGURE 5.7 ▶ In a nuclear chain reaction, the fission of each uranium-235 atom produces three neutrons that cause the nuclear fission of more and more uranium-235 atoms.

Q Why is the fission of uranium-235 called a chain reaction?

Scientists expect less radioactive waste with shorter half-lives from fusion reactors. However, fusion is still in the experimental stage because the extremely high temperatures needed have been difficult to reach and even more difficult to maintain.

▶ SAMPLE PROBLEM 5.9 Identifying Fission and Fusion

TRY IT FIRST

Indicate whether each of the following is characteristic of the fission or fusion process, or both:

a. A large nucleus breaks apart to produce smaller nuclei.
b. Large amounts of energy are released.
c. Extremely high temperatures are needed for reaction.

SOLUTION

a. When a large nucleus breaks apart to produce smaller nuclei, the process is fission.
b. Large amounts of energy are generated in both the fusion and fission processes.
c. An extremely high temperature is required for fusion.

STUDY CHECK 5.9

Classify each of the following nuclear equations as fission or fusion:

a. $^3_2He + ^3_2He \longrightarrow ^4_2He + 2^1_1H$ **b.** $^1_0n + ^{235}_{92}U \longrightarrow ^{140}_{54}Xe + ^{94}_{38}Sr + 2^1_0n$

ANSWER

a. fusion **b.** fission

TEST

Try Practice Problems 5.45 and 5.46

PRACTICE PROBLEMS

5.6 Nuclear Fission and Fusion

5.41 What is nuclear fission?

5.42 How does a chain reaction occur in nuclear fission?

5.43 Complete the following fission reaction:

$${}^{1}_{0}n + {}^{235}_{92}\text{U} \longrightarrow {}^{131}_{50}\text{Sn} + ? + 2\,{}^{1}_{0}n + \text{energy}$$

5.44 In another fission reaction, uranium-235 bombarded with a neutron produces strontium-94, another small nucleus, and three neutrons. Write the balanced nuclear equation for the fission reaction.

5.45 Indicate whether each of the following is characteristic of the fission or fusion process, or both:

a. Neutrons bombard a nucleus.
b. The nuclear process occurs in the Sun.
c. A large nucleus splits into smaller nuclei.
d. Small nuclei combine to form larger nuclei.

5.46 Indicate whether each of the following is characteristic of the fission or fusion process, or both:

a. Very high temperatures are required to initiate the reaction.
b. Less radioactive waste is produced.
c. Hydrogen nuclei are the reactants.
d. Large amounts of energy are released when the nuclear reaction occurs.

CLINICAL UPDATE Cardiac Imaging Using a Radioisotope

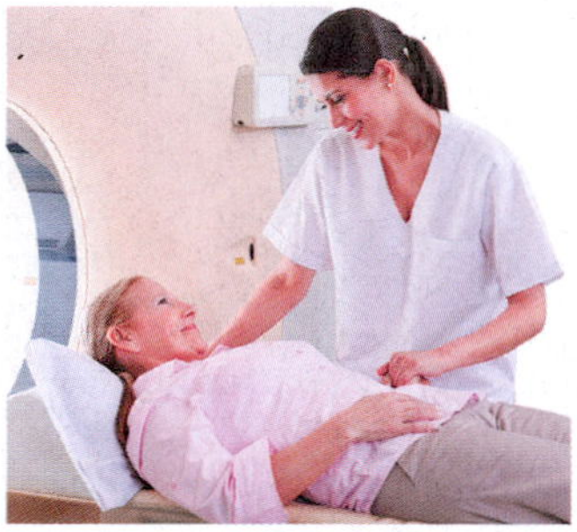

As part of her nuclear stress test, Simone starts to walk on the treadmill. When she reaches the maximum level, Pauline injects a radioactive dye containing Tl-201 with an activity of 74 MBq. The radiation emitted from areas of the heart is detected by a scanner and produces images of her heart muscle. A thallium stress test can determine how effectively coronary arteries provide blood to the heart. If Simone has any damage to her coronary arteries, reduced blood flow during stress would show a narrowing of an artery or a blockage. After Simone rests for 3 h, Pauline injects more dye with Tl-201, and she is placed under the scanner again. A second set of images of her heart muscle at rest are taken. When Simone's doctor reviews her scans, he assures her that she had normal blood flow to her heart muscle, both at rest and under stress.

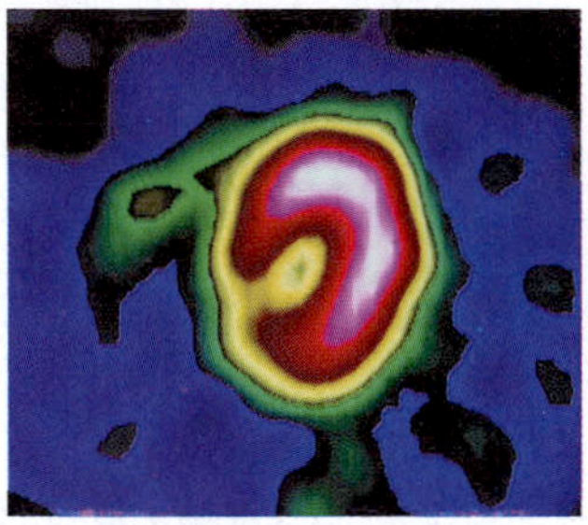

A thallium stress test can show narrowing of an artery during stress.

Clinical Applications

5.47 What is the activity of the radioactive dye injection for Simone

a. in curies? **b.** in millicuries?

5.48 If the half-life of Tl-201 is 3.0 days, what is its activity, in megabecquerels

a. after 3.0 days? **b.** after 6.0 days?

5.49 How many days will it take until the activity of the Tl-201 in Simone's body is one-eighth of the initial activity?

5.50 Radiation from Tl-201 to Simone's kidneys can be 24 mGy. What is this amount of radiation in rads?

CONCEPT MAP

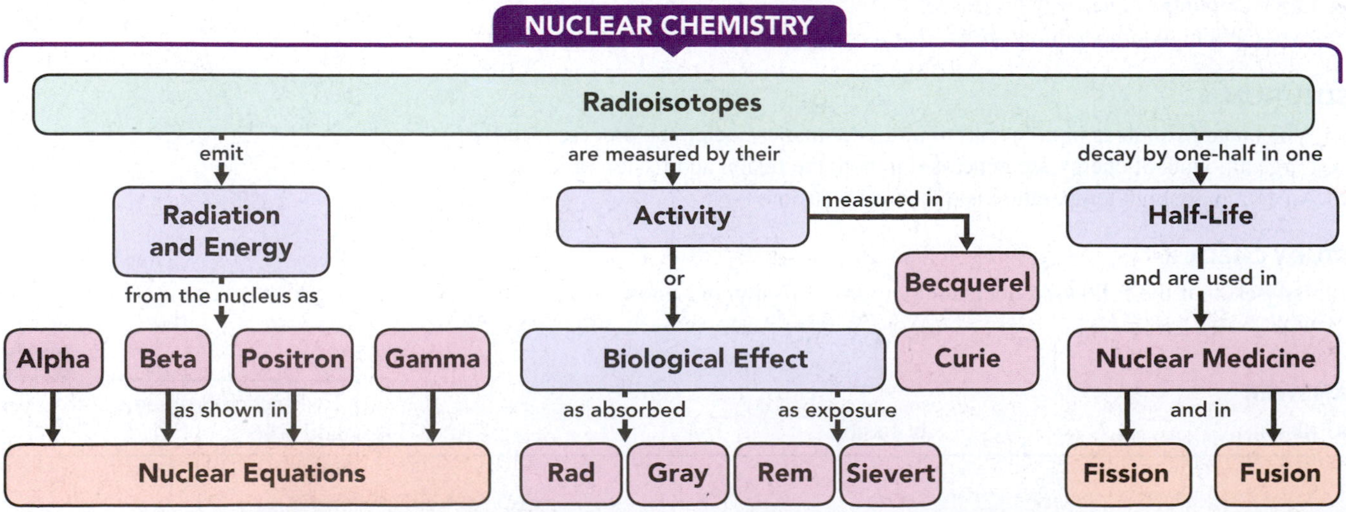

NUCLEAR CHEMISTRY
Radioisotopes
emit
Radiation and Energy
from the nucleus as
Alpha
Beta
Positron
Gamma
as shown in
Nuclear Equations
are measured by their
Activity
measured in
Becquerel
Curie
or
Biological Effect
as absorbed
as exposure
Rad
Gray
Rem
Sievert
decay by one-half in one
Half-Life
and are used in
Nuclear Medicine
and in
Fission
Fusion

CHAPTER REVIEW

5.1 Natural Radioactivity

$^{4}_{2}He$ or α
Alpha particle

LEARNING GOAL Describe alpha, beta, positron, and gamma radiation.

- Radioactive isotopes have unstable nuclei that break down (decay), spontaneously emitting alpha (α), beta (β), positron (β^+), and gamma (γ) radiation.
- Because radiation can damage the cells in the body, proper protection must be used: shielding, limiting the time of exposure, and distance.

5.2 Nuclear Reactions

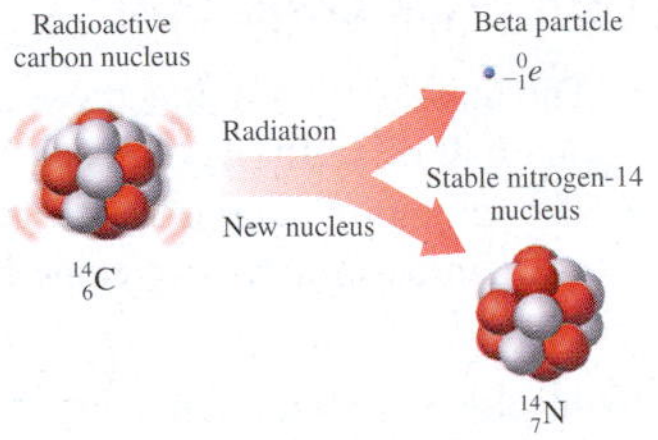

Radioactive carbon nucleus
Beta particle
$^{0}_{-1}e$
Radiation
Stable nitrogen-14 nucleus
New nucleus
$^{14}_{6}C$
$^{14}_{7}N$

LEARNING GOAL Write a balanced nuclear equation for radioactive decay, showing mass numbers and atomic numbers.

- A balanced nuclear equation is used to represent the changes that take place in the nuclei of the reactants and products.
- The new isotopes and the type of radiation emitted can be determined from the symbols that show the mass numbers and atomic numbers of the isotopes in the nuclear equation.
- A radioisotope is produced artificially when a nonradioactive isotope is bombarded by a small particle.

5.3 Radiation Measurement

RADIOACTIVE

LEARNING GOAL Describe the detection and measurement of radiation.

- In a Geiger counter, radiation produces charged particles in the gas contained in a tube, which generates an electrical current.
- The curie (Ci) and the becquerel (Bq) measure the activity, which is the number of nuclear transformations per second.
- The amount of radiation absorbed by a substance is measured in the rad or the gray (Gy).
- The rem and the sievert (Sv) are units used to determine the biological damage from the different types of radiation.

5.4 Half-Life of a Radioisotope

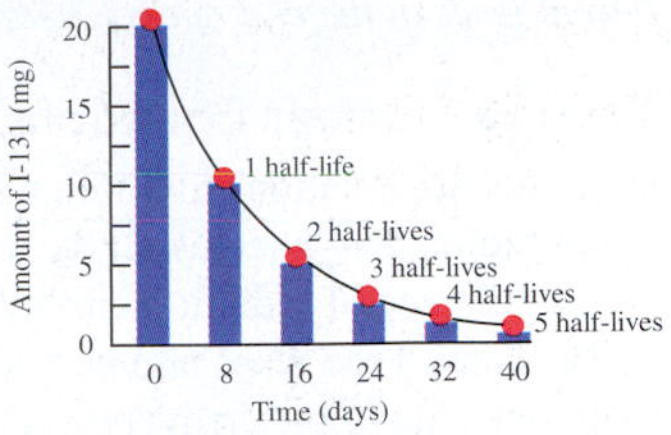

Amount of I-131 (mg)
20
15
10
5
0
1 half-life
2 half-lives
3 half-lives
4 half-lives
5 half-lives
0
8
16
24
32
40
Time (days)

LEARNING GOAL Given the half-life of a radioisotope, calculate the amount of radioisotope remaining after one or more half-lives.

- Every radioisotope has its own rate of emitting radiation.
- The time it takes for one-half of a radioactive sample to decay is called its half-life.
- For many medical radioisotopes, such as Tc-99m and I-131, half-lives are short.
- For other isotopes, usually naturally occurring ones such as C-14, Ra-226, and U-238, half-lives are extremely long.

5.5 Medical Applications Using Radioactivity

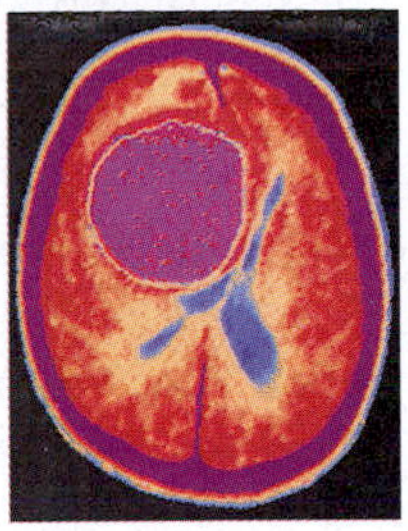

LEARNING GOAL Describe the use of radioisotopes in medicine.

- In nuclear medicine, radioisotopes that go to specific sites in the body are given to the patient.
- By detecting the radiation they emit, an evaluation can be made about the location and extent of an injury, disease, tumor, or the level of function of a particular organ.
- Higher levels of radiation are used to treat or destroy tumors.

5.6 Nuclear Fission and Fusion

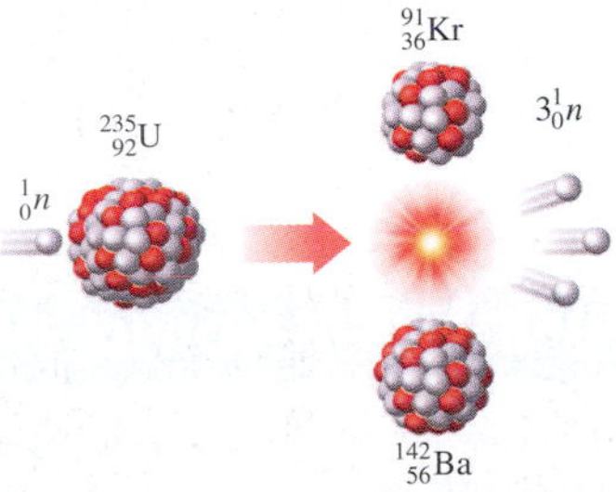

$^{91}_{36}Kr$
$^{235}_{92}U$
$3\,^{1}_{0}n$
$^{1}_{0}n$
$^{142}_{56}Ba$

LEARNING GOAL Describe the processes of nuclear fission and fusion.

- In fission, the bombardment of a large nucleus breaks it apart into smaller nuclei, releasing one or more types of radiation and a great amount of energy.
- In fusion, small nuclei combine to form larger nuclei while great amounts of energy are released.

KEY TERMS

alpha particle A nuclear particle identical to a helium nucleus, symbol α or $^{4}_{2}He$.

becquerel (Bq) A unit of activity of a radioactive sample equal to one disintegration per second.

beta particle A particle identical to an electron, symbol $^{0}_{-1}e$ or β, that forms in the nucleus when a neutron changes to a proton and an electron.

chain reaction A fission reaction that will continue once it has been initiated by a high-energy neutron bombarding a heavy nucleus such as uranium-235.

curie (Ci) A unit of activity of a radioactive sample equal to 3.7×10^{10} disintegrations/s.

decay curve A diagram of the decay of a radioactive element.

equivalent dose The measure of biological damage from an absorbed dose that has been adjusted for the type of radiation.

fission A process in which large nuclei are split into smaller nuclei, releasing large amounts of energy.

fusion A reaction in which large amounts of energy are released when small nuclei combine to form larger nuclei.

gamma ray High-energy radiation, symbol $^{0}_{0}\gamma$, emitted by an unstable nucleus.

gray (Gy) A unit of absorbed dose equal to 100 rad.

half-life The length of time it takes for one-half of a radioactive sample to decay.

positron A particle of radiation with no mass and a positive charge, symbol β^+ or $^{0}_{+1}e$, produced when a proton is transformed into a neutron and a positron.

rad (radiation absorbed dose) A measure of an amount of radiation absorbed by the body.

radiation Energy or particles released by radioactive atoms.

radioactive decay The process by which an unstable nucleus breaks down with the release of high-energy radiation.

radioisotope A radioactive atom of an element.

rem (radiation equivalent in humans) A measure of the biological damage caused by the various kinds of radiation (rad $\times$ radiation biological factor).

sievert (Sv) A unit of biological damage (equivalent dose) equal to 100 rem.

CORE CHEMISTRY SKILLS

The chapter Section containing each Core Chemistry Skill is shown in parentheses at the end of each heading.

Writing Nuclear Equations (5.2)

- A nuclear equation is written with the atomic symbols of the original radioactive nucleus on the left, an arrow, and the new nucleus and the type of radiation emitted on the right.
- The sum of the mass numbers and the sum of the atomic numbers on one side of the arrow must equal the sum of the mass numbers and the sum of the atomic numbers on the other side.
- When an alpha particle is emitted, the mass number of the new nucleus decreases by 4, and its atomic number decreases by 2.
- When a beta particle is emitted, there is no change in the mass number of the new nucleus, but its atomic number increases by one.
- When a positron is emitted, there is no change in the mass number of the new nucleus, but its atomic number decreases by one.
- In gamma emission, there is no change in the mass number or the atomic number of the new nucleus.

Example: **a.** Write a balanced nuclear equation for the alpha decay of Po-210.

b. Write a balanced nuclear equation for the beta decay of Co-60.

Answer: **a.** When an alpha particle is emitted, we calculate the decrease of 4 in the mass number (210) of the polonium, and a decrease of 2 in its atomic number.

$$^{210}_{84}\text{Po} \longrightarrow {}^{206}_{82}? + {}^{4}_{2}\text{He}$$

Because lead has atomic number 82, the new nucleus must be an isotope of lead.

$$^{210}_{84}\text{Po} \longrightarrow {}^{206}_{82}\text{Pb} + {}^{4}_{2}\text{He}$$

b. When a beta particle is emitted, there is no change in the mass number (60) of the cobalt, but there is an increase of 1 in its atomic number.

$$^{60}_{27}\text{Co} \longrightarrow {}^{60}_{28}? + {}^{0}_{-1}e$$

Because nickel has atomic number 28, the new nucleus must be an isotope of nickel.

$$^{60}_{27}\text{Co} \longrightarrow {}^{60}_{28}\text{Ni} + {}^{0}_{-1}e$$

Using Half-Lives (5.4)

- The half-life of a radioisotope is the amount of time it takes for one-half of a sample to decay.
- The remaining amount of a radioisotope is calculated by dividing its quantity or activity by one-half for each half-life that has elapsed.

Example: Co-60 has a half-life of 5.3 yr. If the initial sample of Co-60 has an activity of 1200 mCi, what is its activity after 15.9 yr?

Answer:

$$\text{number of half-lives} = 15.9\ \cancel{\text{yr}} \times \frac{1\ \text{half-life}}{5.3\ \cancel{\text{yr}}} = 3.0\ \text{half-lives}$$

$$1200\ \text{mCi} \xrightarrow{1\ \text{half-life}} 600\ \text{mCi} \xrightarrow{2\ \text{half-lives}} 300\ \text{mCi} \xrightarrow{3\ \text{half-lives}} 150\ \text{mCi}$$

In 15.9 yr, three half-lives have passed. The activity was reduced from 1200 mCi to 150 mCi.

UNDERSTANDING THE CONCEPTS

The chapter Sections to review are shown in parentheses at the end of each problem.

In problems 5.51 to 5.54, a nucleus is shown with protons and neutrons.

proton neutron

5.51 Draw the new nucleus when this isotope emits a positron to complete the following: (5.2)

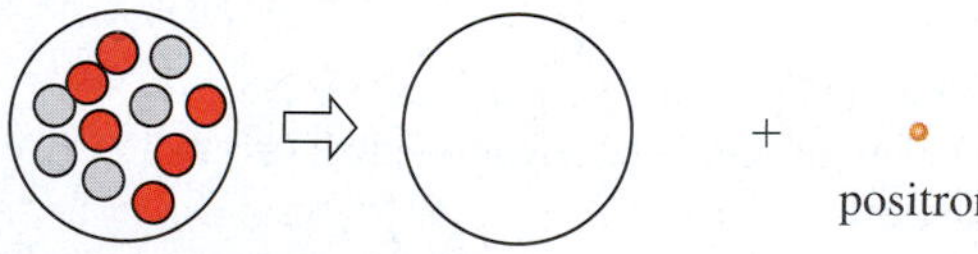
positron

5.52 Draw the nucleus that emits a beta particle to complete the following: (5.2)

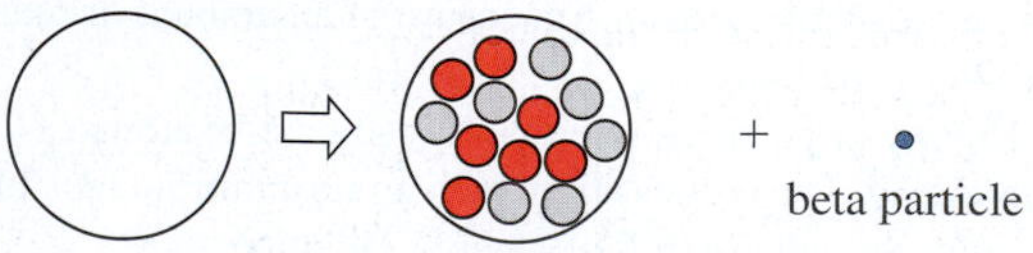
beta particle

5.53 Draw the nucleus of the isotope that is bombarded in the following: (5.2)

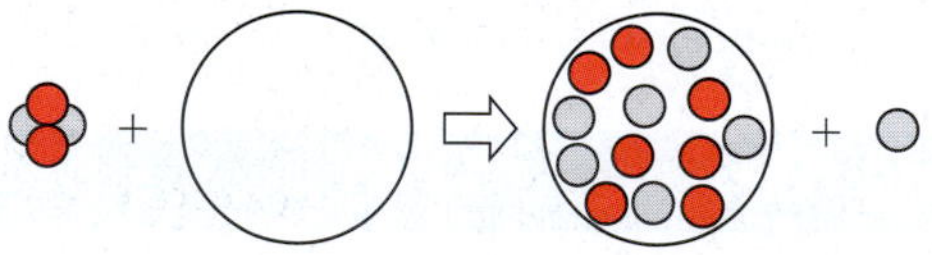

5.54 Complete the bombardment reaction by drawing the nucleus of the new isotope that is produced in the following: (5.2)

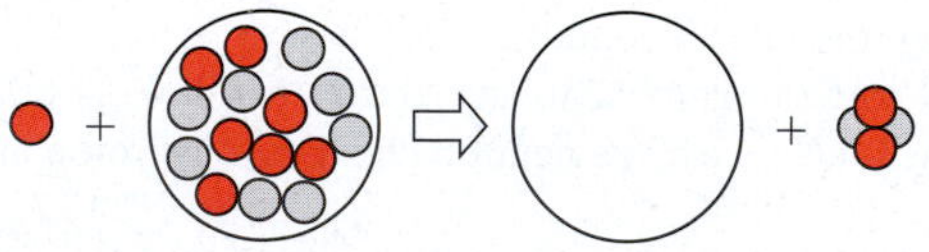

5.55 Carbon dating of small bits of charcoal used in cave paintings has determined that some of the paintings are from 10 000 to 30 000 yr old. Carbon-14 has a half-life of 5730 yr. In a 1-μg sample of carbon from a live tree, the activity of carbon-14 is 6.4 μCi. If researchers determine that 1 μg of charcoal from a prehistoric cave painting in France has an activity of 0.80 μCi, what is the age of the painting? (5.4)

The technique of carbon dating is used to determine the age of ancient cave paintings.

5.56 Use the following decay curve for iodine-131 to answer problems **a** to **c**: (5.4)

a. Complete the values for the mass of radioactive iodine-131 on the vertical axis.

b. Complete the number of days on the horizontal axis.

c. What is the half-life, in days, of iodine-131?

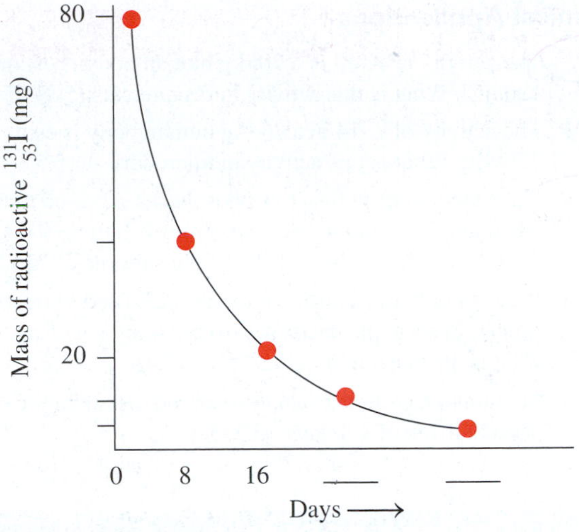
80
20
Mass of radioactive $^{131}_{53}I$ (mg)
0
8
16
Days ⟶

ADDITIONAL PRACTICE PROBLEMS

5.57 Determine the number of protons and number of neutrons in the nucleus of each of the following: (5.1)

a. sodium-25 **b.** nickel-61

c. rubidium-84 **d.** silver-110

5.58 Determine the number of protons and number of neutrons in the nucleus of each of the following: (5.1)

a. boron-10 **b.** zinc-72

c. iron-59 **d.** gold-198

5.59 Identify each of the following as alpha decay, beta decay, positron emission, or gamma emission: (5.2)

a. $^{27m}_{13}Al \longrightarrow ^{27}_{13}Al + ^{0}_{0}\gamma$

b. $^{8}_{5}B \longrightarrow ^{8}_{4}Be + ^{0}_{+1}e$

c. $^{220}_{86}Rn \longrightarrow ^{216}_{84}Po + ^{4}_{2}He$

5.60 Identify each of the following as alpha decay, beta decay, positron emission, or gamma emission: (5.2)

a. $^{127}_{55}Cs \longrightarrow ^{127}_{54}Xe + ^{0}_{+1}e$

b. $^{90}_{38}Sr \longrightarrow ^{90}_{39}Y + ^{0}_{-1}e$

c. $^{218}_{85}At \longrightarrow ^{214}_{83}Bi + ^{4}_{2}He$

5.61 Write the balanced nuclear equation for each of the following: (5.2)

a. Th-225 (α decay) **b.** Bi-210 (α decay)

c. cesium-137 (β decay) **d.** tin-126 (β decay)

e. F-18 (β^+ emission)

5.62 Write the balanced nuclear equation for each of the following: (5.2)

a. potassium-40 (β decay) **b.** sulfur-35 (β decay)

c. platinum-190 (α decay) **d.** Ra-210 (α decay)

e. In-113m (γ emission)

5.63 Complete each of the following nuclear equations: (5.2)

a. $^{4}_{2}He + ^{14}_{7}N \longrightarrow ? + ^{1}_{1}H$

b. $^{4}_{2}He + ^{27}_{13}Al \longrightarrow ^{30}_{14}Si + ?$

c. $^{1}_{0}n + ^{235}_{92}U \longrightarrow ^{90}_{38}Sr + 3^{1}_{0}n + ?$

d. $^{23m}_{12}Mg \longrightarrow ? + ^{0}_{0}\gamma$

5.64 Complete each of the following nuclear equations: (5.2)

a. $? + ^{59}_{27}Co \longrightarrow ^{56}_{25}Mn + ^{4}_{2}He$

b. $? \longrightarrow ^{14}_{7}N + ^{0}_{-1}e$

c. $^{0}_{-1}e + ^{76}_{36}Kr \longrightarrow ?$

d. $^{4}_{2}He + ^{241}_{95}Am \longrightarrow ? + 2^{1}_{0}n$

5.65 Write the balanced nuclear equation for each of the following: (5.2)

a. When two oxygen-16 atoms collide, one of the products is an alpha particle.

b. When californium-249 is bombarded by oxygen-18, a new element, seaborgium-263, and four neutrons are produced.

c. Radon-222 undergoes alpha decay.

d. An atom of strontium-80 emits a positron.

5.66 Write the balanced nuclear equation for each of the following: (5.2)

a. Actinium-225 decays to give francium-221.

b. Bismuth-211 emits an alpha particle.

c. A radioisotope emits a positron to form titanium-48.

d. An atom of germanium-69 emits a positron.

5.67 A 120-mg sample of technetium-99m is used for a diagnostic test. If technetium-99m has a half-life of 6.0 h, how many milligrams of the technetium-99m sample remains active 24 h after the test? (5.4)

5.68 The half-life of oxygen-15 is 124 s. If a sample of oxygen-15 has an activity of 4000 Bq, how many minutes will elapse before it has an activity of 500 Bq? (5.4)

5.69 What is the difference between fission and fusion? (5.6)

5.70 **a.** What are the products in the fission of uranium-235 that make possible a nuclear chain reaction? (5.6)

b. What is the purpose of placing control rods among uranium samples in a nuclear reactor?

5.71 Where does fusion occur naturally? (5.6)

5.72 Why are scientists continuing to try to build a fusion reactor even though the very high temperatures it requires have been difficult to reach and maintain? (5.6)

Clinical Applications

5.73 The activity of K-40 in a 70.-kg human body is estimated to be 120 nCi. What is this activity in becquerels? (5.3)

5.74 The activity of C-14 in a 70.-kg human body is estimated to be 3.7 kBq. What is this activity in microcuries? (5.3)

5.75 If the amount of radioactive phosphorus-32, used to treat leukemia, in a sample decreases from 1.2 mg to 0.30 mg in 28.6 days, what is the half-life of phosphorus-32? (5.4)

5.76 If the amount of radioactive iodine-123, used to treat thyroid cancer, in a sample decreases from 0.4 mg to 0.1 mg in 26.4 h, what is the half-life of iodine-123? (5.4)

5.77 Calcium-47, used to evaluate disorders in calcium metabolism, has a half-life of 4.5 days. (5.2, 5.4)

a. Write the balanced nuclear equation for the beta decay of calcium-47.

b. How many milligrams of a 16-mg sample of calcium-47 remain after 18 days?

c. How many days have passed if 4.8 mg of calcium-47 decayed to 1.2 mg of calcium-47?

5.78 Cesium-137, used in cancer treatment, has a half-life of 30 yr. (5.2, 5.4)

a. Write the balanced nuclear equation for the beta decay of cesium-137.

b. How many milligrams of a 16-mg sample of cesium-137 remain after 90 yr?

c. How many years are required for 28 mg of cesium-137 to decay to 3.5 mg of cesium-137?

CHALLENGE PROBLEMS

The following problems are related to the topics in this chapter. However, they do not all follow the chapter order, and they require you to combine concepts and skills from several Sections. These problems will help you increase your critical thinking skills and prepare for your next exam.

5.79 Write the balanced nuclear equation for each of the following radioactive emissions: (5.2)

a. an alpha particle from Hg-180

b. a beta particle from Au-198

c. a positron from Rb-82

5.80 Write the balanced nuclear equation for each of the following radioactive emissions: (5.2)

a. an alpha particle from Gd-148

b. a beta particle from Sr-90

c. a positron from Al-25

5.81 All the elements beyond uranium, the transuranium elements, have been prepared by bombardment and are not naturally occurring elements. The first transuranium element neptunium, Np, was prepared by bombarding U-238 with neutrons to form a neptunium atom and a beta particle. Complete the following equation: (5.2)

$${}^{1}_{0}n + {}^{238}_{92}\text{U} \longrightarrow ? + ?$$

5.82 One of the most recent transuranium elements, oganesson-294 (Og-294), atomic number 118, was prepared by bombarding californium-249 with another isotope. Complete the following equation for the preparation of this new element: (5.2)

$$? + {}^{249}_{98}\text{Cf} \longrightarrow {}^{294}_{118}\text{Og} + 3\,{}^{1}_{0}n$$

5.83 A 64-μCi sample of Tl-201 decays to 4.0 μCi in 12 days. What is the half-life, in days, of Tl-201? (5.3, 5.4)

5.84 A wooden object from the site of an ancient temple has a carbon-14 activity of 10 counts/min compared with a reference piece of wood cut today that has an activity of 40 counts/min. If the half-life for carbon-14 is 5730 yr, what is the age of the ancient wood object? (5.3, 5.4)

5.85 Element 114 was recently named flerovium, symbol Fl. The reaction for its synthesis involves bombarding Pu-244 with Ca-48. Write the balanced nuclear equation for the synthesis of flerovium. (5.2)

5.86 Element 116 was recently named livermorium, symbol Lv. The reaction for its synthesis involves bombarding Cm-248 with Ca-48. Write the balanced nuclear equation for the synthesis of livermorium. (5.2)

Clinical Applications

5.87 The half-life for the radioactive decay of calcium-47 is 4.5 days. If a sample has an activity of 1.0 μCi after 27 days, what was the initial activity, in microcuries, of the sample? (5.3, 5.4)

5.88 The half-life for the radioactive decay of Ce-141 is 32.5 days. If a sample has an activity of 4.0 μCi after 130. days have elapsed, what was the initial activity, in microcuries, of the sample? (5.3, 5.4)

5.89 A nuclear technician was accidentally exposed to potassium-42 while doing brain scans for possible tumors. The error was not discovered until 36 h later when the activity of the potassium-42 sample was 2.0 μCi. If potassium-42 has a half-life of 12 h, what was the activity of the sample at the time the technician was exposed? (5.3, 5.4)

5.90 The radioisotope sodium-24 is used to determine the levels of electrolytes in the body. A 16-μg sample of sodium-24 decays to 2.0 μg in 45 h. What is the half-life, in hours, of sodium-24? (5.4)

ANSWERS

5.1 **a.** alpha particle **b.** positron
c. gamma radiation

5.3 **a.** ${}^{39}_{19}\text{K}$ ${}^{40}_{19}\text{K}$ ${}^{41}_{19}\text{K}$
b. They all have 19 protons and 19 electrons, but they differ in the number of neutrons.

5.5 **a.** ${}^{0}_{-1}e$ or β **b.** ${}^{4}_{2}\text{He}$ or α **c.** ${}^{1}_{0}n$ or n
d. ${}^{38}_{18}\text{Ar}$ **e.** ${}^{14}_{6}\text{C}$

5.7 **a.** ${}^{64}_{29}\text{Cu}$ **b.** ${}^{75}_{34}\text{Se}$
c. ${}^{24}_{11}\text{Na}$ **d.** ${}^{15}_{7}\text{N}$

5.9

Medical Use	Atomic Symbol	Mass Number	Number of Protons	Number of Neutrons
Heart imaging	${}^{201}_{81}\text{Tl}$	201	81	120
Radiation therapy	${}^{60}_{27}\text{Co}$	60	27	33
Abdominal scan	${}^{67}_{31}\text{Ga}$	67	31	36
Hyperthyroidism	${}^{131}_{53}\text{I}$	131	53	78
Leukemia treatment	${}^{32}_{15}\text{P}$	32	15	17

5.11 **a.** 1, alpha particle **b.** 3, gamma radiation **c.** 1, alpha particle

5.13 **a.** ${}^{208}_{84}\text{Po} \longrightarrow {}^{204}_{82}\text{Pb} + {}^{4}_{2}\text{He}$
b. ${}^{232}_{90}\text{Th} \longrightarrow {}^{228}_{88}\text{Ra} + {}^{4}_{2}\text{He}$
c. ${}^{251}_{102}\text{No} \longrightarrow {}^{247}_{100}\text{Fm} + {}^{4}_{2}\text{He}$
d. ${}^{220}_{86}\text{Rn} \longrightarrow {}^{216}_{84}\text{Po} + {}^{4}_{2}\text{He}$

5.15 **a.** ${}^{25}_{11}\text{Na} \longrightarrow {}^{25}_{12}\text{Mg} + {}^{0}_{-1}e$
b. ${}^{20}_{8}\text{O} \longrightarrow {}^{20}_{9}\text{F} + {}^{0}_{-1}e$
c. ${}^{92}_{38}\text{Sr} \longrightarrow {}^{92}_{39}\text{Y} + {}^{0}_{-1}e$
d. ${}^{60}_{26}\text{Fe} \longrightarrow {}^{60}_{27}\text{Co} + {}^{0}_{-1}e$

5.17 **a.** ${}^{26}_{14}\text{Si} \longrightarrow {}^{26}_{13}\text{Al} + {}^{0}_{+1}e$
b. ${}^{54}_{27}\text{Co} \longrightarrow {}^{54}_{26}\text{Fe} + {}^{0}_{+1}e$
c. ${}^{77}_{37}\text{Rb} \longrightarrow {}^{77}_{36}\text{Kr} + {}^{0}_{+1}e$
d. ${}^{93}_{45}\text{Rh} \longrightarrow {}^{93}_{44}\text{Ru} + {}^{0}_{+1}e$

5.19 **a.** ${}^{28}_{14}\text{Si}$, beta decay
b. ${}^{0}_{0}\gamma$, gamma emission
c. ${}^{0}_{-1}e$, beta decay
d. ${}^{238}_{92}\text{U}$, alpha decay
e. ${}^{188}_{79}\text{Au}$, positron emission

5.21 **a.** ${}^{10}_{4}\text{Be}$ **b.** ${}^{132}_{53}\text{I}$ **c.** ${}^{27}_{13}\text{Al}$ **d.** ${}^{17}_{8}\text{O}$

5.23 **a.** 2, absorbed dose **b.** 3, biological damage **c.** 1, activity **d.** 2, absorbed dose

5.25 The technician exposed to 5 rad received the higher amount of radiation.

5.27 **a.** 294 μCi **b.** 0.5 Gy

5.29 **a.** two half-lives **b.** one half-life **c.** three half-lives

5.31 **a.** 40.0 mg **b.** 20.0 mg **c.** 10.0 mg **d.** 1.25 mg

5.33 **a.** 130 days **b.** 195 days

5.35 **a.** Because the elements Ca and P are part of the bone, the radioactive isotopes of Ca and P will become part of the bony structures of the body, where their radiation can be used to diagnose or treat bone diseases.
b. Strontium (Sr) acts much like calcium (Ca) because both are Group 2A (2) elements. The body will accumulate radioactive strontium in bones in the same way that it incorporates calcium. Radioactive strontium is harmful to children because the radiation it produces causes more damage in cells that are dividing rapidly.

5.37 180 μCi

5.39 **a.** ${}^{68}_{31}\text{Ga} \longrightarrow {}^{68}_{30}\text{Mn} + {}^{0}_{+1}e$ **b.** 16 mcg

5.41 Nuclear fission is the splitting of a large atom into smaller fragments with the release of large amounts of energy.

5.43 ${}^{103}_{42}\text{Mo}$

5.45 **a.** fission **b.** fusion **c.** fission **d.** fusion

5.47 **a.** 2.0×10^{-3} Ci **b.** 2.0 mCi

5.49 9.0 days

5.51

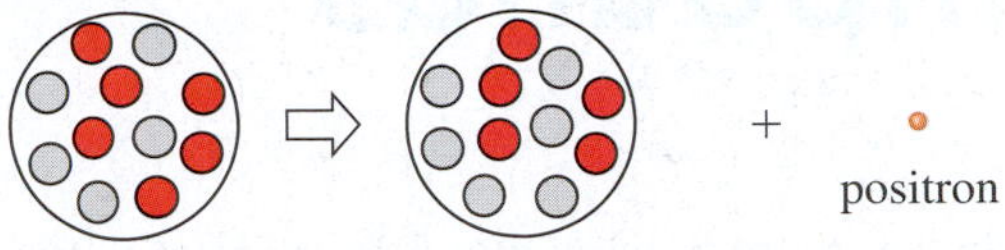
+
positron

5.53

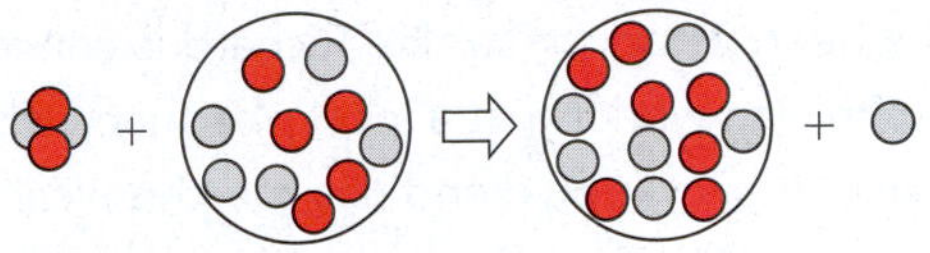
+
+

5.55 17 000 yr old

5.57 **a.** 11 protons and 14 neutrons
b. 28 protons and 33 neutrons
c. 37 protons and 47 neutrons
d. 47 protons and 63 neutrons

5.59 **a.** gamma emission
b. positron emission
c. alpha decay

5.61 **a.** ${}^{225}_{90}\text{Th} \longrightarrow {}^{221}_{88}\text{Ra} + {}^{4}_{2}\text{He}$
b. ${}^{210}_{83}\text{Bi} \longrightarrow {}^{206}_{81}\text{Tl} + {}^{4}_{2}\text{He}$
c. ${}^{137}_{55}\text{Cs} \longrightarrow {}^{137}_{56}\text{Ba} + {}^{0}_{-1}e$
d. ${}^{126}_{50}\text{Sn} \longrightarrow {}^{126}_{51}\text{Sb} + {}^{0}_{-1}e$
e. ${}^{18}_{9}\text{F} \longrightarrow {}^{18}_{8}\text{O} + {}^{0}_{+1}e$

5.63 **a.** ${}^{17}_{8}\text{O}$ **b.** ${}^{1}_{1}\text{H}$ **c.** ${}^{143}_{54}\text{Xe}$ **d.** ${}^{23}_{12}\text{Mg}$

5.65 **a.** ${}^{16}_{8}\text{O} + {}^{16}_{8}\text{O} \longrightarrow {}^{28}_{14}\text{Si} + {}^{4}_{2}\text{He}$
b. ${}^{18}_{8}\text{O} + {}^{249}_{98}\text{Cf} \longrightarrow {}^{263}_{106}\text{Sg} + 4{}^{1}_{0}n$
c. ${}^{222}_{86}\text{Rn} \longrightarrow {}^{218}_{84}\text{PO} + {}^{4}_{2}\text{He}$
d. ${}^{80}_{38}\text{Sr} \longrightarrow {}^{80}_{37}\text{Rb} + {}^{0}_{+1}e$

5.67 7.5 mg of Tc-99m

5.69 In the fission process, an atom splits into smaller nuclei. In fusion, small nuclei combine (fuse) to form a larger nucleus.

5.71 Fusion occurs naturally in the Sun and other stars.

5.73 4.4×10^{3} Bq

5.75 14.3 days

5.77 **a.** ${}^{47}_{20}\text{Ca} \longrightarrow {}^{47}_{21}\text{Sc} + {}^{0}_{-1}e$
b. 1.0 mg of Ca-47
c. 9.0 days

5.79 **a.** ${}^{180}_{80}\text{Hg} \longrightarrow {}^{176}_{78}\text{Pt} + {}^{4}_{2}\text{He}$
b. ${}^{198}_{79}\text{Au} \longrightarrow {}^{198}_{80}\text{Hg} + {}^{0}_{-1}e$
c. ${}^{82}_{37}\text{Rb} \longrightarrow {}^{82}_{36}\text{Kr} + {}^{0}_{+1}e$

5.81 ${}^{1}_{0}n + {}^{238}_{92}\text{U} \longrightarrow {}^{239}_{93}\text{Np} + {}^{0}_{-1}e$

5.83 3.0 days

5.85 ${}^{48}_{20}\text{Ca} + {}^{244}_{94}\text{Pu} \longrightarrow {}^{292}_{114}\text{Fl}$

5.87 64 μCi

5.89 16 μCi

6

Ionic and Molecular Compounds

Richard's doctor recommends that Richard take a low-dose aspirin (81 mg) every day to decrease the chance of a heart attack or stroke. Richard is concerned about taking aspirin and asks Sarah, a pharmacy technician working at a local pharmacy, about the effects of aspirin. Sarah explains that aspirin is acetylsalicylic acid and has the chemical formula $C_9H_8O_4$. Aspirin is a molecular compound, often referred to as an organic molecule because it contains the nonmetal carbon (C) combined with the nonmetals hydrogen (H) and oxygen (O). Sarah explains that aspirin is used to relieve minor pains, to reduce inflammation and fever, and to slow blood clotting. Aspirin is one of several nonsteroidal anti-inflammatory drugs (NSAIDs) that reduce pain and fever by blocking the formation of prostaglandins, which are chemical messengers that transmit pain signals to the brain and cause fever. Some potential side effects of aspirin may include heartburn, upset stomach, nausea, and an increased risk of a stomach ulcer.

CAREER

Pharmacy Technician

Pharmacy technicians work in hospitals, pharmacies, clinics, and long-term care facilities where they are responsible for the preparation and distribution of pharmaceutical medications based on a doctor's orders. They obtain the proper medication, and also calculate, measure, and label the patients' medication. Pharmacy technicians advise clients and health care practitioners on the selection of both prescription and over-the-counter drugs, proper dosages, and geriatric considerations, as well as possible side effects and interactions. They may also administer vaccinations; prepare sterile intravenous solutions; and advise clients about health, diet, and home medical equipment. Pharmacy technicians also prepare insurance claims and create and maintain patient profiles.

CLINICAL UPDATE

Compounds at the Pharmacy

Two weeks later, Richard returns to the pharmacy. Using Sarah's recommendations, he purchases Epsom salts for a sore toe, an antacid for an upset stomach, and an iron supplement. You can see the chemical formulas of these medications in the **CLINICAL UPDATE Compounds at the Pharmacy**, page 210.

6.1 Ions: Transfer of Electrons

LEARNING GOAL Write the symbols for the simple ions of the representative elements.

Most of the elements, except the noble gases, are found in nature combined as compounds. The noble gases are so stable that they form compounds only under extreme conditions. One explanation for the stability of noble gases is that they have a filled valence electron energy level.

Compounds form when electrons are transferred or shared between atoms to give stable electron configurations to the atoms. In the formation of either an *ionic bond* or a *covalent bond*, atoms lose, gain, or share valence electrons to acquire an octet of eight valence electrons. This tendency of atoms to attain a stable electron configuration is known as the **octet rule** and provides a key to our understanding of the ways in which atoms bond and form compounds. A few elements achieve the stability of helium with two valence electrons. However, the octet rule is not used with transition elements.

Ionic bonds occur when the valence electrons of atoms of a metal are transferred to atoms of nonmetals. For example, sodium atoms lose electrons and chlorine atoms gain electrons to form the ionic compound NaCl. **Covalent bonds** form when atoms of nonmetals share valence electrons. In the molecular compounds H_2O and C_3H_8, atoms share electrons (see **TABLE 6.1**).

TABLE 6.1 Types of Particles and Bonds in Compounds

Type	Ionic Compounds	Molecular Compounds	
Particles	Ions	Molecules	
Bonds	Ionic	Covalent	
Examples	Na^+ Cl^- ions	H_2O molecules	C_3H_8 molecules
		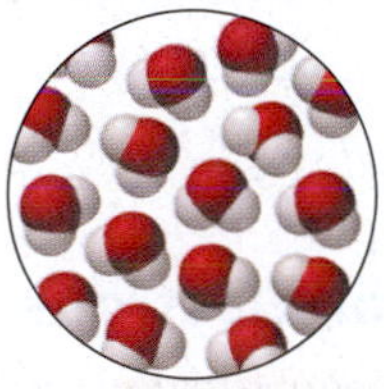	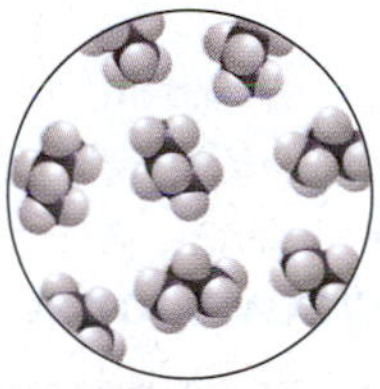

Positive Ions: Loss of Electrons

In ionic bonding, **ions**, which have electrical charges, form when atoms lose or gain electrons to form a stable electron configuration. Because the ionization energies of metals of Groups 1A (1), 2A (2), and 3A (13) are low, metal atoms readily lose their valence electrons. In doing so, they form ions with positive charges. A metal atom obtains the same electron configuration as its nearest noble gas (usually eight valence electrons). For example, when a sodium atom loses its single valence electron, the remaining electrons have a stable configuration. By losing an electron, sodium has 10 negatively charged electrons

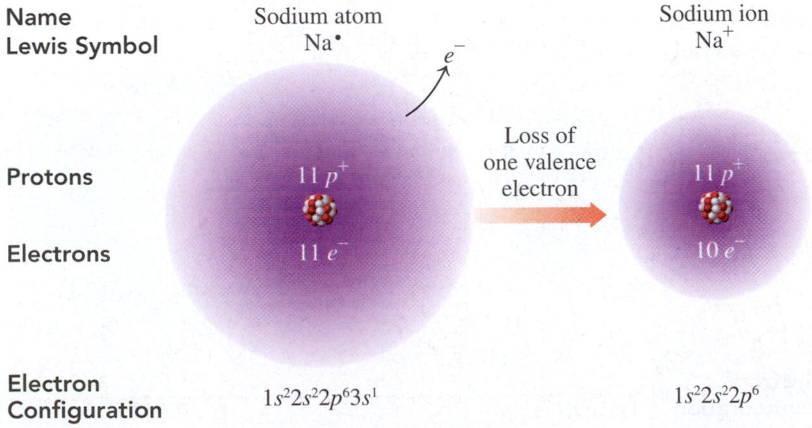

Name
Lewis Symbol
Sodium atom
Na•
Sodium ion
Na+
e−
Loss of one valence electron
Protons
11 p+
11 p+
Electrons
11 e−
10 e−
Electron Configuration
1s²2s²2p⁶3s¹
1s²2s²2p⁶

LOOKING AHEAD

REVIEW

Using Positive and Negative Numbers in Calculations (1.4)

Writing Electron Configurations (4.7)

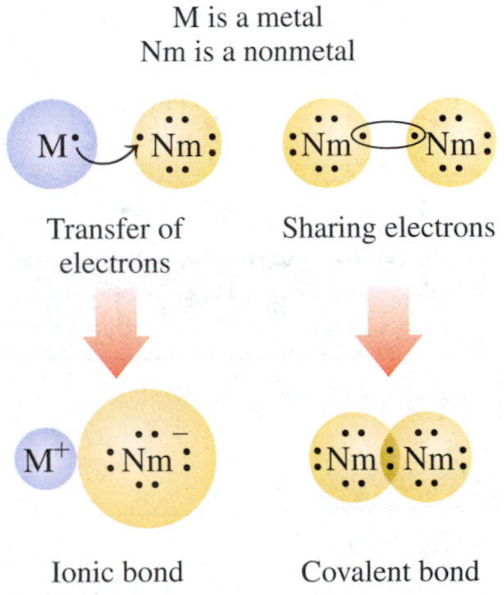

M is a metal
Nm is a nonmetal
M
Nm
Nm
Nm
Transfer of electrons
Sharing electrons
M+
Nm
Nm
Nm
Ionic bond
Covalent bond

ENGAGE

A cesium ion has 55 protons and 54 electrons. Why does the cesium ion have a 1+ charge?

instead of 11. Because there are still 11 positively charged protons in its nucleus, the atom is no longer neutral. It is now a sodium ion with a positive electrical charge, called an **ionic charge**, of 1+. In the Lewis symbol for the sodium ion, the ionic charge of 1+ is written in the upper right corner, Na^+, where the 1 is understood. A metal ion is named by its element name. Thus, Na^+ is named the *sodium ion*. The sodium ion is smaller than the sodium atom because the ion has lost its outermost electron from the third energy level. A positively charged ion of a metal is called a **cation** (pronounced *cat-eye-un*).

$$\text{Ionic charge} = \text{Charge of protons} + \text{Charge of electrons}$$
$$1+ = (11+) + (10-)$$

Magnesium, a metal in Group 2A (2), obtains a stable electron configuration by losing two valence electrons to form a magnesium ion with a 2+ ionic charge, Mg^{2+}. The magnesium ion is smaller than the magnesium atom because the outermost electrons in the third energy level were removed. The octet in the magnesium ion is made up of electrons that fill its second energy level.

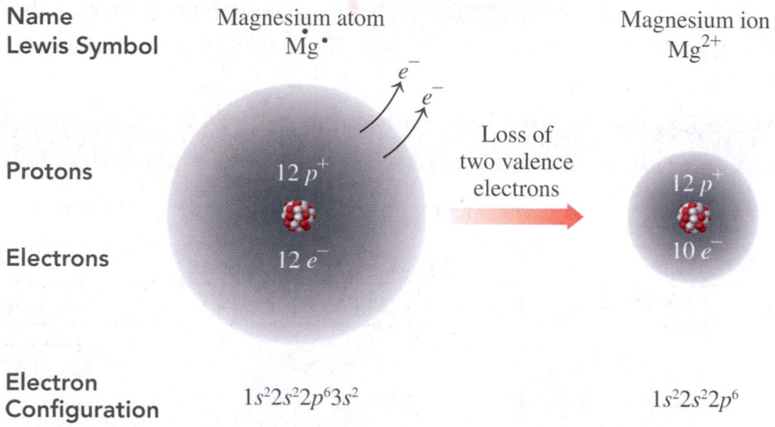

Name
Lewis Symbol
Magnesium atom
Mg
Magnesium ion
Mg2+
e−
e−
Loss of two valence electrons
Protons
12 p+
12 p+
Electrons
12 e−
10 e−
Electron Configuration
1s²2s²2p⁶3s²
1s²2s²2p⁶

Negative Ions: Gain of Electrons

The ionization energy of a nonmetal atom in Groups 5A (15), 6A (16), or 7A (17) is high. In an ionic compound, a nonmetal atom gains one or more valence electrons to obtain a stable electron configuration. By gaining electrons, a nonmetal atom forms a negatively charged ion. For example, an atom of chlorine with seven valence electrons gains one electron to form an octet. Because it now has 18 electrons and 17 protons in its nucleus, the chlorine atom is no longer neutral. It is a *chloride ion* with an ionic charge of 1−, which is written as Cl^-, with the 1 understood. A negatively charged ion, called an **anion** (pronounced *an-eye-un*), is named by using the first syllable of its element name followed by *ide*. The chloride ion is larger than the chlorine atom because the ion has an additional electron, which completes its outermost energy level.

CORE CHEMISTRY SKILL

Writing Positive and Negative Ions

ENGAGE

Why does Li form a positive ion, Li^+, whereas Br forms a negative ion, Br^-?

$$\text{Ionic charge} = \text{Charge of protons} + \text{Charge of electrons}$$
$$1- = (17+) + (18-)$$

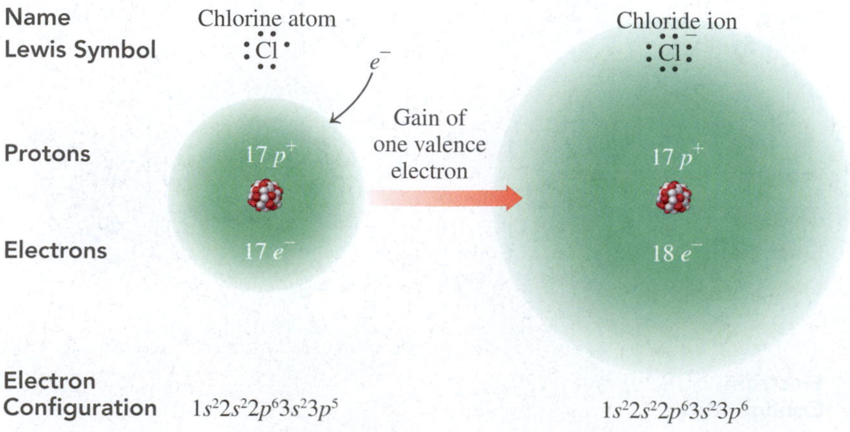

Name
Lewis Symbol
Chlorine atom
Cl
Chloride ion
Cl−
e−
Gain of one valence electron
Protons
17 p+
17 p+
Electrons
17 e−
18 e−
Electron Configuration
1s²2s²2p⁶3s²3p⁵
1s²2s²2p⁶3s²3p⁶

TABLE 6.2 lists the names of some important metal and nonmetal ions.

TABLE 6.2 Formulas and Names of Some Common Ions

Metals			Nonmetals		
Group Number	**Cation**	**Name of Cation**	**Group Number**	**Anion**	**Name of Anion**
1A (1)	Li^+	Lithium	5A (15)	N^{3-}	Nitride
	Na^+	Sodium		P^{3-}	Phosphide
	K^+	Potassium	6A (16)	O^{2-}	Oxide
2A (2)	Mg^{2+}	Magnesium		S^{2-}	Sulfide
	Ca^{2+}	Calcium	7A (17)	F^-	Fluoride
	Ba^{2+}	Barium		Cl^-	Chloride
3A (13)	Al^{3+}	Aluminum		Br^-	Bromide
				I^-	Iodide

▶SAMPLE PROBLEM 6.1 Ions

TRY IT FIRST

a. Write the symbol and name for the ion that has 7 protons and 10 electrons.
b. Write the symbol and name for the ion that has 20 protons and 18 electrons.

SOLUTION

a. The element with 7 protons is nitrogen. In an ion of nitrogen with 10 electrons, the ionic charge would be 3−, [(7+) + (10−) = 3−]. The ion, written as N^{3-}, is the *nitride* ion.

b. The element with 20 protons is calcium. In an ion of calcium with 18 electrons, the ionic charge would be 2+, [(20+) + (18−) = 2+]. The ion, written as Ca^{2+}, is the *calcium* ion.

STUDY CHECK 6.1

How many protons and electrons are in each of the following ions?

a. Sr^{2+} **b.** Cl^- **c.** As^{3+}

ANSWER

a. 38 protons, 36 electrons **b.** 17 protons, 18 electrons **c.** 33 protons, 30 electrons

TEST

Try Practice Problems 6.1 to 6.8

Ionic Charges from Group Numbers

In ionic compounds, representative elements usually lose or gain electrons to give eight valence electrons like their nearest noble gas (or two for helium). We can use the group numbers in the periodic table to determine the charges for the ions of the representative elements. The elements in Group 1A (1) lose one electron to form ions with a 1+ charge. The elements in Group 2A (2) lose two electrons to form ions with a 2+ charge. The elements in Group 3A (13) lose three electrons to form ions with a 3+ charge. In this text, we do not use the group numbers of the transition elements to determine their ionic charges.

In ionic compounds, the elements in Group 7A (17) gain one electron to form ions with a 1− charge. The elements in Group 6A (16) gain two electrons to form ions with a 2− charge. The elements in Group 5A (15) gain three electrons to form ions with a 3− charge.

The nonmetals of Group 4A (14) do not typically form ions. However, the metals Sn and Pb in Group 4A (14) lose electrons to form positive ions. **TABLE 6.3** lists the ionic charges for some common monatomic ions of representative elements.

TABLE 6.3 Examples of Monatomic Ions and Their Nearest Noble Gases

	Metals Lose Valence Electrons			Nonmetals Gain Valence Electrons		
Noble Gas	**1A (1)**	**2A (2)**	**3A (13)**	**5A (15)**	**6A (16)**	**7A (17)**
He	Li^+					
Ne	Na^+	Mg^{2+}	Al^{3+}	N^{3-}	O^{2-}	F^-
Ar	K^+	Ca^{2+}		P^{3-}	S^{2-}	Cl^-
Kr	Rb^+	Sr^{2+}			Se^{2-}	Br^-
Xe	Cs^+	Ba^{2+}				I^-

ENGAGE

Why do all the atoms in Group 2A (2) form ions with 2+ charges?

▶SAMPLE PROBLEM 6.2 Writing Symbols for Ions

TRY IT FIRST

Consider the elements aluminum and oxygen.

a. Identify each as a metal or a nonmetal.
b. State the number of valence electrons for each.
c. State the number of electrons that must be lost or gained for each to achieve an octet.
d. Write the symbol, including its ionic charge, and the name for each ion.

SOLUTION

Aluminum	Oxygen
a. metal	nonmetal
b. 3 valence electrons	6 valence electrons
c. loses 3 e^-	gains 2 e^-
d. Al^{3+}, [(13+) + (10−) = 3+], aluminum ion	O^{2-}, [(8+) + (10−) = 2−], oxide ion

STUDY CHECK 6.2

Consider the elements sulfur and potassium.

a. Identify each as a metal or a nonmetal.
b. State the number of valence electrons for each.
c. State the number of electrons that must be lost or gained for each to achieve an octet.
d. Write the symbol, including its ionic charge, and the name for each ion.

ANSWER

a. Sulfur is a nonmetal; potassium is a metal.
b. Sulfur has six valence electrons; potassium has one valence electron.
c. Sulfur will gain two electrons; potassium will lose one electron.
d. S^{2-}, sulfide ion; K^+, potassium ion

TEST

Try Practice Problems 6.9 to 6.14

PRACTICE PROBLEMS

6.1 Ions: Transfer of Electrons

6.1 State the number of electrons that must be lost by atoms of each of the following to achieve a stable electron configuration:
a. Li **b.** Ca **c.** Ga **d.** Cs **e.** Ba

6.2 State the number of electrons that must be gained by atoms of each of the following to achieve a stable electron configuration:
a. Cl **b.** Se **c.** N **d.** I **e.** S

6.3 State the number of electrons lost or gained when the following elements form ions:
a. Sr **b.** P **c.** Group 7A (17) **d.** Na **e.** Br

6.4 State the number of electrons lost or gained when the following elements form ions:
a. O **b.** Group 2A (2) **c.** F **d.** K **e.** Rb

6.5 Write the symbols for the ions with the following number of protons and electrons:
a. 3 protons, 2 electrons **b.** 9 protons, 10 electrons
c. 12 protons, 10 electrons **d.** 27 protons, 24 electrons

6.6 Write the symbols for the ions with the following number of protons and electrons:
a. 8 protons, 10 electrons **b.** 19 protons, 18 electrons
c. 35 protons, 36 electrons **d.** 50 protons, 46 electrons

6.7 State the number of protons and electrons in each of the following:
a. Cu^{2+} **b.** Se^{2-} **c.** Br^- **d.** Fe^{2+}

6.8 State the number of protons and electrons in each of the following:
a. S^{2-} **b.** Ni^{2+} **c.** Au^{3+} **d.** Ag^+

6.9 Write the symbol for the ion of each of the following:
a. chlorine **b.** cesium **c.** nitrogen **d.** radium

6.10 Write the symbol for the ion of each of the following:
a. fluorine **b.** barium **c.** sodium **d.** iodine

6.11 Write the names for each of the following ions:
a. Li^+ **b.** Ca^{2+} **c.** Ga^{3+} **d.** P^{3-}

6.12 Write the names for each of the following ions:
a. Rb^+ **b.** Sr^{2+} **c.** S^{2-} **d.** F^-

Clinical Applications

6.13 State the number of protons and electrons in each of the following ions:
a. O^{2-}, used to build biomolecules and water
b. K^+, most prevalent positive ion in cells; needed for muscle contraction, nerve impulses
c. I^-, needed for thyroid function
d. Na^+, most prevalent positive ion in extracellular fluid

6.14 State the number of protons and electrons in each of the following ions:
a. P^{3-}, needed for bones and teeth
b. F^-, used to strengthen tooth enamel
c. Mg^{2+}, needed for bones and teeth
d. Ca^{2+}, needed for bones and teeth

Chemistry Link to Health

Some Important Ions in the Body

Several ions in body fluids have important physiological and metabolic functions. Some are listed in **TABLE 6.4**.

Foods such as bananas, milk, cheese, and potatoes provide the body with ions that are important in regulating body functions.

Milk, cheese, bananas, cereal, and potatoes provide ions for the body.

TABLE 6.4 Ions in the Body

Ion	Occurrence	Function	Source	Result of Too Little	Result of Too Much
Na^+	Principal cation outside the cell	Regulates and controls body fluids	Salt, cheese, pickles	Hyponatremia, anxiety, diarrhea, circulatory failure, decrease in body fluid	Hypernatremia, little urine, thirst, edema
K^+	Principal cation inside the cell	Regulates body fluids and cellular functions	Bananas, orange juice, milk, prunes, potatoes	Hypokalemia (hypopotassemia), lethargy, muscle weakness, failure of neurological impulses	Hyperkalemia (hyperpotassemia), irritability, nausea, little urine, cardiac arrest
Ca^{2+}	Cation outside the cell; 90% of calcium in the body in bones	Major cation of bones; needed for muscle contraction	Milk, yogurt, cheese, greens, spinach	Hypocalcemia, tingling fingertips, muscle cramps, osteoporosis	Hypercalcemia, relaxed muscles, kidney stones, deep bone pain
Mg^{2+}	Cation outside the cell; 50% of magnesium in the body in bones	Essential for certain enzymes, muscles, nerve control	Widely distributed (part of chlorophyll of all green plants), nuts, whole grains	Disorientation, hypertension, tremors, slow pulse	Drowsiness
Cl^-	Principal anion outside the cell	Gastric juice, regulates body fluids	Salt	Same as for Na^+	Same as for Na^+

6.2 Ionic Compounds

LEARNING GOAL Using charge balance, write the correct formula for an ionic compound.

We use *ionic compounds* such as salt, NaCl, and baking soda, $NaHCO_3$, every day. Milk of magnesia, $Mg(OH)_2$, or calcium carbonate, $CaCO_3$, may be taken to settle an upset stomach. In a mineral supplement, iron may be present as iron(II) sulfate, $FeSO_4$, iodine as potassium iodide, KI, and manganese as manganese(II) sulfate, $MnSO_4$. Some sunscreens contain zinc oxide, ZnO, while tin(II) fluoride, SnF_2, in toothpaste provides fluoride to help prevent tooth decay. Gemstones are ionic compounds that are cut and polished to make jewelry. For example, sapphires and rubies are aluminum oxide, Al_2O_3. Impurities of chromium ions make rubies red, and iron and titanium ions make sapphires blue.

Rubies and sapphires are the ionic compound aluminum oxide, with chromium ions in rubies, and titanium and iron ions in sapphires.

Properties of Ionic Compounds

In an **ionic compound**, one or more electrons are transferred from metals to nonmetals, which form positive and negative ions. The attraction between these ions is called an ionic bond.

The physical and chemical properties of an ionic compound such as NaCl are very different from those of the original elements. For example, the original elements of NaCl were sodium, a soft, shiny metal, and chlorine, a yellow-green poisonous gas. However, when they react and form positive and negative ions, they produce NaCl, which is ordinary table salt, a hard, white, crystalline substance that is important in our diet.

In a crystal of NaCl, the larger Cl^- ions are arranged in a three-dimensional structure in which the smaller Na^+ ions occupy the spaces between the Cl^- ions (see **FIGURE 6.1**). In this crystal, every Na^+ ion is surrounded by six Cl^- ions, and every Cl^- ion is surrounded by six Na^+ ions. Thus, there are many strong attractions between the positive and negative ions, which account for the high melting points of ionic compounds. For example, the melting point of NaCl is 801 °C. At room temperature, ionic compounds are solids.

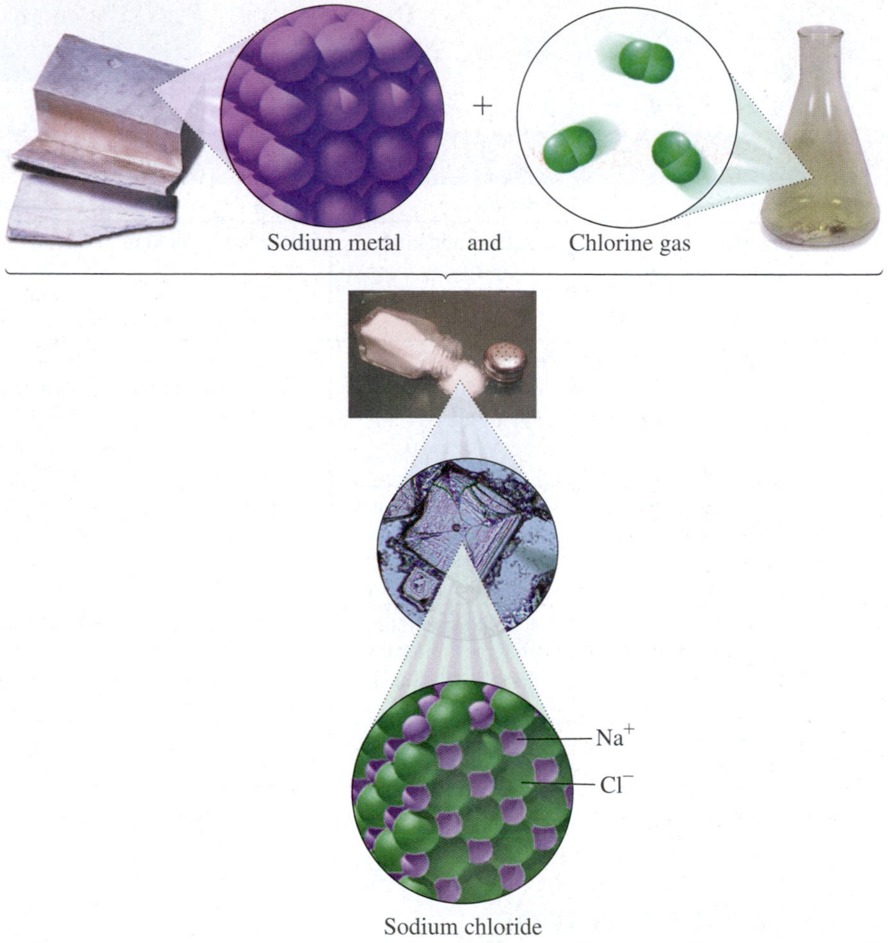

Sodium metal
+
and
Chlorine gas
Na$^+$
Cl$^-$
Sodium chloride

FIGURE 6.1 ▶ The elements sodium and chlorine react to form the ionic compound sodium chloride, the compound that makes up table salt. The magnification of NaCl crystals shows the arrangement of Na^+ and Cl^- ions.

Q What is the type of bonding between Na^+ and Cl^- ions in NaCl?

Chemical Formulas of Ionic Compounds

The **chemical formula** of a compound represents the symbols and subscripts in the lowest whole-number ratio of the atoms or ions. In the formula of an ionic compound, the sum of the ionic charges in the formula is always zero. *Thus, the total amount of positive charge is equal to the total amount of negative charge*. For example, to achieve a stable electron configuration, one Na atom (metal) loses its one valence electron to form Na^+, and one Cl atom (nonmetal) gains one electron to form a Cl^- ion. The formula NaCl indicates that the compound has charge balance because there is one sodium ion, Na^+, for every chloride ion, Cl^-. The ionic charges are not shown in the formula of the compound.

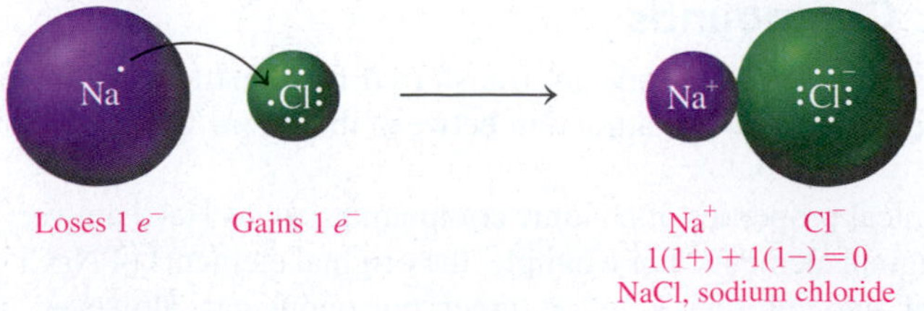

Na
Cl
Na$^+$
Cl
Loses 1 e^-
Gains 1 e^-
Na$^+$ Cl$^-$
$1(1+) + 1(1-) = 0$
NaCl, sodium chloride

Subscripts in Formulas

Consider a compound of magnesium and chlorine. To achieve a stable electron configuration, one Mg atom (metal) loses its two valence electrons to form Mg^{2+}. Two Cl atoms (nonmetals) each gain one electron to form two Cl^- ions. The two Cl^- ions are needed to balance the positive charge of Mg^{2+}. This gives the formula $MgCl_2$, magnesium chloride, in which the subscript 2 shows that two Cl^- ions are needed for charge balance.

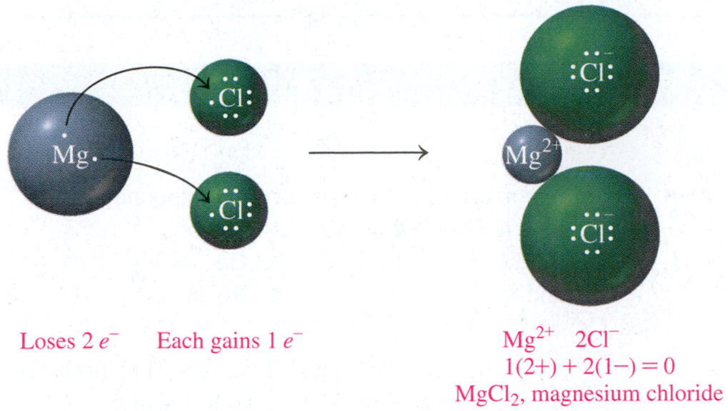

Mg·
:Cl:
:Cl:
Mg2+
:Cl:−
:Cl:−
Loses 2 e−
Each gains 1 e−
Mg2+ 2Cl−
1(2+) + 2(1−) = 0
MgCl2, magnesium chloride

Writing Ionic Formulas from Ionic Charges

The subscripts in the formula of an ionic compound represent the number of positive and negative ions that give an overall charge of zero. Thus, we can now write a formula directly from the ionic charges of the positive and negative ions. Suppose we wish to write the formula for the ionic compound containing Na^+ and S^{2-} ions. To balance the ionic charge of the S^{2-} ion, we will need to place two Na^+ ions in the formula. This gives the formula Na_2S, which has an overall charge of zero. In the formula of an ionic compound, the cation is written first, followed by the anion. Appropriate subscripts are used to show the number of each of the ions. This formula is the lowest ratio of ions in the ionic compound. This lowest ratio of ions is called a *formula unit*.

CORE CHEMISTRY SKILL
Writing Ionic Formulas

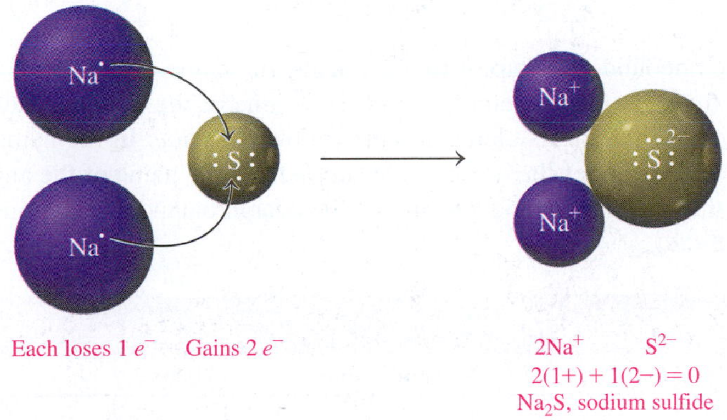

Na·
Na·
:S:
Na+
Na+
:S:2−
Each loses 1 e−
Gains 2 e−
2Na+ S2−
2(1+) + 1(2−) = 0
Na2S, sodium sulfide

ENGAGE
How are the charges of ions used to write an ionic formula?

▶SAMPLE PROBLEM 6.3 Writing Formulas from Ionic Charges

TRY IT FIRST

Write the symbols for the ions, and the correct formula for the ionic compound formed when lithium and nitrogen react.

SOLUTION

Lithium, which is a metal in Group 1A (1), forms Li^+; nitrogen, which is a nonmetal in Group 5A (15), forms N^{3-}. The charge of 3− is balanced by three Li^+ ions.

$$3(1+) + 1(3-) = 0$$

Writing the cation (positive ion) first and the anion (negative ion) second gives the formula Li_3N.

STUDY CHECK 6.3

Write the symbols for the ions, and the correct formula for the ionic compound that would form when each of the following react:

a. calcium and oxygen **b.** magnesium and phosphorus

ANSWER

a. Ca^{2+}, O^{2-}, CaO **b.** Mg^{2+}, P^{3-}, Mg_3P_2

TEST

Try Practice Problems 6.15 to 6.20

PRACTICE PROBLEMS

6.2 Ionic Compounds

6.15 Which of the following pairs of elements are likely to form an ionic compound?
a. lithium and chlorine **b.** oxygen and bromine
c. potassium and oxygen **d.** sodium and neon
e. cesium and magnesium **f.** nitrogen and fluorine

6.16 Which of the following pairs of elements are likely to form an ionic compound?
a. helium and oxygen **b.** magnesium and chlorine
c. chlorine and bromine **d.** potassium and sulfur
e. sodium and potassium **f.** nitrogen and iodine

6.17 Write the correct ionic formula for the compound formed between each of the following pairs of ions:
a. Na^+ and O^{2-} **b.** Al^{3+} and Br^- **c.** Ba^{2+} and N^{3-}
d. Mg^{2+} and F^- **e.** Al^{3+} and S^{2-}

6.18 Write the correct ionic formula for the compound formed between each of the following pairs of ions:
a. Al^{3+} and Cl^- **b.** Ca^{2+} and S^{2-}
c. Li^+ and S^{2-} **d.** Rb^+ and P^{3-}
e. Cs^+ and I^-

6.19 Write the symbols for the ions, and the correct formula for the ionic compound formed by each of the following:
a. potassium and sulfur **b.** sodium and nitrogen
c. aluminum and iodine **d.** gallium and oxygen

6.20 Write the symbols for the ions, and the correct formula for the ionic compound formed by each of the following:
a. calcium and chlorine **b.** rubidium and bromine
c. sodium and phosphorus **d.** magnesium and oxygen

REVIEW

Solving Equations (1.4)

6.3 Naming and Writing Ionic Formulas

LEARNING GOAL Given the formula of an ionic compound, write the correct name; given the name of an ionic compound, write the correct formula.

In the name of an ionic compound made up of two elements, the name of the metal ion, which is written first, is the same as its element name. The name of the nonmetal ion is obtained by using the first syllable of its element name followed by *ide*. In the name of any ionic compound, a space separates the name of the cation from the name of the anion. Subscripts are not used; they are understood because of the charge balance of the ions in the compound (see **TABLE 6.5**).

CORE CHEMISTRY SKILL

Naming Ionic Compounds

IODIZED
SALT

Iodized salt contains KI to prevent iodine deficiency.

TABLE 6.5 Names of Some Ionic Compounds

Compound	Metal Ion	Nonmetal Ion	Name
KI	K^+ Potassium	I^- Iodide	Potassium iodide
$MgBr_2$	Mg^{2+} Magnesium	Br^- Bromide	Magnesium bromide
Al_2O_3	Al^{3+} Aluminum	O^{2-} Oxide	Aluminum oxide

▶SAMPLE PROBLEM 6.4 Naming Ionic Compounds

TRY IT FIRST

Write the name for the ionic compound Mg_3N_2.

ENGAGE

How are the positive and negative ions named in an ionic compound?

SOLUTION

ANALYZE THE PROBLEM	Given	Need	Connect
	Mg_3N_2	name	cation, anion

STEP 1 **Identify the cation and anion.** The cation is Mg^{2+} and the anion is N^{3-}.

STEP 2 **Name the cation by its element name.** The cation Mg^{2+} is magnesium.

STEP 3 **Name the anion by using the first syllable of its element name followed by *ide*.** The anion N^{3-} is nitride.

STEP 4 **Write the name for the cation first and the name for the anion second.** magnesium nitride

STUDY CHECK 6.4

Write the name for each of the following ionic compounds:

a. Ga_2S_3 **b.** Li_3P

ANSWER

a. gallium sulfide **b.** lithium phosphide

TEST

Try Practice Problems 6.21 and 6.22

Metals with Variable Charges

We have seen that the charge of an ion of a representative element can be obtained from its group number. However, we cannot determine the charge of a transition element because it typically forms two or more positive ions. The transition elements lose electrons, but they are lost from the highest energy level and sometimes from a lower energy level as well. This is also true for metals of representative elements in Groups 4A (14) and 5A (15), such as Pb, Sn, and Bi.

In some ionic compounds, iron is in the Fe^{2+} form, but in other compounds, it has the Fe^{3+} form. Copper also forms two different ions, Cu^{+} and Cu^{2+}. When a metal can form two or more types of ions, it has *variable charge*. Then we cannot predict the ionic charge from the group number.

For metals that form two or more ions, a naming system is used to identify the particular cation. A Roman numeral that is equal to the ionic charge is placed in parentheses immediately after the name of the metal. For example, Fe^{2+} is iron(II), and Fe^{3+} is iron(III). **TABLE 6.6** lists the ions of some metals that produce more than one ion.

The transition elements form more than one positive ion except for zinc (Zn^{2+}), cadmium (Cd^{2+}), and silver (Ag^{+}), which form only one ion. Thus, no Roman numerals are used with zinc, cadmium, and silver when naming their cations in ionic compounds. Metals in Groups 4A (14) and 5A (15) also form more than one type of positive ion. For example, lead and tin in Group 4A (14) form cations with charges of 2+ and 4+, and bismuth in Group 5A (15) forms cations with charges of 3+ and 5+.

ENGAGE

Why is a Roman numeral placed after the name of the cations of most transition elements?

TABLE 6.6 Some Metals That Form More Than One Positive Ion

Element	Positive Ions	Name of Ion
Bismuth	Bi^{3+} Bi^{5+}	Bismuth(III) Bismuth(V)
Chromium	Cr^{2+} Cr^{3+}	Chromium(II) Chromium(III)
Cobalt	Co^{2+} Co^{3+}	Cobalt(II) Cobalt(III)
Copper	Cu^{+} Cu^{2+}	Copper(I) Copper(II)
Gold	Au^{+} Au^{3+}	Gold(I) Gold(III)
Iron	Fe^{2+} Fe^{3+}	Iron(II) Iron(III)
Lead	Pb^{2+} Pb^{4+}	Lead(II) Lead(IV)
Manganese	Mn^{2+} Mn^{3+}	Manganese(II) Manganese(III)
Mercury	Hg_2^{2+} Hg^{2+}	Mercury(I)* Mercury(II)
Nickel	Ni^{2+} Ni^{3+}	Nickel(II) Nickel(III)
Tin	Sn^{2+} Sn^{4+}	Tin(II) Tin(IV)

*Mercury(I) ions form an ion pair with a 2+ charge.

Determination of Variable Charge

When you name an ionic compound, you need to determine if the metal is a representative element or a transition element. If it is a transition element, except for zinc, cadmium, or silver, you will need to use its ionic charge as a Roman numeral as part of its name. The calculation of ionic charge depends on the negative charge of the anions in the formula. For example, if we want to name the ionic compound $CuCl_2$, we use charge balance to determine the charge of the copper cation. Because there are two chloride ions, each with a 1− charge, the total negative charge is 2−. To balance this 2− charge, the copper ion must have a charge of 2+, or Cu^{2+}:

$CuCl_2$

$Cu^{?}$ Cl^{-}

Cl^{-}

$$1(?) + 2(1-) = 0$$
$$? = 2+$$

To indicate the 2+ charge for the copper ion Cu^{2+}, we place the Roman numeral (II) immediately after copper when naming this compound: copper(II) chloride. Some ions and their location on the periodic table are seen in **FIGURE 6.2**.

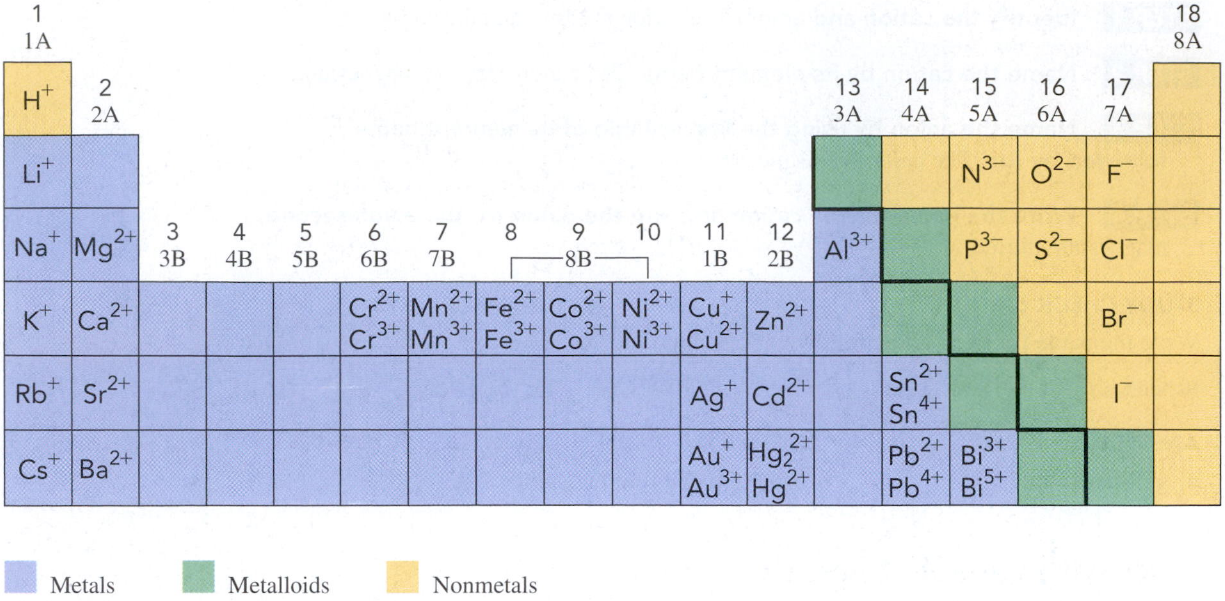
1
1A
18
8A
2
2A
13 14 15 16 17
3A 4A 5A 6A 7A
3 4 5 6 7 8 9 10 11 12
3B 4B 5B 6B 7B 8B 1B 2B
H⁺
Li⁺
Na⁺ Mg²⁺
K⁺ Ca²⁺
Rb⁺ Sr²⁺
Cs⁺ Ba²⁺
Cr²⁺ Cr³⁺
Mn²⁺ Mn³⁺
Fe²⁺ Fe³⁺
Co²⁺ Co³⁺
Ni²⁺ Ni³⁺
Cu⁺ Cu²⁺
Zn²⁺
Ag⁺ Cd²⁺
Au⁺ Au³⁺
Hg₂²⁺ Hg²⁺
Al³⁺
Sn²⁺ Sn⁴⁺
Pb²⁺ Pb⁴⁺
Bi³⁺ Bi⁵⁺
N³⁻ O²⁻ F⁻
P³⁻ S²⁻ Cl⁻
Br⁻
I⁻
Metals
Metalloids
Nonmetals

FIGURE 6.2 ▶ In ionic compounds, metals form positive ions, and nonmetals form negative ions.

Q What are the typical ions of calcium, copper, and oxygen in ionic compounds?

TABLE 6.7 Some Ionic Compounds of Metals That Form Two Compounds

Compound	Systematic Name
$FeCl_2$	Iron(II) chloride
Fe_2O_3	Iron(III) oxide
Cu_3P	Copper(I) phosphide
$CrBr_2$	Chromium(II) bromide
$SnCl_2$	Tin(II) chloride
PbS_2	Lead(IV) sulfide
BiF_3	Bismuth(III) fluoride

The growth of barnacles is prevented by using a paint with Cu_2O on the bottom of a boat.

TABLE 6.7 lists the names of some ionic compounds in which the transition elements and metals from Groups 4A (14) and 5A (15) have more than one positive ion.

▶ SAMPLE PROBLEM 6.5 Naming Ionic Compounds with Variable Charge Metal Ions

TRY IT FIRST

Antifouling paint contains Cu_2O, which prevents the growth of barnacles and algae on the bottoms of boats. What is the name of Cu_2O?

SOLUTION

	Given	Need	Connect
ANALYZE THE PROBLEM	Cu_2O	name	cation, anion, charge balance

STEP 1 Determine the charge of the cation from the anion.

	Metal	Nonmetal
Element	copper (Cu)	oxygen (O)
Group	transition element	6A (16)
Ion	$Cu^?$	O^{2-}
Charge Balance	$2Cu^? \quad +$ $\frac{2Cu^?}{2} = \frac{2+}{2} = 1+$	$2- = 0$
Ion	Cu^+	O^{2-}

STEP 2 Name the cation by its element name, and use a Roman numeral in parentheses for the charge. copper(I)

STEP 3 Name the anion by using the first syllable of its element name followed by *ide*. oxide

STEP 4 Write the name for the cation first and the name for the anion second. copper(I) oxide

STUDY CHECK 6.5

Write the name for each of the following compounds:

a. Mn_2S_3 **b.** SnF_4

ANSWER

a. Mn_2S_3 with the variable charge metal ion Mn^{3+} is named manganese(III) sulfide.
b. SnF_4 with the variable charge metal ion Sn^{4+} is named tin(IV) fluoride.

TEST
Try Practice Problems 6.23 to 6.28

Writing Formulas from the Name of an Ionic Compound

The formula for an ionic compound is written from the first part of the name that describes the metal ion, including its charge, and the second part of the name that specifies the nonmetal ion. Subscripts are added, as needed, to balance the charge. The steps for writing a formula from the name of an ionic compound are shown in Sample Problem 6.6.

INTERACTIVE VIDEO
PEARSON eText 2.0
Naming and Writing Ionic Formulas

▶ SAMPLE PROBLEM 6.6 Writing Formulas for Ionic Compounds

TRY IT FIRST

Write the correct formula for iron(III) chloride.

SOLUTION

ANALYZE THE PROBLEM	Given	Need	Connect
	iron(III) chloride	formula	cation, anion, charge balance

STEP 1 Identify the cation and anion.

Type of Ion	Cation	Anion
Name	iron(III)	chloride
Group	transition element	7A (17)
Symbol of Ion	Fe^{3+}	Cl^-

STEP 2 **Balance the charges.** The charge of 3+ is balanced by three Cl^- ions.

$$\mathbf{1}(3+) + \mathbf{3}(1-) = 0$$

STEP 3 **Write the formula, cation first, using subscripts from the charge balance.** $FeCl_3$

Acrylic
Colour

The pigment chrome green contains chromium(III) oxide.

STUDY CHECK 6.6

Write the correct formula for each of the following paint pigments:

a. chrome green, chromium(III) oxide
b. titanium white, titanium(IV) oxide

ANSWER

a. Cr_2O_3 **b.** TiO_2

TEST
Try Practice Problems 6.29 to 6.34

PRACTICE PROBLEMS

6.3 Naming and Writing Ionic Formulas

6.21 Write the name for each of the following ionic compounds:
a. AlF_3 **b.** $CaCl_2$ **c.** Na_2O
d. Mg_3P_2 **e.** KI **f.** BaF_2

6.22 Write the name for each of the following ionic compounds:
a. $MgCl_2$ **b.** K_3P **c.** Li_2S
d. CsF **e.** MgO **f.** $SrBr_2$

6.23 Write the name for each of the following ions (include the Roman numeral when necessary):
a. Fe^{2+} **b.** Cu^{2+} **c.** Zn^{2+} **d.** Pb^{4+} **e.** Cr^{3+} **f.** Mn^{2+}

6.24 Write the name for each of the following ions (include the Roman numeral when necessary):
a. Ag^+ **b.** Cu^+ **c.** Bi^{3+} **d.** Sn^{2+} **e.** Au^{3+} **f.** Ni^{2+}

6.25 Write the name for each of the following ionic compounds:
a. $SnCl_2$ **b.** FeO **c.** Cu_2S
d. CuS **e.** $CdBr_2$ **f.** $HgCl_2$

6.26 Write the name for each of the following ionic compounds:
a. Ag_3P **b.** PbS **c.** SnO_2
d. $MnCl_3$ **e.** Bi_2O_3 **f.** $CoCl_2$

6.27 Write the symbol for the cation in each of the following ionic compounds:
a. $AuCl_3$ **b.** Fe_2O_3 **c.** PbI_4 **d.** AlP

6.28 Write the symbol for the cation in each of the following ionic compounds:
a. $FeCl_2$ **b.** CrO **c.** Ni_2S_3 **d.** $SnCl_2$

6.29 Write the formula for each of the following ionic compounds:
a. magnesium chloride **b.** sodium sulfide
c. copper(I) oxide **d.** zinc phosphide
e. gold(III) nitride **f.** cobalt(III) fluoride

6.30 Write the formula for each of the following ionic compounds:
a. nickel(III) oxide **b.** barium fluoride
c. tin(IV) chloride **d.** silver sulfide
e. bismuth(V) chloride **f.** potassium nitride

6.31 Write the formula for each of the following ionic compounds:
a. cobalt(III) chloride **b.** lead(IV) oxide
c. silver iodide **d.** calcium nitride
e. copper(I) phosphide **f.** chromium(II) chloride

6.32 Write the formula for each of the following ionic compounds:
a. zinc bromide **b.** iron(III) sulfide
c. manganese(IV) oxide **d.** chromium(III) iodide
e. lithium nitride **f.** gold(I) oxide

Clinical Applications

6.33 The following compounds contain ions that are required in small amounts by the body. Write the formula for each.
a. potassium phosphide **b.** copper(II) chloride
c. iron(III) bromide **d.** magnesium oxide

6.34 The following compounds contain ions that are required in small amounts by the body. Write the formula for each.
a. calcium chloride **b.** nickel(II) iodide
c. manganese(II) oxide **d.** zinc nitride

6.4 Polyatomic Ions

LEARNING GOAL Write the name and formula for an ionic compound containing a polyatomic ion.

An ionic compound may also contain a *polyatomic ion* as one of its cations or anions. A **polyatomic ion** is a group of covalently bonded atoms that has an overall ionic charge. Most polyatomic ions consist of a nonmetal such as phosphorus, sulfur, carbon, or nitrogen covalently bonded to oxygen atoms.

Almost all the polyatomic ions are anions with charges 1−, 2−, or 3−. Only one common polyatomic ion, NH_4^+, has a positive charge. Some models of common polyatomic ions are shown in **FIGURE 6.3**.

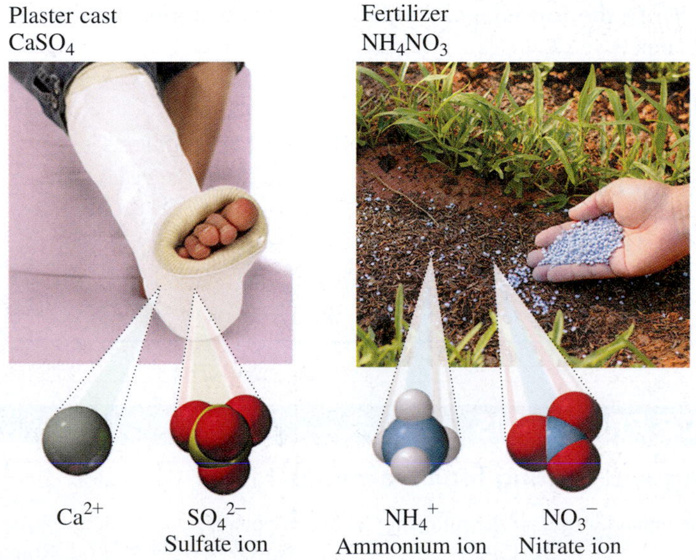
Plaster cast
$CaSO_4$
Fertilizer
NH_4NO_3
Ca^{2+}
SO_4^{2-}
Sulfate ion
NH_4^+
Ammonium ion
NO_3^-
Nitrate ion

FIGURE 6.3 ▶ Many products contain polyatomic ions, which are groups of atoms that have an ionic charge.

Q What is the charge of a sulfate ion?

Names of Polyatomic Ions

The names of the most common polyatomic ions, which end in *ate*, are shown in bold in **TABLE 6.8**. When a related ion has one less O atom, the *ite* ending is used. For the same nonmetal, the *ate* ion and the *ite* ion have the same charge. For example, the sulfate ion is SO_4^{2-}, and the sulfite ion, which has one less oxygen atom, is SO_3^{2-}.

Formula	Charge	Name
SO_4^{2-}	2−	sulf**ate**
SO_3^{2-}	2−	sulf**ite**

Phosphate and phosphite ions each have a 3− charge.

Formula	Charge	Name
PO_4^{3-}	3−	phosph**ate**
PO_3^{3-}	3−	phosph**ite**

The formula of hydrogen carbonate, or *bicarbonate*, is written by placing a hydrogen in front of the polyatomic ion formula for carbonate (CO_3^{2-}), and the charge is decreased from 2− to 1− to give HCO_3^-.

Formula	Charge	Name
CO_3^{2-}	2−	carbon**ate**
HCO_3^-	1−	hydrogen carbon**ate**

The halogens form four different polyatomic ions with oxygen. The prefix *per* is used for one more O than the *ate* ion and the prefix *hypo* for one less O than the *ite* ion.

Formula	Charge	Name
ClO_4^-	1−	**per**chlor**ate**
ClO_3^-	1−	chlor**ate**
ClO_2^-	1−	chlor**ite**
ClO^-	1−	**hypo**chlor**ite**

Recognizing these prefixes and endings will help you identify polyatomic ions in the name of a compound. The hydroxide ion (OH^-) and cyanide ion (CN^-) are exceptions to this naming pattern.

TABLE 6.8 Names and Formulas of Some Common Polyatomic Ions

Nonmetal	Formula of Ion*	Name of Ion
Hydrogen	OH^-	Hydroxide
Nitrogen	NH_4^+ $\mathbf{NO_3^-}$ NO_2^-	Ammonium **Nitrate** Nitrite
Chlorine	ClO_4^- $\mathbf{ClO_3^-}$ ClO_2^- ClO^-	Perchlorate **Chlorate** Chlorite Hypochlorite
Carbon	$\mathbf{CO_3^{2-}}$ HCO_3^- CN^- $C_2H_3O_2^-$	**Carbonate** Hydrogen carbonate (or bicarbonate) Cyanide Acetate
Sulfur	$\mathbf{SO_4^{2-}}$ HSO_4^- SO_3^{2-} HSO_3^-	**Sulfate** Hydrogen sulfate (or bisulfate) Sulfite Hydrogen sulfite (or bisulfite)
Phosphorus	$\mathbf{PO_4^{3-}}$ HPO_4^{2-} $H_2PO_4^-$ PO_3^{3-}	**Phosphate** Hydrogen phosphate Dihydrogen phosphate Phosphite

*Formulas and names in bold type indicate the most common polyatomic ion for that element.

TEST

Try Practice Problems 6.35 to 6.38

Sodium chlorite is used as a disinfectant in contact lens cleaning solutions, mouthwashes, and as a medication to treat amyotrophic lateral sclerosis (ALS).

Writing Formulas for Compounds Containing Polyatomic Ions

No polyatomic ion exists by itself. Like any ion, a polyatomic ion must be associated with ions of opposite charge. The bonding between polyatomic ions and other ions is one of electrical attraction.

To write correct formulas for compounds containing polyatomic ions, we follow the same rules of charge balance that we used for writing the formulas for simple ionic compounds. The total negative and positive charges must equal zero. For example, consider the formula for a compound containing sodium ions and chlorite ions. The ions are written as

Na^+ (Sodium ion) ClO_2^- (Chlorite ion)

$$(1+) + (1-) = 0$$

Because one ion of each balances the charge, the formula is written as

$NaClO_2$
Sodium chlorite

When more than one polyatomic ion is needed for charge balance, parentheses are used to enclose the formula of the ion. A subscript is written outside the right parenthesis of the polyatomic ion to indicate the number needed for charge balance. Consider the formula for magnesium nitrate. The ions in this compound are the magnesium ion and the nitrate ion, a polyatomic ion.

Mg^{2+} (Magnesium ion) NO_3^- (Nitrate ion)

ENGAGE

Why does the formula for magnesium nitrate contain two NO_3^- ions?

To balance the positive charge of 2+ on the magnesium ion, two nitrate ions are needed. In the formula of the compound, parentheses are placed around the nitrate ion, and the subscript 2 is written outside the right parenthesis.

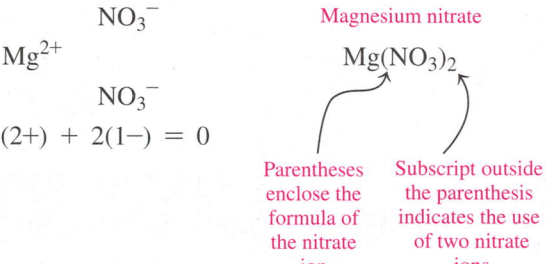

Mg^{2+} with NO_3^- and NO_3^-

$$(2+) + 2(1-) = 0$$

Magnesium nitrate

$Mg(NO_3)_2$

Parentheses enclose the formula of the nitrate ion

Subscript outside the parenthesis indicates the use of two nitrate ions

▶SAMPLE PROBLEM 6.7 Writing Formulas Containing Polyatomic Ions

TRY IT FIRST

Antacid tablets contain aluminum hydroxide, which treats acid indigestion and heartburn. Write the formula for aluminum hydroxide.

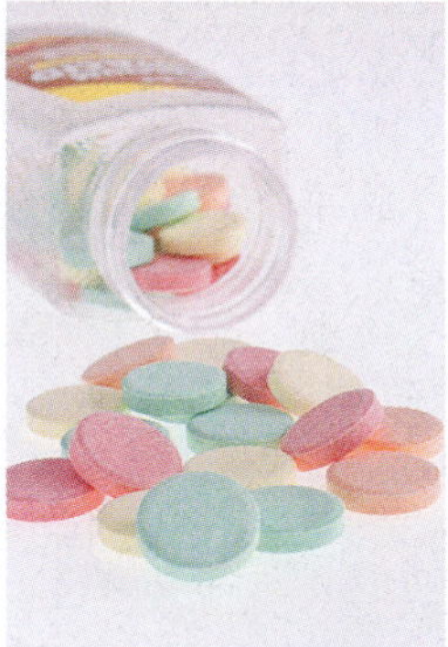

Aluminum hydroxide is an antacid used to treat acid indigestion.

SOLUTION

	Given	Need	Connect
ANALYZE THE PROBLEM	aluminum hydroxide	formula	cation, polyatomic anion, charge balance

STEP 1 Identify the cation and polyatomic ion (anion).

Cation	Polyatomic Ion (anion)
aluminum	hydroxide
Al^{3+}	OH^-

STEP 2 **Balance the charges.** The charge of 3+ is balanced by three OH^- ions.

$$\mathbf{1}(3+) + \mathbf{3}(1-) = 0$$

STEP 3 **Write the formula, cation first, using the subscripts from charge balance.** The formula for the compound is written by enclosing the formula of the hydroxide ion, OH^-, in parentheses and writing the subscript 3 outside the right parenthesis.

$Al(OH)_3$

STUDY CHECK 6.7

Write the formula for a compound containing the following:

a. ammonium ions and phosphate ions **b.** iron(III) ions and bicarbonate ions

ANSWER

a. $(NH_4)_3PO_4$ **b.** $Fe(HCO_3)_3$

Naming Ionic Compounds Containing Polyatomic Ions

When naming ionic compounds containing polyatomic ions, we first write the positive ion, usually a metal, and then we write the name for the polyatomic ion. It is important that you learn to recognize the polyatomic ion in the formula and name it correctly.

Na_2SO_4	$FePO_4$	$Al_2(CO_3)_3$
$Na_2\boxed{SO_4}$	$Fe\boxed{PO_4}$	$Al_2(\boxed{CO_3})_3$
Sodium sulfate	Iron(III) phosphate	Aluminum carbonate

TABLE 6.9 lists the formulas and names of some ionic compounds that include polyatomic ions and also gives their uses in medicine.

TABLE 6.9 Some Ionic Compounds that Contain Polyatomic Ions

Formula	Name	Medical Use
$AlPO_4$	Aluminum phosphate	Antacid
$Al_2(SO_4)_3$	Aluminum sulfate	Antiperspirant, anti-infective
$BaSO_4$	Barium sulfate	Contrast medium for X-rays
$CaCO_3$	Calcium carbonate	Antacid, calcium supplement
$Ca_3(PO_4)_2$	Calcium phosphate	Calcium dietary supplement
$CaSO_4$	Calcium sulfate	Plaster casts
$MgSO_4$	Magnesium sulfate	Cathartic, Epsom salts
K_2CO_3	Potassium carbonate	Alkalizer, diuretic
$AgNO_3$	Silver nitrate	Topical anti-infective
$NaHCO_3$	Sodium hydrogen carbonate or sodium bicarbonate	Antacid
$Zn_3(PO_4)_2$	Zinc phosphate	Dental cement

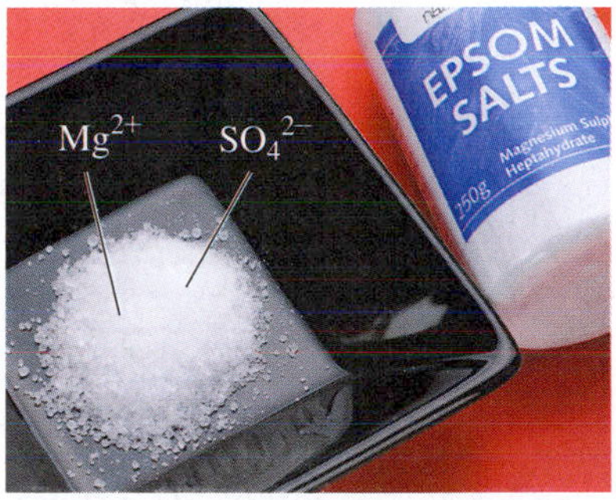

Mg^{2+}
SO_4^{2-}
EPSOM SALTS

A solution of Epsom salts, magnesium sulfate, $MgSO_4$, may be used to soothe sore muscles.

▶ SAMPLE PROBLEM 6.8 Naming Compounds Containing Polyatomic Ions

TRY IT FIRST

Name the following ionic compounds:

a. $Cu(NO_2)_2$ **b.** $KClO_3$

SOLUTION

ANALYZE THE PROBLEM	Given	Need	Connect
	formula	name	cation, polyatomic ion

Formula	Cation	Anion	Name of Cation	Name of Anion	Name of Compound
a. $Cu(NO_2)_2$	Cu^{2+}	NO_2^-	Copper(II) ion	Nitrite ion	Copper(II) nitrite
b. $KClO_3$	K^+	ClO_3^-	Potassium ion	Chlorate ion	Potassium chlorate

STUDY CHECK 6.8

Name each of the following compounds:

a. $Co_3(PO_4)_2$ **b.** $SrSO_3$

ANSWER

a. cobalt(II) phosphate **b.** strontium sulfite

TEST

Try Practice Problems 6.39 to 6.46

PRACTICE PROBLEMS

6.4 Polyatomic Ions

6.35 Write the formula including the charge for each of the following polyatomic ions:
a. hydrogen carbonate (bicarbonate)
b. ammonium
c. phosphite
d. chlorate

6.36 Write the formula including the charge for each of the following polyatomic ions:
a. nitrite **b.** sulfite
c. hydroxide **d.** acetate

6.37 Name the following polyatomic ions:
a. SO_4^{2-} **b.** CO_3^{2-} **c.** HSO_3^- **d.** NO_3^-

6.38 Name the following polyatomic ions:
a. OH^- **b.** PO_4^{3-} **c.** CN^- **d.** NO_2^-

6.39 Complete the following table with the formula and name of the compound that forms between each pair of ions:

	NO_2^-	CO_3^{2-}	HSO_4^-	PO_4^{3-}
Li^+				
Cu^{2+}				
Ba^{2+}				

6.40 Complete the following table with the formula and name of the compound that forms between each pair of ions:

	NO_3^-	HCO_3^-	SO_3^{2-}	HPO_4^{2-}
NH_4^+				
Al^{3+}				
Pb^{4+}				

6.41 Write the correct formula for the following ionic compounds:
a. barium hydroxide **b.** sodium hydrogen sulfate
c. iron(II) nitrite **d.** zinc phosphate
e. iron(III) carbonate

6.42 Write the correct formula for the following ionic compounds:
a. aluminum chlorate **b.** ammonium oxide
c. magnesium bicarbonate **d.** sodium nitrite
e. copper(I) sulfate

6.43 Write the formula for the polyatomic ion and name each of the following compounds:
a. Na_2CO_3 **b.** $(NH_4)_2S$
c. $Ca(OH)_2$ **d.** $Sn(NO_2)_2$

6.44 Write the formula for the polyatomic ion and name each of the following compounds:
a. $MnCO_3$ **b.** Au_2SO_4
c. $Ca_3(PO_4)_2$ **d.** $Fe(HCO_3)_3$

Clinical Applications

6.45 Name each of the following ionic compounds:
a. $Zn(C_2H_3O_2)_2$, cold remedy
b. $Mg_3(PO_4)_2$, antacid
c. NH_4Cl, expectorant
d. $NaHCO_3$, corrects pH imbalance
e. $NaNO_2$, meat preservative

6.46 Name each of the following ionic compounds:
a. Li_2CO_3, antidepressant
b. $MgSO_4$, Epsom salts
c. NaClO, disinfectant
d. Na_3PO_4, laxative
e. $Ba(OH)_2$, component of antacids

6.5 Molecular Compounds: Sharing Electrons

LEARNING GOAL Given the formula of a molecular compound, write its correct name; given the name of a molecular compound, write its formula.

A **molecular compound** consists of atoms of two or more nonmetals that share one or more valence electrons. The atoms are held together by **covalent bonds** that form a *molecule*. There are many more molecular compounds than there are ionic ones. For example, water (H_2O) and carbon dioxide (CO_2) are both molecular compounds. Molecular compounds consist of **molecules**, which are discrete groups of atoms in a definite proportion. A molecule of water (H_2O) consists of two atoms of hydrogen and one atom of oxygen. When you have iced tea, perhaps you add molecules of sugar ($C_{12}H_{22}O_{11}$), which is a molecular compound. Other familiar molecular compounds include propane (C_3H_8), alcohol (C_2H_6O), the antibiotic amoxicillin ($C_{16}H_{19}N_3O_5S$), and the antidepressant Prozac ($C_{17}H_{18}F_3NO$).

Names and Formulas of Molecular Compounds

When naming a molecular compound, the first nonmetal in the formula is named by its element name; the second nonmetal is named using the first syllable of its element name, followed by *ide*. When a subscript indicates two or more atoms of an element, a prefix is shown in front of its name. **TABLE 6.10** lists prefixes used in naming molecular compounds.

The names of molecular compounds need prefixes because several different compounds can be formed from the same two nonmetals. For example, carbon and oxygen can form two different compounds, carbon monoxide, CO, and carbon dioxide, CO_2, in which the number of atoms of oxygen in each compound is indicated by the prefixes *mono* or *di* in their names.

When the vowels *o* and *o* or *a* and *o* appear together, the first vowel is omitted, as in carbon monoxide. In the name of a molecular compound, the prefix *mono* is usually omitted, as in NO, nitrogen oxide. Traditionally, however, CO is named carbon monoxide. **TABLE 6.11** lists the formulas, names, and commercial uses of some molecular compounds.

CORE CHEMISTRY SKILL

Writing the Names and Formulas for Molecular Compounds

TABLE 6.10 Prefixes Used in Naming Molecular Compounds

1 mono	6 hexa
2 di	7 hepta
3 tri	8 octa
4 tetra	9 nona
5 penta	10 deca

TABLE 6.11 Some Common Molecular Compounds

Formula	Name	Commercial Uses
CO_2	Carbon dioxide	Fire extinguishers, dry ice, propellant in aerosols, carbonation of beverages
CS_2	Carbon disulfide	Manufacture of rayon
NO	Nitrogen oxide	Stabilizer, biochemical messenger in cells
N_2O	Dinitrogen oxide	Inhalation anesthetic, "laughing gas"
SF_6	Sulfur hexafluoride	Electrical circuits
SO_2	Sulfur dioxide	Preserving fruits, vegetables; disinfectant in breweries; bleaching textiles
SO_3	Sulfur trioxide	Manufacture of explosives

▶ SAMPLE PROBLEM 6.9 Naming Molecular Compounds

TRY IT FIRST

Name the molecular compound NCl_3.

SOLUTION

ANALYZE THE PROBLEM	Given	Need	Connect
	NCl_3	name	prefixes

STEP 1 **Name the first nonmetal by its element name.** In NCl_3, the first nonmetal (N) is nitrogen.

STEP 2 **Name the second nonmetal by using the first syllable of its element name followed by *ide*.** The second nonmetal (Cl) is chloride.

STEP 3 **Add prefixes to indicate the number of atoms (subscripts).** Because there is one nitrogen atom, no prefix is needed. The subscript 3 for the Cl atoms is shown as the prefix *tri*. The name of NCl_3 is nitrogen trichloride.

STUDY CHECK 6.9

Name each of the following molecular compounds:

a. $SiBr_4$ **b.** Br_2O **c.** S_3N_2

ANSWER

a. silicon tetrabromide **b.** dibromine oxide **c.** trisulfur dinitride

ENGAGE

Why are prefixes used to name molecular compounds?

TEST

Try Practice Problems 6.47 to 6.50

Writing Formulas from the Names of Molecular Compounds

In the name of a molecular compound, the names of two nonmetals are given along with prefixes for the number of atoms of each. To write the formula from the name, we use the symbol for each element and a subscript if a prefix indicates two or more atoms.

▶SAMPLE PROBLEM 6.10 Writing Formulas for Molecular Compounds

TRY IT FIRST

Write the formula for the molecular compound diboron trioxide.

SOLUTION

ANALYZE THE PROBLEM	Given	Need	Connect
	diboron trioxide	formula	subscripts from prefixes

STEP 1 **Write the symbols in the order of the elements in the name.**

Name of Element	Boron	Oxygen
Symbol of Element	B	O

STEP 2 **Write any prefixes as subscripts.** The prefix *di* in *di*boron indicates that there are two atoms of boron, shown as a subscript 2 in the formula. The prefix *tri* in *tri*oxide indicates that there are three atoms of oxygen, shown as a subscript 3 in the formula. B_2O_3

STUDY CHECK 6.10

Write the formula for each of the following molecular compounds:

a. iodine pentafluoride **b.** carbon diselenide

ANSWER

a. IF_5 **b.** CSe_2

TEST

Try Practice Problems 6.51 to 6.54

Summary of Naming Ionic and Molecular Compounds

We have now examined strategies for naming ionic and molecular compounds. In general, compounds having two elements are named by stating the first element name followed by the name of the second element with an *ide* ending. If the first element is a metal, the compound is usually ionic; if the first element is a nonmetal, the compound is usually molecular. For ionic compounds, it is necessary to determine whether the metal can form more than one type of positive ion; if so, a Roman numeral following the name of the metal indicates the particular ionic charge. One exception is the ammonium ion, NH_4^+, which is also written first as a positively charged polyatomic ion. Ionic compounds having three or more elements include some type of polyatomic ion. They are named by ionic rules but have an *ate* or *ite* ending when the polyatomic ion has a negative charge.

In naming molecular compounds having two elements, prefixes are necessary to indicate two or more atoms of each nonmetal as shown in that particular formula (see **FIGURE 6.4**).

▶SAMPLE PROBLEM 6.11 Naming Ionic and Molecular Compounds

TRY IT FIRST

Identify each of the following compounds as ionic or molecular and write its name:

a. $NiSO_4$ **b.** SO_3

ENGAGE

How do you know that sodium phosphate is an ionic compound and diphosphorus pentoxide a molecular compound?

SOLUTION

a. $NiSO_4$, consisting of a cation of a transition element and a polyatomic ion SO_4^{2-}, is an ionic compound. As a transition element, Ni forms more than one type of ion.

In this formula, the 2− charge of SO_4^{2-} is balanced by one nickel ion, Ni^{2+}. In the name, a Roman numeral written after the metal name, nickel(II), specifies the 2+ charge. The anion SO_4^{2-} is a polyatomic ion named sulfate. The compound is named nickel(II) sulfate.

b. SO_3 consists of two nonmetals, which indicates that it is a molecular compound. The first element S is sulfur (no prefix is needed). The second element O, oxide, has subscript 3, which requires a prefix *tri* in the name. The compound is named sulfur trioxide.

STUDY CHECK 6.11

Name each of the following compounds:

a. IF_7 **b.** $Fe(NO_3)_3$

ANSWER

a. iodine heptafluoride **b.** iron(III) nitrate

TEST

Try Practice Problems 6.55 and 6.56

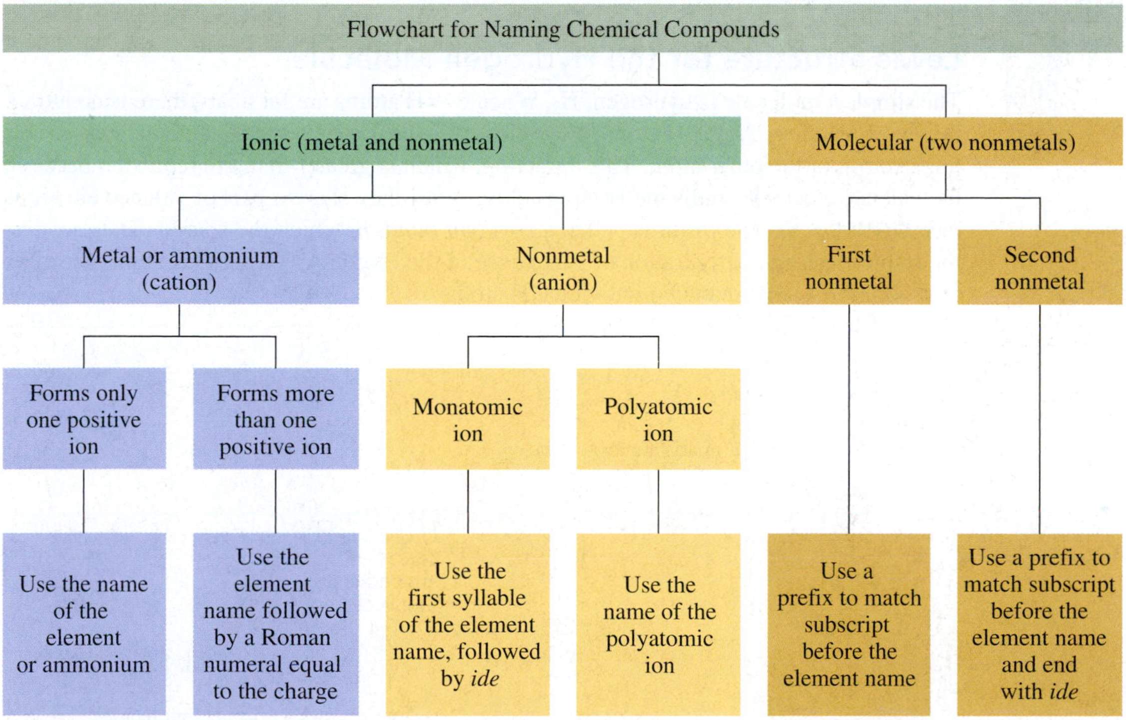

Flowchart for Naming Chemical Compounds
Ionic (metal and nonmetal)
Molecular (two nonmetals)
Metal or ammonium (cation)
Nonmetal (anion)
First nonmetal
Second nonmetal
Forms only one positive ion
Forms more than one positive ion
Monatomic ion
Polyatomic ion
Use the name of the element or ammonium
Use the element name followed by a Roman numeral equal to the charge
Use the first syllable of the element name, followed by ide
Use the name of the polyatomic ion
Use a prefix to match subscript before the element name
Use a prefix to match subscript before the element name and end with ide

FIGURE 6.4 ▶ A flowchart illustrates naming for ionic and molecular compounds.

Q Why are the names of some metal ions followed by a Roman numeral in the name of a compound?

PRACTICE PROBLEMS

6.5 Molecular Compounds: Sharing Electrons

6.47 Name each of the following molecular compounds:
a. PBr_3 **b.** Cl_2O **c.** CBr_4 **d.** HF **e.** NF_3

6.48 Name each of the following molecular compounds:
a. IF_3 **b.** P_2O_5 **c.** SiO_2 **d.** PCl_3 **e.** CO

6.49 Name each of the following molecular compounds:
a. N_2O_3 **b.** Si_2Br_6 **c.** P_4S_3 **d.** PCl_5 **e.** SeF_6

6.50 Name each of the following molecular compounds:
a. SiF_4 **b.** IBr_3 **c.** CO_2 **d.** N_2F_2 **e.** N_2S_3

6.51 Write the formula for each of the following molecular compounds:
a. carbon tetrachloride **b.** carbon monoxide
c. phosphorus trifluoride **d.** dinitrogen tetroxide

6.52 Write the formula for each of the following molecular compounds:
a. sulfur dioxide **b.** silicon tetrachloride
c. iodine trifluoride **d.** dinitrogen oxide

6.53 Write the formula for each of the following molecular compounds:
a. oxygen difluoride **b.** boron trichloride
c. dinitrogen trioxide **d.** sulfur hexafluoride

6.54 Write the formula for each of the following molecular compounds:
a. sulfur dibromide **b.** carbon disulfide
c. tetraphosphorus hexoxide **d.** dinitrogen pentoxide

Clinical Applications

6.55 Name each of the following ionic or molecular compounds:

a. $Al_2(SO_4)_3$, antiperspirant
b. $CaCO_3$, antacid
c. N_2O, "laughing gas," inhaled anesthetic
d. $Mg(OH)_2$, laxative

6.56 Name each of the following ionic or molecular compounds:

a. $Al(OH)_3$, antacid
b. $FeSO_4$, iron supplement in vitamins
c. NO, vasodilator
d. $Cu(OH)_2$, fungicide

REVIEW

Drawing Lewis Symbols (4.8)

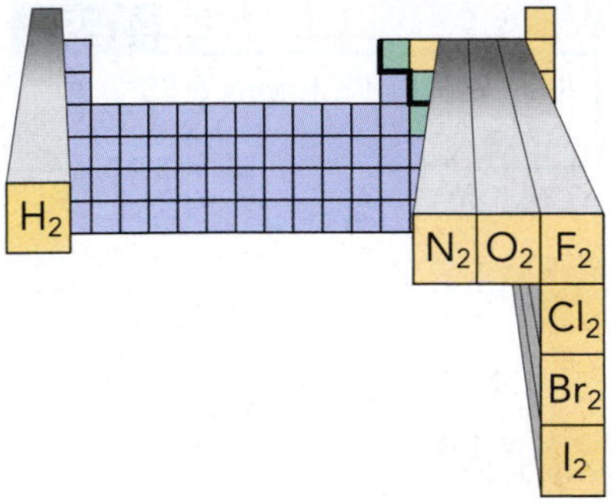
H2
N2 O2 F2
Cl2
Br2
I2

The elements hydrogen, nitrogen, oxygen, fluorine, chlorine, bromine, and iodine exist as diatomic molecules.

6.6 Lewis Structures for Molecules and Polyatomic Ions

LEARNING GOAL Draw the Lewis structures for molecular compounds and polyatomic ions.

Now we can investigate more complex chemical bonds and how they contribute to the structure of a molecule. *Lewis structures* use Lewis symbols to diagram the sharing of valence electrons for single and multiple bonds in molecules.

Lewis Structure for the Hydrogen Molecule

The simplest molecule is hydrogen, H_2. When two H atoms are far apart, there is no attraction between them. As the H atoms move closer, the positive charge of each nucleus attracts the electron of the other atom. This attraction, which is greater than the repulsion between the valence electrons, pulls the H atoms closer until they share a pair of valence electrons (see **FIGURE 6.5**). The result is called a *covalent bond*, in which the shared electrons give the stable electron configuration of He to *each* of the H atoms. When the H atoms form H_2, they are more stable than two individual H atoms.

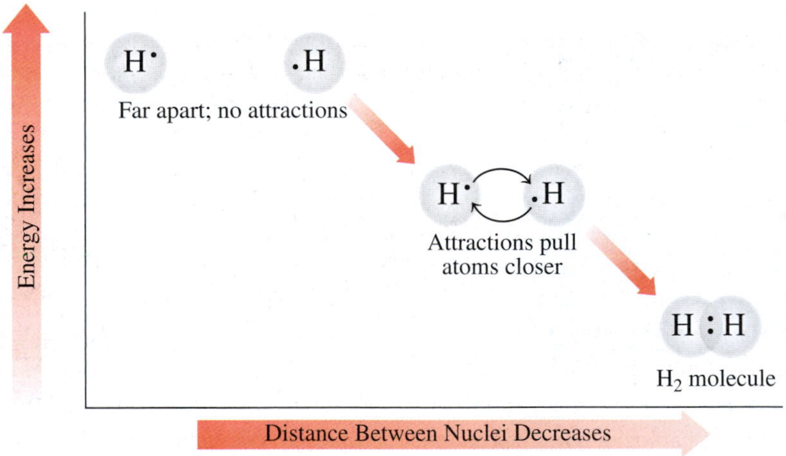
H· ·H
Far apart; no attractions
H· ·H
Attractions pull atoms closer
H:H
H2 molecule
Energy Increases
Distance Between Nuclei Decreases

FIGURE 6.5 ▶ A covalent bond forms as H atoms move close together to share electrons.

Q What determines the attraction between two H atoms?

Lewis Structures for Molecular Compounds

A molecule is represented by a **Lewis structure** in which the valence electrons of all the atoms are arranged to give octets, except for hydrogen, which has two electrons. The shared electrons, or *bonding pairs*, are shown as two dots or a single line between atoms. The nonbonding pairs of electrons, or *lone pairs*, are placed on the outside. For example, a fluorine molecule, F_2, consists of two fluorine atoms, which are in Group 7A (17), each with seven valence electrons. In the Lewis structure for the F_2 molecule, each F atom achieves an octet by sharing its unpaired valence electron.

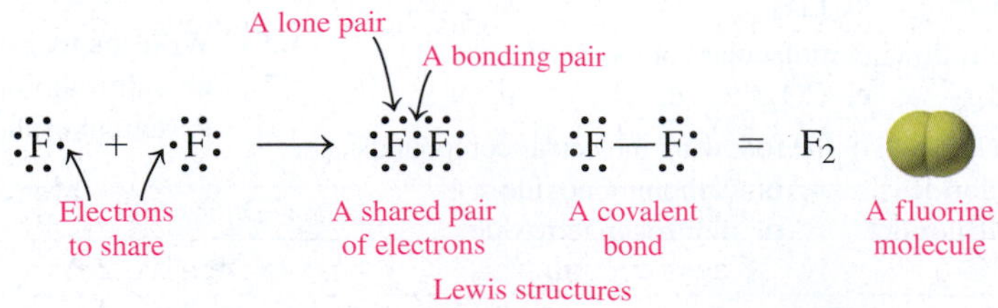
A lone pair
A bonding pair
Electrons to share
A shared pair of electrons
A covalent bond
= F2
A fluorine molecule
Lewis structures

Hydrogen (H_2) and fluorine (F_2) are examples of nonmetal elements whose natural state is diatomic; that is, they contain two like atoms. The elements that exist as diatomic molecules are listed in **TABLE 6.12**.

TABLE 6.12 Elements that Exist as Diatomic Molecules

Diatomic Molecule	Name
H_2	Hydrogen
N_2	Nitrogen
O_2	Oxygen
F_2	Fluorine
Cl_2	Chlorine
Br_2	Bromine
I_2	Iodine

Sharing Electrons Between Atoms of Different Elements

The number of electrons that a nonmetal atom shares and the number of covalent bonds it forms are usually equal to the number of electrons it needs to achieve a stable electron configuration. **TABLE 6.13** gives the most typical bonding patterns for some nonmetals.

TABLE 6.13 Typical Bonding Patterns of Some Nonmetals

1A (1)	3A (13)	4A (14)	5A (15)	6A (16)	7A (17)
*H 1 bond					
	*B 3 bonds	C 4 bonds	N 3 bonds	O 2 bonds	F 1 bond
		Si 4 bonds	P 3 bonds	S 2 bonds	Cl, Br, I 1 bond

*H and B do not form octets. H atoms share one electron pair; B atoms share three electron pairs for a set of six electrons.

Drawing Lewis Structures

To draw the Lewis structure for CH_4, we first draw the Lewis symbols for carbon and hydrogen.

Then we determine the number of valence electrons needed for carbon and hydrogen. When a carbon atom shares its four electrons with four hydrogen atoms, carbon obtains an octet and each hydrogen atom is complete with two shared electrons. The Lewis structure is drawn with the carbon atom as the central atom, with the hydrogen atoms on each of the sides. The bonding pairs of electrons, which are single covalent bonds, may also be shown as single lines between the carbon atom and each of the hydrogen atoms.

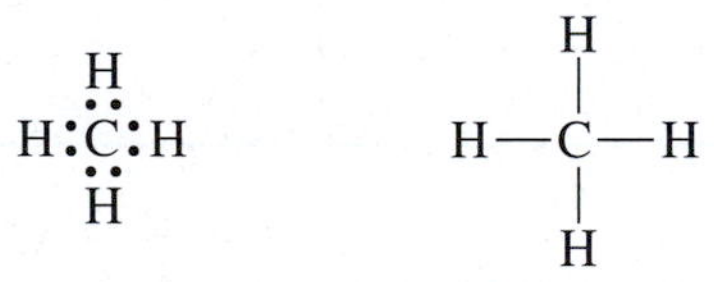

TABLE 6.14 gives examples of Lewis structures and molecular models for some molecules.

TABLE 6.14 Lewis Structures for Some Molecular Compounds

CH_4	NH_3	H_2O
Lewis Structures		
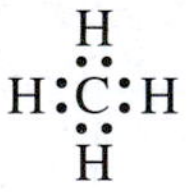	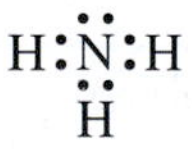	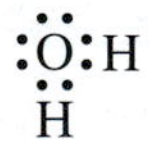
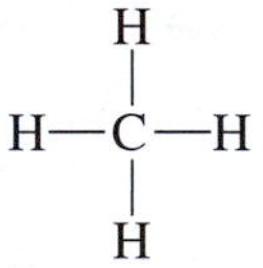	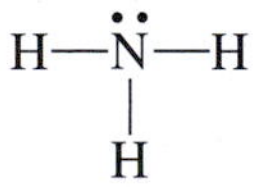	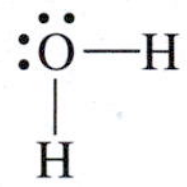
Molecular Models		
 Methane molecule	 Ammonia molecule	 Water molecule

When we draw a Lewis structure for a molecule or polyatomic ion, we show the sequence of atoms, the bonding pairs of electrons shared between atoms, and the nonbonding or *lone pairs* of electrons. From the formula, we identify the central atom, which is the element that has the fewer atoms. Then, the central atom is bonded to the other atoms, as shown in Sample Problem 6.12.

▶ SAMPLE PROBLEM 6.12 Drawing Lewis Structures for Molecules

TRY IT FIRST

Draw the Lewis structures for PCl_3, phosphorus trichloride, used commercially to prepare insecticides and flame retardants.

SOLUTION

ANALYZE THE PROBLEM	Given	Need	Connect
	PCl_3	Lewis structure	total valence electrons

CORE CHEMISTRY SKILL
Drawing Lewis Structures

STEP 1 **Determine the arrangement of atoms.** In PCl_3, the central atom is P because there is only one P atom.

Cl P Cl
  Cl

STEP 2 **Determine the total number of valence electrons.** We use the group number to determine the number of valence electrons for each of the atoms in the molecule.

Element	Group	Atoms		Valence Electrons		Total
P	5A (15)	1 P	×	5 e^-	=	5 e^-
Cl	7A (17)	3 Cl	×	7 e^-	=	21 e^-
				Total valence electrons for PCl_3	=	26 e^-

STEP 3 **Attach each bonded atom to the central atom with a pair of electrons.** Each bonding pair can also be represented by a bond line.

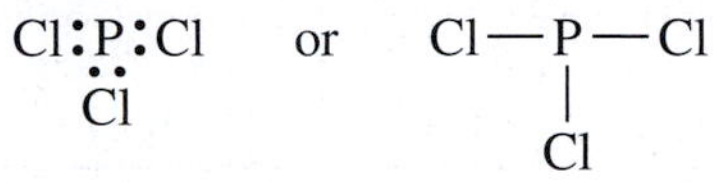

Six electrons ($3 \times 2\ e^-$) are used to bond the central P atom to three Cl atoms. Twenty valence electrons are left.

26 valence e^- − 6 bonding e^- = 20 e^- remaining

The ball-and-stick model of PCl_3 consists of P and Cl atoms connected by single bonds.

STEP 4 **Use the remaining electrons to complete octets.** We use the remaining 20 electrons as lone pairs, which are placed around the outer Cl atoms and on the P atom, such that all the atoms have octets.

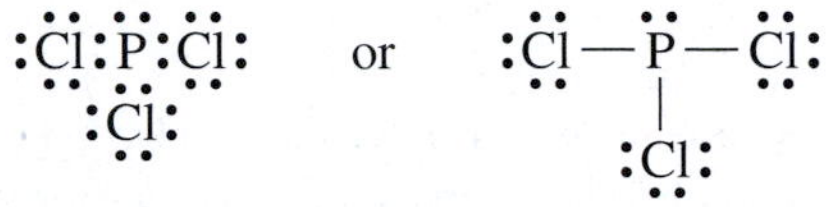

STUDY CHECK 6.12

Draw the Lewis structures for Cl_2O, one with electron dots and the other with lines for bonding pairs.

ANSWER

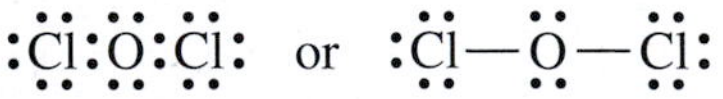

TEST

Try Practice Problems 6.57 to 6.62

▶ SAMPLE PROBLEM 6.13 Drawing Lewis Structures for Polyatomic Ions

TRY IT FIRST

Sodium chlorite, $NaClO_2$, is a component of mouthwashes, toothpastes, and contact lens cleaning solutions. Draw the Lewis structures for the chlorite ion, ClO_2^-.

SOLUTION

STEP 1 **Determine the arrangement of atoms.** For the polyatomic ion ClO_2^-, the central atom is Cl because there is only one Cl atom. For a polyatomic ion, the atoms and electrons are placed in brackets, and the charge is written outside at the upper right.

$[O\ \ Cl\ \ O]^-$

STEP 2 **Determine the total number of valence electrons.** We use the group numbers to determine the number of valence electrons for each of the atoms in the ion. Because the ion has a negative charge, one more electron is added to the valence electrons.

Element	Group	Atoms		Valence Electrons		Total
Cl	7A (17)	1 Cl	×	7 e^-	=	7 e^-
O	6A (16)	2 O	×	6 e^-	=	12 e^-
Ionic charge (negative) add				1 e^-	=	1 e^-
				Total valence electrons for ClO_2^-	=	20 e^-

STEP 3 **Attach each bonded atom to the central atom with a pair of electrons.** Each bonding pair can also be represented by a line, which indicates a single bond.

$[O:Cl:O]^-$ or $[O—Cl—O]^-$

Four electrons (2 × 2 e^-) are used to bond the O atoms to the central Cl atom, which leaves 16 valence electrons.

STEP 4 **Use the remaining electrons to complete octets.** Of the 16 remaining valence electrons, 12 are drawn as lone pairs to complete the octets of the O atoms.

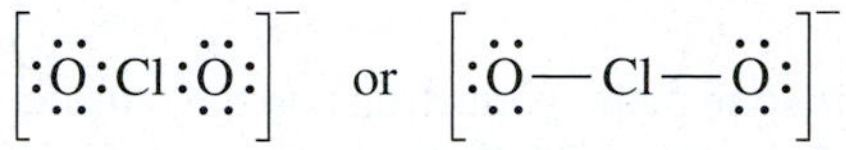

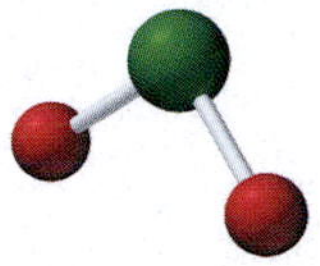

In ClO_2^-, the central Cl atom is bonded to two O atoms.

The remaining four electrons are placed as two lone pairs on the central Cl atom.

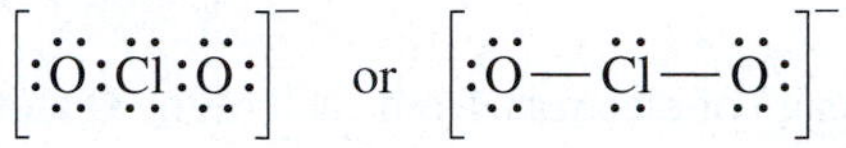

STUDY CHECK 6.13

Draw the Lewis structures for the polyatomic ion NH_2^-, one with electron dots and the other with lines for bonding pairs.

ANSWER

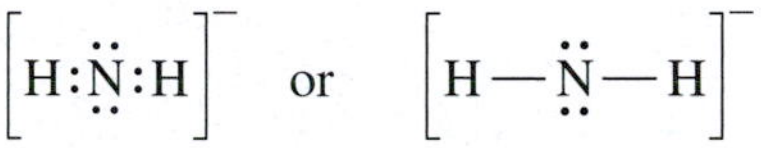

Double and Triple Bonds

Up to now, we have looked at bonding in molecules having only single bonds. In many molecular compounds, atoms share two or three pairs of electrons to complete their octets. Double and triple bonds form when the number of valence electrons is not enough to complete the octets of all the atoms in the molecule. Then one or more lone pairs of electrons from the atoms attached to the central atom are shared with the central atom. A **double bond** occurs when two pairs of electrons are shared; in a **triple bond**, three pairs of electrons are shared. Atoms of carbon, oxygen, nitrogen, and sulfur are most likely to form multiple bonds. Atoms of hydrogen and the halogens do not form double or triple bonds. The process of drawing a Lewis structure with multiple bonds is shown in Sample Problem 6.14.

INTERACTIVE VIDEO

PEARSON eText 2.0 Drawing Lewis Structures with Multiple Bonds

▶ SAMPLE PROBLEM 6.14 Drawing Lewis Structures with Multiple Bonds

TRY IT FIRST

Draw the Lewis structures for carbon dioxide, CO_2.

SOLUTION

ANALYZE THE PROBLEM	Given	Need	Connect
	CO_2	Lewis structure	total valence electrons

STEP 1 **Determine the arrangement of atoms.** In CO_2, the central atom is C because there is only one C atom.

O C O

STEP 2 **Determine the total number of valence electrons.** We use the group number to determine the number of valence electrons for each of the atoms in the molecule.

Element	Group	Atoms	Valence Electrons	Total
C	4A (14)	1 C ×	4 e^- =	4 e^-
O	6A (16)	2 O ×	6 e^- =	12 e^-
			Total valence electrons for CO_2 =	16 e^-

STEP 3 **Attach each bonded atom to the central atom with a pair of electrons.**

O:C:O or O—C—O

We use four valence electrons to attach the central C atom to two O atoms, which leaves 12 valence electrons.

STEP 4 **Use the remaining electrons to complete octets, using multiple bonds if needed.**

The 12 remaining electrons are placed as six lone pairs of electrons on the outside O atoms. However, this does not complete the octet of the C atom.

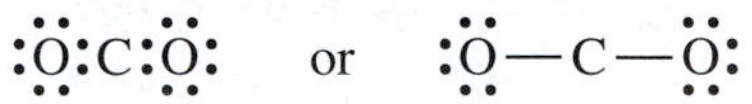

To obtain an octet, the C atom must share pairs of electrons from each of the O atoms. When two bonding pairs occur between atoms, it is known as a double bond.

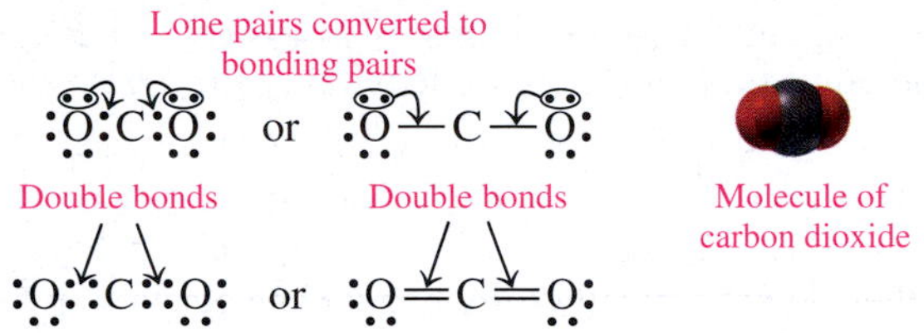
Lone pairs converted to bonding pairs
or
Double bonds
Double bonds
or
Molecule of carbon dioxide

STUDY CHECK 6.14

Draw the Lewis structures for HCN, which has a triple bond. Draw one structure with electron dots and the other with lines for bonding pairs.

ANSWER

H:C⋮⋮N: or H—C≡N:

ENGAGE

How do you know when you need to add one or more multiple bonds to complete a Lewis structure?

TEST

Try Practice Problems 6.63 and 6.64

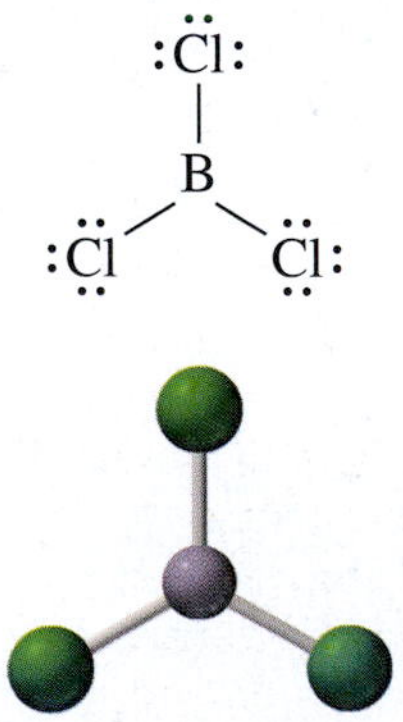

In BCl_3, the central B atom is bonded to three Cl atoms.

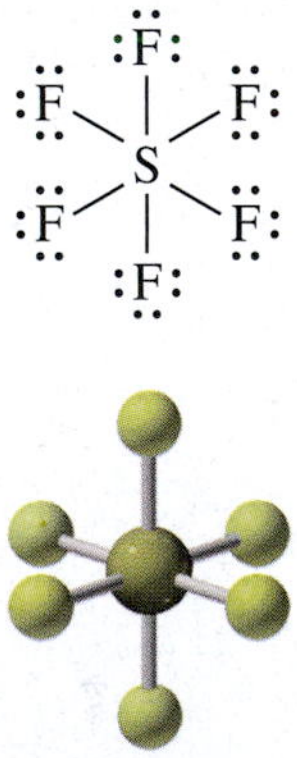

In SF_6, the central S atom is bonded to six F atoms.

Exceptions to the Octet Rule

Although the octet rule is useful for bonding in many compounds, there are exceptions. We have already seen that a hydrogen (H_2) molecule requires just two electrons or a single bond. Usually the nonmetals form octets. However, in BCl_3, the B atom has only three valence electrons to share. Boron compounds typically have six valence electrons on the central B atoms and form just three bonds. Although we will generally see compounds of P, S, Cl, Br, and I with octets, they can form molecules in which they share more of their valence electrons. This expands their valence electrons to 10, 12, or even 14 electrons. For example, we have seen that the P atom in PCl_3 has an octet, but in PCl_5, the P atom has five bonds with 10 valence electrons. In H_2S, the S atom has an octet, but in SF_6, there are six bonds to sulfur with 12 valence electrons.

PRACTICE PROBLEMS

6.6 Lewis Structures for Molecules and Polyatomic Ions

6.57 Determine the total number of valence electrons for each of the following:
a. H_2S **b.** I_2 **c.** CCl_4 **d.** OH^-

6.58 Determine the total number of valence electrons for each of the following:
a. SBr_2 **b.** NBr_3 **c.** CH_3OH **d.** NH_4^+

6.59 Draw the Lewis structures for each of the following molecules or ions:
a. HF **b.** SF_2 **c.** NBr_3 **d.** BH_4^-

6.60 Draw the Lewis structures for each of the following molecules or ions:
a. H_2O **b.** CCl_4 **c.** H_3O^+ **d.** SiF_4

6.61 When is it necessary to draw a multiple bond in a Lewis structure?

6.62 If the available number of valence electrons for a molecule or polyatomic ion does not complete all of the octets in a Lewis structure, what should you do?

6.63 Draw the Lewis structures for each of the following molecules or ions:
a. CO
b. NO_3^-
c. H_2CO (C is the central atom)

6.64 Draw the Lewis structures for each of the following molecules or ions:
a. HCCH
b. CS_2
c. NO^+

6.7 Electronegativity and Bond Polarity

LEARNING GOAL Use electronegativity to determine the polarity of a bond.

CORE CHEMISTRY SKILL
Using Electronegativity

We can learn more about the chemistry of compounds by looking at how bonding electrons are shared between atoms. The bonding electrons are shared equally in a bond between identical nonmetal atoms. However, when a bond is between atoms of different elements, the electron pairs are usually shared unequally. Then the shared pairs of electrons are attracted to one atom in the bond more than the other.

The **electronegativity** of an atom is its ability to attract the shared electrons in a chemical bond. Nonmetals have higher electronegativities than do metals because nonmetals have a greater attraction for electrons than metals. On the electronegativity scale, the nonmetal fluorine located in the upper right corner of the periodic table, was assigned the highest value of 4.0, and the electronegativities for all other elements were determined relative to the attraction of fluorine for shared electrons. The metal cesium located in the lower left corner of the periodic table, has the lowest electronegativity value of 0.7. The electronegativities for the representative elements are shown in **FIGURE 6.6**. Note that there are no electronegativity values for the noble gases because they do not typically form bonds. The electronegativity values for transition elements are also low, but we have not included them in our discussion.

ENGAGE
Why does chlorine have a higher electronegativity than iodine?

TEST
Try Practice Problems 6.65 to 6.68

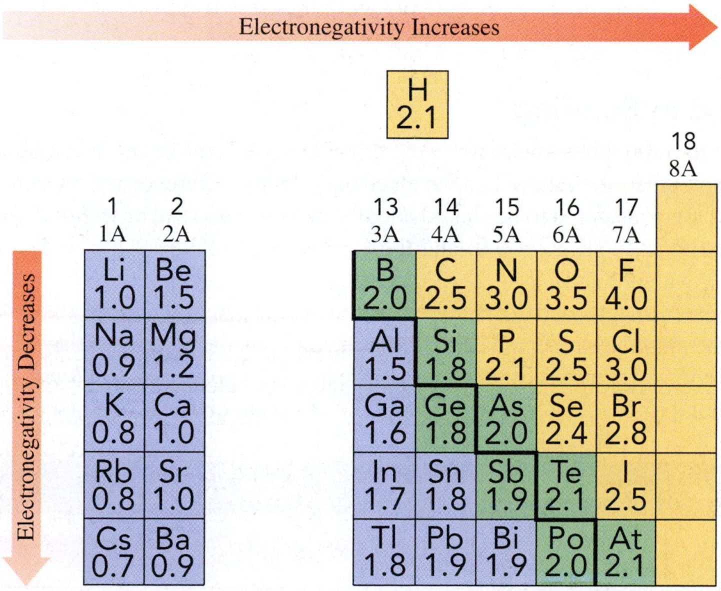

Electronegativity Increases
Electronegativity Decreases
H 2.1
1 1A
2 2A
13 3A
14 4A
15 5A
16 6A
17 7A
18 8A
Li 1.0
Be 1.5
Na 0.9
Mg 1.2
K 0.8
Ca 1.0
Rb 0.8
Sr 1.0
Cs 0.7
Ba 0.9
B 2.0
C 2.5
N 3.0
O 3.5
F 4.0
Al 1.5
Si 1.8
P 2.1
S 2.5
Cl 3.0
Ga 1.6
Ge 1.8
As 2.0
Se 2.4
Br 2.8
In 1.7
Sn 1.8
Sb 1.9
Te 2.1
I 2.5
Tl 1.8
Pb 1.9
Bi 1.9
Po 2.0
At 2.1

FIGURE 6.6 ▶ The electronegativity values of representative elements in Group 1A (1) to Group 7A (17), which indicate the ability of atoms to attract shared electrons, increase going across a period from left to right and decrease going down a group.

Q What element on the periodic table has the strongest attraction for shared electrons?

Polarity of Bonds

The difference in the electronegativity values of two atoms can be used to predict the type of chemical bond, ionic or covalent, that forms. For the H—H bond, the electronegativity difference is zero $(2.1 - 2.1 = 0)$, which means the bonding electrons are shared equally. We illustrate this by drawing a symmetrical electron cloud around the H atoms. A bond between atoms with identical or very similar electronegativity values is a **nonpolar covalent bond**.

However, when covalent bonds are between atoms with different electronegativity values, the electrons are shared unequally; the bond is a **polar covalent bond**. The electron cloud for a polar covalent bond is unsymmetrical. For the H—Cl bond, there is an electronegativity difference of 3.0 (Cl) − 2.1 (H) = 0.9, which means that the H—Cl bond is polar covalent (see **FIGURE 6.7**). When finding the electronegativity difference, the smaller electronegativity is always subtracted from the larger; thus, the difference is always a positive number.

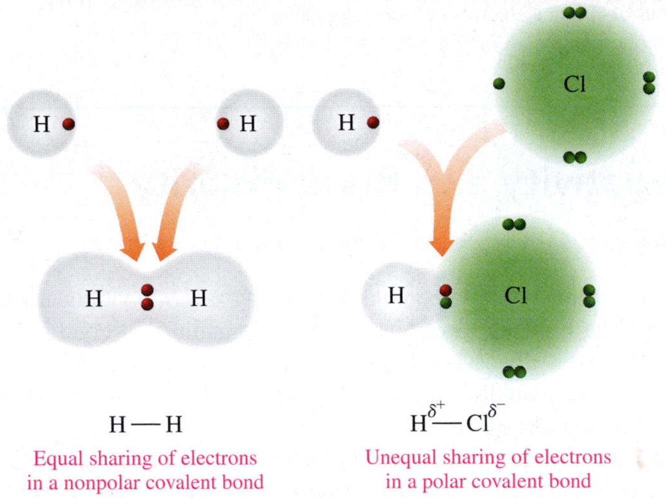
H
H
H
Cl
H H
H Cl
H—H
H^{δ^+}—Cl^{δ^-}
Equal sharing of electrons in a nonpolar covalent bond
Unequal sharing of electrons in a polar covalent bond

FIGURE 6.7 ▶ In the nonpolar covalent bond of H_2, electrons are shared equally. In the polar covalent bond of HCl, electrons are shared unequally.

Q H_2 has a nonpolar covalent bond, but HCl has a polar covalent bond. Explain.

The **polarity** of a bond depends on the difference in the electronegativity values of its atoms. In a polar covalent bond, the shared electrons are attracted to the more electronegative atom, which makes it partially negative, because of the negatively charged electrons around that atom. At the other end of the bond, the atom with the lower electronegativity becomes partially positive because of the lack of electrons at that atom.

A bond becomes more *polar* as the electronegativity difference increases. A polar covalent bond that has a separation of charges is called **a dipole**. The positive and negative ends of the dipole are indicated by the lowercase Greek letter delta with a positive or negative sign, δ^+ and δ^-. Sometimes we use an arrow that points from the positive charge to the negative charge $\longmapsto$ to indicate the dipole.

C^{δ^+}—O^{δ^-} $\quad$ N^{δ^+}—O^{δ^-} $\quad$ Cl^{δ^+}—F^{δ^-}

$\longmapsto$ $\quad$ $\longmapsto$ $\quad$ $\longmapsto$

Dipoles occur in polar covalent bonds containing N, O, or F.

Variations in Bonding

The variations in bonding are continuous; there is no definite point at which one type of bond stops and the next starts. When the electronegativity difference is between 0.0 and 0.4, the electrons are considered to be shared equally in a *nonpolar covalent bond*. For example, the C—C bond (2.5 − 2.5 = 0.0) and the C—H bond (2.5 − 2.1 = 0.4) are classified as nonpolar covalent bonds.

As the electronegativity difference increases, the shared electrons are attracted more strongly to the more electronegative atom, which increases the polarity of the bond. When the electronegativity difference is from 0.5 to 1.8, the bond is a *polar covalent bond*. For example, the O—H bond (3.5 − 2.1 = 1.4) is classified as a polar covalent bond (see **TABLE 6.15**).

ENGAGE

Use electronegativity differences to explain why a Si—S bond is polar covalent and a Si—P bond is nonpolar covalent.

TABLE 6.15 Electronegativity Differences and Types of Bonds

Electronegativity Difference	0.0 to 0.4	0.5 to 1.8	1.9 to 3.3
Bond Type	Nonpolar covalent	Polar covalent	Ionic
Electron Bonding	Electrons shared equally	Electrons shared unequally	Electrons transferred
	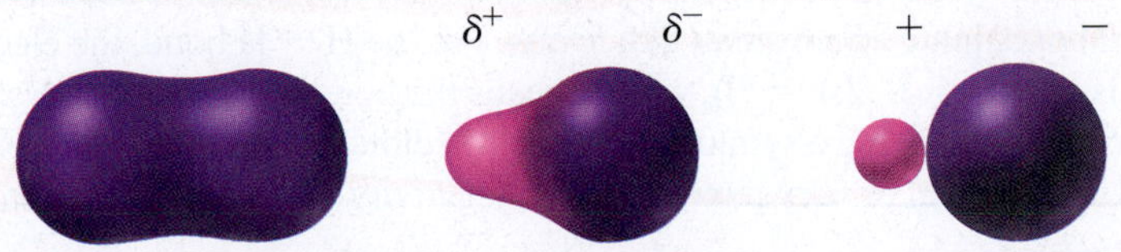	δ^+ δ^-	+ −

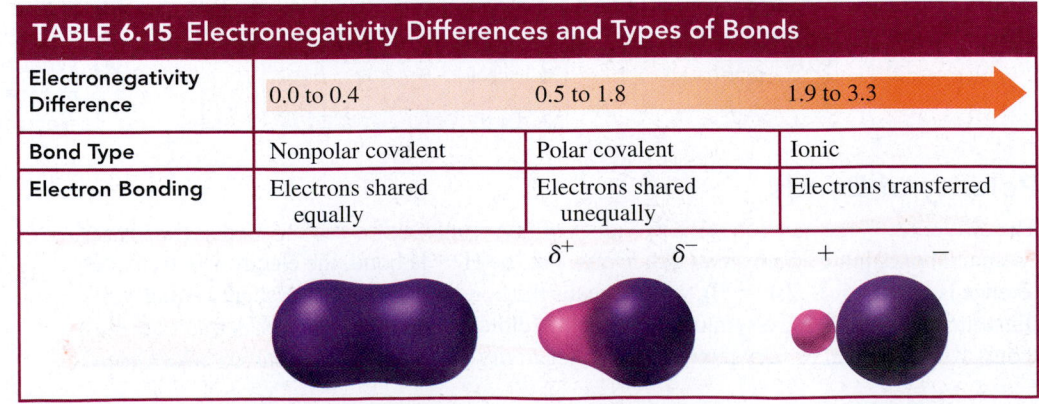

When the electronegativity difference is greater than 1.8, electrons are transferred from one atom to another, which results in an *ionic bond*. For example, the electronegativity difference for the ionic compound NaCl is $3.0 - 0.9 = 2.1$. Thus, for large differences in electronegativity, we would predict an ionic bond (see **TABLE 6.16**).

TABLE 6.16 Predicting Bond Type from Electronegativity Differences

Formula	Bond	Electronegativity Difference*	Type of Bond
H_2	H—H	$2.1 - 2.1 = 0.0$	Nonpolar covalent
BrCl	Br—Cl	$3.0 - 2.8 = 0.2$	Nonpolar covalent
HBr	$\overset{\delta^+}{H}—\overset{\delta^-}{Br}$	$2.8 - 2.1 = 0.7$	Polar covalent
HCl	$\overset{\delta^+}{H}—\overset{\delta^-}{Cl}$	$3.0 - 2.1 = 0.9$	Polar covalent
NaCl	Na^+Cl^-	$3.0 - 0.9 = 2.1$	Ionic
MgO	$Mg^{2+}O^{2-}$	$3.5 - 1.2 = 2.3$	Ionic

*Values are taken from Figure 6.6.

▶SAMPLE PROBLEM 6.15 Bond Polarity

TRY IT FIRST

Using electronegativity values, classify the bond between each of the following pairs of atoms as nonpolar covalent, polar covalent, or ionic and label any polar covalent bond with δ^+ and δ^- and show the direction of the dipole:

a. K and O
b. As and Cl
c. N and N
d. P and Br

SOLUTION

ANALYZE THE PROBLEM	Given	Need	Connect
	pairs of atoms	type of bond	electronegativity values

For each pair of atoms, we obtain the electronegativity values and calculate the difference in electronegativity.

Atoms	Electronegativity Difference	Type of Bond	Dipole
a. K and O	$3.5 - 0.8 = 2.7$	Ionic	
b. As and Cl	$3.0 - 2.0 = 1.0$	Polar covalent	$\overset{\delta^+}{As}—\overset{\delta^-}{Cl}$ $\longmapsto$
c. N and N	$3.0 - 3.0 = 0.0$	Nonpolar covalent	
d. P and Br	$2.8 - 2.1 = 0.7$	Polar covalent	$\overset{\delta^+}{P}—\overset{\delta^-}{Br}$ $\longmapsto$

STUDY CHECK 6.15

Using electronegativity values, classify the bond between each of the following pairs of atoms as nonpolar covalent, polar covalent, or ionic and label any polar covalent bond with δ^+ and δ^- and show the direction of the dipole:

a. P and Cl
b. Br and Br
c. Na and O
d. O and I

ANSWER

a. polar covalent (0.9) $\overset{\delta^+}{P}—\overset{\delta^-}{Cl}$ $\longmapsto$
b. nonpolar covalent (0.0)
c. ionic (2.6)
d. polar covalent (1.0) $\overset{\delta^-}{O}—\overset{\delta^+}{I}$ $\longleftarrow\!\!+$

TEST
Try Practice Problems 6.69 to 6.74

PRACTICE PROBLEMS

6.7 Electronegativity and Bond Polarity

6.65 Describe the trend in electronegativity as *increases* or *decreases* for each of the following:
a. from B to F **b.** from Mg to Ba **c.** from F to I

6.66 Describe the trend in electronegativity as *increases* or *decreases* for each of the following:
a. from Al to Cl **b.** from Br to K **c.** from Li to Cs

6.67 Using the periodic table, arrange the atoms in each of the following sets in order of increasing electronegativity:
a. Li, Na, K **b.** Na, Cl, P **c.** Se, Ca, O

6.68 Using the periodic table, arrange the atoms in each of the following sets in order of increasing electronegativity:
a. Cl, F, Br **b.** B, O, N **c.** Mg, F, S

6.69 Which electronegativity difference (**a**, **b**, or **c**) would you expect for a nonpolar covalent bond?
a. from 0.0 to 0.4 **b.** from 0.5 to 1.8 **c.** from 1.9 to 3.3

6.70 Which electronegativity difference (**a**, **b**, or **c**) would you expect for a polar covalent bond?
a. from 0.0 to 0.4 **b.** from 0.5 to 1.8 **c.** from 1.9 to 3.3

6.71 Predict whether the bond between each of the following pairs of atoms is nonpolar covalent, polar covalent, or ionic:
a. Si and Br **b.** Li and F **c.** Br and F
d. I and I **e.** N and P **f.** C and P

6.72 Predict whether the bond between each of the following pairs of atoms is nonpolar covalent, polar covalent, or ionic:
a. Si and O **b.** K and Cl **c.** S and F
d. P and Br **e.** Li and O **f.** N and S

6.73 For the bond between each of the following pairs of atoms, indicate the positive end with δ^+ and the negative end with δ^-. Draw an arrow to show the dipole for each.
a. N and F **b.** Si and Br **c.** C and O
d. P and Br **e.** N and P

6.74 For the bond between each of the following pairs of atoms, indicate the positive end with δ^+ and the negative end with δ^-. Draw an arrow to show the dipole for each.
a. P and Cl **b.** Se and F **c.** Br and F
d. N and H **e.** B and Cl

6.8 Shapes and Polarity of Molecules

LEARNING GOAL Predict the three-dimensional structure of a molecule, and classify it as polar or nonpolar.

Using the Lewis structures, we can predict the three-dimensional shapes of many molecules. The shape is important in our understanding of how molecules interact with enzymes or certain antibiotics or produce our sense of taste and smell.

The three-dimensional shape of a molecule is determined by drawing its Lewis structure and identifying the number of electron groups (one or more electron pairs) around the central atom. We count lone pairs of electrons, single, double, or triple bonds as *one* electron group. In the **valence shell electron-pair repulsion (VSEPR) theory**, the electron groups are arranged as far apart as possible around the central atom to minimize the repulsion between the groups. Once we have counted the number of electron groups surrounding the central atom, we can determine its specific shape from the number of atoms bonded to the central atom.

CORE CHEMISTRY SKILL
Predicting Shape

Central Atoms with Two Electron Groups

In the Lewis structure for CO_2, there are two electron groups (two double bonds) attached to the central atom. According to VSEPR theory, minimal repulsion occurs when two electron groups are on opposite sides of the central C atom. This gives the CO_2 molecule a *linear* electron-group geometry and a shape that is **linear** with a bond angle of 180°.

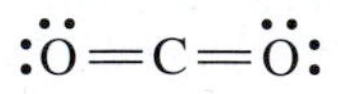

Linear electron-group geometry

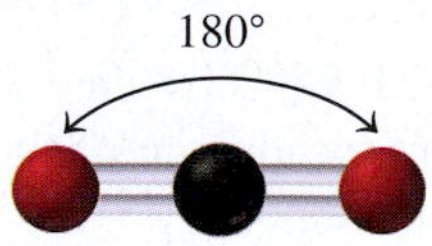
180°

Linear shape

Central Atoms with Three Electron Groups

In the Lewis structure for formaldehyde, H_2CO, the central atom C is attached to two H atoms by single bonds and to the O atom by a double bond. Minimal repulsion occurs when three electron groups are as far apart as possible around the central C atom, which gives 120° bond angles. This type of electron-group geometry is *trigonal planar* and gives a shape for H_2CO, also called **trigonal planar**.

When the central atom has the same number of electron groups as bonded atoms, the shape and the electron-group geometry have the same name.

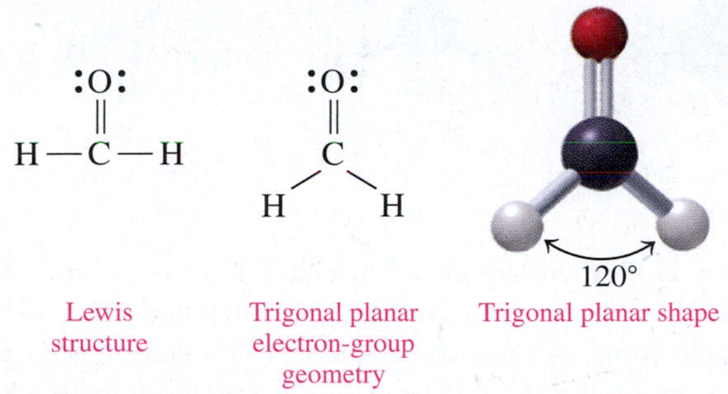
Lewis structure
Trigonal planar electron-group geometry
120°
Trigonal planar shape

In the Lewis structure for SO_2, there are also three electron groups around the central S atom: a single bond to an O atom, a double bond to another O atom, and a lone pair of electrons. As in H_2CO, three electron groups have minimal repulsion when they form trigonal planar electron-group geometry. However, in SO_2, one of the electron groups is a lone pair of electrons. Therefore, the shape of the SO_2 molecule is determined by the two O atoms bonded to the central S atom, which gives the SO_2 molecule a shape that is **bent** with a bond angle of 120°. When the central atom has more electron groups than bonded atoms, the shape and the electron-group geometry have different names.

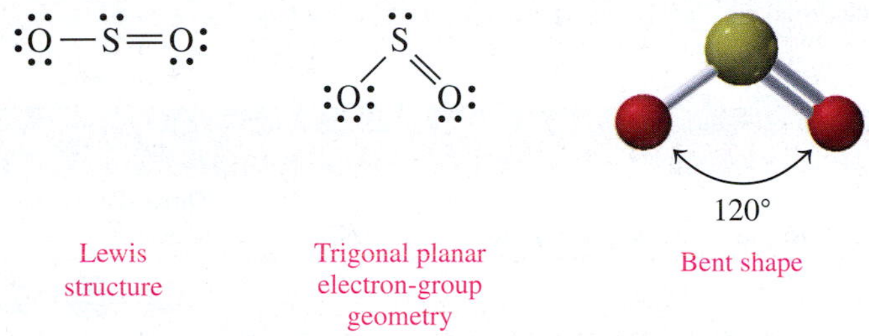
Lewis structure
Trigonal planar electron-group geometry
120°
Bent shape

Central Atom with Four Electron Groups

In a molecule of methane, CH_4, the central C atom is bonded to four H atoms. From the Lewis structure, you may think that CH_4 is planar with 90° bond angles. However, the best geometry for minimal repulsion is *tetrahedral*, giving bond angles of 109°. When there are four atoms attached to four electron groups, the shape of the molecule is also **tetrahedral**.

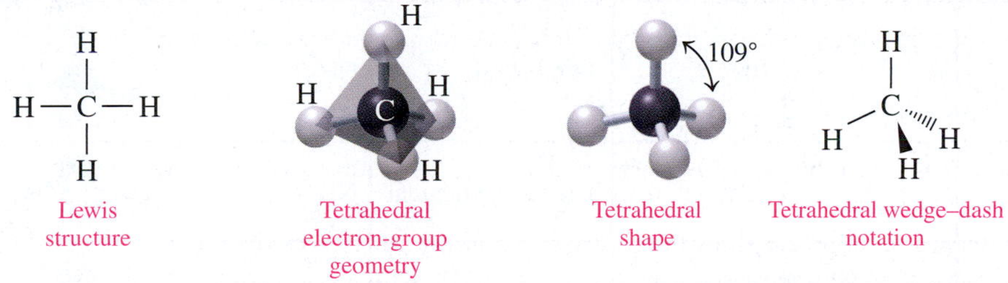
Lewis structure
Tetrahedral electron-group geometry
109°
Tetrahedral shape
Tetrahedral wedge–dash notation

A way to represent the three-dimensional structure of methane is to use the *wedge–dash notation.* In this representation, the two bonds connecting carbon to hydrogen by solid lines are in the plane of the paper. The wedge represents a carbon-to-hydrogen bond coming out of the page toward us, whereas the dash represents a carbon-to-hydrogen bond going into the page away from us.

Now we can look at molecules that have four electron groups, of which one or more are lone pairs of electrons. Then the central atom is attached to only two or three atoms. For example, in the Lewis structure for ammonia, NH_3, four electron groups have a tetrahedral electron-group geometry. However, in NH_3, one of the electron groups is a lone pair of electrons. Therefore, the shape of NH_3 is determined by the three H atoms bonded to the central N atom, which gives the NH_3 molecule a shape that is **trigonal pyramidal**, with a bond angle of 109°. The wedge–dash notation can also represent this three-dimensional structure of ammonia with one N—H bond in the plane, one N—H bond coming toward us, and one N—H bond going away from us.

ENGAGE

If the four electron groups in a PH_3 molecule have a tetrahedral geometry, why does a PH_3 molecule have a trigonal pyramidal shape and not a tetrahedral shape?

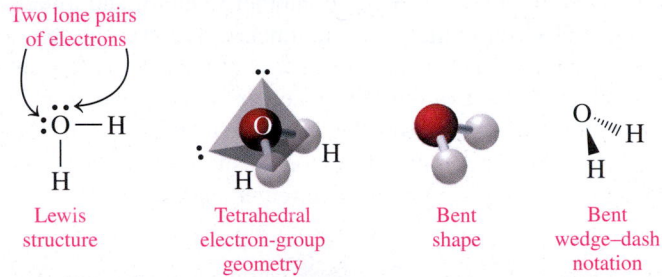

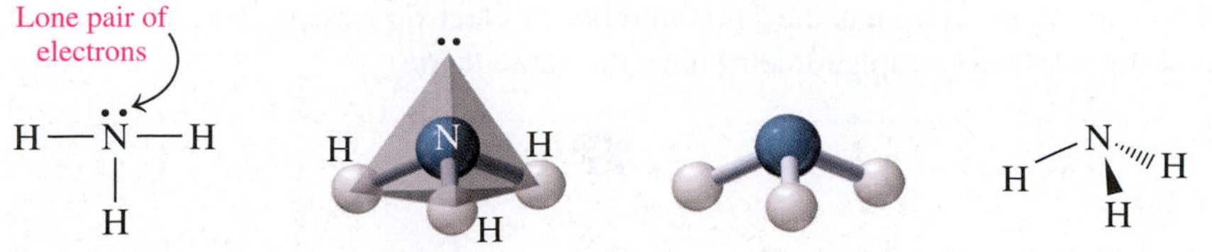
Lone pair of electrons
H
N
H
H
Lewis structure
Tetrahedral electron-group geometry
Trigonal pyramidal shape
Trigonal pyramidal wedge–dash notation

In the Lewis structure for water, H_2O, there are also four electron groups, which have minimal repulsion when the electron-group geometry is tetrahedral. However, in H_2O, two of the electron groups are lone pairs of electrons. Because the shape of H_2O is determined by the two H atoms bonded to the central O atom, the H_2O molecule has a bent shape with a bond angle of 109°. **TABLE 6.17** gives the shapes of molecules with two, three, or four bonded atoms.

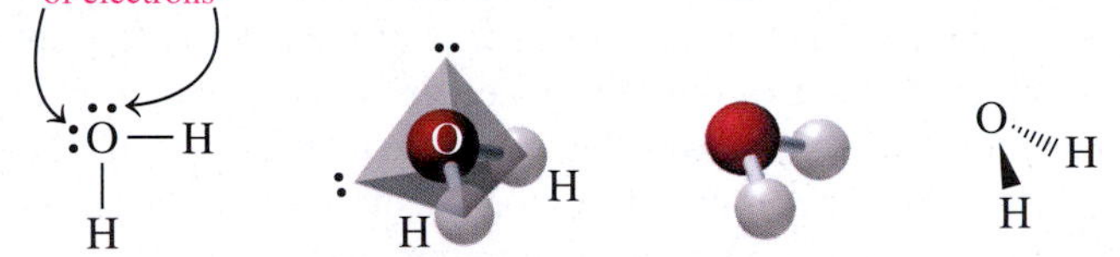
Two lone pairs of electrons
O
H
H
Lewis structure
Tetrahedral electron-group geometry
Bent shape
Bent wedge–dash notation

TABLE 6.17 Molecular Shapes for a Central Atom with Two, Three, and Four Bonded Atoms

Electron Groups	Electron-Group Geometry	Bonded Atoms	Lone Pairs	Bond Angle*	Molecular Shape	Example	Three-Dimensional Model
2	Linear	2	0	180°	Linear	CO_2	
3	Trigonal planar	3	0	120°	Trigonal planar	H_2CO	
3	Trigonal planar	2	1	120°	Bent	SO_2	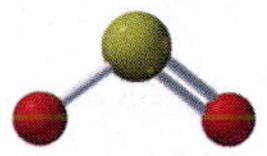
4	Tetrahedral	4	0	109°	Tetrahedral	CH_4	
4	Tetrahedral	3	1	109°	Trigonal pyramidal	NH_3	
4	Tetrahedral	2	2	109°	Bent	H_2O	

*The bond angles in actual molecules may vary slightly.

▶SAMPLE PROBLEM 6.16 Predicting Shapes of Molecules

TRY IT FIRST

Use VSEPR theory to predict the shape of the molecule $SiCl_4$.

SOLUTION

	Given	Need	Connect
ANALYZE THE PROBLEM	$SiCl_4$	shape	Lewis structure, electron groups, bonded atoms

STEP 1 **Draw the Lewis structure.** Using 32 e^-, we draw the bonds and lone pairs for the Lewis structure of $SiCl_4$.

Name of Element	Silicon		Chlorine
Symbol of Element	Si		Cl
Atoms of Element	1 Si		4 Cl
Valence Electrons	4 e^-		7 e^-
Total Electrons	1(4 e^-)	+	4(7 e^-) = 32 e^-

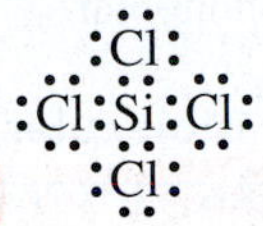

STEP 2 **Arrange the electron groups around the central atom to minimize repulsion.** In the Lewis structure of $SiCl_4$, there are four electron groups around the central Si atom. To minimize repulsion, the electron-group geometry would be tetrahedral.

STEP 3 **Use the atoms bonded to the central atom to determine the shape.** Because the central Si atom has four bonded pairs and no lone pairs of electrons, the $SiCl_4$ molecule has a tetrahedral shape.

STUDY CHECK 6.16

Use VSEPR theory to predict the shape of SCl_2.

ANSWER

The central atom S has four electron groups: two bonded atoms and two lone pairs of electrons. The shape of SCl_2 is bent, with a bond angle of 109°.

▸SAMPLE PROBLEM 6.17 Predicting Shape of an Ion

TRY IT FIRST

Use VSEPR theory to predict the shape of the polyatomic ion NO_3^-.

SOLUTION

STEP 1 **Draw the Lewis structure.** The polyatomic ion NO_3^- contains three electron groups (two single bonds between the central N atom and O atoms, and one double bond between N and O). Note that the double bond can be drawn to any of the O atoms.

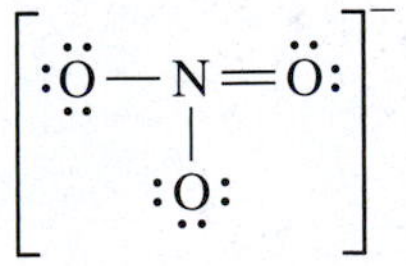

STEP 2 **Arrange the electron groups around the central atom to minimize repulsion.** To minimize repulsion, three electron groups would have a trigonal planar geometry.

STEP 3 **Use the atoms bonded to the central atom to determine the shape.** Because NO_3^- has three bonded atoms, it has a trigonal planar shape.

STUDY CHECK 6.17

Use VSEPR theory to predict the shape of each of the following:

a. ClO_2^- **b.** PO_2^+

ANSWER

a. With two bonded atoms and two lone pairs of electrons on the central Cl atom, the shape of ClO_2^- is bent, with a bond angle of 109°.
b. With two bonded atoms and no lone pairs on the central P atom, the shape of PO_2^+ is linear.

TEST
Try Practice Problems 6.75 to 6.84

CORE CHEMISTRY SKILL
Identifying Polarity of Molecules

Polarity of Molecules

We have seen that covalent bonds in molecules can be polar or nonpolar. Now we will look at how the bonds in a molecule and its shape determine whether that molecule is classified as polar or nonpolar.

Nonpolar Molecules

In a **nonpolar molecule**, all the bonds are nonpolar or the polar bonds cancel each other out. Molecules such as H_2, Cl_2, and CH_4 are nonpolar because they contain only nonpolar covalent bonds. A *nonpolar molecule* also occurs when polar bonds (dipoles) cancel each other because they are in a symmetrical arrangement. For example, CO_2, a linear molecule, contains two equal polar covalent bonds whose dipoles point in opposite directions. As a result, the dipoles cancel out, which makes a CO_2 molecule nonpolar.

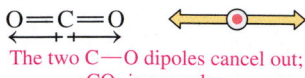

The two C—O dipoles cancel out; CO_2 is nonpolar.

Another example of a nonpolar molecule is the CCl_4 molecule, which has four polar bonds symmetrically arranged around the central C atom. Each of the C—Cl bonds has the same polarity, but because they have a tetrahedral arrangement, their opposing dipoles cancel out. As a result, a molecule of CCl_4 is nonpolar.

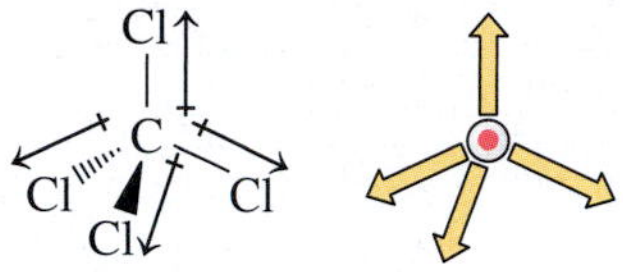
The four C—Cl dipoles cancel out; CCl_4 is nonpolar.

Polar Molecules

In a **polar molecule**, one end of the molecule is more negatively charged than the other end. Polarity in a molecule occurs when the dipoles from the individual polar bonds do not cancel each other. For example, HCl is a polar molecule because it has one covalent bond that is polar.

A single dipole does not cancel out; HCl is polar.

In molecules with two or more electron groups, the shape, such as bent or trigonal pyramidal, determines whether the dipoles cancel. For example, we have seen that H_2O has a bent shape. Thus, a water molecule is polar because the individual dipoles do not cancel.

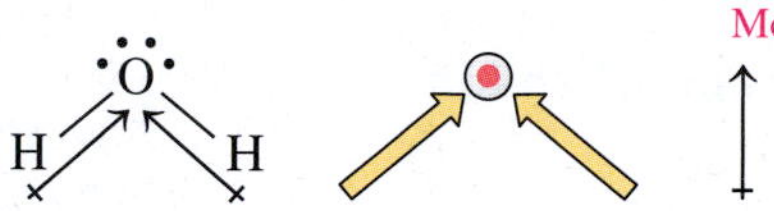
More negative end of molecule
More positive end of molecule
The dipoles do not cancel out; H_2O is polar.

The NH_3 molecule has a tetrahedral electron-group geometry with three bonded atoms, which gives it a trigonal pyramidal shape. Thus, the NH_3 molecule is polar because the individual N—H dipoles do not cancel.

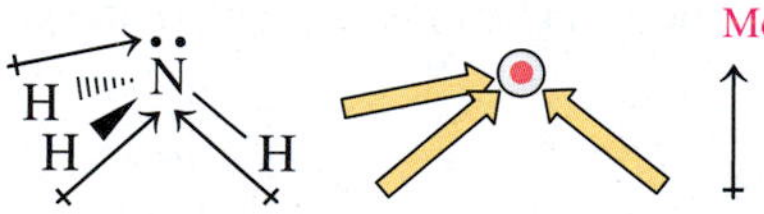
More negative end of molecule
More positive end of molecule
The dipoles do not cancel out; NH_3 is polar.

In the molecule CH_3F, the C—F bond is polar covalent, but the three C—H bonds are nonpolar covalent. Because there is only one dipole in CH_3F, which does not cancel, CH_3F is a polar molecule.

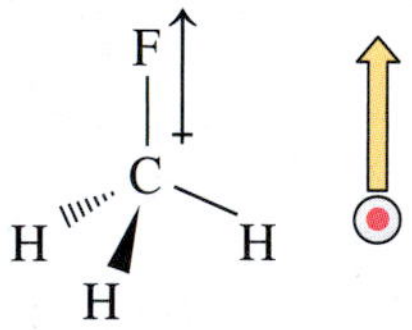
The dipole does not cancel out; CH_3F is polar.

▶SAMPLE PROBLEM 6.18 Polarity of Molecules

TRY IT FIRST

Determine whether a molecule of OF_2 is polar or nonpolar.

SOLUTION

ANALYZE THE PROBLEM	Given	Need	Connect
	OF_2	polarity	Lewis structure, bond polarity

STEP 1 **Determine if the bonds are polar covalent or nonpolar covalent.** From Figure 6.6, F and O have an electronegativity difference of 0.5 (4.0 − 3.5 = 0.5), which makes the O—F bonds polar covalent.

STEP 2 If the bonds are polar covalent, draw the Lewis structure and determine if the dipoles cancel. The Lewis structure for OF_2 has four electron groups and two bonded atoms. The molecule has a bent shape in which the dipoles of the O—F bonds do not cancel. The OF_2 molecule would be polar.

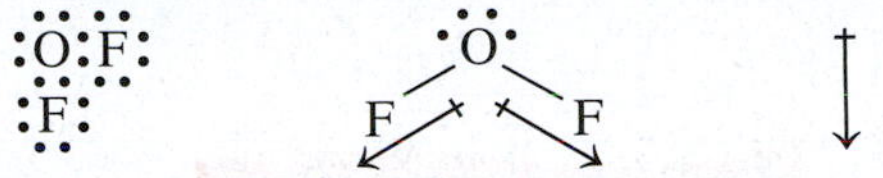

The dipoles do not cancel out; OF_2 is polar.

STUDY CHECK 6.18

Determine if each of the following would be polar or nonpolar:

a. PCl_3 **b.** Cl_2

ANSWER

a. polar **b.** nonpolar

TEST

Try Practice Problems 6.85 and 6.86

PRACTICE PROBLEMS

6.8 Shapes and Polarity of Molecules

6.75 Choose the shape (**1** to **6**) that matches each of the following descriptions (**a** to **c**):

1. linear **2.** bent (109°) **3.** trigonal planar
4. bent (120°) **5.** trigonal pyramidal **6.** tetrahedral

a. a molecule with a central atom that has four electron groups and four bonded atoms
b. a molecule with a central atom that has four electron groups and three bonded atoms
c. a molecule with a central atom that has three electron groups and three bonded atoms

6.76 Choose the shape (**1** to **6**) that matches each of the following descriptions (**a** to **c**):

1. linear **2.** bent (109°) **3.** trigonal planar
4. bent (120°) **5.** trigonal pyramidal **6.** tetrahedral

a. a molecule with a central atom that has four electron groups and two bonded atoms
b. a molecule with a central atom that has two electron groups and two bonded atoms
c. a molecule with a central atom that has three electron groups and two bonded atoms

6.77 Complete each of the following statements for a molecule of SeO_3:

a. There are _____ electron groups around the central Se atom.
b. The electron-group geometry is _____.
c. The number of atoms attached to the central Se atom is _____.
d. The shape of the molecule is _____.

6.78 Complete each of the following statements for a molecule of H_2S:

a. There are _____ electron groups around the central S atom.
b. The electron-group geometry is _____.
c. The number of atoms attached to the central S atom is _____.
d. The shape of the molecule is _____.

6.79 Compare the Lewis structures of PH_3 and NH_3. Why do these molecules have the same shape?

6.80 Compare the Lewis structures of CH_4 and H_2O. Why do these molecules have approximately the same bond angles but different molecular shapes?

6.81 Use VSEPR theory to predict the shape of each of the following:

a. $SeBr_2$ **b.** CCl_4 **c.** OBr_2

6.82 Use VSEPR theory to predict the shape of each of the following:

a. NCl_3 **b.** SeO_2 **c.** SiF_2Cl_2

6.83 Draw the Lewis structure and predict the shape for each of the following:

a. AlH_4^- **b.** SO_4^{2-} **c.** NH_4^+ **d.** NO_2^+

6.84 Draw the Lewis structure and predict the shape for each of the following:

a. NO_2^- **b.** PO_4^{3-} **c.** ClO_4^- **d.** SF_3^+

6.85 Identify each of the following molecules as polar or nonpolar:

a. HBr **b.** NF_3 **c.** CHF_3

6.86 Identify each of the following molecules as polar or nonpolar:

a. SeF_2 **b.** PBr_3 **c.** $SiCl_4$

6.9 Intermolecular Forces in Compounds

LEARNING GOAL Describe the intermolecular forces between ions, polar covalent molecules, and nonpolar covalent molecules.

CORE CHEMISTRY SKILL

Identifying Intermolecular Forces

In gases, the interactions between particles are minimal, which allows gas molecules to move far apart from each other. In solids and liquids, there are sufficient interactions between the particles to hold them close together, although some solids have low melting points whereas others have very high melting points. Such differences in properties are explained by looking at the various kinds of intermolecular forces between particles including *dipole–dipole attractions*, *hydrogen bonding*, *dispersion forces*, as well as ionic bonds.

Ionic compounds have high melting points. For example, solid NaCl melts at 801 °C. Large amounts of energy are needed to overcome the strong intermolecular forces between positive and negative ions. In solids containing molecules with covalent bonds, there are intermolecular forces too, but they are weaker than those of an ionic compound.

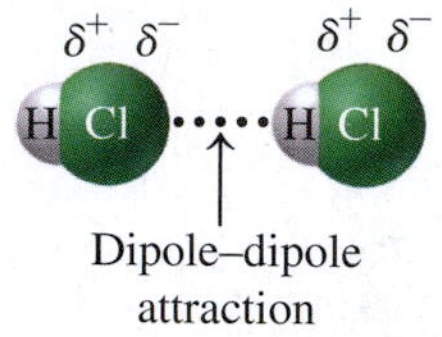
δ+ δ−
δ+ δ−
H Cl
H Cl
Dipole–dipole attraction

Dipole–Dipole Attractions

For polar molecules, intermolecular forces called **dipole–dipole attractions** occur between the positive end of one molecule and the negative end of another. For a polar molecule with a dipole such as HCl, the partially positive H atom of one HCl molecule attracts the partially negative Cl atom in another HCl molecule.

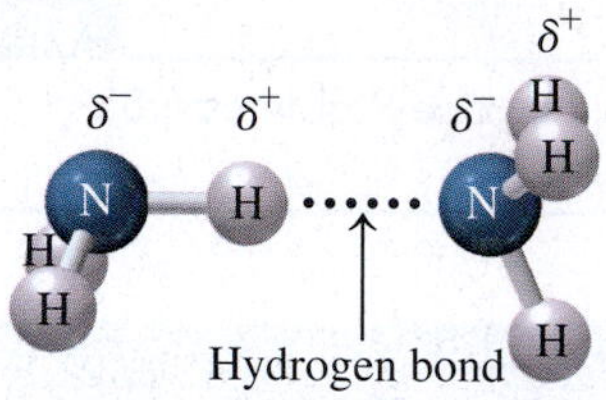
δ−
δ+
δ−
δ+
N
H
H
H
N
H
H
H
Hydrogen bond

Hydrogen Bonds

Polar molecules containing hydrogen atoms bonded to highly electronegative atoms of nitrogen, oxygen, or fluorine form especially strong dipole–dipole attractions. This type of attraction, called a **hydrogen bond**, occurs between the partially positive hydrogen atom in one molecule and the partially negative fluorine, oxygen, or nitrogen atom in another molecule. Hydrogen bonds are the strongest type of intermolecular forces between polar covalent molecules. They are a major factor in the formation and structure of biological molecules such as proteins and DNA.

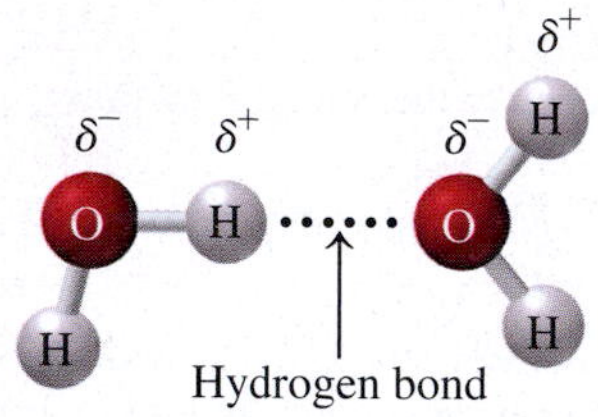
δ−
δ+
δ−
δ+
O
H
H
O
H
H
Hydrogen bond

Dispersion Forces

Very weak attractions called **dispersion forces** are the only intermolecular forces that occur between nonpolar molecules. Usually, the electrons in a nonpolar covalent molecule are distributed symmetrically. However, the movement of the electrons may place more of them in one part of the molecule than another, which forms a *temporary dipole*. These momentary dipoles align the molecules so that the positive end of one molecule is attracted to the negative end of another molecule. Although dispersion forces are very weak, they make it possible for nonpolar molecules to form liquids and solids.

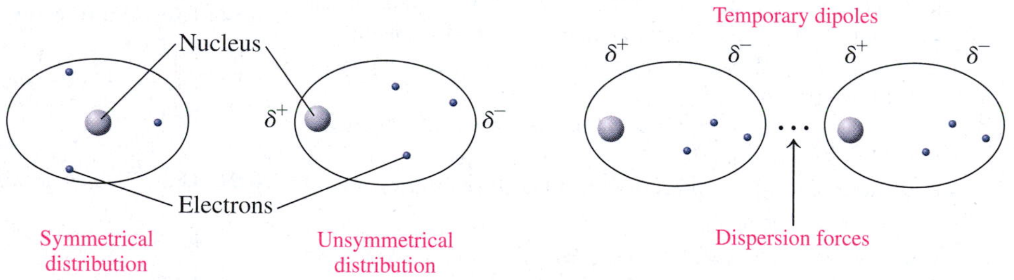
Nucleus
Electrons
δ+
δ−
Symmetrical distribution
Unsymmetrical distribution
Temporary dipoles
δ+
δ−
δ+
δ−
Dispersion forces

Nonpolar covalent molecules have weak attractions when they form temporary dipoles.

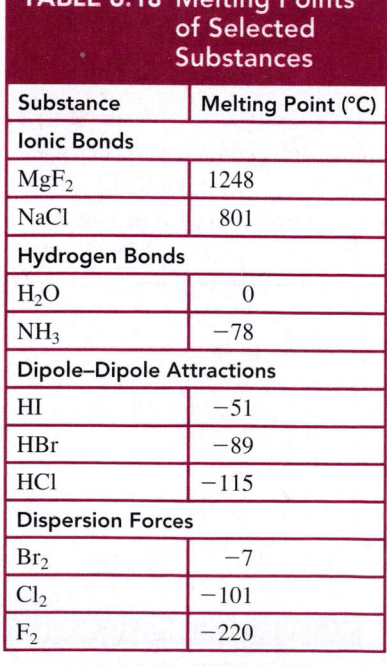

TABLE 6.18 Melting Points of Selected Substances

Substance	Melting Point (°C)
Ionic Bonds	
MgF_2	1248
NaCl	801
Hydrogen Bonds	
H_2O	0
NH_3	−78
Dipole–Dipole Attractions	
HI	−51
HBr	−89
HCl	−115
Dispersion Forces	
Br_2	−7
Cl_2	−101
F_2	−220

Intermolecular Forces and Melting Points

The melting point of a substance is related to the strength of the intermolecular forces between its particles. A compound with weak intermolecular forces, such as dispersion forces, has a low melting point because only a small amount of energy is needed to separate the molecules and form a liquid. A compound with dipole–dipole attractions requires more energy to break the intermolecular forces between the molecules. A compound that can form hydrogen bonds requires even more energy to overcome the intermolecular forces that exist between its molecules. Larger amounts of energy are needed to overcome the strong intermolecular forces between positive and negative ions and to melt an ionic solid. **TABLE 6.18** compares the melting points of some substances with various kinds of intermolecular forces. The various types of attractions between particles in solids and liquids are summarized in **TABLE 6.19**.

TABLE 6.19 Comparison of Bonding and Intermolecular Forces

Type of Force	Particle Arrangement	Example	Strength
Between Atoms or Ions			Strong 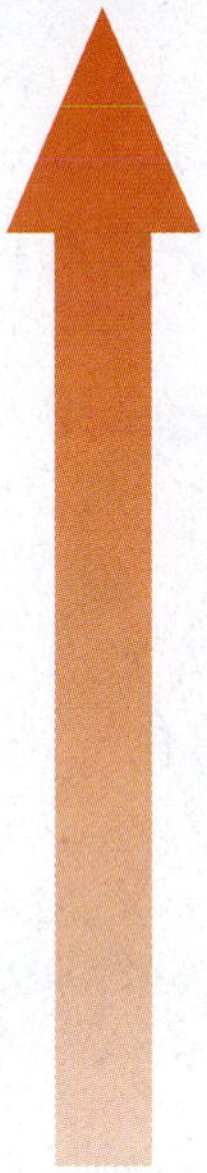 Weak
Ionic bonds	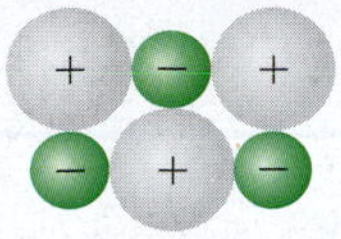 + − + − + −	Na^+Cl^-	
Covalent bonds (X = nonmetal)	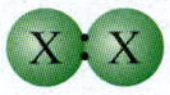 X:X	Cl—Cl	
Between Molecules			
Hydrogen bonds (X = N, O, or F)	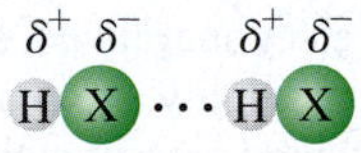 δ+ δ− δ+ δ− H X ··· H X	$H^{\delta+}—F^{\delta-} \cdots H^{\delta+}—F^{\delta-}$	
Dipole–dipole attractions (X and Y = nonmetals)	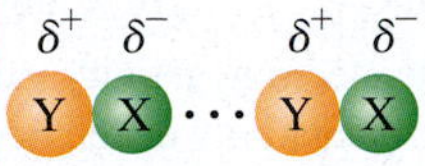 δ+ δ− δ+ δ− Y X ··· Y X	$H^{\delta+}—Cl^{\delta-} \cdots H^{\delta+}—Cl^{\delta-}$	
Dispersion forces (temporary shift of electrons in nonpolar bonds)	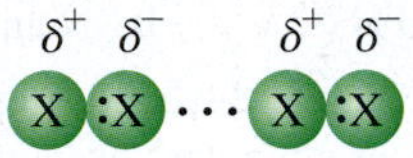 δ+ δ− δ+ δ− X:X ··· X:X (temporary dipoles)	$F^{\delta+}—F^{\delta-} \cdots F^{\delta+}—F^{\delta-}$	

▶SAMPLE PROBLEM 6.19 Intermolecular Forces Between Particles

TRY IT FIRST

Indicate the major type of intermolecular forces—dipole–dipole attractions, hydrogen bonds, or dispersion forces—expected of each of the following:

a. HF **b.** Br_2 **c.** PCl_3

SOLUTION

a. hydrogen bonds
b. dispersion forces
c. dipole–dipole attractions

STUDY CHECK 6.19

Indicate the major type of intermolecular forces in each of the following:

a. H_2S **b.** H_2O

ANSWER

a. dipole–dipole attractions
b. hydrogen bonds

TEST

Try Practice Problems 6.87 to 6.92

PRACTICE PROBLEMS

6.9 Intermolecular Forces in Compounds

6.87 Identify the major type of intermolecular forces between the particles of each of the following:
a. BrF **b.** KCl **c.** NF_3 **d.** Cl_2

6.88 Identify the major type of intermolecular forces between the particles of each of the following:
a. HCl **b.** MgF_2 **c.** PBr_3 **d.** NH_3

6.89 Identify the strongest intermolecular forces between the particles of each of the following:
a. CH_3OH **b.** CO **c.** CF_4 **d.** CH_3CH_3

6.90 Identify the strongest intermolecular forces between the particles of each of the following:
a. O_2 **b.** SiH_4 **c.** CH_3Cl **d.** H_2O_2

6.91 Identify the substance in each of the following pairs that would have the higher melting point and explain your choice:
a. HF or HBr **b.** HF or NaF
c. $MgBr_2$ or PBr_3 **d.** CH_4 or CH_3OH

6.92 Identify the substance in each of the following pairs that would have the higher melting point and explain your choice:
a. NaCl or HCl **b.** H_2O or H_2Se
c. NH_3 or PH_3 **d.** F_2 or HF

CLINICAL UPDATE Compounds at the Pharmacy

Richard returns to the pharmacy to pick up aspirin, $C_9H_8O_4$, and acetaminophen, $C_8H_9NO_2$. He talks to Sarah about a way to treat his sore toe. Sarah recommends soaking his foot in a solution of Epsom salts, which is magnesium sulfate. Richard also asks Sarah to recommend an antacid for his upset stomach and an iron supplement. Sarah suggests an antacid that contains calcium carbonate and aluminum hydroxide, and iron(II) sulfate as an iron supplement. Richard also picks up toothpaste containing tin(II) fluoride, and carbonated water, which contains carbon dioxide.

Clinical Applications

6.93 Using the electronegativity values in Figure 6.6, calculate the electronegativity difference for the bond between each of the following pairs of atoms present in aspirin:
a. C and C **b.** C and H **c.** C and O

6.94 Using the electronegativity values in Figure 6.6, calculate the electronegativity difference for the bond between each of the following pairs of atoms present in acetaminophen:
a. C and N **b.** N and H **c.** H and O

6.95 Classify each of the bonds in problem 6.93 as ionic, polar covalent, or nonpolar covalent.

6.96 Classify each of the bonds in problem 6.94 as ionic, polar covalent, or nonpolar.

CONCEPT MAP

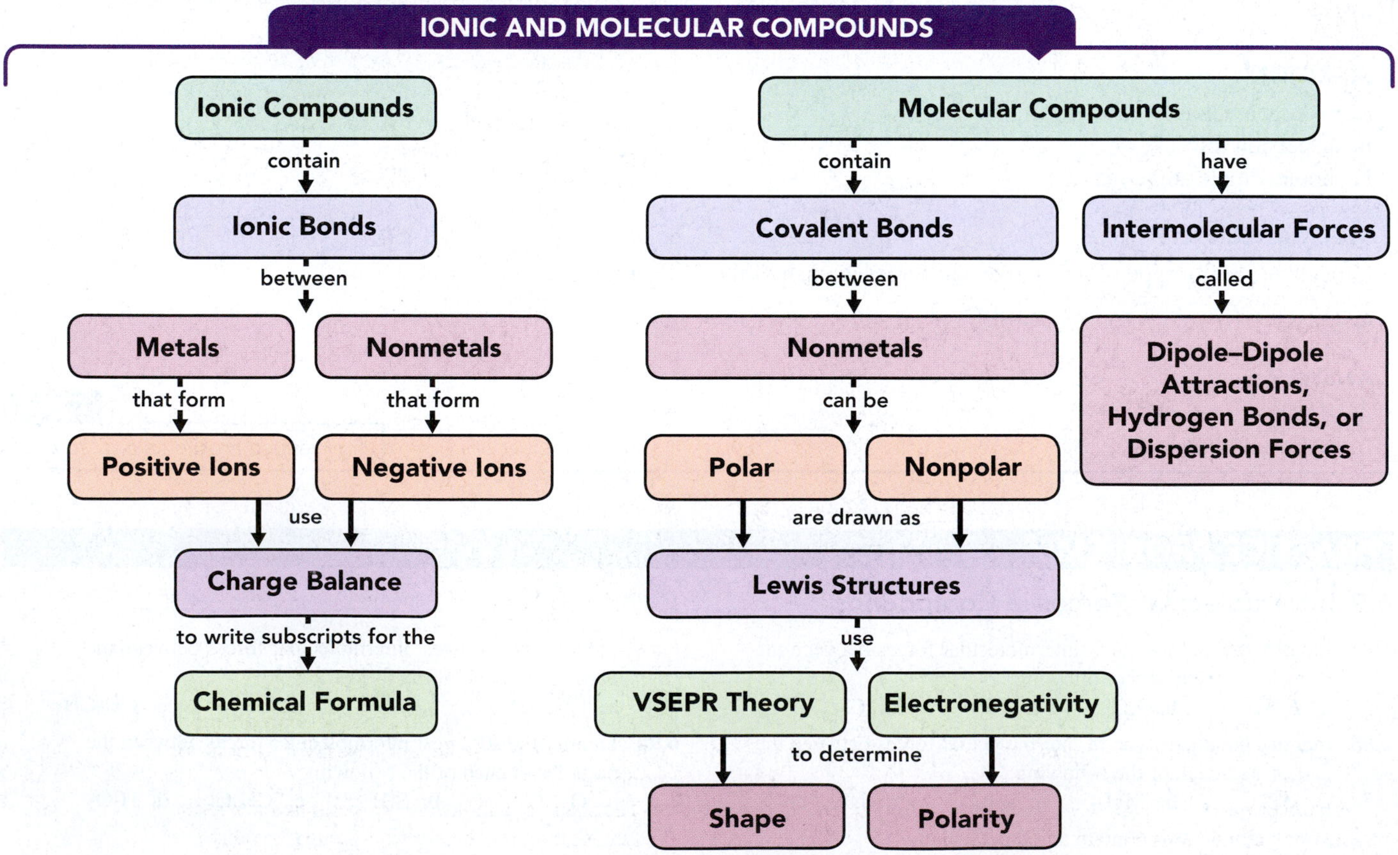
IONIC AND MOLECULAR COMPOUNDS
Ionic Compounds
contain
Ionic Bonds
between
Metals
Nonmetals
that form
that form
Positive Ions
Negative Ions
use
Charge Balance
to write subscripts for the
Chemical Formula
Molecular Compounds
contain
Covalent Bonds
between
Nonmetals
can be
Polar
Nonpolar
are drawn as
Lewis Structures
use
VSEPR Theory
Electronegativity
to determine
Shape
Polarity
have
Intermolecular Forces
called
Dipole–Dipole Attractions, Hydrogen Bonds, or Dispersion Forces

CHAPTER REVIEW

6.1 Ions: Transfer of Electrons

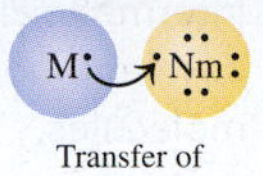
M
Nm
M+
Nm
Transfer of electrons

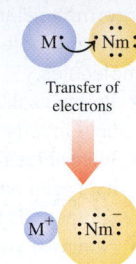

Ionic bond

LEARNING GOAL Write the symbols for the simple ions of the representative elements.

- The stability of the noble gases is associated with a stable electron configuration in the outermost energy level.
- With the exception of helium, which has two electrons, noble gases have eight valence electrons, which is an octet.
- Atoms of elements in Groups 1A to 7A (1, 2, 13 to 17) achieve stability by losing, gaining, or sharing their valence electrons in the formation of compounds.
- Metals of the representative elements lose valence electrons to form positively charged ions (cations): Groups 1A (1), 1+, 2A (2), 2+, and 3A (13), 3+.
- When reacting with metals, nonmetals gain electrons to form octets and form negatively charged ions (anions): Groups 5A (15), 3−, 6A (16), 2−, and 7A (17), 1−.

6.2 Ionic Compounds

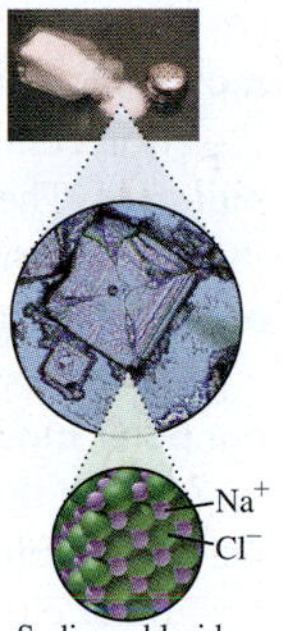
Na^+
Cl^-
Sodium chloride

LEARNING GOAL Using charge balance, write the correct formula for an ionic compound.

- The total positive and negative ionic charge is balanced in the formula of an ionic compound.
- Charge balance in a formula is achieved by using subscripts after each symbol so that the overall charge is zero.

6.3 Naming and Writing Ionic Formulas

LEARNING GOAL Given the formula of an ionic compound, write the correct name; given the name of an ionic compound, write the correct formula.

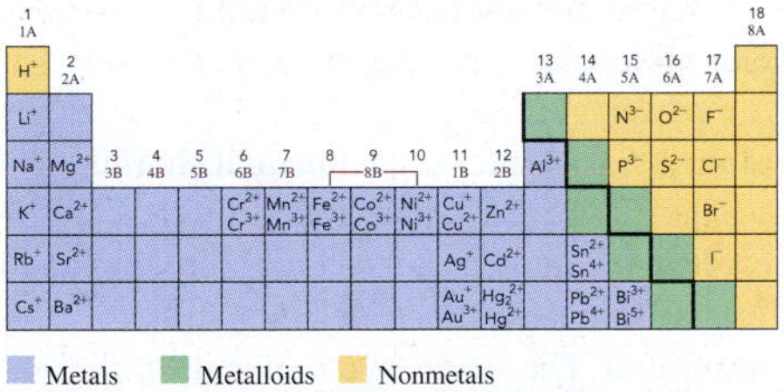
Metals
Metalloids
Nonmetals

- In naming ionic compounds, the positive ion is given first followed by the name of the negative ion.
- The names of ionic compounds containing two elements end with *ide*.
- Except for Ag, Cd, and Zn, transition elements form cations with two or more ionic charges.
- The charge of the cation is determined from the total negative charge in the formula and included as a Roman numeral immediately following the name of the metal that has a variable charge.

6.4 Polyatomic Ions

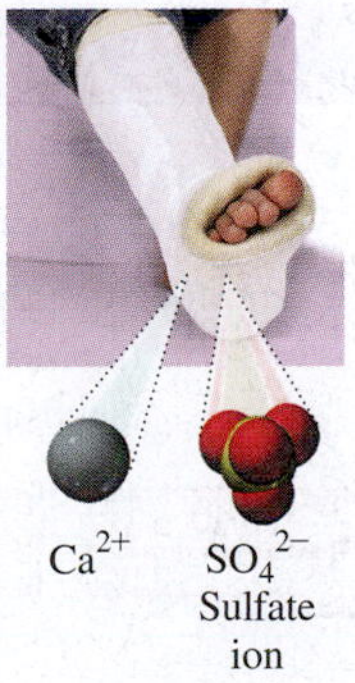
Ca^{2+}
SO_4^{2-}
Sulfate ion

LEARNING GOAL Write the name and formula for an ionic compound containing a polyatomic ion.

- A polyatomic ion is a covalently bonded group of atoms with an electrical charge; for example, the carbonate ion has the formula CO_3^{2-}.
- Most polyatomic ions have names that end with *ate* or *ite*.
- Most polyatomic ions contain a nonmetal and one or more oxygen atoms.
- The ammonium ion, NH_4^+, is a positive polyatomic ion.
- When more than one polyatomic ion is used for charge balance, parentheses enclose the formula of the polyatomic ion.

6.5 Molecular Compounds: Sharing Electrons

LEARNING GOAL Given the formula of a molecular compound, write its correct name; given the name of a molecular compound, write its formula.

1	mono
2	di
3	tri
4	tetra
5	penta

- In a covalent bond, atoms of nonmetals share valence electrons such that each atom has a stable electron configuration.
- The first nonmetal in a molecular compound uses its element name; the second nonmetal uses the first syllable of its element name followed by *ide*.
- The name of a molecular compound with two different atoms uses prefixes to indicate the subscripts in the formula.

6.6 Lewis Structures for Molecules and Polyatomic Ions

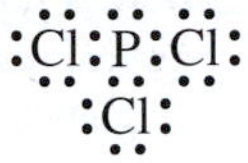

LEARNING GOAL Draw the Lewis structures for molecular compounds and polyatomic ions.

- The total number of valence electrons is determined for all the atoms in the molecule or polyatomic ion.
- Any negative charge is added to the total valence electrons, whereas any positive charge is subtracted.
- In the Lewis structure, a bonding pair of electrons is placed between the central atom and each of the attached atoms.
- Any remaining valence electrons are used as lone pairs to complete the octets of the surrounding atoms and then the central atom.
- If octets are not completed, one or more lone pairs of electrons are placed as bonding pairs forming double or triple bonds.

6.7 Electronegativity and Bond Polarity

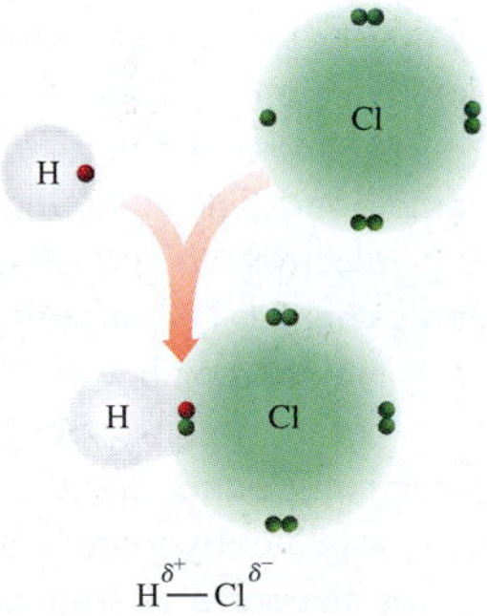
H
Cl
H
Cl
$\overset{\delta^+}{H}—\overset{\delta^-}{Cl}$
Unequal sharing of electrons in a polar covalent bond

LEARNING GOAL Use electronegativity to determine the polarity of a bond.

- Electronegativity is the ability of an atom to attract the electrons it shares with another atom. In general, the electronegativities of metals are low, whereas nonmetals have high electronegativities.
- In a nonpolar covalent bond, atoms share electrons equally.
- In a polar covalent bond, the electrons are unequally shared because they are attracted to the more electronegative atom.
- The atom in a polar bond with the lower electronegativity is partially positive (δ^+), and the atom with the higher electronegativity is partially negative (δ^-).
- Atoms that form ionic bonds have large differences in electronegativities.

6.8 Shapes and Polarity of Molecules

Tetrahedral shape

LEARNING GOAL Predict the three-dimensional structure of a molecule, and classify it as polar or nonpolar.

- The shape of a molecule is determined from the Lewis structure, the electron-group geometry, and the number of bonded atoms.

- The electron-group geometry around a central atom with two electron groups is linear; with three electron groups, the geometry is trigonal planar; and with four electron groups, the geometry is tetrahedral.
- When all the electron groups are bonded to atoms, the shape has the same name as the electron-group geometry.
- A central atom with three electron groups and two bonded atoms has a bent shape, with a bond angle of 120°.
- A central atom with four electron groups and three bonded atoms has a trigonal pyramidal shape.
- A central atom with four electron groups and two bonded atoms has a bent shape, with a bond angle of 109°.
- Nonpolar molecules contain nonpolar covalent bonds or have an arrangement of bonded atoms that causes the dipoles to cancel out.
- In polar molecules, the dipoles do not cancel.

6.9 Intermolecular Forces in Compounds

LEARNING GOAL Describe the intermolecular forces between ions, polar covalent molecules, and nonpolar covalent molecules.

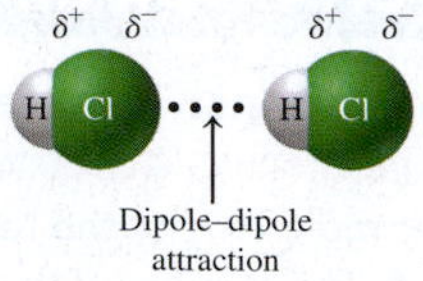

δ+ δ−
δ+ δ−
H Cl
H Cl
Dipole–dipole attraction

- In ionic solids, oppositely charged ions are held in a rigid structure by ionic bonds.
- Intermolecular forces called dipole–dipole attractions and hydrogen bonds hold the solid and liquid states of polar molecular compounds together.
- Nonpolar compounds form solids and liquids by weak attractions between temporary dipoles called dispersion forces.

KEY TERMS

anion A negatively charged ion such as Cl^-, O^{2-}, or SO_4^{2-}.

bent The shape of a molecule with two bonded atoms and one lone pair or two lone pairs.

cation A positively charged ion such as Na^+, Mg^{2+}, Al^{3+}, or NH_4^+.

chemical formula The group of symbols and subscripts that represents the atoms or ions in a compound.

covalent bond A sharing of valence electrons by atoms.

dipole The separation of positive and negative charge in a polar bond indicated by an arrow that is drawn from the more positive atom to the more negative atom.

dipole–dipole attractions Intermolecular forces between oppositely charged ends of polar molecules.

dispersion forces Weak dipole bonding that results from a momentary polarization of nonpolar molecules.

double bond A sharing of two pairs of electrons by two atoms.

electronegativity The relative ability of an element to attract electrons in a bond.

hydrogen bond The attraction between a partially positive H atom and a strongly electronegative atom of N, O, or F.

ion An atom or group of atoms having an electrical charge because of a loss or gain of electrons.

ionic bond The attraction between a positive ion and a negative when electrons are transferred from a metal to a nonmetal.

ionic charge The difference between the number of protons (positive) and the number of electrons (negative) written in the upper right corner of the symbol for the element or polyatomic ion.

ionic compound A compound of positive and negative ions held together by ionic bonds.

Lewis structure A structure drawn in which the valence electrons of all the atoms are arranged to give octets except two electrons for hydrogen.

linear The shape of a molecule that has two bonded atoms and no lone pairs.

molecular compound A combination of atoms in which stable electron configurations are attained by sharing electrons.

molecule The smallest unit of two or more atoms held together by covalent bonds.

nonpolar covalent bond A covalent bond in which the electrons are shared equally between atoms.

nonpolar molecule A molecule that has only nonpolar bonds or in which the bond dipoles cancel.

octet rule Elements in Groups 1A to 7A (1, 2, 13 to 17) react with other elements by forming ionic or covalent bonds to produce a stable electron configuration.

polar covalent bond A covalent bond in which the electrons are shared unequally between atoms.

polar molecule A molecule containing bond dipoles that do not cancel.

polarity A measure of the unequal sharing of electrons, indicated by the difference in electronegativities.

polyatomic ion A group of covalently bonded nonmetal atoms that has an overall electrical charge.

tetrahedral The shape of a molecule with four bonded atoms.

trigonal planar The shape of a molecule with three bonded atoms and no lone pairs.

trigonal pyramidal The shape of a molecule that has three bonded atoms and one lone pair.

triple bond A sharing of three pairs of electrons by two atoms.

valence shell electron-pair repulsion (VSEPR) theory A theory that predicts the shape of a molecule by moving the electron groups on a central atom as far apart as possible to minimize the mutual repulsion of the electrons.

CORE CHEMISTRY SKILLS

The chapter Section containing each Core Chemistry Skill is shown in parentheses at the end of each heading.

Writing Positive and Negative Ions (6.1)

- In the formation of an ionic bond, atoms of a metal lose and atoms of a nonmetal gain valence electrons to acquire a stable electron configuration, usually eight valence electrons.
- This tendency of atoms to attain a stable electron configuration is known as the octet rule.

Example: State the number of electrons lost or gained by atoms and the ion formed for each of the following to obtain a stable electron configuration:

a. Br **b.** Ca **c.** S

Answer: **a.** Br atoms gain one electron to achieve a stable electron configuration, Br^-.
b. Ca atoms lose two electrons to achieve a stable electron configuration, Ca^{2+}.
c. S atoms gain two electrons to achieve a stable electron configuration, S^{2-}.

Writing Ionic Formulas (6.2)

- The chemical formula of a compound represents the lowest whole-number ratio of the atoms or ions.
- In the chemical formula of an ionic compound, the sum of the positive and negative charges is always zero.
- Thus, in a chemical formula of an ionic compound, the total positive charge is equal to the total negative charge.

Example: Write the formula for magnesium phosphide.

Answer: Magnesium phosphide is an ionic compound that contains the ions Mg^{2+} and P^{3-}.
Using charge balance, we determine the number(s) of each type of ion.

$3(2+) + 2(3-) = 0$

$3Mg^{2+}$ and $2P^{3-}$ give the formula Mg_3P_2.

Naming Ionic Compounds (6.3)

- In the name of an ionic compound made up of two elements, the name of the metal ion, which is written first, is the same as its element name.
- For metals that form two or more ions, a Roman numeral that is equal to the ionic charge is placed in parentheses immediately after the name of the metal.
- The name of a nonmetal ion is obtained by using the first syllable of its element name followed by *ide*.
- The name of a polyatomic ion ends in *ate* or *ite*.

Example: What is the name of $PbSO_4$?

Answer: This compound contains the SO_4^{2-} ion, which has a 2− charge.
For charge balance, the positive ion must have a charge of 2+.

$Pb^? + (2-) = 0 \qquad Pb = 2+$

Because lead can form two different positive ions, a Roman numeral (II) is used in the name of the compound: lead(II) sulfate.

Writing the Names and Formulas for Molecular Compounds (6.5)

- When naming a molecular compound, the first nonmetal in the formula is named by its element name; the second nonmetal is named using the first syllable of its element name followed by *ide*.
- When a subscript indicates two or more atoms of an element, a prefix is shown in front of its name.

Example: Name the molecular compound BrF_5.

Answer: Two nonmetals share electrons and form a molecular compound. Br (first nonmetal) is bromine; F (second nonmetal) is fluoride. In the name for a molecular compound, prefixes indicate the subscripts in the formulas. The subscript 1 is understood for Br. The subscript 5 for fluoride is written with the prefix *penta*. The name is bromine pentafluoride.

Drawing Lewis Structures (6.6)

- The Lewis structure for a molecule shows the sequence of atoms, the bonding pairs of electrons shared between atoms, and the nonbonding or *lone pairs* of electrons.
- Double or triple bonds result when a second or third electron pair is shared between the same atoms to complete octets.

Example: Draw the Lewis structures for CS_2.

Answer: The central atom is C.

S C S

Determine the total number of valence electrons.

$$1\text{ C} \times 4\,e^- = 4\,e^-$$
$$2\text{ S} \times 6\,e^- = 12\,e^-$$
$$\text{Total} = 16\,e^-$$

Attach each bonded atom to the central atom using a pair of electrons. Two bonding pairs use four electrons.

S:C:S

Place the 12 remaining electrons as lone pairs around the S atoms.

To complete the octet for C, a lone pair of electrons from each of the S atoms is shared with C, which forms two double bonds.

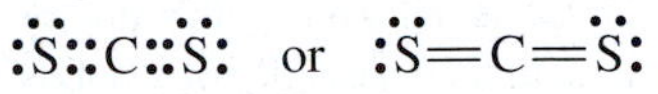

Using Electronegativity (6.7)

- The electronegativity values indicate the ability of atoms to attract shared electrons.
- Electronegativity values increase going across a period from left to right, and decrease going down a group.
- A nonpolar covalent bond occurs between atoms with identical or very similar electronegativity values such that the electronegativity difference is 0.0 to 0.4.
- A polar covalent bond typically occurs when electrons are shared unequally between atoms with electronegativity differences from 0.5 to 1.8.
- An ionic bond typically occurs when the difference in electronegativity for two atoms is greater than 1.8.

Example: Use electronegativity values to classify the bond between each of the following pairs of atoms as nonpolar covalent, polar covalent, or ionic:

a. Sr and Cl **b.** C and S **c.** O and Br

Answer: **a.** An electronegativity difference of 2.0 (Cl 3.0 − Sr 1.0) makes this an ionic bond.
b. An electronegativity difference of 0.0 (C 2.5 − S 2.5) makes this a nonpolar covalent bond.
c. An electronegativity difference of 0.7 (O 3.5 − Br 2.8) makes this a polar covalent bond.

Predicting Shape (6.8)

- The three-dimensional shape of a molecule is determined by drawing a Lewis structure and identifying the number of electron groups (one or more electron pairs) around the central atom and the number of bonded atoms.
- In the valence shell electron-pair repulsion (VSEPR) theory, the electron groups are arranged as far apart as possible around a central atom to minimize the repulsion.

- A central atom with two electron groups bonded to two atoms is linear. A central atom with three electron groups bonded to three atoms is trigonal planar, and to two atoms is bent (120°). A central atom with four electron groups bonded to four atoms is tetrahedral, to three atoms is trigonal pyramidal, and to two atoms is bent (109°).

Example: Predict the shape of $AsCl_3$.

Answer: From its Lewis structure, we see that $AsCl_3$ has four electron groups with three bonded atoms.

The electron-group geometry is tetrahedral, but with the central atom bonded to three Cl atoms, the shape is trigonal pyramidal.

Identifying Polarity of Molecules (6.8)

- A molecule is nonpolar if all of its bonds are nonpolar or it has polar bonds that cancel out. CCl_4 is a nonpolar molecule that consists of four polar bonds that cancel out.

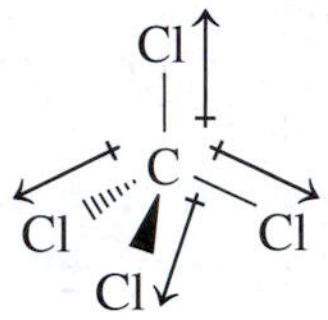

- A molecule is polar if it contains polar bonds that do not cancel out. H_2O is a polar molecule that consists of polar bonds that do not cancel out.

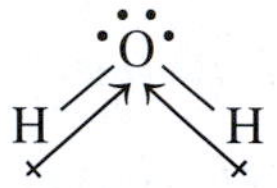

Example: Predict whether $AsCl_3$ is polar or nonpolar.

Answer: From its Lewis structure, we see that $AsCl_3$ has four electron groups with three bonded atoms.

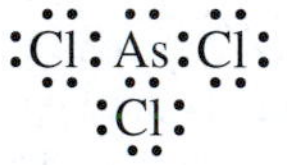

The shape of a molecule of $AsCl_3$ would be trigonal pyramidal with three polar bonds (As—Cl = 3.0 − 2.0 = 1.0) that do not cancel out. Thus, it is a polar molecule.

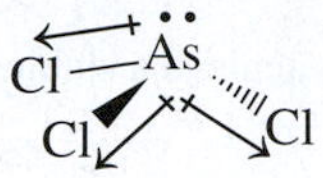

Identifying Intermolecular Forces (6.9)

- Dipole–dipole attractions occur between the dipoles in polar compounds because the positively charged end of one molecule is attracted to the negatively charged end of another molecule.

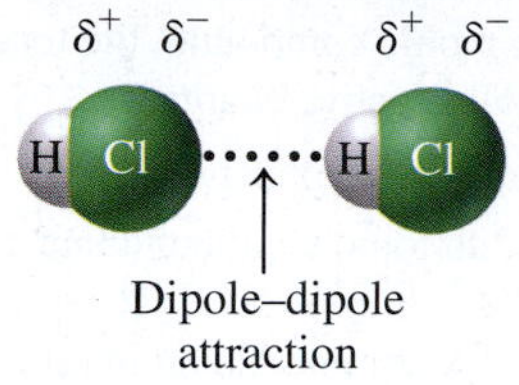
δ+ δ−
δ+ δ−
H Cl
H Cl
Dipole–dipole attraction

- Strong dipole–dipole attractions called hydrogen bonds occur in compounds in which H is bonded to N, O, or F. The partially positive H atom in one molecule has a strong attraction to the partially negative N, O, or F in another molecule.
- Dispersion forces are very weak intermolecular forces between nonpolar molecules that occur when *temporary dipoles* form as electrons are unsymmetrically distributed.

Example: Identify the strongest type of intermolecular forces in each of the following:

a. HF **b.** F_2 **c.** NF_3

Answer:
a. HF molecules, which are polar with H bonded to F, have hydrogen bonding.
b. Nonpolar F_2 molecules have only dispersion forces.
c. NF_3 molecules, which are polar, have dipole–dipole attractions.

UNDERSTANDING THE CONCEPTS

The chapter Sections to review are shown in parentheses at the end of each problem.

6.97 **a.** How does the octet rule explain the formation of a magnesium ion? (6.1)
b. What noble gas has the same electron configuration as the magnesium ion?
c. Why are Group 1A (1) and Group 2A (2) elements found in many compounds, but not Group 8A (18) elements?

6.98 **a.** How does the octet rule explain the formation of a chloride ion? (6.1)
b. What noble gas has the same electron configuration as the chloride ion?
c. Why are Group 7A (17) elements found in many compounds, but not Group 8A (18) elements?

6.99 Identify each of the following atoms or ions: (6.1)

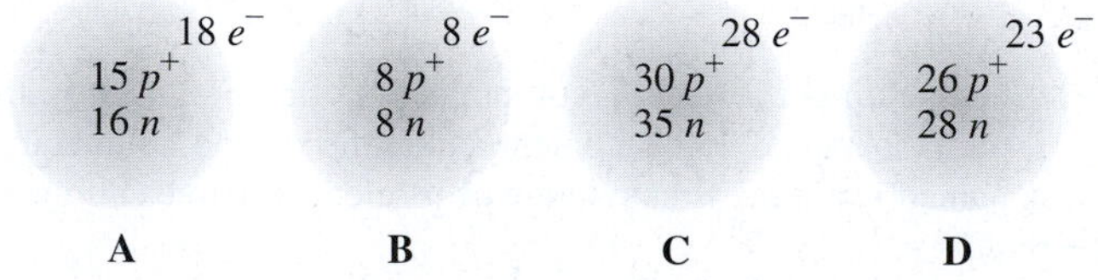
18 e^-
15 p^+
16 n
8 e^-
8 p^+
8 n
28 e^-
30 p^+
35 n
23 e^-
26 p^+
28 n
A
B
C
D

6.100 Identify each of the following atoms or ions: (6.1)

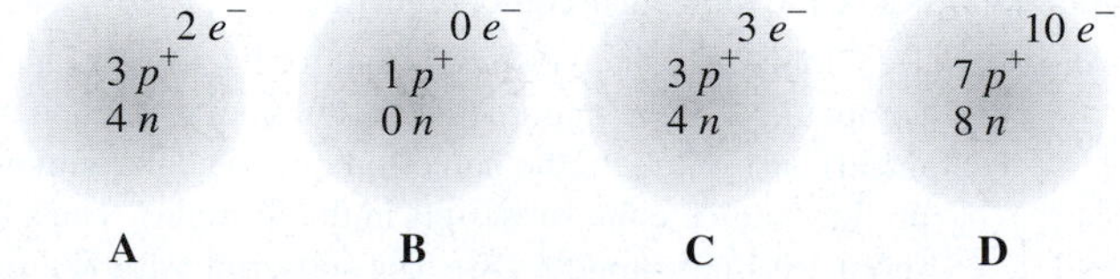
2 e^-
3 p^+
4 n
0 e^-
1 p^+
0 n
3 e^-
3 p^+
4 n
10 e^-
7 p^+
8 n
A
B
C
D

6.101 Consider the following Lewis symbols for elements X and Y: (6.1, 6.2, 6.5)

X· ·Ÿ:

a. What are the group numbers of X and Y?
b. Will a compound of X and Y be ionic or molecular?
c. What ions would be formed by X and Y?
d. What would be the formula of a compound of X and Y?
e. What would be the formula of a compound of X and sulfur?
f. What would be the formula of a compound of Y and chlorine?
g. Is the compound in part **f** ionic or molecular?

6.102 Consider the following Lewis symbols for elements X and Y: (6.1, 6.2, 6.5)

·Ẋ· ·Ÿ·

a. What are the group numbers of X and Y?
b. Will a compound of X and Y be ionic or molecular?
c. What ions would be formed by X and Y?
d. What would be the formula of a compound of X and Y?
e. What would be the formula of a compound of X and sulfur?
f. What would be the formula of a compound of Y and chlorine?
g. Is the compound in part **f** ionic or molecular?

6.103 Using each of the following electron configurations, write the formulas for the cation and anion that form, the formula for the compound they form, and its name. (6.2, 6.3)

Electron Configurations		Cation	Anion	Formula of Compound	Name of Compound
$1s^22s^22p^63s^2$	$1s^22s^22p^3$				
$1s^22s^22p^63s^23p^64s^1$	$1s^22s^22p^4$				
$1s^22s^22p^63s^23p^1$	$1s^22s^22p^63s^23p^5$				

6.104 Using each of the following electron configurations, write the formulas for the cation and anion that form, the formula for the compound they form, and its name. (6.2, 6.3)

Electron Configurations		Cation	Anion	Formula of Compound	Name of Compound
$1s^22s^22p^63s^1$	$1s^22s^22p^5$				
$1s^22s^22p^63s^23p^64s^2$	$1s^22s^22p^63s^23p^4$				
$1s^22s^22p^63s^23p^1$	$1s^22s^22p^63s^23p^3$				

6.105 State the number of valence electrons, bonding pairs, and lone pairs in each of the following Lewis structures: (6.6)

a. H:H b. H:B̤̈r: c. :B̤̈r:B̤̈r:

6.106 State the number of valence electrons, bonding pairs, and lone pairs in each of the following Lewis structures: (6.6)

a. H:Ö: with H above O b. H:N̈:H with H below N c. :B̤̈r:Ö:B̤̈r:

6.107 Match each of the Lewis structures (**a** to **c**) with the correct diagram (**1** to **3**) of its shape, and name the shape; indicate if each molecule is polar or nonpolar. Assume X and Y are nonmetals and all bonds are polar covalent. (6.6, 6.8)

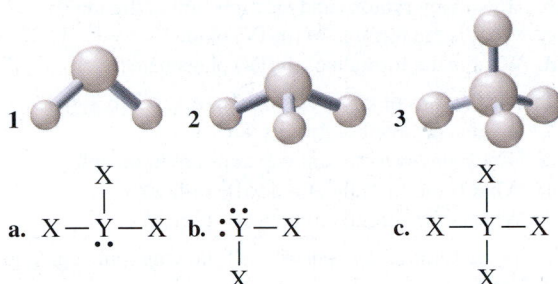

1
2
3

a. X—Y—X with X bonded above Y and two lone pairs on Y b. :Ÿ—X with X bonded below Y c. X—Y—X with X bonded above and below Y

6.108 Match each of the formulas (**a** to **c**) with the correct diagram (**1** to **3**) of its shape, and name the shape; indicate if each molecule is polar or nonpolar. (6.6, 6.8)

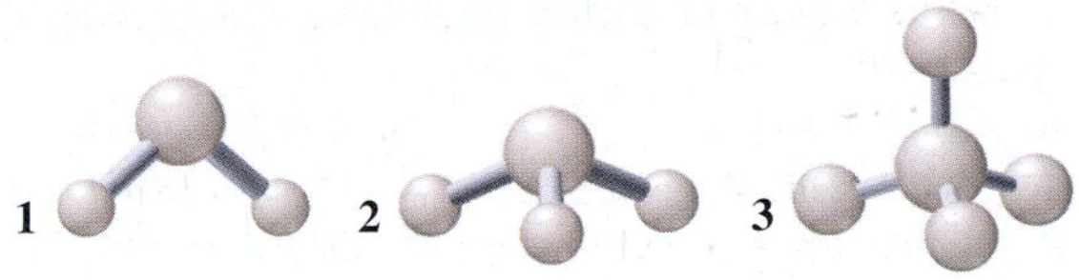
1
2
3

a. PBr_3 b. $SiCl_4$ c. OF_2

6.109 Consider the bond between each of the following pairs of atoms: Ca and O, C and O, K and O, O and O, and N and O. (6.7)
a. Which bonds are polar covalent?
b. Which bonds are nonpolar covalent?
c. Which bonds are ionic?
d. Arrange the covalent bonds in order of decreasing polarity.

6.110 Consider the bond between each of the following pairs of atoms: F and Cl, Cl and Cl, Cs and Cl, O and Cl, and Ca and Cl. (6.7)
a. Which bonds are polar covalent?
b. Which bonds are nonpolar covalent?
c. Which bonds are ionic?
d. Arrange the covalent bonds in order of decreasing polarity.

6.111 Identify the major intermolecular forces between each of the following atoms or molecules: (6.9)
a. PH_3 b. NO_2 c. CH_3NH_2 d. Ar

6.112 Identify the major intermolecular forces between each of the following atoms or molecules: (6.9)
a. He b. HBr
c. SnH_4 d. $CH_3CH_2CH_2OH$

ADDITIONAL PRACTICE PROBLEMS

6.113 Write the name for each of the following ions: (6.1)
a. N^{3-} b. Mg^{2+}
c. O^{2-} d. Al^{3+}

6.114 Write the name for each of the following ions: (6.1)
a. K^+ b. Na^+
c. Ba^{2+} d. Cl^-

6.115 Consider an ion with the symbol X^{2+} formed from a representative element. (6.1, 6.2, 6.3)
a. What is the group number of the element?
b. What is the Lewis symbol of the element?
c. If X is in Period 3, what is the element?
d. What is the formula of the compound formed from X and the nitride ion?

6.116 Consider an ion with the symbol Y^{3-} formed from a representative element. (6.1, 6.2, 6.3)
- **a.** What is the group number of the element?
- **b.** What is the Lewis symbol of the element?
- **c.** If Y is in Period 3, what is the element?
- **d.** What is the formula of the compound formed from the barium ion and Y?

6.117 One of the ions of tin is tin(IV). (6.1, 6.2, 6.3, 6.4)
- **a.** What is the symbol for this ion?
- **b.** How many protons and electrons are in the ion?
- **c.** What is the formula of tin(IV) oxide?
- **d.** What is the formula of tin(IV) phosphate?

6.118 One of the ions of gold is gold(III). (6.1, 6.2, 6.3, 6.4)
- **a.** What is the symbol for this ion?
- **b.** How many protons and electrons are in the ion?
- **c.** What is the formula of gold(III) sulfate?
- **d.** What is the formula of gold(III) nitrate?

6.119 Write the formula for each of the following ionic compounds: (6.2, 6.3)
- **a.** tin(II) sulfide **b.** lead(IV) oxide
- **c.** silver chloride **d.** calcium nitride
- **e.** copper(I) phosphide **f.** chromium(II) bromide

6.120 Write the formula for each of the following ionic compounds: (6.2, 6.3)
- **a.** nickel(III) oxide **b.** iron(III) sulfide
- **c.** lead(II) sulfate **d.** chromium(III) iodide
- **e.** lithium nitride **f.** gold(I) oxide

6.121 Name each of the following molecular compounds: (6.5)
- **a.** NCl_3 **b.** N_2S_3 **c.** N_2O
- **d.** IF **e.** BF_3 **f.** P_2O_5

6.122 Name each of the following molecular compounds: (6.5)
- **a.** CF_4 **b.** SF_6 **c.** BrCl
- **d.** N_2O_4 **e.** SO_2 **f.** CS_2

6.123 Write the formula for each of the following molecular compounds: (6.5)
- **a.** carbon sulfide **b.** diphosphorus pentoxide
- **c.** dihydrogen sulfide **d.** sulfur dichloride

6.124 Write the formula for each of the following molecular compounds: (6.5)
- **a.** silicon dioxide **b.** carbon tetrabromide
- **c.** diphosphorus tetraiodide **d.** dinitrogen trioxide

6.125 Classify each of the following as ionic or molecular, and write its name: (6.3, 6.5)
- **a.** $FeCl_3$ **b.** Na_2SO_4 **c.** NO_2
- **d.** Rb_2S **e.** PF_5 **f.** CF_4

6.126 Classify each of the following as ionic or molecular, and write its name: (6.3, 6.5)
- **a.** $Al_2(CO_3)_3$ **b.** ClF_5 **c.** BCl_3
- **d.** Mg_3N_2 **e.** ClO_2 **f.** $CrPO_4$

6.127 Write the formula for each of the following: (6.3, 6.4, 6.5)
- **a.** tin(II) carbonate **b.** lithium phosphide
- **c.** silicon tetrachloride **d.** manganese(III) oxide
- **e.** tetraphosphorus triselenide **f.** calcium bromide

6.128 Write the formula for each of the following: (6.3, 6.4, 6.5)
- **a.** sodium carbonate **b.** nitrogen dioxide
- **c.** aluminum nitrate **d.** copper(I) nitride
- **e.** potassium phosphate **f.** cobalt(III) sulfate

6.129 Determine the total number of valence electrons in each of the following: (6.6)
- **a.** HNO_2 **b.** CH_3CHO **c.** CH_3NH_2

6.130 Determine the total number of valence electrons in each of the following: (6.6)
- **a.** $COCl_2$ **b.** N_2O **c.** $SeCl_2$

6.131 Draw the Lewis structures for each of the following: (6.6)
- **a.** Cl_2O **b.** H_2NOH (N is the central atom)
- **c.** H_2CCCl_2 **d.** BF_4^-

6.132 Draw the Lewis structures for each of the following: (6.6)
- **a.** H_3COCH_3 (the atoms are in the order C O C)
- **b.** HNO_2 (the atoms are in the order HONO)
- **c.** OBr_2 **d.** NO_2^+

6.133 Use the periodic table to arrange the following atoms in order of increasing electronegativity: (6.7)
- **a.** I, F, Cl **b.** Li, K, S, Cl **c.** Mg, Sr, Ba, Be

6.134 Use the periodic table to arrange the following atoms in order of increasing electronegativity: (6.7)
- **a.** Cl, Br, Se **b.** Na, Cs, O, S **c.** O, F, B, Li

6.135 Select the more polar bond in each of the following pairs: (6.7)
- **a.** C—N or C—O **b.** N—F or N—Br
- **c.** Br—Cl or S—Cl **d.** Br—Cl or Br—I
- **e.** N—F or N—O

6.136 Select the more polar bond in each of the following pairs: (6.7)
- **a.** C—C or C—O **b.** P—Cl or P—Br
- **c.** Si—S or Si—Cl **d.** F—Cl or F—Br
- **e.** P—O or P—S

6.137 Show the dipole arrow for each of the following bonds: (6.7)
- **a.** Si—Cl **b.** C—N **c.** F—Cl
- **d.** C—F **e.** N—O

6.138 Show the dipole arrow for each of the following bonds: (6.7)
- **a.** P—O **b.** N—F **c.** O—Cl
- **d.** S—Cl **e.** P—F

6.139 Calculate the electronegativity difference and classify the bond between each of the following pair of atoms as nonpolar covalent, polar covalent, or ionic: (6.7)
- **a.** Si and Cl **b.** C and C **c.** Na and Cl
- **d.** C and H **e.** F and F

6.140 Calculate the electronegativity difference and classify the bond between each of the following pair of atoms as nonpolar covalent, polar covalent, or ionic: (6.7)
- **a.** C and N **b.** Cl and Cl **c.** K and Br
- **d.** H and H **e.** N and F

6.141 For each of the following, draw the Lewis structures and determine the shape: (6.6, 6.8)
- **a.** NF_3 **b.** $SiBr_4$ **c.** CSe_2

6.142 For each of the following, draw the Lewis structures and determine the shape: (6.6, 6.8)
- **a.** $COCl_2$ (C is the central atom)
- **b.** HCCH
- **c.** SO_2

6.143 Use the Lewis structure to determine the shape for each of the following molecules or polyatomic ions:
- **a.** BrO_2^- **b.** H_2O
- **c.** CBr_4 **d.** PO_3^{3-}

6.144 Use the Lewis structure to determine the shape for each of the following molecules or polyatomic ions:
a. PH_3 **b.** NO_3^-
c. HCN **d.** SO_3^{2-}

6.145 Predict the shape and polarity of each of the following molecules, which have polar covalent bonds: (6.8)
a. A central atom with three identical bonded atoms and one lone pair.
b. A central atom with two bonded atoms and two lone pairs.

6.146 Predict the shape and polarity of each of the following molecules, which have polar covalent bonds: (6.8)
a. A central atom with four identical bonded atoms and no lone pairs.
b. A central atom with four bonded atoms that are not identical and no lone pairs.

6.147 Classify each of the following molecules as polar or nonpolar: (6.7, 6.8)
a. HBr **b.** SiO_2 **c.** NCl_3
d. CH_3Cl **e.** NI_3 **f.** H_2O

6.148 Classify each of the following molecules as polar or nonpolar: (6.7, 6.8)
a. GeH_4 **b.** I_2 **c.** CF_3Cl
d. PCl_3 **e.** BCl_3 **f.** SCl_2

6.149 Indicate the major type of intermolecular forces—(1) ionic bonds, (2) dipole–dipole attractions, (3) hydrogen bonds, (4) dispersion forces—that occurs between particles of the following: (6.9)
a. NF_3 **b.** ClF **c.** Br_2
d. Cs_2O **e.** C_4H_{10} **f.** CH_3OH

6.150 Indicate the major type of intermolecular forces—(1) ionic bonds, (2) dipole–dipole attractions, (3) hydrogen bonds, (4) dispersion forces—that occurs between particles of the following: (6.9)
a. $CHCl_3$ **b.** H_2O **c.** LiCl
d. OBr_2 **e.** HBr **f.** IBr

CHALLENGE PROBLEMS

The following problems are related to the topics in this chapter. However, they do not all follow the chapter order, and they require you to combine concepts and skills from several Sections. These problems will help you increase your critical thinking skills and prepare for your next exam.

6.151 Complete the following table for atoms or ions: (6.1)

Atom or Ion	Number of Protons	Number of Electrons	Electrons Lost/Gained
K^+			
	$12\ p^+$	$10\ e^-$	
	$8\ p^+$		$2\ e^-$ gained
		$10\ e^-$	$3\ e^-$ lost

6.152 Complete the following table for atoms or ions: (6.1)

Atom or Ion	Number of Protons	Number of Electrons	Electrons Lost/Gained
	$30\ p^+$		$2\ e^-$ lost
	$36\ p^+$	$36\ e^-$	
	$16\ p^+$		$2\ e^-$ gained
		$46\ e^-$	$4\ e^-$ lost

6.153 Identify the group number in the periodic table of X, a representative element, in each of the following ionic compounds: (6.2)
a. XCl_3 **b.** Al_2X_3 **c.** XCO_3

6.154 Identify the group number in the periodic table of X, a representative element, in each of the following ionic compounds: (6.2)
a. X_2O_3 **b.** X_2SO_3 **c.** Na_3X

6.155 Classify each of the following as ionic or molecular, and name each: (6.2, 6.3, 6.4, 6.5)
a. Li_2HPO_4 **b.** ClF_3 **c.** $Mg(ClO_2)_2$
d. NF_3 **e.** $Ca(HSO_4)_2$ **f.** $KClO_4$
g. $Au_2(SO_3)_3$

6.156 Classify each of the following as ionic or molecular, and name each: (6.2, 6.3, 6.4, 6.5)
a. $FePO_3$ **b.** Cl_2O_7 **c.** $Ca_3(PO_4)_2$
d. PCl_3 **e.** $Al(ClO_2)_3$ **f.** $Pb(C_2H_3O_2)_2$
g. $MgCO_3$

6.157 Complete the Lewis structure for each of the following: (6.6)

a. 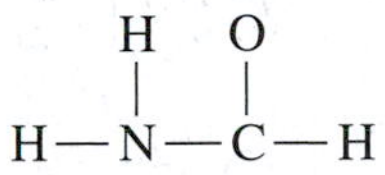

b. 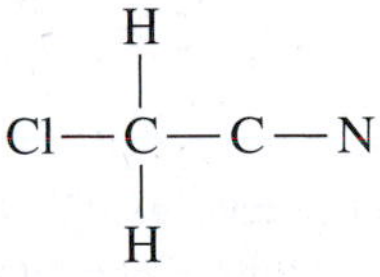

c. 

d. 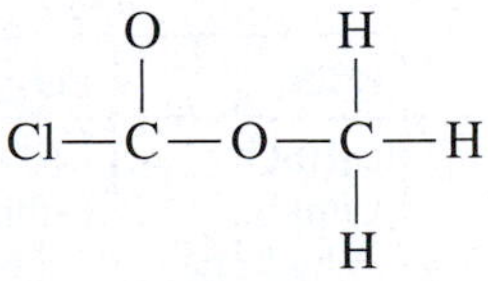

6.158 Identify the errors in each of the following Lewis structures and draw the correct formula: (6.6)

a. 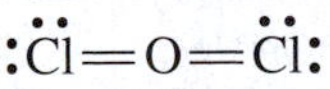

b. 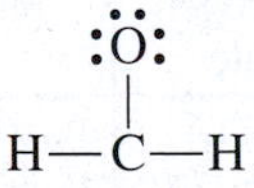

c. 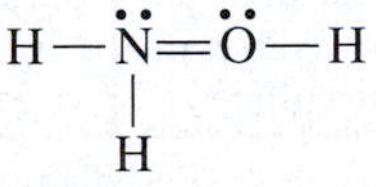

6.159 Predict the shape of each of the following molecules: (6.8)
a. NH_2Cl (N is the central atom) **b.** TeO_2

6.160 Classify each of the following molecules as polar or nonpolar: (6.7, 6.8)
a. N_2 **b.** NH_2Cl (N is the central atom)

ANSWERS

6.1 **a.** 1 **b.** 2 **c.** 3 **d.** 1 **e.** 2

6.3 **a.** 2 e^- lost **b.** 3 e^- gained **c.** 1 e^- gained
d. 1 e^- lost **e.** 1 e^- gained

6.5 **a.** Li^+ **b.** F^- **c.** Mg^{2+} **d.** Co^{3+}

6.7 **a.** 29 protons, 27 electrons **b.** 34 protons, 36 electrons
c. 35 protons, 36 electrons **d.** 26 protons, 24 electrons

6.9 **a.** Cl^- **b.** Cs^+ **c.** N^{3-} **d.** Ra^{2+}

6.11 **a.** lithium **b.** calcium **c.** gallium **d.** phosphide

6.13 **a.** 8 protons, 10 electrons **b.** 19 protons, 18 electrons
c. 53 protons, 54 electrons **d.** 11 protons, 10 electrons

6.15 a and c

6.17 **a.** Na_2O **b.** $AlBr_3$ **c.** Ba_3N_2
d. MgF_2 **e.** Al_2S_3

6.19 **a.** K^+ and S^{2-}, K_2S **b.** Na^+ and N^{3-}, Na_3N
c. Al^{3+} and I^-, AlI_3 **d.** Ga^{3+} and O^{2-}, Ga_2O_3

6.21 **a.** aluminum fluoride **b.** calcium chloride
c. sodium oxide **d.** magnesium phosphide
e. potassium iodide **f.** barium fluoride

6.23 **a.** iron(II) **b.** copper(II) **c.** zinc
d. lead(IV) **e.** chromium(III) **f.** manganese(II)

6.25 **a.** tin(II) chloride **b.** iron(II) oxide
c. copper(I) sulfide **d.** copper(II) sulfide
e. cadmium bromide **f.** mercury(II) chloride

6.27 **a.** Au^{3+} **b.** Fe^{3+} **c.** Pb^{4+} **d.** Al^{3+}

6.29 **a.** $MgCl_2$ **b.** Na_2S **c.** Cu_2O **d.** Zn_3P_2
e. AuN **f.** CoF_3

6.31 **a.** $CoCl_3$ **b.** PbO_2 **c.** AgI **d.** Ca_3N_2
e. Cu_3P **f.** $CrCl_2$

6.33 **a.** K_3P **b.** $CuCl_2$ **c.** $FeBr_3$ **d.** MgO

6.35 **a.** HCO_3^- **b.** NH_4^+ **c.** PO_3^{3-} **d.** ClO_3^-

6.37 **a.** sulfate **b.** carbonate
c. hydrogen sulfite (bisulfite) **d.** nitrate

6.39

	NO_2^-	CO_3^{2-}	HSO_4^-	PO_4^{3-}
Li^+	$LiNO_2$ Lithium nitrite	Li_2CO_3 Lithium carbonate	$LiHSO_4$ Lithium hydrogen sulfate	Li_3PO_4 Lithium phosphate
Cu^{2+}	$Cu(NO_2)_2$ Copper(II) nitrite	$CuCO_3$ Copper(II) carbonate	$Cu(HSO_4)_2$ Copper(II) hydrogen sulfate	$Cu_3(PO_4)_2$ Copper(II) phosphate
Ba^{2+}	$Ba(NO_2)_2$ Barium nitrite	$BaCO_3$ Barium carbonate	$Ba(HSO_4)_2$ Barium hydrogen sulfate	$Ba_3(PO_4)_2$ Barium phosphate

6.41 **a.** $Ba(OH)_2$ **b.** $NaHSO_4$ **c.** $Fe(NO_2)_2$
d. $Zn_3(PO_4)_2$ **e.** $Fe_2(CO_3)_3$

6.43 **a.** CO_3^{2-}, sodium carbonate **b.** NH_4^+, ammonium sulfide
c. OH^-, calcium hydroxide **d.** NO_2^-, tin(II) nitrite

6.45 **a.** zinc acetate
b. magnesium phosphate
c. ammonium chloride
d. sodium hydrogen carbonate or sodium bicarbonate
e. sodium nitrite

6.47 **a.** phosphorus tribromide **b.** dichlorine oxide
c. carbon tetrabromide **d.** hydrogen fluoride
e. nitrogen trifluoride

6.49 **a.** dinitrogen trioxide **b.** disilicon hexabromide
c. tetraphosphorus trisulfide **d.** phosphorus pentachloride
e. selenium hexafluoride

6.51 **a.** CCl_4 **b.** CO **c.** PF_3 **d.** N_2O_4

6.53 **a.** OF_2 **b.** BCl_3 **c.** N_2O_3 **d.** SF_6

6.55 **a.** aluminum sulfate **b.** calcium carbonate
c. dinitrogen oxide **d.** magnesium hydroxide

6.57 **a.** 8 valence electrons **b.** 14 valence electrons
c. 32 valence electrons **d.** 8 valence electrons

6.59 **a.** HF (8 e^-) 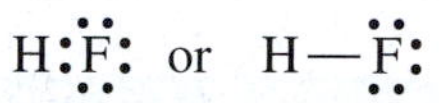

b. SF_2 (20 e^-) 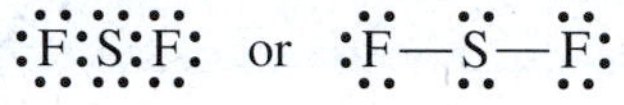

c. NBr_3 (26 e^-) 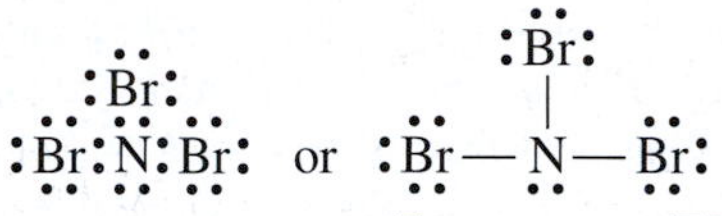

d. BH_4^- (8 e^-) 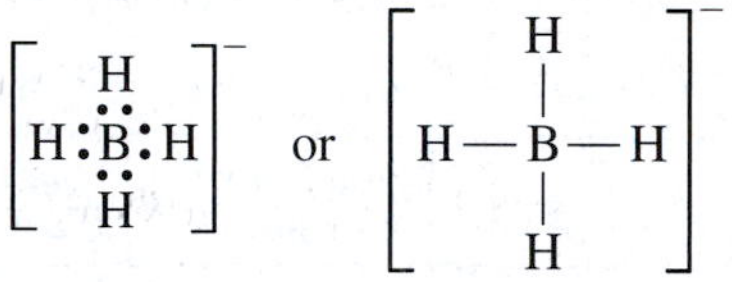

6.61 If complete octets cannot be formed by using all the valence electrons, it is necessary to draw multiple bonds.

6.63 **a.** CO (10 e^-) 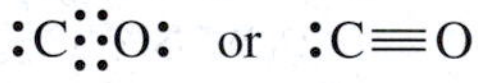

b. NO_3^- (24 e^-) 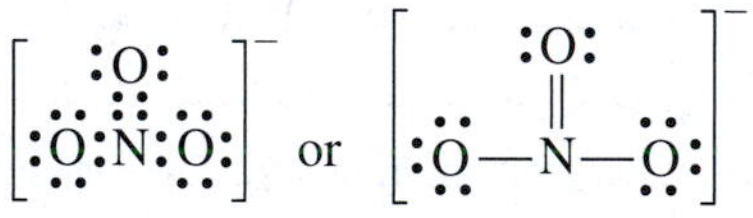

c. H_2CO (12 e^-) 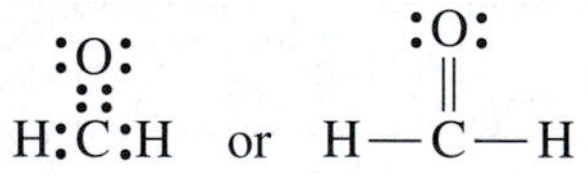

6.65 **a.** increases **b.** decreases **c.** decreases

6.67 from 0.0 and 0.4

6.69 **a.** K, Na, Li **b.** Na, P, Cl **c.** Ca, Se, O

6.71 **a.** polar covalent **b.** ionic
c. polar covalent **d.** nonpolar covalent
e. polar covalent **f.** nonpolar covalent

6.73 **a.** $\overset{\delta+}{N}—\overset{\delta-}{F}$ (dipole arrow ⟶) **b.** $\overset{\delta+}{Si}—\overset{\delta-}{Br}$ (dipole arrow ⟶)
c. $\overset{\delta+}{C}—\overset{\delta-}{O}$ (dipole arrow ⟶) **d.** $\overset{\delta+}{P}—\overset{\delta-}{Br}$ (dipole arrow ⟶)
e. $\overset{\delta-}{N}—\overset{\delta+}{P}$ (dipole arrow ⟵)

6.75 **a.** 6, tetrahedral **b.** 5, trigonal pyramidal
c. 3, trigonal planar

6.77 **a.** three **b.** trigonal planar
c. three **d.** trigonal planar

6.79 In both PH_3, and NH_3, there are three bonded atoms and one lone pair of electrons on the central atoms. The shapes of both are trigonal pyramidal.

6.81 **a.** bent (109°) **b.** trigonal planar **c.** bent (109°)

6.83 **a.** AlH_4^- ($8\ e^-$)

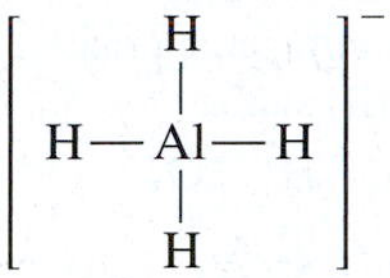

$$\left[H-\underset{\underset{H}{|}}{\overset{\overset{H}{|}}{Al}}-H \right]^-$$ tetrahedral

b. SO_4^{2-} ($32\ e^-$)

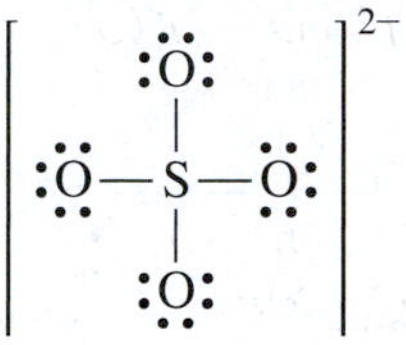

$$\left[:\!\ddot{O}\!:\!-\!\underset{\underset{:\ddot{O}:}{|}}{\overset{\overset{:\ddot{O}:}{}}{S}}\!-\!:\!\ddot{O}\!: \right]^{2-}$$ tetrahedral

c. NH_4^+ ($8\ e^-$)

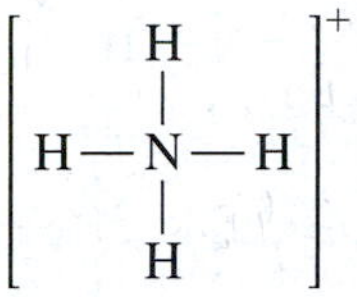

$$\left[H-\underset{\underset{H}{|}}{\overset{\overset{H}{|}}{N}}-H \right]^+$$ tetrahedral

d. NO_2^+ ($16\ e^-$) $\left[:\ddot{O}=N=\ddot{O}: \right]^+$ 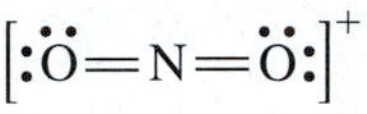 linear

6.85 **a.** polar **b.** polar **c.** polar

6.87 **a.** dipole–dipole attractions **b.** ionic bonds
c. dipole–dipole attractions **d.** dispersion forces

6.89 **a.** hydrogen bonds **b.** dipole–dipole attractions
c. dipole–dipole attractions **d.** dispersion forces

6.91 **a.** HF hydrogen bonds are stronger than dipole-dipole attractions
b. NaF ionic bonds are stronger than hydrogen bonds
c. $MgBr_2$ ionic bonds are stronger than dipole-dipole attractions
d. CH_3OH hydrogen bonds are stronger than dispersion forces

6.93 **a.** 0.0 **b.** 0.4 **c.** 1.0

6.95 **a.** nonpolar covalent **b.** nonpolar covalent
c. polar covalent

6.97 **a.** By losing two valence electrons from the third energy level, magnesium achieves an octet in the second energy level.
b. The magnesium ion Mg^{2+} has the same electron configuration as Ne ($1s^2 2s^2 2p^6$).
c. Group 1A (1) and 2A (2) elements achieve octets by losing electrons to form compounds. Group 8A (18) elements are stable with octets (or two electrons for helium).

6.99 **a.** P^{3-} ion **b.** O atom **c.** Zn^{2+} ion **d.** Fe^{3+} ion

6.101 **a.** X = Group 1A (1), Y = Group 6A (16)
b. ionic **c.** X^+ and Y^{2-} **d.** X_2Y
e. X_2S **f.** YCl_2 **g.** molecular

6.103

Electron Configurations		Cation	Anion	Formula of Compound	Name of Compound
$1s^2 2s^2 2p^6 3s^2$	$1s^2 2s^2 2p^3$	Mg^{2+}	N^{3-}	Mg_3N_2	Magnesium nitride
$1s^2 2s^2 2p^6 3s^2 3p^6 4s^1$	$1s^2 2s^2 2p^4$	K^+	O^{2-}	K_2O	Potassium oxide
$1s^2 2s^2 2p^6 3s^2 3p^1$	$1s^2 2s^2 2p^6 3s^2 3p^5$	Al^{3+}	Cl^-	$AlCl_3$	Aluminum chloride

6.105 **a.** two valence electrons, one bonding pair, no lone pairs
b. eight valence electrons, one bonding pair, three lone pairs
c. 14 valence electrons, one bonding pair, six lone pairs

6.107 **a.** 2, trigonal pyramidal, polar **b.** 1, bent (109°), polar
c. 3, tetrahedral, nonpolar

6.109 **a.** C and O, N and O **b.** O and O
c. Ca and O, K and O **d.** C and O, N and O, O and O

6.111 **a.** dispersion forces **b.** dipole–dipole attractions
c. hydrogen bonds **d.** dispersion forces

6.113 **a.** nitride **b.** magnesium
c. oxide **d.** aluminum

6.115 **a.** 2A (2) **b.** $\dot{\overset{\cdot}{X}}\cdot$
c. Mg **d.** X_3N_2

6.117 **a.** Sn^{4+} **b.** 50 protons and 46 electrons
c. SnO_2 **d.** $Sn_3(PO_4)_4$

6.119 **a.** SnS **b.** PbO_2 **c.** AgCl
d. Ca_3N_2 **e.** Cu_3P **f.** $CrBr_2$

6.121 **a.** nitrogen trichloride **b.** dinitrogen trisulfide
c. dinitrogen oxide **d.** iodine fluoride
e. boron trifluoride **f.** diphosphorus pentoxide

6.123 **a.** CS **b.** P_2O_5 **c.** H_2S **d.** SCl_2

6.125 **a.** ionic, iron(III) chloride
b. ionic, sodium sulfate
c. molecular, nitrogen dioxide
d. ionic, rubidium sulfide
e. molecular, phosphorus pentafluoride
f. molecular, carbon tetrafluoride

6.127 **a.** $SnCO_3$ **b.** Li_3P **c.** $SiCl_4$
d. Mn_2O_3 **e.** P_4Se_3 **f.** $CaBr_2$

6.129 **a.** $1 + 5 + 2(6) = 18$ valence electrons
b. $2(4) + 4(1) + 6 = 18$ valence electrons
c. $4 + 5(1) + 5 = 14$ valence electrons

6.131 **a.** Cl_2O ($20\ e^-$) $:\!\ddot{C}l\!:\!\ddot{O}\!:\!\ddot{C}l\!:$ or $:\!\ddot{C}l\!-\!\ddot{O}\!-\!\ddot{C}l\!:$ 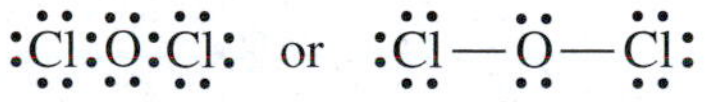

b. H_2NOH ($14\ e^-$) $H\!:\!\ddot{N}\!:\!\ddot{O}\!:\!H$ or $H-\underset{\underset{H}{|}}{N}-\ddot{O}-H$ 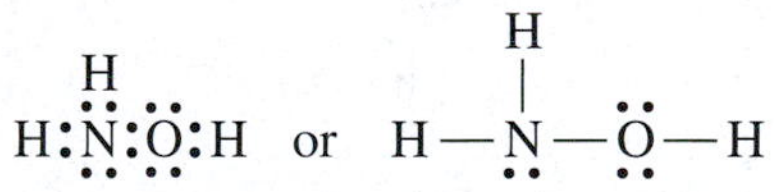

c. H_2CCCl_2 ($24\ e^-$) $H\!:\!\ddot{C}l\!:\!:\!C\!:\!\ddot{C}l\!:$ or $H-\underset{\underset{H}{|}}{C}=\underset{\underset{:\ddot{C}l:}{|}}{C}-\ddot{C}l\!:$ 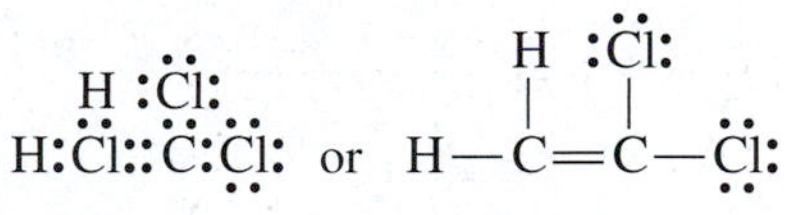

d. BF_4^- ($32\ e^-$) $\left[:\!\ddot{F}\!:\!\overset{\overset{:F:}{}}{\underset{\underset{:F:}{}}{B}}\!:\!\ddot{F}\!: \right]^-$ or $\left[:\!\ddot{F}\!-\!\underset{\underset{:\ddot{F}:}{|}}{\overset{\overset{:\ddot{F}:}{|}}{B}}\!-\!\ddot{F}\!: \right]^-$ 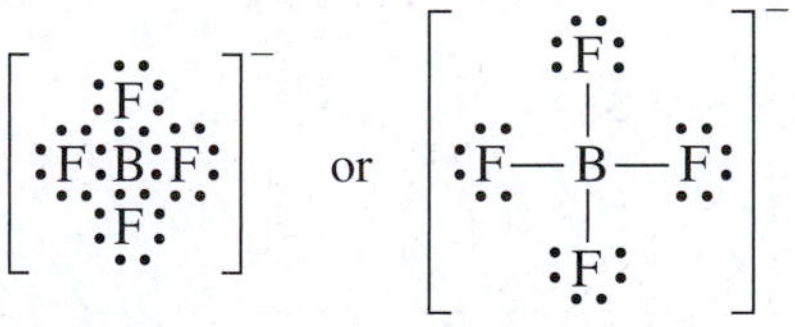

6.133 **a.** I, Cl, F **b.** K, Li, S, Cl **c.** Ba, Sr, Mg, Be

6.135 **a.** C—O **b.** N—F **c.** S—Cl
d. Br—I **e.** N—F

6.137 **a.** $\underset{\longleftrightarrow}{Si-Cl}$ **b.** $\underset{\longleftrightarrow}{C-N}$
c. $\underset{\longleftarrow}{F-Cl}$ **d.** $\underset{\longleftrightarrow}{C-F}$
e. $\underset{\longleftrightarrow}{N-O}$

6.139 **a.** polar covalent **b.** nonpolar covalent
c. ionic **d.** nonpolar covalent
e. nonpolar covalent

6.141 a. NF_3 (26 e^-) 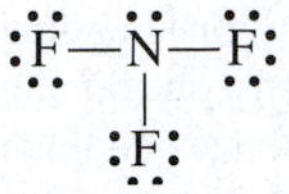 trigonal pyramidal

b. $SiBr_4$ (32 e^-) 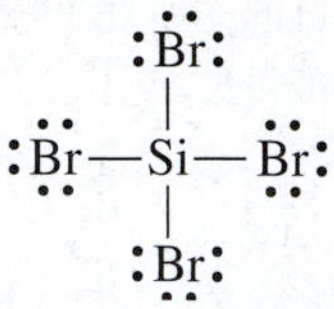 tetrahedral

c. CSe_2 (16 e^-) 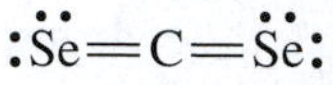 linear

6.143 a. bent (109°) **b.** bent (109°)
c. tetrahedral **d.** trigonal pyramidal

6.145 a. trigonal pyramidal, polar **b.** bent (109°), polar

6.147 a. polar **b.** nonpolar
c. nonpolar **d.** polar
e. polar **f.** polar

6.149 a. (2) dipole–dipole attractions **b.** (2) dipole–dipole attractions
c. (4) dispersion forces **d.** (1) ionic bonds
e. (4) dispersion forces **f.** (3) hydrogen bonds

6.151

Atom or Ion	Number of Protons	Number of Electrons	Electrons Lost/ Gained
K^+	19 p^+	18 e^-	1 e^- lost
Mg^{2+}	12 p^+	10 e^-	2 e^- lost
O^{2-}	8 p^+	10 e^-	2 e^- gained
Al^{3+}	13 p^+	10 e^-	3 e^- lost

6.153 a. Group 3A (13) **b.** Group 6A (16)
c. Group 2A (2)

6.155 a. ionic, lithium hydrogen phosphate
b. molecular, chlorine trifluoride
c. ionic, magnesium chlorite
d. molecular, nitrogen trifluoride
e. ionic, calcium bisulfate or calcium hydrogen sulfate
f. ionic, potassium perchlorate
g. ionic, gold(III) sulfite

6.157 a. (18 e^-) 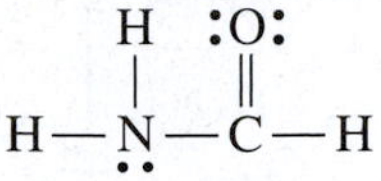

b. (22 e^-) 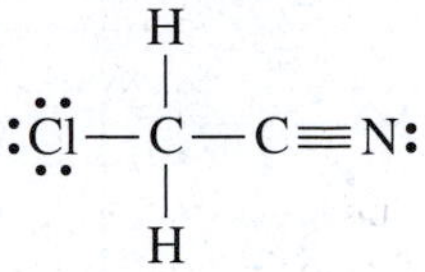

c. (12 e^-) 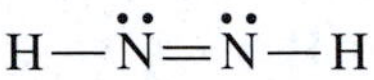

d. (30 e^-) 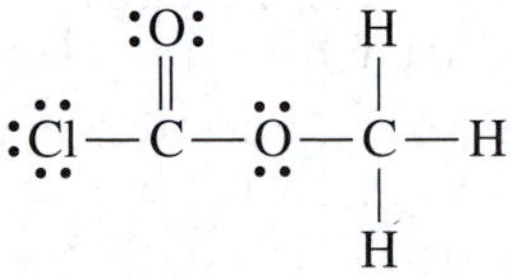

6.159 a. trigonal pyramidal **b.** bent, 120°

COMBINING IDEAS from Chapters 4 to 6

CI.7 For parts **a** to **f**, consider the loss of electrons by atoms of the element X, and a gain of electrons by atoms of the element Y, if X is in Group 2A (2), Period 3, and Y is in Group 7A (17), Period 3. (4.6, 6.1, 6.2, 6.3, 6.8)

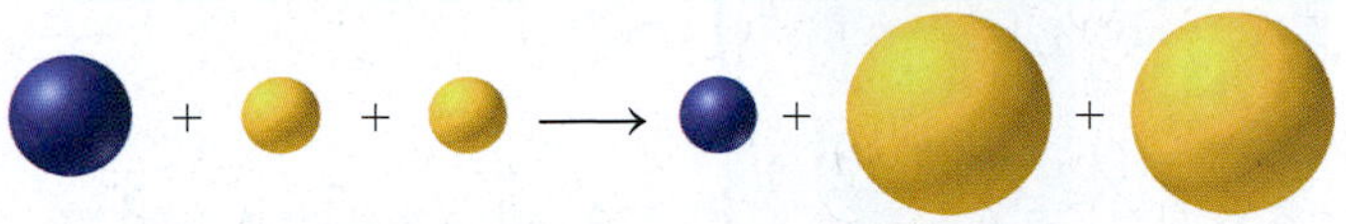

X **Y** **Y**

a. Which reactant has the higher electronegativity?
b. What are the ionic charges of **X** and **Y** in the product?
c. Write the electron configurations for the atoms **X** and **Y**.
d. Write the electron configurations for the ions of **X** and **Y**.
e. Give the names for the noble gases with the same electron configurations as each of these ions.
f. Write the formula and name for the ionic compound formed by the ions of **X** and **Y**.

CI.8 A sterling silver bracelet, which is 92.5% silver by mass, has a volume of 25.6 cm^3 and a density of 10.2 g/cm^3. (2.10, 4.5, 5.2, 5.4)

925

Sterling silver is 92.5% silver by mass.

a. What is the mass, in kilograms, of the bracelet?
b. Silver (Ag) has two naturally occuring isotopes: Ag-107 (106.905 amu) with an abundance of 51.84% and Ag-109 (108.905 amu) with an abundance of 48.16%. Calculate the atomic mass for silver using the weighted average mass method.
c. Determine the number of protons and neutrons in each of the two stable isotopes of silver:

$^{107}_{47}Ag$ $^{109}_{47}Ag$

d. Ag-112 decays by beta emission. Write the balanced nuclear equation for the decay of Ag-112.
e. A 64.0-μCi sample of Ag-112 decays to 8.00 μCi in 9.3 h. What is the half-life of Ag-112?

CI.9 Silicon has three naturally occurring isotopes: Si-28, Si-29, and Si-30. (4.5, 4.6, 5.2, 5.4, 6.6, 6.7)

a. Complete the following table with the number of protons, neutrons, and electrons for each of the naturally occurring isotopes of silicon:

Isotope	Number of Protons	Number of Neutrons	Number of Electrons
$^{28}_{14}Si$			
$^{29}_{14}Si$			
$^{30}_{14}Si$			

b. What is the electron configuration of silicon?
c. Which isotope is the most abundant, if the atomic mass for Si is 28.09?
d. Write the balanced nuclear equation for the beta decay of the radioactive isotope Si-31.
e. How many hours are needed for a sample of Si-31 with an activity of 16 μCi to decay to 2.0 μCi?
f. Draw the Lewis structures and predict the shape of $SiCl_4$.

CI.10 The iceman known as Ötzi was discovered in a high mountain pass on the Austrian–Italian border. Samples of his hair and bones had carbon-14 activity that was 50% of that present in new hair or bone. Carbon-14 undergoes beta decay and has a half-life of 5730 yr. (5.2, 5.4)

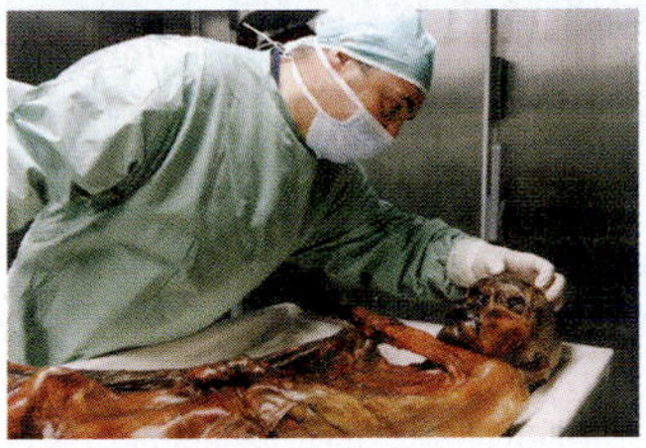

The mummified remains of Ötzi were discovered in 1991.

a. How long ago did Ötzi live?
b. Write a balanced nuclear equation for the decay of carbon-14.

CI.11 K^+ is an electrolyte required by the human body and found in many foods as well as salt substitutes. One of the isotopes of potassium is potassium-40, which has a natural abundance of 0.012% and a half-life of 1.3×10^9 yr. The isotope potassium-40 decays to calcium-40 or to argon-40. A typical activity from potassium-40 is 7.0 μCi per gram of potassium. (5.2, 5.3, 5.4)

Potassium chloride is used as a salt substitute.

a. Write a balanced nuclear equation for each type of decay and identify the particle emitted.
b. A shaker of salt substitute contains 1.6 oz of K. What is the activity, in millicuries and becquerels, of the potassium-40 in the shaker?

CI.12 Of much concern to environmentalists is radon-222, which is a radioactive noble gas that can seep from the ground into basements of homes and buildings. Radon-222 is a product of the decay of radium-226 that occurs naturally in rocks and soil in much of the United States. Radon-222, which has a half-life of 3.8 days, decays by emitting an alpha particle. Radon-222, which is a gas, can be inhaled into the lungs where it is strongly associated with lung cancer. Radon levels in a home can be measured with a home radon-detection kit.

Environmental agencies have set the maximum level of radon-222 in a home at 4 picocuries per liter (pCi/L) of air. (5.2, 5.3, 5.4)

a. Write the balanced nuclear equation for the decay of Ra-226.
b. Write the balanced nuclear equation for the decay of Rn-222.
c. If a room contains 24 000 atoms of radon-222, how many atoms of radon-222 remain after 15.2 days?
d. Suppose a room has a volume of 72 000 L (7.2×10^4 L). If the radon level is the maximum allowed (4 pCi/L), how many alpha particles are emitted from Rn-222 in 1 day? (1 Ci = 3.7×10^{10} disintegrations/s)

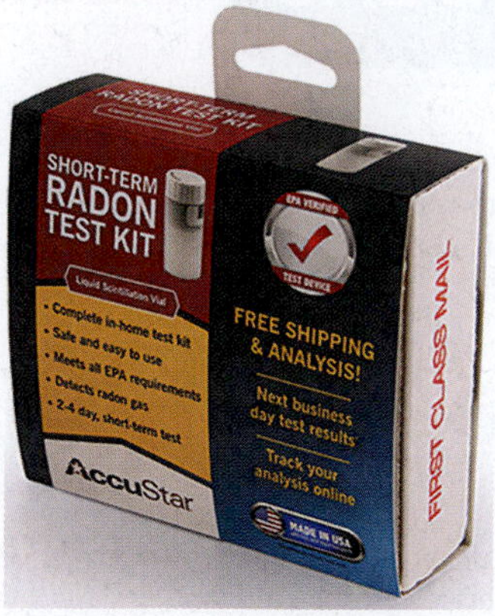
SHORT-TERM
RADON
TEST KIT
FREE SHIPPING & ANALYSIS!
Next business day test results
Track your analysis online
AccuStar
MADE IN USA
FIRST CLASS MAIL

A radon test kit is used to measure radon levels in homes.

ANSWERS

CI.7 **a.** Y has the higher electronegativity.
b. X^{2+}, Y^-
c. X = $1s^22s^22p^63s^2$ Y = $1s^22s^22p^63s^23p^5$
d. X^{2+} = $1s^22s^22p^6$ Y^- = $1s^22s^22p^63s^23p^6$
e. X^{2+} has the same electron configuration as Ne. Y^- has the same electron configuration as Ar.
f. $MgCl_2$, magnesium chloride

CI.9 **a.**

Isotope	Number of Protons	Number of Neutrons	Number of Electrons
$^{28}_{14}Si$	14	14	14
$^{29}_{14}Si$	14	15	14
$^{30}_{14}Si$	14	16	14

b. $1s^22s^22p^63s^23p^2$
c. Si-28
d. $^{31}_{14}Si \longrightarrow ^{31}_{15}P + ^{0}_{-1}e$
e. 7.8 h
f.

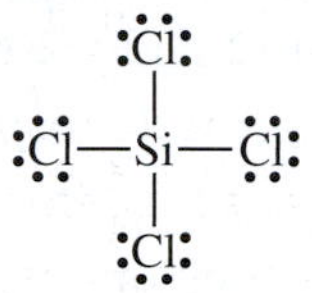
Cl
Cl
Si
Cl
Cl

tetrahedral

CI.11 **a.** $^{40}_{19}K \longrightarrow ^{40}_{20}Ca + ^{0}_{-1}e$ beta particle
$^{40}_{19}K \longrightarrow ^{40}_{18}Ar + ^{0}_{+1}e$ positron
b. 3.8×10^{-5} mCi; 1.4×10^3 Bq

7

Chemical Reactions and Quantities

Natalie was recently diagnosed with mild pulmonary emphysema due to secondhand cigarette smoke. She was referred to Angela, an exercise physiologist, who begins to assess Natalie's condition by connecting her to an electrocardiogram (ECG or EKG), a pulse oximeter, and a blood pressure cuff. The ECG records the electrical activity of Natalie's heart, which is used to measure her heartbeat's rate and rhythm and detect the possible presence of heart damage. The pulse oximeter measures her pulse and the saturation level of oxygen in her arterial blood (the percentage of hemoglobin that is saturated with O_2). The blood pressure cuff determines the pressure exerted by the heart in pumping her blood.

To determine possible heart disease, Natalie has an exercise stress test on a treadmill to measure how her heart rate and blood pressure respond to exertion by walking faster as Angela increases the treadmill's slope. Angela attaches electrical leads to Natalie to measure the heart rate and blood pressure first at rest and then on the treadmill. Angela uses a face mask to collect expired air and measures Natalie's maximal volume of oxygen uptake, or $V_{O_2\ max}$.

CAREER

Exercise Physiologist

Exercise physiologists work with athletes as well as patients diagnosed with diabetes, heart disease, pulmonary disease, or other chronic disabilities or diseases. Patients who have one of these diseases are often prescribed exercise as a form of treatment, and they are referred to an exercise physiologist. The exercise physiologist evaluates the patient's overall health and then creates a customized exercise program for that individual. The program for an athlete might focus on reducing her number of injuries, whereas a program for a cardiac patient would focus on strengthening his heart muscles. The exercise physiologist also monitors the patient for improvement and determines if the exercise is helping to reduce or reverse the progression of the disease.

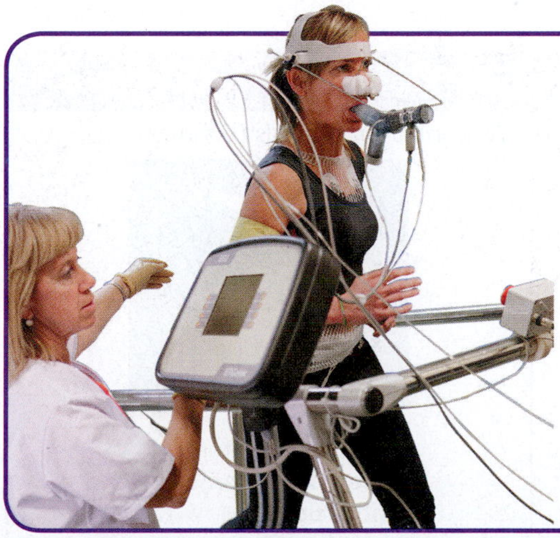

CLINICAL UPDATE

Improving Natalie's Overall Fitness

Natalie's test results indicate that her blood oxygen level is below normal. You can view Natalie's test results and lung function diagnosis in the **CLINICAL UPDATE Improving Natalie's Overall Fitness**, page 262. After reviewing the test results, Angela teaches Natalie how to improve her respiration and her overall fitness.

LOOKING AHEAD

REVIEW

Writing Ionic Formulas (6.2)
Naming Ionic Compounds (6.3)
Writing the Names and Formulas for Molecular Compounds (6.5)

7.1 Equations for Chemical Reactions

LEARNING GOAL Write a balanced chemical equation from the formulas of the reactants and products for a reaction; determine the number of atoms in the reactants and products.

Chemical reactions occur everywhere. The fuel in a car burns with oxygen to make a car move and run the air conditioner. When we cook food or bleach our hair, a chemical reaction takes place. In our bodies, chemical reactions convert food into molecules that build muscles and move them. In plant and tree leaves, carbon dioxide and water are converted into carbohydrates. Chemical equations are used by chemists to describe chemical reactions. In every chemical equation, the atoms in the reacting substances, called *reactants*, are rearranged to give new substances called *products*.

A *chemical change* occurs when a substance is converted into one or more new substances that have different formulas and different properties. For example, when silver tarnishes, the shiny silver metal (Ag) reacts with sulfur (S) to become the dull, black substance we call tarnish (Ag_2S) (see **FIGURE 7.1**).

A chemical change:
the tarnishing of silver

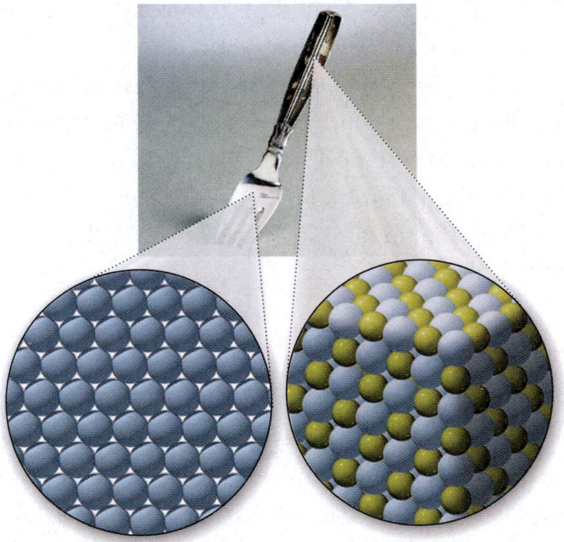

Ag
Ag₂S

FIGURE 7.1 ▶ A chemical change produces new substances with new properties.

Q Why is the formation of tarnish a chemical change?

A *chemical reaction* always involves chemical change because atoms of the reacting substances form new combinations with new properties. For example, a chemical reaction takes place when a piece of iron (Fe) combines with oxygen (O_2) in the air to produce a new substance, rust (Fe_2O_3), which has a reddish-brown color. During a chemical change, new properties become visible, which are an indication that a chemical reaction has taken place (see **TABLE 7.1**).

TABLE 7.1 Types of Evidence of a Chemical Reaction

1. Change in color	**2.** Formation of a gas (bubbles)
 Fe → Fe_2O_3	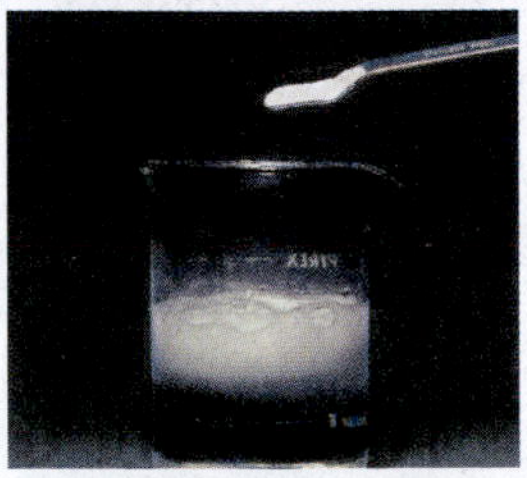
Iron nails change color when they react with oxygen to form rust.	Bubbles (gas) form when $CaCO_3$ reacts with acid.

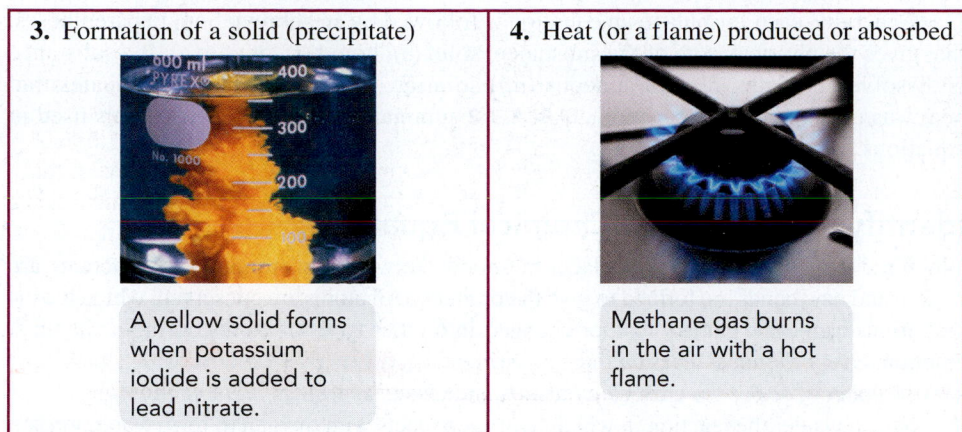

3. Formation of a solid (precipitate)

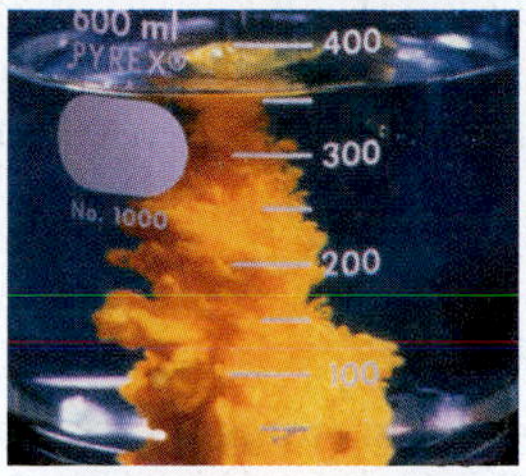
400
300
200

A yellow solid forms when potassium iodide is added to lead nitrate.

4. Heat (or a flame) produced or absorbed

Methane gas burns in the air with a hot flame.

Writing a Chemical Equation

When you build a model airplane, prepare a new recipe, or mix a medication, you follow a set of directions. These directions tell you what materials to use and the products you will obtain. In chemistry, a *chemical equation* tells us the materials we need and the products that will form.

Suppose you work in a bicycle shop, assembling wheels and frames into bicycles. You could represent this process by a simple equation:

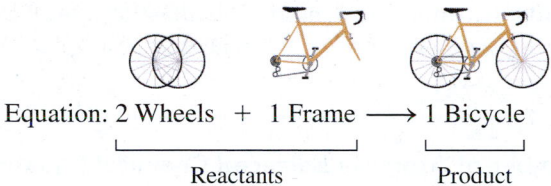

Equation: 2 Wheels + 1 Frame $\longrightarrow$ 1 Bicycle

Reactants (2 Wheels + 1 Frame) — Product (1 Bicycle)

When you burn charcoal in a grill, the carbon in the charcoal combines with oxygen to form carbon dioxide. We can represent this reaction by a chemical equation.

Reactants
Product

Equation: $C(s) + O_2(g) \xrightarrow{\Delta} CO_2(g)$

ENGAGE

What is the evidence for a chemical change in the reaction of carbon and oxygen to form carbon dioxide?

In a chemical equation, the formulas of the reactants are written on the left of the arrow and the formulas of the products on the right.

In a **chemical equation**, the formulas of the **reactants** are written on the left of the arrow and the formulas of the **products** on the right. When there are two or more formulas on the same side, they are separated by plus (+) signs. The chemical equation for burning carbon is *balanced* because there is one carbon atom and two oxygen atoms in both the reactants and the products.

TABLE 7.2 Some Symbols Used in Writing Equations

Symbol	Meaning
+	Separates two or more formulas
$\longrightarrow$	Reacts to form products
(s)	Solid
(l)	Liquid
(g)	Gas
(aq)	Aqueous
$\xrightarrow{\Delta}$	Reactants are heated

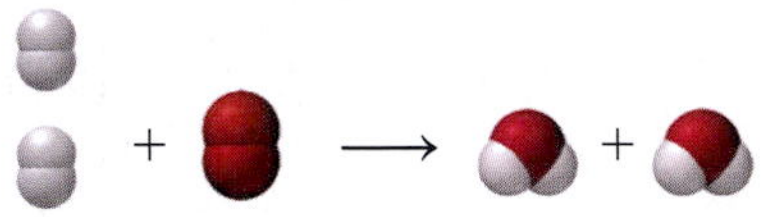

$2H_2(g) + O_2(g) \longrightarrow 2H_2O(g)$
Balanced

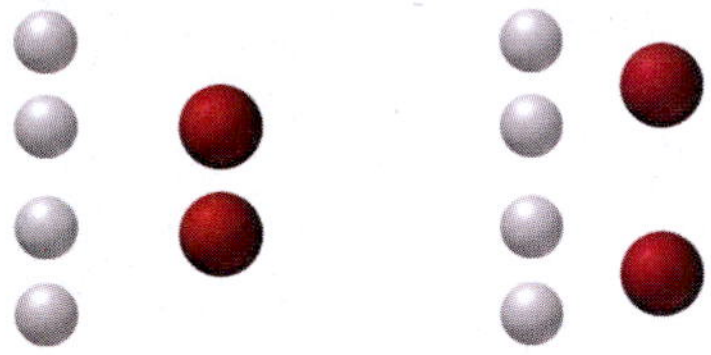

H O H O

Reactant atoms = Product atoms

Generally, each formula in an equation is followed by an abbreviation, in parentheses, that gives the physical state of the substance: solid (s), liquid (l), or gas (g). If a substance is dissolved in water, it is an aqueous (aq) solution. The delta sign (Δ) indicates that heat was used to start the reaction. **TABLE 7.2** summarizes some of the symbols used in equations.

Identifying a Balanced Chemical Equation

When a chemical reaction takes place, the bonds between the atoms of the reactants are broken and new bonds are formed to give the products. All atoms are conserved, which means that atoms cannot be gained, lost, or changed into other types of atoms during a chemical reaction. Every chemical reaction must be written as a **balanced equation**, which shows the same number of atoms for each element in the reactants as well as in the products.

Now consider the reaction in which hydrogen reacts with oxygen to form water, written as follows:

$$H_2(g) + O_2(g) \longrightarrow H_2O(g) \quad \text{Not balanced}$$

In the balanced equation, there are whole numbers called **coefficients** in front of the formulas. On the reactant side, the coefficient of 2 in front of the H_2 formula represents two molecules of hydrogen, which is 4 atoms of H. A coefficient of 1 is understood for O_2, which gives 2 atoms of O. On the product side, the coefficient of 2 in front of the H_2O formula represents 2 molecules of water. Because the coefficient of 2 multiplies all the atoms in H_2O, there are 4 hydrogen atoms and 2 oxygen atoms in the products. Because there are the same number of hydrogen atoms and oxygen atoms in the reactants as in the products, we know that the equation is balanced. This illustrates the *Law of Conservation of Matter*, which states that matter cannot be created or destroyed during a chemical reaction.

▶SAMPLE PROBLEM 7.1 Number of Atoms in Balanced Chemical Equations

TRY IT FIRST

Indicate the number of each type of atom in the following balanced chemical equation:

$$Fe_2S_3(s) + 6HCl(aq) \longrightarrow 2FeCl_3(aq) + 3H_2S(g)$$

	Reactants	Products
Fe		
S		
H		
Cl		

SOLUTION

The total number of atoms in each formula is obtained by multiplying the coefficient by each subscript in a chemical formula.

	Reactants	Products
Fe	2 (1 × 2)	2 (2 × 1)
S	3 (1 × 3)	3 (3 × 1)
H	6 (6 × 1)	6 (3 × 2)
Cl	6 (6 × 1)	6 (2 × 3)

STUDY CHECK 7.1

a. When ethane, C_2H_6, burns in oxygen, the products are carbon dioxide and water. The balanced chemical equation is written as

$$2C_2H_6(g) + 7O_2(g) \xrightarrow{\Delta} 4CO_2(g) + 6H_2O(g)$$

Indicate the number of each type of atom in the reactants and in the products.

b. Iron(III) oxide reacts with carbon monoxide to produce iron and carbon dioxide. The balanced chemical equation is written as

$$Fe_2O_3(s) + 3CO(g) \longrightarrow 2Fe(s) + 3CO_2(g)$$

Indicate the number of each type of atom in the reactants and in the products.

ANSWER

a. In both the reactants and products, there are 4 C atoms, 12 H atoms, and 14 O atoms.
b. In both the reactants and products, there are 2 Fe atoms, 6 O atoms, and 3 C atoms.

TEST
Try Practice Problems 7.1 and 7.2

Balancing a Chemical Equation

CORE CHEMISTRY SKILL
Balancing a Chemical Equation

The chemical reaction that occurs in the flame of a gas burner you use in the laboratory or a gas cooktop is the reaction of methane gas, CH_4, and oxygen to produce carbon dioxide and water. We now show the process of balancing a chemical equation in Sample Problem 7.2.

▶SAMPLE PROBLEM 7.2 Writing and Balancing a Chemical Equation

TRY IT FIRST

The chemical reaction of methane gas (CH_4) and oxygen gas (O_2) produces the gases carbon dioxide (CO_2) and water (H_2O). Write a balanced chemical equation for this reaction.

SOLUTION

	Given	Need	Connect
ANALYZE THE PROBLEM	reactants, products	balanced equation	equal numbers of atoms in reactants and products

STEP 1 **Write an equation using the correct formulas for the reactants and products.**

$$CH_4(g) + O_2(g) \xrightarrow{\Delta} CO_2(g) + H_2O(g)$$

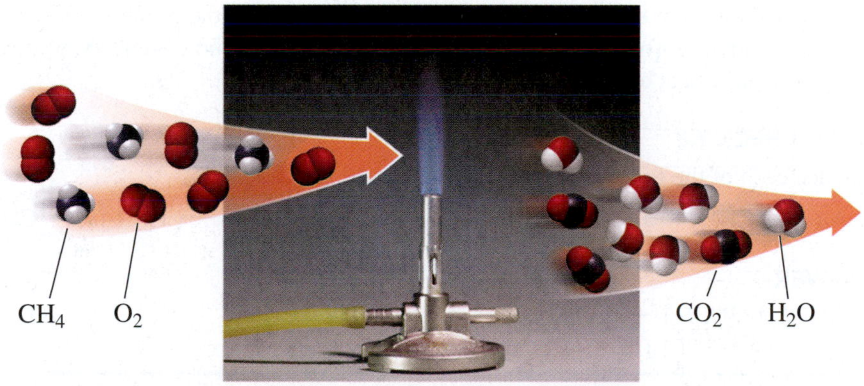

CH4
O2
CO2
H2O

In a balanced chemical equation, the number of each type of atom must be the same in the reactants and in the products.

STEP 2 **Count the atoms of each element in the reactants and products.** When we count the atoms on the reactant side and the atoms on the product side, we see that there are more H atoms in the reactants and more O atoms in the products.

$$CH_4(g) + O_2(g) \xrightarrow{\Delta} CO_2(g) + H_2O(g)$$

Reactants	Products	
1 C atom	1 C atom	Balanced
4 H atoms	2 H atoms	Not balanced
2 O atoms	3 O atoms	Not balanced

STEP 3 **Use coefficients to balance each element.** We will start by balancing the H atoms in CH_4 because it has the most atoms. By placing a coefficient of 2 in front of the formula for H_2O, a total of 4 H atoms in the products is obtained. *Only use*

coefficients to balance an equation. Do not change any of the subscripts: This would alter the chemical formula of a reactant or product.

$$CH_4(g) + O_2(g) \xrightarrow{\Delta} CO_2(g) + 2H_2O(g)$$

Reactants	Products	
1 C atom	1 C atom	Balanced
4 H atoms	4 H atoms	Balanced
2 O atoms	4 O atoms	Not balanced

We can balance the O atoms on the reactant side by placing a coefficient of 2 in front of the formula O_2. There are now 4 O atoms in both the reactants and products.

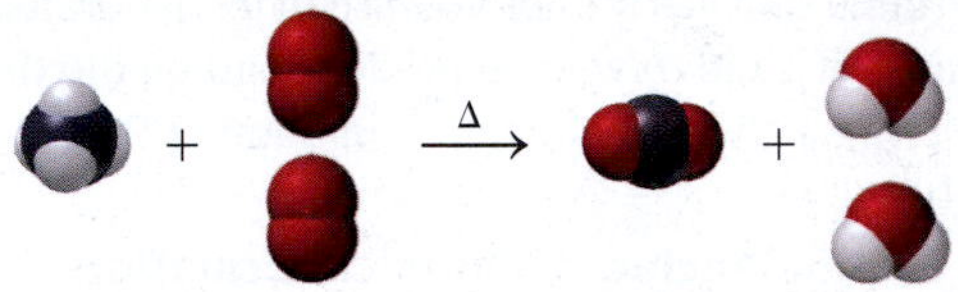

$$CH_4(g) + 2O_2(g) \xrightarrow{\Delta} CO_2(g) + 2H_2O(g)$$ Balanced

ENGAGE

How do you check that a chemical equation is balanced?

STEP 4 Check the final equation to confirm it is balanced.

$$CH_4(g) + 2O_2(g) \xrightarrow{\Delta} CO_2(g) + 2H_2O(g)$$ The equation is balanced.

Reactants	Products	
1 C atom	1 C atom	Balanced
4 H atoms	4 H atoms	Balanced
4 O atoms	4 O atoms	Balanced

In a balanced chemical equation, the coefficients must be the *lowest possible whole numbers*. Suppose you had obtained the following for the balanced equation:

$$2CH_4(g) + 4O_2(g) \xrightarrow{\Delta} 2CO_2(g) + 4H_2O(g)$$ Incorrect

Although there are equal numbers of atoms on both sides of the equation, this is not written correctly. To obtain coefficients that are the lowest whole numbers, we divide all the coefficients by 2.

STUDY CHECK 7.2

Balance each of the following chemical equations:

a. $Al(s) + Cl_2(g) \longrightarrow AlCl_3(s)$ **b.** $Cr_2O_3(s) + CCl_4(l) \longrightarrow CrCl_3(s) + COCl_2(g)$

ANSWER

a. $2Al(s) + 3Cl_2(g) \longrightarrow 2AlCl_3(s)$
b. $Cr_2O_3(s) + 3CCl_4(l) \longrightarrow 2CrCl_3(s) + 3COCl_2(g)$

Equations with Polyatomic Ions

Sometimes an equation contains the same polyatomic ion in both the reactants and the products. Then we can balance the polyatomic ions as a group on both sides of the equation, as shown in Sample Problem 7.3.

▶ SAMPLE PROBLEM 7.3 Balancing Chemical Equations with Polyatomic Ions

TRY IT FIRST

Balance the following chemical equation:

$$Na_3PO_4(aq) + MgCl_2(aq) \longrightarrow Mg_3(PO_4)_2(s) + NaCl(aq)$$

SOLUTION

	Given	Need	Connect
ANALYZE THE PROBLEM	reactants, products	balanced equation	equal numbers of atoms in reactants and products

STEP 1 **Write an equation using the correct formulas for the reactants and products.**

$Na_3PO_4(aq) + MgCl_2(aq) \longrightarrow Mg_3(PO_4)_2(s) + NaCl(aq)$ Not balanced

STEP 2 **Count the atoms of each element in the reactants and products.** When we compare the number of ions in the reactants and products, we find that the equation is not balanced. In this equation, we can balance the phosphate ion as a group of atoms because it appears on both sides of the equation.

ENGAGE

What is the evidence for a chemical reaction when $Na_3PO_4(aq)$ and $MgCl_2(aq)$ are mixed?

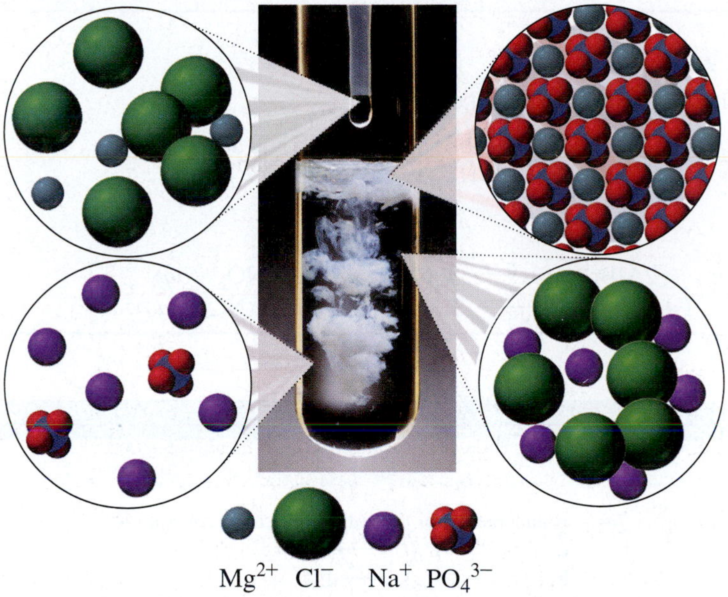
Mg^{2+} Cl^- Na^+ PO_4^{3-}

$Na_3PO_4(aq) + MgCl_2(aq) \longrightarrow Mg_3(PO_4)_2(s) + NaCl(aq)$

Reactants	Products	
3 Na^+	1 Na^+	Not balanced
1 $\mathbf{PO_4^{3-}}$	2 $\mathbf{PO_4^{3-}}$	Not balanced
1 Mg^{2+}	3 Mg^{2+}	Not balanced
2 Cl^-	1 Cl^-	Not balanced

STEP 3 **Use coefficients to balance each element.** We begin with the formula that has the highest subscript values, which in this equation is $Mg_3(PO_4)_2$. The subscript 3 in $Mg_3(PO_4)_2$ is used as a coefficient for $MgCl_2$ to balance magnesium. The subscript 2 in $Mg_3(PO_4)_2$ is used as a coefficient for Na_3PO_4 to balance the phosphate ion.

$2Na_3PO_4(aq) + 3MgCl_2(aq) \longrightarrow Mg_3(PO_4)_2(s) + NaCl(aq)$

Reactants	Products	
6 Na^+	1 Na^+	Not balanced
2 $\mathbf{PO_4^{3-}}$	2 $\mathbf{PO_4^{3-}}$	Balanced
3 Mg^{2+}	3 Mg^{2+}	Balanced
6 Cl^-	1 Cl^-	Not balanced

In the reactants and products, we see that the sodium and chloride ions are not yet balanced. A coefficient of 6 is placed in front of the NaCl to balance the equation.

$$2Na_3PO_4(aq) + 3MgCl_2(aq) \longrightarrow Mg_3(PO_4)_2(s) + 6NaCl(aq)$$

STEP 4 **Check the final equation to confirm it is balanced.**

$$2Na_3PO_4(aq) + 3MgCl_2(aq) \longrightarrow Mg_3(PO_4)_2(s) + 6NaCl(aq)$$ Balanced

Reactants	Products	
6 Na^+	6 Na^+	Balanced
2 $\mathbf{PO_4}^{3-}$	2 $\mathbf{PO_4}^{3-}$	Balanced
3 Mg^{2+}	3 Mg^{2+}	Balanced
6 Cl^-	6 Cl^-	Balanced

STUDY CHECK 7.3

Balance each of the following chemical equations:

a. $Pb(NO_3)_2(aq) + AlBr_3(aq) \longrightarrow PbBr_2(s) + Al(NO_3)_3(aq)$
b. $HNO_3(aq) + Fe_2(SO_4)_3(aq) \longrightarrow H_2SO_4(aq) + Fe(NO_3)_3(aq)$

ANSWER

a. $3Pb(NO_3)_2(aq) + 2AlBr_3(aq) \longrightarrow 3PbBr_2(s) + 2Al(NO_3)_3(aq)$
b. $6HNO_3(aq) + Fe_2(SO_4)_3(aq) \longrightarrow 3H_2SO_4(aq) + 2Fe(NO_3)_3(aq)$

TEST
Try Practice Problems 7.3 to 7.6

PRACTICE PROBLEMS

7.1 Equations for Chemical Reactions

7.1 Determine whether each of the following chemical equations is balanced or not balanced:
a. $S(s) + O_2(g) \longrightarrow SO_3(g)$
b. $2Ga(s) + 3Cl_2(g) \longrightarrow 2GaCl_3(s)$
c. $H_2(g) + O_2(g) \longrightarrow H_2O(g)$
d. $C_3H_8(g) + 5O_2(g) \xrightarrow{\Delta} 3CO_2(g) + 4H_2O(g)$

7.2 Determine whether each of the following chemical equations is balanced or not balanced:
a. $PCl_3(s) + Cl_2(g) \longrightarrow PCl_5(s)$
b. $CO(g) + 2H_2(g) \longrightarrow CH_4O(g)$
c. $2KClO_3(s) \longrightarrow 2KCl(s) + O_2(g)$
d. $Mg(s) + N_2(g) \longrightarrow Mg_3N_2(s)$

7.3 Balance each of the following chemical equations:
a. $N_2(g) + O_2(g) \longrightarrow NO(g)$
b. $HgO(s) \xrightarrow{\Delta} Hg(l) + O_2(g)$
c. $Fe(s) + O_2(g) \longrightarrow Fe_2O_3(s)$
d. $Na(s) + Cl_2(g) \longrightarrow NaCl(s)$

7.4 Balance each of the following chemical equations:
a. $Ca(s) + Br_2(l) \longrightarrow CaBr_2(s)$
b. $P_4(s) + O_2(g) \longrightarrow P_4O_{10}(s)$
c. $Sb_2S_3(s) + HCl(aq) \longrightarrow SbCl_3(aq) + H_2S(g)$
d. $Fe_2O_3(s) + C(s) \longrightarrow Fe(s) + CO(g)$

7.5 Balance each of the following chemical equations:
a. $Mg(s) + AgNO_3(aq) \longrightarrow Mg(NO_3)_2(aq) + Ag(s)$
b. $Al(s) + CuSO_4(aq) \longrightarrow Al_2(SO_4)_3(aq) + Cu(s)$
c. $Pb(NO_3)_2(aq) + NaCl(aq) \longrightarrow PbCl_2(s) + NaNO_3(aq)$
d. $HCl(aq) + Al(s) \longrightarrow H_2(g) + AlCl_3(aq)$

7.6 Balance each of the following chemical equations:
a. $HNO_3(aq) + Zn(s) \longrightarrow H_2(g) + Zn(NO_3)_2(aq)$
b. $H_2SO_4(aq) + Al(s) \longrightarrow H_2(g) + Al_2(SO_4)_3(aq)$
c. $K_2SO_4(aq) + BaCl_2(aq) \longrightarrow BaSO_4(s) + KCl(aq)$
d. $CaCO_3(s) \longrightarrow CaO(s) + CO_2(g)$

7.2 Types of Chemical Reactions

LEARNING GOAL Identify a chemical reaction as a combination, decomposition, single replacement, double replacement, or combustion.

A great number of chemical reactions occur in nature, in biological systems, and in the laboratory. However, there are some general patterns that help us classify most reactions into five general types.

Combination Reactions

In a **combination reaction**, two or more elements or compounds bond to form one product. For example, sulfur and oxygen combine to form the product sulfur dioxide.

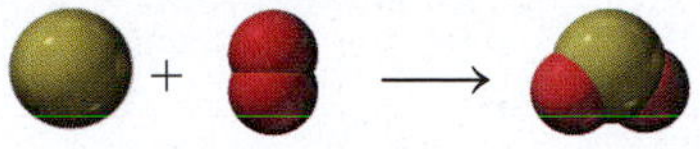

$$S(s) + O_2(g) \longrightarrow SO_2(g)$$

In **FIGURE 7.2**, the elements magnesium and oxygen combine to form a single product, which is the ionic compound magnesium oxide formed from Mg^{2+} and O^{2-} ions.

$$2Mg(s) + O_2(g) \longrightarrow 2MgO(s)$$

In other examples of combination reactions, elements or compounds combine to form a single product.

$$N_2(g) + 3H_2(g) \longrightarrow 2NH_3(g)$$
$$Cu(s) + S(s) \longrightarrow CuS(s)$$
$$MgO(s) + CO_2(g) \longrightarrow MgCO_3(s)$$

Combination

Two or more reactants	combine to yield	a single product
A + B	$\longrightarrow$	AB

CORE CHEMISTRY SKILL
Classifying Types of Chemical Reactions

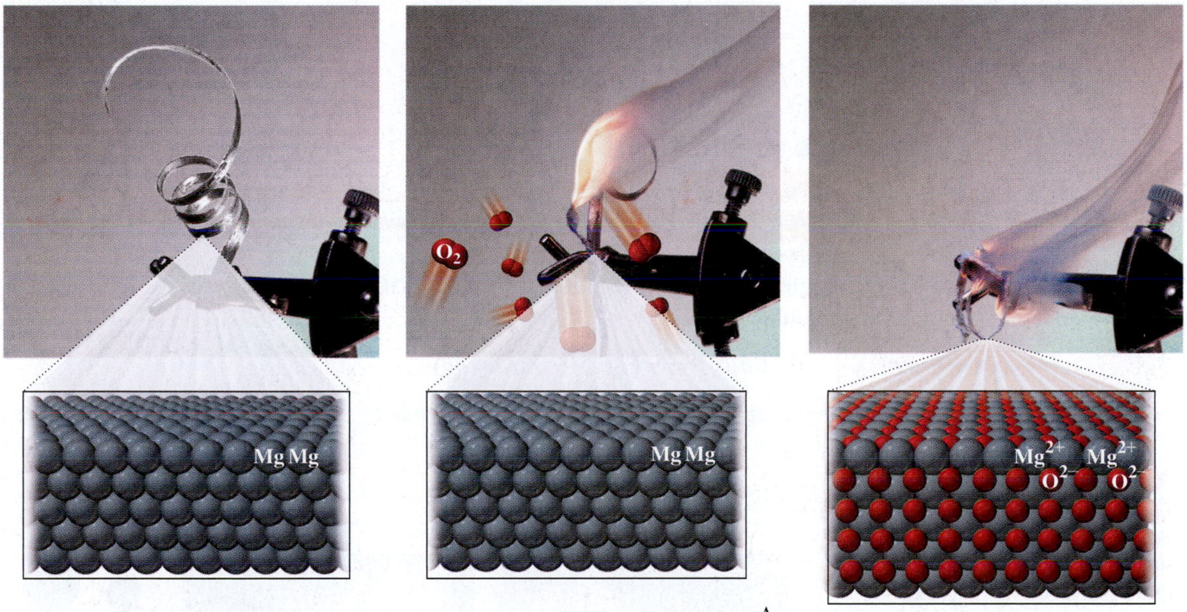

O_2
Mg Mg
Mg Mg
Mg^{2+} Mg^{2+}
O^{2-} O^{2-}

$$\underset{\text{Magnesium}}{2Mg(s)} + \underset{\text{Oxygen}}{O_2(g)} \xrightarrow{\Delta} \underset{\text{Magnesium oxide}}{2MgO(s)}$$

FIGURE 7.2 ▶ In a combination reaction, two or more substances combine to form one substance as product.

Q What happens to the atoms in the reactants in a combination reaction?

Decomposition Reactions

In a **decomposition reaction**, a reactant splits into two or more simpler products. For example, when mercury(II) oxide is heated, the compound breaks apart into mercury atoms and oxygen (see **FIGURE 7.3**).

$$2HgO(s) \xrightarrow{\Delta} 2Hg(l) + O_2(g)$$

In another example of a decomposition reaction, when calcium carbonate is heated, it breaks apart into simpler compounds of calcium oxide and carbon dioxide.

$$CaCO_3(s) \xrightarrow{\Delta} CaO(s) + CO_2(g)$$

Decomposition

A reactant	splits into	two or more products
AB	$\longrightarrow$	A + B

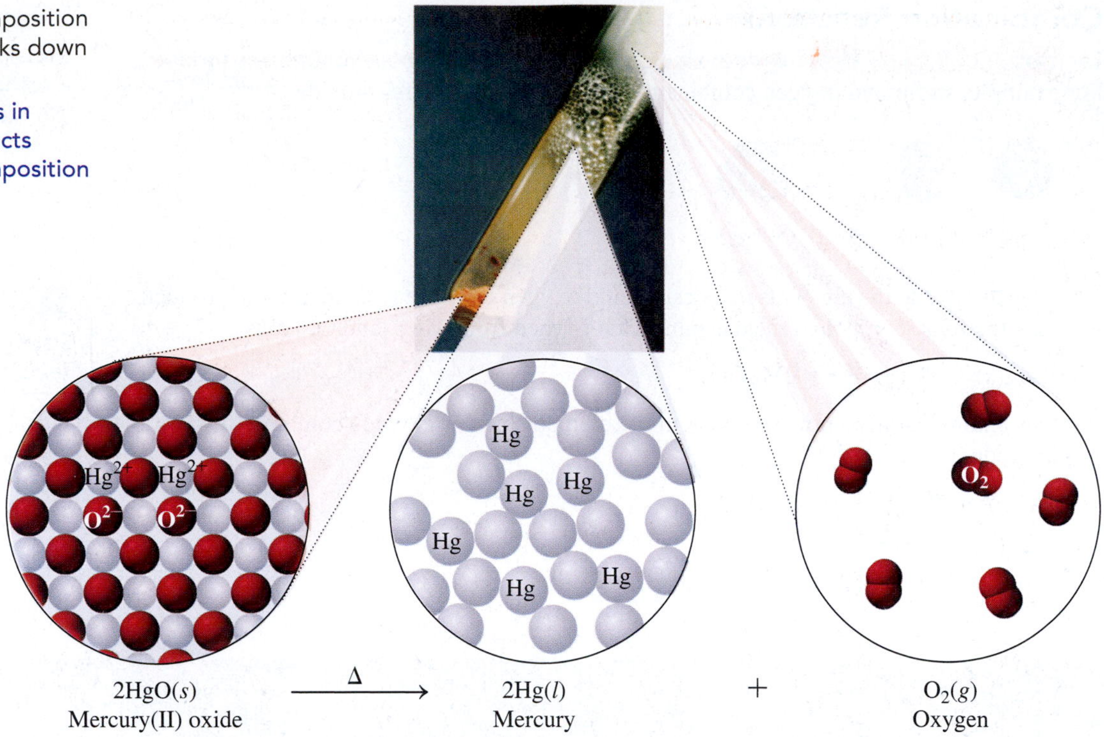

Hg^{2+} Hg^{2+}
O^{2-} O^{2-}
Hg
O_2
$2HgO(s)$ Mercury(II) oxide $\xrightarrow{\Delta}$ $2Hg(l)$ Mercury + $O_2(g)$ Oxygen

FIGURE 7.3 ▶ In a decomposition reaction, one reactant breaks down into two or more products.

Q How do the differences in the reactant and products classify this as a decomposition reaction?

Replacement Reactions

In a replacement reaction, elements in a compound are replaced by other elements. In a **single replacement reaction**, a reacting element switches places with an element in the other reacting compound.

Single replacement

One element replaces another element

A + BC ⟶ AC + B

In the single replacement reaction shown in **FIGURE 7.4**, zinc replaces hydrogen in hydrochloric acid, $HCl(aq)$.

$$Zn(s) + 2HCl(aq) \longrightarrow H_2(g) + ZnCl_2(aq)$$

In another single replacement reaction, chlorine replaces bromine in the compound potassium bromide.

$$Cl_2(g) + 2KBr(s) \longrightarrow 2KCl(s) + Br_2(l)$$

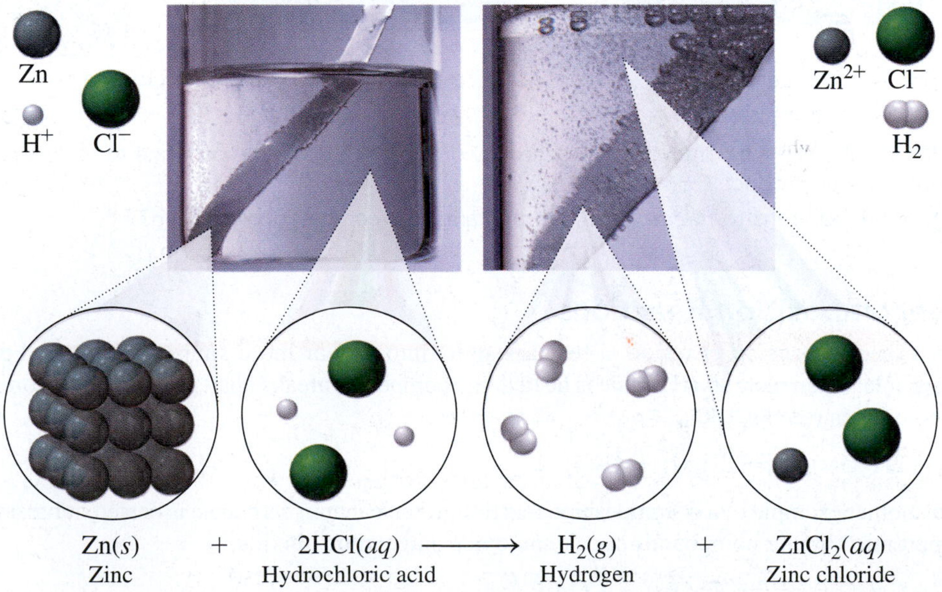

Zn
H^+ Cl^-
Zn^{2+} Cl^-
H_2
$Zn(s)$ Zinc + $2HCl(aq)$ Hydrochloric acid ⟶ $H_2(g)$ Hydrogen + $ZnCl_2(aq)$ Zinc chloride

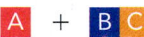

FIGURE 7.4 ▶ In a single replacement reaction, an atom or ion replaces an atom or ion in a compound.

Q What changes in the formulas of the reactants identify this equation as a single replacement?

In a **double replacement reaction**, the positive ions in the reacting compounds switch places.

In the reaction shown in **FIGURE 7.5**, barium ions change places with sodium ions in the reactants to form sodium chloride and a white solid precipitate of barium sulfate. The formulas of the products depend on the charges of the ions.

$$BaCl_2(aq) + Na_2SO_4(aq) \longrightarrow BaSO_4(s) + 2NaCl(aq)$$

When hydrochloric acid (HCl) and sodium hydroxide react, sodium and hydrogen ions switch places, forming water and sodium chloride.

$$HCl(aq) + NaOH(aq) \longrightarrow H_2O(l) + NaCl(aq)$$

Double replacement

Two elements replace each other

A B + C D ⟶ A D + C B

ENGAGE

How can you distinguish a single replacement reaction from a double replacement reaction?

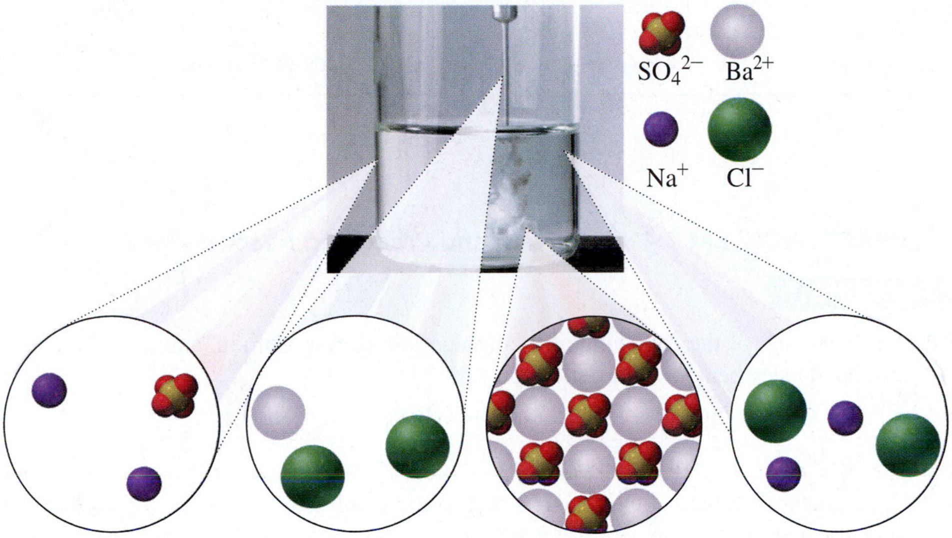
SO_4^{2-}
Ba^{2+}
Na^+
Cl^-

$Na_2SO_4(aq)$ Sodium sulfate + $BaCl_2(aq)$ Barium chloride ⟶ $BaSO_4(s)$ Barium sulfate + $2NaCl(aq)$ Sodium chloride

FIGURE 7.5 ▶ In a double replacement reaction, the positive ions in the reactants replace each other.

Q How do the changes in the formulas of the reactants identify this equation as a double replacement reaction?

Combustion Reactions

The burning of a candle and the burning of fuel in the engine of a car are examples of combustion reactions. In a **combustion reaction**, a carbon-containing compound, usually a fuel, burns in oxygen gas to produce the gases carbon dioxide (CO_2), water (H_2O), and energy in the form of heat or a flame. For example, methane gas (CH_4) undergoes combustion when used to cook our food on a gas cooktop and to heat our homes. In the equation for the combustion of methane, each element in the fuel (CH_4) forms a compound with oxygen.

$$CH_4(g) + 2O_2(g) \xrightarrow{\Delta} CO_2(g) + 2H_2O(g) + \text{energy}$$

Methane

The balanced equation for the combustion of propane (C_3H_8) is

$$C_3H_8(g) + 5O_2(g) \xrightarrow{\Delta} 3CO_2(g) + 4H_2O(g) + \text{energy}$$

Propane is the fuel used in portable heaters and gas barbecues. Gasoline, a mixture of liquid carbon compounds, is the fuel that powers our cars, lawn mowers, and snow blowers.

TABLE 7.3 summarizes the reaction types and gives examples.

In a combustion reaction, a candle burns using the oxygen in the air.

TABLE 7.3 Summary of Reaction Types

Reaction Type	Example
Combination $A + B \longrightarrow AB$	$Ca(s) + Cl_2(g) \longrightarrow CaCl_2(s)$
Decomposition $AB \longrightarrow A + B$	$Fe_2S_3(s) \longrightarrow 2Fe(s) + 3S(s)$
Single Replacement $A + BC \longrightarrow AC + B$	$Cu(s) + 2AgNO_3(aq) \longrightarrow 2Ag(s) + Cu(NO_3)_2(aq)$
Double Replacement $AB + CD \longrightarrow AD + CB$	$BaCl_2(aq) + K_2SO_4(aq) \longrightarrow BaSO_4(s) + 2KCl(aq)$
Combustion $C_XH_Y + ZO_2(g) \xrightarrow{\Delta} XCO_2(g) + \frac{Y}{2}H_2O(g) + \text{energy}$	$CH_4(g) + 2O_2(g) \xrightarrow{\Delta} CO_2(g) + 2H_2O(g) + \text{energy}$

▶SAMPLE PROBLEM 7.4 Identifying and Predicting Reaction Types

TRY IT FIRST

1. Classify each of the following as a combination, decomposition, single replacement, double replacement, or combustion reaction:
   a. $2Fe_2O_3(s) + 3C(s) \longrightarrow 3CO_2(g) + 4Fe(s)$
   b. $2KClO_3(s) \xrightarrow{\Delta} 2KCl(s) + 3O_2(g)$
2. Using Table 7.3, predict the products that would result from the following combustion reaction and balance:

$$C_2H_4(g) + O_2(g) \xrightarrow{\Delta}$$

SOLUTION

1. a. In this single replacement reaction, a C atom replaces Fe in Fe_2O_3 to form the compound CO_2 and Fe atoms.
   b. When one reactant breaks down to produce two products, the reaction is decomposition.
2. In a combustion reaction, a carbon compound reacts with oxygen to produce carbon dioxide, water, and energy.

$$C_2H_4(g) + O_2(g) \xrightarrow{\Delta} 2CO_2(g) + 2H_2O(g) + \text{energy}$$

STUDY CHECK 7.4

1. Write the balanced chemical equation for each of the following reactions, using the correct chemical formulas of the reactants and products, and identify the reaction type:
   a. Nitrogen gas (N_2) and oxygen gas (O_2) react to form nitrogen dioxide gas.
   b. Aqueous sodium sulfide and aqueous nickel(II) chloride react to form solid nickel(II) sulfide and aqueous sodium chloride.
2. Using Table 7.3, predict the products that would result from the following reaction and balance:

   single replacement: $HCl(aq) + Sr(s) \longrightarrow$

ANSWER

1. a. $N_2(g) + 2O_2(g) \longrightarrow 2NO_2(g)$ Combination
   b. $Na_2S(aq) + NiCl_2(aq) \longrightarrow NiS(s) + 2NaCl(aq)$ Double replacement
2. $2HCl(aq) + Sr(s) \longrightarrow H_2(g) + SrCl_2(aq)$

TEST

Try Practice Problems 7.7 to 7.12

PRACTICE PROBLEMS

7.2 Types of Chemical Reactions

7.7 Classify each of the following as a combination, decomposition, single replacement, double replacement, or combustion reaction:

a. $2Al_2O_3(s) \xrightarrow{\Delta} 4Al(s) + 3O_2(g)$
b. $Br_2(l) + BaI_2(s) \longrightarrow BaBr_2(s) + I_2(s)$
c. $2C_2H_2(g) + 5O_2(g) \xrightarrow{\Delta} 4CO_2(g) + 2H_2O(g)$
d. $BaCl_2(aq) + K_2CO_3(aq) \longrightarrow BaCO_3(s) + 2KCl(aq)$
e. $Pb(s) + O_2(g) \longrightarrow PbO_2(s)$

7.8 Classify each of the following as a combination, decomposition, single replacement, double replacement, or combustion reaction:

a. $H_2(g) + Br_2(l) \longrightarrow 2HBr(g)$
b. $AgNO_3(aq) + NaCl(aq) \longrightarrow AgCl(s) + NaNO_3(aq)$
c. $2H_2O_2(aq) \longrightarrow 2H_2O(l) + O_2(g)$
d. $Zn(s) + CuCl_2(aq) \longrightarrow Cu(s) + ZnCl_2(aq)$
e. $C_5H_8(g) + 7O_2(g) \xrightarrow{\Delta} 5CO_2(g) + 4H_2O(g)$

7.9 Classify each of the following as a combination, decomposition, single replacement, double replacement, or combustion reaction:

a. $4Fe(s) + 3O_2(g) \longrightarrow 2Fe_2O_3(s)$
b. $Mg(s) + 2AgNO_3(aq) \longrightarrow 2Ag(s) + Mg(NO_3)_2(aq)$
c. $CuCO_3(s) \xrightarrow{\Delta} CuO(s) + CO_2(g)$
d. $Al_2(SO_4)_3(aq) + 6KOH(aq) \longrightarrow 2Al(OH)_3(s) + 3K_2SO_4(aq)$
e. $C_4H_8(g) + 6O_2(g) \xrightarrow{\Delta} 4CO_2(g) + 4H_2O(g)$

7.10 Classify each of the following as a combination, decomposition, single replacement, double replacement, or combustion reaction:

a. $CuO(s) + 2HCl(aq) \longrightarrow CuCl_2(aq) + H_2O(l)$
b. $2Al(s) + 3Br_2(l) \longrightarrow 2AlBr_3(s)$
c. $C_3H_8(g) + 10O_2(g) \xrightarrow{\Delta} 8CO_2(g) + 4H_2O(g)$
d. $Fe_2O_3(s) + 3C(s) \longrightarrow 2Fe(s) + 3CO(g)$
e. $C_6H_{12}O_6(aq) \longrightarrow 2C_2H_6O(aq) + 2CO_2(g)$

7.11 Using Table 7.3, predict the products that would result from each of the following reactions and balance:

a. combination: $Mg(s) + Cl_2(g) \longrightarrow$
b. decomposition: $HBr(g) \longrightarrow$
c. single replacement: $Mg(s) + Zn(NO_3)_2(aq) \longrightarrow$
d. double replacement: $K_2S(aq) + Pb(NO_3)_2(aq) \longrightarrow$
e. combustion: $C_2H_6(g) + O_2(g) \xrightarrow{\Delta}$

7.12 Using Table 7.3, predict the products that would result from each of the following reactions and balance:

a. combination: $Ca(s) + O_2(g) \longrightarrow$
b. combustion: $C_6H_6(g) + O_2(g) \xrightarrow{\Delta}$
c. decomposition: $PbO_2(s) \xrightarrow{\Delta}$
d. single replacement: $KI(s) + Cl_2(g) \longrightarrow$
e. double replacement: $CuCl_2(aq) + Na_2S(aq) \longrightarrow$

7.3 Oxidation–Reduction Reactions

LEARNING GOAL Define the terms oxidation and reduction; identify the reactants oxidized and reduced.

Rust forms when the oxygen in the air reacts with iron.

Perhaps you have never heard of an oxidation and reduction reaction. However, this type of reaction has many important applications in your everyday life. When you see a rusty nail, tarnish on a silver spoon, or corrosion on metal, you are observing oxidation.

$$4Fe(s) + 3O_2(g) \longrightarrow \underset{\text{Rust}}{2Fe_2O_3(s)} \quad \text{Fe is oxidized}$$

When we turn the lights on in our automobiles, an oxidation–reduction reaction within the car battery provides the electricity. On a cold, wintry day, we might build a fire. As the wood burns, oxygen combines with carbon and hydrogen to produce carbon dioxide, water, and heat. In the previous section, we called this a combustion reaction, but it is also an *oxidation–reduction reaction*. When we eat foods with starches in them, the starches break down to give glucose, which is oxidized in our cells to give us energy along with carbon dioxide and water. Every breath we take provides oxygen to carry out oxidation in our cells.

$$\underset{\text{Glucose}}{C_6H_{12}O_6(aq)} + 6O_2(g) \longrightarrow 6CO_2(g) + 6H_2O(l) + \text{energy}$$

Oxidation–Reduction Reactions

In an **oxidation–reduction reaction** (*redox*), electrons are transferred from one substance to another. If one substance loses electrons, another substance must gain electrons. **Oxidation** is defined as the *loss* of electrons; **reduction** is the *gain* of electrons.

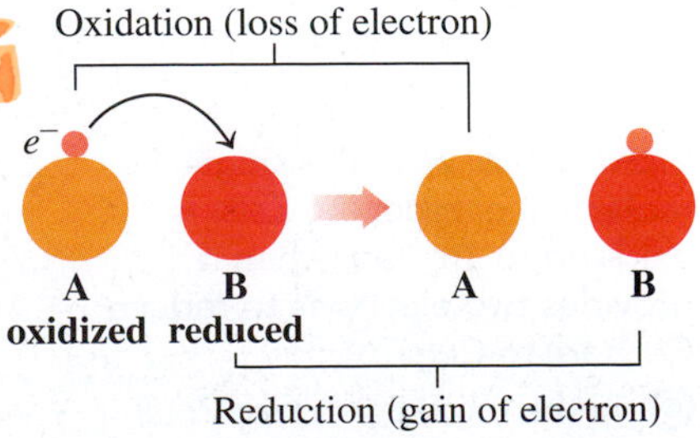
Oxidation (loss of electron)
e⁻
A
B
A
B
oxidized
reduced
Reduction (gain of electron)

One way to remember these definitions is to use the following acronym:

OIL RIG
Oxidation **I**s **L**oss of electrons
Reduction **I**s **G**ain of electrons

The green patina on copper is due to oxidation.

In general, atoms of metals lose electrons to form positive ions, whereas nonmetals gain electrons to form negative ions. Now we can say that metals are oxidized and nonmetals are reduced.

The green color that appears on copper surfaces from weathering, known as *patina*, is a mixture of $CuCO_3$ and CuO. We can now look at the oxidation and reduction reactions that take place when copper metal reacts with oxygen in the air to produce copper(II) oxide.

$$2Cu(s) + O_2(g) \longrightarrow 2CuO(s)$$

The element Cu in the reactants has a charge of 0, but in the CuO product, it is present as Cu^{2+}, which has a 2+ charge. Because the Cu atom lost two electrons, the charge is more positive. This means that Cu was oxidized in the reaction.

$$Cu^0(s) \longrightarrow Cu^{2+}(s) + 2\,e^- \quad \text{Oxidation: loss of electrons by Cu}$$

At the same time, the element O in the reactants has a charge of 0, but in the CuO product, it is present as O^{2-}, which has a 2− charge. Because the O atom has gained two electrons, the charge is more negative. This means that O was reduced in the reaction.

$$O_2{}^0(g) + 4\,e^- \longrightarrow 2O^{2-}(s) \quad \text{Reduction: gain of electrons by O}$$

CORE CHEMISTRY SKILL

Identifying Oxidized and Reduced Substances

Thus, the overall equation for the formation of CuO involves an oxidation and a reduction that occur simultaneously. In every oxidation and reduction, the number of electrons lost must be equal to the number of electrons gained. Therefore, we multiply the oxidation reaction of Cu by 2. Canceling the 4 e^- on each side, we obtain the overall oxidation–reduction equation for the formation of CuO.

$$2Cu(s) \longrightarrow 2Cu^{2+}(s) + \cancel{4\,e^-} \quad \text{Oxidation}$$

$$O_2(g) + \cancel{4\,e^-} \longrightarrow 2O^{2-}(s) \quad \text{Reduction}$$

$$2Cu(s) + O_2(g) \longrightarrow 2CuO(s) \quad \text{Oxidation–reduction equation}$$

As we see in the next reaction between zinc and copper(II) sulfate, there is always an oxidation with every reduction (see **FIGURE 7.6**). We write the equation to show the atoms and ions.

$$\mathbf{Zn}(s) + \mathbf{Cu^{2+}}(aq) + SO_4{}^{2-}(aq) \longrightarrow \mathbf{Zn^{2+}}(aq) + SO_4{}^{2-}(aq) + \mathbf{Cu}(s)$$

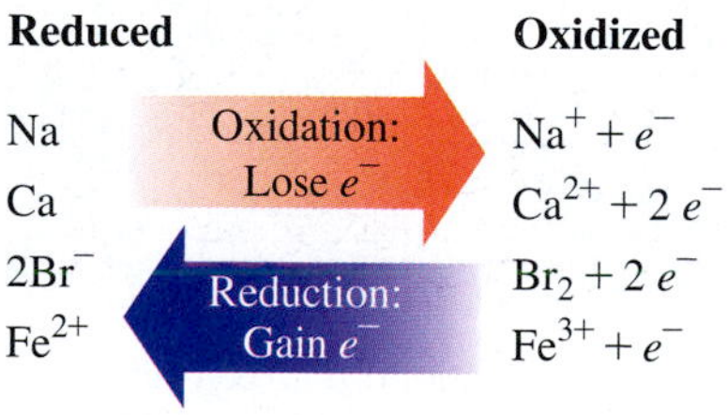

Reduced
Oxidized
Na
Ca
2Br$^-$
Fe^{2+}
Oxidation: Lose e^-
Reduction: Gain e^-
Na$^+$ + e^-
Ca^{2+} + 2 e^-
Br$_2$ + 2 e^-
Fe^{3+} + e^-

Oxidation is a loss of electrons; reduction is a gain of electrons.

In this reaction, Zn atoms lose two electrons to form Zn^{2+}. The increase in positive charge indicates that Zn is oxidized. At the same time, Cu^{2+} gains two electrons. The decrease in charge indicates that Cu is reduced. The $SO_4{}^{2-}$ ions are *spectator ions*, which are present in both the reactants and products and do not change.

$$Zn(s) \longrightarrow Zn^{2+}(aq) + 2\,e^- \quad \text{Oxidation of Zn}$$

$$Cu^{2+}(aq) + 2\,e^- \longrightarrow Cu(s) \quad \text{Reduction of } Cu^{2+}$$

In this single replacement reaction, zinc was oxidized and copper(II) ion was reduced.

ENGAGE

How can you determine that Cu^{2+} is reduced in this reaction?

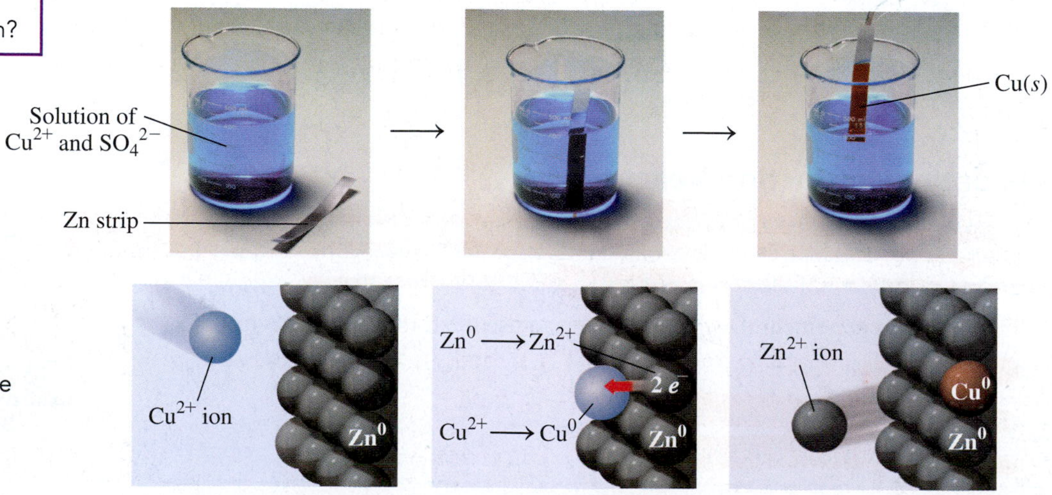

Solution of Cu^{2+} and $SO_4{}^{2-}$
Zn strip
Cu(s)
Cu^{2+} ion
Zn^0
$Zn^0 \longrightarrow Zn^{2+}$
2 e^-
$Cu^{2+} \longrightarrow Cu^0$
Zn^{2+} ion
Cu^0

FIGURE 7.6 ▶ In this single replacement reaction, Zn(s) is oxidized to $Zn^{2+}(aq)$ when it provides two electrons to reduce $Cu^{2+}(aq)$ to Cu(s).

Q In the oxidation, does Zn(s) lose or gain electrons?

Oxidation and Reduction in Biological Systems

Oxidation may also involve the addition of oxygen or the loss of hydrogen, and reduction may involve the loss of oxygen or the gain of hydrogen. In the cells of the body, oxidation of organic (carbon) compounds involves the transfer of hydrogen atoms (H), which are composed of electrons and protons. For example, the oxidation of a typical biochemical molecule can involve the transfer of two hydrogen atoms (or $2H^+$ and $2\ e^-$) to a hydrogen ion acceptor such as the coenzyme FAD (flavin adenine dinucleotide). The coenzyme is reduced to $FADH_2$.

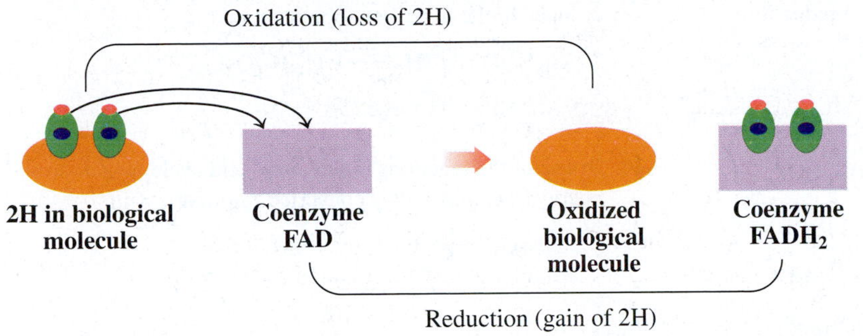

Oxidation (loss of 2H)
2H in biological molecule
Coenzyme FAD
Oxidized biological molecule
Coenzyme FADH₂
Reduction (gain of 2H)

In many biochemical oxidation–reduction reactions, the transfer of hydrogen atoms is necessary for the production of energy in the cells. For example, methyl alcohol (CH_4O), a poisonous substance, is metabolized in the body by the following reactions:

$$\underset{\text{Methyl alcohol}}{CH_4O} \longrightarrow \underset{\text{Formaldehyde}}{CH_2O} + 2H \quad \text{Oxidation: loss of H atoms}$$

The formaldehyde can be oxidized further, this time by the addition of oxygen, to produce formic acid.

$$\underset{\text{Formaldehyde}}{2CH_2O} + O_2 \longrightarrow \underset{\text{Formic acid}}{2CH_2O_2} \quad \text{Oxidation: addition of O atoms}$$

Finally, formic acid is oxidized to carbon dioxide and water.

$$\underset{\text{Formic acid}}{2CH_2O_2} + O_2 \longrightarrow 2CO_2 + 2H_2O \quad \text{Oxidation: addition of O atoms}$$

The intermediate products of the oxidation of methyl alcohol are quite toxic, causing blindness and possibly death, as they interfere with key reactions in the cells of the body.

In summary, we find that the particular definition of oxidation and reduction we use depends on the process that occurs in the reaction. All these definitions are summarized in **TABLE 7.4**. Oxidation always involves a loss of electrons, but it may also be seen as an addition of oxygen or the loss of hydrogen atoms. A reduction always involves a gain of electrons and may also be seen as the loss of oxygen or the gain of hydrogen.

TABLE 7.4 Characteristics of Oxidation and Reduction

Always Involves	May Involve
Oxidation	
Loss of electrons	Addition of oxygen Loss of hydrogen
Reduction	
Gain of electrons	Loss of oxygen Gain of hydrogen

▶ SAMPLE PROBLEM 7.5 Oxidation and Reduction

TRY IT FIRST

For each of the following, determine the number of electrons lost or gained and identify as an oxidation or a reduction:

a. $Mg^{2+}(aq) \longrightarrow Mg(s)$ **b.** $Fe(s) \longrightarrow Fe^{3+}(aq)$

SOLUTION

a. $Mg^{2+}(aq) + 2\ e^- \longrightarrow Mg(s)$ Reduction **b.** $Fe(s) \longrightarrow Fe^{3+}(aq) + 3\ e^-$ Oxidation

STUDY CHECK 7.5

Identify each of the following as an oxidation or a reduction:

a. $Co^{3+}(aq) + e^- \longrightarrow Co^{2+}(aq)$ **b.** $2I^-(aq) \longrightarrow I_2(s) + 2\ e^-$

ANSWER

a. reduction **b.** oxidation

TEST

Try Practice Problems 7.13 to 7.20

PRACTICE PROBLEMS

7.3 Oxidation–Reduction Reactions

7.13 Identify each of the following as an oxidation or a reduction:
- **a.** $Na^+(aq) + e^- \longrightarrow Na(s)$
- **b.** $Ni(s) \longrightarrow Ni^{2+}(aq) + 2\,e^-$
- **c.** $Cr^{3+}(aq) + 3\,e^- \longrightarrow Cr(s)$
- **d.** $2H^+(aq) + 2\,e^- \longrightarrow H_2(g)$

7.14 Identify each of the following as an oxidation or a reduction:
- **a.** $O_2(g) + 4\,e^- \longrightarrow 2O^{2-}(aq)$
- **b.** $Ag(s) \longrightarrow Ag^+(aq) + e^-$
- **c.** $Fe^{3+}(aq) + e^- \longrightarrow Fe^{2+}(aq)$
- **d.** $2Br^-(aq) \longrightarrow Br_2(l) + 2\,e^-$

7.15 In each of the following, identify the reactant that is oxidized and the reactant that is reduced:
- **a.** $Zn(s) + Cl_2(g) \longrightarrow ZnCl_2(s)$
- **b.** $Cl_2(g) + 2NaBr(aq) \longrightarrow 2NaCl(aq) + Br_2(l)$
- **c.** $2PbO(s) \longrightarrow 2Pb(s) + O_2(g)$
- **d.** $2Fe^{3+}(aq) + Sn^{2+}(aq) \longrightarrow 2Fe^{2+}(aq) + Sn^{4+}(aq)$

7.16 In each of the following, identify the reactant that is oxidized and the reactant that is reduced:
- **a.** $2Li(s) + F_2(g) \longrightarrow 2LiF(s)$
- **b.** $Cl_2(g) + 2KI(aq) \longrightarrow I_2(s) + 2KCl(aq)$
- **c.** $2Al(s) + 3Sn^{2+}(aq) \longrightarrow 3Sn(s) + 2Al^{3+}(aq)$
- **d.** $Fe(s) + CuSO_4(aq) \longrightarrow Cu(s) + FeSO_4(aq)$

Clinical Applications

7.17 In the mitochondria of human cells, energy is provided by the oxidation and reduction reactions of the iron ions in the cytochromes in electron transport. Identify each of the following as an oxidation or a reduction:
- **a.** $Fe^{3+}(aq) + e^- \longrightarrow Fe^{2+}(aq)$
- **b.** $Fe^{2+}(aq) \longrightarrow Fe^{3+}(aq) + e^-$

7.18 Chlorine (Cl_2) is a strong germicide used to disinfect drinking water and to kill microbes in swimming pools. If the product is Cl^-, was the elemental chlorine oxidized or reduced?

7.19 When linoleic acid, an unsaturated fatty acid, reacts with hydrogen, it forms a saturated fatty acid. Is linoleic acid oxidized or reduced in the hydrogenation reaction?

$$\underset{\text{Linoleic acid}}{C_{18}H_{32}O_2} + 2H_2 \longrightarrow C_{18}H_{36}O_2$$

7.20 In one of the reactions in the citric acid cycle, which provides energy, succinic acid is converted to fumaric acid.

$$\underset{\text{Succinic acid}}{C_4H_6O_4} \longrightarrow \underset{\text{Fumaric acid}}{C_4H_4O_4} + 2H$$

The reaction is accompanied by a coenzyme, flavin adenine dinucleotide (FAD).

$$FAD + 2H \longrightarrow FADH_2$$

- **a.** Is succinic acid oxidized or reduced?
- **b.** Is FAD oxidized or reduced?
- **c.** Why would the two reactions occur together?

REVIEW
- Writing Numbers in Scientific Notation (1.5)
- Counting Significant Figures (2.2)
- Writing Conversion Factors from Equalities (2.5)
- Using Conversion Factors (2.6)

7.4 The Mole

LEARNING GOAL Use Avogadro's number to determine the number of particles in a given number of moles.

At the grocery store, you buy eggs by the dozen or soda by the case. In an office-supply store, pencils are ordered by the gross and paper by the ream. Common terms *dozen*, *case*, *gross*, and *ream* are used to count the number of items present. For example, when you buy a dozen eggs, you know you will get 12 eggs in the carton.

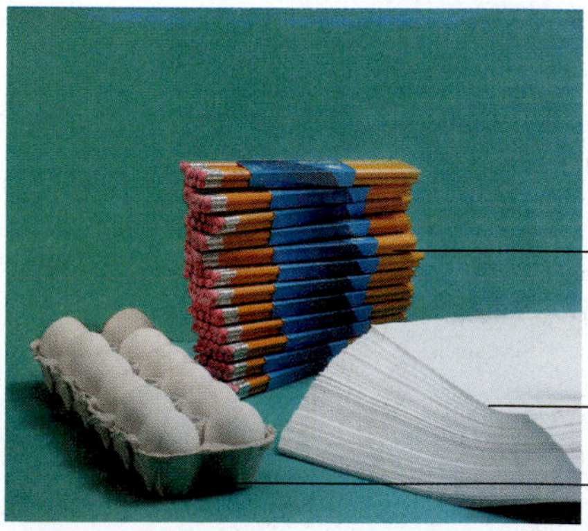
144 pencils = 1 gross
500 sheets = 1 ream
12 eggs = 1 dozen

Collections of items include dozen, gross, ream, and mole.

Avogadro's Number

In chemistry, particles such as atoms, molecules, and ions are counted by the **mole**, which contains 6.02×10^{23} items. This value, known as **Avogadro's number**, is a very big number because atoms are so small that it takes an extremely large number of atoms to provide a sufficient amount to weigh and use in chemical reactions. Avogadro's number is named for Amedeo Avogadro (1776–1856), an Italian physicist.

CORE CHEMISTRY SKILL

Converting Particles to Moles

Avogadro's number

$$6.02 \times 10^{23} = 602\,000\,000\,000\,000\,000\,000\,000$$

One mole of any element always contains Avogadro's number of atoms. For example, 1 mole of carbon contains 6.02×10^{23} carbon atoms; 1 mole of aluminum contains 6.02×10^{23} aluminum atoms; 1 mole of sulfur contains 6.02×10^{23} sulfur atoms.

$$1 \text{ mole of an element} = 6.02 \times 10^{23} \text{ atoms of that element}$$

One mole of sulfur contains 6.02×10^{23} sulfur atoms.

One mole of a molecular compound contains Avogadro's number of molecules. For example, 1 mole of CO_2 contains 6.02×10^{23} molecules of CO_2. One mole of an ionic compound contains Avogadro's number of **formula units**, which are the groups of ions represented by the formula of an ionic compound. One mole of NaCl contains 6.02×10^{23} formula units of NaCl (Na^+, Cl^-). **TABLE 7.5** gives examples of the number of particles in some 1-mole quantities.

TABLE 7.5 Number of Particles in 1-Mole Quantities

Substance	Number and Type of Particles
1 mole of Al	6.02×10^{23} atoms of Al
1 mole of S	6.02×10^{23} atoms of S
1 mole of water (H_2O)	6.02×10^{23} molecules of H_2O
1 mole of vitamin C ($C_6H_8O_6$)	6.02×10^{23} molecules of vitamin C
1 mole of NaCl	6.02×10^{23} formula units of NaCl

We use Avogadro's number as a conversion factor to convert between the moles of a substance and the number of particles it contains.

$$\frac{6.02 \times 10^{23} \text{ particles}}{1 \text{ mole}} \quad \text{and} \quad \frac{1 \text{ mole}}{6.02 \times 10^{23} \text{ particles}}$$

For example, we use Avogadro's number to convert 4.00 moles of sulfur to atoms of sulfur.

$$4.00 \cancel{\text{moles S}} \times \frac{6.02 \times 10^{23} \text{ atoms S}}{1 \cancel{\text{mole S}}} = 2.41 \times 10^{24} \text{ atoms of S}$$

Avogadro's number as a conversion factor

We also use Avogadro's number to convert 3.01×10^{24} molecules of CO_2 to moles of CO_2.

$$3.01 \times 10^{24} \cancel{\text{molecules } CO_2} \times \frac{1 \text{ mole } CO_2}{6.02 \times 10^{23} \cancel{\text{molecules } CO_2}} = 5.00 \text{ moles of } CO_2$$

Avogadro's number as a conversion factor

In calculations that convert between moles and particles, the number of moles will be a small number compared to the number of atoms or molecules, which will be large as shown in Sample Problem 7.6.

ENGAGE

Why is 0.20 mole of aluminum a small number, but the number 1.2×10^{23} atoms of aluminum in 0.20 mole is a large number?

▸SAMPLE PROBLEM 7.6 Calculating the Number of Molecules

TRY IT FIRST

How many molecules are present in 1.75 moles of carbon dioxide, CO_2?

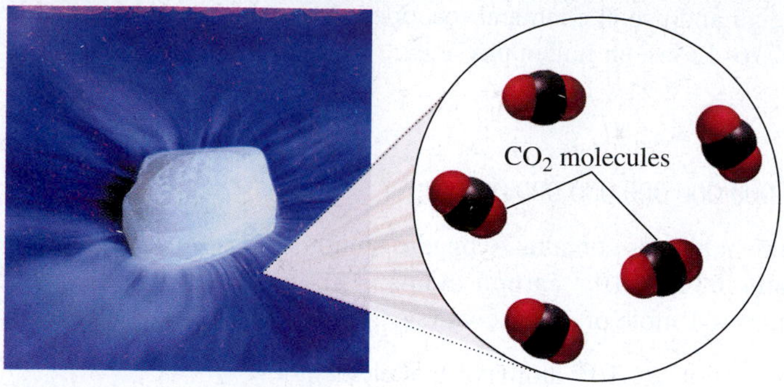
CO2 molecules

The solid form of carbon dioxide is known as "dry ice."

SOLUTION

STEP 1 State the given and needed quantities.

ANALYZE THE PROBLEM	Given	Need	Connect
	1.75 moles of CO_2	molecules of CO_2	Avogadro's number

STEP 2 Write a plan to convert moles to atoms or molecules.

moles of CO_2 → Avogadro's number → molecules of CO_2

STEP 3 Use Avogadro's number to write conversion factors.

$$1 \text{ mole of } CO_2 = 6.02 \times 10^{23} \text{ molecules of } CO_2$$

$$\frac{6.02 \times 10^{23} \text{ molecules } CO_2}{1 \text{ mole } CO_2} \quad \text{and} \quad \frac{1 \text{ mole } CO_2}{6.02 \times 10^{23} \text{ molecules } CO_2}$$

STEP 4 Set up the problem to calculate the number of particles.

$$1.75 \cancel{\text{moles } CO_2} \times \frac{6.02 \times 10^{23} \text{ molecules } CO_2}{1 \cancel{\text{mole } CO_2}} = 1.05 \times 10^{24} \text{ molecules of } CO_2$$

STUDY CHECK 7.6

a. How many moles of water, H_2O, contain 2.60×10^{23} molecules of water?

b. How many molecules are in 2.65 moles of dinitrogen oxide, N_2O, which is an anesthetic called "laughing gas"?

ANSWER

a. 0.432 mole of H_2O

b. 1.60×10^{24} molecules of N_2O

TEST

Try Practice Problems 7.21 to 7.24

Moles of Elements in a Chemical Compound

We have seen that the subscripts in the chemical formula of a compound indicate the number of atoms of each type of element in the compound. For example, aspirin, $C_9H_8O_4$, is a drug used to reduce pain and inflammation in the body. Using the subscripts in the chemical formula of aspirin shows that there are 9 carbon atoms, 8 hydrogen atoms, and

4 oxygen atoms in each molecule of aspirin. The subscripts also tell us the number of moles of each element in 1 mole of aspirin: 9 moles of C atoms, 8 moles of H atoms, and 4 moles of O atoms.

ENGAGE

Why does 1 mole of $Zn(C_2H_3O_2)_2$, a dietary supplement, contain 1 mole of Zn, 4 moles of C, 6 moles of H, and 4 moles of O?

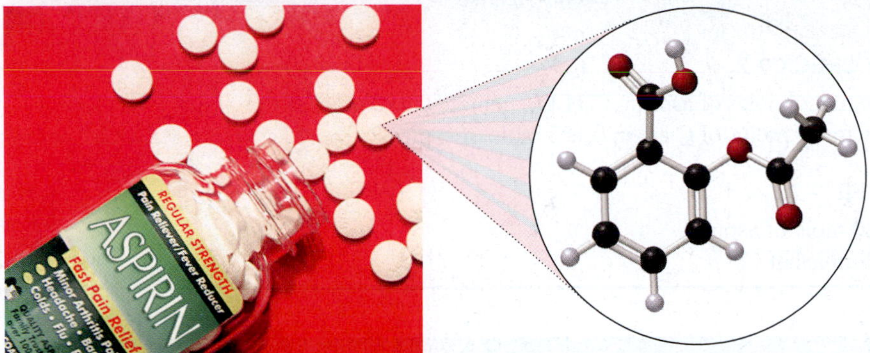

REGULAR STRENGTH
Pain Reliever/Fever Reducer
ASPIRIN
Fast Pain Relief

Aspirin $C_9H_8O_4$

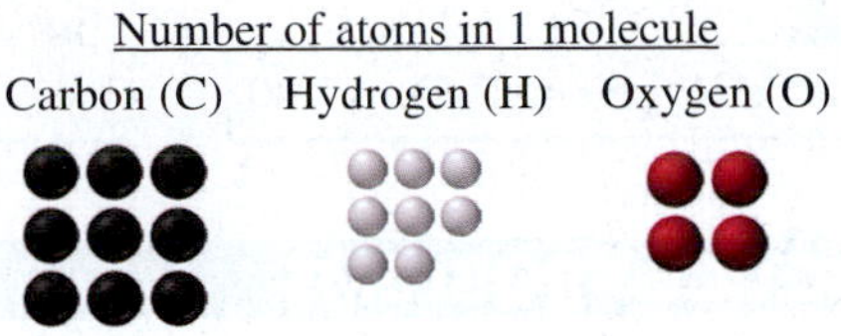

Number of atoms in 1 molecule
Carbon (C)
Hydrogen (H)
Oxygen (O)

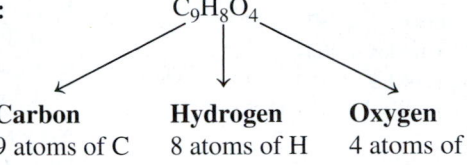

Every molecule of aspirin $C_9H_8O_4$ contains 9 C atoms, 8 H atoms, and 4 O atoms.

The chemical formula subscripts specify the: $C_9H_8O_4$

	Carbon	Hydrogen	Oxygen
Atoms in 1 molecule	9 atoms of C	8 atoms of H	4 atoms of O
Moles of each element in 1 mole	9 moles of C	8 moles of H	4 moles of O

Using a Chemical Formula to Write Conversion Factors

Using the subscripts from the formula, $C_9H_8O_4$, we can write the conversion factors for each of the elements in 1 mole of aspirin:

$$\frac{9\text{ moles C}}{1\text{ mole }C_9H_8O_4} \quad \frac{8\text{ moles H}}{1\text{ mole }C_9H_8O_4} \quad \frac{4\text{ moles O}}{1\text{ mole }C_9H_8O_4}$$

$$\frac{1\text{ mole }C_9H_8O_4}{9\text{ moles C}} \quad \frac{1\text{ mole }C_9H_8O_4}{8\text{ moles H}} \quad \frac{1\text{ mole }C_9H_8O_4}{4\text{ moles O}}$$

▶ SAMPLE PROBLEM 7.7 Calculating the Moles of an Element in a Compound

TRY IT FIRST

How many moles of carbon are present in 1.50 moles of aspirin, $C_9H_8O_4$?

SOLUTION

STEP 1 State the given and needed quantities.

ANALYZE THE PROBLEM	Given	Need	Connect
	1.50 moles of aspirin, $C_9H_8O_4$	moles of C	subscripts in formula

STEP 2 Write a plan to convert moles of a compound to moles of an element.

moles of $C_9H_8O_4$ → Subscript → moles of C

STEP 3 Write the equalities and conversion factors using subscripts.

$$1\text{ mole of }C_9H_8O_4 = 9\text{ moles of C}$$

$$\frac{9\text{ moles C}}{1\text{ mole }C_9H_8O_4} \text{ and } \frac{1\text{ mole }C_9H_8O_4}{9\text{ moles C}}$$

STEP 4 Set up the problem to calculate the moles of an element.

$$1.50 \text{ moles } C_9H_8O_4 \times \frac{9 \text{ moles C}}{1 \text{ mole } C_9H_8O_4} = 13.5 \text{ moles of C}$$

STUDY CHECK 7.7

a. How many moles of aspirin, $C_9H_8O_4$, contain 0.480 mole of O?
b. How many moles of C are in 0.125 mole of aspirin?

ANSWER

a. 0.120 mole of aspirin
b. 1.13 moles of C

TEST

Try Practice Problems 7.25 to 7.30

PRACTICE PROBLEMS

7.4 The Mole

7.21 What is a mole?

7.22 What is Avogadro's number?

7.23 Calculate each of the following:
a. number of C atoms in 0.500 mole of C
b. number of SO_2 molecules in 1.28 moles of SO_2
c. moles of Fe in 5.22×10^{22} atoms of Fe
d. moles of C_2H_6O in 8.50×10^{24} molecules of C_2H_6O

7.24 Calculate each of the following:
a. number of Li atoms in 4.5 moles of Li
b. number of CO_2 molecules in 0.0180 mole of CO_2
c. moles of Cu in 7.8×10^{21} atoms of Cu
d. moles of C_2H_6 in 3.75×10^{23} molecules of C_2H_6

7.25 Calculate each of the following quantities in 2.00 moles of H_3PO_4:
a. moles of H **b.** moles of O **c.** atoms of P **d.** atoms of O

7.26 Calculate each of the following quantities in 0.185 mole of $C_6H_{14}O$:
a. moles of C **b.** moles of O **c.** atoms of H **d.** atoms of C

Clinical Applications

7.27 Quinine, $C_{20}H_{24}N_2O_2$, is a component of tonic water and bitter lemon.
a. How many moles of H are in 1.5 moles of quinine?
b. How many moles of C are in 5.0 moles of quinine?
c. How many moles of N are in 0.020 mole of quinine?

7.28 Aluminum sulfate, $Al_2(SO_4)_3$, is used in some antiperspirants.
a. How many moles of O are present in 3.0 moles of $Al_2(SO_4)_3$?
b. How many moles of aluminum ions (Al^{3+}) are present in 0.40 mole of $Al_2(SO_4)_3$?
c. How many moles of sulfate ions (SO_4^{2-}) are present in 1.5 moles of $Al_2(SO_4)_3$?

7.29 Naproxen, found in Aleve, is used to treat the pain and inflammation caused by arthritis. Naproxen has a formula of $C_{14}H_{14}O_3$.
a. How many moles of C are present in 2.30 moles of naproxen?
b. How many moles of H are present in 0.444 mole of naproxen?
c. How many moles of O are present in 0.0765 mole of naproxen?

7.30 Benadryl is an over-the-counter drug used to treat allergy symptoms. The formula of Benadryl is $C_{17}H_{21}NO$.
a. How many moles of C are present in 0.733 mole of Benadryl?
b. How many moles of H are present in 2.20 moles of Benadryl?
c. How many moles of N are present in 1.54 moles of Benadryl?

7.5 Molar Mass

LEARNING GOAL Given the chemical formula of a substance, calculate its molar mass.

A single atom or molecule is much too small to weigh, even on the most accurate balance. In fact, it takes a huge number of atoms or molecules to make enough of a substance for you to see. An amount of water that contains Avogadro's number of water molecules is only a few sips. However, in the laboratory, we can use a balance to weigh out Avogadro's number of particles for 1 mole of substance.

For any element, the quantity called **molar mass** is the quantity in grams that equals the atomic mass of that element. We are counting 6.02×10^{23} atoms of an element when we weigh out the number of grams equal to its molar mass. For example, carbon has an atomic mass of 12.01 on the periodic table. This means 1 mole of carbon atoms has a mass of 12.01 g. Then to obtain 1 mole of carbon atoms, we would need to weigh out 12.01 g of carbon. The molar mass of carbon is found by looking at its atomic mass on the periodic table.

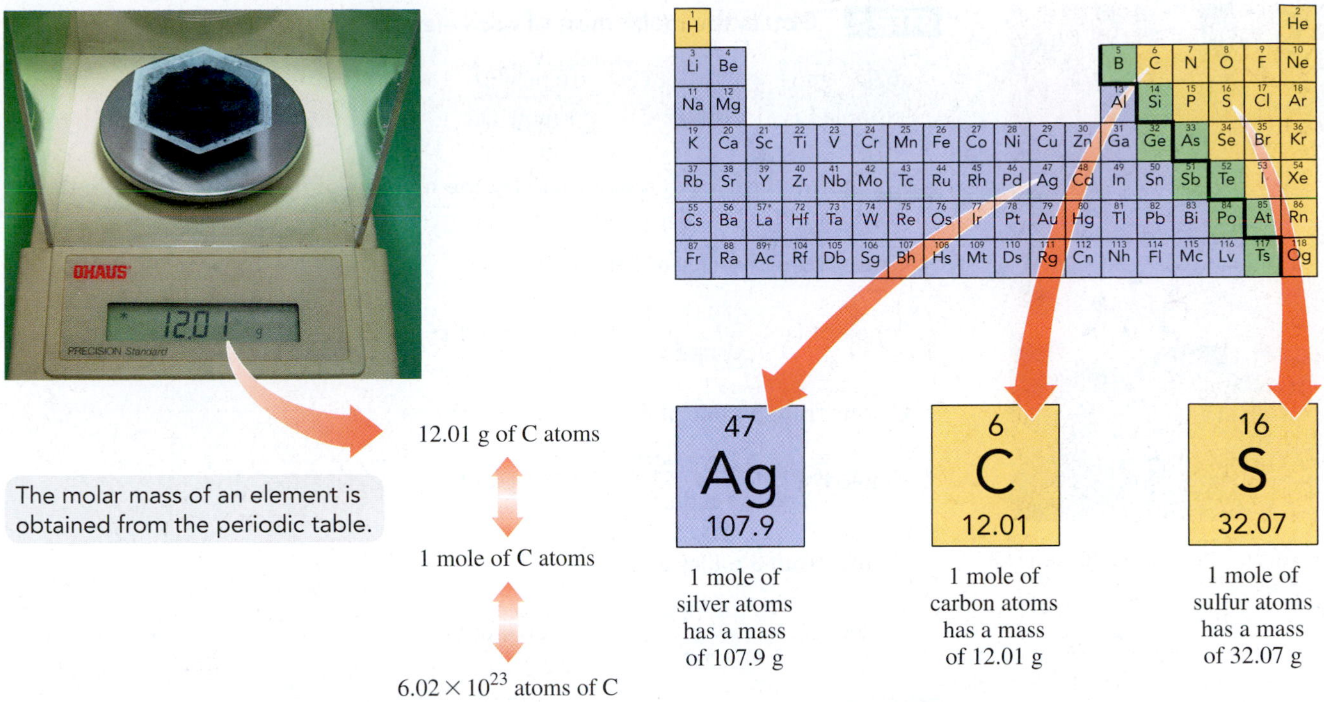

12.01 g
12.01 g of C atoms
1 mole of C atoms
6.02×10^{23} atoms of C
The molar mass of an element is obtained from the periodic table.
47
Ag
107.9
1 mole of silver atoms has a mass of 107.9 g
6
C
12.01
1 mole of carbon atoms has a mass of 12.01 g
16
S
32.07
1 mole of sulfur atoms has a mass of 32.07 g

Molar Mass of a Compound

CORE CHEMISTRY SKILL
Calculating Molar Mass

To determine the molar mass of a compound, multiply the molar mass of each element by its subscript in the formula and add the results as shown in Sample Problem 7.8. **In this text, we round the molar mass of an element to the hundredths place (0.01) or use at least four significant figures for calculations.**

FIGURE 7.7 shows some 1-mole quantities of substances.

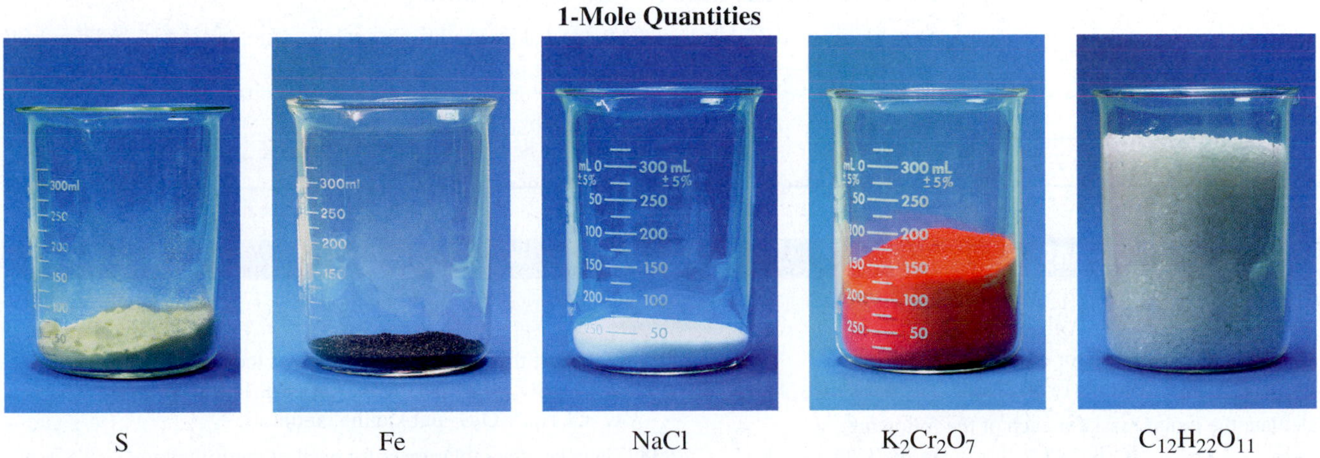

1-Mole Quantities
S
Fe
NaCl
$K_2Cr_2O_7$
$C_{12}H_{22}O_{11}$

FIGURE 7.7 ▶ 1-mole samples: sulfur, S (32.07 g); iron, Fe (55.85 g); salt, NaCl (58.44 g); potassium dichromate, $K_2Cr_2O_7$ (294.2 g); and sucrose, $C_{12}H_{22}O_{11}$ (342.3 g).

Q How is the molar mass for $K_2Cr_2O_7$ obtained?

▶ SAMPLE PROBLEM 7.8 Calculating Molar Mass

TRY IT FIRST

Calculate the molar mass for lithium carbonate, Li_2CO_3, used to treat bipolar disorder.

SOLUTION

	Given	Need	Connect
ANALYZE THE PROBLEM	formula Li_2CO_3	molar mass of Li_2CO_3	periodic table

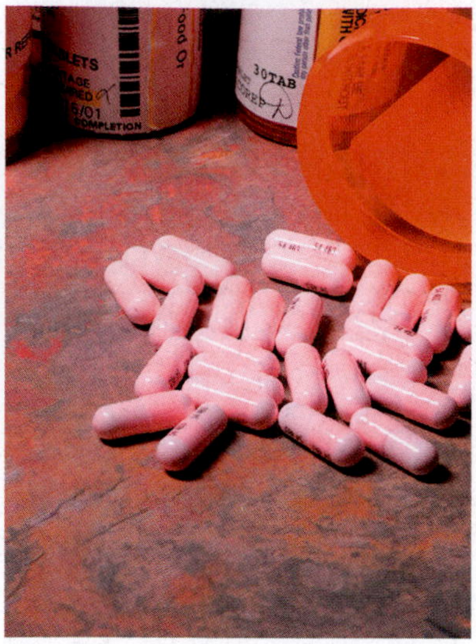

Lithium carbonate is used to treat bipolar disorder.

STEP 1 Obtain the molar mass of each element.

$$\frac{6.941\text{ g Li}}{1\text{ mole Li}} \quad \frac{12.01\text{ g C}}{1\text{ mole C}} \quad \frac{16.00\text{ g O}}{1\text{ mole O}}$$

STEP 2 Multiply each molar mass by the number of moles (subscript) in the formula.

Grams from 2 moles of Li:

$$2\text{ }\cancel{\text{moles Li}} \times \frac{6.941\text{ g Li}}{1\text{ }\cancel{\text{mole Li}}} = 13.88\text{ g of Li}$$

Grams from 1 mole of C:

$$1\text{ }\cancel{\text{mole C}} \times \frac{12.01\text{ g C}}{1\text{ }\cancel{\text{mole C}}} = 12.01\text{ g of C}$$

Grams from 3 moles of O:

$$3\text{ }\cancel{\text{moles O}} \times \frac{16.00\text{ g O}}{1\text{ }\cancel{\text{mole O}}} = 48.00\text{ g of O}$$

STEP 3 Calculate the molar mass by adding the masses of the elements.

2 moles of Li	= 13.88 g of Li
1 mole of C	= 12.01 g of C
3 moles of O	= 48.00 g of O
Molar mass of Li_2CO_3	= 73.89 g

STUDY CHECK 7.8

Calculate the molar mass for each of the following:

a. salicylic acid, $C_7H_6O_3$, used to treat skin conditions such as acne, psoriasis, and dandruff
b. vitamin D_3, $C_{27}H_{44}O$, a vitamin that prevents rickets

ANSWER

a. 138.12 g **b.** 384.6 g

TEST

Try Practice Problems 7.31 to 7.40

PRACTICE PROBLEMS

7.5 Molar Mass

7.31 Calculate the molar mass for each of the following:
a. Cl_2 **b.** $C_3H_6O_3$ **c.** $Mg_3(PO_4)_2$

7.32 Calculate the molar mass for each of the following:
a. O_2 **b.** KH_2PO_4 **c.** $Fe(ClO_4)_3$

7.33 Calculate the molar mass for each of the following:
a. AlF_3 **b.** $C_2H_4Cl_2$ **c.** SnF_2

7.34 Calculate the molar mass for each of the following:
a. $C_4H_8O_4$ **b.** $Ga_2(CO_3)_3$ **c.** $KBrO_4$

Clinical Applications

7.35 Calculate the molar mass for each of the following:
a. KCl, salt substitute
b. $C_6H_5NO_2$, niacin (vitamin B_3) needed for cell metabolism
c. $C_{19}H_{20}FNO_3$, Paxil, antidepressant

7.36 Calculate the molar mass for each of the following:
a. $FeSO_4$, iron supplement **b.** Al_2O_3, absorbent and abrasive
c. $C_7H_5NO_3S$, saccharin, artificial sweetener

7.37 Calculate the molar mass for each of the following:
a. $Al_2(SO_4)_3$, antiperspirant **b.** $KC_4H_5O_6$, cream of tartar
c. $C_{16}H_{19}N_3O_5S$, amoxicillin, antibiotic

7.38 Calculate the molar mass for each of the following:
a. C_3H_8O, rubbing alcohol **b.** $(NH_4)_2CO_3$, baking powder
c. $Zn(C_2H_3O_2)_2$, zinc supplement

7.39 Calculate the molar mass for each of the following:
a. $C_8H_9NO_2$, acetaminophen used in Tylenol
b. $Ca_3(C_6H_5O_7)_2$, calcium supplement
c. $C_{17}H_{18}FN_3O_3$, Cipro, used to treat a range of bacterial infections

7.40 Calculate the molar mass for each of the following:
a. $CaSO_4$, calcium sulfate, used to make casts to protect broken bones
b. $C_6H_{12}N_2O_4Pt$, Carboplatin, used in chemotherapy
c. $C_{12}H_{18}O$, Propofol, used to induce anesthesia during surgery

7.6 Calculations Using Molar Mass

LEARNING GOAL Use molar mass to convert between grams and moles.

The molar mass of an element is one of the most useful conversion factors in chemistry because it converts moles of a substance to grams, or grams to moles. For example, 1 mole of silver has a mass of 107.9 g. To express the molar mass of Ag as an equality, we write

1 mole of Ag = 107.9 g of Ag

From this equality for the molar mass, two conversion factors can be written as

$$\frac{107.9 \text{ g Ag}}{1 \text{ mole Ag}} \quad \text{and} \quad \frac{1 \text{ mole Ag}}{107.9 \text{ g Ag}}$$

Sample Problem 7.9 shows how the molar mass of silver is used as a conversion factor.

CORE CHEMISTRY SKILL
Using Molar Mass as a Conversion Factor

▶SAMPLE PROBLEM 7.9 Converting Between Moles and Grams

TRY IT FIRST

Silver metal is used in the manufacture of tableware, mirrors, jewelry, and dental alloys. If the design for a piece of jewelry requires 0.750 mole of silver, how many grams of silver are needed?

Silver metal is used to make jewelry.

SOLUTION

STEP 1 State the given and needed quantities.

ANALYZE THE PROBLEM	Given	Need	Connect
	0.750 mole of Ag	grams of Ag	molar mass

STEP 2 Write a plan to convert moles to grams.

moles of Ag → Molar mass → grams of Ag

STEP 3 Determine the molar mass and write conversion factors.

1 mole of Ag = 107.9 g of Ag

$$\frac{107.9 \text{ g Ag}}{1 \text{ mole Ag}} \quad \text{and} \quad \frac{1 \text{ mole Ag}}{107.9 \text{ g Ag}}$$

STEP 4 Set up the problem to convert moles to grams.

$$0.750 \cancel{\text{mole Ag}} \times \frac{107.9 \text{ g Ag}}{1 \cancel{\text{mole Ag}}} = 80.9 \text{ g of Ag}$$

STUDY CHECK 7.9

a. A dentist orders 24.4 g of gold (Au) to prepare dental crowns and fillings. Calculate the number of moles of gold in the order.

b. Pencil lead contains graphite, which is the element carbon. Calculate the number of grams in 0.520 mole of carbon.

ANSWER

a. 0.124 mole of Au

b. 6.25 g of C

Writing Conversion Factors for the Molar Mass of a Compound

The conversion factors for a compound are also written from the molar mass. For example, the molar mass of the compound H_2O is written

$$1 \text{ mole of } H_2O = 18.02 \text{ g of } H_2O$$

From this equality, conversion factors for the molar mass of H_2O are written as

$$\frac{18.02 \text{ g } H_2O}{1 \text{ mole } H_2O} \quad \text{and} \quad \frac{1 \text{ mole } H_2O}{18.02 \text{ g } H_2O}$$

We can now change from moles to grams, or grams to moles, using the conversion factors derived from the molar mass of a compound, shown in Sample Problem 7.10. (Remember, you must determine the molar mass first.)

Table salt is sodium chloride, NaCl.

▶ SAMPLE PROBLEM 7.10 Converting Between Mass and Moles of a Compound

TRY IT FIRST

A salt shaker contains 73.7 g of NaCl. How many moles of NaCl are present?

SOLUTION

STEP 1 State the given and needed quantities.

ANALYZE THE PROBLEM	Given	Need	Connect
	73.7 g of NaCl	moles of NaCl	molar mass

STEP 2 Write a plan to convert grams to moles.

grams of NaCl → Molar mass → moles of NaCl

STEP 3 Determine the molar mass and write conversion factors.

$$(1 \times 22.99) + (1 \times 35.45) = 58.44 \text{ g/mole}$$

$$1 \text{ mole of NaCl} = 58.44 \text{ g of NaCl}$$

$$\frac{58.44 \text{ g NaCl}}{1 \text{ mole NaCl}} \quad \text{and} \quad \frac{1 \text{ mole NaCl}}{58.44 \text{ g NaCl}}$$

STEP 4 Set up the problem to convert grams to moles.

$$73.7 \cancel{\text{g NaCl}} \times \frac{1 \text{ mole NaCl}}{58.44 \cancel{\text{g NaCl}}} = 1.26 \text{ moles of NaCl}$$

STUDY CHECK 7.10

a. One tablet of an antacid contains 680. mg of $CaCO_3$. How many moles of $CaCO_3$ are present?

b. A different brand of antacid contains 6.9×10^{-3} mole of $Mg(OH)_2$. How many grams of $Mg(OH)_2$ are present?

ANSWER

a. 0.006 79 or 6.79×10^{-3} mole of $CaCO_3$

b. 0.40 g of $Mg(OH)_2$

TEST

Try Practice Problems 7.41 to 7.54

FIGURE 7.8 gives a summary of the calculations to show the connections between the moles of a compound, its mass in grams, the number of molecules (or formula units if ionic), and the moles and atoms of each element in that compound.

ENGAGE

Why are there more grams of chlorine than grams of fluorine in 1 mole of Freon-12, CCl_2F_2?

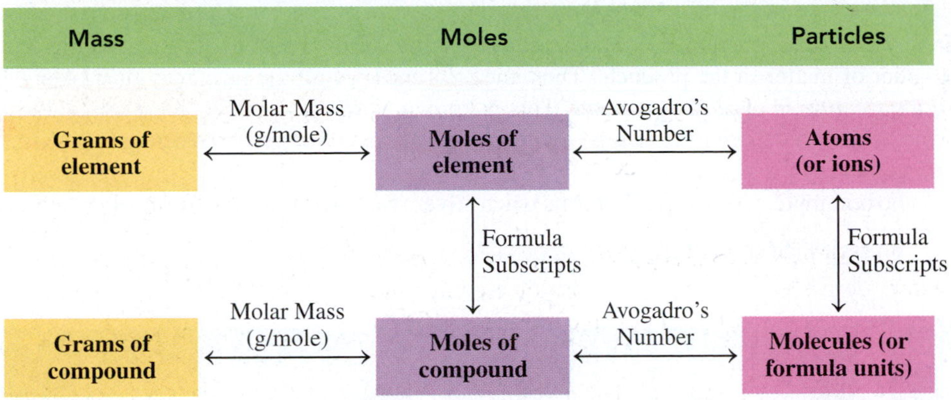

Mass
Moles
Particles
Grams of element
Molar Mass (g/mole)
Moles of element
Avogadro's Number
Atoms (or ions)
Formula Subscripts
Formula Subscripts
Grams of compound
Molar Mass (g/mole)
Moles of compound
Avogadro's Number
Molecules (or formula units)

FIGURE 7.8 ▶ The moles of a compound are related to its mass in grams by molar mass, to the number of molecules (or formula units) by Avogadro's number, and to the moles of each element by the subscripts in the formula.

Q What steps are needed to calculate the number of atoms of H in 5.00 g of CH_4?

PRACTICE PROBLEMS

7.6 Calculations Using Molar Mass

7.41 Calculate the mass, in grams, for each of the following:
a. 1.50 moles of Na **b.** 2.80 moles of Ca
c. 0.125 mole of CO_2 **d.** 0.0485 mole of Na_2CO_3
e. 7.14×10^2 moles of PCl_3

7.42 Calculate the mass, in grams, for each of the following:
a. 5.12 moles of Al **b.** 0.75 mole of Cu
c. 3.52 moles of $MgBr_2$ **d.** 0.145 mole of C_2H_6O
e. 2.08 moles of $(NH_4)_2SO_4$

7.43 Calculate the mass, in grams, in 0.150 mole of each of the following:
a. Ne **b.** I_2 **c.** Na_2O
d. $Ca(NO_3)_2$ **e.** C_6H_{14}

7.44 Calculate the mass, in grams, in 2.28 moles of each of the following:
a. Pd **b.** SO_3 **c.** $C_3H_6O_3$
d. $Mg(HCO_3)_2$ **e.** SF_6

7.45 Calculate the number of moles in each of the following:
a. 82.0 g of Ag **b.** 0.288 g of C
c. 15.0 g of ammonia, NH_3 **d.** 7.25 g of CH_4
e. 245 g of Fe_2O_3

7.46 Calculate the number of moles in each of the following:
a. 85.2 g of Ni **b.** 144 g of K
c. 6.4 g of H_2O **d.** 308 g of $BaSO_4$
e. 252.8 g of fructose, $C_6H_{12}O_6$

7.47 Calculate the number of moles in 25.0 g of each of the following:
a. He **b.** O_2 **c.** $Al(OH)_3$
d. Ga_2S_3 **e.** C_4H_{10}, butane

7.48 Calculate the number of moles in 4.00 g of each of the following:
a. Au **b.** SnO_2 **c.** CS_2
d. Ca_3N_2 **e.** $C_6H_8O_6$, vitamin C

Clinical Applications

7.49 Chloroethane, C_2H_5Cl, is used to diagnose dead tooth nerves.
a. How many moles are in 34.0 g of chloroethane?
b. How many grams are in 1.50 moles of chloroethane?

7.50 Allyl sulfide, $(C_3H_5)_2S$, gives garlic, onions, and leeks their characteristic odor.

The characteristic odor of garlic is due to a sulfur-containing compound.

a. How many moles are in 23.2 g of allyl sulfide?
b. How many grams are in 0.75 mole of allyl sulfide?

7.51 **a.** The compound $MgSO_4$, Epsom salts, is used to soothe sore feet and muscles. How many grams will you need to prepare a bath containing 5.00 moles of Epsom salts?
b. Potassium iodide, KI, is used as an expectorant. How many grams are in 0.450 mole of potassium iodide?

7.52 **a.** Cyclopropane, C_3H_6, is an anesthetic given by inhalation. How many grams are in 0.25 mole of cyclopropane?
b. The sedative Demerol hydrochloride has the formula $C_{15}H_{22}ClNO_2$. How many grams are in 0.025 mole of Demerol hydrochloride?

7.53 Dinitrogen oxide (or nitrous oxide), N_2O, also known as laughing gas, is widely used as an anesthetic in dentistry.
a. How many grams are in 1.50 moles of dinitrogen oxide?
b. How many moles are in 34.0 g of dinitrogen oxide?

7.54 Chloroform, $CHCl_3$, was formerly used as an anesthetic but its use was discontinued due to respiratory and cardiac failure.
a. How many grams are in 0.122 mole of chloroform?
b. How many moles are in 26.7 g of chloroform?

7.7 Mole Relationships in Chemical Equations

LEARNING GOAL Use a mole–mole factor from a balanced chemical equation to calculate the number of moles of another substance in the reaction.

In any chemical reaction, the total amount of matter in the reactants is equal to the total amount of matter in the products. Thus, the total mass of all the reactants must be equal to the total mass of all the products. This is known as the *Law of Conservation of Mass*, which states that there is no change in the total mass of the substances reacting in a chemical reaction.

For example, tarnish (Ag_2S) forms when silver reacts with sulfur to form silver sulfide.

$$2Ag(s) + S(s) \longrightarrow Ag_2S(s)$$

OHAUS
215.8 g
PRECISION Standard
OHAUS
32.1 g
PRECISION Standard
OHAUS
247.9 g
PRECISION Standard

$2Ag(s)$ + $S(s)$ $\longrightarrow$ $Ag_2S(s)$

Mass of reactants = Mass of product

In the chemical reaction of Ag and S, the mass of the reactants is the same as the mass of the product, Ag_2S.

In this reaction, the number of silver atoms that reacts is twice the number of sulfur atoms. When 200 silver atoms react, 100 sulfur atoms are required. However, in the actual chemical reaction, many more atoms of both silver and sulfur would react. If we are dealing with moles of silver and sulfur, then the coefficients in the equation can be interpreted in terms of moles. Thus, 2 moles of silver react with 1 mole of sulfur to form 1 mole of Ag_2S. Because the molar mass of each can be determined, the moles of Ag, S, and Ag_2S can also be stated in terms of mass in grams of each. Thus, 215.8 g of Ag and 32.1 g of S react to form 247.9 g of Ag_2S. The total mass of the reactants (247.9 g) is equal to the mass of product (247.9 g). The various ways in which a chemical equation can be interpreted are seen in **TABLE 7.6**.

TABLE 7.6 Information Available from a Balanced Equation

	Reactants			Products
Equation	$2Ag(s)$	$+ S(s)$	$\longrightarrow$	$Ag_2S(s)$
Atoms	2 Ag atoms	+ 1 S atom	$\longrightarrow$	1 Ag_2S formula unit
	200 Ag atoms	+ 100 S atoms	$\longrightarrow$	100 Ag_2S formula units
Avogadro's Number of Atoms	$2(6.02 \times 10^{23})$ Ag atoms	$+ 1(6.02 \times 10^{23})$ S atoms	$\longrightarrow$	$1(6.02 \times 10^{23})$ Ag_2S formula units
Moles	2 moles of Ag	+ 1 mole of S	$\longrightarrow$	1 mole of Ag_2S
Mass (g)	2(107.9 g) of Ag	+ 1(32.07 g) of S	$\longrightarrow$	1(247.9 g) of Ag_2S
Total Mass (g)	247.9 g		$\longrightarrow$	247.9 g

Mole–Mole Factors from a Balanced Equation

When iron reacts with sulfur, the product is iron(III) sulfide.

$$2Fe(s) + 3S(s) \longrightarrow Fe_2S_3(s)$$

Iron (Fe) — Sulfur (S) — Iron(III) sulfide (Fe_2S_3)

$2Fe(s) + 3S(s) \longrightarrow Fe_2S_3(s)$

In the chemical reaction of Fe and S, the mass of the reactants is the same as the mass of the product, Fe_2S_3.

From the balanced chemical equation, we see that 2 moles of iron reacts with 3 moles of sulfur to form 1 mole of iron(III) sulfide. Actually, any amount of iron or sulfur may be used, but the *ratio* of iron reacting with sulfur will always be the same. From the coefficients, we can write **mole–mole factors** between reactants and between reactants and products. The coefficients used in the mole–mole factors are exact numbers; they do not limit the number of significant figures.

Fe and S: $\frac{2 \text{ moles Fe}}{3 \text{ moles S}}$ and $\frac{3 \text{ moles S}}{2 \text{ moles Fe}}$

Fe and Fe_2S_3: $\frac{2 \text{ moles Fe}}{1 \text{ mole } Fe_2S_3}$ and $\frac{1 \text{ mole } Fe_2S_3}{2 \text{ moles Fe}}$

S and Fe_2S_3: $\frac{3 \text{ moles S}}{1 \text{ mole } Fe_2S_3}$ and $\frac{1 \text{ mole } Fe_2S_3}{3 \text{ moles S}}$

TEST

Try Practice Problems 7.55 and 7.56

Using Mole–Mole Factors in Calculations

CORE CHEMISTRY SKILL

Using Mole–Mole Factors

Whenever you prepare a recipe, adjust an engine for the proper mixture of fuel and air, or prepare medicines in a pharmaceutical laboratory, you need to know the proper amounts of reactants to use and how much of the product will form. Now that we have written all the possible conversion factors for the balanced equation $2Fe(s) + 3S(s) \longrightarrow Fe_2S_3(s)$, we will use those mole–mole factors in a chemical calculation in Sample Problem 7.11.

▶ SAMPLE PROBLEM 7.11 Calculating Moles of a Reactant

TRY IT FIRST

In the chemical reaction of iron and sulfur, how many moles of sulfur are needed to react with 1.42 moles of iron?

$$2Fe(s) + 3S(s) \longrightarrow Fe_2S_3(s)$$

SOLUTION

STEP 1 State the given and needed quantities (moles).

ANALYZE THE PROBLEM	Given	Need	Connect
	1.42 moles of Fe	moles of S	mole–mole factor
	Equation		
	$2Fe(s) + 3S(s) \longrightarrow Fe_2S_3(s)$		

STEP 2 Write a plan to convert the given to the needed quantity (moles).

moles of Fe → Mole–mole factor → moles of S

STEP 3 Use coefficients to write mole–mole factors.

2 moles of Fe = 3 moles of S

$$\frac{2 \text{ moles Fe}}{3 \text{ moles S}} \quad \text{and} \quad \frac{3 \text{ moles S}}{2 \text{ moles Fe}}$$

STEP 4 Set up the problem to give the needed quantity (moles).

$$1.42 \text{ moles Fe} \times \frac{3 \text{ moles S}}{2 \text{ moles Fe}} = 2.13 \text{ moles of S}$$

Three SFs — Exact — Three SFs

STUDY CHECK 7.11

Using the equation in Sample Problem 7.11, calculate each of the following:

a. moles of iron needed to react with 2.75 moles of sulfur
b. moles of iron(III) sulfide produced by the reaction of 0.758 mole of sulfur

ANSWER

a. 1.83 moles of iron
b. 0.253 mole of iron(III) sulfide

TEST

Try Practice Problems 7.57 to 7.60

PRACTICE PROBLEMS

7.7 Mole Relationships in Chemical Equations

7.55 Write all of the mole–mole factors for each of the following chemical equations:

a. $2SO_2(g) + O_2(g) \longrightarrow 2SO_3(g)$
b. $4P(s) + 5O_2(g) \longrightarrow 2P_2O_5(s)$

7.56 Write all of the mole–mole factors for each of the following chemical equations:

a. $2Al(s) + 3Cl_2(g) \longrightarrow 2AlCl_3(s)$
b. $4HCl(g) + O_2(g) \longrightarrow 2Cl_2(g) + 2H_2O(g)$

7.57 The chemical reaction of hydrogen with oxygen produces water.

$$2H_2(g) + O_2(g) \longrightarrow 2H_2O(g)$$

a. How many moles of O_2 are required to react with 2.6 moles of H_2?
b. How many moles of H_2 are needed to react with 5.0 moles of O_2?
c. How many moles of H_2O form when 2.5 moles of O_2 reacts?

7.58 Ammonia is produced by the chemical reaction of nitrogen and hydrogen.

$$N_2(g) + 3H_2(g) \longrightarrow 2NH_3(g)$$

Ammonia

a. How many moles of H_2 are needed to react with 1.8 moles of N_2?
b. How many moles of N_2 reacted if 0.60 mole of NH_3 is produced?
c. How many moles of NH_3 are produced when 1.4 moles of H_2 reacts?

7.59 Carbon disulfide and carbon monoxide are produced when carbon is heated with sulfur dioxide.

$$5C(s) + 2SO_2(g) \xrightarrow{\Delta} CS_2(l) + 4CO(g)$$

a. How many moles of C are needed to react with 0.500 mole of SO_2?
b. How many moles of CO are produced when 1.2 moles of C reacts?
c. How many moles of SO_2 are needed to produce 0.50 mole of CS_2?
d. How many moles of CS_2 are produced when 2.5 moles of C reacts?

7.60 In the acetylene torch, acetylene gas (C_2H_2) burns in oxygen to produce carbon dioxide, water, and energy.

$$2C_2H_2(g) + 5O_2(g) \xrightarrow{\Delta} 4CO_2(g) + 2H_2O(g)$$

a. How many moles of O_2 are needed to react with 2.40 moles of C_2H_2?
b. How many moles of CO_2 are produced when 3.5 moles of C_2H_2 reacts?
c. How many moles of C_2H_2 are needed to produce 0.50 mole of H_2O?
d. How many moles of CO_2 are produced from 0.100 mole of O_2?

7.8 Mass Calculations for Chemical Reactions

LEARNING GOAL Given the mass in grams of a substance in a reaction, calculate the mass in grams of another substance in the reaction.

REVIEW
Using Significant Figures in Calculations (2.3)

When we have the balanced chemical equation for a reaction, we can use the mass of one of the substances (A) in the reaction to calculate the mass of another substance (B) in the reaction. However, the calculations require us to convert the mass of A to moles of A using the molar mass of A. Then we use the mole–mole factor that links substance A to substance B, which we obtain from the coefficients in the balanced equation. This mole–mole factor (B/A) will convert the moles of A to moles of B. Then the molar mass of B is used to calculate the grams of substance B.

CORE CHEMISTRY SKILL
Converting Grams to Grams

Substance A | **Substance B**

grams of A → [Molar mass A] → moles of A → [Mole–mole factor B/A] → moles of B → [Molar mass B] → grams of B

▸SAMPLE PROBLEM 7.12 Calculating Mass of a Product

TRY IT FIRST

When acetylene, C_2H_2, burns in oxygen, high temperatures are produced that are used for welding metals.

$$2C_2H_2(g) + 5O_2(g) \xrightarrow{\Delta} 4CO_2(g) + 2H_2O(g)$$

How many grams of CO_2 are produced when 54.6 g of C_2H_2 is burned?

A mixture of acetylene and oxygen undergoes combustion during the welding of metals.

SOLUTION

STEP 1 State the given and needed quantities (grams).

ANALYZE THE PROBLEM	Given	Need	Connect
	54.6 g of C_2H_2	grams of CO_2	molar masses, mole–mole factor
	Equation		
	$2C_2H_2(g) + 5O_2(g) \xrightarrow{\Delta} 4CO_2(g) + 2H_2O(g)$		

STEP 2 Write a plan to convert the given to the needed quantity (grams).

grams of C_2H_2 → [Molar mass] → moles of C_2H_2 → [Mole–mole factor] → moles of CO_2 → [Molar mass] → grams of CO_2

STEP 3 Use coefficients to write mole–mole factors; write molar masses.

1 mole of C_2H_2 = 26.04 g of C_2H_2

$$\frac{26.04\text{ g }C_2H_2}{1\text{ mole }C_2H_2} \quad \text{and} \quad \frac{1\text{ mole }C_2H_2}{26.04\text{ g }C_2H_2}$$

1 mole of CO_2 = 44.01 g of CO_2

$$\frac{44.01\text{ g }CO_2}{1\text{ mole }CO_2} \quad \text{and} \quad \frac{1\text{ mole }CO_2}{44.01\text{ g }CO_2}$$

2 moles of C_2H_2 = 4 moles of CO_2

$$\frac{2\text{ moles }C_2H_2}{4\text{ moles }CO_2} \quad \text{and} \quad \frac{4\text{ moles }CO_2}{2\text{ moles }C_2H_2}$$

STEP 4 Set up the problem to give the needed quantity (grams).

$$54.6\ \cancel{\text{g }C_2H_2} \times \frac{1\ \cancel{\text{mole }C_2H_2}}{26.04\ \cancel{\text{g }C_2H_2}} \times \frac{4\ \cancel{\text{moles }CO_2}}{2\ \cancel{\text{moles }C_2H_2}} \times \frac{44.01\text{ g }CO_2}{1\ \cancel{\text{mole }CO_2}} = 185\text{ g of }CO_2$$

Three SFs; Exact / Four SFs; Exact / Exact; Four SFs / Exact; Three SFs

INTERACTIVE VIDEO

Problem 7.67

PEARSON eText 2.0

TEST

Try Practice Problems 7.61 to 7.70

STUDY CHECK 7.12

Using the equation in Sample Problem 7.12, calculate each of the following:

a. the grams of CO_2 produced when 25.0 g of O_2 reacts

b. the grams of C_2H_2 needed when 65.0 g of H_2O are produced

ANSWER

a. 27.5 g of CO_2

b. 93.9 g of C_2H_2

PRACTICE PROBLEMS

7.8 Mass Calculations for Chemical Reactions

7.61 Sodium reacts with oxygen to produce sodium oxide.

$$4Na(s) + O_2(g) \longrightarrow 2Na_2O(s)$$

a. How many grams of Na_2O are produced when 57.5 g of Na reacts?

b. If you have 18.0 g of Na, how many grams of O_2 are needed for the reaction?

c. How many grams of O_2 are needed in a reaction that produces 75.0 g of Na_2O?

7.62 Nitrogen reacts with hydrogen to produce ammonia.

$$N_2(g) + 3H_2(g) \longrightarrow 2NH_3(g)$$

a. If you have 3.64 g of H_2, how many grams of NH_3 can be produced?

b. How many grams of H_2 are needed to react with 2.80 g of N_2?

c. How many grams of NH_3 can be produced from 12.0 g of H_2?

7.63 Ammonia and oxygen react to form nitrogen and water.

$$4NH_3(g) + 3O_2(g) \longrightarrow 2N_2(g) + 6H_2O(g)$$

a. How many grams of O_2 are needed to react with 13.6 g of NH_3?

b. How many grams of N_2 can be produced when 6.50 g of O_2 reacts?

c. How many grams of H_2O are formed from the reaction of 34.0 g of NH_3?

7.64 Iron(III) oxide reacts with carbon to give iron and carbon monoxide.

$$Fe_2O_3(s) + 3C(s) \longrightarrow 2Fe(s) + 3CO(g)$$

a. How many grams of C are required to react with 16.5 g of Fe_2O_3?

b. How many grams of CO are produced when 36.0 g of C reacts?

c. How many grams of Fe can be produced when 6.00 g of Fe_2O_3 reacts?

7.65 Nitrogen dioxide and water react to produce nitric acid, HNO_3, and nitrogen oxide.

$$3NO_2(g) + H_2O(l) \longrightarrow 2HNO_3(aq) + NO(g)$$

a. How many grams of H_2O are needed to react with 28.0 g of NO_2?

b. How many grams of NO are produced from 15.8 g of H_2O?

c. How many grams of HNO_3 are produced from 8.25 g of NO_2?

7.66 Calcium cyanamide, $CaCN_2$, reacts with water to form calcium carbonate and ammonia.

$$CaCN_2(s) + 3H_2O(l) \longrightarrow CaCO_3(s) + 2NH_3(g)$$

a. How many grams of H_2O are needed to react with 75.0 g of $CaCN_2$?

b. How many grams of NH_3 are produced from 5.24 g of $CaCN_2$?

c. How many grams of $CaCO_3$ form if 155 g of H_2O reacts?

7.67 When solid lead(II) sulfide reacts with oxygen gas, the products are solid lead(II) oxide and sulfur dioxide gas.

a. Write the balanced chemical equation for the reaction.

b. How many grams of oxygen are required to react with 29.9 g of lead(II) sulfide?

c. How many grams of sulfur dioxide can be produced when 65.0 g of lead(II) sulfide reacts?

d. How many grams of lead(II) sulfide are used to produce 128 g of lead(II) oxide?

7.68 When the gases dihydrogen sulfide and oxygen react, they form the gases sulfur dioxide and water vapor.

a. Write the balanced chemical equation for the reaction.

b. How many grams of oxygen are needed to react with 2.50 g of dihydrogen sulfide?

c. How many grams of sulfur dioxide can be produced when 38.5 g of oxygen reacts?

d. How many grams of oxygen are needed to produce 55.8 g of water vapor?

Clinical Applications

7.69 In the body, the amino acid glycine, $C_2H_5NO_2$, reacts according to the following equation:

$$\underset{\text{Glycine}}{2C_2H_5NO_2(aq)} + 3O_2(g) \longrightarrow 3CO_2(g) + 3H_2O(l) + \underset{\text{Urea}}{CH_4N_2O(aq)}$$

a. How many grams of O_2 are needed to react with 15.0 g of glycine?

b. How many grams of urea are produced from 15.0 g of glycine?

c. How many grams of CO_2 are produced from 15.0 g of glycine?

7.70 In the body, ethanol, C_2H_6O, reacts according to the following equation:

$$\underset{\text{Ethanol}}{C_2H_6O(aq)} + 3O_2(g) \longrightarrow 2CO_2(g) + 3H_2O(l)$$

a. How many grams of O_2 are needed to react with 8.40 g of ethanol?

b. How many grams of H_2O are produced from 8.40 g of ethanol?

c. How many grams of CO_2 are produced from 8.40 g of ethanol?

7.9 Limiting Reactants and Percent Yield

REVIEW

Calculating Percentages (1.4)

LEARNING GOAL Identify a limiting reactant when given the quantities of two reactants; calculate the amount of product formed from the limiting reactant. Given the actual quantity of product, calculate the percent yield for a reaction.

When we make peanut butter sandwiches for lunch, we need 2 slices of bread and 1 tablespoon of peanut butter for each sandwich. As an equation, we could write:

$$2 \text{ slices of bread} + 1 \text{ tablespoon of peanut butter} \longrightarrow 1 \text{ peanut butter sandwich}$$

If we have 8 slices of bread and a full jar of peanut butter, we will run out of bread after we make 4 peanut butter sandwiches. We cannot make any more sandwiches once the bread is used up, even though there is a lot of peanut butter left in the jar. The number of slices of bread has limited the number of sandwiches we can make.

On a different day, we might have 8 slices of bread but only a tablespoon of peanut butter left in the jar. We will run out of peanut butter after we make just 1 sandwich and have 6 slices of bread left over. The amount of peanut butter has limited the number of sandwiches we can make.

 +  ⟶  + 

8 slices of bread + 1 jar of peanut butter make 4 peanut butter sandwiches + peanut butter left over

 +  ⟶  + 

8 slices of bread + 1 tablespoon of peanut butter make 1 peanut butter sandwich + 6 slices of bread left over

The reactant that is completely used up is the **limiting reactant**. The reactant that does not completely react and is left over is called the *excess reactant*.

Bread	Peanut Butter	Sandwiches	Limiting Reactant	Excess Reactant
1 loaf (20 slices)	1 tablespoon	1	peanut butter	bread
4 slices	1 full jar	2	bread	peanut butter
8 slices	1 full jar	4	bread	peanut butter

ENGAGE

For a picnic you have 10 spoons, 8 forks, and 6 knives. If each person requires 1 spoon, 1 fork, and 1 knife, why can you only serve 6 people at the picnic?

Calculating Moles of Product from a Limiting Reactant

CORE CHEMISTRY SKILL

Calculating Quantity of Product from a Limiting Reactant

In a similar way, the reactants in a chemical reaction do not always combine in quantities that allow each to be used up at exactly the same time. In many reactions, there is a reactant that limits the amount of product that can be formed. When we know the quantities of the reactants of a chemical reaction, we calculate the amount of product that is possible from each reactant if it were completely consumed. We are looking for the *limiting reactant*, which is the one that runs out first, producing the smaller amount of product.

▶SAMPLE PROBLEM 7.13 Moles of Product from a Limiting Reactant

TRY IT FIRST

Carbon monoxide and hydrogen are used to produce methanol (CH_4O). The balanced chemical reaction is

$$CO(g) + 2H_2(g) \longrightarrow CH_4O(g)$$

If 3.00 moles of CO and 5.00 moles of H_2 are the initial reactants, what is the limiting reactant, and how many moles of methanol can be produced?

SOLUTION

STEP 1 **State the given and needed quantities (moles).**

	Given	Need	Connect
ANALYZE THE PROBLEM	3.00 moles of CO, 5.00 moles of H_2	moles of CH_4O produced, limiting reactant	mole–mole factors
	Equation		
	$CO(g) + 2H_2(g) \longrightarrow CH_4O(g)$		

STEP 2 **Write a plan to convert the quantity (moles) of each reactant to quantity (moles) of product.**

moles of CO → Mole–mole factor → moles of CH_4O

moles of H_2 → Mole–mole factor → moles of CH_4O

STEP 3 **Use coefficients to write mole–mole factors.**

1 mole of CO = 1 mole of CH_4O

$$\frac{1 \text{ mole CO}}{1 \text{ mole } CH_4O} \text{ and } \frac{1 \text{ mole } CH_4O}{1 \text{ mole CO}}$$

2 moles of H_2 = 1 mole of CH_4O

$$\frac{2 \text{ moles } H_2}{1 \text{ mole } CH_4O} \text{ and } \frac{1 \text{ mole } CH_4O}{2 \text{ moles } H_2}$$

STEP 4 **Calculate the quantity (moles) of product from each reactant, and select the smaller quantity (moles) as the limiting reactant.**

Moles of CH_4O (product) from CO:

$$3.00 \cancel{\text{moles CO}} \times \frac{1 \text{ mole } CH_4O}{1 \cancel{\text{mole CO}}} = 3.00 \text{ moles of } CH_4O$$

Three SFs — Exact — Three SFs

Moles of CH_4O (product) from H_2:

$$5.00 \cancel{\text{moles } H_2} \times \frac{1 \text{ mole } CH_4O}{2 \cancel{\text{moles } H_2}} = 2.50 \text{ moles of } CH_4O$$

Three SFs; Limiting reactant — Exact — Three SFs; Smaller amount of product

The smaller amount, 2.50 moles of CH_4O, is the maximum amount of methanol that can be produced from the limiting reactant, H_2, because it is completely consumed.

STUDY CHECK 7.13

a. If an initial mixture of reactants for Sample Problem 7.13 contains 4.00 moles of CO and 4.00 moles of H_2, what is the limiting reactant, and how many moles of methanol can be produced?

b. If an initial mixture of reactants for Sample Problem 7.13 contains 1.50 moles of CO and 6.00 moles of H_2, what is the limiting reactant, and how many moles of methanol can be produced?

ANSWER

a. H_2 is the limiting reactant; 2.00 moles of methanol can be produced.

b. CO is the limiting reactant; 1.50 moles of methanol can be produced.

TEST

Try Practice Problems 7.71 to 7.74

Calculating Mass of Product from a Limiting Reactant

The quantities of the reactants can also be given in grams. The calculations to identify the limiting reactant are the same as before, but the grams of each reactant must first be converted to moles, then to moles of product, and finally to grams of product. Then select the smaller mass of product, which is from complete use of the limiting reactant. This calculation is shown in Sample Problem 7.14.

▶ SAMPLE PROBLEM 7.14 Mass of Product from a Limiting Reactant

TRY IT FIRST

When silicon dioxide (sand) and carbon are heated, the products are silicon carbide, SiC, and carbon monoxide. Silicon carbide is a ceramic material that tolerates extreme temperatures and is used as an abrasive and in the brake discs of cars. How many grams of CO are formed from 70.0 g of SiO_2 and 50.0 g of C?

$$SiO_2(s) + 3C(s) \xrightarrow{\Delta} SiC(s) + 2CO(g)$$

A ceramic brake disc in a car withstands temperatures of 1400 °C.

SOLUTION

STEP 1 State the given and needed quantities (grams).

ANALYZE THE PROBLEM	Given	Need	Connect
	70.0 g of SiO_2, 50.0 g of C	grams of CO from limiting reactant	molar masses, mole–mole factors
	Equation		
	$SiO_2(s) + 3C(s) \xrightarrow{\Delta} SiC(s) + 2CO(g)$		

STEP 2 Write a plan to convert the quantity (grams) of each reactant to quantity (grams) of product.

grams of SiO_2 → Molar mass → moles of SiO_2 → Mole–mole factor → moles of CO → Molar mass → grams of CO

grams of C → Molar mass → moles of C → Mole–mole factor → moles of CO → Molar mass → grams of CO

STEP 3 Use coefficients to write mole–mole factors; write molar mass factors.

1 mole of SiO_2 = 60.09 g of SiO_2

$$\frac{60.09\text{ g }SiO_2}{1\text{ mole }SiO_2} \text{ and } \frac{1\text{ mole }SiO_2}{60.09\text{ g }SiO_2}$$

1 mole of C = 12.01 g of C

$$\frac{12.01\text{ g C}}{1\text{ mole C}} \text{ and } \frac{1\text{ mole C}}{12.01\text{ g C}}$$

1 mole of CO = 28.01 g of CO

$$\frac{28.01\text{ g CO}}{1\text{ mole CO}} \text{ and } \frac{1\text{ mole CO}}{28.01\text{ g CO}}$$

2 moles of CO = 1 mole of SiO_2

$$\frac{2\text{ moles CO}}{1\text{ mole }SiO_2} \text{ and } \frac{1\text{ mole }SiO_2}{2\text{ moles CO}}$$

2 moles of CO = 3 moles of C

$$\frac{2\text{ moles CO}}{3\text{ moles C}} \text{ and } \frac{3\text{ moles C}}{2\text{ moles CO}}$$

STEP 4 Calculate the quantity (grams) of product from each reactant, and select the smaller quantity (grams) as the limiting reactant.

Grams of CO (product) from SiO_2:

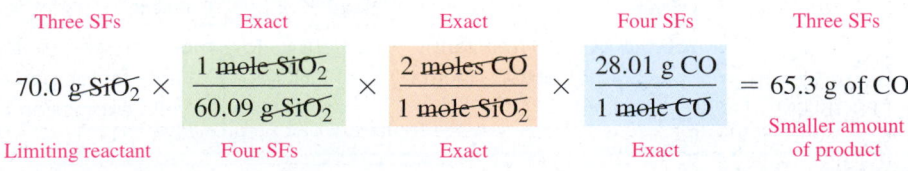

Three SFs | Exact | Exact | Four SFs | Three SFs

$$70.0\ \cancel{\text{g }SiO_2} \times \frac{1\ \cancel{\text{mole }SiO_2}}{60.09\ \cancel{\text{g }SiO_2}} \times \frac{2\ \cancel{\text{moles CO}}}{1\ \cancel{\text{mole }SiO_2}} \times \frac{28.01\text{ g CO}}{1\ \cancel{\text{mole CO}}} = 65.3\text{ g of CO}$$

Limiting reactant | Four SFs | Exact | Exact | Smaller amount of product

Grams of CO (product) from C:

$$\underset{\text{Three SFs}}{50.0\ \cancel{g\ C}} \times \frac{\overset{\text{Exact}}{1\ \cancel{mole\ C}}}{\underset{\text{Four SFs}}{12.01\ \cancel{g\ C}}} \times \frac{\overset{\text{Exact}}{2\ \cancel{moles\ CO}}}{\underset{\text{Exact}}{3\ \cancel{moles\ C}}} \times \frac{\overset{\text{Four SFs}}{28.01\ g\ CO}}{\underset{\text{Exact}}{1\ \cancel{mole\ CO}}} = \overset{\text{Three SFs}}{77.7\ g\ of\ CO}$$

The smaller amount, 65.3 g of CO, is the most CO that can be produced. This also means that the SiO_2 is the limiting reactant.

STUDY CHECK 7.14

Hydrogen sulfide burns with oxygen to give sulfur dioxide and water.

$$2H_2S(g) + 3O_2(g) \xrightarrow{\Delta} 2SO_2(g) + 2H_2O(g)$$

a. How many grams of SO_2 are formed from the reaction of 8.52 g of H_2S and 9.60 g of O_2?
b. How many grams of H_2O are formed from the reaction of 15.0 g of H_2S and 25.0 g of O_2?

ANSWER

a. 12.8 g of SO_2 **b.** 7.93 g of H_2O

TEST

Try Practice Problems 7.75 to 7.78

Percent Yield

In our problems up to now, we assumed that all of the reactants changed completely to product. Thus, we have calculated the amount of product as the maximum quantity possible, or 100%. While this would be an ideal situation, it does not usually happen. As we carry out a reaction and transfer products from one container to another, some product is usually lost. In the lab as well as commercially, the starting materials may not be completely pure, and side reactions may use some of the reactants to give unwanted products. Thus, 100% of the desired product is not actually obtained.

When we do a chemical reaction in the laboratory, we measure out specific quantities of the reactants. We calculate the **theoretical yield** for the reaction, which is the amount of product (100%) we would expect if all the reactants were converted to the desired product. When the reaction ends, we collect and measure the mass of the product, which is the **actual yield** for the product. Because some product is usually lost, the actual yield is less than the theoretical yield. Using the actual yield and the theoretical yield for a product, we can calculate the **percent yield**.

$$\text{Percent yield (\%)} = \frac{\text{actual yield}}{\text{theoretical yield}} \times 100\%$$

ENGAGE

For your chemistry class party, you have prepared cookie dough to make five dozen cookies. However, 12 of the cookies burned and you threw them away. Why is the percent yield of cookies for the chemistry party only 80%?

CORE CHEMISTRY SKILL

Calculating Percent Yield

▶SAMPLE PROBLEM 7.15 Calculating Percent Yield

TRY IT FIRST

On a space shuttle, LiOH is used to absorb exhaled CO_2 from breathing air to form $LiHCO_3$.

$$LiOH(s) + CO_2(g) \longrightarrow LiHCO_3(s)$$

What is the percent yield of $LiHCO_3$ for the reaction if 50.0 g of LiOH gives 72.8 g of $LiHCO_3$?

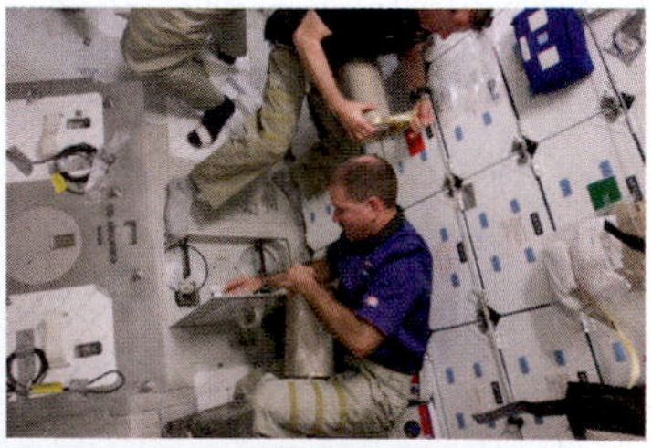

On a space shuttle, the LiOH in the canisters removes CO_2 from the air.

SOLUTION

STEP 1 State the given and needed quantities.

	Given	Need	Connect
ANALYZE THE PROBLEM	50.0 g of LiOH (reactant), 72.8 g of $LiHCO_3$ (actual product)	percent yield of $LiHCO_3$	molar masses, mole–mole factor, percent yield expression
	Equation		
	$LiOH(s) + CO_2(g) \longrightarrow LiHCO_3(s)$		

STEP 2 **Write a plan to calculate the theoretical yield and the percent yield.**

Calculation of theoretical yield:

grams of LiOH → Molar mass → moles of LiOH → Mole–mole factor → moles of $LiHCO_3$ → Molar mass → grams of $LiHCO_3$
Theoretical yield

Calculation of percent yield:

$$\text{Percent yield (\%)} = \frac{\text{actual yield of } LiHCO_3}{\text{theoretical yield of } LiHCO_3} \times 100\%$$

STEP 3 **Use coefficients to write mole–mole factors; write molar mass factors.**

1 mole of LiOH = 23.95 g of LiOH

$$\frac{23.95 \text{ g LiOH}}{1 \text{ mole LiOH}} \text{ and } \frac{1 \text{ mole LiOH}}{23.95 \text{ g LiOH}}$$

1 mole of $LiHCO_3$ = 67.96 g of $LiHCO_3$

$$\frac{67.96 \text{ g } LiHCO_3}{1 \text{ mole } LiHCO_3} \text{ and } \frac{1 \text{ mole } LiHCO_3}{67.96 \text{ g } LiHCO_3}$$

1 mole of $LiHCO_3$ = 1 mole of LiOH

$$\frac{1 \text{ mole } LiHCO_3}{1 \text{ mole LiOH}} \text{ and } \frac{1 \text{ mole LiOH}}{1 \text{ mole } LiHCO_3}$$

STEP 4 **Calculate the percent yield by dividing the actual yield (given) by the theoretical yield and multiplying the result by 100%.**

Calculation of theoretical yield:

$$\underset{\text{Three SFs}}{50.0 \text{ g LiOH}} \times \overset{\text{Exact}}{\underset{\text{Four SFs}}{\frac{1 \text{ mole LiOH}}{23.95 \text{ g LiOH}}}} \times \overset{\text{Exact}}{\underset{\text{Exact}}{\frac{1 \text{ mole } LiHCO_3}{1 \text{ mole LiOH}}}} \times \overset{\text{Four SFs}}{\underset{\text{Exact}}{\frac{67.96 \text{ g } LiHCO_3}{1 \text{ mole } LiHCO_3}}}$$

$$= \underset{\text{Three SFs}}{142 \text{ g of } LiHCO_3}$$

Calculation of percent yield:

$$\frac{\text{actual yield (given)}}{\text{theoretical yield (calculated)}} \times 100\% = \frac{\overset{\text{Three SFs}}{72.8 \text{ g } LiHCO_3}}{\underset{\text{Three SFs}}{142 \text{ g } LiHCO_3}} \times 100\% = \underset{\text{Three SFs}}{51.3\%}$$

STUDY CHECK 7.15

Using the equation in Sample Problem 7.15, calculate each of the following:

a. The percent yield of $LiHCO_3$ if 8.00 g of CO_2 produces 10.5 g of $LiHCO_3$
b. The percent yield of $LiHCO_3$ if 35.0 g of LiOH produces 76.6 g of $LiHCO_3$

ANSWER

a. 84.7% **b.** 77.1%

TEST
Try Practice Problems 7.79 to 7.84

PRACTICE PROBLEMS

7.9 Limiting Reactants and Percent Yield

7.71 A taxi company has 10 taxis.
a. On a certain day, only eight taxi drivers show up for work. How many taxis can be used to pick up passengers?
b. On another day, 10 taxi drivers show up for work but three taxis are in the repair shop. How many taxis can be driven?

7.72 A clock maker has 15 clock faces. Each clock requires one face and two hands.
a. If the clock maker has 42 hands, how many clocks can be produced?
b. If the clock maker has eight hands, how many clocks can be produced?

7.73 Nitrogen and hydrogen react to form ammonia.

$$N_2(g) + 3H_2(g) \longrightarrow 2NH_3(g)$$

Determine the limiting reactant in each of the following mixtures of reactants:

a. 3.0 moles of N_2 and 5.0 moles of H_2
b. 8.0 moles of N_2 and 4.0 moles of H_2
c. 3.0 moles of N_2 and 12.0 moles of H_2

7.74 Iron and oxygen react to form iron(III) oxide.

$$4Fe(s) + 3O_2(g) \longrightarrow 2Fe_2O_3(s)$$

Determine the limiting reactant in each of the following mixtures of reactants:

a. 2.0 moles of Fe and 6.0 moles of O_2
b. 5.0 moles of Fe and 4.0 moles of O_2
c. 16.0 moles of Fe and 20.0 moles of O_2

7.75 For each of the following reactions, 20.0 g of each reactant is present initially. Determine the limiting reactant, and calculate the grams of product in parentheses that would be produced.

a. $2Al(s) + 3Cl_2(g) \longrightarrow 2AlCl_3(s)$ ($AlCl_3$)
b. $4NH_3(g) + 5O_2(g) \xrightarrow{\Delta} 4NO(g) + 6H_2O(g)$ (H_2O)
c. $CS_2(g) + 3O_2(g) \xrightarrow{\Delta} CO_2(g) + 2SO_2(g)$ (SO_2)

7.76 For each of the following reactions, 20.0 g of each reactant is present initially. Determine the limiting reactant, and calculate the grams of product in parentheses that would be produced.

a. $4Al(s) + 3O_2(g) \longrightarrow 2Al_2O_3(s)$ (Al_2O_3)
b. $3NO_2(g) + H_2O(l) \longrightarrow 2HNO_3(aq) + NO(g)$ (HNO_3)
c. $C_2H_6O(l) + 3O_2(g) \xrightarrow{\Delta} 2CO_2(g) + 3H_2O(g)$ (H_2O)

7.77 For each of the following reactions, calculate the grams of indicated product when 25.0 g of the first reactant and 40.0 g of the second reactant is used:

a. $2SO_2(g) + O_2(g) \longrightarrow 2SO_3(g)$ (SO_3)
b. $3Fe(s) + 4H_2O(l) \longrightarrow Fe_3O_4(s) + 4H_2(g)$ (Fe_3O_4)
c. $C_7H_{16}(g) + 11O_2(g) \xrightarrow{\Delta} 7CO_2(g) + 8H_2O(g)$ (CO_2)

7.78 For each of the following reactions, calculate the grams of indicated product when 15.0 g of the first reactant and 10.0 g of the second reactant is used:

a. $4Li(s) + O_2(g) \longrightarrow 2Li_2O(s)$ (Li_2O)
b. $Fe_2O_3(s) + 3H_2(g) \longrightarrow 2Fe(s) + 3H_2O(l)$ (Fe)
c. $Al_2S_3(s) + 6H_2O(l) \longrightarrow 2Al(OH)_3(s) + 3H_2S(g)$ (H_2S)

7.79 Carbon disulfide is produced by the reaction of carbon and sulfur dioxide.

$$5C(s) + 2SO_2(g) \longrightarrow CS_2(g) + 4CO(g)$$

a. What is the percent yield of carbon disulfide if the reaction of 40.0 g of carbon produces 36.0 g of carbon disulfide?
b. What is the percent yield of carbon disulfide if the reaction of 32.0 g of sulfur dioxide produces 12.0 g of carbon disulfide?

7.80 Iron(III) oxide reacts with carbon monoxide to produce iron and carbon dioxide.

$$Fe_2O_3(s) + 3CO(g) \longrightarrow 2Fe(s) + 3CO_2(g)$$

a. What is the percent yield of iron if the reaction of 65.0 g of iron(III) oxide produces 15.0 g of iron?
b. What is the percent yield of carbon dioxide if the reaction of 75.0 g of carbon monoxide produces 85.0 g of carbon dioxide?

7.81 Aluminum reacts with oxygen to produce aluminum oxide.

$$4Al(s) + 3O_2(g) \longrightarrow 2Al_2O_3(s)$$

Calculate the mass of Al_2O_3 that can be produced if the reaction of 50.0 g of aluminum and sufficient oxygen has a 75.0% yield.

7.82 Propane (C_3H_8) burns in oxygen to produce carbon dioxide and water.

$$C_3H_8(g) + 5O_2(g) \xrightarrow{\Delta} 3CO_2(g) + 4H_2O(g)$$

Calculate the mass of CO_2 that can be produced if the reaction of 45.0 g of propane and sufficient oxygen has a 60.0% yield.

7.83 When 30.0 g of carbon is heated with silicon dioxide, 28.2 g of carbon monoxide is produced. What is the percent yield of carbon monoxide for this reaction?

$$SiO_2(s) + 3C(s) \xrightarrow{\Delta} SiC(s) + 2CO(g)$$

7.84 When 56.6 g of calcium is reacted with nitrogen gas, 32.4 g of calcium nitride is produced. What is the percent yield of calcium nitride for this reaction?

$$3Ca(s) + N_2(g) \longrightarrow Ca_3N_2(s)$$

7.10 Energy in Chemical Reactions

LEARNING GOAL Given the heat of reaction, calculate the loss or gain of heat for an exothermic or endothermic reaction.

Almost every chemical reaction involves a loss or gain of energy. To discuss energy change for a reaction, we look at the energy of the reactants before the reaction and the energy of the products after the reaction.

Energy Units for Chemical Reactions

The SI unit for energy is the *joule* (J). Often, the unit of *kilojoules* (kJ) is used to show the energy change in a reaction.

$$1 \text{ kilojoule (kJ)} = 1000 \text{ joules (J)}$$

Heat of Reaction

The **heat of reaction** is the amount of heat absorbed or released during a reaction that takes place at constant pressure. A change of energy occurs as reactants interact, bonds break apart, and products form. We determine a heat of reaction, symbol ΔH, as the difference in the energy of the products and the reactants.

$$\Delta H = H_{\text{products}} - H_{\text{reactants}}$$

Exothermic Reactions

In an **exothermic reaction** (*exo* means "out"), the energy of the products is lower than that of the reactants. This means that heat is released along with the products that form. Let us look at the equation for the exothermic reaction in which 185 kJ of heat is released when 1 mole of hydrogen and 1 mole of chlorine react to form 2 moles of hydrogen chloride. For an exothermic reaction, the heat of reaction can be written as one of the products. It can also be written as a ΔH value with a negative sign (−).

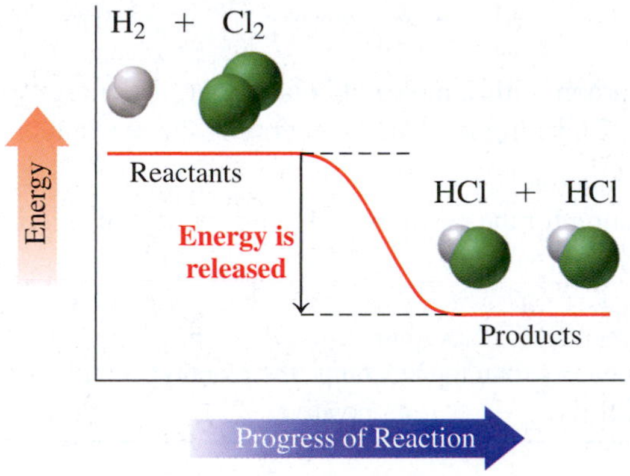

H_2 + Cl_2
Energy
Reactants
Energy is released
HCl + HCl
Products
Progress of Reaction

In an exothermic reaction, the energy of the products is lower than the energy of the reactants.

Exothermic, Heat Released	Heat Is a Product
$H_2(g) + Cl_2(g) \longrightarrow 2HCl(g) + 185\text{ kJ}$	$\Delta H = -185\text{ kJ}$ Negative sign

Endothermic Reactions

In an **endothermic reaction** (*endo* means "within"), the energy of the products is higher than that of the reactants. Heat is required to convert the reactants to products. Let us look at the equation for the endothermic reaction in which 180 kJ of heat is needed to convert 1 mole of nitrogen and 1 mole of oxygen to 2 moles of nitrogen oxide. For an endothermic reaction, the heat of reaction can be written as one of the reactants. It can also be written as a ΔH value with a positive sign (+).

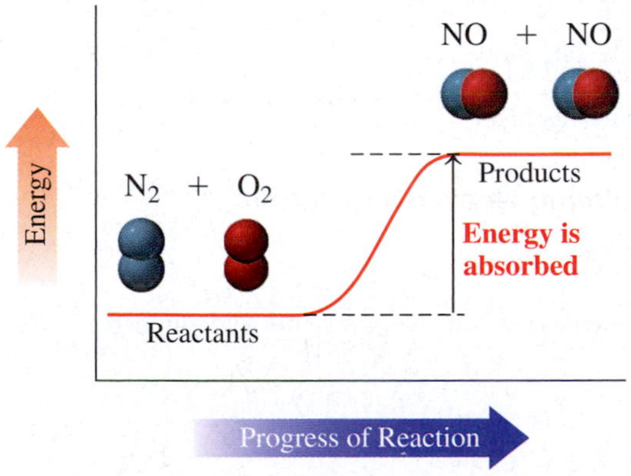

NO + NO
Energy
N_2 + O_2
Products
Energy is absorbed
Reactants
Progress of Reaction

In an endothermic reaction, the energy of the products is higher than the energy of the reactants.

Endothermic, Heat Absorbed	Heat Is a Reactant
$N_2(g) + O_2(g) + 180\text{ kJ} \longrightarrow 2NO(g)$	$\Delta H = +180\text{ kJ}$ Positive sign

Reaction	Energy Change	Heat in the Equation	Sign of ΔH
Exothermic	Heat released	Product side	Negative (−)
Endothermic	Heat absorbed	Reactant side	Positive (+)

▶SAMPLE PROBLEM 7.16 Exothermic and Endothermic Reactions

TRY IT FIRST

In the reaction of 1 mole of solid carbon with oxygen gas, the energy of the carbon dioxide gas produced is 393 kJ lower than the energy of the reactants.

a. Is the reaction exothermic or endothermic?
b. Write the balanced chemical equation for the reaction, including the heat of reaction.
c. Is the sign of ΔH positive or negative?

SOLUTION

a. When the products have a lower energy than the reactants, the reaction is exothermic.
b. $C(s) + O_2(g) \longrightarrow CO_2(g) + 393\text{ kJ}$ **c.** negative

STUDY CHECK 7.16

In the reaction of 1 mole of solid carbon with 2 moles of solid sulfur, the energy of the liquid carbon disulfide produced is 92.0 kJ higher than the energy of the reactants.

a. Is the reaction exothermic or endothermic?
b. Write the balanced chemical equation for the reaction, including the heat of reaction.
c. Is the sign of ΔH positive or negative?

ANSWER

a. When the products have a higher energy than the reactants, the reaction is endothermic.
b. $C(s) + 2S(s) + 92.0\text{ kJ} \longrightarrow CS_2(l)$ **c.** positive

TEST
Try Practice Problems 7.85 to 7.90

Calculations of Heat in Reactions

The value of ΔH refers to the heat change, in kilojoules, for each substance in the balanced equation for the reaction. Consider the following reaction:

$$2H_2O(l) \longrightarrow 2H_2(g) + O_2(g) \quad \Delta H = +572\text{ kJ}$$
$$2H_2O(l) + 572\text{ kJ} \longrightarrow 2H_2(g) + O_2(g)$$

For this reaction, 572 kJ are absorbed by 2 moles of H_2O to produce 2 moles of H_2 and 1 mole of O_2. We can write two heat conversion factors for each substance in this reaction as

$$\frac{+572\text{ kJ}}{2\text{ moles } H_2O} \text{ and } \frac{2\text{ moles } H_2O}{+572\text{ kJ}} \qquad \frac{+572\text{ kJ}}{1\text{ mole } O_2} \text{ and } \frac{1\text{ mole } O_2}{+572\text{ kJ}}$$

Suppose in this reaction that 9.00 g of H_2O undergoes reaction. We can calculate the kilojoules of heat absorbed as

CORE CHEMISTRY SKILL
Using the Heat of Reaction

$$9.00\ \cancel{\text{g } H_2O} \times \frac{1\ \cancel{\text{mole } H_2O}}{18.02\ \cancel{\text{g } H_2O}} \times \frac{+572\text{ kJ}}{2\ \cancel{\text{moles } H_2O}} = +143\text{ kJ}$$

▶SAMPLE PROBLEM 7.17 Calculating Heat in a Reaction

TRY IT FIRST

How much heat, in kilojoules, is released when nitrogen and hydrogen react to form 50.0 g of ammonia?

$$N_2(g) + 3H_2(g) \longrightarrow 2NH_3(g) \qquad \Delta H = -92.2\text{ kJ}$$

SOLUTION

STEP 1 State the given and needed quantities.

	Given	Need	Connect
ANALYZE THE PROBLEM	50.0 g of NH_3	heat released, in kilojoules	$\Delta H = -92.2$ kJ, molar mass
	Equation		
	$N_2(g) + 3H_2(g) \longrightarrow 2NH_3(g)$		

STEP 2 **Write a plan using the heat of reaction and any molar mass needed.**

grams of NH_3 → (Molar mass) → moles of NH_3 → (Heat of reaction) → kilojoules

STEP 3 **Write the conversion factors including heat of reaction.**

1 mole of NH_3 = 17.03 g of NH_3	2 moles of NH_3 = −92.2 kJ
$\frac{17.03\text{ g NH}_3}{1\text{ mole NH}_3}$ and $\frac{1\text{ mole NH}_3}{17.03\text{ g NH}_3}$	$\frac{-92.2\text{ kJ}}{2\text{ moles NH}_3}$ and $\frac{2\text{ moles NH}_3}{-92.2\text{ kJ}}$

STEP 4 **Set up the problem to calculate the heat.**

$$\underset{\text{Three SFs}}{50.0\text{ g NH}_3} \times \overset{\text{Exact}}{\underset{\text{Four SFs}}{\frac{1\text{ mole NH}_3}{17.03\text{ g NH}_3}}} \times \overset{\text{Three SFs}}{\underset{\text{Exact}}{\frac{-92.2\text{ kJ}}{2\text{ moles NH}_3}}} = \underset{\text{Three SFs}}{-135\text{ kJ}}$$

Therefore, 135 kJ are released.

STUDY CHECK 7.17

Mercury(II) oxide decomposes to mercury and oxygen.

$$2HgO(s) \longrightarrow 2Hg(l) + O_2(g) \qquad \Delta H = +182\text{ kJ}$$

a. Is the reaction exothermic or endothermic?
b. How many kilojoules are needed when 25.0 g of mercury(II) oxide reacts?

ANSWER

a. endothermic **b.** 20.5 kJ

TEST
Try Practice Problems 7.91 to 7.94

Chemistry Link to Health

Cold Packs and Hot Packs

In a hospital, at a first-aid station, or at an athletic event, an instant cold pack may be used to reduce swelling from an injury, remove heat from inflammation, or decrease capillary size to lessen the effect of hemorrhaging. Inside the plastic container of a cold pack, there is a compartment containing solid ammonium nitrate (NH_4NO_3) that is separated from a compartment containing water. The pack is activated when it is hit or squeezed hard enough to break the walls between the compartments and cause the ammonium nitrate to mix with the water (shown as H_2O over the reaction arrow). In an endothermic process, 1 mole of NH_4NO_3 that dissolves absorbs 26 kJ of heat. The temperature drops to about 4 to 5 °C to give a cold pack that is ready to use.

Endothermic Reaction in a Cold Pack

$$NH_4NO_3(s) + 26\text{ kJ} \xrightarrow{H_2O} NH_4NO_3(aq)$$

Hot packs are used to relax muscles, lessen aches and cramps, and increase circulation by expanding capillary size. Constructed in the same way as cold packs, a hot pack contains a salt such as $CaCl_2$. When 1 mole of $CaCl_2$ dissolves in water, 82 kJ are released as heat. The temperature increases as much as 66 °C to give a hot pack that is ready to use.

Exothermic Reaction in a Hot Pack

$$CaCl_2(s) \xrightarrow{H_2O} CaCl_2(aq) + 82\text{ kJ}$$

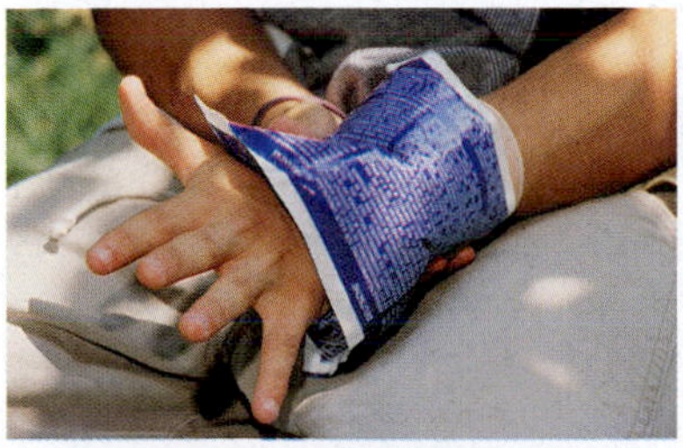

Cold packs use an endothermic reaction.

PRACTICE PROBLEMS

7.10 Energy in Chemical Reactions

7.85 In an exothermic reaction, is the energy of the products higher or lower than that of the reactants?

7.86 In an endothermic reaction, is the energy of the products higher or lower than that of the reactants?

7.87 Classify each of the following as exothermic or endothermic:
a. A reaction releases 550 kJ.
b. The energy level of the products is higher than that of the reactants.
c. The metabolism of glucose in the body provides energy.

7.88 Classify each of the following as exothermic or endothermic:
a. The energy level of the products is lower than that of the reactants.
b. In the body, the synthesis of proteins requires energy.
c. A reaction absorbs 125 kJ.

7.89 Classify each of the following as exothermic or endothermic and give the ΔH for each:
a. $CH_4(g) + 2O_2(g) \xrightarrow{\Delta} CO_2(g) + 2H_2O(g) + 802\text{ kJ}$
b. $Ca(OH)_2(s) + 65.3\text{ kJ} \longrightarrow CaO(s) + H_2O(l)$
c. $2Al(s) + Fe_2O_3(s) \longrightarrow Al_2O_3(s) + 2Fe(l) + 850\text{ kJ}$

The thermite reaction of aluminum and iron(III) oxide produces very high temperatures used to cut or weld railroad tracks.

7.90 Classify each of the following as exothermic or endothermic and give the ΔH for each:
a. $C_3H_8(g) + 5O_2(g) \xrightarrow{\Delta} 3CO_2(g) + 4H_2O(g) + 2220\text{ kJ}$
b. $2Na(s) + Cl_2(g) \longrightarrow 2NaCl(s) + 819\text{ kJ}$
c. $PCl_5(g) + 67\text{ kJ} \longrightarrow PCl_3(g) + Cl_2(g)$

7.91 a. How many kilojoules are released when 125 g of Cl_2 reacts with silicon?
$Si(s) + 2Cl_2(g) \longrightarrow SiCl_4(g) \quad \Delta H = -657\text{ kJ}$
b. How many kilojoules are absorbed when 278 g of PCl_5 reacts?
$PCl_5(g) \longrightarrow PCl_3(g) + Cl_2(g) \quad \Delta H = +67\text{ kJ}$

7.92 a. How many kilojoules are released when 75.0 g of CH_4O reacts?
$2CH_4O(l) + 3O_2(g) \xrightarrow{\Delta} 2CO_2(g) + 4H_2O(l) \quad \Delta H = -726\text{ kJ}$
b. How many kilojoules are absorbed when 315 g of $Ca(OH)_2$ reacts?
$Ca(OH)_2(s) \longrightarrow CaO(s) + H_2O(l) \quad \Delta H = +65.3\text{ kJ}$

Clinical Applications

7.93 In photosynthesis, glucose, $C_6H_{12}O_6$, and O_2 are produced from CO_2 and H_2O. Glucose from starches is the major fuel for the body.
$6CO_2(g) + 6H_2O(l) + 680\text{ kcal} \longrightarrow C_6H_{12}O_6(aq) + 6O_2(g)$
Glucose
a. Is the reaction endothermic or exothermic?
b. How many grams of glucose are produced from 18.0 g of CO_2?
c. How much heat, in kilojoules, is needed to produce 25.0 g of $C_6H_{12}O_6$?

7.94 Ethanol, C_2H_6O, reacts in the body using the following equation:
$C_2H_6O(aq) + 3O_2(g) \longrightarrow 2CO_2(g) + 3H_2O(l) + 327\text{ kcal}$
Ethanol
a. Is the reaction endothermic or exothermic?
b. How much heat, in kilocalories, is produced when 5.00 g of ethanol reacts with O_2 gas?
c. How many grams of ethanol react if 1250 kJ are produced?

CLINICAL UPDATE Improving Natalie's Overall Fitness

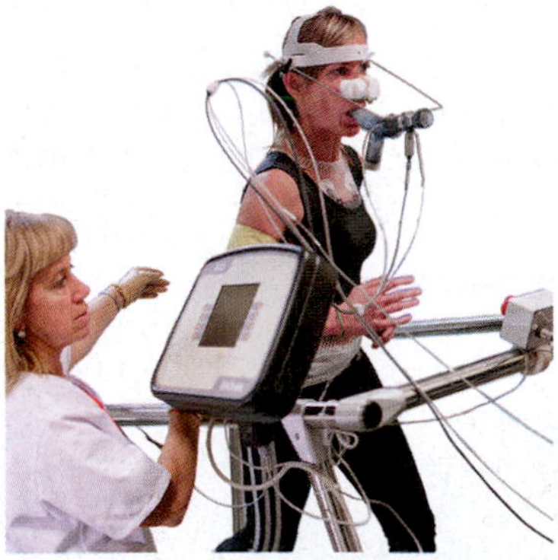

Natalie's test results indicate that she has a blood oxygen level of 89%. The normal values for pulse oximeter readings are 95% to 100%, which means that Natalie's O_2 saturation is low. Thus, Natalie does not have an adequate amount of O_2 in her blood and may be *hypoxic*. This may be the reason she has noticed a shortness of breath and a dry cough. Her doctor diagnoses her with *interstitial lung disease*, which is scarring of the tissue of the lungs.

Angela teaches Natalie to inhale and exhale slower and deeper to fill the lungs with more air and thus more oxygen. Angela also develops a workout program with the goal of increasing Natalie's overall fitness level. During the exercises, Angela continues to monitor Natalie's heart rate, blood O_2 level, and blood pressure to ensure that Natalie is exercising at a level that will enable her to become stronger without breaking down muscle due to a lack of oxygen.

Low-intensity exercises are used at the beginning of Natalie's exercise program.

Clinical Applications

7.95 a. During cellular respiration, aqueous $C_6H_{12}O_6$ (glucose) in the cells undergoes a reaction with oxygen gas to form gaseous carbon dioxide and liquid water. Write and balance the chemical equation for reaction of glucose in the human body.
b. In plants, carbon dioxide gas and liquid water are converted to aqueous glucose ($C_6H_{12}O_6$) and oxygen gas. Write and balance the chemical equation for production of glucose in plants.

7.96 Fatty acids undergo reaction with oxygen gas and form gaseous carbon dioxide and liquid water when utilized for energy in the body.
a. Write and balance the equation for the reaction of the fatty acid aqueous capric acid, $C_{10}H_{20}O_2$.
b. Write and balance the equation for the reaction of the fatty acid aqueous myristic acid, $C_{14}H_{28}O_2$.

CONCEPT MAP

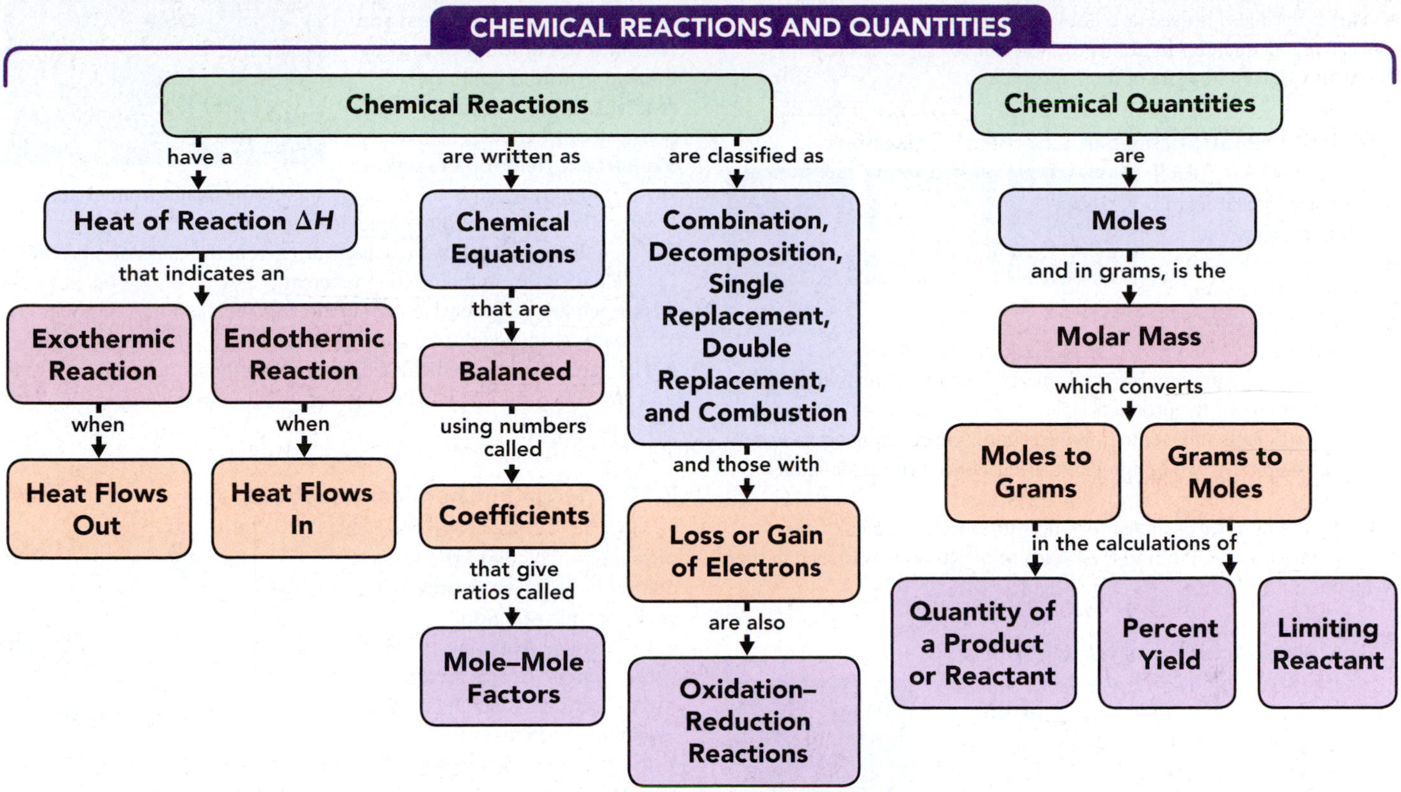
CHEMICAL REACTIONS AND QUANTITIES
Chemical Reactions
have a
Heat of Reaction ΔH
that indicates an
Exothermic Reaction
Endothermic Reaction
when
Heat Flows Out
when
Heat Flows In
are written as
Chemical Equations
that are
Balanced
using numbers called
Coefficients
that give ratios called
Mole–Mole Factors
are classified as
Combination, Decomposition, Single Replacement, Double Replacement, and Combustion
and those with
Loss or Gain of Electrons
are also
Oxidation–Reduction Reactions
Chemical Quantities
are
Moles
and in grams, is the
Molar Mass
which converts
Moles to Grams
Grams to Moles
in the calculations of
Quantity of a Product or Reactant
Percent Yield
Limiting Reactant

CHAPTER REVIEW

7.1 Equations for Chemical Reactions

LEARNING GOAL Write a balanced chemical equation from the formulas of the reactants and products for a reaction; determine the number of atoms in the reactants and products.

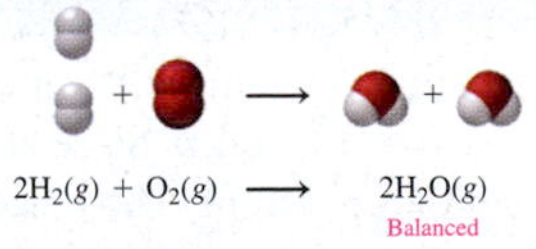
$2H_2(g) + O_2(g) \longrightarrow 2H_2O(g)$
Balanced

- A chemical change occurs when the atoms of the initial substances rearrange to form new substances.
- A chemical equation shows the formulas of the substances that react on the left side of a reaction arrow and the products that form on the right side of the reaction arrow.
- A chemical equation is balanced by writing coefficients, small whole numbers, in front of formulas to equalize the atoms of each of the elements in the reactants and the products.

7.2 Types of Chemical Reactions

LEARNING GOAL Identify a chemical reaction as a combination, decomposition, single replacement, double replacement, or combustion.

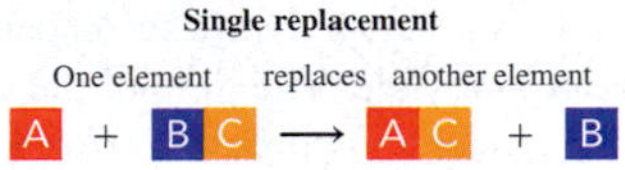
Single replacement
One element replaces another element
A + BC ⟶ AC + B

- Many chemical reactions can be organized by reaction type: combination, decomposition, single replacement, double replacement, or combustion.

7.3 Oxidation–Reduction Reactions

LEARNING GOAL Define the terms oxidation and reduction; identify the reactants oxidized and reduced.

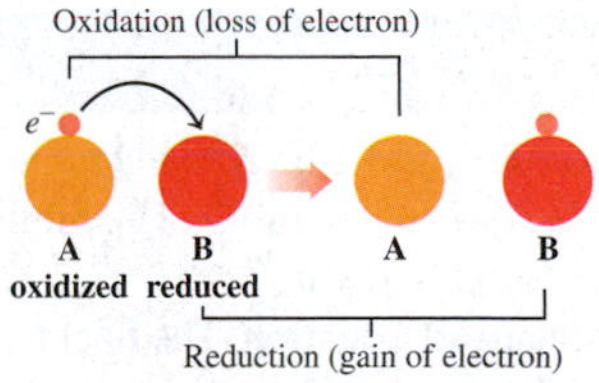
Oxidation (loss of electron)
e⁻
A
B
A
B
oxidized
reduced
Reduction (gain of electron)

- When electrons are transferred in a reaction, it is an oxidation–reduction reaction.
- One reactant loses electrons, and another reactant gains electrons.
- Overall, the number of electrons lost and gained is equal.

7.4 The Mole

LEARNING GOAL Use Avogadro's number to determine the number of particles in a given number of moles.

- One mole of an element contains 6.02×10^{23} atoms.
- One mole of a compound contains 6.02×10^{23} molecules or formula units.

7.5 Molar Mass

LEARNING GOAL Given the chemical formula of a substance, calculate its molar mass.

- The molar mass (g/mole) of any substance is the mass in grams equal numerically to its atomic mass, or the sum of the atomic masses, which have been multiplied by their subscripts in a formula.
- The molar mass is used as a conversion factor to change a quantity in grams to moles or to change a given number of moles to grams.

7.6 Calculations Using Molar Mass

LEARNING GOAL Use molar mass to convert between grams and moles.

- The molar mass is used as a conversion factor to change a quantity in grams to moles or to change a given number of moles to grams.

7.7 Mole Relationships in Chemical Equations

LEARNING GOAL Use a mole–mole factor from a balanced chemical equation to calculate the number of moles of another substance in the reaction.

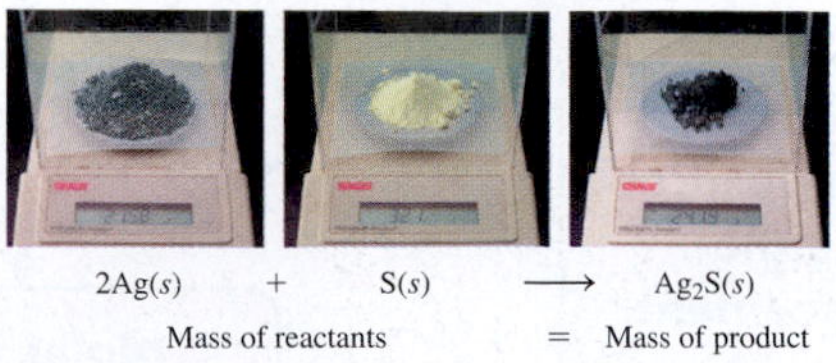
$2Ag(s)$ + $S(s)$ ⟶ $Ag_2S(s)$
Mass of reactants = Mass of product

- In a balanced equation, the total mass of the reactants is equal to the total mass of the products.
- The coefficients in an equation describing the relationship between the moles of any two components are used to write mole–mole factors.
- When the number of moles for one substance is known, a mole–mole factor is used to find the moles of a different substance in the reaction.

7.8 Mass Calculations for Chemical Reactions

LEARNING GOAL Given the mass in grams of a substance in a reaction, calculate the mass in grams of another substance in the reaction.

- In calculations using equations, the molar masses of the substances and their mole–mole factors are used to change the number of grams of one substance to the corresponding grams of a different substance.

7.9 Limiting Reactants and Percent Yield

LEARNING GOAL Identify a limiting reactant when given the quantities of two reactants; and calculate the amount of product formed from the limiting reactant. Given the actual quantity of product, calculate the percent yield for a reaction.

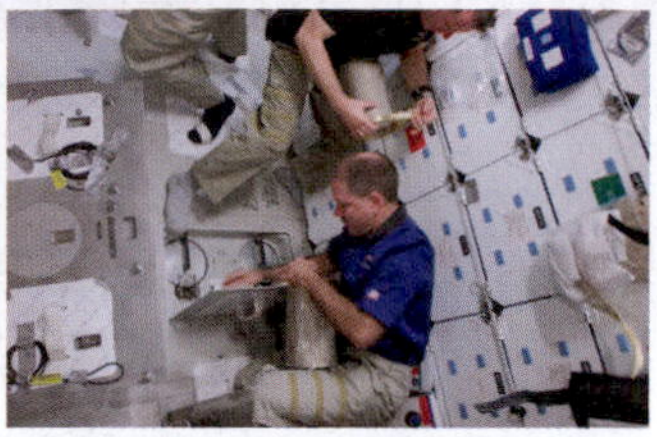

- A limiting reactant is the reactant that produces the smaller amount of product while the other reactant is left over.
- When the masses of two reactants are given, the mass of a product is calculated from the limiting reactant.
- The percent yield for a reaction indicates the percent of product actually produced during a reaction.
- The percent yield is calculated by dividing the actual yield in grams of a product by the theoretical yield in grams and multiplying by 100%.

7.10 Energy in Chemical Reactions

LEARNING GOAL Given the heat of reaction, calculate the loss or gain of heat for an exothermic or endothermic reaction.

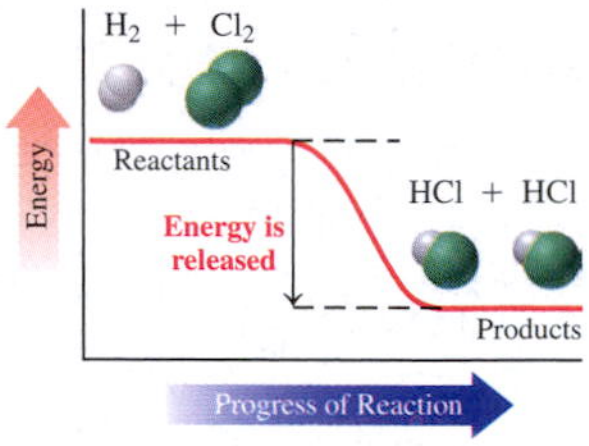
H_2 + Cl_2
Energy
Reactants
Energy is released
HCl + HCl
Products
Progress of Reaction

- In chemical reactions, the heat of reaction (ΔH) is the energy difference between the products and the reactants.
- In an exothermic reaction, the energy of the products is lower than that of the reactants. Heat is released, and ΔH is negative.
- In an endothermic reaction, the energy of the products is higher than that of the reactants. Heat is absorbed, and ΔH is positive.

KEY TERMS

actual yield The actual amount of product produced by a reaction.

Avogadro's number The number of items in a mole, equal to 6.02×10^{23}.

balanced equation The final form of a chemical equation that shows the same number of atoms of each element in the reactants and products.

chemical equation A shorthand way to represent a chemical reaction using chemical formulas to indicate the reactants and products and coefficients to show reacting ratios.

coefficients Whole numbers placed in front of the formulas to balance the number of atoms or moles of atoms of each element on both sides of an equation.

combination reaction A chemical reaction in which reactants combine to form a single product.

combustion reaction A chemical reaction in which a fuel containing carbon and hydrogen reacts with oxygen to produce CO_2, H_2O, and energy.

decomposition reaction A reaction in which a single reactant splits into two or more simpler substances.

double replacement reaction A reaction in which the positive ions in the reacting compounds exchange places.

endothermic reaction A reaction that requires heat; the energy of the products is higher than the energy of the reactants.

exothermic reaction A reaction that releases heat; the energy of the products is lower than the energy of the reactants.

formula unit The group of ions represented by the formula of an ionic compound.

heat of reaction The heat (symbol ΔH) absorbed or released when a reaction takes place at constant pressure.

limiting reactant The reactant used up during a chemical reaction, which limits the amount of product that can form.

molar mass The mass in grams of 1 mole of an element equal numerically to its atomic mass. The molar mass of a compound is equal to the sum of the masses of the elements in the formula.

mole A group of atoms, molecules, or formula units that contains 6.02×10^{23} of these items.

mole–mole factor A conversion factor that relates the number of moles of two compounds in an equation derived from the coefficients.

oxidation The loss of electrons by a substance. Biological oxidation may involve the addition of oxygen or the loss of hydrogen.

oxidation–reduction reaction A reaction in which the oxidation of one reactant is always accompanied by the reduction of another reactant.

percent yield The ratio of the actual yield for a reaction to the theoretical yield possible for the reaction.

products The substances formed as a result of a chemical reaction.

reactants The initial substances that undergo change in a chemical reaction.

reduction The gain of electrons by a substance. Biological reduction may involve the loss of oxygen or the gain of hydrogen.

single replacement reaction A reaction in which an element replaces a different element in a compound.

theoretical yield The maximum amount of product that a reaction can produce from a given amount of reactant.

CORE CHEMISTRY SKILLS

The chapter Section containing each Core Chemistry Skill is shown in parentheses at the end of each heading.

Balancing a Chemical Equation (7.1)

- In a *balanced* chemical equation, whole numbers called coefficients multiply each of the atoms in the chemical formulas so that the number of each type of atom in the reactants is equal to the number of the same type of atom in the products.

Example: Balance the following chemical equation:

$$SnCl_4(s) + H_2O(l) \longrightarrow Sn(OH)_4(s) + HCl(aq) \quad \text{Not balanced}$$

Answer: When we compare the atoms on the reactant side and the product side, we see that there are more Cl atoms in the reactants and more O and H atoms in the products.

To balance the equation, we need to use coefficients in front of the formulas containing the Cl atoms, H atoms, and O atoms.

- Place a 4 in front of the formula HCl to give 8 H atoms and 4 Cl atoms in the products.

$$SnCl_4(s) + H_2O(l) \longrightarrow Sn(OH)_4(s) + 4HCl(aq)$$

- Place a 4 in front of the formula H_2O to give 8 H atoms and 4 O atoms in the reactants.

$$SnCl_4(s) + 4H_2O(l) \longrightarrow Sn(OH)_4(s) + 4HCl(aq)$$

- The total number of Sn (1), Cl (4), H (8), and O (4) atoms is now equal on both sides of the equation. Thus, this equation is balanced.

Classifying Types of Chemical Reactions (7.2)

- Chemical reactions are classified by identifying general patterns in their equations.
- In a combination reaction, two or more elements or compounds bond to form one product.
- In a decomposition reaction, a single reactant splits into two or more products.
- In a single replacement reaction, an uncombined element takes the place of an element in a compound.
- In a double replacement reaction, the positive ions in the reacting compounds switch places.
- In combustion reaction, a fuel containing carbon and hydrogen reacts with oxygen from the air to produce carbon dioxide (CO_2), water (H_2O), and energy.

Example: Classify the type of the following reaction:

$$2Al(s) + Fe_2O_3(s) \xrightarrow{\Delta} Al_2O_3(s) + 2Fe(l)$$

Answer: The iron in iron(III) oxide is replaced by aluminum, which makes this a single replacement reaction.

Identifying Oxidized and Reduced Substances (7.3)

- In an oxidation–reduction reaction (abbreviated *redox*), one reactant is oxidized when it loses electrons, and another reactant is reduced when it gains electrons.
- Oxidation is the *loss* of electrons; reduction is the *gain* of electrons.

Example: For the following redox reaction, identify the reactant that is oxidized, and the reactant that is reduced:

$$Fe(s) + Cu^{2+}(aq) \longrightarrow Fe^{2+}(aq) + Cu(s)$$

Answer: $Fe^0(s) \longrightarrow Fe^{2+}(aq) + 2\,e^-$

Fe loses electrons; it is oxidized.

$Cu^{2+}(aq) + 2\,e^- \longrightarrow Cu^0(s)$

Cu^{2+} gains electrons; it is reduced.

Converting Particles to Moles (7.4)

- In chemistry, atoms, molecules, and ions are counted by the mole, a unit that contains 6.02×10^{23} items, which is Avogadro's number.
- For example, 1 mole of carbon contains 6.02×10^{23} atoms of carbon and 1 mole of H_2O contains 6.02×10^{23} molecules of H_2O.
- Avogadro's number is used to convert between particles and moles.

Example: How many moles of nickel contain 2.45×10^{24} Ni atoms?

Answer: $2.45 \times 10^{24}\ \cancel{\text{Ni atoms}} \times \dfrac{1\ \text{mole Ni}}{6.02 \times 10^{23}\ \cancel{\text{Ni atoms}}}$ (Three SFs; Exact / Three SFs)

$= 4.07$ moles of Ni (Three SFs)

Calculating Molar Mass (7.5)

Pinene is a component of pine sap.

- The molar mass of an element is its mass in grams equal numerically to its atomic mass.
- The molar mass of a compound is the sum of the molar mass of each element in its chemical formula multiplied by its subscript in the formula.

Example: Pinene, $C_{10}H_{16}$, which is found in pine tree sap and essential oils, has anti-inflammatory properties. Calculate the molar mass for pinene.

Answer:

$$10\ \cancel{\text{moles C}} \times \frac{12.01\ \text{g C}}{1\ \cancel{\text{mole C}}} = 120.1\ \text{g of C}$$

$$16\ \cancel{\text{moles H}} \times \frac{1.008\ \text{g H}}{1\ \cancel{\text{mole H}}} = 16.13\ \text{g of H}$$

$$\text{Molar mass of } C_{10}H_{16} = 136.2\ \text{g}$$

Using Molar Mass as a Conversion Factor (7.6)

A racing bicycle has an aluminum frame.

- Molar mass is used as a conversion factor to convert between the moles and grams of a substance.

Example: The frame of a bicycle contains 6500 g of aluminum. How many moles of aluminum are in the bicycle frame?

Equality: 1 mole of Al = 26.98 g of Al

Conversion Factors: $\dfrac{26.98\ \text{g Al}}{1\ \text{mole Al}}$ and $\dfrac{1\ \text{mole Al}}{26.98\ \text{g Al}}$

Answer: $6500\ \cancel{\text{g Al}} \times \dfrac{1\ \text{mole Al}}{26.98\ \cancel{\text{g Al}}} = 240$ moles of Al

(Two SFs; Exact / Four SFs; Two SFs)

Using Mole–Mole Factors (7.7)

Consider the balanced chemical equation

$$4Na(s) + O_2(g) \longrightarrow 2Na_2O(s)$$

- The coefficients in a balanced chemical equation represent the moles of reactants and the moles of products. Thus, 4 moles of Na react with 1 mole of O_2 to form 2 moles of Na_2O.

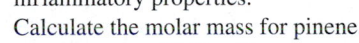

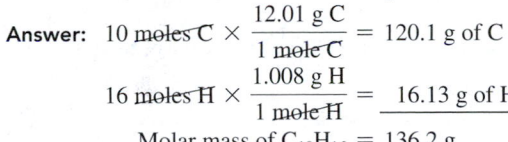

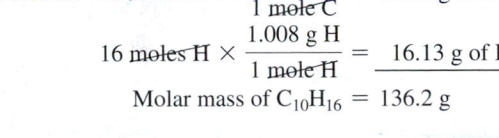

- From the coefficients, mole–mole factors can be written for any two substances as follows:

Na and O_2 $\frac{\text{4 moles Na}}{\text{1 mole } O_2}$ and $\frac{\text{1 mole } O_2}{\text{4 moles Na}}$

Na and Na_2O $\frac{\text{4 moles Na}}{\text{2 moles } Na_2O}$ and $\frac{\text{2 moles } Na_2O}{\text{4 moles Na}}$

O_2 and Na_2O $\frac{\text{2 moles } Na_2O}{\text{1 mole } O_2}$ and $\frac{\text{1 mole } O_2}{\text{2 moles } Na_2O}$

- A mole–mole factor is used to convert the number of moles of one substance in the reaction to the number of moles of another substance in the reaction.

Example: How many moles of sodium are needed to produce 3.5 moles of sodium oxide?

Answer:

Given	Need	Connect
3.5 moles of Na_2O	moles of Na	mole–mole factor

$$3.5 \text{ moles } Na_2O \times \frac{\text{4 moles Na}}{\text{2 moles } Na_2O} = 7.0 \text{ moles of Na}$$

Two SFs; Exact; Two SFs

Converting Grams to Grams (7.8)

When we have the balanced chemical equation for a reaction, we can use the mass of substance A and then calculate the mass of substance B. The process is as follows:

- Use the molar mass factor of A to convert the mass, in grams, of A to moles of A.
- Use the mole–mole factor that converts moles of A to moles of B.
- Use the molar mass factor of B to calculate the mass, in grams, of B.

grams of A $\xrightarrow{\text{Molar mass A}}$ moles of A $\xrightarrow{\text{Mole–mole factor}}$ moles of B $\xrightarrow{\text{Molar mass B}}$ grams of B

Example: How many grams of O_2 are needed to completely react with 14.6 g of Na?

$$4Na(s) + O_2(g) \longrightarrow 2Na_2O(s)$$

Answer:

$$14.6 \text{ g Na} \times \frac{\text{1 mole Na}}{22.99 \text{ g Na}} \times \frac{\text{1 mole } O_2}{\text{4 moles Na}} \times \frac{32.00 \text{ g } O_2}{\text{1 mole } O_2} = 5.08 \text{ g of } O_2$$

Three SFs; Exact / Four SFs; Exact / Exact; Four SFs / Exact; Three SFs

Calculating Quantity of Product from a Limiting Reactant (7.9)

Often in reactions, the reactants are not present in quantities that allow both reactants to be completely used up. Then one of the reactants, called the *limiting reactant*, determines the maximum amount of product that can form.

- To determine the limiting reactant, we calculate the amount of product that is possible from each reactant.
- The limiting reactant is the one that produces the smaller amount of product.

Example: If 12.5 g of S reacts with 17.2 g of O_2, what is the limiting reactant and the mass, in grams, of SO_3 produced?

$$2S(s) + 3O_2(g) \longrightarrow 2SO_3(g)$$

Answer:

Mass of SO_3 from S:

$$12.5 \text{ g S} \times \frac{\text{1 mole S}}{32.07 \text{ g S}} \times \frac{\text{2 moles } SO_3}{\text{2 moles S}} \times \frac{80.07 \text{ g } SO_3}{\text{1 mole } SO_3} = 31.2 \text{ g of } SO_3$$

Three SFs; Exact / Four SFs; Exact / Exact; Four SFs / Exact; Three SFs

Mass of SO_3 from O_2:

$$17.2 \text{ g } O_2 \times \frac{\text{1 mole } O_2}{32.00 \text{ g } O_2} \times \frac{\text{2 moles } SO_3}{\text{3 moles } O_2} \times \frac{80.07 \text{ g } SO_3}{\text{1 mole } SO_3}$$

Three SFs; Exact / Four SFs; Exact / Exact; Four SFs / Exact

Limiting reactant $= 28.7 \text{ g of } SO_3$ (Three SFs) Smaller amount of SO_3

Calculating Percent Yield (7.9)

- The *theoretical yield* for a reaction is the amount of product (100%) formed if all the reactants were converted to desired product.
- The *actual yield* for the reaction is the mass, in grams, of the product obtained at the end of the experiment. Because some product is usually lost, the actual yield is less than the theoretical yield.
- The *percent yield* is calculated from the actual yield divided by the theoretical yield and multiplied by 100%.

$$\text{Percent yield (\%)} = \frac{\text{actual yield}}{\text{theoretical yield}} \times 100\%$$

Example: If 22.6 g of Al reacts completely with O_2, and 37.8 g of Al_2O_3 is obtained, what is the percent yield of Al_2O_3 for the reaction?

$$4Al(s) + 3O_2(g) \longrightarrow 2Al_2O_3(s)$$

Answer:

Calculation of theoretical yield:

$$22.6 \text{ g Al} \times \frac{\text{1 mole Al}}{26.98 \text{ g Al}} \times \frac{\text{2 moles } Al_2O_3}{\text{4 moles Al}} \times \frac{101.96 \text{ g } Al_2O_3}{\text{1 mole } Al_2O_3}$$

Three SFs; Exact / Four SFs; Exact / Exact; Five SFs / Exact

$= 42.7 \text{ g of } Al_2O_3$ Theoretical yield (Three SFs)

Calculation of percent yield:

$$\frac{\text{actual yield (given)}}{\text{theoretical yield (calculated)}} \times 100\%$$

$$= \frac{37.8 \text{ g } Al_2O_3}{42.7 \text{ g } Al_2O_3} \times 100\%$$

Three SFs / Three SFs

$= 88.5\%$

Three SFs

Using the Heat of Reaction (7.10)

- The heat of reaction is the amount of heat, in kJ or kcal, that is absorbed or released during a reaction.
- The heat of reaction, symbol ΔH, is the difference in the energy of the products and the reactants.

$$\Delta H = H_{\text{products}} - H_{\text{reactants}}$$

- In an exothermic reaction (*exo* means "out"), the energy of the products is lower than that of the reactants. This means that heat is released along with the products that form. Then the sign for the heat of reaction, ΔH, is negative.
- In an endothermic reaction (*endo* means "within"), the energy of the products is higher than that of the reactants. The heat is required to convert the reactants to products. Then the sign for the heat of reaction, ΔH, is positive.

Example: How many kilojoules are released when 3.50 g of CH_4 undergoes combustion?

$$CH_4(g) + 2O_2(g) \xrightarrow{\Delta} CO_2(g) + 2H_2O(g) \quad \Delta H = -802\text{ kJ}$$

Answer: $$3.50\ \cancel{g\ CH_4} \times \frac{1\ \cancel{mole\ CH_4}}{16.04\ \cancel{g\ CH_4}} \times \frac{-802\text{ kJ}}{1\ \cancel{mole\ CH_4}} = -175\text{ kJ}$$

Three SFs · Exact / Four SFs · Three SFs / Exact · Three SFs

Therefore 175 kJ are released.

UNDERSTANDING THE CONCEPTS

The chapter Sections to review are shown in parentheses at the end of each problem.

7.97 Balance each of the following by adding coefficients, and identify the type of reaction for each: (7.1, 7.2)

a. 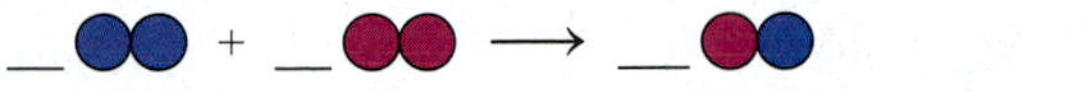

b. 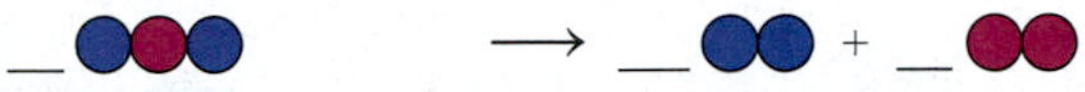

7.98 Balance each of the following by adding coefficients, and identify the type of reaction for each: (7.1, 7.2)

a. 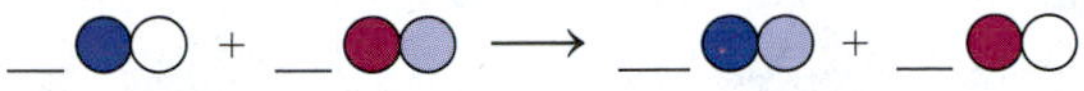

b. 

7.99 If red spheres represent oxygen atoms, blue spheres represent nitrogen atoms, and all the molecules are gases, (7.1, 7.2)

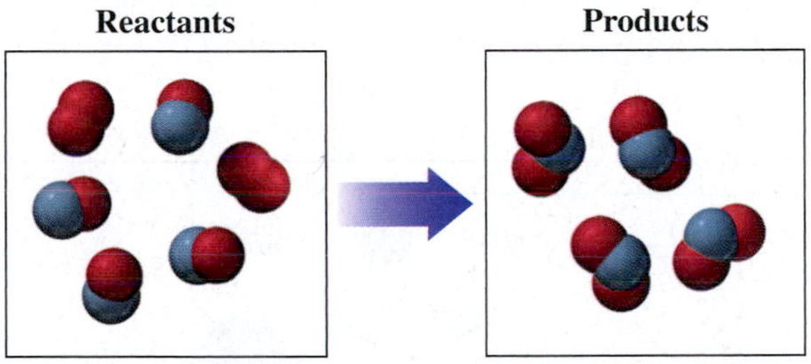
Reactants
Products

a. write the formula for each of the reactants and products.
b. write a balanced equation for the reaction.
c. indicate the type of reaction as combination, decomposition, single replacement, double replacement, or combustion.

7.100 If purple spheres represent iodine atoms, white spheres represent hydrogen atoms, and all the molecules are gases, (7.1, 7.2)

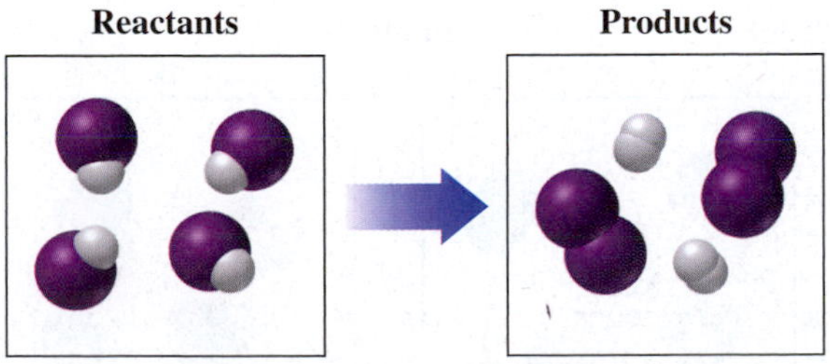
Reactants
Products

a. write the formula for each of the reactants and products.
b. write a balanced equation for the reaction.
c. indicate the type of reaction as combination, decomposition, single replacement, double replacement, or combustion.

7.101 If blue spheres represent nitrogen atoms, purple spheres represent iodine atoms, the reacting molecules are solid, and the products are gases, (7.1, 7.2)

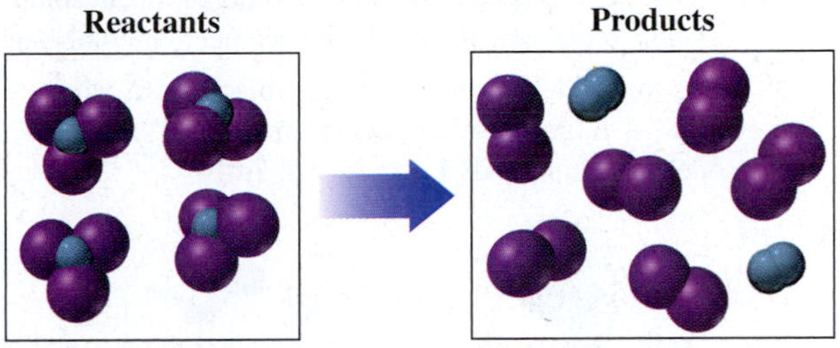
Reactants
Products

a. write the formula for each of the reactants and products.
b. write a balanced equation for the reaction.
c. indicate the type of reaction as combination, decomposition, single replacement, double replacement, or combustion.

7.102 If green spheres represent chlorine atoms, yellow-green spheres represent fluorine atoms, white spheres represent hydrogen atoms, and all the molecules are gases, (7.1, 7.2)

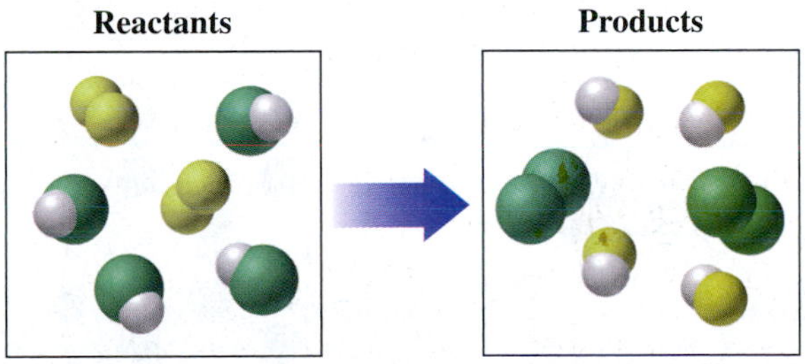
Reactants
Products

a. write the formula for each of the reactants and products.
b. write a balanced equation for the reaction.
c. indicate the type of reaction as combination, decomposition, single replacement, double replacement, or combustion.

7.103 If green spheres represent chlorine atoms, red spheres represent oxygen atoms, and all the molecules are gases, (7.1, 7.2)

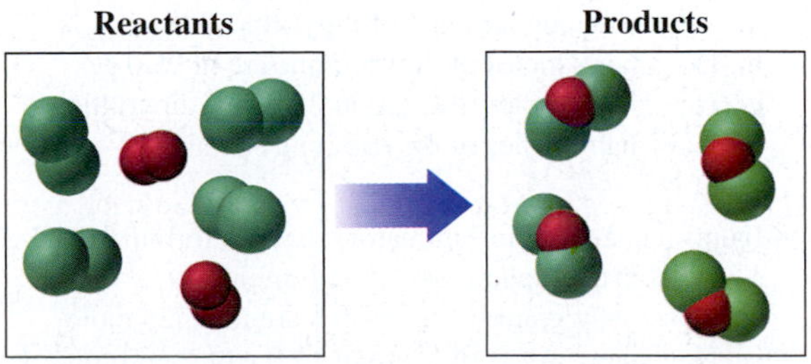
Reactants
Products

a. write the formula for each of the reactants and products.
b. write a balanced equation for the reaction.
c. indicate the type of reaction as combination, decomposition, single replacement, double replacement, or combustion.

7.104 If blue spheres represent nitrogen atoms, purple spheres represent iodine atoms, the reacting molecules are gases, and the products are solid, (7.1, 7.2)

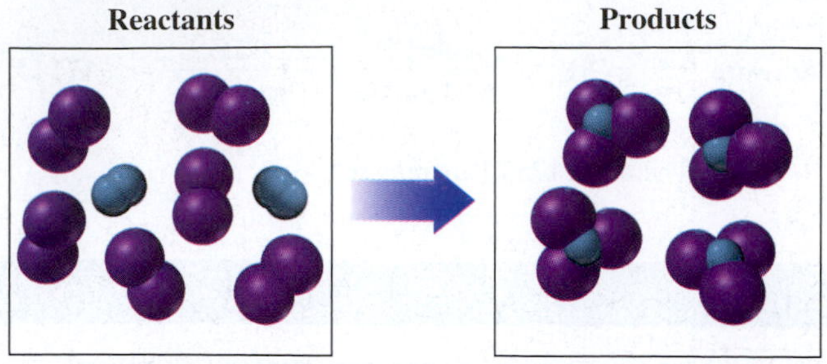
Reactants
Products

a. write the formula for each of the reactants and products.
b. write a balanced equation for the reaction.
c. indicate the type of reaction as combination, decomposition, single replacement, double replacement, or combustion.

7.105 Using the models of the molecules (black = C, white = H, yellow = S, green = Cl), determine each of the following for models of compounds **1** and **2**: (7.5, 7.6)

1.

2.

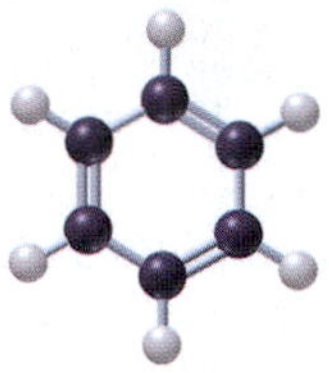

a. molecular formula
b. molar mass
c. number of moles in 10.0 g

7.106 Using the models of the molecules (black = C, white = H, yellow = S, red = O), determine each of the following for models of compounds **1** and **2**: (7.5, 7.6)

1.

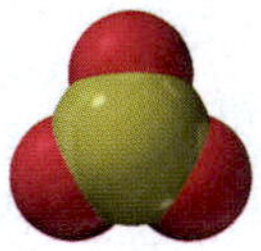

2.

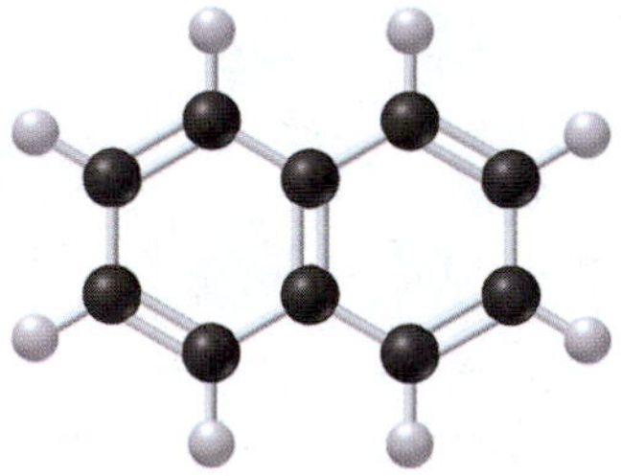

a. molecular formula
b. molar mass
c. number of moles in 10.0 g

7.107 A dandruff shampoo contains dipyrithione, $C_{10}H_8N_2O_2S_2$, which acts as an antibacterial and antifungal agent. (7.4, 7.5, 7.6)
a. What is the molar mass of dipyrithione?
b. How many moles of dipyrithione are in 25.0 g?
c. How many moles of C are in 25.0 g of dipyrithione?
d. How many moles of dipyrithione contain 8.2×10^{24} atoms of N?

7.108 Ibuprofen, an anti-inflammatory, has the formula $C_{13}H_{18}O_2$.
a. What is the molar mass of ibuprofen?
b. How many grams of ibuprofen are in 0.525 mole?
c. How many moles of C are in 12.0 g of ibuprofen?
d. How many moles of ibuprofen contain 1.22×10^{23} atoms of C?

7.109 If red spheres represent oxygen atoms and blue spheres represent nitrogen atoms, and all the molecules are gases, (7.1, 7.9)

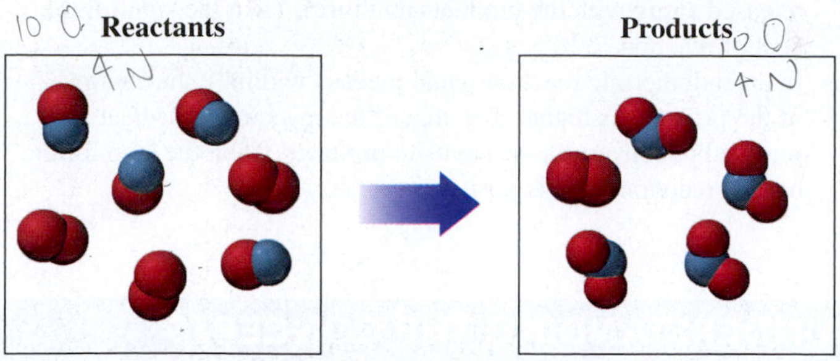
Reactants
Products

a. write a balanced equation for the reaction.
b. identify the limiting reactant.

7.110 If green spheres represent chlorine atoms, yellow-green spheres represent fluorine atoms, and white spheres represent hydrogen atoms, and all the molecules are gases, (7.1, 7.9)

Reactants
Products

a. write a balanced equation for the reaction.
b. identify the limiting reactant.

7.111 If blue spheres represent nitrogen atoms and white spheres represent hydrogen atoms, and all the molecules are gases, (7.1, 7.9)

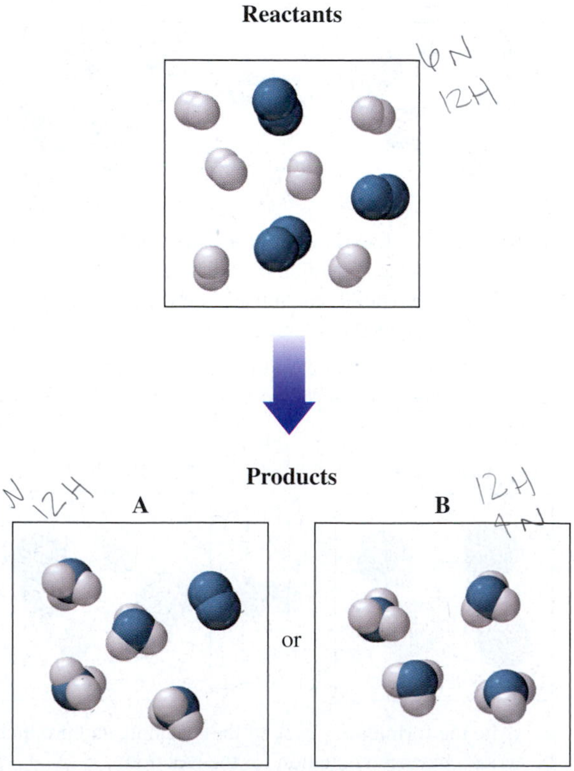
Reactants
Products
A
or
B

a. write a balanced equation for the reaction.
b. identify the diagram that shows the products.

7.112 If purple spheres represent iodine atoms and white spheres represent hydrogen atoms, and all the molecules are gases, (7.1, 7.9)

a. write a balanced equation for the reaction.

b. identify the diagram that shows the products.

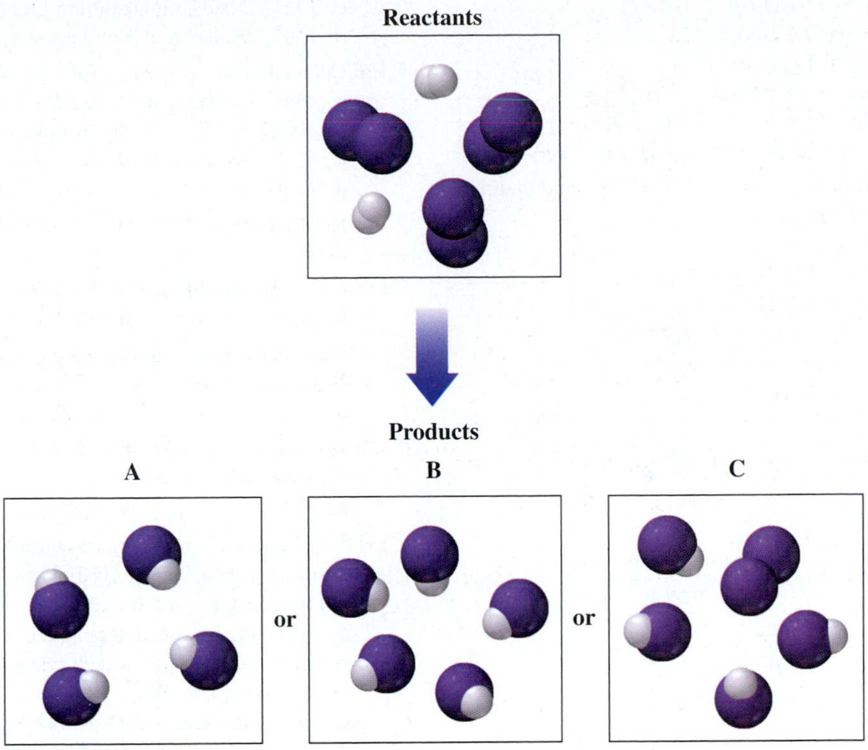
Reactants
Products
A
B
C
or
or

7.113 If blue spheres represent nitrogen atoms and purple spheres represent iodine atoms, and the reacting molecules are solid and the product molecules are gases, (7.1, 7.9)

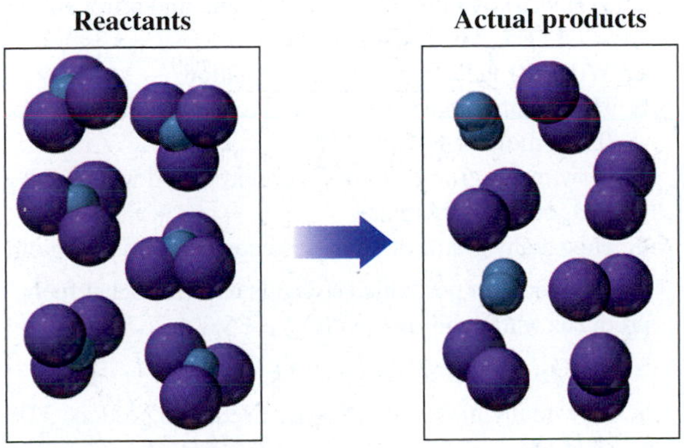
Reactants
Actual products

a. write a balanced equation for the reaction.

b. from the diagram of the actual products that result, calculate the percent yield for the reaction.

7.114 If green spheres represent chlorine atoms and red spheres represent oxygen atoms, and all the molecules are gases, (7.1, 7.9)

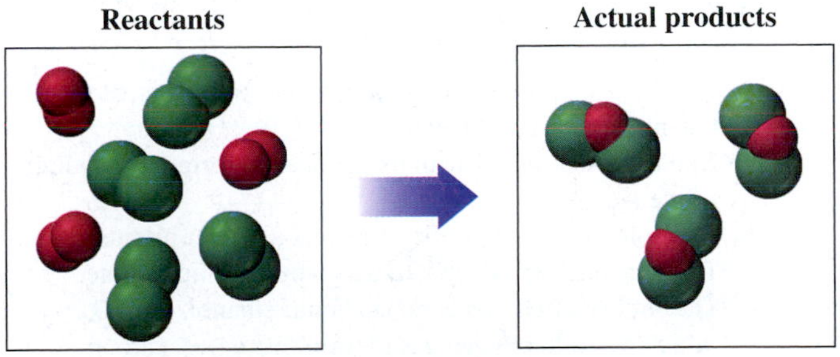
Reactants
Actual products

a. write a balanced equation for the reaction.

b. identify the limiting reactant.

c. from the diagram of the actual products that result, calculate the percent yield for the reaction.

ADDITIONAL PRACTICE PROBLEMS

7.115 Identify the type of reaction for each of the following as combination, decomposition, single replacement, double replacement, or combustion: (7.2)

a. A metal and a nonmetal form an ionic compound.

b. A compound of hydrogen and carbon reacts with oxygen to produce carbon dioxide, water, and energy.

c. Heating magnesium carbonate produces magnesium oxide and carbon dioxide.

d. Zinc replaces copper in $Cu(NO_3)_2$.

7.116 Identify the type of reaction for each of the following as combination, decomposition, single replacement, double replacement, or combustion: (7.2)

a. A compound breaks apart into its elements.

b. Copper and bromine form copper(II) bromide.

c. Iron(II) sulfite breaks down to iron(II) oxide and sulfur dioxide.

d. Silver ion from $AgNO_3(aq)$ forms a solid with bromide ion from $KBr(aq)$.

7.117 Balance each of the following chemical equations, and identify the type of reaction: (7.1, 7.2)

a. $NH_3(g) + HCl(g) \longrightarrow NH_4Cl(s)$
b. $C_4H_8(g) + O_2(g) \xrightarrow{\Delta} CO_2(g) + H_2O(g)$
c. $Sb(s) + Cl_2(g) \longrightarrow SbCl_3(s)$
d. $NI_3(s) \longrightarrow N_2(g) + I_2(g)$
e. $KBr(aq) + Cl_2(aq) \longrightarrow KCl(aq) + Br_2(l)$
f. $Fe(s) + H_2SO_4(aq) \longrightarrow H_2(g) + Fe_2(SO_4)_3(aq)$
g. $Al_2(SO_4)_3(aq) + NaOH(aq) \longrightarrow Al(OH)_3(s) + Na_2SO_4(aq)$

7.118 Balance each of the following chemical equations, and identify the type of reaction: (7.1, 7.2)

a. $Si_3N_4(s) \longrightarrow Si(s) + N_2(g)$
b. $Mg(s) + N_2(g) \longrightarrow Mg_3N_2(s)$
c. $Al(s) + H_3PO_4(aq) \longrightarrow H_2(g) + AlPO_4(aq)$
d. $C_3H_4(g) + O_2(g) \xrightarrow{\Delta} CO_2(g) + H_2O(g)$
e. $Cr_2O_3(s) + H_2(g) \longrightarrow Cr(s) + H_2O(g)$
f. $Al(s) + Cl_2(g) \longrightarrow AlCl_3(s)$
g. $MgCl_2(aq) + AgNO_3(aq) \longrightarrow AgCl(s) + Mg(NO_3)_2(aq)$

7.119 Predict the products and write a balanced equation for each of the following: (7.1, 7.2)

a. single replacement:
$HCl(aq) + Zn(s) \longrightarrow$ ________ + ________
b. decomposition: $BaCO_3(s) \xrightarrow{\Delta}$ ________ + ________
c. double replacement:
$HCl(aq) + NaOH(aq) \longrightarrow$ ________ + ________
d. combination: $Al(s) + F_2(g) \longrightarrow$ ________

7.120 Predict the products and write a balanced equation for each of the following: (7.1, 7.2)

a. decomposition: $NaCl(s) \xrightarrow{\text{Electricity}}$ ________ + ________
b. combination: $Ca(s) + Br_2(g) \longrightarrow$ ________
c. combustion:
$C_2H_4(g) + O_2(g) \xrightarrow{\Delta}$ ________ + ________
d. double replacement:
$NiCl_2(aq) + NaOH(aq) \longrightarrow$ ________ + ________

7.121 Write a balanced equation for each of the following reactions and identify the type of reaction: (7.1, 7.2)

a. Sodium metal reacts with oxygen gas to form solid sodium oxide.
b. Aqueous sodium chloride and aqueous silver nitrate react to form solid silver chloride and aqueous sodium nitrate.
c. Gasohol is a fuel that contains liquid ethanol, C_2H_6O, which burns in oxygen gas to form two gases, carbon dioxide and water.

7.122 Write a balanced equation for each of the following reactions and identify the type of reaction: (7.1, 7.2)

a. Solid potassium chlorate is heated to form solid potassium chloride and oxygen gas.
b. Carbon monoxide gas and oxygen gas combine to form carbon dioxide gas.
c. Ethene gas, C_2H_4, reacts with chlorine gas, Cl_2, to form dichloroethane, $C_2H_4Cl_2$.

7.123 In each of the following reactions, identify the reactant that is oxidized and the reactant that is reduced: (7.3)

a. $N_2(g) + 2O_2(g) \longrightarrow 2NO_2(g)$
b. $CO(g) + 3H_2(g) \longrightarrow CH_4(g) + H_2O(g)$
c. $Mg(s) + Br_2(l) \longrightarrow MgBr_2(s)$

7.124 In each of the following reactions, identify the reactant that is oxidized and the reactant that is reduced: (7.3)

a. $2Al(s) + 3F_2(g) \longrightarrow 2AlF_3(s)$
b. $ZnO(s) + H_2(g) \longrightarrow Zn(s) + H_2O(g)$
c. $2CuS(s) + 3O_2(g) \longrightarrow 2CuO(s) + 2SO_2(g)$

7.125 Calculate the molar mass for each of the following: (7.5)

a. $ZnSO_4$, zinc sulfate, zinc supplement
b. $Ca(IO_3)_2$, calcium iodate, iodine source in table salt
c. $C_5H_8NNaO_4$, monosodium glutamate, flavor enhancer
d. $C_6H_{12}O_2$, isoamyl formate, used to make artificial fruit syrups

7.126 Calculate the molar mass for each of the following: (7.5)

a. $MgCO_3$, magnesium carbonate, used in antacids
b. $Au(OH)_3$, gold(III) hydroxide, used in gold plating
c. $C_{18}H_{34}O_2$, oleic acid, from olive oil
d. $C_{21}H_{26}O_5$, prednisone, anti-inflammatory

7.127 How many grams are in 0.150 mole of each of the following? (7.6)

a. K b. Cl_2 c. Na_2CO_3

7.128 How many grams are in 2.25 moles of each of the following? (7.6)

a. N_2 b. NaBr c. C_6H_{14}

7.129 How many moles are in 25.0 g of each of the following compounds? (7.6)

a. CO_2 b. Al_2O_3 c. $MgCl_2$

7.130 How many moles are in 4.00 g of each of the following compounds? (7.6)

a. NH_3 b. $Ca(NO_3)_2$ c. SO_3

7.131 When ammonia (NH_3) reacts with fluorine, the products are dinitrogen tetrafluoride (N_2F_4) and hydrogen fluoride (HF). All the reactants and products are gases. (7.1, 7.5, 7.6, 7.7)

a. Write the balanced chemical equation.
b. How many moles of each reactant are needed to produce 4.00 moles of HF?
c. How many grams of F_2 are needed to react with 25.5 g of NH_3?
d. How many grams of N_2F_4 can be produced when 3.40 g of NH_3 reacts?

7.132 When gaseous nitrogen dioxide (NO_2) from car exhaust combines with water vapor in the air, it forms aqueous nitric acid (HNO_3), which produces acid rain, and nitrogen oxide gas. (7.1, 7.5, 7.6, 7.7, 7.8)

a. Write the balanced chemical equation.
b. How many moles of each product are produced from 0.250 mole of H_2O?
c. How many grams of HNO_3 are produced when 60.0 g of NO_2 completely reacts?
d. How many grams of NO_2 are needed to form 75.0 g of HNO_3?

7.133 When hydrogen peroxide (H_2O_2) is used in rocket fuels, it produces water and oxygen (O_2). (7.5, 7.6, 7.7, 7.8)

$$2H_2O_2(l) \longrightarrow 2H_2O(l) + O_2(g)$$

a. How many moles of H_2O_2 are needed to produce 3.00 moles of H_2O?
b. How many grams of H_2O_2 are required to produce 36.5 g of O_2?
c. How many grams of H_2O can be produced when 12.2 g of H_2O_2 reacts?

7.134 Propane gas (C_3H_8) reacts with oxygen to produce carbon dioxide and water. (7.5, 7.6, 7.7, 7.8)

$$C_3H_8(g) + 5O_2(g) \xrightarrow{\Delta} 3CO_2(g) + 4H_2O(l)$$

a. How many moles of H_2O form when 5.00 moles of C_3H_8 completely reacts?
b. How many grams of CO_2 are produced from 18.5 g of oxygen gas?
c. How many grams of H_2O can be produced when 46.3 g of C_3H_8 reacts?

7.135 When 12.8 g of Na and 10.2 g of Cl_2 react, what is the mass, in grams, of NaCl that is produced? (7.5, 7.6, 7.7, 7.8, 7.9)

$$2Na(s) + Cl_2(g) \longrightarrow 2NaCl(s)$$

7.136 If 35.8 g of CH_4 and 75.5 g of S react, how many grams of H_2S are produced? (7.5, 7.6, 7.7, 7.8, 7.9)

$$CH_4(g) + 4S(g) \longrightarrow CS_2(g) + 2H_2S(g)$$

7.137 Pentane gas (C_5H_{12}) reacts with oxygen to produce carbon dioxide and water. (7.5, 7.6, 7.7 7.8, 7.9)

$$C_5H_{12}(g) + 8O_2(g) \xrightarrow{\Delta} 5CO_2(g) + 6H_2O(g)$$

a. How many moles of C_5H_{12} must react to produce 4.00 moles of water?
b. How many grams of CO_2 are produced from 32.0 g of O_2?
c. How many grams of CO_2 are formed if 44.5 g of C_5H_{12} is reacted with 108 g of O_2?

7.138 Gasohol is a fuel that contains ethanol (C_2H_6O) that burns in oxygen (O_2) to give carbon dioxide and water. (7.5, 7.6, 7.7, 7.8, 7.9)

$$C_2H_6O(g) + 3O_2(g) \xrightarrow{\Delta} 2CO_2(g) + 3H_2O(g)$$

a. How many moles of O_2 are needed to completely react with 4.0 moles of C_2H_6O?
b. If a car produces 88 g of CO_2, how many grams of O_2 are used up in the reaction?
c. How many grams of CO_2 are produced if 50.0 g of C_2H_6O are reacted with 75.0 g of O_2?

7.139 The gaseous hydrocarbon acetylene (C_2H_2), used in welders' torches, burns according to the following equation: (7.5, 7.6, 7.7, 7.8, 7.9)

$$2C_2H_2(g) + 5O_2(g) \xrightarrow{\Delta} 4CO_2(g) + 2H_2O(g)$$

a. What is the theoretical yield, in grams, of CO_2, if 22.0 g of C_2H_2 completely reacts?
b. If the actual yield in part **a** is 64.0 g of CO_2, what is the percent yield of CO_2 for the reaction?

7.140 The equation for the decomposition of potassium chlorate is written as (7.5, 7.6, 7.7, 7.8, 7.9)

$$2KClO_3(s) \xrightarrow{\Delta} 2KCl(s) + 3O_2(g)$$

a. When 46.0 g of $KClO_3$ is completely decomposed, what is the theoretical yield, in grams, of O_2?
b. If the actual yield in part a is 12.1 g of O_2, what is the percent yield of O_2?

7.141 When 28.0 g of acetylene reacts with hydrogen, 24.5 g of ethane is produced. What is the percent yield of C_2H_6 for the reaction? (7.5, 7.6, 7.7, 7.8, 7.9)

$$C_2H_2(g) + 2H_2(g) \xrightarrow{Pt} C_2H_6(g)$$

7.142 When 50.0 g of iron(III) oxide reacts with carbon monoxide, 32.8 g of iron is produced. What is the percent yield of Fe for the reaction? (7.5, 7.6, 7.7, 7.8, 7.9)

$$Fe_2O_3(s) + 3CO(g) \longrightarrow 2Fe(s) + 3CO_2(g)$$

7.143 The equation for the reaction of nitrogen and oxygen to form nitrogen oxide is written as (7.5, 7.6, 7.7, 7.9, 7.10)

$$N_2(g) + O_2(g) \longrightarrow 2NO(g) \qquad \Delta H = +90.2 \text{ kJ}$$

a. How many kilojoules are required to form 3.00 g of NO?
b. What is the complete equation (including heat) for the decomposition of NO?
c. How many kilojoules are released when 5.00 g of NO decomposes to N_2 and O_2?

7.144 The equation for the reaction of iron and oxygen gas to form rust (Fe_2O_3) is written as (7.5, 7.6, 7.7, 7.9, 7.10)

$$4Fe(s) + 3O_2(g) \longrightarrow 2Fe_2O_3(s) \qquad \Delta H = -1.7 \times 10^3 \text{ kJ}$$

a. How many kilojoules are released when 2.00 g of Fe reacts?
b. How many grams of rust form when 150 kJ are released?

Clinical Applications

7.145 Each of the following is a reaction that occurs in the cells of the body. Identify which is exothermic and endothermic. (7.10)
a. Succinyl CoA + $H_2O \longrightarrow$ succinate + CoA + 37 kJ
b. GDP + P_i + 34 kJ $\longrightarrow$ GTP + H_2O

7.146 Each of the following is a reaction that is needed by the cells of the body. Identify which is exothermic and endothermic. (7.10)
a. Phosphocreatine + $H_2O \longrightarrow$ creatine + P_i + 42.7 kJ
b. Fructose-6-phosphate + P_i + 16 kJ $\longrightarrow$ fructose-1,6-bisphosphate

CHALLENGE PROBLEMS

The following problems are related to the topics in this chapter. However, they do not all follow the chapter order, and they require you to combine concepts and skills from several Sections. These problems will help you increase your critical thinking skills and prepare for your next exam.

7.147 Balance each of the following chemical equations, and identify the type of reaction: (7.1, 7.2)
a. $K_2O(s) + H_2O(g) \longrightarrow KOH(aq)$
b. $C_8H_{18}(l) + O_2(g) \xrightarrow{\Delta} CO_2(g) + H_2O(g)$
c. $Fe(OH)_3(s) \longrightarrow Fe_2O_3(s) + H_2O(g)$
d. $CuS(s) + HCl(aq) \longrightarrow CuCl_2(aq) + H_2S(g)$

7.148 Balance each of the following chemical equations, and identify the type of reaction: (7.1, 7.2)
a. $TiCl_4(s) + Mg(s) \longrightarrow MgCl_2(s) + Ti(s)$
b. $P_4O_{10}(s) + H_2O(g) \xrightarrow{\Delta} H_3PO_4(aq)$
c. $KClO_3(s) \xrightarrow{\Delta} KCl(s) + O_2(g)$
d. $C_3H_6(g) + O_2(g) \xrightarrow{\Delta} CO_2(g) + H_2O(g)$

7.149 Complete and balance each of the following chemical equations: (7.1, 7.2)
a. single replacement: $Fe_3O_4(s) + H_2(g) \longrightarrow$
b. combustion: $C_4H_{10}(g) + O_2(g) \xrightarrow{\Delta}$
c. combination: $Al(s) + O_2(g) \longrightarrow$
d. double replacement: $NaOH(aq) + ZnSO_4(aq) \longrightarrow$

7.150 Complete and balance each of the following chemical equations: (7.1, 7.2)
a. decomposition: $HgO(s) \xrightarrow{\Delta}$
b. double replacement: $BaCl_2(aq) + AgNO_3(aq) \longrightarrow$
c. single replacement: $Ca(s) + AlCl_3(s) \longrightarrow$
d. combination: $Mg(s) + N_2(g) \longrightarrow$

7.151 Write the correct formulas for the reactants and products, the balanced equation for each of the following reaction descriptions, and identify each type of reaction: (7.1, 7.2)
a. An aqueous solution of lead(II) nitrate is mixed with aqueous sodium phosphate to produce solid lead(II) phosphate and aqueous sodium nitrate.
b. Gallium metal heated in oxygen gas forms solid gallium(III) oxide.
c. When solid sodium nitrate is heated, solid sodium nitrite and oxygen gas are produced.

7.152 Write the correct formulas for the reactants and products, the balanced equation for each of the following reaction descriptions, and identify each type of reaction: (7.1, 7.2)

a. Solid bismuth(III) oxide and solid carbon react to form bismuth metal and carbon monoxide gas.
b. Solid sodium bicarbonate is heated and forms solid sodium carbonate, gaseous carbon dioxide, and liquid water.
c. Liquid hexane, C_6H_{14}, reacts with oxygen gas to form two gaseous products: carbon dioxide and water.

7.153 At a winery, glucose ($C_6H_{12}O_6$) in grapes undergoes fermentation to produce ethanol (C_2H_6O) and carbon dioxide. (7.6, 7.7, 7.8)

$$\underset{\text{Glucose}}{C_6H_{12}O_6} \longrightarrow \underset{\text{Ethanol}}{2C_2H_6O} + 2CO_2$$

Glucose in grapes ferments to produce ethanol.

a. How many grams of glucose are required to form 124 g of ethanol?
b. How many grams of ethanol would be formed from the reaction of 0.240 kg of glucose?

7.154 Gasohol is a fuel containing liquid ethanol (C_2H_6O) that burns in oxygen gas to give carbon dioxide gas and water vapor. (7.1, 7.6, 7.7, 7.8)

a. Write the balanced chemical equation.
b. How many moles of O_2 are needed to completely react with 8.0 moles of C_2H_6O?
c. If a car produces 4.4 g of CO_2, how many grams of O_2 are used up in the reaction?
d. If you burn 125 g of C_2H_6O, how many grams of CO_2 and H_2O can be produced?

7.155 Consider the following *unbalanced* equation: (7.1, 7.2, 7.6, 7.7, 7.9)

$$Al(s) + O_2(g) \longrightarrow Al_2O_3(s)$$

a. Write the balanced chemical equation.
b. Identify the type of reaction.
c. How many moles of O_2 are needed to react with 4.50 moles of Al?
d. How many grams of Al_2O_3 are produced when 50.2 g of Al reacts?
e. When Al is reacted with 8.00 g of O_2, how many grams of Al_2O_3 can form?

7.156 A toothpaste contains 0.240% by mass sodium fluoride, used to prevent cavities, and 5.0% by mass KNO_3, which decreases pain sensitivity. One tube contains 119 g of toothpaste. (7.4, 7.5, 7.6)

a. How many moles of NaF are in the tube of toothpaste?
b. How many fluoride ions, F^-, are in the tube of toothpaste?
c. How many grams of sodium ion, Na^+, are in 1.50 g of toothpaste?
d. How many KNO_3 formula units are in the tube of toothpaste?

7.157 Chromium and oxygen combine to form chromium(III) oxide. (7.5, 7.6, 7.7, 7.8, 7.9)

$$4Cr(s) + 3O_2(g) \longrightarrow 2Cr_2O_3(s)$$

a. How many moles of O_2 react with 4.50 moles of Cr?
b. How many grams of Cr_2O_3 are produced when 24.8 g of Cr reacts?
c. When 26.0 g of Cr reacts with 8.00 g of O_2, how many grams of Cr_2O_3 can form?
d. If 74.0 g of Cr and 62.0 g of O_2 are mixed, and 87.3 g of Cr_2O_3 is actually obtained, what is the percent yield of Cr_2O_3 for the reaction?

7.158 Aluminum and chlorine combine to form aluminum chloride. (7.5, 7.6, 7.7, 7.8, 7.9)

$$2Al(s) + 3Cl_2(g) \longrightarrow 2AlCl_3(s)$$

a. How many moles of Cl_2 are needed to react with 2.50 moles of Al?
b. How many grams of $AlCl_3$ are produced when 37.7 g of Al reacts?
c. When 13.5 g of Al reacts with 8.00 g of Cl_2, how many grams of $AlCl_3$ can form?
d. If 45.0 g of Al and 62.0 g of Cl_2 are mixed, and 66.5 g of $AlCl_3$ is actually obtained, what is the percent yield of $AlCl_3$ for the reaction?

7.159 The combustion of propyne (C_3H_4) releases heat when it burns according to the following equation: (7.5, 7.6, 7.7, 7.8, 7.9)

$$C_3H_4(g) + 4O_2(g) \xrightarrow{\Delta} 3CO_2(g) + 2H_2O(g)$$

a. How many moles of O_2 are needed to react completely with 0.225 mole of C_3H_4?
b. How many grams of H_2O are produced from the complete reaction of 64.0 g of O_2?
c. How many grams of CO_2 are produced from the complete reaction of 78.0 g of C_3H_4?
d. If the reaction in part **c** produces 186 g of CO_2, what is the percent yield of CO_2 for the reaction?

7.160 The gaseous hydrocarbon butane (C_4H_{10}) burns according to the following equation: (7.5, 7.6, 7.7, 7.8, 7.9)

$$2C_4H_{10}(g) + 13O_2(g) \xrightarrow{\Delta} 8CO_2(g) + 10H_2O(g)$$

a. How many moles of H_2O are produced from the complete reaction of 2.50 moles of C_4H_{10}?
b. How many grams of O_2 are needed to react completely with 22.5 g of C_4H_{10}?
c. How many grams of CO_2 are produced from the complete reaction of 55.0 g of C_4H_{10}?
d. If the reaction in part **c** produces 145 g of CO_2, what is the percent yield of CO_2 for the reaction?

7.161 Sulfur trioxide decomposes to sulfur and oxygen. (7.5, 7.6, 7.7, 7.10)

$$2SO_3(g) \longrightarrow 2S(s) + 3O_2(g) \quad \Delta H = +790\text{ kJ}$$

a. Is the reaction endothermic or exothermic?
b. How many kilojoules are required when 1.5 moles of SO_3 reacts?
c. How many kilojoules are required when 150 g of O_2 is formed?

7.162 When hydrogen peroxide (H_2O_2) is used in rocket fuels, it produces water, oxygen, and heat. (7.5, 7.6, 7.7, 7.10)

$$2H_2O_2(l) \longrightarrow 2H_2O(l) + O_2(g) \quad \Delta H = -196\text{ kJ}$$

a. Is the reaction endothermic or exothermic?
b. How many kilojoules are released when 2.50 moles of H_2O_2 reacts?
c. How many kilojoules are released when 275 g of O_2 is produced?

ANSWERS

7.1 **a.** not balanced **b.** balanced
c. not balanced **d.** balanced

7.3 **a.** $N_2(g) + O_2(g) \longrightarrow 2NO(g)$
b. $2HgO(s) \longrightarrow Hg(l) + O_2(g)$
c. $4Fe(s) + 3O_2(g) \longrightarrow 2Fe_2O_3(s)$
d. $2Na(s) + Cl_2(g) \longrightarrow 2NaCl(s)$

7.5 **a.** $Mg(s) + 2AgNO_3(aq) \longrightarrow Mg(NO_3)_2(aq) + 2Ag(s)$
b. $2Al(s) + 3CuSO_4(aq) \longrightarrow 3Cu(s) + Al_2(SO_4)_3(aq)$
c. $Pb(NO_3)_2(aq) + 2NaCl(aq) \longrightarrow PbCl_2(s) + 2NaNO_3(aq)$
d. $6HCl(aq) + 2Al(s) \longrightarrow 3H_2(g) + 2AlCl_3(aq)$

7.7 **a.** decomposition **b.** single replacement
c. combustion **d.** double replacement
e. combination

7.9 **a.** combination **b.** single replacement
c. decomposition **d.** double replacement
e. combustion

7.11 **a.** $Mg(s) + Cl_2(g) \longrightarrow MgCl_2(s)$
b. $2HBr(g) \longrightarrow H_2(g) + Br_2(l)$
c. $Mg(s) + Zn(NO_3)_2(aq) \longrightarrow Zn(s) + Mg(NO_3)_2(aq)$
d. $K_2S(aq) + Pb(NO_3)_2(aq) \longrightarrow PbS(s) + 2KNO_3(aq)$
e. $2C_2H_6(g) + 7O_2(g) \xrightarrow{\Delta} 4CO_2(g) + 6H_2O(g)$

7.13 **a.** reduction **b.** oxidation
c. reduction **d.** reduction

7.15 **a.** Zn is oxidized; Cl_2 is reduced.
b. The Br^- in NaBr is oxidized; Cl_2 is reduced.
c. The O^{2-} in PbO is oxidized; the Pb^{2+} in PbO is reduced.
d. Sn^{2+} is oxidized; Fe^{3+} is reduced.

7.17 **a.** reduction **b.** oxidation

7.19 Linoleic acid gains hydrogen atoms and is reduced.

7.21 One mole contains 6.02×10^{23} atoms of an element, molecules of a molecular substance, or formula units of an ionic substance.

7.23 **a.** 3.01×10^{23} atoms of C
b. 7.71×10^{23} molecules of SO_2
c. 0.0867 mole of Fe
d. 14.1 moles of C_2H_6O

7.25 **a.** 6.00 moles of H **b.** 8.00 moles of O
c. 1.20×10^{24} atoms of P **d.** 4.82×10^{24} atoms of O

7.27 **a.** 36 moles of H **b.** 1.0×10^2 moles of C
c. 0.040 mole of N

7.29 **a.** 32.2 moles of C **b.** 6.22 moles of H
c. 0.230 mole of O

7.31 **a.** 70.90 g **b.** 90.08 g **c.** 262.9 g

7.33 **a.** 83.98 g **b.** 98.95 g **c.** 156.7 g

7.35 **a.** 74.55 g **b.** 123.11 g **c.** 329.4 g

7.37 **a.** 342.2 g **b.** 188.18 g **c.** 365.5 g

7.39 **a.** 151.16 g **b.** 489.4 g **c.** 331.4 g

7.41 **a.** 34.5 g **b.** 112 g **c.** 5.50 g
d. 5.14 g **e.** 9.80×10^4 g

7.43 **a.** 3.03 g **b.** 38.1 g **c.** 9.30 g
d. 24.6 g **e.** 12.9 g

7.45 **a.** 0.760 mole of Ag **b.** 0.0240 mole of C
c. 0.881 mole of NH_3 **d.** 0.452 mole of CH_4
e. 1.53 moles of Fe_2O_3

7.47 **a.** 6.25 moles of He **b.** 0.781 mole of O_2
c. 0.321 mole of $Al(OH)_3$ **d.** 0.106 mole of Ga_2S_3
e. 0.430 mole of C_4H_{10}

7.49 **a.** 0.527 mole **b.** 96.8 g

7.51 **a.** 602 g **b.** 74.7 g

7.53 **a.** 66.0 g **b.** 0.772 mole

7.55 **a.** $\frac{2 \text{ moles } SO_2}{1 \text{ mole } O_2}$ and $\frac{1 \text{ mole } O_2}{2 \text{ moles } SO_2}$

$\frac{2 \text{ moles } SO_2}{2 \text{ moles } SO_3}$ and $\frac{2 \text{ moles } SO_3}{2 \text{ moles } SO_2}$

$\frac{2 \text{ moles } SO_3}{1 \text{ mole } O_2}$ and $\frac{1 \text{ mole } O_2}{2 \text{ moles } SO_3}$

b. $\frac{4 \text{ moles P}}{5 \text{ moles } O_2}$ and $\frac{5 \text{ moles } O_2}{4 \text{ moles P}}$

$\frac{4 \text{ moles P}}{2 \text{ moles } P_2O_5}$ and $\frac{2 \text{ moles } P_2O_5}{4 \text{ moles P}}$

$\frac{5 \text{ moles } O_2}{2 \text{ moles } P_2O_5}$ and $\frac{2 \text{ moles } P_2O_5}{5 \text{ moles } O_2}$

7.57 **a.** 1.3 moles of O_2 **b.** 10. moles of H_2
c. 5.0 moles of H_2O

7.59 **a.** 1.25 moles of C **b.** 0.96 mole of CO
c. 1.0 mole of SO_2 **d.** 0.50 mole of CS_2

7.61 **a.** 77.5 g of Na_2O **b.** 6.26 g of O_2 **c.** 19.4 g of O_2

7.63 **a.** 19.2 g of O_2 **b.** 3.79 g of N_2 **c.** 54.0 g of H_2O

7.65 **a.** 3.66 g of H_2O **b.** 26.3 g of NO **c.** 7.53 g of HNO_3

7.67 **a.** $2PbS(s) + 3O_2(g) \longrightarrow 2PbO(s) + 2SO_2(g)$
b. 6.00 g of O_2 **c.** 17.4 g of SO_2 **d.** 137 g of PbS

7.69 **a.** 9.59 g of O_2 **b.** 6.00 g of urea
c. 13.2 g of CO_2

7.71 **a.** Eight taxis can be used to pick up passengers.
b. Seven taxis can be driven.

7.73 **a.** 5.0 moles of H_2 **b.** 4.0 moles of H_2
c. 3.0 moles of N_2

7.75 **a.** 25.1 g of $AlCl_3$ **b.** 13.5 g of H_2O
c. 26.7 g of SO_2

7.77 **a.** 31.2 g of SO_3 **b.** 34.6 g of Fe_3O_4
c. 35.0 g of CO_2

7.79 **a.** 71.0% **b.** 63.2%

7.81 70.9 g of Al_2O_3

7.83 60.5%

7.85 In exothermic reactions, the energy of the products is lower than that of the reactants.

7.87 **a.** exothermic **b.** endothermic
c. exothermic

7.89 **a.** Heat is released, exothermic, $\Delta H = -802$ kJ
b. Heat is absorbed, endothermic, $\Delta H = +65.3$ kJ
c. Heat is released, exothermic, $\Delta H = -850$ kJ

7.91 **a.** 579 kJ **b.** 89 kJ

7.93 **a.** endothermic **b.** 12.3 g of glucose
c. 390 kJ

7.95 **a.** $C_6H_{12}O_6(aq) + 6O_2(g) \longrightarrow 6CO_2(g) + 6H_2O(l)$
b. $6CO_2(g) + 6H_2O(l) \longrightarrow C_6H_{12}O_6(aq) + 6O_2(g)$

7.97 **a.** 1,1,2 combination **b.** 2,2,1 decomposition

7.99 **a.** reactants NO and O_2; product NO_2
b. $2NO(g) + O_2(g) \longrightarrow 2NO_2(g)$
c. combination products

7.101 **a.** reactant NI_3; products N_2 and I_2
b. $2NI_3(s) \longrightarrow N_2(g) + 3I_2(g)$
c. decomposition

7.103 **a.** reactants Cl_2 and O_2; product OCl_2
b. $2Cl_2(g) + O_2(g) \longrightarrow 2OCl_2(g)$
c. combination

7.105 **1.** **a.** S_2Cl_2
b. 135.04 g/mole
c. 0.0741 mole
2. **a.** C_6H_6
b. 78.11 g/mole
c. 0.128 mole

7.107 **a.** 252.3 g/mole
b. 0.0991 mole of dipyrithione
c. 0.991 mole of C
d. 6.8 moles of dipyrithione

7.109 **a.** $2NO(g) + O_2(g) \longrightarrow 2NO_2(g)$
b. NO is the limiting reactant.

7.111 **a.** $N_2(g) + 3H_2(g) \longrightarrow 2NH_3(g)$ **b.** A

7.113 **a.** $2NI_3(s) \longrightarrow N_2(g) + 3I_2(s)$ **b.** 67%

7.115 **a.** combination **b.** combustion
c. decomposition **d.** single replacement

7.117 **a.** $NH_3(g) + HCl(g) \longrightarrow NH_4Cl(s)$ combination
b. $C_4H_8(g) + 6O_2(g) \xrightarrow{\Delta} 4CO_2(g) + 4H_2O(g)$ combustion
c. $2Sb(s) + 3Cl_2(g) \longrightarrow 2SbCl_3(s)$ combination
d. $2NI_3(s) \longrightarrow N_2(g) + 3I_2(g)$ decomposition
e. $2KBr(aq) + Cl_2(aq) \longrightarrow 2KCl(aq) + Br_2(l)$ single replacement
f. $2Fe(s) + 3H_2SO_4(aq) \longrightarrow 3H_2(g) + Fe_2(SO_4)_3(aq)$ single replacement
g. $Al_2(SO_4)_3(aq) + 6NaOH(aq) \longrightarrow 2Al(OH)_3(s) + 3Na_2SO_4(aq)$ double replacement

7.119 **a.** $Zn(s) + 2HCl(aq) \longrightarrow H_2(g) + ZnCl_2(aq)$
b. $BaCO_3(s) \xrightarrow{\Delta} BaO(s) + CO_2(g)$
c. $HCl(aq) + NaOH(aq) \longrightarrow NaCl(aq) + H_2O(l)$
d. $2Al(s) + 3F_2(g) \longrightarrow 2AlF_3(s)$

7.121 **a.** $4Na(s) + O_2(g) \longrightarrow 2Na_2O(s)$ combination
b. $NaCl(aq) + AgNO_3(aq) \longrightarrow AgCl(s) + NaNO_3(aq)$ double replacement
c. $C_2H_6O(l) + 3O_2(g) \xrightarrow{\Delta} 2CO_2(g) + 3H_2O(g)$ combustion

7.123 **a.** N_2 is oxidized; O_2 is reduced.
b. The C in CO is oxidized; H_2 is reduced.
c. Mg is oxidized; Br_2 is reduced.

7.125 **a.** 161.48 g/mole **b.** 389.9 g/mole
c. 169.11 g/mole **d.** 116.16 g/mole

7.127 **a.** 5.87 g **b.** 10.6 g
c. 15.9 g

7.129 **a.** 0.568 mole **b.** 0.245 mole
c. 0.263 mole

7.131 **a.** $2NH_3(g) + 5F_2(g) \longrightarrow N_2F_4(g) + 6HF(g)$
b. 1.33 moles of NH_3 and 3.33 moles of F_2
c. 142 g of F_2
d. 10.4 g of N_2F_4

7.133 **a.** 3.00 moles of H_2O_2 **b.** 77.6 g of H_2O_2
c. 6.46 g of H_2O

7.135 16.8 g of NaCl

7.137 **a.** 0.667 mole of C_5H_{12}
b. 27.5 g of CO_2
c. 92.8 g of CO_2

7.139 **a.** 74.4 g of CO_2
b. 86.0%

7.141 75.9%

7.143 **a.** 4.51 kJ
b. $2NO(g) \longrightarrow N_2(g) + O_2(g) + 90.2$ kJ
c. 7.51 kJ

7.145 **a.** exothermic **b.** endothermic

7.147 **a.** $K_2O(s) + H_2O(g) \longrightarrow 2KOH(s)$ combination
b. $2C_8H_{18}(l) + 25O_2(g) \xrightarrow{\Delta} 16CO_2(g) + 18H_2O(g)$ combustion
c. $2Fe(OH)_3(s) \longrightarrow Fe_2O_3(s) + 3H_2O(g)$ decomposition
d. $CuS(s) + 2HCl(aq) \longrightarrow CuCl_2(aq) + H_2S(g)$ double replacement

7.149 **a.** $Fe_3O_4(s) + 4H_2(g) \longrightarrow 3Fe(s) + 4H_2O(g)$
b. $2C_4H_{10}(g) + 13O_2(g) \xrightarrow{\Delta} 8CO_2(g) + 10H_2O(g)$
c. $4Al(s) + 3O_2(g) \longrightarrow 2Al_2O_3(s)$
d. $2NaOH(aq) + ZnSO_4(aq) \longrightarrow Zn(OH)_2(s) + Na_2SO_4(aq)$

7.151 **a.** $3Pb(NO_3)_2(aq) + 2Na_3PO_4(aq) \longrightarrow Pb_3(PO_4)_2(s) + 6NaNO_3(aq)$ double replacement
b. $4Ga(s) + 3O_2(g) \xrightarrow{\Delta} 2Ga_2O_3(s)$ combination
c. $2NaNO_3(s) \xrightarrow{\Delta} 2NaNO_2(s) + O_2(g)$ decomposition

7.153 **a.** 242 g of glucose **b.** 123 g of ethanol

7.155 **a.** $4Al(s) + 3O_2(g) \longrightarrow 2Al_2O_3(s)$
b. combination **c.** 3.38 moles of O_2
d. 94.9 g of Al_2O_3 **e.** 17.0 g of Al_2O_3

7.157 **a.** 3.38 moles of O_2
b. 36.2 g of Cr_2O_3
c. 25.3 g of Cr_2O_3
d. 80.8%

7.159 **a.** 0.900 mole of O_2
b. 18.0 g of H_2O
c. 257 g of CO_2
d. 72.4%

7.161 **a.** endothermic
b. 590 kJ
c. 1200 kJ

8 Gases

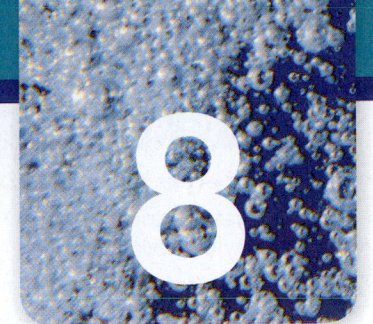

After soccer practice, Whitney complains that she is having difficulty breathing. Her father takes her to the emergency room, where she is seen by Sam, a respiratory therapist, who listens to Whitney's chest and tests her breathing capacity using a spirometer. Based on her limited breathing capacity and the wheezing noise in her chest, Whitney is diagnosed with asthma.

Sam gives Whitney a nebulizer containing a bronchodilator that opens the airways and allows more air to go into the lungs. During the breathing treatment, he measures the amount of oxygen (O_2) in her blood and explains to Whitney and her father that air is a mixture of gases containing 78% nitrogen (N_2) gas and 21% O_2 gas. Because Whitney has difficulty obtaining sufficient oxygen, Sam gives her supplemental oxygen through an oxygen mask. Within a short period of time, Whitney's breathing returns to normal. Sam explains that the lungs work according to Boyle's law: The volume of the lungs increases upon inhalation, and the pressure decreases to allow air to flow in. However, during an asthma attack, the airways become restricted, and it becomes more difficult to expand the volume of the lungs.

CAREER

Respiratory Therapist

Respiratory therapists assess and treat a range of patients, including premature infants whose lungs have not developed, asthmatics, and patients with emphysema or cystic fibrosis. In assessing patients, they perform a variety of diagnostic tests including breathing capacity and concentrations of oxygen and carbon dioxide in a patient's blood, as well as blood pH. In order to treat patients, therapists provide oxygen or aerosol medications to the patient, as well as chest physiotherapy to remove mucus from their lungs. Respiratory therapists also educate patients on how to correctly use their inhalers.

CLINICAL UPDATE

Exercise-Induced Asthma

Whitney's doctor prescribes an inhaled medication that opens up her airways before she starts exercise. In the **CLINICAL UPDATE Exercise-Induced Asthma**, page 301, you can view the impact of Whitney's medication and learn about other treatments that help prevent exercise-induced asthma.

LOOKING AHEAD

REVIEW

Using Significant Figures in Calculations (2.3)

Writing Conversion Factors from Equalities (2.5)

Using Conversion Factors (2.6)

INTERACTIVE VIDEO

Kinetic Molecular Theory

PEARSON eText 2.0

ENGAGE

Use the kinetic molecular theory to explain why a gas completely fills a container of any size and shape.

TEST

Try Practice Problems 8.1 and 8.2

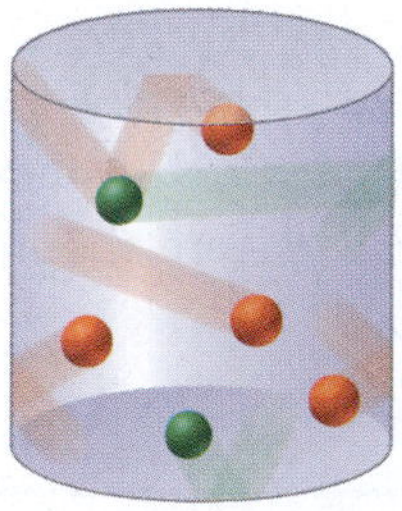

FIGURE 8.1 ▶ Gas particles moving in straight lines within a container exert pressure when they collide with the walls of the container.

Q Why does heating the container increase the pressure of the gas within it?

8.1 Properties of Gases

LEARNING GOAL Describe the kinetic molecular theory of gases and the units of measurement used for gases.

We all live at the bottom of a sea of gases called the atmosphere. The most important of these gases is oxygen, which constitutes about 21% of the atmosphere. Without oxygen, life on this planet would be impossible: Oxygen is vital to all life processes of plants and animals. Ozone (O_3), formed in the upper atmosphere by the interaction of oxygen with ultraviolet light, absorbs some of the harmful radiation before it can strike Earth's surface. The other gases in the atmosphere include nitrogen (78%), argon, carbon dioxide (CO_2), and water vapor. Carbon dioxide gas, a product of combustion and metabolism, is used by plants in photosynthesis, which produces the oxygen that is essential for humans and animals.

The behavior of gases is quite different from that of liquids and solids. Gas particles are far apart, whereas particles of both liquids and solids are held close together. A gas has no definite shape or volume and will completely fill any container. Because there are great distances between gas particles, a gas is less dense than a solid or liquid, and easy to compress. A model for the behavior of a gas, called the **kinetic molecular theory of gases**, helps us understand gas behavior.

Kinetic Molecular Theory of Gases

1. **A gas consists of small particles (atoms or molecules) that move randomly with high velocities.** Gas molecules moving in random directions at high speeds cause a gas to fill the entire volume of a container.
2. **The attractive forces between the particles of a gas are usually very small.** Gas particles are far apart and fill a container of any size and shape.
3. **The actual volume occupied by gas molecules is extremely small compared to the volume that the gas occupies.** The volume of the gas is considered equal to the volume of the container. Most of the volume of a gas is empty space, which allows gases to be easily compressed.
4. **Gas particles are in constant motion, moving rapidly in straight paths.** When gas particles collide, they rebound and travel in new directions. Every time they hit the walls of the container, they exert pressure. An increase in the number or force of collisions against the walls of the container causes an increase in the pressure of the gas.
5. **The average kinetic energy of gas molecules is proportional to the Kelvin temperature.** Gas particles move faster as the temperature increases. At higher temperatures, gas particles hit the walls of the container more often and with more force, producing higher pressures.

The kinetic molecular theory helps explain some of the characteristics of gases. For example, you can smell perfume when a bottle is opened on the other side of a room because its particles move rapidly in all directions. At room temperature, the molecules in the air are moving at about 450 m/s, which is 1000 mi/h. They move faster at higher temperatures and more slowly at lower temperatures. Sometimes tires and gas-filled containers explode when temperatures are too high. From the kinetic molecular theory, you know that gas particles move faster when heated, hit the walls of a container with more force, and cause a buildup of pressure inside a container.

When we talk about a gas, we describe it in terms of four properties: pressure, volume, temperature, and the amount of gas.

Pressure (*P*)

Gas particles are extremely small and move rapidly. When they hit the walls of a container, they exert a **pressure** (see FIGURE 8.1). If we heat the container, the molecules move faster and smash into the walls of the container more often and with increased force, thus increasing the pressure. The gas particles in the air, mostly oxygen and nitrogen, exert a pressure on us called **atmospheric pressure** (see FIGURE 8.2). As you go to higher altitudes, the

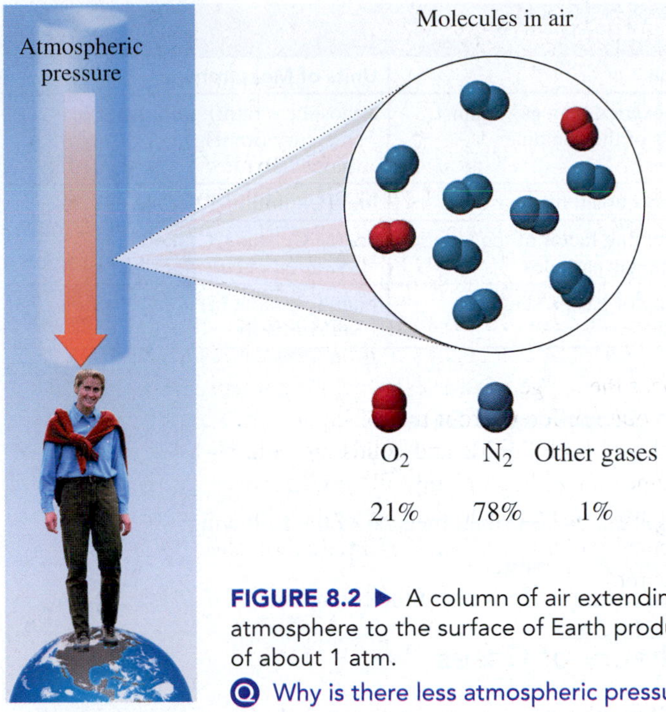

Atmospheric pressure
Molecules in air
O_2
N_2
Other gases
21%
78%
1%

FIGURE 8.2 ▶ A column of air extending from the top of the atmosphere to the surface of Earth produces pressure on each of us of about 1 atm.

Q Why is there less atmospheric pressure at higher altitudes?

atmospheric pressure is less because there are fewer particles in the air. The most common units used to measure gas pressure are the *atmosphere* (atm) and *millimeters of mercury* (mmHg). On a TV weather report, the atmospheric pressure may be reported in inches of mercury, or in kilopascals in countries other than the United States. In a hospital, the unit torr or pounds per square inch (psi) may be used.

Volume (*V*)

The volume of gas equals the size of the container in which the gas is placed. When you inflate a tire or a balloon, you are adding more gas particles. The increase in the number of particles hitting the walls of the tire or balloon increases the volume. Sometimes, on a cold morning, a tire looks flat. The volume of the tire has decreased because a lower temperature decreases the speed of the molecules, which in turn reduces the force of their impacts on the walls of the tire. The most common units for volume measurement are liters (L) and milliliters (mL).

ENGAGE

Why does the volume of a gas increase when the number of gas particles increases?

Temperature (*T*)

The temperature of a gas is related to the kinetic energy of its particles. For example, if we have a gas at 200 K and heat it to a temperature of 400 K, the gas particles will have twice the kinetic energy that they did at 200 K. This also means that the gas at 400 K exerts twice the pressure of the gas at 200 K, if the volume and amount of gas do not change. Although we measure gas temperature using a Celsius thermometer, all comparisons of gas behavior and all calculations related to temperature must use the Kelvin temperature scale. No one has created the conditions for absolute zero (0 K), but scientists predict that the particles will have zero kinetic energy and exert zero pressure at absolute zero.

Amount of Gas (*n*)

When you add air to a bicycle tire, you increase the amount of gas, which results in a higher pressure in the tire. Usually, we measure the amount of gas by its mass, in grams. In gas law calculations, we need to change the grams of gas to moles.

A summary of the four properties of a gas is given in **TABLE 8.1**.

TABLE 8.1 Properties That Describe a Gas		
Property	**Description**	**Units of Measurement**
Pressure (*P*)	The force exerted by a gas against the walls of the container	atmosphere (atm); millimeter of mercury (mmHg); torr (Torr); pascal (Pa)
Volume (*V*)	The space occupied by a gas	liter (L); milliliter (mL)
Temperature (*T*)	The determining factor of the kinetic energy of gas particles	degree Celsius (°C); kelvin (K) *is required in calculations*
Amount (*n*)	The quantity of gas present in a container	gram (g); mole (*n*) *is required in calculations*

▶ SAMPLE PROBLEM 8.1 Properties of Gases

TRY IT FIRST

Identify the property of a gas that is described by each of the following:

a. increases the kinetic energy of gas particles
b. the force of the gas particles hitting the walls of the container
c. the space that is occupied by a gas

SOLUTION

a. temperature **b.** pressure **c.** volume

STUDY CHECK 8.1

What property of a gas is described in each of the following?

a. Helium gas is added to a balloon. **b.** A balloon expands as it rises higher.

ANSWER

a. amount **b.** volume

TEST

Try Practice Problems 8.3 and 8.4

Measurement of Gas Pressure

When billions and billions of gas particles hit against the walls of a container, they exert pressure, which is a force acting on a certain area.

$$\text{Pressure } (P) = \frac{\text{force}}{\text{area}}$$

The atmospheric pressure can be measured using a barometer (see **FIGURE 8.3**). At a pressure of exactly 1 atmosphere (atm), a mercury column in an inverted tube would be *exactly* 760 mm high. One **atmosphere (atm)** is defined as *exactly* 760 mmHg (millimeters of mercury). One atmosphere is also 760 Torr, a pressure unit named to honor Evangelista Torricelli, the inventor of the barometer. Because the units of Torr and mmHg are equal, they are used interchangeably. One atmosphere is also equivalent to 29.9 in. of mercury (inHg).

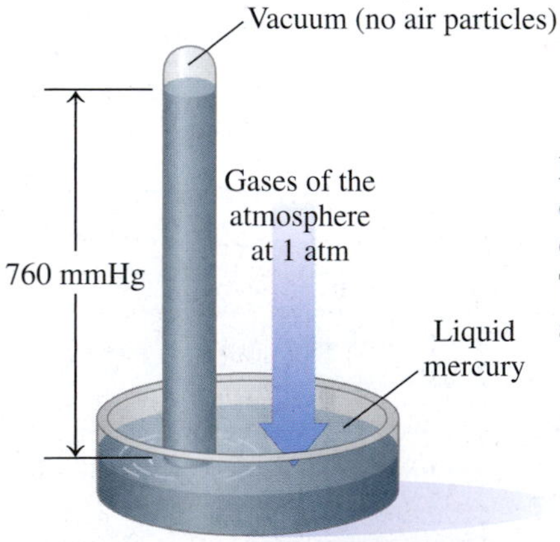

Vacuum (no air particles)
Gases of the atmosphere at 1 atm
760 mmHg
Liquid mercury

FIGURE 8.3 ▶ A barometer: The pressure exerted by the gases in the atmosphere is equal to the downward pressure of a mercury column in a closed glass tube. The height of the mercury column measured in mmHg is called atmospheric pressure.

Q Why does the height of the mercury column change from day to day?

$$1 \text{ atm} = 760 \text{ mmHg} = 760 \text{ Torr (exact)}$$
$$1 \text{ atm} = 29.9 \text{ inHg}$$
$$1 \text{ mmHg} = 1 \text{ Torr (exact)}$$

In SI units, pressure is measured in pascals (Pa); 1 atm is equal to 101 325 Pa. Because a pascal is a very small unit, pressures are usually reported in kilopascals.

$$1 \text{ atm} = 101\,325 \text{ Pa} = 101.325 \text{ kPa}$$

The U.S. equivalent of 1 atm is 14.7 lb/in.^{2} (psi). When you use a pressure gauge to check the air pressure in the tires of a car, it may read 30 to 35 psi. This measurement is actually 30 to 35 psi above the pressure that the atmosphere exerts on the outside of the tire.

$$1 \text{ atm} = 14.7 \text{ lb/in.}^2$$

TABLE 8.2 summarizes the various units used in the measurement of pressure.

TABLE 8.2 Units for Measuring Pressure

Unit	Abbreviation	Unit Equivalent to 1 atm
Atmosphere	atm	1 atm (exact)
Millimeters of Hg	mmHg	760 mmHg (exact)
Torr	Torr	760 Torr (exact)
Inches of Hg	inHg	29.9 inHg
Pounds per square inch	lb/in.^{2} (psi)	14.7 lb/in.^{2}
Pascal	Pa	101 325 Pa
Kilopascal	kPa	101.325 kPa

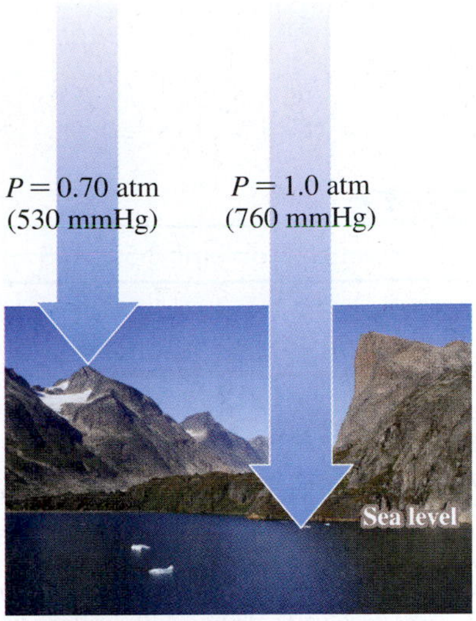

P = 0.70 atm (530 mmHg)
P = 1.0 atm (760 mmHg)
Sea level

The atmospheric pressure decreases as the altitude increases.

Atmospheric pressure changes with variations in weather and altitude. On a hot, sunny day, the mercury column rises, indicating a higher atmospheric pressure. On a rainy day, the atmosphere exerts less pressure, which causes the mercury column to fall. In a weather report, this type of weather is called a *low-pressure system*. Above sea level, the density of the gases in the air decreases, which causes lower atmospheric pressures; the atmospheric pressure is greater than 760 mmHg at the Dead Sea because it is below sea level (see **TABLE 8.3**).

TABLE 8.3 Altitude and Atmospheric Pressure

Location	Altitude (km)	Atmospheric Pressure (mmHg)
Dead Sea	−0.40	800
Sea level	0.00	760
Los Angeles	0.09	752
Las Vegas	0.70	700
Denver	1.60	630
Mount Whitney	4.50	440
Mount Everest	8.90	253

Divers must be concerned about increasing pressures on their ears and lungs when they dive below the surface of the ocean. Because water is more dense than air, the pressure on a diver increases rapidly as the diver descends. At a depth of 33 ft below the surface of the ocean, an additional 1 atm of pressure is exerted by the water on a diver, which gives a total pressure of 2 atm. At 100 ft, there is a total pressure of 4 atm on a diver. The regulator that a diver uses continuously adjusts the pressure of the breathing mixture to match the increase in pressure.

▶ SAMPLE PROBLEM 8.2 Units of Pressure

TRY IT FIRST

The oxygen in a tank in the hospital respiratory unit has a pressure of 4820 mmHg. Calculate the pressure, in atmospheres, of the oxygen gas.

SOLUTION

The equality 1 atm = 760 mmHg can be written as two conversion factors:

$$\frac{760\text{ mmHg}}{1\text{ atm}} \quad \text{and} \quad \frac{1\text{ atm}}{760\text{ mmHg}}$$

Using the conversion factor that cancels mmHg and gives atm, we can set up the problem as

$$4820\ \cancel{\text{mmHg}} \times \frac{1\text{ atm}}{760\ \cancel{\text{mmHg}}} = 6.34\text{ atm}$$

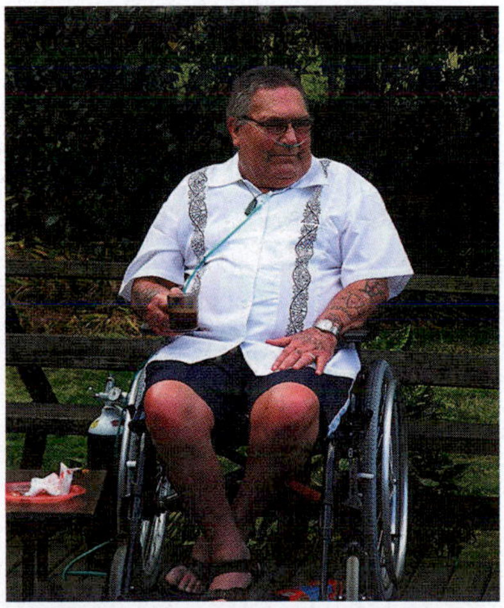

A patient with severe COPD obtains oxygen from an oxygen tank.

STUDY CHECK 8.2

A tank of nitrous oxide (N_2O) used as an anesthetic has a pressure of 48 psi.

a. What is that pressure in atmospheres?
b. What is the pressure in torr?

TEST

Try Practice Problems 8.5 to 8.8

ANSWER

a. 3.3 atm **b.** 2500 (2.5×10^3) Torr

Chemistry Link to Health

Measuring Blood Pressure

Your blood pressure is one of the vital signs a doctor or nurse checks during a physical examination. It actually consists of two separate measurements. Acting like a pump, the heart contracts to create the pressure that pushes blood through the circulatory system. During contraction, the blood pressure is at its highest; this is your *systolic* pressure. When the heart muscles relax, the blood pressure falls; this is your *diastolic* pressure. The normal range for systolic pressure is 100 to 200 mmHg. For diastolic pressure, it is 60 to 80 mmHg. These two measurements are usually expressed as a ratio such as 100/80. These values are somewhat higher in older people. When blood pressures are elevated, such as 140/90, there is a greater risk of stroke, heart attack, or kidney damage. Low blood pressure prevents the brain from receiving adequate oxygen, causing dizziness and fainting.

The blood pressures are measured by a *sphygmomanometer*, an instrument consisting of a stethoscope and an inflatable cuff connected to a tube of mercury called a manometer. After the cuff is wrapped around the upper arm, it is pumped up with air until it cuts off the flow of blood. With the stethoscope over the artery, the air is slowly released from the cuff, decreasing the pressure on the artery. When the blood flow first starts again in the artery, a noise can be heard through the stethoscope signifying the systolic blood pressure as the pressure shown on the manometer. As air continues to be released, the cuff deflates until no sound is heard in the artery. A second pressure reading is taken at the moment of silence and denotes the diastolic pressure, the pressure when the heart is not contracting.

The use of digital blood pressure monitors is becoming more common. However, they have not been validated for use in all situations and can sometimes give inaccurate readings.

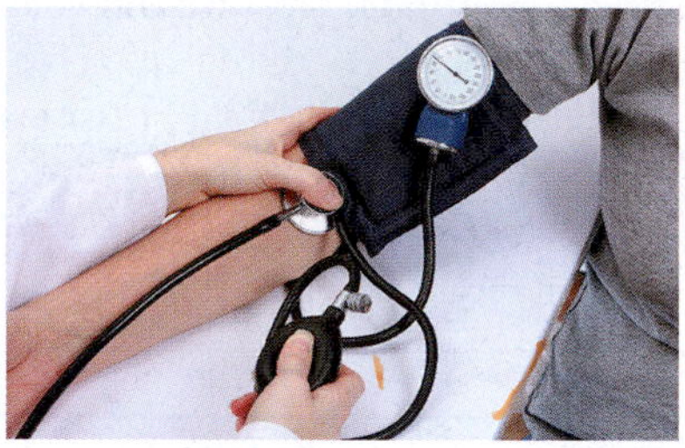

The measurement of blood pressure is part of a routine physical exam.

PRACTICE PROBLEMS

8.1 Properties of Gases

8.1 Use the kinetic molecular theory of gases to explain each of the following:
- **a.** Gases move faster at higher temperatures.
- **b.** Gases can be compressed much more easily than liquids or solids.
- **c.** Gases have low densities.

8.2 Use the kinetic molecular theory of gases to explain each of the following:
- **a.** A container of nonstick cooking spray explodes when thrown into a fire.
- **b.** The air in a hot-air balloon is heated to make the balloon rise.
- **c.** You can smell the odor of cooking onions from far away.

8.3 Identify the property of a gas that is measured in each of the following:
- **a.** 350 K
- **b.** 125 mL
- **c.** 2.00 g of O_2
- **d.** 755 mmHg

8.4 Identify the property of a gas that is measured in each of the following:
- **a.** 425 K
- **b.** 1.0 atm
- **c.** 10.0 L
- **d.** 0.50 mole of He

8.5 Which of the following statements describe the pressure of a gas?
- **a.** the force of the gas particles on the walls of the container
- **b.** the number of gas particles in a container
- **c.** 4.5 L of helium gas
- **d.** 750 Torr
- **e.** 28.8 lb/in.^{2}

8.6 Which of the following statements describe the pressure of a gas?
- **a.** the temperature of a gas
- **b.** the volume of the container
- **c.** 3.00 atm
- **d.** 0.25 mole of O_2
- **e.** 101 kPa

Clinical Applications

8.7 A tank contains oxygen (O_2) at a pressure of 2.00 atm. What is the pressure in the tank in terms of the following units?
- **a.** torr
- **b.** lb/in.^{2}
- **c.** mmHg
- **d.** kPa

8.8 On a climb up Mount Whitney, the atmospheric pressure drops to 467 mmHg. What is the pressure in terms of the following units?
- **a.** atm
- **b.** torr
- **c.** inHg
- **d.** Pa

8.2 Pressure and Volume (Boyle's Law)

LEARNING GOAL Use the pressure–volume relationship (Boyle's law) to calculate the unknown pressure or volume when the temperature and amount of gas do not change.

REVIEW

Solving Equations (1.4)

Imagine that you can see air particles hitting the walls inside a bicycle tire pump. What happens to the pressure inside the pump as you push down on the handle? As the volume decreases, there is a decrease in the surface area of the container. The air particles are crowded together, more collisions occur, and the pressure increases within the container.

When a change in one property (in this case, volume) causes a change in another property (in this case, pressure), the properties are related. If the change occurs in opposite directions, the properties have an **inverse relationship**. The inverse relationship between the pressure and volume of a gas is known as **Boyle's law**. The law states that the volume (V) of a sample of gas changes inversely with the pressure (P) of the gas as long as there is no change in the temperature (T) and amount of gas (n), as illustrated in **FIGURE 8.4**.

If the volume or pressure of a gas changes without any change occurring in the temperature or in the amount of the gas, then the final pressure and volume will give the same PV product as the initial pressure and volume. Then we can set the initial and final PV products equal to each other. In the equation for Boyle's law, the initial pressure and volume are written as P_1 and V_1 and the final pressure and volume are written as P_2 and V_2.

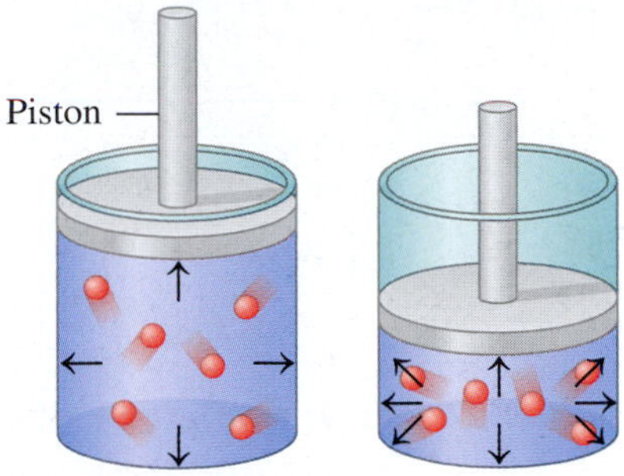
Piston
$V_1 = 4$ L
$P_1 = 1$ atm
$V_2 = 2$ L
$P_2 = 2$ atm

FIGURE 8.4 ▶ **Boyle's law**: As volume decreases, gas molecules become more crowded, which causes the pressure to increase. Pressure and volume are inversely related.

Q If the volume of a gas increases, what will happen to its pressure?

Boyle's Law

$P_1V_1 = P_2V_2$ No change in temperature and amount of gas

▶SAMPLE PROBLEM 8.3 Calculating Volume When Pressure Changes

CORE CHEMISTRY SKILL

Using the Gas Laws

TRY IT FIRST

When Whitney had her asthma attack, she was given oxygen through a face mask. The gauge on a 12-L tank of compressed oxygen reads 3800 mmHg. How many liters would this same gas occupy at a final pressure of 570 mmHg when temperature and amount of gas do not change?

SOLUTION

STEP 1 **State the given and needed quantities.** We place the gas data in a table by writing the initial pressure and volume as P_1 and V_1 and the final pressure and volume as P_2 and V_2. We see that the pressure decreases from 3800 mmHg to 570 mmHg. Using Boyle's law, we predict that the volume increases.

	Given	Need	Connect
ANALYZE THE PROBLEM	$P_1 = 3800$ mmHg $P_2 = 570$ mmHg $V_1 = 12$ L **Factors that do not change:** T and n	V_2	Boyle's law, $P_1V_1 = P_2V_2$ **Predict:** P decreases, V increases

Oxygen therapy increases the oxygen available to the tissues of the body.

STEP 2 **Rearrange the gas law equation to solve for the unknown quantity.** For a PV relationship, we use Boyle's law and solve for V_2 by dividing both sides by P_2.

$$P_1V_1 = P_2V_2$$

$$\frac{P_1V_1}{P_2} = \frac{\cancel{P_2}V_2}{\cancel{P_2}}$$

$$V_2 = V_1 \times \frac{P_1}{P_2}$$

ENGAGE

Why can we predict that the pressure of a gas decreases in a closed container when the volume of the container increases?

STEP 3 **Substitute values into the gas law equation and calculate.** When we substitute in the values with pressures in units of mmHg, the ratio of pressures (pressure factor) is greater than 1, which increases the volume as predicted.

$$V_2 = 12\text{ L} \times \frac{3800\ \cancel{\text{mmHg}}}{570\ \cancel{\text{mmHg}}} = 80.\text{ L}$$

Pressure factor increases volume

STUDY CHECK 8.3

In an underground gas reserve, a bubble of methane gas (CH_4) has a volume of 45.0 mL at 1.60 atm pressure.

a. What volume, in milliliters, will the bubble occupy when it reaches the surface where the atmospheric pressure is 744 mmHg, if there is no change in the temperature and amount of gas?

b. What is the new pressure, in atmospheres, when the volume of the methane gas bubble expands to 125 mL, if there is no change in the temperature and amount of gas?

ANSWER

a. 73.5 mL **b.** 0.576 atm

TEST

Try Practice Problems 8.9 to 8.24

Chemistry Link to Health

Pressure–Volume Relationship in Breathing

The importance of Boyle's law becomes apparent when you consider the mechanics of breathing. Our lungs are elastic, balloon-like structures contained within an airtight chamber called the thoracic cavity. The diaphragm, a muscle, forms the flexible floor of the cavity.

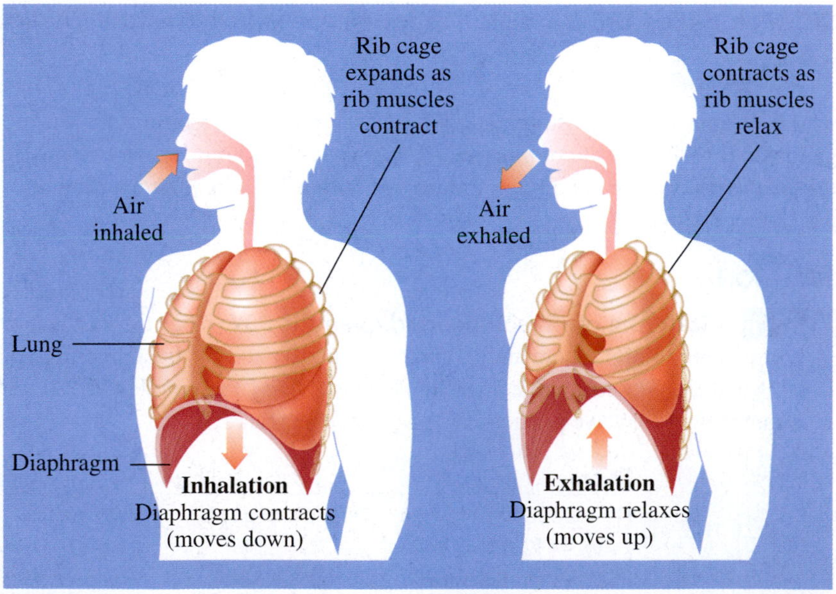

Rib cage expands as rib muscles contract
Rib cage contracts as rib muscles relax
Air inhaled
Air exhaled
Lung
Diaphragm
Inhalation
Diaphragm contracts (moves down)
Exhalation
Diaphragm relaxes (moves up)

Inspiration

The process of taking a breath of air begins when the diaphragm contracts and the rib cage expands, causing an increase in the volume of the thoracic cavity. The elasticity of the lungs allows them to expand when the thoracic cavity expands. According to Boyle's law, the pressure inside the lungs decreases when their volume increases, causing the pressure inside the lungs to fall below the pressure of the atmosphere. This difference in pressures produces a *pressure gradient* between the lungs and the atmosphere. In a pressure gradient, molecules flow from an area of higher pressure to an area of lower pressure. During the inhalation phase of breathing, air flows into the lungs (*inspiration*), until the pressure within the lungs becomes equal to the pressure of the atmosphere.

Expiration

Expiration, the exhalation phase of breathing, occurs when the diaphragm relaxes and moves back up into the thoracic cavity to its resting position. The volume of the thoracic cavity decreases, which squeezes the lungs and decreases their volume. Now the pressure in the lungs is higher than the pressure of the atmosphere, so air flows out of the lungs. Thus, breathing is a process in which pressure gradients are continuously created between the lungs and the environment because of the changes in the volume and pressure.

PRACTICE PROBLEMS

8.2 Pressure and Volume (Boyle's Law)

8.9 Why do scuba divers need to exhale air when they ascend to the surface of the water?

8.10 Why does a sealed bag of chips expand when you take it to a higher altitude?

8.11 The air in a cylinder with a piston has a volume of 220 mL and a pressure of 650 mmHg.

a. To obtain a higher pressure inside the cylinder when the temperature and amount of gas do not change, would the cylinder change as shown in **A** or **B**? Explain your choice.

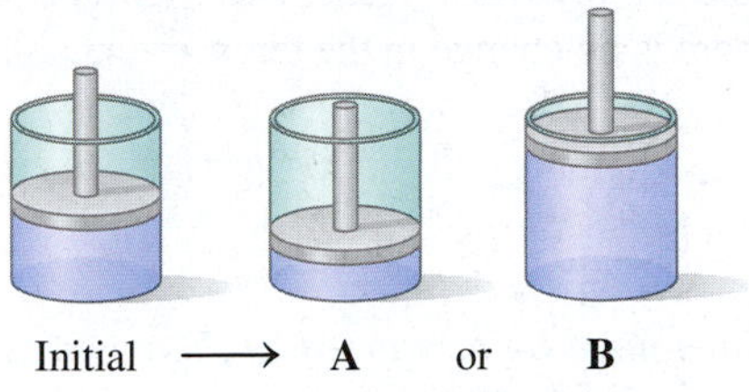
Initial ⟶ A or B

b. If the pressure inside the cylinder increases to 1.2 atm, what is the final volume, in milliliters, of the cylinder?

8.12 A balloon is filled with helium gas. When each of the following changes are made with no change in temperature and amount of gas, which of these diagrams (**A**, **B**, or **C**) shows the final volume of the balloon?

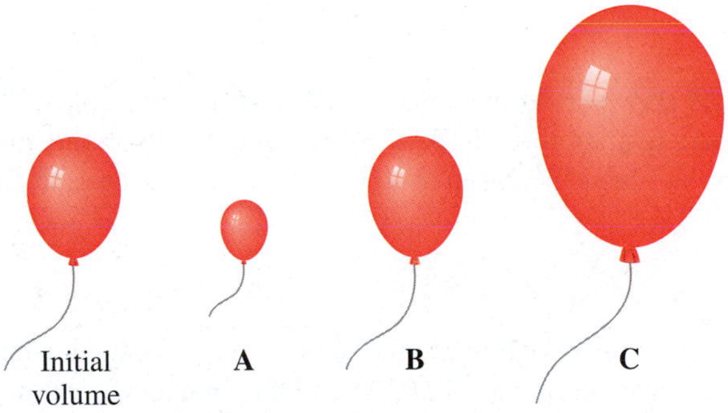
Initial volume
A
B
C

a. The balloon floats to a higher altitude where the outside pressure is lower.
b. The balloon is taken inside the house, but the atmospheric pressure does not change.
c. The balloon is put in a hyperbaric chamber in which the pressure is increased.

8.13 A gas with a volume of 4.0 L is in a closed container. Indicate the changes (*increases*, *decreases*, *does not change*) in its pressure when the volume undergoes the following changes at the same temperature and amount of gas:

a. The volume is compressed to 2.0 L.
b. The volume expands to 12 L.
c. The volume is compressed to 0.40 L.

8.14 A gas at a pressure of 2.0 atm is in a closed container. Indicate the changes (*increases*, *decreases*, *does not change*) in its volume when the pressure undergoes the following changes at the same temperature and amount of gas:

a. The pressure increases to 6.0 atm.
b. The pressure remains at 2.0 atm.
c. The pressure drops to 0.40 atm.

8.15 A 10.0-L balloon contains helium gas at a pressure of 655 mmHg. What is the final pressure, in millimeters of mercury, when the helium is placed in tanks that have the following volumes, if there is no change in temperature and amount of gas?

a. 20.0 L **b.** 2.50 L
c. 13 800 mL **d.** 1250 mL

8.16 The air in a 5.00-L tank has a pressure of 1.20 atm. What is the final pressure, in atmospheres, when the air is placed in tanks that have the following volumes, if there is no change in temperature and amount of gas?

a. 1.00 L **b.** 2500. mL
c. 750. mL **d.** 8.00 L

8.17 A sample of nitrogen (N_2) has a volume of 50.0 L at a pressure of 760. mmHg. What is the final volume, in liters, of the gas at each of the following pressures, if there is no change in temperature and amount of gas?

a. 725 mmHg **b.** 1520 mmHg
c. 0.500 atm **d.** 850 Torr

8.18 A sample of methane (CH_4) has a volume of 25 mL at a pressure of 0.80 atm. What is the final volume, in milliliters, of the gas at each of the following pressures, if there is no change in temperature and amount of gas?

a. 0.40 atm **b.** 2.00 atm
c. 2500 mmHg **d.** 80.0 Torr

8.19 A sample of Ar gas has a volume of 5.40 L with an unknown pressure. The gas has a volume of 9.73 L when the pressure is 3.62 atm, with no change in temperature and amount of gas. What was the initial pressure, in atmospheres, of the gas?

8.20 A sample of Ne gas has a pressure of 654 mmHg with an unknown volume. The gas has a pressure of 345 mmHg when the volume is 495 mL, with no change in temperature and amount of gas. What was the initial volume, in milliliters, of the gas?

Clinical Applications

8.21 Cyclopropane, C_3H_6, is a general anesthetic. A 5.0-L sample has a pressure of 5.0 atm. What is the final volume, in liters, of this gas given to a patient at a pressure of 1.0 atm with no change in temperature and amount of gas?

8.22 A patient's oxygen tank holds 20.0 L of oxygen (O_2) at a pressure of 15.0 atm. What is the final volume, in liters, of this gas when it is released at a pressure of 1.00 atm with no change in temperature and amount of gas?

8.23 Use the words *inspiration* and *expiration* to describe the part of the breathing cycle that occurs as a result of each of the following:

a. The diaphragm contracts.
b. The volume of the lungs decreases.
c. The pressure within the lungs is less than that of the atmosphere.

8.24 Use the words *inspiration* and *expiration* to describe the part of the breathing cycle that occurs as a result of each of the following:

a. The diaphragm relaxes, moving up into the thoracic cavity.
b. The volume of the lungs expands.
c. The pressure within the lungs is higher than that of the atmosphere.

As the gas in a hot-air balloon is heated, it expands.

8.3 Temperature and Volume (Charles's Law)

LEARNING GOAL Use the temperature–volume relationship (Charles's law) to calculate the unknown temperature or volume when the pressure and amount of gas do not change.

Suppose that you are going to take a ride in a hot-air balloon. The captain turns on a propane burner to heat the air inside the balloon. As the air is heated, it expands and becomes less dense than the air outside, causing the balloon and its passengers to lift off. In 1787, Jacques Charles, a balloonist as well as a physicist, proposed that the volume of a gas is related to the temperature. This proposal became **Charles's law**, which states that the volume (V) of a gas is directly related to the temperature (T) when there is no change in the pressure (P) and amount (n) of gas. A **direct relationship** is one in which the related properties increase or decrease together. For two conditions, initial and final, we can write Charles's law as follows:

Charles's Law

$$\frac{V_1}{T_1} = \frac{V_2}{T_2} \quad \text{No change in pressure and amount of gas}$$

All temperatures used in gas law calculations must be converted to their corresponding Kelvin (K) temperatures.

To determine the effect of changing temperature on the volume of a gas, the pressure and the amount of gas are not changed. If we increase the temperature of a gas sample, the volume of the container must increase (see **FIGURE 8.5**). If the temperature of the gas is decreased, the volume of the container must also decrease when pressure and the amount of gas do not change.

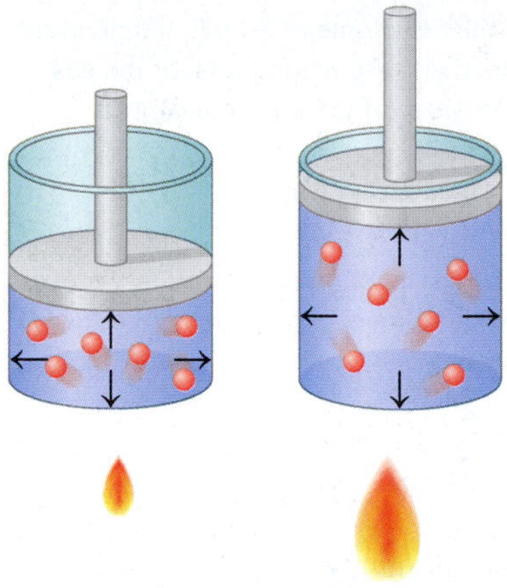
$T_1 = 200$ K
$V_1 = 1$ L
$T_2 = 400$ K
$V_2 = 2$ L

FIGURE 8.5 ▶ Charles's law: The Kelvin temperature of a gas is directly related to the volume of the gas when there is no change in the pressure and amount of gas.

Q If the temperature of a gas decreases, how will the volume change if the pressure and amount of gas do not change?

▶ SAMPLE PROBLEM 8.4 Calculating Volume When Temperature Changes

TRY IT FIRST

Helium gas is used to inflate the abdomen during laparoscopic surgery. A sample of helium gas has a volume of 5.40 L and a temperature of 15 °C. What is the final volume, in liters, of the gas after the temperature has been increased to 42 °C when the pressure and amount of gas do not change?

SOLUTION

STEP 1 **State the given and needed quantities.** We place the gas data in a table by writing the initial temperature and volume as T_1 and V_1 and the final temperature and volume as T_2 and V_2. We see that the temperature increases from 15 °C to 42 °C. Using Charles's law, we predict that the volume increases.

$$T_1 = 15\,°\text{C} + 273 = 288\text{ K}$$
$$T_2 = 42\,°\text{C} + 273 = 315\text{ K}$$

	Given	Need	Connect
ANALYZE THE PROBLEM	$T_1 = 288$ K $T_2 = 315$ K $V_1 = 5.40$ L **Factors that do not change:** P and n	V_2	Charles's law, $\frac{V_1}{T_1} = \frac{V_2}{T_2}$ **Predict:** T increases, V increases

ENGAGE

Why can we predict that the volume of the helium gas will increase when the temperature increases?

STEP 2 **Rearrange the gas law equation to solve for the unknown quantity.** In this problem, we want to know the final volume (V_2) when the temperature increases. Using Charles's law, we solve for V_2 by multiplying both sides by T_2.

$$\frac{V_1}{T_1} = \frac{V_2}{T_2}$$

$$\frac{V_1}{T_1} \times T_2 = \frac{V_2}{\cancel{T_2}} \times \cancel{T_2}$$

$$V_2 = V_1 \times \frac{T_2}{T_1}$$

STEP 3 **Substitute values into the gas law equation and calculate.** From the table, we see that the temperature has increased. Because temperature is directly related to volume, the volume must increase. When we substitute in the values, we see that the ratio of the temperatures (temperature factor) is greater than 1, which increases the volume as predicted.

$$V_2 = 5.40\ \text{L} \times \frac{315\ \cancel{\text{K}}}{288\ \cancel{\text{K}}} = 5.91\ \text{L}$$

Temperature factor increases volume

STUDY CHECK 8.4

a. A mountain climber inhales air that has a temperature of −8 °C. If the final volume of air in the lungs is 569 mL at a body temperature of 37 °C, what was the initial volume of air, in milliliters, inhaled by the climber, if the pressure and amount of gas do not change?

b. A hiker inhales 598 mL of air. If the final volume of air in the lungs is 612 mL at a body temperature of 37 °C, what was the initial temperature of the air in degrees Celsius?

ANSWER

a. 486 mL **b.** 30. °C

TEST

Try Practice Problems 8.25 to 8.32

PRACTICE PROBLEMS

8.3 Temperature and Volume (Charles's Law)

8.25 Select the diagram that shows the final volume of a balloon when each of the following changes are made when the pressure and amount of gas do not change:

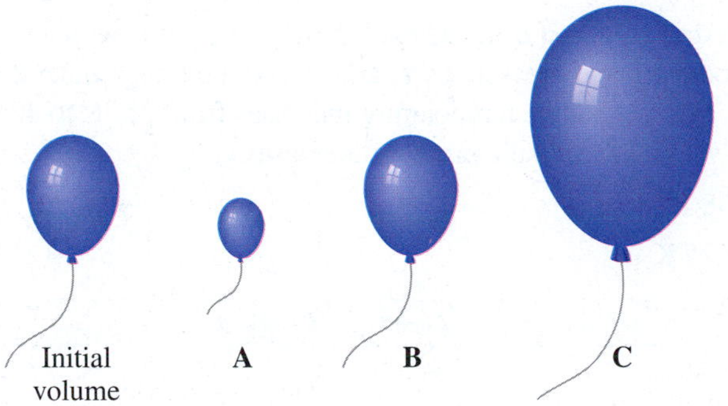
Initial volume
A
B
C

a. The temperature is changed from 100 K to 300 K.
b. The balloon is placed in a freezer.
c. The balloon is first warmed and then returned to its starting temperature.

8.26 Indicate whether the final volume of gas in each of the following is the *same*, *larger*, or *smaller* than the initial volume, if pressure and amount of gas do not change:
a. A volume of 505 mL of air on a cold winter day at −15 °C is breathed into the lungs, where body temperature is 37 °C.
b. The heater used to heat the air in a hot-air balloon is turned off.
c. A balloon filled with helium at the amusement park is left in a car on a hot day.

8.27 A sample of neon initially has a volume of 2.50 L at 15 °C. What final temperature, in degrees Celsius, is needed to change the volume of the gas to each of the following, if pressure and amount of gas do not change?
a. 5.00 L **b.** 1250 mL **c.** 7.50 L **d.** 3550 mL

8.28 A gas has a volume of 4.00 L at 0 °C. What final temperature, in degrees Celsius, is needed to change the volume of the gas to each of the following, if pressure and amount of gas do not change?
a. 1.50 L **b.** 1200 mL **c.** 10.0 L **d.** 50.0 mL

8.29 A balloon contains 2500 mL of helium gas at 75 °C. What is the final volume, in milliliters, of the gas when the temperature changes to each of the following, if pressure and amount of gas do not change?
a. 55 °C **b.** 680. K **c.** −25 °C **d.** 240. K

8.30 An air bubble has a volume of 0.500 L at 18 °C. What is the final volume, in liters, of the gas when the temperature changes to each of the following, if pressure and amount of gas do not change?
a. 0 °C **b.** 425 K **c.** −12 °C **d.** 575 K

8.31 A gas sample has a volume of 0.256 L with an unknown temperature. The same gas has a volume of 0.198 L when the temperature is 32 °C, with no change in the pressure and amount of gas. What was the initial temperature, in degrees Celsius, of the gas?

8.32 A gas sample has a temperature of 22 °C with an unknown volume. The same gas has a volume of 456 mL when the temperature is 86 °C, with no change in the pressure and amount of gas. What was the initial volume, in milliliters, of the gas?

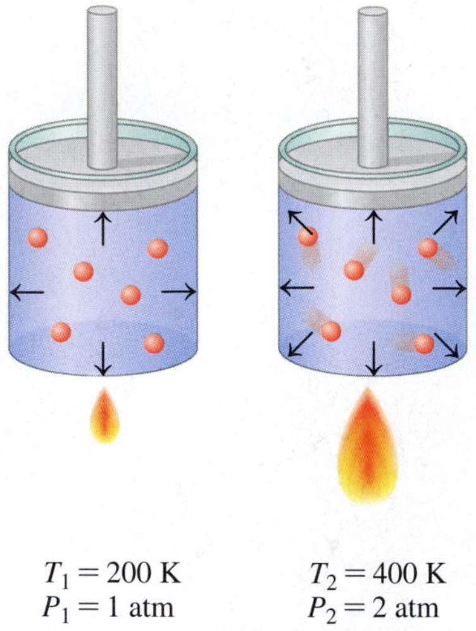

$T_1 = 200$ K
$P_1 = 1$ atm
$T_2 = 400$ K
$P_2 = 2$ atm

FIGURE 8.6 ▶ **Gay-Lussac's law:** When the Kelvin temperature of a gas is doubled and the volume and amount of gas do not change, the pressure also doubles.

Q How does a decrease in the temperature of a gas affect its pressure when the volume and amount of gas do not change?

8.4 Temperature and Pressure (Gay-Lussac's Law)

LEARNING GOAL Use the temperature–pressure relationship (Gay-Lussac's law) to calculate the unknown temperature or pressure when the volume and amount of gas do not change.

If we could observe the molecules of a gas as the temperature rises, we would notice that they move faster and hit the sides of the container more often and with greater force. If volume and amount of gas do not change, the pressure would increase. In the temperature–pressure relationship known as **Gay-Lussac's law**, the pressure of a gas is directly related to its Kelvin temperature. This means that an increase in temperature increases the pressure of a gas, and a decrease in temperature decreases the pressure of the gas as long as the volume and amount of gas do not change (see **FIGURE 8.6**).

Gay-Lussac's Law

$$\frac{P_1}{T_1} = \frac{P_2}{T_2} \quad \text{No change in volume and amount of gas}$$

All temperatures used in gas law calculations must be converted to their corresponding Kelvin (K) temperatures.

ENGAGE

Why can we predict that the pressure in the oxygen tank will increase when the temperature increases?

▶ SAMPLE PROBLEM 8.5 Calculating Pressure When Temperature Changes

TRY IT FIRST

Home oxygen tanks can be dangerous if they are heated, because they can explode. Suppose an oxygen tank has a pressure of 120 atm at a room temperature of 25 °C. If a fire in the room causes the temperature of the gas inside the oxygen tank to reach 402 °C, what is its pressure, in atmospheres, if the volume and amount of gas do not change? The oxygen tank may rupture if the pressure inside exceeds 180 atm. Would you expect it to rupture?

SOLUTION

STEP 1 State the given and needed quantities. We place the gas data in a table by writing the initial temperature and pressure as T_1 and P_1 and the final temperature and pressure as T_2 and P_2. We see that the temperature increases from 25 °C to 402 °C. Using Gay-Lussac's law, we predict that the pressure increases.

$T_1 = 25\text{ °C} + 273 = 298$ K

$T_2 = 402\text{ °C} + 273 = 675$ K

ANALYZE THE PROBLEM	Given	Need	Connect
	$P_1 = 120$ atm $T_1 = 298$ K $\quad T_2 = 675$ K **Factors that do not change:** V and n	P_2	Gay-Lussac's law, $\frac{P_1}{T_1} = \frac{P_2}{T_2}$ **Predict:** T increases, P increases

STEP 2 Rearrange the gas law equation to solve for the unknown quantity. Using Gay-Lussac's law, we solve for P_2 by multiplying both sides by T_2.

$$\frac{P_1}{T_1} = \frac{P_2}{T_2}$$

$$\frac{P_1}{T_1} \times T_2 = \frac{P_2}{\cancel{T_2}} \times \cancel{T_2}$$

$$P_2 = P_1 \times \frac{T_2}{T_1}$$

STEP 3 **Substitute values into the gas law equation and calculate.** When we substitute in the values, we see that the ratio of the temperatures (temperature factor) is greater than 1, which increases the pressure as predicted.

$$P_2 = 120 \text{ atm} \times \frac{675 \cancel{\text{K}}}{298 \cancel{\text{K}}} = 270 \text{ atm}$$

Temperature factor increases pressure

Because the calculated pressure of 270 atm exceeds the limit of 180 atm, we would expect the oxygen tank to rupture.

Cylinders of oxygen gas are placed in a hospital storage room.

STUDY CHECK 8.5

In a storage area of a hospital where the temperature has reached 55 °C, the pressure of oxygen gas in a 15.0-L steel cylinder is 965 Torr.

a. To what temperature, in degrees Celsius, would the gas have to be cooled to reduce the pressure to 850. Torr, when the volume and the amount of the gas do not change?

b. What is the final pressure, in millimeters of mercury, when the temperature of the oxygen gas drops to 24 °C, and the volume and the amount of the gas do not change?

ANSWER

a. 16 °C **b.** 874 mmHg

TEST

Try Practice Problems 8.33 to 8.40

Vapor Pressure and Boiling Point

When liquid molecules with sufficient kinetic energy break away from the surface, they become gas particles or vapor. In an open container, all the liquid will eventually evaporate. In a closed container, the vapor accumulates and creates pressure called **vapor pressure**. Each liquid exerts its own vapor pressure at a given temperature. As temperature increases, more vapor forms, and vapor pressure increases. **TABLE 8.4** lists the vapor pressure of water at various temperatures.

A liquid reaches its boiling point when its vapor pressure becomes equal to the external pressure. As boiling occurs, bubbles of the gas form within the liquid and quickly rise to the surface. For example, at an atmospheric pressure of 760 mmHg, water will boil at 100 °C, the temperature at which its vapor pressure reaches 760 mmHg.

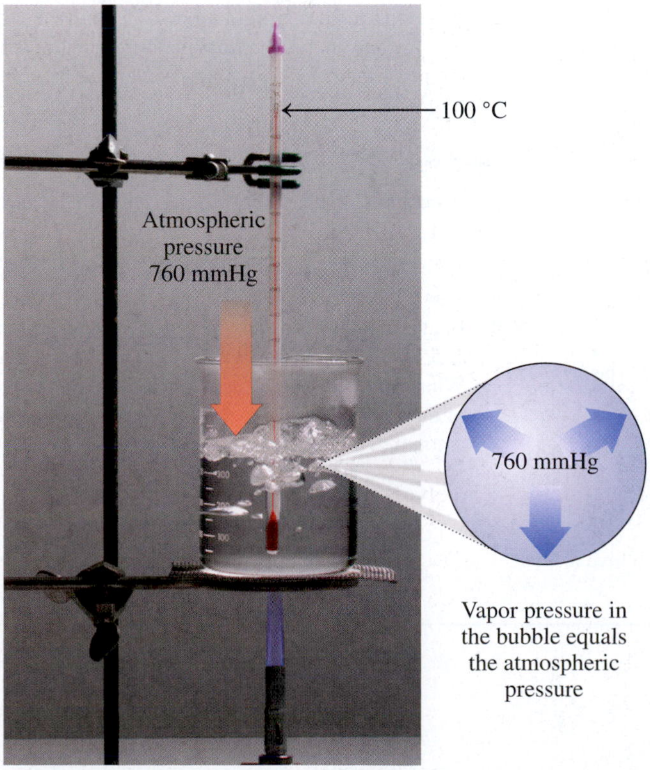
100 °C
Atmospheric pressure 760 mmHg
760 mmHg
Vapor pressure in the bubble equals the atmospheric pressure

TABLE 8.4 Vapor Pressure of Water

Temperature (°C)	Vapor Pressure (mmHg)
0	5
10	9
20	18
30	32
37*	47
40	55
50	93
60	149
70	234
80	355
90	528
100	760

*Normal body temperature

TABLE 8.5 Pressure and the Boiling Point of Water

Pressure (mmHg)	Boiling Point (°C)
270	70
467	87
630	95
752	99
760	100
800	100.4
1075	110
1520 (2 atm)	120
3800 (5 atm)	160
7600 (10 atm)	180

At high altitudes, where atmospheric pressures are lower than 760 mmHg, the boiling point of water is lower than 100 °C. For example, a typical atmospheric pressure in Denver is 630 mmHg. This means that water in Denver boils when the vapor pressure is 630 mmHg. From **TABLE 8.5**, we see that water has a vapor pressure of 630 mmHg at 95 °C, which means that water boils at 95 °C in Denver.

In a closed container such as a pressure cooker, a pressure greater than 1 atm can be obtained, which means that water boils at a temperature higher than 100 °C. Laboratories and hospitals use closed containers called *autoclaves* to sterilize laboratory and surgical equipment.

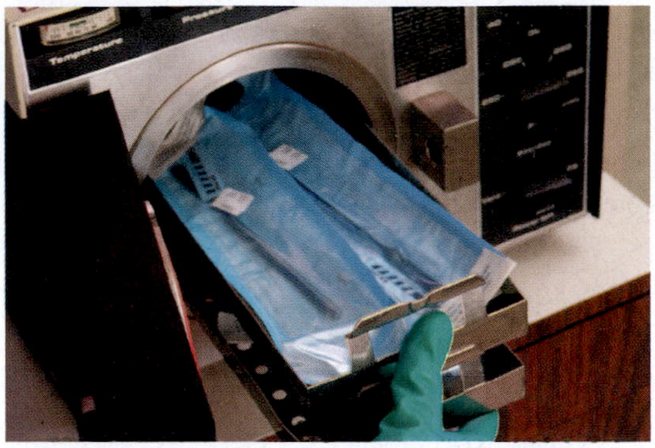

An autoclave used to sterilize equipment attains a temperature higher than 100 °C.

PRACTICE PROBLEMS

8.4 Temperature and Pressure (Gay-Lussac's Law)

8.33 Calculate the final pressure, in millimeters of mercury, for each of the following, if volume and amount of gas do not change:
- **a.** A gas with an initial pressure of 1200 Torr at 155 °C is cooled to 0 °C.
- **b.** A gas in an aerosol can at an initial pressure of 1.40 atm at 12 °C is heated to 35 °C.

8.34 Calculate the final pressure, in atmospheres, for each of the following, if volume and amount of gas do not change:
- **a.** A gas with an initial pressure of 1.20 atm at 75 °C is cooled to −32 °C.
- **b.** A sample of N_2 with an initial pressure of 780. mmHg at −75 °C is heated to 28 °C.

8.35 Calculate the final temperature, in degrees Celsius, for each of the following, if volume and amount of gas do not change:
- **a.** A sample of xenon gas at 25 °C and 740. mmHg is cooled to give a pressure of 620. mmHg.
- **b.** A tank of argon gas with a pressure of 0.950 atm at −18 °C is heated to give a pressure of 1250 Torr.

8.36 Calculate the final temperature, in degrees Celsius, for each of the following, if volume and amount of gas do not change:
- **a.** A sample of helium gas with a pressure of 250 Torr at 0 °C is heated to give a pressure of 1500 Torr.
- **b.** A sample of air at 40 °C and 740. mmHg is cooled to give a pressure of 680. mmHg.

8.37 A gas sample has a pressure of 744 mmHg when the temperature is 22 °C. What is the final temperature, in degrees Celsius, when the pressure is 766 mmHg, with no change in the volume and amount of gas?

8.38 A gas sample has a pressure of 2.35 atm when the temperature is −15 °C. What is the final pressure, in atmospheres, when the temperature is 46 °C, with no change in the volume and amount of gas?

Clinical Applications

8.39 A tank contains isoflurane, an inhaled anesthetic, at a pressure of 1.8 atm and 5 °C. What is the pressure, in atmospheres, if the gas is warmed to a temperature of 22 °C, if volume and amount of gas do not change?

8.40 Bacteria and viruses are inactivated by temperatures above 135 °C. An autoclave contains steam at 1.00 atm and 100 °C. What is the pressure, in atmospheres, when the temperature of the steam in the autoclave reaches 135 °C, if volume and amount of gas do not change?

8.5 The Combined Gas Law

LEARNING GOAL Use the combined gas law to calculate the unknown pressure, volume, or temperature of a gas when changes in two of these properties are given and the amount of gas does not change.

All of the pressure–volume–temperature relationships for gases that we have studied may be combined into a single relationship called the **combined gas law**. This expression is useful for studying the effect of changes in two of these variables on the third as long as the amount of gas (number of moles) does not change.

Combined Gas Law

$$\frac{P_1V_1}{T_1} = \frac{P_2V_2}{T_2} \quad \text{No change in number of moles of gas}$$

By using the combined gas law, we can derive any of the gas laws by omitting those properties that do not change, as seen in **TABLE 8.6**.

TABLE 8.6 Summary of Gas Laws

Combined Gas Law	Properties That Do Not Change	Relationship	Name of Gas Law
$\frac{P_1V_1}{\cancel{T_1}} = \frac{P_2V_2}{\cancel{T_2}}$	T, n	$P_1V_1 = P_2V_2$	Boyle's law
$\frac{\cancel{P_1}V_1}{T_1} = \frac{\cancel{P_2}V_2}{T_2}$	P, n	$\frac{V_1}{T_1} = \frac{V_2}{T_2}$	Charles's law
$\frac{P_1\cancel{V_1}}{T_1} = \frac{P_2\cancel{V_2}}{T_2}$	V, n	$\frac{P_1}{T_1} = \frac{P_2}{T_2}$	Gay-Lussac's law

ENGAGE

Why does the pressure of a gas decrease to one-fourth of its initial pressure when the volume of the gas doubles and the Kelvin temperature decreases by half, if the amount of gas does not change?

▶ SAMPLE PROBLEM 8.6 Using the Combined Gas Law

TRY IT FIRST

A 25.0-mL bubble is released from a diver's air tank at a pressure of 4.00 atm and a temperature of 11 °C. What is the volume, in milliliters, of the bubble when it reaches the ocean surface where the pressure is 1.00 atm and the temperature is 18 °C? (Assume the amount of gas in the bubble does not change.)

SOLUTION

STEP 1 **State the given and needed quantities.** We list the properties that change, which are the pressure, volume, and temperature. The temperatures in degrees Celsius must be changed to kelvins.

$$T_1 = 11\text{ °C} + 273 = 284\text{ K}$$
$$T_2 = 18\text{ °C} + 273 = 291\text{ K}$$

	Given	Need	Connect
ANALYZE THE PROBLEM	$P_1 = 4.00$ atm $P_2 = 1.00$ atm $V_1 = 25.0$ mL $T_1 = 284$ K $T_2 = 291$ K **Factor that does not change:** n	V_2	combined gas law, $\frac{P_1V_1}{T_1} = \frac{P_2V_2}{T_2}$

SCUBAPRO

Under water, the pressure on a diver is greater than the atmospheric pressure.

STEP 2 **Rearrange the gas law equation to solve for the unknown quantity.** Using the combined gas law, we solve for V_2 by multiplying both sides by T_2 and dividing both sides by P_2.

$$\frac{P_1V_1}{T_1} = \frac{P_2V_2}{T_2}$$

$$\frac{P_1V_1}{T_1} \times \frac{T_2}{P_2} = \frac{\cancel{P_2}V_2}{\cancel{T_2}} \times \frac{\cancel{T_2}}{\cancel{P_2}}$$

$$V_2 = V_1 \times \frac{P_1}{P_2} \times \frac{T_2}{T_1}$$

ENGAGE

Rearrange the combined gas law to solve for T_1.

STEP 3 **Substitute values into the gas law equation and calculate.** From the data table, we determine that both the pressure decrease and the temperature increase will increase the volume.

$$V_2 = 25.0\text{ mL} \times \frac{4.00\ \cancel{\text{atm}}}{1.00\ \cancel{\text{atm}}} \times \frac{291\ \cancel{\text{K}}}{284\ \cancel{\text{K}}} = 102\text{ mL}$$

Pressure factor increases volume; Temperature factor increases volume

However, in situations where the unknown value is decreased by one change but increased by the second change, it is difficult to predict the overall change for the unknown.

STUDY CHECK 8.6

A weather balloon is filled with 15.0 L of helium at a temperature of 25 °C and a pressure of 685 mmHg.

a. What is the pressure, in millimeters of mercury, of the helium in the balloon in the upper atmosphere when the final temperature is −35 °C and the final volume becomes 34.0 L, if the amount of gas does not change?

b. What is the temperature, in degrees Celsius, if the helium in the balloon has a final volume of 22.0 L and a final pressure of 426 mmHg, if the amount of gas does not change?

ANSWER

a. 241 mmHg **b.** −1 °C

TEST

Try Practice Problems 8.41 to 8.46

PRACTICE PROBLEMS

8.5 The Combined Gas Law

8.41 Rearrange the variables in the combined gas law to solve for T_2.

8.42 Rearrange the variables in the combined gas law to solve for P_2.

8.43 A sample of helium gas has a volume of 6.50 L at a pressure of 845 mmHg and a temperature of 25 °C. What is the final pressure of the gas, in atmospheres, when the volume and temperature of the gas sample are changed to the following, if the amount of gas does not change?
a. 1850 mL and 325 K **b.** 2.25 L and 12 °C
c. 12.8 L and 47 °C

8.44 A sample of argon gas has a volume of 735 mL at a pressure of 1.20 atm and a temperature of 112 °C. What is the final volume of the gas, in milliliters, when the pressure and temperature of the gas sample are changed to the following, if the amount of gas does not change?
a. 658 mmHg and 281 K **b.** 0.55 atm and 75 °C
c. 15.4 atm and −15 °C

Clinical Applications

8.45 A 124-mL bubble of hot gas initially at 212 °C and 1.80 atm is emitted from an active volcano. What is the final temperature, in degrees Celsius, of the gas in the bubble outside the volcano when the final volume of the bubble is 138 mL and the pressure is 0.800 atm, if the amount of gas does not change?

8.46 A scuba diver 60 ft below the ocean surface inhales 50.0 mL of compressed air from a scuba tank at a pressure of 3.00 atm and a temperature of 8 °C. What is the final pressure of air, in atmospheres, in the lungs when the gas expands to 150.0 mL at a body temperature of 37 °C, if the amount of gas does not change?

8.6 Volume and Moles (Avogadro's Law)

LEARNING GOAL Use Avogadro's law to calculate the unknown amount or volume of a gas when the pressure and temperature do not change.

REVIEW

Using Molar Mass as a Conversion Factor (7.3)

In our study of the gas laws, we have looked at changes in properties for a specified amount (n) of gas. Now we consider how the properties of a gas change when there is a change in the number of moles or grams of the gas.

When you blow up a balloon, its volume increases because you add more air molecules. If the balloon has a small hole in it, air leaks out, causing its volume to decrease. In 1811, Amedeo Avogadro formulated **Avogadro's law**, which states that the volume of a gas is directly related to the number of moles of a gas when temperature and pressure do not change. For example, if the number of moles of a gas is doubled, then the volume will also double as long as we do not change the pressure or the temperature (see **FIGURE 8.7**). When pressure and temperature do not change, we can write Avogadro's law as follows:

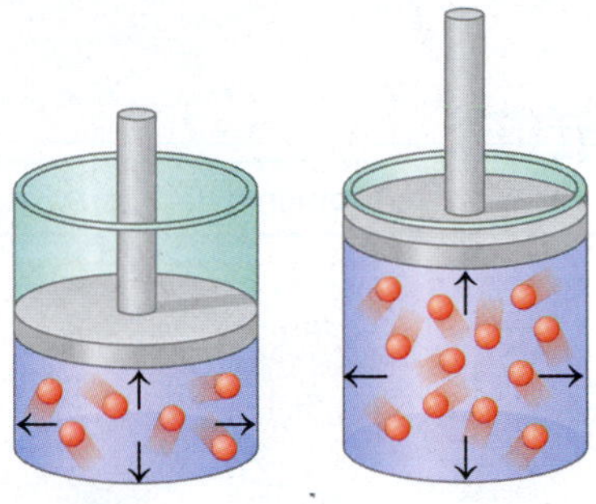

$n_1 = 1$ mole $V_1 = 1$ L $n_2 = 2$ moles $V_2 = 2$ L

FIGURE 8.7 ▶ **Avogadro's law:** The volume of a gas is directly related to the number of moles of the gas. If the number of moles is doubled, the volume must double when temperature and pressure do not change.

Q If a balloon has a leak, what happens to its volume?

Avogadro's Law

$$\frac{V_1}{n_1} = \frac{V_2}{n_2}$$ No change in pressure and temperature

▶ SAMPLE PROBLEM 8.7 Calculating Volume for a Change in Moles

TRY IT FIRST

A weather balloon with a volume of 44 L is filled with 2.0 moles of helium. What is the final volume, in liters, if helium are added to give a total of 5.0 moles of helium, if the pressure and temperature do not change?

SOLUTION

STEP 1 **State the given and needed quantities.** We list those properties that change, which are volume and amount (moles). Because there is an increase in the number of moles of gas, we can predict that the volume increases.

	Given	Need	Connect
ANALYZE THE PROBLEM	$V_1 = 44$ L $n_1 = 2.0$ moles $n_2 = 5.0$ moles **Factors that do not change:** P and T	V_2	Avogadro's law, $\frac{V_1}{n_1} = \frac{V_2}{n_2}$ **Predict:** n increases, V increases

ENGAGE

Why can we predict that the volume of a balloon increases when the number of moles of gas increases, if the pressure and temperature do not change?

STEP 2 **Rearrange the gas law equation to solve for the unknown quantity.** Using Avogadro's law, we can solve for V_2 by multiplying both sides of the equation by n_2.

$$\frac{V_1}{n_1} = \frac{V_2}{n_2}$$

$$\frac{V_1}{n_1} \times n_2 = \frac{V_2}{\cancel{n_2}} \times \cancel{n_2}$$

$$V_2 = V_1 \times \frac{n_2}{n_1}$$

STEP 3 **Substitute values into the gas law equation and calculate.** When we substitute in the values, we see that the ratio of the moles (mole factor) is greater than 1, which increases the volume as predicted.

$$V_2 = 44 \text{ L} \times \frac{5.0 \cancel{\text{moles}}}{2.0 \cancel{\text{moles}}} = 110 \text{ L}$$

Mole factor increases volume

STUDY CHECK 8.7

A balloon containing 8.00 g of oxygen gas has a volume of 5.00 L.

a. What is the volume, in liters, after 4.00 g of oxygen gas is added to the 8.00 g of oxygen in the balloon, if the temperature and pressure do not change?

b. More oxygen gas is added to the balloon until the volume is 12.6 L. How many grams of oxygen gas are in the balloon, if the pressure and temperature do not change?

ANSWER

a. 7.50 L **b.** 20.2 g of oxygen

TEST

Try Practice Problems 8.47 to 8.50

The molar volume of a gas at STP is about the same as the volume of three basketballs.

STP and Molar Volume

Using Avogadro's law, we can say that any two gases will have equal volumes if they contain the same number of moles of gas at the same temperature and pressure. To help us make comparisons between gases, arbitrary conditions called *standard temperature* (273 K) and *standard pressure* (1 atm), together abbreviated **STP**, were selected by scientists:

STP Conditions

Standard temperature is *exactly* 0 °C (273 K).

Standard pressure is *exactly* 1 atm (760 mmHg).

At STP, one mole of any gas occupies a volume of 22.4 L, which is about the same as the volume of three basketballs. This volume, 22.4 L, of any gas is called the **molar volume** (see **FIGURE 8.8**). When a gas is at STP conditions (0 °C and 1 atm), its molar volume can be used to write conversion factors between the number of moles of gas and its volume, in liters.

Molar Volume Conversion Factors

1 mole of gas = 22.4 L (STP)

$$\frac{22.4\ \text{L (STP)}}{1\ \text{mole gas}} \quad \text{and} \quad \frac{1\ \text{mole gas}}{22.4\ \text{L (STP)}}$$

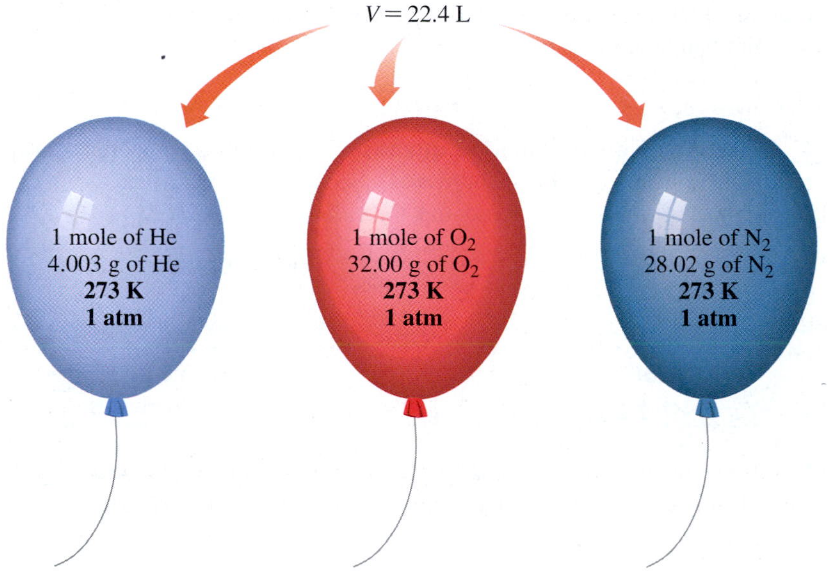
V = 22.4 L
1 mole of He
4.003 g of He
273 K
1 atm
1 mole of O2
32.00 g of O2
273 K
1 atm
1 mole of N2
28.02 g of N2
273 K
1 atm

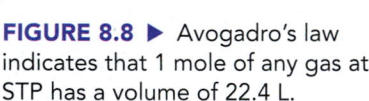

FIGURE 8.8 ▶ Avogadro's law indicates that 1 mole of any gas at STP has a volume of 22.4 L.

Q What volume of gas is occupied by 16.0 g of methane gas, CH_4, at STP?

▶SAMPLE PROBLEM 8.8 Using Molar Volume

TRY IT FIRST

What is the volume, in liters, of 64.0 g of O_2 gas at STP?

ENGAGE

Why would the molar volume of a gas be greater than 22.4 L, if the pressure is 1 atm and the temperature is 100 °C?

SOLUTION

STEP 1 State the given and needed quantities.

	Given	Need	Connect
ANALYZE THE PROBLEM	64.0 g of $O_2(g)$ at STP	liters of O_2 gas at STP	molar mass, molar volume (STP)

STEP 2 Write a plan to calculate the needed quantity.

grams of O_2 → Molar mass → moles of O_2 → Molar volume → liters of O_2

STEP 3 Write the equalities and conversion factors including 22.4 L/mole at STP.

1 mole of O_2 = 32.00 g of O_2

$$\frac{32.00 \text{ g } O_2}{1 \text{ mole } O_2} \text{ and } \frac{1 \text{ mole } O_2}{32.00 \text{ g } O_2}$$

1 mole of O_2 = 22.4 L of O_2 (STP)

$$\frac{22.4 \text{ L } O_2 \text{ (STP)}}{1 \text{ mole } O_2} \text{ and } \frac{1 \text{ mole } O_2}{22.4 \text{ L } O_2 \text{ (STP)}}$$

STEP 4 Set up the problem with factors to cancel units.

$$64.0 \cancel{\text{g } O_2} \times \frac{1 \cancel{\text{mole } O_2}}{32.00 \cancel{\text{g } O_2}} \times \frac{22.4 \text{ L } O_2 \text{ (STP)}}{1 \cancel{\text{mole } O_2}} = 44.8 \text{ L of } O_2 \text{ (STP)}$$

STUDY CHECK 8.8

a. How many grams of $Cl_2(g)$ are in 5.00 L of $Cl_2(g)$ at STP?
b. What is the volume, in liters, of 42.4 g of $O_2(g)$ at STP?

ANSWER

a. 15.8 g of Cl_2 **b.** 29.7 L of O_2

TEST

Try Practice Problems 8.51 and 8.52

PRACTICE PROBLEMS

8.6 Volume and Moles (Avogadro's Law)

8.47 What happens to the volume of a bicycle tire or a basketball when you use an air pump to add air?

8.48 Sometimes when you blow up a balloon and release it, it flies around the room. What is happening to the air in the balloon and its volume?

8.49 A sample containing 1.50 moles of Ne gas has an initial volume of 8.00 L. What is the final volume, in liters, when each of the following occurs and pressure and temperature do not change?
a. A leak allows one-half of Ne atoms to escape.
b. A sample of 3.50 moles of Ne is added to the 1.50 moles of Ne gas in the container.
c. A sample of 25.0 g of Ne is added to the 1.50 moles of Ne gas in the container.

8.50 A sample containing 4.80 g of O_2 gas has an initial volume of 15.0 L. What is the final volume, in liters, when each of the following occurs and pressure and temperature do not change?
a. A sample of 0.500 mole of O_2 is added to the 4.80 g of O_2 in the container.
b. A sample of 2.00 g of O_2 is removed.
c. A sample of 4.00 g of O_2 is added to the 4.80 g of O_2 gas in the container.

8.51 Use the molar volume to calculate each of the following at STP:
a. the number of moles of O_2 in 2.24 L of O_2 gas
b. the volume, in liters, occupied by 2.50 moles of N_2 gas
c. the volume, in liters, occupied by 50.0 g of Ar gas
d. the number of grams of H_2 in 1620 mL of H_2 gas

8.52 Use the molar volume to calculate each of the following at STP:
a. the number of moles of CO_2 in 4.00 L of CO_2 gas
b. the volume, in liters, occupied by 0.420 mole of He gas
c. the volume, in liters, occupied by 6.40 g of O_2 gas
d. the number of grams of Ne contained in 11.2 L of Ne gas

8.7 The Ideal Gas Law

LEARNING GOAL Use the ideal gas law equation to solve for *P*, *V*, *T*, or *n* of a gas when given three of the four values in the ideal gas law equation. Calculate mass or volume of a gas in a chemical reaction.

The **ideal gas law** is the combination of the four properties used in the measurement of a gas—pressure (P), volume (V), temperature (T), and amount of a gas (n)—to give a single expression, which is written as

Ideal Gas Law

$$PV = nRT$$

Rearranging the ideal gas law equation shows that the four gas properties equal the **ideal gas constant, *R*.**

$$\frac{PV}{nT} = R$$

To calculate the value of R, we substitute the STP conditions for molar volume into the expression: 1.00 mole of any gas occupies 22.4 L at STP (273 K and 1.00 atm).

$$R = \frac{(1.00 \text{ atm})(22.4 \text{ L})}{(1.00 \text{ mole})(273 \text{ K})} = \frac{0.0821 \text{ L} \cdot \text{atm}}{\text{mole} \cdot \text{K}}$$

The value for R is 0.0821 L · atm per mole · K. If we use 760. mmHg for the pressure, we obtain another useful value for R of 62.4 L · mmHg per mole · K.

$$R = \frac{(760.\ \text{mmHg})\ (22.4\ \text{L})}{(1.00\ \text{mole})\ (273\ \text{K})} = \frac{62.4\ \text{L} \cdot \text{mmHg}}{\text{mole} \cdot \text{K}}$$

ENGAGE

If you use the R value of 0.0821, what unit must you use for pressure?

The ideal gas law is a useful expression when you are given the quantities for any three of the four properties of a gas. Although real gases show some deviations in behavior, the ideal gas law closely approximates the behavior of real gases at typical conditions. In working problems using the ideal gas law, the units of each variable must match the units in the R you select.

Ideal Gas Constant (R)	$\frac{0.0821\ \text{L} \cdot \text{atm}}{\text{mole} \cdot \text{K}}$	$\frac{62.4\ \text{L} \cdot \text{mmHg}}{\text{mole} \cdot \text{K}}$
Pressure (P)	atm	mmHg
Volume (V)	L	L
Amount (n)	mole	mole
Temperature (T)	K	K

CORE CHEMISTRY SKILL

Using the Ideal Gas Law

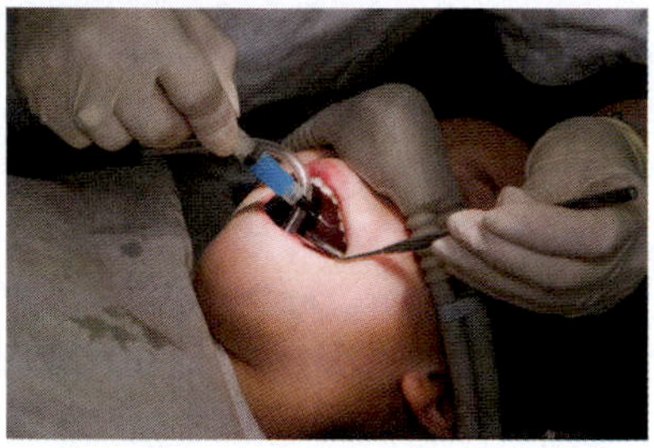

Dinitrogen oxide is used as an anesthetic in dentistry.

▶SAMPLE PROBLEM 8.9 Using the Ideal Gas Law

TRY IT FIRST

Dinitrogen oxide, N_2O, which is used in dentistry, is an anesthetic called laughing gas. What is the pressure, in atmospheres, of 0.350 mole of N_2O gas at 22 °C in a 5.00-L container?

SOLUTION

STEP 1 **State the given and needed quantities.** When three of the four quantities (P, V, n, and T) are known, we use the ideal gas law equation to solve for the unknown quantity. It is helpful to organize the data in a table. The temperature is converted from degrees Celsius to kelvins so that the units of V, n, and T match the unit of the gas constant, R.

	Given	Need	Connect
ANALYZE THE PROBLEM	$V = 5.00$ L $n = 0.350$ mole $T = 22\ °\text{C} + 273 = 295$ K	P	ideal gas law, $PV = nRT$ $R = \frac{0.0821\ \text{L} \cdot \text{atm}}{\text{mole} \cdot \text{K}}$

STEP 2 **Rearrange the ideal gas law equation to solve for the needed quantity.** By dividing both sides of the ideal gas law equation by V, we solve for pressure, P.

$$PV = nRT \quad \text{Ideal gas law equation}$$

$$\frac{P\cancel{V}}{\cancel{V}} = \frac{nRT}{V}$$

$$P = \frac{nRT}{V}$$

STEP 3 **Substitute the gas data into the equation, and calculate the needed quantity.**

$$P = \frac{0.350\ \cancel{\text{mole}} \times \frac{0.0821\ \cancel{\text{L}} \cdot \text{atm}}{\cancel{\text{mole}} \cdot \cancel{\text{K}}} \times 295\ \cancel{\text{K}}}{5.00\ \cancel{\text{L}}} = 1.70\ \text{atm}$$

STUDY CHECK 8.9

a. Chlorine gas, Cl_2, is used to purify water. How many moles of chlorine gas are in a 7.00-L tank, if the gas has a pressure of 865 mmHg and a temperature of 24 °C?

b. Isobutane gas, C_4H_{10}, is used as a propellant in aerosol spray cans of whipping cream, shaving cream, and suntan spray. What is the pressure, in atmospheres, of 0.125 mole of isobutane in an aerosol can at 21 °C with a volume of 465 mL?

An aerosol spray can contains a gas that propels the contents.

ANSWER

a. 0.327 mole of Cl_2 **b.** 6.49 atm

Many times we need to know the amount of gas, in grams, involved in a reaction. Then the ideal gas law equation can be rearranged to solve for the amount (n) of gas, which is converted to mass in grams using its molar mass as shown in Sample Problem 8.10.

▶ SAMPLE PROBLEM 8.10 Calculating Mass Using the Ideal Gas Law

TRY IT FIRST

Butane, C_4H_{10}, is used as a fuel for camping stoves. If you have 108 mL of butane gas at 715 mmHg and 25 °C, what is the mass, in grams, of butane?

When camping, butane is used as a fuel for a portable stove.

SOLUTION

STEP 1 **State the given and needed quantities.** When three of the quantities (P, V, and T) are known, we use the ideal gas law equation to solve for the unknown quantity, moles (n). Because the pressure is given in mmHg, we will use R in mmHg. The volume given in milliliters (mL) is converted to a volume in liters (L). The temperature is converted from degrees Celsius to kelvins.

	Given	Need	Connect
ANALYZE THE PROBLEM	$P = 715$ mmHg $V = 108$ mL (0.108 L) $T = 25\ °C + 273 = 298$ K	n	ideal gas law, $PV = nRT$, molar mass

STEP 2 **Rearrange the ideal gas law equation to solve for the needed quantity.** By dividing both sides of the ideal gas law equation by RT, we solve for moles, n.

$$PV = nRT \quad \text{Ideal gas law equation}$$

$$\frac{PV}{RT} = \frac{n\cancel{RT}}{\cancel{RT}}$$

$$n = \frac{PV}{RT}$$

STEP 3 **Substitute the gas data into the equation, and calculate the needed quantity.**

$$n = \frac{715\ \cancel{\text{mmHg}} \times 0.108\ \cancel{\text{L}}}{\dfrac{62.4\ \cancel{\text{L}} \cdot \cancel{\text{mmHg}}}{\text{mole} \cdot \cancel{\text{K}}} \times 298\ \cancel{\text{K}}} = 0.004\ 15\ \text{mole}\ (4.15 \times 10^{-3}\ \text{mole})$$

Now we can convert the moles of butane to grams using its molar mass of 58.12 g/mole:

$$0.004\ 15\ \cancel{\text{mole } C_4H_{10}} \times \frac{58.12\ \text{g } C_4H_{10}}{1\ \cancel{\text{mole } C_4H_{10}}} = 0.241\ \text{g of } C_4H_{10}$$

STUDY CHECK 8.10

a. What is the volume, in liters, of 1.20 g of carbon monoxide gas at 8 °C, if it has a pressure of 724 mmHg?

b. What is the pressure, in atmospheres, of 1.23 g of CH_4 gas in a 3.22-L container at 37 °C?

ANSWER

a. 1.04 L

b. 0.606 atm

TEST

Try Practice Problems 8.53 to 8.58, 8.65 and 8.66

Chemistry Link to Health

Hyperbaric Chambers

A burn patient may undergo treatment for burns and infections in a *hyperbaric chamber*, a device in which pressures can be obtained that are two to three times greater than atmospheric pressure. A greater oxygen pressure increases the level of dissolved oxygen in the blood and tissues, where it fights bacterial infections. High levels of oxygen are toxic to many strains of bacteria. The hyperbaric chamber may also be used during surgery, to help counteract carbon monoxide (CO) poisoning, and to treat some cancers.

The blood is normally capable of dissolving up to 95% of the oxygen. Thus, if the partial pressure of the oxygen in the hyperbaric chamber is 2280 mmHg (3 atm), about 2170 mmHg of oxygen can dissolve in the blood, saturating the tissues. In the treatment for carbon monoxide poisoning, oxygen at high pressure is used to displace the CO from the hemoglobin faster than breathing pure oxygen at 1 atm.

A patient undergoing treatment in a hyperbaric chamber must also undergo decompression (reduction of pressure) at a rate that slowly reduces the concentration of dissolved oxygen in the blood. If decompression is too rapid, the oxygen dissolved in the blood may form gas bubbles in the circulatory system.

Similarly, if a scuba diver does not decompress slowly, a condition called the "bends" may occur. While below the surface of the ocean, a diver uses a breathing mixture with higher pressures. If there is nitrogen in the mixture, higher quantities of nitrogen gas will dissolve in the blood. If the diver ascends to the surface too quickly, the dissolved nitrogen forms gas bubbles that can block a blood vessel and cut off the flow of blood in the joints and tissues of the body and be quite painful. A diver suffering from the bends is placed immediately into a hyperbaric chamber where pressure is first increased and then slowly decreased. The dissolved nitrogen can then diffuse through the lungs until atmospheric pressure is reached.

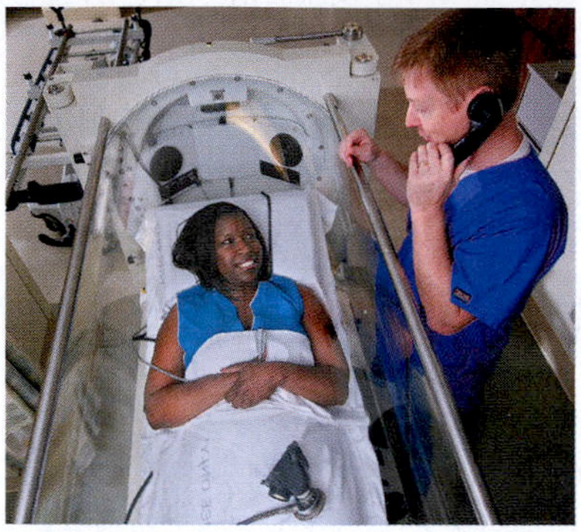

A hyperbaric chamber is used in the treatment of certain diseases.

CORE CHEMISTRY SKILL
Calculating Mass or Volume of a Gas in a Chemical Reaction

Gas Laws and Chemical Reactions

Gases are involved as reactants and products in many chemical reactions. Typically, the information given for a gas in a reaction is its pressure (P), volume (V), and temperature (T). Then we can use the ideal gas law equation to determine the moles of a gas in a reaction. If we know the number of moles for one of the gases in a reaction, we can use mole–mole factors to determine the moles of any other substance.

▶SAMPLE PROBLEM 8.11 Gases in Chemical Reactions

TRY IT FIRST

Calcium carbonate, $CaCO_3$, in antacids reacts with HCl in the stomach to reduce acid reflux.

$$2HCl(aq) + CaCO_3(s) \longrightarrow CO_2(g) + H_2O(l) + CaCl_2(aq)$$

How many liters of CO_2 are produced at 752 mmHg and 24 °C from a 25.0-g sample of calcium carbonate?

SOLUTION

STEP 1 State the given and needed quantities.

	Given	Need	Connect
ANALYZE THE PROBLEM	25.0 g of $CaCO_3$ $P = 752$ mmHg $T = 24\ °C + 273 = 297$ K	V of $CO_2(g)$	ideal gas law, $PV = nRT$, molar mass
	Equation		
	$2HCl(aq) + CaCO_3(s) \longrightarrow CO_2(g) + H_2O(l) + CaCl_2(aq)$		

STEP 2 Write a plan to convert the given quantity to the needed moles.

grams of $CaCO_3$ → Molar mass → moles of $CaCO_3$ → Mole–mole factor → moles of CO_2

STEP 3 **Write the equalities and conversion factors for molar mass and mole–mole factors.**

$$1 \text{ mole of } CaCO_3 = 100.09 \text{ g of } CaCO_3$$
$$\frac{100.09 \text{ g } CaCO_3}{1 \text{ mole } CaCO_3} \text{ and } \frac{1 \text{ mole } CaCO_3}{100.09 \text{ g } CaCO_3}$$

$$1 \text{ mole of } CaCO_3 = 1 \text{ mole of } CO_2$$
$$\frac{1 \text{ mole } CaCO_3}{1 \text{ mole } CO_2} \text{ and } \frac{1 \text{ mole } CO_2}{1 \text{ mole } CaCO_3}$$

STEP 4 **Set up the problem to calculate moles of needed quantity.**

$$25.0 \cancel{\text{g } CaCO_3} \times \frac{1 \cancel{\text{mole } CaCO_3}}{100.09 \cancel{\text{g } CaCO_3}} \times \frac{1 \text{ mole } CO_2}{1 \cancel{\text{mole } CaCO_3}} = 0.250 \text{ mole of } CO_2$$

TEST

Try Practice Problems 8.59 to 8.64

STEP 5 **Convert the moles of needed quantity to volume using the ideal gas law equation.**

$$V = \frac{nRT}{P}$$

$$V = \frac{0.250 \cancel{\text{mole}} \times \dfrac{62.4 \text{ L} \cdot \cancel{\text{mmHg}}}{\cancel{\text{mole}} \cdot \cancel{\text{K}}} \times 297 \cancel{\text{K}}}{752 \cancel{\text{mmHg}}} = 6.16 \text{ L of } CO_2$$

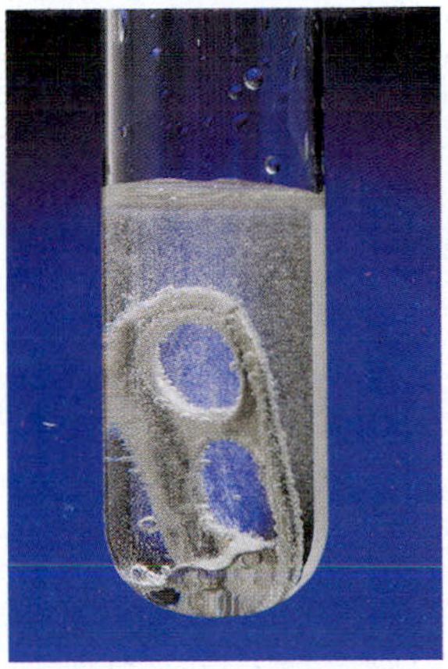

When HCl reacts with aluminum, bubbles of H_2 gas form.

STUDY CHECK 8.11

HCl reacts with aluminum metal to produce hydrogen gas.

$$6HCl(aq) + 2Al(s) \longrightarrow 3H_2(g) + 2AlCl_3(aq)$$

a. If 12.8 g of aluminum reacts with HCl, how many liters of H_2 would be formed at 715 mmHg and 19 °C?

b. If 9.28 L of H_2 is produced at 24 °C and 1.20 atm, how many grams of aluminum react?

ANSWER

a. 18.1 L of H_2

b. 8.22 g of Al

PRACTICE PROBLEMS

8.7 The Ideal Gas Law

8.53 Calculate the pressure, in atmospheres, of 2.00 moles of helium gas in a 10.0-L container at 27 °C.

8.54 What is the volume, in liters, of 4.00 moles of methane gas, CH_4, at 18 °C and 1.40 atm?

8.55 An oxygen gas container has a volume of 20.0 L. How many grams of oxygen are in the container, if the gas has a pressure of 845 mmHg at 22 °C?

8.56 A 10.0-g sample of krypton has a temperature of 25 °C at 575 mmHg. What is the volume, in milliliters, of the krypton gas?

8.57 A 25.0-g sample of nitrogen, N_2, has a volume of 50.0 L and a pressure of 630. mmHg. What is the temperature, in kelvins and degrees Celsius, of the gas?

8.58 A 0.226-g sample of carbon dioxide, CO_2, has a volume of 525 mL and a pressure of 455 mmHg. What is the temperature, in kelvins and degrees Celsius, of the gas?

8.59 HCl reacts with magnesium metal to produce hydrogen gas.

$$2HCl(aq) + Mg(s) \longrightarrow H_2(g) + MgCl_2(aq)$$

a. What volume, in liters, of hydrogen at 0 °C and 1.00 atm (STP) is released when 8.25 g of Mg reacts?

b. How many grams of magnesium are needed to prepare 5.00 L of H_2 at 735 mmHg and 18 °C?

8.60 When heated to 350 °C at 0.950 atm, ammonium nitrate decomposes to produce nitrogen, water, and oxygen gases.

$$2NH_4NO_3(s) \xrightarrow{\Delta} 2N_2(g) + 4H_2O(g) + O_2(g)$$

a. How many liters of water vapor are produced when 25.8 g of NH_4NO_3 decomposes?

b. How many grams of NH_4NO_3 are needed to produce 10.0 L of oxygen?

8.61 Butane undergoes combustion when it reacts with oxygen to produce carbon dioxide and water. What volume, in liters, of oxygen is needed to react with 55.2 g of butane at 0.850 atm and 25 °C?

$$2C_4H_{10}(g) + 13O_2(g) \xrightarrow{\Delta} 8CO_2(g) + 10H_2O(g)$$

8.62 Potassium nitrate decomposes to potassium nitrite and oxygen. What volume, in liters, of O_2 can be produced from the decomposition of 50.0 g of KNO_3 at 35 °C and 1.19 atm?

$$2KNO_3(s) \longrightarrow 2KNO_2(s) + O_2(g)$$

8.63 Aluminum and oxygen react to form aluminum oxide. How many liters of oxygen at 0 °C and 760 mmHg (STP) are required to completely react with 5.4 g of aluminum?

$$4Al(s) + 3O_2(g) \longrightarrow 2Al_2O_3(s)$$

8.64 Nitrogen dioxide reacts with water to produce oxygen and ammonia. How many grams of NH_3 can be produced when 4.00 L of NO_2 reacts at 415 °C and 725 mmHg?

$$4NO_2(g) + 6H_2O(g) \longrightarrow 7O_2(g) + 4NH_3(g)$$

Clinical Applications

8.65 A single-patient hyperbaric chamber has a volume of 640 L. At a temperature of 24 °C, how many grams of oxygen are needed to give a pressure of 1.6 atm?

8.66 A multipatient hyperbaric chamber has a volume of 3400 L. At a temperature of 22 °C, how many grams of oxygen are needed to give a pressure of 2.4 atm?

8.8 Partial Pressures (Dalton's Law)

LEARNING GOAL Use Dalton's law of partial pressures to calculate the total pressure of a mixture of gases.

Many gas samples are a mixture of gases. For example, the air you breathe is a mixture of mostly oxygen and nitrogen gases. In gas mixtures, scientists observed that all gas particles behave in the same way. Therefore, the total pressure of the gases in a mixture is a result of the collisions of the gas particles regardless of what type of gas they are.

In a gas mixture, each gas exerts its **partial pressure**, which is the pressure it would exert if it were the only gas in the container. **Dalton's law** states that the total pressure of a gas mixture is the sum of the partial pressures of the gases in the mixture.

CORE CHEMISTRY SKILL
Calculating Partial Pressure

ENGAGE
Why does the combination of helium with a pressure of 2.0 atm and argon with a pressure of 4.0 atm produce a gas mixture with a total pressure of 6.0 atm?

Dalton's Law

$$P_{total} = P_1 + P_2 + P_3 + \cdots$$

Total pressure of a gas mixture = Sum of the partial pressures of the gases in the mixture

Suppose we have two separate tanks, one filled with helium at a pressure of 2.0 atm and the other filled with argon at a pressure of 4.0 atm. When the gases are combined in a single tank with the same volume and temperature, the number of gas molecules, not the type of gas, determines the pressure in the container. The total pressure of the gas mixture would be 6.0 atm, which is the sum of their individual or partial pressures.

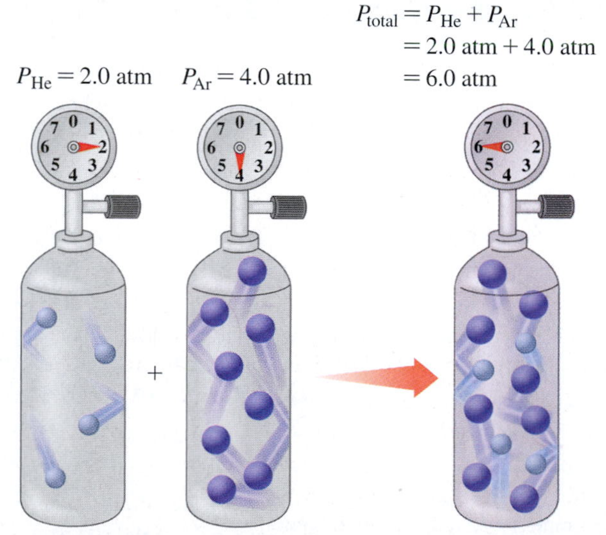
$P_{He} = 2.0$ atm
$P_{Ar} = 4.0$ atm
$P_{total} = P_{He} + P_{Ar}$
$= 2.0$ atm $+ 4.0$ atm
$= 6.0$ atm
+

The total pressure of two gases is the sum of their partial pressures.

TABLE 8.7 Typical Composition of Air

Gas	Partial Pressure (mmHg)	Percentage (%)
Nitrogen, N_2	594	78.2
Oxygen, O_2	160.	21.0
Carbon dioxide, CO_2; Argon, Ar; Water vapor, H_2O	6	0.8
Total air	760.	100

Air Is a Gas Mixture

The air you breathe is a mixture of gases. What we call the *atmospheric pressure* is actually the sum of the partial pressures of all the gases in the air. **TABLE 8.7** lists partial pressures for the gases in air on a typical day.

Chemistry Link to Health

Blood Gases

Our cells continuously use oxygen and produce carbon dioxide. Both gases move in and out of the lungs through the membranes of the alveoli, the tiny air sacs at the ends of the airways in the lungs. An exchange of gases occurs in which oxygen from the air diffuses into the lungs and into the blood, while carbon dioxide produced in the cells is carried to the lungs to be exhaled. In **TABLE 8.8**, partial pressures are given for the gases in air that we inhale (inspired air), air in the alveoli, and air that we exhale (expired air).

At sea level, oxygen normally has a partial pressure of 100 mmHg in the alveoli of the lungs. Because the partial pressure of oxygen in venous blood is 40 mmHg, oxygen diffuses from the alveoli into the bloodstream. The oxygen combines with hemoglobin, which carries it to the tissues of the body where the partial pressure of oxygen can be very low, less than 30 mmHg. Oxygen diffuses from the blood, where the partial pressure of O_2 is high, into the tissues where O_2 pressure is low.

As oxygen is used in the cells of the body during metabolic processes, carbon dioxide is produced, so the partial pressure of CO_2 may be as high as 50 mmHg or more. Carbon dioxide diffuses from the tissues into the bloodstream and is carried to the lungs. There it diffuses out of the blood, where CO_2 has a partial pressure of 46 mmHg, into the alveoli, where the CO_2 is at 40 mmHg and is exhaled. **TABLE 8.9** gives the partial pressures of blood gases in the tissues and in oxygenated and deoxygenated blood.

TABLE 8.8 Partial Pressures of Gases During Breathing

	Partial Pressure (mmHg)		
Gas	**Inspired Air**	**Alveolar Air**	**Expired Air**
Nitrogen, N_2	594	573	569
Oxygen, O_2	160.	100.	116
Carbon dioxide, CO_2	0.3	40.	28
Water vapor, H_2O	5.7	47	47
Total	760.	760.	760.

TABLE 8.9 Partial Pressures of Oxygen and Carbon Dioxide in Blood and Tissues

	Partial Pressure (mmHg)		
Gas	**Oxygenated Blood**	**Deoxygenated Blood**	**Tissues**
O_2	100	40	30 or less
CO_2	40	46	50 or greater

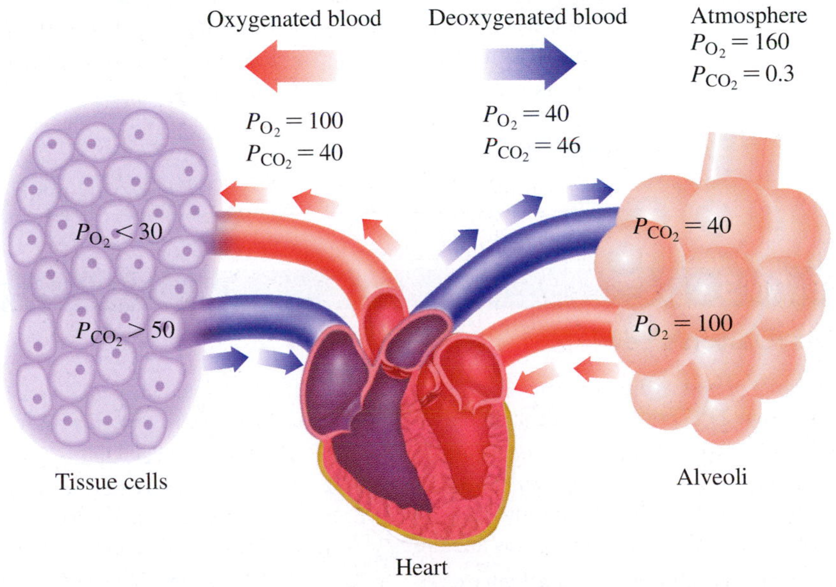

Oxygenated blood
Deoxygenated blood
Atmosphere
$P_{O_2} = 160$
$P_{CO_2} = 0.3$
$P_{O_2} = 100$
$P_{CO_2} = 40$
$P_{O_2} = 40$
$P_{CO_2} = 46$
$P_{O_2} < 30$
$P_{CO_2} > 50$
$P_{CO_2} = 40$
$P_{O_2} = 100$
Tissue cells
Alveoli
Heart

▶ SAMPLE PROBLEM 8.12 Calculating the Partial Pressure of a Gas in a Mixture

TRY IT FIRST

A heliox breathing mixture of oxygen and helium is prepared for a patient with *chronic obstructive pulmonary disease* (COPD). The gas mixture has a total pressure of 7.00 atm. If the partial pressure of the oxygen in the tank is 1140 mmHg, what is the partial pressure, in atmospheres, of the helium in the breathing mixture?

SOLUTION

	Given	Need	Connect
ANALYZE THE PROBLEM	$P_{total} = 7.00$ atm $P_{O_2} = 1140$ mmHg	partial pressure of He	Dalton's law

STEP 1 **Write the equation for the sum of the partial pressures.**

$P_{total} = P_{O_2} + P_{He}$ Dalton's law

STEP 2 **Rearrange the equation to solve for the unknown pressure.** To solve for the partial pressure of helium (P_{He}), we rearrange the equation to give the following:

$P_{He} = P_{total} - P_{O_2}$

Convert units to match.

$$P_{O_2} = 1140 \cancel{\text{mmHg}} \times \frac{1 \text{ atm}}{760 \cancel{\text{mmHg}}} = 1.50 \text{ atm}$$

STEP 3 **Substitute known pressures into the equation, and calculate the unknown pressure.**

$P_{He} = P_{total} - P_{O_2}$

$P_{He} = 7.00 \text{ atm} - 1.50 \text{ atm} = 5.50 \text{ atm}$

STUDY CHECK 8.12

a. An anesthetic consists of a mixture of cyclopropane gas, C_3H_6, and oxygen gas, O_2. If the mixture has a total pressure of 1.09 atm, and the partial pressure of the cyclopropane is 73 mmHg, what is the partial pressure, in millimeters of mercury, of the oxygen in the anesthetic?

b. Another anesthetic contains nitrous oxide gas, N_2O, and oxygen gas, O_2. If the mixture has a total pressure of 786 mmHg and the partial pressure of the nitrous oxide is 0.110 atm, what is the partial pressure, in millimeters of mercury, of the oxygen in the mixture?

ANSWER

a. 755 mmHg **b.** 702 mmHg

PRACTICE PROBLEMS

8.8 Partial Pressures (Dalton's Law)

8.67 In a gas mixture, the partial pressures are nitrogen 425 Torr, oxygen 115 Torr, and helium 225 Torr. What is the total pressure, in torr, exerted by the gas mixture?

8.68 In a gas mixture, the partial pressures are argon 415 mmHg, neon 75 mmHg, and nitrogen 125 mmHg. What is the total pressure, in millimeters of mercury, exerted by the gas mixture?

8.69 A gas mixture containing oxygen, nitrogen, and helium exerts a total pressure of 925 Torr. If the partial pressures are oxygen 425 Torr and helium 75 Torr, what is the partial pressure, in torr, of the nitrogen in the mixture?

8.70 A gas mixture containing oxygen, nitrogen, and neon exerts a total pressure of 1.20 atm. If helium added to the mixture increases the pressure to 1.50 atm, what is the partial pressure, in atmospheres, of the helium?

Clinical Applications

8.71 In certain lung ailments such as emphysema, there is a decrease in the ability of oxygen to diffuse into the blood.

a. How would the partial pressure of oxygen in the blood change?

b. Why does a person with severe emphysema sometimes use a portable oxygen tank?

8.72 An accident to the head can affect the ability of a person to ventilate (breathe in and out).

a. What would happen to the partial pressures of oxygen and carbon dioxide in the blood if a person cannot properly ventilate?

b. When a person who cannot breathe properly is placed on a ventilator, an air mixture is delivered at pressures that are alternately above the air pressure in the person's lung, and then below. How will this move oxygen gas into the lungs, and carbon dioxide out?

8.73 An air sample in the lungs contains oxygen at 93 mmHg, nitrogen at 565 mmHg, carbon dioxide at 38 mmHg, and water vapor at 47 mmHg. What is the total pressure, in atmospheres, exerted by the gas mixture?

8.74 A nitrox II gas mixture for scuba diving contains oxygen gas at 53 atm and nitrogen gas at 94 atm. What is the total pressure, in torr, of the scuba gas mixture?

CLINICAL UPDATE Exercise-Induced Asthma

Vigorous exercise can induce asthma, particularly in children. When Whitney had her asthma attack, her breathing became more rapid, the temperature within her airways increased, and the muscles around the bronchi contracted, causing a narrowing of the airways. Whitney's symptoms, which may occur within 5 to 20 min after the start of vigorous exercise, include shortness of breath, wheezing, and coughing.

Whitney now does several things to prevent exercise-induced asthma. She uses a pre-exercise inhaled medication before she starts her activity. The medication relaxes the muscles that surround the airways and opens up the airways. Then she does a warm-up set of exercises. If pollen counts are high, she avoids exercising outdoors.

Clinical Applications

8.75 Whitney's lung capacity was measured as 3.2 L at a body temperature of 37 °C and a pressure of 745 mmHg. What is her lung capacity, in liters, at STP?

8.76 Using the answer from problem 8.75, how many grams of nitrogen are in Whitney's lungs at STP if air contains 78% nitrogen?

CONCEPT MAP

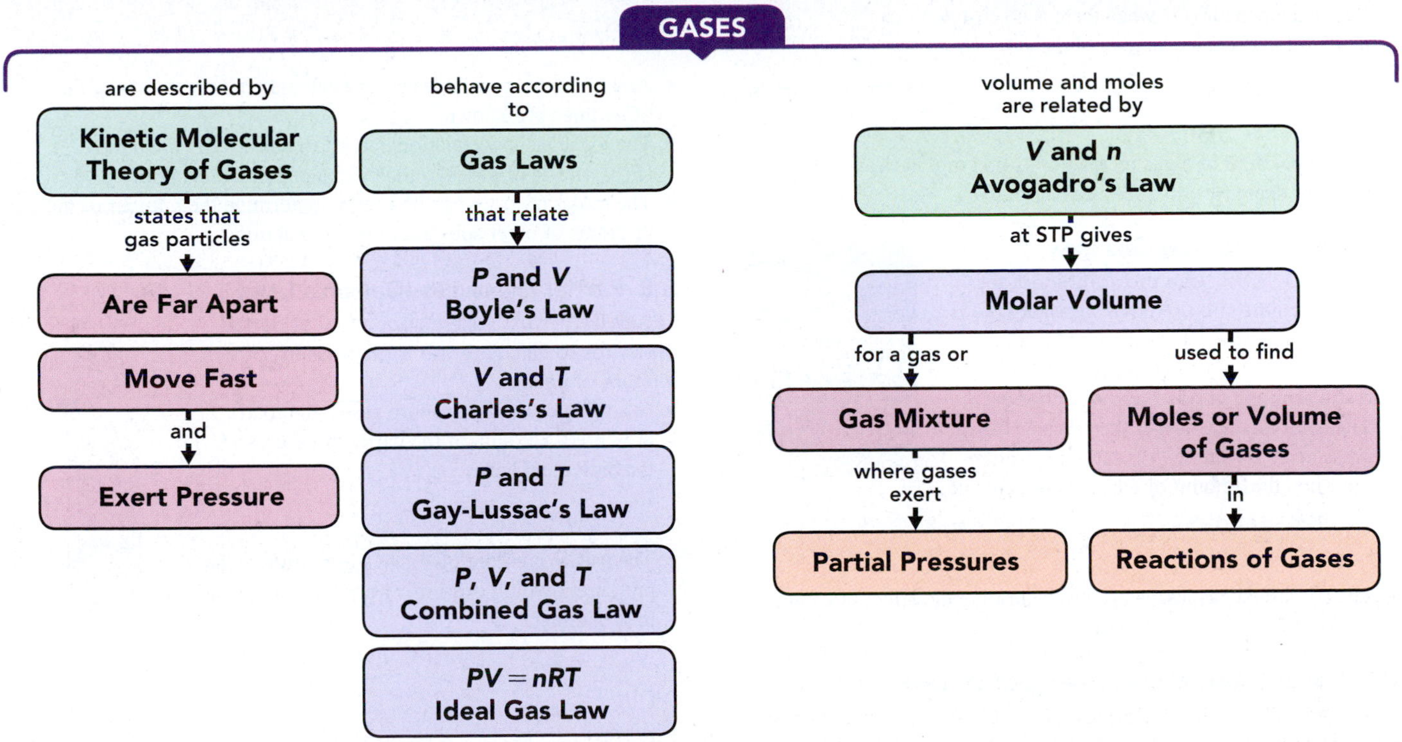
GASES
are described by
Kinetic Molecular Theory of Gases
states that gas particles
Are Far Apart
Move Fast
and
Exert Pressure
behave according to
Gas Laws
that relate
P and V
Boyle's Law
V and T
Charles's Law
P and T
Gay-Lussac's Law
P, V, and T
Combined Gas Law
PV = nRT
Ideal Gas Law
volume and moles are related by
V and n
Avogadro's Law
at STP gives
Molar Volume
for a gas or
Gas Mixture
where gases exert
Partial Pressures
used to find
Moles or Volume of Gases
in
Reactions of Gases

CHAPTER REVIEW

8.1 Properties of Gases

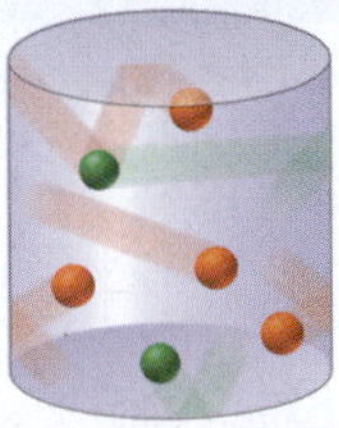

LEARNING GOAL Describe the kinetic molecular theory of gases and the units of measurement used for gases.

- In a gas, particles are so far apart and moving so fast that their attractions are negligible.
- A gas is described by the physical properties of pressure (P), volume (V), temperature (T), and amount in moles (n).
- A gas exerts pressure, the force of the gas particles striking the surface of a container.
- Gas pressure is measured in units such as torr, mmHg, atm, and Pa.

8.2 Pressure and Volume (Boyle's Law)

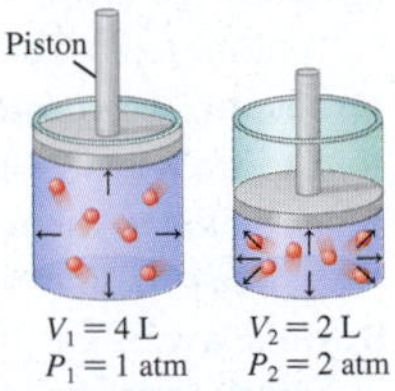
Piston
V1 = 4 L
P1 = 1 atm
V2 = 2 L
P2 = 2 atm

LEARNING GOAL Use the pressure–volume relationship (Boyle's law) to calculate the unknown pressure or volume when the temperature and amount of gas do not change.

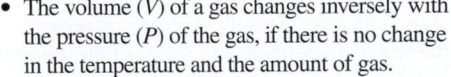

- The volume (V) of a gas changes inversely with the pressure (P) of the gas, if there is no change in the temperature and the amount of gas.

$$P_1V_1 = P_2V_2$$

- The pressure increases if volume decreases; its pressure decreases if the volume increases.

8.3 Temperature and Volume (Charles's Law)

LEARNING GOAL Use the temperature–volume relationship (Charles's law) to calculate the unknown temperature or volume when the pressure and amount of gas do not change.

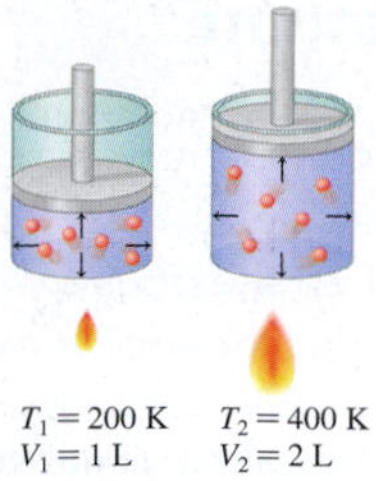
$T_1 = 200$ K
$V_1 = 1$ L
$T_2 = 400$ K
$V_2 = 2$ L

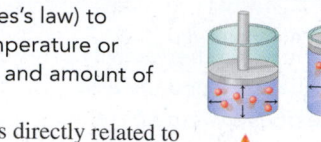

- The volume (V) of a gas is directly related to its Kelvin temperature (T) when there is no change in the pressure and the amount of gas.

$$\frac{V_1}{T_1} = \frac{V_2}{T_2}$$

- If the temperature of a gas increases, its volume increases; if its temperature decreases, the volume decreases.

8.4 Temperature and Pressure (Gay-Lussac's Law)

LEARNING GOAL Use the temperature–pressure relationship (Gay-Lussac's law) to calculate the unknown temperature or pressure when the volume and amount of gas do not change.

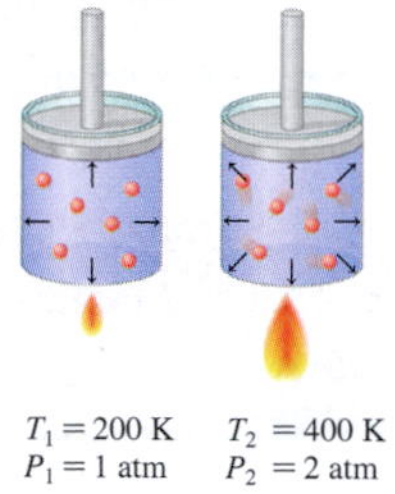
$T_1 = 200$ K
$P_1 = 1$ atm
$T_2 = 400$ K
$P_2 = 2$ atm

- The pressure (P) of a gas is directly related to its Kelvin temperature (T) when there is no change in the volume and the amount of the gas.

$$\frac{P_1}{T_1} = \frac{P_2}{T_2}$$

- As temperature of a gas increases, its pressure increases; if its temperature decreases, its pressure decreases.

8.5 The Combined Gas Law

LEARNING GOAL Use the combined gas law to calculate the unknown pressure, volume, or temperature of a gas when changes in two of these properties are given and the amount of gas does not change.

- The combined gas law is the relationship of pressure (P), volume (V), and temperature (T) when the amount of gas does not change.

$$\frac{P_1V_1}{T_1} = \frac{P_2V_2}{T_2}$$

- The combined gas law is used to determine the effect of changes in two of the variables on the third.

8.6 Volume and Moles (Avogadro's Law)

LEARNING GOAL Use Avogadro's law to calculate the unknown amount or volume of a gas when the pressure and temperature do not change.

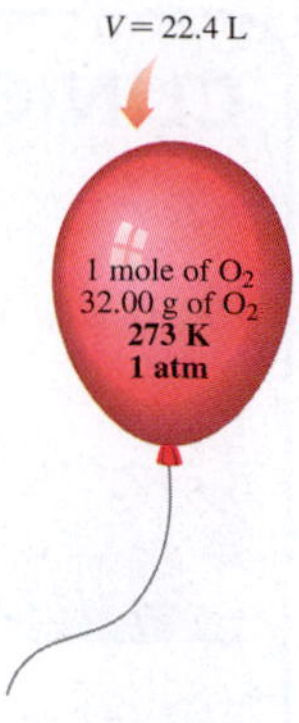
$V = 22.4$ L
1 mole of O_2
32.00 g of O_2
273 K
1 atm

- The volume (V) of a gas is directly related to the number of moles (n) of the gas when the pressure and temperature of the gas do not change.

$$\frac{V_1}{n_1} = \frac{V_2}{n_2}$$

- If the moles of gas increase, the volume must increase; if the moles of gas decrease, the volume must decrease.
- At standard temperature (273 K) and standard pressure (1 atm), abbreviated STP, 1 mole of any gas has a volume of 22.4 L.

8.7 The Ideal Gas Law

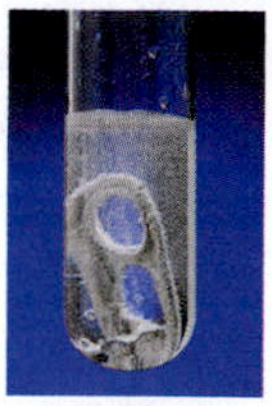

LEARNING GOAL Use the ideal gas law equation to solve for P, V, T, or n of a gas when given three of the four values in the ideal gas law equation. Calculate mass or volume of a gas in a chemical reaction.

- The ideal gas law gives the relationship of the quantities P, V, n, and T that describe and measure a gas.

$$PV = nRT$$

- Any of the four variables can be calculated if the values of the other three are known.
- The ideal gas law equation is used to convert the quantities (P, V, and T) of gases to moles in a chemical reaction.
- The moles of gases can be used to determine the number of moles or grams of other substances in the reaction.

8.8 Partial Pressures (Dalton's Law)

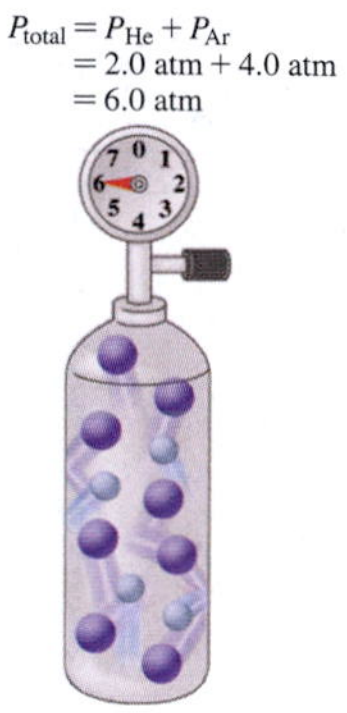
$P_{total} = P_{He} + P_{Ar}$
$= 2.0$ atm $+ 4.0$ atm
$= 6.0$ atm

LEARNING GOAL Use Dalton's law of partial pressures to calculate the total pressure of a mixture of gases.

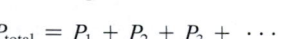

- In a mixture of two or more gases, the total pressure is the sum of the partial pressures of the individual gases.

$$P_{total} = P_1 + P_2 + P_3 + \cdots$$

- The partial pressure of a gas in a mixture is the pressure it would exert if it were the only gas in the container.

KEY TERMS

atmosphere (atm) A unit equal to the pressure exerted by a column of mercury 760 mm high.

atmospheric pressure The pressure exerted by the atmosphere.

Avogadro's law A gas law stating that the volume of a gas is directly related to the number of moles of gas when pressure and temperature do not change.

Boyle's law A gas law stating that the pressure of a gas is inversely related to the volume when temperature and moles of the gas do not change.

Charles's law A gas law stating that the volume of a gas is directly related to the Kelvin temperature when pressure and moles of the gas do not change.

combined gas law A relationship that combines several gas laws relating pressure, volume, and temperature when the amount of gas does not change.

$$\frac{P_1V_1}{T_1} = \frac{P_2V_2}{T_2}$$

Dalton's law A gas law stating that the total pressure exerted by a mixture of gases in a container is the sum of the partial pressures that each gas would exert alone.

direct relationship A relationship in which two properties increase or decrease together.

Gay-Lussac's law A gas law stating that the pressure of a gas is directly related to the Kelvin temperature when the number of moles of a gas and its volume do not change.

ideal gas constant, *R* A numerical value that relates the quantities P, V, n, and T in the ideal gas law, $PV = nRT$.

ideal gas law A law that combines the four measured properties of a gas.

$$PV = nRT$$

inverse relationship A relationship in which two properties change in opposite directions.

kinetic molecular theory of gases A model used to explain the behavior of gases.

molar volume A volume of 22.4 L occupied by 1 mole of a gas at STP conditions of 0 °C (273 K) and 1 atm.

partial pressure The pressure exerted by a single gas in a gas mixture.

pressure The force exerted by gas particles that hit the walls of a container.

STP Standard conditions of exactly 0 °C (273 K) temperature and 1 atm pressure used for the comparison of gases.

vapor pressure The pressure exerted by the particles of vapor above a liquid.

CORE CHEMISTRY SKILLS

The chapter Section containing each Core Chemistry Skill is shown in parentheses at the end of each heading.

Using the Gas Laws (8.2)

- Boyle's law is one of the gas laws that shows the relationships between two properties of a gas.

$$P_1V_1 = P_2V_2 \quad \text{Boyle's law}$$

- When two of the four properties of a gas (P, V, T, or n) vary and the other two do not change, we list the initial and final conditions of each property.

Example: A sample of helium gas (He) has a volume of 6.8 L and a pressure of 2.5 atm. What is the final volume, in liters, if it has a final pressure of 1.2 atm with no change in temperature and amount of gas?

Answer: Using Boyle's law, we can write the relationship for V_2, which we predict will increase.

$P_1 = 2.5$ atm $\qquad P_2 = 1.2$ atm

$V_1 = 6.8$ L $\qquad V_2$ Need

$$V_2 = V_1 \times \frac{P_1}{P_2}$$

$$V_2 = 6.8\text{ L} \times \frac{2.5\text{ }\cancel{\text{atm}}}{1.2\text{ }\cancel{\text{atm}}} = 14\text{ L}$$

Using the Ideal Gas Law (8.7)

- The ideal gas law equation combines the relationships of the four properties of a gas into one equation.

$$PV = nRT$$

- When three of the four properties are given, we rearrange the ideal gas law equation for the needed quantity.

Example: What is the volume, in liters, of 0.750 mole of CO_2 at a pressure of 1340 mmHg and a temperature of 295 K?

Answer: $V = \dfrac{nRT}{P}$

$$= \frac{0.750\text{ }\cancel{\text{mole}} \times \dfrac{62.4\text{ L}\cdot\cancel{\text{mmHg}}}{\cancel{\text{mole}}\cdot\cancel{\text{K}}} \times 295\text{ }\cancel{\text{K}}}{1340\text{ }\cancel{\text{mmHg}}} = 10.3\text{ L}$$

Calculating Mass or Volume of a Gas in a Chemical Reaction (8.7)

- The ideal gas law equation is used to calculate the pressure, volume, or moles (or grams) of a gas in a chemical reaction.

Example: What is the volume, in liters, of N_2 required to react with 18.5 g of magnesium at a pressure of 1.20 atm and a temperature of 303 K?

$$3Mg(s) + N_2(g) \longrightarrow Mg_3N_2(s)$$

Answer: Initially, we convert the grams of Mg to moles and use a mole–mole factor from the balanced equation to calculate the moles of N_2 gas.

$$18.5\text{ }\cancel{\text{g Mg}} \times \frac{1\text{ }\cancel{\text{mole Mg}}}{24.31\text{ }\cancel{\text{g Mg}}} \times \frac{1\text{ mole }N_2}{3\text{ }\cancel{\text{moles Mg}}} = 0.254\text{ mole of }N_2$$

Now, we use the moles of N_2 in the ideal gas law equation and solve for liters, the needed quantity.

$$V = \frac{nRT}{P} = \frac{0.254\text{ }\cancel{\text{mole }N_2} \times \dfrac{0.0821\text{ L}\cdot\cancel{\text{atm}}}{\cancel{\text{mole}}\cdot\cancel{\text{K}}} \times 303\text{ }\cancel{\text{K}}}{1.20\text{ }\cancel{\text{atm}}} = 5.27\text{ L}$$

Calculating Partial Pressure (8.8)

- In a gas mixture, each gas exerts its partial pressure, which is the pressure it would exert if it were the only gas in the container.
- Dalton's law states that the total pressure of a gas mixture is the sum of the partial pressures of the gases in the mixture.

$$P_{total} = P_1 + P_2 + P_3 + \cdots$$

Example: A gas mixture with a total pressure of 1.18 atm contains helium gas at a partial pressure of 465 mmHg and nitrogen gas. What is the partial pressure, in atmospheres, of the nitrogen gas?

Answer: Initially, we convert the partial pressure of helium gas from mmHg to atm.

$$465\text{ }\cancel{\text{mmHg}} \times \frac{1\text{ atm}}{760\text{ }\cancel{\text{mmHg}}} = 0.612\text{ atm of He gas}$$

Using Dalton's law, we solve for the needed quantity, P_{N_2} in atm.

$$P_{total} = P_{N_2} + P_{He}$$
$$P_{N_2} = P_{total} - P_{He}$$
$$P_{N_2} = 1.18\text{ atm} - 0.612\text{ atm} = 0.57\text{ atm}$$

UNDERSTANDING THE CONCEPTS

The chapter Sections to review are shown in parentheses at the end of each problem.

8.77 Two flasks of equal volume and at the same temperature contain different gases. One flask contains 10.0 g of Ne, and the other flask contains 10.0 g of He. Is each of the following statements *true* or *false*? Explain. (8.1)

a. The flask that contains He has a higher pressure than the flask that contains Ne.

b. The densities of the gases are the same.

8.78 Two flasks of equal volume and at the same temperature contain different gases. One flask contains 5.0 g of O_2, and the other flask contains 5.0 g of H_2. Is each of the following statements *true* or *false*? Explain. (8.1)

a. Both flasks contain the same number of molecules.

b. The pressures in the flasks are the same.

8.79 At 100 °C, which of the following diagrams (**1**, **2**, or **3**) represents a gas sample that exerts the: (8.1)

a. lowest pressure? **b.** highest pressure?

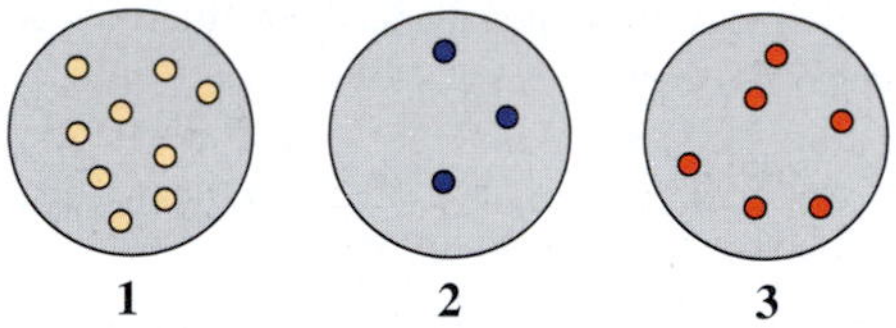
1
2
3

8.80 Indicate which diagram (**1**, **2**, or **3**) represents the volume of the gas sample in a flexible container when each of the following changes (**a** to **d**) takes place: (8.2, 8.3)

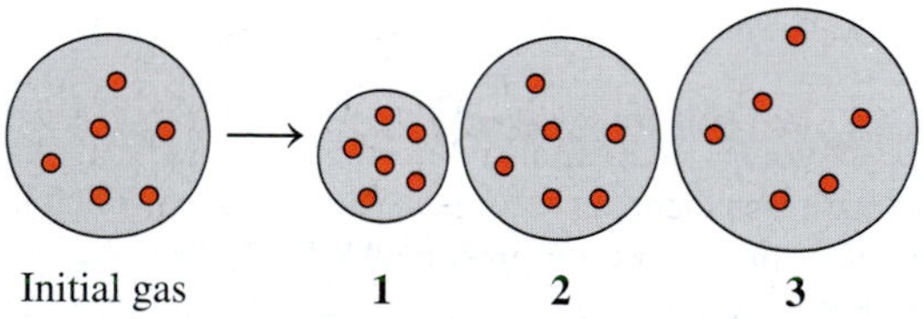
Initial gas
1
2
3

a. Temperature increases, if pressure does not change.

b. Temperature decreases, if pressure does not change.

c. Atmospheric pressure decreases, if temperature does not change.

d. Doubling the atmospheric pressure and doubling the Kelvin temperature.

8.81 A balloon is filled with helium gas with a partial pressure of 1.00 atm and neon gas with a partial pressure of 0.50 atm. For each of the following changes (**a** to **e**) of the initial balloon, select the diagram (**A**, **B**, or **C**) that shows the final volume of the balloon: (8.2, 8.3, 8.6)

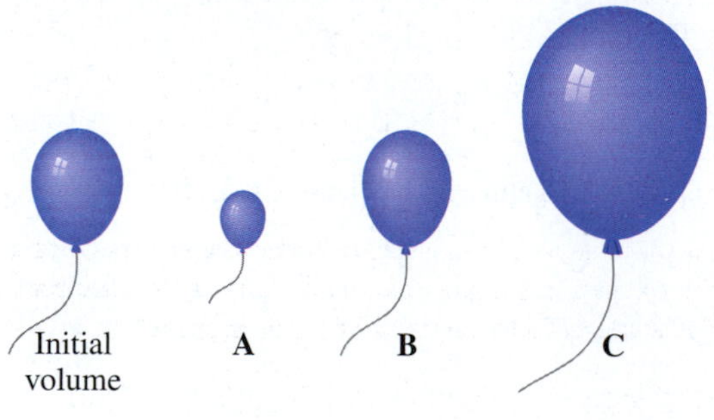
Initial volume
A
B
C

a. The balloon is put in a cold storage unit (pressure and amount of gas do not change).

b. The balloon floats to a higher altitude where the pressure is less (temperature and amount of gas do not change).

c. All of the neon gas is removed (temperature and pressure do not change).

d. The Kelvin temperature doubles and half of the gas atoms leak out (pressure does not change).

e. 2.0 moles of O_2 gas is added (temperature and pressure do not change).

8.82 Indicate if pressure *increases*, *decreases*, or *stays the same* in each of the following: (8.2, 8.4, 8.6)

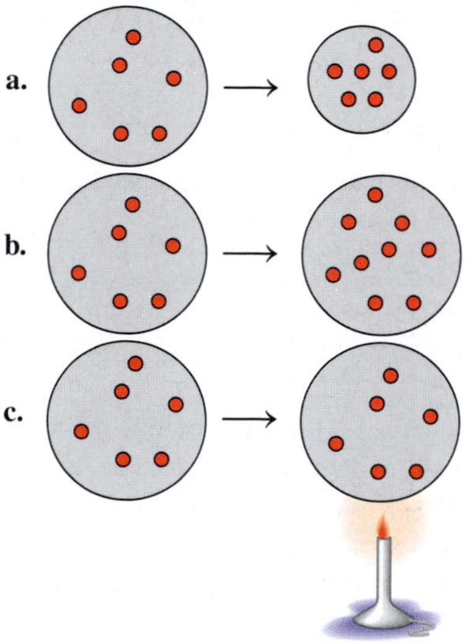
a.
b.
c.

8.83 At a restaurant, a customer chokes on a piece of food. You put your arms around the person's waist and use your fists to push up on the person's abdomen, an action called the Heimlich maneuver. (8.2)

a. How would this action change the volume of the chest and lungs?

b. Why does it cause the person to expel the food item from the airway?

8.84 An airplane is pressurized with air to 650. mmHg. (8.8)

a. If air is 21% oxygen, what is the partial pressure of oxygen on the plane?

b. If the partial pressure of oxygen drops below 100. mmHg, passengers become drowsy. If this happens, oxygen masks are released. What is the total cabin pressure at which oxygen masks are dropped?

ADDITIONAL PRACTICE PROBLEMS

8.85 In 1783, Jacques Charles launched his first balloon filled with hydrogen gas, which he chose because it was lighter than air. If the balloon had a volume of 31 000 L, how many kilograms of hydrogen were needed to fill the balloon at STP? (8.6)

Jacques Charles used hydrogen to launch his balloon in 1783.

8.86 In problem 8.85, the balloon reached an altitude of 1000 m, where the pressure was 658 mmHg and the temperature was −8 °C. What was the volume, in liters, of the balloon at these conditions, if the amount of gas does not change? (8.5)

8.87 A sample of hydrogen (H_2) gas at 127 °C has a pressure of 2.00 atm. At what final temperature, in degrees Celsius, will the pressure of the H_2 decrease to 0.25 atm, if volume and amount of gas do not change? (8.4)

8.88 A fire extinguisher has a pressure of 10. atm at 25 °C. What is the final pressure, in atmospheres, when the fire extinguisher is used at a temperature of 75 °C, if volume and amount of gas do not change? (8.4)

8.89 A weather balloon has a volume of 750 L when filled with helium at 8 °C at a pressure of 380 Torr. What is the final volume, in liters, of the balloon, where the pressure is 0.20 atm and the temperature is −45 °C, when the amount of gas does not change? (8.5)

8.90 During laparoscopic surgery, carbon dioxide gas is used to expand the abdomen to help create a larger working space. If 4.80 L of CO_2 gas at 18 °C at 785 mmHg is used, what is the final volume, in liters, of the gas at 37 °C and a pressure of 745 mmHg, if the amount of CO_2 does not change? (8.5)

8.91 A 2.00-L container is filled with methane gas, CH_4, at a pressure of 2500. mmHg and a temperature of 18 °C. How many grams of methane are in the container? (8.7)

8.92 A steel cylinder with a volume of 15.0 L is filled with 50.0 g of nitrogen gas at 25 °C. What is the pressure, in atmospheres, of the N_2 gas in the cylinder? (8.7)

8.93 How many grams of CO_2 are in 35.0 L of $CO_2(g)$ at 1.20 atm and 5 °C? (8.7)

8.94 A container is filled with 0.644 g of O_2 at 5 °C and 845 mmHg. What is the volume, in milliliters, of the container? (8.7)

8.95 How many liters of H_2 gas can be produced at 0 °C and 1.00 atm (STP) from 25.0 g of Zn? (8.7)

$$2HCl(aq) + Zn(s) \longrightarrow H_2(g) + ZnCl_2(aq)$$

8.96 In the formation of smog, nitrogen and oxygen gas react to form nitrogen dioxide. How many grams of NO_2 will be produced when 2.0 L of nitrogen at 840 mmHg and 24 °C are completely reacted? (8.7)

$$N_2(g) + 2O_2(g) \longrightarrow 2NO_2(g)$$

8.97 Nitrogen dioxide reacts with water to produce oxygen and ammonia. A 5.00-L sample of $H_2O(g)$ reacts at a temperature of 375 °C and a pressure of 725 mmHg. How many grams of NH_3 can be produced? (8.7)

$$4NO_2(g) + 6H_2O(g) \longrightarrow 7O_2(g) + 4NH_3(g)$$

8.98 Hydrogen gas can be produced in the laboratory through the reaction of hydrochloric acid with magnesium metal. When 12.0 g of Mg reacts, what volume, in liters, of H_2 gas is produced at 24 °C and 835 mmHg? (8.7)

$$2HCl(aq) + Mg(s) \longrightarrow H_2(g) + MgCl_2(aq)$$

8.99 A gas mixture contains oxygen and argon at partial pressures of 0.60 atm and 425 mmHg. If nitrogen gas added to the sample increases the total pressure to 1250 Torr, what is the partial pressure, in torr, of the nitrogen added? (8.8)

8.100 What is the total pressure, in millimeters of mercury, of a gas mixture containing argon gas at 0.25 atm, helium gas at 350 mmHg, and nitrogen gas at 360 Torr? (8.8)

CHALLENGE PROBLEMS

The following problems are related to the topics in this chapter. However, they do not all follow the chapter order, and they require you to combine concepts and skills from several Sections. These problems will help you increase your critical thinking skills and prepare for your next exam.

8.101 Your spaceship has docked at a space station above Mars. The temperature inside the space station is a carefully controlled 24 °C at a pressure of 745 mmHg. A balloon with a volume of 425 mL drifts into the airlock where the temperature is −95 °C and the pressure is 0.115 atm. What is the final volume, in milliliters, of the balloon, if the amount of gas does not change and the balloon is very elastic? (8.5)

8.102 You are doing research on planet X. The temperature inside the space station is a carefully controlled 24 °C and the pressure is 755 mmHg. Suppose that a balloon, which has a volume of 850. mL inside the space station, is placed into the airlock, and floats out to planet X. If planet X has an atmospheric pressure of 0.150 atm and the volume of the balloon changes to 3.22 L, what is the temperature, in degrees Celsius, on planet X (the amount of gas does not change)? (8.5)

8.103 A gas sample has a volume of 4250 mL at 15 °C and 745 mmHg. What is the final temperature, in degrees Celsius, after the sample is transferred to a different container with a volume of 2.50 L and a pressure of 1.20 atm, when the amount of gas does not change? (8.5)

8.104 In the fermentation of glucose (wine making), 780 mL of CO_2 gas was produced at 37 °C and 1.00 atm. What is the final volume, in liters, of the gas when measured at 22 °C and 675 mmHg, when the amount of gas does not change? (8.5)

8.105 The propane, C_3H_8, in a fuel cylinder, undergoes combustion with oxygen in the air. How many liters of CO_2 are produced at STP, if the cylinder contains 881 g of propane? (8.7)

$$C_3H_8(g) + 5O_2(g) \xrightarrow{\Delta} 3CO_2(g) + 4H_2O(g)$$

8.106 When sensors in a car detect a collision, they cause the reaction of sodium azide, NaN_3, which generates nitrogen gas to fill the airbags within 0.03 s. How many liters of N_2 are produced at STP, if the airbag contains 132 g of NaN_3? (8.7)

$$2NaN_3(s) \longrightarrow 2Na(s) + 3N_2(g)$$

8.107 Glucose, $C_6H_{12}O_6$, is metabolized in living systems. How many grams of water can be produced from the reaction of 18.0 g of glucose and 7.50 L of O_2 at 1.00 atm and 37 °C? (8.7)

$$C_6H_{12}O_6(s) + 6O_2(g) \longrightarrow 6CO_2(g) + 6H_2O(l)$$

8.108 2.00 L of N_2, at 25 °C and 1.08 atm, is mixed with 4.00 L of O_2, at 25 °C and 0.118 atm, and the mixture is allowed to react. How much NO, in grams, is produced? (8.7)

$$N_2(g) + O_2(g) \longrightarrow 2NO(g)$$

Clinical Applications

8.109 Hyperbaric therapy uses 100% oxygen at pressure to help heal wounds and infections, and to treat carbon monoxide poisoning. If the pressure inside a hyperbaric chamber is 3.0 atm, what is the volume, in liters, of the chamber containing 2400 g of O_2 at 28 °C? (8.7)

8.110 A hyperbaric chamber has a volume of 1510 L. How many kilograms of O_2 gas are needed to give an oxygen pressure of 2.04 atm at 25 °C? (8.7)

8.111 Laparoscopic surgery involves inflating the abdomen with carbon dioxide gas to separate the internal organs and the abdominal wall. If the CO_2 injected into the abdomen produces a pressure of 20. mmHg and a volume of 4.00 L at 32 °C, how many grams of CO_2 were used? (8.7)

8.112 In another laparoscopic surgery, helium is used to inflate the abdomen to separate the internal organs and the abdominal wall. If the He injected into the abdomen produces a pressure of 15 mmHg and a volume of 3.1 L at 28 °C, how many grams of He were used? (8.7)

8.113 A gas mixture with a total pressure of 2400 Torr is used by a scuba diver. If the mixture contains 2.0 moles of helium and 6.0 moles of oxygen, what is the partial pressure, in torr, of each gas in the sample? (8.8)

8.114 A gas mixture with a total pressure of 4.6 atm is used in a hospital. If the mixture contains 5.4 moles of nitrogen and 1.4 moles of oxygen, what is the partial pressure, in atmospheres, of each gas in the sample? (8.8)

ANSWERS

8.1 **a.** At a higher temperature, gas particles have greater kinetic energy, which makes them move faster.
b. Because there are great distances between the particles of a gas, they can be pushed closer together and still remain a gas.
c. Gas particles are very far apart, which means that the mass of a gas in a certain volume is very small, resulting in a low density.

8.3 **a.** temperature **b.** volume **c.** amount **d.** pressure

8.5 Statements **a**, **d**, and **e** describe the pressure of a gas.

8.7 **a.** 1520 Torr **b.** 29.4 $lb/in.^2$ **c.** 1520 mmHg **d.** 203 kPa

8.9 As a diver ascends to the surface, external pressure decreases. If the air in the lungs were not exhaled, its volume would expand and severely damage the lungs. The pressure in the lungs must adjust to changes in the external pressure.

8.11 **a.** The pressure is greater in cylinder **A**. According to Boyle's law, a decrease in volume pushes the gas particles closer together, which will cause an increase in the pressure.
b. 160 mL

8.13 **a.** increases **b.** decreases **c.** increases

8.15 **a.** 328 mmHg **b.** 2620 mmHg **c.** 475 mmHg **d.** 5240 mmHg

8.17 **a.** 52.4 L **b.** 25.0 L **c.** 100. L **d.** 45 L

8.19 6.52 atm

8.21 25 L of cyclopropane

8.23 **a.** inspiration **b.** expiration **c.** inspiration

8.25 **a.** C **b.** A **c.** B

8.27 **a.** 303 °C **b.** −129 °C **c.** 591 °C **d.** 136 °C

8.29 **a.** 2400 mL **b.** 4900 mL **c.** 1800 mL **d.** 1700 mL

8.31 121 °C

8.33 **a.** 770 mmHg **b.** 1150 mmHg

8.35 **a.** −23 °C **b.** 168 °C

8.37 31 °C

8.39 1.9 atm

8.41 $T_2 = T_1 \times \frac{P_2}{P_1} \times \frac{V_2}{V_1}$

8.43 **a.** 4.26 atm **b.** 3.07 atm **c.** 0.606 atm

8.45 −33 °C

8.47 The volume increases because the number of gas particles is increased.

8.49 **a.** 4.00 L **b.** 26.7 L **c.** 14.6 L

8.51 **a.** 0.100 mole of O_2 **b.** 56.0 L **c.** 28.0 L **d.** 0.146 g of H_2

8.53 4.93 atm

8.55 29.4 g of O_2

8.57 566 K (293 °C)

8.59 **a.** 7.60 L of H_2 **b.** 4.92 g of Mg

8.61 178 L of O_2

8.63 3.4 L of O_2

8.65 1300 g of O_2

8.67 765 Torr

8.69 425 Torr

8.71 **a.** The partial pressure of oxygen will be lower than normal.
b. Breathing a higher concentration of oxygen will help to increase the supply of oxygen in the lungs and blood and raise the partial pressure of oxygen in the blood.

8.73 0.978 atm

8.75 2.8 L

8.77 **a.** True. The flask containing helium has more moles of helium and thus more helium atoms.
b. True. The mass and volume of each are the same, which means the mass/volume ratio or density is the same in both flasks.

8.79 **a.** 2 **b.** 1

8.81 **a.** A **b.** C **c.** A **d.** B **e.** C

8.83 **a.** The volume of the chest and lungs is decreased.
b. The decrease in volume increases the pressure, which can dislodge the food in the trachea.

8.85 2.8 kg of H_2

8.87 −223 °C

8.89 1500 L of He

8.91 4.41 g of CH_4

8.93 81.0 g of CO_2

8.95 8.56 L of H_2

8.97 1.02 g of NH_3

8.99 370 Torr

8.101 2170 mL

8.103 −66 °C

8.105 1340 L of CO_2

8.107 5.31 g of water

8.109 620 L

8.111 0.18 g of CO_2

8.113 He 600 Torr, O_2 1800 Torr

COMBINING IDEAS from Chapters 7 and 8

CI.13 In the following diagram, blue spheres represent the element **A** and yellow spheres represent the element **B**: (6.5, 7.1, 7.2)

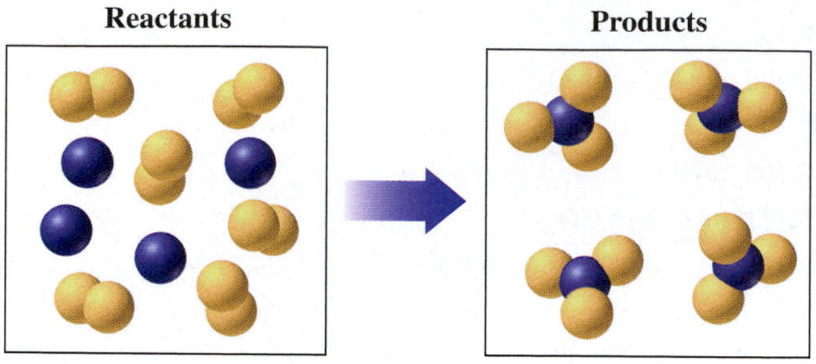

Reactants
Products

a. Write the formulas for each of the reactants and products.
b. Write a balanced chemical equation for the reaction.
c. Indicate the type of reaction as combination, decomposition, single replacement, double replacement, or combustion.

CI.14 The active ingredient in Tums is calcium carbonate. One Tums tablet contains 500. mg of calcium carbonate. (6.3, 6.4, 7.4, 7.5)
a. What is the formula of calcium carbonate?
b. What is the molar mass of calcium carbonate?
c. How many moles of calcium carbonate are in one roll of Tums that contains 12 tablets?
d. If a person takes two Tums tablets, how many grams of calcium are obtained?
e. If the daily recommended quantity of Ca^{2+} to maintain bone strength in older women is 1500 mg, how many Tums tablets are needed each day to supply the needed calcium?

CI.15 Tamiflu (Oseltamivir), $C_{16}H_{28}N_2O_4$, is an antiviral drug used to treat influenza. The preparation of Tamiflu begins with the extraction of shikimic acid from the seedpods of star anise. From 2.6 g of star anise, 0.13 g of shikimic acid can be obtained and used to produce one capsule containing 75 mg of Tamiflu. The usual adult dosage for treatment of influenza is two capsules of Tamiflu daily for 5 days. (7.4, 7.5)

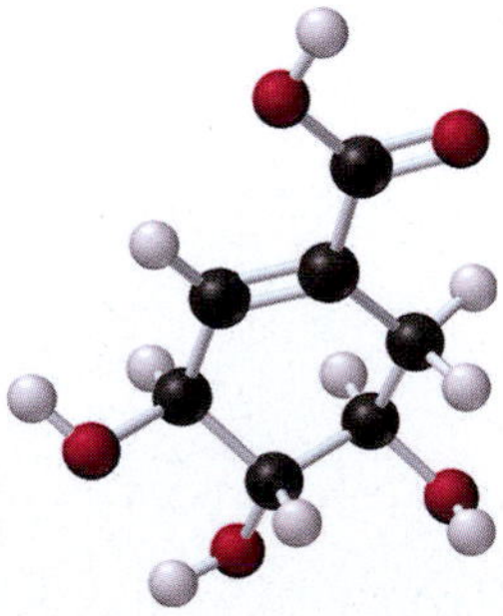

Shikimic acid is the basis for the antiviral drug in Tamiflu.

The spice called star anise is a plant source of shikimic acid.

a. If black spheres are carbon atoms, white spheres are hydrogen atoms, and red spheres are oxygen atoms, what is the formula of shikimic acid?
b. What is the molar mass of shikimic acid?
c. How many moles of shikimic acid are contained in 130 g of shikimic acid?
d. How many capsules containing 75 mg of Tamiflu could be produced from 155 g of star anise?

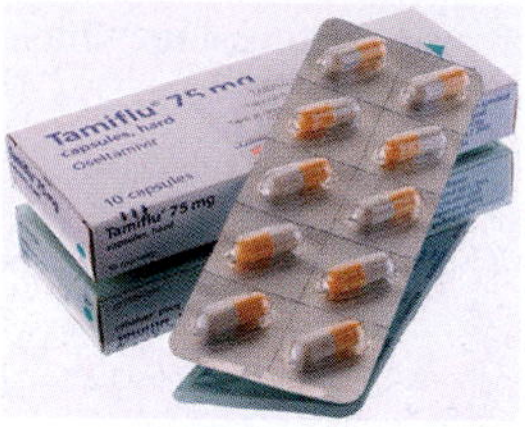

Each capsule contains 75 mg of Tamiflu.

e. What is the molar mass of Tamiflu?
f. How many kilograms of Tamiflu would be needed to treat all the people in a city with a population of 500 000 if each person consumes two Tamiflu capsules a day for 5 days?

CI.16 The compound butyric acid gives rancid butter its characteristic odor. (6.7, 7.4, 7.5)

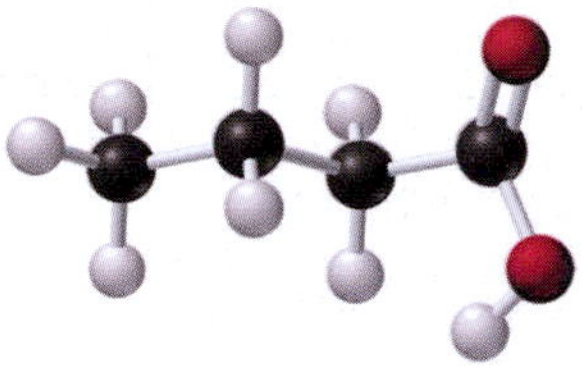

Butyric acid produces the characteristic odor of rancid butter.

a. If black spheres are carbon atoms, white spheres are hydrogen atoms, and red spheres are oxygen atoms, what is the molecular formula of butyric acid?
b. What is the molar mass of butyric acid?
c. How many grams of butyric acid contain 3.28×10^{23} atoms of oxygen?
d. How many grams of carbon are in 5.28 g of butyric acid?
e. Butyric acid has a density of 0.959 g/mL at 20 °C. How many moles of butyric acid are contained in 1.56 mL of butyric acid?
f. Identify the bonds C—C, C—H, and C—O in a molecule of butyric acid as polar covalent or nonpolar covalent.

CI.17 Methane is a major component of purified natural gas used for heating and cooking. When 1.0 mole of methane gas burns with oxygen to produce carbon dioxide and water, 883 kJ is produced. Methane gas has a density of 0.715 g/L at STP. For transport, the natural gas is cooled to −163 °C to form liquefied natural gas (LNG) with a density of 0.45 g/mL. A tank on a ship can hold 7.0 million gallons of LNG. (2.7, 6.6, 7.1, 7.5, 7.7, 7.9, 8.6)

An LNG carrier transports liquefied natural gas.

a. Draw the Lewis structures for methane, which has the formula CH_4.
b. What is the mass, in kilograms, of LNG (assume that LNG is all methane) transported in one tank on a ship?
c. What is the volume, in liters, of LNG (methane) from one tank when the LNG is converted to gas at STP?
d. Write a balanced chemical equation for the combustion of methane in a gas burner, including the heat of reaction.

Methane is the fuel burned in a gas cooktop.

e. How many kilograms of oxygen are needed to react with all of the methane provided by one tank of LNG?
f. How much heat, in kilojoules, is released from burning all of the methane in one tank of LNG?

CI.18 Automobile exhaust is a major cause of air pollution. One pollutant is nitrogen oxide, which forms from nitrogen and oxygen gases in the air at the high temperatures in an automobile engine. Once emitted into the air, nitrogen oxide reacts with oxygen to produce nitrogen dioxide, a reddish brown gas with a sharp, pungent odor that makes up smog. One component of gasoline is heptane, C_7H_{16}, which has a density of 0.684 g/mL. In one year, a typical automobile uses 550 gal of gasoline and produces 41 lb of nitrogen oxide. (7.1, 7.5, 7.7, 8.6)

Two gases found in automobile exhaust are carbon dioxide and nitrogen oxide.

a. Write balanced chemical equations for the production of nitrogen oxide and nitrogen dioxide.
b. If all the nitrogen oxide emitted by one automobile is converted to nitrogen dioxide in the atmosphere, how many kilograms of nitrogen dioxide are produced in one year by a single automobile?
c. Write a balanced chemical equation for the combustion of heptane.
d. How many moles of CO_2 are produced from the gasoline used by the typical automobile in one year, assuming the gasoline is all heptane?
e. How many liters of carbon dioxide at STP are produced in one year from the gasoline used by the typical automobile?

ANSWERS

CI.13 a. reactants A and B_2; product AB_3
b. $2A + 3B_2 \longrightarrow 2AB_3$
c. combination

CI.15 a. $C_7H_{10}O_5$
b. 174.15 g/mole
c. 0.75 mole
d. 59 capsules
e. 312.4 g/mole
f. 400 kg

CI.17 a. 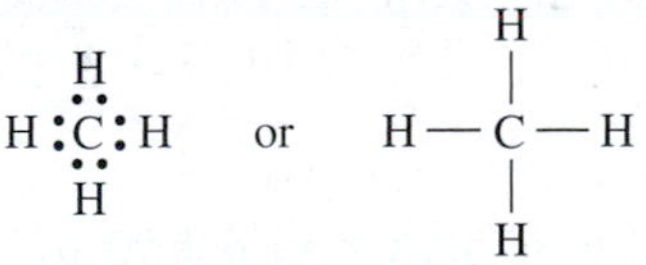

b. 1.2×10^7 kg of LNG (methane)
c. 1.7×10^{10} L of LNG (methane) at STP
d. $CH_4(g) + 2O_2(g) \xrightarrow{\Delta} CO_2(g) + 2H_2O(g) + 883\ kJ$
e. 4.8×10^7 kg of O_2
f. 6.6×10^{11} kJ

Solutions

9

Our kidneys produce urine, which carries waste products and excess fluid from the body. They also reabsorb electrolytes such as potassium and produce hormones that regulate blood pressure and the levels of calcium in the blood. Diseases such as diabetes and high blood pressure can cause a decrease in kidney function. Symptoms of kidney malfunction include protein in the urine, an abnormal level of urea in the blood, frequent urination, and swollen feet. If kidney failure occurs, it may be treated with dialysis or transplantation.

Michelle suffers from kidney disease because of severe strep throat she contracted as a child. When her kidneys stopped functioning, Michelle was placed on dialysis three times a week. As she enters the dialysis unit, her dialysis nurse, Amanda, asks Michelle how she is feeling. Michelle indicates that she feels tired today and has considerable swelling around her ankles. Amanda informs her that these side effects occur because of her body's inability to regulate the amount of water in her cells. Amanda explains that the retention of water is regulated by the concentration of electrolytes in her body fluids and the rate at which waste products are removed from her body. Amanda explains that although water is essential for the many chemical reactions that occur in the body, the amount of water can become too high or too low because of various diseases and conditions. Because Michelle's kidneys no longer function properly, she cannot regulate the amount of electrolytes or waste in her body fluids. As a result, she has an electrolyte imbalance and a buildup of waste products, so her body retains water. Amanda then explains that the dialysis machine does the work of her kidneys to reduce the high levels of electrolytes and waste products.

CAREER

Dialysis Nurse

A dialysis nurse specializes in assisting patients with kidney disease undergoing dialysis. This requires monitoring the patient before, during, and after dialysis for any complications such as a drop in blood pressure or cramping. The dialysis nurse connects the patient to the dialysis unit via a dialysis catheter that is inserted into the neck or chest. A dialysis nurse must have considerable knowledge about how the dialysis machine functions to ensure that it is operating correctly at all times.

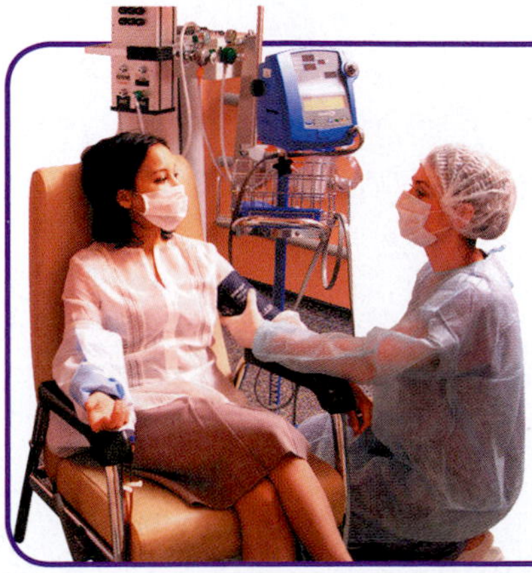

CLINICAL UPDATE

Using Dialysis for Renal Failure

Michelle continues to have dialysis treatment three times a week. You can see more details of Michelle's dialysis in the **CLINICAL UPDATE** **Using Dialysis for Renal Failure**, page 346, and discover that dialysis requires 120 L of fluid containing electrolytes to adjust Michelle's blood level to that of normal serum.

9.1 Solutions

LEARNING GOAL Identify the solute and solvent in a solution; describe the formation of a solution.

Solutions are everywhere around us. Most of the gases, liquids, and solids we see are mixtures of at least one substance dissolved in another. There are different types of solutions. The air we breathe is a solution that is primarily oxygen and nitrogen gases. Carbon dioxide gas dissolved in water makes carbonated drinks. When we make solutions of coffee or tea, we use hot water to dissolve substances from coffee beans or tea leaves. The ocean is also a solution, consisting of many ionic compounds such as sodium chloride dissolved in water. In your medicine cabinet, the antiseptic tincture of iodine is a solution of iodine dissolved in ethanol.

A **solution** is a homogeneous mixture in which one substance, called the **solute**, is uniformly dispersed in another substance called the **solvent**. Because the solute and the solvent do not react with each other, they can be mixed in varying proportions. A solution of a little salt dissolved in water tastes slightly salty. When a large amount of salt is dissolved in water, the solution tastes very salty. Usually, the solute (in this case, salt) is the substance present in the lesser amount, whereas the solvent (in this case, water) is present in the greater amount. For example, in a solution composed of 5.0 g of salt and 50. g of water, salt is the solute and water is the solvent. In a solution, the particles of the solute are evenly dispersed among the molecules within the solvent (see **FIGURE 9.1**).

REVIEW

Identifying Polarity of Molecules (6.8)

Identifying Intermolecular Forces (6.9)

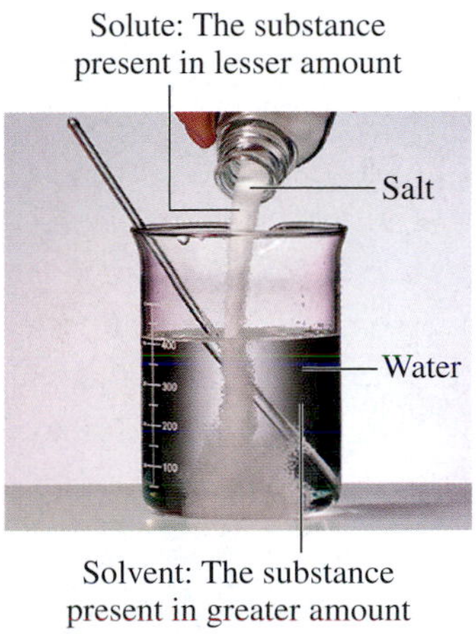

Solute: The substance present in lesser amount
Salt
Water
Solvent: The substance present in greater amount

A solution has at least one solute dispersed in a solvent.

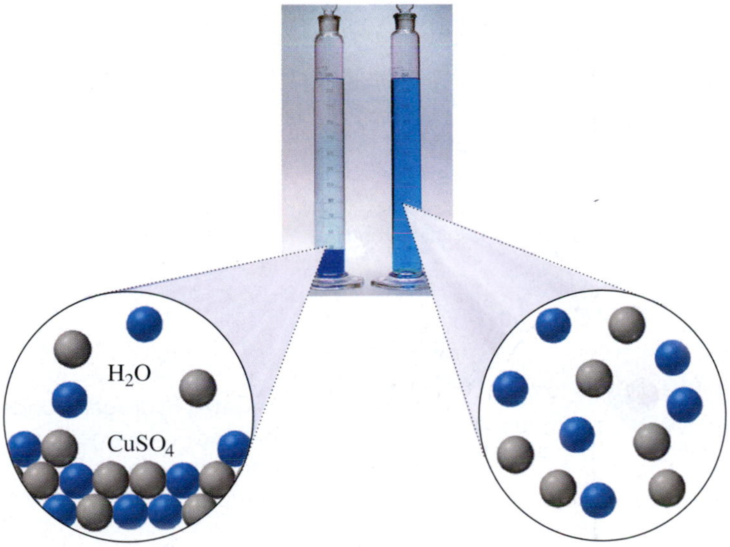

H_2O
$CuSO_4$

FIGURE 9.1 ▶ A solution of copper(II) sulfate ($CuSO_4$) forms as particles of solute dissolve, move away from the crystal, and become evenly dispersed among the solvent (water) molecules.

Q What does the uniform blue color in the graduated cylinder on the right indicate?

Types of Solutes and Solvents

Solutes and solvents may be solids, liquids, or gases. The solution that forms has the same physical state as the solvent. When sugar crystals are dissolved in water, the resulting sugar solution is liquid. Sugar is the solute, and water is the solvent. Soda water and soft drinks are prepared by dissolving carbon dioxide gas in water. The carbon dioxide gas is the solute, and water is the solvent. **TABLE 9.1** lists some solutions and their solutes and solvents.

Water as a Solvent

Water is one of the most common solvents in nature. In the H_2O molecule, an oxygen atom shares electrons with two hydrogen atoms. Because oxygen is much more electronegative than hydrogen, the O—H bonds are polar. In each polar bond, the oxygen atom has a partial

TABLE 9.1 Some Examples of Solutions

Type	Example	Solute	Solvent
Gas Solutions			
Gas in a gas	Air	$O_2(g)$	$N_2(g)$
Liquid Solutions			
Gas in a liquid	Soda water	$CO_2(g)$	$H_2O(l)$
	Household ammonia	$NH_3(g)$	$H_2O(l)$
Liquid in a liquid	Vinegar	$HC_2H_3O_2(l)$	$H_2O(l)$
Solid in a liquid	Seawater	$NaCl(s)$	$H_2O(l)$
	Tincture of iodine	$I_2(s)$	$C_2H_6O(l)$
Solid Solutions			
Solid in a solid	Brass	$Zn(s)$	$Cu(s)$
	Steel	$C(s)$	$Fe(s)$

TEST

Try Practice Problems 9.1 and 9.2

negative (δ^-) charge, and the hydrogen atom has a partial positive (δ^+) charge. Because the shape of a water molecule is bent, its dipoles do not cancel out. Thus, water is polar and is a *polar solvent*.

Attractive forces known as *hydrogen bonds* occur between molecules where partially positive hydrogen atoms are attracted to the partially negative atoms N, O, or F. As seen in the diagram, the hydrogen bonds are shown as a series of dots. Although hydrogen bonds are much weaker than covalent or ionic bonds, there are many of them linking water molecules together. Hydrogen bonds are important in the properties of biological compounds such as proteins, carbohydrates, and DNA.

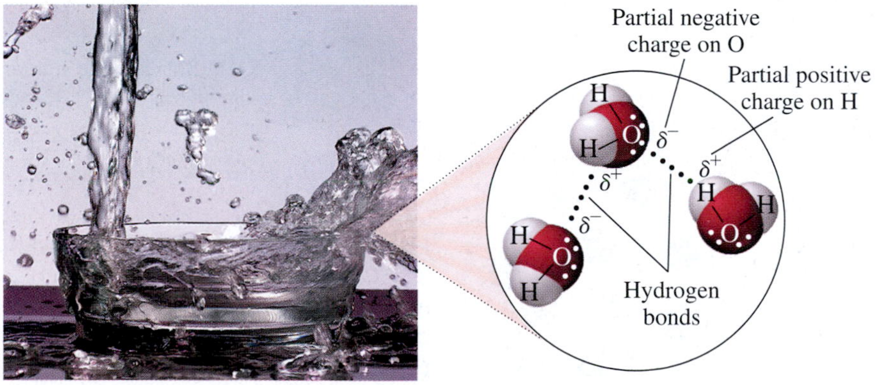

Partial negative charge on O
Partial positive charge on H
Hydrogen bonds
H
O
δ−
δ+

In water, hydrogen bonds form between an oxygen atom in one water molecule and the hydrogen atom in another.

Chemistry Link to Health

Water in the Body

The average adult is about 60% water by mass, and the average infant about 75%. About 60% of the body's water is contained within the cells as intracellular fluids; the other 40% makes up extracellular fluids, which include the interstitial fluid in tissue and the plasma in the blood. These external fluids carry nutrients and waste materials between the cells and the circulatory system.

Every day you lose between 1500 and 3000 mL of water from the kidneys as urine, from the skin as perspiration, from the lungs as you exhale, and from the gastrointestinal tract. Serious dehydration

Typical water gain and loss during 24 hours

Water Gain	
Liquid	1000 mL
Food	1200 mL
Metabolism	300 mL
Total	2500 mL

Water Loss	
Urine	1500 mL
Perspiration	300 mL
Breath	600 mL
Feces	100 mL
Total	2500 mL

(continued)

Chemistry Link to Health (continued)

The water lost from the body is replaced by the intake of fluids.

can occur in an adult if there is a 10% net loss in total body fluid; a 20% loss of fluid can be fatal. An infant suffers severe dehydration with only a 5 to 10% loss in body fluid.

Water loss is continually replaced by the liquids and foods in the diet and from metabolic processes that produce water in the cells of the body. **TABLE 9.2** lists the percentage by mass of water contained in some foods.

TABLE 9.2 Percentage of Water in Some Foods

Food	Water (% by mass)	Food	Water (% by mass)
Vegetables		**Meats/Fish**	
Carrot	88	Chicken, cooked	71
Celery	94	Hamburger, broiled	60
Cucumber	96	Salmon	71
Tomato	94		
Fruits		**Milk Products**	
Apple	85	Cottage cheese	78
Cantaloupe	91	Milk, whole	87
Orange	86	Yogurt	88
Strawberry	90		
Watermelon	93		

Formation of Solutions

The interactions between solute and solvent will determine whether a solution will form. Initially, energy is needed to separate the particles in the solute and the solvent particles. Then energy is released as solute particles move between the solvent particles to form a solution. However, there must be attractions between the solute and the solvent particles to provide the energy for the initial separation. These attractions occur when the solute and the solvent have similar polarities. The expression "like dissolves like" is a way of saying that the polarities of a solute and a solvent must be similar for a solution to form (see **FIGURE 9.2**). In the absence of attractions between a solute and a solvent, there is insufficient energy to form a solution (see **TABLE 9.3**).

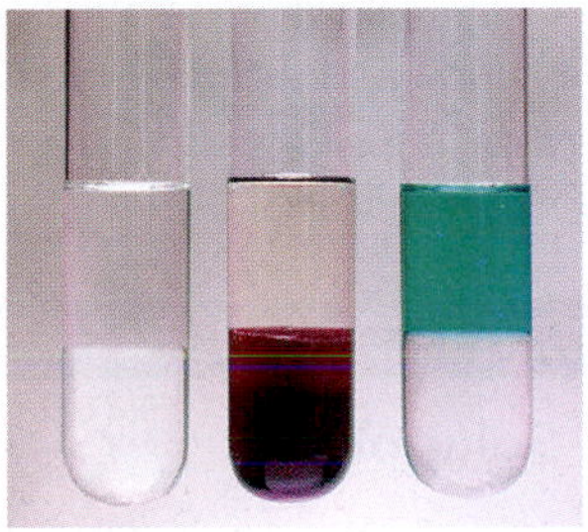

(a) (b) (c)

FIGURE 9.2 ▶ Like dissolves like. In each test tube, the lower layer is CH_2Cl_2 (more dense), and the upper layer is water (less dense). **(a)** CH_2Cl_2 is nonpolar and water is polar; the two layers do not mix. **(b)** The nonpolar solute I_2 (purple) is soluble in the nonpolar solvent CH_2Cl_2. **(c)** The ionic solute $Ni(NO_3)_2$ (green) is soluble in the polar solvent water.

Q In which layer would polar molecules of sucrose ($C_{12}H_{22}O_{11}$) be soluble?

TABLE 9.3 Possible Combinations of Solutes and Solvents

Solutions Will Form		Solutions Will Not Form	
Solute	**Solvent**	**Solute**	**Solvent**
Polar	Polar	Polar	Nonpolar
Nonpolar	Nonpolar	Nonpolar	Polar

Solutions with Ionic and Polar Solutes

In ionic solutes such as sodium chloride, NaCl, there are strong ionic bonds between positively charged Na^+ ions and negatively charged Cl^- ions. In water, a polar solvent, the hydrogen bonds provide strong solvent–solvent attractions. When NaCl crystals are placed in water, partially negative oxygen atoms in water molecules attract positive Na^+ ions, and the partially positive hydrogen atoms in other water molecules attract negative Cl^- ions (see **FIGURE 9.3**). As soon as the Na^+ ions and the Cl^- ions form a solution, they undergo **hydration** as water molecules surround each ion. Hydration of the ions diminishes their attraction to other ions and keeps them in solution.

In the equation for the formation of the NaCl solution, the solid and aqueous NaCl are shown with the formula H_2O over the arrow, which indicates that water is needed for the dissociation process but is not a reactant.

$$NaCl(s) \xrightarrow[\text{Dissociation}]{H_2O} Na^+(aq) + Cl^-(aq)$$

ENGAGE

Why does KCl form a solution with water, but nonpolar hexane (C_6H_{14}) does not form a solution with water?

TEST

Try Practice Problems 9.3 to 9.6

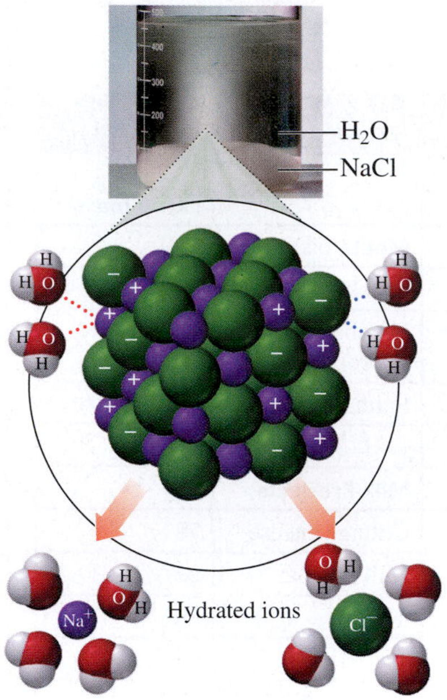

H_2O
NaCl
Hydrated ions
Na^+
Cl^-

FIGURE 9.3 ▶ Ions on the surface of a crystal of NaCl dissolve in water as they are attracted to the polar water molecules that pull the ions into solution and surround them.

Q What helps keep Na^+ and Cl^- ions in solution?

In another example, we find that a polar molecular compound such as methanol, CH_4O, is soluble in water because methanol has a polar —OH group that forms hydrogen bonds with water (see **FIGURE 9.4**). Polar solutes require polar solvents for a solution to form.

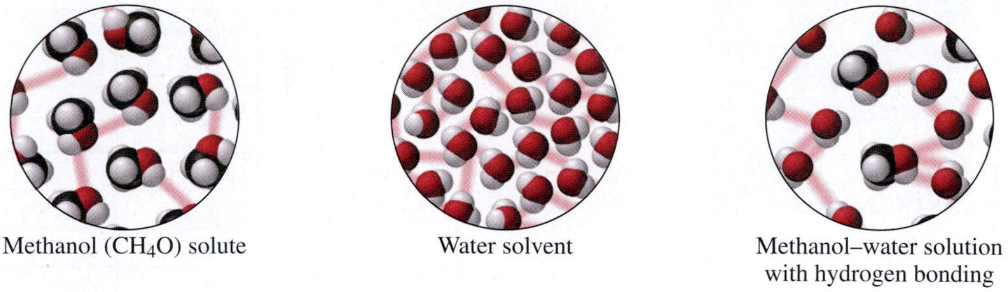

Methanol (CH_4O) solute
Water solvent
Methanol–water solution with hydrogen bonding

FIGURE 9.4 ▶ Polar molecules of methanol, CH_4O, form hydrogen bonds with polar water molecules to form a methanol–water solution.

Q Why are there attractions between the solute methanol and the solvent water?

Solutions with Nonpolar Solutes

Compounds containing nonpolar molecules, such as iodine (I_2), oil, or grease, do not dissolve in water because there are no attractions between the particles of a nonpolar solute and the polar solvent. Nonpolar solutes require nonpolar solvents for a solution to form.

PRACTICE PROBLEMS

9.1 Solutions

9.1 Identify the solute and the solvent in each solution composed of the following:
a. 10.0 g of NaCl and 100.0 g of H_2O
b. 50.0 mL of ethanol, C_2H_6O, and 10.0 mL of H_2O
c. 0.20 L of O_2 and 0.80 L of N_2

9.2 Identify the solute and the solvent in each solution composed of the following:
a. 10.0 mL of acetic acid, $HC_2H_3O_2$, and 200. mL of H_2O
b. 100.0 mL of H_2O and 5.0 g of sugar, $C_{12}H_{22}O_{11}$
c. 1.0 g of Br_2 and 50.0 mL of methylene chloride, CH_2Cl_2

9.3 Describe the formation of an aqueous KI solution, when solid KI dissolves in water.

9.4 Describe the formation of an aqueous LiBr solution, when solid LiBr dissolves in water.

Clinical Applications

9.5 Water is a polar solvent and carbon tetrachloride (CCl_4) is a nonpolar solvent. In which solvent is each of the following, which is found or used in the body, more likely to be soluble?
a. $CaCO_3$ (calcium supplement), ionic
b. retinol (vitamin A), nonpolar
c. sucrose (table sugar), polar
d. cholesterol (lipid), nonpolar

9.6 Water is a polar solvent, and hexane (C_6H_{14}) is a nonpolar solvent. In which solvent is each of the following, which is found or used in the body, more likely to be soluble?
a. vegetable oil, nonpolar
b. oleic acid (lipid), nonpolar
c. niacin (vitamin B_3), polar
d. $FeSO_4$ (iron supplement), ionic

REVIEW

Writing Conversion Factors from Equalities (2.5)
Using Conversion Factors (2.6)
Writing Positive and Negative Ions (6.1)

9.2 Electrolytes and Nonelectrolytes

LEARNING GOAL Identify solutes as electrolytes or nonelectrolytes.

Solutes can be classified by their ability to conduct an electrical current. When **electrolytes** dissolve in water, the process of *dissociation* separates them into ions forming solutions that conduct electricity. When **nonelectrolytes** dissolve in water, they do not separate into ions, and their solutions do not conduct electricity.

To test solutions for the presence of ions, we can use an apparatus that consists of a battery and a pair of electrodes connected by wires to a light bulb. The light bulb glows when electricity can flow, which can only happen when electrolytes provide ions that move between the electrodes to complete the circuit.

Types of Electrolytes

Electrolytes can be further classified as *strong electrolytes* or *weak electrolytes*. For a **strong electrolyte**, such as sodium chloride (NaCl), there is 100% dissociation of the solute into ions. When the electrodes from the light bulb apparatus are placed in the NaCl solution, the light bulb glows very brightly.

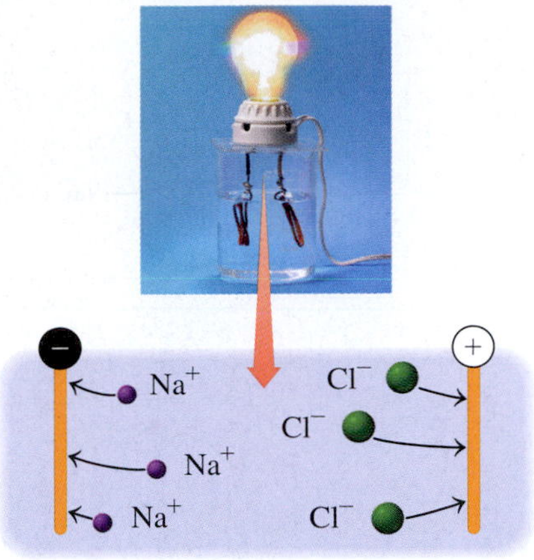

Na$^+$
Cl$^-$
Cl$^-$
Na$^+$
Na$^+$
Cl$^-$

(a) Strong electrolyte

In an equation for dissociation of a compound in water, the charges must balance. For example, magnesium nitrate dissociates to give one magnesium ion for every two nitrate ions. However, only the ionic bonds between Mg^{2+} and NO_3^- are broken, not the covalent bonds within the polyatomic ion. The equation for the dissociation of $Mg(NO_3)_2$ is written as follows:

$$Mg(NO_3)_2(s) \xrightarrow[\text{Dissociation}]{H_2O} Mg^{2+}(aq) + 2NO_3^-(aq)$$

A **weak electrolyte** is a compound that dissolves in water mostly as molecules. Only a few of the dissolved solute molecules undergo *dissociation*, producing a small number of ions in solution. Thus, solutions of weak electrolytes do not conduct electrical current as well as solutions of strong electrolytes. When the electrodes are placed in a solution of a weak electrolyte, the glow of the light bulb is very dim. In an aqueous solution of the weak electrolyte HF, a few HF molecules dissociate to produce H^+ and F^- ions. As more H^+ and F^- ions form, some recombine to give HF molecules. These forward and reverse reactions of molecules to ions and back again are indicated by two arrows between reactant and products that point in opposite directions:

$$HF(aq) \underset{\text{Recombination}}{\overset{\text{Dissociation}}{\rightleftarrows}} H^+(aq) + F^-(aq)$$

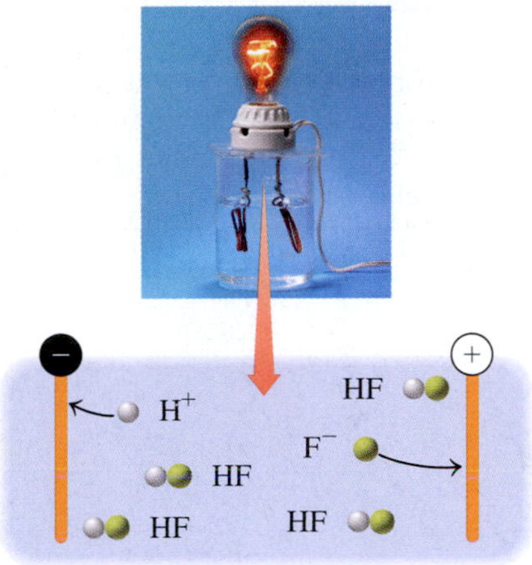

H$^+$
HF
F$^-$
HF
HF
HF

(b) Weak electrolyte

A nonelectrolyte such as methanol (CH_4O) dissolves in water only as molecules, which do not dissociate. When electrodes of the light bulb apparatus are placed in a solution of a nonelectrolyte, the light bulb does not glow because the solution does not contain ions and cannot conduct electricity.

$$CH_4O(l) \xrightarrow{H_2O} CH_4O(aq)$$

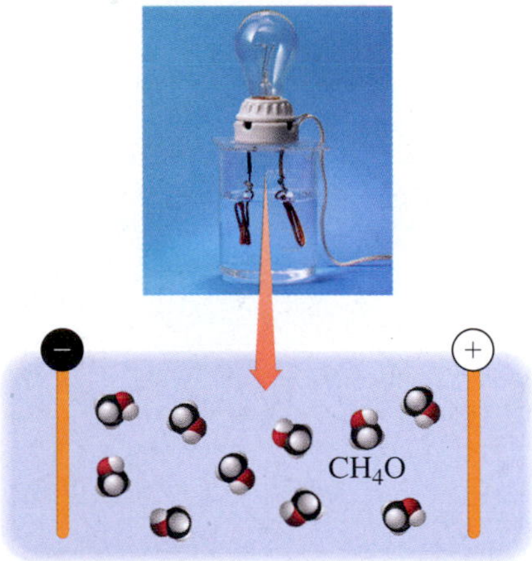

CH$_4$O

(c) Nonelectrolyte

TABLE 9.4 summarizes the classification of solutes in aqueous solutions.

TABLE 9.4 Classification of Solutes in Aqueous Solutions

Type of Solute	In Solution	Type(s) of Particles in Solution	Conducts Electricity?	Examples
Strong electrolyte	Dissociates completely	Only ions	Yes	Ionic compounds such as NaCl, KBr, $MgCl_2$, $NaNO_3$; bases such as NaOH, KOH; acids such as HCl, HBr, HI, HNO_3, $HClO_4$, H_2SO_4
Weak electrolyte	Dissociates partially	Mostly molecules and a few ions	Weakly	HF, H_2O, NH_3, $HC_2H_3O_2$ (acetic acid)
Nonelectrolyte	No dissociation	Only molecules	No	Carbon compounds such as CH_4O (methanol), C_2H_6O (ethanol), $C_{12}H_{22}O_{11}$ (sucrose), CH_4N_2O (urea)

▸SAMPLE PROBLEM 9.1 Solutions of Electrolytes and Nonelectrolytes

TRY IT FIRST

Indicate whether solutions of each of the following contain only ions, only molecules, or mostly molecules and a few ions. Write the equation for the formation of a solution for each of the following:

a. $Na_2SO_4(s)$, a strong electrolyte
b. sucrose, $C_{12}H_{22}O_{11}(s)$, a nonelectrolyte
c. acetic acid, $HC_2H_3O_2(l)$, a weak electrolyte

SOLUTION

a. An aqueous solution of $Na_2SO_4(s)$ contains only the ions Na^+ and SO_4^{2-}.

$$Na_2SO_4(s) \xrightarrow{H_2O} 2Na^+(aq) + SO_4^{2-}(aq)$$

b. A nonelectrolyte such as sucrose, $C_{12}H_{22}O_{11}(s)$, produces only molecules when it dissolves in water.

$$C_{12}H_{22}O_{11}(s) \xrightarrow{H_2O} C_{12}H_{22}O_{11}(aq)$$

c. A weak electrolyte such as $HC_2H_3O_2(l)$ produces mostly molecules and a few ions when it dissolves in water.

$$HC_2H_3O_2(l) \overset{H_2O}{\rightleftarrows} H^+(aq) + C_2H_3O_2^-(aq)$$

STUDY CHECK 9.1

a. Boric acid, $H_3BO_3(s)$, is a weak electrolyte. Would you expect a boric acid solution to contain only ions, only molecules, or mostly molecules and a few ions?
b. Hydrochloric acid, HCl, is a strong electrolyte. Would you expect a solution of hydrochloric acid to contain only ions, only molecules, or mostly molecules and a few ions?

ANSWER

a. A solution of a weak electrolyte would contain mostly molecules and a few ions.
b. A solution of a strong electrolyte would contain only ions.

ENGAGE
Why does a solution of $LiNO_3$, a strong electrolyte, contain only ions whereas a solution of urea, CH_4N_2O, a nonelectrolyte, contains only molecules?

TEST
Try Practice Problems 9.7 to 9.14

Equivalents

Body fluids contain a mixture of electrolytes, such as Na^+, Cl^-, K^+, and Ca^{2+}. We measure each individual ion in terms of an **equivalent (Eq)**, which is the amount of that ion equal to 1 mole of positive or negative electrical charge. For example, 1 mole of Na^+ ions and 1 mole of Cl^- ions are each 1 equivalent because they each contain 1 mole of charge. For an ion with a charge of 2+ or 2−, there are 2 equivalents for each mole. Some examples of ions and equivalents are shown in **TABLE 9.5**.

ENGAGE
Why does one mole of magnesium ion have two equivalents, whereas one mole of phosphate ion has three equivalents?

TABLE 9.5 Equivalents of Electrolytes in Clinical Intravenous (IV) Solutions

Ion	Ionic Charge	Number of Equivalents in 1 Mole
Na^+, K^+, Li^+, NH_4^+	1+	1 Eq
Ca^{2+}, Mg^{2+}	2+	2 Eq
Fe^{3+}	3+	3 Eq
Cl^-, $C_2H_3O_2^-$ (acetate), $H_2PO_4^-$, $C_3H_5O_3^-$ (lactate)	1−	1 Eq
CO_3^{2-}, HPO_4^{2-}	2−	2 Eq
PO_4^{3-}, $C_6H_5O_7^{3-}$ (citrate)	3−	3 Eq

In any solution, the charge of the positive ions is always balanced by the charge of the negative ions. The concentrations of electrolytes in intravenous fluids are expressed in milliequivalents per liter (mEq/L); 1 Eq = 1000 mEq. For example, a solution containing 25 mEq/L of Na^+ and 4 mEq/L of K^+ has a total positive charge of 29 mEq/L. If Cl^- is the only anion, its concentration must be 29 mEq/L.

▶ SAMPLE PROBLEM 9.2 Electrolyte Concentration

TRY IT FIRST

The laboratory tests for Michelle indicate that she has hypercalcemia with a blood calcium level of 8.8 mEq/L. How many moles of calcium ion are in 0.50 L of her blood?

SOLUTION

STEP 1 State the given and needed quantities.

	Given	Need	Connect
ANALYZE THE PROBLEM	0.50 L, 8.8 mEq of Ca^{2+}/L	moles of Ca^{2+}	1 Eq = 1000 mEq, 1 mole of Ca^{2+} = 2 Eq of Ca^{2+}

STEP 2 Write a plan to calculate the moles.

liters of solution → **Electrolyte concentration** → milliequivalents of Ca^{2+} → **Metric factor** → equivalents of Ca^{2+} → **Eq/mole** → moles of Ca^{2+}

STEP 3 State the equalities and conversion factors.

1 L of solution = 8.8 mEq of Ca^{2+}

$$\frac{8.8 \text{ mEq Ca}^{2+}}{1 \text{ L solution}} \quad \text{and} \quad \frac{1 \text{ L solution}}{8.8 \text{ mEq Ca}^{2+}}$$

1 Eq = 1000 mEq

$$\frac{1000 \text{ mEq Ca}^{2+}}{1 \text{ Eq Ca}^{2+}} \quad \text{and} \quad \frac{1 \text{ Eq Ca}^{2+}}{1000 \text{ mEq Ca}^{2+}}$$

1 mole of Ca^{2+} = 2 Eq of Ca^{2+}

$$\frac{2 \text{ Eq Ca}^{2+}}{1 \text{ mole Ca}^{2+}} \quad \text{and} \quad \frac{1 \text{ mole Ca}^{2+}}{2 \text{ Eq Ca}^{2+}}$$

STEP 4 Set up the problem to calculate the number of moles.

$$0.50 \cancel{\text{L}} \times \frac{8.8 \cancel{\text{mEq Ca}^{2+}}}{1 \cancel{\text{L}}} \times \frac{1 \cancel{\text{Eq Ca}^{2+}}}{1000 \cancel{\text{mEq Ca}^{2+}}} \times \frac{1 \text{ mole Ca}^{2+}}{2 \cancel{\text{Eq Ca}^{2+}}} = 0.0022 \text{ mole of Ca}^{2+}$$

Two SFs (0.50 L); Two SFs / Exact (8.8 mEq / 1 L); Exact / Exact (1 Eq / 1000 mEq); Exact / Exact (1 mole / 2 Eq); Two SFs (answer)

STUDY CHECK 9.2

a. A lactated Ringer's solution for intravenous fluid replacement contains 109 mEq/L of Cl^-. If a patient received 1250 mL of Ringer's solution, how many moles of chloride ion were given?

b. If a patient receives 680 mL of a maintenance solution that contains 35 mEq/L of K^+, how many moles of potassium ion were given?

ANSWER

a. 0.136 mole of Cl^-
b. 0.024 mole of K^+

TEST

Try Practice Problems 9.15 to 9.20

Chemistry Link to Health

Electrolytes in Body Fluids

Electrolytes in the body play an important role in maintaining the proper function of the cells and organs in the body. Typically, the electrolytes sodium, potassium, chloride, and bicarbonate are measured in a blood test. Sodium ions regulate the water content in the body and are important in carrying electrical impulses through the nervous system. Potassium ions are also involved in the transmission of electrical impulses and play a role in the maintenance of a regular heartbeat. Chloride ions balance the charges of the positive ions and also control the balance of fluids in the body. Bicarbonate is important in maintaining the proper pH of the blood. Sometimes when vomiting, diarrhea, or sweating is excessive, the concentrations of certain electrolytes may decrease. Then fluids such as Pedialyte may be given to return electrolyte levels to normal.

The concentrations of electrolytes present in body fluids and in intravenous fluids given to a patient are expressed in milliequivalents per liter (mEq/L) of solution. For example, one liter of Pedialyte contains the following electrolytes: Na^+ 45 mEq, Cl^- 35 mEq, K^+ 20 mEq, and citrate^{3-} 30 mEq.

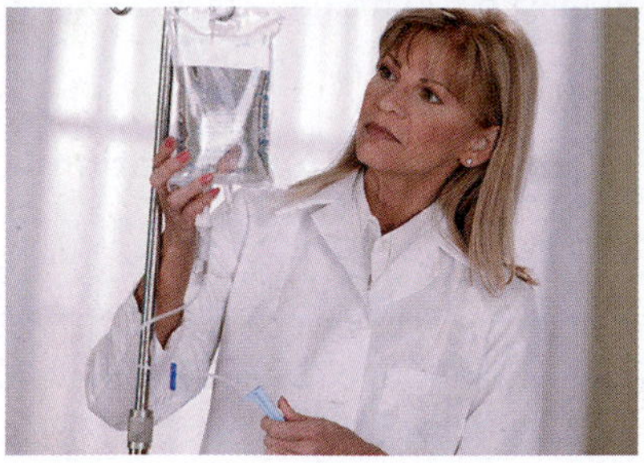

An intravenous solution is used to replace electrolytes in the body.

TABLE 9.6 gives the concentrations of some typical electrolytes in blood plasma and various types of solutions. The use of a specific intravenous solution depends on the nutritional, electrolyte, and fluid needs of the individual patient.

TABLE 9.6 Electrolytes in Blood Plasma and Selected Intravenous Solutions

Normal Concentrations of Ions (mEq/L)

	Blood Plasma	Normal Saline (0.9% NaCl)	Lactated Ringer's Solution	Maintenance Solution	Replacement Solution (Extracellular)
Purpose		Replaces fluid loss	Hydration	Maintains electrolytes and fluids	Replaces electrolytes
Cations					
Na^+	135–145	154	130	40	140
K^+	3.5–5.5		4	35	10
Ca^{2+}	4.5–5.5		3		5
Mg^{2+}	1.5–3.0				3
Total		154	137	75	158
Anions					
Acetate$^-$					47
Cl^-	95–105	154	109	40	103
HCO_3^-	22–28				
Lactate$^-$			28	20	
HPO_4^{2-}	1.8–2.3			15	
Citrate^{3-}					8
Total		154	137	75	158

PRACTICE PROBLEMS

9.2 Electrolytes and Nonelectrolytes

9.7 KF is a strong electrolyte, and HF is a weak electrolyte. How is the solution of KF different from that of HF?

9.8 NaOH is a strong electrolyte, and CH_4O is a nonelectrolyte. How is the solution of NaOH different from that of CH_4O?

9.9 Write a balanced equation for the dissociation of each of the following strong electrolytes in water:
a. KCl **b.** $CaCl_2$ **c.** K_3PO_4 **d.** $Fe(NO_3)_3$

9.10 Write a balanced equation for the dissociation of each of the following strong electrolytes in water:
a. LiBr **b.** $NaNO_3$ **c.** $CuCl_2$ **d.** K_2CO_3

9.11 Indicate whether aqueous solutions of each of the following solutes contain only ions, only molecules, or mostly molecules and a few ions:
a. acetic acid, $HC_2H_3O_2$, a weak electrolyte
b. NaBr, a strong electrolyte
c. fructose, $C_6H_{12}O_6$, a nonelectrolyte

9.12 Indicate whether aqueous solutions of each of the following solutes contain only ions, only molecules, or mostly molecules and a few ions:
a. NH_4Cl, a strong electrolyte
b. ethanol, C_2H_6O, a nonelectrolyte
c. hydrocyanic acid, HCN, a weak electrolyte

9.13 Classify the solute represented in each of the following equations as a strong, weak, or nonelectrolyte:
a. $K_2SO_4(s) \xrightarrow{H_2O} 2K^+(aq) + SO_4^{2-}(aq)$
b. $NH_3(g) + H_2O(l) \rightleftarrows NH_4^+(aq) + OH^-(aq)$
c. $C_6H_{12}O_6(s) \xrightarrow{H_2O} C_6H_{12}O_6(aq)$

9.14 Classify the solute represented in each of the following equations as a strong, weak, or nonelectrolyte:
a. $C_4H_{10}O(l) \xrightarrow{H_2O} C_4H_{10}O(aq)$
b. $MgCl_2(s) \xrightarrow{H_2O} Mg^{2+}(aq) + 2Cl^-(aq)$
c. $HClO(aq) \overset{H_2O}{\rightleftarrows} H^+(aq) + ClO^-(aq)$

9.15 Calculate the number of equivalents in each of the following:
a. 1 mole of K^+
b. 2 moles of OH^-
c. 1 mole of Ca^{2+}
d. 3 moles of CO_3^{2-}

9.16 Calculate the number of equivalents in each of the following:
a. 1 mole of Mg^{2+}
b. 0.5 mole of H^+
c. 4 moles of Cl^-
d. 2 moles of Fe^{3+}

Clinical Applications

9.17 An intravenous saline solution contains 154 mEq/L each of Na^+ and Cl^-. How many moles each of Na^+ and Cl^- are in 1.00 L of the saline solution?

9.18 An intravenous solution to replace potassium loss contains 40 mEq/L each of K^+ and Cl^-. How many moles each of K^+ and Cl^- are in 1.5 L of the solution?

9.19 An intravenous solution contains 40. mEq/L of Cl^- and 15 mEq/L of HPO_4^{2-}. If Na^+ is the only cation in the solution, what is the Na^+ concentration, in milliequivalents per liter?

9.20 A Ringer's solution contains the following concentrations (mEq/L) of cations: Na^+ 147, K^+ 4, and Ca^{2+} 4. If Cl^- is the only anion in the solution, what is the Cl^- concentration, in milliequivalents per liter?

9.3 Solubility

REVIEW
Interpreting Graphs (1.4)

LEARNING GOAL Define solubility; distinguish between an unsaturated and a saturated solution. Identify an ionic compound as soluble or insoluble.

The term *solubility* is used to describe the amount of a solute that can dissolve in a given amount of solvent. Many factors, such as the type of solute, the type of solvent, and the temperature, affect the solubility of a solute. **Solubility**, usually expressed in grams of solute in 100. g of solvent, is the maximum amount of solute that can dissolve at a certain temperature. If a solute readily dissolves when added to the solvent, the solution does not contain the maximum amount of solute. We call this solution an **unsaturated solution**.

A solution that contains all the solute that can dissolve is a **saturated solution**. When a solution is saturated, the rate at which the solute dissolves becomes equal to the rate at which solid forms, a process known as *crystallization*. Then there is no further change in the amount of dissolved solute in solution.

$$\text{Solute} + \text{solvent} \underset{\text{Solute recrystallizes}}{\overset{\text{Solute dissolves}}{\rightleftarrows}} \text{saturated solution}$$

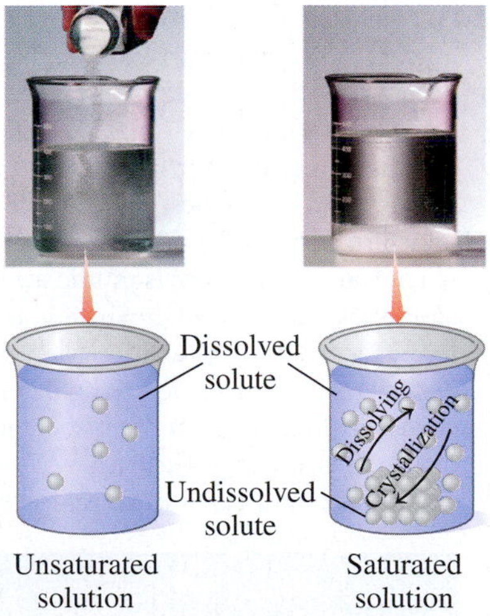
Dissolved solute
Undissolved solute
Dissolving
Crystallization
Unsaturated solution
Saturated solution

Additional solute can dissolve in an unsaturated solution, but not in a saturated solution.

We can prepare a saturated solution by adding an amount of solute greater than that needed to reach solubility. Stirring the solution will dissolve the maximum amount of solute and leave the excess on the bottom of the container. The addition of more solute to the saturated solution will only increase the amount of undissolved solute.

▶SAMPLE PROBLEM 9.3 Saturated Solutions

TRY IT FIRST

At 20 °C, the solubility of KCl is 34 g/100. g of H_2O. In the laboratory, a student mixes 75 g of KCl with 200. g of H_2O at a temperature of 20 °C.

a. How much of the KCl will dissolve?
b. Is the solution saturated or unsaturated?
c. What is the mass, in grams, of any solid KCl left undissolved on the bottom of the container?

SOLUTION

a. At 20 °C, KCl has a solubility of 34 g of KCl in 100. g of water. Using the solubility as a conversion factor, we can calculate the maximum amount of KCl that can dissolve in 200. g of water as follows:

$$200.\ \text{g}\ \cancel{H_2O} \times \frac{34\ \text{g KCl}}{100.\ \text{g}\ \cancel{H_2O}} = 68\ \text{g of KCl}$$

b. Because 75 g of KCl exceeds the maximum amount (68 g) that can dissolve in 200. g of water, the KCl solution is saturated.

c. If we add 75 g of KCl to 200. g of water and only 68 g of KCl can dissolve, there is 7 g (75 g − 68 g) of solid (undissolved) KCl on the bottom of the container.

STUDY CHECK 9.3

At 40 °C, the solubility of KNO_3 is 65 g/100. g of H_2O.

a. How many grams of KNO_3 will dissolve in 120 g of H_2O at 40 °C?
b. How many grams of H_2O are needed to completely dissolve 110 g of KNO_3 at 20 °C?

ANSWER

a. 78 g of KNO_3 **b.** 170 g of H_2O

TEST

Try Practice Problems 9.21 to 9.24

Chemistry Link to Health

Gout and Kidney Stones: Saturation in Body Fluids

The conditions of gout and kidney stones involve compounds in the body that exceed their solubility levels and form solid products. Gout affects adults, primarily men, over the age of 40. Attacks of gout may occur when the concentration of uric acid in blood plasma exceeds its solubility, which is 7 mg/100 mL of plasma at 37 °C. Insoluble deposits of needle-like crystals of uric acid can form in the cartilage, tendons, and soft tissues, where they cause painful gout attacks. They may also form in the tissues of the kidneys, where they can cause renal damage. High levels of uric acid in the body can be caused by an increase in uric acid production, failure of the kidneys to remove uric acid, or a diet with an overabundance of foods containing purines, which are metabolized to uric acid in the body. Foods in the diet that contribute to high levels of uric acid include certain meats, sardines, mushrooms, asparagus, and beans. Drinking alcoholic beverages may also significantly increase uric acid levels and bring about gout attacks.

Treatment for gout involves diet changes and drugs. Medications, such as probenecid, which helps the kidneys eliminate uric acid, or allopurinol, which blocks the production of uric acid by the body, may be useful.

Kidney stones are solid materials that form in the urinary tract. Most kidney stones are composed of calcium phosphate and calcium oxalate, although they can be solid uric acid. Insufficient water intake and high levels of calcium, oxalate, and phosphate in the urine can lead to the formation of kidney stones. When a kidney stone passes through the urinary tract, it causes considerable pain and discomfort, necessitating the use of painkillers and surgery. Sometimes ultrasound is used to break up kidney stones. Persons prone to kidney stones are advised to drink six to eight glasses of water every day to prevent saturation levels of minerals in the urine.

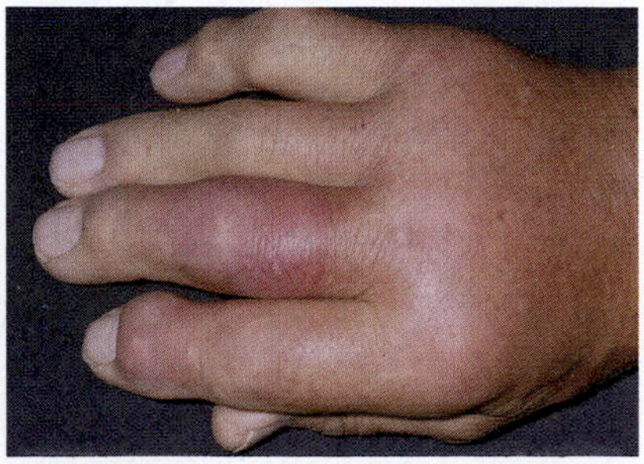

Gout occurs when uric acid exceeds its solubility in blood plasma.

Kidney stones form when calcium phosphate exceeds its solubility.

Effect of Temperature on Solubility

The solubility of most solids is greater as temperature increases, which means that solutions usually contain more dissolved solute at higher temperature. A few substances show little change in solubility at higher temperatures, and a few are less soluble (see **FIGURE 9.5**). For example, when you add sugar to iced tea, some undissolved sugar may form on the bottom of the glass. But if you add sugar to hot tea, many teaspoons of sugar are needed before solid sugar appears. Hot tea dissolves more sugar than does cold tea because the solubility of sugar is much greater at a higher temperature.

When a saturated solution is carefully cooled, it becomes a *supersaturated solution* because it contains more solute than the solubility allows. Such a solution is unstable, and if the solution is agitated or if a solute crystal is added, the excess solute will recrystallize to give a saturated solution again.

Conversely, the solubility of a gas in water decreases as the temperature increases. At higher temperatures, more gas molecules have the energy to escape from the solution. Perhaps you have observed the bubbles escaping from a cold carbonated soft drink as it warms. At high temperatures, bottles containing carbonated solutions may burst as more gas molecules leave the solution and increase the gas pressure inside the bottle. Biologists have found that increased temperatures in rivers and lakes cause the amount of dissolved oxygen to decrease until the warm water can no longer support a biological community. Electricity-generating plants are required to have their own ponds to use with their cooling towers to lessen the threat of thermal pollution in surrounding waterways.

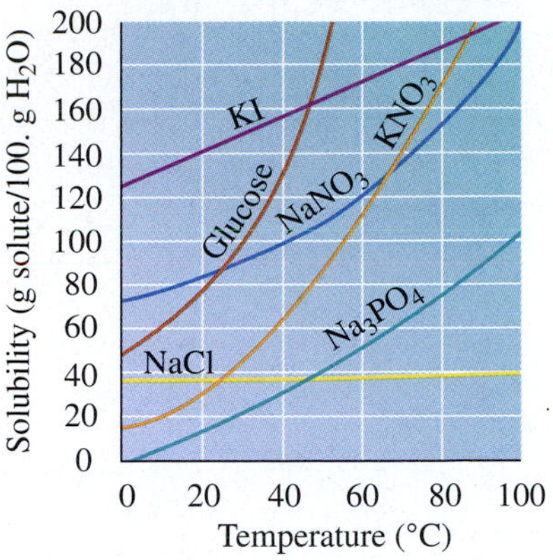
Solubility (g solute/100. g H_2O)
200
180
160
140
120
100
80
60
40
20
0
KI
Glucose
$NaNO_3$
KNO_3
Na_3PO_4
NaCl
0
20
40
60
80
100
Temperature (°C)

FIGURE 9.5 ▶ In water, most common solids are more soluble as the temperature increases.

Q Compare the solubility of $NaNO_3$ at 20 °C and 60 °C.

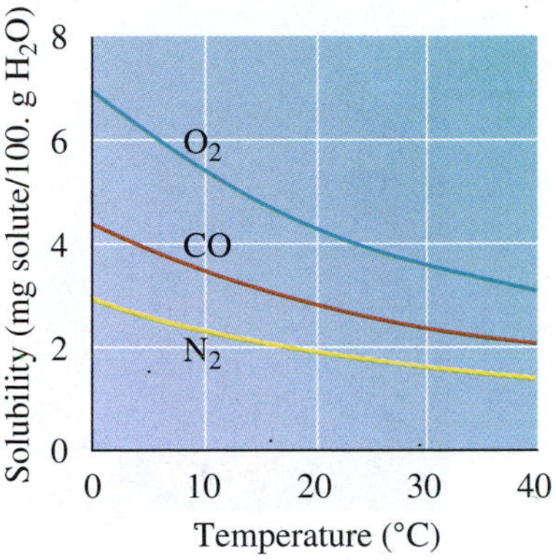
Solubility (mg solute/100. g H_2O)
8
6
4
2
0
O_2
CO
N_2
0
10
20
30
40
Temperature (°C)

In water, gases are less soluble as the temperature increases.

Henry's Law

Henry's law states that the solubility of gas in a liquid is directly related to the pressure of that gas above the liquid. At higher pressures, there are more gas molecules available to enter and dissolve in the liquid. A can of soda is carbonated by using CO_2 gas at high pressure to increase the solubility of the CO_2 in the beverage. When you open the can at atmospheric pressure, the pressure on the CO_2 drops, which decreases the solubility of CO_2. As a result, bubbles of CO_2 rapidly escape from the solution. The burst of bubbles is even more noticeable when you open a warm can of soda.

TEST

Try Practice Problems 9.25 to 9.28

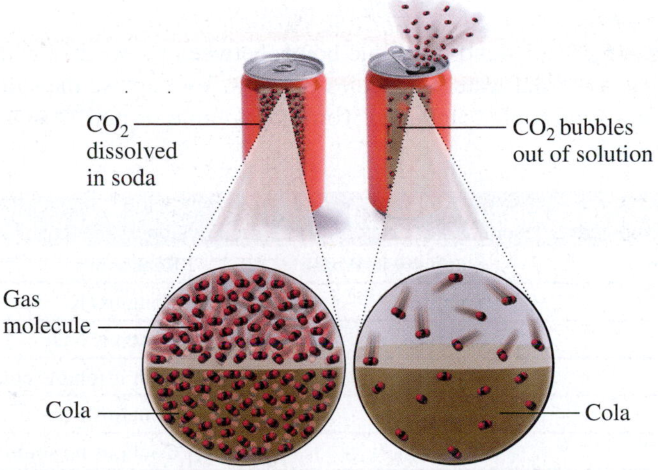
CO_2 dissolved in soda
CO_2 bubbles out of solution
Gas molecule
Cola
Cola

When the pressure of a gas above a solution decreases, the solubility of that gas in the solution also decreases.

Soluble and Insoluble Ionic Compounds

In our discussion up to now, we have considered ionic compounds that dissolve in water. However, some ionic compounds do not dissociate into ions and remain as solids even in contact with water. The **solubility rules** give some guidelines about the solubility of ionic compounds in water.

Ionic compounds that are soluble in water typically contain at least one of the ions in **TABLE 9.7**. *Only an ionic compound containing a soluble cation or anion will dissolve in water.* Most ionic compounds containing Cl^- are soluble, but AgCl, $PbCl_2$, and Hg_2Cl_2 are insoluble. Similarly, most ionic compounds containing SO_4^{2-} are soluble, but a few are insoluble. Most other ionic compounds are insoluble (see **FIGURE 9.6**).

TABLE 9.7 Solubility Rules for Ionic Compounds in Water

An ionic compound is soluble in water if it contains one of the following:	
Positive Ions	Li^+, Na^+, K^+, Rb^+, Cs^+, NH_4^+
Negative Ions	NO_3^-, $C_2H_3O_2^-$ Cl^-, Br^-, I^- except when combined with Ag^+, Pb^{2+}, or Hg_2^{2+} SO_4^{2-} except when combined with Ba^{2+}, Pb^{2+}, Ca^{2+}, Sr^{2+}, or Hg_2^{2+}
Ionic compounds that do not contain at least one of these ions are usually insoluble.	

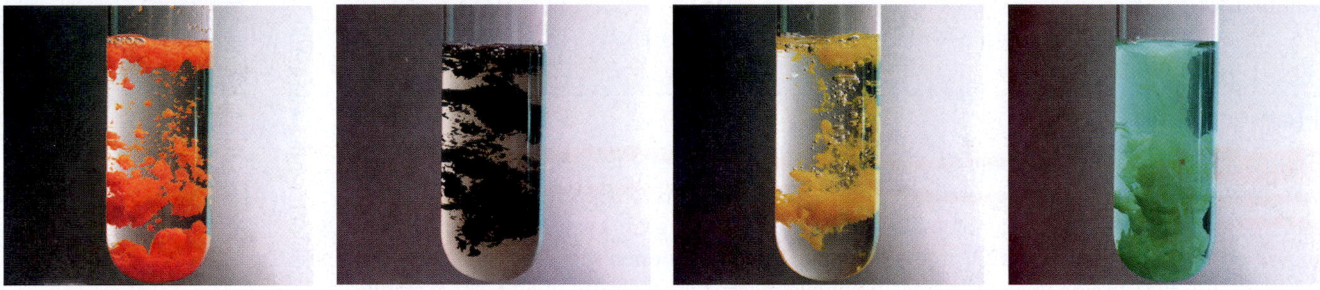

CdS   FeS PbI_2 $Ni(OH)_2$

FIGURE 9.6 ▶ If an ionic compound contains a combination of a cation and an anion that are not soluble, that ionic compound is insoluble. For example, combinations of cadmium and sulfide, iron and sulfide, lead and iodide, and nickel and hydroxide do not contain any soluble ions. Thus, they form insoluble ionic compounds.

Q Why are each of these ionic compounds insoluble in water?

CORE CHEMISTRY SKILL
Using Solubility Rules

In an insoluble ionic compound, the ionic bonds between its positive and negative ions are too strong for the polar water molecules to break. We can use the solubility rules to predict whether a solid ionic compound would be soluble or not. **TABLE 9.8** illustrates the use of these rules.

ENGAGE
Why is K_2S soluble in water whereas $PbCl_2$ is not soluble?

TABLE 9.8 Using Solubility Rules

Ionic Compound	Solubility in Water	Reasoning
K_2S	Soluble	Contains K^+
$Ca(NO_3)_2$	Soluble	Contains NO_3^-
$PbCl_2$	Insoluble	Is an insoluble chloride
NaOH	Soluble	Contains Na^+
$AlPO_4$	Insoluble	Contains no soluble ions

In medicine, insoluble $BaSO_4$ is used as an opaque substance to enhance X-rays of the gastrointestinal tract (see **FIGURE 9.7**). $BaSO_4$ is so insoluble that it does not dissolve in gastric fluids. Other ionic barium compounds cannot be used because they would dissolve in water, releasing Ba^{2+}, which is poisonous.

▶ SAMPLE PROBLEM 9.4 Soluble and Insoluble Ionic Compounds

TRY IT FIRST

Predict whether each of the following ionic compounds is soluble in water or not and explain your answer:

a. Na_3PO_4 **b.** $CaCO_3$

SOLUTION

a. The ionic compound Na_3PO_4 is soluble in water because any compound that contains Na^+ is soluble.

b. The ionic compound $CaCO_3$ is not soluble because it does not contain a soluble positive or negative ion.

STUDY CHECK 9.4

a. In some electrolyte drinks, $MgCl_2$ is added to provide magnesium. Why would you expect $MgCl_2$ to be soluble in water?

b. Even though barium ions are toxic if ingested, why can barium sulfate be given to a patient for an X-ray of the lower gastrointestinal tract?

ANSWER

a. $MgCl_2$ is soluble in water because ionic compounds that contain chloride are soluble unless they contain Ag^+, Pb^{2+}, or Hg_2^{2+}.

b. Barium sulfate is insoluble in water; it does not produce barium ions as it passes through the GI tract during an X-ray.

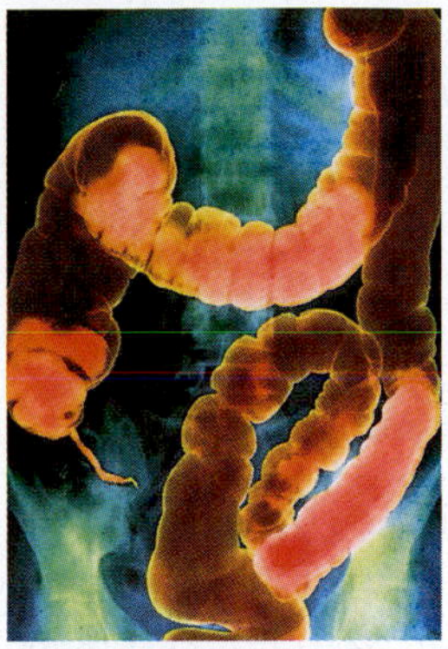

FIGURE 9.7 ▶ A barium sulfate-enhanced X-ray of the abdomen shows the lower gastrointestinal (GI) tract.

Q Is $BaSO_4$ a soluble or an insoluble substance?

TEST

Try Practice Problems 9.29 and 9.30

Formation of a Solid

We can use solubility rules to predict whether a solid, called a *precipitate*, forms when two solutions containing soluble reactants are mixed, as shown in Sample Problem 9.5.

▶ SAMPLE PROBLEM 9.5 Writing Equations for the Formation of an Insoluble Ionic Compound

TRY IT FIRST

When solutions of NaCl and $AgNO_3$ are mixed, a white solid forms. Write the ionic and net ionic equations for the reaction.

SOLUTION

STEP 1 **Write the ions of the reactants.**

Reactants
(initial combinations)

$Ag^+(aq) + NO_3^-(aq)$

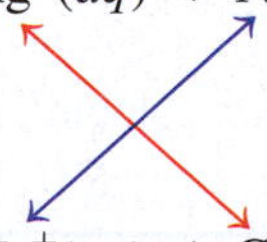

$Na^+(aq) + Cl^-(aq)$

STEP 2 **Write the combinations of ions, and determine if any are insoluble.**

When we look at the ions of each solution, we see that the combination of Ag^+ and Cl^- forms an insoluble ionic compound.

Mixture (new combinations)	Product	Soluble
$Ag^+(aq) + Cl^-(aq)$	AgCl	No
$Na^+(aq) + NO_3^-(aq)$	$NaNO_3$	Yes

STEP 3 **Write the ionic equation including any solid.** In the *ionic equation*, we show all the ions of the reactants. The products include the solid AgCl that forms along with the remaining ions Na^+ and NO_3^-.

$$Ag^+(aq) + NO_3^-(aq) + Na^+(aq) + Cl^-(aq) \longrightarrow AgCl(s) + Na^+(aq) + NO_3^-(aq)$$

STEP 4 **Write the net ionic equation.** We remove the Na^+ and NO_3^- ions, known as *spectator ions*, which are unchanged. This gives the *net ionic equation*, which only shows the ions that form a solid precipitate.

ENGAGE

When mixing solutions of $Pb(NO_3)_2$ and NaBr, how do we know that $PbBr_2(s)$ is the solid that forms?

$$Ag^+(aq) + \underbrace{\cancel{NO_3^-(aq)} + \cancel{Na^+(aq)}}_{\text{Spectator ions}} + Cl^-(aq) \longrightarrow AgCl(s) + \underbrace{\cancel{Na^+(aq)} + \cancel{NO_3^-(aq)}}_{\text{Spectator ions}}$$

$$Ag^+(aq) + Cl^-(aq) \longrightarrow AgCl(s) \quad \text{Net ionic equation}$$

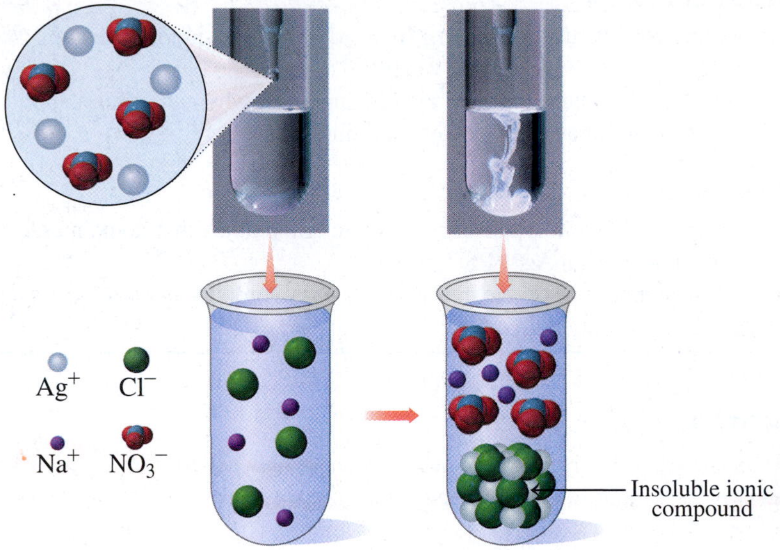
Ag+
Cl−
Na+
NO3−
Insoluble ionic compound

Type of Equation	
Chemical	$AgNO_3(aq) + NaCl(aq) \longrightarrow AgCl(s) + NaNO_3(aq)$
Ionic	$Ag^+(aq) + NO_3^-(aq) + Na^+(aq) + Cl^-(aq) \rightarrow AgCl(s) + Na^+(aq) + NO_3^-(aq)$
Net Ionic	$Ag^+(aq) + Cl^-(aq) \longrightarrow AgCl(s)$

STUDY CHECK 9.5

Predict whether a solid might form in each of the following mixtures of solutions. If so, write the net ionic equation for the reaction.

a. $NH_4Cl(aq) + Ca(NO_3)_2(aq)$ **b.** $Pb(NO_3)_2(aq) + KCl(aq)$

ANSWER

a. No solid forms because the products, $NH_4NO_3(aq)$ and $CaCl_2(aq)$, are soluble.
b. $Pb^{2+}(aq) + 2Cl^-(aq) \longrightarrow PbCl_2(s)$

TEST

Try Practice Problems 9.31 and 9.32

PRACTICE PROBLEMS

9.3 Solubility

9.21 State whether each of the following refers to a saturated or an unsaturated solution:
a. A crystal added to a solution does not change in size.
b. A sugar cube completely dissolves when added to a cup of coffee.
c. A uric acid concentration of 4.6 mg/100 mL in the kidney does not cause gout.

9.22 State whether each of the following refers to a saturated or an unsaturated solution:
a. A spoonful of salt added to boiling water dissolves.
b. A layer of sugar forms on the bottom of a glass of tea as ice is added.
c. A kidney stone of calcium phosphate forms in the kidneys when urine becomes concentrated.

Use the following table for problems 9.23 to 9.26:

	Solubility (g/100. g H_2O)	
Substance	20 °C	50 °C
KCl	34	43
$NaNO_3$	88	110
$C_{12}H_{22}O_{11}$ (sugar)	204	260

9.23 Determine whether each of the following solutions will be saturated or unsaturated at 20 °C:
a. adding 25 g of KCl to 100. g of H_2O
b. adding 11 g of $NaNO_3$ to 25 g of H_2O
c. adding 400. g of sugar to 125 g of H_2O

9.24 Determine whether each of the following solutions will be saturated or unsaturated at 50 °C:
a. adding 25 g of KCl to 50. g of H_2O
b. adding 150. g of $NaNO_3$ to 75 g of H_2O
c. adding 80. g of sugar to 25 g of H_2O

9.25 A solution containing 80. g of KCl in 200. g of H_2O at 50 °C is cooled to 20 °C.
a. How many grams of KCl remain in solution at 20 °C?
b. How many grams of solid KCl crystallized after cooling?

9.26 A solution containing 80. g of $NaNO_3$ in 75 g of H_2O at 50 °C is cooled to 20 °C.
a. How many grams of $NaNO_3$ remain in solution at 20 °C?
b. How many grams of solid $NaNO_3$ crystallized after cooling?

9.27 Explain the following observations:
a. More sugar dissolves in hot tea than in iced tea.
b. Champagne in a warm room goes flat.
c. A warm can of soda has more spray when opened than a cold one.

9.28 Explain the following observations:
a. An open can of soda loses its "fizz" faster at room temperature than in the refrigerator.
b. Chlorine gas in tap water escapes as the sample warms to room temperature.
c. Less sugar dissolves in iced coffee than in hot coffee.

9.29 Predict whether each of the following ionic compounds is soluble in water:
a. LiCl
b. PbS
c. $BaCO_3$
d. K_2O
e. $Fe(NO_3)_3$

9.30 Predict whether each of the following ionic compounds is soluble in water:
a. AgCl
b. KI
c. Na_2S
d. Ag_2O
e. $CaSO_4$

9.31 Determine whether a solid forms when solutions containing the following ionic compounds are mixed. If so, write the ionic equation and the net ionic equation.
a. $KCl(aq)$ and $Na_2S(aq)$
b. $AgNO_3(aq)$ and $K_2S(aq)$
c. $CaCl_2(aq)$ and $Na_2SO_4(aq)$
d. $CuCl_2(aq)$ and $Li_3PO_4(aq)$

9.32 Determine whether a solid forms when solutions containing the following ionic compounds are mixed. If so, write the ionic equation and the net ionic equation.
a. $Na_3PO_4(aq)$ and $AgNO_3(aq)$
b. $K_2SO_4(aq)$ and $Na_2CO_3(aq)$
c. $Pb(NO_3)_2(aq)$ and $Na_2CO_3(aq)$
d. $BaCl_2(aq)$ and $KOH(aq)$

9.4 Solution Concentrations and Reactions

LEARNING GOAL Calculate the concentration of a solute in a solution; use concentration units to calculate the amount of solute or solution. Given the volume and concentration of a solution, calculate the amount of another reactant or product in a reaction.

REVIEW
Calculating Percentages (1.4)
Using Molar Mass as a Conversion Factor (7.6)
Using Mole–Mole Factors (7.7)

The amount of solute dissolved in a certain amount of solution is called the **concentration of the solution.** We will look at ways to express a concentration as a ratio of a certain amount of solute in a given amount of solution as shown in **TABLE 9.9**. The amount of a solute may be expressed in units of grams, milliliters, or moles. The amount of a solution may be expressed in units of grams, milliliters, or liters.

CORE CHEMISTRY SKILL
Calculating Concentration

$$\text{Concentration of a solution} = \frac{\text{amount of solute}}{\text{amount of solution}}$$

TABLE 9.9 Summary of Types of Concentration Expressions and Their Units

	Mass Percent (m/m)	Volume Percent (v/v)	Mass/Volume Percent (m/v)	Molarity (M)
Solute Unit	g	mL	g	mole
Solvent Unit	g	mL	mL	L

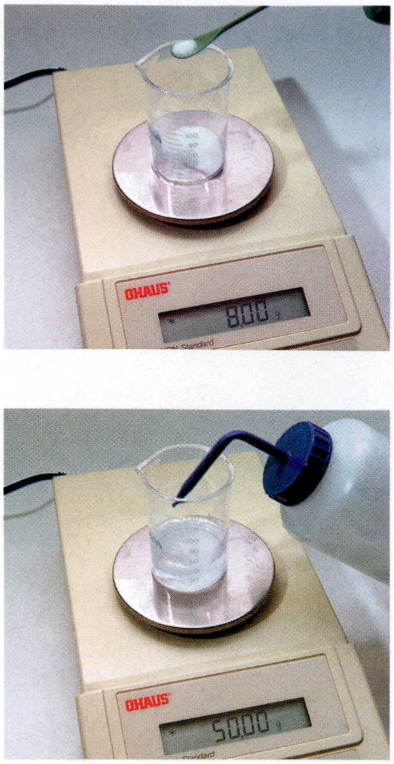
OHAUS
8.00
OHAUS
50.00

When water is added to 8.00 g of KCl to form 50.00 g of a KCl solution, the mass percent concentration is 16.0% (m/m).

Mass Percent (m/m) Concentration

Mass percent (m/m) describes the mass of the solute in grams for 100. g of solution. The mass percent is calculated by dividing the mass of a solute by the mass of the solution multiplied by 100% to give the percentage. In the calculation of mass percent (m/m), the units of mass of the solute and solution must be the same. If the mass of the solute is given as grams, then the mass of the solution must also be grams. The mass of the solution is the sum of the mass of the solute and the mass of the solvent.

$$\text{Mass percent (m/m)} = \frac{\text{mass of solute (g)}}{\text{mass of solute (g)} + \text{mass of solvent (g)}} \times 100\%$$
$$= \frac{\text{mass of solute (g)}}{\text{mass of solution (g)}} \times 100\%$$

Suppose we prepared a solution by mixing 8.00 g of KCl (solute) with 42.00 g of water (solvent). Together, the mass of the solute and mass of solvent give the mass of the solution (8.00 g + 42.00 g = 50.00 g). Mass percent is calculated by substituting the mass of the solute and the mass of the solution into the mass percent expression.

$$\frac{8.00 \text{ g KCl}}{50.00 \text{ g solution}} \times 100\% = 16.0\% \text{ (m/m) KCl solution}$$

$$\underbrace{8.00 \text{ g KCl}}_{\text{Solute}} + \underbrace{42.00 \text{ g } H_2O}_{\text{Solvent}}$$

▶SAMPLE PROBLEM 9.6 Calculating Mass Percent (m/m) Concentration

TRY IT FIRST

What is the mass percent of NaOH in a solution prepared by dissolving 30.0 g of NaOH in 120.0 g of H_2O?

SOLUTION

STEP 1 **State the given and needed quantities.**

ANALYZE THE PROBLEM	Given	Need	Connect
	30.0 g of NaOH, 120.0 g of H_2O	mass percent (m/m)	$\frac{\text{mass of solute}}{\text{mass of solution}} \times 100\%$

STEP 2 **Write the concentration expression.**

$$\text{Mass percent (m/m)} = \frac{\text{grams of solute}}{\text{grams of solution}} \times 100\%$$

STEP 3 **Substitute solute and solution quantities into the expression and calculate.** The mass of the solution is obtained by adding the mass of the solute and the mass of the solution.

$$\text{mass of solution} = 30.0 \text{ g NaOH} + 120.0 \text{ g } H_2O = 150.0 \text{ g of NaOH solution}$$

$$\text{Mass percent (m/m)} = \frac{\overset{\text{Three SFs}}{30.0 \text{ g NaOH}}}{\underset{\text{Four SFs}}{150.0 \text{ g solution}}} \times 100\%$$

$$= \underset{\text{Three SFs}}{20.0\%} \text{ (m/m) NaOH solution}$$

STUDY CHECK 9.6

a. What is the mass percent (m/m) of NaCl in a solution made by dissolving 2.0 g of NaCl in 56.0 g of H_2O?

b. What is the mass percent (m/m) of $MgCl_2$ in a solution prepared by dissolving 1.8 g of $MgCl_2$ in 18.5 g of H_2O?

ANSWER

a. 3.4% (m/m) NaCl solution **b.** 8.9% (m/m) $MgCl_2$ solution

TEST

Try Practice Problems 9.33 and 9.34

Using Mass Percent Concentration as a Conversion Factor

CORE CHEMISTRY SKILL
Using Concentration as a Conversion Factor

In the preparation of solutions, we often need to calculate the amount of solute or solution. Then the concentration of a solution is useful as a conversion factor as shown in Sample Problem 9.7.

▶ SAMPLE PROBLEM 9.7 Using Mass Percent to Calculate Mass of Solute

TRY IT FIRST

The topical antibiotic ointment Neosporin is 3.5% (m/m) neomycin solution. How many grams of neomycin are in a tube containing 64 g of ointment?

SOLUTION

STEP 1 **State the given and needed quantities.**

	Given	Need	Connect
ANALYZE THE PROBLEM	64 g of 3.5% (m/m) neomycin solution	grams of neomycin	mass percent factor $\frac{\text{g of solute}}{\text{100. g of solution}}$

STEP 2 **Write a plan to calculate the mass.**

grams of ointment → % (m/m) factor → grams of neomycin

STEP 3 **Write equalities and conversion factors.** The mass percent (m/m) indicates the grams of a solute in every 100. g of a solution. The mass percent (3.5% m/m) can be written as two conversion factors.

$$3.5 \text{ g of neomycin} = 100. \text{ g of ointment}$$
$$\frac{3.5 \text{ g neomycin}}{100. \text{ g ointment}} \quad \text{and} \quad \frac{100. \text{ g ointment}}{3.5 \text{ g neomycin}}$$

STEP 4 **Set up the problem to calculate the mass.**

$$\underset{\text{Two SFs}}{64 \cancel{\text{g ointment}}} \times \underset{\text{Exact}}{\overset{\text{Two SFs}}{\frac{3.5 \text{ g neomycin}}{100. \cancel{\text{g ointment}}}}} = \underset{\text{Two SFs}}{2.2 \text{ g of neomycin}}$$

STUDY CHECK 9.7

a. Calculate the grams of KCl in 225 g of an 8.00% (m/m) KCl solution.
b. Calculate the grams of KNO_3 in 45.0 g of a 1.85% (m/m) KNO_3 solution.

ANSWER

a. 18.0 g of KCl **b.** 0.833 g of KNO_3

ENGAGE
How is the mass percent (m/m) of a solution used to convert the mass of the solution to the grams of solute?

Volume Percent (v/v) Concentration

Because the volumes of liquids or gases are easily measured, the concentrations of their solutions are often expressed as **volume percent (v/v)**. The units of volume used in the ratio must be the same, for example, both in milliliters or both in liters.

$$\text{Volume percent (v/v)} = \frac{\text{volume of solute}}{\text{volume of solution}} \times 100\%$$

We interpret a volume percent as the volume of solute in 100. mL of solution. On a bottle of extract of vanilla, a label that reads alcohol 35% (v/v) means 35 mL of alcohol solute in 100. mL of vanilla solution.

PURE
Vanilla
EXTRACT
INGREDIENTS: VANILLA BEAN EXTRACTIVES IN WATER, ALCOHOL (35%), AND CORN SYRUP.
The label indicates that vanilla extract contains 35% (v/v) alcohol.

LEMON EXTRACT
2 FL OZ
2 FL OZ (59mL)

Lemon extract is a solution of lemon flavor and alcohol.

▶ SAMPLE PROBLEM 9.8 Calculating Volume Percent (v/v) Concentration

TRY IT FIRST

A bottle contains 59 mL of lemon extract solution. If the extract contains 49 mL of alcohol, what is the volume percent (v/v) of the alcohol in the solution?

SOLUTION

STEP 1 State the given and needed quantities.

ANALYZE THE PROBLEM	Given	Need	Connect
	49 mL of alcohol, 59 mL of solution	volume percent (v/v)	$\frac{\text{volume of solute}}{\text{volume of solution}} \times 100\%$

STEP 2 Write the concentration expression.

$$\text{Volume percent (v/v)} = \frac{\text{volume of solute}}{\text{volume of solution}} \times 100\%$$

STEP 3 Substitute solute and solution quantities into the expression and calculate.

$$\text{Volume percent (v/v)} = \frac{\overset{\text{Two SFs}}{49 \text{ mL alcohol}}}{\underset{\text{Two SFs}}{59 \text{ mL solution}}} \times 100\% = \underset{\text{Two SFs}}{83\%} \text{ (v/v) alcohol solution}$$

STUDY CHECK 9.8

a. What is the volume percent (v/v) of Br_2 in a solution prepared by dissolving 12 mL of liquid bromine (Br_2) in the solvent carbon tetrachloride (CCl_4) to make 250 mL of solution?

b. What is the volume percent (v/v) of a "rubbing alcohol" solution that contains 14 mL of isopropyl alcohol in 20. mL of solution?

ANSWER

a. 4.8% (v/v) Br_2 in CCl_4 **b.** 70.% (v/v)

TEST

Try Practice Problems 9.35 and 9.36

Mass/Volume Percent (m/v) Concentration

Mass/volume percent (m/v) describes the mass of the solute in grams for 100. mL of solution. In the calculation of mass/volume percent, the unit of mass of the solute is grams, and the unit of the solution volume is milliliters.

$$\text{Mass/volume percent (m/v)} = \frac{\text{grams of solute}}{\text{milliliters of solution}} \times 100\%$$

The mass/volume percent is widely used in hospitals and pharmacies for the preparation of intravenous solutions and medicines. For example, a 5% (m/v) glucose solution contains 5 g of glucose in 100. mL of solution. The volume of solution represents the combined volumes of the glucose and H_2O.

ENGAGE

What units should be used for a 3.0% (m/v) solution?

▶ SAMPLE PROBLEM 9.9 Calculating Mass/Volume Percent (m/v) Concentration

TRY IT FIRST

A potassium iodide solution may be used in a diet that is low in iodine. A KI solution is prepared by dissolving 5.0 g of KI in enough water to give a final volume of 250 mL. What is the mass/volume percent (m/v) of KI in the solution?

SOLUTION

STEP 1 State the given and needed quantities.

ANALYZE THE PROBLEM	Given	Need	Connect
	5.0 g of KI solute, 250 mL of KI solution	mass/volume percent (m/v)	$\frac{\text{mass of solute}}{\text{volume of solution}} \times 100\%$

STEP 2 **Write the concentration expression.**

$$\text{Mass/volume percent (m/v)} = \frac{\text{mass of solute}}{\text{volume of solution}} \times 100\%$$

STEP 3 **Substitute solute and solution quantities into the expression and calculate.**

$$\text{Mass/volume percent (m/v)} = \frac{\overset{\text{Two SFs}}{5.0\text{ g KI}}}{\underset{\text{Two SFs}}{250\text{ mL solution}}} \times 100\% = \underset{\text{Two SFs}}{2.0\%}\text{ (m/v) KI solution}$$

Water added to make a solution
250 mL
5.0 g of KI
250 mL of KI solution

STUDY CHECK 9.9

a. What is the mass/volume percent (m/v) of NaOH in a solution prepared by dissolving 12 g of NaOH in enough water to make 220 mL of solution?

b. What is the mass/volume percent (m/v) of $MgCl_2$ in a solution prepared by dissolving 3.25 g of $MgCl_2$ in enough water to make 125 mL of solution?

ANSWER

a. 5.5% (m/v) NaOH solution **b.** 2.60% (m/v) $MgCl_2$ solution

▶SAMPLE PROBLEM 9.10 Using Mass/Volume Percent to Calculate Mass of Solute

TRY IT FIRST

A topical antibiotic is 1.0% (m/v) clindamycin. How many grams of clindamycin are in 60. mL of the 1.0% (m/v) clindamycin solution?

SOLUTION

STEP 1 **State the given and needed quantities.**

	Given	Need	Connect
ANALYZE THE PROBLEM	60. mL of 1.0% (m/v) clindamycin solution	grams of clindamycin	% (m/v) factor

STEP 2 **Write a plan to calculate the mass.**

milliliters of solution → % (m/v) factor → grams of clindamycin

STEP 3 **Write equalities and conversion factors.** The percent (m/v) indicates the grams of a solute in every 100. mL of a solution. The 1.0% (m/v) can be written as two conversion factors.

1.0 g of clindamycin = 100. mL of solution

$$\frac{1.0\text{ g clindamycin}}{100.\text{ mL solution}} \quad\text{and}\quad \frac{100.\text{ mL solution}}{1.0\text{ g clindamycin}}$$

STEP 4 **Set up the problem to calculate the mass.** The volume of the solution is converted to mass of solute using the conversion factor that cancels mL.

$$\underset{\text{Two SFs}}{60.\ \cancel{\text{mL solution}}} \times \frac{\overset{\text{Two SFs}}{1.0\text{ g clindamycin}}}{\underset{\text{Exact}}{100.\ \cancel{\text{mL solution}}}} = \underset{\text{Two SFs}}{0.60}\text{ g of clindamycin}$$

STUDY CHECK 9.10

a. In 2010, the FDA approved a 2.0% (m/v) morphine oral solution to treat severe or chronic pain. How many grams of morphine does a patient receive if 0.60 mL of 2.0% (m/v) morphine solution was ordered?

b. For a skin infection, the doctor orders 300 mg of ampicillin. How many milliliters of 5.0% (m/v) ampicillin should be given?

ANSWER

a. 0.012 g of morphine **b.** 6 mL

TEST

Try Practice Problems 9.37 to 9.42, 9.53 to 9.60

Molarity (M) Concentration

When chemists work with solutions, they often use **molarity (M)**, a concentration that states the number of moles of solute in exactly 1 L of solution.

$$\text{Molarity (M)} = \frac{\text{moles of solute}}{\text{liters of solution}}$$

The molarity of a solution can be calculated when we know the moles of solute and the volume of solution in liters. For example, if 1.0 mole of NaCl were dissolved in enough water to prepare 1.0 L of solution, the resulting NaCl solution has a molarity of 1.0 M. The abbreviation M indicates the units of mole per liter (mole/L).

$$\text{M} = \frac{\text{moles of solute}}{\text{liters of solution}} = \frac{1.0 \text{ mole NaCl}}{1 \text{ L solution}} = 1.0 \text{ M NaCl solution}$$

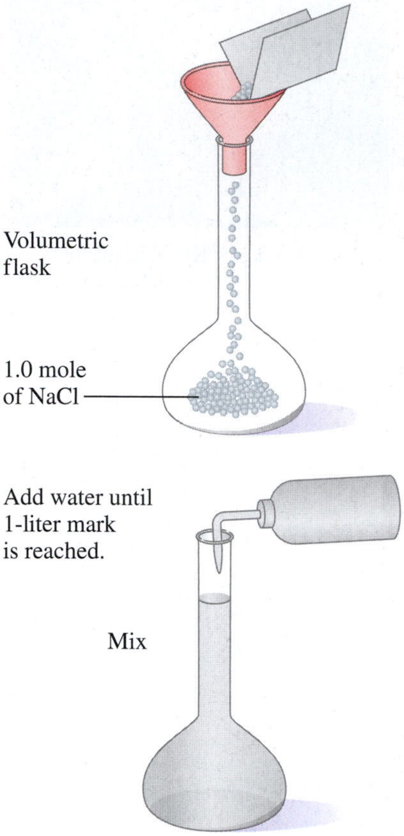
Volumetric flask
1.0 mole of NaCl
Add water until 1-liter mark is reached.
Mix

A 1.0 molar (M) NaCl solution

ENGAGE

Why does a solution containing 2.0 moles of KCl in 4.0 L of KCl solution have a molarity of 0.50 M?

▶SAMPLE PROBLEM 9.11 Calculating Molarity

TRY IT FIRST

What is the molarity (M) of 60.0 g of NaOH in 0.250 L of NaOH solution?

SOLUTION

STEP 1 State the given and needed quantities.

	Given	Need	Connect
ANALYZE THE PROBLEM	60.0 g of NaOH, 0.250 L of NaOH solution	molarity (mole/L)	molar mass of NaOH, $\frac{\text{moles of solute}}{\text{liters of solution}}$

To calculate the moles of NaOH, we need to write the equality and conversion factors for the molar mass of NaOH. Then the moles in 60.0 g of NaOH can be determined.

$$1 \text{ mole of NaOH} = 40.00 \text{ g of NaOH}$$

$$\frac{40.00 \text{ g NaOH}}{1 \text{ mole NaOH}} \quad \text{and} \quad \frac{1 \text{ mole NaOH}}{40.00 \text{ g NaOH}}$$

$$\text{moles of NaOH} = 60.0 \cancel{\text{g NaOH}} \times \frac{1 \text{ mole NaOH}}{40.00 \cancel{\text{g NaOH}}}$$

$$= 1.50 \text{ moles of NaOH}$$

$$\text{volume of solution} = 0.250 \text{ L of NaOH solution}$$

STEP 2 Write the concentration expression.

$$\text{Molarity (M)} = \frac{\text{moles of solute}}{\text{liters of solution}}$$

STEP 3 Substitute solute and solution quantities into the expression and calculate.

$$\text{M} = \frac{\overset{\text{Three SFs}}{1.50 \text{ moles NaOH}}}{\underset{\text{Three SFs}}{0.250 \text{ L solution}}} = \frac{\overset{\text{Three SFs}}{6.00 \text{ moles NaOH}}}{\underset{\text{Exact}}{1 \text{ L solution}}} = \underset{\text{Three SFs}}{6.00 \text{ M NaOH solution}}$$

STUDY CHECK 9.11

a. What is the molarity of a solution that contains 75.0 g of KNO_3 dissolved in 0.350 L of solution?

b. What is the molarity of a mouthwash solution that contains 6.40 g of the disinfectant thymol, $C_{10}H_{14}O$, in 2.00 L of solution?

ANSWER

a. 2.12 M KNO_3 solution

b. 0.0213 M thymol solution

▶ SAMPLE PROBLEM 9.12 Using Molarity to Calculate Volume of Solution

TRY IT FIRST

How many liters of a 2.00 M NaCl solution are needed to provide 67.3 g of NaCl?

SOLUTION

STEP 1 State the given and needed quantities.

ANALYZE THE PROBLEM	Given	Need	Connect
	67.3 g of NaCl, 2.00 M NaCl solution	liters of NaCl solution	molar mass of NaCl, molarity

STEP 2 Write a plan to calculate the volume.

grams of NaCl → Molar mass → moles of NaCl → Molarity → liters of NaCl solution

STEP 3 Write equalities and conversion factors.

1 mole of NaCl = 58.44 g of NaCl

$$\frac{58.44 \text{ g NaCl}}{1 \text{ mole NaCl}} \text{ and } \frac{1 \text{ mole NaCl}}{58.44 \text{ g NaCl}}$$

1 L of NaCl solution = 2.00 moles of NaCl

$$\frac{2.00 \text{ moles NaCl}}{1 \text{ L NaCl solution}} \text{ and } \frac{1 \text{ L NaCl solution}}{2.00 \text{ moles NaCl}}$$

STEP 4 Set up the problem to calculate the volume.

$$67.3 \text{ g NaCl} \times \frac{1 \text{ mole NaCl}}{58.44 \text{ g NaCl}} \times \frac{1 \text{ L NaCl solution}}{2.00 \text{ moles NaCl}} = 0.576 \text{ L of NaCl solution}$$

Three SFs; Exact / Four SFs; Exact / Three SFs; Three SFs

STUDY CHECK 9.12

a. How many milliliters of a 6.0 M HCl solution will provide 164 g of HCl?

b. What volume, in milliliters, of a 2.50 M NaOH solution contains 12.5 g of NaOH?

ANSWER

a. 750 mL of HCl solution

b. 125 mL of NaOH solution

TEST

Try Practice Problems 9.43 to 9.48

A summary of percent concentrations and molarity, their meanings, and conversion factors is given in **TABLE 9.10**.

ENGAGE

What conversion factors are provided by the concentration of a 12% (v/v) solution?

TABLE 9.10 Conversion Factors from Concentrations

Percent Concentration	Meaning	Conversion Factors
10% (m/m) KCl solution	10 g of KCl in 100. g of KCl solution	$\frac{10\text{ g KCl}}{100.\text{ g solution}}$ and $\frac{100.\text{ g solution}}{10\text{ g KCl}}$
12% (v/v) ethanol solution	12 mL of ethanol in 100. mL of ethanol solution	$\frac{12\text{ mL ethanol}}{100.\text{ mL solution}}$ and $\frac{100.\text{ mL solution}}{12\text{ mL ethanol}}$
5% (m/v) glucose solution	5 g of glucose in 100. mL of glucose solution	$\frac{5\text{ g glucose}}{100.\text{ mL solution}}$ and $\frac{100.\text{ mL solution}}{5\text{ g glucose}}$
Molarity	**Meaning**	**Conversion Factors**
6.0 M HCl solution	6.0 moles of HCl in 1 L of HCl solution	$\frac{6.0\text{ moles HCl}}{1\text{ L solution}}$ and $\frac{1\text{ L solution}}{6.0\text{ moles HCl}}$

INTERACTIVE VIDEO

Solutions

PEARSON eText 2.0

CORE CHEMISTRY SKILL

Calculating the Quantity of a Reactant or Product for a Chemical Reaction in Solution

Concentrations in Chemical Reactions in Solution

When chemical reactions involve aqueous solutions, we use their concentrations along with the balanced chemical equation to determine the concentrations or quantities of other reactants or products. For example, we can calculate the volume of a solution from the molarity and volume of the other reactant as shown in Sample Problem 9.13.

When a $BaCl_2$ solution is added to a Na_2SO_4 solution, a white solid of $BaSO_4$ forms.

▶ SAMPLE PROBLEM 9.13 Calculations Involving Solutions in Reactions

TRY IT FIRST

How many milliliters of a 0.250 M $BaCl_2$ solution are needed to react with 0.0325 L of a 0.160 M Na_2SO_4 solution?

$$Na_2SO_4(aq) + BaCl_2(aq) \longrightarrow BaSO_4(s) + 2NaCl(aq)$$

SOLUTION

STEP 1 State the given and needed quantities.

	Given	Need	Connect
ANALYZE THE PROBLEM	0.0325 L of 0.160 M Na_2SO_4 solution, 0.250 M $BaCl_2$ solution	milliliters of $BaCl_2$ solution	mole–mole factor
	Equation		
	$Na_2SO_4(aq) + BaCl_2(aq) \longrightarrow BaSO_4(s) + 2NaCl(aq)$		

STEP 2 Write a plan to calculate the needed quantity.

liters of Na_2SO_4 solution → Molarity → moles of Na_2SO_4 → Mole–mole factor → moles of $BaCl_2$

→ Molarity → liters of $BaCl_2$ solution → Metric factor → milliliters of $BaCl_2$ solution

STEP 3 Write equalities and conversion factors including mole–mole and concentration factors.

1 L of solution = 0.160 mole of Na_2SO_4

$$\frac{0.160\text{ mole }Na_2SO_4}{1\text{ L solution}} \text{ and } \frac{1\text{ L solution}}{0.160\text{ mole }Na_2SO_4}$$

1 mole of Na_2SO_4 = 1 mole of $BaCl_2$

$$\frac{1\text{ mole }Na_2SO_4}{1\text{ mole }BaCl_2} \text{ and } \frac{1\text{ mole }BaCl_2}{1\text{ mole }Na_2SO_4}$$

1 L of solution = 0.250 mole of $BaCl_2$

$$\frac{0.250\text{ mole }BaCl_2}{1\text{ L solution}} \text{ and } \frac{1\text{ L solution}}{0.250\text{ mole }BaCl_2}$$

1 L = 1000 mL

$$\frac{1000\text{ mL}}{1\text{ L}} \text{ and } \frac{1\text{ L}}{1000\text{ mL}}$$

STEP 4 **Set up the problem to calculate the needed quantity.**

$$0.0325\ \cancel{\text{L solution}} \times \frac{0.160\ \cancel{\text{mole } Na_2SO_4}}{1\ \cancel{\text{L solution}}} \times \frac{1\ \cancel{\text{mole } BaCl_2}}{1\ \cancel{\text{mole } Na_2SO_4}} \times \frac{1\ \cancel{\text{L solution}}}{0.250\ \cancel{\text{mole } BaCl_2}} \times \frac{1000\ \text{mL } BaCl_2 \text{ solution}}{1\ \cancel{\text{L solution}}}$$

$$= 20.8\ \text{mL of } BaCl_2 \text{ solution}$$

STUDY CHECK 9.13

a. For the reaction in Sample Problem 9.13, how many milliliters of a 0.330 M Na_2SO_4 solution are needed to react with 26.8 mL of a 0.216 M $BaCl_2$ solution?

b. For the reaction in Sample Problem 9.13, what is the molarity of a $BaCl_2$ solution if 19.4 mL of the $BaCl_2$ solution is needed to react with 15.0 mL of a 0.225 M Na_2SO_4 solution?

INTERACTIVE VIDEO

PEARSON eText 2.0

Calculations Involving Solutions in Reactions

ANSWER

a. 17.5 mL of Na_2SO_4 solution

b. 0.174 M $BaCl_2$ solution

TEST

Try Practice Problems 9.49 to 9.52

FIGURE 9.8 gives a summary of the pathways and conversion factors needed for substances including solutions involved in chemical reactions.

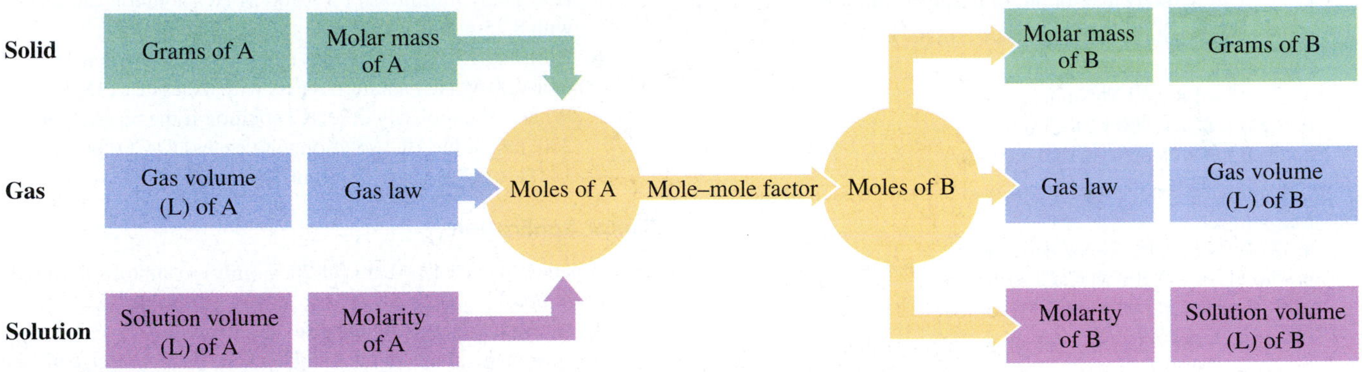
Solid
Grams of A
Molar mass of A
Gas
Gas volume (L) of A
Gas law
Solution
Solution volume (L) of A
Molarity of A
Moles of A
Mole–mole factor
Moles of B
Molar mass of B
Grams of B
Gas law
Gas volume (L) of B
Molarity of B
Solution volume (L) of B

FIGURE 9.8 ▶ In calculations involving chemical reactions, substance A is converted to moles of A using molar mass (if solid), gas laws (if gas), or molarity (if solution). Then moles of A are converted to moles of substance B, which are converted to grams of solid, liters of gas, or liters of solution, as needed.

Q What sequence of conversion factors would you use to calculate the number of grams of $CaCO_3$ needed to react with 1.50 L of a 2.00 M HCl solution in the reaction $2HCl(aq) + CaCO_3(s) \longrightarrow CO_2(g) + H_2O(l) + CaCl_2(aq)$?

PRACTICE PROBLEMS

9.4 Solution Concentrations and Reactions

9.33 Calculate the mass percent (m/m) for the solute in each of the following:
- **a.** 25 g of KCl and 125 g of H_2O
- **b.** 12 g of sucrose in 225 g of tea solution
- **c.** 8.0 g of $CaCl_2$ in 80.0 g of $CaCl_2$ solution

9.34 Calculate the mass percent (m/m) for the solute in each of the following:
- **a.** 75 g of NaOH in 325 g of NaOH solution
- **b.** 2.0 g of KOH and 20.0 g of H_2O
- **c.** 48.5 g of Na_2CO_3 in 250.0 g of Na_2CO_3 solution

9.35 A mouthwash contains 22.5% (v/v) alcohol. If the bottle of mouthwash contains 355 mL, what is the volume, in milliliters, of alcohol?

9.36 A bottle of champagne is 11% (v/v) alcohol. If there are 750 mL of champagne in the bottle, what is the volume, in milliliters, of alcohol?

9.37 Calculate the mass/volume percent (m/v) for the solute in each of the following:
- **a.** 75 g of Na_2SO_4 in 250 mL of Na_2SO_4 solution
- **b.** 39 g of sucrose in 355 mL of a carbonated drink

9.38 Calculate the mass/volume percent (m/v) for the solute in each of the following:
- **a.** 2.50 g of LiCl in 40.0 mL of LiCl solution
- **b.** 7.5 g of casein in 120 mL of low-fat milk

9.39 Calculate the grams or milliliters of solute needed to prepare the following:
- **a.** 50. g of a 5.0% (m/m) KCl solution
- **b.** 1250 mL of a 4.0% (m/v) NH_4Cl solution
- **c.** 250. mL of a 10.0% (v/v) acetic acid solution

9.40 Calculate the grams or milliliters of solute needed to prepare the following:
- **a.** 150. g of a 40.0% (m/m) $LiNO_3$ solution
- **b.** 450 mL of a 2.0% (m/v) KOH solution
- **c.** 225 mL of a 15% (v/v) isopropyl alcohol solution

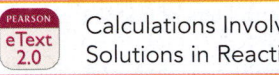

9.41 For each of the following solutions, calculate the:
a. grams of a 25% (m/m) $LiNO_3$ solution that contains 5.0 g of $LiNO_3$
b. milliliters of a 10.0% (m/v) KOH solution that contains 40.0 g of KOH
c. milliliters of a 10.0% (v/v) formic acid solution that contains 2.0 mL of formic acid

9.42 For each of the following solutions, calculate the:
a. grams of a 2.0% (m/m) NaCl solution that contains 7.50 g of NaCl
b. milliliters of a 25% (m/v) NaF solution that contains 4.0 g of NaF
c. milliliters of an 8.0% (v/v) ethanol solution that contains 20.0 mL of ethanol

9.43 Calculate the molarity of each of the following:
a. 2.00 moles of glucose in 4.00 L of a glucose solution
b. 4.00 g of KOH in 2.00 L of a KOH solution
c. 5.85 g of NaCl in 400. mL of a NaCl solution

9.44 Calculate the molarity of each of the following:
a. 0.500 mole of glucose in 0.200 L of a glucose solution
b. 73.0 g of HCl in 2.00 L of a HCl solution
c. 30.0 g of NaOH in 350. mL of a NaOH solution

9.45 Calculate the grams of solute needed to prepare each of the following:
a. 2.00 L of a 1.50 M NaOH solution
b. 4.00 L of a 0.200 M KCl solution
c. 25.0 mL of a 6.00 M HCl solution

9.46 Calculate the grams of solute needed to prepare each of the following:
a. 2.00 L of a 6.00 M NaOH solution
b. 5.00 L of a 0.100 M $CaCl_2$ solution
c. 175 mL of a 3.00 M $NaNO_3$ solution

9.47 For each of the following solutions, calculate the:
a. liters of a 2.00 M KBr solution to obtain 3.00 moles of KBr
b. liters of a 1.50 M NaCl solution to obtain 15.0 moles of NaCl
c. milliliters of a 0.800 M $Ca(NO_3)_2$ solution to obtain 0.0500 mole of $Ca(NO_3)_2$

9.48 For each of the following solutions, calculate the:
a. liters of a 4.00 M KCl solution to obtain 0.100 mole of KCl
b. liters of a 6.00 M HCl solution to obtain 5.00 moles of HCl
c. milliliters of a 2.50 M K_2SO_4 solution to obtain 1.20 moles of K_2SO_4

9.49 Calculate the volume, in milliliters, for each of the following that provides the given amount of solute:
a. 12.5 g of Na_2CO_3 from a 0.120 M Na_2CO_3 solution
b. 0.850 mole of $NaNO_3$ from a 0.500 M $NaNO_3$ solution
c. 30.0 g of LiOH from a 2.70 M LiOH solution

9.50 Calculate the volume, in liters, for each of the following that provides the given amount of solute:
a. 5.00 moles of NaOH from a 12.0 M NaOH solution
b. 15.0 g of Na_2SO_4 from a 4.00 M Na_2SO_4 solution
c. 28.0 g of $NaHCO_3$ from a 1.50 M $NaHCO_3$ solution

9.51 Answer the following for the reaction:

$$Pb(NO_3)_2(aq) + 2KCl(aq) \longrightarrow PbCl_2(s) + 2KNO_3(aq)$$

a. How many grams of $PbCl_2$ will be formed from 50.0 mL of a 1.50 M KCl solution?
b. How many milliliters of a 2.00 M $Pb(NO_3)_2$ solution will react with 50.0 mL of a 1.50 M KCl solution?
c. What is the molarity of 20.0 mL of a KCl solution that reacts completely with 30.0 mL of a 0.400 M $Pb(NO_3)_2$ solution?

9.52 Answer the following for the reaction:

$$NiCl_2(aq) + 2NaOH(aq) \longrightarrow Ni(OH)_2(s) + 2NaCl(aq)$$

a. How many milliliters of a 0.200 M NaOH solution are needed to react with 18.0 mL of a 0.500 M $NiCl_2$ solution?
b. How many grams of $Ni(OH)_2$ are produced from the reaction of 35.0 mL of a 1.75 M NaOH solution and excess $NiCl_2$?
c. What is the molarity of 30.0 mL of a $NiCl_2$ solution that reacts completely with 10.0 mL of a 0.250 M NaOH solution?

9.53 Answer the following for the reaction:

$$2HCl(aq) + Mg(s) \longrightarrow H_2(g) + MgCl_2(aq)$$

a. How many milliliters of a 6.00 M HCl solution are required to react with 15.0 g of magnesium?
b. How many liters of hydrogen gas can form at STP when 0.500 L of a 2.00 M HCl solution reacts with excess magnesium?
c. What is the molarity of a HCl solution if the reaction of 45.2 mL of the HCl solution with excess magnesium produces 5.20 L of H_2 gas at 735 mmHg and 25 °C?

9.54 Answer the following for the reaction:

$$2HCl(aq) + CaCO_3(s) \longrightarrow CO_2(g) + H_2O(l) + CaCl_2(aq)$$

a. How many milliliters of a 0.200 M HCl solution can react with 8.25 g of $CaCO_3$?
b. How many liters of CO_2 gas can form at STP when 15.5 mL of a 3.00 M HCl solution reacts with excess $CaCO_3$?
c. What is the molarity of a HCl solution if the reaction of 200. mL of the HCl solution with excess $CaCO_3$ produces 12.0 L of CO_2 gas at 725 mmHg and 18 °C?

Clinical Applications

9.55 A patient receives 100. mL of 20.% (m/v) mannitol solution every hour.
a. How many grams of mannitol are given in 1 h?
b. How many grams of mannitol does the patient receive in 12 h?

9.56 A patient receives 250 mL of a 4.0% (m/v) amino acid solution twice a day.
a. How many grams of amino acids are in 250 mL of solution?
b. How many grams of amino acids does the patient receive in 1 day?

9.57 Calculate the volume required for each of the following:
a. A patient needs 100. g of glucose in the next 12 h. How many liters of a 5% (m/v) glucose solution must be given?
b. A patient received 2.0 g of NaCl in 8 h. How many milliliters of a 0.90% (m/v) NaCl (saline) solution were delivered?

9.58 Calculate the volume required for each of the following:
a. A doctor orders 0.075 g of chlorpromazine, which is used to treat schizophrenia. If the stock solution is 2.5% (m/v), how many milliliters are administered to the patient?
b. A doctor orders 5 mg of compazine, which is used to treat nausea, vertigo, and migraine headaches. If the stock solution is 2.5% (m/v), how many milliliters are administered to the patient?

9.59 A $CaCl_2$ solution is given to increase blood levels of calcium. If a patient receives 5.0 mL of a 10.% (m/v) $CaCl_2$ solution, how many grams of $CaCl_2$ were given?

9.60 An intravenous solution of mannitol is used as a diuretic to increase the loss of sodium and chloride by a patient. If a patient receives 30.0 mL of a 25% (m/v) mannitol solution, how many grams of mannitol were given?

9.5 Dilution of Solutions

REVIEW

Solving Equations (1.4)

LEARNING GOAL Describe the dilution of a solution; calculate the unknown concentration or volume when a solution is diluted.

In chemistry and biology, we often prepare diluted solutions from more concentrated solutions. In a process called **dilution**, a solvent, usually water, is added to a solution, which increases the volume. As a result, the concentration of the solution decreases. In an everyday example, you are making a dilution when you add three cans of water to a can of concentrated orange juice.

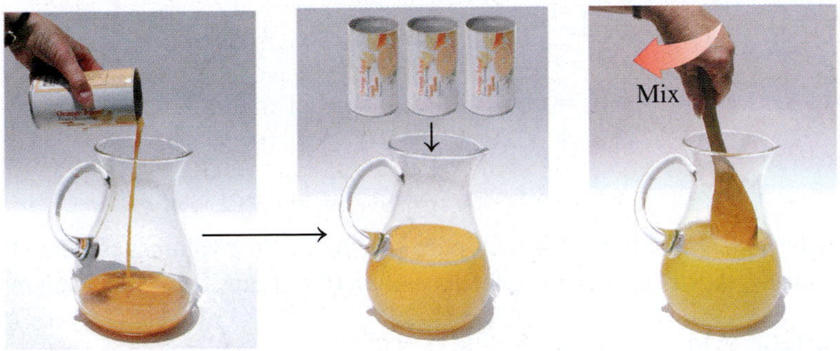
Mix

1 can of orange juice concentrate + 3 cans of water = 4 cans of orange juice

In the dilution of a solution, solvent is added to increase the volume and decrease the concentration.

Although the addition of solvent increases the volume, the amount of solute does not change; it is the same in the concentrated solution and the diluted solution (see **FIGURE 9.9**).

$$\underset{\text{Concentrated solution}}{\text{Grams or moles of solute}} = \underset{\text{Diluted solution}}{\text{grams or moles of solute}}$$

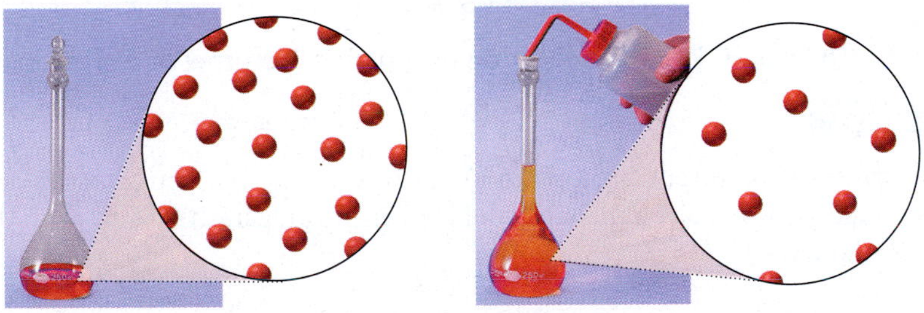

FIGURE 9.9 ▶ When water is added to a concentrated solution, there is no change in the number of particles. However, the solute particles spread out as the volume of the diluted solution increases.

Q What is the concentration of the diluted solution after an equal volume of water is added to a sample of a 6 M HCl solution?

We can write this equality in terms of the concentration, C, and the volume, V. The concentration, C, may be percent concentration or molarity.

$$\underset{\text{Concentrated solution}}{C_1V_1} = \underset{\text{Diluted solution}}{C_2V_2}$$

If we are given any three of the four variables (C_1, C_2, V_1, or V_2) we can rearrange the dilution expression to solve for the unknown quantity as seen in Sample Problems 9.14 and 9.15.

TEST

Try Practice Problems 9.61 and 9.62

▶ SAMPLE PROBLEM 9.14 Calculations Involving Concentration of Solutions

TRY IT FIRST

What is the molarity of a solution when 75.0 mL of a 4.00 M KCl solution is diluted to a volume of 500. mL?

SOLUTION

STEP 1 Prepare a table of the concentrations and volumes of the solutions.

	Given	Need	Connect
ANALYZE THE PROBLEM	$C_1 = 4.00$ M $V_1 = 75.0$ mL $V_2 = 500.$ mL	C_2	$C_1V_1 = C_2V_2$ **Predict:** V_2 increases, C_2 decreases

STEP 2 Rearrange the dilution expression to solve for the unknown quantity.

$$C_1V_1 = C_2V_2$$

$$\frac{C_1V_1}{V_2} = \frac{C_2\cancel{V_2}}{\cancel{V_2}} \quad \text{Divide both sides by } V_2$$

$$C_2 = C_1 \times \frac{V_1}{V_2}$$

STEP 3 Substitute the known quantities into the dilution expression and calculate.

$$C_2 = 4.00\ \text{M} \times \frac{75.0\ \cancel{\text{mL}}}{500.\ \cancel{\text{mL}}} = 0.600\ \text{M (diluted KCl solution)}$$

Three SFs (4.00 M); Three SFs (75.0 mL); Three SFs (500. mL); Three SFs (0.600 M)

Volume factor decreases concentration

When the initial molarity (C_1) is multiplied by a ratio of the volumes (volume factor) that is less than 1, the molarity of the diluted solution (C_2) decreases as predicted in Step 1.

STUDY CHECK 9.14

a. What is the molarity of a solution when 50.0 mL of a 4.00 M KOH solution is diluted to 200. mL?

b. What is the final concentration (m/v) when 18 mL of a 12% (m/v) NaCl solution is diluted to a volume of 85 mL?

ANSWER

a. 1.00 M KOH solution **b.** 2.5% (m/v) NaCl solution

TEST

Try Practice Problems 9.63 to 9.66

▶ SAMPLE PROBLEM 9.15 Calculations Involving Volume of Solutions

TRY IT FIRST

A doctor orders 1000. mL of a 35.0% (m/v) dextrose solution. If you have a 50.0% (m/v) dextrose solution, how many milliliters would you use to prepare 1000. mL of 35.0% (m/v) dextrose solution?

SOLUTION

STEP 1 Prepare a table of the concentrations and volumes of the solutions.

	Given	Need	Connect
ANALYZE THE PROBLEM	C_1 = 50.0% (m/v) C_2 = 35.0% (m/v) V_2 = 1000. mL	V_1	$C_1V_1 = C_2V_2$ **Predict:** C_1 increases, V_1 decreases

STEP 2 Rearrange the dilution expression to solve for the unknown quantity.

$$C_1V_1 = C_2V_2$$

$$\frac{\cancel{C_1}V_1}{\cancel{C_1}} = \frac{C_2V_2}{C_1} \quad \text{Divide both sides by } C_1$$

$$V_1 = V_2 \times \frac{C_2}{C_1}$$

STEP 3 Substitute the known quantities into the dilution expression and calculate.

$$V_1 = 1000.\ \text{mL} \times \frac{35.0\cancel{\%}}{50.0\cancel{\%}} = 700.\ \text{mL of dextrose solution}$$

Four SFs (1000. mL); Three SFs (35.0%); Three SFs (50.0%); Three SFs (700. mL)

Concentration factor decreases volume

When the final volume (V_2) is multiplied by a ratio of the percent concentrations (concentration factor) that is less than 1, the initial volume (V_1) is less than the final volume (V_2) as predicted in Step 1.

STUDY CHECK 9.15

a. What initial volume, in milliliters, of a 15% (m/v) mannose solution is needed to prepare 125 mL of a 3.0% (m/v) mannose solution?

b. What initial volume, in milliliters, of a 2.40 M NaCl solution is needed to prepare 80.0 mL of a 0.600 M NaCl solution?

ANSWER

a. 25 mL of a 15% (m/v) mannose solution

b. 20.0 mL of a 2.40 M NaCl solution

TEST

Try Practice Problems 9.67 to 9.70

PRACTICE PROBLEMS

9.5 Dilution of Solutions

9.61 To make tomato soup, you add one can of water to the condensed soup. Why is this a dilution?

9.62 A can of frozen lemonade calls for the addition of three cans of water to make a pitcher of the beverage. Why is this a dilution?

9.63 Calculate the final concentration of each of the following:

a. 2.0 L of a 6.0 M HCl solution is added to water so that the final volume is 6.0 L.

b. Water is added to 0.50 L of a 12 M NaOH solution to make 3.0 L of a diluted NaOH solution.

c. A 10.0-mL sample of a 25% (m/v) KOH solution is diluted with water so that the final volume is 100.0 mL.

d. A 50.0-mL sample of a 15% (m/v) H_2SO_4 solution is added to water to give a final volume of 250 mL.

9.64 Calculate the final concentration of each of the following:

a. 1.0 L of a 4.0 M HNO_3 solution is added to water so that the final volume is 8.0 L.

b. Water is added to 0.25 L of a 6.0 M NaF solution to make 2.0 L of a diluted NaF solution.

c. A 50.0-mL sample of an 8.0% (m/v) KBr solution is diluted with water so that the final volume is 200.0 mL.

d. A 5.0-mL sample of a 50.0% (m/v) acetic acid ($HC_2H_3O_2$) solution is added to water to give a final volume of 25 mL.

9.65 Determine the final volume, in milliliters, of each of the following:

a. a 1.5 M HCl solution prepared from 20.0 mL of a 6.0 M HCl solution

b. a 2.0% (m/v) LiCl solution prepared from 50.0 mL of a 10.0% (m/v) LiCl solution

c. a 0.500 M H_3PO_4 solution prepared from 50.0 mL of a 6.00 M H_3PO_4 solution

d. a 5.0% (m/v) glucose solution prepared from 75 mL of a 12% (m/v) glucose solution

9.66 Determine the final volume, in milliliters, of each of the following:

a. a 1.00% (m/v) H_2SO_4 solution prepared from 10.0 mL of a 20.0% H_2SO_4 solution

b. a 0.10 M HCl solution prepared from 25 mL of a 6.0 M HCl solution

c. a 1.0 M NaOH solution prepared from 50.0 mL of a 12 M NaOH solution

d. a 1.0% (m/v) $CaCl_2$ solution prepared from 18 mL of a 4.0% (m/v) $CaCl_2$ solution

9.67 Determine the initial volume, in milliliters, required to prepare each of the following:

a. 255 mL of a 0.200 M HNO_3 solution using a 4.00 M HNO_3 solution

b. 715 mL of a 0.100 M $MgCl_2$ solution using a 6.00 M $MgCl_2$ solution

c. 0.100 L of a 0.150 M KCl solution using an 8.00 M KCl solution

9.68 Determine the initial volume, in milliliters, required to prepare each of the following:

a. 20.0 mL of a 0.250 M KNO_3 solution using a 6.00 M KNO_3 solution

b. 25.0 mL of a 2.50 M H_2SO_4 solution using a 12.0 M H_2SO_4 solution

c. 0.500 L of a 1.50 M NH_4Cl solution using a 10.0 M NH_4Cl solution

Clinical Applications

9.69 You need 500. mL of a 5.0% (m/v) glucose solution. If you have a 25% (m/v) glucose solution on hand, how many milliliters do you need?

9.70 A doctor orders 100. mL of 2.0% (m/v) ibuprofen. If you have 8.0% (m/v) ibuprofen on hand, how many milliliters do you need?

9.6 Properties of Solutions

LEARNING GOAL Identify a mixture as a solution, a colloid, or a suspension. Describe how the number of particles in a solution affects the freezing point, the boiling point, and the osmotic pressure of a solution.

The size and number of solute particles in different types of mixtures play an important role in determining the properties of that mixture.

Solutions

In the solutions discussed up to now, the solute was dissolved as small particles that are uniformly dispersed throughout the solvent to give a homogeneous solution. When you observe a solution, such as salt water, you cannot visually distinguish the solute from the solvent. The solution appears transparent, although it may have a color. The particles are so small that they go through filters and through *semipermeable membranes*. A semipermeable membrane allows solvent molecules such as water and very small solute particles to pass through, but does not allow the passage of large solute molecules.

Colloids

The particles in a **colloid** are much larger than the solute particles in a solution. Colloidal particles are large molecules, such as proteins, or groups of molecules or ions. Colloids, similar to solutions, are homogeneous mixtures that do not separate or settle out. Colloidal particles are small enough to pass through filters, but too large to pass through semipermeable membranes. **TABLE 9.11** lists several examples of colloids.

TABLE 9.11 Examples of Colloids

Colloid	Substance Dispersed	Dispersing Medium
Fog, clouds, hairsprays	Liquid	Gas
Dust, smoke	Solid	Gas
Shaving cream, whipped cream, soap suds	Gas	Liquid
Styrofoam, marshmallows	Gas	Solid
Mayonnaise, homogenized milk	Liquid	Liquid
Cheese, butter	Liquid	Solid
Blood plasma, paints (latex), gelatin	Solid	Liquid

Suspensions

Suspensions are heterogeneous, nonuniform mixtures that are very different from solutions or colloids. The particles of a suspension are so large that they can often be seen with the naked eye. They are trapped by filters and semipermeable membranes.

The weight of the suspended solute particles causes them to settle out soon after mixing. If you stir muddy water, it mixes but then quickly separates as the suspended particles settle to the bottom and leave clear liquid at the top. You can find suspensions among the medications in a hospital or in your medicine cabinet. These include Kaopectate, calamine lotion, antacid mixtures, and liquid penicillin. It is important to follow the instructions on the label that states "shake well before using" so that the particles form a suspension.

Water-treatment plants make use of the properties of suspensions to purify water. When chemicals such as aluminum sulfate or iron(III) sulfate are added to untreated water, they react with impurities to form large suspended particles called *floc*. In the water-treatment plant, a system of filters traps the suspended particles, but clean water passes through. **TABLE 9.12** compares the different types of mixtures, and **FIGURE 9.10** illustrates some properties of solutions, colloids, and suspensions.

Suspensions separate into solids and water that can be removed by filtration.

TABLE 9.12 Comparison of Solutions, Colloids, and Suspensions

Type of Mixture	Type of Particle	Settling	Separation
Solution	Small particles such as atoms, ions, or small molecules	Particles do not settle	Particles cannot be separated by filters or semipermeable membranes
Colloid	Larger molecules or groups of molecules or ions	Particles do not settle	Particles can be separated by semipermeable membranes but not by filters
Suspension	Very large particles that may be visible	Particles settle rapidly	Particles can be separated by filters

TEST

Try Practice Problems 9.71 and 9.72

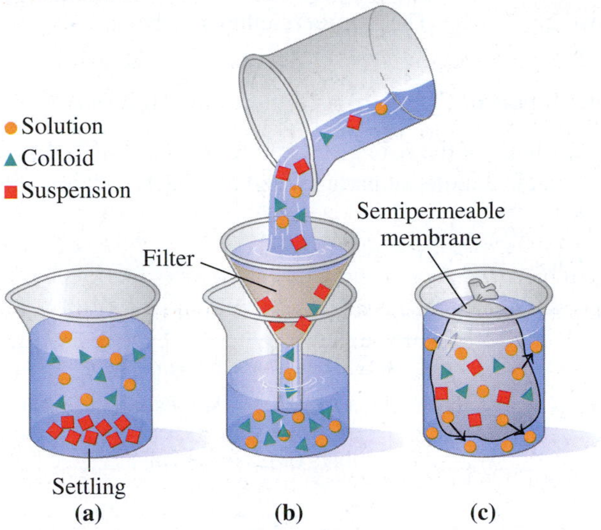
Solution
Colloid
Suspension
Filter
Semipermeable membrane
Settling
(a)
(b)
(c)

FIGURE 9.10 ▶ Properties of different types of mixtures: **(a)** suspensions settle out; **(b)** suspensions are separated by a filter; **(c)** solution particles go through a semipermeable membrane, but colloids and suspensions do not.

Q A filter can be used to separate suspension particles from a solution, but a semipermeable membrane is needed to separate colloids from a solution. Explain.

Boiling Point Elevation and Freezing Point Lowering

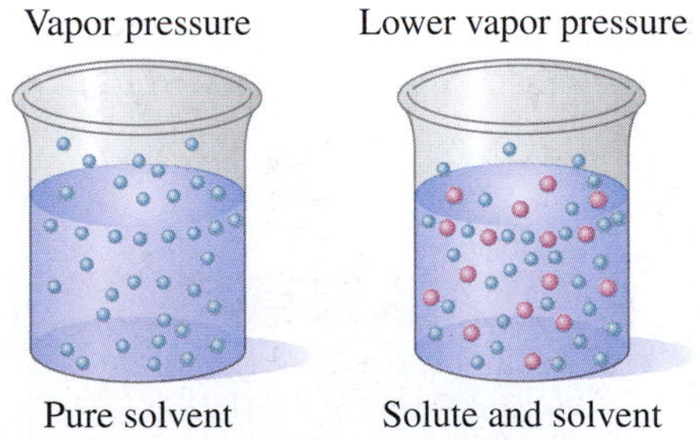
Vapor pressure
Lower vapor pressure
Pure solvent
Solute and solvent

When we add a solute to water, it changes the vapor pressure, boiling point, freezing point, and osmotic pressure of pure water. These types of changes in physical properties, known as *colligative properties*, depend only on the concentration of solute particles in the solution.

We can illustrate how these changes in physical properties occur by comparing the number of evaporating solvent molecules in pure solvent with those in a solution with a nonvolatile solute in the solvent. In the solution, there are fewer solvent molecules at the surface because the solute that has been added takes up some of the space at the surface. As a result, fewer solvent molecules can evaporate compared to the pure solvent. Vapor pressure is lowered for the solution. If we add more solute molecules, the vapor pressure will be lowered even more.

The boiling point of a solvent is raised when a nonvolatile solute is added. The vapor pressure of a solvent must reach atmospheric pressure before it begins boiling. However, because the solute lowers the vapor pressure of the solvent, a temperature higher than the normal boiling point of the pure solvent is needed to cause the solution to boil.

The freezing point of a solvent is lowered when a nonvolatile solute is added. In this case, the solute particles prevent the organization of solvent molecules needed to form the solid state. Thus, a lower temperature is required before the molecules of the solvent can become organized enough to freeze.

Thus, when you spread salt on an icy sidewalk when temperatures drop below freezing, the particles from the salt combine with water to lower the freezing point, which causes the ice to melt. Another example is the addition of antifreeze, such as ethylene glycol, $C_2H_6O_2$, to the water in a car radiator. Ethylene glycol forms many hydrogen

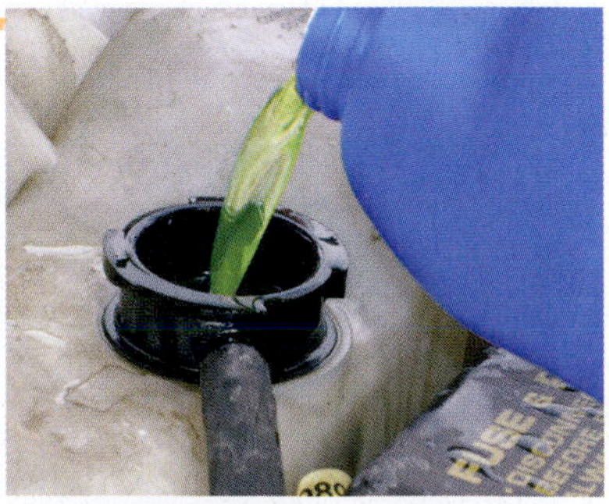

Ethylene glycol is added to a radiator to form an aqueous solution that has a lower freezing point and a higher boiling point than water.

The Alaska Upis beetle produces biological antifreeze to survive subfreezing temperatures.

bonds, which makes it very soluble in water. If an ethylene glycol and water mixture is about 50–50% by mass, it does not freeze until the temperature drops to about −30 °F, and does not boil unless the temperature reaches about 220 °F. The solution in the radiator prevents the water in the radiator from forming ice in cold weather or boiling over on a hot desert highway.

Insects and fish in climates with subfreezing temperatures control ice formation by producing biological antifreezes made of glycerol, proteins, and sugars, such as glucose, within their bodies. Some insects can survive temperatures below −60 °C. These forms of biological antifreezes may one day be applied to the long-term preservation of human organs.

Particles in Solution

A solute that is a nonelectrolyte dissolves as molecules, whereas a solute that is a strong electrolyte dissolves entirely as ions. The antifreeze ethylene glycol, $C_2H_6O_2$, a nonelectrolyte, dissolves as molecules.

$$\text{Nonelectrolyte: 1 mole of } C_2H_6O_2(l) = \text{1 mole of } C_2H_6O_2(aq)$$

However, when 1 mole of the strong electrolyte NaCl or $CaCl_2$ dissolves in water, the NaCl solution will contain 2 moles of particles and the $CaCl_2$ solution will contain 3 moles of particles.

ENGAGE

Why does 1 mole of KBr, a strong electrolyte, lower the freezing point of a solution more than 1 mole of urea, a nonelectrolyte?

Strong electrolytes:

$$\text{1 mole of NaCl}(s) = \underbrace{\text{1 mole of } Na^{+}(aq) + \text{1 mole of } Cl^{-}(aq)}_{\text{2 moles of particles } (aq)}$$

$$\text{1 mole of } CaCl_2(s) = \underbrace{\text{1 mole of } Ca^{2+}(aq) + \text{2 moles of } Cl^{-}(aq)}_{\text{3 moles of particles } (aq)}$$

Change in Boiling Point and Freezing Point

The change in boiling point or freezing point depends on the number of particles in the solution. One mole of an electrolyte such as glucose raises the boiling point of one kilogram of pure water (100.00 °C) by 0.51 °C, which gives a boiling point of 100.51 °C. If the solute is a strong electrolyte such as NaCl, one mole produces two moles of particles (ions). Then the boiling point of one kilogram of water is raised by 1.02 °C (2 × 0.51 °C), to give a boiling point of 101.02 °C.

CORE CHEMISTRY SKILL

Calculating the Boiling Point/Freezing Point of a Solution

The presence of solute particles also changes the freezing point. One mole of a nonelectrolyte such as glucose lowers the freezing point of one kilogram of pure water (0.00 °C) by 1.86 °C, to give a freezing point of −1.86 °C. For NaCl, one mole dissociates to form two moles of particles (ions). Then the freezing point of one kilogram of water is lowered by 3.72 °C (2 × 1.86 °C), which gives a freezing point of −3.72 °C.

The effect of some solutes on freezing and boiling point is summarized in **TABLE 9.13**.

TABLE 9.13 Effect of Solute Concentration on Freezing and Boiling Points of 1 kg of Water

Substance in One Kilogram of Water	Type of Solute	Moles of Solute Particles in One Kilogram of Water	Freezing Point	Boiling Point
Pure water	None	0	0.00 °C	100.00 °C
1 mole of $C_2H_6O_2$	Nonelectrolyte	1	−1.86 °C	100.51 °C
1 mole of NaCl	Strong electrolyte	2	−3.72 °C	101.02 °C
1 mole of $CaCl_2$	Strong electrolyte	3	−5.58 °C	101.53 °C

TEST

Try Practice Problems 9.73 and 9.74

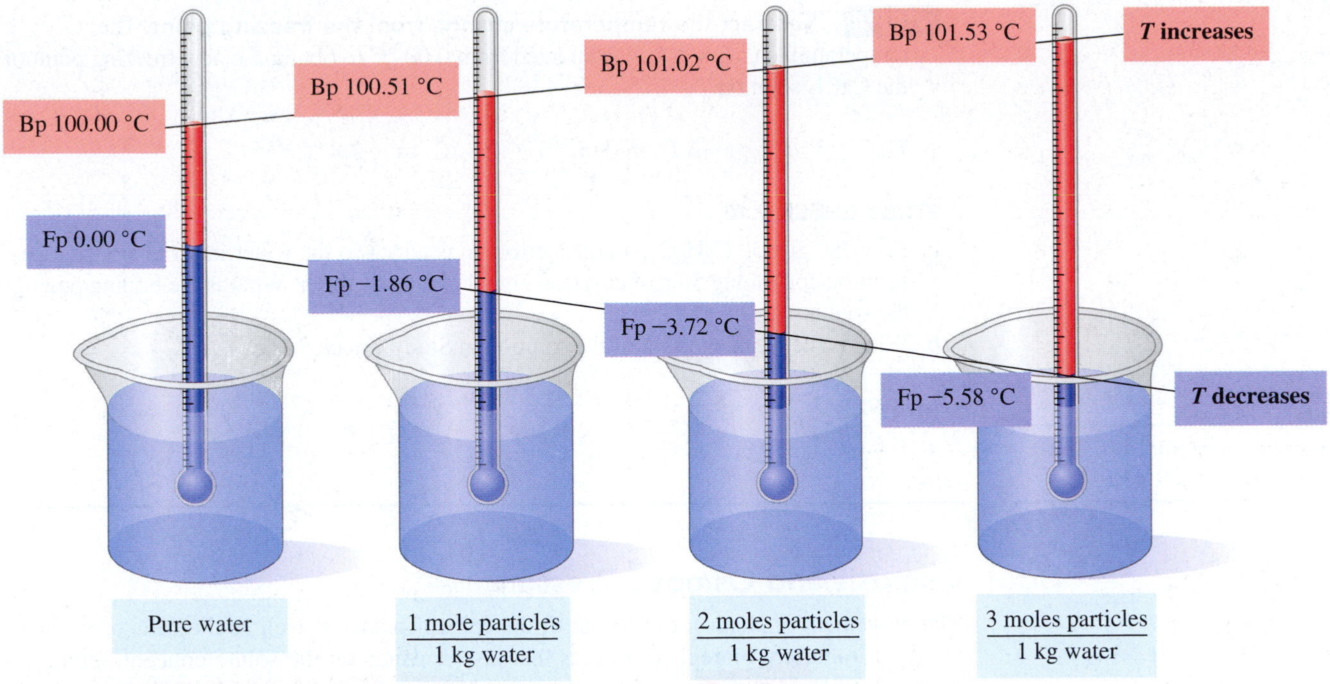
Bp 100.00 °C
Bp 100.51 °C
Bp 101.02 °C
Bp 101.53 °C
T increases
Fp 0.00 °C
Fp −1.86 °C
Fp −3.72 °C
Fp −5.58 °C
T decreases
Pure water
1 mole particles / 1 kg water
2 moles particles / 1 kg water
3 moles particles / 1 kg water

As the concentration of an aqueous solution increases, the freezing point is lowered and the boiling point is raised.

▶ SAMPLE PROBLEM 9.16 Calculating the Freezing Point of a Solution

A truck spreads calcium chloride on the road to melt ice and snow.

TRY IT FIRST

In the northeastern United States during freezing weather, $CaCl_2$ is spread on icy highways to melt the ice. Calculate the freezing point of a solution containing 0.50 mole of $CaCl_2$ in 1 kg of water.

SOLUTION

STEP 1 State the given and needed quantities.

	Given	Need	Connect
ANALYZE THE PROBLEM	0.50 mole of $CaCl_2$, 1 kg of water	freezing point of solution	$CaCl_2$ strong electrolyte, 1 mole/kg lowers freezing point 1.86 °C

STEP 2 Determine the number of moles of solute particles.

$$CaCl_2(s) \longrightarrow Ca^{2+}(aq) + 2Cl^-(aq) = 3 \text{ moles of solute particles}$$

$$1 \text{ mole of } CaCl_2 = 3 \text{ moles of solute particles}$$

$$\frac{3 \text{ moles solute particles}}{1 \text{ mole } CaCl_2} \text{ and } \frac{1 \text{ mole } CaCl_2}{3 \text{ moles solute particles}}$$

STEP 3 Determine the temperature change using the moles of solute particles and the degrees Celsius change per mole of particles.

$$1 \text{ mole of solute particles} = 1.86 \text{ °C}$$

$$\frac{1.86 \text{ °C}}{1 \text{ mole solute particles}} \text{ and } \frac{1 \text{ mole solute particles}}{1.86 \text{ °C}}$$

$$\text{Temperature change} = 0.50 \cancel{\text{mole } CaCl_2} \times \frac{3 \cancel{\text{moles solute particles}}}{1 \cancel{\text{mole } CaCl_2}} \times \frac{1.86 \text{ °C}}{1 \cancel{\text{mole solute particles}}}$$

$$= 2.8 \text{ °C}$$

STEP 4 **Subtract the temperature change from the freezing point.** The temperature change, ΔT, is subtracted from 0.00 °C to obtain the new freezing point of the $CaCl_2$ solution.

$$T_{solution} = T_{water} - \Delta T = 0.00\ °C - 2.8\ °C = -2.8\ °C$$

STUDY CHECK 9.16

a. Ethylene glycol, $C_2H_6O_2$, a nonelectrolyte, is added to the water in a radiator to give a solution containing 56 g of ethylene glycol in 1 kg of water. What is the boiling point of the solution?

b. What is the freezing point of the solution in Study Check 9.16a?

ANSWER

a. 100.46 °C
b. −1.7 °C

TEST

Try Practice Problems 9.75 and 9.76, and 9.81 and 9.82

Osmosis and Osmotic Pressure

The movement of water into and out of the cells of plants as well as the cells of our bodies is an important biological process that also depends on the solute concentration. In a process called **osmosis**, water molecules move through a semipermeable membrane from the solution with the lower concentration of solute into a solution with the higher solute concentration. In an osmosis apparatus, water is placed on one side of a semipermeable membrane and a sucrose (sugar) solution on the other side. The semipermeable membrane allows water molecules to flow back and forth but blocks the sucrose molecules because they cannot pass through the membrane. Because the sucrose solution has a higher solute concentration, more water molecules flow into the sucrose solution than out of the sucrose solution. The volume level of the sucrose solution rises as the volume level on the water side falls. The increase of water dilutes the sucrose solution to equalize (or attempt to equalize) the concentrations on both sides of the membrane.

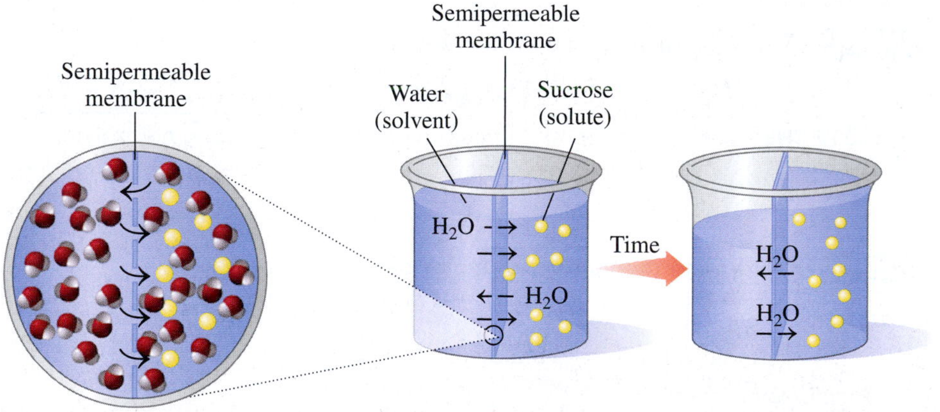
Semipermeable membrane
Semipermeable membrane
Water (solvent)
Sucrose (solute)
H_2O
H_2O
Time
H_2O
H_2O

Water flows into the solution with a higher solute concentration until the flow of water becomes equal in both directions.

Eventually the height of the sucrose solution creates sufficient pressure to equalize the flow of water between the two compartments. This pressure, called **osmotic pressure**, prevents the flow of additional water into the more concentrated solution. Then there is no further change in the volumes of the two solutions. The osmotic pressure depends on the concentration of solute particles in the solution. The greater the number of particles dissolved, the higher its osmotic pressure. In this example, the sucrose solution has a higher osmotic pressure than pure water, which has an osmotic pressure of zero.

In a process called *reverse osmosis*, a pressure greater than the osmotic pressure is applied to a solution so that it is forced through a purification membrane. The flow of water is reversed because water flows from an area of lower water concentration to an area of higher water concentration. The molecules and ions in solution stay behind, trapped by the membrane, while water passes through the membrane. This process of reverse osmosis is used in desalination plants to obtain pure water from sea (salt) water. However, the pressure that must be applied requires so much energy that reverse osmosis is not yet an economical method for obtaining pure water in most parts of the world.

TEST

Try Practice Problems 9.77 to 9.80

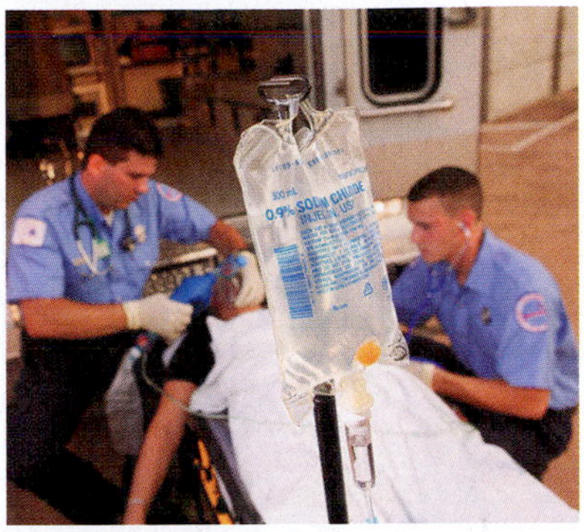

A 0.9% NaCl solution is isotonic with the solute concentration of the blood cells of the body.

Isotonic Solutions

Because the cell membranes in biological systems are semipermeable, osmosis is an ongoing process. The solutes in body solutions such as blood, tissue fluids, lymph, and plasma all exert osmotic pressure. Most intravenous (IV) solutions used in a hospital are **isotonic solutions**, which exert the same osmotic pressure as body fluids such as blood. The percent concentration typically used in IV solutions is mass/volume percent (m/v), which is a type of percent concentration we have already discussed. The most typical isotonic solutions are 0.9% (m/v) NaCl solution, or 0.9 g of NaCl/100. mL of solution, and 5% (m/v) glucose, or 5 g of glucose/100. mL of solution. Although they do not contain the same kinds of particles, a 0.9% (m/v) NaCl solution as well as a 5% (m/v) glucose solution both have the same osmotic pressure. A red blood cell placed in an isotonic solution retains its volume because there is an equal flow of water into and out of the cell (see **FIGURE 9.11a**).

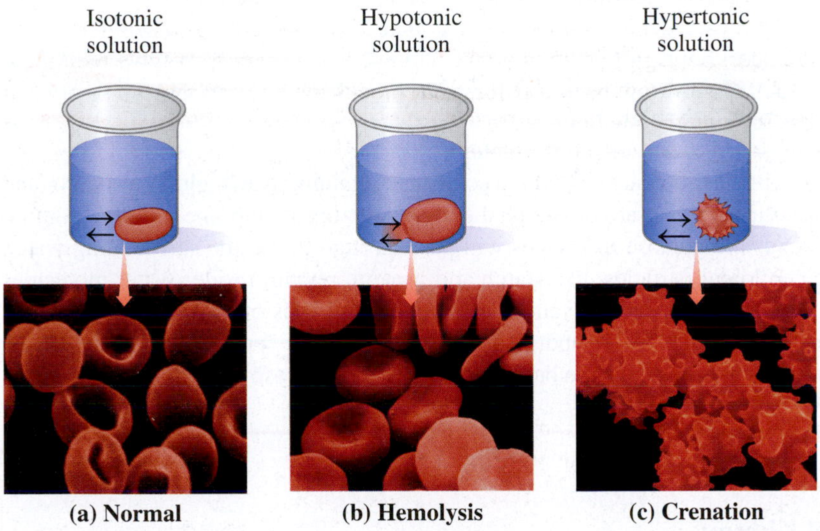
Isotonic solution
Hypotonic solution
Hypertonic solution
(a) Normal
(b) Hemolysis
(c) Crenation

FIGURE 9.11 ▶ **(a)** In an isotonic solution, a red blood cell retains its normal volume. **(b)** Hemolysis: In a hypotonic solution, water flows into a red blood cell, causing it to swell and burst. **(c)** Crenation: In a hypertonic solution, water leaves the red blood cell, causing it to shrink.

Q What happens to a red blood cell placed in a 4% NaCl solution?

Hypotonic and Hypertonic Solutions

If a red blood cell is placed in a solution that is not isotonic, the differences in osmotic pressure inside and outside the cell can alter the volume of the cell. When a red blood cell is placed in a **hypotonic solution**, which has a lower solute concentration (*hypo* means "lower than"), water flows into the cell by osmosis. The increase in fluid causes the cell to swell, and possibly burst—a process called **hemolysis** (see **FIGURE 9.11b**). A similar process occurs when you place dehydrated food, such as raisins or dried fruit, in water. The water enters the cells, and the food becomes plump and smooth.

If a red blood cell is placed in a **hypertonic solution**, which has a higher solute concentration (*hyper* means "greater than"), water flows out of the cell into the hypertonic solution by osmosis. Suppose a red blood cell is placed in a 10% (m/v) NaCl solution. Because the osmotic pressure in the red blood cell is the same as a 0.9% (m/v) NaCl solution, the 10% (m/v) NaCl solution has a much greater osmotic pressure. As water leaves the cell, it shrinks, a process called **crenation** (see **FIGURE 9.11c**). A similar process occurs when making pickles, which uses a hypertonic salt solution that causes the cucumbers to shrivel as they lose water.

▶SAMPLE PROBLEM 9.17 Isotonic, Hypotonic, and Hypertonic Solutions

TRY IT FIRST

Describe each of the following solutions as isotonic, hypotonic, or hypertonic. Indicate whether a red blood cell placed in each solution will undergo hemolysis, crenation, or no change.

a. a 5% (m/v) glucose solution
b. a 0.2% (m/v) NaCl solution

SOLUTION

a. A 5% (m/v) glucose solution is isotonic. A red blood cell will not undergo any change.
b. A 0.2% (m/v) NaCl solution is hypotonic. A red blood cell will undergo hemolysis.

STUDY CHECK 9.17

a. What will happen to a red blood cell placed in a 10% (m/v) glucose solution?
b. What will happen to a red blood cell placed in a 0.2% (m/v) glucose solution?

ANSWER

a. The red blood cell will shrink (crenate).
b. The red blood cell will expand (hemolyze).

TEST

Try Practice Problems 9.83 and 9.84

Dialysis

Dialysis is a process that is similar to osmosis. In dialysis, a semipermeable membrane, called a dialyzing membrane, permits small solute molecules and ions as well as solvent water molecules to pass through, but it retains large particles, such as colloids. Dialysis is a way to separate solution particles from colloids.

Suppose we fill a cellophane bag with a solution containing NaCl, glucose, starch, and protein and place it in pure water. Cellophane is a dialyzing membrane, and the sodium ions, chloride ions, and glucose molecules will pass through it into the surrounding water. However, large colloidal particles, like starch and protein, remain inside. Water molecules will flow into the cellophane bag. Eventually the concentrations of sodium ions, chloride ions, and glucose molecules inside and outside the dialysis bag become equal. To remove more NaCl or glucose, the cellophane bag must be placed in a fresh sample of pure water.

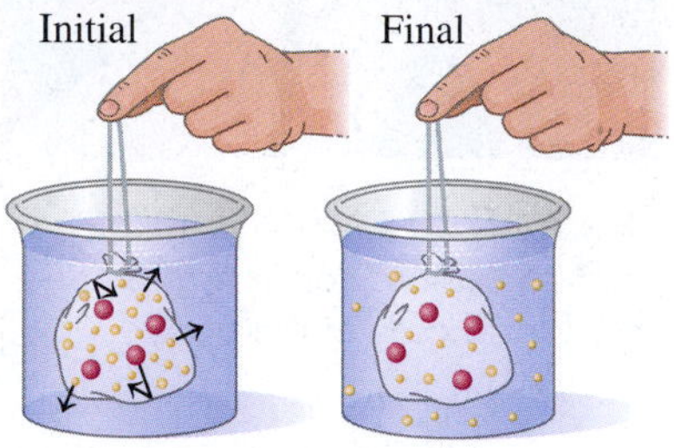
Initial
Final

- Solution particles such as Na^+, Cl^-, glucose
- Colloidal particles such as protein, starch

Solution particles pass through a dialyzing membrane, but colloidal particles are retained.

TEST

Try Practice Problems 9.85 and 9.86

Chemistry Link to Health

Dialysis by the Kidneys and the Artificial Kidney

The fluids of the body undergo dialysis by the membranes of the kidneys, which remove waste materials, excess salts, and water. In an adult, each kidney contains about 2 million nephrons. At the top of each nephron, there is a network of arterial capillaries called the *glomerulus*.

As blood flows into the glomerulus, small particles, such as amino acids, glucose, urea, water, and certain ions, will move through the capillary membranes into the nephron. As this solution moves through the nephron, substances still of value to the body (such as amino acids, glucose, certain ions, and 99% of the water) are reabsorbed. The major waste product, urea, is excreted in the urine.

Hemodialysis

If the kidneys fail to dialyze waste products, increased levels of urea can become life-threatening in a relatively short time. A person with kidney failure must use an artificial kidney, which cleanses the blood by *hemodialysis*.

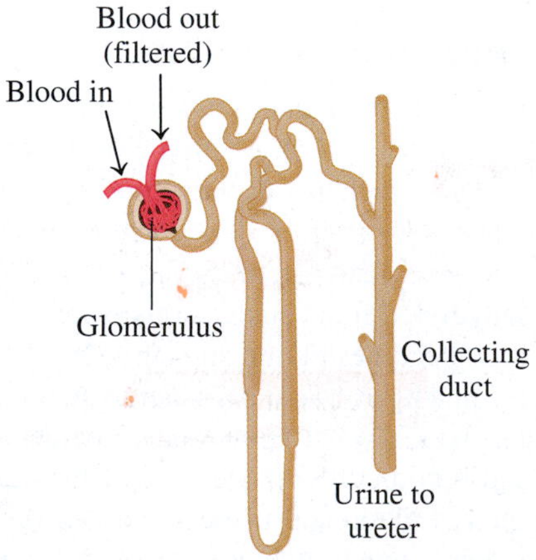
Blood out (filtered)
Blood in
Glomerulus
Collecting duct
Urine to ureter

In the kidneys, each nephron contains a glomerulus where urea and waste products are removed to form urine.

(continued)

Chemistry Link to Health (*continued*)

A typical artificial kidney machine contains a tank filled with water containing selected electrolytes. In the center of this dialyzing bath (dialysate), there is a dialyzing coil or membrane made of cellulose tubing. As the patient's blood flows through the dialyzing coil, the highly concentrated waste products dialyze out of the blood. No blood is lost because the membrane is not permeable to large particles such as red blood cells.

Dialysis patients do not produce much urine. As a result, they retain large amounts of water between dialysis treatments, which produces a strain on the heart. The intake of fluids for a dialysis patient may be restricted to as little as a few teaspoons of water a day. In the dialysis procedure, the pressure of the blood is increased as it circulates through the dialyzing coil so water can be squeezed out of the blood. For some dialysis patients, 2 to 10 L of water may be removed during one treatment. Dialysis patients have from two to three treatments a week, each treatment requiring about 5 to 7 h. Some of the newer treatments require less time. For many patients, dialysis is done at home with a home dialysis unit.

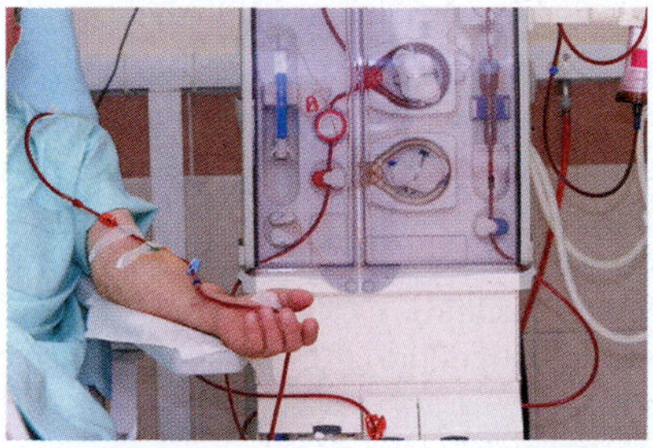

During dialysis, waste products and excess water are removed from the blood.

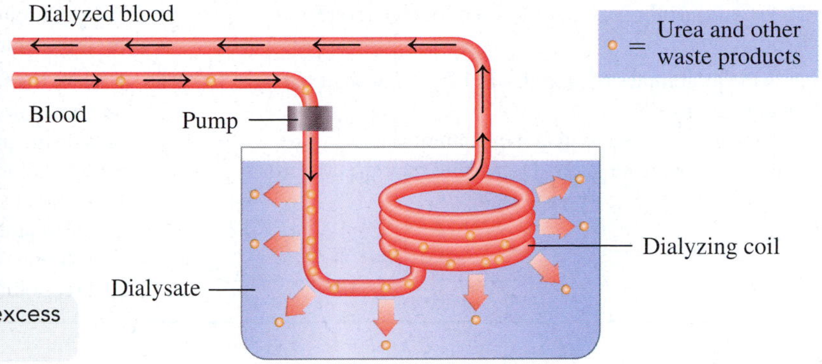
Dialyzed blood
Blood
Pump
Urea and other waste products
Dialyzing coil
Dialysate

PRACTICE PROBLEMS

9.6 Properties of Solutions

9.71 Identify the following as characteristic of a solution, a colloid, or a suspension:

a. a mixture that cannot be separated by a semipermeable membrane

b. a mixture that settles out upon standing

9.72 Identify the following as characteristic of a solution, a colloid, or a suspension:

a. particles of this mixture remain inside a semipermeable membrane but pass through filters

b. particles of solute in this solution are very large and visible

9.73 In each pair, identify the solution that will have a lower freezing point. Explain.

a. 1.0 mole of glycerol (nonelectrolyte) and 2.0 moles of ethylene glycol (nonelectrolyte) each in 1.0 kg of water

b. 0.50 mole of KCl (strong electrolyte) and 0.50 mole of $MgCl_2$ (strong electrolyte) each in 1.0 kg of water

9.74 In each pair, identify the solution that will have a higher boiling point. Explain.

a. 1.50 moles of LiOH (strong electrolyte) and 3.00 moles of KOH (strong electrolyte) each in 1.0 kg of water

b. 0.40 mole of $Al(NO_3)_3$ (strong electrolyte) and 0.40 mole of CsCl (strong electrolyte) each in 1.0 kg of water

9.75 Calculate the freezing point of each of the following solutions:

a. 1.36 moles of methanol, CH_4O, a nonelectrolyte, added to 1.00 kg of water

b. 640. g of the antifreeze propylene glycol, $C_3H_8O_2$, a nonelectrolyte, dissolved in 1.00 kg of water

c. 111 g of KCl, a strong electrolyte, dissolved in 1.00 kg of water

9.76 Calculate the boiling point of each of the following solutions:

a. 2.12 moles of glucose, $C_6H_{12}O_6$, a nonelectrolyte, added to 1.00 kg of water

b. 110. g of sucrose, $C_{12}H_{22}O_{11}$, a nonelectrolyte, dissolved in 1.00 kg of water

c. 146 g of $NaNO_3$, a strong electrolyte, dissolved in 1.00 kg of water

9.77 A 10% (m/v) starch solution is separated from a 1% (m/v) starch solution by a semipermeable membrane. (Starch is a colloid.)

a. Which compartment has the higher osmotic pressure?

b. In which direction will water flow initially?

c. In which compartment will the volume level rise?

9.78 A 0.1% (m/v) albumin solution is separated from a 2% (m/v) albumin solution by a semipermeable membrane. (Albumin is a colloid.)

a. Which compartment has the higher osmotic pressure?

b. In which direction will water flow initially?

c. In which compartment will the volume level rise?

9.79 Indicate the compartment (**A** or **B**) that will increase in volume for each of the following pairs of solutions separated by a semipermeable membrane:

A	B
a. 5% (m/v) sucrose	10% (m/v) sucrose
b. 8% (m/v) albumin	4% (m/v) albumin
c. 0.1% (m/v) starch	10% (m/v) starch

9.80 Indicate the compartment (**A** or **B**) that will increase in volume for each of the following pairs of solutions separated by a semipermeable membrane:

A	B
a. 20% (m/v) starch	10% (m/v) starch
b. 10% (m/v) albumin	2% (m/v) albumin
c. 0.5% (m/v) sucrose	5% (m/v) sucrose

Clinical Applications

9.81 What is the total moles of particles in 1.0 L of each of the following solutions?

a. half-normal NaCl, 0.45% (m/v), used to provide electrolytes

b. 50.% (m/v) dextrose, $C_6H_{12}O_6$, used to treat severe hypoglycemia

9.82 What is the total moles of particles in 1.0 L of each of the following solutions?

a. 3% (m/v) NaCl used to treat hyponatremia

b. 2.5% (m/v) dextrose, $C_6H_{12}O_6$, used to treat cystic fibrosis

9.83 Are the following solutions isotonic, hypotonic, or hypertonic compared with a red blood cell?

a. distilled H_2O **b.** 1% (m/v) glucose

c. 0.9% (m/v) NaCl **d.** 15% (m/v) glucose

9.84 Will a red blood cell undergo crenation, hemolysis, or no change in each of the following solutions?

a. 1% (m/v) glucose **b.** 2% (m/v) NaCl

c. 5% (m/v) glucose **d.** 0.1% (m/v) NaCl

9.85 Each of the following mixtures is placed in a dialyzing bag and immersed in distilled water. Which substances will be found outside the bag in the distilled water?

a. NaCl solution

b. starch solution (colloid) and alanine (an amino acid) solution

c. NaCl solution and starch solution (colloid)

d. urea solution

9.86 Each of the following mixtures is placed in a dialyzing bag and immersed in distilled water. Which substances will be found outside the bag in the distilled water?

a. KCl solution and glucose solution

b. albumin solution (colloid)

c. an albumin solution (colloid), KCl solution, and glucose solution

d. urea solution and NaCl solution

CLINICAL UPDATE Using Dialysis for Renal Failure

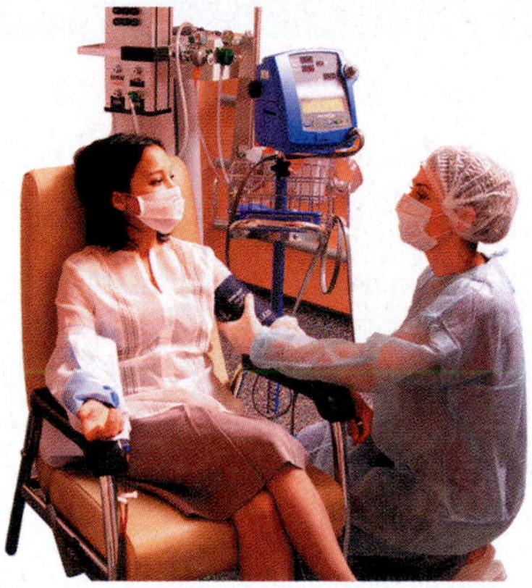

As a dialysis patient, Michelle has a 4-h dialysis treatment three times a week. When she arrives at the dialysis clinic, her weight, temperature, and blood pressure are taken, and blood tests are done to determine the level of electrolytes and urea in her blood. In the dialysis center, tubes to the dialyzer are connected to the catheter she has had implanted. Blood is then pumped out of her body, through the dialyzer, where it is filtered and returned to her body. As Michelle's blood flows through the dialyzer, electrolytes from the dialysate move into her blood, and waste products in her blood move into the dialysate, which is continually renewed. To achieve normal serum electrolyte levels, dialysate fluid contains sodium, chloride, and magnesium levels that are equal to serum concentrations. These electrolytes are removed from the blood only if their concentrations are higher than normal. Typically, in dialysis patients, the potassium ion level is higher than normal. Therefore, initial dialysis may start with a low concentration of potassium ion in the dialysate. During dialysis, excess fluid is removed by osmosis. A 4-h dialysis session requires at least 120 L of dialysis fluid. During dialysis, the electrolytes in the dialysate are adjusted until the electrolytes have the same levels as normal serum. Michelle's blood tests prior to dialysis show that the electrolyte levels in her blood are: HCO_3^- 21.0 mEq/L, K^+ 6.0 mEq/L, Na^+ 148 mEq/L, Ca^{2+} 3.0 mEq/L, Mg^{2+} 1.0 mEq/L, Cl^- 111.0 mEq/L.

A dialysate solution is prepared for Michelle that contains the following: HCO_3^- 35.0 mEq/L, K^+ 2.0 mEq/L, Na^+ 130 mEq/L, Ca^{2+} 5.0 mEq/L, Mg^{2+} 3.0 mEq/L, Cl^- 105.0 mEq/L, glucose 5.0% (m/v).

Clinical Applications

9.87 When Michelle's blood was tested, the chloride level was 0.45 g/dL.

a. What is this value in milliequivalents per liter?

b. According to Table 9.6, is this value above, below, or within the normal range?

9.88 After dialysis, the level of magnesium in Michelle's blood was 0.0026 g/dL.

a. What is this value in milliequivalents per liter?

b. According to Table 9.6, is this value above, below, or within the normal range?

9.89 **a.** Which electrolytes in Michelle's blood serum need to be increased by dialysis (see Table 9.6)?

b. Which electrolytes in Michelle's blood serum need to be decreased by dialysis (see Table 9.6)?

9.90 **a.** What is the total positive charge, in milliequivalents/L, of the electrolytes in the dialysate fluid?

b. What is the total negative charge, in milliequivalents/L, of the electrolytes in the dialysate fluid?

CONCEPT MAP

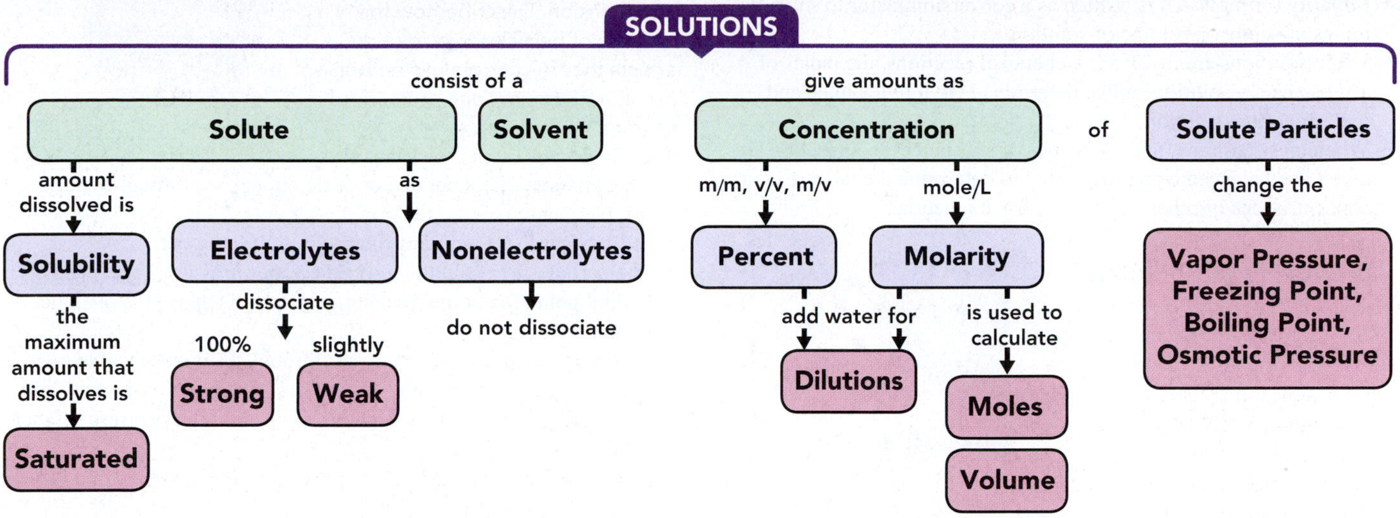

SOLUTIONS
consist of a
Solute
Solvent
give amounts as
Concentration
of
Solute Particles
amount dissolved is
Solubility
the maximum amount that dissolves is
Saturated
as
Electrolytes
dissociate
100%
slightly
Strong
Weak
Nonelectrolytes
do not dissociate
m/m, v/v, m/v
Percent
mole/L
Molarity
add water for
Dilutions
is used to calculate
Moles
Volume
change the
Vapor Pressure, Freezing Point, Boiling Point, Osmotic Pressure

CHAPTER REVIEW

9.1 Solutions

LEARNING GOAL Identify the solute and solvent in a solution; describe the formation of a solution.

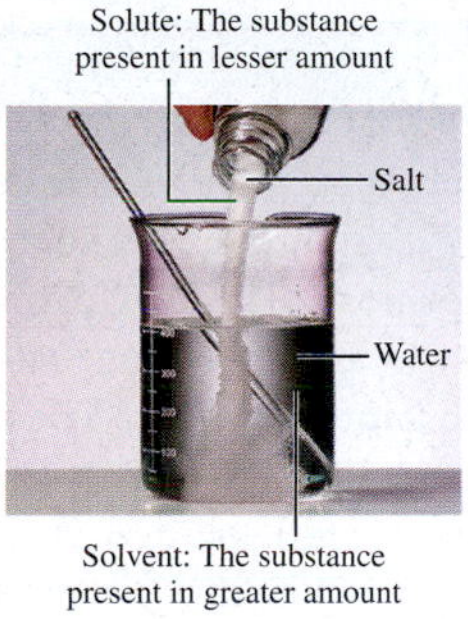

Solute: The substance present in lesser amount
Salt
Water
Solvent: The substance present in greater amount

- A solution forms when a solute dissolves in a solvent.
- In a solution, the particles of solute are evenly dispersed in the solvent.
- The solute and solvent may be solid, liquid, or gas.
- The polar O—H bond leads to hydrogen bonding between water molecules.
- An ionic solute dissolves in water, a polar solvent, because the polar water molecules attract and pull the ions into solution, where they become hydrated.
- The expression *like dissolves like* means that a polar or an ionic solute dissolves in a polar solvent while a nonpolar solute dissolves in a nonpolar solvent.

9.2 Electrolytes and Nonelectrolytes

LEARNING GOAL Identify solutes as electrolytes or nonelectrolytes.

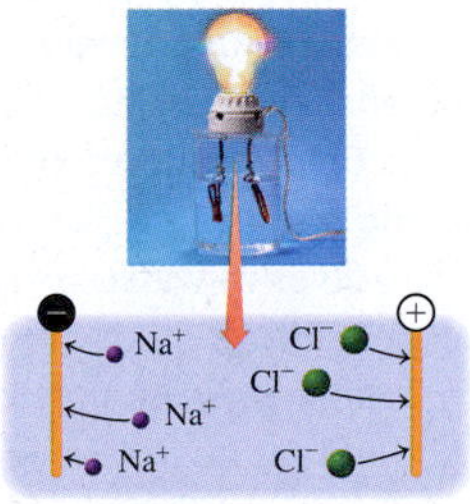

Na+
Cl−

Strong electrolyte

- Substances that produce ions in water are called electrolytes because their solutions will conduct an electrical current.
- Strong electrolytes are completely dissociated, whereas weak electrolytes are only partially dissociated.
- Nonelectrolytes are substances that dissolve in water to produce only molecules and cannot conduct electrical currents.
- An equivalent (Eq) is the amount of an electrolyte that carries one mole of positive or negative charge.

9.3 Solubility

LEARNING GOAL Define solubility; distinguish between an unsaturated and a saturated solution. Identify an ionic compound as soluble or insoluble.

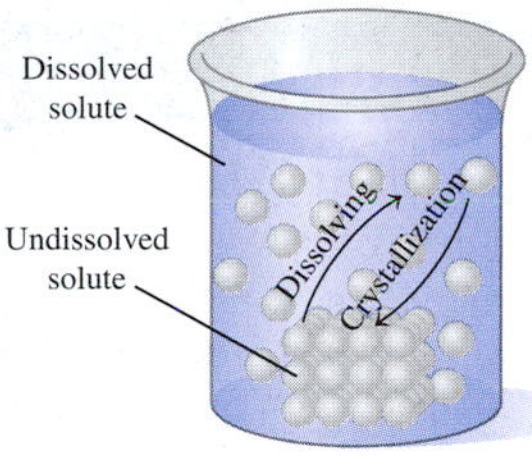

Dissolved solute
Undissolved solute
Dissolving
Crystallization

Saturated solution

- The solubility of a solute is the maximum amount of a solute that can dissolve in 100. g of solvent.
- A solution that contains the maximum amount of dissolved solute is a saturated solution.
- A solution containing less than the maximum amount of dissolved solute is unsaturated.
- An increase in temperature increases the solubility of most solids in water, but decreases the solubility of gases in water.
- Ionic compounds that are soluble in water usually contain Li^+, Na^+, K^+, NH_4^+, NO_3^-, or acetate, $C_2H_3O_2^-$.
- When two solutions are mixed, solubility rules can be used to predict whether a precipitate will form.

9.4 Solution Concentrations and Reactions

LEARNING GOAL Calculate the concentration of a solute in a solution; use concentration units to calculate the amount of solute or solution. Given the volume and concentration of a solution, calculate the amount of another reactant or product in a reaction.

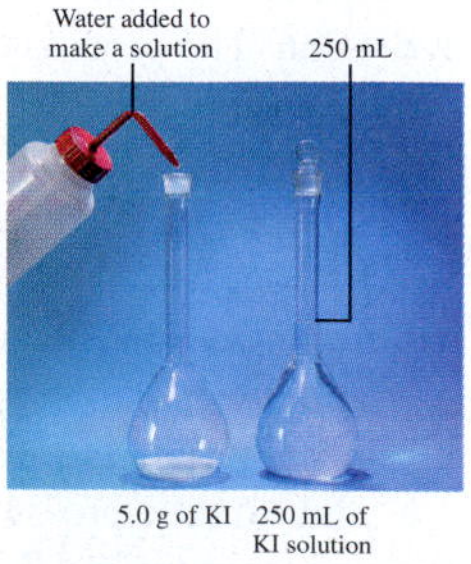

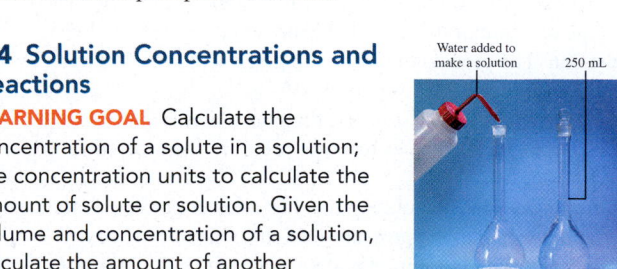

Water added to make a solution
250 mL
5.0 g of KI
250 mL of KI solution

- Mass percent expresses the mass/mass (m/m) ratio of the mass of solute to the mass of solution multiplied by 100%.
- Percent concentration can also be expressed as volume/volume (v/v) and mass/volume (m/v) ratios.

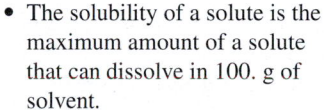

- Molarity is the moles of solute per liter of solution.
- In calculations of grams or milliliters of solute or solution, the concentration is used as a conversion factor.
- Molarity (or moles/L) is written as a conversion factor to solve for moles of solute or volume of solution.
- When solutions are involved in chemical reactions, the moles of a substance in solution can be determined from the volume and molarity of the solution.
- When mass, volume, and molarities of substances in a reaction are given, the balanced equation is used to determine the quantities or concentrations of other substances in the reaction.

9.5 Dilution of Solutions

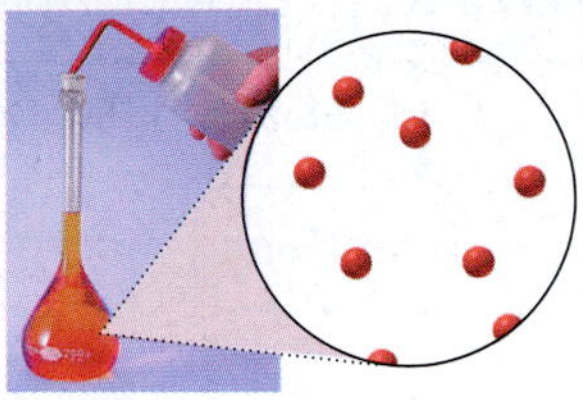

LEARNING GOAL Describe the dilution of a solution; calculate the unknown concentration or volume when a solution is diluted.

- In dilution, a solvent such as water is added to a solution, which increases its volume and decreases its concentration.

9.6 Properties of Solutions

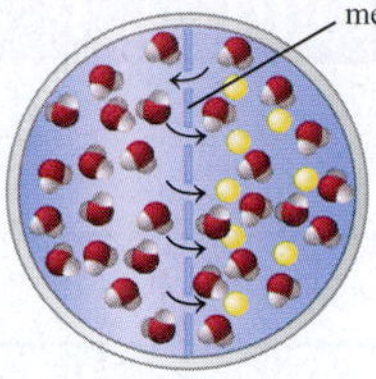
Semipermeable
membrane

LEARNING GOAL Identify a mixture as a solution, a colloid, or a suspension. Describe how the number of particles in a solution affects the freezing point, the boiling point, and the osmotic pressure of a solution.

- Colloids contain particles that pass through most filters but do not settle out or pass through semipermeable membranes.
- Suspensions have very large particles that settle out.
- The particles in a solution lower the vapor pressure, raise the boiling point, lower the freezing point, and increase the osmotic pressure.
- In osmosis, solvent (water) passes through a semipermeable membrane from a solution with a lower osmotic pressure (lower solute concentration) to a solution with a higher osmotic pressure (higher solute concentration).
- Isotonic solutions have osmotic pressures equal to that of body fluids.
- A red blood cell maintains its volume in an isotonic solution but swells in a hypotonic solution, and shrinks in a hypertonic solution.
- In dialysis, water and small solute particles pass through a dialyzing membrane, while larger particles are retained.

KEY TERMS

colloid A mixture having particles that are moderately large. Colloids pass through filters but cannot pass through semipermeable membranes.

concentration A measure of the amount of solute that is dissolved in a specified amount of solution.

crenation The shriveling of a cell because water leaves the cell when the cell is placed in a hypertonic solution.

dialysis A process in which water and small solute particles pass through a semipermeable membrane.

dilution A process by which water (solvent) is added to a solution to increase the volume and decrease (dilute) the concentration of the solute.

electrolyte A substance that produces ions when dissolved in water; its solution conducts electricity.

equivalent (Eq) The amount of a positive or negative ion that supplies 1 mole of electrical charge.

hemolysis A swelling and bursting of red blood cells in a hypotonic solution because of an increase in fluid volume.

Henry's law The solubility of a gas in a liquid is directly related to the pressure of that gas above the liquid.

hydration The process of surrounding dissolved ions by water molecules.

hypertonic solution A solution that has a higher particle concentration and higher osmotic pressure than the cells of the body.

hypotonic solution A solution that has a lower particle concentration and lower osmotic pressure than the cells of the body.

isotonic solution A solution that has the same particle concentration and osmotic pressure as that of the cells of the body.

mass percent (m/m) The grams of solute in 100. g of solution.

mass/volume percent (m/v) The grams of solute in 100. mL of solution.

molarity (M) The number of moles of solute in exactly 1 L of solution.

nonelectrolyte A substance that dissolves in water as molecules; its solution does not conduct an electrical current.

osmosis The flow of a solvent, usually water, through a semipermeable membrane into a solution of higher solute concentration.

osmotic pressure The pressure that prevents the flow of water into the more concentrated solution.

saturated solution A solution containing the maximum amount of solute that can dissolve at a given temperature. Any additional solute will remain undissolved in the container.

solubility The maximum amount of solute that can dissolve in 100. g of solvent, usually water, at a given temperature.

solubility rules A set of guidelines that states whether an ionic compound is soluble or insoluble in water.

solute The component in a solution that is present in the lesser amount.

solution A homogeneous mixture in which the solute is made up of small particles (ions or molecules) that can pass through filters and semipermeable membranes.

solvent The substance in which the solute dissolves; usually the component present in greater amount.

strong electrolyte A compound that ionizes completely when it dissolves in water. Its solution is a good conductor of electricity.

suspension A mixture in which the solute particles are large enough and heavy enough to settle out and be retained by both filters and semipermeable membranes.

unsaturated solution A solution that contains less solute than can be dissolved.

volume percent (v/v) A percent concentration that relates the volume of the solute in 100. mL of solution.

weak electrolyte A substance that produces only a few ions along with many molecules when it dissolves in water. Its solution is a weak conductor of electricity.

CORE CHEMISTRY SKILLS

The chapter Section containing each Core Chemistry Skill is shown in parentheses at the end of each heading.

Using Solubility Rules (9.3)

- Ionic compounds that are soluble in water contain Li^+, Na^+, K^+, NH_4^+, NO_3^-, or $C_2H_3O_2^-$ (acetate).
- Ionic compounds containing Cl^-, Br^-, or I^- are soluble, but if they contain Ag^+, Pb^{2+}, or Hg_2^{2+}, they are insoluble.
- Ionic compounds containing SO_4^{2-} are soluble, but if they contain Ba^{2+}, Pb^{2+}, Ca^{2+}, Sr^{2+}, or Hg_2^{2+}, they are insoluble.
- Most other ionic compounds, including those containing the anions CO_3^{2-}, S^{2-}, PO_4^{3-}, or OH^-, are insoluble.

Example: Is K_2SO_4 soluble or insoluble in water?
Answer: K_2SO_4 is soluble in water because it contains K^+.

Calculating Concentration (9.4)

The amount of solute dissolved in a certain amount of solution is called the concentration of the solution.

- $\text{Mass percent (m/m)} = \dfrac{\text{mass of solute}}{\text{mass of solution}} \times 100\%$
- $\text{Volume percent (v/v)} = \dfrac{\text{volume of solute}}{\text{volume of solution}} \times 100\%$
- $\text{Mass/volume percent (m/v)} = \dfrac{\text{grams of solute}}{\text{milliliters of solution}} \times 100\%$
- $\text{Molarity (M)} = \dfrac{\text{moles of solute}}{\text{liters of solution}}$

Example: What is the mass/volume percent (m/v) and the molarity (M) of 225 mL (0.225 L) of a LiCl solution that contains 17.1 g of LiCl?

Answer:

$$\text{Mass/volume percent (m/v)} = \frac{\text{grams of solute}}{\text{milliliters of solution}} \times 100\%$$
$$= \frac{17.1 \text{ g LiCl}}{225 \text{ mL solution}} \times 100\%$$
$$= 7.60\% \text{ (m/v) LiCl solution}$$

$$\text{moles of LiCl} = 17.1 \cancel{\text{g LiCl}} \times \frac{1 \text{ mole LiCl}}{42.39 \cancel{\text{g LiCl}}}$$
$$= 0.403 \text{ mole of LiCl}$$

$$\text{Molarity (M)} = \frac{\text{moles of solute}}{\text{liters of solution}} = \frac{0.403 \text{ mole LiCl}}{0.225 \text{ L solution}}$$
$$= 1.79 \text{ M LiCl solution}$$

Using Concentration as a Conversion Factor (9.4)

- When we need to calculate the amount of solute or solution, we use the concentration as a conversion factor.
- For example, the concentration of a 4.50 M HCl solution means there are 4.50 moles of HCl in 1 L of HCl solution, which gives two conversion factors written as

$$\frac{4.50 \text{ moles HCl}}{1 \text{ L solution}} \quad \text{and} \quad \frac{1 \text{ L solution}}{4.50 \text{ moles HCl}}$$

Example: How many milliliters of a 4.50 M HCl solution will provide 1.13 moles of HCl?

Answer: $1.13 \cancel{\text{moles HCl}} \times \dfrac{1 \cancel{\text{L solution}}}{4.50 \cancel{\text{moles HCl}}} \times \dfrac{1000 \text{ mL solution}}{1 \cancel{\text{L solution}}}$

$= 251$ mL of HCl solution

Calculating the Quantity of a Reactant or Product for a Chemical Reaction in Solution (9.4)

- When chemical reactions involve aqueous solutions of reactants or products, we use the balanced chemical equation, the molarity, and the volume to determine the moles or grams of the reactants or products.

Example: How many grams of zinc metal will react with 0.315 L of a 1.20 M HCl solution?

$$2HCl(aq) + Zn(s) \longrightarrow H_2(g) + ZnCl_2(aq)$$

Answer:

$$0.315 \cancel{\text{L solution}} \times \frac{1.20 \cancel{\text{moles HCl}}}{1 \cancel{\text{L solution}}} \times \frac{1 \cancel{\text{mole Zn}}}{2 \cancel{\text{moles HCl}}} \times \frac{65.41 \text{ g Zn}}{1 \cancel{\text{mole Zn}}}$$
$$= 12.4 \text{ g of Zn}$$

Calculating the Boiling Point / Freezing Point of a Solution (9.6)

- The particles in a solution raise the boiling point, lower the freezing point, and increase the osmotic pressure.
- The boiling point elevation is determined from the moles of particles in one kilogram of water.
- The freezing point lowering is determined from the moles of particles in one kilogram of water.

Example: What is the boiling point of a solution that contains 1.5 moles of the strong electrolyte KCl in 1.0 kg of water?

Answer: A solution of 1.5 moles of KCl in 1.0 kg of water contains 3.0 moles of solute particles (1.5 moles of K^+ and 1.5 moles of Cl^-) in 1.0 kg of water and has a boiling point change of

$$3.0 \cancel{\text{moles solute particles}} \times \frac{0.51 \text{ °C}}{1 \cancel{\text{mole solute particles}}} = 1.53 \text{ °C}$$

The boiling point would be
100.00 °C + 1.53 °C = 101.53 °C.

UNDERSTANDING THE CONCEPTS

The chapter Sections to review are shown in parentheses at the end of each problem.

9.91 Match the diagrams with the following: (9.1)

a. a polar solute and a polar solvent
b. a nonpolar solute and a polar solvent
c. a nonpolar solute and a nonpolar solvent

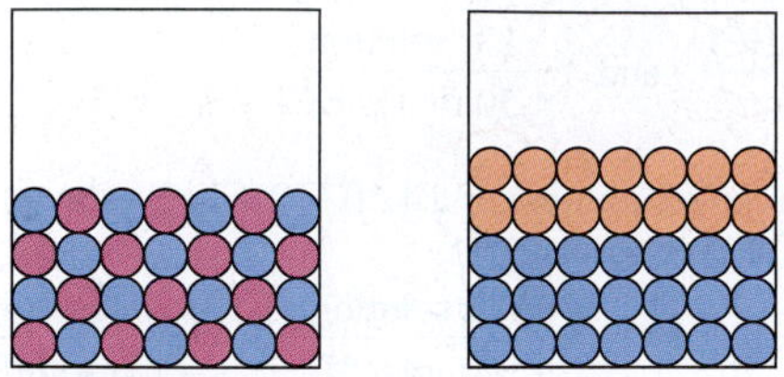

1 2

9.92 If all the solute is dissolved in diagram **1**, how would heating or cooling the solution cause each of the following changes? (9.3)

a. **2** to **3** **b.** **2** to **1**

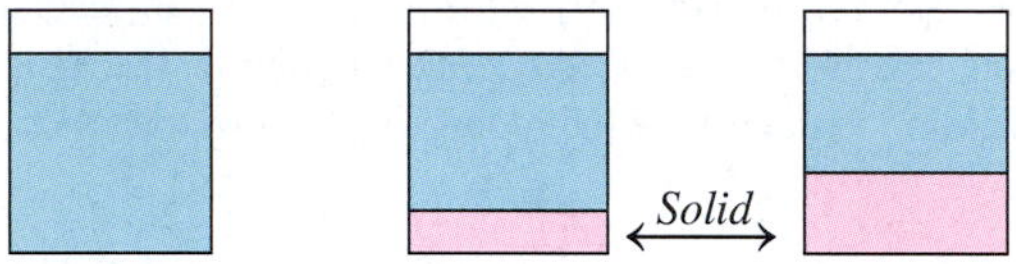
Solid

1 2 3

9.93 Select the diagram that represents the solution formed by a solute that is a (9.2)

a. nonelectrolyte **b.** weak electrolyte
c. strong electrolyte

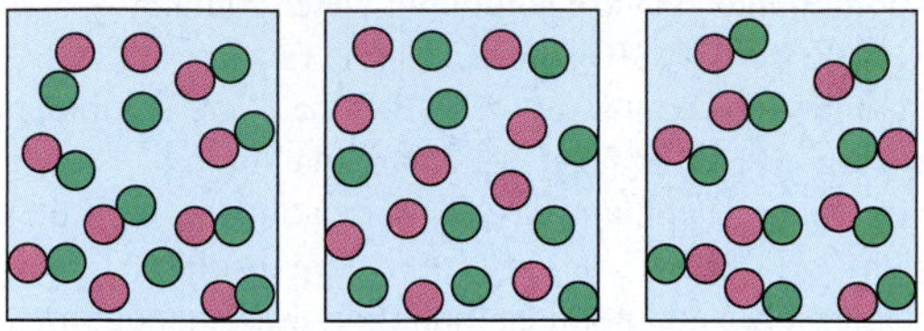

1 2 3

9.94 Select the container that represents the dilution of a 4% (m/v) KCl solution to give each of the following: (9.5)

a. a 2% (m/v) KCl solution
b. a 1% (m/v) KCl solution

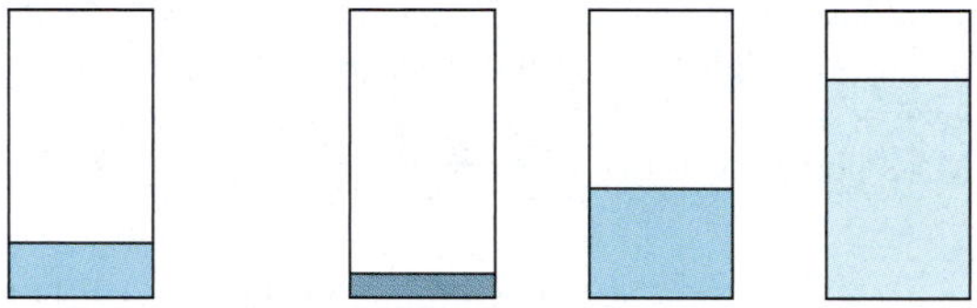

4% (m/v) KCl 1 2 3

Use the following illustration of beakers and solutions for problems 9.95 and 9.96:

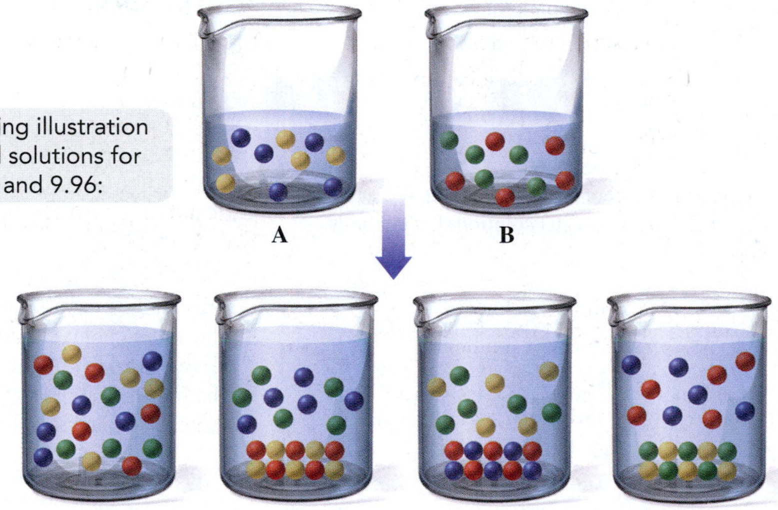
A
B

1 2 3 4

9.95 Use the following: (9.3)

Na^+ Cl^- Ag^+ NO_3^-

a. Select the beaker (**1**, **2**, **3**, or **4**) that contains the products after the solutions in beakers **A** and **B** are mixed.
b. If an insoluble ionic compound forms, write the ionic equation.
c. If a reaction occurs, write the net ionic equation.

9.96 Use the following: (9.3)

K^+ NO_3^- NH_4^+ Br^-

a. Select the beaker (**1**, **2**, **3**, or **4**) that contains the products after the solutions in beakers **A** and **B** are mixed.
b. If an insoluble ionic compound forms, write the ionic equation.
c. If a reaction occurs, write the net ionic equation.

9.97 A pickle is made by soaking a cucumber in brine, a salt-water solution. What makes the smooth cucumber become wrinkled like a prune? (9.6)

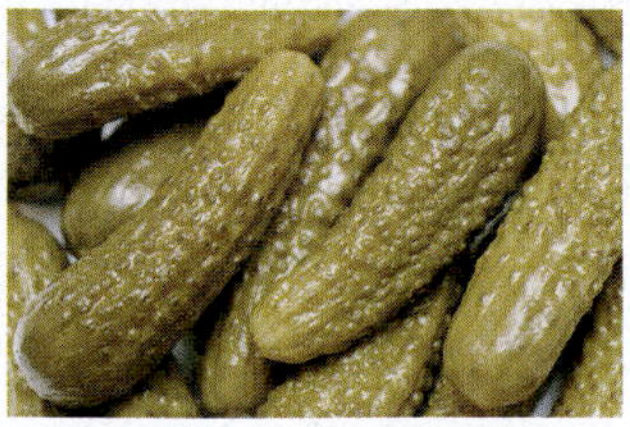

9.98 Why do lettuce leaves in a salad wilt after a vinaigrette dressing containing salt is added? (9.6)

9.99 A semipermeable membrane separates two compartments, **A** and **B**. If the levels in **A** and **B** are equal initially, select the diagram that illustrates the final levels in **a** to **d**: (9.6)

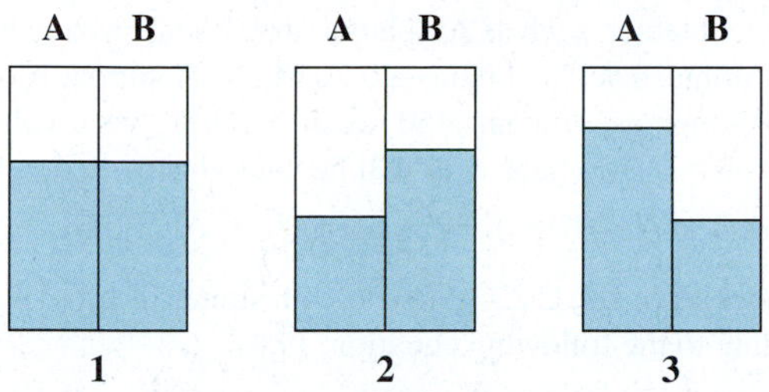
A B
A B
A B
1
2
3

Solution in A	Solution in B
a. 2% (m/v) starch	8% (m/v) starch
b. 1% (m/v) starch	1% (m/v) starch
c. 5% (m/v) sucrose	1% (m/v) sucrose
d. 0.1% (m/v) sucrose	1% (m/v) sucrose

9.100 Select the diagram that represents the shape of a red blood cell when placed in each of the following **a** to **e**: (9.6)

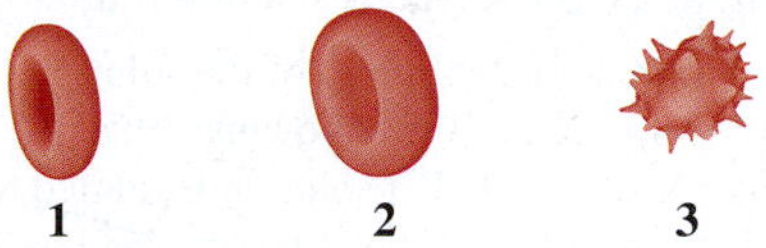
1
2
3
Normal red blood cell

a. 0.9% (m/v) NaCl solution
b. 10% (m/v) glucose solution
c. 0.01% (m/v) NaCl solution
d. 5% (m/v) glucose solution
e. 1% (m/v) glucose solution

ADDITIONAL PRACTICE PROBLEMS

9.101 Why does iodine dissolve in hexane, but not in water? (9.1)

9.102 How do temperature and pressure affect the solubility of solids and gases in water? (9.3)

9.103 Potassium nitrate has a solubility of 32 g of KNO_3 in 100. g of H_2O at 20 °C. Determine if each of the following forms an unsaturated or saturated solution at 20 °C: (9.3)
a. adding 32 g of KNO_3 to 200. g of H_2O
b. adding 19 g of KNO_3 to 50. g of H_2O
c. adding 68 g of KNO_3 to 150. g of H_2O

9.104 Potassium chloride has a solubility of 43 g of KCl in 100. g of H_2O at 50 °C. Determine if each of the following forms an unsaturated or saturated solution at 50 °C: (9.3)
a. adding 25 g of KCl to 100. g of H_2O
b. adding 15 g of KCl to 25 g of H_2O
c. adding 86 g of KCl to 150. g of H_2O

9.105 Indicate whether each of the following ionic compounds is soluble or insoluble in water: (9.3)
a. KCl
b. $MgSO_4$
c. CuS
d. $AgNO_3$
e. $Ca(OH)_2$

9.106 Indicate whether each of the following ionic compounds is soluble or insoluble in water: (9.3)
a. $CuCO_3$
b. FeO
c. $Mg_3(PO_4)_2$
d. $(NH_4)_2SO_4$
e. $NaHCO_3$

9.107 Write the net ionic equation to show the formation of a solid (insoluble ionic compound) when the following solutions are mixed. Write *none* if no solid forms. (9.3)
a. $AgNO_3(aq)$ and $LiBr(aq)$
b. $NaCl(aq)$ and $KNO_3(aq)$
c. $Na_2SO_4(aq)$ and $BaCl_2(aq)$

9.108 Write the net ionic equation to show the formation of a solid (insoluble ionic compound) when the following solutions are mixed. Write *none* if no solid forms. (9.3)
a. $Ca(NO_3)_2(aq)$ and $Na_2S(aq)$
b. $Na_3PO_4(aq)$ and $Pb(NO_3)_2(aq)$
c. $FeCl_3(aq)$ and $NH_4NO_3(aq)$

9.109 Calculate the mass percent (m/m) of a solution containing 15.5 g of Na_2SO_4 and 75.5 g of H_2O. (9.4)

9.110 Calculate the mass percent (m/m) of a solution containing 26 g of K_2CO_3 and 724 g of H_2O. (9.4)

9.111 How many milliliters of a 12% (v/v) propyl alcohol solution would you need to obtain 4.5 mL of propyl alcohol? (9.4)

9.112 An 80-proof brandy is a 40.% (v/v) ethanol solution. The "proof" is twice the percent concentration of alcohol in the beverage. How many milliliters of alcohol are present in 750 mL of brandy? (9.4)

9.113 How many liters of a 12% (m/v) KOH solution would you need to obtain 86.0 g of KOH? (9.4)

9.114 How many liters of a 5.0% (m/v) glucose solution would you need to obtain 75 g of glucose? (9.4)

9.115 What is the molarity of a solution containing 8.0 g of NaOH in 400. mL of NaOH solution? (9.4)

9.116 What is the molarity of a solution containing 15.6 g of KCl in 274 mL of KCl solution? (9.4)

9.117 How many grams of solute are in each of the following solutions? (9.4)
a. 2.5 L of a 3.0 M $Al(NO_3)_3$ solution
b. 75 mL of a 0.50 M $C_6H_{12}O_6$ solution
c. 235 mL of a 1.80 M LiCl solution

9.118 How many grams of solute are in each of the following solutions? (9.4)
a. 0.428 L of a 0.450 M K_2CO_3 solution
b. 10.5 mL of a 2.50 M $AgNO_3$ solution
c. 28.4 mL of a 6.00 M H_3PO_4 solution

9.119 Calculate the final concentration of the solution when water is added to prepare each of the following: (9.5)
a. 25.0 mL of a 0.200 M NaBr solution is diluted to 50.0 mL
b. 15.0 mL of a 12.0% (m/v) K_2SO_4 solution is diluted to 40.0 mL
c. 75.0 mL of a 6.00 M NaOH solution is diluted to 255 mL

9.120 Calculate the final concentration of the solution when water is added to prepare each of the following: (9.5)
a. 25.0 mL of a 18.0 M HCl solution is diluted to 500. mL
b. 50.0 mL of a 15.0% (m/v) NH_4Cl solution is diluted to 125 mL
c. 4.50 mL of a 8.50 M KOH solution is diluted to 75.0 mL

9.121 What is the initial volume, in milliliters, needed to prepare each of the following diluted solutions? (9.5)
a. 250 mL of a 3.0% (m/v) HCl solution using a 10.0% (m/v) HCl solution
b. 500. mL of a 0.90% (m/v) NaCl solution using a 5.0% (m/v) NaCl solution
c. 350. mL of a 2.00 M NaOH solution using a 6.00 M NaOH solution

9.122 What is the initial volume, in milliliters, needed to prepare each of the following diluted solutions? (9.5)
a. 250 mL of a 5.0% (m/v) glucose solution using a 20.% (m/v) glucose solution
b. 45.0 mL of a 1.0% (m/v) $CaCl_2$ solution using a 5.0% (m/v) $CaCl_2$ solution
c. 100. mL of a 6.00 M H_2SO_4 solution using a 18.0 M H_2SO_4 solution

9.123 Calculate the freezing point of each of the following solutions: (9.6)
a. 0.580 mole of lactose, a nonelectrolyte, dissolved in 1.00 kg of water
b. 45.0 g of KCl, a strong electrolyte, dissolved in 1.00 kg of water
c. 1.5 moles of K_3PO_4, a strong electrolyte, dissolved in 1.00 kg of water

9.124 Calculate the boiling point of each of the following solutions: (9.6)
a. 175 g of glucose, $C_6H_{12}O_6$, a nonelectrolyte, dissolved in 1.00 kg of water
b. 1.8 moles of $CaCl_2$, a strong electrolyte, dissolved in 1.00 kg of water
c. 50.0 g of $LiNO_3$, a strong electrolyte, dissolved in 1.00 kg of water

Clinical Applications

9.125 An antacid tablet, such as Amphojel, may be taken to reduce excess stomach acid, which is a 0.20 M HCl solution. If one dose of Amphojel contains 450 mg of $Al(OH)_3$, what volume, in milliliters, of stomach acid will be neutralized? (9.4)

$$3HCl(aq) + Al(OH)_3(s) \longrightarrow 3H_2O(l) + AlCl_3(aq)$$

9.126 Calcium carbonate, $CaCO_3$, reacts with stomach acid, HCl, according to the following equation: (9.4)

$$2HCl(aq) + CaCO_3(s) \longrightarrow CO_2(g) + H_2O(l) + CaCl_2(aq)$$

Tums, an antacid, contains $CaCO_3$. If Tums is added to 20.0 mL of a 0.400 M HCl solution, how many grams of CO_2 gas are produced?

9.127 A patient on dialysis has a high level of urea, a high level of sodium, and a low level of potassium in the blood. Why is the dialyzing solution prepared with a high level of potassium but no sodium or urea? (9.6)

9.128 Why would a dialysis unit (artificial kidney) use isotonic concentrations of NaCl, KCl, $NaHCO_3$, and glucose in the dialysate? (9.6)

CHALLENGE PROBLEMS

The following problems are related to the topics in this chapter. However, they do not all follow the chapter order, and they require you to combine concepts and skills from several Sections. These problems will help you increase your critical thinking skills and prepare for your next exam.

9.129 In a laboratory experiment, a 10.0-mL sample of NaCl solution is poured into an evaporating dish with a mass of 24.10 g. The combined mass of the evaporating dish and NaCl solution is 36.15 g. After heating, the evaporating dish and dry NaCl have a combined mass of 25.50 g. (9.4)
a. What is the mass percent (m/m) of the NaCl solution?
b. What is the molarity (M) of the NaCl solution?
c. If water is added to 10.0 mL of the initial NaCl solution to give a final volume of 60.0 mL, what is the molarity of the diluted NaCl solution?

9.130 In a laboratory experiment, a 15.0-mL sample of KCl solution is poured into an evaporating dish with a mass of 24.10 g. The combined mass of the evaporating dish and KCl solution is 41.50 g. After heating, the evaporating dish and dry KCl have a combined mass of 28.28 g. (9.4)
a. What is the mass percent (m/m) of the KCl solution?
b. What is the molarity (M) of the KCl solution?
c. If water is added to 10.0 mL of the initial KCl solution to give a final volume of 60.0 mL, what is the molarity of the diluted KCl solution?

9.131 Potassium fluoride has a solubility of 92 g of KF in 100. g of H_2O at 18 °C. Determine if each of the following mixtures forms an unsaturated or saturated solution at 18 °C: (9.3)
a. adding 35 g of KF to 25 g of H_2O
b. adding 42 g of KF to 50. g of H_2O
c. adding 145 g of KF to 150. g of H_2O

9.132 Lithium chloride has a solubility of 55 g of LiCl in 100. g of H_2O at 25 °C. Determine if each of the following mixtures forms an unsaturated or saturated solution at 25 °C: (9.3)
a. adding 10 g of LiCl to 15 g of H_2O
b. adding 25 g of LiCl to 50. g of H_2O
c. adding 75 g of LiCl to 150. g of H_2O

9.133 A solution is prepared with 70.0 g of HNO_3 and 130.0 g of H_2O. The HNO_3 solution has a density of 1.21 g/mL. (9.4)
a. What is the mass percent (m/m) of the HNO_3 solution?
b. What is the total volume, in milliliters, of the solution?
c. What is the mass/volume percent (m/v) of the solution?
d. What is the molarity (M) of the solution?

9.134 A solution is prepared by dissolving 22.0 g of NaOH in 118.0 g of water. The NaOH solution has a density of 1.15 g/mL. (9.4)
a. What is the mass percent (m/m) of the NaOH solution?
b. What is the total volume, in milliliters, of the solution?
c. What is the mass/volume percent (m/v) of the solution?
d. What is the molarity (M) of the solution?

Clinical Applications

9.135 A Pedialyte solution contains Na^+ 45 mEq/L, K^+ 20. mEq/L, Cl^- 35 mEq/L, $citrate^{3-}$ 30. mEq/L, and glucose ($C_6H_{12}O_6$) 25 g/L. (9.6)
a. What is the total positive charge of the electrolytes?
b. What is the total negative charge of the electrolytes?

9.136 A lactated Ringer's solution with 5% glucose contains Na^+ 130. mEq/L, K^+ 4 mEq/L, Ca^{2+} 3 mEq/L, Cl^- 109 mEq/L, $lactate^-$ 28 mEq/L, and glucose ($C_6H_{12}O_6$) 50. g/L. (9.6)
a. What is the total positive charge of the electrolytes?
b. What is the total negative charge of the electrolytes?

ANSWERS

9.1 **a.** NaCl, solute; H_2O, solvent
b. H_2O, solute; C_2H_6O, solvent
c. O_2, solute; N_2, solvent

9.3 The polar water molecules pull the K^+ and I^- ions away from the solid and into solution, where they are hydrated.

9.5 **a.** water **b.** CCl_4
c. water **d.** CCl_4

9.7 In a solution of KF, only the ions of K^+ and F^- are present in the solvent. In an HF solution, there are a few ions of H^+ and F^- present but mostly dissolved HF molecules.

9.9 **a.** $KCl(s) \xrightarrow{H_2O} K^+(aq) + Cl^-(aq)$
b. $CaCl_2(s) \xrightarrow{H_2O} Ca^{2+}(aq) + 2Cl^-(aq)$
c. $K_3PO_4(s) \xrightarrow{H_2O} 3K^+(aq) + PO_4^{3-}(aq)$
d. $Fe(NO_3)_3(s) \xrightarrow{H_2O} Fe^{3+}(aq) + 3NO_3^-(aq)$

9.11 **a.** mostly molecules and a few ions
b. ions only
c. molecules only

9.13 **a.** strong electrolyte
b. weak electrolyte
c. nonelectrolyte

9.15 **a.** 1 Eq **b.** 2 Eq
c. 2 Eq **d.** 6 Eq

9.17 0.154 mole of Na^+, 0.154 mole of Cl^-

9.19 55 mEq/L

9.21 **a.** saturated **b.** unsaturated **c.** unsaturated

9.23 **a.** unsaturated **b.** unsaturated **c.** saturated

9.25 **a.** 68 g of KCl **b.** 12 g of KCl

9.27 **a.** The solubility of solid solutes typically increases as temperature increases.
b. The solubility of a gas is less at a higher temperature.
c. Gas solubility is less at a higher temperature and the CO_2 pressure in the can is increased.

9.29 **a.** soluble **b.** insoluble **c.** insoluble
d. soluble **e.** soluble

9.31 **a.** No solid forms.
b. $2Ag^+(aq) + 2NO_3^-(aq) + 2K^+(aq) + S^{2-}(aq) \longrightarrow Ag_2S(s) + 2K^+(aq) + 2NO_3^-(aq)$
$2Ag^+(aq) + S^{2-}(aq) \longrightarrow Ag_2S(s)$
c. $Ca^{2+}(aq) + 2Cl^-(aq) + 2Na^+(aq) + SO_4^{2-}(aq) \longrightarrow CaSO_4(s) + 2Na^+(aq) + 2Cl^-(aq)$
$Ca^{2+}(aq) + SO_4^{2-}(aq) \longrightarrow CaSO_4(s)$
d. $3Cu^{2+}(aq) + 6Cl^-(aq) + 6Li^+(aq) + 2PO_4^{3-}(aq) \longrightarrow Cu_3(PO_4)_2(s) + 6Li^+(aq) + 6Cl^-(aq)$
$3Cu^{2+}(aq) + 2PO_4^{3-}(aq) \longrightarrow Cu_3(PO_4)_2(s)$

9.33 **a.** 17% (m/m) KCl solution
b. 5.3% (m/m) sucrose solution
c. 10.% (m/m) $CaCl_2$ solution

9.35 79.9 mL of alcohol

9.37 **a.** 30.% (m/v) Na_2SO_4 solution
b. 11% (m/v) sucrose solution

9.39 **a.** 2.5 g of KCl **b.** 50. g of NH_4Cl
c. 25.0 mL of acetic acid

9.41 **a.** 20. g of $LiNO_3$ solution **b.** 400. mL of KOH solution
c. 20. mL of formic acid solution

9.43 **a.** 0.500 M glucose solution **b.** 0.0356 M KOH solution
c. 0.250 M NaCl solution

9.45 **a.** 120. g of NaOH **b.** 59.6 g of KCl
c. 5.47 g of HCl

9.47 **a.** 1.50 L of KBr solution **b.** 10.0 L of NaCl solution
c. 62.5 mL of $Ca(NO_3)_2$ solution

9.49 **a.** 983 mL **b.** 1.70×10^3 mL **c.** 464 mL

9.51 **a.** 10.4 g of $PbCl_2$ **b.** 18.8 mL of $Pb(NO_3)_2$ solution
c. 1.20 M KCl solution

9.53 **a.** 206 mL of HCl solution **b.** 11.2 L of H_2 gas
c. 9.12 M HCl solution

9.55 **a.** 20. g of mannitol **b.** 240 g of mannitol

9.57 **a.** 2 L of glucose solution **b.** 220 mL of NaCl solution

9.59 0.50 g of $CaCl_2$

9.61 Adding water (solvent) to the soup increases the volume and decreases the tomato soup concentration.

9.63 **a.** 2.0 M HCl solution
b. 2.0 M NaOH solution
c. 2.5% (m/v) KOH solution
d. 3.0% (m/v) H_2SO_4 solution

9.65 **a.** 80. mL of HCl solution
b. 250 mL of LiCl solution
c. 600. mL of H_3PO_4 solution
d. 180 mL of glucose solution

9.67 **a.** 12.8 mL of the HNO_3 solution
b. 11.9 mL of the $MgCl_2$ solution
c. 1.88 mL of the KCl solution

9.69 1.0×10^2 mL of glucose solution

9.71 **a.** solution **b.** suspension

9.73 **a.** 2.0 moles of ethylene glycol in 1.0 kg of water will have a lower freezing point because it has more solute particles in solution.
b. 0.50 mole of $MgCl_2$ in 1.0 kg of water has a lower freezing point because each $MgCl_2$ dissociates in water to give three solute particles, whereas each KCl dissociates to give only two solute particles.

9.75 **a.** -2.53 °C **b.** -15.6 °C **c.** -5.54 °C

9.77 **a.** 10% (m/v) starch solution
b. from the 1% (m/v) starch solution into the 10% (m/v) starch solution
c. 10% (m/v) starch solution

9.79 **a.** B 10% (m/v) sucrose solution
b. A 8% (m/v) albumin solution
c. B 10% (m/v) starch solution

9.81 **a.** 0.15 mole **b.** 2.8 moles

9.83 **a.** hypotonic **b.** hypotonic
c. isotonic **d.** hypertonic

9.85 **a.** NaCl **b.** alanine
c. NaCl **d.** urea

9.87 **a.** 130 mEq/L **b.** above normal level

9.89 **a.** Ca^{2+}, Mg^{2+}, HCO_3^- **b.** K^+, Na^+, Cl^-

9.91 **a.** 1 **b.** 2 **c.** 1

9.93 **a.** 3 **b.** 1 **c.** 2

9.95 **a.** beaker 3
b. $Na^+(aq) + Cl^-(aq) + Ag^+(aq) + NO_3^-(aq) \longrightarrow AgCl(s) + Na^+(aq) + NO_3^-(aq)$
c. $Ag^+(aq) + Cl^-(aq) \longrightarrow AgCl(s)$

9.97 The skin of the cucumber acts like a semipermeable membrane, and the more dilute solution inside flows into the brine.

9.99 **a.** 2 **b.** 1 **c.** 3 **d.** 2

9.101 Because iodine is a nonpolar molecule, it will dissolve in hexane, a nonpolar solvent. Iodine does not dissolve in water because water is a polar solvent.

9.103 **a.** unsaturated solution **b.** saturated solution
c. saturated solution

9.105 **a.** soluble **b.** soluble **c.** insoluble
d. soluble **e.** insoluble

9.107 **a.** $Ag^+(aq) + Br^-(aq) \longrightarrow AgBr(s)$
b. none
c. $Ba^{2+}(aq) + SO_4^{2-}(aq) \longrightarrow BaSO_4(s)$

9.109 17.0% (m/m) Na_2SO_4 solution

9.111 38 mL of propyl alcohol solution

9.113 0.72 L of KOH solution

9.115 0.50 M NaOH solution

9.117 **a.** 1600 g of $Al(NO_3)_3$ **b.** 6.8 g of $C_6H_{12}O_6$
c. 17.9 g of LiCl

9.119 **a.** 0.100 M NaBr solution
b. 4.50% (m/v) K_2SO_4 solution
c. 1.76 M NaOH solution

9.121 **a.** 75 mL of 10.0% (m/v) HCl solution
b. 90. mL of 5.0% (m/v) NaCl solution
c. 117 mL of 6.00 M NaOH solution

9.123 **a.** -1.08 °C **b.** -2.25 °C **c.** -11 °C

9.125 87 mL of HCl solution

9.127 A dialysis solution contains no sodium or urea so that these substances will flow out of the blood into the solution. High levels of potassium are maintained in the dialyzing solution so that the potassium will dialyze into the blood.

9.129 **a.** 11.6% (m/m) NaCl solution
b. 2.40 M NaCl solution
c. 0.400 M NaCl solution

9.131 **a.** saturated **b.** unsaturated
c. saturated

9.133 **a.** 35.0% (m/m) HNO_3 solution
b. 165 mL
c. 42.4% (m/v) HNO_3 solution
d. 6.73 M HNO_3 solution

9.135 **a.** 65 mEq/L **b.** 65 mEq/L

10 Reaction Rates and Chemical Equilibrium

Andrew is born 12 weeks early with a breathing problem called *infant respiratory distress syndrome* (IRDS). He is limp, purple, and not breathing. IRDS is caused by a lack of surfactant in underdeveloped lungs. Normally produced in the pulmonary alveoli, surfactants help the lungs inflate and prevent the air sacs from collapsing. Without surfactants, a premature infant has difficulty breathing, which leads to decreased ventilation and hypoxia. Lack of oxygen can lead to the injury or death of cells and ultimately to irreversible brain damage. To prevent this, Andrew's metabolism needs to be slowed down.

At the hospital, Judy, a neonatal nurse, wraps Andrew in a cooling blanket that lowers his body temperature to 33.5 °C. She attaches electrodes to his head to monitor his brain activity. With a decrease in body temperature, Andrew's demand for oxygen is reduced. Artificial surfactants are administered to help him breathe more easily. When Andrew's oxygen levels in his bloodstream fall too low, Judy inserts a baby cannula in his nose that provides an air–oxygen mixture. Then more O_2 is picked up by Andrew's hemoglobin (Hb) at the lungs and released from the hemoglobin in the tissues.

$$Hb(aq) + O_2(g) \rightleftarrows HbO_2(aq)$$

After Andrew successfully manages the three-day danger period, his body temperature is slowly restored to normal. Two weeks later, he is discharged from the hospital.

CAREER

Neonatal Nurse

A neonatal nurse works with newborns that are premature, have birth defects, cardiac malformations, and surgical problems. Most neonatal nurses care for infants from the time of birth until they are discharged from the hospital.

A neonatal nurse monitors life support machines, administers medications, operates ventilators, and resuscitates a premature infant who stops breathing. Many neonatal nurses take a national test to become part of a neonatal team or participate on an oxygenation team that provides heart–lung bypass for critically ill infants.

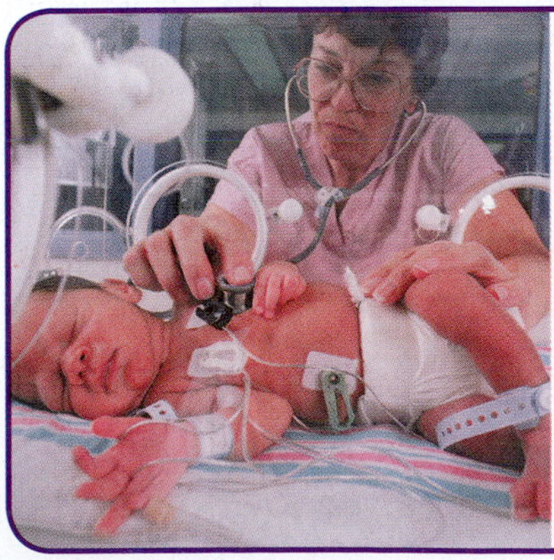

CLINICAL UPDATE

An Iron-Rich Diet for Children's Anemia

Andrew has grown to be a toddler and is in his age group for height and weight. However, he shows signs of fatigue with headaches. Read about the decreased hemoglobin level in Andrew's blood and how his mother changes his diet to include more iron-containing foods in the **CLINICAL UPDATE An Iron-Rich Diet for Children's Anemia**, page 376.

LOOKING AHEAD

10.1 Rates of Reactions

LEARNING GOAL Describe how temperature, concentration, and catalysts affect the rate of a reaction.

Earlier we looked at chemical reactions and determined the amounts of substances that react and the products that form. Now we are interested in how fast a reaction goes. If we know how fast a medication acts on the body, we can adjust the time over which the medication is taken. Some reactions such as explosions or the formation of precipitates in a solution are very fast. When we roast a turkey or bake a cake, the reaction is slower. Some reactions such as the tarnishing of silver and the aging of the body are much slower (see **FIGURE 10.1**). We have seen that some reactions need energy while other reactions produce energy. In this chapter, we will also look at the effect of changing the concentrations of reactants or products on the rate of reaction.

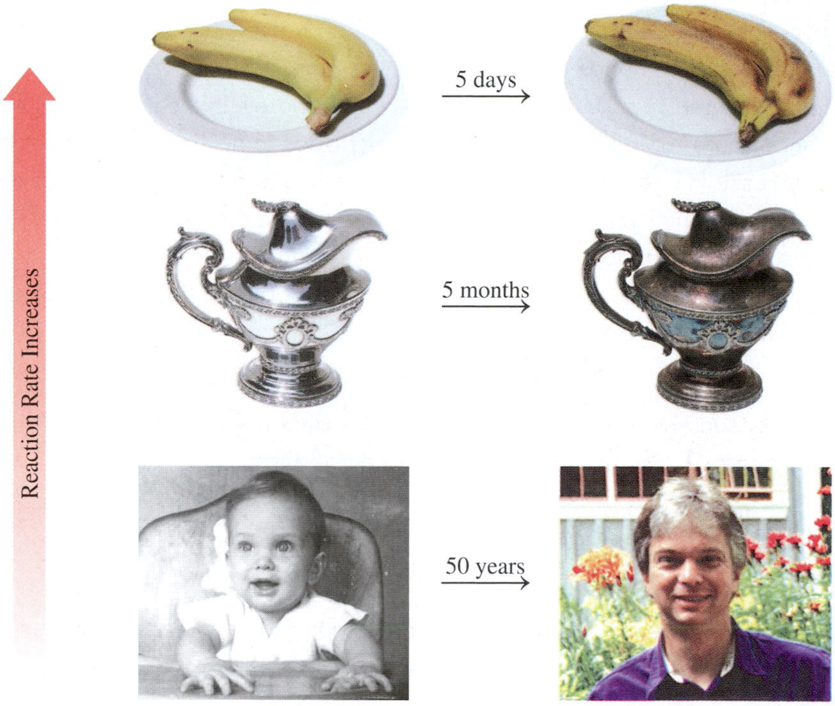
Reaction Rate Increases
5 days
5 months
50 years

FIGURE 10.1 ▶ Reaction rates vary greatly for everyday processes. A banana ripens in a few days, silver tarnishes in a few months, while the aging process of humans takes many years.

Q How would you compare the rates of the reaction that forms sugars in plants by photosynthesis with the reactions that digest sugars in the body?

For a chemical reaction to take place, the molecules of the reactants must come in contact with each other. The **collision theory** indicates that a reaction takes place only when molecules collide with the proper orientation and sufficient energy. Many collisions can occur, but only a few actually lead to the formation of product. For example, consider the reaction of nitrogen and oxygen molecules (see **FIGURE 10.2**). To form nitrogen oxide (NO), the collisions between N_2 and O_2 molecules must place the atoms in the proper alignment. If the molecules are not aligned properly, no reaction takes place.

Activation Energy

Even when a collision has the proper orientation, there still must be sufficient energy to break the bonds between the atoms of the reactants. The **activation energy** is the minimum amount of energy required to break the bonds between atoms of the reactants. The concept of activation energy is analogous to climbing a hill. To reach a destination on the other side, we must have the energy needed to climb to the top of the hill. Once we are at the top, we can run down the other side. The energy needed to get us from our starting point to the top of the hill would be our activation energy.

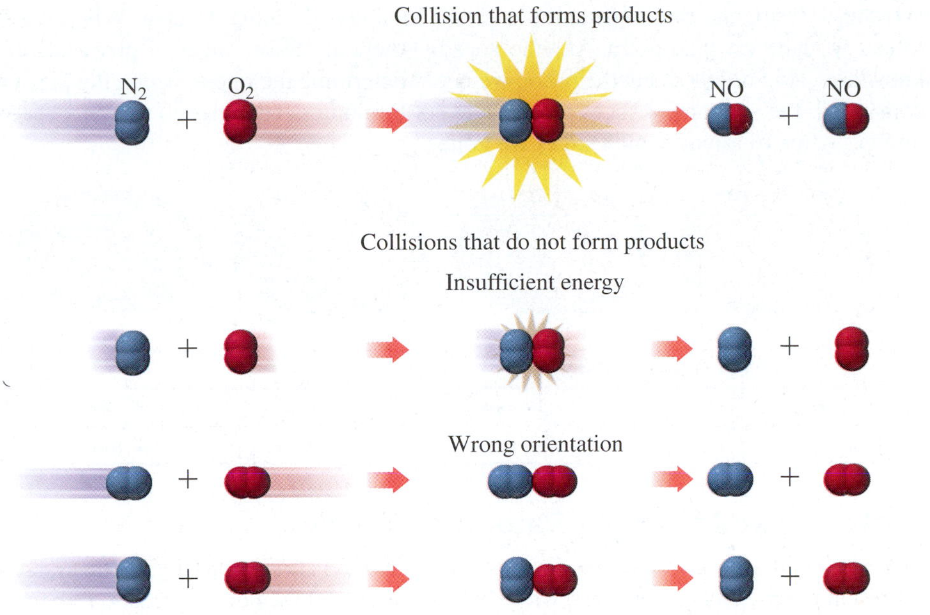
Collision that forms products
N_2
O_2
NO
NO
Collisions that do not form products
Insufficient energy
Wrong orientation

FIGURE 10.2 ▶ Reacting molecules must collide, have a minimum amount of energy, and have the proper orientation to form product.

Q What happens when reacting molecules collide with the minimum energy but don't have the proper orientation?

In the same way, a collision must provide enough energy to push the reactants to the top of the energy hill. Then the reactants may be converted to products. If the energy provided by the collision is less than the activation energy, the molecules simply bounce apart, and no reaction occurs. The features that lead to a successful reaction are summarized next (see **FIGURE 10.3**).

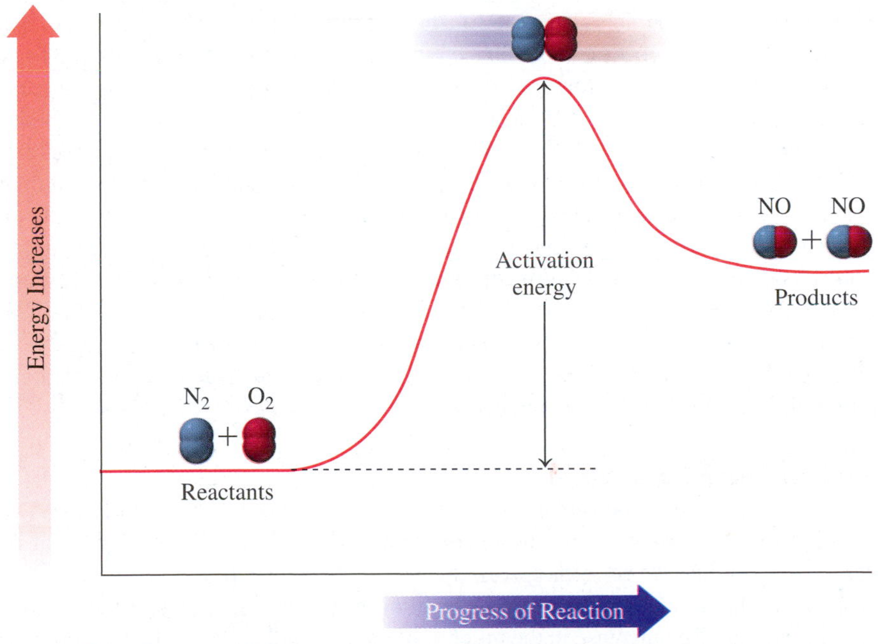
Energy Increases
NO
NO
Activation energy
Products
N_2
O_2
Reactants
Progress of Reaction

FIGURE 10.3 ▶ The activation energy is the minimum energy needed to convert the colliding molecules into product.

Q What happens in a collision of reacting molecules that have the proper orientation, but not the energy of activation?

Three Conditions Required for a Reaction to Occur

1. **Collision** The reactants must collide.
2. **Orientation** The reactants must align properly to break and form bonds.
3. **Energy** The collision must provide the energy of activation.

Rate of Reaction

The **rate** (or speed) **of reaction** is determined by measuring the amount of a reactant used up, or the amount of a product formed, in a certain period of time.

$$\text{Rate of reaction} = \frac{\text{change in concentration of reactant or product}}{\text{change in time}}$$

We can describe the rate of reaction with the analogy of eating a pizza. When we start to eat, we have a whole pizza. As time goes by, there are fewer slices of pizza left. If we know how long it took to eat the pizza, we could determine the rate at which the pizza was consumed. Let's assume 4 slices are eaten every 8 minutes. That gives a rate of $\frac{1}{2}$ slice per minute. After 16 minutes, all 8 slices are gone.

Rate at Which Pizza Slices Are Eaten				
Slices Eaten	0	4 slices	6 slices	8 slices
Time (min)	0	8 min	12 min	16 min

Factors that Affect the Rate of a Reaction

Reactions with low activation energies go faster than reactions with high activation energies. Some reactions go very fast, while others are very slow. For any reaction, the rate is affected by changes in temperature, changes in the concentration of the reactants, and the addition of catalysts.

ENGAGE

Why would an increase in temperature increase the rate of a reaction?

Temperature

At higher temperatures, the increase in kinetic energy of the reactants makes them move faster and collide more often, and it provides more collisions with the required energy of activation. Reactions almost always go faster at higher temperatures. For every 10 °C increase in temperature, most reaction rates approximately double. If we want food to cook faster, we increase the temperature. When body temperature rises, there is an increase in the pulse rate, rate of breathing, and metabolic rate. If we are exposed to extreme heat, we may experience *heat stroke*, which is a condition that occurs when body temperature goes above 40.5 °C (105 °F). If the body loses its ability to regulate temperature, body temperature continues to rise and may cause damage to the brain and internal organs. On the other hand, we slow down a reaction by decreasing the temperature. For example, we refrigerate perishable foods to make them last longer. In some cardiac surgeries, body temperature is decreased to 28 °C so the heart can be stopped and less oxygen is required by the brain. This is also the reason why some people have survived submersion in icy lakes for long periods of time. Cool water or an ice blanket may also be used to decrease the body temperature of a person with hyperthermia or heat stroke.

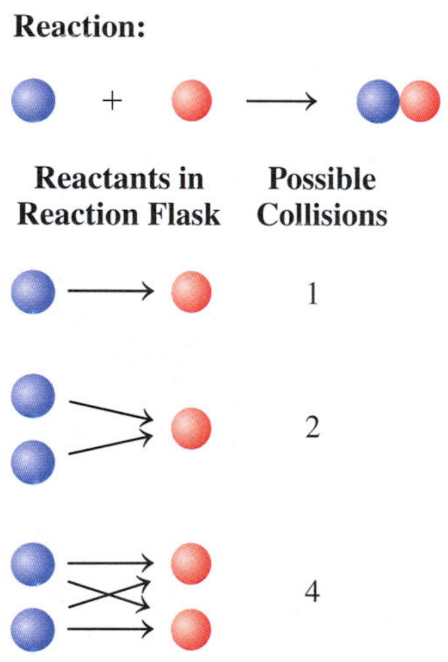
Reaction:
+
Reactants in Reaction Flask
Possible Collisions
1
2
4

FIGURE 10.4 ▶ Increasing the concentration of a reactant increases the number of collisions that are possible.

Q Why does doubling the reactants increase the rate of reaction?

Concentrations of Reactants

For virtually all reactions, the rate of a reaction increases when the concentration of the reactants increases. When there are more reacting molecules, more collisions that form products can occur, and the reaction goes faster (see **FIGURE 10.4**). For example, a patient having difficulty breathing may be given a breathing mixture with a higher oxygen content than the atmosphere. The increase in the number of oxygen molecules in the lungs increases the rate at which oxygen combines with hemoglobin. The increased rate of oxygenation of the blood means that the patient can breathe more easily.

$$\underset{\text{Hemoglobin}}{Hb(aq)} + \underset{\text{Oxygen}}{O_2(g)} \longrightarrow \underset{\text{Oxyhemoglobin}}{HbO_2(aq)}$$

Catalysts

Another way to speed up a reaction is to lower the *energy of activation*. The energy of activation is the minimum energy needed to break apart the bonds of the reacting molecules. If a collision provides less than the activation energy, the bonds do not break and the reactant molecules bounce apart. A **catalyst** speeds up a reaction by providing an alternative pathway that has a lower energy of activation. When activation energy is lowered, more collisions provide sufficient energy for reactants to form product. During a reaction, a catalyst is not changed or consumed.

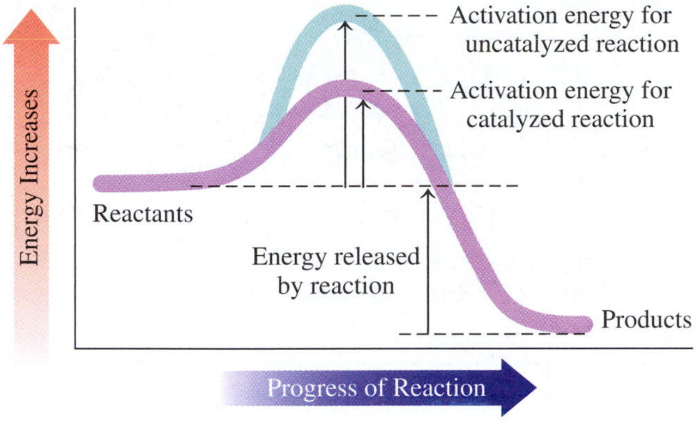
Energy Increases
Activation energy for uncatalyzed reaction
Activation energy for catalyzed reaction
Reactants
Energy released by reaction
Products
Progress of Reaction

When a catalyst lowers the activation energy, the reaction occurs at a faster rate.

Catalysts have many uses in industry. In the manufacturing of margarine, hydrogen (H_2) is added to vegetable oils. Normally, the reaction is very slow because it has a high activation energy. However, when platinum (Pt) is used as a catalyst, the reaction occurs rapidly. In the body, biocatalysts called *enzymes* make metabolic reactions proceed at rates necessary for proper cellular activity. Enzymes are added to laundry detergents to break down proteins (proteases), starches (amylases), or greases (lipases) that have stained clothes. Such enzymes function at the low temperatures that are used in home washing machines, and they are biodegradable as well.

The factors affecting reaction rates are summarized in **TABLE 10.1**.

TABLE 10.1 Factors That Increase Reaction Rate

Factor	Reason
Increasing temperature	More collisions, more collisions with energy of activation
Increasing reactant concentration	More collisions
Adding a catalyst	Lowers energy of activation

▶ SAMPLE PROBLEM 10.1 Factors That Affect the Rate of Reaction

TRY IT FIRST

Indicate whether each of the following changes will increase, decrease, or have no effect on the rate of reaction:

a. increasing the temperature
b. decreasing the number of reacting molecules
c. adding a catalyst

SOLUTION

a. A higher temperature increases the kinetic energy of the particles, which increases the number of collisions and makes more collisions effective, causing an increase in the rate of reaction.
b. Decreasing the number of reacting molecules decreases the number of collisions and the rate of the reaction.
c. Adding a catalyst increases the rate of reaction by lowering the activation energy, which increases the number of collisions that form product.

STUDY CHECK 10.1

a. How does using an ice blanket on a patient affect the rate of metabolism in the body?
b. How does adding more reacting molecules affect the rate of the reaction?

ANSWER

a. Lowering the temperature will decrease the rate of metabolism.
b. Adding more reacting molecules will increase the rate of the reaction.

TEST

Try Practice Problems 10.1 to 10.6

PRACTICE PROBLEMS

10.1 Rates of Reactions

10.1 **a.** What is meant by the rate of a reaction?
b. Why does bread grow mold more quickly at room temperature than in the refrigerator?

10.2 **a.** How does a catalyst affect the activation energy?
b. Why is pure oxygen used in respiratory distress?

10.3 In the following reaction, what happens to the number of collisions when more $Br_2(g)$ molecules are added?

$$H_2(g) + Br_2(g) \longrightarrow 2HBr(g)$$

10.4 In the following reaction, what happens to the number of collisions when the temperature of the reaction is decreased?

$$2H_2(g) + CO(g) \longrightarrow CH_4O(g)$$

10.5 How would each of the following change the rate of the reaction shown here?

$$2SO_2(g) + O_2(g) \longrightarrow 2SO_3(g)$$

a. adding some $SO_2(g)$ **b.** increasing the temperature
c. adding a catalyst **d.** removing some $O_2(g)$

10.6 How would each of the following change the rate of the reaction shown here?

$$2NO(g) + 2H_2(g) \longrightarrow N_2(g) + 2H_2O(g)$$

a. adding some $NO(g)$ **b.** decreasing the temperature
c. removing some $H_2(g)$ **d.** adding a catalyst

10.2 Chemical Equilibrium

LEARNING GOAL Use the concept of reversible reactions to explain chemical equilibrium.

We consider the *forward reaction* in an equation and assumed that all of the reactants were converted to products. However, most of the time reactants are not completely converted to products because a *reverse reaction* takes place in which products collide to form the reactants. When a reaction proceeds in both a forward and reverse direction, it is said to be *reversible*. We have looked at other reversible processes. For example, the melting of solids to form liquids and the freezing of liquids to solids is a reversible physical change. Even in our daily life we have reversible events. We go from home to school and we return from school to home. We go up an escalator and we come back down. We put money in our bank account and we take money out.

An analogy for a forward and reverse reaction can be found in the phrase "We are going to the grocery store." Although we mention our trip in one direction, we know that we will also return home from the store. Because our trip has both a forward and reverse direction, we can say the trip is reversible. It is not very likely that we would stay at the store forever.

A trip to the grocery store can be used to illustrate another aspect of reversible reactions. Perhaps the grocery store is nearby and we usually walk. However, we can change our rate. Suppose that one day we drive to the store, which increases our rate and gets us to the store faster. Correspondingly, a car also increases the rate at which we return home.

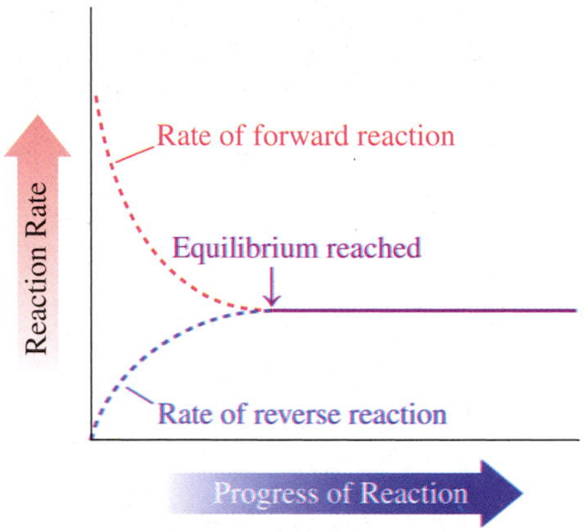
Rate of forward reaction
Equilibrium reached
Rate of reverse reaction
Reaction Rate
Progress of Reaction

As a reaction progresses, the rate of the forward reaction decreases and that of the reverse reaction increases. At equilibrium, the rates of the forward and reverse reactions are equal.

Reversible Chemical Reactions

A **reversible reaction** proceeds in both the forward and reverse direction. That means there are two reaction rates: one is the rate of the forward reaction, and the other is the rate of the reverse reaction. When molecules begin to react, the rate of the forward reaction is faster than the rate of the reverse reaction. As reactants are consumed and products accumulate, the rate of the forward reaction decreases, and the rate of the reverse reaction increases.

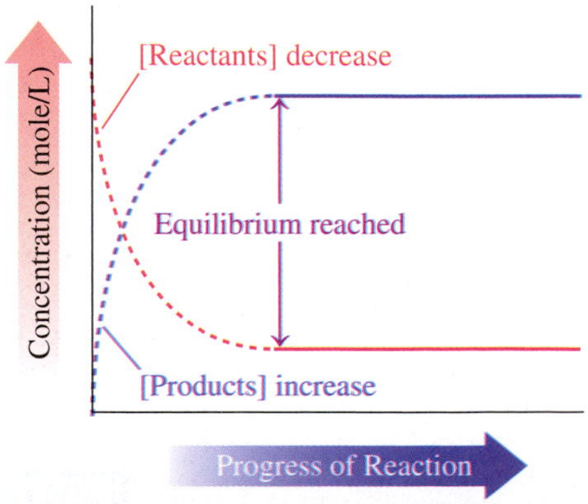
[Reactants] decrease
Equilibrium reached
[Products] increase
Concentration (mole/L)
Progress of Reaction

Equilibrium is reached when there are no further changes in the concentrations of reactants and products.

Equilibrium

Eventually, the rates of the forward and reverse reactions become equal; the reactants form products at the same rate that the products form reactants. A reaction reaches **chemical equilibrium** when no further change takes place in the concentrations of the reactants and products, even though the two reactions continue at equal but opposite rates.

At Equilibrium:

- The rate of the forward reaction is equal to the rate of the reverse reaction.
- No further changes occur in the concentrations of reactants and products, even though the two reactions continue at equal but opposite rates.

Let us look at the process as the reaction of H_2 and I_2 proceeds to equilibrium. Initially, only the reactants H_2 and I_2 are present. Soon, a few molecules of HI are produced by the forward reaction. With more time, additional HI molecules are produced. As the concentration of HI increases, more HI molecules collide and react in the reverse direction. As HI product builds up, the rate of the reverse reaction increases, while the rate of the forward reaction decreases. Eventually, the rates become equal, which means the reaction has reached equilibrium. Even though the concentrations remain constant at equilibrium, the forward and reverse reactions continue to occur. The forward and reverse reactions are usually shown together in a single equation by using a double arrow. A reversible reaction is two opposing reactions that occur at the same time (see **FIGURE 10.5**).

ENGAGE

Why do the concentrations of the reactants decrease before equilibrium is reached?

$$H_2(g) + I_2(g) \underset{\text{Reverse reaction}}{\overset{\text{Forward reaction}}{\rightleftarrows}} 2HI(g)$$

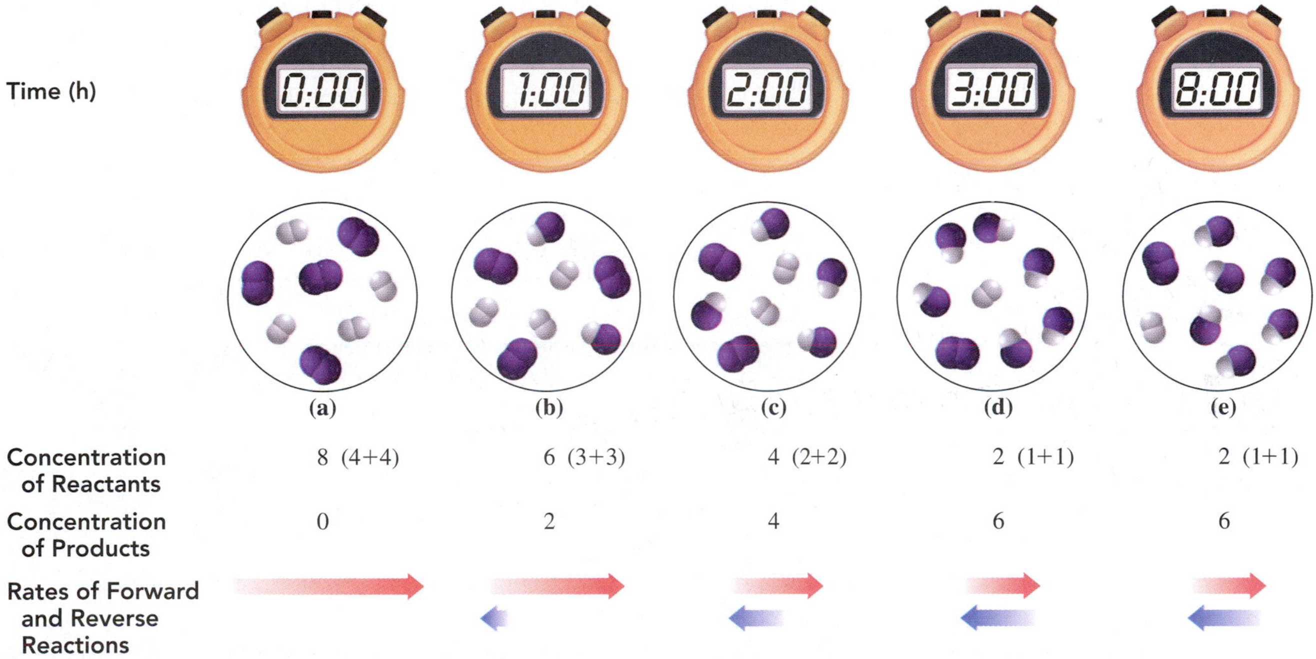
Time (h)
0:00
1:00
2:00
3:00
8:00
(a)
(b)
(c)
(d)
(e)
Concentration of Reactants
8 (4+4)
6 (3+3)
4 (2+2)
2 (1+1)
2 (1+1)
Concentration of Products
0
2
4
6
6
Rates of Forward and Reverse Reactions

FIGURE 10.5 ▶ **(a)** Initially, the reaction flask contains only the reactants H_2 (white) and I_2 (purple). **(b)** The forward reaction between H_2 and I_2 begins to produce HI. **(c)** As the reaction proceeds, there are fewer molecules of H_2 and I_2 and more molecules of HI, which increases the rate of the reverse reaction. **(d)** At equilibrium, the concentrations of reactants H_2 and I_2 and product HI are constant. **(e)** The reaction continues with the rate of the forward reaction equal to the rate of the reverse reaction.

Q How do the rates of the forward and reverse reactions compare once a chemical reaction reaches equilibrium?

▶ SAMPLE PROBLEM 10.2 Reaction Rates and Equilibrium

TRY IT FIRST

Complete each of the following with *change* or *do not change*, *faster* or *slower*, *equal* or *not equal*:

a. Before equilibrium is reached, the concentrations of the reactants and products _____.

b. Initially, reactants placed in a container have a _____ rate of reaction than the rate of reaction of the products.

c. At equilibrium, the rate of the forward reaction is _____ to the rate of the reverse reaction.

SOLUTION

a. Before equilibrium is reached, the concentrations of the reactants and products *change*.
b. Initially, reactants placed in a container have a *faster* rate of reaction than the rate of reaction of the products.
c. At equilibrium, the rate of the forward reaction is *equal* to the rate of the reverse reaction.

STUDY CHECK 10.2

Complete the following statements with *changes* or *does not change*:

a. At equilibrium, the concentration of the reactant _____.
b. The rate of the forward reaction _____ as the reaction proceeds toward equilibrium.

TEST

Try Practice Problems 10.7 to 10.12

ANSWER

a. At equilibrium, the concentration of the reactant *does not change*.
b. The rate of the forward reaction *changes* as the reaction proceeds toward equlibrium.

We can set up a reaction starting with only reactants or with only products. Let's look at the initial reactions in each, the forward and reverse reactions, and the equilibrium mixture that forms (see **FIGURE 10.6**).

$$2SO_2(g) + O_2(g) \rightleftarrows 2SO_3(g)$$

If we start with only the reactants SO_2 and O_2 in the container, the reaction to form SO_3 takes place until equilibrium is reached. However, if we start with only the product SO_3 in the container, the reaction to form SO_2 and O_2 takes place until equilibrium is reached. In both containers, the equilibrium mixture contains the same concentrations of SO_2, O_2, and SO_3.

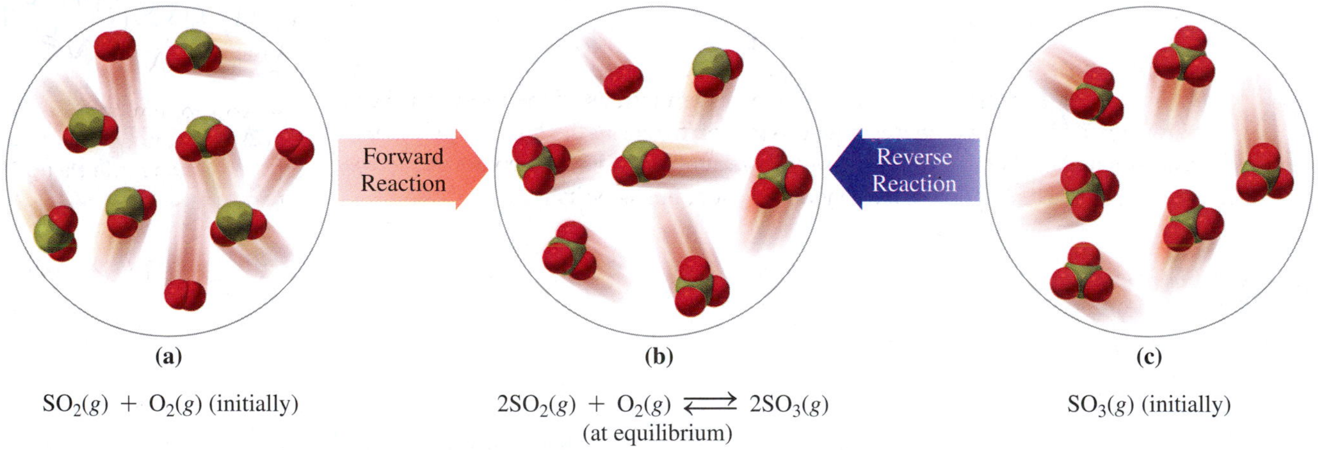

Forward Reaction
Reverse Reaction
(a)
$SO_2(g) + O_2(g)$ (initially)
(b)
$2SO_2(g) + O_2(g) \rightleftarrows 2SO_3(g)$
(at equilibrium)
(c)
$SO_3(g)$ (initially)

FIGURE 10.6 ▶ Sample **(a)** initially contains $SO_2(g)$ and $O_2(g)$. At equilibrium, sample **(b)** contains mostly $SO_3(g)$ and only small amounts of $SO_2(g)$ and $O_2(g)$, whereas sample **(c)** contains only $SO_3(g)$.

Q Why is the same equilibrium mixture obtained from $SO_2(g)$ and $O_2(g)$ as from $SO_3(g)$?

PRACTICE PROBLEMS

10.2 Chemical Equilibrium

10.7 What is meant by the term reversible reaction?

10.8 When does a reversible reaction reach equilibrium?

10.9 Which of the following are at equilibrium?
a. The rate of the forward reaction is twice as fast as the rate of the reverse reaction.
b. The concentrations of the reactants and the products do not change.
c. The rate of the reverse reaction does not change.

10.10 Which of the following are *not* at equilibrium?
a. The rates of the forward and reverse reactions are equal.
b. The rate of the forward reaction does not change.
c. The concentrations of reactants and the products are not constant.

10.11 The following diagrams show the chemical reaction with time:

$$A \rightleftarrows B$$

If A is blue and B is orange, state whether or not the reaction has reached equilibrium in this time period and explain why.

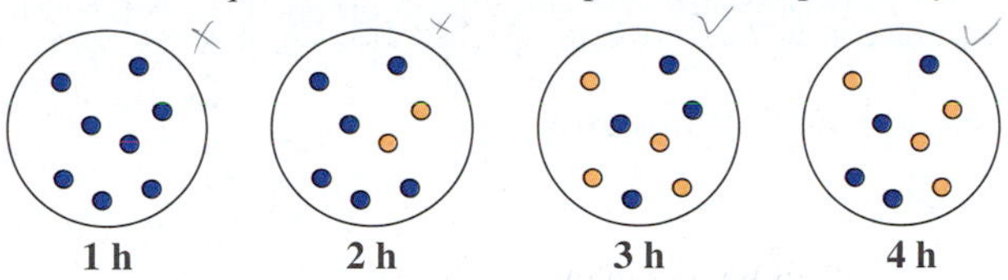
1 h
2 h
3 h
4 h

10.12 The following diagrams show the chemical reaction with time:

$$C \rightleftarrows D$$

If C is blue and D is yellow, state whether or not the reaction has reached equilibrium in this time period and explain why.

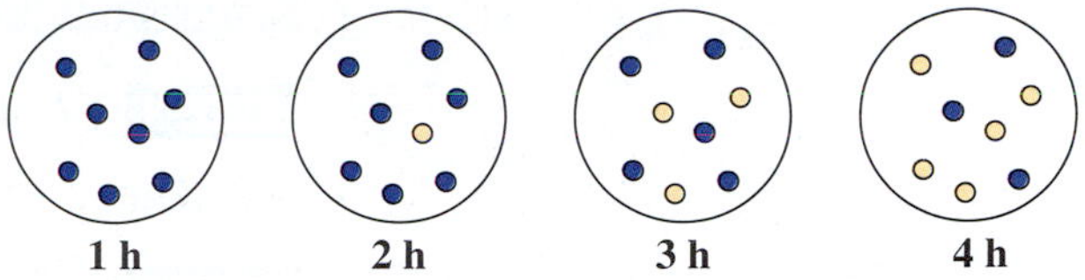
1 h
2 h
3 h
4 h

10.3 Equilibrium Constants

LEARNING GOAL Calculate the equilibrium constant for a reversible reaction given the concentrations of reactants and products at equilibrium.

REVIEW

Using Significant Figures in Calculations (2.3)

Balancing a Chemical Equation (7.1)

Calculating Concentration (9.4)

At equilibrium, the concentrations of the reactants and products are constant. We can use a ski lift as an analogy. Early in the morning, skiers at the bottom of the mountain begin to ride the ski lift up to the slopes. After the skiers reach the top of the mountain, they ski down. Eventually, the number of people riding up the ski lift becomes equal to the number of people skiing down the mountain. There is no further change in the number of skiers on the slopes; the system is at equilibrium.

At equilibrium, the number of people riding up the lift and the number of people skiing down the slope are constant.

Equilibrium Expression

At equilibrium, the concentrations can be used to set up a relationship between the products and the reactants. Suppose we write a general equation for reactants A and B that form products C and D. The small italic letters are the coefficients in the balanced equation.

$$aA + bB \rightleftarrows cC + dD$$

An **equilibrium expression** for a reversible chemical reaction multiplies the concentrations of the products together and divides by the concentrations of the reactants. Each concentration is raised to a power that is equal to its coefficient in the balanced chemical equation. The square bracket around each substance indicates that the concentration is expressed in moles per liter (M). For our general reaction, this is written as:

$$K_c = \frac{[\text{Products}]}{[\text{Reactants}]} = \frac{[C]^c\,[D]^d}{[A]^a\,[B]^b}$$

Coefficients (labels pointing to the exponents)

Equilibrium expression (label under the fraction)

CORE CHEMISTRY SKILL

Writing the Equilibrium Expression

We can now describe how to write the equilibrium expression for the reaction of H_2 and I_2 that forms HI. The balanced chemical equation is written with a double arrow between the reactants and the products.

$$H_2(g) + I_2(g) \rightleftarrows 2HI(g)$$

We show the concentration of the products using brackets in the numerator and the concentrations of the reactants in brackets in the denominator and write any coefficient as an exponent of its concentration (a coefficient 1 is understood).

$$K_c = \frac{[HI]^2}{[H_2][I_2]}$$

Coefficient of HI (label pointing to the exponent 2)

▶ **SAMPLE PROBLEM 10.3** Writing Equilibrium Expressions

TRY IT FIRST

Write the equilibrium expression for the following reaction:

$$2SO_2(g) + O_2(g) \rightleftarrows 2SO_3(g)$$

SOLUTION

ANALYZE THE PROBLEM	Given	Need	Connect
	equation	equilibrium expression	$\frac{[\text{products}]}{[\text{reactants}]}$

STEP 1 Write the balanced chemical equation.

$$2SO_2(g) + O_2(g) \rightleftharpoons 2SO_3(g)$$

STEP 2 Write the concentrations of the products as the numerator and the reactants as the denominator.

$$\frac{[\text{Products}]}{[\text{Reactants}]} \longrightarrow \frac{[SO_3]}{[SO_2][O_2]}$$

STEP 3 Write any coefficient in the equation as an exponent.

$$K_c = \frac{[SO_3]^2}{[SO_2]^2[O_2]}$$

STUDY CHECK 10.3

Write the equilibrium expression for each of the following reactions:

a. $2NO(g) + O_2(g) \rightleftharpoons 2NO_2(g)$ **b.** $2N_2O_5(g) \rightleftharpoons 4NO_2(g) + O_2(g)$

ANSWER

a. $K_c = \frac{[NO_2]^2}{[NO]^2[O_2]}$ **b.** $K_c = \frac{[NO_2]^4[O_2]}{[N_2O_5]^2}$

TEST

Try Practice Problems 10.13 to 10.16

CORE CHEMISTRY SKILL

Calculating an Equilibrium Constant

Calculating Equilibrium Constants

The **equilibrium constant, K_c,** is the numerical value obtained by substituting experimentally measured molar concentrations at equilibrium into the equilibrium expression. For example, the equilibrium expression for the reaction of H_2 and I_2 is written

$$H_2(g) + I_2(g) \rightleftharpoons 2HI(g) \qquad K_c = \frac{[HI]^2}{[H_2][I_2]}$$

In the first experiment, the molar concentrations for the reactants and products at equilibrium are found to be $[H_2] = 0.10$ M, $[I_2] = 0.20$ M, and $[HI] = 1.04$ M. When we substitute these values into the equilibrium expression, we obtain the numerical value of K_c.

In additional experiments 2 and 3, the mixtures have different equilibrium concentrations for the system at equilibrium at the same temperature. However, when these concentrations are used to calculate the equilibrium constant, we obtain the same value of K_c for each (see **TABLE 10.2**). *Thus, a reaction at a specific temperature can have only one value for the equilibrium constant.*

TABLE 10.2 Equilibrium Constant for $H_2(g) + I_2(g) \rightleftharpoons 2HI(g)$ at 427 °C

Experiment	Concentrations at Equilibrium			Equilibrium Constant
	$[H_2]$	$[I_2]$	$[HI]$	$K_c = \frac{[HI]^2}{[H_2][I_2]}$
1	0.10 M	0.20 M	1.04 M	$K_c = \frac{[1.04]^2}{[0.10][0.20]} = 54$
2	0.20 M	0.20 M	1.47 M	$K_c = \frac{[1.47]^2}{[0.20][0.20]} = 54$
3	0.30 M	0.17 M	1.66 M	$K_c = \frac{[1.66]^2}{[0.30][0.17]} = 54$

The units of K_c depend on the specific equation. In this example, the units of $[M]^2/[M]^2$ cancel out to give a value of 54. In other equations, the concentration units do not cancel. However, in this text, the numerical value will be given without any units as shown in Sample Problem 10.4.

▶SAMPLE PROBLEM 10.4 Calculating an Equilibrium Constant

TRY IT FIRST

The decomposition of dinitrogen tetroxide forms nitrogen dioxide.

$$N_2O_4(g) \rightleftharpoons 2NO_2(g)$$

What is the numerical value of K_c at 100 °C if a reaction mixture at equilibrium contains 0.45 M N_2O_4 and 0.31 M NO_2?

SOLUTION

STEP 1 State the given and needed quantities.

	Given	Need	Connect
ANALYZE THE PROBLEM	0.45 M N_2O_4, 0.31 M NO_2	K_c	equilibrium expression
	Equation		
	$N_2O_4(g) \rightleftharpoons 2NO_2(g)$		

STEP 2 Write the equilibrium expression.

$$K_c = \frac{[NO_2]^2}{[N_2O_4]}$$

STEP 3 Substitute equilibrium (molar) concentrations and calculate K_c.

$$K_c = \frac{[0.31]^2}{[0.45]} = 0.21$$

STUDY CHECK 10.4

Calculate the numerical value of K_c if an equilibrium mixture contains 0.040 M NH_3, 0.60 M H_2, and 0.20 M N_2.

$$2NH_3(g) \rightleftharpoons 3H_2(g) + N_2(g)$$

ANSWER

$K_c = 27$

TEST

Try Practice Problems 10.17 to 10.20

PRACTICE PROBLEMS

10.3 Equilibrium Constants

10.13 Write the equilibrium expression for each of the following reactions:

a. $CH_4(g) + 2H_2S(g) \rightleftharpoons CS_2(g) + 4H_2(g)$
b. $2NO(g) \rightleftharpoons N_2(g) + O_2(g)$
c. $2SO_3(g) + CO_2(g) \rightleftharpoons CS_2(g) + 4O_2(g)$
d. $CH_4(g) + H_2O(g) \rightleftharpoons 3H_2(g) + CO(g)$

10.14 Write the equilibrium expression for each of the following reactions:

a. $2HBr(g) \rightleftharpoons H_2(g) + Br_2(g)$
b. $2BrNO(g) \rightleftharpoons Br_2(g) + 2NO(g)$
c. $CH_4(g) + Cl_2(g) \rightleftharpoons CH_3Cl(g) + HCl(g)$
d. $Br_2(g) + Cl_2(g) \rightleftharpoons 2BrCl(g)$

10.15 Write the equilibrium expression for the reaction in the diagram, and calculate the numerical value of K_c. In the diagram, X atoms are orange and Y atoms are blue.

$$X_2(g) + Y_2(g) \rightleftharpoons 2XY(g)$$

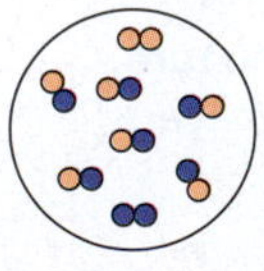

Equilibrium mixture

10.16 Write the equilibrium expression for the reaction in the diagram, and calculate the numerical value of K_c. In the diagram, A atoms are red and B atoms are green.

$$2AB(g) \rightleftharpoons A_2(g) + B_2(g)$$

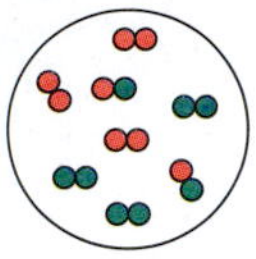

Equilibrium mixture

10.17 What is the numerical value of K_c for the following reaction if the equilibrium mixture contains 0.030 M N_2O_4 and 0.21 M NO_2?

$$N_2O_4(g) \rightleftharpoons 2NO_2(g)$$

10.18 What is the numerical value of K_c for the following reaction if the equilibrium mixture contains 0.30 M CO_2, 0.033 M H_2, 0.20 M CO, and 0.30 M H_2O?

$$CO_2(g) + H_2(g) \rightleftharpoons CO(g) + H_2O(g)$$

10.19 What is the numerical value of K_c for the following reaction if the equilibrium mixture contains 0.51 M CO, 0.30 M H_2, 1.8 M CH_4, and 2.0 M H_2O?

$$CO(g) + 3H_2(g) \rightleftharpoons CH_4(g) + H_2O(g)$$

10.20 What is the numerical value of K_c for the following reaction if the equilibrium mixture contains 0.44 M N_2, 0.40 M H_2, and 2.2 M NH_3?

$$N_2(g) + 3H_2(g) \rightleftharpoons 2NH_3(g)$$

REVIEW

Solving Equations (1.4)

10.4 Using Equilibrium Constants

LEARNING GOAL Use an equilibrium constant to predict the extent of reaction and to calculate equilibrium concentrations.

The values of K_c can be large or small. The size of the equilibrium constant depends on whether equilibrium is reached with more products than reactants, or more reactants than products. However, the size of an equilibrium constant does not affect how fast equilibrium is reached.

Equilibrium with a Large K_c

When a reaction has a large equilibrium constant, it means that the forward reaction produced a large amount of products when equilibrium was reached. Then the equilibrium mixture contains mostly products, which makes the concentrations of the products in the numerator higher than the concentrations of the reactants in the denominator. Thus at equilibrium, this reaction has a large K_c. Consider the reaction of SO_2 and O_2, which has a large K_c. At equilibrium, the reaction mixture contains mostly product and few reactants (see **FIGURE 10.7**).

$$2SO_2(g) + O_2(g) \rightleftharpoons 2SO_3(g)$$

$$K_c = \frac{[SO_3]^2}{[SO_2]^2[O_2]} \quad \frac{\text{Mostly product}}{\text{Few reactants}} = 3.4 \times 10^2$$

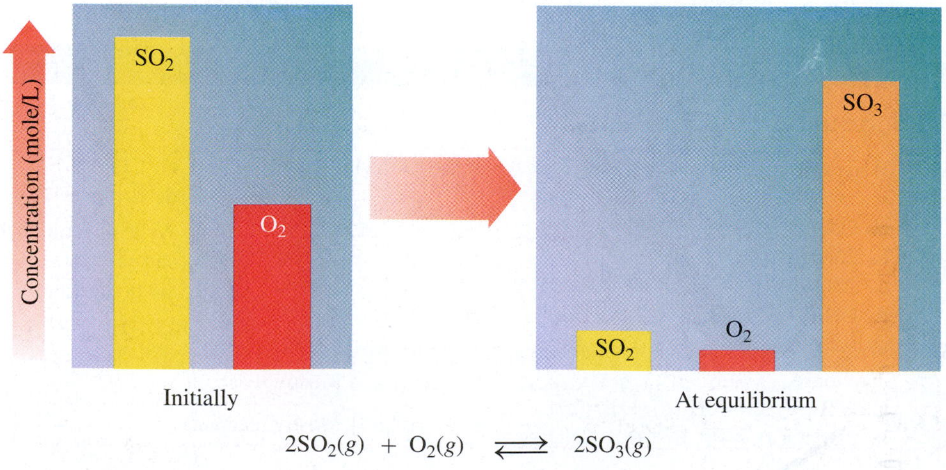

Concentration (mole/L)
SO_2
O_2
Initially
SO_2
O_2
SO_3
At equilibrium
$2SO_2(g) + O_2(g) \rightleftharpoons 2SO_3(g)$

FIGURE 10.7 ▶ In the reaction of $SO_2(g)$ and $O_2(g)$, the equilibrium mixture contains mostly product $SO_3(g)$, which results in a large K_c.

Q Why does an equilibrium mixture containing mostly product have a large K_c?

Equilibrium with a Small K_c

When a reaction has a small equilibrium constant, the equilibrium mixture contains a high concentration of reactants and a low concentration of products. Then the equilibrium expression has a small number in the numerator and a large number in the denominator. Thus at equilibrium, this reaction has a small K_c. Consider the reaction for the formation of $NO(g)$ from $N_2(g)$ and $O_2(g)$, which has a small K_c (see **FIGURE 10.8**).

$$N_2(g) + O_2(g) \rightleftharpoons 2NO(g)$$

$$K_c = \frac{[NO]^2}{[N_2][O_2]} \quad \frac{\text{Few products}}{\text{Mostly reactants}} = 2 \times 10^{-9}$$

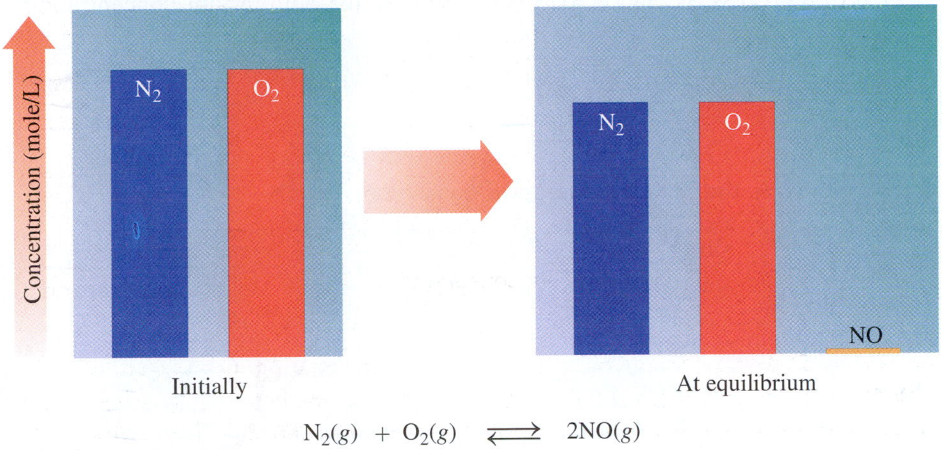
Concentration (mole/L)
N_2
O_2
N_2
O_2
NO
Initially
At equilibrium
$N_2(g) + O_2(g) \rightleftharpoons 2NO(g)$

FIGURE 10.8 ▶ The equilibrium mixture contains a very small amount of the product NO and a large amount of the reactants N_2 and O_2, which results in a small K_c.

Q Does a reaction with a $K_c = 3.2 \times 10^{-5}$ contain mostly reactants or products at equilibrium?

A few reactions have equilibrium constants close to 1, which means they have about equal concentrations of reactants and products. Moderate amounts of reactants have been converted to products upon reaching equilibrium (see **FIGURE 10.9**).

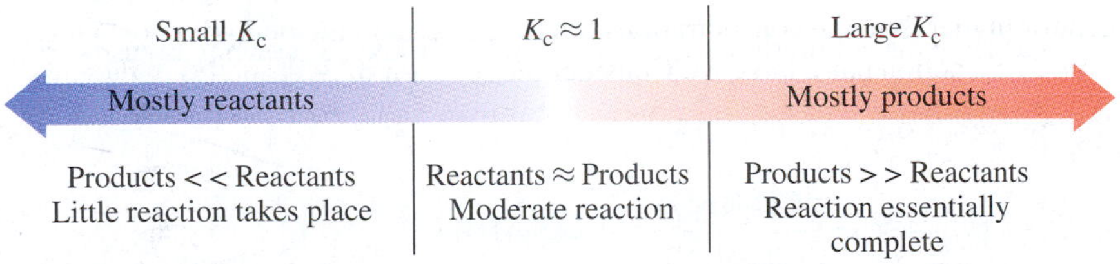
Small K_c
$K_c \approx 1$
Large K_c
Mostly reactants
Mostly products
Products < < Reactants
Little reaction takes place
Reactants ≈ Products
Moderate reaction
Products > > Reactants
Reaction essentially complete

FIGURE 10.9 ▶ At equilibrium, a reaction with a large K_c contains mostly products, whereas a reaction with a small K_c contains mostly reactants.

Q Does a reaction with a $K_c = 1.2$ contain mostly reactants, mostly products, or about equal amounts of both reactants and products at equilibrium?

TABLE 10.3 lists some equilibrium constants and the extent of their reaction.

TABLE 10.3 Examples of Reactions with Large and Small K_c Values

Reactants		Products	K_c	Equilibrium Mixture Contains
$2CO(g) + O_2(g)$	$\rightleftharpoons$	$2CO_2(g)$	2×10^{11}	Mostly product
$2H_2(g) + S_2(g)$	$\rightleftharpoons$	$2H_2S(g)$	1.1×10^{7}	Mostly product
$N_2(g) + 3H_2(g)$	$\rightleftharpoons$	$2NH_3(g)$	1.6×10^{2}	Mostly product
$PCl_5(g)$	$\rightleftharpoons$	$PCl_3(g) + Cl_2(g)$	1.2×10^{-2}	Mostly reactant
$N_2(g) + O_2(g)$	$\rightleftharpoons$	$2NO(g)$	2×10^{-9}	Mostly reactants

TEST

Try Practice Problems 10.21 to 10.24

ENGAGE

Why does the equilibrium mixture for a reaction with a $K_c = 2 \times 10^{-9}$ contain mostly reactants and only a few products?

CORE CHEMISTRY SKILL
Calculating Equilibrium Concentrations

Calculating Concentrations at Equilibrium

When we know the numerical value of the equilibrium constant and all the equilibrium concentrations except one, we can calculate the unknown concentration as shown in Sample Problem 10.5.

▶ SAMPLE PROBLEM 10.5 Calculating Concentration Using an Equilibrium Constant

TRY IT FIRST

For the reaction of carbon dioxide and hydrogen, the equilibrium concentrations are 0.25 M CO_2, 0.80 M H_2, and 0.50 M H_2O. What is the equilibrium concentration of $CO(g)$?

$$CO_2(g) + H_2(g) \rightleftharpoons CO(g) + H_2O(g) \qquad K_c = 0.11$$

SOLUTION

STEP 1 State the given and needed quantities.

ANALYZE THE PROBLEM	Given	Need	Connect
	0.25 M CO_2, 0.80 M H_2, 0.50 M H_2O	[CO]	equilibrium expression, value of K_c
	Equation		
	$CO_2(g) + H_2(g) \rightleftharpoons CO(g) + H_2O(g)$ $K_c = 0.11$		

STEP 2 Write the equilibrium expression and solve for the needed concentration.

$$K_c = \frac{[CO][H_2O]}{[CO_2][H_2]}$$

We rearrange the equilibrium expression to solve for the unknown [CO] as follows:

Multiply both sides by $[CO_2][H_2]$.

$$K_c \times [CO_2][H_2] = \frac{[CO][H_2O]}{\cancel{[CO_2][H_2]}} \times \cancel{[CO_2][H_2]}$$

$$K_c \times [CO_2][H_2] = [CO][H_2O]$$

Divide both sides by $[H_2O]$.

$$K_c \times \frac{[CO_2][H_2]}{[H_2O]} = \frac{[CO]\cancel{[H_2O]}}{\cancel{[H_2O]}}$$

$$[CO] = K_c \times \frac{[CO_2][H_2]}{[H_2O]}$$

STEP 3 Substitute the equilibrium (molar) concentrations and calculate the needed concentration.

$$[CO] = K_c \times \frac{[CO_2][H_2]}{[H_2O]} = 0.11 \times \frac{[0.25][0.80]}{[0.50]} = 0.044\ M$$

STUDY CHECK 10.5

When ethene (C_2H_4) reacts with water vapor, ethanol (C_2H_6O) is produced. If an equilibrium mixture contains 0.020 M C_2H_4 and 0.015 M H_2O, what is the equilibrium concentration of C_2H_6O? At 327 °C, the K_c is 9.0×10^3.

$$C_2H_4(g) + H_2O(g) \rightleftharpoons C_2H_6O(g)$$

ANSWER

$[C_2H_6O] = 2.7$ M

TEST

Try Practice Problems 10.25 to 10.28

PRACTICE PROBLEMS

10.4 Using Equilibrium Constants

10.21 If the K_c for this reaction is 4, which of the following diagrams represents the molecules in an equilibrium mixture? In the diagrams, X atoms are orange and Y atoms are blue.

$$X_2(g) + Y_2(g) \rightleftharpoons 2XY(g)$$

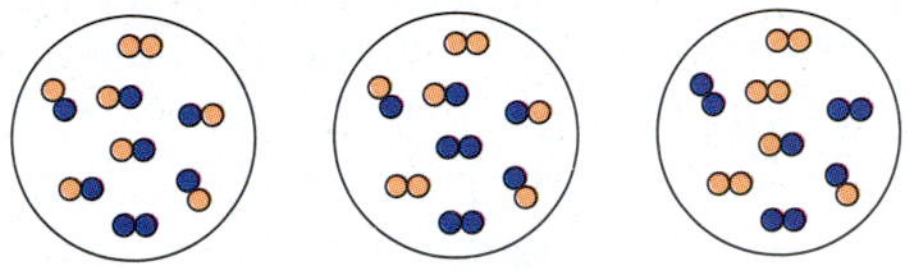
A
B
C

10.22 If the K_c for this reaction is 2, which of the following diagrams represents the molecules in an equilibrium mixture? In the diagrams, A atoms are orange and B atoms are green.

$$2AB(g) \rightleftharpoons A_2(g) + B_2(g)$$

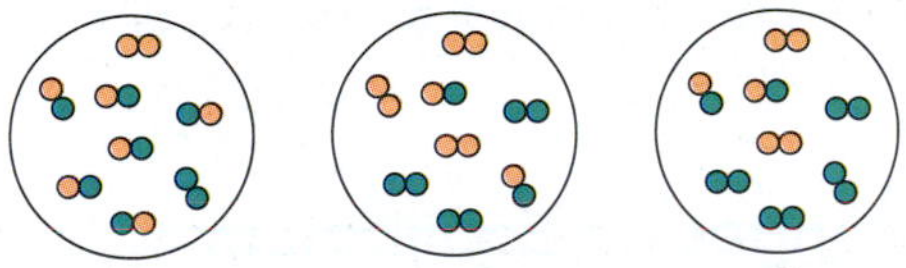
A
B
C

10.23 Indicate whether each of the following equilibrium mixtures contains mostly products or mostly reactants:

a. $Cl_2(g) + NO(g) \rightleftharpoons 2NOCl(g)$ $K_c = 3.7 \times 10^8$
b. $2H_2(g) + S_2(g) \rightleftharpoons 2H_2S(g)$ $K_c = 1.1 \times 10^7$
c. $3O_2(g) \rightleftharpoons 2O_3(g)$ $K_c = 1.7 \times 10^{-56}$

10.24 Indicate whether each of the following equilibrium mixtures contains mostly products or mostly reactants:

a. $CO(g) + Cl_2(g) \rightleftharpoons COCl_2(g)$ $K_c = 5.0 \times 10^{-9}$
b. $2HF(g) \rightleftharpoons H_2(g) + F_2(g)$ $K_c = 1.0 \times 10^{-95}$
c. $2NO(g) + O_2(g) \rightleftharpoons 2NO_2(g)$ $K_c = 6.0 \times 10^{13}$

10.25 The equilibrium constant, K_c, for this reaction is 54.

$$H_2(g) + I_2(g) \rightleftharpoons 2HI(g)$$

If the equilibrium mixture contains 0.015 M I_2 and 0.030 M HI, what is the molar concentration of H_2?

10.26 The equilibrium constant, K_c, for the following reaction is 4.6×10^{-3}. If the equilibrium mixture contains 0.050 M NO_2, what is the molar concentration of N_2O_4?

$$N_2O_4(g) \rightleftharpoons 2NO_2(g)$$

10.27 The equilibrium constant, K_c, for the following reaction is 2.0. If the equilibrium mixture contains 2.0 M NO and 1.0 M Br_2, what is the molar concentration of NOBr?

$$2NOBr(g) \rightleftharpoons 2NO(g) + Br_2(g)$$

10.28 The equilibrium constant, K_c, for the following reaction is 1.7×10^2. If the equilibrium mixture contains 0.18 M H_2 and 0.020 M N_2, what is the molar concentration of NH_3?

$$3H_2(g) + N_2(g) \rightleftharpoons 2NH_3(g)$$

10.5 Changing Equilibrium Conditions: Le Châtelier's Principle

LEARNING GOAL Use Le Châtelier's principle to describe the changes made in equilibrium concentrations when reaction conditions change.

We have seen that when a reaction reaches equilibrium, the rates of the forward and reverse reactions are equal and the concentrations remain constant. Now we will look at what happens to a system at equilibrium when changes occur in reaction conditions, such as changes in concentration, volume, or temperature.

Le Châtelier's Principle

When we alter any of the conditions of a system at equilibrium, the rates of the forward and reverse reactions may no longer be equal. We say that a *stress* is placed on the equilibrium. Then the system responds by changing the rate of the forward or reverse reaction in the direction that relieves that stress to reestablish equilibrium. We can use **Le Châtelier's principle**, which states that when a system at equilibrium is disturbed, the system will shift in the direction that will reduce that stress.

Le Châtelier's Principle

When a stress (change in conditions) is placed on a reaction at equilibrium, the equilibrium will shift in the direction that relieves the stress.

Suppose we have two water tanks connected by a pipe. When the water levels in the tanks are equal, water flows in the forward direction from Tank A to Tank B at the same rate as it flows in the reverse direction from Tank B to Tank A. Suppose we add more water to Tank A. With a higher level of water in Tank A, more water flows in the forward direction from Tank A to Tank B than in the reverse direction from Tank B to Tank A, which is shown with a longer arrow. Eventually, equilibrium is reached as the levels in both tanks become equal, but higher than before. Then the rate of water flows equally between Tank A and Tank B.

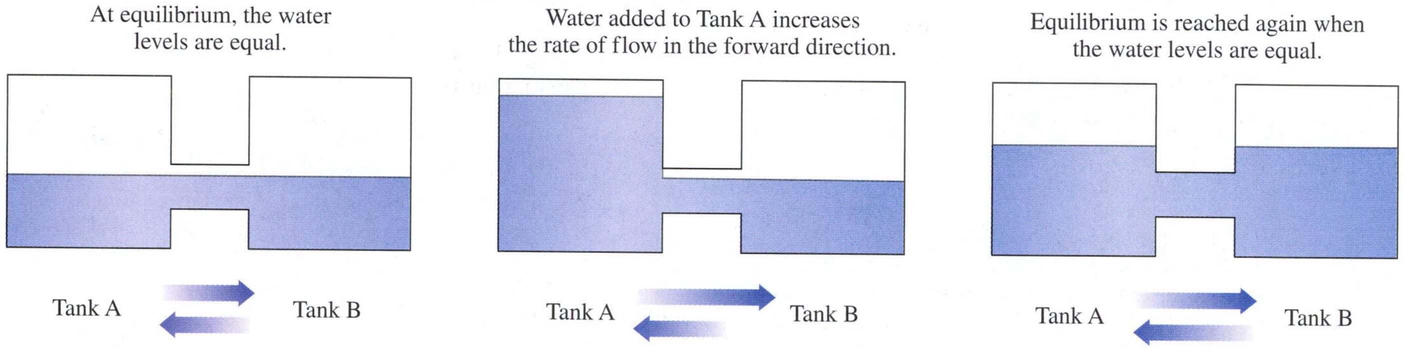
At equilibrium, the water levels are equal.
Water added to Tank A increases the rate of flow in the forward direction.
Equilibrium is reached again when the water levels are equal.
Tank A
Tank B
Tank A
Tank B
Tank A
Tank B

The stress of adding water to Tank A increases the rate of the forward direction to reestablish equal water levels and equilibrium.

CORE CHEMISTRY SKILL
Using Le Châtelier's Principle

Effect of Concentration Changes on Equilibrium

We will now use the reaction of H_2 and I_2 to illustrate how a change in concentration disturbs the equilibrium and how the system responds to that stress.

$$H_2(g) + I_2(g) \rightleftharpoons 2HI(g)$$

Suppose that more of the reactant H_2 is added to the equilibrium mixture, which increases the concentration of H_2. Because K_c cannot change for a reaction at a given temperature, adding more H_2 places a stress on the system (see **FIGURE 10.10**). Then the system

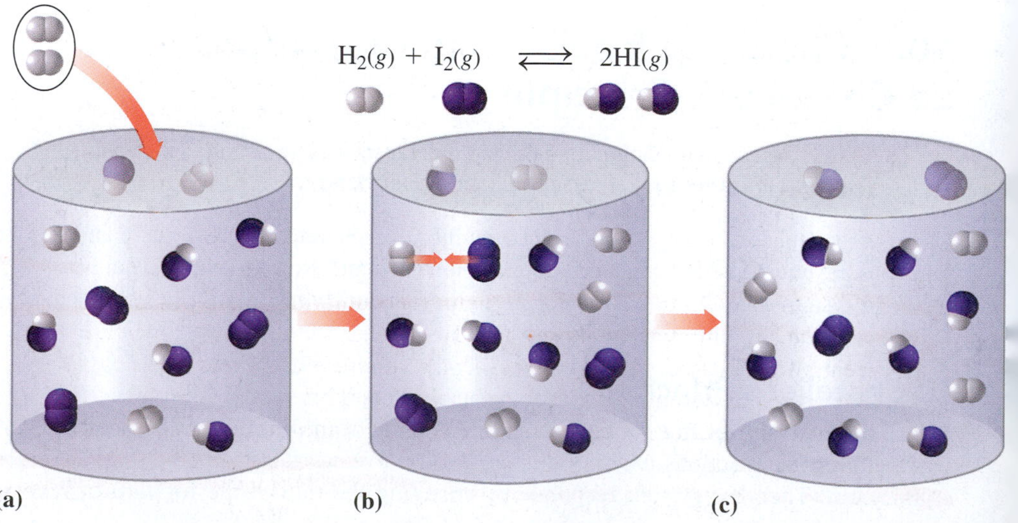
H2(g) + I2(g) ⇌ 2HI(g)
(a)
(b)
(c)

FIGURE 10.10 ▶ **(a)** The addition of H_2 places stress on the equilibrium system of $H_2(g) + I_2(g) \rightleftharpoons 2HI(g)$. **(b)** To relieve the stress, the forward reaction converts some reactants H_2 and I_2 to product HI. **(c)** A new equilibrium is established when the rates of the forward reaction and the reverse reaction become equal.

Q If more product HI is added, will the equilibrium shift in the direction of the product or reactants? Why?

relieves this stress by increasing the rate of the forward reaction, which is indicated by the direction of the large arrow. Thus, more product is formed until the system is again at equilibrium. According to Le Châtelier's principle, adding more reactant causes the system to *shift* in the direction of the product until equilibrium is reestablished.

Add H_2

$$H_2(g) + I_2(g) \rightleftarrows 2HI(g)$$

Suppose now that some H_2 is removed from the reaction mixture at equilibrium, which lowers the concentration of H_2 and slows the rate of the forward reaction. Using Le Châtelier's principle, we know that when some of the reactants are removed, the system will *shift* in the direction of the reactants until equilibrium is reestablished.

Remove H_2

$$H_2(g) + I_2(g) \rightleftarrows 2HI(g)$$

The concentrations of the products of an equilibrium mixture can also increase or decrease. For example, if more HI is added, there is an increase in the rate of the reaction in the reverse direction, which converts some of the product to reactants. The concentration of the products decreases and the concentration of the reactants increases until equilibrium is reestablished. Using Le Châtelier's principle, we see that the addition of a product causes the system to *shift* in the direction of the reactants.

Add HI

$$H_2(g) + I_2(g) \rightleftarrows 2HI(g)$$

In another example, some HI is removed from an equilibrium mixture, which decreases the concentration of the product. Then there is a *shift* in the direction of the product to reestablish equilibrium.

Remove HI

$$H_2(g) + I_2(g) \rightleftarrows 2HI(g)$$

In summary, Le Châtelier's principle indicates that a stress caused by adding a substance at equilibrium is relieved when the equilibrium system shifts the reaction away from that substance. Adding more reactant causes an increase in the forward reaction to products. Adding more product causes an increase in the reverse reaction to reactants. When some of a substance is removed, the equilibrium system shifts in the direction of that substance. These features of Le Châtelier's principle are summarized in **TABLE 10.4**.

TABLE 10.4 Effect of Concentration Changes on Equilibrium $H_2(g) + I_2(g) \rightleftarrows 2HI(g)$

Stress	Shift in the Direction of
Increasing [H_2]	Product
Decreasing [H_2]	Reactants
Increasing [I_2]	Product
Decreasing [I_2]	Reactants
Increasing [HI]	Reactants
Decreasing [HI]	Product

Effect of a Catalyst on Equilibrium

Sometimes a catalyst is added to a reaction to speed up a reaction by lowering the activation energy. As a result, the rates of both the forward and reverse reactions increase. The time required to reach equilibrium is shorter, but the same ratios of products and reactants are attained. Therefore, a catalyst speeds up the forward and reverse reactions, but it has no effect on the equilibrium mixture.

Effect of Volume Change on Equilibrium

If there is a change in the volume of a gas mixture at equilibrium, there will also be a change in the concentrations of those gases. Decreasing the volume will increase the concentration of gases, whereas increasing the volume will decrease their concentration. Then the system responds to reestablish equilibrium.

Let's look at the effect of decreasing the volume of the equilibrium mixture of the following reaction:

$$2CO(g) + O_2(g) \rightleftarrows 2CO_2(g)$$

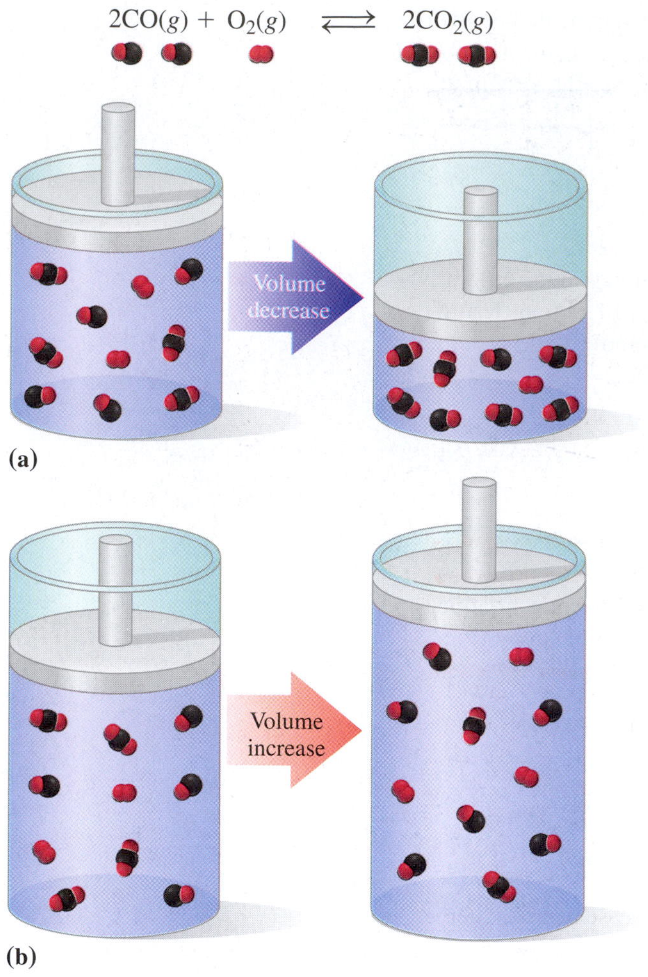

2CO(g) + O2(g) ⇌ 2CO2(g)
Volume decrease
(a)
Volume increase
(b)

If we decrease the volume, all the concentrations increase. According to Le Châtelier's principle, the increase in concentration is relieved when the system shifts in the direction of the smaller number of moles.

Decrease V

$$2CO(g) + O_2(g) \rightleftharpoons 2CO_2(g)$$

3 moles of gas — 2 moles of gas

On the other hand, when the volume of the equilibrium gas mixture increases, the concentrations of all the gases decrease. Then the system shifts in the direction of the greater number of moles to reestablish equilibrium (see **FIGURE 10.11**).

Increase V

$$2CO(g) + O_2(g) \rightleftharpoons 2CO_2(g)$$

3 moles of gas — 2 moles of gas

When a reaction has the same number of moles of reactants as products, a volume change does not affect the equilibrium mixture because the concentrations of the reactants and products change in the same way.

$$H_2(g) + I_2(g) \rightleftharpoons 2HI(g)$$

2 moles of gas — 2 moles of gas

FIGURE 10.11 ▶ **(a)** A decrease in the volume of the container causes the system to shift in the direction of fewer moles of gas. **(b)** An increase in the volume of the container causes the system to shift in the direction of more moles of gas.

Q If you want to increase the product, should you increase or decrease the volume of the container?

Chemistry Link to Health

Oxygen–Hemoglobin Equilibrium and Hypoxia

The transport of oxygen involves an equilibrium between hemoglobin (Hb), oxygen, and oxyhemoglobin (HbO_2).

$$Hb(aq) + O_2(g) \rightleftharpoons HbO_2(aq)$$

When the O_2 level is high in the alveoli of the lung, the reaction shifts in the direction of the product HbO_2. In the tissues where O_2 concentration is low, the reverse reaction releases the oxygen from the hemoglobin. The equilibrium expression is written

$$K_c = \frac{[HbO_2]}{[Hb][O_2]}$$

At normal atmospheric pressure, oxygen diffuses into the blood because the partial pressure of oxygen in the alveoli is higher than that in the blood. At an altitude above 8000 ft, a decrease in the atmospheric pressure results in a significant reduction in the partial pressure of oxygen, which means that less oxygen is available for the blood and body tissues. The fall in atmospheric pressure at higher altitudes decreases the partial pressure of inhaled oxygen, and there is less driving pressure for gas exchange in the lungs. At an altitude of 18 000 ft, a person will obtain 29% less oxygen. When oxygen levels are lowered, a person may experience *hypoxia*, characterized by increased respiratory rate, headache, decreased mental acuteness, fatigue, decreased physical coordination, nausea, vomiting, and cyanosis. A similar problem occurs in persons with a history of lung disease that impairs gas diffusion in the alveoli or in persons with a reduced number of red blood cells, such as smokers.

According to Le Châtelier's principle, we see that a decrease in oxygen will shift the equilibrium in the direction of the reactants. Such a shift depletes the concentration of HbO_2 and causes the hypoxia.

Remove O_2

$$Hb(aq) + O_2(g) \rightleftharpoons HbO_2(aq)$$

Immediate treatment of altitude sickness includes hydration, rest, and if necessary, descending to a lower altitude. The adaptation to lowered oxygen levels requires about 10 days. During this time, the bone marrow increases red blood cell production, providing more red blood cells and more hemoglobin. A person living at a high altitude can have 50% more red blood cells than someone at sea level. This increase in hemoglobin causes a shift in the equilibrium back in the

(continued)

Chemistry Link to Health (*continued*)

direction of HbO_2 product. Eventually, the higher concentration of HbO_2 will provide more oxygen to the tissues and the symptoms of hypoxia will lessen.

$$\text{Hb}(aq) + \text{O}_2(g) \xrightleftharpoons{\text{Add O}_2} \text{HbO}_2(aq)$$

For some who climb high mountains, it is important to stop and acclimatize for several days at increasing altitudes. At very high altitudes, it may be necessary to use an oxygen tank.

Hypoxia may occur at high altitudes where the oxygen concentration is lower.

Effect of a Change in Temperature on Equilibrium

We can think of heat as a reactant or a product in a reaction. For example, in the equation for an endothermic reaction, heat is written on the reactant side. When the temperature of an endothermic reaction increases, the system responds by shifting in the direction of the products to remove heat.

$$\text{N}_2(g) + \text{O}_2(g) + \text{heat} \xrightleftharpoons{\text{Increase } T} 2\text{NO}(g)$$

If the temperature is decreased for an endothermic reaction, there is a decrease in heat. Then the system shifts in the direction of the reactants to add heat.

$$\text{N}_2(g) + \text{O}_2(g) + \text{heat} \xrightleftharpoons{\text{Decrease } T} 2\text{NO}(g)$$

In the equation for an exothermic reaction, heat is written on the product side. When the temperature of an exothermic reaction increases, the system responds by shifting in the direction of the reactants to remove heat.

$$2\text{SO}_2(g) + \text{O}_2(g) \xrightleftharpoons{\text{Increase } T} 2\text{SO}_3(g) + \text{heat}$$

If the temperature is decreased for an exothermic reaction, there is a decrease in heat. Then the system shifts in the direction of the products to add heat.

$$2\text{SO}_2(g) + \text{O}_2(g) \xrightleftharpoons{\text{Decrease } T} 2\text{SO}_3(g) + \text{heat}$$

TABLE 10.5 summarizes the ways we can use Le Châtelier's principle to determine the shift in equilibrium that relieves a stress caused by the change in a condition.

TABLE 10.5 Effects of Condition Changes on Equilibrium

Condition	Change (Stress)	Shift in the Direction of
Concentration	Adding a reactant Removing a reactant Adding a product Removing a product	Products (forward reaction) Reactants (reverse reaction) Reactants (reverse reaction) Products (forward reaction)

(*continued*)

TABLE 10.5 Effects of Condition Changes on Equilibrium (*continued*)

Condition	Change (Stress)	Shift in the Direction of
Volume (container)	Decreasing the volume Increasing the volume	Fewer moles of gas More moles of gas
Temperature	**Endothermic Reaction** Increasing the temperature Decreasing the temperature **Exothermic Reaction** Increasing the temperature Decreasing the temperature	 Products (forward reaction) Reactants (reverse reaction) Reactants (reverse reaction) Products (forward reaction)
Catalyst	Increasing the rates equally	No effect

▶SAMPLE PROBLEM 10.6 Using Le Châtelier's Principle

TRY IT FIRST

Methanol, CH_4O, is finding use as a fuel additive. Describe the effect of each of the following changes on the equilibrium mixture for the combustion of methanol:

$$2CH_4O(g) + 3O_2(g) \rightleftharpoons 2CO_2(g) + 4H_2O(g) + 1450\text{ kJ}$$

a. adding more CO_2
b. adding more O_2
c. increasing the volume of the container
d. increasing the temperature
e. adding a catalyst

ENGAGE

Why does adding $H_2O(g)$ to this reaction at equilibrium cause a shift in the direction of the reactants?

SOLUTION

a. When the concentration of the product CO_2 increases, the equilibrium shifts in the direction of the reactants.
b. When the concentration of the reactant O_2 increases, the equilibrium shifts in the direction of the products.
c. When the volume increases, the equilibrium shifts in the direction of the greater number of moles of gas, which is the products.
d. When the temperature is increased for an exothermic reaction, the equilibrium shifts in the direction of the reactants.
e. When a catalyst is added, there is no change in the equilibrium mixture.

STUDY CHECK 10.6

Describe the effect of each of the following changes on the equilibrium mixture for the following reaction:

$$2HF(g) + Cl_2(g) + 357\text{ kJ} \rightleftharpoons 2HCl(g) + F_2(g)$$

a. adding more Cl_2
b. decreasing the volume of the container
c. decreasing the temperature

ANSWER

a. When the concentration of the reactant Cl_2 increases, the equilibrium shifts in the direction of the products.
b. There is no change in equilibrium mixture because the moles of reactants are equal to the moles of products.
c. When the temperature for an endothermic reaction decreases, the equilibrium shifts in the direction of the reactants.

TEST

Try Practice Problems 10.29 to 10.34

Chemistry Link to Health

Homeostasis: Regulation of Body Temperature

In a physiological system of equilibrium called *homeostasis*, changes in our environment are balanced by changes in our bodies. It is crucial to our survival that we balance heat gain with heat loss. If we do not lose enough heat, our body temperature rises. At high temperatures, the body can no longer regulate our metabolic reactions. If we lose too much heat, body temperature drops. At low temperatures, essential functions proceed too slowly.

The skin plays an important role in the maintenance of body temperature. When the outside temperature rises, receptors in the skin send signals to the brain. The temperature-regulating part of the brain stimulates the sweat glands to produce perspiration. As perspiration evaporates from the skin, heat is removed and the body temperature is decreased.

In cold temperatures, epinephrine is released, causing an increase in metabolic rate, which increases the production of heat. Receptors on the skin signal the brain to constrict the blood vessels. Less blood flows through the skin, and heat is conserved. The production of perspiration stops, thereby lessening the heat lost by evaporation.

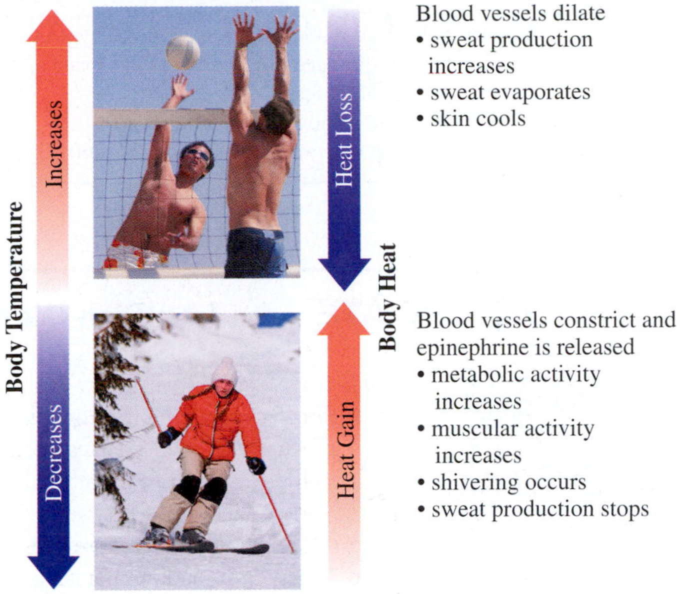
Body Temperature
Increases
Decreases
Heat Loss
Heat Gain
Body Heat
Blood vessels dilate
• sweat production increases
• sweat evaporates
• skin cools
Blood vessels constrict and epinephrine is released
• metabolic activity increases
• muscular activity increases
• shivering occurs
• sweat production stops

PRACTICE PROBLEMS

10.5 Changing Equilibrium Conditions: Le Châtelier's Principle

10.29 In the lower atmosphere, oxygen is converted to ozone (O_3) by the energy provided from lightning.

$$3O_2(g) + \text{heat} \rightleftharpoons 2O_3(g)$$

For each of the following changes at equilibrium, indicate whether the equilibrium shifts in the direction of the product, the reactant, or does not change:

a. adding more $O_2(g)$
b. adding more $O_3(g)$
c. increasing the temperature
d. increasing the volume of the container
e. adding a catalyst

10.30 Ammonia is produced by reacting nitrogen gas and hydrogen gas.

$$N_2(g) + 3H_2(g) \rightleftharpoons 2NH_3(g) + 92\text{ kJ}$$

For each of the following changes at equilibrium, indicate whether the equilibrium shifts in the direction of the product, the reactants, or does not change:

a. removing some $N_2(g)$
b. decreasing the temperature
c. adding more $NH_3(g)$
d. adding more $H_2(g)$
e. increasing the volume of the container

10.31 Hydrogen chloride can be made by reacting hydrogen gas and chlorine gas.

$$H_2(g) + Cl_2(g) + \text{heat} \rightleftharpoons 2HCl(g)$$

For each of the following changes at equilibrium, indicate whether the equilibrium shifts in the direction of the product, the reactants, or does not change:

a. adding more $H_2(g)$
b. increasing the temperature
c. removing some $HCl(g)$
d. adding a catalyst
e. removing some $Cl_2(g)$

10.32 When heated, carbon monoxide reacts with water to produce carbon dioxide and hydrogen.

$$CO(g) + H_2O(g) \rightleftharpoons CO_2(g) + H_2(g) + \text{heat}$$

For each of the following changes at equilibrium, indicate whether the equilibrium shifts in the direction of the products, the reactants, or does not change:

a. decreasing the temperature
b. adding more $H_2(g)$
c. removing some $CO_2(g)$
d. adding more $H_2O(g)$
e. decreasing the volume of the container

Clinical Applications

Use the following equation for the equilibrium of hemoglobin in the blood to answer problems 10.33 and 10.34:

$$\mathrm{Hb}(aq) + \mathrm{O_2}(g) \rightleftharpoons \mathrm{HbO_2}(aq)$$

10.33 Athletes who train at high altitudes initially experience hypoxia, which causes their body to produce more hemoglobin. When an athlete first arrives at high altitude,
- **a.** What is the stress on the hemoglobin equilibrium?
- **b.** In what direction does the hemoglobin equilibrium shift?

10.34 A person who has been a smoker and has a low oxygen blood saturation uses an oxygen tank for supplemental oxygen. When oxygen is first supplied,
- **a.** What is the stress on the hemoglobin equilibrium?
- **b.** In what direction does the hemoglobin equilibrium shift?

CLINICAL UPDATE An Iron-Rich Diet for Children's Anemia

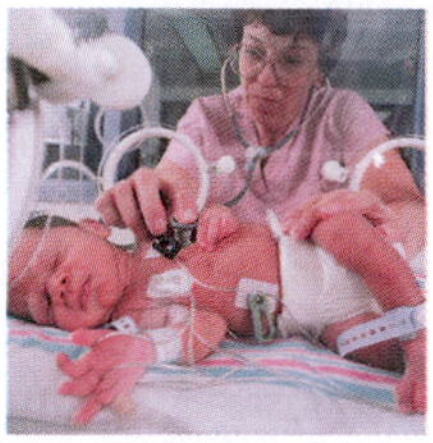

Today, Andrew is a thriving two-year old toddler. However, his mother is concerned that Andrew looks pale, seems to have some trouble breathing, has a lot of headaches, and generally seems tired. At the clinic, Andrew and his mother see Judy, his neonatal nurse when he was an infant. Judy notes his symptoms and takes a blood sample, which shows that Andrew has *iron-deficiency anemia.* Judy explains to his mother that Andrew's red-blood cell count is below normal, which means that he has a low concentration of hemoglobin in his blood.

Hemoglobin, present in our red blood cells, is an iron-containing protein that carries oxygen from our lungs to the muscles and tissues of our body. Each hemoglobin molecule contains four iron-containing heme groups. Because each heme group can attach to one O_2 in the lungs, one hemoglobin can carry as many as four O_2 molecules. The transport of oxygen involves an equilibrium reaction between the reactants Hb and O_2 and the product HbO_2.

$$\mathrm{Hb}(aq) + \mathrm{O_2}(g) \rightleftharpoons \mathrm{HbO_2}(aq)$$

The equilibrium expression can be written

$$K_c = \frac{[\mathrm{HbO_2}]}{[\mathrm{Hb}][\mathrm{O_2}]}$$

In the alveoli of the lungs where the O_2 concentration is high, the forward reaction is faster, which shifts the equilibrium in the direction of the product HbO_2. As a result, O_2 binds to hemoglobin in the lungs.

$$\mathrm{Hb}(aq) + \mathrm{O_2}(g) \longrightarrow \mathrm{HbO_2}(aq)$$

In the tissues and the muscles where the O_2 concentration is low, the reverse reaction is faster, which shifts the equilibrium in the direction of the reactant Hb and releases the oxygen from the hemoglobin.

$$\mathrm{Hb}(aq) + \mathrm{O_2}(g) \longleftarrow \mathrm{HbO_2}(aq)$$

Judy recommends that Andrew's diet include more high iron-containing foods such as chicken, fish, spinach, cereal, eggs, peas, beans, peanut butter, and whole-grain bread. After three months of an iron-rich diet, Andrew's hemoglobin is within normal range.

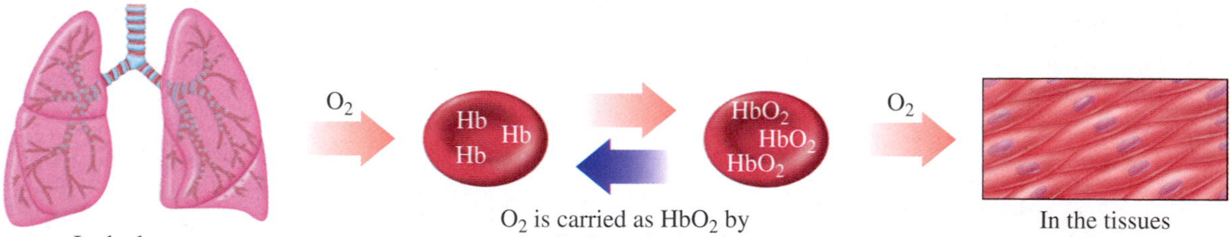
O2
Hb
Hb
Hb
HbO2
HbO2
HbO2
O2

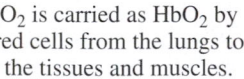

In the lungs, Hb binds O_2.

O_2 is carried as HbO_2 by red cells from the lungs to the tissues and muscles.

In the tissues and muscles, HbO_2 releases O_2.

Clinical Applications

10.35 For the hemoglobin equilibrium, indicate if each of the following changes shifts the equilibrium in the direction of the product, the reactants, or does not change:
- **a.** increasing $[O_2]$
- **b.** increasing $[HbO_2]$
- **c.** decreasing [Hb]

10.36 For the hemoglobin equilibrium, indicate if each of the following changes shifts the equilibrium in the direction of the product, the reactants, or does not change:
- **a.** decreasing $[O_2]$
- **b.** increasing [Hb]
- **c.** decreasing $[HbO_2]$

10.37 Andrew is diagnosed as being anemic. Why does Andrew not obtain sufficient oxygen in his tissues?

10.38 Andrew's diet is changed to include more iron-rich foods. How would an increase in iron affect his hemoglobin equilibrium?

CONCEPT MAP

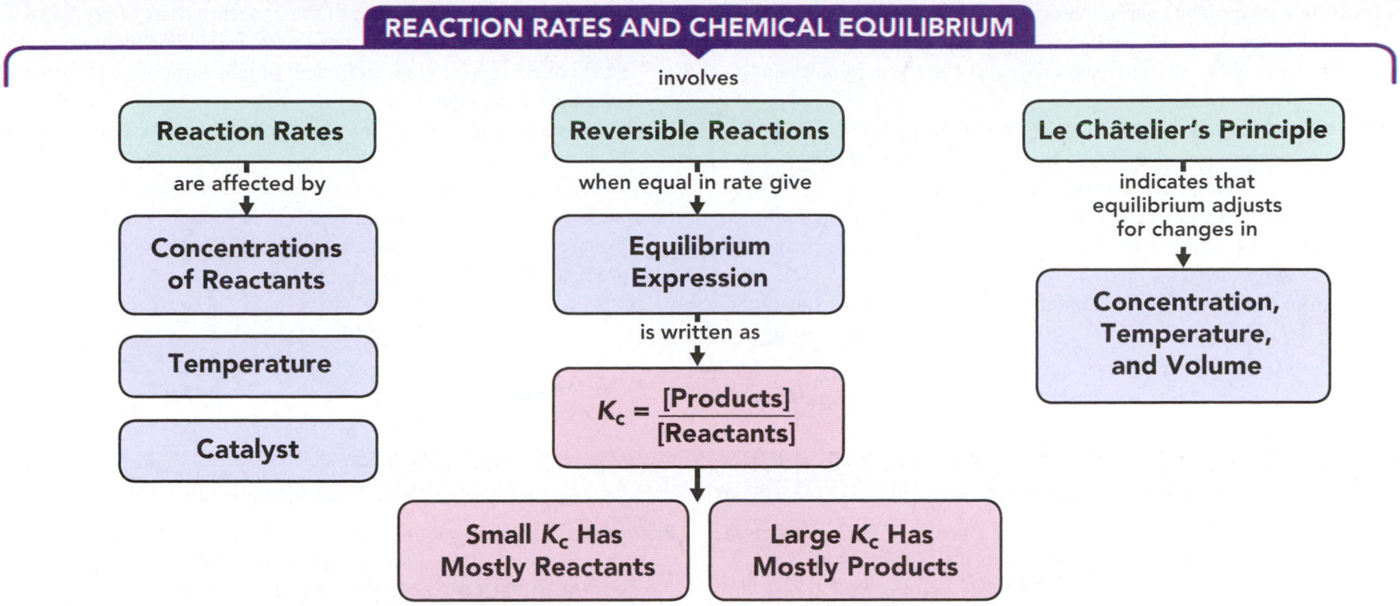

REACTION RATES AND CHEMICAL EQUILIBRIUM
involves
Reaction Rates
are affected by
Concentrations of Reactants
Temperature
Catalyst
Reversible Reactions
when equal in rate give
Equilibrium Expression
is written as
$K_c = \frac{[Products]}{[Reactants]}$
Small K_c Has Mostly Reactants
Large K_c Has Mostly Products
Le Châtelier's Principle
indicates that equilibrium adjusts for changes in
Concentration, Temperature, and Volume

CHAPTER REVIEW

10.1 Rates of Reactions

LEARNING GOAL Describe how temperature, concentration, and catalysts affect the rate of a reaction.

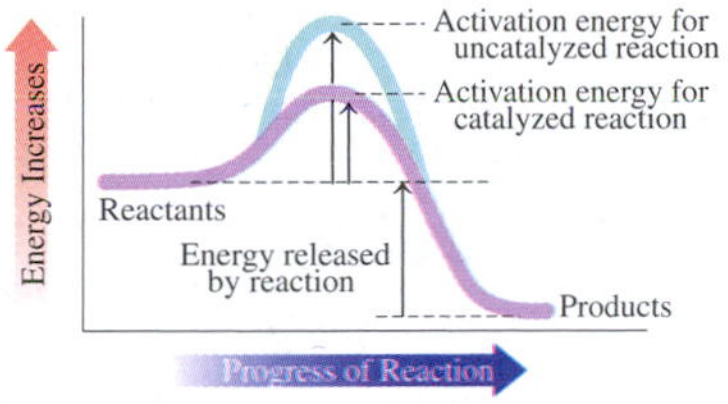

Activation energy for uncatalyzed reaction
Activation energy for catalyzed reaction
Energy Increases
Reactants
Energy released by reaction
Products
Progress of Reaction

- The rate of a reaction is the speed at which the reactants are converted to products.
- Increasing the concentrations of reactants, increasing the temperature, or adding a catalyst can increase the rate of a reaction.

10.2 Chemical Equilibrium

LEARNING GOAL Use the concept of reversible reactions to explain chemical equilibrium.

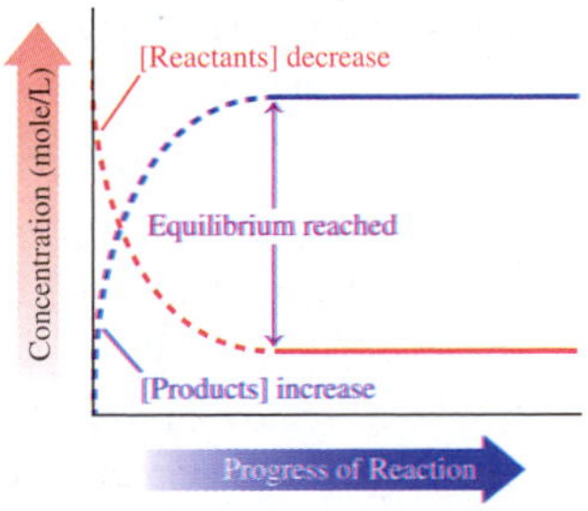

[Reactants] decrease
Concentration (mole/L)
Equilibrium reached
[Products] increase
Progress of Reaction

- Chemical equilibrium occurs in a reversible reaction when the rate of the forward reaction becomes equal to the rate of the reverse reaction.
- At equilibrium, no further change occurs in the concentrations of the reactants and products as the forward and reverse reactions continue.

10.3 Equilibrium Constants

LEARNING GOAL Calculate the equilibrium constant for a reversible reaction given the concentrations of reactants and products at equilibrium.

- An equilibrium constant, K_c, is the ratio of the concentrations of the products to the concentrations of the reactants, with each concentration raised to a power equal to its coefficient in the balanced chemical equation.

10.4 Using Equilibrium Constants

LEARNING GOAL Use an equilibrium constant to predict the extent of reaction and to calculate equilibrium concentrations.

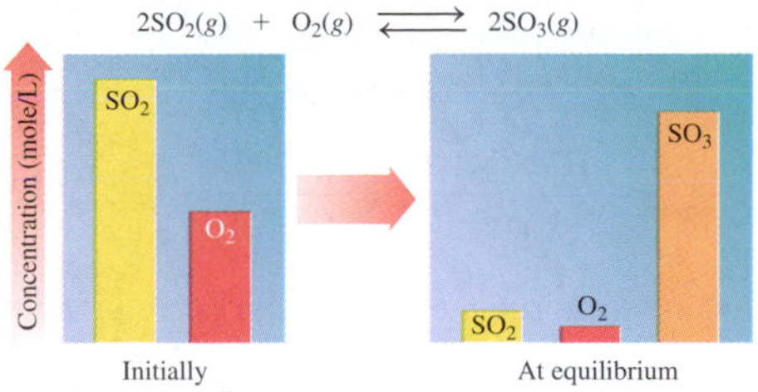

$2SO_2(g) + O_2(g) \rightleftarrows 2SO_3(g)$
Concentration (mole/L)
SO_2
O_2
Initially
SO_3
SO_2
O_2
At equilibrium

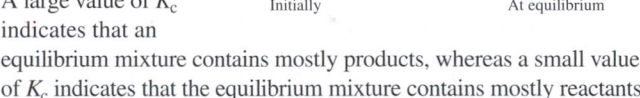

- A large value of K_c indicates that an equilibrium mixture contains mostly products, whereas a small value of K_c indicates that the equilibrium mixture contains mostly reactants.
- Equilibrium constants can be used to calculate the concentration of a component in the equilibrium mixture.

10.5 Changing Equilibrium Conditions: Le Châtelier's Principle

LEARNING GOAL Use Le Châtelier's principle to describe the changes made in equilibrium concentrations when reaction conditions change.

- When reactants are removed or products are added to an equilibrium mixture, the system shifts in the direction of the reactants.
- When reactants are added or products are removed from an equilibrium mixture, the system shifts in the direction of the products.
- A decrease in the volume of a reaction container causes a shift in the direction of the smaller number of moles of gas.
- An increase in the volume of a reaction container causes a shift in the direction of the greater number of moles of gas.
- Increasing the temperature of an endothermic reaction or decreasing the temperature of an exothermic reaction will cause the system to shift in the direction of the products.
- Decreasing the temperature of an endothermic reaction or increasing the temperature of an exothermic reaction will cause the system to shift in the direction of the reactants.

KEY TERMS

activation energy The energy that must be provided by a collision to break apart the bonds of the reacting molecules.

catalyst A substance that increases the rate of reaction by lowering the activation energy.

chemical equilibrium The point at which the rate of forward and reverse reactions are equal so that no further change in concentrations of reactants and products takes place.

collision theory A model for a chemical reaction stating that molecules must collide with sufficient energy and proper orientation to form products.

equilibrium constant, K_c The numerical value obtained by substituting the equilibrium concentrations of the components into the equilibrium expression.

equilibrium expression The ratio of the concentrations of products to the concentrations of reactants, with each component raised to an exponent equal to the coefficient of that compound in the balanced chemical equation.

Le Châtelier's principle When a stress is placed on a system at equilibrium, the equilibrium shifts to relieve that stress.

rate of reaction The speed at which reactants form products.

reversible reaction A reaction in which a forward reaction occurs from reactants to products, and a reverse reaction occurs from products back to reactants.

CORE CHEMISTRY SKILLS

The chapter Section containing each Core Chemistry Skill is shown in parentheses at the end of each heading.

Writing the Equilibrium Expression (10.3)

- An equilibrium expression for a reversible reaction is written by multiplying the concentrations of the products in the numerator and dividing by the product of the concentrations of the reactants in the denominator.
- Each concentration is raised to a power equal to its coefficient in the balanced chemical equation:

$$K_c = \frac{[\text{Products}]}{[\text{Reactants}]} = \frac{[C]^c\,[D]^d}{[A]^a\,[B]^b} \quad \leftarrow \text{Coefficients}$$

Example: Write the equilibrium expression for the following chemical reaction:

$$2NO_2(g) \rightleftharpoons N_2O_4(g)$$

Answer: $K_c = \frac{[N_2O_4]}{[NO_2]^2}$

Calculating an Equilibrium Constant (10.3)

- The equilibrium constant, K_c, is the numerical value obtained by substituting experimentally measured molar concentrations at equilibrium into the equilibrium expression.

Example: Calculate the numerical value of K_c for the following reaction when the equilibrium mixture contains 0.025 M NO_2 and 0.087 M N_2O_4:

$$2NO_2(g) \rightleftharpoons N_2O_4(g)$$

Answer: Write the equilibrium expression, substitute the molar concentrations, and calculate.

$$K_c = \frac{[N_2O_4]}{[NO_2]^2} = \frac{[0.087]}{[0.025]^2} = 140$$

Calculating Equilibrium Concentrations (10.4)

- To determine the concentration of a product or a reactant at equilibrium, we use the equilibrium expression to solve for the unknown concentration.

Example: Calculate the equilibrium concentration for CF_4 if $K_c = 2.0$, and the equilibrium mixture contains 0.10 M COF_2 and 0.050 M CO_2.

$$2COF_2(g) \rightleftharpoons CO_2(g) + CF_4(g)$$

Answer: Write the equilibrium expression.

$$K_c = \frac{[CO_2][CF_4]}{[COF_2]^2}$$

Solve the equation for the unknown concentration, substitute the molar concentrations, and calculate.

$$[CF_4] = K_c \times \frac{[COF_2]^2}{[CO_2]} = 2.0 \times \frac{[0.10]^2}{[0.050]} = 0.40\ \text{M}$$

Using Le Châtelier's Principle (10.5)

- Le Châtelier's principle states that when a system at equilibrium is disturbed by changes in concentration, volume, or temperature, the system will shift in the direction that will reduce that stress.

Example: Nitrogen and oxygen form dinitrogen pentoxide in an exothermic reaction.

$$2N_2(g) + 5O_2(g) \rightleftharpoons 2N_2O_5(g) + \text{heat}$$

For each of the following changes at equilibrium, indicate whether the equilibrium shifts in the direction of the product, the reactants, or does not change:

a. removing some $N_2(g)$
b. decreasing the temperature
c. increasing the volume of the container

Answer:
a. Removing a reactant shifts the equilibrium in the direction of the reactants.
b. Decreasing the temperature shifts the equilibrium of an exothermic reaction in the direction of the product.
c. Increasing the volume of the container shifts the equilibrium in the direction of the greater number of moles of gas, which is in the direction of the reactants.

UNDERSTANDING THE CONCEPTS

The chapter Sections to review are shown in parentheses at the end of each problem.

10.39 Write the equilibrium expression for each of the following reactions: (10.3)

a. $CH_4(g) + 2O_2(g) \rightleftharpoons CO_2(g) + 2H_2O(g)$

b. $4NH_3(g) + 3O_2(g) \rightleftharpoons 2N_2(g) + 6H_2O(g)$

10.40 Write the equilibrium expression for each of the following reactions: (10.3)

a. $2C_2H_6(g) + 7O_2(g) \rightleftharpoons 4CO_2(g) + 6H_2O(g)$

b. $4NH_3(g) + 5O_2(g) \rightleftharpoons 4NO(g) + 6H_2O(g)$

10.41 Would the equilibrium constant, K_c, for the reaction in the diagrams have a large or small value? (10.4)

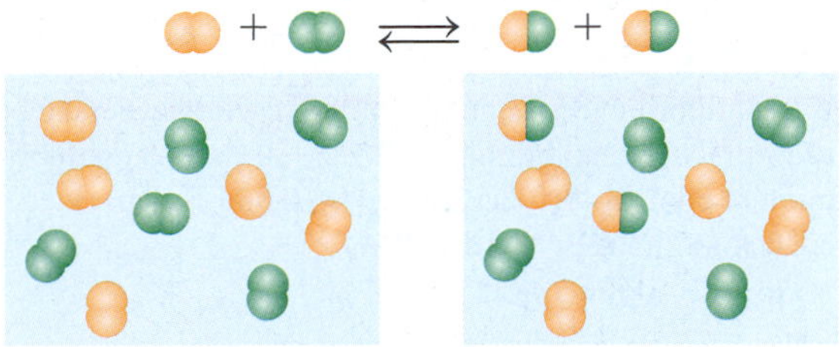

Initially | At equilibrium

10.42 Would the equilibrium constant, K_c, for the reaction in the diagrams have a large or small value? (10.4)

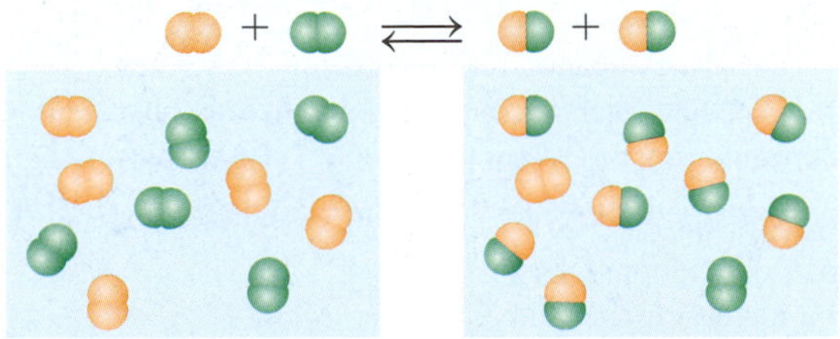

Initially | At equilibrium

10.43 Would T_2 be higher or lower than T_1 for the reaction shown in the diagrams? (10.5)

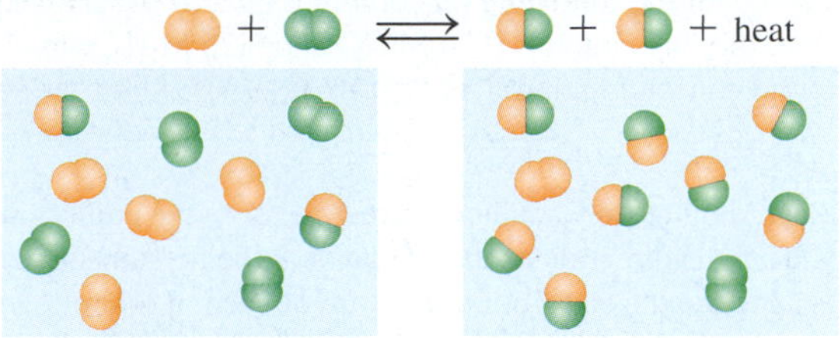
+ heat

$T_1 = 300\ °C$ | $T_2 = ?$

10.44 Would the reaction shown in the diagrams be exothermic or endothermic? (10.5)

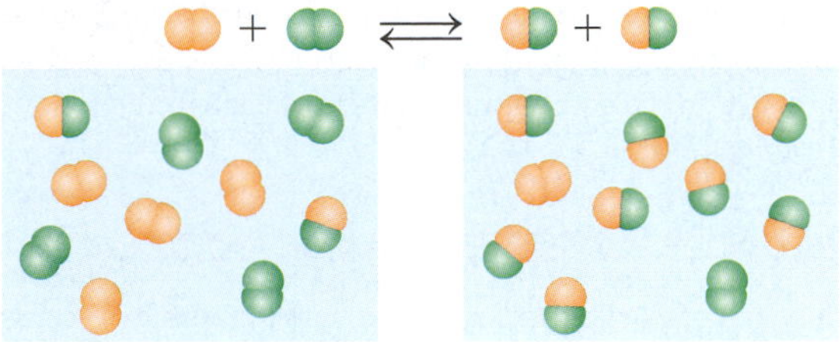

$T_1 = 100\ °C$ | $T_2 = 200\ °C$

ADDITIONAL PRACTICE PROBLEMS

10.45 For each of the following changes at equilibrium, indicate whether the equilibrium shifts in the direction of the product, the reactants, or does not change: (10.5)

$$C_2H_4(g) + Cl_2(g) \rightleftharpoons C_2H_4Cl_2(g) + \text{heat}$$

a. increasing the temperature

b. decreasing the volume of the container

c. adding a catalyst

d. adding more $Cl_2(g)$

10.46 For each of the following changes at equilibrium, indicate whether the equilibrium shifts in the direction of the product, the reactants, or does not change: (10.5)

$$N_2(g) + O_2(g) + \text{heat} \rightleftharpoons 2NO(g)$$

a. increasing the temperature

b. decreasing the volume of the container

c. adding a catalyst

d. adding more $N_2(g)$

10.47 For each of the following reactions, indicate if the equilibrium mixture contains mostly products, mostly reactants, or both reactants and products: (10.4)

a. $H_2(g) + Cl_2(g) \rightleftharpoons 2HCl(g)\ \ K_c = 1.3 \times 10^{34}$

b. $2NOBr(g) \rightleftharpoons 2NO(g) + Br_2(g)\ \ K_c = 2.0$

c. $2H_2S(g) + CH_4(g) \rightleftharpoons CS_2(g) + 4H_2(g)\ \ K_c = 5.3 \times 10^{-8}$

10.48 For each of the following reactions, indicate if the equilibrium mixture contains mostly products, mostly reactants, or both reactants and products: (10.4)

a. $2H_2O(g) \rightleftharpoons 2H_2(g) + O_2(g)\ \ K_c = 4 \times 10^{-48}$

b. $N_2(g) + 3H_2(g) \rightleftharpoons 2NH_3(g)\ \ K_c = 0.30$

c. $2SO_2(g) + O_2(g) \rightleftharpoons 2SO_3(g)\ \ K_c = 1.2 \times 10^9$

10.49 Consider the reaction: (10.3)

$$2NH_3(g) \rightleftharpoons N_2(g) + 3H_2(g)$$

a. Write the equilibrium expression.

b. What is the numerical value of K_c for the reaction if the concentrations at equilibrium are 0.20 M NH_3, 3.0 M N_2, and 0.50 M H_2?

10.50 Consider the reaction: (10.3)

$$2SO_2(g) + O_2(g) \rightleftharpoons 2SO_3(g)$$

a. Write the equilibrium expression.

b. What is the numerical value of K_c for the reaction if the concentrations at equilibrium are 0.10 M SO_2, 0.12 M O_2, and 0.60 M SO_3?

10.51 The K_c for the following reaction is 5.0 at 100 °C. If an equilibrium mixture contains 0.50 M NO_2, what is the molar concentration of N_2O_4? (10.3, 10.4)

$$2NO_2(g) \rightleftharpoons N_2O_4(g)$$

10.52 The K_c for the following reaction is 15 at 220 °C. If an equilibrium mixture contains 0.40 M CO and 0.20 M H_2, what is the molar concentration of CH_4O? (10.3, 10.4)

$$CO(g) + 2H_2(g) \rightleftharpoons CH_4O(g)$$

10.53 According to Le Châtelier's principle, does the equilibrium shift in the direction of the products or the reactants when O_2 is added to the equilibrium mixture of each of the following reactions? (10.5)

a. $3O_2(g) \rightleftharpoons 2O_3(g)$
b. $2CO_2(g) \rightleftharpoons 2CO(g) + O_2(g)$
c. $2SO_2(g) + O_2(g) \rightleftharpoons 2SO_3(g)$
d. $2SO_2(g) + 2H_2O(g) \rightleftharpoons 2H_2S(g) + 3O_2(g)$

10.54 According to Le Châtelier's principle, does the equilibrium shift in the direction of the products or the reactants when N_2 is added to the equilibrium mixture of each of the following reactions? (10.5)

a. $2NH_3(g) \rightleftharpoons 3H_2(g) + N_2(g)$
b. $N_2(g) + O_2(g) \rightleftharpoons 2NO(g)$
c. $2NO_2(g) \rightleftharpoons N_2(g) + 2O_2(g)$
d. $4NH_3(g) + 3O_2(g) \rightleftharpoons 2N_2(g) + 6H_2O(g)$

10.55 Would decreasing the volume of the container for each of the following reactions cause the equilibrium to shift in the direction of the products, the reactants, or not change? (10.5)

a. $3O_2(g) \rightleftharpoons 2O_3(g)$
b. $2CO_2(g) \rightleftharpoons 2CO(g) + O_2(g)$
c. $2SO_2(g) + 2H_2O(g) \rightleftharpoons 2H_2S(g) + 3O_2(g)$

10.56 Would increasing the volume of the container for each of the following reactions cause the equilibrium to shift in the direction of the products, the reactants, or not change? (10.5)

a. $2NH_3(g) \rightleftharpoons 3H_2(g) + N_2(g)$
b. $N_2(g) + O_2(g) \rightleftharpoons 2NO(g)$
c. $N_2(g) + 2O_2(g) \rightleftharpoons 2NO_2(g)$

10.57 The equilibrium constant, K_c, for the decomposition of $COCl_2$ to CO and Cl_2 is 0.68. If an equilibrium mixture contains 0.40 M CO and 0.74 M Cl_2, what is the molar concentration of $COCl_2$? (10.3, 10.4)

$$COCl_2(g) \rightleftharpoons CO(g) + Cl_2(g)$$

10.58 The equilibrium constant, K_c, for the reaction of iodine bromide to form iodine and bromine is 2.5×10^{-3}. If an equilibrium mixture contains 0.40 M IBr and 0. 010 M I_2, what is the molar concentration of Br_2? (10.3, 10.4)

$$2IBr(g) \rightleftharpoons I_2(g) + Br_2(g)$$

CHALLENGE PROBLEMS

The following problems are related to the topics in this chapter. However, they do not all follow the chapter order, and they require you to combine concepts and skills from several Sections. These problems will help you increase your critical thinking skills and prepare for your next exam.

10.59 The K_c at 100 °C is 2.0 for the decomposition reaction of NOBr. (10.3, 10.4, 10.5)

$$2NOBr(g) \rightleftharpoons 2NO(g) + Br_2(g)$$

In an experiment, 1.0 mole of NOBr, 1.0 mole of NO, and 1.0 mole of Br_2 were placed in a 1.0-L container.

a. Write the equilibrium expression for the reaction.
b. Is the system at equilibrium?
c. If not, will the rate of the forward or reverse reaction initially speed up?
d. At equilibrium, which concentration(s) will be greater than 1.0 mole/L, and which will be less than 1.0 mole/L?

10.60 Consider the following reaction: (10.3, 10.4, 10.5)

$$PCl_5(g) \rightleftharpoons PCl_3(g) + Cl_2(g)$$

a. Write the equilibrium expression for the reaction.
b. Initially, 0.60 mole of PCl_5 is placed in a 1.0-L flask. At equilibrium, there is 0.16 mole of PCl_3 in the flask. What are the equilibrium concentrations of PCl_5 and Cl_2?
c. What is the numerical value of the equilibrium constant, K_c, for the reaction?
d. If 0.20 mole of Cl_2 is added to the equilibrium mixture, will the concentration of PCl_5 increase or decrease?

10.61 Indicate how each of the following will affect the equilibrium concentration of H_2O in the reaction: (10.3, 10.5)

$$O_2(g) + 2HF(g) + 76\text{ kcal} \rightleftharpoons OF_2(g) + H_2O(g)$$

a. adding more $OF_2(g)$
b. increasing the temperature
c. increasing the volume of the container
d. decreasing the volume of the container
e. adding a catalyst

10.62 Indicate how each of the following will affect the equilibrium concentration of NH_3 in the reaction: (10.3, 10.5)

$$4NH_3(g) + 5O_2(g) \rightleftharpoons 4NO(g) + 6H_2O(g) + 906\text{ kJ}$$

a. adding more $O_2(g)$
b. increasing the temperature
c. increasing the volume of the container
d. adding more $NO(g)$
e. removing some $H_2O(g)$

10.63 Indicate if you would increase or decrease the volume of the container to *increase* the yield of the products in each of the following: (10.5)

a. $2H_2S(g) + CH_4(g) \rightleftharpoons CS_2(g) + 4H_2(g)$
b. $2CH_4(g) \rightleftharpoons C_2H_2(g) + 3H_2(g)$
c. $2H_2(g) + O_2(g) \rightleftharpoons 2H_2O(g)$

10.64 Indicate if you would increase or decrease the volume of the container to *increase* the yield of the products in each of the following: (10.5)

a. $Cl_2(g) + 2NO(g) \rightleftharpoons 2NOCl(g)$
b. $N_2(g) + 2H_2(g) \rightleftharpoons N_2H_4(g)$
c. $N_2O_4(g) \rightleftharpoons 2NO_2(g)$

ANSWERS

10.1 a. The rate of the reaction indicates how fast the products form or how fast the reactants are used up.
b. At room temperature, the reactions involved in the growth of bread mold will proceed at a faster rate than at the lower temperature of the refrigerator.

10.3 The number of collisions will increase when the number of Br_2 molecules is increased.

10.5 a. increase **b.** increase
c. increase **d.** decrease

10.7 A reversible reaction is one in which a forward reaction converts reactants to products, whereas a reverse reaction converts products to reactants.

10.9 a. not at equilibrium **b.** at equilibrium
c. at equilibrium

10.11 The reaction has reached equilibrium because the number of reactants and products does not change.

10.13 a. $K_c = \frac{[CS_2][H_2]^4}{[CH_4][H_2S]^2}$ **b.** $K_c = \frac{[N_2][O_2]}{[NO]^2}$
c. $K_c = \frac{[CS_2][O_2]^4}{[SO_3]^2[CO_2]}$ **d.** $K_c = \frac{[H_2]^3[CO]}{[CH_4][H_2O]}$

10.15 $K_c = \frac{[XY]^2}{[X_2][Y_2]} = 36$

10.17 $K_c = 1.5$

10.19 $K_c = 260$

10.21 Diagram **B** represents the equilibrium mixture.

10.23 a. mostly products
b. mostly products
c. mostly reactants

10.25 $[H_2] = 1.1 \times 10^{-3}$ M

10.27 $[NOBr] = 1.4$ M

10.29 a. Equilibrium shifts in the direction of the product.
b. Equilibrium shifts in the direction of the reactant.
c. Equilibrium shifts in the direction of the product.
d. Equilibrium shifts in the direction of the reactant.
e. No shift in equilibrium occurs.

10.31 a. Equilibrium shifts in the direction of the product.
b. Equilibrium shifts in the direction of the product.
c. Equilibrium shifts in the direction of the product.
d. No shift in equilibrium occurs.
e. Equilibrium shifts in the direction of the reactants.

10.33 a. The oxygen concentration is lowered.
b. Equilibrium shifts in the direction of the reactants.

10.35 a. Equilibrium shifts in the direction of the products.
b. Equilibrium shifts in the direction of the reactants.
c. Equilibrium shifts in the direction of the reactants.

10.37 Without enough iron, Andrew will not have sufficient hemoglobin. When Hb is decreased, the equilibrium shifts in the direction of the reactants.

10.39 a. $K_c = \frac{[CO_2][H_2O]^2}{[CH_4][O_2]^2}$ **b.** $K_c = \frac{[N_2]^2[H_2O]^6}{[NH_3]^4[O_2]^3}$

10.41 The equilibrium constant for the reaction would have a small value.

10.43 T_2 is lower than T_1.

10.45 a. Equilibrium shifts in the direction of the reactants.
b. Equilibrium shifts in the direction of the product.
c. There is no change in equilibrium.
d. Equilibrium shifts in the direction of the product.

10.47 a. mostly products
b. both reactants and products
c. mostly reactants

10.49 a. $K_c = \frac{[N_2][H_2]^3}{[NH_3]^2}$ **b.** $K_c = 9.4$

10.51 $[N_2O_4] = 1.3$ M

10.53 a. Equilibrium shifts in the direction of the product.
b. Equilibrium shifts in the direction of the reactant.
c. Equilibrium shifts in the direction of the product.
d. Equilibrium shifts in the direction of the reactants.

10.55 a. Equilibrium shifts in the direction of the product.
b. Equilibrium shifts in the direction of the reactant.
c. Equilibrium shifts in the direction of the reactants.

10.57 $[COCl_2] = 0.44$ M

10.59 a. $K_c = \frac{[NO]^2[Br_2]}{[NOBr]^2}$
b. When the concentrations are placed in the equilibrium expression, the result is 1.0, which is not equal to K_c. The system is not at equilibrium.
c. The rate of the forward reaction will increase.
d. The $[Br_2]$ and [NO] will increase and [NOBr] will decrease.

10.61 a. decrease **b.** increase **c.** decrease
d. increase **e.** no change

10.63 a. increase **b.** increase **c.** decrease

11

Acids and Bases

Larry, a 30-year-old man, is brought to the emergency room after an automobile accident where he is unresponsive. One of the emergency room nurses takes a blood sample, which is then sent to Brianna, a clinical laboratory technician, who begins the process of analyzing the pH, the partial pressures of O_2 and CO_2, and the concentrations of glucose and electrolytes.

Brianna determines that Larry's blood pH is 7.30 and the partial pressure of CO_2 gas is above the desired level. Blood pH is typically in the range of 7.35 to 7.45, and a value less than 7.35 indicates a state of acidosis. Respiratory acidosis occurs because an increase in the partial pressure of CO_2 gas in the bloodstream prevents the biochemical buffers in blood from making a change in the pH.

Brianna recognizes these signs and immediately contacts the emergency room to inform them that Larry's airway may be blocked. In the emergency room, they provide Larry with an IV containing bicarbonate to increase the blood pH and begin the process of unblocking his airway. Shortly afterward, Larry's airway is cleared, and his blood pH and partial pressure of CO_2 gas return to normal.

CAREER

Clinical Laboratory Technician

Clinical laboratory technicians, also known as medical laboratory technicians, perform a wide variety of tests on body fluids and cells that help in the diagnosis and treatment of patients. These tests range from determining blood concentrations of glucose and cholesterol to determining drug levels in the blood for transplant patients or a patient undergoing treatment. Clinical laboratory technicians also prepare specimens in the detection of cancerous tumors and type blood samples for transfusions. Clinical laboratory technicians must also interpret and analyze the test results, which are then passed on to the physician.

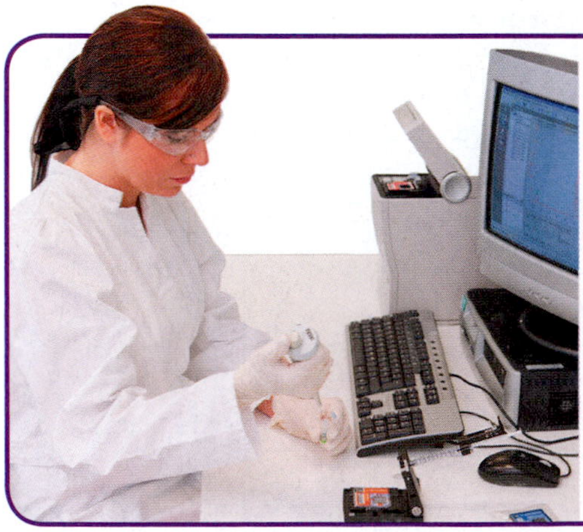

CLINICAL UPDATE

Acid Reflux Disease

After Larry was discharged from the hospital, he complained of a sore throat and dry cough, which his doctor diagnosed as acid reflux. You can view the symptoms of acid reflux disease (GERD) in the **CLINICAL UPDATE** **Acid Reflux Disease**, pages 414–415, and learn about the pH changes in the stomach and how the condition is treated.

11.1 Acids and Bases

LEARNING GOAL Describe and name acids and bases.

Acids and bases are important substances in health, industry, and the environment. One of the most common characteristics of acids is their sour taste. Lemons and grapefruits taste sour because they contain acids such as citric and ascorbic acid (vitamin C). Vinegar tastes sour because it contains acetic acid. We produce lactic acid in our muscles when we exercise. Acid from bacteria turns milk sour in the production of yogurt and cottage cheese. We have hydrochloric acid in our stomachs that helps us digest food. Sometimes we take antacids, which are bases such as sodium bicarbonate or milk of magnesia, to neutralize the effects of too much stomach acid.

The term *acid* comes from the Latin word *acidus*, which means "sour." You are probably already familiar with the sour tastes of vinegar and lemons.

In 1887, Swedish chemist Svante Arrhenius was the first to describe **acids** as substances that produce hydrogen ions (H^+) when they dissolve in water. Because acids produce ions in water, they are electrolytes. For example, hydrogen chloride dissociates in water to give hydrogen ions, H^+, and chloride ions, Cl^-. The hydrogen ions give acids a sour taste, change the blue litmus indicator to red, and corrode some metals.

$$HCl(g) \xrightarrow{H_2O} H^+(aq) + Cl^-(aq)$$

Polar molecular compound — Dissociation — Hydrogen ion

Naming Acids

Acids dissolve in water to produce hydrogen ions, along with a negative ion that may be a simple nonmetal anion or a polyatomic ion. When an acid dissolves in water to produce a hydrogen ion and a simple nonmetal anion, the prefix *hydro* is used before the name of the nonmetal, and its *ide* ending is changed to *ic acid*. For example, hydrogen chloride (HCl) dissolves in water to form HCl(*aq*), which is named hydrochloric acid. An exception is hydrogen cyanide (HCN), which as an acid is named hydrocyanic acid.

When an acid contains oxygen, it dissolves in water to produce a hydrogen ion and an oxygen-containing polyatomic anion. The most common form of an oxygen-containing acid has a name that ends with *ic acid*. The name of its polyatomic anion ends in *ate*. If the acid contains a polyatomic ion with an *ite* ending, its name ends in *ous acid*. When the acid has one oxygen atom less than the common form, the suffix *ous* is used and the polyatomic ion is named with an *ite* ending. The names of some common acids and their anions are listed in **TABLE 11.1**.

TABLE 11.1 Names of Common Acids and Their Anions

Acid	Name of Acid	Anion	Name of Anion
HF	**Hydro**fluor**ic acid**	F^-	Fluor**ide**
HCl	**Hydro**chlor**ic acid**	Cl^-	Chlor**ide**
HBr	**Hydro**brom**ic acid**	Br^-	Brom**ide**
HI	**Hydro**iod**ic acid**	I^-	Iod**ide**
HCN	**Hydro**cyan**ic acid**	CN^-	Cyan**ide**
HNO_3	Nitr**ic acid**	NO_3^-	Nitr**ate**
HNO_2	Nitr**ous acid**	NO_2^-	Nitr**ite**
H_2SO_4	Sulfur**ic acid**	SO_4^{2-}	Sulf**ate**
H_2SO_3	Sulfur**ous acid**	SO_3^{2-}	Sulf**ite**
H_2CO_3	Carbon**ic acid**	CO_3^{2-}	Carbon**ate**
$HC_2H_3O_2$	Acet**ic acid**	$C_2H_3O_2^-$	Acet**ate**
H_3PO_4	Phosphor**ic acid**	PO_4^{3-}	Phosph**ate**
H_3PO_3	Phosphor**ous acid**	PO_3^{3-}	Phosph**ite**

LOOKING AHEAD

REVIEW

Writing Ionic Formulas (6.2)

Citrus fruits are sour because of the presence of acids.

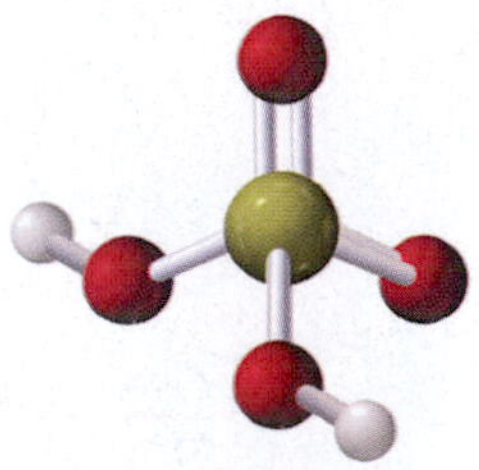

Sulfuric acid dissolves in water to produce one or two H^+ and an anion.

ENGAGE

Why is HBr named hydrobromic acid but $HBrO_3$ is named bromic acid?

The halogens in Group 7A (17) can form more than two oxygen-containing acids. For chlorine, the common form is chloric acid, $HClO_3$, which contains the polyatomic ion chlorate, ClO_3^-. For the acid that contains one more oxygen atom than the common form, the prefix *per* is used; $HClO_4$ is *perchloric acid*. When the polyatomic ion in the acid has one oxygen atom less than the common form, the suffix *ous* is used. Thus, $HClO_2$ is *chlorous acid*; it contains the chlorite ion, ClO_2^-. The prefix *hypo* is used for the acid that has two oxygen atoms less than the common form; HClO is *hypochlorous acid*.

Names of Common Acids and Their Anions

Acid	Name of Acid	Anion	Name of Anion
$HClO_4$	**Perchlor**ic **acid**	ClO_4^-	Perchlor**ate**
$HClO_3$	Chlor**ic acid**	ClO_3^-	Chlor**ate**
$HClO_2$	Chlor**ous acid**	ClO_2^-	Chlor**ite**
HClO	**Hypo**chlor**ous acid**	ClO^-	**Hypo**chlor**ite**

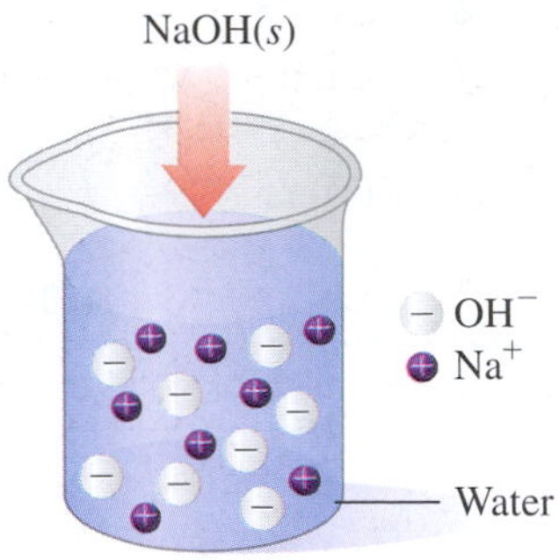
NaOH(*s*)
OH^-
Na^+
Water

$$NaOH(s) \xrightarrow{H_2O} Na^+(aq) + OH^-(aq)$$

Ionic compound   Dissociation   Hydroxide ion

An Arrhenius base produces cations and OH^- anions in an aqueous solution.

Bases

You may be familiar with household bases such as antacids, drain openers, and oven cleaners. According to the Arrhenius theory, **bases** are ionic compounds that dissociate into cations and hydroxide ions (OH^-) when they dissolve in water. They are strong electrolytes. For example, sodium hydroxide is an Arrhenius base that dissociates completely in water to give sodium ions Na^+, and hydroxide ions OH^-.

Most Arrhenius bases are formed from Groups 1A (1) and 2A (2) metals, such as NaOH, KOH, LiOH, and $Ca(OH)_2$. The hydroxide ions (OH^-) give Arrhenius bases common characteristics, such as a bitter taste and a slippery feel. A base turns litmus indicator blue and phenolphthalein indicator pink. **TABLE 11.2** compares some characteristics of acids and bases.

TABLE 11.2 Some Characteristics of Acids and Bases

Characteristic	Acids	Bases
Arrhenius	Produce H^+	Produce OH^-
Electrolytes	Yes	Yes
Taste	Sour	Bitter, chalky
Feel	May sting	Soapy, slippery
Litmus	Red	Blue
Phenolphthalein	Colorless	Pink
Neutralization	Neutralize bases	Neutralize acids

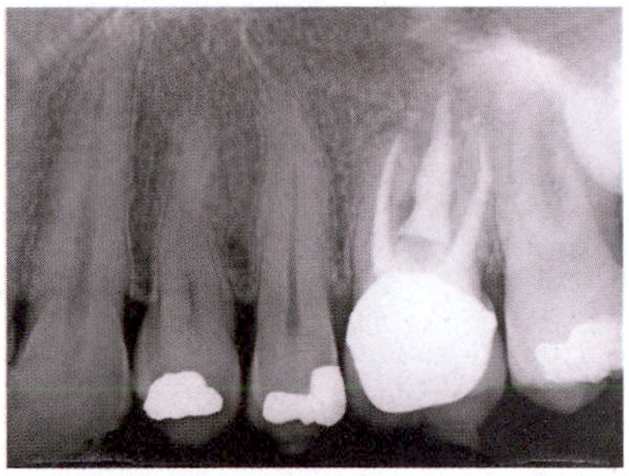

Calcium hydroxide, $Ca(OH)_2$, is used in the food industry to produce beverages, and in dentistry as a filler for root canals.

Naming Bases

Typical Arrhenius bases are named as *hydroxides*.

Base	Name
LiOH	Lithium **hydroxide**
NaOH	Sodium **hydroxide**
KOH	Potassium **hydroxide**
$Ca(OH)_2$	Calcium **hydroxide**
$Al(OH)_3$	Aluminum **hydroxide**

A soft drink contains H_3PO_4 and H_2CO_3.

▶SAMPLE PROBLEM 11.1 Names and Formulas of Acids and Bases

TRY IT FIRST

a. Identify each of the following as an acid or a base, and give its name:

1. H_3PO_4, ingredient in soft drinks   **2.** NaOH, ingredient in oven cleaner

b. Write the formula for each of the following:

1. magnesium hydroxide, ingredient in antacids

2. hydrobromic acid, used industrially to prepare bromide compounds

SOLUTION

a. 1. acid, phosphoric acid **2.** base, sodium hydroxide

b. 1. $Mg(OH)_2$ **2.** HBr

STUDY CHECK 11.1

a. Identify as an acid or a base and give the name for H_2CO_3.

b. Write the formula for iron(III) hydroxide.

ANSWER

a. acid, carbonic acid **b.** $Fe(OH)_3$

TEST

Try Practice Problems 11.1 to 11.6

PRACTICE PROBLEMS

11.1 Acids and Bases

11.1 Indicate whether each of the following statements is characteristic of an acid, a base, or both:

a. has a sour taste
b. neutralizes bases
c. produces H^+ ions in water
d. is named barium hydroxide
e. is an electrolyte

11.2 Indicate whether each of the following statements is characteristic of an acid, a base, or both:

a. neutralizes acids
b. produces OH^- ions in water
c. has a slippery feel
d. conducts an electrical current in solution
e. turns litmus red

11.3 Name each of the following acids or bases:

a. HCl **b.** $Ca(OH)_2$ **c.** $HClO_4$
d. $Sr(OH)_2$ **e.** H_2SO_3 **f.** $HBrO_2$

11.4 Name each of the following acids or bases:

a. $Al(OH)_3$ **b.** HBr **c.** H_2SO_4
d. KOH **e.** HNO_2 **f.** $HClO_2$

11.5 Write formulas for each of the following acids and bases:

a. rubidium hydroxide
b. hydrofluoric acid
c. phosphoric acid
d. lithium hydroxide
e. ammonium hydroxide
f. periodic acid

11.6 Write formulas for each of the following acids and bases:

a. barium hydroxide
b. hydroiodic acid
c. nitric acid
d. strontium hydroxide
e. acetic acid
f. hypochlorous acid

11.2 Brønsted–Lowry Acids and Bases

LEARNING GOAL Identify conjugate acid–base pairs for Brønsted–Lowry acids and bases.

In 1923, J. N. Brønsted in Denmark and T. M. Lowry in Great Britain expanded the definition of acids and bases to include bases that do not contain OH^- ions. A **Brønsted–Lowry acid** can donate a hydrogen ion, H^+, and a **Brønsted–Lowry base** can accept a hydrogen ion.

A Brønsted–Lowry acid is a substance that donates H^+.

A Brønsted–Lowry base is a substance that accepts H^+.

A free hydrogen ion, H^+, does not actually exist in water. Its attraction to polar water molecules is so strong that the H^+ bonds to a water molecule and forms a **hydronium ion, H_3O^+**.

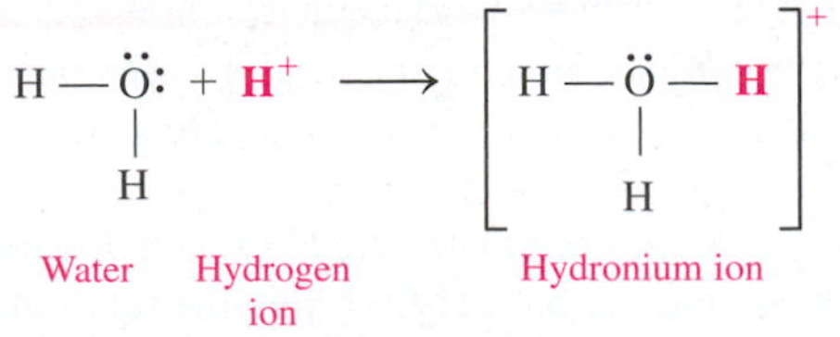
Water
Hydrogen ion
Hydronium ion

We can write the formation of a hydrochloric acid solution as a transfer of H^+ from hydrogen chloride to water. By accepting an H^+ in the reaction, water is acting as a base according to the Brønsted–Lowry concept.

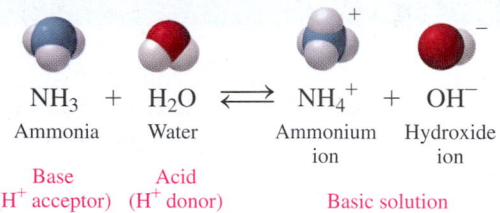

$$HCl + H_2O \longrightarrow H_3O^+ + Cl^-$$

Hydrogen chloride — Water — Hydronium ion — Chloride ion

Acid (H^+ donor) — Base (H^+ acceptor) — Acidic solution

In another reaction, ammonia, NH_3, acts as a base by accepting H^+ when it reacts with water. Because the nitrogen atom of NH_3 has a stronger attraction for H^+ than oxygen, water acts as an acid by donating H^+.

$$NH_3 + H_2O \rightleftarrows NH_4^+ + OH^-$$

Ammonia — Water — Ammonium ion — Hydroxide ion

Base (H^+ acceptor) — Acid (H^+ donor) — Basic solution

▶SAMPLE PROBLEM 11.2 Acids and Bases

TRY IT FIRST

In each of the following equations, identify the reactant that is a Brønsted–Lowry acid and the reactant that is a Brønsted–Lowry base:

a. $HBr(aq) + H_2O(l) \longrightarrow H_3O^+(aq) + Br^-(aq)$
b. $CN^-(aq) + H_2O(l) \rightleftarrows HCN(aq) + OH^-(aq)$

SOLUTION

a. HBr, Brønsted–Lowry acid; H_2O, Brønsted–Lowry base
b. H_2O, Brønsted–Lowry acid; CN^-, Brønsted–Lowry base

STUDY CHECK 11.2

a. When HNO_3 reacts with water, water acts as a Brønsted–Lowry base. Write the equation for the reaction.
b. When hypochlorite ion, ClO^-, reacts with water, water acts as a Brønsted–Lowry acid. Write the equation for the reaction.

ANSWER

a. $HNO_3(aq) + H_2O(l) \rightleftarrows H_3O^+(aq) + NO_3^-(aq)$
b. $ClO^-(aq) + H_2O(l) \rightleftarrows HClO(aq) + OH^-(aq)$

TEST

Try Practice Problems 11.7 and 11.8

CORE CHEMISTRY SKILL

Identifying Conjugate Acid–Base Pairs

Conjugate Acid–Base Pairs

According to the Brønsted–Lowry theory, a **conjugate acid–base pair** consists of molecules or ions related by the loss of one H^+ by an acid, and the gain of one H^+ by a base. Every acid–base reaction contains two conjugate acid–base pairs because an H^+ is transferred in both the forward and reverse directions. When an acid such as HF loses one H^+, the conjugate base F^- is formed. When the base H_2O gains an H^+, its conjugate acid, H_3O^+, is formed.

Because the overall reaction of HF is *reversible*, the conjugate acid H_3O^+ can donate H^+ to the conjugate base F^- and re-form the acid HF and the base H_2O. Using the relationship of loss and gain of one H^+, we can now identify the conjugate acid–base pairs as HF/F^- along with H_3O^+/H_2O.

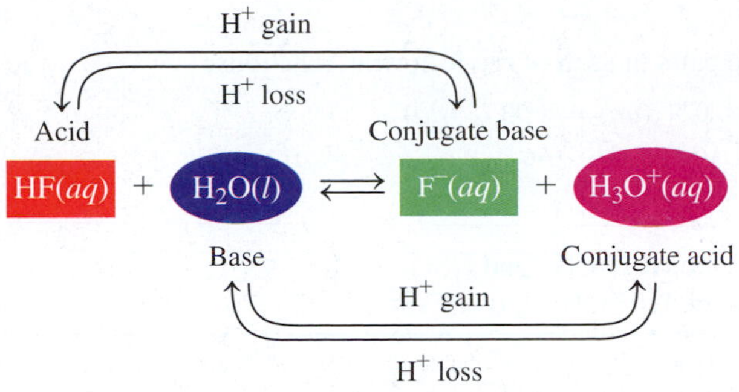

H$^+$ gain
H$^+$ loss
Acid
Conjugate base
HF(aq) + H$_2$O(l) ⇌ F$^-$(aq) + H$_3$O$^+$(aq)
Base
Conjugate acid
H$^+$ gain
H$^+$ loss

HF, an acid, loses one H^+ to form its conjugate base F^-. Water acts as a base by gaining one H^+ to form its conjugate acid H_3O^+.

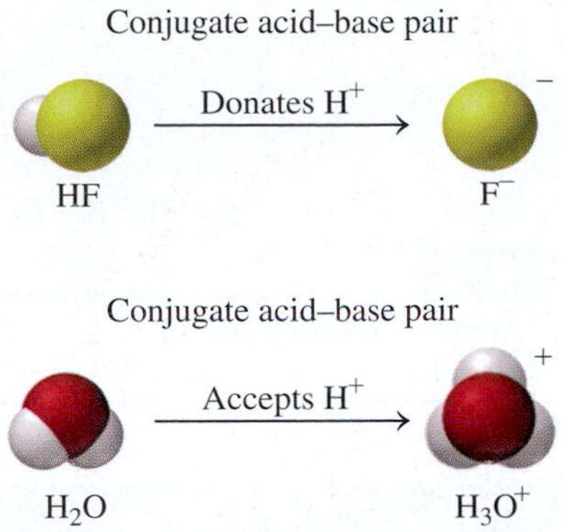

Conjugate acid–base pair
Donates H$^+$
HF
F$^-$
Conjugate acid–base pair
Accepts H$^+$
H$_2$O
H$_3$O$^+$

In another reaction, ammonia (NH_3) accepts H^+ from H_2O to form the conjugate acid NH_4^+ and conjugate base OH^-. Each of these conjugate acid–base pairs, NH_4^+/NH_3 and H_2O/OH^-, is related by the loss and gain of one H^+.

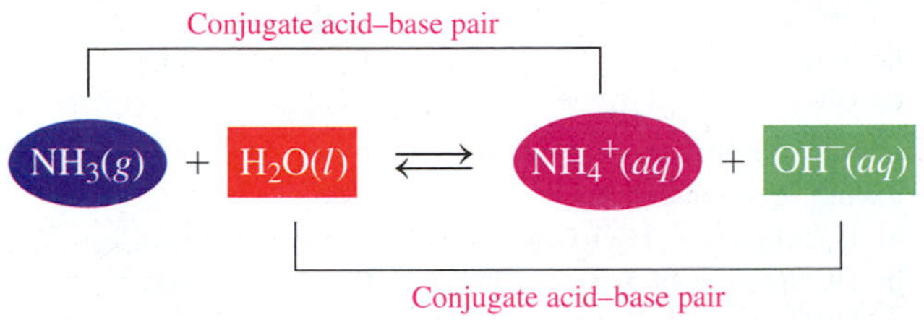

Conjugate acid–base pair
NH$_3$(g) + H$_2$O(l) ⇌ NH$_4^+$(aq) + OH$^-$(aq)
Conjugate acid–base pair

Ammonia, NH_3, acts as a base when it gains one H^+ to form its conjugate acid NH_4^+. Water acts as an acid by losing one H^+ to form its conjugate base OH^-.

ENGAGE

Why is $HBrO_2$ the conjugate acid of BrO_2^-?

In these two examples, we see that water can act as an acid when it donates H^+ or as a base when it accepts H^+. Substances that can act as both acids and bases are **amphoteric** or *amphiprotic*. For water, the most common amphoteric substance, the acidic or basic behavior depends on the other reactant. Water donates H^+ when it reacts with a stronger base, and it accepts H^+ when it reacts with a stronger acid. Another example of an amphoteric substance is bicarbonate (HCO_3^-). With a base, HCO_3^- acts as an acid and donates one H^+ to give CO_3^{2-}. However, when HCO_3^- reacts with an acid, it acts as a base and accepts one H^+ to form H_2CO_3.

ENGAGE

Why can H_2O be both the conjugate base of H_3O^+ and the conjugate acid of OH^-?

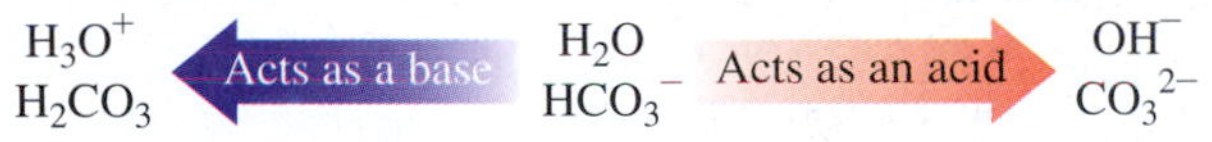

H$_3$O$^+$
H$_2$CO$_3$
Acts as a base
H$_2$O
HCO$_3^-$
Acts as an acid
OH$^-$
CO$_3^{2-}$

Amphoteric substances act as both acids and bases.

▶ SAMPLE PROBLEM 11.3 Identifying Conjugate Acid–Base Pairs

TRY IT FIRST

Identify the conjugate acid–base pairs in the following reaction:

$$HBr(aq) + NH_3(aq) \longrightarrow Br^-(aq) + NH_4^+(aq)$$

SOLUTION

ANALYZE THE PROBLEM	Given		Need	Connect
	HBr NH$_3$	Br$^-$ NH$_4^+$	conjugate acid–base pairs	lose/gain one H$^+$

STEP 1 **Identify the reactant that loses H^+ as the acid.** In the reaction, HBr loses H^+ to form the product Br^-. Thus HBr is the acid and Br^- is its conjugate base.

STEP 2 **Identify the reactant that gains H^+ as the base.** In the reaction, NH_3 gains H^+ to form the product NH_4^+. Thus, NH_3 is the base and NH_4^+ is its conjugate acid.

STEP 3 **Write the conjugate acid–base pairs.**

HBr/Br^- and NH_4^+/NH_3

STUDY CHECK 11.3

Identify the conjugate acid–base pairs in each of the following reactions:

a. $HCN(aq) + SO_4^{2-}(aq) \rightleftharpoons CN^-(aq) + HSO_4^-(aq)$
b. $H_2O(l) + S^{2-}(aq) \rightleftharpoons OH^-(aq) + HS^-(aq)$

TEST

Try Practice Problems 11.9 to 11.16

ANSWER

a. The conjugate acid–base pairs are HCN/CN^- and HSO_4^-/SO_4^{2-}.
b. The conjugate acid–base pairs are H_2O/OH^- and HS^-/S^{2-}.

PRACTICE PROBLEMS

11.2 Brønsted–Lowry Acids and Bases

11.7 Identify the reactant that is a Brønsted–Lowry acid and the reactant that is a Brønsted–Lowry base in each of the following:
- **a.** $HI(aq) + H_2O(l) \longrightarrow I^-(aq) + H_3O^+(aq)$
- **b.** $F^-(aq) + H_2O(l) \rightleftharpoons HF(aq) + OH^-(aq)$
- **c.** $H_2S(aq) + C_2H_5{-}NH_2(aq) \rightleftharpoons HS^-(aq) + C_2H_5{-}NH_3^+(aq)$

11.8 Identify the reactant that is a Brønsted–Lowry acid and the reactant that is a Brønsted–Lowry base in each of the following:
- **a.** $CO_3^{2-}(aq) + H_2O(l) \rightleftharpoons HCO_3^-(aq) + OH^-(aq)$
- **b.** $H_2SO_4(aq) + H_2O(l) \longrightarrow HSO_4^-(aq) + H_3O^+(aq)$
- **c.** $C_2H_3O_2^-(aq) + H_3O^+(aq) \rightleftharpoons HC_2H_3O_2(aq) + H_2O(l)$

11.9 Write the formula for the conjugate base for each of the following acids:
a. HF **b.** H_2O **c.** $H_2PO_3^-$
d. HSO_4^- **e.** $HClO_2$

11.10 Write the formula for the conjugate base for each of the following acids:
a. HCO_3^- **b.** $CH_3{-}NH_3^+$ **c.** HPO_4^{2-}
d. HNO_2 **e.** HBrO

11.11 Write the formula for the conjugate acid for each of the following bases:
a. CO_3^{2-} **b.** H_2O **c.** $H_2PO_4^-$
d. Br^- **e.** ClO_4^-

11.12 Write the formula for the conjugate acid for each of the following bases:
a. SO_4^{2-} **b.** CN^- **c.** NH_3
d. ClO_2^- **e.** HS^-

11.13 Identify the Brønsted–Lowry acid–base pairs in each of the following equations:
- **a.** $H_2CO_3(aq) + H_2O(l) \rightleftharpoons HCO_3^-(aq) + H_3O^+(aq)$
- **b.** $HCN(aq) + NO_2^-(aq) \rightleftharpoons CN^-(aq) + HNO_2(aq)$
- **c.** $CHO_2^-(aq) + HF(aq) \rightleftharpoons HCHO_2(aq) + F^-(aq)$

11.14 Identify the Brønsted–Lowry acid–base pairs in each of the following equations:
- **a.** $H_3PO_4(aq) + H_2O(l) \rightleftharpoons H_2PO_4^-(aq) + H_3O^+(aq)$
- **b.** $H_3PO_4(aq) + NH_3(aq) \rightleftharpoons H_2PO_4^-(aq) + NH_4^+(aq)$
- **c.** $HNO_2(aq) + C_2H_5{-}NH_2(aq) \rightleftharpoons C_2H_5{-}NH_3^+(aq) + NO_2^-(aq)$

11.15 When ammonium chloride dissolves in water, the ammonium ion, NH_4^+, acts as an acid. Write a balanced equation for the reaction of the ammonium ion with water.

11.16 When sodium carbonate dissolves in water, the carbonate ion, CO_3^{2-}, acts as a base. Write a balanced equation for the reaction of the carbonate ion with water.

11.3 Strengths of Acids and Bases

LEARNING GOAL Write equations for the dissociation of strong and weak acids; identify the direction of reaction.

In the process called **dissociation**, an acid or a base separates into ions in water. The *strength* of an acid is determined by the moles of H_3O^+ that are produced for each mole of acid that dissolves. The *strength* of a base is determined by the moles of OH^- that are produced for each mole of base that dissolves. Strong acids and strong bases dissociate completely in water, whereas weak acids and weak bases dissociate only slightly, leaving most of the initial acid or base undissociated.

Strong and Weak Acids

Strong acids are examples of strong electrolytes because they donate H^+ so easily that their dissociation in water is essentially complete. For example, when HCl, a strong acid, dissociates in water, H^+ is transferred to H_2O; the resulting solution contains essentially only the ions H_3O^+ and Cl^-. We consider the reaction of HCl in H_2O as going 100% to products. Thus, one mole of a strong acid dissociates in water to yield one mole of H_3O^+ and one mole of its conjugate base. We write the equation for a strong acid such as HCl with a single arrow.

$$HCl(g) + H_2O(l) \longrightarrow H_3O^+(aq) + Cl^-(aq)$$

There are only six common strong acids, which are stronger acids than H_3O^+. All other acids are weak. **TABLE 11.3** lists the relative strengths of acids and bases. **Weak acids** are weak electrolytes because they dissociate slightly in water, forming only a small amount of H_3O^+ ions. A weak acid has a strong conjugate base, which is why the reverse reaction is more prevalent. Even at high concentrations, weak acids produce low concentrations of H_3O^+ ions (see **FIGURE 11.1**).

Many of the products you use at home contain weak acids. Citric acid is a weak acid found in fruits and fruit juices such as lemons, oranges, and grapefruit. The vinegar used in salad dressings is typically a 5% (m/v) acetic acid, $HC_2H_3O_2$, solution. In water, a few $HC_2H_3O_2$ molecules donate H^+ to H_2O to form H_3O^+ ions and acetate ions $C_2H_3O_2^-$. The reverse reaction also takes place, which converts the H_3O^+ ions and acetate ions $C_2H_3O_2^-$ back to reactants. The formation of hydronium ions from vinegar is the reason we notice the sour taste of vinegar. We write the equation for a weak acid in an aqueous solution with a double arrow to indicate that the forward and reverse reactions are at equilibrium.

$$\underset{\text{Acetic acid}}{HC_2H_3O_2(aq)} + H_2O(l) \rightleftharpoons \underset{\text{Acetate ion}}{C_2H_3O_2^-(aq)} + H_3O^+(aq)$$

Weak acids are found in foods and household products.

ENGAGE

Which is the weaker acid: H_2SO_4 or H_2S?

TABLE 11.3 Relative Strengths of Acids and Bases

(Acid Strength Increases ↑ ; Base Strength Increases ↓)

Acid		Conjugate Base	
Strong Acids		**Weak Bases**	
Hydroiodic acid	HI	I^-	Iodide ion
Hydrobromic acid	HBr	Br^-	Bromide ion
Perchloric acid	$HClO_4$	ClO_4^-	Perchlorate ion
Hydrochloric acid	HCl	Cl^-	Chloride ion
Sulfuric acid	H_2SO_4	HSO_4^-	Hydrogen sulfate ion
Nitric acid	HNO_3	NO_3^-	Nitrate ion
Hydronium ion	H_3O^+	H_2O	Water
Weak Acids		**Strong Bases**	
Hydrogen sulfate ion	HSO_4^-	SO_4^{2-}	Sulfate ion
Phosphoric acid	H_3PO_4	$H_2PO_4^-$	Dihydrogen phosphate ion
Nitrous acid	HNO_2	NO_2^-	Nitrite ion
Hydrofluoric acid	HF	F^-	Fluoride ion
Acetic acid	$HC_2H_3O_2$	$C_2H_3O_2^-$	Acetate ion
Carbonic acid	H_2CO_3	HCO_3^-	Bicarbonate ion
Hydrosulfuric acid	H_2S	HS^-	Hydrogen sulfide ion
Dihydrogen phosphate ion	$H_2PO_4^-$	HPO_4^{2-}	Hydrogen phosphate ion
Ammonium ion	NH_4^+	NH_3	Ammonia
Hydrocyanic acid	HCN	CN^-	Cyanide ion
Bicarbonate ion	HCO_3^-	CO_3^{2-}	Carbonate ion
Methylammonium ion	$CH_3{-}NH_3^+$	$CH_3{-}NH_2$	Methylamine
Hydrogen phosphate ion	HPO_4^{2-}	PO_4^{3-}	Phosphate ion
Water	H_2O	OH^-	Hydroxide ion

TEST

Try Practice Problems 11.17 to 11.22

HCl
1 M
$HC_2H_3O_2$
1 M
HCl: Completely dissociated
$HC_2H_3O_2$: Mostly molecules and a few ions

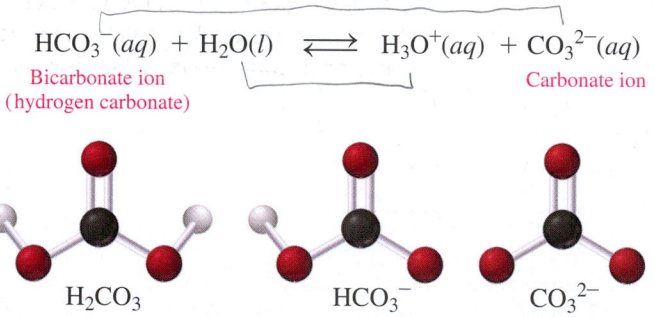

FIGURE 11.1 ▶ A strong acid such as HCl is completely dissociated (≈100%), whereas a weak acid such as $HC_2H_3O_2$ contains mostly molecules and a few ions.

Q What is the difference between a strong acid and a weak acid?

Diprotic Acids

Some weak acids, such as carbonic acid, are *diprotic acids* that have two H^+, which dissociate one at a time. For example, carbonated soft drinks are prepared by dissolving CO_2 in water to form carbonic acid, H_2CO_3. A weak acid such as H_2CO_3 reaches equilibrium between the mostly undissociated H_2CO_3 molecules and the ions H_3O^+ and HCO_3^-.

$$H_2CO_3(aq) + H_2O(l) \rightleftharpoons H_3O^+(aq) + HCO_3^-(aq)$$

Carbonic acid — Bicarbonate ion (hydrogen carbonate)

Because HCO_3^- is also a weak acid, a second dissociation can take place to produce another hydronium ion and the carbonate ion, CO_3^{2-}.

$$HCO_3^-(aq) + H_2O(l) \rightleftharpoons H_3O^+(aq) + CO_3^{2-}(aq)$$

Bicarbonate ion (hydrogen carbonate) — Carbonate ion

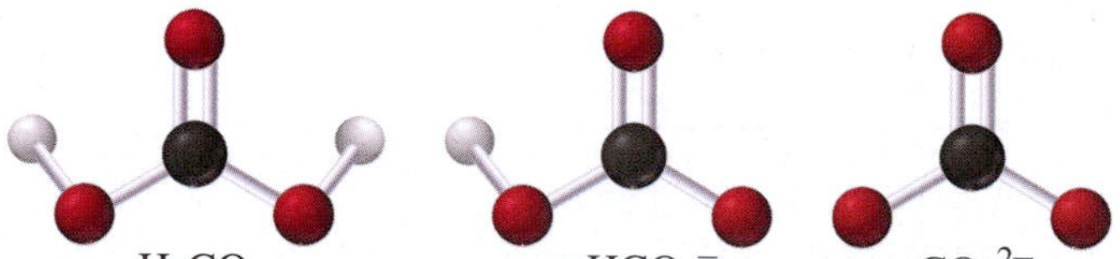

H_2CO_3 — HCO_3^- — CO_3^{2-}

Carbonic acid, a weak acid, loses one H^+ to form hydrogen carbonate ion, which loses a second H^+ to form carbonate ion.

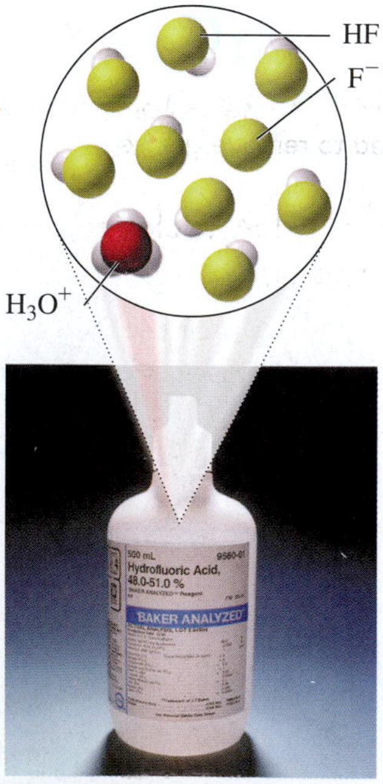
HF
F^-
H_3O^+
500 mL
Hydrofluoric Acid,
48.0-51.0 %
BAKER ANALYZED

Hydrofluoric acid is the only halogen acid that is a weak acid.

Sulfuric acid, H_2SO_4, is also a diprotic acid. However, its first dissociation is complete (100%), which means H_2SO_4 is a strong acid. The product, hydrogen sulfate HSO_4^-, can dissociate again but only slightly, which means that the hydrogen sulfate ion is a weak acid.

$$H_2SO_4(aq) + H_2O(l) \longrightarrow H_3O^+(aq) + HSO_4^-(aq)$$

Sulfuric acid — Bisulfate ion (hydrogen sulfate)

$$HSO_4^-(aq) + H_2O(l) \rightleftharpoons H_3O^+(aq) + SO_4^{2-}(aq)$$

Bisulfate ion (hydrogen sulfate) — Sulfate ion

In summary, a strong acid such as HI in water dissociates completely to form an aqueous solution of the ions H_3O^+ and I^-. A weak acid such as HF dissociates only slightly in water to form an aqueous solution that consists mostly of HF molecules and a few H_3O^+ and F^- ions (see **FIGURE 11.2**).

Strong acid: $HI(aq) + H_2O(l) \longrightarrow H_3O^+(aq) + I^-(aq)$ Completely dissociated

Weak acid: $HF(aq) + H_2O(l) \rightleftarrows H_3O^+(aq) + F^-(aq)$ Slightly dissociated

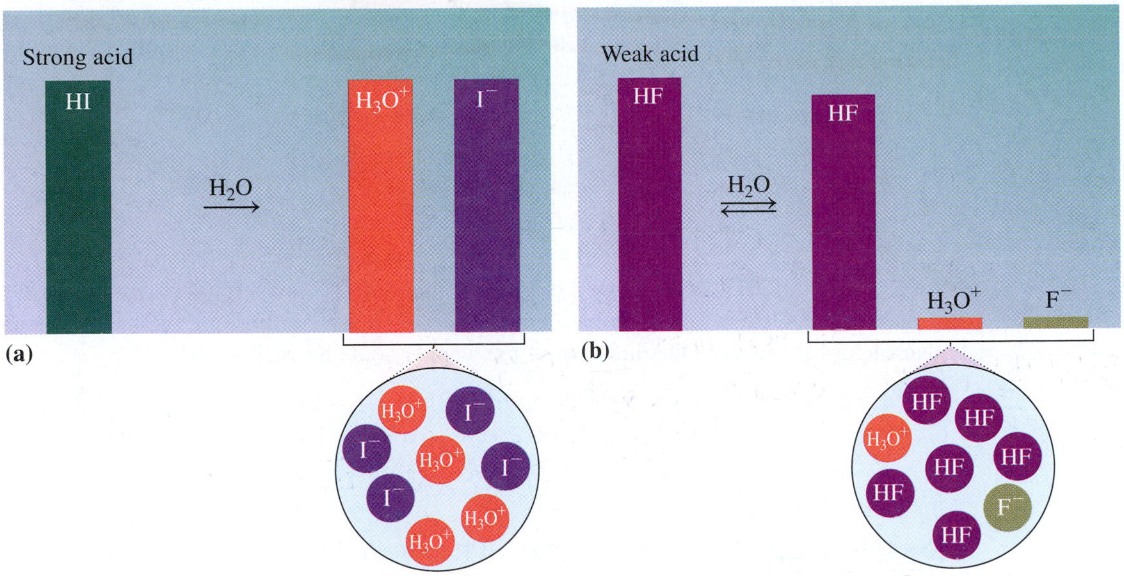

Strong acid
HI
H2O
H3O+
I−
(a)
Weak acid
HF
H2O
HF
H3O+
F−
(b)

FIGURE 11.2 ▶ After dissociation in water, **(a)** the strong acid HI has high concentrations of H_3O^+ and I^-, and **(b)** the weak acid HF has a high concentration of HF and low concentrations of H_3O^+ and F^-.

Q How do the heights of H_3O^+ and F^- compare to the height of the weak acid HF in the bar diagram for HF?

Strong and Weak Bases

As strong electrolytes, **strong bases** dissociate completely in water. Because these strong bases are ionic compounds, they dissociate in water to give an aqueous solution of metal ions and hydroxide ions. The Group 1A (1) hydroxides are very soluble in water, which can give high concentrations of OH^- ions. A few strong bases are less soluble in water, but what does dissolve dissociates completely as ions. For example, when KOH forms a KOH solution, it contains only the ions K^+ and OH^-.

$$KOH(s) \xrightarrow{H_2O} K^+(aq) + OH^-(aq)$$

Strong Bases	
Lithium hydroxide	LiOH
Sodium hydroxide	NaOH
Potassium hydroxide	KOH
Rubidium hydroxide	RbOH
Cesium hydroxide	CsOH
Calcium hydroxide	$Ca(OH)_2$*
Strontium hydroxide	$Sr(OH)_2$*
Barium hydroxide	$Ba(OH)_2$*
*Low solubility, but dissociates completely	

Weak bases are weak electrolytes that are poor acceptors of hydrogen ions and produce very few ions in solution. A typical weak base, ammonia, NH_3, is found in window cleaners. In an aqueous solution, only a few ammonia molecules accept hydrogen ions to form NH_4^+ and OH^-.

$$NH_3(g) + H_2O(l) \rightleftarrows NH_4^+(aq) + OH^-(aq)$$

Ammonia   Ammonium hydroxide

Bases in household products are used to remove grease.

Bases in Household Products

Weak Bases

Window cleaner, ammonia, NH_3
Bleach, NaOCl
Laundry detergent, Na_2CO_3, Na_3PO_4
Toothpaste and baking soda, $NaHCO_3$
Baking powder, scouring powder, Na_2CO_3
Lime for lawns and agriculture, $CaCO_3$
Laxatives, antacids, $Mg(OH)_2$, $Al(OH)_3$

Strong Bases

Drain cleaner, oven cleaner, NaOH

Direction of Reaction

There is a relationship between the components in each conjugate acid–base pair. Strong acids have weak conjugate bases that do not readily accept H^+. As the strength of the acid decreases, the strength of its conjugate base increases.

In any acid–base reaction, there are two acids and two bases. However, one acid is stronger than the other acid, and one base is stronger than the other base. By comparing their relative strengths, we can determine the direction of the reaction. For example, the strong acid H_2SO_4 readily gives up H^+ to water. The hydronium ion H_3O^+ produced is a weaker acid than H_2SO_4, and the conjugate base HSO_4^- is a weaker base than water.

$$\underset{\text{Stronger acid}}{H_2SO_4(aq)} + \underset{\text{Stronger base}}{H_2O(l)} \longrightarrow \underset{\text{Weaker acid}}{H_3O^+(aq)} + \underset{\text{Weaker base}}{HSO_4^-(aq)} \quad \text{Mostly products}$$

Let's look at another reaction in which water donates one H^+ to carbonate, CO_3^{2-}, to form HCO_3^- and OH^-. From Table 11.3, we see that HCO_3^- is a stronger acid than H_2O. We also see that OH^- is a stronger base than CO_3^{2-}. To reach equilibrium, the stronger acid and stronger base react in the direction of the weaker acid and weaker base.

$$\underset{\text{Weaker base}}{CO_3^{2-}(aq)} + \underset{\text{Weaker acid}}{H_2O(l)} \rightleftarrows \underset{\text{Stronger acid}}{HCO_3^-(aq)} + \underset{\text{Stronger base}}{OH^-(aq)} \quad \text{Mostly reactants}$$

▶ SAMPLE PROBLEM 11.4 Direction of Reaction

TRY IT FIRST

Does the equilibrium mixture of the following reaction contain mostly reactants or products?

$$HF(aq) + H_2O(l) \rightleftarrows H_3O^+(aq) + F^-(aq)$$

SOLUTION

From Table 11.3, we see that HF is a weaker acid than H_3O^+ and that H_2O is a weaker base than F^-. Thus, the equilibrium mixture contains mostly reactants.

$$\underset{\text{Weaker acid}}{HF(aq)} + \underset{\text{Weaker base}}{H_2O(l)} \rightleftarrows \underset{\text{Stronger acid}}{H_3O^+(aq)} + \underset{\text{Stronger base}}{F^-(aq)}$$

STUDY CHECK 11.4

Does the equilibrium mixture for the reaction of nitric acid and water contain mostly reactants or products?

ANSWER

$$HNO_3(aq) + H_2O(l) \longrightarrow H_3O^+(aq) + NO_3^-(aq)$$

The equilibrium mixture contains mostly products because HNO_3 is a stronger acid than H_3O^+, and H_2O is a stronger base than NO_3^-.

TEST

Try Practice Problems 11.23 to 11.26

PRACTICE PROBLEMS

11.3 Strengths of Acids and Bases

11.17 What is meant by the phrase "A strong acid has a weak conjugate base"?

11.18 What is meant by the phrase "A weak acid has a strong conjugate base"?

11.19 Identify the stronger acid in each of the following pairs:
a. HBr or HNO_2 **b.** H_3PO_4 or HSO_4^- **c.** HCN or H_2CO_3

11.20 Identify the stronger acid in each of the following pairs:
a. NH_4^+ or H_3O^+ **b.** H_2SO_4 or HCl **c.** H_2O or H_2CO_3

11.21 Identify the weaker acid in each of the following pairs:
a. HCl or HSO_4^- **b.** HNO_2 or HF **c.** HCO_3^- or NH_4^+

11.22 Identify the weaker acid in each of the following pairs:
a. HNO_3 or HCO_3^- **b.** HSO_4^- or H_2O **c.** H_2SO_4 or H_2CO_3

11.23 Predict whether each of the following reactions contains mostly reactants or products at equilibrium:

a. $H_2CO_3(aq) + H_2O(l) \rightleftarrows HCO_3^-(aq) + H_3O^+(aq)$

b. $NH_4^+(aq) + H_2O(l) \rightleftarrows NH_3(aq) + H_3O^+(aq)$

c. $HNO_2(aq) + NH_3(aq) \rightleftarrows NO_2^-(aq) + NH_4^+(aq)$

11.24 Predict whether each of the following reactions contains mostly reactants or products at equilibrium:

a. $H_3PO_4(aq) + H_2O(l) \rightleftarrows H_3O^+(aq) + H_2PO_4^-(aq)$

b. $CO_3^{2-}(aq) + H_2O(l) \rightleftarrows OH^-(aq) + HCO_3^-(aq)$

c. $HS^-(aq) + F^-(aq) \rightleftarrows HF(aq) + S^{2-}(aq)$

11.25 Write an equation for the acid–base reaction between ammonium ion and sulfate ion. Why does the equilibrium mixture contain mostly reactants?

11.26 Write an equation for the acid–base reaction between nitrous acid and hydroxide ion. Why does the equilibrium mixture contain mostly products?

11.4 Dissociation of Weak Acids and Bases

LEARNING GOAL Write the expression for the dissociation of a weak acid or weak base.

REVIEW
Balancing a Chemical Equation (7.1)

As we have seen, acids have different strengths depending on how much they dissociate in water. Because the dissociation of strong acids in water is essentially complete, the reaction is not considered to be an equilibrium situation. However, because weak acids in water dissociate only slightly, the ion products reach equilibrium with the undissociated weak acid molecules. For example, formic acid $HCHO_2$, the acid found in bee and ant stings, is a weak acid. Formic acid is a weak acid that dissociates in water to form hydronium ion, H_3O^+, and formate ion, CHO_2^-.

$$HCHO_2(aq) + H_2O(l) \rightleftarrows H_3O^+(aq) + CHO_2^-(aq)$$

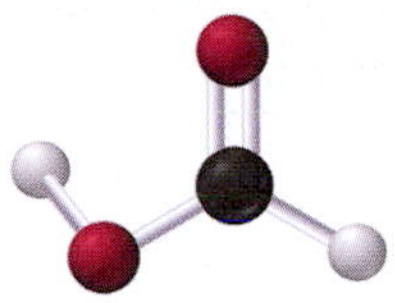
$HCHO_2$

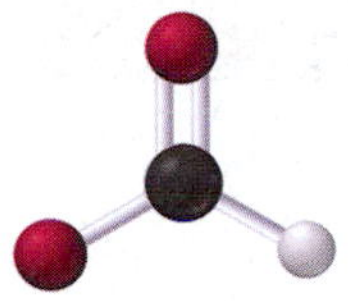
CHO_2^-

Formic acid, a weak acid, loses one H^+ to form formate ion.

Writing Dissociation Expressions

An acid dissociation expression can be written for weak acids that gives the ratio of the concentrations of products to the reactants. As with other dissociation expressions, the molar concentration of the products is divided by the molar concentration of the reactants. (Recall that the brackets in the acid dissociation expression represent the molar concentrations of the reactants and products.) Because water is a pure liquid with a constant concentration, it is omitted.

$$K_a = \frac{[H_3O^+][CHO_2^-]}{[HCHO_2]} \quad \text{Acid dissociation expression}$$

The numerical value of the acid dissociation expression is the **acid dissociation constant, K_a.** The value of the K_a for formic acid at 25 °C is determined by experiment to be 1.8×10^{-4}. Thus, for the weak acid $HCHO_2$, the K_a is written

$$K_a = \frac{[H_3O^+][CHO_2^-]}{[HCHO_2]} = 1.8 \times 10^{-4} \quad \text{Acid dissociation constant}$$

The K_a for formic acid is small, which means that the equilibrium mixture of formic acid in water contains mostly reactants and only small amounts of the products. Weak acids have small K_a values. However, strong acids, which are essentially 100% dissociated, have very large K_a values, but these values are not usually given. **TABLE 11.4** gives K_a and K_b values for selected weak acids and bases.

Let us now consider the dissociation of the weak base methylamine:

$$CH_3—NH_2(aq) + H_2O(l) \rightleftarrows CH_3—NH_3^+(aq) + OH^-(aq)$$

TABLE 11.4 K_a and K_b Values for Selected Weak Acids and Bases

Acids		K_a
Phosphoric acid	H_3PO_4	7.5×10^{-3}
Nitrous acid	HNO_2	4.5×10^{-4}
Hydrofluoric acid	HF	3.5×10^{-4}
Formic acid	$HCHO_2$	1.8×10^{-4}
Acetic acid	$HC_2H_3O_2$	1.8×10^{-5}
Carbonic acid	H_2CO_3	4.3×10^{-7}
Hydrosulfuric acid	H_2S	9.1×10^{-8}
Dihydrogen phosphate	$H_2PO_4^-$	6.2×10^{-8}
Hypochlorous acid	$HClO$	3.0×10^{-8}
Hydrocyanic acid	HCN	4.9×10^{-10}
Hydrogen carbonate	HCO_3^-	5.6×10^{-11}
Hydrogen phosphate	HPO_4^{2-}	2.2×10^{-13}
Bases		K_b
Methylamine	$CH_3—NH_2$	4.4×10^{-4}
Carbonate	CO_3^{2-}	2.2×10^{-4}
Ammonia	NH_3	1.8×10^{-5}

As we did with the acid dissociation expression, the concentration of water is omitted from the base dissociation expression. The **base dissociation constant**, K_b, for methylamine is written

$$K_b = \frac{[CH_3—NH_3^+][OH^-]}{[CH_3—NH_2]} = 4.4 \times 10^{-4}$$

TABLE 11.5 summarizes the characteristics of acids and bases in terms of strength and equilibrium position.

TABLE 11.5 Characteristics of Acids and Bases

Characteristic	Strong Acids	Weak Acids
Equilibrium Position	Toward products	Toward reactants
K_a	Large	Small
$[H_3O^+]$ and $[A^-]$	100% of [HA] dissociates	Small percent of [HA] dissociates
Conjugate Base	Weak	Strong
Characteristic	**Strong Bases**	**Weak Bases**
Equilibrium Position	Toward products	Toward reactants
K_b	Large	Small
$[BH^+]$ and $[OH^-]$	100% of [B] reacts	Small percent of [B] reacts
Conjugate Acid	Weak	Strong

▶SAMPLE PROBLEM 11.5 Writing an Acid Dissociation Expression

TRY IT FIRST

Write the acid dissociation expression for the weak acid, nitrous acid.

SOLUTION

The equation for the dissociation of nitrous acid is written

$$HNO_2(aq) + H_2O(l) \rightleftharpoons H_3O^+(aq) + NO_2^-(aq)$$

The acid dissociation expression is written as the concentration of the products divided by the concentration of the undissociated weak acid.

$$K_a = \frac{[H_3O^+][NO_2^-]}{[HNO_2]}$$

STUDY CHECK 11.5

Write the acid dissociation expression for each of the following:

a. hydrogen phosphate **b.** hypochlorous acid

ANSWER

a. $K_a = \frac{[H_3O^+][PO_4^{3-}]}{[HPO_4^{2-}]}$

b. $K_a = \frac{[H_3O^+][ClO^-]}{[HClO]}$

TEST

Try Practice Problems 11.27 to 11.32

PRACTICE PROBLEMS

11.4 Dissociation of Weak Acids and Bases

11.27 Answer *True* or *False* for each of the following: A strong acid

a. is completely dissociated in aqueous solution
b. has a small value of K_a
c. has a strong conjugate base
d. has a weak conjugate base
e. is slightly dissociated in aqueous solution

11.28 Answer *True* or *False* for each of the following: A weak acid

a. is completely dissociated in aqueous solution
b. has a small value of K_a
c. has a strong conjugate base
d. has a weak conjugate base
e. is slightly dissociated in aqueous solution

11.29 Consider the following acids and their dissociation constants:

$$H_2SO_3(aq) + H_2O(l) \rightleftharpoons H_3O^+(aq) + HSO_3^-(aq) \quad K_a = 1.2 \times 10^{-2}$$

$$HS^-(aq) + H_2O(l) \rightleftharpoons H_3O^+(aq) + S^{2-}(aq) \quad K_a = 1.3 \times 10^{-19}$$

a. Which is the stronger acid, H_2SO_3 or HS^-?
b. What is the conjugate base of H_2SO_3?
c. Which acid has the weaker conjugate base?
d. Which acid has the stronger conjugate base?
e. Which acid produces more ions?

11.30 Consider the following acids and their dissociation constants:

$$HPO_4^{2-}(aq) + H_2O(l) \rightleftharpoons H_3O^+(aq) + PO_4^{3-}(aq) \quad K_a = 2.2 \times 10^{-13}$$

$$HCHO_2(aq) + H_2O(l) \rightleftharpoons H_3O^+(aq) + CHO_2^-(aq) \quad K_a = 1.8 \times 10^{-4}$$

a. Which is the weaker acid, HPO_4^{2-} or $HCHO_2$?
b. What is the conjugate base of HPO_4^{2-}?
c. Which acid has the weaker conjugate base?
d. Which acid has the stronger conjugate base?
e. Which acid produces more ions?

11.31 Phosphoric acid dissociates to form hydronium ion and dihydrogen phosphate. Phosphoric acid has a K_a of 7.5×10^{-3}. Write the equation for the reaction and the acid dissociation expression for phosphoric acid.

11.32 Aniline, $C_6H_5{-}NH_2$, a weak base with a K_b of 4.0×10^{-10}, reacts with water to form $C_6H_5{-}NH_3^+$ and hydroxide ion. Write the equation for the reaction and the base dissociation expression for aniline.

11.5 Dissociation of Water

LEARNING GOAL Use the water dissociation expression to calculate the $[H_3O^+]$ and $[OH^-]$ in an aqueous solution.

In many acid–base reactions, water is *amphoteric*, which means that it can act either as an acid or as a base. In pure water, there is a forward reaction between two water molecules that transfers H^+ from one water molecule to the other. One molecule acts as an acid by losing H^+, and the water molecule that gains H^+ acts as a base. Every time H^+ is transferred between two water molecules, the products are one H_3O^+ and one OH^-, which react in the reverse direction to re-form two water molecules. Thus, equilibrium is reached between the conjugate acid–base pairs of water molecules.

ENGAGE

Why is the $[H_3O^+]$ equal to the $[OH^-]$ in pure water?

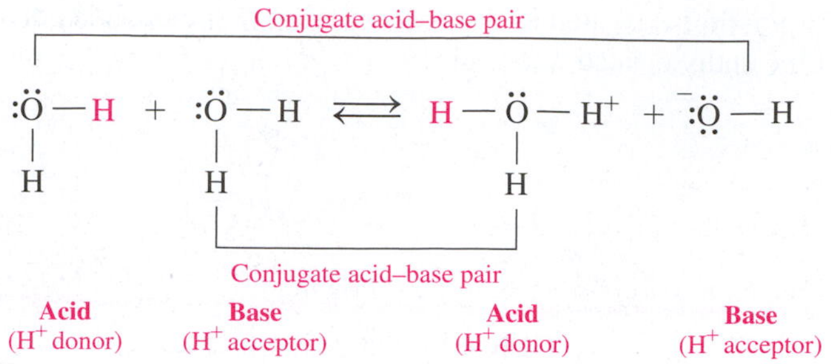

Conjugate acid–base pair
Conjugate acid–base pair
Acid
(H+ donor)
Base
(H+ acceptor)
Acid
(H+ donor)
Base
(H+ acceptor)

Writing the Water Dissociation Expression

Using the equation for water at equilibrium, we can write the equilibrium expression that shows the concentrations of the products divided by the concentrations of the reactants. Recall that square brackets around the symbols indicate their concentrations in moles per liter (M).

$$H_2O(l) + H_2O(l) \rightleftharpoons H_3O^+(aq) + OH^-(aq)$$

$$K = \frac{[H_3O^+][OH^-]}{[H_2O][H_2O]}$$

By omitting the constant concentration of pure water, we can write the **water dissociation expression**.

$$K_w = [H_3O^+][OH^-]$$

Experiments have determined that, in pure water, the concentration of H_3O^+ and OH^- at 25 °C are each 1.0×10^{-7} M.

Pure water $\quad [H_3O^+] = [OH^-] = 1.0 \times 10^{-7}\,M$

When we place the $[H_3O^+]$ and $[OH^-]$ into the water dissociation expression, we obtain the numerical value of the **water dissociation constant, K_w**, which is 1.0×10^{-14} at 25 °C. As before, the concentration units are omitted in the K_w value.

$$K_w = [H_3O^+][OH^-]$$
$$= [1.0 \times 10^{-7}][1.0 \times 10^{-7}] = 1.0 \times 10^{-14}$$

Neutral, Acidic, and Basic Solutions

TEST

Try Practice Problems 11.33 and 11.34

The K_w value (1.0×10^{-14}) applies to any aqueous solution at 25 °C because all aqueous solutions contain both H_3O^+ and OH^- (see **FIGURE 11.3**). When the $[H_3O^+]$ and $[OH^-]$ in a solution are equal, the solution is **neutral**. However, most solutions are not neutral; they have different concentrations of H_3O^+ and OH^-. If acid is added to water, there is an increase

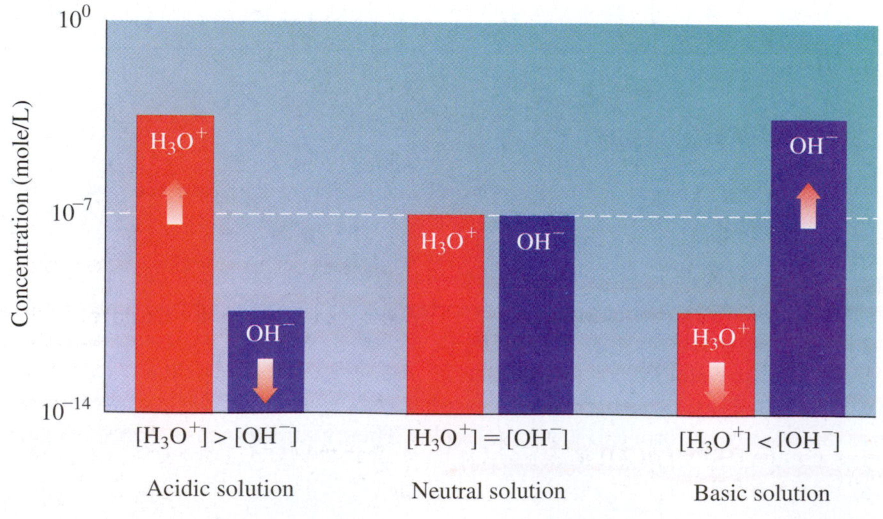

Concentration (mole/L)
10^0
10^-7
10^-14
H3O+
OH−
H3O+
OH−
H3O+
OH−
[H3O+] > [OH−]
[H3O+] = [OH−]
[H3O+] < [OH−]
Acidic solution
Neutral solution
Basic solution

FIGURE 11.3 ▶ In a neutral solution, $[H_3O^+]$ and $[OH^-]$ are equal. In acidic solutions, the $[H_3O^+]$ is greater than the $[OH^-]$. In basic solutions, the $[OH^-]$ is greater than the $[H_3O^+]$.

Q Is a solution that has a $[H_3O^+]$ of 1.0×10^{-3} M acidic, basic, or neutral?

in $[H_3O^+]$ and a decrease in $[OH^-]$, which makes an acidic solution. If base is added, $[OH^-]$ increases and $[H_3O^+]$ decreases, which gives a basic solution. However, for any aqueous solution, whether it is neutral, acidic, or basic, the product $[H_3O^+][OH^-]$ is equal to K_w (1.0×10^{-14} at 25 °C) (see **TABLE 11.6**).

TABLE 11.6 Examples of $[H_3O^+]$ and $[OH^-]$ in Neutral, Acidic, and Basic Solutions

Type of Solution	$[H_3O^+]$	$[OH^-]$	K_w (25 °C)
Neutral	1.0×10^{-7} M	1.0×10^{-7} M	1.0×10^{-14}
Acidic	1.0×10^{-2} M	1.0×10^{-12} M	1.0×10^{-14}
Acidic	2.5×10^{-5} M	4.0×10^{-10} M	1.0×10^{-14}
Basic	1.0×10^{-8} M	1.0×10^{-6} M	1.0×10^{-14}
Basic	5.0×10^{-11} M	2.0×10^{-4} M	1.0×10^{-14}

TEST

Try Practice Problems 11.35 and 11.36

Using the K_w to Calculate $[H_3O^+]$ and $[OH^-]$ in a Solution

If we know the $[H_3O^+]$ of a solution, we can use the K_w to calculate $[OH^-]$. If we know the $[OH^-]$ of a solution, we can calculate $[H_3O^+]$ from their relationship in the K_w, as shown in Sample Problem 11.6.

$$K_w = [H_3O^+][OH^-]$$

$$[OH^-] = \frac{K_w}{[H_3O^+]} \qquad [H_3O^+] = \frac{K_w}{[OH^-]}$$

ENGAGE

If you know the $[H_3O^+]$ of a solution, how do you use the K_w to calculate the $[OH^-]$?

▶SAMPLE PROBLEM 11.6 Calculating the $[H_3O^+]$ of a Solution

CORE CHEMISTRY SKILL

Calculating $[H_3O^+]$ and $[OH^-]$ in Solutions

TRY IT FIRST

A vinegar solution has a $[OH^-] = 5.0 \times 10^{-12}$ M at 25 °C. What is the $[H_3O^+]$ of the vinegar solution? Is the solution acidic, basic, or neutral?

SOLUTION

STEP 1 State the given and needed quantities.

ANALYZE THE PROBLEM	Given	Need	Connect
	$[OH^-] = 5.0 \times 10^{-12}$ M	$[H_3O^+]$	$K_w = [H_3O^+][OH^-]$

STEP 2 Write the K_w for water and solve for the unknown $[H_3O^+]$.

$$K_w = [H_3O^+][OH^-] = 1.0 \times 10^{-14}$$

Solve for $[H_3O^+]$ by dividing both sides by $[OH^-]$.

$$\frac{K_w}{[OH^-]} = \frac{[H_3O^+]\cancel{[OH^-]}}{\cancel{[OH^-]}}$$

$$[H_3O^+] = \frac{1.0 \times 10^{-14}}{[OH^-]}$$

STEP 3 Substitute the known $[OH^-]$ into the equation and calculate.

$$[H_3O^+] = \frac{1.0 \times 10^{-14}}{[5.0 \times 10^{-12}]} = 2.0 \times 10^{-3}\ \text{M}$$

Because the $[H_3O^+]$ of 2.0×10^{-3} M is larger than the $[OH^-]$ of 5.0×10^{-12} M, the solution is acidic.

ENGAGE

Why does the $[H_3O^+]$ of an aqueous solution increase if the $[OH^-]$ decreases?

STUDY CHECK 11.6

a. What is the $[H_3O^+]$ of an ammonia cleaning solution with $[OH^-] = 4.0 \times 10^{-4}$ M? Is the solution acidic, basic, or neutral?

b. The $[H_3O^+]$ of tomato juice is 6.3×10^{-5} M. What is the $[OH^-]$ of the juice? Is the tomato juice acidic, basic, or neutral?

ANSWER

a. $[H_3O^+] = 2.5 \times 10^{-11}$ M, basic **b.** $[OH^-] = 1.6 \times 10^{-10}$ M, acidic

PRACTICE PROBLEMS

11.5 Dissociation of Water

11.33 Why are the concentrations of H_3O^+ and OH^- equal in pure water?

11.34 What is the meaning and value of K_w at 25 °C?

11.35 In an acidic solution, how does the concentration of H_3O^+ compare to the concentration of OH^-?

11.36 If a base is added to pure water, why does the $[H_3O^+]$ decrease?

11.37 Indicate whether each of the following solutions is acidic, basic, or neutral:
- **a.** $[H_3O^+] = 2.0 \times 10^{-5}$ M
- **b.** $[H_3O^+] = 1.4 \times 10^{-9}$ M
- **c.** $[OH^-] = 8.0 \times 10^{-3}$ M
- **d.** $[OH^-] = 3.5 \times 10^{-10}$ M

11.38 Indicate whether each of the following solutions is acidic, basic, or neutral:
- **a.** $[H_3O^+] = 6.0 \times 10^{-12}$ M
- **b.** $[H_3O^+] = 1.4 \times 10^{-4}$ M
- **c.** $[OH^-] = 5.0 \times 10^{-12}$ M
- **d.** $[OH^-] = 4.5 \times 10^{-2}$ M

11.39 Calculate the $[H_3O^+]$ of each aqueous solution with the following $[OH^-]$:
- **a.** coffee, 1.0×10^{-9} M
- **b.** soap, 1.0×10^{-6} M
- **c.** cleanser, 2.0×10^{-5} M
- **d.** lemon juice, 4.0×10^{-13} M

11.40 Calculate the $[H_3O^+]$ of each aqueous solution with the following $[OH^-]$:
- **a.** NaOH solution, 1.0×10^{-2} M
- **b.** milk of magnesia, 1.0×10^{-5} M
- **c.** aspirin, 1.8×10^{-11} M
- **d.** seawater, 2.5×10^{-6} M

Clinical Applications

11.41 Calculate the $[OH^-]$ of each aqueous solution with the following $[H_3O^+]$:
- **a.** stomach acid, 4.0×10^{-2} M
- **b.** urine, 5.0×10^{-6} M
- **c.** orange juice, 2.0×10^{-4} M
- **d.** bile, 7.9×10^{-9} M

11.42 Calculate the $[OH^-]$ of each aqueous solution with the following $[H_3O^+]$:
- **a.** baking soda, 1.0×10^{-8} M
- **b.** blood, 4.2×10^{-8} M
- **c.** milk, 5.0×10^{-7} M
- **d.** pancreatic juice, 4.0×10^{-9} M

11.6 The pH Scale

LEARNING GOAL Calculate pH from $[H_3O^+]$; given the pH, calculate the $[H_3O^+]$ and $[OH^-]$ of a solution.

The proper level of acidity is necessary to evaluate the functioning of the lungs and kidneys, to control bacterial growth in foods, and to prevent the growth of pests in food crops. In the environment, the acidity, or pH, of rain, water, and soil can have significant effects. When rain becomes too acidic, it can dissolve marble statues and accelerate the corrosion of metals. In lakes and ponds, the acidity of water can affect the ability of plants and fish to survive. The acidity of soil around plants affects their growth. If the soil pH is too acidic or too basic, the roots of the plant cannot take up some nutrients. Most plants thrive in soil with a nearly neutral pH, although certain plants, such as orchids, camellias, and blueberries, require a more acidic soil.

Although we have expressed H_3O^+ and OH^- as molar concentrations, it is more convenient to describe the acidity of solutions using the *pH scale*. On this scale, a number between 0 and 14 represents the H_3O^+ concentration for common solutions. A neutral solution has a pH of 7.0 at 25 °C. An acidic solution has a pH less than 7.0; a basic solution has a pH greater than 7.0 (see **FIGURE 11.4**).

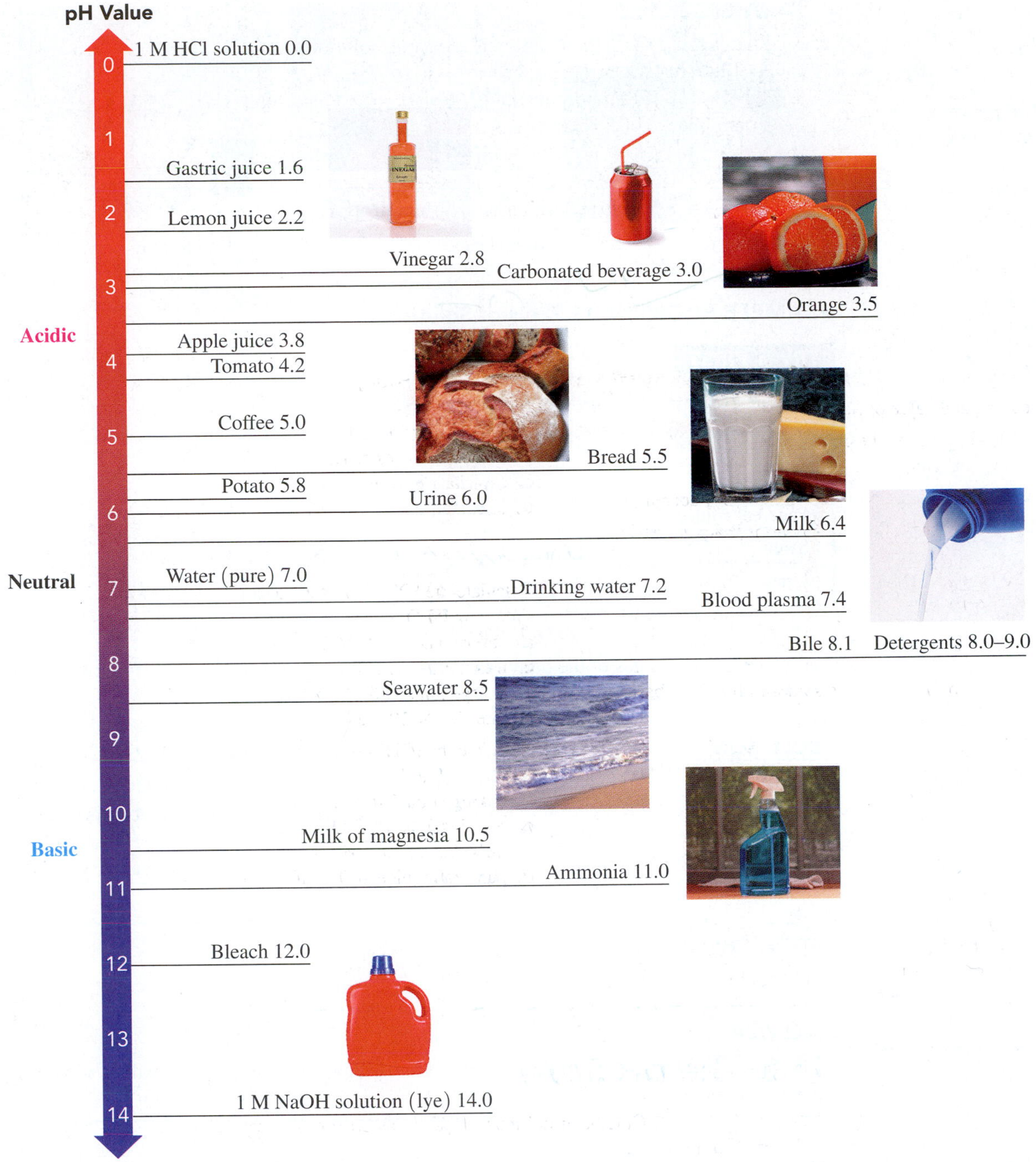

pH Value
1 M HCl solution 0.0
Gastric juice 1.6
Lemon juice 2.2
Vinegar 2.8
Carbonated beverage 3.0
Orange 3.5
Acidic
Apple juice 3.8
Tomato 4.2
Coffee 5.0
Bread 5.5
Potato 5.8
Urine 6.0
Milk 6.4
Neutral
Water (pure) 7.0
Drinking water 7.2
Blood plasma 7.4
Bile 8.1
Detergents 8.0–9.0
Seawater 8.5
Basic
Milk of magnesia 10.5
Ammonia 11.0
Bleach 12.0
1 M NaOH solution (lye) 14.0
0
1
2
3
4
5
6
7
8
9
10
11
12
13
14

FIGURE 11.4 ▶ On the pH scale, values below 7.0 are acidic, a value of 7.0 is neutral, and values above 7.0 are basic.

Q Is apple juice an acidic, a basic, or a neutral solution?

When we relate acidity and pH, we are using an inverse relationship, which is when one component increases while the other component decreases. When an acid is added to pure water, the $[H_3O^+]$ (acidity) of the solution increases but its pH decreases. When a base is added to pure water, it becomes more basic, which means its acidity decreases and the pH increases.

In the laboratory, a pH meter is commonly used to determine the pH of a solution. There are also various indicators and pH papers that turn specific colors when placed in solutions of different pH values. The pH is found by comparing the color on the test paper or the color of the solution to a color chart (see **FIGURE 11.5**).

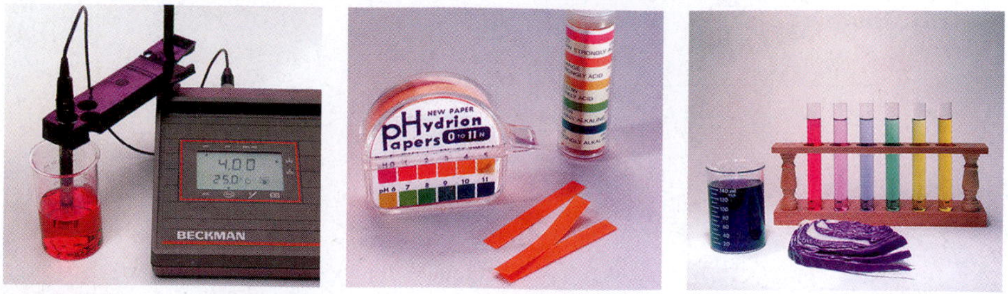
4.00
BECKMAN

(a) pH meter (b) pH paper (c) pH indicator

FIGURE 11.5 ▶ The pH of a solution can be determined using different methods.

Q If a pH meter reads 4.00, is the solution acidic, basic, or neutral?

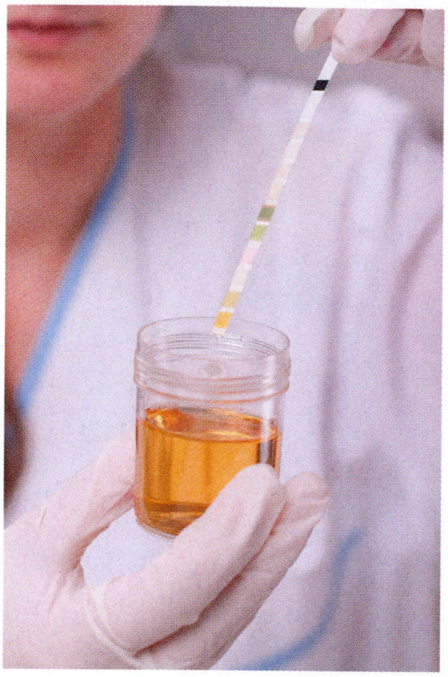

A dipstick is used to measure the pH of a urine sample.

▶ SAMPLE PROBLEM 11.7 pH of Solutions

TRY IT FIRST

Consider the pH of the following body fluids:

Body Fluid	pH
Stomach acid	1.4
Pancreatic juice	8.4
Sweat	4.8
Urine	5.3
Cerebrospinal fluid	7.3

a. Place the pH values of the body fluids on the list in order of most acidic to most basic.
b. Which body fluid has the highest $[H_3O^+]$?

SOLUTION

a. The most acidic body fluid is the one with the lowest pH, and the most basic is the body fluid with the highest pH: stomach acid (1.4), sweat (4.8), urine (5.3), cerebrospinal fluid (7.3), pancreatic juice (8.4).
b. The body fluid with the highest $[H_3O^+]$ would have the lowest pH value, which is stomach acid.

STUDY CHECK 11.7

Which body fluid has the highest $[OH^-]$?

ANSWER

The body fluid with the highest $[OH^-]$ would have the highest pH value, which is pancreatic juice.

TEST

Try Practice Problems 11.43 to 11.46

KEY MATH SKILL

Calculating pH from $[H_3O^+]$

If soil is too acidic, nutrients are not absorbed by crops. Then lime ($CaCO_3$), which acts as a base, may be added to increase the soil pH.

Calculating the pH of Solutions

The pH scale is a logarithmic scale that corresponds to the $[H_3O^+]$ of aqueous solutions. Mathematically, **pH is** the negative logarithm (base 10) of the $[H_3O^+]$.

$$pH = -\log[H_3O^+]$$

Essentially, the negative powers of 10 in the molar concentrations are converted to positive numbers. For example, a lemon juice solution with $[H_3O^+] = 1.0 \times 10^{-2}$ M has a pH of 2.00. This can be calculated using the pH equation:

$$pH = -\log[1.0 \times 10^{-2}]$$
$$pH = -(-2.00)$$
$$= 2.00$$

The number of *decimal places* in the pH value is the same as the number of significant figures in the $[H_3O^+]$. The number to the left of the decimal point in the pH value is the power of 10.

$[H_3O^+] = \mathbf{1.0} \times 10^{-2}$ pH $= 2.\mathbf{00}$

Two SFs → Two SFs

Because pH is a log scale, a change of one pH unit corresponds to a tenfold change in $[H_3O^+]$. It is important to note that the pH decreases as the $[H_3O^+]$ increases. For example, a solution with a pH of 2.00 has a $[H_3O^+]$ that is ten times greater than a solution with a pH of 3.00 and 100 times greater than a solution with a pH of 4.00. The pH of a solution is calculated from the $[H_3O^+]$ by using the *log* key and changing the sign as shown in Sample Problem 11.8.

ENGAGE

Explain why 6.00 but not 6.0 is the correct pH for $[H_3O^+] = 1.0 \times 10^{-6}$ M.

▸SAMPLE PROBLEM 11.8 Calculating pH from $[H_3O^+]$

TRY IT FIRST

Aspirin, which is acetylsalicylic acid, was the first nonsteroidal anti-inflammatory drug (NSAID) used to alleviate pain and fever. If a solution of aspirin has a $[H_3O^+] = 1.7 \times 10^{-3}$ M, what is the pH of the solution?

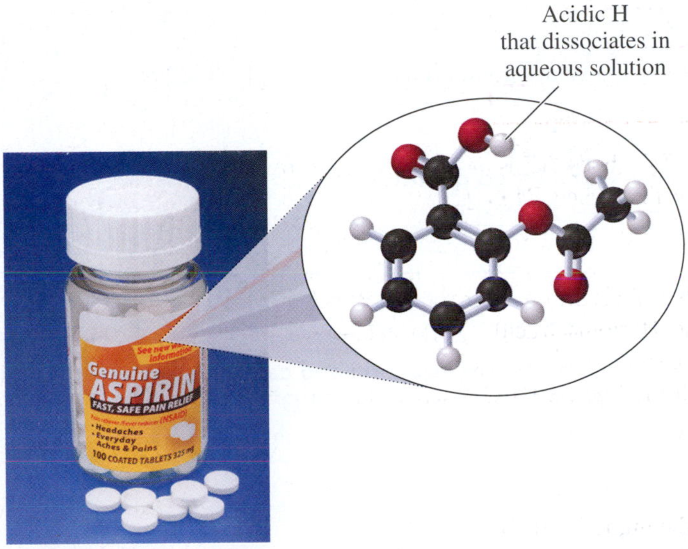
Acidic H that dissociates in aqueous solution
Genuine ASPIRIN
FAST, SAFE PAIN RELIEF
• Headaches
• Everyday Aches & Pains
100 COATED TABLETS

Aspirin, acetylsalicylic acid, is a weak acid.

SOLUTION

STEP 1 **State the given and needed quantities.**

ANALYZE THE PROBLEM	Given	Need	Connect
	$[H_3O^+] = 1.7 \times 10^{-3}$ M	pH	pH equation

STEP 2 **Enter the $[H_3O^+]$ into the pH equation and calculate.**

$$pH = -\log[H_3O^+] = -\log[1.7 \times 10^{-3}]$$

Calculator Procedure — **Calculator Display**

1.7 [EE or EXP] [+/−] 3 [log] [+/−] [=] or [+/−] [log] 1.7 [EE or EXP] [+/−] 3 [=] — 2.769551079

Be sure to check the instructions for your calculator.

STEP 3 **Adjust the number of SFs on the *right* of the decimal point.** In a pH value, the number to the *left* of the decimal point is an *exact* number derived from the power of 10. Thus, the two SFs in the coefficient determine that there are two SFs after the decimal point in the pH value.

Coefficient		Power of ten	
1.7	$\times$	10^{-3} M	$pH = -\log[1.7 \times 10^{-3}] = 2.77$
Two SFs		Exact	Exact, Two SFs

STUDY CHECK 11.8

a. What is the pH of bleach with $[H_3O^+] = 4.2 \times 10^{-12}$ M?
b. What is the pH of a borax solution with $[H_3O^+] = 6.1 \times 10^{-10}$ M?

ANSWER

a. pH = 11.38 **b.** pH = 9.21

When we need to calculate the pH from $[OH^-]$, we use the K_w to calculate $[H_3O^+]$, place it in the pH equation, and calculate the pH of the solution as shown in Sample Problem 11.9.

▶ SAMPLE PROBLEM 11.9 Calculating pH from $[OH^-]$

TRY IT FIRST

What is the pH of an ammonia solution with $[OH^-] = 3.7 \times 10^{-3}$ M?

SOLUTION

STEP 1 **State the given and needed quantities.**

ANALYZE THE PROBLEM	Given	Need	Connect
	$[OH^-] = 3.7 \times 10^{-3}$ M	$[H_3O^+]$, pH	$K_w = [H_3O^+][OH^-]$, pH equation

STEP 2 **Enter the $[H_3O^+]$ into the pH equation and calculate.** Because $[OH^-]$ is given for the ammonia solution, we have to calculate $[H_3O^+]$. Using the water dissociation expression, we divide both sides by $[OH^-]$ to obtain $[H_3O^+]$.

$$K_w = [H_3O^+][OH^-] = 1.0 \times 10^{-14}$$

$$\frac{K_w}{[OH^-]} = \frac{[H_3O^+]\cancel{[OH^-]}}{\cancel{[OH^-]}}$$

$$[H_3O^+] = \frac{1.0 \times 10^{-14}}{[3.7 \times 10^{-3}]} = 2.7 \times 10^{-12}\ \text{M}$$

Now, we enter the $[H_3O^+]$ into the pH equation.

$$pH = -\log[H_3O^+] = -\log[2.7 \times 10^{-12}]$$

Calculator Procedure **Calculator Display**

2.7 [EE or EXP] [+/−] 12 [log] [+/−] [=] or [+/−] [log] 2.7 [EE or EXP] [+/−] 12 [=] 11.56863624

STEP 3 **Adjust the number of SFs on the *right* of the decimal point.**

2.7×10^{-12} M — Two SFs

pH = 11.57 — Two SFs to the *right* of the decimal point

STUDY CHECK 11.9

a. Calculate the pH of a sample of bile that has $[OH^-] = 1.3 \times 10^{-6}$ M.
b. What is the pH of egg whites with $[OH^-] = 1.2 \times 10^{-5}$ M?

ANSWER

a. pH = 8.11 **b.** pH = 9.08

A comparison of $[H_3O^+]$, $[OH^-]$, and their corresponding pH values is given in **TABLE 11.7**.

TABLE 11.7 A Comparison of pH Values at 25 °C, $[H_3O^+]$, and $[OH^-]$

pH	$[H_3O^+]$	$[OH^-]$	
0	10^0	10^{-14}	
1	10^{-1}	10^{-13}	
2	10^{-2}	10^{-12}	
3	10^{-3}	10^{-11}	
4	10^{-4}	10^{-10}	Acidic
5	10^{-5}	10^{-9}	
6	10^{-6}	10^{-8}	
7	10^{-7}	10^{-7}	Neutral
8	10^{-8}	10^{-6}	
9	10^{-9}	10^{-5}	
10	10^{-10}	10^{-4}	Basic
11	10^{-11}	10^{-3}	
12	10^{-12}	10^{-2}	
13	10^{-13}	10^{-1}	
14	10^{-14}	10^0	

pH = 3.9
pH = 3.4
pH = 2.0
pH = 2.4
Acids produce the sour taste of the fruits we eat.

Calculating $[H_3O^+]$ from pH

KEY MATH SKILL
Calculating $[H_3O^+]$ from pH

If we are given the pH of the solution and asked to determine the $[H_3O^+]$, we need to reverse the calculation of pH.

$$[H_3O^+] = 10^{-pH}$$

For example, if the pH of a solution is 3.0, we can substitute it into this equation. The number of significant figures in $[H_3O^+]$ is equal to the number of decimal places in the pH value.

$$[H_3O^+] = 10^{-pH} = 10^{-3.0} = 1 \times 10^{-3}\text{ M}$$

For pH values that are not whole numbers, the calculation requires the use of the 10^x key, which is usually a *2nd function* key. On some calculators, this operation is done using the inverse log equation as shown in Sample Problem 11.10.

▶ SAMPLE PROBLEM 11.10 Calculating $[H_3O^+]$ from pH

TRY IT FIRST

Calculate $[H_3O^+]$ for a urine sample, which has a pH of 7.5.

SOLUTION

STEP 1 State the given and needed quantities.

ANALYZE THE PROBLEM	Given	Need	Connect
	pH = 7.5	$[H_3O^+]$	$[H_3O^+] = 10^{-pH}$

STEP 2 Enter the pH value into the inverse log equation and calculate.

$$[H_3O^+] = 10^{-pH} = 10^{-7.5}$$

Calculator Procedure

[2nd] [log] [+/−] 7.5 [=] or 7.5 [+/−] [2nd] [log] [=]

Calculator Display

3.16227766E−08

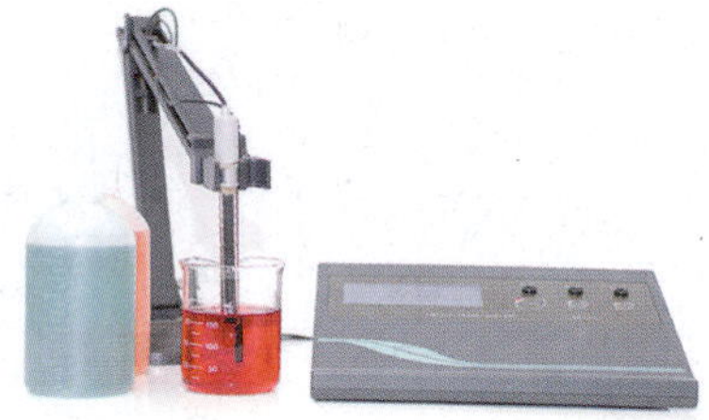

The pH of a liquid is measured using a pH meter.

STEP 3 **Adjust the SFs for the coefficient.** Because the pH value 7.5 has one digit to the *right* of the decimal point, the coefficient for $[H_3O^+]$ is written with one SF.

$$[H_3O^+] = 3 \times 10^{-8}\ M$$

One SF

STUDY CHECK 11.10

What are the $[H_3O^+]$ and $[OH^-]$ of cola that has a pH of 3.17?

ANSWER

$[H_3O^+] = 6.8 \times 10^{-4}$ M, $[OH^-] = 1.5 \times 10^{-11}$ M

TEST

Try Practice Problems 11.47 to 11.54

Chemistry Link to Health

Stomach Acid, HCl

Gastric acid, which contains HCl, is produced by parietal cells that line the stomach. When the stomach expands with the intake of food, the gastric glands begin to secrete a strongly acidic solution of HCl. In a single day, a person may secrete 2000 mL of gastric juice, which contains hydrochloric acid, mucins, and the enzymes pepsin and lipase.

The HCl in the gastric juice activates a digestive enzyme from the chief cells called *pepsinogen* to form *pepsin*, which breaks down proteins in food entering the stomach. The secretion of HCl continues until the stomach has a pH of about 2, which is the optimum for activating the digestive enzymes without ulcerating the stomach lining. In addition, the low pH destroys bacteria that reach the stomach. Normally, large quantities of viscous mucus are secreted within the stomach to protect its lining from acid and enzyme damage. Gastric acid may also form under conditions of stress when the nervous system activates the production of HCl. As the contents of the stomach move into the small intestine, cells produce bicarbonate that neutralizes the gastric acid until the pH is about 5.

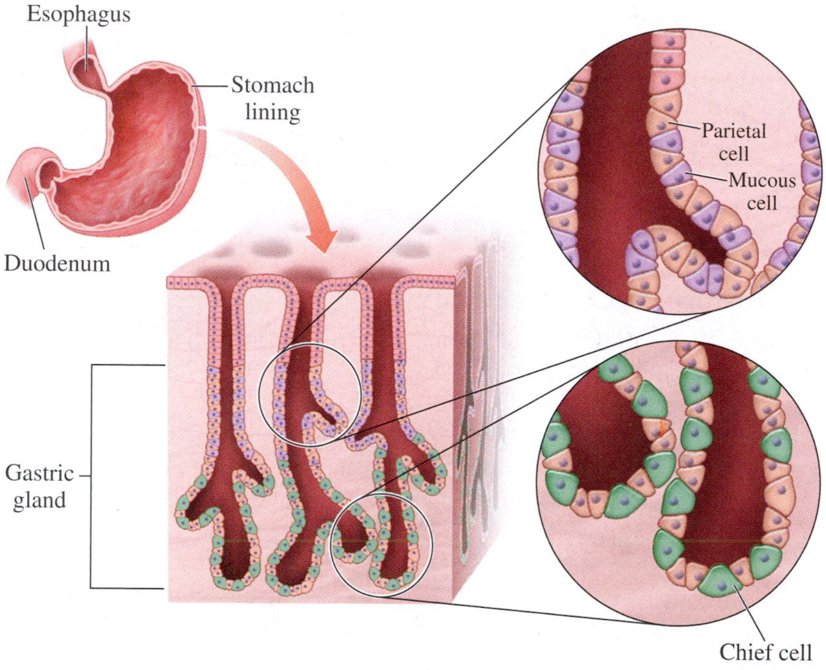
Esophagus
Stomach lining
Duodenum
Gastric gland
Parietal cell
Mucous cell
Chief cell

Parietal cells in the lining of the stomach secrete gastric acid HCl.

PRACTICE PROBLEMS

11.6 The pH Scale

11.43 Why does a neutral solution have a pH of 7.0?

11.44 If the $[OH^-]$ of a solution is greater than the $[H_3O^+]$, is the solution acidic or basic?

11.45 State whether each of the following solutions is acidic, basic, or neutral:

a. blood plasma, pH 7.38
b. vinegar, pH 2.8
c. coffee, pH 5.52
d. tomatoes, pH 4.2
e. chocolate cake, pH 7.6

11.46 State whether each of the following solutions is acidic, basic, or neutral:

a. soda, pH 3.22
b. shampoo, pH 5.7
c. rain, pH 5.8
d. honey, pH 3.9
e. cheese, pH 5.2

11.47 A solution with a pH of 3 is 10 times more acidic than a solution with pH 4. Explain.

11.48 A solution with a pH of 10 is 100 times more basic than a solution with pH 8. Explain.

11.49 Calculate the pH of each solution given the following:

a. $[H_3O^+] = 1 \times 10^{-4}$ M **b.** $[H_3O^+] = 3 \times 10^{-9}$ M
c. $[OH^-] = 1 \times 10^{-5}$ M **d.** $[OH^-] = 2.5 \times 10^{-11}$ M
e. $[H_3O^+] = 6.7 \times 10^{-8}$ M **f.** $[OH^-] = 8.2 \times 10^{-4}$ M

11.50 Calculate the pH of each solution given the following:

a. $[H_3O^+] = 1 \times 10^{-8}$ M **b.** $[H_3O^+] = 5 \times 10^{-6}$ M
c. $[OH^-] = 1 \times 10^{-2}$ M **d.** $[OH^-] = 8.0 \times 10^{-3}$ M
e. $[H_3O^+] = 4.7 \times 10^{-2}$ M **f.** $[OH^-] = 3.9 \times 10^{-6}$ M

Clinical Applications

11.51 Complete the following table for a selection of foods:

Food	$[H_3O^+]$	$[OH^-]$	pH	Acidic, Basic, or Neutral?
Rye bread		6.3×10^{-6} M		
Tomatoes			4.64	
Peas	6.2×10^{-7} M			

11.52 Complete the following table for a selection of foods:

Food	$[H_3O^+]$	$[OH^-]$	pH	Acidic, Basic, or Neutral?
Strawberries			3.90	
Soy milk				Neutral
Tuna fish (canned)	6.3×10^{-7} M			

11.53 A patient with severe metabolic acidosis has a blood plasma pH of 6.92. What is the $[H_3O^+]$ of the blood plasma?

11.54 A patient with respiratory alkalosis has a blood plasma pH of 7.58. What is the $[H_3O^+]$ of the blood plasma?

11.7 Reactions of Acids and Bases

LEARNING GOAL Write balanced equations for reactions of acids with metals, carbonates or bicarbonates, and bases; calculate the molarity or volume of an acid from titration information.

Magnesium reacts rapidly with acid and forms H_2 gas and a salt of magnesium.

Typical reactions of acids and bases include the reactions of acids with metals, carbonates or bicarbonates, and bases. For example, when you drop an antacid tablet in water, the bicarbonate ion and citric acid in the tablet react to produce carbon dioxide bubbles, water, and a salt. A **salt** is an ionic compound that does not have H^+ as the cation or OH^- as the anion.

Acids and Metals

Acids react with certain metals to produce hydrogen gas (H_2) and a salt. Active metals include potassium, sodium, calcium, magnesium, aluminum, zinc, iron, and tin. In these single replacement reactions, the metal ion replaces the hydrogen in the acid.

$$\underset{\text{Acid}}{2HCl(aq)} + \underset{\text{Metal}}{Mg(s)} \longrightarrow \underset{\text{Hydrogen}}{H_2(g)} + \underset{\text{Salt}}{MgCl_2(aq)}$$

$$\underset{\text{Acid}}{2HNO_3(aq)} + \underset{\text{Metal}}{Zn(s)} \longrightarrow \underset{\text{Hydrogen}}{H_2(g)} + \underset{\text{Salt}}{Zn(NO_3)_2(aq)}$$

Acids React with Carbonates or Bicarbonates

When an acid is added to a carbonate or bicarbonate, the products are carbon dioxide gas, water, and a salt. The acid reacts with CO_3^{2-} or HCO_3^- to produce carbonic acid, H_2CO_3, which breaks down rapidly to CO_2 and H_2O.

$$\underset{\text{Acid}}{2HCl(aq)} + \underset{\text{Carbonate}}{Na_2CO_3(aq)} \longrightarrow \underset{\text{Carbon dioxide}}{CO_2(g)} + \underset{\text{Water}}{H_2O(l)} + \underset{\text{Salt}}{2NaCl(aq)}$$

$$\underset{\text{Acid}}{HBr(aq)} + \underset{\text{Bicarbonate}}{NaHCO_3(aq)} \longrightarrow \underset{\text{Carbon dioxide}}{CO_2(g)} + \underset{\text{Water}}{H_2O(l)} + \underset{\text{Salt}}{NaBr(aq)}$$

VINEGAR

When sodium bicarbonate (baking soda) reacts with an acid (vinegar), the products are carbon dioxide gas, water, and a salt.

Acids and Hydroxides: Neutralization

CORE CHEMISTRY SKILL

Writing Equations for Reactions of Acids and Bases

Neutralization is a reaction between a strong or weak acid with a strong base to produce a water and a salt. The H^+ of the acid and the OH^- of the base combine to form water. The salt is the combination of the cation from the base and the anion from the acid. We can write the following equation for the neutralization reaction between HCl and NaOH:

$$\underset{\text{Acid}}{HCl(aq)} + \underset{\text{Base}}{NaOH(aq)} \longrightarrow \underset{\text{Water}}{H_2O(l)} + \underset{\text{Salt}}{NaCl(aq)}$$

If we write the strong acid HCl and the strong base NaOH as ions, we see that H^+ combines with OH^- to form water, leaving the ions Na^+ and Cl^- in solution.

$$H^+(aq) + Cl^-(aq) + Na^+(aq) + OH^-(aq) \longrightarrow H_2O(l) + Na^+(aq) + Cl^-(aq)$$

When we omit the ions that do not change during the reaction (*spectator ions*), we obtain the *net ionic equation.*

$$H^+(aq) + \cancel{Cl^-(aq)} + \cancel{Na^+(aq)} + OH^-(aq) \longrightarrow H_2O(l) + \cancel{Na^+(aq)} + \cancel{Cl^-(aq)}$$

The net ionic equation for the neutralization of H^+ and OH^- to form H_2O is

$$H^+(aq) + OH^-(aq) \longrightarrow H_2O(l) \quad \text{Net ionic equation}$$

ENGAGE

What salt is produced when KOH completely neutralizes H_2SO_4?

Balancing Neutralization Equations

In a neutralization reaction, one H^+ always reacts with one OH^-. Therefore, a neutralization equation may need coefficients to balance the H^+ from the acid with the OH^- from the base as shown in Sample Problem 11.11.

▶ SAMPLE PROBLEM 11.11 Balancing Equations for Acids

TRY IT FIRST

Write the balanced equation for the neutralization of $HCl(aq)$ and $Ba(OH)_2(s)$.

SOLUTION

STEP 1 **Write the reactants and products.**

$$HCl(aq) + Ba(OH)_2(s) \longrightarrow H_2O(l) + \text{salt}$$

STEP 2 **Balance the H^+ in the acid with the OH^- in the base.** Placing a coefficient of 2 in front of the HCl provides $2H^+$ for the $2OH^-$ from $Ba(OH)_2$.

$$2HCl(aq) + Ba(OH)_2(s) \longrightarrow H_2O(l) + \text{salt}$$

STEP 3 **Balance the H_2O with the H^+ and the OH^-.** Use a coefficient of 2 in front of H_2O to balance $2H^+$ and $2OH^-$.

$$2HCl(aq) + Ba(OH)_2(s) \longrightarrow 2H_2O(l) + \text{salt}$$

STEP 4 **Write the salt from the remaining ions.** Use the ions Ba^{2+} and $2Cl^-$ and write the formula for the salt as $BaCl_2$.

$$2HCl(aq) + Ba(OH)_2(s) \longrightarrow 2H_2O(l) + BaCl_2(aq)$$

STUDY CHECK 11.11

Write the balanced equation for the reaction between $H_2SO_4(aq)$ and $K_2CO_3(s)$.

ANSWER

$$H_2SO_4(aq) + K_2CO_3(s) \longrightarrow CO_2(g) + H_2O(l) + K_2SO_4(aq)$$

TEST

Try Practice Problems 11.55 to 11.60

Chemistry Link to Health

Antacids

Antacids are substances used to neutralize excess stomach acid (HCl). Some antacids are mixtures of aluminum hydroxide and magnesium hydroxide. These hydroxides are not very soluble in water, so the levels of available OH^- are not damaging to the intestinal tract. However, aluminum hydroxide has the side effects of producing constipation and binding phosphate in the intestinal tract, which may cause weakness and loss of appetite. Magnesium hydroxide has a laxative effect. These side effects are less likely when a combination of the antacids is used.

$$3HCl(aq) + Al(OH)_3(s) \longrightarrow 3H_2O(l) + AlCl_3(aq)$$

$$2HCl(aq) + Mg(OH)_2(s) \longrightarrow 2H_2O(l) + MgCl_2(aq)$$

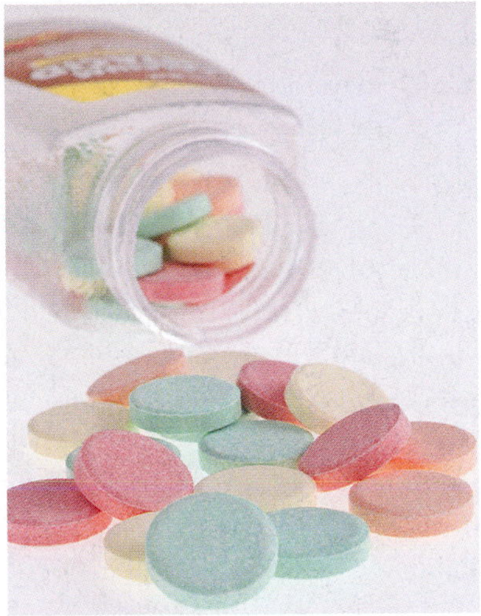

Antacids neutralize excess stomach acid.

Some antacids use calcium carbonate to neutralize excess stomach acid. About 10% of the calcium is absorbed into the bloodstream, where it elevates the level of serum calcium. Calcium carbonate is not recommended for patients who have peptic ulcers or a tendency to form kidney stones, which typically consist of an insoluble calcium salt.

$$2HCl(aq) + CaCO_3(s) \longrightarrow CO_2(g) + H_2O(l) + CaCl_2(aq)$$

Still other antacids contain sodium bicarbonate. This type of antacid neutralizes excess gastric acid, increases blood pH, but also elevates sodium levels in the body fluids. It also is not recommended in the treatment of peptic ulcers.

$$HCl(aq) + NaHCO_3(s) \longrightarrow CO_2(g) + H_2O(l) + NaCl(aq)$$

The neutralizing substances in some antacid preparations are given in **TABLE 11.8**.

TABLE 11.8 Basic Compounds in Some Antacids

Antacid	Base(s)
Amphojel	$Al(OH)_3$
Milk of magnesia	$Mg(OH)_2$
Mylanta, Maalox, Di-Gel, Gelusil, Riopan	$Mg(OH)_2$, $Al(OH)_3$
Bisodol, Rolaids	$CaCO_3$, $Mg(OH)_2$
Titralac, Tums, Pepto-Bismol	$CaCO_3$
Alka-Seltzer	$NaHCO_3$, $KHCO_3$

Acid–Base Titration

Suppose we need to find the molarity of a solution of HCl, which has an unknown concentration. We can do this by a laboratory procedure called **titration** in which we neutralize an acid sample with a known amount of base. In a titration, we place a measured volume of the acid in a flask and add a few drops of an **indicator**, such as phenolphthalein. An indicator is a compound that dramatically changes color when pH of the solution changes. In an acidic solution, phenolphthalein is colorless. Then we fill a buret with a NaOH solution of known molarity and carefully add NaOH solution to neutralize the acid in the flask (see **FIGURE 11.6**). We know that neutralization has taken place when the phenolphthalein in the solution changes from colorless to pink. This is called the neutralization **endpoint**. From the measured volume of the NaOH solution and its molarity, we calculate the number of moles of NaOH, the moles of acid, and use the measured volume of acid to calculate its concentration.

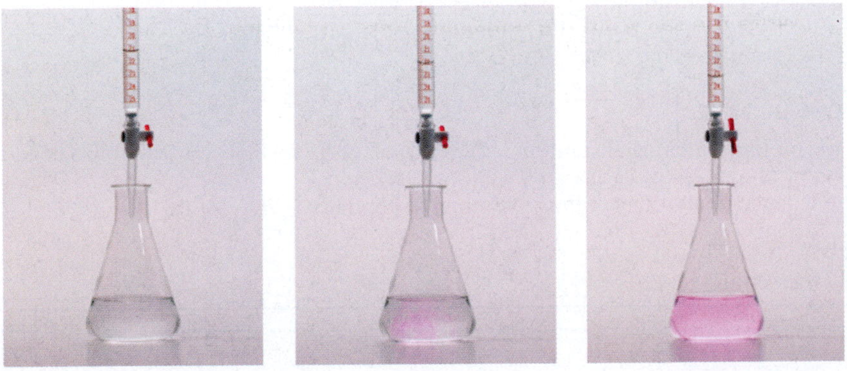

FIGURE 11.6 ▶ The titration of an acid. A known volume of an acid is placed in a flask with an indicator and titrated with a measured volume of a base solution, such as NaOH, to the neutralization endpoint.

Q What data is needed to determine the molarity of the acid in the flask?

CORE CHEMISTRY SKILL
Calculating Molarity or Volume of an Acid or Base in a Titration

ENGAGE
Why do you need to use a balanced equation for a neutralization problem?

▶SAMPLE PROBLEM 11.12 Titration of an Acid

TRY IT FIRST

If 16.3 mL of a 0.185 M $Sr(OH)_2$ solution is used to titrate the HCl in a 25.0-mL (0.0250 L) sample of gastric juice, what is the molarity of the HCl solution?

$$2HCl(aq) + Sr(OH)_2(aq) \longrightarrow 2H_2O(l) + SrCl_2(aq)$$

SOLUTION

STEP 1 State the given and needed quantities and concentrations.

ANALYZE THE PROBLEM	Given	Need	Connect
	25.0 mL (0.0250 L) of HCl solution, 16.3 mL of 0.185 M $Sr(OH)_2$ solution	molarity of the HCl solution	molarity, mole–mole factor
	Neutralization Equation		
	$2HCl(aq) + Sr(OH)_2(aq) \longrightarrow 2H_2O(l) + SrCl_2(aq)$		

STEP 2 Write a plan to calculate the molarity.

mL of $Sr(OH)_2$ solution → Metric factor → L of $Sr(OH)_2$ solution → Molarity → moles of $Sr(OH)_2$ → Mole–mole factor → moles of HCl → Divide by liters → molarity of HCl solution

STEP 3 State equalities and conversion factors, including concentrations.

$$1000 \text{ mL of } Sr(OH)_2 \text{ solution} = 1 \text{ L of } Sr(OH)_2 \text{ solution}$$
$$\frac{1000 \text{ mL } Sr(OH)_2 \text{ solution}}{1 \text{ L } Sr(OH)_2 \text{ solution}} \text{ and } \frac{1 \text{ L } Sr(OH)_2 \text{ solution}}{1000 \text{ mL } Sr(OH)_2 \text{ solution}}$$

$$0.185 \text{ mole of } Sr(OH)_2 = 1 \text{ L of } Sr(OH)_2 \text{ solution}$$
$$\frac{0.185 \text{ mole } Sr(OH)_2}{1 \text{ L } Sr(OH)_2 \text{ solution}} \text{ and } \frac{1 \text{ L } Sr(OH)_2 \text{ solution}}{0.185 \text{ mole } Sr(OH)_2}$$

$$2 \text{ moles of HCl} = 1 \text{ mole of } Sr(OH)_2$$
$$\frac{2 \text{ moles HCl}}{1 \text{ mole } Sr(OH)_2} \text{ and } \frac{1 \text{ mole } Sr(OH)_2}{2 \text{ moles HCl}}$$

STEP 4 Set up the problem to calculate the needed quantity.

$$16.3 \cancel{\text{mL } Sr(OH)_2 \text{ solution}} \times \frac{1 \cancel{\text{L } Sr(OH)_2 \text{ solution}}}{1000 \cancel{\text{mL } Sr(OH)_2 \text{ solution}}} \times \frac{0.185 \cancel{\text{mole } Sr(OH)_2}}{1 \cancel{\text{L } Sr(OH)_2 \text{ solution}}} \times \frac{2 \text{ moles HCl}}{1 \cancel{\text{mole } Sr(OH)_2}} = 0.006\,03 \text{ mole of HCl}$$

$$\text{molarity of HCl solution} = \frac{0.006\,03 \text{ mole HCl}}{0.0250 \text{ L HCl solution}} = 0.241 \text{ M HCl solution}$$

INTERACTIVE VIDEO
Acid–Base Titration
PEARSON eText 2.0

TEST
Try Practice Problems 11.61 to 11.66

STUDY CHECK 11.12

What is the molarity of an HCl solution if 28.6 mL of a 0.175 M NaOH solution is needed to titrate a 25.0-mL sample of the HCl solution?

ANSWER

0.200 M HCl solution

PRACTICE PROBLEMS

11.7 Reactions of Acids and Bases

11.55 Complete and balance the equation for each of the following reactions:

a. $HBr(aq) + ZnCO_3(s) \longrightarrow$
b. $HCl(aq) + Zn(s) \longrightarrow$
c. $HCl(aq) + NaHCO_3(s) \longrightarrow$
d. $H_2SO_4(aq) + Mg(OH)_2(s) \longrightarrow$

11.56 Complete and balance the equation for each of the following reactions:

a. $HBr(aq) + KHCO_3(s) \longrightarrow$
b. $H_2SO_4(aq) + Ca(s) \longrightarrow$
c. $H_2SO_4(aq) + Ca(OH)_2(s) \longrightarrow$
d. $H_2SO_4(aq) + Na_2CO_3(s) \longrightarrow$

11.57 Balance each of the following neutralization reactions:

a. $HCl(aq) + Mg(OH)_2(s) \longrightarrow H_2O(l) + MgCl_2(aq)$
b. $H_3PO_4(aq) + LiOH(aq) \longrightarrow H_2O(l) + Li_3PO_4(aq)$

11.58 Balance each of the following neutralization reactions:

a. $HNO_3(aq) + Ba(OH)_2(s) \longrightarrow H_2O(l) + Ba(NO_3)_2(aq)$
b. $H_2SO_4(aq) + Al(OH)_3(s) \longrightarrow H_2O(l) + Al_2(SO_4)_3(aq)$

11.59 Write a balanced equation for the neutralization of each of the following:

a. $H_2SO_4(aq)$ and $NaOH(aq)$
b. $HCl(aq)$ and $Fe(OH)_3(s)$
c. $H_2CO_3(aq)$ and $Mg(OH)_2(s)$

11.60 Write a balanced equation for the neutralization of each of the following:

a. $H_3PO_4(aq)$ and $NaOH(aq)$
b. $HI(aq)$ and $LiOH(aq)$
c. $HNO_3(aq)$ and $Ca(OH)_2(s)$

11.61 What is the molarity of a solution of HCl if 5.00 mL of the HCl solution is titrated with 28.6 mL of a 0.145 NaOH solution?

$$HCl(aq) + NaOH(aq) \longrightarrow H_2O(l) + NaCl(aq)$$

11.62 What is the molarity of the acetic acid solution if 29.7 mL of a 0.205 M KOH solution is required to titrate 25.0 mL of a solution of $HC_2H_3O_2$?

$$HC_2H_3O_2(aq) + KOH(aq) \longrightarrow H_2O(l) + KC_2H_3O_2(aq)$$

11.63 If 38.2 mL of a 0.163 M KOH solution is required to titrate 25.0 mL of a H_2SO_4 solution, what is the molarity of the H_2SO_4 solution?

$$H_2SO_4(aq) + 2KOH(aq) \longrightarrow 2H_2O(l) + K_2SO_4(aq)$$

11.64 If 32.8 mL of a 0.162 M NaOH solution is required to titrate 25.0 mL of a solution of H_2SO_4, what is the molarity of the H_2SO_4 solution?

$$H_2SO_4(aq) + 2NaOH(aq) \longrightarrow 2H_2O(l) + Na_2SO_4(aq)$$

11.65 A solution of 0.204 M NaOH is used to titrate 50.0 mL of a 0.0224 M H_3PO_4 solution. What volume, in milliliters, of the NaOH solution is required?

$$H_3PO_4(aq) + 3NaOH(aq) \longrightarrow 3H_2O(l) + Na_3PO_4(aq)$$

11.66 A solution of 0.312 M KOH is used to titrate 15.0 mL of a 0.186 M H_3PO_4 solution. What volume, in milliliters, of the KOH solution is required?

$$H_3PO_4(aq) + 3KOH(aq) \longrightarrow 3H_2O(l) + K_3PO_4(aq)$$

11.8 Buffers

LEARNING GOAL Describe the role of buffers in maintaining the pH of a solution; calculate the pH of a buffer.

The pH of water and most solutions changes drastically when a small amount of acid or base is added. However, when an acid or a base is added to a *buffer* solution, there is little change in pH. A **buffer solution** maintains the pH of a solution by neutralizing small amounts of added acid or base. In the human body, whole blood contains plasma, white blood cells and platelets, and red blood cells. Blood plasma contains buffers that maintain a consistent pH of about 7.4. If the pH of the blood plasma goes slightly above or below 7.4, changes in our oxygen levels and our metabolic processes can be drastic enough to cause death. Even though we obtain acids and bases from foods and cellular reactions, the buffers in the body absorb those compounds so effectively that the pH of our blood plasma remains essentially unchanged (see **FIGURE 11.7**).

In a buffer, an acid must be present to react with any OH^- that is added, and a base must be available to react with any added H_3O^+. However, that acid and base must not neutralize each other. Therefore, a combination of an acid–base conjugate pair is used in buffers. Most buffer solutions consist of nearly equal concentrations of a weak acid and a salt containing its conjugate base. Buffers may also contain a weak base and the salt of the weak base, which contains its conjugate acid.

ENGAGE

Why does a buffer require the presence of a weak acid or weak base and the salt of that weak acid or weak base?

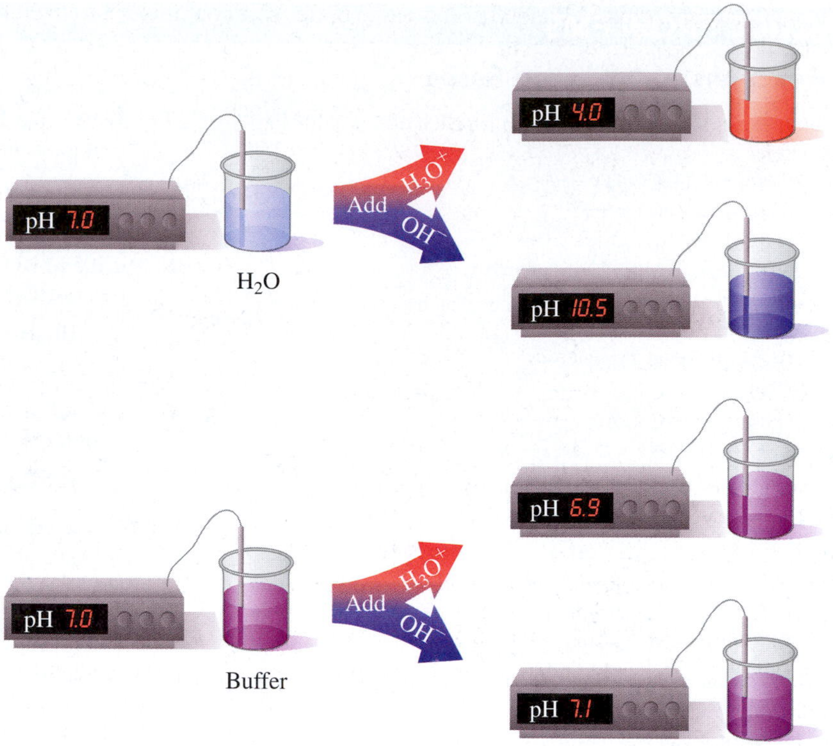
pH 4.0
pH 7.0
Add
H_3O^+
OH^-
H_2O
pH 10.5
pH 6.9
pH 7.0
Add
H_3O^+
OH^-
Buffer
pH 7.1

FIGURE 11.7 ▶ Adding an acid or a base to water changes the pH drastically, but a buffer resists pH change when small amounts of acid or base are added.

Q Why does the pH change several pH units when acid is added to water, but not when acid is added to a buffer?

For example, a typical buffer can be made from the weak acid acetic acid ($HC_2H_3O_2$) and its salt, sodium acetate ($NaC_2H_3O_2$). As a weak acid, acetic acid dissociates slightly in water to form H_3O^+ and a very small amount of $C_2H_3O_2^-$. The addition of its salt, sodium acetate, provides a much larger concentration of acetate ion ($C_2H_3O_2^-$), which is necessary for its buffering capability.

$$\underset{\text{Large amount}}{HC_2H_3O_2(aq)} + H_2O(l) \rightleftharpoons H_3O^+(aq) + \underset{\text{Large amount}}{C_2H_3O_2^-(aq)}$$

ENGAGE

Which part of a buffer neutralizes any H_3O^+ that is added?

We can now describe how this buffer solution maintains the $[H_3O^+]$. When a small amount of acid is added, the additional H_3O^+ combines with the acetate ion, $C_2H_3O_2^-$, causing the equilibrium to shift in the direction of the reactants, acetic acid and water. There will be a slight decrease in the $[C_2H_3O_2^-]$ and a slight increase in the $[HC_2H_3O_2]$, but both the $[H_3O^+]$ and pH are maintained (see **FIGURE 11.8**).

$$HC_2H_3O_2(aq) + H_2O(l) \longleftarrow H_3O^+(aq) + C_2H_3O_2^-(aq)$$

Equilibrium shifts in the direction of the reactants

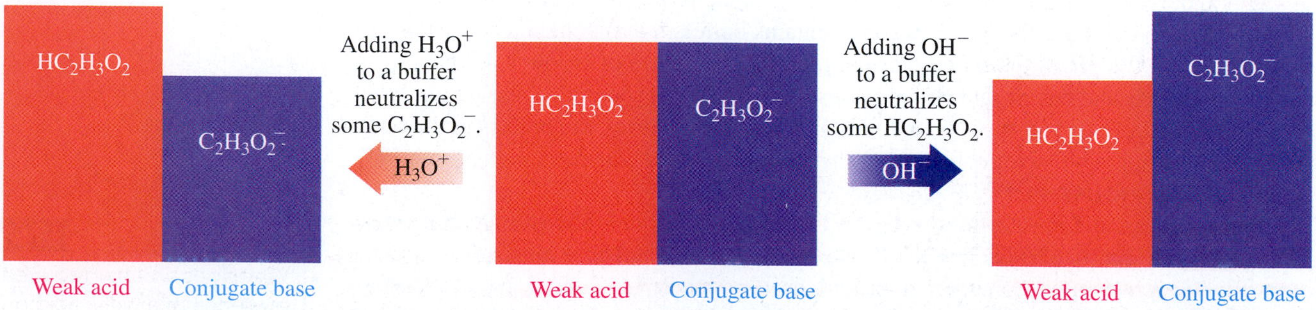
$HC_2H_3O_2$
$C_2H_3O_2^-$
Weak acid
Conjugate base
Adding H_3O^+ to a buffer neutralizes some $C_2H_3O_2^-$.
H_3O^+
$HC_2H_3O_2$
$C_2H_3O_2^-$
Weak acid
Conjugate base
Adding OH^- to a buffer neutralizes some $HC_2H_3O_2$.
OH^-
$HC_2H_3O_2$
$C_2H_3O_2^-$
Weak acid
Conjugate base

FIGURE 11.8 ▶ The buffer described here consists of about equal concentrations of acetic acid ($HC_2H_3O_2$) and its conjugate base acetate ion ($C_2H_3O_2^-$). The pH of the solution is maintained as long as the added amount of acid or base is small compared to the concentrations of the buffer components.

Q How does this acetic acid–acetate ion buffer maintain pH?

If a small amount of base is added to this same buffer solution, it is neutralized by the acetic acid, $HC_2H_3O_2$, which shifts the equilibrium in the direction of the products, acetate ion and water. The $[HC_2H_3O_2]$ decreases slightly and the $[C_2H_3O_2^-]$ increases slightly, but again the $[H_3O^+]$ and thus the pH of the solution are maintained.

$$HC_2H_3O_2(aq) + OH^-(aq) \longrightarrow H_2O(l) + C_2H_3O_2^-(aq)$$

Equilibrium shifts in the direction of the products

TEST

Try Practice Problems 11.67 to 11.70

Calculating the pH of a Buffer

CORE CHEMISTRY SKILL

Calculating the pH of a Buffer

INTERACTIVE VIDEO

PEARSON eText 2.0 Calculating the pH of a Buffer

By rearranging the K_a expression to give $[H_3O^+]$, we can obtain the ratio of the acetic acid/acetate buffer.

$$K_a = \frac{[H_3O^+][C_2H_3O_2^-]}{[HC_2H_3O_2]}$$

Solving for $[H_3O^+]$ gives:

$$[H_3O^+] = K_a \times \frac{[HC_2H_3O_2]}{[C_2H_3O_2^-]} \begin{array}{l} \longleftarrow \text{Weak acid} \\ \longleftarrow \text{Conjugate base} \end{array}$$

In this rearrangement of K_a, the weak acid is in the numerator and the conjugate base in the denominator. We can now calculate the $[H_3O^+]$ and pH for an acetic acid buffer as shown in Sample Problem 11.13.

▶ SAMPLE PROBLEM 11.13 Calculating the pH of a Buffer

TRY IT FIRST

The K_a for acetic acid, $HC_2H_3O_2$, is 1.8×10^{-5}. What is the pH of a buffer prepared with 1.0 M $HC_2H_3O_2$ and 1.0 M $C_2H_3O_2^-$?

$$HC_2H_3O_2(aq) + H_2O(l) \rightleftarrows H_3O^+(aq) + C_2H_3O_2^-(aq)$$

SOLUTION

STEP 1 **State the given and needed quantities.**

ANALYZE THE PROBLEM	Given	Need	Connect
	1.0 M $HC_2H_3O_2$, 1.0 M $C_2H_3O_2^-$	pH	K_a expression
	Equation		
	$HC_2H_3O_2(aq) + H_2O(l) \rightleftarrows H_3O^+(aq) + C_2H_3O_2^-(aq)$		

STEP 2 **Write the K_a expression and rearrange for $[H_3O^+]$.**

$$K_a = \frac{[H_3O^+][C_2H_3O_2^-]}{[HC_2H_3O_2]}$$

$$[H_3O^+] = K_a \times \frac{[HC_2H_3O_2]}{[C_2H_3O_2^-]}$$

STEP 3 **Substitute [HA] and $[A^-]$ into the K_a expression.**

$$[H_3O^+] = 1.8 \times 10^{-5} \times \frac{[1.0]}{[1.0]}$$

$$[H_3O^+] = 1.8 \times 10^{-5}\ \text{M}$$

STEP 4 **Use $[H_3O^+]$ to calculate pH.** Placing the $[H_3O^+]$ into the pH equation gives the pH of the buffer.

$$\text{pH} = -\log[1.8 \times 10^{-5}] = 4.74$$

STUDY CHECK 11.13

a. One of the conjugate acid–base pairs that buffers the blood plasma is $H_2PO_4^-/HPO_4^{2-}$. The K_a for $H_2PO_4^-$ is 6.2×10^{-8}. What is the pH of a buffer that is prepared from 0.10 M $H_2PO_4^-$ and 0.50 M HPO_4^{2-}?

b. What is the pH of a buffer made from 0.10 M formic acid ($HCHO_2$) and 0.010 M formate (CHO_2^-)? The K_a for formic acid is 1.8×10^{-4}.

ANSWER

a. pH = 7.91 **b.** pH = 2.74

TEST

Try Practice Problems 11.71 to 11.76

Because K_a is a constant at a given temperature, the $[H_3O^+]$ is determined by the $[HC_2H_3O_2]/[C_2H_3O_2^-]$ ratio. As long as the addition of small amounts of either acid or base changes the ratio of $[HC_2H_3O_2]/[C_2H_3O_2^-]$ only slightly, the changes in $[H_3O^+]$ will be small and the pH will be maintained. If a large amount of acid or base is added, the *buffering capacity* of the system may be exceeded. Buffers can be prepared from conjugate acid–base pairs such as $H_2PO_4^-/HPO_4^{2-}$, HPO_4^{2-}/PO_4^{3-}, HCO_3^-/CO_3^{2-}, or NH_4^+/NH_3. The pH of the buffer solution will depend on the conjugate acid–base pair chosen.

ENGAGE

How would a solution composed of $H_2PO_4^-$ and HPO_4^{2-} act as a buffer?

Using a common phosphate buffer for biological specimens, we can look at the effect of using different ratios of $[H_2PO_4^-]/[HPO_4^{2-}]$ on the $[H_3O^+]$ and pH. The K_a of $H_2PO_4^-$ is 6.2×10^{-8}. The equation and the $[H_3O^+]$ are written as follows:

$$H_2PO_4^-(aq) + H_2O(l) \rightleftharpoons H_3O^+(aq) + HPO_4^{2-}(aq)$$

$$[H_3O^+] = K_a \times \frac{[H_2PO_4^-]}{[HPO_4^{2-}]}$$

K_a	$\frac{[H_2PO_4^-]}{[HPO_4^{2-}]}$	Ratio	$[H_3O^+]$	pH
6.2×10^{-8}	$\frac{1.0\ M}{0.10\ M}$	$\frac{10}{1}$	6.2×10^{-7}	6.21
6.2×10^{-8}	$\frac{1.0\ M}{1.0\ M}$	$\frac{1}{1}$	6.2×10^{-8}	7.21
6.2×10^{-8}	$\frac{0.10\ M}{1.0\ M}$	$\frac{1}{10}$	6.2×10^{-9}	8.21

To prepare a phosphate buffer with a pH close to the pH of a biological sample, 7.4, we would choose concentrations that are about equal, such as 1.0 M $H_2PO_4^-$ and 1.0 M HPO_4^{2-}.

Chemistry Link to Health

Buffers in the Blood Plasma

The arterial blood plasma has a normal pH of 7.35 to 7.45. If changes in H_3O^+ lower the pH below 6.8 or raise it above 8.0, cells cannot function properly, and death may result. In our cells, CO_2 is continually produced as an end product of cellular metabolism. Some CO_2 is carried to the lungs for elimination, and the rest dissolves in body fluids such as plasma and saliva, forming carbonic acid, H_2CO_3. As a weak acid, carbonic acid dissociates to give bicarbonate, HCO_3^-, and H_3O^+. More of the anion HCO_3^- is supplied by the kidneys to give an important buffer system in the body fluid—the H_2CO_3/HCO_3^- buffer.

$$CO_2(g) + H_2O(l) \rightleftharpoons H_2CO_3(aq) \rightleftharpoons H_3O^+(aq) + HCO_3^-(aq)$$

Excess H_3O^+ entering the body fluids reacts with the HCO_3^-, and excess OH^- reacts with the carbonic acid.

$$H_2CO_3(aq) + H_2O(l) \longleftarrow H_3O^+(aq) + HCO_3^-(aq)$$

Equilibrium shifts in the direction of the reactants

$$H_2CO_3(aq) + OH^-(aq) \longrightarrow H_2O(l) + HCO_3^-(aq)$$

Equilibrium shifts in the direction of the products

For carbonic acid, we can write the equilibrium expression as

$$K_a = \frac{[H_3O^+][HCO_3^-]}{[H_2CO_3]}$$

(continued)

Chemistry Link to Health (*continued*)

To maintain the normal blood plasma pH (7.35 to 7.45), the ratio of $[H_2CO_3]/[HCO_3^-]$ needs to be about 1 to 10, which is obtained by the concentrations in the blood plasma of 0.0024 M H_2CO_3 and 0.024 M HCO_3^-.

$$[H_3O^+] = K_a \times \frac{[H_2CO_3]}{[HCO_3^-]}$$

$$[H_3O^+] = 4.3 \times 10^{-7} \times \frac{[0.0024]}{[0.024]}$$

$$= 4.3 \times 10^{-7} \times 0.10 = 4.3 \times 10^{-8}\text{ M}$$

$$\text{pH} = -\log[4.3 \times 10^{-8}] = 7.37$$

In the body, the concentration of carbonic acid is closely associated with the partial pressure of CO_2, P_{CO_2}. **TABLE 11.9** lists the normal values for arterial blood. If the CO_2 level rises, increasing $[H_2CO_3]$, the equilibrium shifts to produce more H_3O^+, which lowers the pH. This condition is called *acidosis*. Difficulty with ventilation or gas diffusion can lead to respiratory acidosis, which can happen in emphysema or when an accident or depressive drugs affect the medulla of the brain.

A lowering of the CO_2 level leads to a high blood pH, a condition called *alkalosis*. Excitement, trauma, or a high temperature may cause a person to hyperventilate, which expels large amounts of CO_2. As the partial pressure of CO_2 in the blood falls below normal, the equilibrium shifts from H_2CO_3 to CO_2 and H_2O. This shift decreases the $[H_3O^+]$ and raises the pH. The kidneys also regulate H_3O^+ and HCO_3^-, but they do so more slowly than the adjustment made by the lungs during ventilation.

TABLE 11.10 lists some of the conditions that lead to changes in the blood pH and some possible treatments.

TABLE 11.9 Normal Values for Blood Buffer in Arterial Blood

P_{CO_2}	40 mmHg
H_2CO_3	2.4 mmoles/L of plasma
HCO_3^-	24 mmoles/L of plasma
pH	7.35 to 7.45

TABLE 11.10 Acidosis and Alkalosis: Symptoms, Causes, and Treatments

Respiratory Acidosis: $CO_2\uparrow$ pH$\downarrow$	
Symptoms	Failure to ventilate, suppression of breathing, disorientation, weakness, coma
Causes	Lung disease blocking gas diffusion (e.g., emphysema, pneumonia, bronchitis, asthma); depression of respiratory center by drugs, cardiopulmonary arrest, stroke, poliomyelitis, or nervous system disorders
Treatment	Correction of disorder, infusion of bicarbonate
Metabolic Acidosis: $H^+\uparrow$ pH$\downarrow$	
Symptoms	Increased ventilation, fatigue, confusion
Causes	Renal disease, including hepatitis and cirrhosis; increased acid production in diabetes mellitus, hyperthyroidism, alcoholism, and starvation; loss of alkali in diarrhea; acid retention in renal failure
Treatment	Sodium bicarbonate given orally, dialysis for renal failure, insulin treatment for diabetic ketosis
Respiratory Alkalosis: $CO_2\downarrow$ pH$\uparrow$	
Symptoms	Increased rate and depth of breathing, numbness, light-headedness, tetany
Causes	Hyperventilation because of anxiety, hysteria, fever, exercise; reaction to drugs such as salicylate, quinine, and antihistamines; conditions causing hypoxia (e.g., pneumonia, pulmonary edema, heart disease)
Treatment	Elimination of anxiety-producing state, rebreathing into a paper bag
Metabolic Alkalosis: $H^+\downarrow$ pH$\uparrow$	
Symptoms	Depressed breathing, apathy, confusion
Causes	Vomiting, diseases of the adrenal glands, ingestion of excess alkali
Treatment	Infusion of saline solution, treatment of underlying diseases

PRACTICE PROBLEMS

11.8 Buffers

11.67 Which of the following represents a buffer system? Explain.

a. NaOH and NaCl
b. H_2CO_3 and $NaHCO_3$
c. HF and KF
d. KCl and NaCl

11.68 Which of the following represents a buffer system? Explain.

a. $HClO_2$
b. $NaNO_3$
c. $HC_2H_3O_2$ and $NaC_2H_3O_2$
d. HCl and NaOH

11.69 Consider the buffer system of hydrofluoric acid, HF, and its salt, NaF.

$$HF(aq) + H_2O(l) \rightleftharpoons H_3O^+(aq) + F^-(aq)$$

a. The purpose of this buffer system is to:
1. maintain [HF] **2.** maintain $[F^-]$
3. maintain pH

b. The salt of the weak acid is needed to:
1. provide the conjugate base **2.** neutralize added H_3O^+
3. provide the conjugate acid

c. If OH^- is added, it is neutralized by:
1. the salt **2.** H_2O **3.** H_3O^+

d. When H_3O^+ is added, the equilibrium shifts in the direction of the:
1. reactants **2.** products **3.** does not change

11.70 Consider the buffer system of nitrous acid, HNO_2, and its salt, $NaNO_2$.

$$HNO_2(aq) + H_2O(l) \rightleftharpoons H_3O^+(aq) + NO_2^-(aq)$$

a. The purpose of this buffer system is to:
1. maintain $[HNO_2]$ **2.** maintain $[NO_2^-]$
3. maintain pH

b. The weak acid is needed to:
1. provide the conjugate base **2.** neutralize added OH^-
3. provide the conjugate acid

c. If H_3O^+ is added, it is neutralized by:
1. the salt **2.** H_2O **3.** OH^-

d. When OH^- is added, the equilibrium shifts in the direction of the:
1. reactants **2.** products **3.** does not change

11.71 Nitrous acid has a K_a of 4.5×10^{-4}. What is the pH of a buffer solution containing 0.10 M HNO_2 and 0.10 M NO_2^-?

11.72 Acetic acid has a K_a of 1.8×10^{-5}. What is the pH of a buffer solution containing 0.15 M $HC_2H_3O_2$ and 0.15 M $C_2H_3O_2^-$?

11.73 Using Table 11.4 for K_a values, compare the pH of a HF buffer that contains 0.10 M HF and 0.10 M NaF with another HF buffer that contains 0.060 M HF and 0.120 M NaF.

11.74 Using Table 11.4 for K_a values, compare the pH of a H_2CO_3 buffer that contains 0.10 M H_2CO_3 and 0.10 M $NaHCO_3$ with another H_2CO_3 buffer that contains 0.15 M H_2CO_3 and 0.050 M $NaHCO_3$.

Clinical Applications

11.75 Why would the pH of your blood plasma increase if you breathe fast?

11.76 Why would the pH of your blood plasma decrease if you hold your breath?

11.77 Someone with kidney failure excretes urine with large amounts of HCO_3^-. How would this loss of HCO_3^- affect the pH of the blood plasma?

11.78 Someone with severe diabetes obtains energy by the breakdown of fats, which produce large amounts of acidic substances. How would this affect the pH of the blood plasma?

CLINICAL UPDATE Acid Reflux Disease

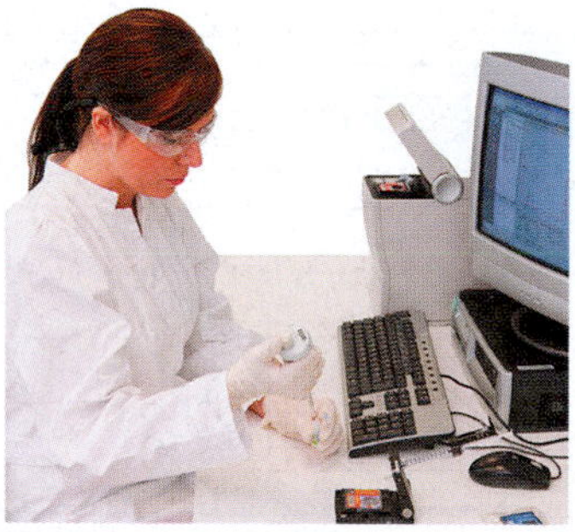

Larry has not been feeling well lately. He tells his doctor that he has discomfort and a burning feeling in his chest, and a sour taste in his throat and mouth. At times, Larry says he feels bloated after a big meal, has a dry cough, is hoarse, and sometimes has a sore throat. He has tried antacids, but they do not bring any relief.

The doctor tells Larry that he thinks he has acid reflux. At the top of the stomach there is a valve, the lower esophageal sphincter, that normally closes after food passes through it. However, if the valve does not close completely, acid produced in the stomach to digest food can move up into the esophagus, a condition called *acid reflux*. The acid, which is hydrochloric acid, HCl, is produced in the stomach to kill microorganisms, and activate the enzymes we need to break down food.

If acid reflux occurs, the strong acid HCl comes in contact with the lining of the esophagus, where it causes irritation and produces a burning feeling in the chest. Sometimes the pain in the chest is called *heartburn*. If the HCl reflux goes high enough to reach the throat, a sour taste may be noticed in the mouth. If Larry's symptoms occur three or more times a week, he may have a chronic condition known as *acid reflux disease* or *gastroesophageal reflux disease* (GERD).

Larry's doctor orders an *esophageal pH test* in which the amount of acid entering the esophagus from the stomach is measured over 24 h. A probe that measures the pH is inserted into the lower esophagus above the esophageal sphincter. The pH measurements indicate a reflux episode each time the pH drops to 4 or less.

In the 24-h period, Larry has several reflux episodes, and his doctor determines that he has chronic GERD. He and Larry discuss treatment for GERD, which includes eating smaller meals, not lying down for 3 h after eating, making dietary changes, and losing weight. Antacids may be used to neutralize the acid coming up from the stomach. Other medications known as *proton pump inhibitors* (PPIs), such as Prilosec and Nexium, may be used to suppress the production of HCl in the stomach (gastric parietal cells), which raises the pH in the stomach to between 4 and 5, and gives the esophagus time to heal. Nexium may be given in oral doses of 40 mg once a day for 4 weeks. In severe GERD cases, an artificial valve may be created at the top of the stomach to strengthen the lower esophageal sphincter.

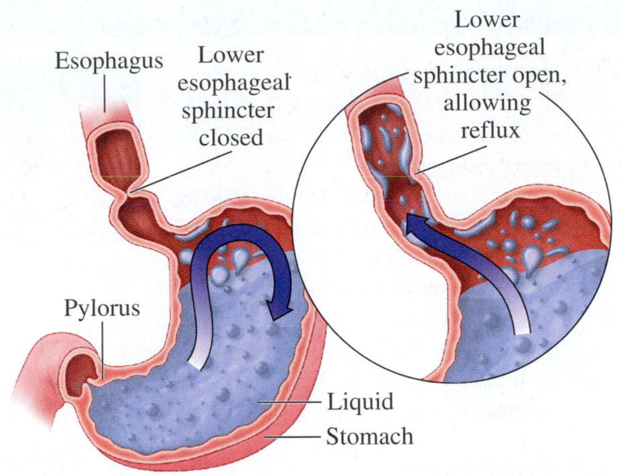
Esophagus
Lower esophageal sphincter closed
Lower esophageal sphincter open, allowing reflux
Pylorus
Liquid
Stomach

In acid reflux disease, the lower esophageal sphincter opens, allowing acidic fluid from the stomach to enter the esophagus.

Clinical Applications

11.79 At rest, the $[H_3O^+]$ of the stomach fluid is 2.0×10^{-4} M. What is the pH of the stomach fluid?

11.80 When food enters the stomach, HCl is released and the $[H_3O^+]$ of the stomach fluid rises to 4.2×10^{-2} M. What is the pH of the stomach fluid while eating?

11.81 In Larry's esophageal pH test, a pH value of 3.60 was recorded in the esophagus. What is the $[H_3O^+]$ in his esophagus?

11.82 After Larry had taken Nexium for 4 weeks, the pH in his stomach was raised to 4.52. What is the $[H_3O^+]$ in his stomach?

11.83 Write the balanced chemical equation for the neutralization reaction of stomach acid HCl with $CaCO_3$, an ingredient in some antacids.

11.84 Write the balanced chemical equation for the neutralization reaction of stomach acid HCl with $Al(OH)_3$, an ingredient in some antacids.

11.85 How many grams of $CaCO_3$ are required to neutralize 100. mL of stomach acid HCl, which is 0.0400 M HCl?

11.86 How many grams of $Al(OH)_3$ are required to neutralize 150. mL of stomach acid HCl with a pH of 1.50?

CONCEPT MAP

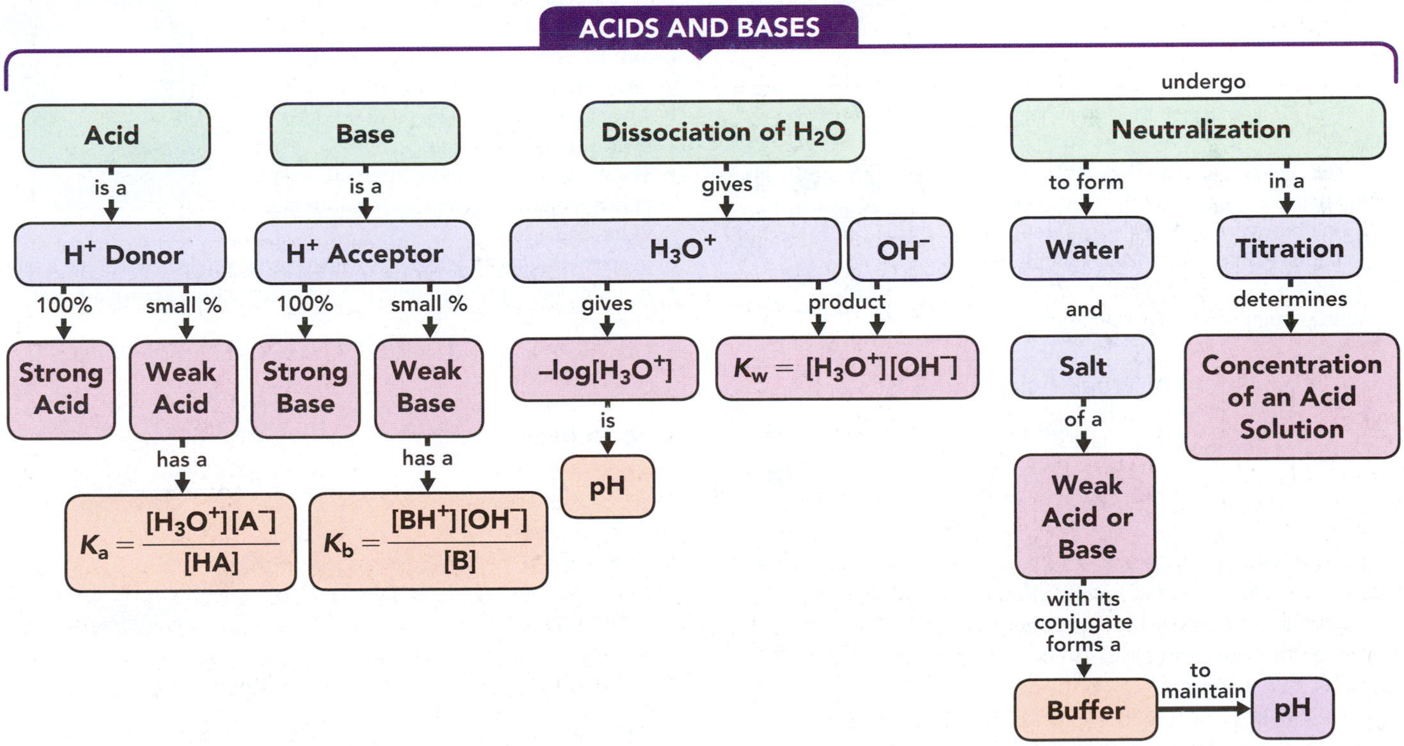
ACIDS AND BASES
Acid
is a
H⁺ Donor
100%
small %
Strong Acid
Weak Acid
has a
$K_a = \frac{[H_3O^+][A^-]}{[HA]}$
Base
is a
H⁺ Acceptor
100%
small %
Strong Base
Weak Base
has a
$K_b = \frac{[BH^+][OH^-]}{[B]}$
Dissociation of H_2O
gives
H_3O^+
OH^-
gives
$-\log[H_3O^+]$
is
pH
product
$K_w = [H_3O^+][OH^-]$
undergo
Neutralization
to form
Water
and
Salt
of a
Weak Acid or Base
with its conjugate forms a
Buffer
to maintain
pH
in a
Titration
determines
Concentration of an Acid Solution

CHAPTER REVIEW

11.1 Acids and Bases

LEARNING GOAL Describe and name acids and bases.

NaOH(s)
OH−
Na+
Water
NaOH(s) —H2O→ Na+(aq) + OH−(aq)
Ionic compound
Dissociation
Hydroxide ion

- An Arrhenius acid produces H^+ and an Arrhenius base produces OH^- in aqueous solutions.
- Acids taste sour, may sting, and neutralize bases.
- Bases taste bitter, feel slippery, and neutralize acids.
- Acids containing a simple anion use a *hydro* prefix, whereas acids with oxygen-containing polyatomic anions are named as *ic* or *ous acids*.

11.2 Brønsted–Lowry Acids and Bases

LEARNING GOAL Identify conjugate acid–base pairs for Brønsted–Lowry acids and bases.

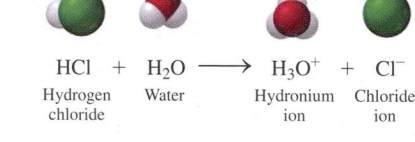

$$HCl + H_2O \longrightarrow H_3O^+ + Cl^-$$

Hydrogen chloride (Acid, H^+ donor) · Water (Base, H^+ acceptor) · Hydronium ion · Chloride ion — Acidic solution

- According to the Brønsted–Lowry theory, acids are H^+ donors and bases are H^+ acceptors.
- A conjugate acid–base pair is related by the loss or gain of one H^+.
- For example, when the acid HF donates H^+, the F^- is its conjugate base. The other acid–base pair would be H_3O^+/H_2O.

$$HF(aq) + H_2O(l) \rightleftharpoons H_3O^+(aq) + F^-(aq)$$

11.3 Strengths of Acids and Bases

LEARNING GOAL Write equations for the dissociation of strong and weak acids; identify the direction of reaction.

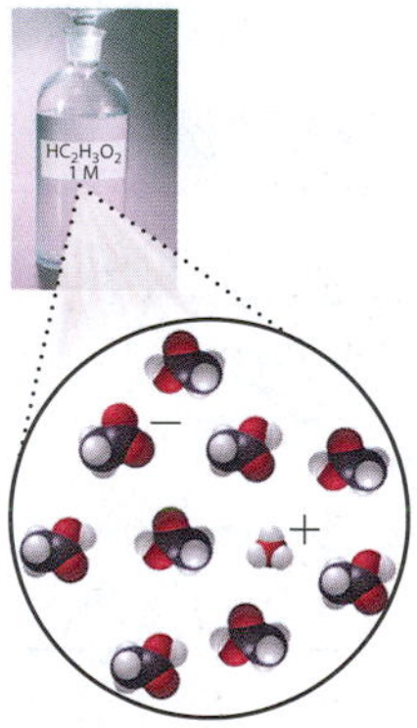
HC2H3O2
1 M

- Strong acids dissociate completely in water, and the H^+ is accepted by H_2O acting as a base.
- A weak acid dissociates slightly in water, producing only a small percentage of H_3O^+.
- Strong bases are hydroxides of Groups 1A (1) and 2A (2) that dissociate completely in water.
- An important weak base is ammonia, NH_3.

11.4 Dissociation of Weak Acids and Bases

LEARNING GOAL Write the expression for the dissociation of a weak acid or weak base.

$HCHO_2$ CHO_2^-

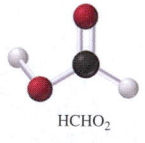

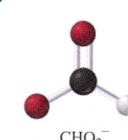

- In water, weak acids and weak bases produce only a few ions when equilibrium is reached.
- Weak acids have small K_a values whereas strong acids, which are essentially 100% dissociated, have very large K_a values.
- The reaction for a weak acid can be written as $HA + H_2O \rightleftharpoons H_3O^+ + A^-$. The acid dissociation expression is written as

$$K_a = \frac{[H_3O^+][A^-]}{[HA]}.$$

- For a weak base, $B + H_2O \rightleftharpoons BH^+ + OH^-$, the base dissociation expression is written as

$$K_b = \frac{[BH^+][OH^-]}{[B]}.$$

11.5 Dissociation of Water

LEARNING GOAL Use the water dissociation expression to calculate the $[H_3O^+]$ and $[OH^-]$ in an aqueous solution.

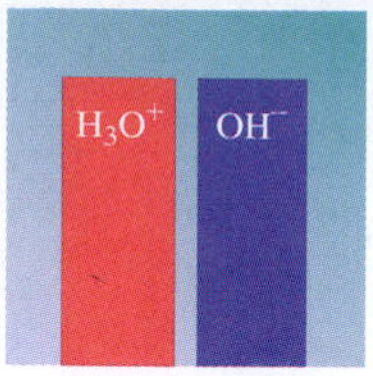
H3O+
OH−

$[H_3O^+] = [OH^-]$

- In pure water, a few water molecules transfer H^+ to other water molecules, producing small, but equal, $[H_3O^+]$ and $[OH^-]$.
- In pure water, the molar concentrations of H_3O^+ and OH^- are each 1.0×10^{-7} mole/L.
- The water dissociation expression, is written as $K_w = [H_3O^+][OH^-]$.
- In acidic solutions, the $[H_3O^+]$ is greater than the $[OH^-]$.
- In basic solutions, the $[OH^-]$ is greater than the $[H_3O^+]$.

11.6 The pH Scale

LEARNING GOAL Calculate pH from $[H_3O^+]$; given the pH, calculate the $[H_3O^+]$ and $[OH^-]$ of a solution.

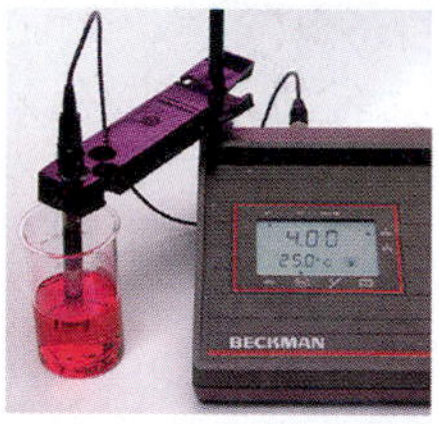
4.00
BECKMAN

- The pH scale is a range of numbers typically from 0 to 14, which represents the $[H_3O^+]$ of the solution.
- A neutral solution has a pH of 7.0. In acidic solutions, the pH is below 7.0; in basic solutions, the pH is above 7.0.
- Mathematically, pH is the negative logarithm of the hydronium ion concentration,

$$pH = -\log[H_3O^+].$$

11.7 Reactions of Acids and Bases

LEARNING GOAL Write balanced equations for reactions of acids with metals, carbonates or bicarbonates, and bases; calculate the molarity or volume of an acid from titration information.

- An acid reacts with a metal to produce hydrogen gas and a salt.
- The reaction of an acid with a carbonate or bicarbonate produces carbon dioxide, water, and a salt.
- In neutralization, an acid reacts with a base to produce water and a salt.
- In a titration, an acid sample is neutralized with a known amount of a base.
- From the volume and molarity of the base, the concentration of the acid is calculated.

11.8 Buffers

LEARNING GOAL Describe the role of buffers in maintaining the pH of a solution; calculate the pH of a buffer.

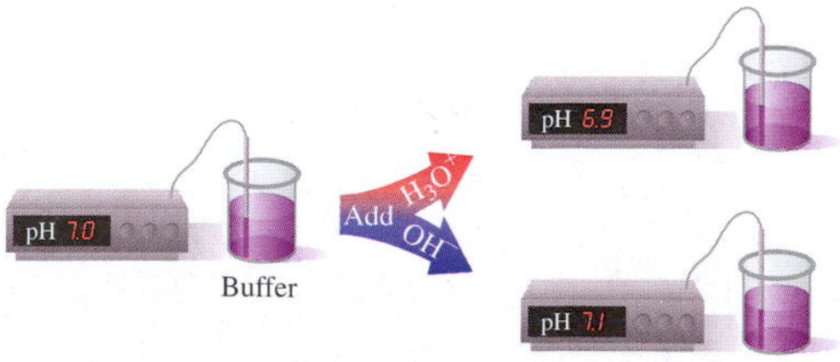
pH 7.0
Buffer
Add H3O+
OH−
pH 6.9
pH 7.1

- A buffer solution resists changes in pH when small amounts of an acid or a base are added.
- A buffer contains either a weak acid and its salt or a weak base and its salt.
- In a buffer, the weak acid reacts with added OH^-, and the anion of the salt reacts with added H_3O^+.
- Most buffer solutions consist of nearly equal concentrations of a weak acid and a salt containing its conjugate base.
- The pH of a buffer is calculated by solving the K_a expression for $[H_3O^+]$.

KEY TERMS

acid A substance that dissolves in water and produces hydrogen ions (H^+), according to the Arrhenius theory. All acids are hydrogen ion donors, according to the Brønsted–Lowry theory.

acid dissociation constant, K_a The numerical value of the product of the ions from the dissociation of a weak acid divided by the concentration of the weak acid.

amphoteric Substances that can act as either an acid or a base in water.

base A substance that dissolves in water and produces hydroxide ions (OH^-), according to the Arrhenius theory. All bases are hydrogen ion acceptors, according to the Brønsted–Lowry theory.

base dissociation constant, K_b The numerical value of the product of the ions from the dissociation of a weak base divided by the concentration of the weak base.

Brønsted–Lowry acids and bases An acid is a hydrogen ion donor; a base is a hydrogen ion acceptor.

buffer solution A solution of a weak acid and its conjugate base or a weak base and its conjugate acid that maintains the pH by neutralizing added acid or base.

conjugate acid base pair An acid and a base that differ by one H^+. When an acid donates a hydrogen ion, the product is its conjugate base, which is capable of accepting a hydrogen ion in the reverse reaction.

dissociation The separation of an acid or a base into ions in water.

endpoint The point at which an indicator changes color. For the indicator phenolphthalein, the color change occurs when the number of moles of OH^- is equal to the number of moles of H_3O^+ in the sample.

hydronium ion, H_3O^+ The ion formed by the attraction of a hydrogen ion, H^+, to a water molecule.

indicator A substance added to a titration sample that changes color when the pH of the solution changes.

neutral The term that describes a solution with equal concentrations of $[H_3O^+]$ and $[OH^-]$.

neutralization A reaction between an acid and a base to form water and a salt.

pH A measure of the $[H_3O^+]$ in a solution; $pH = -\log[H_3O^+]$.

salt An ionic compound that contains a metal ion or NH_4^+ and a nonmetal or polyatomic ion other than OH^-.

strong acid An acid that completely dissociates in water.

strong base A base that completely dissociates in water.

titration The addition of base to an acid sample to determine the concentration of the acid.

water dissociation constant, K_w The numerical value of the product of $[H_3O^+]$ and $[OH^-]$ in solution; $K_w = 1.0 \times 10^{-14}$.

water dissociation expression The product of $[H_3O^+]$ and $[OH^-]$ in solution; $K_w = [H_3O^+][OH^-]$.

weak acid An acid that is a poor donor of H^+ and dissociates only slightly in water.

weak base A base that is a poor acceptor of H^+ and produces only a small number of ions in water.

KEY MATH SKILLS

The chapter Section containing each Key Math Skill is shown in parentheses at the end of each heading.

Calculating pH from $[H_3O^+]$ (11.6)

- The pH of a solution is calculated from the negative log of the $[H_3O^+]$.

$$pH = -\log[H_3O^+]$$

Example: What is the pH of a solution that has $[H_3O^+] = 2.4 \times 10^{-11}$ M?

Answer: We substitute the given $[H_3O^+]$ into the pH equation and calculate the pH.

$$pH = -\log[H_3O^+]$$
$$= -\log[2.4 \times 10^{-11}]$$

Calculator Display

[+/−] [log] 2.4 [EE or EXP] [+/−] 11 [=] 10.61978876

$$= 10.62$$

Two decimal places in the pH equal the two SFs in the $[H_3O^+]$ coefficient.

Calculating $[H_3O^+]$ from pH (11.6)

- The calculation of $[H_3O^+]$ from the pH is done by reversing the pH calculation using the negative pH.

$$[H_3O^+] = 10^{-pH}$$

Example: What is the $[H_3O^+]$ of a solution with a pH of 4.80?

Answer: $[H_3O^+] = 10^{-pH}$

$$= 10^{-4.80}$$

Calculator Display

[2nd] [log] [+/−] 4.80 [=] 1.584893192−05

$$= 1.6 \times 10^{-5}\ M$$

Two SFs in the $[H_3O^+]$ equal the two decimal places in the pH.

CORE CHEMISTRY SKILLS

The chapter Section containing each Core Chemistry Skill is shown in parentheses at the end of each heading.

Identifying Conjugate Acid–Base Pairs (11.2)

- According to the Brønsted–Lowry theory, a conjugate acid–base pair consists of molecules or ions related by the loss of one H^+ by an acid, and the gain of one H^+ by a base.
- Every acid–base reaction contains two conjugate acid–base pairs because an H^+ is transferred in both the forward and reverse directions.
- When an acid such as HF loses one H^+, the conjugate base F^- is formed. When H_2O acts as a base, it gains one H^+, which forms its conjugate acid, H_3O^+.

Example: Identify the conjugate acid–base pairs in the following reaction:

$$H_2SO_4(aq) + H_2O(l) \rightleftharpoons HSO_4^-(aq) + H_3O^+(aq)$$

Answer:

$$\underset{\text{Acid}}{H_2SO_4(aq)} + \underset{\text{Base}}{H_2O(l)} \rightleftharpoons \underset{\text{Conjugate base}}{HSO_4^-(aq)} + \underset{\text{Conjugate acid}}{H_3O^+(aq)}$$

Conjugate acid–base pairs: H_2SO_4/HSO_4^- and H_3O^+/H_2O

Calculating $[H_3O^+]$ and $[OH^-]$ in Solutions (11.5)

- For all aqueous solutions, the product of $[H_3O^+]$ and $[OH^-]$ is equal to the water dissociation constant, K_w.

$$K_w = [H_3O^+][OH^-]$$

- Because pure water contains equal numbers of OH^- ions and H_3O^+ ions, each with molar concentrations of 1.0×10^{-7} M, the numerical value of K_w is 1.0×10^{-14} at 25 °C.

$$K_w = [H_3O^+][OH^-] = [1.0 \times 10^{-7}][1.0 \times 10^{-7}]$$
$$= 1.0 \times 10^{-14}$$

- If we know the $[H_3O^+]$ of a solution, we can use the K_w expression to calculate the $[OH^-]$. If we know the $[OH^-]$ of a solution, we can calculate the $[H_3O^+]$ using the K_w expression.

$$[OH^-] = \frac{K_w}{[H_3O^+]} \quad [H_3O^+] = \frac{K_w}{[OH^-]}$$

Example: What is the $[OH^-]$ in a solution that has $[H_3O^+] = 2.4 \times 10^{-11}$ M? Is the solution acidic or basic?

Answer: We solve the K_w expression for $[OH^-]$ and substitute in the known values of K_w and $[H_3O^+]$.

$$[OH^-] = \frac{K_w}{[H_3O^+]} = \frac{1.0 \times 10^{-14}}{[2.4 \times 10^{-11}]} = 4.2 \times 10^{-4}\text{ M}$$

Because the $[OH^-]$ is greater than the $[H_3O^+]$, this is a basic solution.

Writing Equations for Reactions of Acids and Bases (11.7)

- Acids react with certain metals to produce hydrogen gas (H_2) and a salt.

$$2HCl(aq) + Mg(s) \longrightarrow H_2(g) + MgCl_2(aq)$$

Acid | Metal | Hydrogen | Salt

- When an acid is added to a carbonate or bicarbonate, the products are carbon dioxide gas, water, and a salt.

$$2HCl(aq) + Na_2CO_3(aq) \longrightarrow CO_2(g) + H_2O(l) + 2NaCl(aq)$$

Acid | Carbonate | Carbon dioxide | Water | Salt

- Neutralization is a reaction between a strong or weak acid with a strong base to produce water and a salt.

$$HCl(aq) + NaOH(aq) \longrightarrow H_2O(l) + NaCl(aq)$$

Acid | Base | Water | Salt

Example: Write the balanced chemical equation for the reaction of hydrobromic acid $HBr(aq)$ and $ZnCO_3(s)$.

Answer:

$$2HBr(aq) + ZnCO_3(s) \longrightarrow CO_2(g) + H_2O(l) + ZnBr_2(aq)$$

Calculating Molarity or Volume of an Acid or Base in a Titration (11.7)

- In a titration, a measured volume of acid is neutralized by a strong base solution of known molarity.
- From the measured volume of the strong base solution required for titration and its molarity, the number of moles of the strong base, the moles of acid, and the concentration of the acid are calculated.

Example: A 15.0-mL sample of a H_2SO_4 solution is titrated with 24.0 mL of a 0.245 M NaOH solution. What is the molarity of the H_2SO_4 solution?

$$H_2SO_4(aq) + 2NaOH(aq) \longrightarrow 2H_2O(l) + Na_2SO_4(aq)$$

Answer:

$$24.0\text{ mL NaOH solution} \times \frac{1\text{ L NaOH solution}}{1000\text{ mL NaOH solution}}$$
$$\times \frac{0.245\text{ mole NaOH}}{1\text{ L NaOH solution}} \times \frac{1\text{ mole }H_2SO_4}{2\text{ moles NaOH}} = 0.002\,94\text{ mole of }H_2SO_4$$

$$\text{Molarity (M)} = \frac{0.002\,94\text{ mole }H_2SO_4}{0.0150\text{ L }H_2SO_4\text{ solution}} = 0.196\text{ M }H_2SO_4\text{ solution}$$

Calculating the pH of a Buffer (11.8)

- A buffer solution maintains pH by neutralizing small amounts of added acid or base.
- Most buffer solutions consist of nearly equal concentrations of a weak acid and a salt containing its conjugate base such as acetic acid, $HC_2H_3O_2$, and its salt, $NaC_2H_3O_2$.
- The $[H_3O^+]$ is calculated by solving the K_a expression for $[H_3O^+]$, then substituting the values of $[H_3O^+]$, [HA], and K_a into the equation.

$$K_a = \frac{[H_3O^+][C_2H_3O_2^-]}{[HC_2H_3O_2]}$$

Solving for $[H_3O^+]$ gives:

$$[H_3O^+] = K_a \times \frac{[HC_2H_3O_2]}{[C_2H_3O_2^-]} \begin{matrix} \longleftarrow \text{Weak acid} \\ \longleftarrow \text{Conjugate base} \end{matrix}$$

- The pH of the buffer is calculated from the $[H_3O^+]$.

$$pH = -\log[H_3O^+]$$

Example: What is the pH of a buffer prepared with 0.40 M $HC_2H_3O_2$ and 0.20 M $C_2H_3O_2^-$, if the K_a of acetic acid is 1.8×10^{-5}?

Answer:

$$[H_3O^+] = K_a \times \frac{[HC_2H_3O_2]}{[C_2H_3O_2^-]} = 1.8 \times 10^{-5} \times \frac{[0.40]}{[0.20]}$$
$$= 3.6 \times 10^{-5}\text{ M}$$
$$pH = -\log[3.6 \times 10^{-5}] = 4.44$$

UNDERSTANDING THE CONCEPTS

The chapter Sections to review are shown in parentheses at the end of each problem.

11.87 Determine if each of the following diagrams represents a strong acid or a weak acid. The acid has the formula HX. (11.3)

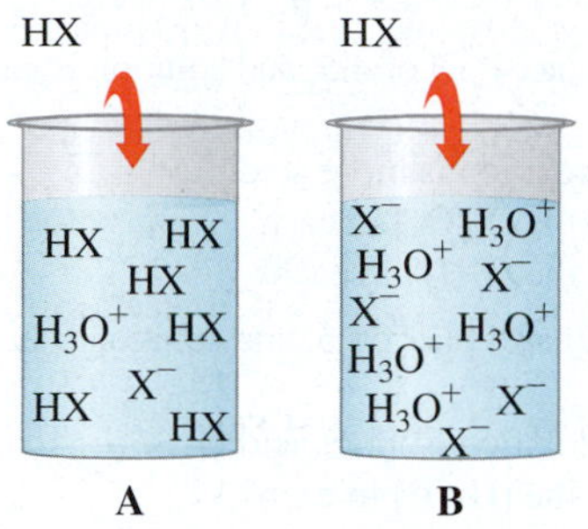
HX
HX
HX HX
HX
H_3O^+ HX
X^-
HX HX
X^- H_3O^+
H_3O^+ X^-
X^- H_3O^+
H_3O^+
H_3O^+ X^-
X^-
A
B

11.88 Adding a few drops of a strong acid to water will lower the pH appreciably. However, adding the same number of drops to a buffer does not appreciably alter the pH. Why? (11.8)

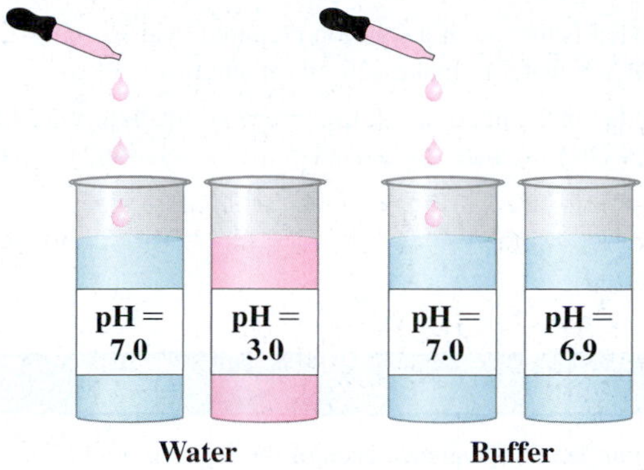
pH = 7.0
pH = 3.0
pH = 7.0
pH = 6.9
Water
Buffer

11.89 Identify each of the following as an acid or a base: (11.1)

a. H_2SO_4 **b.** RbOH
c. $Ca(OH)_2$ **d.** HI

11.90 Identify each of the following as an acid or a base: (11.1)

a. $Sr(OH)_2$ **b.** H_2SO_3
c. $HC_2H_3O_2$ **d.** CsOH

11.91 Complete the following table: (11.2)

Acid	Conjugate Base
H_2O	
	CN^-
HNO_2	
	$H_2PO_4^-$

11.92 Complete the following table: (11.2)

Base	Conjugate Acid
	HS^-
	H_3O^+
NH_3	
HCO_3^-	

Clinical Applications

11.93 Sometimes, during stress or trauma, a person can start to hyperventilate. Then the person might breathe into a paper bag to avoid fainting. (11.8)

a. What changes occur in the blood pH during hyperventilation?

b. How does breathing into a paper bag help return blood pH to normal?

Breathing into a paper bag can help a person who is hyperventilating.

11.94 In the blood plasma, pH is maintained by the carbonic acid–bicarbonate buffer system. (11.8)

a. How is pH maintained when acid is added to the buffer system?

b. How is pH maintained when base is added to the buffer system?

11.95 State whether each of the following solutions is acidic, basic, or neutral: (11.5)

a. sweat, pH 5.2 **b.** tears, pH 7.5
c. bile, pH 8.1 **d.** stomach acid, pH 2.5

11.96 State whether each of the following solutions is acidic, basic, or neutral: (11.5)

a. saliva, pH 6.8 **b.** urine, pH 5.9
c. pancreatic juice, pH 8.0 **d.** blood, pH 7.45

ADDITIONAL PRACTICE PROBLEMS

11.97 Identify each of the following as an acid, base, or salt, and give its name: (11.1)

a. $HBrO_2$ **b.** CsOH
c. $Mg(NO_3)_2$ **d.** $HClO_4$

11.98 Identify each of the following as an acid, base, or salt, and give its name: (11.1)

a. HNO_2 **b.** $MgBr_2$
c. NH_3 **d.** Li_2SO_3

11.99 Complete the following table: (11.2)

Acid	Conjugate Base
HI	
	Cl^-
NH_4^+	
	HS^-

11.100 Complete the following table: (11.2)

Base	Conjugate Acid
F^-	
	$HC_2H_3O_2$
	HSO_3^-
ClO^-	

11.101 Using Table 11.3, identify the stronger acid in each of the following pairs: (11.3)
a. HF or HCN
b. H_3O^+ or H_2S
c. HNO_2 or $HC_2H_3O_2$
d. H_2O or HCO_3^-

11.102 Using Table 11.3, identify the stronger base in each of the following pairs: (11.3)
a. H_2O or Cl^-
b. OH^- or NH_3
c. SO_4^{2-} or NO_2^-
d. CO_3^{2-} or H_2O

11.103 Determine the pH for each of the following solutions: (11.6)
a. $[H_3O^+] = 2.0 \times 10^{-8}$ M
b. $[H_3O^+] = 5.0 \times 10^{-2}$ M
c. $[OH^-] = 3.5 \times 10^{-4}$ M
d. $[OH^-] = 0.0054$ M

11.104 Determine the pH for each of the following solutions: (11.6)
a. $[OH^-] = 1.0 \times 10^{-7}$ M
b. $[H_3O^+] = 4.2 \times 10^{-3}$ M
c. $[H_3O^+] = 0.0001$ M
d. $[OH^-] = 8.5 \times 10^{-9}$ M

11.105 Are the solutions in problem 11.103 acidic, basic, or neutral? (11.6)

11.106 Are the solutions in problem 11.104 acidic, basic, or neutral? (11.6)

11.107 Calculate the $[H_3O^+]$ and $[OH^-]$ for a solution with each of the following pH values: (11.6)
a. 3.00
b. 6.2
c. 8.85
d. 11.00

11.108 Calculate the $[H_3O^+]$ and $[OH^-]$ for a solution with each of the following pH values: (11.6)
a. 10.00
b. 5.0
c. 6.54
d. 1.82

11.109 Solution A has a pH of 4.5, and solution B has a pH of 6.7. (11.6)
a. Which solution is more acidic?
b. What is the $[H_3O^+]$ in each?
c. What is the $[OH^-]$ in each?

11.110 Solution X has a pH of 9.5, and solution Y has a pH of 7.5. (11.6)
a. Which solution is more acidic?
b. What is the $[H_3O^+]$ in each?
c. What is the $[OH^-]$ in each?

11.111 What is the $[OH^-]$ in a solution that contains 0.225 g of NaOH in 0.250 L of solution? (11.7)

11.112 What is the $[H_3O^+]$ in a solution that contains 1.54 g of HNO_3 in 0.500 L of solution? (11.7)

11.113 What is the pH of a solution prepared by dissolving 2.5 g of HCl in water to make 425 mL of solution? (11.6)

11.114 What is the pH of a solution prepared by dissolving 1.0 g of $Ca(OH)_2$ in water to make 875 mL of solution? (11.6)

CHALLENGE PROBLEMS

The following problems are related to the topics in this chapter. However, they do not all follow the chapter order, and they require you to combine concepts and skills from several Sections. These problems will help you increase your critical thinking skills and prepare for your next exam.

11.115 For each of the following: (11.2, 11.3)
1. H_2S
2. H_3PO_4
a. Write the formula for the conjugate base.
b. Write the K_a expression.
c. Which is the weaker acid?

11.116 For each of the following: (11.2, 11.3)
1. HCO_3^-
2. $HC_2H_3O_2$
a. Write the formula for the conjugate base.
b. Write the K_a expression.
c. Which is the stronger acid?

11.117 Using Table 11.3, identify the conjugate acid–base pairs in each of the following equations and whether the equilibrium mixture contains mostly products or mostly reactants: (11.2, 11.3)
a. $NH_3(aq) + HNO_3(aq) \rightleftharpoons NH_4^+(aq) + NO_3^-(aq)$
b. $HBr(aq) + H_2O(l) \rightleftharpoons H_3O^+(aq) + Br^-(aq)$

11.118 Using Table 11.3, identify the conjugate acid–base pairs in each of the following equations and whether the equilibrium mixture contains mostly products or mostly reactants: (11.2, 11.3)
a. $HNO_2(aq) + HS^-(aq) \rightleftharpoons H_2S(g) + NO_2^-(aq)$
b. $Cl^-(aq) + H_2O(l) \rightleftharpoons OH^-(aq) + HCl(aq)$

11.119 Complete and balance each of the following: (11.7)
a. $H_2SO_4(aq) + ZnCO_3(s) \longrightarrow$
b. $HNO_3(aq) + Al(s) \longrightarrow$

11.120 Complete and balance each of the following: (11.7)
a. $H_3PO_4(aq) + Ca(OH)_2(s) \longrightarrow$
b. $HNO_3(aq) + KHCO_3(s) \longrightarrow$

11.121 Determine each of the following for a 0.050 M KOH solution: (11.6, 11.7)
a. $[H_3O^+]$
b. pH
c. the balanced equation for the reaction with H_2SO_4
d. milliliters of KOH solution required to neutralize 40.0 mL of a 0.035 M H_2SO_4 solution

11.122 Determine each of the following for a 0.10 M HBr solution: (11.6, 11.7)
a. $[H_3O^+]$
b. pH
c. the balanced equation for the reaction with LiOH
d. milliliters of HBr solution required to neutralize 36.0 mL of a 0.25 M LiOH solution

11.123 Calculate the volume, in milliliters, of a 0.150 M NaOH solution that will completely neutralize each of the following: (11.7)
a. 25.0 mL of a 0.288 M HCl solution
b. 10.0 mL of a 0.560 M H_2SO_4 solution

11.124 Calculate the volume, in milliliters, of a 0.215 M NaOH solution that will completely neutralize each of the following: (11.7)

a. 3.80 mL of a 1.25 M HNO_3 solution

b. 8.50 mL of a 0.825 M H_3PO_4 solution

11.125 A solution of 0.205 M NaOH is used to titrate 20.0 mL of a H_2SO_4 solution. If 45.6 mL of the NaOH solution is required to reach the endpoint, what is the molarity of the H_2SO_4 solution? (11.7)

$$H_2SO_4(aq) + 2NaOH(aq) \longrightarrow 2H_2O(l) + Na_2SO_4(aq)$$

11.126 A 10.0-mL sample of vinegar, which is an aqueous solution of acetic acid, $HC_2H_3O_2$, requires 16.5 mL of a 0.500 M NaOH solution to reach the endpoint in a titration. What is the molarity of the acetic acid solution? (11.7)

$$HC_2H_3O_2(aq) + NaOH(aq) \longrightarrow H_2O(l) + NaC_2H_3O_2(aq)$$

11.127 A buffer solution is made by dissolving H_3PO_4 and NaH_2PO_4 in water. (11.8)

a. Write an equation that shows how this buffer neutralizes added acid.

b. Write an equation that shows how this buffer neutralizes added base.

c. Calculate the pH of this buffer if it contains 0.50 M H_3PO_4 and 0.20 M $H_2PO_4^-$. The K_a for H_3PO_4 is 7.5×10^{-3}.

11.128 A buffer solution is made by dissolving $HC_2H_3O_2$ and $NaC_2H_3O_2$ in water. (11.8)

a. Write an equation that shows how this buffer neutralizes added acid.

b. Write an equation that shows how this buffer neutralizes added base.

c. Calculate the pH of this buffer if it contains 0.20 M $HC_2H_3O_2$ and 0.40 M $C_2H_3O_2^-$. The K_a for $HC_2H_3O_2$ is 1.8×10^{-5}.

11.129 One of the most acidic lakes in the United States is Little Echo Pond in the Adirondacks in New York. Recently, this lake had a pH of 4.2, well below the recommended pH of 6.5. (11.6, 11.7)

a. What are the $[H_3O^+]$ and $[OH^-]$ of Little Echo Pond?

b. What are the $[H_3O^+]$ and $[OH^-]$ of a lake that has a pH of 6.5?

A helicopter drops calcium carbonate on an acidic lake to increase its pH.

c. One way to raise the pH of an acidic lake (and restore aquatic life) is to add limestone ($CaCO_3$). How many grams of $CaCO_3$ are needed to neutralize 1.0 kL of the acidic water from the lake if the acid is sulfuric acid?

$$H_2SO_4(aq) + CaCO_3(s) \longrightarrow CO_2(g) + H_2O(l) + CaSO_4(aq)$$

Clinical Applications

11.130 The daily output of stomach acid (gastric juice) is 1000 mL to 2000 mL. Prior to a meal, stomach acid (HCl) typically has a pH of 1.42. (11.6, 11.7)

a. What is the $[H_3O^+]$ of the stomach acid?

b. One chewable tablet of the antacid Maalox contains 600. mg of $CaCO_3$. Write the neutralization equation, and calculate the milliliters of stomach acid neutralized by two tablets of Maalox.

c. The antacid milk of magnesia contains 400. mg of $Mg(OH)_2$ per teaspoon. Write the neutralization equation, and calculate the number of milliliters of stomach acid that are neutralized by 1 tablespoon of milk of magnesia (1 tablespoon = 3 teaspoons).

ANSWERS

11.1 **a.** acid **b.** acid **c.** acid **d.** base **e.** both

11.3 **a.** hydrochloric acid **b.** calcium hydroxide **c.** perchloric acid **d.** strontium hydroxide **e.** sulfurous acid **f.** bromous acid

11.5 **a.** RbOH **b.** HF **c.** H_3PO_4 **d.** LiOH **e.** NH_4OH **f.** HIO_4

11.7 **a.** HI is the acid (hydrogen ion donor), and H_2O is the base (hydrogen ion acceptor).
b. H_2O is the acid (hydrogen ion donor), and F^- is the base (hydrogen ion acceptor).
c. H_2S is the acid (hydrogen ion donor), and $C_2H_5—NH_2$ is the base (hydrogen ion acceptor).

11.9 **a.** F^- **b.** OH^- **c.** HPO_3^{2-} **d.** SO_4^{2-} **e.** ClO_2^-

11.11 **a.** HCO_3^- **b.** H_3O^+ **c.** H_3PO_4 **d.** HBr **e.** $HClO_4$

11.13 **a.** The conjugate acid–base pairs are H_2CO_3/HCO_3^- and H_3O^+/H_2O.
b. The conjugate acid–base pairs are HCN/CN^- and HNO_2/NO_2^-.
c. The conjugate acid–base pairs are HF/F^- and $HCHO_2/CHO_2^-$.

11.15 $NH_4^+(aq) + H_2O(l) \rightleftharpoons NH_3(aq) + H_3O^+(aq)$

11.17 A strong acid is a good hydrogen ion donor, whereas its conjugate base is a poor hydrogen ion acceptor.

11.19 **a.** HBr **b.** HSO_4^- **c.** H_2CO_3

11.21 a. HSO_4^- **b.** HF **c.** HCO_3^-

11.23 a. reactants **b.** reactants **c.** products

11.25 $NH_4^+(aq) + SO_4^{2-}(aq) \rightleftarrows NH_3(aq) + HSO_4^-(aq)$
The equilibrium mixture contains mostly reactants because NH_4^+ is a weaker acid than HSO_4^-, and SO_4^{2-} is a weaker base than NH_3.

11.27 a. True **b.** False **c.** False **d.** True **e.** False

11.29 a. H_2SO_3 **b.** HSO_3^- **c.** H_2SO_3 **d.** HS^- **e.** H_2SO_3

11.31 $H_3PO_4(aq) + H_2O(l) \rightleftarrows H_3O^+(aq) + H_2PO_4^-(aq)$

$$K_a = \frac{[H_3O^+][H_2PO_4^-]}{[H_3PO_4]}$$

11.33 In pure water, $[H_3O^+] = [OH^-]$ because one of each is produced every time a hydrogen ion is transferred from one water molecule to another.

11.35 In an acidic solution, the $[H_3O^+]$ is greater than the $[OH^-]$.

11.37 a. acidic **b.** basic **c.** basic **d.** acidic

11.39 a. 1.0×10^{-5} M **b.** 1.0×10^{-8} M **c.** 5.0×10^{-10} M **d.** 2.5×10^{-2} M

11.41 a. 2.5×10^{-13} M **b.** 2.0×10^{-9} M **c.** 5.0×10^{-11} M **d.** 1.3×10^{-6} M

11.43 In a neutral solution, the $[H_3O^+]$ is 1.0×10^{-7} M and the pH is 7.00, which is the negative value of the power of 10.

11.45 a. basic **b.** acidic **c.** acidic **d.** acidic **e.** basic

11.47 An increase or decrease of one pH unit changes the $[H_3O^+]$ by a factor of 10. Thus a pH of 3 is 10 times more acidic than a pH of 4.

11.49 a. 4.0 **b.** 8.5 **c.** 9.0 **d.** 3.40 **e.** 7.17 **f.** 10.92

11.51

Food	$[H_3O^+]$	$[OH^-]$	pH	Acidic, Basic, or Neutral?
Rye bread	1.6×10^{-9} M	6.3×10^{-6} M	8.80	Basic
Tomatoes	2.3×10^{-5} M	4.3×10^{-10} M	4.64	Acidic
Peas	6.2×10^{-7} M	1.6×10^{-8} M	6.21	Acidic

11.53 1.2×10^{-7} M

11.55 a. $2HBr(aq) + ZnCO_3(s) \longrightarrow CO_2(g) + H_2O(l) + ZnBr_2(aq)$
b. $2HCl(aq) + Zn(s) \longrightarrow H_2(g) + ZnCl_2(aq)$
c. $HCl(aq) + NaHCO_3(s) \longrightarrow CO_2(g) + H_2O(l) + NaCl(aq)$
d. $H_2SO_4(aq) + Mg(OH)_2(s) \longrightarrow 2H_2O(l) + MgSO_4(aq)$

11.57 a. $2HCl(aq) + Mg(OH)_2(s) \longrightarrow 2H_2O(l) + MgCl_2(aq)$
b. $H_3PO_4(aq) + 3LiOH(aq) \longrightarrow 3H_2O(l) + Li_3PO_4(aq)$

11.59 a. $H_2SO_4(aq) + 2NaOH(aq) \longrightarrow 2H_2O(l) + Na_2SO_4(aq)$
b. $3HCl(aq) + Fe(OH)_3(s) \longrightarrow 3H_2O(l) + FeCl_3(aq)$
c. $H_2CO_3(aq) + Mg(OH)_2(s) \longrightarrow 2H_2O(l) + MgCO_3(s)$

11.61 0.829 M HCl solution

11.63 0.124 M H_2SO_4 solution

11.65 16.5 mL

11.67 **b** and **c** are buffer systems. **b** contains the weak acid H_2CO_3 and its salt $NaHCO_3$. **c** contains HF, a weak acid, and its salt, KF.

11.69 a. 3 **b.** 1 and 2 **c.** 3 **d.** 1

11.71 pH = 3.35

11.73 The pH of the 0.10 M HF/0.10 M NaF buffer is 3.46.
The pH of the 0.060 M HF/0.120 M NaF buffer is 3.76.

11.75 If you breathe fast, CO_2 is expelled and the equilibrium shifts to lower H_3O^+, which raises the pH.

11.77 If large amounts of HCO_3^- are lost, equilibrium shifts to higher H_3O^+, which lowers the pH.

11.79 pH = 3.70

11.81 2.5×10^{-4} M

11.83 $2HCl(aq) + CaCO_3(s) \longrightarrow CO_2(g) + H_2O(l) + CaCl_2(aq)$

11.85 0.200 g of $CaCO_3$

11.87 a. This diagram represents a weak acid; only a few HX molecules separate into H_3O^+ and X^- ions.
b. This diagram represents a strong acid; all the HX molecules separate into H_3O^+ and X^- ions.

11.89 a. acid **b.** base **c.** base **d.** acid

11.91

Acid	Conjugate Base
H_2O	OH^-
HCN	CN^-
HNO_2	NO_2^-
H_3PO_4	$H_2PO_4^-$

11.93 a. During hyperventilation, a person will lose CO_2 and the blood pH will rise.
b. Breathing into a paper bag will increase the CO_2 concentration and lower the blood pH.

11.95 a. acidic **b.** basic **c.** basic **d.** acidic

11.97 a. acid, bromous acid **b.** base, cesium hydroxide **c.** salt, magnesium nitrate **d.** acid, perchloric acid

11.99

Acid	Conjugate Base
HI	I^-
HCl	Cl^-
NH_4^+	NH_3
H_2S	HS^-

11.101 a. HF **b.** H_3O^+
c. HNO_2 **d.** HCO_3^-

11.103 a. pH = 7.70 **b.** pH = 1.30
c. pH = 10.54 **d.** pH = 11.73

11.105 a. basic **b.** acidic
c. basic **d.** basic

11.107 a. $[H_3O^+] = 1.0 \times 10^{-3}$ M; $[OH^-] = 1.0 \times 10^{-11}$ M
b. $[H_3O^+] = 6 \times 10^{-7}$ M; $[OH^-] = 2 \times 10^{-8}$ M
c. $[H_3O^+] = 1.4 \times 10^{-9}$ M; $[OH^-] = 7.1 \times 10^{-6}$ M
d. $[H_3O^+] = 1.0 \times 10^{-11}$ M; $[OH^-] = 1.0 \times 10^{-3}$ M

11.109 a. Solution A
b. Solution A $[H_3O^+] = 3 \times 10^{-5}$ M; Solution B $[H_3O^+] = 2 \times 10^{-7}$ M
c. Solution A $[OH^-] = 3 \times 10^{-10}$ M; Solution B $[OH^-] = 5 \times 10^{-8}$ M

11.111 $[OH^-] = 0.0225$ M

11.113 pH = 0.80

11.115 a. 1. HS^-
2. $H_2PO_4^-$
b. 1. $\dfrac{[H_3O^+][HS^-]}{[H_2S]}$
2. $\dfrac{[H_3O^+][H_2PO_4^-]}{[H_3PO_4]}$
c. H_2S

11.117 a. HNO_3/NO_3^- and NH_4^+/NH_3; equilibrium mixture contains mostly products
b. HBr/Br^- and H_3O^+/H_2O; equilibrium mixture contains mostly products

11.119 a. $H_2SO_4(aq) + ZnCO_3(s) \longrightarrow CO_2(g) + H_2O(l) + ZnSO_4(aq)$
b. $6HNO_3(aq) + 2Al(s) \longrightarrow 3H_2(g) + 2Al(NO_3)_3(aq)$

11.121 a. $[H_3O^+] = 2.0 \times 10^{-13}$ M
b. pH = 12.70
c. $2KOH(aq) + H_2SO_4(aq) \longrightarrow 2H_2O(l) + K_2SO_4(aq)$
d. 56 mL of the KOH solution

11.123 a. 48.0 mL of NaOH solution
b. 74.7 mL of NaOH solution

11.125 0.234 M H_2SO_4

11.127 a. acid: $H_2PO_4^-(aq) + H_3O^+(aq) \longrightarrow H_3PO_4(aq) + H_2O(l)$
b. base: $H_3PO_4(aq) + OH^-(aq) \longrightarrow H_2PO_4^-(aq) + H_2O(l)$
c. pH = 1.72

11.129 a. $[H_3O^+] = 6 \times 10^{-5}$ M; $[OH^-] = 2 \times 10^{-10}$ M
b. $[H_3O^+] = 3 \times 10^{-7}$ M; $[OH^-] = 3 \times 10^{-8}$ M
c. 3 g of $CaCO_3$

COMBINING IDEAS from Chapters 9 to 11

CI.19 Consider the following reaction at equilibrium:

$$2H_2(g) + S_2(g) \rightleftharpoons 2H_2S(g) + \text{heat}$$

In a 10.0-L container, an equilibrium mixture contains 2.02 g of H_2, 10.3 g of S_2, and 68.2 g of H_2S. (7.5, 10.2, 10.3, 10.4, 10.5)

a. What is the numerical value of K_c for this equilibrium mixture?
b. If more H_2 is added to the equilibrium mixture, how will the equilibrium shift?
c. How will the equilibrium shift if the mixture is placed in a 5.00-L container with no change in temperature?
d. If a 5.00-L container has an equilibrium mixture of 0.300 mole of H_2 and 2.50 moles of H_2S, what is the $[S_2]$ if temperature remains constant?

CI.20 In wine-making, glucose ($C_6H_{12}O_6$) from grapes undergoes fermentation in the absence of oxygen to produce ethanol, C_2H_6O, and carbon dioxide. A bottle of vintage port wine has a volume of 750 mL and contains 135 mL of ethanol, which has a density of 0.789 g/mL. In 1.5 lb of grapes, there are 26 g of glucose. (2.7, 7.1, 7.5, 9.4)

a. Calculate the volume percent (v/v) of ethanol in the port wine.
b. What is the molarity (M) of ethanol in the port wine?
c. Write the balanced chemical equation for the fermentation reaction of glucose in grapes.
d. How many grams of glucose from grapes are required to produce one bottle of port wine?
e. How many bottles of port wine can be produced from 1.0 ton of grapes (1 ton = 2000 lb)?

When the glucose in grapes is fermented, ethanol is produced.

CI.21 A metal M with a mass of 0.420 g completely reacts with 34.8 mL of a 0.520 M HCl solution to form H_2 gas and aqueous MCl_3. (7.1, 7.5, 8.7, 11.7)

a. Write a balanced chemical equation for the reaction of HCl(*aq*) and the metal (M).
b. What volume, in milliliters, of H_2 at 720. mmHg and 24 °C is produced?
c. How many moles of metal M reacted?
d. Using your results from part **c**, determine the molar mass and name of metal M.
e. Write the balanced chemical equation for the reaction.

When a metal reacts with a strong acid, bubbles of hydrogen gas form.

CI.22 A solution of HCl is prepared by diluting 15.0 mL of a 12.0 M HCl solution with enough water to make 750. mL of HCl solution. (9.4, 11.6, 11.7)

a. What is the molarity of the HCl solution?
b. What is the $[H_3O^+]$ and pH of the HCl solution?
c. Write the balanced chemical equation for the reaction of HCl and $MgCO_3$.
d. How many milliliters of the diluted HCl solution is required to completely react with 350. mg of $MgCO_3$?

CI.23 A KOH solution is prepared by dissolving 8.57 g of KOH in enough water to make 850. mL of KOH solution. (9.4, 11.6, 11.7)

a. What is the molarity of the KOH solution?
b. What is the $[H_3O^+]$ and pH of the KOH solution?
c. Write the balanced chemical equation for the neutralization of KOH by H_2SO_4.
d. How many milliliters of a 0.250 M H_2SO_4 solution is required to neutralize 10.0 mL of the KOH solution?

Clinical Applications

CI.24 In a teaspoon (5.0 mL) of a liquid antacid, there are 400. mg of $Mg(OH)_2$ and 400. mg of $Al(OH)_3$. A 0.080 M HCl solution, which is similar to stomach acid, is used to neutralize 5.0 mL of the liquid antacid. (9.4, 11.6, 11.7)

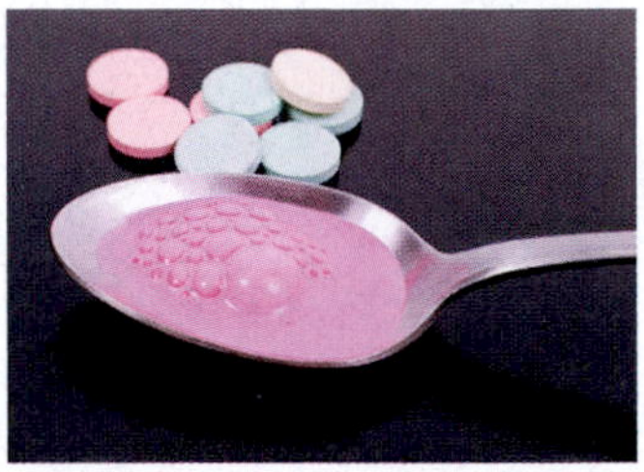

An antacid neutralizes stomach acid and raises the pH.

a. What is the pH of the HCl solution?
b. Write the balanced chemical equation for the neutralization of HCl and $Mg(OH)_2$.
c. Write the balanced chemical equation for the neutralization of HCl and $Al(OH)_3$.
d. How many milliliters of the HCl solution are needed to neutralize the $Mg(OH)_2$?
e. How many milliliters of the HCl solution are needed to neutralize the $Al(OH)_3$?

CI.25 A volume of 200. mL of a carbonic acid buffer for blood plasma is prepared that contains 0.403 g of $NaHCO_3$ and 0.0149 g of H_2CO_3. At body temperature (37 °C), the K_a of carbonic acid is 7.9×10^{-7}. (9.4, 11.4, 11.6, 11.7, 11.8)

$$H_2CO_3(aq) + H_2O(l) \rightleftharpoons HCO_3^-(aq) + H_3O^+(aq)$$

a. What is the $[H_2CO_3]$? **b.** What is the $[HCO_3^-]$?
c. What is the $[H_3O^+]$? **d.** What is the pH of the buffer?
e. Write a balanced chemical equation that shows how this buffer neutralizes added acid.
f. Write a balanced chemical equation that shows how this buffer neutralizes added base.

CI.26 In the kidneys, the ammonia buffer system buffers high H_3O^+. Ammonia, which is produced in renal tubules from amino acids, combines with H^+ to be excreted as NH_4Cl. At body temperature (37 °C), the $K_a = 5.6 \times 10^{-10}$. A buffer solution with a volume of 125 mL contains 3.34 g of NH_4Cl and 0.0151 g of NH_3. (9.4, 11.4, 11.6, 11.7, 11.8)

$$NH_4^+(aq) + H_2O(l) \rightleftharpoons NH_3(aq) + H_3O^+(aq)$$

a. What is the $[NH_4^+]$? **b.** What is the $[NH_3]$?
c. What is the $[H_3O^+]$? **d.** What is the pH of the buffer?
e. Write a balanced chemical equation that shows how this buffer neutralizes added acid.
f. Write a balanced chemical equation that shows how this buffer neutralizes added base.

ANSWERS

CI.19 a. $K_c = 248$

b. If H_2 is added, the equilibrium will shift in the direction of the products.

c. If the volume decreases, the equilibrium will shift in the direction of the products.

d. $[S_2] = 0.280$ M

CI.21 a. $6HCl(aq) + 2M(s) \longrightarrow 3H_2(g) + 2MCl_3(aq)$

b. 233 mL of H_2

c. 6.03×10^{-3} mole of M

d. 69.7 g/mole; gallium

e. $6HCl(aq) + 2Ga(s) \longrightarrow 3H_2(g) + 2GaCl_3(aq)$

CI.23 a. 0.180 M

b. $[H_3O^+] = 5.56 \times 10^{-14}$ M; pH = 13.255

c. $H_2SO_4(aq) + 2KOH(aq) \longrightarrow 2H_2O(l) + K_2SO_4(aq)$

d. 3.60 mL

CI.25 a. 1.20×10^{-3} M

b. 0.0240 M

c. 4.0×10^{-8} M

d. 7.40

e. $HCO_3^-(aq) + H_3O^+(aq) \longrightarrow H_2CO_3(aq) + H_2O(l)$

f. $H_2CO_3(aq) + OH^-(aq) \longrightarrow HCO_3^-(aq) + H_2O(l)$

12

Introduction to Organic Chemistry: Hydrocarbons

At 4:35 A.M., a rescue crew responds to a call about a house fire. At the scene, Jack, a firefighter/emergency medical technician (EMT), finds Diane, a 62-year old woman, lying in the front yard of her house. In his assessment, Jack reports that Diane has second- and third-degree burns over 40% of her body as well as a broken leg. He places an oxygen re-breather mask on Diane to provide a high concentration of oxygen. Another firefighter/EMT, Nancy, begins dressing the burns with sterile water and cling film, a first aid material made of polyvinyl chloride, which does not stick to the skin and is protective. Jack and his crew transport Diane to the burn center for further treatment.

At the scene of the fire, arson investigators use trained dogs to find traces of accelerants and fuel. Gasoline, which is often found at arson scenes, is a mixture of organic molecules called alkanes. Alkanes or hydrocarbons are chains of carbon and hydrogen atoms. The alkanes present in gasoline consist of a mixture of compounds with five to eight carbon atoms in a chain. Alkanes are extremely combustible; they react with oxygen to form carbon dioxide, water, and large amounts of heat. Because alkanes undergo combustion reactions, they can be used to start arson fires.

CAREER

Firefighter/Emergency Medical Technician

Firefighters/emergency medical technicians are first responders to fires, accidents, and other emergency situations. They are required to have an emergency medical technician certification in order to be able to treat seriously injured people. By combining the skills of a firefighter and an emergency medical technician, they increase the survival rates of the injured. The physical demands of firefighters are extremely high as they fight, extinguish, and prevent fires while wearing heavy protective clothing. They also train for and participate in firefighting drills, and maintain fire equipment so that it is always working and ready. Firefighters must also be knowledgeable about fire codes, arson, and the handling and disposal of hazardous materials. Because firefighters also provide emergency care for sick and injured people, they need to be aware of emergency medical and rescue procedures, as well as the proper methods for controlling the spread of infectious disease.

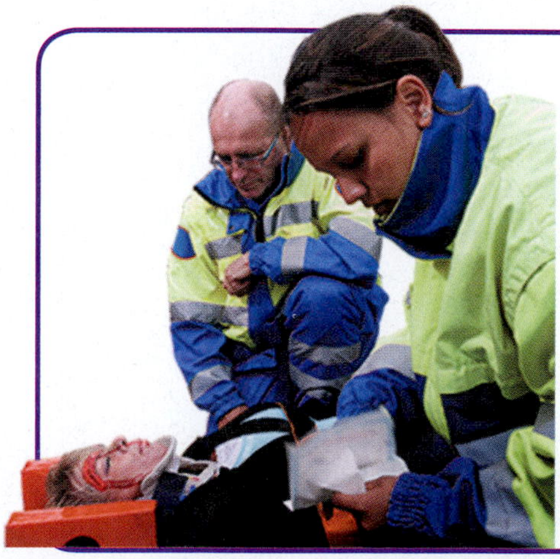

CLINICAL UPDATE

Diane's Treatment in the Burn Unit

When Diane arrives at the hospital, she is diagnosed with second- and third-degree burns. You can see Diane's treatment in the **CLINICAL UPDATE Diane's Treatment in the Burn Unit**, pages 456–457, and see the results of the arson investigation into the house fire.

12.1 Organic Compounds

LEARNING GOAL Identify the properties of organic or inorganic compounds.

At the beginning of the nineteenth century, scientists classified chemical compounds as either inorganic or organic. An inorganic compound was a substance that was composed of minerals, and an organic compound was a substance that came from an organism, thus the use of the word *organic*. Early scientists thought that some type of "vital force," which could only be found in living cells, was required to synthesize an organic compound. This idea was shown to be incorrect in 1828, when German chemist Friedrich Wöhler synthesized urea, a product of protein metabolism, by heating an inorganic compound, ammonium cyanate.

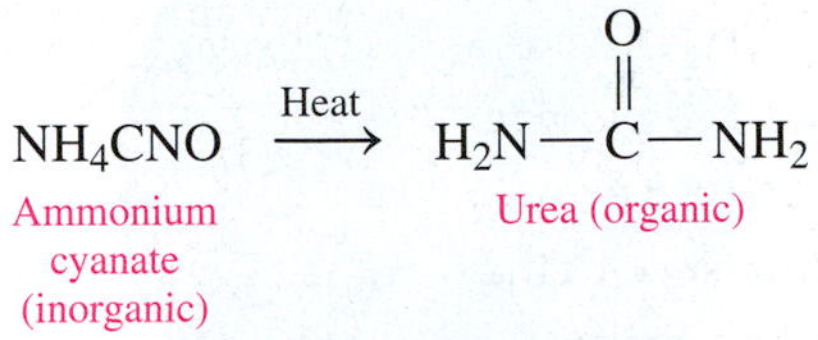
NH_4CNO
Heat
$H_2N-C(=O)-NH_2$
Ammonium cyanate (inorganic)
Urea (organic)

Organic chemistry is the study of carbon compounds. The element carbon has a special role because many carbon atoms can bond together to give a vast array of molecular compounds. **Organic compounds** always contain carbon and hydrogen, and sometimes other nonmetals such as oxygen, sulfur, nitrogen, phosphorus, or a halogen. We find organic compounds in many common products we use every day, such as gasoline, medicines, shampoos, plastics, and perfumes. The food we eat is composed of organic compounds such as carbohydrates, fats, and proteins that supply us with fuel for energy and the carbon atoms needed to build and repair the cells of our bodies.

The formulas of organic compounds are written with carbon first, followed by hydrogen, and then any other elements. Organic compounds typically have low melting and boiling points, are not soluble in water, and are less dense than water. For example, vegetable oil, which is a mixture of organic compounds, does not dissolve in water but floats on top. Many organic compounds undergo combustion and burn vigorously in air. By contrast, many inorganic compounds have high melting and boiling points. Inorganic compounds that are ionic are usually soluble in water, and most do not burn in air. **TABLE 12.1** contrasts some of the properties associated with organic and inorganic compounds, such as propane, C_3H_8, and sodium chloride, NaCl (see **FIGURE 12.1**).

TABLE 12.1 Some Properties of Organic and Inorganic Compounds

Property	Organic	Example: C_3H_8	Inorganic	Example: NaCl
Elements Present	C and H, sometimes O, S, N, P, or Cl (F, Br, I)	C and H	Most metals and nonmetals	Na and Cl
Particles	Molecules	C_3H_8	Mostly ions	Na^+ and Cl^-
Bonding	Mostly covalent	Covalent	Many are ionic, some covalent	Ionic
Polarity of Bonds	Nonpolar unless a strongly electronegative atom is present	Nonpolar	Most are ionic or polar covalent, a few are nonpolar covalent	Ionic
Melting Point	Usually low	−188 °C	Usually high	801 °C
Boiling Point	Usually low	−42 °C	Usually high	1413 °C
Flammability	High	Burns in air	Low	Does not burn
Solubility in Water	Not soluble unless a polar group is present	No	Most are soluble unless nonpolar	Yes

LOOKING AHEAD

REVIEW

Drawing Lewis Structures (6.6)
Predicting Shape (6.8)

Vegetable oil, a mixture of organic compounds, is not soluble in water.

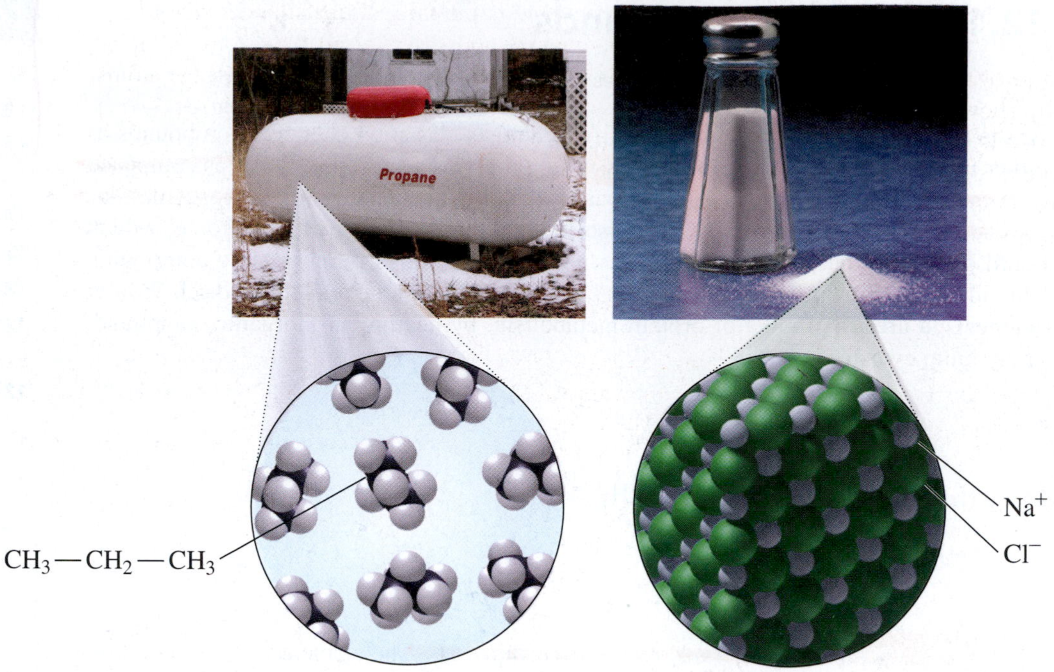
Propane
$CH_3—CH_2—CH_3$
Na^+
Cl^-

FIGURE 12.1 ▶ Propane, C_3H_8, is an organic compound, whereas sodium chloride, NaCl, is an inorganic compound.

Q Why is propane used as a fuel?

▶SAMPLE PROBLEM 12.1 Properties of Organic Compounds

TRY IT FIRST

Indicate whether the following properties are more typical of organic or inorganic compounds:

a. is not soluble in water
b. has a high melting point
c. burns in air

SOLUTION

a. Many organic compounds are not soluble in water.
b. Inorganic compounds are more likely to have high melting points.
c. Organic compounds are more likely to burn in air.

STUDY CHECK 12.1

a. What elements are always found in organic compounds?
b. If butane in a lighter is used to start a fire, is butane an inorganic or organic compound?

ANSWER

a. C and H
b. organic compound

Representations of Carbon Compounds

Hydrocarbons are organic compounds that consist of only carbon and hydrogen. In organic molecules, every carbon atom has four bonds. In the simplest hydrocarbon, methane (CH_4), the carbon atom forms an octet by sharing its four valence electrons with four hydrogen atoms.

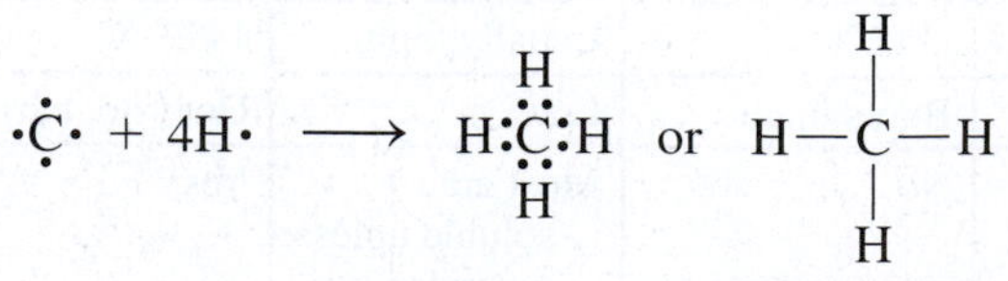

Methane

The most accurate representation of methane is the three-dimensional *space-filling model* **(a)** in which spheres show the relative size and shape of all the atoms. Another type of three-dimensional representation is the *ball-and-stick model* **(b)**, where the atoms are shown as balls and the bonds between them are shown as sticks. In the ball-and-stick model of methane, CH_4, the covalent bonds from the carbon atom to each hydrogen atom are directed to the corners of a tetrahedron with bond angles of 109°. In the *wedge–dash model* **(c)**, the three-dimensional shape is represented by symbols of the atoms with lines for bonds in the plane of the page, wedges for bonds that project out from the page, and dashes for bonds that are behind the page.

However, the three-dimensional models are awkward to draw and view for more complex molecules. Therefore, it is more practical to use their corresponding two-dimensional formulas. The **expanded structural formula (d)** shows all of the atoms and the bonds connected to each atom. A **condensed structural formula (e)** shows the carbon atoms each grouped with the attached number of hydrogen atoms.

ENGAGE

Why does methane have a tetrahedral shape?

Three-Dimensional and Two-Dimensional Representations of Methane

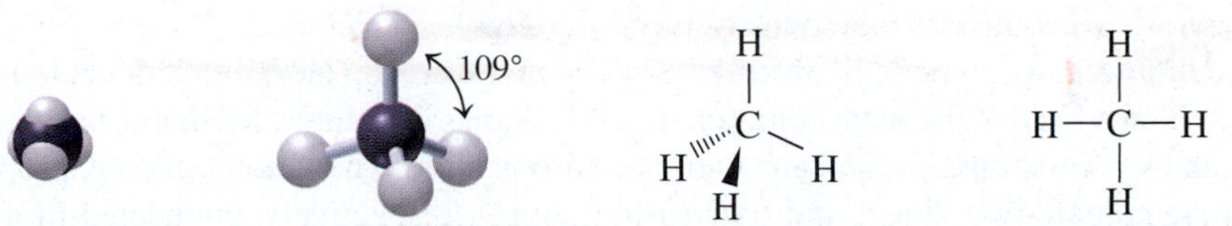
109°

CH_4

(a) Space-filling model **(b)** Ball-and-stick model **(c)** Wedge–dash model **(d)** Expanded structural formula **(e)** Condensed structural formula

The hydrocarbon ethane with two carbon atoms and six hydrogen atoms can be represented by a similar set of three- and two-dimensional models and formulas in which each carbon atom is bonded to another carbon and three hydrogen atoms. As in methane, each carbon atom in ethane retains a tetrahedral shape. A hydrocarbon is referred to as a *saturated hydrocarbon* when all the bonds in the molecule are single bonds.

TEST

Try Practice Problems 12.1 to 12.6

Three-Dimensional and Two-Dimensional Representations of Ethane

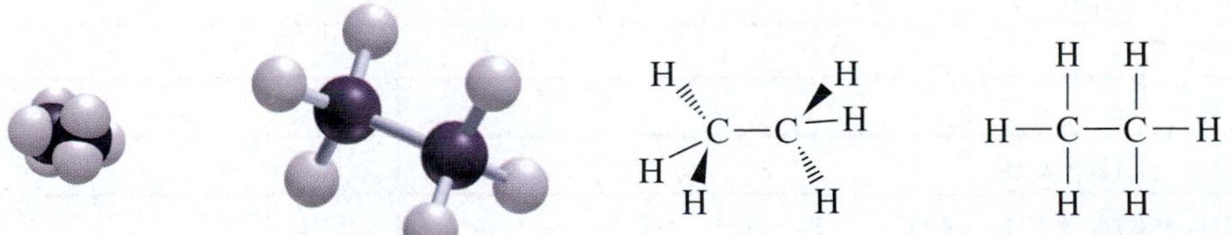

$CH_3—CH_3$

(a) Space-filling model **(b)** Ball-and-stick model **(c)** Wedge–dash model **(d)** Expanded structural formula **(e)** Condensed structural formula

PRACTICE PROBLEMS

12.1 Organic Compounds

12.1 Identify each of the following as a formula of an organic or inorganic compound. For an organic compound, indicate if represented as molecular formula, expanded, or condensed structural formula:

a. KCl **b.** $CH_3—CH_2—CH_2—CH_3$

c. 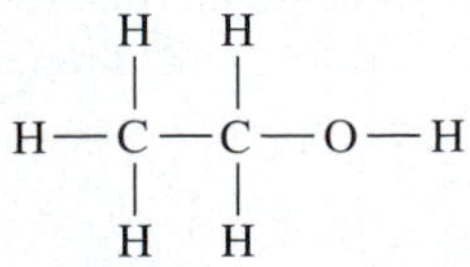 **d.** H_2SO_4

e. $CaCl_2$ **f.** C_3H_7Cl

12.2 Identify each of the following as a formula of an organic or inorganic compound. For an organic compound, indicate if represented as molecular formula, expanded, or condensed structural formula:

a. $C_6H_{12}O_6$ **b.** K_3PO_4

c. I_2 **d.** 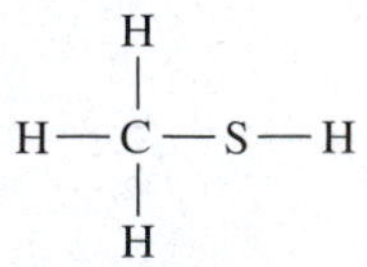

e. $CH_3—CH_2—CH_2—CH_2—CH_3$

f. C_4H_9Br

12.3 Identify each of the following properties as more typical of an organic or inorganic compound:
a. is soluble in water **b.** has a low boiling point
c. contains carbon and hydrogen **d.** contains ionic bonds

12.4 Identify each of the following properties as more typical of an organic or inorganic compound:
a. contains Li and F **b.** is a gas at room temperature
c. contains covalent bonds **d.** produces ions in water

12.5 Match each of the following physical and chemical properties with ethane, C_2H_6, or sodium bromide, NaBr:
a. boils at −89 °C **b.** burns vigorously in air
c. is a solid at 250 °C **d.** dissolves in water

12.6 Match each of the following physical and chemical properties with cyclohexane, C_6H_{12}, or calcium nitrate, $Ca(NO_3)_2$:
a. melts at 500 °C **b.** is insoluble in water
c. does not burn in air **d.** is a liquid at room temperature

12.2 Alkanes

LEARNING GOAL Write the IUPAC names and draw the condensed structural and line-angle formulas for alkanes and cycloalkanes.

More than 90% of the compounds in the world are organic compounds. The large number of carbon compounds is possible because the covalent bond between carbon atoms (C—C) is very strong, allowing carbon atoms to form long, stable chains.

The **alkanes** are a type of hydrocarbon in which the carbon atoms are connected only by single bonds. One of the most common uses of alkanes is as fuels. Methane, used in gas heaters and gas cooktops, is an alkane with one carbon atom. The alkanes ethane, propane, and butane contain two, three, and four carbon atoms, respectively, connected in a row or a *continuous chain*. As we can see, the names for alkanes end in *ane*. Such names are part of the **IUPAC system** (International Union of Pure and Applied Chemistry) used by chemists to name organic compounds. Alkanes with five or more carbon atoms in a chain are named using Greek prefixes: *pent* (5), *hex* (6), *hept* (7), *oct* (8), *non* (9), and *dec* (10) (see **TABLE 12.2**).

TABLE 12.2 IUPAC Names and Formulas of the First 10 Alkanes

Number of Carbon Atoms	IUPAC Name	Molecular Formula	Condensed Structural Formula	Line-Angle Structural Formula
1	Methane	CH_4	CH_4	
2	Ethane	C_2H_6	$CH_3—CH_3$	
3	Propane	C_3H_8	$CH_3—CH_2—CH_3$	
4	Butane	C_4H_{10}	$CH_3—CH_2—CH_2—CH_3$	
5	Pentane	C_5H_{12}	$CH_3—CH_2—CH_2—CH_2—CH_3$	
6	Hexane	C_6H_{14}	$CH_3—CH_2—CH_2—CH_2—CH_2—CH_3$	
7	Heptane	C_7H_{16}	$CH_3—CH_2—CH_2—CH_2—CH_2—CH_2—CH_3$	
8	Octane	C_8H_{18}	$CH_3—CH_2—CH_2—CH_2—CH_2—CH_2—CH_2—CH_3$	
9	Nonane	C_9H_{20}	$CH_3—CH_2—CH_2—CH_2—CH_2—CH_2—CH_2—CH_2—CH_3$	
10	Decane	$C_{10}H_{22}$	$CH_3—CH_2—CH_2—CH_2—CH_2—CH_2—CH_2—CH_2—CH_2—CH_3$	

CORE CHEMISTRY SKILL
Naming and Drawing Alkanes

Condensed Structural and Line-Angle Formulas

In a condensed structural formula, each carbon atom and its attached hydrogen atoms are written as a group. A subscript indicates the number of hydrogen atoms bonded to each carbon atom.

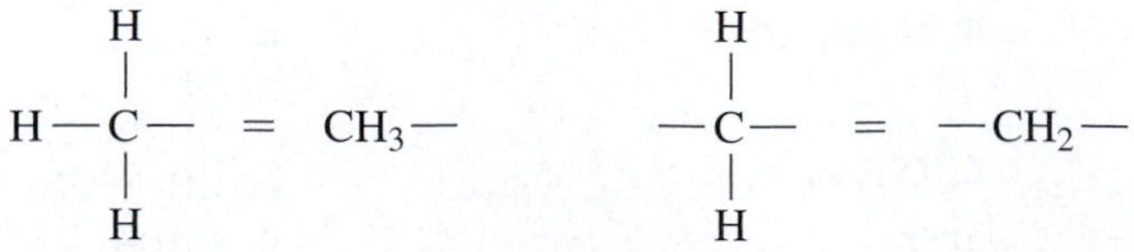
H
H—C—
H
= CH₃—
H
—C—
H
= —CH₂—

Expanded Condensed Expanded Condensed

When an organic molecule consists of a chain of three or more carbon atoms, the carbon atoms do not lie in a straight line. Rather, they are arranged in a zigzag pattern.

A simplified formula called the **line-angle formula** shows a zigzag line in which carbon atoms are represented as the ends of each line and as corners. For example, in the line-angle formula of pentane, each line in the zigzag drawing represents a single bond. The carbon atoms on the ends are bonded to three hydrogen atoms. However, the carbon atoms in the middle of the carbon chain are each bonded to two carbons and two hydrogen atoms as shown in Sample Problem 12.2.

ENGAGE

How does a line-angle formula represent an organic compound of carbon and hydrogen with single bonds?

▶SAMPLE PROBLEM 12.2 Drawing Formulas for an Alkane

TRY IT FIRST

Draw the expanded, condensed structural, and line-angle formulas for pentane.

SOLUTION

ANALYZE THE PROBLEM	Given	Need	Connect
	pentane	expanded, condensed structural, and line-angle formulas	carbon chain, zigzag line

STEP 1 **Draw the carbon chain.** A molecule of pentane has five carbon atoms in a continuous chain.

C—C—C—C—C

STEP 2 **Draw the expanded structural formula by adding the hydrogen atoms using single bonds to each of the carbon atoms.**

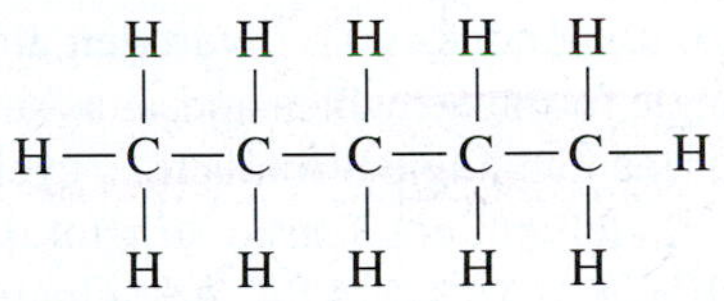

STEP 3 **Draw the condensed structural formula by combining the H atoms with each C atom.**

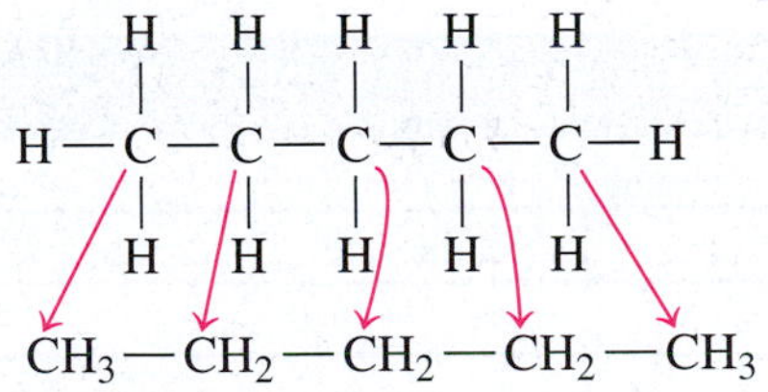

Expanded structural formula

Condensed structural formula

STEP 4 **Draw the line-angle formula as a zigzag line in which the ends and corners represent C atoms.**

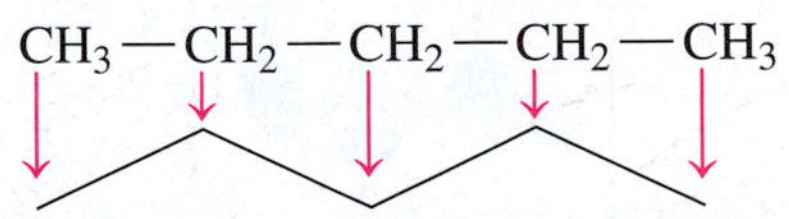

Condensed structural formula

Line-angle formula

STUDY CHECK 12.2

Draw the condensed structural formula and write the name for each of the following line-angle formulas:

a. 

b. 

ANSWER

a. $CH_3—CH_2—CH_2—CH_2—CH_2—CH_2—CH_3$ heptane

b. $CH_3—CH_2—CH_2—CH_3$ butane

2-2-di-methyl, 4-propyl 5-propyl

Because an alkane has only single carbon–carbon bonds, the groups attached to each C are not in fixed positions. They can rotate freely around the bond connecting the carbon atoms. For example, butane can be drawn using a variety of structural formulas as shown in **TABLE 12.3**. All of these formulas represent the same compound with four carbon atoms.

TABLE 12.3 Some Structural Formulas for Butane, C_4H_{10}

Expanded Structural Formula	Line-Angle Formulas
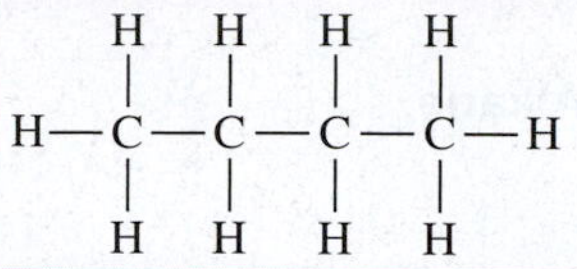	

Condensed Structural Formulas
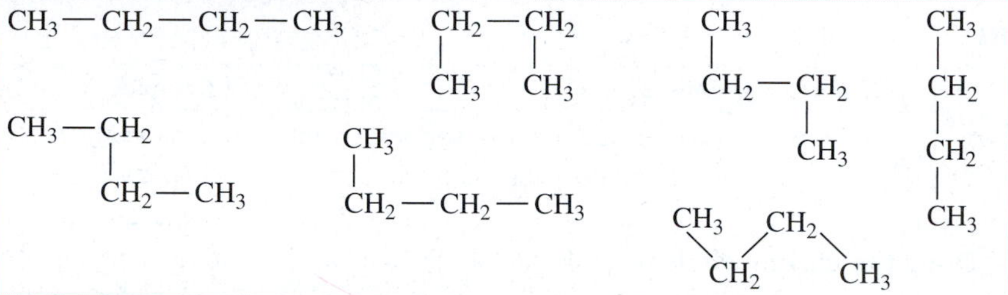

Cycloalkanes

Hydrocarbons can also form cyclic or ring structures called **cycloalkanes**, which have two fewer hydrogen atoms than the corresponding alkanes. The simplest cycloalkane, cyclopropane (C_3H_6), has a ring of three carbon atoms bonded to six hydrogen atoms. Most often cycloalkanes are drawn using their line-angle formulas, which appear as simple geometric figures. As seen for alkanes, each corner of the line-angle formula for a cycloalkane represents a carbon atom. The ball-and-stick model and condensed structural and line-angle formulas for several cycloalkanes are shown in **TABLE 12.4**. A cycloalkane is named by adding the prefix *cyclo* to the name of the alkane with the same number of carbon atoms.

TABLE 12.4 Formulas of Some Common Cycloalkanes

Name	Cyclopropane	Cyclobutane	Cyclopentane	Cyclohexane
Ball-and-Stick Model	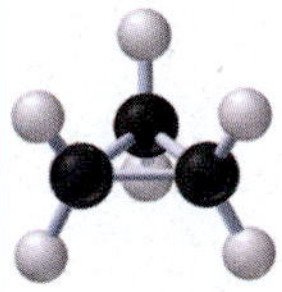	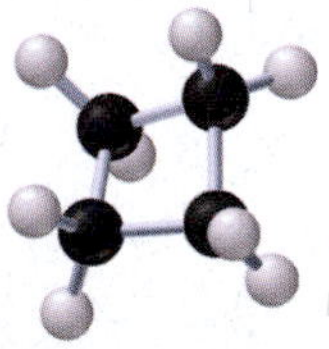	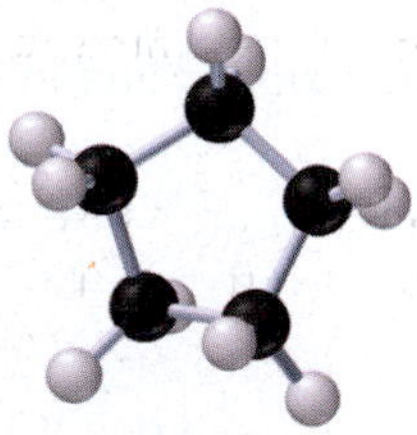	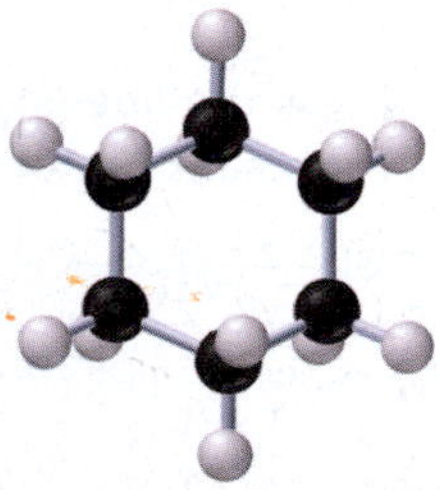
Condensed Structural Formula	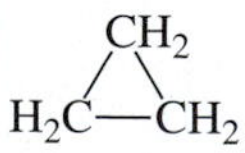	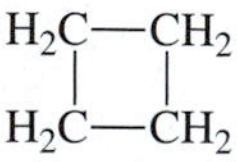	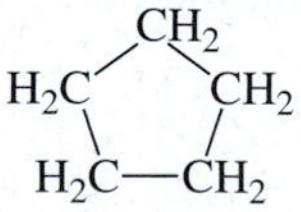	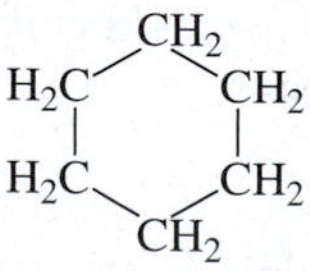
Line-Angle Formula				

▶ SAMPLE PROBLEM 12.3 Naming Alkanes and Cycloalkanes

TRY IT FIRST

Write the IUPAC name for each of the following:

a.  **b.** 

SOLUTION

a. A chain with eight carbon atoms is octane.
b. The ring of six carbon atoms is named cyclohexane.

STUDY CHECK 12.3

Write the IUPAC name for each of the following compounds:

a.  **b.** 

ANSWER

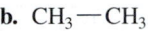

a. cyclobutane **b.** decane

TEST

Try Practice Problems 12.7 to 12.10

PRACTICE PROBLEMS

12.2 Alkanes

12.7 Write the IUPAC name for each of the following alkanes and cycloalkanes:

a. 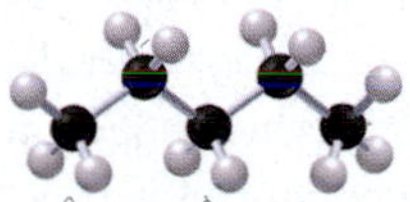
b. $CH_3—CH_3$
c. 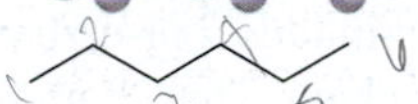
d. 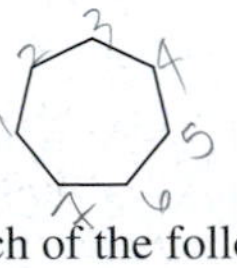

12.8 Write the IUPAC name for each of the following alkanes and cycloalkanes:

a. CH_4
b. 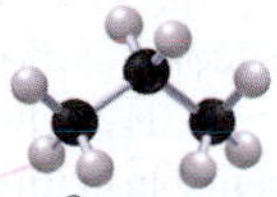
c. 
d. 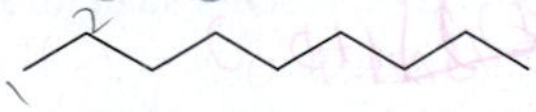

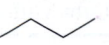

12.9 Draw the condensed structural formula for alkanes or the line-angle formula for cycloalkanes for each of the following:

a. methane **b.** ethane
c. butane **d.** cyclopropane

12.10 Draw the condensed structural formula for alkanes or the line-angle formula for cycloalkanes for each of the following:

a. propane **b.** hexane
c. heptane **d.** cyclohexane

12.3 Alkanes with Substituents

LEARNING GOAL Write the IUPAC names for alkanes with substituents and draw their condensed structural and line-angle formulas.

When an alkane has four or more carbon atoms, the atoms can be arranged so that a side group called a *branch* or **substituent** is attached to a carbon chain. For example, **FIGURE 12.2** shows two different ball-and-stick models for two compounds that have the same molecular

ENGAGE

How can two or more structural isomers have the same molecular formula?

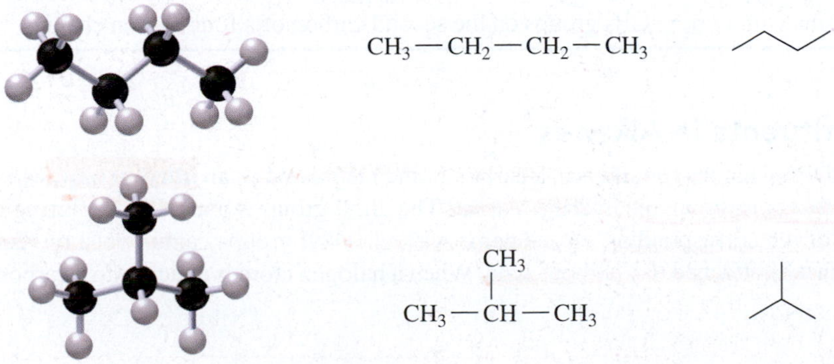
$CH_3—CH_2—CH_2—CH_3$
CH_3
$CH_3—CH—CH_3$

FIGURE 12.2 ▶ The structural isomers of C_4H_{10} have the same number and type of atoms but are bonded in a different order.

Q What makes these molecules structural isomers?

formula, C_4H_{10}. One model is shown as a chain of four carbon atoms. In the other model, a carbon atom is attached as a branch or substituent to a carbon in a chain of three atoms. An alkane with at least one branch is called a *branched alkane*. When the two compounds have the same molecular formula but different arrangements of atoms, they are **structural isomers**.

In another example, we can draw the condensed structural and line-angle formulas for three different isomers with the molecular formula C_5H_{12} as follows:

<table>
<tr><th></th><th colspan="3">Isomers of C_5H_{12}</th></tr>
<tr><td>Condensed Structural Formula</td><td>$CH_3-CH_2-CH_2-CH_2-CH_3$</td><td>CH_3
|
$CH_3-CH-CH_2-CH_3$</td><td>CH_3
|
CH_3-C-CH_3
|
CH_3</td></tr>
<tr><td>Line-Angle Formula</td><td></td><td>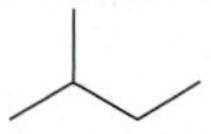</td><td></td></tr>
</table>

▶SAMPLE PROBLEM 12.4 Structural Isomers

TRY IT FIRST

Identify each pair of formulas as structural isomers or the same molecule.

a. CH_3 CH_3
| |
CH_2-CH_2 and $CH_2-CH_2-CH_3$
|
CH_3

b. 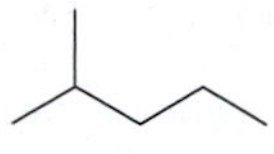 and 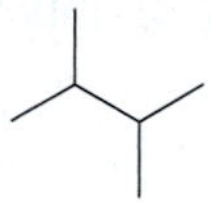

SOLUTION

a. Both have the same molecular formula C_4H_{10}. Both have continuous four-carbon chains even though the $-CH_3$ ends are drawn above or below the chain. Thus, both formulas represent the same molecule.

b. Both have the same molecular formula C_6H_{14}. The formula on the left has a five-carbon chain with a $-CH_3$ substituent on the second carbon of the chain. The formula on the right has a four-carbon chain with two $-CH_3$ substituents. Thus, there is a different order of bonding of atoms, which represents structural isomers.

STUDY CHECK 12.4

Why do each of the following line-angle formulas represent a different structural isomer of the molecules in Sample Problem 12.4, part **b**?

a. 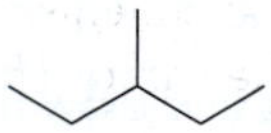

b. 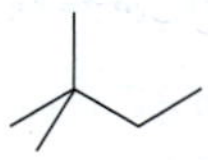

ANSWER

a. This isomer with the molecular formula C_6H_{14} has a different arrangement of carbon atoms with a $-CH_3$ group on the third carbon of a five-carbon chain.

b. This isomer with the molecular formula C_6H_{14} has a different arrangement of carbon atoms with two $-CH_3$ groups on the second carbon of a four-carbon chain.

TEST

Try Practice Problems 12.11 and 12.12

Substituents in Alkanes

In the IUPAC names for alkanes, a carbon branch is named as an **alkyl group**, which is an alkane that is missing one hydrogen atom. The alkyl group is named by replacing the *ane* ending of the corresponding alkane name with *yl*. Alkyl groups cannot exist on their own: They must be attached to a carbon chain. When a halogen atom is attached to a carbon chain, it is named as a *halo* group: *fluoro*, *chloro*, *bromo*, or *iodo*. Some of the common groups attached to carbon chains are illustrated in **TABLE 12.5**.

TABLE 12.5 Formulas and Names of Some Common Substituents

Formula	CH_3-	CH_3-CH_2-		
Name	methyl	ethyl		
Formula	$CH_3-CH_2-CH_2-$	$CH_3-\underset{\vert}{CH}-CH_3$		
Name	propyl	isopropyl		
Formula	$CH_3-CH_2-CH_2-CH_2-$	$CH_3-\underset{CH_3}{\underset{\vert}{CH}}-CH_2-$	$CH_3-\underset{\vert}{CH}-CH_2-CH_3$	$CH_3-\overset{CH_3}{\overset{\vert}{\underset{\vert}{C}}}-CH_3$
Name	butyl	isobutyl	*sec*-butyl	*tert*-butyl
Formula	$F-$ $Cl-$	$Br-$ $I-$		
Name	fluoro chloro	bromo iodo		

Naming Alkanes with Substituents

In the IUPAC system of naming, a carbon chain is numbered to give the location of the substituents. Let's take a look at how we use the IUPAC system to name the alkane shown in Sample Problem 12.5.

INTERACTIVE VIDEO
PEARSON eText 2.0 Naming Alkanes

▶ SAMPLE PROBLEM 12.5 Writing IUPAC Names for Alkanes with Substituents

TRY IT FIRST

Write the IUPAC name for the following alkane:

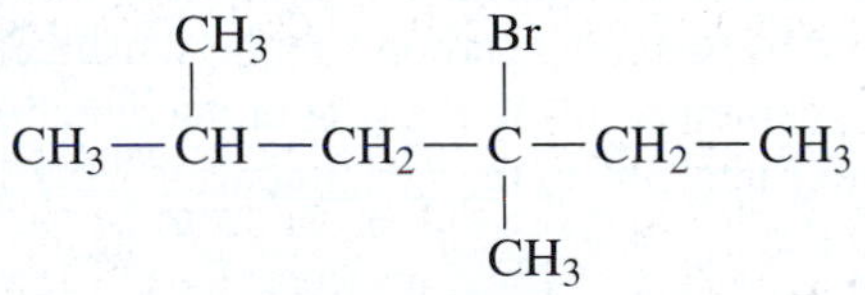

SOLUTION

	Given	Need	Connect
ANALYZE THE PROBLEM	six-carbon chain, two methyl groups, one bromo group	IUPAC name	position of substituents on the carbon chain

STEP 1 **Write the alkane name for the longest chain of carbon atoms.**

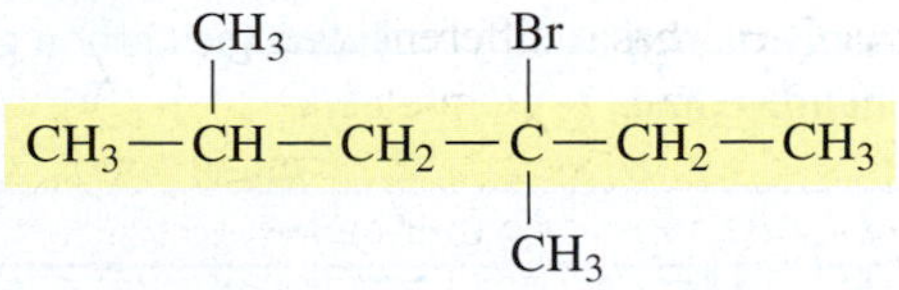

hexane

STEP 2 **Number the carbon atoms from the end nearer a substituent.**

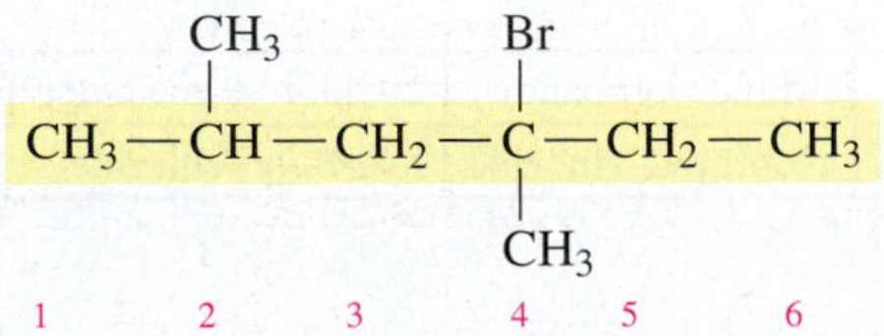

hexane

STEP 3 **Give the location and name for each substituent (alphabetical order) as a prefix to the name of the main chain.** The substituents are listed in alphabetical order (bromo first, then methyl). A hyphen is placed between the number and the substituent name. When there are two or more of the same substituent, a prefix (*di*, *tri*, *tetra*) is used and commas separate the numbers. However, prefixes are not used to determine the alphabetical order of the substituents.

ENGAGE

What part of the IUPAC name gives (a) the number of carbon atoms in the carbon chain and (b) the substituents on the carbon chain?

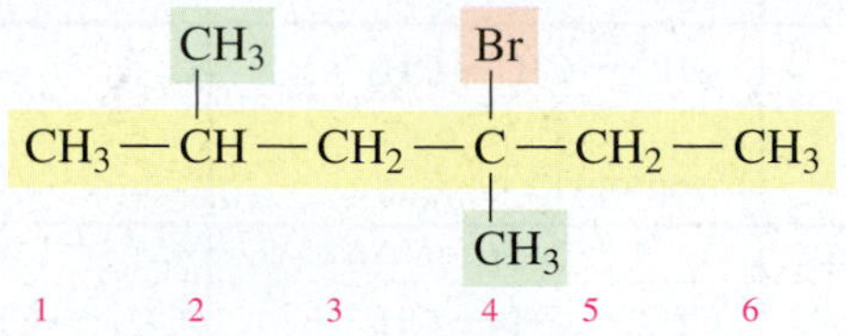

4-bromo-2,4-dimethylhexane

STUDY CHECK 12.5

Write the IUPAC name for each of the following compounds:

a. 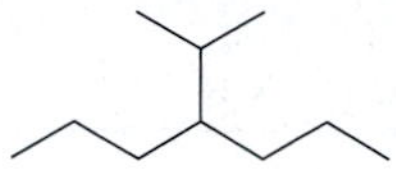

b. 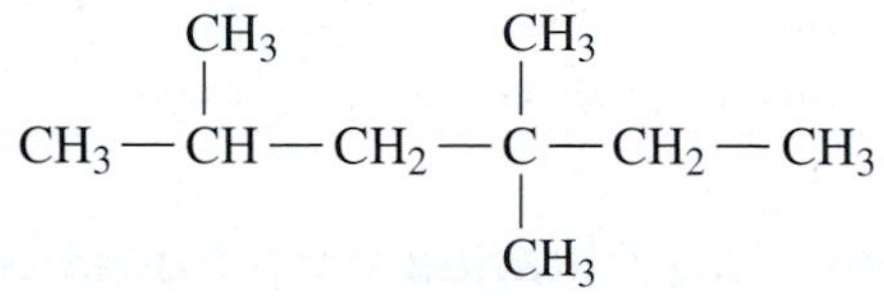

TEST

Try Practice Problems 12.13 and 12.14

ANSWER

a. 4-isopropylheptane

b. 2,4,4-trimethylhexane

Naming Cycloalkanes with Substituents

When one substituent is attached to a carbon atom in a cycloalkane, the name of the substituent is placed in front of the cycloalkane name. No number is needed for a single alkyl group or halogen atom because the carbon atoms in the cycloalkane are equivalent. However, if two or more substituents are attached, the ring is numbered by assigning carbon 1 to the substituent that comes first alphabetically. Then we count the carbon atoms in the ring in the direction (clockwise or counterclockwise) that gives the lower numbers to the substituents.

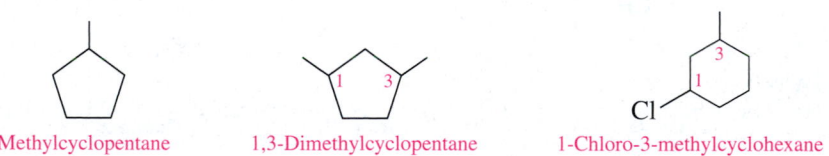

Methylcyclopentane

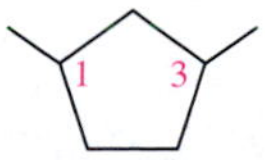

1,3-Dimethylcyclopentane

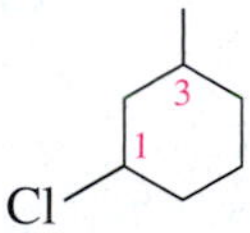

1-Chloro-3-methylcyclohexane

Haloalkanes

In a *haloalkane*, halogen atoms replace hydrogen atoms in an alkane. The halo substituents are numbered and arranged alphabetically, just as we did with the alkyl groups. Many times chemists use the common, traditional name for these compounds rather than the systematic IUPAC name. Simple haloalkanes are commonly named as alkyl halides; the carbon group is named as an alkyl group followed by the halide name.

Examples of Haloalkanes

Formula	$CH_3—Cl$	$CH_3—CH_2—Br$	$CH_3—CH(F)—CH_3$	$CH_3—C(Cl)(CH_3)—CH_3$
IUPAC	Chloromethane	Bromoethane	2-Fluoropropane	2-Chloro-2-methylpropane
Common	Methyl chloride	Ethyl bromide	Isopropyl fluoride	*tert*-Butyl chloride

Drawing Structural Formulas for Alkanes with Substituents

The IUPAC name gives all the information needed to draw the condensed structural formula for an alkane. Suppose you are asked to draw the condensed structural formula for 2,3-dimethylbutane. The alkane name gives the number of carbon atoms in the longest chain. The other names indicate the substituents and where they are attached. We can break down the name in the following way:

2,3-Dimethylbutane				
2,3-	**Di**	**methyl**	**but**	**ane**
Substituents on carbons 2 and 3	two identical groups	$-CH_3$ alkyl groups	four C atoms in the main chain	single (C—C) bonds

▶SAMPLE PROBLEM 12.6 Drawing Condensed Structural and Line-Angle Formulas from IUPAC Names

TRY IT FIRST

Draw the condensed structural and line-angle formulas for 2,3-dimethylbutane.

SOLUTION

	Given	Need	Connect
ANALYZE THE PROBLEM	2,3-dimethylbutane	condensed structural and line-angle formulas	four-carbon chain, two methyl groups

STEP 1 **Draw the main chain of carbon atoms.** For butane, we draw a chain and a zigzag line of four carbon atoms.

C—C—C—C

STEP 2 **Number the chain and place the substituents on the carbons indicated by the numbers.** The first part of the name indicates two methyl groups $-CH_3$: one on carbon 2 and one on carbon 3.

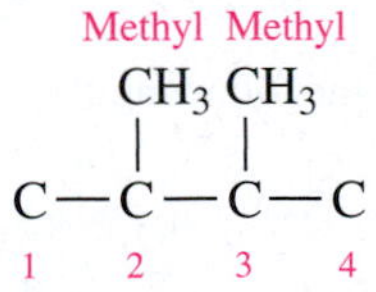
Methyl Methyl
CH3 CH3
C—C—C—C
1 2 3 4

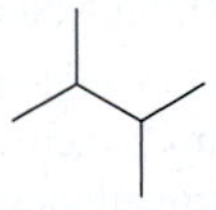
1 2 3 4

STEP 3 **For the condensed structural formula, add the correct number of hydrogen atoms to give four bonds to each C atom.**

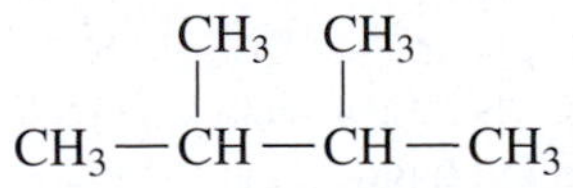

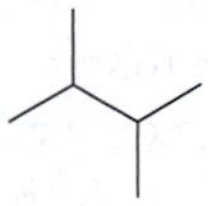

STUDY CHECK 12.6

Draw the condensed structural and line-angle formulas for each of the following:

a. 2-bromo-4-methylpentane
b. 2-chloro-2-methylpentane

ANSWER

a. 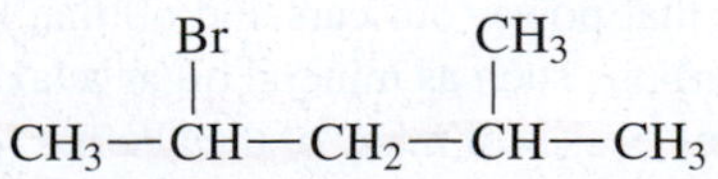

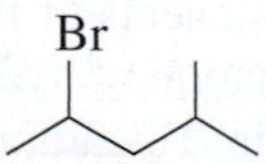

b. 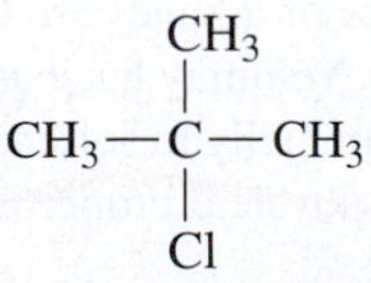

TEST

Try Practice Problems 12.15 to 12.18

PRACTICE PROBLEMS

12.3 Alkanes with Substituents

12.11 Indicate whether each of the following pairs represent structural isomers or the same molecule:

a. 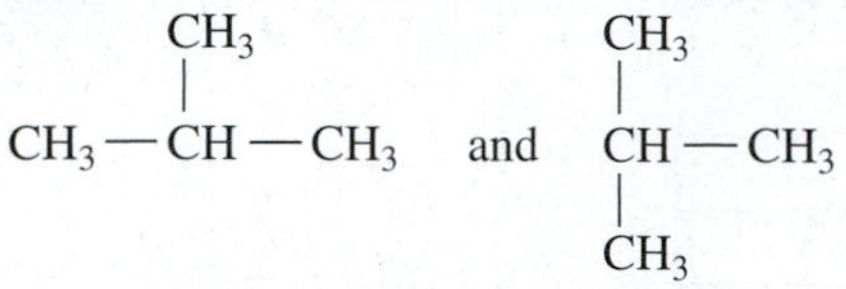

b. 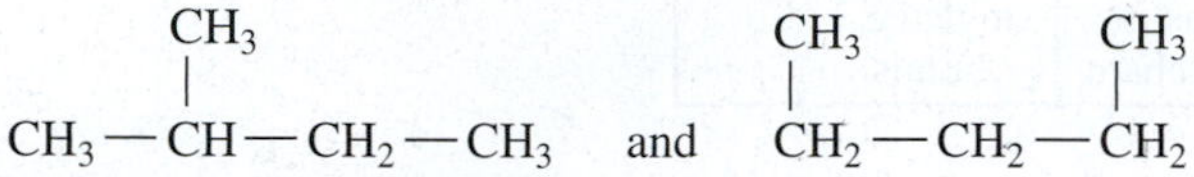

c. 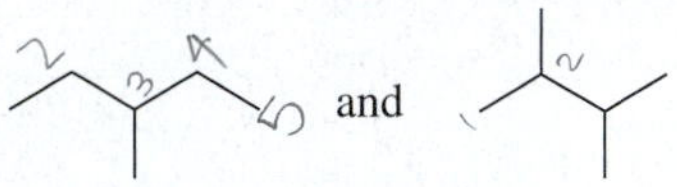

12.12 Indicate whether each of the following pairs represent structural isomers or the same molecule:

a. 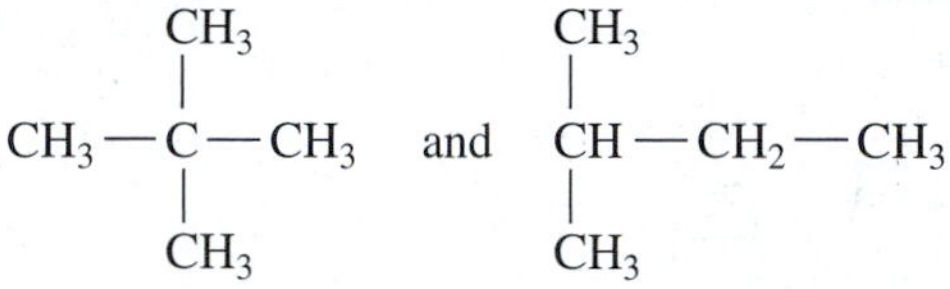

b. 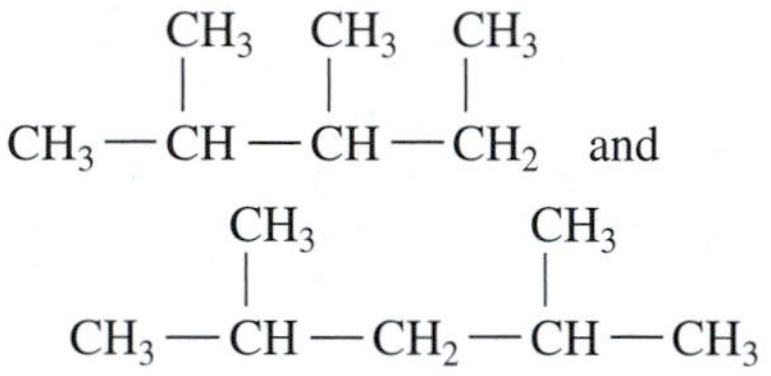

c. 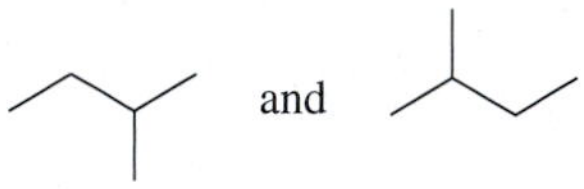

12.13 Write the IUPAC name for each of the following:

a. 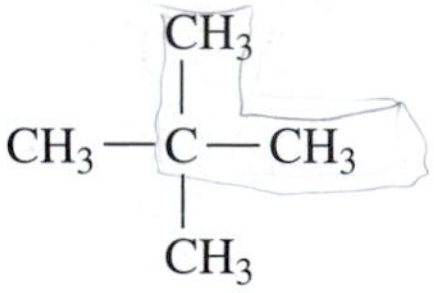

b. 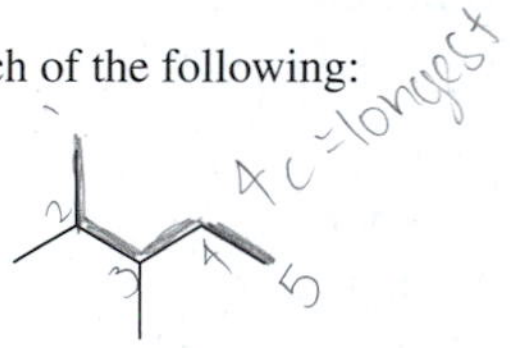

c. 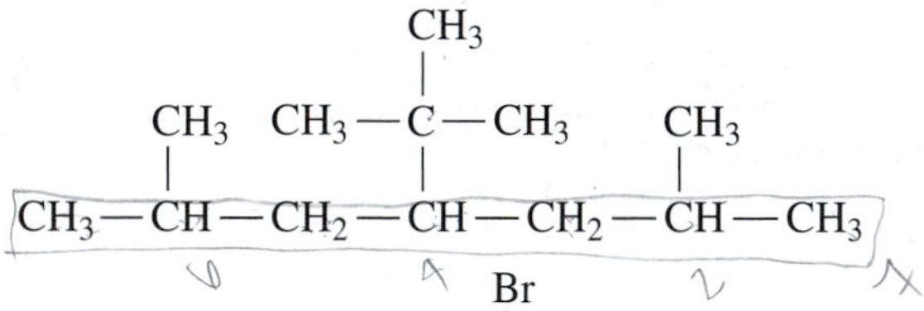

d. 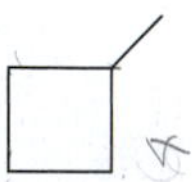

e. 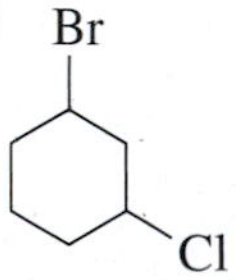

12.14 Write the IUPAC name for each of the following:

a. 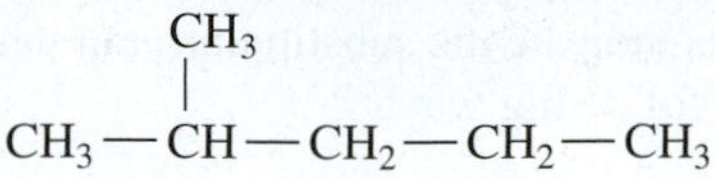

b. 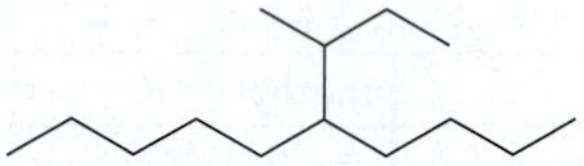

c. 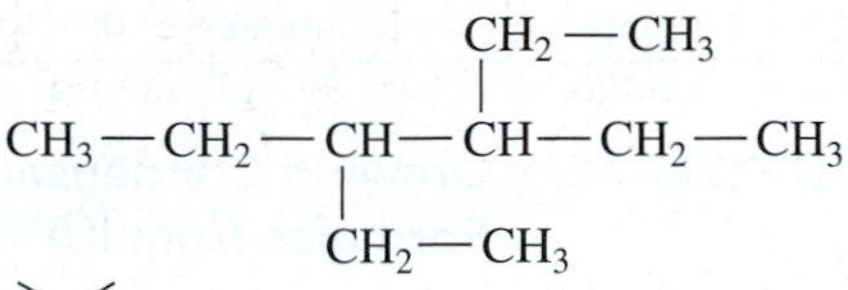

d. 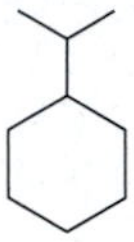

e. 

12.15 Draw the condensed structural formula for each of the following alkanes:

a. 3,3-dimethylpentane
b. 2,3,5-trimethylhexane
c. 3-ethyl-2,5-dimethyloctane
d. 1-bromo-2-chloroethane

12.16 Draw the condensed structural formula for each of the following alkanes:

a. 3-ethylpentane
b. 4-isopropyl-3-methylheptane
c. 4-ethyl-2,2-dimethyloctane
d. 2-bromopropane

12.17 Draw the line-angle formula for each of the following:

a. 3-methylheptane
b. 1-chloro-3-ethylcyclopentane
c. bromocyclobutane
d. 2,3-dichlorohexane

12.18 Draw the line-angle formula for each of the following:

a. 1-bromo-2-methylpentane
b. 1,2,3-trimethylcyclopropane
c. ethylcyclohexane
d. 4-chlorooctane

12.4 Properties of Alkanes

LEARNING GOAL Identify the properties of alkanes and write a balanced chemical equation for combustion.

Many types of alkanes are the components of fuels that power our cars and oil that heats our homes. You may have used a mixture of hydrocarbons such as mineral oil as a laxative or petrolatum jelly to soften your skin. The differences in uses of many of the alkanes result from their physical properties, including solubility and density.

Some Uses of Alkanes

The first four alkanes—methane, ethane, propane, and butane—are gases at room temperature and are widely used as heating fuels.

Alkanes having five to eight carbon atoms (pentane, hexane, heptane, and octane) are liquids at room temperature. They are highly volatile, which makes them useful in fuels such as gasoline.

Liquid alkanes with 9 to 17 carbon atoms have higher boiling points and are found in kerosene, diesel, and jet fuels. Motor oil is a mixture of high-molecular-weight liquid hydrocarbons and is used to lubricate the internal components of engines. Mineral oil is a mixture of liquid hydrocarbons and is used as a laxative and a lubricant. Alkanes with 18 or more carbon atoms are waxy solids at room temperature. Known as paraffins, they are used in waxy coatings added to fruits and vegetables to retain moisture, inhibit mold, and enhance appearance. Petrolatum jelly, or Vaseline, is a semisolid mixture of hydrocarbons with more than 25 carbon atoms used in ointments and cosmetics and as a lubricant.

The solid alkanes that make up waxy coatings on fruits and vegetables help retain moisture, inhibit mold, and enhance appearance.

Melting and Boiling Points

Alkanes have the lowest melting and boiling points of all the organic compounds. This occurs because alkanes contain only the nonpolar bonds of C—C and C—H. Therefore, the attractions that occur between alkane molecules in the solid and liquid states are because of the relatively weak dispersion forces. As the number of carbon atoms increases, there is also an increase in the number of electrons, which increases the attraction because of the dispersion forces. Thus, alkanes with higher masses have higher melting and boiling points.

CH_4	$CH_3—CH_3$	$CH_3—CH_2—CH_3$	$CH_3—CH_2—CH_2—CH_3$	$CH_3—CH_2—CH_2—CH_2—CH_3$
Methane	Ethane	Propane	Butane	Pentane
bp = −164 °C	bp = −89 °C	bp = −42 °C	bp = 0.5 °C	bp = 36 °C

Number of Carbon Atoms Increases
Boiling Point Increases

The boiling points of branched alkanes are generally lower than the straight-chain isomers. The branched-chain alkanes tend to be more compact, which reduces the points of contact between the molecules. In an analogy, we can think of the carbon chains of straight-chain alkanes as pieces of licorice in a package. Because they have linear shapes, they can line up very close to each other, which gives many points of contact between the surface of the molecules. In our analogy, we can also think of branched alkanes as tennis balls in a can that, because of their spherical shapes, have only a small area of contact. One tennis ball represents an entire molecule. Because branched alkanes have fewer attractions, they have lower melting and boiling points.

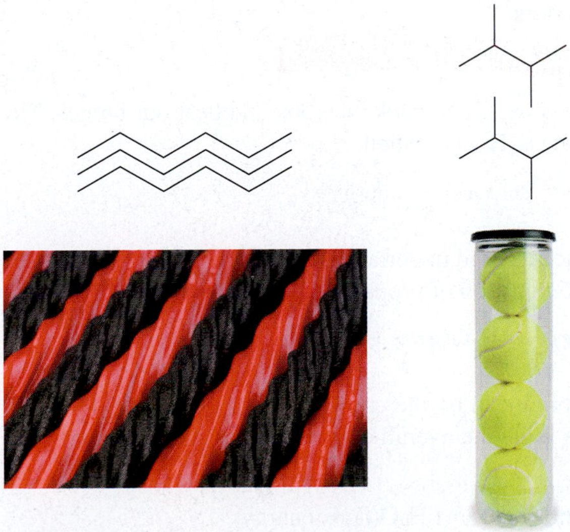

There is more contact between the surfaces of hexane molecules (red and black licorice) than between the surfaces of 2,3-dimethylbutane molecules (tennis balls).

ENGAGE

Why does pentane have a higher boiling point than dimethylpropane?

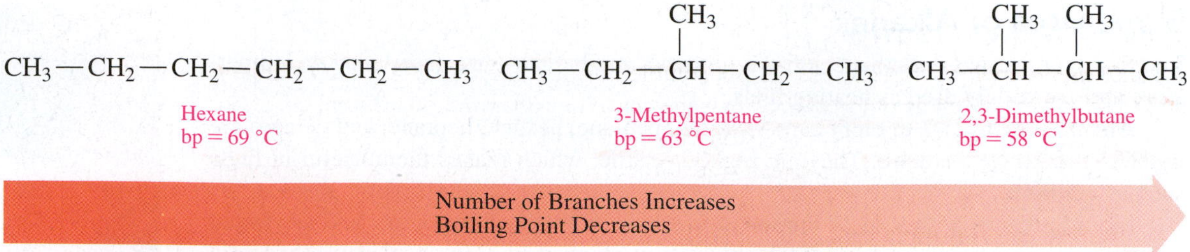
Hexane
bp = 69 °C
3-Methylpentane
bp = 63 °C
2,3-Dimethylbutane
bp = 58 °C
Number of Branches Increases
Boiling Point Decreases

Cycloalkanes have higher boiling points than the straight-chain alkanes with the same number of carbon atoms. Because rotation of carbon bonds is restricted, cycloalkanes maintain a rigid structure. Cycloalkanes with their rigid structures can be stacked closely together, which gives them many points of contact and therefore many attractions to each other.

TEST
Try Practice Problems 12.19 and 12.20

We can compare the boiling points of straight-chain alkanes, branched-chain alkanes, and cycloalkanes with five carbon atoms as shown in **TABLE 12.6**.

TABLE 12.6 Comparison of Boiling Points of Alkanes and Cycloalkanes with Five Carbons

Formula	Name	Boiling Point (°C)
Straight-Chain Alkane		
$CH_3-CH_2-CH_2-CH_2-CH_3$	Pentane	36
Branched-Chain Alkanes		
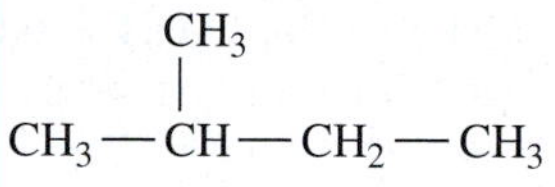	2-Methylbutane	28
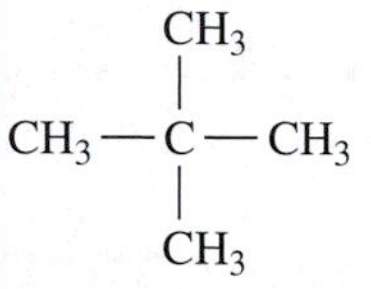	Dimethylpropane	10
Cycloalkane		
	Cyclopentane	49

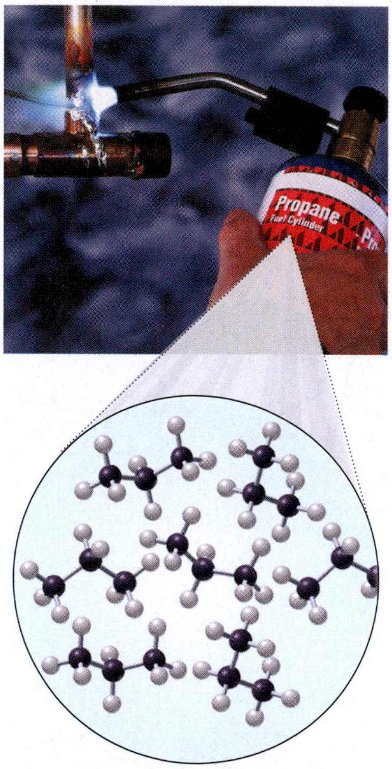
Propane
Fuel Cylinder

FIGURE 12.3 ▶ The propane fuel in the tank undergoes combustion, which provides energy.

Q What is the balanced equation for the combustion of propane?

Combustion of Alkanes

The carbon–carbon single bonds in alkanes are difficult to break, which makes them the least reactive family of organic compounds. However, alkanes burn readily in oxygen to produce carbon dioxide, water, and energy.

$$\text{Alkane}(g) + O_2(g) \xrightarrow{\Delta} CO_2(g) + H_2O(g) + \text{energy}$$

For example, methane is the natural gas we use to cook our food and heat our homes. The equation for the combustion of methane (CH_4) is written:

$$\underset{\text{Methane}}{CH_4(g)} + 2O_2(g) \xrightarrow{\Delta} CO_2(g) + 2H_2O(g) + \text{energy}$$

In another example, propane is the gas used in portable heaters and gas barbecues (see **FIGURE 12.3**). The equation for the combustion of propane (C_3H_8) is written:

$$\underset{\text{Propane}}{C_3H_8(g)} + 5O_2(g) \xrightarrow{\Delta} 3CO_2(g) + 4H_2O(g) + \text{energy}$$

In the cells of our bodies, energy is produced by the combustion of glucose. Although a series of reactions is involved, we can write the overall combustion of glucose in our cells as follows:

$$\underset{\text{Glucose}}{C_6H_{12}O_6(aq)} + 6O_2(g) \xrightarrow{\text{Enzymes}} 6CO_2(g) + 6H_2O(l) + \text{energy}$$

TEST
Try Practice Problems 12.21 to 12.24

Solubility and Density

Alkanes are nonpolar, which makes them insoluble in water. However, they are soluble in nonpolar solvents such as other alkanes. Alkanes have densities from 0.62 g/mL to about 0.79 g/mL, which is less than the density of water (1.0 g/mL).

If there is an oil spill in the ocean, the alkanes in the oil, which do not mix with water, form a thin layer on the surface that spreads over a large area. In April 2010, an explosion on an oil-drilling rig in the Gulf of Mexico caused the largest oil spill in U.S. history. At its maximum, an estimated 10 million liters of oil was leaked every day (see **FIGURE 12.4**). If the crude oil reaches land, there can be considerable damage to beaches, shellfish, birds, and wildlife habitats. When animals such as birds are covered with oil, they must be cleaned quickly because ingestion of the hydrocarbons when they try to clean themselves is fatal.

FIGURE 12.4 ▶ In oil spills, large quantities of oil spread out to form a thin layer on top of the ocean surface.

Q What physical properties cause oil to remain on the surface of water?

PRACTICE PROBLEMS

12.4 Properties of Alkanes

12.19 In each of the following pairs of hydrocarbons, which one would you expect to have the higher boiling point?

a. pentane or heptane
b. propane or cyclopropane
c. hexane or 2-methylpentane

12.20 In each of the following pairs of hydrocarbons, which one would you expect to have the higher boiling point?

a. propane or butane
b. hexane or cyclohexane
c. 2,2-dimethylpentane or heptane

12.21 Heptane, used as a solvent for rubber cement, has a density of 0.68 g/mL and boils at 98 °C.

a. Draw the condensed structural and line-angle formulas for heptane.
b. Is heptane a solid, liquid, or gas at room temperature?
c. Is heptane soluble in water?
d. Will heptane float on water or sink?
e. Write the balanced chemical equation for the complete combustion of heptane.

12.22 Nonane has a density of 0.79 g/mL and boils at 151 °C.

a. Draw the condensed structural and line-angle formulas for nonane.
b. Is nonane a solid, liquid, or gas at room temperature?
c. Is nonane soluble in water?
d. Will nonane float on water or sink?
e. Write the balanced chemical equation for the complete combustion of nonane.

12.23 Write the balanced chemical equation for the complete combustion of each of the following compounds:

a. ethane
b. cyclopropane
c. 2,3-dimethylhexane

12.24 Write the balanced chemical equation for the complete combustion of each of the following compounds:

a. butane
b. cyclopentane
c. 2,3-dimethylbutane

Butane in a portable burner undergoes combustion.

12.5 Alkenes and Alkynes

LEARNING GOAL Write the IUPAC names and draw the condensed structural and line-angle formulas for alkenes and alkynes.

We organize organic compounds into *classes* or *families* by their *functional groups*, which are groups of specific atoms. Compounds that contain the same functional group have similar physical and chemical properties. Identifying functional groups allows us to classify organic compounds according to their structure, to name compounds, predict their chemical reactions, and draw the structures for their products.

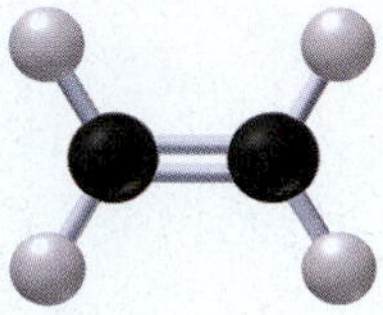

H   H
 C=C
H   H

Bond angles = 120°
Ethene

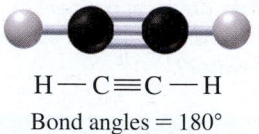

H—C≡C—H

Bond angles = 180°
Ethyne

FIGURE 12.5 ▶ Ball-and-stick models of ethene and ethyne show the double or triple bonds and the bond angles.

Q Why are these compounds called unsaturated hydrocarbons?

Alkenes and *alkynes* are families of hydrocarbons that contain double and triple bonds, respectively. They are called *unsaturated hydrocarbons* because they do not contain the maximum number of hydrogen atoms, as do alkanes. They react with hydrogen gas to increase the number of hydrogen atoms to become alkanes, which are *saturated hydrocarbons*.

Identifying Alkenes and Alkynes

Alkenes contain one or more carbon–carbon double bonds that form when adjacent carbon atoms share two pairs of valence electrons. Recall that *a carbon atom always forms four covalent bonds*. In the simplest alkene, ethene (C_2H_4), two carbon atoms are connected by a double bond and each is also attached to two H atoms. This gives each carbon atom in the double bond a trigonal planar arrangement with bond angles of 120°. As a result, the ethene molecule is flat because the carbon and hydrogen atoms all lie in the same plane (see **FIGURE 12.5**).

In an **alkyne**, a triple bond forms when two carbon atoms share three pairs of valence electrons. In the simplest alkyne, ethyne (C_2H_2), the two carbon atoms of the triple bond are each attached to one hydrogen atom, which gives a triple bond a linear geometry.

Ethene, more commonly called ethylene, is an important plant hormone involved in promoting the ripening of fruit. Commercially grown fruit, such as avocados, bananas, and tomatoes, are often picked before they are ripe. Before the fruit is brought to market, it is exposed to ethylene to accelerate the ripening process. Ethylene also accelerates the breakdown of cellulose in plants, which causes flowers to wilt and leaves to fall from trees. Ethyne, commonly called acetylene, is used in welding, where it reacts with oxygen to produce flames with temperatures above 3300 °C.

Fruit is ripened with ethene, a plant hormone.

Naming Alkenes and Alkynes

The IUPAC names for alkenes and alkynes are similar to those of alkanes. Using the alkane name with the same number of carbon atoms, the *ane* ending is replaced with *ene* for an alkene and *yne* for an alkyne (see **TABLE 12.7**). Cyclic alkenes are named as *cycloalkenes*.

TABLE 12.7 Comparison of Names for Alkanes, Alkenes, and Alkynes

Alkane	Alkene	Alkyne
$CH_3—CH_3$	$H_2C═CH_2$	HC≡CH
Ethane	Ethene (ethylene)	Ethyne (acetylene)
$CH_3—CH_2—CH_2$	$CH_3—CH═CH_2$	$CH_3—C≡CH$
	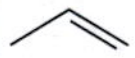	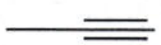
Propane	Propene	Propyne

An example of naming an alkene is seen in Sample Problem 12.7.

▶SAMPLE PROBLEM 12.7 Naming Alkenes and Alkynes

TRY IT FIRST

Write the IUPAC name for the following:

```
       CH3
       |
CH3—CH—CH═CH—CH3
```

SOLUTION

	Given	Need	Connect
ANALYZE THE PROBLEM	five-carbon chain, double bond, methyl group	IUPAC name	replace the *ane* of the alkane name with *ene*

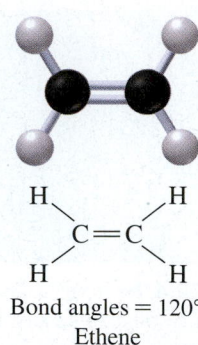

STEP 1 **Name the longest carbon chain that contains the double bond.** There are five carbon atoms in the longest carbon chain containing the double bond. Replace the *ane* in the corresponding alkane name with *ene* to give pentene.

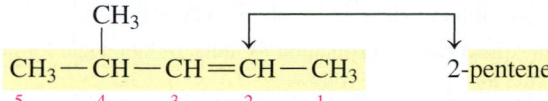

$$\begin{array}{l} \quad\; CH_3 \\ \quad\;\; | \\ CH_3-CH-CH=CH-CH_3 \end{array} \qquad \text{pentene}$$

STEP 2 **Number the carbon chain starting from the end nearer the double bond.** Place the number of the first carbon in the double bond in front of the alkene name.

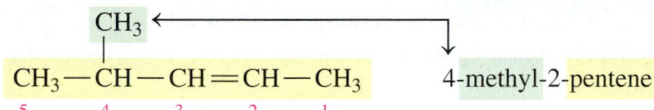

$$\begin{array}{l} \quad\; CH_3 \\ \quad\;\; | \\ CH_3-CH-CH=CH-CH_3 \\ \;5 \qquad\; 4 \qquad 3 \qquad 2 \qquad 1 \end{array} \qquad \text{2-pentene}$$

Alkenes with two or three carbons do not need numbers.

STEP 3 **Give the location and name for each substituent (alphabetical order) as a prefix to the alkene name.**

$$\begin{array}{l} \quad\; CH_3 \\ \quad\;\; | \\ CH_3-CH-CH=CH-CH_3 \\ \;5 \qquad\; 4 \qquad 3 \qquad 2 \qquad 1 \end{array} \qquad \text{4-methyl-2-pentene}$$

ENGAGE

Why is the name 2-methyl-3-pentene incorrect for this compound?

STUDY CHECK 12.7

Draw the condensed structural formula for each of the following:

a. 1-chloro-3-hexyne
b. 1-bromo-1-pentene

ANSWER

a. $Cl-CH_2-CH_2-C\equiv C-CH_2-CH_3$ **b.** $Br-CH=CH-CH_2-CH_2-CH_3$

Naming Cycloalkenes

Some alkenes called *cycloalkenes* have a double bond within a ring structure. If there is no substituent, the double bond does not need a number. If there is a substituent, the carbons in the double bond are numbered as 1 and 2, and the ring is numbered in the direction that will give the lower number to the substituent.

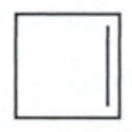

Cyclobutene

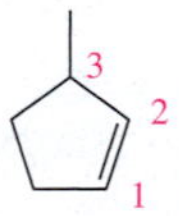
3
2
1

3-Methylcyclopentene

TEST

Try Practice Problems 12.25 to 12.30

PRACTICE PROBLEMS

12.5 Alkenes and Alkynes

12.25 Identify the following as alkanes, alkenes, cycloalkenes, or alkynes:

a. 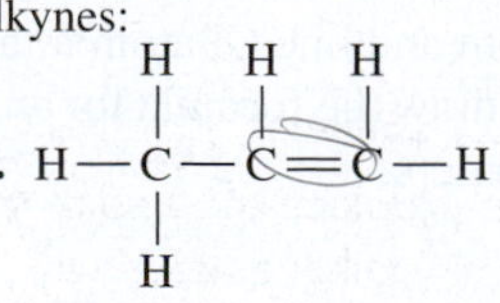
H H H
H—C—C=C—H
H

b. $CH_3-CH_2-C\equiv CH$

c. 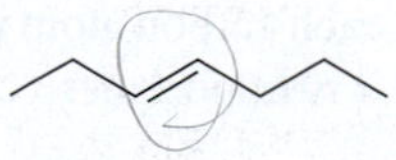

d. 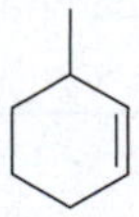

12.26 Identify the following as alkanes, alkenes, cycloalkenes, or alkynes:

a. 

b. 

c. 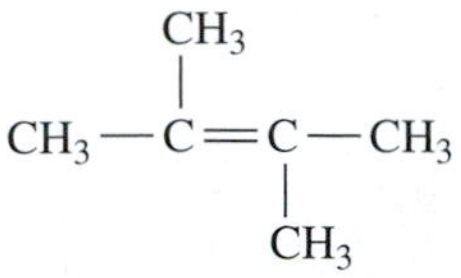

d. 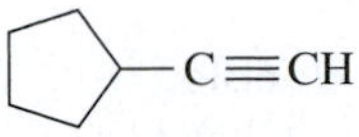

12.27 Write the IUPAC name for each of the following:

a. 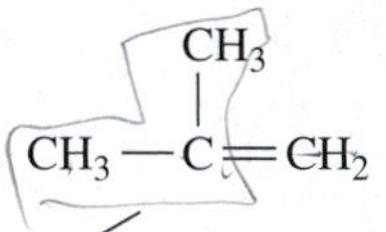

b. 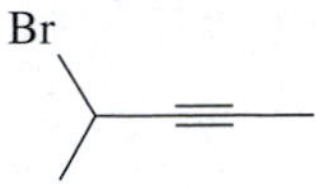

c. 

d. 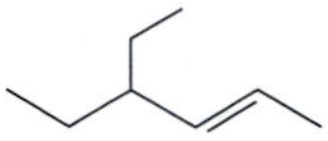

12.28 Write the IUPAC name for each of the following:

a. $H_2C{=}CH{-}CH_2{-}CH_2{-}CH_2{-}CH_3$

b. 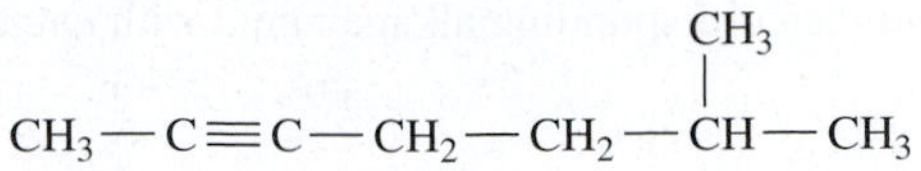

c. 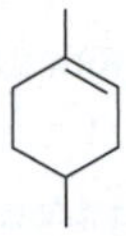

d. 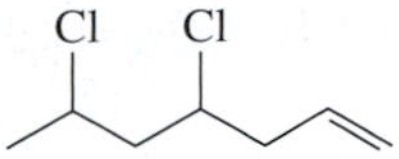

12.29 Draw the condensed structural formula, or line-angle formula, if cyclic, for each of the following:

a. 1-pentene **b.** 2-methyl-1-butene
c. 3-methylcyclohexene **d.** 3-chloro-1-butyne

12.30 Draw the condensed structural formula, or line-angle formula, if cyclic, for each of the following:

a. 1-methylcyclopentene **b.** 1-bromo-3-hexyne
c. 3,4-dimethyl-1-pentene **d.** 2-methyl-2-hexene

12.6 Cis–Trans Isomers

LEARNING GOAL Draw the condensed structural formulas and write the names for the cis–trans isomers of alkenes.

In any alkene, the double bond is rigid, which means there is no rotation around the double bond. As a result, the atoms or groups are attached to the carbon atoms in the double bond on one side or the other, which gives two different structures called *geometric isomers* or *cis–trans isomers*.

For example, the formula for 2-butene can be drawn as two different molecules, which are cis–trans isomers. In the ball-and-stick models, the atoms bonded to the carbon atoms in the double bond have bond angles of 120° (see **FIGURE 12.6**). We add the prefix *cis* or *trans* to denote whether the atoms bonded to the double bond are on the same side or opposite sides. In the **cis isomer**, the CH_3- groups are on the same side of the double bond. In the **trans isomer**, the CH_3- groups are on opposite sides. *Trans* means "across," as in transcontinental; *cis* means "on this side."

As with any pair of cis–trans isomers, *cis*-2-butene and *trans*-2-butene are different compounds with different physical and chemical properties.

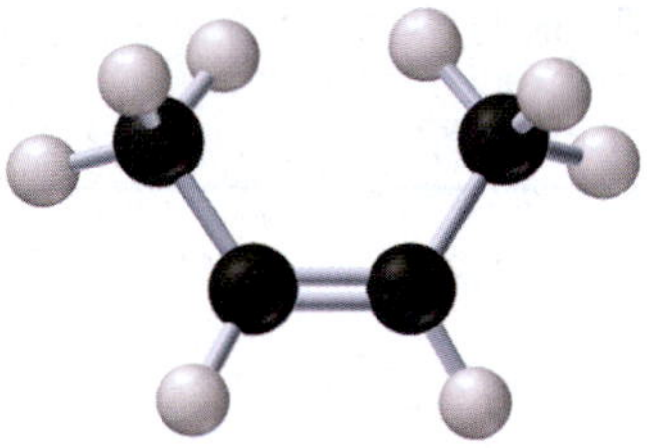

cis-2-Butene

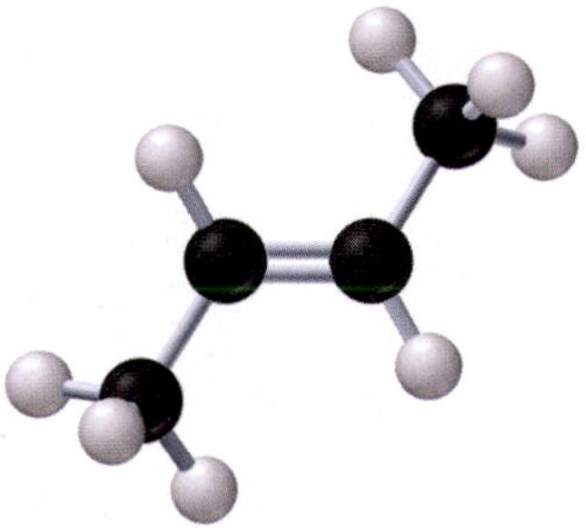

trans-2-Butene

FIGURE 12.6 ▶ Ball-and-stick models of the cis and trans isomers of 2-butene.

Q What feature in 2-butene accounts for the cis and trans isomers?

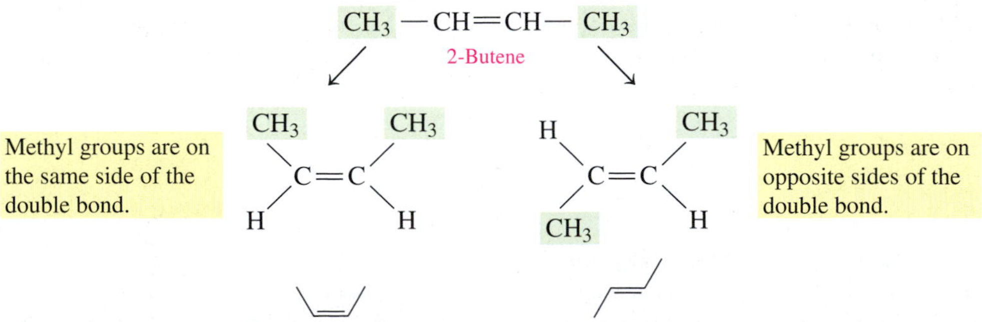

CH3 —CH=CH— CH3
2-Butene
Methyl groups are on the same side of the double bond.
Methyl groups are on opposite sides of the double bond.

cis-2-Butene (mp −139 °C; bp 3.7 °C)

trans-2-Butene (mp −106 °C; bp 0.3 °C)

When the carbon atoms in the double bond are attached to two different atoms or groups of atoms, an alkene can have cis–trans isomers. For example, 1,2-dichloroethene can be drawn with cis and trans isomers because there is one H atom and one Cl atom attached to each carbon atom in the double bond. When you are asked to draw the formula for an alkene, it is important to consider the possibility of cis and trans isomers.

ENGAGE

Explain how an alkene can have two geometric isomers.

Cl—CH=CH—Cl

1,2-Dichloroethene

Chlorine atoms are on the same side of the double bond.

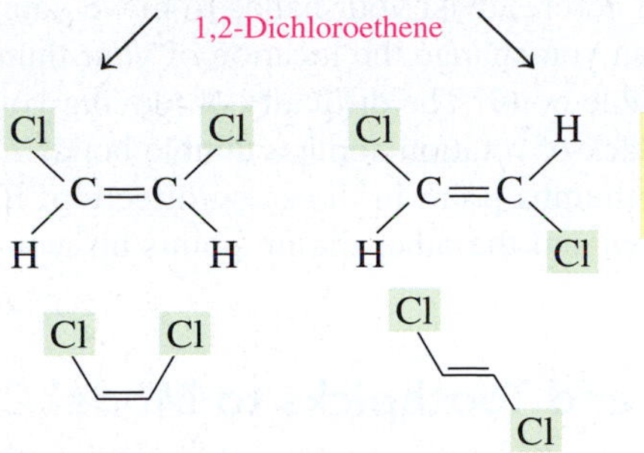

Chlorine atoms are on opposite sides of the double bond.

cis-1,2-Dichloroethene *trans*-1,2-Dichloroethene

When an alkene has identical groups on the same carbon atom of the double bond, cis and trans isomers cannot be drawn. For example, 1,1-dichloropropene has just one condensed structural formula without cis and trans isomers.

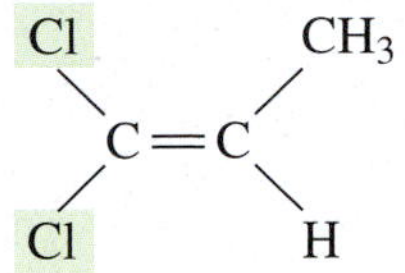

1,1-Dichloropropene

▶SAMPLE PROBLEM 12.8 Identifying Cis–Trans Isomers

TRY IT FIRST

Identify each of the following as the cis or trans isomer and write its name:

a. 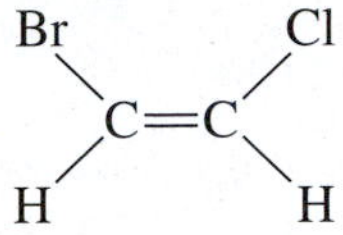

b. 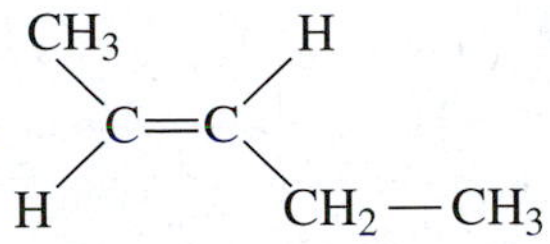

SOLUTION

a. This is a cis isomer because the two halogen atoms attached to the carbon atoms of the double bond are on the same side. The name of the two-carbon alkene, starting with the bromo group on carbon 1, is *cis*-1-bromo-2-chloroethene.

b. This is a trans isomer because the two alkyl groups attached to the carbon atoms of the double bond are on opposite sides of the double bond. This isomer of the five-carbon alkene is named *trans*-2-pentene.

STUDY CHECK 12.8

Write the name for each of the following compounds, including cis or trans:

a. 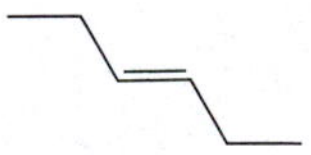

b. 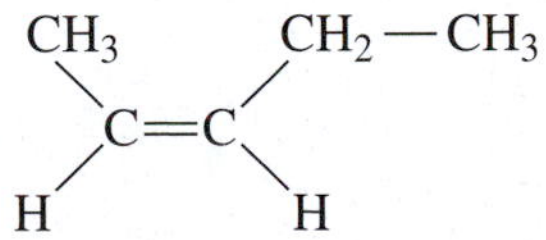

ANSWER

a. *trans*-3-hexene **b.** *cis*-2-pentene

INTERACTIVE VIDEO

PEARSON eText 2.0 Cis–Trans Isomers

TEST

Try Practice Problems 12.31 to 12.34

Modeling Cis–Trans Isomers

Because cis–trans isomerism is not easy to visualize, here are some things you can do to understand the difference in rotation around a single bond compared to a double bond and how it affects groups that are attached to the carbon atoms in the double bond.

Put the tips of your index fingers together. This is a model of a single bond. Consider the index fingers as a pair of carbon atoms, and think of your thumbs and other fingers as other parts of a carbon chain. While your index fingers are touching, twist your hands and change the position of your thumbs relative to each other. Notice how the relationship of your other fingers changes.

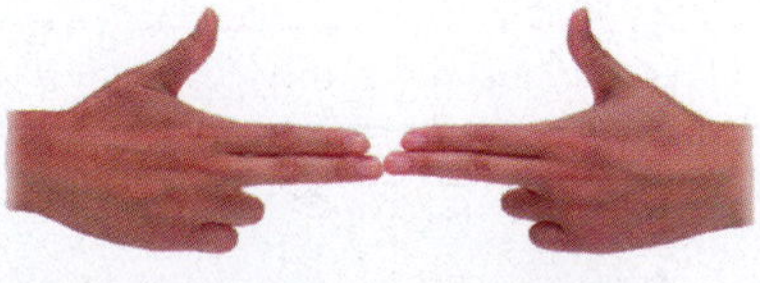

Cis-hands (cis-thumbs/fingers)

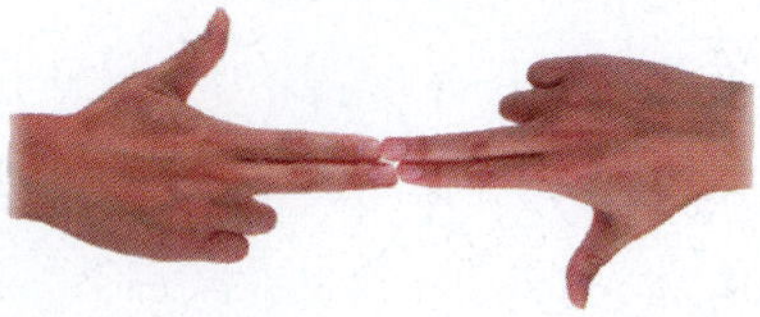

Trans-hands (trans-thumbs/fingers)

Now place the tips of your index fingers and middle fingers together in a model of a double bond. As you did before, twist your hands to move your thumbs away from each other. What happens? Can you change the location of your thumbs relative to each other without breaking the double bond? The difficulty of moving your hands with two fingers touching represents the lack of rotation about a double bond. You have made a model of a cis isomer when both thumbs point in the same direction. If you turn one hand over so one thumb points down and the other thumb points up, you have made a model of a trans isomer.

Using Gumdrops and Toothpicks to Model Cis–Trans Isomers

Obtain some toothpicks and yellow, green, and black gumdrops. The black gumdrops represent C atoms, the yellow gumdrops represent H atoms, and the green gumdrops represent Cl atoms. Place a toothpick between two black gumdrops. Use three more toothpicks to attach two yellow gumdrops and one green gumdrop to each black gumdrop carbon atom. Move one of the black gumdrops to show the rotation of the attached H and Cl atoms.

Remove a toothpick and yellow gumdrop from each black gumdrop. Place a second toothpick between the carbon atoms, which makes a double bond. Try to twist the double bond of toothpicks. Can you do it? When you observe the location of the green gumdrops, does the model you made represent a cis or trans isomer? Why? If your model is a cis isomer, how would you change it to a trans isomer? If your model is a trans isomer, how could you change it to a cis isomer?

Models from gumdrops represent cis and trans isomers.

PRACTICE PROBLEMS

12.6 Cis–Trans Isomers

12.31 Write the IUPAC name for each of the following, using cis or trans prefixes, if needed:

a. 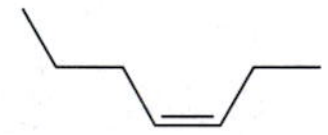

b. 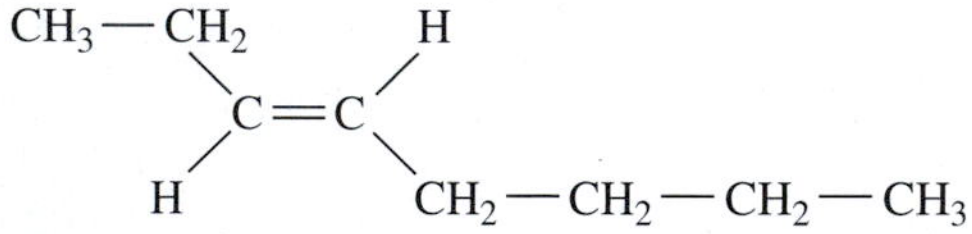

CH3—CH2
H
C=C
H
CH2—CH2—CH2—CH3

c. 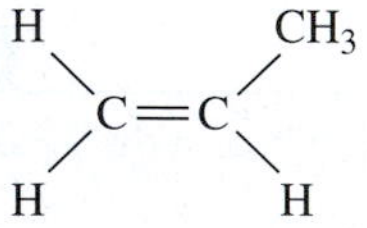

H
CH3
C=C
H
H

12.32 Write the IUPAC name for each of the following, using cis or trans prefixes, if needed:

a. 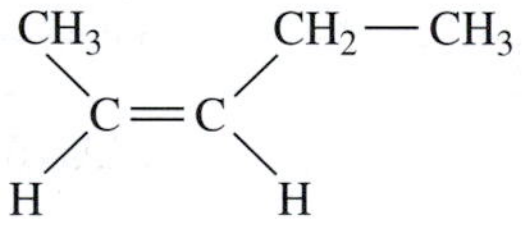

CH3
CH2—CH3
C=C
H
H

b. 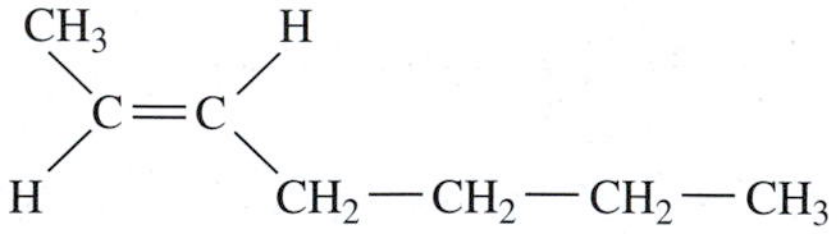

CH3
H
C=C
H
CH2—CH2—CH2—CH3

c. 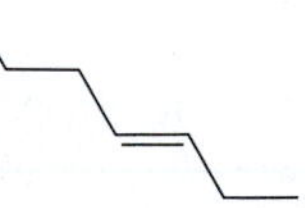

12.33 Draw the condensed structural formula for each of the following:
a. *trans*-1-bromo-2-chloroethene **b.** *cis*-2-hexene
c. *cis*-4-octene

12.34 Draw the condensed structural formula for each of the following:
a. *cis*-1,2-difluoroethene **b.** *trans*-2-pentene
c. *trans*-2-heptene

Chemistry Link to the Environment

Pheromones in Insect Communication

Many insects emit minute quantities of chemicals called *pheromones* to send messages to individuals of the same species. Some pheromones warn of danger, others call for defense, mark a trail, or attract the opposite sex. One of the most studied is bombykol, the sex pheromone produced by the female silkworm moth. The bombykol molecule contains one cis double bond and one trans double bond. Even a few nanograms of bombykol will attract male silkworm moths from distances of over 1 km. The effectiveness of many of these pheromones depends on the cis or trans configuration of the double bonds in the molecules. A certain species will respond to one isomer but not the other.

Pheromones allow insects to attract mates from a great distance.

Scientists are interested in synthesizing pheromones to use as nontoxic alternatives to pesticides. When placed in a trap, bombykol can be used to capture male silkworm moths. When a synthetic pheromone is released in a field, the males cannot locate the females, which disrupts the reproductive cycle. This technique has been successful with controlling the oriental fruit moth, the grapevine moth, and the pink bollworm.

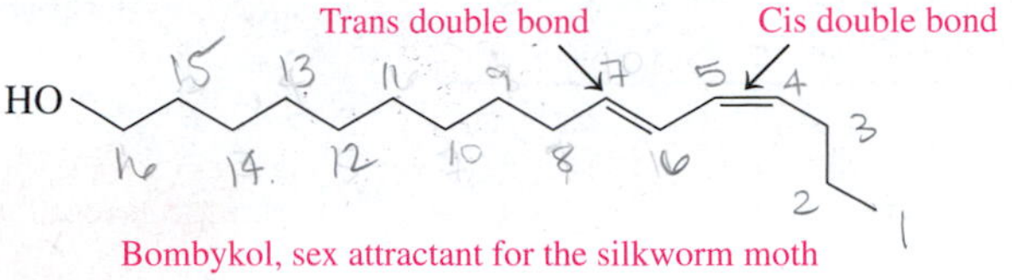
Trans double bond
Cis double bond
HO

Bombykol, sex attractant for the silkworm moth

Chemistry Link to Health

Cis–Trans Isomers for Night Vision

The retinas of the eyes consist of two types of cells: rods and cones. The rods on the edge of the retina allow us to see in dim light, and the cones, in the center, produce our vision in bright light. In the rods, there is a substance called *rhodopsin* that absorbs light. Rhodopsin is composed of *cis*-11-retinal, an unsaturated compound, attached to a protein. When rhodopsin absorbs light, the *cis*-11-retinal isomer is converted to its trans isomer, which changes its shape. The trans form no longer fits the protein, and it separates from the protein. The change from the cis to trans isomer and its separation from the protein generate an electrical signal that the brain converts into an image.

An enzyme (isomerase) converts the trans isomer back to the *cis*-11-retinal isomer and the rhodopsin re-forms. If there is a deficiency of rhodopsin in the rods of the retina, *night blindness* may occur. One common cause is a lack of vitamin A in the diet. In our diet, we obtain vitamin A from plant pigments containing β-carotene, which is found in foods such as carrots, squash, and spinach. In the small intestine, the β-carotene is converted to vitamin A, which can be converted to *cis*-11-retinal or stored in the liver for future use. Without a sufficient quantity of retinal, not enough rhodopsin is produced to enable us to see adequately in dim light.

Cis–Trans Isomers of Retinal

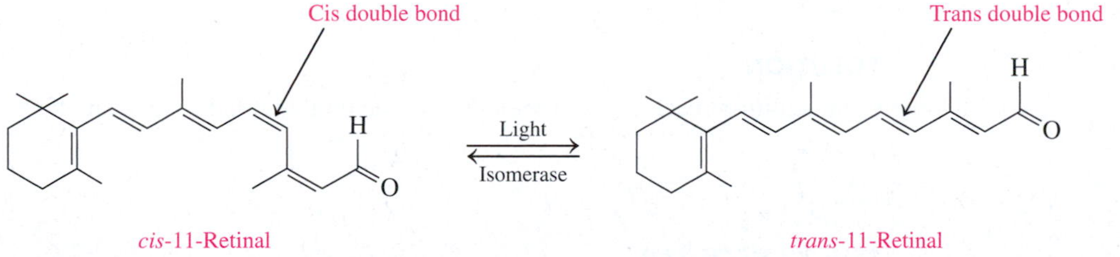
Cis double bond
H
O
Light
Isomerase
Trans double bond
H
O
cis-11-Retinal
trans-11-Retinal

12.7 Addition Reactions for Alkenes

LEARNING GOAL Draw the condensed structural and line-angle formulas and write the names for the organic products of addition reactions of alkenes.

The most characteristic reaction of alkenes is the **addition** of atoms or groups of atoms to the carbon atoms in a double bond. Addition occurs because double bonds are easily broken, providing electrons to form new single bonds.

The addition reactions have different names that depend on the type of reactant we add to the alkene, as **TABLE 12.8** shows.

TABLE 12.8 Summary of Addition Reactions

Name of Addition Reaction	Reactants	Catalysts	Product
Hydrogenation	Alkene + H_2	Pt, Ni, or Pd	Alkane
Hydration	Alkene + H_2O	H^+ (strong acid)	Alcohol
Polymerization	Alkenes	High temperature, pressure	Polymer

Hydrogenation

CORE CHEMISTRY SKILL
Writing Equations for Hydrogenation, Hydration, and Polymerization

In a reaction called **hydrogenation**, H atoms add to each of the carbon atoms in a double bond of an alkene. During hydrogenation, the double bonds are converted to single bonds in alkanes. A catalyst such as finely divided platinum (Pt), nickel (Ni), or palladium (Pd) is used to speed up the reaction. The general equation for hydrogenation can be written as follows:

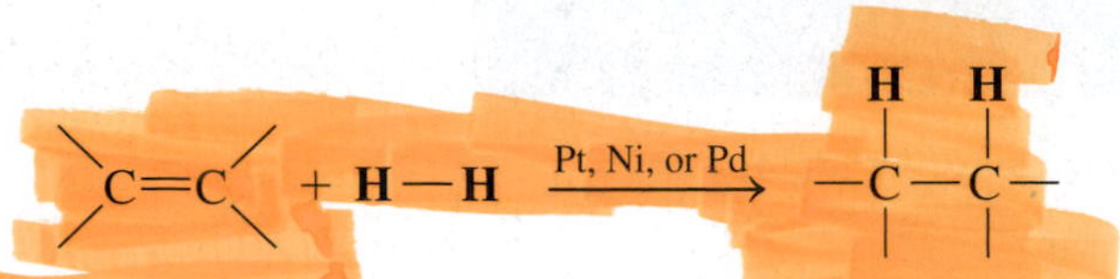
C=C
+ H—H
Pt, Ni, or Pd
H H
—C—C—

Double bond (alkene) Single bond (alkane)

Some examples of the hydrogenation of alkenes follow:

$$CH_3-CH=CH-CH_3 + H_2 \xrightarrow{Pt} CH_3-CH_2-CH_2-CH_3$$

2-Butene Butane

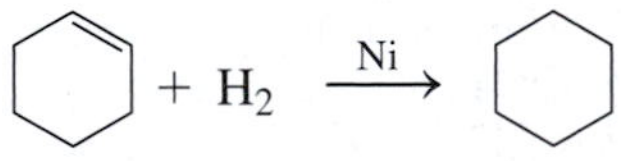
+ H2
Ni

Cyclohexene Cyclohexane

▶SAMPLE PROBLEM 12.9 Writing Equations for Hydrogenation

TRY IT FIRST

Draw the line-angle formula for the product of the following hydrogenation reaction:

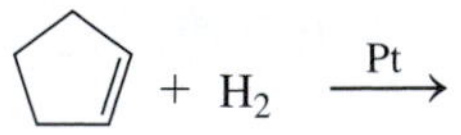
+ H2
Pt

SOLUTION

In an addition reaction, hydrogen adds to the double bond to give an alkane.

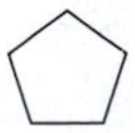

STUDY CHECK 12.9

a. Draw the condensed structural formula for the product of the hydrogenation of 2-methyl-1-butene, using a platinum catalyst.

b. Draw the line-angle formula for the product of the hydrogenation of 3-methylcyclohexene, using a platinum catalyst.

ANSWER

a.

$$\begin{array}{c} \quad CH_3 \\ \quad | \\ CH_3-CH-CH_2-CH_3 \end{array}$$

b. 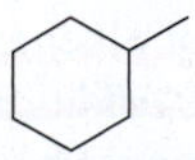

Chemistry Link to Health

Hydrogenation of Unsaturated Fats

Vegetable oils such as corn oil or safflower oil contain unsaturated fats composed of fatty acids that contain double bonds. The process of hydrogenation is used commercially to convert the double bonds in the unsaturated fats in vegetable oils to saturated fats such as margarine, which are more solid. Adjusting the amount of hydrogen added produces partially hydrogenated fats such as soft margarine, solid margarine in sticks, and shortenings, which are used in cooking. For example, oleic acid is a typical unsaturated fatty acid in olive oil and has a cis double bond at carbon 9. When oleic acid is hydrogenated, it is converted to stearic acid, a saturated fatty acid.

Crisco
All-Vegetable
Jif
CREAMY
Blue Bonnet

The unsaturated fats in vegetable oils are converted to saturated fats to make a more solid product.

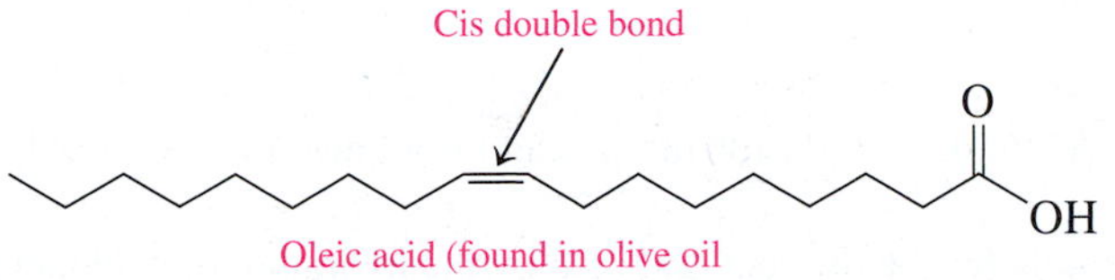
Cis double bond
O
OH

Oleic acid (found in olive oil and other unsaturated fats)

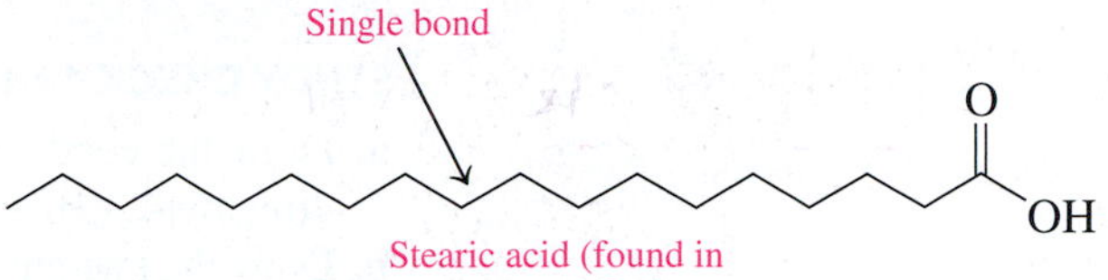
Single bond
O
OH

Stearic acid (found in saturated fats)

Hydration

In **hydration**, an alkene reacts with water (H—OH). A hydrogen atom (H—) from water forms a bond with one carbon atom in the double bond, and the oxygen atom in —OH forms a bond with the other carbon. The reaction is catalyzed by a strong acid such as H_2SO_4, written as H^+. Hydration is used to prepare alcohols, which have the hydroxyl (—OH) functional group. When water (H—OH) adds to a symmetrical alkene, such as ethene, a single product is formed.

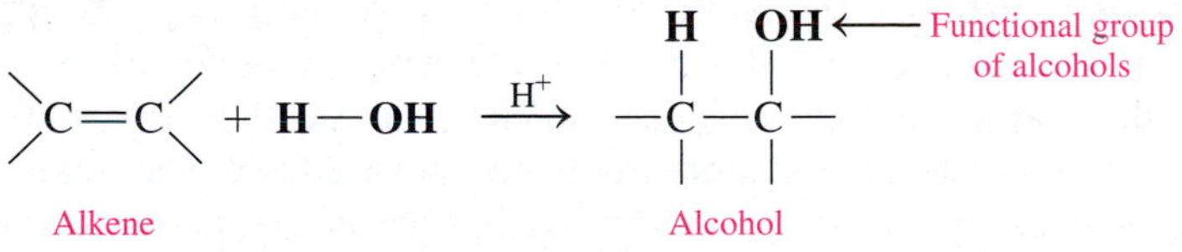
H OH ← Functional group of alcohols
C=C + H—OH
H+
—C—C—
Alkene
Alcohol

$$H_2C{=}CH_2 + \mathbf{H_2O} \xrightarrow{H^+} CH_3{-}CH_2{-}OH$$

Ethene — Ethanol (ethyl alcohol)

However, when H_2O adds to a double bond in an asymmetrical alkene, two products are possible. We can write the prevalent product by attaching the H— from H_2O to the carbon that has the *greater* number of H atoms and the —OH from H_2O to the other carbon atom from the double bond. In the following example, the H— from H_2O attaches to the end carbon of the double bond, which has more hydrogen atoms, and the —OH adds to the middle carbon atom:

ENGAGE

Why do you need to determine if the double bond in an alkene is symmetrical or asymmetrical?

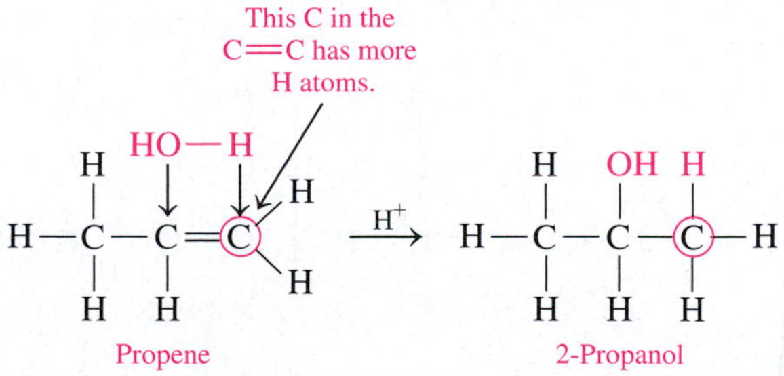
This C in the C=C has more H atoms.
HO—H
H+
OH
Propene
2-Propanol

▶ SAMPLE PROBLEM 12.10 Hydration

TRY IT FIRST

Draw the condensed structural formula for the product that forms in the following hydration reaction:

$$CH_3-CH_2-CH_2-CH=CH_2 + H_2O \xrightarrow{H^+}$$

SOLUTION

The H— and —OH from water add to the carbon atoms in the double bond. The H— adds to the carbon with the *greater* number of hydrogen atoms, and the —OH bonds to the carbon with fewer H atoms.

$$CH_3-CH_2-CH_2-\overset{\overset{OH}{\downarrow}}{CH}=\overset{\overset{H}{\downarrow}}{CH_2} \xrightarrow{H^+} CH_3-CH_2-CH_2-\overset{\overset{OH}{|}}{CH}-CH_3$$

STUDY CHECK 12.10

a. Draw the condensed structural formula for the product obtained by the hydration of 2-methyl-2-pentene.

b. Draw the line-angle formula for the product obtained by the hydration of cyclopentene.

ANSWER

a. $CH_3-\overset{\overset{OH}{|}}{\underset{\underset{CH_3}{|}}{C}}-CH_2-CH_2-CH_3$

b. 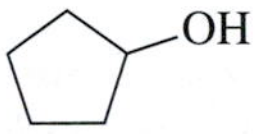
OH

INTERACTIVE VIDEO

Addition to an Asymmetric Bond

PEARSON eText 2.0

TEST

Try Practice Problems 12.35 and 12.36

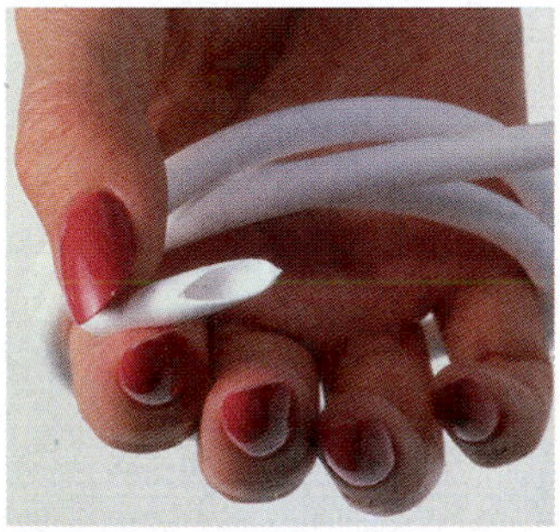

FIGURE 12.7 ▶ Synthetic polymers are used to replace diseased veins and arteries.

Q Why are the substances in these plastic devices called polymers?

Addition of Alkenes: Polymerization

A **polymer** is a large molecule that consists of small repeating units called **monomers**. In the past hundred years, the plastics industry has made synthetic polymers that are in many of the materials we use every day, such as carpeting, plastic wrap, nonstick pans, plastic cups, and rain gear. In medicine, synthetic polymers are used to replace diseased or damaged body parts such as hip joints, pacemakers, teeth, heart valves, and blood vessels (see **FIGURE 12.7**). There are about 100 billion kg of plastics produced every year, which is about 15 kg for every person on Earth.

Many of the synthetic polymers are made by addition reactions of small alkene monomers. Many polymerization reactions require high temperature, a catalyst, and high pressure (over 1000 atm). In an addition reaction, a polymer grows longer as each monomer is added at the end of the chain. A polymer may contain as many as 1000 monomers. Polyethylene, a polymer made from ethylene monomers, is used in plastic bottles, film, and plastic dinnerware. More polyethylene is produced worldwide than any other polymer. Low-density polyethylene (LDPE) is flexible, breakable, less dense, and more branched than high-density polyethylene (HDPE). High-density polyethylene is stronger, is more dense, and melts at a higher temperature than LDPE.

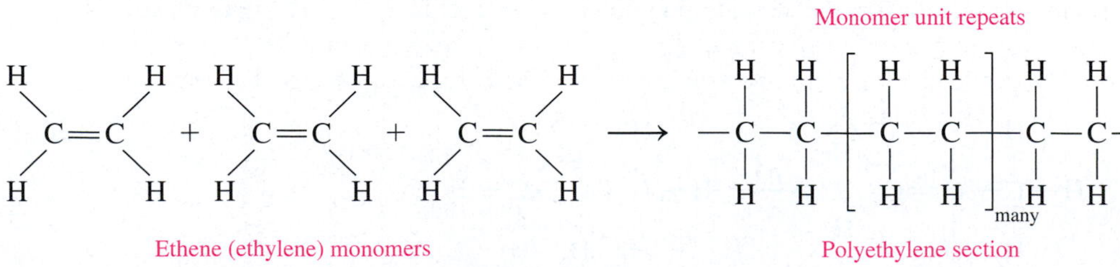
Monomer unit repeats
many
Ethene (ethylene) monomers
Polyethylene section

TABLE 12.9 lists several alkene monomers that are used to produce common synthetic polymers, and **FIGURE 12.8** shows examples of each. The alkane-like nature of these plastic synthetic polymers makes them unreactive. Thus, they do not decompose easily (they are not biodegradable). As a result, they have become significant contributors to pollution, on land and in the oceans. Efforts are being made to make them more degradable.

TABLE 12.9 Some Alkenes and Their Polymers

Recycling Code	Monomer	Polymer Section	Common Uses
2 HDPE, 4 LDPE	$H_2C{=}CH_2$ Ethene (ethylene)	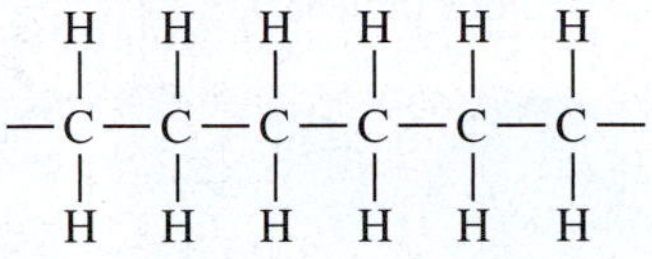 Polyethylene (PE)	Plastic bottles, film, insulation materials, shopping bags
3 V	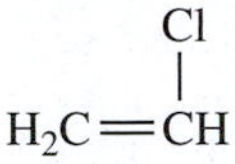 Chloroethene (vinyl chloride)	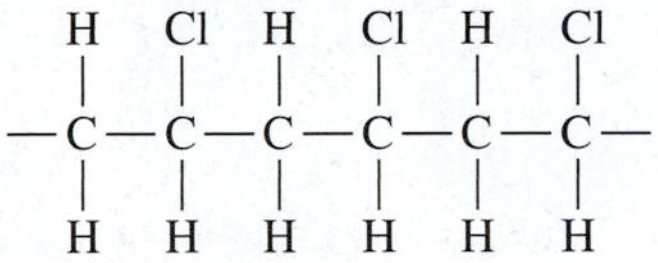 Polyvinyl chloride (PVC)	Plastic pipes and tubing, garden hoses, garbage bags, shower curtains, credit cards, medical containers
5 PP	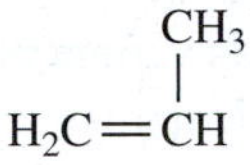 Propene (propylene)	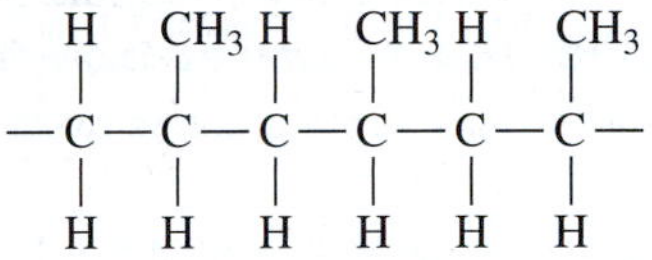 Polypropylene (PP)	Ski and hiking clothing, carpets, artificial joints, plastic bottles, food containers, medical face masks, tubing
	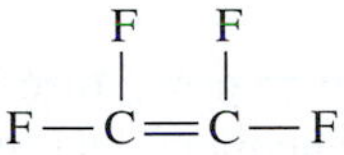 Tetrafluoroethene	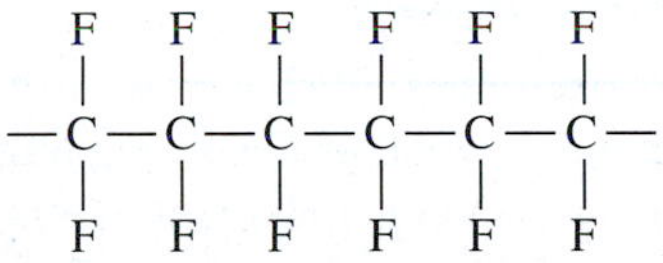 Polytetrafluoroethylene (PTFE)	Nonstick coatings, Teflon
	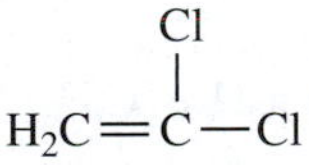 1,1-Dichloroethene	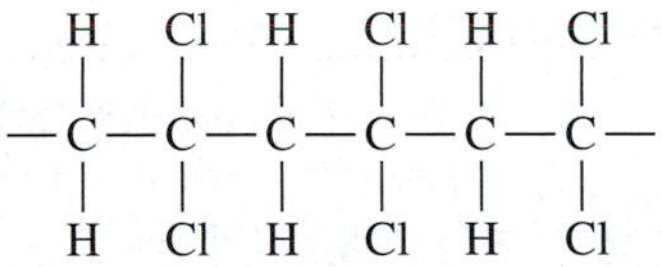 Polydichloroethylene (Saran)	Plastic film and wrap, Gore-Tex rainwear, medical implants
6 PS	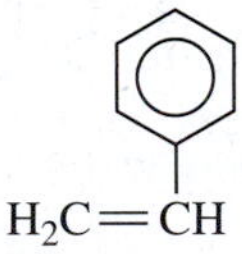 Phenylethene (styrene)	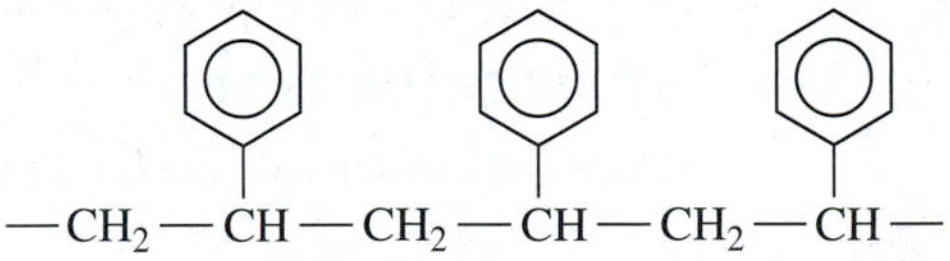 Polystyrene (PS)	Plastic coffee cups and cartons, insulation

You can identify the type of polymer used to manufacture a plastic item by looking for the recycling symbol (arrows in a triangle) found on the label or on the bottom of the plastic container. For example, the number 5 or the letters PP inside the triangle is the code for a polypropylene plastic. There are now many cities that maintain recycling programs that reduce the amount of plastic materials that are transported to landfills.

Today, products such as lumber, tables and benches, trash receptacles, and pipes used for irrigation systems are made from recycled plastics.

Tables, benches, and trash receptacles can be manufactured from recycled plastics.

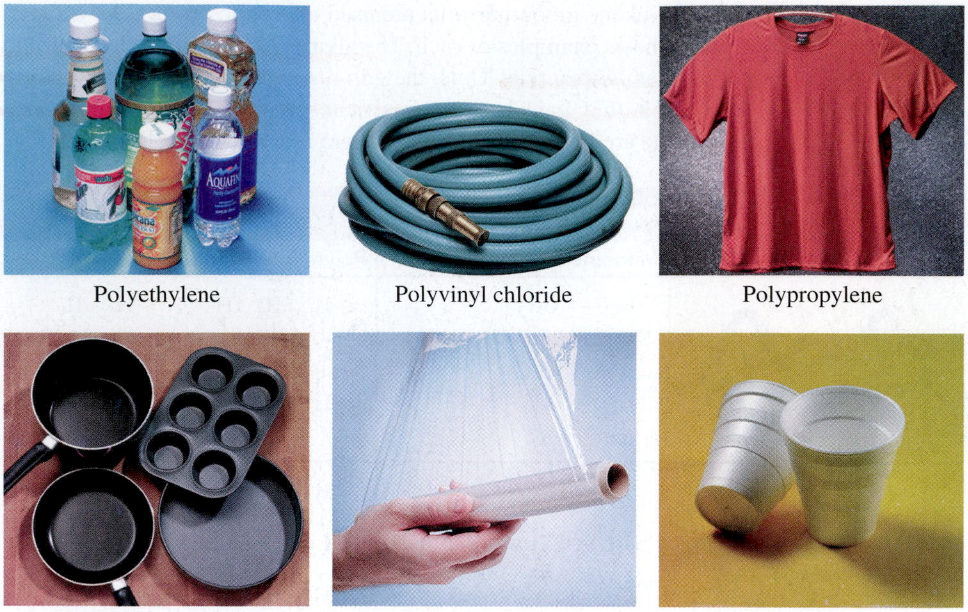
Polyethylene

Polyvinyl chloride

Polypropylene

Polytetrafluoroethylene
(Teflon)

Polydichloroethylene
(Saran)

Polystyrene

FIGURE 12.8 ▶ Synthetic polymers provide a wide variety of items that we use every day.
Q What are some alkenes used to make the polymers in these plastic items?

▶**SAMPLE PROBLEM 12.11** Polymers

TRY IT FIRST

A firefighter/EMT arrives at a home where a premature baby has been delivered. To prevent hypothermia during transport to the neonatal facility, she wraps the baby in cling wrap, which is polydichloroethylene. Draw and name the monomer unit and draw a portion of the polymer formed from three monomer units.

SOLUTION

$$H_2C = \overset{\overset{\displaystyle Cl}{|}}{C} - Cl$$

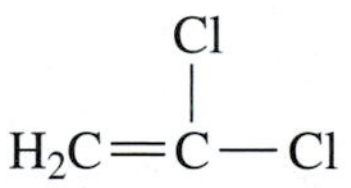

1,1-Dichloroethene

$$-\overset{\overset{\displaystyle H}{|}}{\underset{\underset{\displaystyle H}{|}}{C}} - \overset{\overset{\displaystyle Cl}{|}}{\underset{\underset{\displaystyle Cl}{|}}{C}} - \overset{\overset{\displaystyle H}{|}}{\underset{\underset{\displaystyle H}{|}}{C}} - \overset{\overset{\displaystyle Cl}{|}}{\underset{\underset{\displaystyle Cl}{|}}{C}} - \overset{\overset{\displaystyle H}{|}}{\underset{\underset{\displaystyle H}{|}}{C}} - \overset{\overset{\displaystyle Cl}{|}}{\underset{\underset{\displaystyle Cl}{|}}{C}} -$$

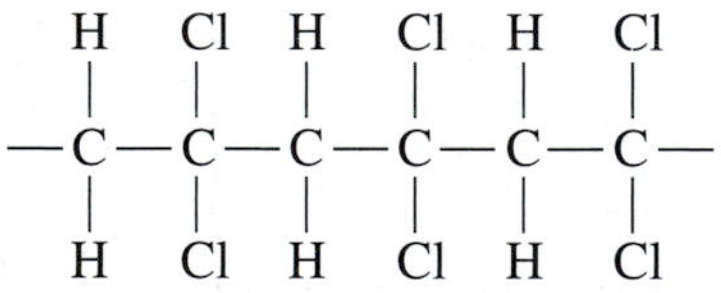

STUDY CHECK 12.11

Draw the condensed structural formula for the monomer used in the manufacturing of PVC.

ANSWER

$$H_2C = \overset{\overset{\displaystyle Cl}{|}}{CH}$$

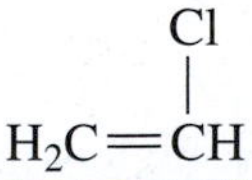

TEST
Try Practice Problems 12.37
to 12.42

PRACTICE PROBLEMS

12.7 Addition Reactions for Alkenes

12.35 Draw the structural formula for the product in each of the following reactions:

a. $CH_3 - CH_2 - CH_2 - CH = CH_2 + H_2 \xrightarrow{Pt}$

b. $H_2C = \overset{\overset{\displaystyle CH_3}{|}}{C} - CH_2 - CH_3 + H_2O \xrightarrow{H^+}$ 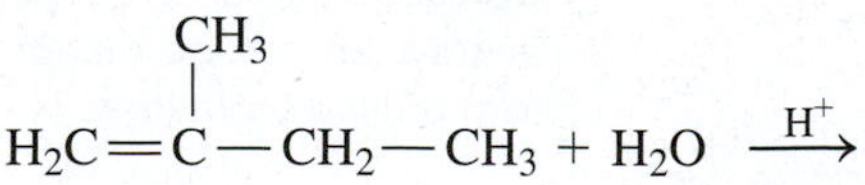

c. ⌇ $+ H_2 \xrightarrow{Pt}$ 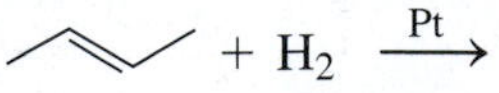

d. ⬡ $+ H_2O \xrightarrow{H^+}$ 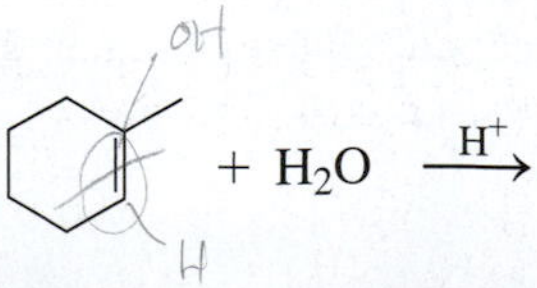

12.36 Draw the structural formula for the product in each of the following reactions:

a. $CH_3—CH_2—CH=CH_2 + H_2O \xrightarrow{H^+}$

b. 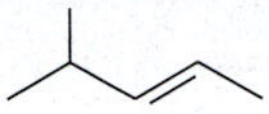 $+ H_2 \xrightarrow{Pt}$

c. 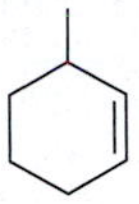 $+ H_2 \xrightarrow{Pt}$

d. 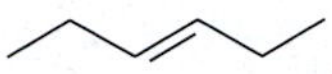 $+ H_2O \xrightarrow{H^+}$

12.37 What is a polymer?

12.38 What is a monomer?

12.39 Write an equation that represents the formation of a portion of polypropylene from three of its monomers.

12.40 Write an equation that represents the formation of a portion of polystyrene from three of its monomers.

12.41 The plastic polyvinylidene difluoride, PVDF, is made from monomers of 1,1-difluoroethene. Draw the expanded structural formula for a portion of the polymer formed from three monomers of 1,1-difluoroethene.

12.42 The polymer polyacrylonitrile, PAN, used in the fabric material Orlon is made from monomers of acrylonitrile. Draw the expanded structural formula for a portion of the polymer formed from three monomers of acrylonitrile.

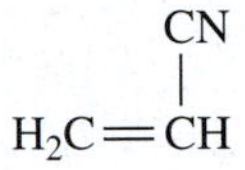

Acrylonitrile

12.8 Aromatic Compounds

LEARNING GOAL Describe the bonding in benzene; name aromatic compounds and draw their line-angle formulas.

In 1825, Michael Faraday isolated a hydrocarbon called **benzene,** which consists of a ring of six carbon atoms with one hydrogen atom attached to each carbon. Because many compounds containing benzene had fragrant odors, the family of benzene compounds became known as **aromatic compounds**. Some common examples of aromatic compounds that we use for flavor are anisole from anise, estragole from tarragon, and thymol from thyme.

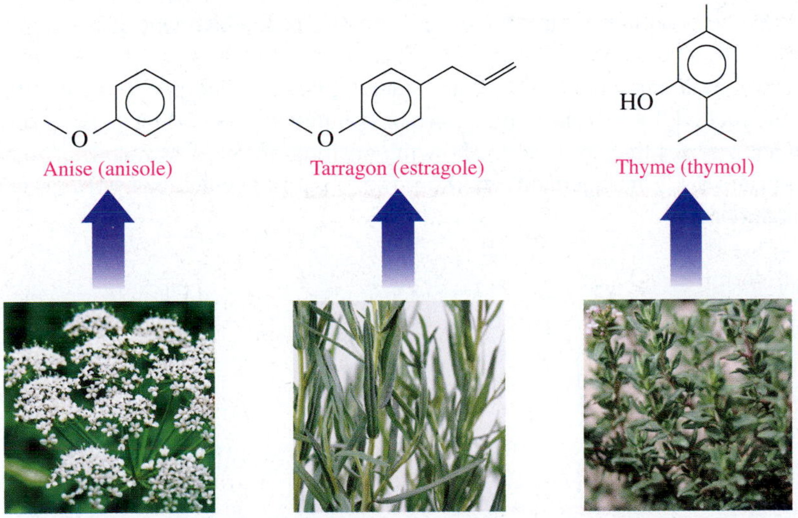
Anise (anisole)
Tarragon (estragole)
Thyme (thymol)
HO
O
O

The aroma and flavor of the herbs anise, tarragon, and thyme are because of aromatic compounds.

In benzene, each carbon atom uses three valence electrons to bond to the hydrogen atom and two adjacent carbons. That leaves one valence electron, which scientists first thought was shared in a double bond with an adjacent carbon. In 1865, August Kekulé proposed that the carbon atoms in benzene were arranged in a flat ring with alternating single and double bonds between the adjacent carbon atoms. There are two possible structural representations of benzene in which the double bonds can form between two different carbon atoms. If there were double bonds as in alkenes, then benzene should be much more reactive than it is.

However, unlike the alkenes and alkynes, aromatic hydrocarbons do not easily undergo addition reactions. If their reaction behavior is quite different, they must also differ in how the atoms are bonded in their structures. Today, we know that the six electrons are shared equally among the six carbon atoms. This unique feature of benzene makes it especially stable. Benzene is most often represented as a line-angle structural formula, which shows a hexagon with a circle in the center. Some of the ways to represent benzene are shown as follows:

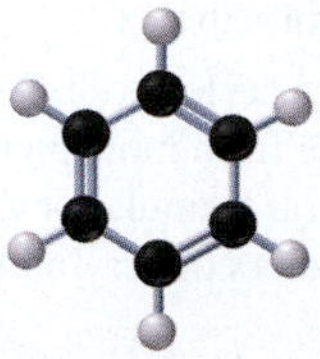

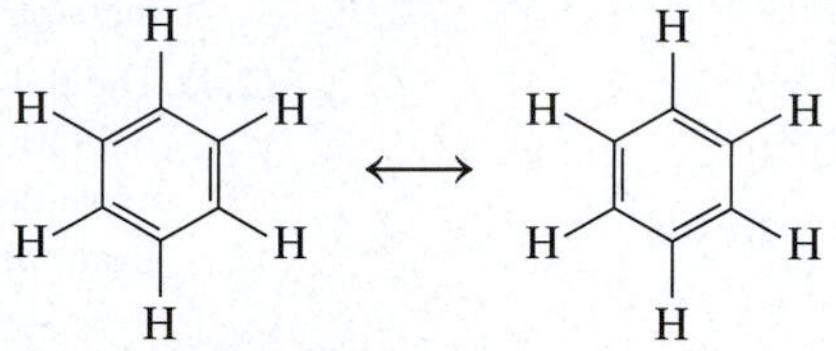

Equivalent structures for benzene

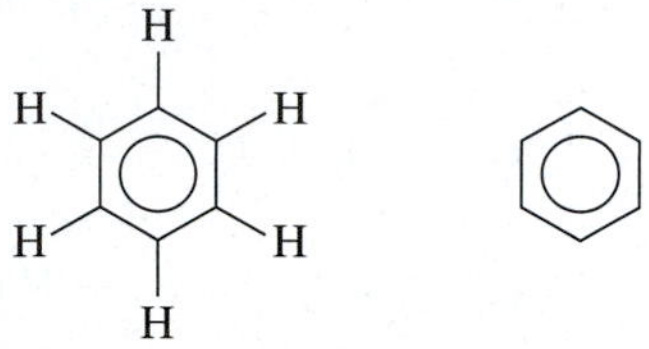

Structural formulas for benzene

Naming Aromatic Compounds

Many compounds containing benzene have been important in chemistry for many years and still use their common names. Toluene consists of benzene with a methyl group ($—CH_3$), aniline is benzene with an amino group ($—NH_2$), and phenol is benzene with a hydroxyl group (—OH). The names toluene, aniline, and phenol are allowed by IUPAC rules.

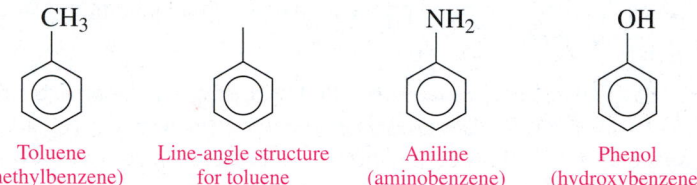

ENGAGE

How is the formula of toluene similar to and different from that of phenol?

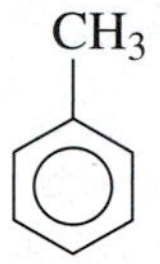

Toluene (methylbenzene)

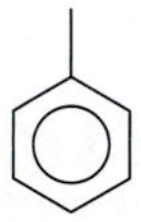

Line-angle structure for toluene

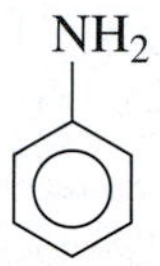

Aniline (aminobenzene)

Phenol (hydroxybenzene)

When benzene has only one substituent, the ring is not numbered. When there are two or more substituents, the benzene ring is numbered to give the lowest numbers to the substituents. In common names, when the benzene ring is a substituent, it is named as a phenyl group.

When a common name is used such as toluene, phenol, or aniline, the carbon atom attached to the methyl, hydroxyl, or amine group is numbered as carbon 1. In a common name, there are prefixes that are used to show the position of two substituents. The prefix **ortho** (*o*) indicates a 1,2-arrangement, **meta** (*m*) is a 1,3-arrangement, and **para** (*p*) is used for 1,4-arrangements.

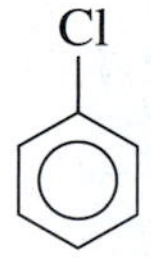

Chlorobenzene (phenyl chloride)

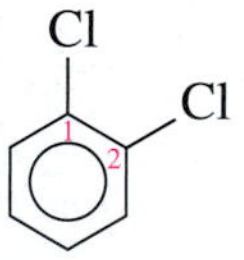

1,2-Dichlorobenzene (*o*-dichlorobenzene)

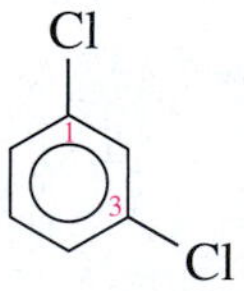

1,3-Dichlorobenzene (*m*-dichlorobenzene)

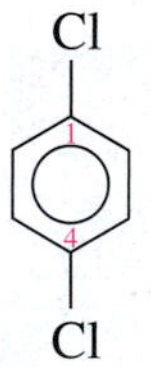

1,4-Dichlorobenzene (*p*-dichlorobenzene)

When aniline, phenol, or toluene has substituents, the carbon atom attached to the amine, hydroxyl, or methyl group is numbered as carbon 1 and the substituents are named alphabetically.

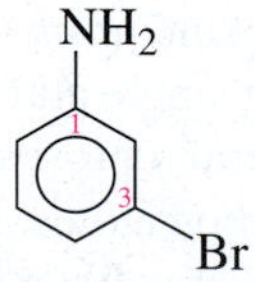

3-Bromoaniline

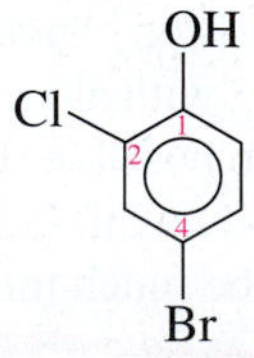

4-Bromo-2-chlorophenol

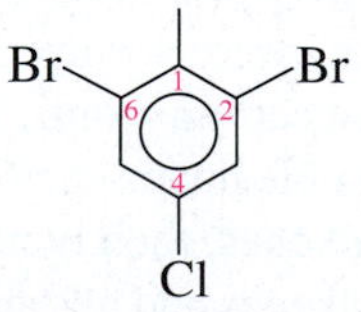

2,6-Dibromo-4-chlorotoluene

▶ SAMPLE PROBLEM 12.12 Naming Aromatic Compounds

TRY IT FIRST

Write the IUPAC name for the following:

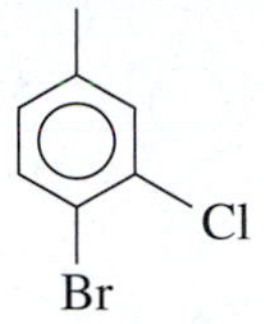

SOLUTION

	Given	Need	Connect
ANALYZE THE PROBLEM	structural formula	IUPAC name	name of aromatic compound, number and list substituents

STEP 1 **Write the name for the aromatic compound.** A benzene ring with a methyl group is named toluene.

STEP 2 **If there are two or more substituents, number the aromatic ring starting from the substituent.** The methyl group of toluene is attached to carbon 1, and the ring is numbered to give the lower numbers.

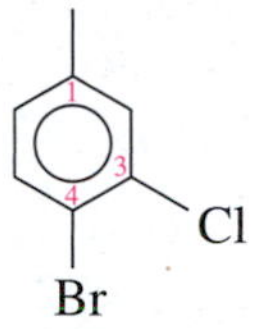

STEP 3 **Name each substituent as a prefix.** Naming the substituents in alphabetical order, this aromatic compound is 4-bromo-3-chlorotoluene.

STUDY CHECK 12.12

Write the IUPAC name for each of the following:

a. 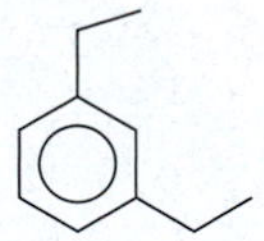

b. 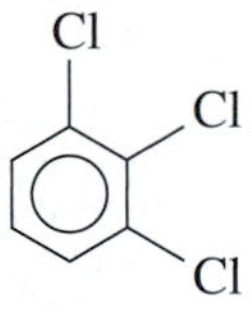

ANSWER

a. 1,3-diethylbenzene **b.** 1,2,3-trichlorobenzene

TEST

Try Practice Problems 12.43 to 12.46

Chemistry Link to Health

Some Common Aromatic Compounds

Aromatic compounds are common in nature and in medicine. Toluene is used as a reactant to make drugs, dyes, and explosives such as TNT (trinitrotoluene). The benzene ring is found in pain relievers such as aspirin and acetaminophen, and in flavorings such as vanillin.

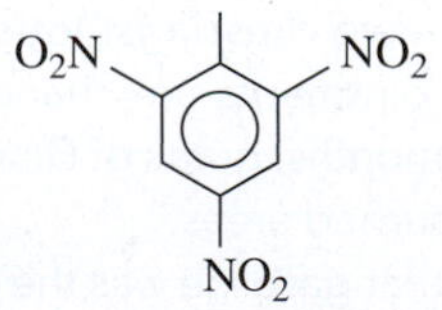

TNT (2,4,6-trinitrotoluene)

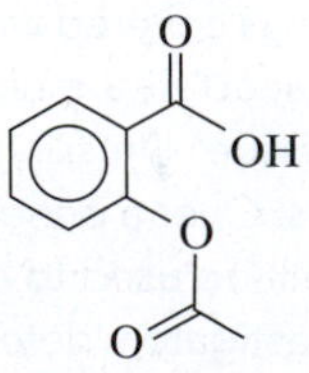

Aspirin

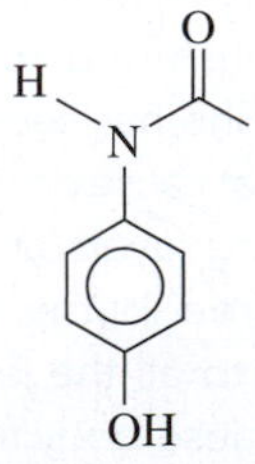

Acetaminophen

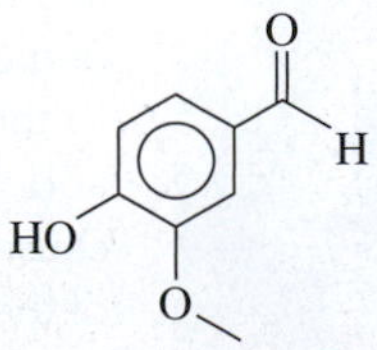

Vanillin

PRACTICE PROBLEMS

12.8 Aromatic Compounds

12.43 Write the IUPAC name and any common name for each of the following:

a. 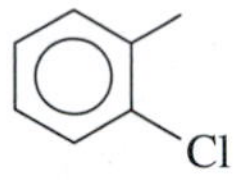
b. 
c. 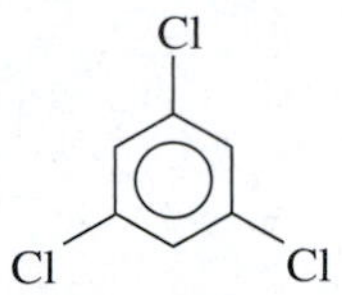
d. 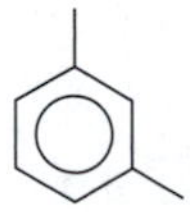
e. 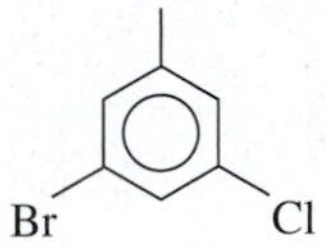
f. 

12.44 Write the IUPAC name and any common name for each of the following:

a. 
b. 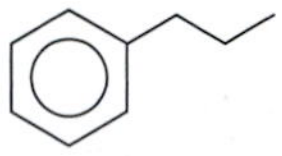
c. 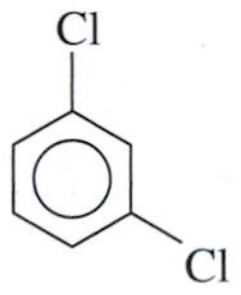
d. 
e. 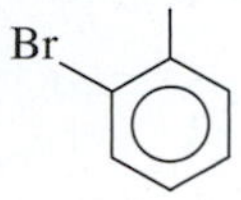
f. 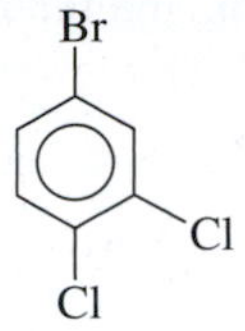

12.45 Draw the line-angle formula for each of the following:

a. aniline
b. 1-bromo-3-chlorobenzene
c. 1-ethyl-4-methylbenzene
d. *p*-chlorotoluene

12.46 Draw the line-angle formula for each of the following:

a. *m*-dibromobenzene
b. *o*-chloroethylbenzene
c. butylbenzene
d. 1,2,4-trichlorobenzene

Chemistry Link to Health

Polycyclic Aromatic Hydrocarbons (PAHs)

Large aromatic compounds known as *polycyclic aromatic hydrocarbons* (PAHs) are formed by fusing together two or more benzene rings edge to edge. In a fused-ring compound, neighboring benzene rings share two carbon atoms. Naphthalene, with two benzene rings, is well known for its use in mothballs; anthracene, with three rings, is used in the manufacture of dyes.

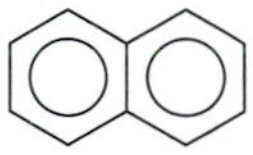
Naphthalene

Anthracene

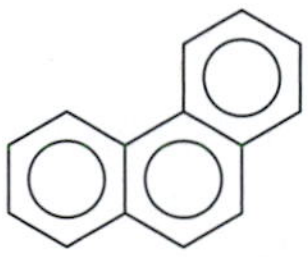
Phenanthrene

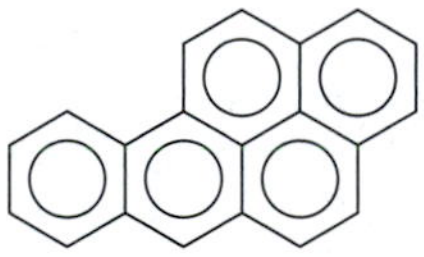
Benzo[*a*]pyrene

When a polycyclic compound contains phenanthrene, it may act as a carcinogen, a substance known to cause cancer.

Compounds containing five or more fused benzene rings such as benzo[*a*]pyrene are potent carcinogens. The molecules interact with the DNA in the cells, causing abnormal cell growth and cancer. Increased exposure to carcinogens increases the chance of DNA alterations in the cells. Benzo[*a*]pyrene, a product of combustion, has been identified in coal tar, tobacco smoke, barbecued meats, and automobile exhaust.

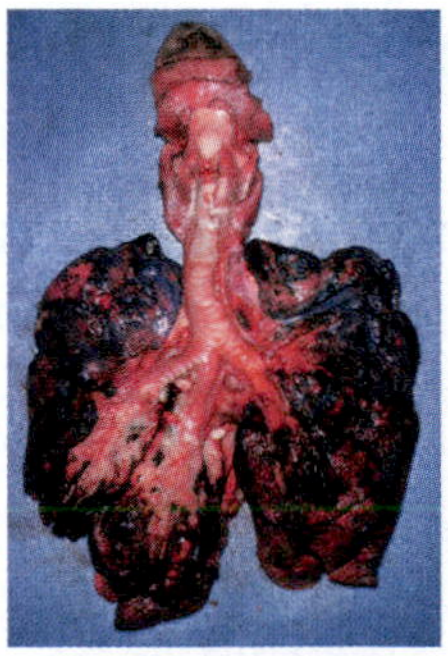
Aromatic compounds such as benzo[*a*]pyrene are strongly associated with lung cancers.

CLINICAL UPDATE Diane's Treatment in the Burn Unit

When Diane arrives at the hospital, she is transferred to the ICU burn unit. She has second-degree burns that cause damage to the underlying layers of skin and third-degree burns that cause damage to all the layers of the skin. Because body fluids are lost when deep burns occur, a lactated Ringer's solution is administered. The most common complications of burns are related to infection. To prevent infection, her skin is covered with a topical antibiotic. The next day, Diane is placed in a tank to remove dressings, lotions, and damaged tissue. Dressings and ointments are changed every eight hours. Over a period of 3 months, grafts of Diane's unburned skin will be used to cover burned areas.

The arson investigators determine that gasoline was the primary accelerant used to start the fire at Diane's house. Because there was a lot of paper and dry wood in the area, the fire spread quickly. Some of the hydrocarbons found in gasoline include hexane, heptane, octane, nonane, decane, and cyclohexane. Some other hydrocarbons in gasoline are 3-ethyltoluene, isopentane, and toluene.

Clinical Applications

12.47 The medications for Diane are contained in blister packs made of polychlorotrifluoroethylene (PCTFE). Draw the expanded structural formula for a portion of PCTFE polymer from three monomers of chlorotrifluoroethylene shown below:

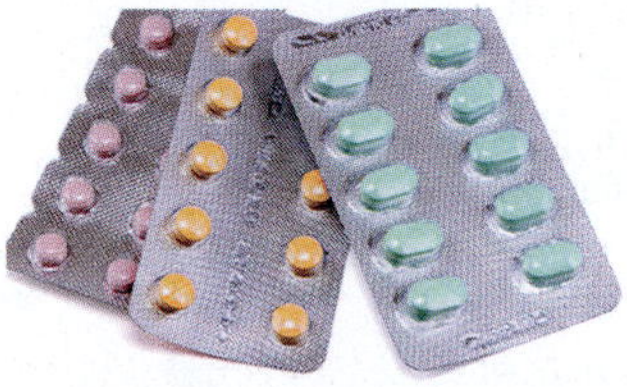

Tablets are contained in blister packs made from polychlorotrifluoroethylene (PCTFE).

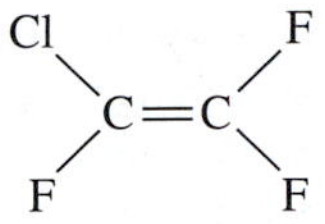

Chlorotrifluoroethylene

12.48 New polymers have been synthesized to replace PVC in IV bags and cling film. One of these is PEVA, polyethylene vinyl acetate. Draw the expanded structural formula for a portion of PEVA using an alternating sequence that has two of each monomer shown below:

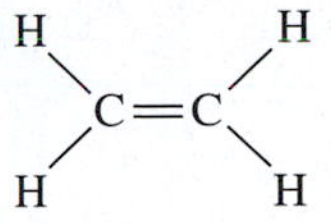

Ethylene

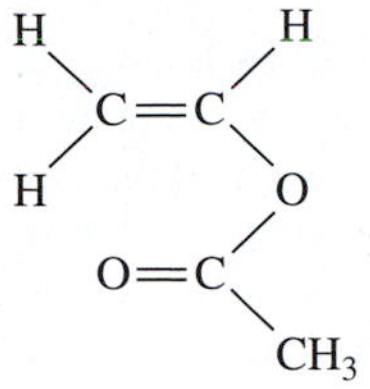

Vinyl acetate

12.49 Write the balanced chemical equation for the complete combustion of each of the following hydrocarbons found in gasoline:

a. octane
b. isopentane (2-methylbutane)
c. 3-ethyltoluene

12.50 Write the balanced chemical equation for the complete combustion of each of the following hydrocarbons found in gasoline:

a. decane **b.** cyclohexane **c.** toluene

CONCEPT MAP

INTRODUCTION TO ORGANIC CHEMISTRY: HYDROCARBONS

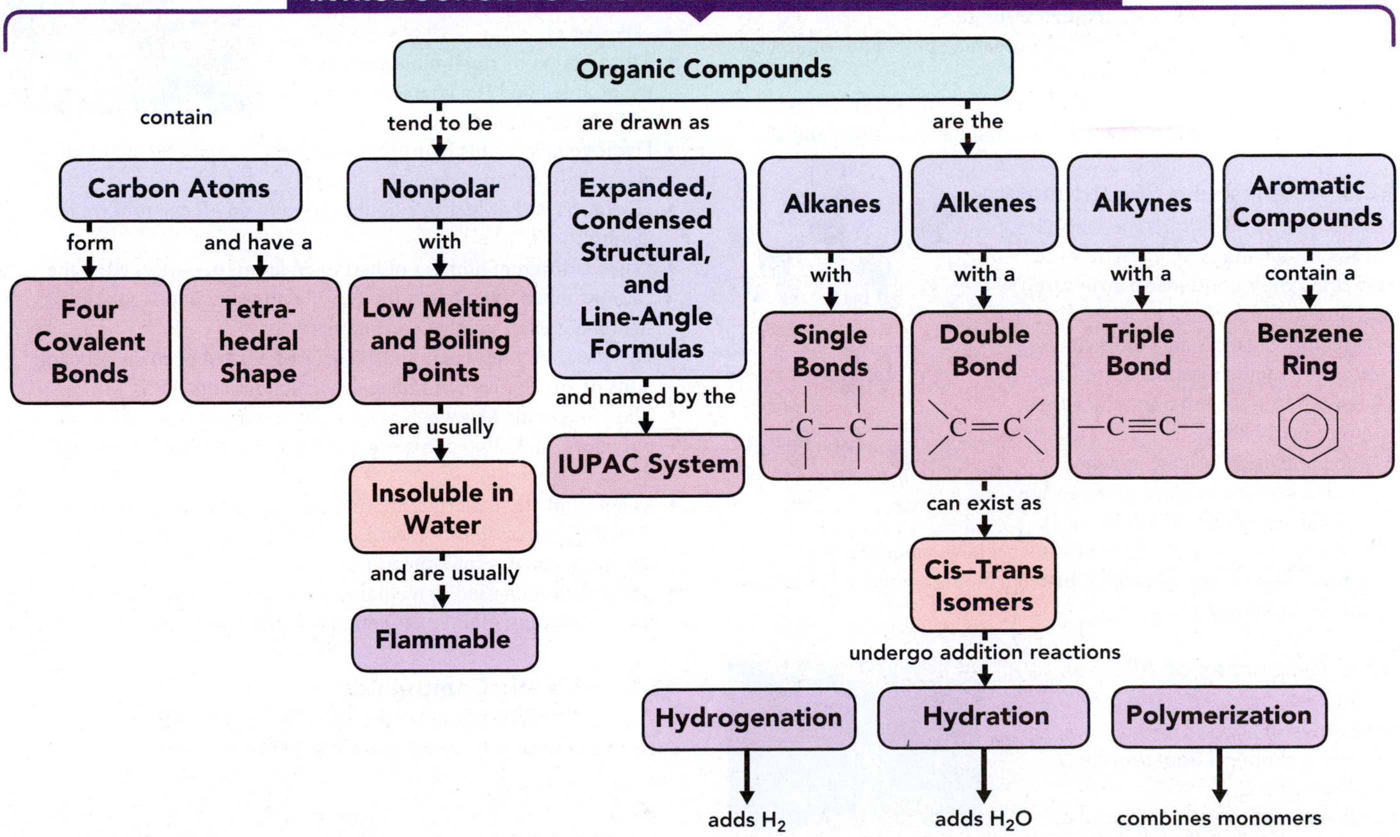
Organic Compounds
contain
Carbon Atoms
form
Four Covalent Bonds
and have a
Tetrahedral Shape
tend to be
Nonpolar
with
Low Melting and Boiling Points
are usually
Insoluble in Water
and are usually
Flammable
are drawn as
Expanded, Condensed Structural, and Line-Angle Formulas
and named by the
IUPAC System
are the
Alkanes
with
Single Bonds
Alkenes
with a
Double Bond
can exist as
Cis–Trans Isomers
undergo addition reactions
Hydrogenation
adds H2
Hydration
adds H2O
Polymerization
combines monomers
Alkynes
with a
Triple Bond
Aromatic Compounds
contain a
Benzene Ring

CHAPTER REVIEW

12.1 Organic Compounds

LEARNING GOAL Identify properties characteristic of organic or inorganic compounds.

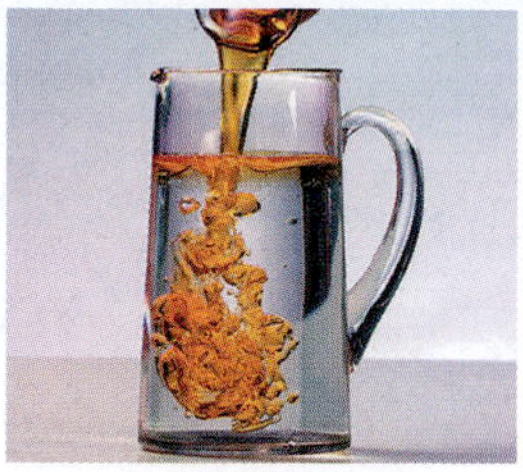

- Organic compounds have covalent bonds, most form nonpolar molecules, have low melting points and low boiling points, are not very soluble in water, dissolve as molecules in solutions, and burn vigorously in air.
- Inorganic compounds are often ionic or contain polar covalent bonds and form polar molecules, have high melting and boiling points, are usually soluble in water, produce ions in water, and do not burn in air.
- In the simplest organic molecule, methane, CH_4, the C—H bonds that attach four hydrogen atoms to the carbon atom are directed to the corners of a tetrahedron with bond angles of 109°.
- In the expanded structural formula, a separate line is drawn for every bond.

12.2 Alkanes

LEARNING GOAL Write the IUPAC names and draw the condensed structural and line-angle formulas for alkanes and cycloalkanes.

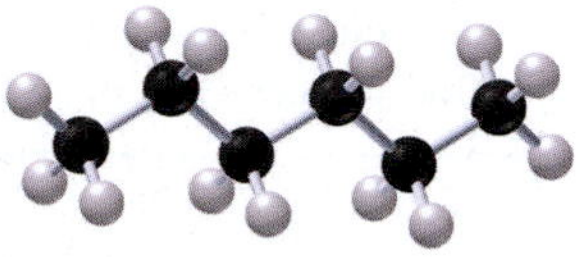

- Alkanes are hydrocarbons that have only C—C single bonds.
- A condensed structural formula depicts groups composed of each carbon atom and its attached hydrogen atoms.
- A line-angle formula represents the carbon skeleton as ends and corners of a zigzag line or geometric figure.
- The IUPAC system is used to name organic compounds by indicating the number of carbon atoms.
- The name of a cycloalkane is written by placing the prefix *cyclo* before the alkane name with the same number of carbon atoms.

12.3 Alkanes with Substituents

LEARNING GOAL Write the IUPAC names for alkanes with substituents and draw their condensed structural and line-angle formulas.

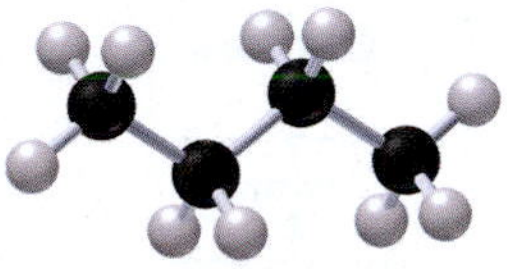

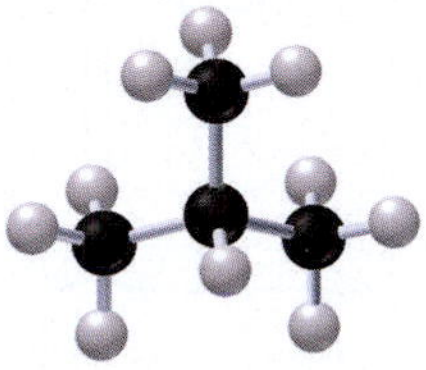

- Structural isomers are compounds with the same molecular formulas that differ in the order in which their atoms are bonded.
- Substituents, which are attached to an alkane chain, include alkyl groups and halogen atoms (F, Cl, Br, or I).
- In the IUPAC system, alkyl substituents have names such as methyl, propyl, and isopropyl; halogen atoms are named as fluoro, chloro, bromo, or iodo.

12.4 Properties of Alkanes

LEARNING GOAL Identify the properties of alkanes and write a balanced chemical equation for combustion.

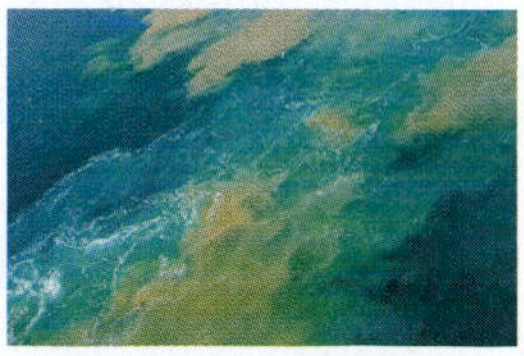

- Alkanes, which are nonpolar molecules, are not soluble in water, and usually less dense than water.
- Alkanes undergo combustion in which they react with oxygen to produce carbon dioxide, water, and energy.

12.5 Alkenes and Alkynes

LEARNING GOAL Write the IUPAC names and draw the condensed structural and line-angle formulas for alkenes and alkynes.

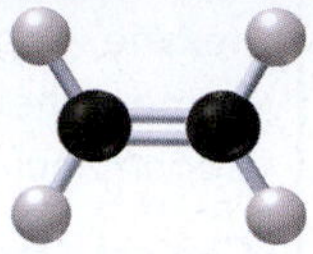

Ethene

- Alkenes are unsaturated hydrocarbons that contain carbon–carbon double bonds.
- Alkynes contain a carbon–carbon triple bond.
- The IUPAC names of alkenes end with *ene*, while alkyne names end with *yne*. The main chain is numbered from the end nearer the double or triple bond.

12.6 Cis–Trans Isomers

LEARNING GOAL Draw the condensed structural formulas and write the names for the cis–trans isomers of alkenes.

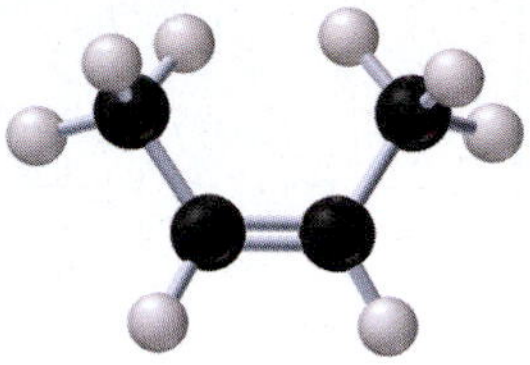

cis-2-Butene

- Geometric or cis–trans isomers of alkenes occur when the carbon atoms in the double bond are connected to different atoms or groups.
- In the cis isomer, the similar groups are on the same side of the double bond, whereas in the trans isomer they are connected on opposite sides of the double bond.

12.7 Addition Reactions for Alkenes

LEARNING GOAL Draw the condensed structural and line-angle formulas and write the names for the organic products of addition reactions of alkenes.

Crisco
Jif

- The addition of small molecules to the double bond is a characteristic reaction of alkenes.
- Hydrogenation adds hydrogen atoms to the double bond of an alkene to yield an alkane.
- Hydration adds water to a double bond of an alkene to form an alcohol.
- When a different number of hydrogen atoms are attached to the carbons in a double bond, the H— from water adds to the carbon with the greater number of H atoms.
- Polymers are long-chain molecules that consist of many repeating units of smaller carbon molecules called monomers.
- Many materials that we use every day are made from synthetic polymers, including carpeting, plastic wrap, nonstick pans, and nylon.
- Many synthetic materials are made using addition reactions in which a catalyst links the carbon atoms from various kinds of alkene molecules (monomers).
- Many materials used in medicine are made from synthetic polymers, including tubing, IV containers, and face masks.

12.8 Aromatic Compounds

LEARNING GOAL Describe the bonding in benzene; name aromatic compounds and draw their line-angle formulas.

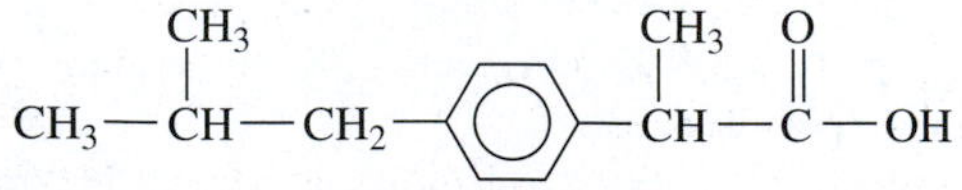
CH3
CH3—CH—CH2
CH3 O
CH—C—OH

Ibuprofen

- Most aromatic compounds contain benzene, C_6H_6, a cyclic structure containing six carbon atoms and six hydrogen atoms.
- The structure of benzene is represented as a hexagon with a circle in the center.
- The IUPAC system uses the names of benzene, toluene, aniline, and phenol.
- In the IUPAC name, two or more substituents are numbered and listed in alphabetical order. In the common name, with two substituents, the positions are shown by the prefixes *ortho* (1,2-), *meta* (1,3-), and *para* (1,4-).

SUMMARY OF NAMING

Family	Structure	IUPAC Name	Common Name
Alkane	$CH_3-CH_2-CH_3$	Propane	
	$CH_3-\underset{\vert}{\overset{CH_3}{}}CH-CH_2-CH_3$ (with CH_3 on CH)	2-Methylbutane	Isopentane
Haloalkane	$CH_3-CH_2-CH_2-Cl$	1-Chloropropane	Propyl chloride
Cycloalkane		Cyclohexane	
Alkene	$CH_3-CH=CH_2$	Propene	Propylene
Cycloalkene		Cyclopropene	
Alkyne	$CH_3-C\equiv CH$	Propyne	
Aromatic	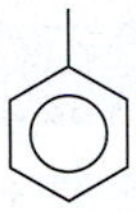	Methylbenzene; toluene	

SUMMARY OF REACTIONS

The chapter Sections to review are shown after the name of each reaction.

Combustion (12.4)

$$\underset{\text{Propane}}{CH_3-CH_2-CH_3(g)} + 5O_2(g) \xrightarrow{\Delta} 3CO_2(g) + 4H_2O(g) + \text{energy}$$

Hydrogenation (12.7)

$$\underset{\text{Propene}}{H_2C=CH-CH_3} + H_2 \xrightarrow{Pt} \underset{\text{Propane}}{CH_3-CH_2-CH_3}$$

Hydration (12.7)

$$\underset{\text{Propene}}{H_2C=CH-CH_3} + H_2O \xrightarrow{H^+} \underset{\text{2-Propanol}}{CH_3-\overset{\overset{OH}{|}}{CH}-CH_3}$$

Polymerization (12.7)

$$\underset{\text{Ethene monomers}}{H_2C=CH_2 + H_2C=CH_2 + H_2C=CH_2} \xrightarrow[\text{catalyst}]{\text{Heat, pressure,}} \underset{\text{Polyethylene}}{-CH_2-CH_2-CH_2-CH_2-CH_2-CH_2-}$$

KEY TERMS

addition A reaction in which atoms or groups of atoms bond to a carbon–carbon double bond. Addition reactions include the addition of hydrogen (hydrogenation), water (hydration), and monomers (polymerization).

alkane A hydrocarbon that contains only single bonds between carbon atoms.

alkene A hydrocarbon that contains one or more carbon–carbon double bonds (C=C).

alkyl group An alkane minus one hydrogen atom. Alkyl groups are named like the alkanes except a *yl* ending replaces *ane*.

alkyne A hydrocarbon that contains one or more carbon–carbon triple bonds ($C\equiv C$).

aromatic compound A compound that contains the ring structure of benzene.

benzene A ring of six carbon atoms, each of which is attached to a hydrogen atom, C_6H_6.

cis isomer An isomer of an alkene in which similar groups in the double bond are on the same side.

condensed structural formula A structural formula that shows the arrangement of the carbon atoms in a molecule but groups each carbon atom with its bonded hydrogen atoms

$-CH_3$, $-CH_2-$, or $-\overset{|}{C}H-$.

cycloalkane An alkane that is a ring or cyclic structure.

expanded structural formula A type of structural formula that shows the arrangement of the atoms by drawing each bond in the hydrocarbon.

hydration An addition reaction in which the components of water, H— and —OH, bond to the carbon–carbon double bond to form an alcohol.

hydrocarbon An organic compound that contains only carbon and hydrogen.

hydrogenation The addition of hydrogen (H_2) to the double bond of alkenes to yield alkanes.

IUPAC system A system for naming organic compounds devised by the International Union of Pure and Applied Chemistry.

line-angle formula A simplified structure that shows a zigzag line in which carbon atoms are represented as the ends of each line and as corners.

meta A method of naming that indicates substituents at carbons 1 and 3 of a benzene ring.

monomer The small organic molecule that is repeated many times in a polymer.

organic compound A compound made of carbon that typically has covalent bonds, is nonpolar, has low melting and boiling points, is insoluble in water, and is flammable.

ortho A method of naming that indicates substituents at carbons 1 and 2 of a benzene ring.

para A method of naming that indicates substituents at carbons 1 and 4 of a benzene ring.

polymer A very large molecule that is composed of many small, repeating monomer units.

structural isomers Organic compounds in which identical molecular formulas have different arrangements of atoms.

substituent Groups of atoms such as an alkyl group or a halogen bonded to the main carbon chain or ring of carbon atoms.

trans isomer An isomer of an alkene in which similar groups in the double bond are on opposite sides.

CORE CHEMISTRY SKILLS

The chapter Section containing each Core Chemistry Skill is shown in parentheses at the end of each heading.

Naming and Drawing Alkanes (12.2)

- The alkanes ethane, propane, and butane contain two, three, and four carbon atoms, respectively, connected in a row or a *continuous* chain.
- Alkanes with five or more carbon atoms in a chain are named using the prefixes *pent* (5), *hex* (6), *hept* (7), *oct* (8), *non* (9), and *dec* (10).
- In the condensed structural formula, the carbon and hydrogen atoms on the ends are written as $-CH_3$ and the carbon and hydrogen atoms in the middle are written as $-CH_2-$.

Example: **a.** What is the name of $CH_3-CH_2-CH_2-CH_3$?
b. Draw the condensed structural formula for pentane.

Answer: **a.** An alkane with a four-carbon chain is named butane.
b. Pentane is an alkane with a five-carbon chain. The carbon atoms on the ends are attached to three H atoms each, and the carbon atoms in the middle are attached to two H each. $CH_3-CH_2-CH_2-CH_2-CH_3$

Writing Equations for Hydrogenation, Hydration, and Polymerization (12.7)

- The addition of small molecules to the double bond is a characteristic reaction of alkenes.
- Hydrogenation adds hydrogen atoms to the double bond of an alkene to form an alkane.
- Hydration adds water to a double bond of an alkene to form an alcohol.
- When a different number of hydrogen atoms are attached to the carbons in a double bond, the H— from H_2O adds to the carbon atom with the greater number of H atoms, and the —OH adds to the other carbon atom from the double bond.
- In polymerization, many small molecules (monomers) join together to form a polymer.

Example: **a.** Draw the condensed structural formula for 2-methyl-2-butene.
b. Draw the condensed structural formula for the product of the hydrogenation of 2-methyl-2-butene.
c. Draw the condensed structural formula for the product of the hydration of 2-methyl-2-butene.
d. Draw the condensed structural formula of a portion of the polymer polybutene (polybutylene), used in synthetic rubber, formed from three monomers of 1-butene.

Answer: **a.**

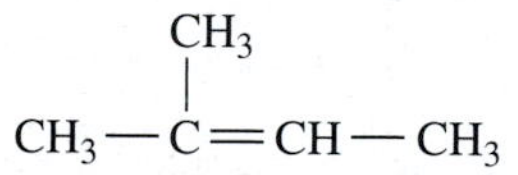

b. When H_2 adds to a double bond, the product is an alkane with the same number of carbon atoms.

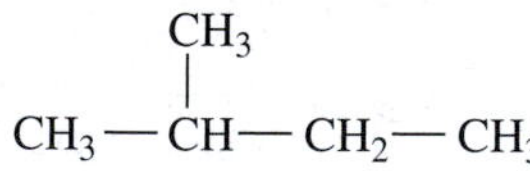

c. When water adds to a double bond, the product is an alcohol, which has an —OH group. The H— from water adds to the carbon atom in the double bond that has the greater number of H atoms.

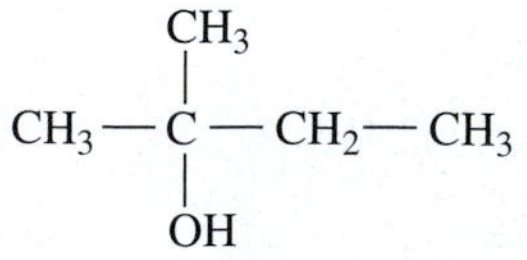

d.

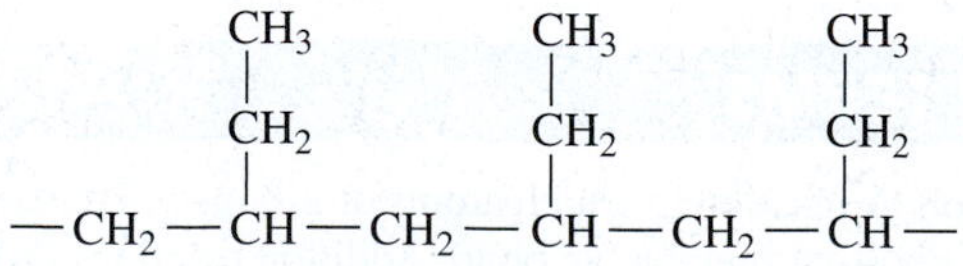

UNDERSTANDING THE CONCEPTS

The chapter Sections to review are shown in parentheses at the end of each problem.

12.51 Match the following physical and chemical properties with potassium chloride, KCl, used in salt substitutes, or butane, C_4H_{10}, used in lighters: (12.1)

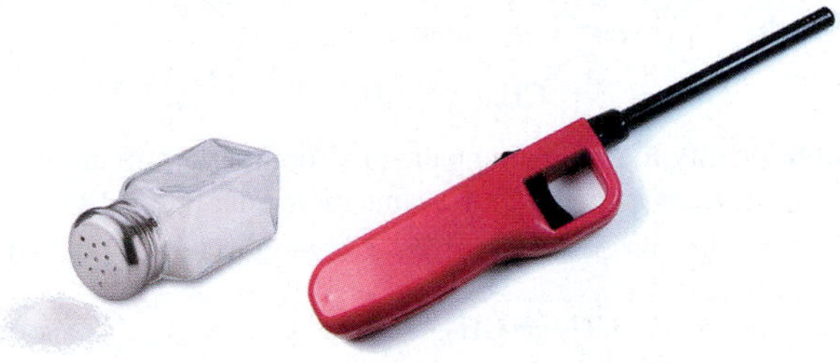

a. melts at −138 °C
b. burns vigorously in air
c. melts at 770 °C
d. contains ionic bonds
e. is a gas at room temperature

12.52 Match the following physical and chemical properties with octane, C_8H_{18}, found in gasoline, or magnesium sulfate, $MgSO_4$, also called Epsom salts: (12.1)

a. contains only covalent bonds
b. melts at 1124 °C
c. is insoluble in water
d. is a liquid at room temperature
e. is a strong electrolyte

12.53 Identify the compounds in each of the following pairs as structural isomers or not structural isomers: (12.3)

a.  and 

b.  and 

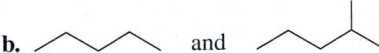

12.54 Identify the compounds in each of the following pairs as structural isomers or not structural isomers: (12.3)

a.
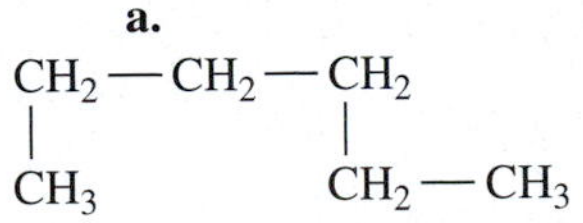 and 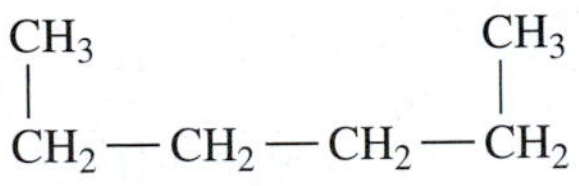

b. 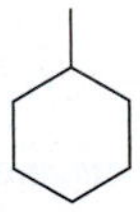 and 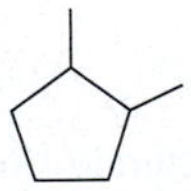

12.55 Convert each of the following line-angle formulas to a condensed structural formula and write its IUPAC name: (12.3)

a. 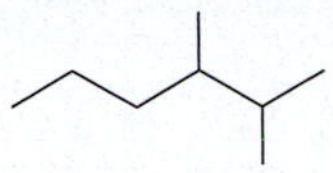 **b.** 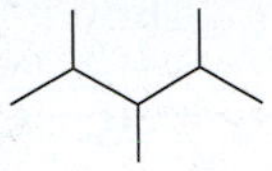

12.56 Convert each of the following line-angle formulas to a condensed structural formula and write its IUPAC name: (12.3)

a. 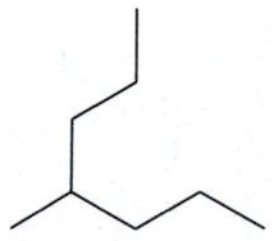 **b.** 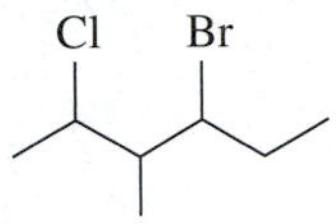

12.57 Draw a portion of the polymer of Teflon, which is made from 1,1,2,2-tetrafluoroethene (use three monomers). (12.7)

Teflon is used as a nonstick coating on cooking pans.

12.58 A garden hose is made of polyvinyl chloride (PVC) from chloroethene (vinyl chloride). Draw a portion of the polymer (use three monomers) for PVC. (12.7)

A garden hose is made of polyvinyl chloride (PVC).

12.59 Give the number of carbon atoms and the types of carbon–carbon bonds for each of the following: (12.2, 12.3, 12.5)

a. propane
b. cyclopropane
c. propene
d. propyne

12.60 Give the number of carbon atoms and the types of carbon–carbon bonds for each of the following: (12.2, 12.3, 12.5)

a. butane
b. cyclobutane
c. cyclobutene
d. 2-butyne

ADDITIONAL PRACTICE PROBLEMS

12.61 Write the IUPAC name for each of the following: (12.3)

a. 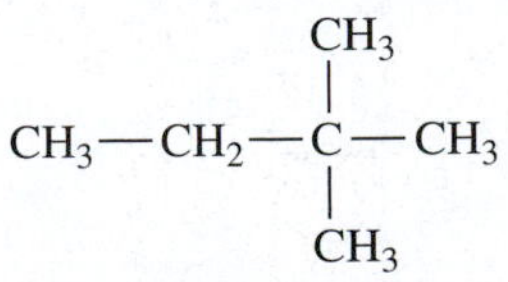

b. CH_3-CH_2-Cl

c. 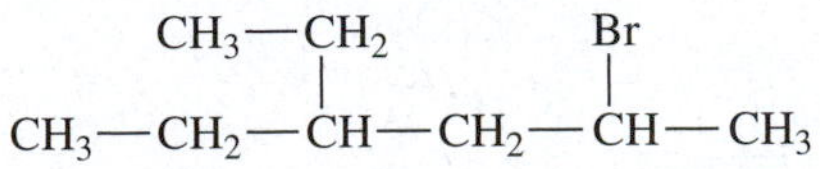

d. 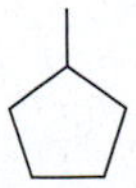

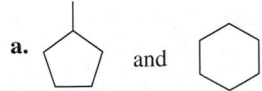

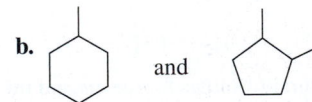

12.62 Write the IUPAC name for each of the following: (12.3)

a. 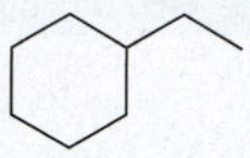

b. 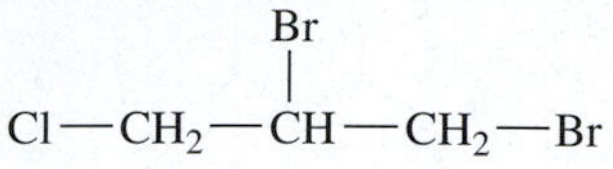

c. 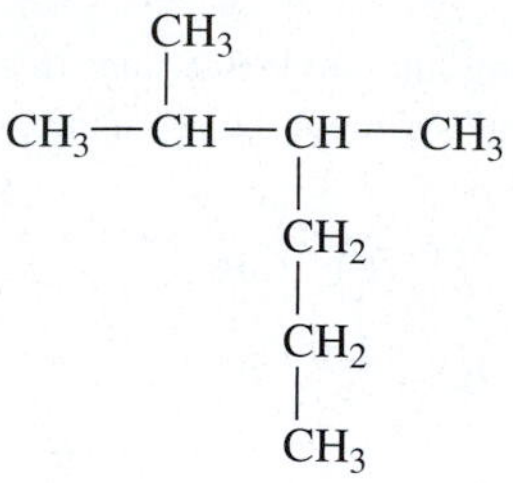

d. 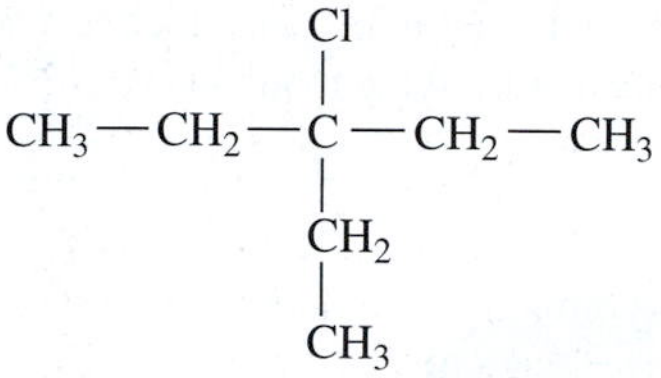

12.63 Draw the condensed structural formula for alkanes or the line-angle formula for cycloalkanes for each of the following: (12.3, 12.5, 12.6)

a. 1,2-dibromocyclopentene
b. 2-hexyne
c. *cis*-2-heptene

12.64 Draw the condensed structural formula for alkanes or the line-angle formula for cycloalkanes for each of the following: (12.3, 12.5, 12.6)

a. *trans*-3-hexene
b. 2-bromo-3-chlorocyclohexene
c. 3-chloro-1-butyne

12.65 Write the IUPAC name (including cis or trans, if needed) for each of the following: (12.5, 12.6)

a. 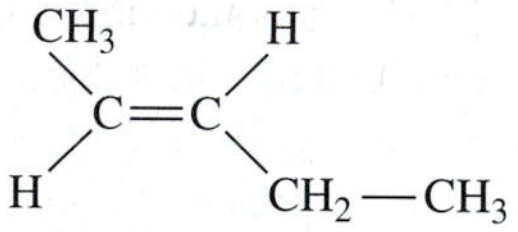

b. 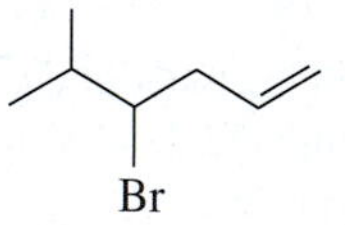

c. 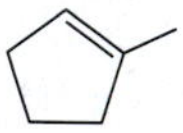

12.66 Write the IUPAC name (including cis or trans, if needed) for each of the following: (12.5, 12.6)

a. 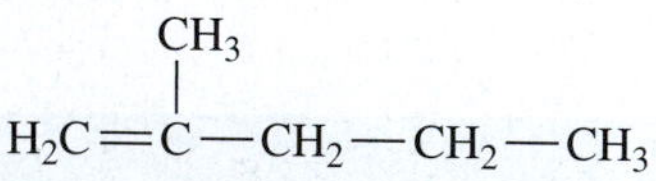

b. 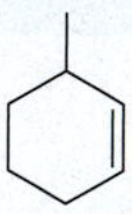

c. 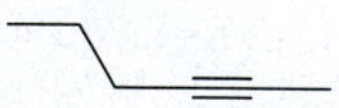

12.67 Identify the following pairs of structures as structural isomers, cis–trans isomers, or the same molecule: (12.5, 12.6)

a. 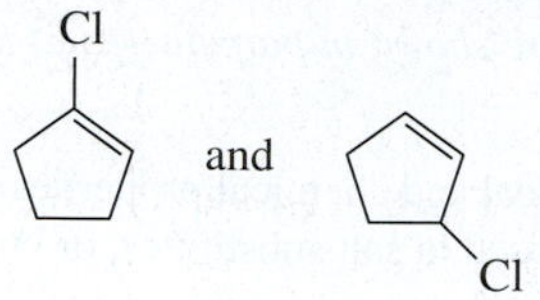

b. 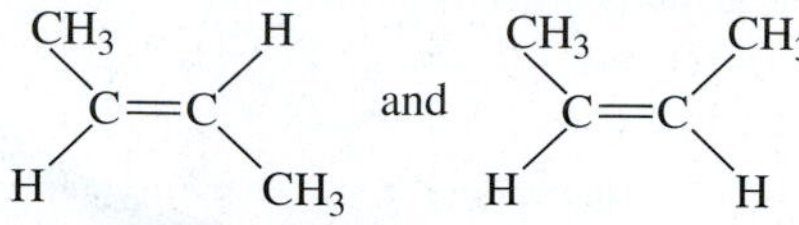

12.68 Identify the following pairs of structures as structural isomers, cis–trans isomers, or the same molecule: (12.5, 12.6)

a. 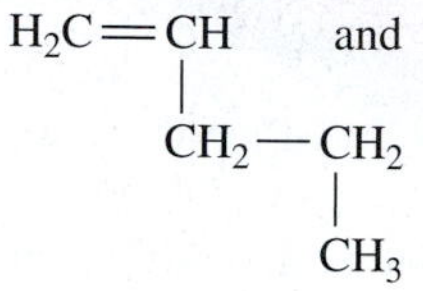 and $CH_3—CH_2—CH_2—CH{=}CH_2$

b. 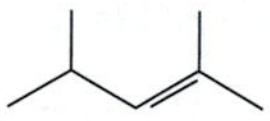 and 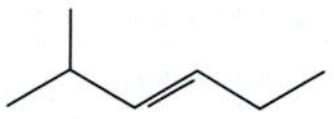

12.69 Write the IUPAC name, including cis or trans, for each of the following: (12.5, 12.6)

a. 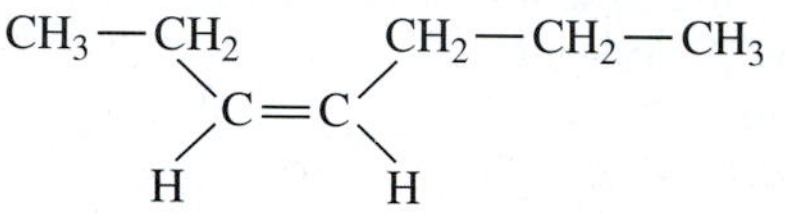

b. 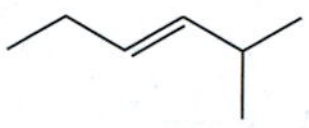

c. 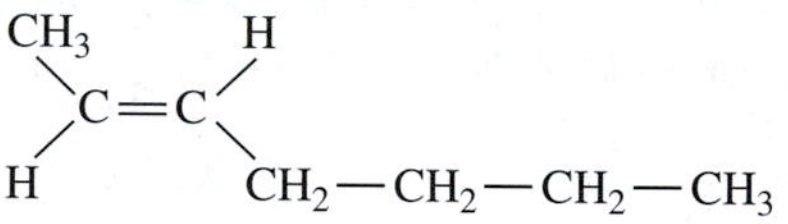

12.70 Write the IUPAC name, including cis or trans, for each of the following: (12.5, 12.6)

a. 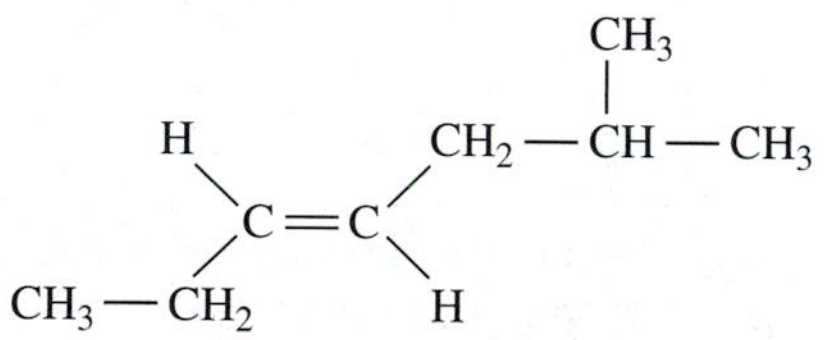

b. 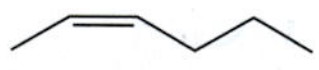

c. 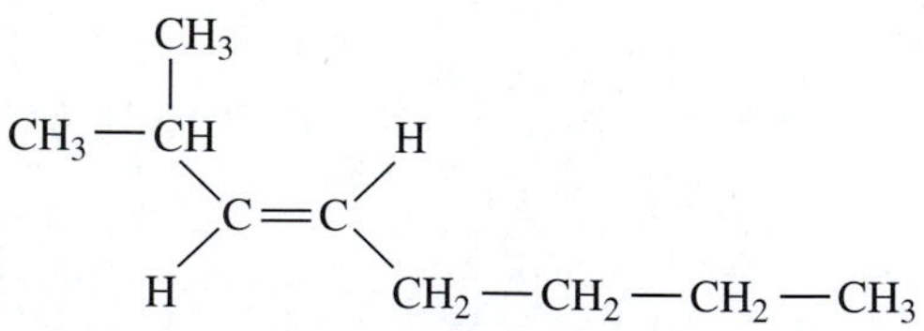

12.71 Draw the condensed structural formulas for the cis and trans isomers for each of the following: (12.5, 12.6)

a. 2-pentene
b. 3-hexene

12.72 Draw the condensed structural formulas for the cis and trans isomers for each of the following: (12.5, 12.6)

a. 2-butene
b. 2-hexene

12.73 Write a balanced chemical equation for the complete combustion of each of the following: (12.4, 12.5)
a. dimethylpropane
b. cyclobutane
c. 2-heptene

12.74 Write a balanced chemical equation for the complete combustion of each of the following: (12.4, 12.5)
a. cycloheptane
b. 2-methyl-1-pentene
c. 3,3-dimethylhexane

12.75 Write the name for the product from the hydrogenation of each of the following: (12.7)
a. 3-methyl-2-pentene
b. 3-methylcyclohexene
c. 2-butene

12.76 Write the name for the product from the hydrogenation of each of the following: (12.7)
a. 1-hexene
b. 2-methyl-2-butene
c. 1-ethylcyclopentene

12.77 Draw the line-angle formula for the product of each of the following: (12.7)

a.  $+ H_2 \xrightarrow{Ni}$

b.  $+ H_2O \xrightarrow{H^+}$

c. $CH_3—CH_2—\overset{CH_3}{\overset{|}{C}}{=}CH_2 + H_2O \xrightarrow{H^+}$

12.78 Draw the line-angle formula for the product of each of the following: (12.7)

a.  $+ H_2O \xrightarrow{H^+}$

b. $CH_3—CH_2—CH_2—CH{=}CH_2 + H_2 \xrightarrow{Ni}$

c. 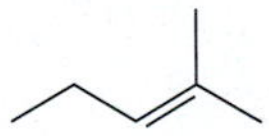 $+ H_2O \xrightarrow{H^+}$

12.79 Copolymers contain more than one type of monomer. One copolymer used in medicine is made of alternating units of styrene and acrylonitrile. Draw the expanded structural formula for a portion of the copolymer SAN that would have three each of these alternating units. (12.7)

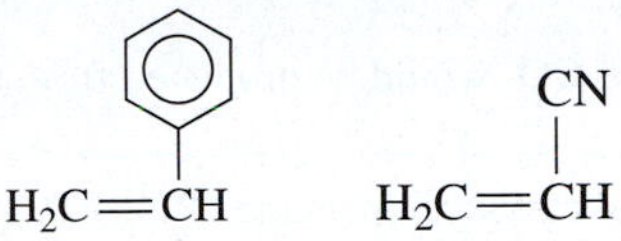

H₂C=CH
CN
H₂C=CH

Styrene   Acrylonitrile

12.80 Lucite, or Plexiglas, is a polymer of methyl methacrylate. Draw the expanded structural formula for a portion of the polymer that is made from the addition of three of these monomers. (12.7)

$$H_2C{=}\overset{CH_3}{\overset{|}{C}}—\overset{O}{\overset{\|}{C}}—O—CH_3$$

Methyl methacrylate

12.81 Name each of the following aromatic compounds: (12.8)

a. 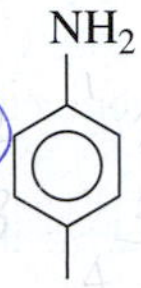

NH₂

b. 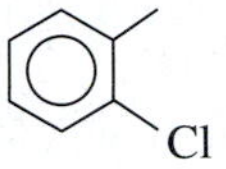

Cl

c. 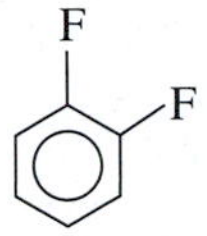

F
F

12.82 Name each of the following aromatic compounds: (12.8)

para-(1,4-) dichloro benzene **a.** 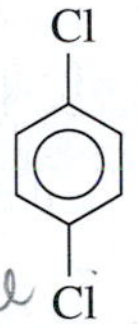

Cl
Cl

b. 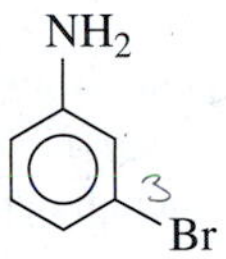

NH₂
Br

3-bromoanaline

c. 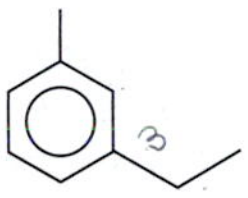
3-ethyl tolmene

12.83 Draw the line-angle formula for each of the following: (12.8)
a. *p*-bromotoluene
b. 2,6-dimethylaniline
c. 1,4-dimethylbenzene

12.84 Draw the line-angle formula for each of the following: (12.8)
a. ethylbenzene
b. *m*-chloroaniline
c. 1,2,4-trimethylbenzene

CHALLENGE PROBLEMS

The following problems are related to the topics in this chapter. However, they do not all follow the chapter order, and they require you to combine concepts and skills from several Sections. These problems will help you increase your critical thinking skills and prepare for your next exam.

12.85 The density of pentane, a component of gasoline, is 0.63 g/mL. The heat of combustion for pentane is 845 kcal/mole. (7.2, 7.4, 7.7, 7.8, 8.6, 12.2, 12.4)
a. Write the balanced chemical equation for the complete combustion of pentane.
b. What is the molar mass of pentane?
c. How much heat is produced when 1 gal of pentane is burned (1 gal = 4 qt)?
d. How many liters of CO_2 at STP are produced from the complete combustion of 1 gal of pentane?

12.86 Acetylene (ethyne) gas reacts with oxygen and burns at 3300 °C in an acetylene torch. (7.4, 7.7, 8.6, 12.5)
a. Write the balanced chemical equation for the complete combustion of acetylene.
b. What is the molar mass of acetylene?
c. How many grams of oxygen are needed to react with 8.5 L of acetylene gas at STP?
d. How many liters of CO_2 gas at STP are produced when 30.0 g of acetylene undergoes combustion?

12.87 Draw the condensed structural formulas for all the possible alkane isomers that have a total of six carbon atoms and a four-carbon chain. (12.3)

12.88 Draw the condensed structural formulas for all the possible haloalkane isomers that have four carbon atoms and a bromine. (12.3)

12.89 Consider the compound propane. (7.4, 7.7, 12.2, 12.4)
- **a.** Draw the condensed structural formula for propane.
- **b.** Write the balanced chemical equation for the complete combustion of propane.
- **c.** How many grams of O_2 are needed to react with 12.0 L of propane gas at STP?
- **d.** How many grams of CO_2 would be produced from the reaction in part **c**?

12.90 Consider the compound ethylcyclopentane. (7.4, 7.7, 8.6, 12.3, 12.4)
- **a.** Draw the line-angle formula for ethylcyclopentane.
- **b.** Write the balanced chemical equation for the complete combustion of ethylcyclopentane.
- **c.** Calculate the grams of O_2 required for the combustion of 25.0 g of ethylcyclopentane.
- **d.** How many liters of CO_2 would be produced at STP from the reaction in part **c**?

12.91 If a female silkworm moth secretes 50 ng of bombykol, a sex attractant, how many molecules did she secrete (see Chemistry Link to the Environment: Pheromones in Insect Communication)? (7.2, 12.6)

12.92 How many grams of hydrogen are needed to hydrogenate 30.0 g of 2-butene? (12.7)

12.93 Draw the condensed structural formula for and write the name of all the possible alkenes with molecular formula C_5H_{10} that have a five-carbon chain, including those with cis and trans isomers. (12.5, 12.6)

12.94 Draw the condensed structural formula for and write the name of all the possible alkenes with molecular formula C_6H_{12} that have a six-carbon chain, including those with cis and trans isomers. (12.5, 12.6)

12.95 Explosives used in mining contain TNT or trinitrotoluene. (12.8)
- **a.** If the functional group *nitro* is $—NO_2$, draw the line-angle formula for 2,4,6-trinitrotoluene, one isomer of TNT.
- **b.** TNT is actually a mixture of structural isomers of trinitrotoluene. Draw the line-angle formulas for two other possible structural isomers.

Explosives containing TNT are used in mining.

12.96 Margarine is produced from the hydrogenation of vegetable oils, which contain unsaturated fatty acids. How many grams of hydrogen are required to completely saturate 75.0 g of oleic acid, $C_{18}H_{34}O_2$, which has one double bond? (12.7)

Margarines are produced by the hydrogenation of unsaturated fats.

ANSWERS

12.1
- **a.** inorganic
- **b.** organic, condensed structural formula
- **c.** organic, expanded structural formula
- **d.** inorganic
- **e.** inorganic
- **f.** organic, molecular formula

12.3 **a.** inorganic **b.** organic **c.** organic **d.** inorganic

12.5 **a.** ethane **b.** ethane **c.** NaBr **d.** NaBr

12.7 **a.** pentane **b.** ethane **c.** hexane **d.** cycloheptane

12.9
- **a.** CH_4
- **b.** $CH_3—CH_3$
- **c.** $CH_3—CH_2—CH_2—CH_2—CH_3$
- **d.** 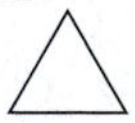

12.11
- **a.** same molecule
- **b.** structural isomers of C_5H_{12}
- **c.** structural isomers of C_6H_{14}

12.13
- **a.** dimethylpropane
- **b.** 2,3-dimethylpentane
- **c.** 4-*tert*-butyl-2,6-dimethylheptane
- **d.** methylcyclobutane
- **e.** 1-bromo-3-chlorocyclohexane

12.15
- **a.** 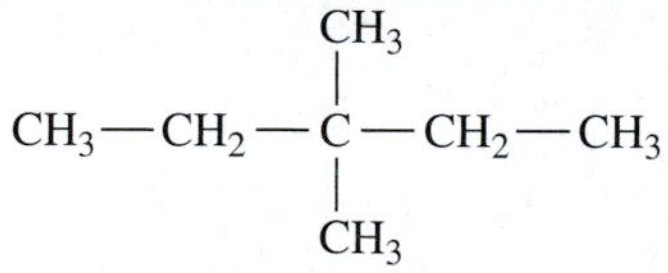
- **b.** 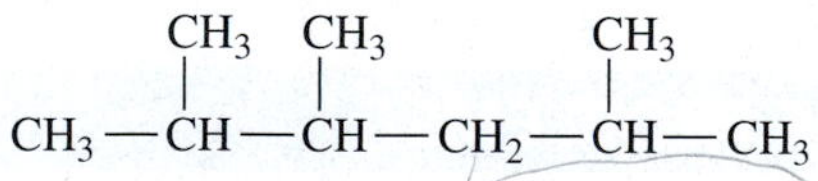
- **c.** 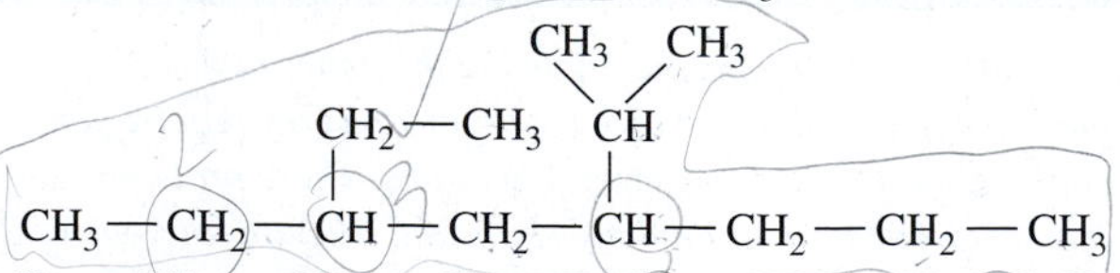
- **d.** $Br—CH_2—CH_2—Cl$

12.17
- **a.** 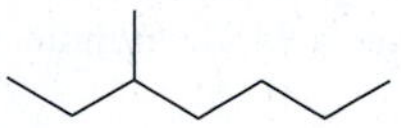
- **b.** 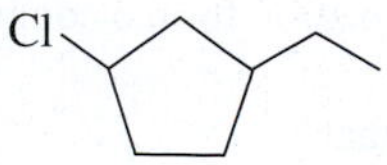
- **c.** 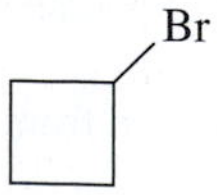
- **d.** 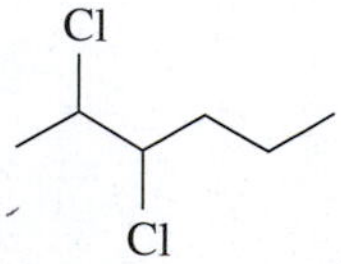

12.19 a. heptane **b.** cyclopropane **c.** hexane

12.21 a. $CH_3-CH_2-CH_2-CH_2-CH_2-CH_2-CH_3$

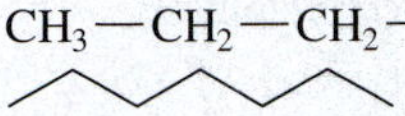

b. liquid
c. no
d. float
e. $C_7H_{16}(g) + 11O_2(g) \xrightarrow{\Delta} 7CO_2(g) + 8H_2O(g) + \text{energy}$

12.23 a. $2C_2H_6(g) + 7O_2(g) \xrightarrow{\Delta} 4CO_2(g) + 6H_2O(g) + \text{energy}$
b. $2C_3H_6(g) + 9O_2(g) \xrightarrow{\Delta} 6CO_2(g) + 6H_2O(g) + \text{energy}$
c. $2C_8H_{18}(g) + 25O_2(g) \xrightarrow{\Delta} 16CO_2(g) + 18H_2O(g) + \text{energy}$

12.25 a. alkene **b.** alkyne **c.** alkene **d.** cycloalkene

12.27 a. 2-methylpropene **b.** 4-bromo-2-pentyne
c. 4-ethylcyclopentene **d.** 4-ethyl-2-hexene

12.29 a. $H_2C{=}CH-CH_2-CH_2-CH_3$

b. 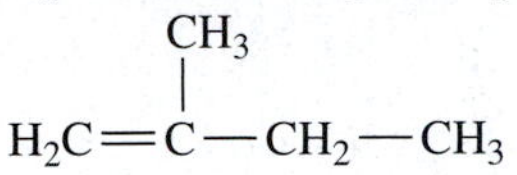

c. 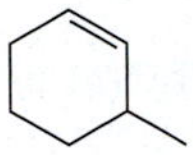

d. 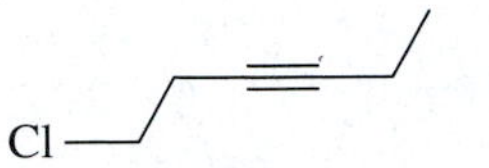

12.31 a. *cis*-3-heptene **b.** *trans*-3-octene **c.** propene

12.33 a. 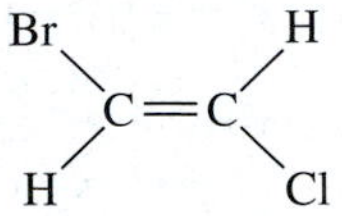

b. 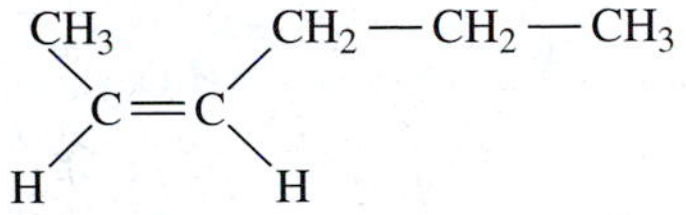

c. 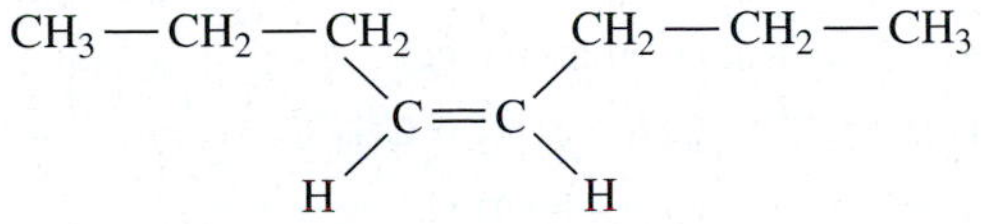

12.35 a. $CH_3-CH_2-CH_2-CH_2-CH_3$

b. 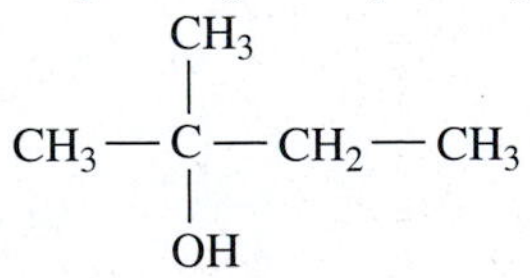

c. 

d. 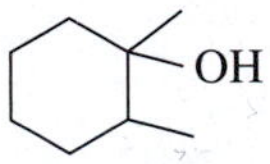

12.37 A polymer is a very large molecule composed of small units (monomers) that are repeated many times.

12.39 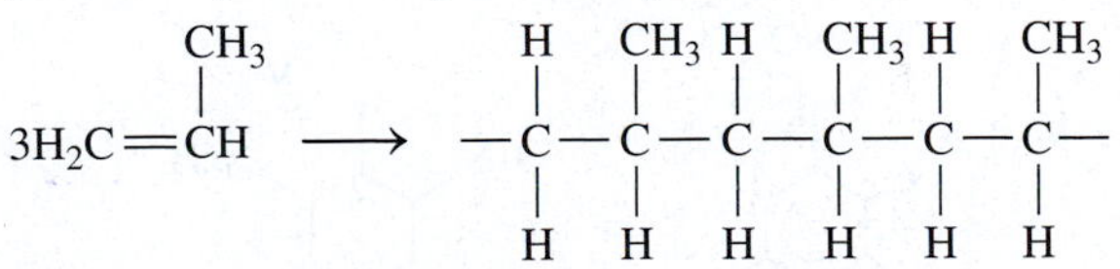

12.41 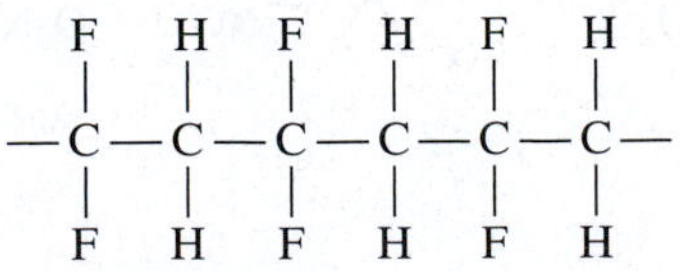

12.43 a. 2-chlorotoluene
b. ethylbenzene
c. 1,3,5-trichlorobenzene
d. 1,3-dimethylbenzene (*m*-dimethylbenzene)
e. 3-bromo-5-chlorotoluene
f. isopropylbenzene

12.45 a. 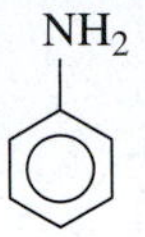

b. 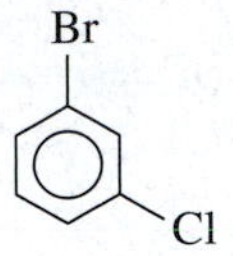

c. 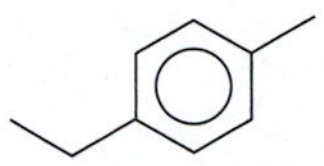

d. 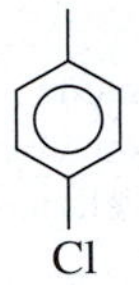

12.47 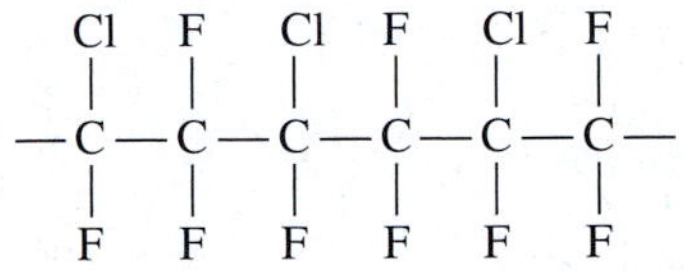

12.49
a. $C_9H_{20}(l) + 14O_2(g) \xrightarrow{\Delta} 9CO_2(g) + 10H_2O(g) + \text{energy}$
b. $C_5H_{12}(l) + 8O_2(g) \xrightarrow{\Delta} 5CO_2(g) + 6H_2O(g) + \text{energy}$
c. $C_9H_{12}(l) + 12O_2(g) \xrightarrow{\Delta} 9CO_2(g) + 6H_2O(g) + \text{energy}$

12.51 a. butane
b. butane
c. potassium chloride
d. potassium chloride
e. butane

12.53 a. structural isomers
b. not structural isomers

12.55 a. 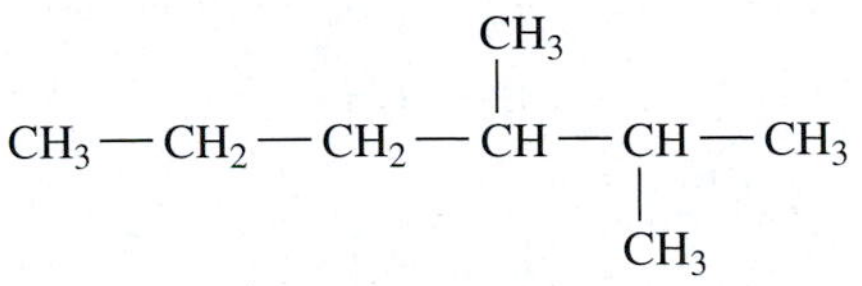

2,3-Dimethylhexane

b. 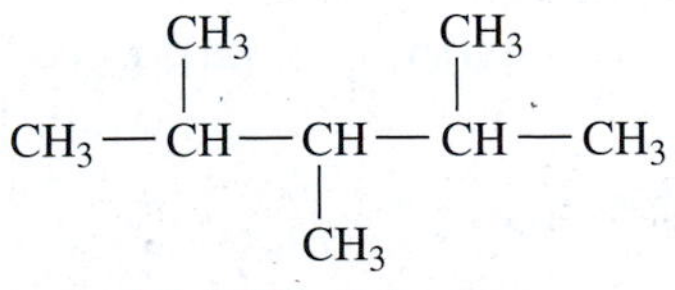

2,3,4-Trimethylpentane

12.57 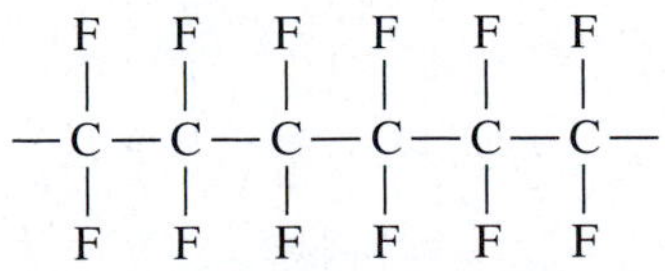

12.59 a. three carbon atoms; two carbon–carbon single bonds
b. three carbon atoms; three carbon–carbon single bonds in a ring
c. three carbon atoms; one carbon–carbon single bond; one carbon–carbon double bond
d. three carbon atoms; one carbon–carbon single bond; one carbon–carbon triple bond

12.61 a. 2,2-dimethylbutane
b. chloroethane
c. 2-bromo-4-ethylhexane
d. methylcyclopentane

12.63 a. 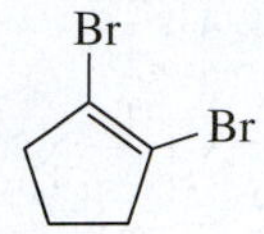

b. $CH_3—C{\equiv}C—CH_2—CH_2—CH_3$

c. 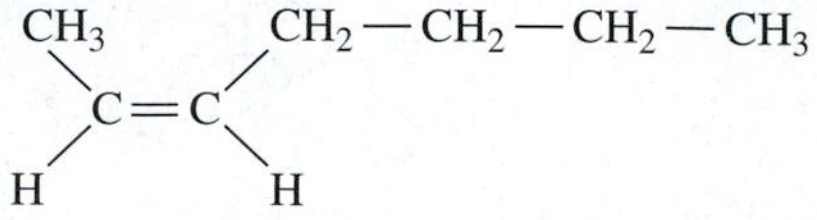

12.65 a. *trans*-2-pentene
b. 4-bromo-5-methyl-1-hexene
c. 1-methylcyclopentene

12.67 a. structural isomers
b. cis-trans isomers

12.69 a. *cis*-3-heptene
b. *trans*-2-methyl-3-hexene
c. *trans*-2-heptene

12.71 a. 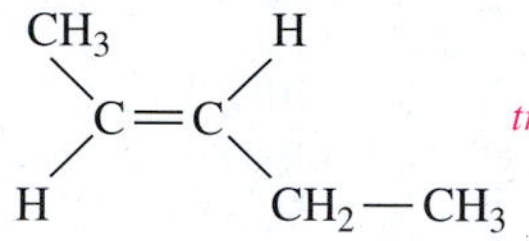 *trans*-2-Pentene

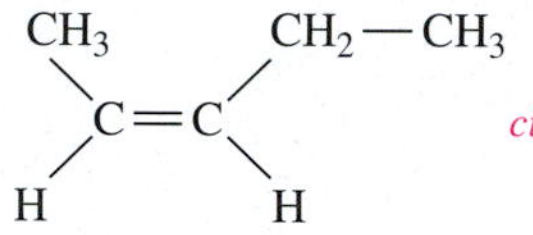 *cis*-2-Pentene

b. 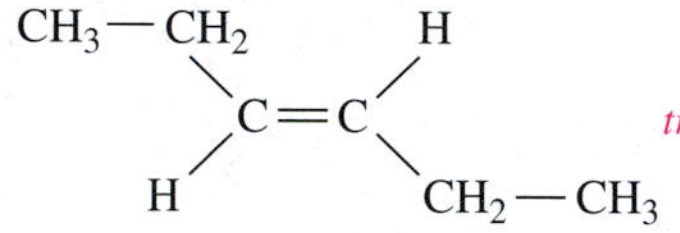 *trans*-3-Hexene

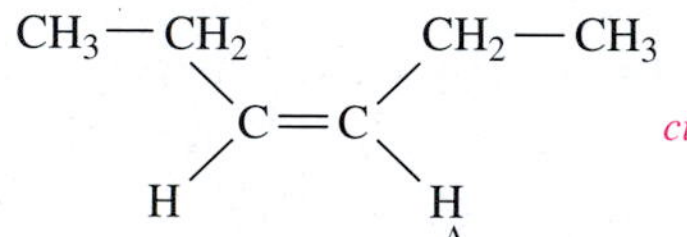 *cis*-3-Hexene

12.73 a. $C_5H_{12}(g) + 8O_2(g) \xrightarrow{\Delta} 5CO_2(g) + 6H_2O(g) + \text{energy}$
b. $C_4H_8(g) + 6O_2(g) \xrightarrow{\Delta} 4CO_2(g) + 4H_2O(g) + \text{energy}$
c. $2C_7H_{14}(l) + 21O_2(g) \xrightarrow{\Delta} 14CO_2(g) + 14H_2O(g) + \text{energy}$

12.75 a. 3-methylpentane
b. methylcyclohexane
c. butane

12.77 a. 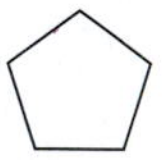

b. 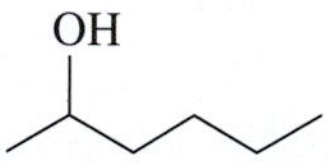

c. 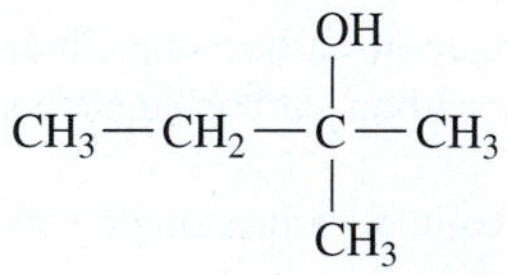

12.79 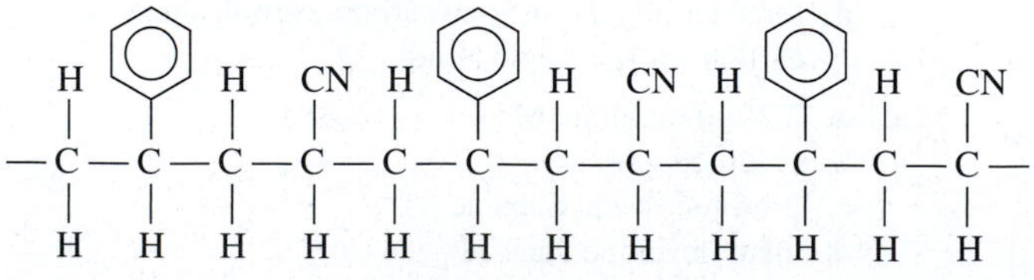

12.81 a. 3-methylaniline (*p*-methylaniline)
b. 1-chloro-2-methylbenzene, 2-chlorotoluene (*o*-chlorotoluene)
c. 1,2-difluorobenzene (*o*-difluorobenzene)

12.83 a. 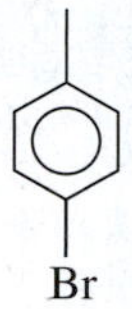

b. 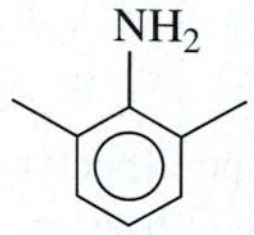

c. 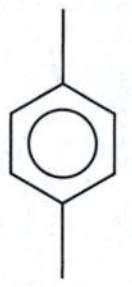

12.85 a. $C_5H_{12}(g) + 8O_2(g) \xrightarrow{\Delta} 5CO_2(g) + 6H_2O(g) + \text{energy}$
b. 72.15 g/mole
c. 2.8×10^4 kcal
d. 3.7×10^3 L of CO_2 at STP

12.87 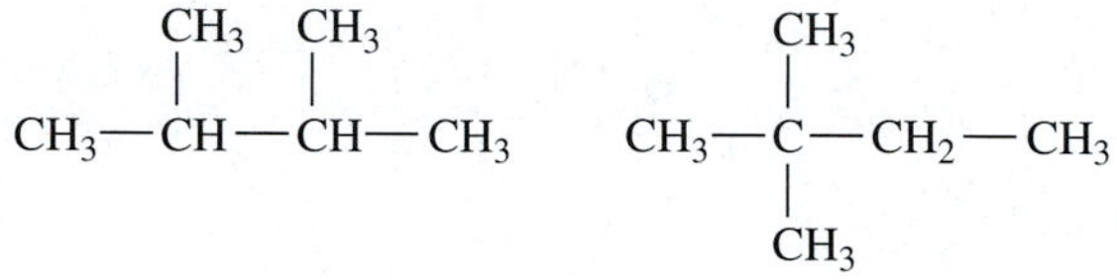

12.89 a. $CH_3—CH_2—CH_3$
b. $C_3H_8(g) + 5O_2(g) \xrightarrow{\Delta} 3CO_2(g) + 4H_2O(g) + \text{energy}$
c. 85.7 g of O_2
d. 70.7 g of CO_2

12.91 1×10^{14} molecules of bombykol

12.93 a. $H_2C{=}CH—CH_2—CH_2—CH_3$ 1-Pentene

b. 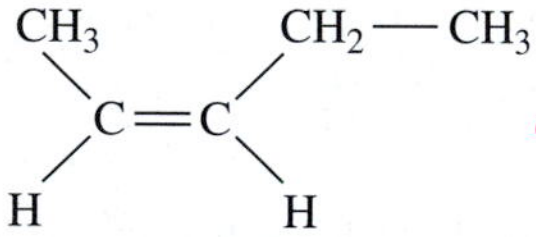 *cis*-2-Pentene

c. 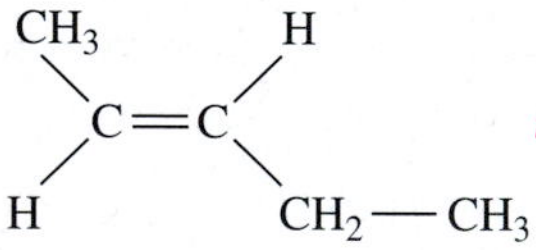 *trans*-2-Pentene

12.95 a. 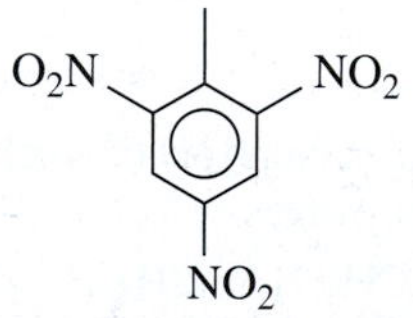

b. 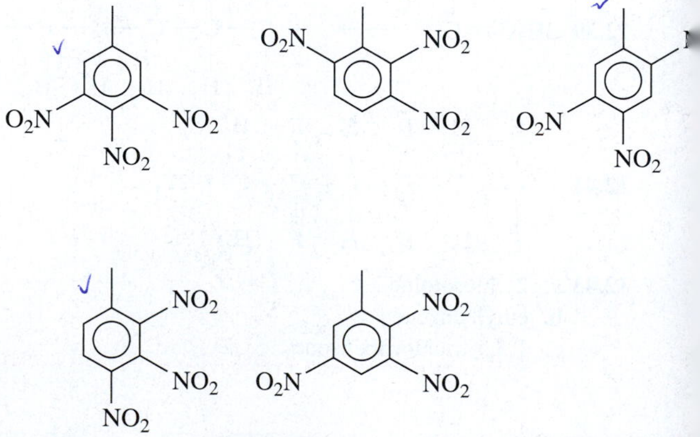

13

Alcohols, Phenols, Thiols, and Ethers

Janet injured her knee in a ski accident six months ago. She has tried physical therapy as well as anti-inflammatory medications. However, her knee is still very painful, and her doctor recommends surgery to repair the torn ACL (anterior cruciate ligament). Liz, a nurse anesthetist, prepares an injection of 50 mg of propofol (Diprivan) used to relax Janet before surgery. During surgery, more propofol is give by an IV drip. After surgery, Janet's doctor suggests that losing weight would help her to recover more quickly from the knee surgery. She selects a diet plan that she likes, which includes eating more fruits and vegetables.

One of the early anesthetics was diethyl ether, commonly referred to as ether. It contains a functional group that has an oxygen atom bonded by single bonds to two carbon groups. Today, it is not used because it is extremely flammable and produces undesirable side effects. Today, more modern inhaled anesthetics, such as Sevoflurane, that do not cause nausea and vomiting are used.

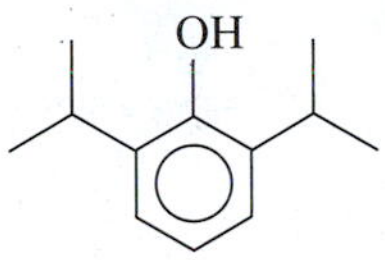
OH

Propofol (Diprivan)

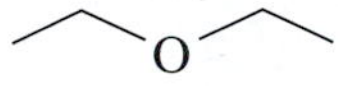
O

Ethoxyethane (diethyl ether)

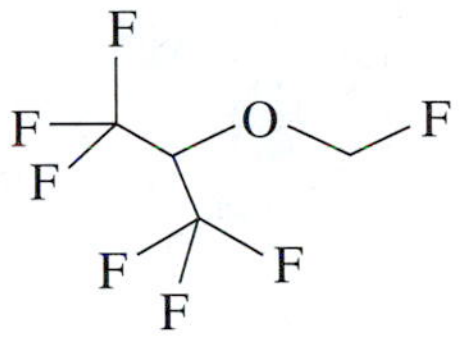
F
F
F
O
F
F
F
F

Sevoflurane

CAREER

Nurse Anesthetist

Each year, more than 26 million people in the United States undergo medical procedures that require anesthesia. Anesthesia is typically administered by a nurse anesthetist who provides care before, during, and after the procedure by giving medications to keep a patient asleep and pain-free while monitoring the patient's vital signs. A nurse anesthetist also obtains supplies, equipment, and determines how the anesthesia will affect the patient. A nurse anesthetist may also be required to insert artificial airways, administer oxygen, or work to prevent surgical shock during a procedure, working under the direction of the attending surgeon, dentist, or anesthesiologist.

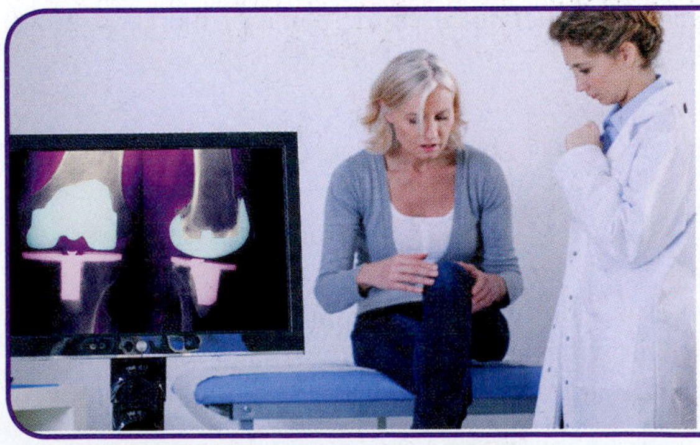

CLINICAL UPDATE

Janet's New Diet Plan

Janet has started on her diet plan to lose weight. You can see details of her diet in the **CLINICAL UPDATE Janet's New Diet Plan**, page 487, including the functional groups in the line-angle structures of some of the antioxidants found in foods.

LOOKING AHEAD

REVIEW

Naming and Drawing Alkanes (12.2)

CORE CHEMISTRY SKILL

Identifying Alcohols, Phenols, and Thiols

13.1 Alcohols, Phenols, and Thiols

LEARNING GOAL Write the IUPAC and common names for alcohols, phenols, and thiols. Draw their condensed structural and line-angle formulas.

Alcohols, which contain the *hydroxyl group* (—OH), are commonly found in nature and are used in industry and at home. For centuries, grains, vegetables, and fruits have been fermented to produce the ethanol present in alcoholic beverages. The hydroxyl group is important in biomolecules such as sugars and starches as well as in steroids such as cholesterol and estradiol. The phenols contain the hydroxyl group attached to a benzene ring. Thiols, which contain the —SH group, give the strong odors we associate with garlic and onions.

In an **alcohol**, the hydroxyl functional group (—OH) replaces a hydrogen atom in a hydrocarbon. Oxygen (O) atoms are shown in red in the ball-and-stick models (see **FIGURE 13.1**). In a **phenol**, the hydroxyl group replaces a hydrogen atom attached to a benzene ring. A **thiol** contains a sulfur atom, shown in yellow-green in the ball-and-stick model, which makes a thiol similar to an alcohol except that —OH is replaced by an —SH group. Molecules of alcohols, phenols, and thiols have bent shapes around the oxygen or sulfur atom, similar to water.

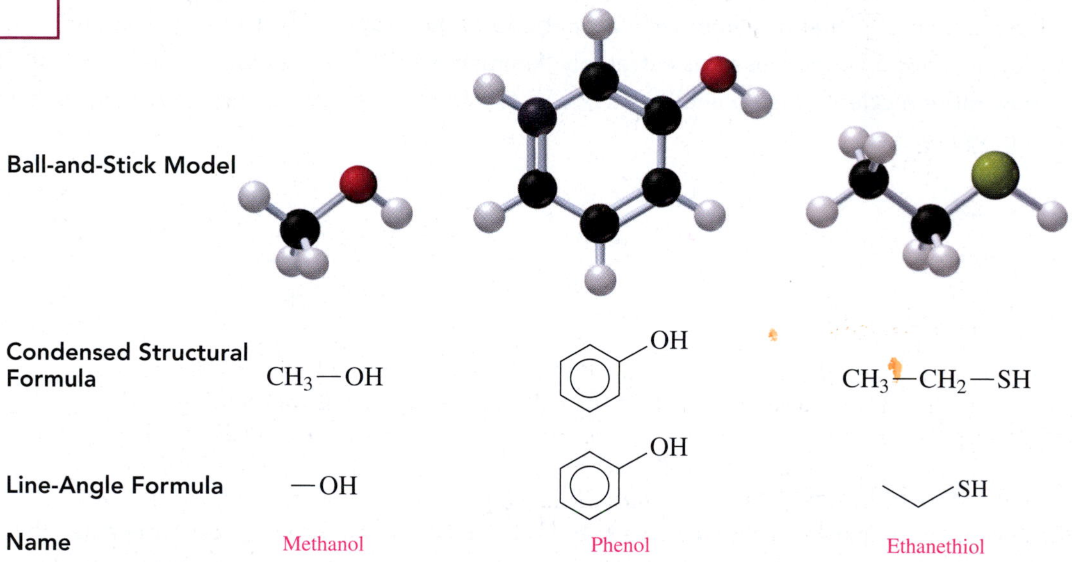

Ball-and-Stick Model
Condensed Structural Formula
CH3—OH
OH
CH3—CH2—SH
Line-Angle Formula
—OH
OH
SH
Name
Methanol
Phenol
Ethanethiol

FIGURE 13.1 ▶ An alcohol has a hydroxyl group (—OH) attached to carbon, and a phenol has a hydroxyl group (—OH) attached to a benzene ring. A thiol has a thiol group (—SH) attached to carbon.

Q How is an alcohol different from a thiol?

CORE CHEMISTRY SKILL

Naming Alcohols and Phenols

Naming Alcohols

In the IUPAC system, an alcohol is named by replacing the *e* of the corresponding alkane name with *ol*. The common name of a simple alcohol uses the name of the alkyl group followed by *alcohol*.

$CH_3—H$	$CH_3—OH$	$CH_3—CH_2—H$	$CH_3—CH_2—OH$
Methane	Methanol (methyl alcohol)	Ethane	Ethanol (ethyl alcohol)

Alcohols with one or two carbon atoms do not require a number for the hydroxyl group. When an alcohol consists of a chain with three or more carbon atoms, the chain is numbered to give the position of the —OH group and any substituents on the chain. An alcohol with two —OH groups is named as a *diol*, and an alcohol with three —OH groups is named as a *triol*.

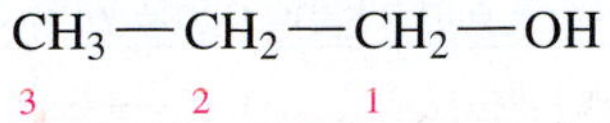
1-Propanol
(propyl alcohol)

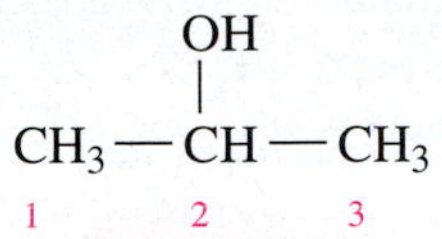
2-Propanol
(isopropyl alcohol)

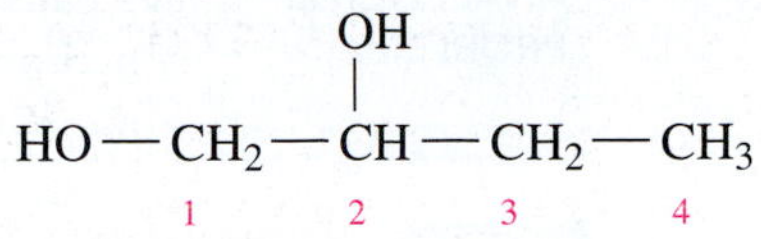
1,2-Butanediol

We can also draw the line-angle formulas for alcohols as shown for 2-propanol, 2-butanol, and 1,2-butanediol.

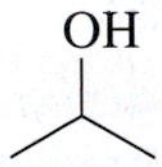
2-Propanol

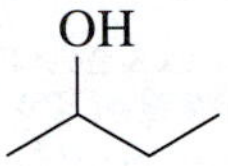
2-Butanol

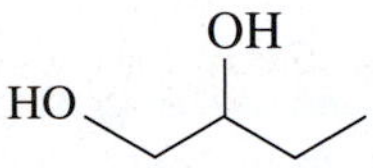
1,2-Butanediol

A cyclic alcohol is named as a *cycloalkanol*. If there are substituents, the ring is numbered from carbon 1, which is the carbon attached to the —OH group. Compounds with no substituents on the ring do not require a number for the hydroxyl group.

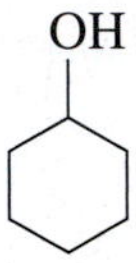
Cyclohexanol

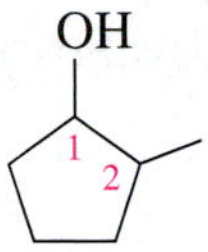
2-Methylcyclopentanol

Naming Phenols

The term *phenol* is the IUPAC name when a hydroxyl group (—OH) is bonded to a benzene ring. When there is a second substituent, the benzene ring is numbered starting from carbon 1, which is the carbon bonded to the —OH group. As we have seen for other aromatic compounds with 2 substituents, the terms *ortho*, *meta*, and *para* (abbreviated *o*-, *m*-, and *p*-) are used for the common names of simple phenols. The common name *cresol* is also used for methylphenols.

Phenol

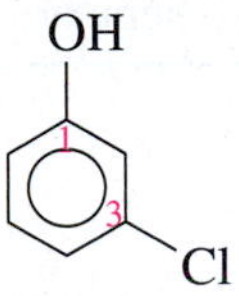
3-Chlorophenol
(*m*-chlorophenol)

4-Methylphenol
(*p*-methylphenol, *p*-cresol)

ENGAGE

If there are six carbon atoms in 2-hexanol, phenol, and cyclohexanol, how do the structures of the compounds differ?

▶SAMPLE PROBLEM 13.1 Naming Alcohols

TRY IT FIRST

Write the IUPAC name for the following:

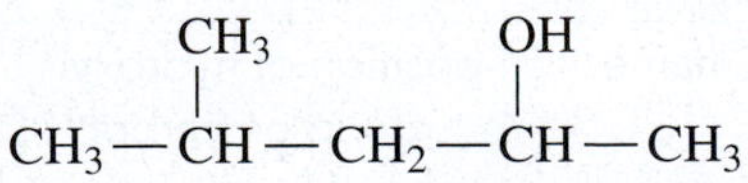

SOLUTION

	Given	Need	Connect
ANALYZE THE PROBLEM	five carbon chain, hydroxyl group, methyl group	IUPAC name	position of methyl and hydroxyl groups, replace e in alkane name with *ol*

STEP 1 **Name the longest carbon chain attached to the —OH group by replacing the *e* in the corresponding alkane name with *ol*.** To name the alcohol, the *e* in alkane name pentane is replaced by *ol*.

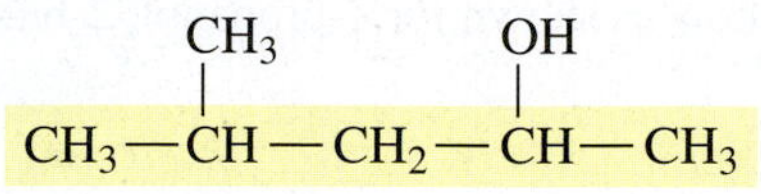

pentanol

STEP 2 **Number the chain starting at the end nearer to the —OH group.** This carbon chain is numbered from right to left to give the position of the —OH group as carbon 2, which is shown as a prefix in the name 2-pentanol.

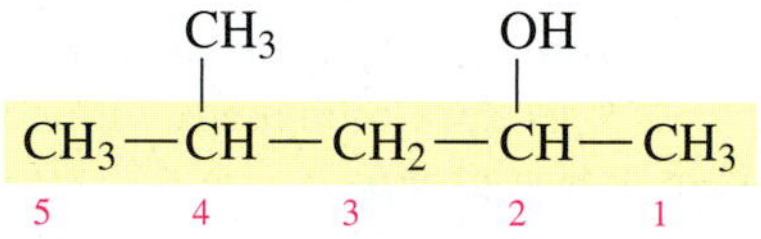

2-pentanol

STEP 3 **Give the location and name of each substituent relative to the —OH group.**

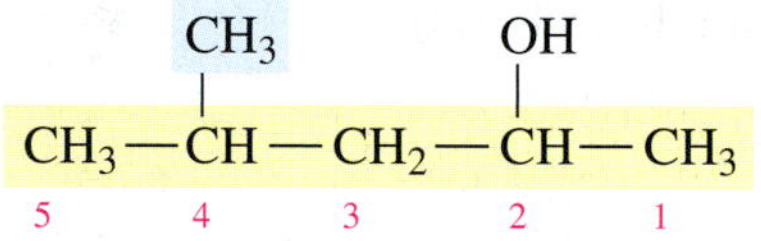

4-methyl-2-pentanol

STUDY CHECK 13.1

Write the IUPAC name for each of the following:

a. 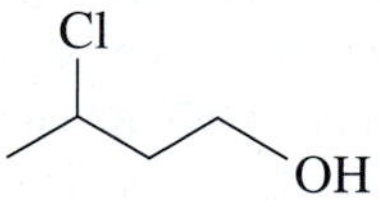

b. 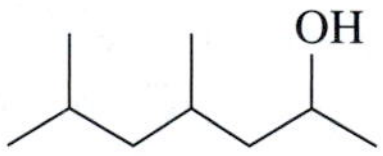

ANSWER

a. 3-chloro-1-butanol

b. 4,6-dimethyl-2-heptanol

▶SAMPLE PROBLEM 13.2 Naming Phenols

TRY IT FIRST

Write the IUPAC and common names for the following:

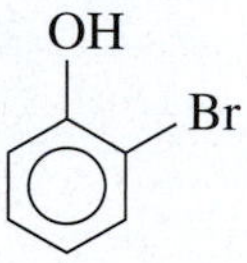

SOLUTION

	Given	Need	Connect
ANALYZE THE PROBLEM	hydroxyl group bonded to benzene ring	IUPAC name	position of hydroxyl group on carbon 1, bromo group

STEP 1 **Name an aromatic alcohol as a *phenol*.**

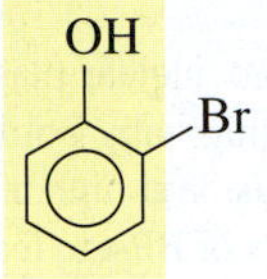
OH
Br

phenol

STEP 2 **Number the chain starting at the end nearer to the —OH group.** For a phenol, the carbon atom attached to the —OH group is carbon 1.

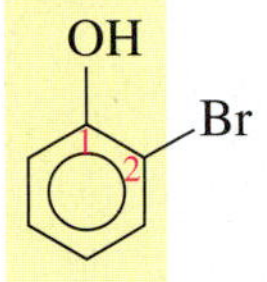
OH
Br
1
2

phenol

STEP 3 **Give the location and name for each substituent relative to the —OH group.** For the common name, use the prefix *ortho* (*o*-).

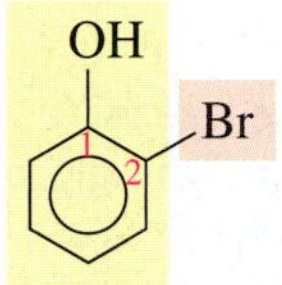
OH
Br
1
2

2-bromophenol (*o*-bromophenol)

STUDY CHECK 13.2

Write the IUPAC and common names for each of the following:

a. 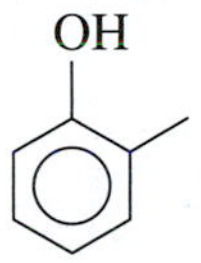
OH

b. 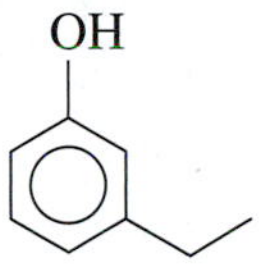
OH

ANSWER

a. 2-methylphenol (*o*-methylphenol, *o*-cresol) **b.** 3-ethylphenol (*m*-ethylphenol)

Chemistry Link to Health

Some Important Alcohols and Phenols

Methanol (methyl alcohol), the simplest alcohol, is found in many solvents and paint removers. If ingested, methanol is oxidized to formaldehyde, which can cause headaches, blindness, and death. Methanol is used to make plastics, medicines, and fuels. In car racing, it is used as a fuel because it is less flammable and has a higher octane rating than does gasoline.

Ethanol (ethyl alcohol) has been known since prehistoric times as an intoxicating product formed by the fermentation of grains, sugars, and starches.

$$\underset{\text{Glucose}}{C_6H_{12}O_6} \xrightarrow{\text{Fermentation}} \underset{\text{Ethanol}}{2CH_3{-}CH_2{-}OH} + 2CO_2$$

Today, ethanol for commercial use is produced by reacting ethene and water at high temperatures and pressures. Ethanol is used as a solvent for perfumes, varnishes, and some medicines, such as tincture of iodine. Recent interest in alternative fuels has led to increased production of ethanol by the fermentation of sugars from grains such as corn, wheat, and rice. "Gasohol" is a mixture of ethanol and gasoline used as a fuel.

$$\underset{\text{Ethene}}{H_2C{=}CH_2} + H_2O \xrightarrow{300\,°C,\ 200\ \text{atm},\ H^+} \underset{\text{Ethanol}}{CH_3{-}CH_2{-}OH}$$

1,2-Ethanediol (ethylene glycol) is used as an antifreeze in heating and cooling systems. It is also a solvent for paints, inks, and plastics, and it is used in the production of synthetic fibers such

(continued)

Chemistry Link to Health (*continued*)

as Dacron. If ingested, it is extremely toxic. In the body, it is oxidized to oxalic acid, which forms insoluble salts in the kidneys that cause renal damage, convulsions, and death. Because its sweet taste is attractive to pets and children, ethylene glycol solutions must be carefully stored.

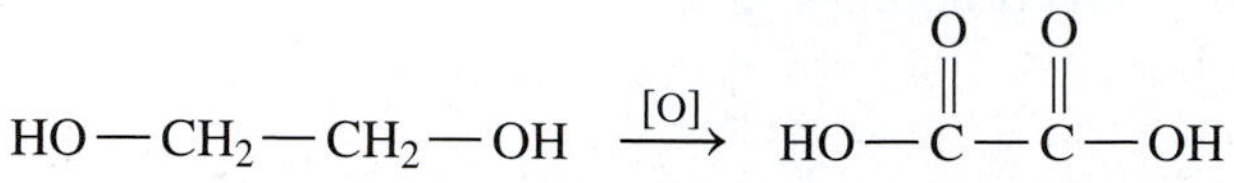

1,2-Ethanediol (ethylene glycol) → Oxalic acid

1,2,3-Propanetriol (*glycerol* or *glycerin*), a trihydroxy alcohol, is a viscous liquid obtained from oils and fats during the production of soaps. The presence of several polar —OH groups makes it strongly attracted to water, a feature that makes glycerol useful as a skin softener in products such as skin lotions, cosmetics, shaving creams, and liquid soaps.

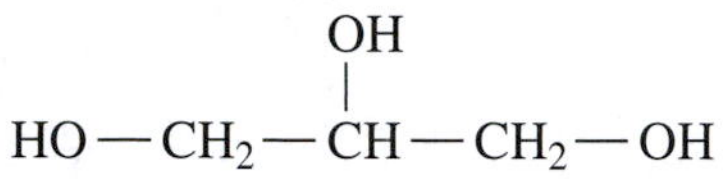

1,2,3-Propanetriol (glycerol)

Bisphenol A (BPA) is used to make polycarbonate, a clear plastic that is used to manufacture beverage bottles, including baby bottles. Washing polycarbonate bottles with certain detergents or at high temperatures disrupts the polymer, causing small amounts of BPA to leach from the bottles. Because BPA is an estrogen mimic, there are concerns about the harmful effects from low levels of BPA. In 2008, Canada banned the use of polycarbonate baby bottles, which are now BPA free.

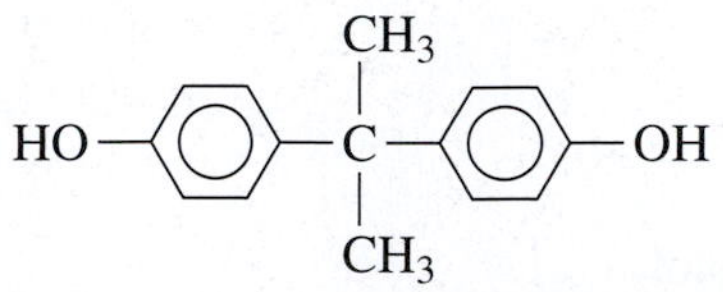

Bisphenol A (BPA)

Antifreeze (ethylene glycol) raises the boiling point and decreases the freezing point of water in a radiator.

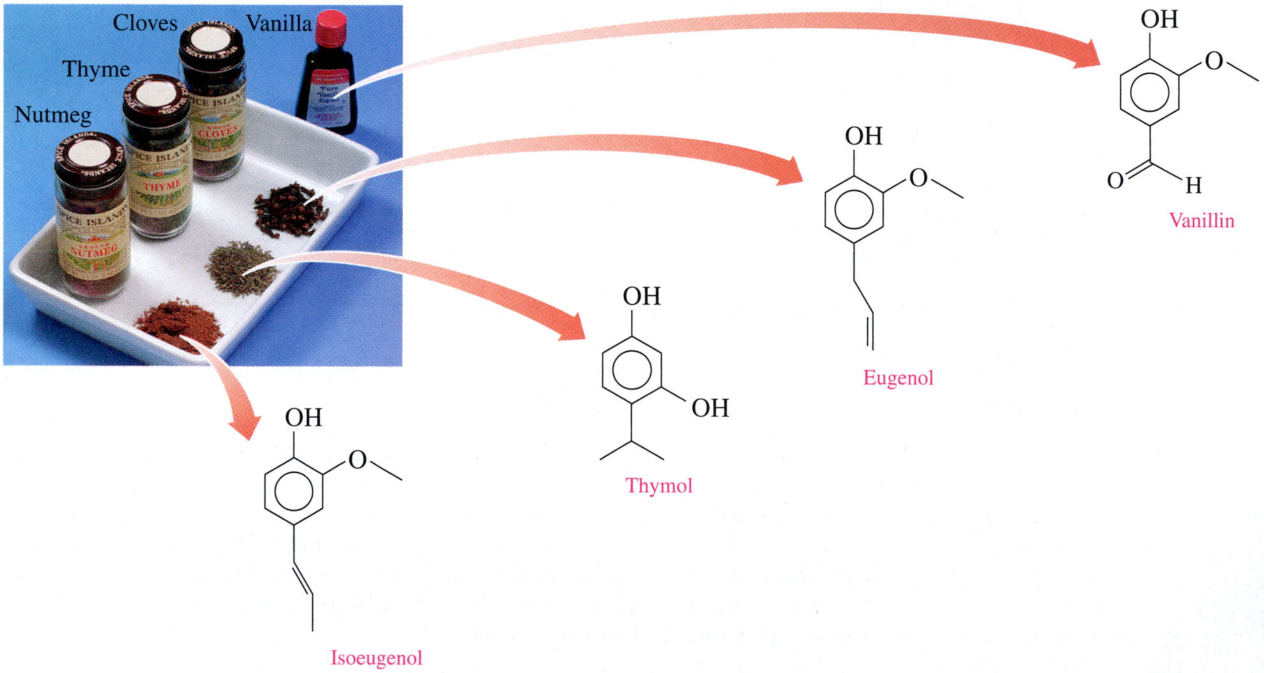
Cloves
Vanilla
Thyme
Nutmeg
Vanillin
Eugenol
Thymol
Isoeugenol

Phenols found in essential oils of plants produce the odor or flavor of the plant.

Thiols

Thiols, also known as *mercaptans*, are a family of sulfur-containing organic compounds that have a *thiol* group (—SH). In the IUPAC system, thiols are named by adding *thiol* to the alkane name of the longest carbon chain and numbering the carbon chain from the end nearer the —SH group.

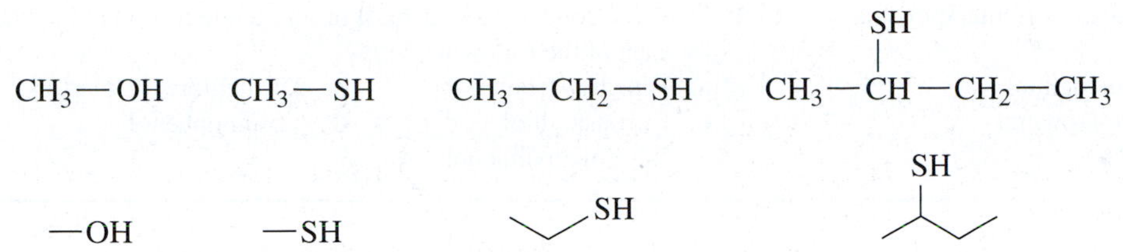

CH3—OH
CH3—SH
CH3—CH2—SH
SH
CH3—CH—CH2—CH3
—OH
—SH
SH
SH

Methanol | Methanethiol | Ethanethiol | 2-Butanethiol

The spray of a skunk contains a mixture of thiols.

An important property of thiols is a strong, sometimes disagreeable, odor. To help us detect natural gas (methane) leaks, a small amount of *tert*-butylthiol is added to the gas supply, which is normally odorless. Thiols such as *trans*-2-butene-1-thiol are in the spray emitted when a skunk senses danger.

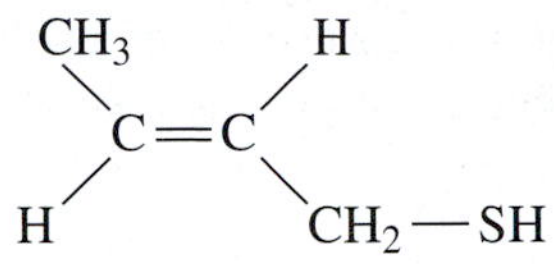

CH3
H
C=C
H
CH2—SH

trans-2-Butene-1-thiol
(in skunk spray)

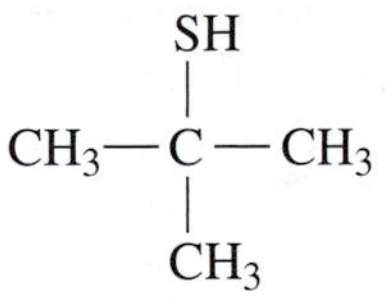

SH
CH3—C—CH3
CH3

tert-Butylthiol
(odorant in gas)

Methanethiol is the characteristic odor of oysters, cheddar cheese, and garlic. Garlic also contains 2-propene-1-thiol. The odor of onions is because of 1-propanethiol, which is also a lachrymator, a substance that makes eyes tear (see **FIGURE 13.2**).

TEST

Try Practice Problems 13.1 to 13.4

CH3—SH
Methanethiol
(oysters, cheese, and garlic)
CH3—CH2—CH2—SH
1-Propanethiol
(onions)
H2C=CH—CH2—SH
2-Propene-1-thiol
(garlic)

FIGURE 13.2 ▶ Thiols are sulfur-containing compounds and often have strong odors.

Q How are the structures of thiols similar to alcohols?

PRACTICE PROBLEMS

13.1 Alcohols, Phenols, and Thiols

13.1 Write the IUPAC and common names for each of the following:

a. $CH_3—CH_2—OH$

b. $CH_3—CH_2—CH(OH)—CH_3$

c. 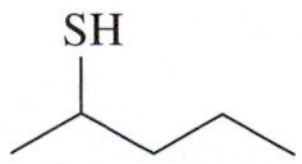

d. 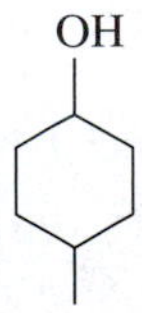

e. 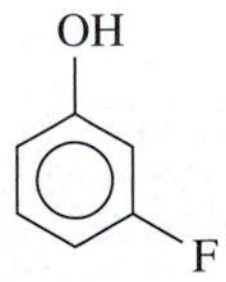

OH
F

13.2 Write the IUPAC and common names for each of the following:

a. 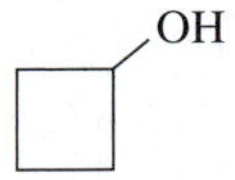

OH

b. $CH_3—CH_2—CH(CH_3)—CH_2—OH$

c. 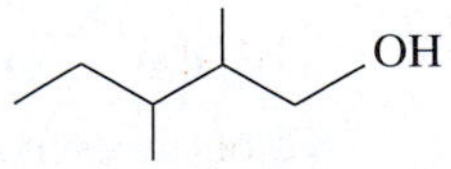

OH

d. 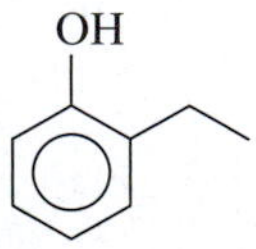

OH

e. 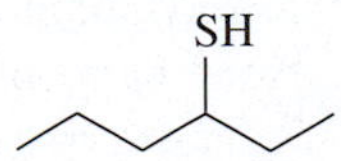

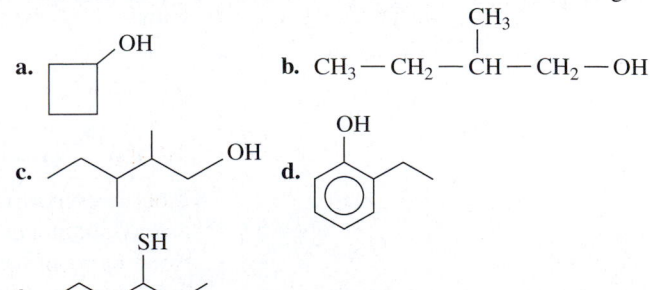

SH

13.3 Draw the condensed structural or line-angle formula, if cyclic, for each of the following:
a. 1-propanol **b.** 3-pentanethiol
c. 2-methyl-2-butanol **d.** *p*-chlorophenol
e. 2-bromo-5-chlorophenol

13.4 Draw the condensed structural or line-angle formula, if cyclic, for each of the following:
a. 3-methyl-1-butanol **b.** 2,4-dichlorocyclohexanol
c. 1-propanethiol **d.** *o*-bromophenol
e. 2,4-dimethylphenol

13.2 Ethers

LEARNING GOAL Write the IUPAC and common names for ethers; draw their condensed structural and line-angle formulas.

An **ether** consists of the functional group that has an oxygen atom (—O—) attached by single bonds to two carbon groups that are alkyl or aromatic groups. Molecules of alcohols, phenols, thiols, and ethers have bent shapes around the oxygen or sulfur atom, similar to water.

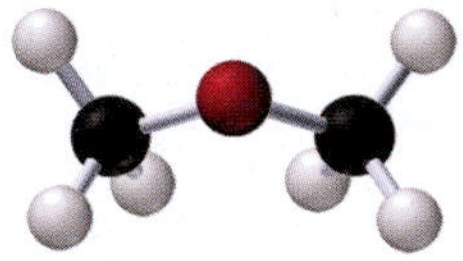

$CH_3—O—CH_3$

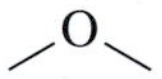

Methoxymethane
(dimethyl ether)

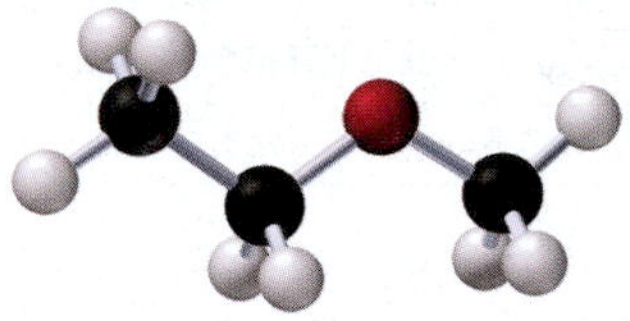

$CH_3—CH_2—O—CH_3$

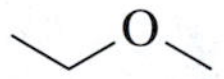

Methoxyethane
(ethyl methyl ether)

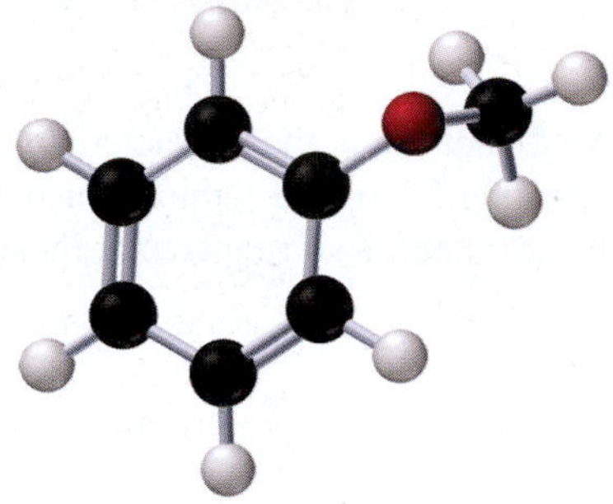

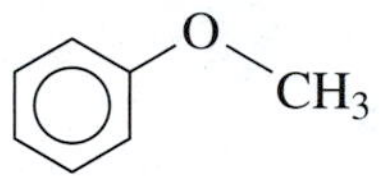
O
CH3

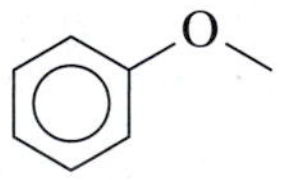
O

Methoxybenzene
(methyl phenyl ether)

ENGAGE

How does the functional group of an ether differ from that of an alcohol?

Naming Ethers

In the common name of an ether, the names of the alkyl or aromatic groups attached to the oxygen atom are written in alphabetical order, followed by the word *ether*.

Methyl (↓ CH_3) Propyl (↓ $CH_2—CH_2—CH_3$)

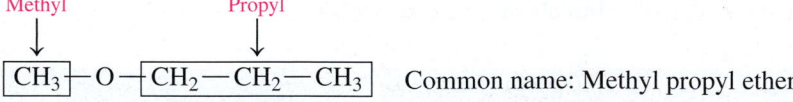

$\boxed{CH_3}—O—\boxed{CH_2—CH_2—CH_3}$ Common name: Methyl propyl ether

In the IUPAC system, an ether is named with an *alkoxy* group made up of the smaller alkyl group and the oxygen atom, followed by the alkane name of the longer carbon chain.

Methoxy (↓ $CH_3—O$) Propane (↓ $CH_2—CH_2—CH_3$, carbons numbered 1, 2, 3)

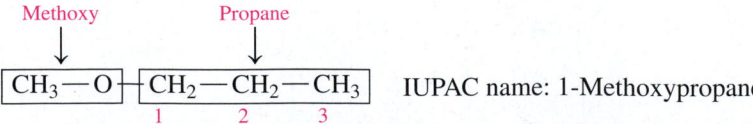

$\boxed{CH_3—O}—\boxed{CH_2—CH_2—CH_3}$ IUPAC name: 1-Methoxypropane

More examples of naming ethers with both IUPAC and common names follow:

$CH_3—CH_2—O—CH_2—CH_3$

Ethoxyethane
(diethyl ether)

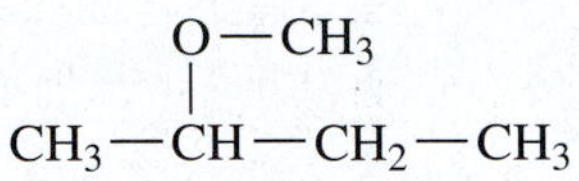

2-Methoxybutane

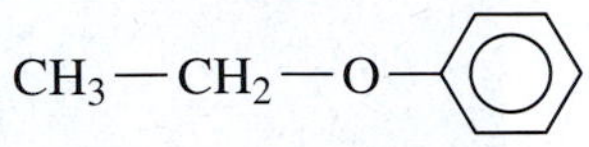

Ethoxybenzene
(ethyl phenyl ether)

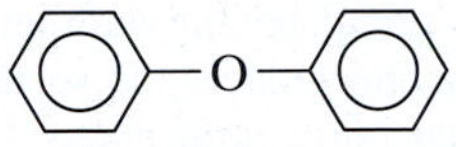

Phenoxybenzene
(diphenyl ether)

▶SAMPLE PROBLEM 13.3 Naming Ethers

TRY IT FIRST

Write the IUPAC name for the following:

$CH_3—CH_2—O—CH_2—CH_2—CH_2—CH_3$

SOLUTION

ANALYZE THE PROBLEM	Given	Need	Connect
	ether	IUPAC name	alkoxy group, carbon chain

STEP 1 **Write the alkane name of the longer carbon chain.**

$CH_3—CH_2—O—\underbrace{CH_2—CH_2—CH_2—CH_3}_{\text{Longer carbon chain}}$ butane

STEP 2 **Name the oxygen and smaller alkyl group as an alkoxy group.**

$CH_3—CH_2—O—CH_2—CH_2—CH_2—CH_3$ ethoxybutane

↑ Ethoxy group

STEP 3 **Number the longer carbon chain from the end nearer the alkoxy group and give its location.**

$CH_3—CH_2—O—\overset{}{\underset{1}{CH_2}}—\underset{2}{CH_2}—\underset{3}{CH_2}—\underset{4}{CH_3}$ 1-ethoxybutane

STUDY CHECK 13.3

Write the IUPAC name for each of the following:

a. methyl phenyl ether **b.** ethyl propyl ether

ANSWER

a. methoxybenzene **b.** 1-ethoxypropane

TEST

Try Practice Problems 13.5 to 13.8

Chemistry Link to Health

Ethers as Anesthetics

Anesthesia is the loss of sensation and consciousness. A general anesthetic is a substance that blocks signals to the awareness centers in the brain so the person has a loss of memory, a loss of feeling pain, and an artificial sleep. The term *ether* has been associated with anesthesia because diethyl ether was the most widely used anesthetic for more than a hundred years. Although it is easy to administer, ether is very volatile and highly flammable. A small spark in the operating room could cause an explosion. Since the 1950s, anesthetics such as Forane (isoflurane), Ethrane (enflurane), Suprane (desflurane), and Sevoflurane have been developed that are not as flammable. Most of these anesthetics retain the ether group, but the addition of halogen atoms reduces the volatility and flammability of the ethers.

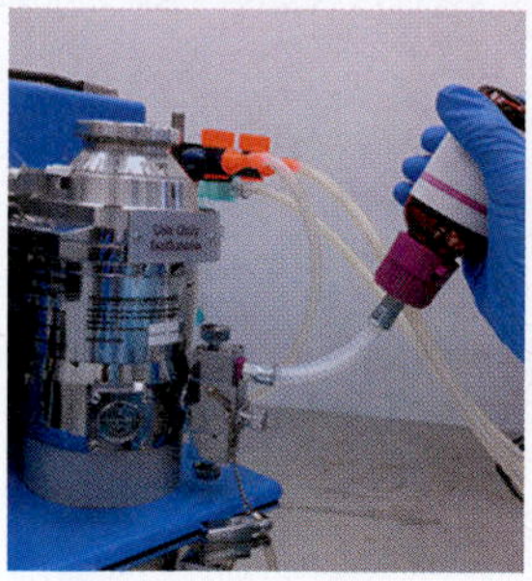

Isoflurane (Forane) is an inhaled anesthetic.

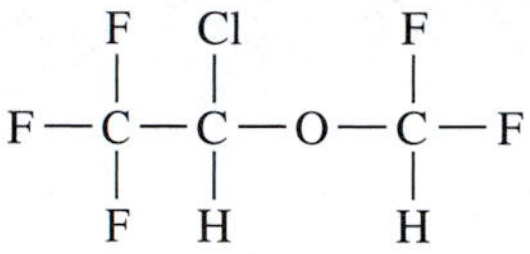

Forane (isoflurane)

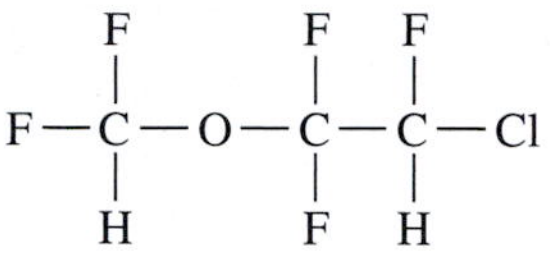

Ethrane (enflurane)

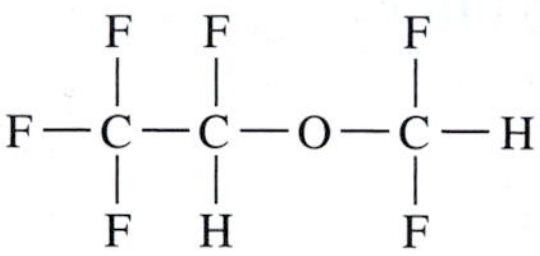

Suprane (desflurane)

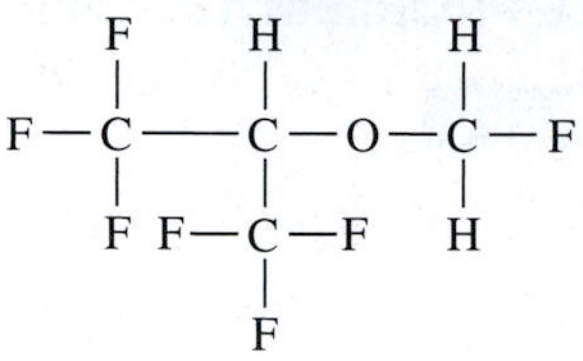

Sevoflurane

PRACTICE PROBLEMS

13.2 Ethers

13.5 Write the IUPAC name and any common name for each of the following ethers:

a. $CH_3—O—CH_2—CH_3$ **b.**

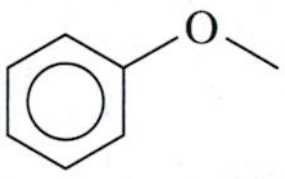

c.

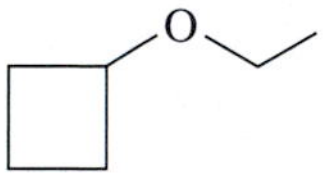

d. $CH_3—CH_2—CH_2—O—CH_3$

13.6 Write the IUPAC name and any common name for each of the following ethers:

a. $CH_3—CH_2—CH_2—O—CH_2—CH_3$

b.

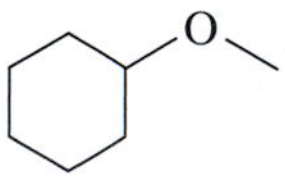

c.

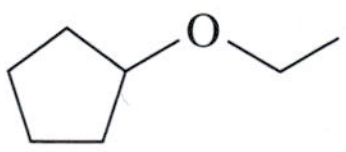

d. $CH_3—O—CH_3$

13.7 Draw the condensed structural or line-angle formula for each of the following:

a. ethyl propyl ether
b. cyclopropyl ethyl ether
c. methoxycyclopentane
d. 1-ethoxy-2-methylbutane
e. 2,3-dimethoxypentane

13.8 Draw the condensed structural or line-angle formula for each of the following:

a. diethyl ether
b. diphenyl ether
c. ethoxycyclohexane
d. 2-methoxy-2,3-dimethylbutane
e. 1,2-dimethoxybenzene

13.3 Physical Properties of Alcohols, Phenols, and Ethers

LEARNING GOAL Describe the classification of alcohols; describe the boiling points and solubility of alcohols, phenols, and ethers.

Alcohols are classified by the number of alkyl groups attached to the carbon atom bonded to the hydroxyl group (—OH). A **primary (1°) alcohol** has one alkyl group attached to the carbon atom bonded to the —OH group. The simplest alcohol, methanol (CH_3OH), which has a carbon attached to three H atoms but no alkyl group, is considered a primary alcohol; a **secondary (2°) alcohol** has two alkyl groups, and a **tertiary (3°) alcohol** has three alkyl groups.

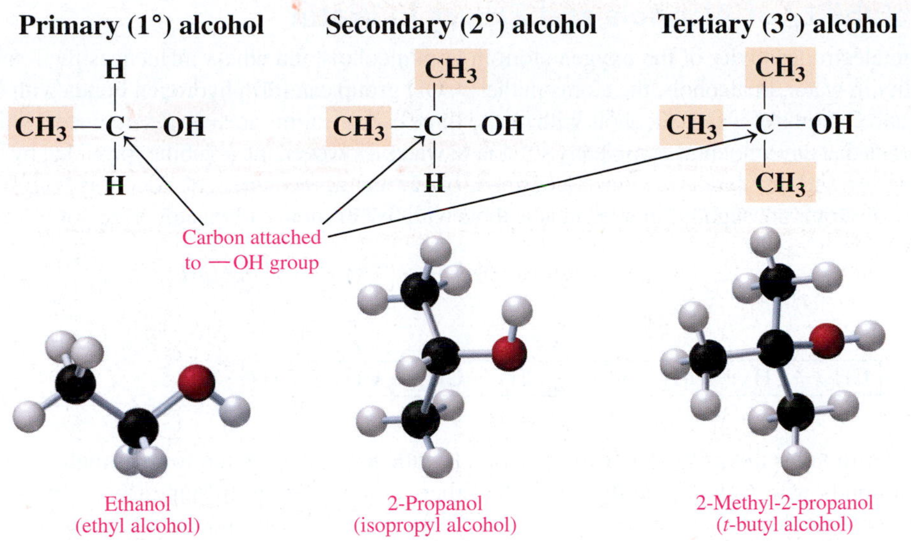

Primary (1°) alcohol
Secondary (2°) alcohol
Tertiary (3°) alcohol
Carbon attached to —OH group
Ethanol (ethyl alcohol)
2-Propanol (isopropyl alcohol)
2-Methyl-2-propanol (t-butyl alcohol)

▸SAMPLE PROBLEM 13.4 Classifying Alcohols

TRY IT FIRST

Classify each of the following alcohols as primary (1°), secondary (2°), or tertiary (3°):

a. $CH_3—CH_2—CH_2—OH$

b.

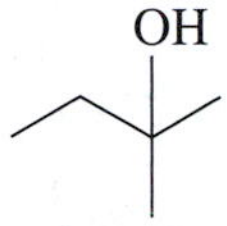

OH

SOLUTION

a. The carbon atom bonded to the —OH group is attached to one alkyl group, which makes this a primary (1°) alcohol.

b. The carbon atom bonded to the —OH group is attached to three alkyl groups, which makes this a tertiary (3°) alcohol.

STUDY CHECK 13.4

Classify each of the following as primary (1°), secondary (2°), or tertiary (3°):

a.

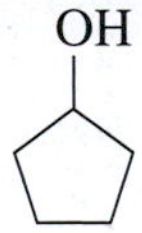

OH

b.

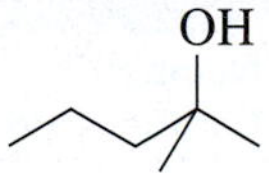

OH

ANSWER

a. secondary (2°) **b.** tertiary (3°)

TEST

Try Practice Problems 13.9 and 13.10

Boiling Points

Because there is a large electronegativity difference between the oxygen and hydrogen atoms in the —OH group, the oxygen has a partially negative charge, and the hydrogen has a partially positive charge. As a result, hydrogen bonds form between the oxygen of one alcohol and hydrogen in the —OH group of another alcohol. Hydrogen bonds cannot form between ether molecules because there are not any polar —OH groups.

Alcohols have higher boiling points than do ethers of the same molar mass because alcohols require higher temperatures to provide sufficient energy to break the many hydrogen bonds. The boiling points of ethers are similar to those of alkanes because neither can form hydrogen bonds.

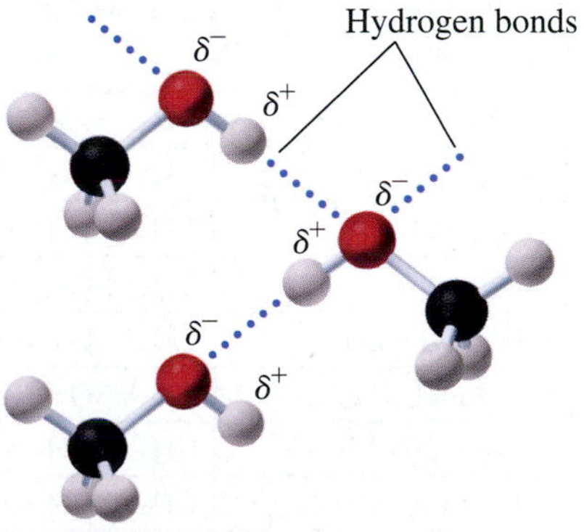

Hydrogen bonds
δ−
δ+

Methanol
(methyl alcohol)

Because they can form hydrogen bonds, alcohols have higher boiling points than ethers of the same molar mass.

Solubility of Alcohols and Ethers in Water

The electronegativity of the oxygen atom in both alcohols and ethers influences their solubility in water. In alcohols, the atoms in the —OH group can form hydrogen bonds with the H and O atoms of water. Alcohols with one to three carbon atoms are *miscible* in water, which means that any amount is completely soluble in water. However, the solubility provided by the polar —OH group decreases as the number of carbon atoms increases. Alcohols with four carbon atoms are slightly soluble, and alcohols with five or more carbon atoms are not soluble.

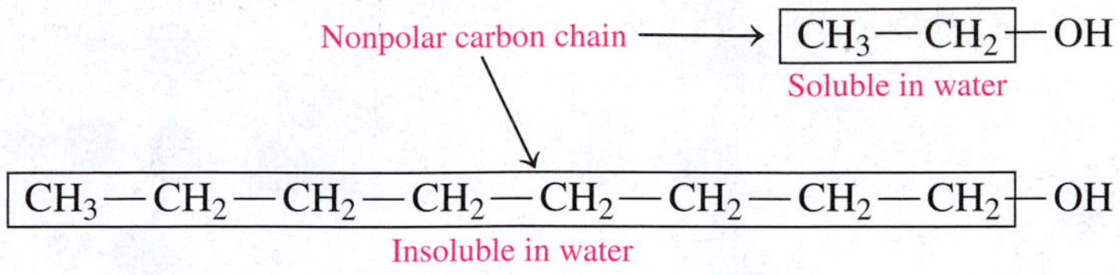
Nonpolar carbon chain
$CH_3—CH_2$—OH
Soluble in water
$CH_3—CH_2—CH_2—CH_2—CH_2—CH_2—CH_2—CH_2$—OH
Insoluble in water

Although ethers can form hydrogen bonds with water, they do not form as many hydrogen bonds with water as do the alcohols. Ethers containing up to four carbon atoms are slightly soluble in water. **TABLE 13.1** compares the boiling points and the solubility in water of some alcohols and ethers of similar molar mass.

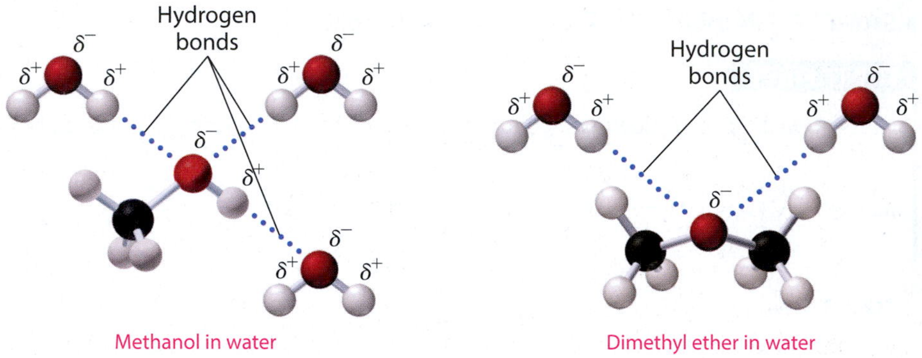
Hydrogen bonds
δ−
δ+
Hydrogen bonds
Methanol in water
Dimethyl ether in water

(a) Small alcohol molecules are soluble in water because they form hydrogen bonds with water.

(b) Small ether molecules are slightly soluble in water because they form hydrogen bonds with water.

TABLE 13.1 Boiling Points and Solubility of Some Alcohols and Ethers

Compound	Condensed Structural Formula	Number of Carbon Atoms	Boiling Point (°C)	Solubility in Water
Methanol	$CH_3—OH$	1	65	Soluble
Ethanol	$CH_3—CH_2—OH$	2	78	Soluble
1-Propanol	$CH_3—CH_2—CH_2—OH$	3	97	Soluble
1-Butanol	$CH_3—CH_2—CH_2—CH_2—OH$	4	118	Slightly soluble
1-Pentanol	$CH_3—CH_2—CH_2—CH_2—CH_2—OH$	5	138	Insoluble
Dimethyl ether	$CH_3—O—CH_3$	2	−23	Slightly soluble
Ethyl methyl ether	$CH_3—O—CH_2—CH_3$	3	8	Slightly soluble
Diethyl ether	$CH_3—CH_2—O—CH_2—CH_3$	4	35	Slightly soluble
Ethyl propyl ether	$CH_3—CH_2—O—CH_2—CH_2—CH_3$	5	64	Insoluble

Solubility and Boiling Point of Phenol

Phenol has a high boiling point (182 °C) because the —OH group allows phenol molecules to hydrogen bond with other phenol molecules. Phenol is slightly soluble in water because the —OH group can form hydrogen bonds with water molecules. In water, the —OH group of phenol ionizes slightly, which makes it a weak acid ($K_a = 1 \times 10^{-10}$). In fact, an early name for phenol was *carbolic acid*.

TEST

Try Practice Problems 13.11 to 13.16

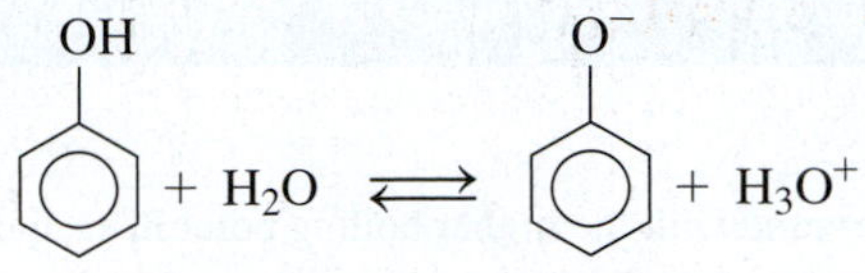

Phenol Phenoxide ion

Phenol and Antiseptics

An *antiseptic* is a substance applied to the skin to kill microorganisms that cause infection. At one time, dilute solutions of phenol (carbolic acid) were used in hospitals as antiseptics. Joseph Lister (1827–1912) is considered a pioneer in antiseptic surgery and was the first to sterilize surgical instruments and dressings with phenol. Phenol was also used to disinfect wounds to prevent postsurgical infections such as gangrene. However, phenol is very corrosive and highly irritating to the skin; it can cause severe burns, and ingestion can be fatal. Soon phenol solutions were replaced with other disinfectants. 4-Hexylresorcinol is a form of phenol used in topical antiseptics, throat lozenges, mouthwash and throat sprays. Lysol, used to disinfect surfaces in a home or hospital, contains the antiseptics 2-phenylphenol and 2-benzyl-4-chlorophenol.

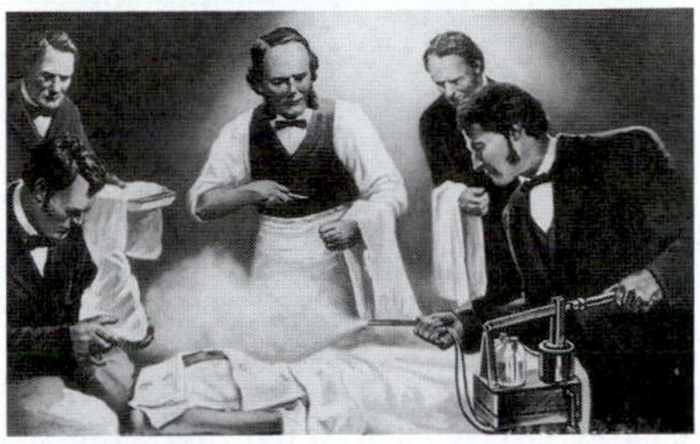

Joseph Lister was the first to use phenol to sterilize surgical instruments.

Lysol
KILLS FLU VIRUS

Lysol, used as a disinfectant, contains phenol compounds.

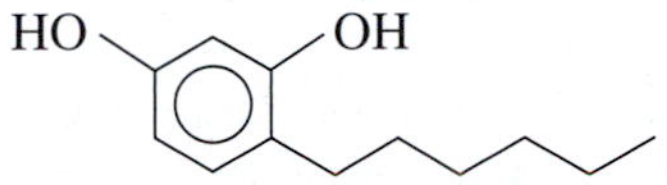

4-Hexylresorcinol

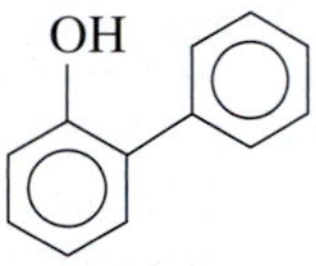

2-Phenylphenol (*o*-phenylphenol)

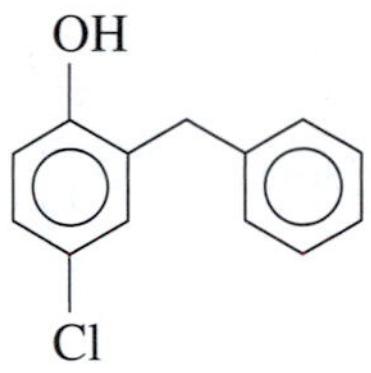

2-Benzyl-4-chlorophenol

Chemistry Link to Health

Hand Sanitizers

When soap and water are not available, hand sanitizers may be used to kill most bacteria and viruses that spread colds and flu. As a gel or liquid solution, many hand sanitizers use ethanol or propanol as their active ingredient. Sanitizers also contain glycerin and propylene glycol to prevent the skin from drying. When children use hand sanitizers, they must be carefully supervised because the ingestion of a small amount can cause alcohol poisoning.

In an alcohol-containing sanitizer, the amount of ethanol is typically 60% (v/v) but can be as high as 85% (v/v). This amount of ethanol can make hand sanitizers a fire hazard in the home because ethanol is highly flammable. When ethanol undergoes combustion, it produces a transparent blue flame. When using an ethanol-containing sanitizer, it is important to rub hands until they are completely dry. It is also recommended that sanitizers containing ethanol be placed in storage areas that are away from heat sources in the home.

Some sanitizers are alcohol-free, but often the active ingredient is triclosan, which contains aromatic, ether, and phenol functional groups. The Food and Drug Administration has banned triclosan in personal care products because its use may promote the growth of antibiotic-resistant bacteria. Recent reports indicate that triclosan may disrupt the endocrine system and interfere with the function of estrogens, androgens, and thyroid hormones.

Hand Sanitizer

Hand sanitizers that contain ethanol or propanol are used to kill bacteria on the hands.

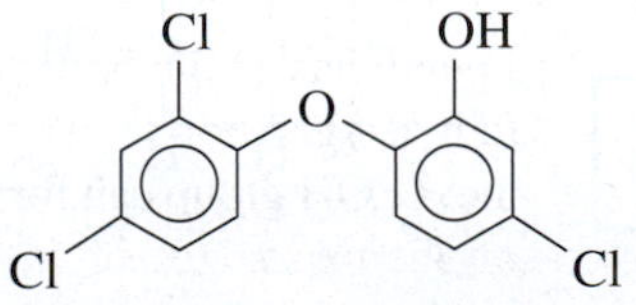

Triclosan, an antibacterial compound, is now banned in personal-care products.

PRACTICE PROBLEMS

13.3 Physical Properties of Alcohols, Phenols, and Ethers

13.9 Classify each of the following alcohols as primary (1°), secondary (2°), or tertiary (3°):

a. 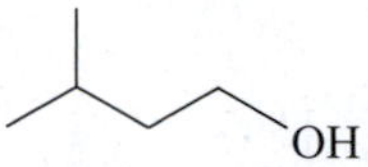

b. $CH_3-CH_2-CH_2-CH_2-OH$

c. 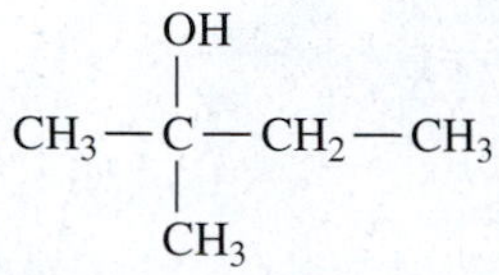

d. 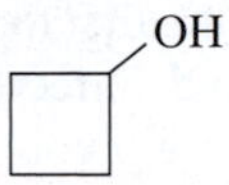

13.10 Classify each of the following alcohols as primary (1°), secondary (2°), or tertiary (3°):

a. 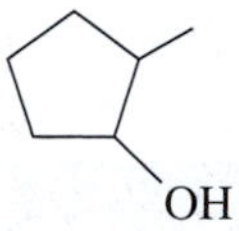

b. 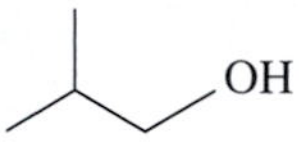

c. 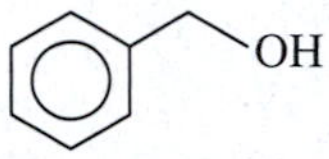

d. 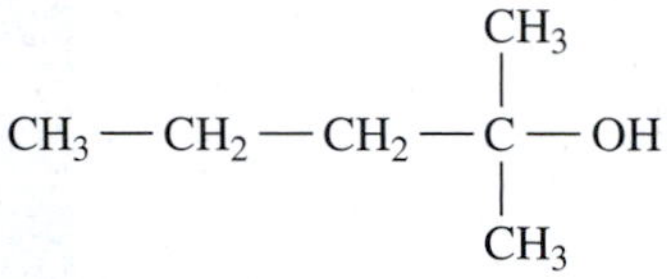

13.11 Predict the compound with the higher boiling point in each of the following pairs:
a. ethane or methanol **b.** diethyl ether or 1-butanol
c. 1-butanol or pentane

13.12 Predict the compound with the higher boiling point in each of the following pairs:
a. 2-propanol and 2-butanol **b.** dimethyl ether or ethanol
c. dimethyl ether or diethyl ether

13.13 Are each of the following soluble, slightly soluble, or insoluble in water? Explain.
a. CH_3-CH_2-OH
b. CH_3-O-CH_3
c. $CH_3-CH_2-CH_2-CH_2-CH_2-CH_2-OH$

13.14 Are each of the following soluble, slightly soluble, or insoluble in water? Explain.
a. $CH_3-CH_2-CH_2-OH$
b. 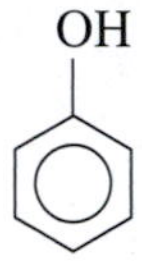
c. $CH_3-CH_2-O-CH_2-CH_3$

13.15 Give an explanation for each of the following observations:
a. Methanol is soluble in water, but ethane is not.
b. 2-Propanol is soluble in water, but 1-butanol is only slightly soluble.
c. 1-Propanol is soluble in water, but ethyl methyl ether is only slightly soluble.

13.16 Give an explanation for each of the following observations:
a. Ethanol is soluble in water, but propane is not.
b. Dimethyl ether is slightly soluble in water, but pentane is not.
c. 1-Propanol is soluble in water, but 1-hexanol is not.

13.4 Reactions of Alcohols and Thiols

LEARNING GOAL Write balanced chemical equations for the combustion, dehydration, and oxidation of alcohols and thiols.

FIGURE 13.3 ▶ A flaming dessert is prepared using an alcohol that undergoes combustion.

Q What is the equation for the complete combustion of the ethanol in the liquor?

TEST
Try Practice Problems 13.17 and 13.18

Alcohols, similar to hydrocarbons, undergo combustion in the presence of oxygen. For example, in a restaurant, a flaming dessert may be prepared by pouring a liquor on fruit or ice cream and lighting it (see **FIGURE 13.3**). The combustion of the ethanol in the liquor proceeds as follows:

$$CH_3-CH_2-OH(g) + 3O_2(g) \xrightarrow{\Delta} 2CO_2(g) + 3H_2O(g) + \text{energy}$$

$C_2H_6O + O_2 \xrightarrow{\Delta} 2CO_2 + 3H_2O$ + energy

Dehydration of Alcohols to Form Alkenes

In a **dehydration** reaction, alcohols lose a water molecule when they are heated at a high temperature (180 °C) with an acid catalyst such as H_2SO_4. During the dehydration of an alcohol, H— and —OH are removed from *adjacent carbon atoms of the same alcohol* to produce a water molecule. A double bond forms between the same two carbon atoms to produce an alkene product.

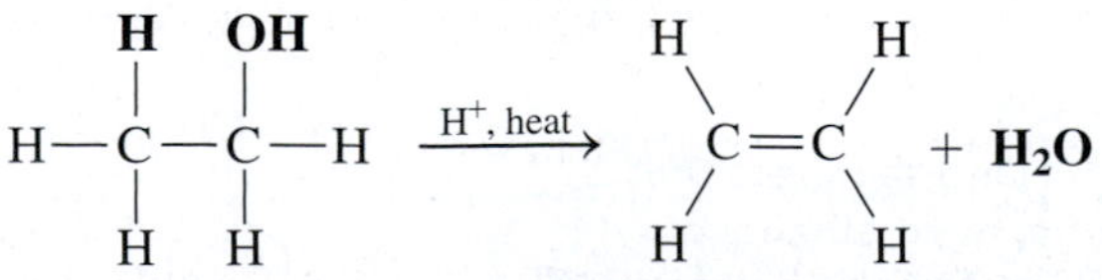
H⁺, heat
+ H₂O

Ethanol Ethene

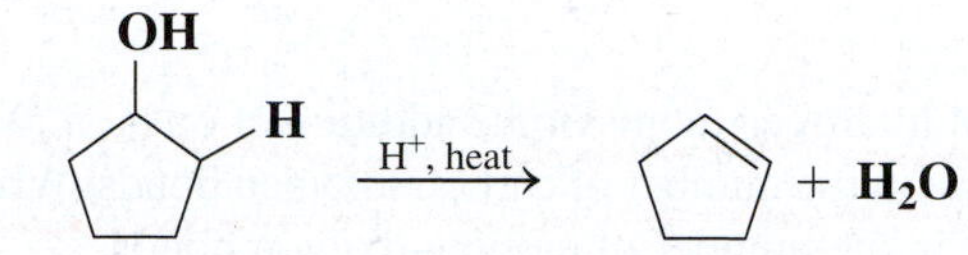
OH
H
H+, heat
+ H2O

Cyclopentanol Cyclopentene

The dehydration of a secondary alcohol can result in the formation of two products. The major product is the one that forms by removing the hydrogen from the carbon atom that has the smaller number of hydrogen atoms. A hydrogen atom is easier to remove from the carbon atom that has fewer hydrogen atoms.

CORE CHEMISTRY SKILL
Writing Equations for the Dehydration of Alcohols

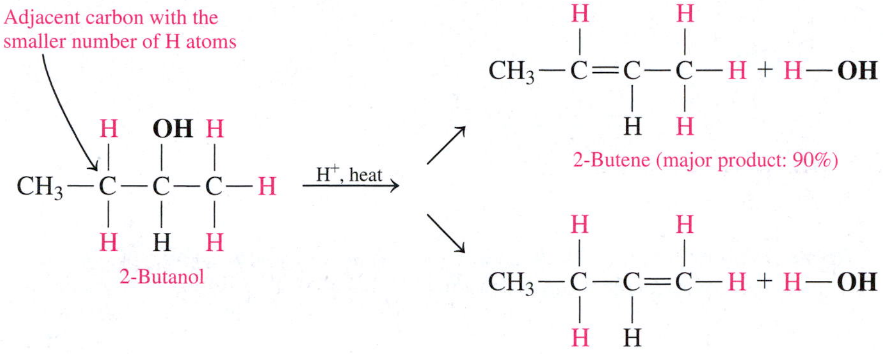
Adjacent carbon with the smaller number of H atoms
H+, heat

2-Butanol

2-Butene (major product: 90%)

1-Butene (minor product: 10%)

ENGAGE
If 2-pentanol is dehydrated, what type of organic product forms?

▶ SAMPLE PROBLEM 13.5 Dehydration of Alcohols

TRY IT FIRST

Draw the condensed structural formula for the major alkene produced by the dehydration of each of the following alcohols:

a. $CH_3—CH_2—CH_2—CH_2—OH \xrightarrow{H^+,\ heat}$

b. $CH_3—CH(OH)—CH_2—CH_2—CH_3 \xrightarrow{H^+,\ heat}$

SOLUTION

a. The 1-butanol loses —OH from carbon 1 and H— from carbon 2 to form 1-butene. This is the only possible product.

$CH_3—CH_2—CH{=}CH_2$

b. For the dehydration of an asymmetrical alcohol, we remove —OH from carbon 2 and H— from carbon 3, which has the smaller number of H atoms. The major product is 2-pentene.

$CH_3—CH{=}CH—CH_2—CH_3$

STUDY CHECK 13.5

Write the name of the alkene produced by the dehydration of each of the following:

a. cyclopentanol **b.** 2-methyl-3-hexanol

ANSWER

a. cyclopentene **b.** 2-methyl-2-hexene

TEST
Try Practice Problems 13.19 to 13.22

CORE CHEMISTRY SKILL

Writing Equations for the Oxidation of Alcohols

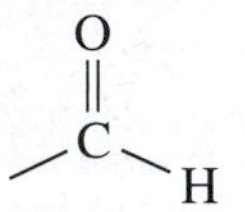

The aldehyde functional group at the end of a chain has a C=O bonded to one carbon group and to H.

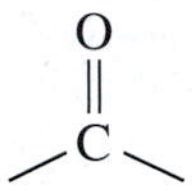

The ketone functional group in the middle of a chain has a C=O bonded to two carbon groups.

Oxidation of Alcohols

In organic chemistry, **oxidation** is a loss of hydrogen atoms or the addition of oxygen. When a compound is oxidized, there is an increase in the number of carbon–oxygen bonds. When a compound is reduced, there is a decrease in the number of carbon–oxygen bonds.

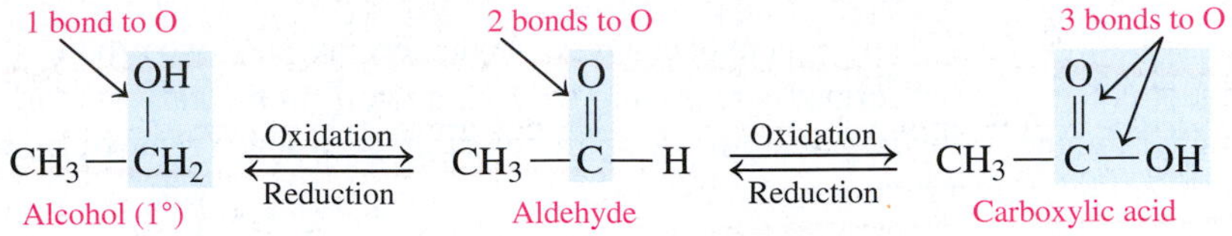
1 bond to O
2 bonds to O
3 bonds to O
Oxidation
Reduction
Alcohol (1°)
Aldehyde
Carboxylic acid

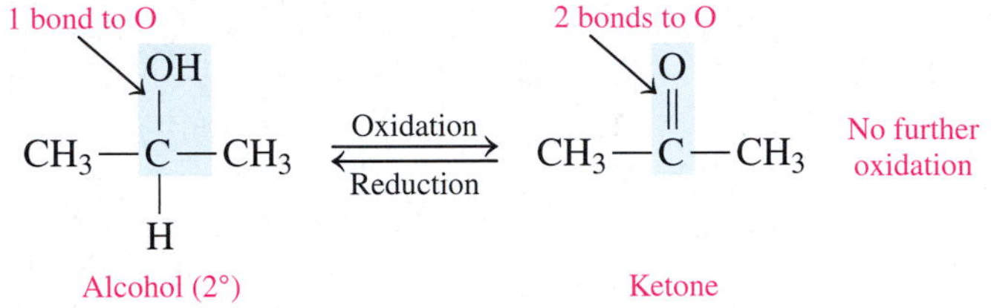
1 bond to O
2 bonds to O
Oxidation
Reduction
No further oxidation
Alcohol (2°)
Ketone

An aldehyde or ketone is more oxidized than an alcohol; a carboxylic acid is more oxidized than an aldehyde.

Oxidation of Primary and Secondary Alcohols

The oxidation of a primary alcohol produces an *aldehyde*, which contains a double bond between oxygen and the carbon atom at the end of the chain. For example, the oxidation of methanol and ethanol occurs by removing two hydrogen atoms, one from the —OH group and another from the carbon that is bonded to the —OH group. The oxidized product contains the same number of carbon atoms as the reactant. The reaction is written with the symbol [O] over the arrow to indicate that O is obtained from an oxidizing agent, such as $KMnO_4$ or $K_2Cr_2O_7$.

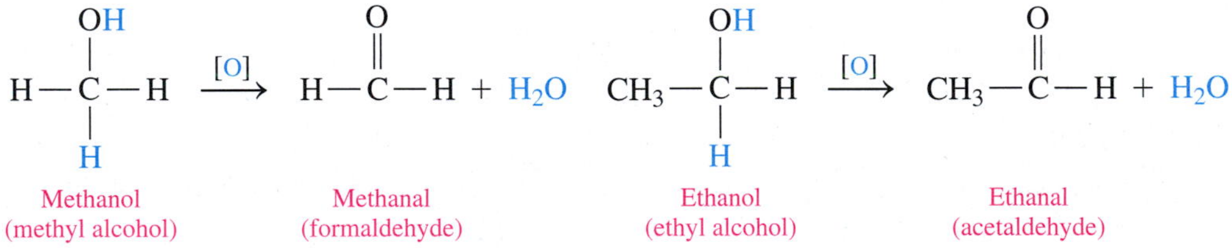
[O]
Methanol (methyl alcohol)
Methanal (formaldehyde)
[O]
Ethanol (ethyl alcohol)
Ethanal (acetaldehyde)

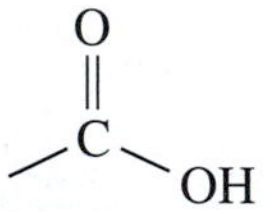

The carboxylic acid functional group at the end of a chain has a C=O bonded to one carbon group and to OH.

Aldehydes oxidize further by the addition of oxygen to form *carboxylic acids*, which have three carbon–oxygen bonds. This step occurs so readily that it is often difficult to isolate the aldehyde product during oxidation.

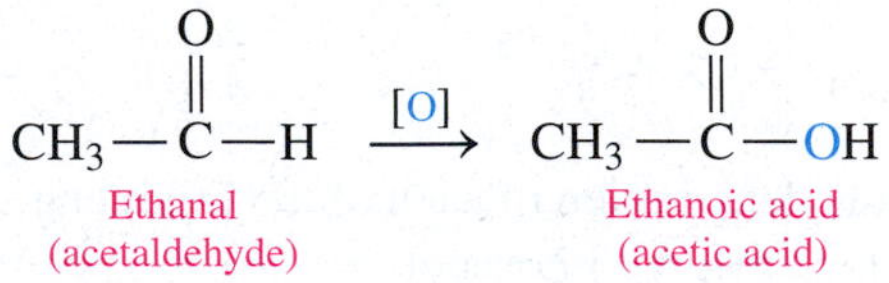
[O]
Ethanal (acetaldehyde)
Ethanoic acid (acetic acid)

In the oxidation of secondary alcohols, the products are *ketones*. Two hydrogen atoms are removed, one from the —OH and the other from the carbon bonded to the —OH group. The result is a ketone that has the carbon–oxygen double bond attached to two alkyl or aromatic groups. There is no further oxidation of a ketone because there are no hydrogen atoms attached to the carbon of the ketone group.

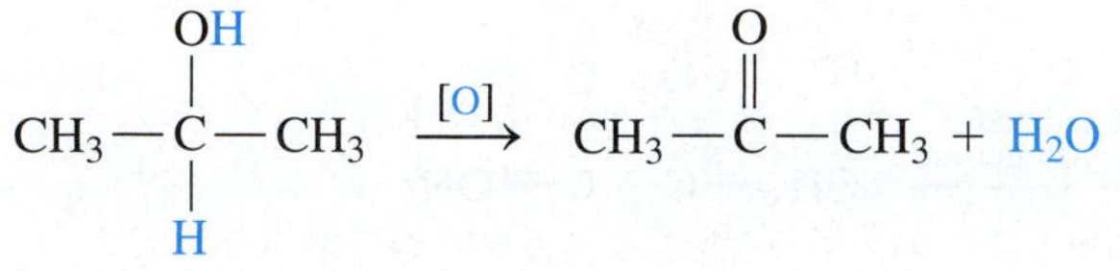

2-Propanol (isopropyl alcohol) — Propanone (dimethyl ketone; acetone)

ENGAGE

Why is it more difficult to oxidize a ketone than to oxidize an aldehyde?

Tertiary alcohols do not oxidize readily because there is no hydrogen atom on the carbon bonded to the —OH group. Because C—C bonds are usually too strong to oxidize, tertiary alcohols resist oxidation.

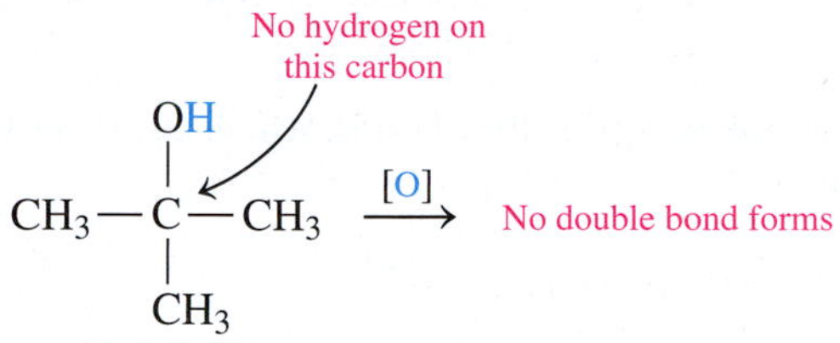
No hydrogen on this carbon
[O]
No double bond forms

Alcohol (3°)

▶SAMPLE PROBLEM 13.6 Oxidation of Alcohols

TRY IT FIRST

Classify each of the following alcohols as primary (1°), secondary (2°), or tertiary (3°); draw the condensed structural or line-angle formula for the aldehyde or ketone formed by the oxidation of each of the following:

a. 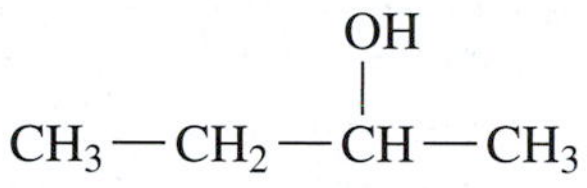

b. 

SOLUTION

a. This is a secondary (2°) alcohol, which can oxidize to a ketone.

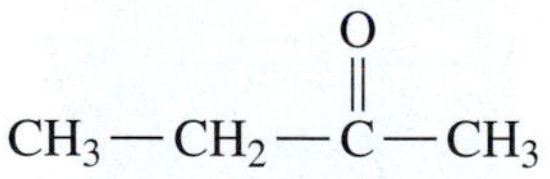

b. This is a primary (1°) alcohol, which can oxidize to an aldehyde.

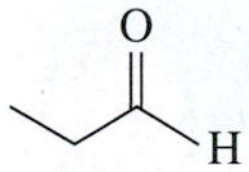

STUDY CHECK 13.6

Draw the condensed structural formula for the product formed by the oxidation of each of the following:

a. 2-pentanol **b.** 1-pentanol

ANSWER

a. 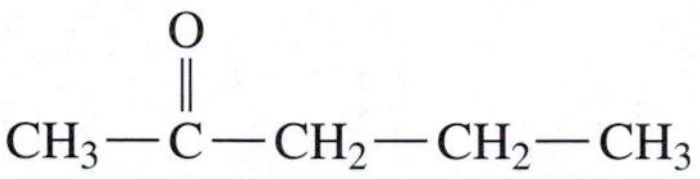

b. 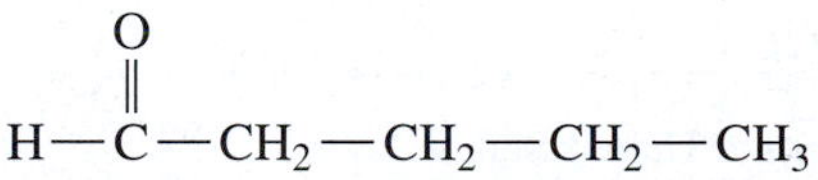

INTERACTIVE VIDEO

PEARSON eText 2.0 Oxidation of Alcohols

TEST

Try Practice Problems 13.23 to 13.26

During vigorous exercise, lactic acid accumulates in the muscles and causes fatigue. When the activity level is decreased, oxygen enters the muscles. The secondary —OH group in lactic acid is oxidized to a ketone group in pyruvic acid, which eventually is oxidized to CO_2 and H_2O. The muscles in highly trained athletes are capable of taking up greater quantities of oxygen so that vigorous exercise can be maintained for longer periods of time.

When oxygen levels fall during intense exercise, lactic acid (lactate) is produced that may cause muscle pain and cramps, rapid breathing, and nausea.

$$CH_3-\underset{\text{Secondary alcohol}}{CH(OH)}-\overset{O}{\overset{\|}{C}}-OH \xrightarrow{\text{Lactic acid dehydrogenase}} CH_3-\overset{O}{\overset{\|}{C}}-\overset{O}{\overset{\|}{C}}-OH$$

Lactic acid (lactate) — Secondary alcohol; Pyruvic acid (pyruvate) — Ketone group

Oxidation of Thiols

Thiols also undergo oxidation by a loss of hydrogen atoms from each of two —SH groups. The oxidized product is a **disulfide** that contains a —S—S— bond.

$$\underset{\text{Methanethiol}}{CH_3-S-H + H-S-CH_3} \xrightarrow{[O]} \underset{\text{Dimethyl disulfide}}{CH_3-S-S-CH_3} + H_2O$$

Much of the protein in the hair is cross-linked by disulfide bonds, which occur between the thiol groups of the amino acid cysteine:

$$\text{Protein Chain}-\underset{\text{Cysteine side groups}}{CH_2-SH + HS-CH_2}-\text{Protein Chain} \xrightarrow{[O]}$$

$$\text{Protein Chain}-\underset{\text{Disulfide bond}}{CH_2-S-S-CH_2}-\text{Protein Chain} + H_2O$$

When a person has his or her hair permanently curled or straightened, a reducing substance is used to break the disulfide bonds. While the hair is wrapped around curlers or straightened, an oxidizing substance is applied that causes new disulfide bonds to form between different parts of the protein hair strands, which gives the hair a new shape.

$$\underset{\text{Methanethiol}}{CH_3-S-H + H-S-CH_3} \xrightarrow{[O]} \underset{\text{Dimethyl disulfide}}{CH_3-S-S-CH_3} + H_2O$$

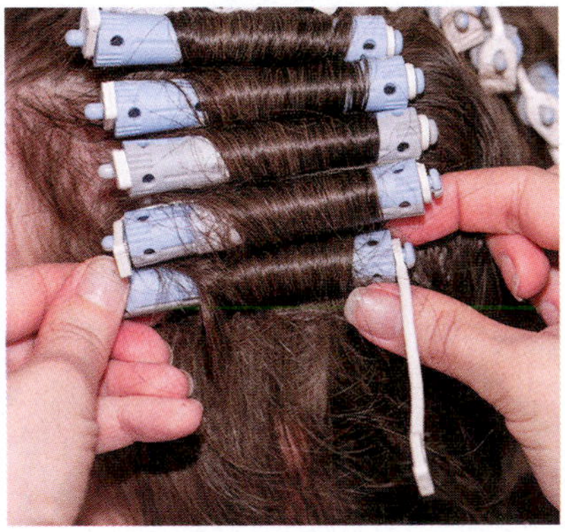

Proteins in the hair take new shapes when disulfide bonds are reduced and oxidized.

Chemistry Link to Health

Oxidation of Alcohol in the Body

Ethanol is the most commonly abused drug in the United States. When ingested in small amounts, ethanol may produce a feeling of euphoria in the body despite the fact that it is a depressant. In the liver, enzymes such as alcohol dehydrogenase oxidize ethanol to acetaldehyde, a substance that impairs mental and physical coordination. If the blood alcohol concentration exceeds 0.4%, coma or death may occur. **TABLE 13.2** gives some of the typical behaviors exhibited at various blood alcohol levels.

$$\underset{\substack{\text{Ethanol}\\\text{(ethyl alcohol)}}}{CH_3-CH_2-OH} \xrightarrow{[O]} \underset{\substack{\text{Ethanal}\\\text{(acetaldehyde)}}}{CH_3-\overset{O}{\overset{\|}{C}}-H} \xrightarrow{[O]} 2CO_2 + H_2O$$

(continued)

Chemistry Link to Health (*continued*)

TABLE 13.2 Typical Behaviors Exhibited by a 150-lb Person Consuming Alcohol

Number of Beers (12 oz) or Glasses of Wine (5 oz) in 1 h	Blood Alcohol Level (% m/v)	Typical Behavior
1	0.025	Slightly dizzy, talkative
2	0.050	Euphoria, loud talking and laughing
4	0.10	Loss of inhibition, loss of coordination, drowsiness, legally intoxicated in most states
8	0.20	Intoxicated, quick to anger, exaggerated emotions
12	0.30	Unconscious
16 to 20	0.40 to 0.50	Coma and death

The acetaldehyde produced from ethanol in the liver is further oxidized to acetic acid, which is converted to carbon dioxide and water in the citric acid cycle. Thus, the enzymes in the liver can eventually break down ethanol, but the aldehyde and carboxylic acid intermediates can cause considerable damage while they are present within the cells of the liver.

A person weighing 150 lb requires about one hour to metabolize the alcohol in 12 ounces of beer. However, the rate of metabolism of ethanol varies between nondrinkers and drinkers. Typically, nondrinkers and social drinkers can metabolize 12 to 15 mg of ethanol/dL of blood in one hour, but an alcoholic can metabolize as much as 30 mg of ethanol/dL in one hour. Some effects of alcohol metabolism include an increase in liver lipids (fatty liver), gastritis, pancreatitis, ketoacidosis, alcoholic hepatitis, and psychological disturbances.

When alcohol is present in the blood, it evaporates through the lungs. Thus, the percentage of alcohol in the lungs can be used to calculate the blood alcohol concentration (BAC). Several devices are used to measure the BAC. When a Breathalyzer is used, a suspected drunk driver exhales through a mouthpiece into a solution containing the orange Cr^{6+} ion. Any alcohol present in the exhaled air is oxidized, which reduces the orange Cr^{6+} to a green Cr^{3+}.

$$\underset{\text{Ethanol}}{CH_3-CH_2-OH} + \underset{\text{Orange}}{Cr^{6+}} \xrightarrow{[O]} \underset{\text{Ethanoic acid}}{CH_3-\overset{\overset{\displaystyle O}{\|}}{C}-OH} + \underset{\text{Green}}{Cr^{3+}}$$

The Alcosensor uses the oxidation of alcohol in a fuel cell to generate an electric current that is measured. The Intoxilyzer measures the amount of light absorbed by the alcohol molecules.

Sometimes alcoholics are treated with a drug called Antabuse (disulfiram), which prevents the oxidation of acetaldehyde to acetic acid. As a result, acetaldehyde accumulates in the blood, which causes nausea, profuse sweating, headache, dizziness, vomiting, and respiratory difficulties. Because of these unpleasant side effects, the person is less likely to use alcohol.

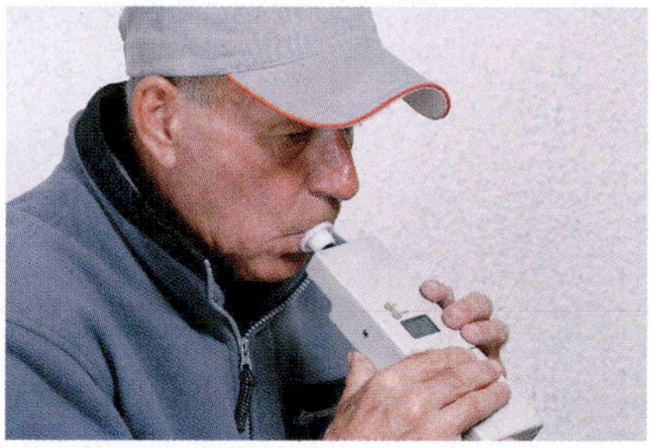

A Breathalyzer test is used to determine blood alcohol level.

Methanol Poisoning

Methanol (methyl alcohol), CH_3OH, is a highly toxic alcohol present in products such as windshield washer fluid, Sterno, and paint strippers. Methanol is rapidly absorbed in the gastrointestinal tract. In the liver, it is oxidized to formaldehyde and then formic acid, a substance that causes nausea, severe abdominal pain, and blurred vision. Blindness can occur because the intermediate products destroy the retina of the eye. As little as 4 mL of methanol can produce blindness. The formic acid, which is not readily eliminated from the body, lowers blood pH so severely that just 30 mL of methanol can lead to coma and death.

$$\underset{\substack{\text{Methanol}\\\text{(methyl alcohol)}}}{CH_3-OH} \xrightarrow{[O]} \underset{\substack{\text{Methanal}\\\text{(formaldehyde)}}}{H-\overset{\overset{\displaystyle O}{\|}}{C}-H} \xrightarrow{[O]} \underset{\substack{\text{Methanoic acid}\\\text{(formic acid)}}}{H-\overset{\overset{\displaystyle O}{\|}}{C}-OH}$$

The treatment for methanol poisoning involves giving sodium bicarbonate to neutralize the formic acid in the blood. In some cases, ethanol is given intravenously to the patient. The enzymes in the liver pick up ethanol molecules to oxidize instead of methanol molecules. This process gives time for the methanol to be eliminated via the lungs without the formation of its dangerous oxidation products.

PRACTICE PROBLEMS

13.4 Reactions of Alcohols and Thiols

13.17 Write the balanced chemical equation for the complete combustion of each of the following:

a. methanol **b.** 2-butanol

13.18 Write the balanced chemical equation for the complete combustion of each of the following:

a. cyclopentanol **b.** 3-hexanol

13.19 Draw the condensed structural or line-angle formula for the alkene that is the major product from each of the following dehydration reactions:

a. $CH_3-\underset{\displaystyle CH_3}{CH}-CH_2-CH_2-OH \xrightarrow{H^+,\ heat}$

b. 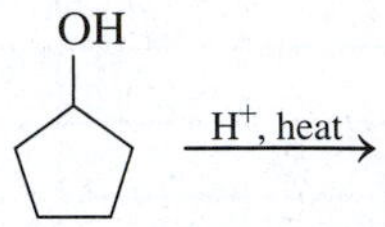 c. 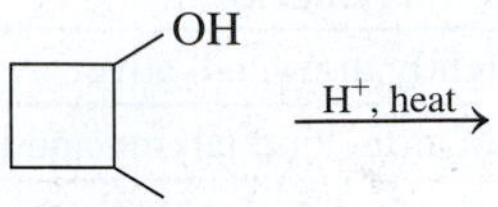

d. $CH_3-\underset{\displaystyle CH_3}{CH}-CH_2-CH_2-\underset{\displaystyle OH}{CH}-CH_3 \xrightarrow{H^+,\ heat}$

13.20 Draw the condensed structural or line-angle formula for the alkene that is the major product from each of the following dehydration reactions:

a. $CH_3-\underset{\displaystyle CH_3}{CH}-CH_2-OH \xrightarrow{H^+,\ heat}$

b. $CH_3-\underset{\displaystyle OH}{CH}-\underset{\displaystyle CH_3}{CH}-CH_2-CH_3 \xrightarrow{H^+,\ heat}$

c. 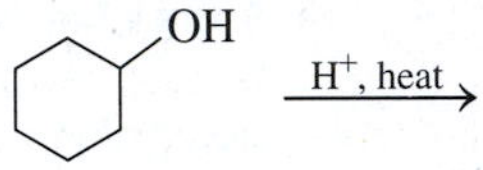

d. 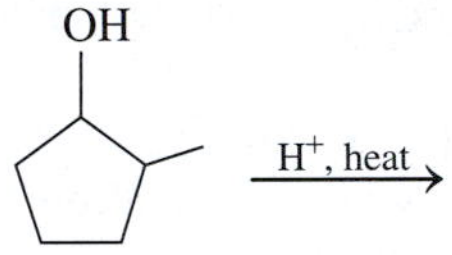

13.21 What alcohol(s) could be used to produce each of the following compounds?

a. $H_2C{=}CH_2$ b. 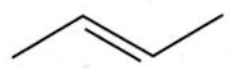 c. 

13.22 What alcohol(s) could be used to produce each of the following compounds?

a. 

b. $CH_3-CH_2-\underset{\displaystyle CH_3}{C}{=}CH-CH_3$

c. 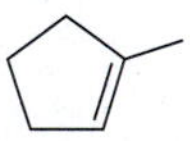

13.23 Draw the condensed structural or line-angle formula for the aldehyde or ketone produced when each of the following alcohols is oxidized [O] (if no reaction, write *none*):

a. $CH_3-CH_2-CH_2-CH_2-CH_2-OH$

b. $CH_3-CH_2-\underset{\displaystyle OH}{CH}-CH_3$

c. 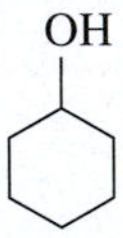

d. 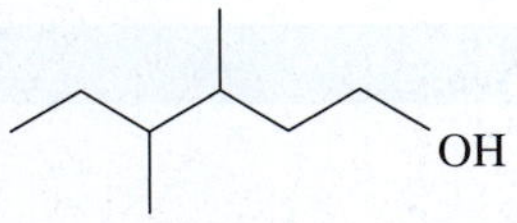

e. $CH_3-\underset{\displaystyle CH_3}{CH}-CH_2-CH_2-OH$

13.24 Draw the condensed structural or line-angle formula for the aldehyde or ketone produced when each of the following alcohols is oxidized [O] (if no reaction, write *none*):

a. 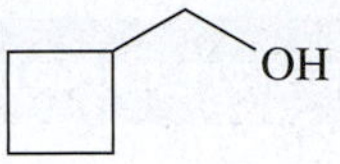

b. 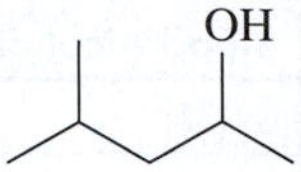

c. $CH_3-CH_2-\overset{\displaystyle OH}{\underset{\displaystyle CH_3}{C}}-CH_3$

d. 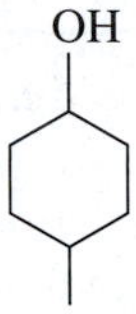

e. 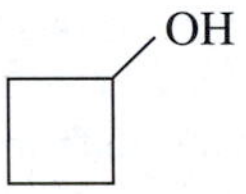

13.25 Draw the condensed structural or line-angle formula for the alcohol needed to give each of the following oxidation products:

a. $H-\overset{\displaystyle O}{\overset{\|}{C}}-H$ b. 

c. 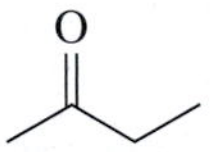 d. 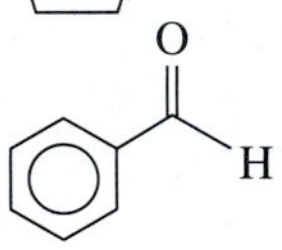

e. 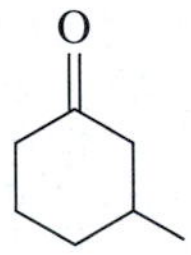

13.26 Draw the condensed structural or line-angle formula for the alcohol needed to give each of the following oxidation products:

a. $CH_3-\overset{\displaystyle O}{\overset{\|}{C}}-H$

b. 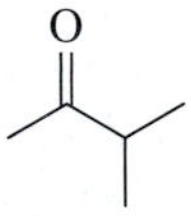

c. 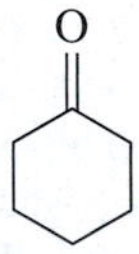

d. $CH_3-CH_2-\overset{\displaystyle O}{\overset{\|}{C}}-H$

e. $CH_3-\underset{\displaystyle CH_3}{CH}-CH_2-\overset{\displaystyle O}{\overset{\|}{C}}-H$

CLINICAL UPDATE Janet's New Diet Plan

At her doctor's suggestion, Janet decided to lose some weight after her surgery to repair her ACL. For her new diet, she decided to eat more fruits and vegetables, including peppers and onions. Vegetables are low in calories, have no fat or cholesterol, contain fiber, vitamins, especially vitamin B_3 (niacin), and minerals. Peppers contain antioxidants, such as capsaicin, which gives hot peppers a burning taste. Onions contain fiber, folic acid, calcium, and antioxidants, which inhibit oxidation reactions in the body and act as protective agents. 2-Propene-1-thiol gives the taste and odor to garlic. Janet added ginger, which contains pungent compounds, such as gingerol and shogaol, which is a bioactive compound that has anti-nausea, anti-inflammatory, and anti-carcinogenic properties. Resveratrol, obtained from the skin of red grapes, may improve oxygenation of muscles.

Clinical Applications

13.27 **a.** Which of the functional groups alkene, alcohol, phenol, thiol, and ether are found in capsaicin?

b. Which of the functional groups alkene, alcohol, phenol, thiol, and ether found in 2-propene-1-thiol from garlic?

c. Which of the functional groups alkene, alcohol, phenol, thiol, and ether are found in resveratrol from grapes?

13.28 **a.** Which of the functional groups alkene, alcohol, phenol, thiol, and ether are found in both gingerol and capsaicin?

b. Which of the functional groups alkene, alcohol, phenol, thiol, and ether are found in both gingerol and shogaol?

c. What type of reaction would convert gingerol to shogaol?

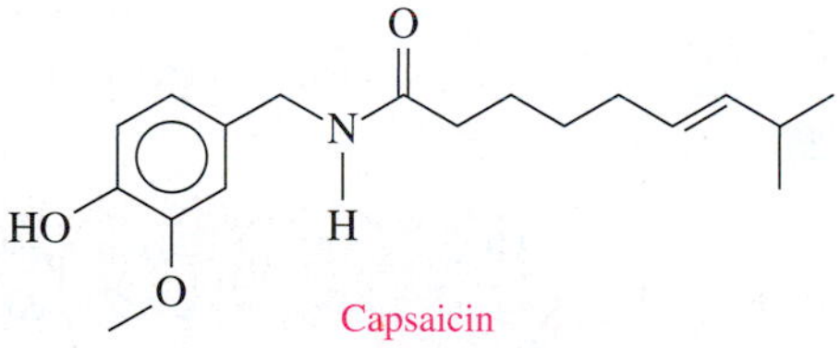
O
N
H
HO
O

Capsaicin

SH

2-Propene-1-thiol

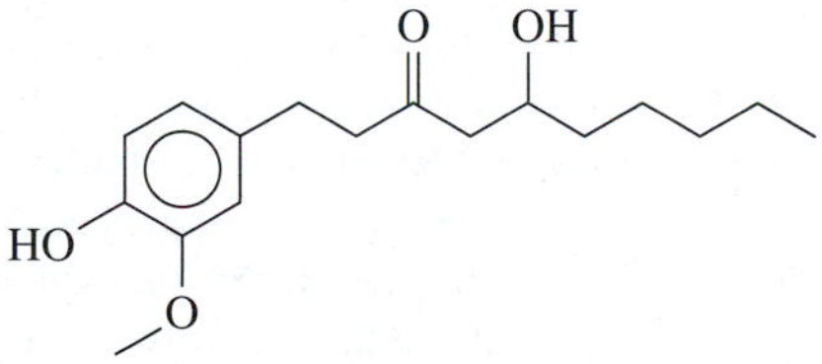
O
OH
HO
O

Gingerol

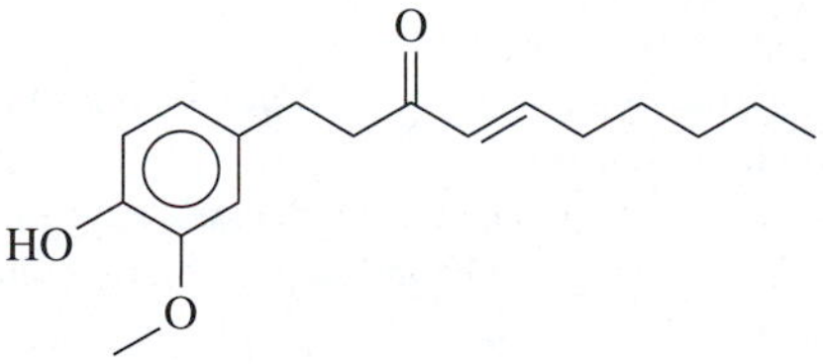
O
HO
O

Shogaol

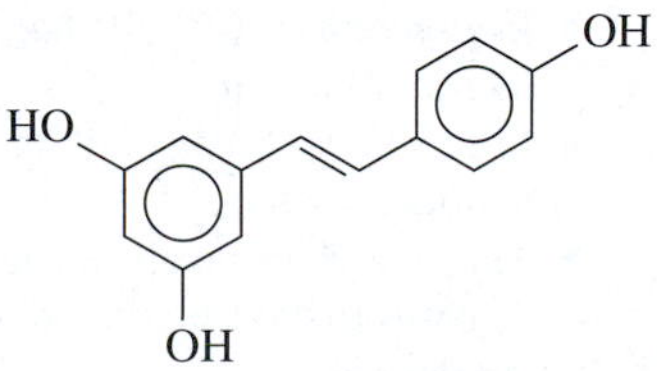
OH
HO
OH

Resveratrol

CONCEPT MAP

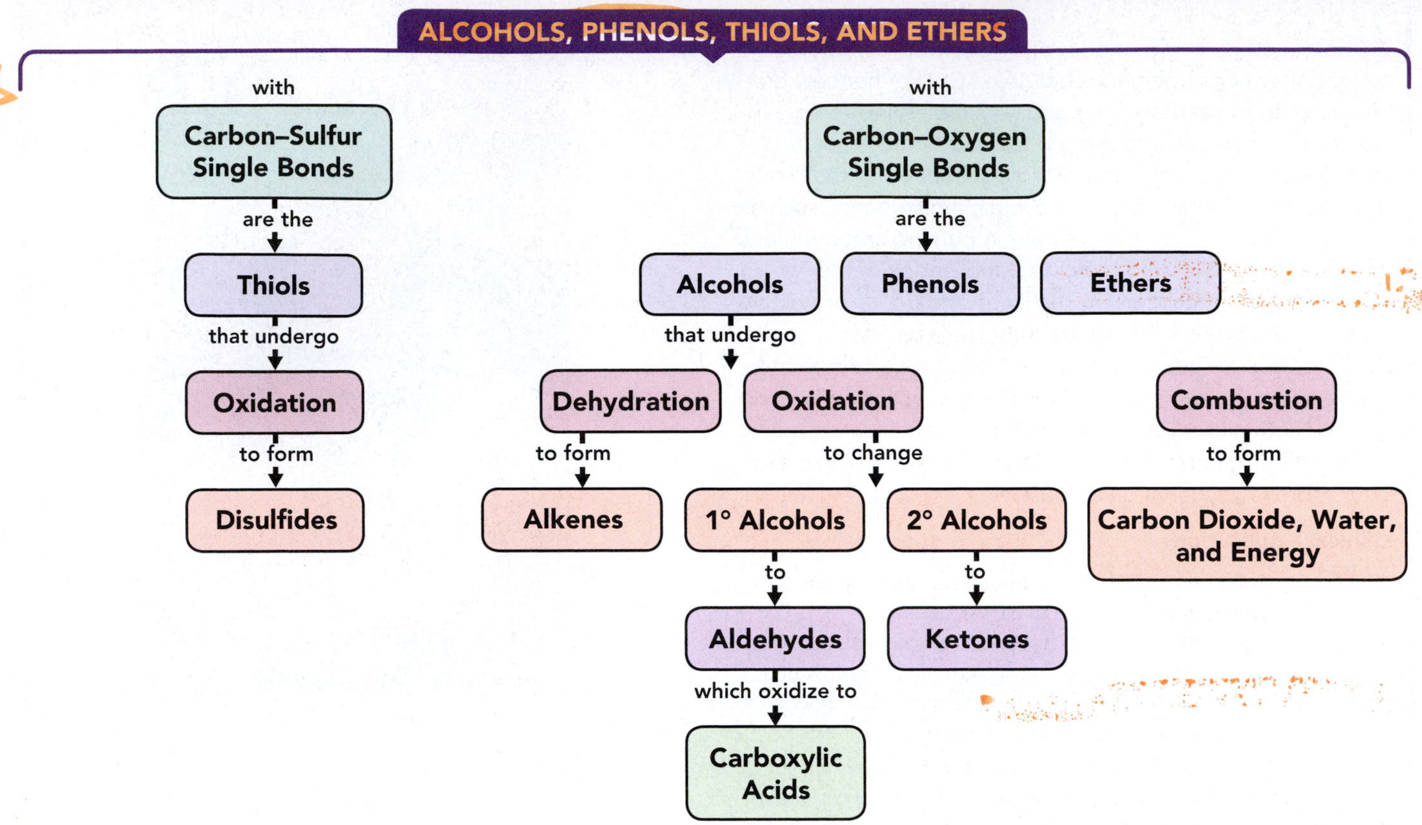

ALCOHOLS, PHENOLS, THIOLS, AND ETHERS
with
Carbon–Sulfur Single Bonds
are the
Thiols
that undergo
Oxidation
to form
Disulfides
with
Carbon–Oxygen Single Bonds
are the
Alcohols
Phenols
Ethers
that undergo
Dehydration
to form
Alkenes
Oxidation
to change
1° Alcohols
to
Aldehydes
which oxidize to
Carboxylic Acids
2° Alcohols
to
Ketones
Combustion
to form
Carbon Dioxide, Water, and Energy

CHAPTER REVIEW

13.1 Alcohols, Phenols, and Thiols

LEARNING GOAL Write the IUPAC and common names for alcohols, phenols, and thiols. Draw their condensed structural and line-angle formulas.

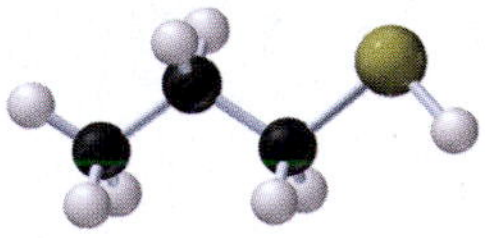

$CH_3—CH_2—CH_2—SH$

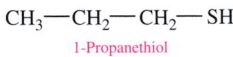

1-Propanethiol (onions)

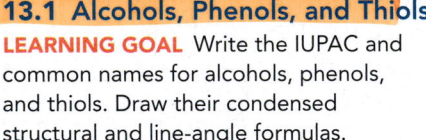

- The functional group of an alcohol is the hydroxyl group (—OH) bonded to a carbon chain.
- In a phenol, the hydroxyl group is bonded to an aromatic ring.
- In thiols, the functional group is —SH, which is analogous to the —OH group of alcohols.
- In the IUPAC system, the names of alcohols have *ol* endings, and the location of the —OH group is given by numbering the carbon chain.
- A cyclic alcohol is named as a cycloalkanol.
- Simple alcohols are generally named by their common names, with the alkyl name preceding the term *alcohol*.
- An aromatic alcohol is named as a phenol.

13.2 Ethers

LEARNING GOAL Write the IUPAC and common names for ethers; draw their condensed structural and line-angle formulas.

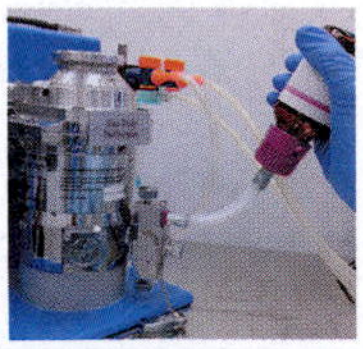

- In an ether, an oxygen atom is connected by single bonds to two alkyl or aromatic groups.
- In the common names of ethers, the alkyl groups are listed alphabetically, followed by the word *ether*.
- In the IUPAC name of an ether, the smaller alkyl group with the oxygen is named as an alkoxy group and is attached to the longer alkane chain, which is numbered to give the location of the alkoxy group.

13.3 Physical Properties of Alcohols, Phenols, and Ethers

LEARNING GOAL Describe the classification of alcohols; describe the boiling points and solubility of alcohols, phenols, and ethers.

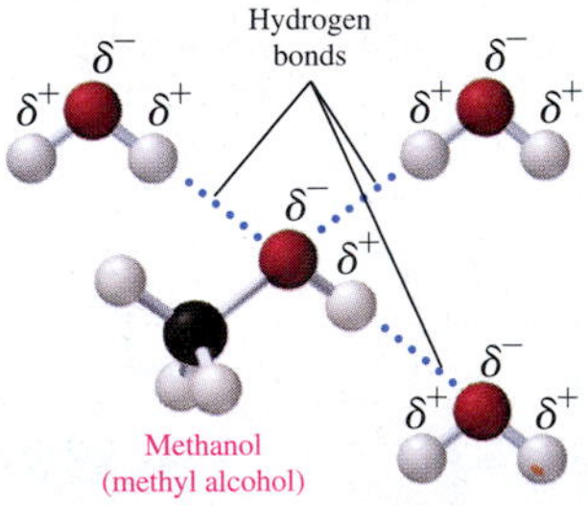

Hydrogen bonds
δ−
δ+

Methanol (methyl alcohol)

- Alcohols are classified according to the number of alkyl groups bonded to the carbon that holds the —OH group.
- In a primary (1°) alcohol, one group is attached to the hydroxyl carbon.
- In a secondary (2°) alcohol, two groups are attached to the hydroxyl carbon.
- In a tertiary (3°) alcohol, there are three groups bonded to the hydroxyl carbon.
- The —OH group allows alcohols to hydrogen bond, which causes alcohols to have higher boiling points than alkanes and ethers of similar mass.
- Short-chain alcohols and ethers can hydrogen bond with water, which makes them soluble.

13.4 Reactions of Alcohols and Thiols

LEARNING GOAL Write balanced chemical equations for the combustion, dehydration, and oxidation of alcohols and thiols.

- Alcohols undergo combustion with O_2 to form CO_2, H_2O, and energy.
- At high temperatures, alcohols dehydrate in the presence of an acid to yield alkenes.
- Primary alcohols are oxidized to aldehydes, which can oxidize further to carboxylic acids.
- Secondary alcohols are oxidized to ketones.
- Tertiary alcohols do not oxidize.
- Thiols undergo oxidation to form disulfides.

SUMMARY OF NAMING

Family	Structure	IUPAC Name	Common Name
Alcohol	CH_3-**OH**	Methanol	Methyl alcohol
Phenol	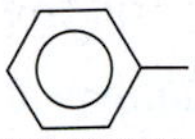 —**OH**	Phenol	Phenol
Thiol	CH_3-**SH**	Methanethiol	
Ether	CH_3-**O**$-CH_3$	Methoxymethane	Dimethyl ether

SUMMARY OF REACTIONS

The chapter Sections to review are shown after the name of each reaction.

Combustion of Alcohols (13.4)

$$\underset{\text{Ethanol}}{CH_3-CH_2-OH} + \underset{\text{Oxygen}}{3O_2} \xrightarrow{\Delta} \underset{\text{Carbon dioxide}}{2CO_2} + \underset{\text{Water}}{3H_2O} + \text{energy}$$

Dehydration of Alcohols to Form Alkenes (13.4)

$$\underset{\text{1-Propanol}}{CH_3-CH_2-CH_2-OH} \xrightarrow{H^+,\ \text{heat}} \underset{\text{Propene}}{CH_3-CH=CH_2} + H_2O$$

Oxidation of Primary Alcohols to Form Aldehydes (13.4)

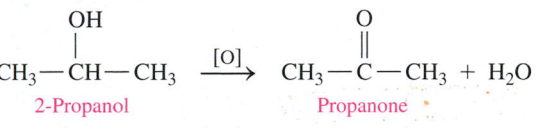

$$\underset{\text{Ethanol}}{CH_3-\overset{\overset{\displaystyle OH}{|}}{CH_2}} \xrightarrow{[O]} \underset{\text{Ethanal}}{CH_3-\overset{\overset{\displaystyle O}{\|}}{C}-H} + H_2O$$

Oxidation of Secondary Alcohols to Form Ketones (13.4)

$$\underset{\text{2-Propanol}}{CH_3-\overset{\overset{\displaystyle OH}{|}}{CH}-CH_3} \xrightarrow{[O]} \underset{\text{Propanone}}{CH_3-\overset{\overset{\displaystyle O}{\|}}{C}-CH_3} + H_2O$$

Oxidation of Aldehydes to Form Carboxylic Acids (13.4)

$$\underset{\text{Ethanal}}{CH_3-\overset{\overset{\displaystyle O}{\|}}{C}-H} \xrightarrow{[O]} \underset{\text{Ethanoic acid}}{CH_3-\overset{\overset{\displaystyle O}{\|}}{C}-OH}$$

KEY TERMS

alcohol An organic compound that contains the hydroxyl functional group ($-$OH) attached to a carbon chain.

dehydration A reaction that removes water from an alcohol in the presence of an acid to form alkenes at high temperatures.

disulfide A compound formed from thiols; disulfides contain the $-S-S-$ functional group.

ether An organic compound in which an oxygen atom is bonded to two carbon groups that are alkyl or aromatic.

oxidation The loss of two hydrogen atoms from a reactant to give a more oxidized compound: primary alcohols oxidize to aldehydes, secondary alcohols oxidize to ketones. An oxidation can also be the addition of an oxygen atom, as in the oxidation of aldehydes to carboxylic acids.

phenol An organic compound that has a hydroxyl group ($-$OH) attached to a benzene ring.

primary (1°) alcohol An alcohol that has one alkyl group bonded to the alcohol's carbon atom.

secondary (2°) alcohol An alcohol that has two alkyl groups bonded to the carbon atom with the $-$OH group.

tertiary (3°) alcohol An alcohol that has three alkyl groups bonded to the carbon atom with the $-$OH group.

thiol An organic compound that contains a thiol group ($-$SH).

CORE CHEMISTRY SKILLS

The chapter Section containing each Core Chemistry Skill is shown in parentheses at the end of each heading.

Identifying Alcohols, Phenols, and Thiols (13.1)

- Functional groups are specific groups of atoms in organic compounds, which undergo characteristic chemical reactions.
- Organic compounds with the same functional group have similar properties and reactions.

Example: Identify the functional group and classify the following molecule:

$$CH_3-CH_2-CH_2-CH_2-OH$$

Answer: The hydroxyl group ($-OH$) makes it an alcohol.

Naming Alcohols and Phenols (13.1)

- In the IUPAC system, an alcohol is named by replacing the *e* in the alkane name with *ol*.
- Simple alcohols are named with the alkyl name preceding the term *alcohol*.
- The carbon chain is numbered from the end nearer the $-OH$ group and the location of the $-OH$ is given in front of the name.
- A cyclic alcohol is named as a cycloalkanol.
- An aromatic alcohol is named as a phenol.

Example: Write the IUPAC name for the following:

$$\underset{}{CH_3-\overset{\overset{\displaystyle CH_3}{|}}{CH}-CH_2-OH}$$

Answer: 2-methyl-1-propanol

Writing Equations for the Dehydration of Alcohols (13.4)

- At high temperatures, alcohols dehydrate in the presence of an acid to yield alkenes.
- The major product from the dehydration of an asymmetrical alcohol is the one in which the $-H$ atom is removed from the carbon atom that has the smaller number of H atoms.

Example: Draw the condensed structural formula for the organic product from the dehydration of 3-methyl-2-butanol.

Answer: $CH_3-\overset{\overset{\displaystyle CH_3}{|}}{C}=CH-CH_3$

Writing Equations for the Oxidation of Alcohols (13.4)

- Primary alcohols oxidize to aldehydes, which can oxidize further to carboxylic acids.
- Secondary alcohols oxidize to ketones.
- Tertiary alcohols do not oxidize.

Example: Draw the condensed structural formula for the product from the oxidation of 2-butanol.

Answer: $CH_3-\overset{\overset{\displaystyle O}{\|}}{C}-CH_2-CH_3$

UNDERSTANDING THE CONCEPTS

The chapter Sections to review are shown in parentheses at the end of each problem.

13.29 Identify each of the following as an alcohol, a phenol, an ether, or a thiol: (13.1, 13.2)

a. 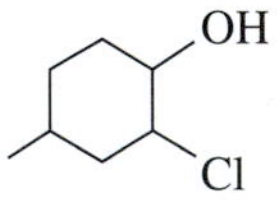

b. 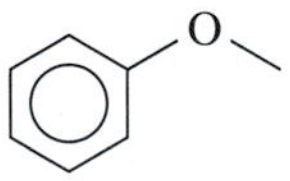

c. $CH_3-\overset{\overset{\displaystyle SH}{|}}{CH}-CH_3$

d. $CH_3-\overset{\overset{\displaystyle OH}{|}}{\underset{\underset{\displaystyle CH_3}{|}}{C}}-CH_2-\overset{\overset{\displaystyle CH_3}{|}}{CH}-CH_3$

13.30 Identify each of the following as an alcohol, a phenol, an ether, or a thiol: (13.1, 13.2)

a. 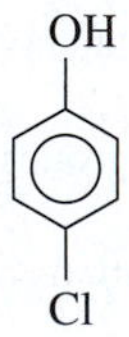

b. $CH_3-CH_2-CH_2-SH$

c. 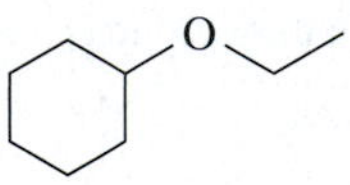

d. 

13.31 Write the IUPAC and common names (if any) for each of the compounds in problem 13.29. (13.1, 13.2)

13.32 Write the IUPAC and common names (if any) for each of the compounds in problem 13.30. (13.1, 13.2)

13.33 Identify each of the following as an alcohol, a phenol, an ether, or a thiol: (13.1, 13.2)

a. $CH_3-CH_2-CH_2-O-CH_3$

b. $CH_3-CH_2-\overset{\overset{\displaystyle SH}{|}}{CH}-CH_3$

c. $CH_3-\overset{\overset{\displaystyle Br}{|}}{CH}-CH_2-\overset{\overset{\displaystyle OH}{|}}{CH}-CH_3$

d. 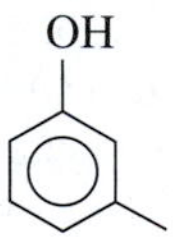

13.34 Identify each of the following as an alcohol, a phenol, an ether, or a thiol: (13.1, 13.2)

a. 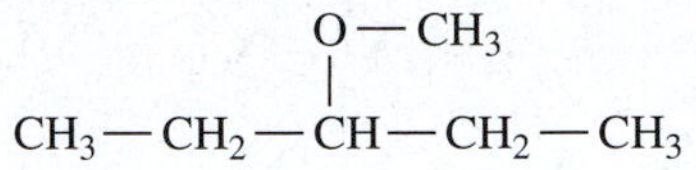

b. 

c. 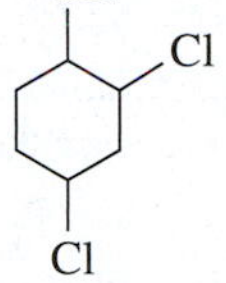

d. 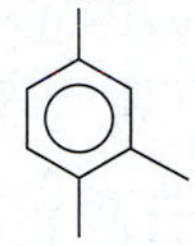

13.35 Write the IUPAC and common names (if any) for each of the compounds in problem 13.33. (13.1, 13.2)

13.36 Write the IUPAC and common names (if any) for each of the compounds in problem 13.34. (13.1, 13.2)

ADDITIONAL PRACTICE PROBLEMS

13.37 Draw the condensed structural or line-angle formula, if cyclic, for each of the following compounds: (13.1, 13.2)

a. 3-methylcyclopentanol **b.** 4-chlorophenol
c. 2-methyl-3-pentanol **d.** ethyl phenyl ether

13.38 Draw the condensed structural or line-angle formula, if cyclic, for each of the following compounds: (13.1, 13.2)

a. 3-methoxypentane **b.** *m*-chlorophenol
c. 2,3-pentanediol **d.** methyl propyl ether

13.39 Draw the condensed structural or line-angle formula, if cyclic, for each of the following compounds: (13.1, 13.2)

a. 3-pentanethiol **b.** 2-methoxypentane
c. 2,4-dibromophenol **d.** 2,3-dimethyl-2-butanol

13.40 Draw the condensed structural or line-angle formula, if cyclic, for each of the following compounds: (13.1, 13.2)

a. methanethiol **b.** 3-methyl-2-butanol
c. 3,4-dichlorocyclohexanol **d.** 1-ethoxypropane

13.41 Write the IUPAC name for each of the following alcohols, phenols, and ethers: (13.1, 13.2)

a. 

b. 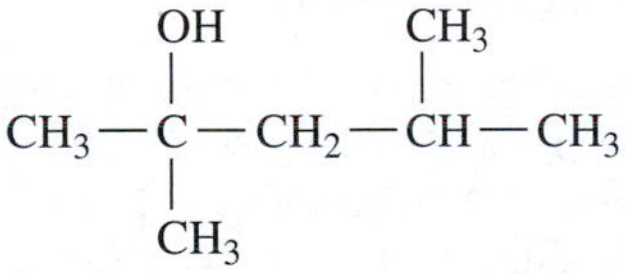

c. 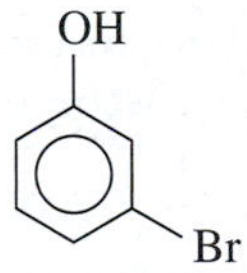

d. 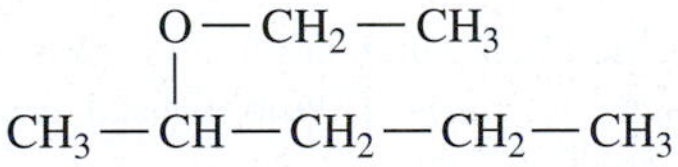

e. 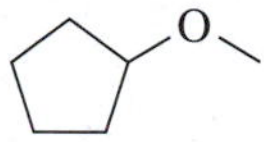

13.42 Write the IUPAC name for each of the following alcohols, phenols, and ethers: (13.1, 13.2)

a. 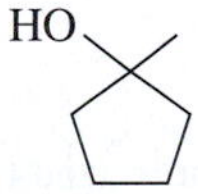

b. 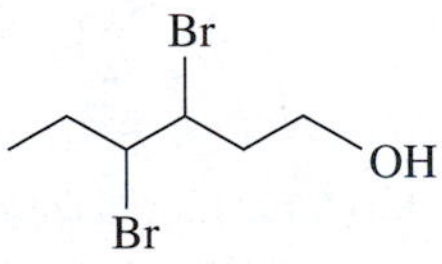

c. 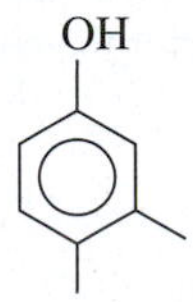

d. 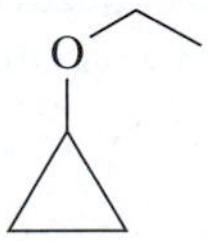

e. 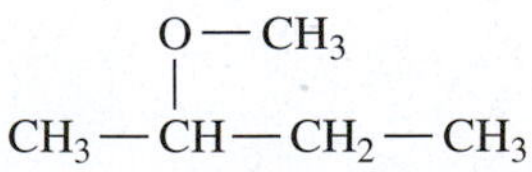

13.43 Draw the condensed structural formulas for all the alcohols with a molecular formula $C_4H_{10}O$. (13.1)

13.44 Draw the condensed structural formulas for all the ethers with a molecular formula $C_4H_{10}O$. (13.2)

13.45 Classify each of the following alcohols as primary (1°), secondary (2°), or tertiary (3°): (13.3)

a. 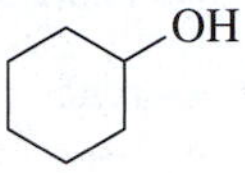

b. 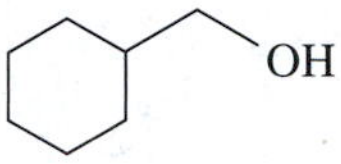

c. 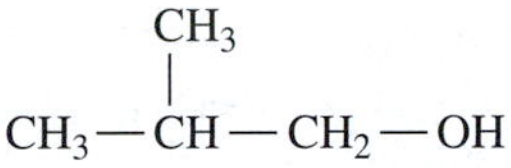

d. 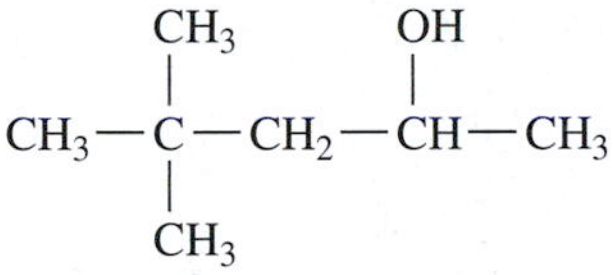

e. $HO-CH_2-CH_2-CH_3$

f. 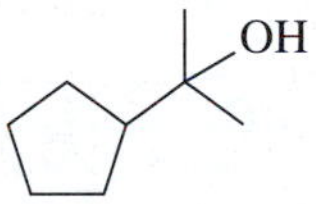

13.46 Classify each of the following alcohols as primary (1°), secondary (2°), or tertiary (3°): (13.3)

a. 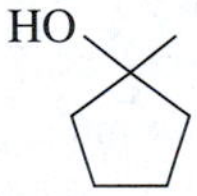

b. 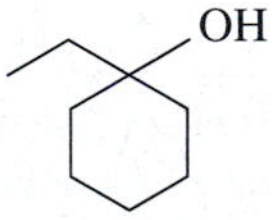

c. 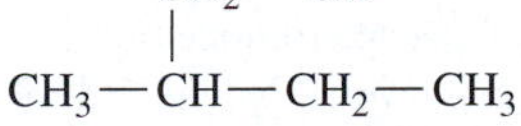

d. 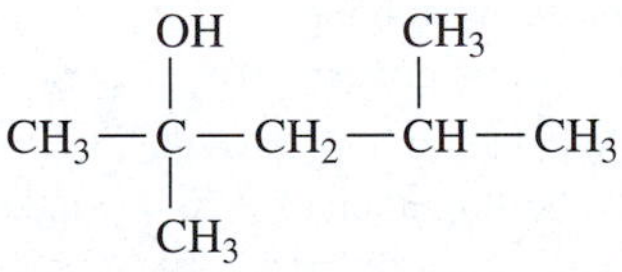

e. $CH_3-CH_2-CH_2-CH_2-OH$

f. 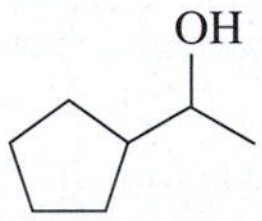

13.47 Which compound in each of the following pairs would you expect to have the higher boiling point? Explain. (13.3)

a. butane or 1-propanol
b. 1-propanol or ethyl methyl ether
c. ethanol or 1-butanol

13.48 Which compound in each of the following pairs would you expect to have the higher boiling point? Explain. (13.3)
- **a.** propane or ethyl alcohol
- **b.** 2-propanol or 2-pentanol
- **c.** methyl propyl ether or 1-butanol

13.49 Explain why each of the following compounds would be soluble or insoluble in water: (13.3)
- **a.** 2-propanol
- **b.** dipropyl ether
- **c.** 1-hexanol

13.50 Explain why each of the following compounds would be soluble or insoluble in water: (13.3)
- **a.** glycerol
- **b.** butane
- **c.** 1,3-hexanediol

13.51 Which compound in each pair would be more soluble in water? Explain. (13.3)
- **a.** butane or 1-propanol
- **b.** 1-propanol or ethyl methyl ether
- **c.** ethanol or 1-hexanol

13.52 Which compound in each pair would be more soluble in water? Explain. (13.3)
- **a.** ethane or ethanol
- **b.** 2-propanol or 2-pentanol
- **c.** methyl propyl ether or 1-butanol

13.53 Write the balanced chemical equation for the complete combustion of each of the following: (13.4)

a. $CH_3-CH_2-CH_2-CH_2-OH + O_2 \xrightarrow{\Delta}$

b.  $+ O_2 \xrightarrow{\Delta}$

c. 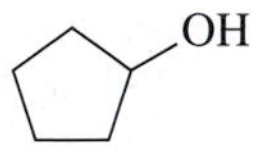 $+ O_2 \xrightarrow{\Delta}$

13.54 Write the balanced chemical equation for the complete combustion of each of the following: (13.4)

a. $CH_3-CH_2-CH_2-CH_2-OH + O_2 \xrightarrow{\Delta}$

b.  $+ O_2 \xrightarrow{\Delta}$

c. 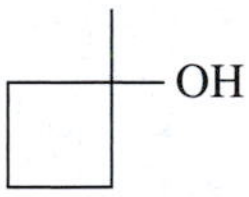 $+ O_2 \xrightarrow{\Delta}$

13.55 Draw the condensed structural or line-angle formula, if cyclic, for each of the following naturally occurring compounds: (13.1, 13.2)
- **a.** 2,5-dichlorophenol, a defense pheromone of a grasshopper
- **b.** 3-methyl-1-butanethiol and *trans*-2-butene-1-thiol, a mixture that gives skunk scent its odor
- **c.** pentachlorophenol, a wood preservative

13.56 Dimethyl ether and ethyl alcohol both have the molecular formula C_2H_6O. One has a boiling point of −24 °C, and the other, 79 °C. (13.1, 13.2, 13.3)
- **a.** Draw the condensed structural formula for each compound.
- **b.** Decide which boiling point goes with which compound and explain.

13.57 Draw the condensed structural or line-angle formula for the alkene (major product), aldehyde, ketone, or *none* produced in each of the following: (13.4)

a. $CH_3-CH_2-CH_2-OH \xrightarrow{H^+,\ heat}$

b. $CH_3-CH_2-CH_2-OH \xrightarrow{[O]}$

c. 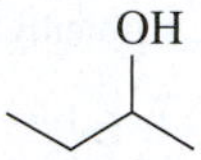 $\xrightarrow{H^+,\ heat}$

d. 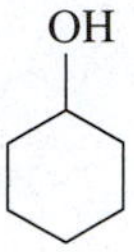 $\xrightarrow{H^+,\ heat}$

e. 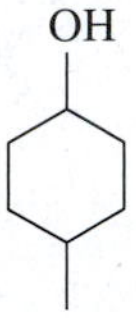 $\xrightarrow{[O]}$

13.58 Draw the condensed structural or line-angle formula for the alkene (major product), aldehyde, ketone, or *none* produced in each of the following: (13.4)

a. 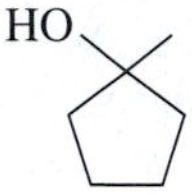 $\xrightarrow{H^+,\ heat}$

b. $CH_3-CH(CH_3)-CH(OH)-CH_3 \xrightarrow{H^+,\ heat}$

c. 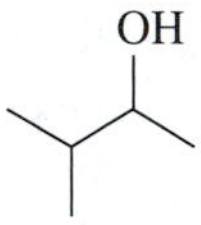 $\xrightarrow{[O]}$

d. 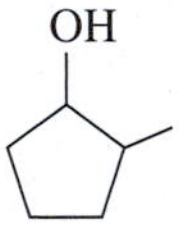 $\xrightarrow{[O]}$

e. $CH_3-CH_2-CH_2-CH(OH)-CH_3 \xrightarrow{H^+,\ heat}$

Clinical Applications

13.59 Hexylresorcinol, an antiseptic ingredient used in mouthwashes and throat lozenges, has the IUPAC name of 4-hexyl-1,3-benzenediol. Draw its condensed structural formula. (13.1)

13.60 Menthol, which has a minty flavor, is used in throat sprays and lozenges. Thymol is used as a topical antiseptic to destroy mold. (13.1)
- **a.** Write the IUPAC name for each.
- **b.** What is similar and what is different about their structures?

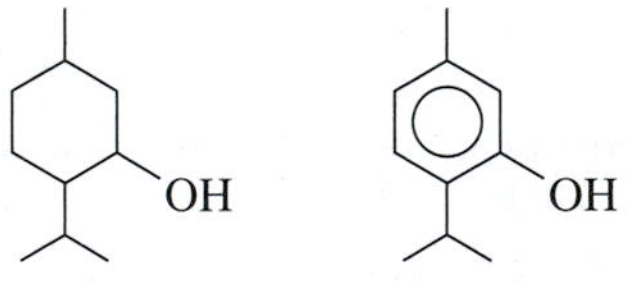

OH
OH

Menthol Thymol

13.61 A compound with the formula C_4H_8O is synthesized from 2-methyl-1-propanol and oxidizes easily to give a carboxylic acid. Draw the condensed structural formula for the compound. (13.4)

13.62 Methyl *tert*-butyl ether (MTBE), or 2-methoxy-2-methylpropane, has been used as a fuel additive for gasoline to boost the octane rating and to reduce CO emissions. (2.7, 7.5, 7.6, 7.7, 8.6, 13.4)

- **a.** If fuel mixtures are required to contain 2.7% oxygen by mass, how many grams of MTBE must be present in each 100. g of gasoline?
- **b.** How many liters of MTBE would be in 1.0 L of fuel if the density of both gasoline and MTBE is 0.740 g/mL?
- **c.** Write the balanced chemical equation for the complete combustion of MTBE.
- **d.** How many liters of air containing 21% (v/v) O_2 are required at STP to completely react (combust) 1.00 L of liquid MTBE?

Methyl *tert*-butyl ether (MTBE) is a gasoline additive.

CHALLENGE PROBLEMS

The following problems are related to the topics in this chapter. However, they do not all follow the chapter order, and they require you to combine concepts and skills from several Sections. These problems will help you increase your critical thinking skills and prepare for your next exam.

13.63 Compound **A** is a primary alcohol whose formula is C_3H_8O. When compound **A** is heated with strong acid, it dehydrates to form compound **B** (C_3H_6). When compound **A** is oxidized, compound **C** (C_3H_6O) forms. Draw the condensed structural formulas and write the IUPAC names for compounds **A**, **B**, and **C**. (13.4)

13.64 Compound **X** is a secondary alcohol whose formula is C_3H_8O. When compound **X** is heated with strong acid, it dehydrates to form compound **Y** (C_3H_6). When compound **X** is oxidized, compound **Z** (C_3H_6O) forms, which cannot be oxidized further. Draw the condensed structural formulas and write the IUPAC names for compounds **X**, **Y**, and **Z**. (13.4)

13.65 Sometimes several steps are needed to prepare a compound. Using a combination of the reactions we have studied, indicate how you might prepare the following from the starting substance given. For example, 2-propanol could be prepared from 1-propanol by first dehydrating the alcohol to give propene and then hydrating it again to give 2-propanol as follows: (13.4)

$$\underset{\text{1-Propanol}}{CH_3-CH_2-CH_2-OH} \xrightarrow{H^+,\ \text{heat}} \underset{\text{Propene}}{CH_3-CH=CH_2} + H_2O$$

$$\xrightarrow{H^+} \underset{\text{2-Propanol}}{CH_3-\overset{\overset{OH}{|}}{CH}-CH_3}$$

- **a.** prepare 2-methylpropane from 2-methyl-2-propanol
- **b.** prepare $CH_3-\overset{\overset{O}{\|}}{C}-CH_3$ from 1-propanol

13.66 As in problem 13.65, indicate how you might prepare the following from the starting substance given: (13.4)

- **a.** prepare 1-pentene from 1-pentanol
- **b.** prepare cyclohexane from cyclohexanol

13.67 One of the molecules that gives the flavor and odor to mushrooms is 1-octanol. (13.1, 13.3, 13.4)

- **a.** Draw the condensed structural and line-angle formulas for 1-octanol.
- **b.** Classify 1-octanol as a 1°, 2°, or 3° alcohol.
- **c.** Draw the line-angle formula for the product from the oxidation of 1-octanol.

13.68 One of the molecules that gives the flavor and odor to mushrooms is 3-octanol. (13.1, 13.3, 13.4)

- **a.** Draw the condensed structural and line-angle formulas for 3-octanol.
- **b.** Classify 3-octanol as a 1°, 2°, or 3° alcohol.
- **c.** Draw the line-angle formula for the product from the oxidation of 3-octanol.

ANSWERS

13.1
- **a.** ethanol (ethyl alcohol)
- **b.** 2-butanol (*sec*-butyl alcohol)
- **c.** 2-pentanethiol
- **d.** 4-methylcyclohexanol
- **e.** 3-fluorophenol (*m*-fluorophenol)

13.3
- **a.** $CH_3-CH_2-CH_2-OH$
- **b.** $CH_3-CH_2-\overset{\overset{SH}{|}}{CH}-CH_2-CH_3$
- **c.** $CH_3-\underset{\underset{CH_3}{|}}{\overset{\overset{OH}{|}}{C}}-CH_2-CH_3$
- **d.** 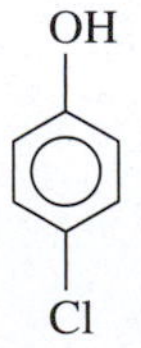
- **e.** 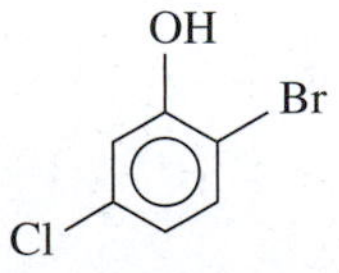

13.5 **a.** methoxyethane (ethyl methyl ether)
b. methoxybenzene (methyl phenyl ether)
c. ethoxycyclobutane (cyclobutyl ethyl ether)
d. 1-methoxypropane (methyl propyl ether)

13.7 **a.** $CH_3-CH_2-O-CH_2-CH_2-CH_3$

b. 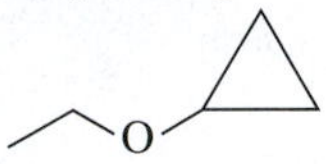

c. 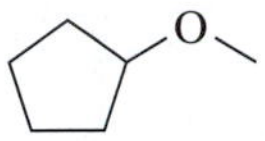

d. $CH_3-CH_2-O-CH_2-\underset{}{\overset{\overset{\large CH_3}{|}}{CH}}-CH_2-CH_3$

e. $CH_3-\overset{\overset{\large O-CH_3}{|}}{CH}-\underset{\underset{\large O-CH_3}{|}}{CH}-CH_2-CH_3$

13.9 **a.** 1° **b.** 1° **c.** 3° **d.** 2°

13.11 **a.** methanol **b.** 1-butanol **c.** 1-butanol

13.13 **a.** Soluble; ethanol with a short carbon chain is soluble because the hydroxyl group forms hydrogen bonds with water.
b. Slightly soluble; ethers with up to four carbon atoms are slightly soluble in water because they can form a few hydrogen bonds with water.
c. Insoluble; an alcohol with a carbon chain of five or more carbon atoms is not soluble in water.

13.15 **a.** Methanol can form hydrogen bonds with water, but ethane cannot.
b. 2-Propanol is more soluble because it has a shorter carbon chain.
c. 1-Propanol is more soluble because it can form more hydrogen bonds.

13.17 **a.** $2CH_3-OH + 3O_2 \xrightarrow{\Delta} 2CO_2 + 4H_2O + \text{energy}$

b. $CH_3-\overset{\overset{\large OH}{|}}{CH}-CH_2-CH_3 + 6O_2 \xrightarrow{\Delta} 4CO_2 + 5H_2O + \text{energy}$

13.19 **a.** $CH_3-\overset{\overset{\large CH_3}{|}}{CH}-CH=CH_2$

b. 

c. 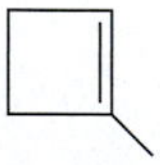

d. $CH_3-\overset{\overset{\large CH_3}{|}}{CH}-CH_2-CH=CH-CH_3$

13.21 **a.** CH_3-CH_2-OH

b. 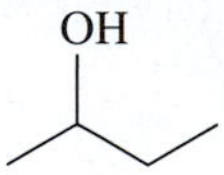

c. 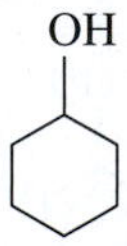

13.23 **a.** $CH_3-CH_2-CH_2-CH_2-\overset{\overset{\large O}{\|}}{C}-H$

b. $CH_3-CH_2-\overset{\overset{\large O}{\|}}{C}-CH_3$

c. 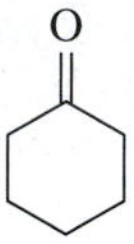

d. 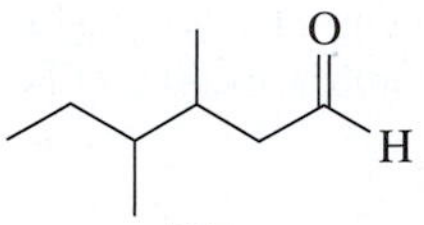

e. $CH_3-\overset{\overset{\large CH_3}{|}}{CH}-CH_2-\overset{\overset{\large O}{\|}}{C}-H$

13.25 **a.** CH_3-OH

b. 

c. 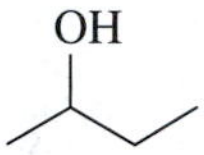

d. 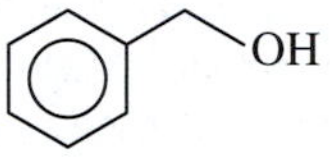

e. 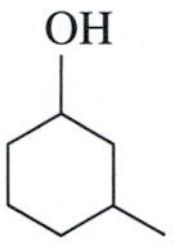

13.27 **a.** Capsaicin contains alkene, phenol, and ether functional groups.
b. Resveratrol contains phenol and alkene functional groups.
c. 2-Propene-1-thiol contains alkene and thiol function groups.

13.29 **a.** alcohol **b.** ether
c. thiol **d.** alcohol

13.31 **a.** 2-chloro-4-methylcyclohexanol
b. methoxybenzene (methyl phenyl ether)
c. 2-propanethiol
d. 2,4-dimethyl-2-pentanol

13.33 **a.** ether **b.** thiol
c. alcohol **d.** phenol

13.35 **a.** 1-methoxypropane (methyl propyl ether)
b. 2-butanethiol
c. 4-bromo-2-pentanol
d. 3-methylphenol (*m*-methylphenol, *m*-cresol)

13.37 **a.** 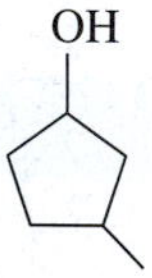

b. 

c. $CH_3-\overset{\overset{\large CH_3}{|}}{CH}-\overset{\overset{\large OH}{|}}{CH}-CH_2-CH_3$

d. 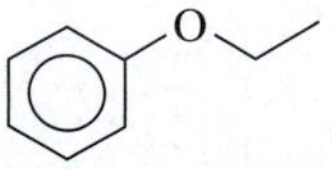

13.39 a. 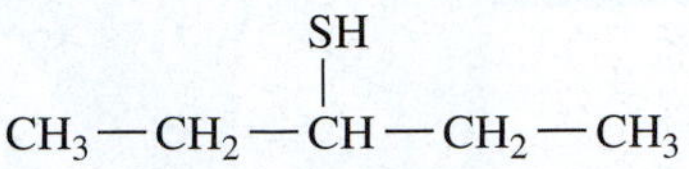

b. 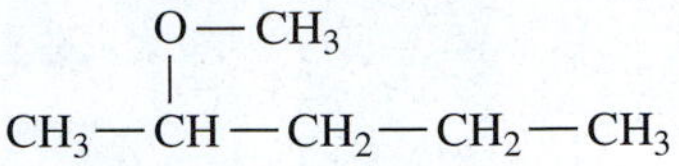

c. 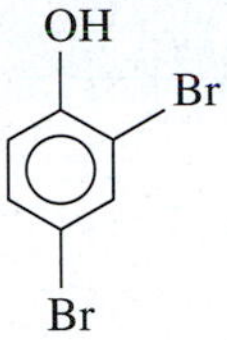

d. 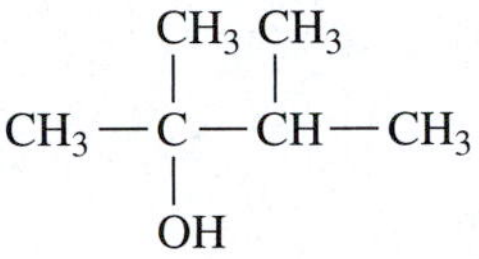

13.41 a. 2-methyl-2-propanol
b. 2,4-dimethyl-2-pentanol
c. 3-bromophenol
d. 2-ethoxypentane
e. methoxycyclopentane

13.43 $CH_3-CH_2-CH_2-CH_2-OH$

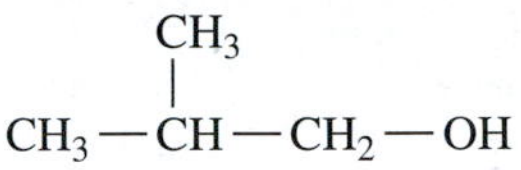

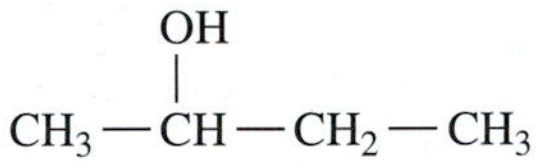

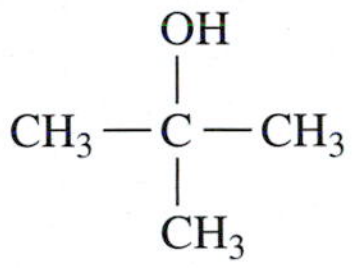

13.45 a. 2°	**b.** 1°	**c.** 1°
d. 2°	**e.** 1°	**f.** 3°

13.47 a. 1-propanol, hydrogen bonding
b. 1-propanol, hydrogen bonding
c. 1-butanol, greater molar mass

13.49 a. soluble, hydrogen bonding
b. insoluble, long carbon chain diminishes effect of hydrogen bonding of water to —O—
c. insoluble, long carbon chain diminishes effect of polar —OH group on hydrogen bonding

13.51 a. 1-Propanol can form hydrogen bonds with its polar hydroxyl group.
b. An alcohol forms more hydrogen bonds than an ether.
c. Ethanol has a smaller number of carbon atoms.

13.53 a. $CH_3-CH_2-CH_2-CH_2-OH + 6O_2 \xrightarrow{\Delta} 4CO_2 + 5H_2O + \text{energy}$

b. 2 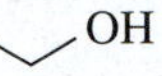 $+ 9O_2 \xrightarrow{\Delta} 6CO_2 + 8H_2O + \text{energy}$

c. 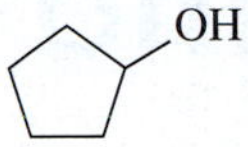 $+ 7O_2 \xrightarrow{\Delta} 5CO_2 + 5H_2O + \text{energy}$

13.55 a. 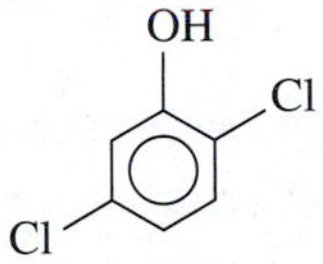

b. 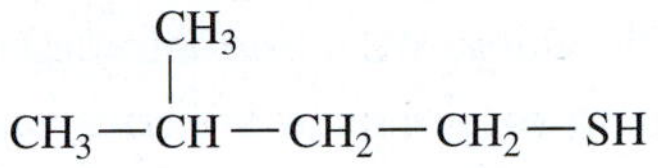 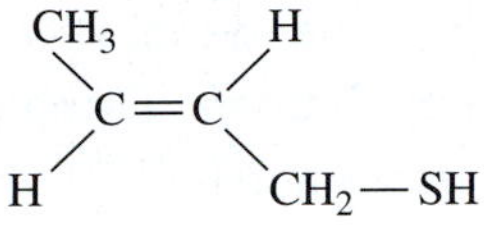

c. 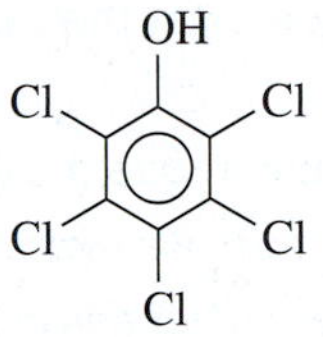

13.57 a. $CH_3-CH{=}CH_2$ **b.** 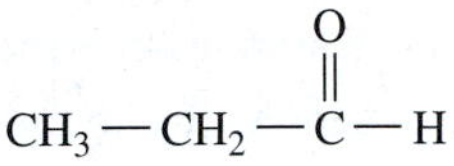

c.  **d.** 

e. 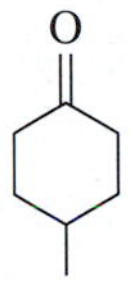

13.59 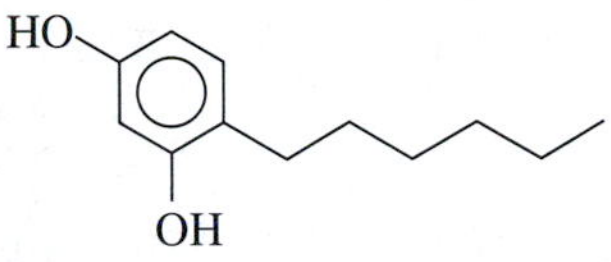

13.61 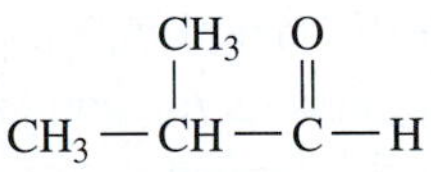

13.63 $CH_3-CH_2-CH_2-OH$ 1-Propanol
A

$CH_3-CH{=}CH_2$ Propene
B

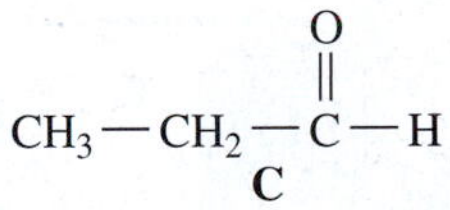 Propanal
C

13.65 a. 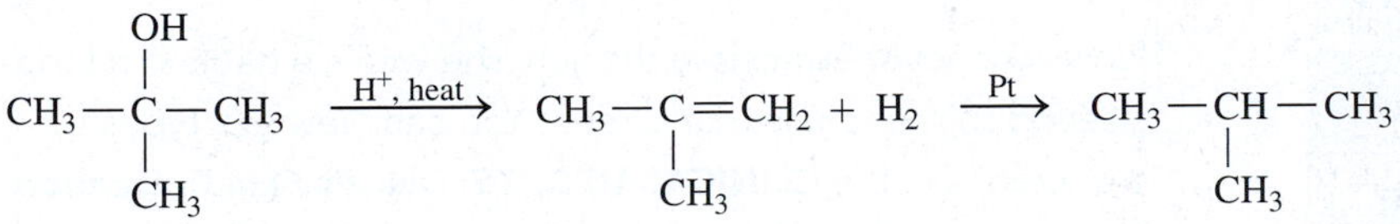

b. 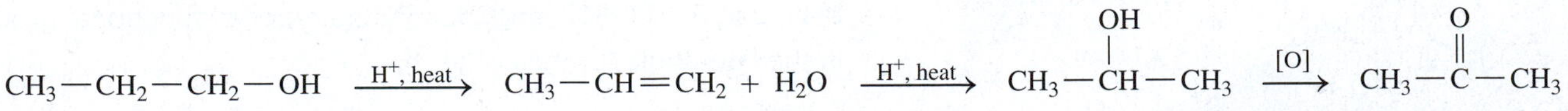

13.67 a. $CH_3-CH_2-CH_2-CH_2-CH_2-CH_2-CH_2-CH_2-OH$ 
b. 1°
c. 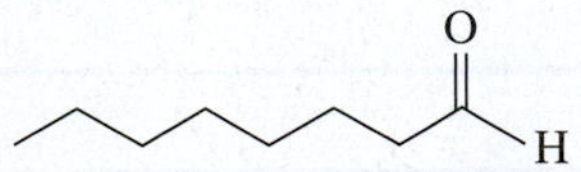

14

Aldehydes and Ketones

Recently, Diana noticed that a mole on her arm changed appearance. For many years, it was light brown in color, with a flat, circular appearance. But over the last few weeks, the mole has become raised, with irregular borders, and has darkened. She calls her dermatologist for an appointment. Diana tells Margaret, a dermatology nurse, that she has been going to tanning salons and had 20 tanning sessions the previous year. She loves being outside but does not always apply sunscreen while in the sun or at the beach. Diana says she has no family history of malignant melanoma.

The risk factors for melanoma include frequent exposure to sun, severe sunburns at an early age, skin type, and family history. During Diana's skin exam, Margaret looks for other suspicious moles with nonuniform borders, changes in color, and changes in size. To treat the mole on Diana's arm, Margaret numbs the area, then removes a sample of skin tissue, which she sends to a lab for evaluation. Because the results indicate the presence of malignant melanoma cells, a surgeon excises the entire mole including subcutaneous fat. Fortunately, the mole is not very large and no further treatment is needed. Margaret suggests that Diana return in six months for a follow-up skin check.

The number of cases of melanoma is rising, unlike many other types of cancer. Doctors think this change may be because of unprotected sun exposure, an increase in the use of tanning salons, and perhaps an increased awareness and detection of the disease.

CAREER

Dermatology Nurse

A dermatology nurse performs many of the duties of dermatologists, including treating skin conditions, assisting in surgeries, performing biopsies and excisions, writing prescriptions, freezing skin lesions, and screening patients for skin cancer. To become a dermatology nurse, you first become a nurse or physician assistant and then specialize in dermatology. To be certified, the RN must have an additional two years of dermatology experience, with a minimum of 2000 h of work experience in dermatology, and pass an examination. Advanced training is available through colleges and universities, allied health schools, and medical schools.

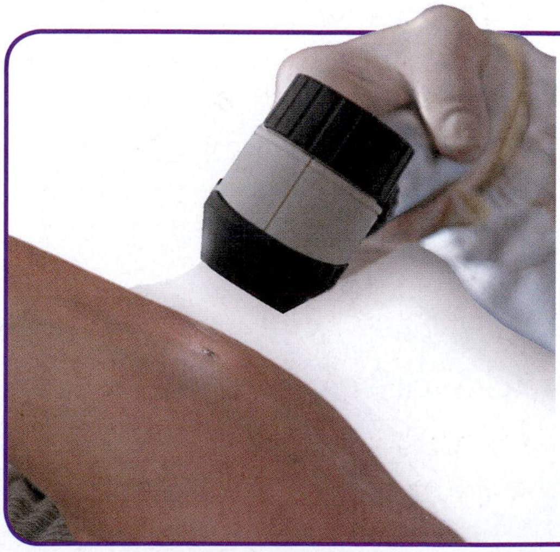

CLINICAL UPDATE

Diana's Skin Protection Plan

Now, whenever Diana is in the sun, she wears a hat and a long-sleeved shirt and uses sunscreen. You can view the types of sunscreens in the **CLINICAL UPDATE Diana's Skin Protection Plan**, pages 511–512, and identify the types of functional groups in the line-angle formula of each.

14.1 Aldehydes and Ketones

LEARNING GOAL Identify compounds with carbonyl groups as aldehydes and ketones. Write the IUPAC and common names for aldehydes and ketones; draw their condensed structural and line-angle formulas.

LOOKING AHEAD

Aldehydes and ketones contain a **carbonyl group** that consists of a carbon–oxygen double bond with two groups of atoms attached to the carbon at angles of 120°. The oxygen atom with two lone pairs of electrons is much more electronegative than the carbon atom. Therefore, the carbonyl group has a strong dipole with a partial negative charge (δ^-) on the oxygen and a partial positive charge (δ^+) on the carbon. The polarity of the carbonyl group strongly influences the physical and chemical properties of aldehydes and ketones. Aldehydes and ketones with the same number of carbon atoms are *structural isomers*; they have the same molecular formula, but the atoms are arranged differently.

In an **aldehyde**, the carbon of the carbonyl group is bonded to at least one hydrogen atom. That carbon may also be bonded to another hydrogen atom, a carbon of an alkyl group, or an aromatic ring (see **FIGURE 14.1**). The aldehyde group may be written as separate atoms or as —CHO, with the double bond understood. In a **ketone**, the carbonyl group is bonded to two alkyl groups or aromatic rings. The keto group (C═O) can sometimes be written as CO. A line-angle formula may also be used to represent an aldehyde or ketone.

REVIEW

Naming and Drawing Alkanes (12.2)

Structural Isomers of C_3H_6O

Aldehyde

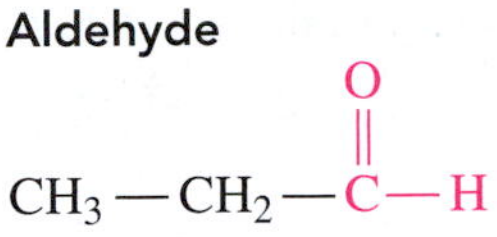
$= CH_3-CH_2-CHO =$
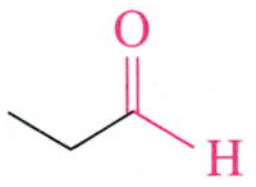

Ketone

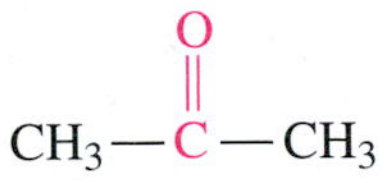
$= CH_3-CO-CH_3 =$

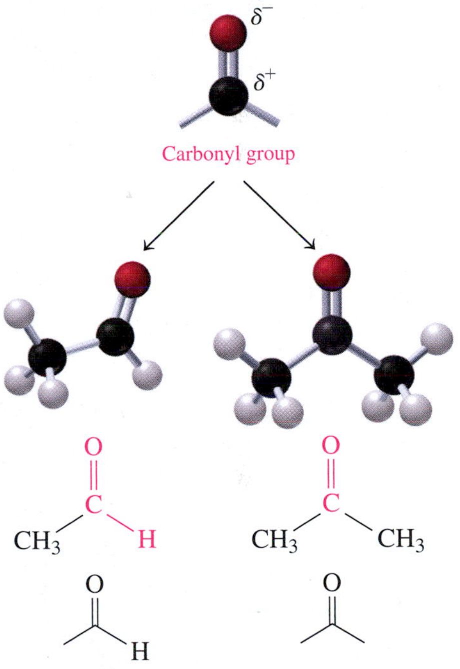

δ−
δ+
Carbonyl group
Aldehyde
Ketone

FIGURE 14.1 ▶ The carbonyl group is found in aldehydes and ketones.

Q If aldehydes and ketones both contain a carbonyl group, how can you differentiate between compounds from each family?

▶ SAMPLE PROBLEM 14.1 Identifying Aldehydes and Ketones

TRY IT FIRST

Identify each of the following compounds as an aldehyde or ketone:

a. 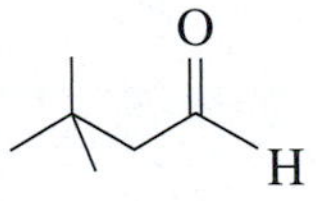 **b.** 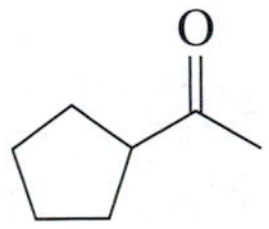 **c.** 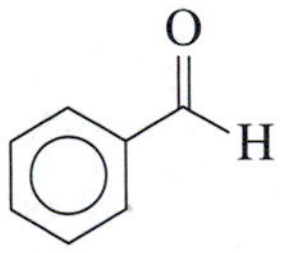

SOLUTION

a. aldehyde **b.** ketone **c.** aldehyde

STUDY CHECK 14.1

Identify each of the following compounds as an aldehyde or ketone:

a. 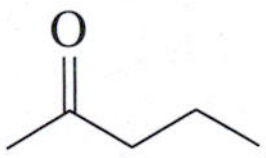

b. 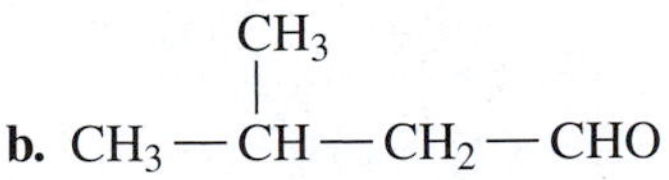

ANSWER

a. ketone **b.** aldehyde

TEST

Try Practice Problems 14.1 to 14.4

Naming Aldehydes

In the IUPAC system, an aldehyde is named by replacing the *e* of the corresponding alkane name with *al*. No number is needed for the aldehyde group because it always appears at the end of the chain. The aldehydes with carbon chains of one to four carbons are often referred

ENGAGE

How does the structure of an aldehyde differ from that of a ketone?

to by their common names, which end in *aldehyde* (see **FIGURE 14.2**). The roots (*form*, *acet*, *propion*, and *butyr*) of these common names are derived from Latin or Greek words.

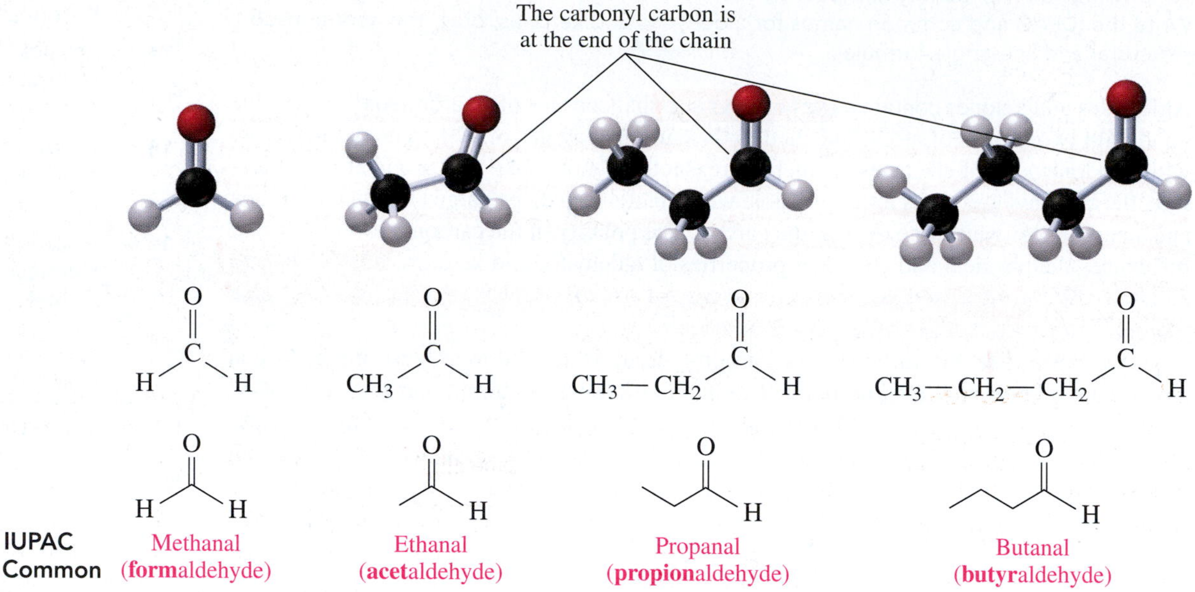

The carbonyl carbon is at the end of the chain

	O‖ H—C—H	O‖ CH₃—C—H	O‖ CH₃—CH₂—C—H	O‖ CH₃—CH₂—CH₂—C—H
IUPAC	Methanal	Ethanal	Propanal	Butanal
Common	(**form**aldehyde)	(**acet**aldehyde)	(**propion**aldehyde)	(**butyr**aldehyde)

FIGURE 14.2 ▶ In aldehydes, the carbonyl group is always the end carbon.
ⓠ Why is the carbon in the carbonyl group in aldehydes always at the end of the chain?

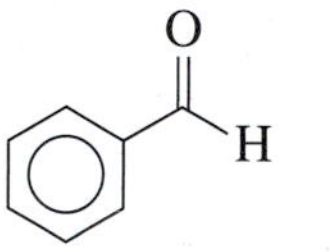

The IUPAC system names the aldehyde of benzene as benzaldehyde.

Any substituents are numbered from the carbonyl group in benzaldehyde, which is counted as carbon 1.

Benzaldehyde

▶**SAMPLE PROBLEM 14.2** Naming Aldehydes

TRY IT FIRST

Write the IUPAC name for the following:

$$CH_3—CH_2—\underset{\underset{CH_3}{|}}{CH}—CH_2—\overset{\overset{O}{\|}}{C}—H$$

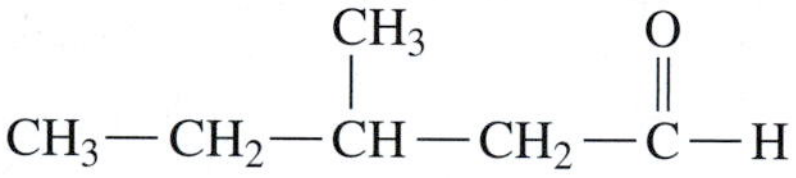

SOLUTION

	Given	Need	Connect
ANALYZE THE PROBLEM	five-carbon chain, carbonyl group, methyl substituent	IUPAC name	position of methyl group, replace *e* in alkane name with *al*

STEP 1 **Name the longest carbon chain by replacing the e in the alkane name with al.** The longest carbon chain containing the carbonyl group has five carbon atoms, which is named pentanal in the IUPAC system.

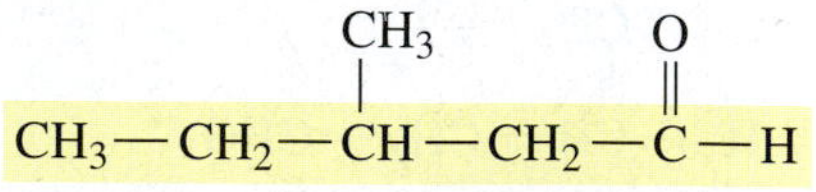

$$CH_3—CH_2—\underset{\underset{CH_3}{|}}{CH}—CH_2—\overset{\overset{O}{\|}}{C}—H \qquad \text{pentanal}$$

STEP 2 **Name and number any substituents by counting the carbonyl group as carbon 1.** The substituent, which is on carbon 3, is methyl. The IUPAC name for this compound is 3-methylpentanal.

$$\underset{5\quad\quad 4\quad\quad 3\quad\quad 2\quad\quad 1}{CH_3—CH_2—\underset{\underset{CH_3}{|}}{CH}—CH_2—\overset{\overset{O}{\|}}{C}—H} \qquad \text{3-methylpentanal}$$

ENGAGE

How is the carbon chain in an aldehyde numbered to indicate the position of a substituent?

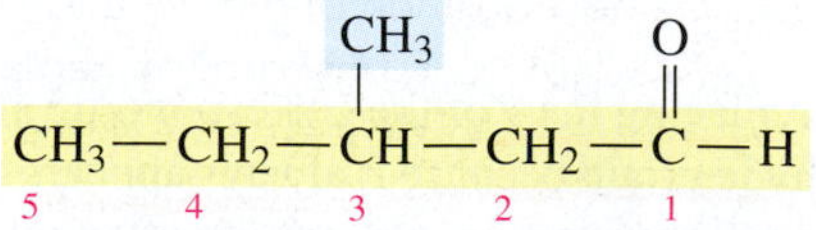

STUDY CHECK 14.2

a. What is the IUPAC name of the following compound?

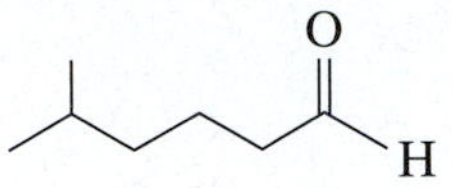

b. Draw the line-angle formula for 4-chlorobenzaldehyde.

ANSWER

a. 5-methylhexanal

b.

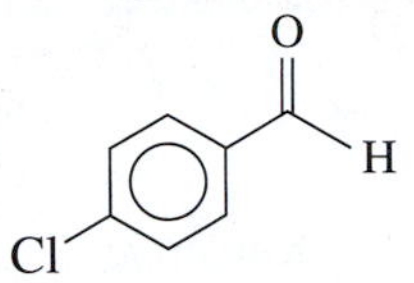

Naming Ketones

Because ketones play a major role in organic chemistry, they are still often referred to by their common names. In the common names for unbranched ketones, the alkyl groups bonded to the carbonyl group are named as substituents and are listed alphabetically, followed by *ketone*. Acetone, which is another name for propanone, has been retained by the IUPAC system.

In the IUPAC system, the name of a ketone is obtained by replacing the *e* in the corresponding alkane name with *one*. Carbon chains with five carbon atoms or more are numbered from the end nearer the carbonyl group.

ENGAGE

Why is ethyl propyl ketone the same compound as 3-hexanone?

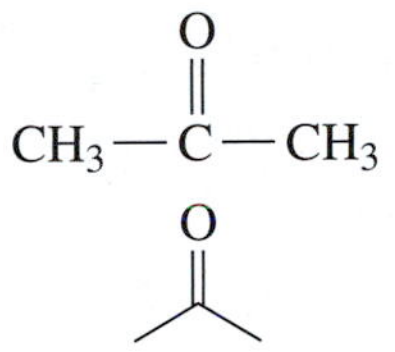

Propanone
(dimethyl ketone; acetone)

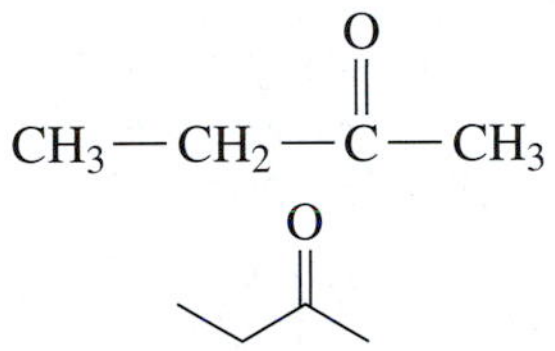

Butanone
(ethyl methyl ketone)

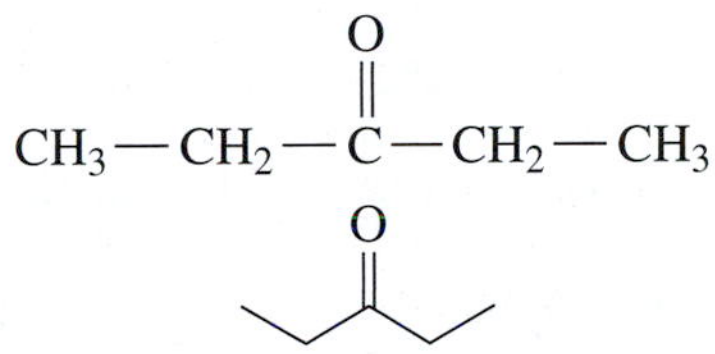

3-Pentanone
(diethyl ketone)

For cyclic ketones, the prefix *cyclo* is used in front of the ketone name. Any substituent is located by numbering the ring starting with the carbonyl carbon as carbon 1. The ring is numbered in the direction to give substituents the lower possible numbers.

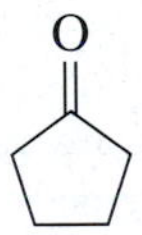

Cyclopentanone

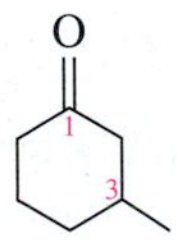

3-Methylcyclohexanone

▶ SAMPLE PROBLEM 14.3 Naming Ketones

TRY IT FIRST

Write the IUPAC name for the following ketone:

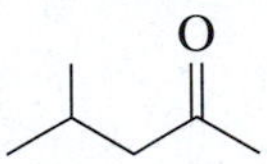

SOLUTION

	Given	Need	Connect
ANALYZE THE PROBLEM	five-carbon chain, methyl substituent	IUPAC name	position of methyl and carbonyl groups, replace *e* in alkane name with *one*

STEP 1 **Name the longest carbon chain by replacing the *e* in the alkane name with *one*.** The longest chain has five carbon atoms, which is named pentanone.

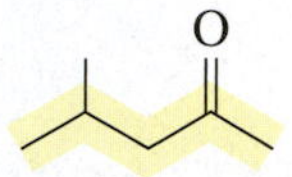

pentanone

ENGAGE

How is the carbon chain of a ketone numbered to indicate the position of a substituent?

STEP 2 **Number the carbon chain starting from the end nearer the carbonyl group and indicate its location.** Counting from the right, the carbonyl group is on carbon 2.

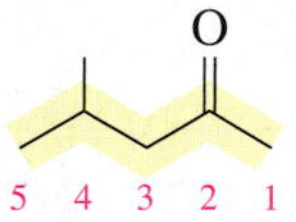
5 4 3 2 1

2-pentanone

STEP 3 **Name and number any substituents on the carbon chain.** Counting from the right, the methyl group is on carbon 4. The IUPAC name is 4-methyl-2-pentanone.

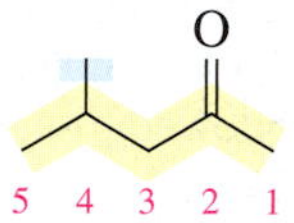
5 4 3 2 1

4-methyl-2-pentanone

STUDY CHECK 14.3

a. What is the IUPAC name of the following compound?

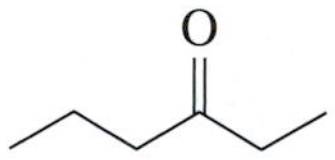

b. Draw the line-angle formula for 3-chlorobutanone.

ANSWER

a. 3-hexanone

b.

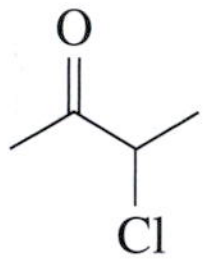

TEST

Try Practice Problems 14.5 to 14.12

Chemistry Link to Health

Some Important Aldehydes and Ketones

Formaldehyde, the simplest aldehyde, is a colorless gas with a pungent odor. An aqueous solution called *formalin*, which contains 40% formaldehyde, is used as a germicide and to preserve biological specimens. Industrially, it is a reactant in the synthesis of polymers used to make fabrics, insulation materials, carpeting, pressed wood products such as plywood, and plastics for kitchen counters. Exposure to formaldehyde fumes can irritate the eyes, nose, and upper respiratory tract and cause skin rashes, headaches, dizziness, and general fatigue.

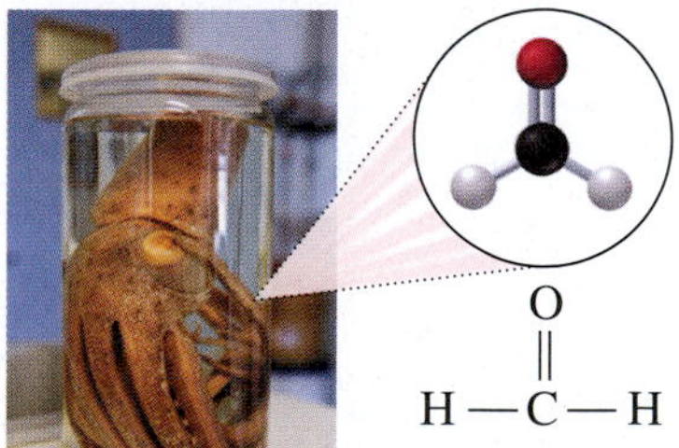

Methanal (formaldehyde)

Several naturally occurring aromatic aldehydes are used to flavor food and as fragrances in perfumes. Benzaldehyde is found in almonds, vanillin in vanilla beans, and cinnamaldehyde in cinnamon.

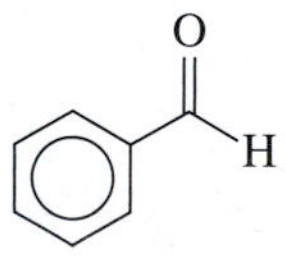

Benzaldehyde (almond)

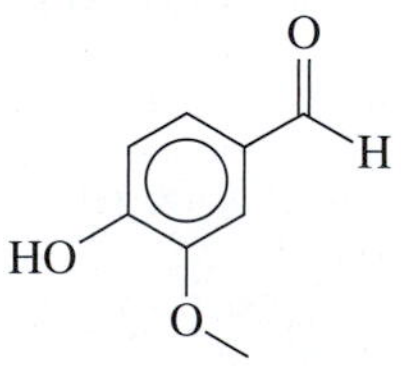

Vanillin (vanilla)

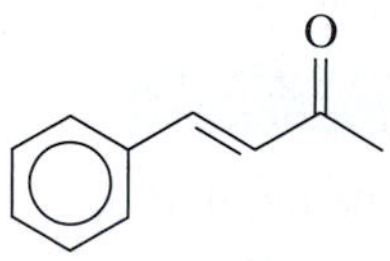

Cinnamaldehyde (cinnamon)

The simplest ketone, known as *acetone* or propanone (dimethyl ketone), is a colorless liquid with a mild odor that has wide use as a solvent in cleaning fluids, paint and nail polish removers, and rubber cement.

(*continued*)

Chemistry Link to Health (*continued*)

Acetone is extremely flammable, and care must be taken when using it. In the body, acetone may be produced in uncontrolled diabetes, fasting, and high-protein diets when large amounts of fats are metabolized for energy (see **FIGURE 14.3**).

Muscone is a ketone used to make musk perfumes, and oil of spearmint contains carvone.

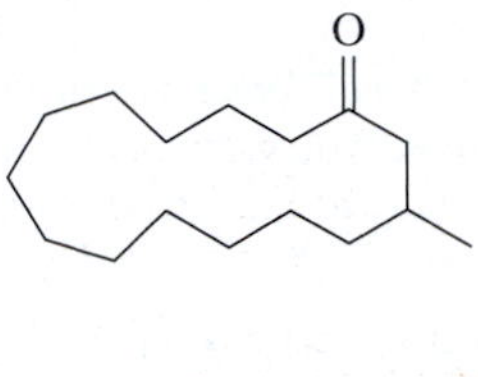

Muscone (musk)

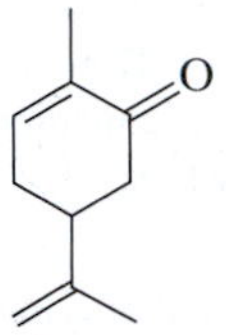

Carvone (spearmint oil)

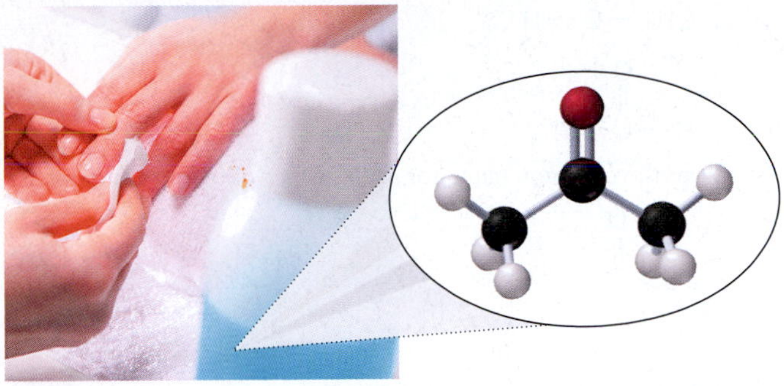

FIGURE 14.3 ▶ Acetone is used as a solvent in paint and nail polish removers.

Q What is the IUPAC name for acetone?

PRACTICE PROBLEMS

14.1 Aldehydes and Ketones

14.1 Identify each of the following compounds as an aldehyde or a ketone:

a. $CH_3-CH_2-\overset{O}{\overset{\|}{C}}-CH_3$

b. 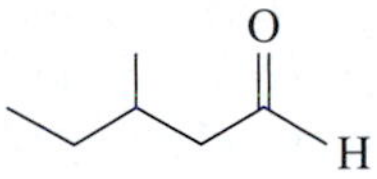

c. 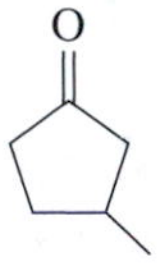

d. 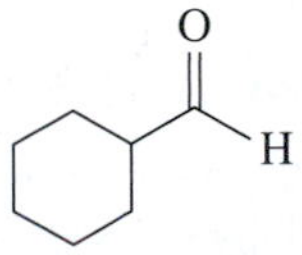

14.2 Identify each of the following compounds as an aldehyde or a ketone:

a. 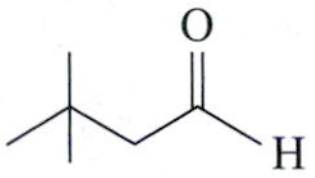

b. $CH_3-\overset{CH_3}{\overset{|}{CH}}-\overset{O}{\overset{\|}{C}}-H$

c. 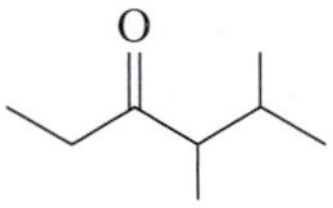

d. 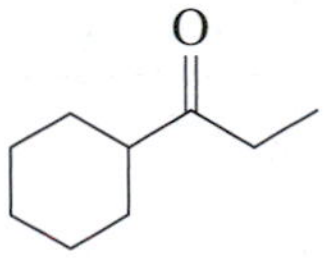

14.3 Indicate if each of the following pairs represents structural isomers or not:

a. $CH_3-\overset{O}{\overset{\|}{C}}-CH_3$ and $CH_3-CH_2-\overset{O}{\overset{\|}{C}}-H$

b. 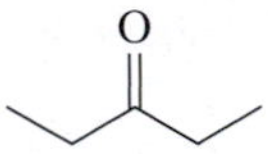 and 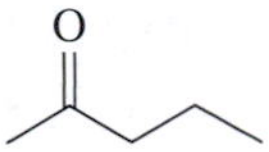

c. $CH_3-\overset{O}{\overset{\|}{C}}-CH_2-CH_3$ and $CH_3-\overset{O}{\overset{\|}{C}}-CH_2-CH_2-CH_3$

14.4 Indicate if each of the following pairs represents structural isomers or not:

a. $CH_3-CH_2-CH_2-\overset{O}{\overset{\|}{C}}-H$ and 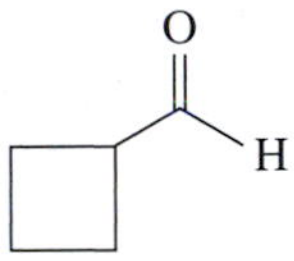

b. 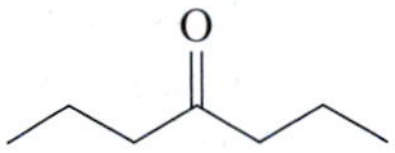 and 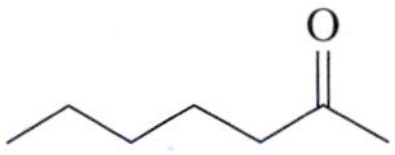

c. 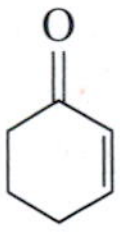 and 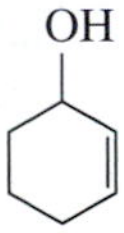

14.5 Write the IUPAC name for each of the following compounds:

a. $CH_3-\overset{Br}{\overset{|}{CH}}-CH_2-\overset{O}{\overset{\|}{C}}-H$

b. 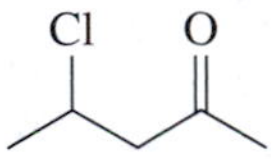

c. 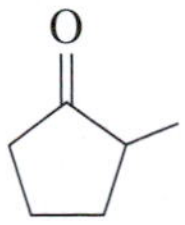

d. 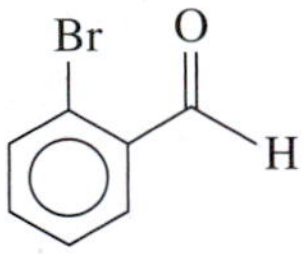

14.6 Write the IUPAC name for each of the following compounds:

a. 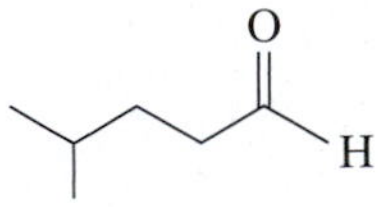

b. 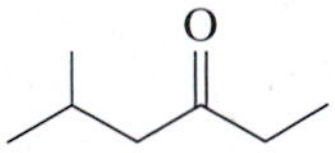

c. 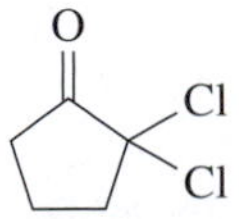

d. 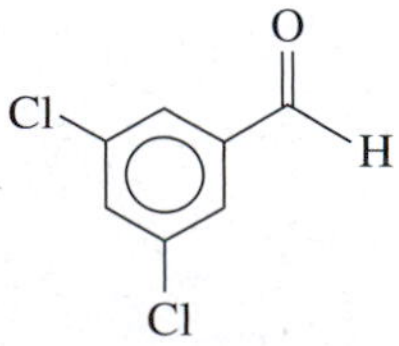

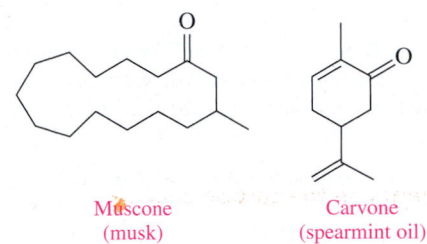

14.7 Write the common name for each of the following compounds:

a. $CH_3-\overset{O}{\overset{\|}{C}}-H$ **b.** 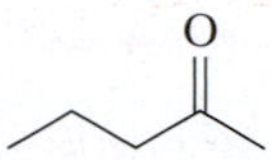

c. $H-\overset{O}{\overset{\|}{C}}-H$

14.8 Write the common name for each of the following compounds:

a. 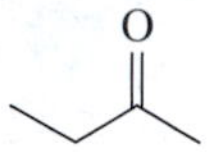

b. $CH_3-CH_2-\overset{O}{\overset{\|}{C}}-CH_2-CH_3$

c. $CH_3-CH_2-\overset{O}{\overset{\|}{C}}-H$

14.9 Draw the condensed structural and line-angle formulas for each of the following compounds:

a. ethanal **b.** 2-methyl-3-pentanone
c. butyl methyl ketone **d.** 3-methylhexanal

14.10 Draw the condensed structural and line-angle formulas for each of the following compounds:

a. butyraldehyde **b.** 3,4-dichloropentanal
c. 4-bromobutanone **d.** acetone

Clinical Applications

14.11 Anisaldehyde, from Korean mint or blue licorice, is a medicinal herb used in Chinese medicine. The IUPAC name of anisaldehyde is 4-methoxybenzaldehyde. Draw the line-angle formula for anisaldehyde.

14.12 The IUPAC name of ethyl vanillin, a synthetic compound used as a flavoring, is 3-ethoxy-4-hydroxybenzaldehyde. Draw the line-angle formula for ethyl vanillin.

14.2 Physical Properties of Aldehydes and Ketones

LEARNING GOAL Describe the boiling points and solubilities of aldehydes and ketones.

At room temperature, methanal (formaldehyde) and ethanal (acetaldehyde) are gases. Aldehydes and ketones containing 3 to 10 carbon atoms are liquids. The polar carbonyl group with a partially negative oxygen atom and a partially positive carbon atom has an influence on the boiling points and the solubility of aldehydes and ketones in water.

ENGAGE

If an aldehyde and an alkane have the same molar mass, why would the aldehyde have the higher boiling point?

Boiling Points of Aldehydes and Ketones

The polar carbonyl group in aldehydes and ketones provides dipole–dipole attractions, which alkanes do not have. Thus, aldehydes and ketones have higher boiling points than alkanes. However, aldehydes and ketones cannot form hydrogen bonds with each other as do alcohols. Thus, alcohols have higher boiling points than aldehydes and ketones of similar molar mass.

Dipole–dipole attractions in formaldehyde

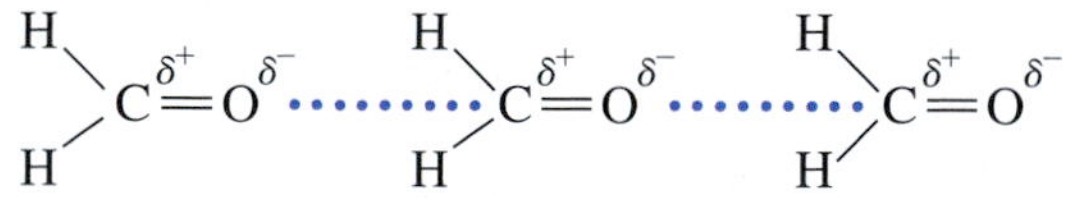

	$CH_3-CH_2-CH_2-CH_3$	$CH_3-CH_2-\overset{O}{\overset{\|}{C}}-H$	$CH_3-\overset{O}{\overset{\|}{C}}-CH_3$	$CH_3-CH_2-CH_2-OH$
Name	Butane	Propanal	Propanone	1-Propanol
Molar Mass	58	58	58	60
Family	Alkane	Aldehyde	Ketone	Alcohol
bp	0 °C	49 °C	56 °C	97 °C

Boiling Point Increases →

For aldehydes and ketones, the boiling points increase as the number of carbon atoms in the chain increases. As the molecules become larger, there are more electrons and more temporary dipoles (dispersion forces), which give higher boiling points (see **TABLE 14.1**).

TABLE 14.1 Boiling Points and Solubility of Selected Aldehydes and Ketones

Compound	Condensed Structural Formula	Number of Carbon Atoms	Boiling Point (°C)	Solubility in Water
Methanal (formaldehyde)	$H-CHO$	1	−21	Soluble
Ethanal (acetaldehyde)	CH_3-CHO	2	21	Soluble
Propanal (propionaldehyde)	CH_3-CH_2-CHO	3	49	Soluble
Propanone (acetone)	$CH_3-CO-CH_3$	3	56	Soluble
Butanal (butyraldehyde)	$CH_3-CH_2-CH_2-CHO$	4	75	Soluble
Butanone	$CH_3-CO-CH_2-CH_3$	4	80	Soluble
Pentanal	$CH_3-CH_2-CH_2-CH_2-CHO$	5	103	Slightly soluble
2-Pentanone	$CH_3-CO-CH_2-CH_2-CH_3$	5	102	Slightly soluble
Hexanal	$CH_3-CH_2-CH_2-CH_2-CH_2-CHO$	6	129	Insoluble
2-Hexanone	$CH_3-CO-CH_2-CH_2-CH_2-CH_3$	6	127	Insoluble

Solubility of Aldehydes and Ketones in Water

Aldehydes and ketones contain a polar carbonyl group (carbon–oxygen double bond), which has a partially negative oxygen atom and a partially positive carbon atom. Because the electronegative oxygen atom forms hydrogen bonds with water molecules, aldehydes and ketones with one to four carbons are soluble (see **FIGURE 14.4**). However, aldehydes and ketones with five or more carbon atoms are not soluble because longer hydrocarbon chains, which are nonpolar, diminish the solubility effect of the polar carbonyl group (see **TABLE 14.1**).

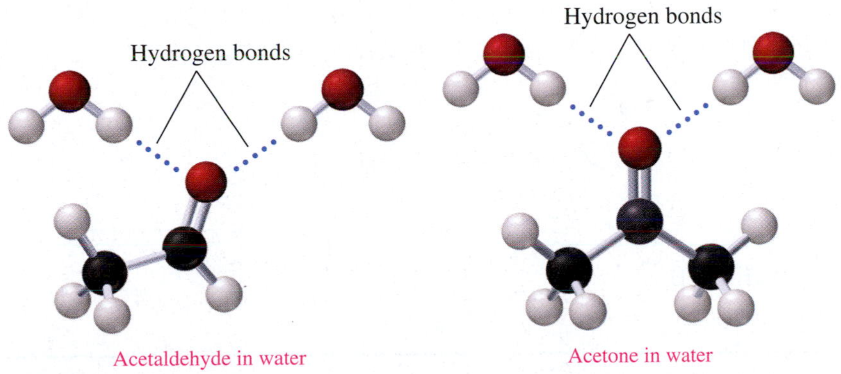
Hydrogen bonds
Hydrogen bonds
Acetaldehyde in water
Acetone in water

FIGURE 14.4 ▶ Acetaldehyde and acetone form hydrogen bonds with water.

Q Would you expect propanal to be soluble in water?

▶SAMPLE PROBLEM 14.4 Boiling Point and Solubility

TRY IT FIRST

a. Arrange pentane, 2-butanol, and butanone, which have similar molar masses, in order of increasing boiling points. Explain.

b. Why is butanone soluble in water, but 2-hexanone is not?

SOLUTION

a. Pentane with only dispersion forces has the lowest boiling point of the three compounds. Butanone has a higher boiling point than pentane because its carbonyl group forms dipole–dipole interactions. 2-Butanol can form hydrogen bonds with other butanol molecules: It has the highest boiling point of the three compounds. The actual boiling points are pentane (36 °C), butanone (80 °C), and 2-butanol (100 °C).

b. Butanone contains a carbonyl group with an electronegative oxygen atom that forms hydrogen bonds with water. 2-Hexanone also contains a carbonyl group, but its longer hydrocarbon chain reduces the impact on the solubility of the polar carbonyl group.

STUDY CHECK 14.4

a. If acetone molecules cannot hydrogen bond with each other, why is acetone soluble in water?

b. Arrange propanone, 2-pentanone, and butanone in order of decreasing solubility in water.

ANSWER

a. The oxygen atom in the carbonyl group of acetone forms hydrogen bonds with water molecules.

b. The increase in molar mass decreases solubility in water: propanone, butanone, 2-pentanone.

TEST

Try Practice Problems 14.13 to 14.18

PRACTICE PROBLEMS

14.2 Physical Properties of Aldehydes and Ketones

14.13 Which compound in each of the following pairs would have the higher boiling point? Explain.

a. $CH_3-CH_2-CH_3$ or $CH_3-\overset{\overset{\displaystyle O}{\|}}{C}-H$

b. propanal or pentanal **c.** butanal or 1-butanol

14.14 Which compound in each of the following pairs would have the higher boiling point? Explain.

a. 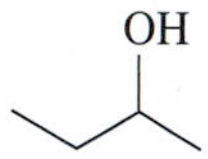 or 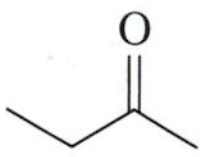

b. hexane or 3-pentanone **c.** propanone or pentanone

14.15 Which compound in each of the following pairs would be more soluble in water? Explain.

a. $CH_3-\overset{\overset{\displaystyle O}{\|}}{C}-CH_2-CH_2-CH_3$ or $CH_3-\overset{\overset{\displaystyle O}{\|}}{C}-\overset{\overset{\displaystyle O}{\|}}{C}-CH_2-CH_3$

b. propanal or pentanal **c.** acetone or 2-pentanone

14.16 Which compound in each of the following pairs would be more soluble in water? Explain.

a. $CH_3-CH_2-CH_3$ or CH_3-CH_2-CHO

b. propanone or 3-hexanone

c. butane or propanone

14.17 Would you expect an aldehyde with a formula of $C_8H_{16}O$ to be soluble in water? Explain.

14.18 Would you expect an aldehyde with a formula of C_3H_6O to be soluble in water? Explain.

REVIEW

Writing Equations for the Oxidation of Alcohols (13.4)

14.3 Oxidation and Reduction of Aldehydes and Ketones

LEARNING GOAL Draw the condensed structural and line-angle formulas for the reactants and products in the oxidation or reduction of aldehydes and ketones.

Aldehydes, which have an H bonded to the carbonyl C, oxidize readily to form carboxylic acids. In contrast, ketones, which do not have an H that can be removed easily, do not usually undergo further oxidation.

$$\underset{\text{Ethanal}}{CH_3-\overset{\overset{\displaystyle O}{\|}}{C}-H} \xrightarrow{\text{Oxidation}} \underset{\text{Ethanoic acid}}{CH_3-\overset{\overset{\displaystyle O}{\|}}{C}-OH}$$

$$\underset{\text{Propanone}}{CH_3-\overset{\overset{\displaystyle O}{\|}}{C}-CH_3} \xrightarrow{\text{Oxidation}} \text{no reaction}$$

Tollens' Test

The ease of oxidation of aldehydes allows certain mild oxidizing agents to oxidize the aldehyde functional group without oxidizing other functional groups. In the laboratory, **Tollens' test** may be used to distinguish between aldehydes and ketones. Tollens' reagent, which is a solution of Ag^+ ($AgNO_3$) and ammonia, oxidizes aldehydes, but not ketones. The silver ion is reduced and forms a silver mirror on the inside of the container.

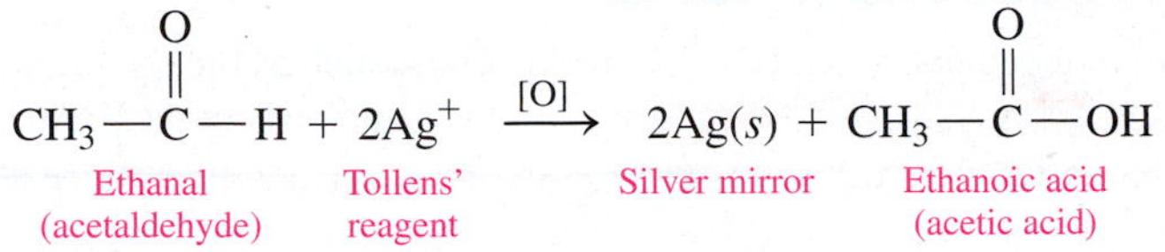
[O]
Ethanal
(acetaldehyde)
Tollens'
reagent
Silver mirror
Ethanoic acid
(acetic acid)

$$CH_3-\overset{O}{\overset{\|}{C}}-H + 2Ag^+ \xrightarrow{[O]} 2Ag(s) + CH_3-\overset{O}{\overset{\|}{C}}-OH$$

Commercially, a similar process is used to make the silvery surfaces of mirrors by applying a solution of $AgNO_3$ and ammonia on glass with a spray gun (see **FIGURE 14.5**).

ENGAGE

Why does a silver mirror form when Tollens' reagent is added to propanal?

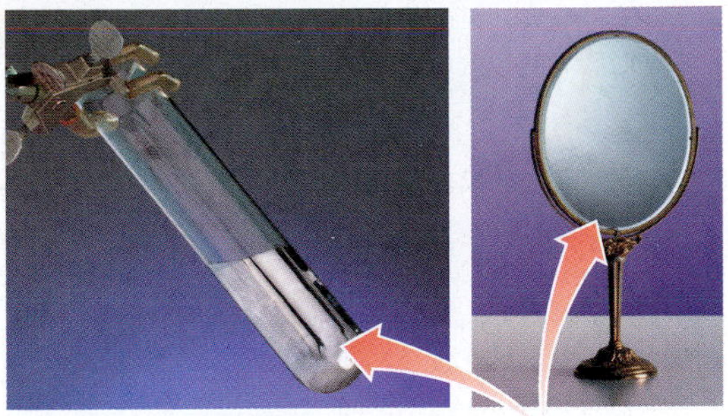

$$Ag^+ + 1\,e^- \longrightarrow Ag(s)$$

FIGURE 14.5 ▶ In Tollens' test, a silver mirror forms when the oxidation of an aldehyde reduces silver ions to metallic silver.

Q What is the product of the oxidation of an aldehyde?

▶ SAMPLE PROBLEM 14.5 Oxidation of Aldehydes

TRY IT FIRST

Draw the condensed structural formula for the product of oxidation, if any, when Tollens' reagent is added to each of the following compounds:

a. propanal **b.** propanone **c.** 2-methylbutanal

SOLUTION

Tollens' reagent will oxidize aldehydes but not ketones.

a. $CH_3-CH_2-\overset{O}{\overset{\|}{C}}-OH$ **b.** no reaction **c.** $CH_3-CH_2-\overset{CH_3}{\overset{|}{CH}}-\overset{O}{\overset{\|}{C}}-OH$

STUDY CHECK 14.5

Why does a silver mirror form when Tollens' reagent is added to a test tube containing benzaldehyde?

ANSWER

The oxidation of benzaldehyde reduces Ag^+ to metallic silver, which forms a silvery coating on the inside of the test tube.

TEST

Try Practice Problems 14.19 to 14.22

Benedict's test also gives a positive test with aldehydes. When Benedict's solution containing Cu^{2+} ($CuSO_4$) is added to an aldehyde and heated, a brick-red solid of Cu_2O forms (see **FIGURE 14.6**). The test is negative for ketones. Because sugars such as glucose contain an aldehyde group, Benedict's reagent can be used to determine the presence of glucose in blood or urine.

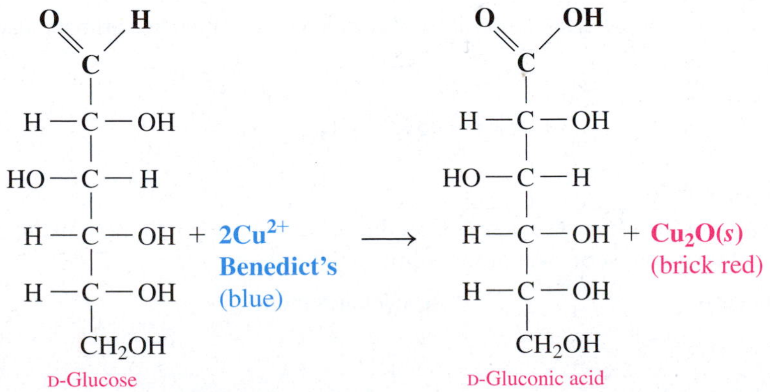
2Cu2+
Benedict's
(blue)
Cu2O(s)
(brick red)
D-Glucose
D-Gluconic acid

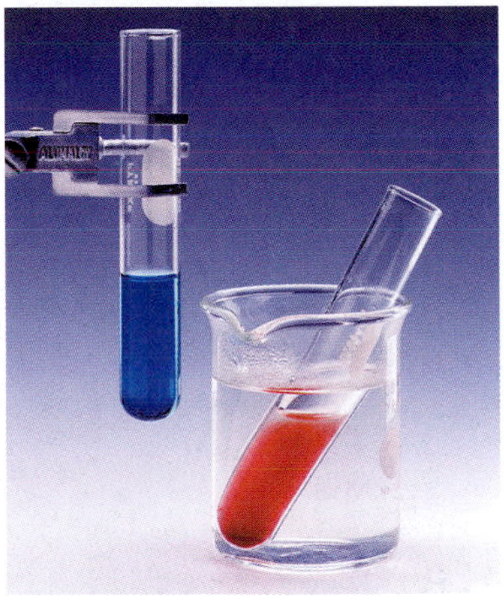

Cu^{2+} $Cu_2O(s)$

FIGURE 14.6 ▶ The blue Cu^{2+} in Benedict's solution forms a brick-red solid of Cu_2O in a positive test for sugars and aldehydes when heated.

Q Which test tube indicates that glucose is present?

ENGAGE

When Benedict's solution is added to propanone and heated, will a brick-red solid appear or not?

A summary of the Tollens' and Benedict's tests is shown below:

Test	Aldehydes	Ketones
Tollens'	form silver mirror	no change
Benedict's	form brick-red solid	no change

Reduction of Aldehydes and Ketones

Aldehydes and ketones are reduced hydrogen (H_2), using a catalyst such as nickel, platinum, or palladium. In the **reduction** of an aldehyde or ketone, there is a decrease in the number of carbon–oxygen bonds. Aldehydes are reduced to primary alcohols, and ketones are reduced to secondary alcohols.

Aldehydes Reduce to Primary Alcohols

$$CH_3-CH_2-\overset{\overset{\displaystyle O}{\|}}{C}-H + \mathbf{H_2} \xrightarrow{Pt} CH_3-CH_2-\underset{\underset{\displaystyle \mathbf{H}}{|}}{\overset{\overset{\displaystyle \mathbf{OH}}{|}}{C}}-H$$

Propanal (propionaldehyde) → 1-Propanol (1° alcohol) (propyl alcohol)

Ketones Reduce to Secondary Alcohols

$$CH_3-\overset{\overset{\displaystyle O}{\|}}{C}-CH_3 + \mathbf{H_2} \xrightarrow{Ni} CH_3-\underset{\underset{\displaystyle \mathbf{H}}{|}}{\overset{\overset{\displaystyle \mathbf{OH}}{|}}{C}}-CH_3$$

Propanone (dimethyl ketone) → 2-Propanol (2° alcohol) (isopropyl alcohol)

ENGAGE

Why does 2-butanol form when butanone is reduced?

▶ SAMPLE PROBLEM 14.6 Reduction of Carbonyl Groups

TRY IT FIRST

Write the balanced chemical equation for the reduction of each of the following:

a. cyclopentanone

b. ethanal, using hydrogen in the presence of a nickel catalyst

SOLUTION

a. The reacting molecule is a cyclic ketone that has five carbon atoms. During the reduction, the ketone is converted to the corresponding secondary alcohol.

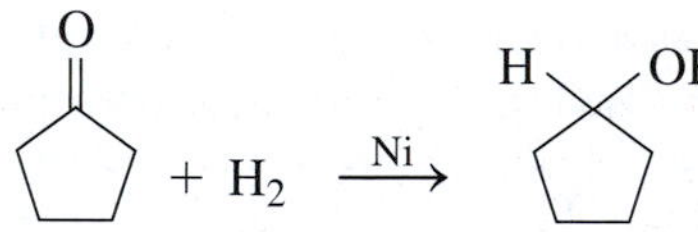
O
+ H2
Ni
H
OH

b. The reacting molecule is a two-carbon aldehyde, which is converted to a primary alcohol.

$$CH_3-\overset{\overset{\displaystyle O}{\|}}{C}-H + H_2 \xrightarrow{Ni} CH_3-CH_2-OH$$

STUDY CHECK 14.6

What is the name of the product obtained from the reduction using hydrogen in the presence of a palladium catalyst of each of the following?

a. 2-methylbutanal **b.** cyclohexanone

ANSWER

a. 2-methyl-1-butanol **b.** cyclohexanol

TEST

Try Practice Problems 14.23 and 14.24

PRACTICE PROBLEMS

14.3 Oxidation and Reduction of Aldehydes and Ketones

14.19 Draw the condensed structural or line-angle formula, if cyclic, for the product of oxidation (if any) for each of the following compounds:
- **a.** methanal
- **b.** butanone
- **c.** benzaldehyde
- **d.** 3-methylcyclohexanone

14.20 Draw the condensed structural or line-angle formula, if cyclic, for the product of oxidation (if any) for each of the following compounds:
- **a.** acetaldehyde
- **b.** 3-methylbutanone
- **c.** cyclohexanone
- **d.** 3-methylbutanal

14.21 Will each of the following compounds react with Tollens' reagent or not?

a. 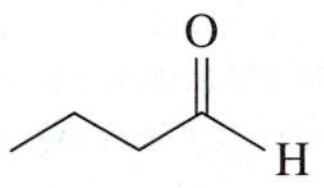

b. $CH_3-\overset{O}{\overset{\|}{C}}-CH_3$

c. $CH_3-CH_2-\overset{O}{\overset{\|}{C}}-H$

14.22 Will each of the following compounds react with Benedict's reagent or not?

a. 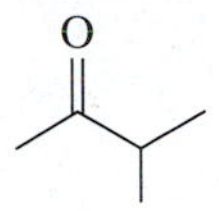

b. $CH_3-CH_2-CH_2-\overset{O}{\overset{\|}{C}}-H$

c. $CH_3-\overset{O}{\overset{\|}{C}}-H$

14.23 Draw the condensed structural formula for the product formed when each of the following is reduced by hydrogen in the presence of a nickel catalyst:
- **a.** butyraldehyde
- **b.** acetone
- **c.** 3-bromohexanal
- **d.** 2-methyl-3-pentanone

14.24 Draw the condensed structural formula for the product formed when each of the following is reduced by hydrogen in the presence of a nickel catalyst:
- **a.** ethyl propyl ketone
- **b.** formaldehyde
- **c.** 3-chloropentanal
- **d.** 2-pentanone

14.4 Addition of Alcohols: Hemiacetals and Acetals

LEARNING GOAL Draw the condensed structural and line-angle formulas for the products of the addition of alcohols to the carbonyl group of aldehydes and ketones.

Aldehydes and ketones are very reactive because the carbonyl group has a partially negatively charged O atom and a partially positively charged C atom. In a typical *addition reaction*, one or two molecules of an alcohol add to the C and O of the carbonyl. This addition reaction usually requires an acid catalyst.

Hemiacetal Formation

The formation of a *hemiacetal* occurs when one alcohol adds to the carbonyl group of an aldehyde or ketone. **Hemiacetals** contain two functional groups, a hydroxyl group (—OH) and an alkoxy group (—OR), bonded to the same C atom.

In general, aldehydes are more reactive than ketones because the carbonyl carbon is more positive in aldehydes. Also, the presence of two alkyl groups in ketones makes it more difficult for an alcohol to bond with the carbon in the carbonyl group.

CORE CHEMISTRY SKILL
Forming Hemiacetals and Acetals

General Equation for the Formation of a Hemiacetal

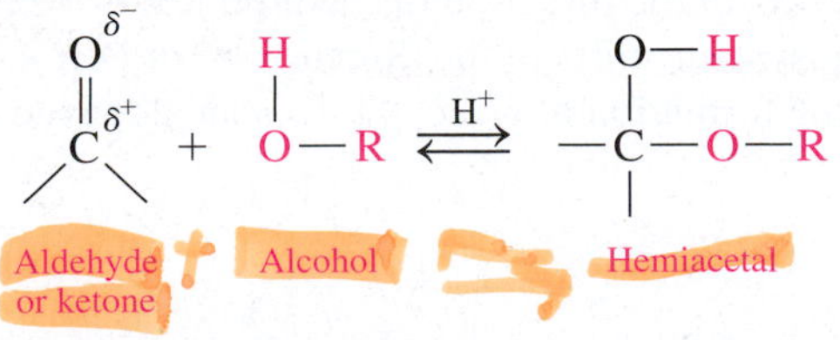
Aldehyde or ketone
Alcohol
Hemiacetal

Examples of the Formation of a Hemiacetal

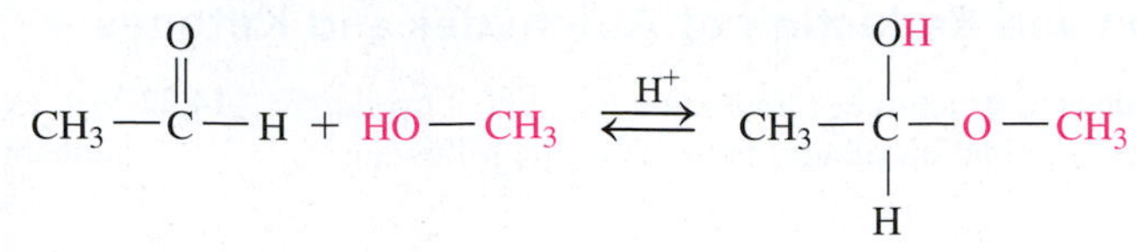

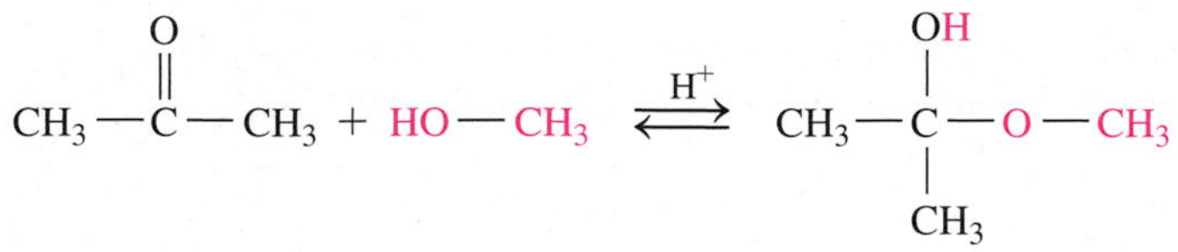

$$CH_3 - \overset{\overset{\displaystyle O}{\|}}{C} - H \ + \ HO - CH_3 \ \underset{}{\overset{H^+}{\rightleftharpoons}} \ CH_3 - \overset{\overset{\displaystyle OH}{|}}{\underset{\underset{\displaystyle H}{|}}{C}} - O - CH_3$$

Ethanal   Methanol   Hemiacetal
(acetaldehyde)   (methyl alcohol)

$$CH_3 - \overset{\overset{\displaystyle O}{\|}}{C} - CH_3 \ + \ HO - CH_3 \ \underset{}{\overset{H^+}{\rightleftharpoons}} \ CH_3 - \overset{\overset{\displaystyle OH}{|}}{\underset{\underset{\displaystyle CH_3}{|}}{C}} - O - CH_3$$

Propanone   Methanol   Hemiacetal
(dimethyl ketone)   (methyl alcohol)

Acetal Formation

An *acetal* forms when two alcohols add to the carbonyl group of an aldehyde or ketone in the presence of an acid catalyst. An **acetal** contains two alkoxy groups (—OR) attached to the initial carbonyl carbon atom. Acetals form readily because hemiacetals are generally unstable and react with a second molecule of the alcohol to form a stable acetal and water.

General Equation for the Formation of an Acetal

$$\overset{\overset{\displaystyle O}{\|}}{C} \ + \ \overset{\overset{\displaystyle H}{|}}{\underset{}{O}} - R \ \underset{}{\overset{H^+}{\rightleftharpoons}} \ -\overset{\overset{\displaystyle OH}{|}}{\underset{}{C}} - O - R \ + \ \overset{\overset{\displaystyle H}{|}}{\underset{}{O}} - R \ \underset{}{\overset{H^+}{\rightleftharpoons}} \ -\overset{\overset{\displaystyle O - R}{|}}{\underset{}{C}} - O - R \ + \ H - O - H$$

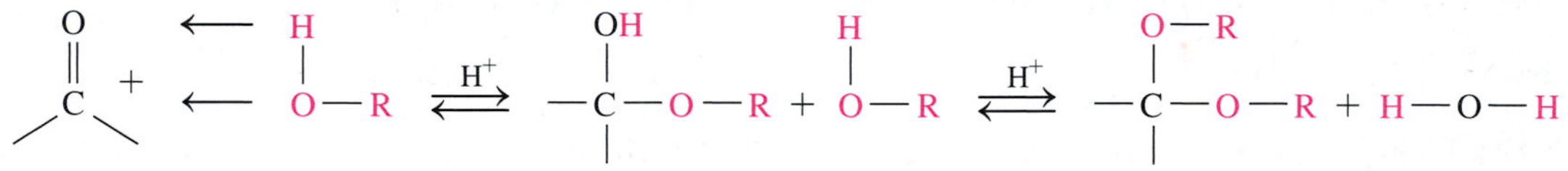

Aldehyde or ketone   Alcohol   Hemiacetal   Alcohol   Acetal
                                      unstable

Commercially, compounds that are acetals are used to produce vitamins, dyes, pharmaceuticals, and perfumes. For example, rose acetal is commonly used to give fruity and woody fragrances to perfumes. The flavor of grapefruit comes from pamplemousse acetal.

ENGAGE

What is the alcohol used in rose acetal and pamplemousse acetal?

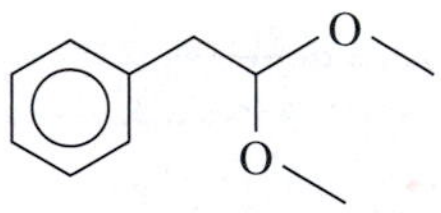

Rose acetal

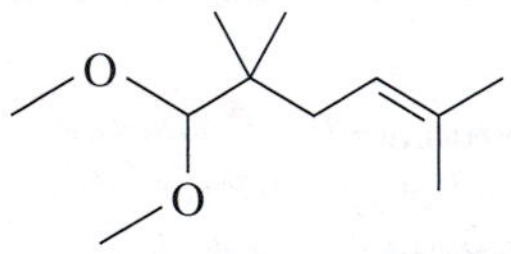

Pamplemousse acetal

The reactions in the formation of hemiacetals and acetals are reversible. As predicted by Le Châtelier's principle, the system shifts in the direction of the products by removing water from the reaction mixture. The system shifts in the direction of the ketone or aldehyde by adding water. Examples of the formation of an acetal from an aldehyde and a ketone follow:

Examples of the Formation of an Acetal

$$CH_3-\overset{\overset{\displaystyle O}{\|}}{C}-H + HO-CH_3 \underset{}{\overset{H^+}{\rightleftharpoons}} CH_3-\overset{\overset{\displaystyle OH}{|}}{\underset{\underset{\displaystyle OH}{|}}{C}}-O-CH_3 + HO-CH_3 \underset{}{\overset{H^+}{\rightleftharpoons}} CH_3-\overset{\overset{\displaystyle O-CH_3}{|}}{\underset{\underset{\displaystyle H}{|}}{C}}-O-CH_3 + H_2O$$

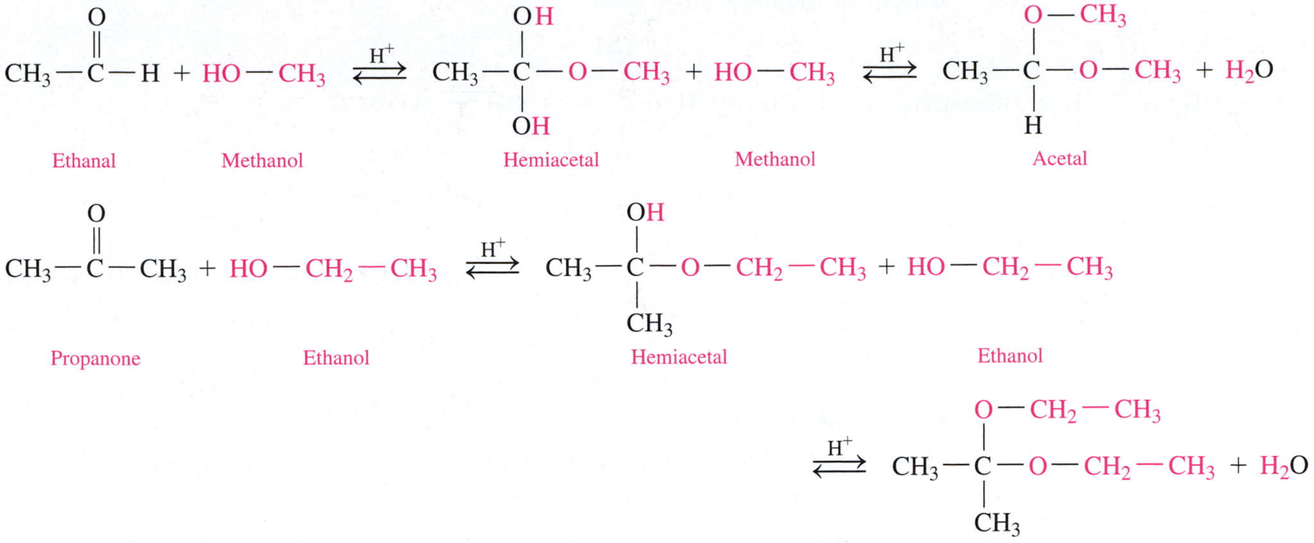

| Ethanal | Methanol | Hemiacetal | Methanol | Acetal |

$$CH_3-\overset{\overset{\displaystyle O}{\|}}{C}-CH_3 + HO-CH_2-CH_3 \underset{}{\overset{H^+}{\rightleftharpoons}} CH_3-\overset{\overset{\displaystyle OH}{|}}{\underset{\underset{\displaystyle CH_3}{|}}{C}}-O-CH_2-CH_3 + HO-CH_2-CH_3$$

| Propanone | Ethanol | Hemiacetal | Ethanol |

$$\underset{}{\overset{H^+}{\rightleftharpoons}} CH_3-\overset{\overset{\displaystyle O-CH_2-CH_3}{|}}{\underset{\underset{\displaystyle CH_3}{|}}{C}}-O-CH_2-CH_3 + H_2O$$

▶ **SAMPLE PROBLEM 14.7 Hemiacetals and Acetals**

TRY IT FIRST

Identify each of the following as a hemiacetal or acetal:

a. $CH_3-CH_2-\overset{\overset{\displaystyle OH}{|}}{\underset{\underset{\displaystyle CH_3}{|}}{C}}-O-CH_3$ 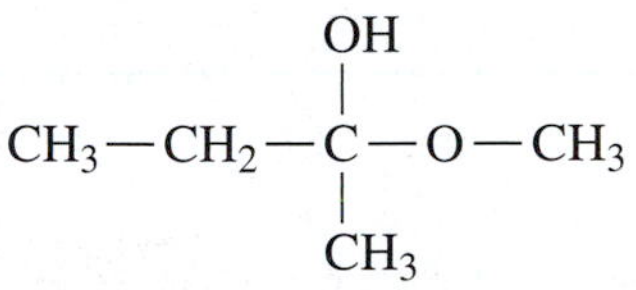
     b. $H-\overset{\overset{\displaystyle O-CH_3}{|}}{\underset{\underset{\displaystyle CH_3}{|}}{C}}-O-CH_3$ 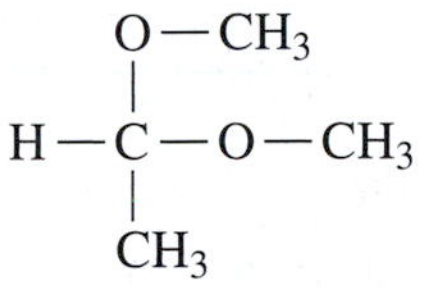

c. $CH_3-CH_2-\overset{\overset{\displaystyle OH}{|}}{\underset{\underset{\displaystyle CH_3}{|}}{C}}-O-CH_2-CH_3$ 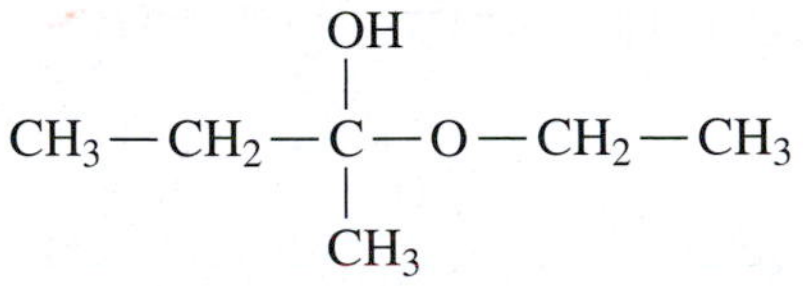

SOLUTION

a. A hemiacetal has a carbon bonded to one alkoxy group and one hydroxyl group.
b. An acetal has a carbon bonded to two alkoxy groups.
c. A hemiacetal has a carbon bonded to one alkoxy group and one hydroxyl group.

STUDY CHECK 14.7

From each of the following descriptions, identify the compound as a hemiacetal or an acetal:

a. a molecule that contains a carbon atom attached to a hydroxyl group and an ethoxy group
b. a molecule that contains a carbon atom attached to two ethoxy groups

ANSWER

a. hemiacetal   **b.** acetal

▶ **SAMPLE PROBLEM 14.8 Formation of Hemiacetals and Acetals**

TRY IT FIRST

Use condensed structural formulas to write the balanced chemical equation for the formation of the hemiacetal and acetal when methanol reacts with propanal.

SOLUTION

To draw the hemiacetal, the hydrogen from methanol is added to the oxygen of the carbonyl group to form a new hydroxyl group. The remaining part of the methanol adds to the carbon atom in the carbonyl group. The acetal forms when a second molecule of methanol adds to the carbonyl carbon atom.

$$CH_3-CH_2-\overset{\overset{\displaystyle O}{\|}}{C}-H + HO-CH_3 \underset{}{\overset{H^+}{\rightleftarrows}} CH_3-CH_2-\overset{\overset{\displaystyle OH}{|}}{\underset{\underset{\displaystyle H}{|}}{C}}-O-CH_3 + HO-CH_3$$

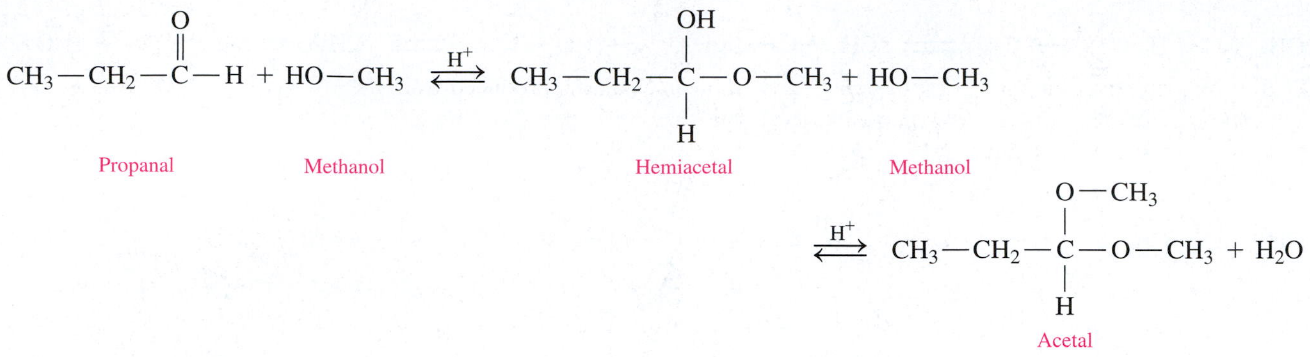

Propanal   Methanol   Hemiacetal   Methanol

$$\overset{H^+}{\rightleftarrows} CH_3-CH_2-\overset{\overset{\displaystyle O-CH_3}{|}}{\underset{\underset{\displaystyle H}{|}}{C}}-O-CH_3 + H_2O$$

Acetal

STUDY CHECK 14.8

In the addition of methanol to butanone, a hemiacetal and an acetal are formed. Draw the condensed structural formulas for each of the following:

a. the hemiacetal   **b.** the acetal

ANSWER

$$\textbf{a. } CH_3-\overset{\overset{\displaystyle OH}{|}}{\underset{\underset{\displaystyle CH_2-CH_3}{|}}{C}}-O-CH_3 \qquad \textbf{b. } CH_3-\overset{\overset{\displaystyle O-CH_3}{|}}{\underset{\underset{\displaystyle CH_2-CH_3}{|}}{C}}-O-CH_3$$

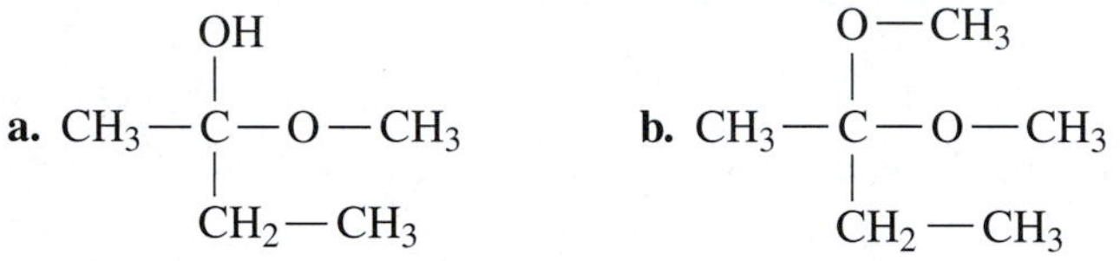

Cyclic Hemiacetals

One very important type of hemiacetal that is stable and can be isolated is a *cyclic hemiacetal* that forms when the carbonyl group and the —OH group are in the *same* molecule.

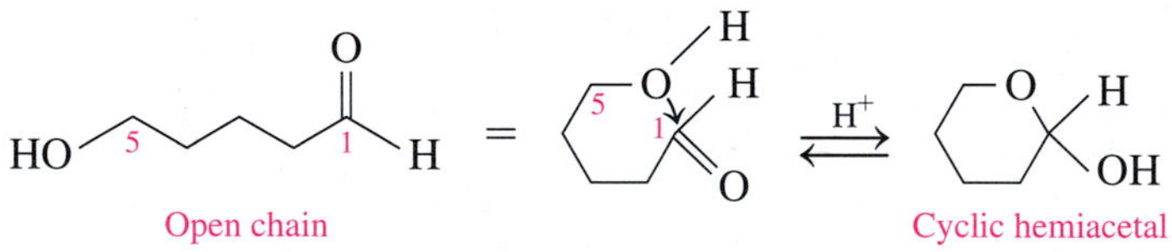

Open chain   Cyclic hemiacetal

The five- and six-atom cyclic hemiacetals and acetals are more stable than their open-chain isomers. The importance of acetals is shown for glucose, a carbohydrate, which has both carbonyl and hydroxyl groups that can form acetal bonds. Glucose forms a cyclic hemiacetal when the hydroxyl group on carbon 5 bonds with the carbonyl group on carbon 1. The hemiacetal of glucose is so stable that almost all the glucose (99%) exists as the cyclic hemiacetal in aqueous solution.

ENGAGE

Why is carbon 1 in the cyclic structure of glucose a hemiacetal?

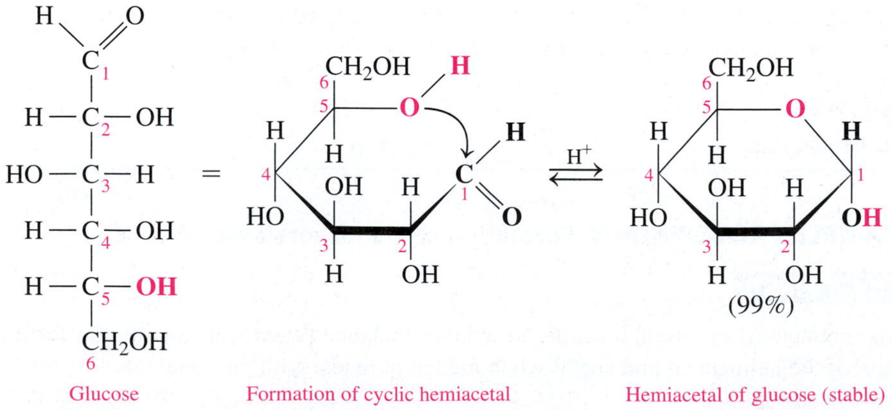

Glucose   Formation of cyclic hemiacetal   Hemiacetal of glucose (stable)

An alcohol adds to a cyclic hemiacetal to form a cyclic acetal. This reaction is very important in carbohydrate chemistry. It is the linkage that bonds glucose molecules to other glucose molecules in the formation of disaccharides and polysaccharides. In the reaction, the oxygen in the —OH in the hemiacetal of one glucose bonds with an —OH group in a second glucose.

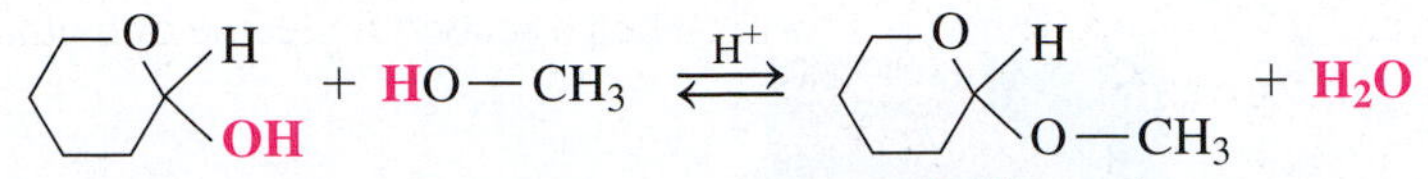
Cyclic hemiacetal
Cyclic acetal

Maltose is a disaccharide consisting of two glucose molecules and is produced from the hydrolysis of starch from grains. In maltose, an acetal bond (shown in red) links two glucose molecules. One glucose retains the cyclic hemiacetal bond (shown in green).

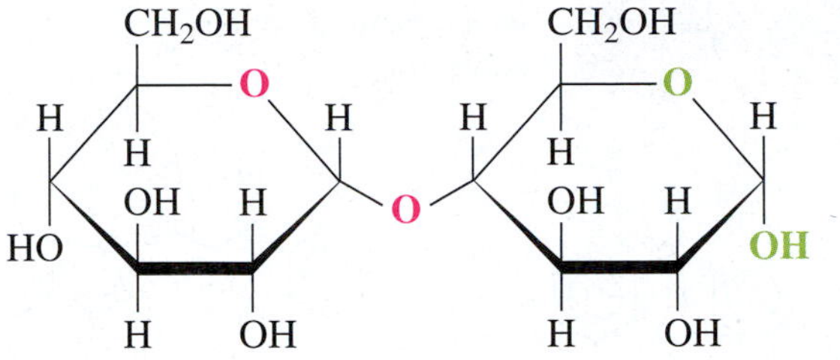

α-Maltose

Maltose, a disaccharide, is produced from the hydrolysis of the starches in grains, such as barley.

TEST

Try Practice Problems 14.25 to 14.30

PRACTICE PROBLEMS

14.4 Addition of Alcohols: Hemiacetals and Acetals

14.25 Indicate whether each of the following is a hemiacetal, acetal, or neither:

a. $CH_3—CH_2—O—CH_2—CH_3$

b. 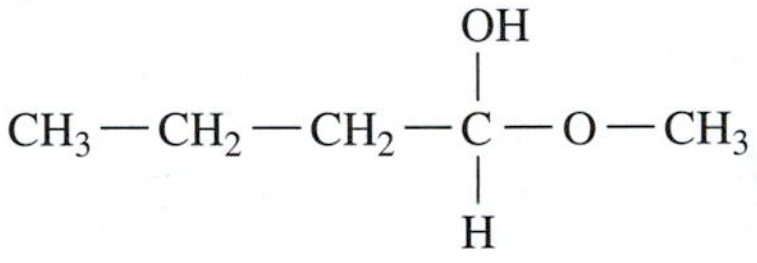

c. 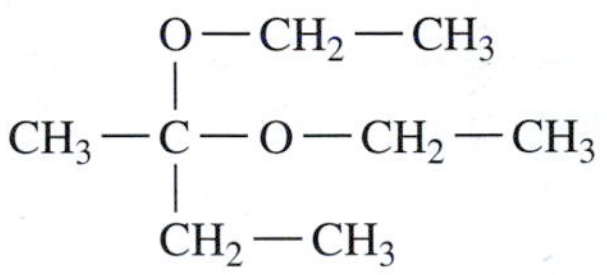

d. 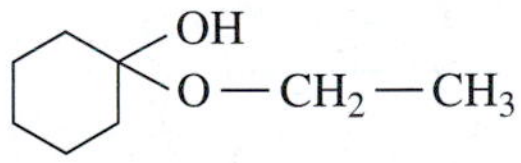

e. 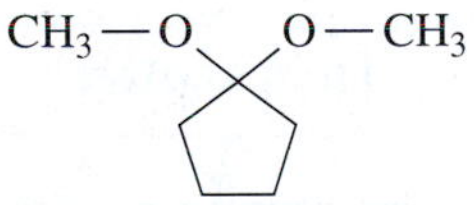

14.26 Indicate whether each of the following is a hemiacetal, acetal, or neither:

a. $CH_3—CH_2—O—CH_2—OH$

b. $HO—CH_2—CH_2—O—CH_2—CH_2—O—CH_3$

c. 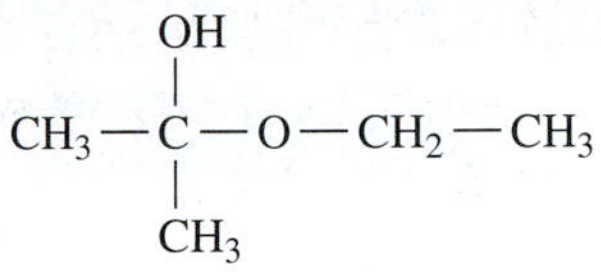

d. 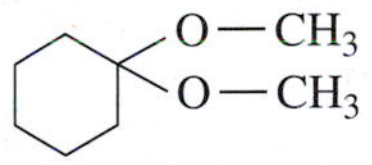

e. 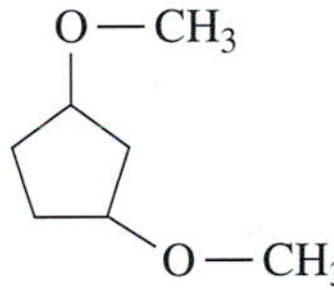

14.27 Draw the condensed structural formula for the hemiacetal formed by adding one methanol molecule to each of the following:

a. ethanal **b.** propanone **c.** butanal

14.28 Draw the condensed structural formula for the hemiacetal formed by adding one ethanol molecule to each of the following:

a. propanal **b.** butanone **c.** methanal

14.29 Draw the condensed structural formula for the acetal formed by adding a second methanol molecule to the compounds in problem 14.27.

14.30 Draw the condensed structural formula for the acetal formed by adding a second ethanol molecule to the compounds in problem 14.28.

CLINICAL UPDATE Diana's Skin Protection Plan

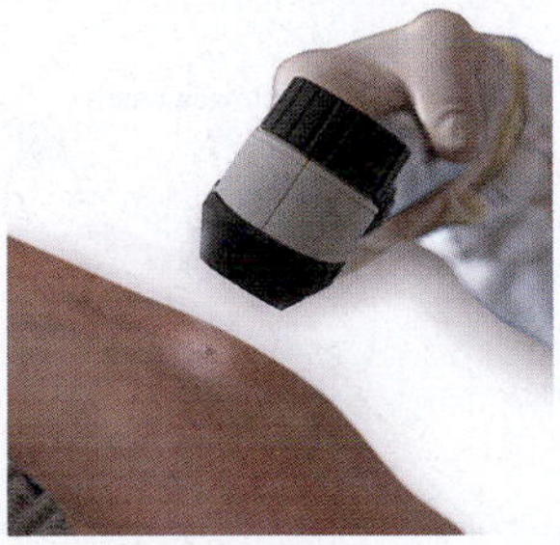

After six months, Diana returns to the dermatology office for a follow-up skin check. There has been no change in the skin where the mole was excised and no other moles are found that were suspicious. Margaret reminds Diana to limit her exposure to the sun, especially between 10 A.M. and 3 P.M., and to wear protective clothing, including a hat, a long-sleeved shirt, and pants that cover her legs. On all exposed skin, Margaret tells Diana to use broad-spectrum sunscreen with an SPF of at least 15 every day. Sunscreen absorbs UV light radiation when the skin is exposed to sunlight and thus helps protect against sunburn. The SPF (sun protection factor) number, which ranges from 2 to 100, gives the amount of time for protected skin to sunburn compared to the time for unprotected skin to sunburn. The principal ingredients in sunscreens, such as oxybenzone and avobenzone, are usually aromatic molecules with carbonyl groups.

Clinical Applications

14.31 Oxybenzone is an effective sunscreen whose structural formula is shown below.
- **a.** What functional groups are in oxybenzone?
- **b.** What is the molecular formula and molar mass of oxybenzone?
- **c.** If a bottle of sunscreen containing 178 mL has 6.0% (m/v) oxybenzone, how many grams of oxybenzone are present?

14.32 Avobenzone is a common ingredient in sunscreen. Its structural formula is shown below.
- **a.** What functional groups are in avobenzone?
- **b.** What is the molecular formula and molar mass of avobenzone?
- **c.** If a bottle of sunscreen containing 236 mL has 3.0% (m/v) avobenzone, how many grams of avobenzone are present?

Sunscreen absorbs UV light, which protects against sunburn.

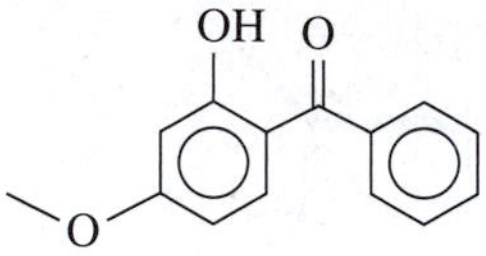

Oxybenzone

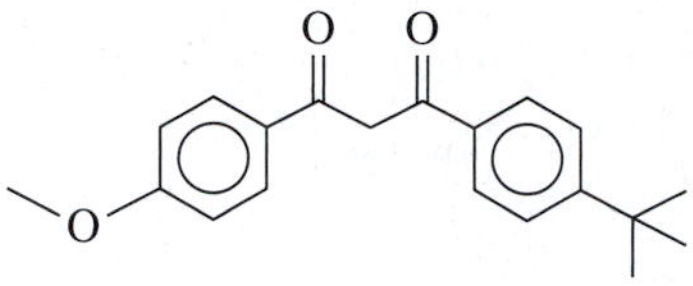

Avobenzone

CONCEPT MAP

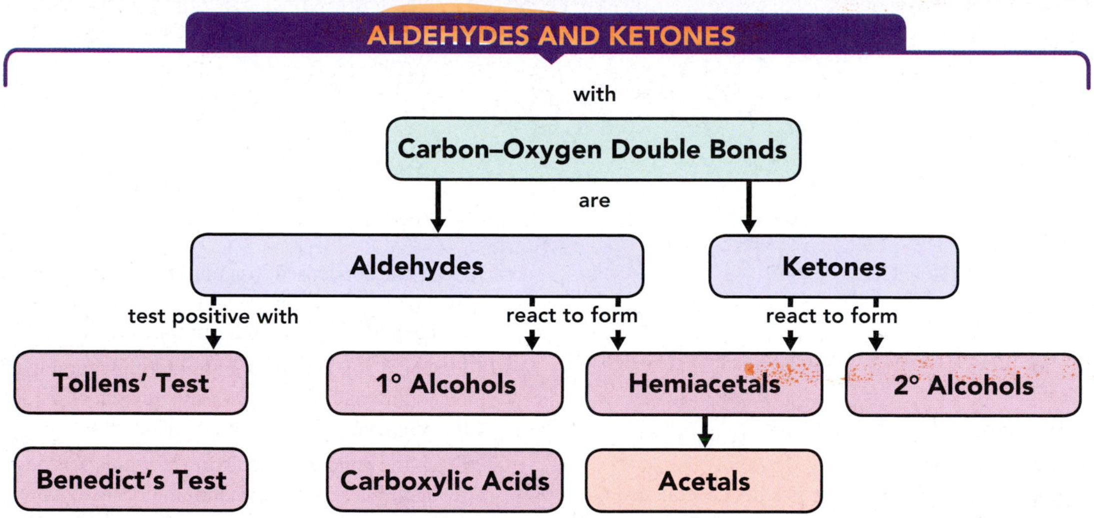

ALDEHYDES AND KETONES
with
Carbon–Oxygen Double Bonds
are
Aldehydes
Ketones
test positive with
react to form
react to form
Tollens' Test
Benedict's Test
1° Alcohols
Carboxylic Acids
Hemiacetals
Acetals
2° Alcohols

CHAPTER REVIEW

14.1 Aldehydes and Ketones

LEARNING GOAL Identify compounds with carbonyl groups as aldehydes and ketones. Write the IUPAC and common names for aldehydes and ketones; draw their condensed structural and line-angle formulas.

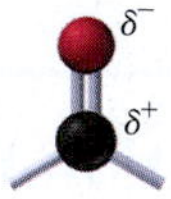

δ−
δ+

Carbonyl group

- Aldehydes and ketones contain a carbonyl group (C=O), which is strongly polar.
- In aldehydes, the carbonyl group appears at the end of carbon chains attached to at least one hydrogen atom.
- In ketones, the carbonyl group occurs between two alkyl or aromatic groups.
- In the IUPAC system, the *e* in the corresponding alkane name is replaced with *al* for aldehydes and *one* for ketones. For ketones with more than four carbon atoms in the main chain, the carbonyl group is numbered to show its location.
- Many of the simple aldehydes and ketones use common names.
- Many aldehydes and ketones are found in biological systems, flavorings, and drugs.

14.2 Physical Properties of Aldehydes and Ketones

LEARNING GOAL Describe the boiling points and solubilites of aldehydes and ketones.

- The polar carbonyl group in aldehydes and ketones gives higher boiling points than alkanes.

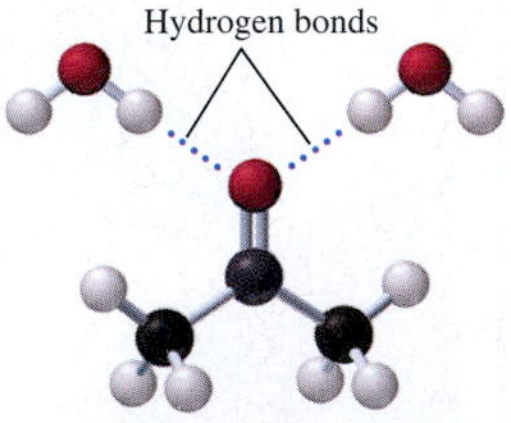

Hydrogen bonds

Acetone

- The boiling points of aldehydes and ketones are lower than alcohols because aldehydes and ketones cannot hydrogen bond with each other.
- Aldehydes and ketones form hydrogen bonds with water molecules, which makes carbonyl compounds with one to four carbon atoms soluble in water.

- Aldehydes, but not ketones, react with Tollens' reagent to give silver mirrors.
- In Benedict's test, aldehydes reduce blue Cu^{2+} to give a brick-red Cu_2O solid.
- The reduction of aldehydes with hydrogen produces primary alcohols, whereas ketones are reduced to secondary alcohols.

14.3 Oxidation and Reduction of Aldehydes and Ketones

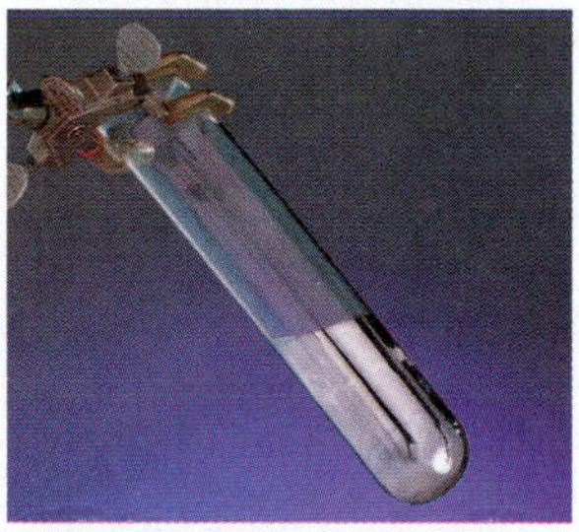

LEARNING GOAL Draw the condensed structural and line-angle formulas for the reactants and products in the oxidation or reduction of aldehydes and ketones.

- Primary alcohols can be oxidized to aldehydes, whereas secondary alcohols can oxidize to ketones.
- Aldehydes are easily oxidized to carboxylic acids, but ketones do not oxidize further.

14.4 Addition of Alcohols: Hemiacetals and Acetals

LEARNING GOAL Draw the condensed structural and line-angle formulas for the products of the addition of alcohols to the carbonyl group aldehydes and ketones.

- Alcohols can add to the carbonyl group of aldehydes and ketones.
- The addition of one alcohol molecule forms a hemiacetal, whereas the addition of two alcohol molecules forms an acetal.
- Hemiacetals are not usually stable, except for cyclic hemiacetals, which are the most common form of simple sugars such as glucose.

SUMMARY OF NAMING

Family	Structure	IUPAC Name	Common Name
Aldehyde	$\overset{\displaystyle O}{\overset{\|}{H-C-H}}$	Methanal	Formaldehyde
Ketone	$\overset{\displaystyle O}{\overset{\|}{CH_3-C-CH_3}}$	Propanone	Acetone; dimethyl ketone

SUMMARY OF REACTIONS

The chapter Sections to review are shown after the name of each reaction.

Oxidation of Aldehydes to Form Carboxylic Acids (14.3)

$$CH_3-\overset{\overset{\displaystyle O}{\|}}{C}-H \xrightarrow{[O]} CH_3-\overset{\overset{\displaystyle O}{\|}}{C}-OH$$

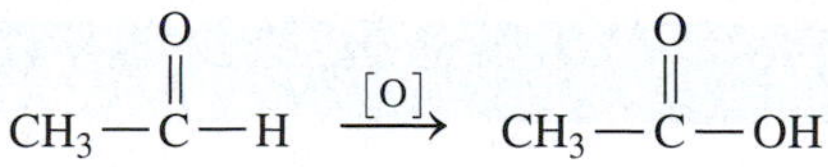

Ethanal → Ethanoic acid

Reduction of Aldehydes to Form Primary Alcohols (14.3)

$$CH_3-\overset{\overset{\displaystyle O}{\|}}{C}-H + H_2 \xrightarrow{Ni} CH_3-\overset{\overset{\displaystyle OH}{\|}}{CH_2}$$

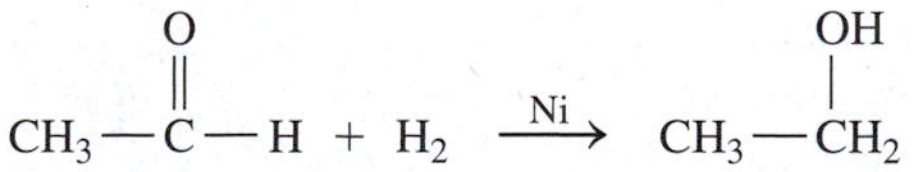

Ethanal → Ethanol

Reduction of Ketones to Form Secondary Alcohols (14.3)

$$CH_3-\overset{\overset{\displaystyle O}{\|}}{C}-CH_3 + H_2 \xrightarrow{Ni} CH_3-\overset{\overset{\displaystyle OH}{\|}}{CH}-CH_3$$

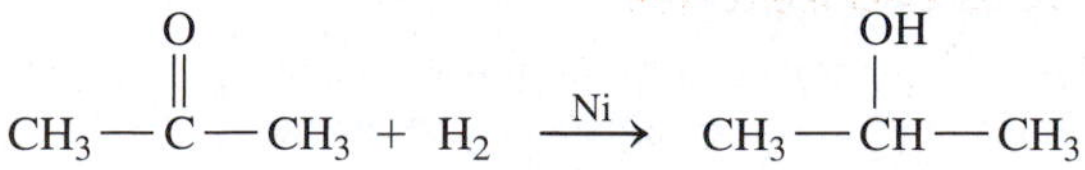

Propanone → 2-Propanol

Addition of Alcohols to Form Hemiacetals and Acetals (14.4)

$$H-\overset{\overset{\displaystyle O}{\|}}{C}-H + HO-CH_3 \overset{H^+}{\rightleftharpoons} H-\overset{\overset{\displaystyle OH}{|}}{\underset{\displaystyle H}{C}}-O-CH_3 + H_2O$$

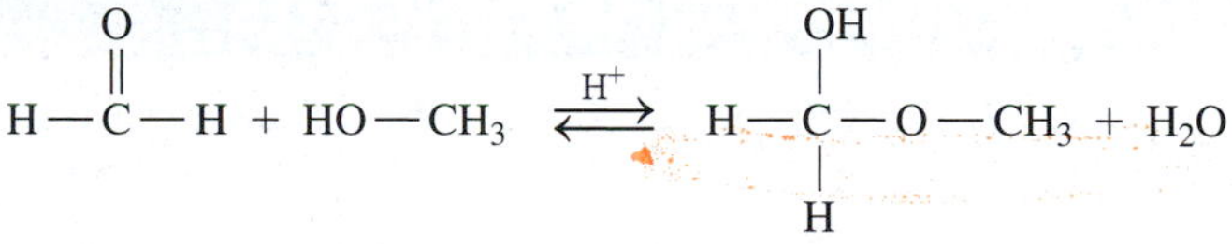

Methanal   Methanol   Hemiacetal

$$H-\overset{\overset{\displaystyle O-CH_3}{|}}{\underset{\displaystyle OH}{C}}-H + HO-CH_3 \overset{H^+}{\rightleftharpoons} H-\overset{\overset{\displaystyle O-CH_3}{|}}{\underset{\displaystyle H}{C}}-O-CH_3 + H_2O$$

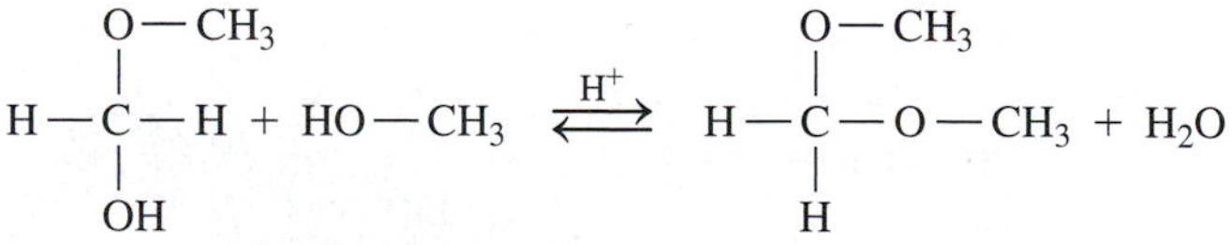

Hemiacetal   Methanol   Acetal

KEY TERMS

acetal The product of the addition of two alcohols to an aldehyde or ketone.

aldehyde An organic compound with a carbonyl functional group and at least one hydrogen attached to the carbon in the carbonyl group.

$$\overset{\overset{\displaystyle O}{\|}}{-C}-H \ = \ -CHO$$

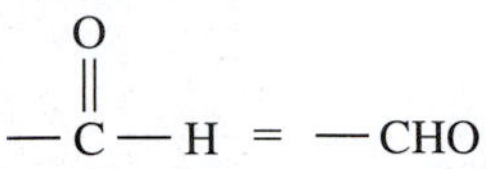

Benedict's test A test for aldehydes in which Cu^{2+} ($CuSO_4$) ions in Benedict's reagent are reduced to a brick-red solid of Cu_2O.

carbonyl group A functional group that contains a carbon–oxygen double bond ($C=O$).

hemiacetal The product of the addition of one alcohol to the double bond of the carbonyl group in aldehydes and ketones.

ketone An organic compound in which the carbonyl functional group is bonded to two alkyl or aromatic groups.

$$\overset{\overset{\displaystyle O}{\|}}{-C}- \ = \ -CO-$$

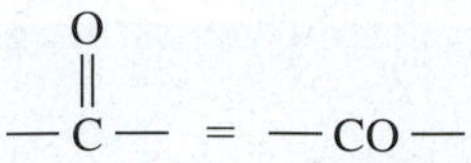

reduction A decrease in the number of carbon–oxygen bonds by the addition of hydrogen to a carbonyl bond. Aldehydes are reduced to primary alcohols; ketones to secondary alcohols.

Tollens' test A test for aldehydes in which the Ag^+ in Tollens' reagent is reduced to metallic silver, which forms a silver mirror on the walls of the container.

CORE CHEMISTRY SKILLS

The chapter Section containing each Core Chemistry Skill is shown in parentheses at the end of each heading.

Naming Aldehydes and Ketones (14.1)

- In the IUPAC system, an aldehyde is named by replacing the *e* in the alkane name with *al* and a ketone by replacing the *e* with *one*.
- The position of a substituent on an aldehyde is indicated by numbering the carbon chain from the carbonyl group and given in front of the name.
- For a ketone, the carbon chain is numbered from the end nearer the carbonyl group.

Example: Write the IUPAC name for the following:

$$CH_3 - \overset{\overset{\displaystyle CH_3}{|}}{CH} - CH_2 - \overset{\overset{\displaystyle O}{\|}}{C} - CH_3$$

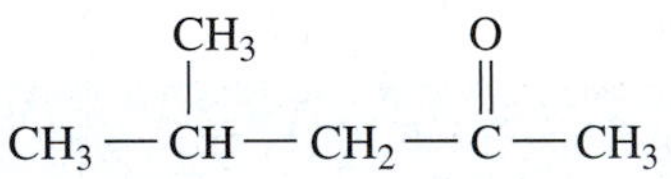

Answer: 4-methyl-2-pentanone

Forming Hemiacetals and Acetals (14.4)

- A hemiacetal forms when an alcohol adds to the carbonyl group of an aldehyde or ketone.
- An acetal forms when two alcohols add to the carbonyl group of an aldehyde or ketone.

Example: Draw the condensed structural formulas of the hemiacetal and acetal formed when 1-propanol reacts with butanone.

Answer:
$$CH_3 - CH_2 - \overset{\overset{\displaystyle OH}{|}}{\underset{\underset{\displaystyle CH_3}{|}}{C}} - O - CH_2 - CH_2 - CH_3$$
Hemiacetal

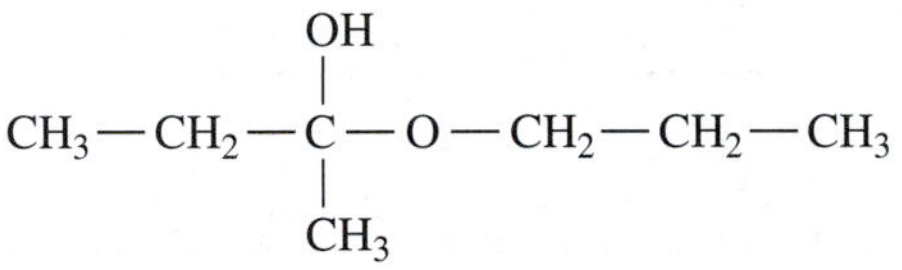

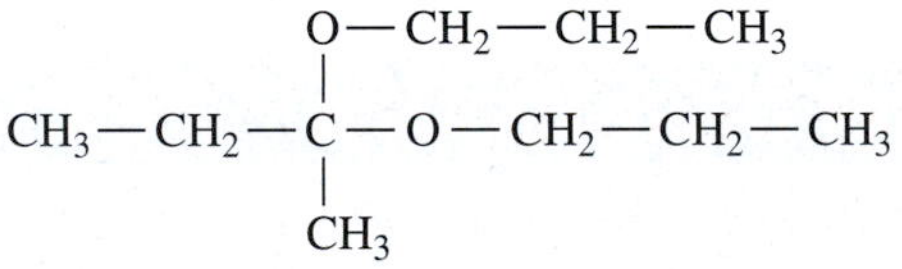

$$CH_3 - CH_2 - \overset{\overset{\displaystyle O-CH_2-CH_2-CH_3}{|}}{\underset{\underset{\displaystyle CH_3}{|}}{C}} - O - CH_2 - CH_2 - CH_3$$
Acetal

UNDERSTANDING THE CONCEPTS

The chapter Sections to review are shown in parentheses at the end of each problem.

14.33 The compound cinnamaldehyde gives the flavor to cinnamon. Identify the functional groups in cinnamaldehyde. (14.1)

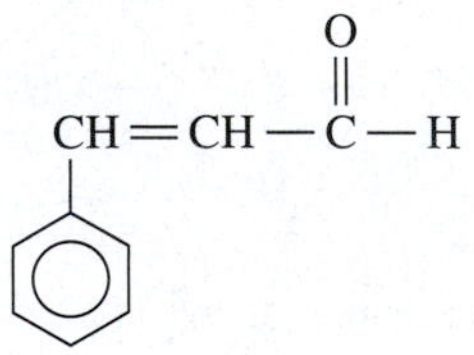

Cinnamaldehyde   Cinnamon sticks

14.34 The compound frambinone has the taste of raspberries and has been used in weight loss. Identify the functional groups in frambinone. (14.1)

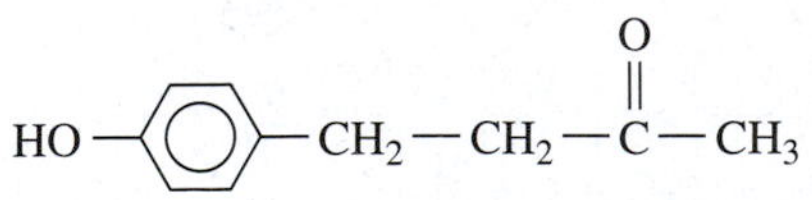

Frambinone

14.35 Draw the condensed structural and line-angle formulas for each of the following: (14.1)
  a. *trans*-2-hexenal, alarm pheromone of ants
  b. 2,6-dimethyl-5-heptenal, communication pheromone of ants

14.36 Draw the condensed structural and line-angle formulas for each of the following: (14.1)
  a. 4-methyl-3-heptanone, ant trail pheromone
  b. 2-nonanone, moth sex attractant pheromone

14.37 Why does the $C=O$ double bond have a dipole, whereas the $C=C$ double bond does not? (14.1)

14.38 Why are aldehydes and ketones with one to four carbon atoms soluble in water? (14.2)

14.39 Which of the following will give a positive Tollens' test? (14.3)

a. 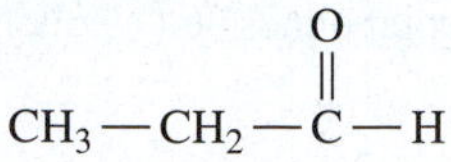

b. 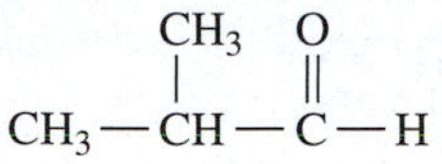

c. $CH_3—O—CH_2—CH_3$

14.40 Which of the following will give a positive Tollens' test? (14.3)

a. $CH_3—CH_2—CH_2—OH$ **b.** 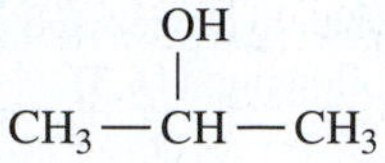

c. 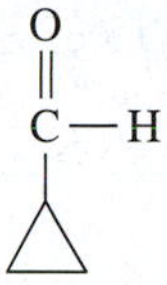

ADDITIONAL PRACTICE PROBLEMS

14.41 Write the IUPAC name for each of the following compounds: (14.1)

a. 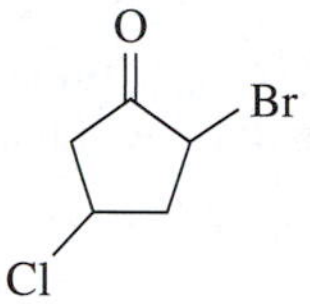 **b.** 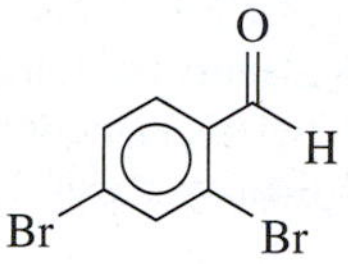

c. 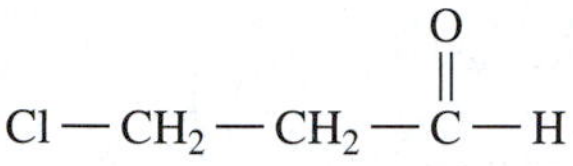 **d.** 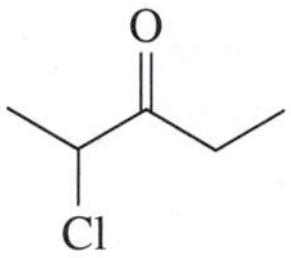

e. 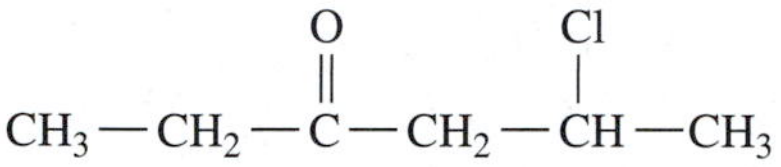

14.42 Write the IUPAC name for each of the following compounds: (14.1)

a. 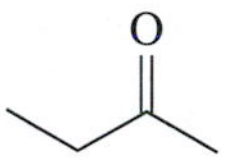 **b.** 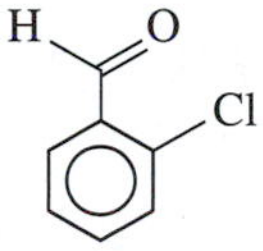

c. 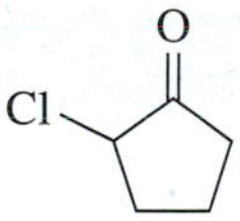 **d.** 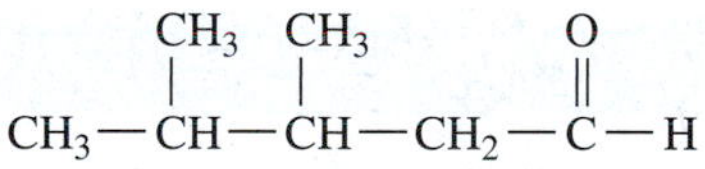

e. 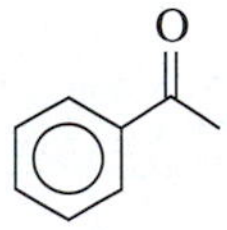

14.43 Draw the condensed structural formula or line-angle formula, if cyclic, for each of the following: (14.1)

a. 3-methylcyclopentanone **b.** pentanal
c. ethyl methyl ketone **d.** 4-methylhexanal

14.44 Draw the condensed structural formula or line-angle formula, if cyclic, for each of the following: (14.1)

a. 2-chlorobutanal
b. 2-methylcyclohexanone
c. 3,5-dimethylhexanal
d. 3-bromocyclopentanone

14.45 Which of the following aldehydes or ketones are soluble in water? (14.2)

a. 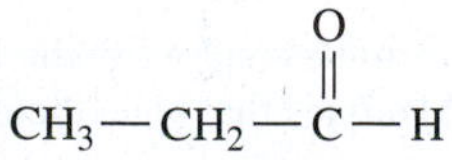 **b.** 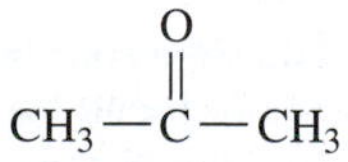

c. 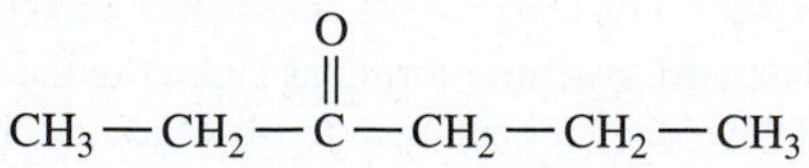

14.46 Which of the following aldehydes or ketones are soluble in water? (14.2)

a. 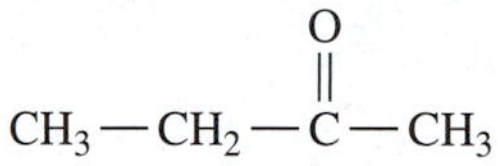 **b.** 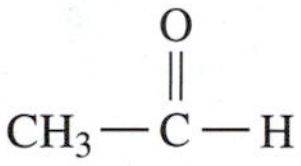

c. 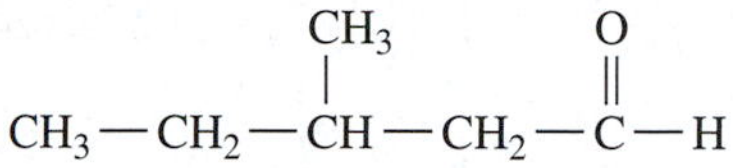

14.47 In each of the following pairs of compounds, select the compound with the higher boiling point: (14.2)

a. $CH_3—CH_2—CH_2—OH$ or 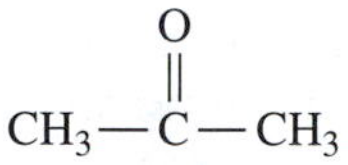

b. $CH_3—CH_2—CH_2—CH_3$ or 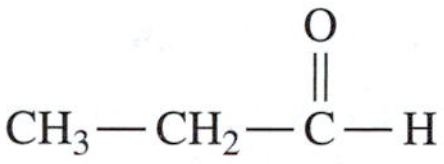

c. $CH_3—CH_2—OH$ or 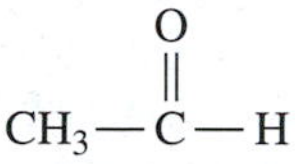

14.48 In each of the following pairs of compounds, select the compound with the higher boiling point: (14.2)

a. 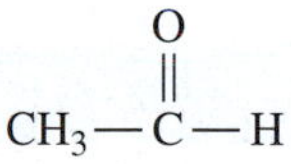 or 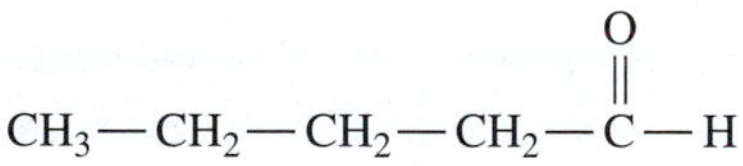

b. 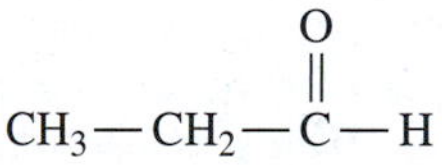 or 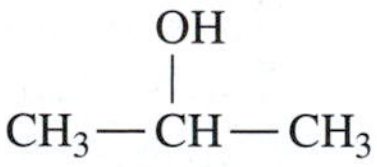

c. $CH_3—CH_2—CH_2—CH_3$ or 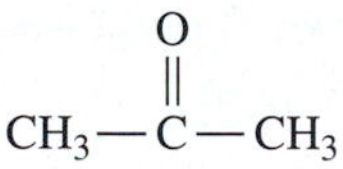

14.49 Draw the condensed structural formula for the product, if any, when each of the following is oxidized: (14.3)

a. 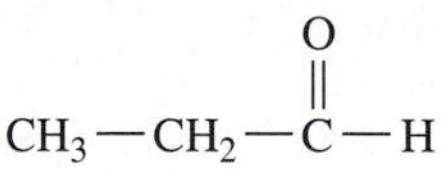 **b.** 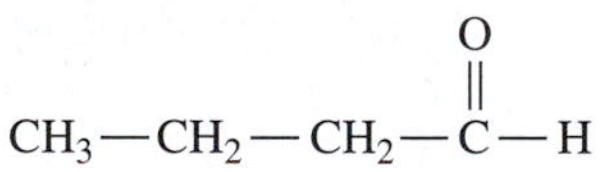

c. 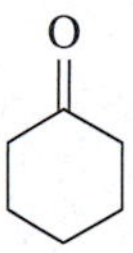

14.50 Draw the condensed structural formula for the product, if any, when each of the following is oxidized: (14.3)

a. 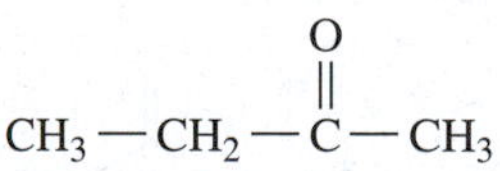 **b.** 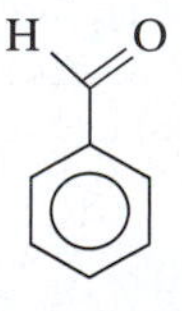

c. 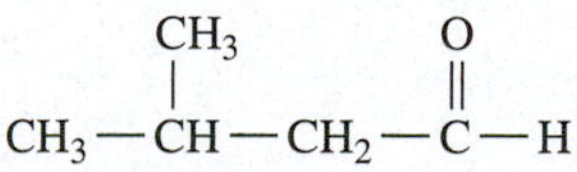

14.51 Draw the condensed structural formula for the product when hydrogen and a nickel catalyst reduce each of the following: (14.3)

a. 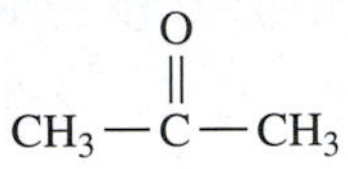

b. 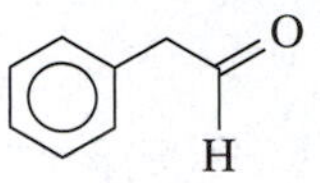

c. 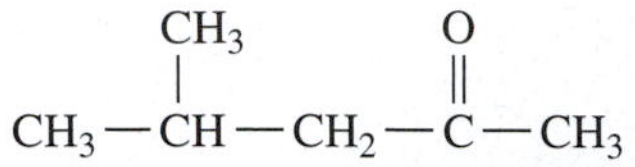

14.52 Draw the condensed structural formula for the product when hydrogen and a platinum catalyst reduce each of the following: (14.3)

a. 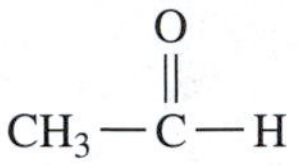

b. 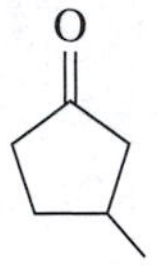

c. 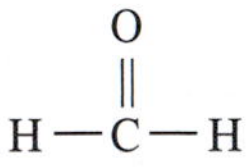

14.53 Write the name of the alcohol, aldehyde, or ketone produced from each of the following reactions: (14.3)

a. oxidation of 1-propanol
b. oxidation of 2-pentanol
c. reduction of butanone
d. oxidation of cyclohexanol

14.54 Write the name of the alcohol, aldehyde, or ketone produced from each of the following reactions: (14.3)

a. reduction of butyraldehyde
b. oxidation of 3-methyl-2-pentanol
c. reduction of 4-methyl-2-hexanone
d. oxidation of 3-methylcyclopentanol

14.55 Identify the following as hemiacetals or acetals. Write the IUPAC names of the carbonyl compounds and alcohols used in their synthesis. (14.4)

a. 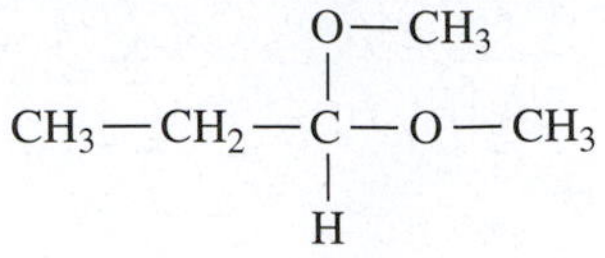

b. 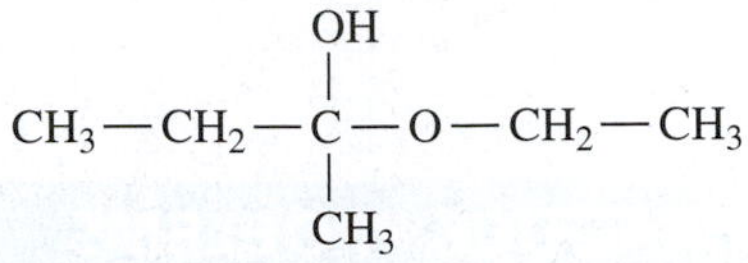

c. 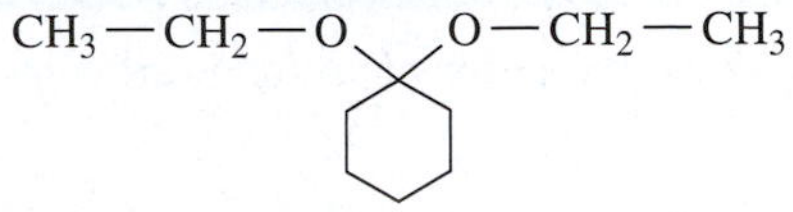

14.56 Identify the following as hemiacetals or acetals. Write the IUPAC names of the carbonyl compounds and alcohols used in their synthesis. (14.4)

a. 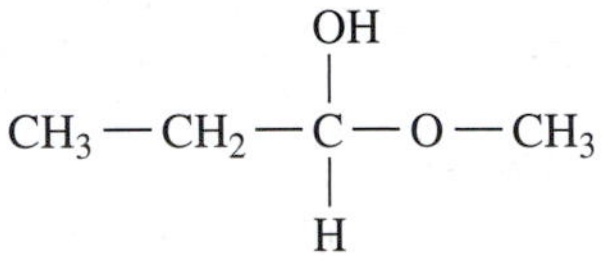

b. 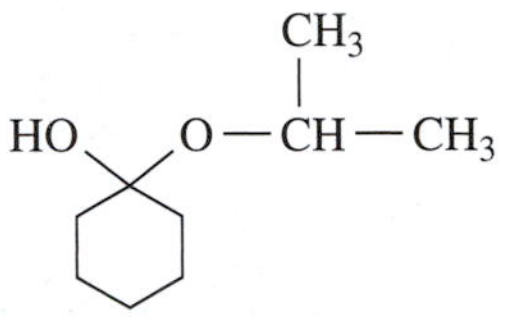

c. 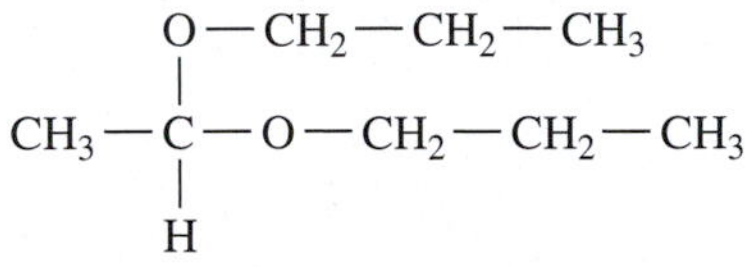

CHALLENGE PROBLEMS

The following problems are related to the topics in this chapter. However, they do not all follow the chapter order, and they require you to combine concepts and skills from several Sections. These problems will help you increase your critical thinking skills and prepare for your next exam.

Use the following condensed structural and line-angle formulas **A** to **F** to answer problems 14.57 and 14.58: (14.1, 14.5)

A 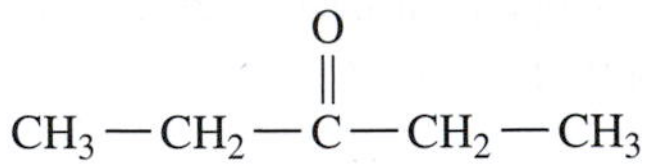

B 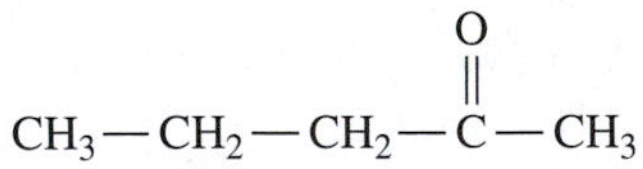

C 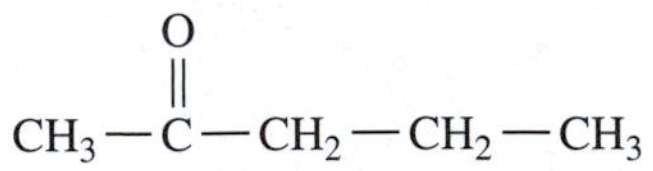

D 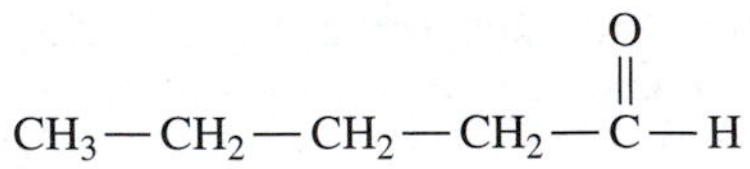

E 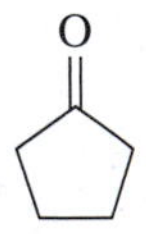

F 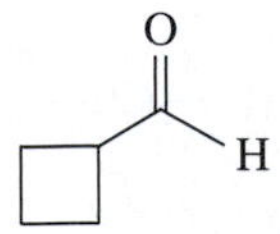

14.57 *True* or *False?*

a. **A** and **B** are structural isomers.
b. **D** and **F** are aldehydes.
c. **B** and **C** are the same compound.
d. **C** and **D** are structural isomers.

14.58 *True* or *False?*

a. **E** and **F** are structural isomers.
b. **A** is chiral.
c. **A** and **C** are the same compound.
d. **B** and **E** are ketones.

14.59 A compound with the formula C_4H_8O is made by oxidation of 2-butanol and cannot be oxidized further. Draw the condensed structural formula and write the IUPAC name for the compound. (14.1, 14.3)

14.60 A compound with the formula C_4H_8O is made by oxidation of 2-methyl-1-propanol and oxidizes easily to give a carboxylic acid. Draw the condensed structural formula and give the IUPAC name for the compound. (14.1, 14.3)

14.61 Draw the condensed structural formulas and write the IUPAC names for all the aldehydes and ketones that have the molecular formula C_4H_8O. (14.1)

14.62 Draw the condensed structural formulas and write the IUPAC names for all the aldehydes and ketones that have the molecular formula $C_5H_{10}O$. (14.1)

ANSWERS

14.1 **a.** ketone **b.** aldehyde
c. ketone **d.** aldehyde

14.3 **a.** structural isomers of C_3H_6O
b. structural isomers of $C_5H_{10}O$
c. not structural isomers

14.5 **a.** 3-bromobutanal **b.** 4-chloro-2-pentanone
c. 2-methylcyclopentanone **d.** 2-bromobenzaldehyde

14.7 **a.** acetaldehyde **b.** methyl propyl ketone
c. formaldehyde

14.9 **a.** $CH_3-\overset{\overset{O}{\|}}{C}-H$ 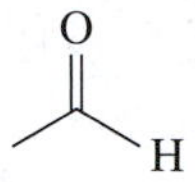

b. $CH_3-\overset{\overset{CH_3}{|}}{CH}-\overset{\overset{O}{\|}}{C}-CH_2-CH_3$ 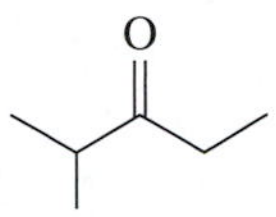

c. $CH_3-\overset{\overset{O}{\|}}{C}-CH_2-CH_2-CH_2-CH_3$

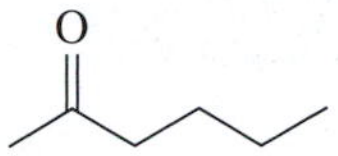

d. $CH_3-CH_2-CH_2-\overset{\overset{CH_3}{|}}{CH}-CH_2-\overset{\overset{O}{\|}}{C}-H$

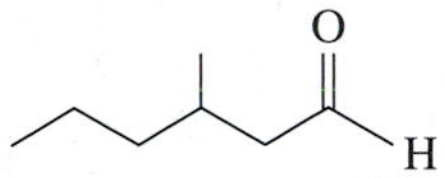

14.11 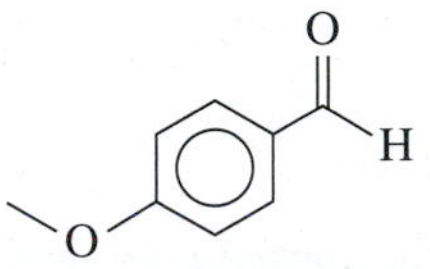

14.13 **a.** $CH_3-\overset{\overset{O}{\|}}{C}-H$
Ethanal has a higher boiling point than propane because the polar carbonyl group in ethanal forms dipole–dipole attractions.
b. Pentanal has a longer carbon chain, more electrons, and more dispersion forces, which give it a higher boiling point.
c. 1-Butanol has a higher boiling point because it can hydrogen bond with other 1-butanol molecules.

14.15 **a.** $CH_3-\overset{\overset{O}{\|}}{C}-\overset{\overset{O}{\|}}{C}-CH_2-CH_3$ has two polar carbonyl groups and can form more hydrogen bonds with water.
b. Propanal has a shorter carbon chain than pentanal, in which the larger hydrocarbon chain reduces the impact on solubility of the polar carbonyl group.
c. Acetone has a shorter carbon chain than 2-pentanone, in which the larger hydrocarbon chain reduces the impact on solubility of the polar carbonyl group.

14.17 No. A hydrocarbon chain of eight carbon atoms reduces the impact on solubility of the polar carbonyl group.

14.19 **a.** $H-\overset{\overset{O}{\|}}{C}-OH$ **b.** none
c. 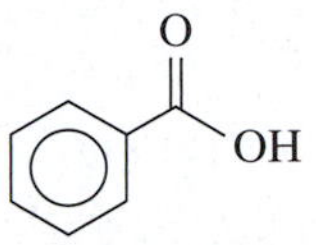 **d.** none

14.21 **a.** yes **b.** no **c.** yes

14.23 **a.** $CH_3-CH_2-CH_2-CH_2-OH$
b. $CH_3-\overset{\overset{OH}{|}}{CH}-CH_3$
c. $CH_3-CH_2-CH_2-\overset{\overset{Br}{|}}{CH}-CH_2-CH_2-OH$
d. $CH_3-\overset{\overset{CH_3}{|}}{CH}-\overset{\overset{OH}{|}}{CH}-CH_2-CH_3$

14.25 **a.** neither **b.** hemiacetal **c.** acetal
d. hemiacetal **e.** acetal

14.27 **a.** $CH_3-\underset{\underset{H}{|}}{\overset{\overset{OH}{|}}{C}}-O-CH_3$
b. $CH_3-\underset{\underset{CH_3}{|}}{\overset{\overset{OH}{|}}{C}}-O-CH_3$
c. $CH_3-CH_2-CH_2-\underset{\underset{H}{|}}{\overset{\overset{OH}{|}}{C}}-O-CH_3$

14.29 **a.** $CH_3-\underset{\underset{H}{|}}{\overset{\overset{O-CH_3}{|}}{C}}-O-CH_3$
b. $CH_3-\underset{\underset{CH_3}{|}}{\overset{\overset{O-CH_3}{|}}{C}}-O-CH_3$
c. $CH_3-CH_2-CH_2-\underset{\underset{H}{|}}{\overset{\overset{O-CH_3}{|}}{C}}-O-CH_3$

14.31 **a.** ether, phenol, ketone, aromatic
b. $C_{14}H_{12}O_3$, 228.2 g/mole
c. 10.7 g of oxybenzone

14.33 aromatic, alkene, aldehyde

14.35 a. 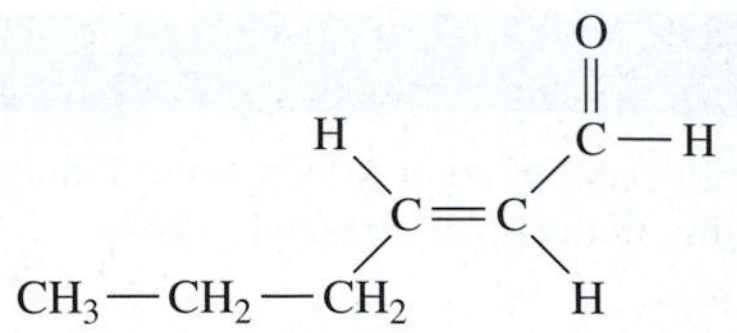

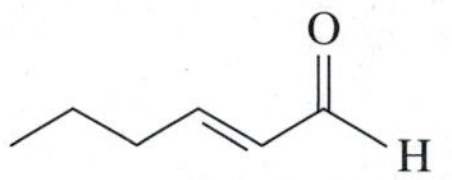

b. $CH_3-\overset{\overset{\large CH_3}{|}}{C}=CH-CH_2-CH_2-\overset{\overset{\large CH_3}{|}}{CH}-\overset{\overset{\large O}{\|}}{C}-H$

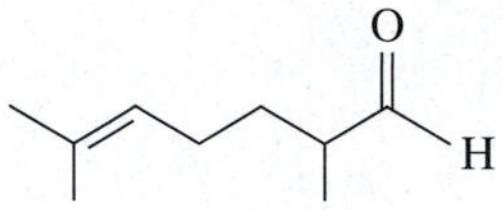

14.37 The C=O double bond has a dipole because the oxygen atom is highly electronegative compared to the carbon atom. In the C=C double bond, both atoms have the same electronegativity, and there is no dipole.

14.39 **a** and **b**

14.41 a. 2-bromo-4-chlorocyclopentanone
b. 2,4-dibromobenzaldehyde
c. 3-chloropropanal
d. 2-chloro-3-pentanone
e. 5-chloro-3-hexanone

14.43 a. 

b. $CH_3-CH_2-CH_2-CH_2-\overset{\overset{\large O}{\|}}{C}-H$

c. $CH_3-CH_2-\overset{\overset{\large O}{\|}}{C}-CH_3$

d. $CH_3-CH_2-\overset{\overset{\large CH_3}{|}}{CH}-CH_2-CH_2-\overset{\overset{\large O}{\|}}{C}-H$

14.45 **a** and **b**

14.47 a. $CH_3-CH_2-CH_2-OH$ **b.** $CH_3-CH_2-\overset{\overset{\large O}{\|}}{C}-H$
c. CH_3-CH_2-OH

14.49 a. $CH_3-CH_2-\overset{\overset{\large O}{\|}}{C}-OH$

b. $CH_3-CH_2-CH_2-\overset{\overset{\large O}{\|}}{C}-OH$

c. no reaction

14.51 a. $CH_3-\overset{\overset{\large OH}{|}}{CH}-CH_3$

b. 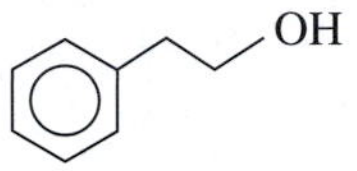

c. $CH_3-\overset{\overset{\large CH_3}{|}}{CH}-CH_2-\overset{\overset{\large OH}{|}}{CH}-CH_3$

14.53 a. propanal **b.** 2-pentanone
c. 2-butanol **d.** cyclohexanone

14.55 a. acetal; propanal and methanol
b. hemiacetal; butanone and ethanol
c. acetal; cyclohexanone and ethanol

14.57 a. true **b.** true **c.** true **d.** true

14.59 $CH_3-\overset{\overset{\large O}{\|}}{C}-CH_2-CH_3$ Butanone

14.61 $CH_3-CH_2-CH_2-\overset{\overset{\large O}{\|}}{C}-H$ Butanal

$CH_3-\overset{\overset{\large CH_3}{|}}{CH}-\overset{\overset{\large O}{\|}}{C}-H$ 2-Methylpropanal

$CH_3-\overset{\overset{\large O}{\|}}{C}-CH_2-CH_3$ Butanone

COMBINING IDEAS from Chapters 12 to 14

CI.27 A compound called butylated hydroxytoluene, or BHT, has been added to cereal and other foods as an antioxidant. A common name for BHT is 2,6-di-*tert*-butyl-4-methylphenol. (2.6, 7.5, 7.6, 12.3, 13.1)

a. Draw the line-angle formula for BHT.
b. BHT is produced from 4-methylphenol and 2-methylpropene. Draw the line-angle formulas for these reactants.
c. What are the molecular formula and molar mass of BHT?
d. The U.S. Food and Drug Administration (FDA) allows a maximum of 50. ppm of BHT added to cereal. How many milligrams of BHT could be added to a box of cereal that contains 15 oz of dry cereal?

CI.28 Used in sunless tanning lotions, the compound 1,3-dihydroxypropanone, or dihydroxyacetone (DHA), darkens the skin without exposure to sunlight. DHA reacts with amino acids in the outer surface of the skin. A typical drugstore lotion contains 4.0% (m/v) DHA. (2.6, 7.4, 7.5, 14.1)

A sunless tanning lotion contains DHA to darken the skin.

a. Draw the condensed structural formula for DHA.
b. What are the functional groups in DHA?
c. What are the molecular formula and molar mass of DHA?
d. A bottle of sunless tanning lotion contains 200 mL of lotion. How many milligrams of DHA are in a bottle?

CI.29 Acetone (propanone), a clear liquid solvent with an acrid odor, is used to remove nail polish, paints, and resins. It has a low boiling point and is highly flammable. (7.4, 7.5, 13.4, 14.1)

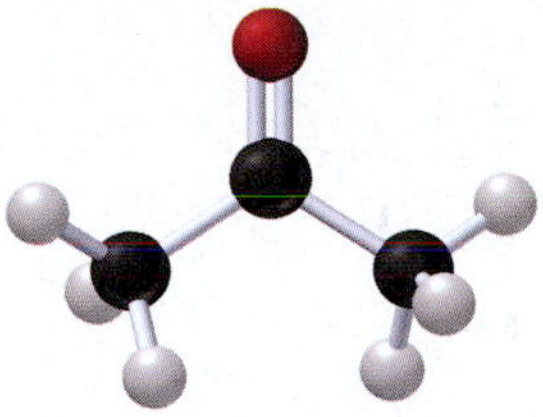

Acetone is a solvent used in polish remover.

a. Draw the condensed structural formula for propanone.
b. What are the molecular formula and molar mass of propanone?
c. Draw the condensed structural formula for the alcohol that can be oxidized to produce propanone.

CI.30 Acetone (propanone) has a density of 0.786 g/mL and a heat of combustion of 428 kcal/mole. Use your answers to problem CI.29 to answer the following: (7.1, 7.6, 7.9, 8.6, 14.1)
a. Write the balanced chemical equation for the complete combustion of propanone.
b. How much heat, in kilojoules, is released if 2.58 g of propanone reacts with oxygen?
c. How many grams of oxygen gas are needed to react with 15.0 mL of propanone?
d. How many liters of carbon dioxide gas are produced at STP in part **c**?

CI.31 One of the components of gasoline is octane, C_8H_{18}, which has a density of 0.803 g/mL. The combustion of 1 mole of octane provides 5510 kJ. A hybrid car has a fuel tank with a capacity of 11.9 gal and a gas mileage of 45 mi/gal. (2.6, 7.1, 7.5, 7.7, 7.9, 12.2)

Octane is one of the components of gasoline.

a. Write the balanced chemical equation for the combustion of octane.
b. Is the combustion of octane endothermic or exothermic?
c. How many moles of octane are in one tank of fuel, assuming it is all octane?
d. If the total mileage of this hybrid car for one year is 24 500 mi, how many kilograms of carbon dioxide would be produced from the combustion of the fuel, assuming it is all octane?

CI.32 Ionone is a compound that gives violets their aroma. The small, purple flowers of violets are used in salads and to make teas. Liquid ionone has a density of 0.935 g/mL. (7.4, 7.7, 7.9, 8.6, 12.6, 13.1, 14.1)

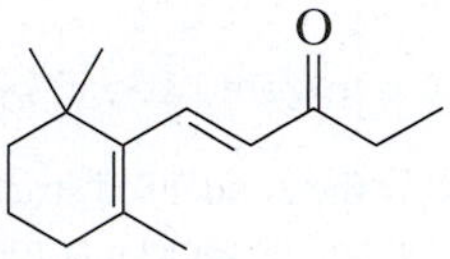

Ionone

The aroma of violets is due to ionone.

a. What functional groups are present in ionone?
b. Is the double bond on the side chain cis or trans?
c. What are the molecular formula and molar mass of ionone?
d. How many moles are in 2.00 mL of ionone?
e. When ionone reacts with hydrogen in the presence of a platinum catalyst, hydrogen adds to the double bonds and converts the ketone group to an alcohol. Draw the condensed structural formula and give the molecular formula for the product.
f. How many milliliters of hydrogen gas at STP are needed to completely react with 5.0 mL of ionone?

CI.33 Butyraldehyde is a clear liquid solvent with an unpleasant odor. It has a low boiling point and is highly flammable. (14.1, 14.3)

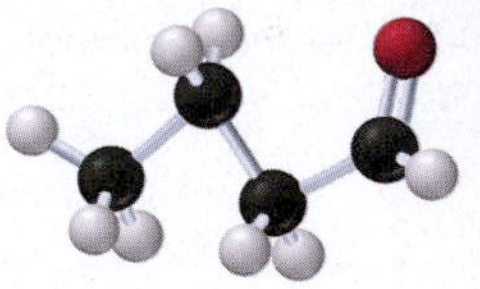

The unpleasant odor of old gym socks is because of butyraldehyde.

a. Draw the condensed structural formula for butyraldehyde.
b. Draw the line-angle formula for butyraldehyde.
c. What is the IUPAC name of butyraldehyde?
d. Draw the condensed structural formula for the alcohol that is produced when butyraldehyde is reduced.

CI.34 Butyraldehyde has a density of 0.802 g/mL and a heat of combustion of 1520 kJ/mole. Use your answers to problem CI.33 to answer the following: (7.1, 7.7, 7.9, 8.6, 14.1)
a. Write the balanced chemical equation for the complete combustion of butyraldehyde.
b. How many grams of oxygen gas are needed to completely react with 15.0 mL of butyraldehyde?
c. How many liters of carbon dioxide gas are produced at STP in part **b**?
d. Calculate the heat, in kilojoules, that is released from the combustion of butyraldehyde in part **b**.

ANSWERS

CI.27 a.

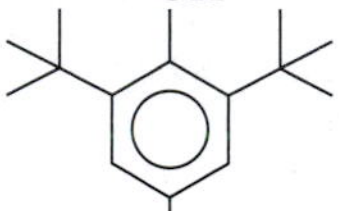

b.

4-Methylphenol

2-Methylpropene

c. $C_{15}H_{24}O$, 220.4 g/mole **d.** 21 mg

CI.29 a. $CH_3-\overset{\overset{\displaystyle O}{\|}}{C}-CH_3$
b. C_3H_6O, 58.08 g/mole
c. $CH_3-\overset{\overset{\displaystyle OH}{|}}{CH}-CH_3$

CI.31 a.
$$2C_8H_{18}(l) + 25O_2(g) \xrightarrow{\Delta} 16CO_2(g) + 18H_2O(g) + 11\,020$$
b. exothermic
c. 317 moles of octane
d. 5.10×10^3 kg of CO_2

CI.33 a. $CH_3-CH_2-CH_2-\overset{\overset{\displaystyle O}{\|}}{C}-H$
b.

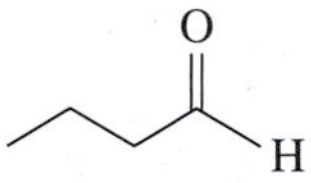

c. butanal
d. $CH_3-CH_2-CH_2-CH_2-OH$

15

Carbohydrates

During her annual physical examination, Kate, a 64-year old woman, complains that her vision is blurry, she feels a frequent need to urinate, and she has gained 22 lb over the past year. She tried to lose weight and increase her exercise for the past 6 months without success. Her diet is high in carbohydrates. For dinner, Kate typically eats two cups of pasta and three to four slices of bread with butter or olive oil. She also eats eight to ten pieces of fresh fruit per day at meals and as snacks. A lab test shows that her fasting blood glucose level is 178 mg/dL, indicating type 2 diabetes. She is referred to the diabetes specialty clinic, where she meets Paula, a diabetes nurse. Paula explains to Kate that foods like pasta, bread, and fruit can raise blood glucose levels because they contain large amounts of molecules called carbohydrates, which are broken down in the body into glucose molecules.

CAREER

Diabetes Nurse

Diabetes nurses teach patients about diabetes so they can self-manage and control their condition. This includes education on proper diets and nutrition for both diabetic and pre-diabetic patients. Diabetes nurses help patients learn to monitor their medication and blood sugar levels and look for symptoms like diabetic nerve damage and vision loss. Diabetes nurses may also work with patients who have been hospitalized because of complications from their disease. Working with patients with diabetes requires a thorough knowledge of the endocrine system, the system that regulates metabolism, so there is some overlap between diabetes and endocrinology nursing.

CLINICAL UPDATE

Kate's Program for Type 2 Diabetes

Now Kate uses a glucose meter to check her blood glucose level. You can read about changes she makes to her lifestyle in the **CLINICAL UPDATE Kate's Program for Type 2 Diabetes**, page 550, and see how a change in eating habits and an increase in exercise helps Kate to decrease her blood sugar and lose weight.

LOOKING AHEAD

Carbohydrates contained in foods such as pasta and bread provide energy for the body.

15.1 Carbohydrates

LEARNING GOAL Classify a monosaccharide as an aldose or a ketose, and indicate the number of carbon atoms.

We have many carbohydrates in our food. There are polysaccharides called starches in bread and pasta. The table sugar used to sweeten cereal, tea, or coffee is sucrose, a disaccharide that consists of two simple sugars, glucose and fructose. **Carbohydrates** such as table sugar, lactose in milk, and cellulose are all made of carbon, hydrogen, and oxygen. Simple sugars, which have formulas of $C_n(H_2O)_n$, were once thought to be hydrates of carbon, thus the name *carbohydrate*.

Glucose ($C_6H_{12}O_6$) is the most important simple carbohydrate in metabolism. In a series of reactions called *photosynthesis*, energy from the Sun is used to combine the carbon atoms from carbon dioxide (CO_2) and the hydrogen and oxygen atoms of water (H_2O) to form glucose.

$$6CO_2 + 6H_2O + \text{energy} \underset{\text{Respiration}}{\overset{\text{Photosynthesis}}{\rightleftharpoons}} \underset{\text{Glucose}}{C_6H_{12}O_6} + 6O_2$$

In the body, glucose is oxidized in a series of metabolic reactions known as *respiration* to release chemical energy to do work in the cells. Carbon dioxide and water are produced and returned to the atmosphere. The combination of photosynthesis and respiration is called the *carbon cycle*, in which energy from the Sun is stored in plants by photosynthesis and made available to us when the carbohydrates in our diets are metabolized (see **FIGURE 15.1**).

Energy
RESPIRATION
Energy
$CO_2 + H_2O$
Carbohydrates + O_2
PHOTOSYNTHESIS

FIGURE 15.1 ▶ During photosynthesis, energy from the Sun combines CO_2 and H_2O to form glucose ($C_6H_{12}O_6$) and O_2. During respiration in the body, carbohydrates are oxidized to CO_2 and H_2O, while energy is released.

Q What are the reactants and products of respiration?

TEST

Try Practice Problems 15.1 and 15.2

Types of Carbohydrates

The simplest carbohydrates are the **monosaccharides**. A monosaccharide cannot be split or hydrolyzed into smaller carbohydrates. Glucose, $C_6H_{12}O_6$, is a monosaccharide. A **disaccharide** consists of two monosaccharide units joined together. For example, ordinary table sugar, sucrose, $C_{12}H_{22}O_{11}$, is a disaccharide that can be split by water (hydrolysis) in the presence of an acid or an enzyme to give one molecule of glucose and one molecule of another monosaccharide, fructose.

$$\underset{\text{Sucrose}}{C_{12}H_{22}O_{11}} + H_2O \xrightarrow{H^+ \text{ or enzyme}} \underset{\text{Glucose}}{C_6H_{12}O_6} + \underset{\text{Fructose}}{C_6H_{12}O_6}$$

TEST

Try Practice Problems 15.3 and 15.4

A **polysaccharide** is a carbohydrate that consists of many linked monosaccharide units. In the presence of an acid or an enzyme, a polysaccharide can be completely hydrolyzed to yield many monosaccharide molecules.

Example	Type of Carbohydrate	Products of Hydrolysis
 Honey contains the monosaccharides fructose and glucose.	Monosaccharide + H_2O $\xrightarrow{H^+ \text{ or enzyme}}$ 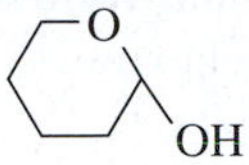	no hydrolysis 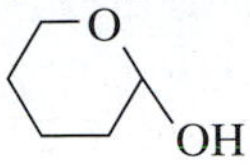
 Molasses contains maltose, a disaccharide.	Disaccharide + H_2O $\xrightarrow{H^+ \text{ or enzyme}}$ 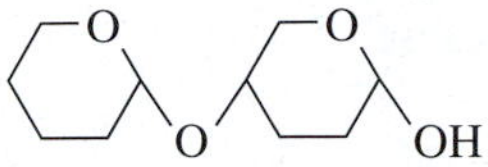	two monosaccharide molecules 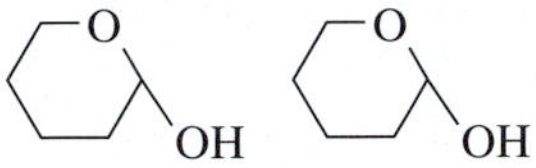
 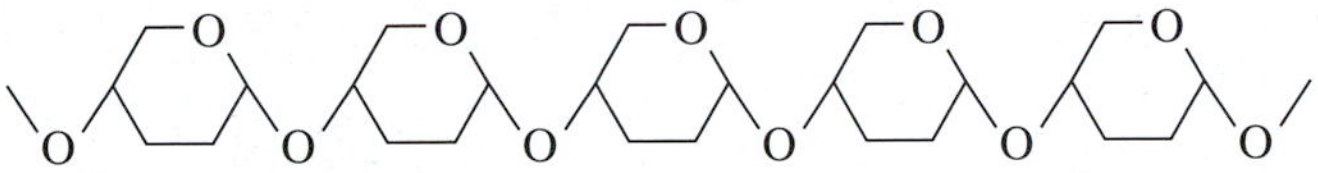 Potatoes contain the polysaccharide amylose.	Polysaccharide + many H_2O $\xrightarrow{H^+ \text{ or enzyme}}$	many monosaccharide molecules 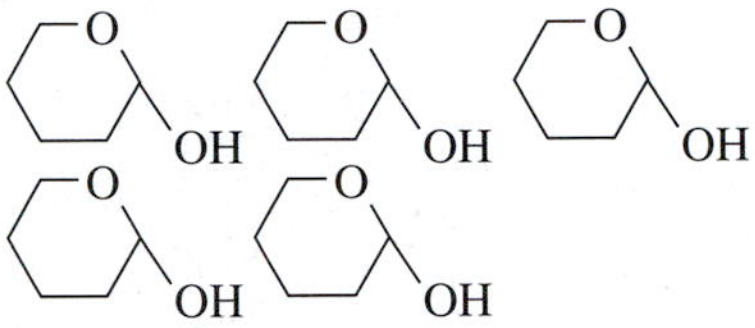

Monosaccharides

A monosaccharide contains several hydroxyl groups attached to a chain of three to seven carbon atoms that also contains an aldehyde or a ketone group. Thus, a monosaccharide is known as a polyhydroxy aldehyde or polyhydroxy ketone. In an **aldose**, the first carbon in the chain is an aldehyde, whereas a **ketose** has a ketone as the second carbon atom.

TEST

Try Practice Problems 15.5 and 15.6

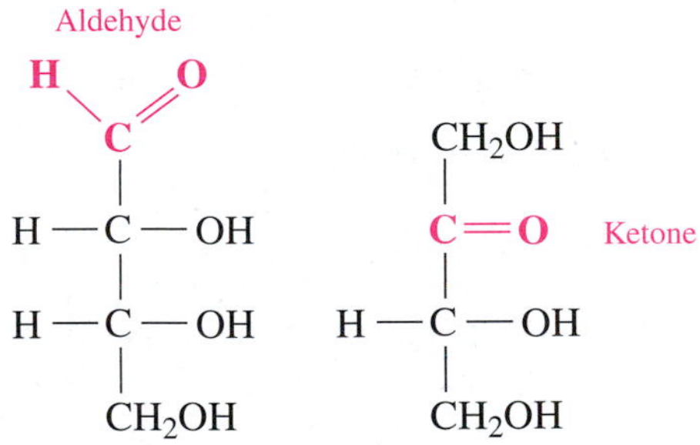
Aldehyde
Ketone

Erythrose, an aldose — Erythrulose, a ketose

Erythrulose, used in sunless tanning lotions, reacts with amino acids or proteins in the skin to produce a bronze color.

Monosaccharides are also classified by the number of carbon atoms. A monosaccharide with three carbon atoms is a *triose*, and one with four carbon atoms is a *tetrose*; a *pentose* has five carbons, and a *hexose* contains six carbons. We can use both classification systems to indicate the aldehyde or ketone group and the number of carbon atoms. An aldopentose is a five-carbon monosaccharide that is an aldehyde; a ketohexose is a six-carbon monosaccharide that is a ketone.

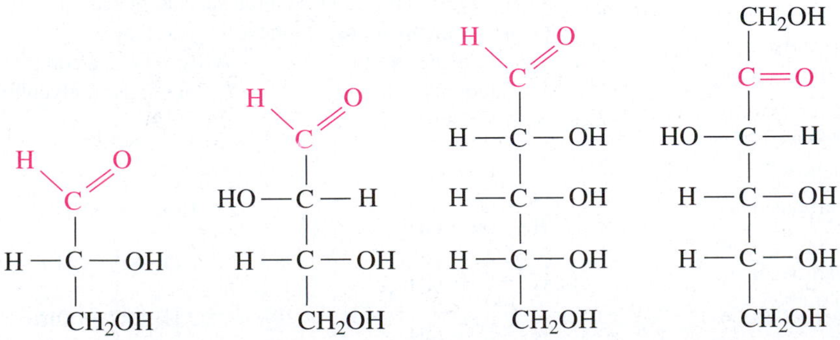

Glyceraldehyde (aldotriose) — Threose (aldotetrose) — Ribose (aldopentose) — Fructose (ketohexose)

ENGAGE

Why is fructose classified as a ketohexose?

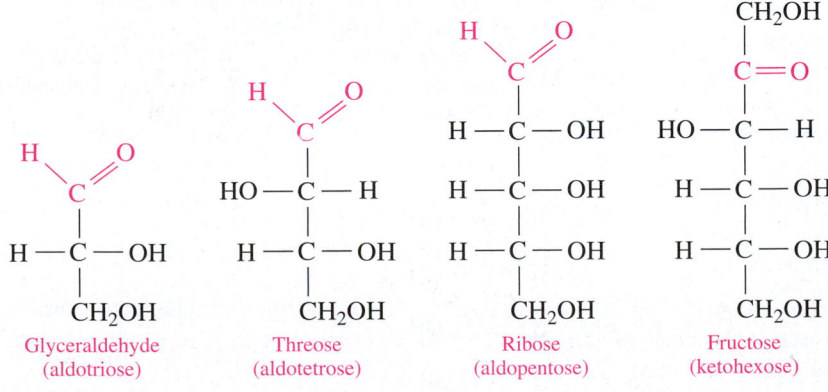

▶ SAMPLE PROBLEM 15.1 Monosaccharides

TRY IT FIRST

Classify each of the following monosaccharides as an aldopentose, ketopentose, aldohexose, or ketohexose:

a.

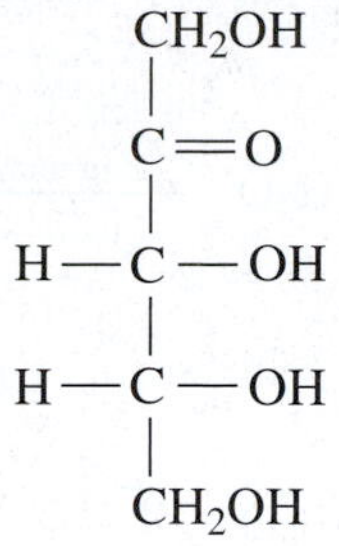

Ribulose

b.

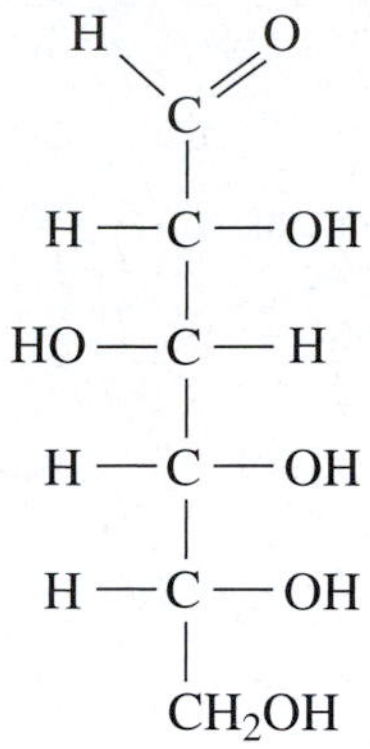

Glucose

SOLUTION

a. Ribulose has five carbon atoms (pentose) and is a ketone, which makes it a ketopentose.
b. Glucose has six carbon atoms (hexose) and is an aldehyde, which makes it an aldohexose.

STUDY CHECK 15.1

Classify each of the following monosaccharides as an aldotetrose, ketotetrose, aldopentose, or ketopentose:

a.

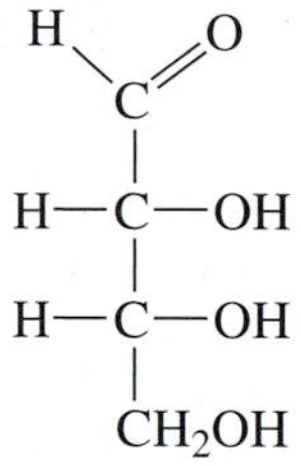

Erythrose

b.

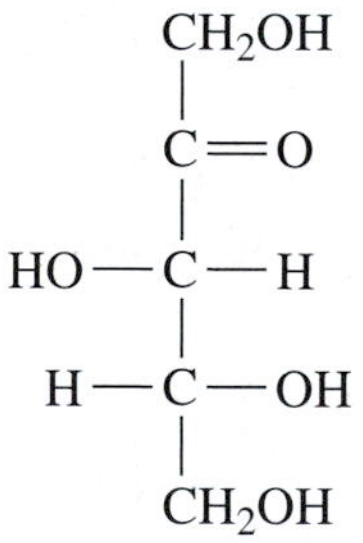

Xylulose

ANSWER

a. aldotetrose **b.** ketopentose

TEST

Try Practice Problems 15.7 to 15.10

PRACTICE PROBLEMS

15.1 Carbohydrates

15.1 What reactants are needed for photosynthesis and respiration?

15.2 What is the relationship between photosynthesis and respiration?

15.3 What is a monosaccharide? A disaccharide?

15.4 What is a polysaccharide?

15.5 What functional groups are found in all monosaccharides?

15.6 What is the difference between an aldose and a ketose?

15.7 What are the functional groups and number of carbons in a ketopentose?

15.8 What are the functional groups and number of carbons in an aldohexose?

Clinical Applications

15.9 Classify each of the following monosaccharides as an aldopentose, ketopentose, aldohexose, or ketohexose:

a. Psicose is present in low amounts in foods.

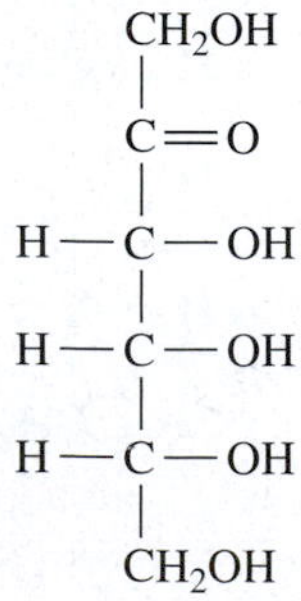

Psicose

b. Lyxose is a component of bacterial glycolipids.

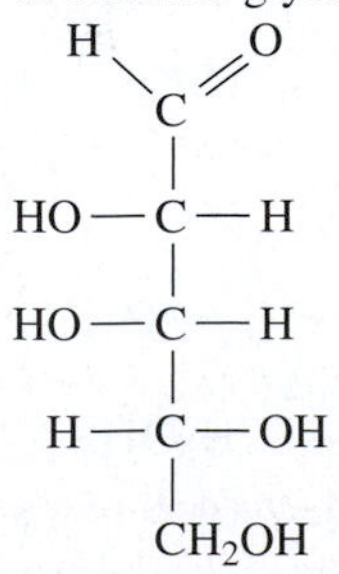

Lyxose

15.10 Classify each of the following monosaccharides as an aldopentose, ketopentose, aldohexose, or ketohexose:

a. A solution of xylose is given to test its absorption by the intestines.

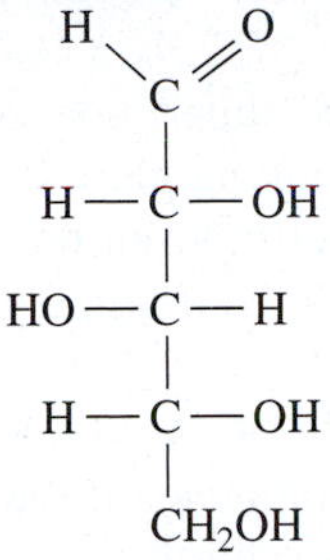

Xylose

b. Tagatose, found in fruit, is similar in sweetness to sugar.

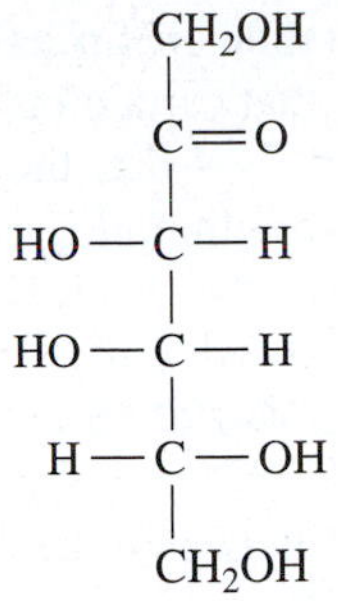

Tagatose

15.2 Chiral Molecules

LEARNING GOAL Identify chiral and achiral carbon atoms in an organic molecule. Use Fischer projections to identify the D and L enantiomers of monosaccharides.

Everything has a mirror image, and molecules are no exception. Sometimes, the mirror image of a molecule has a different three-dimensional shape than the original molecule. Because the three-dimensional shape of a molecule determines what the molecule can do, a molecule and its mirror image with different three-dimensional shapes can have different properties and functions. Many carbohydrates have mirror images with different three-dimensional shapes and, thus, different properties. Therefore, it is important to become familiar with mirror images.

We will start by thinking about a familiar example of a mirror image. If you hold your right hand up to a mirror, you see its mirror image, which matches your left hand (see **FIGURE 15.2**). If you turn your palms toward each other, one hand is the mirror image of the other. If you look at the palms of your hands, your thumbs are on opposite sides. If you then place your right hand over your left hand, you cannot match up all the parts of the hands: palms, backs, thumbs, and little fingers.

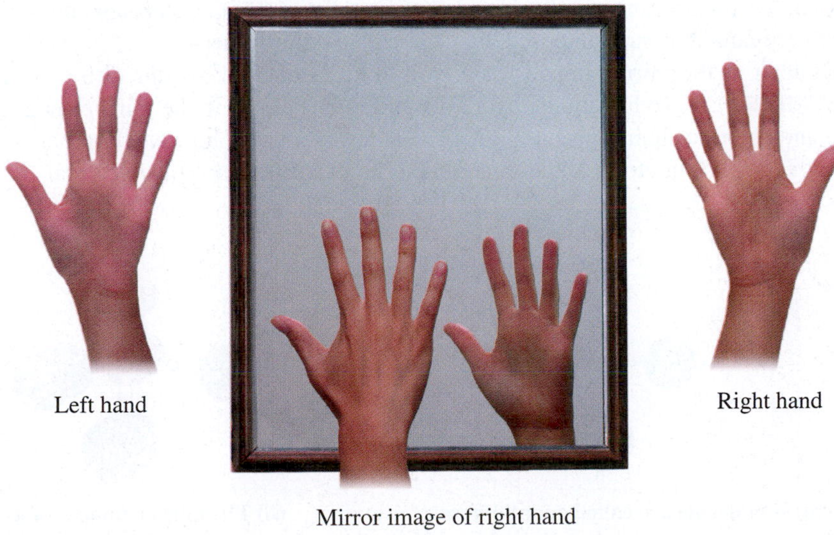
Left hand
Right hand
Mirror image of right hand

FIGURE 15.2 ▶ The left and right hands are chiral because they have mirror images that cannot be superimposed on each other.

Q Why are your shoes chiral objects?

The thumbs and little fingers can be matched, but then the palms or backs of your hands are facing each other. Your hands are mirror images that cannot be superimposed on each other. When the mirror images cannot be completely matched, they are *nonsuperimposable*. In other words, the mirror images have different three-dimensional shapes because we cannot line them up exactly with each other.

INTERACTIVE VIDEO
Chirality

Objects such as hands that have nonsuperimposable mirror images are **chiral** (pronunciation *kai-ral*). Left and right shoes are chiral; left- and right-handed golf clubs are chiral. When we think of how difficult it is to put a left-hand glove on our right hand, put a right shoe on our left foot, or use left-handed scissors if we are right-handed, we begin to realize that certain properties of mirror images are very different.

When the mirror image of an object is identical and can be superimposed on the original, it is *achiral*. For example, the mirror image of a plain drinking glass is identical to the original glass, which means the mirror image can be superimposed on the glass and that the mirror image has the same properties and functions as the original glass (see **FIGURE 15.3**).

Many important compounds in medicine are chiral, including medications such as ibuprofen, penicillin, epinephrine, and morphine. Most of the compounds in biochemistry—carbohydrates, fats, amino acids, proteins, and DNA—are also chiral.

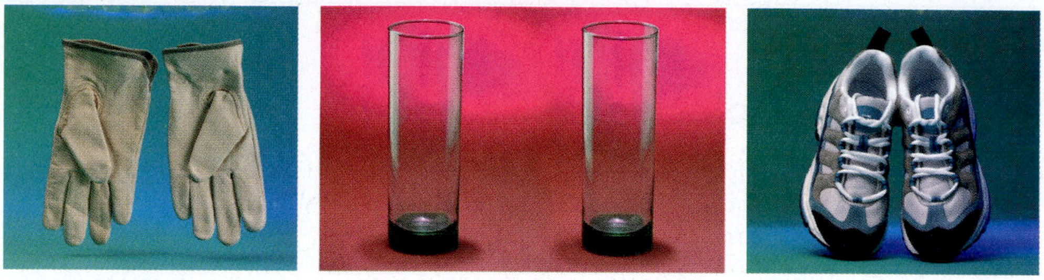

Chiral Achiral Chiral

FIGURE 15.3 ▶ Everyday objects such as gloves and shoes are chiral, but a plain glass is achiral.

Q Why are some of the objects chiral and others achiral?

CORE CHEMISTRY SKILL
Identifying Chiral Molecules

Chiral Carbon Atoms

While it is common to think of the mirror images of objects we see in our everyday lives, it is not as easy to think about the mirror images of the molecules we cannot see. Fortunately, there is a simple way to determine if a carbon compound is chiral. A carbon compound is chiral if it has at least one carbon atom bonded to *four different atoms or groups*. This type of carbon atom is called a **chiral carbon** because there are two different ways that it can bond to four atoms or groups of atoms. The resulting structures are nonsuperimposable mirror images. When two or more chiral structures have the same molecular formula, but differ in the three-dimensional arrangements of atoms in space, they are called **stereoisomers**. When stereoisomers cannot be superimposed, they are called **enantiomers**.

Let's look at the mirror images of a carbon atom bonded to four different atoms (see **FIGURE 15.4**). If we line up the hydrogen and iodine atoms in the mirror images, the bromine and chlorine atoms appear on opposite sides. No matter how we turn the models, we cannot align all four atoms at the same time. Therefore, this molecule is chiral.

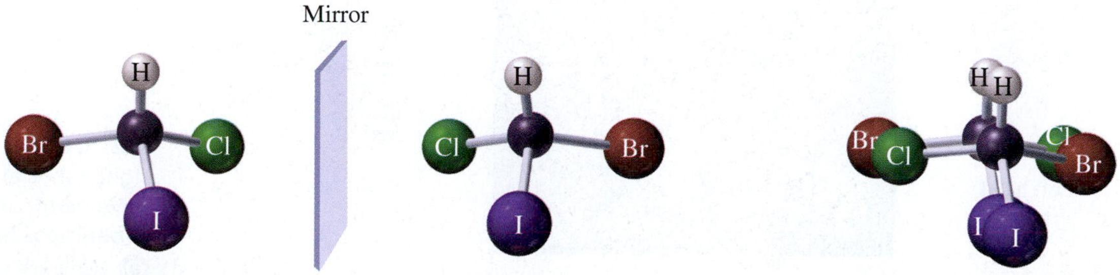
Mirror
H
Br
Cl
I
H
Cl
Br
I
H H
Br Cl
Cl Br
I I

(a) The mirror images of a chiral molecule are called enantiomers.

(b) The mirror images of a chiral molecule cannot be superimposed on each other.

FIGURE 15.4 ▶ Enantiomers are chiral molecules that are mirror images of each other, but that cannot be superimposed.

Q Why is the carbon atom in this compound chiral?

Now let's look at the mirror images of a carbon atom that has two identical atoms bonded to it. If we rotate the mirror image, the atoms in the original molecule align with the atoms in the rotated mirror image. Therefore, the molecule is achiral, and the mirror images represent the same structure (see **FIGURE 15.5**).

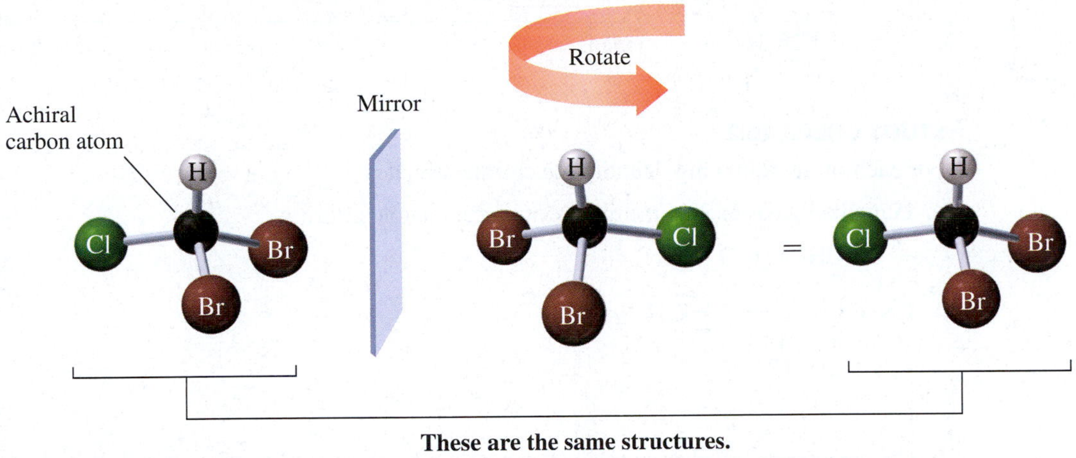
Rotate
Achiral
carbon atom
Mirror
H
Cl
Br
Br
H
Br
Cl
Br
=
H
Cl
Br
Br
These are the same structures.

FIGURE 15.5 ▶ The mirror images of an achiral compound can be superimposed on each other.

Q Why can the mirror images of this compound be superimposed?

▶ SAMPLE PROBLEM 15.2 Chiral Carbons

TRY IT FIRST

For each of the following, indicate whether the carbon in red is chiral or achiral:

a. Glycerol is used in soaps, creams, and hair care products.

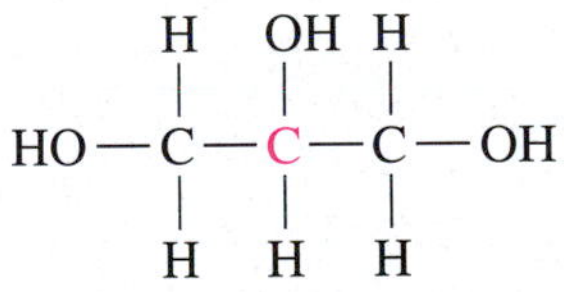

Glycerol

b. Ibuprofen is used to relieve fever and pain.

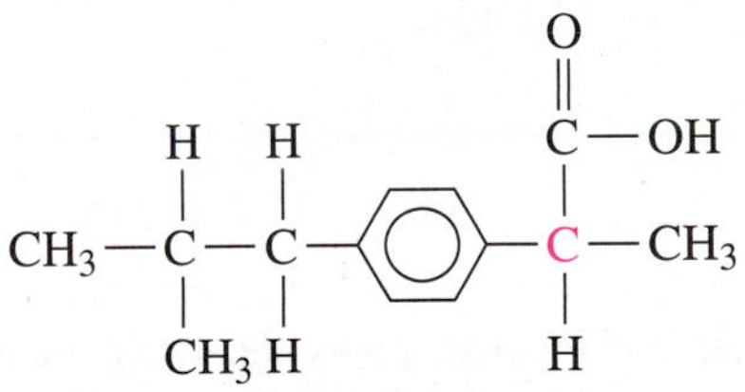

Ibuprofen

SOLUTION

a. Achiral. Boxes are drawn around the groups of atoms attached to the red C to determine chirality. Two of the substituents on the carbon in red are the same.

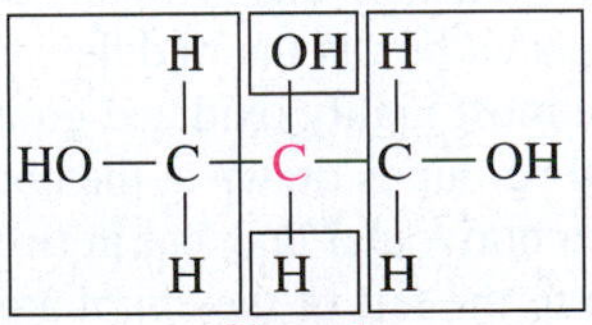

Glycerol

b. Chiral. The carbon in red is bonded to four different groups.

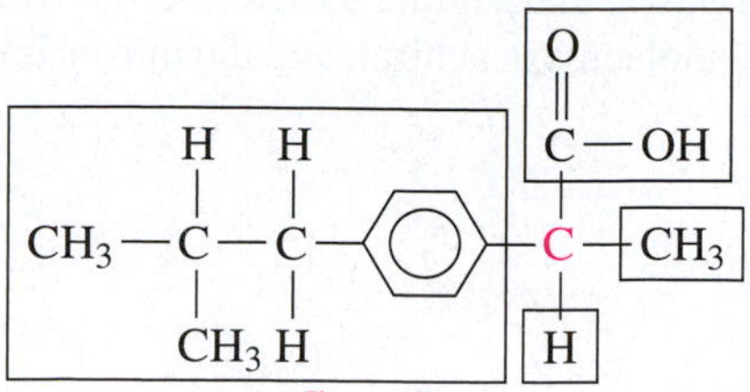

Ibuprofen

ENGAGE

Why does 3-methylhexane have a chiral carbon whereas 2-methylhexane does not?

STUDY CHECK 15.2

For each of the following, identify the chiral carbon(s):

a. Penicillamine is used in the treatment of rheumatoid arthritis.

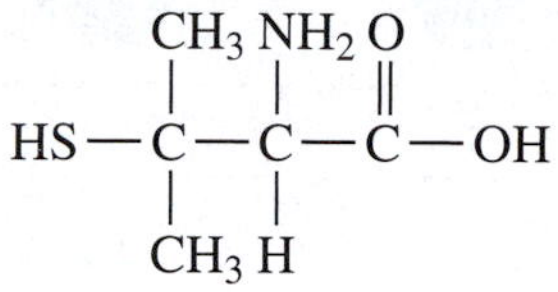

Penicillamine

b. One enantiomer of ethambutol is used to treat tuberculosis. The other enantiomer causes blindness.

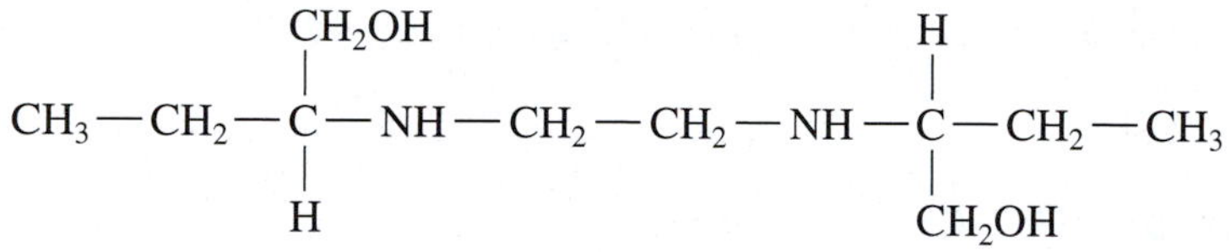

Ethambutol

ANSWER

a.

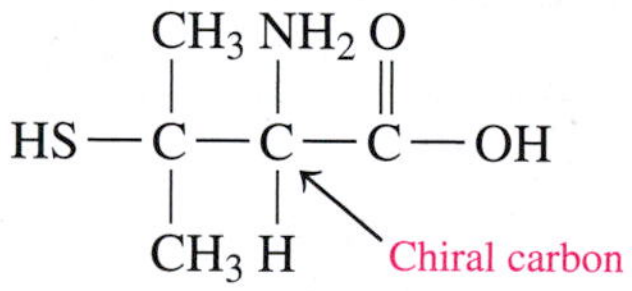
Chiral carbon

Penicillamine

b.

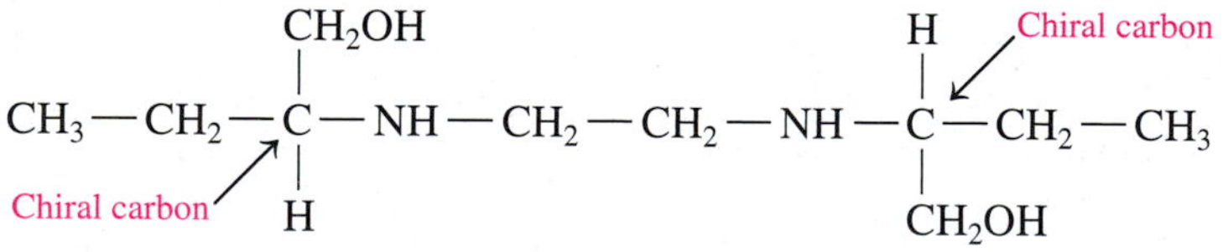
Chiral carbon
Chiral carbon

Ethambutol

TEST

Try Practice Problems 15.11, 15.12, 15.19, and 15.20

Drawing Fischer Projections

Emil Fischer devised a simplified system for drawing isomers that shows the arrangements of the atoms around the chiral carbons. Fischer received the Nobel Prize in Chemistry in 1902 for his contributions to carbohydrate and protein chemistry. Now we use his model, called a **Fischer projection**, to represent three-dimensional structures of enantiomers. Vertical lines represent bonds that project backward from a carbon atom, and horizontal lines represent bonds that project forward. In this model, the most highly oxidized carbon is placed at the top, and the intersections of vertical and horizontal lines represent a carbon atom that is usually chiral.

For the simplest aldose, glyceraldehyde, the only chiral carbon is the middle carbon. In the Fischer projection, the carbonyl group, which is the most highly oxidized group, is drawn at the top above the chiral carbon and the $-CH_2OH$ group is drawn at the bottom. The $-H$ and $-OH$ groups can be drawn at each end of a horizontal line, but in two different ways. The isomer that has the $-OH$ group drawn to the left of the chiral atom is designated as the L isomer. The isomer with the $-OH$ group drawn to the right of the chiral carbon represents the D isomer (see **FIGURE 15.6**).

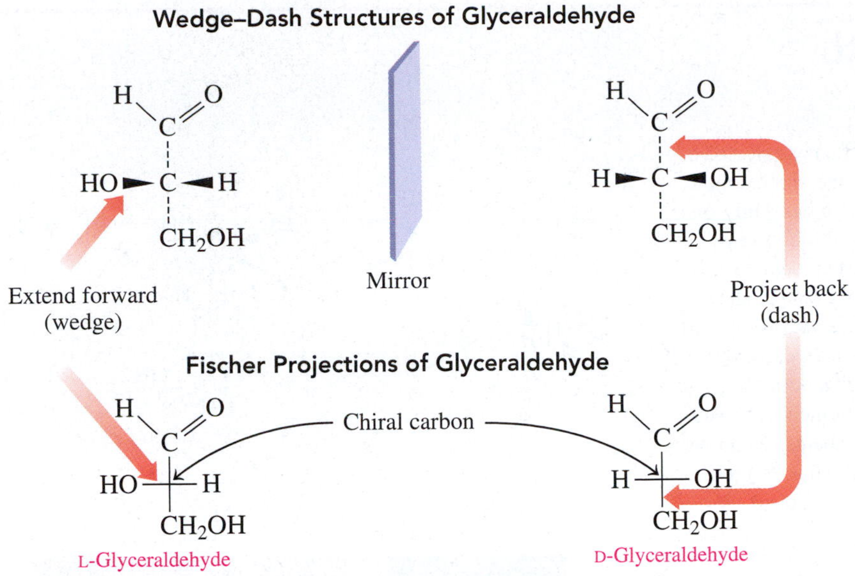
Wedge–Dash Structures of Glyceraldehyde
Mirror
Extend forward (wedge)
Project back (dash)
Fischer Projections of Glyceraldehyde
Chiral carbon
L-Glyceraldehyde
D-Glyceraldehyde

FIGURE 15.6 ▶ In the Fischer projection for glyceraldehyde, the chiral carbon atom is at the center, with horizontal lines for bonds that project forward and vertical lines for bonds that point away.

Q Why does glyceraldehyde have only one chiral carbon atom?

Fischer projections can also be drawn for compounds that have two or more chiral carbons. For example, in erythrose, both of the carbon atoms at the intersections are chiral. Then the designation as a D or L isomer is determined by the position of the —OH group attached to the chiral carbon *farthest from the carbonyl group*.

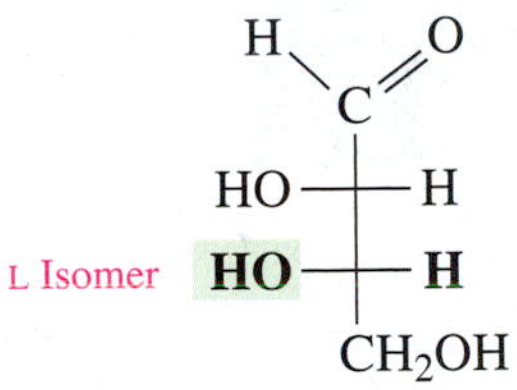
L Isomer

L-Erythrose

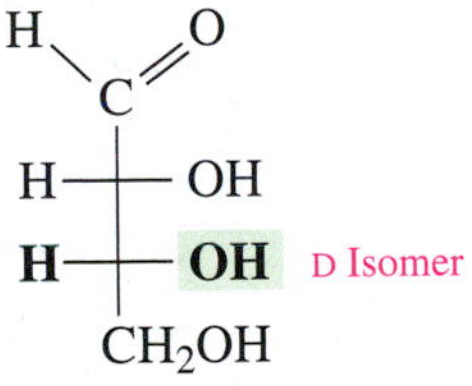
D Isomer

D-Erythrose

▶ SAMPLE PROBLEM 15.3 Fischer Projections

TRY IT FIRST

Identify each of the following as the D or L isomer:

a. 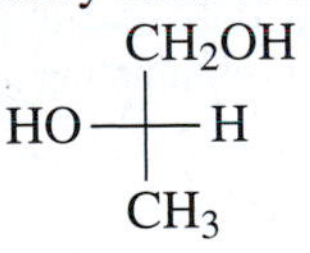

b. 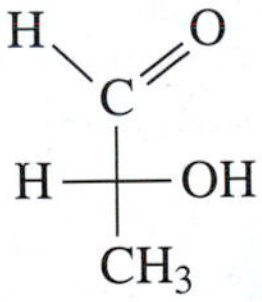

c. 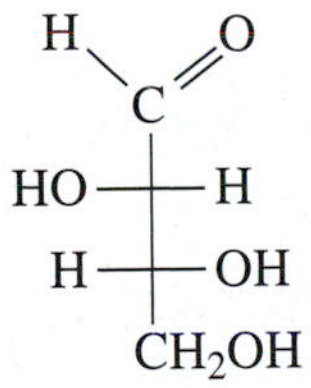

SOLUTION

a. When the —OH group is drawn to the left of the chiral carbon, it is the L isomer.

b. When the —OH group is drawn to the right of the chiral carbon, it is the D isomer.

c. When the —OH group on the chiral carbon farthest from the top of the Fischer projection is drawn to the right, it is the D isomer.

STUDY CHECK 15.3

Identify each of the following as the D or L isomer:

a. 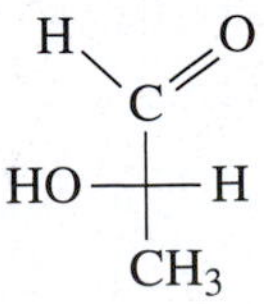

b. 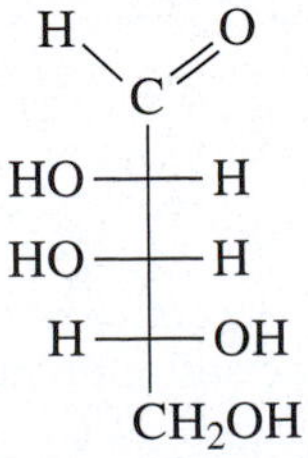

c. 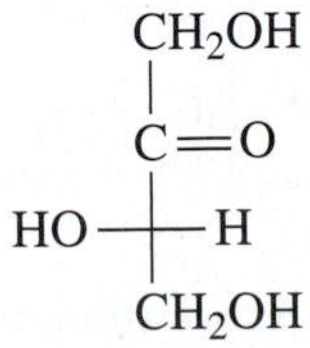

ANSWER

a. L isomer **b.** D isomer **c.** L isomer

TEST

Try Practice Problems 15.13 to 15.18

Chemistry Link to Health

Enantiomers in Biological Systems

Molecules in nature also have mirror images, and often one stereoisomer has a different biological effect than the other one. For example, one enantiomer of nicotine is more toxic than the other. Only one enantiomer of epinephrine causes the constriction of blood vessels.

For some compounds, one enantiomer has a certain odor, and the other enantiomer has a completely different odor. For example, the oils that give the scent of spearmint and caraway seeds are both composed of carvone. However, carvone has one chiral carbon in the carbon ring, indicated by an asterisk, which gives carvone two enantiomers.

Olfactory receptors in the nose detect these enantiomers as two different odors. One smells and tastes like spearmint, whereas its mirror image has the odor and taste of caraway in rye bread. Thus, our senses of smell and taste are responsive to the chirality of molecules.

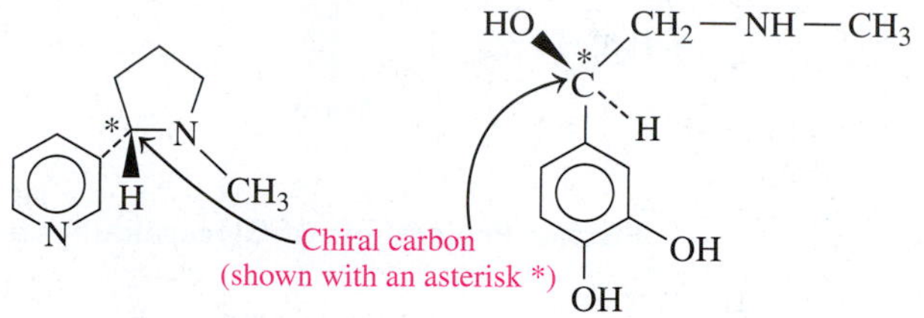
Chiral carbon
(shown with an asterisk *)

Nicotine

Adrenaline (epinephrine)

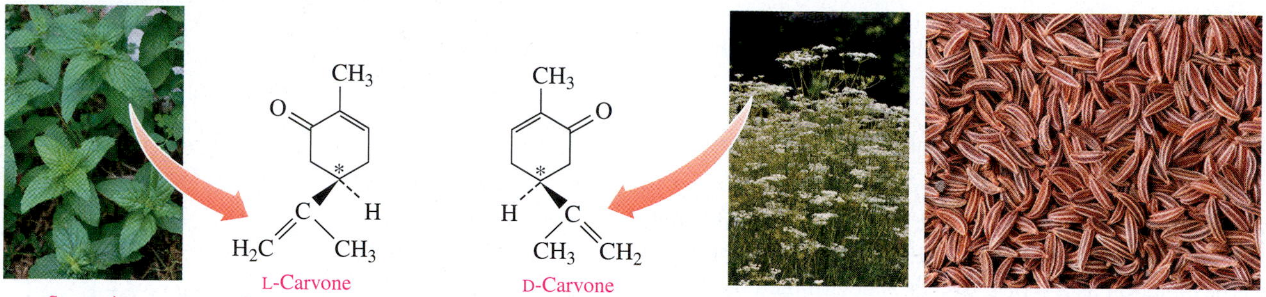

Spearmint

L-Carvone from spearmint

D-Carvone from caraway plant and seeds

Caraway plant and seeds

Many compounds in biological systems have only one active enantiomer because the enzymes and cell surface receptors on which metabolic reactions take place are themselves chiral. The chiral receptor fits the arrangement of the substituents in only one enantiomer; its mirror image does not fit properly (see **FIGURE 15.7**).

Today, drug researchers are using *chiral technology* to produce the active enantiomers of chiral drugs. Chiral catalysts are being designed that direct the formation of just one enantiomer rather than both. The active forms of several enantiomers are now being produced, such as naproxen and ibuprofen, which are nonsteroidal anti-inflammatory drugs used to relieve pain, fever, and inflammation caused by osteoarthritis and tendinitis. The benefits of producing only the active enantiomer include using a lower dose, enhancing activity, reducing interactions with other drugs, and eliminating possible harmful side effects from the enantiomer.

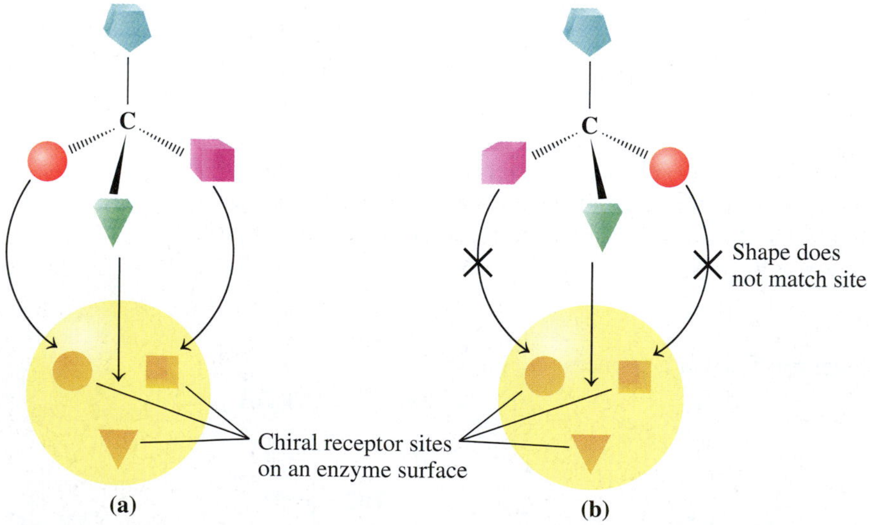
Shape does
not match site
Chiral receptor sites
on an enzyme surface
(a)
(b)

FIGURE 15.7 ▶ **(a)** The substituents on the biologically active enantiomer bind to all the sites on a chiral receptor; **(b)** its enantiomer does not bind properly and is not active biologically.

Q Why don't all the substituents of the mirror image of the active enantiomer fit into a chiral receptor site?

PRACTICE PROBLEMS

15.2 Chiral Molecules

15.11 Identify each of the following structures as chiral or achiral. If chiral, indicate the chiral carbon.

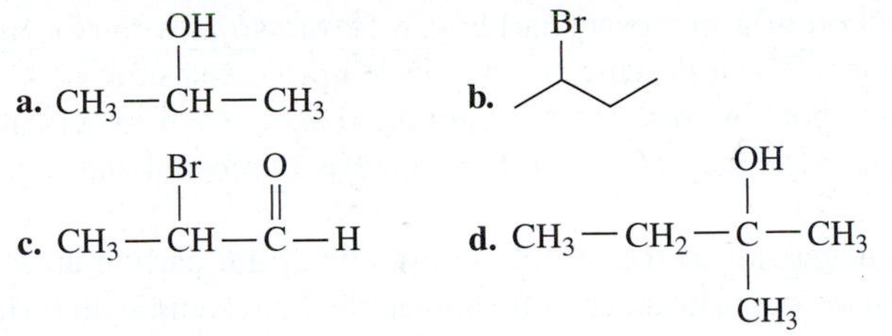

15.12 Identify each of the following structures as chiral or achiral. If chiral, indicate the chiral carbon.

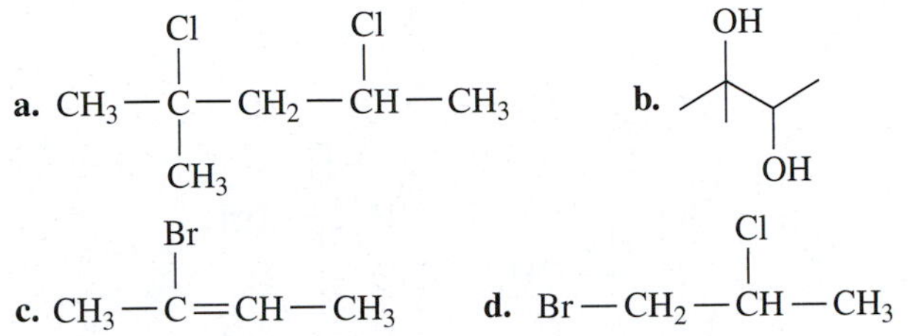

15.13 Draw the Fischer projection for each of the following wedge–dash structures:

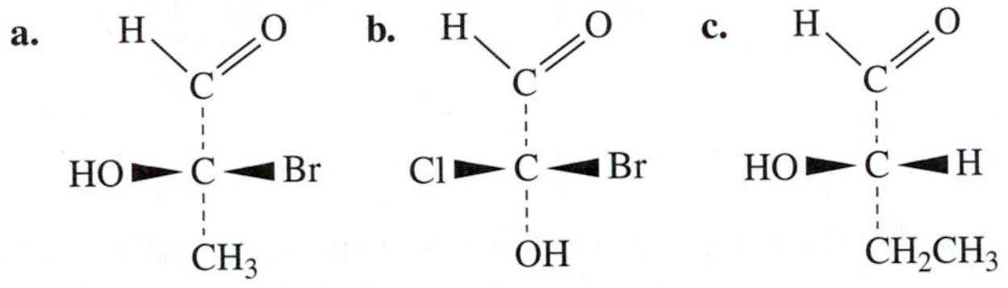

15.14 Draw the Fischer projection for each of the following wedge–dash structures:

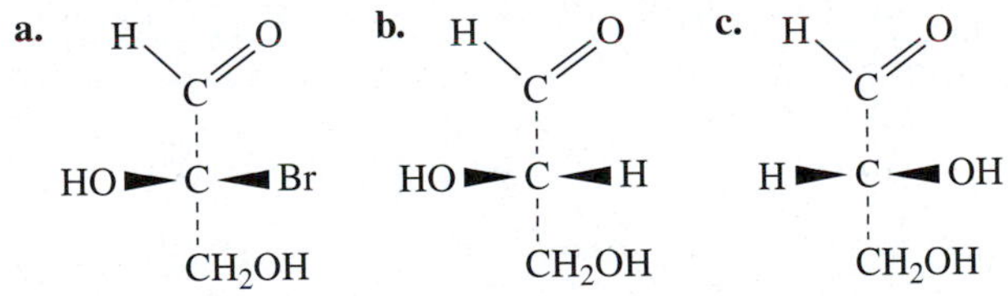

15.15 Indicate whether each pair of Fischer projections represents enantiomers or identical structures.

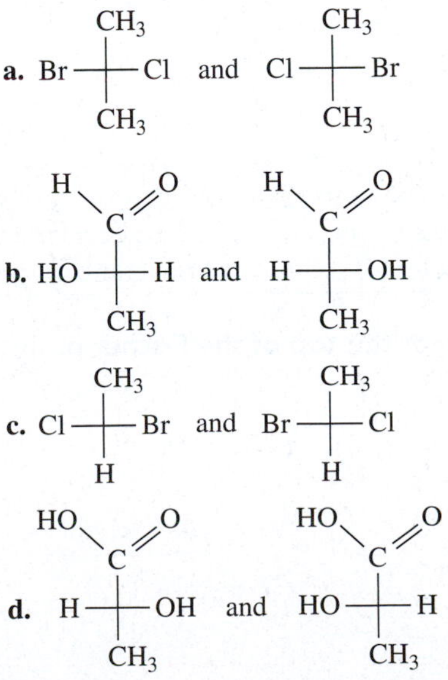

15.16 Indicate whether each pair of Fischer projections represents enantiomers or identical structures.

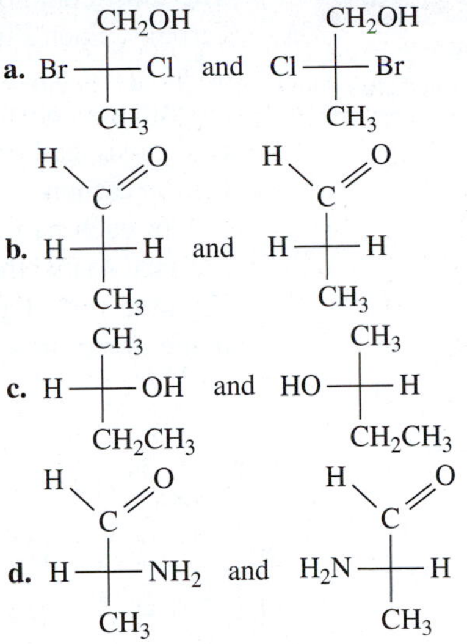

15.17 Identify each of the following as D or L:

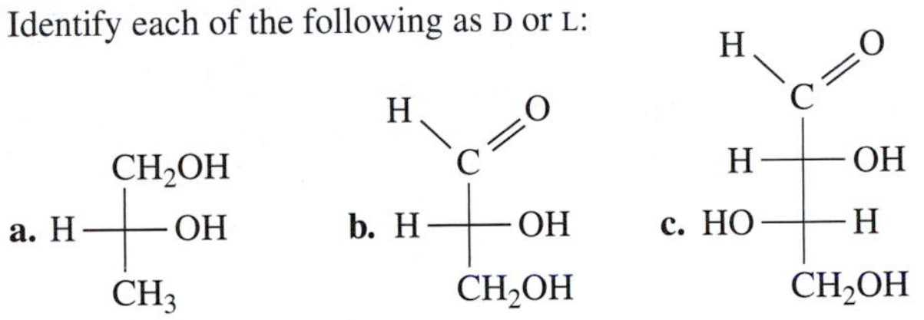

15.18 Identify each of the following as D or L:

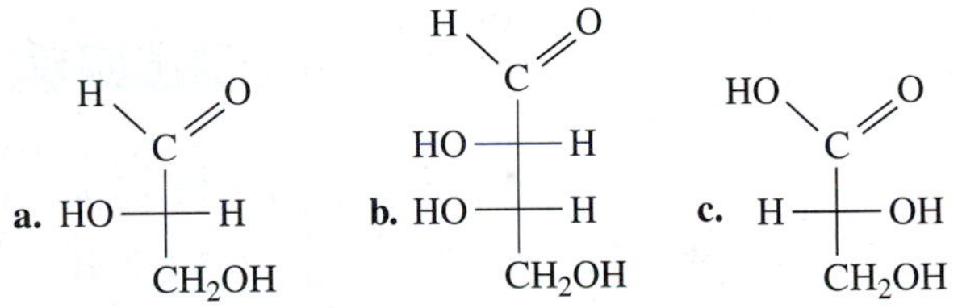

Clinical Applications

15.19 Identify the chiral carbon in each of the following compounds:

a. citronellol; one enantiomer has the odor of geranium

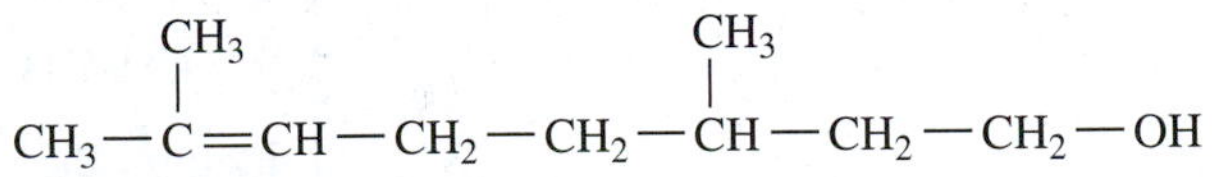

b. alanine, an amino acid

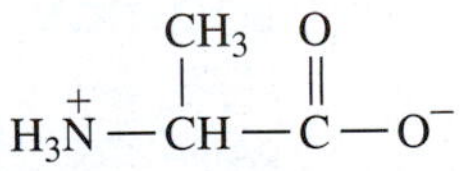

15.20 Identify the chiral carbon in each of the following compounds:

a. amphetamine (Benzedrine), stimulant, used in the treatment of hyperactivity

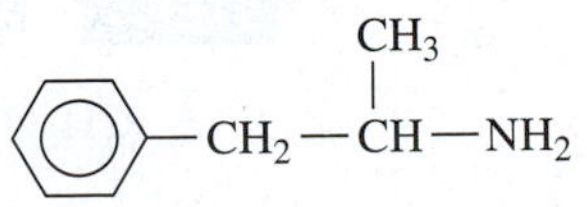

b. norepinephrine, increases blood pressure and nerve transmission

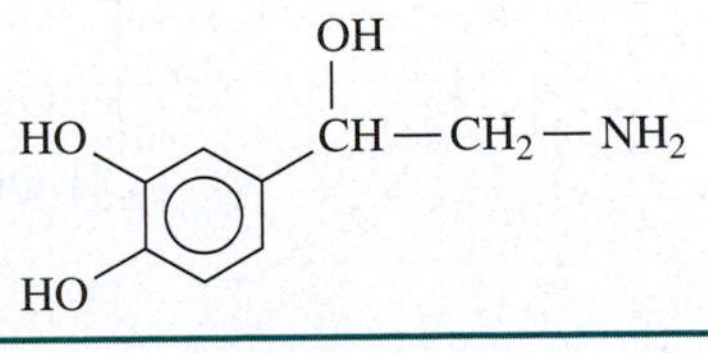

15.3 Fischer Projections of Monosaccharides

LEARNING GOAL Identify or draw the D and L configurations of the Fischer projections for common monosaccharides.

CORE CHEMISTRY SKILL
Identifying D and L Fischer Projections for Carbohydrates

The most common monosaccharides contain five or six carbon atoms with several chiral carbons. Each Fischer projection of a monosaccharide can be drawn as a mirror image, which gives a pair of enantiomers. The following are the Fischer projections for the D and L enantiomers of ribose, a five-carbon monosaccharide, and the D and L enantiomers of glucose, a six-carbon monosaccharide. The vertical carbon chain is numbered starting from the top carbon.

In each pair of mirror images, all of the —OH groups on chiral carbon atoms are reversed so that they appear on the opposite sides of the molecule. For example, in L-ribose, all of the —OH groups drawn on the left side of the vertical line are drawn on the right side in the mirror image D-ribose.

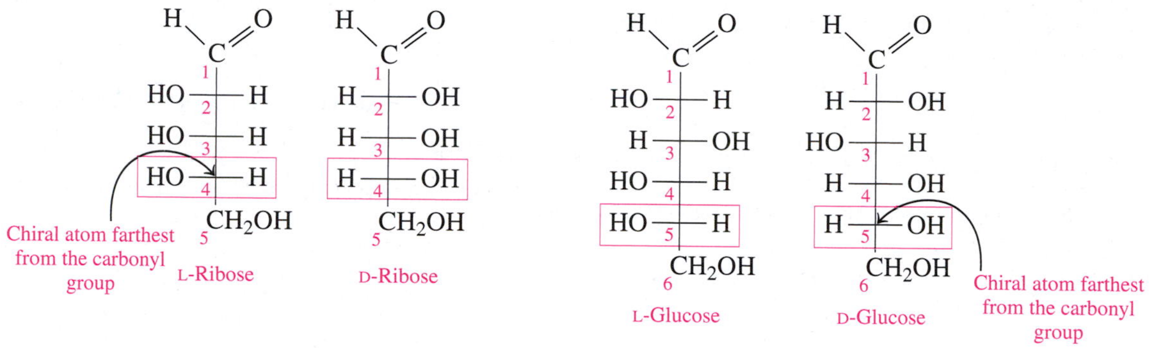
Chiral atom farthest from the carbonyl group
L-Ribose
D-Ribose
L-Glucose
D-Glucose
Chiral atom farthest from the carbonyl group

▶ SAMPLE PROBLEM 15.4 Fischer Projections for Monosaccharides

TRY IT FIRST

Ribulose, which is used in various brands of artificial sweeteners, has the following Fischer projection:

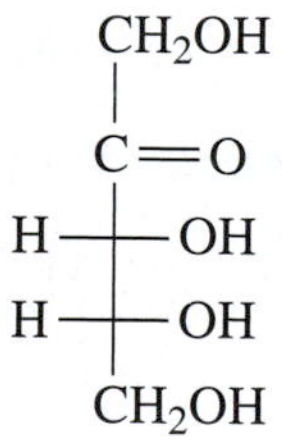

Identify the compound as D- or L-ribulose.

SOLUTION

ANALYZE THE PROBLEM	Given	Need	Connect
	Fischer projection of ribulose	identify as D- or L-ribulose	chiral carbon farthest from carbonyl group

STEP 1 Number the carbon chain starting at the top of the Fischer projection.

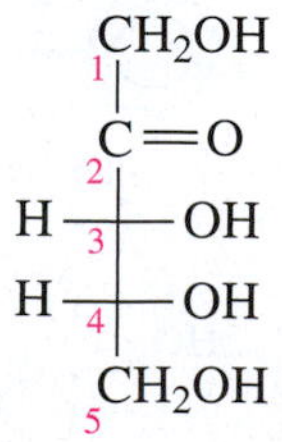

STEP 2 **Locate the chiral carbon farthest from the top of the Fischer projection.** The chiral carbon farthest from the top is carbon 4.

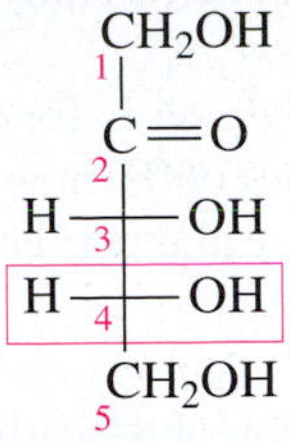

STEP 3 **Identify the position of the —OH group as D or L.** In this Fischer projection, the —OH group is on the right of carbon 4, which makes it D-ribulose.

ENGAGE

Which of the chiral carbon atoms in a carbohydrate determines the D or L isomer?

STUDY CHECK 15.4

a. Draw and name the Fischer projection for the mirror image of the ribulose in Sample Problem 15.4.

b. The structure of one of the enantiomers of the carbohydrate mannose is shown below. Draw and name the Fischer projection for its mirror image.

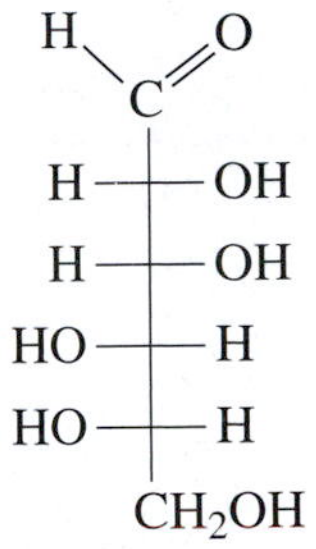

ANSWER

a.

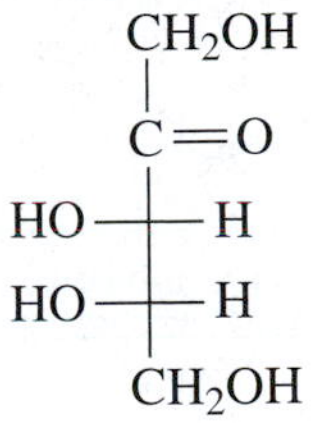

L-Ribulose

b.

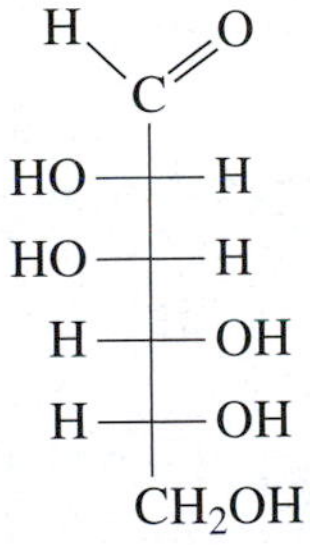

D-Mannose

INTERACTIVE VIDEO

PEARSON eText 2.0 Fischer Projections of Monosaccharides

TEST

Try Practice Problems 15.21 to 15.28

Some Important Monosaccharides

D-Glucose, D-galactose, and D-fructose are important monosaccharides. They are all hexoses with the molecular formula $C_6H_{12}O_6$ and are isomers of each other. Although we can draw Fischer projections for their D and L enantiomers, the D enantiomers are more commonly found in nature and used in the cells of the body. The Fischer projections for the D enantiomers are drawn as follows:

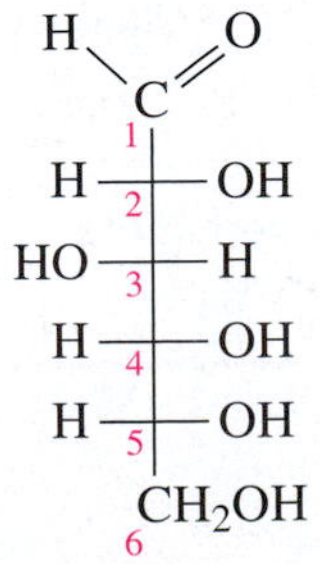

D-Glucose

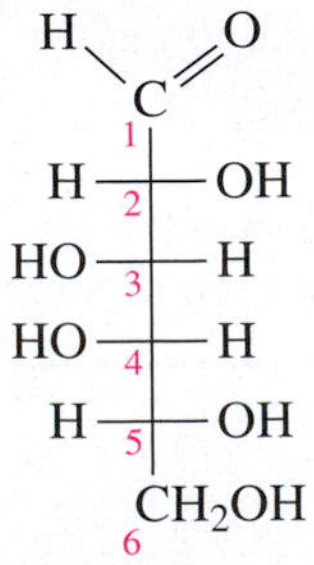

D-Galactose

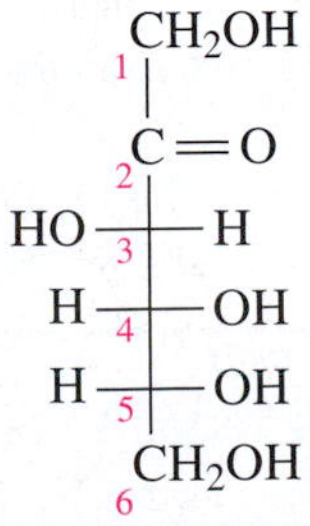

D-Fructose

The sweet taste of honey is due to the monosaccharides D-glucose and D-fructose.

ENGAGE

What are some differences in the Fischer projections of D-glucose, D-galactose, and D-fructose?

TEST

Try Practice Problems 15.29 and 15.30

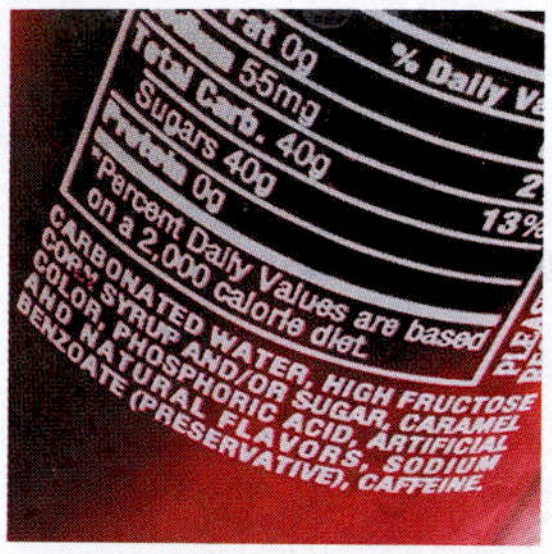

% Daily Va
Total Carb. 40g
Sugars 40g
13%
Percent Daily Values are based on a 2,000 calorie diet.
CARBONATED WATER, HIGH FRUCTOSE CORN SYRUP AND/OR SUGAR, CARAMEL COLOR, PHOSPHORIC ACID, ARTIFICIAL AND NATURAL FLAVORS, SODIUM BENZOATE (PRESERVATIVE), CAFFEINE.

A food label shows that high-fructose corn syrup is the sweetener in this product.

The most common hexose, **D-glucose**, $C_6H_{12}O_6$, also known as dextrose and blood sugar, is found in fruits, vegetables, corn syrup, and honey. D-Glucose is a building block of the disaccharides sucrose, lactose, and maltose, and polysaccharides such as amylose, cellulose, and glycogen.

D-Galactose, $C_6H_{12}O_6$, is obtained from the disaccharide lactose, which is found in milk and milk products. D-Galactose is important in the cellular membranes of the brain and nervous system. The only difference in the Fischer projections of D-glucose and D-galactose is the arrangement of the —OH group on carbon 4.

In contrast to D-glucose and D-galactose, **D-fructose**, $C_6H_{12}O_6$, is a ketohexose. The structure of D-fructose differs from glucose at carbons 1 and 2 by the location of the carbonyl group. D-Fructose is the sweetest of the carbohydrates, almost twice as sweet as sucrose (table sugar). This characteristic makes D-fructose popular with dieters because less fructose, and therefore fewer calories, is needed to provide a pleasant taste. D-Fructose, also called levulose and fruit sugar, is found in fruit juices and honey.

D-Fructose is also obtained as one of the hydrolysis products of sucrose, the disaccharide known as table sugar. High-fructose corn syrup (HFCS) is a sweetener produced when enzymes convert the glucose in corn syrup to fructose. When the fructose is mixed with corn syrup containing only glucose, the sweetener HFCS is produced. HFCS with 42% fructose is used in bakery goods, and HFCS with 55% fructose is used in soft drinks.

Chemistry Link to Health

Hyperglycemia and Hypoglycemia

Kate's doctor ordered an oral glucose tolerance test (OGTT) to evaluate her body's ability to return to normal glucose concentrations (70 to 99 mg/dL) in response to the ingestion of a specified amount of glucose (dextrose). After Kate fasts for 12 h, she drinks a solution containing 75 g of glucose. A blood sample is taken immediately, followed by more blood samples each half-hour for 2 h, and then every hour for a total of 5 h. After her test, Kate is told that her blood glucose was 178 mg/dL, which indicates hyperglycemia. The prefix *hyper* means above or over; *hypo* means below or under. The term *glyc* or *gluco* refers to "sugar." Thus, the blood sugar level in *hyperglycemia* is above normal and in *hypoglycemia* it is below normal.

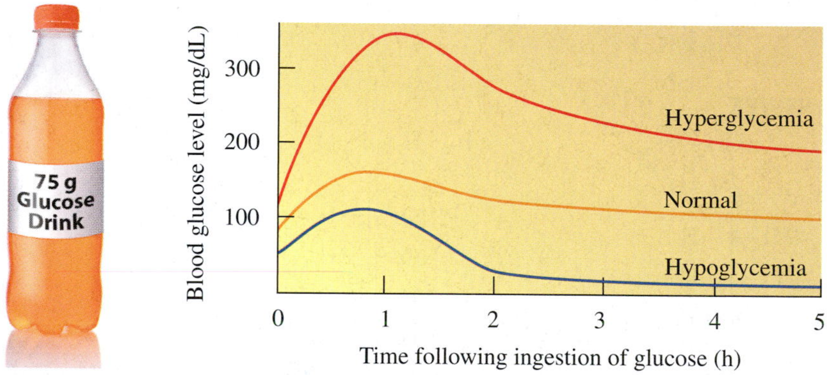

75 g
Glucose
Drink
Blood glucose level (mg/dL)
300
200
100
Hyperglycemia
Normal
Hypoglycemia
0
1
2
3
4
5
Time following ingestion of glucose (h)

A glucose solution is given to determine blood glucose levels.

A disease that can cause hyperglycemia is *type 2 diabetes*, which occurs when the pancreas is unable to produce sufficient quantities of insulin. As a result, glucose levels in the body fluids can rise as high as 350 mg/dL. Kate's symptoms of type 2 diabetes include thirst, excessive urination, and increased appetite. In older adults, type 2 diabetes is sometimes a consequence of excessive weight gain.

When a person is hypoglycemic, the blood glucose level rises and then decreases rapidly to levels as low as 40 mg/dL. In some cases, hypoglycemia is caused by overproduction of insulin by the pancreas. Low blood glucose can cause dizziness, general weakness, and muscle tremors. A diet may be prescribed that consists of several small meals high in protein and low in carbohydrate. Some hypoglycemic patients are finding success with diets that include more complex carbohydrates rather than simple sugars.

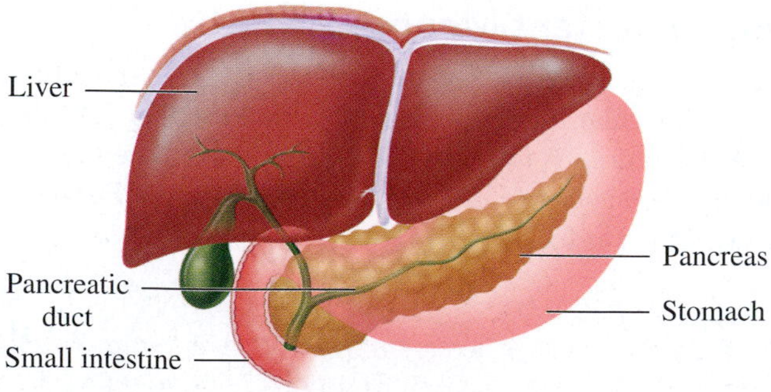

Liver
Pancreatic duct
Small intestine
Pancreas
Stomach

The insulin produced in the pancreas is needed in the digestive system for the metabolism of glucose.

PRACTICE PROBLEMS

15.3 Fischer Projections of Monosaccharides

15.21 Identify each of the following as the D or L enantiomer:

a.

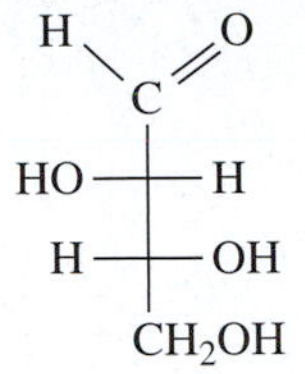

Threose

b.

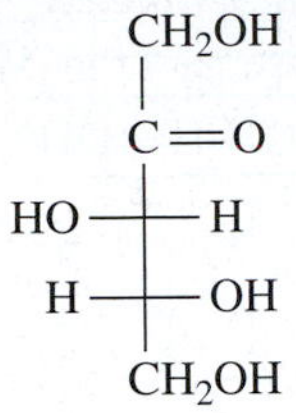

Xylulose

c.

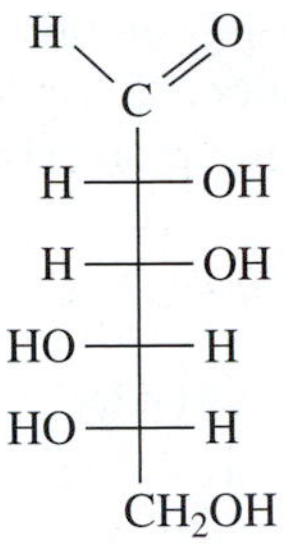

Mannose

d.

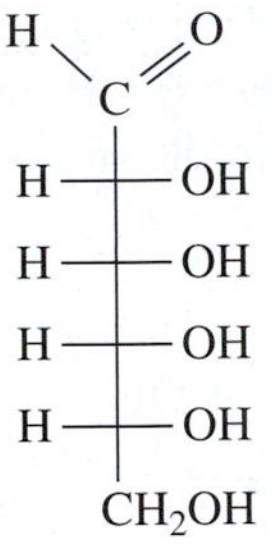

Allose

15.22 Identify each of the following as the D or L enantiomer:

a.

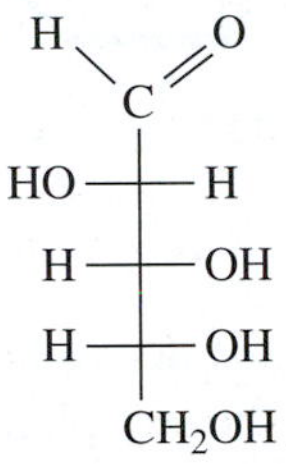

Arabinose

b.

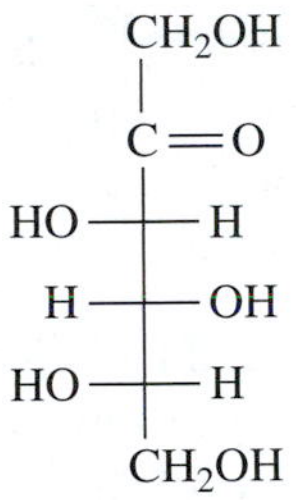

Sorbose

c.

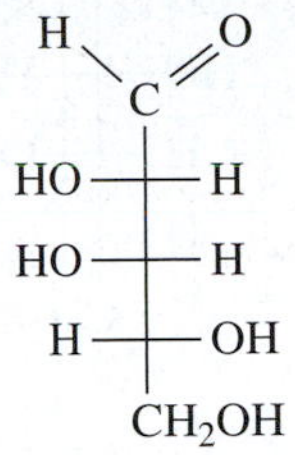

Lyxose

d.

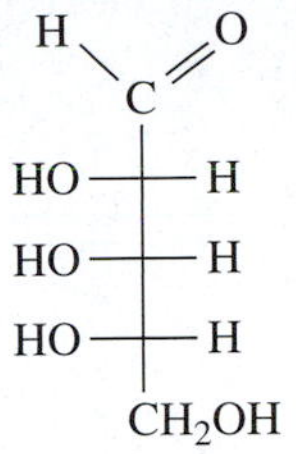

Ribose

15.23 Draw the Fischer projection for the other enantiomer of **a** to **d** in problem 15.21.

15.24 Draw the Fischer projection for the other enantiomer of **a** to **d** in problem 15.22.

15.25 Draw the Fischer projections for D-glucose and L-glucose.

15.26 Draw the Fischer projections for D-fructose and L-fructose.

Clinical Applications

15.27 An infant with galactosemia can utilize D-glucose in milk but not D-galactose. How does the Fischer projection of D-galactose differ from that of D-glucose?

15.28 D-Fructose is the sweetest monosaccharide. How does the Fischer projection of D-fructose differ from that of D-glucose?

15.29 Identify the monosaccharide that fits each of the following descriptions:

a. is also called blood sugar

b. is not metabolized in a condition known as galactosemia

15.30 Identify a monosaccharide that fits each of the following descriptions:

a. found in high blood levels in diabetes

b. is also called fruit sugar

15.4 Haworth Structures of Monosaccharides

LEARNING GOAL Draw and identify the Haworth structures for monosaccharides.

Up to now, we have drawn the Fischer projections for monosaccharides as open chains. However, the most stable form of pentoses and hexoses are five- or six-atom rings. These rings are produced from the reaction of a carbonyl group and a hydroxyl group in the *same* molecule and are represented by **Haworth structures**. We will now show how to draw the Haworth structure for D-glucose from its Fischer projection.

CORE CHEMISTRY SKILL
Drawing Haworth Structures

Drawing Haworth Structures

To convert a Fischer projection to a Haworth structure, turn the Fischer projection clockwise by 90°. The —H and —OH groups on the right of the vertical carbon chain are now below the horizontal carbon chain. Those on the left of the open chain are now above the horizontal carbon chain.

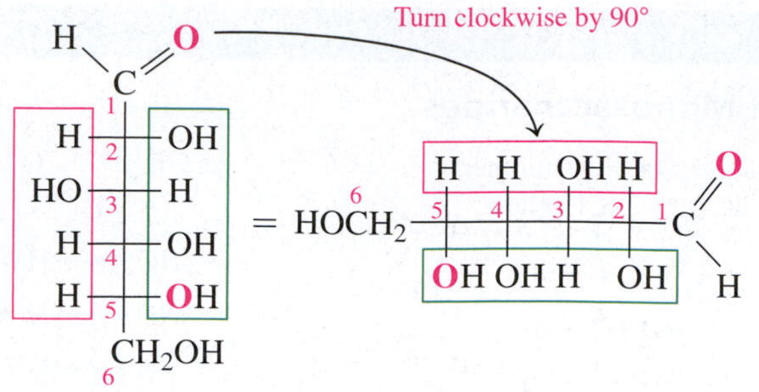
Turn clockwise by 90°
H
O
C
1
H
OH
2
HO
H
3
H
OH
4
H
OH
5
CH2OH
6
= HOCH2
6
H H OH H
5 4 3 2 1
C
O
OH OH H OH
H

D-Glucose (open chain)

ENGAGE

Why is an —OH group drawn on carbon 1 in the Haworth structure of an aldohexose?

With carbons 2 and 3 as the base of a hexagon, move the remaining carbons upward. Rotate the groups on carbon 5 so that the —OH group is close to carbon 1. To complete the Haworth structure, draw a bond from the oxygen of the —OH group on carbon 5 to carbon 1.

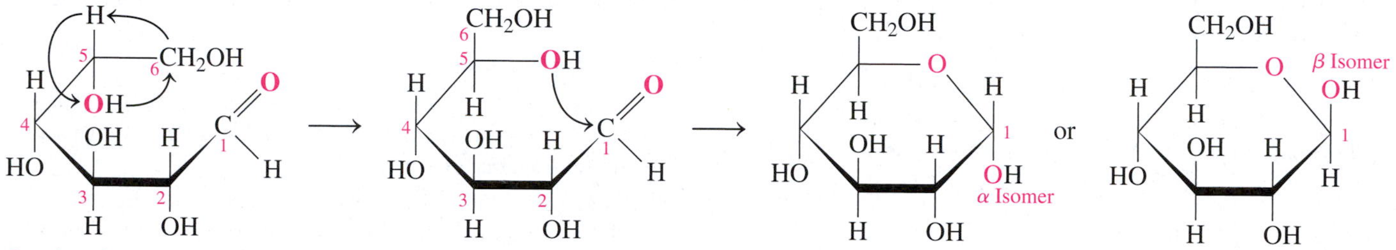
H
5
6
CH2OH
H
OH
O
4
OH
H
C
1
HO
H
3
2
H
OH
Rotation of groups on carbon 5
CH2OH
6
5
OH
O
H
C
1
Oxygen on carbon 5 bonds to carbon 1
O
α Isomer
α-D-Glucose
or
β Isomer
β-D-Glucose

Because the —OH group on the new chiral carbon (1) can be drawn above or below the plane of the Haworth structure, there are two isomers of D-glucose. In the α (alpha) isomer, the —OH group is drawn below the plane of the ring. In the β (beta) isomer, the —OH group is drawn above the plane of the ring. The carbon atoms in the ring are drawn as corners.

Mutarotation of α- and β-D-Glucose

In an aqueous solution, the Haworth structure of α-D-glucose opens to give the open chain of D-glucose, which has an aldehyde group. At any given time, there is only a trace amount of the open chain because it closes quickly to form a stable cyclic structure. However, when the open chain closes again, it can form β-D-glucose. In this process called *mutarotation*, each isomer converts to the open chain and back again. As the ring opens and closes, the —OH group on carbon 1 can form either the α or β isomer. An aqueous glucose solution contains a mixture of 36% α-D-glucose and 64% β-D-glucose.

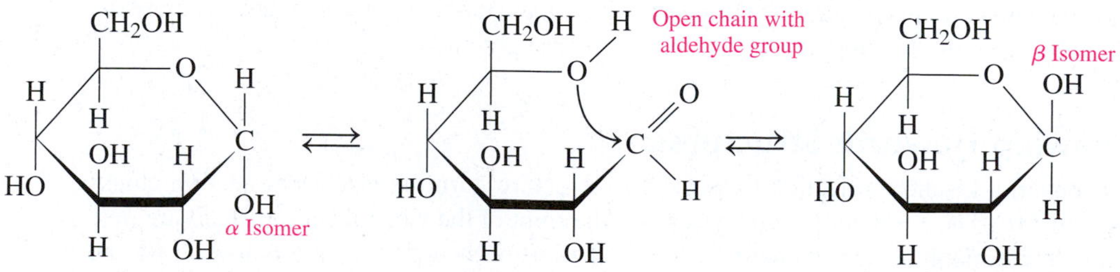
CH2OH
O
H
C
OH
α Isomer
Open chain with
aldehyde group
β Isomer

α-D-Glucose (36%)

D-Glucose open chain (trace)

β-D-Glucose (64%)

Haworth Structures of Galactose

Galactose is an aldohexose that differs from glucose only in the arrangement of the —OH group on carbon 4. Thus, its Haworth structure is similar to that of glucose, except that the —OH group on carbon 4 is drawn above the ring. Galactose also exists as α and β isomers.

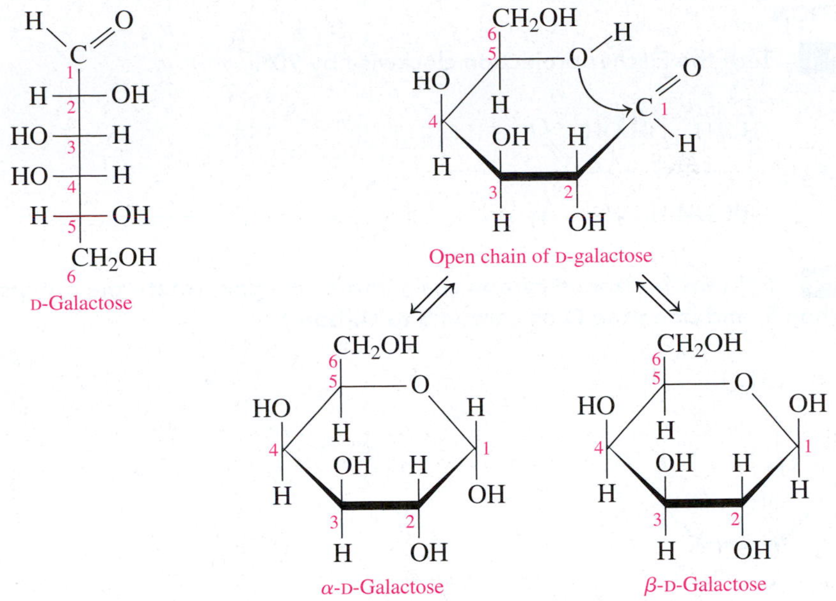
D-Galactose
Open chain of D-galactose
α-D-Galactose
β-D-Galactose

Haworth Structures of Fructose

In contrast to glucose and galactose, fructose is a ketohexose. The Haworth structure for fructose is a five-atom ring with carbon 2 at the right corner of the ring. The cyclic structure forms when the —OH group on carbon 5 bonds to carbon 2 in the carbonyl group. The new —OH group on carbon 2 gives the α and β isomers of fructose.

INTERACTIVE VIDEO

PEARSON eText 2.0 Haworth Structures of Monosaccharides

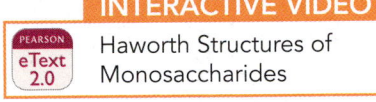

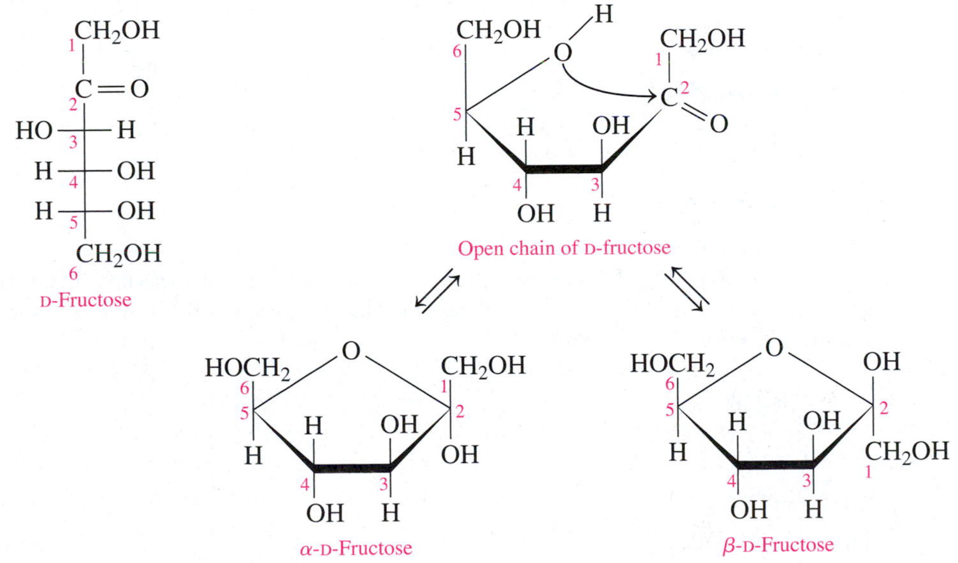
D-Fructose
Open chain of D-fructose
α-D-Fructose
β-D-Fructose

▸SAMPLE PROBLEM 15.5 Drawing Haworth Structures for Sugars

TRY IT FIRST

D-Mannose, a carbohydrate found in immunoglobulins, has the following Fischer projection. Draw the Haworth structure for β-D-mannose.

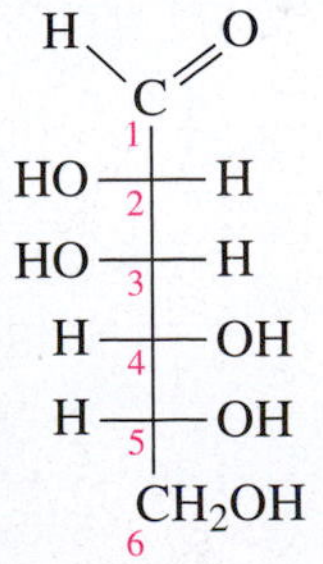

D-Mannose

ENGAGE

If a hydroxyl group on carbon 3 is drawn to the left in the Fischer projection for a sugar, will the hydroxyl group appear above or below the ring in the Haworth structure for the sugar?

SOLUTION

STEP 1 Turn the Fischer projection clockwise by 90°.

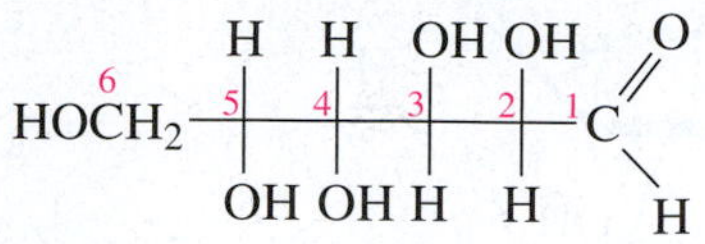

H H OH OH
O
6
HOCH2
5 4 3 2 1
C
OH OH H H
H

STEP 2 Fold the horizontal carbon chain into a hexagon, rotate the groups on carbon 5, and bond the O on carbon 5 to carbon 1.

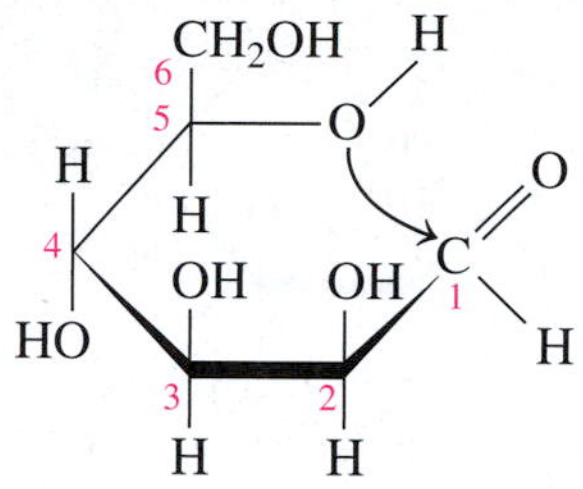

CH2OH
H
6
5
O
H
O
4
H
OH
OH
C
1
HO
H
3
2
H
H

STEP 3 Draw the new —OH group on carbon 1 above the ring to give the β isomer.

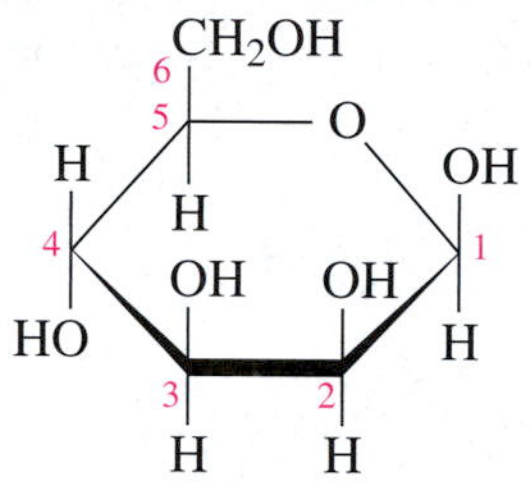

CH2OH
6
5
O
H
OH
H
4
1
OH
OH
HO
H
3
2
H
H

STUDY CHECK 15.5

a. Draw the Haworth structure for α-D-mannose.

b. The Fischer projection of D-sorbose, a monosaccharide with a sweetness level that is similar to that of sucrose, is given below. Draw the Haworth structures for both α-D-sorbose and β-D-sorbose.

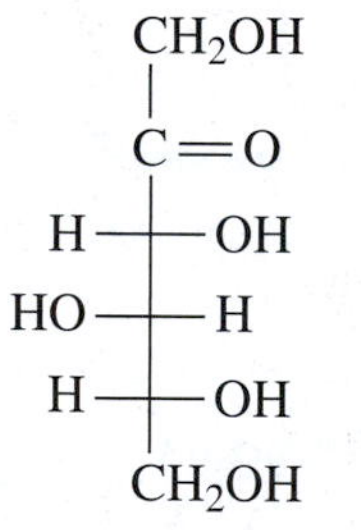

CH2OH
C=O
H OH
HO H
H OH
CH2OH

D-Sorbose

ANSWER

a.

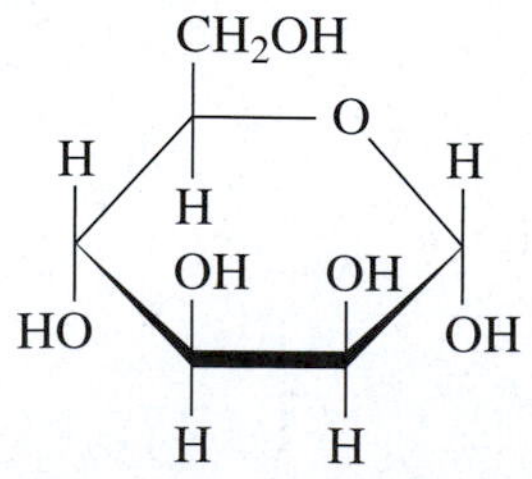

CH2OH
O
H
H
H
OH
OH
HO
OH
H
H

b.

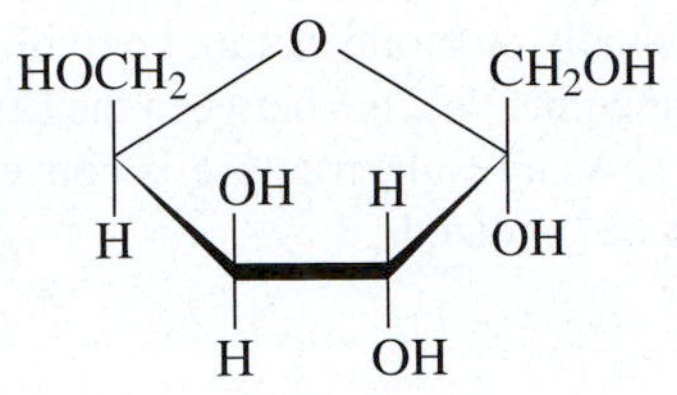

α-D-Sorbose

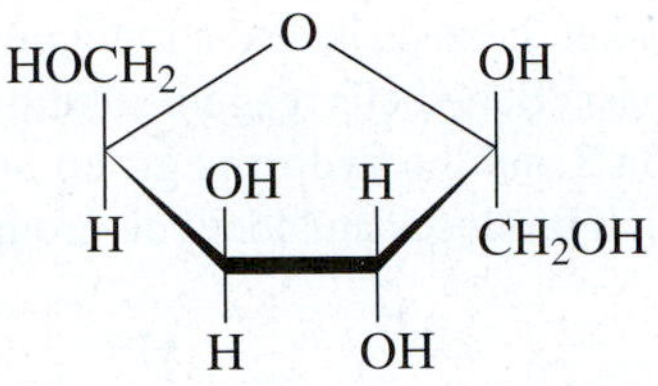

β-D-Sorbose

TEST

Try Practice Problems 15.31 to 15.36

PRACTICE PROBLEMS

15.4 Haworth Structures of Monosaccharides

15.31 What are the kind and number of atoms in the ring portion of the Haworth structure of glucose?

15.32 What are the kind and number of atoms in the ring portion of the Haworth structure of fructose?

15.33 Draw the Haworth structures for α- and β-D-glucose.

15.34 Draw the Haworth structures for α- and β-D-fructose.

15.35 Identify each of the following as the α or β isomer:

a.

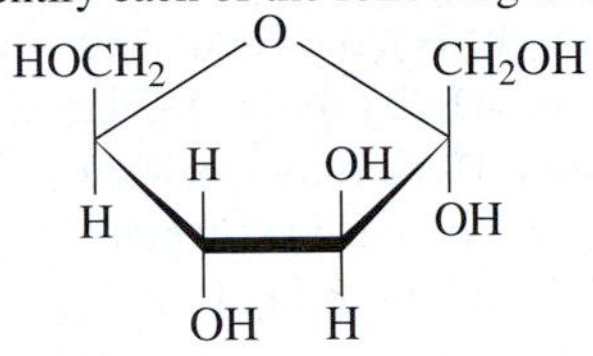

b.

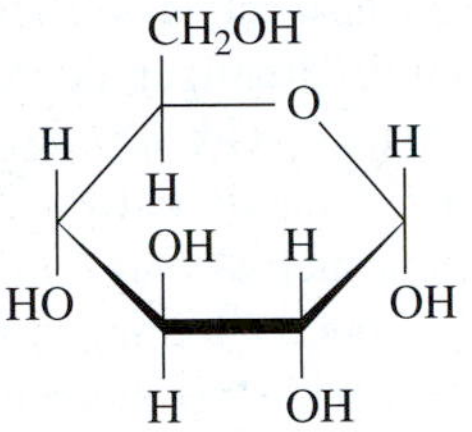

15.36 Identify each of the following as the α or β isomer:

a.

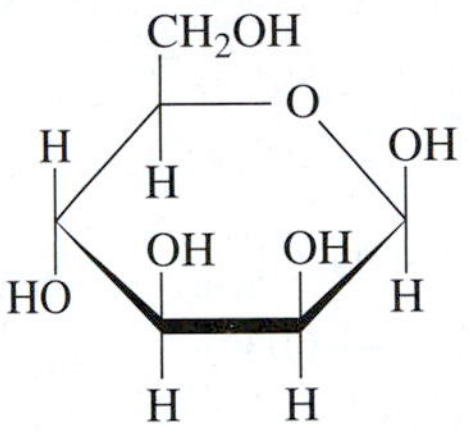

b.

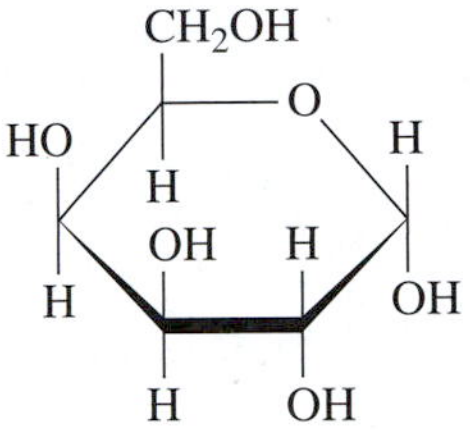

15.5 Chemical Properties of Monosaccharides

REVIEW

Writing Equations for the Oxidation of Alcohols (13.4)

LEARNING GOAL Identify the products of oxidation or reduction of monosaccharides; determine if a carbohydrate is a reducing sugar.

Monosaccharides contain functional groups that can undergo chemical reactions. In an aldose, the aldehyde group can be oxidized to a carboxylic acid. The carbonyl group in both an aldose and a ketose can be reduced to give a hydroxyl group. The hydroxyl groups can react with other compounds to form a variety of derivatives that are important in biological structures.

Oxidation of Monosaccharides

Although monosaccharides exist mostly in cyclic forms, a small amount of the open-chain form is always present. The open-chain forms can participate in chemical reactions. For example, the aldehyde group in aldose monosaccharides can be oxidized to a carboxylic acid (a "sugar acid") by an oxidizing agent such as Benedict's reagent. The sugar acids are named by replacing the *ose* ending of the monosaccharide with *onic acid*. The sugar reduces the Cu^{2+} in the Benedict's reagent to Cu^{+}, which forms a brick-red precipitate of Cu_2O. A carbohydrate, such as the glucose shown below, that reduces another substance is called a **reducing sugar**.

ENGAGE

When is a monosaccharide called a reducing sugar?

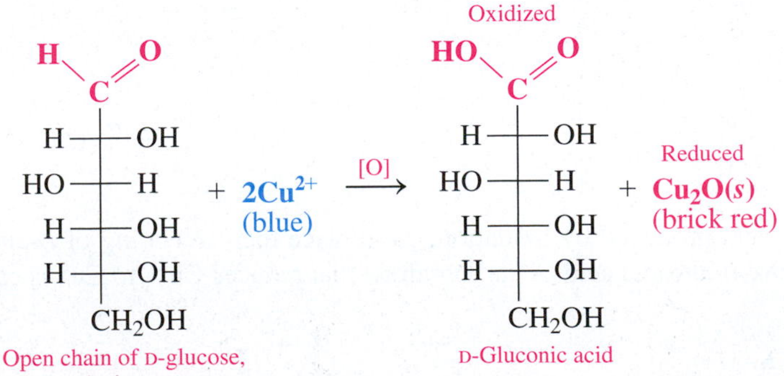

Oxidized
+ 2Cu2+ (blue)
[O]
Reduced
+ Cu2O(s) (brick red)

Open chain of D-glucose, a reducing sugar

D-Gluconic acid

Benedict's Reagent (no glucose)

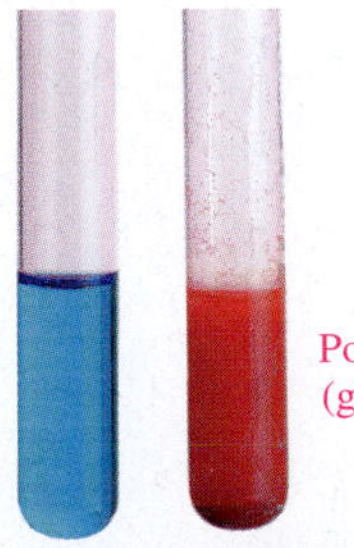

Positive Reaction (glucose present)

Benedict's reagent gives a brick-red precipitate with reducing sugars.

Fructose, a ketohexose, is also a reducing sugar. Usually a ketone cannot be oxidized. However, in a basic Benedict's reagent solution, a rearrangement occurs between the ketone group on carbon 2 and the hydroxyl group on carbon 1. As a result, fructose is converted to glucose, which produces an aldehyde group that can be oxidized.

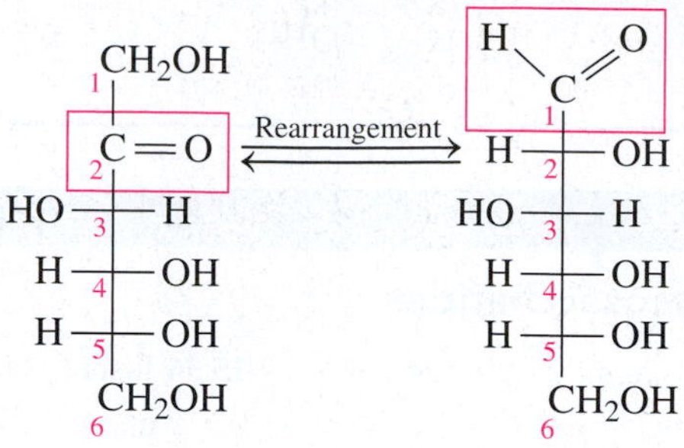
Rearrangement

D-Fructose (ketose) D-Glucose (aldose)

Reduction of Monosaccharides

The reduction of the carbonyl group in monosaccharides produces sugar alcohols, which are also called *alditols*. D-Glucose is reduced to D-glucitol, better known as D-sorbitol. The sugar alcohols are named by replacing the *ose* ending of the monosaccharide with *itol*. Sugar alcohols such as D-sorbitol, D-xylitol from D-xylose, and D-mannitol from D-mannose are used as sweeteners in many sugar-free products such as diet drinks and sugarless gum, as well as products for people with diabetes. Some people experience discomfort such as gas and diarrhea from the ingestion of sugar alcohols. Accumulation of D-sorbitol in the lens of the eye can cause cataracts in diabetics.

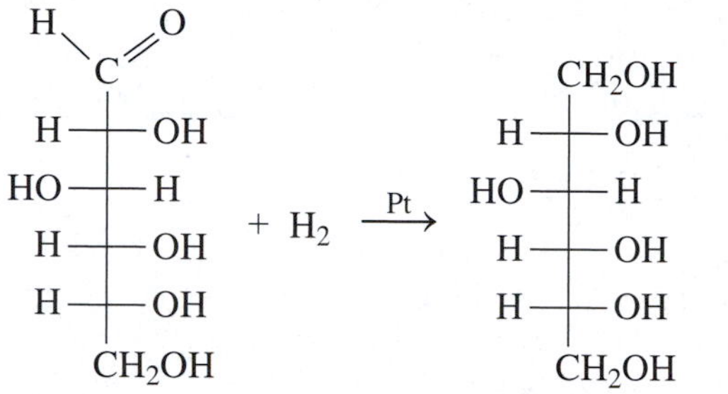

D-Glucose D-Glucitol or D-Sorbitol

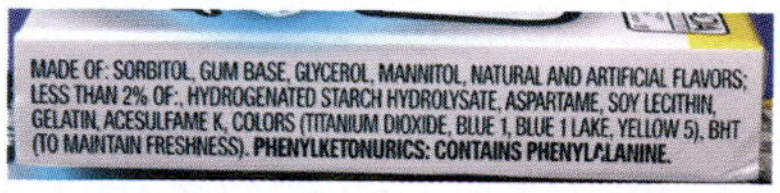
MADE OF: SORBITOL, GUM BASE, GLYCEROL, MANNITOL, NATURAL AND ARTIFICIAL FLAVORS; LESS THAN 2% OF:, HYDROGENATED STARCH HYDROLYSATE, ASPARTAME, SOY LECITHIN, GELATIN, ACESULFAME K, COLORS (TITANIUM DIOXIDE, BLUE 1, BLUE 1 LAKE, YELLOW 5), BHT (TO MAINTAIN FRESHNESS). PHENYLKETONURICS: CONTAINS PHENYLALANINE.

D-glucitol or D-sorbitol is used as a low-calorie sweetener in diet drinks and sugarless gum.

▶SAMPLE PROBLEM 15.6 Reducing Sugars

TRY IT FIRST

Draw and name the oxidation product of D-altrose. Why is D-altrose a reducing sugar?

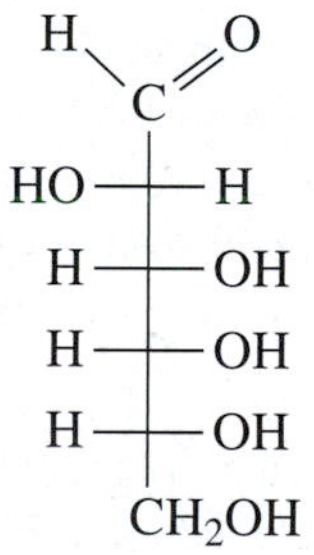

D-Altrose

SOLUTION

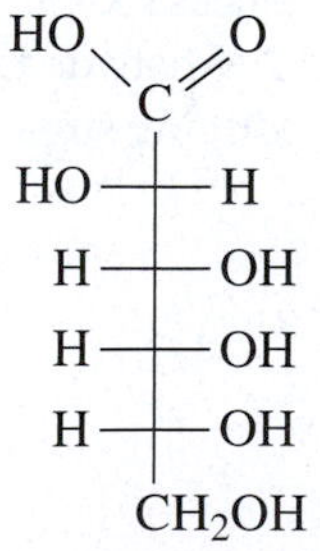

D-Altronic acid

To name the sugar acid produced by oxidation, we replace the *ose* ending of D-altrose with *onic acid* to give D-altronic acid. A carbohydrate that reduces Cu^{2+} to Cu^{+} is called a reducing sugar.

STUDY CHECK 15.6

a. Draw and name the reduction product of D-altrose.

b. A solution containing Benedict's reagent turns brick red with a urine sample. What might this result indicate?

ANSWER

a.

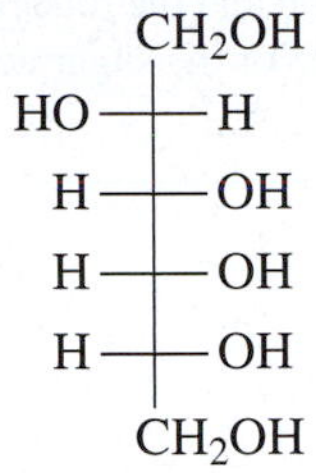

D-Altritol

b. The brick-red color of the Benedict's reagent shows a high level of reducing sugar (probably glucose) in the urine, which may indicate type 2 diabetes.

TEST

Try Practice Problems 15.37 to 15.40

Chemistry Link to Health

Dental Cavities and Xylitol Gum

According to the U.S. Centers for Disease Control and Prevention (CDC), *dental caries* (cavities) is the most common chronic disease in children and adolescents. It also affects adults, with 9 out of 10 adults having some form of tooth decay.

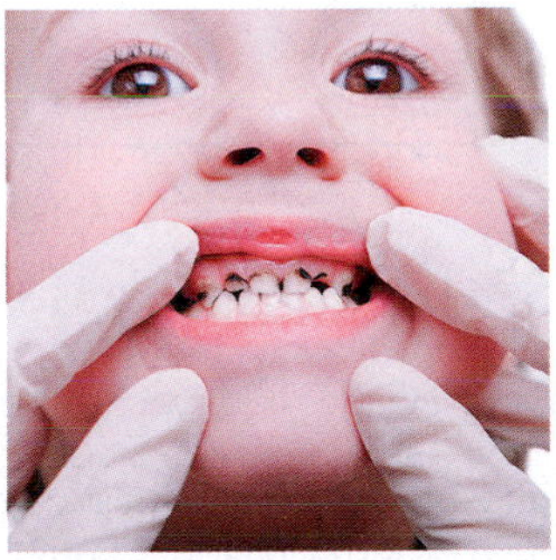

Dental cavities is the most common chronic disease in children.

One of the major contributors to tooth decay is the bacterium *Streptococcus mutans*. This bacterium is found in the mouth, where it converts residual food particles to lactic acid via metabolic processes that do not require oxygen (see Chapter 22). If sufficient amounts of lactic acid are produced to overcome the natural buffers in the mouth, the pH in the mouth decreases. When the pH decreases below 5.0, the enamel on the teeth begins to dissolve. As the acid dissolves the underlying dentin to expose the pulp of the tooth, inflammation and pain result.

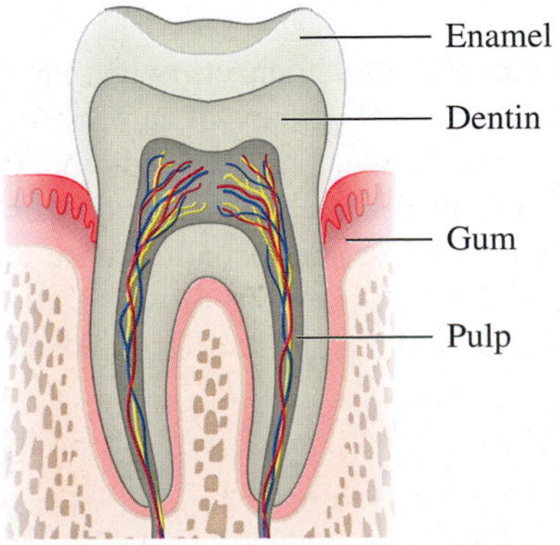
Enamel
Dentin
Gum
Pulp

When acid dissolves the enamel and dentin of the tooth, the pulp is exposed and pain results.

The preferred food for *Streptococcus mutans* are 6-carbon sugars (hexoses). The bacterium is not able to convert 5-carbon sugars (pentoses) or 5-carbon sugar alcohols into lactic acid. Therefore, many dentists suggest that their patients use gums and candies sweetened with xylitol, a 5-carbon sugar alcohol. Xylitol is produced by the reduction of xylose, a naturally occurring aldopentose that can be isolated from many woody plant sources.

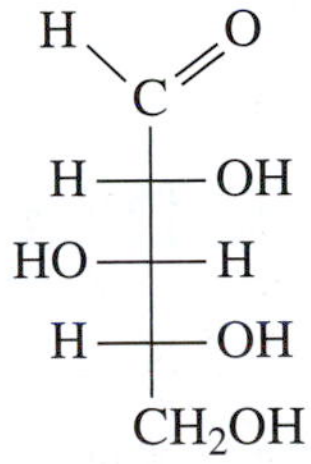

Xylose

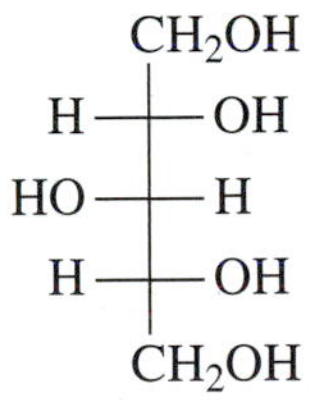

Xylitol

A number of different products contain xylitol, including gum, mints, mouthwash, and candies. Because bacteria in the mouth cannot convert xylitol into the lactic acid that contributes to tooth decay, the U.S. Food and Drug Administration (FDA) regulations state that products containing xylitol, such as chewing gum, may include a label that the product "does not promote dental caries." Proponents of the use of xylitol suggest that patients "strive for five" exposures of xylitol daily to decrease the incidence of dental cavities. The optimal intake of xylitol is approximately 5 g daily. Ingestion of more than the optimal amount of xylitol can lead to intestinal discomfort. Although xylitol is generally well tolerated by humans, it can be toxic to animals. Therefore, it is best to keep xylitol-containing products away from pets.

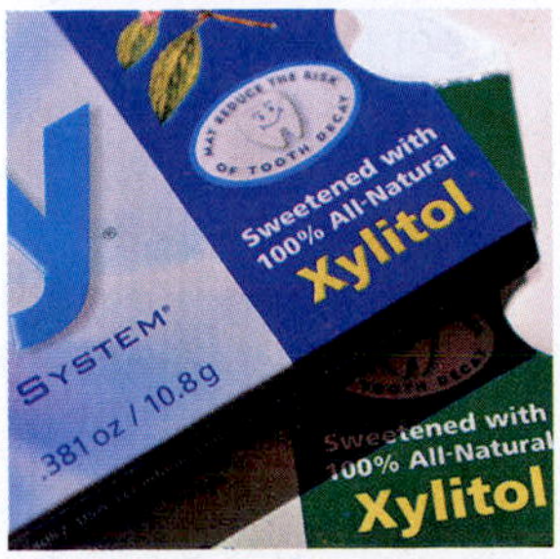
Sweetened with 100% All-Natural Xylitol
SYSTEM
.381 oz / 10.8g
Sweetened with 100% All-Natural Xylitol

Five exposures to xylitol-containing products a day can reduce the incidence of dental cavities.

PRACTICE PROBLEMS

15.5 Chemical Properties of Monosaccharides

15.37 Draw the Fischer projection for the oxidation and the reduction products of D-xylose. What are the names of the sugar acid and the sugar alcohol produced?

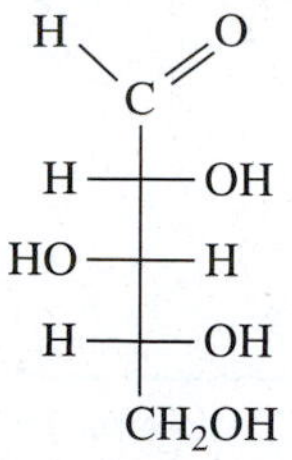

D-Xylose

15.38 Draw the Fischer projection for the oxidation and the reduction products of D-mannose. What are the names of the sugar acid and the sugar alcohol produced?

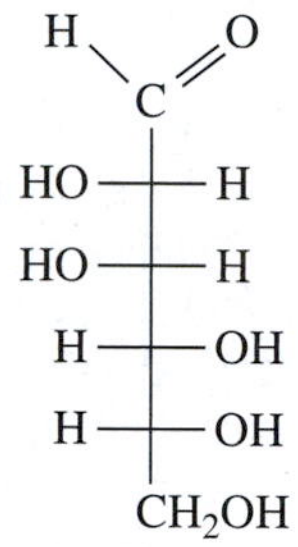

D-Mannose

15.39 Draw the Fischer projection for the oxidation and the reduction products of D-arabinose. What are the names of the sugar acid and the sugar alcohol produced?

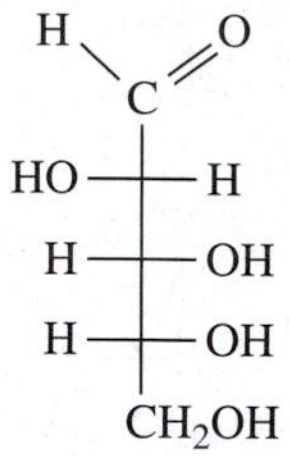

D-Arabinose

15.40 Draw the Fischer projection for the oxidation and the reduction products of D-ribose. What are the names of the sugar acid and the sugar alcohol produced?

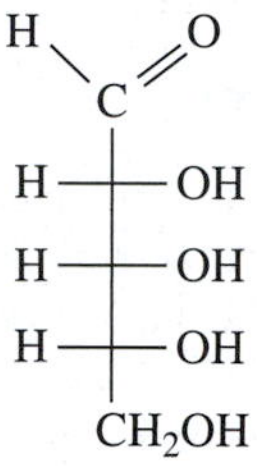

D-Ribose

15.6 Disaccharides

LEARNING GOAL Describe the monosaccharide units and linkages in disaccharides.

A disaccharide is composed of two monosaccharides linked together. The most common disaccharides are maltose, lactose, and sucrose. When two monosaccharides combine in a dehydration reaction, the product is a disaccharide. The reaction occurs between the hydroxyl group on carbon 1 and one of the hydroxyl groups on a second monosaccharide.

ENGAGE
Which disaccharide contains only glucose?

Glucose + glucose $\longrightarrow$ maltose + H_2O

Glucose + galactose $\longrightarrow$ lactose + H_2O

Glucose + fructose $\longrightarrow$ sucrose + H_2O

Maltose, or malt sugar, is obtained from starch and is found in germinating grains. When maltose in barley and other grains is hydrolyzed by yeast enzymes, glucose is obtained, which can undergo fermentation to give ethanol. Maltose is used in cereals, candies, and the brewing of beverages.

In the Haworth structure of a disaccharide, a **glycosidic bond** connects two monosaccharides. In maltose, a glycosidic bond forms between the —OH groups of carbons 1 and 4 of two α-D-glucose molecules with a loss of a water molecule. The glycosidic bond in maltose is designated as an $\alpha(1 \rightarrow 4)$ linkage to show that an alpha —OH group on carbon 1 is joined to carbon 4 of the second glucose molecule. Because the second glucose molecule still has a free —OH group on carbon 1, it can form an open chain, which allows maltose to form both α and β isomers. The open chain provides an aldehyde group that can be oxidized, making maltose a reducing sugar.

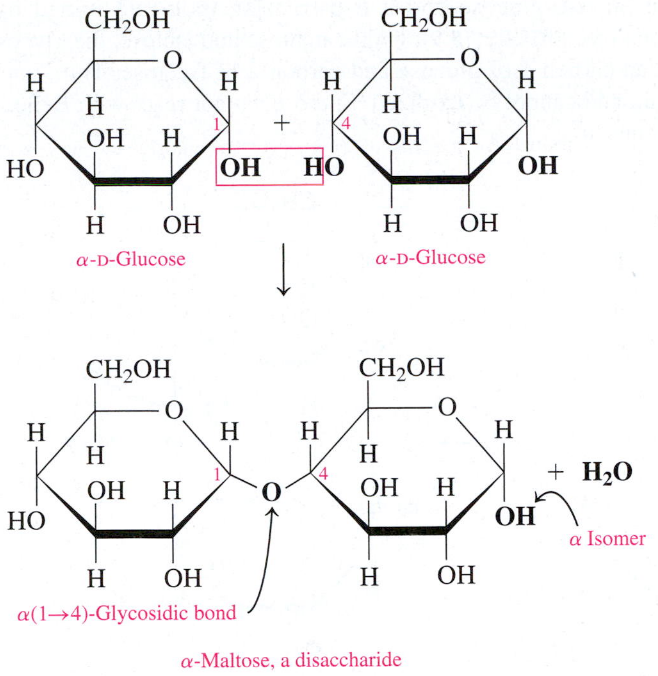
α-D-Glucose
α-D-Glucose
+ H_2O
α Isomer
α(1→4)-Glycosidic bond
α-Maltose, a disaccharide

ENGAGE

How is α-maltose different from β-maltose?

Lactose, milk sugar, is a disaccharide found in milk and milk products (see **FIGURE 15.8**). The bond in lactose is a β(1→4)-glycosidic bond because the —OH group on carbon 1 of β-D-galactose forms a glycosidic bond with the —OH group on carbon 4 of a D-glucose molecule. Because D-glucose still has a free —OH group on carbon 1, it can form an open chain, which allows lactose to form both α and β isomers. The open chain provides an aldehyde group that can be oxidized, making lactose a reducing sugar.

Lactose makes up 6 to 8% of human milk and about 4 to 5% of cow's milk, and it is used in products that attempt to duplicate mother's milk. When a person does not produce sufficient quantities of the enzyme lactase, which is needed to hydrolyze lactose, it remains undigested when it enters the colon. Then bacteria in the colon digest the lactose in a fermentation process that creates large amounts of gas, including carbon dioxide and methane, which cause bloating and abdominal cramps. In some commercial milk products, lactase has already been added to break down lactose.

FIGURE 15.8 ▶ Lactose, a disaccharide found in milk and milk products, contains galactose and glucose.

Q What type of glycosidic bond links galactose and glucose in lactose?

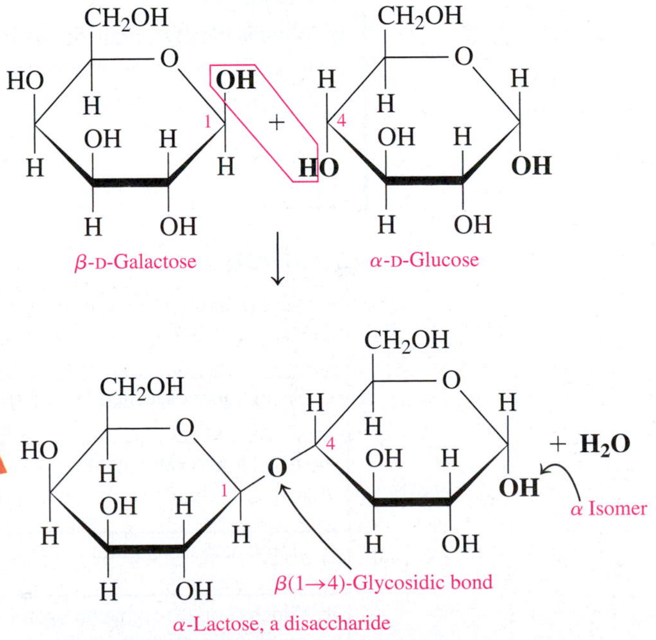
β-D-Galactose
α-D-Glucose
+ H_2O
α Isomer
β(1→4)-Glycosidic bond
α-Lactose, a disaccharide

Sucrose consists of an α-D-glucose and a β-D-fructose molecule joined by an $\alpha,\beta(1\rightarrow2)$-glycosidic bond (see **FIGURE 15.9**). Unlike maltose and lactose, the glycosidic bond in sucrose is between carbon 1 of glucose and carbon 2 of fructose. Thus, sucrose cannot form an open chain and cannot be oxidized. Sucrose cannot react with Benedict's reagent and is not a reducing sugar.

PURE CANE SUGAR
GRANULATED

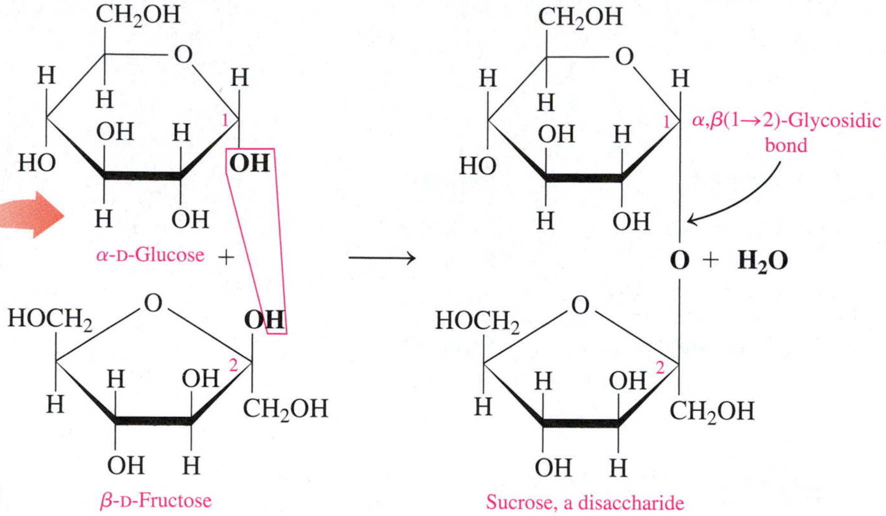
α-D-Glucose
β-D-Fructose
Sucrose, a disaccharide
α,β(1→2)-Glycosidic bond
O + H2O

FIGURE 15.9 ▶ Sucrose, a disaccharide obtained from sugar beets and sugar cane, contains glucose and fructose.

Q Why is sucrose not a reducing sugar?

The sugar we use to sweeten our cereal, coffee, or tea is sucrose. Most of the sucrose for table sugar comes from sugar cane (20% by mass) or sugar beets (15% by mass). Both the raw and refined forms of sugar are sucrose. Some estimates indicate that each person in the United States consumes an average of 68 kg (150 lb) of sucrose every year, either by itself or in a variety of food products.

▶ SAMPLE PROBLEM 15.7 Glycosidic Bonds in Disaccharides

TRY IT FIRST

Melibiose is a disaccharide that is 30 times sweeter than sucrose.

a. What are the monosaccharide units in melibiose?

b. What is the glycosidic link in melibiose?

c. Is this the α or β isomer of melibiose?

d. Is melibiose a reducing sugar?

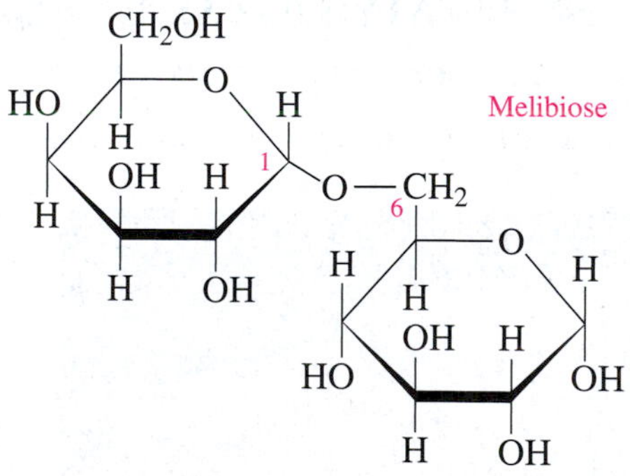
Melibiose

SOLUTION

a. First monosaccharide (left)	When the —OH group on carbon 4 is above the plane, it is D-galactose. When the —OH group on carbon 1 is below the plane, it is α-D-galactose.
Second monosaccharide (right)	When the —OH group on carbon 4 is below the plane, it is α-D-glucose.
b. Type of glycosidic bond	The —OH group at carbon 1 of α-D-galactose bonds with the —OH group on carbon 6 of glucose, which makes it an $\alpha(1\rightarrow6)$-glycosidic bond.
c. Name of disaccharide	The —OH group on carbon 1 of glucose is below the plane, which is α-melibiose.
d. Glucose on the right can open to form an aldehyde that can be oxidized, which makes melibiose a reducing sugar.	

STUDY CHECK 15.7

a. Isomaltulose is a sugar substitute that is digested more slowly than sucrose. What are the monosaccharide units in isomaltulose, and what is the glycosidic link in isomaltulose?

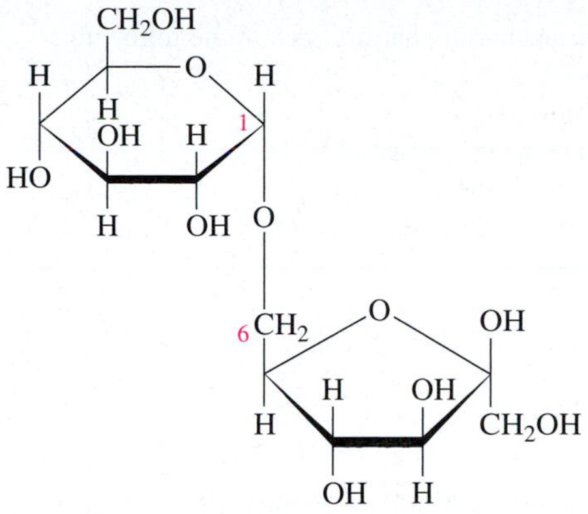
CH2OH
H
O
H
H
OH
H
1
HO
H
OH
O
6
CH2
O
OH
H
OH
H
CH2OH
OH
H

Isomaltulose

b. Cellobiose is a disaccharide composed of two D-glucose molecules connected by a $\beta(1\rightarrow4)$-glycosidic linkage. Draw the Haworth structure for β-cellobiose.

ANSWER

a. The two monosaccharides in isomaltulose are glucose and fructose. They are linked by an $\alpha(1\rightarrow6)$-glycosidic bond.

b.

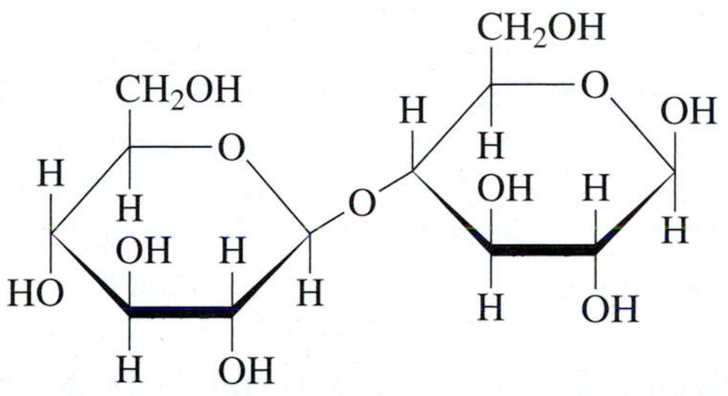

TEST

Try Practice Problems 15.41 to 15.46

PRACTICE PROBLEMS

15.6 Disaccharides

15.41 For each of the following, give the monosaccharide units produced by hydrolysis, the type of glycosidic bond, and the name of the disaccharide, including α or β:

a.

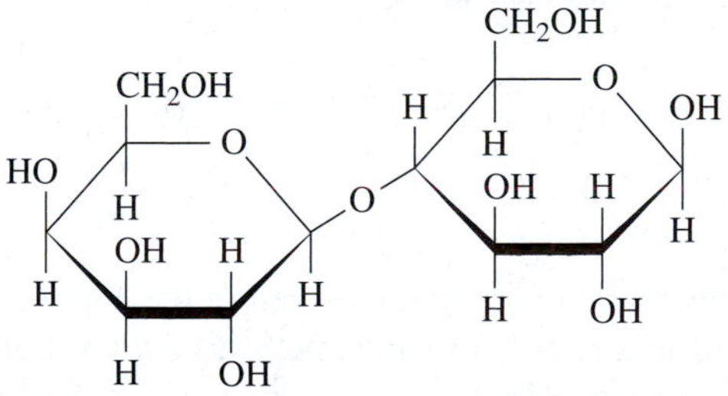

b.

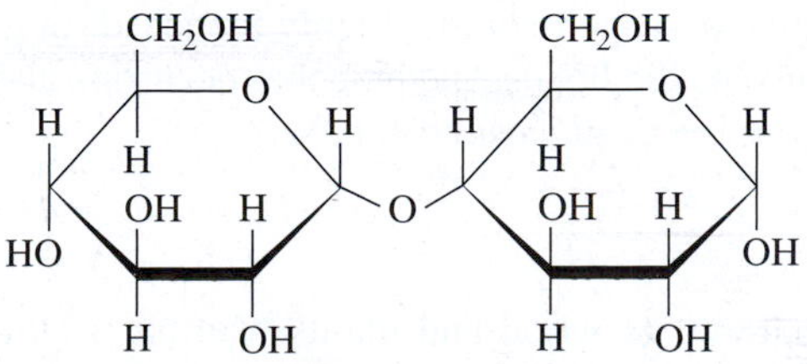

15.42 For each of the following, give the monosaccharide units produced by hydrolysis, the type of glycosidic bond, and the name of the disaccharide, including α or β:

a.

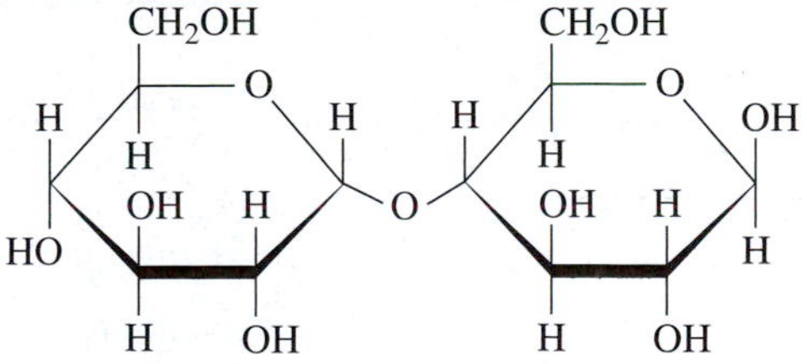

b.

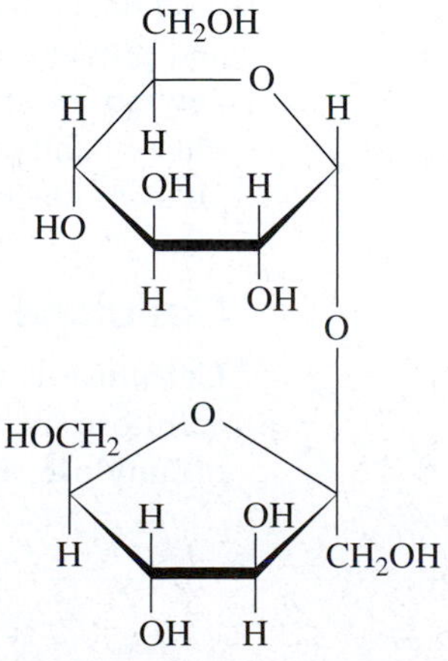

15.43 Indicate whether each disaccharide in Problem 15.41 is a reducing sugar or not.

15.44 Indicate whether each disaccharide in Problem 15.42 is a reducing sugar or not.

Clinical Applications

15.45 Identify the disaccharide that fits each of the following descriptions:
- **a.** ordinary table sugar
- **b.** found in milk and milk products
- **c.** also called malt sugar
- **d.** hydrolysis gives galactose and glucose

15.46 Identify the disaccharide that fits each of the following descriptions:
- **a.** not a reducing sugar
- **b.** composed of two glucose units
- **c.** also called milk sugar
- **d.** hydrolysis gives glucose and fructose

15.7 Polysaccharides

LEARNING GOAL Describe the structural features of amylose, amylopectin, glycogen, and cellulose.

A polysaccharide is a polymer of many monosaccharides joined together. Four important polysaccharides—*amylose*, *amylopectin*, *glycogen*, and *cellulose*—are polymers of D-glucose that differ only in the type of glycosidic bonds and the amount of branching in the molecule.

Starch

Starch, a storage form of glucose in plants, is found as insoluble granules in rice, wheat, potatoes, beans, and cereals. Starch is composed of two kinds of polysaccharides, amylose and amylopectin. **Amylose**, which makes up about 20% of starch, consists of 250 to 4000 α-D-glucose molecules connected by $\alpha(1\rightarrow4)$-glycosidic bonds in a continuous chain. Sometimes called a straight-chain polymer, polymers of amylose are actually coiled in helical fashion (see **FIGURE 15.10a**).

ENGAGE

What types of glycosidic bonds are present in amylose and amylopectin?

Amylopectin, which makes up as much as 80% of starch, is a branched-chain polysaccharide. Like amylose, glucose molecules are connected by $\alpha(1\rightarrow4)$-glycosidic bonds. However, at about every 25 glucose units, there is a branch of glucose molecules attached by an $\alpha(1\rightarrow6)$-glycosidic bond between carbon 1 of the branch and carbon 6 in the main chain (see **FIGURE 15.10b**).

Starches hydrolyze in water and acid to give smaller saccharides, called *dextrins*, which then hydrolyze to maltose and finally glucose. In our bodies, these complex carbohydrates are digested by the enzymes amylase (in saliva) and maltase (in the intestine). The glucose obtained provides about 50% of our nutritional calories.

$$\text{Amylose, amylopectin} \xrightarrow{H^+ \text{ or amylase}} \text{dextrins} \xrightarrow{H^+ \text{ or amylase}} \text{maltose} \xrightarrow{H^+ \text{ or maltase}} \text{many D-glucose units}$$

Glycogen

Glycogen, or animal starch, is a polymer of glucose that is stored in the liver and muscle of animals. It is hydrolyzed in our cells at a rate that maintains the blood level of glucose and provides energy between meals. The structure of glycogen is very similar to that of amylopectin found in plants, except that glycogen is more highly branched. In glycogen, glucose units are joined by $\alpha(1\rightarrow4)$-glycosidic bonds, and branches occurring about every 10 to 15 glucose units are attached by $\alpha(1\rightarrow6)$-glycosidic bonds.

Cellulose

Cellulose is the major structural material of wood and plants. Cotton is almost pure cellulose. In cellulose, glucose molecules form a long unbranched chain similar to that of amylose. However, the glucose units in cellulose are linked by $\beta(1\rightarrow4)$-glycosidic

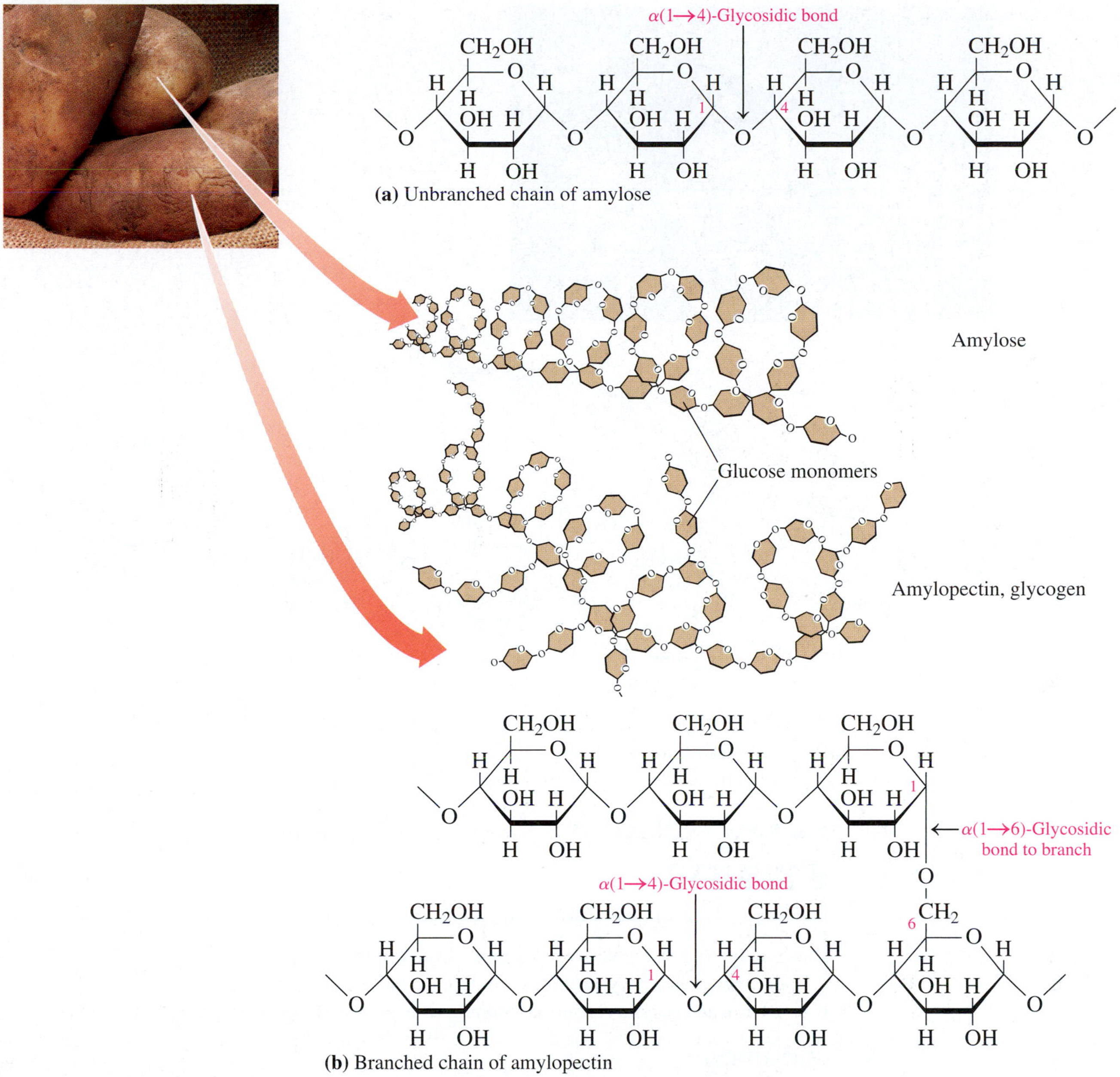
α(1→4)-Glycosidic bond
(a) Unbranched chain of amylose
Amylose
Glucose monomers
Amylopectin, glycogen
α(1→6)-Glycosidic bond to branch
α(1→4)-Glycosidic bond
(b) Branched chain of amylopectin

FIGURE 15.10 ▶ The structure of amylose **(a)** is a straight-chain polysaccharide of glucose units, and the structure of amylopectin **(b)** is a branched chain of glucose.

Q What are the two types of glycosidic bonds that link glucose molecules in amylopectin?

bonds. The cellulose chains do not form coils like amylose but are aligned in parallel rows that are held in place by hydrogen bonds between hydroxyl groups in adjacent chains, making cellulose insoluble in water. This gives a rigid structure to the cell walls in wood and fiber that is more resistant to hydrolysis than are the starches (see **FIGURE 15.11**).

Humans have an enzyme called α-amylase in saliva and pancreatic juices that hydrolyzes the $\alpha(1\rightarrow4)$-glycosidic bonds of starches, but not the $\beta(1\rightarrow4)$-glycosidic bonds of cellulose. Thus, humans cannot digest cellulose. Animals such as horses, cows, and goats can obtain glucose from cellulose because their digestive systems contain bacteria that provide enzymes such as cellulase to hydrolyze $\beta(1\rightarrow4)$-glycosidic bonds.

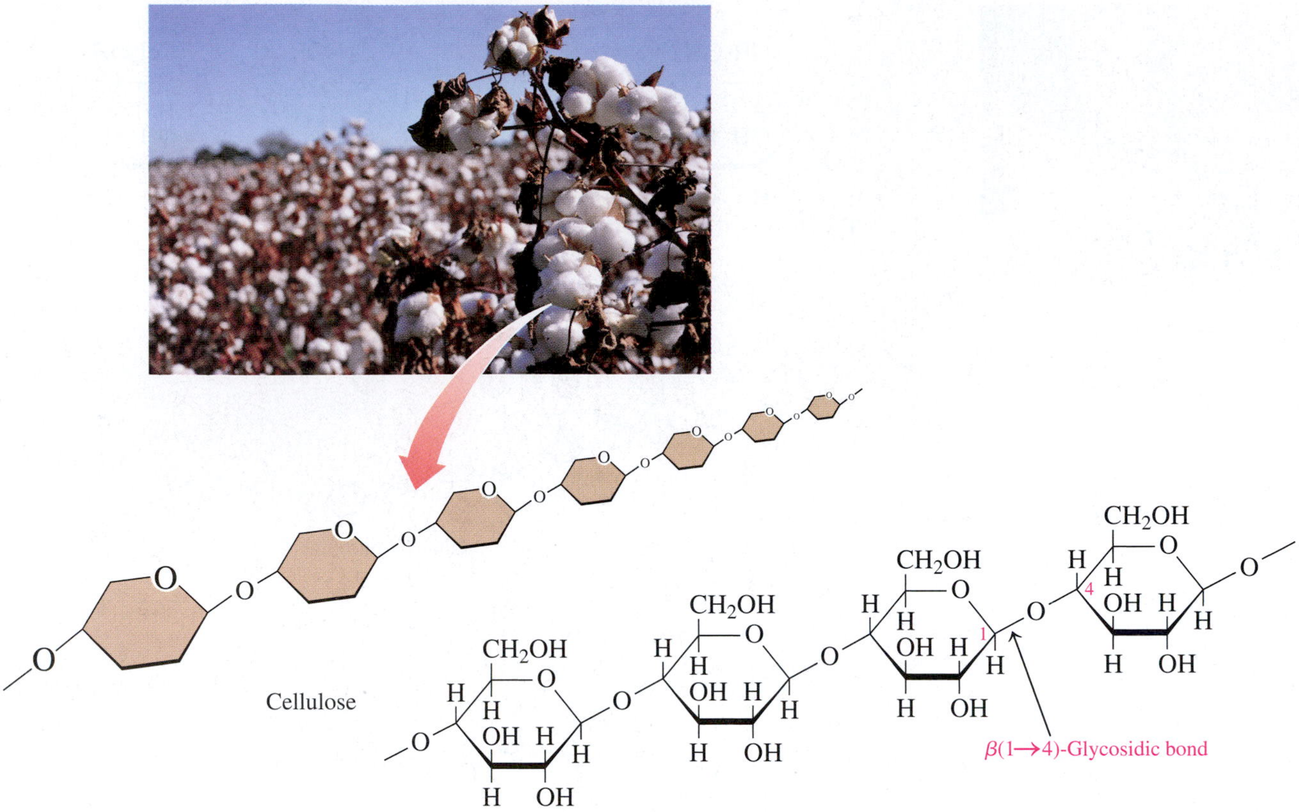
Cellulose
CH2OH
O
H
OH
$\beta(1\rightarrow4)$-Glycosidic bond

FIGURE 15.11 ▶ The polysaccharide cellulose is composed of glucose units connected by $\beta(1\rightarrow4)$-glycosidic bonds.

Q Why are humans unable to digest cellulose?

▶ SAMPLE PROBLEM 15.8 Structures of Polysaccharides

TRY IT FIRST

Give the name of one or more polysaccharides described by each of the following:

a. a polysaccharide that is stored in the liver and muscle tissues
b. an unbranched polysaccharide containing $\beta(1\rightarrow4)$-glycosidic bonds
c. a branched polysaccharide containing $\alpha(1\rightarrow4)$- and $\alpha(1\rightarrow6)$-glycosidic bonds

SOLUTION

a. glycogen **b.** cellulose **c.** amylopectin, glycogen

STUDY CHECK 15.8

Cellulose and amylose are both unbranched glucose polymers. How do they differ?

ANSWER

Cellulose contains glucose units connected by $\beta(1\rightarrow4)$-glycosidic bonds, whereas the glucose units in amylose are connected by $\alpha(1\rightarrow4)$-glycosidic bonds.

TEST

Try Practice Problems 15.47 to 15.50

Chemistry Link to Health

Varied Biological Roles of Carbohydrate Polymers: The Case of Glycosaminoglycans

Although we primarily think of them as providing fuel and energy for our bodies, carbohydrates play a number of biological roles. For example, *glycosaminoglycans* are a type of unbranched polysaccharide that consist of repeating disaccharide units. Each disaccharide unit is composed of one amino sugar molecule and either one uronic acid or one galactose molecule. *Amino sugars* are monosaccharides in which the hydroxyl group on carbon 2 has been replaced by a nitrogen-containing group like an amine group ($-NH_2$) or an N-acetyl group ($-NHCOCH_3$), while *uronic acids* are monosaccharides that include a carboxylate group ($-COO^-$) in addition

(continued)

Chemistry Link to Health (*continued*)

to the carbonyl group that all aldose and ketose sugars have. Some examples of amino sugars and uronic acids are shown below.

Amino Sugars

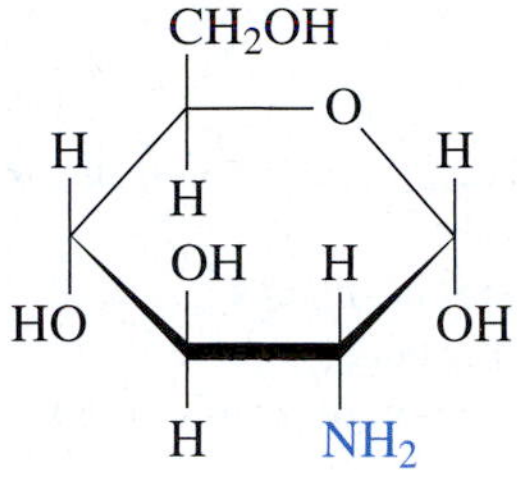

α-D-Glucosamine

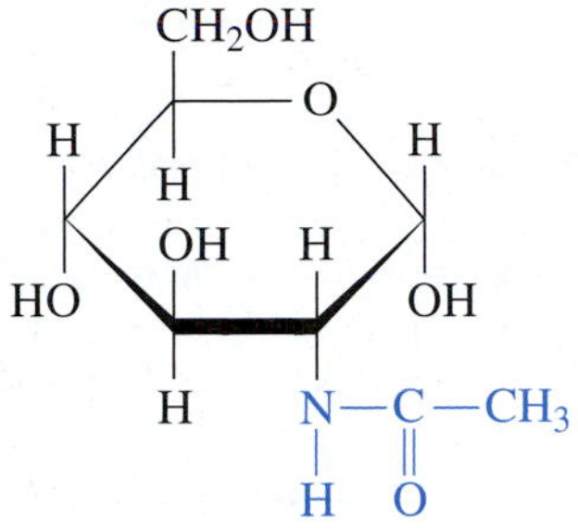

α-D-*N*-Acetylglucosamine

Amino sugars have a nitrogen-containing group on carbon 2.

Uronic Acids

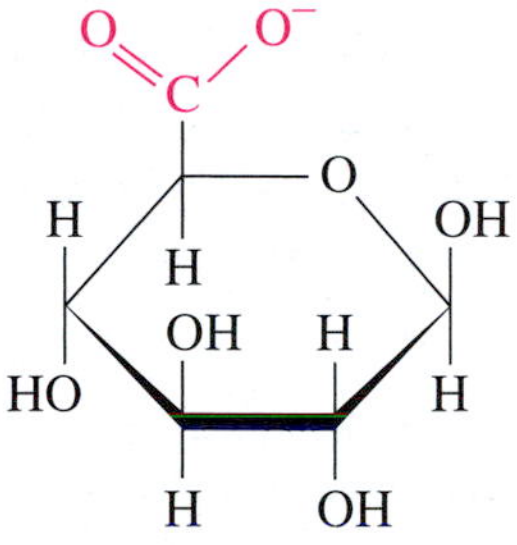

α-D-Glucuronic Acid

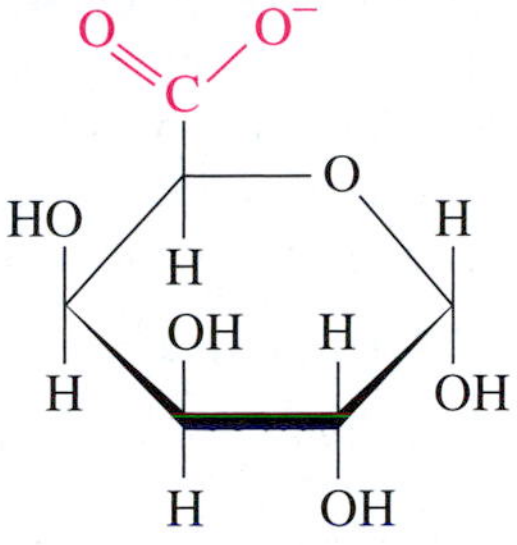

α-D-Galacturonic Acid

Uronic acids are monosaccharides that include both carbonyl and carboxylic acid groups.

There are many different glycosaminoglycans. Overall, they are very polar and negatively charged. As a consequence, they can attract water and can cushion or lubricate structures in the body. For example, keratan sulfate is found in the cornea of the eye, cartilage, bone, and the horns of animals.

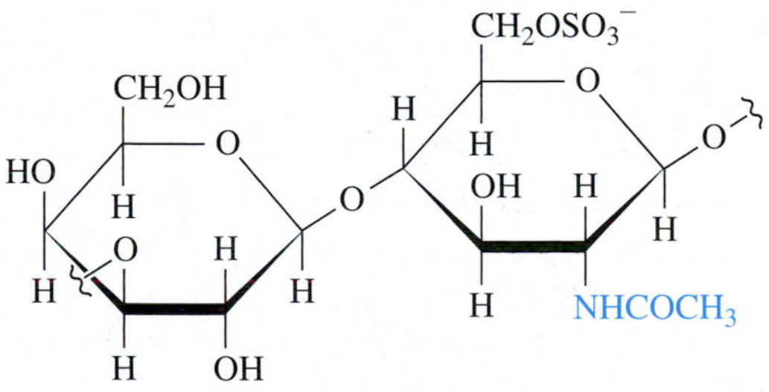

Keratan sulfate

In the eye, keratan sulfate helps to maintain corneal hydration. It also maintains the transparency of the cornea, the clear outer coat on the front of the eye. Inherited mutations in a gene can affect the biosynthesis of keratan sulfate and lead to *macular corneal dystrophy*. In these cases, the sulfate group does not become attached to the keratan. The unsulfonated keratan forms deposits in corneal cells, causing cloudiness in the cornea that becomes worse with age.

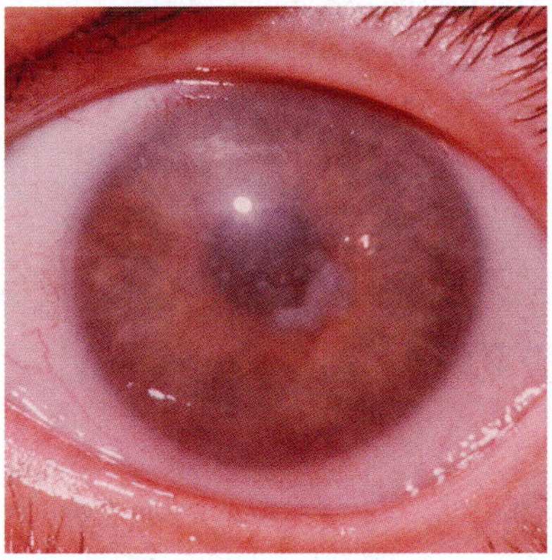

Unsulfonated keratan deposits cause cloudiness to develop in the cornea.

The glycosaminoglycan hyaluronic acid is found throughout connective, epithelial, and neural tissues. It can act as a lubricant for joints, or as a biological "glue," holding together gel-like connective tissue.

Like other glycosaminoglycans, hyaluronic acid attracts water. In fact, it can bind 1000 times its weight in water. Because of this ability, hyaluronic acid is included in many skin care products. It can also be injected subcutaneously as a filler for facial wrinkles. The effects of hyaluronic acid fillers typically last 6 to 12 months before the body naturally absorbs the injected molecules.

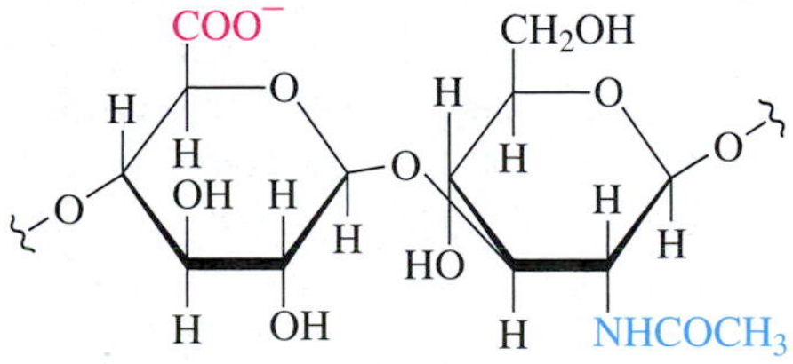

Hyaluronic acid

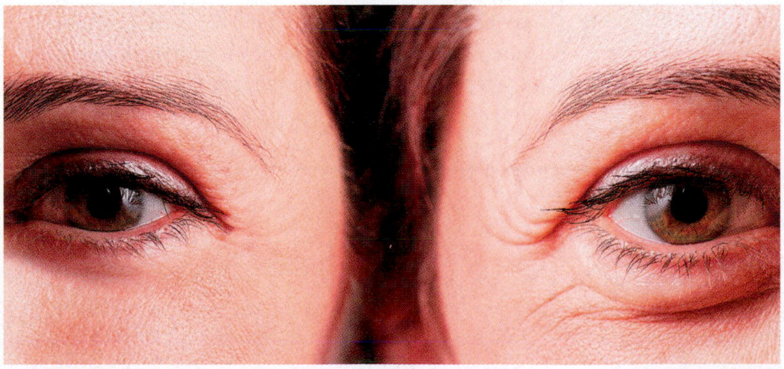

The subcutaneous injection of hyaluronic acid fillers decreases the appearance of facial lines and wrinkles.

PRACTICE PROBLEMS

15.7 Polysaccharides

15.47 Describe the similarities and differences in the following:
- **a.** amylose and amylopectin
- **b.** amylopectin and glycogen

15.48 Describe the similarities and differences in the following:
- **a.** amylose and cellulose
- **b.** cellulose and glycogen

Clinical Applications

15.49 Give the name of one or more polysaccharides that matches each of the following descriptions:
- **a.** not digestible by humans
- **b.** the storage form of carbohydrates in plants
- **c.** contains only $\alpha(1\rightarrow4)$-glycosidic bonds
- **d.** the most highly branched polysaccharide

15.50 Give the name of one or more polysaccharides that matches each of the following descriptions:
- **a.** the storage form of carbohydrates in animals
- **b.** contains only $\beta(1\rightarrow4)$-glycosidic bonds
- **c.** contains both $\alpha(1\rightarrow4)$- and $\alpha(1\rightarrow6)$-glycosidic bonds
- **d.** produces maltose during digestion

CLINICAL UPDATE Kate's Program for Type 2 Diabetes

At Kate's next appointment, Paula shows Kate how to use a glucose meter. She instructs Kate to measure her blood glucose level before and after breakfast and dinner each day. Paula explains to Kate that her pre-meal blood glucose level should be 110 mg/dL or less, and if it increases by more than 50 mg/dL after the meal, she needs to lower the amount of carbohydrates she consumes.

Kate and Paula plan several meals. They combine fruits and vegetables that have high and low levels of carbohydrates in the same meal to stay within the recommended range of about 45 to 60 g of carbohydrate per meal. Kate and Paula also discuss the fact that complex carbohydrates in the body take longer to break down into glucose and, therefore, raise the blood sugar level more gradually.

After her appointment, Kate increases her exercise to walking 30 minutes twice a day. She begins to change her diet by eating six small meals a day consisting mostly of small amounts of fruits, vegetables without starch such as green beans and broccoli, small servings of whole grains, and chicken or fish. Kate also decreases the amounts of breads and pasta she eats because she knows carbohydrates will raise her blood sugar.

After three months, Kate reports that she lost 10 lb and that her blood glucose dropped to 146 mg/dL. Her blurry vision improved, and her need to urinate decreased.

Clinical Applications

15.51 Kate's blood volume is 3.9 L. Before treatment, if her blood glucose was 178 mg/dL, how many grams of glucose were in her blood?

15.52 Kate's blood volume is 3.9 L. After three months of diet and exercise, if her blood glucose is 146 mg/dL, how many grams of glucose are in her blood?

15.53 For breakfast, Kate had 1 cup of orange juice (23 g carbohydrate), 2 slices of wheat toast (24 g carbohydrate), 2 tablespoons of grape jam (26 g carbohydrate), and coffee with sugar substitute (0 g carbohydrate).
- **a.** Has Kate remained within the limit of 45 to 60 g of carbohydrate?
- **b.** Using the energy value of 4 kcal/g for carbohydrate, calculate the total kilocalories from carbohydrates in Kate's breakfast, rounded to the tens place.

15.54 The next day, Kate had 1 cup of cereal (15 g carbohydrate) with skim milk (7 g carbohydrate), 1 banana (17 g carbohydrate), and 1/2 cup of orange juice (12 g carbohydrate) for breakfast.
- **a.** Has Kate remained within the limit of 45 to 60 g of carbohydrate?
- **b.** Using the energy value of 4 kcal/g for carbohydrate, calculate the total kilocalories from carbohydrates in Kate's breakfast, rounded to the tens place.

CONCEPT MAP

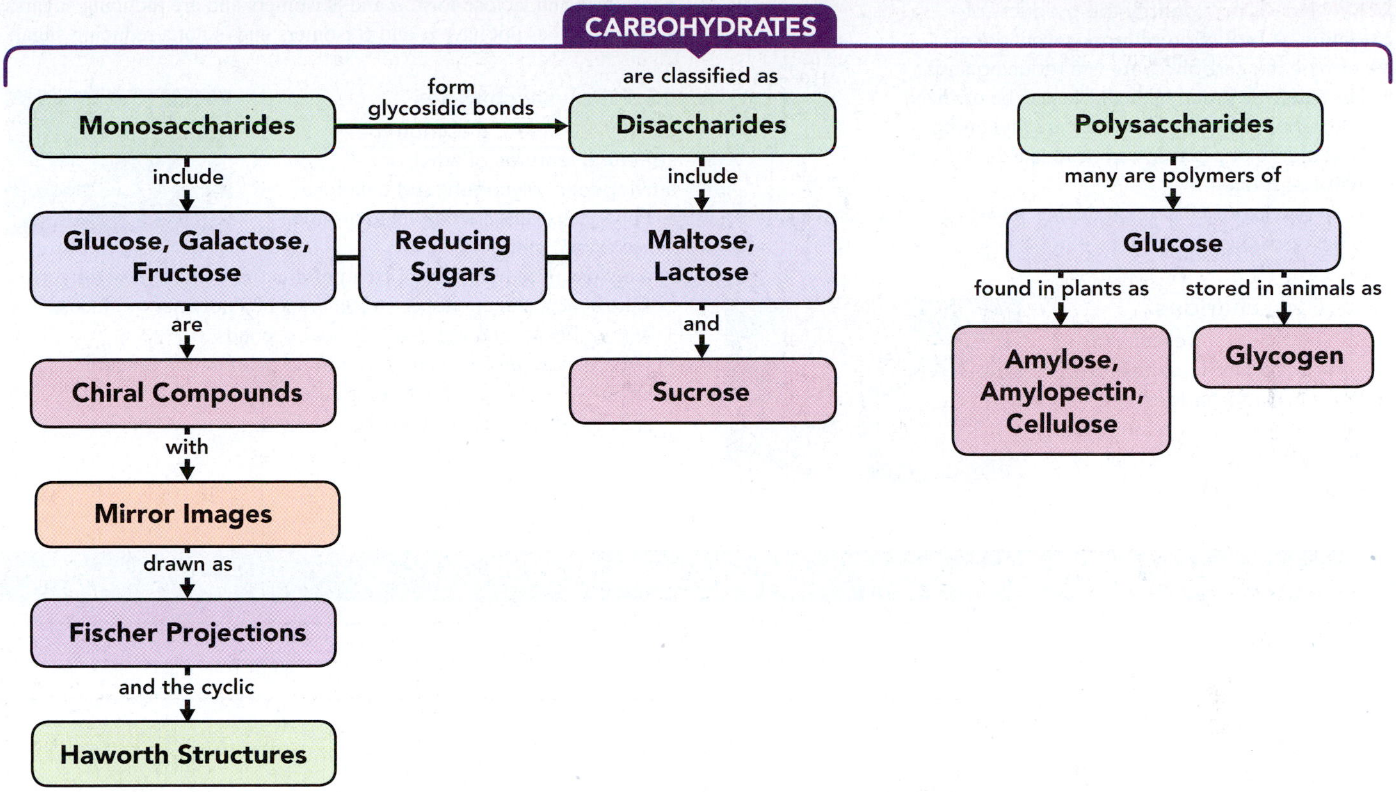
CARBOHYDRATES
are classified as
Monosaccharides
form
glycosidic bonds
Disaccharides
Polysaccharides
include
include
many are polymers of
Glucose, Galactose, Fructose
Reducing Sugars
Maltose, Lactose
Glucose
found in plants as
stored in animals as
are
and
Chiral Compounds
Sucrose
Amylose, Amylopectin, Cellulose
Glycogen
with
Mirror Images
drawn as
Fischer Projections
and the cyclic
Haworth Structures

CHAPTER REVIEW

15.1 Carbohydrates

LEARNING GOAL Classify a monosaccharide as an aldose or a ketose, and indicate the number of carbon atoms.

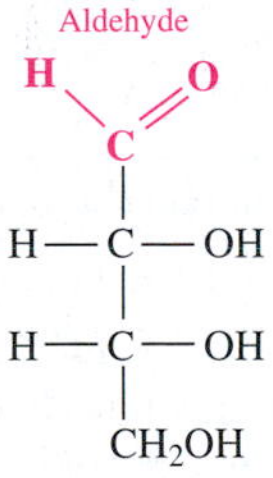
Aldehyde

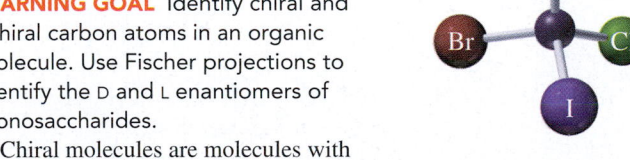

Erythrose, an aldose

- Carbohydrates are classified as monosaccharides (simple sugars), disaccharides (two monosaccharide units), or polysaccharides (many monosaccharide units).
- Monosaccharides are polyhydroxy aldehydes (aldoses) or ketones (ketoses).
- Monosaccharides are also classified by their number of carbon atoms: triose, tetrose, pentose, or hexose.

15.2 Chiral Molecules

LEARNING GOAL Identify chiral and achiral carbon atoms in an organic molecule. Use Fischer projections to identify the D and L enantiomers of monosaccharides.

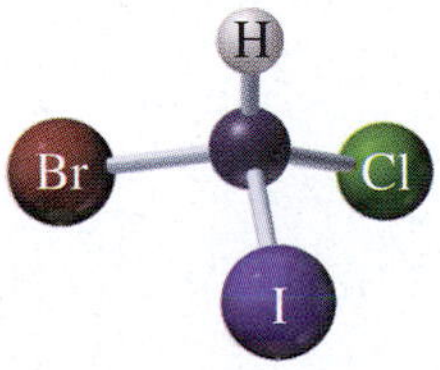
H
Br
Cl
I

- Chiral molecules are molecules with mirror images that cannot be superimposed on each other. These types of stereoisomers are called enantiomers.
- A chiral molecule must have at least one chiral carbon, which is a carbon bonded to four different atoms or groups of atoms.
- The Fischer projection is a simplified way to draw the arrangements of atoms by placing the carbon atoms at the intersection of vertical and horizontal lines.
- The mirror images in Fischer projections are labeled D or L to differentiate between enantiomers.

15.3 Fischer Projections of Monosaccharides

LEARNING GOAL Identify or draw the D and L configurations of the Fischer projections for common monosaccharides.

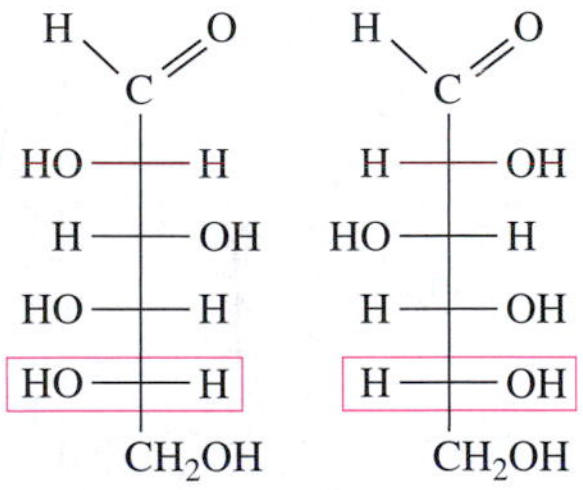

L-Glucose D-Glucose

- In a D enantiomer, the —OH group is on the right of the chiral carbon farthest from the carbonyl group; the —OH group is on the left in the L enantiomer.
- Important monosaccharides are the aldohexoses glucose and galactose, and the ketohexose fructose.

15.4 Haworth Structures of Monosaccharides

LEARNING GOAL Draw and identify the Haworth structures for monosaccharides.

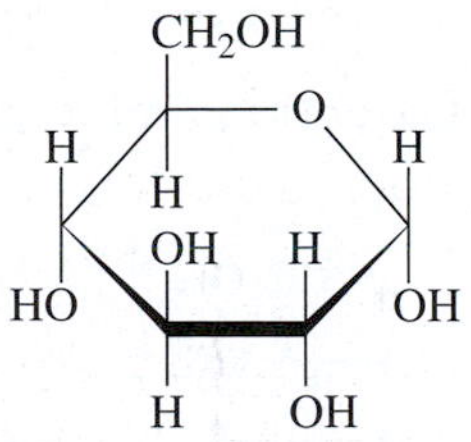

α-D-Glucose

- The predominant form of a monosaccharide is a ring of five or six atoms.
- The cyclic structure forms when an —OH group (usually the one on carbon 5 in hexoses) reacts with the carbonyl group of the same molecule.
- The formation of a new hydroxyl group on carbon 1 gives α and β isomers of the cyclic monosaccharide.

15.5 Chemical Properties of Monosaccharides

LEARNING GOAL Identify the products of oxidation or reduction of monosaccharides; determine if a carbohydrate is a reducing sugar.

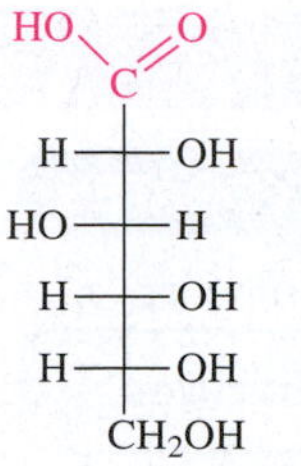

D-Gluconic acid

- The aldehyde group in an aldose can be oxidized to a carboxylic acid, while the carbonyl group in an aldose or a ketose can be reduced to give a hydroxyl group.
- Monosaccharides that are reducing sugars have an aldehyde group in the open chain that can be oxidized.

15.6 Disaccharides

LEARNING GOAL Describe the monosaccharide units and linkages in disaccharides.

PURE
CANE
SUGAR
GRANULATED

- Disaccharides are two monosaccharide units joined together by a glycosidic bond.
- In the common disaccharides maltose, lactose, and sucrose, there is at least one glucose unit.
- Maltose and lactose form α and β isomers and are reducing sugars.
- Sucrose does not have α and β isomers and is not a reducing sugar.

15.7 Polysaccharides

LEARNING GOAL Describe the structural features of amylose, amylopectin, glycogen, and cellulose.

- Polysaccharides are polymers of monosaccharide units.
- Amylose is an unbranched chain of glucose with $\alpha(1\rightarrow4)$-glycosidic bonds, and amylopectin is a branched polymer of glucose with $\alpha(1\rightarrow4)$- and $\alpha(1\rightarrow6)$-glycosidic bonds.
- Glycogen is similar to amylopectin but with more branching.
- Cellulose is also a polymer of glucose, but in cellulose, the glycosidic bonds are $\beta(1\rightarrow4)$ bonds.

SUMMARY OF CARBOHYDRATES

Carbohydrate	Found In	
Monosaccharides		
Glucose	Fruit juices, honey, corn syrup	
Galactose	Lactose hydrolysis	
Fructose	Fruit juices, honey, sucrose hydrolysis	
Disaccharides		**Monosaccharide Components**
Maltose	Germinating grains, starch hydrolysis	Glucose + glucose
Lactose	Milk, yogurt, ice cream	Glucose + galactose
Sucrose	Sugar cane, sugar beets	Glucose + fructose
Polysaccharides		
Amylose	Rice, wheat, grains, cereals	Unbranched polymer of glucose joined by $\alpha(1\rightarrow4)$-glycosidic bonds
Amylopectin	Rice, wheat, grains, cereals	Branched polymer of glucose joined by $\alpha(1\rightarrow4)$- and $\alpha(1\rightarrow6)$-glycosidic bonds
Glycogen	Liver, muscles	Highly branched polymer of glucose joined by $\alpha(1\rightarrow4)$- and $\alpha(1\rightarrow6)$-glycosidic bonds
Cellulose	Plant fiber, bran, beans, celery	Unbranched polymer of glucose joined by $\beta(1\rightarrow4)$-glycosidic bonds

SUMMARY OF REACTIONS

The chapter Sections to review are shown after the name of each reaction.

Oxidation and Reduction of Monosaccharides (15.5)

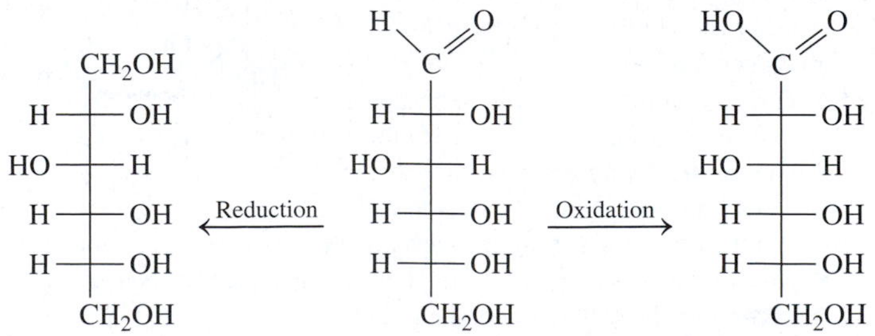
Reduction
Oxidation

D-Glucitol D-Glucose D-Gluconic acid

Formation of Disaccharides (15.6)

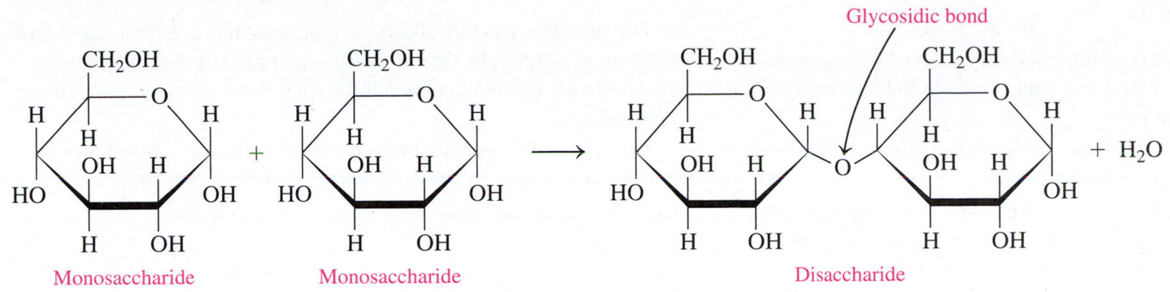
Glycosidic bond
Monosaccharide
Monosaccharide
Disaccharide
+ H_2O

Hydrolysis of Polysaccharides (15.7)

Amylose, amylopectin $\xrightarrow{H^+ \text{ or enzymes}}$ many D-glucose units

KEY TERMS

aldose A monosaccharide that contains an aldehyde group.

amylopectin A branched-chain polymer of starch composed of glucose units joined by $\alpha(1\rightarrow4)$- and $\alpha(1\rightarrow6)$-glycosidic bonds.

amylose An unbranched polymer of starch composed of glucose units joined by $\alpha(1\rightarrow4)$-glycosidic bonds.

carbohydrate A simple or complex sugar composed of carbon, hydrogen, and oxygen.

cellulose An unbranched polysaccharide composed of glucose units linked by $\beta(1\rightarrow4)$-glycosidic bonds that cannot be hydrolyzed by the human digestive system.

chiral An object or molecule that has a nonsuperimposable mirror image.

chiral carbon A carbon atom that is bonded to four different atoms or groups.

disaccharide A carbohydrate composed of two monosaccharides joined by a glycosidic bond.

enantiomers Chiral compounds that are mirror images that cannot be superimposed.

Fischer projection A system for drawing stereoisomers; an intersection of a vertical and horizontal line represents a carbon atom. A vertical line represents bonds that project backward from a carbon atom and a horizontal line represents bonds that project forward. The most highly oxidized carbon is at the top.

fructose A monosaccharide that is also called levulose and fruit sugar and is found in honey and fruit juices; it is combined with glucose in sucrose.

galactose A monosaccharide that occurs combined with glucose in lactose.

glucose An aldohexose found in fruits, vegetables, corn syrup, and honey that is also known as blood sugar and dextrose. The most prevalent monosaccharide in the diet. Many polysaccharides are polymers of glucose.

glycogen A polysaccharide formed in the liver and muscles for the storage of glucose as an energy reserve. It is composed of glucose in a highly branched polymer joined by $\alpha(1\rightarrow4)$- and $\alpha(1\rightarrow6)$-glycosidic bonds.

glycosidic bond The bond that forms when the hydroxyl group of one monosaccharide reacts with the hydroxyl group of another monosaccharide. It is the type of bond that links monosaccharides in di- or polysaccharides.

Haworth structure The ring structure of a monosaccharide.

ketose A monosaccharide that contains a ketone group.

lactose A disaccharide consisting of glucose and galactose found in milk and milk products.

maltose A disaccharide consisting of two glucose units; it is obtained from the hydrolysis of starch and is found in germinating grains.

monosaccharide A polyhydroxy compound that contains an aldehyde or ketone group.

polysaccharide A polymer of many monosaccharide units, usually glucose. Polysaccharides differ in the types of glycosidic bonds and the amount of branching in the polymer.

reducing sugar A carbohydrate with an aldehyde group capable of reducing the Cu^{2+} in Benedict's reagent.

stereoisomers Isomers that have atoms bonded in the same order, but with different arrangements in space.

sucrose A disaccharide composed of glucose and fructose; a nonreducing sugar, commonly called table sugar or "sugar."

CORE CHEMISTRY SKILLS

The chapter Section containing each Core Chemistry Skill is shown in parentheses at the end of each heading.

Identifying Chiral Molecules (15.2)

- Chiral molecules are molecules with mirror images that cannot be superimposed. These types of stereoisomers are called enantiomers.
- A chiral molecule must have at least one chiral carbon, which is a carbon bonded to four different atoms or groups of atoms.

Example: Identify each of the following as chiral or achiral:

a. $CH_3-CH_2-\underset{}{\overset{Br}{\overset{|}{C}}}H-CH_3$

b. $CH_3-CH_2-CH_2-Br$

c. $CH_3-\overset{Br}{\overset{|}{C}}H-CH_3$

Answer: **a** is chiral; **b** and **c** are achiral.

Identifying D and L Fischer Projections for Carbohydrates (15.3)

- The Fischer projection is a simplified way to draw the arrangements of atoms by placing the carbon atoms at the intersection of vertical and horizontal lines.
- The names of the mirror images are labeled D or L to differentiate between enantiomers of carbohydrates.

Example: Identify each of the following as the D or L enantiomer:

a.

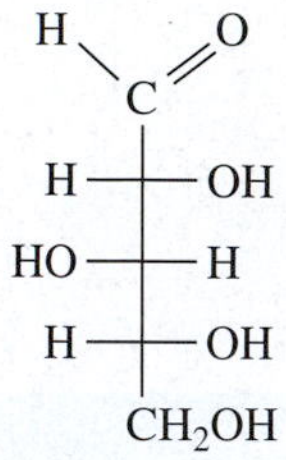
H
O
C
H OH
HO H
H OH
CH₂OH

b.

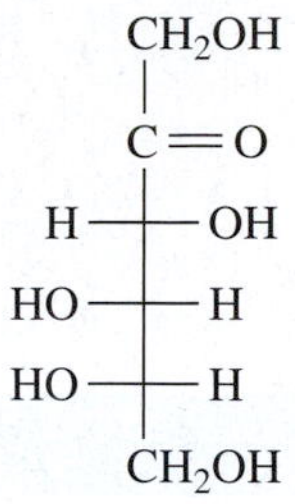
CH₂OH
C=O
H OH
HO H
HO H
CH₂OH

Answer: In **a**, the —OH group on the chiral carbon farthest from the carbonyl group is on the right; it is the D enantiomer.

In **b**, the —OH group on the chiral carbon farthest from the carbonyl group is on the left; it is the L enantiomer.

Drawing Haworth Structures (15.4)

- The Haworth structure shows the ring structure of a monosaccharide.
- Groups on the right side of the Fischer projection of the monosaccharide are below the plane of the ring, those on the left are above the plane.
- The new —OH group that forms is the α isomer if it is below the plane of the ring, or the β isomer if it is above the plane.

Example: Draw the Haworth structure for β-D-idose.

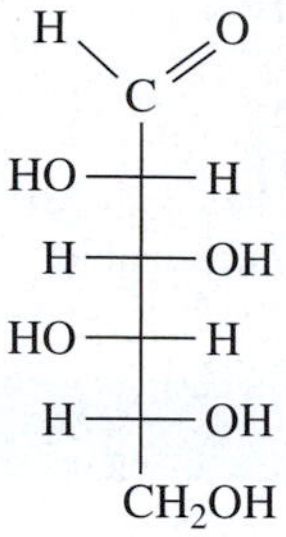
H
O
C
HO H
H OH
HO H
H OH
CH₂OH

D-Idose

Answer: In the β-isomer, the new —OH group is above the plane of the ring.

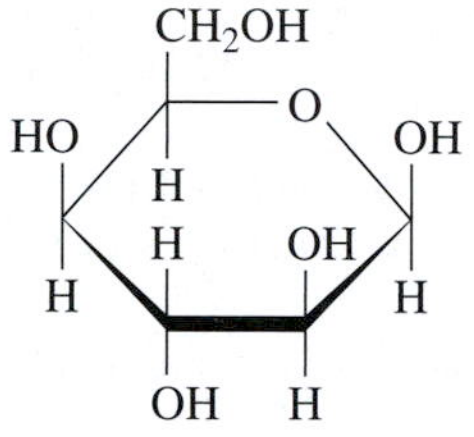
CH₂OH
O
HO
OH
H
H
OH
H
H
OH
H

UNDERSTANDING THE CONCEPTS

The chapter Sections to review are shown in parentheses at the end of each problem.

15.55 Isomaltose, obtained from the breakdown of starch, has the following Haworth structure: (15.4, 15.5, 15.6)

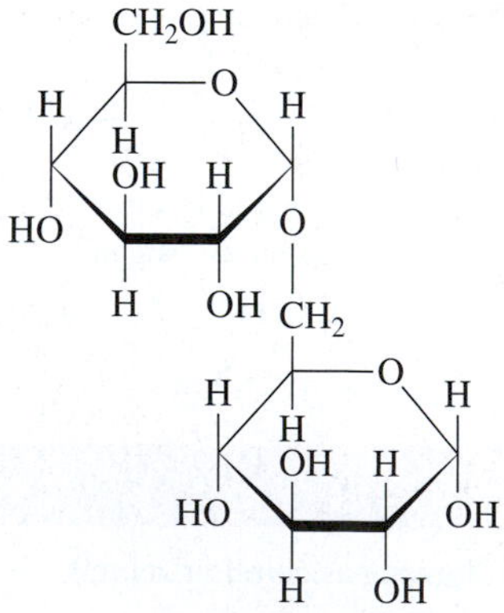
CH₂OH
O
H
H
H
OH H
HO
O
H OH
CH₂
O
H
H
H
OH H
HO
OH
H OH

Isomaltose

a. Is isomaltose a mono-, di-, or polysaccharide?
b. What are the monosaccharides in isomaltose?
c. What is the glycosidic link in isomaltose?
d. Is this the α or β isomer of isomaltose?
e. Is isomaltose a reducing sugar?

15.56 Sophorose, a carbohydrate found in certain types of beans, has the following Haworth structure: (15.4, 15.5, 15.6)

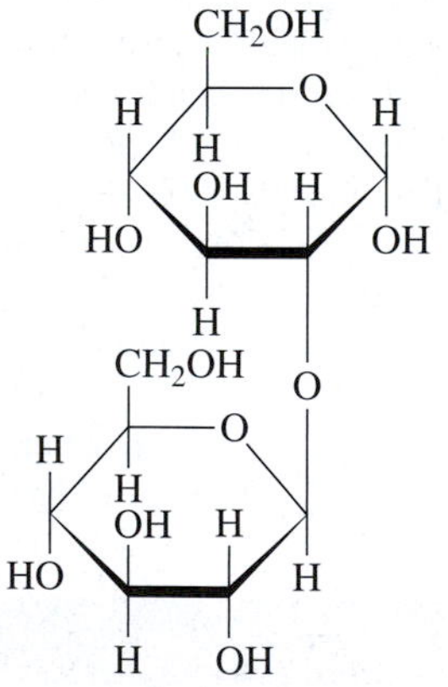
CH₂OH
O
H
H
H
OH H
HO
OH
H
CH₂OH
O
O
H
H
OH H
HO
H
H OH

Sophorose

a. Is sophorose a mono-, di-, or polysaccharide?
b. What are the monosaccharides in sophorose?
c. What is the glycosidic link in sophorose?
d. Is this the α or β isomer of sophorose?
e. Is sophorose a reducing sugar?

15.57 Melezitose, a carbohydrate secreted by insects, has the following Haworth structure: (15.4, 15.5, 15.6)

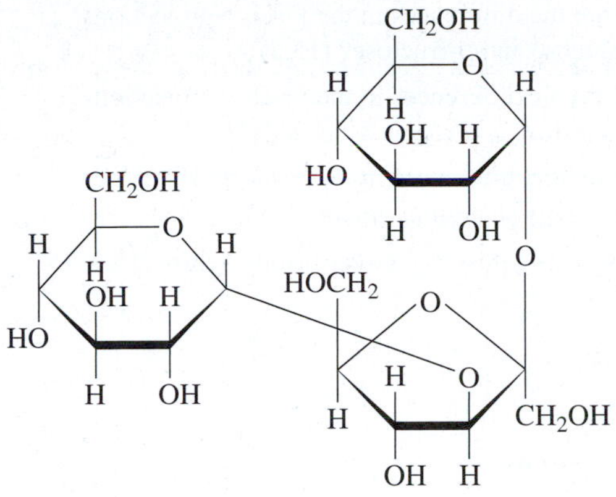

Melezitose

a. Is melezitose a mono-, di-, or trisaccharide?
b. What monosaccharides are present in melezitose?
c. Is melezitose a reducing sugar?

15.58 Raffinose, found in Australian manna and in cottonseed meal, has the following Haworth structure: (15.4, 15.5, 15.6)

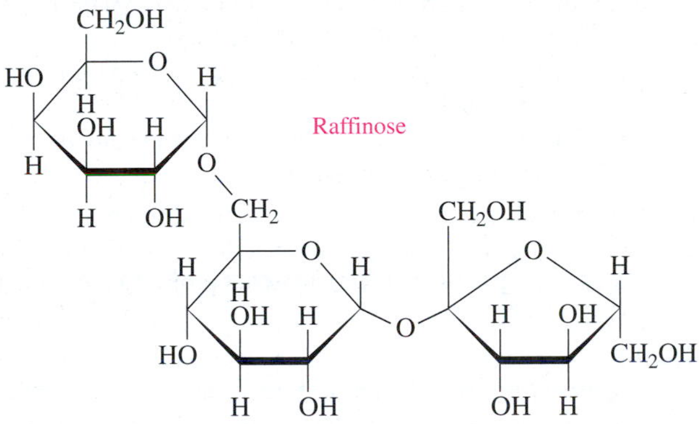

Raffinose

a. Is raffinose a mono-, di-, or trisaccharide?
b. What monosaccharides are present in raffinose?
c. Is raffinose a reducing sugar?

15.59 What are the disaccharides and polysaccharides present in each of the following? (15.6, 15.7)

PURE
CANE SUGAR
GRANULATED

(a)

(b)

15.60 What are the disaccharides and polysaccharides present in each of the following? (15.6, 15.7)

(a)

(b)

ADDITIONAL PRACTICE PROBLEMS

15.61 Identify the chiral carbons, if any, in each of the following compounds: (15.2)

a. 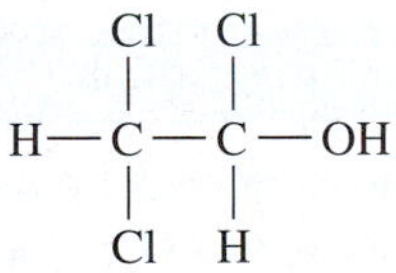

b. 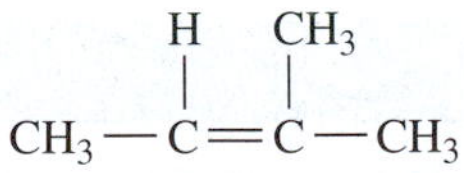

c. 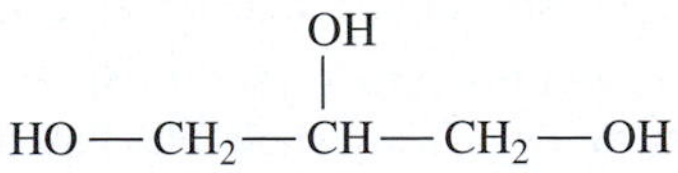

d. 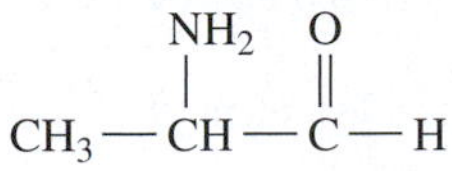

e. 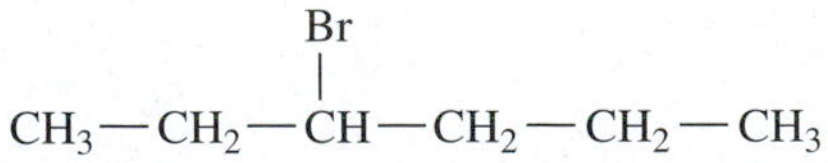

15.62 Identify the chiral carbons, if any, in each of the following compounds: (15.2)

a. 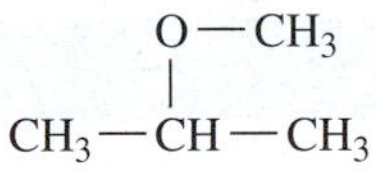

b. 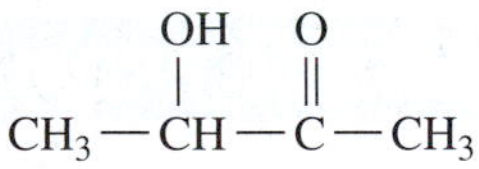

c. 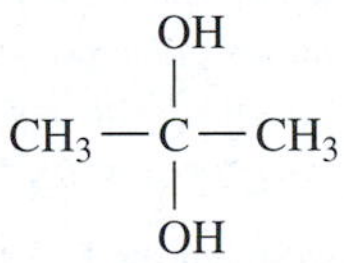

d. 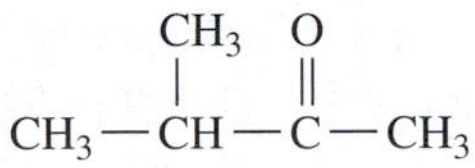

e. 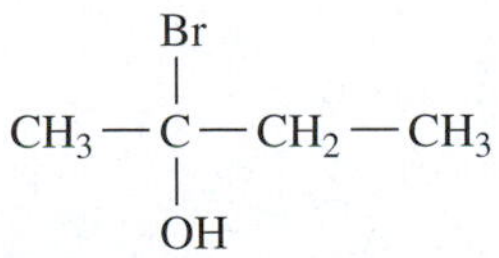

15.63 Identify each of the following pairs of Fischer projections as enantiomers or identical compounds: (15.2, 15.3)

a. 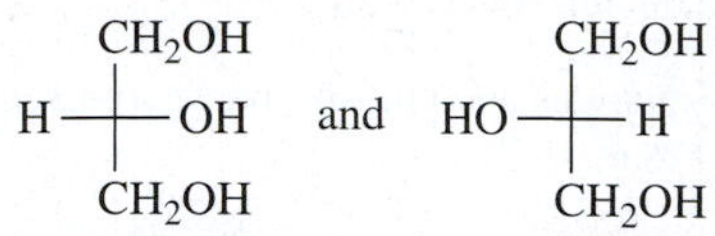

b. 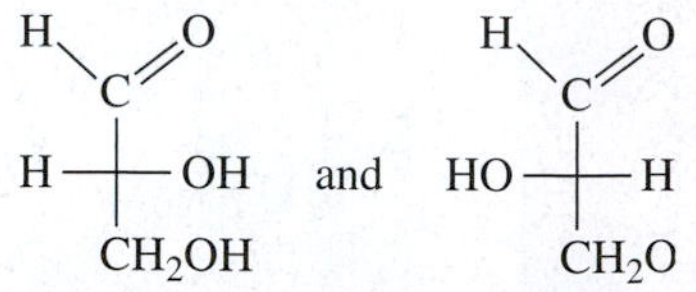

c. 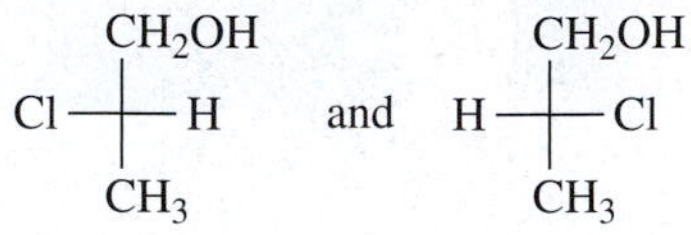

d. 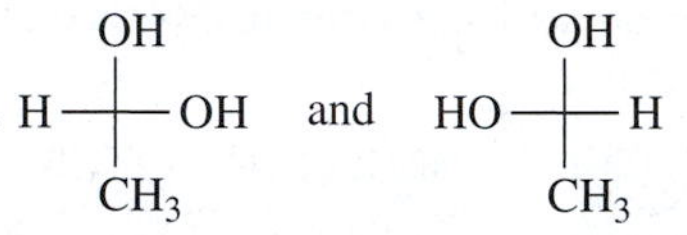

15.64 Identify each of the following pairs of Fischer projections as enantiomers or identical compounds: (15.2, 15.3)

a. 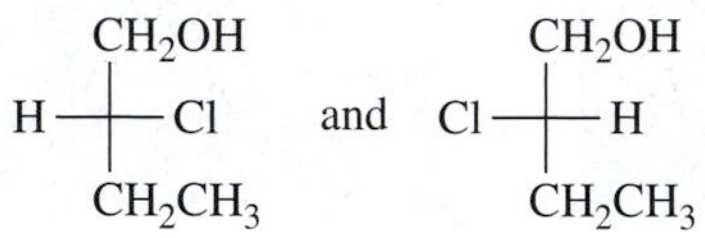

b. 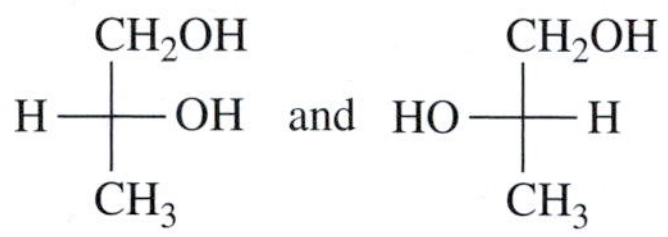

c. 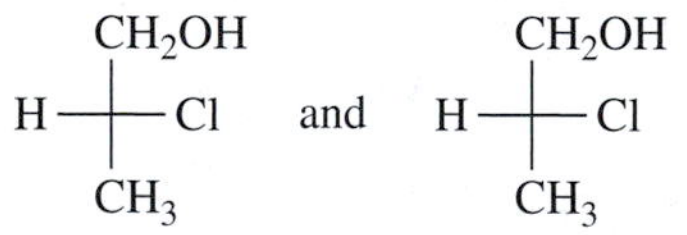

d. 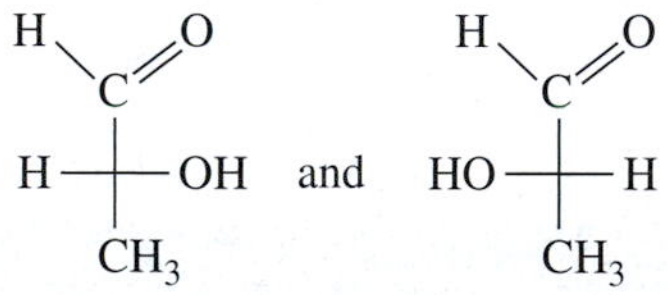

15.65 What are the differences in the Fischer projections of D-fructose and D-galactose? (15.3)

15.66 What are the differences in the Fischer projections of D-glucose and D-fructose? (15.3)

15.67 What are the differences in the Fischer projections of D-galactose and L-galactose? (15.3)

15.68 What are the differences in the Haworth structures of α-D-glucose and β-D-glucose? (15.4)

15.69 The sugar D-gulose is a sweet-tasting syrup. (15.3, 15.4)

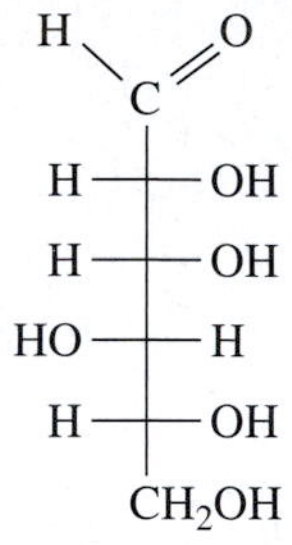

D-Gulose

a. Draw the Fischer projection for L-gulose.

b. Draw the Haworth structures for α- and β-D-gulose.

15.70 Use the Fischer projection for D-gulose in problem 15.69 to answer each of the following: (15.3, 15.5)

a. Draw the Fischer projection and name the product formed by the reduction of D-gulose.

b. Draw the Fischer projection and name the product formed by the oxidation of D-gulose.

15.71 D-Sorbitol, a sweetener found in seaweed and berries, contains only hydroxyl functional groups. When D-sorbitol is oxidized, it forms D-glucose. Draw the Fischer projection for D-sorbitol. (15.3, 15.5)

15.72 D-Erythritol is 70% as sweet as sucrose and contains only hydroxyl functional groups. When D-erythritol is oxidized, it forms D-erythrose. Draw the Fischer projection for D-erythritol. (15.3, 15.5)

15.73 If α-galactose is dissolved in water, β-galactose is eventually present. Explain how this occurs. (15.5)

15.74 Why are lactose and maltose reducing sugars, but sucrose is not? (15.6)

CHALLENGE PROBLEMS

The following problems are related to the topics in this chapter. However, they do not all follow the chapter order, and they require you to combine concepts and skills from several Sections. These problems will help you increase your critical thinking skills and prepare for your next exam.

15.75 α-Cellobiose is a disaccharide obtained from the hydrolysis of cellulose. It is quite similar to maltose except it has a $\beta(1\rightarrow4)$-glycosidic bond. Draw the Haworth structure for α-cellobiose. (15.4, 15.6)

15.76 The disaccharide trehalose found in mushrooms is composed of two α-D-glucose molecules joined by an $\alpha(1\rightarrow1)$-glycosidic bond. Draw the Haworth structure for trehalose. (15.4, 15.6)

15.77 Gentiobiose is found in saffron. (15.4, 15.5, 15.6)

a. Gentiobiose contains two glucose molecules linked by a $\beta(1\rightarrow6)$-glycosidic bond. Draw the Haworth structure for β-gentiobiose.

b. Is gentiobiose a reducing sugar? Explain.

15.78 Identify the Fischer projection **A** to **D** that matches each of the following: (15.1, 15.3)

a. the L enantiomer of mannose
b. a ketopentose
c. an aldopentose
d. a ketohexose

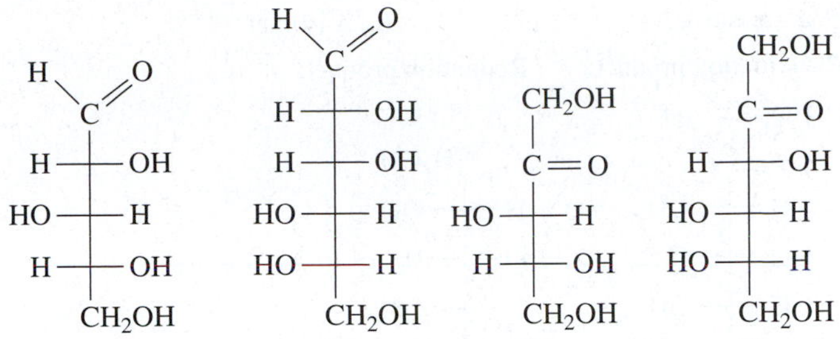
A
B
C
D

ANSWERS

15.1 Photosynthesis requires CO_2, H_2O, and the energy from the Sun. Respiration requires O_2 from the air and glucose from foods.

15.3 Monosaccharides can be a chain of three to eight carbon atoms, one in a carbonyl group as an aldehyde or ketone, and the rest attached to hydroxyl groups. A monosaccharide cannot be split or hydrolyzed into smaller carbohydrates. A disaccharide consists of two monosaccharide units joined together by a glycosidic bond.

15.5 Hydroxyl groups are found in all monosaccharides along with a carbonyl on the first or second carbon that gives an aldehyde or ketone functional group.

15.7 A ketopentose contains hydroxyl and ketone functional groups and has five carbon atoms.

15.9 a. ketohexose **b.** aldopentose

15.11 a. achiral **b.** chiral

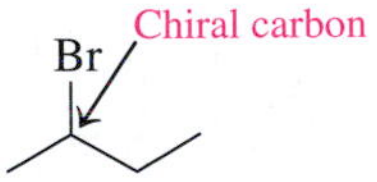
Chiral carbon
Br

c. chiral

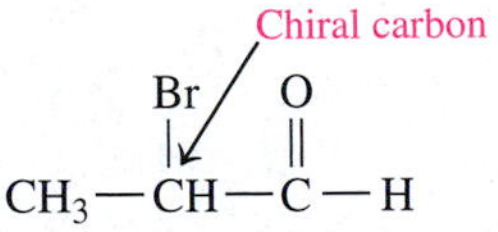
Chiral carbon

d. achiral

15.13 a.

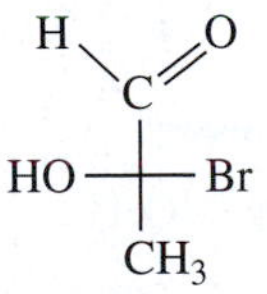

b.

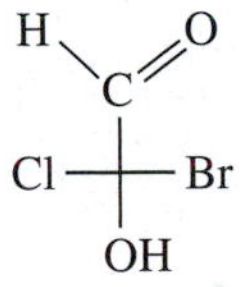

c.

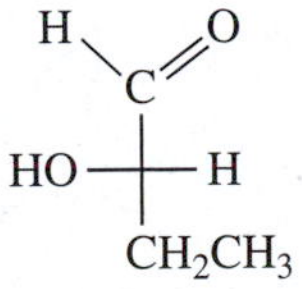

15.15 a. identical **b.** enantiomers
c. enantiomers **d.** enantiomers

15.17 a. D **b.** D **c.** L

15.19 a.

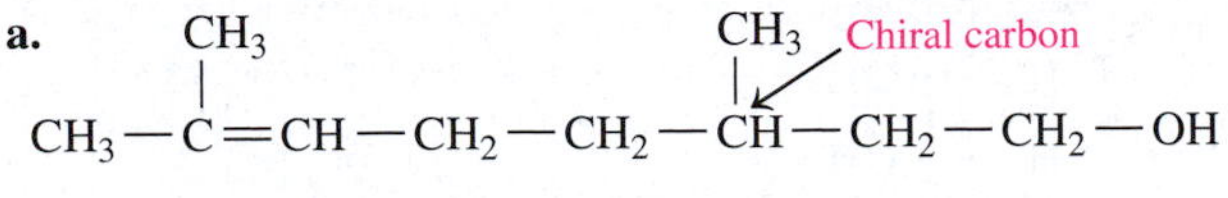
Chiral carbon

b.

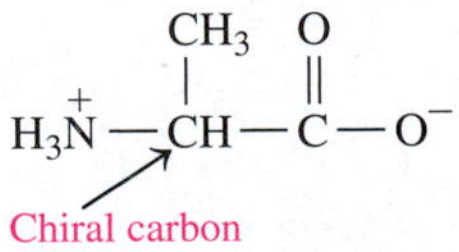
Chiral carbon

15.21 a. D **b.** D **c.** L **d.** D

15.23 a.

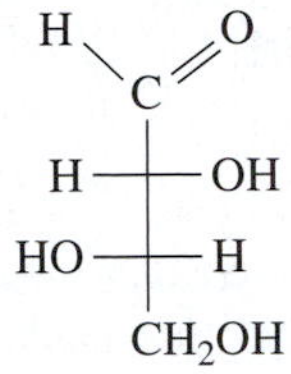

b.

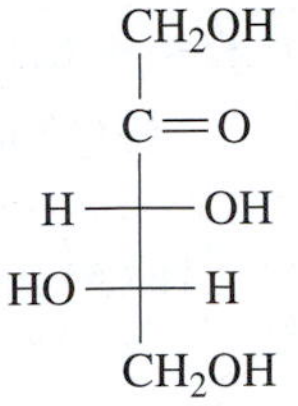

c.

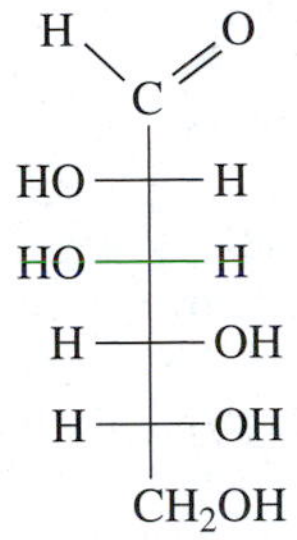

d.

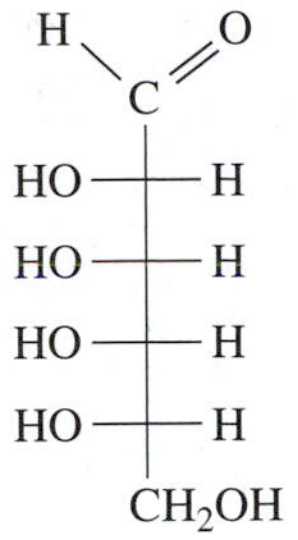

15.25

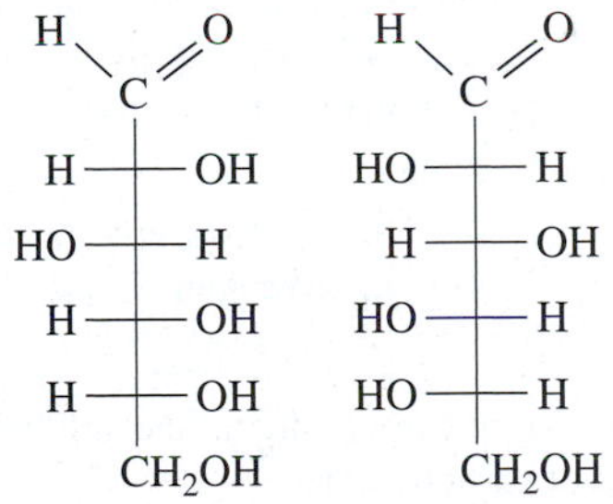

D-Glucose L-Glucose

15.27 In D-galactose, the —OH group on carbon 4 extends to the left. In D-glucose, this —OH group extends to the right.

15.29 a. glucose **b.** galactose

15.31 In the cyclic structure of glucose, there are five carbon atoms and an oxygen atom.

15.33

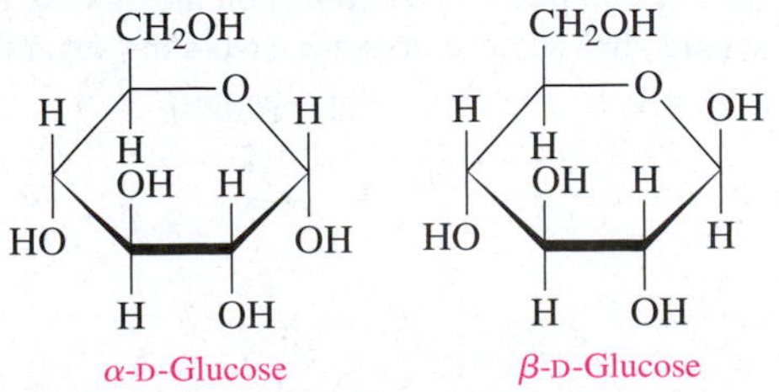

α-D-Glucose β-D-Glucose

15.35 **a.** α isomer **b.** α isomer

15.37 Oxidation product: Reduction product:

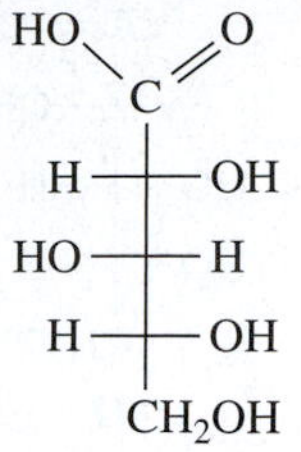

D-Xylonic acid

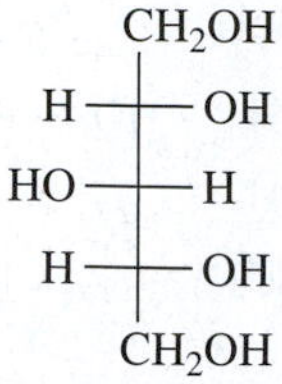

D-Xylitol

15.39 Oxidation product: Reduction product:

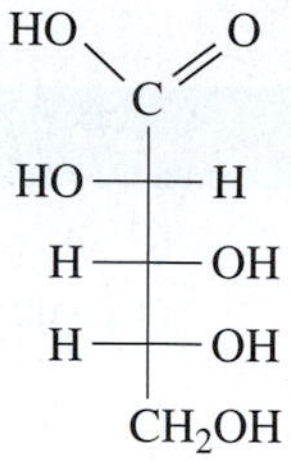

D-Arabinonic acid

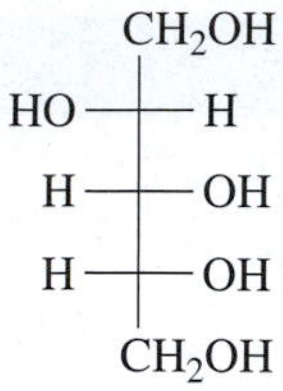

D-Arabitol

15.41 **a.** galactose and glucose; β(1→4)-glycosidic bond; β-lactose
b. glucose and glucose; α(1→4)-glycosidic bond; α-maltose

15.43 **a.** is a reducing sugar **b.** is a reducing sugar

15.45 **a.** sucrose **b.** lactose
c. maltose **d.** lactose

15.47 **a.** Amylose is an unbranched polymer of glucose units joined by α(1→4)-glycosidic bonds; amylopectin is a branched polymer of glucose joined by α(1→4)- and α(1→6)-glycosidic bonds.
b. Amylopectin, which is produced in plants, is a branched polymer of glucose, joined by α(1→4)- and α(1→6)-glycosidic bonds. The branches in amylopectin occur about every 25 glucose units. Glycogen, which is produced in animals, is a highly branched polymer of glucose, joined by α(1→4)- and α(1→6)-glycosidic bonds. The branches in glycogen occur about every 10 to 15 glucose units.

15.49 **a.** cellulose **b.** amylose, amylopectin
c. amylose **d.** glycogen

15.51 6.9 g of glucose

15.53 **a.** Kate's breakfast had 73 g of carbohydrate. She still needs to cut down the amount of carbohydrate.
b. 290 kcal

15.55 **a.** disaccharide **b.** α-D-glucose
c. α(1→6)-glycosidic bond **d.** α
e. is a reducing sugar

15.57 **a.** trisaccharide
b. two glucose and one fructose
c. Melezitose has no free —OH groups on the glucose or fructose units; like sucrose, it is not a reducing sugar.

15.59 **a.** sucrose **b.** cellulose

15.61 **a.**

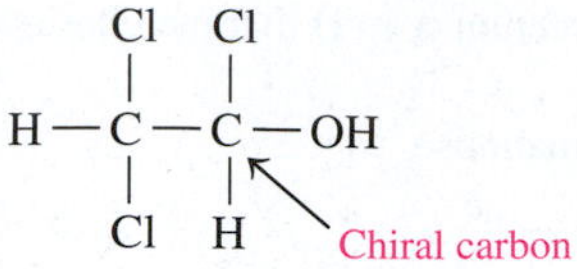
Chiral carbon

b. none

c. none

d.

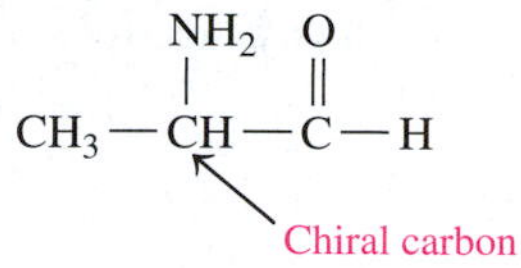
Chiral carbon

e.

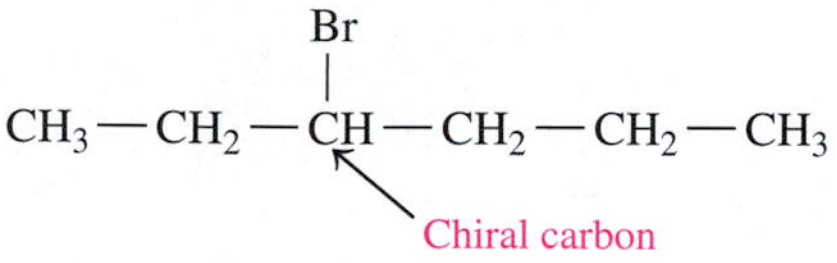
Chiral carbon

15.63 **a.** identical **b.** enantiomers
c. enantiomers **d.** identical

15.65 D-Fructose is a ketohexose, whereas D-galactose is an aldohexose. In the Fischer projection of D-galactose, the —OH group on carbon 4 is drawn on the left; in fructose, the —OH group is drawn on the right.

15.67 L-Galactose is the mirror image of D-galactose. In the Fischer projection of D-galactose, the —OH groups on carbon 2 and carbon 5 are drawn on the right side, but they are drawn on the left for carbon 3 and carbon 4. In L-galactose, the —OH groups are reversed; carbons 2 and 5 have —OH groups drawn on the left, and carbons 3 and 4 have —OH groups drawn on the right.

15.69 **a.**

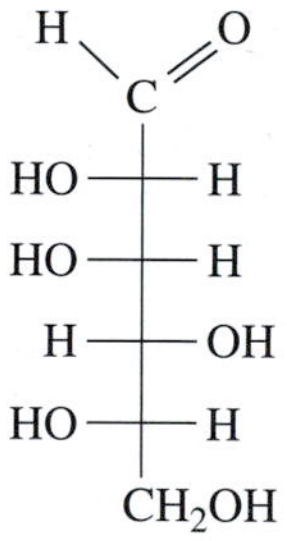

L-Gulose

b.

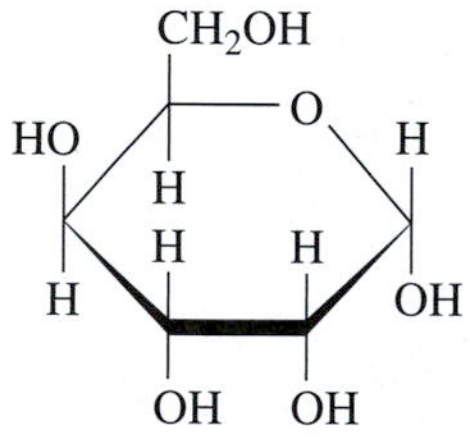

α-D-Gulose

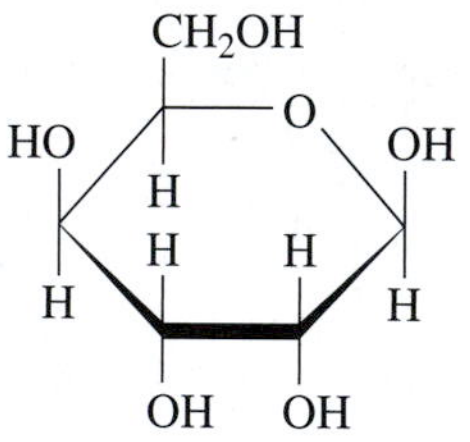

β-D-Gulose

15.71

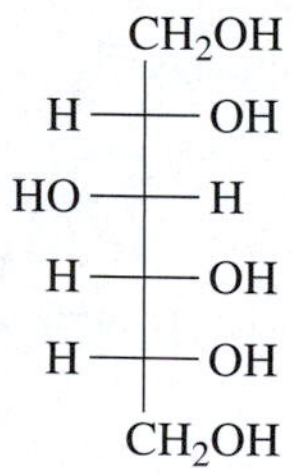

15.73 When α-galactose forms an open-chain structure, it can close to form either α- or β-galactose.

15.75

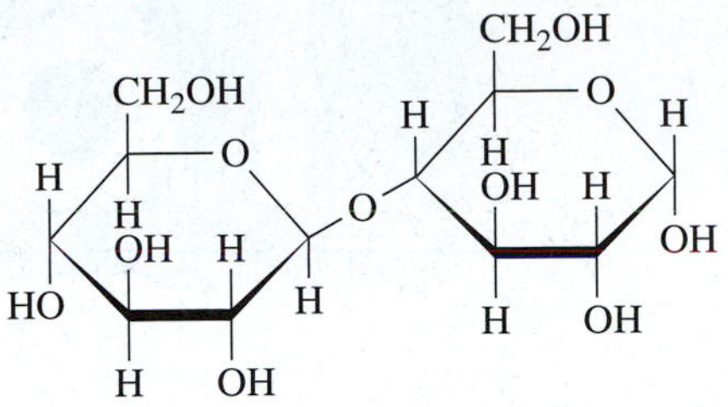
CH₂OH
O
H
H
OH H
HO
H OH
H
O
CH₂OH
O
H
H
H
OH H
OH
H OH

15.77 a.

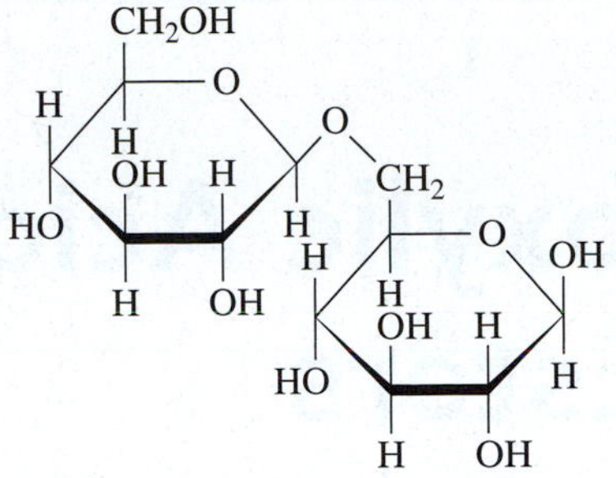
CH₂OH
O
H
H
OH H
HO
H OH
H
O
CH₂
H
O
OH
H
OH H
HO
H
H OH

b. Yes. Gentiobiose is a reducing sugar. The ring on the right can open up to form an aldehyde that can be oxidized.

16

Carboxylic Acids and Esters

Maureen, a surgical technician, begins preparing for Robert's heart surgery. After Maureen places all of the surgical instruments into an autoclave for sterilization, she prepares the room by ensuring that all of the equipment is working properly. She also determines that the room is sterile, as this will minimize the chance of an infection. Right before surgery begins, Maureen shaves Robert's chest and disinfects the incision sites.

After the surgery, Maureen applies Indermil, which is a type of liquid bandage, to Robert's incision sites. Liquid bandages are tissue adhesives that seal surgical or wound incisions on a patient. Stitches and staples are not required, and scarring is minimal. The polymer in a liquid bandage is typically dissolved in an alcohol-based solvent. The alcohol also acts as an antiseptic.

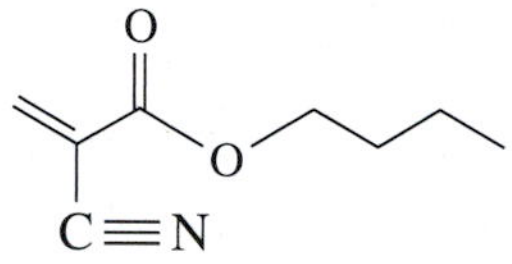
O
O
C≡N
Indermil

Indermil is a cyanoester or cyanoacrylate, as it contains a cyano group (C≡N) and an ester functional group. Esters have a carbonyl group (C═O), which has a single bond to a carbon group on one side of the carbonyl and an oxygen atom on the other side. The oxygen atom is then bonded to a carbon group by a single bond. Cyanoacrylates are a group of adhesives that include "Super Glue" and Indermil, which is butyl cyanoacrylate. The *ate* in butyl cyanoacrylate indicates that an ester is present in the molecule, and the *butyl* indicates the four-carbon group that is bonded to the oxygen atom. In water and body fluids, these adhesives rapidly form strong chains that join the edges of surfaces and hold them in place.

CAREER

Surgical Technician

Surgical technicians prepare the operating room by creating a sterile environment. This includes setting up surgical instruments and equipment and ensuring that all of the equipment is working properly. A sterile environment is critical to the patient's recovery, as it helps lower the chance of an infection. They also prepare patients for surgery by washing, shaving, and disinfecting incision sites. During the surgery, a surgical technician provides the sterile instruments and supplies to the surgeons and surgical assistants.

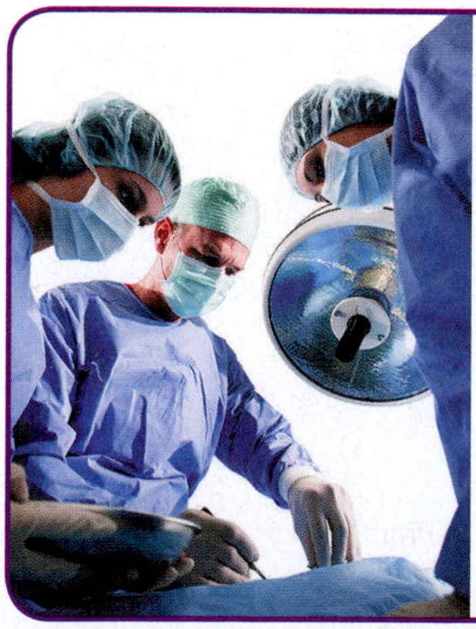

CLINICAL UPDATE

Liquid Bandages

At home, Robert's wound healed with the liquid bandage, which eventually came off. You can read more about liquid bandages in the **CLINICAL UPDATE Liquid Bandages**, pages 577–578, to see how Robert's wife also uses liquid bandage.

16.1 Carboxylic Acids

LEARNING GOAL Write the IUPAC and common names for carboxylic acids; draw their condensed structural and line-angle formulas.

Carboxylic acids are weak acids. They have a sour or tart taste, produce hydronium ions in water, and neutralize bases. You encounter carboxylic acids when you use a salad dressing containing vinegar, which is a solution of acetic acid and water, or experience the sour taste of citric acid in a grapefruit or lemon.

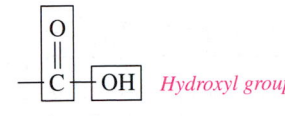

In a **carboxylic acid**, the carbon atom of a carbonyl group is attached to a hydroxyl group that forms a **carboxyl group**. The carboxyl functional group may be attached to an alkyl group or an aromatic group. Some ways to represent the carboxyl group in carboxylic acids are shown for propanoic acid.

$$CH_3-CH_2-\overset{\displaystyle O}{\overset{\|}{C}}-OH$$

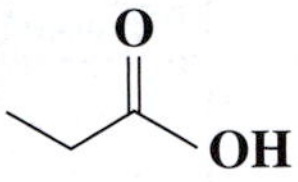

$$CH_3-CH_2-COOH$$

Propanoic acid
(propionic acid)

IUPAC Names of Carboxylic Acids

The IUPAC name of a carboxylic acid replaces the *e* of the corresponding alkane name with *oic acid.* If there are substituents, the carbon chain is numbered beginning with the carboxyl carbon.

$$H-\overset{\displaystyle O}{\overset{\|}{C}}-OH \qquad CH_3-CH_2-\overset{\displaystyle O}{\overset{\|}{C}}-OH \qquad CH_3-\overset{\displaystyle CH_3}{\overset{|}{C}H}-\overset{\displaystyle O}{\overset{\|}{C}}-OH$$

Methanoic acid — Propanoic acid — 2-Methylpropanoic acid

The simplest aromatic carboxylic acid is named benzoic acid. With the carboxyl carbon bonded to carbon 1, the ring is numbered in the direction that gives substituents the smallest possible numbers. The prefixes *ortho*, *meta*, and *para* may be used to show the position of one substituent on the aromatic ring.

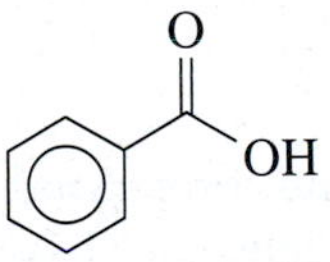

Benzoic acid

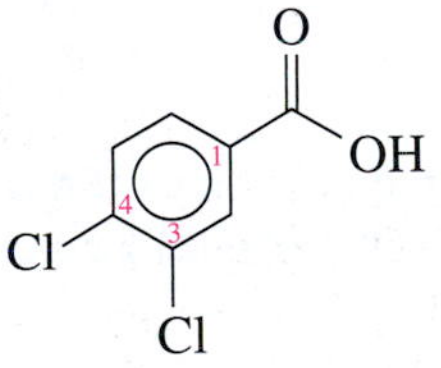

3,4-Dichlorobenzoic acid

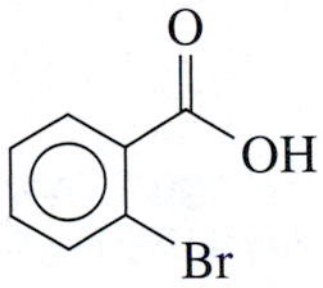

2-Bromobenzoic acid
(*o*-bromobenzoic acid)

Common Names of Carboxylic Acids

Many carboxylic acids are still named by their common names, which use prefixes: *form, acet, propion, butyr*. When using the common names, the Greek letters alpha (α), beta (β), and gamma (γ) are assigned to the carbons adjacent to the carboxyl carbon.

$$CH_3-\overset{\displaystyle CH_3}{\overset{|}{C}H}-CH_2-\overset{\displaystyle O}{\overset{\|}{C}}-OH$$

IUPAC	4	3	2	1	3-Methylbutanoic acid
(common)	γ	β	α		(β-methylbutyric acid)

Formic acid is injected under the skin from bee or red ant stings and other insect bites. Acetic acid is the oxidation product of the ethanol in wines and apple cider. The resulting solution of acetic acid and water is known as vinegar. Propionic acid is obtained from the fats of dairy products. Butyric acid gives the foul odor to rancid butter (see **TABLE 16.1**).

LOOKING AHEAD

REVIEW

Naming and Drawing Alkanes (12.2)

Carbonyl group

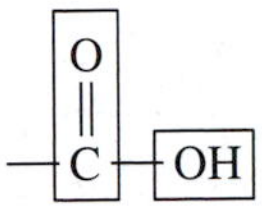

Hydroxyl group

Carboxyl group

CORE CHEMISTRY SKILL

Naming Carboxylic Acids

A red ant sting contains formic acid, that irritates the skin.

TEST

Try Practice Problems 16.1 and 16.2

TABLE 16.1 IUPAC and Common Names of Selected Carboxylic Acids

Condensed Structural Formula	Line-Angle Formula	IUPAC Name	Common Name	Ball-and-Stick Model
$\mathrm{H-\overset{O}{\overset{\Vert}{C}}-OH}$	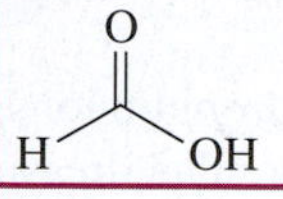	Methanoic acid	Formic acid	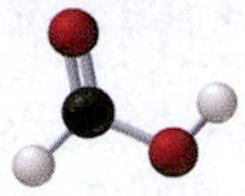
$\mathrm{CH_3-\overset{O}{\overset{\Vert}{C}}-OH}$	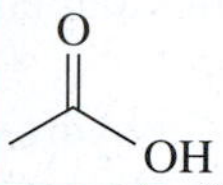	Ethanoic acid	Acetic acid	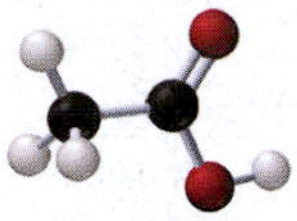
$\mathrm{CH_3-CH_2-\overset{O}{\overset{\Vert}{C}}-OH}$	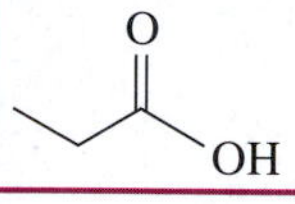	Propanoic acid	Propionic acid	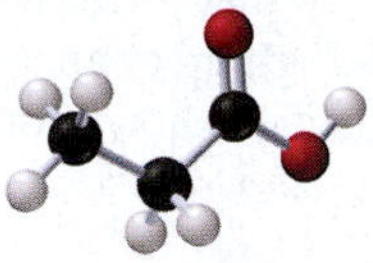
$\mathrm{CH_3-CH_2-CH_2-\overset{O}{\overset{\Vert}{C}}-OH}$	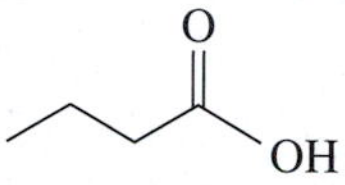	Butanoic acid	Butyric acid	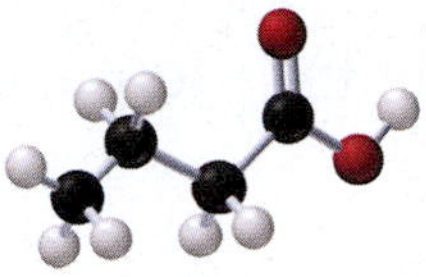

▸SAMPLE PROBLEM 16.1 Naming Carboxylic Acids

TRY IT FIRST

Write the IUPAC name for the following:

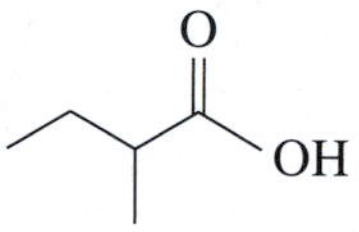

SOLUTION GUIDE

	Given	Need	Connect
ANALYZE THE PROBLEM	four-carbon chain, methyl substituent	IUPAC name	position of methyl group, replace *e* in alkane name with *oic acid*

STEP 1 **Identify the longest carbon chain and replace the *e* in the corresponding alkane name with *oic acid*.** The longest chain contains four carbon atoms, which is butanoic acid.

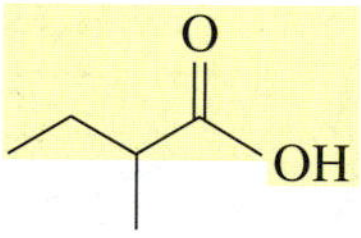

butanoic acid

STEP 2 **Name and number any substituents by counting the carboxyl group as carbon 1.** The substituent on carbon 2 is methyl. The IUPAC name for this compound is 2-methylbutanoic acid.

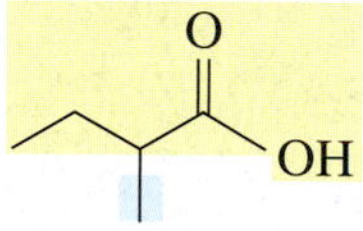

4 3 2 1

2-methylbutanoic acid

STUDY CHECK 16.1

a. Write the IUPAC name for the following:

$$\mathrm{CH_3-\overset{Cl}{\overset{|}{C}H}-CH_2-\overset{Cl}{\overset{|}{C}H}-\overset{O}{\overset{\Vert}{C}}-OH}$$

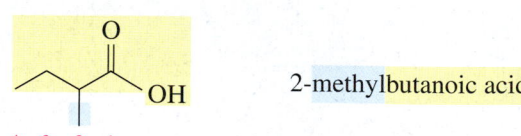

b. Draw the line-angle formula for 3-chlorobenzoic acid.

ANSWER

a. 2,4-dichloropentanoic acid

b. 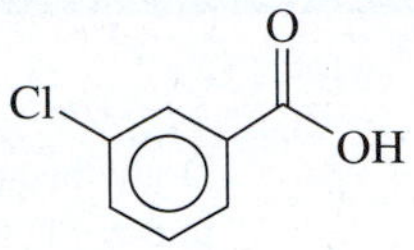

TEST

Try Practice Problems 16.3 to 16.8

The sour taste of vinegar is because of ethanoic acid (acetic acid).

Preparation of Carboxylic Acids

Carboxylic acids can be prepared from primary alcohols or aldehydes. For example, when ethyl alcohol in wine is exposed to oxygen in the air, vinegar is produced. The oxidation process converts the ethyl alcohol (1° alcohol) to acetaldehyde, and then to acetic acid, the carboxylic acid in vinegar.

$$\underset{\text{Ethanol (ethyl alcohol)}}{CH_3-\overset{\overset{\displaystyle OH}{|}}{C}H_2} \xrightarrow{[O]} \underset{\text{Ethanal (acetaldehyde)}}{CH_3-\overset{\overset{\displaystyle O}{\|}}{C}-H} \xrightarrow{[O]} \underset{\text{Ethanoic acid (acetic acid)}}{CH_3-\overset{\overset{\displaystyle O}{\|}}{C}-OH}$$

TEST

Try Practice Problems 16.9 and 16.10

Chemistry Link to Health

Alpha Hydroxy Acids

Alpha hydroxy acids (AHAs), found in fruits, milk, and sugar cane, are naturally occurring carboxylic acids with a hydroxyl group (—OH) on the carbon atom that is adjacent to the carboxyl group. Cleopatra, Queen of Egypt, reportedly bathed in sour milk to smooth her skin. Dermatologists have been using products with high concentrations (20 to 70%) of AHAs to remove acne scars and in skin peels to reduce irregular pigmentation and age spots. Lower concentrations (8 to 10%) of AHAs are added to skin care products for the purpose of smoothing fine lines, improving skin texture, and cleansing pores. Several different AHAs may be found in skin care products singly or in combination. Glycolic acid and lactic acid are most frequently used.

Recent studies indicate that products with AHAs increase sensitivity of the skin to sun and UV radiation. It is recommended that a sunscreen with a sun protection factor (SPF) of at least 15 be used when treating the skin with products that include AHAs. Products containing AHAs at concentrations under 10% and pH values greater than 3.5 are generally considered safe. However, the Food and Drug Administration (FDA) has received reports of AHAs causing skin irritation including blisters, rashes, and discoloration of the skin. The FDA does not require product safety reports from cosmetic manufacturers, although they are responsible for marketing safe products. The FDA advises that you test any product containing AHAs on a small area of skin before you use it on a large area.

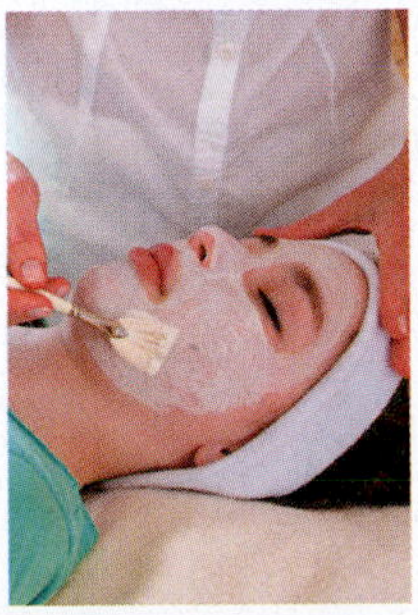

Alpha hydroxy carboxylic acids are used in many skin care products.

Alpha Hydroxy Acid (Source)	Condensed Structural Formula
Glycolic acid (sugar cane)	$HO-CH_2-\overset{\overset{\displaystyle O}{\|}}{C}-OH$
Lactic acid (sour milk)	$CH_3-\overset{\overset{\displaystyle OH}{\vert}}{C}H-\overset{\overset{\displaystyle O}{\|}}{C}-OH$
Tartaric acid (grapes)	$HO-\overset{\overset{\displaystyle O}{\|}}{C}-\overset{\overset{\displaystyle OH}{\vert}}{C}H-\overset{\overset{\displaystyle OH}{\vert}}{C}H-\overset{\overset{\displaystyle O}{\|}}{C}-OH$
Malic acid (apples)	$HO-\overset{\overset{\displaystyle O}{\|}}{C}-CH_2-\overset{\overset{\displaystyle OH}{\vert}}{C}H-\overset{\overset{\displaystyle O}{\|}}{C}-OH$
Citric acid (citrus fruits)	$HO-\overset{\overset{\displaystyle CH_2-COOH}{\vert}}{\underset{\underset{\displaystyle CH_2-COOH}{\vert}}{C}}-COOH$

PRACTICE PROBLEMS

16.1 Carboxylic Acids

16.1 What carboxylic acid is responsible for the pain of an ant sting?

16.2 What carboxylic acid is found in vinegar?

16.3 Draw the condensed structural formula and write the IUPAC name for each of the following:
- **a.** a carboxylic acid that has the formula $C_6H_{12}O_2$, with no substituents
- **b.** a carboxylic acid that has the formula $C_6H_{12}O_2$, with one ethyl substituent

16.4 Draw the condensed structural formula and write the IUPAC name for each of the following:
- **a.** a carboxylic acid that has the formula $C_5H_{10}O_2$, with no substituents
- **b.** a carboxylic acid that has the formula $C_5H_{10}O_2$, with two methyl substituents

16.5 Write the IUPAC and common name, if any, for each of the following carboxylic acids:

a. $CH_3-\overset{\overset{\large O}{\|}}{C}-OH$

b. 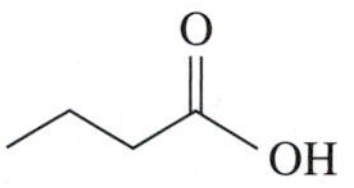

c. 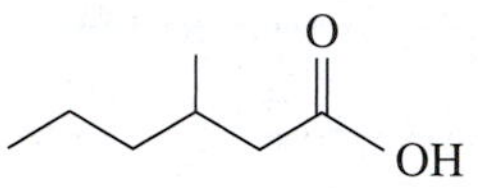

d. 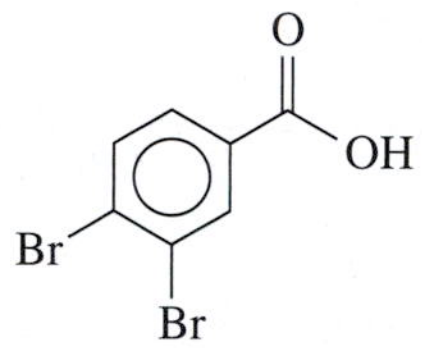

16.6 Write the IUPAC and common name, if any, for each of the following carboxylic acids:

a. $H-\overset{\overset{\large O}{\|}}{C}-OH$

b. 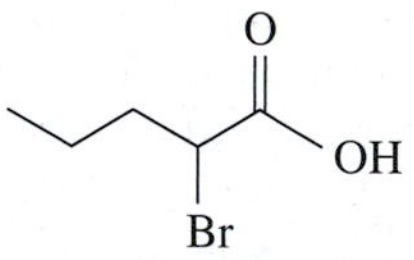

c. 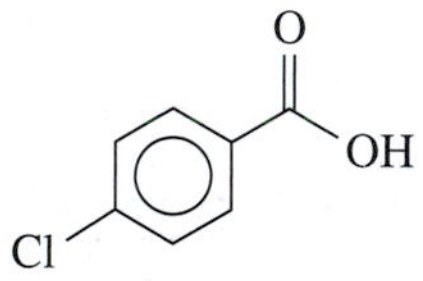

d. $CH_3-\underset{\underset{\large CH_3}{|}}{\overset{\overset{\large CH_3}{|}}{C}}-CH_2-\overset{\overset{\large O}{\|}}{C}-OH$

16.7 Draw the condensed structural formulas for **a** and **b** and the line-angle formulas for **c** and **d**.
- **a.** 2-chloroethanoic acid
- **b.** 3-hydroxypropanoic acid
- **c.** α-methylbutyric acid
- **d.** 3,5-dibromoheptanoic acid

16.8 Draw the condensed structural formulas for **a** and **b** and the line-angle formulas for **c** and **d**.
- **a.** pentanoic acid
- **b.** 3-ethylbenzoic acid
- **c.** α-hydroxyacetic acid
- **d.** 2,4-dibromobutanoic acid

16.9 Draw the condensed structural or line-angle formula for the carboxylic acid formed by the oxidation of each of the following:

a. CH_3-OH

b. $CH_3-\overset{\overset{\large O}{\|}}{C}-H$

c. 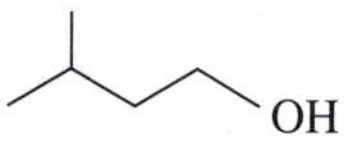

d. 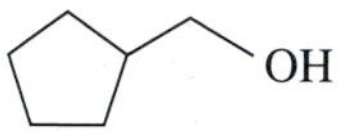

16.10 Draw the condensed structural or line-angle formula for the carboxylic acid formed by the oxidation of each of the following:

a. 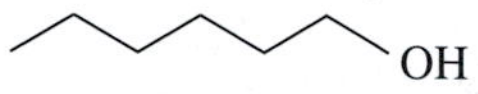

b. 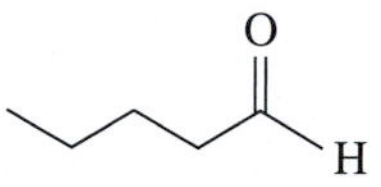

c. $CH_3-\overset{\overset{\large CH_3}{|}}{CH}-CH_2-\overset{\overset{\large O}{\|}}{C}-H$

d. 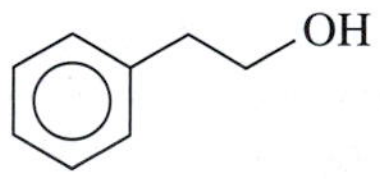

REVIEW

Writing Equations for Reactions of Acids and Bases (11.7)

16.2 Properties of Carboxylic Acids

LEARNING GOAL Describe the boiling points, solubility, dissociation, and neutralization of carboxylic acids.

Carboxylic acids are among the most polar organic compounds because their functional group consists of two polar groups: a hydroxyl group ($-OH$) and a carbonyl group ($C=O$). The $-OH$ group is similar to the functional group in alcohols, and the $C=O$ is similar to the functional group of aldehydes and ketones.

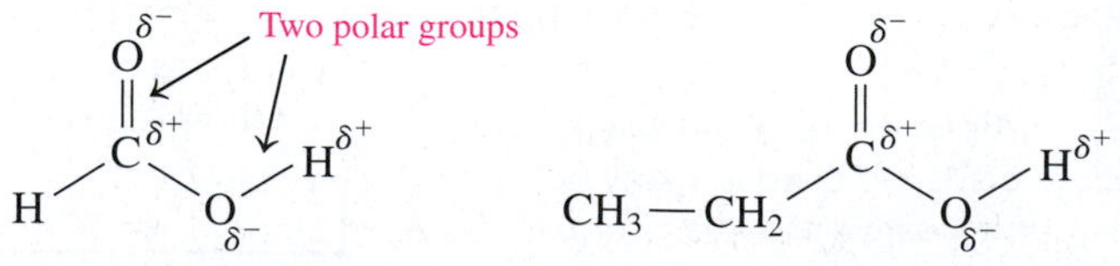
Two polar groups

Boiling Points

The polar carboxyl groups allow carboxylic acids to form several hydrogen bonds with other carboxylic acid molecules. This effect of hydrogen bonds gives carboxylic acids higher boiling points than alcohols, ketones, and aldehydes of similar molar mass.

	$CH_3-CH_2-C(=O)-H$	$CH_3-CH_2-CH_2-OH$	$CH_3-C(=O)-OH$
Name	Propanal	1-Propanol	Ethanoic acid
Family	Aldehyde	Alcohol	Carboxylic acid
Molar Mass	58	60	60
bp	49 °C	97 °C	118 °C

Boiling Point Increases →

An important reason for the higher boiling points of carboxylic acids is that two carboxylic acids form hydrogen bonds between their carboxyl groups, resulting in a *dimer*. As a dimer, the mass of the carboxylic acid is effectively doubled, which means that a higher temperature is required to reach the boiling point. **TABLE 16.2** lists the boiling points for some selected carboxylic acids.

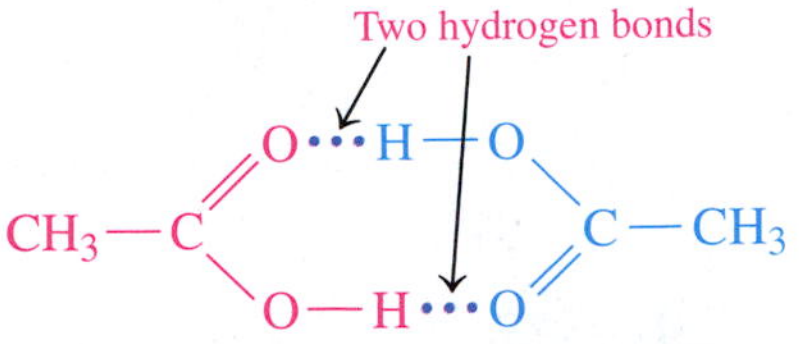
Two hydrogen bonds
CH3—C
C—CH3
O···H—O
O—H···O
A dimer of two ethanoic acid molecules

TABLE 16.2 Boiling Points, Solubilities, and Acid Dissociation Constants for Selected Carboxylic Acids

IUPAC Name	Condensed Structural Formula	Boiling Point (°C)	Solubility in Water	Acid Dissociation Constant (at 25 °C)
Methanoic acid	$H-C(=O)-OH$	101	Soluble	1.8×10^{-4}
Ethanoic acid	$CH_3-C(=O)-OH$	118	Soluble	1.8×10^{-5}
Propanoic acid	$CH_3-CH_2-C(=O)-OH$	141	Soluble	1.3×10^{-5}
Butanoic acid	$CH_3-CH_2-CH_2-C(=O)-OH$	164	Soluble	1.5×10^{-5}
Pentanoic acid	$CH_3-CH_2-CH_2-CH_2-C(=O)-OH$	187	Soluble	1.5×10^{-5}
Hexanoic acid	$CH_3-CH_2-CH_2-CH_2-CH_2-C(=O)-OH$	205	Slightly soluble	1.4×10^{-5}
Benzoic acid	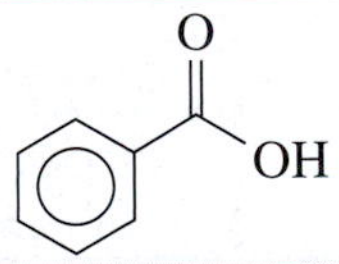	250	Slightly soluble	6.4×10^{-5}

▶ SAMPLE PROBLEM 16.2 Boiling Points of Carboxylic Acids

TRY IT FIRST

Match each of the compounds 2-butanol, pentane, and propanoic acid with a boiling point of 141 °C, 100 °C, or 36 °C. (They have about the same molar mass.)

SOLUTION

The boiling point increases when the molecules of a compound can form hydrogen bonds or have dipole–dipole attractions. Pentane has the lowest boiling point, 36 °C, because alkanes cannot form hydrogen bonds or dipole–dipole attractions. 2-Butanol has a higher boiling point, 100 °C, than pentane because an alcohol can form hydrogen bonds. Propanoic acid has the highest boiling point, 141 °C, because carboxylic acids can form stable dimers through hydrogen bonding to increase their effective molar mass and therefore their boiling points.

STUDY CHECK 16.2

Why would methanoic acid (molar mass 46, bp 101 °C) have a higher boiling point than ethanol (molar mass 46, bp 78 °C)?

ANSWER

Two methanoic acid molecules form a dimer, which gives an effective molar mass that is double that of a single acid molecule. Thus, a higher boiling point is required than for ethanol.

TEST

Try Practice Problems 16.11 and 16.12

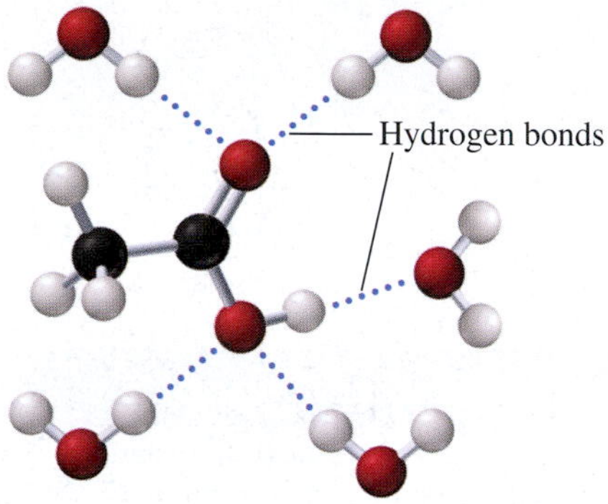
Hydrogen bonds

FIGURE 16.1 ▶ Acetic acid forms hydrogen bonds with water molecules.

Q Why do the atoms in the carboxyl group form hydrogen bonds with water molecules?

Solubility in Water

Carboxylic acids with one to five carbons are soluble in water because the carboxyl group forms hydrogen bonds with several water molecules (see **FIGURE 16.1**). However, as the length of the hydrocarbon chain increases, the nonpolar portion reduces the solubility of the carboxylic acid in water. Carboxylic acids having more than five carbons are not very soluble in water. **TABLE 16.2** lists the solubility for some selected carboxylic acids.

Acidity of Carboxylic Acids

An important property of carboxylic acids is their dissociation in water. When a carboxylic acid dissociates in water, H^+ is transferred to a water molecule to form a negatively charged **carboxylate ion** and a positively charged hydronium ion (H_3O^+). Carboxylic acids are more acidic than most other organic compounds including phenols. However, they are weak acids because only 1% of the carboxylic acid molecules dissociate in water. The acid dissociation constants of some carboxylic acids are given in **TABLE 16.2**.

Carboxylic Acid **Carboxylate Ion**

$$\underset{\text{Ethanoic acid (acetic acid)}}{CH_3-\overset{\overset{\displaystyle O}{\|}}{C}-OH} + H_2O \rightleftharpoons \underset{\text{Ethanoate ion (acetate ion)}}{CH_3-\overset{\overset{\displaystyle O}{\|}}{C}-O^-} + \underset{\text{Hydronium ion}}{H_3O^+}$$

▶SAMPLE PROBLEM 16.3 Dissociation of Carboxylic Acids in Water

TRY IT FIRST

Write the balanced chemical equation for the dissociation of propanoic acid in water.

SOLUTION

ANALYZE THE PROBLEM	Given	Need	Connect
	propanoic acid, H_2O	dissociation equation	products: propanoate ion, H_3O^+

The dissociation of propanoic acid produces a carboxylate ion and a hydronium ion.

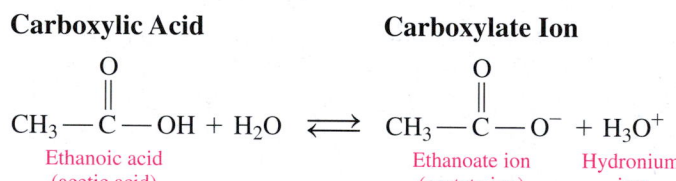

$$\underset{\text{Propanoic acid (propionic acid)}}{CH_3-CH_2-\overset{\overset{\displaystyle O}{\|}}{C}-OH} + H_2O \rightleftharpoons \underset{\text{Propanoate ion (propionate ion)}}{CH_3-CH_2-\overset{\overset{\displaystyle O}{\|}}{C}-O^-} + H_3O^+$$

STUDY CHECK 16.3

a. Use condensed structural formulas to write the balanced chemical equation for the dissociation of formic acid in water.

b. Use line-angle formulas to write the balanced chemical equation for the dissociation of 3-bromobenzoic acid in water.

ANSWER

a. $H{-}\overset{O}{\overset{\|}{C}}{-}OH + H_2O \rightleftarrows H{-}\overset{O}{\overset{\|}{C}}{-}O^- + H_3O^+$

b. 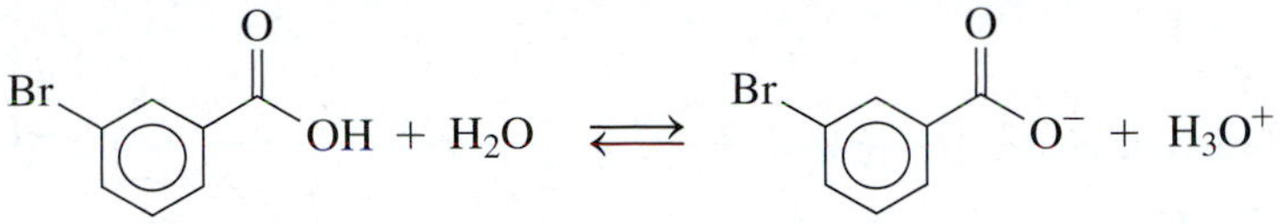

TEST

Try Practice Problems 16.13 to 16.16

Neutralization of Carboxylic Acids

Because carboxylic acids are weak acids, they are completely neutralized by strong bases such as NaOH and KOH. The products are a **carboxylate salt** and water. The carboxylate ion is named by replacing the *ic acid* ending of the acid name with *ate*.

$$H{-}\overset{O}{\overset{\|}{C}}{-}OH + NaOH \longrightarrow H{-}\overset{O}{\overset{\|}{C}}{-}O^-\ Na^+ + H_2O$$

Methanoic acid (formic acid) — Sodium methanoate (sodium formate)

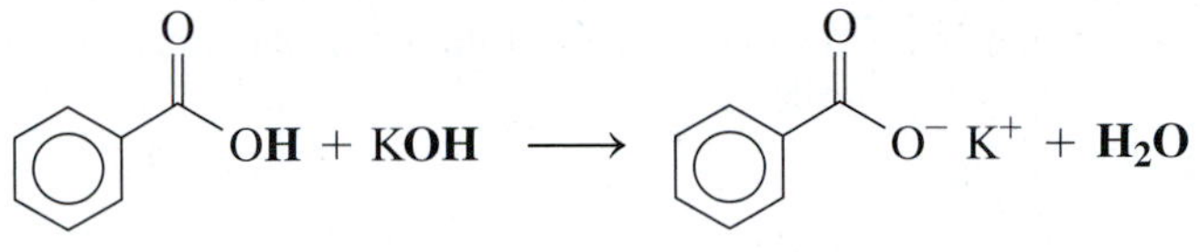

Benzoic acid — Potassium benzoate

CUP NOODLES
Chicken
Chef BOYARDEE
BEEF
Ravioli
Maruchan Ramen
Oriental Flavor

FIGURE 16.2 ▶ Carboxylate salts are often used as preservatives and flavor enhancers in soups and seasonings.

Sodium propionate, a preservative, is added to bread, cheeses, and bakery items to inhibit the spoilage of the food by microorganisms. Sodium benzoate, an inhibitor of mold and bacteria, is added to juices, margarine, relishes, salads, and jams. Monosodium glutamate (MSG) is added to meats, fish, vegetables, and bakery items to enhance flavor, although it causes headache in some people.

$$CH_3{-}CH_2{-}\overset{O}{\overset{\|}{C}}{-}O^-\ Na^+$$

Sodium propanoate (sodium propionate)

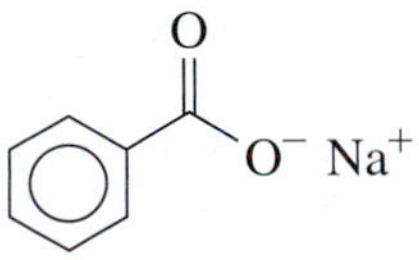

Sodium benzoate

$$HO{-}\overset{O}{\overset{\|}{C}}{-}CH_2{-}CH_2{-}\overset{NH_2}{\overset{|}{CH}}{-}\overset{O}{\overset{\|}{C}}{-}O^-\ Na^+$$

Monosodium glutamate

Carboxylate salts are ionic compounds with strong attractions between positively charged metal ions such as Li^+, Na^+, and K^+ and the negatively charged carboxylate ion. Like most salts, the carboxylate salts are solids at room temperature, have high melting points, and are usually soluble in water.

▶SAMPLE PROBLEM 16.4 Neutralization of a Carboxylic Acid

TRY IT FIRST

Write the balanced chemical equation for the neutralization of propanoic acid (propionic acid) with sodium hydroxide.

SOLUTION

ANALYZE THE PROBLEM	Given	Need	Connect
	propanoic acid, NaOH	neutralization equation	products: sodium propanoate, H_2O

A chemical equation for the neutralization of carboxylic acid includes the reactants, a carboxylic acid and a base, and the products, a carboxylate salt and water.

$$CH_3{-}CH_2{-}\overset{O}{\overset{\|}{C}}{-}OH + NaOH \longrightarrow CH_3{-}CH_2{-}\overset{O}{\overset{\|}{C}}{-}O^-\ Na^+ + H_2O$$

Propanoic acid (propionic acid) — Sodium hydroxide — Sodium propanoate (sodium propionate)

STUDY CHECK 16.4

a. What carboxylic acid will give potassium butanoate (potassium butyrate) when it is neutralized by KOH?

b. What carboxylic acid will give sodium benzoate when it is neutralized by NaOH?

ANSWER

a. butanoic acid (butyric acid)

b. benzoic acid

TEST

Try Practice Problems 16.17 to 16.20

Chemistry Link to Health

Carboxylic Acids in Metabolism

Several carboxylic acids are part of the metabolic processes within our cells. For example, during glycolysis, a molecule of glucose is broken down into two molecules of pyruvic acid, or actually, its carboxylate ion, pyruvate. During strenuous exercise, when oxygen levels are low (anaerobic), pyruvic acid is reduced to give lactic acid or the lactate ion.

Pyruvic acid   Lactic acid

During exercise, pyruvic acid is converted to lactic acid in the muscles.

In the *citric acid cycle*, also called the *Krebs cycle*, di- and tricarboxylic acids are oxidized and decarboxylated (loss of CO_2) to produce energy for the cells of the body. These carboxylic acids are normally referred to by their common names. At the start of the citric acid cycle, citric acid with six carbons is converted to five-carbon α-ketoglutaric acid. Citric acid is also the acid that gives the sour taste to citrus fruits such as lemons and grapefruits.

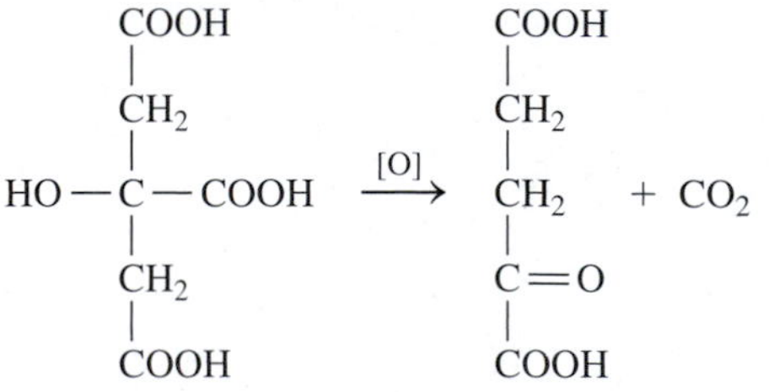

Citric acid α-Ketoglutaric acid

The citric acid cycle continues as α-ketoglutaric acid loses CO_2 to give a four-carbon succinic acid. Then a series of reactions converts succinic acid to oxaloacetic acid. We see that some of the functional groups we have studied, along with reactions such as hydration and oxidation, are part of the metabolic processes that take place in our cells.

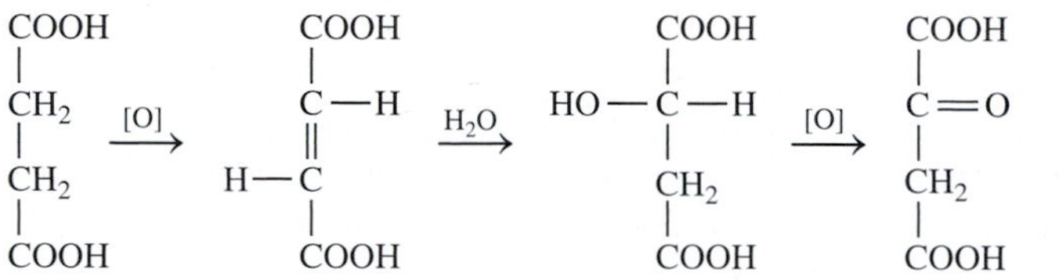

Succinic acid   Fumaric acid   Malic acid   Oxaloacetic acid

At the pH of the aqueous environment in the cells, the carboxylic acids are dissociated, which means it is actually the carboxylate ions that take part in the reactions of the citric acid cycle. For example, in water, succinic acid is in equilibrium with its carboxylate ion, succinate.

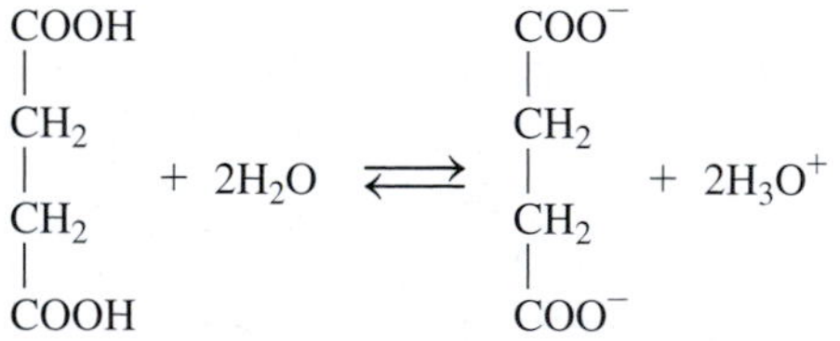

Succinic acid   Succinate ion

Citric acid gives the sour taste to citrus fruits.

PRACTICE PROBLEMS

16.2 Properties of Carboxylic Acids

16.11 Identify the compound in each of the following pairs that has the higher boiling point. Explain.
a. ethanoic acid (acetic acid) or butanoic acid
b. 1-propanol or propanoic acid
c. butanone or butanoic acid

16.12 Identify the compound in each of the following pairs that has the higher boiling point. Explain.
a. propanone (acetone) or propanoic acid
b. propanoic acid or hexanoic acid
c. ethanol or ethanoic acid (acetic acid)

16.13 Identify the compound in each of the following groups that is most soluble in water. Explain.
a. propanoic acid, hexanoic acid, benzoic acid
b. pentane, 1-hexanol, propanoic acid

16.14 Identify the compound in each of the following groups that is most soluble in water. Explain.
a. butanone, butanoic acid, butane
b. acetic acid, hexanoic acid, octanoic acid

16.15 Write the balanced chemical equation for the dissociation of each of the following carboxylic acids in water:
a. butanoic acid
b. $CH_3-\underset{\displaystyle CH_3}{\overset{|}{CH}}-\overset{\displaystyle O}{\overset{\|}{C}}-OH$

16.16 Write the balanced chemical equation for the dissociation of each of the following carboxylic acids in water:
a. 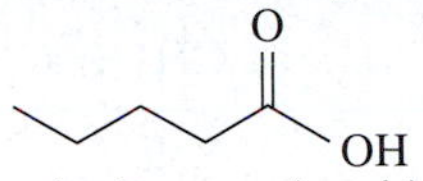
O
OH
b. α-hydroxyacetic acid

16.17 Use condensed structural formulas for **a** and **b** and the line-angle formula for **c** to write the balanced chemical equation for the reaction of each of the following carboxylic acids with NaOH:
a. pentanoic acid
b. 2-chloropropanoic acid
c. benzoic acid

16.18 Use condensed structural formulas for **a** and **b** and the line-angle formula for **c** to write the balanced chemical equation for the reaction of each of the following carboxylic acids with KOH:
a. hexanoic acid
b. 2-methylbutanoic acid
c. *p*-chlorobenzoic acid

16.19 Write the IUPAC and common names, if any, of the carboxylate salts produced in problem 16.17.

16.20 Write the IUPAC and common names, if any, of the carboxylate salts produced in problem 16.18.

16.3 Esters

LEARNING GOAL Write the IUPAC and common names for esters; draw condensed structural and line-angle formulas.

When a carboxylic acid reacts with an alcohol, an **ester** and water are produced when the —H of the carboxylic acid is replaced by an alkyl group. Fats and oils are esters of glycerol and fatty acids, which are long-chain carboxylic acids. Esters produce the pleasant aromas and flavors of many fruits, such as bananas, strawberries, and oranges.

Carboxylic Acid	**Ester**
$CH_3-\overset{\displaystyle O}{\overset{\|}{C}}-OH$	$CH_3-\overset{\displaystyle O}{\overset{\|}{C}}-O-CH_3$
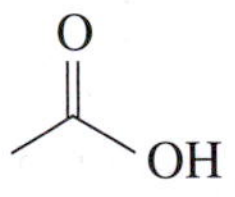	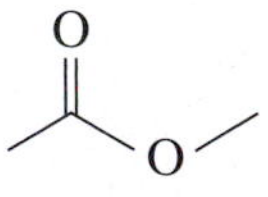
Ethanoic acid (acetic acid)	Methyl ethanoate (methyl acetate)

Esterification

In a reaction called **esterification**, an ester is produced when a carboxylic acid and an alcohol react in the presence of an acid catalyst (usually H_2SO_4) and heat. In esterification, the —OH group from the carboxylic acid and the —H from the alcohol are removed and combine to form water. An excess of the alcohol is used to shift the equilibrium in the direction of the formation of the ester product.

$$\underset{\text{Ethanoic acid (acetic acid)}}{CH_3-\overset{\displaystyle O}{\overset{\|}{C}}-OH} + \underset{\text{Methanol (methyl alcohol)}}{H-O-CH_3} \overset{H^+,\text{ heat}}{\rightleftarrows} \underset{\text{Methyl ethanoate (methyl acetate)}}{CH_3-\overset{\displaystyle O}{\overset{\|}{C}}-O-CH_3} + H-OH$$

Pentyl ethanoate provides the flavor and odor of bananas.

For example, pentyl ethanoate, which is the ester responsible for the flavor and odor of bananas, can be prepared using ethanoic acid and 1-pentanol. The equation for this esterification is written as

$$\underset{\text{Ethanoic acid (acetic acid)}}{CH_3-\overset{\overset{\displaystyle O}{\|}}{C}-\mathbf{OH}} + \underset{\text{1-Pentanol (pentyl alcohol)}}{\mathbf{H}-O-CH_2-CH_2-CH_2-CH_2-CH_3} \xrightleftharpoons{H^+,\ \text{heat}}$$

$$\underset{\text{Pentyl ethanoate (pentyl acetate)}}{CH_3-\overset{\overset{\displaystyle O}{\|}}{C}-O-CH_2-CH_2-CH_2-CH_2-CH_3} + \mathbf{H_2O}$$

▶SAMPLE PROBLEM 16.5 Writing Esterification Equations

TRY IT FIRST

An ester that has the smell of pineapple can be synthesized from butanoic acid and methanol. Use condensed structural formulas to write the balanced chemical equation for the formation of this ester.

SOLUTION

ANALYZE THE PROBLEM	Given	Need	Connect
	butanoic acid, methanol	esterification equation	products: methyl butanoate, H_2O

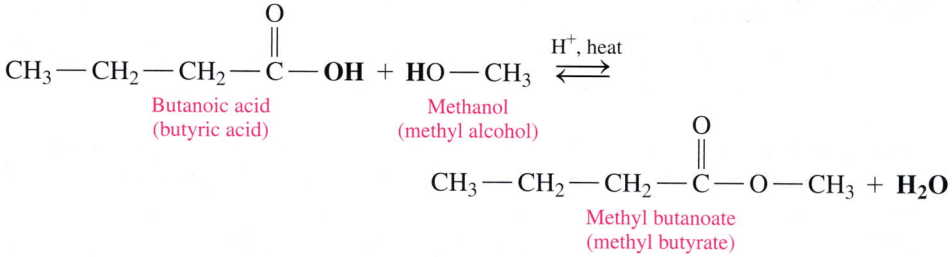

$$\underset{\text{Butanoic acid (butyric acid)}}{CH_3-CH_2-CH_2-\overset{\overset{\displaystyle O}{\|}}{C}-\mathbf{OH}} + \underset{\text{Methanol (methyl alcohol)}}{\mathbf{HO}-CH_3} \xrightleftharpoons{H^+,\ \text{heat}}$$

$$\underset{\text{Methyl butanoate (methyl butyrate)}}{CH_3-CH_2-CH_2-\overset{\overset{\displaystyle O}{\|}}{C}-O-CH_3} + \mathbf{H_2O}$$

The flavor and odor of plums is provided by an ester from methanoic acid and 1-butanol.

STUDY CHECK 16.5

The ester that smells like plums can be synthesized from methanoic acid and 1-butanol. Write the balanced chemical equation using line-angle formulas for the formation of this ester.

ANSWER

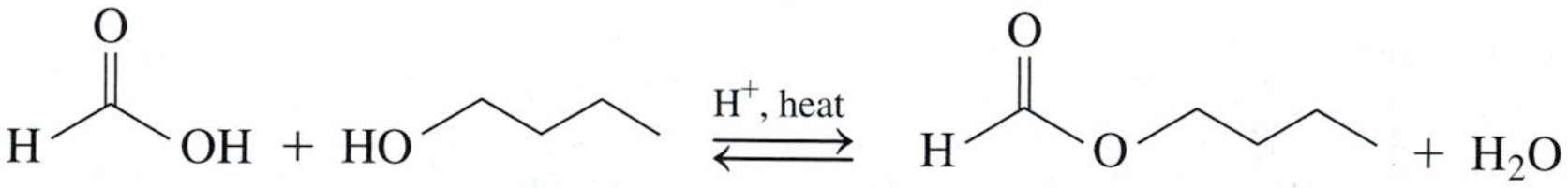

O
H
OH
+ HO
H⁺, heat
O
H
O
+ H₂O

TEST

Try Practice Problems 16.21 to 16.24

Naming Esters

The name of an ester consists of two words, which are derived from the names of the alcohol and the acid in that ester. The first word indicates the *alkyl* part from the alcohol. The second word is the name of the *carboxylate* from the carboxylic acid. The IUPAC names of esters use the IUPAC names of the acids, while the common names of esters use the common names of the acids. Let's take a look at the following ester, which has a pleasant, fruity odor. We start by separating the ester bond to identify the alkyl part from the alcohol and the carboxylate part from the acid. Then we name the ester as an alkyl carboxylate.

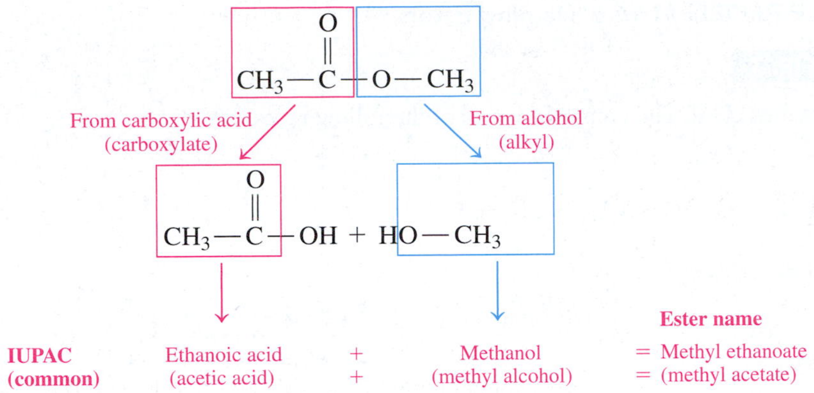
From carboxylic acid
(carboxylate)
From alcohol
(alkyl)
Ester name
IUPAC
(common)
Ethanoic acid
(acetic acid)
+
+
Methanol
(methyl alcohol)
= Methyl ethanoate
= (methyl acetate)

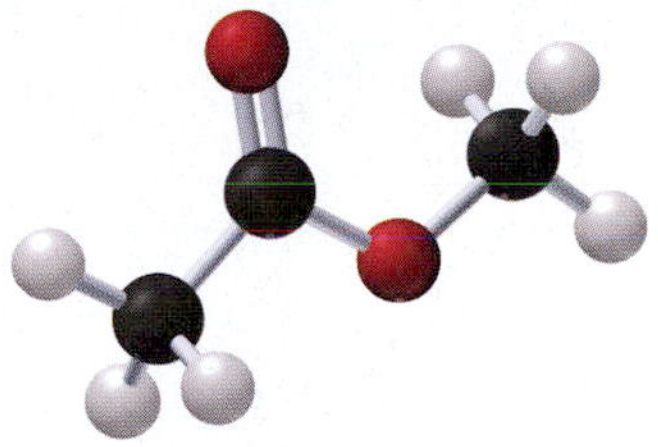

Methyl ethanoate
(methyl acetate)

The following examples of some typical esters show the IUPAC names, as well as the common names, of esters:

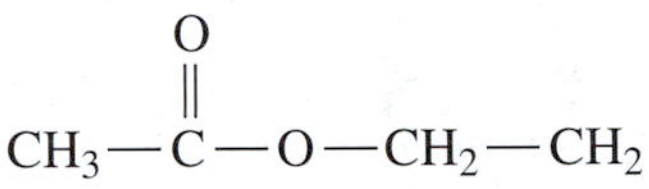

Ethyl ethanoate
(ethyl acetate)

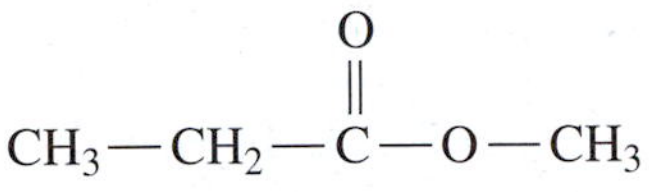

Methyl propanoate
(methyl propionate)

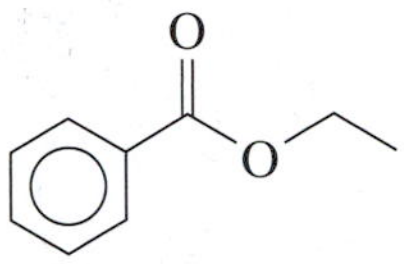

Ethyl benzoate

Chemistry Link to Health

Salicylic Acid from a Willow Tree

For many centuries, relief from pain and fever was obtained by chewing on the leaves or a piece of bark from the willow tree. By the 1800s, chemists discovered that salicin was the agent in the bark responsible for the relief of pain. However, the body converts salicin to salicylic acid, which has a carboxyl group and a hydroxyl group that irritates the stomach lining. In 1899, the Bayer chemical company in Germany produced an ester of salicylic acid and acetic acid, called acetylsalicylic acid (aspirin), which is less irritating. In some aspirin preparations, a buffer is added to neutralize the carboxylic acid group. Today, aspirin is used as an analgesic (pain reliever), antipyretic (fever reducer), and anti-inflammatory agent. Many people take a daily low-dose aspirin, which has been found to lower the risk of heart attack and stroke.

The discovery of salicin in the leaves and bark of the willow tree led to the development of aspirin.

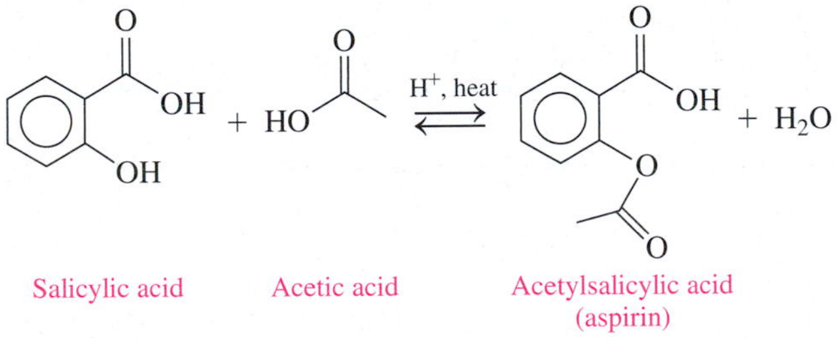
Salicylic acid
Acetic acid
Acetylsalicylic acid
(aspirin)

Oil of wintergreen, or methyl salicylate, has a pungent, minty odor and flavor. Because it can pass through the skin, methyl salicylate is used in skin ointments, where it acts as a counterirritant, producing heat to soothe sore muscles.

ICY
HOT

Ointments containing methyl salicylate are used to soothe sore muscles.

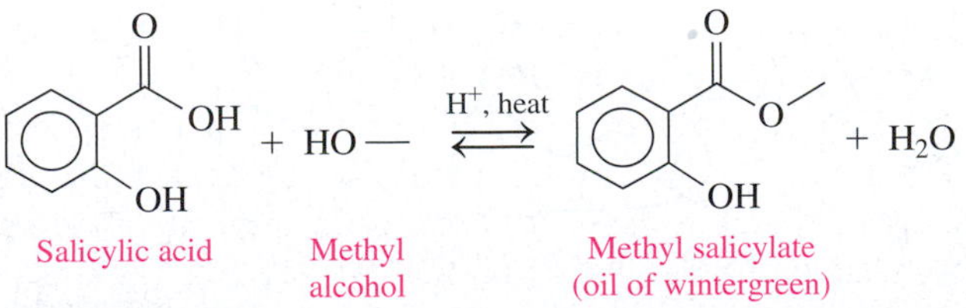
Salicylic acid
Methyl
alcohol
Methyl salicylate
(oil of wintergreen)

▶SAMPLE PROBLEM 16.6 Naming Esters

TRY IT FIRST

What are the IUPAC and common names of the following ester?

$$CH_3-CH_2-\overset{\overset{\displaystyle O}{\|}}{C}-O-CH_2-CH_2-CH_3$$

SOLUTION GUIDE

ANALYZE THE PROBLEM	Given	Need	Connect
	ester	IUPAC name, common name	write the alkyl name for the alcohol, change *ic acid* to *ate*

STEP 1 **Write the name for the carbon chain from the alcohol as an *alkyl* group.** The alcohol part of the ester is from propanol (propyl alcohol). The alkyl group is propyl.

$$CH_3-CH_2-\overset{\overset{\displaystyle O}{\|}}{C}-O-CH_2-CH_2-CH_3 \qquad \text{propyl}$$

STEP 2 **Change the *ic acid* of the acid name to *ate*.** The carboxylic acid part of the ester is from propanoic (propionic) acid, which becomes propanoate (propionate). Replacing the *ic acid* in the common name propionic acid with *ate* gives propionate.

$$CH_3-CH_2-\overset{\overset{\displaystyle O}{\|}}{C}-O-CH_2-CH_2-CH_3 \qquad \text{propyl propanoate}$$

(propyl propionate)

The odor of grapes is due to an ester.

TEST

Try Practice Problems 16.25 to 16.28

STUDY CHECK 16.6

a. Write the IUPAC name for the following ester, which gives the odor and flavor to grapes:

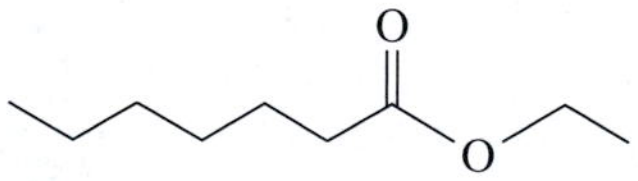
O
O

b. Draw the line-angle formula for ethyl methanoate, which is present in bee stings.

ANSWER

a. ethyl heptanoate

b. 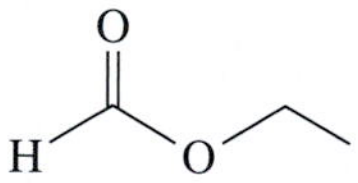
O
H
O

Chemistry Link to the Environment

Plastics

Terephthalic acid (an aromatic acid with two carboxyl groups) is produced in large quantities for the manufacture of polyesters such as Dacron. When terephthalic acid reacts with ethylene glycol, ester bonds form on both ends of the molecules, allowing many molecules to combine into a long molecule.

Dacron, a synthetic material first produced by DuPont in the 1960s, is a polyester used to make permanent press fabrics, carpets, and clothes. Permanent press is a chemical process in which fabrics are permanently shaped and treated for wrinkle resistance. In medicine, artificial blood vessels and valves are made of Dacron, which is biologically inert and does not clot the blood.

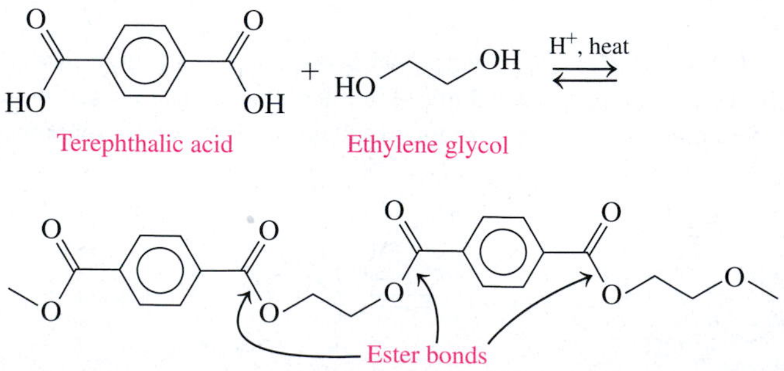
HO
OH
+
HO
OH
H+, heat
Terephthalic acid
Ethylene glycol
Ester bonds

A section of the polyester Dacron

(continued)

Chemistry Link to the Environment (*continued*)

The polyester can also be made into a film called Mylar and a plastic known as PETE (polyethyleneterephthalate). PETE is used for plastic soft drink and water bottles as well as for peanut butter jars, containers of salad dressings, shampoos, and dishwashing liquids. Today, PETE (recycling symbol "1") is the most widely recycled of all the plastics. Every year, more than 1.5×10^9 lb (6.8×10^8 kg) of PETE is being recycled. After PETE is separated from other plastics, it is used to make useful items, including polyester fabric for T-shirts and coats, carpets, fill for sleeping bags, doormats, and containers for tennis balls.

Dacron is a polyester used in permanent press clothing.

Polyester, in the form of the plastic PETE, is used to make soft drink bottles.

Esters in Plants

Many of the fragrances of perfumes and flowers and the flavors of fruits are due to esters. Small esters are volatile, so we can smell them, and they are soluble in water, so we can taste them. Several of these, with their flavor and odor, are listed in **TABLE 16.3**.

TEST

Try Practice Problems 16.29 to 16.32

TABLE 16.3 Some Esters in Fruits and Flavorings

Condensed Structural Formula and Name	Flavor/Odor
$CH_3-\overset{\overset{O}{\Vert}}{C}-O-CH_2-CH_2-CH_3$ Propyl ethanoate (propyl acetate)	Pears
$CH_3-\overset{\overset{O}{\Vert}}{C}-O-CH_2-CH_2-CH_2-CH_2-CH_3$ Pentyl ethanoate (pentyl acetate)	Bananas
$CH_3-\overset{\overset{O}{\Vert}}{C}-O-CH_2-CH_2-CH_2-CH_2-CH_2-CH_2-CH_2-CH_3$ Octyl ethanoate (octyl acetate)	Oranges
$CH_3-CH_2-CH_2-\overset{\overset{O}{\Vert}}{C}-O-CH_2-CH_3$ Ethyl butanoate (ethyl butyrate)	Pineapples
$CH_3-CH_2-CH_2-\overset{\overset{O}{\Vert}}{C}-O-CH_2-CH_2-CH_2-CH_2-CH_3$ Pentyl butanoate (pentyl butyrate)	Apricots

Esters such as ethyl butanoate provide the odor and flavor of many fruits such as pineapples.

PRACTICE PROBLEMS

16.3 Esters

16.21 Draw the condensed structural formula for the ester formed when each of the following reacts with ethyl alcohol:
a. acetic acid **b.** butyric acid
c. benzoic acid

16.22 Draw the condensed structural formula for the ester formed when each of the following reacts with methyl alcohol:
a. formic acid **b.** propionic acid
c. 2-methylpentanoic acid

16.23 Draw the line-angle formula for the ester formed in each of the following reactions:

a. 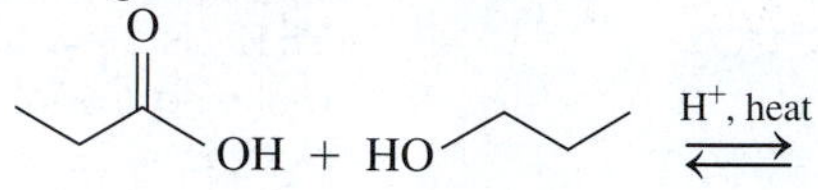

b. 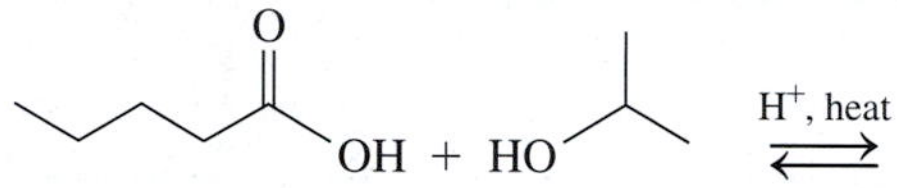

16.24 Draw the line-angle formula for the ester formed in each of the following reactions:

a. 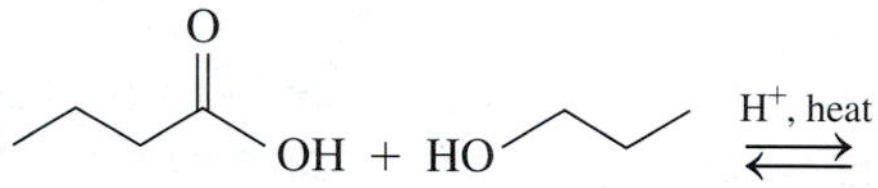

b. 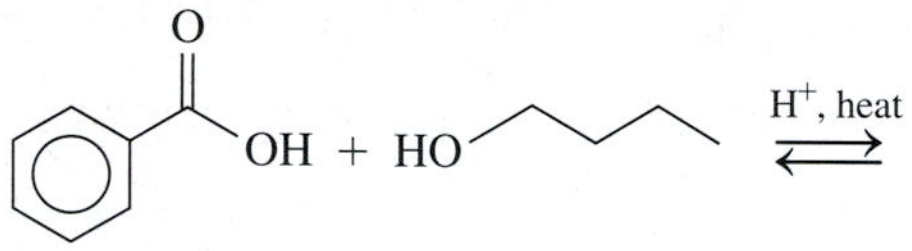

16.25 Write the IUPAC and common names, if any, of the carboxylic acid and alcohol needed to produce each of the following esters:

a. 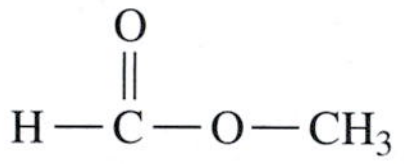

b. 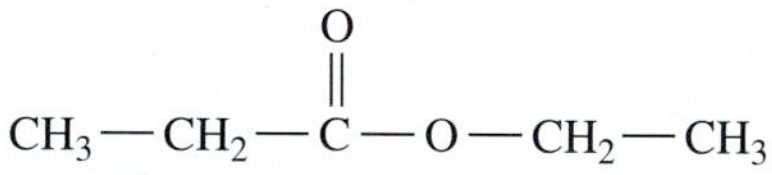

c. 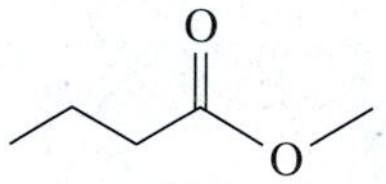

d. 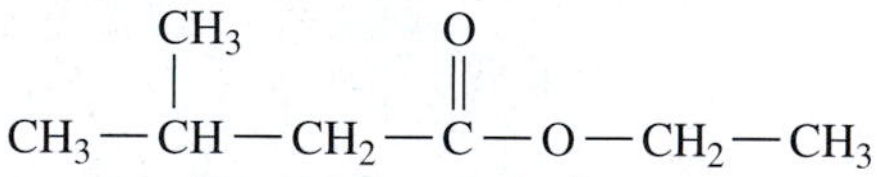

16.26 Write the IUPAC and common names, if any, of the carboxylic acid and alcohol needed to produce each of the following esters:

a. 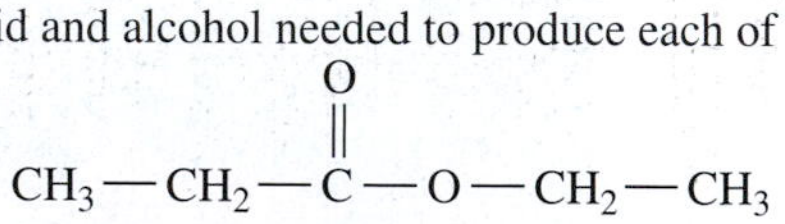

b. 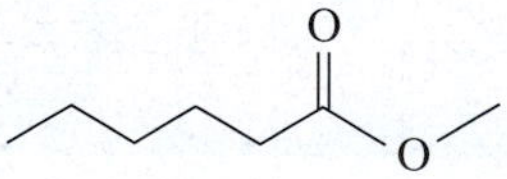

c. 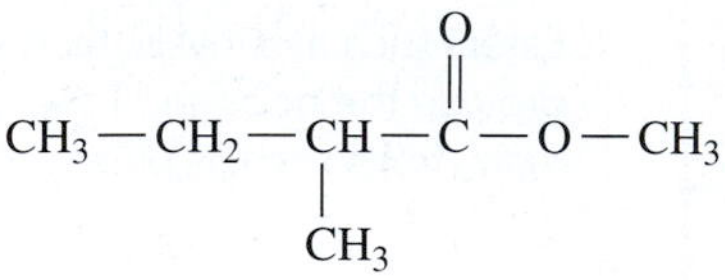

d. 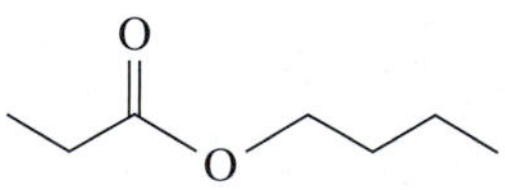

16.27 Write the IUPAC name and common names, if any, for each of the following esters:

a. 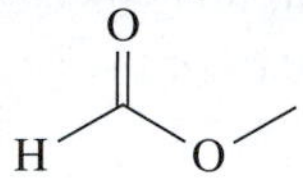

b. 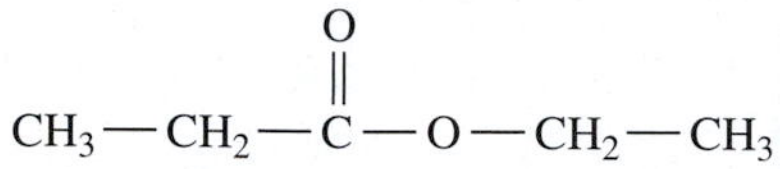

c. 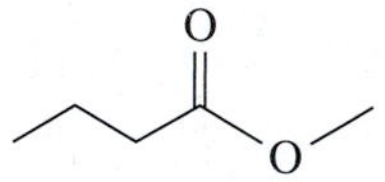

d. 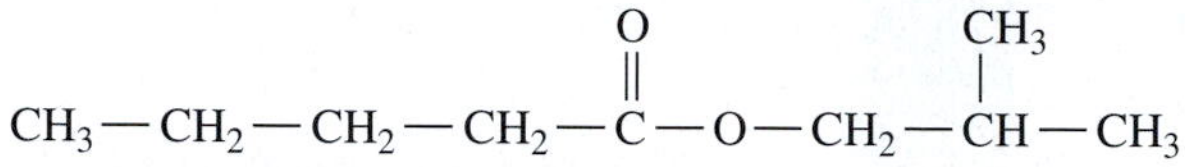

16.28 Write the IUPAC name and common names, if any, for each of the following esters:

a. 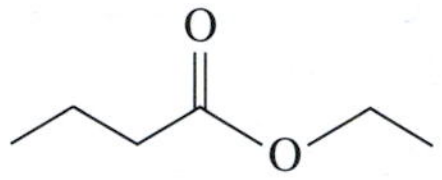

b. 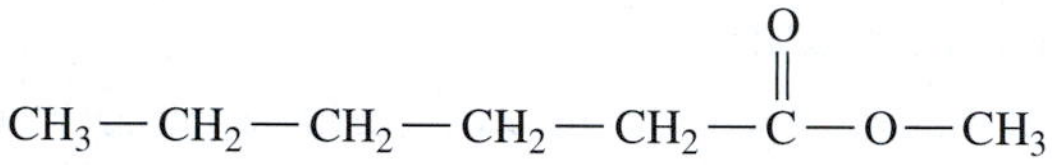

c. 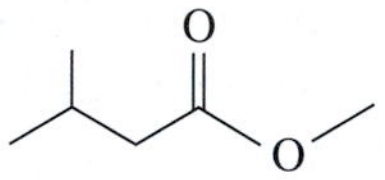

d. 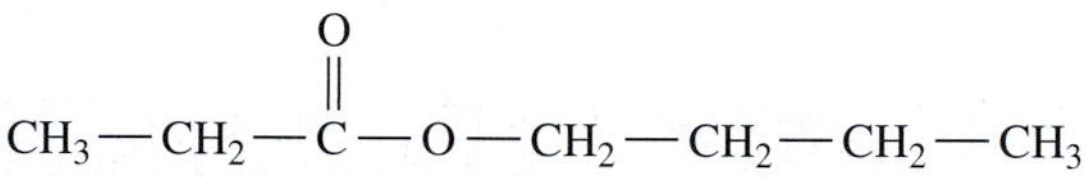

16.29 Draw the condensed structural formulas for **a** and **b** and line-angle formulas for **c** and **d**:
a. propyl butyrate
b. butyl formate
c. ethyl pentanoate
d. methyl propanoate

16.30 Draw the condensed structural formulas for **a** and **b** and line-angle formulas for **c** and **d**:
a. hexyl acetate
b. ethyl methanoate
c. propyl benzoate
d. methyl octanoate

16.31 What is the ester responsible for the flavor and odor of the following fruit?
a. banana
b. orange
c. apricot

16.32 What flavor would you notice if you smelled or tasted the following?
a. ethyl butanoate
b. propyl acetate
c. pentyl acetate

16.4 Properties of Esters

LEARNING GOAL Describe the boiling points and solubility of esters; draw the condensed structural and line-angle formulas for the products from acid and base hydrolysis of esters.

TEST
Try Practice Problems 16.33 to 16.36

Esters have boiling points higher than those of alkanes and ethers, but lower than those of alcohols and carboxylic acids of similar mass. Because ester molecules do not have hydroxyl groups, they cannot hydrogen bond to each other.

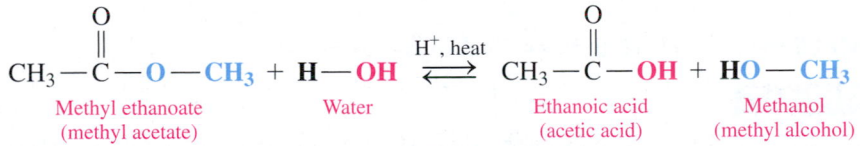

	$CH_3-CH_2-CH_2-CH_3$	$H-\overset{\overset{\large O}{\|}}{C}-O-CH_3$	$CH_3-CH_2-CH_2-OH$	$CH_3-\overset{\overset{\large O}{\|}}{C}-OH$
Name	Butane	Methyl methanoate	1-Propanol	Ethanoic acid
Family	Alkane	Ester	Alcohol	Carboxylic acid
Molar Mass	58	60	60	60
bp	0 °C	32 °C	97 °C	118 °C

Boiling Point Increases →

Solubility in Water

Esters with two to five carbon atoms are soluble in water. The partially negative oxygen of the carbonyl group forms hydrogen bonds with the partially positive hydrogen atoms of water molecules. The solubility of esters decreases as the number of carbon atoms increases.

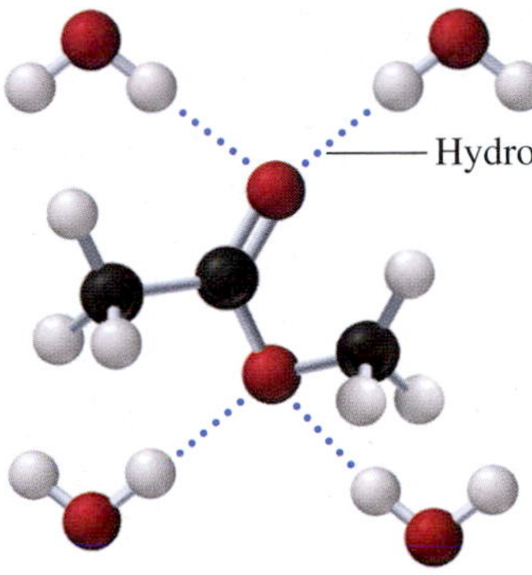
Hydrogen bonds
Esters with two to five carbon atoms are soluble because they form hydrogen bonds with water.

Acid Hydrolysis of Esters

In **acid hydrolysis**, water reacts with an ester in the presence of a strong acid, usually H_2SO_4 or HCl, to form a carboxylic acid and an alcohol. Therefore, hydrolysis is the reverse of the esterification reaction. During acid hydrolysis, a water molecule provides the —OH group to convert the carbonyl group of the ester to a carboxyl group. A large quantity of water is used to shift the equilibrium in the direction of the carboxylic acid and alcohol products. When hydrolysis of biological esters occurs in the cells, an enzyme replaces the acid as the catalyst.

CORE CHEMISTRY SKILL
Hydrolyzing Esters

$$CH_3-\overset{\overset{\large O}{\|}}{C}-O-CH_3 + H-OH \underset{}{\overset{H^+,\text{ heat}}{\rightleftharpoons}} CH_3-\overset{\overset{\large O}{\|}}{C}-OH + HO-CH_3$$

Methyl ethanoate (methyl acetate) + Water ⇌ Ethanoic acid (acetic acid) + Methanol (methyl alcohol)

▸SAMPLE PROBLEM 16.7 Acid Hydrolysis of Esters

TRY IT FIRST

Aspirin (acetylsalicylic acid) that has been stored for a long time may undergo acid hydrolysis in the presence of water and heat. Use line-angle formulas to write the balanced chemical equation for the acid hydrolysis of aspirin.

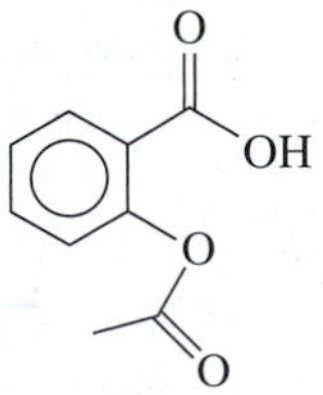
O
OH
O
O
Aspirin (acetylsalicylic acid)

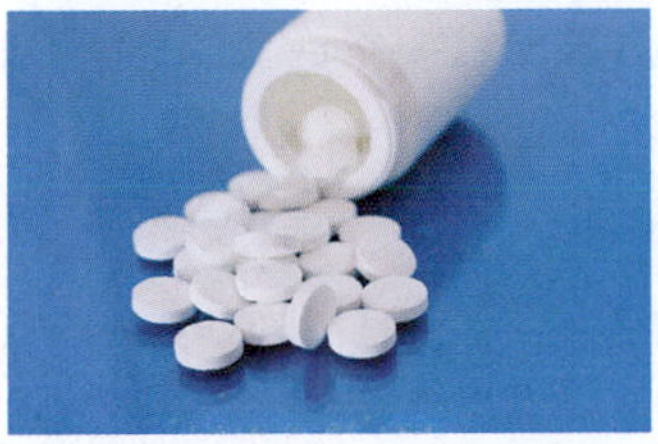
Aspirin stored in a warm, humid place may undergo hydrolysis.

SOLUTION

ANALYZE THE PROBLEM	Given	Need	Connect
	aspirin and water	acid hydrolysis equation	products: carboxylic acid, alcohol

For the hydrolysis products, draw the line-angle formula of the carboxylic acid by adding —OH to the carbonyl group and —H to complete the alcohol group. The acetic acid in the products gives the odor of vinegar to the aspirin that has hydrolyzed.

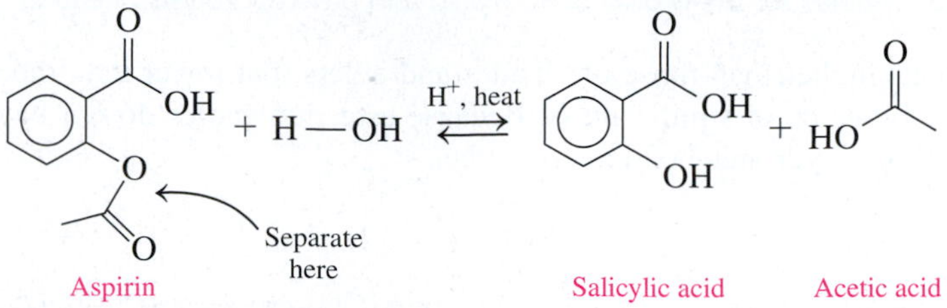

OH
+ H—OH
H+, heat
OH
OH
+ HO
Separate here

Aspirin — Salicylic acid — Acetic acid

STUDY CHECK 16.7

Write the IUPAC and common names of the products from the acid hydrolysis of each of the following:

a. ethyl propanoate (ethyl propionate) **b.** methyl benzoate

ANSWER

a. propanoic acid (propionic acid) and ethanol (ethyl alcohol)
b. benzoic acid and methanol (methyl alcohol)

Base Hydrolysis of Esters (Saponification)

When an ester undergoes hydrolysis with a strong base such as NaOH or KOH, the products are the carboxylate salt and the corresponding alcohol. This **base hydrolysis** is also called **saponification**.

$$CH_3—\overset{\overset{\displaystyle O}{\|}}{C}—O—CH_3 + \mathbf{NaOH} \xrightarrow{\text{Heat}} CH_3—\overset{\overset{\displaystyle O}{\|}}{C}—O^- \, Na^+ + \mathbf{HO—CH_3}$$

Methyl ethanoate (methyl acetate) + Sodium hydroxide → Sodium ethanoate (sodium acetate) + Methanol (methyl alcohol)

Ethyl acetate is the solvent in fingernail polish.

▶ SAMPLE PROBLEM 16.8 Base Hydrolysis of Esters

TRY IT FIRST

Ethyl acetate is a solvent used for fingernail polish, plastics, and lacquers. Write the balanced chemical equation for the hydrolysis of ethyl acetate with NaOH.

SOLUTION

ANALYZE THE PROBLEM	Given	Need	Connect
	ethyl acetate, NaOH	base hydrolysis equation	products: carboxylate ion, alcohol

The hydrolysis of ethyl acetate with NaOH gives the carboxylate salt, sodium acetate, and ethyl alcohol.

$$CH_3—\overset{\overset{\displaystyle O}{\|}}{C}—O—CH_2—CH_3 + NaOH \xrightarrow{\text{Heat}} CH_3—\overset{\overset{\displaystyle O}{\|}}{C}—O^- \, Na^+ + HO—CH_2—CH_3$$

Ethyl ethanoate (ethyl acetate) → Sodium ethanoate (sodium acetate) + Ethanol (ethyl alcohol)

STUDY CHECK 16.8

Draw the line-angle formulas for the products from

a. the hydrolysis of methyl benzoate with KOH
b. the hydrolysis of ethyl propionate with NaOH

INTERACTIVE VIDEO

Study Check 16.8 — PEARSON eText 2.0

ANSWER

a. 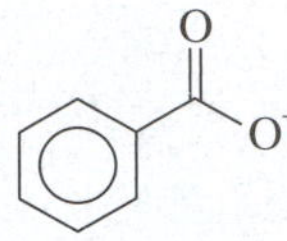 $O^- K^+ + HO-$

b. 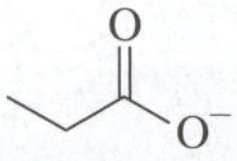 $O^- Na^+ + HO$

TEST

Try Practice Problems 16.37 and 16.38

PRACTICE PROBLEMS

16.4 Properties of Esters

16.33 For each of the following pairs of compounds, select the compound that has the higher boiling point:

a. $CH_3-\overset{O}{\overset{\|}{C}}-O-CH_3$ or $CH_3-CH_2-\overset{O}{\overset{\|}{C}}-OH$

b. $CH_3-\overset{O}{\overset{\|}{C}}-O-CH_3$ or $CH_3-CH_2-CH_2-CH_2-OH$

c. 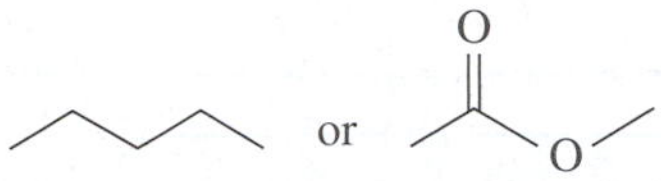

16.34 For each of the following pairs of compounds, select the compound that has the higher boiling point:

a. $H-\overset{O}{\overset{\|}{C}}-O-CH_3$ or $CH_3-CH_2-CH_2-OH$

b. 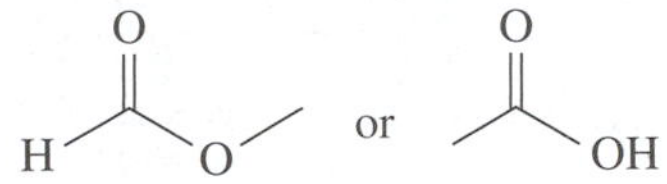

c. $CH_3-O-CH_2-CH_3$ or $H-\overset{O}{\overset{\|}{C}}-O-CH_3$

16.35 Predict if each of the following is water soluble or not:

a. ethyl acetate b. pentyl butanoate

16.36 Predict if each of the following is water soluble or not:

a. octyl acetate b. propyl formate

16.37 Draw the condensed structural or line-angle formulas for the products from the acid- or base-catalyzed hydrolysis of each of the following compounds:

a. $CH_3-CH_2-\overset{O}{\overset{\|}{C}}-O-CH_3 + NaOH \xrightarrow{\text{Heat}}$

b. 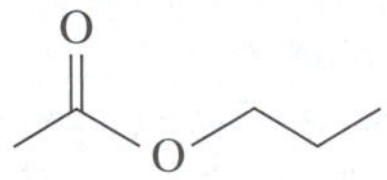 $+ H_2O \overset{H^+,\ heat}{\rightleftarrows}$

c. $CH_3-CH_2-CH_2-\overset{O}{\overset{\|}{C}}-O-CH_2-CH_3 + H_2O \overset{H^+,\ heat}{\rightleftarrows}$

d. 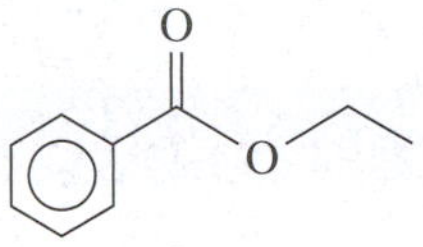 $+ NaOH \xrightarrow{\text{Heat}}$

e. 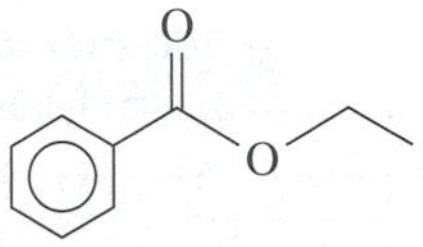 $+ H_2O \overset{H^+,\ heat}{\rightleftarrows}$

16.38 Draw the condensed structural or line-angle formulas for the products from the acid- or base-catalyzed hydrolysis of each of the following compounds:

a. $CH_3-CH_2-\overset{O}{\overset{\|}{C}}-O-CH_2-CH_2-CH_2-CH_3 + H_2O \overset{H^+,\ heat}{\rightleftarrows}$

b. 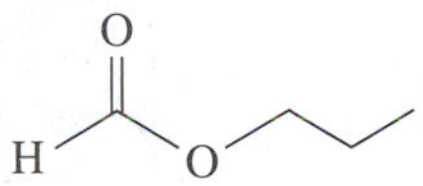 $+ NaOH \xrightarrow{\text{Heat}}$

c. $CH_3-CH_2-\overset{O}{\overset{\|}{C}}-O-CH_3 + H_2O \overset{H^+,\ heat}{\rightleftarrows}$

d. 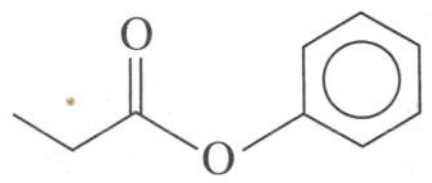 $+ H_2O \overset{H^+,\ heat}{\rightleftarrows}$

e. 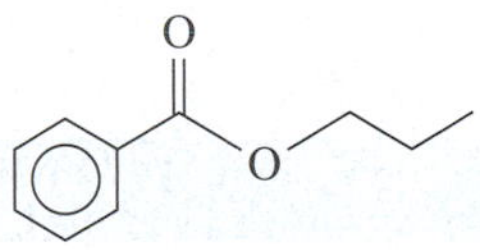 $+ NaOH \xrightarrow{\text{Heat}}$

CLINICAL UPDATE Liquid Bandages

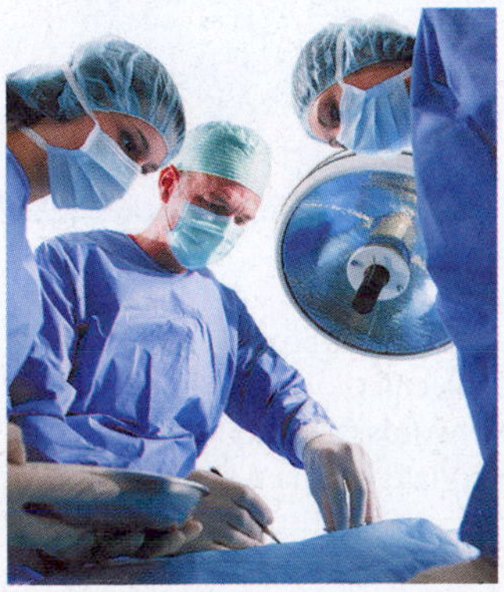

The liquid bandage used to close Robert's surgical incisions is also available in pharmacies for home use in closing wounds. The liquid hardens quickly when it comes in contact with blood and tissue, stops bleeding, and holds the edges of the wound together as skin heals. In a few days, the plastic bandage sloughs off, and there is no need to remove stitches or staples. Anyone applying liquid bandage needs to be aware that contact of the liquid with gloves or other surfaces will cause them to adhere. Persons with an allergy to plastic should not use liquid bandage products.

The polymer in the liquid bandage consists of repeating units of Indermil. Indermil can be produced by the esterification reaction of 2-cyano-2-propenoic acid and 1-butanol.

$$H_2C=\overset{CN}{\overset{|}{C}}-\overset{O}{\overset{\|}{C}}-OH$$

2-Cyano-2-propenoic acid

Clinical Applications

16.39 a. Use condensed structural formulas to write the balanced chemical equation for the reaction between 2-cyano-2-propenoic acid and 1-butanol.

b. What is the IUPAC name of the ester formed?

16.40 a. Draw the condensed structural formulas for the products from the acid hydrolysis of the ester in problem 16.39.

b. Draw the condensed structural formulas for the products from the base hydrolysis with NaOH of the ester in problem 16.39.

A spray containing a liquid bandage applied to a wound closes the wound quickly and stops bleeding.

CONCEPT MAP

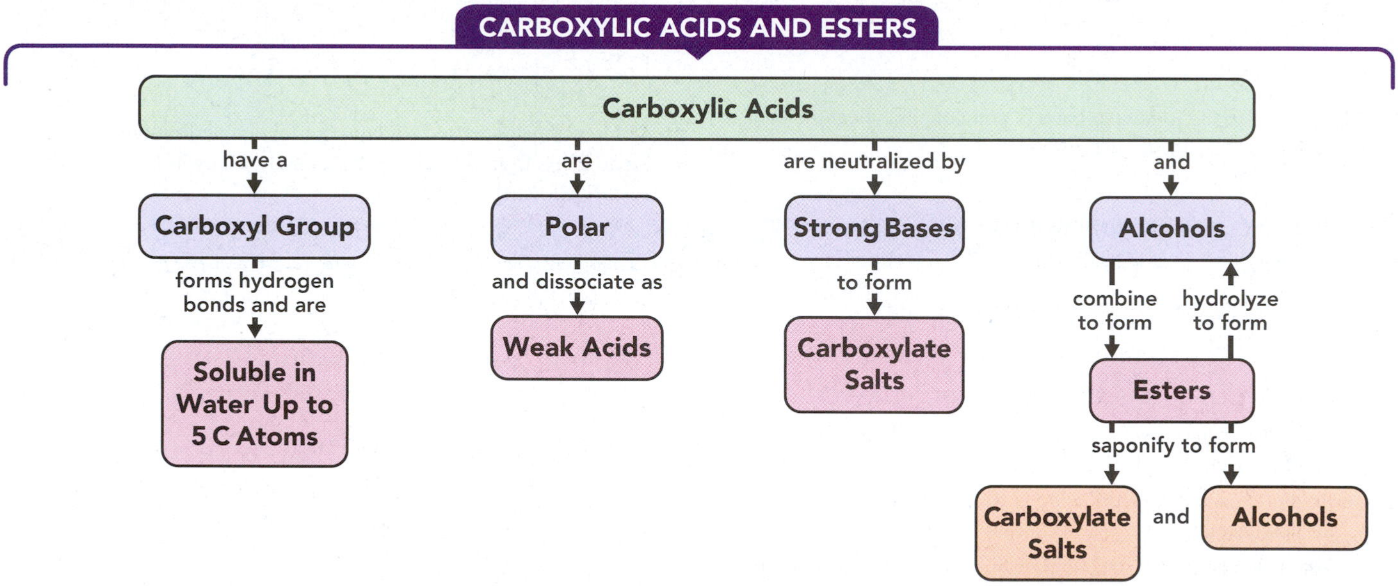
CARBOXYLIC ACIDS AND ESTERS
Carboxylic Acids
have a
Carboxyl Group
forms hydrogen bonds and are
Soluble in Water Up to 5 C Atoms
are
Polar
and dissociate as
Weak Acids
are neutralized by
Strong Bases
to form
Carboxylate Salts
and
Alcohols
combine to form
hydrolyze to form
Esters
saponify to form
Carboxylate Salts
and
Alcohols

CHAPTER REVIEW

16.1 Carboxylic Acids

HEINZ
VINEGAR

LEARNING GOAL Write the IUPAC and common names for carboxylic acids; draw their condensed structural and line-angle formulas.

- A carboxylic acid contains the carboxyl functional group, which is a hydroxyl group connected to a carbonyl group.
- The IUPAC name of a carboxylic acid is obtained by replacing the *e* in the alkane name with *oic acid*.
- The common names of carboxylic acids with one to four carbon atoms are formic acid, acetic acid, propionic acid, and butyric acid.

16.2 Properties of Carboxylic Acids

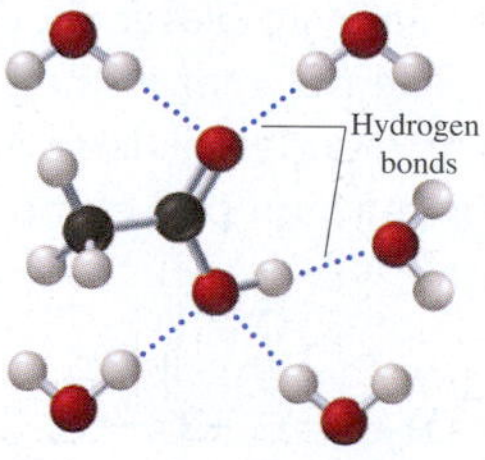
Hydrogen bonds

LEARNING GOAL Describe the boiling points, solubility, dissociation, and neutralization of carboxylic acids.

- The carboxyl group contains polar bonds of O—H and C═O, which makes a carboxylic acid with one to five carbon atoms soluble in water.
- As weak acids, carboxylic acids dissociate slightly by donating H^+ to water to form carboxylate and hydronium ions.
- Carboxylic acids are neutralized by base, producing a carboxylate salt and water.

16.3 Esters

LEARNING GOAL Write the IUPAC and common names for esters; draw condensed structural and line-angle formulas.

- In an ester, an alkyl or aromatic group replaces the H of the hydroxyl group of a carboxylic acid.
- In the presence of a strong acid, a carboxylic acid reacts with an alcohol to produce an ester and a molecule of water from the —OH removed from the carboxylic acid, and the —H removed from the alcohol molecule.
- The names of esters consist of two words: the alkyl part from the alcohol and the name of the carboxylate obtained by replacing *ic acid* with *ate*.

16.4 Properties of Esters

LEARNING GOAL Describe the boiling points and solubility of esters; draw the condensed structural and line-angle formulas for the products from acid and base hydrolysis of esters.

- Esters have boiling points that are lower than alcohols and carboxylic acids, but higher than alkanes.

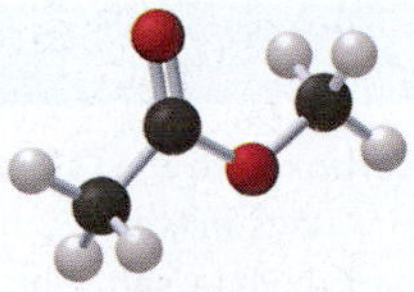

Methyl ethanoate (methyl acetate)

- Esters with two to five carbon atoms are soluble in water.
- Esters undergo acid hydrolysis by adding water to yield the carboxylic acid and alcohol.
- Base hydrolysis, or saponification, of an ester produces the carboxylate salt and an alcohol.

SUMMARY OF NAMING

Family	Structure	IUPAC Name	Common Name
Carboxylic acid	$CH_3-\overset{\overset{\displaystyle O}{\|}}{C}-OH$	Ethanoic acid	Acetic acid
Carboxylate salt	$CH_3-\overset{\overset{\displaystyle O}{\|}}{C}-O^- \, Na^+$	Sodium ethanoate	Sodium acetate
Ester	$CH_3-\overset{\overset{\displaystyle O}{\|}}{C}-O-CH_3$	Methyl ethanoate	Methyl acetate

SUMMARY OF REACTIONS

The chapter Sections to review are shown after the name of each reaction.

Dissociation of a Carboxylic Acid in Water (16.2)

$$\underset{\text{Ethanoic acid (acetic acid)}}{CH_3-\overset{\overset{\displaystyle O}{\|}}{C}-OH} + H_2O \rightleftarrows \underset{\text{Ethanoate ion (acetate ion)}}{CH_3-\overset{\overset{\displaystyle O}{\|}}{C}-O^-} + \underset{\text{Hydronium ion}}{H_3O^+}$$

Neutralization of a Carboxylic Acid (16.2)

$$\underset{\text{Propanoic acid (propionic acid)}}{CH_3-CH_2-\overset{\overset{\displaystyle O}{\|}}{C}-OH} + \underset{\text{Sodium hydroxide}}{NaOH} \longrightarrow \underset{\text{Sodium propanoate (sodium propionate)}}{CH_3-CH_2-\overset{\overset{\displaystyle O}{\|}}{C}-O^- \, Na^+} + H_2O$$

Esterification: Carboxylic Acid and an Alcohol (16.3)

$$\underset{\text{Ethanoic acid (acetic acid)}}{CH_3-\overset{\overset{\displaystyle O}{\|}}{C}-OH} + \underset{\text{Methanol (methyl alcohol)}}{HO-CH_3} \overset{H^+,\text{ heat}}{\rightleftarrows} \underset{\text{Methyl ethanoate (methyl acetate)}}{CH_3-\overset{\overset{\displaystyle O}{\|}}{C}-O-CH_3} + H_2O$$

Acid Hydrolysis of an Ester (16.4)

$$\underset{\text{Methyl ethanoate (methyl acetate)}}{CH_3-\overset{\overset{\displaystyle O}{\|}}{C}-O-CH_3} + H_2O \overset{H^+,\text{ heat}}{\rightleftarrows} \underset{\text{Ethanoic acid (acetic acid)}}{CH_3-\overset{\overset{\displaystyle O}{\|}}{C}-OH} + \underset{\text{Methanol (methyl alcohol)}}{HO-CH_3}$$

Base Hydrolysis of an Ester (Saponification) (16.4)

$$\underset{\text{Methyl propanoate (methyl propionate)}}{CH_3-CH_2-\overset{\overset{\displaystyle O}{\|}}{C}-O-CH_3} + \underset{\text{Sodium hydroxide}}{NaOH} \xrightarrow{\text{Heat}} \underset{\text{Sodium propanoate (sodium propionate)}}{CH_3-CH_2-\overset{\overset{\displaystyle O}{\|}}{C}-O^- \, Na^+} + \underset{\text{Methanol (methyl alcohol)}}{HO-CH_3}$$

KEY TERMS

acid hydrolysis The splitting of an ester molecule in the presence of a strong acid to produce a carboxylic acid and an alcohol.

base hydrolysis (saponification) The splitting of an ester molecule by a strong base to produce a carboxylate salt and an alcohol.

carboxyl group A functional group found in carboxylic acids composed of carbonyl and hydroxyl groups.

$$\begin{array}{c} O \\ \parallel \\ -C-OH \end{array}$$

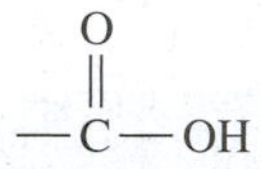

Carboxyl group

carboxylate ion The anion produced when a carboxylic acid dissociates in water.

carboxylate salt The product of neutralization of a carboxylic acid, which is a carboxylate ion and a metal ion from the base.

carboxylic acid An organic compound containing the carboxyl group.

ester An organic compound in which an alkyl or aromatic group replaces the hydrogen atom in a carboxylic acid.

esterification The formation of an ester from a carboxylic acid and an alcohol with the elimination of a molecule of water in the presence of an acid catalyst.

CORE CHEMISTRY SKILLS

The chapter Section containing each Core Chemistry Skill is shown in parentheses at the end of each heading.

Naming Carboxylic Acids (16.1)

- The IUPAC name of a carboxylic acid is obtained by replacing the *e* in the corresponding alkane name with *oic acid*.
- The common names of carboxylic acids with one to four carbon atoms are formic acid, acetic acid, propionic acid, and butyric acid.

Example: Write the IUPAC name for the following:

$$\begin{array}{ccccccc} & CH_3 & & Cl & & O \\ & | & & | & & \parallel \\ CH_3-CH-CH_2-&C&-CH_2-&C&-OH \\ & & & | & & \\ & & & CH_3 & & \end{array}$$

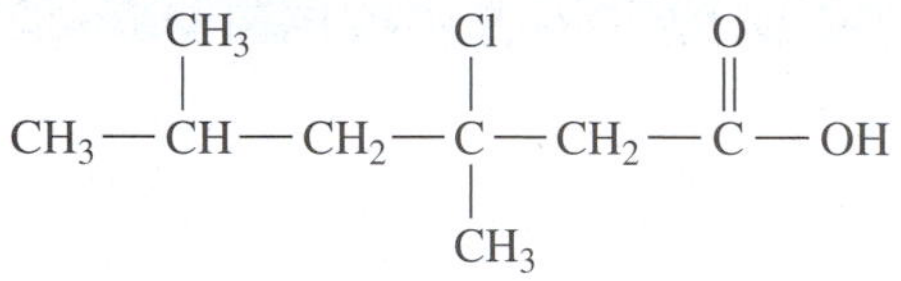

Answer: 3-chloro-3,5-dimethylhexanoic acid

Hydrolyzing Esters (16.4)

- Esters undergo acid hydrolysis by adding water to produce a carboxylic acid and an alcohol.
- Esters undergo base hydrolysis, or saponification, to produce a carboxylate salt and an alcohol.

Example: Draw the condensed structural formulas for the following products from the (**a**) acid and (**b**) base hydrolysis (NaOH) of ethyl butanoate:

Answer:

$$\textbf{a.}\ CH_3-CH_2-CH_2-\overset{\overset{\displaystyle O}{\parallel}}{C}-OH + HO-CH_2-CH_3$$

$$\textbf{b.}\ CH_3-CH_2-CH_2-\overset{\overset{\displaystyle O}{\parallel}}{C}-O^-\ Na^+ + HO-CH_2-CH_3$$

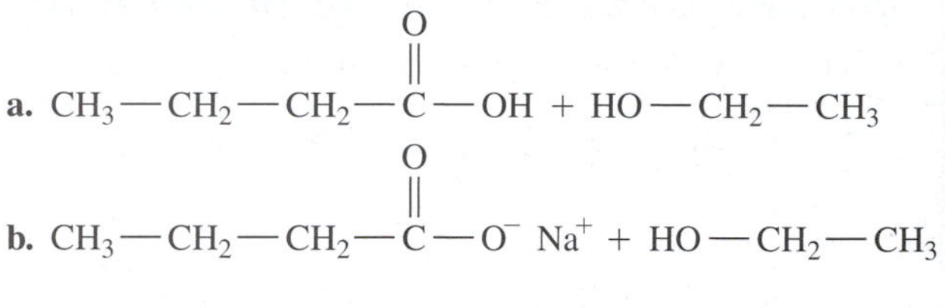

UNDERSTANDING THE CONCEPTS

The chapter Sections to review are shown in parentheses at the end of each problem.

16.41 Draw the condensed structural formulas and write the IUPAC names of two structural isomers of the carboxylic acids that have the molecular formula $C_4H_8O_2$. (16.1)

16.42 Draw the condensed structural formulas and write the IUPAC names of four structural isomers of the carboxylic acids that have the molecular formula $C_5H_{10}O_2$. (16.1)

16.43 Draw the condensed structural formulas and write the IUPAC names of two structural isomers of the esters that have the molecular formula $C_3H_6O_2$. (16.3)

16.44 Draw the condensed structural formulas and write the IUPAC names of four structural isomers of the esters that have the molecular formula $C_4H_8O_2$. (16.3)

Clinical Applications

16.45 A strawberry nutritional drink used for a liquid diet is flavored with methyl butanoate. (16.1, 16.3, 16.4)
   a. Draw the condensed structural formula for methyl butanoate.

   b. Write the IUPAC name of the carboxylic acid and the alcohol used to prepare methyl butanoate.
   c. Use condensed structural formulas to write the balanced chemical equation for the acid hydrolysis of methyl butanoate.
   d. Use condensed structural formulas to write the balanced chemical equation for the base hydrolysis of methyl butanoate with NaOH.

16.46 Methyl benzoate, which smells like pineapple guava, is used to train detection dogs. (16.1, 16.3, 16.4)
   a. Draw the line-angle formula for methyl benzoate.
   b. Write the name of the carboxylic acid and alcohol used to prepare methyl benzoate.
   c. Use line-angle formulas to write the balanced chemical equation for the acid hydrolysis of methyl benzoate.
   d. Use line-angle formulas to write the balanced chemical equation for the base hydrolysis of methyl benzoate with NaOH.

ADDITIONAL PRACTICE PROBLEMS

16.47 Write the IUPAC and common names, if any, for each of the following compounds: (16.1, 16.3)

a. $CH_3-CH(CH_3)-CH_2-C(=O)-OH$

b. 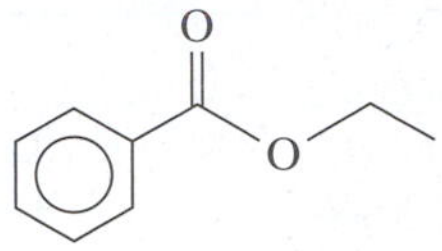

c. $CH_3-CH_2-C(=O)-O-CH_2-CH_3$

d. 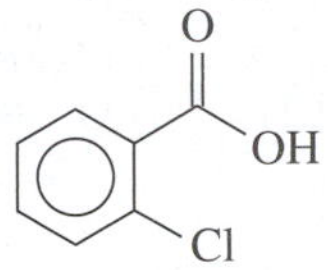

e. 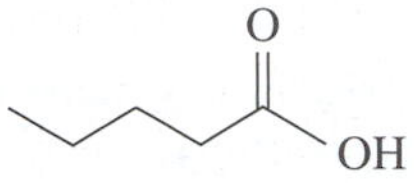

f. $CH_3-C(=O)-O-CH(CH_3)-CH_3$

16.48 Write the IUPAC and common names, if any, for each of the following compounds: (16.1, 16.3)

a. $CH_3-CH(CH_3)-CH_2-CH_2-C(=O)-OH$

b. 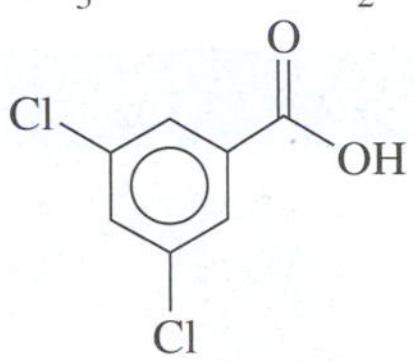

c. 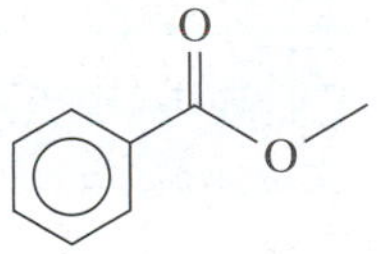

d. $CH_3-CH_2-CH_2-C(=O)-O-CH_3$

e. 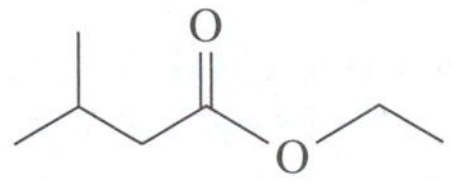

f. $CH_3-CH(CH_3)-CH_2-CH_2-C(=O)-OH$

16.49 Draw the line-angle formulas for two methyl esters that have the molecular formula $C_5H_{10}O_2$. (16.3)

16.50 Draw the line-angle formulas for the carboxylic acid and the ester that have the molecular formula $C_2H_4O_2$. (16.1, 16.3)

16.51 Draw the condensed structural formula, or line-angle formula if aromatic, for each of the following: (16.1, 16.3)

a. methyl hexanoate
b. *p*-chlorobenzoic acid
c. β-chloropropionic acid
d. ethyl butanoate
e. 3-methylpentanoic acid
f. ethyl benzoate

16.52 Draw the condensed structural formula, or line-angle formula if aromatic, for each of the following: (16.1, 16.3)

a. α-bromobutyric acid
b. ethyl butyrate
c. 2-methyloctanoic acid
d. 3,5-dimethylhexanoic acid
e. propyl acetate
f. 3,4-dibromobenzoic acid

16.53 For each of the following pairs, identify the compound that would have the higher boiling point. Explain. (16.2, 16.4)

a. $CH_3-CH_2-CH_2-OH$ or $CH_3-C(=O)-OH$

b. $CH_3-CH_2-CH_2-CH_3$ or $CH_3-C(=O)-OH$

16.54 For each of the following pairs, identify the compound that would have the higher boiling point. Explain. (16.2, 16.4)

a. $CH_3-CH(OH)-CH_3$ or $H-C(=O)-O-CH_3$

b. $H-C(=O)-O-CH_3$ or $CH_3-CH_2-CH_2-CH_3$

16.55 Acetic acid, methyl formate, and 1-propanol all have the same molar mass. The possible boiling points are 32 °C, 97 °C, and 118 °C. Match the compounds with the boiling points and explain your choice. (16.2, 16.4)

16.56 Propionic acid, 1-butanol, and butyraldehyde all have the same molar mass. The possible boiling points are 76 °C, 118 °C, and 141 °C. Match the compounds with the boiling points and explain your choice. (16.2, 16.4)

16.57 Which of the following compounds are soluble in water? (16.2, 16.4)

a. $CH_3-CH_2-C(=O)-O^-\ Na^+$

b. 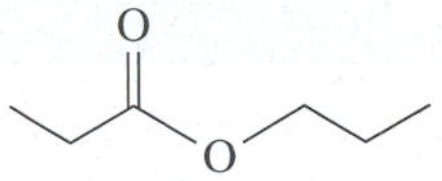

c. $CH_3-CH_2-CH_2-OH$

d. $CH_3-CH_2-C(=O)-OH$

16.58 Which of the following compounds are soluble in water? (16.2, 16.4)

a. $CH_3-CH_2-CH_2-C(=O)-OH$

b. $CH_3-C(=O)-O-CH_3$

c. $CH_3-CH_2-CH_2-CH_3$

d. 

16.59 Draw the condensed structural or line-angle formulas for the products from each of the following reactions: (16.2, 16.3)

a. $CH_3-CH_2-C(=O)-OH + H_2O \rightleftarrows$

b. $CH_3-CH_2-C(=O)-OH + KOH \longrightarrow$

c. 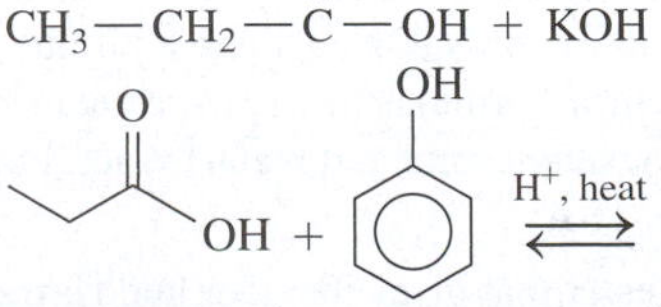

d. 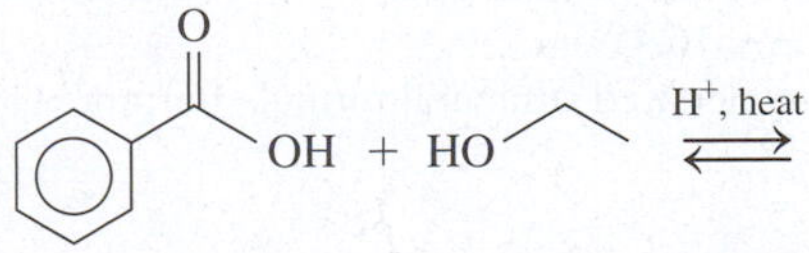

16.60 Draw the condensed structural or line-angle formulas for the products from each of the following reactions: (16.2, 16.3)

a. $CH_3-\overset{\overset{O}{\|}}{C}-OH + NaOH \longrightarrow$

b. $CH_3-\overset{\overset{O}{\|}}{C}-OH + H_2O \rightleftarrows$

c. 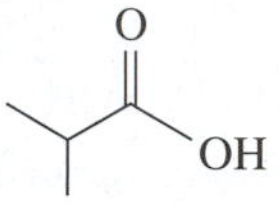 $+ KOH \longrightarrow$

d. $CH_3-\overset{\overset{CH_3}{|}}{CH}-\overset{\overset{O}{\|}}{C}-OH + HO-CH_3 \xrightleftharpoons{H^+, \text{heat}}$

16.61 Write the IUPAC names of the carboxylic acid and alcohol needed to produce each of the following esters: (16.3)

a. $CH_3-\overset{\overset{CH_3}{|}}{CH}-CH_2-\overset{\overset{O}{\|}}{C}-O-CH_3$

b. 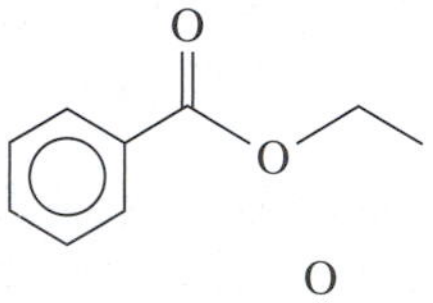

c. 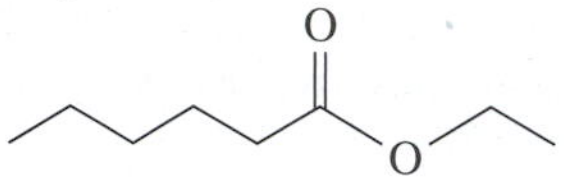

16.62 Write the IUPAC names of the carboxylic acid and alcohol needed to produce each of the following esters: (16.3)

a. $CH_3-CH_2-CH_2-\overset{\overset{O}{\|}}{C}-O-CH_2-CH_3$

b. 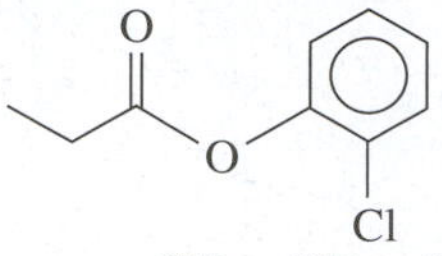

c. $CH_3-\overset{\overset{CH_3}{|}}{CH}-\overset{\overset{CH_3}{|}}{CH}-\overset{\overset{O}{\|}}{C}-O-CH_3$

16.63 Draw the condensed structural formulas for the products from each of the following reactions: (16.4)

a. $CH_3-CH_2-\overset{\overset{O}{\|}}{C}-O-\overset{\overset{CH_3}{|}}{CH}-CH_3 + H_2O \xrightleftharpoons{H^+, \text{heat}}$

b. $CH_3-\overset{\overset{CH_3}{|}}{CH}-\overset{\overset{O}{\|}}{C}-O-CH_2-CH_2-CH_3 + NaOH \xrightarrow{\text{Heat}}$

16.64 Draw the condensed structural or line-angle formulas for the products from each of the following reactions: (16.4)

a. $CH_3-CH_2-\overset{\overset{O}{\|}}{C}-O-\overset{\overset{CH_3}{|}}{CH}-CH_3 + NaOH \xrightarrow{\text{Heat}}$

b. 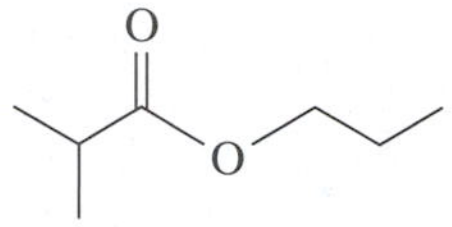 $+ H_2O \xrightleftharpoons{H^+, \text{heat}}$

CHALLENGE PROBLEMS

The following problems are related to the topics in this chapter. However, they do not all follow the chapter order, and they require you to combine concepts and skills from several Sections. These problems will help you increase your critical thinking skills and prepare for your next exam.

16.65 Using the reactions we have studied, indicate how you might prepare the following from the starting substance given: (12.7, 16.1, 16.3)
a. acetic acid from ethene
b. butyric acid from 1-butanol

16.66 Using the reactions we have studied, indicate how you might prepare the following from the starting substance given: (16.1, 16.3)
a. pentanoic acid from 1-pentanol
b. ethyl acetate from two molecules of ethanol

16.67 Methyl benzoate is not soluble in water; however, when it is heated with KOH, the ester forms soluble products. When HCl is added to neutralize the basic solution, a white solid forms. Use condensed structural formulas to write the balanced chemical equation when methyl benzoate and KOH react. Explain what happens. (16.3, 16.4)

16.68 Hexanoic acid is not soluble in water. However, when hexanoic acid is added to a solution of NaOH, a soluble product forms. Use condensed structural formulas to write the balanced chemical equation when hexanoic acid and NaOH react. Explain what happens. (16.3, 16.4)

16.69 Propyl acetate is an ester that gives the odor and flavor of pears. (9.4, 16.3, 16.4)
a. Draw the condensed structural formula for propyl acetate.

b. Use condensed structural formulas to write the balanced chemical equation for the formation of propyl acetate.
c. Use condensed structural formulas to write the balanced chemical equation for the hydrolysis of propyl acetate with HCl.
d. Use condensed structural formulas to write the balanced chemical equation for the hydrolysis of propyl acetate with NaOH.
e. How many milliliters of a 0.208 M NaOH solution is needed to completely hydrolyze (saponify) 1.58 g of propyl acetate?

16.70 Ethyl octanoate is a flavor component of mangos. (9.4, 16.3, 16.4)

a. Draw the condensed structural formula for ethyl octanoate.
b. Use condensed structural formulas to write the balanced chemical equation for the formation of ethyl octanoate.
c. Use condensed structural formulas to write the balanced chemical equation for the hydrolysis of ethyl octanoate with HCl.
d. Use condensed structural formulas to write the balanced chemical equation for the hydrolysis of ethyl octanoate with NaOH.
e. How many milliliters of a 0.315 M NaOH solution is needed to completely hydrolyze (saponify) 2.84 g of ethyl octanoate?

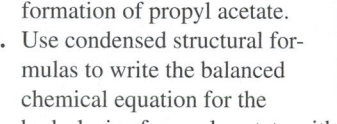

ANSWERS

16.1 methanoic acid (formic acid)

16.3 **a.** $CH_3-CH_2-CH_2-CH_2-CH_2-\overset{O}{\overset{\|}{C}}-OH$, hexanoic acid

b. $CH_3-CH_2-\underset{}{\overset{CH_3-CH_2}{\overset{|}{CH}}}-\overset{O}{\overset{\|}{C}}-OH$, 2-ethylbutanoic acid

16.5 **a.** ethanoic acid (acetic acid)
b. butanoic acid (butyric acid)
c. 3-methylhexanoic acid
d. 3,4-dibromobenzoic acid

16.7 **a.** $Cl-CH_2-\overset{O}{\overset{\|}{C}}-OH$

b. $HO-CH_2-CH_2-\overset{O}{\overset{\|}{C}}-OH$

c. 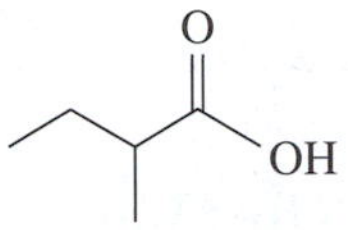

d. 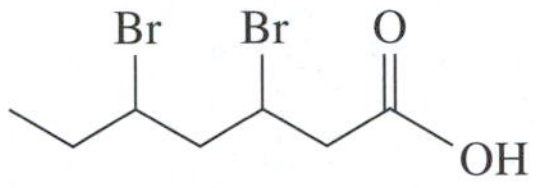

16.9 **a.** $H-\overset{O}{\overset{\|}{C}}-OH$ **b.** $CH_3-\overset{O}{\overset{\|}{C}}-OH$

c. 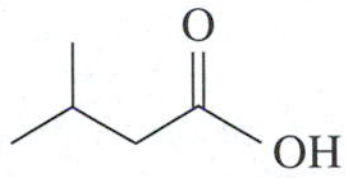 **d.** 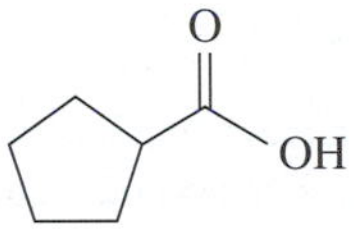

16.11 **a.** Butanoic acid has a greater molar mass and would have a higher boiling point.
b. Propanoic acid can form dimers, effectively doubling the molar mass, which gives propanoic acid a higher boiling point.
c. Butanoic acid can form dimers, effectively doubling the molar mass, which gives butanoic acid a higher boiling point.

16.13 **a.** Propanoic acid has the smallest alkyl group, which makes it most soluble.
b. Propanoic acid forms more hydrogen bonds with water, which makes it most soluble.

16.15 **a.** $CH_3-CH_2-CH_2-\overset{O}{\overset{\|}{C}}-OH + H_2O \rightleftarrows$

$CH_3-CH_2-CH_2-\overset{O}{\overset{\|}{C}}-O^- + H_3O^+$

b. $CH_3-\overset{CH_3}{\overset{|}{CH}}-\overset{O}{\overset{\|}{C}}-OH + H_2O \rightleftarrows$

$CH_3-\overset{CH_3}{\overset{|}{CH}}-\overset{O}{\overset{\|}{C}}-O^- + H_3O^+$

16.17 **a.** $CH_3-CH_2-CH_2-CH_2-\overset{O}{\overset{\|}{C}}-OH + NaOH \longrightarrow$

$CH_3-CH_2-CH_2-CH_2-\overset{O}{\overset{\|}{C}}-O^-\ Na^+ + H_2O$

b. $CH_3-\overset{Cl}{\overset{|}{CH}}-\overset{O}{\overset{\|}{C}}-OH + NaOH \longrightarrow$

$CH_3-\overset{Cl}{\overset{|}{CH}}-\overset{O}{\overset{\|}{C}}-O^-\ Na^+ + H_2O$

c. 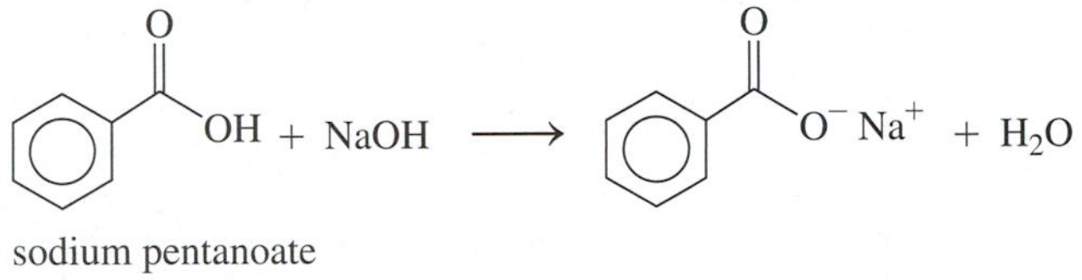

16.19 **a.** sodium pentanoate
b. sodium 2-chloropropanoate (sodium α-chloropropionate)
c. sodium benzoate

16.21 **a.** $CH_3-\overset{O}{\overset{\|}{C}}-O-CH_2-CH_3$

b. $CH_3-CH_2-CH_2-\overset{O}{\overset{\|}{C}}-O-CH_2-CH_3$

c. 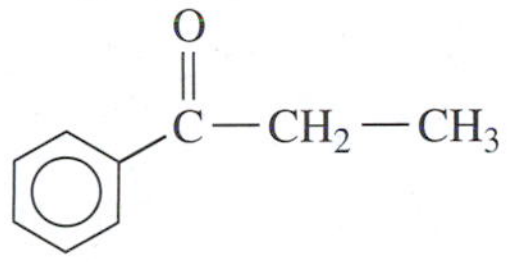

16.23 **a.** 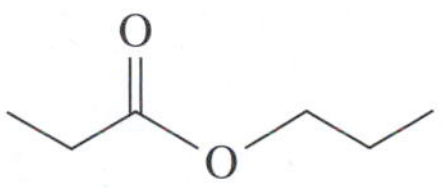

b. 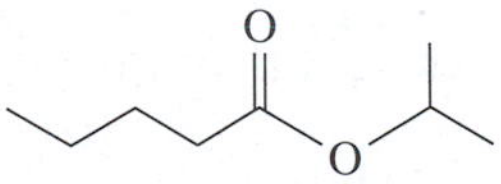

16.25 **a.** methanoic acid (formic acid) and methanol (methyl alcohol)
b. propanoic acid (propionic acid) and ethanol (ethyl alcohol)
c. butanoic acid (butyric acid) and methanol (methyl alcohol)
d. 3-methylbutanoic acid (β-methylbutyric acid) and ethanol (ethyl alcohol)

16.27 **a.** methyl methanoate (methyl formate)
b. ethyl propanoate (ethyl propionate)
c. methyl butanoate (methyl butyrate)
d. 2-methylpropyl pentanoate

16.29 **a.** $CH_3-CH_2-CH_2-\overset{O}{\overset{\|}{C}}-O-CH_2-CH_2-CH_3$

b. $H-\overset{O}{\overset{\|}{C}}-O-CH_2-CH_2-CH_2-CH_3$

c. 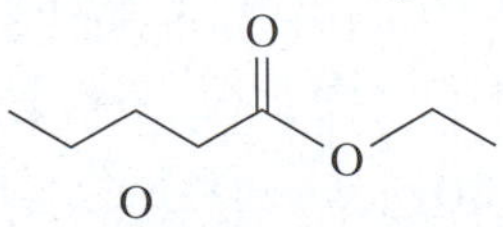

d. 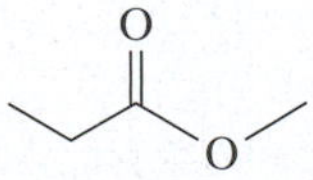

16.31 a. pentyl ethanoate (pentyl acetate)
b. octyl ethanoate (octyl acetate)
c. pentyl butanoate (pentyl butyrate)

16.33 a. $CH_3-CH_2-C(=O)-OH$
b. $CH_3-CH_2-CH_2-CH_2-OH$
c. 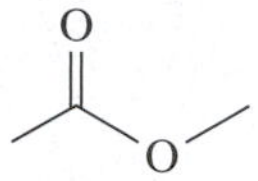

16.35 a. soluble
b. not soluble

16.37 a. $CH_3-CH_2-C(=O)-O^- \ Na^+ + HO-CH_3$
b. 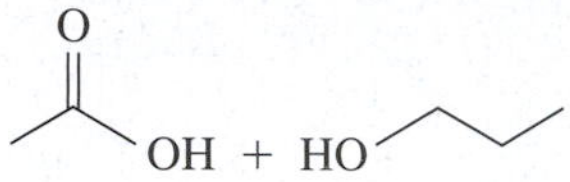
c. $CH_3-CH_2-CH_2-C(=O)-OH + HO-CH_2-CH_3$
d. $H-C(=O)-O^- \ Na^+ + HO-CH_3$
e. 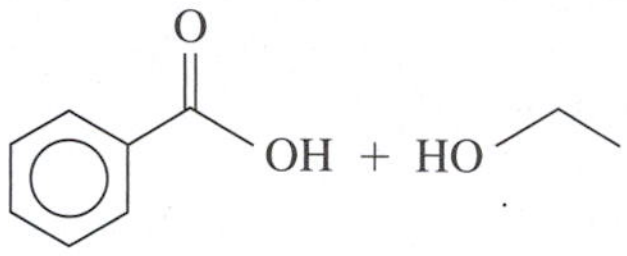

16.39 a.
$$H_2C{=}C(CN)-C(=O)-OH + HO-CH_2-CH_2-CH_2-CH_3 \xrightleftharpoons{H^+,\ heat} H_2C{=}C(CN)-C(=O)-O-CH_2-CH_2-CH_2-CH_3 + H_2O$$
b. butyl 2-cyano-2-propenoate

16.41 $CH_3-CH_2-CH_2-C(=O)-OH$ Butanoic acid; $CH_3-CH(CH_3)-C(=O)-OH$ 2-Methylpropanoic acid

16.43 $CH_3-C(=O)-O-CH_3$ Methyl ethanoate; $H-C(=O)-O-CH_2-CH_3$ Ethyl methanoate

16.45 a. $CH_3-CH_2-CH_2-C(=O)-O-CH_3$
b. butanoic acid and methanol
c.
$$CH_3-CH_2-CH_2-C(=O)-O-CH_3 + H_2O \xrightleftharpoons{H^+,\ heat} CH_3-CH_2-CH_2-C(=O)-OH + HO-CH_3$$
d.
$$CH_3-CH_2-CH_2-C(=O)-O-CH_3 + NaOH \xrightarrow{Heat} CH_3-CH_2-CH_2-C(=O)-O^- \ Na^+ + HO-CH_3$$

16.47 a. 3-methylbutanoic acid (β-methylbutyric acid)
b. ethyl benzoate
c. ethyl propanoate (ethyl propionate)
d. 2-chlorobenzoic acid (*o*-chlorobenzoic acid)
e. pentanoic acid
f. 2-propyl ethanoate (isopropyl acetate)

16.49 

16.51 a. $CH_3-CH_2-CH_2-CH_2-CH_2-C(=O)-O-CH_3$
b. 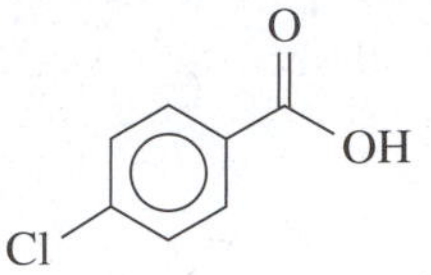
c. $Cl-CH_2-CH_2-C(=O)-OH$
d. $CH_3-CH_2-CH_2-C(=O)-O-CH_2-CH_3$
e. $CH_3-CH_2-CH(CH_3)-CH_2-C(=O)-OH$
f. 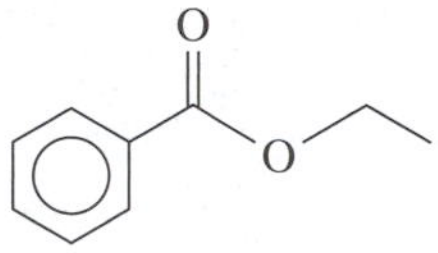

16.53 a. Ethanoic acid has a higher boiling point than 1-propanol because two molecules of ethanoic acid hydrogen bond to form a dimer, which effectively doubles the molar mass and requires a higher temperature to reach the boiling point.
b. Ethanoic acid forms hydrogen bonds, but butane does not.

16.55 Of the three compounds, methyl formate would have the lowest boiling point because it only has dipole–dipole attractions. Both acetic acid and 1-propanol can form hydrogen bonds, but because acetic acid can form dimers and double the effective molar mass, it has the highest boiling point. Methyl formate, 32 °C; 1-propanol, 97 °C; acetic acid, 118 °C.

16.57 **a**, **c**, and **d** are soluble in water.

16.59 a. $CH_3-CH_2-C(=O)-O^- + H_3O^+$
b. $CH_3-CH_2-C(=O)-O^- \ K^+ + H_2O$
c. 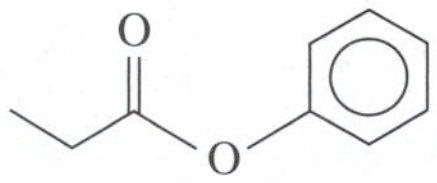 $+ H_2O$
d. 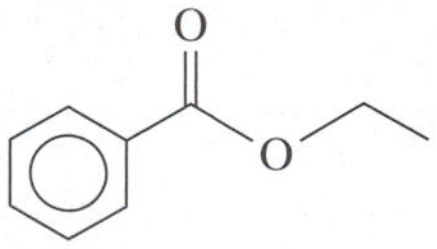

16.61 a. 3-methylbutanoic acid and methanol

b. 3-chlorobenzoic acid and ethanol

c. hexanoic acid and ethanol

16.63 a. $CH_3-CH_2-C(=O)-OH + HO-CH(CH_3)-CH_3$

b. $CH_3-CH(CH_3)-C(=O)-O^- Na^+ + HO-CH_2-CH_2-CH_3$

16.65 a. $H_2C{=}CH_2 + H_2O \xrightarrow{H^+}$

$CH_3-CH_2-OH \xrightarrow{[O]} CH_3-C(=O)-OH$

b. $CH_3-CH_2-CH_2-CH_2-OH \xrightarrow{[O]}$

$CH_3-CH_2-CH_2-C(=O)-OH$

16.67 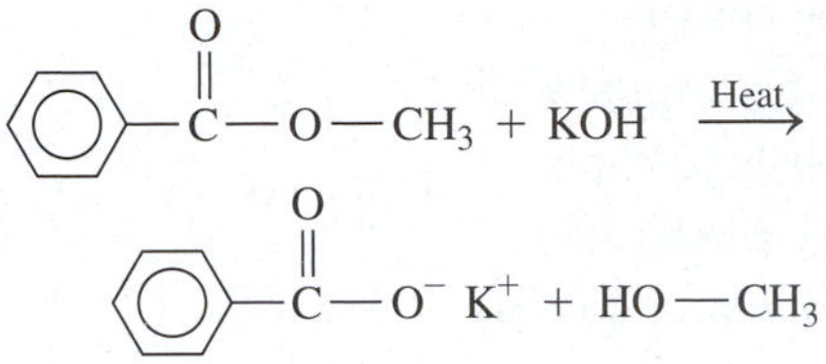
Heat

In KOH solution, the ester undergoes saponification to form the carboxylate salt, potassium benzoate, and methanol, which are soluble in water. When HCl is added, the salt is converted to benzoic acid, which is insoluble.

16.69 a. $CH_3-C(=O)-O-CH_2-CH_2-CH_3$

b. $CH_3-C(=O)-OH + HO-CH_2-CH_2-CH_3 \overset{H^+,\ heat}{\rightleftarrows}$

$CH_3-C(=O)-O-CH_2-CH_2-CH_3 + H_2O$

c. $CH_3-C(=O)-O-CH_2-CH_2-CH_3 + H_2O \overset{H^+,\ heat}{\rightleftarrows}$

$CH_3-C(=O)-OH + HO-CH_2-CH_2-CH_3$

d. $CH_3-C(=O)-O-CH_2-CH_2-CH_3 + NaOH \xrightarrow{Heat}$

$CH_3-C(=O)-O^- Na^+ + HO-CH_2-CH_2-CH_3$

e. 74.4 mL of a 0.208 M NaOH solution

Lipids

17

Rebecca learned of her family's hypercholesterolemia after her mother died of a heart attack at 44. Her mother had a total cholesterol level that was three times the normal level. Rebecca learned that the lumps under her mother's skin and around the corneas of her eyes, called xanthomas, were caused by cholesterol that was stored throughout her body. A cholesterol level of less than 200 mg/dL is desirable. In adults with familial hypercholesterolemia (FH), cholesterol levels in the blood may be greater than 300 mg/dL.

At the lipid clinic, Rebecca's blood tests show that she has a total cholesterol level of 420 mg/dL. Rebecca is diagnosed with FH. She learns that from birth, cholesterol has been accumulating throughout her arteries. FH is caused by a genetic mutation that blocks the removal of cholesterol from the blood, resulting in the formation of deposits on the arterial walls.

After her diagnosis, Rebecca meets with Susan, a clinical lipid specialist at the lipid clinic, who tells her about managing risk factors that can lead to heart attack and stroke. Susan asks Rebecca to change her diet, increase her exercise, and return to the clinic one year later so they can determine if these changes have lowered her cholesterol levels sufficiently.

CAREER

Clinical Lipid Specialist

Clinical lipid specialists work with patients who have lipid disorders such as high cholesterol, high triglycerides, coronary heart disease, obesity, and FH. A clinical lipid specialist reviews a patient's lipid profile, which includes total cholesterol, high-density lipoprotein (HDL), and low-density lipoprotein (LDL) levels. If a lipid disorder is identified, the lipid specialist assesses the patient's current diet and exercise program. The lipid specialist diagnoses and determines treatment including dietary changes such as reducing salt intake, increasing the amount of fiber in the diet, and lowering the amount of fat. Lipid specialists also discuss drug therapy using lipid-lowering medications that remove LDL-cholesterol to help patients achieve and maintain good health.

Allied health professionals such as nurses, nurse practitioners, pharmacists, and dietitians are certified by a clinical lipid specialist program for specialized care of patients with lipid disorders.

CLINICAL UPDATE

Rebecca's Program to Lower Cholesterol

When Rebecca returns to the lipid clinic, Susan tells Rebecca that her diet and exercise are not lowering her cholesterol level sufficiently. You can view the medications that Susan prescribes and determine their effects on Rebecca's cholesterol levels in the **CLINICAL UPDATE** **Rebecca's Program to Lower Cholesterol**, pages 615–616.

17.1 Lipids

LEARNING GOAL Describe the classes of lipids.

Lipids are a family of biomolecules that have the common property of being soluble in organic solvents, like ether or chloroform, but not in water. They serve multiple purposes in the body, such as storing energy, protecting and insulating internal organs, and acting as chemical messengers. Because they are not soluble in water, lipids are also important components of cellular membranes that function to separate the internal contents of cells from the external environment.

Types of Lipids

Within the lipid family, there are specific structures that distinguish the different types of lipids. Lipids such as waxes, triacylglycerols, glycerophospholipids, and sphingolipids are esters that can be hydrolyzed to give fatty acids along with other molecules. Triacylglycerols and glycerophospholipids contain the alcohol *glycerol*, whereas sphingolipids contain the amino alcohol *sphingosine*. Steroids, which have a completely different structure, do not contain fatty acids and cannot be hydrolyzed. Steroids are characterized by the *steroid nucleus* of four fused carbon rings. **FIGURE 17.1** illustrates the types and general structure of lipids we discuss in this chapter.

LOOKING AHEAD

ENGAGE

What is one feature that is characteristic of all lipids?

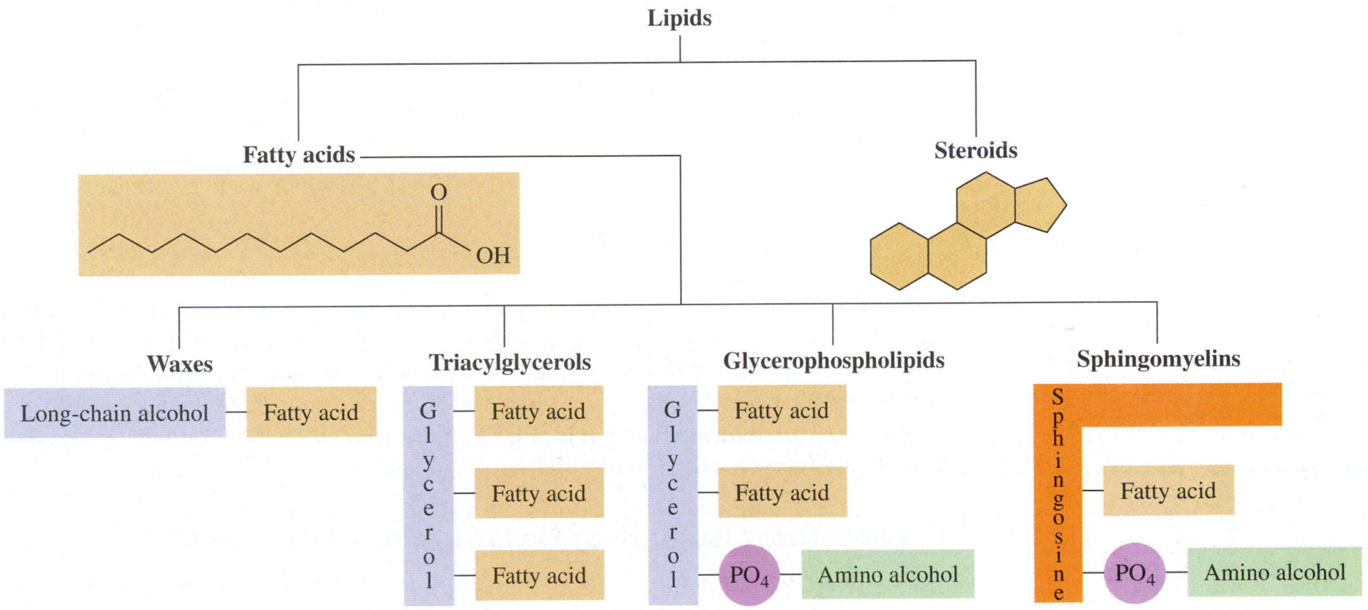
Lipids
Fatty acids
O
OH
Steroids
Waxes
Long-chain alcohol
Fatty acid
Triacylglycerols
Glycerol
Fatty acid
Fatty acid
Fatty acid
Glycerophospholipids
Glycerol
Fatty acid
Fatty acid
PO4
Amino alcohol
Sphingomyelins
Sphingosine
Fatty acid
PO4
Amino alcohol

FIGURE 17.1 ▶ Lipids are naturally occurring biomolecules which are soluble in organic solvents but not in water.

Q What chemical structure do waxes, triacylglycerols, and glycerophospholipids have in common?

TEST

Try Practice Problems 17.1 to 17.4

PRACTICE PROBLEMS

17.1 Lipids

17.1 Lipids are not soluble in water. Are lipids polar or nonpolar molecules?

17.2 Which of the following solvents might be used to dissolve an oil stain?

a. water **b.** CCl_4 **c.** diethyl ether
d. benzene **e.** NaCl solution

Clinical Applications

17.3 What are some functions of lipids in the body?

17.4 What are some of the different kinds of lipids?

17.2 Fatty Acids

LEARNING GOAL Draw the condensed structural and line-angle formulas for a fatty acid, and identify it as saturated or unsaturated.

Many lipids include fatty acids as part of their structures. A **fatty acid** contains a long, unbranched carbon chain with a carboxylic acid group at one end. Although the carboxylic acid part is hydrophilic, the long hydrophobic carbon chain makes fatty acids insoluble in water.

CORE CHEMISTRY SKILL
Identifying Fatty Acids

Most naturally occurring fatty acids have an even number of carbon atoms, usually between 12 and 20. An example of a fatty acid is lauric acid, a 12-carbon acid found in coconut oil. The shorthand notation description for lauric acid is 12:0; 12 for the number of carbon atoms and 0 for the number of double bonds.

Drawing Structural Formulas for Lauric Acid (12:0)

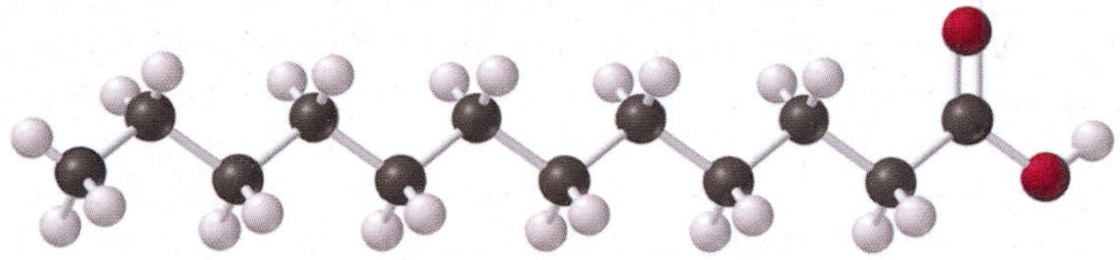

Ball-and-stick model of lauric acid

$CH_3-(CH_2)_{10}-\overset{\overset{\displaystyle O}{\|}}{C}-OH$ $\qquad$ $CH_3-(CH_2)_{10}-COOH$

$CH_3-CH_2-CH_2-CH_2-CH_2-CH_2-CH_2-CH_2-CH_2-CH_2-CH_2-C(=O)OH$

Condensed structural formulas of lauric acid

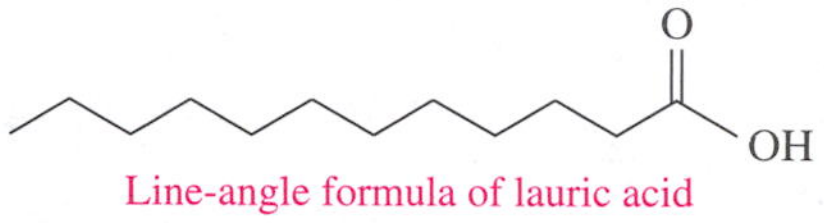
O
OH

Line-angle formula of lauric acid

A **saturated fatty acid (SFA)** contains only carbon–carbon single bonds, which make the properties of a long-chain fatty acid similar to those of an alkane. An **unsaturated fatty acid (UFA)** contains one or more carbon–carbon double bonds. In a **monounsaturated fatty acid (MUFA)**, the long carbon chain has one double bond, which makes its properties similar to those of an alkene. A **polyunsaturated fatty acid (PUFA)** has at least two carbon–carbon double bonds. **TABLE 17.1** lists some of the typical fatty acids in lipids.

TEST
Try Practice Problems 17.5 and 17.6

Cis and Trans Isomers of Unsaturated Fatty Acids

Unsaturated fatty acids can be drawn as *cis* and *trans* isomers. For example, oleic acid, an 18-carbon monounsaturated fatty acid found in olives, has one double bond starting at carbon 9. We can show its cis and trans structures using line-angle formulas. In the cis isomer, the carbon chain has a "kink" at the double bond site. The trans isomer of oleic acid, *elaidic acid*, is a straight chain without a kink at the double bond site. The cis structure is the more prevalent isomer found in naturally occurring unsaturated fatty acids.

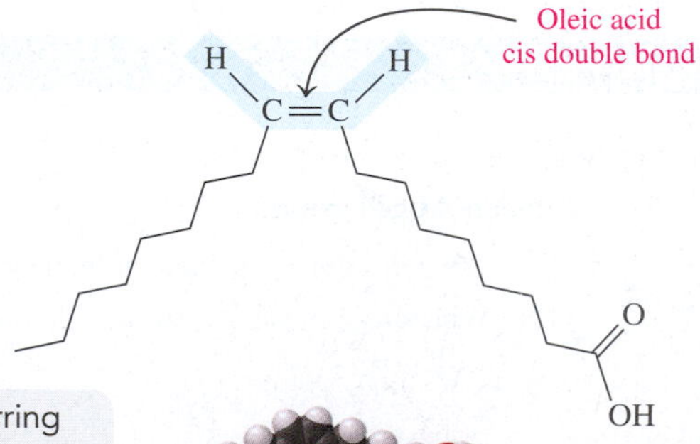
Oleic acid
cis double bond
H
H
C=C
O
OH

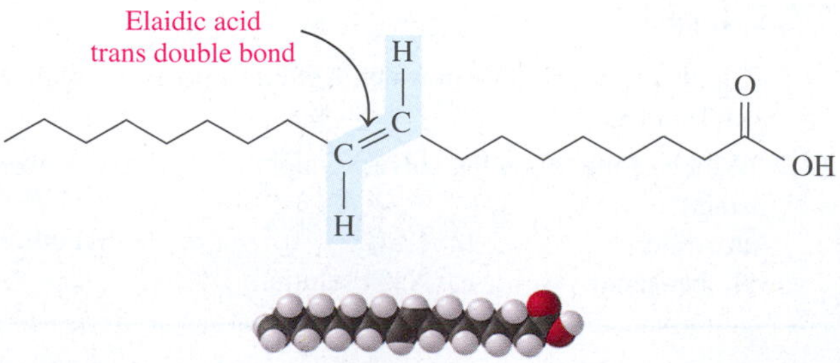
Elaidic acid
trans double bond
H
C
C
H
O
OH

Almost all naturally occurring unsaturated fatty acids have one or more cis double bonds.

TABLE 17.1 Structures and Melting Points of Common Fatty Acids

Name	Carbon Atoms: Double Bonds	Present in	Melting Point (°C)	Structures
Saturated Fatty Acids				
Lauric acid	12:0	Coconut	44	$CH_3-(CH_2)_{10}-COOH$ 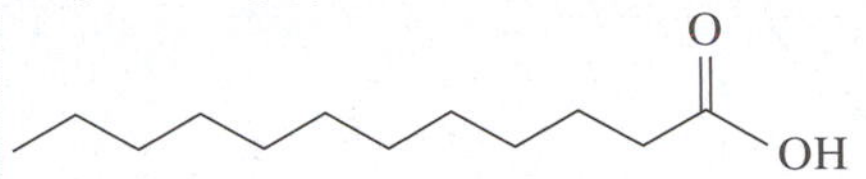
Myristic acid	14:0	Nutmeg	55	$CH_3-(CH_2)_{12}-COOH$ 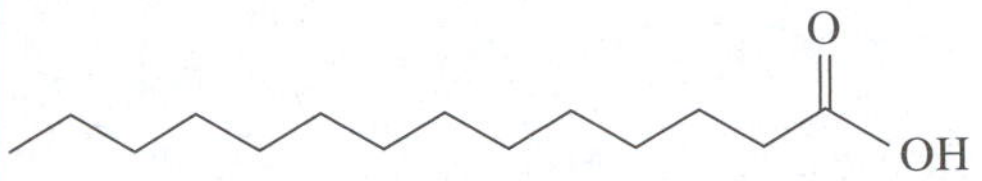
Palmitic acid	16:0	Palm	63	$CH_3-(CH_2)_{14}-COOH$ 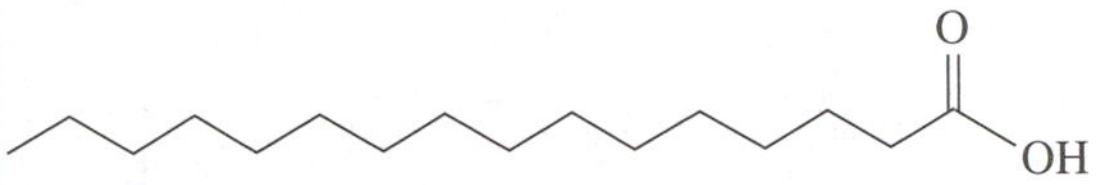
Stearic acid	18:0	Animal fat	69	$CH_3-(CH_2)_{16}-COOH$ 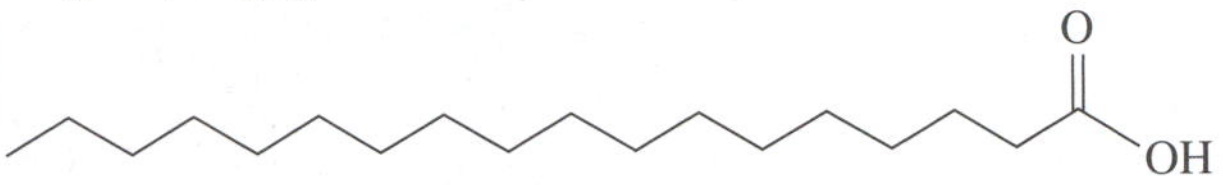
Monounsaturated Fatty Acids				
Palmitoleic acid	16:1	Butter	0	$CH_3-(CH_2)_5-CH=CH-(CH_2)_7-COOH$ 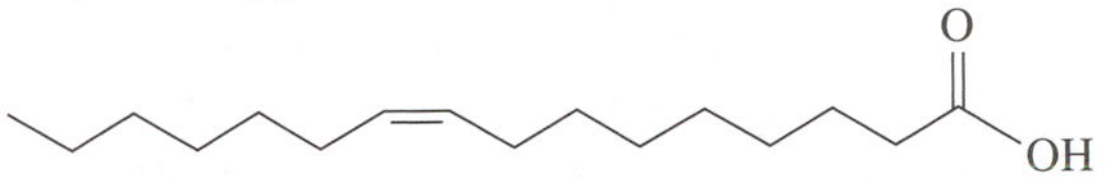
Oleic acid	18:1	Olives, pecan, grapeseed	14	$CH_3-(CH_2)_7-CH=CH-(CH_2)_7-COOH$ 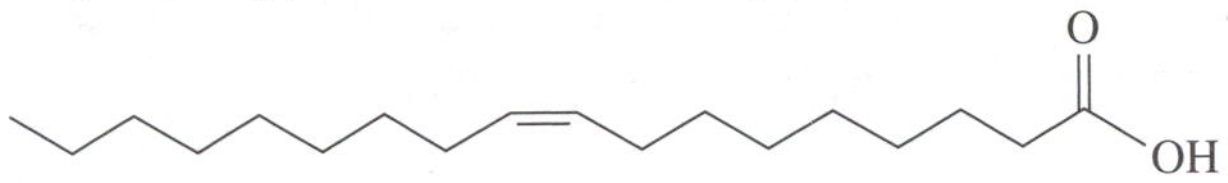
Polyunsaturated Fatty Acids				
Linoleic acid	18:2	Soybean, safflower, sunflower	−5	$CH_3-(CH_2)_4-CH=CH-CH_2-CH=CH-(CH_2)_7-COOH$ 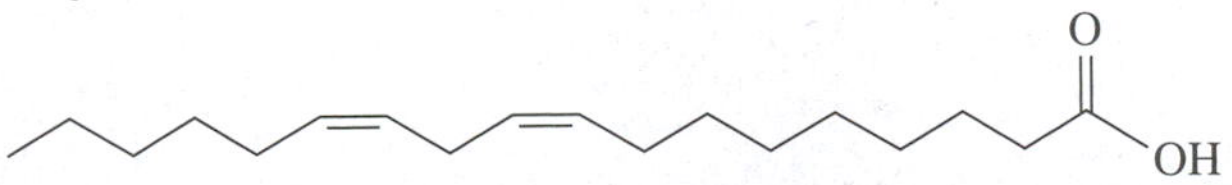
Linolenic acid	18:3	Corn	−11	$CH_3-CH_2-CH=CH-CH_2-CH=CH-CH_2-CH=CH-(CH_2)_7-COOH$ 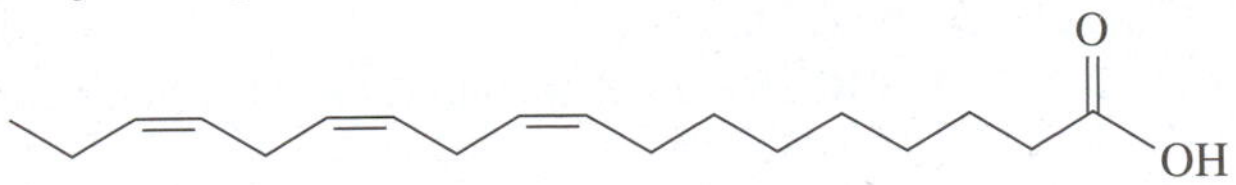
Arachidonic acid	20:4	Meat, eggs, fish	−50	$CH_3-(CH_2)_3-(CH_2-CH=CH)_4-(CH_2)_3-COOH$ 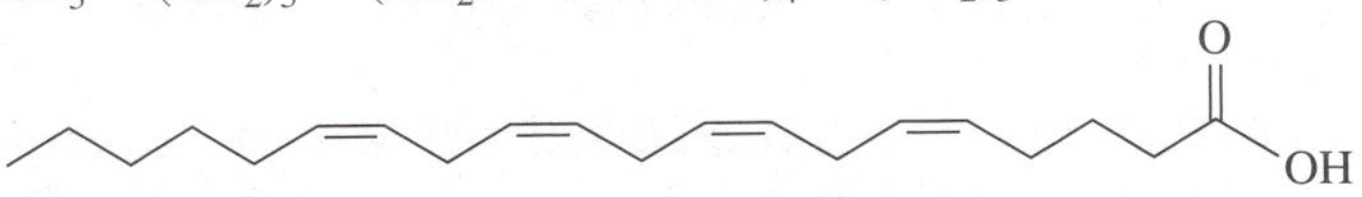

Physical Properties of Fatty Acids

The saturated fatty acids fit closely together in a regular pattern, which allows many dispersion forces between the carbon chains. These normally weak forces of attraction become important when molecules of fatty acids are close together. As a result, a significant amount of energy and higher temperatures are required to separate the fatty acids before melting occurs. As the length of the carbon chain increases, more interactions occur between the fatty acids, requiring higher temperatures to melt. Saturated fatty acids are usually solids at room temperature (see **FIGURE 17.2**).

In unsaturated fatty acids, the cis double bonds cause the carbon chain to bend or kink, giving the molecules an irregular shape. As a result, unsaturated fatty acids cannot stack

ENGAGE

Why do unsaturated fatty acids have lower melting points than saturated fatty acids?

as closely as saturated fatty acids, and thus have fewer dispersion forces between carbon chains. We might think of saturated fatty acids as regularly shaped chips that stack closely together in a can. Similarly, unsaturated fatty acids would be like irregularly shaped chips that do not fit closely together. Consequently, less energy is required to separate these molecules, making the melting points of unsaturated fatty acids lower than those of saturated fatty acids. Most unsaturated fatty acids are liquids at room temperature.

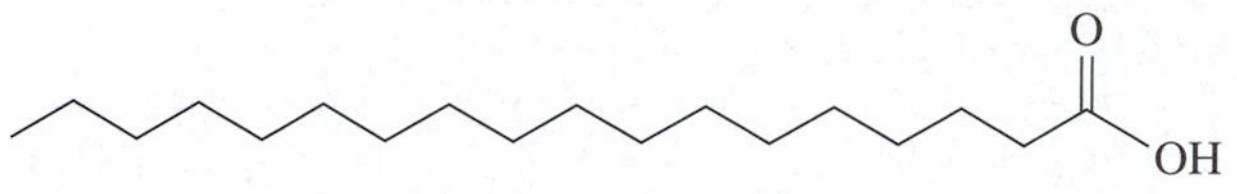
O
OH

Stearic acid, mp 69 °C

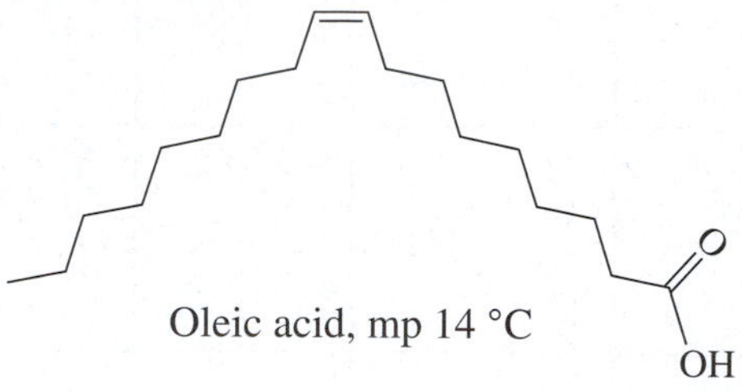
O
OH

Oleic acid, mp 14 °C

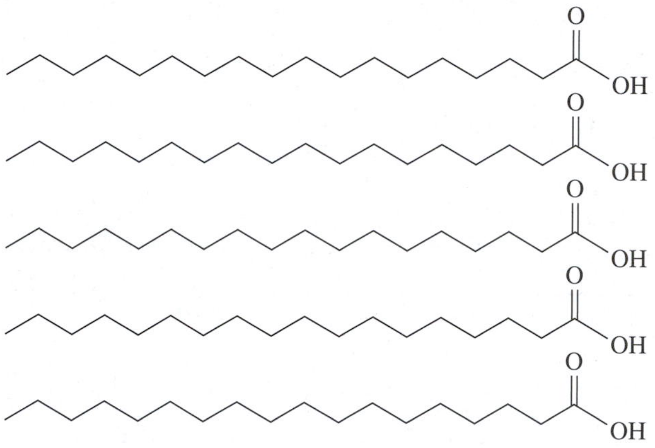
O
OH
O
OH
O
OH
O
OH
O
OH

(a) In saturated fatty acids, the molecules fit closely together, resulting in higher melting points.

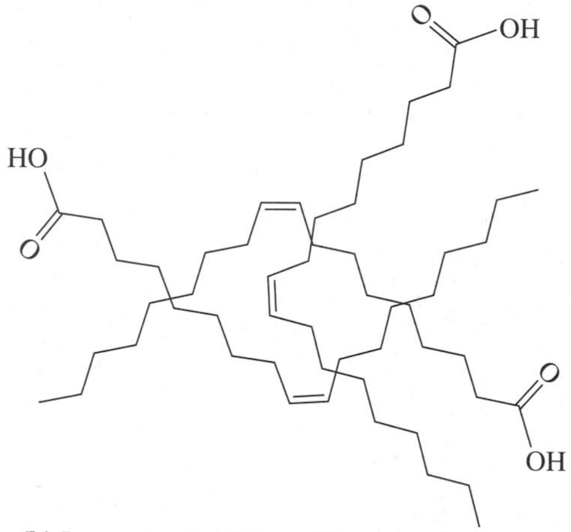
O
OH
HO
O
O
OH

(b) In unsaturated fatty acids, molecules cannot fit closely together, resulting in lower melting points.

FIGURE 17.2 ▶ Saturated fatty acids can fit together more closely than can unsaturated fatty acids.

Q Why does the cis double bond cause unsaturated fatty acids to have lower melting points than saturated fatty acids?

▶ SAMPLE PROBLEM 17.1 Structures and Properties of Fatty Acids

TRY IT FIRST

Consider the line-angle formula for vaccenic acid, a fatty acid found in dairy products.

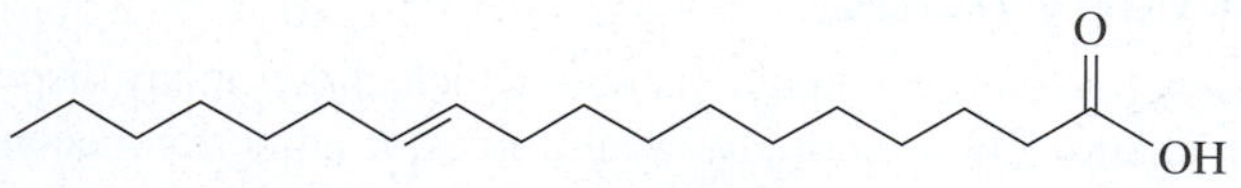
O
OH

a. Why is this substance an acid?
b. How many carbon atoms are in vaccenic acid?
c. Is the fatty acid saturated, monounsaturated, or polyunsaturated?
d. Give the shorthand notation for the number of carbon atoms and double bonds in vaccenic acid.
e. Would it be soluble in water?

SOLUTION

a. Vaccenic acid contains a carboxylic acid group.
b. It contains 18 carbon atoms.
c. It is a monounsaturated fatty acid (MUFA).
d. 18:1
e. No. Its long hydrocarbon chain makes it insoluble in water.

STUDY CHECK 17.1

Palmitoleic acid is a fatty acid with the following condensed structural formula:

$$CH_3-(CH_2)_5-CH=CH-(CH_2)_7-\overset{\overset{\displaystyle O}{\|}}{C}-OH$$

a. How many carbon atoms are in palmitoleic acid?
b. Is the fatty acid saturated, monounsaturated, or polyunsaturated?
c. Give the shorthand notation for the number of carbon atoms and double bonds in palmitoleic acid.
d. Is it most likely to be solid or liquid at room temperature?

ANSWER

a. 16
b. monounsaturated
c. 16:1
d. liquid

TEST
Try Practice Problems 17.7 to 17.12

Chemistry Link to Health

Omega-3 Fatty Acids in Fish Oils

Because unsaturated fats are more beneficial to health than saturated fats, American diets should include more unsaturated fats and less saturated fatty acids. Research indicates that atherosclerosis and heart disease are associated with high levels of fats in the diet. However, the Inuit people of Alaska have a diet with high levels of unsaturated fats as well as high levels of blood cholesterol, but a very low occurrence of atherosclerosis and heart attacks. The fats in the Inuit diet are primarily unsaturated fats from fish, rather than saturated fats from land animals.

Both vegetable and fish oils have high levels of unsaturated fats. The fatty acids in vegetable oils are *omega-6 fatty acids* (ω-6 fatty acids), in which the first double bond occurs at carbon 6 counting from the methyl end of the carbon chain. Omega is the last letter in the Greek alphabet and is used to denote the end. Two common omega-6 acids are linoleic acid (LA) and arachidonic acid (AA). However, the fatty acids in fish oils are mostly the omega-3 type, in which the first double bond occurs at the third carbon counting from the methyl group. Three common *omega-3 fatty acids* (or ω-3 fatty acids) in fish are linolenic acid (ALA), eicosapentaenoic acid (EPA), and docosahexaenoic acid (DHA).

In atherosclerosis and heart disease, cholesterol forms plaques that adhere to the walls of the blood vessels. Blood pressure rises as blood has to squeeze through a smaller opening in the blood vessel. As more plaque forms, there is also a possibility of blood clots blocking the blood vessels and causing a heart attack. Omega-3 fatty acids lower the tendency of blood platelets to stick together, thereby reducing the possibility of blood clots. It does seem that a diet that includes fish such as salmon, tuna, and herring can provide higher amounts of the omega-3 fatty acids, which help lessen the possibility of developing heart disease. However, high levels of omega-3 fatty acids can increase bleeding if the ability of the platelets to form blood clots is reduced too much.

Cold water fish are a good source of omega-3 fatty acids.

(continued)

Chemistry Link to Health (*continued*)

Omega-6 Fatty Acids

Linoleic acid (LA)

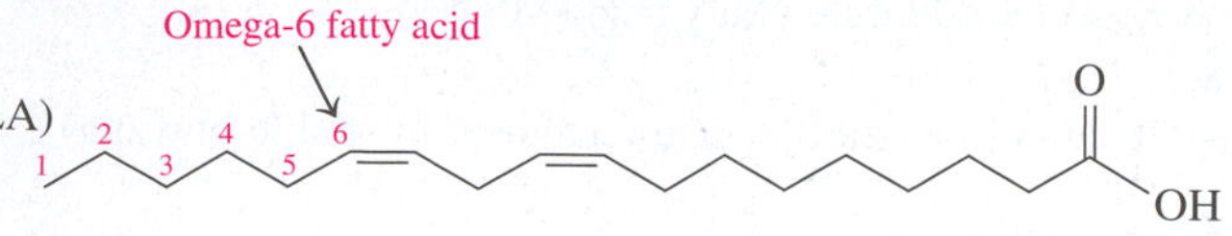
Omega-6 fatty acid
1
2
3
4
5
6
O
OH

Arachidonic acid (AA)

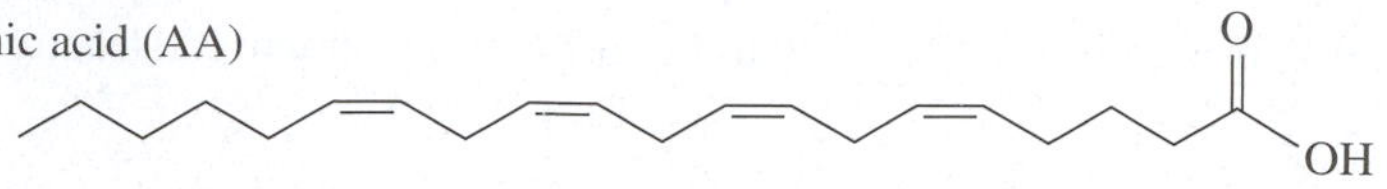
O
OH

Omega-3 Fatty Acids

Linolenic acid (ALA)

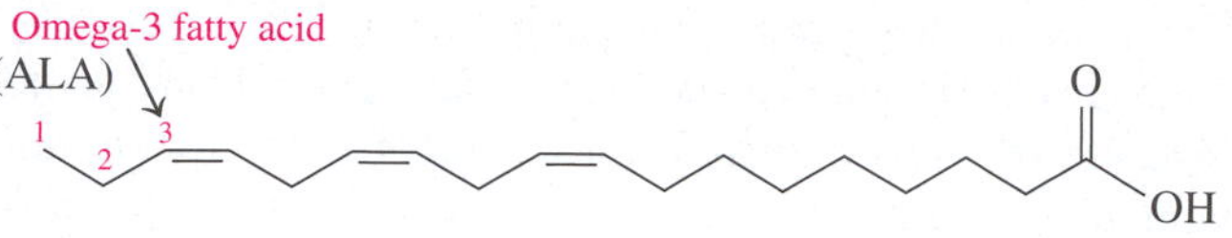
Omega-3 fatty acid
1
2
3
O
OH

Eicosapentaenoic acid (EPA)

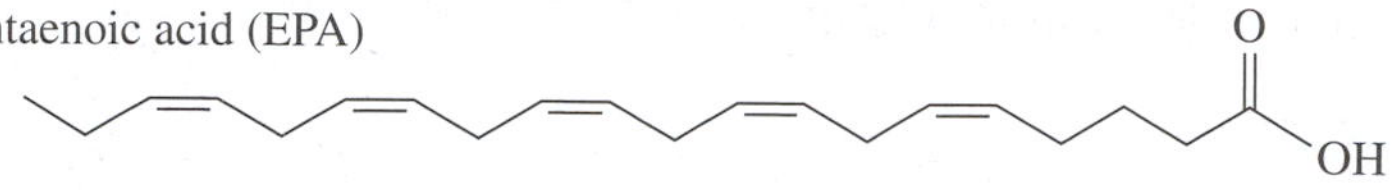
O
OH

Docosahexaenoic acid (DHA)

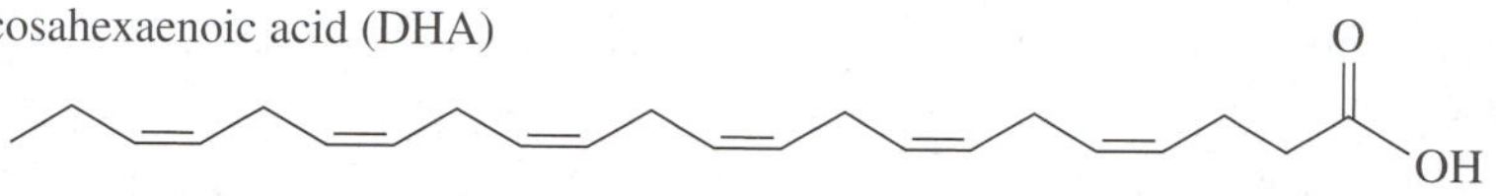
O
OH

TEST

Try Practice Problems 17.13 and 17.14

Prostaglandins

The human body is capable of synthesizing some fatty acids from carbohydrates or other fatty acids. However, humans cannot synthesize the polyunsaturated fatty acids linoleic acid and linolenic acid. Because they must be obtained from the diet, they are known as *essential fatty acids* (EFAs).

The EFA linoleic acid plays an important role in the body because it is a precursor of arachidonic acid, the polyunsaturated fatty acid with 20 carbon atoms. The prostaglandins, also known as *eicosanoids*, are formed from arachidonic acid (*eicos* is the Greek word for 20). **Prostaglandins** are hormone-like substances produced in small amounts in most cells of the body.

Swedish chemists first discovered prostaglandins and named them "prostaglandin E" (soluble in ether) and "prostaglandin F" (soluble in phosphate buffer or *fosfat* in Swedish). The various kinds of prostaglandins differ by the substituents attached to

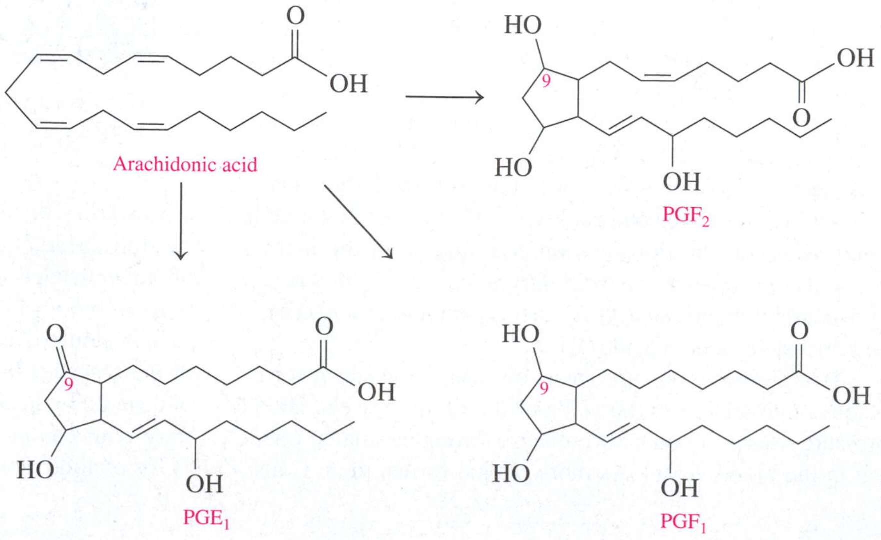
Arachidonic acid
O
OH
HO
9
OH
O
HO
OH
PGF2
O
9
O
OH
HO
OH
PGE1
HO
9
O
OH
HO
OH
PGF1

the five-carbon ring. Prostaglandin E (PGE) has a ketone group on carbon 9, whereas prostaglandin F (PGF) has a hydroxyl group. The number of double bonds is shown as a subscript 1 or 2.

Although prostaglandins are broken down quickly, they have potent physiological effects. Some prostaglandins increase blood pressure, and others lower blood pressure. Other prostaglandins stimulate contraction and relaxation in the smooth muscle of the uterus. When tissues are injured, arachidonic acid is converted to prostaglandins that produce inflammation and pain in the area.

The treatment of pain, fever, and inflammation is based on inhibiting the enzymes that convert arachidonic acid to prostaglandins. Several nonsteroidal anti-inflammatory drugs (NSAIDs), such as aspirin, block the production of prostaglandins and in doing so decrease pain, inflammation, and fever. Ibuprofen has similar anti-inflammatory and pain relief effects. Other NSAIDs include naproxen (Aleve and Naprosyn), ketoprofen (Actron), and nabumetone (Relafen). Although NSAIDs are helpful, their long-term use can result in liver, kidney, and gastrointestinal damage.

ENGAGE

What are some functions of prostaglandins in the body?

TEST

Try Practice Problems 17.15 to 17.18

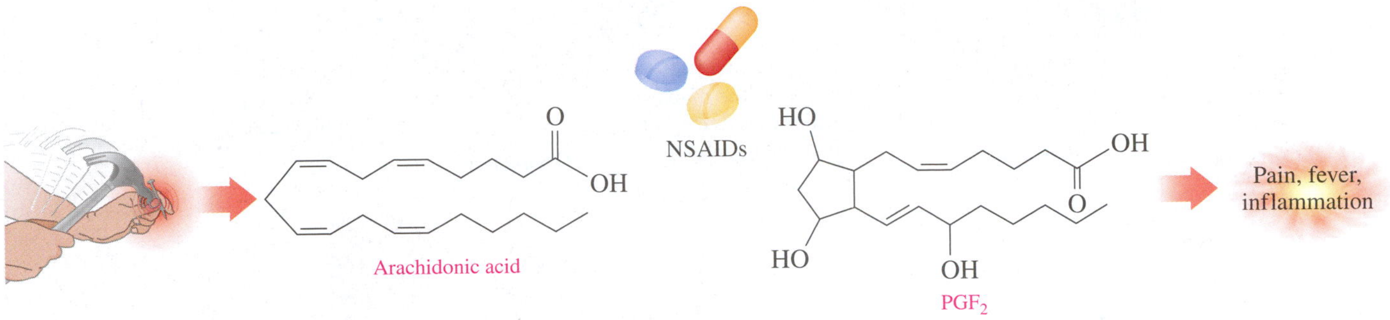
O
OH
Arachidonic acid
NSAIDs
HO
OH
O
HO
OH
PGF_2
Pain, fever, inflammation

When tissues are injured, prostaglandins are produced, which cause pain and inflammation.

PRACTICE PROBLEMS

17.2 Fatty Acids

17.5 Describe some similarities and differences in the structures of a saturated fatty acid and an unsaturated fatty acid.

17.6 Stearic acid and linoleic acid each have 18 carbon atoms. Why does stearic acid melt at 69 °C but linoleic acid melts at −5 °C?

17.7 Draw the line-angle formula for each of the following fatty acids:
- **a.** palmitic acid
- **b.** oleic acid

17.8 Draw the line-angle formula for each of the following fatty acids:
- **a.** stearic acid
- **b.** linoleic acid

17.9 For each of the following fatty acids, give the shorthand notation for the number of carbon atoms and double bonds, and classify as saturated, monounsaturated, or polyunsaturated:
- **a.** lauric acid
- **b.** linolenic acid
- **c.** palmitoleic acid
- **d.** stearic acid

17.10 For each of the following fatty acids, give the shorthand notation for the number of carbon atoms and double bonds, and classify as saturated, monounsaturated, or polyunsaturated:
- **a.** linoleic acid
- **b.** palmitic acid
- **c.** myristic acid
- **d.** oleic acid

17.11 How does the structure of a fatty acid with a cis double bond differ from the structure of a fatty acid with a trans double bond?

17.12 How does the double bond influence the dispersion forces that can form between the hydrocarbon chains of fatty acids?

Clinical Applications

17.13 What is the difference in the location of the first double bond in an omega-3 and an omega-6 fatty acid (see Chemistry Link to Health: Omega-3 Fatty Acids in Fish Oils)?

17.14 What are some sources of omega-3 and omega-6 fatty acids (see Chemistry Link to Health: Omega-3 Fatty Acids in Fish Oils)?

17.15 Compare the structures and functional groups of arachidonic acid and prostaglandin PGE_1.

17.16 Compare the structures and functional groups of PGF_1 and PGF_2.

17.17 What are some effects of prostaglandins in the body?

17.18 How does an anti-inflammatory drug reduce inflammation?

Chemistry Link to Health

A Prostaglandin-like Medication for Glaucoma That Also Thickens Eyelashes

Glaucoma is a group of diseases in which an increase in fluid pressure in the eye can damage the optic nerve. It is one of the leading causes of irreversible blindness. The increased fluid pressure typically results when fluid is unable to drain or doesn't drain fast enough from the eye.

Glaucoma is usually treated through the application of eye drops that either decrease the amount of fluid that the eye produces or cause more fluid to drain from the eye. The medication bimatoprost (Lumigan) decreases pressure in the eye by increasing fluid drainage. Bimatoprost is a prostaglandin *analog*, meaning that it has a structure that is very similar to that of the prostaglandins.

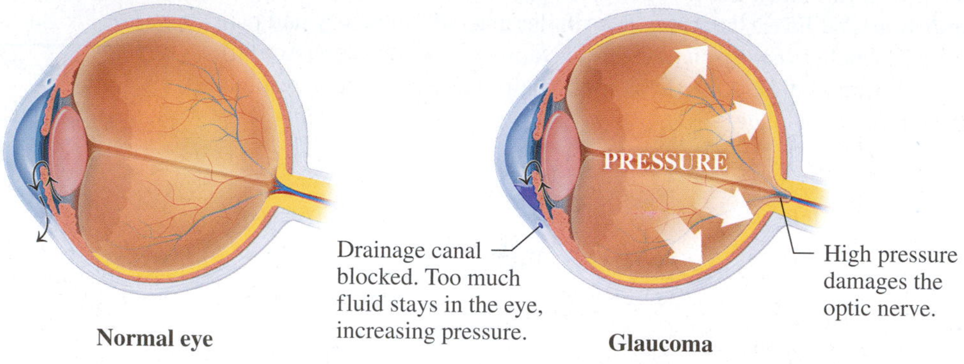
PRESSURE
Drainage canal blocked. Too much fluid stays in the eye, increasing pressure.
High pressure damages the optic nerve.
Normal eye
Glaucoma

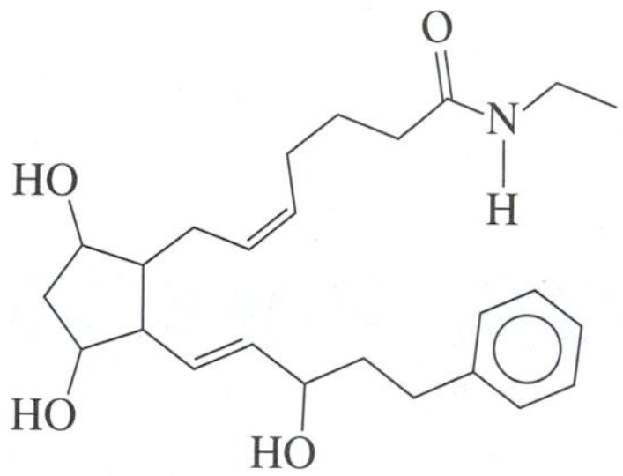
O
N
H
HO
HO
HO

The structure of bimatoprost is similar to that of prostaglandins.

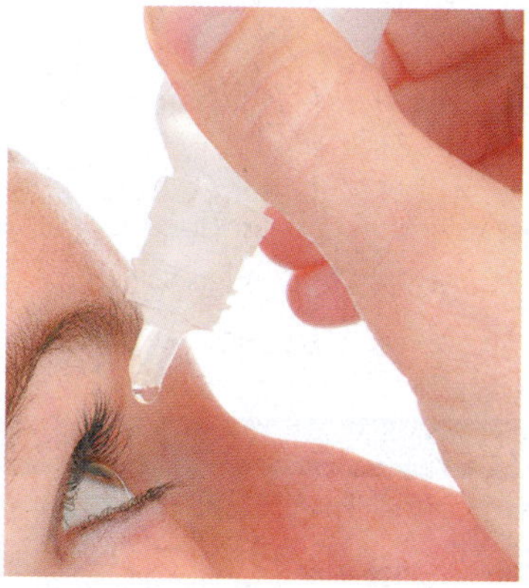

In the treatment of glaucoma, bimatoprost is applied as eye drops.

Doctors noticed that many of their patients that used bimatoprost for glaucoma developed longer, thicker, and darker eyelashes than they had previously. Therefore, Allergan, one of the companies that markets bimatoprost for glaucoma treatment, investigated bimatoprost's potential cosmetic uses. In 2008, the U.S. Food and Drug Administration approved the use of bimatoprost for the lengthening of eyelashes (bimatoprost is marketed under the name Latisse when intended for cosmetic use). When used cosmetically, the bimatoprost solution is applied with a brush to the upper lid of the eye, close to the eyelashes, once a day. Within 16 weeks, most patients using bimatoprost cosmetically notice longer, thicker, and darker eyelashes. If patients discontinue applying the bimatoprost solution, the eyelashes return to their original state.

17.3 Waxes and Triacylglycerols

LEARNING GOAL Draw the condensed structural and line-angle formulas for a wax or triacylglycerol produced by the reaction of a fatty acid and an alcohol or glycerol.

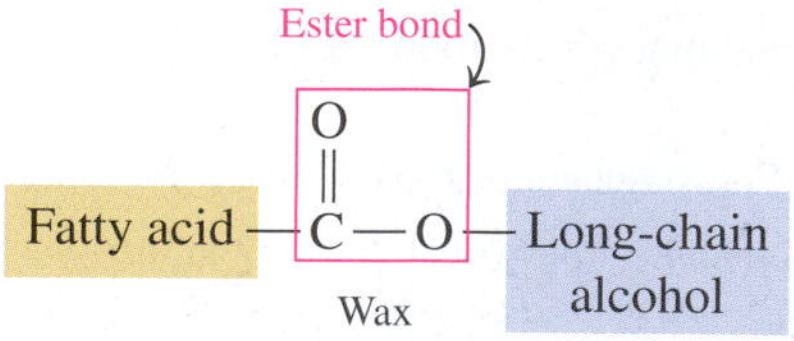
Ester bond
O
Fatty acid
C
O
Long-chain alcohol
Wax

Fatty acids are the building blocks of other compounds that are important in living organisms, like waxes and triacylglycerols. *Waxes* are found in many plants and animals. Natural waxes are found on the surface of fruits, and on the leaves and stems of plants where they help prevent loss of water and damage from pests. Waxes on the skin, fur, and feathers of animals provide a waterproof coating. A **wax** is an ester of a long-chain fatty acid and a long-chain alcohol, each containing from 14 to 30 carbon atoms.

The formulas of some common waxes are given in **TABLE 17.2**. Beeswax obtained from honeycomb and carnauba wax from palm trees are used to give a protective coating to furniture, cars, and floors. Jojoba wax is used in making candles and cosmetics such as lipstick. Lanolin, a mixture of waxes obtained from wool, is used in hand and facial lotions to aid retention of water, which softens the skin.

Waxes are esters of long-chain alcohols and fatty acids.

TABLE 17.2 Some Typical Waxes

Type	Condensed Structural Formula	Source	Uses
Beeswax	$CH_3-(CH_2)_{14}-\overset{\displaystyle O}{\overset{\|}{C}}-O-(CH_2)_{29}-CH_3$	Honeycomb	Candles, shoe polish, wax paper
Carnauba wax	$CH_3-(CH_2)_{24}-\overset{\displaystyle O}{\overset{\|}{C}}-O-(CH_2)_{29}-CH_3$	Brazilian palm tree	Waxes for furniture, cars, floors, shoes
Jojoba wax	$CH_3-(CH_2)_{18}-\overset{\displaystyle O}{\overset{\|}{C}}-O-(CH_2)_{19}-CH_3$	Jojoba bush	Candles, soaps, cosmetics

TEST

Try Practice Problems 17.19 and 17.20

Triacylglycerols

In the body, fatty acids are stored as **triacylglycerols**, also called *triglycerides*, which are triesters of glycerol (a trihydroxy alcohol) and fatty acids. The general structure of a triacylglycerol follows:

O
O
O
CH_2-O
$CH-O$
CH_2-O

Triacylglycerol

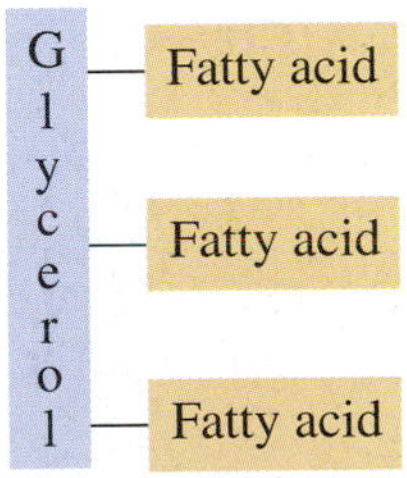
Glycerol
Fatty acid
Fatty acid
Fatty acid

ENGAGE

What functional group is present in all fats (triacylglycerols)?

In a triacylglycerol, three hydroxyl groups of glycerol form ester bonds with the carboxyl groups of three fatty acids. For example, glycerol and three molecules of stearic acid form a triacylglycerol. In the name, glycerol is named *glyceryl* and the fatty acids are named as carboxylates. For example, stearic acid is named as stearate, which gives the name glyceryl tristearate. The common name of this compound is tristearin.

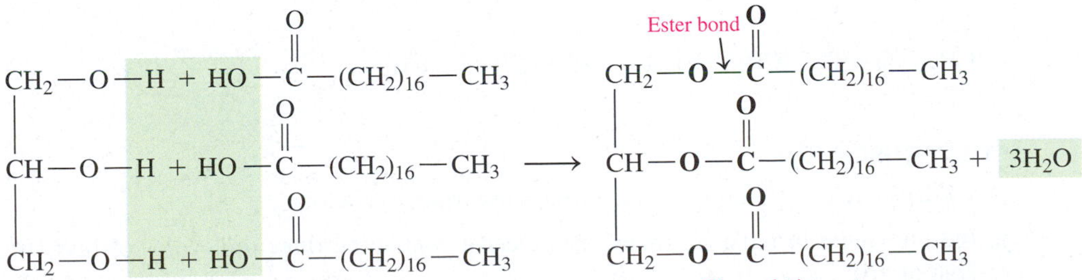
$CH_2-O-H + HO-\overset{O}{\overset{\|}{C}}-(CH_2)_{16}-CH_3$
$CH-O-H + HO-\overset{O}{\overset{\|}{C}}-(CH_2)_{16}-CH_3$
$CH_2-O-H + HO-\overset{O}{\overset{\|}{C}}-(CH_2)_{16}-CH_3$
Ester bond
$CH_2-O-\overset{O}{\overset{\|}{C}}-(CH_2)_{16}-CH_3$
$CH-O-\overset{O}{\overset{\|}{C}}-(CH_2)_{16}-CH_3 + 3H_2O$
$CH_2-O-\overset{O}{\overset{\|}{C}}-(CH_2)_{16}-CH_3$
Glycerol
3 Stearic acid molecules
Glyceryl tristearate (tristearin)

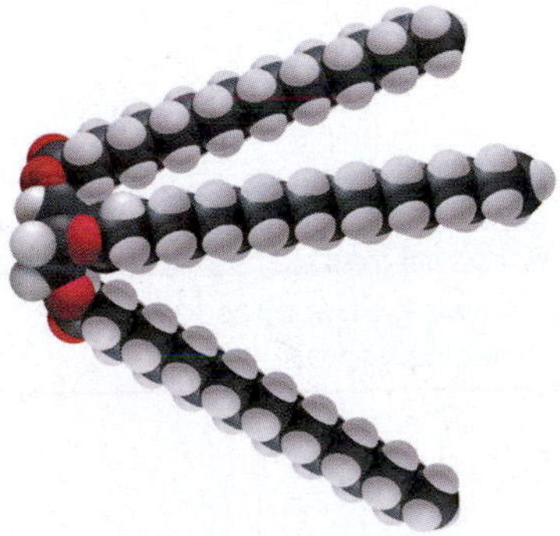

Glyceryl tristearate consists of glycerol with three ester bonds to stearic acid molecules.

Most naturally occurring triacylglycerols contain glycerol bonded to two or three different fatty acids, typically palmitic acid, oleic acid, linoleic acid, and stearic acid. For example, a mixed triacylglycerol might be made from stearic acid, oleic acid, and palmitic acid. One possible structure for this mixed triacylglycerol follows:

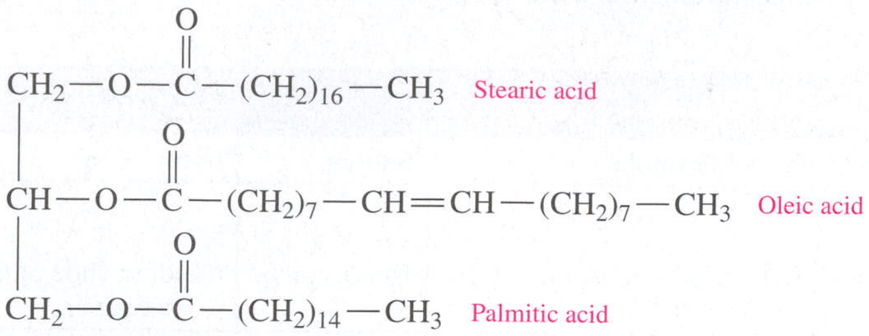

$$
\begin{array}{l}
\underset{}{CH_2}-O-\overset{\overset{\textstyle O}{\|}}{C}-(CH_2)_{16}-CH_3 \quad \text{Stearic acid}\\[4pt]
\underset{}{CH}-O-\overset{\overset{\textstyle O}{\|}}{C}-(CH_2)_7-CH=CH-(CH_2)_7-CH_3 \quad \text{Oleic acid}\\[4pt]
\underset{}{CH_2}-O-\overset{\overset{\textstyle O}{\|}}{C}-(CH_2)_{14}-CH_3 \quad \text{Palmitic acid}
\end{array}
$$

A mixed triacylglycerol

▶ **SAMPLE PROBLEM 17.2** Drawing the Structure for a Triacylglycerol

TRY IT FIRST

Draw the condensed structural formula for glyceryl tripalmitoleate (tripalmitolein), which is used in cosmetic creams and lotions.

SOLUTION

ANALYZE THE PROBLEM	Given	Need	Connect
	glyceryl tripalmitoleate (tripalmitolein)	condensed structural formula	ester bonds of glycerol and three fatty acids

STEP 1　Draw the condensed structural formulas for glycerol and the fatty acids.

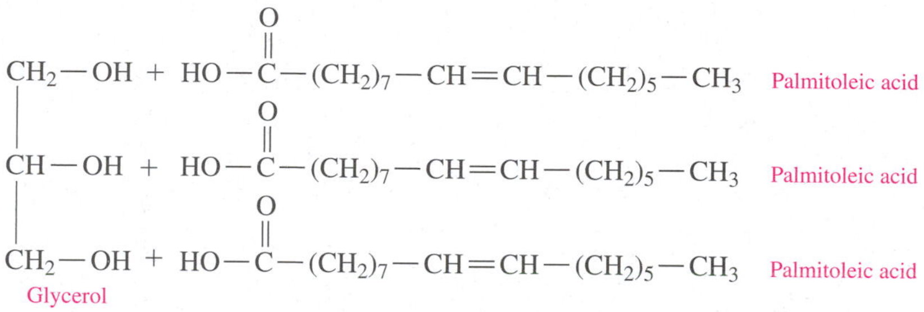

$$
\begin{array}{l}
CH_2-OH \; + \; HO-\overset{\overset{\textstyle O}{\|}}{C}-(CH_2)_7-CH=CH-(CH_2)_5-CH_3 \quad \text{Palmitoleic acid}\\[4pt]
CH-OH \; + \; HO-\overset{\overset{\textstyle O}{\|}}{C}-(CH_2)_7-CH=CH-(CH_2)_5-CH_3 \quad \text{Palmitoleic acid}\\[4pt]
CH_2-OH \; + \; HO-\overset{\overset{\textstyle O}{\|}}{C}-(CH_2)_7-CH=CH-(CH_2)_5-CH_3 \quad \text{Palmitoleic acid}\\
\text{Glycerol}
\end{array}
$$

STEP 2　Form ester bonds between the hydroxyl groups on glycerol and the carboxyl groups on each fatty acid.

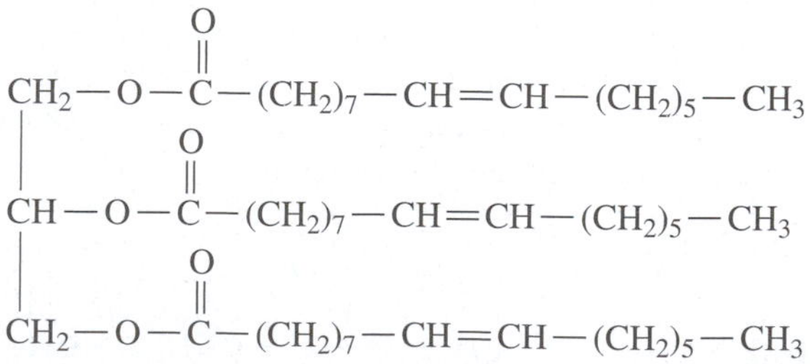

$$
\begin{array}{l}
CH_2-O-\overset{\overset{\textstyle O}{\|}}{C}-(CH_2)_7-CH=CH-(CH_2)_5-CH_3\\[4pt]
CH-O-\overset{\overset{\textstyle O}{\|}}{C}-(CH_2)_7-CH=CH-(CH_2)_5-CH_3\\[4pt]
CH_2-O-\overset{\overset{\textstyle O}{\|}}{C}-(CH_2)_7-CH=CH-(CH_2)_5-CH_3\\
\text{Glyceryl tripalmitoleate (tripalmitolein)}
\end{array}
$$

STUDY CHECK 17.2

Draw the indicated structures for the following triacylglycerols:

a. the line-angle formula for the triacylglycerol containing three molecules of myristic acid (14:0)

b. the condensed structural formula for the triacylglycerol containing three molecules of linoleic acid (18:2)

Triacylglycerols are used to thicken creams and lotions.

CORE CHEMISTRY SKILL

Drawing Structures for Triacylglycerols

ANSWER

a.

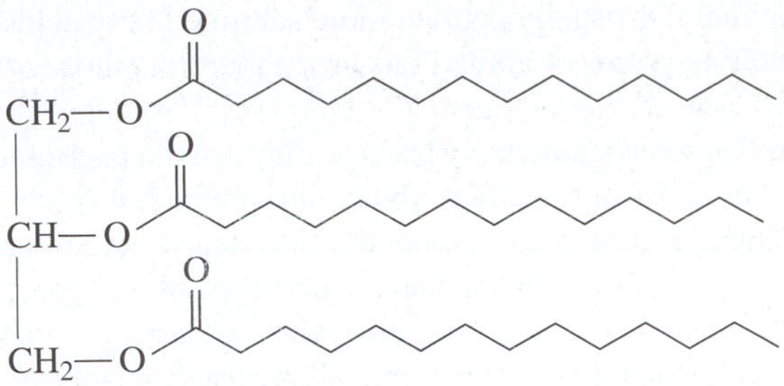

b.

$$\begin{array}{l} \overset{\displaystyle O}{\overset{\|}{CH_2-O-C-(CH_2)_7-CH=CH-CH_2-CH=CH-(CH_2)_4-CH_3}} \\ \quad | \\ \overset{\displaystyle O}{\overset{\|}{CH-O-C-(CH_2)_7-CH=CH-CH_2-CH=CH-(CH_2)_4-CH_3}} \\ \quad | \\ \overset{\displaystyle O}{\overset{\|}{CH_2-O-C-(CH_2)_7-CH=CH-CH_2-CH=CH-(CH_2)_4-CH_3}} \end{array}$$

TEST

Try Practice Problems 17.21 to 17.24

Triacylglycerols are the major form of energy storage for animals. Animals that hibernate eat large quantities of plants, seeds, and nuts that are high in calories. Prior to hibernation, these animals, such as polar bears, gain as much as 14 kg per week. As the external temperature drops, the animal goes into hibernation. The body temperature drops to nearly freezing, and cellular activity, respiration, and heart rate are drastically reduced. Animals that live in extremely cold climates hibernate for 4 to 7 months. During this time, stored fat is the only source of energy.

Prior to hibernation, a polar bear eats food with a high caloric content.

Melting Points of Fats and Oils

A **fat** is a triacylglycerol that is solid at room temperature, and it usually comes from animal sources such as meat, whole milk, butter, and cheese. An **oil** is a triacylglycerol that is usually a liquid at room temperature and is obtained from a plant source. Olive oil and peanut oil are monounsaturated because they contain large amounts of oleic acid. Oils from corn, cottonseed, safflower seed, and sunflower seed are polyunsaturated because they contain large amounts of fatty acids with two or more double bonds (see **FIGURE 17.3**).

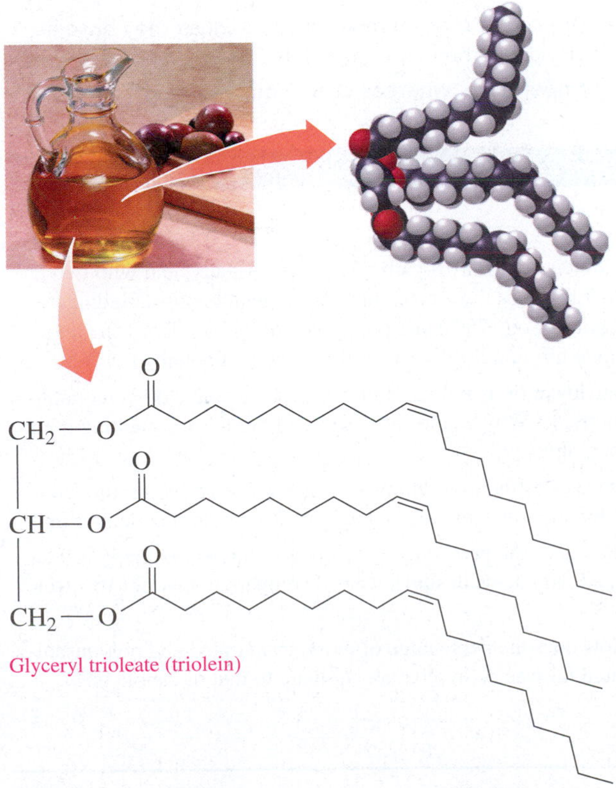

Glyceryl trioleate (triolein)

FIGURE 17.3 ▶ Vegetable oils such as olive oil, corn oil, and safflower oil contain unsaturated fats.

Q Why is olive oil a liquid at room temperature?

Saturated fatty acids have higher melting points than unsaturated fatty acids because they pack together more tightly. Animal fats usually contain more saturated fatty acids than do vegetable oils. Therefore, the melting points of animal fats are higher than those of vegetable oils.

Palm oil and coconut oil are solids at room temperature because they contain large amounts of saturated fatty acids. Coconut oil is 92% saturated fats, about half of which is lauric acid, which has 12 carbon atoms rather than 18 carbon atoms found in the stearic acid of animal sources. Thus coconut oil has a melting point that is higher than typical vegetable oils, but not as high as fats from animal sources that contain stearic acid. The percentages of saturated, monounsaturated, and polyunsaturated fatty acids in some typical fats and oils are listed in **FIGURE 17.4**.

ENGAGE

Why would you expect arachidic acid (20:0) to have a higher melting point than arachidonic acid (20:4)?

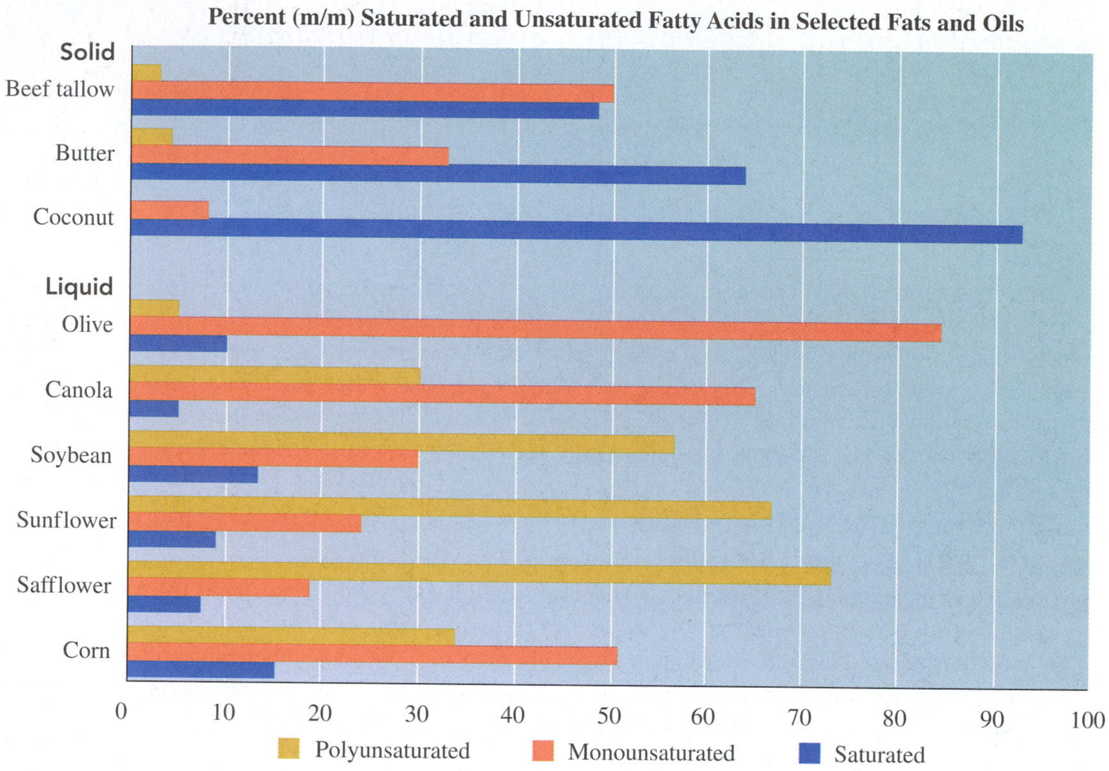
Percent (m/m) Saturated and Unsaturated Fatty Acids in Selected Fats and Oils
Solid
Beef tallow
Butter
Coconut
Liquid
Olive
Canola
Soybean
Sunflower
Safflower
Corn
0
10
20
30
40
50
60
70
80
90
100
Polyunsaturated
Monounsaturated
Saturated

FIGURE 17.4 ▶ Vegetable oils are liquids at room temperature because they have a higher percentage of unsaturated fatty acids than do animal fats.

Q Why is butter a solid at room temperature, whereas canola oil is a liquid?

TEST

Try Practice Problems 17.25 to 17.28

PRACTICE PROBLEMS

17.3 Waxes and Triacylglycerols

17.19 Draw the condensed structural formula for the ester in beeswax that is formed from myricyl alcohol, $CH_3—(CH_2)_{29}—OH$, and palmitic acid.

17.20 Draw the condensed structural formula for the ester in jojoba wax that is formed from arachidic acid, a 20-carbon saturated fatty acid, and 1-docosanol, $CH_3—(CH_2)_{21}—OH$.

17.21 Draw the condensed structural formula for a triacylglycerol that contains stearic acid and glycerol.

17.22 Draw the condensed structural formula for a mixed triacylglycerol that contains two palmitic acid molecules and one oleic acid molecule on the center carbon of glycerol.

Clinical Applications

17.23 Caprylic acid is an 8-carbon saturated fatty acid that occurs in coconut oil (10%) and palm kernel oil (4%). Draw the line-angle formula for glyceryl tricaprylate (tricaprylin).

17.24 Linoleic acid is a 18-carbon unsaturated fatty acid with two double bonds that occurs in many vegetable oils including safflower oil (75%) and poppyseed oil (75%). Draw the line-angle formula for glyceryl trilinoleate (trilinolein).

17.25 Safflower oil is polyunsaturated, whereas olive oil is monounsaturated. Why would safflower oil have a lower melting point than olive oil?

17.26 Olive oil is monounsaturated, whereas butter fat is saturated. Why does olive oil have a lower melting point than butter fat?

17.27 How does the percentage of monounsaturated and polyunsaturated fatty acids in sunflower oil compare to that of safflower oil?

17.28 How does the percentage of monounsaturated and polyunsaturated fatty acids in olive oil compare to that of canola oil?

17.4 Chemical Properties of Triacylglycerols

LEARNING GOAL Draw the condensed structural and line-angle formulas for the products of a triacylglycerol that undergoes hydrogenation, hydrolysis, or saponification.

REVIEW

Writing Equations for Hydrogenation, Hydration, and Polymerization (12.7)

Hydrolyzing Esters (16.4)

The chemical reactions of triacylglycerols involve the hydrogenation of the double bonds in the fatty acids, and the hydrolysis and saponification of the ester bonds between glycerol and the fatty acids.

Hydrogenation

In the **hydrogenation** reaction, hydrogen gas is bubbled through a heated oil, typically in the presence of a nickel catalyst. As a result, H atoms add to one or more carbon–carbon double bonds to form carbon–carbon single bonds.

$$-CH=CH- + H_2 \xrightarrow{\text{Ni}} \begin{matrix} H & H \\ | & | \\ -C-C- \\ | & | \\ H & H \end{matrix}$$

For example, when hydrogen adds to all of the double bonds of glyceryl trioleate (triolein) using a nickel catalyst, the product is the saturated fat glyceryl tristearate (tristearin).

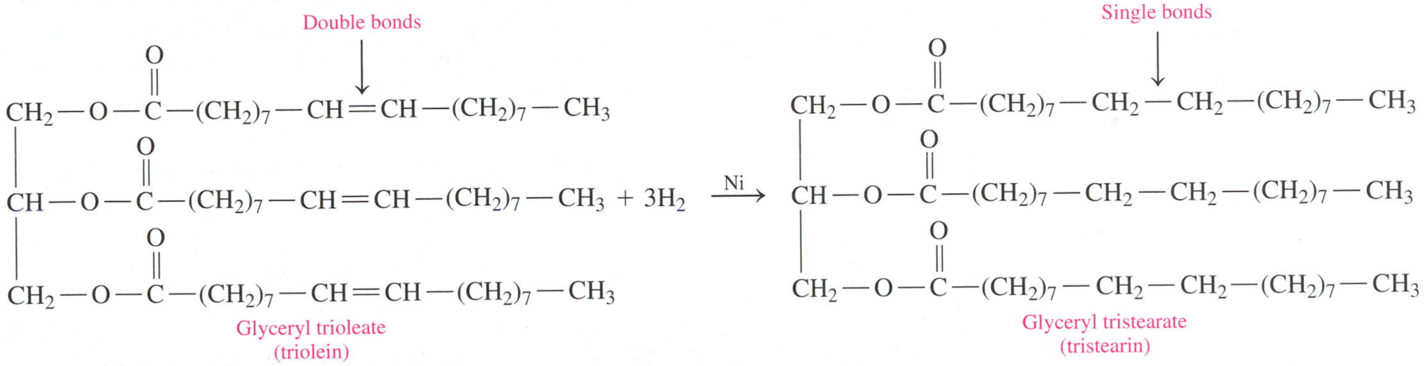

Double bonds

$$CH_2-O-\overset{\overset{\displaystyle O}{\|}}{C}-(CH_2)_7-CH=CH-(CH_2)_7-CH_3$$
$$CH-O-\overset{\overset{\displaystyle O}{\|}}{C}-(CH_2)_7-CH=CH-(CH_2)_7-CH_3 + 3H_2 \xrightarrow{\text{Ni}}$$
$$CH_2-O-\overset{\overset{\displaystyle O}{\|}}{C}-(CH_2)_7-CH=CH-(CH_2)_7-CH_3$$

Glyceryl trioleate
(triolein)

Single bonds

$$CH_2-O-\overset{\overset{\displaystyle O}{\|}}{C}-(CH_2)_7-CH_2-CH_2-(CH_2)_7-CH_3$$
$$CH-O-\overset{\overset{\displaystyle O}{\|}}{C}-(CH_2)_7-CH_2-CH_2-(CH_2)_7-CH_3$$
$$CH_2-O-\overset{\overset{\displaystyle O}{\|}}{C}-(CH_2)_7-CH_2-CH_2-(CH_2)_7-CH_3$$

Glyceryl tristearate
(tristearin)

In commercial hydrogenation, the addition of hydrogen is stopped before all the double bonds in a liquid vegetable oil become saturated. Complete hydrogenation gives a brittle product, whereas the partial hydrogenation of a liquid vegetable oil changes it to a soft, semisolid fat. By controlling the amount of hydrogen, manufacturers can produce various types of products such as soft margarines, solid stick margarines, and solid shortenings (see **FIGURE 17.5**). Although these products now contain more saturated fatty acids than the original oils, they contain no cholesterol, unlike similar products from animal sources, such as butter and lard.

CORE CHEMISTRY SKILL

Drawing the Products for the Hydrogenation, Hydrolysis, and Saponification of a Triacylglycerol

FIGURE 17.5 ▶ Many soft margarines, stick margarines, and solid shortenings are produced by the partial hydrogenation of vegetable oils.

Q How does hydrogenation change the structure of the fatty acids in the vegetable oils?

ENGAGE

What is the function of a lipase?

Hydrolysis

Triacylglycerols are hydrolyzed (split by water) in the presence of strong acids such as HCl or H_2SO_4, or digestive enzymes called *lipases*. The products of hydrolysis of the ester bonds are glycerol and three fatty acids. The polar glycerol is soluble in water, but the fatty acids with their long hydrocarbon chains are not.

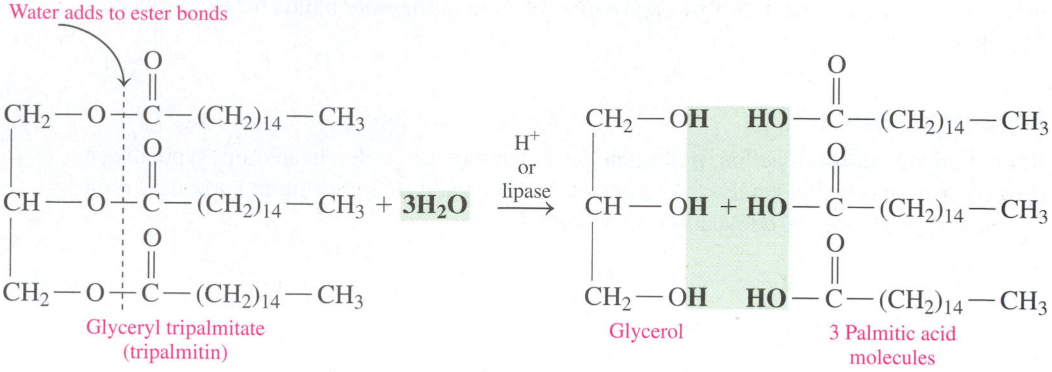

Water adds to ester bonds
H+ or lipase
Glyceryl tripalmitate (tripalmitin)
Glycerol
3 Palmitic acid molecules

Saponification

Saponification occurs when a fat is heated with a strong base such as NaOH to form glycerol and the salts of the fatty acids. This is a process by which soap can be made. When NaOH is used as the base, a solid soap is produced that can be molded into a desired shape; KOH produces a softer, liquid soap. An oil that is polyunsaturated produces a softer soap. Names like "coconut" or "avocado shampoo" tell you the sources of the oil used in the reaction.

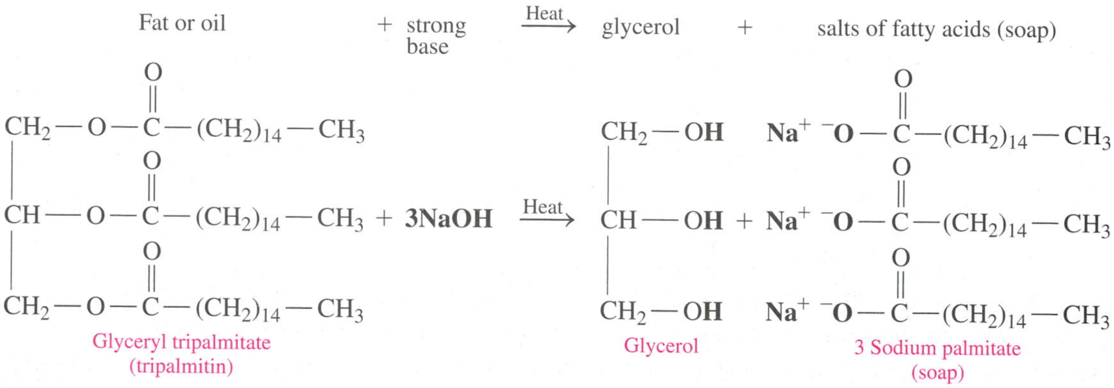

Fat or oil + strong base Heat glycerol + salts of fatty acids (soap)
Heat
Glyceryl tripalmitate (tripalmitin)
Glycerol
3 Sodium palmitate (soap)

The reactions for fatty acids and triacylglycerols are summarized in **TABLE 17.3**.

TABLE 17.3 Summary of Lipid Reactions

Reaction	Reactants and Products
Hydrogenation	Unsaturated fat (double bonds) + hydrogen $\xrightarrow{\text{Ni}}$ saturated fat (single bonds)
Hydrolysis	Triacylglycerol (fat) + 3 water $\xrightarrow{\text{Enzyme}}$ 3 fatty acids + glycerol
Saponification	Triacylglycerol (fat) + 3 sodium hydroxide $\xrightarrow{\text{Heat}}$ 3 sodium salts of fatty acid (soap) + glycerol

▶SAMPLE PROBLEM 17.3 Lipid Reactions

TRY IT FIRST

Use condensed structural formulas to write the balanced chemical equation for the reaction catalyzed by lipase enzymes that hydrolyze glyceryl trilaurate (trilaurin) during the digestion process.

SOLUTION

	Given	Need	Connect
ANALYZE THE PROBLEM	glyceryl trilaurate, lipases	chemical equation for hydrolysis	products: glycerol and three fatty acids

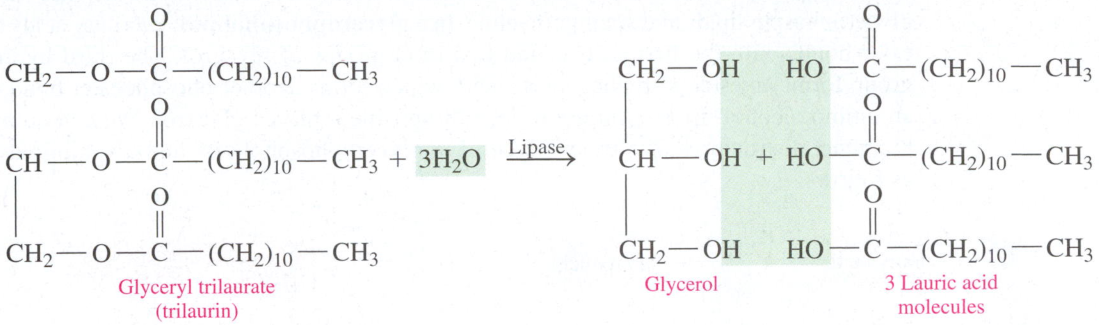
Lipase
Glyceryl trilaurate
(trilaurin)
Glycerol
3 Lauric acid
molecules

STUDY CHECK 17.3

a. What is the name of the product formed when a triacylglycerol containing oleic acid (18:1) and linoleic acid (18:2) is completely hydrogenated?

b. Use condensed structural formulas to write the balanced equation for the saponification of glyceryl tristearate (tristearin) by KOH.

TEST

Try Practice Problems 17.29 to 17.38

ANSWER

a. glyceryl tristearate (tristearin)

b.

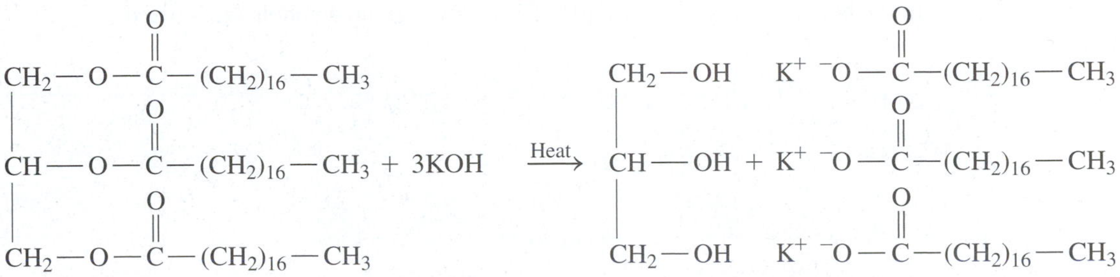
Heat

PRACTICE PROBLEMS

17.4 Chemical Properties of Triacylglycerols

17.29 Identify each of the following processes as hydrogenation, hydrolysis, or saponification and give the products:

a. the reaction of palm oil with KOH

b. the reaction of glyceryl trilinoleate from safflower oil with water and HCl

17.30 Identify each of the following processes as hydrogenation, hydrolysis, or saponification and give the products:

a. the reaction of corn oil and hydrogen (H_2) with a nickel catalyst

b. the reaction of glyceryl tristearate with water in the presence of lipase enzyme

17.31 Use condensed structural formulas to write the balanced chemical equation for the hydrogenation of glyceryl tripalmitoleate, a fat containing glycerol and three palmitoleic acid molecules.

17.32 Use condensed structural formulas to write the balanced chemical equation for the hydrogenation of glyceryl trilinolenate, a fat containing glycerol and three linolenic acid molecules.

17.33 Use line-angle formulas to write the balanced chemical equation for the acid hydrolysis of glyceryl trimyristate (trimyristin).

17.34 Use line-angle formulas to write the balanced chemical equation for the acid hydrolysis of glyceryl trioleate (triolein).

17.35 Use condensed structural formulas to write the balanced chemical equation for the NaOH saponification of glyceryl trimyristate (trimyristin).

17.36 Use condensed structural formulas to write the balanced chemical equation for the NaOH saponification of glyceryl trioleate (triolein).

17.37 Draw the condensed structural formula for the product of the hydrogenation of the following triacylglycerol:

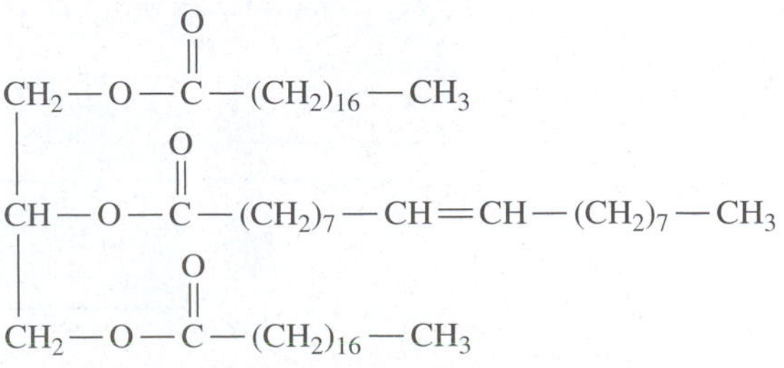

17.38 Draw the condensed structural formulas for all the products that would be obtained when the triacylglycerol in problem 17.37 undergoes complete hydrolysis.

17.5 Phospholipids

LEARNING GOAL Draw the structure of a phospholipid that contains glycerol or sphingosine.

The **phospholipids** are a family of lipids similar in structure to triacylglycerols; they include glycerophospholipids and sphingomyelins. In a **glycerophospholipid**, two fatty acids form ester bonds with the first and second hydroxyl groups of glycerol. The third hydroxyl group forms an ester with phosphoric acid, which forms another phosphoester bond with an amino alcohol. In a *sphingomyelin*, sphingosine replaces glycerol. We can compare the general structures of a triacylglycerol, a glycerophospholipid, and a sphingomyelin as follows:

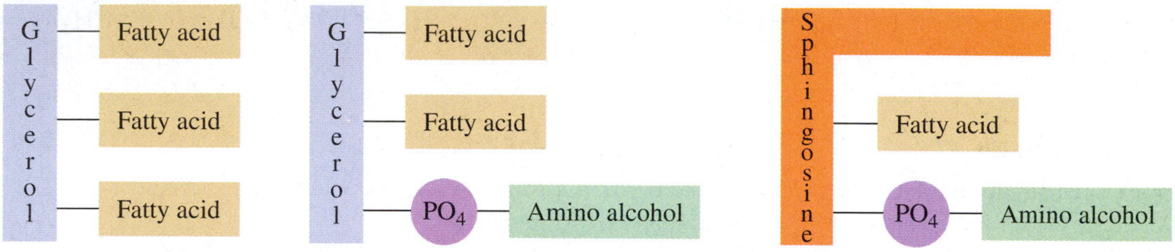
Glycerol
Fatty acid
Fatty acid
Fatty acid
Glycerol
Fatty acid
Fatty acid
PO_4
Amino alcohol
Sphingosine
Fatty acid
PO_4
Amino alcohol

Triacylglycerol   Glycerophospholipid   Sphingomyelin

Amino Alcohols

Three amino alcohols found in glycerophospholipids are choline, serine, and ethanolamine. In the body, at a physiological pH of 7.4, these amino alcohols are ionized.

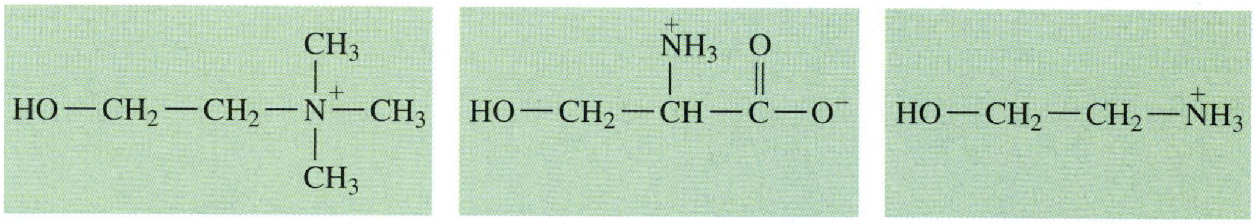

Choline   Serine   Ethanolamine

TEST

Try Practice Problems 17.39 and 17.40

Lecithins and *cephalins* are two types of glycerophospholipids that are particularly abundant in brain and nerve tissue as well as in egg yolk, wheat germ, and yeast. Lecithins contain choline, and cephalins usually contain ethanolamine and sometimes serine. In the following structural formulas, the fatty acid that is used in each example is palmitic acid:

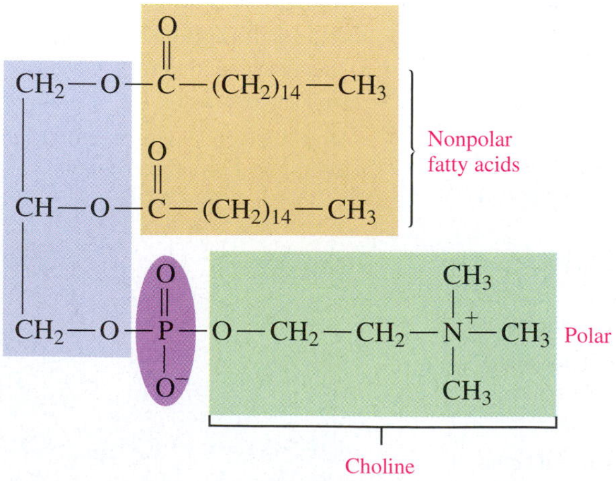
Nonpolar
fatty acids
Polar
Choline

A lecithin

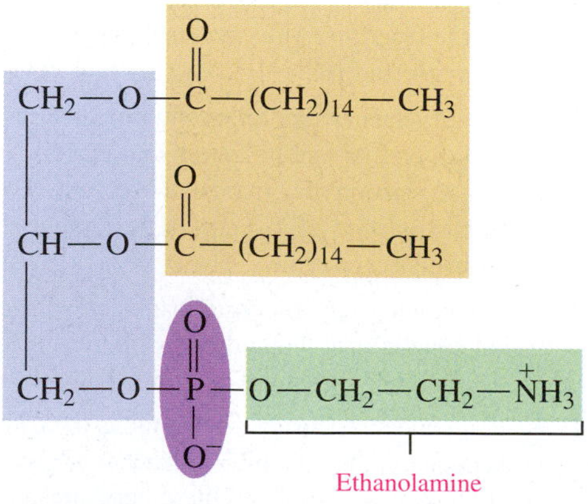
Ethanolamine

A cephalin

Glycerophospholipids contain both polar and nonpolar regions, which allow them to interact with both polar and nonpolar substances. The ionized amino alcohol and phosphate portion, called "the head," is polar and strongly attracted to water (see **FIGURE 17.6**). The hydrocarbon chains of the two fatty acids are the nonpolar "tails" of the glycerophospholipid, which are soluble in other nonpolar substances, mostly lipids.

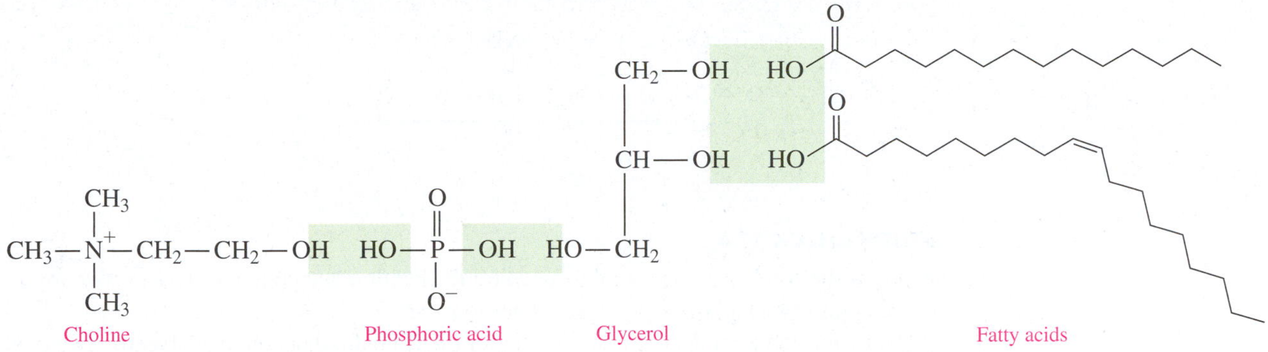
Choline
Phosphoric acid
Glycerol
Fatty acids

(a) Components of a typical glycerophospholipid

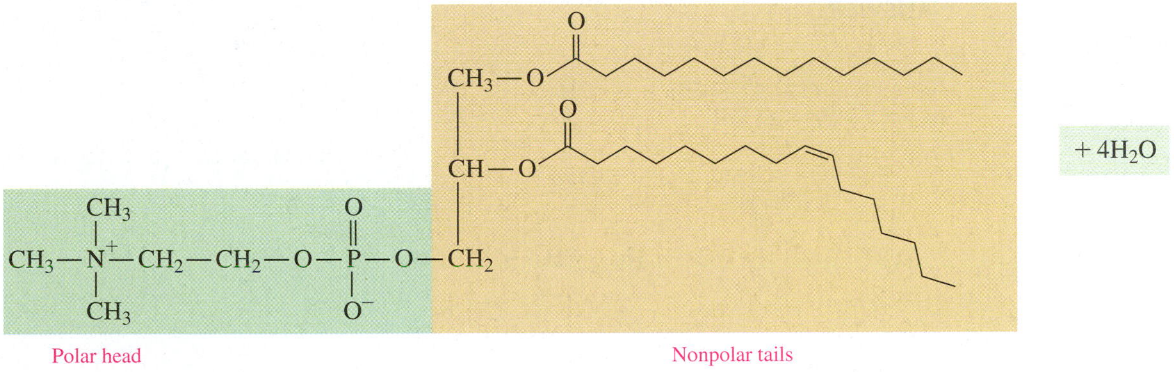
Polar head
Nonpolar tails

(b) Glycerophospholipid

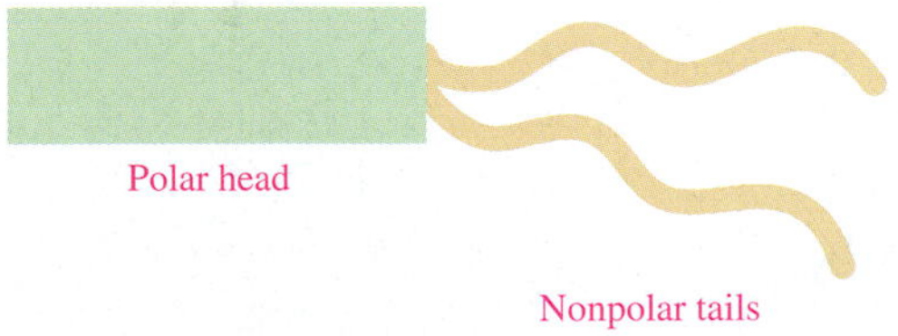
Polar head
Nonpolar tails

(c) Simplified way to draw a glycerophospholipid

FIGURE 17.6 ▶ **(a)** The components of a typical glycerophospholipid: an amino alcohol, phosphoric acid, glycerol, and two fatty acids. **(b)** In a glycerophospholipid, a polar "head" contains the ionized amino alcohol and phosphate, while the hydrocarbon chains of two fatty acids make up the nonpolar "tails." **(c)** A simplified drawing indicates the polar and nonpolar regions.

Q Which components make the glycerophospholipid "head" polar?

▶ SAMPLE PROBLEM 17.4 Drawing Glycerophospholipid Structures

TRY IT FIRST

Draw the condensed structural formula for the cephalin that contains glycerol, two stearic acids (18:0), phosphate, and serine (ionized).

SOLUTION

	Given	Need	Connect
ANALYZE THE PROBLEM	cephalin, two stearic acids (18:0), phosphate, serine (ionized)	condensed structural formula	ester bonds with fatty acids, phosphoester bond with amino alcohol

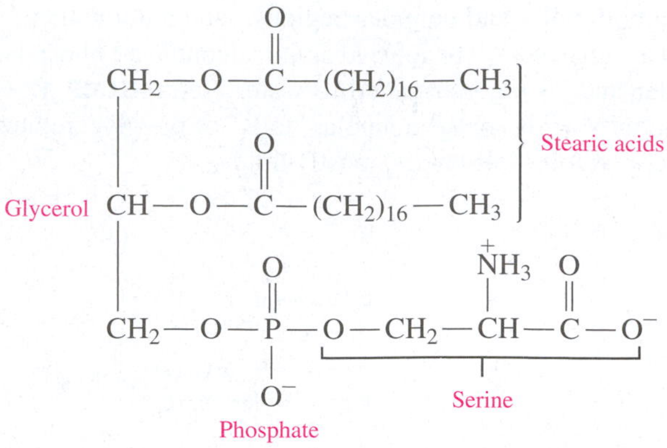
Stearic acids
Glycerol
Serine
Phosphate

STUDY CHECK 17.4

a. Draw the condensed structural formula for the lecithin that contains glycerol, two myristic acids (14:0), phosphate, and choline (ionized).

b. Draw the condensed structural formula for the cephalin that contains glycerol, two oleic acids (18:1), phosphate, and ethanolamine (ionized).

ANSWER

a.

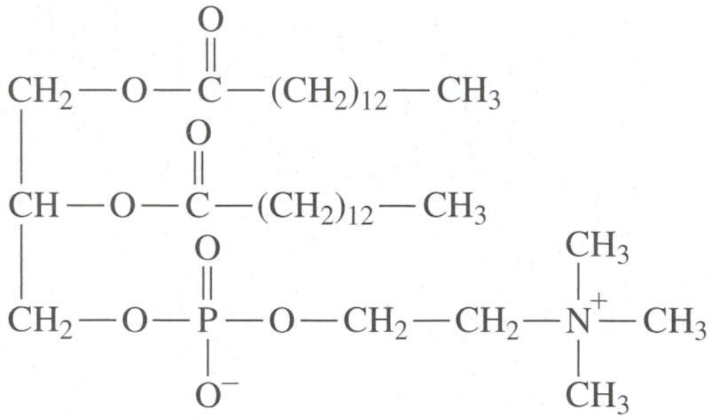

b.

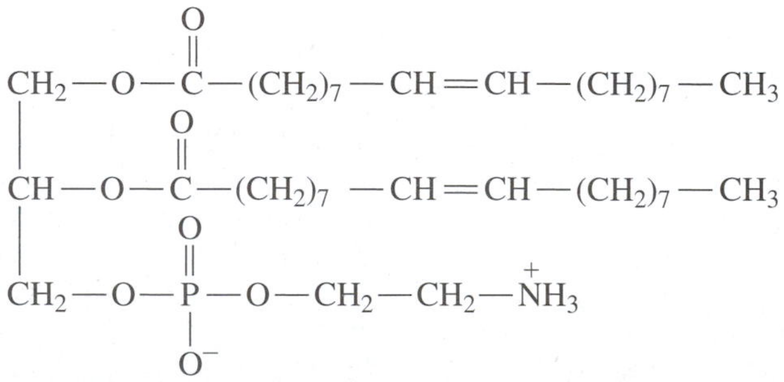

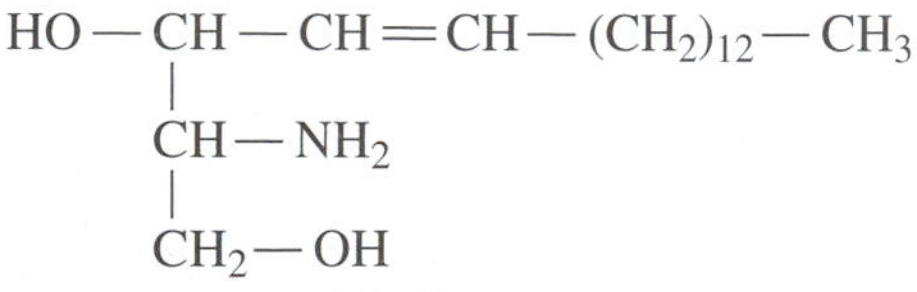

Sphingosine

Sphingosine, found in sphingomyelins and other sphingolipids, is a long-chain amino alcohol. In a **sphingomyelin**, the amine group of sphingosine forms an amide bond to a fatty acid and the hydroxyl group forms an ester bond with phosphate, which forms another phosphoester bond to choline or ethanolamine. The sphingomyelins are abundant in the white matter of the myelin sheath, a coating surrounding the nerve cells that increases the speed of nerve impulses and insulates and protects the nerve cells.

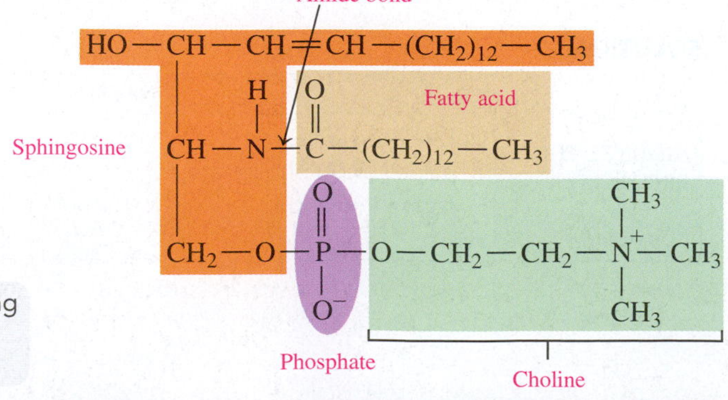
Amide bond
Fatty acid
Sphingosine
Phosphate
Choline

A sphingomyelin containing sphingosine, myristic acid, and choline

In *multiple sclerosis*, sphingomyelin is lost from the myelin sheath. As the disease progresses, the myelin sheath deteriorates. Scars form on the neurons and impair the transmission of nerve signals. The symptoms of multiple sclerosis include various levels of muscle weakness with loss of coordination and vision, depending on the amount of damage. The cause of multiple sclerosis is not known, although some researchers suggest that a virus is involved. Several studies also suggest that adequate levels of vitamin D may lessen the severity or lower the risk of developing multiple sclerosis.

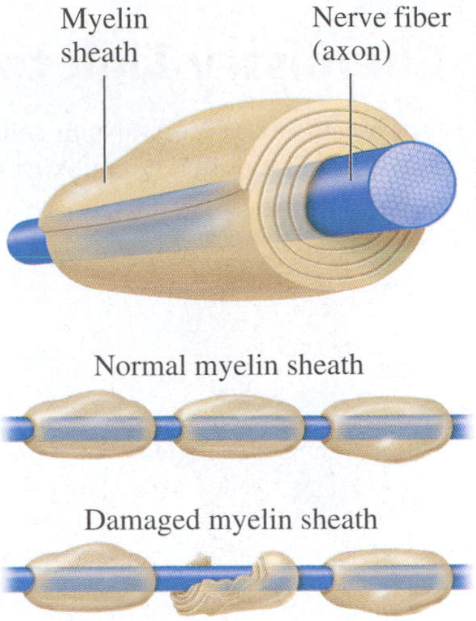
Myelin sheath
Nerve fiber (axon)
Normal myelin sheath
Damaged myelin sheath

When the myelin sheath loses sphingomyelin, it deteriorates and the transmission of nerve signals is impaired.

▶ SAMPLE PROBLEM 17.5 Sphingomyelin

TRY IT FIRST

A sphingomyelin found in eggs contains sphingosine, palmitic acid (16:0), phosphate, and choline (ionized). Draw the condensed structural formula for this sphingomyelin.

SOLUTION

	Given	Need	Connect
ANALYZE THE PROBLEM	sphingosine, palmitic acid (16:0), phosphate, choline (ionized)	condensed structural formula of the sphingomyelin	amide bond with fatty acid, phosphoester bond with amino alcohol

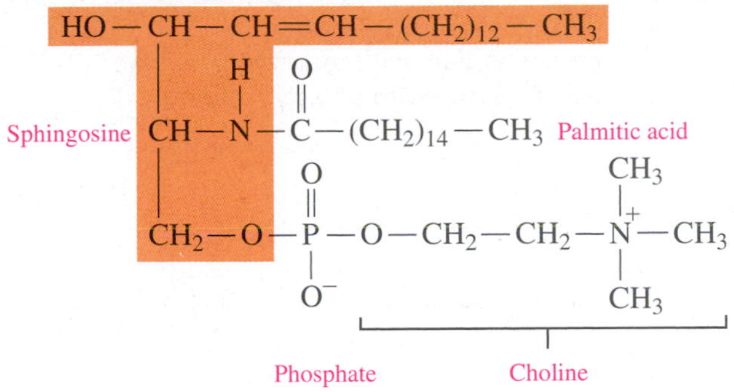
Sphingosine
Palmitic acid
Phosphate
Choline

STUDY CHECK 17.5

Stearic acid (18:0) is found in sphingomyelin in the brain. Draw the condensed structural formula for this sphingomyelin using ethanolamine (ionized).

ANSWER

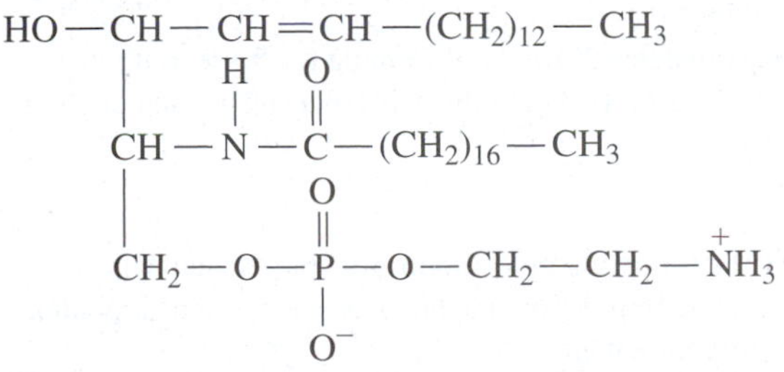

ENGAGE

How is the polarity of a sphingomyelin different from that of a triacylglycerol?

TEST

Try Practice Problems 17.41 to 17.46

Chemistry Link to Health

Infant Respiratory Distress Syndrome (IRDS)

When an infant is born, an important key to its survival is proper lung function. In the lungs, there are many tiny air sacs called *alveoli*, where the exchange of O_2 and CO_2 takes place. Upon birth of a mature infant, surfactant is released into the lung tissues where it lowers the surface tension in the alveoli, which helps the air sacs inflate. The production of a *pulmonary surfactant*, which is a mixture of phospholipids including lecithin and sphingomyelin produced by specific lung cells, occurs in a fetus after 24 to 28 weeks of pregnancy. If an infant is born before 28 weeks of gestation, the low level of surfactant and immature lung development lead to a high risk of *infant respiratory distress syndrome* (IRDS). Without sufficient surfactant, the air sacs collapse and have to reopen with

(*continued*)

Chemistry Link to Health (*continued*)

each breath. As a result, alveoli cells are damaged, less oxygen is taken in, and more carbon dioxide is retained, which can lead to hypoxia and acidosis.

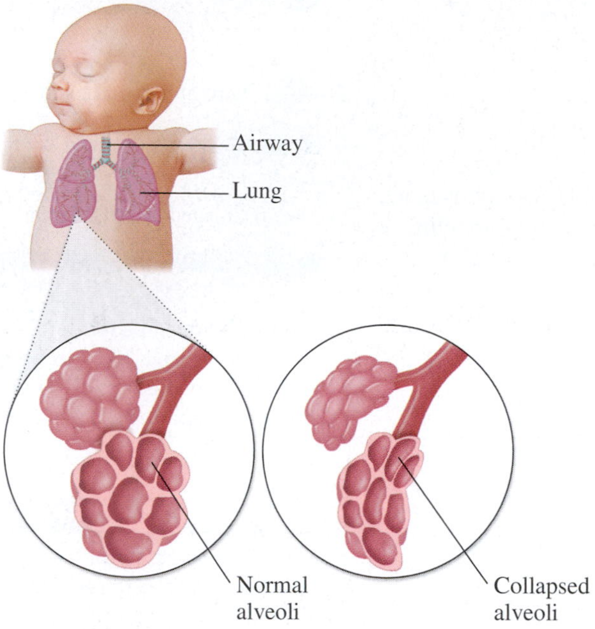

Airway

Lung

Normal alveoli

Collapsed alveoli

Without sufficient surfactant in the lungs of a premature infant, the alveoli collapse, which decreases pulmonary function.

One way to determine the maturity of the lungs of a fetus is to measure the *lecithin–sphingomyelin (L/S) ratio*. A ratio of 2.5 indicates mature fetal lung function, an L/S ratio of 2.4 to 1.6 indicates a low risk, and a ratio of less than 1.5 indicates a high risk of IRDS. Before the initiation of an early delivery, the L/S ratio of the amniotic fluid is measured. If the L/S ratio is low, steroids may be given to the mother to assist the lung development and production of surfactant in the fetus. Once a premature infant is born, treatment includes the use of steroids to help maturation of the lungs, the application of surfactants, and the administration of supplemental oxygen with ventilation to help minimize damage to the lungs.

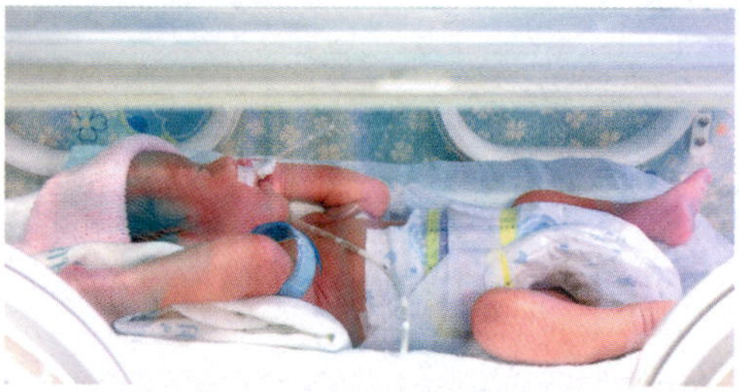

A premature infant with respiratory distress is treated with a surfactant and oxygen.

PRACTICE PROBLEMS

17.5 Phospholipids

17.39 Describe the similarities and differences between triacylglycerols and glycerophospholipids.

17.40 Describe the similarities and differences between lecithins and cephalins.

Clinical Applications

17.43 Identify the following glycerophospholipid, which is found in the nerves and spinal cord in the body, as a lecithin or cephalin, and list its components:

$$
\begin{array}{l}
\quad\quad\quad\quad\;\; O \\
\quad\quad\quad\quad\;\; \| \\
CH_2-O-C-(CH_2)_7-CH=CH-(CH_2)_7-CH_3 \\
|\quad\quad\quad\; O \\
|\quad\quad\quad\; \| \\
CH-O-C-(CH_2)_{16}-CH_3 \\
|\quad\quad\quad\; O \\
|\quad\quad\quad\; \| \\
CH_2-O-P-O-CH_2-CH_2-\overset{+}{N}H_3 \\
\quad\quad\quad\;\; | \\
\quad\quad\quad\;\; O^-
\end{array}
$$

17.41 Draw the condensed structural formula for the cephalin that contains glycerol, two palmitic acids, phosphate, and ethanolamine (ionized).

17.42 Draw the condensed structural formula for the lecithin that contains glycerol, two palmitic acids, phosphate, and choline (ionized).

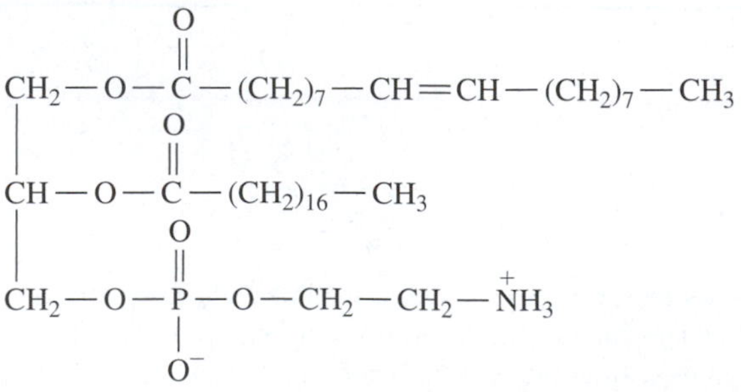

17.44 Identify the following glycerophospholipid, which helps conduct nerve impulses in the body, as a lecithin or cephalin, and list its components:

$$
\begin{array}{l}
\quad\quad\quad\quad\;\; O \\
\quad\quad\quad\quad\;\; \| \\
CH_2-O-C-(CH_2)_{14}-CH_3 \\
|\quad\quad\quad\; O \\
|\quad\quad\quad\; \| \\
CH-O-C-(CH_2)_{16}-CH_3 \\
|\quad\quad\quad\; O\quad\quad\quad\quad\quad\;\; CH_3 \\
|\quad\quad\quad\; \|\quad\quad\quad\quad\quad\quad | \\
CH_2-O-P-O-CH_2-CH_2-\overset{+}{N}-CH_3 \\
\quad\quad\quad\;\; |\quad\quad\quad\quad\quad\quad\quad | \\
\quad\quad\quad\;\; O^-\quad\quad\quad\quad\quad\quad CH_3
\end{array}
$$

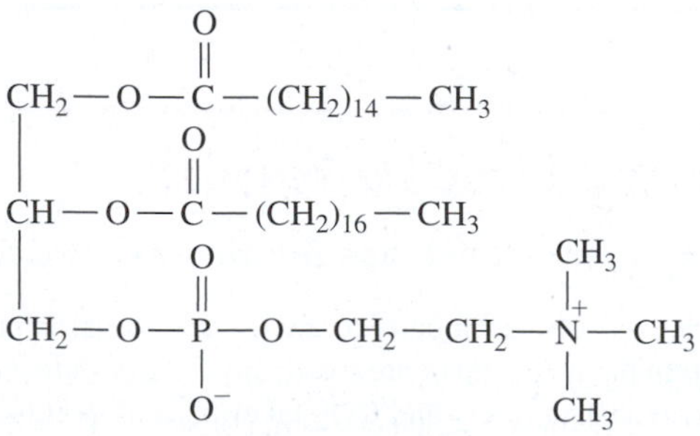

17.45 Identify the following features of this phospholipid, which is abundant in the myelin sheath that surrounds nerve cells:

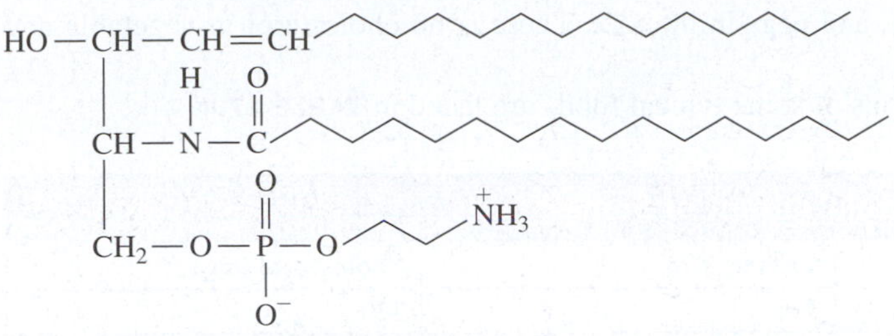
HO—CH—CH=CH
H O
CH—N—C
O
NH3
CH2—O—P—O
O−

a. Is the phospholipid formed from glycerol or sphingosine?
b. What is the fatty acid?
c. What type of bond connects the fatty acid?
d. What is the amino alcohol?

17.46 Identify the following features of this phospholipid, which is needed for the brain and nerve tissues:

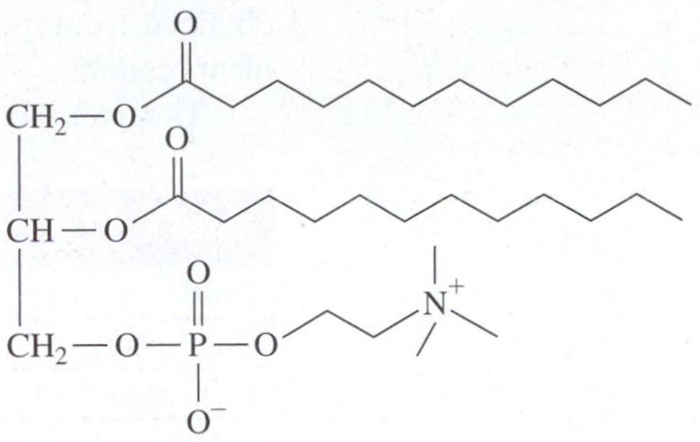
O
CH2—O
O
CH—O
O
N+
CH2—O—P—O
O−

a. Is the phospholipid formed from glycerol or sphingosine?
b. What is the fatty acid?
c. What type of bond connects the fatty acid?
d. What is the amino alcohol?

17.6 Steroids: Cholesterol, Bile Salts, and Steroid Hormones

LEARNING GOAL Draw the structure of the steroid nucleus; compare the structures and functions of steroid compounds.

While many lipids include fatty acids in their structures, the steroids do not. **Steroids** are compounds containing the *steroid nucleus*, which consists of three cyclohexane rings and one cyclopentane ring fused together. The four rings in the steroid nucleus are designated A, B, C, and D. The carbon atoms are numbered beginning with the carbons in ring A, and in steroids like cholesterol, ending with two methyl groups.

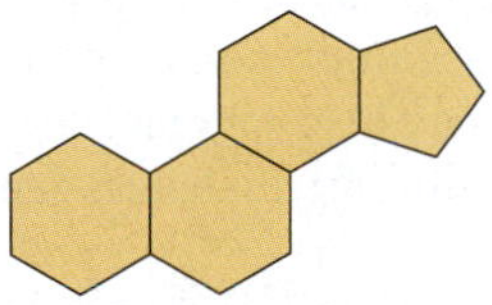

Steroid nucleus

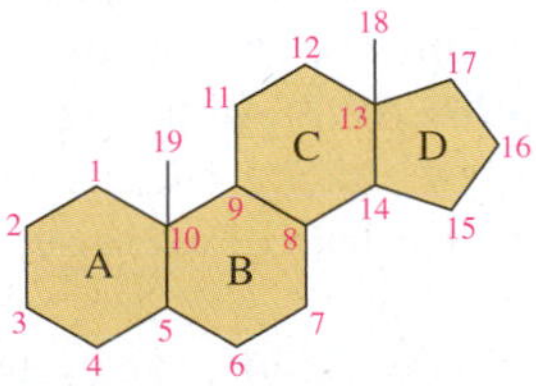
18
12
17
11
13
19
C
D
16
1
9
14
2
10
8
15
A
B
3
5
7
4
6

Steroid numbering system

Cholesterol

CORE CHEMISTRY SKILL
Identifying the Steroid Nucleus

Cholesterol, which is one of the most important and abundant steroids in the body, is a *sterol* because it contains an oxygen atom as a hydroxyl group (—OH) on carbon 3. Like many steroids, cholesterol has a double bond between carbon 5 and carbon 6, methyl groups at carbon 10 and carbon 13, and a carbon chain at carbon 17. In other steroids, the oxygen atom forms a carbonyl group (C=O) at carbon 3.

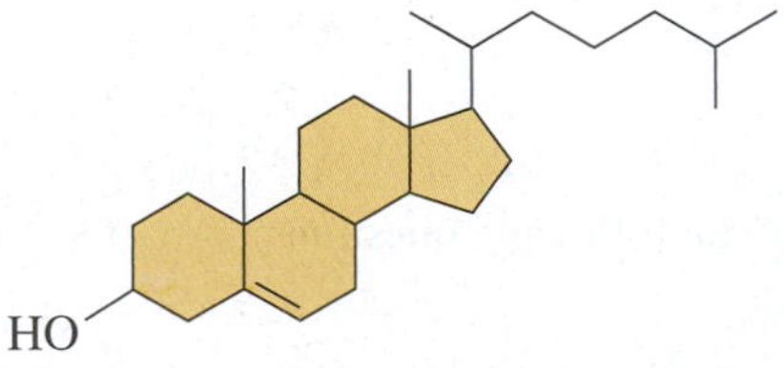
HO

Cholesterol

Cholesterol is a component of cellular membranes, the myelin sheath, and brain and nerve tissue. It is also found in the liver and bile salts; large quantities of it are found in the skin, and some of it becomes vitamin D when the skin is exposed to direct sunlight. In the adrenal gland, cholesterol is used to synthesize steroid hormones. The liver synthesizes cholesterol for the body from fats, carbohydrates, and proteins. Additional cholesterol is obtained from meat, milk, and eggs in the diet. There is no cholesterol in vegetable and plant products.

The cholesterol contents of some typical foods are listed in **TABLE 17.4**.

TABLE 17.4 Cholesterol Content of Some Foods

Food	Serving Size	Cholesterol (mg)
Liver (beef)	3 oz	370
Large egg	1	200
Lobster	3 oz	175
Fried chicken	$3\frac{1}{2}$ oz	130
Hamburger	3 oz	85
Chicken (no skin)	3 oz	75
Fish (salmon)	3 oz	40
Whole milk	1 cup	35
Butter	1 tablespoon	30
Skim milk	1 cup	5
Margarine	1 tablespoon	0

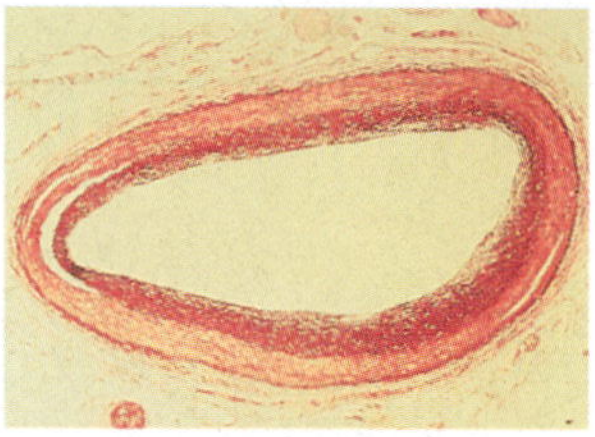

(a) A cross section of a normal, open artery shows no buildup of plaque.

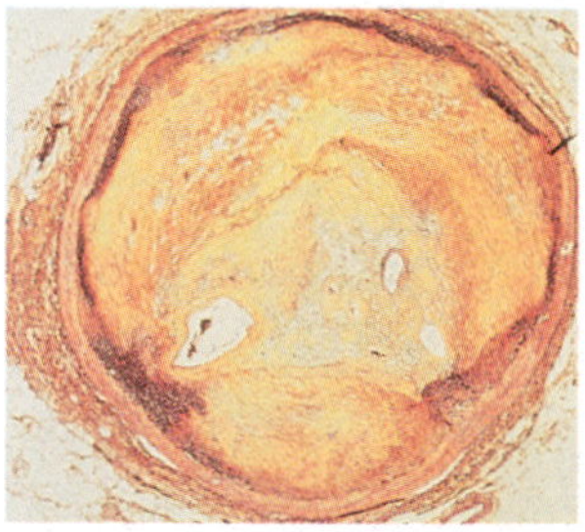

(b) A cross section of an artery that is almost completely clogged by atherosclerotic plaque.

FIGURE 17.7 ▶ Excess cholesterol forms plaque that can block an artery, resulting in a heart attack.

Q What property of cholesterol would cause it to form deposits along the coronary arteries?

Cholesterol in the Body

If a diet is high in cholesterol, the liver produces less cholesterol. A typical daily American diet includes 400 to 500 mg of cholesterol, one of the highest in the world. The American Heart Association has recommended that we consume no more than 300 mg of cholesterol a day. Researchers suggest that saturated fats and cholesterol are associated with diseases such as diabetes; breast, pancreas, and colon cancers; and atherosclerosis. In *atherosclerosis*, deposits of a protein–lipid complex (plaque) accumulate in the coronary blood vessels, restricting the flow of blood to the tissue and causing necrosis (death) of the tissue (see **FIGURE 17.7**). In the heart, plaque accumulation could result in a *myocardial infarction* (heart attack). Other factors that may also increase the risk of heart disease are family history, lack of exercise, smoking, obesity, diabetes, gender, and age.

Clinically, cholesterol levels are considered elevated if the total plasma cholesterol level exceeds 200 mg/dL. Saturated fats in the diet may stimulate the production of cholesterol by the liver. A diet that is low in foods containing cholesterol and saturated fats appears to be helpful in reducing the plasma cholesterol level. The American Institute for Cancer Research (AICR) has recommended that our diet contain more fiber and starch by adding more vegetables, fruits, whole grains, and moderate amounts of foods with low levels of fat and cholesterol such as fish, poultry, lean meats, and low-fat dairy products. The AICR also suggests that we limit our intake of foods high in cholesterol such as eggs, nuts, French fries, fatty or organ meats, cheeses, butter, and coconut and palm oil.

▶SAMPLE PROBLEM 17.6 Cholesterol

TRY IT FIRST

Refer to the structure of cholesterol for each of the following questions:

a. What part of cholesterol is the steroid nucleus?
b. What features have been added to the steroid nucleus in cholesterol?
c. What classifies cholesterol as a sterol?

SOLUTION

a. The four fused rings form the steroid nucleus.
b. The cholesterol molecule contains a hydroxyl group ($-OH$) on the first ring, one double bond in the second ring, methyl groups ($-CH_3$) at carbons 10 and 13, and a branched carbon chain at carbon 17.
c. The hydroxyl group determines the sterol classification.

STUDY CHECK 17.6

Why is cholesterol in the lipid family?

ANSWER

Cholesterol is soluble in organic solvents but not in water, the common characteristic of all lipids.

TEST

Try Practice Problems 17.47 and 17.48

Bile Salts

The *bile salts* in the body are synthesized from cholesterol in the liver and stored in the gallbladder. When bile is secreted into the small intestine, the bile salts mix with the water-insoluble fats and oils in our diets. The bile salts, with their nonpolar and polar regions, act much like soaps, breaking down large globules of fat into smaller droplets. The smaller droplets containing fat have a larger surface area to react with lipases, which are the enzymes that digest fat. The bile salts also help with the absorption of cholesterol into the intestinal mucosa.

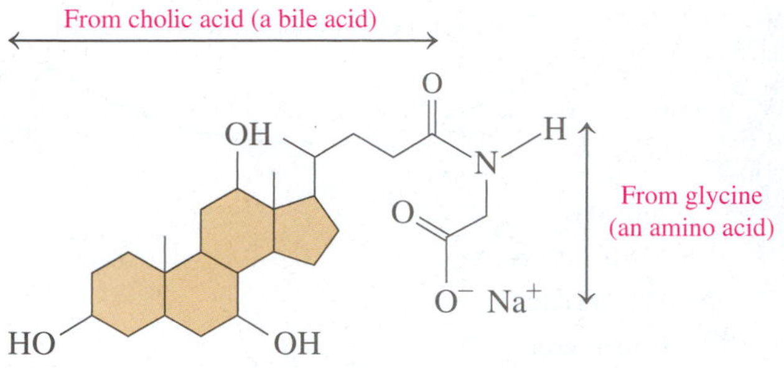
From cholic acid (a bile acid)
OH
O
H
N
O
From glycine
(an amino acid)
O⁻ Na⁺
HO
OH

Sodium glycocholate (a bile salt)

When large amounts of cholesterol accumulate in the gallbladder, cholesterol can become solid, which forms gallstones (see **FIGURE 17.8**). Gallstones are composed of almost 100% cholesterol, with some calcium salts, fatty acids, and glycerophospholipids. Normally, small stones pass through the bile duct into the duodenum, the first part of the small intestine immediately beyond the stomach. If a large stone passes into the bile duct, it can get stuck, and the pain can be severe. If the gallstone obstructs the duct, bile cannot be excreted. Then bile pigments known as bilirubin will not be able to pass through the bile duct into the duodenum. They will back up into the liver and be excreted via the blood, causing jaundice (*hyperbilirubinemia*), which gives a yellow color to the skin and the whites of the eyes.

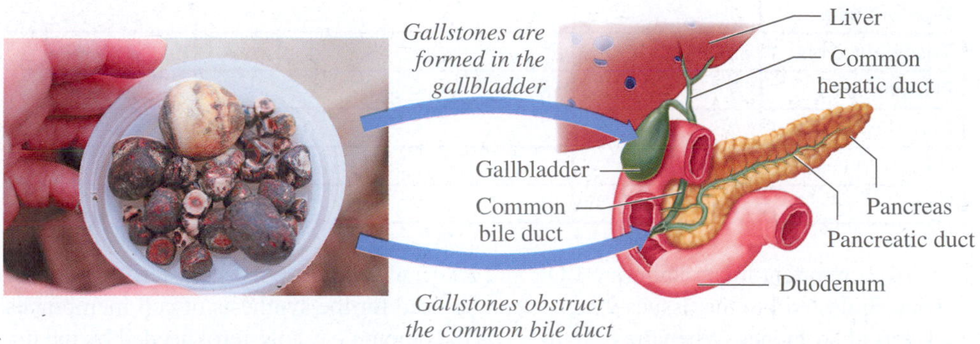
Gallstones are formed in the gallbladder
Liver
Common hepatic duct
Gallbladder
Common bile duct
Pancreas
Pancreatic duct
Duodenum
Gallstones obstruct the common bile duct

FIGURE 17.8 ▶ When cholesterol levels are higher, gallstones form in the gallbladder. They usually pass through the bile duct into the duodenum. If too large, they may obstruct the bile duct, causing pain and blocking bile.

Q What type of steroid is stored in the gallbladder?

Lipoproteins: Transporting Lipids

In the body, lipids must be moved through the bloodstream to tissues where they are stored, used for energy, or used to make hormones. However, most lipids are nonpolar and insoluble in the aqueous environment of blood. They are made more soluble by combining them with

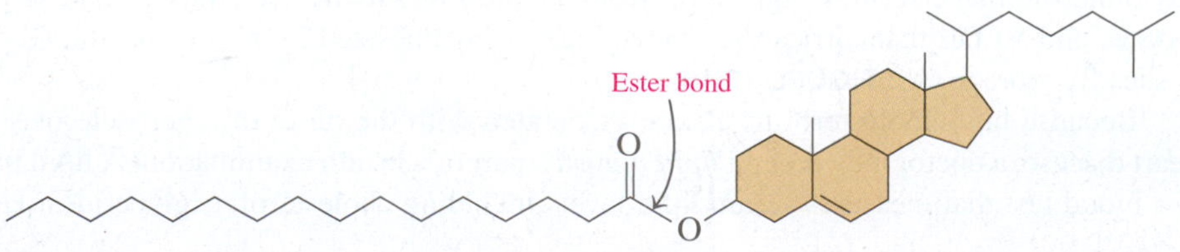
Ester bond
O
O

Cholesteryl ester

phospholipids and proteins to form water-soluble complexes called **lipoproteins**. In general, lipoproteins are spherical particles with an outer surface of polar proteins and phospholipids that surround hundreds of nonpolar molecules of triacylglycerols and cholesteryl esters (see **FIGURE 17.9**). Cholesteryl esters are the prevalent form of cholesterol in the blood. They are formed by the esterification of the hydroxyl group in cholesterol with a fatty acid.

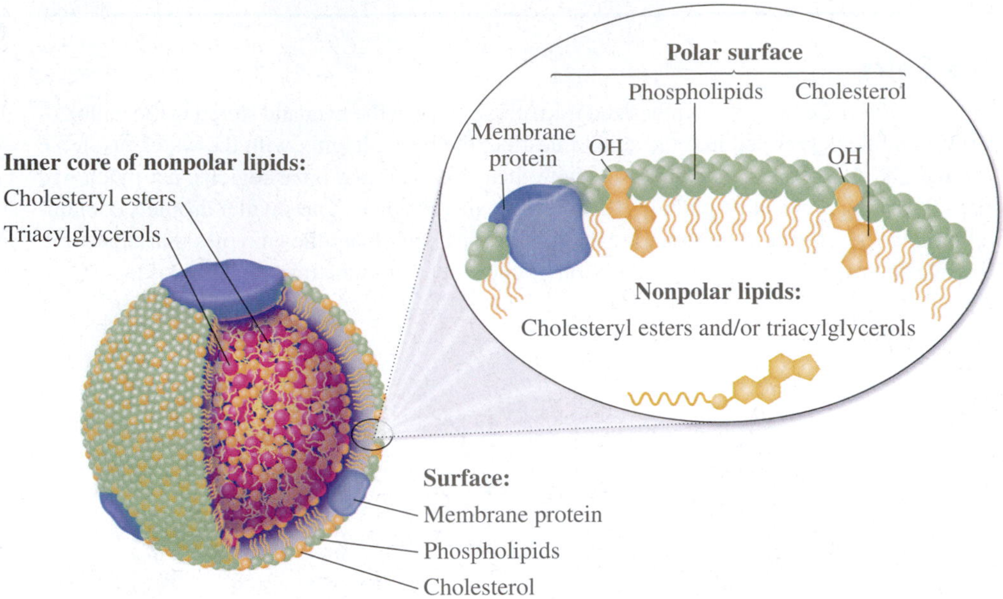
Polar surface
Phospholipids
Cholesterol
Membrane protein
OH
OH
Nonpolar lipids:
Cholesteryl esters and/or triacylglycerols
Inner core of nonpolar lipids:
Cholesteryl esters
Triacylglycerols
Surface:
Membrane protein
Phospholipids
Cholesterol

FIGURE 17.9 ▶ A spherical lipoprotein particle surrounds nonpolar lipids with polar lipids and protein for transport to body cells.

Q Why are the polar components on the surface of a lipoprotein particle and the nonpolar components at the center?

There are a variety of lipoproteins, which differ in density, lipid composition, and function. They include chylomicrons, very-low-density lipoproteins (VLDLs), low-density lipoproteins (LDLs), and high-density lipoproteins (HDLs). The density of the lipoproteins increases as the percentage of protein in each type increases (see **TABLE 17.5**).

ENGAGE

How does the percentage of protein in HDL compare with the percentage of protein in LDL?

TABLE 17.5 Composition and Properties of Plasma Lipoproteins

	Chylomicrons	VLDL	LDL	HDL
Density (g/mL)	0.940	0.940–1.006	1.006–1.063	1.063–1.210
	Composition (% by mass)			
Type of Lipid				
Triacylglycerols	86	55	6	4
Phospholipids	7	18	22	24
Cholesterol	2	7	8	2
Cholesteryl esters	3	12	42	15
Protein	2	8	22	55

Two important lipoproteins are LDL and HDL, which transport cholesterol. The LDL carries cholesterol to the tissues where it can be used for the synthesis of cell membranes and steroid hormones. When the LDL exceeds the amount of cholesterol needed by the tissues, the LDL deposits cholesterol in the arteries (plaque), which can restrict blood flow and increase the risk of developing heart disease and/or myocardial infarctions (heart attacks). This is why LDL is called "bad" cholesterol. The HDL picks up cholesterol from the tissues and carries it to the liver, where it can be converted to bile salts, which are eliminated from the body. This is why HDL is called "good" cholesterol. Other lipoproteins include chylomicrons that carry triacylglycerols from the intestines to the liver, muscle, and adipose tissues, and VLDL that carries the triacylglycerols synthesized in the liver to the adipose tissues for storage (see **FIGURE 17.10**).

Because high cholesterol levels are associated with the onset of atherosclerosis and heart disease, a doctor may order a *lipid panel* as part of a health examination. A lipid panel is a blood test that measures serum lipid levels including cholesterol, triglycerides, HDL,

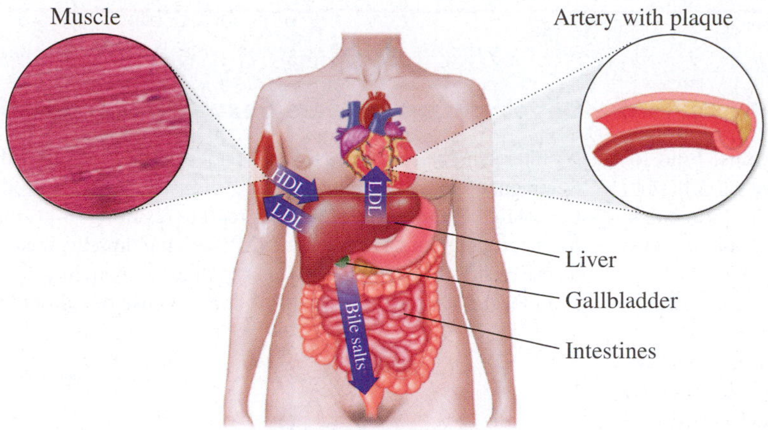
Muscle
Artery with plaque
HDL
LDL
LDL
Liver
Gallbladder
Bile salts
Intestines

FIGURE 17.10 ▶ High- and low-density lipoproteins transport cholesterol between the tissues and the liver.

Q What type of lipoprotein transports cholesterol to the liver?

and LDL. The results of a lipid panel are used to evaluate a patient's risk of heart disease and to help a doctor determine the type of treatment needed.

Lipid Panel	Recommended Level	Greater Risk of Heart Disease
Total Cholesterol	Less than 200 mg/dL	Greater than 240 mg/dL
Triglycerides (triacylglycerols)	Less than 150 mg/dL	Greater than 200 mg/dL
HDL ("good" cholesterol)	Greater than 60 mg/dL	Less than 40 mg/dL
LDL ("bad" cholesterol)	Less than 100 mg/dL	Greater than 160 mg/dL
Cholesterol/HDL Ratio	Less than 4	Greater than 7

TEST

Try Practice Problems 17.49 to 17.54

A lipid panel may also be useful in the diagnosis of other diseases or conditions. For example, elevated serum triglyceride levels are often seen in alcoholics. Alcohol metabolism, which mainly occurs in the liver, results in decreased metabolism of fatty acids and an increase in triglyceride synthesis. The excess triglycerides are packaged into VLDL and excreted into the blood, raising serum triglyceride levels. In chronic alcoholics, the ability to excrete VLDL becomes impaired, and fats begin to accumulate in the liver, leading to *fatty liver disease*.

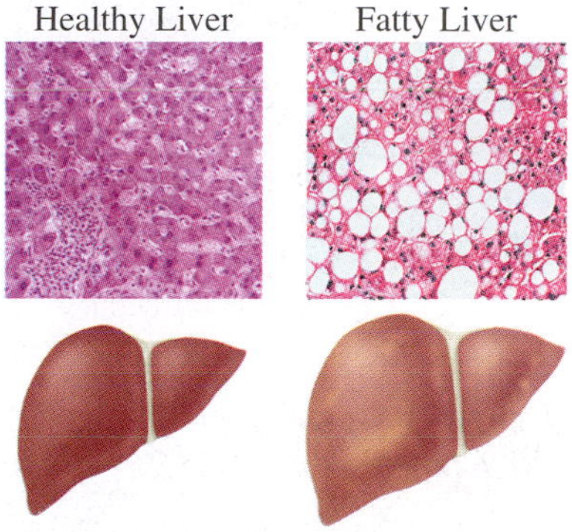
Healthy Liver
Fatty Liver

In chronic alcoholics, VLDL excretion from the liver is impaired, resulting in an accumulation of fat in the liver.

Steroid Hormones

The word *hormone* comes from the Greek "to arouse" or "to excite." Hormones are chemical messengers that serve as a communication system from one part of the body to another. The *steroid* hormones, which include the sex hormones and the adrenocortical hormones, are closely related in structure to cholesterol and depend on cholesterol for their synthesis.

Two of the male sex hormones, testosterone and androsterone, promote the growth of muscle and facial hair, and the maturation of the male sex organs and of sperm.

The *estrogens*, a group of female sex hormones, direct the development of female sexual characteristics: the uterus increases in size, fat is deposited in the breasts, and the pelvis broadens. Progesterone prepares the uterus for the implantation of a fertilized egg. If an egg is not fertilized, the levels of progesterone and estrogen drop sharply, and menstruation follows. Synthetic forms of the female sex hormones are used in birth control pills. As with other kinds of steroids, side effects include weight gain and a greater risk of forming blood clots. The structures of some steroid hormones are shown below:

ENGAGE

How does testosterone differ from estradiol?

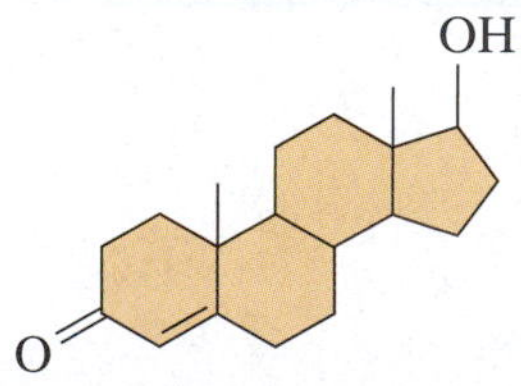
OH
O

Testosterone (androgen)
(produced in testes)

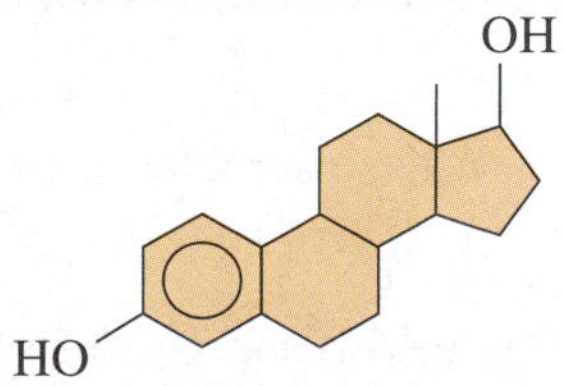
OH
HO

Estradiol (estrogen)
(produced in ovaries)

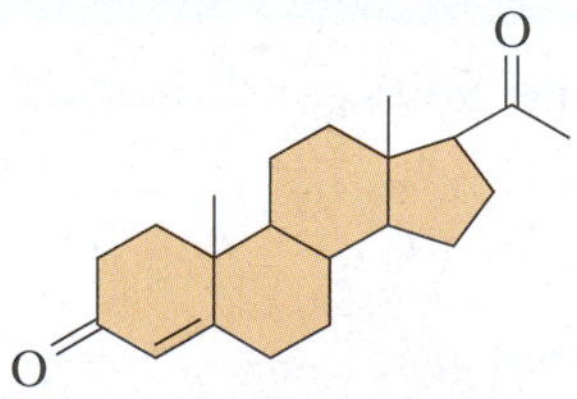
O
O

Progesterone
(produced in ovaries)

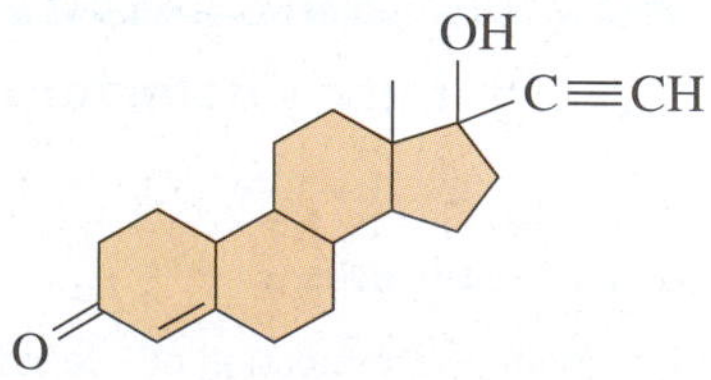
OH
C≡CH
O

Norethindrone
(synthetic progestin)

Chemistry Link to Health

A Steroid Receptor Antagonist That Prevents the Development of Male Sexual Characteristics

In order to elicit their cellular effects, hormones must bind to *receptors*, specific molecules that are either bound to cell surfaces or that are found in the cytoplasm of the cells. Molecules that block the binding of a hormone to its receptor are known as *antagonists*.

For example, to promote the development of male sexual characteristics, testosterone and andosterone must first bind to an *androgen receptor* that is found in the cytoplasm of cells. In cases where the development of male sexual characteristics is not desirable, a doctor may prescribe an *androgen receptor antagonist*. One such example is spironolactone. Spironolactone not only blocks the binding of male sexual hormones to androgen receptors, but inhibits the synthesis of these hormones as well. For this reason, spironolactone can be used as part of hormone replacement therapy for transgender women. Dermatologists also often prescribe it for hormonal acne and hair loss in females, both conditions that can be exacerbated by increased amounts of testosterone.

Interestingly, spironolactone was not originally developed for the treatment of the conditions listed above. Instead, it was introduced to treat fluid buildup and high blood pressure resulting from conditions like heart failure. It acts as a potassium-sparing diuretic, meaning that its use results in the removal of excess fluid, without lowering potassium levels. This remains the most common use of spironolactone.

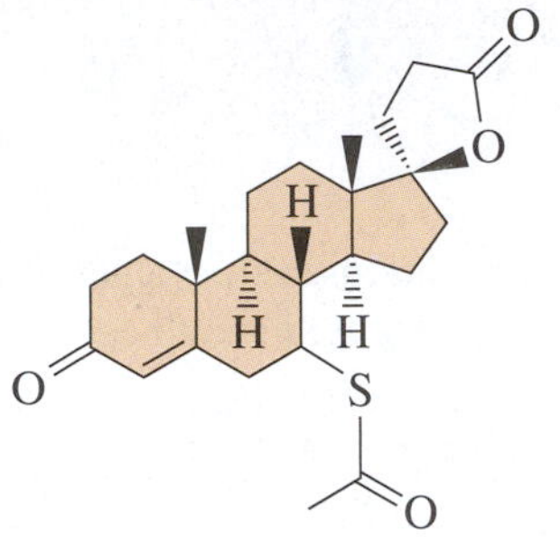

The structure of spironolactone includes the steroid nucleus.

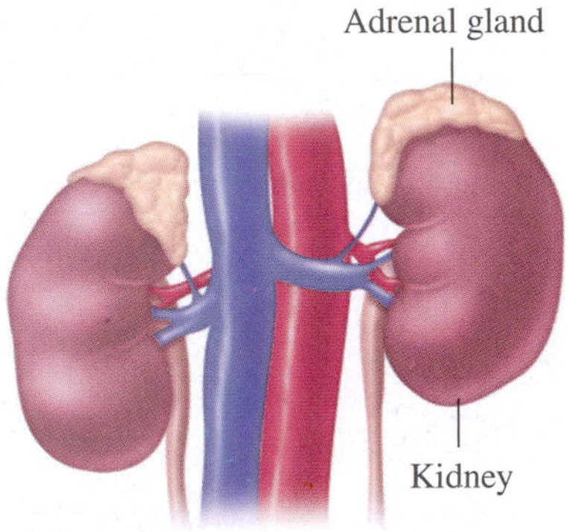
Adrenal gland
Kidney

The adrenal glands on the kidneys produce corticosteroids.

Adrenal Corticosteroids

The adrenal glands, located on the top of each kidney, produce a large number of compounds known as the *corticosteroids*. Cortisone increases the blood glucose level and stimulates the synthesis of glycogen in the liver. Aldosterone is responsible for the regulation of electrolytes and water balance by the kidneys. Cortisol is released under stress to increase blood sugar and regulate carbohydrate, fat, and protein metabolism. Synthetic corticosteroid drugs such as prednisone are derived from cortisone and used medically for reducing inflammation and treating asthma and rheumatoid arthritis, although health problems can result from long-term use.

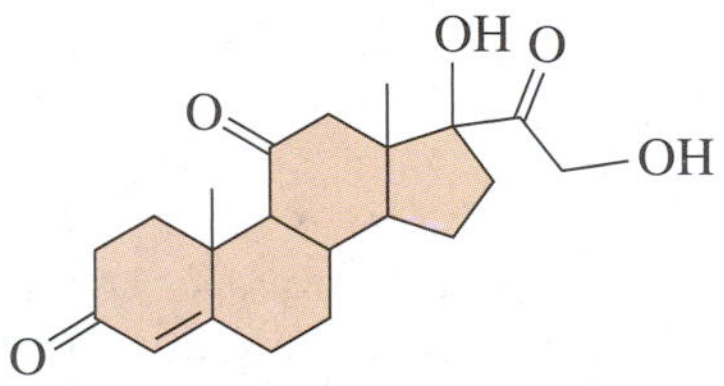

Cortisone
(produced in adrenal gland)

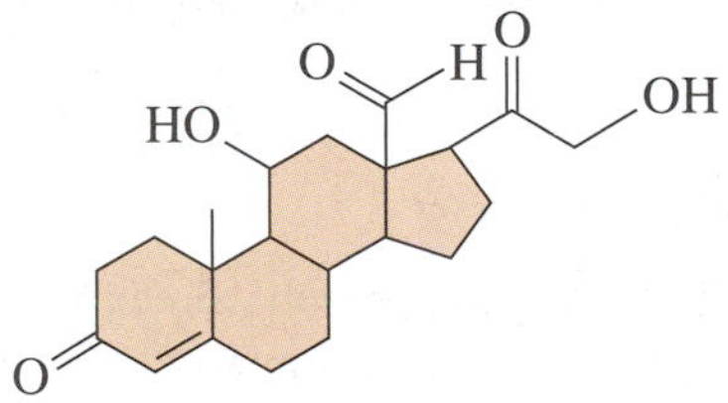

Aldosterone (mineralocorticoid)
(produced in adrenal gland)

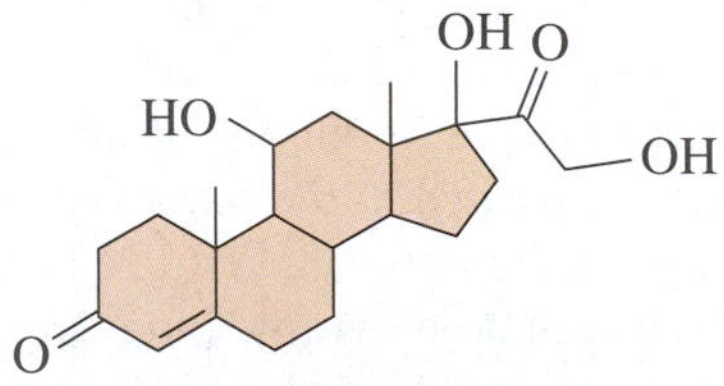

Cortisol
(produced in adrenal cortex)

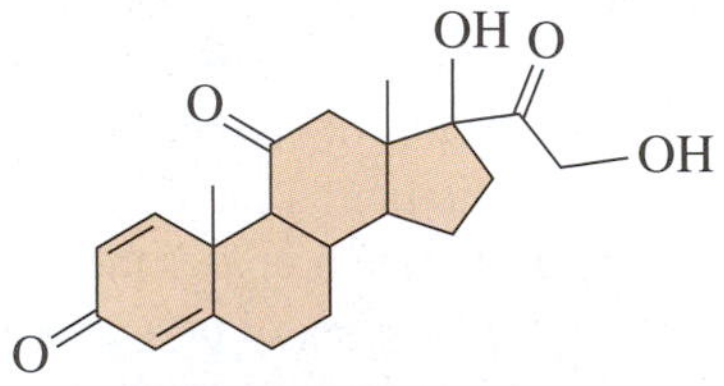

Prednisone
(synthetic corticoid)

TEST

Try Practice Problems 17.55 to 17.58

PRACTICE PROBLEMS

17.6 Steroids: Cholesterol, Bile Salts, and Steroid Hormones

17.47 Draw the structure for the steroid nucleus.

17.48 Draw the structure for cholesterol.

Clinical Applications

17.49 What is the function of bile salts in digestion?

17.50 Why are lipoproteins needed to transport lipids in the bloodstream?

17.51 How do chylomicrons differ from VLDL?

17.52 How does LDL differ from HDL?

17.53 Why is LDL called "bad" cholesterol?

17.54 Why is HDL called "good" cholesterol?

17.55 What are the similarities and differences between the steroid hormones estradiol and testosterone?

17.56 What are the similarities and differences between the adrenal hormone cortisone and the synthetic corticoid prednisone?

17.57 Which of the following are steroid hormones?

a. cholesterol **b.** cortisol
c. estradiol **d.** testosterone

17.58 Which of the following are adrenal corticosteroids?

a. prednisone **b.** aldosterone
c. cortisol **d.** testosterone

17.7 Cell Membranes

LEARNING GOAL Describe the composition and function of the lipid bilayer in cell membranes.

The membrane of a cell separates the contents of a cell from external fluids. It is *semipermeable* so that nutrients can enter the cell and waste products can leave. The main components of a cell membrane are the glycerophospholipids and sphingolipids. Earlier in this chapter, we saw that the phospholipids consist of a nonpolar region, or hydrocarbon "tail," with two long-chain fatty acids and a polar region, or ionic "head" of phosphate and an ionized amino alcohol.

In a cell (plasma) membrane, two layers of phospholipids are arranged with their hydrophilic heads at the outer and inner surfaces of the membrane, and their hydrophobic tails in the center. This double layer arrangement of phospholipids is called a **lipid bilayer** (see **FIGURE 17.11**). The outer layer of phospholipids is in contact with the external fluids, and the inner layer is in contact with the internal contents of the cell.

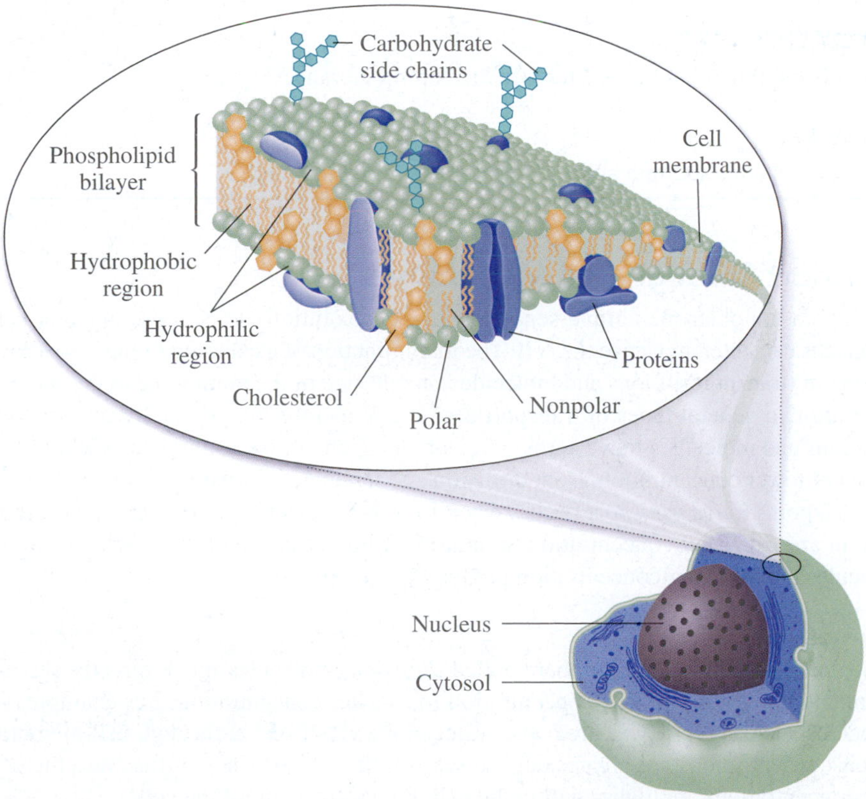
Carbohydrate side chains
Cell membrane
Phospholipid bilayer
Hydrophobic region
Hydrophilic region
Cholesterol
Polar
Nonpolar
Proteins
Nucleus
Cytosol

FIGURE 17.11 ▶ In the fluid mosaic model of a cell membrane, proteins and cholesterol are embedded in a lipid bilayer of phospholipids. The bilayer forms a barrier with polar heads at the membrane surfaces and the nonpolar tails in the center, away from the water.

Q What types of fatty acids are found in the phospholipids of the lipid bilayer?

ENGAGE

Which components of the lipid bilayer give it fluid-like characteristics?

Most of the phospholipids in the lipid bilayer contain unsaturated fatty acids. Because of the kinks in the carbon chains at the cis double bonds, the phospholipids do not fit closely together. As a result, the lipid bilayer is not a rigid, fixed structure, but one that is dynamic and fluid-like. For this reason, the model of biological membranes is referred to as the **fluid mosaic model** of membranes. It is called a *mosaic* because the bilayer also contains proteins, carbohydrates, and cholesterol molecules.

In the fluid mosaic model, proteins known as *peripheral proteins* emerge on just one of the surfaces, outer or inner. The *integral proteins* extend through the entire lipid bilayer and appear on both surfaces of the membrane. Some proteins and lipids on the outer surface of the cell membrane are attached to carbohydrates. These carbohydrate chains project into the surrounding fluid environment where they are responsible for cell recognition and communication with chemical messengers such as hormones and neurotransmitters. In animals, cholesterol molecules embedded among the phospholipids make up 20 to 25% of the lipid bilayer. Because cholesterol molecules are large and rigid, they reduce the flexibility of the lipid bilayer and add strength to the cell membrane.

INTERACTIVE VIDEO

Membrane Structure

PEARSON eText 2.0

▶ SAMPLE PROBLEM 17.7 Lipid Bilayer in Cell Membranes

TRY IT FIRST

Describe the role of phospholipids in the lipid bilayer.

SOLUTION

Phospholipids consist of polar and nonpolar parts. In a cell membrane, an alignment of the nonpolar sections toward the center with the polar sections on the outside produces a barrier that prevents the contents of a cell from mixing with the fluids on the outside of the cell.

STUDY CHECK 17.7

What is the function of cholesterol in the cell membrane?

ANSWER

Cholesterol adds strength and rigidity to the cell membrane.

TEST

Try Practice Problems 17.59 to 17.64

Transport Through Cell Membranes

Although a nonpolar membrane separates aqueous solutions, it is necessary that certain substances can enter and leave the cell. The main function of a cell membrane is to allow the movement (transport) of ions and molecules on one side of the membrane to the other side. There are two general types of transport across cell membranes. *Passive transport* occurs when ions and molecules move across a membrane from an area of higher concentration to an area of lower concentration. Both diffusion and facilitated diffusion are examples of passive transport. *Active transport* occurs when ions and molecules move across a membrane from an area of lower concentration to an area of higher concentration. Moving ions and molecules *against* their concentration gradients requires energy.

Diffusion

In the simplest transport mechanism called *diffusion*, molecules move directly across the cell membrane from a higher concentration to a lower concentration. For example, small molecules such as O_2, CO_2, urea, and water move via diffusion through cell membranes. If their concentrations are greater outside the cell than inside, they diffuse into the cell. If their concentrations are higher within the cell, they diffuse out of the cell.

Facilitated Diffusion

In *facilitated diffusion*, proteins that extend from one side of the bilayer membrane to the other provide a channel through which certain substances can diffuse more rapidly to meet cellular needs. These protein channels allow transport of chloride ion (Cl^-), bicarbonate ion (HCO_3^-), and glucose molecules in and out of the cell.

Active Transport

Certain ions such as K^+, Na^+, and Ca^{2+} move across a cell membrane *against* their concentration gradients. For example, the K^+ concentration is greater inside a cell, and the Na^+ concentration is greater outside. However, in the conduction of nerve impulses and contraction of muscles, K^+ moves into the cell, and Na^+ moves out by an energy-requiring process known as *active transport*.

ENGAGE

Why are ions like Na^+ or K^+ unable to pass through the lipid bilayer?

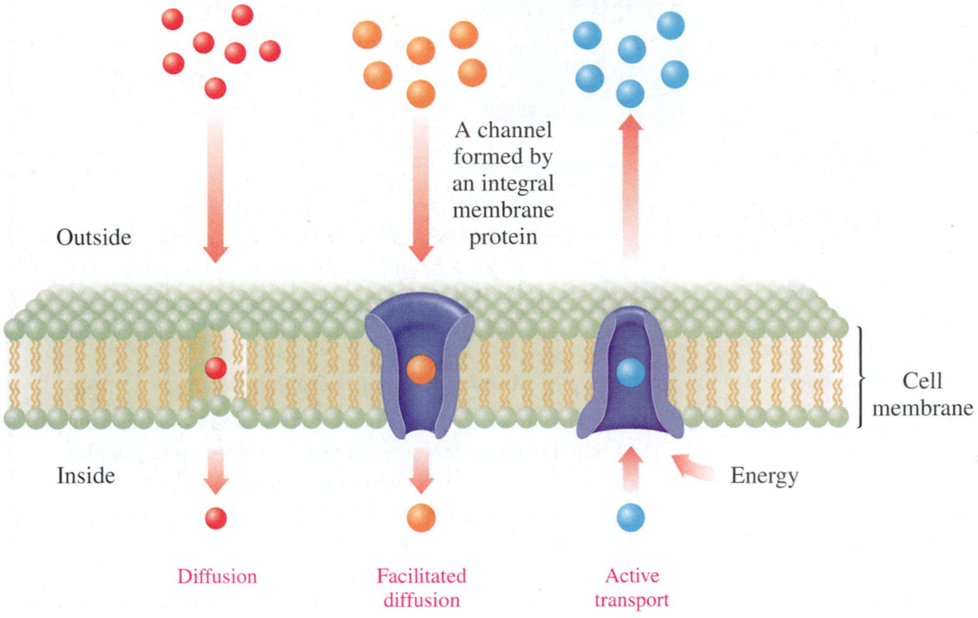

Outside
A channel formed by an integral membrane protein
Cell membrane
Inside
Energy
Diffusion
Facilitated diffusion
Active transport

Substances are transported across a cell membrane by either diffusion, facilitated diffusion, or active transport.

TEST

Try Practice Problems 17.65 and 17.66

PRACTICE PROBLEMS

17.7 Cell Membranes

Clinical Applications

17.59 What is the function of the lipid bilayer in a cell membrane?

17.60 Describe the structure of a lipid bilayer.

17.61 How do molecules of cholesterol affect the structure of cell membranes?

17.62 How do the unsaturated fatty acids in the phospholipids affect the structure of cell membranes?

17.63 Where are proteins located in cell membranes?

17.64 What is the difference between peripheral and integral proteins?

17.65 Identify the type of transport described by each of the following:

a. A molecule moves through a protein channel.

b. O_2 moves into the cell from a higher concentration outside the cell.

17.66 Identify the type of transport described by each of the following:

a. An ion moves from low to high concentration in the cell.

b. Carbon dioxide moves through a cell membrane.

CLINICAL UPDATE Rebecca's Program to Lower Cholesterol

During their first visit, Susan asked Rebecca to maintain a diet with less beef and chicken, more fish with omega-3 oils, low-fat dairy products, no egg yolks, and no coconut or palm oils. Rebecca maintained her new diet containing lower quantities of fats and increased quantities of fiber for the next year. She also increased her exercise and lost 35 lb. When Rebecca returns to the lipid clinic to check her progress, a new set of blood tests indicates that her total cholesterol had dropped from 420 to 390 mg/dL.

Because Rebecca's changes in diet and exercise did not sufficiently lower her cholesterol level, Susan prescribes a medication to help lower blood cholesterol levels further. The most common medications used are the statins lovastatin (Mevacor), pravastatin (Pravachol), simvastatin (Zocor), atorvastatin (Lipitor), and rosuvastatin (Crestor). For some FH patients, statin therapy may be combined with fibrates such as gemfibrozil (Lopid) or

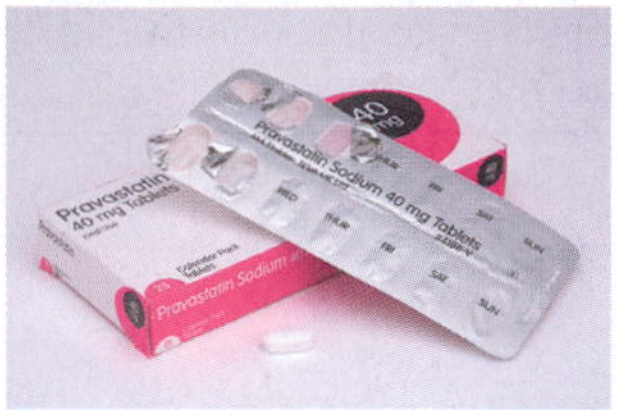
Pravastatin Sodium 40 mg Tablets

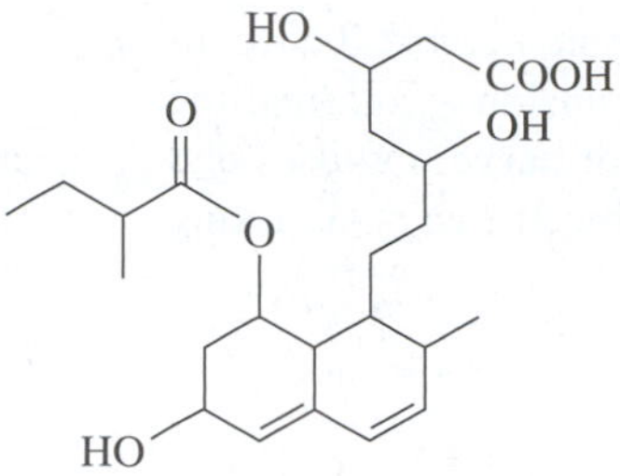
HO
COOH
OH
O
O
HO

Pravastatin (Pravachol)

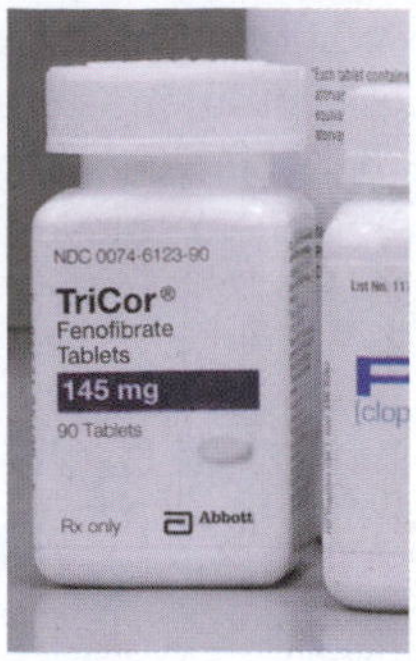
NDC 0074-6123-90
TriCor®
Fenofibrate
Tablets
145 mg
90 Tablets
Rx only
Abbott

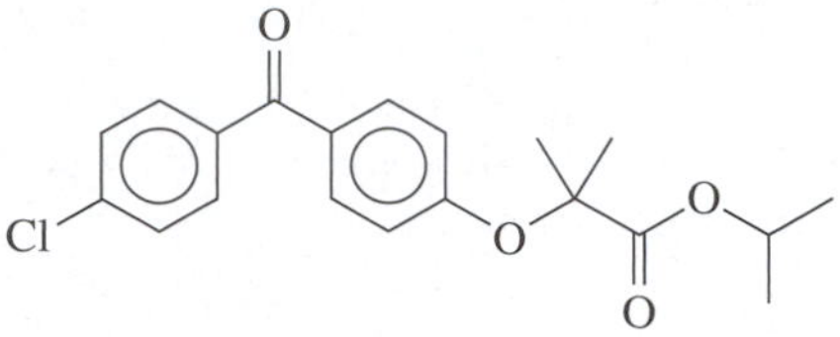
O
Cl
O
O
O

Fenofibrate (TriCor)

fenofibrate (TriCor) that activate enzymes that break down fats in the blood.

Susan prescribes pravastatin (Pravachol), 80 mg once a day, which is effective. Later, Susan adds fenofibrate (TriCor), one 145-mg tablet each day, to the Pravachol. Rebecca understands that her medications, diet, and exercise plan are part of a lifelong process. Because Rebecca was diagnosed with FH, both her children are tested for FH. Her older son is diagnosed with FH and is also being treated with a statin. Her younger son does not have FH.

Clinical Applications

17.67 Identify the functional groups in Pravachol.

17.68 Identify the functional groups in TriCor.

17.69 Six months after Rebecca started using Pravachol, her blood test showed that a 5.0-mL blood sample contained a total cholesterol of 18 mg.

a. How many grams of Pravachol did Rebecca consume in 1 week?

b. What was her total cholesterol in mg/dL?

17.70 Five months after Rebecca added the fibrate TriCor to the statin, her blood test showed that a 5.0-mL blood sample contained a total cholesterol of 14 mg.

a. How many grams of TriCor did Rebecca consume in 1 week?

b. What was her total cholesterol in mg/dL?

CONCEPT MAP

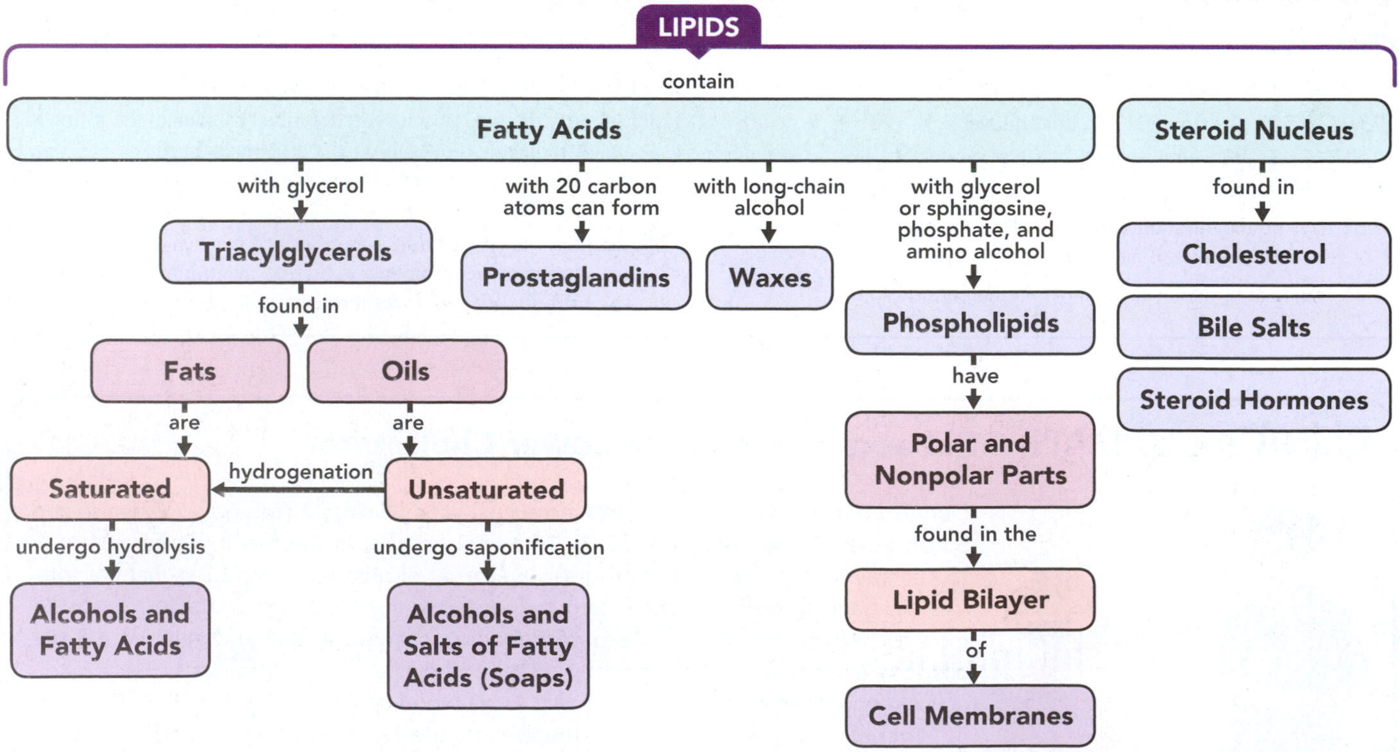
LIPIDS
contain
Fatty Acids
Steroid Nucleus
with glycerol
with 20 carbon atoms can form
with long-chain alcohol
with glycerol or sphingosine, phosphate, and amino alcohol
found in
Triacylglycerols
Prostaglandins
Waxes
Phospholipids
Cholesterol
Bile Salts
Steroid Hormones
found in
Fats
Oils
have
are
are
Saturated
hydrogenation
Unsaturated
Polar and Nonpolar Parts
undergo hydrolysis
undergo saponification
found in the
Alcohols and Fatty Acids
Alcohols and Salts of Fatty Acids (Soaps)
Lipid Bilayer
of
Cell Membranes

CHAPTER REVIEW

17.1 Lipids

LEARNING GOAL Describe the classes of lipids.

- Lipids are biomolecules that are not soluble in water.
- Classes of lipids include waxes, triacylglycerols, glycerophospholipids, sphingolipids, and steroids.

17.2 Fatty Acids

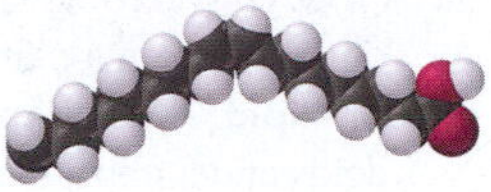

LEARNING GOAL Draw the condensed structural and line-angle formulas for a fatty acid, and identify it as saturated or unsaturated.

- Fatty acids are unbranched carboxylic acids that typically contain an even number (12 to 20) of carbon atoms.
- Fatty acids may be saturated, monounsaturated with one double bond, or polyunsaturated with two or more double bonds.
- The double bonds in unsaturated fatty acids are almost always cis.

17.3 Waxes and Triacylglycerols

LEARNING GOAL Draw the condensed structural and line-angle formulas for a wax or triacylglycerol produced by the reaction of a fatty acid and an alcohol or glycerol.

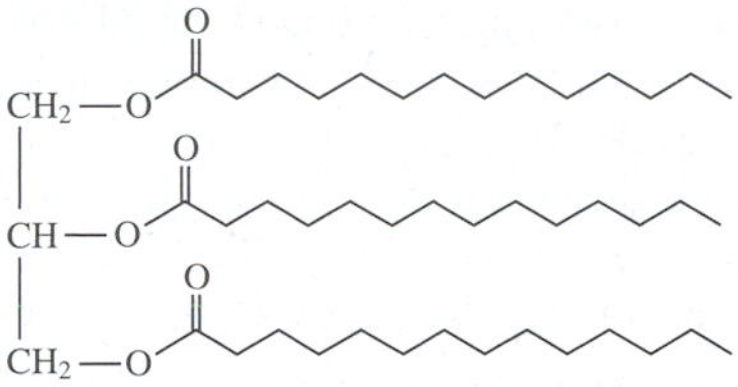

Triacylglycerol

- A wax is an ester of a long-chain fatty acid and a long-chain alcohol.
- The triacylglycerols are esters of glycerol with three long-chain fatty acids.
- Fats contain more saturated fatty acids and have higher melting points than most vegetable oils.

17.4 Chemical Properties of Triacylglycerols

LEARNING GOAL Draw the condensed structural and line-angle formulas for the products of a triacylglycerol that undergoes hydrogenation, hydrolysis, or saponification.

Shortening (solid)
Crisco
Blue Bonnet
Margarine (soft)
Stick margarine (solid)

- The hydrogenation of unsaturated fatty acids of a triacylglycerol converts double bonds to single bonds.
- The hydrolysis of the ester bonds in triacylglycerols in the presence of a strong acid produces glycerol and fatty acids.
- In saponification, a triacylglycerol heated with a strong base produces glycerol and the salts of the fatty acids (soap).

17.5 Phospholipids

LEARNING GOAL Draw the structure of a phospholipid that contains glycerol or sphingosine.

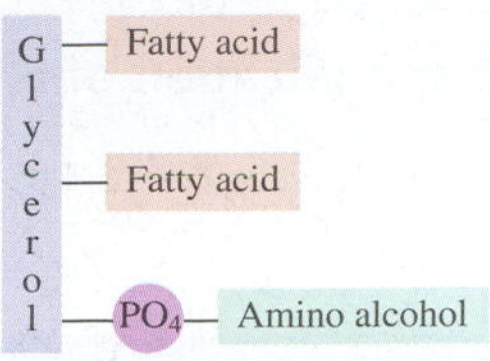

Phospholipid
Glycerol
Fatty acid
Fatty acid
PO4
Amino alcohol

- Glycerophospholipids are esters of glycerol with two fatty acids and a phosphate group attached to an amino alcohol.
- In a sphingomyelin, the amino alcohol sphingosine forms an amide bond with a fatty acid, and phosphoester bonds to phosphate and an amino alcohol.

17.6 Steroids: Cholesterol, Bile Salts, and Steroid Hormones

LEARNING GOAL Draw the structure of the steroid nucleus; compare the structures and functions of steroid compounds.

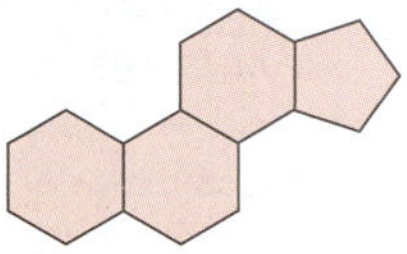

Steroid nucleus

- Steroids are lipids containing the steroid nucleus, which is a fused structure of four rings.
- Steroids include cholesterol, bile salts, and steroid hormones.
- Bile salts, synthesized from cholesterol, mix with water-insoluble fats and break them apart during digestion.
- Lipoproteins, such as chylomicrons and LDL, transport triacylglycerols from the intestines and the liver to fat cells and muscles for storage and energy.
- HDL transports cholesterol from the tissues to the liver for elimination.
- The steroid hormones are closely related in structure to cholesterol and depend on cholesterol for their synthesis.
- The sex hormones, such as estrogen and testosterone, are responsible for sexual characteristics and reproduction.
- The adrenal corticosteroids, such as aldosterone and cortisone, regulate water balance and glucose levels in the cells, respectively.

17.7 Cell Membranes

LEARNING GOAL Describe the composition and function of the lipid bilayer in cell membranes.

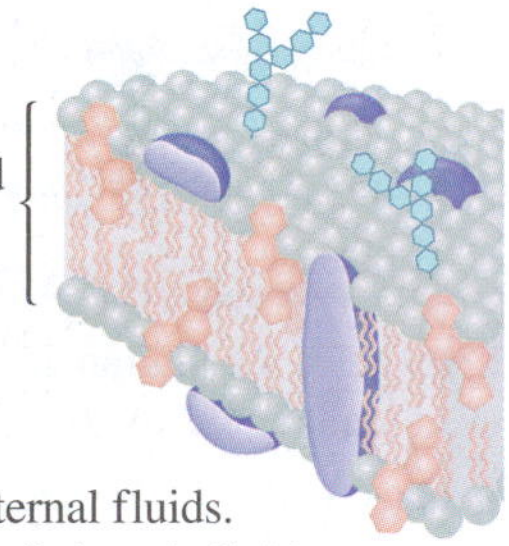

Phospholipid bilayer

- All animal cells are surrounded by a semipermeable membrane that separates the cellular contents from the external fluids.
- The membrane is composed of two rows of phospholipids in a lipid bilayer.
- Proteins and cholesterol are embedded in the lipid bilayer, and carbohydrates are attached to its surface.
- Nutrients and waste products move through the cell membrane by diffusion, facilitated diffusion, or active transport.

SUMMARY OF REACTIONS

The chapter Sections to review are shown after the name of each reaction.

Hydrogenation of a Triacylglycerol (17.4)

$$\text{Triacylglycerol (unsaturated)} + H_2 \xrightarrow{Ni} \text{triacylglycerol (saturated)}$$

Hydrolysis of a Triacylglycerol (17.4)

Triacylglycerol + $3H_2O \xrightarrow{H^+ \text{ or lipase}}$ glycerol + 3 fatty acid molecules

Saponification of a Triacylglycerol (17.4)

Triacylglycerol + 3NaOH $\xrightarrow{\text{Heat}}$ glycerol + 3 sodium salts of fatty acids (soaps)

KEY TERMS

cholesterol The most prevalent of the steroid compounds; needed for cellular membranes and the synthesis of vitamin D, hormones, and bile salts.

fat A triacylglycerol that is solid at room temperature and usually comes from animal sources.

fatty acid A long-chain carboxylic acid found in many lipids.

fluid mosaic model The concept that cell membranes are lipid bilayer structures that contain an assortment of polar lipids and proteins in a dynamic, fluid arrangement.

glycerophospholipid A polar lipid of glycerol attached to two fatty acids and a phosphate group connected to an amino alcohol such as choline, serine, or ethanolamine.

hydrogenation The addition of hydrogen to unsaturated fats.

lipid bilayer A model of a cell membrane in which phospholipids are arranged in two rows.

lipids A family of biomolecules that is nonpolar in nature and not soluble in water; includes prostaglandins, waxes, triacylglycerols, phospholipids, and steroids.

lipoprotein A polar complex composed of a combination of nonpolar lipids with glycerophospholipids and proteins to form a polar complex that can be transported through body fluids.

monounsaturated fatty acid (MUFA) A fatty acid with one double bond.

oil A triacylglycerol that is a liquid at room temperature and is obtained from a plant source.

phospholipid A polar lipid of glycerol or sphingosine attached to fatty acids and a phosphate group connected to an amino alcohol.

polyunsaturated fatty acid (PUFA) A fatty acid that contains two or more double bonds.

prostaglandin A compound derived from arachidonic acid that regulates several physiological processes.

saturated fatty acid (SFA) A fatty acid that has no double bonds, which has a higher melting point than unsaturated lipids, and is usually solid at room temperature.

sphingomyelin A compound in which the amine group of sphingosine forms an amide bond to a fatty acid and the hydroxyl group forms an ester bond with phosphate, which forms another phosphoester bond to an amino alcohol.

steroid Type of lipid containing a multicyclic ring system.

triacylglycerol A lipid composed of three fatty acids bonded through ester bonds to glycerol, a trihydroxy alcohol.

unsaturated fatty acid (UFA) A fatty acid that contains one or more carbon–carbon double bonds, which has a lower melting point than saturated fatty acids, and is usually liquid at room temperature.

wax The ester of a long-chain alcohol and a long-chain fatty acid.

CORE CHEMISTRY SKILLS

The chapter Section containing each Core Chemistry Skill is shown in parentheses at the end of each heading.

Identifying Fatty Acids (17.2)

Fatty acids are unbranched carboxylic acids that typically contain an even number (12 to 20) of carbon atoms.

- Fatty acids may be saturated, monounsaturated with one double bond, or polyunsaturated with two or more double bonds.
- The double bonds in unsaturated fatty acids are almost always cis.

Example: State the number of carbon atoms, saturated or unsaturated, and name of the following:

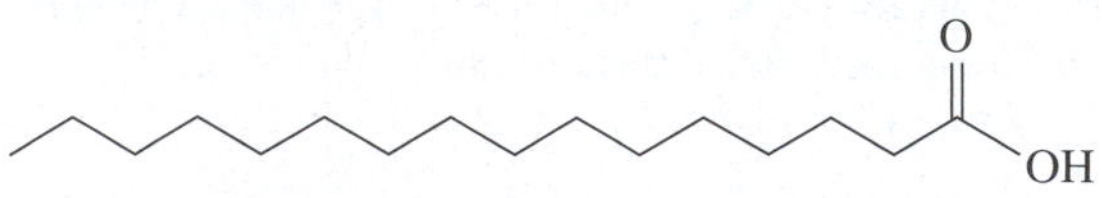
O
OH

Answer: 16 carbon atoms, saturated, palmitic acid

Drawing Structures for Triacylglycerols (17.3)

- Triacylglycerols are esters of glycerol with three long-chain fatty acids.

Example: Draw the line-angle formula and name the triacylglycerol formed from glycerol and palmitic acid.

Answer:

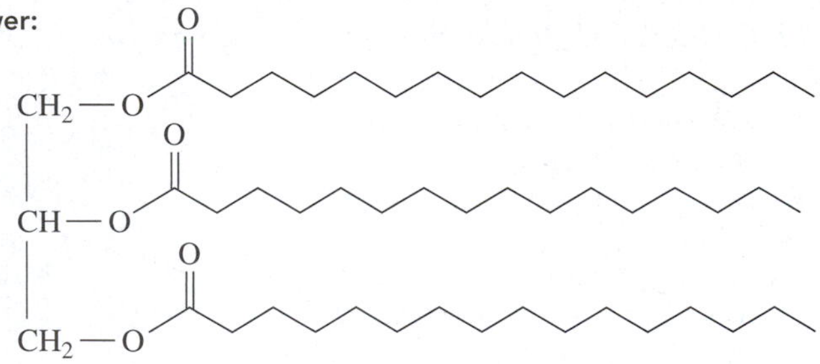
O
CH2—O
O
CH—O
O
CH2—O

Glyceryl tripalmitate (tripalmitin)

Drawing the Products for the Hydrogenation, Hydrolysis, and Saponification of a Triacylglycerol (17.4)

- The hydrogenation of unsaturated fatty acids of a triacylglycerol in the presence of a Ni catalyst converts double bonds to single bonds.
- The hydrolysis of the ester bonds in triacylglycerols in the presence of a strong acid or digestive enzyme produces glycerol and fatty acids.
- In saponification, a triacylglycerol heated with a strong base produces glycerol and the salts of the fatty acids (soap).

Example: Identify each of the following as hydrogenation, hydrolysis, or saponification and state the products:

a. the reaction of palm oil with NaOH
b. the reaction of glyceryl trilinoleate from corn oil with water in the presence of an acid catalyst
c. the reaction of corn oil and H_2 using a nickel catalyst

Answer: **a.** The reaction of palm oil with NaOH is saponification, and the products are glycerol and the sodium salts of the fatty acids, which is soap.
b. In hydrolysis, glyceryl trilinoleate reacts with water in the presence of an acid catalyst, which splits the ester bonds to produce glycerol and three molecules of linoleic acid.
c. In hydrogenation, H_2 adds to double bonds in corn oil, which produces a more saturated, and thus more solid, fat.

Identifying the Steroid Nucleus (17.6)

- The steroid nucleus consists of three cyclohexane rings and one cyclopentane ring fused together.

Example: Why are cholesterol, sodium glycocholate (a bile salt), and cortisone (a corticosteroid) all considered steroids?

Answer: They all contain the steroid nucleus of three six-carbon rings and one five-carbon ring fused together.

UNDERSTANDING THE CONCEPTS

The chapter Sections to review are shown in parentheses at the end of each problem.

17.71 Palm oil has a high level of glyceryl tripalmitate (tripalmitin). Draw the condensed structural formula for glyceryl tripalmitate. (17.2, 17.3)

The fruit from palm trees are a source of palm oil.

17.72 Jojoba wax in candles consists of a 20-carbon saturated fatty acid and a 20-carbon saturated alcohol. Draw the condensed structural formula for jojoba wax. (17.2, 17.3)

Candles contain jojoba wax.

Clinical Applications

17.73 Identify each of the following as a saturated, monounsaturated, polyunsaturated, omega-3, or omega-6 fatty acid: (17.2)

a. $CH_3-(CH_2)_7-CH{=}CH-(CH_2)_7-\overset{\overset{\displaystyle O}{\|}}{C}-OH$

b. linoleic acid

c. $CH_3-CH_2-(CH{=}CH-CH_2)_5-CH_2-CH_2-\overset{\overset{\displaystyle O}{\|}}{C}-OH$

17.74 Identify each of the following as a saturated, monounsaturated, polyunsaturated, omega-3, or omega-6 fatty acid: (17.2)

a.

$CH_3-(CH_2)_4-CH{=}CH-CH_2-CH{=}CH-(CH_2)_7-\overset{\overset{\displaystyle O}{\|}}{C}-OH$

b. linolenic acid

c. $CH_3-(CH_2)_{14}-\overset{\overset{\displaystyle O}{\|}}{C}-OH$

ADDITIONAL PRACTICE PROBLEMS

17.75 Among the ingredients in lipstick are beeswax, carnauba wax, hydrogenated vegetable oils, and glyceryl tricaprate (tricaprin). (17.1, 17.2, 17.3)

a. What types of lipids are these?
b. Draw the condensed structural formula for glyceryl tricaprate (tricaprin). Capric acid is a saturated 10-carbon fatty acid.

17.76 Because peanut oil floats on the top of peanut butter, the peanut oil in many brands of peanut butter is hydrogenated and the solid is mixed into the peanut butter to give a product that does not separate. If a triacylglycerol in peanut oil that contains one oleic acid and two linoleic acids is completely hydrogenated, draw the condensed structural formula for the product. (17.3, 17.4)

Clinical Applications

17.77 The total kilocalories and grams of fat for some typical meals at fast-food restaurants are listed here. Calculate the number of kilocalories and the percentage of total kilocalories from fat (1 gram of fat = 9 kcal). Round answers to the tens place. (17.2, 17.3)

a. a chicken dinner, 830 kcal, 46 g of fat
b. a quarter-pound cheeseburger, 520 kcal, 29 g of fat
c. pepperoni pizza (three slices), 560 kcal, 18 g of fat

17.78 The total kilocalories and grams of fat for some typical meals at fast-food restaurants are listed here. Calculate the number of kilocalories and the percentage of total kilocalories from fat (1 gram of fat = 9 kcal). Round answers to the tens place. (17.2, 17.3)

a. a beef burrito, 470 kcal, 21 g of fat
b. deep-fried fish (three pieces), 480 kcal, 28 g of fat
c. a jumbo hot dog, 180 kcal, 18 g of fat

17.79 Identify each of the following as a fatty acid, soap, triacylglycerol, wax, glycerophospholipid, sphingolipid, or steroid: (17.1, 17.2, 17.3, 17.5, 17.6)

a. beeswax **b.** cholesterol
c. lecithin **d.** glyceryl tripalmitate (tripalmitin)
e. sodium stearate **f.** safflower oil

17.80 Identify each of the following as a fatty acid, soap, triacylglycerol, wax, glycerophospholipid, sphingolipid, or steroid: (17.1, 17.2, 17.3, 17.5, 17.6)

a. sphingomyelin **b.** whale blubber
c. adipose tissue **d.** progesterone
e. cortisone **f.** stearic acid

17.81 Identify the components (**1** to **6**) contained in each of the following lipids (**a** to **d**): (17.1, 17.2, 17.3, 17.5, 17.6)

1. glycerol **2.** fatty acid
3. phosphate **4.** amino alcohol
5. steroid nucleus **6.** sphingosine

a. estrogen **b.** cephalin
c. wax **d.** triacylglycerol

17.82 Identify the components (**1** to **6**) contained in each of the following lipids (**a** to **d**): (17.1, 17.2, 17.3, 17.5, 17.6)

1. glycerol **2.** fatty acid
3. phosphate **4.** amino alcohol
5. steroid nucleus **6.** sphingosine

a. glycerophospholipid **b.** sphingomyelin
c. aldosterone **d.** linoleic acid

17.83 Which of the following are found in cell membranes? (17.7)

a. cholesterol **b.** triacylglycerols **c.** carbohydrates

17.84 Which of the following are found in cell membranes? (17.7)

a. proteins **b.** waxes **c.** phospholipids

CHALLENGE PROBLEMS

The following problems are related to the topics in this chapter. However, they do not all follow the chapter order, and they require you to combine concepts and skills from several Sections. These problems will help you increase your critical thinking skills and prepare for your next exam.

17.85 Draw the condensed structural formula for a glycerophospholipid that contains glycerol, two stearic acids, phosphate, and ethanolamine (ionized). (17.2, 17.5)

17.86 Sunflower seed oil can be used to make margarine. A triacylglycerol in sunflower seed oil contains two linoleic acids and one oleic acid. (17.2, 17.3, 17.5)

Sunflower oil is obtained from the seeds of the sunflower.

a. Draw the condensed structural formulas for two isomers of the triacylglycerol in sunflower seed oil.
b. Using one of the isomers, write the reaction that takes place when sunflower seed oil is used to make solid margarine.

Clinical Applications

17.87 Match the lipoprotein (**1** to **4**) with its description (**a** to **d**). (17.6)

1. chylomicrons **2.** VLDL
3. LDL **4.** HDL

a. "good" cholesterol
b. transports most of the cholesterol to the cells
c. carries triacylglycerols from the intestine to the fat cells
d. transports cholesterol to the liver

17.88 Match the lipoprotein (**1** to **4**) with its description (**a** to **d**). (17.6)

1. chylomicrons **2.** VLDL
3. LDL **4.** HDL

a. has the greatest abundance of protein
b. "bad" cholesterol
c. carries triacylglycerols synthesized in the liver to the muscles
d. has the lowest density

17.89 A sink drain can become clogged with solid fat such as glyceryl tristearate (tristearin). (7.1, 7.7, 7.8, 9.4, 17.3, 17.4)

A sink drain can become clogged with saturated fats.

a. How would adding lye (NaOH) to the sink drain remove the blockage?
b. Write a balanced chemical equation for the reaction that occurs.
c. How many milliliters of a 0.500 M NaOH solution are needed to completely react with 10.0 g of glyceryl tristearate (tristearin)?

17.90 One of the triacylglycerols in olive oil is glyceryl tripalmitoleate (tripalmitolein). (7.1, 7.7, 7.8, 8.6, 9.4, 17.3, 17.4)

Olive oil contains glyceryl tripalmitoleate (tripalmitolein).

a. Draw the condensed structural formula for glyceryl tripalmitoleate (tripalmitolein).
b. How many liters of H_2 gas at STP are needed to completely saturate 100. g of glyceryl tripalmitoleate (tripalmitolein)?
c. How many milliliters of a 0.250 M NaOH solution are needed to completely react with 100. g of glyceryl tripalmitoleate (tripalmitolein)?

17.91 1.00 mole of glyceryl trioleate (triolein) is completely hydrogenated. (7.5, 7.6, 8.6, 17.3, 17.4)
a. Draw the condensed structural formula for the product.
b. How many moles of hydrogen are required?
c. How many grams of hydrogen are required?
d. How many liters of hydrogen gas are needed if the reaction is run at STP?

ANSWERS

17.1 Because lipids are not soluble in water, a polar solvent, they are nonpolar molecules.

17.3 Lipids can store energy, protect and insulate internal organs, and act as chemical messengers. They are also important components of cell membranes.

17.5 All fatty acids contain a long chain of carbon atoms with a carboxylic acid group. Saturated fatty acids contain only carbon–carbon single bonds; unsaturated fatty acids contain one or more double bonds.

17.7 a. palmitic acid

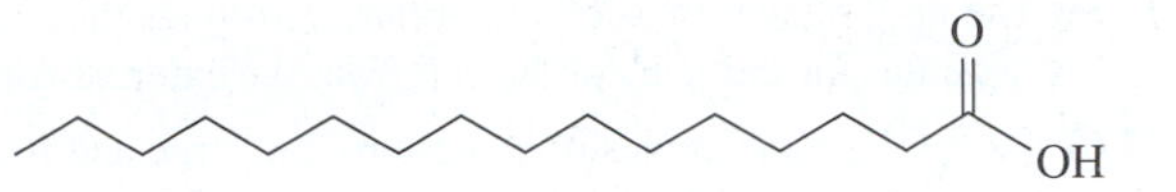

b. oleic acid

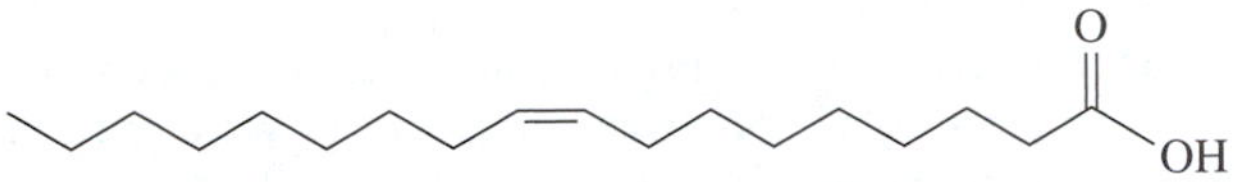

17.9 a. 12:0, saturated b. 18:3, polyunsaturated
c. 16:1, monounsaturated d. 18:0, saturated

17.11 In a cis fatty acid, the hydrogen atoms are on the same side of the double bond, which produces a kink in the carbon chain. In a trans fatty acid, the hydrogen atoms are on opposite sides of the double bond, which gives a carbon chain without any kink.

17.13 In an omega-3 fatty acid, there is a double bond beginning at carbon 3, counting from the methyl group, whereas in an omega-6 fatty acid, there is a double bond beginning at carbon 6, counting from the methyl group.

17.15 Arachidonic acid and PGE_1 are both carboxylic acids with 20 carbon atoms. The differences are that arachidonic acid has four cis double bonds and no other functional groups, whereas PGE_1 has one trans double bond, one ketone functional group, and two hydroxyl functional groups. In addition, a part of the PGE_1 chain forms a cyclopentane ring.

17.17 Prostaglandins raise or lower blood pressure, stimulate contraction and relaxation of smooth muscle, and may cause inflammation and pain.

17.19 $CH_3-(CH_2)_{14}-\overset{\displaystyle O}{\overset{\|}{C}}-O-(CH_2)_{29}-CH_3$

17.21
$$CH_2-O-\overset{O}{\overset{\|}{C}}-(CH_2)_{16}-CH_3$$
$$CH-O-\overset{O}{\overset{\|}{C}}-(CH_2)_{16}-CH_3$$
$$CH_2-O-\overset{O}{\overset{\|}{C}}-(CH_2)_{16}-CH_3$$

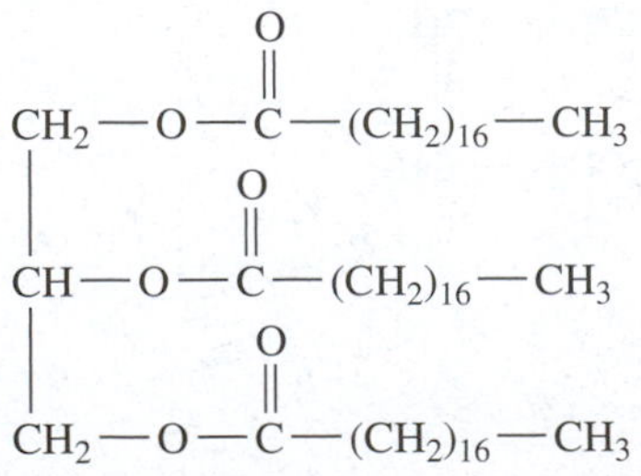

17.23

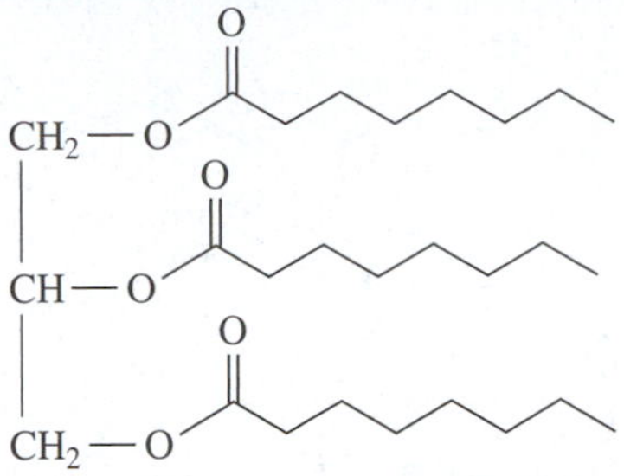

17.25 Safflower oil has a lower melting point because it contains fatty acids with two or three double bonds; olive oil contains a large amount of oleic acid, which has only one double bond (monounsaturated).

17.27 Sunflower oil has about 24% monounsaturated fats, whereas safflower oil has about 18%. Sunflower oil has about 66% polyunsaturated fats, whereas safflower oil has about 73%.

17.29 a. The reaction of palm oil with KOH is saponification; the products are glycerol and the potassium salts of the fatty acids, which are soaps.
b. The reaction of glyceryl trilinoleate from safflower oil with water and HCl is hydrolysis, which splits the ester bonds to produce glycerol and three molecules of linoleic acid.

17.31
$$CH_2-O-\overset{O}{\overset{\|}{C}}-(CH_2)_7-CH=CH-(CH_2)_5-CH_3$$
$$CH-O-\overset{O}{\overset{\|}{C}}-(CH_2)_7-CH=CH-(CH_2)_5-CH_3 + 3H_2$$
$$CH_2-O-\overset{O}{\overset{\|}{C}}-(CH_2)_7-CH=CH-(CH_2)_5-CH_3$$

$\xrightarrow{\text{Ni}}$

$$CH_2-O-\overset{O}{\overset{\|}{C}}-(CH_2)_{14}-CH_3$$
$$CH-O-\overset{O}{\overset{\|}{C}}-(CH_2)_{14}-CH_3$$
$$CH_2-O-\overset{O}{\overset{\|}{C}}-(CH_2)_{14}-CH_3$$

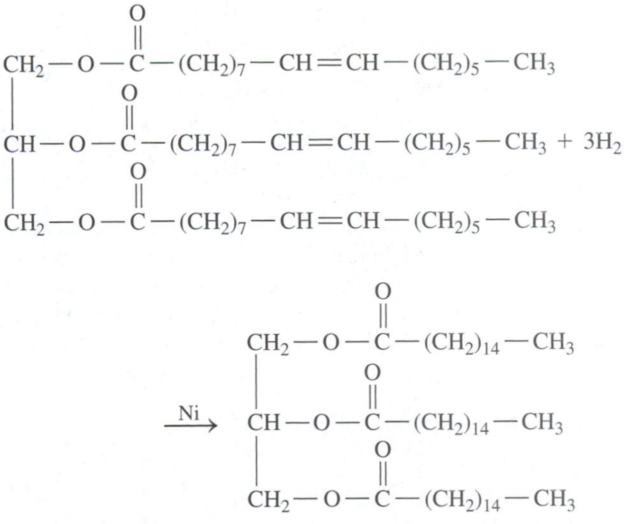

17.33

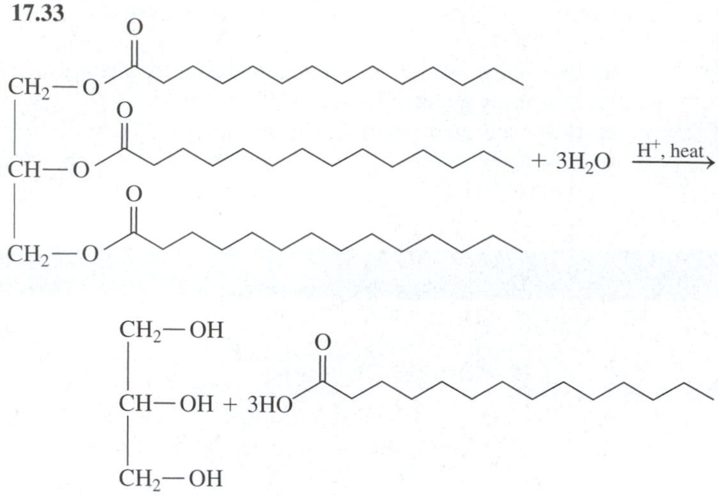

17.35

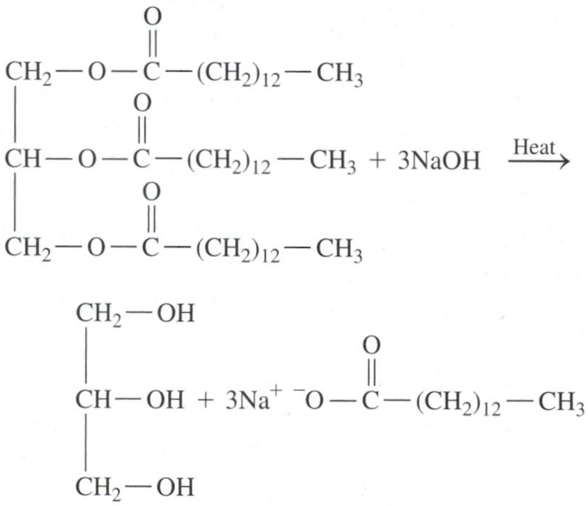

17.37

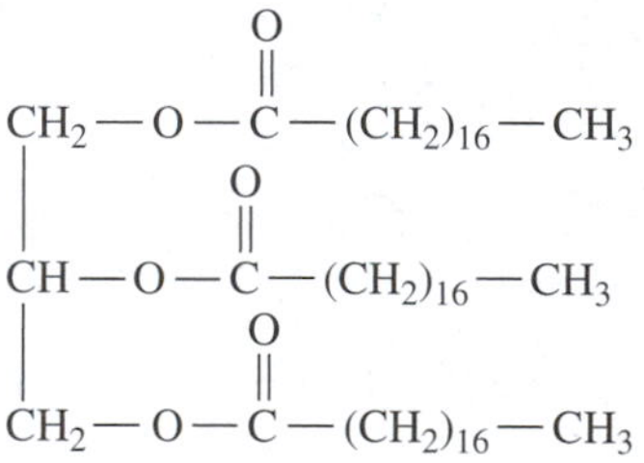

17.39 A triacylglycerol consists of glycerol and three fatty acids. A glycerophospholipid also contains glycerol, but has only two fatty acids. The hydroxyl group on the third carbon of glycerol is attached by a phosphoester bond to an amino alcohol.

17.41

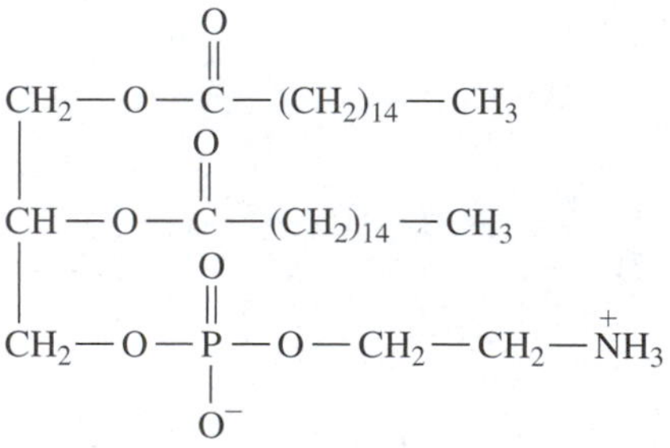

17.43 This glycerophospholipid is a cephalin. It contains glycerol, oleic acid, stearic acid, phosphate, and ethanolamine.

17.45 **a.** sphingosine **b.** palmitic acid **c.** an amide bond **d.** ethanolamine

17.47

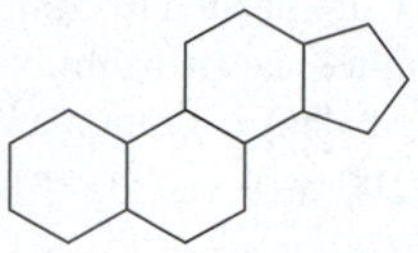

17.49 Bile salts act to emulsify fat globules, allowing the fat to be more easily digested.

17.51 Chylomicrons have a lower density than VLDLs. They pick up triacylglycerols from the intestine, whereas VLDLs transport triacylglycerols synthesized in the liver.

17.53 "Bad" cholesterol is the cholesterol carried by LDL that can form deposits in the arteries called plaque, which narrows the arteries.

17.55 Both estradiol and testosterone contain the steroid nucleus and a hydroxyl group. Testosterone has a ketone group, a double bond, and two methyl groups. Estradiol has an aromatic ring, a hydroxyl group in place of the ketone, and a methyl group.

17.57 **b**, **c**, and **d**

17.59 The lipid bilayer in a cell membrane surrounds the cell and separates the contents of the cell from the external fluids.

17.61 Because the molecules of cholesterol are large and rigid, they reduce the flexibility of the lipid bilayer and add strength to the cell membrane.

17.63 The peripheral proteins in the membrane emerge on the inner or outer surface only, whereas the integral proteins extend through the membrane to both surfaces.

17.65 Substances move through cell membranes by diffusion, facilitated diffusion, and active transport.
a. facilitated diffusion **b.** diffusion

17.67 alkene, alcohol, ester, carboxylic acid

17.69 **a.** 0.56 g of Pravachol **b.** 360 mg/dL

17.71

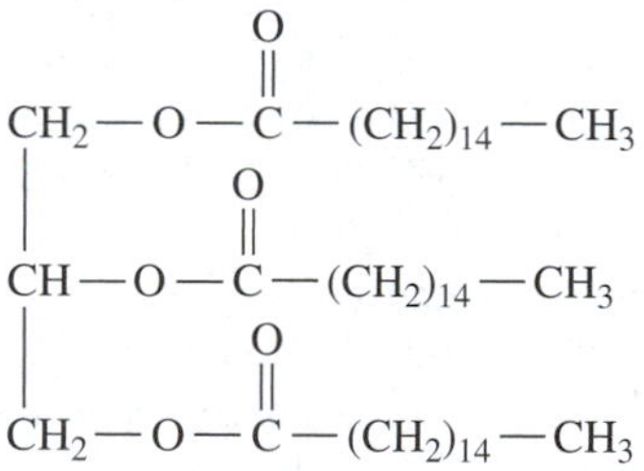

17.73 **a.** monounsaturated **b.** polyunsaturated, omega-6 **c.** polyunsaturated, omega-3

17.75 **a.** Beeswax and carnauba are waxes. Vegetable oil and glyceryl tricaprate (tricaprin) are triacylglycerols.

b.

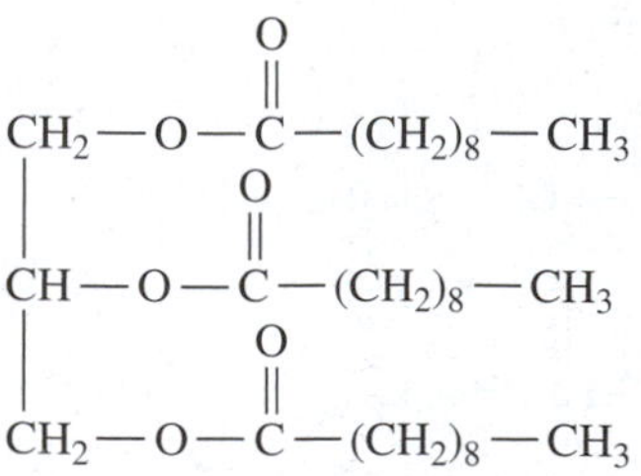

Glyceryl tricaprate (tricaprin)

17.77 a. 410 kcal from fat; 49% fat
  b. 260 kcal from fat; 50.% fat
  c. 160 kcal from fat; 29% fat

17.79 a. wax   **b.** steroid
  c. glycerophospholipid   **d.** triacylglycerol
  e. soap   **f.** triacylglycerol

17.81 a. 5   **b.** 1, 2, 3, 4
  c. 2   **d.** 1, 2

17.83 a and c

17.85

$$CH_2-O-\overset{\overset{\displaystyle O}{\|}}{C}-(CH_2)_{16}-CH_3$$
$$CH-O-\overset{\overset{\displaystyle O}{\|}}{C}-(CH_2)_{16}-CH_3$$
$$CH_2-O-\overset{\overset{\displaystyle O}{\|}}{P}-O-CH_2-CH_2-\overset{+}{N}H_3$$
$$\underset{O^-}{|}$$

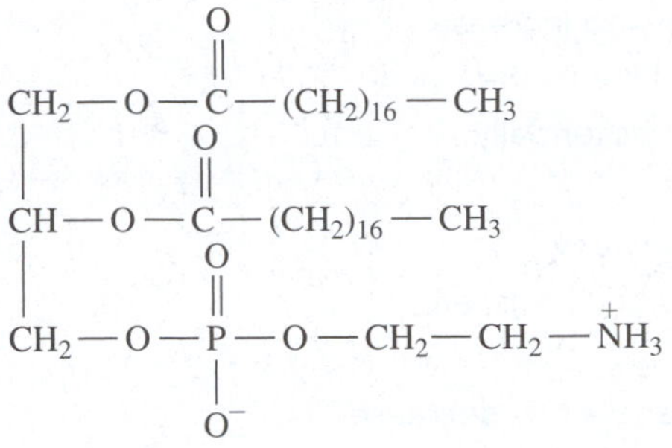

17.87 a. (4) HDL   **b.** (3) LDL
  c. (1) chylomicrons   **d.** (4) HDL

17.89 a. Adding NaOH would saponify lipids such as glyceryl tristearate (tristearin), forming glycerol and salts of the fatty acids that are soluble in water and would wash down the drain.

b.

$$CH_2-O-\overset{\overset{\displaystyle O}{\|}}{C}-(CH_2)_{16}-CH_3$$
$$CH-O-\overset{\overset{\displaystyle O}{\|}}{C}-(CH_2)_{16}-CH_3 + 3NaOH \xrightarrow{\text{Heat}}$$
$$CH_2-O-\overset{\overset{\displaystyle O}{\|}}{C}-(CH_2)_{16}-CH_3$$

$$CH_2-OH$$
$$CH-OH + 3Na^+ \ {}^-O-\overset{\overset{\displaystyle O}{\|}}{C}-(CH_2)_{16}-CH_3$$
$$CH_2-OH$$

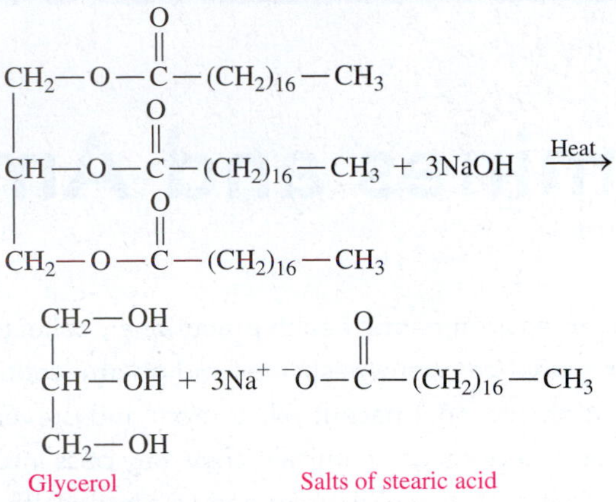

Glycerol   Salts of stearic acid

c. 67.3 mL of a 0.500 M NaOH solution

17.91 a.

$$CH_2-O-\overset{\overset{\displaystyle O}{\|}}{C}-(CH_2)_{16}-CH_3$$
$$CH-O-\overset{\overset{\displaystyle O}{\|}}{C}-(CH_2)_{16}-CH_3$$
$$CH_2-O-\overset{\overset{\displaystyle O}{\|}}{C}-(CH_2)_{16}-CH_3$$

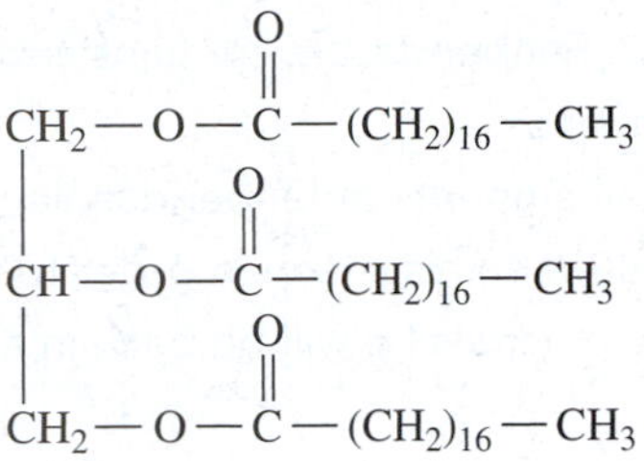

b. 3.00 moles of H_2
c. 6.05 g of H_2
d. 67.2 L of H_2

18 Amines and Amides

Lance, an environmental health practitioner, collects soil and water samples at a ranch to test for the presence of any pesticides and pharmaceuticals. Ranchers use pesticides to increase production and pharmaceuticals to treat and prevent animal-related diseases. Due to the common use of these chemicals, they may pass into the soil and water supply, potentially contaminating the environment and causing health problems.

Recently, the sheep on the ranch were treated with fenbendazole to destroy any gastrointestinal worms. Fenbendazole contains several functional groups: aromatic rings, an ester, an amide, and amines.

Lance detects small amounts of fenbendazole in the soil. He advises the rancher to decrease the dosage he administers to his sheep in order to reduce the amounts currently being detected in the soil. Lance then indicates he will be back in a month to retest the soil and water.

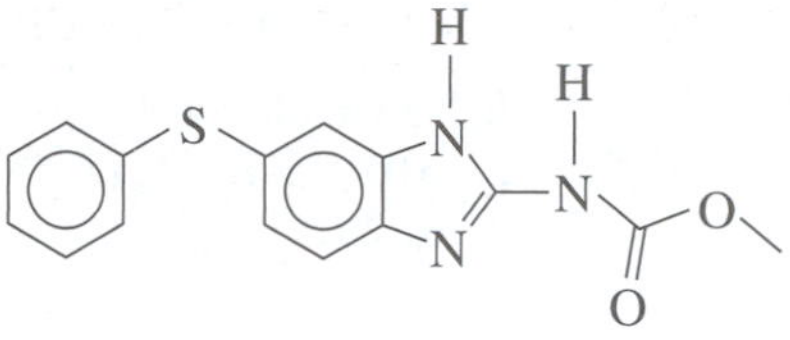

Fenbendazole

CAREER

Environmental Health Practitioner

Environmental health practitioners (EHPs) monitor environmental pollution to protect the public's health. By using specialized equipment, EHPs measure pollution levels in soil, air, and water, as well as noise and radiation levels. EHPs can specialize in a specific area, such as air quality or hazardous and solid waste. For instance, air quality experts monitor indoor air for allergens, mold, and toxins; they measure outdoor air pollutants created by businesses, vehicles, and agriculture. Because EHPs obtain samples with potentially hazardous materials, they must be knowledgeable about safety protocols and wear personal protective equipment. EHPs also recommend methods to diminish various pollutants and may assist in clean-up and remediation efforts.

CLINICAL UPDATE

Testing Soil and Water Samples for Chemicals

When Lance returns to the ranch, the owners are using insecticide spray to treat an infestation of flies and maggots. You can see the chemicals used in the **CLINICAL UPDATE Testing Soil and Water Samples for Chemicals**, page 650, and see the structures of these insecticides and how Lance will analyze their levels in the environment.

18.1 Amines

LEARNING GOAL Write the IUPAC and common names for amines; draw the condensed structural and line-angle formulas when given their names. Classify amines as primary (1°), secondary (2°), or tertiary (3°).

LOOKING AHEAD

REVIEW

Naming and Drawing Alkanes (12.2)

Amines and amides are organic compounds that contain nitrogen. Many nitrogen-containing compounds are important to life as components of amino acids, proteins, and nucleic acids (DNA and RNA). Many amines exhibit strong physiological activity and are used in medicine as decongestants, anesthetics, and sedatives. Examples include dopamine, histamine, epinephrine, and amphetamine. Alkaloids such as caffeine, nicotine, cocaine, and digitalis, which have powerful physiological activity, are naturally occurring amines obtained from plants.

Amines are derivatives of ammonia (NH_3) in which the nitrogen atom, which has one lone pair of electrons, has three bonds to hydrogen atoms. In an amine, the nitrogen atom is bonded to one, two, or three alkyl or aromatic groups.

Naming Amines

In the IUPAC names for amines, the *e* in the corresponding alkane name is replaced with *amine*. For the common names of amines, the alkyl or aromatic groups bonded to the nitrogen are listed in alphabetical order followed by *amine*. The prefixes *di* and *tri* are used to indicate two and three identical substituents.

CH_4	$CH_3—NH_2$	$CH_3—CH_3$	$CH_3—CH_2—NH_2$
Methane	Methanamine (methylamine)	Ethane	Ethanamine (ethylamine)

In the IUPAC name, a chain of three or more carbon atoms is numbered to show the position of the $—NH_2$ group and any other substituents.

$$\underset{3}{CH_3}—\underset{2}{CH_2}—\underset{1}{CH_2}—NH_2$$

1-Propan**amine**
(propylamine)

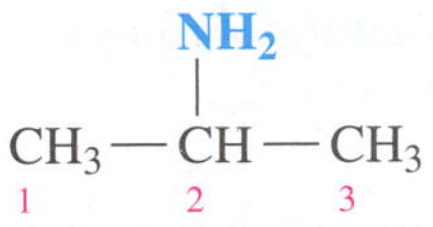

2-Propan**amine**
(isopropylamine)

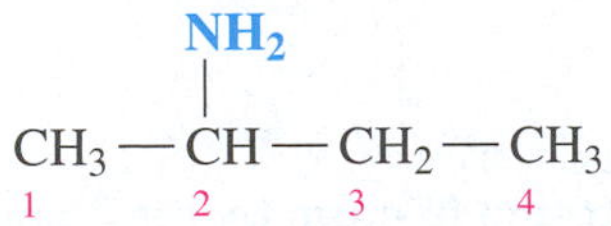

2-Butan**amine**

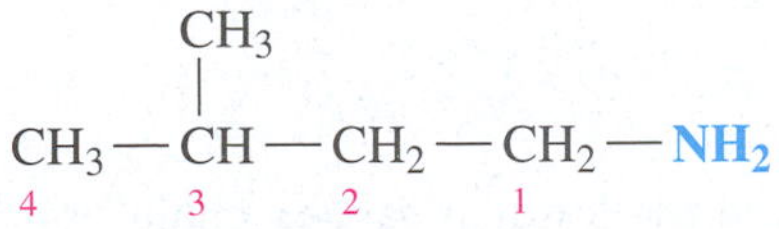

3-Methyl-1-butan**amine**

When an alkyl group is attached to the nitrogen atom, the prefix *N*- and the alkyl name are placed in front of the amine name. If there are two alkyl groups bonded to the N atom, the prefix *N*- is used for each, and they are listed alphabetically.

ENGAGE

When is the prefix *N*- used in naming amines?

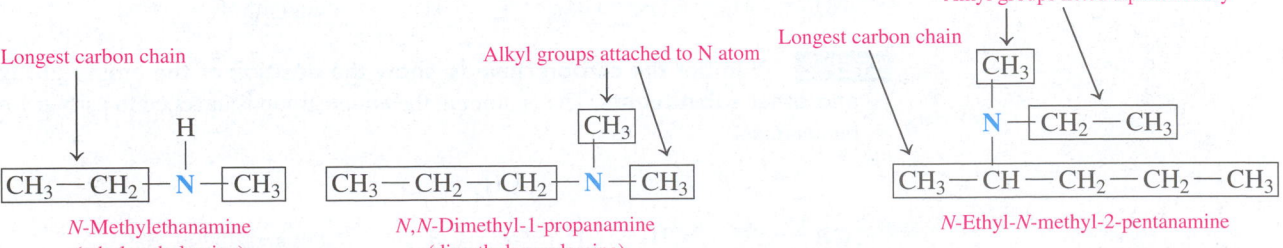

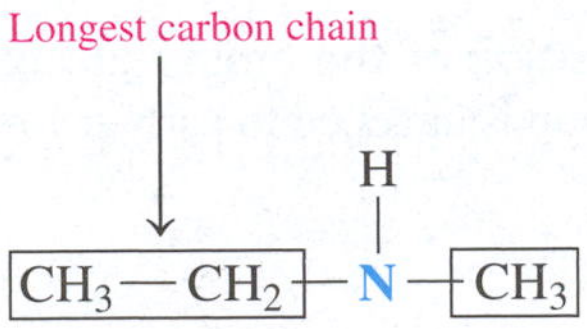
Longest carbon chain

N-Methylethanamine
(ethylmethylamine)

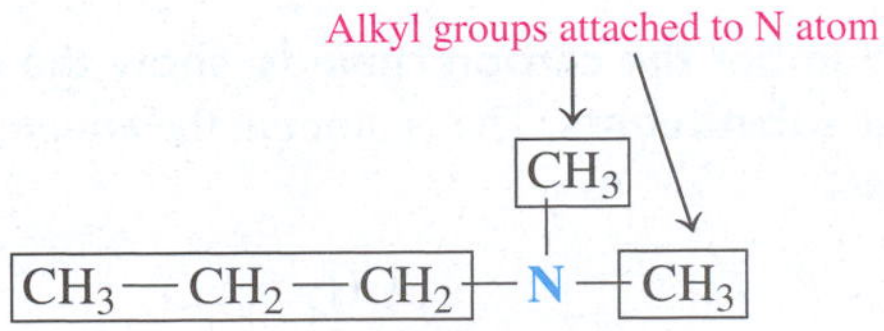
Alkyl groups attached to N atom

N,*N*-Dimethyl-1-propanamine
(dimethylpropylamine)

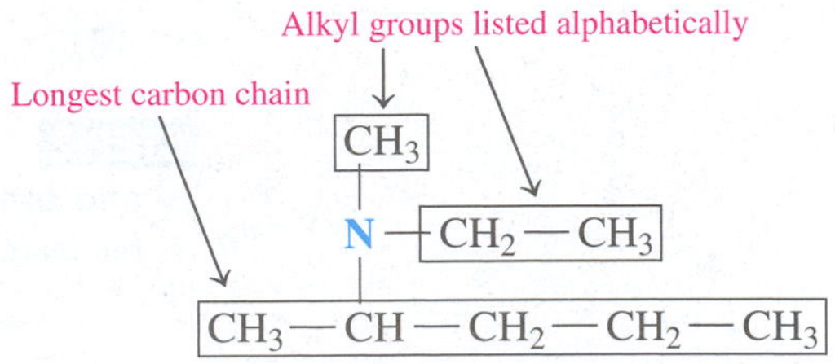
Alkyl groups listed alphabetically
Longest carbon chain

N-Ethyl-*N*-methyl-2-pentanamine

Aromatic Amines

The aromatic amines use the name *aniline*, which is approved by IUPAC. Aniline is the simplest aromatic amine; it is used to make many industrial chemicals.

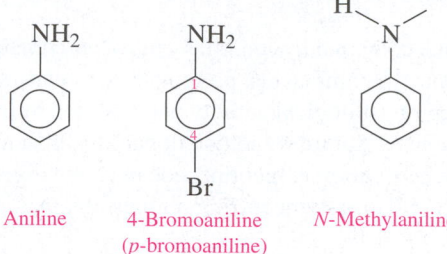

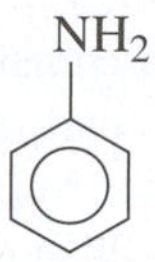

Aniline

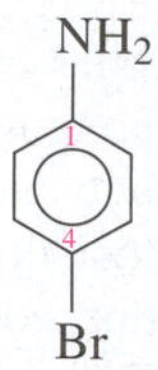

4-Bromoaniline
(*p*-bromoaniline)

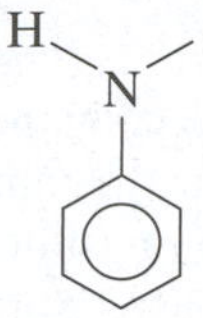

N-Methylaniline

TEST

Try Practice Problems 18.1 to 18.4

Aniline, which was discovered in 1826 when it was first isolated from indigo plants, is used to make many dyes, which give color to wool, cotton, and silk fibers, as well as blue jeans. It is also used to make the polymer polyurethane and in the synthesis of the pain reliever acetaminophen.

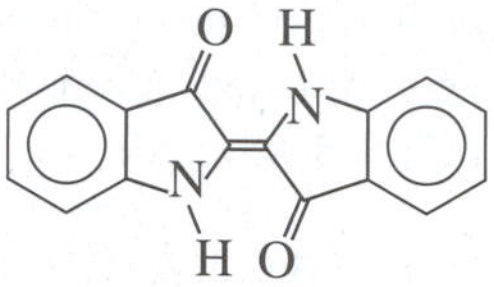

Indigo

Indigo used in blue dyes can be obtained from tropical plants such as *Indigofera tinctoria*.

▶SAMPLE PROBLEM 18.1 Naming Amines

TRY IT FIRST

Write the IUPAC and common names for the following amine:

$$CH_3-CH_2-CH_2-CH_2-\underset{}{\overset{\displaystyle CH_3}{\overset{|}{N}}}-CH_3$$

SOLUTION

STEP 1 **Name the longest carbon chain bonded to the N atom by replacing the *e* of its alkane name with *amine*.** The longest carbon chain bonded to the N atom has four carbon atoms, which is named by replacing the *e* in the alkane name with *amine* to give butanamine.

$$CH_3-CH_2-CH_2-CH_2-\overset{\displaystyle CH_3}{\overset{|}{N}}-CH_3 \qquad \text{butanamine}$$

STEP 2 **Number the carbon chain to show the position of the amine group and other substituents.** The N atom in the amine group is attached to carbon 1 of butanamine.

$$\underset{4}{CH_3}-\underset{3}{CH_2}-\underset{2}{CH_2}-\underset{1}{CH_2}-\overset{\displaystyle CH_3}{\overset{|}{N}}-CH_3 \qquad \text{1-butanamine}$$

STEP 3 **Any alkyl group attached to the nitrogen atom is indicated by the prefix *N*- and the alkyl name, which is placed in front of the amine name.** Alkyl groups attached to the N atom are listed alphabetically.

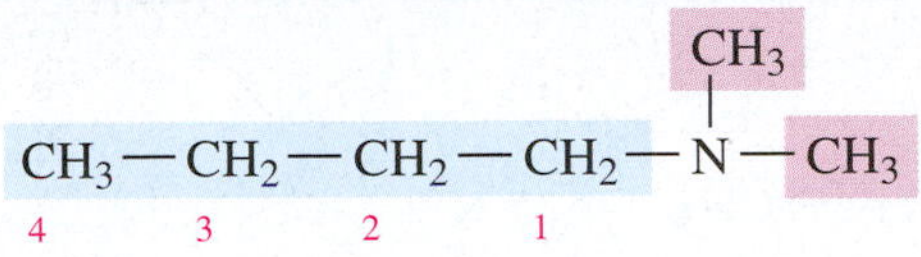

N,N-dimethyl-1-butanamine

In the common name, the alkyl groups, butyl and methyl, bonded to the nitrogen atom are listed in alphabetical order.

butyldimethylamine

STUDY CHECK 18.1

a. Draw the condensed structural formula for *N*-ethyl-1-propanamine.
b. Write the common name for the structure in part **a**.

ANSWER

a. 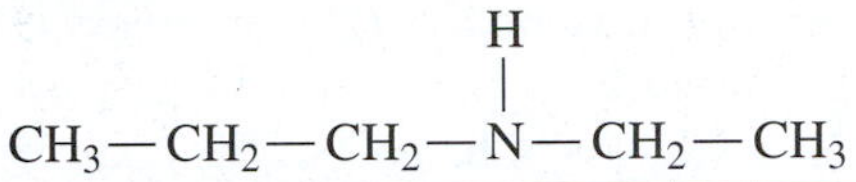

b. ethylpropylamine

Classification of Amines

Amines are classified by counting the number of carbon atoms directly bonded to the nitrogen atom. In a primary (1°) amine, the nitrogen atom is bonded to one alkyl group. In a secondary (2°) amine, the nitrogen atom is bonded to two alkyl groups. In a tertiary (3°) amine, the nitrogen atom is bonded to three alkyl groups. In each of the following models of ammonia and amines, the atoms are arranged around the nitrogen atom (blue) in a trigonal pyramidal shape:

Ammonia | **Primary (1°) amine** | **Secondary (2°) amine** | **Tertiary (3°) amine**

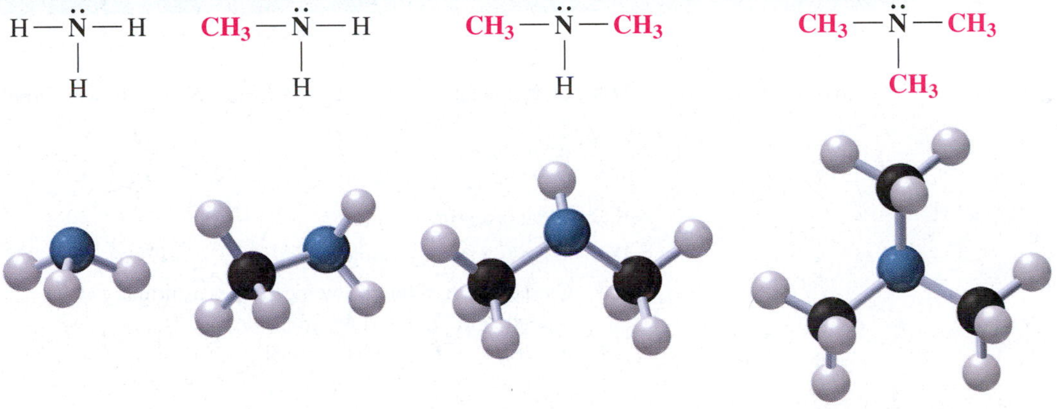

Ammonia | Methanamine (methylamine) | *N*-Methylmethanamine (dimethylamine) | *N,N*-Dimethylmethanamine (trimethylamine)

Line-Angle Formulas for Amines

We can draw line-angle formulas for amines just as we have for other organic compounds. In the line-angle formula for an amine, we show the hydrogen atoms bonded to the N atom. For example, we can draw the following line-angle formulas and classify each:

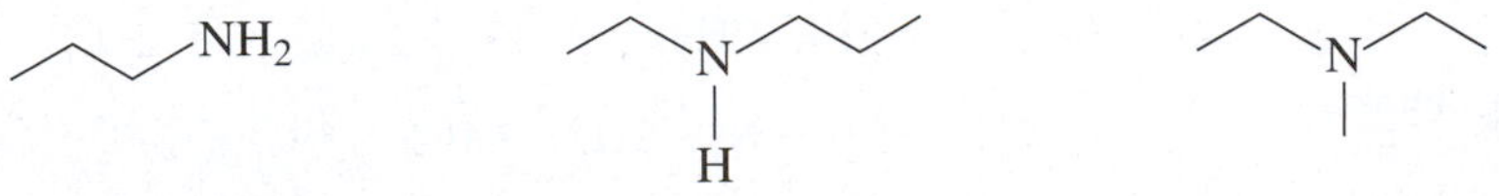

1-Propanamine (propylamine) Primary (1°) amine | *N*-Ethyl-1-propanamine (ethylpropylamine) Secondary (2°) amine | *N*-Ethyl-*N*-methylethanamine (diethylmethylamine) Tertiary (3°) amine

▸SAMPLE PROBLEM 18.2 Classifying Amines

TRY IT FIRST

Classify each of the following amines as primary (1°), secondary (2°), or tertiary (3°):

a. 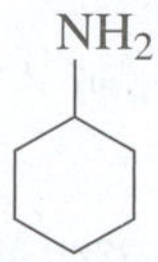

b. $CH_3-N(CH_3)-CH_2-CH_3$

c. 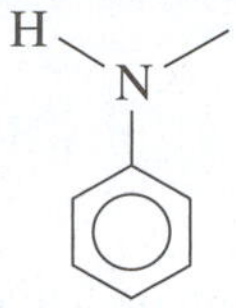

SOLUTION

a. This is a primary (1°) amine because there is one carbon atom bonded to the nitrogen atom.

b. This is a tertiary (3°) amine. There are three carbon atoms bonded to the nitrogen atom.

c. This is a secondary (2°) amine with two carbon atoms bonded to the nitrogen atom.

STUDY CHECK 18.2

Classify each of the following amines as primary (1°), secondary (2°), or tertiary (3°):

a. $CH_3-CH_2-N(CH_3)-CH_2-CH_2-CH_3$

b. 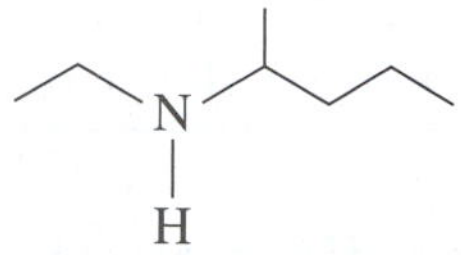

ANSWER

a. tertiary (3°) **b.** secondary (2°)

TEST

Try Practice Problems 18.5 to 18.8

PRACTICE PROBLEMS

18.1 Amines

18.1 Write the IUPAC and common names for each of the following:

a. $CH_3-CH_2-NH_2$

b. $CH_3-N(H)-CH_2-CH_2-CH_3$

c. 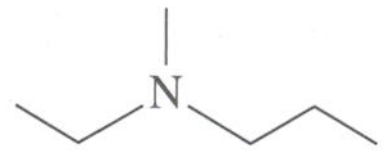

d. $CH_3-CH(NH_2)-CH_3$

18.2 Write the IUPAC and common names for each of the following:

a. $CH_3-CH_2-CH_2-NH_2$

b. $CH_3-N(H)-CH_2-CH_3$

c. 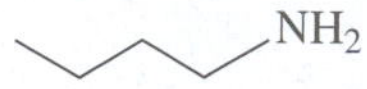

d. $CH_3-CH_2-N(CH_2-CH_3)-CH_2-CH_3$

18.3 Draw the line-angle formula for each of the following amines:

a. 2-chloroethanamine

b. *N*-methylaniline

c. butylpropylamine

18.4 Draw the line-angle formula for each of the following amines:

a. dimethylamine

b. *p*-chloroaniline

c. *N*,*N*-diethylaniline

18.5 What is a primary amine?

18.6 What is a tertiary amine?

18.7 Classify each of the following amines as primary (1°), secondary (2°), or tertiary (3°):

a. $CH_3-CH_2-CH_2-NH_2$

b. $CH_3-N(H)-CH_2-CH_3$

c. 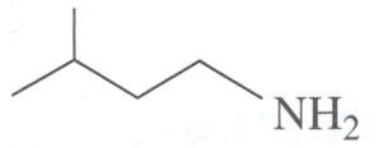

d. 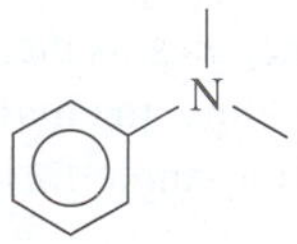

e. $CH_3-CH(CH_3)-N(CH_3)-CH_2-CH_3$

18.8 Classify each of the following amines as primary (1°), secondary (2°), or tertiary (3°):

a. $CH_3-CH_2-CH(NH_2)-CH_3$

b. $CH_3-CH_2-N(CH_2-CH_2-CH_3)-CH_2-CH_3$

c. 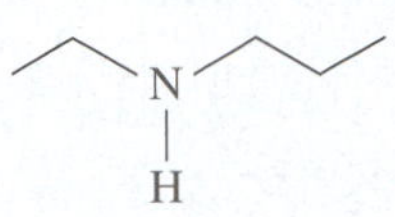

d. 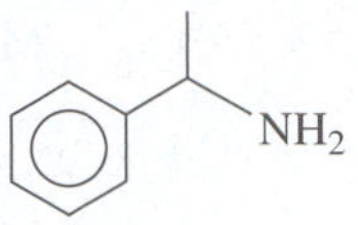

e. $CH_3-NH-C(CH_3)_2-CH_3$

18.2 Properties of Amines

REVIEW
Writing Equations for Reactions of Acids and Bases (11.7)

LEARNING GOAL Describe the boiling points and solubility of amines; write balanced chemical equations for the reaction of amines with water and with acids.

Amines contain polar N—H bonds, which allow primary and secondary amines to form hydrogen bonds with each other, while all amines can form hydrogen bonds with water. However, nitrogen is not as electronegative as oxygen, which means that the hydrogen bonds in amines are weaker than the hydrogen bonds in alcohols.

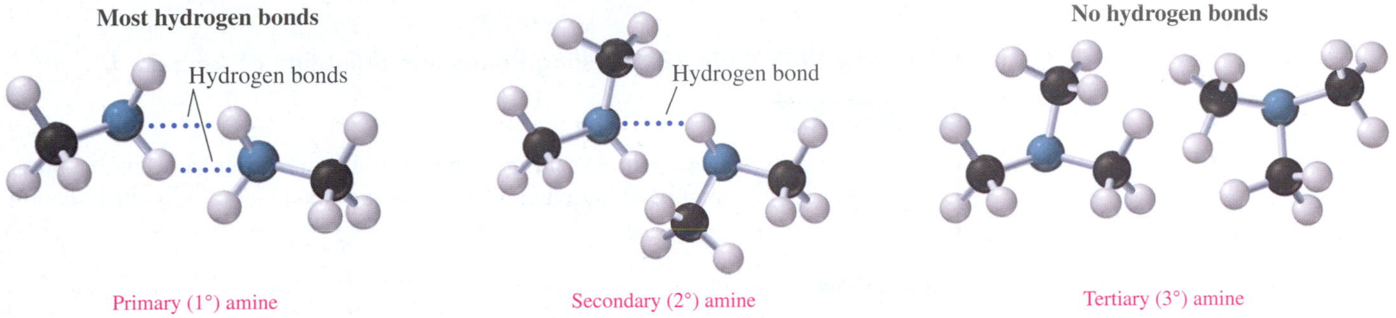
Most hydrogen bonds
Hydrogen bonds
Hydrogen bond
No hydrogen bonds
Primary (1°) amine
Secondary (2°) amine
Tertiary (3°) amine

Primary and secondary amines form hydrogen bonds that produce higher boiling points than tertiary amines that cannot form hydrogen bonds.

Boiling Points of Amines

Amines have boiling points that are higher than alkanes but lower than alcohols. Primary (1°) amines, with two N—H bonds, can form more hydrogen bonds, and thus have higher boiling points than secondary (2°) amines of the same mass. It is not possible for tertiary (3°) amines to hydrogen bond with each other because they have no N—H bonds. Tertiary amines have lower boiling points than primary or secondary amines of the same mass.

$CH_3-CH_2-CH_2-NH_2$ Propylamine (1°) bp 48 °C

$CH_3-CH_2-NH-CH_3$ Ethylmethylamine (2°) bp 37 °C

$CH_3-N(CH_3)-CH_3$ Trimethylamine (3°) bp 3 °C

Boiling Point Decreases →

TEST
Try Practice Problems 18.9 to 18.12

Solubility of Amines in Water

Because amines contain a polar N—H bond, they form hydrogen bonds with water. In primary (1°) amines, $-NH_2$ can form more hydrogen bonds than the secondary (2°) amines. A tertiary (3°) amine, which has no hydrogen on the nitrogen atom, can form only hydrogen bonds with water from the N atom in the amine to the H atom of a water molecule. Like alcohols, the smaller amines, including tertiary ones, are soluble because they form hydrogen bonds with water. However, in amines with more than six carbon atoms the effect of hydrogen bonding is diminished. Then the nonpolar hydrocarbon chains of the amine decreases its solubility in water (see **FIGURE 18.1**).

ENGAGE
Why can methylamine form more hydrogen bonds than dimethylamine?

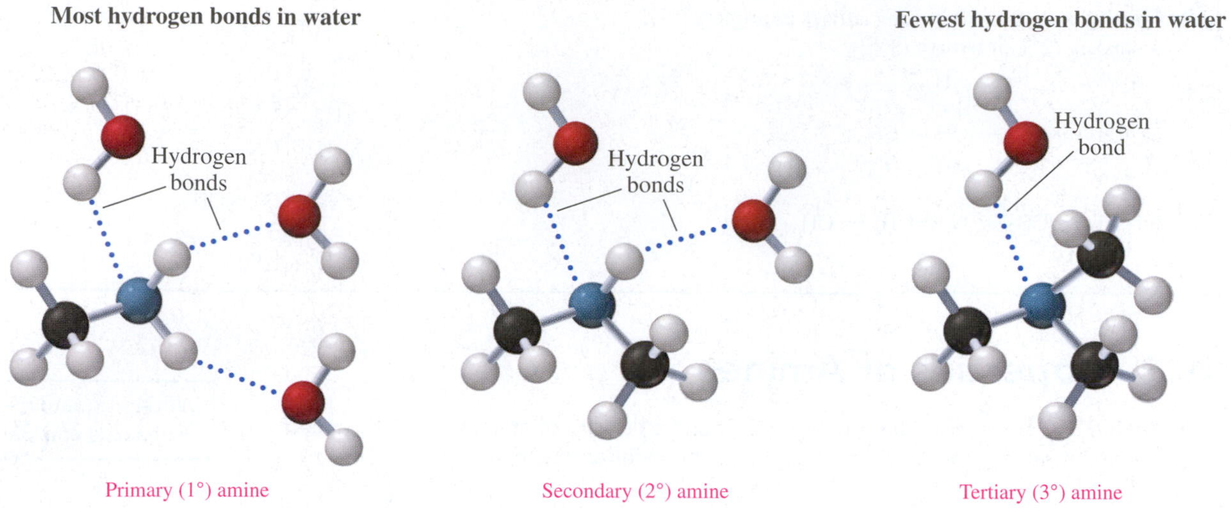
Most hydrogen bonds in water
Fewest hydrogen bonds in water
Hydrogen bonds
Hydrogen bonds
Hydrogen bond
Primary (1°) amine
Secondary (2°) amine
Tertiary (3°) amine

FIGURE 18.1 ▶ Primary, secondary, and tertiary amines form hydrogen bonds with water molecules, but primary amines form the most and tertiary amines form the least.

Q Why do tertiary (3°) amines form fewer hydrogen bonds with water than primary amines?

▶ SAMPLE PROBLEM 18.3 Boiling Points and Solubility of Amines

TRY IT FIRST

The compounds trimethylamine and ethylmethylamine have the same molar mass. Why is the boiling point of trimethylamine (3 °C) lower than that of ethylmethylamine (37 °C)?

SOLUTION

With polar N—H bonds, ethylmethylamine molecules form hydrogen bonds with each other. Trimethylamine, which is a tertiary amine, does not have N—H bonds and cannot hydrogen bond with other trimethylamine molecules. Thus, a higher temperature is required to break the hydrogen bonds in ethylmethylamine than in trimethylamine.

STUDY CHECK 18.3

Why is diethylamine more soluble in water than butylpropylamine?

ANSWER

Longer nonpolar hydrocarbon chains diminish the solubility effect of the polar amine group. Diethylamine, a secondary amine, has four carbon atoms, which makes it more soluble than butylpropylamine, a secondary amine that has seven carbon atoms.

TEST

Try Practice Problems 18.13 and 18.14

Amines React as Bases in Water

Ammonia (NH_3) acts as a Brønsted–Lowry base by accepting H^+ from water to produce an ammonium ion (NH_4^+) and a hydroxide ion (OH^-).

$$\ddot{N}H_3 + H_2O \rightleftharpoons NH_4^+ + OH^-$$

Ammonia; Ammonium ion; Hydroxide ion

Amines are also Brønsted–Lowry bases because the lone pair of electrons on the nitrogen atom accepts H^+ from water. In the reaction of an amine with water, the products are a positively charged alkylammonium ion and a negatively charged hydroxide ion. The organic product is named by adding *ammonium ion* to the name of its alkyl group.

Reaction of a Primary Amine with Water

$$CH_3-\ddot{N}H_2 + H_2O \rightleftarrows CH_3-\overset{+}{N}H_3 + OH^-$$

Methylamine — Methylammonium ion — Hydroxide ion

Reaction of a Secondary Amine with Water

$$CH_3-\underset{\displaystyle CH_3}{\underset{|}{\ddot{N}H}} + H_2O \rightleftarrows CH_3-\underset{\displaystyle CH_3}{\underset{|}{\overset{+}{N}H_2}} + OH^-$$

Dimethylamine — Dimethylammonium ion — Hydroxide ion

Reaction of a Tertiary Amine with Water

$$CH_3-\underset{\displaystyle CH_3}{\underset{|}{\ddot{N}}}-CH_3 + H_2O \rightleftarrows CH_3-\underset{\displaystyle CH_3}{\underset{|}{\overset{+}{N}H}}-CH_3 + OH^-$$

Trimethylamine — Trimethylammonium ion — Hydroxide ion

ENGAGE

Is methylamine a Brønsted–Lowry acid or base?

TEST

Try Practice Problems 18.15 and 18.16

Neutralization of an Amine

When you squeeze lemon juice on fish, the "fishy" odor is removed by converting the amines to their ammonium salts. In a *neutralization reaction*, an amine acts as a base and reacts with an acid to form an **ammonium salt**. The lone pair of electrons on the nitrogen atom accepts H^+ from an acid to give an ammonium salt; no water is formed. An ammonium salt is named by using its alkylammonium ion name followed by the name of the negative ion.

The amines in fish react with the acid in lemon to neutralize the "fishy" odor.

Amine **Acid** **Ammonium Salt**

$$CH_3-\ddot{N}H_2 + HCl \longrightarrow CH_3-\overset{+}{N}H_3\ Cl^-$$

Methylamine — Methylammonium chloride

$$CH_3-\underset{\displaystyle CH_3}{\underset{|}{\ddot{N}H}} + HCl \longrightarrow CH_3-\underset{\displaystyle CH_3}{\underset{|}{\overset{+}{N}H_2}}\ Cl^-$$

Dimethylamine — Dimethylammonium chloride

In a **quaternary ammonium salt**, a nitrogen atom is bonded to four carbon groups, which classifies it as a quaternary (4°) amine. As in other ammonium salts, the nitrogen atom has a positive charge. Choline, a component of glycerophospholipids, is a quaternary ammonium ion. The quaternary salts differ from other ammonium salts because the nitrogen atom is not bonded to an H atom.

$$CH_3-\underset{\displaystyle CH_3}{\underset{|}{\overset{\displaystyle CH_3}{\overset{|}{N^+}}}}-CH_3\ Cl^- \qquad HO-CH_2-CH_2-\underset{\displaystyle CH_3}{\underset{|}{\overset{\displaystyle CH_3}{\overset{|}{N^+}}}}-CH_3$$

Tetramethylammonium chloride — Choline

Properties of Ammonium Salts

Ammonium salts are ionic compounds with strong attractions between the positively charged ammonium ion and an anion, usually chloride. Like most salts, ammonium salts are solid at room temperature, odorless, and soluble in water and body fluids. For this reason, amines used as drugs are usually converted to their ammonium salts, which are soluble in

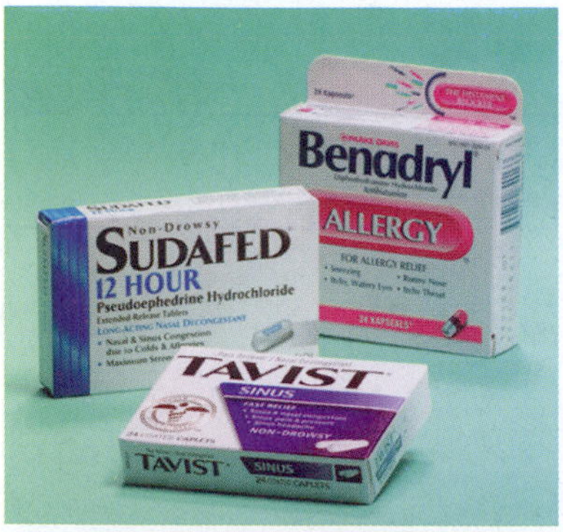
Benadryl
ALLERGY
Non-Drowsy
SUDAFED
12 HOUR
Pseudoephedrine Hydrochloride
TAVIST
SINUS
TAVIST SINUS

FIGURE 18.2 ▶ Decongestants and products that relieve itchy skin often contain ammonium salts.

Q Why are ammonium salts used in drugs rather than the biologically active amines?

water and body fluids. The ammonium salt of ephedrine is used as a bronchodilator and in decongestant products such as Sudafed. The ammonium salt of diphenhydramine is used in products such as Benadryl for relief of itching and pain from skin irritations and rashes (see **FIGURE 18.2**). In pharmaceuticals, the naming of the ammonium salt follows an older method of giving the amine name followed by the name of the acid.

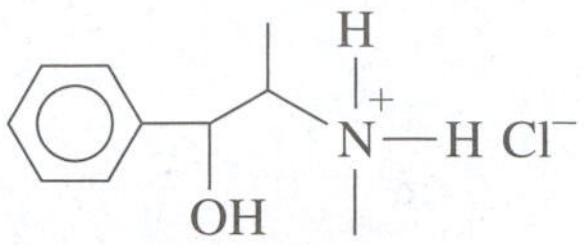

Ephedrine hydrochloride
Ephedrine HCl
Sudafed

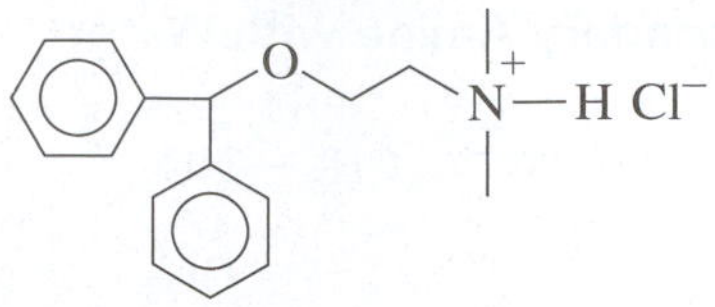

Diphenhydramine hydrochloride
Diphenylhydramine HCl
Benadryl

When an ammonium salt reacts with a strong base such as NaOH, it is converted back to the amine, which is also called the free amine or free base.

$$CH_3{-}\overset{+}{N}H_3\ Cl^- + NaOH \longrightarrow CH_3{-}NH_2 + NaCl + H_2O$$

Coca leaves are a source of cocaine.

Cocaine is typically extracted from coca leaves using an acidic HCl solution to give a white, solid ammonium salt, which is cocaine hydrochloride. It is the salt of cocaine (cocaine hydrochloride) that is smuggled and used illegally on the street. "Crack cocaine" is the free amine or free base of the amine obtained by treating the cocaine hydrochloride with NaOH and ether, a process known as "free-basing." The solid product is known as "crack cocaine" because it makes a cracking noise when heated. The free amine is rapidly absorbed when smoked and gives stronger highs than the cocaine hydrochloride, which makes crack cocaine more addictive.

Crack cocaine is obtained by treating cocaine hydrochloride with NaOH and ether.

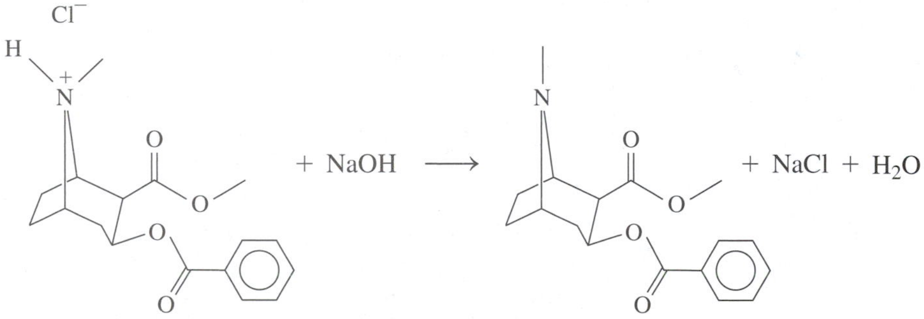

Cocaine hydrochloride　　Cocaine ("free base"; crack cocaine)

▶ SAMPLE PROBLEM 18.4 Reactions of Amines

TRY IT FIRST

Write the balanced chemical equation that shows ethylamine

a. acting as a weak base in water

b. neutralized by HCl

SOLUTION

a. In water, ethylamine acts as a weak base by accepting a hydrogen ion from water to produce ethylammonium and hydroxide ions.

$$CH_3{-}CH_2{-}NH_2 + H_2O \rightleftharpoons CH_3{-}CH_2{-}\overset{+}{N}H_3 + OH^-$$

b. In a reaction with an acid, ethylamine acts as a weak base by accepting a hydrogen ion from HCl to produce ethylammonium chloride.

$$CH_3{-}CH_2{-}NH_2 + HCl \longrightarrow CH_3{-}CH_2{-}\overset{+}{N}H_3\ Cl^-$$

INTERACTIVE VIDEO

Reactions of Amines　PEARSON eText 2.0

STUDY CHECK 18.4

a. Draw the condensed structural formula for the ammonium salt formed by the reaction of trimethylamine and HCl.

b. Draw the condensed structural formula for the products formed by the reaction of diethylamine and water.

ANSWER

a. $CH_3-\overset{CH_3}{\underset{CH_3}{N^+}}-H\ Cl^-$ **b.** $CH_3-CH_2-\overset{H}{\underset{H}{N^+}}-CH_2-CH_3 + OH^-$

TEST

Try Practice Problems 18.17 to 18.20

PRACTICE PROBLEMS

18.2 Properties of Amines

18.9 Identify the compound in each pair that has the higher boiling point. Explain.

a. $CH_3-CH_2-NH_2$ or CH_3-CH_2-OH

b. CH_3-NH_2 or $CH_3-CH_2-CH_2-NH_2$

c. $CH_3-\overset{CH_3}{N}-CH_3$ or $CH_3-CH_2-CH_2-NH_2$

18.10 Identify the compound in each pair that has the higher boiling point. Explain.

a. $CH_3-CH_2-CH_2-CH_3$ or $CH_3-CH_2-CH_2-NH_2$

b. CH_3-NH_2 or $CH_3-CH_2-NH_2$

c. $CH_3-CH_2-CH_2-OH$ or $CH_3-\overset{NH_2}{CH}-CH_3$

18.11 Propylamine (59 g/mole) has a boiling point of 48 °C, and ethylmethylamine (59 g/mole) has a boiling point of 37 °C. Butane (58 g/mole) has a much lower boiling point of −1 °C. Explain.

18.12 Assign the boiling point of 3 °C, 48 °C, or 97 °C to the appropriate compound: 1-propanol, propylamine, and trimethylamine.

18.13 Indicate if each of the following is soluble in water. Explain.

a. $CH_3-CH_2-NH_2$ **b.** $CH_3-\overset{H}{N}-CH_3$

c. $CH_3-CH_2-CH_2-\overset{CH_2-CH_2-CH_3}{N}-CH_2-CH_2-CH_3$

d. $CH_3-\overset{NH_2}{CH}-CH_2-CH_3$

18.14 Indicate if each of the following is soluble in water. Explain.

a. $CH_3-CH_2-CH_2-NH_2$

b. $CH_3-CH_2-CH_2-\overset{H}{N}-CH_2-CH_3$

c. $CH_3-\overset{NH_2}{CH}-CH_3$ **d.**

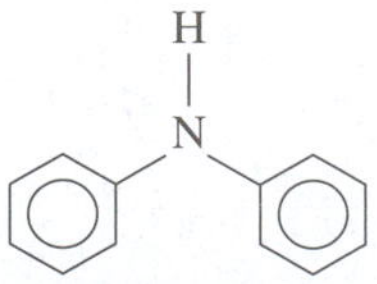

18.15 Write the balanced chemical equation for the reaction of each of the following amines with water:

a. methylamine **b.** dimethylamine

c. aniline

18.16 Write the balanced chemical equation for the reaction of each of the following amines with water:

a. *N*-methylethanamine **b.** propylamine

c. *N*-methylaniline

18.17 Draw the condensed structural or line-angle formula for the ammonium salt obtained when each of the amines in problem 18.15 reacts with HCl.

18.18 Draw the line-angle formula for the ammonium salt obtained when each of the amines in problem 18.16 reacts with HCl.

Clinical Applications

18.19 Novocaine, a local anesthetic, is the ammonium salt of procaine.

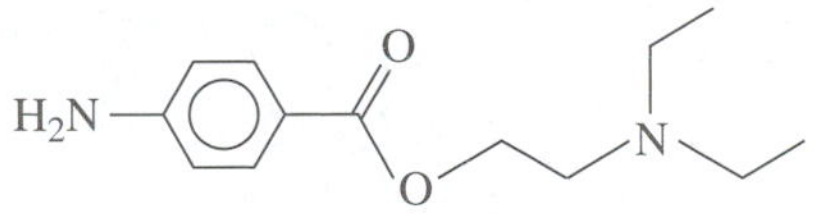

Procaine

a. Draw the line-angle formula for the ammonium salt (procaine hydrochloride) formed when procaine reacts with HCl. (*Hint*: The tertiary amine reacts with HCl.)

b. Why is procaine hydrochloride used rather than procaine?

18.20 Lidocaine (Xylocaine) is used as a local anesthetic and cardiac depressant.

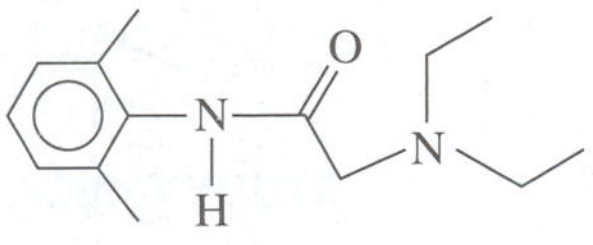

Lidocaine (Xylocaine)

a. Draw the line-angle formula for the ammonium salt formed when lidocaine reacts with HCl.

b. Why is the ammonium salt of lidocaine used rather than the amine?

18.3 Heterocyclic Amines

LEARNING GOAL Identify heterocyclic amines; distinguish between the types of heterocyclic amines.

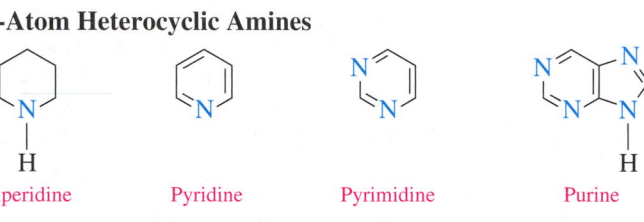

A **heterocyclic amine** is a cyclic organic compound that consists of a ring of five or six atoms, of which one or two are nitrogen atoms. Of the five-atom rings, the simplest one is pyrrolidine, which is a ring of four carbon atoms and one nitrogen atom, all with single bonds. Pyrrole is a five-atom ring with one nitrogen atom and two double bonds. Imidazole is a five-atom ring that contains two nitrogen atoms. Piperidine is a six-atom heterocyclic ring with a nitrogen atom. Some of the pungent aroma and taste of black pepper that we use to season our food is due to piperidine. Purine and pyrimidine rings are found in DNA and RNA. In purine, the structures of 6-atom pyrimidine and 5-atom imidazole are combined. Heterocyclic amines with two or three double bonds have aromatic properties similar to benzene.

ENGAGE

How does pyrrole differ from pyrrolidine?

The aroma of pepper is due to piperidine, a heterocyclic amine.

5-Atom Heterocyclic Amines

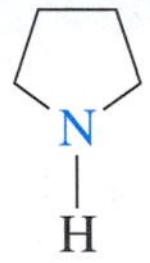

Pyrrolidine

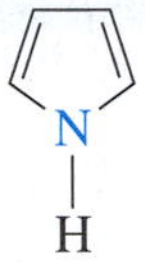

Pyrrole

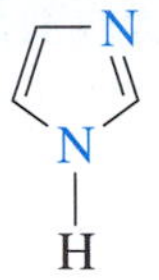

Imidazole

6-Atom Heterocyclic Amines

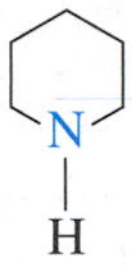

Piperidine

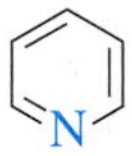

Pyridine

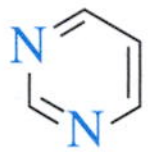

Pyrimidine

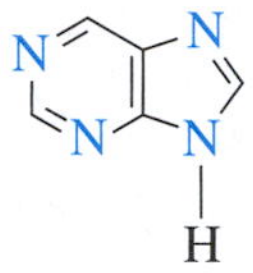

Purine

▶SAMPLE PROBLEM 18.5 Heterocyclic Amines

TRY IT FIRST

Identify the heterocyclic amines that are part of the structure of nicotine.

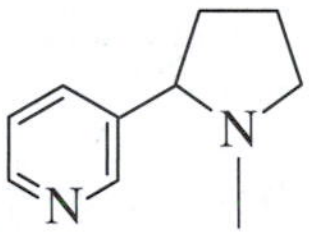

Nicotine

SOLUTION

Nicotine contains two heterocyclic rings. The 6-atom ring with one N atom and three double bonds is pyridine, and the 5-atom ring with one N atom and no double bonds is pyrrolidine. The N atom in the pyrrolidine ring is bonded to a methyl group.

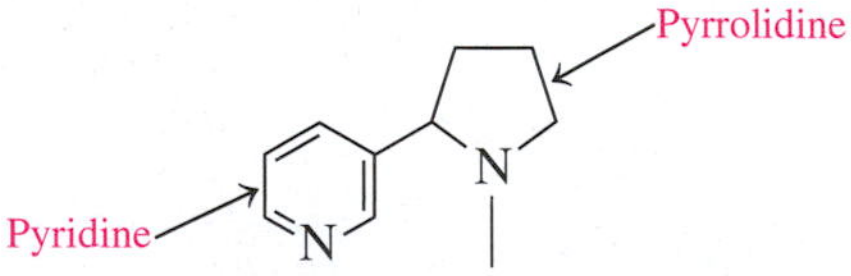
Pyrrolidine
Pyridine

STUDY CHECK 18.5

Sedamine is used as a sleep aid. Identify the heterocyclic amine that is part of the structure of sedamine.

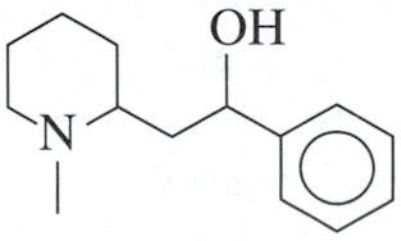

Sedamine

ANSWER

piperidine

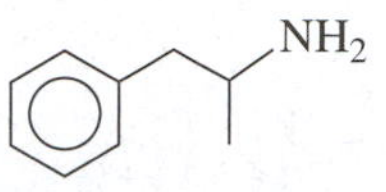
NH2

Amphetamine
(Benzedrine, Adderall)

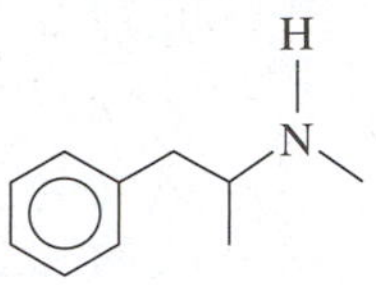
H
N

Methamphetamine
(Methedrine)

Catecholamines

The word catecholamine refers to the catechol part (3,4-dihydroxyphenyl group) of these aromatic amines. The most important catecholamine neurotransmitters are dopamine, norepinephrine, and epinephrine, which are closely related in structure and all are synthesized from the amino acid tyrosine. In the diet, tyrosine is found in meats, nuts, eggs, and cheese.

Amphetamine and methamphetamine are synthetic central nervous system stimulants that increase the levels of excitatory catecholamine neurotransmitters, particularly dopamine and norepinephrine. Amphetamine is used in the treatment of ADHD to improve cognition and decrease hyperactivity.

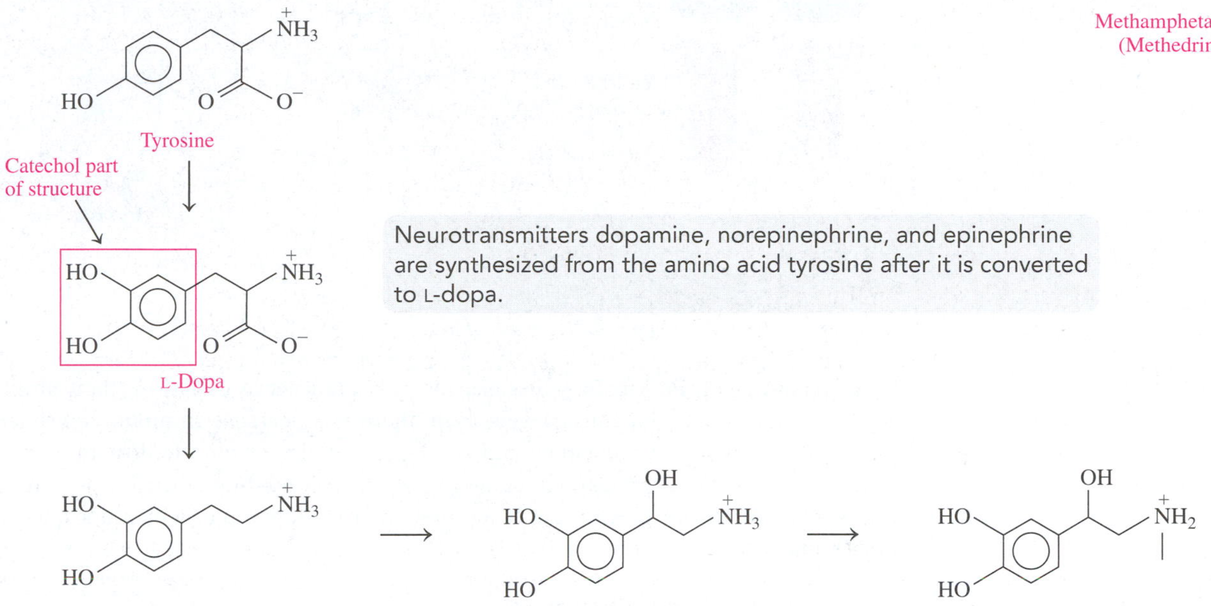
HO
NH3+
O
O−
Tyrosine
Catechol part
of structure
HO
HO
L-Dopa
Dopamine
OH
Norepinephrine (noradrenaline)
NH2+
Epinephrine (adrenaline)

Neurotransmitters dopamine, norepinephrine, and epinephrine are synthesized from the amino acid tyrosine after it is converted to L-dopa.

Dopamine

Dopamine, which is produced in the nerve cells of the midbrain, works as a natural stimulant, to give us energy and feelings of enjoyment. It plays a role in controlling muscle movement, regulation of the sleep–wake cycle, and helps to improve cognition, attention, memory, and learning. High levels of dopamine may be involved in addictive behavior and schizophrenia. Cocaine and amphetamine block the reuptake of dopamine into the vesicles of the nerve cell. As a result, dopamine remains in the synapse longer. The addiction to cocaine may be a result of the extended exposure to high levels of dopamine in the synapses.

▶SAMPLE PROBLEM 18.6

TRY IT FIRST

Place the following in order of occurrence for nerve impulses:

a. Neurotransmitters diffuse across synapse to receptors on dendrites.
b. Neurotransmitters move away from receptors.
c. An electrical signal reaches the axon terminal of a nerve cell.
d. Reuptake moves neurotransmitters into the vesicles for storage.
e. Vesicles release neurotransmitters into the synapse of nearby nerve cells.
f. Neurotransmitters stimulate receptors to send new nerve impulse.

SOLUTION

c, e, a, f, b, d

STUDY CHECK 18.6

How does an excitatory neurotransmitter send a nerve impulse at the receptor?

ANSWER

When the neurotransmitter attaches to the receptor, ion channels open and positive ions flow to nearby nerve cells, creating new electrical impulses.

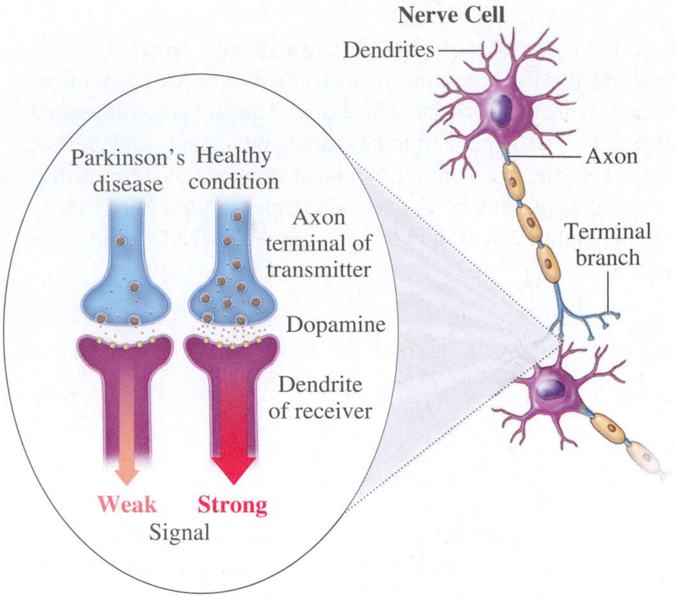
Nerve Cell
Dendrites
Axon
Terminal branch
Parkinson's disease
Healthy condition
Axon terminal of transmitter
Dopamine
Dendrite of receiver
Weak
Strong
Signal

L-Dopa is used to increase dopamine levels in the brain.

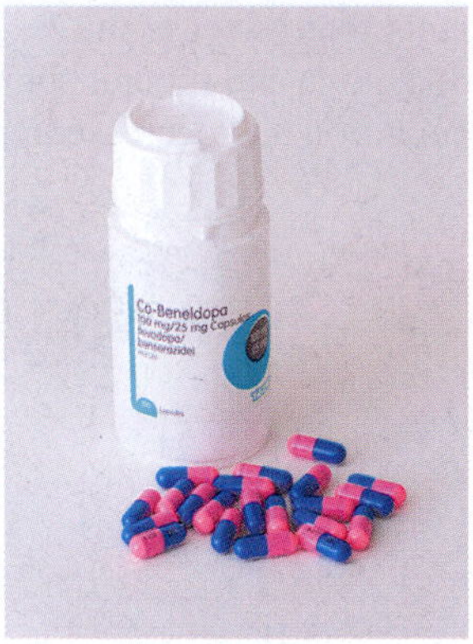
Co-Beneldopa

In persons with Parkinson's disease, the midbrain nerve cells lose their ability to produce dopamine. As dopamine levels drop, there is a decrease in motor coordination, resulting in a slowing of movement and shuffling, rigidity of muscle, loss of cognition, and dementia. Although dopamine cannot cross the blood–brain barrier, persons with Parkinson's disease are given L-dopa, the precursor of dopamine, which does cross the blood–brain barrier.

Norepinephrine and Epinephrine

Norepinephrine (noradrenaline) and epinephrine (adrenaline) are hormonal neurotransmitters that play a role in sleep, attention and focus, and alertness. Epinephrine is synthesized from norepinephrine by the addition of a methyl group to the amine group. Norepinephrine (noradrenaline) and epinephrine (adrenaline) are normally produced in the adrenal glands and are produced in large quantities when the stress of physical threat causes the fight-or-flight response. Then they cause an increase in blood pressure and heart rate, constrict blood vessels, dilate airways, and stimulate the breakdown of glycogen, which provides glucose and energy for the body. Because of its physiological effects, epinephrine is administered during cardiac arrest and used as a bronchodilator during allergy or asthma attacks. As a neurotransmitter, low levels of norepinephrine as well as dopamine contribute to *attention deficit disorder* (ADD). Medications such as Ritalin or Dexedrine may be prescribed to increase levels of norepinephrine and dopamine.

Serotonin

Serotonin (5-hydroxytryptamine) helps us to relax, sleep deeply and peacefully, think rationally, and gives us a feeling of well-being and calmness. Serotonin is synthesized from the amino acid tryptophan, which can cross the blood–brain barrier. A diet that contains foods such as eggs, fish, cheese, turkey, chicken, and beef, which have high levels of tryptophan, will increase serotonin levels. Foods with a low level of tryptophan, such as whole wheat, will lower serotonin levels. Psychedelic drugs such as LSD and mescaline stimulate the action of serotonin at its receptors.

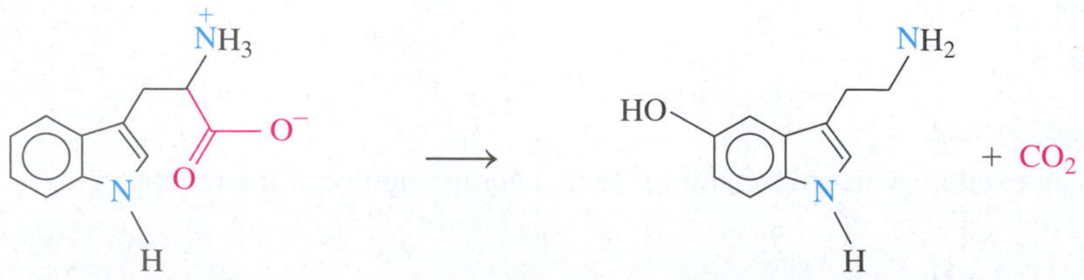

Tryptophan (amino acid) Serotonin

Low levels of serotonin in the brain may be associated with depression, anxiety disorders, obsessive–compulsive disorder, and eating disorders. Many antidepressant drugs, such as fluoxetine (Prozac) and paroxetine (Paxil), are selective serotonin reuptake inhibitors (SSRIs). When the reuptake of serotonin is slowed, it remains longer at the receptors, where it continues its action; the net effect is as if additional quantities of serotonin were taken.

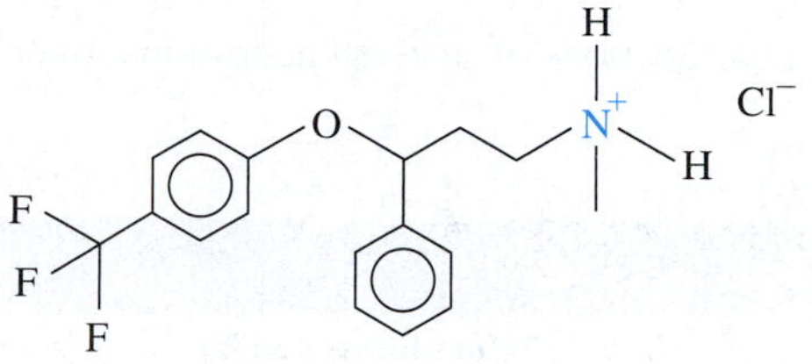

Fluoxetine hydrochloride (Prozac)

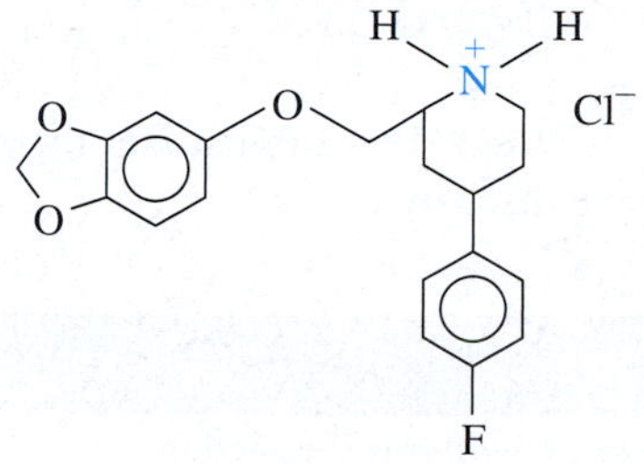

Paroxetine hydrochloride (Paxil)

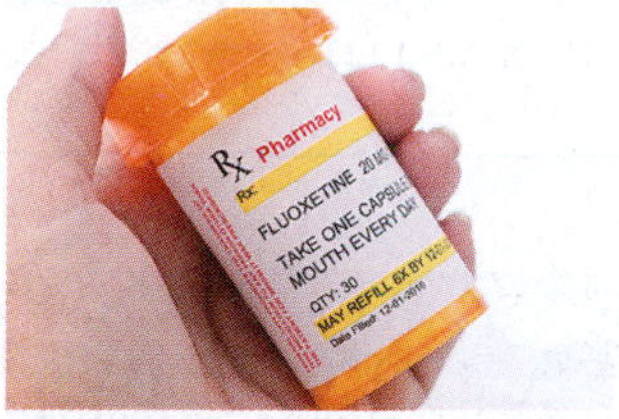
Pharmacy
FLUOXETINE

Prozac is one of the selective serotonin reuptake inhibitors (SSRIs) used to slow the reuptake of serotonin.

Histamine

Histamine is synthesized in the nerve cells in the hypothalamus from the amino acid histidine. Histamine is produced by the immune system in response to pathogens and invaders, or injury. When histamine combines with histamine receptors, it causes allergic reactions, which may include inflammation, watery eyes, itchy skin, and hay fever. Histamine can also cause smooth muscle constriction, such as the closing of the trachea in persons allergic to shellfish. Histamine is also stored and released in the cells of the stomach, where it stimulates acid production. Antihistamines, such as Benadryl, Zantac, and Tagamet, are used to block the histamine receptors and stop the allergic reactions.

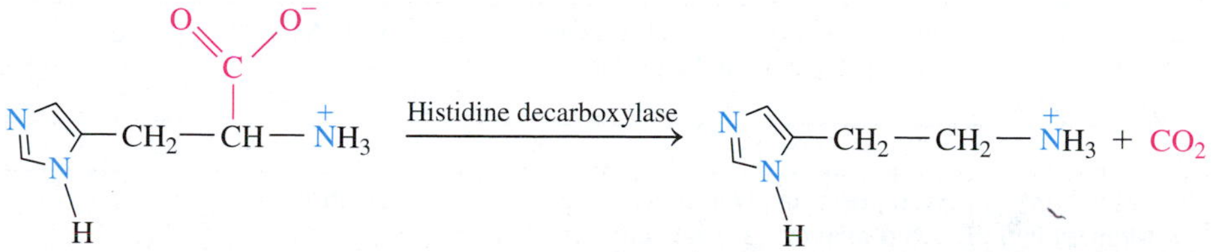
Histidine decarboxylase

Histidine　　　　Histamine

Glutamate

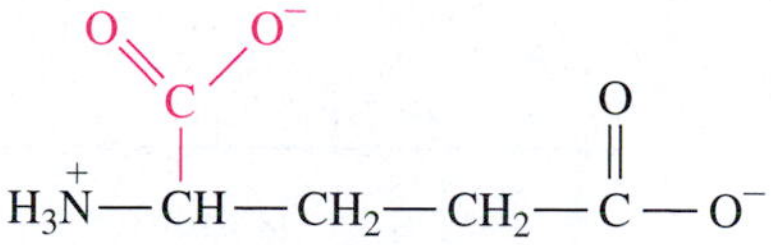

Glutamate

Glutamate is the most abundant neurotransmitter in the nervous system, where it is used to stimulate over 90% of the synapses. When glutamate binds to its receptor cells, it stimulates the synthesis of nitrogen oxide (NO), also a neurotransmitter in the brain. As NO reaches the transmitting nerve cells, more glutamate is released. Glutamate and NO are thought to be involved in learning and memory. Glutamate is used in many fast excitatory synapses in the brain and spinal cord. The reuptake of glutamate out of the synapse and back into the nerve cell occurs rapidly, which keeps glutamate levels low. If the reuptake of glutamate does not take place fast enough, a condition called *excitotoxicity* occurs in which excess glutamate at the receptors can destroy brain cells. In Lou Gehrig's disease (ALS), the production of an excessive amount of glutamate causes the degeneration of nerve cells in the spinal cord. As a result, a person with Lou Gehrig's disease suffers increasing weakness and muscular atrophy. When the reuptake of glutamate is too rapid, the levels of glutamate fall too low in the synapse, which may result in mental illness such as schizophrenia.

Gamma(γ)-Aminobutyric Acid or GABA

Gamma(γ)-aminobutyric acid or GABA, which is produced from glutamate, is the most common inhibitory neurotransmitter in the brain. GABA produces a calming effect and reduces anxiety by inhibiting the ability of nerve cells to send electrical signals to nearby nerve cells. It is involved in the regulation of muscle tone, sleep, and anxiety. GABA can be obtained as a nutritional supplement. Medications such as benzodiazepines, and barbiturates such as phenobarbital, are used to increase GABA levels at the GABA receptors. Alcohol,

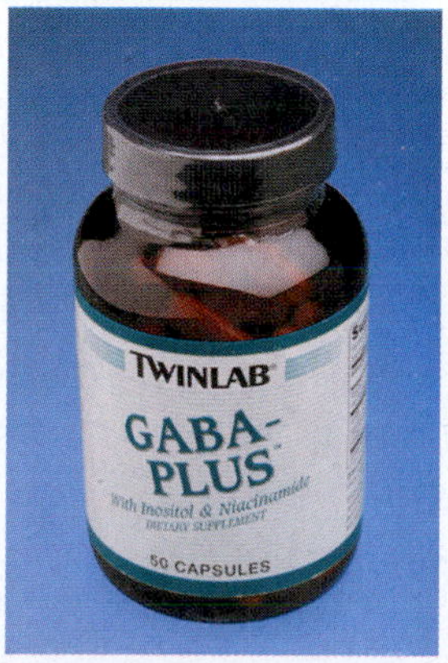
TWINLAB
GABA-PLUS
with Inositol & Niacinamide
DIETARY SUPPLEMENT
50 CAPSULES

GABA is the major inhibitory neurotransmitter.

sedatives, and tranquilizers increase the inhibitory effects of GABA. Caffeine decreases the GABA levels in the synapses, leading to conditions of anxiety and sleep problems.

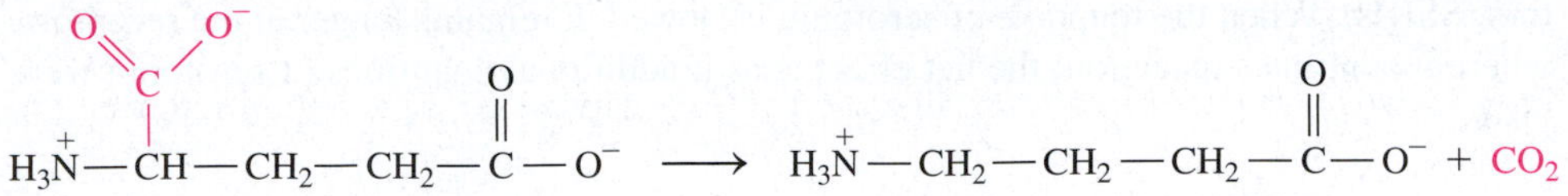

Glutamate → Gamma(γ)-aminobutyric acid (GABA)

TEST

Try Practice Problems 18.33 to 18.44

TABLE 18.1 summarizes the properties and functions of selected neurotransmitters we have discussed.

TABLE 18.1 Selected Neurotransmitters

Amine Transmitters	Synthesized from	Site of Synthesis	Function	Effect Enhanced by
Acetylcholine	Acetyl CoA and choline	Central nervous system	Excitatory: regulates muscle activation, learning, and short-term memory	Nicotine
Dopamine	Tyrosine	Central nervous system	Excitatory and inhibitory: regulates muscle movement, cognition, sleep, mood, and learning; deficiency leads to Parkinson's disease and schizophrenia	L-dopa, amphetamines, cocaine
Norepinephrine	Tyrosine	Central nervous system	Excitatory and inhibitory: sleep, focus, and alertness	Ritalin, Dexedrine
Epinephrine	Tyrosine	Adrenal glands	Excitatory: plays a role in sleep and being alert, and is involved in the fight-or-flight response	Ritalin, Dexedrine
Serotonin	Tryptophan	Central nervous system	Inhibitory: regulates anxiety, eating, mood, sleep, learning, and memory	LSD; Prozac and Paxil block its action to relieve anxiety
Histamine	Histidine	Central nervous system: hypothalamus	Inhibitory: involved in allergic reactions, inflammation, and hay fever	Antihistamines
Amino Acid Transmitters				
Glutamate		Central nervous system: spinal cord	Excitatory: involved in learning and memory; main excitatory neurotransmitter in brain	Alcohol
GABA	Glutamate	Central nervous system: brain, hypothalamus	Inhibitory: regulates muscle tone, sleep, and anxiety	Alcohol, sedatives, tranquilizers; synthesis blocked by antianxiety drugs (benzodiazepines); caffeine decreases GABA levels in synapse

PRACTICE PROBLEMS

18.4 Neurotransmitters

18.27 What is a neurotransmitter?

18.28 Where are the neurotransmitters stored in a neuron?

18.29 When is a neurotransmitter released into the synapse?

18.30 What happens to a neurotransmitter once it is in the synapse?

18.31 Why is it important to remove a neurotransmitter from its receptor?

18.32 What are three ways in which a neurotransmitter can be separated from a receptor?

Clinical Applications

18.33 What is the role of acetylcholine?

18.34 What are some physiological effects of low levels of acetylcholine?

18.35 What is the function of dopamine in the body?

18.36 What are some physiological effects of low levels of dopamine?

18.37 What is the function of serotonin in the body?

18.38 What are the physiological effects of low serotonin?

18.39 What is the function of histamine in the body?

18.40 How do antihistamines stop the action of histamine?

18.41 Why is it important that the levels of glutamate in the synapse remain low?

18.42 What happens if there is an excess of glutamate in the synapse?

18.43 What is the function of GABA in the body?

18.44 What is the effect of caffeine on GABA levels?

18.5 Amides

LEARNING GOAL Draw the amide product from amidation, and write the IUPAC and common names.

The **amides** are derivatives of carboxylic acids in which an amino group replaces the hydroxyl group.

Carboxylic Acid

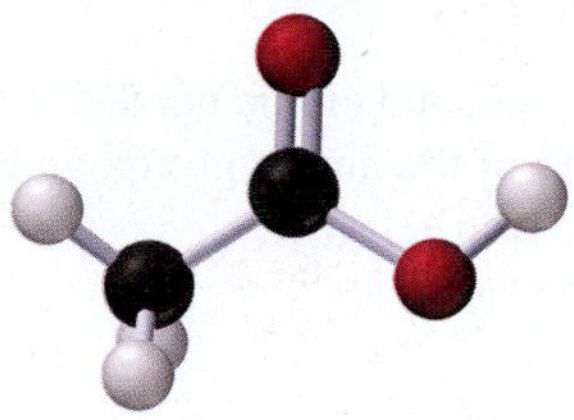

Ethanoic acid
(acetic acid)

Amide

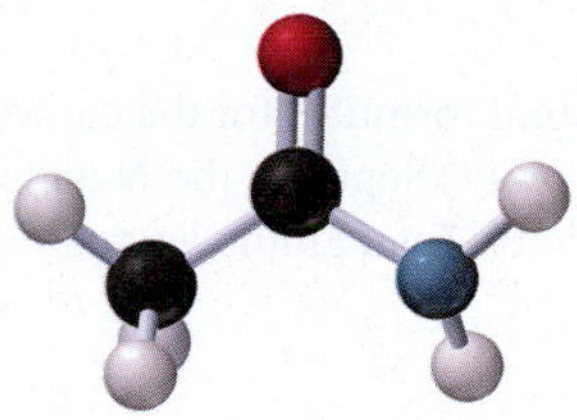

Ethanamide
(acetamide)

In amides, the functional group consists of a carbonyl group attached to an amine. In biochemistry, the amide bond that links amino acids in a protein is called a *peptide bond*. Some medically important amides include acetaminophen (Tylenol) used to reduce fever; phenobarbital, a sedative and anticonvulsant medication; and penicillin, an antibiotic.

Preparation of Amides

CORE CHEMISTRY SKILL
Forming Amides

An amide is produced in a reaction called **amidation**, or *condensation*, in which a carboxylic acid reacts with ammonia or a primary (1°) or secondary (2°) amine. A molecule of water is eliminated, and the fragments of the carboxylic acid and amine molecules join to form the amide, much like the formation of an ester. Because a tertiary (3°) amine does not contain a hydrogen atom, it cannot undergo amidation.

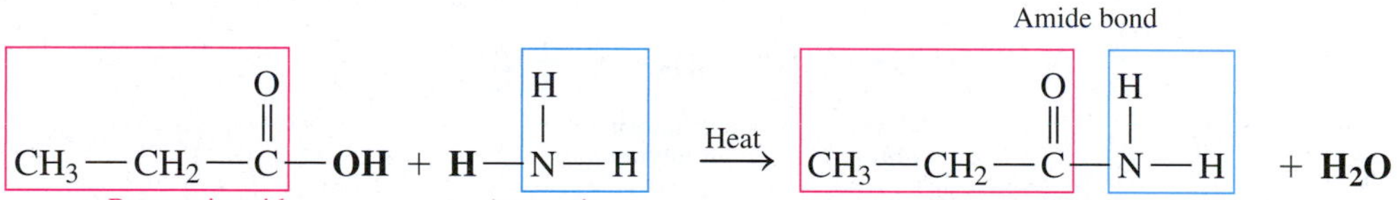

Amide bond
Heat

Propanoic acid (propionic acid) + Ammonia → Propanamide (propionamide) + H_2O

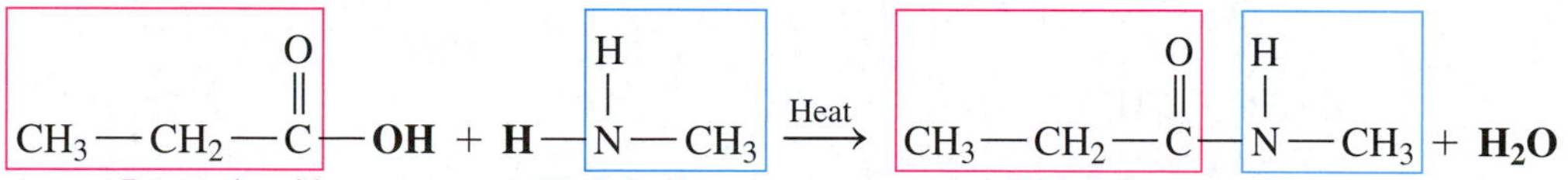

Heat

Propanoic acid (propionic acid) + Methylamine → *N*-Methylpropanamide (*N*-methylpropionamide) + H_2O

▶SAMPLE PROBLEM 18.7 Amidation

TRY IT FIRST

Draw the line-angle formula in **a** and the condensed structural formula in **b** for the amide product of each of the following reactions:

a.

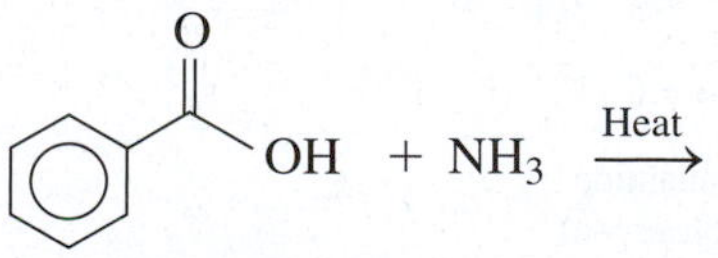

Heat

b. $CH_3-\overset{\overset{\large O}{\|}}{C}-OH + H_2N-CH_2-CH_3 \xrightarrow{\text{Heat}}$

SOLUTION

The line-angle or condensed structural formula for the amide product can be drawn by attaching the carbonyl group from the acid to the nitrogen atom of the amine. The —OH group is removed from the acid and —H from the amine to form water.

a. 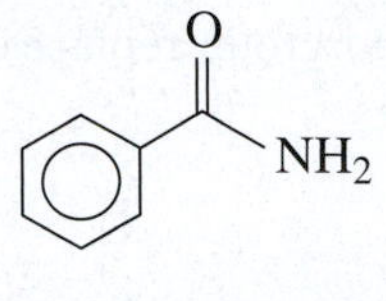

b. 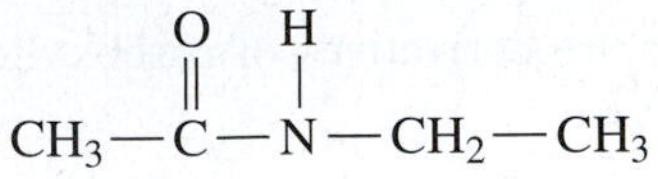

STUDY CHECK 18.7

Draw the condensed structural formulas for the carboxylic acid and amine needed to prepare the following amide. (*Hint*: Separate the N and C=O of the amide group, and add —H and —OH to give the original amine and carboxylic acid.)

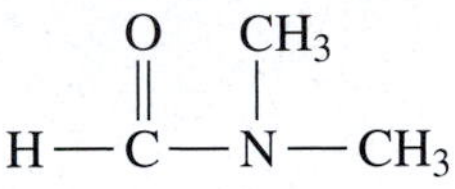

ANSWER

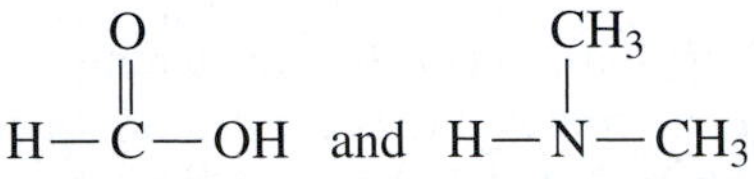

TEST

Try Practice Problems 18.45 and 18.46

Naming Amides

In the IUPAC and common names for amides, the *oic acid* or *ic acid* from the corresponding carboxylic acid name is replaced with *amide*. We can diagram the name of an amide in the following way:

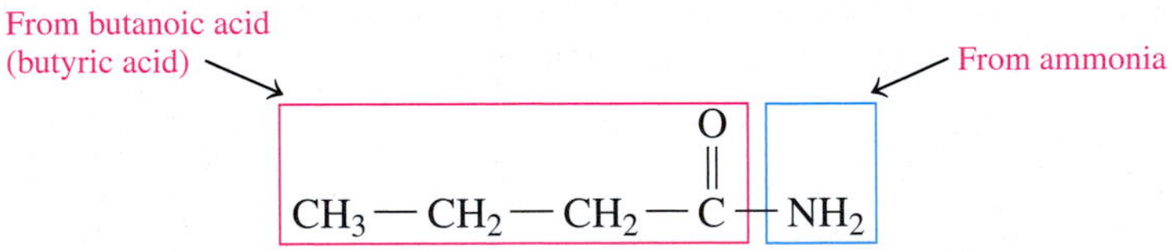
From butanoic acid
(butyric acid)
From ammonia

IUPAC Butanamide
(common) (butyramide)

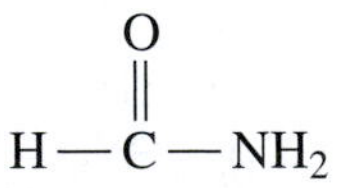

Methanamide
(formamide)

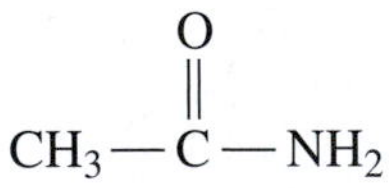

Ethanamide
(acetamide)

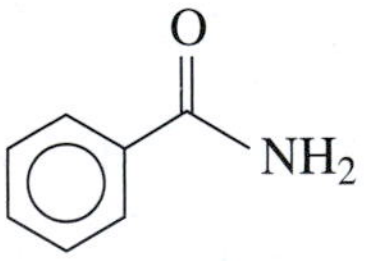

Benzamide

When alkyl groups are attached to the nitrogen atom, the prefix *N*- or *N,N*- precedes the name of the amide, depending on whether there are one or two groups. We can diagram the name of a substituted amide in the following way:

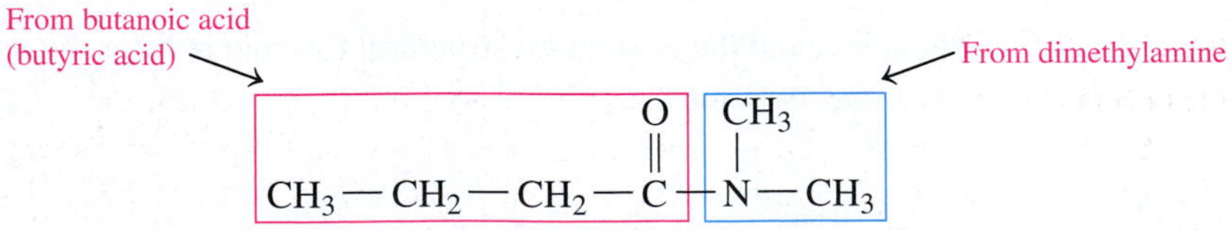
From butanoic acid
(butyric acid)
From dimethylamine

IUPAC *N,N*-Dimethylbutanamide
(common) (*N,N*-dimethylbutyramide)

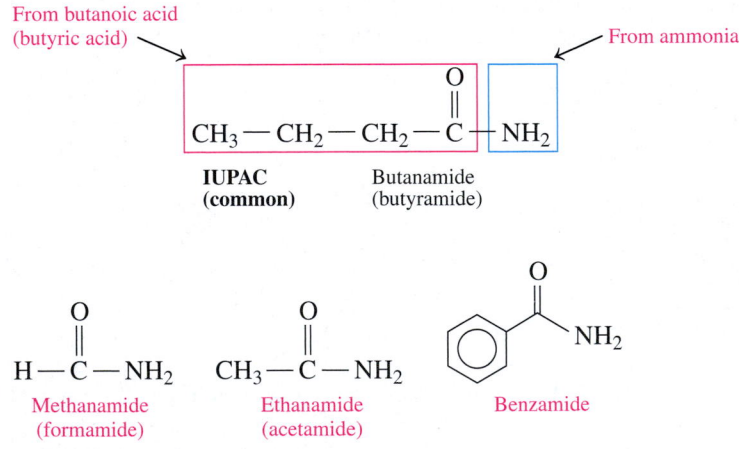

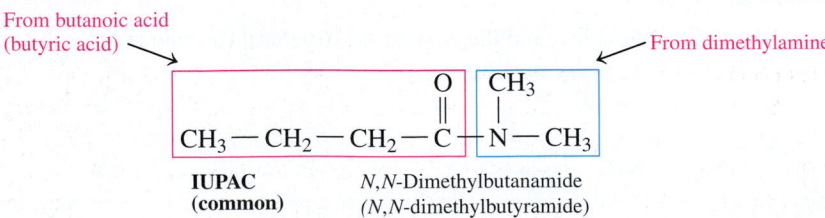

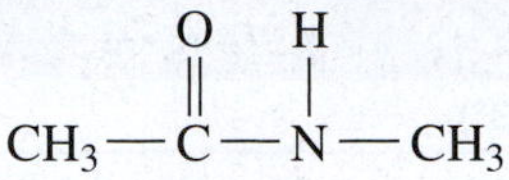

N-Methylethanamide
(N-methylacetamide)

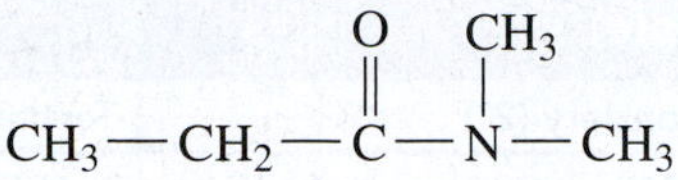

N,N-Dimethylpropanamide
(N,N-dimethylpropionamide)

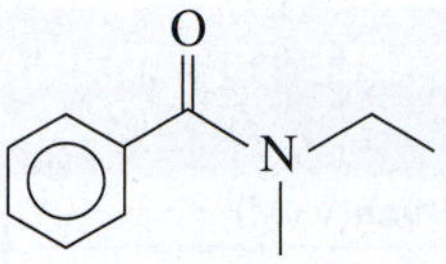

N-Ethyl-N-methylbenzamide

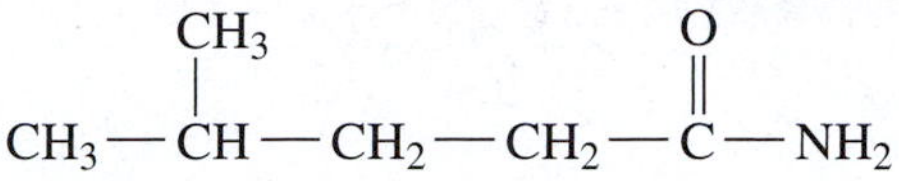

4-Methylpentanamide

▶ SAMPLE PROBLEM 18.8 Naming Amides

TRY IT FIRST

Write the IUPAC name for the following amide:

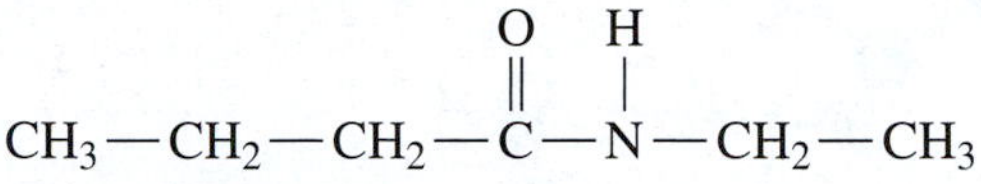

SOLUTION

ANALYZE THE PROBLEM	Given	Need	Connect
	amide	IUPAC name	replace *oic acid* with *amide*, alkyl substituent has prefix *N*-

STEP 1 Replace the *oic acid* in the carboxylic acid name with *amide*.

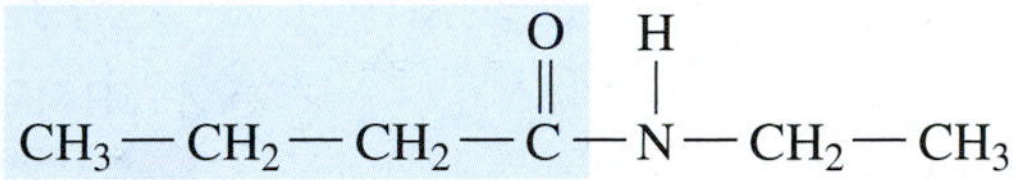

butanamide

STEP 2 Name each substituent on the N atom using the prefix *N*- and the alkyl name.

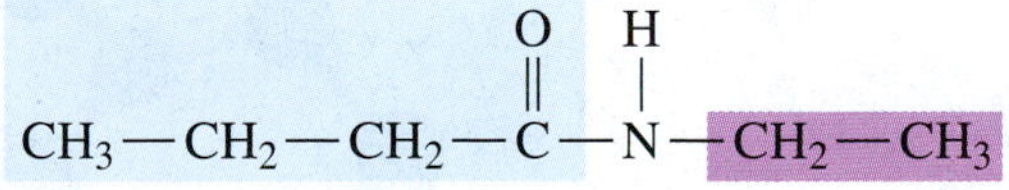

N-ethyl butanamide

STUDY CHECK 18.8

Draw the line-angle formula for *N*,*N*-dimethylbenzamide.

ANSWER

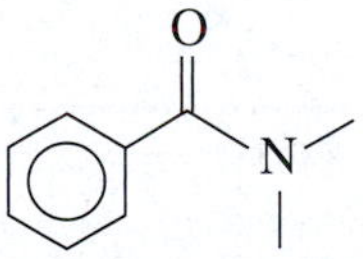

TEST

Try Practice Problems 18.47 to 18.50

Physical Properties of Amides

The amides do not have the properties of bases that we saw for the amines. Only methanamide is a liquid at room temperature, while the other amides are solids. For primary (1°) amides, the $-NH_2$ group can form more hydrogen bonds, which gives primary amides the highest melting points. The melting points of the secondary (2°) amides are lower because they form fewer hydrogen bonds. Because tertiary (3°) amides cannot form hydrogen bonds, they have the lowest melting points (see **TABLE 18.2**).

ENGAGE

Which amide forms more hydrogen bonds, ethanamide or N-methylethanamide?

TABLE 18.2 Melting Points of Selected Amides with the Same Molar Mass (73.0 g/mole)

Primary (1°)	Secondary (2°)	Tertiary (3°)
Propanamide 80. °C	*N*-Methylethanamide 28 °C	*N*,*N*-Dimethylmethanamide −61 °C

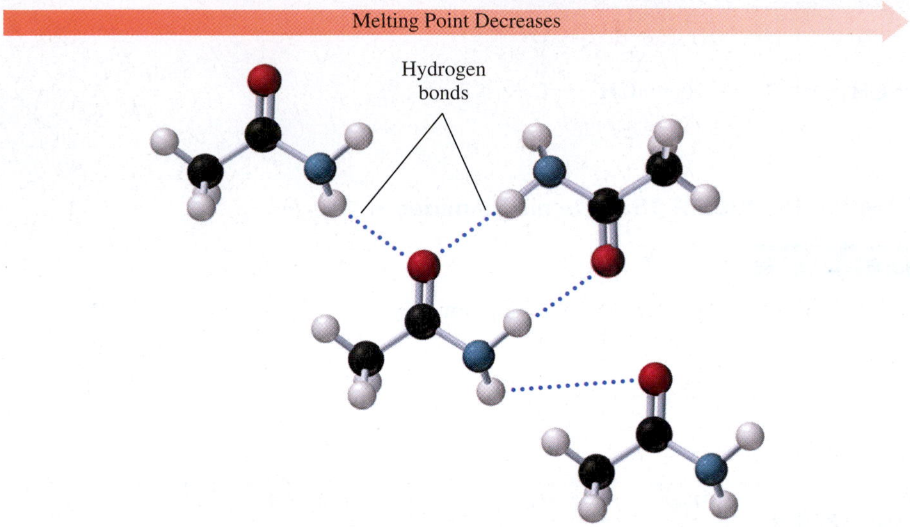
Melting Point Decreases
Hydrogen
bonds

Hydrogen bonding between molecules of a primary amide increases the melting point.

TEST

Try Practice Problems 18.51 and 18.52

The amides with one to five carbon atoms are soluble in water because they can hydrogen bond with water molecules.

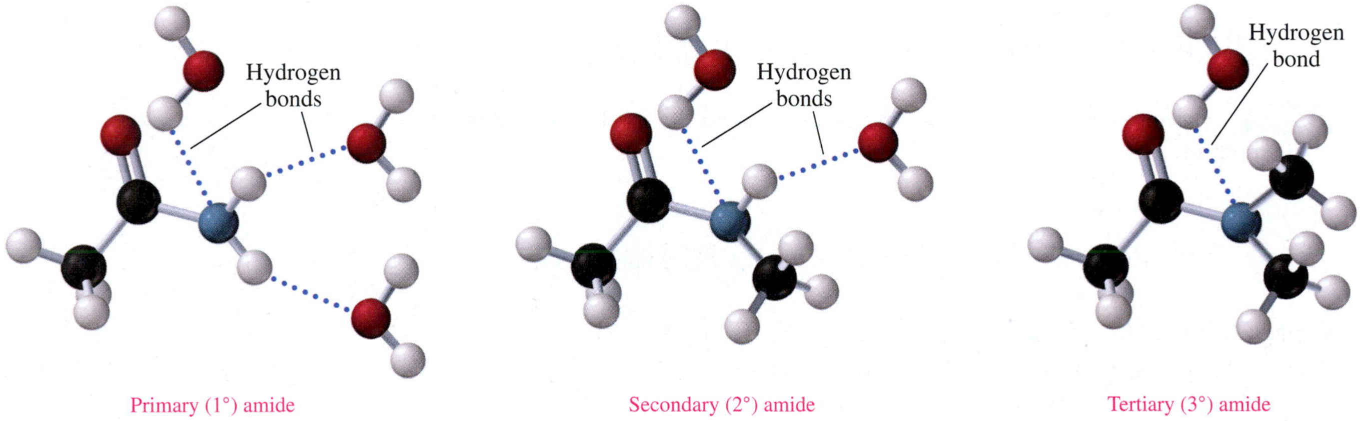
Hydrogen
bonds
Hydrogen
bonds
Hydrogen
bond
Primary (1°) amide
Secondary (2°) amide
Tertiary (3°) amide

In water, primary amides form more hydrogen bonds than secondary and tertiary amides.

Chemistry Link to Health

Amides in Health and Medicine

The simplest natural amide is urea, an end product of protein metabolism in the body. The kidneys remove urea from the blood and excrete it in urine. If the kidneys malfunction, urea is not removed and builds to a toxic level, a condition called *uremia*. Urea is also used as a component of fertilizer to increase nitrogen in the soil.

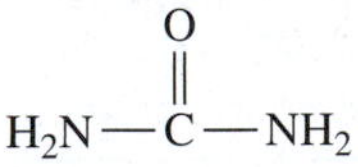
O
H_2N—C—NH_2

Urea

Synthetic amides are used as substitutes for sugar. Saccharin is a very powerful sweetener and is used as a sugar substitute. The sweetener aspartame is made from two amino acids, aspartic acid and phenylalanine, joined by an amide bond.

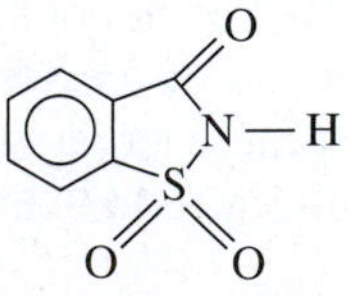
O
N—H
S
O
O

Saccharin

(continued)

Chemistry Link to Health (continued)

VICKS
NyQuil
SWEET'N LOW
TYLENOL
Extra Strength

Amides are found in synthetic sweeteners, such as aspartame, and pain relievers, such as acetaminophen.

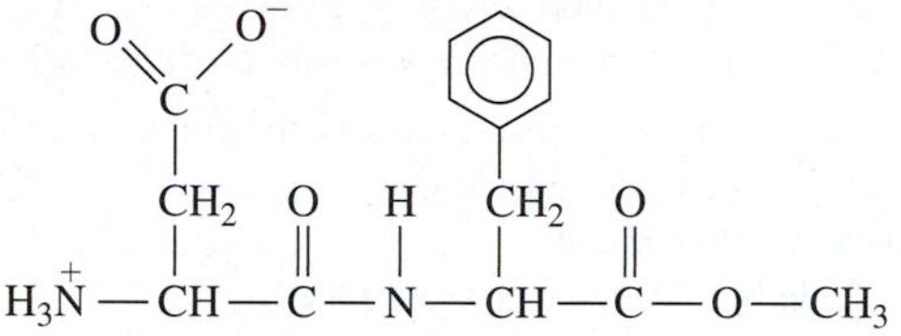

Aspartic acid Phenylalanine Methyl ester

Aspartame

The compounds phenacetin and acetaminophen, which is used in Tylenol, reduce fever and pain, but they have little anti-inflammatory effect.

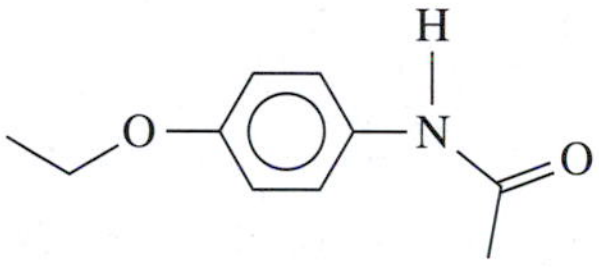

Phenacetin

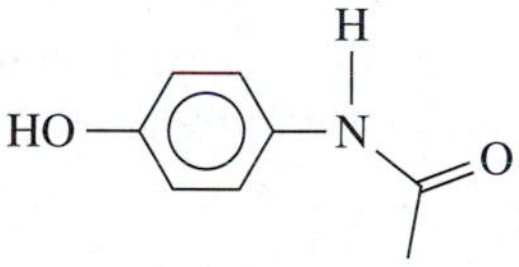

Acetaminophen

Many barbiturates are cyclic amides of barbituric acid that act as sedatives in small dosages or sleep inducers in larger dosages. They are often habit forming. Barbiturate drugs include phenobarbital (Luminal), pentobarbital (Nembutal), and secobarbital (Seconal). Other amides, such as meprobamate and diazepam, act as sedatives and tranquilizers.

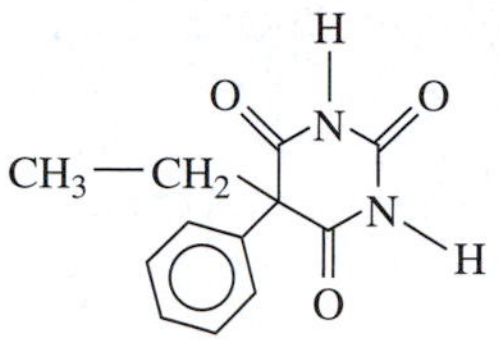

Phenobarbital
(Luminal)

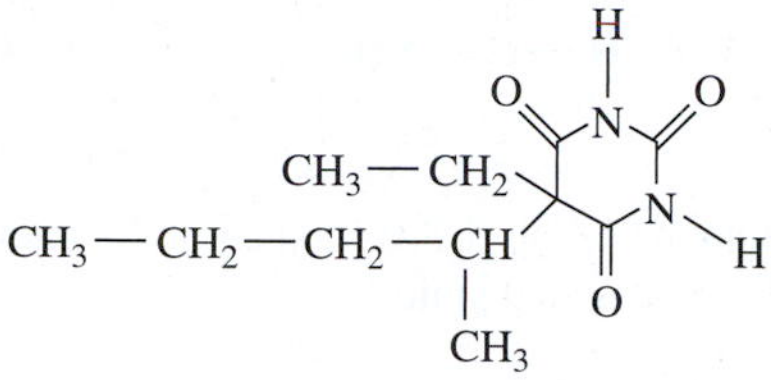

Pentobarbital
(Nembutal)

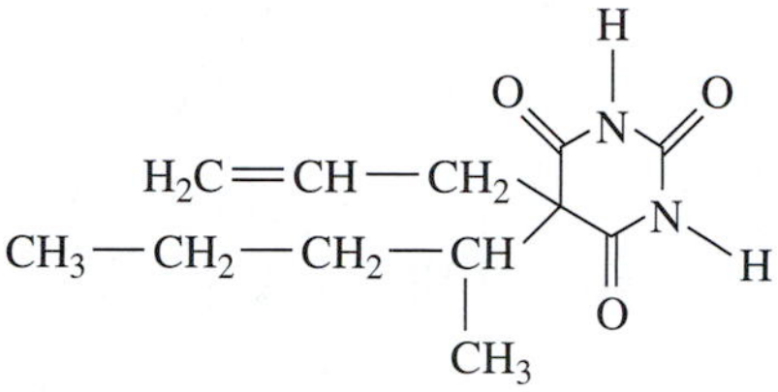

Secobarbital
(Seconal)

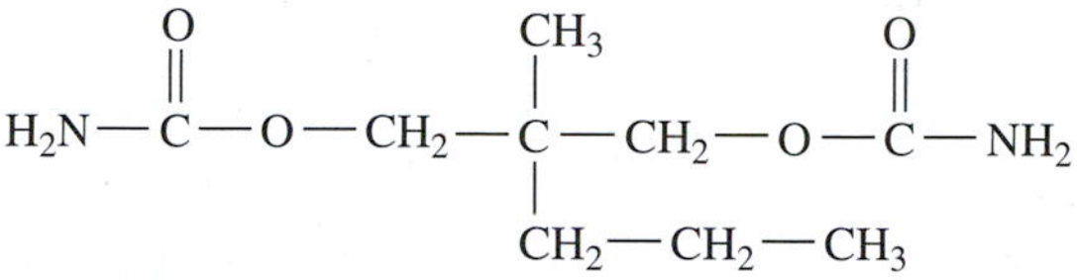

Meprobamate
(Miltown)

Diazepam
(Valium)

PRACTICE PROBLEMS

18.5 Amides

18.45 Draw the condensed structural or line-angle formula for the amide formed in each of the following reactions:

a. $CH_3-\overset{\overset{\large O}{\|}}{C}-OH + NH_3 \xrightarrow{\text{Heat}}$

b. $CH_3-\overset{\overset{\large O}{\|}}{C}-OH + H_2N-CH_2-CH_3 \xrightarrow{\text{Heat}}$

c. 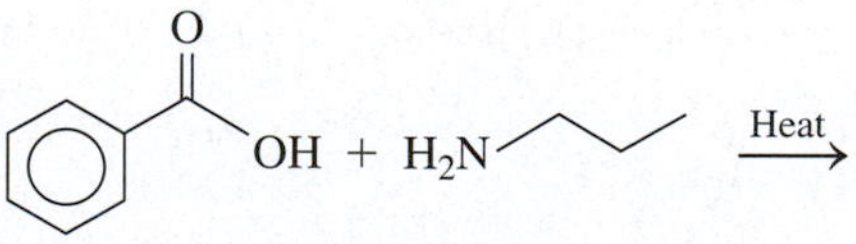

18.46 Draw the condensed structural or line-angle formula for the amide formed in each of the following reactions:

a. $CH_3-CH_2-CH_2-CH_2-\overset{\overset{\large O}{\|}}{C}-OH + NH_3 \xrightarrow{\text{Heat}}$

b. $CH_3-\underset{}{\overset{\overset{\large CH_3}{|}}{CH}}-\overset{\overset{\large O}{\|}}{C}-OH + H_2N-CH_2-CH_3 \xrightarrow{\text{Heat}}$

c. 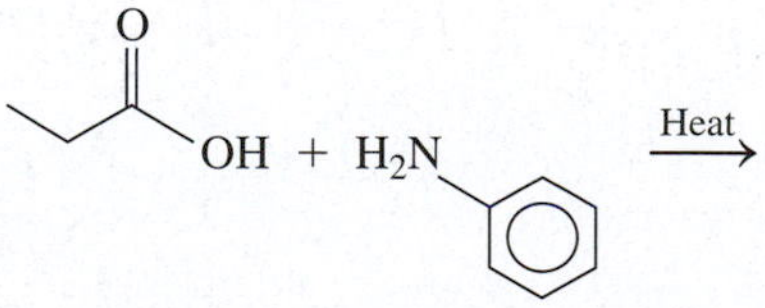

18.47 Write the IUPAC and common names (if any) for each of the following amides:

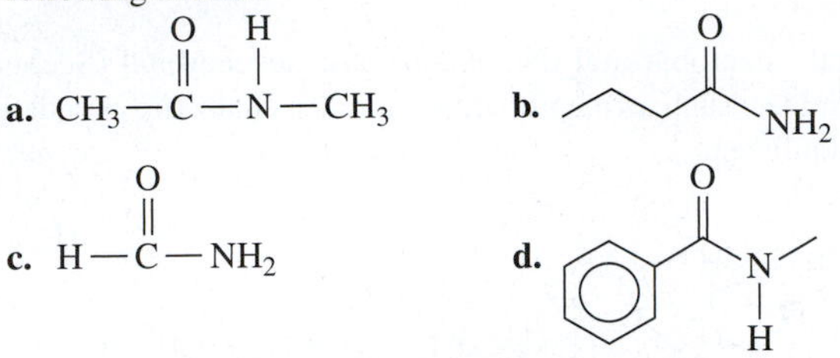

18.48 Write the IUPAC and common names (if any) for each of the following amides:

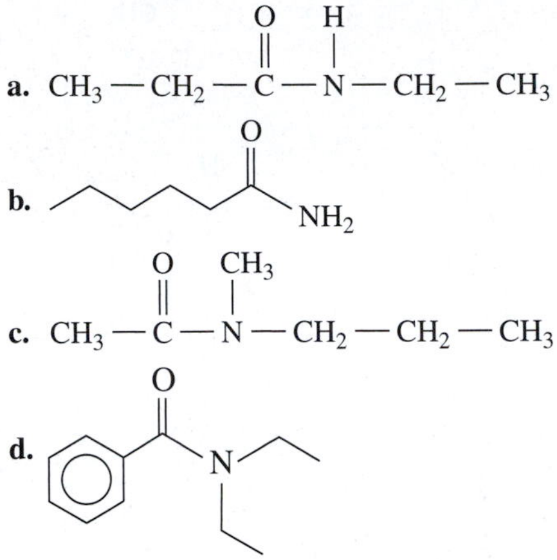

18.49 Draw the condensed structural formula for **a** and **b** and the line-angle formula for **c** and **d**.

a. propionamide
b. pentanamide
c. *N*-ethylbenzamide
d. *N*-ethylbutyramide

18.50 Draw the condensed structural formula for **a** and **b** and the line-angle formula for **c** and **d**.

a. hexanamide
b. 3-methylbutanamide
c. *N*-ethyl-*N*-methylbenzamide
d. *N*-propylpentanamide

18.51 For each of the following pairs, identify the compound that has the higher melting point. Explain.

a. ethanamide or *N*-methylethanamide
b. butane or propionamide
c. *N*,*N*-dimethylpropanamide or *N*-methylpropanamide

18.52 For each of the following pairs, identify the compound that has the higher melting point. Explain.

a. propane or ethanamide
b. *N*-methylethanamide or propanamide
c. *N*-ethylethanamide or *N*,*N*-dimethylethanamide

18.6 Hydrolysis of Amides

LEARNING GOAL Write balanced chemical equations for the hydrolysis of amides.

CORE CHEMISTRY SKILL
Hydrolyzing Amides

In a reverse reaction of amidation, **hydrolysis** occurs when water and an acid or a base split an amide. In acid hydrolysis of an amide, the products are the carboxylic acid and the ammonium salt. In base hydrolysis, the products are the carboxylate salt and the amine or ammonia.

Acid Hydrolysis of Amides

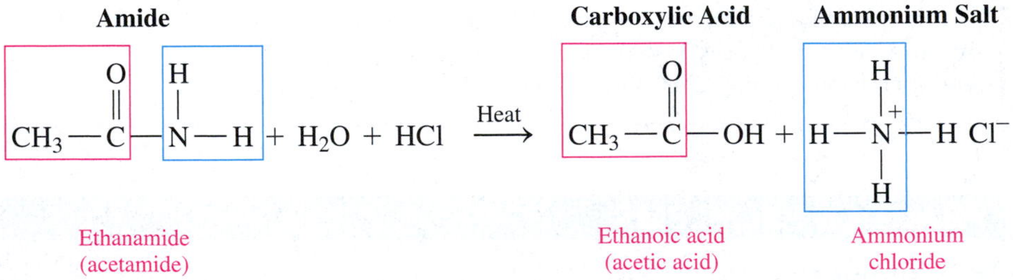
Amide
Carboxylic Acid
Ammonium Salt
Heat
Ethanamide
(acetamide)
Ethanoic acid
(acetic acid)
Ammonium
chloride

Base Hydrolysis of Amides

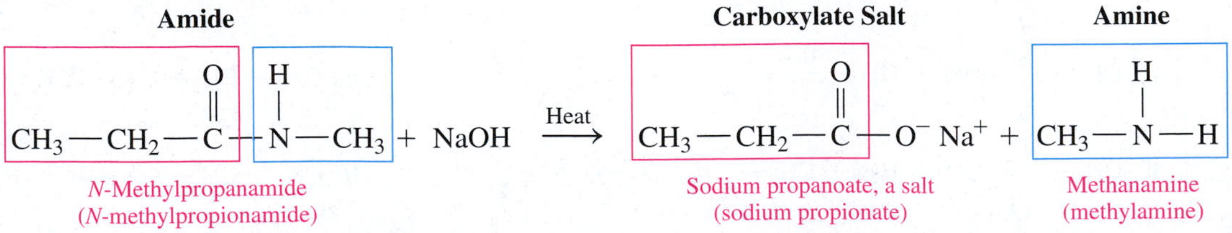
Amide
Carboxylate Salt
Amine
Heat
N-Methylpropanamide
(N-methylpropionamide)
Sodium propanoate, a salt
(sodium propionate)
Methanamine
(methylamine)

▶ SAMPLE PROBLEM 18.9 Hydrolysis of Amides

TRY IT FIRST

Draw the condensed structural formulas and write the IUPAC names for the products from the hydrolysis of *N*-methylpentanamide with NaOH.

SOLUTION

ANALYZE THE PROBLEM	Given	Need	Connect
	N-methylpentanamide, NaOH	products: carboxylate salt, amine	base hydrolysis of an amide

In base hydrolysis, the amide bond is broken between the carboxyl carbon atom and the nitrogen atom. When a base such as NaOH is used, the products are the carboxylate salt and an amine.

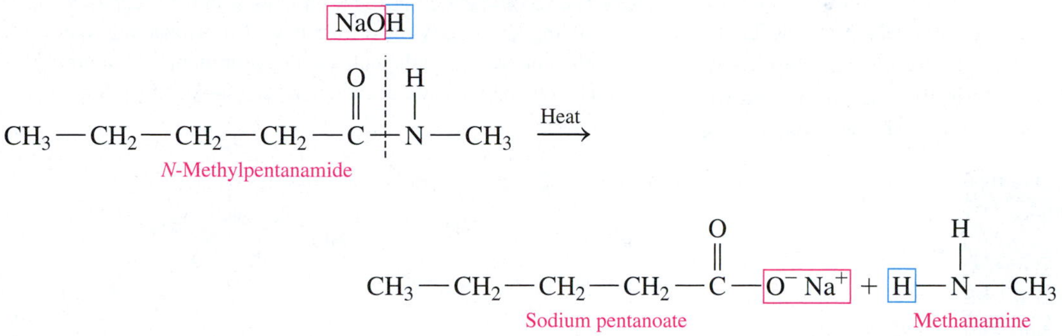
NaOH
Heat
N-Methylpentanamide
Sodium pentanoate
Methanamine

STUDY CHECK 18.9

Draw the condensed structural formulas for the products obtained from the hydrolysis of *N*-methylbutyramide with HBr.

ANSWER

$CH_3-CH_2-CH_2-\overset{\overset{O}{\|}}{C}-OH$ and $CH_3-\overset{+}{N}H_3\ Br^-$

TEST

Try Practice Problems 18.53 and 18.54

PRACTICE PROBLEMS

18.6 Hydrolysis of Amides

18.53 Draw the condensed structural or line-angle formulas for the products from the hydrolysis of each of the following with HCl:

a. 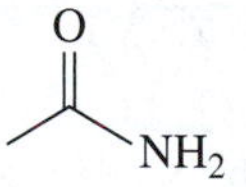
O
NH2

b. $CH_3-CH_2-\overset{\overset{O}{\|}}{C}-NH_2$

c. $CH_3-CH_2-CH_2-\overset{\overset{O}{\|}}{C}-\overset{\overset{H}{|}}{N}-CH_3$

d. 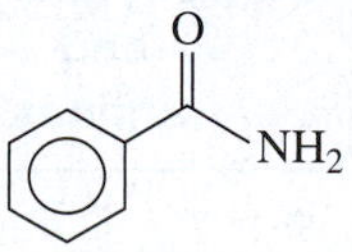
O
NH2

18.54 Draw the condensed structural or line-angle formulas for the products from the hydrolysis of each of the following with NaOH:

a. $CH_3-CH_2-\overset{\overset{CH_3}{|}}{CH}-\overset{\overset{O}{\|}}{C}-NH_2$

b. $CH_3-CH_2-CH_2-\overset{\overset{O}{\|}}{C}-\overset{\overset{CH_2-CH_3}{|}}{N}-CH_2-CH_3$

c. 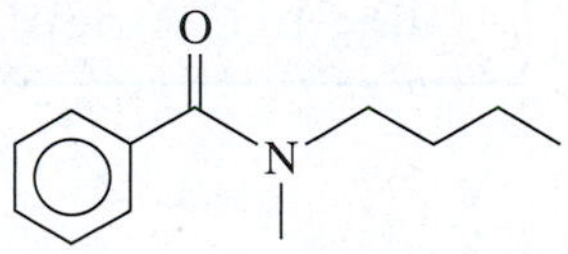
O
N

d. $CH_3-\overset{\overset{Cl}{|}}{CH}-\overset{\overset{O}{\|}}{C}-\overset{\overset{CH_3}{|}}{N}-CH_2-CH_3$

CLINICAL UPDATE Testing Soil and Water Samples for Chemicals

Lance, an environmental health practitioner, returns to the sheep ranch to obtain more soil and water samples. When he arrives at the ranch, he notices that the owners are spraying the sheep because they had an infestation of flies and maggots. Lance is told that the insecticide they were spraying on the sheep was called dicyclanil, a potent insect growth regulator. Lance is also informed that the sheep are being treated with enrofloxacin to counteract a respiratory infection.

Because high levels of these chemicals could be hazardous if they exceed acceptable environmental standards, Lance collects samples of soil and water to test for contaminants or breakdown products to evaluate in the laboratory. He will send the test results to the rancher, who will use those results to adjust the levels of the drugs, order protective equipment, and do a cleanup, if necessary.

Clinical Applications

18.55 Name the functional groups in dicyclanil. Identify the heterocyclic amine in dicyclanil.

18.56 Name the functional groups in enrofloxacin.

18.57 The recommended application for dicyclanil for an adult sheep is 65 mg/kg of body mass. If dicyclanil is supplied in a spray with a concentration of 50. mg/mL, how many milliliters of the spray are required to treat a 70.-kg adult sheep?

18.58 The recommended dose for enrofloxacin for sheep is 30. mg/kg of body mass for 5 days. Enrofloxacin is supplied in 50.-mL vials with a concentration of 100. mg/mL. How many vials are needed to treat a 64-kg sheep for 5 days?

A soil bag is filled with soil from areas where sheep were sprayed with dicyclanil.

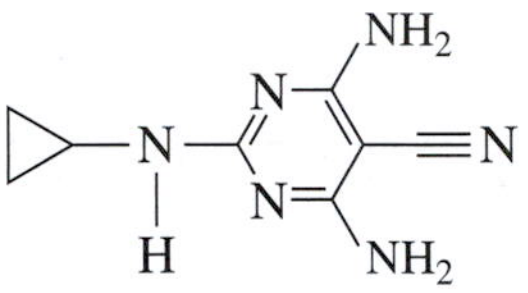

Dicyclanil

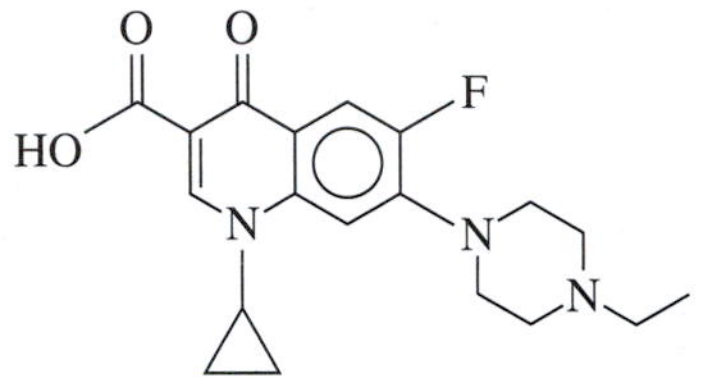

Enrofloxacin

CONCEPT MAP

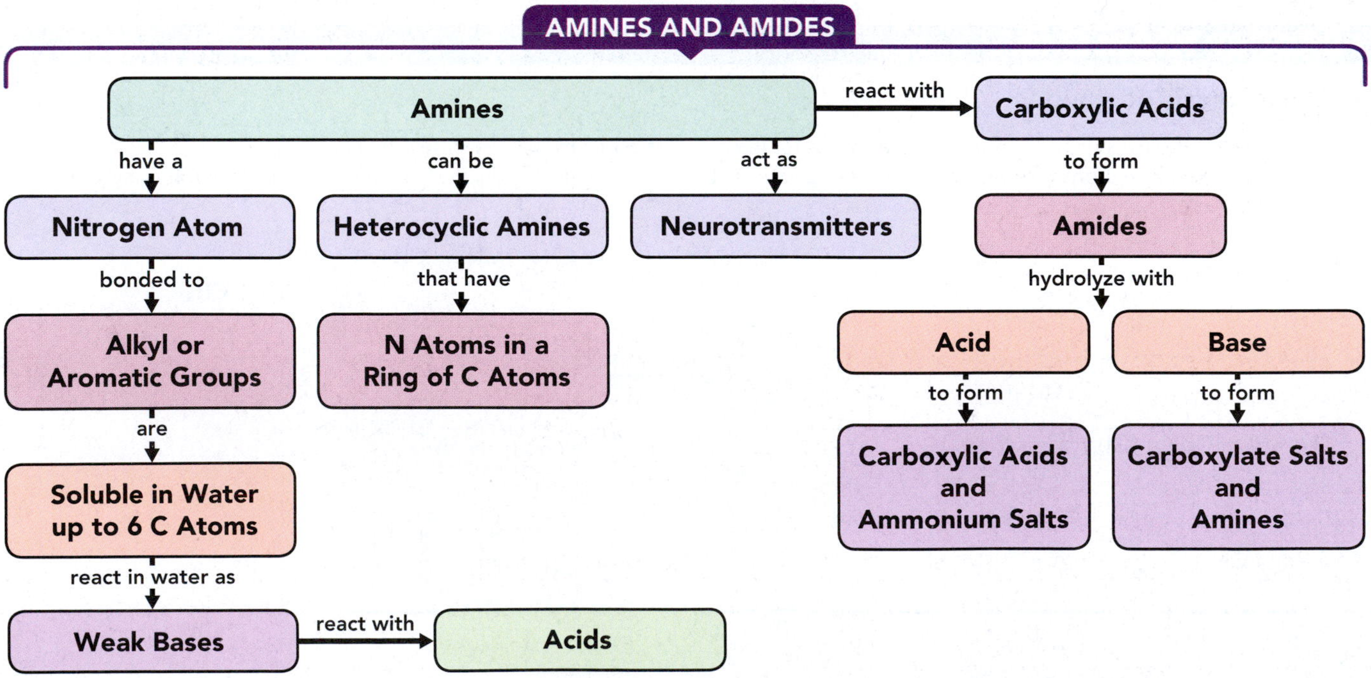
AMINES AND AMIDES
Amines
react with
Carboxylic Acids
have a
can be
act as
to form
Nitrogen Atom
Heterocyclic Amines
Neurotransmitters
Amides
bonded to
that have
hydrolyze with
Alkyl or Aromatic Groups
N Atoms in a Ring of C Atoms
Acid
Base
are
to form
to form
Soluble in Water up to 6 C Atoms
Carboxylic Acids and Ammonium Salts
Carboxylate Salts and Amines
react in water as
Weak Bases
react with
Acids

CHAPTER REVIEW

18.1 Amines

LEARNING GOAL Write the IUPAC and common names for amines; draw the condensed structural and line-angle formulas when given their names. Classify amines as primary (1°), secondary (2°), or tertiary (3°).

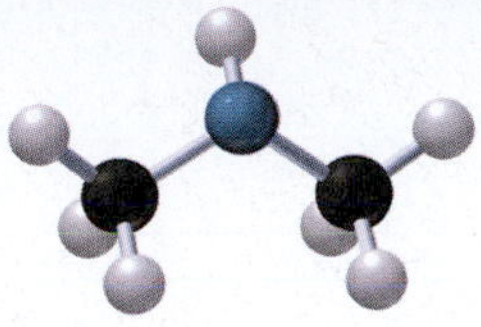

Dimethylamine

- In the IUPAC system, name the longest carbon chain bonded to the N atom by replacing the *e* of its alkane name with *amine*. Groups attached to the nitrogen atom use the *N*- prefix.
- When other functional groups are present, the $-NH_2$ is named as an amino group.
- In the common names of simple amines, the alkyl groups are listed alphabetically followed by *amine*.
- A nitrogen atom attached to one, two, or three alkyl or aromatic groups forms a primary (1°), secondary (2°), or tertiary (3°) amine.

18.2 Properties of Amines

LEARNING GOAL Describe the boiling points and solubility of amines; write balanced chemical equations for the reaction of amines with water and with acids.

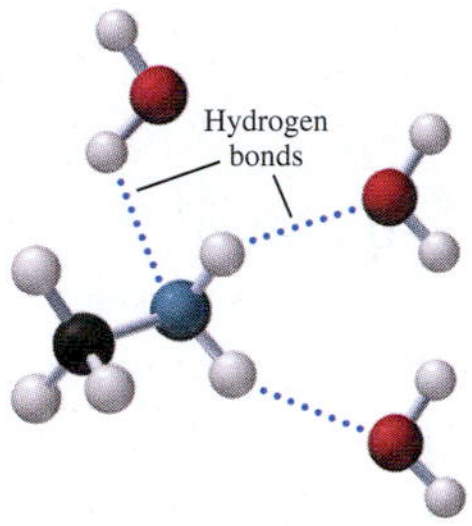
Hydrogen
bonds

- Primary and secondary amines form hydrogen bonds, which make their boiling points higher than those of alkanes of similar mass, but lower than those of alcohols.
- Amines with up to six carbon atoms are soluble in water.
- In water, amines act as weak bases because the nitrogen atom accepts H^+ from water to produce ammonium and hydroxide ions.
- When amines react with acids, they form ammonium salts. As ionic compounds, ammonium salts are solids, soluble in water, and odorless.
- Quaternary (4°) ammonium salts contain four carbon groups bonded to the nitrogen atom.

18.3 Heterocyclic Amines

LEARNING GOAL Identify heterocyclic amines; distinguish between the types of heterocyclic amines.

- Heterocyclic amines are cyclic organic compounds that contain one or more nitrogen atoms in the ring.
- Heterocyclic amines typically consist of five or six atoms and one or more nitrogen atoms.
- Many heterocyclic compounds are known for their physiological activity.
- Alkaloids such as caffeine and nicotine are physiologically active amines derived from plants.

18.4 Neurotransmitters

LEARNING GOAL Describe the role of amines as neurotransmitters.

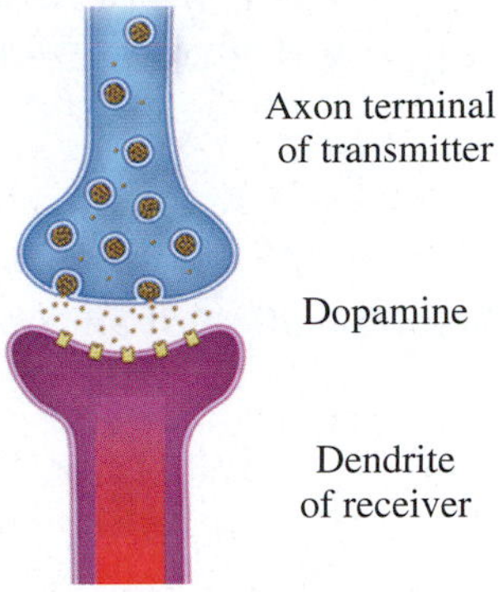
Axon terminal
of transmitter
Dopamine
Dendrite
of receiver

- Neurotransmitters are chemicals that transfer a signal between nerve cells.

18.5 Amides

LEARNING GOAL Draw the amide product from amidation, and write the IUPAC and common names.

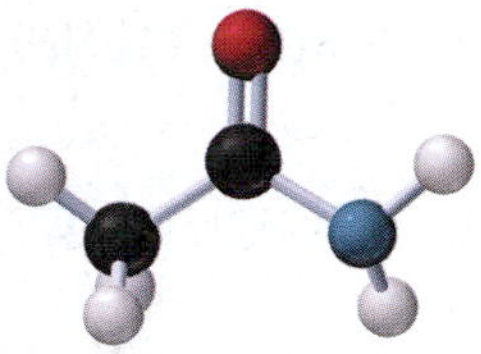

Ethanamide (acetamide)

- Amides are derivatives of carboxylic acids in which the hydroxyl group is replaced by $-NH_2$ or a primary or secondary amine group.
- Amides are formed when carboxylic acids react with ammonia or primary or secondary amines in the presence of heat.
- Amides are named by replacing the *oic acid* or *ic acid* from the carboxylic acid name with *amide*. Any carbon group attached to the nitrogen atom is named using the *N*- prefix.

18.6 Hydrolysis of Amides

LEARNING GOAL Write balanced chemical equations for the hydrolysis of amides.

$$\underset{\text{Ethanamide (acetamide)}}{CH_3-\overset{\overset{\displaystyle O}{\|}}{C}-NH_2} + H_2O + HCl \xrightarrow{\text{Heat}} \underset{\text{Ethanoic acid (acetic acid)}}{CH_3-\overset{\overset{\displaystyle O}{\|}}{C}-OH} + \underset{\text{Ammonium chloride}}{NH_4^+Cl^-}$$

- Hydrolysis of an amide by an acid produces a carboxylic acid and an ammonium salt.
- Hydrolysis of an amide by a base produces a carboxylate salt and an amine.

SUMMARY OF NAMING

Family	Structure	IUPAC Name	Common Name
Amine	$CH_3-CH_2-NH_2$ $CH_3-CH_2-NH-CH_3$	Ethanamine *N*-Methylethanamine	Ethylamine Ethylmethylamine
Ammonium salt	$CH_3-CH_2-\overset{+}{N}H_3\ Cl^-$	Ethylammonium chloride	Ethylammonium chloride
Amide	$CH_3-\overset{\overset{\displaystyle O}{\|}}{C}-NH_2$	Ethanamide	Acetamide

SUMMARY OF REACTIONS

The chapter Sections to review are shown after the name of each reaction.

Reaction of an Amine with Water (18.2)

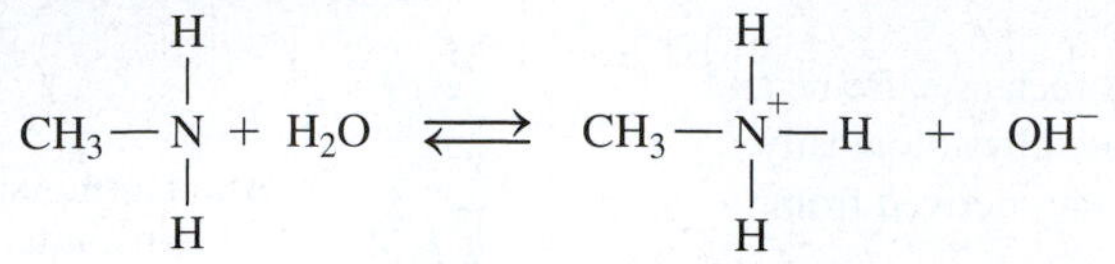

$$CH_3-\underset{\underset{\displaystyle H}{|}}{\overset{\overset{\displaystyle H}{|}}{N}} + H_2O \rightleftharpoons CH_3-\underset{\underset{\displaystyle H}{|}}{\overset{\overset{\displaystyle H}{|}}{\overset{+}{N}}}-H + OH^-$$

Methylamine   Methylammonium   Hydroxide
                            ion   ion

Neutralization of an Amine (18.2)

$$CH_3-\underset{\underset{\displaystyle H}{|}}{\overset{\overset{\displaystyle H}{|}}{N}} + HCl \longrightarrow CH_3-\underset{\underset{\displaystyle H}{|}}{\overset{\overset{\displaystyle H}{|}}{\overset{+}{N}}}-H\ Cl^-$$

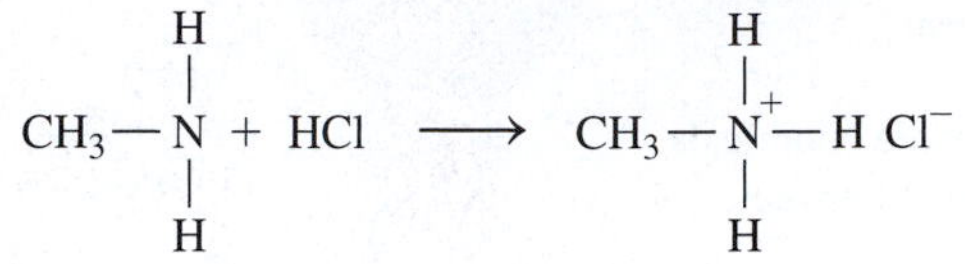

Methylamine   Methylammonium chloride

Amidation: Carboxylic Acid and an Amine (18.5)

$$CH_3-CH_2-\overset{\overset{\displaystyle O}{||}}{C}-OH + H-\underset{\underset{\displaystyle H}{|}}{\overset{\overset{\displaystyle H}{|}}{N}}-H \xrightarrow{\text{Heat}} CH_3-CH_2-\overset{\overset{\displaystyle O}{||}}{C}-\underset{\underset{\displaystyle H}{|}}{N}-H + H_2O$$

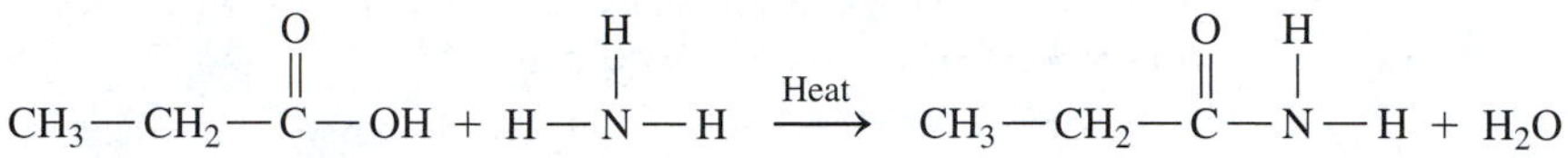

Propanoic acid   Ammonia   Propanamide
(propionic acid)   (propionamide)

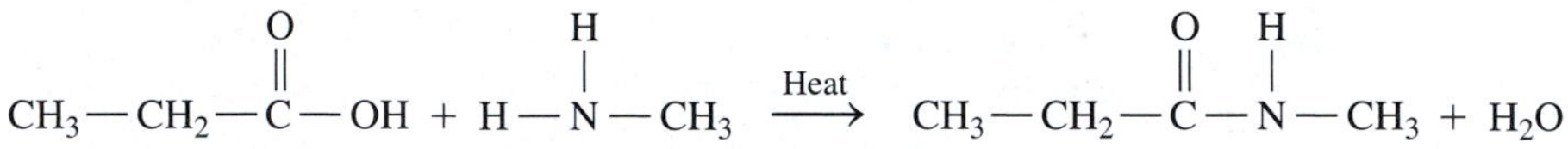

$$CH_3-CH_2-\overset{\overset{\displaystyle O}{||}}{C}-OH + H-\underset{\underset{\displaystyle H}{|}}{\overset{\overset{\displaystyle H}{|}}{N}}-CH_3 \xrightarrow{\text{Heat}} CH_3-CH_2-\overset{\overset{\displaystyle O}{||}}{C}-\underset{\underset{\displaystyle H}{|}}{N}-CH_3 + H_2O$$

Propanoic acid   Methanamine   N-Methylpropanamide
(propionic acid)   (methylamine)   (N-methylpropionamide)

Acid Hydrolysis of an Amide (18.6)

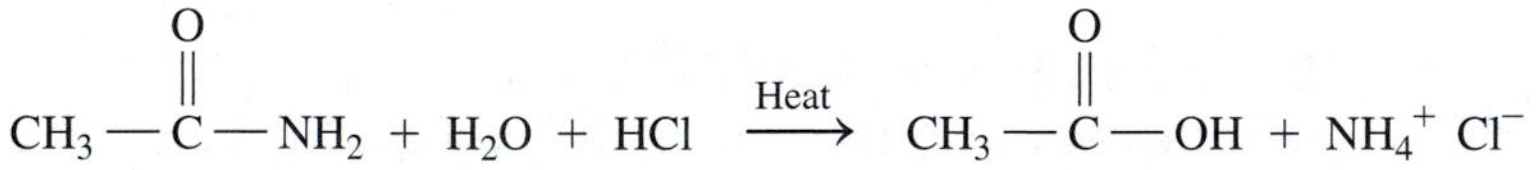

$$CH_3-\overset{\overset{\displaystyle O}{||}}{C}-NH_2 + H_2O + HCl \xrightarrow{\text{Heat}} CH_3-\overset{\overset{\displaystyle O}{||}}{C}-OH + NH_4^+\ Cl^-$$

Ethanamide   Ethanoic acid   Ammonium
(acetamide)   (acetic acid)   chloride

Base Hydrolysis of an Amide (18.6)

$$CH_3-CH_2-\overset{\overset{\displaystyle O}{||}}{C}-\underset{\underset{\displaystyle H}{|}}{N}-CH_3 + NaOH \xrightarrow{\text{Heat}} CH_3-CH_2-\overset{\overset{\displaystyle O}{||}}{C}-O^-\ Na^+ + H_2N-CH_3$$

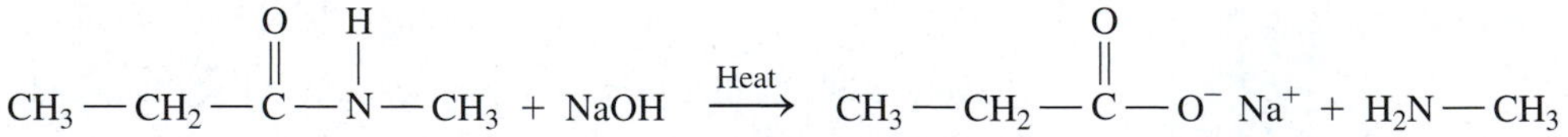

N-Methylpropanamide   Sodium propanoate   Methanamine
(N-methylpropionamide)   (sodium propionate)   (methylamine)

KEY TERMS

alkaloid An amine having physiological activity that is derived from plants.

amidation The formation of an amide from a carboxylic acid and ammonia or an amine.

amide An organic compound containing the carbonyl group attached to an amino group or a substituted nitrogen atom.

amine An organic compound containing a nitrogen atom attached to one, two, or three alkyl or aromatic groups.

ammonium salt An ionic compound produced from an amine and an acid.

heterocyclic amine A cyclic organic compound that contains one or more nitrogen atoms in the ring.

hydrolysis The splitting of a molecule by the addition of water. Amides yield the corresponding carboxylic acid and amine, or their salts.

neurotransmitter A chemical compound that transmits an impulse from a nerve cell (neuron) to a target cell such as another nerve cell, a muscle cell, or a gland cell.

quaternary ammonium salt An ammonium salt in which the nitrogen atom is bonded to four carbon groups.

CORE CHEMISTRY SKILLS

The chapter Section containing each Core Chemistry Skill is shown in parentheses at the end of each heading.

Forming Amides (18.5)

- Amides are formed when carboxylic acids react with ammonia or primary or secondary amines in the presence of heat.

Example: Draw the condensed structural formula for the amide product from the reaction of 3-methylbutanoic acid and ethylamine.

Answer:

$$CH_3-\underset{\displaystyle CH_3}{\overset{|}{CH}}-CH_2-\overset{\displaystyle O}{\overset{\|}{C}}-\overset{\displaystyle H}{\overset{|}{N}}-CH_2-CH_3$$

Hydrolyzing Amides (18.6)

- An amide undergoes acid hydrolysis to produce a carboxylic acid and an ammonium salt.
- An amide undergoes base hydrolysis to produce the carboxylate salt and an amine.

Example: Draw the condensed structural formulas for the products from the **(a)** acid (HCl) and **(b)** base (NaOH) hydrolysis of *N*-ethylbutanamide.

Answer:

a. $CH_3-CH_2-CH_2-\overset{\displaystyle O}{\overset{\|}{C}}-OH + CH_3-CH_2-\overset{+}{N}H_3\ Cl^-$

b. $CH_3-CH_2-CH_2-\overset{\displaystyle O}{\overset{\|}{C}}-O^-\ Na^+ + CH_3-CH_2-NH_2$

UNDERSTANDING THE CONCEPTS

The chapter Sections to review are shown in parentheses at the end of each problem.

Clinical Applications

18.59 The noncaloric sweetener Equal contains aspartame, which is made from two amino acids: aspartic acid and phenylalanine. Identify the functional groups in aspartame. (16.1, 16.3, 18.2, 18.5)

Equal
0 Calorie Sweetener

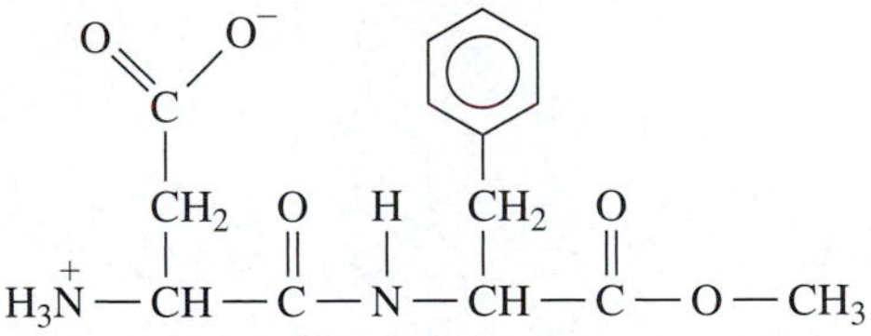

Aspartame

18.60 Some aspirin substitutes contain phenacetin to reduce fever. Identify the functional groups in phenacetin. (13.2, 18.5)

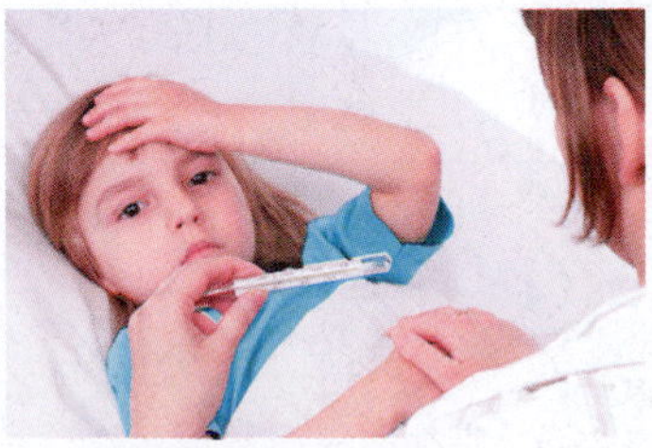

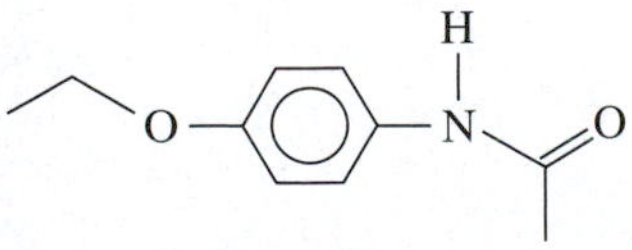

Phenacetin

18.61 Neo-Synephrine is the active ingredient in some nose sprays used to reduce the swelling of nasal membranes. Identify the functional groups in Neo-Synephrine. (13.1, 18.1)

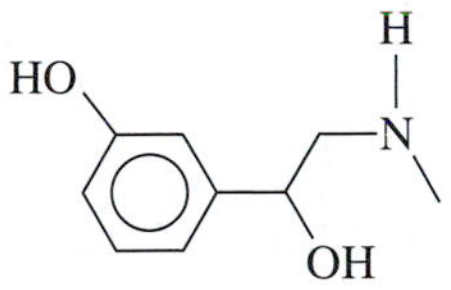

Phenylephrine

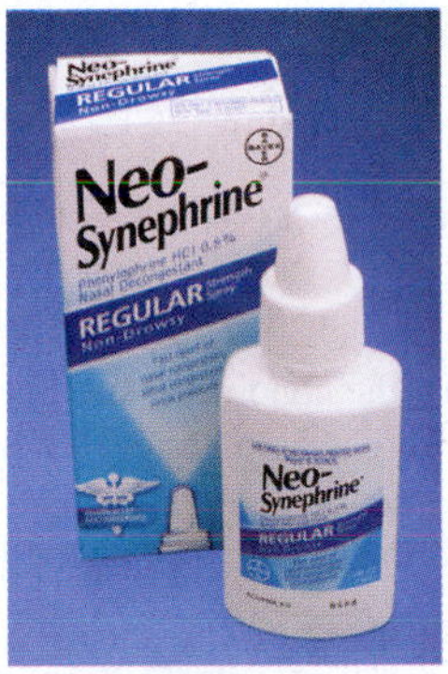
Neo-Synephrine
REGULAR

18.62 Melatonin is a naturally occurring compound in plants and animals, where it regulates the biological time clock. Melatonin is sometimes used to counteract jet lag. Identify the functional groups in melatonin. (13.2, 18.1, 18.5)

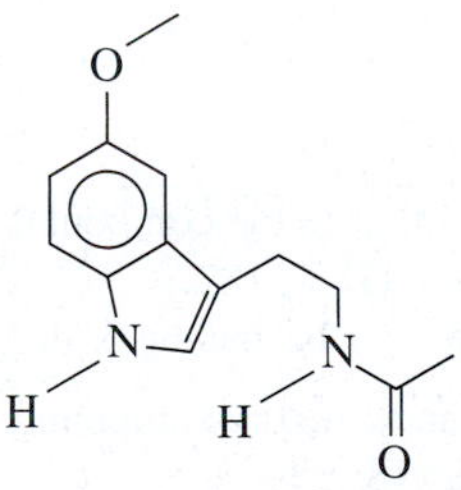

Melatonin

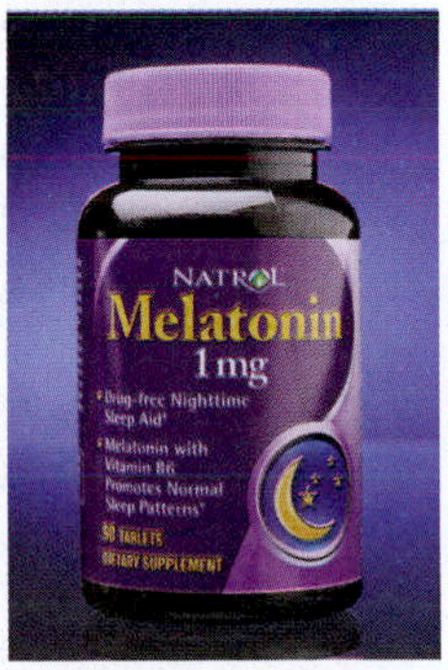
NATROL
Melatonin
1mg

18.63 What is an excitatory neurotransmitter? (18.4)

18.64 What is an inhibitory neurotransmitter? (18.4)

18.65 Identify the structural components that are the same in dopamine, norepinephrine, and epinephrine. (18.4)

18.66 Identify the structural components that are different in dopamine, norepinephrine, and epinephrine. (18.4)

ADDITIONAL PRACTICE PROBLEMS

18.67 Write the IUPAC and common names (if any) and classify each of the following compounds as a primary (1°), secondary (2°), tertiary (3°) amine, or as a quaternary (4°) ammonium salt: (18.1, 18.2)

a. 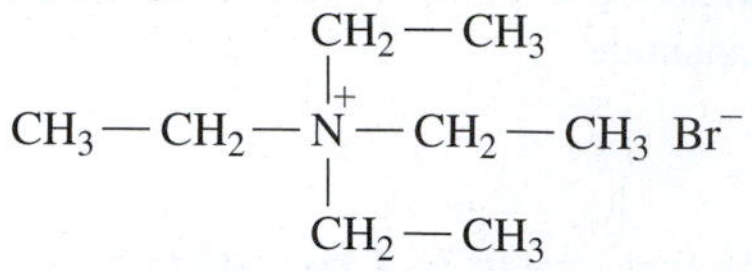

b. $CH_3-CH_2-CH_2-CH_2-CH_2-NH_2$

c. 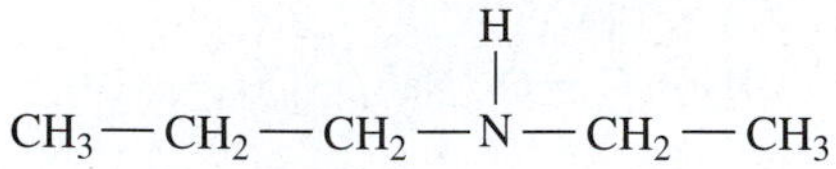

18.68 Write the IUPAC and common names (if any), and classify each of the following compounds as a primary (1°), secondary (2°), tertiary (3°) amine, or as a quaternary (4°) ammonium salt: (18.1, 18.2)

a. 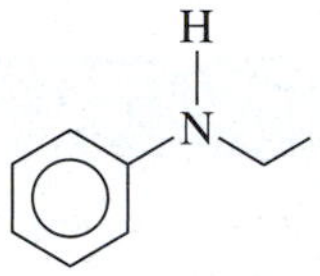

b. 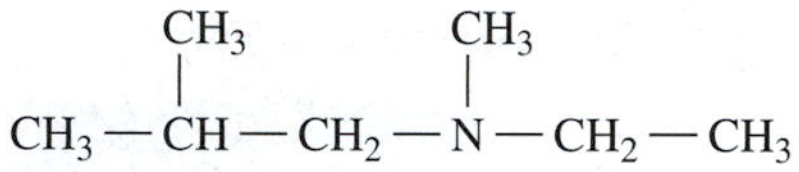

c. 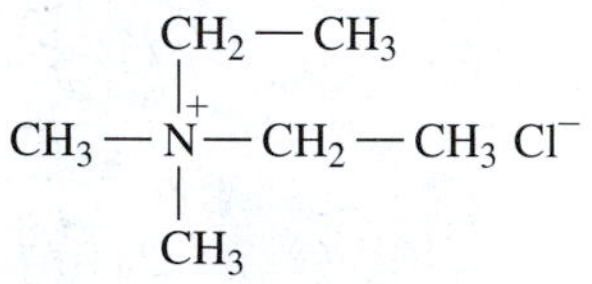

18.69 Draw the condensed structural formula for **a**, **b**, and **c** and the line-angle formula for **d**. (18.1, 18.2)

a. 3-pentanamine **b.** triethylamine
c. dimethylammonium chloride **d.** cyclohexylamine

18.70 Draw the condensed structural formula for **a**, **b**, and **c**, and the line-angle fomula for **d**. (18.1, 18.2)

a. 3-amino-2-hexanol
b. tetramethylammonium bromide
c. butylethylmethylamine
d. *N,N*-dimethylaniline

18.71 In each of the following pairs, indicate the compound that has the higher boiling point. Explain. (18.1, 18.2)

a. 1-butanol or butanamine
b. ethylamine or dimethylamine

18.72 In each of the following pairs, indicate the compound that has the higher boiling point. Explain. (18.1, 18.2)

a. butylamine or diethylamine **b.** butane or propylamine

18.73 In each of the following pairs, indicate the compound that is more soluble in water. Explain. (18.1, 18.2)

a. ethylamine or dibutylamine
b. trimethylamine or *N*-ethylcyclohexylamine

18.74 In each of the following pairs, indicate the compound that is more soluble in water. Explain. (18.1, 18.2, 18.5)

a. butylamine or pentane **b.** butyramide or hexane

18.75 Write the IUPAC name for each of the following amides: (18.5)

a. 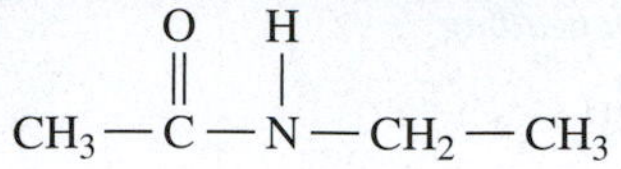

b. 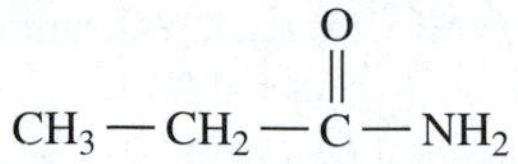

c. 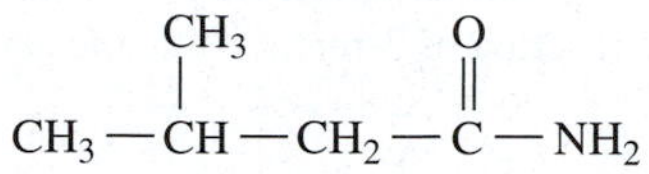

18.76 Write the IUPAC name for each of the following amides: (18.5)

a. 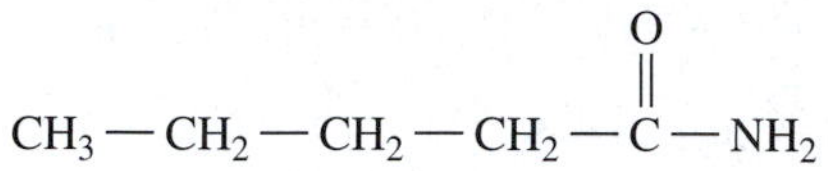

b. 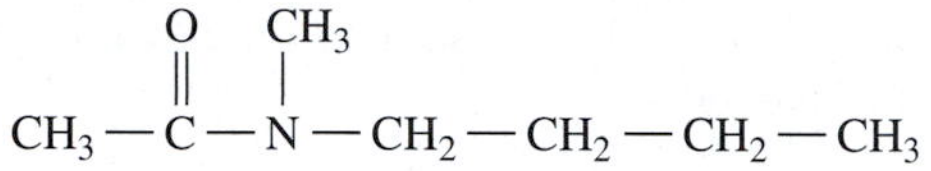

c. 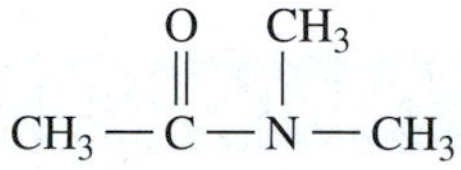

18.77 Draw the condensed structural formulas for the products of the following reactions: (18.2)

a. 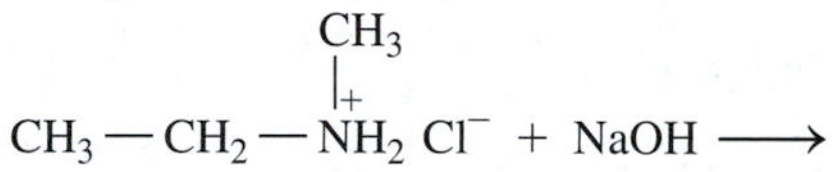

b. 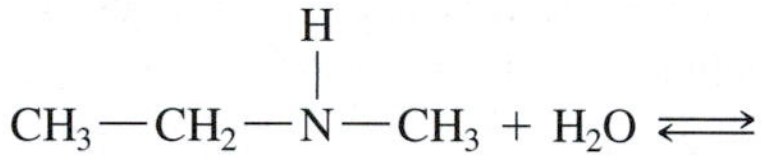

18.78 Draw the condensed structural formulas for the products of the following reactions: (18.2)

a. 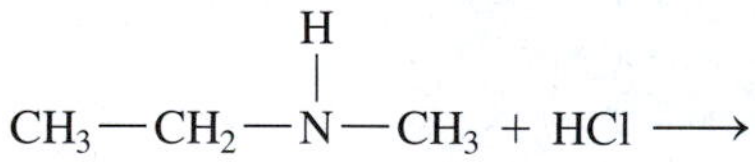

b. $CH_3-CH_2-CH_2-\overset{+}{N}H_3\ Cl^- + NaOH \longrightarrow$

Clinical Applications

18.79 Voltaren (diclofenac) is indicated for acute and chronic treatment of the symptoms of rheumatoid arthritis. Name the functional groups in voltaren. (18.1)

Voltaren

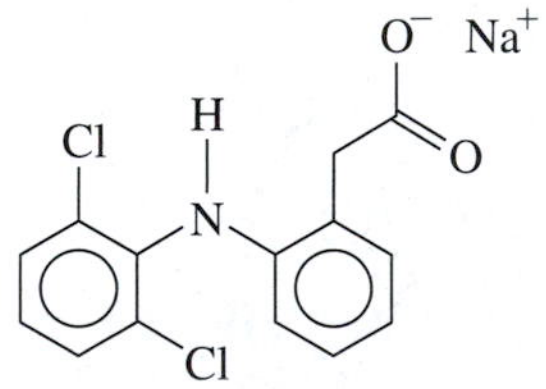

18.80 Toradol (ketorolac) is used in dentistry to relieve pain. Name the functional groups in toradol. (18.3, 18.5)

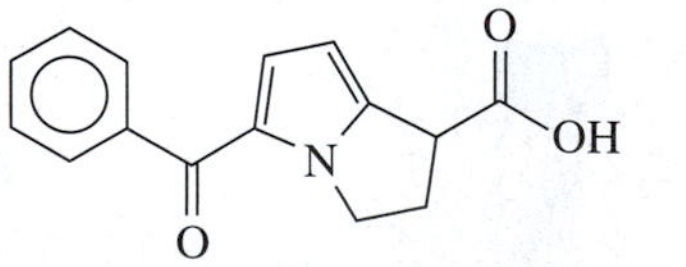

Toradol

18.81 Write the name of the alkaloid described in each of the following: (18.3)
- **a.** from the bark of the cinchona tree and used in malaria treatment
- **b.** found in tobacco
- **c.** found in coffee and tea
- **d.** a painkiller found in the opium poppy plant

18.82 Identify the heterocyclic amine(s) in each of the following: (18.3)
- **a.** caffeine
- **b.** Demerol (meperidine)
- **c.** coniine
- **d.** quinine

18.83 What is the structural difference between tryptophan and serotonin? (18.4)

18.84 What is the structural difference between histidine and histamine? (18.4)

18.85 What does the abbreviation SSRI signify? (18.4)

18.86 Give an example of an SSRI and its role in treating depression. (18.4)

CHALLENGE PROBLEMS

The following problems are related to the topics in this chapter. However, they do not all follow the chapter order, and they require you to combine concepts and skills from several Sections. These problems will help you increase your critical thinking skills and prepare for your next exam.

18.87 There are four amine isomers with the molecular formula C_3H_9N. Draw their condensed structural formulas. Write the common name, and classify each as a primary (1°), secondary (2°), or tertiary (3°) amine. (18.1)

18.88 There are four amide isomers with the molecular formula C_3H_7NO. Draw their condensed structural formulas and write the IUPAC name for each. (18.5)

Clinical Applications

18.89 Use the Internet to look up the structural formula for the following medicinal drugs. List the functional groups in each compound. (18.1, 18.3, 18.5)
- **a.** Keflex, an antibiotic (cefalexin)
- **b.** Inderal, a β-channel blocker used to treat heart irregularities (propranolol)
- **c.** Ibuprofen, an anti-inflammatory agent
- **d.** Aldomet (methyldopa)
- **e.** Carfentanil, a synthetic opioid
- **f.** Triamterene, a diuretic

18.90 Kevlar is a polymer used in tires and bulletproof vests. Part of the strength of Kevlar is due to hydrogen bonds between polymer chains. The polymer chain is: (16.1, 18.1, 18.5)

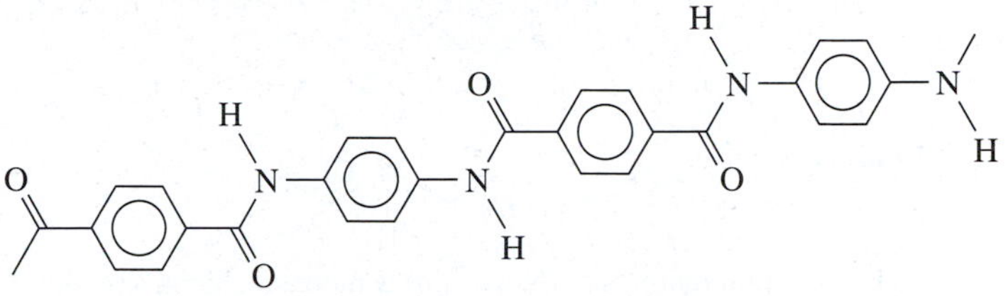

Polyparaphenylene terephthalamide (Kevlar)

- **a.** Draw the line-angle formulas for the carboxylic acid and amine that polymerize to make Kevlar.
- **b.** What feature of Kevlar will give the hydrogen bonds between the polymer chains?

18.91 Why is L-dopa, not dopamine, given to persons with low dopamine levels? (18.4)

18.92 How does cocaine increase the dopamine levels in the synapse? (18.4)

ANSWERS

18.1
- **a.** ethanamine (ethylamine)
- **b.** *N*-methyl-1-propanamine (methylpropylamine)
- **c.** *N*-ethyl-*N*-methylpropanamine (ethylmethylpropylamine)
- **d.** 2-propanamine (isopropylamine)

18.3 **a.** 

b. 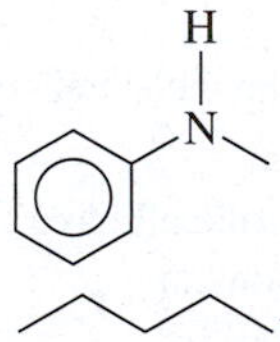

c. 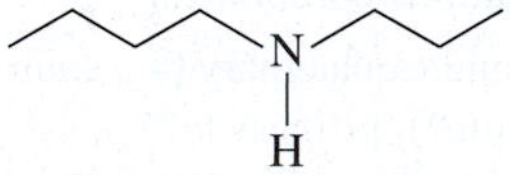

18.5 In a primary amine, there is one alkyl group (and two hydrogens) attached to a nitrogen atom.

18.7
- **a.** primary (1°)
- **b.** secondary (2°)
- **c.** primary (1°)
- **d.** tertiary (3°)
- **e.** tertiary (3°)

18.9
- **a.** CH_3-CH_2-OH has a higher boiling point because the $-OH$ group forms stronger hydrogen bonds than the $-NH_2$ group.
- **b.** $CH_3-CH_2-CH_2-NH_2$ has the higher boiling point because it has a greater molar mass.
- **c.** $CH_3-CH_2-CH_2-NH_2$ has the higher boiling point because it is a primary amine that forms hydrogen bonds. A tertiary amine cannot form hydrogen bonds with other tertiary amines.

18.11 As a primary amine, propylamine can form two hydrogen bonds, which gives it the highest boiling point. Ethylmethylamine, a secondary amine, can form one hydrogen bond, and butane cannot form hydrogen bonds. Thus, butane has the lowest boiling point of the three compounds.

18.13
- **a.** Yes, amines with fewer than seven carbon atoms are soluble in water.
- **b.** Yes, amines with fewer than seven carbon atoms are soluble in water.
- **c.** No, an amine with nine carbon atoms is not soluble in water.
- **d.** Yes, amines with fewer than seven carbon atoms are soluble in water.

18.15 a. $CH_3-NH_2 + H_2O \rightleftharpoons CH_3-\overset{+}{N}H_3 + OH^-$

b. $CH_3-\overset{\displaystyle H}{\overset{|}{N}}-CH_3 + H_2O \rightleftharpoons CH_3-\overset{\displaystyle H}{\overset{|}{\underset{\underset{\displaystyle H}{|}}{\overset{+}{N}}}}-CH_3 + OH^-$

c. 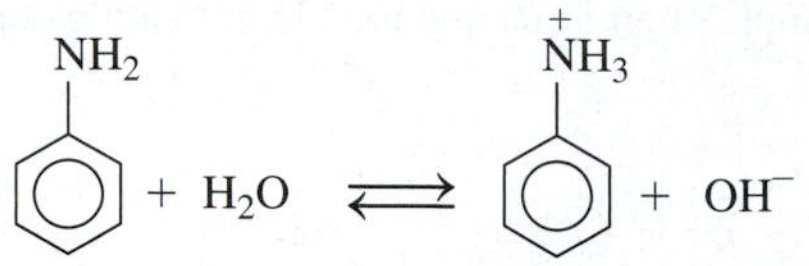

18.17 a. $CH_3-\overset{+}{N}H_3\ Cl^-$

b. $CH_3-\overset{+}{N}H_2-CH_3\ Cl^-$

c. 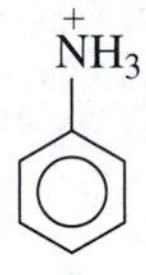

18.19 a. 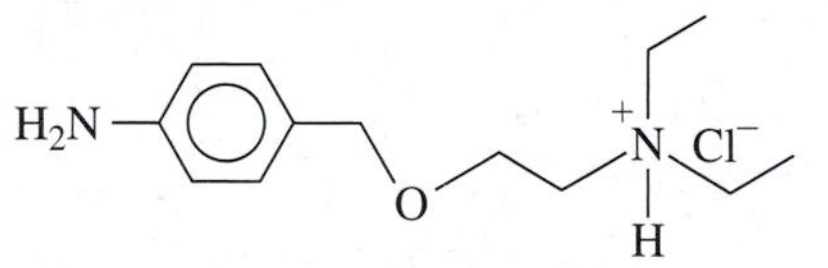

b. The ammonium salt (Novocain) is more soluble in aqueous body fluids than procaine.

18.21 a. piperidine **b.** pyrimidine **c.** pyrrole

18.23 piperidine

18.25 pyrrole

18.27 A neurotransmitter is a chemical compound that transmits an impulse from a nerve cell to a target cell.

18.29 When a nerve impulse reaches the axon terminal, it stimulates the release of neurotransmitters into the synapse.

18.31 A neurotransmitter must be removed from its receptor so that new signals can come from the nerve cells.

18.33 Acetylcholine is a neurotransmitter that communicates between the nervous system and muscle cells.

18.35 Dopamine is a neurotransmitter that controls muscle movement, regulates the sleep–wake cycle, and helps to improve cognition, attention, memory, and learning.

18.37 Serotonin is a neurotransmitter that helps to decrease anxiety, improve mood, learning, and memory; it also reduces appetite, and induces sleep.

18.39 Histamine is a neurotransmitter that causes allergic reactions, which may include inflammation, watery eyes, itchy skin, and hay fever.

18.41 Excess glutamate in the synapse can lead to destruction of brain cells.

18.43 GABA is a neurotransmitter that regulates muscle tone, sleep, and anxiety.

18.45 a. $CH_3-\overset{\displaystyle O}{\overset{\|}{C}}-NH_2$

b. $CH_3-\overset{\displaystyle O}{\overset{\|}{C}}-\overset{\displaystyle H}{\overset{|}{N}}-CH_2-CH_3$

c. 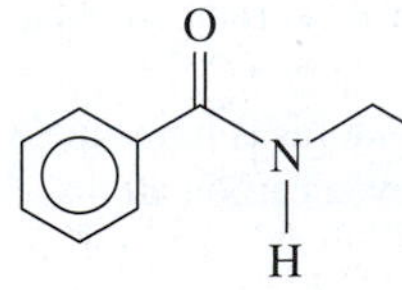

18.47 a. *N*-methylethanamide (*N*-methylacetamide)
b. butanamide (butyramide)
c. methanamide (formamide)
d. *N*-methylbenzamide

18.49 a. $CH_3-CH_2-\overset{\displaystyle O}{\overset{\|}{C}}-NH_2$

b. $CH_3-CH_2-CH_2-CH_2-\overset{\displaystyle O}{\overset{\|}{C}}-NH_2$

c. 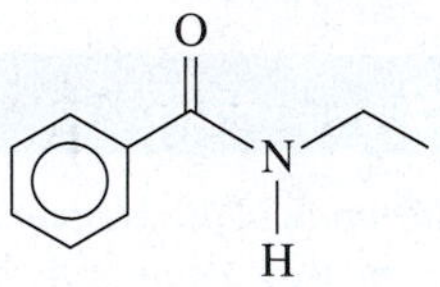

d. 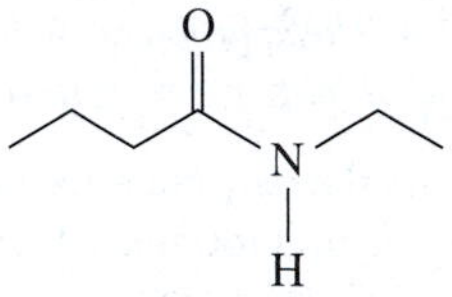

18.51 a. Ethanamide has the higher melting point because it forms more hydrogen bonds as a primary amide than *N*-methylethanamide, which is a secondary amide.
b. Propionamide has the higher melting point because it forms hydrogen bonds, but butane does not.
c. *N*-methylpropanamide, a secondary amide, has a higher melting point because it can form hydrogen bonds, whereas *N,N*-dimethylpropanamide, a tertiary amide, cannot form hydrogen bonds.

18.53 a. 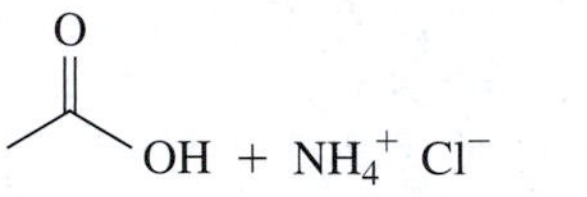

b. $CH_3-CH_2-\overset{\displaystyle O}{\overset{\|}{C}}-OH + NH_4^+\ Cl^-$

c. $CH_3-CH_2-CH_2-\overset{\displaystyle O}{\overset{\|}{C}}-OH + CH_3-\overset{+}{N}H_3\ Cl^-$

d. 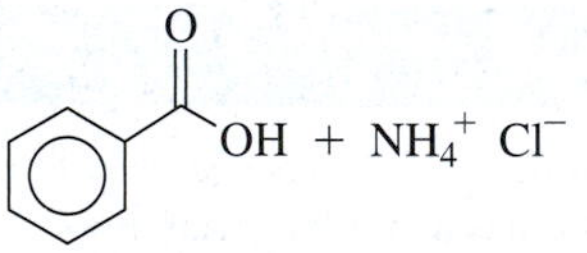

18.55 amine, pyrimidine

18.57 91 mL

18.59 amine, carboxylic acid, amide, aromatic, ester

18.61 phenol, alcohol, amine

18.63 Excitatory neurotransmitters open ion channels and stimulate the receptors to send more signals.

18.65 Dopamine, norepinephrine, and epinephrine all have catechol (3,4-dihydroxyphenyl) and amine components.

18.67 a. tetraethylammonium bromide; quaternary (4°) ammonium salt
b. 1-pentanamine (pentylamine); primary (1°)
c. *N*-ethyl-1-propanamine (ethylpropylamine); secondary (2°)

18.69 a. $CH_3-CH_2-\overset{\displaystyle NH_2}{\overset{|}{CH}}-CH_2-CH_3$

b. $CH_3-CH_2-\overset{\displaystyle CH_2-CH_3}{\overset{|}{N}}-CH_2-CH_3$

c. $CH_3-\overset{+}{N}H_2(CH_3)\ Cl^-$ (CH₃ drawn above N)

d. 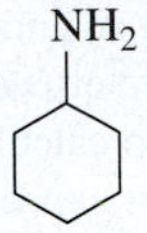

18.71 a. An alcohol with an —OH group such as 1-butanol forms stronger hydrogen bonds than an amine, and has a higher boiling point than an amine.

b. Ethylamine, a primary amine, forms more hydrogen bonds and has a higher boiling point than dimethylamine, which forms fewer hydrogen bonds as a secondary amine.

18.73 a. Ethylamine is a small amine that is soluble because it forms hydrogen bonds with water. Dibutylamine has two large nonpolar alkyl groups that decrease its solubility in water.

b. Trimethylamine is a small tertiary amine that is soluble because it hydrogen bonds with water. *N*-ethylcyclohexylamine has a large nonpolar cycloalkyl group that decreases its solubility in water.

18.75 a. *N*-ethylethanamide

b. propanamide

c. 3-methylbutanamide

18.77 a. $CH_3-CH_2-\underset{}{\overset{H}{N}}-CH_3 + NaCl + H_2O$

b. $CH_3-CH_2-\overset{+}{N}H_2-CH_3 + OH^-$

18.79 aromatic, amine, carboxylate salt

18.81 a. quinine

b. nicotine

c. caffeine

d. morphine, codeine

18.83 Tryptophan contains a carboxylic acid group that is not present in serotonin. Serotonin has a hydroxyl group (—OH) on the aromatic ring that is not present in tryptophan.

18.85 SSRI stands for selective serotonin reuptake inhibitor.

18.87 $CH_3-CH_2-CH_2-NH_2$

Propylamine (1°)

$CH_3-CH_2-\overset{H}{N}-CH_3$

Ethylmethylamine (2°)

$CH_3-\overset{CH_3}{N}-CH_3$

Trimethylamine (3°)

$CH_3-\overset{CH_3}{CH}-NH_2$

Isopropylamine (1°)

18.89 a. aromatic, amine, amide, carboxylic acid, cycloalkene

b. aromatic, ether, alcohol, amine

c. aromatic, carboxylic acid

d. phenol, amine, carboxylic acid

e. aromatic, amine, ester, amide

f. aromatic, amine

18.91 Dopamine is needed in the brain, where it is important in controlling muscle movement. Because dopamine cannot cross the blood–brain barrier, persons with low levels of dopamine are given L-dopa, which can cross the blood–brain barrier, where it is converted to dopamine.

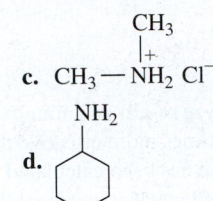

COMBINING IDEAS from Chapters 15 to 18

CI.35 The plastic known as PETE (**p**oly**e**thylene**te**rephthalate) is used to make plastic soft drink bottles and containers for salad dressing, shampoos, and dishwashing liquids. PETE is a polymer of terephthalic acid and ethylene glycol. Today, PETE is the most widely recycled of all the plastics. After it is separated from other plastics, PETE can be used in polyester fabric, door mats, and tennis ball containers. In 2015, 1.80×10^9 lb of PETE bottles were recycled in the U.S. The density of PETE is 1.38 g/mL. (2.5, 2.7, 16.3)

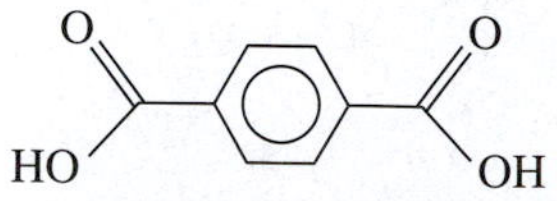
Terephthalic acid

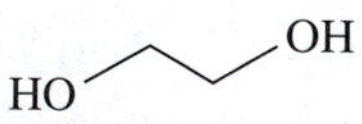
Ethylene glycol

Plastic bottles made of PETE are ready to be recycled.

a. Draw the line-angle formula for the ester formed from one molecule of terephthalic acid and one molecule of ethylene glycol.
b. Draw the line-angle formula for the product formed when a second molecule of ethylene glycol reacts with the ester you drew in part **a**.
c. How many kilograms of PETE bottles were recycled in 2015 in the U.S.?
d. If 2.71×10^9 kg of PETE bottles were sold in 2015, what percentage of those bottles were recycled?
e. What volume, in liters, of PETE bottles were recycled in 2015 in the U.S.?
f. Suppose a landfill holds 2.7×10^7 L of recycled PETE. If all the PETE bottles recycled in 2015 in the U.S. were placed instead in landfills, how many landfills would be needed?

CI.36 Epibatidine is one of the alkaloids that the Ecuadorian poison dart frog (*Epipedobates tricolor*) secretes through its skin. Natives of the rainforest prepare poison darts by rubbing the tips on the skin of the poison dart frogs. The effect of a very small amount of the poison can paralyze or kill an animal. As a pain reliever, epibatidine is 200 times more effective than morphine. Although a therapeutic dose has been calculated as 2.5 mcg/kg, epibatidine has adverse effects. It is expected that more research to chemically modify the epibatidine molecule will produce an important pain reliever. (7.4, 7.5, 18.3)

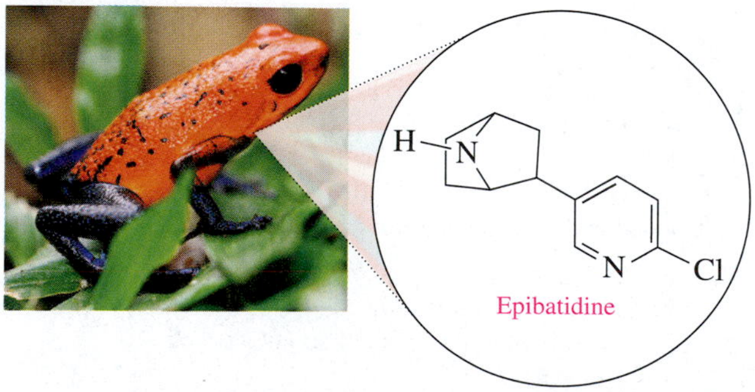
Epibatidine

Natives in the South American rainforests obtain poison for blow darts from the alkaloids secreted by poison dart frogs.

a. What two heterocyclic amines are in the structure of epibatidine?
b. What is the molecular formula of epibatidine?
c. What is the molar mass of epibatidine?
d. How many grams of epibatidine would be given to a 60.-kg person if the dose is 2.5 mcg/kg?
e. How many molecules of epibatidine would be given to a 60.-kg person for the dose in part **d**?

CI.37 A person with type 2 diabetes develops hyperglycemia (high blood sugar) because their body resists the effects of insulin, which causes glucose to be absorbed from the blood into the cells. For some type 2 diabetics, changes in diet and exercise are combined with medication to lower blood sugar levels. The most commonly prescribed medication for this purpose is metformin (Glucophage). (7.5, 12.2, 18.1, 18.2)

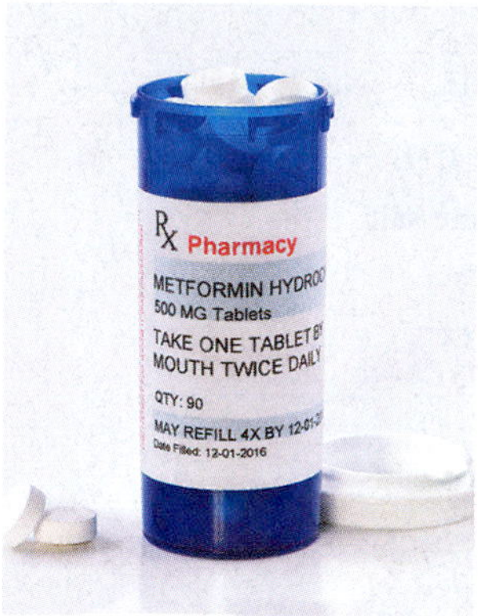
Rx Pharmacy
METFORMIN HYDRO
500 MG Tablets
TAKE ONE TABLET B
MOUTH TWICE DAIL
QTY: 90
MAY REFILL 4X BY 12-01-2
Date Filled: 12-01-2016
Medication may be used to control high blood sugar.

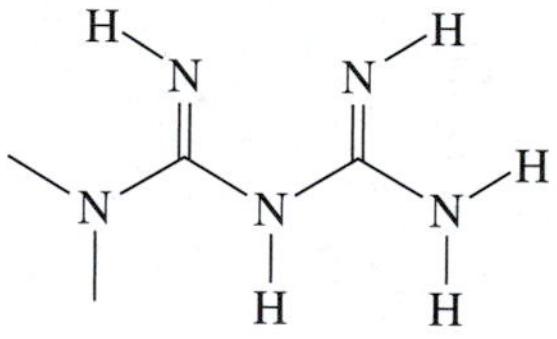
Metformin

a. Identify the functional group in metformin.
b. Would you expect metformin to act like an acid or like a base in water?
c. What is the molecular formula of metformin?
d. What is the molar mass of metformin?
e. The recommended initial dose for adult patients is one 500.-mg tablet of metformin twice a day. How many moles of metformin would an adult take during one week of metformin treatment?

CI.38 Glyceryl trimyristate (trimyristin) is found in the seeds of the nutmeg (*Myristica fragrans*). Trimyristin is used as a lubricant and fragrance in soaps and shaving creams. Isopropyl myristate is used to increase absorption of skin creams. Draw the condensed structural formula for each of the following: (16.1, 16.2, 16.3, 16.4, 16.5)

Nutmeg contains high levels of glyceryl trimyristate.

a. myristic acid (14:0)
b. glyceryl trimyristate (trimyristin)
c. isopropyl myristate
d. products from the hydrolysis of glyceryl trimyristate with an acid catalyst
e. products from the saponification of glyceryl trimyristate with KOH
f. reactant and product for oxidation of myristyl alcohol to myristic acid

CI.39 Panose is a trisaccharide that is being considered as a possible sweetener by the food industry. (15.4, 15.5, 15.6)

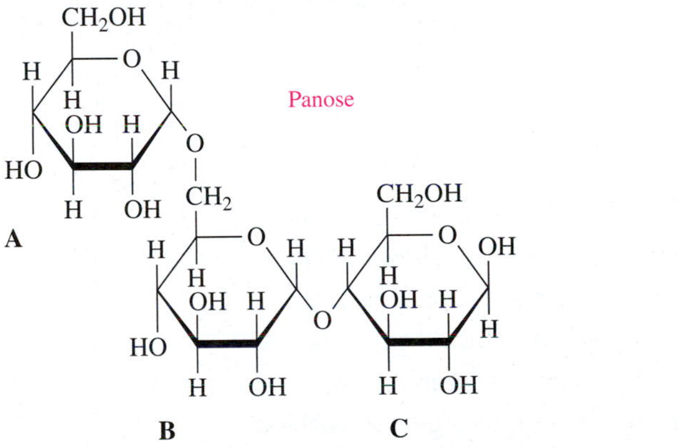

A
B
C

Panose

a. What are the monosaccharide units **A**, **B**, and **C** in panose?
b. What type of glycosidic bond connects monosaccharides **A** and **B**?
c. What type of glycosidic bond connects monosaccharides **B** and **C**?
d. Is the structure drawn as α- or β-panose?
e. Why would panose be a reducing sugar?

CI.40 Hyaluronic acid (HA), a polymer of about 25 000 disaccharide units, is a natural component of eye and joint fluid as well as skin and cartilage. Due to the ability of HA to absorb water, it is used in skin care products and injections to smooth wrinkles and for treatment of arthritis. The repeating disaccharide units in HA consist of D-gluconic acid and *N*-acetyl-D-glucosamine. *N*-Acetyl-D-glucosamine is an amide derived from acetic acid and D-glucosamine, in which an amine group ($—NH_2$) replaces the hydroxyl on carbon 2 of D-glucose. Another natural polymer called chitin is found in the shells of crabs and lobsters. Chitin is made of repeating units of *N*-acetyl-D-glucosamine connected by $\beta(1\rightarrow4)$-glycosidic bonds. (15.4, 15.5, 15.6, 18.5)

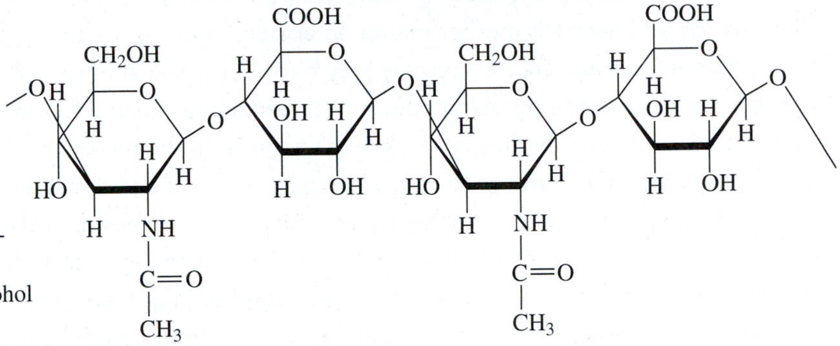

Hyaluronic acid

The shells of crabs and lobsters contain chitin.

a. Draw the Haworth structure for the product from the oxidation reaction of the hydroxyl group on carbon 6 in β-D-glucose to form β-D-gluconic acid.
b. Draw the Haworth structure for β-D-glucosamine.
c. Draw the Haworth structure for the amide of β-D-glucosamine and acetic acid.
d. What are the two types of glycosidic bonds that link the monosaccharides in hyaluronic acid?
e. Draw the structure for a section of chitin with two *N*-acetyl-D-glucosamine units linked by $\beta(1\rightarrow4)$-glycosidic bonds.

ANSWERS

CI.35 a.

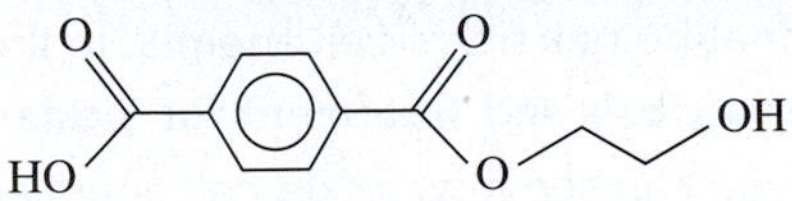

b.

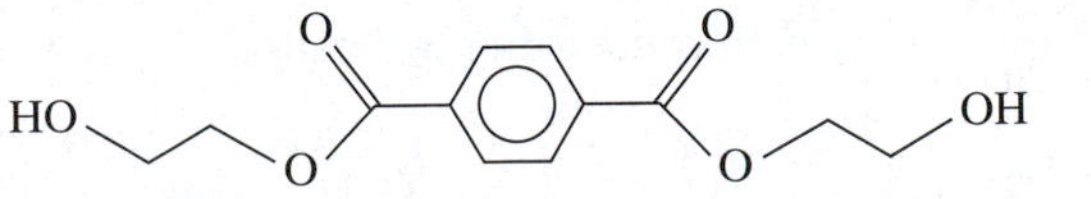

c. 30.2%
d. 8.18×10^8 kg of PETE
e. 5.93×10^8 L of PETE
f. 22 landfills

CI.37 a. amine
b. base
c. $C_4H_{11}N_5$
d. 129.18 g/mole
e. 0.0542 mole of metformin

CI.39 a. **A**, **B**, and **C** are all glucose.
b. An $\alpha(1\rightarrow6)$-glycosidic bond links **A** and **B**.
c. An $\alpha(1\rightarrow4)$-glycosidic bond links **B** and **C**.
d. β-panose
e. Panose is a reducing sugar because it has a free hydroxyl group on carbon 1 of structure **C**, which allows glucose **C** to form an aldehyde.

19

Amino Acids and Proteins

Jeremy, a 9-month-old boy recently adopted from Africa, has a fever, along with painful swelling of his hands and feet. His mother makes an appointment to see Samantha, a hematology nurse. Samantha suspects that Jeremy may have sickle-cell anemia, a disease that results from a defective form of hemoglobin in the blood. Samantha draws a blood sample, which will be tested to determine if Jeremy has sickle-cell anemia. In the meantime, Samantha suggests that Jeremy's mother give him over-the-counter pain relievers and that she increase his fluid intake.

Hemoglobin is a protein that transports oxygen in the blood. Proteins are large biological molecules that carry out many different functions in living organisms. Proteins are made of smaller molecules, called amino acids, linked together in a chain. When the order of amino acids in a protein is altered, the result may be a defective protein that is no longer able to effectively carry out its original function. In the case of sickle-cell anemia, an alteration in the amino acids of the hemoglobin protein leads to patients experiencing anemia (a deficiency of red blood cells) and its accompanying symptoms.

CAREER

Hematology Nurse

A hematology nurse specializes in treating patients with blood disorders like hemophilia, sickle-cell anemia, leukemia, and lymphoma. They take medical histories, perform exams, work with physicians to develop and initiate patient treatment plans, and educate patients about how to manage their disorders. They are often called on to start IVs and to assist with blood transfusions, to take blood samples, or to administer drugs for pain management or chemotherapy.

Hematology nursing is closely associated with oncology (cancer) nursing, so a person wishing to specialize in hematology nursing often gains experience in a cancer clinic after becoming a licensed RN but before becoming certified as an Oncology Certified Nurse (OCN) or as a Certified Pediatric Hematology Oncology Nurse (CPHON).

CLINICAL UPDATE

Jeremy's Diagnosis and Treatment for Sickle-Cell Anemia

Jeremy's blood test indicates that he has sickle-cell anemia. In the **CLINICAL UPDATE Jeremy's Diagnosis and Treatment for Sickle-Cell Anemia**, pages 681–682, you can see how sickle-cell anemia is diagnosed, and how Jeremy is treated with medication.

LOOKING AHEAD

19.1 Proteins and Amino Acids

LEARNING GOAL Classify proteins by their functions. Give the name and abbreviations for an amino acid and draw its structure at physiological pH.

Proteins are one of the most prevalent types of molecules in living organisms. In fact, researchers derived the name "protein" from the Greek word *proteios*, meaning "first," to indicate the central roles that proteins play in living organisms. For example, there are proteins that form structural components such as cartilage, muscles, hair, and nails. Wool, silk, feathers, and horns in animals are made of proteins (see **FIGURE 19.1**). Proteins that function as enzymes regulate biological reactions such as digestion and cellular metabolism. Other proteins, such as hemoglobin and myoglobin, transport oxygen in the blood and muscle. **TABLE 19.1** gives examples of proteins that are classified by their functions.

FIGURE 19.1 ▶ The horns of animals are made of proteins.

Q What class of protein would be in horns?

TEST

Try Practice Problems 19.1 and 19.2

TABLE 19.1 Classification of Some Proteins and Their Functions

Class of Protein	Function	Examples
Structural	Provide structural components	*Collagen* is in tendons and cartilage. *Keratin* is in hair, skin, wool, and nails.
Contractile	Make muscles move	*Myosin* and *actin* contract muscle fibers.
Transport	Carry essential substances throughout the body	*Hemoglobin* transports oxygen. *Lipoproteins* transport lipids.
Storage	Store nutrients	*Casein* stores protein in milk. *Ferritin* stores iron in the spleen and liver.
Hormone	Regulate body metabolism and the nervous system	*Insulin* regulates blood glucose level. *Growth hormone* regulates body growth.
Enzyme	Catalyze biochemical reactions in the cells	*Sucrase* catalyzes the hydrolysis of sucrose. *Trypsin* catalyzes the hydrolysis of proteins.
Protection	Recognize and destroy foreign substances	*Immunoglobulins* stimulate immune responses.

Although there are many types of proteins with many different functions, all proteins are made from the same building blocks. Proteins are formed when smaller molecules called amino acids link together in a chain. The specific order of the amino acids in the chain determines how the protein will fold into its three-dimensional shape, which determines the protein's function.

Just how important are the amino acids in determining the function of a protein? Sickle-cell anemia is a disease caused by an abnormality in one of the subunits of the hemoglobin protein. This abnormal protein causes red blood cells (RBCs) to change from a rounded shape to a crescent shape, like a sickle, which interferes with their ability to transport adequate quantities of oxygen. Patients who suffer from sickle-cell anemia experience fatigue, shortness of breath, dizziness, headaches, coldness in the hands and feet, and even jaundice. All of these problems are caused by the substitution of *one* amino acid out of more than 800 in the hemoglobin molecule. To understand why one amino acid can have such a significant effect on the structure and function of a large protein molecule, we need to look at the structures of amino acids.

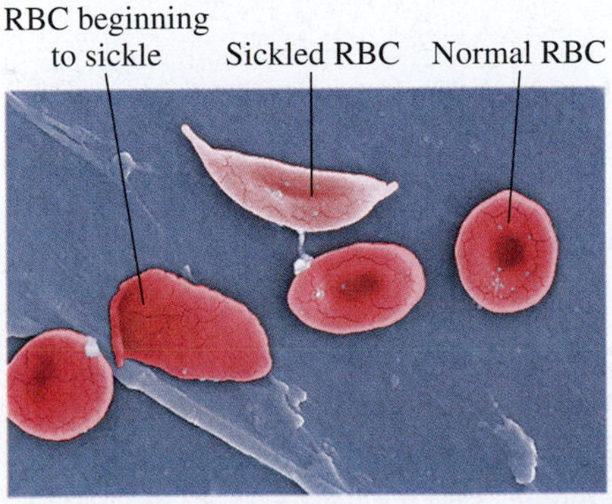

RBC beginning to sickle
Sickled RBC
Normal RBC

Red blood cells containing normal hemoglobin have a rounded shape, whereas red blood cells containing sickled hemoglobin have a crescent or sickle shape.

Amino Acids

Proteins are composed of molecular building blocks called *amino acids*. There are many amino acids in nature; however, there are only 20 amino acids commonly found in the proteins of living organisms. Every **amino acid** has a central carbon atom, called the α carbon, bonded to two α functional groups: an ammonium group ($-NH_3^+$) and a carboxylate group ($-COO^-$). The α carbon is also bonded to a hydrogen atom and a side chain called an R group. The differences in the 20 amino acids present in human proteins are due to the unique characteristics of the R groups. For example, alanine has a methyl, $-CH_3$, as its R group.

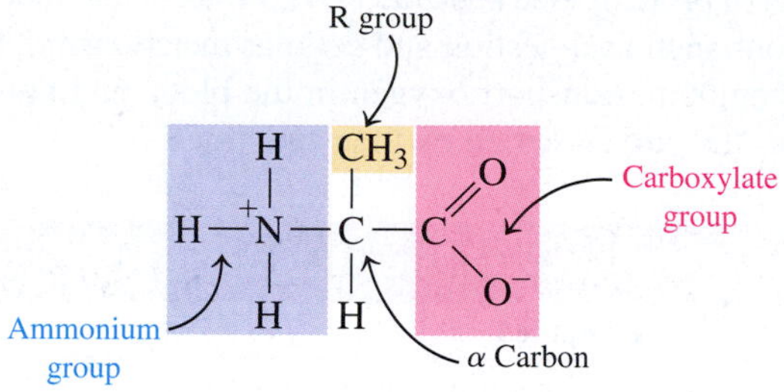
R group
Carboxylate group
Ammonium group
α Carbon

Ionized form of alanine

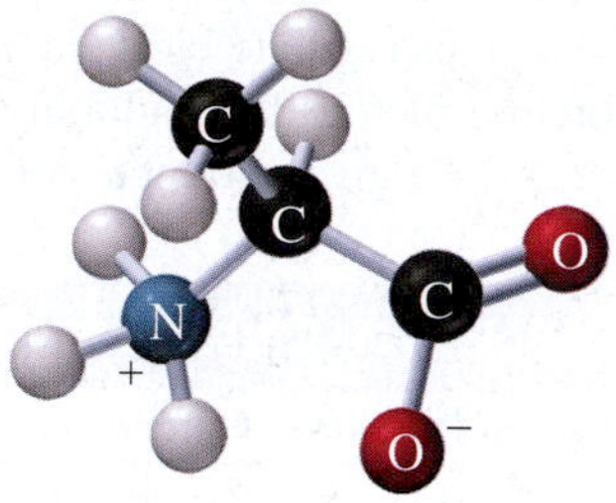

Ball-and-stick model of alanine

Classification of Amino Acids

We classify amino acids using their specific R groups, which determine their properties in aqueous solution. The **nonpolar amino acids** have hydrogen, alkyl, or aromatic R groups, which make them *hydrophobic* ("water fearing"). The **polar amino acids** have R groups that interact with water, which makes them *hydrophilic* ("water loving"). There are three groups of polar amino acids. The R groups of *polar neutral amino acids* contain hydroxyl ($-OH$), thiol ($-SH$), or amide ($-CONH_2$) groups. The R group of a polar **acidic amino acid** contains a carboxylate group ($-COO^-$). The R group of a polar **basic amino acid** contains an amino group, which ionizes to give an ammonium ion. The names and structures of the 20 alpha amino acids commonly found in proteins, along with their three-letter and one-letter abbreviations, are shown at physiological pH (7.4) in **TABLE 19.2**.

ENGAGE

Why is leucine classified as a nonpolar amino acid, whereas threonine is classified as a polar amino acid?

▶ SAMPLE PROBLEM 19.1 Polarity of Amino Acids

TRY IT FIRST

Use the information in Table 19.2 to classify each of the following amino acids as nonpolar or polar. If polar, indicate if the R group is neutral, acidic, or basic. Indicate if each would be hydrophobic or hydrophilic.

a. valine **b.** asparagine

SOLUTION

a. The R group in valine is an alkyl group consisting of C atoms and H atoms, which makes valine a nonpolar amino acid that is hydrophobic.

b. The R group in asparagine contains an amide group ($-CONH_2$), which makes asparagine a polar amino acid that is neutral and hydrophilic.

STUDY CHECK 19.1

Use the information in Table 19.2 to classify each of the following amino acids as nonpolar or polar. If polar, indicate if the R group is neutral, acidic, or basic and if it would be hydrophobic or hydrophilic.

a. lysine **b.** aspartate

ANSWER

a. The R group in lysine contains an amino group, which makes lysine a polar amino acid that is basic and hydrophilic.

b. The R group in aspartate contains a carboxylate group ($-COO^-$), which makes aspartate a polar amino acid that is acidic and hydrophilic.

TEST

Try Practice Problems 19.3, 19.4, 19.7, and 19.8

TABLE 19.2 Structures, Names, and Abbreviations of 20 Common Amino Acids at Physiological pH (7.4)

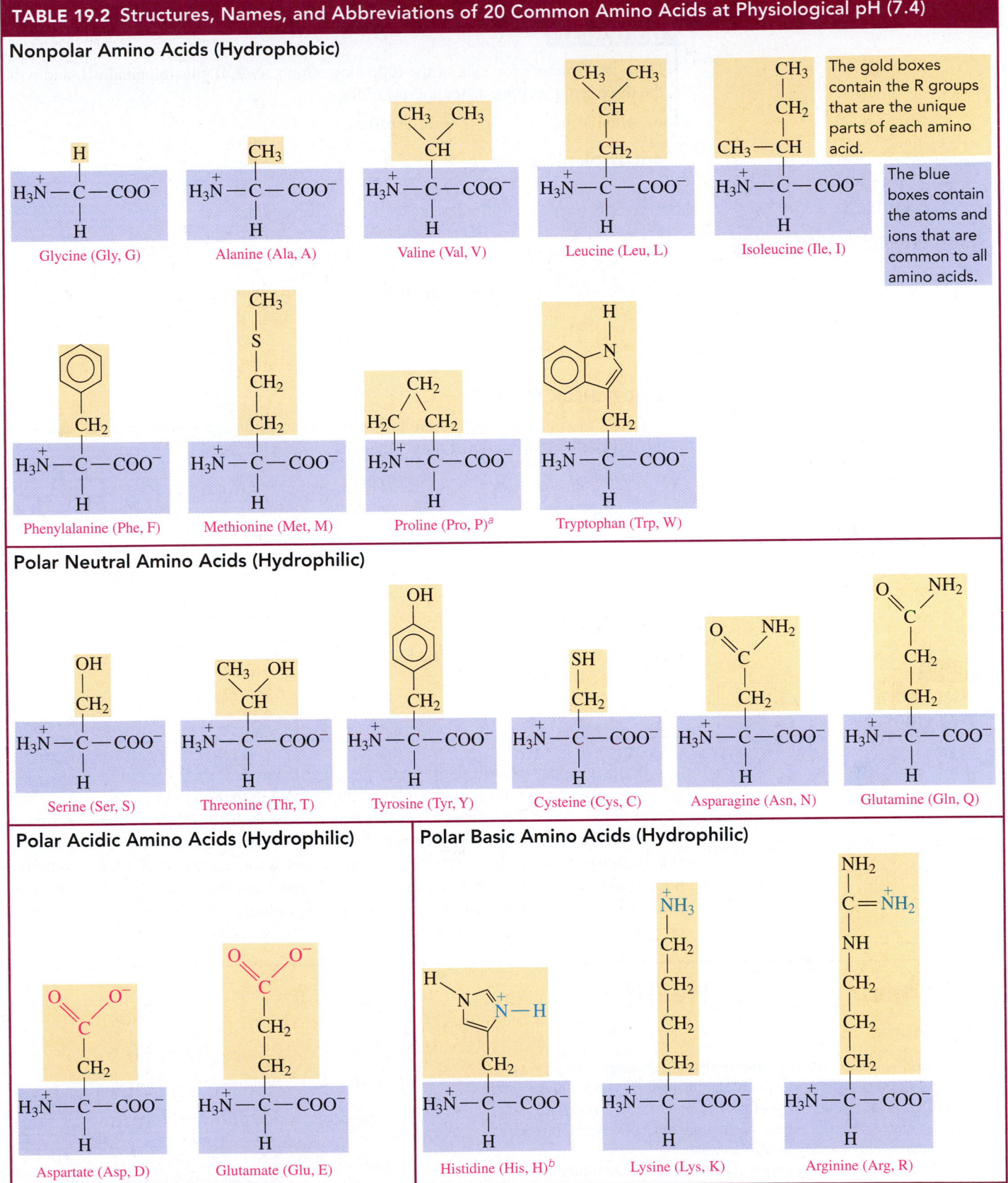

Nonpolar Amino Acids (Hydrophobic)

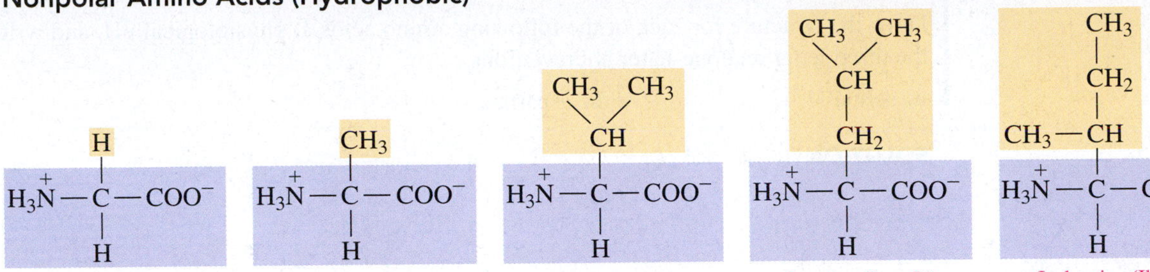

Glycine (Gly, G) · Alanine (Ala, A) · Valine (Val, V) · Leucine (Leu, L) · Isoleucine (Ile, I)

The gold boxes contain the R groups that are the unique parts of each amino acid.

The blue boxes contain the atoms and ions that are common to all amino acids.

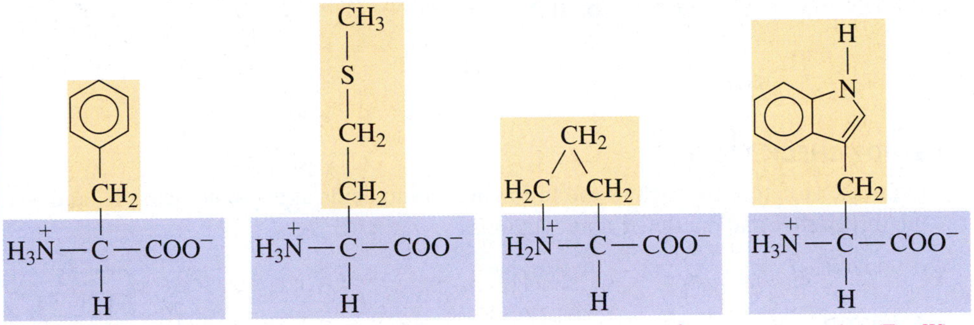

Phenylalanine (Phe, F) · Methionine (Met, M) · Proline (Pro, P)[a] · Tryptophan (Trp, W)

Polar Neutral Amino Acids (Hydrophilic)

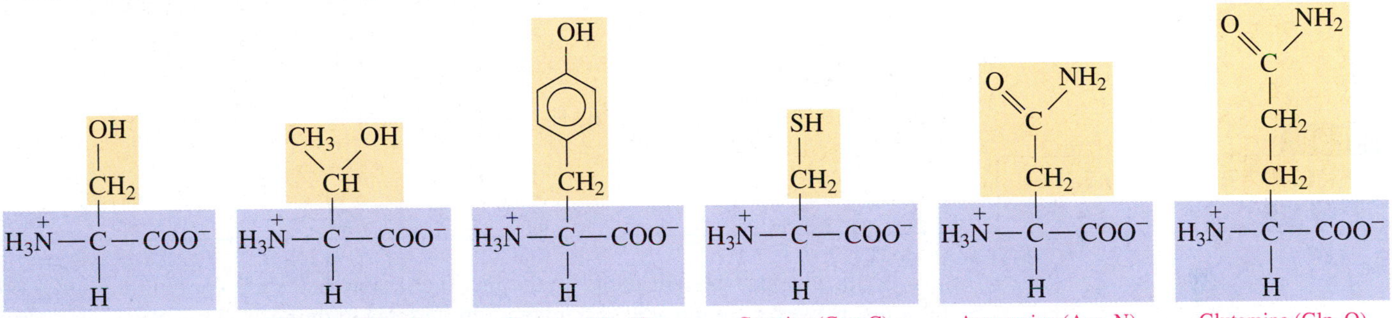

Serine (Ser, S) · Threonine (Thr, T) · Tyrosine (Tyr, Y) · Cysteine (Cys, C) · Asparagine (Asn, N) · Glutamine (Gln, Q)

Polar Acidic Amino Acids (Hydrophilic)

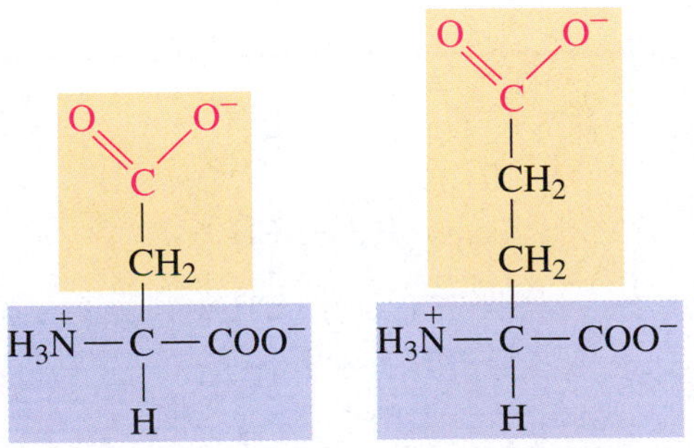

Aspartate (Asp, D) · Glutamate (Glu, E)

Polar Basic Amino Acids (Hydrophilic)

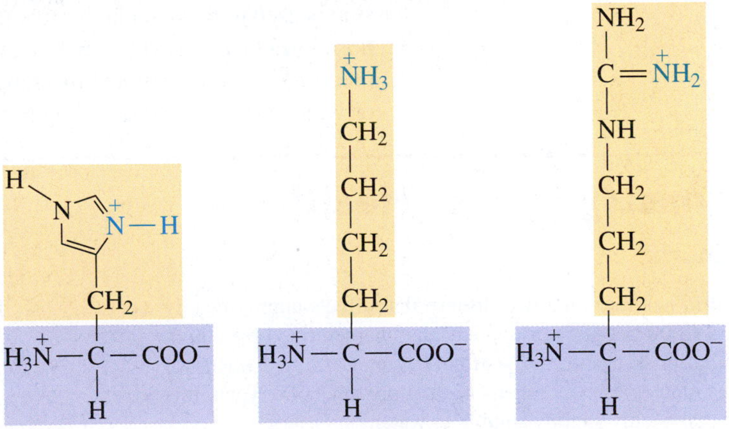

Histidine (His, H)[b] · Lysine (Lys, K) · Arginine (Arg, R)

[a] Proline is a cyclic amino acid because its R group bonds to the nitrogen atom attached to the α carbon.
[b] At physiological pH, some histidine molecules have a positively charged R group, while other histidine molecules have a neutral R group. The molecule with the positively charged R group is shown here.

▸SAMPLE PROBLEM 19.2 Structural Formulas of Amino Acids

TRY IT FIRST

Draw the structure for each of the following amino acids at physiological pH, and write the three-letter and one-letter abbreviations:

a. serine **b.** aspartate

SOLUTION

CORE CHEMISTRY SKILL

Drawing the Structure for an Amino Acid at Physiological pH

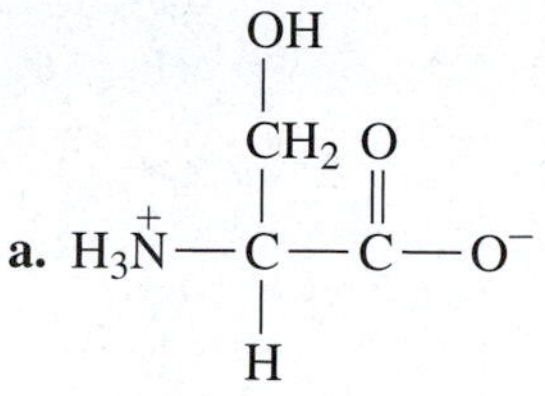

Serine (Ser, S)

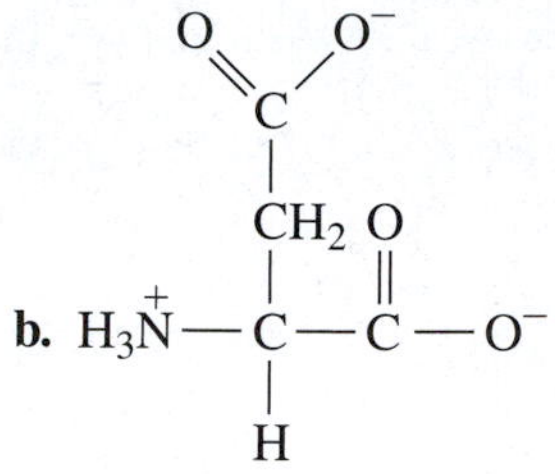

Aspartate (Asp, D)

STUDY CHECK 19.2

Draw the structure for each of the following amino acids at physiological pH, and write the three-letter and one-letter abbreviations:

a. leucine **b.** cysteine

INTERACTIVE VIDEO

Amino Acids at Physiological pH

PEARSON eText 2.0

ANSWER

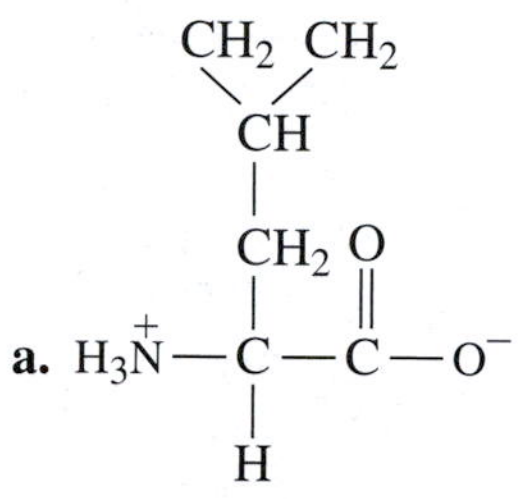

Leucine (Leu, L)

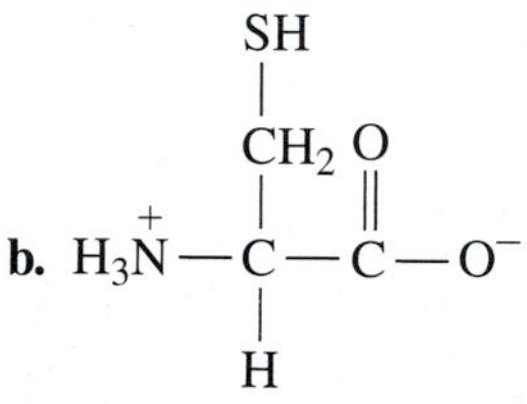

Cysteine (Cys, C)

TEST

Try Practice Problems 19.5, 19.6, 19.9, and 19.10

In this chapter, we will focus on the fact that amino acids are the building blocks of proteins. However, amino acids can play multiple roles in living systems. Yes, amino acids can be incorporated into proteins, but they can also be metabolized for energy or used as precursors for the synthesis of other biologically important compounds. For example, tyrosine is used to make melanins, the pigments responsible for hair and skin color, and thyroid hormones like thyroxin, which regulate the body's basal metabolic rate. Tyrosine is also the precursor for the chemical messengers epinephrine (adrenaline) and norepinephrine, both of which are involved in the fight-or-flight response to a perceived threat.

Chemistry Link to Health

Cystinuria

Cystinuria is a rare inherited disease that affects approximately 1 out of 7000 people worldwide. It is characterized by a high concentration of cystine in the urine. *Cystine* is formed when two cysteine molecules react together in the presence of oxygen. It is a solid that is only slightly soluble in water.

When the amount of cystine in the urine exceeds its solubility, hexagonal crystals form. Ultimately, the crystals form jagged stones that locate in the kidneys, ureter, or bladder. If the stones remain small, they may be passed through urination. Larger stones may need to be removed surgically. People who suffer from cystinuria experience nausea, blood in the urine, and pain that

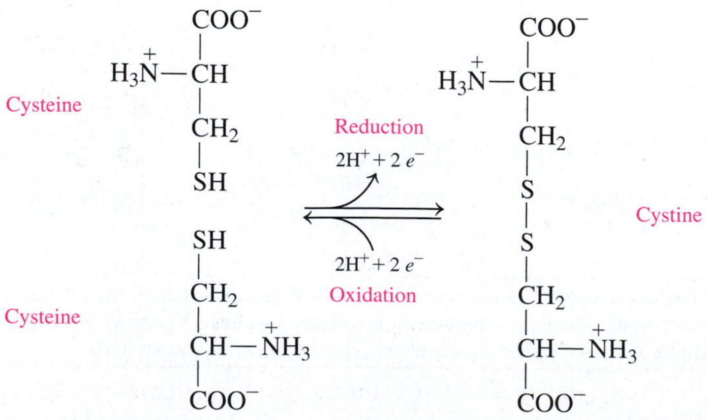
Cysteine
Cysteine
Reduction
2H⁺ + 2 e⁻
2H⁺ + 2 e⁻
Oxidation
Cystine

(continued)

Chemistry Link to Health (*continued*)

is typically located in the kidney, abdominal, and groin areas. They also often develop urinary tract infections. Because cystine stones are jagged, they can cause scarring in the kidneys.

Patients with cystinuria develop cystine stones in their urine.

The kidneys constantly filter blood to create urine. Usually, the amino acids in this filtered fluid are reabsorbed into the blood. However, in people with cystinuria, a genetic mutation prevents the proper reabsorption of certain amino acids, including lysine, arginine, and cystine. Of these, cystine is the only one that forms crystals in the urine.

Currently, there is no cure for cystinuria. Certain thiol-containing medications can keep cystine in the urine from forming stones; but many patients experience significant adverse reactions to these medications, so the management of this disease is focused on actions that increase the solubility of cystine. For example, patients with cystinuria are cautioned to remain properly hydrated to increase the volume of their urine and to decrease the amount of cystine that crystallizes. They can also take substances that increase the pH of their urine because cystine is more soluble at higher pH.

PRACTICE PROBLEMS

19.1 Proteins and Amino Acids

19.1 Classify each of the following proteins according to its function:
- **a.** hemoglobin, oxygen carrier in the blood
- **b.** collagen, a major component of tendons and cartilage
- **c.** keratin, a protein found in hair
- **d.** amylases that catalyze the hydrolysis of starch

19.2 Classify each of the following proteins according to its function:
- **a.** insulin, a protein needed for glucose utilization
- **b.** antibodies that disable foreign proteins
- **c.** casein, milk protein
- **d.** lipases that catalyze the hydrolysis of lipids

19.3 What functional groups are found in all α-amino acids?

19.4 How does the polarity of the R group in leucine compare to that of the R group in serine?

19.5 Draw the structure for each of the following amino acids at physiological pH:
a. glycine **b.** T **c.** glutamate **d.** Phe

19.6 Draw the structure for each of the following amino acids at physiological pH:
a. lysine **b.** proline **c.** V **d.** Tyr

19.7 Classify each of the amino acids in problem 19.5 as polar or nonpolar. If polar, indicate if the R group is neutral, acidic, or basic. Indicate if each is hydrophobic or hydrophilic.

19.8 Classify each of the amino acids in problem 19.6 as polar or nonpolar. If polar, indicate if the R group is neutral, acidic, or basic. Indicate if each is hydrophobic or hydrophilic.

19.9 Give the name for the amino acid represented by each of the following abbreviations:
a. Ala **b.** Q **c.** K **d.** Cys

19.10 Give the name for the amino acid represented by each of the following abbreviations:
a. Trp **b.** M **c.** Pro **d.** G

19.2 Proteins: Primary Structure

REVIEW
Forming Amides (18.5)

LEARNING GOAL Draw the condensed structural formula for a peptide and give its name. Describe the primary structure for a protein.

Now that we know about the structures of individual amino acids, we can look at how amino acids link together to form a protein. A **peptide bond** is an amide bond that forms when the $-COO^-$ group of one amino acid reacts with the $-NH_3^+$ group of the next amino acid. The linking of two or more amino acids by peptide bonds forms a **peptide**. An O atom is removed from the carboxylate end of one amino acid, and two H atoms are removed from the ammonium end of the other amino acid, producing water. Two amino acids form a *dipeptide*, three amino acids form a *tripeptide*, and four amino acids form a *tetrapeptide*. A chain of five amino acids is a *pentapeptide*, and longer chains of amino acids are *polypeptides*.

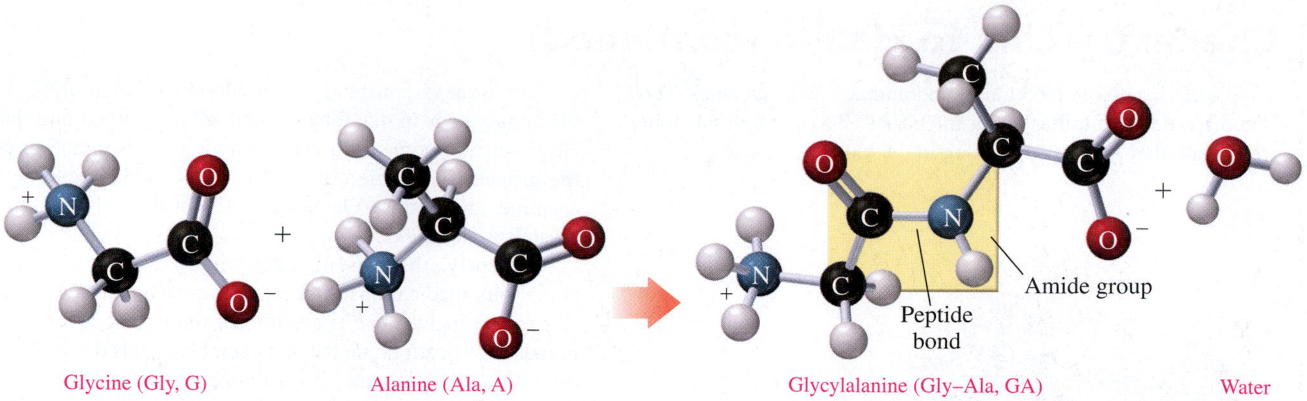
Glycine (Gly, G)
Alanine (Ala, A)
Peptide bond
Amide group
Glycylalanine (Gly–Ala, GA)
Water

FIGURE 19.2 ▶ A peptide bond between glycine and alanine forms the dipeptide glycylalanine.

Q What functional groups in glycine and alanine form the peptide bond?

We can write the amidation reaction for the formation of a dipeptide formed between glycine and alanine, glycylalanine (Gly–Ala, GA) (see **FIGURE 19.2**). During the amidation, the O atom removed from the carboxylate group of glycine combines with two H atoms from the $-NH_3^+$ group of alanine to produce H_2O. The amino acid written on the left, glycine, has a free (unbonded) $-NH_3^+$ group. Therefore, it is the amino acid at the **N-terminus** of the peptide. The amino acid written on the right, alanine, has a free (unbonded) $-COO^-$ group. Therefore, it is the amino acid at the **C-terminus**. The dipeptide forms when the carbonyl group in glycine bonds to the N atom in the $-NH_3^+$ group of alanine.

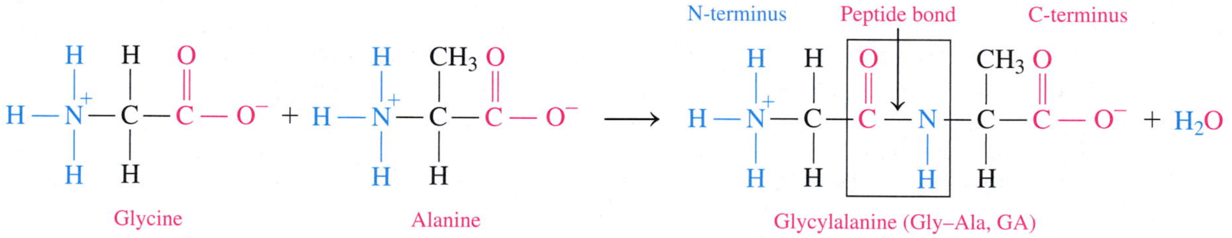
N-terminus
Peptide bond
C-terminus
Glycine
Alanine
Glycylalanine (Gly–Ala, GA)
+ H₂O

Naming Peptides

By convention, peptides are drawn and named from N-terminus to C-terminus. With the exception of the amino acid at the C-terminus, the names of all the other amino acids in a peptide end with *yl*. For example, a tripeptide consisting of alanine at the N-terminus, glycine, and serine at the C-terminus is named as one word: alan**yl**glyc**yl**serine. For convenience, the order of amino acids in the peptide is often written as the sequence of three-letter or one-letter abbreviations.

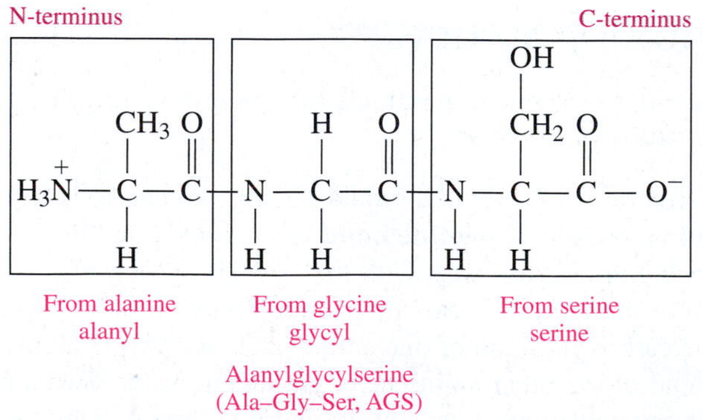
N-terminus
C-terminus
From alanine
alanyl
From glycine
glycyl
From serine
serine
Alanylglycylserine
(Ala–Gly–Ser, AGS)

▶SAMPLE PROBLEM 19.3 Identifying a Tripeptide

TRY IT FIRST

Answer the questions for the tripeptide that is shown below:

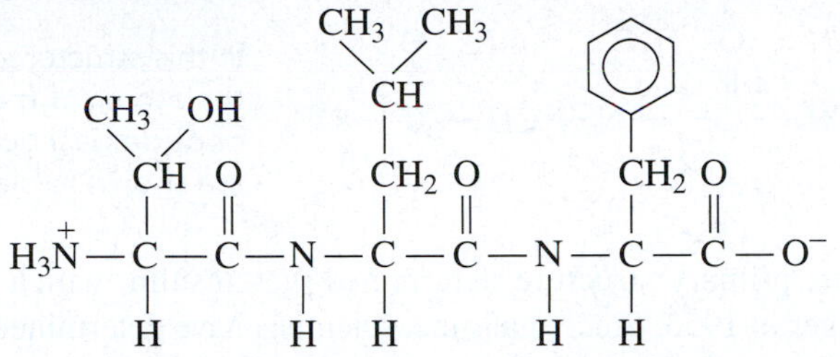

a. What is the amino acid at the N-terminus? What is the amino acid at the C-terminus?

b. Use the three-letter and one-letter abbreviations of amino acids to give the amino acid order in the tripeptide.

c. What is the name of the tripeptide?

SOLUTION

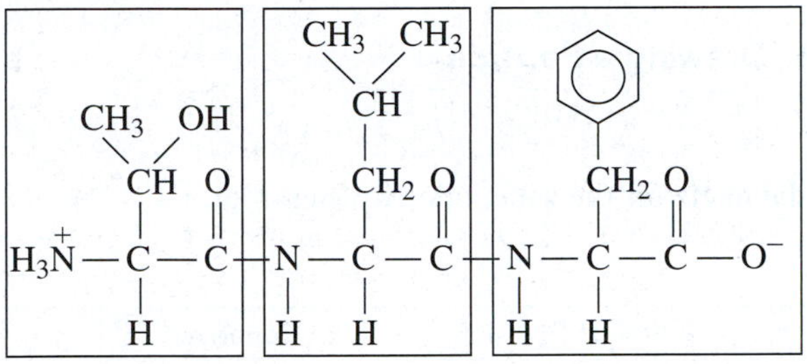

Threonine (Thr, T) N-terminus | Leucine (Leu, L) | Phenylalanine (Phe, F) C-terminus

a. Threonine is the amino acid at the N-terminus; phenylalanine is the amino acid at the C-terminus.

b. Thr–Leu–Phe; TLF

c. threonylleucylphenylalanine

STUDY CHECK 19.3

A tripeptide has the abbreviation Pro–His–Met.

a. What is the amino acid at the N-terminus?

b. What is the amino acid at the C-terminus?

c. What is the name of the tripeptide?

ANSWER

a. proline **b.** methionine **c.** prolylhistidylmethionine

Primary Structure of a Protein

A **protein** is a polypeptide of 50 or more amino acids that has biological activity. The **primary structure** of a protein is the particular sequence of amino acids held together by peptide bonds from N- to C-terminus. For example, a hormone that stimulates the thyroid to release thyroxin is a tripeptide with the amino acid sequence Glu–His–Pro, EHP. Although five other amino acid sequences of the same three amino acids are possible, such as His–Pro–Glu or Pro–His–Glu, they do not produce hormonal activity. Thus the biological function of peptides and proteins depends on the specific sequence of the amino acids.

In the following structure, the R groups of the amino acids are colored red. The atoms colored black are known as the **backbone** of the peptide or protein, which is the repeating sequence of the **N** in the ammonium group, the **C** from α carbon, and the **C** from the carboxylate group (**—N—C—C—N—C—C—N—C—C**).

ENGAGE

How can two peptides with exactly the same number and types of amino acids have different primary structures?

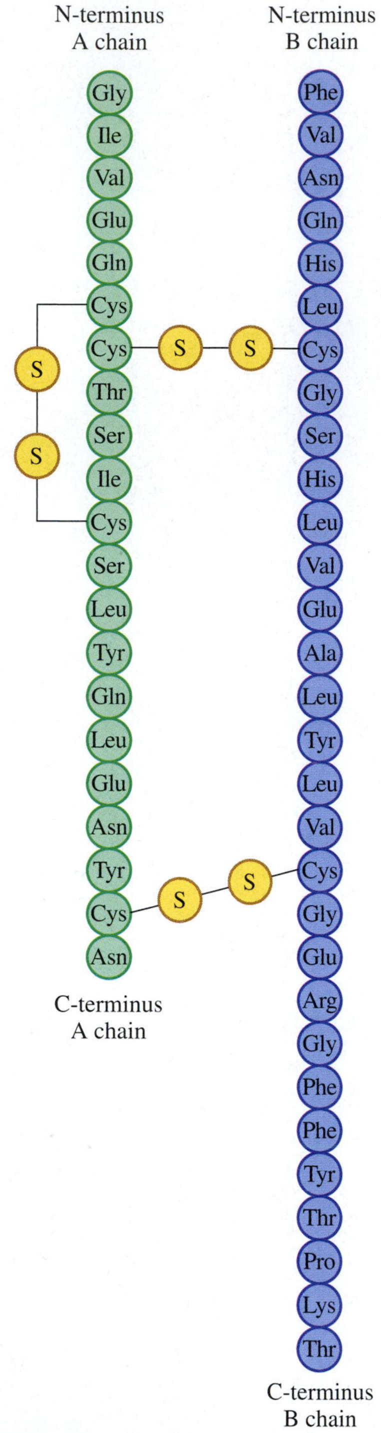
N-terminus
A chain
N-terminus
B chain
Gly Ile Val Glu Gln Cys Cys Thr Ser Ile Cys Ser Leu Tyr Gln Leu Glu Asn Tyr Cys Asn
Phe Val Asn Gln His Leu Cys Gly Ser His Leu Val Glu Ala Leu Tyr Leu Val Cys Gly Glu Arg Gly Phe Phe Tyr Thr Pro Lys Thr
S
C-terminus
A chain
C-terminus
B chain

FIGURE 19.3 ▶ The sequence of amino acids in human insulin is its primary structure.

Q What kinds of bonds occur in the primary structure of a protein?

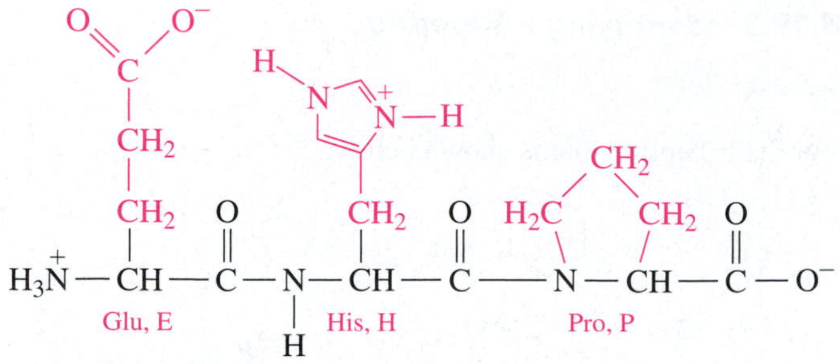
Glu, E
His, H
Pro, P

In this structure, the atoms of the peptide backbone are shown in black.

The first protein to have its primary structure determined was insulin, which was accomplished by Frederick Sanger in 1953. Since that time, scientists have determined the amino acid sequences of thousands of proteins. Insulin is a hormone that regulates the glucose level in the blood. In the primary structure of human insulin, there are two polypeptide chains. In chain A, there are 21 amino acids; chain B has 30 amino acids. The polypeptide chains are held together by *disulfide bonds* formed by the thiol groups of the cysteine amino acids in each of the chains (see **FIGURE 19.3**). Today, human insulin with this exact structure is produced through genetic engineering.

▶ SAMPLE PROBLEM 19.4 Drawing a Peptide

TRY IT FIRST

Draw the structure and give the name for the tripeptide Gly–Ser–Met.

SOLUTION

ANALYZE THE PROBLEM	Given	Need	Connect
	Gly–Ser–Met	tripeptide structure, name	peptide bonds, amino acid names

STEP 1 **Draw the structure for each amino acid in the peptide, starting with the N-terminus.**

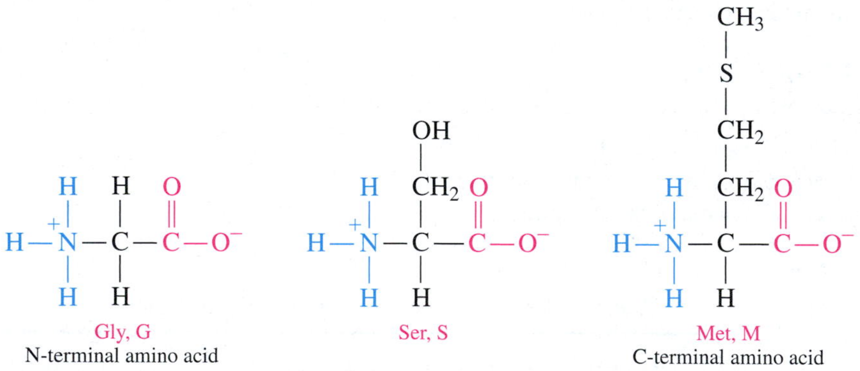
Gly, G
N-terminal amino acid
Ser, S
Met, M
C-terminal amino acid

STEP 2 **Remove the O atom from the carboxylate group of the N-terminal amino acid and two H atoms from the ammonium group in the adjacent amino acid.** Repeat this process until the C-terminus is reached.

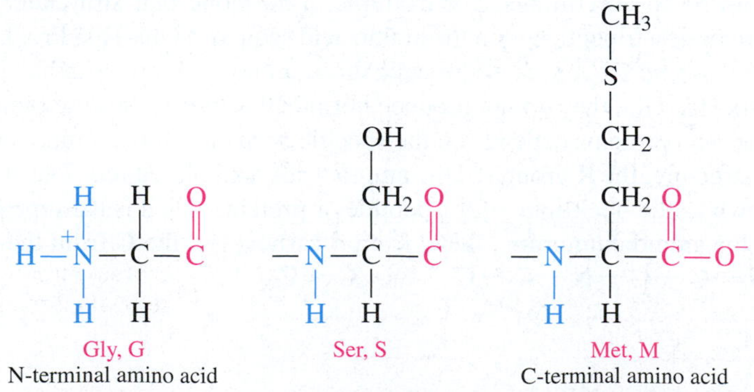
Gly, G
N-terminal amino acid
Ser, S
Met, M
C-terminal amino acid

STEP 3 **Use peptide bonds to connect the amino acids.**

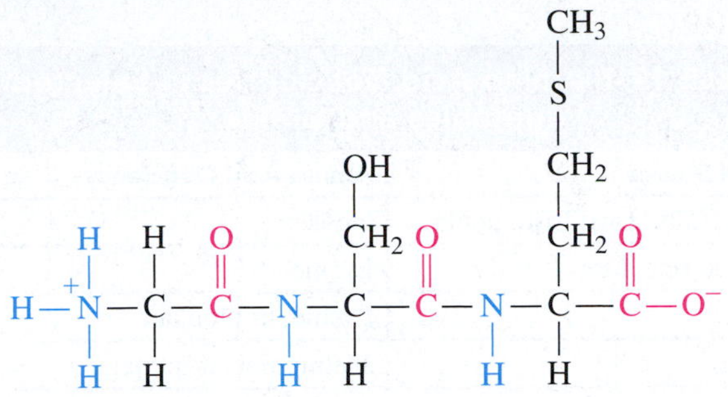

N-terminus C-terminus

Gly–Ser–Met, GSM

The tripeptide is named by replacing the last syllable of each amino acid name with *yl*, starting at the N-terminus. The C-terminal amino acid retains its complete name.

N-terminus glycine is named glycyl
serine is named seryl
C-terminus methionine keeps its complete name

The tripeptide is named glycylserylmethionine.

STUDY CHECK 19.4

a. Draw the structure and give the name for Phe–Thr, a section in glucagon, which is a peptide hormone that increases blood glucose levels.

b. Draw the structure and give the name for Gln–Asn, a section in oxytocin, a nonapeptide used to induce labor.

ANSWER

a. 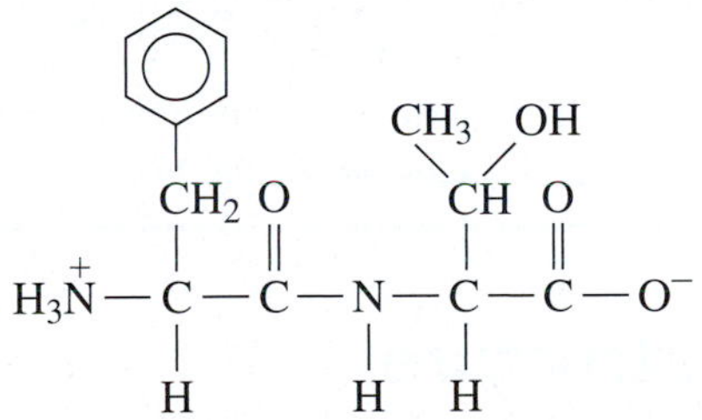

Phenylalanylthreonine

b. 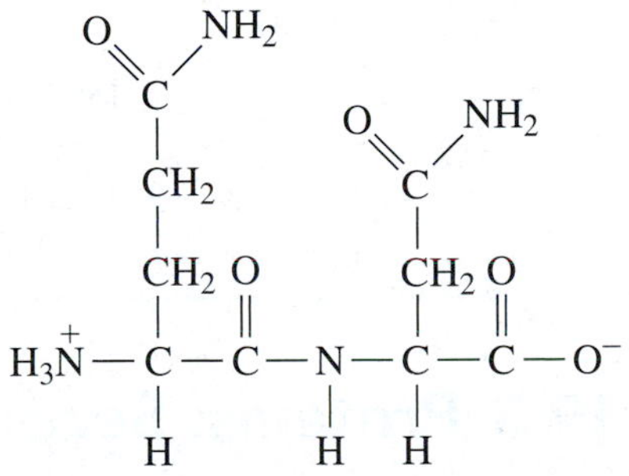

Glutaminylasparagine

TEST

Try Practice Problems 19.11 to 19.14

Chemistry Link to Health

Essential Amino Acids and Complete Proteins

Of the 20 common amino acids used to build the proteins in the body, only 11 can be synthesized in the body. The other 9 amino acids, listed in **TABLE 19.3**, are **essential amino acids** that must be obtained from the proteins in the diet.

Complete proteins, which contain all of the essential amino acids, are found in most animal products such as eggs, milk, meat, fish, and

TABLE 19.3 Essential Amino Acids for Adults

Histidine (His, H)	Phenylalanine (Phe, F)
Isoleucine (Ile, I)	Threonine (Thr, T)
Leucine (Leu, L)	Tryptophan (Trp, W)
Lysine (Lys, K)	Valine (Val, V)
Methionine (Met, M)	

Complete proteins such as eggs, milk, meat, and fish contain all of the essential amino acids. Incomplete proteins from plants such as grains, beans, and nuts are deficient in one or more essential amino acids.

(continued)

Chemistry Link to Health (*continued*)

poultry. However, gelatin and plant proteins such as grains, beans, and nuts are *incomplete proteins* because they are deficient in one or more of the essential amino acids. Diets that rely on plant foods for protein must contain a variety of protein sources to obtain all the essential amino acids. For example, a diet of rice and beans contains all the essential amino acids because they are *complementary protein* sources. Rice contains the methionine and tryptophan that are deficient in beans, while beans contain the lysine that is lacking in rice (see **TABLE 19.4**).

TABLE 19.4 Amino Acid Deficiencies in Selected Vegetables and Grains

Food Source	Amino Acid Deficiency
Eggs, milk, meat, fish, poultry	None
Wheat, rice, oats	Lysine
Corn	Lysine, tryptophan
Beans	Methionine, tryptophan
Peas, peanuts	Methionine
Almonds, walnuts	Lysine, tryptophan
Soy	Methionine

TEST

Try Practice Problems 19.15 and 19.16

PRACTICE PROBLEMS

19.2 Proteins: Primary Structure

19.11 Draw the condensed structural formula for each of the following peptides, and give its three-letter and one-letter abbreviations:
- **a.** alanylcysteine
- **b.** serylphenylalanine
- **c.** glycylalanylvaline
- **d.** valylisoleucyltryptophan

19.12 Draw the condensed structural formula for each of the following peptides, and give its three-letter and one-letter abbreviations:
- **a.** prolylaspartate
- **b.** threonylleucine
- **c.** methionylglutaminyllysine
- **d.** histidylglycylglutamylisoleucine

Clinical Applications

19.13 Peptides isolated from rapeseed that may lower blood pressure have the following sequence of amino acids. Draw the structure for each peptide and write the one-letter abbreviations.
a. Arg–Ile–Tyr **b.** Val–Trp–Ile–Ser

19.14 Peptides from sweet potato with antioxidant properties have the following sequence of amino acids. Draw the structure for each peptide and write the one-letter abbreviations.
a. Asp–Cys–Gly–Tyr **b.** Asn–Tyr–Asp–Glu–Tyr

19.15 Explain why each of the following pairs are complementary proteins:
a. corn and peas **b.** rice and soy

19.16 Explain why each of the following pairs are complementary proteins:
a. beans and oats **b.** almonds and peanuts

19.3 Proteins: Secondary Structure

LEARNING GOAL Describe the two most common types of secondary structures for a protein. Describe the structure of collagen.

The **secondary structure** of a protein describes the type of structure that forms when the atoms in the backbone of a protein or peptide form hydrogen bonds within a single polypeptide chain or between polypeptide chains. The most common types of secondary structure are the *alpha helix* and the *beta-pleated sheet.*

Alpha Helix

In an **alpha helix (α helix)**, hydrogen bonds form between the oxygen atoms of the C═O groups and the hydrogen atoms of the N—H groups of the amide bonds in the next turn of the α helix (see **FIGURE 19.4**). The formation of many hydrogen bonds along the polypeptide chain gives the characteristic helical shape of a spiral staircase. All of the R groups of the different amino acids extend to the outside of the helix.

Beta-Pleated Sheet

Another type of secondary structure found in proteins is the **beta-pleated sheet (β-pleated sheet)**. In a β-pleated sheet, hydrogen bonds form between the oxygen atoms in the carbonyl groups in one section of the polypeptide chain, and the hydrogen atoms in the N—H groups of the amide bonds in a nearby section of the polypeptide chain. A beta-pleated sheet

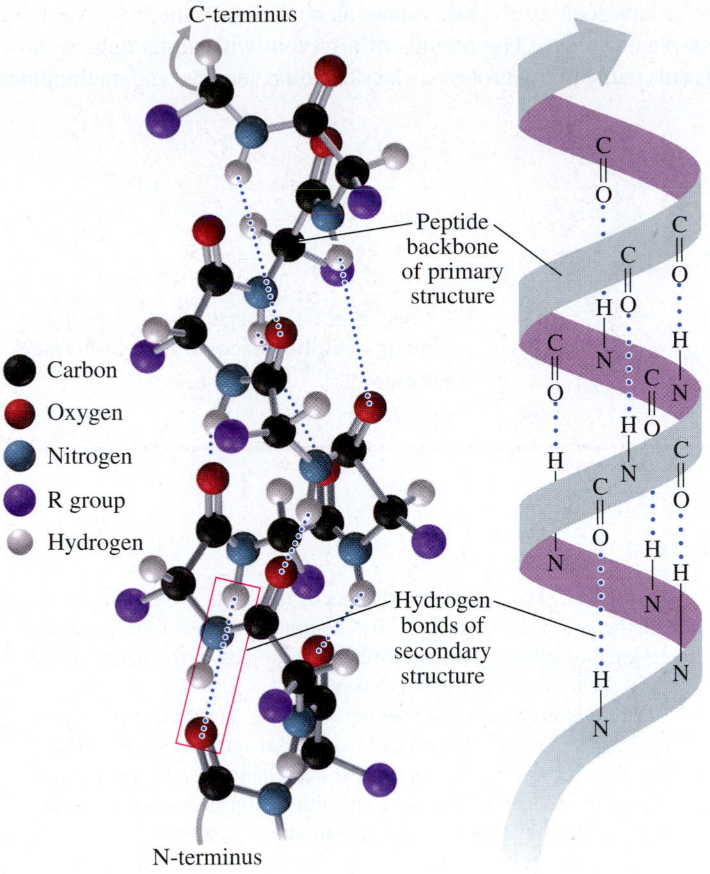
C-terminus
Peptide backbone of primary structure
Carbon
Oxygen
Nitrogen
R group
Hydrogen
Hydrogen bonds of secondary structure
N-terminus

FIGURE 19.4 ▶ The α helix acquires a coiled shape from hydrogen bonds between the oxygen of the C=O group and the hydrogen of the N—H group in the next turn.

Q What are the partial charges of the H in N—H and the O in C=O that permit hydrogen bonds to form?

can form between adjacent polypeptide chains or within the same polypeptide chain. The hydrogen bonds holding the β-pleated sheets tightly in place account for the strength and durability of proteins such as silk (see **FIGURE 19.5**).

The tendency to form various kinds of secondary structures depends on the amino acids in a particular segment of the polypeptide chain. Typically, beta-pleated sheets contain mostly

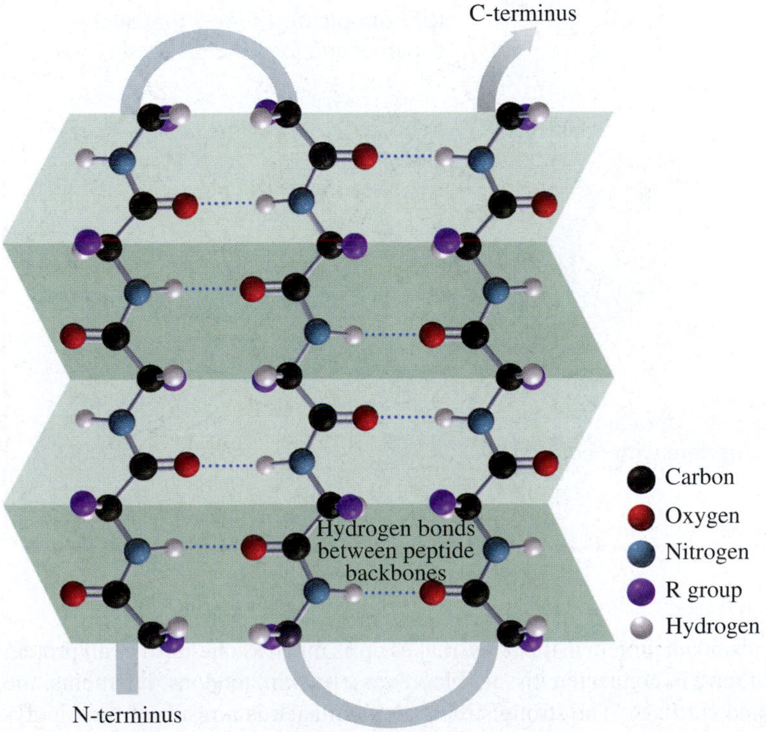
C-terminus
Hydrogen bonds between peptide backbones
Carbon
Oxygen
Nitrogen
R group
Hydrogen
N-terminus

FIGURE 19.5 ▶ In the secondary structure of a β-pleated sheet, hydrogen bonds form between the peptide chains.

Q How do the hydrogen bonds in a β-pleated sheet differ from the hydrogen bonds in an α helix?

amino acids with small R groups such as glycine, valine, alanine, and serine, which extend above and below the beta-pleated sheet. The regions of a protein with alpha helices have higher amounts of amino acids with large R groups such as histidine, leucine, and methionine.

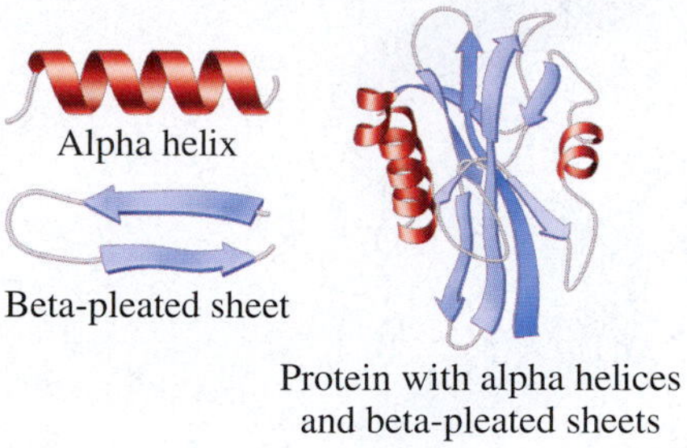
Alpha helix
Beta-pleated sheet
Protein with alpha helices and beta-pleated sheets

A ribbon model of a protein shows the regions of alpha helices and beta-pleated sheets.

TEST

Try Practice Problems 19.17 to 19.20

Chemistry Link to Health

Protein Secondary Structures and Alzheimer's Disease

Alzheimer's disease is a form of dementia in which a person has increasing memory loss and inability to handle daily tasks. Alzheimer's usually occurs after age 60 and is irreversible. Although researchers are still investigating its causes, Alzheimer's patients have distinctly different brain tissue from people who do not have the disease. In the brain of a normal person, small beta-amyloid proteins, made up of 42 amino acids, exist in the alpha-helical form. In the brain of a person with Alzheimer's, the beta-amyloid proteins change shape from the normal alpha helices that are soluble, to sticky beta-pleated sheets, forming clusters of insoluble protein fragments called *plaques.* The diagnosis of Alzheimer's disease is based on the presence of plaques and neurofibrillary tangles in the neurons that affect the transmission of nerve signals.

There is no cure for Alzheimer's disease, but there are medications that can slow its progression or lessen symptoms for a limited time. Medications like donepezil (Aricept) and rivastigmine (Exelon) help keep the levels of nerve transmitters high in the brain, which improves learning and memory in patients.

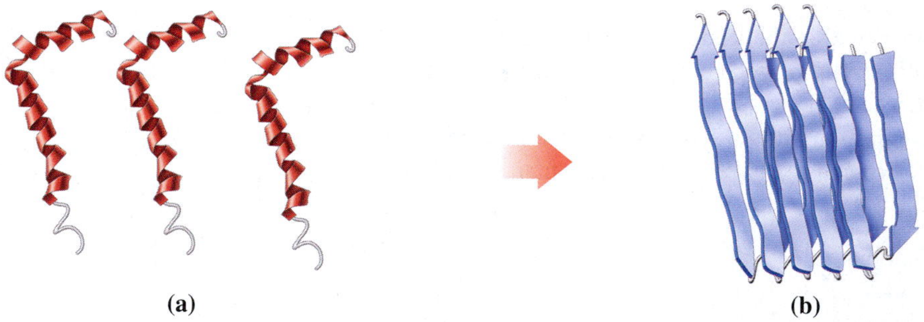
(a)
(b)

In patients with Alzheimer's disease, beta-amyloid proteins change from **(a)** a normal alpha-helical shape to **(b)** beta-pleated sheets that stick together and form plaques in the brain.

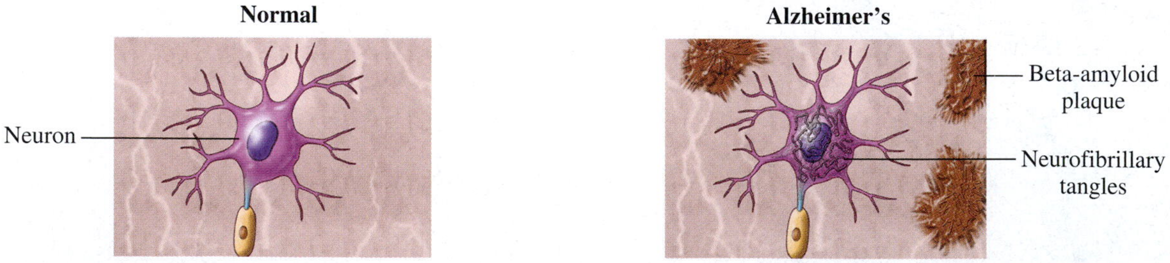
Normal
Neuron
Alzheimer's
Beta-amyloid plaque
Neurofibrillary tangles

In an Alzheimer's brain, beta-amyloid plaques and neurofibrillary tangles damage the neurons and interfere with nerve signals.

TEST

Try Practice Problems 19.21 and 19.22

Collagen

Collagen, the most abundant protein in the body, makes up as much as one-third of all protein in vertebrates. It is found in connective tissue, blood vessels, skin, tendons, ligaments, the cornea of the eye, and cartilage. The strong structure of collagen is a result of three polypeptides twisted together to form a **triple helix** (see **FIGURE 19.6**).

Collagen has a high content of glycine (33%), proline (22%), alanine (12%), and smaller amounts of hydroxyproline and hydroxylysine, which are modified forms of proline and lysine. The —OH groups on these modified amino acids provide additional hydrogen bonds between the peptide chains to give strength to the collagen triple helix. When many triple helices bundle together, they form the fibrils that make up connective tissues and tendons. When a diet is deficient in vitamin C, collagen triple helices are weakened because the enzymes needed to form hydroxyproline and hydroxylysine require vitamin C. Because there are fewer —OH groups, there is less hydrogen bonding between the polypeptide chains in collagen. Collagen becomes less elastic as a person ages because additional cross-links form between the fibrils. Bones, cartilage, and tendons become more brittle, and wrinkles are seen as the skin loses elasticity.

TEST

Try Practice Problems 19.23 and 19.24

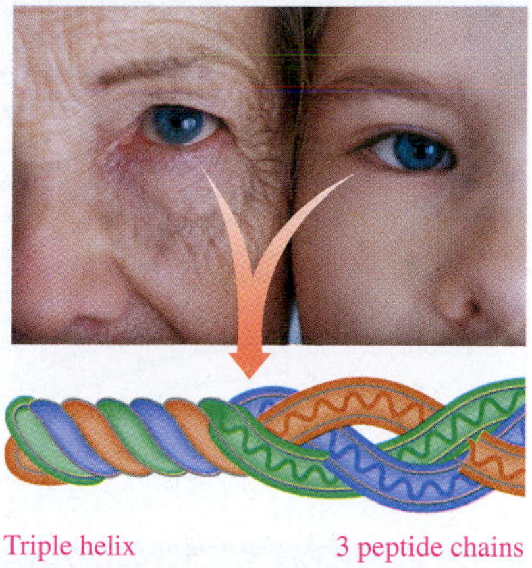
Triple helix
3 peptide chains

FIGURE 19.6 ▶ Collagen triple helices are composed of three polypeptide chains twisted and held together by hydrogen bonds.

Q What are some of the amino acids in collagen that form hydrogen bonds between the polypeptide chains?

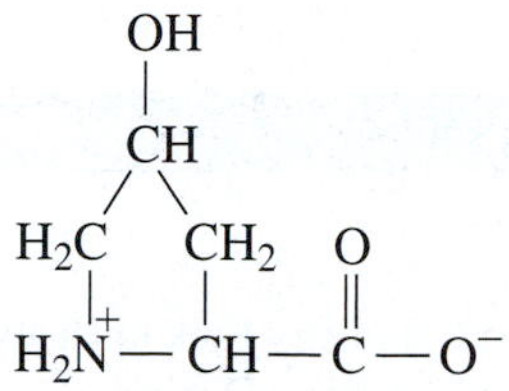

Hydroxyproline

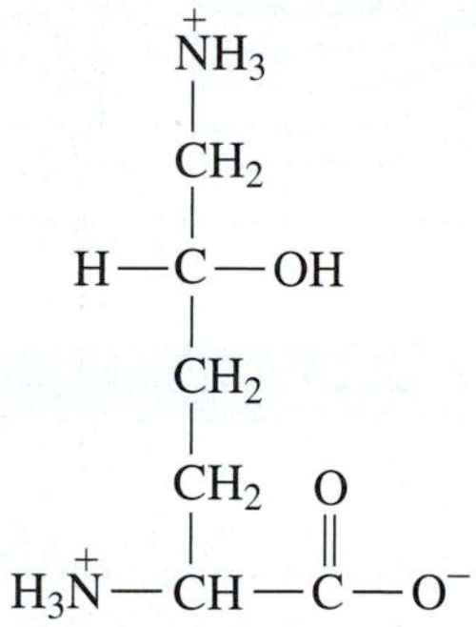

Hydroxylysine

Hydroxyproline and hydroxylysine provide additional hydrogen bonds in the triple helices of collagen.

Chemistry Link to Health

Keratoconus

Keratoconus is a noninflammatory, progressive eye disorder that affects approximately 1 out of 2000 people worldwide. The specific cause of the disease is currently unknown, but it is characterized by a thinning of the cornea, the transparent front layer of the eye. Layers of collagen fibers help to maintain the shape of the cornea. However, as the cornea thins, it loses its biomechanical strength. As a consequence, the cornea begins to bulge until it resembles a cone. In fact, the word *keratoconus* comes from the Greek *kéras* (cornea) and the Latin *cōnus* (cone).

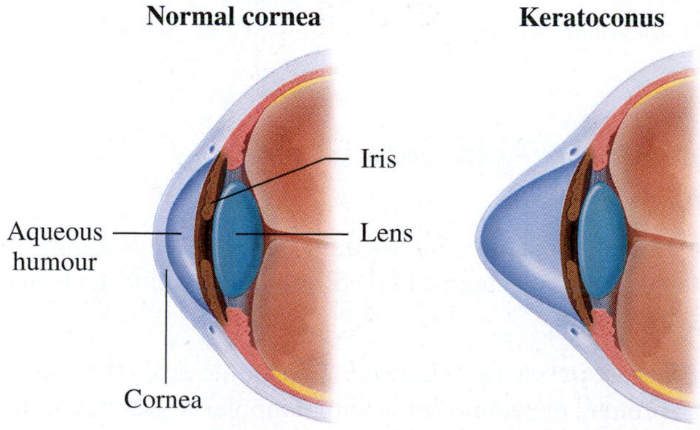
Normal cornea
Keratoconus
Iris
Aqueous humour
Lens
Cornea

In keratoconus, the cornea takes on a cone-like shape.

Because the cornea is responsible for refracting the light that comes into the eye, any change in the shape of the cornea affects the way we see. People that suffer from keratoconus typically experience blurry vision, double vision, and sensitivity to light and glare.

The bulging of the cornea in keratoconus causes vision distortion.

The disease typically presents in late childhood or early adulthood. In the early stages of keratoconus, resulting vision distortions can be corrected with soft contact lenses or glasses. In severe, late stages of the disease, corneal transplants may be required to correct for the vision distortion that results from the bulging of the cornea.

In April 2016, the U.S. Food and Drug Administration (FDA) approved a treatment that appears to halt the progression of keratoconus. The disease is often associated with a loss of adhesion between the collagen fibers in the cornea. Corneal Collagen Cross-Linking (CXL) is a treatment that increases the cross-linking between these fibers, thus increasing the biomechanical strength of the cornea.

(*continued*)

Chemistry Link to Health (*continued*)

CXL is an outpatient treatment. A solution of riboflavin (vitamin B_2) is applied to the eye. After the surgeon has verified that there is a sufficient amount of riboflavin in the cornea, the eye is exposed to ultraviolet (UV) light for approximately 30 minutes. The riboflavin acts as a *photosensitizer*, inducing the formation of an increased number of cross-links between the collagen fibers in the cornea.

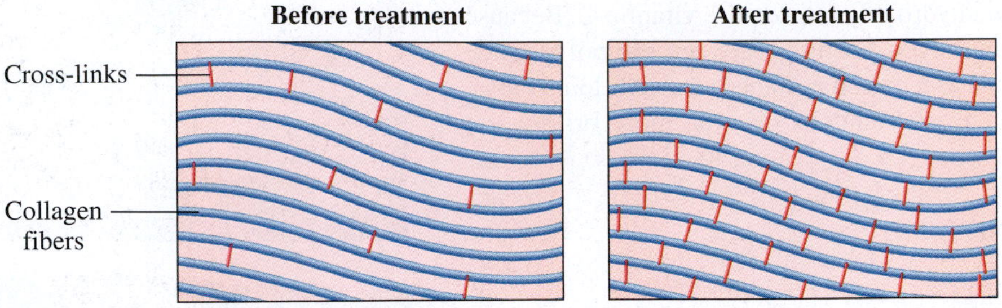
Before treatment
After treatment
Cross-links
Collagen fibers

Increasing the number of cross-links between collagen fibers increases the biomechanical strength of the cornea.

PRACTICE PROBLEMS

19.3 Proteins: Secondary Structure

19.17 What happens when a primary structure forms a secondary structure?

19.18 What are two types of secondary protein structure?

19.19 What is the difference in hydrogen bonding between an α helix and a β-pleated sheet?

19.20 In an α helix, how does hydrogen bonding occur between the amino acids in the polypeptide chain?

Clinical Applications

19.21 What is the difference between the secondary structures of the beta-amyloid protein in a normal adult brain and the brain of an adult with Alzheimer's disease (see Chemistry Link to Health: Protein Secondary Structures and Alzheimer's Disease)?

19.22 What causes the formation of beta-amyloid plaques in Alzheimer's disease (see Chemistry Link to Health: Protein Secondary Structures and Alzheimer's Disease)?

19.23 Collagen is one of the most prevalent proteins in the cornea. How does the structure of collagen differ from that of a typical α helix?

19.24 What type of bond holds the three polypeptide chains together in an individual collagen triple helix?

19.4 Proteins: Tertiary and Quaternary Structures

LEARNING GOAL Describe the tertiary and quaternary structures of a protein.

The **tertiary structure** of a protein involves attractions and repulsions between the R groups of the amino acids in the polypeptide chain. As interactions occur between different parts of the peptide chain, segments of the chain twist and bend until the protein acquires a specific three-dimensional shape.

CORE CHEMISTRY SKILL
Identifying the Primary, Secondary, Tertiary, and Quaternary Structures of Proteins

Interactions Between Amino Acid Side Chains in Tertiary Structures

The tertiary structure of a protein is stabilized by interactions between the R groups of the amino acids in one region of the polypeptide chain and the R groups of amino acids in other regions of the protein (see **FIGURE 19.7**).

1. **Hydrophobic interactions** are interactions between two amino acids that have nonpolar R groups. Within a protein, the amino acids with nonpolar R groups are pushed away from the aqueous environment to form a hydrophobic center at the interior of the protein molecule.

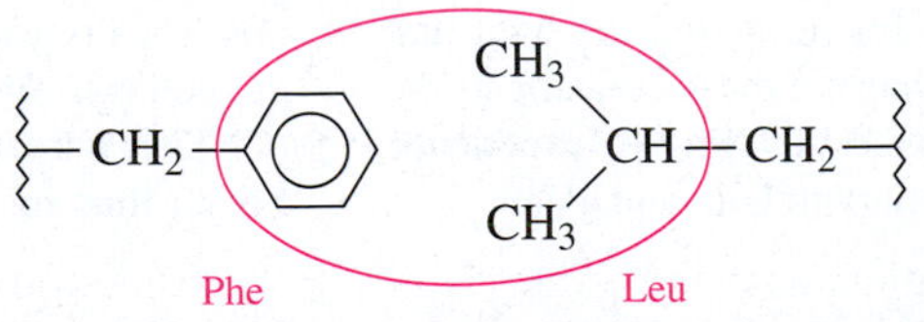
Phe
Leu

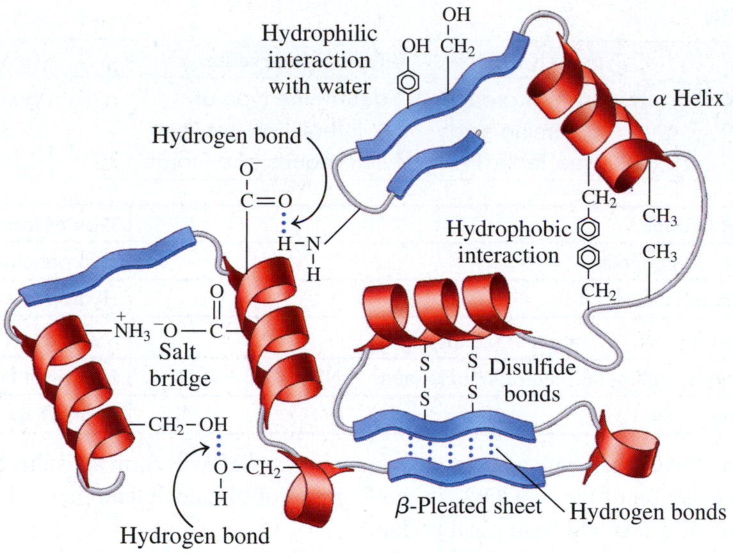

Hydrophilic interaction with water
OH
OH CH2
α Helix
Hydrogen bond
O−
C=O
H−N
H
Hydrophobic interaction
CH2
CH3
CH3
CH2
O
NH3 −O—C
Salt bridge
S S
S S
Disulfide bonds
CH2−OH
O−CH2
H
Hydrogen bond
β-Pleated sheet
Hydrogen bonds

FIGURE 19.7 ▶ Interactions between amino acid R groups fold a protein into a specific three-dimensional shape called its tertiary structure.

Q Why would one section of a protein move to the center while another section remains on the surface of the tertiary structure?

2. **Hydrophilic interactions** are attractions between the external aqueous environment and the R groups of polar amino acids that are pulled to the outer surface of most proteins, where they form hydrogen bonds with water.

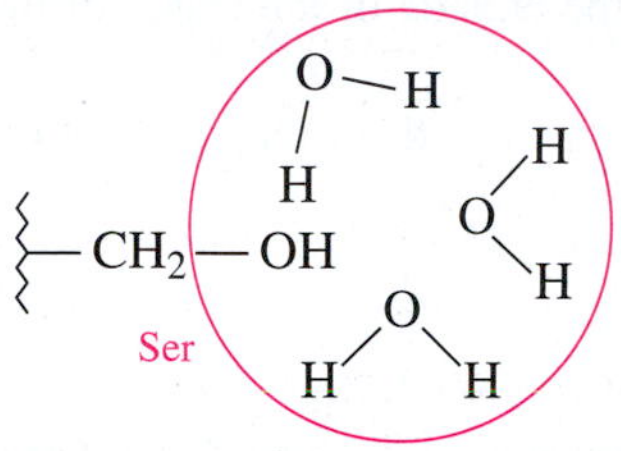

CH2—OH
Ser

3. **Salt bridges** are ionic attractions between ionized R groups of polar basic and polar acidic amino acids. For example, at physiological pH, the ionized R group of lysine, which has a positive charge, can form a salt bridge with the R group in aspartate, which has a negative charge.

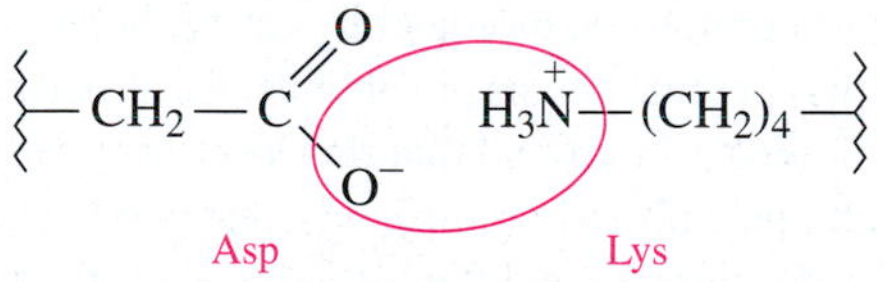

Asp
Lys

ENGAGE

Why is the interaction between arginine and aspartate in a tertiary structure called a salt bridge?

4. **Hydrogen bonds** form between the H of a polar R group and the O or N of another polar amino acid. For example, a hydrogen bond can form between the —OH groups of two serines or between the —OH of serine and the $—NH_2$ in the R group of asparagine.

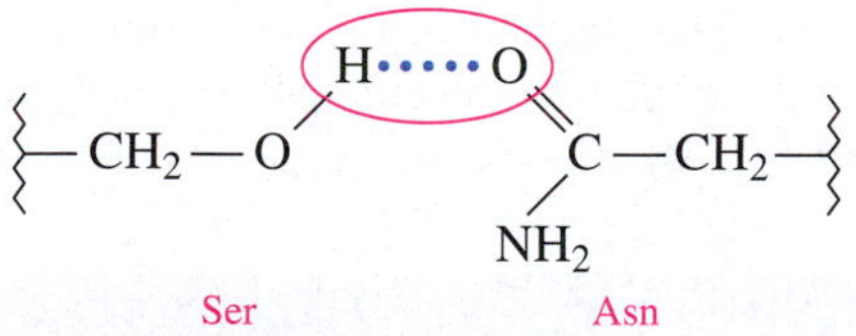

Ser
Asn

5. **Disulfide bonds** (—S—S—) are covalent bonds that form between the —SH groups of cysteines in a polypeptide chain.

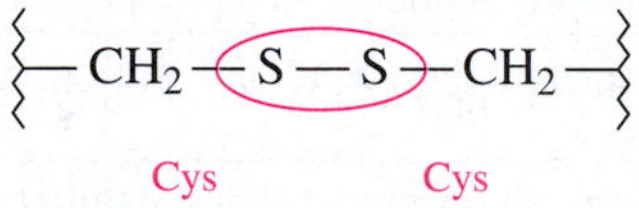

Cys
Cys

▶ **SAMPLE PROBLEM 19.5** **Interaction Between R Groups in Tertiary Structures**

TRY IT FIRST

What type of interaction would you expect between the R groups of each of the following amino acids in a tertiary structure?

a. cysteine and cysteine **b.** glutamate and lysine **c.** tyrosine and water

SOLUTION

	Type of R Groups	Type of Interaction	Connect
ANALYZE THE PROBLEM	identify R groups on amino acids (see Table 19.2)	determine type of interaction between R groups (see Figure 19.7)	polarity of R groups

Type of R Groups	Type of Interaction
Nonpolar and nonpolar	Hydrophobic
Polar (neutral) and water	Hydrophilic
Polar (basic) $-NH_3^+$ and polar (acidic) $-COO^-$	Salt bridges
Polar (neutral) and polar (neutral) $-OH$ and $-NH-$ or $-NH_2$	Hydrogen bonds
$-SH$ and $-SH$	Disulfide bonds

a. Two cysteines, each with an R group containing $-SH$, will form a disulfide bond.
b. The interaction of the $-COO^-$ in the R group of glutamate and the $-NH_3^+$ in the R group of lysine will form a salt bridge.
c. The R group in tyrosine has an $-OH$ group that is attracted to water by hydrophilic interactions.

STUDY CHECK 19.5

a. Would you expect to find valine and leucine on the outside or the inside of the tertiary structure? Why?
b. Would you expect to find threonine and serine on the outside or the inside of the tertiary structure? Why?

ANSWER

a. Both are nonpolar and would be found on the inside of the tertiary structure.
b. Both are polar and would likely be found on the outside of the tertiary structure.

TEST

Try Practice Problems 19.25 to 19.28

INTERACTIVE VIDEO

Different Levels of Protein Structure

PEARSON eText 2.0

Quaternary Structure: Hemoglobin

While many proteins are biologically active as tertiary structures, some proteins require two or more tertiary structures, working together as a unit, to be biologically active. When a biologically active protein consists of two or more polypeptide chains or subunits, the structural level is referred to as a **quaternary structure**. Hemoglobin, a protein that transports oxygen in blood, consists of four polypeptide chains: two α-chains with 141 amino acids, and two β-chains with 146 amino acids. Although the α-chains and β-chains have different sequences of amino acids, they both form similar tertiary structures with similar shapes (see **FIGURE 19.8**).

In the quaternary structure, the subunits are held together by the same interactions that stabilize tertiary structures, such as hydrogen bonds, salt bridges, disulfide links, and hydrophobic interactions between R groups. Each subunit of the hemoglobin contains a heme group that binds oxygen. In the adult hemoglobin molecule, all four subunits ($\alpha_2\beta_2$) *must* be combined for hemoglobin to properly function as an oxygen carrier. Therefore,

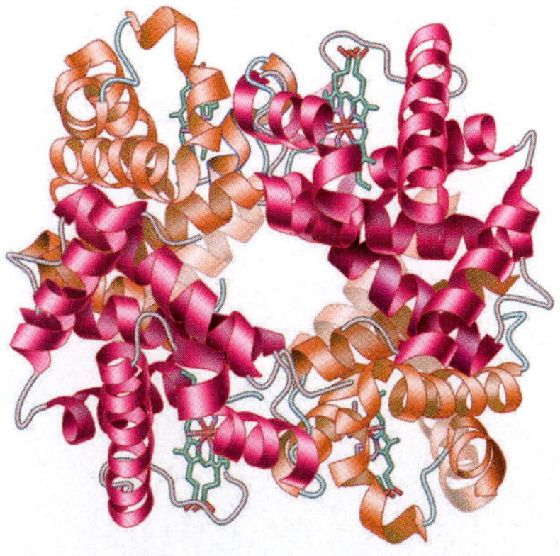

FIGURE 19.8 ▶ In the ribbon model of hemoglobin, the quaternary structure is made up of four polypeptide subunits, two (orange) are α-chains and two (red) are β-chains. The heme groups (green) in the four subunits bind oxygen.

Q What is the difference between a tertiary structure and a quaternary structure?

TABLE 19.5 Summary of Structural Levels in Proteins

Structural Level	Characteristics
Primary	Peptide bonds join amino acids in a specific sequence in a polypeptide.
Secondary	The α helix or β-pleated sheet forms by hydrogen bonding between the atoms in the peptide backbone.
Tertiary	A polypeptide folds into a compact, three-dimensional shape stabilized by interactions (hydrogen bonds, salt bridges, hydrophobic, hydrophilic, disulfide) between R groups of amino acids to form a biologically active protein.
Quaternary	Two or more protein subunits combine and are stabilized by interactions (hydrogen bonds, salt bridges, hydrophobic, hydrophilic, disulfide) to form a biologically active protein.

the complete quaternary structure of hemoglobin can bind and transport four molecules of oxygen.

TABLE 19.5 and **FIGURE 19.9** summarize the structural levels of proteins.

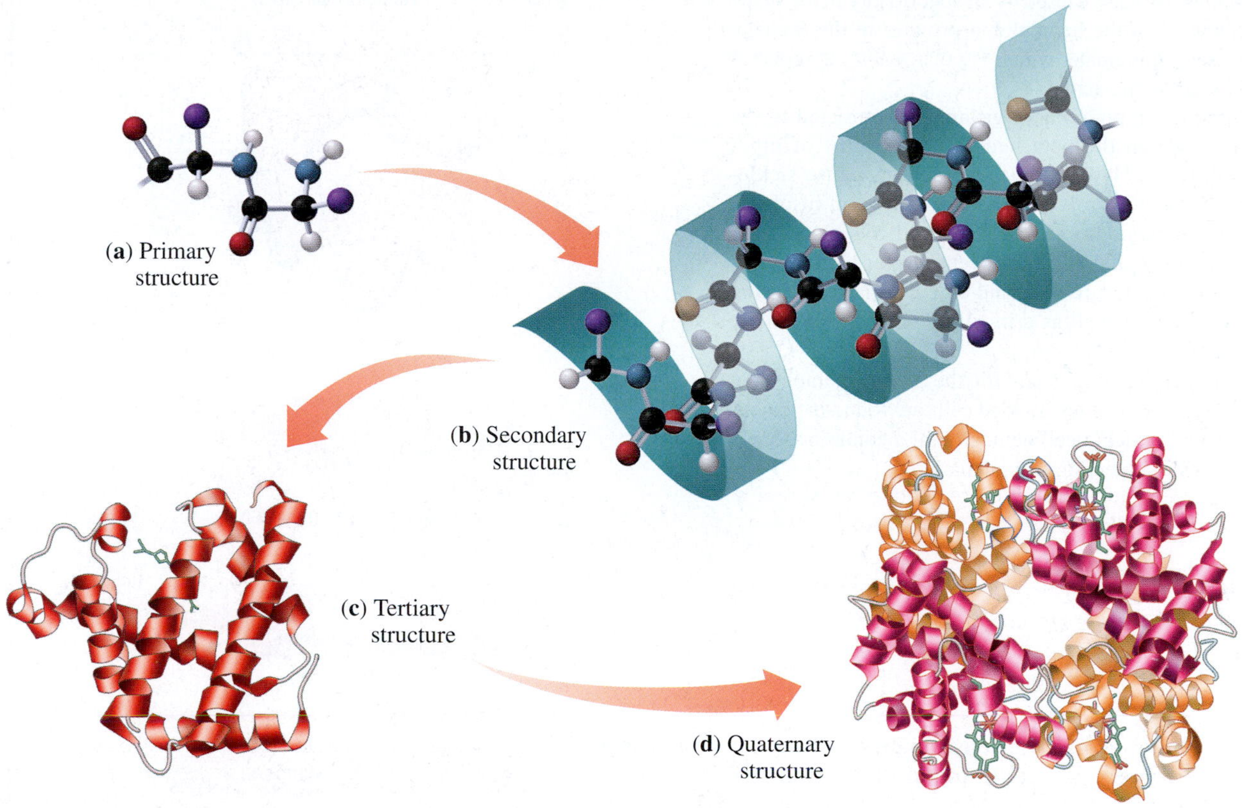
(a) Primary structure
(b) Secondary structure
(c) Tertiary structure
(d) Quaternary structure

FIGURE 19.9 ▶ The structural levels of protein are **(a)** primary, **(b)** secondary, **(c)** tertiary, and **(d)** quaternary.

Q What is the difference between a primary structure and a tertiary structure?

▶ SAMPLE PROBLEM 19.6 Identifying Protein Structure

TRY IT FIRST

Indicate whether the following conditions are responsible for primary, secondary, tertiary, or quaternary protein structures:

a. disulfide bonds that form between portions of a protein chain
b. peptide bonds that form a chain of amino acids
c. hydrogen bonds between the H of a peptide bond and the O of a peptide bond four amino acids away

SOLUTION

a. Disulfide bonds are a type of interaction between R groups found in the tertiary and quaternary levels of protein structure.
b. The peptide bonds in the sequence of amino acids in a polypeptide form the primary level of protein structure.
c. Hydrogen bonding between peptide bonds in the protein backbone forms the secondary level of protein structure.

STUDY CHECK 19.6

What structural level is represented by the interactions of the two subunits in insulin?

ANSWER

quaternary

TEST

Try Practice Problems 19.29 and 19.30

Chemistry Link to Health

Sickle-Cell Anemia

Sickle-cell anemia is a disease caused by an abnormality in the shape of one of the subunits of the hemoglobin protein. In the β-chain, the sixth amino acid, glutamate, which is polar acidic, is replaced by valine, a nonpolar amino acid.

Because valine has a nonpolar R group, it is attracted to the nonpolar regions within the beta hemoglobin chains of other hemoglobin molecules. Hydrophobic interactions cause sickle-cell hemoglobin molecules to stick together and form insoluble fibers of sickle-cell hemoglobin. The affected red blood cells change from a rounded shape to a crescent shape, like a sickle. The sickled red blood cells are rigid and clog capillaries, causing inflammation and pain as well as critically low oxygen levels in the affected tissues.

In sickle-cell anemia, both genes for the altered hemoglobin must be inherited. However, a few sickled cells are found in persons who carry one gene for sickle-cell hemoglobin, a condition that is also known to provide protection from malaria.

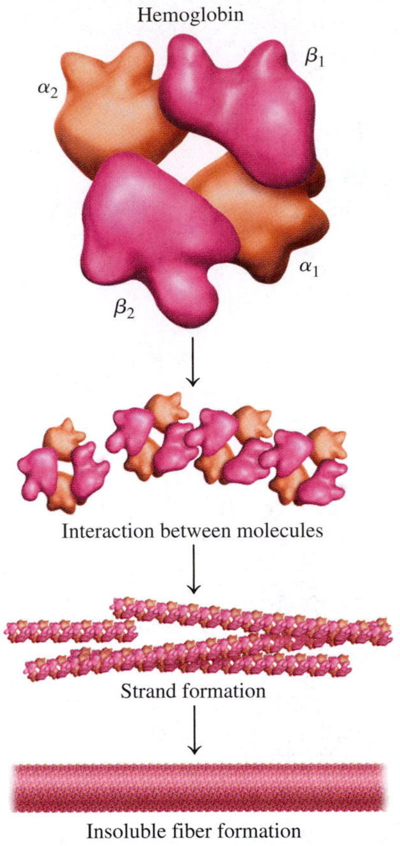
Hemoglobin
α_2
β_1
α_1
β_2
Interaction between molecules
Strand formation
Insoluble fiber formation

Polar acidic amino acid

Normal β-chain: Val–His–Leu–Thr–Pro–Glu–Glu–Lys–

Sickled β-chain: Val–His–Leu–Thr–Pro–Val–Glu–Lys–

Nonpolar amino acid

In sickle-cell hemoglobin, the substitution of a nonpolar amino acid causes hydrophobic interactions that produce insoluble fibers that change the shape of the red blood cell and block blood flow.

PRACTICE PROBLEMS

19.4 Proteins: Tertiary and Quaternary Structures

19.25 What type of interaction would you expect between the R groups of the following amino acids in a tertiary structure?
- **a.** cysteine and cysteine
- **b.** aspartate and lysine
- **c.** serine and aspartate
- **d.** leucine and leucine

19.26 What type of interaction would you expect between the R groups of the following amino acids in a tertiary structure?
- **a.** phenylalanine and isoleucine
- **b.** glutamate and arginine
- **c.** asparagine and tyrosine
- **d.** alanine and proline

19.27 A portion of a polypeptide chain contains the following sequence of amino acids:

–Leu–Val–Cys–Asp–

- **a.** Which amino acids are likely to be found on the inside of the protein structure? Why?
- **b.** Which amino acids would be found on the outside of the protein? Why?
- **c.** How does the primary structure of a protein affect its tertiary structure?

19.28 In myoglobin, about one-half of the 153 amino acids have nonpolar R groups.
- **a.** Where would you expect those amino acids to be located in the tertiary structure?
- **b.** Where would you expect the polar R groups to be in the tertiary structure?
- **c.** Why is myoglobin more soluble in water than silk or wool?

19.29 Indicate whether each of the following statements describes primary, secondary, tertiary, or quaternary protein structure:
- **a.** R groups interact to form disulfide bonds or salt bridges.
- **b.** Peptide bonds join amino acids in a polypeptide chain.
- **c.** Several polypeptides in a beta-pleated sheet are held together by hydrogen bonds between adjacent chains.
- **d.** Hydrogen bonding between amino acids in the same polypeptide gives a coiled shape to the protein.

19.30 Indicate whether each of the following statements describes primary, secondary, tertiary, or quaternary protein structure:
- **a.** Hydrophobic R groups seeking a nonpolar environment move toward the inside of the folded protein.
- **b.** Protein chains of collagen form a triple helix.
- **c.** An active protein contains four tertiary subunits.
- **d.** In sickle-cell anemia, valine replaces glutamate in the β-chain.

19.5 Protein Hydrolysis and Denaturation

LEARNING GOAL Describe the hydrolysis and denaturation of proteins.

The amidation reactions that produce peptide bonds also produce water. Therefore, peptide bonds can be broken through hydrolysis reactions, yielding individual amino acids. This process occurs in the stomach when enzymes such as pepsin or trypsin catalyze the hydrolysis of proteins to give amino acids. This hydrolysis breaks up the primary structure by breaking the covalent peptide bonds that link the amino acids. In the digestion of proteins, the amino acids are absorbed through the intestinal walls and carried to the cells, where they can be used to synthesize new proteins.

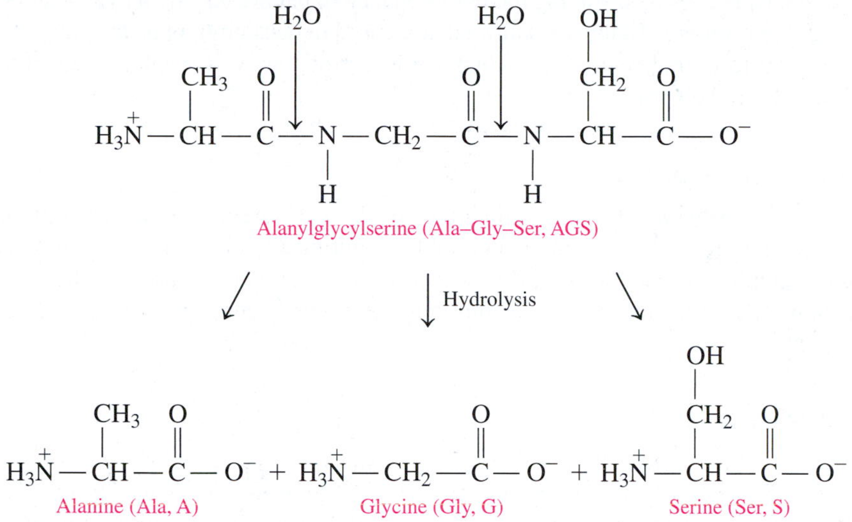

Alanylglycylserine (Ala–Gly–Ser, AGS)
Hydrolysis
Alanine (Ala, A)
Glycine (Gly, G)
Serine (Ser, S)

TEST

Try Practice Problems 19.31 to 19.35

Denaturation of Proteins

ENGAGE

What is the difference between the hydrolysis and the denaturation of proteins?

Denaturation of a protein occurs when there is a change that disrupts the interactions that stabilize the secondary, tertiary, or quaternary structure. However, the covalent amide bonds of the primary structure are not affected.

The loss of secondary and tertiary structures occurs when conditions change, such as increasing the temperature or making the pH very acidic or basic. If the pH changes, the charges on the R groups of basic and acidic amino acids can change in such a way that they can no longer form salt bridges, causing a change in the shape of the protein. Denaturation can also occur by adding certain organic compounds or heavy metal ions, or through mechanical agitation. When the interactions between the R groups are disrupted, a protein unfolds like a wet spaghetti noodle. With the loss of its overall shape (tertiary structure), the protein is no longer biologically active (see **FIGURE 19.10**).

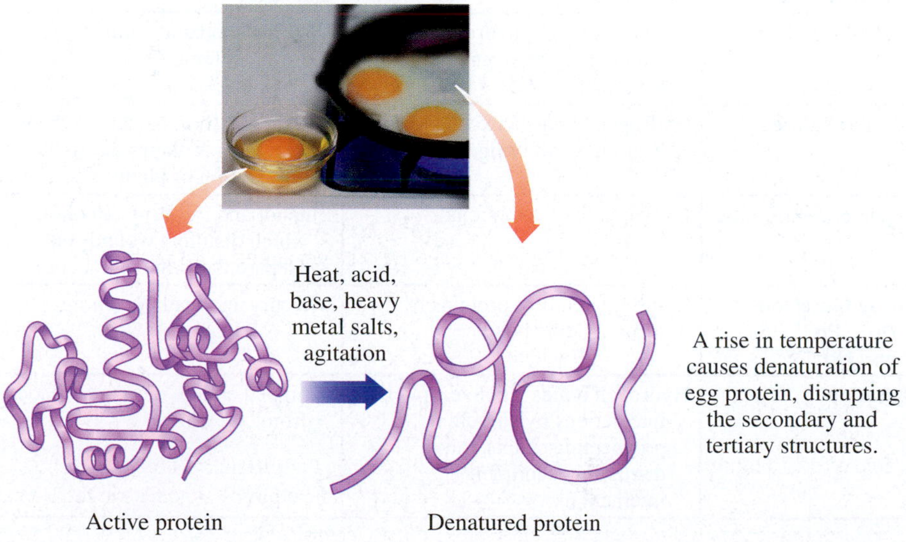

Heat, acid, base, heavy metal salts, agitation
A rise in temperature causes denaturation of egg protein, disrupting the secondary and tertiary structures.
Active protein
Denatured protein

FIGURE 19.10 ▶ Denaturation of a protein occurs when the interactions that stabilize secondary, tertiary, or quaternary structures are disrupted, which destroys the shape and renders the protein biologically inactive.

Q What are some ways in which proteins are denatured?

Heat

Whenever you cook food, you are using heat to denature protein. Hydrogen bonds and hydrophobic interactions between nonpolar R groups are disrupted. Few proteins can remain biologically active above 50 °C. The nutritional value of the proteins in food is not changed, but they are more digestible. High temperatures are also used to disinfect surgical instruments and gowns by denaturing the proteins of any bacteria present.

Acids and Bases

Like a cooked egg, the denatured proteins in milk form more of a semisolid when denatured. In the preparation of yogurt and cheese, bacteria that produce lactic acid are added to denature the proteins by changing the pH. A change in pH breaks down hydrogen bonds and disrupts salt bridges. Tannic acid, a weak acid used in burn ointments, is applied to the site of the burn to coagulate proteins, which forms a protective cover and prevents further loss of fluid from the burn.

Organic Compounds

Ethanol and isopropyl alcohol act as disinfectants by exchanging the bacterial protein's hydrogen bonds to water with their own and disrupting the side chain intramolecular hydrogen bonding. An alcohol swab is used to clean wounds or to prepare the skin for an injection because the alcohol passes through the cell walls and coagulates the proteins inside the bacteria.

Heavy Metal Ions

In hospitals, a dilute (1%) solution of $AgNO_3$ is placed in the eyes of newborn babies to destroy the bacteria that cause gonorrhea. Heavy metal ions such as Ag^+, Pb^{2+}, and Hg^{2+} denature protein by forming bonds with ionic R groups or reacting with disulfide (—S—S—) bonds. If heavy metals are ingested, they act as poisons by severely denaturing body proteins and disrupting metabolic reactions. An antidote is a high-protein food, such as milk, eggs, or cheese, that combines with the heavy metal ions until the stomach can be pumped.

Agitation

The whipping of cream and the beating of egg whites are examples of using mechanical agitation to denature protein. The whipping action stretches the polypeptide chains until the stabilizing interactions are disrupted. **TABLE 19.6** summarizes protein denaturation.

TEST

Try Practice Problems 19.36 to 19.38

TABLE 19.6 Protein Denaturation

Denaturing Agent	Bonds Disrupted	Examples
Heat Above 50 °C	Hydrogen bonds; hydrophobic interactions between nonpolar R groups	Cooking food and autoclaving surgical items
Acids and Bases	Hydrogen bonds between polar R groups; salt bridges	Lactic acid from bacteria, which denatures milk protein in the preparation of yogurt and cheese
Organic Compounds	Hydrophobic interactions	Ethanol and isopropyl alcohol, which disinfect wounds and prepare the skin for injections
Heavy Metal Ions Ag^+, Pb^{2+}, and Hg^{2+}	Disulfide bonds in proteins by forming ionic bonds	Mercury and lead poisoning
Agitation	Hydrogen bonds and hydrophobic interactions by stretching polypeptide chains and disrupting stabilizing interactions	Whipped cream, meringue made from egg whites

▶ SAMPLE PROBLEM 19.7 Denaturation of Proteins

TRY IT FIRST

Describe the denaturation process in each of the following:

a. An appetizer known as ceviche is prepared without heat by placing slices of raw fish in a solution of lemon or lime juice. After 3 or 4 h, the fish appears to be "cooked."

b. When baking scalloped potatoes, the added milk curdles (forms solids).

SOLUTION

a. The acids in lemon or lime juice break down the hydrogen bonds between polar R groups and disrupt salt bridges, which denature the proteins of the fish.

b. The heat during baking breaks apart hydrogen bonds and hydrophobic interactions between nonpolar R groups in the milk proteins. When the milk denatures, the proteins become insoluble and form solids called curds.

STUDY CHECK 19.7

a. Why is a dilute solution of $AgNO_3$ used to disinfect the eyes of newborn infants?

b. Why is tannic acid included in many burn ointments?

ANSWER

a. The heavy metal Ag^+ denatures the proteins in bacteria that cause gonorrhea.

b. The tannic acid causes proteins to coagulate and form a protective layer over the burn.

PRACTICE PROBLEMS

19.5 Protein Hydrolysis and Denaturation

19.31 What products would result from the complete hydrolysis of Gly–Ala–Ser?

19.32 Would the hydrolysis products of the tripeptide Ala–Ser–Gly be the same or different from the products in problem 19.31? Explain.

19.33 What dipeptides could be produced from the partial hydrolysis of His–Met–Gly–Val?

19.34 What tripeptides could be produced from the partial hydrolysis of Ser–Leu–Gly–Gly–Ala?

19.35 What structural level of a protein is affected by hydrolysis?

19.36 What structural level of a protein is affected by denaturation?

Clinical Applications

19.37 Indicate the changes in the secondary and tertiary structural levels of proteins for each of the following:

a. An egg placed in water at 100 °C is soft boiled after 3 minutes.

b. Prior to giving an injection, the skin is wiped with an alcohol swab.

c. Surgical instruments are placed in a 120 °C autoclave.

d. During surgery, a wound is closed by cauterization (heat).

19.38 Indicate the changes in the secondary and tertiary structural levels of proteins for each of the following:

a. Tannic acid is placed on a burn.

b. Milk is heated to 60 °C to make yogurt.

c. To avoid spoilage, seeds are treated with a solution of $HgCl_2$.

d. Hamburger is cooked at high temperatures to destroy *E. coli* bacteria that may cause intestinal illness.

CLINICAL UPDATE Jeremy's Diagnosis and Treatment for Sickle-Cell Anemia

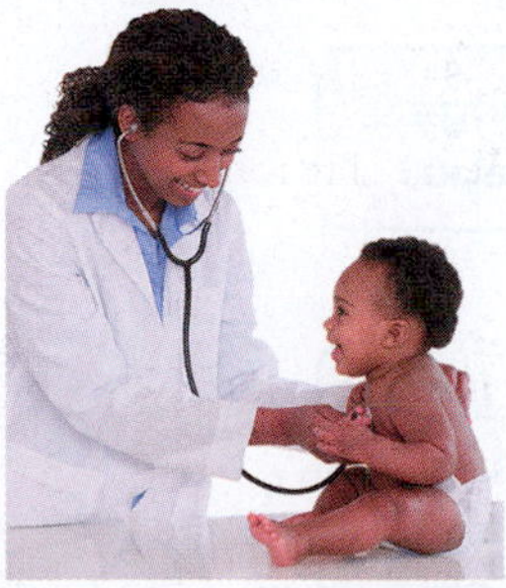

Diagnosis of Sickle-Cell Anemia

When Jeremy and his mother return to the clinic, the hematology nurse explains the result of the blood test. Proteins are used to identify and diagnose diseases such as sickle-cell anemia by *electrophoresis*. For Jeremy's diagnosis, normal hemoglobin, sickle-cell hemoglobin, and Jeremy's hemoglobin are placed on a gel and an electric current is applied. The proteins move along the gel based on their charge, size, and shape.

Treatment of Sickle-Cell Anemia

By comparing the movement of Jeremy's hemoglobin with that of sickle-cell hemoglobin and normal hemoglobin, electrophoresis indicates that he does have sickle-cell anemia. After discussing the test results with Samantha, the doctor prescribes a daily dose of hydroxyurea for Jeremy. Hydroxyurea is a medication that stimulates the production of fetal hemoglobin, a variation of the hemoglobin molecule that binds more tightly to oxygen molecules than normal adult hemoglobin. Fetal hemoglobin is produced by a fetus *in utero*, but production of this molecule typically stops after birth.

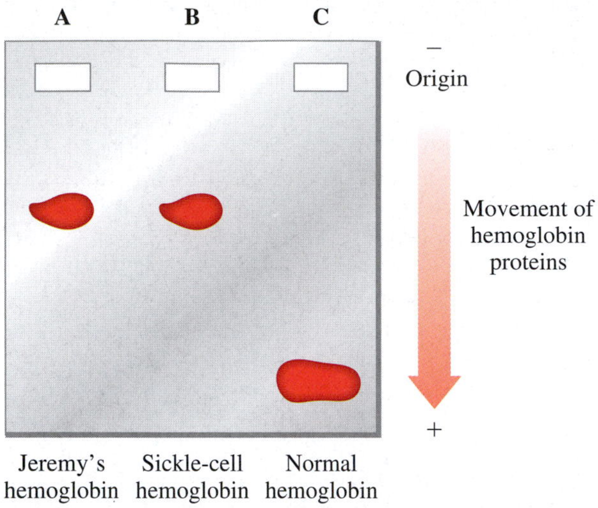
A
B
C
–
Origin
Movement of hemoglobin proteins
+
Jeremy's hemoglobin
Sickle-cell hemoglobin
Normal hemoglobin

In electrophoresis, hemoglobin samples from Jeremy **(A)**, a person with sickle-cell anemia **(B)**, and a person with normal hemoglobin **(C)** are placed on a gel and an electrical field is applied. The differences in the movement of the abnormal and normal proteins allow the diagnosis of sickle-cell anemia.

Patients with sickle-cell anemia who take hydroxyurea daily experience less pain, fewer events in which blood vessels are obstructed by sickled red blood cells, and increased blood flow. The suggested daily dose for children is 20. mg of hydroxyurea per kg of body weight. Samantha will monitor Jeremy regularly to determine if his hydroxyurea dose needs to be adjusted.

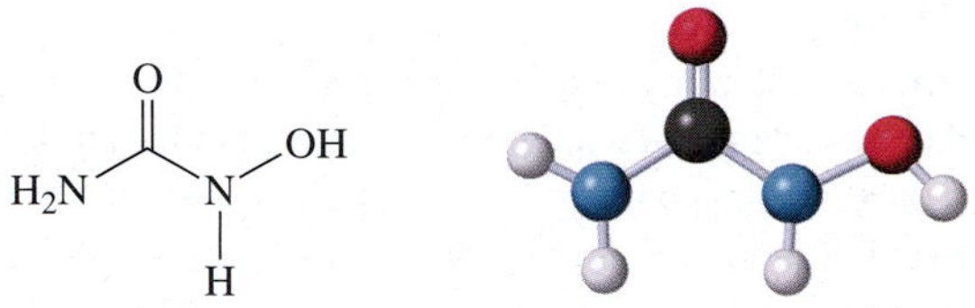
O
H_2N
N
OH
H

Hydroxyurea

Clinical Applications

19.39 What is the chemical formula for hydroxyurea?

19.40 What is the molar mass of hydroxyurea?

19.41 If Jeremy's current weight is 21 lbs, how many milligrams of hydroxyurea should Jeremy be given each day?

19.42 Fetal hemoglobin binds oxygen more tightly than does adult hemoglobin. Why do you think fetal hemoglobin functions differently than adult hemoglobin?

CONCEPT MAP

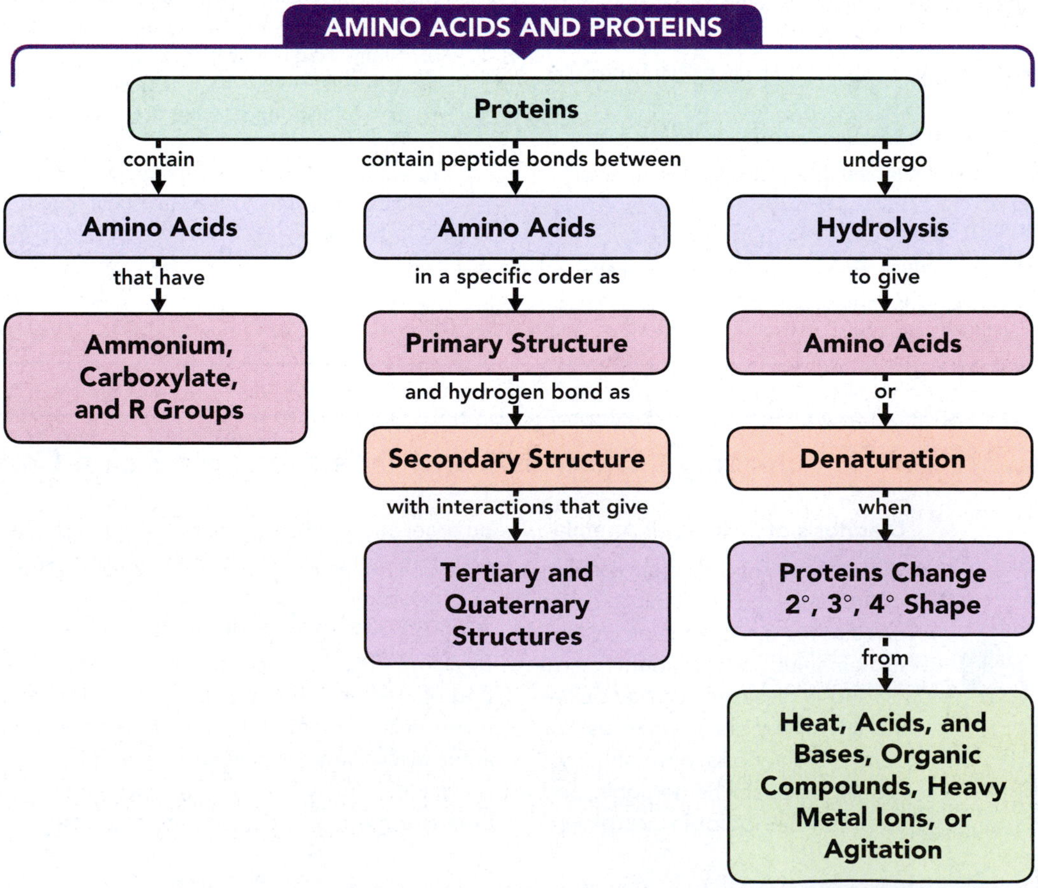
AMINO ACIDS AND PROTEINS
Proteins
contain
contain peptide bonds between
undergo
Amino Acids
Amino Acids
Hydrolysis
that have
in a specific order as
to give
Ammonium, Carboxylate, and R Groups
Primary Structure
Amino Acids
and hydrogen bond as
or
Secondary Structure
Denaturation
with interactions that give
when
Tertiary and Quaternary Structures
Proteins Change 2°, 3°, 4° Shape
from
Heat, Acids, and Bases, Organic Compounds, Heavy Metal Ions, or Agitation

CHAPTER REVIEW

19.1 Proteins and Amino Acids

LEARNING GOAL Classify proteins by their functions. Give the name and abbreviations for an amino acid and draw its structure at physiological pH.

- Some proteins are enzymes or hormones, whereas others are important in structure, transport, protection, storage, and muscle contraction.
- A group of 20 amino acids provides the molecular building blocks of proteins.
- Attached to the central α carbon of each amino acid are an ammonium group, a carboxylate group, and a unique R group.
- The R group gives an amino acid the property of being nonpolar, polar, acidic, or basic.

19.2 Proteins: Primary Structure

LEARNING GOAL Draw the condensed structural formula for a peptide and give its name. Describe the primary structure for a protein.

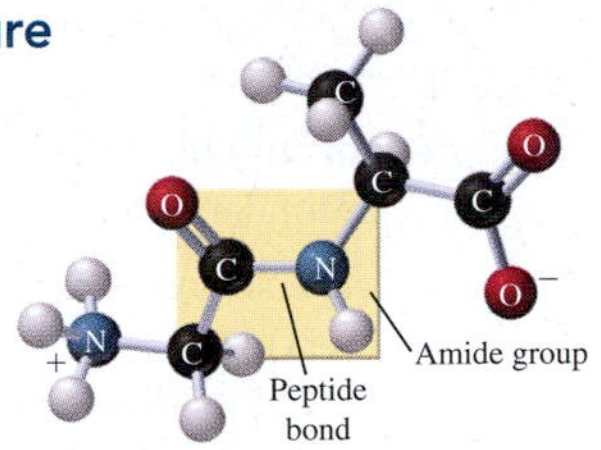

Amide group
Peptide bond

- Peptides form when an amide bond links the carboxylate group of one amino acid and the ammonium group of a second amino acid.
- Long chains of amino acids that are biologically active are called proteins.
- The primary structure of a protein is its sequence of amino acids joined by peptide bonds.
- Peptides are named from the N-terminus by replacing the *ine* or *ate* of each amino acid name with *yl* followed by the name of the amino acid at the C-terminus.
- Essential amino acids must be obtained from the proteins in the diet; they cannot be made in the body.

19.3 Proteins: Secondary Structure

LEARNING GOAL Describe the two most common types of secondary structures for a protein. Describe the structure of collagen.

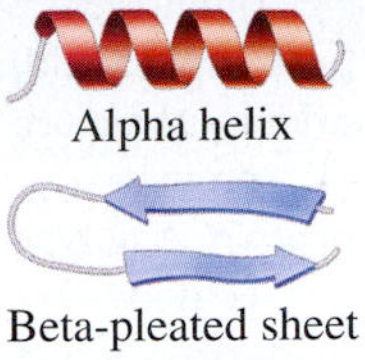

Alpha helix
Beta-pleated sheet

- In secondary structures, hydrogen bonds between atoms in the peptide bonds produce a characteristic shape.
- The two most common secondary structures are α helices and β-pleated sheets.
- In the structure of collagen, three helices twist around each other, with hydrogen bonds forming links between the individual helices.

19.4 Proteins: Tertiary and Quaternary Structures

LEARNING GOAL Describe the tertiary and quaternary structures of a protein.

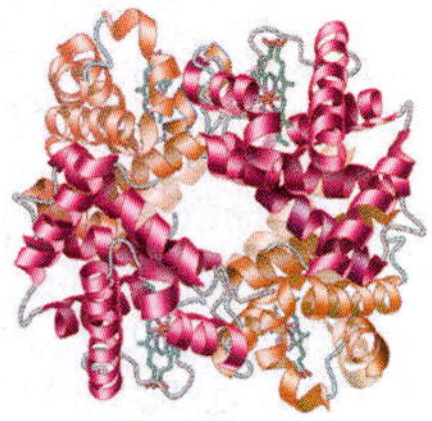

- A tertiary structure is stabilized by interactions that push amino acids with hydrophobic R groups to the center and pull amino acids with hydrophilic R groups to the surface, and by interactions between amino acids with R groups that form hydrogen bonds, disulfide bonds, and salt bridges.
- In a quaternary structure, two or more tertiary subunits are joined together for biological activity, held by the same interactions found in tertiary structures.

19.5 Protein Hydrolysis and Denaturation

LEARNING GOAL Describe the hydrolysis and denaturation of proteins.

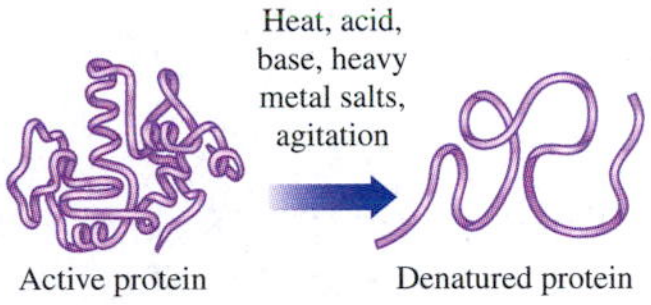

Heat, acid, base, heavy metal salts, agitation
Active protein
Denatured protein

- Hydrolysis of a protein occurs when peptide bonds in the primary structure are broken, producing amino acids.
- Denaturation of a protein destroys the secondary, tertiary, or quaternary structures of a protein, causing a loss of biological activity.
- Denaturation is caused by high temperatures, acids or bases, organic compounds, metal ions, or agitation.

KEY TERMS

acidic amino acid An amino acid that has an R group with a carboxylate group ($-COO^-$).

α (alpha) helix A secondary level of protein structure, in which hydrogen bonds connect the C=O of one peptide bond with the N—H of a peptide bond farther along in the chain to form a coiled or corkscrew structure.

amino acid The building block of proteins, consisting of an ammonium group, a carboxylate group, and a unique R group attached to the α carbon.

backbone The repeating sequence of N—C—C atoms from the amino acids in a protein.

basic amino acid An amino acid that contains an amine R group.

β (beta)-pleated sheet A secondary level of protein structure that consists of hydrogen bonds between peptide links in parallel polypeptide chains.

C-terminus The end of a peptide chain with a free $-COO^-$ group.

collagen The most abundant form of protein in the body, composed of three polypeptide chains twisted together into a triple helix and with hydrogen bonding between —OH groups of hydroxyproline and hydroxylysine.

denaturation The loss of secondary, tertiary, and quaternary protein structures caused by heat, acids, bases, organic compounds, heavy metals, and/or agitation.

disulfide bond Covalent —S—S— bond that forms between the —SH groups of cysteines in a protein, which stabilizes the tertiary and quaternary structures.

essential amino acid An amino acid that must be supplied by the diet because it is not synthesized by the body.

hydrogen bond Attraction between polar R groups, such as —OH, $-NH_2$, and —COOH, of amino acids in a polypeptide chain.

hydrophilic interaction The attraction between water and polar R groups on the outside of the protein.

hydrophobic interaction The attraction between nonpolar R groups on the inside of a protein.

N-terminus The end of a peptide chain with a free $-NH_3^+$ group.

nonpolar amino acid An amino acid with a nonpolar R group.

peptide The combination of two or more amino acids joined by peptide bonds; dipeptide, tripeptide, and so on.

peptide bond The amide bond in peptides that joins the carboxylate group of one amino acid with the ammonium group in the next amino acid.

polar amino acid An amino acid with a polar R group.

primary structure The specific sequence of the amino acids in a protein.

protein A polypeptide of 50 or more amino acids that has biological activity.

quaternary structure A protein structure in which two or more protein subunits form an active protein.

salt bridge The attraction between ionized R groups of basic and acidic amino acids in the tertiary structure of a protein.

secondary structure The structure that forms when the atoms in the backbone of a protein or peptide form hydrogen bonds within a polypeptide or between polypeptide chains.

tertiary structure The folding of the secondary structure of a protein into a compact structure that is stabilized by the interactions of R groups.

triple helix The protein structure found in collagen, consisting of three polypeptide chains twisted together.

CORE CHEMISTRY SKILLS

The chapter Section containing each Core Chemistry Skill is shown in parentheses at the end of each heading.

Drawing the Structure for an Amino Acid at Physiological pH (19.1)

- The central α carbon of each amino acid is bonded to an ammonium group ($-NH_3^+$), a carboxylate group ($-COO^-$), a hydrogen atom, and a unique R group.
- The R group gives an amino acid the property of being nonpolar, polar neutral, polar acidic, or polar basic.

Example: Draw the structure for cysteine at physiological pH.

Answer:

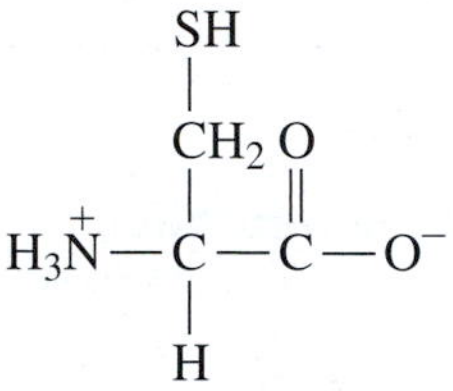

Identifying the Primary, Secondary, Tertiary, and Quaternary Structures of Proteins (19.2, 19.3, 19.4)

- The primary structure of a protein is the sequence of amino acids joined by peptide bonds.
- In the secondary structures of proteins, hydrogen bonds between atoms in the peptide bonds produce an alpha helix or a beta-pleated sheet.
- The tertiary structure of a protein is stabilized by R groups that form hydrogen bonds, disulfide bonds, and salt bridges, and by hydrophobic R groups that move to the center and hydrophilic R groups that move to the surface.
- In a quaternary structure, two or more tertiary subunits are combined for biological activity, held by the same interactions found in tertiary structures.

Example: Identify the following as characteristic of the primary, secondary, tertiary, or quaternary structure of a protein:

a. The R groups of two amino acids interact to form a salt bridge.
b. Eight amino acids form peptide bonds.
c. A polypeptide forms an alpha helix.
d. Two amino acids with hydrophobic R groups move toward the inside of the folded protein.
e. A protein with biological activity contains four tertiary polypeptide subunits.

Answer: **a.** tertiary, quaternary **b.** primary **c.** secondary **d.** tertiary **e.** quaternary

UNDERSTANDING THE CONCEPTS

The chapter Sections to review are shown in parentheses at the end of each problem.

19.43 Aspartame, which is used in artificial sweeteners, contains the following dipeptide: (19.1, 19.2)

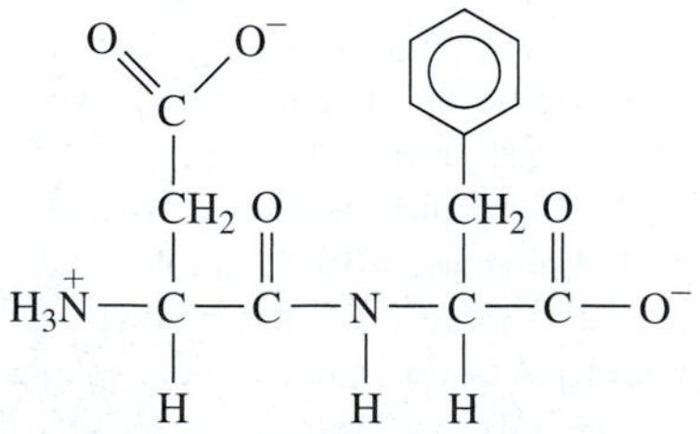

Some artificial sweeteners contain aspartame, which is an ester of a dipeptide.

a. What are the amino acids in aspartame?
b. How would you name the dipeptide in aspartame?

19.44 Fresh pineapple contains the enzyme bromelain, which hydrolyzes peptide bonds in proteins. (19.2, 19.3, 19.4, 19.5)

Fresh pineapple contains the enzyme bromelain.

a. The directions for a gelatin (protein) dessert say not to add fresh pineapple. However, canned pineapple where pineapple is heated to high temperatures can be added. Why?
b. Fresh pineapple is used in a marinade to tenderize tough meat. Why?
c. What structural level of a protein does the bromelain enzyme destroy?

19.45 Identify the amino acids and type of interaction that occurs between the following R groups in tertiary protein structures: (19.1, 19.4)

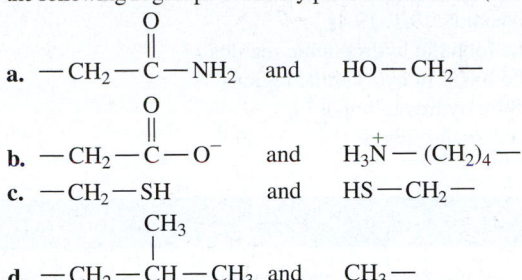

a. $-CH_2-\overset{\overset{\large O}{\|}}{C}-NH_2$ and $HO-CH_2-$

b. $-CH_2-\overset{\overset{\large O}{\|}}{C}-O^-$ and $H_3\overset{+}{N}-(CH_2)_4-$

c. $-CH_2-SH$ and $HS-CH_2-$

d. $-CH_2-\underset{}{\overset{\overset{\large CH_3}{|}}{CH}}-CH_3$ and CH_3-

19.46 What type of interaction would you expect between the following in a tertiary structure? (19.1, 19.4)

a. threonine and glutamine **b.** valine and alanine

c. arginine and glutamate

19.47 Seeds and vegetables are often deficient in one or more essential amino acids. Using the following table, state whether each combination provides all of the essential amino acids: (19.2)

Source	Lysine	Tryptophan	Methionine
Oatmeal	No	Yes	Yes
Rice	No	Yes	Yes
Garbanzo beans	Yes	No	Yes
Lima beans	Yes	No	No
Cornmeal	No	No	Yes

a. rice and garbanzo beans

b. lima beans and cornmeal

c. garbanzo beans and lima beans

19.48 Seeds and vegetables are often deficient in one or more essential amino acids. Using the table in problem 19.47, state whether each combination provides all of the essential amino acids. (19.2)

Oatmeal is deficient in the essential amino acid lysine.

a. rice and lima beans

b. rice and oatmeal

c. oatmeal and lima beans

ADDITIONAL PRACTICE PROBLEMS

19.49 Each of three peptides contains one molecule of valine and two molecules of serine. Use the three-letter and one-letter abbreviations of these amino acids to describe the primary structures of the three possible peptides. (19.1, 19.2)

19.50 Each of three peptides contains one molecule of glycine, one molecule of alanine, and two molecules of isoleucine. In these peptides, the amino acid at the N-terminus is glycine. Use the three-letter and one-letter abbreviations of these amino acids to describe the primary structures of the three possible peptides. (19.1, 19.2)

19.51 a. Draw the condensed structural formula for Ser–Lys–Asp. (19.1, 19.2, 19.3)

b. Would you expect to find this segment at the center or at the surface of a protein? Why?

19.52 a. Draw the condensed structural formula for Val–Ala–Leu. (19.1, 19.2, 19.3)

b. Would you expect to find this segment at the center or at the surface of a protein? Why?

19.53 a. Where in the body is collagen found? (19.3)

b. What type of structure is used to form collagen?

19.54 a. What are some functions of collagen? (19.3)

b. What amino acids give strength to collagen?

19.55 Would you expect a polypeptide with a high content of His, Met, and Leu to have more α-helical sections or β-pleated sheet sections? (19.3)

19.56 Why would a polypeptide with a high content of Val, Pro, and Ser be likely to have more α-helical sections than β-pleated sheet sections? (19.3)

19.57 If serine was replaced by valine in a protein, how might the tertiary structure be affected? (19.1, 19.4)

19.58 If glycine was replaced by alanine in a protein, how might the tertiary structure be affected? (19.1, 19.4)

19.59 In the preparation of meringue for a pie, a few drops of lemon juice are added and the egg whites are whipped. What causes the meringue to form? (19.5)

19.60 How does denaturation of a protein differ from its hydrolysis? (19.5)

CHALLENGE PROBLEMS

The following problems are related to the topics in this chapter. However, they do not all follow the chapter order, and they require you to combine concepts and skills from several Sections. These problems will help you increase your critical thinking skills and prepare for your next exam.

19.61 What are some differences between each of the following pairs? (19.1, 19.2, 19.3)

a. secondary and tertiary protein structures

b. essential and nonessential amino acids

c. polar and nonpolar amino acids

d. dipeptides and tripeptides

19.62 What are some differences between each of the following pairs? (19.1, 19.2, 19.3)

a. a salt bridge and a disulfide bond

b. an α helix and collagen

c. an α helix and a β-pleated sheet

d. tertiary and quaternary structures of proteins

19.63 Consider the amino acids lysine, valine, and aspartate in an enzyme. State which of these amino acids have R groups that would: (19.1, 19.4)

a. be found in hydrophobic regions
b. be found in hydrophilic regions
c. form hydrogen bonds
d. form salt bridges

19.64 Consider the amino acids histidine, phenylalanine, and serine in an enzyme. State which of these amino acids have R groups that would: (19.1, 19.4)

a. be found in hydrophobic regions
b. be found in hydrophilic regions
c. form hydrogen bonds
d. form salt bridges

UNDERSTANDING PROTEIN STRUCTURES

In order to understand how proteins function, we need to know about their tertiary structures; but predicting the tertiary structures of proteins is a very challenging task. Proteins are large molecules consisting of hundreds to thousands of different amino acids. They can fold in many, many different ways.

Researchers at the University of Washington built an online game called Foldit that allows students to fold proteins into three-dimensional shapes. Game players learn about the rules that guide protein folding and then use these rules to fold provided primary structures into three-dimensional shapes. To try your hand at solving for the three-dimensional structure of a protein, go to https://fold.it/portal/.

ANSWERS

19.1 **a.** transport **b.** structural
c. structural **d.** enzyme

19.3 All amino acids contain a carboxylate group and an ammonium group on the α carbon.

19.5 **a.** 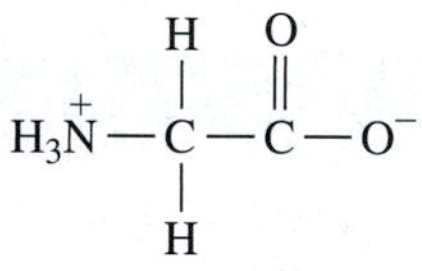

b. 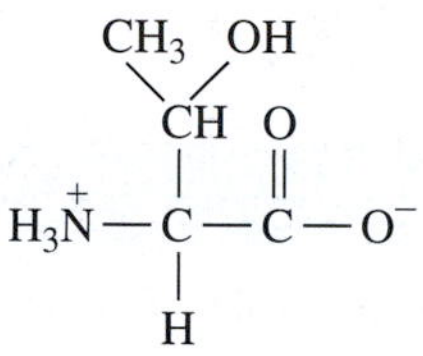

c. 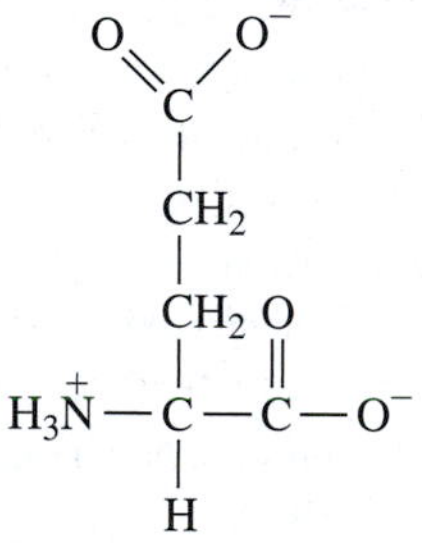

d. 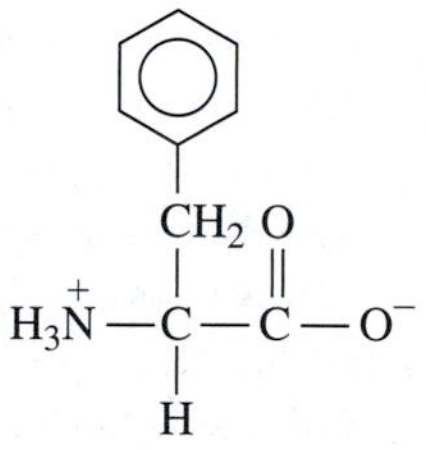

19.7 **a.** nonpolar, hydrophobic **b.** polar neutral, hydrophilic
c. polar acidic, hydrophilic **d.** nonpolar, hydrophobic

19.9 **a.** alanine **b.** glutamine
c. lysine **d.** cysteine

19.11 **a.** 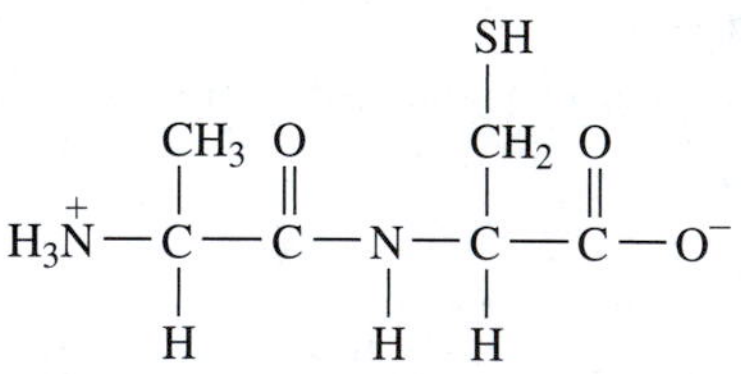

Ala–Cys, AC

b. 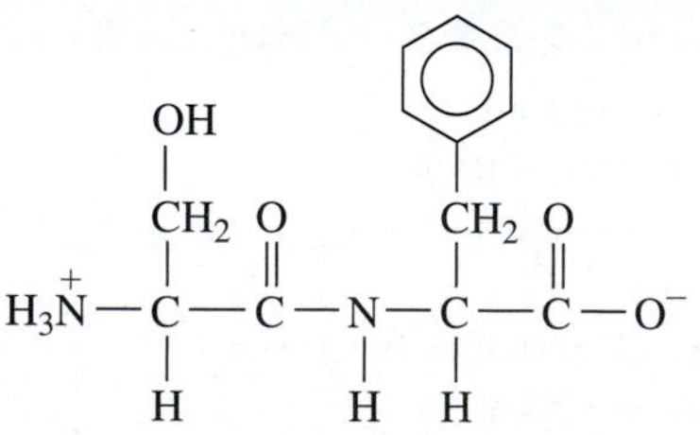

Ser–Phe, SF

c. 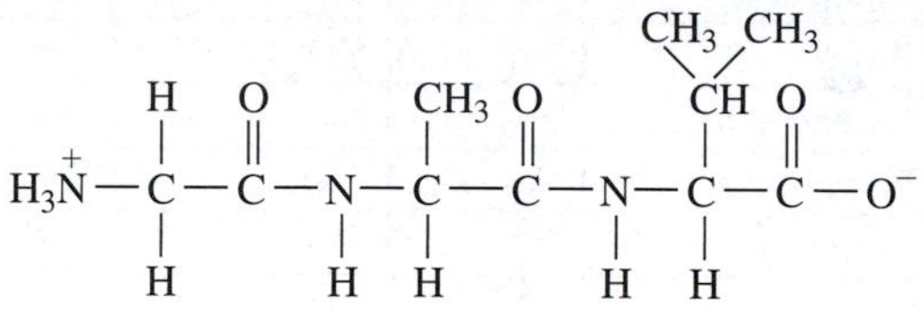

Gly–Ala–Val, GAV

d. 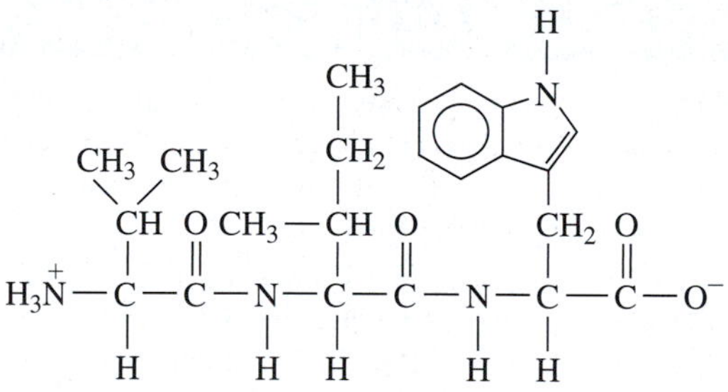

Val–Ile–Trp, VIW

19.13 **a.** 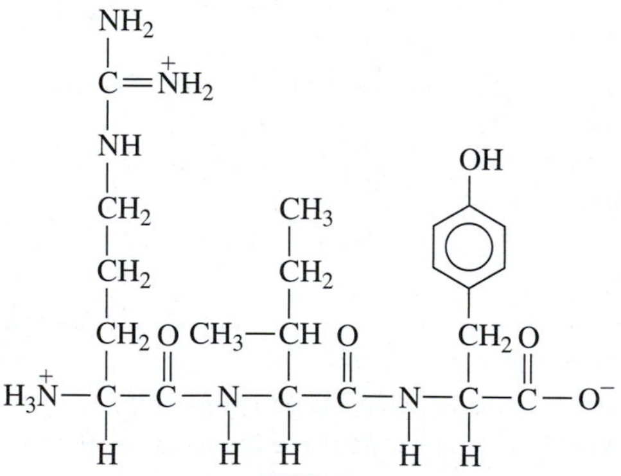

RIY

b. 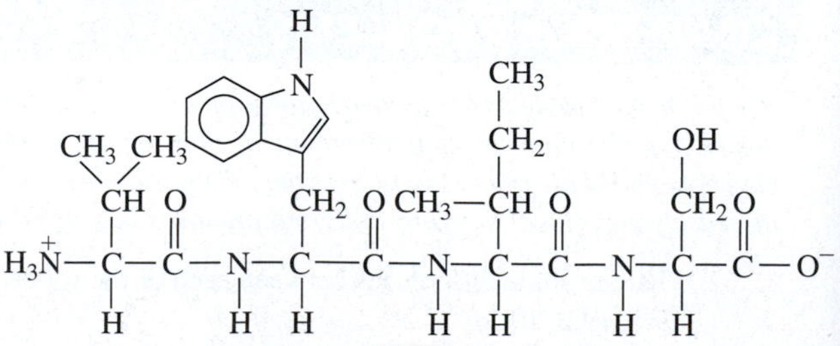

VWIS

19.15 a. Lysine and tryptophan are lacking in corn but supplied by peas; methionine is lacking in peas but supplied by corn.
b. Lysine is lacking in rice but supplied by soy; methionine is lacking in soy but supplied by rice.

19.17 The oxygen atoms of the carbonyl groups and the hydrogen atoms attached to the nitrogen atoms form alpha helices or beta-pleated sheets.

19.19 In the α helix, hydrogen bonds form between the carbonyl oxygen atom and the amino hydrogen atom in the next turn of the helical chain. In the β-pleated sheet, hydrogen bonds occur between parallel sections of a long polypeptide chain.

19.21 The secondary structure of beta-amyloid protein in the brain of a normal person is alpha-helical, whereas the secondary structure in the brain of a person with Alzheimer's changes to a beta-pleated sheet.

19.23 An α helix consists of only one peptide chain twisted into a helix. The structure of collagen consists of three helices twisted around each other.

19.25 a. disulfide bond **b.** salt bridge
c. hydrogen bond **d.** hydrophobic interaction

19.27 a. Leucine and valine will be found on the inside of the protein because they have R groups that are hydrophobic.
b. The cysteine and aspartate would be on the outside of the protein because they have R groups that are polar.
c. The order of the amino acids (the primary structure) provides R groups that interact to determine the tertiary structure of the protein.

19.29 a. tertiary, quaternary **b.** primary
c. secondary **d.** secondary

19.31 The products would be the amino acids glycine, alanine, and serine.

19.33 His–Met, Met–Gly, Gly–Val

19.35 Hydrolysis splits the peptide bonds in the primary structure.

19.37 a. Placing an egg in boiling water coagulates the proteins of the egg because the heat disrupts hydrogen bonds and hydrophobic interactions.
b. The alcohol on the swab coagulates the proteins of any bacteria present by forming hydrogen bonds and disrupting hydrophobic interactions.
c. The heat from an autoclave will coagulate the proteins of any bacteria on the surgical instruments by disrupting hydrogen bonds and hydrophobic interactions.
d. Heat will coagulate the surrounding proteins to close the wound by disrupting hydrogen bonds and hydrophobic interactions.

19.39 $CH_4N_2O_2$

19.41 76 g

19.43 a. aspartate and phenylalanine **b.** aspartylphenylalanine

19.45 a. asparagine and serine; hydrogen bond
b. aspartate and lysine; salt bridge
c. cysteine and cysteine; disulfide bond
d. leucine and alanine; hydrophobic interaction

19.47 a. yes **b.** no **c.** no

19.49 Val–Ser–Ser (VSS), Ser–Ser–Val (SSV), Ser–Val–Ser (SVS)

19.51 a.

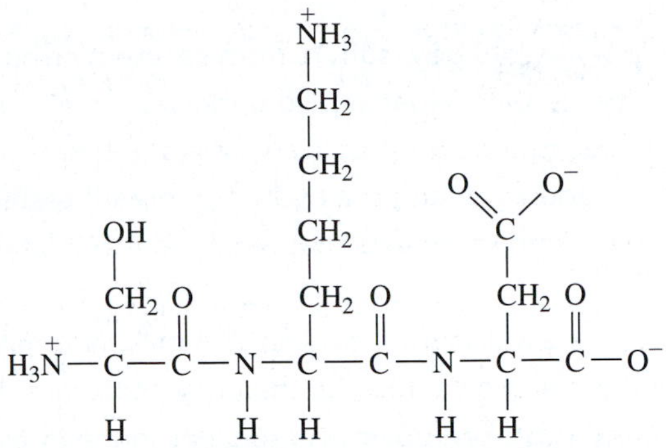

b. This segment contains polar R groups, which would be found on the surface of a protein where they hydrogen bond with water.

19.53 a. Collagen is found in connective tissue, blood vessels, skin, tendons, ligaments, the cornea of the eye, and cartilage.
b. A triple helix is the structure found in collagen.

19.55 α helical

19.57 Serine is a polar amino acid, whereas valine is a nonpolar amino acid. Serine would move to the outside surface of the protein where it can form hydrogen bonds with water. However, valine, which is nonpolar, would be pushed to the center of the tertiary structure where it is stabilized by forming hydrophobic interactions.

19.59 The acid from the lemon juice and the mechanical whipping (agitation) denature the proteins of the egg white, which turn into solids as meringue.

19.61 a. In the secondary structure of proteins, hydrogen bonds form a helix or a pleated sheet; the tertiary structure is determined by the interactions of R groups such as disulfide bonds, hydrogen bonds, and salt bridges.
b. Nonessential amino acids can be synthesized by the body; essential amino acids must be supplied by the diet.
c. Polar amino acids have hydrophilic R groups, whereas nonpolar amino acids have hydrophobic R groups.
d. Dipeptides contain two amino acids, whereas tripeptides contain three.

19.63 a. valine **b.** lysine, aspartate
c. lysine, aspartate **d.** lysine, aspartate

20

Enzymes and Vitamins

Noah, a 6-year-old boy, suffers from severe diarrhea, abdominal pain, and intestinal growling two hours after he eats. Noah is also underweight for his age. His mother makes an appointment for Noah to see Emma, his physician assistant. Emma suspects that Noah may be lactose intolerant, so she schedules Noah for a hydrogen breath test (HBT) in a few days to confirm her suspicion. Until then, Emma suggests that Noah's mother limit the amount of milk and milk products in Noah's diet.

Lactose is a disaccharide found in milk. Normally, it is broken down in the intestinal tract into glucose and galactose by the enzyme lactase. However, when there is a lack of this enzyme, lactose is not broken down and remains in the intestinal tract. There, the lactose undergoes bacterial fermentation, which results in the production of gases such as H_2, CH_4, and CO_2. The presence of these gases in the intestine causes nausea, abdominal cramps, and diarrhea. The severity of the symptoms depends on how much lactose is present in the food that has been consumed and by the amount of lactase a person produces. Noah's HBT will determine if excess hydrogen gas is being produced in his intestines when he consumes lactose.

CAREER

Physician Assistant

A physician assistant, commonly referred to as a PA, helps a doctor by examining and treating patients, as well as prescribing medications. Many physician assistants take on the role of the primary caregiver. Their duties also include obtaining patient medical records and histories, diagnosing illnesses, educating and counseling patients, and referring patients, when needed, to a specialist. Physician assistants may also help the doctor during major surgery. Because of the potential diversity in their duties, physician assistants must be knowledgeable about a variety of medical conditions.

Physician assistants can work in clinics, hospitals, health maintenance organizations, or private practices. They can also take on a more administrative role that involves hiring new PAs and acting as the representative of the hospital to the patient or their family.

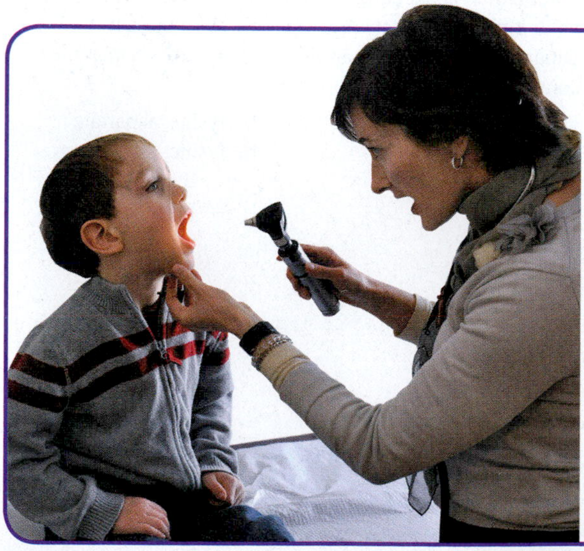

CLINICAL UPDATE

Noah's Diet for Lactose Intolerance

In the **CLINICAL UPDATE Noah's Diet for Lactose Intolerance**, pages 713–714, you can see how an HBT test is used to diagnose lactose intolerance and read about potential treatments for this condition.

20.1 Enzymes and Enzyme Action

LEARNING GOAL Describe enzymes and their roles in enzyme-catalyzed reactions.

LOOKING AHEAD

Every second, thousands of chemical reactions occur in the cells of our body. For example, many reactions occur to digest the food we eat, convert the digestion products to chemical energy, and synthesize proteins and other macromolecules in our cells. These same reactions can occur outside of the cells in our body, but they occur at rates that are much too slow to meet our physiological and metabolic needs. In our bodies, catalysts known as **enzymes** increase the rates at which most biological reactions occur.

As discussed in section 10.1, collision theory states that three conditions must be met in order for a reaction to occur: (1) the reactants must collide, (2) the reactants must align properly to break and form bonds, and (3) the collision between the reactants must provide enough energy to overcome the energy of activation. Enzymes act as *catalysts* by increasing the probability that these conditions are met and, thus, that a reaction will occur. For example, enzymes can hold reactants in place so that they align properly. Enzymes also lower the activation energy required for a chemical reaction to occur (see **FIGURE 20.1**). Less energy is required to convert reactant molecules to products, which increases the rate of an enzyme-catalyzed reaction compared to the rate of the uncatalyzed reaction.

REVIEW

Rates of Reaction (10.1)

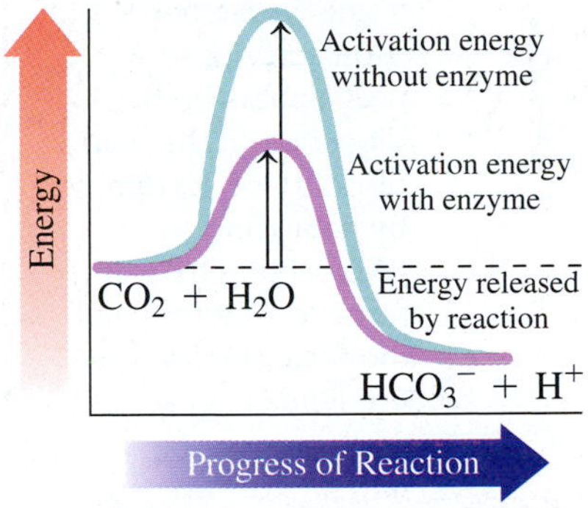

Activation energy without enzyme
Activation energy with enzyme
Energy
$CO_2 + H_2O$
Energy released by reaction
$HCO_3^- + H^+$
Progress of Reaction

FIGURE 20.1 ▶ The enzyme carbonic anhydrase lowers the activation energy for the reversible reaction that converts CO_2 and H_2O to bicarbonate and H^+.

Q Why are enzymes needed in biological reactions?

The rates of enzyme-catalyzed reactions are much faster than the rates of the uncatalyzed reactions. Some enzymes can increase the rate of a biological reaction by a factor of a billion, a trillion, or even a hundred million trillion compared to the rate of the uncatalyzed reaction. For example, an enzyme in the blood called carbonic anhydrase catalyzes the rapid interconversion of carbon dioxide and water to bicarbonate and H^+. In one second, one molecule of carbonic anhydrase can catalyze the reaction of about one million molecules of carbon dioxide. Carbonic anhydrase also catalyzes the reverse reaction, converting bicarbonate and H^+ to carbon dioxide and water.

$$CO_2 + H_2O \xrightleftharpoons{\text{Carbonic anhydrase}} HCO_3^- + H^+$$

TEST

Try Practice Problems 20.1 and 20.2

Chemistry Link to Health

Fabry Disease

Without enzymes, most reactions in the body do not occur quickly enough to meet our physiological needs. Therefore, a deficiency in even one enzyme can have serious consequences. Such is the case with *Fabry disease*, a rare, inherited disease that is caused by a deficiency in the enzyme α-galactosidase A.

α-Galactosidase A is an enzyme that is present in the lysosomes of the cell, where it participates in the digestion of a particular lipid, globotriaosylceramide, referred to as Gb3. Specifically, α-galactosidase A catalyzes the reaction that removes the terminal sugar group (galactose) from this molecule.

When α-galactosidase A is not present or is deficient, Gb3 begins to accumulate in the body and particularly in the blood vessels. The accumulation of the lipid leads to decreased blood flow and, consequently, to decreased nourishment to the tissues supplied by those blood vessels.

The symptoms of Fabry disease usually first manifest in early childhood and include pain and burning in the hands and feet; small, dark red spots (angiokeratomas) on the skin; decreased ability to sweat; gastrointestinal problems; and the development of a cloudiness in the front of the eye that does not affect vision. The disease

(continued)

Chemistry Link to Health (*continued*)

Globotriaosylceramide (Gb3)

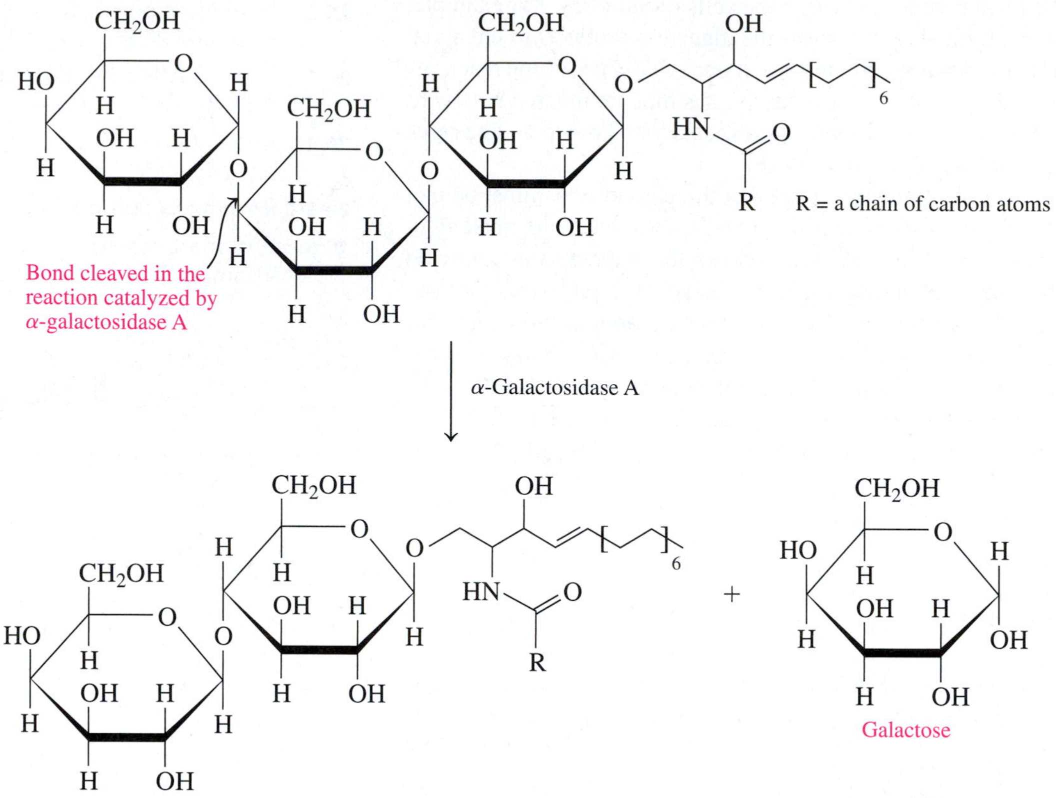
R = a chain of carbon atoms
Bond cleaved in the reaction catalyzed by α-galactosidase A
α-Galactosidase A
Galactose

In the lysosomes, α-galactosidase A participates in the digestion of the lipid globotriaosylceramide by catalyzing the reaction that removes the terminal monosaccharide from the lipid.

is progressive and ultimately leads to renal failure, heart attack, and stroke when patients reach 30 to 45 years of age.

Fabry disease is an X-linked disease and thus affects more males than females. It is rare, affecting approximately 1 out of every 40 000 to 60 000 males, and pan-ethnic, meaning that it does not seem to affect the members of one ethnic group more than those of another ethnic group. The life expectancy for people who suffer from Fabry disease is approximately 50 years.

In 2003, the U.S. Food and Drug Administration (FDA) approved an enzyme replacement therapy (ERT) for the treatment of Fabry disease. In this treatment, a recombinant form of α-galactosidase A is administered to patients. The administered enzyme breaks down the Gb3 so it does not accumulate in the blood vessels. This treatment appears to be more successful in curtailing the symptoms of Fabry disease if it is administered early in the lifetime of the patient.

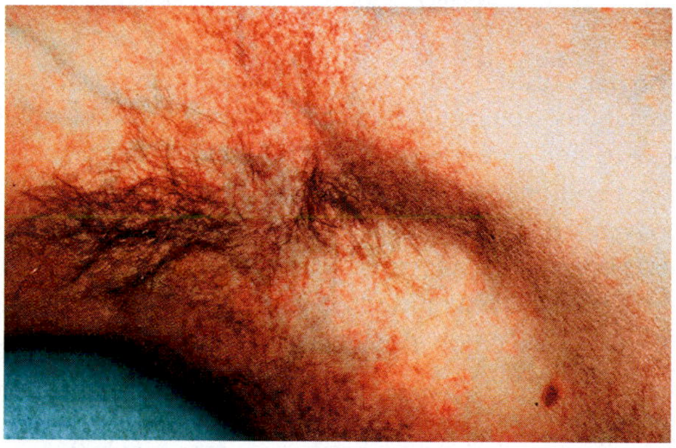

Patients with Fabry disease develop small, dark red spots, or angiokeratomas, on their skin.

Enzymes and Active Sites

Nearly all enzymes are proteins. Each has a unique three-dimensional shape that recognizes and binds a small group of reacting molecules, which are called **substrates**. The three-dimensional tertiary structure of an enzyme plays an important role in how that enzyme catalyzes reactions.

A typical enzyme is much larger than its substrate. However, within the enzyme's tertiary structure is a region called the **active site**, in which the substrate or substrates are held while the reaction takes place. The active site has a structure that closely fits the substrate, like a hand fits in a glove. Within the active site, specific amino acid R groups that are part of the enzyme interact with functional groups on the substrate through hydrogen bonds, salt bridges, and hydrophobic interactions. These interactions hold the substrate in proper alignment for a reaction to occur.

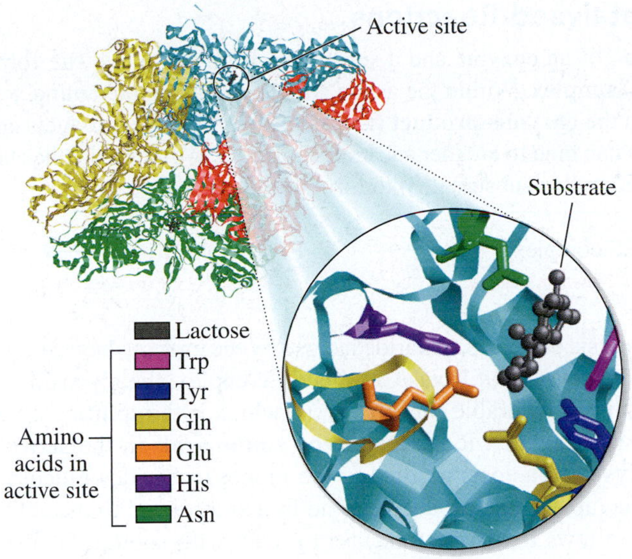
Active site
Substrate
Lactose
Trp
Tyr
Gln
Glu
His
Asn
Amino acids in active site

FIGURE 20.2 ▶ Enzymes like lactase have an active site where the substrate fits for catalysis to occur. The quaternary structure of lactase consists of four subunits. The substrate, lactose (gray), is held in place in the active site by hydrogen bonds with amino acid R groups.

Q Why does an enzyme catalyze a reaction for only certain substrates?

FIGURE 20.2 shows the three-dimensional structure of the enzyme lactase. Lactase hydrolyzes the disaccharide lactose to the monosaccharides glucose and galactose. The protein backbone of lactase is represented by a ribbon. As lactase folds into its three-dimensional structure, the pocket that forms in one part of its surface is its active site. This is where the hydrolysis of lactose occurs. At least six amino acid R groups are involved in the binding of the lactose substrate to the lactase enzyme.

Lactose fits into the active site of lactase, but other disaccharides will not fit into lactase's active site. As a consequence, lactase catalyzes only the specific reaction in which lactose is hydrolyzed into galactose and glucose. Some enzymes demonstrate this type of absolute *substrate specificity*, in which they catalyze only one type of reaction for one specific substrate. Other enzymes catalyze only one type of reaction, but do so for a group of similar substrates. Three categories of substrate specificity are outlined in **TABLE 20.1**.

TABLE 20.1 Types of Substrate Specificity

Type	Reaction Type	Example
Absolute	Catalyzes one type of reaction for one substrate	Urease catalyzes only the hydrolysis of urea.
Group	Catalyzes one type of reaction for similar substrates	Hexokinase adds a phosphate group to hexoses.
Linkage	Catalyzes one type of reaction for a specific type of bond	Chymotrypsin catalyzes the hydrolysis of peptide bonds.

TEST

Try Practice Problems 20.7 and 20.8

▶ SAMPLE PROBLEM 20.1 The Enzyme Active Site

TRY IT FIRST

What is the function of the active site in an enzyme?

SOLUTION

The R groups of amino acids within the active site of an enzyme bind the substrate by forming hydrogen bonds, salt bridges, and hydrophobic interactions, holding the substrate in correct alignment so the reaction can occur.

STUDY CHECK 20.1

Which is a larger molecule, glucose or hexokinase?

ANSWER

Hexokinase. Typically, enzymes are large protein molecules containing many amino acids, and substrates are small molecules such as the monosaccharide glucose.

Enzyme-Catalyzed Reactions

The combination of an enzyme and a substrate within the active site forms an **enzyme–substrate (ES) complex**. Within the active site, R groups on the amino acids catalyze the reaction to form the **enzyme–product (EP) complex**. Then the products are released from the enzyme so it can bind to another substrate molecule. We can write the catalyzed reaction of an enzyme (E) with a substrate (S) to form product (P) as follows:

$$E + S \rightleftarrows \text{ES complex} \longrightarrow \text{EP complex} \longrightarrow E + P$$

Enzyme and Substrate | Enzyme–Substrate Complex | Enzyme–Product Complex | Enzyme and Product

In the hydrolysis of the disaccharide lactose by the enzyme lactase, a molecule of lactose binds to the active site of lactase. In this ES complex, the glycosidic bond of lactose is in a position that is favorable for *hydrolysis*, which is the splitting by water of a large molecule into smaller parts. R groups on the amino acids in the active site then catalyze the hydrolysis of lactose, producing the monosaccharides glucose and galactose. Because the structures of the products are no longer attracted to the active site, they are released, which allows the enzyme lactase to react with another lactose molecule (see **FIGURE 20.3**).

TEST

Try Practice Problems 20.3 and 20.4

$$E + S \rightleftarrows \text{ES complex} \longrightarrow \text{EP complex} \longrightarrow E + P$$

Lactase and Lactose | Lactase–Lactose Complex | Lactase–Glucose–Galactose Complex | Lactase and Glucose and Galactose

CORE CHEMISTRY SKILL

Describing Enzyme Action

Models of Enzyme Action

An early theory of enzyme action, called the *lock-and-key model*, described the active site as having a rigid, nonflexible shape. According to the lock-and-key model, the shape of the active site was analogous to a lock, and its substrate was the key that specifically fit that lock. However, this model was a static one that did not account for the flexibility of the tertiary structure of an enzyme and the way we now know that the active site can adjust to the shape of a substrate.

In the dynamic model of enzyme action, called the **induced-fit model**, the flexibility of the active site allows it to adapt to the shape of the substrate. At the same time, the shape of the substrate may be modified to better fit the geometry of the active site. The adjusted fit of the active site with the substrate provides the best alignment for the catalysis of the substrate. In the induced-fit model, substrate and enzyme work together to acquire a geometrical arrangement that lowers the activation energy of the reaction (see **FIGURE 20.3**).

ENGAGE

How can the induced-fit model be used to explain why some enzymes interact with a range of substances?

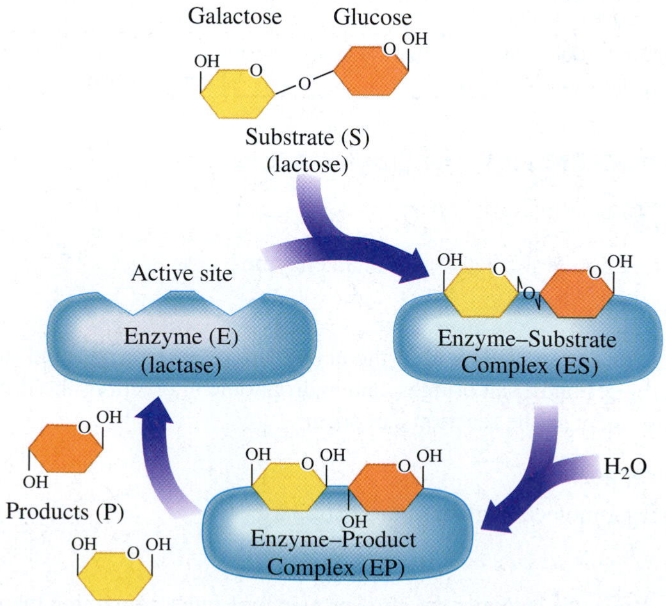

Galactose
Glucose
OH
O
O
O
OH
Substrate (S) (lactose)
Active site
Enzyme (E) (lactase)
Enzyme–Substrate Complex (ES)
H_2O
Products (P)
Enzyme–Product Complex (EP)

FIGURE 20.3 ▶ A flexible active site in lactase and the flexible substrate lactose adjust to provide the best fit for the hydrolysis reaction. Once the disaccharide is hydrolyzed, the monosaccharide products are released from the enzyme, which is ready to bind another lactose.

Q Why does the enzyme-catalyzed hydrolysis of lactose proceed at a faster rate than the hydrolysis of lactose in a test tube?

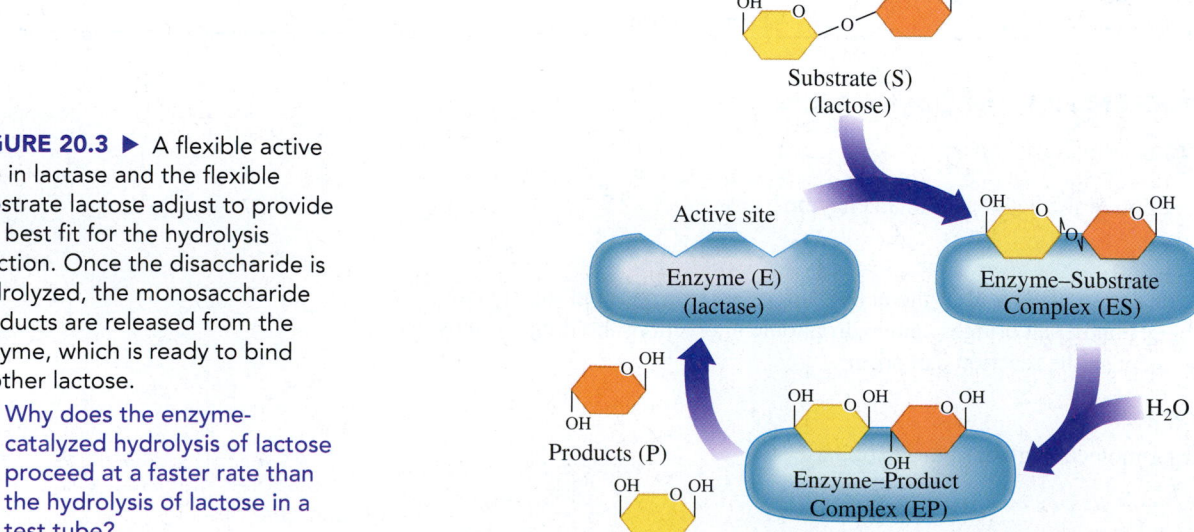

▶SAMPLE PROBLEM 20.2 The Induced-Fit Model

TRY IT FIRST

How does the induced-fit model explain the binding of the substrate at the active site?

SOLUTION

In the induced-fit model, the shapes of the substrate and the active site adjust so that the substrate is in the optimum position needed for the enzyme to catalyze the reaction.

STUDY CHECK 20.2

How does the lock-and-key model explain the binding of the substrate at an active site?

ANSWER

In the lock-and-key model, the substrate fits perfectly into the rigid active site of the enzyme, similar to a key fitting into a lock. We know today that most enzymes do not have a rigid active site and that the active site flexes to accommodate the substrate.

TEST

Try Practice Problems 20.5 and 20.6

PRACTICE PROBLEMS

20.1 Enzymes and Enzyme Action

20.1 Why do chemical reactions in the body require enzymes?

20.2 How do enzymes make chemical reactions in the body proceed at faster rates?

20.3 **a.** Write an equation that represents an enzyme-catalyzed reaction.
b. How is the active site different from the whole enzyme structure?

20.4 **a.** How does an enzyme speed up the reaction of a substrate?
b. After the products have formed, what happens to the enzyme?

20.5 Match the terms, (1) enzyme–substrate complex, (2) enzyme, and (3) substrate, with each of the following:
a. has a tertiary structure that recognizes the substrate
b. the combination of an enzyme with the substrate
c. has a structure that fits the active site of an enzyme

20.6 Match the terms, (1) active site, (2) induced-fit model, and (3) lock-and-key model with each of the following:
a. an active site that has a rigid shape
b. the portion of an enzyme where catalytic activity occurs
c. an active site that adapts to the shape of a substrate

Clinical Applications

20.7 Which type of substrate specificity is demonstrated by the digestive enzyme α-amylase, which catalyzes the hydrolysis of $\alpha(1\rightarrow 4)$-glycosidic bonds in both saliva and in the pancreas?

20.8 Which type of substrate specificity is demonstrated by the digestive enzyme α-galactosidase, which catalyzes the hydrolysis of a terminal galactose molecule from glycolipids and glycoproteins in lysosomes?

20.2 Classification of Enzymes

LEARNING GOAL Classify enzymes and give their names.

CORE CHEMISTRY SKILL

Classifying Enzymes

The names of enzymes describe the substrate or the reaction that is catalyzed. The names are derived by replacing the end of the name of the reaction or substrate with the suffix *ase*. For example, an *oxidase* is an enzyme that catalyzes an oxidation reaction, and a *dehydrogenase* removes hydrogen atoms. These enzymes are classified as *oxidoreductases* because they catalyze the loss or gain of hydrogen or oxygen.

The Enzyme Commission of the International Union of Biochemistry and Molecular Biology systematically classifies enzymes according to the six general types of reactions they catalyze (see **TABLE 20.2**).

Often the name of the enzyme can be used to predict its class and function. The enzyme *sucrase* is a catalyst in the hydrolysis of sucrose, and *lipase* catalyzes reactions that hydrolyze lipids. Some enzymes use names that end in the suffix *in*, such as *papain* found in papaya; *rennin* found in milk; and *pepsin* and *trypsin*, enzymes that catalyze the hydrolysis of proteins. These enzymes are classified as *hydrolases* because they use water to split large molecules into smaller ones.

ENGAGE

What does the name of an enzyme describe?

TABLE 20.2 Classification of Enzymes

Class	Examples of Reaction Catalyzed
1. Oxidoreductases Catalyze oxidation–reduction reactions	*Oxidases* catalyze the oxidation of a substrate. *Dehydrogenases* catalyze the removal or addition of two H atoms and utilize a coenzyme.

$$CH_3-CH_2-OH + NAD^+ \xrightarrow{\text{Alcohol dehydrogenase}} CH_3-\overset{\displaystyle O}{\overset{\|}{C}}-H + NADH + H^+$$

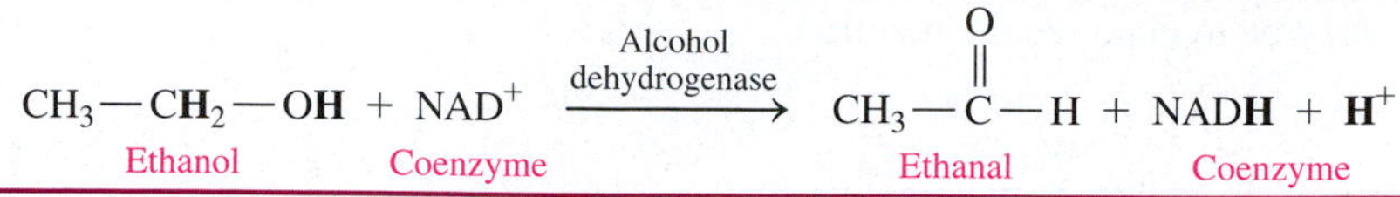

Ethanol   Coenzyme   Ethanal   Coenzyme

Class	Examples of Reaction Catalyzed
2. Transferases Catalyze the transfer of a functional group between two compounds	*Transaminases* catalyze the transfer of an amino group from one substrate to another. *Kinases* catalyze the transfer of phosphate groups from one substrate to another.

$$CH_3-\overset{\displaystyle \overset{+}{N}H_3}{\overset{|}{C}H}-COO^- + {}^-OOC-\overset{\displaystyle O}{\overset{\|}{C}}-CH_2-CH_2-COO^- \underset{}{\overset{\text{Alanine transaminase}}{\rightleftharpoons}} CH_3-\overset{\displaystyle O}{\overset{\|}{C}}-COO^- + {}^-OOC-\overset{\displaystyle \overset{+}{N}H_3}{\overset{|}{C}H}-CH_2-CH_2-COO^-$$

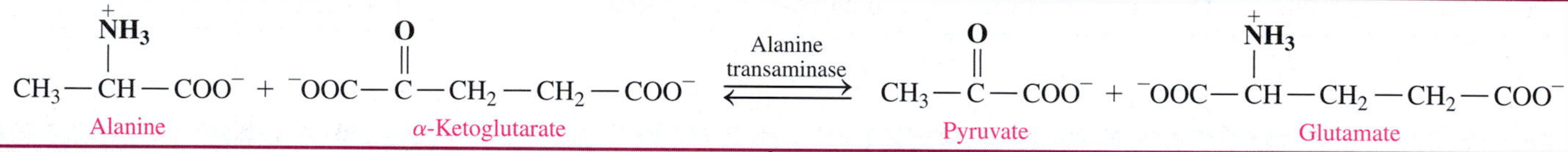

Alanine   α-Ketoglutarate   Pyruvate   Glutamate

Class	Examples of Reaction Catalyzed
3. Hydrolases Catalyze hydrolysis reactions, which add water to split a compound into two products	*Proteases* catalyze the hydrolysis of peptide bonds in proteins. *Lipases* catalyze the hydrolysis of ester bonds in lipids. *Nucleases* catalyze the hydrolysis of the phosphate ester bonds in nucleic acids.

$$-NH-\overset{\displaystyle R}{\overset{|}{C}H}-\overset{\displaystyle O}{\overset{\|}{C}}-NH-\overset{\displaystyle R}{\overset{|}{C}H}-COO^- + H_2O \xrightarrow{\text{Protease}} -NH-\overset{\displaystyle R}{\overset{|}{C}H}-\overset{\displaystyle O}{\overset{\|}{C}}-O^- + H_3\overset{+}{N}-\overset{\displaystyle R}{\overset{|}{C}H}-COO^-$$

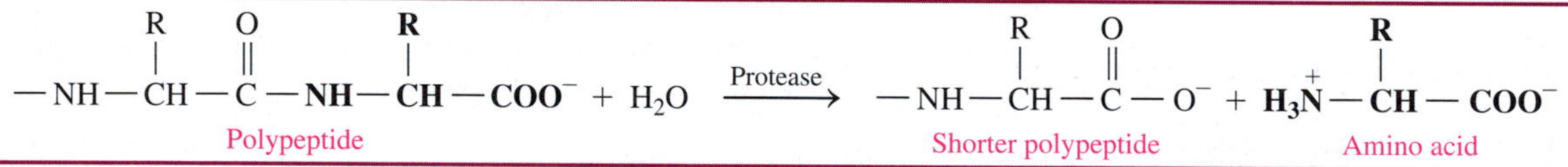

Polypeptide   Shorter polypeptide   Amino acid

Class	Examples of Reaction Catalyzed
4. Lyases Catalyze the addition or removal of a group without hydrolysis	*Decarboxylases* catalyze the removal of CO_2 from a substrate. *Deaminases* catalyze the removal of NH_3 from a substrate. *Dehydratases* catalyze the removal of H_2O from a substrate. *Hydratases* catalyze the addition of H_2O to a substrate.

$$CH_3-\overset{\displaystyle O}{\overset{\|}{C}}-COO^- + H^+ \xrightarrow{\text{Pyruvate decarboxylase}} CH_3-\overset{\displaystyle O}{\overset{\|}{C}}-H + CO_2$$

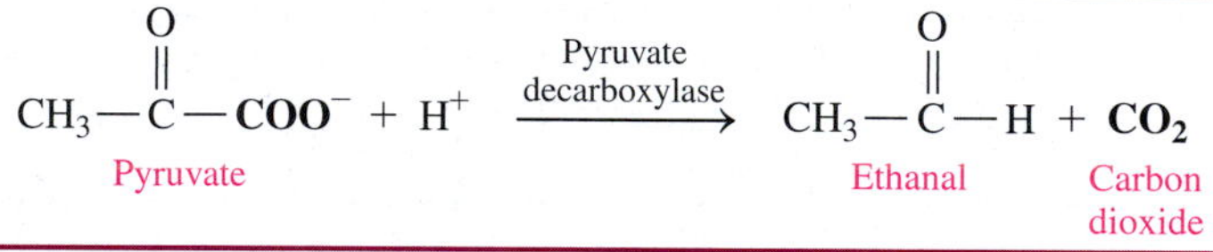

Pyruvate   Ethanal   Carbon dioxide

Class	Examples of Reaction Catalyzed
5. Isomerases Catalyze the rearrangement (isomerization) of atoms within a substrate	*Isomerases* catalyze rearrangement reactions. *Epimerases* catalyze rearrangements of carbohydrates.

$$\underset{H}{\overset{{}^-OOC}{}}C=C\underset{H}{\overset{COO^-}{}} \rightleftharpoons \underset{H}{\overset{{}^-OOC}{}}C=C\underset{COO^-}{\overset{H}{}}$$

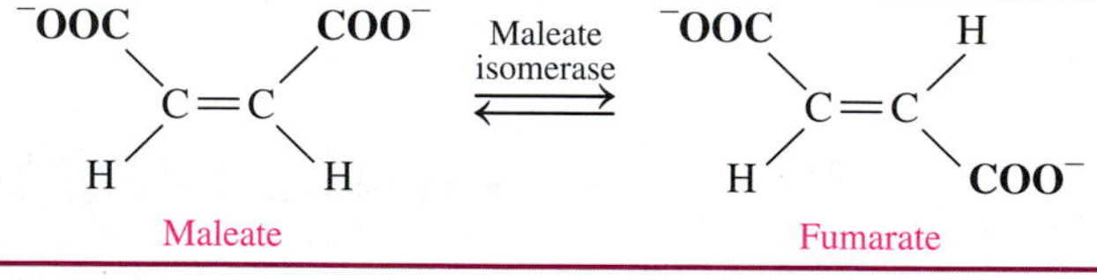

Maleate isomerase

Maleate   Fumarate

Class	Examples of Reaction Catalyzed
6. Ligases Catalyze the joining of two substrates, using ATP energy	*Synthetases* catalyze the formation of a bond between two substrates, utilizing the energy of ATP. *Carboxylases* catalyze the formation of a bond between CO_2 and a substrate, utilizing the energy of ATP.

$${}^-OOC-\overset{\displaystyle O}{\overset{\|}{C}}-CH_3 + CO_2 + ATP \xrightarrow{\text{Pyruvate carboxylase}} {}^-OOC-\overset{\displaystyle O}{\overset{\|}{C}}-CH_2-COO^- + ADP + P_i + H^+$$

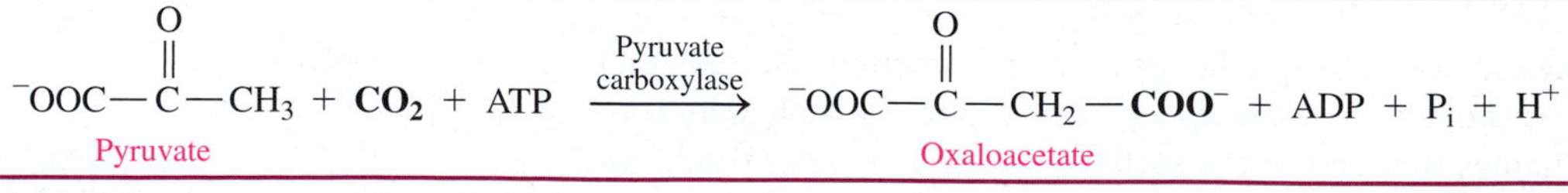

Pyruvate   Oxaloacetate

▶SAMPLE PROBLEM 20.3 Classifying Enzymes

TRY IT FIRST

Using Table 20.2, identify the chemical reaction each of the following enzymes catalyzes and classify the enzyme:

a. aminotransferase **b.** lactate dehydrogenase

SOLUTION

ANALYZE THE PROBLEM	Given	Need
	name of the enzyme	type of reaction catalyzed, class of enzyme

a. An aminotransferase catalyzes the transfer of an amino group between two substrates and is classified as a transferase.

b. Lactate dehydrogenase catalyzes the removal of two H atoms from lactate and is classified as an oxidoreductase.

STUDY CHECK 20.3

a. What is the name and classification of the enzyme that catalyzes the hydrolysis of lipids?

b. What is the name and classification of an enzyme that catalyzes rearrangements of carbohydrates?

ANSWER

a. An enzyme that hydrolyzes a lipid is a lipase, which is classified as a hydrolase.

b. An enzyme that catalyzes rearrangements of carbohydrates is an epimerase, which is classified as an isomerase.

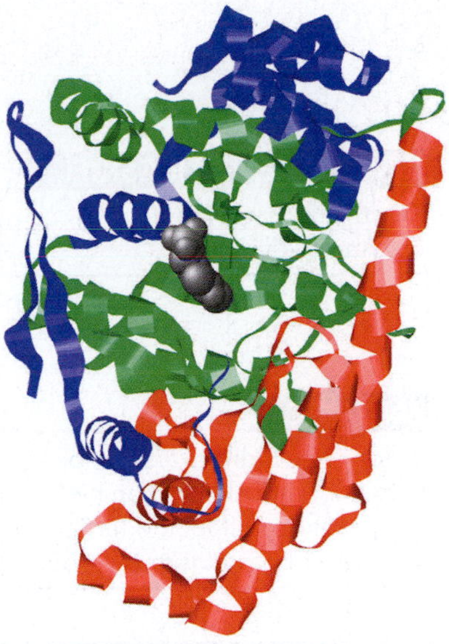

This ribbon structure represents alanine transaminase, an aminotransferase that contains 495 amino acids.

TEST

Try Practice Problems 20.9 to 20.16

PRACTICE PROBLEMS

20.2 Classification of Enzymes

20.9 What type of reaction is catalyzed by each of the following classes of enzymes?
a. oxidoreductases **b.** transferases **c.** hydrolases

20.10 What type of reaction is catalyzed by each of the following classes of enzymes?
a. lyases **b.** isomerases **c.** ligases

20.11 What is the name of the class of enzymes that catalyzes each of the following reactions?
a. hydrolyzing sucrose
b. adding oxygen
c. converting glucose-6-phosphate to fructose-6-phosphate
d. moving an amino group from one molecule to another

20.12 What is the name of the class of enzymes that catalyzes each of the following reactions?
a. adding water to a double bond
b. removing hydrogen atoms from a substrate
c. splitting peptide bonds in proteins
d. converting a tertiary alcohol to a secondary alcohol

20.13 Classify the enzyme that catalyzes each of the following reactions:

a. $CH_3{-}\overset{\overset{\displaystyle O}{\|}}{C}{-}COO^- + H^+ \longrightarrow CH_3{-}\overset{\overset{\displaystyle O}{\|}}{C}{-}H + CO_2$

b. $CH_3{-}\underset{}{\overset{\overset{\displaystyle \overset{+}{N}H_3}{|}}{C}H}{-}COO^- + {}^-OOC{-}\overset{\overset{\displaystyle O}{\|}}{C}{-}CH_2{-}CH_3 \rightleftarrows$

$CH_3{-}\overset{\overset{\displaystyle O}{\|}}{C}{-}COO^- + {}^-OOC{-}\overset{\overset{\displaystyle \overset{+}{N}H_3}{|}}{C}H{-}CH_2{-}CH_3$

20.14 Classify the enzyme that catalyzes each of the following reactions:

a. $CH_3{-}\overset{\overset{\displaystyle O}{\|}}{C}{-}COO^- + CO_2 + ATP \longrightarrow$

${}^-OOC{-}CH_2{-}\overset{\overset{\displaystyle O}{\|}}{C}{-}COO^- + ADP + P_i + H^+$

b. $CH_3{-}CH_2{-}OH + NAD^+ \longrightarrow$

$CH_3{-}\overset{\overset{\displaystyle O}{\|}}{C}{-}H + NADH + H^+$

20.15 Predict the name of the enzyme that catalyzes each of the following reactions:
a. oxidizes succinate
b. combines glutamate and ammonia to form glutamine
c. removes 2H from an alcohol

20.16 Predict the name of the enzyme that catalyzes each of the following reactions:
a. hydrolyzes maltose
b. transfers an amino group from aspartate
c. removes a carboxylate group from pyruvate

20.3 Factors Affecting Enzyme Activity

LEARNING GOAL Describe the effects that changes of temperature, pH, enzyme concentration, and substrate concentration have on enzyme activity.

CORE CHEMISTRY SKILL
Identifying Factors Affecting Enzyme Activity

The **activity** of an enzyme describes how fast an enzyme catalyzes the reaction that converts a substrate to product. This activity is strongly affected by reaction conditions, which include the temperature, pH, concentration of the enzyme, and concentration of the substrate.

Temperature

ENGAGE
Excessively high fevers can be very dangerous. How can high temperatures affect the tertiary structure of proteins in the body?

Enzymes are very sensitive to temperature. At low temperatures, most enzymes show little activity because there is not a sufficient amount of energy for the catalyzed reaction to take place. At higher temperatures, enzyme activity increases as reacting molecules move faster, causing more collisions with enzymes. Enzymes are most active at **optimum temperature**, which is 37 °C, or body temperature, for most enzymes (see **FIGURE 20.4**). At temperatures above 50 °C, there is enough energy to disrupt many of the interactions between the R groups in the protein. When too many of those interactions are disrupted, the tertiary structure, and thus the shape of most proteins, is destroyed, causing a loss in enzyme activity. For this reason, equipment in hospitals and laboratories is sterilized in autoclaves where the high temperatures denature the enzymes in harmful bacteria.

Certain organisms, known as thermophiles, live in environments where temperatures range from 50 to 120 °C. In order to survive in these extreme conditions, thermophiles have enzymes with tertiary structures that are not destroyed by such high temperatures. Some research shows that their enzymes are very similar to ordinary enzymes except that they contain more hydrogen bonds and salt bridges that stabilize their tertiary structures at high temperatures.

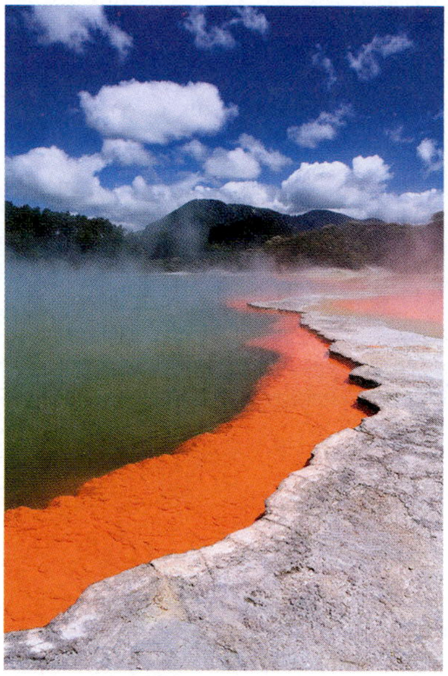

Thermophiles survive in the high temperatures (50 to 120 °C) of a hot spring.

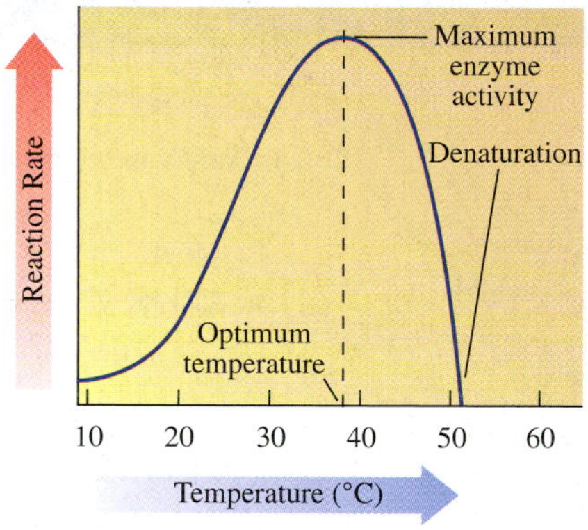
Maximum enzyme activity
Denaturation
Reaction Rate
Optimum temperature
10
20
30
40
50
60
Temperature (°C)

FIGURE 20.4 ▶ An enzyme attains maximum activity at its optimum temperature, usually 37 °C. Lower temperatures slow the rate of reaction, and temperatures above 50 °C denature most enzymes, with a loss of catalytic activity.

Q Why is 37 °C the optimum temperature for many enzymes?

pH

Enzymes are most active at their **optimum pH**, the pH that maintains the proper tertiary structure of the protein (see **FIGURE 20.5**). Changes in the pH of the solution surrounding an enzyme can change the charges on certain amino acid R groups in the enzyme. Any change in the charges of amino acid R groups can disrupt the interactions that maintain the

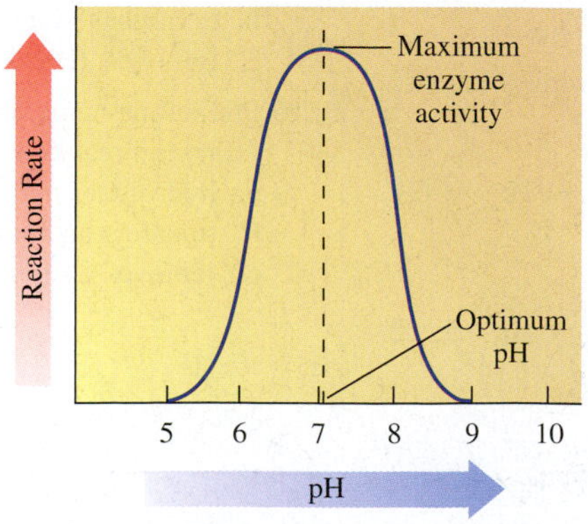
Maximum enzyme activity
Reaction Rate
Optimum pH
5
6
7
8
9
10
pH

FIGURE 20.5 ▶ Enzymes are most active at their optimum pH. At a higher or lower pH, denaturation of the enzyme causes a loss of catalytic activity.

Q Why would we expect the digestive enzyme pepsin to have an optimum pH of 2.0?

tertiary structure of the enzyme. For example, at a pH of 7.4, the R group of aspartate has a negative charge. With a negative charge, aspartate can participate in salt bridges to stabilize the tertiary structure of an enzyme. However, if the pH drops below 4.4, the R group of an aspartate can gain a hydrogen ion (H^+) and lose its negative charge and its ability to participate in a salt bridge. When enough of the interactions between amino acid R groups are disrupted, the enzyme will no longer be able to maintain its three-dimensional tertiary structure. As a result, the enzyme no longer binds substrate, and no catalysis occurs. Small changes in pH are reversible, meaning that an enzyme can regain its tertiary or quaternary structure and biological activity when the pH returns to its optimum value. However, large variations from optimum pH permanently destroy the structure of the enzyme.

Enzymes in most cells have optimum pH values at physiological pH, around 7.4. However, enzymes in the stomach have a low optimum pH because they hydrolyze proteins at the acidic pH in the stomach. For example, pepsin, a digestive enzyme in the stomach, has an optimum pH of 1.5 to 2.0. Between meals, the pH in the stomach is 4 to 5, and pepsin shows little or no digestive activity. When food enters the stomach, the secretion of HCl lowers the pH to about 2, which activates the digestive enzyme pepsin. **TABLE 20.3** lists the optimum pH values for selected enzymes.

TABLE 20.3 Optimum pH for Selected Enzymes

Enzyme	Location	Substrate	Optimum pH
Pepsin	Stomach	Peptide bonds	1.5–2.0
Lactase	GI tract	Lactose	6.0
Sucrase	Small intestine	Sucrose	6.2
Amylase	Pancreas	Amylose	6.7–7.0
Urease	Liver	Urea	7.0
Trypsin	Small intestine	Peptide bonds	7.7–8.0
Lipase	Pancreas	Lipid (ester bonds)	8.0
Arginase	Liver	Arginine	9.7

Enzyme and Substrate Concentration

In any catalyzed reaction, the substrate must first bind with the enzyme to form the enzyme–substrate complex. For a particular substrate concentration, an increase in enzyme concentration increases the rate of the catalyzed reaction. At higher enzyme concentrations, more molecules are available to bind substrate and catalyze the reaction. As long as the substrate concentration is greater than the enzyme concentration, there is a direct relationship between the enzyme concentration and enzyme activity (see **FIGURE 20.6a**). In most enzyme-catalyzed reactions, the concentration of the substrate is much greater than the concentration of the enzyme.

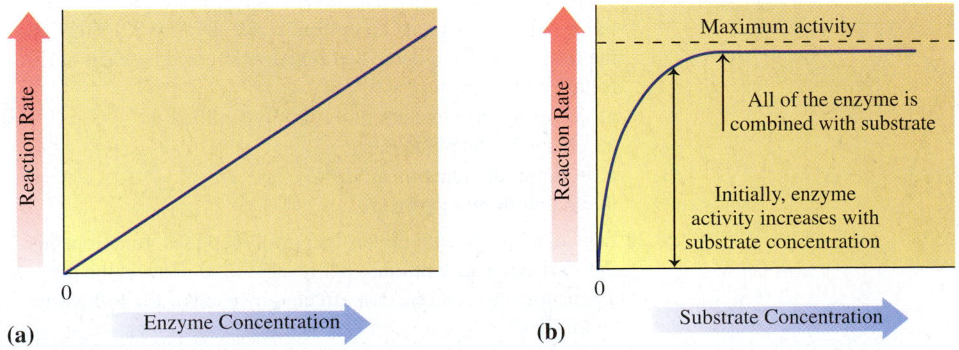
Reaction Rate
0
Enzyme Concentration
(a)
Maximum activity
All of the enzyme is combined with substrate
Initially, enzyme activity increases with substrate concentration
Reaction Rate
0
Substrate Concentration
(b)

FIGURE 20.6 ▶ **(a)** Increasing the enzyme concentration increases the rate of reaction. **(b)** Increasing the substrate concentration increases the rate of reaction until all the enzyme molecules are combined with substrate.

Q What happens to the rate of reaction when all the available enzymes are attached to substrate molecules?

ENGAGE

Why does adding more substrate not change the rate of reaction when the substrate concentration is already high?

When the enzyme concentration is kept constant, the addition of more substrate will increase the rate of the reaction. If the substrate concentration is high, the substrate can bind with all of the available enzyme molecules. Then the rate of the reaction reaches its maximum, and the addition of more substrate does not increase the rate further (see **FIGURE 20.6b**).

▶ SAMPLE PROBLEM 20.4 Factors Affecting Enzymatic Activity

TRY IT FIRST

Describe the effect each of the following changes would have on the rate of the reaction that is catalyzed by urease:

$$H_2N-\overset{\overset{\displaystyle O}{\|}}{C}-NH_2 + H_2O \xrightarrow{\text{Urease}} 2NH_3 + CO_2$$

Urea

a. increasing the urea concentration **b.** lowering the temperature to 10 °C

SOLUTION

a. An increase in urea concentration will increase the rate of reaction until all the enzyme molecules bind to urea. Then no further increase in rate occurs.

b. Because 10 °C is lower than the optimum temperature of 37 °C, there is a decrease in the rate of the reaction.

STUDY CHECK 20.4

Urease has an optimum pH of 7.0. Describe the effect each of the following changes would have on the rate of the reaction that is catalyzed by urease:

a. lowering the pH of the reaction mixture to 3.0

b. increasing the concentration of urease

ANSWER

a. At a pH lower than the optimum pH, the hydrogen bonds and salt bridges of urease will be disrupted, resulting in the denaturation of the enzyme, loss of tertiary and quaternary structure, and a decrease in the rate of reaction.

b. An increase in the concentration of urease will increase the rate of reaction.

TEST

Try Practice Problems 20.17 to 20.20

PRACTICE PROBLEMS

20.3 Factors Affecting Enzyme Activity

Clinical Applications

20.17 The following graph shows the activity versus pH curves for pepsin, sucrase, and trypsin. Estimate the optimum pH for each.

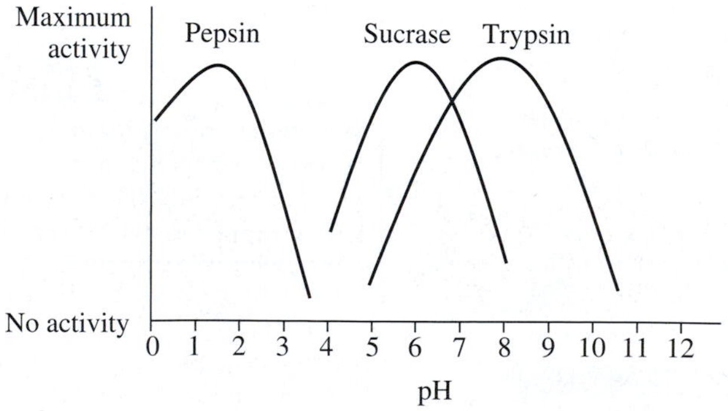
Maximum activity
No activity
Pepsin
Sucrase
Trypsin
0 1 2 3 4 5 6 7 8 9 10 11 12
pH

20.18 Refer to the graph in problem 20.17 to determine if the reaction rate in each condition will be at the optimum rate or not.

a. trypsin, pH 5.0
b. sucrase, pH 5.0
c. pepsin, pH 4.0
d. trypsin, pH 8.0
e. pepsin, pH 2.0

20.19 Trypsin, a protease that hydrolyzes polypeptides, functions in the small intestine at an optimum pH of 7.7 to 8.0. How is the rate of a trypsin-catalyzed reaction affected by each of the following conditions?

a. lowering the concentration of polypeptides
b. changing the pH to 3.0
c. running the reaction at 75 °C
d. adding more trypsin

20.20 Pepsin, a protease that hydrolyzes polypeptides, functions in the stomach at an optimum pH of 1.5 to 2.0. How is the rate of a pepsin-catalyzed reaction affected by each of the following conditions?

a. increasing the concentration of polypeptides
b. changing the pH to 5.0
c. running the reaction at 0 °C
d. using less pepsin

20.4 Regulation of Enzyme Activity

LEARNING GOAL Describe the role of allosteric enzymes, feedback control, and covalent modification in regulating enzyme activity.

In enzyme-catalyzed reactions, substances are produced in the amounts and at the times they are needed in a cell. Therefore, the rates of enzyme-catalyzed reactions are controlled by regulatory enzymes that increase the reaction rate when more of a particular substance is needed, and decrease the reaction rate when that substance is not needed. Enzyme activity can be regulated by allosteric enzymes, feedback control, and covalent modifications.

Allosteric Enzymes

Many enzymes are **allosteric enzymes**, which bind with a regulator molecule that is different from the substrate. The regulator binds to the enzyme at an *allosteric site*, which is a different location than the active site. As a result, the shape of the enzyme changes, which causes a change in the shape of the active site. There are both positive and negative regulators. A *positive regulator* changes the shape of the active site to allow the substrate to bind more effectively to the active site, which increases the rate of the catalyzed reaction. A *negative regulator* changes the shape of the active site to prevent the proper binding of the substrate to the active site, which decreases the rate of the catalyzed reaction (see **FIGURE 20.7**).

TEST

Try Practice Problems 20.21 and 20.22

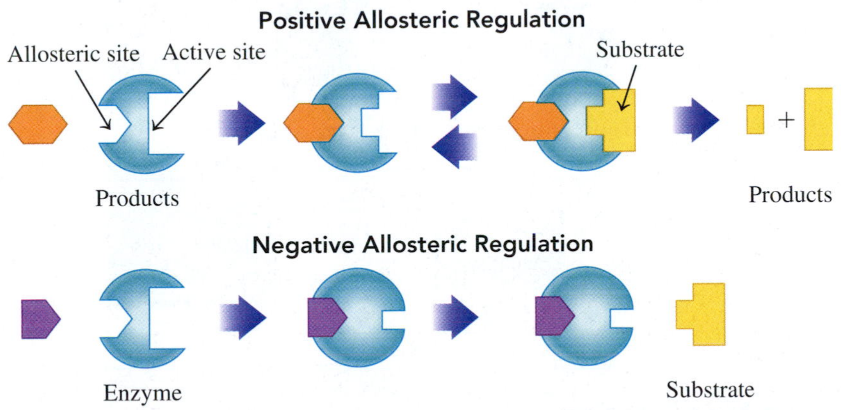

Positive Allosteric Regulation
Allosteric site
Active site
Substrate
+
Products
Products
Negative Allosteric Regulation
Enzyme
Substrate

FIGURE 20.7 ▶ A positive regulator changes the shape of the active site, allowing the substrate to bind more effectively to the active site and increasing the reaction rate. A negative regulator changes the shape of the active site, preventing the proper binding of the substrate to the active site and decreasing the reaction rate.

Feedback Control

In **feedback control**, the end product of a series of reactions acts as a negative regulator of an enzyme that appears earlier in the reaction series (see **FIGURE 20.8**). When the end product is present in sufficient amounts for the needs of the cell, end product molecules bind to the allosteric site of an enzyme at or near the beginning of the reaction series (E_1 in Figure 20.8). The binding of an end product molecule to the enzyme's allosteric site causes

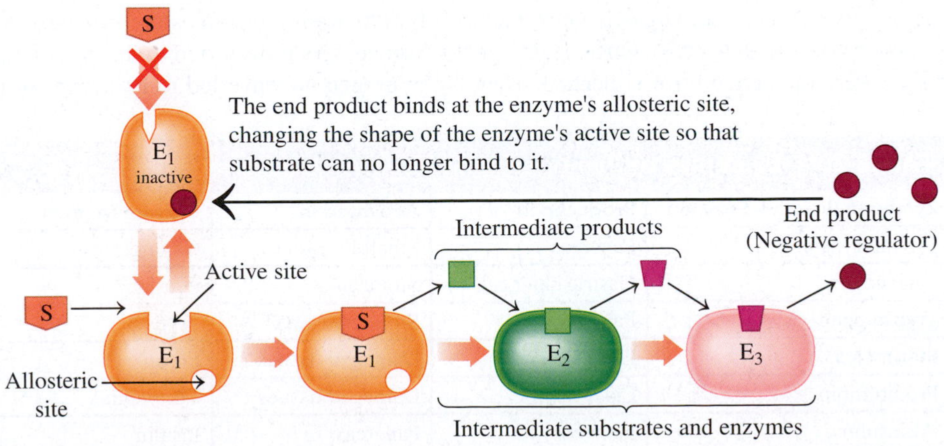

S
E1
inactive
The end product binds at the enzyme's allosteric site, changing the shape of the enzyme's active site so that substrate can no longer bind to it.
End product
(Negative regulator)
Intermediate products
Active site
S
E1
S
E1
E2
E3
Allosteric site
Intermediate substrates and enzymes

FIGURE 20.8 ▶ In feedback control, the end product binds to an allosteric site on an enzyme that appears at or near the beginning of the reaction sequence. The binding of the end product inactivates the enzyme and prevents the formation of all intermediate compounds needed in the synthesis of the end product.

Q Do the intermediate enzymes in a reaction sequence have regulatory sites?

the shape of the enzyme's active site to change in such a way that the enzyme's substrate can no longer bind to it. As a result, the production of all the intermediate compounds in the subsequent reaction sequence stops.

When the level of end product is low, the regulator dissociates from the allosteric enzyme. The enzyme returns to its original, active three-dimensional shape and is able to bind with a substrate molecule. Feedback control allows the enzyme-catalyzed reactions to operate only when the end product is needed by the cell. Feedback control also prevents the accumulation of intermediate products as well as end product, thereby conserving energy and materials in the cell.

TEST

Try Practice Problems 20.23 and 20.24

An example of feedback control is the reaction sequence in which the amino acid threonine is converted to the amino acid isoleucine. When the level of isoleucine is high in the cell, some of the isoleucine binds to the allosteric enzyme threonine deaminase (E_1). The binding of isoleucine changes the shape of the threonine deaminase, which prevents threonine from binding to its active site. As a result, the entire reaction sequence does not function. As isoleucine is utilized in the cell and its concentration decreases, isoleucine molecules are released from the allosteric site on threonine deaminase. The shape of the deaminase returns to its active form, which allows the enzyme-catalyzed reaction sequence to convert threonine to isoleucine.

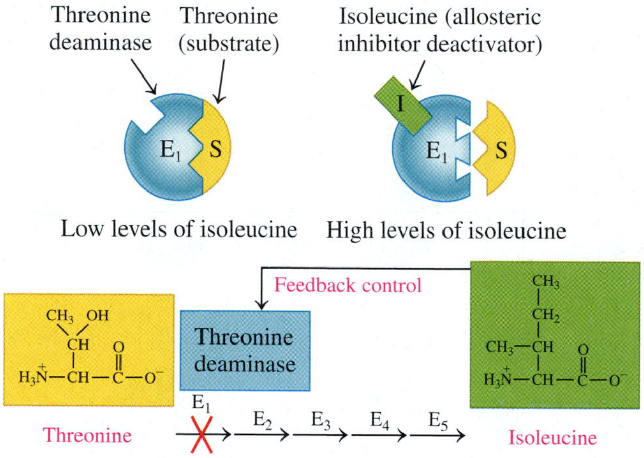

Threonine deaminase
Threonine (substrate)
Isoleucine (allosteric inhibitor deactivator)
E1
S
I
Low levels of isoleucine
High levels of isoleucine
Feedback control
Threonine deaminase
Threonine
Isoleucine
E1
E2
E3
E4
E5

Covalent Modification

The activity of some enzymes is regulated through the addition or cleavage of bonds on the enzyme's polypeptide chain. This type of modification is called **covalent modification**. There are multiple types of covalent modifications that can be used to regulate enzyme activity. Two of these are discussed below.

Zymogens (Proenzymes)

Many enzymes are active as soon as they are synthesized and fold into their tertiary or quaternary structures. However, **zymogens**, or *proenzymes*, are produced as inactive forms and stored for later use. Digestive enzymes and blood clotting enzymes are produced as zymogens (see **TABLE 20.4**). Once a zymogen is formed, it is transported to the part of the body where the active form is needed. Then the zymogen is converted to its active form

TABLE 20.4 Examples of Zymogens and Their Active Forms

Zymogen (Inactive Enzyme)	Produced in	Activated in	Enzyme (Active)
Chymotrypsinogen	Pancreas	Small intestine	Chymotrypsin
Pepsinogen	Gastric chief cells	Stomach	Pepsin
Trypsinogen	Pancreas	Small intestine	Trypsin
Fibrinogen	Liver	Damaged tissues	Fibrin
Prothrombin	Liver	Damaged tissues	Thrombin
Proinsulin	Pancreas	Pancreas	Insulin

by a covalent modification, typically the removal of a polypeptide section that had been covering the enzyme's active site.

In the case of digestive enzymes specifically, the production of zymogens protects cells from being digested by the enzymes they produce. For example, the pancreatic zymogens are stored in *zymogen granules*, which have membranes that are resistant to enzymatic digestion while in the pancreas. If zymogen activation should occur in the pancreas, the proteins within the tissues of the pancreas would undergo digestion, which would cause inflammation and could result in a painful condition called *pancreatitis*.

The proteases trypsin and chymotrypsin are produced as zymogens and stored in the pancreas. After food is ingested, hormones trigger the release of the zymogens from the pancreas for transport to the small intestine. There the zymogens are converted into active digestive enzymes by proteases that remove peptide sections from the protein chains. A hexapeptide from the zymogen trypsinogen is removed by a peptidase to form the active digestive enzyme trypsin. As more trypsin forms, it in turn activates additional trypsinogen molecules by removing the hexapeptide from those molecules. Trypsin also activates the zymogen chymotrypsinogen by removing two dipeptides to give the active enzyme chymotrypsin, which has three polypeptide sections held together by disulfide bonds.

ENGAGE

How does the production of zymogens protect the cells in which digestive enzymes are produced?

The ribbon representation of chymotrypsin shows six cysteines (green) that form three disulfide bonds (yellow).

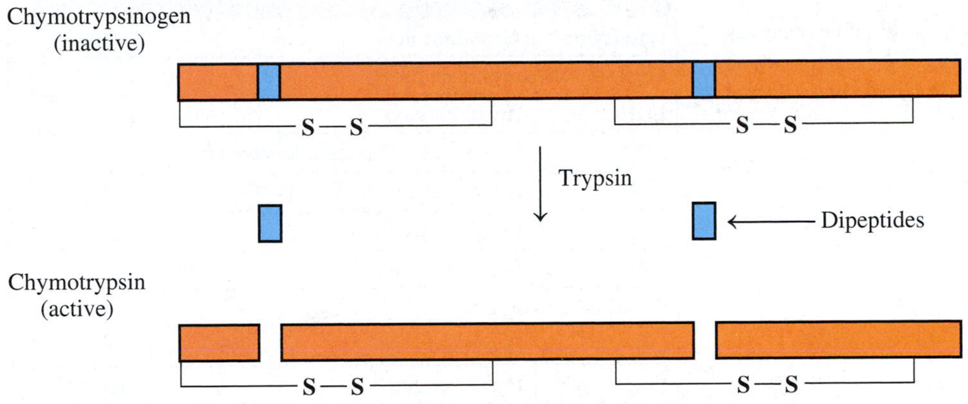
Chymotrypsinogen (inactive)
S—S
S—S
Trypsin
Dipeptides
Chymotrypsin (active)
S—S
S—S

Once formed, trypsin catalyzes the removal of dipeptides from inactive chymotrypsinogen to give the active protease chymotrypsin.

Phosphorylation

Another common covalent modification is the addition or removal of a phosphate group, which can induce a structural change in an enzyme and change its catalytic activity. Almost one-third of all enzymes are regulated through addition or removal of a phosphate group. An inactive enzyme can be activated by the addition of a phosphate from ATP in a reaction catalyzed by a kinase (see **FIGURE 20.9a**). For enzymes that are inactive when phosphorylated, a phosphatase catalyzes the hydrolysis of phosphate in order to activate that enzyme (see **FIGURE 20.9b**).

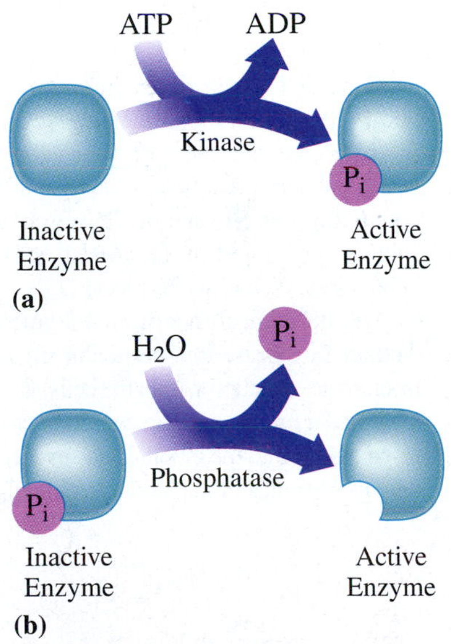
ATP
ADP
Kinase
P_i
Inactive Enzyme
Active Enzyme
(a)
H_2O
P_i
Phosphatase
P_i
Inactive Enzyme
Active Enzyme
(b)

FIGURE 20.9 ▶ Phosphorylation is a type of covalent modification through which an enzyme is deactivated or activated. **(a)** A kinase can activate an inactive enzyme by phosphorylation. **(b)** A phosphatase can activate an inactive enzyme by removal of a phosphate group.

Q What is the difference between a kinase and a phosphatase if they both transfer phosphates?

▶ SAMPLE PROBLEM 20.5 Enzyme Regulation

TRY IT FIRST

How is the rate of a reaction sequence regulated in feedback control?

SOLUTION

When the end product of a reaction sequence is produced at sufficient levels for the cell, some product molecules bind to the allosteric enzyme in the sequence, which shuts down all the reactions that follow and stops the synthesis of intermediate products.

STUDY CHECK 20.5

a. Why is pepsin, a digestive enzyme, produced as a zymogen?

b. What effect does the binding of a positive regulator to an enzyme's allosteric site have on the active site of the enzyme?

TEST

Try Practice Problems 20.25 to 20.28

ANSWER

a. Pepsin hydrolyzes proteins in the foods we ingest. It is synthesized as a zymogen, pepsinogen, to prevent its digestion of the proteins that make up the organs in the body.
b. The binding of a positive regulator to an enzyme's allosteric site causes the shape of the active site to change in such a way to make the binding of substrate more effective.

Chemistry Link to Health

Isoenzymes as Diagnostic Tools

Not only do enzymes need to be regulated so that they produce substances in the amounts and at the times they are needed in a cell, but they also need to be regulated so that they produce the substances in the *locations* in which they are needed. **Isoenzymes** are different forms of an enzyme that catalyze the same reaction in different cells or tissues of the body. Because they have different structures, isoenzymes can be regulated differently, so that a substance can be produced in one type of cells but not in another.

Isoenzymes consist of quaternary structures with slight variations in the amino acids in the polypeptide subunits. For example, there are five isoenzymes of *lactate dehydrogenase* (LDH) that catalyze the conversion between lactate and pyruvate.

$$\underset{\text{Lactate}}{CH_3-\underset{|}{\overset{OH}{CH}}-COO^-} + NAD^+ \xrightleftharpoons[]{\text{Lactate dehydrogenase}} \underset{\text{Pyruvate}}{CH_3-\overset{O}{\overset{\|}{C}}-COO^-} + NADH + H^+$$

Each LDH isoenzyme contains a mix of polypeptide subunits, M and H. In the liver and muscle, lactate is converted to pyruvate by the LDH_5 isoenzyme with four M subunits designated M_4. In the heart, the same reaction is catalyzed by the LDH_1 isoenzyme (H_4) containing four H subunits. Different combinations of the M and H subunits are found in the LDH isoenzymes of the brain, red blood cells, kidneys, and white blood cells.

The different forms of an enzyme allow a medical diagnosis of damage or disease to a particular organ or tissue. In healthy tissues, isoenzymes function within the cells. However, when a disease damages a particular organ, cells die, which releases their contents, including the isoenzymes, into the blood. Measurements of the elevated levels of specific isoenzymes in the blood serum help to identify the disease and its location in the body. For example, an elevation in serum LDH_5 (M_4) indicates liver damage or disease. When a *myocardial infarction* or heart attack damages heart muscle, an increase in the level of LDH_1 (H_4) isoenzyme is detected in the blood serum (see **TABLE 20.5**).

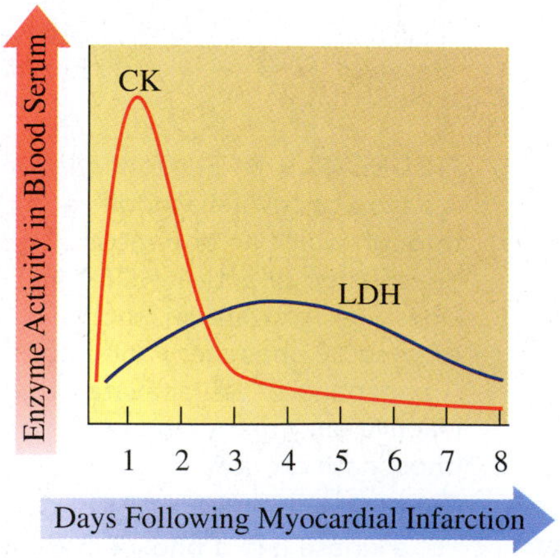

CK
LDH
Enzyme Activity in Blood Serum
1 2 3 4 5 6 7 8
Days Following Myocardial Infarction

A myocardial infarction may be indicated by an increase in the levels of creatine kinase (CK) and lactate dehydrogenase (LDH).

TABLE 20.5 Isoenzymes of Lactate Dehydrogenase and Creatine Kinase

Isoenzyme	Abundant in	Subunits
Lactate Dehydrogenase (LDH)		
LDH_1	Heart, kidneys	H_4
LDH_2	Red blood cells, heart, kidney, brain	H_3M
LDH_3	Brain, lung, white blood cells	H_2M_2
LDH_4	Lung, skeletal muscle	HM_3
LDH_5	Skeletal muscle, liver	M_4
Creatine Kinase (CK)		
CK_1	Brain, lung	BB
CK_2	Heart muscle	MB
CK_3	Skeletal muscle, red blood cells	MM

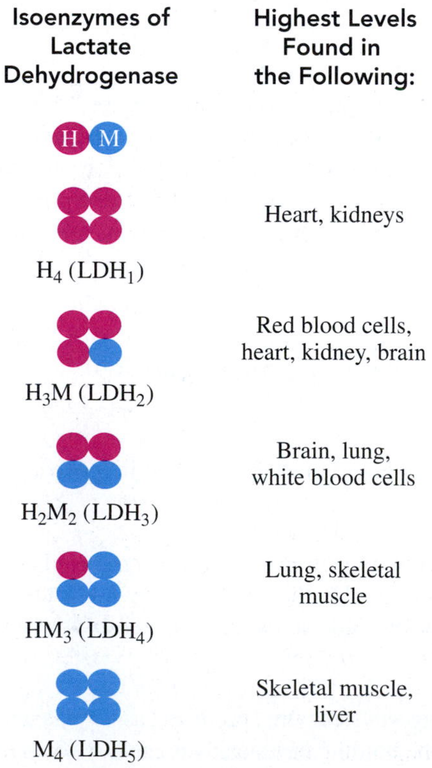

Isoenzymes of Lactate Dehydrogenase
Highest Levels Found in the Following:
H M
Heart, kidneys
H4 (LDH1)
Red blood cells, heart, kidney, brain
H3M (LDH2)
Brain, lung, white blood cells
H2M2 (LDH3)
Lung, skeletal muscle
HM3 (LDH4)
Skeletal muscle, liver
M4 (LDH5)

The different isoenzymes of lactate dehydrogenase (LDH) indicate damage to different organs in the body.

(continued)

Chemistry Link to Health (*continued*)

TABLE 20.6 Serum Enzymes Used in Diagnosis of Tissue Damage

Condition	Diagnostic Enzymes Elevated
Heart attack or liver disease (cirrhosis, hepatitis)	Lactate dehydrogenase (LDH) Aspartate transaminase (AST)
Heart attack	Creatine kinase (CK)
Hepatitis	Alanine transaminase (ALT)
Liver (carcinoma) or bone disease (rickets)	Alkaline phosphatase (ALP)
Pancreatic disease	Pancreatic amylase (PA) Cholinesterase (CE) Lipase (LPS)
Prostate carcinoma	Acid phosphatase (ACP) Prostate-specific antigen (PSA)

Another isoenzyme used diagnostically is creatine kinase (CK), which consists of two types of polypeptide subunits. Subunit B is prevalent in the brain, and subunit M predominates in muscle. In a patient who has suffered a myocardial infarction, the level of CK_2 (subunits MB) is elevated in the blood within 4 to 6 h and reaches a peak in about 24 h. **TABLE 20.6** lists some enzymes used to diagnose tissue damage and diseases of certain organs.

TEST

Try Practice Problems 20.29 to 20.34

PRACTICE PROBLEMS

20.4 Regulation of Enzyme Activity

20.21 How does an allosteric enzyme function as a regulatory enzyme?

20.22 What is the difference between a negative regulator and a positive regulator?

20.23 In feedback control, how does the end product of a reaction sequence regulate enzyme activity?

20.24 Why is it important that, in feedback control, the enzyme that is regulated appears at or near the beginning of the reaction sequence?

Clinical Applications

20.25 Why is the active form of thrombin, which helps blood clot, produced from prothrombin only when injury and bleeding occur?

20.26 The zymogen trypsinogen produced in the pancreas is activated in the small intestine, where it catalyzes the digestion of proteins. Explain how the activation of the zymogen in the pancreas can lead to an inflammation of the pancreas called pancreatitis.

20.27 Indicate if the following statements describe enzyme regulation by (1) an allosteric enzyme, (2) a zymogen, or (3) covalent modification:
- **a.** The enzyme activity increases due to phosphorylation.
- **b.** Fibrinogen in the blood forms fibrin in damaged tissues.
- **c.** A positive regulator stimulates enzyme action.

20.28 Indicate if the following statements describe enzyme regulation by (1) an allosteric enzyme, (2) a zymogen, or (3) covalent modification:
- **a.** An end product attaches to the regulatory site of the first enzyme in the reaction sequence.
- **b.** Trypsin removes two dipeptides from chymotrypsinogen to convert it into chymotrypsin.
- **c.** Phosphorylase kinase deactivates pyruvate dehydrogenase.

20.29 Name the enzyme that is indicated by each of the following abbreviations:
a. PSA **b.** CK **c.** CE

20.30 Name the enzyme that is indicated by each of the following abbreviations:
a. ALP **b.** AST **c.** PA

20.31 What are isoenzymes?

20.32 How is the LDH isoenzyme in the heart different from the LDH isoenzyme in the liver?

20.33 A patient arrives in the emergency room, complaining of chest pains. What enzymes might be present in the patient's blood serum?

20.34 A patient who is an alcoholic has elevated levels of LDH and AST. What condition might be indicated?

20.5 Enzyme Inhibition

LEARNING GOAL Describe reversible and irreversible inhibition. Describe competitive and noncompetitive inhibition.

In the previous Section, we discussed the regulation of allosteric enzymes; however, not all enzymes are allosteric enzymes. The activity of many of the nonallosteric enzymes can be inhibited by the binding of specific small molecules or ions called **inhibitors**. Although inhibitors act differently, they all prevent a substrate from properly binding with the active site. There are two general types of inhibitors. An enzyme with a *reversible inhibitor* can regain enzymatic activity, but an *irreversible inhibitor* causes the permanent loss of enzymatic activity.

Reversible Inhibition

ENGAGE

What happens to the rate of an enzyme-catalyzed reaction when an inhibitor binds to the enzyme?

In **reversible inhibition**, the inhibitor causes a loss of enzymatic activity that can be reversed. A reversible inhibitor can act in different ways but does not form covalent bonds with the enzyme. Reversible inhibition can be competitive or noncompetitive. In *competitive inhibition*, an inhibitor competes for the active site, whereas in *noncompetitive inhibition*, the inhibitor acts on another site that is not the active site.

Competitive Inhibitors

A **competitive inhibitor** has a chemical structure and polarity that is similar to that of the substrate. Thus, a competitive inhibitor competes with the substrate for the active site on the enzyme. When the inhibitor occupies the active site, the substrate cannot bind to the enzyme and no reaction can occur (see **FIGURE 20.10**). As long as the concentration of the inhibitor is substantial, there is a loss of enzyme activity. The effects of competitive enzyme inhibition can be overcome by increasing the substrate concentration. As enzyme molecules bind to substrate instead of to the inhibitor, enzyme activity will be restored.

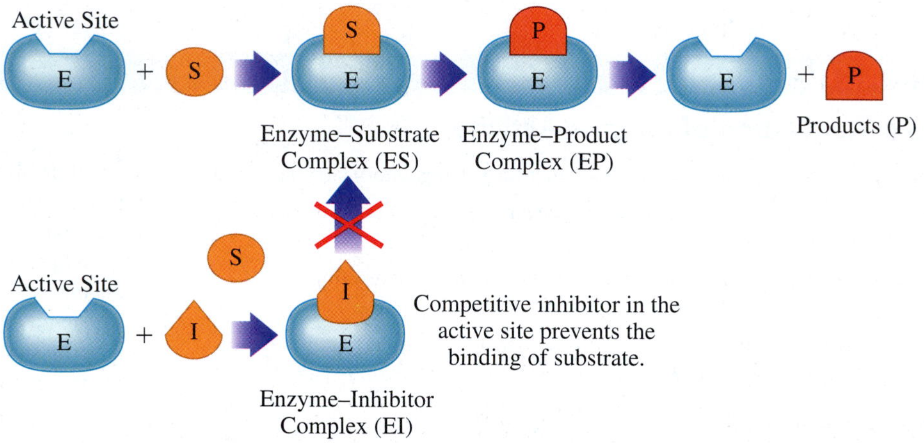
Active Site
E
+
S
S
E
P
E
E
+
P
Enzyme–Substrate Complex (ES)
Enzyme–Product Complex (EP)
Products (P)
S
Active Site
E
+
I
I
E
Competitive inhibitor in the active site prevents the binding of substrate.
Enzyme–Inhibitor Complex (EI)

FIGURE 20.10 ▶ With a structure similar to the substrate of an enzyme, a competitive inhibitor (I) also fits the active site and competes with the substrate when both are present.

Q Why does increasing the substrate concentration reverse the inhibition by a competitive inhibitor?

Malonate, which has a structure and polarity similar to that of succinate, competes for the active site on the enzyme succinate dehydrogenase. As long as malonate, a competitive inhibitor, occupies the active site, no reaction occurs. When more of the substrate succinate is added, it becomes more likely for the enzyme to bind substrate than inhibitor, and the inhibition is reversed.

Some bacterial infections are treated with competitive inhibitors called *antimetabolites*. Sulfanilamide, one of the first sulfa drugs, competes with *p*-aminobenzoic acid (PABA), which is an essential substance (metabolite) in the growth cycle of bacteria.

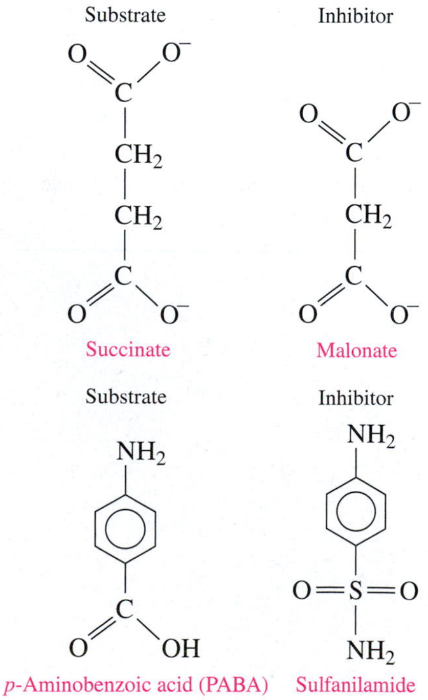
Substrate
Inhibitor
Succinate
Malonate
Substrate
Inhibitor
p-Aminobenzoic acid (PABA)
Sulfanilamide

Noncompetitive Inhibitors

The structure of a **noncompetitive inhibitor** does not resemble the substrate and does not compete for the active site. Instead, a noncompetitive inhibitor binds to a site on the enzyme that is not the active site. When bonded, the noncompetitive inhibitor distorts the shape of the enzyme. Inhibition occurs because the substrate cannot fit properly in the altered active site. Without the proper binding of the substrate, no catalysis can take place (see **FIGURE 20.11**). Because a noncompetitive inhibitor is not competing for the active site, the addition of more substrate does not reverse the effect of a noncompetitive inhibitor. Examples of noncompetitive inhibitors are the heavy metal ions Pb^{2+}, Ag^{+}, and Hg^{2+} that bond with functional groups such as $-COO^-$ or $-OH$ in the protein chains of enzymes, causing a change in the tertiary structure of the enzyme and a loss of catalytic activity. The effect of these metallic noncompetitive inhibitors can be reversed by using chemical reagents known as *chelators*, which bind to the heavy metal ions more tightly than the enzyme does. Once the heavy metal ions are no longer bound to the enzyme, the enzyme can regain its three-dimensional tertiary structure and, thus, its biological activity.

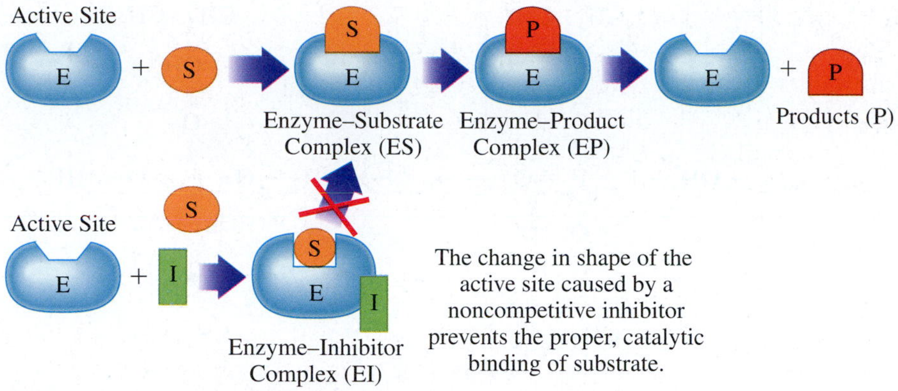

Active Site
E
S
E
Enzyme–Substrate Complex (ES)
P
E
Enzyme–Product Complex (EP)
E
P
Products (P)
Active Site
S
E
I
S
E
I
Enzyme–Inhibitor Complex (EI)
The change in shape of the active site caused by a noncompetitive inhibitor prevents the proper, catalytic binding of substrate.

FIGURE 20.11 ▶ A noncompetitive inhibitor (I) binds to an enzyme at a site other than the active site, which distorts the enzyme and active site and prevents the proper binding and catalysis of the substrate at the active site.

Q Why does an increase in the substrate concentration not reverse inhibition by a noncompetitive inhibitor?

Chemistry Link to Health

Taking Advantage of Enzyme Inhibition to Treat Cancer: Imatinib

Chronic myeloid leukemia (CML) is a cancer of the bone marrow and blood. It affects 5000 to 8000 people each year. Before the year 2001, less than 30% of people with CML survived more than five years after their diagnosis.

CML results from a genetic mutation that causes the body to produce an abnormally active enzyme. The enzyme is a tyrosine kinase. Tyrosine kinases are enzymes that transfer a phosphate group from ATP onto a tyrosine R group in a protein. They play important roles in cell growth, cell differentiation, metabolism, and apoptosis (programmed cell death). This particular tyrosine kinase stimulates uncontrolled cell growth in white blood cells, or CML.

Once scientists realized that CML was caused by a specific tyrosine kinase, they reasoned that if they could inhibit the activity of that enzyme, they could also inhibit the uncontrolled growth of white blood cells. They started looking for compounds that inhibit tyrosine kinases. After developing and testing several tyrosine kinase inhibitors (TKIs), the scientists identified a compound (initially called "ST1571") that strongly inhibited the CML tyrosine kinase but did not seem to affect healthy cells. In 2001, the compound, now known as imatinib (Gleevec), was approved for use as a cancer treatment by the U.S. Food and Drug Administration (FDA).

Imatinib acts as a competitive inhibitor of the tyrosine kinase affected in CML.

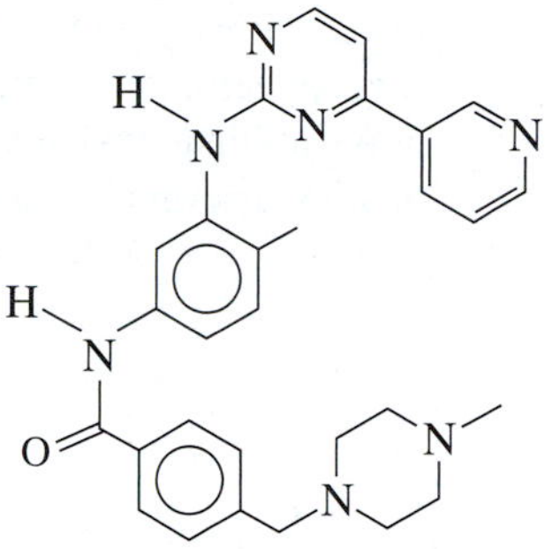

Imatinib was the first tyrosine kinase inhibitor used in a clinical setting.

In 2017, the results of a follow-up study showed that 83% of patients who were taking imatinib had survived at least 10 years after their diagnosis. Moreover, because imatinib is a targeted cancer therapy, meaning that it targets a particular gene or protein that is involved in cancer growth, there were fewer side effects associated with its use than there are with typical chemotherapy drugs.

Although the use of imatinib has shown very positive results, not all patients with CML respond to imatinib treatment. Additionally, some patients with advanced CML can develop resistance to imatinib. Imatinib was the first tyrosine kinase inhibitor used in a clinical setting. Since its development, a number of other TKIs have been developed and approved as cancer treatments.

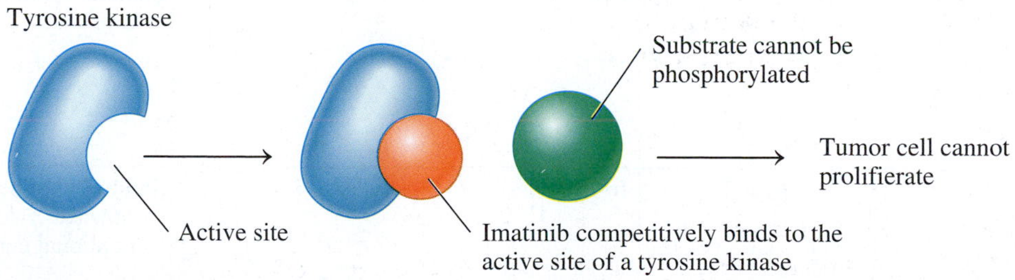

Tyrosine kinase
Active site
Imatinib competitively binds to the active site of a tyrosine kinase
Substrate cannot be phosphorylated
Tumor cell cannot prolifierate

Irreversible Inhibition

In irreversible inhibition, a molecule causes an enzyme to lose all enzymatic activity. Most **irreversible inhibitors** are toxic substances that destroy enzymes. Usually an irreversible inhibitor forms a covalent bond with an R group of an amino acid within the active site, preventing the substrate from binding to the active site and causing a loss of catalytic activity.

For example, insecticides and nerve gases act as irreversible inhibitors of acetylcholinesterase, an enzyme needed for nerve conduction. The compound diisopropyl fluorophosphate (DFP) forms a covalent bond with the $-CH_2OH$ group of a serine in the active site. When acetylcholinesterase is inhibited, the transmission of nerve impulses is blocked, and paralysis occurs.

ENGAGE

What effect does an irreversible inhibitor have on an enzyme?

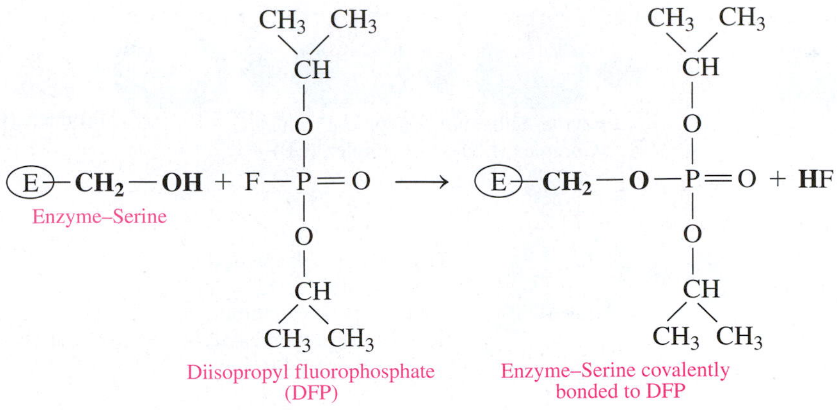
Enzyme–Serine
Diisopropyl fluorophosphate (DFP)
Enzyme–Serine covalently bonded to DFP

Some **antibiotics** act as irreversible inhibitors to stop bacterial growth. For example, penicillin inhibits an enzyme needed for the formation of cell walls in bacteria, but not human cell membranes. With an incomplete cell wall, bacteria cannot survive, and the infection is stopped. Some bacteria are resistant to penicillin because they produce penicillinase, an enzyme that breaks down penicillin. The penicillinase hydrolyzes the four-atom ring converting penicillin to penicillinoic acid, which is inactive. Over the years, derivatives of penicillin to which bacteria have not yet become resistant have been produced. Examples of some irreversible enzyme inhibitors are listed in **TABLE 20.7**.

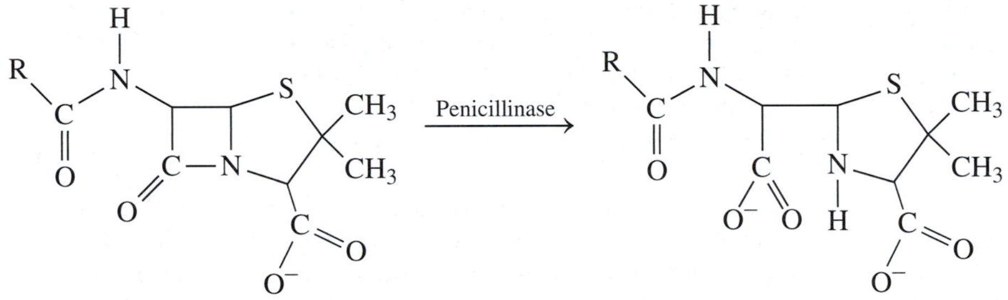
Penicillinase

Penicillin   Penicillinoic acid

TABLE 20.7 Examples of Irreversible Enzyme Inhibitors

Name	Structure	Source	Inhibitory Action
Cyanide	CN^-	Bitter almonds	Bonds to iron in the heme group of cytochrome *c* oxidase, an enzyme in electron transport
Allopurinol	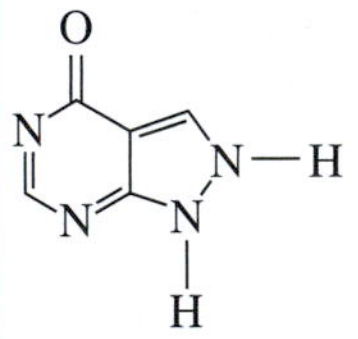	Drug	Inhibits xanthine oxidase from forming uric acid
Parathion	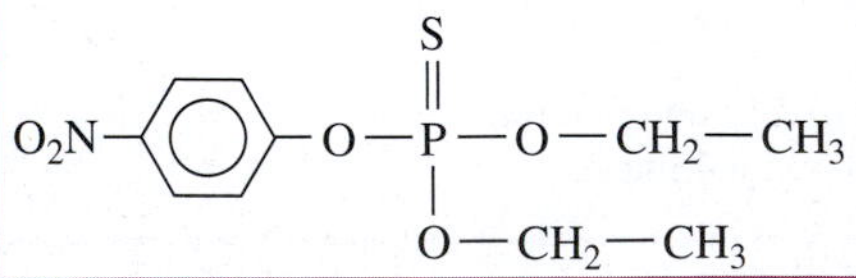	Insecticide	Inhibits cholinesterase from breaking down acetylcholine, resulting in continual nerve transmission
Penicillin	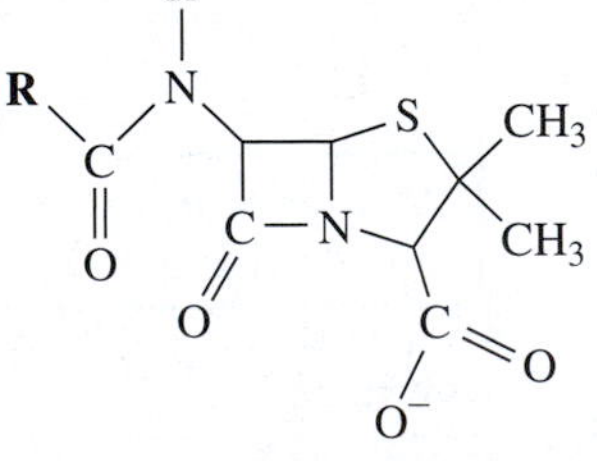	*Penicillium* fungus	Inhibits enzymes that build cell walls in bacteria
		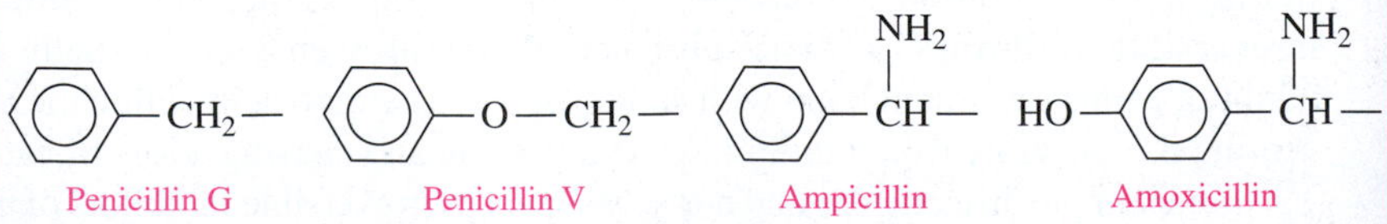 Penicillin G Penicillin V Ampicillin Amoxicillin **R Groups for Penicillin Derivatives**	

A summary of competitive, noncompetitive, and irreversible inhibitor characteristics is shown in **TABLE 20.8**.

TABLE 20.8 Summary of Competitive, Noncompetitive, and Irreversible Inhibitors

Characteristics	Competitive	Noncompetitive	Irreversible
Shape of Inhibitor	Similar shape to the substrate	Does not have a similar shape to the substrate	May or may not have a similar shape to the substrate
Binding to Enzyme	Competes for and binds at the active site	Binds away from the active site to change the shape of the enzyme and its activity	Forms a covalent bond with the enzyme
Reversibility	Adding more substrate reverses the inhibition	Not reversed by adding more substrate, but by a chemical change that removes the inhibitor	Permanent, not reversible

▶ SAMPLE PROBLEM 20.6 Enzyme Inhibition

TRY IT FIRST

State the type of inhibition in each of the following:

a. The inhibitor has a structure similar to that of the substrate.

b. The inhibitor binds to the surface of the enzyme, changing its shape in such a way that it cannot bind to substrate.

SOLUTION

a. STEP 1 Compare the structure of the inhibitor to that of the substrate. This inhibitor has a structure similar to that of the substrate.

STEP 2 Describe the characteristics of the inhibitor. This inhibitor would compete with the substrate for the active site.

STEP 3 Assign the type of inhibition based on the characteristics. This type of inhibition is competitive inhibition, which is reversed by increasing the concentration of the substrate.

b. STEP 1 Compare the structure of the inhibitor to that of the substrate. This inhibitor does not have a structure similar to the substrate.

STEP 2 Describe the characteristics of the inhibitor. This inhibitor binds to the surface of the enzyme where it changes the shape of the enzyme and the active site.

STEP 3 Assign the type of inhibition based on the characteristics. This type of inhibition is noncompetitive inhibition.

STUDY CHECK 20.6

a. What type of inhibition occurs when aspirin, also called acetylsalicylic acid, forms a covalent bond with serine in the active site of the enzyme cyclooxygenase?

b. What type of inhibition occurs when the chemotherapy drug methotrexate binds reversibly to the active site of the enzyme dihydrofolate reductase?

ANSWER

a. irreversible inhibition

b. competitive inhibition

TEST

Try Practice Problems 20.35 to 20.38

PRACTICE PROBLEMS

20.5 Enzyme Inhibition

20.35 Indicate whether each of the following describes a competitive or a noncompetitive enzyme inhibitor:

a. The inhibitor has a structure similar to the substrate.
b. The effect of the inhibitor cannot be reversed by adding more substrate.
c. The inhibitor competes with the substrate for the active site.
d. The structure of the inhibitor is not similar to the substrate.
e. The addition of more substrate reverses the inhibition.

20.36 Oxaloacetate is an inhibitor of succinate dehydrogenase.

a. Would you expect oxaloacetate to be a competitive or a noncompetitive inhibitor? Why?
b. Would oxaloacetate bind to the active site or elsewhere on the enzyme?
c. How would you reverse the effect of the inhibitor?

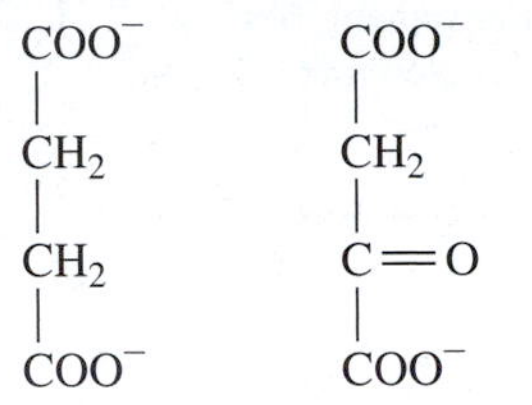

Succinate Oxaloacetate

Clinical Applications

20.37 Methanol and ethanol are oxidized by alcohol dehydrogenase. In methanol poisoning, ethanol is given intravenously to prevent the formation of formaldehyde, which has toxic effects.

a. Draw the condensed structural formulas for methanol and ethanol.
b. Would ethanol compete with methanol for the active site or bind to a different site on the enzyme?
c. Would ethanol be a competitive or noncompetitive inhibitor of alcohol dehydrogenase?

20.38 In humans, the antibiotic amoxicillin (a type of penicillin) is used to treat certain bacterial infections.

a. Does the antibiotic inhibit human enzymes?
b. Why does the antibiotic kill bacteria but not humans?
c. Is amoxicillin a reversible or irreversible inhibitor?

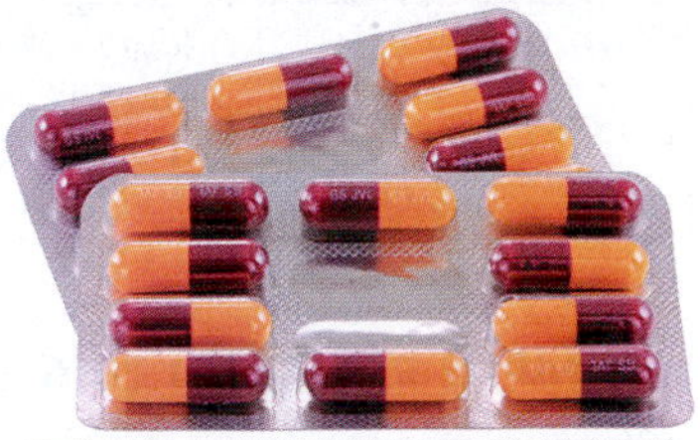

Amoxicillin is a derivative of the antibiotic penicillin.

20.6 Enzyme Cofactors and Vitamins

LEARNING GOAL Describe the types of cofactors found in enzymes.

CORE CHEMISTRY SKILL
Describing the Role of Cofactors

Although some enzymes are biologically active as folded polypeptide chains, many enzymes require cofactors for activity. **Cofactors** are metal ions or small molecules that must associate with an enzyme in order for the enzyme to catalyze a given reaction (see **FIGURE 20.12**). Many cofactors do not remain bonded to a particular enzyme, but are used repeatedly by different enzyme molecules to facilitate an enzyme-catalyzed reaction. Thus, only small amounts of those cofactors are required in the cells.

TEST
Try Practice Problems 20.39 and 20.40

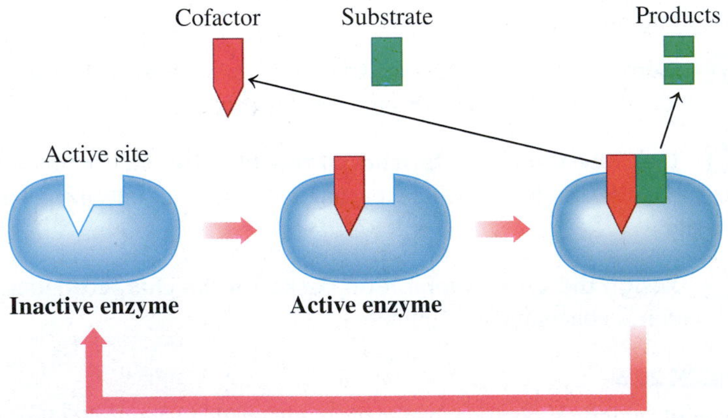
Cofactor
Substrate
Products
Active site
Inactive enzyme
Active enzyme

FIGURE 20.12 ▶ A cofactor is required so that many enzymes can become active.
Q What are the two types of cofactors?

TABLE 20.9 Enzymes and the Metal Ions Required as Cofactors

Metal Ion	Enzymes Requiring Metal Ion Cofactors
Cu^{2+}/Cu^{+}	Cytochrome oxidase
Fe^{2+}/Fe^{3+}	Catalase Cytochrome oxidase
Zn^{2+}	Alcohol dehydrogenase Carbonic anhydrase Carboxypeptidase A
Mg^{2+}	Glucose-6-phosphatase Hexokinase
Mn^{2+}	Arginase
Ni^{2+}	Urease

Metal Ions as Cofactors

The metal ions from the minerals that we obtain from foods in our diet have various functions in enzyme catalysis. Metal ion cofactors such as Fe^{2+} and Cu^{2+} are often required by enzymes involved in the loss or gain of electrons (oxidoreductases). Other metal ions such as Zn^{2+} stabilize the amino acid R groups in the active site of hydrolases. Some of the metal ion cofactors that are required by enzymes are listed in **TABLE 20.9**.

The enzyme carboxypeptidase is produced in the pancreas and moves into the small intestine where it catalyzes the hydrolysis of the C-terminal amino acid with an aromatic residue in a protein (see **FIGURE 20.13**).

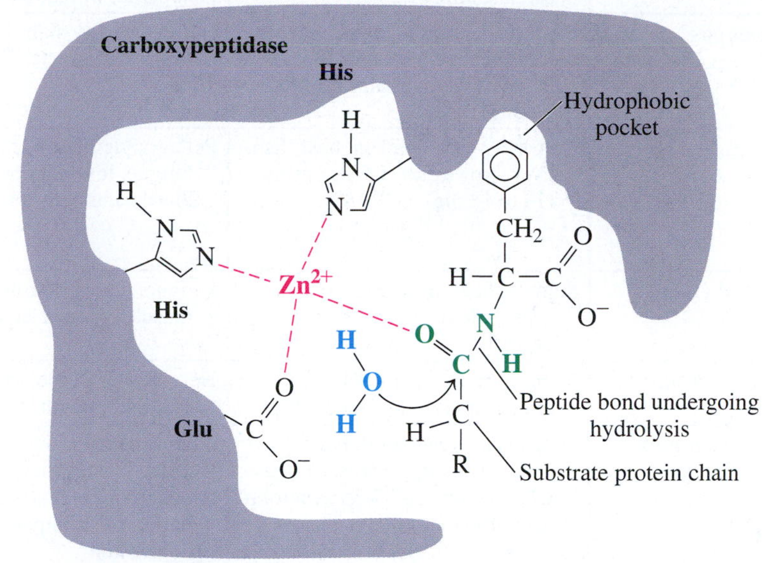
Carboxypeptidase
His
Hydrophobic pocket
His
Zn^{2+}
Glu
Peptide bond undergoing hydrolysis
Substrate protein chain

FIGURE 20.13 ▶ Carboxypeptidase requires a Zn^{2+} cofactor for the hydrolysis of the peptide bond of a C-terminal aromatic amino acid.

Q Why is Zn^{2+} a cofactor for carboxypeptidase?

The active site includes a pocket that fits the bulky hydrophobic amino acid R group on a protein substrate. In the center of the active site, a Zn^{2+} cofactor is bonded to the nitrogen atom in each of two histidine R groups, and an oxygen atom in the R group of glutamate. The Zn^{2+} promotes the hydrolysis of the peptide bond by stabilizing the partial negative charge of the oxygen in the carbonyl group of the peptide bond undergoing hydrolysis.

Coenzymes

If the cofactor is a small organic molecule, it is known as a **coenzyme**. Many coenzymes are derived from vitamins. **Vitamins** are organic molecules that are essential for normal health and growth. They are required in trace amounts and need to be obtained from the diet because vitamins are not synthesized in the body.

Vitamins are classified into two groups by solubility: water-soluble and fat-soluble. **Water-soluble vitamins** have polar groups such as —OH and —COOH, which make them soluble in the aqueous environment of the cells. The **fat-soluble vitamins** are nonpolar compounds, which are soluble in the fat (lipid) components of the body such as fat deposits and cell membranes.

Water-Soluble Vitamins

Water-soluble vitamins are precursors of many of the coenzymes required by enzymes to carry out certain aspects of catalytic action (see **TABLE 20.10**). Because water-soluble vitamins cannot be stored in the body, any excess amounts are excreted in the urine each day. Therefore, the water-soluble vitamins must be in the foods of our daily diets (see **FIGURE 20.14**). Because many water-soluble vitamins are easily destroyed by heat, oxygen, and ultraviolet light, care must be taken in food preparation, processing, and storage. Because refining grains such as wheat causes a loss of vitamins, during the 1940s, the Committee on Food and Nutrition of the National Research Council began to recommend dietary enrichment of cereal grains. Vitamin B_1 (thiamine), vitamin B_2 (riboflavin), and iron were in the first group of added nutrients recommended. The Recommended Daily Allowance (RDA) for many vitamins and

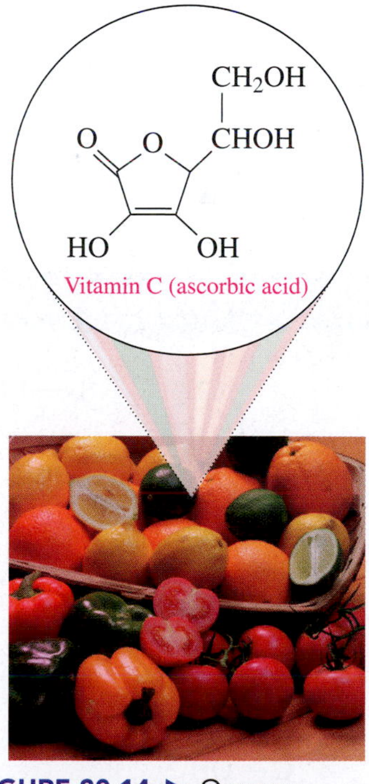
Vitamin C (ascorbic acid)

FIGURE 20.14 ▶ Oranges, lemons, peppers, and tomatoes contain the water-soluble vitamin, vitamin C (ascorbic acid), which is an antioxidant.

Q What happens to the excess vitamin C that may be consumed in one day?

TABLE 20.10 Water-Soluble Vitamins, Coenzymes, Functions, Sources and RDA, and Deficiency Symptoms

Vitamin	Coenzyme	Transfer Function	Sources and RDA (Adults)	Deficiency Symptoms
B_1 (Thiamine)	Thiamine pyrophosphate (TPP)	Aldehyde groups	Liver, yeast, whole grain bread, cereals, milk (1.2 mg)	Beriberi: fatigue, poor appetite, weight loss
B_2 (Riboflavin)	Flavin adenine dinucleotide (FAD); flavin mononucleotide (FMN)	Electrons	Beef, liver, chicken, eggs, green leafy vegetables, dairy foods, peanuts, whole grains (1.2 to 1.8 mg)	Dermatitis; dry skin; red, sore tongue; cataracts
B_3 (Niacin)	Nicotinamide adenine dinucleotide (NAD^+); nicotinamide adenine dinucleotide phosphate ($NADP^+$)	Electrons	Brewer's yeast, chicken, beef, fish, liver, brown rice, whole grains (14 to 18 mg)	Pellagra: dermatitis, muscle fatigue, loss of appetite, diarrhea, mouth sores
B_5 (Pantothenic acid)	Coenzyme A	Acetyl groups	Salmon, beef, liver, eggs, brewer's yeast, whole grains, fresh vegetables (5 mg)	Fatigue, retarded growth, muscle cramps, anemia
B_6 (Pyridoxine)	Pyridoxal phosphate (PLP)	Amino groups	Meat, liver, fish, nuts, whole grains, spinach (1.3 to 2.0 mg)	Dermatitis, fatigue, anemia, retarded growth
B_9 (Folic acid)	Tetrahydrofolate (THF)	Methyl groups	Green leafy vegetables, beans, meat, seafood, yeast, asparagus, whole grains enriched with folic acid (400 mcg)	Abnormal red blood cells, anemia, intestinal tract disturbances, loss of hair, growth impairment, depression
B_{12} (Cobalamin)	Methylcobalamin	Methyl groups, hydrogen	Liver, beef, kidney, chicken, fish, milk products (2.0 to 2.6 mcg)	Pernicious anemia, malformed red blood cells, nerve damage
C (Ascorbic acid)	Ascorbic acid	Electrons	Blueberries, citrus fruits, strawberries, cantaloupe, tomatoes, peppers, broccoli, cabbage, spinach (75 to 90 mg)	Scurvy: bleeding gums, weakened connective tissues, slow-healing wounds, anemia
H (Biotin)	Biocytin	Carbon dioxide	Liver, yeast, nuts, eggs (30 mcg)	Dermatitis, loss of hair, fatigue, anemia, depression

minerals appears on food product labels. Individuals eating balanced meals including fresh produce and whole grains can readily achieve the U.S. RDAs. **TABLE 20.11** gives the structures of the water-soluble vitamins.

TABLE 20.11 Structures of Water-Soluble Vitamins

B_1 (Thiamine)

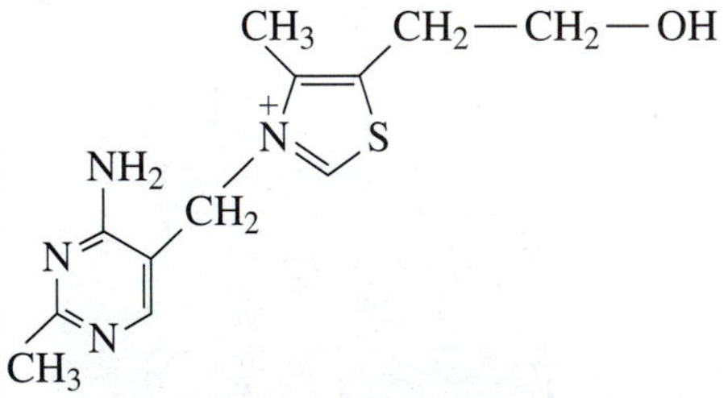

B_2 (Riboflavin)

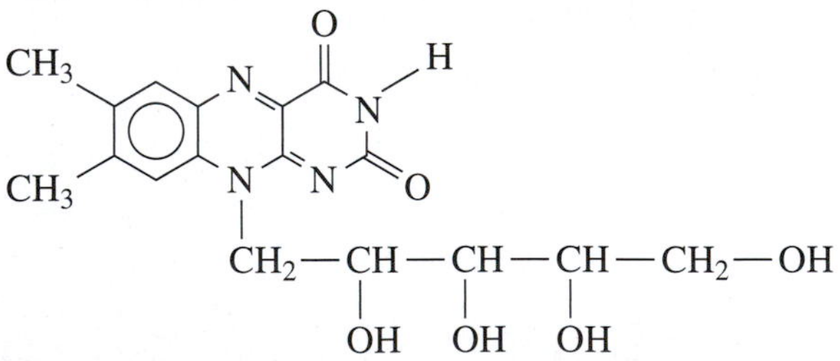

B_3 (Niacin)

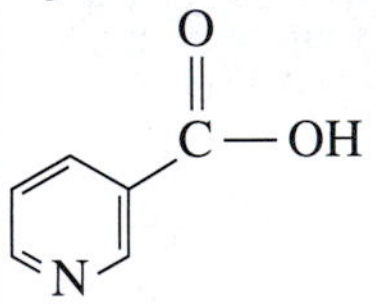

B_5 (Pantothenic acid)

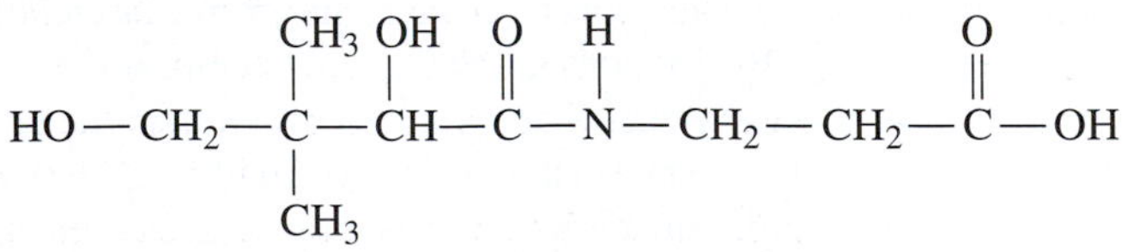

TABLE 20.11 Structures of Water-Soluble Vitamins (*continued*)

B_6 (Pyridoxine)

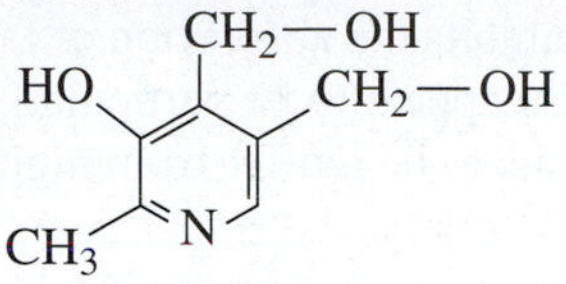

C (Ascorbic acid)

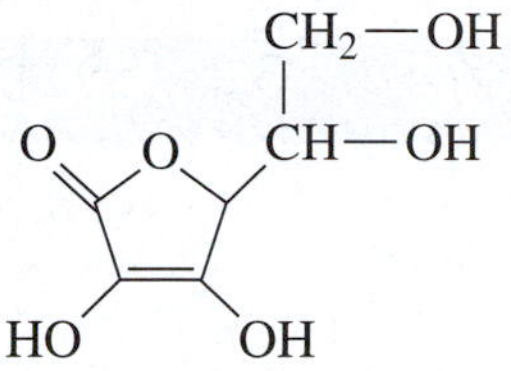

H (Biotin)

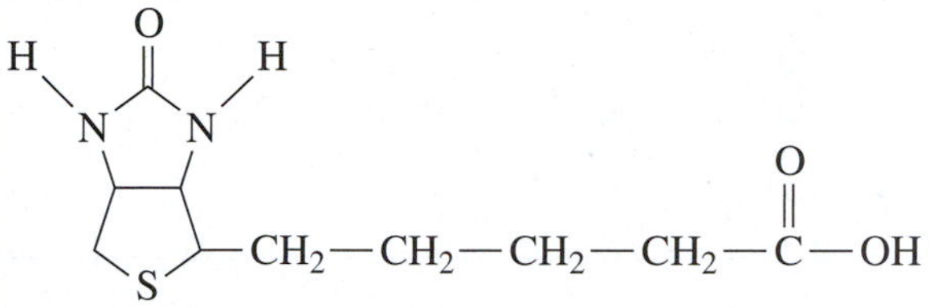

B_{12} (Cobalamin)

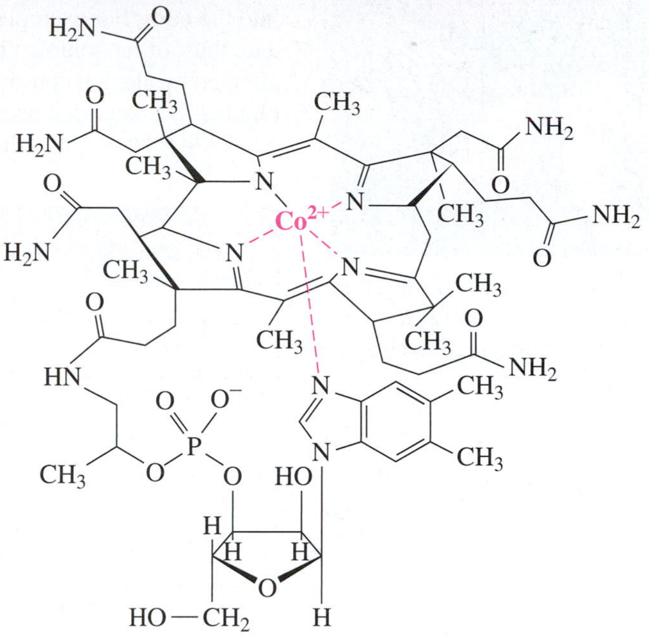

B_9 (Folic acid)

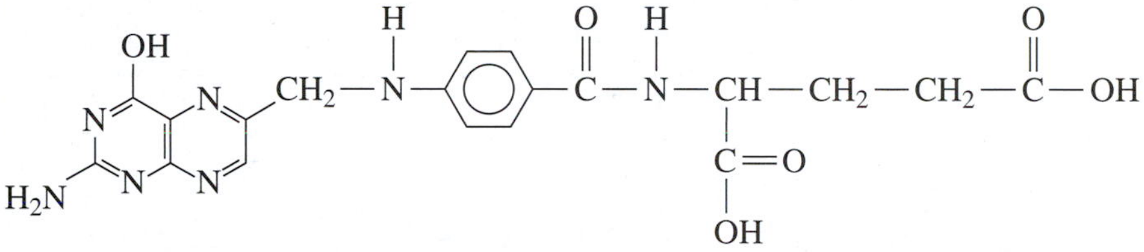

Fat-Soluble Vitamins

The fat-soluble vitamins—A, D, E, and K—are not involved as coenzymes in catalytic reactions, but they are important in processes such as vision, formation of bone, and proper blood clotting (see **TABLE 20.12**). They also provide protection from oxidation. The production of energy by cells occurs through chemical oxidation. During this process, reactive oxygen compounds called *free radicals*, such as NO•, OH•, and O_2•, are formed. These molecules are called free radicals because they have an unpaired electron. Because electrons are more stable when they are paired, free radicals react readily. An overabundance of free radicals can cause cellular and DNA damage such as aging, inflammation, cataracts, and cancer. The body uses *antioxidants* to defend against free radicals. These antioxidants include vitamin E and beta-carotene, a precursor to vitamin A. The water-soluble vitamin C can also act as an antioxidant.

ENGAGE

What type of molecules does the body use to defend against free radicals?

TABLE 20.12 Functions, Sources and RDA, and Deficiency Symptoms of Fat-Soluble Vitamins

Vitamin	Function	Sources and RDA (Adults)	Deficiency Symptoms
A (Retinol)	Formation of visual pigments, synthesis of RNA	Yellow and green fruits and vegetables (800 mcg)	Night blindness, immune system repression, slowed growth, rickets
D (Cholecalciferol)	Regulation of absorption of P and Ca during bone growth	Sunlight, cod liver oil, enriched milk, eggs (5 to 10 mcg)	Rickets, weak bone structure, osteomalacia
E (Tocopherol)	Antioxidant; prevents oxidation of vitamin A and unsaturated fatty acids	Meats, whole grains, vegetables (15 mg)	Hemolysis, anemia
K (Menaquinone)	Synthesis of zymogen prothrombin for blood clotting	Liver, spinach, cauliflower (9 to 120 mcg)	Prolonged bleeding time, bruising

The orange pigment (carotene) in carrots is used to synthesize vitamin A (retinol), an antioxidant, in the body.

Because the fat-soluble vitamins are stored in the body, and not eliminated, it is possible to take too much, which could be toxic, primarily in the liver and fatty tissues. There are, however, conditions that cause insufficient amounts of fat-soluble vitamins to be absorbed into the body. For example, excessive consumption of alcohol inhibits the absorption of fat and, thus, of fat-soluble vitamins, into the body. Vitamin A levels appear to be particularly affected by alcohol consumption, and as a consequence, many alcoholics suffer from night blindness, a condition caused by a deficiency in vitamin A.

TABLE 20.13 gives the structures of the fat-soluble vitamins.

TABLE 20.13 Structures of Fat-Soluble Vitamins

Vitamin A (Retinol)	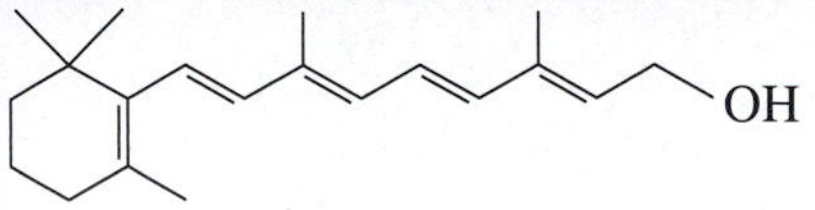
Vitamin D (Cholecalciferol)	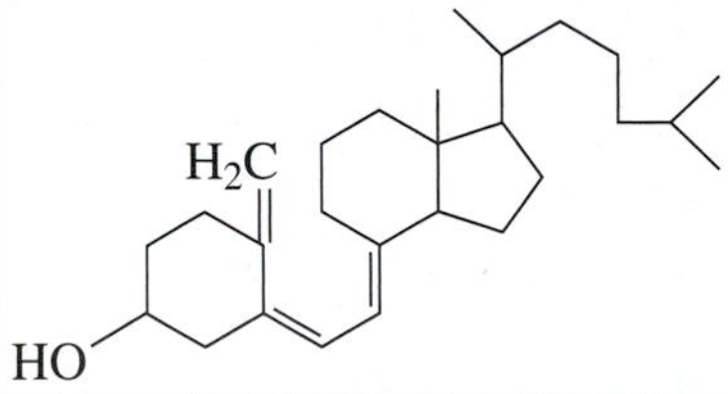
Vitamin E (Tocopherol)	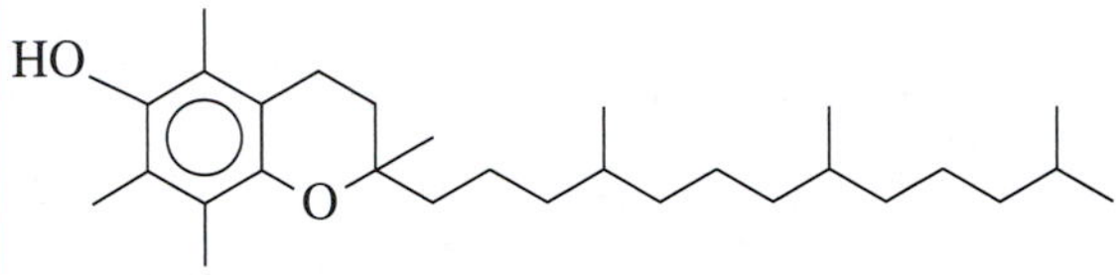
Vitamin K (Menaquinone)	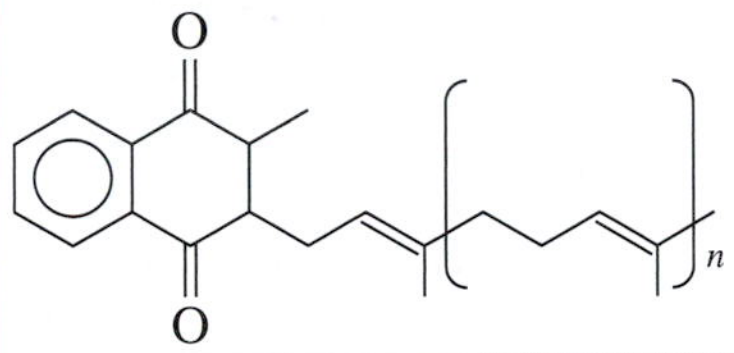

TEST

Try Practice Problems 20.41 to 20.46

▶ SAMPLE PROBLEM 20.7 Cofactors

TRY IT FIRST

Indicate whether each of the following enzymes is active with or without a cofactor:

a. an enzyme that needs Mg^{2+} for catalytic activity
b. a polypeptide chain that is biologically active
c. an enzyme that binds to vitamin B_6 to become active

SOLUTION

a. The enzyme would be active with the metal ion Mg^{2+} cofactor.
b. An active enzyme that is only a polypeptide chain does not require a cofactor.
c. An enzyme that requires vitamin B_6 is active with a cofactor.

STUDY CHECK 20.7

Which cofactor for the enzymes in Sample Problem 20.7 would be called a coenzyme?

ANSWER

vitamin B_6

▶ SAMPLE PROBLEM 20.8 Vitamins

TRY IT FIRST

Why is it more important to get regular, daily amounts of vitamins B_1 and B_2, but not vitamins A or D?

SOLUTION

Water-soluble vitamins like vitamin B_1 (thiamine) and vitamin B_2 (riboflavin) are not stored in the body, whereas fat-soluble vitamins such as vitamin A (retinol) and vitamin D (cholecalciferol) are stored in the liver and body fat. Any excess of thiamine or riboflavin is eliminated in the urine and must be replenished each day from the diet.

STUDY CHECK 20.8

a. Why are fresh fruits rather than cooked fruits recommended as a source of vitamin C?
b. Why is it unlikely that a person can consume a toxic amount of vitamin C?

ANSWER

a. Water-soluble vitamins such as vitamin C are easily destroyed by heat.
b. Vitamin C is a water-soluble vitamin and is not stored in the body. Any excess vitamin C is excreted in the urine. Therefore, it is difficult for toxic levels of vitamin C to accumulate in the body.

PRACTICE PROBLEMS

20.6 Enzyme Cofactors and Vitamins

20.39 Is the enzyme described in each of the following statements active with or without a cofactor?
a. requires vitamin B_1 (thiamine)
b. needs Zn^{2+} for catalytic activity
c. its active form consists of two polypeptide chains

20.40 Is the enzyme described in each of the following statements active with or without a cofactor?
a. requires vitamin B_2 (riboflavin)
b. its active form is composed of 155 amino acids
c. uses Cu^{2+} during catalysis

20.41 Identify the vitamin that is a component of each of the following coenzymes:
a. coenzyme A
b. tetrahydrofolate (THF)
c. NAD^+

20.42 Identify the vitamin that is a component of each of the following coenzymes:
a. thiamine pyrophosphate **b.** FAD
c. pyridoxal phosphate

Clinical Applications

20.43 What vitamin may be deficient in the following conditions?
a. rickets **b.** scurvy **c.** pellagra

20.44 What vitamin may be deficient in the following conditions?
a. poor night vision **b.** pernicious anemia **c.** beriberi

20.45 The RDA for vitamin B_6 (pyridoxine) is 1.3 to 2.0 mg. Why will it not improve your nutrition to take 100 mg of pyridoxine daily?

20.46 The RDA for vitamin C (ascorbic acid) is 75 to 90 mg. If you take 1000 mg of vitamin C daily, what happens to the vitamin C you do not need?

CLINICAL UPDATE Noah's Diet for Lactose Intolerance

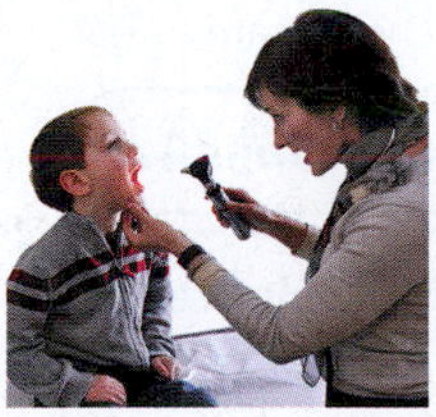

When Noah and his mother return to the clinic, Emma administers the hydrogen breath test (HBT) after Noah has fasted for 8 h. First, Emma has Noah breathe into a balloon to determine a base level of hydrogen (H_2). Then Noah drinks a solution containing lactose, and more breath samples are collected and measured for hydrogen over the next 3 h.

Hydrogen is present in the breath when the bacteria in the colon are fermenting lactose and producing hydrogen (H_2). Lactose intolerance is confirmed if the hydrogen levels rise 20 ppm above baseline levels over a 3-h period after ingesting the lactose solution. The results of the test, shown as the black line on the graph, indicate that Noah's levels were 130 ppm after ingestion of lactose, so Emma and the attending physician conclude that Noah has a problem digesting lactose, which means he has *lactose intolerance*.

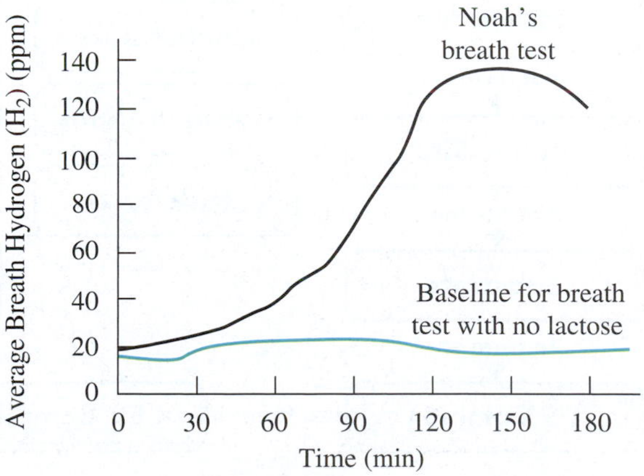
Noah's breath test
Baseline for breath test with no lactose
Average Breath Hydrogen (H2) (ppm)
0
20
40
60
80
100
120
140
0
30
60
90
120
150
180
Time (min)

Treatment of Lactose Intolerance

Emma tells Noah's mother that he will need to avoid products that contain lactose, including milk and milk products such as cheese, butter, and ice cream. A person who is lactose intolerant should also know that some foods that are not dairy products contain lactose. For example, many baked goods, cereals, breakfast drinks, salad dressings, and even lunchmeat can contain lactose. Food labels must be read carefully to see if the ingredients include "milk" or "lactose."

Another way to combat lactose intolerance is to use the enzyme lactase with meals. Lactase is available in many forms, such as tablets that are taken with meals, or drops that are added to milk. When lactase is added to milk that is left in the refrigerator for 24 h, the lactose level is reduced by 70 to 90%. Lactase pills or chewable tablets are taken when a person begins to eat a meal that contains dairy foods.

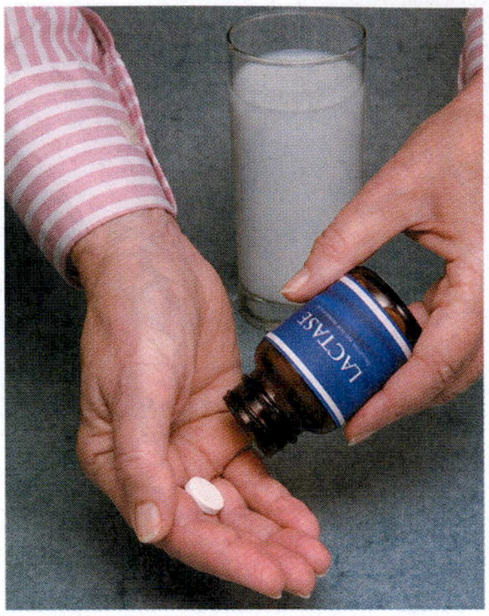
LACTASE

People who are lactose intolerant may use lactase tablets to decrease the amount of gastrointestinal distress they feel after consuming milk.

Clinical Applications

20.47 Using the graph for the hydrogen breath test, estimate
   a. the baseline ppm at 60 min
   b. the hydrogen ppm for Noah at 60 min

20.48 Using the graph for the hydrogen breath test, estimate
   a. the baseline ppm at 90 min
   b. the hydrogen ppm for Noah at 90 min

20.49 Noah's mother makes him some hot chocolate, to which she adds a tablet of lactase. A few hours after Noah drinks the hot chocolate, he has symptoms of lactose intolerance. Explain why the lactase was ineffective.

20.50 Noah's mother adds a tablet of lactase to a glass of milk and places it in the refrigerator for 24 h. The next day, Noah drinks the milk and has no symptoms of lactose intolerance. Explain why the lactase was effective.

CONCEPT MAP

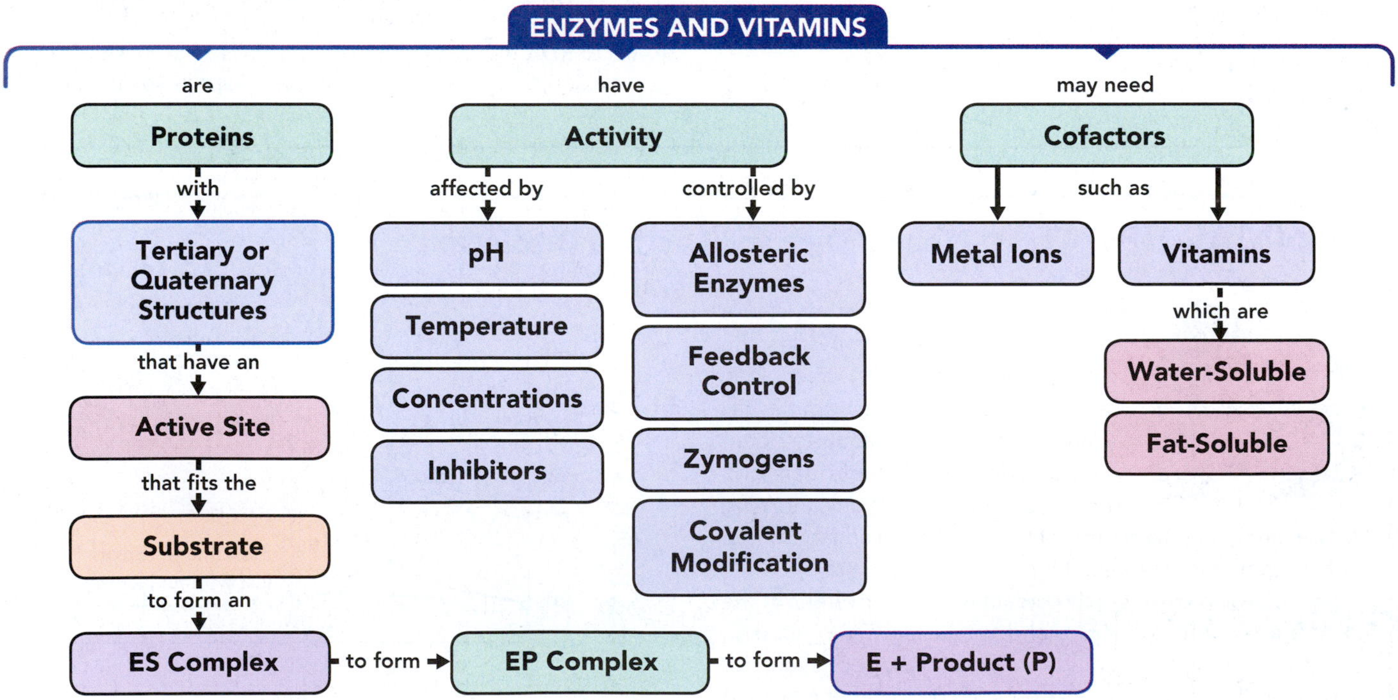
ENZYMES AND VITAMINS
are
have
may need
Proteins
with
Tertiary or Quaternary Structures
that have an
Active Site
that fits the
Substrate
to form an
ES Complex
to form
EP Complex
to form
E + Product (P)
Activity
affected by
pH
Temperature
Concentrations
Inhibitors
controlled by
Allosteric Enzymes
Feedback Control
Zymogens
Covalent Modification
Cofactors
such as
Metal Ions
Vitamins
which are
Water-Soluble
Fat-Soluble

CHAPTER REVIEW

20.1 Enzymes and Enzyme Action

LEARNING GOAL Describe enzymes and their roles in enzyme-catalyzed reactions.

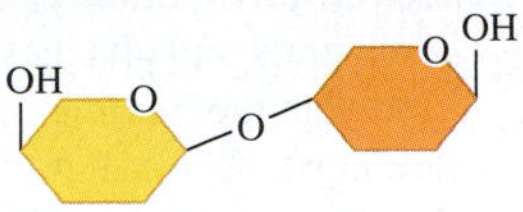
OH
O
O
O
OH

- Enzymes are biological catalysts, usually proteins, that accelerate the rates of cellular reactions by lowering activation energy.
- Within the tertiary structure of an enzyme, a small pocket called the active site binds the substrate.
- In the enzyme–substrate complex, catalysis takes place when the R groups of amino acids in the active site of an enzyme react with a substrate.
- When the products of catalysis are released, the enzyme can bind to another substrate molecule.
- In the lock-and-key model, an early theory of enzyme action, a substrate precisely fits the shape of the active site.
- In the induced-fit model, both the active site and the substrate undergo changes in their shapes to give the best fit for efficient catalysis.

20.2 Classification of Enzymes

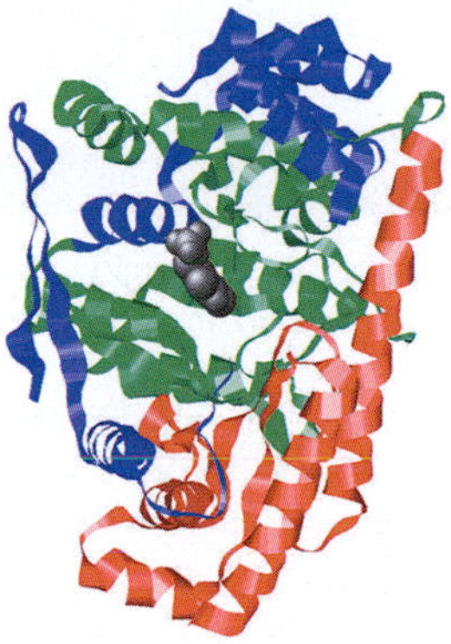

LEARNING GOAL Classify enzymes and give their names.

- The names of most enzymes ending in *ase* describe the substrate or reaction catalyzed by the enzyme.
- Enzymes are classified by the type of reaction they catalyze, such as oxidoreductase, transferase, or isomerase.

20.3 Factors Affecting Enzyme Activity

LEARNING GOAL Describe the effects that changes of temperature, pH, enzyme concentration, and substrate concentration have on enzyme activity.

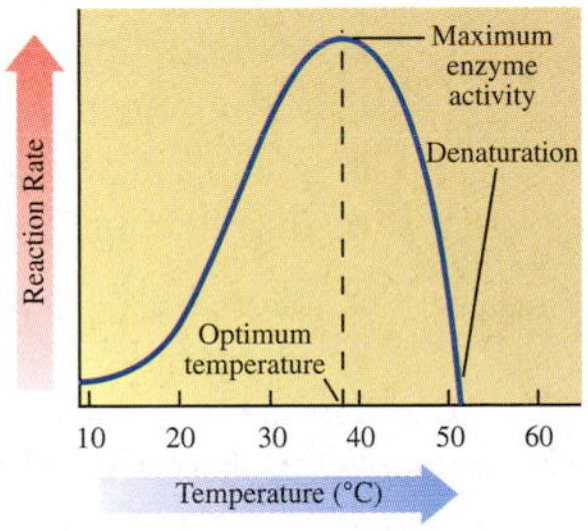
Maximum enzyme activity
Denaturation
Reaction Rate
Optimum temperature
10
20
30
40
50
60
Temperature (°C)

- The optimum temperature at which most enzymes are effective is usually 37 °C, and the optimum pH is usually 7.4.
- The rate of an enzyme-catalyzed reaction decreases as temperature and pH go above or below the optimum temperature and pH values.
- An increase in the enzyme concentration increases the rate of reaction.
- An increase in substrate concentration increases the reaction rate of an enzyme-catalyzed reaction, but a maximum rate is reached when all of the enzyme molecules are combined with substrate.

20.4 Regulation of Enzyme Activity

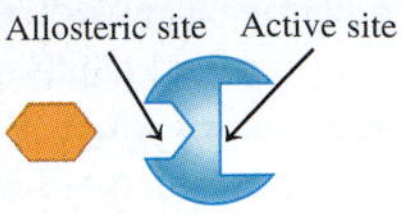
Allosteric site
Active site

LEARNING GOAL Describe the role of allosteric enzymes, feedback control, and covalent modification in regulating enzyme activity.

- The activity of allosteric enzymes is altered when regulator molecules bind to their allosteric sites.
- In feedback control, the end product of a sequence of enzyme-catalyzed reactions acts as a negative regulator of an enzyme at or near the beginning of the reaction sequence.
- Covalent modification regulates enzymes by adding or removing a covalently bonded group.
- Production of zymogens and phosphorylation are types of covalent modification.

20.5 Enzyme Inhibition

S
E
Enzyme–Substrate Complex

LEARNING GOAL Describe reversible and irreversible inhibition. Describe competitive and noncompetitive inhibition.

- An inhibitor reduces the activity of an enzyme or makes it inactive.
- An inhibitor can be reversible or irreversible.
- A competitive inhibitor has a structure similar to the substrate and competes for the active site. When the active site is occupied, the enzyme cannot catalyze the reaction of the substrate.
- A noncompetitive inhibitor attaches to the enzyme away from the active site, changing the shape of both the enzyme and its active site.
- An irreversible inhibitor forms a covalent bond within the active site that permanently prevents catalytic activity.

20.6 Enzyme Cofactors and Vitamins

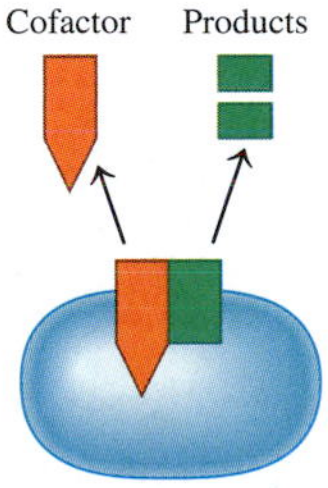
Cofactor
Products

LEARNING GOAL Describe the types of cofactors found in enzymes.

- Some enzymes are biologically active as proteins only, whereas other enzymes require a nonprotein component called a cofactor.
- A cofactor may be a metal ion, such as Cu^{2+} or Fe^{2+}, or an organic molecule called a coenzyme. Many coenzymes are derived from vitamins.
- A vitamin is a small organic molecule needed for health and normal growth that is obtained in small amounts from the diet.
- The water-soluble vitamins are B and C, and they function as coenzymes. Vitamin B is essential for the workings of certain enzymes in the body, and vitamin C is an antioxidant.
- The fat-soluble vitamins are A, D, E, and K. Vitamin A is important for vision, vitamin D for proper bone growth, vitamin E is an antioxidant, and vitamin K is required for proper blood clotting.
- Antioxidants neutralize the highly reactive oxygen free radicals generated during oxidative processes.

KEY TERMS

active site A pocket in a part of the tertiary enzyme structure that binds to a substrate and catalyzes a reaction.

activity The rate at which an enzyme catalyzes the reaction that converts a substrate to a product.

allosteric enzyme An enzyme that regulates the rate of a reaction when a regulator molecule attaches to a site other than the active site.

antibiotic A substance that inhibits the growth of bacteria.

coenzyme An organic molecule, usually derived from a vitamin, required as a cofactor in enzyme action.

cofactor A metal ion or an organic molecule that is necessary for a biologically functional enzyme.

competitive inhibitor A molecule that has a structure similar to a substrate and that inhibits enzyme action by competing for the active site.

covalent modification Regulation of an enzyme that involves forming or breaking covalent bonds to a group on the polypeptide chain.

enzyme A molecule, usually a protein and sometimes with a cofactor, that catalyzes a biological reaction.

enzyme–product (EP) complex An intermediate consisting of an enzyme that binds to a product in an enzyme-catalyzed reaction.

enzyme–substrate (ES) complex An intermediate consisting of an enzyme that binds to a substrate in an enzyme-catalyzed reaction.

fat-soluble vitamin A vitamin that is not soluble in water and can be stored in the liver and body fat.

feedback control A type of inhibition in which an end product inhibits an enzyme at or near the beginning of a sequence of enzyme-catalyzed reactions.

induced-fit model A model of enzyme action in which the shape of a substrate and the active site of the enzyme adjust to give an optimal fit.

inhibitor A substance that makes an enzyme inactive by interfering with its ability to react with a substrate.

irreversible inhibitor A compound or metal ion that causes the loss of enzymatic activity by forming a covalent bond near or at the active site.

isoenzymes Enzymes with different combinations of polypeptide subunits that catalyze the same reaction in different tissues of the body.

noncompetitive inhibitor An inhibitor that does not resemble the substrate and attaches to the enzyme away from the active site to prevent the binding of the substrate.

optimum pH The pH at which an enzyme is most active.

optimum temperature The temperature at which an enzyme is most active.

reversible inhibition The loss of enzymatic activity by an inhibitor whose effect can be reversed.

substrate The molecule that reacts in the active site in an enzyme-catalyzed reaction.

vitamin An organic molecule that is essential for normal health and growth, and obtained in small amounts from the diet.

water-soluble vitamin A vitamin that is soluble in water; cannot be stored in the body; is easily destroyed by heat, ultraviolet light, and oxygen; and functions as a coenzyme.

zymogen An inactive form of an enzyme that is activated by the removal of a peptide portion from the protein.

CORE CHEMISTRY SKILLS

Describing Enzyme Action (20.1)

- Enzymes are biological catalysts that lower the activation energy and accelerate the rate of cellular reactions.
- Within the tertiary structure of an enzyme, a small pocket called the active site binds the substrate.
- In the induced-fit model of enzyme action, the active site and the substrate change their shapes for efficient catalysis.
- When the products of catalysis are released, the enzyme can bind to another substrate molecule.

Example: Match the terms (1) enzyme, (2) substrate, (3) active site, and (4) induced-fit model with each of the following:

a. an active site that adapts to the shape of a substrate
b. has a structure that fits the active site of an enzyme
c. the portion of an enzyme where catalytic activity occurs
d. has a tertiary structure that recognizes the substrate

Answer: **a.** (4) induced-fit model **b.** (2) substrate
c. (3) active site **d.** (1) enzyme

Classifying Enzymes (20.2)

An enzyme can be classified from its name or the type of reaction it catalyzes. Enzymes have six general classes: oxidoreductase, transferase, hydrolase, lyase, isomerase, and ligase.

Example: Match the enzyme names (1) DNA ligase, (2) HIV protease, (3) protein kinase, and (4) glucose oxidase with one of the following classifications:

a. oxidoreductase **b.** transferase
c. hydrolase **d.** ligase

Answer: **a.** (4) glucose oxidase **b.** (3) protein kinase
c. (2) HIV protease **d.** (1) DNA ligase

Identifying Factors Affecting Enzyme Activity (20.3)

- The activity of many enzymes is regulated by allosteric enzymes, feedback control, and covalent modification, including production of zymogens and phosphorylation.
- The rate of an enzyme-catalyzed reaction is affected by changes in temperature, pH, enzyme concentration, and substrate concentration.
- Molecules called inhibitors decrease enzyme activity by binding in the active site (competitive inhibition) or at another site on the enzyme (noncompetitive inhibition).

Example: Lactase, the enzyme that hydrolyzes lactose, operates at an optimal pH of 6.0 and an optimal temperature of 37 °C. Would the following changes increase, decrease, or have no effect on the activity of lactase?

a. lowering the pH to 2
b. raising the temperature above 65 °C
c. greatly increasing the amount of lactose available
d. introducing a competitive inhibitor of lactase

Answer: **a.** decrease **b.** decrease
c. increase **d.** decrease

Describing the Role of Cofactors (20.6)

- Metal ions can be cofactors that stabilize the active site of enzymes and are vital for catalysis.
- Many vitamins are coenzymes necessary for catalysis by some enzymes.

Example: Match the coenzymes (1) NAD^+, (2) FAD, (3) coenzyme A, and (4) THF with one of the following vitamins:

a. riboflavin (B_2) **b.** niacin (B_3)
c. folic acid (B_9) **d.** pantothenic acid (B_5)

Answer: **a.** (2) FAD **b.** (1) NAD^+
c. (4) THF **d.** (3) coenzyme A

UNDERSTANDING THE CONCEPTS

The chapter Sections to review are shown in parentheses at the end of each problem.

20.51 Chymosin is the main enzyme used in the production of cheese. It is a protease that causes the curdling of milk. (20.2)

During cheese making, the enzyme chymosin is added to curdle milk.

a. What type of reaction does chymosin catalyze?
b. Into which class of enzymes could chymosin be categorized?

20.52 Vitamin C, also called ascorbic acid, can act as a coenzyme and an antioxidant. (20.6)

Rose hips are rich in vitamin C.

a. Is vitamin C a water-soluble or a fat-soluble vitamin? How do you know?
b. How do antioxidants protect cells from damage?

Clinical Applications

20.53 Ethylene glycol (HO—CH_2—CH_2—OH) is a major component of antifreeze. If ingested, it is first converted to HOOC—CHO (oxoethanoic acid) and then to HOOC—COOH (oxalic acid), which is toxic. (20.1, 20.4)

Ethylene glycol is added to a radiator to prevent freezing and boiling.

a. What class of enzyme catalyzes both of the reactions of ethylene glycol?
b. The treatment for the ingestion of ethylene glycol is an intravenous solution of ethanol. How might this help prevent toxic levels of oxalic acid in the body?

20.54 Adults who are lactose intolerant cannot break down the disaccharide in milk products. To help digest dairy food, a product known as Lactaid can be ingested prior to eating dairy products. (20.1, 20.3)

The disaccharide lactose is present in milk products.

a. What is the name of the enzyme present in Lactaid, and what is the major class of this enzyme?
b. What might happen to the enzyme if the digestion product were stored in a warm area?
c. A label on a bottle of Lactaid recommends it is stored at room temperature (68 to 77 °F) and kept away from heat and moisture. Why should Lactaid be stored at cool temperatures and dry conditions?

ADDITIONAL PRACTICE PROBLEMS

20.55 Why do the cells in the body have so many enzymes? (20.1)

20.56 Are all the possible enzymes present at the same time in a cell? (20.1)

20.57 How are enzymes different from the catalysts used in chemistry laboratories? (20.1)

20.58 Why do enzymes function only under mild conditions? (20.1)

20.59 How does an enzyme change the activation energy for a reaction in a cell? (20.1)

20.60 Why do most enzyme-catalyzed reactions proceed quickly? (20.1)

20.61 How does the lock-and-key model explain that sucrose hydrolyzes sucrose but not lactose? (20.1)

20.62 How does the induced-fit model of enzyme action allow an enzyme to catalyze a reaction of a group of substrates? (20.1)

20.63 Indicate whether each of the following would be a substrate (S) or an enzyme (E): (20.2)
a. lactose **b.** lactase **c.** lipase
d. trypsin **e.** pyruvate **f.** transaminase

20.64 Indicate whether each of the following would be a substrate (S) or an enzyme (E): (20.2)
a. glucose **b.** hydrolase **c.** maleate isomerase
d. alanine **e.** amylose **f.** amylase

20.65 Give the substrate for each of the following enzymes: (20.2)
a. urease **b.** succinate dehydrogenase
c. aspartate transaminase **d.** phenylalanine hydroxylase

20.66 Give the substrate for each of the following enzymes: (20.2)
a. maltase **b.** fructose oxidase
c. phenolase **d.** sucrase

20.67 Predict the major class for each of the following enzymes: (20.2)
a. acyltransferase **b.** oxidase
c. lipase **d.** decarboxylase

20.68 Predict the major class for each of the following enzymes: (20.2)
a. cis–trans isomerase **b.** reductase
c. carboxylase **d.** peptidase

20.69 What is meant by the optimum temperature for an enzyme? (20.3)

20.70 What is meant by the optimum pH for an enzyme? (20.3)

20.71 Indicate how each of the following will affect an enzyme-catalyzed reaction if the enzyme has an optimum temperature of 37 °C and an optimum pH of 7.0: (20.3)
a. heating the reaction mixture to 100 °C
b. placing the reaction mixture in ice
c. adjusting the pH of the reaction mixture to pH 2.0

20.72 Indicate how each of the following will affect an enzyme-catalyzed reaction if the enzyme has an optimum temperature of 37 °C and an optimum pH of 8.0: (20.3)
a. decreasing the temperature of the reaction mixture from 37 °C to 15 °C
b. adjusting the pH of the reaction mixture to pH 5.0
c. adjusting the pH of the reaction mixture to pH 10.0

20.73 Indicate whether the enzyme is saturated or not saturated with substrate in each of the following conditions: (20.3)
a. adding more substrate does not increase the rate of reaction
b. doubling the substrate concentration doubles the rate of reaction

20.74 Indicate whether each of the following enzymes would be functional: (20.3)
a. pepsin, a digestive enzyme, at pH 2.0
b. an enzyme at 37 °C, if the enzyme is from a type of thermophilic bacteria that has an optimum temperature of 100 °C

20.75 What is an allosteric enzyme? (20.4)

20.76 Why can some regulator molecules speed up a reaction, whereas others slow it down? (20.4)

20.77 In feedback control, what type of regulator slows down the catalytic activity of the reaction series? (20.4)

20.78 Why is covalent modification used to regulate the activity of an enzyme? (20.4)

20.79 How does reversible inhibition differ from irreversible inhibition? (20.5)

20.80 How does competitive reversible inhibition differ from noncompetitive reversible inhibition? (20.5)

20.81 Match the type of inhibitor (**1** to **3**) with the following statements (**a** to **d**): (20.5)
1. competitive inhibitor
2. noncompetitive inhibitor
3. irreversible inhibitor

a. forms a covalent bond with an amino acid R group in the active site
b. has a structure similar to the substrate
c. the addition of more substrate reverses the inhibition
d. bonds to the surface of the enzyme, causing a change in the shape of the enzyme and active site

20.82 Match the type of inhibitor (**1** to **3**) with the following statements (**a** to **d**): (20.5)
1. competitive inhibitor
2. noncompetitive inhibitor
3. irreversible inhibitor

a. has a structure that is not similar to the substrate
b. the addition of more substrate does not reverse the inhibition, but the removal of the inhibitor by chemical reaction can return activity to the enzyme
c. the inhibition is permanent, and it cannot be reversed
d. competes with the substrate for the active site

20.83 Which of the following statements describe an enzyme that requires a cofactor? (20.6)
a. contains Mg^{2+} in the active site
b. has catalytic activity as a tertiary protein structure
c. requires folic acid for catalytic activity

20.84 Which of the following statements describes an enzyme that requires a cofactor? (20.6)
a. contains riboflavin (vitamin B_2)
b. has four subunits of polypeptide chains
c. requires Fe^{3+} in the active site for catalytic activity

Clinical Applications

20.85 The zymogen pepsinogen is produced in the gastric chief cells of the stomach. (20.4)
a. How and where does pepsinogen become the active form, pepsin?
b. Why are proteases such as pepsin produced in inactive forms?

20.86 Thrombin is an enzyme that helps produce blood clotting when an injury and bleeding occur. (20.4)
a. What would be the name of the zymogen of thrombin?
b. Why would the active form of thrombin be produced only when an injury to tissue occurs?

20.87 If a blood test indicates a high level of LDH and CK, what could be the cause? (20.4)

20.88 If a blood test indicates a high level of ALT, what could be the cause? (20.4)

20.89 Many drugs are competitive inhibitors of enzymes. When scientists design inhibitors to serve as drugs, why do you suppose they choose to design competitive inhibitors instead of noncompetitive inhibitors? (20.5)

20.90 When lead acts as a poison, it can do so by either replacing another ion (such as zinc) in the active site of an enzyme or it can react with the R group of cysteine to form covalent bonds. Which of these is irreversible and why? (20.5)

20.91 **a.** What type of an inhibitor is the antibiotic ampicillin? (20.5)
b. Why is ampicillin used to treat bacterial infections?

20.92 A gardener using the insecticide Parathion develops a headache, dizziness, nausea, blurred vision, excessive salivation, and muscle twitching. (20.5)
a. What might be happening to the gardener?
b. Why must humans be careful when using insecticides?

20.93 Match the following coenzymes (**1** to **3**) with their vitamins (**a** to **c**): (20.6)
1. NAD^+
2. thiamine pyrophosphate (TPP)
3. coenzyme A

a. pantothenic acid (B_5)
b. niacin (B_3)
c. thiamine (B_1)

20.94 Match the following coenzymes (**1** to **3**) with their vitamins (**a** to **c**): (20.6)
1. pyridoxal phosphate
2. tetrahydrofolate (THF)
3. FAD

a. folic acid
b. riboflavin (B_2)
c. pyridoxine

20.95 Why are only small amounts of vitamins needed in the cells when there are several enzymes that require coenzymes? (20.6)

20.96 Why is there a daily requirement for vitamins? (20.6)

20.97 Match each of the following symptoms or conditions (**1** to **3**) with a vitamin deficiency (**a** to **c**): (20.6)
1. night blindness
2. weak bone structure
3. pellagra

a. niacin
b. vitamin A
c. vitamin D

20.98 Match each of the following symptoms or conditions (**1** to **3**) with a vitamin deficiency (**a** to **c**): (20.6)
1. bleeding
2. anemia
3. scurvy

a. cobalamin
b. vitamin C
c. vitamin K

CHALLENGE PROBLEMS

The following problems are related to the topics in this chapter. However, they do not all follow the chapter order, and they require you to combine concepts and skills from several Sections. These problems will help you increase your critical thinking skills and prepare for your next exam.

20.99 Lactase is an enzyme that hydrolyzes lactose to glucose and galactose. (20.1, 20.2)
   a. What are the reactants and products of the reaction?
   b. Draw an energy diagram for the reaction with and without lactase.
   c. How does lactase make the reaction go faster?

20.100 Maltase is an enzyme that hydrolyzes maltose into two glucose molecules. (20.1, 20.2)
   a. What are the reactants and products of the reaction?
   b. Draw an energy diagram for the reaction with and without maltase.
   c. How does maltase make the reaction go faster?

20.101 What class of enzyme would catalyze each of the following reactions? (20.2)

   a. $CH_3-\overset{\overset{\displaystyle O}{\|}}{C}-H \longrightarrow CH_3-\overset{\overset{\displaystyle O}{\|}}{C}-OH$ 

   b. $\overset{+}{H_3N}-CH_2-\overset{\overset{\displaystyle O}{\|}}{C}-\overset{\overset{\displaystyle H}{|}}{N}-\overset{\overset{\displaystyle CH_3}{|}}{CH}-\overset{\overset{\displaystyle O}{\|}}{C}-O^- + H_2O \longrightarrow$ 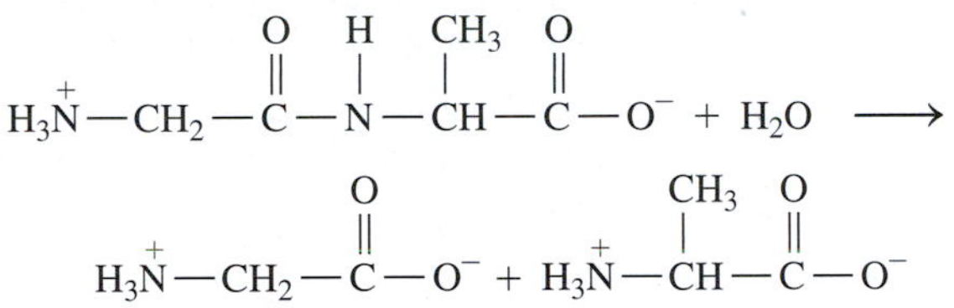

   $\overset{+}{H_3N}-CH_2-\overset{\overset{\displaystyle O}{\|}}{C}-O^- + \overset{+}{H_3N}-\overset{\overset{\displaystyle CH_3}{|}}{CH}-\overset{\overset{\displaystyle O}{\|}}{C}-O^-$

   c. $CH_3-CH=CH-CH_3 + H_2O \longrightarrow$ 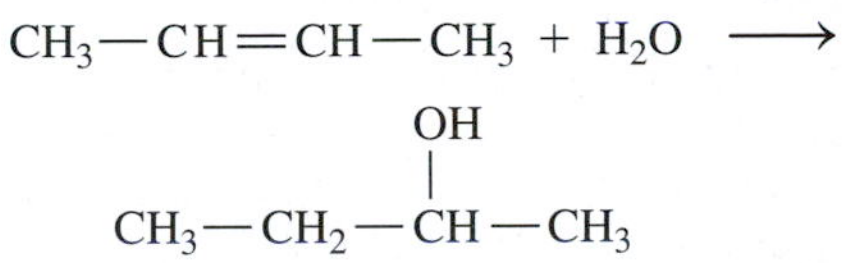

   $CH_3-CH_2-\overset{\overset{\displaystyle OH}{|}}{CH}-CH_3$

20.102 What class of enzyme would catalyze each of the following reactions? (20.2)

   a. $CH_3-\overset{\overset{\displaystyle O}{\|}}{C}-\overset{\overset{\displaystyle O}{\|}}{C}-OH \longrightarrow$ 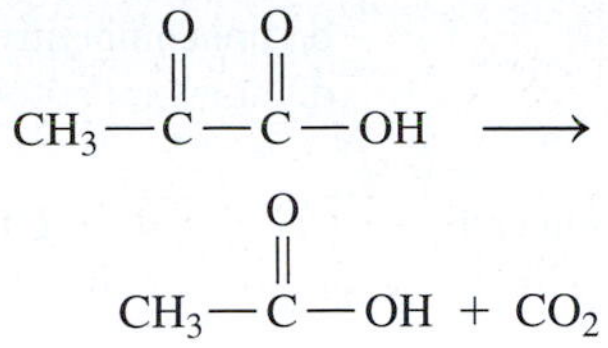

   $CH_3-\overset{\overset{\displaystyle O}{\|}}{C}-OH + CO_2$

   b. $CH_3-\overset{\overset{\displaystyle O}{\|}}{C}-\overset{\overset{\displaystyle O}{\|}}{C}-OH + CO_2 + ATP \longrightarrow$ 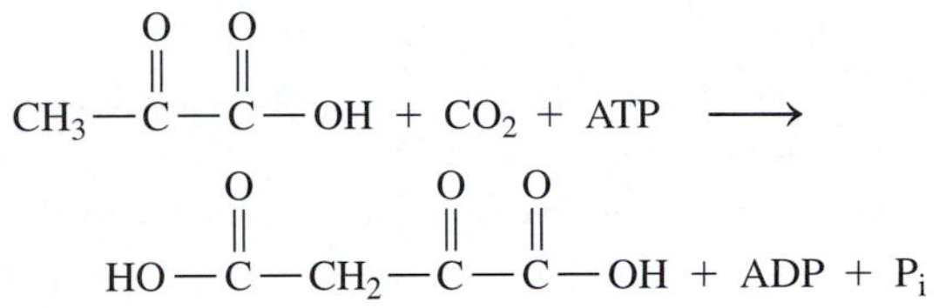

   $HO-\overset{\overset{\displaystyle O}{\|}}{C}-CH_2-\overset{\overset{\displaystyle O}{\|}}{C}-\overset{\overset{\displaystyle O}{\|}}{C}-OH + ADP + P_i$

   c. glucose-6-phosphate $\longrightarrow$ fructose-6-phosphate

Clinical Applications

20.103 Cadmium is a poisonous metal used in industries that produce batteries and plastics. Cadmium ions (Cd^{2+}) are inhibitors of the enzyme hexokinase. Increasing the concentration of glucose or ATP, the substrates of hexokinase, or Mg^{2+}, the cofactor of hexokinase, does not change the rate of the cadmium-inhibited reaction. Is cadmium a competitive or noncompetitive inhibitor? Explain. (20.5, 20.6)

20.104 Beano contains an enzyme that breaks down polysaccharides into smaller, more digestible sugars, which diminishes the intestinal gas formation that can occur after eating foods such as vegetables and beans. (20.1, 20.2, 20.5)
   a. The label says "contains alpha-galactosidase." What class of enzyme is this?
   b. What is the substrate for the enzyme?
   c. The directions indicate you should not cook with or heat Beano. Why?

ANSWERS

20.1 Chemical reactions can occur without enzymes, but the rates are too slow. Catalyzed reactions, which are many times faster, provide the amounts of products needed by the cell at a particular time.

20.3 a. $E + S \rightleftharpoons ES \longrightarrow EP \longrightarrow E + P$
   b. The active site is a region or pocket within the tertiary structure of an enzyme that accepts the substrate, aligns the substrate for reaction, and catalyzes the reaction.

20.5 a. (2) enzyme
   b. (1) enzyme–substrate complex
   c. (3) substrate

20.7 linkage specificity

20.9 a. oxidation–reduction
   b. transfer of a group from one substance to another
   c. hydrolysis (splitting) of molecules with the addition of water

20.11 a. hydrolase   **b.** oxidoreductase
   c. isomerase   **d.** transferase

20.13 a. lyase   **b.** transferase

20.15 a. succinate oxidase   **b.** glutamine synthetase
   c. alcohol dehydrogenase

20.17 pepsin, pH 2; sucrase, pH 6; trypsin, pH 8

20.19 a. The reaction will be slower.
   b. The reaction will slow or stop because the enzyme will be denatured at low pH.
   c. The reaction will slow or stop because the high temperature will denature the enzyme.
   d. The reaction will go faster as long as there are polypeptides to react.

20.21 When a regulator molecule binds to an allosteric site, the shape of the enzyme is altered, which makes the active site more reactive or less reactive, and thereby increases or decreases the rate of the reaction.

20.23 In feedback control, the product binds to the first enzyme in a series and changes the shape of the active site. If the active site can no longer bind the substrate effectively, the reaction will stop.

20.25 In order to conserve material and energy and to prevent clotting in the bloodstream, thrombin is produced from prothrombin only at a wound site.

20.27 a. (3) covalent modification   **b.** (2) a zymogen
   c. (1) an allosteric enzyme

20.29 a. prostate-specific antigen   **b.** creatine kinase
   c. cholinesterase

20.31 Isoenzymes are slightly different forms of an enzyme that catalyze the same reaction in different organs and tissues of the body.

20.33 A doctor might run tests for the enzymes CK and LDH to determine if the patient had a heart attack.

20.35 **a.** competitive **b.** noncompetitive
c. competitive **d.** noncompetitive
e. competitive

20.37 **a.** methanol, $CH_3—OH$; ethanol, $CH_3—CH_2—OH$
b. Ethanol competes with methanol for the active site in the enzyme.
c. Ethanol is a competitive inhibitor.

20.39 **a.** active with a cofactor **b.** active with a cofactor
c. does not require a cofactor

20.41 **a.** pantothenic acid (vitamin B_5) **b.** folic acid
c. niacin (vitamin B_3)

20.43 **a.** vitamin D (cholecalciferol) **b.** vitamin C (ascorbic acid)
c. vitamin B_3 (niacin)

20.45 Vitamin B_6 is a water-soluble vitamin, which means that each day, any excess of vitamin B_6 is eliminated from the body.

20.47 **a.** 20 ppm H_2 **b.** 40 ppm H_2

20.49 Heating the milk in the hot chocolate will denature the protein enzyme, lactase. Because the enzyme is no longer active, Noah will suffer from the symptoms of lactose intolerance.

20.51 **a.** As a protease, chymosin catalyzes reactions in which the peptide bonds in a protein are hydrolyzed.
b. Chymosin fits into the hydrolase category of enzymes.

20.53 **a.** oxidoreductase
b. Ethanol competes with ethylene glycol for the active site of the alcohol dehydrogenase enzyme, saturating the active site, which allows ethylene glycol to be removed from the body without producing oxalic acid.

20.55 The many different reactions that take place in cells require different enzymes because enzymes react with only a certain type of substrate.

20.57 Enzymes are catalysts that are usually proteins and that function only at mild temperature and pH. Catalysts used in chemistry laboratories are usually inorganic materials that function at high temperatures and in strongly acidic or basic conditions.

20.59 An enzyme lowers the activation energy for a reaction.

20.61 Sucrose fits the shape of the active site in sucrose, but lactose does not.

20.63 **a.** S **b.** E **c.** E **d.** E **e.** S **f.** E

20.65 **a.** urea **b.** succinate
c. aspartate **d.** phenylalanine

20.67 **a.** transferase **b.** oxidoreductase
c. hydrolase **d.** lyase

20.69 The optimum temperature for an enzyme is the temperature at which the enzyme is fully active and most effective.

20.71 **a.** The rate of catalysis will slow and stop as a high temperature denatures the enzyme.
b. The rate of the catalyzed reaction will slow as temperature is lowered.
c. The enzyme will not be functional at pH 2.0.

20.73 **a.** saturated **b.** unsaturated

20.75 An allosteric enzyme contains sites for regulators that alter the structure of the active site of an enzyme, which speeds up or slows down the rate of the catalyzed reaction.

20.77 In feedback control, a negative regulator slows down the catalytic activity.

20.79 In reversible inhibition, the inhibitor can dissociate from the enzyme, whereas in irreversible inhibition, the inhibitor forms a strong covalent bond with the enzyme and does not dissociate. Irreversible inhibitors act as poisons to enzymes.

20.81 **a.** (3) irreversible inhibitor **b.** (1) competitive inhibitor
c. (1) competitive inhibitor **d.** (2) noncompetitive inhibitor

20.83 **a.** requires a cofactor **b.** does not require a cofactor
c. requires a cofactor (coenzyme)

20.85 **a.** When pepsinogen enters the stomach, the low pH cleaves a peptide from its protein chain to form pepsin.
b. An active protease would digest the proteins of the stomach rather than the proteins in foods.

20.87 A heart attack may be the cause.

20.89 When designing an inhibitor, the substrate of an enzyme usually is known even if the structure of the enzyme is not. It is easier to design a molecule that resembles the substrate (competitive) than to find an inhibitor that binds to a second site on an enzyme (noncompetitive).

20.91 **a.** Antibiotics such as ampicillin are irreversible inhibitors.
b. Antibiotics inhibit enzymes needed to form cell walls in bacteria, not humans.

20.93 **a.** (3) coenzyme A **b.** (1) NAD^+
c. (2) thiamine pyrophosphate (TPP)

20.95 A vitamin combines with an enzyme only when the enzyme and coenzyme are needed to catalyze a reaction. When the enzyme is not needed, the vitamin dissociates for use by other enzymes in the cell.

20.97 **a.** niacin, (3) pellagra
b. vitamin A, (1) night blindness
c. vitamin D, (2) weak bone structure

20.99 **a.** The reactants are lactose and water and the products are glucose and galactose.
b.

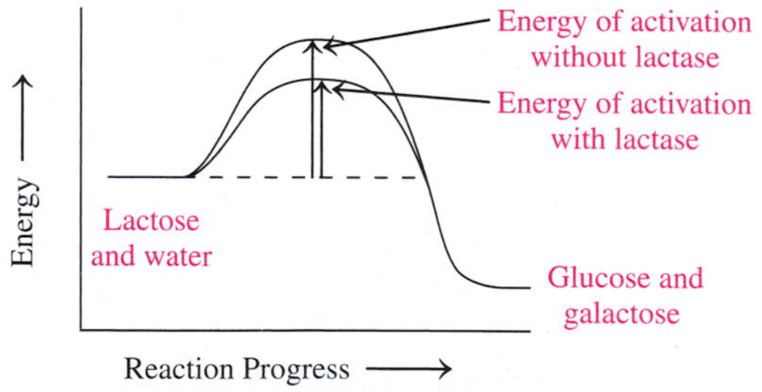
Energy of activation without lactase
Energy of activation with lactase
Energy
Lactose and water
Glucose and galactose
Reaction Progress

c. By lowering the energy of activation, the enzyme furnishes a lower energy pathway by which the reaction can take place.

20.101 **a.** oxidoreductase **b.** hydrolase **c.** lyase

20.103 Cadmium would be a noncompetitive inhibitor because increasing the substrate and cofactor has no effect on the rate. This implies that the cadmium is binding to another site on the enzyme.

21

Nucleic Acids and Protein Synthesis

Ellen found a pea-sized lump in her breast. A needle biopsy confirms that Ellen has breast cancer. She undergoes a lumpectomy, during which the surgeon removes the tumor along with a small margin of surrounding normal tissue. The surgeon also makes an incision under her arm and removes the sentinel lymph node. Because cancer cells are found in the sentinel node, more lymph nodes are removed. He sends the excised tumor and lymph nodes to Lisa, a histology technician.

Lisa prepares the tissue sample to be viewed by a pathologist. To prepare the tissue, Lisa cuts the tissue into very thin sections, about 0.001 mm, and mounts them onto microscope slides. She treats the tissue on the slides with a dye to stain the cells that enables the pathologist to distinguish abnormal cells more easily.

When a person's DNA (deoxyribonucleic acid) is damaged, mutations that promote the abnormal cell growth found in cancer may result. Cancer, as well as genetic diseases, can be a result of mutations caused by environmental and hereditary factors.

CAREER

Histology Technician

A histology technician studies the microscopic makeup of tissues, cells, and bodily fluids with the purpose of detecting and identifying the presence of a specific disease. They determine blood types and the concentrations of drugs and other substances in the blood. Sample preparation is a critical component of a histologist's job. They cut tissue samples, using specialized equipment, into extremely thin sections, which are then mounted and stained using various dyes. The dyes provide contrast for the cells to be viewed and help highlight any abnormalities that may exist. Utilization of various dyes requires the histologist be familiar with solution preparation and the handling of potentially hazardous chemicals.

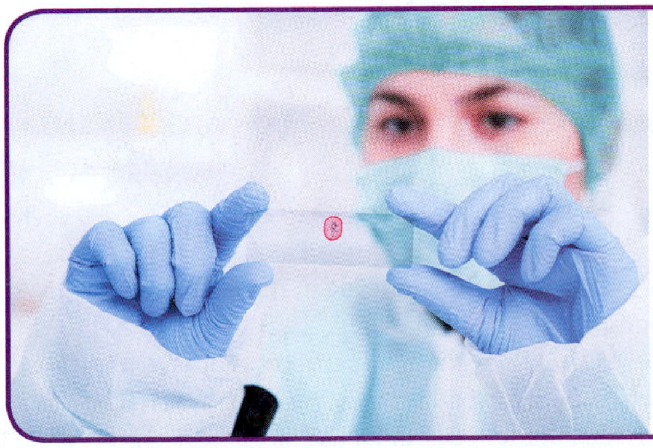

CLINICAL UPDATE

Ellen's Medical Treatment Following Breast Cancer Surgery

You can learn more about the type of breast cancer that Ellen has, as well as some of the drugs that are available to treat this type of breast cancer, by reading the **CLINICAL UPDATE** **Ellen's Medical Treatment Following Breast Cancer Surgery**, pages 754–755.

LOOKING AHEAD

21.1 Components of Nucleic Acids

LEARNING GOAL Describe the bases and ribose sugars that make up the nucleic acids DNA and RNA.

Nucleic acids are large molecules, found in the nuclei of cells, that store information and direct activities for cellular growth and reproduction. There are two closely related types of nucleic acids: *deoxyribonucleic acid* (**DNA**) and *ribonucleic acid* (**RNA**). Deoxyribonucleic acid, the genetic material in the nucleus of a cell, contains all the information needed for the development of a complete living organism. Ribonucleic acid interprets the genetic information in DNA for the synthesis of protein.

Both DNA and RNA are composed of smaller units known as *nucleotides*, linked together in unbranched chains. Each nucleotide has three components: a base that contains nitrogen, a five-carbon sugar, and a phosphate group (see **FIGURE 21.1**). A DNA molecule may contain several million nucleotides; smaller RNA molecules may contain up to several thousand.

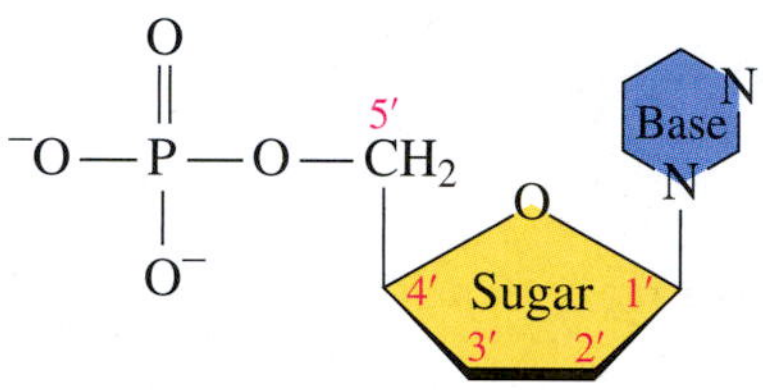

Base
Sugar
5′
4′
3′
2′
1′

FIGURE 21.1 ▶ The general structure of a nucleotide includes a nitrogen-containing base, a sugar, and a phosphate group.

Q In a nucleotide, what types of groups are bonded to a five-carbon sugar?

Bases

The nitrogen-containing **bases** in nucleic acids are derivatives of the heterocyclic amines *pyrimidine* or *purine*. A pyrimidine has a single ring with two nitrogen atoms, and a purine has two rings each with two nitrogen atoms. They are basic because the nitrogen atoms are H^+ acceptors. In DNA, the pyrimidine bases with single rings are cytosine (C) and thymine (T), and the purine bases with double rings are adenine (A) and guanine (G). RNA contains the same bases, except thymine (5-methyluracil) is replaced by uracil (U) (see **FIGURE 21.2**).

Pyrimidine Bases in Nucleic Acids

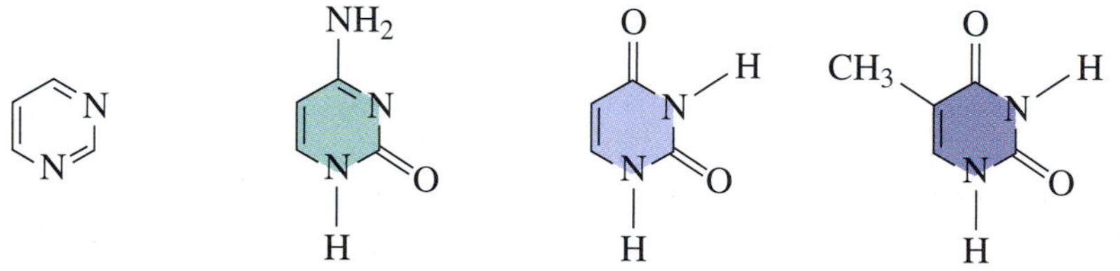

Pyrimidine
Cytosine (C)
(DNA and RNA)
Uracil (U)
(RNA only)
Thymine (T)
(DNA only)

Purine Bases in Nucleic Acids

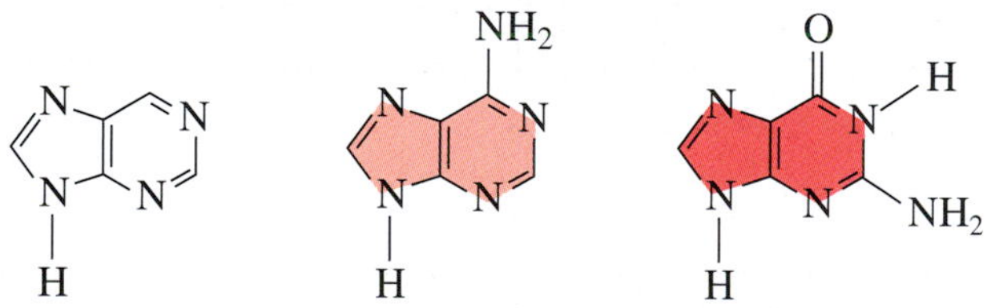

Purine
Adenine (A)
(DNA and RNA)
Guanine (G)
(DNA and RNA)

FIGURE 21.2 ▶ DNA contains the bases A, G, C, and T; RNA contains A, G, C, and U.

Q Which bases are found in DNA?

TEST

Try Practice Problems 21.1 to 21.4

Pentose Sugars in Nucleic Acids

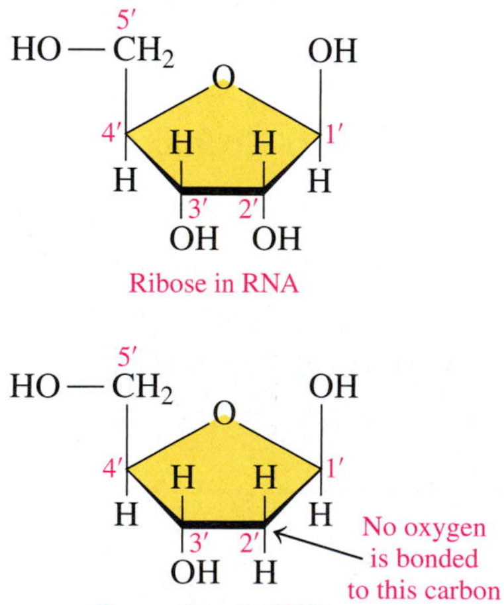

Ribose in RNA
Deoxyribose in DNA
No oxygen is bonded to this carbon

The five-carbon pentose sugar found in RNA is ribose and in DNA, deoxyribose.

Pentose Sugars

In RNA, the five-carbon sugar is *ribose*, which gives the letter R in the abbreviation RNA. The atoms in the pentose sugars are numbered with primes (1′, 2′, 3′, 4′, and 5′) to differentiate them from the atoms in the bases. In DNA, the five-carbon sugar is *deoxyribose*, which is similar to ribose except that there is no hydroxyl group (—OH) on C2′. The *deoxy* prefix means "without oxygen" and provides the letter D in DNA.

Nucleosides and Nucleotides

A **nucleoside** is composed of one of the nitrogen-containing bases and one of the sugars, either ribose or deoxyribose. A nitrogen atom of the base is connected by a β-*N*-glycosidic bond to the C1′ of the sugar. For example, the combination of adenine, a purine, and ribose forms the nucleoside adenosine.

Sugar + Base ⟶ Nucleoside + H_2O

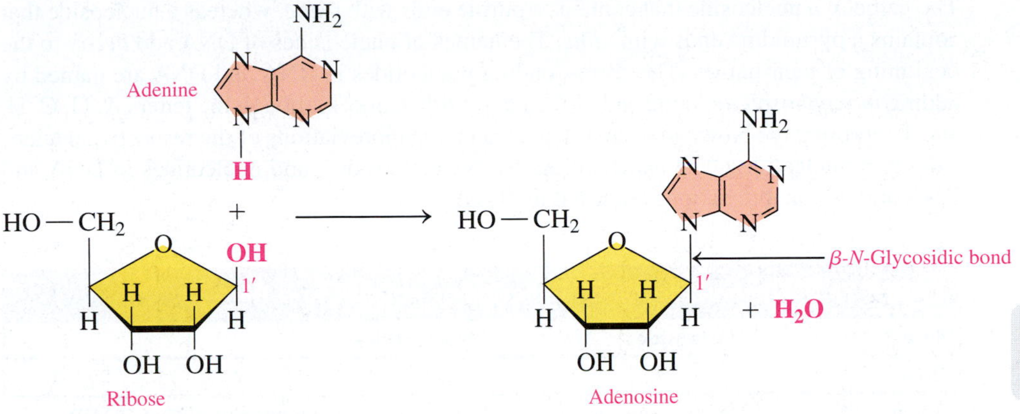

NH2
Adenine
H
+
OH
1′
HO—CH2
O
H H
H H
OH OH
Ribose
HO—CH2
β-N-Glycosidic bond
+ H2O
Adenosine

A base forms a β-*N*-glycosidic bond with a pentose sugar to form a nucleoside and water.

Nucleotides are produced when the C5′ hydroxyl group of ribose or deoxyribose in a nucleoside forms a phosphate ester. All the nucleotides in RNA and DNA are shown in **FIGURE 21.3**.

Phosphate + Nucleoside ⟶ Nucleotide + H_2O

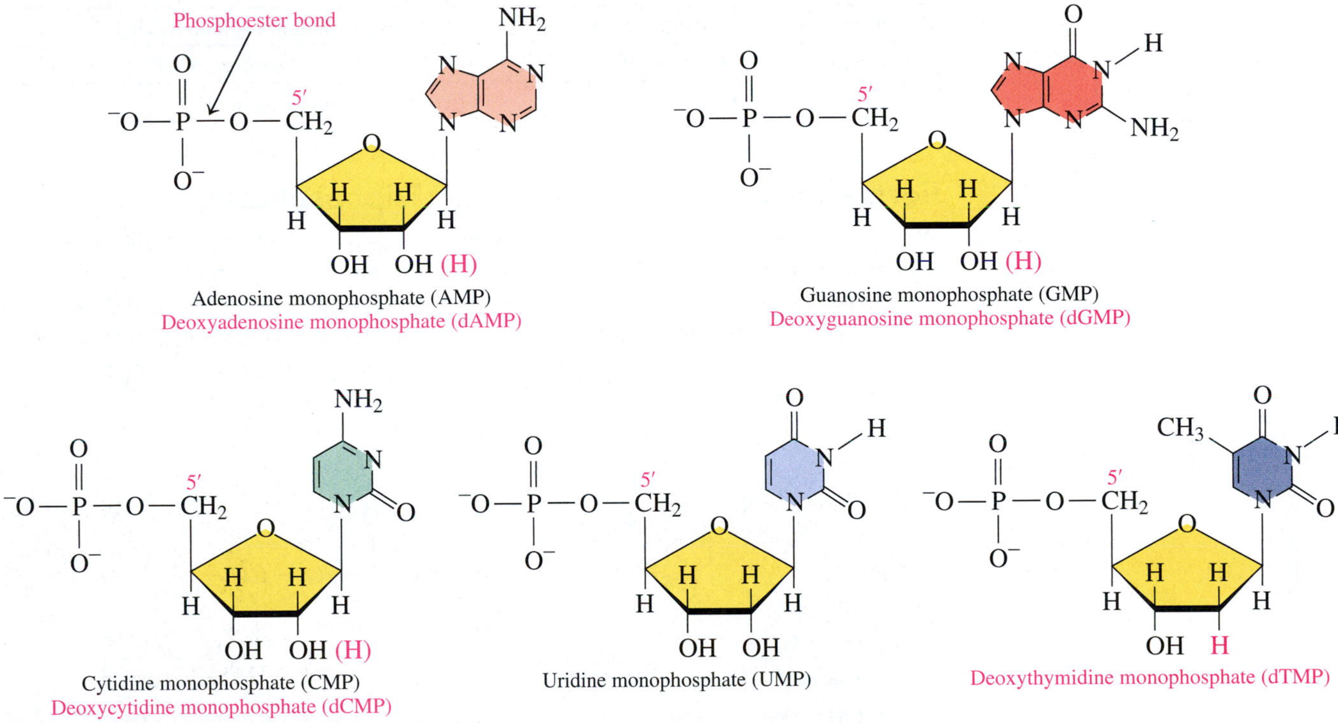

Phosphoester bond
5′
OH OH (H)
Adenosine monophosphate (AMP)
Deoxyadenosine monophosphate (dAMP)
Guanosine monophosphate (GMP)
Deoxyguanosine monophosphate (dGMP)
Cytidine monophosphate (CMP)
Deoxycytidine monophosphate (dCMP)
Uridine monophosphate (UMP)
CH3
OH H
Deoxythymidine monophosphate (dTMP)

FIGURE 21.3 ▶ The nucleotides of RNA (shown in black) are similar to those of DNA (shown in magenta), except in DNA the sugar is deoxyribose and deoxythymidine replaces uridine.

Q What are two differences in the nucleotides of RNA and DNA?

TABLE 21.1 summarizes the components in DNA and RNA.

TABLE 21.1 Components in DNA and RNA

Component	DNA	RNA
Bases	A, G, C, and T	A, G, C, and U
Sugar	Deoxyribose	Ribose
Nucleoside	Base + deoxyribose	Base + ribose
Nucleotide	Base + deoxyribose + phosphate	Base + ribose + phosphate
Nucleic Acid	Linear chain of deoxyribose nucleotides	Linear chain of ribose nucleotides

Naming Nucleosides and Nucleotides

The name of a nucleoside that contains a purine ends with *osine*, whereas a nucleoside that contains a pyrimidine ends with *idine*. The names of nucleosides of DNA add *deoxy* to the beginning of their names. The corresponding nucleotides in RNA and DNA are named by adding *monophosphate* to the end of the nucleoside name. Although the letters A, G, C, U, and T represent the bases, they are often used in the abbreviations of the respective nucleosides and nucleotides. The names of the bases, nucleosides, and nucleotides in DNA and RNA and their abbreviations are listed in **TABLE 21.2**.

ENGAGE

Is uridine monophosphate (UMP) found in DNA or RNA?

TABLE 21.2 Nucleosides and Nucleotides in DNA and RNA

Base	Nucleosides	Nucleotides
DNA		
Adenine (A)	Deoxyadenosine (A)	Deoxyadenosine monophosphate (dAMP)
Guanine (G)	Deoxyguanosine (G)	Deoxyguanosine monophosphate (dGMP)
Cytosine (C)	Deoxycytidine (C)	Deoxycytidine monophosphate (dCMP)
Thymine (T)	Deoxythymidine (T)	Deoxythymidine monophosphate (dTMP)
RNA		
Adenine (A)	Adenosine (A)	Adenosine monophosphate (AMP)
Guanine (G)	Guanosine (G)	Guanosine monophosphate (GMP)
Cytosine (C)	Cytidine (C)	Cytidine monophosphate (CMP)
Uracil (U)	Uridine (U)	Uridine monophosphate (UMP)

Formation of Nucleoside Di- and Triphosphates

When one or two more phosphate groups add to a nucleotide, a nucleoside di- or triphosphate is produced. For example, adding one phosphate group to AMP gives ADP (*adenosine diphosphate*). Adding another phosphate group to ADP gives ATP (*adenosine triphosphate*) (see **FIGURE 21.4**). Of the triphosphates, ATP is of particular interest because it is the major source of energy for most energy-requiring activities in the cell.

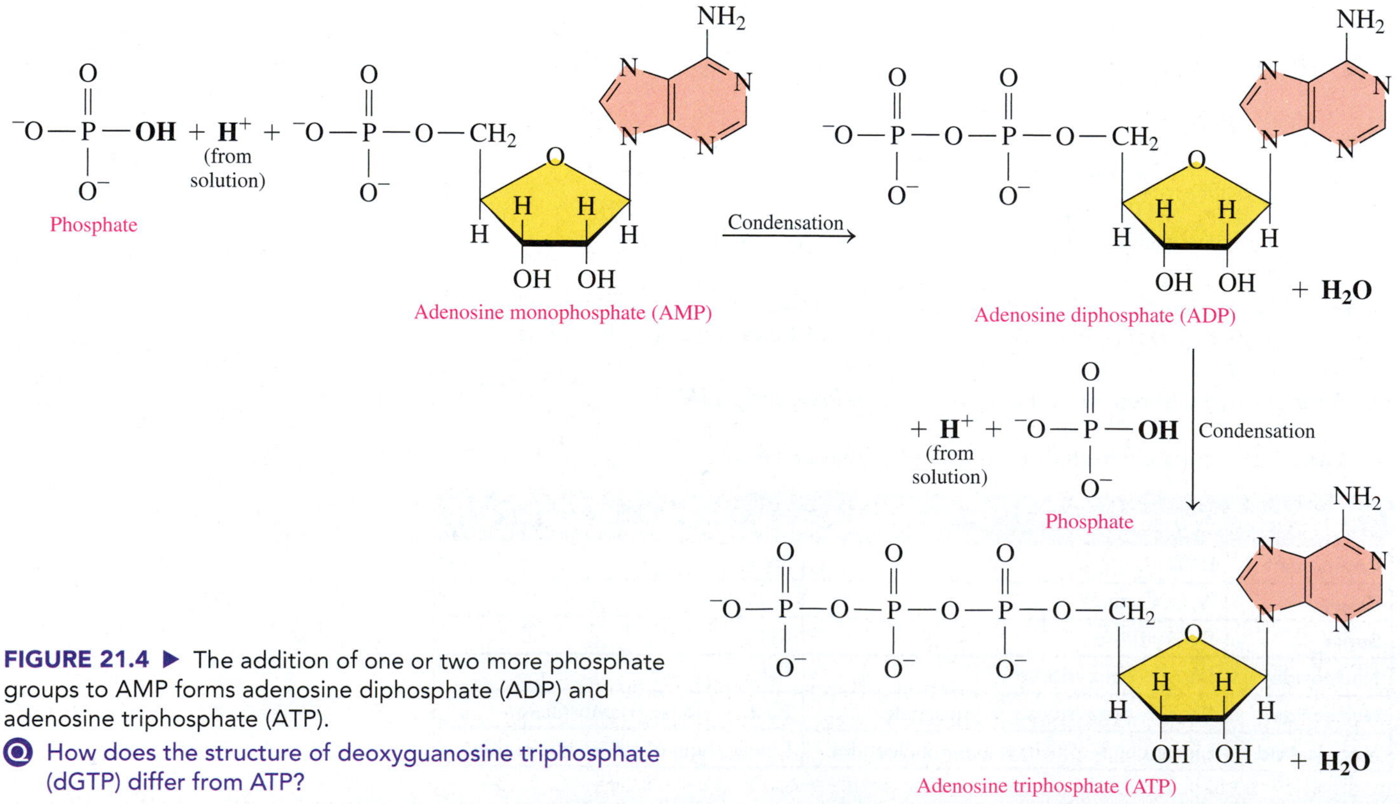

Phosphate
(from solution)
Adenosine monophosphate (AMP)
Condensation
Adenosine diphosphate (ADP)
$+ H_2O$
(from solution)
Phosphate
Condensation
Adenosine triphosphate (ATP)
$+ H_2O$

FIGURE 21.4 ▶ The addition of one or two more phosphate groups to AMP forms adenosine diphosphate (ADP) and adenosine triphosphate (ATP).

Q How does the structure of deoxyguanosine triphosphate (dGTP) differ from ATP?

▶ SAMPLE PROBLEM 21.1 Nucleotides

TRY IT FIRST

For each of the following nucleotides, identify the components and whether the nucleotide is found in DNA only, RNA only, or both DNA and RNA:

a. deoxyguanosine monophosphate (dGMP) **b.** adenosine monophosphate (AMP)

SOLUTION

a. This nucleotide of deoxyribose, guanine, and a phosphate group is only found in DNA.
b. This nucleotide of ribose, adenine, and a phosphate group is only found in RNA.

STUDY CHECK 21.1

a. What is the name and abbreviation of the DNA nucleotide of cytosine?
b. What is the name and abbreviation of the nucleotide that contains a base that is only found in RNA?

ANSWER

a. deoxycytidine monophosphate (dCMP) **b.** uridine monophosphate (UMP)

TEST

Try Practice Problems 21.5 to 21.14

PRACTICE PROBLEMS

21.1 Components of Nucleic Acids

21.1 Identify each of the following bases as a purine or a pyrimidine:
a. thymine **b.**

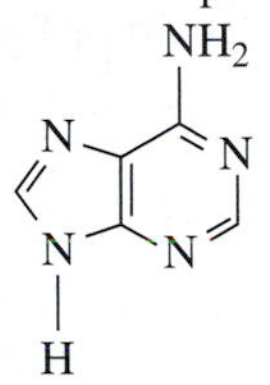

NH_2
N
N
N
N
H

21.2 Identify each of the following bases as a purine or a pyrimidine:
a. guanine **b.**

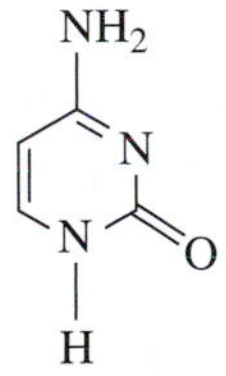

NH_2
N
N
O
H

21.3 Identify each of the bases in problem 21.1 as a component of DNA only, RNA only, or both DNA and RNA.

21.4 Identify each of the bases in problem 21.2 as a component of DNA only, RNA only, or both DNA and RNA.

21.5 What are the names and abbreviations of the four nucleotides in DNA?

21.6 What are the names and abbreviations of the four nucleotides in RNA?

21.7 Identify each of the following as a nucleoside or a nucleotide:
a. adenosine **b.** deoxycytidine
c. uridine **d.** cytidine monophosphate

21.8 Identify each of the following as a nucleoside or a nucleotide:
a. deoxythymidine **b.** guanosine
c. deoxyadenosine monophosphate **d.** uridine monophosphate

21.9 State whether each of the following components is present in DNA only, RNA only, or both DNA and RNA:
a. phosphate **b.** ribose
c. deoxycytidine monophosphate **d.** UMP

21.10 State whether each of the following components is present in DNA only, RNA only, or both DNA and RNA:
a. deoxyribose **b.** guanosine monophosphate
c. uracil **d.** adenine

Clinical Applications

21.11 In the genetic disease *adenosine deaminase deficiency*, there is an accumulation of adenosine. Draw the condensed structural formula for deoxyadenosine monophosphate.

21.12 In the genetic disease *uridine monophosphate synthase deficiency*, symptoms include anemia, cardiac malformations, and infections. Draw the condensed structural formula for uridine monophosphate.

21.13 In *Lesch–Nyhan syndrome*, there is a deficiency in the enzyme guanine transferase, which catalyzes the conversion of guanine to guanosine monophosphate. Draw the condensed structural formula for guanosine monophosphate.

21.14 A deficiency of the enzyme adenine transferase causes a lack of adenine for purine synthesis and a high level of adenine in the urine. Draw the condensed structural formula for adenosine monophosphate.

21.2 Primary Structure of Nucleic Acids

LEARNING GOAL Describe the primary structures of RNA and DNA.

The **nucleic acids** are unbranched chains of many nucleotides in which the 3′ hydroxyl group of the sugar in one nucleotide bonds to the phosphate group on the 5′ carbon atom in the sugar of the next nucleotide. This connection between a phosphate and sugars in adjacent nucleotides is referred to as a **phosphodiester linkage**. As more nucleotides are added, a backbone forms that consists of alternating sugar and phosphate groups. The bases, which are attached to each sugar, extend out from the sugar–phosphate backbone.

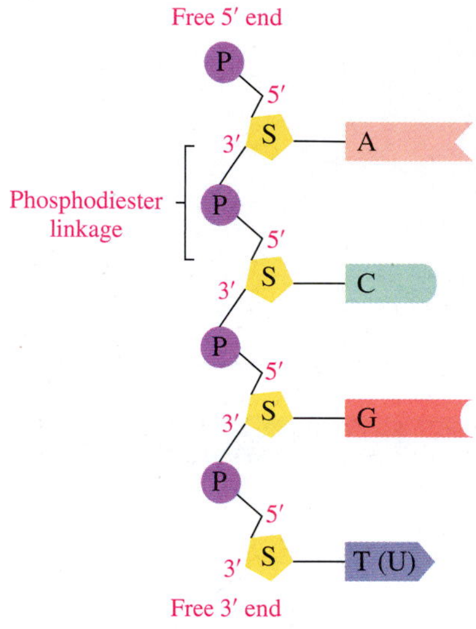

Free 5′ end
P
5′
3′
S
A
Phosphodiester linkage
P
5′
S
C
3′
P
5′
3′
S
G
P
5′
3′
S
T (U)
Free 3′ end

In the primary structure of nucleic acids, each sugar in a sugar–phosphate backbone is attached to a base.

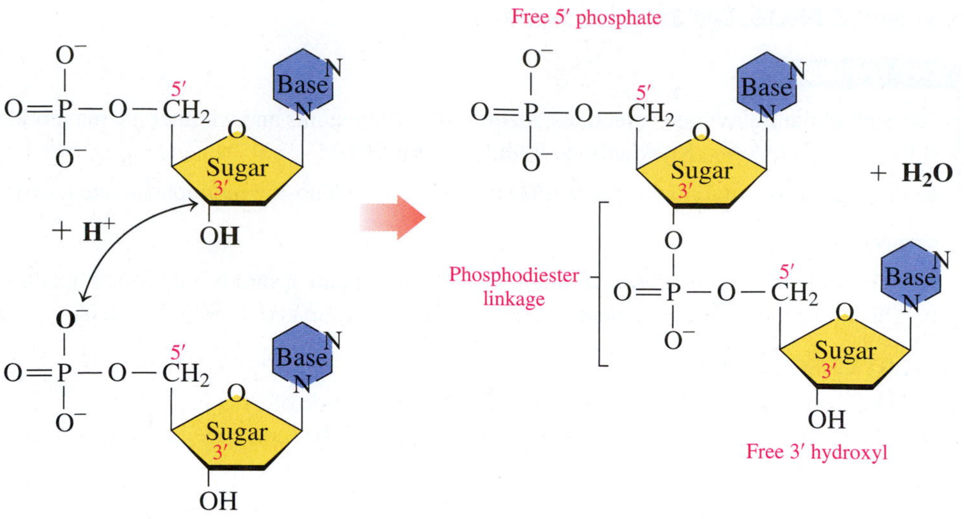

Free 5′ phosphate
Base
Sugar
+ H+
OH
+ H2O
Phosphodiester linkage
Free 3′ hydroxyl

A phosphodiester linkage forms between the 3′ hydroxyl group in the sugar of one nucleotide and the phosphate group on the 5′ carbon atom in the sugar of the next nucleotide.

Each nucleic acid has its own unique sequence of bases, which is known as its **primary structure**. It is this sequence of bases that carries the genetic information. In any nucleic acid, the sugar at one end has an unreacted or free 5′ phosphate terminal end, and the sugar at the other end has a free 3′ hydroxyl group.

A nucleic acid sequence is read from the sugar with the free 5′ phosphate to the sugar with the free 3′ hydroxyl group. The order of nucleotides in a nucleic acid is often written using the letters of the bases. For example, the nucleotide sequence starting with adenine (free 5′ phosphate end) in the section of RNA shown in **FIGURE 21.5** is A C G U.

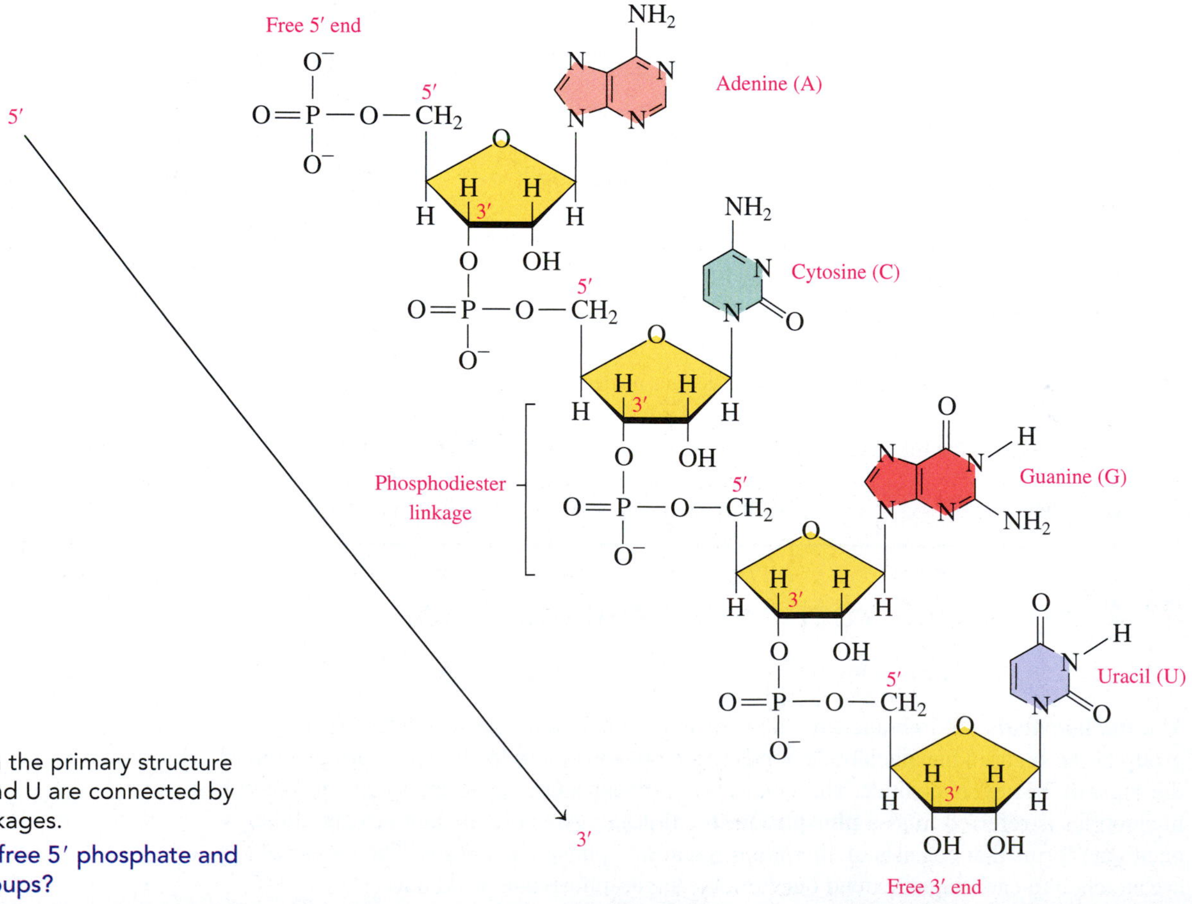

Free 5′ end
5′
Adenine (A)
Cytosine (C)
Phosphodiester linkage
Guanine (G)
Uracil (U)
3′
Free 3′ end

FIGURE 21.5 ▶ In the primary structure of RNA, A, C, G, and U are connected by phosphodiester linkages.

Q Where are the free 5′ phosphate and 3′ hydroxyl groups?

▶SAMPLE PROBLEM 21.2 Bonding of Nucleotides

TRY IT FIRST

Draw the condensed structural formula for an RNA dinucleotide formed by joining the 3′ OH group of adenosine monophosphate and the 5′ phosphate group of cytidine monophosphate.

SOLUTION

The dinucleotide is drawn by connecting the 3′ hydroxyl group on the adenosine monophosphate with the 5′ phosphate group on the cytidine monophosphate.

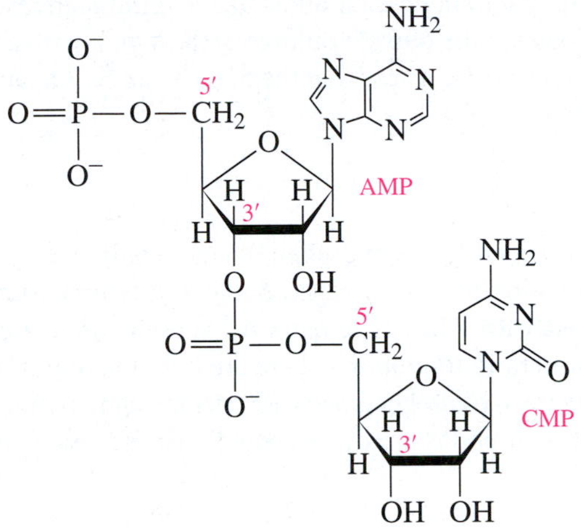
AMP
CMP

ENGAGE

How many phosphodiester linkages are in a dinucleotide?

STUDY CHECK 21.2

What type of linkage connects the two nucleotides in Sample Problem 21.2?

ANSWER

A phosphodiester linkage connects the 3′ OH group in the ribose of AMP and the 5′ carbon in the ribose of CMP.

TEST

Try Practice Problems 21.15 to 21.22

PRACTICE PROBLEMS

21.2 Primary Structure of Nucleic Acids

21.15 What nucleic acid subunits are connected in a phosphodiester linkage in a polynucleotide?

21.16 What is the difference between the 3′ end and the 5′ end of a polynucleotide chain?

21.17 What components join together to form the backbone of a nucleic acid?

21.18 What component in the backbone of a nucleic acid is bonded to a nitrogen base?

21.19 What component in a nucleic acid determines the free 5′ end?

21.20 What component in a nucleic acid determines the free 3′ end?

21.21 Draw the condensed structural formula for the dinucleotide 5′ G C 3′ that would be in RNA.

21.22 Draw the condensed structural formula for the dinucleotide 5′ A T 3′ that would be in DNA.

21.3 DNA Double Helix and Replication

LEARNING GOAL Describe the double helix of DNA; describe the process of DNA replication.

During the 1940s, biologists determined that the bases in DNA from a variety of organisms had a specific relationship: the amount of adenine (A) was equal to the amount of thymine (T), and the amount of guanine (G) was equal to the amount of cytosine (C) (see **TABLE 21.3**). Eventually, scientists determined that adenine is paired (1:1) with thymine, and guanine is paired (1:1) with cytosine.

Number of purine molecules = Number of pyrimidine molecules
Adenine (A) = Thymine (T)
Guanine (G) = Cytosine (C)

TABLE 21.3 Percentages of Bases in the DNA of Selected Organisms

Organism	%A	%T	%G	%C
Human	30	30	20	20
Chicken	28	28	22	22
Salmon	28	28	22	22
Corn (maize)	27	27	23	23

In 1953, James Watson and Francis Crick proposed that DNA was a **double helix** that consisted of two polynucleotide strands winding about each other like a spiral staircase. The sugar–phosphate backbones are analogous to the outside railings of the stairs, with the bases arranged like steps along the inside. One strand goes from the 5′ to 3′ direction, and the other strand goes in the 3′ to 5′ direction.

Complementary Base Pairs

ENGAGE

Why does it require more energy to separate the strands of a DNA double helix containing GC pairs than it does to separate the strands containing AT pairs?

Each of the bases along one polynucleotide strand forms hydrogen bonds to only one specific base on the opposite DNA strand. Adenine forms hydrogen bonds to thymine only, and guanine bonds to cytosine only (see **FIGURE 21.6**). The pairs AT and GC are called **complementary base pairs**. Because of structural limitations, there are only two kinds of stable base pairs. The bases that bind utilizing two hydrogen bonds are adenine and thymine, and the bases that bind utilizing three hydrogen bonds are cytosine and guanine. *No other*

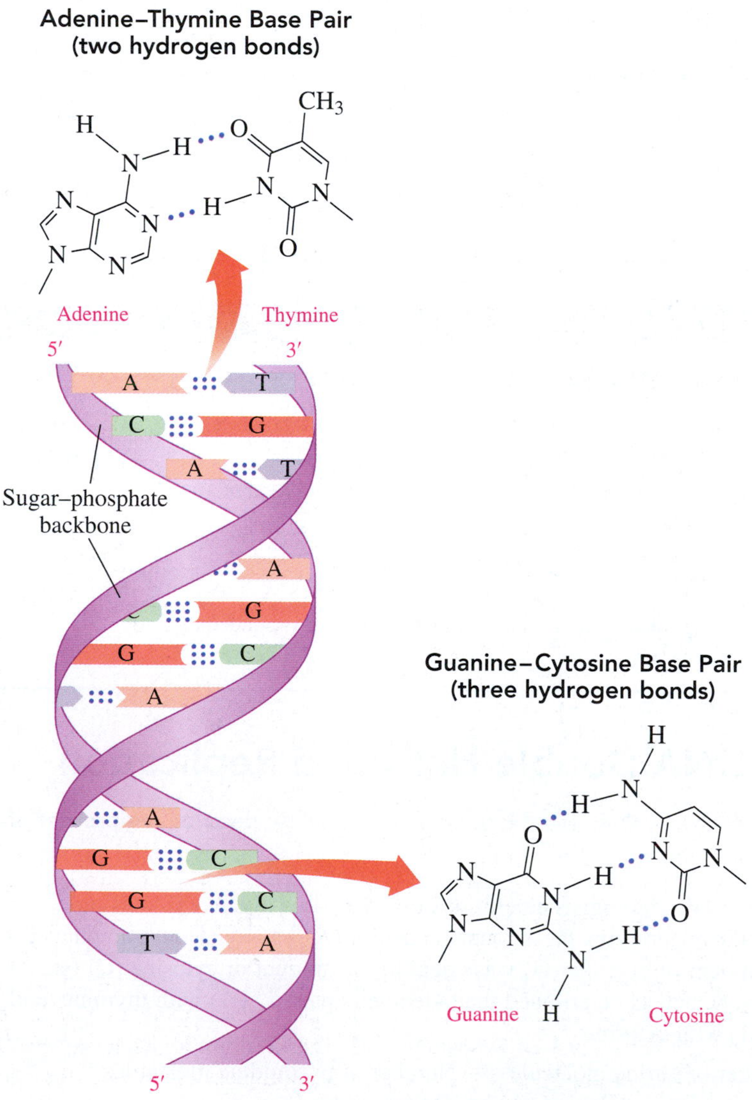

Adenine–Thymine Base Pair
(two hydrogen bonds)
Adenine
Thymine
5′
3′
Sugar–phosphate backbone
Guanine–Cytosine Base Pair
(three hydrogen bonds)
Guanine
Cytosine
5′
3′

FIGURE 21.6 ▶ In the model shown, the sugar–phosphate backbone is represented by a ribbon with hydrogen bonds between complementary base pairs.

Q Why are GC base pairs more stable than AT base pairs?

stable base pairs occur. For example, adenine does not form hydrogen bonds with cytosine or guanine; cytosine does not form hydrogen bonds with adenine or thymine. This explains why DNA has equal amounts of A and T bases and equal amounts of G and C.

▶SAMPLE PROBLEM 21.3 Complementary Base Pairs

TRY IT FIRST

Write the complementary base sequence for the following segment of a strand of DNA: 5′ A C G A T C T 3′

SOLUTION

The complementary base pairs are AT and GC. The complementary strand is written in the opposite direction, from the 3′ end to the 5′.

Original segment of DNA:	5′	A	C	G	A	T	C	T	3′
		⋮	⋮	⋮	⋮	⋮	⋮	⋮	
Complementary segment:	3′	T	G	C	T	A	G	A	5′

STUDY CHECK 21.3

a. What sequence of bases is complementary to a DNA segment with a base sequence of 5′ G G T T A A C C 3′?

b. What sequence of bases is complementary to a DNA segment with a base sequence of 5′ G A T T A C A C G 3′?

ANSWER

a. 3′ C C A A T T G G 5′ **b.** 3′ C T A A T G T G C 5′

CORE CHEMISTRY SKILL

Writing the Complementary DNA Strand

TEST

Try Practice Problems 21.23 to 21.28

DNA Replication

The function of DNA in cells of animals and plants as well as in bacteria is to preserve genetic information. As cells divide, copies of DNA are produced that transfer genetic information to the new cells.

In DNA **replication**, the strands in the original or *parent* DNA molecule separate to allow the synthesis of complementary DNA strands. The process begins with the unwinding of a portion of the double helix by breaking the hydrogen bonds between the complementary bases. The resulting single strands act as templates for the synthesis of new complementary strands of DNA (see **FIGURE 21.7**).

Within the nucleus, nucleoside triphosphates of the four types of bases (dATP, dTTP, dGTP, and dCTP) are available so that each exposed base on the template strand can form hydrogen bonds with its complementary base. For example, T in the template strand hydrogen bonds with A, and G on the template strand hydrogen bonds with C. As the hydrogen bonds form base pairs, phosphodiester linkages are formed between the nucleotides.

Eventually, the entire double helix of the parent DNA is copied. In each new DNA molecule, one strand of the double helix is from the parent DNA, and one is a newly synthesized DNA strand. This process produces two new DNAs called *daughter DNAs* that are identical to each other and exact copies of the original parent DNA. In DNA replication, complementary base pairing ensures the correct placement of bases in the new daughter DNA strands.

Direction of Replication

Now that we have seen the overall process, we can take a look at some of the details of DNA replication, which requires the interaction of many enzymes and proteins as well as the parent DNA strands (see **FIGURE 21.8**). The unwinding of DNA by *helicase* occurs simultaneously in several sections along the parent DNA molecule. The separated strands are held open by a small protein called *single-strand binding protein* so that *DNA polymerase* can bind to the exposed bases. As a result, DNA polymerase can catalyze the replication process at each of these open DNA sections called **replication forks**. However, DNA polymerase only moves in the 5′ to 3′ direction, which means it catalyzes the formation of

ENGAGE

In which direction does DNA polymerase move?

FIGURE 21.7 ▶ In DNA replication, the separate strands of the parent DNA are the templates for the synthesis of complementary strands, producing two exact copies of DNA.

Q How many strands of the parent DNA are contained in each of the daughter DNAs?

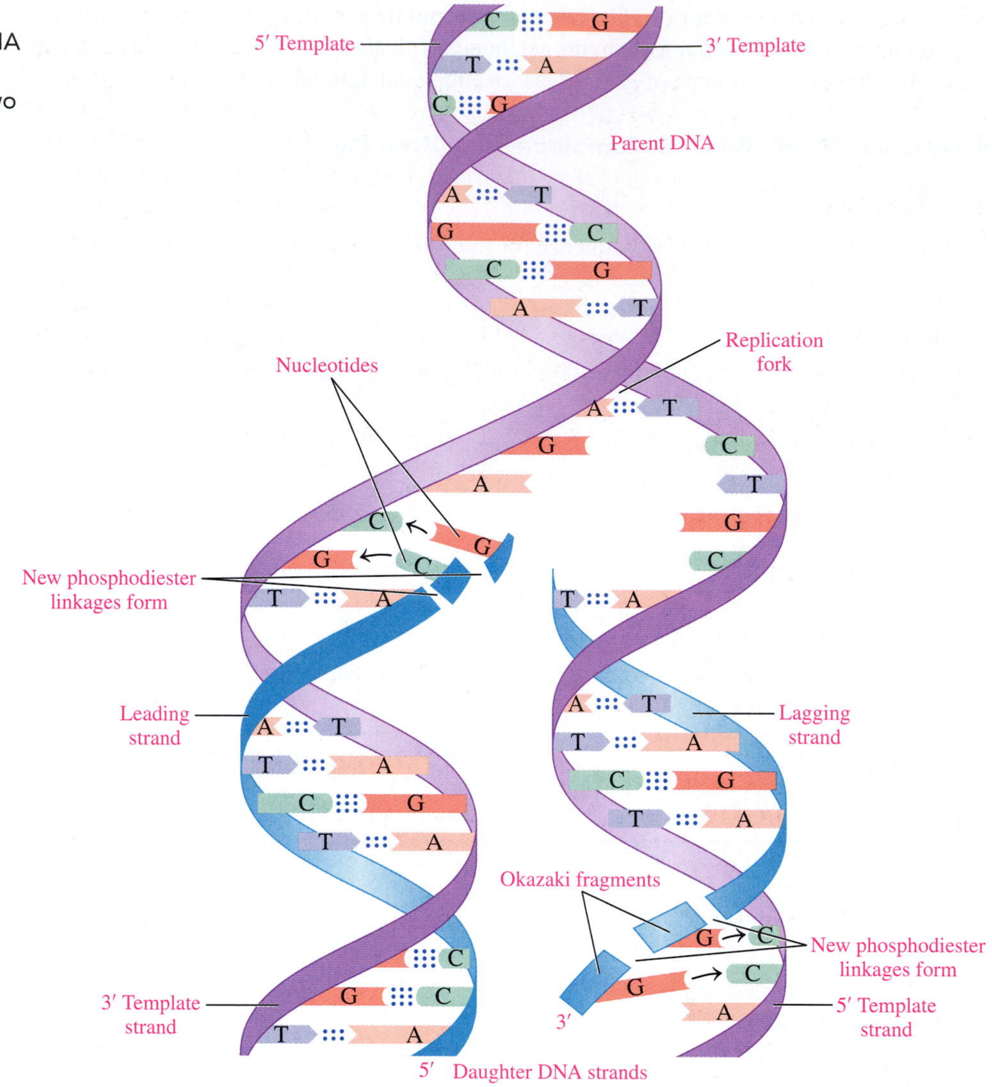
5′ Template
3′ Template
Parent DNA
Replication fork
Nucleotides
New phosphodiester linkages form
Leading strand
Lagging strand
Okazaki fragments
New phosphodiester linkages form
3′ Template strand
5′ Template strand
5′ Daughter DNA strands

FIGURE 21.8 ▶ DNA replication occurs continuously in the 5′ to 3′ direction on the leading strand and discontinuously on the lagging strand in the 5′ to 3′ direction. Smaller Okazaki fragments are joined by DNA ligase.

Q Why are the Okazaki fragments formed only for the lagging strand?

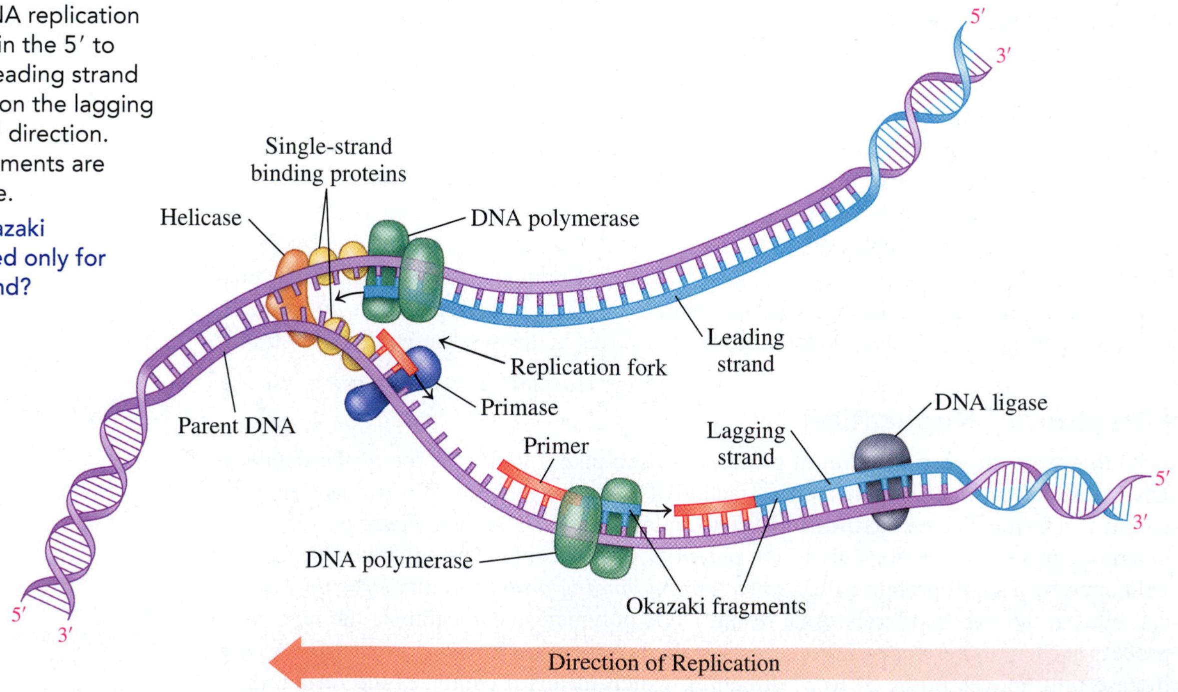
Single-strand binding proteins
Helicase
DNA polymerase
Leading strand
Replication fork
Primase
Parent DNA
Primer
Lagging strand
DNA ligase
DNA polymerase
Okazaki fragments
Direction of Replication

phosphodiester linkages between the hydroxyl group at the end of the growing nucleic acid and the phosphate group of a nucleoside triphosphate. The new DNA strand that grows in the 5′ to 3′ direction, the *leading strand*, is synthesized continuously.

The other new DNA strand, which is the *lagging strand*, is synthesized discontinuously in short, separate segments. To initiate synthesis on the lagging strand, the enzyme *primase* produces short complementary RNA fragments, called *primers*. These RNA primers serve as starting points for DNA polymerase to begin extending the complementary DNA fragments at the 3′ end until the polymerase reaches the next primer. As a result, DNA replication on the lagging strand consists of DNA segments called **Okazaki fragments**. Eventually, DNA polymerase replaces the RNA primers in each segment with the corresponding DNA. Finally, *DNA ligase* connects the Okazaki fragments by phosphodiester linkages, forming a daughter lagging strand. The enzymes and proteins required for replication are summarized in **TABLE 21.4**.

TABLE 21.4 Enzymes and Proteins in DNA Replication

Enzymes and Proteins		
1. Helicase	Helicase breaks the hydrogen bonds of the parent DNA strands at the *replication forks*, giving two separate DNA strands.	
2. Single-Strand Binding Protein	Single-strand binding proteins attach to the separated parent strands to keep them apart and the bases exposed.	
	Leading Strand 5′ to 3′	**Lagging Strand 3′ to 5′**
3. Primase		Primase synthesizes short RNA segments called primers that are starting points for DNA polymerase.
4. DNA Polymerase	DNA polymerase catalyzes the formation of phosphodiester linkages on the new DNA strands between the 3′ ends of the growing chains and available complementary nucleotides (dATP, dTTP, dGTP, and dCTP).	
	DNA polymerase adds nucleotides continuously in the 5′ to 3′ direction to a growing DNA chain.	At each primer, DNA polymerase forms short, separate segments (Okazaki fragments) until it reaches the next primer and stops.
5. DNA Ligase		DNA ligase joins the Okazaki fragments.

TEST

Try Practice Problems 21.29 to 21.36

▶ SAMPLE PROBLEM 21.4 Direction of DNA Replication

TRY IT FIRST

In an original DNA strand, a segment has the base sequence 5′ A G T C C G 3′.

a. What is the sequence of nucleotides in the daughter DNA strand that is complementary to this segment?

b. Why would the complementary sequence in the daughter DNA strand be synthesized as Okazaki fragments that require a DNA ligase?

SOLUTION

a. Only one possible nucleotide can pair with each base in the original segment. Thymine will pair only with adenine, whereas cytosine pairs only with guanine to give the complementary base sequence: 3′ T C A G G C 5′.

b. Because DNA polymerase only adds nucleotides in the 5′ to 3′ direction, DNA on the lagging strand is synthesized as short Okazaki fragments, which are joined by DNA ligase.

STUDY CHECK 21.4

a. How many daughter strands are formed during the replication of DNA?

b. What type of reaction is catalyzed by DNA ligase during the replication of DNA?

ANSWER

a. Two daughter strands are formed, one from each strand of the DNA double helix.

b. DNA ligase catalyzes the formation of phosphodiester bonds to join Okazaki fragments on the lagging strand.

PRACTICE PROBLEMS

21.3 DNA Double Helix and Replication

21.23 List three structural characteristics of DNA.

21.24 What is meant by double helix?

21.25 How are the two strands of nucleic acid in DNA held together?

21.26 What is meant by complementary base pairing?

21.27 Write the base sequence in a complementary DNA segment if each original segment has the following base sequence:
- **a.** 5′ A A A A A A 3′
- **b.** 5′ G G G G G G 3′
- **c.** 5′ A G T C C A G G T 3′
- **d.** 5′ C T G T A T A C G T T A 3′

21.28 Write the base sequence in a complementary DNA segment if each original segment has the following base sequence:
- **a.** 5′ T T T T T T 3′
- **b.** 5′ C C C C C C C C C 3′
- **c.** 5′ A T G G C A 3′
- **d.** 5′ A T A T G C G C T A A A 3′

21.29 What is the function of the enzyme helicase in DNA replication?

21.30 What is the function of the enzyme DNA polymerase in DNA replication?

21.31 What process ensures that the replication of DNA produces identical copies?

21.32 Why are Okazaki fragments formed in the synthesis of the lagging strand?

21.33 Match each component with one of the following descriptions:
1. lagging strand **2.** helicase
3. primase **4.** replication fork
- **a.** unwinds the DNA helix
- **b.** synthesizes primers at the replication fork
- **c.** point in DNA where nucleotides add to daughter DNA strand
- **d.** daughter DNA synthesized from short sections called Okazaki fragments

21.34 Match each component with one of the following descriptions:
1. leading strand **2.** DNA ligase
3. single-strand binding protein **4.** DNA polymerase
- **a.** catalyzes the formation of phosphodiester linkages between nucleotides
- **b.** daughter DNA synthesized continuously in 5′ to 3′ direction
- **c.** attaches to unwound strands to keep them open
- **d.** combines Okazaki fragments to form daughter DNA strand

21.35 Answer each of the following for a segment of a template strand 3′ G C T C C A T G 5′:
- **a.** Write the sequence of the new DNA segment.
- **b.** Will the production of this new DNA segment require the synthesis of Okazaki fragments?

21.36 Answer each of the following for a segment of a template strand 5′ A A G T C G T G 3′:
- **a.** Write the sequence of the new DNA segment.
- **b.** Will the production of this new DNA segment require the synthesis of Okazaki fragments?

21.4 RNA and Transcription

LEARNING GOAL Identify the different types of RNA; describe the synthesis of mRNA.

ENGAGE
What are some ways in which RNA is different from DNA?

Ribonucleic acid, RNA, which makes up most of the nucleic acid found in the cell, is involved with transmitting the genetic information needed to operate the cell. Similar to DNA, RNA molecules are unbranched chains of nucleotides. However, RNA differs from DNA in several important ways:

1. The sugar in RNA is ribose rather than the deoxyribose found in DNA.
2. In RNA, the base uracil replaces thymine.
3. RNA molecules are single stranded, not double stranded.
4. RNA molecules are much smaller than DNA molecules.

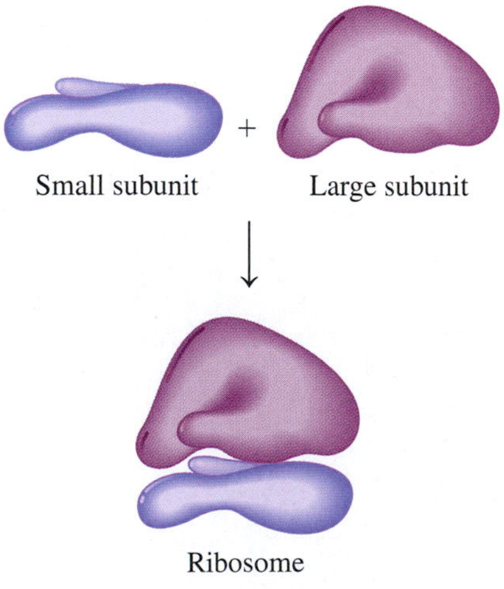

Small subunit
+
Large subunit
Ribosome

FIGURE 21.9 ▶ A typical ribosome consists of a small subunit and a large subunit. The subunit shapes shown contain both protein and rRNA.

Q Why would there be many thousands of ribosomes in a cell?

Types of RNA

There are three major types of RNA in the cells: *messenger RNA*, *ribosomal RNA*, and *transfer RNA*. Ribosomal RNA (**rRNA**), the most abundant type of RNA, is combined with proteins to form ribosomes. Ribosomes, which are the sites for protein synthesis, consist of two subunits: a large subunit and a small subunit (see **FIGURE 21.9**). Cells that synthesize large numbers of proteins have thousands of ribosomes.

Messenger RNA (**mRNA**) carries genetic information from the DNA, located in the nucleus of the cell, to the ribosomes, located in the *cytosol*, the liquid outside the nucleus. A *gene* is a segment of DNA that produces a separate mRNA used to synthesize a protein needed in the cell.

Transfer RNA (**tRNA**), the smallest of the RNA molecules, interprets the genetic information in mRNA and brings specific amino acids to the ribosome for protein synthesis. Only tRNA can translate the genetic information in the mRNA into the amino acid sequence that makes a protein. The major types of RNA molecules in humans are summarized in **TABLE 21.5**.

TABLE 21.5 Types of RNA Molecules in Humans

Type	Abbreviation	Percentage of Total RNA	Function in the Cell
Ribosomal RNA	rRNA	80	Major component of the ribosomes; site of protein synthesis
Messenger RNA	mRNA	5	Carries information for protein synthesis from the DNA to the ribosomes
Transfer RNA	tRNA	15	Brings specific amino acids to the site of protein synthesis

TEST

Try Practice Problems 21.37 to 21.40

The structure of all tRNA molecules is similar. Although tRNA structures are complex in three dimensions, we can draw tRNA as a two-dimensional cloverleaf (see **FIGURE 21.10a**). In the three-dimensional model, the RNA chain has more twists that shows the L-shape of tRNA (see **FIGURE 21.10b**). All tRNA molecules have a 3′ end with the nucleotide sequence ACC, which is known as the *acceptor stem*. An enzyme attaches an amino acid to the 3′ end of the acceptor stem by forming an ester bond with the free OH group of the acceptor stem. Each tRNA contains an **anticodon**, which is a series of three bases that complements three bases on mRNA.

ENGAGE

Which components are part of the structure of every tRNA molecule?

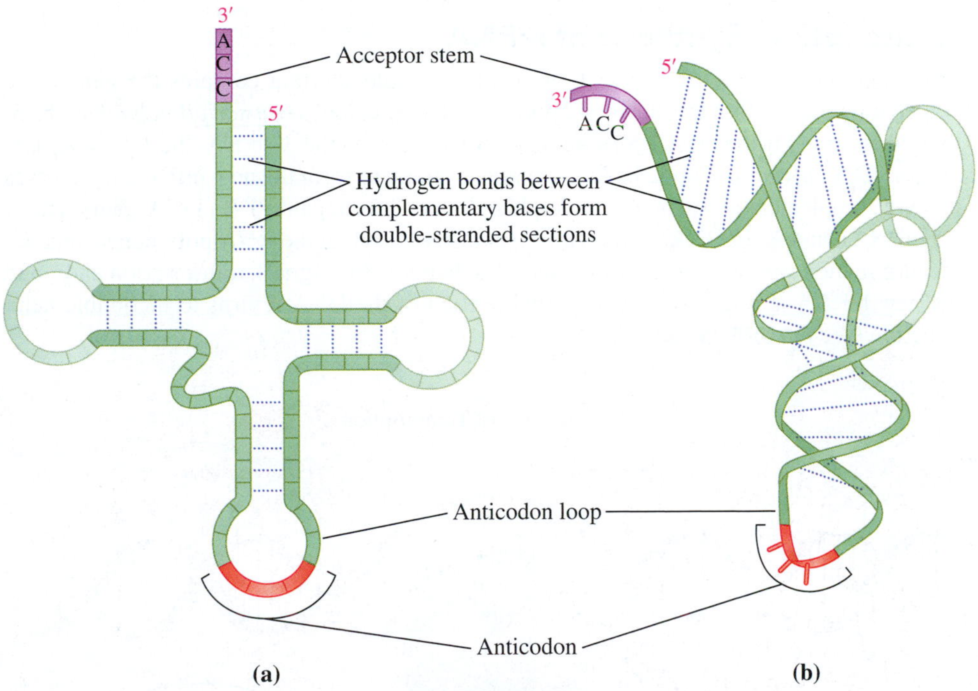

3′
A
C
C
5′
Acceptor stem
3′
ACC
5′
Hydrogen bonds between complementary bases form double-stranded sections
Anticodon loop
Anticodon
(a)
(b)

FIGURE 21.10 ▶ A typical tRNA molecule has an acceptor stem at the 3′ end that attaches to an amino acid and an anticodon loop that complements a codon on mRNA.

Q Why will different tRNAs have different bases in the anticodon loop?

RNA and Protein Synthesis

We now look at the overall processes involved in transferring genetic information encoded in the DNA to the production of proteins. In the nucleus, genetic information for the synthesis of a protein is copied from a gene in DNA to make mRNA, a process called **transcription**. The mRNA molecules move out of the nucleus into the cytosol, where they bind with the ribosomes. Then, in a process called **translation**, tRNA molecules convert the information in the mRNA into amino acids, which are placed in the proper sequence to synthesize a protein (see **FIGURE 21.11**). In this section, we focus on transcription. In the next, we focus on translation.

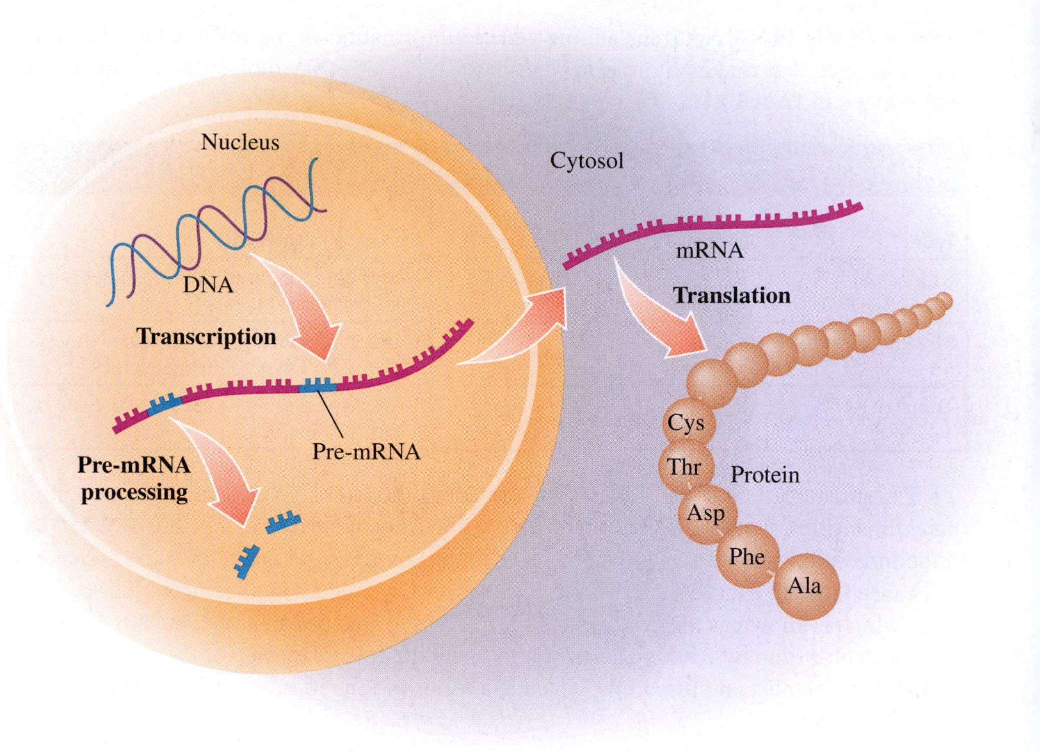

Nucleus
Cytosol
DNA
Transcription
mRNA
Translation
Pre-mRNA
Pre-mRNA processing
Cys
Thr
Protein
Asp
Phe
Ala

FIGURE 21.11 ▶ The genetic information in DNA is replicated in cell division and used to produce messenger RNA that codes for amino acids used in protein synthesis at the ribosomes.

Q What is the difference between transcription and translation?

Transcription: Synthesis of mRNA

Transcription begins when the section of a DNA molecule that contains the gene to be copied unwinds. Within this unwound portion of DNA, called a *transcription bubble*, the 5′ to 3′ open strand is the *DNA informational strand*, whereas the 3′ to 5′ is the *DNA template strand*. *RNA polymerase* uses the DNA template strand to form the new mRNA using bases that are complementary to the DNA template: C and G form pairs, T (in DNA) pairs with A (in mRNA), and A (in DNA) pairs with U (in mRNA). When the RNA polymerase reaches the termination site (a sequence of nucleotides that is a stop signal), transcription ends, and the new mRNA is released. The unwound portion of the DNA returns to its double helix structure (see **FIGURE 21.12**).

ENGAGE

How do the nucleotides found in the DNA template strand differ from the nucleotides found in a transcribed mRNA strand?

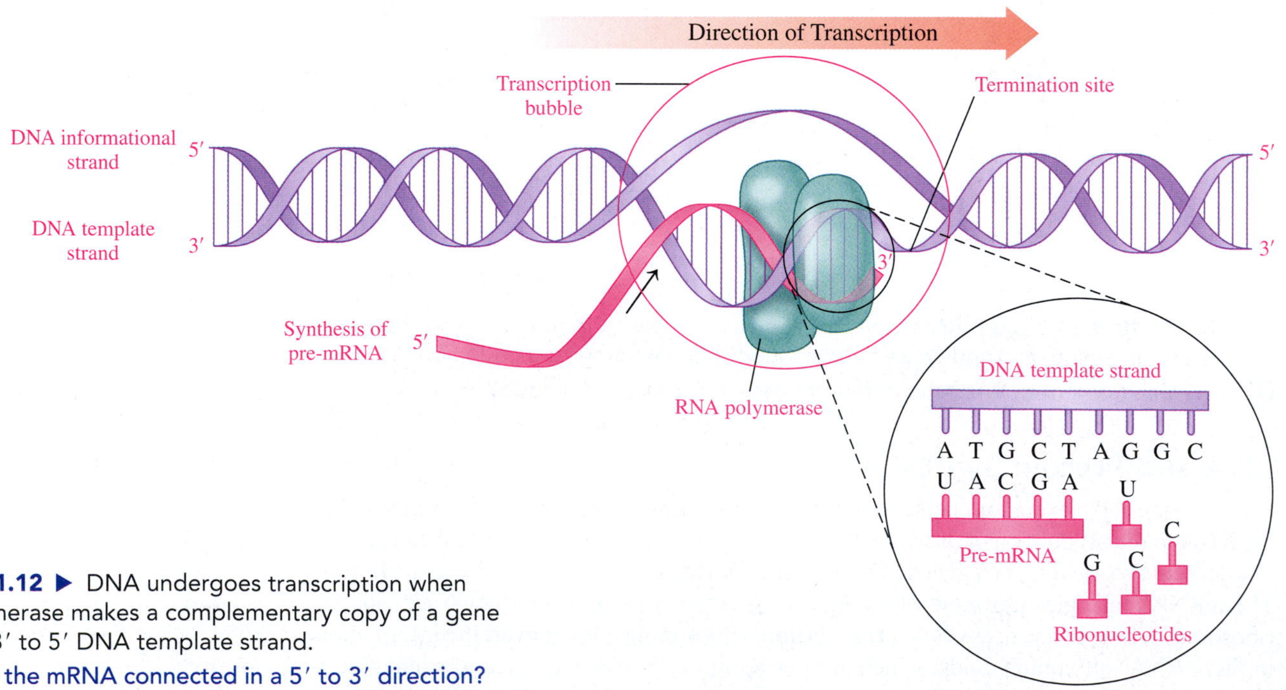

Direction of Transcription
Transcription bubble
Termination site
DNA informational strand
5′
DNA template strand
3′
5′
3′
3′
Synthesis of pre-mRNA
5′
RNA polymerase
DNA template strand
A T G C T A G G C
U A C G A
U
Pre-mRNA
C
G
C
Ribonucleotides

FIGURE 21.12 ▶ DNA undergoes transcription when RNA polymerase makes a complementary copy of a gene using the 3′ to 5′ DNA template strand.

Q Why is the mRNA connected in a 5′ to 3′ direction?

▶ SAMPLE PROBLEM 21.5 RNA Synthesis

CORE CHEMISTRY SKILL

Writing the mRNA Segment for a DNA Template

TRY IT FIRST

The sequence of bases in a part of the DNA template strand is 3′ C G A T C A 5′. What corresponding mRNA is produced?

SOLUTION

To form the mRNA, the bases in the DNA template are paired with their complementary bases: G with C, C with G, T with A, and A with U.

DNA template strand: 3′ C G A T C A 5′

Transcription ↓ ↓ ↓ ↓ ↓ ↓

Complementary base sequence in mRNA: 5′ G C U A G U 3′

STUDY CHECK 21.5

What are the DNA template strand segments that code for the mRNA segments with the following nucleotide sequences?

a. 5′ G G G U U U A A A 3′
b. 5′ C C G G A U A U C 3′

ANSWER

a. 3′ C C C A A A T T T 5′ **b.** 3′ G G C C T A T A G 5′

TEST

Try Practice Problems 21.41 to 21.44

Processing of mRNA

The DNA in plants and animals (eukaryotes) contains sections known as *exons* and *introns*. **Exons**, which code for proteins, are mixed in with sections called **introns** that do not code for proteins. A newly formed mRNA called a *pre-mRNA* or *heterogeneous nuclear RNA* (hnRNA) is a copy of the entire DNA template, including the noncoding introns. However, the pre-mRNA exists for only a short time before it is processed. The introns are cut out and the remaining exons spliced together by a large molecular complex called a *spliceosome*. This processing of pre-mRNA produces a mature mRNA that leaves the nucleus to deliver the genetic information to the ribosomes for the synthesis of protein (see **FIGURE 21.13**). In other words, the exons contain the mRNA sequences that are expressed as protein.

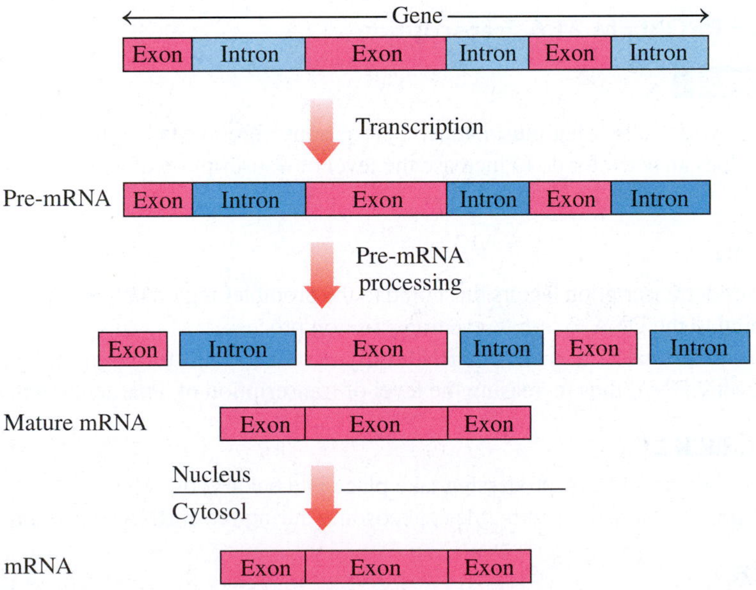
Gene
Exon
Intron
Exon
Intron
Exon
Intron
Transcription
Pre-mRNA
Pre-mRNA processing
Mature mRNA
Nucleus
Cytosol
mRNA

FIGURE 21.13 ▶ A pre-mRNA, containing copies of the exons and introns from the gene, is processed to remove the introns to form a mature mRNA that codes for a protein.

Q What is the difference between exons and introns?

TEST

Try Practice Problems 21.45 to 21.48

Regulation of Transcription: Transcription Factors

The synthesis of proteins is regulated so that proteins are produced when they are needed by a cell. Although regulation of the translation process can occur, protein synthesis is primarily regulated at the transcription level. In other words, in most cases, mRNA is only synthesized when cells require a particular protein. In eukaryotes (plants and animals), the control of transcription occurs mainly through the regulation of the binding of RNA polymerase to DNA.

In eukaryotes, a series of proteins called **transcription factors** must bind to a segment of DNA called a *promoter region* before the RNA polymerase can bind and start the transcription process. The promoter region is a sequence of nucleotides that is present on the DNA before the start site for a gene (see **FIGURE 21.14**). Multiple transcription factors (at least six) must bind to the promoter region for a particular gene or that gene will not be transcribed. For genes that will be transcribed, the level at which transcription occurs is regulated by the binding of *activator* or *repressor* proteins to separate sections of the DNA called *enhancers* or *silencers*. Activators make it easier for RNA polymerase to bind to the promoter region, increasing the level of transcription of a particular gene. Repressors make it harder for RNA polymerase to bind to the promoter region, decreasing the level of transcription.

ENGAGE

How does the binding of activators and repressors affect the level of transcription of a particular gene?

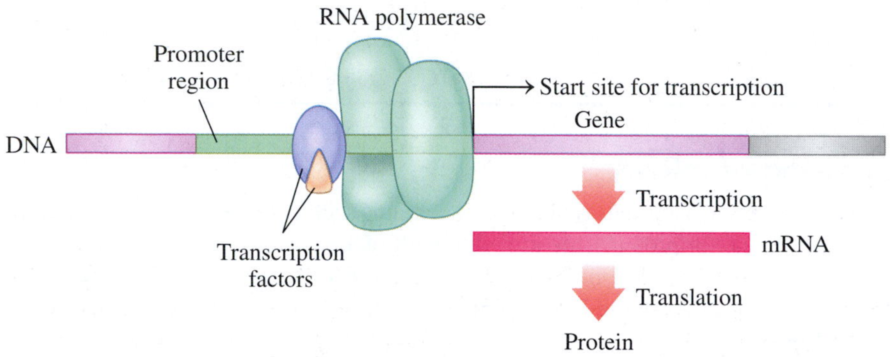
RNA polymerase
Promoter region
Start site for transcription
Gene
DNA
Transcription
mRNA
Transcription factors
Translation
Protein

FIGURE 21.14 ▶ Transcription factors bind at the promoter region of eukaryotic DNA, making it easier for RNA polymerase to bind to DNA and initiate the transcription of a gene. (For simplicity, only two transcription factors are shown.)

Q Where is the promoter region located in relation to the gene that will be transcribed?

TEST

Try Practice Problems 21.49 to 21.52

▶ SAMPLE PROBLEM 21.6 Transcription

TRY IT FIRST

a. In eukaryotic cells, what must happen before transcription can begin?
b. What does an activator do to increase the level of transcription of a gene in a eukaryotic cell?

SOLUTION

a. A series of transcription factors must bind to the promoter region before RNA polymerase can bind to the DNA to initiate the transcription process.
b. An activator makes it easier for RNA polymerase to bind to the promoter region of eukaryotic DNA, thus increasing the level of transcription of a particular gene.

STUDY CHECK 21.6

a. Where does pre-mRNA processing take place in a eukaryotic cell?
b. What process is carried out by the spliceosome during pre-mRNA processing?

ANSWER

a. The processing of pre-mRNA to mature mRNA takes place in the nucleus before the mRNA can move to the ribosomes in the cytosol.
b. The spliceosome links the exons together after the introns have been removed from the pre-mRNA.

PRACTICE PROBLEMS

21.4 RNA and Transcription

21.37 What are the three different types of RNA?

21.38 What are the functions of each type of RNA?

21.39 What is the composition of a ribosome?

21.40 What is the smallest RNA?

21.41 What is meant by the term "transcription"?

21.42 What bases in mRNA are used to complement the bases A, T, G, and C in DNA?

21.43 Write the corresponding section of mRNA produced from the following section of DNA template strand:

3′ C C G A A G G T T C A C 5′

21.44 Write the corresponding section of mRNA produced from the following section of DNA template strand:

3′ T A C G G C A A G C T A 5′

21.45 What are introns and exons?

21.46 What kind of processing do pre-mRNA molecules undergo before they leave the nucleus?

21.47 A pre-mRNA segment has the following base sequence. If underlined sections **A**, **C**, and **D** are introns, and **B**, **E**, and **F** are exons, what is the base sequence of the mature mRNA produced after pre-mRNA processing?

5′ ACC	GGC	ACA	UUC	GGA	UCG 3′
A	B	C	D	E	F

21.48 A pre-mRNA segment has the following base sequence. If underlined sections **B**, **D**, and **F** are introns, and **A**, **C**, and **E** are exons, what is the base sequence of the mature mRNA produced after pre-mRNA processing?

5′ GAC	UAU	GGC	AAC	GGC	GUC 3′
A	B	C	D	E	F

21.49 What is a transcription factor?

21.50 How do transcription factors control protein synthesis at the transcription level?

21.51 What is an activator?

21.52 What is a repressor?

21.5 The Genetic Code and Protein Synthesis

LEARNING GOAL Use the genetic code to write the amino acid sequence for a segment of mRNA.

REVIEW
Forming Amides (18.5)

After the genetic information encoded in DNA is transcribed, mRNA molecules move out of the nucleus to the ribosomes in the cytosol. At the ribosomes, the genetic information in the mRNA is translated into a sequence of amino acids in protein.

Genetic Code

The **genetic code** consists of a series of three nucleotides (triplets) in mRNA called **codons** that specify the amino acids and their sequence in a protein. Early work on protein synthesis showed that repeating triplets of uracil (UUU) produced a polypeptide that contained only phenylalanine. Therefore, a sequence of UUU UUU UUU codes for three phenylalanines.

Codons in mRNA: 5′ UUU UUU UUU 3′

↓ ↓ ↓

Amino acid sequence: Phe — Phe — Phe

Codons have been determined for all 20 amino acids. A total of 64 codons are possible from the triplet combinations of A, G, C, and U. Three of these, UGA, UAA, and UAG, are stop signals that code for the termination of protein synthesis. All the other three-base codons shown in **TABLE 21.6** specify amino acids. Thus, one amino acid can have several codons. For example, glycine has four codons: GGU, GGC, GGA, and GGG. The triplet AUG has two roles in protein synthesis. At the beginning of an mRNA, the codon AUG signals the start of protein synthesis. In the middle of a series of codons, the AUG codon specifies the amino acid methionine.

ENGAGE
Why is it possible for one amino acid to have several codons?

TEST
Try Practice Problems 21.53 to 21.58

Translation

In the *translation* process, tRNA molecules, amino acids, and enzymes convert the mRNA codons into amino acids to build a protein.

Activation of tRNA

Each tRNA molecule contains a loop called the *anticodon*, which is a triplet of bases that complements a codon in mRNA. An amino acid is attached to the acceptor stem of each tRNA by an enzyme called *aminoacyl–tRNA synthetase*. Each amino acid has a different

TABLE 21.6 Codons in mRNA: The Genetic Code for Amino Acids

First Letter	Second Letter: U	Second Letter: C	Second Letter: A	Second Letter: G	Third Letter
U	UUU Phe (F)	UCU Ser (S)	UAU Tyr (Y)	UGU Cys (C)	U
U	UUC Phe (F)	UCC Ser (S)	UAC Tyr (Y)	UGC Cys (C)	C
U	UUA Leu (L)	UCA Ser (S)	UAA STOP[b]	UGA STOP[b]	A
U	UUG Leu (L)	UCG Ser (S)	UAG STOP[b]	UGG Trp (W)	G
C	CUU Leu (L)	CCU Pro (P)	CAU His (H)	CGU Arg (R)	U
C	CUC Leu (L)	CCC Pro (P)	CAC His (H)	CGC Arg (R)	C
C	CUA Leu (L)	CCA Pro (P)	CAA Gln (Q)	CGA Arg (R)	A
C	CUG Leu (L)	CCG Pro (P)	CAG Gln (Q)	CGG Arg (R)	G
A	AUU Ile (I)	ACU Thr (T)	AAU Asn (N)	AGU Ser (S)	U
A	AUC Ile (I)	ACC Thr (T)	AAC Asn (N)	AGC Ser (S)	C
A	AUA Ile (I)	ACA Thr (T)	AAA Lys (K)	AGA Arg (R)	A
A	AUG START[a]/ Met (M)	ACG Thr (T)	AAG Lys (K)	AGG Arg (R)	G
G	GUU Val (V)	GCU Ala (A)	GAU Asp (D)	GGU Gly (G)	U
G	GUC Val (V)	GCC Ala (A)	GAC Asp (D)	GGC Gly (G)	C
G	GUA Val (V)	GCA Ala (A)	GAA Glu (E)	GGA Gly (G)	A
G	GUG Val (V)	GCG Ala (A)	GAG Glu (E)	GGG Gly (G)	G

START[a] codon signals the initiation of a peptide chain.
STOP[b] codons signal the end of a peptide chain.

CORE CHEMISTRY SKILL

Writing the Amino Acid for an mRNA Codon

synthetase. Activation of tRNA occurs when aminoacyl–tRNA synthetase forms an ester bond between the carboxylate group of its amino acid and the hydroxyl group on the acceptor stem (see **FIGURE 21.15**). Each synthetase then checks the tRNA–amino acid combination and hydrolyzes any incorrect combinations.

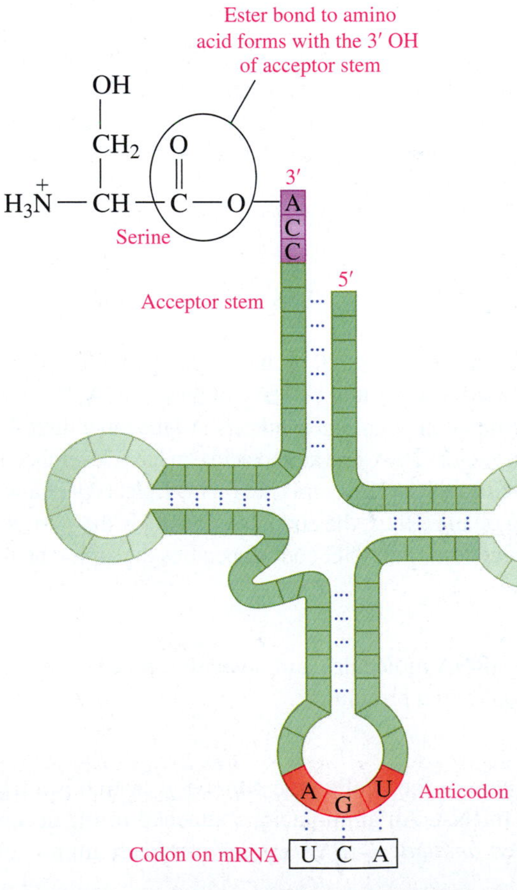

Ester bond to amino acid forms with the 3′ OH of acceptor stem
OH
CH2
O
H3N+ — CH — C — O
Serine
3′
A
C
C
5′
Acceptor stem
A G U
Anticodon
Codon on mRNA
U C A

FIGURE 21.15 ▶ An activated tRNA with anticodon AGU bonds to serine at the acceptor stem.

Q What would be an anticodon on a methionine–tRNA?

Initiation and Chain Elongation

Protein synthesis begins when mRNA binds to a ribosome. The first codon in an mRNA is a *start codon*, AUG, which forms hydrogen bonds with methionine–tRNA. Another tRNA hydrogen bonds to the next codon, placing a second amino acid adjacent to methionine. A peptide bond forms between the C terminus of methionine and the N terminus of the second amino acid (see **FIGURE 21.16**). The initial tRNA detaches from the ribosome, which shifts to the next

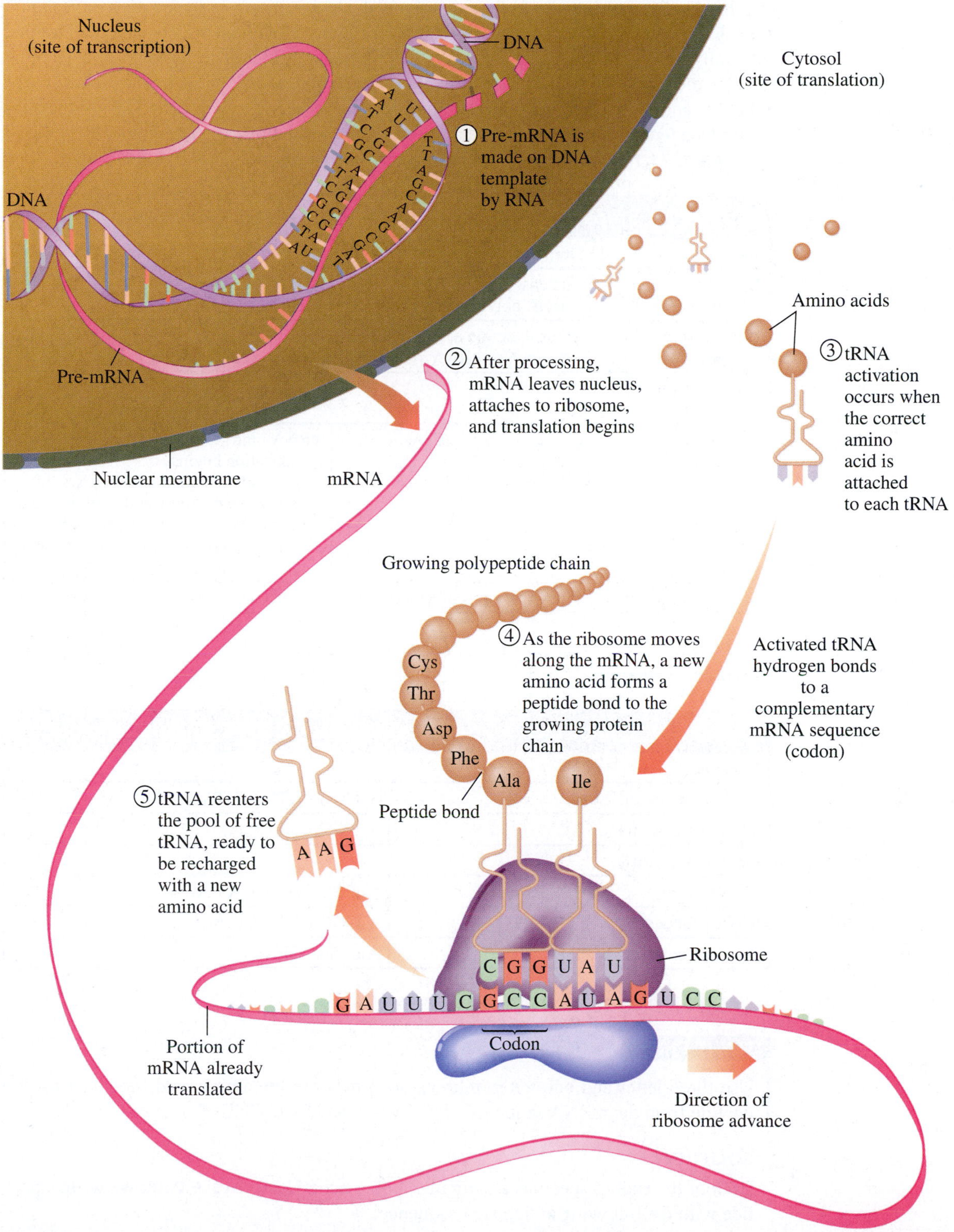
Nucleus
(site of transcription)
DNA
Cytosol
(site of translation)
① Pre-mRNA is made on DNA template by RNA
DNA
Pre-mRNA
② After processing, mRNA leaves nucleus, attaches to ribosome, and translation begins
Nuclear membrane
mRNA
Amino acids
③ tRNA activation occurs when the correct amino acid is attached to each tRNA
Growing polypeptide chain
④ As the ribosome moves along the mRNA, a new amino acid forms a peptide bond to the growing protein chain
Activated tRNA hydrogen bonds to a complementary mRNA sequence (codon)
Cys
Thr
Asp
Phe
Ala
Ile
Peptide bond
⑤ tRNA reenters the pool of free tRNA, ready to be recharged with a new amino acid
A A G
Ribosome
C G G U A U
G A U U U C G C C A U A G U C C
Codon
Portion of mRNA already translated
Direction of ribosome advance

FIGURE 21.16 ▶ In the translation process, the mRNA synthesized by transcription attaches to a ribosome, and tRNAs pick up their amino acids, bind to the appropriate codon, and place them in a growing peptide chain.

Q How is the correct amino acid placed in the peptide chain?

INTERACTIVE VIDEO

Pearson eText 2.0 Protein Synthesis

TEST

Try Practice Problems 21.59 to 21.62

available codon, a process called *translocation*. During *chain elongation*, the ribosome moves along the mRNA from codon to codon so that the tRNAs can attach new amino acids to the growing polypeptide chain. Sometimes, a group of several ribosomes, called a *polysome*, translate the same strand of mRNA to produce several copies of the polypeptide at the same time.

Chain Termination

Eventually, a ribosome encounters a codon—UAA, UGA, or UAG—that has no corresponding tRNAs. These are *stop codons*, which signal the termination of polypeptide synthesis and the release of the polypeptide chain from the ribosome. The initial amino acid, methionine, is usually removed from the beginning of the polypeptide chain. The R groups on the amino acids in the new polypeptide chain form hydrogen bonds to give the secondary structures of α helices and β-pleated sheets and form interactions such as salt bridges and disulfide bonds to produce tertiary and quaternary structures, which make the protein biologically active.

TABLE 21.7 summarizes the steps in protein synthesis.

TABLE 21.7 Steps in Protein Synthesis

Step	Site: Materials	Process
1. DNA Transcription	Nucleus: nucleotides, RNA polymerase	A DNA template is used to produce mRNA.
2. Activation of tRNA	Cytosol: amino acids, tRNAs, aminoacyl–tRNA synthetase	Molecules of tRNA pick up specific amino acids according to their anticodons.
3. Translation of mRNA		
3a. and 3b. Initiation and Chain Elongation	Ribosome: Met–tRNA, mRNA, aminoacyl–tRNAs	mRNA binds to ribosomes where translation begins. A start codon binds the first tRNA carrying the amino acid methionine to the mRNA. Successive tRNAs bind to and detach from the ribosome as they add an amino acid to the polypeptide.
3c. Chain Termination	Ribosome: stop codon on mRNA	The protein is released from the ribosome.

TEST

Try Practice Problems 21.63 and 21.64

TABLE 21.8 gives an example of corresponding nucleotide and amino acid sequences in protein synthesis.

TABLE 21.8 Complementary Sequences in DNA, mRNA, tRNA, and Peptides

Nucleus	
DNA informational strand	5′ GCG AGT GGA TAC 3′
DNA template strand	3′ CGC TCA CCT ATG 5′
Ribosome (cytosol)	
mRNA	5′ GCG AGU GGA UAC 3′
tRNA anticodons	3′ CGC UCA CCU AUG 5′
Polypeptide amino acids	Ala–Ser–Gly–Tyr

ENGAGE

What is one possible mRNA sequence that would code for the peptide with the amino acid sequence Cys–Ala–Arg?

▶ SAMPLE PROBLEM 21.7 Protein Synthesis

TRY IT FIRST

Use three-letter and one-letter abbreviations to write the amino acid sequence for the peptide from the mRNA sequence of 5′ UCA AAA GCC CUU 3′.

SOLUTION

Each of the codons specifies a particular amino acid. Using **TABLE 21.6**, we write a peptide with the following amino acid sequence:

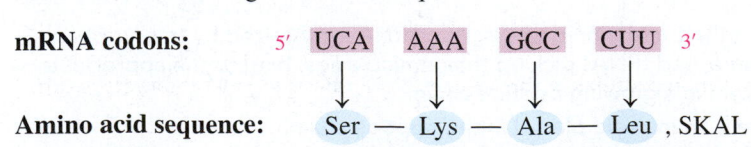

mRNA codons: 5′ UCA AAA GCC CUU 3′

↓ ↓ ↓ ↓

Amino acid sequence: Ser — Lys — Ala — Leu , SKAL

STUDY CHECK 21.7

Use three-letter and one-letter abbreviations to write the amino acid sequences for the peptides produced from the mRNA sequences listed below.

a. 5′ GGG AGC UGC GAG GUU 3′
b. 5′ GCA AGG UGU AUU GAA 3′

ANSWER

a. Gly–Ser–Cys–Glu–Val, GSCEV
b. Ala–Arg–Cys–Ile–Glu, ARCIE

TEST

Try Practice Problems 21.65 to 21.70

Chemistry Link to Health

Cataracts

Even before they are completely synthesized and released from the ribosome, proteins begin to fold into the three-dimensional structures that determine their biological functions. Molecules known as *chaperones* interact with newly synthesized proteins, helping them to fold into their correct, three-dimensional structures. Chaperones can also interact with proteins that have unfolded so that the proteins do not aggregate or clump together.

An important example of a chaperone can be seen in the lens of the eye. The lens is transparent, flexible tissue that is located behind the iris and the pupil of the eye. It is mainly composed of water and proteins. There are some unique characteristics of proteins in the lens:

1. There are three main types of proteins in the eye: alpha-crystallins, beta-crystallins, and gamma-crystallins.
2. These proteins are present in fairly high concentrations in the eye. The high concentrations of proteins are necessary to focus the light coming into our eyes.
3. For the lens to remain transparent, the proteins must remain in their properly folded structures and not aggregate.
4. Unlike other proteins in the body, proteins in the lens are not turned over. In other words, if the proteins in the lens are damaged, they will not be replaced.

As we age, the beta- and gamma-crystallins are subjected to oxidative stresses. As a consequence, these proteins begin to unfold. The damaged, unfolded proteins tend to stick together, forming aggregates that decrease the transparency of the lens. Eventually, the aggregates cause a clouding of the lens that results in blurry vision. This clouding of the lens is known as *cataracts*.

The alpha-crystallins act as chaperones. Once a beta- or gamma-crystallin molecule begins to unfold, an alpha-crystallin binds to the unfolded protein so that it cannot aggregate with other unfolded molecules. In this way, the alpha-crystallins maintain the transparency of the lens. Unfortunately, because the proteins of the lens are not replaced, the number of alpha-crystallins in the lens is finite. Eventually, there are not enough alpha-crystallins remaining to counteract the effects of oxidative stress on the other crystallin molecules. At this point, cataracts start to form.

Because there is a limited number of alpha-crystallins in our lenses, every person who lives long enough will eventually develop cataracts. In fact, the most common surgery in the word is cataract surgery, in which the cloudy lens is replaced with a clear artificial lens. This surgery results in vision improvement for almost all patients.

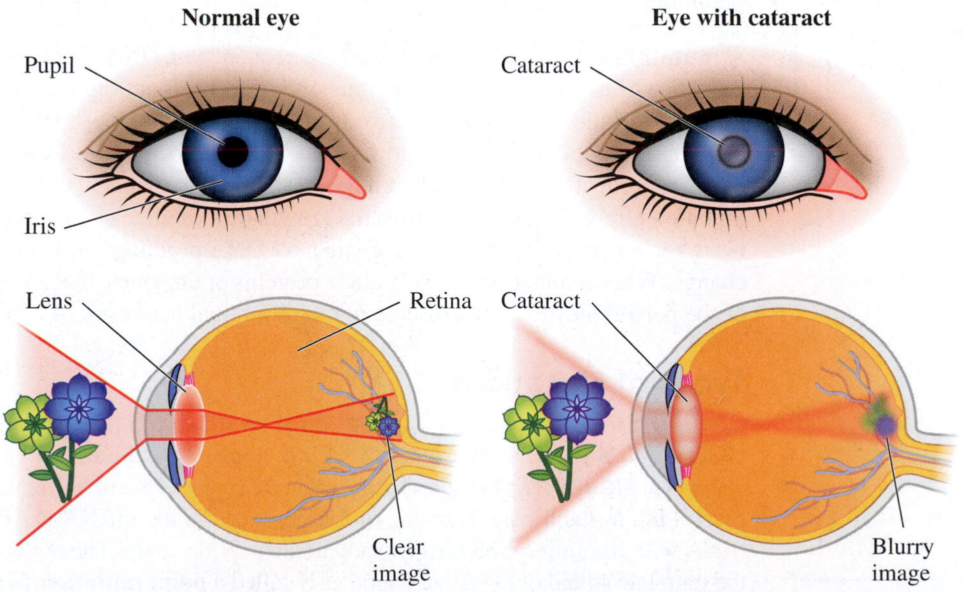
Normal eye
Eye with cataract
Pupil
Iris
Cataract
Lens
Retina
Cataract
Clear image
Blurry image

When proteins in the lens unfold and aggregate, cataracts result.

PRACTICE PROBLEMS

21.5 The Genetic Code and Protein Synthesis

21.53 What is a codon?

21.54 What is the genetic code?

21.55 What amino acid is coded for by each of the following mRNA codons?
a. CCA **b.** AAC **c.** GGU **d.** AGG

21.56 What amino acid is coded for by each of the following mRNA codons?
a. UCU **b.** UUC **c.** CGG **d.** GCA

21.57 When does the codon AUG signal the start of a protein? When does it code for the amino acid methionine?

21.58 The codons UGA, UAA, and UAG do not code for amino acids. What is their role as codons in mRNA?

21.59 What is the difference between a *codon* and an *anticodon*?

21.60 Why are there at least 20 different tRNAs?

21.61 How is a peptide chain extended?

21.62 What is meant by "translocation"?

21.63 What are the three steps of translation?

21.64 Where does protein synthesis take place?

21.65 Use three-letter and one-letter abbreviations to write the amino acid sequence for the peptide from each of the following mRNA sequences:
a. 5′ ACC ACA ACU 3′ **b.** 5′ UUU CCG UUC CCA 3′
c. 5′ UAC GGG AGA UGU 3′

21.66 Use three-letter and one-letter abbreviations to write the amino acid sequence for the peptide from each of the following mRNA sequences:
a. 5′ AAA CCC UUG GCC 3′
b. 5′ CCU CGC AGC GGC UGA 3′
c. 5′ AUG CAC AAG GAA GUA CUG 3′

21.67 The following sequence is a portion of the DNA template strand:
3′ GCT ATA CCA AAA 5′
a. Write the corresponding mRNA segment.
b. What are the anticodons of the tRNAs?
c. Write the three-letter and one-letter abbreviations for this segment in the peptide chain.

21.68 The following sequence is a portion of the DNA template strand:
3′ TGT GGG GTT ATT 5′
a. Write the corresponding mRNA segment.
b. What are the anticodons of the tRNAs?
c. Write the three-letter and one-letter abbreviations for this segment in the peptide chain.

Clinical Applications

21.69 The following is a segment of the DNA template that codes for human insulin:
3′ TTT GTG AAC CAA CAC CTG 5′
a. Write the corresponding mRNA segment.
b. Write the three-letter and one-letter abbreviations for the corresponding peptide segment.

21.70 The following is a segment of the DNA template that codes for human insulin:
3′ TGC GGC TCA CAC CTG GTG 5′
a. Write the corresponding mRNA segment.
b. Write the three-letter and one-letter abbreviations for the corresponding peptide segment.

21.6 Genetic Mutations

LEARNING GOAL Identify the type of change in DNA for a point mutation, a deletion mutation, and an insertion mutation.

REVIEW
Identifying the Primary, Secondary, Tertiary, and Quaternary Structures of Proteins (19.2, 19.3, 19.4)
Describing Enzyme Action (20.1)

A **mutation** is a change in the nucleotide sequence of DNA. Such a change may alter the sequence of amino acids, affecting the structure and function of a protein in a cell. Mutations may result from X-rays, overexposure to sun (ultraviolet [UV] light), chemicals called *mutagens*, and possibly some viruses. If a mutation occurs in a somatic cell (a cell other than a reproductive cell), the altered DNA is limited to that cell and its daughter cells. If the mutation affects DNA that controls the growth of the cell, cancer could result. If a mutation occurs in a germ cell (egg or sperm), then all DNA produced will contain the same genetic change. When a mutation severely alters proteins or enzymes, the new cells may not survive or the person may exhibit a disease or condition that is a result of a genetic defect.

Types of Mutations

Consider a triplet of bases CCG in the template strand of DNA, which produces the codon GGC in mRNA. At the ribosome, tRNA would place the amino acid glycine in the peptide chain (see **FIGURE 21.17a**). Now, suppose that T replaces the first C in the DNA triplet, which gives TCG as the triplet. Then the codon produced in the mRNA is AGC, which brings the tRNA with the amino acid serine to add to the peptide chain. The replacement of one base in the template strand of DNA with another is called a **point mutation**. When there is a change of a nucleotide in the codon, a different amino acid may be inserted into the polypeptide. However, if a point mutation does not change the amino acid, it is a *silent mutation*. A point mutation is the most common way in which mutations occur (see **FIGURE 21.17b**).

ENGAGE
Why does replacing the U in CGU with an A not change the primary structure of a protein?

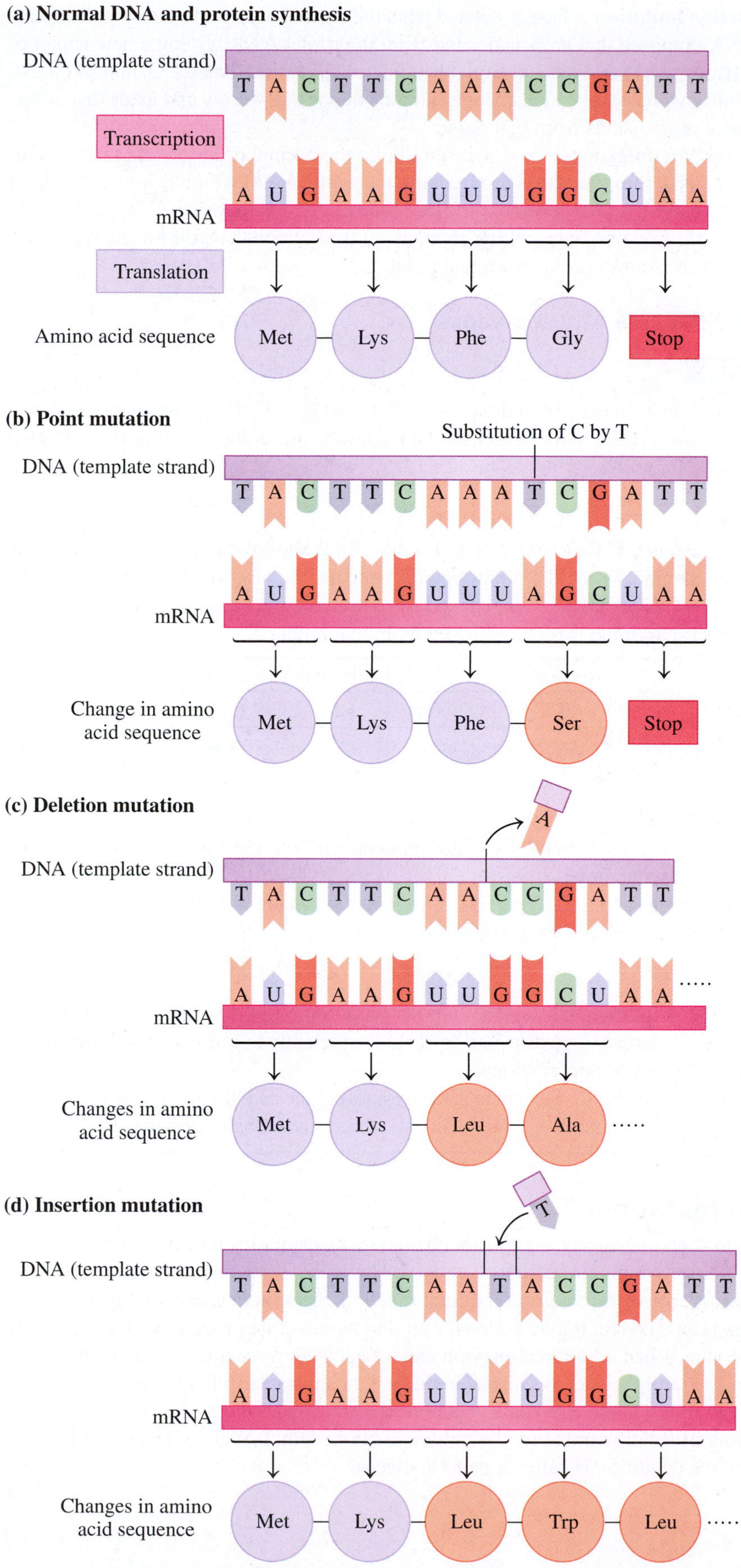
(a) Normal DNA and protein synthesis
DNA (template strand)
T A C T T C A A A C C G A T T
Transcription
A U G A A G U U U G G C U A A
mRNA
Translation
Amino acid sequence
Met Lys Phe Gly Stop
(b) Point mutation
Substitution of C by T
DNA (template strand)
T A C T T C A A A T C G A T T
A U G A A G U U U A G C U A A
mRNA
Change in amino acid sequence
Met Lys Phe Ser Stop
(c) Deletion mutation
A
DNA (template strand)
T A C T T C A A C C G A T T
A U G A A G U U G G C U A A .....
mRNA
Changes in amino acid sequence
Met Lys Leu Ala .....
(d) Insertion mutation
T
DNA (template strand)
T A C T T C A A T A C C G A T T
A U G A A G U U A U G G C U A A
mRNA
Changes in amino acid sequence
Met Lys Leu Trp Leu .....

FIGURE 21.17 ▶ An alteration in the DNA template strand produces a change in the sequence of amino acids in the protein, which may result in a mutation. **(a)** A normal DNA leads to the correct amino acid order in a protein. **(b)** In a point mutation, the change of a base in DNA leads to a change in the mRNA codon and possibly a change in one amino acid. **(c)** The deletion of a base causes a deletion mutation, which changes the mRNA codons that follow the mutation and produces a different amino acid sequence. **(d)** The insertion of a base causes an insertion mutation, which changes the mRNA codons that follow the mutation and produces a different amino acid sequence.

Q When would a point mutation cause protein synthesis to stop?

In a **deletion mutation**, a base is deleted from the normal order of bases in the template strand of DNA. Suppose that an A is deleted from the triplet AAA, giving a new triplet of AAC (see **FIGURE 21.17c**). The next triplet becomes CGA rather than CCG, and so on. All the triplets shift by one base, which changes all the codons that follow and leads to a different sequence of amino acids from that point.

In an **insertion mutation**, a base is inserted into the normal order of bases in the template strand of DNA. Suppose a T is inserted into the triplet AAA, which gives a new triplet of AAT (see **FIGURE 21.17d**). The next triplet becomes ACC rather than CCG, and so on. All the triplets shift by one base, which changes all the codons that follow and leads to a different sequence of amino acids from that point.

▶SAMPLE PROBLEM 21.8 Mutations

TRY IT FIRST

An mRNA has the sequence of codons 5′ CCC AGA GCC 3′. If a point mutation in the DNA changes the mRNA codon of AGA to GGA, how is the amino acid sequence affected in the resulting protein?

SOLUTION

The mRNA sequence 5′ CCC AGA GCC 3′ codes for the following amino acids: proline, arginine, and alanine. When the point mutation occurs, the new sequence of the mRNA codons is 5′ CCC GGA GCC 3′, which codes for proline, glycine, and alanine. The polar basic amino acid arginine is replaced by the nonpolar amino acid glycine.

	Normal	After Point Mutation
mRNA codons	5′ CCC AGA GCC 3′	5′ CCC GGA GCC 3′
Amino acid sequence	Pro—Arg—Ala	Pro—Gly—Ala

STUDY CHECK 21.8

a. How might the protein made from the mutated mRNA in Sample Problem 21.8 be affected by this mutation?

b. How might the protein be affected if a point mutation changed a codon in the middle of the corresponding mRNA from UAC to UAG?

ANSWER

a. Because the point mutation replaces a polar basic amino acid with a nonpolar neutral amino acid, the tertiary structure may be altered sufficiently to cause the resulting protein to be less effective or nonfunctional.

b. The codon UAC codes for the amino acid tyrosine, but the codon UAG is a stop codon. In this case, the point mutation would result in a shortened, and probably nonfunctional, protein.

TEST

Try Practice Problems 21.71 to 21.78

Effect of Mutations

Some mutations do not cause a significant change in the primary structure of a protein and the protein is able to maintain biological activity. However, if the mutation causes a change to an amino acid critical to protein structure or function, the protein loses biological activity. If the protein is an enzyme, it may no longer bind to its substrate or react with the substrate at the active site. When an altered enzyme cannot catalyze a reaction, certain substances may accumulate until they act as poisons in the cell, or substances vital to survival may not be synthesized. If a defective enzyme occurs in a major metabolic pathway or is involved in the building of a cell membrane, the mutation can be lethal. When a protein deficiency is hereditary, the condition is called a **genetic disease**.

ENGAGE

Under what conditions can a DNA mutation result in a genetic disease?

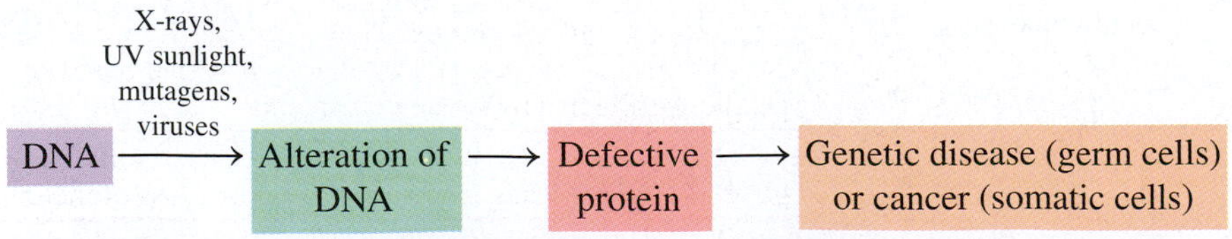
X-rays, UV sunlight, mutagens, viruses
DNA
Alteration of DNA
Defective protein
Genetic disease (germ cells) or cancer (somatic cells)

Genetic Diseases

A genetic disease is the result of a defective enzyme caused by a mutation in its genetic code. For example, *phenylketonuria* (PKU) results when DNA cannot direct the synthesis of the enzyme phenylalanine hydroxylase, required for the conversion of phenylalanine to tyrosine. In an attempt to break down the phenylalanine, other enzymes in the cells convert it to phenylpyruvate. If phenylalanine and phenylpyruvate accumulate in the blood of an infant, it can lead to severe brain damage and mental retardation. If PKU is detected in a newborn baby, a diet is prescribed that eliminates all foods that contain phenylalanine. Preventing the buildup of phenylpyruvate ensures normal growth and development.

The amino acid tyrosine is needed in the formation of melanin, the pigment that gives the color to our skin and hair. If the enzyme that converts tyrosine to melanin is defective, no melanin is produced, and a genetic disease known as *albinism* results. People and animals with no melanin have no skin, eye, or hair pigment (see **FIGURE 21.18**). **TABLE 21.9** lists some other common genetic diseases and the type of metabolism or area affected.

FIGURE 21.18 ▶ A peacock with albinism does not produce the melanin needed to make bright colors for its feathers.

Why are traits such as albinism related to the gene?

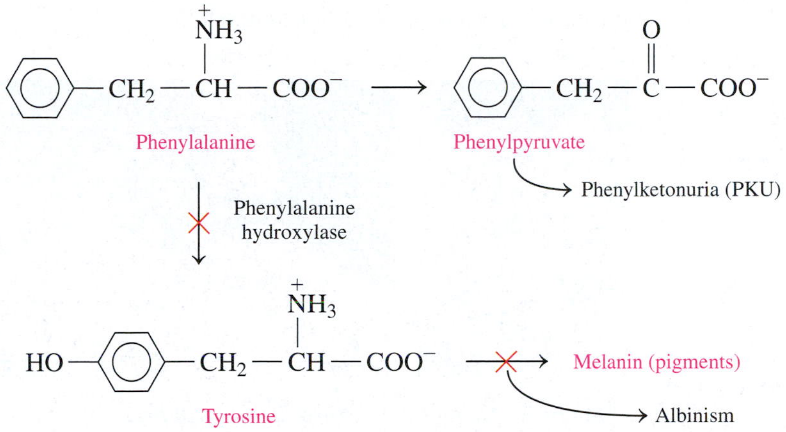

Phenylalanine
Phenylpyruvate
Phenylketonuria (PKU)
Phenylalanine hydroxylase
Tyrosine
Melanin (pigments)
Albinism

TEST

Try Practice Problems 21.79 and 21.80

TABLE 21.9 Some Genetic Diseases

Genetic Disease	Result
Cystic fibrosis (CF)	Cystic fibrosis is caused by a mutation in the gene for the protein that regulates the production of stomach fluids and mucus. CF is one of the most common inherited diseases in children, in which thick mucus secretions make breathing difficult and block pancreatic function.
Down syndrome	Down syndrome is the leading cause of mental retardation, occurring in about 1 of every 800 live births; the mother's age strongly influences its occurrence. Mental and physical problems, including heart and eye defects, are the result of the formation of three chromosomes (trisomy), usually number 21, instead of a pair.
Familial hypercholesterolemia (FH)	Familial hypercholesterolemia occurs when there is a mutation of a gene on chromosome 19 which produces high cholesterol levels that lead to early coronary heart disease in people 30 to 40 years old.
Galactosemia	In galactosemia, the transferase enzyme required for the metabolism of galactose-1-phosphate is absent, resulting in the accumulation of galactose-1-phosphate, which leads to cataracts and mental retardation. Galactosemia occurs in about 1 in every 50 000 births.
Hemophilia	Hemophilia is the result of one or more defective blood clotting factors that lead to poor coagulation, excessive bleeding, and internal hemorrhages. There are about 20 000 hemophilia patients in the United States.
Huntington's disease (HD)	Huntington's disease affects the nervous system, leading to total physical impairment. It is the result of a mutation in a gene on chromosome 4, which can now be mapped to test people in families with a history of HD. There are about 30 000 people with Huntington's disease in the United States.
Muscular dystrophy (MD) (Duchenne)	Muscular dystrophy, Duchenne form, is caused by a mutation in the X chromosome. This muscle-destroying disease appears at about age 5, with death by age 20, and occurs in about 1 of 10 000 males.
Sickle-cell anemia	Sickle-cell anemia is caused by a defective form of hemoglobin resulting from a mutation in a gene on chromosome 11. It decreases the oxygen-carrying ability of red blood cells, which take on a sickled shape, causing anemia and plugged capillaries from red blood cell aggregation. In the United States, about 72 000 people are affected by sickle-cell anemia.
Tay–Sachs disease	Tay–Sachs disease is the result of a defective hexosaminidase A, which causes an accumulation of gangliosides and leads to mental retardation, loss of motor control, and early death.

Chemistry Link to Health

Ehlers–Danlos Syndrome

Collagen is the most abundant protein in the body. It plays a number of important roles in the human body, including providing strength and structure to the skin and connecting bones across joints.

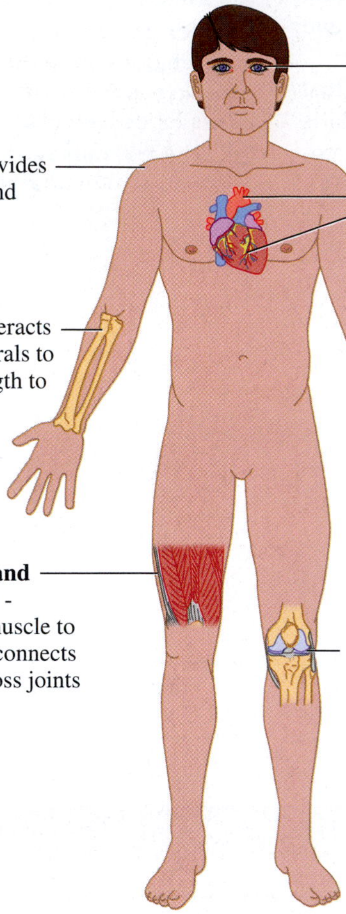

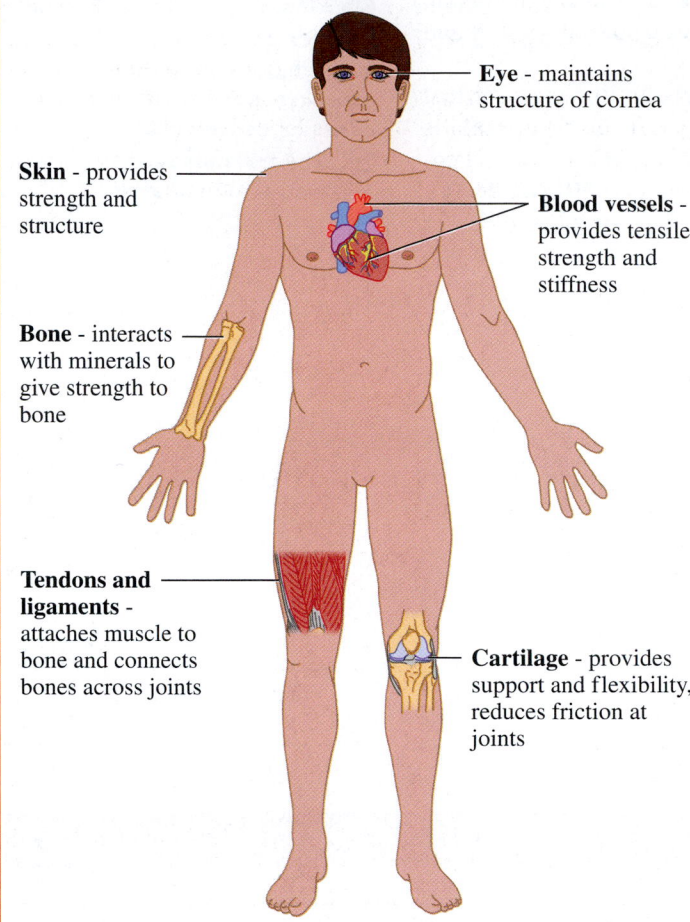

Eye - maintains structure of cornea
Skin - provides strength and structure
Blood vessels - provides tensile strength and stiffness
Bone - interacts with minerals to give strength to bone
Tendons and ligaments - attaches muscle to bone and connects bones across joints
Cartilage - provides support and flexibility, reduces friction at joints

Collagen plays a number of important roles in the human body.

There are many varieties of collagen in the human body. Many different enzymes are involved in the production and processing of collagen itself or in the production and processing of proteins that interact with collagen. A mutation in the gene that codes for any one of these enzymes can either reduce the amount of collagen or weaken the collagen that is present in the body, thus weakening the connective tissue in skin, bones, blood vessels, and other organs.

Ehlers–Danlos syndrome is a group of connective tissue disorders that result from mutations in the genes that affect the production, processing, or function of collagen. As a result of reduced amounts of functional collagen in their bodies, patients with Ehlers–Danlos syndrome can have joints that are overly flexible and skin that is extremely elastic. They suffer from joint pain and experience frequent dislocation of their joints. Their skin tends to bruise easily, and their skin wounds often do not heal well.

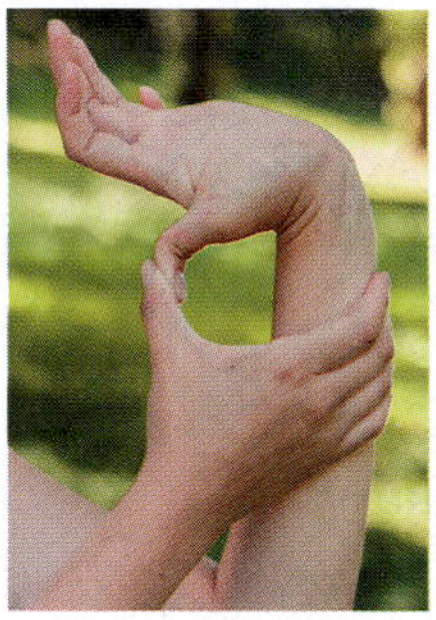

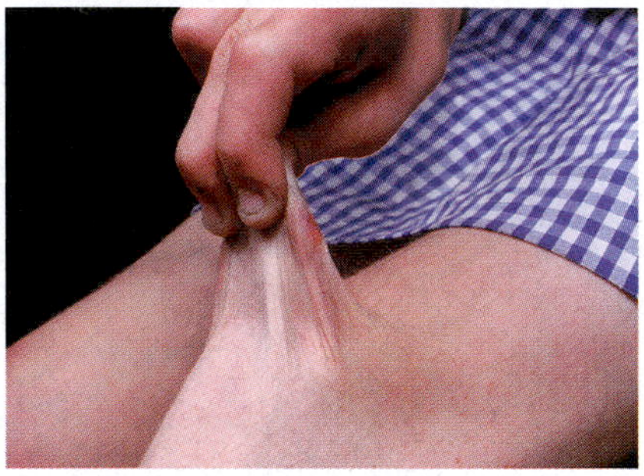

Patients suffering from Ehlers–Danlos syndrome disorders often exhibit joint hypermobility and skin hyperextensibility.

There is no known cure for Ehlers–Danlos syndrome disorders. Management of the disorders typically focuses on strengthening the muscles around the joints, stabilizing the joints, and/or protecting the joints and skin from damage.

PRACTICE PROBLEMS

21.6 Genetic Mutations

21.71 What is a point mutation?

21.72 How does a point mutation for an enzyme affect the order of amino acids in that protein?

21.73 What is the effect of a deletion mutation on the amino acid sequence of a polypeptide?

21.74 How can a mutation decrease the activity of a protein?

21.75 How is protein synthesis affected if the normal base sequence TTT in the DNA template strand is changed to TTC?

21.76 How is protein synthesis affected if the normal base sequence CCC in the DNA template strand is changed to ACC?

21.77 Consider the following segment of mRNA produced by the normal order of DNA nucleotides:

5′ ACA UCA CGG GUA 3′

a. What is the amino acid order produced from this mRNA?
b. What is the amino acid order if a point mutation changes UCA to ACA?
c. What is the amino acid order if a point mutation changes CGG to GGG?
d. What happens to protein synthesis if a point mutation changes UCA to UAA?

e. What is the amino acid order if an insertion mutation adds a G to the beginning of the mRNA segment?

f. What is the amino acid order if a deletion mutation removes the A at the beginning of the mRNA segment?

21.78 Consider the following segment of mRNA produced by the normal order of DNA nucleotides:

5′ CUU AAA CGA CAU 3′

a. What is the amino acid order produced from this mRNA?

b. What is the amino acid order if a point mutation changes CUU to CCU?

c. What is the amino acid order if a point mutation changes CGA to AGA?

d. What happens to protein synthesis if a point mutation changes AAA to UAA?

e. What is the amino acid order if an insertion mutation adds a G to the beginning of the mRNA segment?

f. What is the amino acid order if a deletion mutation removes the C at the beginning of the mRNA segment?

Clinical Applications

21.79 a. A point mutation changes a codon in the mRNA for an enzyme from GCC to GCA. Why is there no change in the amino acid order in the protein?

b. In sickle-cell anemia, a point mutation in the mRNA for hemoglobin results in the replacement of glutamate with valine in the resulting hemoglobin molecule. Why does the replacement of one amino acid cause such a drastic change in biological function?

21.80 a. A point mutation in the mRNA for an enzyme results in the replacement of leucine with alanine in the resulting enzyme molecule. Why does this change in amino acids have little effect on the biological activity of the enzyme?

b. A point mutation in mRNA replaces cytosine in the codon UCA with adenine. How would this substitution affect the amino acid order in the protein?

21.7 Recombinant DNA

LEARNING GOAL Describe the preparation and uses of recombinant DNA.

Techniques in the field of genetic engineering permit scientists to cut and recombine DNA fragments to form **recombinant DNA**. The technology of recombinant DNA is used to produce important medicines like human insulin for diabetics, the antiviral substance interferon, blood clotting factor VIII, and human growth hormone.

Preparing Recombinant DNA

Much of the work with recombinant DNA is done with *Escherichia coli* (*E. coli*) bacteria, which contain small circular plasmids of DNA that can be easily isolated and replicated. Initially, *E. coli* cells are soaked in a detergent solution to disrupt the plasma membrane, releasing the plasmids. A *restriction enzyme* is used to cut the double strands of DNA between specific bases in the DNA sequence (see **FIGURE 21.19**). The same restriction enzymes are

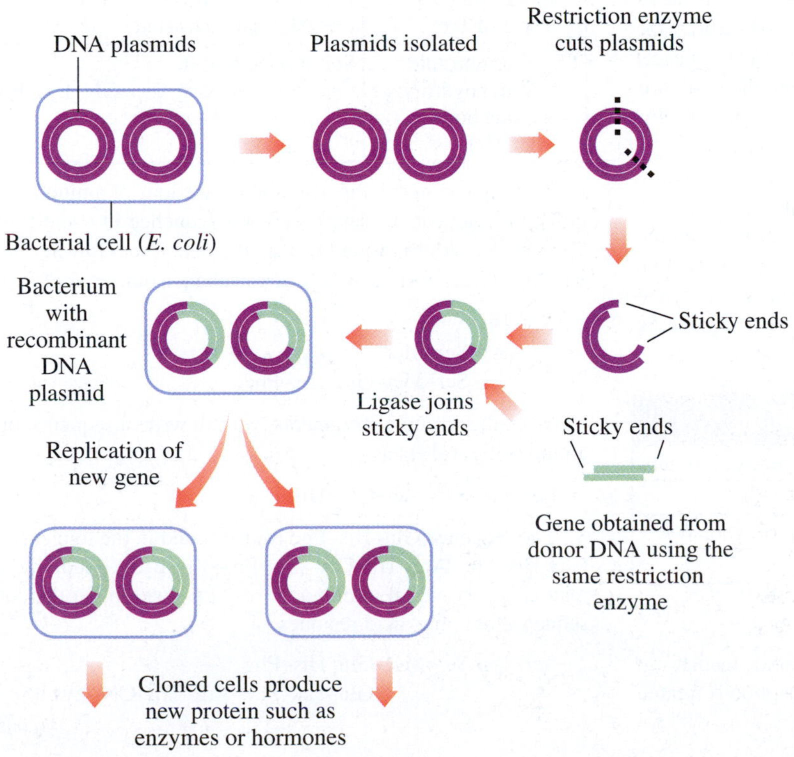

DNA plasmids
Plasmids isolated
Restriction enzyme cuts plasmids
Bacterial cell (E. coli)
Bacterium with recombinant DNA plasmid
Sticky ends
Ligase joins sticky ends
Sticky ends
Replication of new gene
Gene obtained from donor DNA using the same restriction enzyme
Cloned cells produce new protein such as enzymes or hormones

FIGURE 21.19 ▶ Recombinant DNA is formed by placing a gene from another organism in a plasmid DNA of the bacterium, which causes the bacterium to produce a nonbacterial protein such as insulin or growth hormone.

How can recombinant DNA help a person with a genetic disease?

used to cut a piece of DNA called donor DNA from a gene of a different organism, such as the gene that produces insulin or growth hormone. When the donor DNA is mixed with the cut plasmids, the nucleotides in their "sticky ends" join to make *recombinant DNA*. The resulting altered plasmids containing the recombinant DNA are placed in a fresh culture of *E. coli* bacteria. The *E. coli* that take up the plasmids are then selected. As the recombined cells divide and replicate, they produce the protein from the inserted gene in the plasmids. **TABLE 21.10** lists some of the products developed through recombinant DNA technology that are now used therapeutically.

TABLE 21.10 Therapeutic Products of Recombinant DNA

Product	Therapeutic Use
Epidermal growth factor (EGF)	Stimulates healing of wounds and burns
Erythropoietin (EPO)	Treats anemia; stimulates production of erythrocytes
Hepatitis B virus (HBV) vaccine	Prevents viral hepatitis
Human blood clotting factor VIII	Treats hemophilia; allows blood to clot normally
Human growth hormone (HGH)	Stimulates growth
Human insulin	Treats diabetes
Influenza vaccine	Prevents influenza
Interferon	Treats cancer and viral disease
Interleukin	Stimulates immune system; treats cancer
Monoclonal antibody	Transports drugs needed to treat cancer and transplant rejection
Prourokinase	Destroys blood clots; treats myocardial infarctions
Tumor necrosis factor (TNF)	Destroys tumor cells

TEST

Try Practice Problems 21.81 to 21.86

Chemistry Link to Health

Protein Sequencing

Recombinant DNA techniques are used to produce a variety of therapeutic proteins. It is important to verify that the primary structure of the proteins produced through recombinant techniques is the same as the primary structure of natural proteins. Methods exist for determining the order of amino acids in short peptides, but most proteins are very large molecules consisting of hundreds of amino acids. Using a method called *protein sequencing*, scientists determine the order of amino acids in large protein molecules. An enzyme or reagent is used to hydrolyze the protein into smaller fragments for which the amino acid order (sequence) is obtained. Then using a different enzyme or reagent, the protein is hydrolyzed into different fragments, which are also sequenced. Finally, all the different fragments are arranged to find where they overlap.

Digestive enzymes are often used to produce fragments because they hydrolyze specific peptide bonds. For example, the digestive enzyme trypsin only hydrolyzes peptide bonds after the basic amino acids arginine and lysine, whereas chymotrypsin hydrolyzes peptide bonds after the aromatic amino acids phenylalanine, tyrosine, and tryptophan. **TABLE 21.11** shows some common protein sequencing reagents and the amino acids after which hydrolysis occurs.

TABLE 21.11 Protein Sequencing Reagents

Reagent	Amino Acids Where Hydrolysis Occurs
Trypsin	After Arg and Lys (R or K)
Chymotrypsin	After Phe, Tyr, and Trp (F, Y, or W)
Cyanogen bromide (CNBr)	After Met (M)

Example of Protein Sequencing

The hormone α-melanotropin is a 13-amino acid peptide that stimulates cells to produce the pigment melanin. If the peptide is treated separately with chymotrypsin and cyanogen bromide, the following fragments are produced.

Reagent	Fragments Produced
Chymotrypsin hydrolyzes peptide bonds after Phe, Tyr, and Trp.	Arg–Trp Ser–Tyr Gly–Lys–Pro–Val Ser–Met–Glu–His–Phe
Cyanogen bromide (CNBr) hydrolyzes peptide bonds after Met.	Ser–Tyr–Ser–Met Glu–His–Phe–Arg–Trp–Gly–Lys–Pro–Val

Now we can match the overlapping sections of amino acids from different fragments to determine their sequence in α-melanotropin. The pair Ser–Met is found in the fragments Ser–Tyr–Ser–Met and Ser–Met–Glu–His–Phe, which we can align in our series of fragments.

Ser–Tyr
Ser–Tyr–Ser–Met
Ser–Met–Glu–His–Phe

From this group of fragments, we can write a sequence of seven amino acid residues as

Ser–Tyr–Ser–Met–Glu–His–Phe

The sequence Glu–His–Phe is also found in the longer fragment Glu–His–Phe–Arg–Trp–Gly–Lys–Pro–Val. By overlapping the two matching sections of amino acid residues, we obtain the complete sequence of amino acid residues.

Ser–Tyr–Ser–Met–Glu–His–Phe
Glu–His–Phe–Arg–Trp–Gly–Lys–Pro–Val

(*continued*)

Chemistry Link to Health (*continued*)

Thus, the primary structure of the 13 amino acid residues in α-melanotropin is written as

Ser–Tyr–Ser–Met–Glu–His–Phe–Arg–Trp–Gly–Lys–Pro–Val

Now we can use this primary structure of α-melanotropin to see where chymotrypsin and CNBr hydrolyzed peptide bonds to give the fragments we used to determine its sequence.

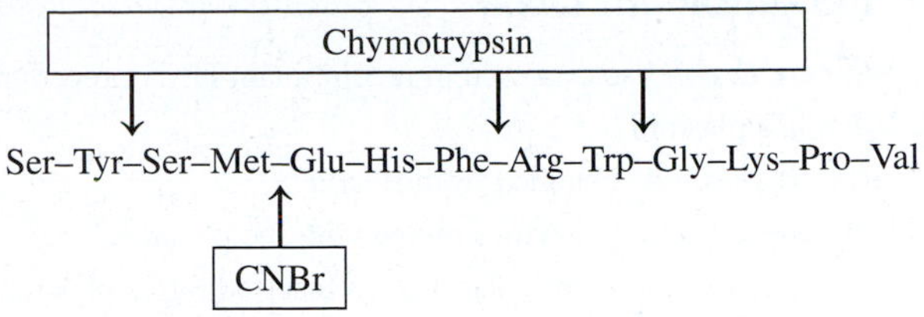
Chymotrypsin
Ser–Tyr–Ser–Met–Glu–His–Phe–Arg–Trp–Gly–Lys–Pro–Val
CNBr

A similar process is followed to determine the primary structure of unknown proteins.

TEST

Try Practice Problems 21.89 and 21.90

Polymerase Chain Reaction

Before gene cloning can occur, a purified gene has to be isolated. In 1987, a process called the **polymerase chain reaction (PCR)** was developed. PCR made it possible to produce multiple copies of a gene (amplify the gene) in a short time. In the PCR technique, a sequence of a DNA molecule is selected to copy, and the DNA is heated to separate the strands. The DNA strands are mixed with a heat-stable DNA polymerase and the four deoxyribonucleotides, and undergo repeated cycles of heating and cooling to produce complementary strands of the DNA section. After several cycles of the PCR process, millions of copies of the initial DNA section are produced (see **FIGURE 21.20**).

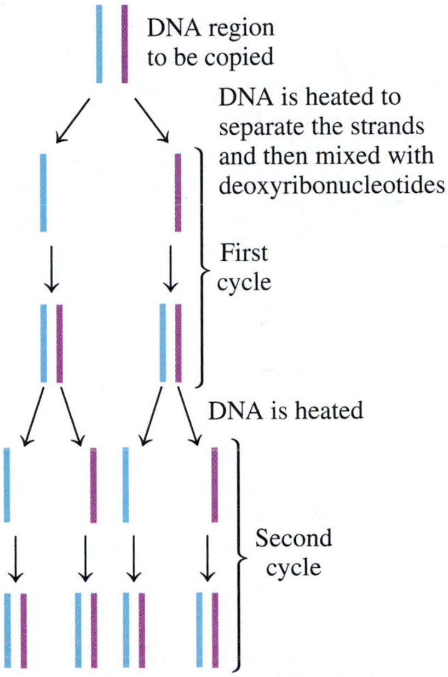
DNA region to be copied
DNA is heated to separate the strands and then mixed with deoxyribonucleotides
First cycle
DNA is heated
Second cycle

FIGURE 21.20 ▶ Each cycle of the polymerase chain reaction doubles the number of copies of the DNA section.

Q Why are the DNA strands heated at the start of each cycle?

Genetic Testing

PCR allows screening for defective genes whose sequences are known. For example, there are several defects in two known breast cancer genes, called BRCA1 and BRCA2, that correlate to a higher risk of breast cancer. Patients are screened for the defects in these genes by using a DNA sample from their blood or saliva. PCR amplifies the defective genes while incorporating a fluorescent label that is visible if the test is positive.

DNA Fingerprinting

DNA fingerprinting or *DNA profiling* uses PCR to identify individuals based on a set of 13 hereditary traits. The chance that two individuals of the same ethnic background have the same 13 genetic traits is 1 in 575 trillion unless they are identical twins. Only a very small sample size is needed from blood, skin, saliva, or semen. Fluorescent or radioactive isotopes are incorporated into the amplified DNA during the PCR process. The DNA is cut into smaller pieces by restriction enzymes, which are placed on a gel and separated using electrophoresis. The banding pattern on the gel is called a DNA fingerprint.

One application of DNA fingerprinting is in forensic science, where DNA samples are used to connect a suspect with a crime (see **FIGURE 21.21**). Recently, DNA fingerprinting has been used to gain the release of individuals who were wrongly convicted. Other applications of DNA fingerprinting are determining the biological parents of a child, establishing the identity of a deceased person, and matching recipients with organ donors.

ENGAGE

How many cycles of PCR would be required to produce 16 copies of an original segment of DNA?

FIGURE 21.21 ▶ Bands on film represent DNA fingerprints that can be used to identify a person involved in a crime.

Q What causes DNA fragments to appear on X-ray film?

TEST

Try Practice Problems 21.87 and 21.88

PRACTICE PROBLEMS

21.7 Recombinant DNA

21.81 Why are *E. coli* bacteria used in recombinant DNA procedures?

21.82 What is a plasmid?

21.83 How are plasmids obtained from *E. coli*?

21.84 Why are restriction enzymes mixed with the plasmids?

21.85 How is a gene for a particular protein inserted into a plasmid?

21.86 What beneficial proteins are produced from recombinant DNA technology?

21.87 What is a DNA fingerprint?

21.88 Why is DNA polymerase useful in criminal investigations?

Clinical Applications

21.89 Write the three-letter sequence for the amino acid residues in the octapeptide angiotensin II, a hormone involved in regulating blood pressure in humans, if treatment with chymotrypsin and trypsin gives the following fragments (see Chemistry Link to Health: Protein Sequencing):

Reagent	Fragments Produced
Chymotrypsin	Ile–His–Pro–Phe Asp–Arg–Val–Tyr
Trypsin	Asp–Arg Val–Tyr–Ile–His–Pro–Phe

21.90 Write the three-letter sequence for the amino acid residues in somatostatin, a peptide that inhibits the release of growth hormone, if treatment with chymotrypsin and trypsin gives the following fragments (see Chemistry Link to Health: Protein Sequencing):

Reagent	Fragments Produced
Chymotrypsin	Phe Trp Lys–Thr Phe Thr–Ser–Cys Ala–Gly–Cys–Lys–Asn–Phe
Trypsin	Ala–Gly–Cys–Lys Asn–Phe–Phe–Trp–Lys Thr–Phe–Thr–Ser–Cys

21.8 Viruses

LEARNING GOAL Describe the methods by which a virus infects a cell.

Viruses are small particles of 3 to 200 genes that cannot replicate without a host cell. A typical virus contains a nucleic acid, DNA or RNA, but not both, inside a protein coat. A virus does not have the necessary material such as nucleotides and enzymes to make proteins and grow. The only way a virus can replicate (make additional copies of itself) is to invade a host cell and take over the machinery and materials necessary for protein synthesis and growth. Some infections caused by viruses invading human cells are listed in **TABLE 21.12**. There are also viruses that attack bacteria, plants, and animals.

ENGAGE

What materials do viruses need from the host cell in order to replicate?

TABLE 21.12 Some Diseases Caused by Viral Infection

Disease	Virus
Chicken pox (shingles)	*Varicella zoster* virus (VZV)
Common cold	Coronavirus (over 100 types), rhinovirus (over 110 types)
Epstein–Barr	Epstein–Barr virus (EBV)
Hepatitis	Hepatitis A virus (HAV), hepatitis B virus (HBV), hepatitis C virus (HCV)
Herpes	Herpesvirus
HPV	Human papilloma virus
Influenza	Orthomyxovirus
Leukemia, cancers, AIDS	Retrovirus
Mumps	Paramyxovirus
Warts	Papovavirus

A viral infection begins when an enzyme in the protein coat of the virus makes a hole in the outside of the host cell, allowing the viral nucleic acid to enter and mix with the materials in the host cell (see **FIGURE 21.22**). If the virus contains DNA, the host cell

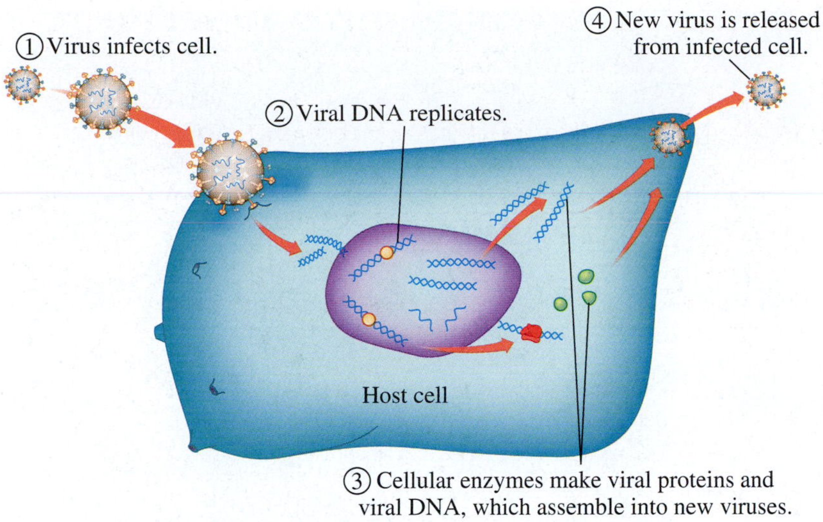

① Virus infects cell.
② Viral DNA replicates.
④ New virus is released from infected cell.
Host cell
③ Cellular enzymes make viral proteins and viral DNA, which assemble into new viruses.

FIGURE 21.22 ▶ After a virus attaches to the host cell, it injects its viral DNA and uses the host cell's machinery and materials to make viral mRNA, new viral DNA, and viral proteins. The newly assembled viruses are released to infect other cells.

Q Why does a virus need a host cell for replication?

begins to replicate the viral DNA in the same way it would replicate normal DNA. Viral DNA produces viral RNA, and a protease processes proteins to produce a protein coat to form a viral particle that leaves the cell. The cell synthesizes so many virus particles that it eventually releases new viruses to infect more cells.

Vaccines are inactive forms of viruses that boost the immune response by causing the body to produce antibodies to the virus. Several childhood diseases, such as polio, mumps, chicken pox, and measles, can be prevented through the use of vaccines.

TEST

Try Practice Problems 21.91 and 21.92

Reverse Transcription

Most viruses contain RNA rather than DNA as the genetic material. A subset of these RNA-containing viruses are **retroviruses**. Once inside the host cell, a retrovirus makes viral DNA using a process known as *reverse transcription*. A retrovirus contains a polymerase enzyme called *reverse transcriptase* that uses the viral RNA template to synthesize complementary strands of DNA. Once produced, the single DNA strands form double-stranded DNA using the nucleotides present in the host cell. This newly formed viral DNA, called a *provirus*, integrates with the DNA of the host cell.

AIDS

During the early 1980s, a disease called *acquired immune deficiency syndrome*, commonly known as AIDS, began to claim an alarming number of lives. We now know that the HIV virus (human immunodeficiency virus) causes the disease (see **FIGURE 21.23**). HIV is a retrovirus that infects and destroys T4 lymphocyte cells, which are involved in the immune response. After the HIV binds to receptors on the surface of a T4 cell, the virus injects viral RNA into the host cell. As a retrovirus, the genes of the viral RNA direct the formation of viral DNA, which is then incorporated into the host's genome so it can replicate as part of the host cell's DNA. The gradual depletion of T4 cells reduces the ability of the immune system to destroy harmful organisms. The AIDS syndrome is characterized by opportunistic infections such as *Pneumocystis carinii*, which causes pneumonia, and *Kaposi's sarcoma*, a skin cancer.

Treatment for AIDS is based on attacking the HIV at different points in its life cycle, including cell entry, reverse transcription, and protein synthesis (see **FIGURE 21.24**). Nucleoside analogs mimic the structures of the nucleosides used for DNA synthesis and are able to successfully inhibit the reverse transcriptase enzyme. For example, the drug AZT (3′-azido-2′-deoxythymidine) is similar to thymidine, and ddI (2′,3′-dideoxyinosine) is similar to guanosine. Two other drugs are 2′,3′-dideoxycytidine (ddC) and 2′,3′-didehydro-2′,3′-dideoxythymidine (d4T). Such compounds are found in the "cocktails" that are providing extended remission of HIV infections. When a nucleoside analog is incorporated into viral DNA, the lack of a hydroxyl group on the 3′ carbon in the sugar prevents the formation of the sugar–phosphate bonds and stops the replication of the virus.

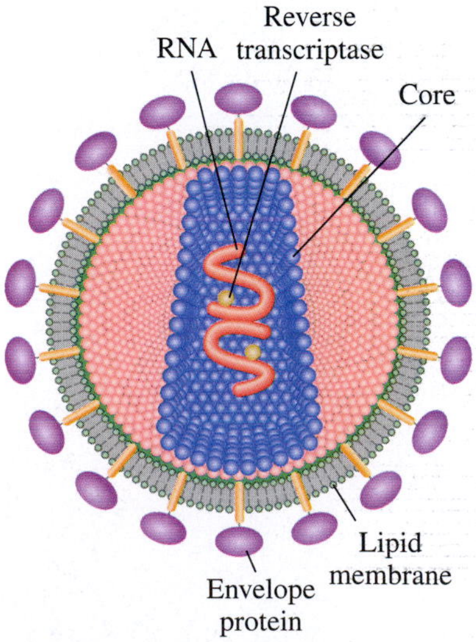

RNA
Reverse transcriptase
Core
Lipid membrane
Envelope protein

FIGURE 21.23 ▶ The HIV virus causes AIDS, which destroys the immune system in the body.

Q Is HIV a DNA virus or an RNA retrovirus?

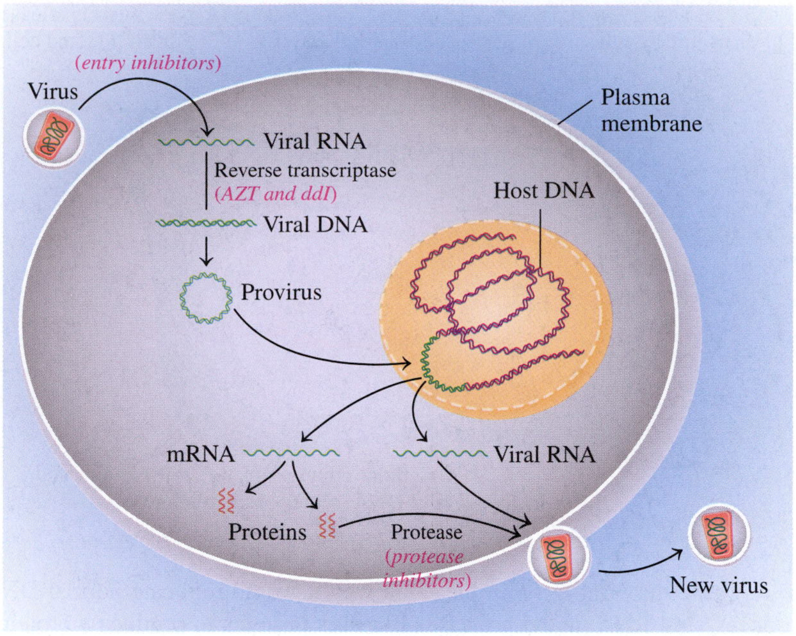
(entry inhibitors)
Virus
Plasma membrane
Viral RNA
Reverse transcriptase
(AZT and ddI)
Viral DNA
Host DNA
Provirus
mRNA
Viral RNA
Proteins
Protease
(protease inhibitors)
New virus

FIGURE 21.24 ▶ After a retrovirus injects its viral RNA into a cell, it forms a DNA strand by reverse transcription. The single-stranded DNA forms a double-stranded DNA called a provirus, which is incorporated into the host cell DNA. When the cell replicates, the provirus produces the viral RNA needed to produce more virus particles. The magenta text shows the locations where inhibitors can affect the life cycle of the HIV virus.

Q What is reverse transcription?

Today, people with HIV and AIDS are treated with a combination of drugs that include entry inhibitors, reverse transcriptase inhibitors, and protease inhibitors. Entry inhibitors attach to the surface of either the lymphocyte or HIV virus, which blocks entry into the cell. Entry inhibitors include enfuvirtide (Fuzeon) and maraviroc (Selzentry). Protease inhibitors prevent the proper cutting and formation of proteins used by the virus to make more copies of its own proteins. Protease inhibitors include saquinavir (Invirase), ritonavir (Norvir), fosamprenavir (Lexiva), and several others. When patients become resistant to certain drugs, different combinations are used to prolong life expectancy.

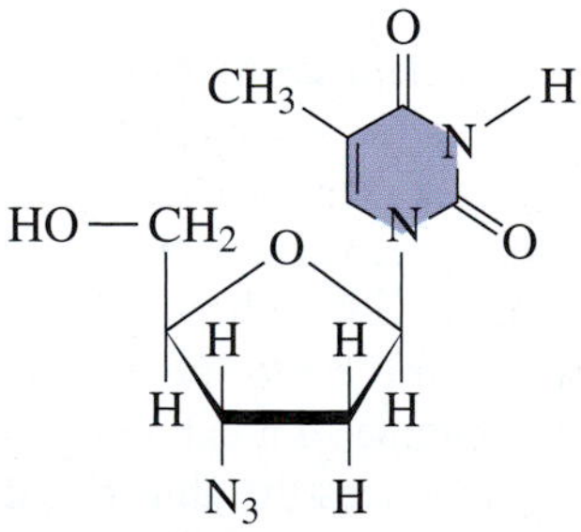

3′-Azido-2′-deoxythymidine (AZT)

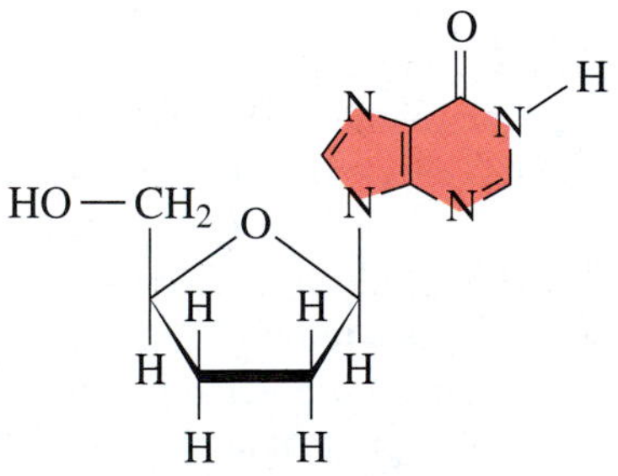

2′,3′-Dideoxyinosine (ddI)

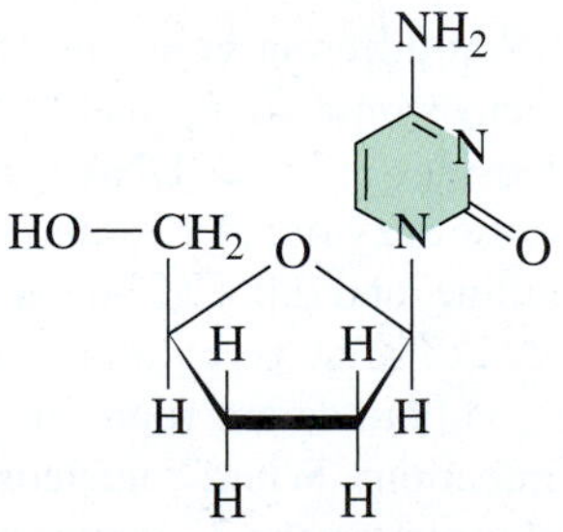

2′,3′-Dideoxycytidine (ddC)

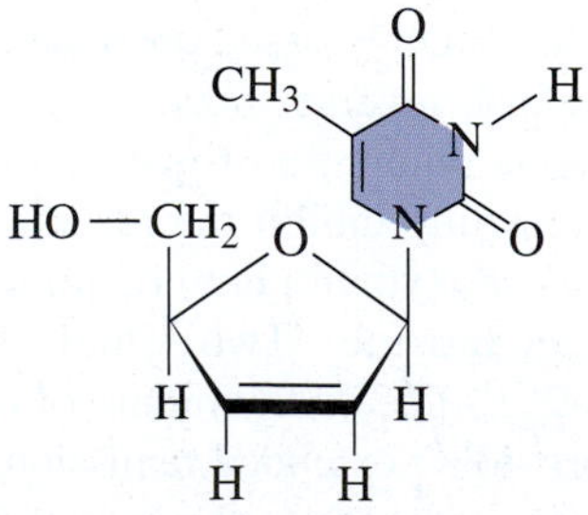

2′,3′-Didehydro-2′,3′-dideoxythymidine (d4T)

Nucleoside analogs inhibit the reverse transcriptase enzyme used by the HIV virus.

▶ SAMPLE PROBLEM 21.9 Viruses

TRY IT FIRST

Why are viruses unable to replicate on their own?

SOLUTION

Viruses contain only packets of DNA or RNA, but not the necessary replication machinery that includes enzymes and nucleosides.

STUDY CHECK 21.9

What are the essential parts of a virus?

ANSWER

nucleic acid (DNA or RNA) and a protein coat

TEST

Try Practice Problems 21.93 to 21.96

PRACTICE PROBLEMS

21.8 Viruses

21.91 What type of genetic information is found in a virus?

21.92 Why do viruses need to invade a host cell?

Clinical Applications

21.93 A specific virus contains RNA as its genetic material.

a. Why might reverse transcription be used in the life cycle of this type of virus?

b. What is the name of the type of virus that uses reverse transcription?

21.94 What is the purpose of a vaccine?

21.95 How do nucleoside analogs disrupt the life cycle of the HIV virus?

21.96 How do protease inhibitors disrupt the life cycle of the HIV virus?

Chemistry Link to Health

Cancer

Normally, cells in the body undergo orderly and controlled cell division. When cells begin to grow and multiply without control, they can invade neighboring cells and appear as a tumor. When they interfere with normal functions of the body, the tumors are cancerous. Cancer can be caused by chemical and environmental substances, by ultraviolet or medical radiation, or by *oncogenic viruses*, which are associated with human cancers.

Some reports estimate that 70 to 80% of all human cancers are initiated by chemical and environmental substances. A *carcinogen* is any substance that can cause cancer by reacting with molecules in a cell, probably DNA, and altering the growth of that cell. Some known carcinogens are listed in **TABLE 21.13**.

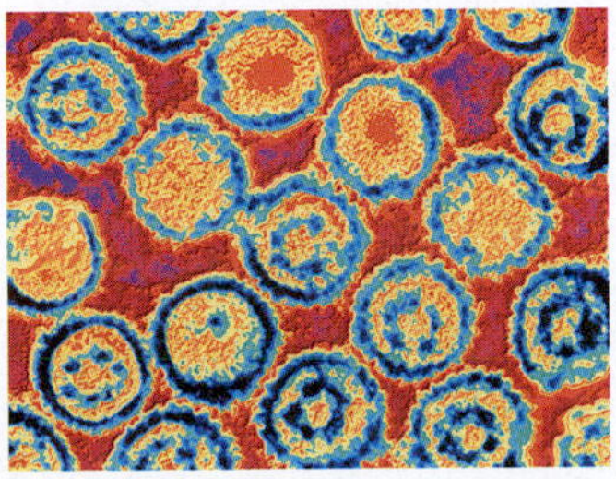

Epstein–Barr virus (EBV), herpesvirus 4, causes cancer in humans.

TABLE 21.13 Some Chemical and Environmental Carcinogens

Carcinogen	Tumor Site
Aflatoxin	Liver
Aniline dyes	Bladder
Arsenic	Skin, lungs
Asbestos	Lungs, respiratory tract
Cadmium	Prostate, kidneys
Chromium	Lungs
Nickel	Lungs, sinuses
Nitrites	Stomach
Vinyl chloride	Liver

Skin cancer has become one of the most prevalent forms of cancer. It appears that DNA damage in the areas of the skin exposed to ultraviolet radiation may eventually cause mutations. The cells lose their ability to control protein synthesis. This type of uncontrolled cell division becomes skin cancer.

Oncogenic viruses cause cancer when cells are infected. Several viruses associated with human cancers are listed in **TABLE 21.14**.

(*continued*)

Chemistry Link to Health (*continued*)

TABLE 21.14 Human Cancers Caused by Oncogenic Viruses

Virus	Disease
RNA Viruses	
Human T-cell lymphotropic virus–type 1 (HTLV-1)	Leukemia
DNA Viruses	
Epstein–Barr virus (EBV)	Burkitt's lymphoma (cancer of white blood B cells) Hodgkin's disease Nasopharyngeal carcinoma
Hepatitis B virus (HBV)	Liver cancer
Herpes simplex virus (HSV, type 2)	Cervical and uterine cancer
Human papilloma virus (HPV)	Cervical and colon cancer, genital warts

CLINICAL UPDATE Ellen's Medical Treatment Following Breast Cancer Surgery

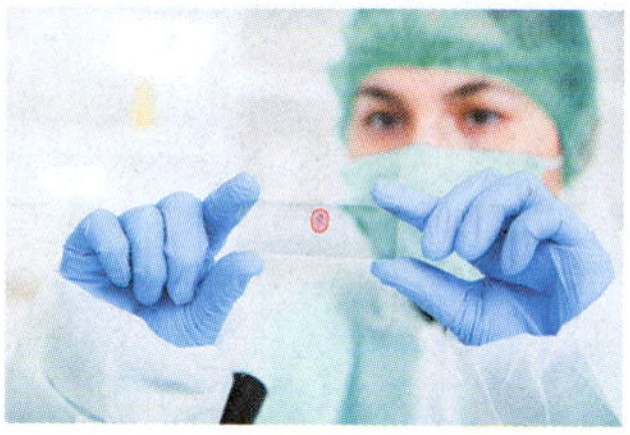

Ellen's oncologist tells her that the cells in her breast tumor and surrounding tissue tested positive for estrogen receptor (ER), which verified that she has breast cancer. Tumors that are estrogen positive require estrogen for their growth. Estrogen is a hormone that travels through the bloodstream and activates estrogen receptors in the cells of the breast and ovaries. The bonding of estrogen and the estrogen receptor to DNA increase the production of mammary cells and the number of potential mutations that can lead to cancer. High levels of the estrogen receptor appear in more than 60% of all breast cancer cases.

Because Ellen is 45 years old and has a family history of breast and ovarian cancer, she was also tested for altered genes BRCA1 and BRCA2. Normally, these genes suppress tumor growth by repairing DNA defects. However, if the mutated genes are inherited, a person's cells lose the ability to suppress tumor growth, and the risk for breast and ovarian cancer as well as other cancers becomes much greater. Both parents can carry a BRCA mutation and pass it on to their sons and daughters. Ellen is relieved that her test results for mutated BRCA1 and BRCA2 were negative.

After Ellen completed a radiation series, she discusses the choice of drugs available for the treatment of breast cancer. Drugs such as tamoxifen and raloxifene block the binding of estrogen to the estrogen receptor, which prevents the growth

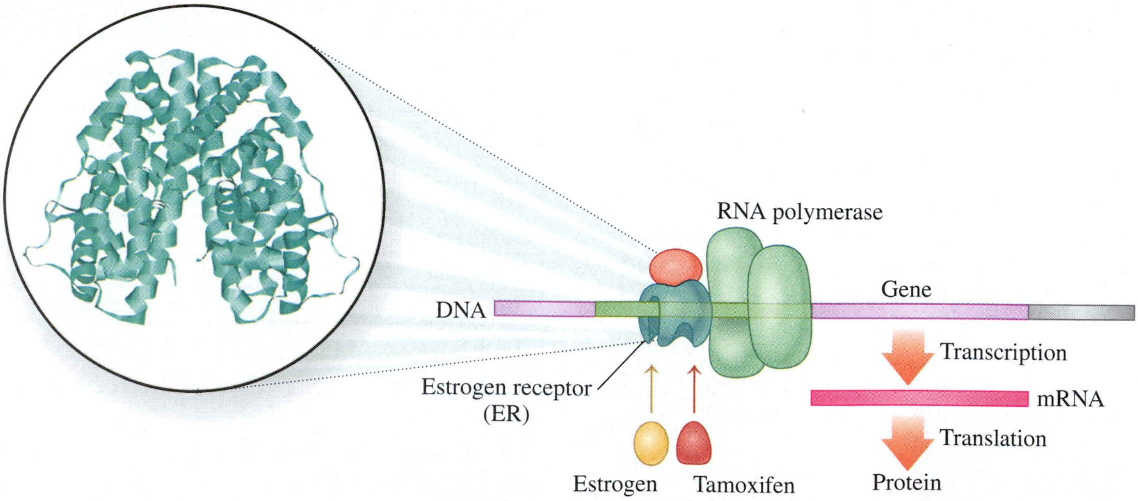
RNA polymerase
Gene
DNA
Transcription
mRNA
Estrogen receptor (ER)
Translation
Estrogen
Tamoxifen
Protein

The estrogen receptor has a binding region for either estrogen or tamoxifen. Tamoxifen slows down transcription by blocking the binding of estrogen to the estrogen receptor.

of cancers that are estrogen positive. Other drugs such as anastrozole (Arimidex) and letrozole (Femara) are aromatase inhibitors (AIs) that prevent the growth of new tumors by blocking the synthesis of aromatase (estrogen synthetase), an enzyme that produces estrogen. Aromatase inhibitors are used primarily for postmenopausal women whose ovaries no longer produce estrogen. Ellen and her oncologist agree upon the use of tamoxifen, which she will take for the next five years to prevent breast cancer from recurring after her surgery.

Clinical Applications

21.97 What are estrogen receptors?

21.98 How do estrogen and the estrogen receptor influence cancer growth of mammary cells?

21.99 How does tamoxifen reduce the potential of breast cancer?

21.100 How does letrozole reduce the potential of breast cancer?

21.101 The following is a segment of the template strand of human BRCA1 gene:

3′ TGG AAT TAT CT**G** CTC TTC GCG 5′

a. Write the corresponding mRNA segment.
b. Write the three-letter and one-letter abbreviations for the corresponding peptide segment.
c. If there is a point mutation in the fourth nucleotide triplet and A replaces **G**, what is the change, if any, in the amino acid sequence?

21.102 The following is a segment of the template strand of human BRCA1 gene:

3′ ACA T**A**T TTT GCA AAT TTT GCA 5′

a. Write the corresponding mRNA segment.
b. Write the three-letter and one-letter abbreviations for the corresponding peptide segment.
c. If there is a point mutation in the second nucleotide triplet and C replaces **A**, what is the change, if any, in the amino acid sequence?

CONCEPT MAP

NUCLEIC ACIDS AND PROTEIN SYNTHESIS

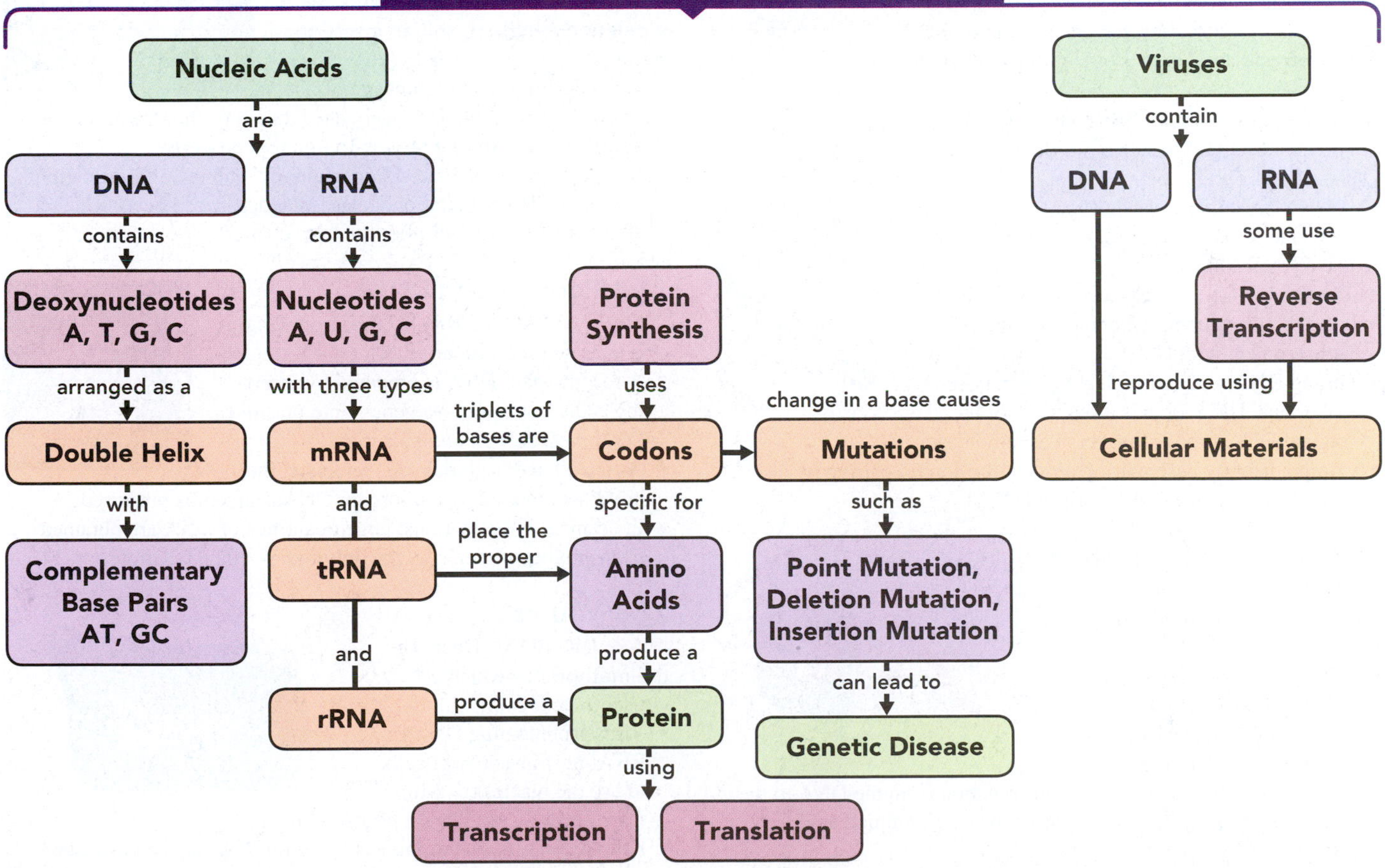

Nucleic Acids
are
DNA
RNA
contains
contains
Deoxynucleotides A, T, G, C
Nucleotides A, U, G, C
arranged as a
with three types
Double Helix
mRNA
with
and
Complementary Base Pairs AT, GC
tRNA
and
rRNA
triplets of bases are
place the proper
produce a
Protein Synthesis
uses
Codons
specific for
Amino Acids
produce a
Protein
using
Transcription
Translation
change in a base causes
Mutations
such as
Point Mutation, Deletion Mutation, Insertion Mutation
can lead to
Genetic Disease
Viruses
contain
DNA
RNA
some use
Reverse Transcription
reproduce using
Cellular Materials

CHAPTER REVIEW

21.1 Components of Nucleic Acids

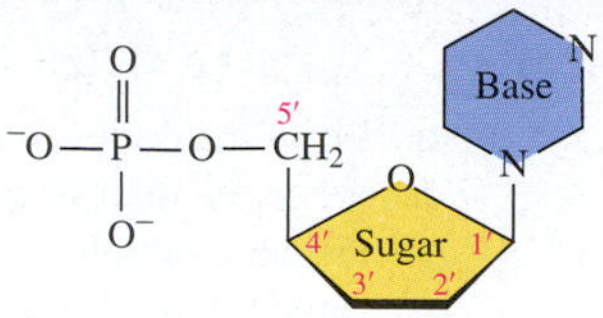
O
$^{-}$O—P—O—CH$_2$
O^{-}
5′
Base
N
N
O
Sugar
4′
1′
3′
2′

LEARNING GOAL Describe the bases and ribose sugars that make up the nucleic acids DNA and RNA.

- Nucleic acids, such as deoxyribonucleic acid (DNA) and ribonucleic acid (RNA), are unbranched chains of nucleotides.
- A nucleoside is a combination of a pentose sugar and a base.
- A nucleotide is composed of three parts: a pentose sugar, a base, and a phosphate group.
- In DNA, the sugar is deoxyribose and the base can be adenine, thymine, guanine, or cytosine.
- In RNA, the sugar is ribose, and the base can be adenine, uracil, guanine, or cytosine.

21.2 Primary Structure of Nucleic Acids

LEARNING GOAL Describe the primary structures of RNA and DNA.

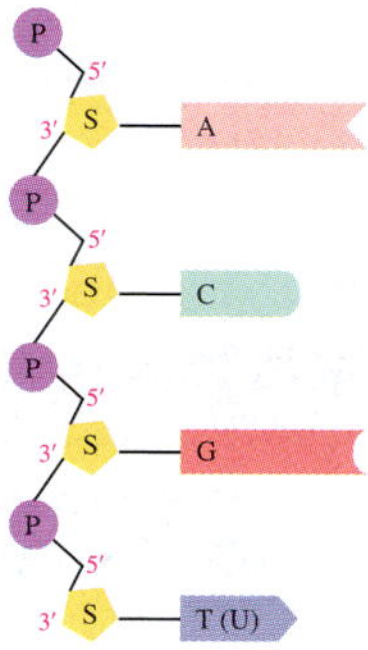
P
5′
3′
S
A
C
G
T (U)

- Each nucleic acid has its own unique sequence of bases known as its primary structure.
- In a nucleic acid, the 3′ OH group of each ribose in RNA or deoxyribose in DNA forms a phosphodiester linkage to the phosphate group of the 5′ carbon atom of the sugar in the next nucleotide to give a backbone of alternating sugar and phosphate groups.
- There is a free 5′ phosphate at one end of the nucleic acid and a free 3′ OH group at the other end.

21.3 DNA Double Helix and Replication

LEARNING GOAL Describe the double helix of DNA; describe the process of DNA replication.

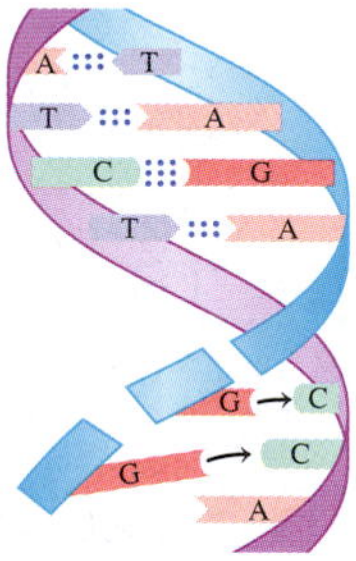
A
T
T
A
C
G
T
A
G
C
C
G
A

- A DNA molecule consists of two strands of nucleotides that are wound around each other like a spiral staircase.
- The two strands are held together by hydrogen bonds between complementary base pairs, A with T, and G with C.
- During DNA replication, DNA polymerase makes new DNA strands along each of the original DNA strands that serve as templates.
- Complementary base pairing ensures the correct pairing of bases to give identical copies of the original DNA.

21.4 RNA and Transcription

LEARNING GOAL Identify the different types of RNA; describe the synthesis of mRNA.

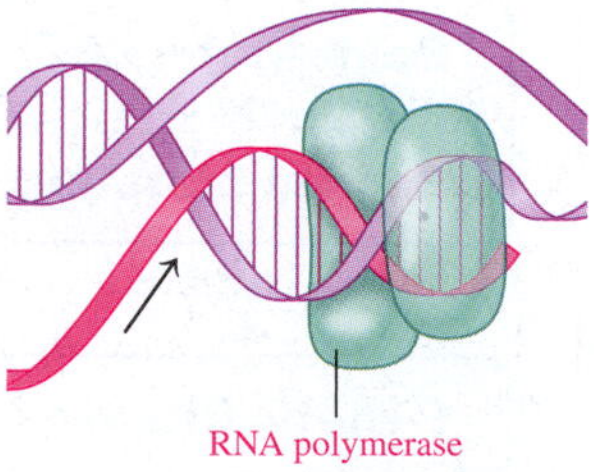
RNA polymerase

- The three main types of RNA differ by function in the cell: ribosomal RNA makes up most of the structure of the ribosomes, messenger RNA carries genetic information from the DNA to the ribosomes, and transfer RNA places the correct amino acids in a growing peptide chain.
- Transcription is the process by which RNA polymerase produces mRNA from one strand of DNA.
- The bases in the mRNA are complementary to the DNA, except A in DNA is paired with U in RNA.
- The production of mRNA occurs when certain proteins are needed in the cell.
- In eukaryotes, transcription factors at the promoter region bind RNA polymerase to DNA, which initiates the transcription of a gene.
- In eukaryotes, newly transcribed mRNA is processed to remove introns.

21.5 The Genetic Code and Protein Synthesis

LEARNING GOAL Use the genetic code to write the amino acid sequence for a segment of mRNA.

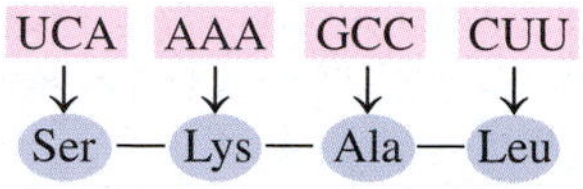
UCA
AAA
GCC
CUU
Ser—Lys—Ala—Leu

- The genetic code consists of a series of codons, which are sequences of three bases that specify the order for the amino acids in a protein.
- There are 64 codons for the 20 amino acids, which means there are multiple codons for most amino acids.
- The codon AUG signals the start of transcription, and codons UAG, UGA, and UAA signal it to stop.
- Proteins are synthesized at the ribosomes in a translation process that includes three steps: initiation, chain elongation, and termination.
- During translation, tRNAs bring the appropriate amino acids to the ribosome, and peptide bonds form to join the amino acids in a peptide chain.
- When the polypeptide is released, it takes on its secondary and tertiary structures and becomes a functional protein in the cell.

21.6 Genetic Mutations

LEARNING GOAL Identify the type of change in DNA for a point mutation, a deletion mutation, and an insertion mutation.

- A genetic mutation is a change of one or more bases in the DNA sequence that may alter the structure and ability of the resulting protein to function properly.
- In a point mutation, one codon is altered, and in an insertion mutation or a deletion mutation, a base is added or removed, which changes all the codons after the base change.

21.7 Recombinant DNA

LEARNING GOAL Describe the preparation and uses of recombinant DNA.

- A recombinant DNA is prepared by inserting a DNA segment—a gene—into plasmid DNA present in *E. coli* bacteria.
- As the altered bacterial cells replicate, the protein expressed by the foreign DNA segment is produced.
- In criminal investigations, large quantities of DNA are obtained from smaller amounts by the polymerase chain reaction.

21.8 Viruses

LEARNING GOAL Describe the methods by which a virus infects a cell.

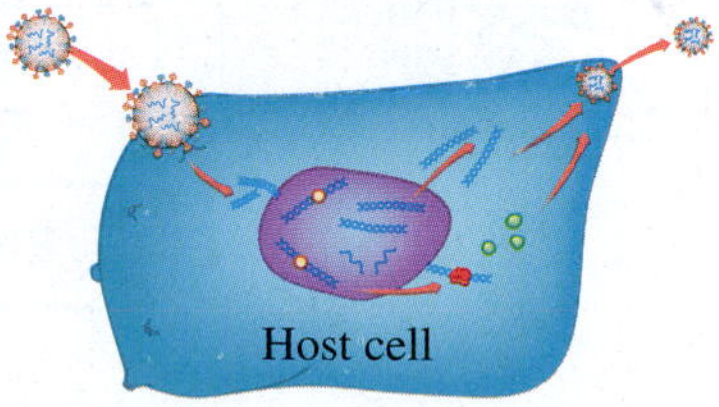
Host cell

- Viruses containing DNA or RNA must invade host cells to use the machinery within the cell for the synthesis of more viruses.
- For a retrovirus containing RNA, a viral DNA is synthesized by reverse transcription using the nucleotides and enzymes in the host cell.
- In the treatment of AIDS, entry inhibitors block the virus from entering the cell, nucleoside analogs inhibit the reverse transcriptase of the HIV virus, and protease inhibitors disrupt the catalytic activity of protease needed to produce proteins for the synthesis of more viruses.

KEY TERMS

anticodon The triplet of bases in the center loop of tRNA that is complementary to a codon on mRNA.

base A nitrogen-containing compound found in DNA and RNA: adenine (A), thymine (T), cytosine (C), guanine (G), and uracil (U).

codon A sequence of three bases in mRNA that specifies a certain amino acid to be placed in a protein. A few codons signal the start or stop of protein synthesis.

complementary base pairs In DNA, adenine is always paired with thymine (A and T or T and A), and guanine is always paired with cytosine (G and C or C and G). In forming RNA, adenine is always paired with uracil (A and U or U and A).

deletion mutation A mutation that deletes a base from a DNA sequence.

DNA Deoxyribonucleic acid; the genetic material of all cells containing nucleotides with deoxyribose, phosphate, and the four bases: adenine, thymine, guanine, and cytosine.

double helix The helical shape of the double chain of DNA that is like a spiral staircase with a sugar–phosphate backbone on the outside and base pairs like stair steps on the inside.

exons The sections in a DNA template that code for proteins.

genetic code The sequence of codons in mRNA that specifies the amino acid order for the synthesis of protein.

genetic disease A physical malformation or metabolic dysfunction caused by a mutation in the base sequence of DNA.

insertion mutation A mutation that inserts a base in a DNA sequence.

introns The sections in DNA that do not code for proteins.

mRNA Messenger RNA; produced in the nucleus from DNA to carry the genetic information to the ribosomes for the construction of a protein.

mutation A change in the DNA base sequence that may alter the formation of a protein in the cell.

nucleic acid A large molecule composed of nucleotides; found as a double helix in DNA and as the single strands of RNA.

nucleoside The combination of a pentose sugar and a base.

nucleotide Building block of a nucleic acid consisting of a base, a pentose sugar (ribose or deoxyribose), and a phosphate group.

Okazaki fragment A short segment formed by DNA polymerase in the daughter DNA strand that runs in the 3′ to 5′ direction.

phosphodiester linkage The phosphate link that joins the 3′ hydroxyl group in one nucleotide to the phosphate group on the 5′ carbon atom in the next nucleotide.

point mutation A mutation that replaces one base in a DNA with a different base.

polymerase chain reaction (PCR) A procedure in which a strand of DNA is copied many times by mixing it with DNA polymerase and a mixture of deoxyribonucleotides, and subjecting it to repeated cycles of heating and cooling.

primary structure The sequence of nucleotides in nucleic acids.

recombinant DNA DNA combined from different organisms to form new, synthetic DNA.

replication The process of duplicating DNA by pairing the bases on each parent strand with their complementary bases.

replication forks The open sections in unwound DNA strands where DNA polymerase begins the replication process.

retrovirus A virus that contains RNA as its genetic material and that synthesizes a complementary DNA strand inside a cell.

RNA Ribonucleic acid; a type of nucleic acid that is a single strand of nucleotides containing ribose, phosphate, and the four bases adenine, cytosine, guanine, and uracil.

rRNA Ribosomal RNA; the most prevalent type of RNA and a major component of the ribosomes.

transcription The transfer of genetic information from DNA by the formation of mRNA.

transcription factor A protein that binds to the promoter region of DNA and makes it easier for RNA polymerase to bind.

translation The interpretation of the codons in mRNA as amino acids in a peptide.

tRNA Transfer RNA; an RNA that places a specific amino acid into a peptide chain at the ribosome so that a protein can be made. There is one or more tRNA for each of the 20 different amino acids.

virus A small particle containing DNA or RNA in a protein coat that requires a host cell for replication.

CORE CHEMISTRY SKILLS

The chapter Section containing each Core Chemistry Skill is shown in parentheses at the end of each heading.

Writing the Complementary DNA Strand (21.3)

- During DNA replication, new DNA strands are made along each of the original DNA strands.
- The new strand of DNA is made by forming hydrogen bonds with the bases in the template strand: A with T and T with A; C with G and G with C.

Example: Write the complementary base sequence for the following DNA segment:
5′ A T T C G G T A C 3′

Answer: 3′ T A A G C C A T G 5′

Writing the mRNA Segment for a DNA Template (21.4)

- Transcription is the process that produces mRNA from one strand of DNA.
- The bases in the mRNA are complementary to the DNA, except that A in DNA is paired with U in RNA.

Example: What is the mRNA produced from the DNA segment 5′ C G C A T G T C A 3′ ?

Answer: 3′ G C G U A C A G U 5′

Writing the Amino Acid for an mRNA Codon (21.5)

- The genetic code consists of a sequence of three bases (codons) that specifies the order for the amino acids in a protein.
- The codon AUG signals the start of transcription and codons UAG, UGA, and UAA signal it to stop.

Example: Use three-letter and one-letter abbreviations to write the amino acid sequence you would expect in a peptide for the mRNA sequence 5′ CCG UAU GGG 3′.

Answer: Pro–Tyr–Gly, PYG

UNDERSTANDING THE CONCEPTS

The chapter Sections to review are shown in parentheses at the end of each problem.

21.103 Answer the following questions for the given section of DNA: (21.3, 21.4, 21.5)

a. Complete the bases in the parent and new strands.

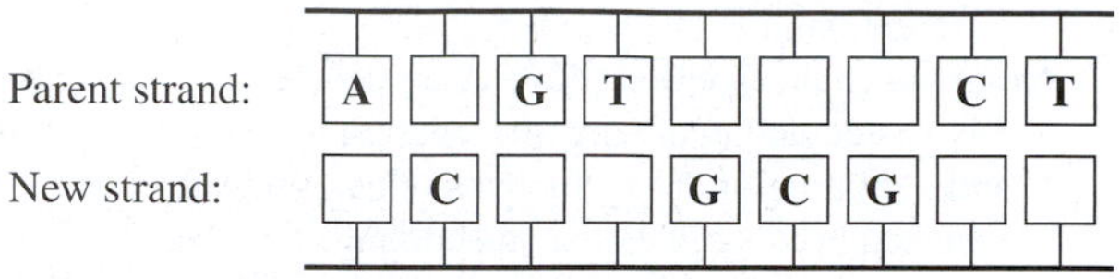
Parent strand:
A G T C T
New strand:
C G C G

b. Using the new strand as a template, write the mRNA sequence.

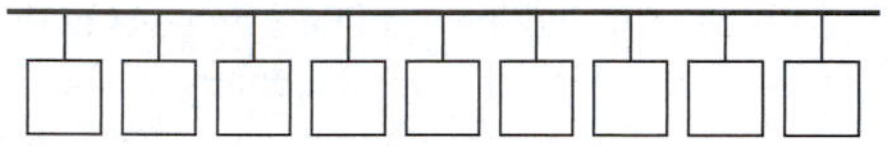

c. Write the three-letter symbols for the amino acids that would go into the peptide from the mRNA you wrote in part **b**.

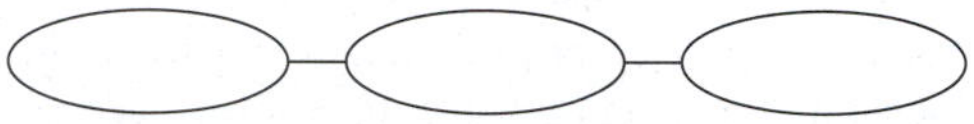

21.104 Suppose a mutation occurs in the DNA section in problem 21.103, and the first base in the parent chain, adenine, is replaced by guanine. (21.3, 21.4, 21.5, 21.6)

a. What type of mutation has occurred?

b. Using the new strand that results from this mutation, write the order of bases in the altered mRNA.

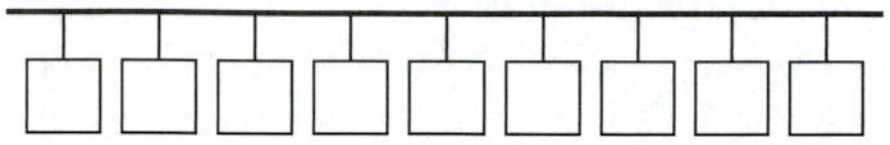

c. Write the three-letter symbols for the amino acids that would go into the peptide from the mRNA you wrote in part **b**.

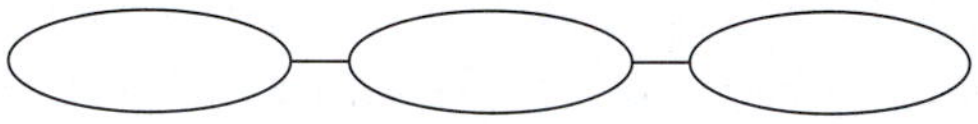

d. What effect, if any, might this mutation have on the structure and/or function of the resulting protein?

ADDITIONAL PRACTICE PROBLEMS

21.105 Identify each of the following bases as a pyrimidine or a purine: (21.1)

a. cytosine **b.** adenine **c.** uracil
d. thymine **e.** guanine

21.106 Indicate if each of the bases in problem 21.105 is found in DNA only, RNA only, or both DNA and RNA. (21.1)

21.107 Identify the base and sugar in each of the following nucleosides: (21.1)

a. deoxythymidine **b.** adenosine
c. cytidine **d.** deoxyguanosine

21.108 Identify the base and sugar in each of the following nucleotides: (21.1)

a. CMP **b.** dAMP **c.** dTMP **d.** UMP

21.109 How do the bases thymine and uracil differ? (21.1)

21.110 How do the bases cytosine and uracil differ? (21.1)

21.111 Draw the condensed structural formula for CMP. (21.1)

21.112 Draw the condensed structural formula for dGMP. (21.1)

21.113 What is similar about the primary structures of RNA and DNA? (21.2)

21.114 What is different about the primary structures of RNA and DNA? (21.2)

21.115 If the DNA double helix in salmon contains 28% adenine, what are the percentages of thymine, guanine, and cytosine? (21.3)

21.116 If the DNA double helix in humans contains 20% cytosine, what are the percentages of guanine, adenine, and thymine? (21.3)

21.117 In DNA, how many hydrogen bonds form between adenine and thymine? (21.3)

21.118 In DNA, how many hydrogen bonds form between guanine and cytosine? (21.3)

21.119 Write the complementary base sequence for each of the following DNA segments: (21.3)

a. 5′ G A C T T A G G C 3′
b. 3′ T G C A A A C T A G C T 5′
c. 5′ A T C G A T C G A T C G 3′

21.120 Write the complementary base sequence for each of the following DNA segments: (21.3)

a. 5′ T T A C G G A C C G C 3′
b. 5′ A T A G C C C T T A C T G G 3′
c. 3′ G G C C T A C C T T A A C G A C G 5′

21.121 In DNA replication, what is the difference between the synthesis of the leading strand and the synthesis of the lagging strand? (21.3)

21.122 How are the Okazaki fragments joined to the growing DNA strand? (21.3)

21.123 After the replication of a DNA, where are the original DNA strands located in the daughter DNA molecules? (21.3)

21.124 How can replication occur at several places along a DNA double helix? (21.3)

21.125 Match the following statements with rRNA, mRNA, or tRNA: (21.4)

a. is the smallest type of RNA
b. makes up the highest percentage of RNA in the cell
c. carries genetic information from the nucleus to the ribosomes

21.126 Match the following statements with rRNA, mRNA, or tRNA: (21.4)

a. combines with proteins to form ribosomes
b. brings amino acids to the ribosomes for protein synthesis
c. acts as a template for protein synthesis

21.127 What are the possible codons for each of the following amino acids? (21.5)

a. threonine **b.** serine **c.** cysteine

21.128 What are the possible codons for each of the following amino acids? (21.5)
a. valine **b.** arginine **c.** histidine

21.129 What is the amino acid for each of the following codons? (21.5)
a. AAG **b.** AUU **c.** CGA

21.130 What is the amino acid for each of the following codons? (21.5)
a. CAA **b.** GGC **c.** UGG

21.131 What is the anticodon on tRNA for each of the following codons in an mRNA? (21.5)
a. AGC **b.** UAU **c.** CCA

21.132 What is the anticodon on tRNA for each of the following codons in an mRNA? (21.5)
a. GUG **b.** CCC **c.** GAA

Clinical Applications

21.133 Endorphins are polypeptides that reduce pain. What is the amino acid order for the endorphin leucine enkephalin (leu-enkephalin), which has the following mRNA? (21.5)

5′ AUG UAC GGU GGA UUU CUA UAA 3′

21.134 Endorphins are polypeptides that reduce pain. What is the amino acid order for the endorphin methionine enkephalin (met-enkephalin), which has the following mRNA? (21.5)

5′ AUG UAC GGU GGA UUU AUG UAA 3′

21.135 Beta-amyloids are proteins that are found in the brains of Alzheimer's patients. Determine the amino acid sequence of a peptide section of beta-amyloid containing 10 amino acid residues when it is treated with chymotrypsin and trypsin to give the following fragments: (21.7)

Reagent	Fragments Produced from Hydrolysis
Chymotrypsin	Glu–Phe Glu–Val Arg–His–Asp–Ser–Gly–Tyr
Trypsin	Glu–Phe–Arg His–Asp–Ser–Gly–Tyr–Glu–Val

21.136 In the brains of Alzheimer's patients, beta-amyloid proteins change from a normal alpha-helical shape to beta-pleated sheets that stick together and form plaques. Determine the amino acid sequence of a peptide section of beta-amyloid containing 10 amino acid residues when it is treated with chymotrypsin and trypsin to give the following fragments: (21.7)

Reagent	Fragments Produced from Hydrolysis
Chymotrypsin	Phe Ala–Glu–Asp His–His–Gln–Lys–Leu–Val–Phe
Trypsin	His–His–Gln–Lys Leu–Val–Phe–Phe–Ala–Glu–Asp

CHALLENGE PROBLEMS

The following problems are related to the topics in this chapter. However, they do not all follow the chapter order, and they require you to combine concepts and skills from several Sections. These problems will help you increase your critical thinking skills and prepare for your next exam.

21.137 Oxytocin is a peptide that contains nine amino acids. How many nucleotides would be found in the mRNA for this peptide? (21.5)

21.138 A polypeptide contains 36 amino acids. How many nucleotides would be found in the mRNA for this polypeptide? (21.5)

21.139 What is the difference between a DNA virus and a retrovirus? (21.8)

21.140 Why are there no base pairs in DNA between adenine and guanine or thymine and cytosine? (21.3)

21.141 Match each of the following processes (**1** to **5**) with one of the items (**a** to **e**): (21.3, 21.4, 21.5, 21.7, 21.8)
1. replication of DNA **2.** transcription **3.** translation
4. recombinant DNA **5.** reverse transcription

a. DNA polymerase
b. mRNA is synthesized from nuclear DNA
c. viruses
d. restriction enzymes
e. tRNA molecules bond to codons

21.142 Match each of the following processes (**1** to **5**) with one of the items (**a** to **e**): (21.3, 21.4, 21.5, 21.7, 21.8)
1. replication of DNA **2.** transcription **3.** translation
4. recombinant DNA **5.** reverse transcription

a. amino acids are linked together
b. RNA template is used to synthesize DNA
c. helicase unwinds DNA
d. genetic information is transferred from DNA
e. sticky ends join new DNA segment

ANSWERS

21.1 **a.** pyrimidine **b.** purine

21.3 **a.** DNA **b.** both DNA and RNA

21.5 deoxyadenosine monophosphate (dAMP), deoxythymidine monophosphate (dTMP), deoxycytidine monophosphate (dCMP), and deoxyguanosine monophosphate (dGMP)

21.7 **a.** nucleoside **b.** nucleoside
c. nucleoside **d.** nucleotide

21.9 **a.** both DNA and RNA **b.** RNA only
c. DNA only **d.** RNA only

21.11

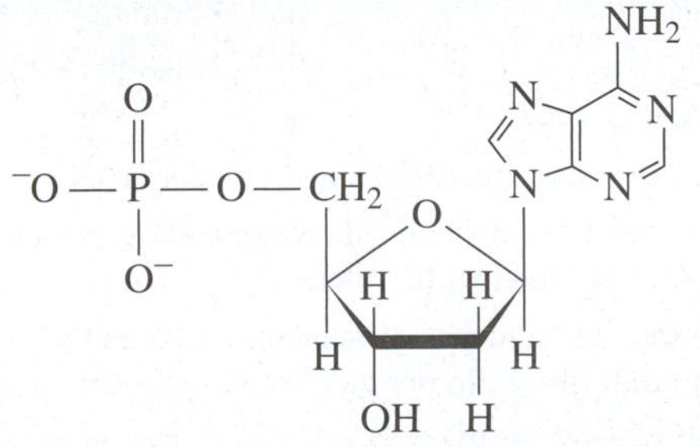
NH2
O
N
N
N
N
⁻O — P — O — CH2
O
O⁻
H H
H H
OH H

21.13

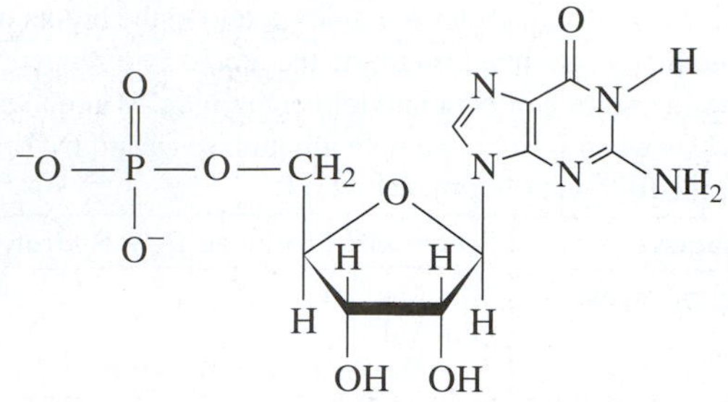

21.15 The nucleotides in nucleic acids are held together by phosphodiester linkages between the 3′ OH group of a sugar (ribose or deoxyribose) and a phosphate group on the 5′ carbon of another sugar.

21.17 –sugar–phosphate–sugar–phosphate–

21.19 the free phosphate on the 5′ carbon of ribose or deoxyribose of a nucleic acid

21.21

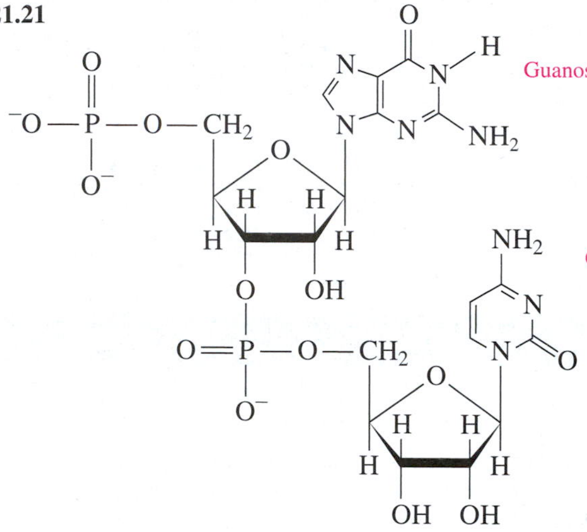

Guanosine (G)
Cytidine (C)

21.23 Structural features of DNA include that it is shaped like a double helix, it contains a sugar–phosphate backbone, the nitrogen-containing bases are hydrogen bonded between strands, and the strands run in opposite directions. A forms two hydrogen bonds to T, and G forms three hydrogen bonds to C.

21.25 The two DNA strands are held together by hydrogen bonds between the complementary bases in each strand.

21.27 **a.** 3′ T T T T T T 5′
b. 3′ C C C C C C 5′
c. 3′ T C A G G T C C A 5′
d. 3′ G A C A T A T G C A A T 5′

21.29 The enzyme helicase unwinds the DNA helix so that the parent DNA strands can be replicated into daughter DNA strands.

21.31 Once the DNA strands separate, the DNA polymerase pairs each of the bases with its complementary base and produces two exact copies of the original DNA.

21.33 **a.** (2) helicase **b.** (3) primase
c. (4) replication fork **d.** (1) lagging strand

21.35 **a.** 5′ C G A G G T A C 3′ **b.** No

21.37 ribosomal RNA, messenger RNA, and transfer RNA

21.39 A ribosome consists of a small subunit and a large subunit that contain rRNA combined with proteins.

21.41 In transcription, the sequence of nucleotides on a DNA template (one strand) is used to produce the base sequence of a messenger RNA.

21.43 5′ GGC UUC CAA GUG 3′

21.45 In eukaryotic cells, genes contain sections called exons that code for proteins and sections called introns that do not code for proteins.

21.47 5′ GGC GGA UCG 3′

21.49 A transcription factor is a molecule that binds to DNA and regulates transcription.

21.51 An activator is a protein that binds to eukaryotic DNA, making it easier for RNA polymerase to bind to the promoter region for a gene and increasing the level of transcription of that gene.

21.53 A codon is a three-base sequence in mRNA that codes for a specific amino acid in a protein.

21.55 **a.** proline (Pro) **b.** asparagine (Asn)
c. glycine (Gly) **d.** arginine (Arg)

21.57 When AUG is the first codon, it signals the start of protein synthesis. Thereafter, AUG codes for methionine.

21.59 A codon is a base triplet in the mRNA. An anticodon is the complementary triplet on a tRNA for a specific amino acid.

21.61 The new amino acid is joined by a peptide bond to the growing peptide chain. The ribosome moves to the next codon, which attaches to a tRNA carrying the next amino acid.

21.63 initiation, chain elongation, and termination

21.65 **a.** Thr–Thr–Thr, TTT **b.** Phe–Pro–Phe–Pro, FPFP
c. Tyr–Gly–Arg–Cys, YGRC

21.67 **a.** 5′ CGA UAU GGU UUU 3′ **b.** GCU, AUA, CCA, AAA
c. Arg–Tyr–Gly–Phe, RYGF

21.69 **a.** 5′ AAA CAC UUG GUU GUG GAC 3′
b. Lys–His–Leu–Val–Val–Asp, KHLVVD

21.71 In a point mutation, a base in DNA is replaced by a different base.

21.73 In a mutation caused by the deletion of a base, all the codons from the mutation onward are changed, which changes the order of amino acids in the rest of the polypeptide chain.

21.75 The normal triplet TTT forms a codon AAA, which codes for lysine. The mutation TTC forms a codon AAG, which also codes for lysine. There is no effect on the amino acid sequence.

21.77 **a.** Thr–Ser–Arg–Val
b. Thr–Thr–Arg–Val
c. Thr–Ser–Gly–Val
d. Thr–STOP. Protein synthesis would terminate early. If this occurs early in the formation of the polypeptide, the resulting protein will probably be nonfunctional.
e. The new protein will contain the sequence Asp–Ile–Thr–Gly.
f. The new protein will contain the sequence His–His–Gly.

21.79 **a.** GCC and GCA both code for alanine.
b. A vital ionic interaction in the tertiary structure of hemoglobin cannot be formed when the polar acidic glutamate is replaced by valine, which is nonpolar. The resulting hemoglobin is malformed and less capable of carrying oxygen.

21.81 *E. coli* bacterial cells contain several small circular plasmids of DNA that can be isolated easily. After the recombinant DNA is formed, *E. coli* multiply rapidly, producing many copies of the recombinant DNA in a relatively short time.

21.83 *E. coli* can be soaked in a detergent solution that disrupts the plasma membrane and releases the cell contents, including the plasmids, which are collected.

21.85 When a gene has been obtained using a restriction enzyme, it is mixed with plasmids that have been cut by the same enzyme. When mixed together, the sticky ends of the DNA fragments bond with the sticky ends of the plasmid DNA to form a recombinant DNA.

21.87 In DNA fingerprinting, restriction enzymes cut a sample DNA into fragments, which are sorted by size using gel electrophoresis. A radioactive probe that adheres to specific DNA sequences exposes an X-ray film and creates a pattern of dark and light bands called a DNA fingerprint.

21.89 Asp–Arg–Val–Tyr–Ile–His–Pro–Phe

21.91 DNA or RNA, but not both

21.93 **a.** A viral RNA is used to synthesize a viral DNA to produce the proteins for the protein coat, which allows the virus to replicate and leave the cell.
b. retrovirus

21.95 Nucleoside analogs such as AZT and ddI are similar to the nucleosides required to make viral DNA in reverse transcription. However, they interfere with the ability of the DNA to form and thereby disrupt the life cycle of the HIV virus.

21.97 Estrogen receptors bind to DNA and increase the production of mammary cells, which can lead to mutations and cancer.

21.99 Tamoxifen blocks the binding of estrogen to the estrogen receptor, which prevents the activation of the gene.

21.101 **a.** 5′ ACC UUA AUA GAC GAG AAG CGC 3′
b. Thr–Leu–Ile–Asp–Glu–Lys–Arg, TLIDEKR
c. There is no change; Asp is still the fourth amino acid.

21.103 **a.**

Parent strand:
New strand:

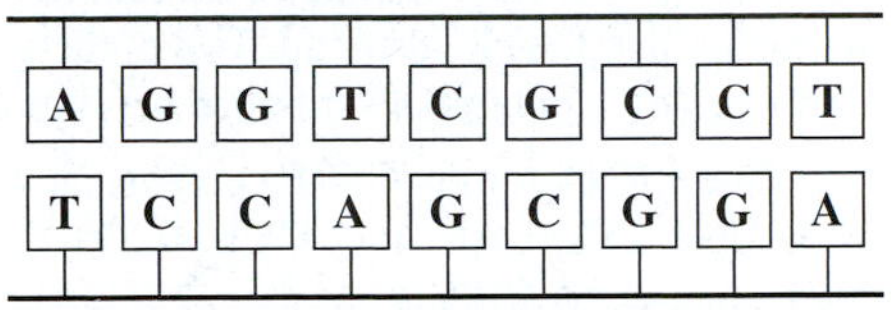
A G G T C G C C T
T C C A G C G G A

b.

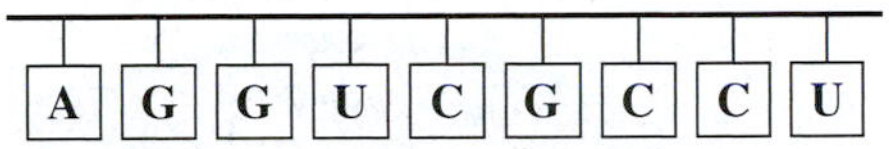
A G G U C G C C U

c.

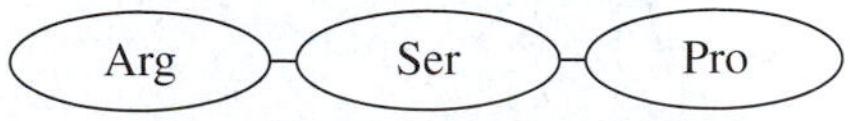
Arg
Ser
Pro

21.105 **a.** pyrimidine **b.** purine **c.** pyrimidine
d. pyrimidine **e.** purine

21.107 **a.** thymine and deoxyribose **b.** adenine and ribose
c. cytosine and ribose **d.** guanine and deoxyribose

21.109 They are both pyrimidines, but thymine has a methyl group.

21.111

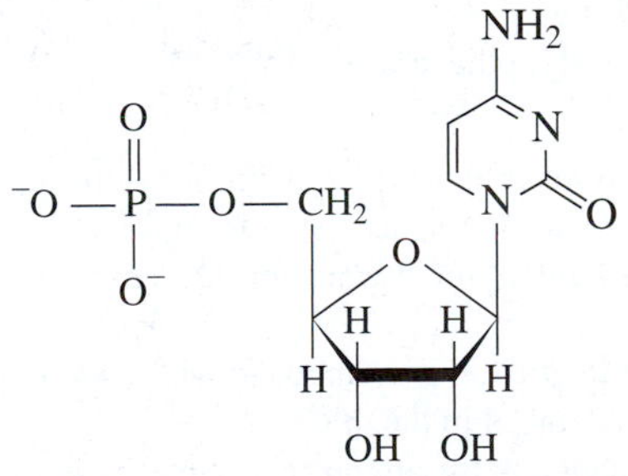

21.113 They are both unbranched chains of nucleotides connected through phosphodiester linkages between alternating sugar and phosphate groups, with bases extending out from each sugar.

21.115 28% T, 22% G, and 22% C

21.117 two

21.119 **a.** 3′ C T G A A T C C G 5′
b. 5′ A C G T T T G A T C G A 3′
c. 3′ T A G C T A G C T A G C 5′

21.121 DNA polymerase synthesizes the leading strand continuously in the 5′ to 3′ direction. The lagging strand is synthesized in small segments called Okazaki fragments because it must grow in the 3′ to 5′ direction and DNA polymerase can only work in the 5′ to 3′ direction.

21.123 One strand of the parent DNA is found in each of the two copies of the daughter DNA molecule.

21.125 **a.** tRNA **b.** rRNA **c.** mRNA

21.127 **a.** ACU, ACC, ACA, and ACG
b. UCU, UCC, UCA, UCG, AGU, and AGC
c. UGU and UGC

21.129 **a.** lysine **b.** isoleucine **c.** arginine

21.131 **a.** UCG **b.** AUA **c.** GGU

21.133 START–Tyr–Gly–Gly–Phe–Leu–STOP

21.135 Glu–Phe–Arg–His–Asp–Ser–Gly–Tyr–Glu–Val

21.137 Three nucleotides are needed to code for each amino acid, plus the start and stop codons consisting of three nucleotides each, which makes a minimum total of 33 nucleotides.

21.139 A DNA virus attaches to a cell and injects viral DNA that uses the host cell to produce copies of the DNA to make viral RNA. A retrovirus injects viral RNA from which complementary DNA is produced by reverse transcription.

21.141 **a.** (1) replication of DNA
b. (2) transcription
c. (5) reverse transcription
d. (4) recombinant DNA
e. (3) translation

COMBINING IDEAS from Chapters 19 to 21

CI.41 Aromasin is a drug that is used to inhibit the activity of the enzyme aromatase (estrogen synthetase). In the ovaries and mammary glands, aromatase converts testosterone to estrogen. Aromasin is used to treat breast cancer because it binds irreversibly to the active site of the enzyme, preventing the synthesis of estrogen and slowing the growth of estrogen-related cancers. Aromasin has the molecular formula $C_{20}H_{24}O_2$. (7.5, 12.5, 13.1, 14.1, 17.6, 20.2, 20.5)

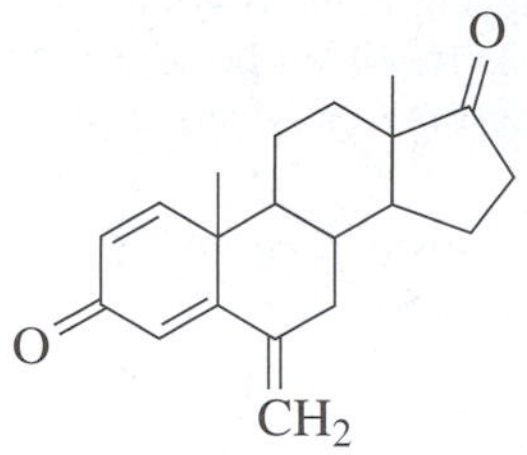
O
O
CH_2
Aromasin

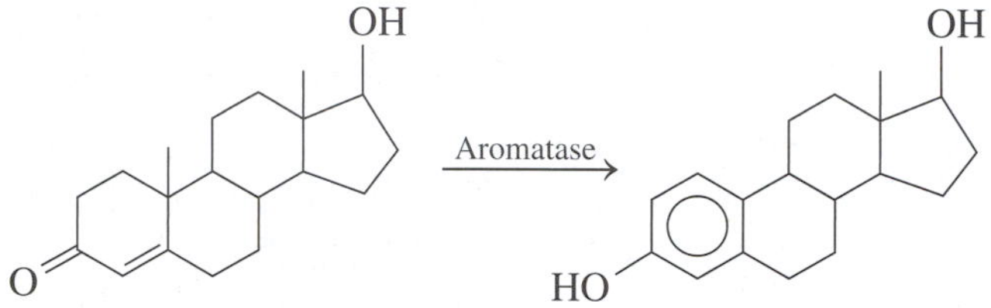
OH
O
Aromatase
OH
HO
Testosterone Estrogen

a. What are the functional groups in aromasin?
b. What are the functional groups in testosterone?
c. What are the functional groups in estrogen?
d. Why is the enzyme that converts testosterone to estrogen called aromatase?
e. Why would aromasin compete with testosterone, the precursor of estrogen?
f. What is the molar mass of aromasin?
g. If one tablet contains 25 mg of aromasin and there are 30 tablets in a package, how many moles of aromasin are in the package?

CI.42 Thalassemia is an inherited genetic mutation that limits the production of the beta chain needed for the formation of hemoglobin. If low levels of the beta chain are produced, there is a shortage of red blood cells (anemia). As a result, the body does not have sufficient amounts of oxygen. In one form of thalassemia, a single nucleotide is deleted in the DNA that codes for the beta chain. This mutation involves the deletion of thymine (T) from section 91 in the following informational strand of normal DNA: (21.3, 21.4, 21.5, 21.6)

89	90	91	92	93	94	
AGT	GAG	CTG	CAC	TGT	GAC	A . . .

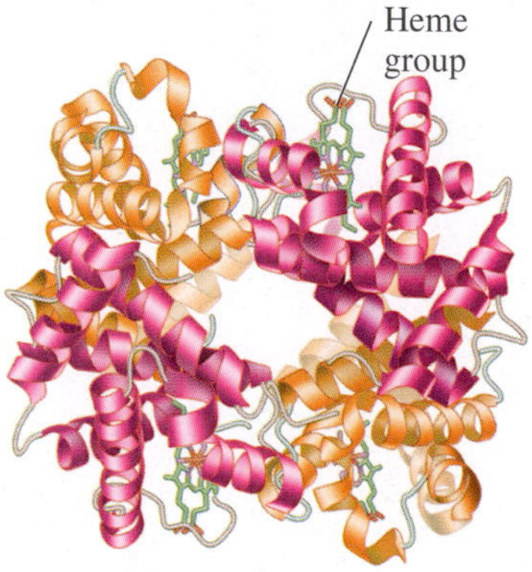
Heme group
The ribbon structure of hemoglobin shows the four polypeptide subunits; two (orange) are α-chains and two (red) are β-chains. The heme groups (green) in the four subunits bind oxygen.

a. Write the complementary (template) strand for this normal DNA segment.
b. Write the mRNA sequence from normal DNA using the template strand in part **a**.
c. What amino acids are placed in the beta chain by this portion of mRNA?
d. What is the order of nucleotides in the mutation?
e. Write the template strand for the mutated DNA segment.
f. Write the mRNA sequence from the mutated DNA segment using the template strand in part **e**.
g. What amino acids are placed in the beta chain by the mutated DNA segment?
h. What type of mutation occurs in this form of thalassemia?
i. How might the properties of this segment of the beta chain be different from the properties of the normal protein?
j. How might the level of structure in hemoglobin be affected if beta chains are not produced?

Use the following information for CI.43 and CI.44:

In response to signals from the nervous system, the hypothalamus secretes a polypeptide hormone known as gonadotropin-releasing factor (GnRF), which stimulates the pituitary gland to release other hormones into the bloodstream. Two of these hormones are luteinizing hormone (LH) and follicle-stimulating hormone (FSH).

The mRNA for the GnRF contains the codon sequence: 5′ GAA CAC UGG UCC UAU GGC CUU AGG CCA GGA 3′ (19.1, 19.3, 19.4, 20.5, 21.5, 21.6)

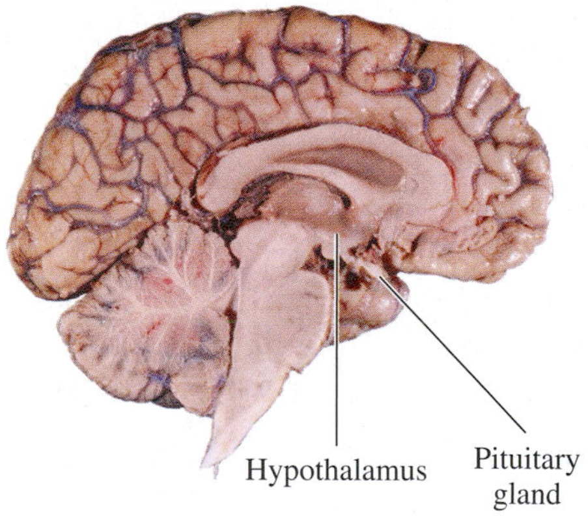
Hypothalamus
Pituitary gland
The hypothalamus secretes GnRF.

CI.43 **a.** Write the three-letter abbreviations for the sequence of amino acid residues in the decapeptide GnRF.
b. Write the one-letter abbreviations for the decapeptide GnRF.
c. If the fourth triplet UCC is changed to UUC, what type of mutation has occurred in the mRNA?
d. What is the change in the amino acid sequence in the mutated GnRF?
e. How could the tertiary structure and biological behavior of the mutated GnRF change?
f. When the level of LH or FSH is high in the bloodstream, the hypothalamus stops secreting GnRF. What type of regulation of proteins does this represent?

CI.44 Use the amino acid sequence you determined in CI.43 for the GnRF to answer the following: (19.1, 19.2)

- **a.** What amino acid is at the N-terminus?
- **b.** What amino acid is at the C-terminus?
- **c.** Which amino acids in GnRF are nonpolar?
- **d.** Which amino acids in GnRF are polar neutral?
- **e.** Draw the condensed structural formulas for the acidic amino acids at physiological pH.
- **f.** Draw the condensed structural formulas for the basic amino acids at physiological pH.

CI.45 A segment of the DNA template from BRCA1 mutated gene has the nucleotide sequence: 3′ A C A T A T T T T G C A A A T T T T G C A T G C 5′ (21.4, 21.5, 21.6)

- **a.** Write the corresponding section of mRNA produced from the segment of the BRCA1 mutated gene.
- **b.** What are the tRNA anticodons?
- **c.** Write the three-letter abbreviations for the amino acid residues in the peptide for this segment.

CI.46 Acyclovir (Zovirax) is an antiviral medication that is used to treat infections by the herpes simplex virus, chickenpox, and shingles. Acyclovir is a nucleoside analog, meaning that it has a structure that is very similar to a nucleoside. In the body, a series of kinase enzymes converts acyclovir into its corresponding nucleotide. This biologically active form of acyclovir is incorporated into viral DNA by specific herpes simplex virus DNA polymerases. Once the biologically active form of acyclovir is incorporated into viral DNA, no additional nucleotides can be added to that strand. (20.4, 21.1, 21.2, 21.8)

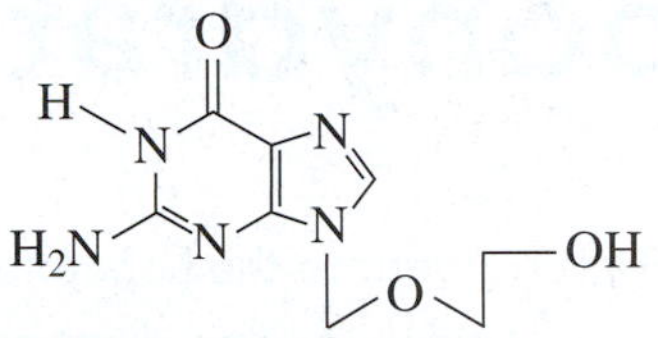
O
H
N
N
N
H2N
N
OH
O

Acyclovir is a nucleoside analog that is used to inhibit the synthesis of viral DNA.

- **a.** Which nucleoside has a structure that is most similar to that of acyclovir?
- **b.** What functional group is added to the acyclovir through the reactions catalyzed by the kinase enzymes?
- **c.** Why can't additional nucleotides be added to viral DNA after the acyclovir-based nucleotide has been incorporated?

ANSWERS

CI.41 a. ketone, alkene, cycloalkene
- **b.** alkene, cycloalkene, alcohol, ketone
- **c.** phenol, alcohol
- **d.** The enzyme aromatase converts a six-carbon cycloalkene into an aromatic ring.
- **e.** Aromasin has a steroid structure similar to that of testosterone, which binds with aromatase at the active site.
- **f.** 296.4 g/mole
- **g.** 2.5×10^{-3} mole

CI.43 a. Glu–His–Trp–Ser–Tyr–Gly–Leu–Arg–Pro–Gly
- **b.** EHWSYGLRPG
- **c.** a point mutation
- **d.** serine is replaced with phenylalanine
- **e.** Serine has a polar —OH group, whereas phenylalanine is nonpolar. The polar serine would have been on the outside of the normal peptide, whereas the phenylalanine will move to the inside of the mutated GnRF. This could change the tertiary structure of the decapeptide, and possibly cause a loss in biological activity.
- **f.** feedback control

CI.45 a. 5′ UGU AUA AAA CGU UUA AAA CGU ACG 3′
- **b.** ACA UAU UUU GCA AAU UUU GCA UGC
- **c.** Cys–Ile–Lys–Arg–Leu–Lys–Arg–Thr

22

Metabolic Pathways for Carbohydrates

Philip, who is 8 months old, is brought to the hepatology unit with symptoms that include convulsions, lethargy, difficulty breathing, and diarrhea. Blood tests indicate that Philip has very low blood glucose, low blood pH, high lactate levels, and high liver glycogen. During examination, Barbara, a hepatology nurse, notes that Philip is small for his age, has a doll-like face from fat deposits, and has a very large liver causing an extended abdomen. Because of his characteristic doll-like face and the large amount of glycogen in his liver, Philip is diagnosed with glycogen storage disease type I (GSD I), also known as von Gierke's disease. Patients with this disorder have a genetic mutation that results in a deficiency in glucose-6-phosphatase. Without this enzyme, the liver cannot properly break down glycogen into free glucose in order to maintain blood sugar levels. As a consequence, patients with von Gierke's disease often have large accumulations of glycogen in their livers and experience severe episodes of hypoglycemia.

CAREER

Hepatology Nurse

Hepatology nursing is a specialty of nursing that emphasizes liver health, prevention of illness, and the care of clients with liver-related health problems that can have viral, genetic, and metabolic causes. The hepatology nurse provides care for patients with liver-related conditions in clinics, outreach centers, or hospitals. Nursing care involves diagnosing illnesses, ordering and analyzing laboratory tests, developing appropriate treatment plans, providing patient education, and giving instruction in preventative care. In addition, a hepatology nurse may be responsible for documenting patient medical records, clinical research, and drug testing.

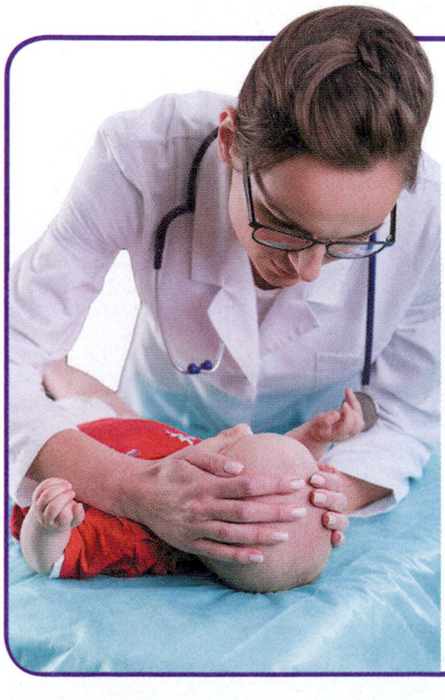

CLINICAL UPDATE

Philip's Diet for von Gierke's Disease

In the **CLINICAL UPDATE** **Philip's Diet for von Gierke's Disease**, page 795, you can read more about how diet can be adjusted to maintain blood glucose levels in patients with this type of glycogen storage disease.

22.1 Metabolism and Energy

LEARNING GOAL Describe the three stages of catabolism, the structure of ATP, and the role of ATP.

When we eat food, the polysaccharides, fats, and proteins are digested to smaller molecules that can be absorbed into the cells of our body. As the glucose, fatty acids, and amino acids are broken down further, energy is released. Because we do not use all the energy from our foods at one time, we store energy in the cells as high-energy adenosine triphosphate (ATP). Our cells use the energy stored in ATP when they do work such as contracting muscles, synthesizing large molecules, sending nerve impulses, and moving substances across cell membranes.

The term **metabolism** refers to all the chemical reactions that provide energy and the substances required for continued cell growth. There are two types of metabolic reactions: *catabolic* and *anabolic*. In **catabolic reactions**, complex molecules are broken down to small ones with an accompanying release of energy. **Anabolic reactions** utilize energy available in the cell to build large molecules from simple ones.

In general, we can think of catabolism as consisting of three stages (see **FIGURE 22.1**).

LOOKING AHEAD

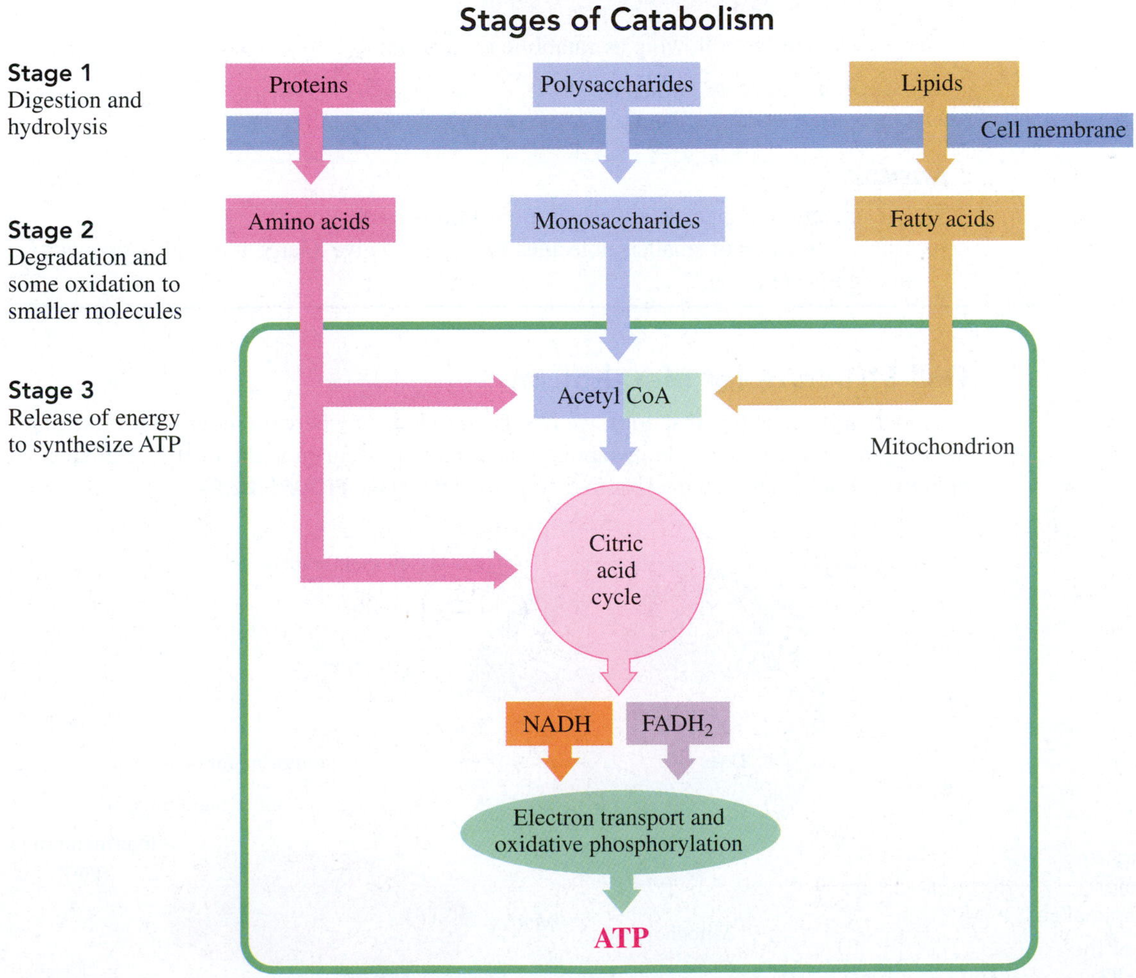
Stages of Catabolism
Stage 1
Digestion and hydrolysis
Stage 2
Degradation and some oxidation to smaller molecules
Stage 3
Release of energy to synthesize ATP
Proteins
Polysaccharides
Lipids
Cell membrane
Amino acids
Monosaccharides
Fatty acids
Acetyl CoA
Mitochondrion
Citric acid cycle
NADH
FADH2
Electron transport and oxidative phosphorylation
ATP

FIGURE 22.1 ▶ In the three stages of catabolism, large molecules from foods are digested and degraded to give smaller molecules that can be oxidized to produce ATP.

Where is most of the ATP produced in the cells?

Stage 1 Catabolism begins with the processes of **digestion** in which enzymes in the digestive tract hydrolyze large molecules into smaller ones. The polysaccharides break down to monosaccharides, fats break down to glycerol and fatty acids, and the proteins yield amino acids. These digestion products diffuse into the bloodstream for transport to cells.

Stage 2 Within the cells, catabolic reactions continue as the digestion products are broken down further to yield smaller groups, such as the two-carbon acetyl group.

Stage 3 The major release of energy takes place in the mitochondria, as the two-carbon acetyl group is oxidized in the citric acid cycle and the reduced coenzymes NADH and $FADH_2$ are produced. As long as the cells have oxygen, NADH and $FADH_2$ can be reoxidized via electron transport to release the energy needed to synthesize ATP.

TEST

Try Practice Problems 22.1 to 22.4

▶ SAMPLE PROBLEM 22.1 Metabolism

TRY IT FIRST

Identify each of the following as catabolic or anabolic:

a. digestion of lipids
b. synthesis of lactase
c. hydrolysis of sucrose
d. synthesis of adenine monophosphate from adenine, ribose, and phosphate

SOLUTION

a. The breakdown of large molecules involves catabolic reactions.
b. The synthesis of large molecules involves anabolic reactions.
c. The breakdown of sucrose to smaller molecules is a catabolic reaction.
d. The combination of small molecules to give a large molecule is an anabolic reaction.

STUDY CHECK 22.1

Identify each of the following as catabolic or anabolic:

a. oxidation of glucose to CO_2 and H_2O
b. synthesis of glycogen from glucose molecules

ANSWER

a. The oxidation of glucose to smaller molecules is catabolic.
b. The combination of smaller molecules (glucose) to give a large molecule (glycogen) is an anabolic reaction.

TEST

Try Practice Problems 22.5 and 22.6

Cell Structure for Metabolism

To understand metabolic reactions, we first need to look at where these reactions take place in cells. We will focus on the metabolic reactions that occur in animal cells. Animal cells are *eukaryotic* and have a nucleus that contains DNA (see **FIGURE 22.2**).

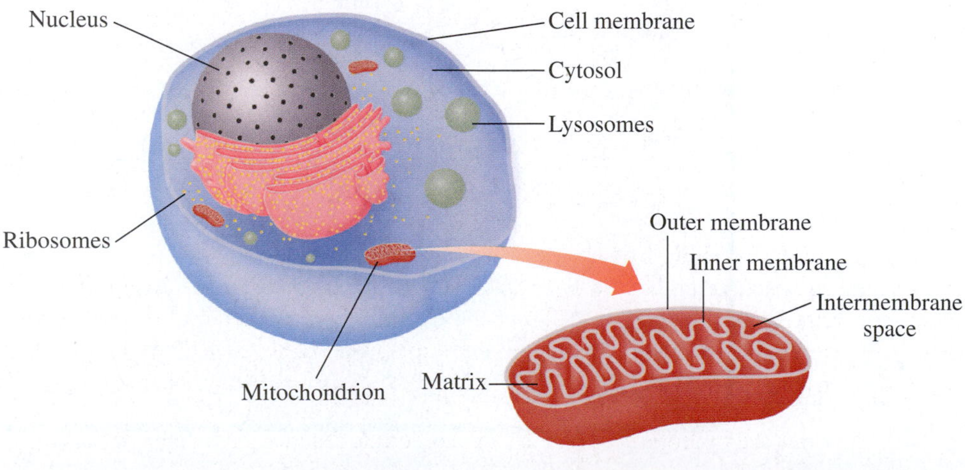
Nucleus
Cell membrane
Cytosol
Lysosomes
Ribosomes
Outer membrane
Inner membrane
Intermembrane space
Mitochondrion
Matrix

FIGURE 22.2 ▶ The diagram illustrates some of the major components of a typical animal cell.

Q What is the function of the mitochondrion in a cell?

In eukaryotic cells, a *cell membrane* separates the materials inside the cell from the aqueous environment surrounding the cell. The *nucleus* contains the genes that control DNA replication and protein synthesis. The **cytosol**, the fluid part of the cell, is an aqueous solution of electrolytes and enzymes that catalyze many of the chemical reactions in the cell.

In the cytosol are small structures called *organelles* that carry out specific functions in the cell. The *ribosomes* are the sites of protein synthesis. *Lysosomes* contain enzymes that break down recyclable cellular structures that are no longer needed by the cell. The **mitochondria** are the energy-releasing factories of the cells. A mitochondrion has an *outer membrane* and an *inner membrane*, with an *intermembrane space* between them. The fluid section surrounded by the inner membrane is called the *matrix*. Enzymes located in the matrix and along the inner membrane catalyze the oxidation of carbohydrates, fats, and amino acids. All of these oxidation pathways eventually produce CO_2 and H_2O, and release energy, which is used to form energy-rich compounds. **TABLE 22.1** summarizes some of the functions of these components in animal cells.

ENGAGE

Which organelle contains the enzymes that oxidize carbohydrates, fats, and amino acids?

TABLE 22.1 Functions of Some Major Components in Animal Cells

Component	Description and Function
Cell membrane	Separates the contents of a cell from the external environment and contains structures that communicate with other cells
Cytosol	Fluid part of the cytoplasm that contains enzymes for many of the cell's chemical reactions
Lysosome	Contains hydrolytic enzymes that digest and recycle old cell structures
Mitochondrion	Contains the structures for the synthesis of ATP from energy-releasing reactions
Nucleus	Contains genetic information for the replication of DNA and the synthesis of protein
Ribosome	Site of protein synthesis based on mRNA templates

ATP and Energy

In our cells, the energy released from the oxidation of the food we eat is stored in the form of a "high-energy" compound called *adenosine triphosphate* (ATP). The **ATP** molecule is composed of the base adenine, a ribose sugar, and three phosphate groups (see **FIGURE 22.3**).

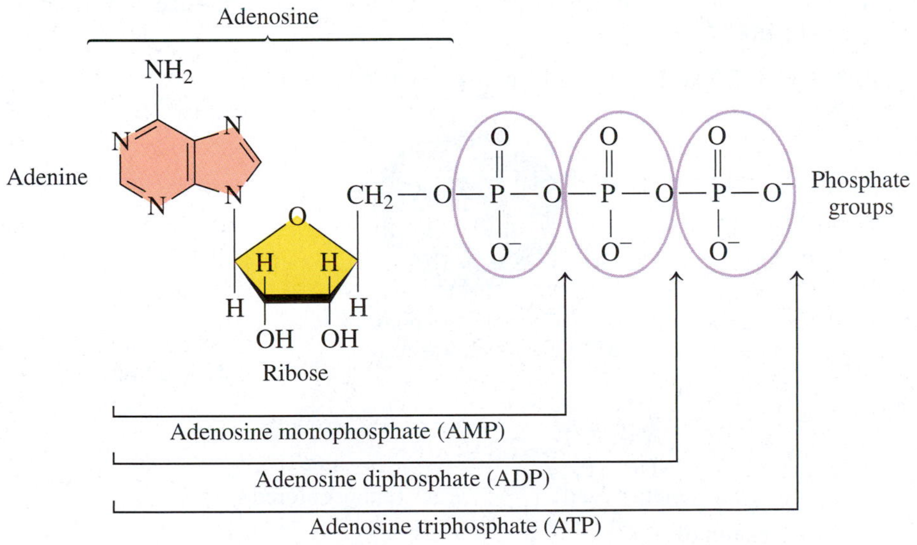
Adenosine
NH_2
Adenine
N
N
N
N
O
H
H
H
H
OH
OH
Ribose
CH_2—O—P—O—P—O—P—O^-
O
O
O
O^-
O^-
O^-
Phosphate groups
Adenosine monophosphate (AMP)
Adenosine diphosphate (ADP)
Adenosine triphosphate (ATP)

FIGURE 22.3 ▶ Adenosine triphosphate (ATP), a "high-energy" compound, is composed of adenine, ribose, and three phosphate groups.

Q How many phosphate groups are in adenosine monophosphate (AMP)?

Hydrolysis of ATP Yields Energy

Adenosine triphosphate undergoes hydrolysis to yield adenosine diphosphate (**ADP**), an inorganic phosphate group HPO_4^{2-}, abbreviated P_i, and energy needed for cellular metabolism. The hydrolysis of ATP molecules can provide 7.3 kcal/mole (31 kJ/mole).

$$ATP + H_2O \longrightarrow ADP + P_i + 7.3\ \text{kcal/mole}\ (31\ \text{kJ/mole})$$

We can draw the hydrolysis of ATP as an exothermic reaction on an energy curve.

ENGAGE

How much energy is released when one mole of ATP molecules are hydrolyzed?

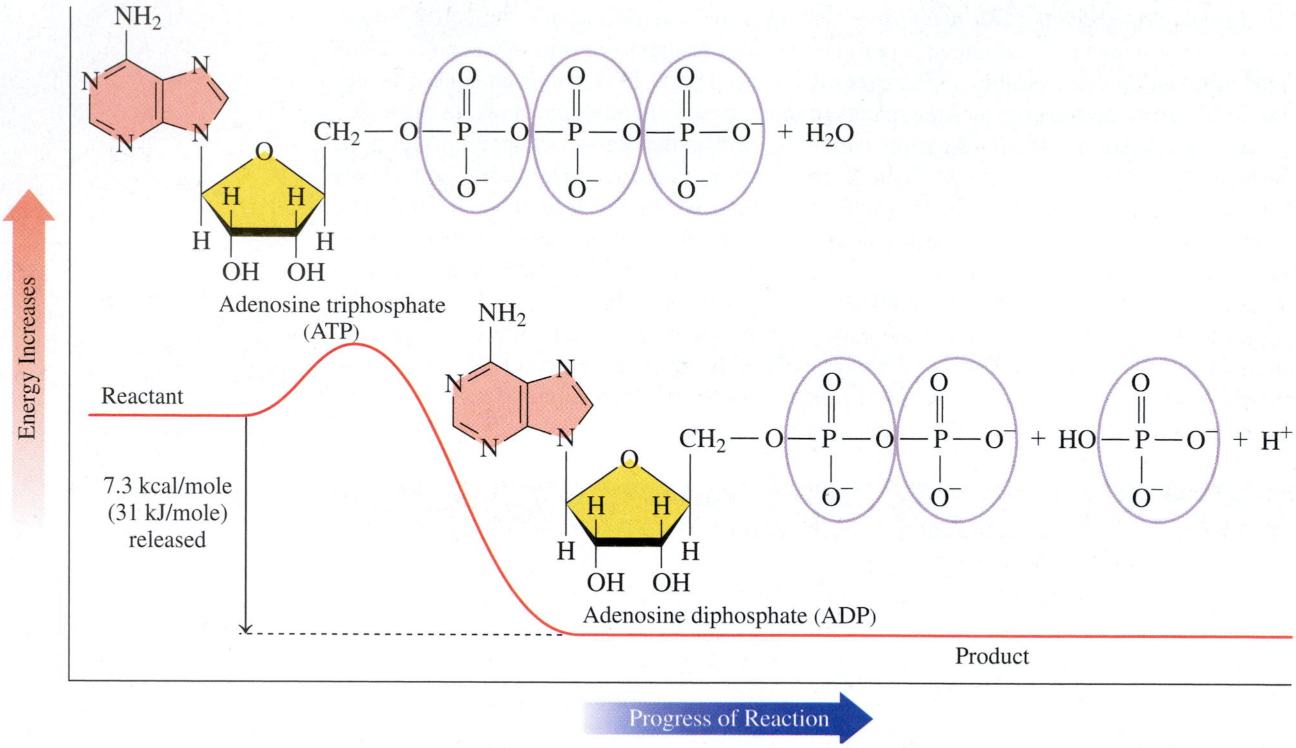
NH2
N
N
N
N
O
H H
H H
OH OH
CH2—O—P—O—P—O—P—O− + H2O
Adenosine triphosphate (ATP)
Energy Increases
Reactant
7.3 kcal/mole (31 kJ/mole) released
CH2—O—P—O—P—O− + HO—P—O− + H+
Adenosine diphosphate (ADP)
Product
Progress of Reaction

The ADP can also hydrolyze to form adenosine monophosphate (AMP) and an inorganic phosphate (P_i).

$$ADP + H_2O \longrightarrow AMP + P_i + 7.3\ \text{kcal/mole}\ (31\ \text{kJ/mole})$$

Every time we contract muscles, move substances across cellular membranes, send nerve signals, or synthesize an enzyme, we use energy from ATP hydrolysis. In a cell that is doing work (anabolic processes), 1 to 2 million ATP molecules may be hydrolyzed in one second. The amount of ATP hydrolyzed in one day can be as much as our body mass, even though only about 1 g of ATP is present in all our cells at any given time.

When we take in food, the resulting catabolic reactions provide energy to regenerate ATP in our cells. Then 7.3 kcal/mole (31 kJ/mole) is used to synthesize ATP from ADP and P_i (see **FIGURE 22.4**).

$$ADP + P_i + 7.3\ \text{kcal/mole}\ (31\ \text{kJ/mole}) \longrightarrow ATP + H_2O$$

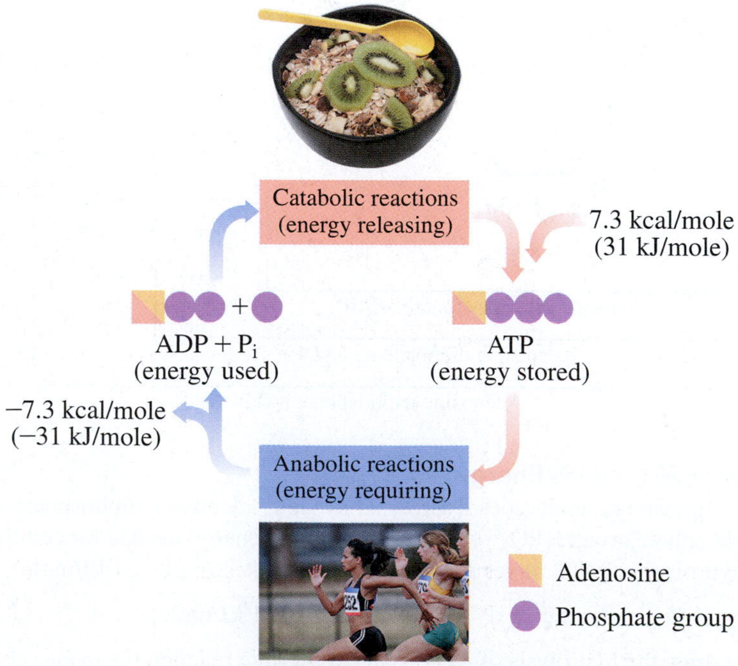
Catabolic reactions
(energy releasing)
7.3 kcal/mole
(31 kJ/mole)
ADP + Pi
(energy used)
ATP
(energy stored)
−7.3 kcal/mole
(−31 kJ/mole)
Anabolic reactions
(energy requiring)
Adenosine
Phosphate group

FIGURE 22.4 ▶ ATP, the energy-storage molecule, links energy-releasing reactions with energy-requiring reactions in the cells.

Q What type of reaction provides energy for ATP synthesis?

▶ SAMPLE PROBLEM 22.2 Energy from ATP Hydrolysis

TRY IT FIRST

A person might burn 2600 kcal running a marathon (26.2 miles). If the molar mass of ATP is 507 g/mole, how many grams of ATP are hydrolyzed to provide this amount of energy?

SOLUTION

STEP 1 **State the given and needed quantities.**

ANALYZE THE PROBLEM	Given	Need	Connect
	2600 kcal	grams of ATP	molar energy factor, molar mass

STEP 2 **Write a plan to convert the given unit to the needed unit.**

kilocalories → Molar energy factor → moles of ATP → Molar mass → grams of ATP

STEP 3 **State the equalities and conversion factors.**

1 mole of ATP = 7.3 kcal

$$\frac{7.3\text{ kcal}}{1\text{ mole ATP}} \text{ and } \frac{1\text{ mole ATP}}{7.3\text{ kcal}}$$

1 mole of ATP = 507 g of ATP

$$\frac{507\text{ g ATP}}{1\text{ mole ATP}} \text{ and } \frac{1\text{ mole ATP}}{507\text{ g ATP}}$$

STEP 4 **Set up the problem to calculate the needed quantity.**

$$2600\ \cancel{\text{kcal}} \times \frac{1\ \cancel{\text{mole ATP}}}{7.3\ \cancel{\text{kcal}}} \times \frac{507\text{ g ATP}}{1\ \cancel{\text{mole ATP}}} = 1.8 \times 10^5\text{ g of ATP}$$

STUDY CHECK 22.2

A 130-lb person might burn 945 kcal in a sprint distance triathlon, which includes a 500-yard swim, a 13-mile bike ride, and a 3.1-mile run. If the molar mass of ATP is 507 g/mole, how many grams of ATP are hydrolyzed to provide this amount of energy?

ANSWER

6.56×10^4 g of ATP

TEST

Try Practice Problems 22.11 and 22.12

ATP Hydrolysis Drives Energy-Requiring Reactions

Many of the reactions essential to a cell require energy to proceed. Often this energy comes from the hydrolysis of ATP or other high-energy compounds. For example, the first step in breaking down glucose in the cell is to add a phosphate group. Adding a phosphate group to glucose requires 3.3 kcal/mole (14 kJ/mole). When the hydrolysis of ATP, which releases 7.3 kcal/mole, is coupled with the reaction, the addition of a phosphate group to glucose can take place.

Note that, below, the reaction for the hydrolysis of ATP is simplified by leaving the water out. This is a common simplification when the hydrolysis of ATP is coupled with an energy-requiring reaction.

ENGAGE

Why are anabolic reactions that require energy always linked with the hydrolysis of a high-energy compound like ATP?

TEST

Try Practice Problems 22.7 and 22.8

ATP $\longrightarrow$ ADP + P_i + 7.3 kcal/mole (31 kJ/mole)		Releases energy
Glucose + P_i + 3.3 kcal/mole (14 kJ/mole) $\longrightarrow$ glucose-6-phosphate		Requires energy
Overall: ATP + glucose $\longrightarrow$ ADP + glucose-6-phosphate + 4.0 kcal/mole (17 kJ/mole)		

▶SAMPLE PROBLEM 22.3 ATP and Energy

TRY IT FIRST

The reaction of glutamate (Glu) with ammonia (NH_3) produces the amino acid glutamine (Gln). The reaction requires 3.4 kcal/mole.

a. Is the formation of glutamine a catabolic or an anabolic reaction?
b. Write the equation for the formation of glutamine (Gln), including the energy.
c. Write the equation for the hydrolysis of ATP, including the energy.
d. Write the overall equation for the combined reactions, including the net energy change.

SOLUTION

a. anabolic
b. $\text{Glu} + NH_3 + 3.4\ \text{kcal/mole} \longrightarrow \text{Gln}$
c. $\text{ATP} \longrightarrow \text{ADP} + P_i + 7.3\ \text{kcal/mole}$
d. $\text{Glu} + NH_3 + \text{ATP} \longrightarrow \text{Gln} + \text{ADP} + P_i + 3.9\ \text{kcal/mole}$

STUDY CHECK 22.3

Phosphocreatine is a high-energy compound that releases 43.1 kJ/mole of energy when it hydrolyzes to creatine and P_i in muscle cells. This reaction can be coupled with the synthesis of ATP from ADP and P_i.

a. Write an equation for the energy-releasing reaction of phosphocreatine.
b. Write an equation for the energy-requiring reaction that forms ATP.
c. Write the overall equation for the combined reaction, including the net energy change (in kilojoules).

ANSWER

a. $\text{phosphocreatine} \longrightarrow \text{creatine} + P_i + 43.1\ \text{kJ/mole}$
b. $\text{ADP} + P_i + 31\ \text{kJ/mole} \longrightarrow \text{ATP}$
c. $\text{phosphocreatine} + \text{ADP} \longrightarrow \text{ATP} + \text{creatine} + 12.1\ \text{kJ/mole}$

TEST

Try Practice Problems 22.9 and 22.10

PRACTICE PROBLEMS

22.1 Metabolism and Energy

22.1 What stage of catabolism involves the digestion of polysaccharides?

22.2 What stage of catabolism involves the conversion of small molecules to CO_2, H_2O, and energy for the synthesis of ATP?

22.3 What is meant by a catabolic reaction in metabolism?

22.4 What is meant by an anabolic reaction in metabolism?

22.5 Identify each of the following as a catabolic or an anabolic reaction:
a. synthesis of lipids from glycerol and fatty acids
b. glucose adds P_i to form glucose-6-phosphate
c. hydrolysis of ATP to ADP and P_i
d. digestion of proteins in the stomach

22.6 Identify each of the following as a catabolic or an anabolic reaction:
a. digestion of fats to fatty acids and glycerol
b. hydrolysis of proteins into amino acids
c. synthesis of nucleic acids from nucleotides
d. glucose and galactose form the disaccharide lactose

22.7 Why is ATP considered a high-energy compound?

22.8 What is meant when we say that the hydrolysis of ATP is used to "drive" a reaction?

22.9 Phosphoenolpyruvate (PEP) is a high-energy compound that releases 14.8 kcal/mole of energy when it hydrolyzes to pyruvate and P_i. This reaction can be coupled with the synthesis of ATP from ADP and P_i.
a. Write an equation for the energy-releasing reaction of PEP.
b. Write an equation for the energy-requiring reaction that forms ATP.
c. Write the overall equation for the combined reaction, including the net energy change.

22.10 The phosphorylation of glycerol to glycerol-3-phosphate requires 2.2 kcal/mole and is driven by the hydrolysis of ATP.
a. Write an equation for the energy-releasing reaction of ATP.
b. Write an equation for the energy-requiring reaction that forms glycerol-3-phosphate.
c. Write the overall equation for the combined reaction, including the net energy change.

Clinical Applications

22.11 A 185-lb person bicycling vigorously on a stationary bicycle uses 2100 kJ. If the molar mass of ATP is 507 g/mole, how many grams of ATP are hydrolyzed to provide this amount of energy?

22.12 A 185-lb person bicycling moderately on a stationary bicycle uses 1300 kJ. If the molar mass of ATP is 507 g/mole, how many grams of ATP are hydrolyzed to provide this amount of energy?

22.2 Important Coenzymes in Metabolic Pathways

REVIEW
Describing the Role of Cofactors (20.6)

LEARNING GOAL Describe the components and functions of the coenzymes NAD^+, $NADP^+$, FAD, and coenzyme A.

To occur at rates that meet our physiological needs, most metabolic reactions are catalyzed by enzymes. Because several reactions that extract energy from our food involve oxidation and reduction reactions, many metabolic reactions require coenzymes that can participate in the transfer of electrons. Therefore, we will review important coenzymes in their oxidized and reduced forms. Each of the four coenzymes we will discuss participates in a particular type of reaction.

In chemistry, oxidation is often associated with the loss of H atoms, whereas reduction is associated with the gain of H atoms. We can represent two H atoms as two hydrogen ions ($2H^+$) and two electrons ($2\ e^-$). In both types of reactions, coenzymes are required to carry the hydrogen ions and electrons from or to the reacting substrate. A coenzyme that gains hydrogen ions and electrons is reduced, whereas a coenzyme that loses hydrogen ions and electrons to a substrate is oxidized.

In general, oxidation reactions release energy, and reduction reactions require energy. **TABLE 22.2** summarizes the characteristics of oxidation and reduction.

TABLE 22.2 Characteristics of Oxidation and Reduction in Metabolic Pathways

Oxidation	Reduction
Loss of electrons (e^-)	Gain of electrons (e^-)
Loss of hydrogen (H or H^+ and e^-)	Gain of hydrogen (H or H^+ and e^-)
Gain of oxygen	Loss of oxygen
Release of energy	Input of energy

NAD^+

CORE CHEMISTRY SKILL
Identifying Important Coenzymes in Metabolism

NAD^+ (nicotinamide adenine dinucleotide) is an important coenzyme in which the vitamin *niacin* provides the *nicotinamide* group, which is bonded to ribose and the nucleotide adenosine diphosphate (ADP) (see **FIGURE 22.5**). The oxidized form of NAD^+ undergoes

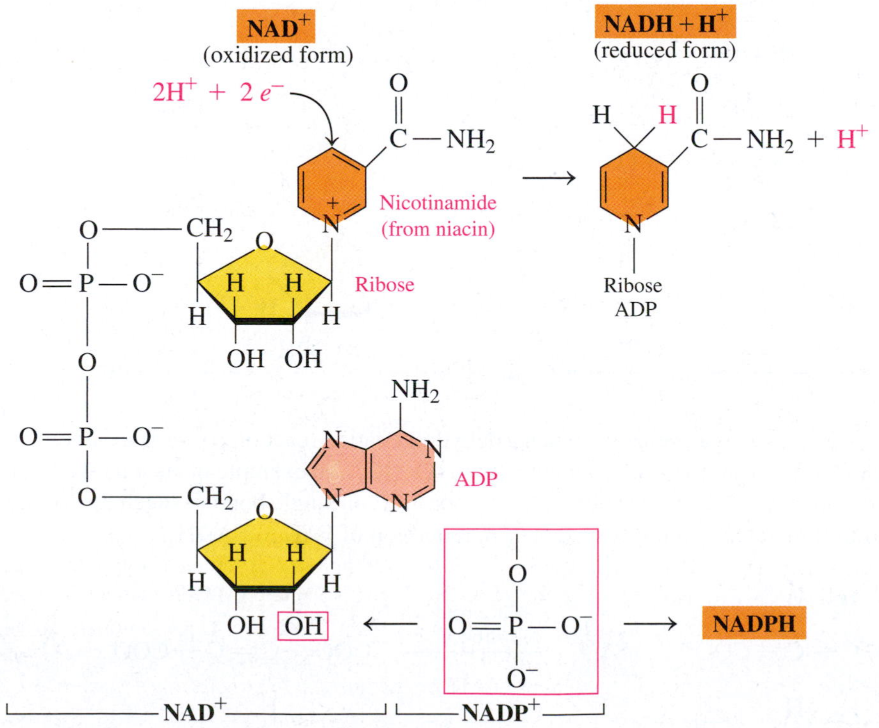
NAD+
(oxidized form)
2H+ + 2 e−
Nicotinamide
(from niacin)
Ribose
ADP
NADH + H+
(reduced form)
Ribose
ADP
NADPH
NAD+
NADP+

FIGURE 22.5 ▶ The coenzyme NAD^+(nicotinamide adenine dinucleotide), which consists of adenosine diphosphate, nicotinamide from the vitamin niacin, and ribose, is reduced to $NADH + H^+$ by adding a hydrogen ion and two electrons to NAD^+. The coenzyme $NADP^+$ is similar to NAD^+ except that a 2′ OH group is replaced by a phosphate. The reduced form of $NADP^+$ is NADPH.

Q Why is the conversion of NAD^+ to NADH and H^+ called a reduction?

reduction when a carbon atom in the nicotinamide ring reacts with 2H (two hydrogen ions and two electrons), leaving one H^+.

The NAD^+ coenzyme is required in dehydrogenation reactions that produce carbon–oxygen double bonds (C=O), such as the oxidation of alcohols to aldehydes and ketones. An example of such an oxidation–reduction reaction is the oxidation of ethanol in the liver to ethanal and the reduction of NAD^+ to $NADH + H^+$.

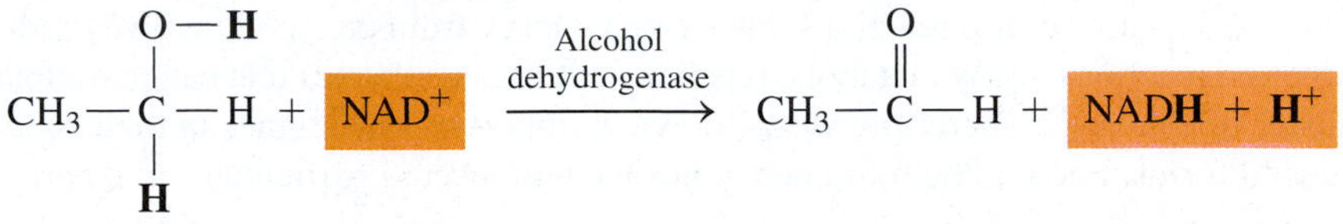
NAD+
Alcohol dehydrogenase
NADH + H+

Ethanol — Ethanal

ENGAGE

In which type of reactions does $NADP^+$ participate?

$NADP^+$ (nicotinamide adenine dinucleotide phosphate) is a coenzyme used in anabolic reactions, such as lipid and nucleic acid synthesis, which require a source of hydrogen ions and electrons. The coenzyme $NADP^+$ is similar to NAD^+ except that a 2′ OH group is replaced by a phosphate group (see **FIGURE 22.5**). The reduced form of $NADP^+$ is NADPH.

FAD

FAD (flavin adenine dinucleotide) is a coenzyme that contains the nucleotide adenosine diphosphate and riboflavin. Riboflavin, also known as vitamin B_2, consists of ribitol (a sugar alcohol) and flavin. The oxidized form of FAD undergoes reduction when the two nitrogen atoms in the flavin part of the FAD coenzyme react with two hydrogen atoms ($2H^+ + 2\,e^-$), reducing FAD to $FADH_2$ (see **FIGURE 22.6**).

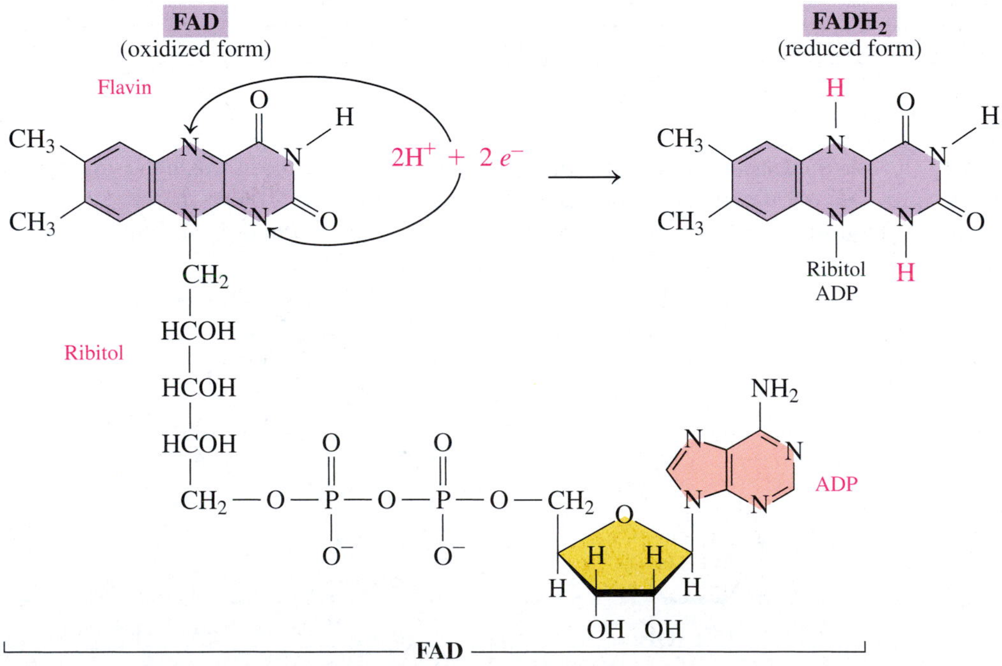
FAD
(oxidized form)
Flavin
Ribitol
2H+ + 2 e−
FADH2
(reduced form)
Ribitol
ADP
ADP
FAD

FIGURE 22.6 ▶ The coenzyme FAD (flavin adenine dinucleotide), made from riboflavin (vitamin B_2) and adenosine diphosphate, is reduced to $FADH_2$ by adding two hydrogen atoms.

Q What is the type of reaction in which FAD accepts hydrogen?

FAD is used as a coenzyme when a dehydrogenation reaction converts a carbon–carbon single bond to a carbon–carbon double bond (C=C). An example of such a reaction in the citric acid cycle is the conversion of the carbon–carbon single bond in succinate to a double bond in fumarate, with the corresponding reduction of FAD to $FADH_2$.

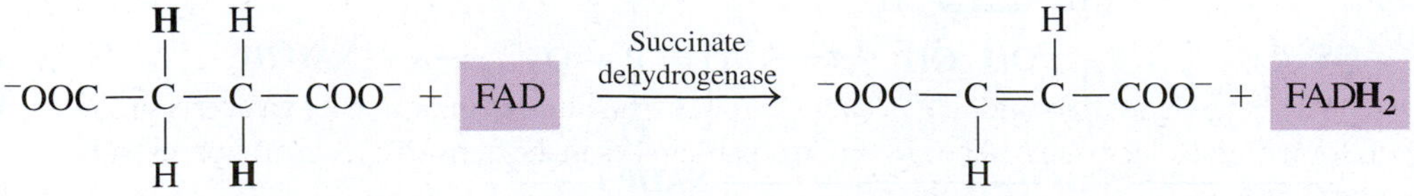
FAD
Succinate dehydrogenase
FADH2

Succinate — Fumarate

Coenzyme A

Coenzyme A (CoA) is made up of several components: pantothenic acid (vitamin B_5), phosphorylated ADP, and aminoethanethiol (see **FIGURE 22.7**).

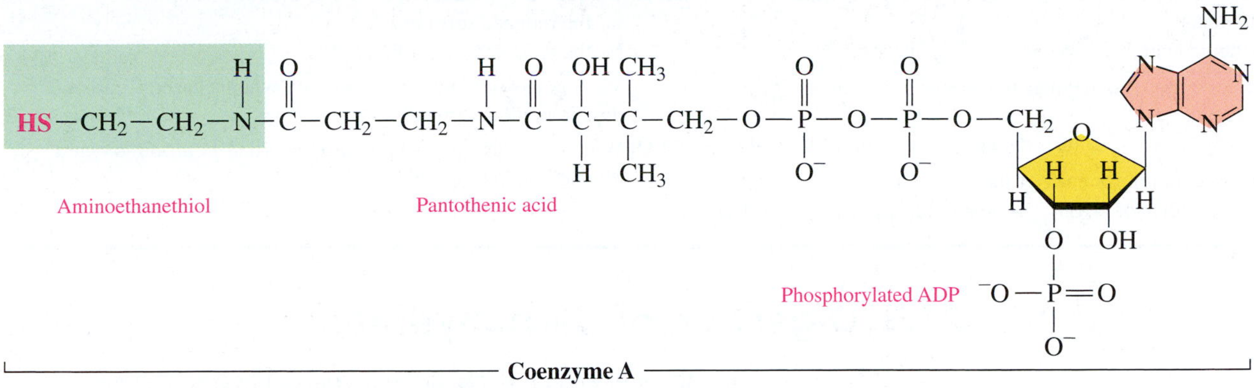
Aminoethanethiol
Pantothenic acid
Phosphorylated ADP
Coenzyme A

FIGURE 22.7 ▶ Coenzyme A is derived from a phosphorylated ADP and pantothenic acid bonded by an amide bond to aminoethanethiol, which contains the —SH reactive part of the molecule.

Q What part of coenzyme A reacts with a two-carbon acetyl group?

An important function of coenzyme A is to prepare small acyl groups (represented by the letter A in the name), such as acetyl, for reactions with enzymes. The reactive feature of coenzyme A is the thiol group (—SH), which bonds to a two-carbon acetyl group to produce the energy-rich thioester **acetyl CoA**.

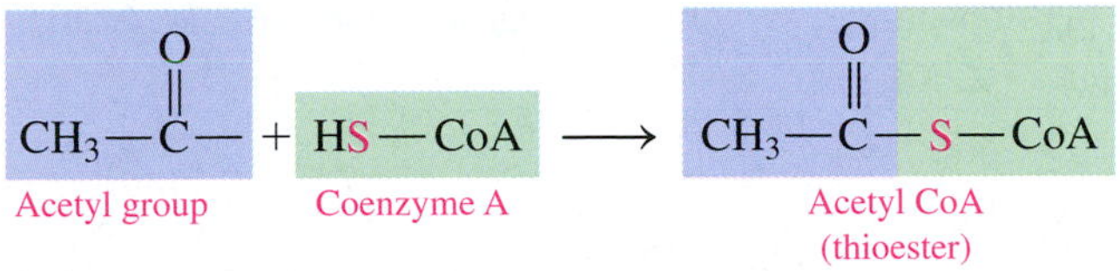
Acetyl group
Coenzyme A
Acetyl CoA
(thioester)

ENGAGE

Is coenzyme A oxidized or reduced when an acetyl group is added?

In biochemistry, several abbreviations are used for coenzyme A and acetyl coenzyme A. For discussions in this text, we will use CoA for coenzyme A and acetyl CoA when the acetyl group is bonded to the sulfur atom (—S—) in coenzyme A. In equations, we will show the —SH group in coenzyme A as HS—CoA.

▶ SAMPLE PROBLEM 22.4 Coenzymes

TRY IT FIRST

Describe the reactive part of each of the following coenzymes and the way each participates in metabolic pathways:

a. FAD **b.** NAD^+

SOLUTION

a. The oxidized form of FAD undergoes reduction when two nitrogen atoms in the flavin part react with $2H^+$ and $2\,e^-$ to give $FADH_2$. The coenzyme FAD is utilized in oxidation reactions that produce carbon–carbon double bonds (C=C).

b. The oxidized form of NAD^+ undergoes reduction when a carbon atom in the nicotinamide part reacts with $2H^+$ and $2\,e^-$, to give NADH and H^+. The coenzyme NAD^+ is utilized in oxidation reactions that produce carbon–oxygen double bonds (C=O).

STUDY CHECK 22.4

Describe the reactive part of coenzyme A and how it participates in metabolic reactions.

ANSWER

In coenzyme A, the thiol group (—SH) of aminoethanethiol combines with an acetyl group to form acetyl coenzyme A. The HS—CoA participates in the transfer of acyl groups, usually acetyl groups.

TEST

Try Practice Problems 22.13 to 22.18

PRACTICE PROBLEMS

22.2 Important Coenzymes in Metabolic Pathways

22.13 Identify one or more coenzymes that contain each of the following components:
- **a.** pantothenic acid **b.** niacin **c.** ribitol

22.14 Identify one or more coenzymes that contain each of the following components:
- **a.** riboflavin **b.** adenine **c.** aminoethanethiol

22.15 Give the abbreviation for each of the following:
- **a.** the reduced form of NAD^+ **b.** the oxidized form of $FADH_2$

22.16 Give the abbreviation for each of the following:
- **a.** the reduced form of FAD
- **b.** the oxidized form of NADPH

22.17 What coenzyme picks up hydrogen when a carbon–carbon double bond is formed?

22.18 What coenzyme picks up hydrogen when a carbon–oxygen double bond is formed?

22.3 Digestion of Carbohydrates

LEARNING GOAL Give the sites and products of the digestion of carbohydrates.

There are different metabolic reactions associated with the different foods we eat. In this chapter, we focus on the reactions associated with carbohydrates.

Digestion of Carbohydrates

We begin the digestion of carbohydrates as soon as we chew food. Enzymes produced in the salivary glands hydrolyze some of the α-glycosidic bonds in amylose and amylopectin, producing maltose, glucose, and smaller polysaccharides called dextrins, which contain three to eight glucose units. After swallowing, the partially digested starches enter the acidic environment of our stomach, where the low pH stops further carbohydrate digestion (see **FIGURE 22.8**).

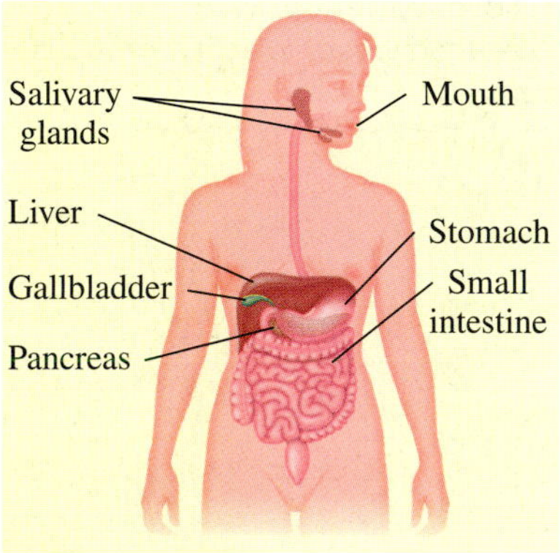
Salivary glands
Mouth
Liver
Stomach
Gallbladder
Small intestine
Pancreas

Carbohydrate digestion begins in the mouth and continues in the small intestine.

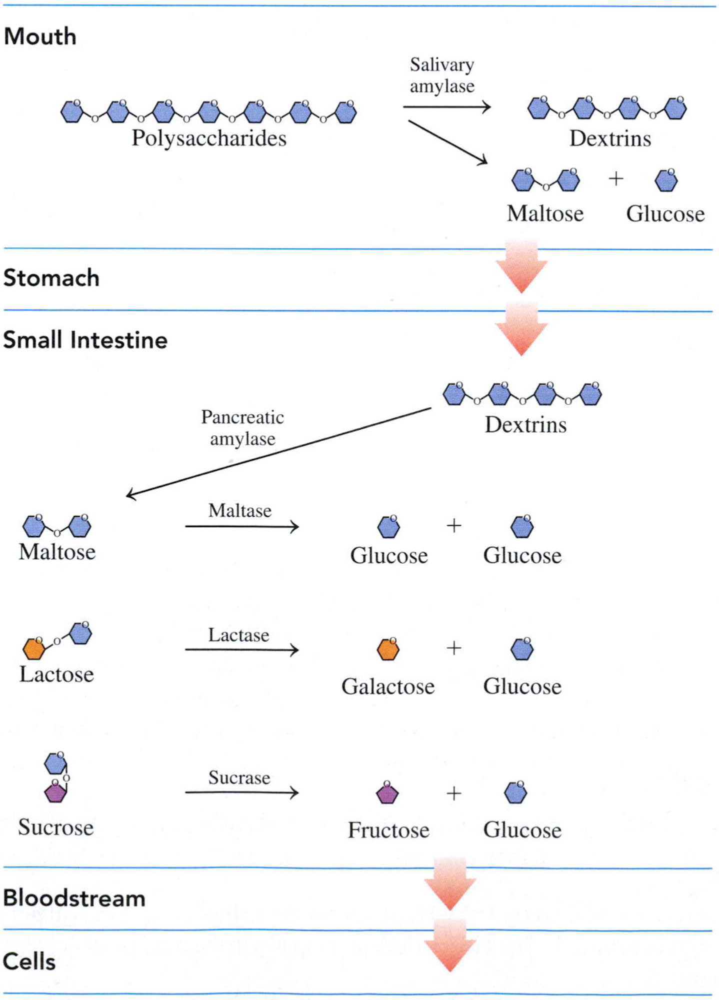
Mouth
Salivary amylase
Polysaccharides
Dextrins
Maltose
+
Glucose
Stomach
Small Intestine
Dextrins
Pancreatic amylase
Maltase
Maltose
Glucose
+
Glucose
Lactase
Lactose
Galactose
+
Glucose
Sucrase
Sucrose
Fructose
+
Glucose
Bloodstream
Cells

FIGURE 22.8 ▶ In stage 1 of catabolism, the digestion of carbohydrates begins in the mouth and is completed in the small intestine.

Q Why is there little or no digestion of carbohydrates in the stomach?

In the small intestine, which has a pH of about 8, enzymes produced in the pancreas hydrolyze the remaining dextrins to maltose and glucose. Then enzymes produced in the mucosal cells that line the small intestine hydrolyze maltose, lactose, and sucrose. The resulting monosaccharides are absorbed through the intestinal wall into the bloodstream, which carries them to the liver, where the hexoses fructose and galactose are converted to glucose. Glucose is the primary energy source for muscle contractions, red blood cells, and the brain.

▶SAMPLE PROBLEM 22.5 Digestion of Carbohydrates

TRY IT FIRST

Indicate the carbohydrate that undergoes digestion in each of the following sites:

a. mouth
b. stomach
c. small intestine

SOLUTION

a. starches amylose and amylopectin ($\alpha(1\rightarrow4)$-glycosidic bonds only)
b. essentially no digestion of carbohydrates
c. dextrins, maltose, sucrose, and lactose

STUDY CHECK 22.5

Describe the digestion of amylose, a polymer of glucose molecules joined by $\alpha(1\rightarrow4)$-glycosidic bonds.

ANSWER

The digestion of amylose begins in the mouth when salivary amylase hydrolyzes some of the $\alpha(1\rightarrow4)$-glycosidic bonds. In the small intestine, pancreatic amylase hydrolyzes more glycosidic bonds, and finally maltose is hydrolyzed by maltase to yield glucose.

TEST

Try Practice Problems 22.19 to 22.22

PRACTICE PROBLEMS

22.3 Digestion of Carbohydrates

22.19 What is the general type of reaction that occurs during the digestion of carbohydrates?

22.20 Why is α-amylase produced in the salivary glands and in the pancreas?

22.21 Complete the following equations by filling in the missing words:

a. ____________ + $H_2O \longrightarrow$ galactose + glucose
b. Sucrose + $H_2O \longrightarrow$ ____________ + ____________
c. Maltose + $H_2O \longrightarrow$ glucose + ____________

22.22 Give the site and the enzyme for each of the reactions in problem 22.21.

22.4 Glycolysis: Oxidation of Glucose

LEARNING GOAL Describe the conversion of glucose to pyruvate in glycolysis.

REVIEW

Classifying Enzymes (20.2)
Identifying Factors Affecting Enzyme Activity (20.3)

The major source of energy for the body is the glucose produced when we digest the carbohydrates in our food or from glycogen, a polysaccharide stored in the liver and skeletal muscle. Glucose in the bloodstream enters our cells where it undergoes degradation in a pathway called *glycolysis*. Early organisms used glycolysis to produce energy from simple nutrients long before there was any oxygen in Earth's atmosphere. Glycolysis is an **anaerobic** process; no oxygen is required.

CORE CHEMISTRY SKILL

Identifying the Compounds in Glycolysis

In **glycolysis**, a six-carbon glucose molecule is broken down to two molecules of three-carbon pyruvate (see **FIGURE 22.9**). All the reactions in glycolysis take place in the cytosol of the cell. In the first five reactions (1 to 5), called the *energy-investing phase*, energy from the hydrolysis of two ATP is used to form two three-carbon, high-energy phosphate compounds. In the last five reactions (6 to 10), called the *energy-generating phase*, energy from the hydrolysis of the high-energy phosphate compounds is used to synthesize four ATP.

ENGAGE

In which steps of glycolysis is ADP phosphorylated to become ATP?

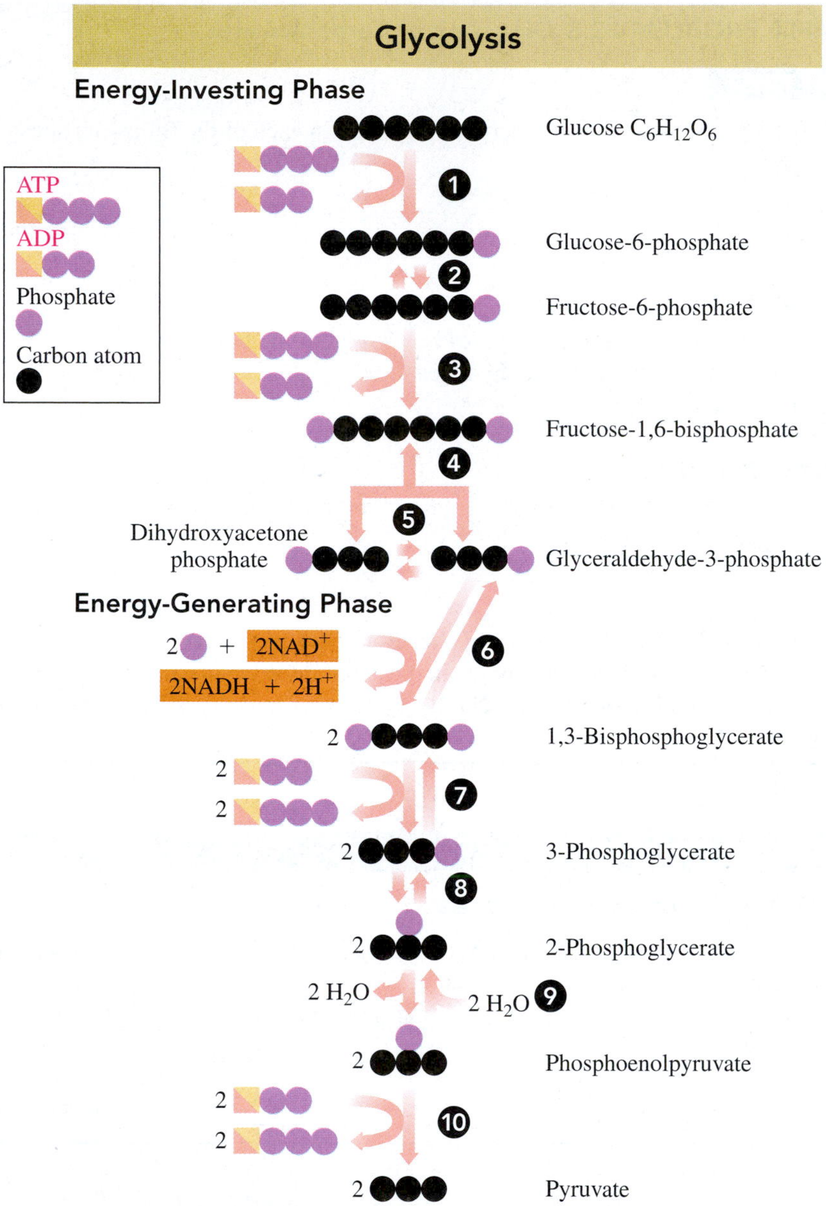

Glycolysis
Energy-Investing Phase
ATP
ADP
Phosphate
Carbon atom
Glucose $C_6H_{12}O_6$
1
Glucose-6-phosphate
2
Fructose-6-phosphate
3
Fructose-1,6-bisphosphate
4
5
Dihydroxyacetone phosphate
Glyceraldehyde-3-phosphate
Energy-Generating Phase
2 + $2NAD^+$
$2NADH + 2H^+$
6
2
1,3-Bisphosphoglycerate
2
2
7
2
3-Phosphoglycerate
8
2
2-Phosphoglycerate
2 H_2O
2 H_2O
9
2
Phosphoenolpyruvate
2
2
10
2
Pyruvate

FIGURE 22.9 ▶ In glycolysis, the six-carbon glucose molecule is degraded to yield two three-carbon pyruvate molecules. A net of two ATP is produced along with two NADH.

Q Where in the glycolysis pathway is glucose cleaved to yield two three-carbon compounds?

Energy-Investing Reactions: 1 to 5

Reaction 1 Phosphorylation

In the initial reaction, a phosphate group from ATP is added to glucose to form glucose-6-phosphate and ADP. The enzyme that catalyzes this reaction is called a *hexokinase*, *hexo* because glucose is a hexose sugar and *kinase* because it uses the energy of ATP to transfer the phosphate group.

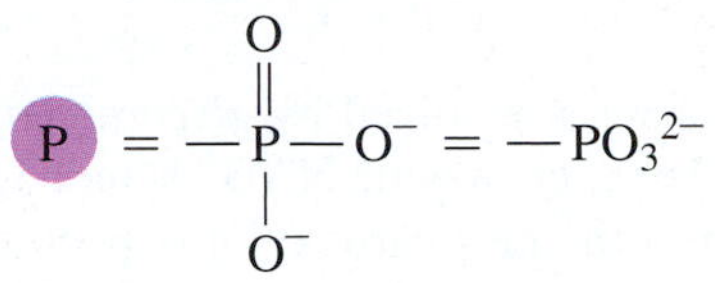
P
$= -P-O^- = -PO_3^{2-}$

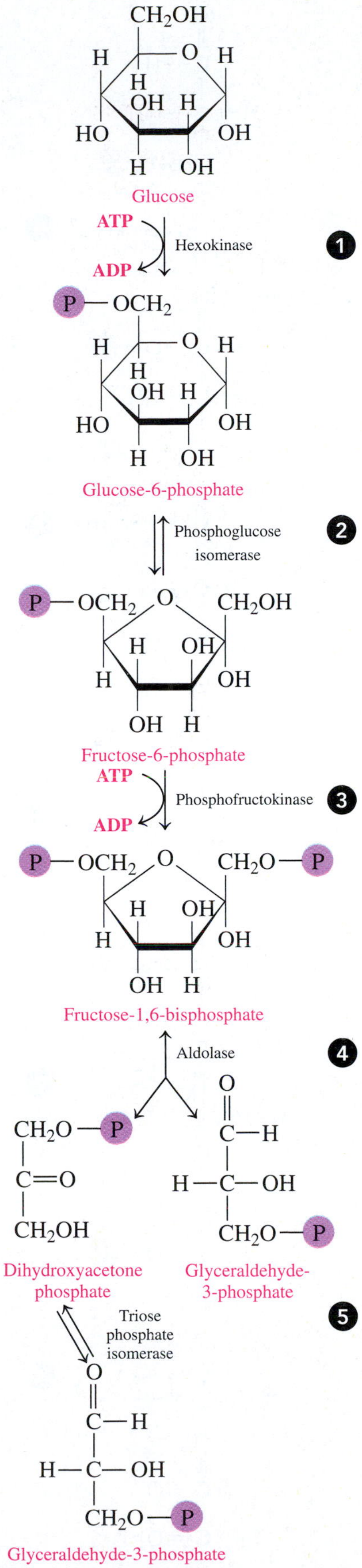
Glucose
ATP
Hexokinase
ADP
1
Glucose-6-phosphate
Phosphoglucose isomerase
2
Fructose-6-phosphate
ATP
Phosphofructokinase
ADP
3
Fructose-1,6-bisphosphate
Aldolase
4
Dihydroxyacetone phosphate
Glyceraldehyde-3-phosphate
Triose phosphate isomerase
5
Glyceraldehyde-3-phosphate

Reaction 2 Isomerization

The glucose-6-phosphate, the aldose from reaction 1, undergoes isomerization to fructose-6-phosphate, which is a ketose. The enzyme that catalyzes this reaction is *phosphoglucose isomerase*.

Reaction 3 Phosphorylation

The hydrolysis of another ATP provides a second phosphate group, which converts fructose-6-phosphate to fructose-1,6-bisphosphate. The word *bisphosphate* is used to show that the two phosphate groups are on different carbons in fructose and not connected to each other. This reaction is catalyzed by a second kinase enzyme called *phosphofructokinase*.

Reaction 4 Cleavage

Fructose-1,6-bisphosphate is split by *aldolase* into two three-carbon phosphate isomers: dihydroxyacetone phosphate and glyceraldehyde-3-phosphate.

Reaction 5 Isomerization

Because dihydroxyacetone phosphate is a ketone, it cannot undergo further oxidation. However, it undergoes isomerization by *triose phosphate isomerase* to provide a second molecule of glyceraldehyde-3-phosphate, which can be oxidized. Now all six carbon atoms from glucose are contained in two identical triose phosphates.

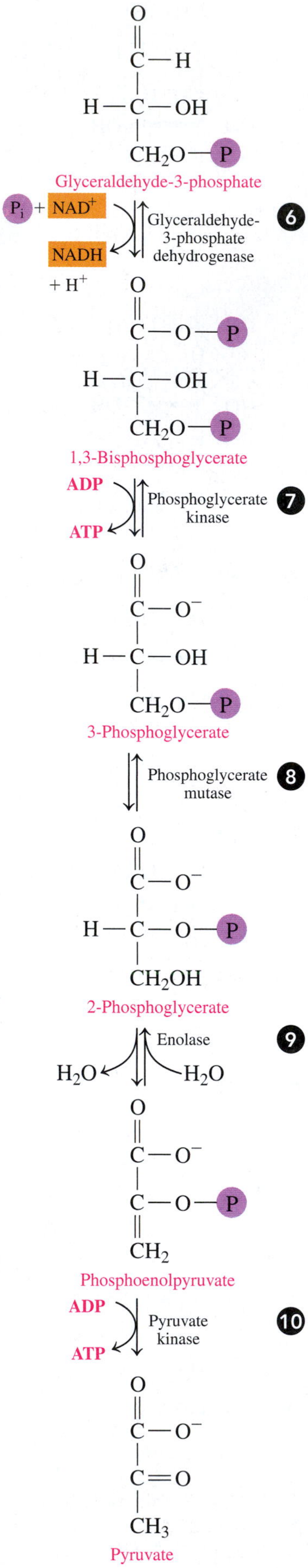
Glyceraldehyde-3-phosphate
Pi + NAD+
NADH + H+
Glyceraldehyde-3-phosphate dehydrogenase
6
1,3-Bisphosphoglycerate
ADP
ATP
Phosphoglycerate kinase
7
3-Phosphoglycerate
Phosphoglycerate mutase
8
2-Phosphoglycerate
Enolase
H2O
H2O
9
Phosphoenolpyruvate
ADP
ATP
Pyruvate kinase
10
Pyruvate

Energy-Generating Reactions: 6 to10

In our discussion of glycolysis from this point, the two molecules of glyceraldehyde-3-phosphate produced in step 5 are undergoing the same reactions. For simplicity, we show the structures and reactions for only one three-carbon molecule for reactions 6 to 10.

Reaction 6 Oxidation and Phosphorylation

The aldehyde group of each glyceraldehyde-3-phosphate is oxidized by *glyceraldehyde-3-phosphate dehydrogenase* to a carboxyl group, while the coenzyme NAD^+ is reduced to NADH and H^+. A phosphate group (P_i) adds to each of the new carboxyl groups to form two molecules of the high-energy compound 1,3-bisphosphoglycerate.

Reaction 7 Phosphate Transfer

A phosphate group from each 1,3-bisphosphoglycerate is transferred to two ADP molecules by *phosphoglycerate kinase*, yielding two molecules of the high-energy compound ATP. At this point in glycolysis, two ATP are produced, which balance the two ATP consumed in reactions 1 and 3.

Reaction 8 Isomerization

Two 3-phosphoglycerate molecules undergo isomerization by *phosphoglycerate mutase*, which moves the phosphate group from carbon 3 to carbon 2, yielding two molecules of 2-phosphoglycerate. The enzyme here is a specific type of isomerase called a *mutase* because the location of the functional group merely shifts from one carbon to another.

Reaction 9 Dehydration

Each of the phosphoglycerate molecules undergoes dehydration (loss of water) by the enzyme *enolase*, producing two molecules of phosphoenolpyruvate, a high-energy compound.

Reaction 10 Phosphate Transfer

In a second direct phosphorylation, phosphate groups from two phosphoenolpyruvate molecules are transferred by *pyruvate kinase* to two ADP to form two pyruvate and two ATP. Here, a fourth kinase enzyme is involved because a phosphate is transferred with ATP production.

Summary of Glycolysis

In the glycolysis pathway, a six-carbon glucose molecule is converted to two three-carbon pyruvates. Initially, two ATP are required to form fructose-1,6-bisphosphate. In later reactions (7 and 10), phosphate transfers produce a total of four ATP. Overall, glycolysis yields two ATP and two NADH when a glucose molecule is converted to two pyruvates.

TEST

Try Practice Problems 22.23 to 22.36

$$\underset{\text{Glucose}}{C_6H_{12}O_6} + 2\,NAD^+ + 2\,ADP + 2P_i \longrightarrow \underset{\text{Pyruvate}}{2CH_3-\overset{\overset{\displaystyle O}{\|}}{C}-COO^-} + 2\,NADH + 2\,ATP + 4H^+ + 2H_2O$$

▶SAMPLE PROBLEM 22.6 Reactions in Glycolysis

TRY IT FIRST

Identify each of the following reactions as an isomerization, phosphorylation, or dehydration:

a. A phosphate group is transferred to ADP to form ATP.
b. 3-Phosphoglycerate is converted to 2-phosphoglycerate.
c. A water molecule is removed from 2-phosphoglycerate.

SOLUTION

a. phosphorylation **b.** isomerization **c.** dehydration

STUDY CHECK 22.6

Indicate the reaction and enzyme that catalyzes each of the reactions in Sample Problem 22.6.

ANSWER

a. Phosphorylation in reaction 7 is catalyzed by phosphoglycerate kinase and in reaction 10 by pyruvate kinase.
b. Isomerization in reaction 8 is catalyzed by phosphoglycerate mutase.
c. Dehydration in reaction 9 is catalyzed by enolase.

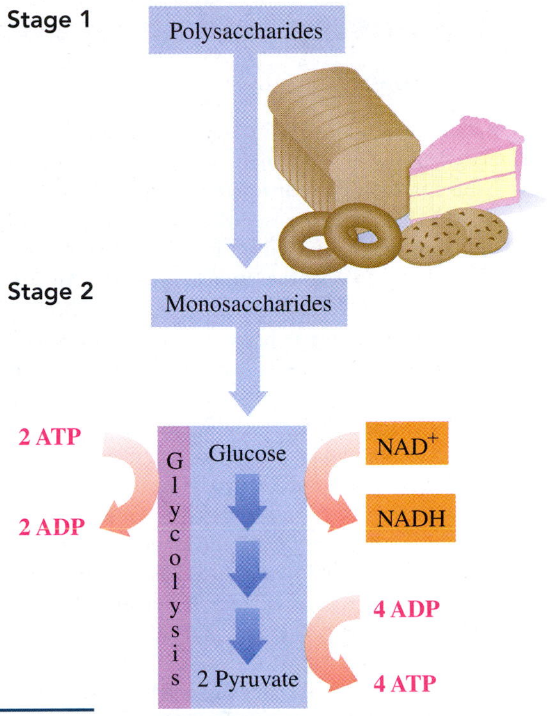

Stage 1
Polysaccharides
Stage 2
Monosaccharides
2 ATP
2 ADP
Glycolysis
Glucose
2 Pyruvate
NAD+
NADH
4 ADP
4 ATP

Glucose obtained from the digestion of polysaccharides is degraded in glycolysis to pyruvate.

Fructose and Galactose Enter Glycolysis

Other monosaccharides in our diet such as fructose, a monosaccharide found in table sugar, and galactose, a monosaccharide found in lactose, also enter glycolysis, but first they must be converted to intermediates that can enter into the glycolytic pathway.

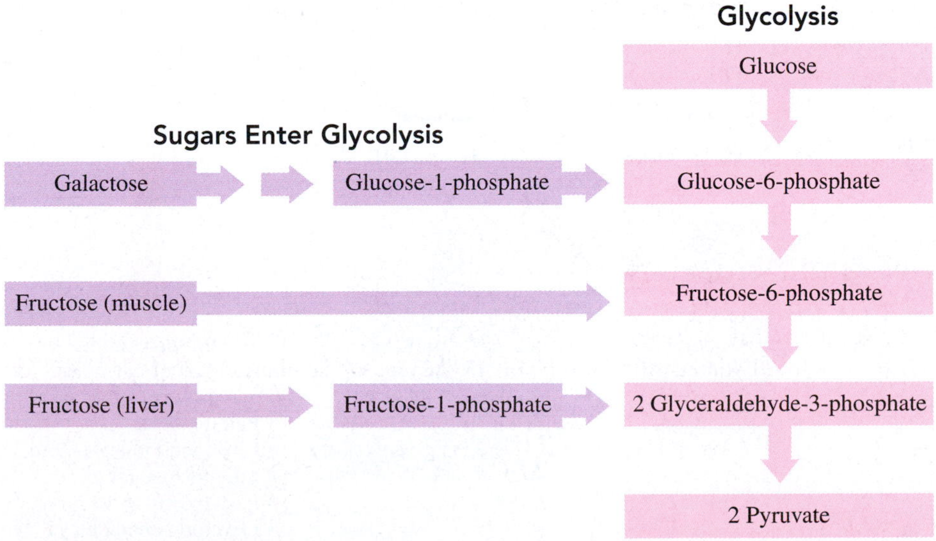

Glycolysis
Glucose
Sugars Enter Glycolysis
Galactose
Glucose-1-phosphate
Glucose-6-phosphate
Fructose (muscle)
Fructose-6-phosphate
Fructose (liver)
Fructose-1-phosphate
2 Glyceraldehyde-3-phosphate
2 Pyruvate

Galactose and fructose form intermediates that enter the glycolysis pathway to be metabolized.

TEST

Try Practice Problems 22.37, 22.45, and 22.46

ENGAGE

Where do fructose and galactose enter glycolysis?

Fructose

In the muscles and kidneys, fructose is phosphorylated to fructose-6-phosphate, which enters glycolysis in reaction 3. In the liver, fructose is converted to glyceraldehyde-3-phosphate, which enters glycolysis at reaction 6.

Galactose

In a series of four enzymatic steps, galactose reacts with ATP to yield glucose-1-phosphate, which is converted to glucose-6-phosphate. Glucose-6-phosphate enters glycolysis at reaction 2.

Chemistry Link to Health

Galactosemia

Adults meet their caloric needs by eating a variety of foods, but the main source of calories for babies in the first four to six months of their lives is milk. The most prevalent carbohydrate in milk is the disaccharide lactose, which is broken down in the body to give the monosaccharides glucose and galactose. Glucose enters glycolysis directly. Galactose is converted, via a series of enzymatic reactions, to glucose-6-phosphate, which then enters glycolysis.

In the case of a rare inherited disease called *classic galactosemia*, a genetic mutation results in a deficiency of one of the enzymes needed to metabolize galactose. As a result, galactose and two of its toxic by-products begin to build up in the tissues and in the blood. Newborns with galactosemia have difficulty with feeding and fail to gain weight as expected (*failure to thrive*). They are typically lethargic and jaundiced, show signs of liver and kidney failure, have an increased likelihood of developing infantile cataracts, and are particularly susceptible to *E. coli* infections (*sepsis*).

The symptoms of galactosemia are severe enough that, left untreated, the disease results in the death of up to 75% of affected infants. Therefore, the U.S. Department of Health and Human Services recommends that all newborns be screened for galactosemia soon after birth. Babies with this disease are placed on a lactose-restricted diet, which must be followed for the remainder of their lives. Even with adequate dietary treatment from an early age, children with the disease often show an increased risk of developmental delays in learning, speech, and motor function.

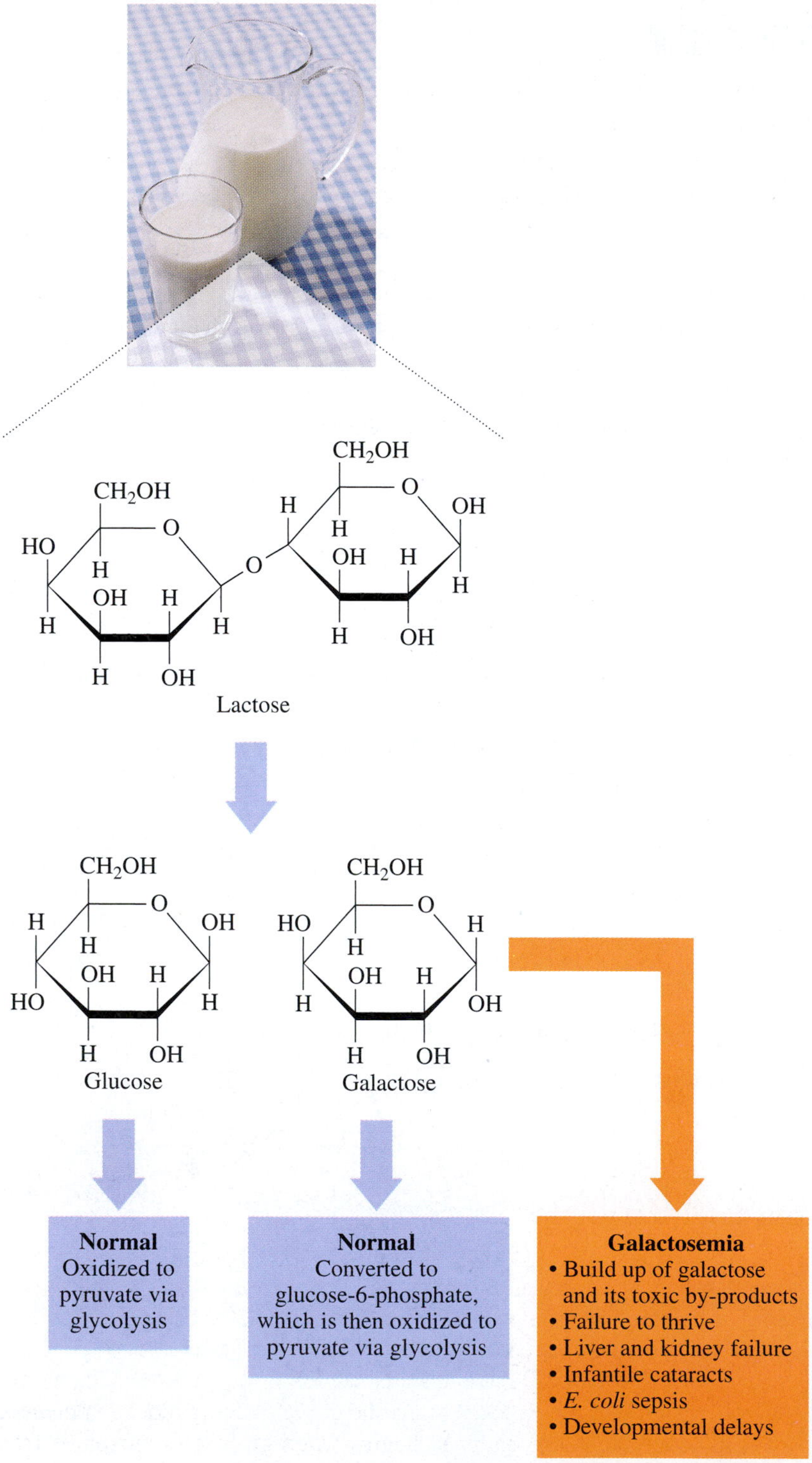
Lactose
Glucose
Galactose
Normal
Oxidized to pyruvate via glycolysis
Normal
Converted to glucose-6-phosphate, which is then oxidized to pyruvate via glycolysis
Galactosemia
• Build up of galactose and its toxic by-products
• Failure to thrive
• Liver and kidney failure
• Infantile cataracts
• E. coli sepsis
• Developmental delays

TEST

Try Practice Problems 22.47 and 22.48

Regulation of Glycolysis

Metabolic pathways such as glycolysis do not run at the same rates all the time. The amount of glucose that is broken down is controlled by the requirements in the cells for pyruvate, ATP, and other intermediates of glycolysis. Within the 10-step pathway of glycolysis, reactions 1, 3, and 10 are irreversible and can be regulated so they are more active when the cell needs ATP or the end products of glycolysis and less active when these compounds are plentiful.

ENGAGE

Which enzymes are regulated in glycolysis?

Reaction 1 Hexokinase

The amount of glucose entering the glycolysis pathway decreases when high levels of glucose-6-phosphate are present in the cell. This phosphorylation product inhibits hexokinase, which prevents glucose from reacting with ATP. This inhibition of the first enzyme in a pathway is an example of feedback control, which is a type of enzyme regulation.

Reaction 3 Phosphofructokinase

The reaction catalyzed by phosphofructokinase is the major control point for glycolysis. Both glucose and glucose-6-phosphate are substrates for multiple metabolic reactions. However, fructose-1,6-bisphosphate is only a substrate for the reaction catalyzed by phosphofructokinase. Therefore, once fructose-1,6-bisphosphate is formed, it must continue through the remaining reactions to pyruvate. Phosphofructokinase is an allosteric enzyme and is inhibited by high levels of ATP and activated by high levels of ADP and AMP. In other words, phosphofructokinase is less active when ATP levels are high and more active when the cell needs to replenish ATP (when ADP and AMP levels are high).

Reaction 10 Pyruvate Kinase

The final enzyme that can be regulated in glycolysis is pyruvate kinase, another allosteric enzyme. This enzyme is inhibited by either high levels of ATP or acetyl CoA.

▶SAMPLE PROBLEM 22.7 Regulation of Glycolysis

TRY IT FIRST

How is glycolysis regulated by each of the following enzymes?

a. hexokinase **b.** phosphofructokinase **c.** pyruvate kinase

SOLUTION

a. High levels of glucose-6-phosphate inhibit hexokinase, which stops the addition of a phosphate group to glucose in reaction 1.
b. Phosphofructokinase, which catalyzes the formation of fructose-1,6-bisphosphate, is inhibited by high levels of ATP, and activated by high levels of ADP and AMP.
c. High levels of ATP or acetyl CoA inhibit pyruvate kinase, which stops the formation of pyruvate in reaction 10.

STUDY CHECK 22.7

Name the molecule that inhibits hexokinase by feedback regulation in step 1 of glycolysis.

ANSWER

glucose-6-phosphate

TEST

Try Practice Problems 22.38 to 22.40

Pentose Phosphate Pathway

The **pentose phosphate pathway** is an alternative pathway for the oxidation of glucose. Whereas glycolysis is a catabolic pathway, the pentose phosphate pathway oxidizes glucose for anabolic purposes, producing the coenzyme NADPH and five-carbon pentoses. The NADPH, which is the reduced form of $NADP^+$, is an important coenzyme required in the pathways that synthesize nucleic acids, cholesterol, and fatty acids in the liver and fat (adipose) cells. Five-carbon pentoses are one of the building blocks of nucleotides.

ENGAGE

Which products from the pentose phosphate pathway are used in anabolic reactions?

The pentose phosphate pathway begins when glucose-6-phosphate from reaction 1 in glycolysis is converted to ribulose-5-phosphate in a process that requires two $NADP^+$.

Ribulose-5-phosphate is isomerized by *phosphopentose isomerase* to ribose-5-phosphate, an important component of nucleotides.

$$\text{Glucose-6-phosphate} + 2\text{NADP}^+ + \text{H}_2\text{O} \longrightarrow \text{ribose-5-phosphate} + 2\text{NADPH} + 2\text{H}^+ + \text{CO}_2$$

A series of subsequent steps convert three ribose-5-phosphate molecules to two hexose molecules and one triose molecule, each of which can participate in glycolytic reactions.

$$3\ \text{Ribose-5-phosphate} \longrightarrow 2\ \text{fructose-6-phosphate} + \text{glyceraldehyde-3-phosphate}$$

TEST

Try Practice Problems 22.41 to 22.44

We can write the overall reaction of the pentose phosphate pathway as

$$3\ \text{Glucose-6-phosphate} + 6\text{NADP}^+ + 3\text{H}_2\text{O} \longrightarrow 2\ \text{fructose-6-phosphate} + \text{glyceraldehyde-3-phosphate} + 6\text{NADPH} + 6\text{H}^+ + 3\text{CO}_2$$

Pentose Phosphate Pathway

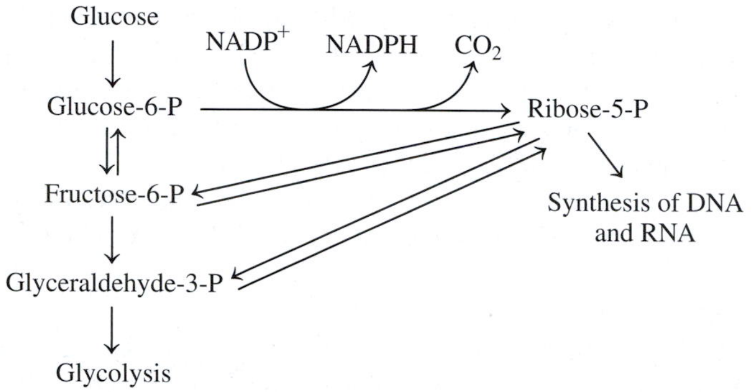
Glucose
NADP+
NADPH
CO2
Glucose-6-P
Ribose-5-P
Fructose-6-P
Synthesis of DNA and RNA
Glyceraldehyde-3-P
Glycolysis

The pentose phosphate pathway converts glucose-6-P to ribose-5-P, which is needed for the synthesis of DNA and RNA. Ribose-5-P can also return to glycolysis as needed. NADPH is produced as a result of these reactions.

PRACTICE PROBLEMS

22.4 Glycolysis: Oxidation of Glucose

22.23 What is the starting compound of glycolysis?

22.24 What is the three-carbon product of glycolysis?

22.25 How is ATP used in the initial steps of glycolysis?

22.26 How many ATP are used in the initial steps of glycolysis?

22.27 What three-carbon intermediates are obtained when fructose-1,6-bisphosphate splits?

22.28 Why does one of the three-carbon intermediates undergo isomerization?

22.29 How does substrate phosphorylation account for the production of ATP in glycolysis?

22.30 Why are there two ATP molecules formed for one molecule of glucose?

22.31 Indicate the enzyme(s) that catalyze(s) each of the following reactions in glycolysis:
a. phosphorylation **b.** direct transfer of a phosphate group

22.32 Indicate the enzyme(s) that catalyze(s) each of the following reactions in glycolysis:
a. isomerization
b. formation of a three-carbon ketone and a three-carbon aldehyde

22.33 How many ATP or NADH are produced (or required) in each of the following steps in glycolysis?
a. glucose to glucose-6-phosphate
b. glyceraldehyde-3-phosphate to 1,3-bisphosphoglycerate
c. glucose to pyruvate

22.34 How many ATP or NADH are produced (or required) in each of the following steps in glycolysis?
a. 1,3-bisphosphoglycerate to 3-phosphoglycerate
b. fructose-6-phosphate to fructose-1,6-bisphosphate
c. phosphoenolpyruvate to pyruvate

22.35 Which step(s) in glycolysis involve(s) the following?
a. The first ATP molecule is hydrolyzed.
b. Direct substrate phosphorylation occurs.
c. Six-carbon sugar splits into two three-carbon molecules.

22.36 Which step(s) in glycolysis involve(s) the following?
a. Isomerization takes place. **b.** NAD^+ is reduced.
c. A second ATP molecule is synthesized.

22.37 How do galactose and fructose, obtained from the digestion of carbohydrates, enter glycolysis?

22.38 What are three enzymes that regulate glycolysis?

22.39 Indicate whether each of the following would activate or inhibit phosphofructokinase:
a. low levels of ATP **b.** high levels of ATP

22.40 Indicate whether each of the following would activate or inhibit pyruvate kinase:
a. low levels of ATP **b.** high levels of ATP

22.41 What is the initial substrate for the pentose phosphate pathway?

22.42 What are the end products of the pentose phosphate pathway?

22.43 What is the purpose of producing NADPH in the pentose phosphate pathway?

22.44 What is the purpose of producing ribose-5-phosphate in the pentose phosphate pathway?

Clinical Applications

22.45 Table sugar, which is the disaccharide sucrose, contains the monosaccharides glucose and fructose. After digestion of sucrose, how do each of the monosaccharides enter glycolysis?

22.46 Milk sugar, which is the disaccharide lactose, contains the monosaccharides glucose and galactose. After digestion of lactose, how do each of the monosaccharides enter glycolysis?

22.47 Why must patients with galactosemia follow a lactose-restricted diet?

22.48 What are some of the symptoms associated with galactosemia?

22.5 Pathways for Pyruvate

LEARNING GOAL Give the conditions for the conversion of pyruvate to lactate, ethanol, and acetyl coenzyme A.

The pyruvate produced from glucose can now enter pathways that continue to extract energy. The available pathway depends on whether there is sufficient oxygen in the cell. Under **aerobic** conditions, oxygen is available to convert pyruvate to acetyl coenzyme A (acetyl CoA) and CO_2. When oxygen levels are low, pyruvate is reduced to lactate. In yeast cells, pyruvate is reduced to ethanol and CO_2 (see **FIGURE 22.10**).

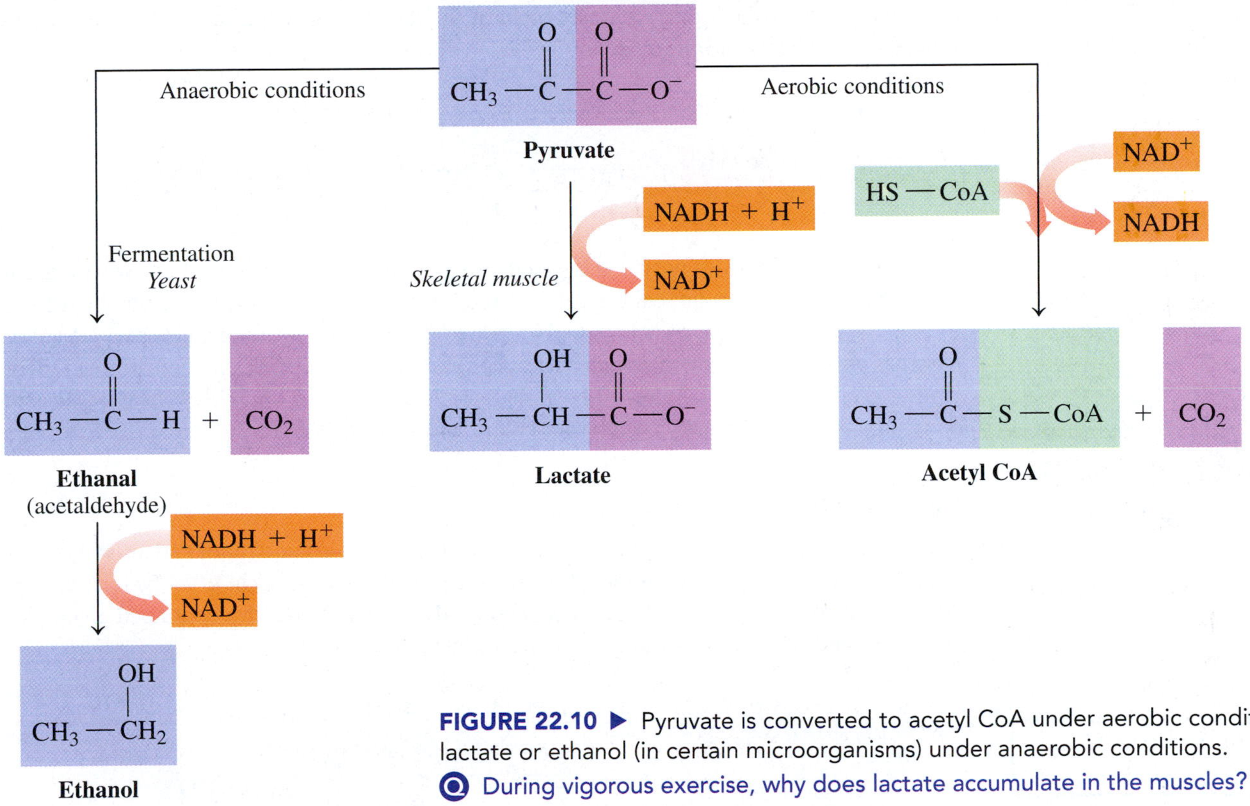
Anaerobic conditions
Aerobic conditions
Pyruvate
NAD+
HS—CoA
NADH + H+
NADH
Fermentation
Yeast
Skeletal muscle
NAD+
CO2
Ethanal
(acetaldehyde)
Lactate
Acetyl CoA
NADH + H+
NAD+
Ethanol

FIGURE 22.10 ▶ Pyruvate is converted to acetyl CoA under aerobic conditions and to lactate or ethanol (in certain microorganisms) under anaerobic conditions.

Q During vigorous exercise, why does lactate accumulate in the muscles?

Aerobic Conditions

In glycolysis, two ATP are generated when one glucose molecule is converted to two pyruvate molecules. However, much more energy is obtained from glucose when oxygen levels are high in the cells. Under aerobic conditions, pyruvate moves from the cytosol into the mitochondria to be oxidized further. In a complex reaction, pyruvate is oxidized, and a carbon atom is removed from pyruvate as CO_2. The coenzyme NAD^+ is reduced during the oxidation. The resulting two-carbon acetyl compound is attached to CoA, producing acetyl CoA, an important intermediate in many metabolic pathways.

ENGAGE

Which coenzyme is produced when pyruvate is oxidized under aerobic conditions?

TEST

Try Practice Problems 22.49 to 22.51

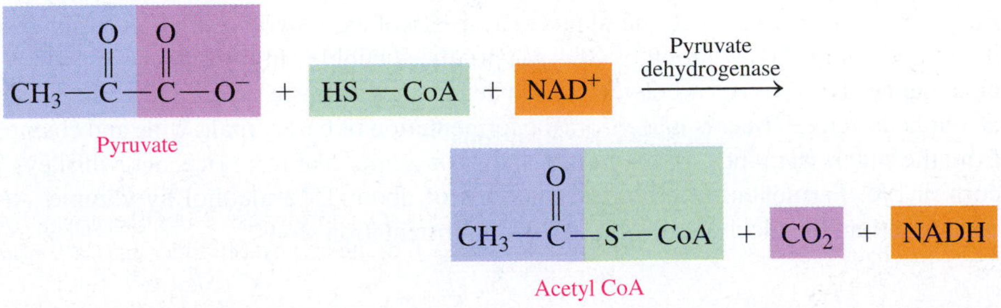
HS—CoA
NAD+
Pyruvate dehydrogenase
Pyruvate
CO2
NADH
Acetyl CoA

Anaerobic Conditions

When we engage in strenuous exercise, the oxygen stored in our muscle cells is quickly depleted. Under anaerobic conditions, pyruvate remains in the cytosol where it is reduced to lactate. The product NAD^+ is used to oxidize glyceraldehyde-3-phosphate in the glycolysis pathway, which produces a small but needed amount of ATP.

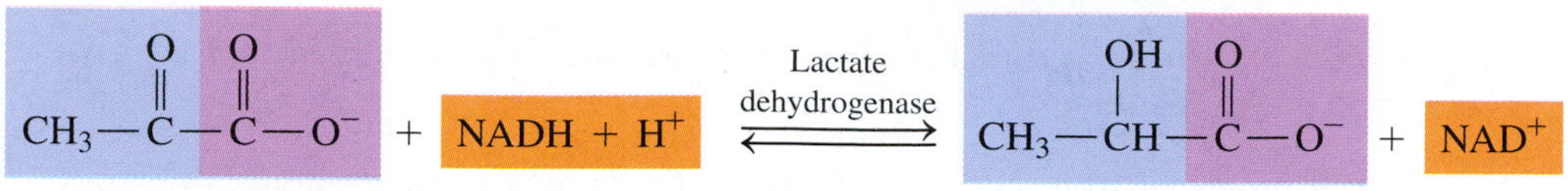
CH3—C—C—O⁻
NADH + H+
Lactate dehydrogenase
CH3—CH—C—O⁻
NAD+

Pyruvate (oxidized) Lactate (reduced)

After vigorous exercise, rapid breathing helps to repay the oxygen debt.

The accumulation of lactate and the corresponding drop in pH causes the muscles to tire and become sore. After exercise, a person continues to breathe rapidly to repay the *oxygen debt* incurred during exercise. Most of the lactate is transported to the liver, where it is converted back into pyruvate.

$$\underset{\text{Glucose}}{C_6H_{12}O_6} + 2\,ADP + 2P_i \longrightarrow \underset{\text{Lactate}}{2CH_3-\overset{\displaystyle OH}{\overset{|}{C}H}-COO^-} + 2\,ATP$$

Olives are cured in a vinegar and salt brine.

Bacteria also convert pyruvate to lactate under anaerobic conditions. In the preparation of kimchee and sauerkraut, cabbage is covered with salt brine. The glucose obtained from the starches in the cabbage is converted to lactate. The acid environment acts as a preservative that prevents the growth of other bacteria. The pickling of olives and cucumbers gives similar products. When cultures of bacteria that produce lactate are added to milk, the acid denatures the milk proteins to give sour cream and yogurt.

TEST

Try Practice Problems 22.52 to 22.54

Fermentation

Some microorganisms, particularly yeast, convert sugars to ethanol under anaerobic conditions by a process called **fermentation**. After pyruvate is formed in glycolysis, a carbon atom is removed in the form of CO_2 (**decarboxylation**). The NAD^+ for continued glycolysis is regenerated when the resulting ethanal is reduced to ethanol.

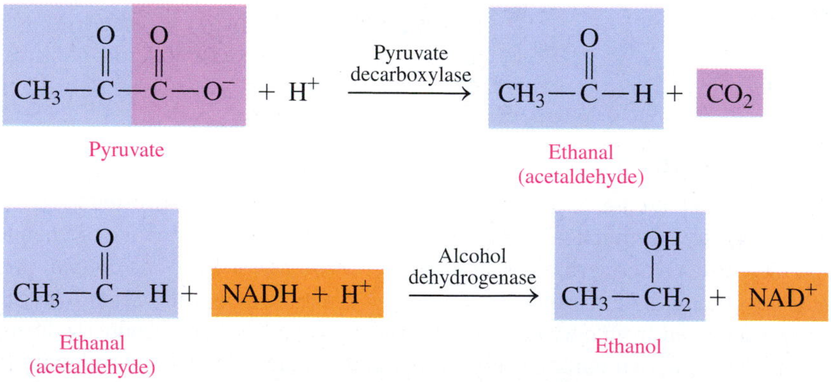
CH3—C—C—O⁻ + H+
Pyruvate decarboxylase
CH3—C—H + CO2
Pyruvate
Ethanal (acetaldehyde)
CH3—C—H + NADH + H+
Alcohol dehydrogenase
CH3—CH2 + NAD+
Ethanal (acetaldehyde)
Ethanol

Beer is produced by the fermentation of pyruvate from barley malt, which gives carbon dioxide and ethanol.

The process of fermentation by yeast is one of the oldest known chemical reactions. Enzymes in the yeast convert the sugars in a variety of carbohydrate sources to glucose and then to ethanol. The evolution of CO_2 gas produces bubbles in beer, sparkling wines, and champagne. The type of carbohydrate used determines the taste associated with a particular alcoholic beverage. Beer is made from the fermentation of barley malt, wine and champagne from the sugars in grapes, vodka from potatoes or grain, sake from rice, and whiskeys from corn or rye. Fermentation produces solutions of about 15% alcohol by volume. At this concentration, the alcohol kills the yeast and fermentation stops.

▶SAMPLE PROBLEM 22.8 Pathways for Pyruvate

TRY IT FIRST

Identify the pathway that has each of the following characteristics:

a. In this pathway, NAD^+ is reduced to $NADH + H^+$.
b. The product of this pathway contains three carbon atoms.
c. NADH is the oxidizing agent in this anaerobic pathway that produces a two-carbon molecule.
d. CO_2 is a product of this aerobic pathway.

SOLUTION

a. NAD^+ is used to oxidize pyruvate to acetyl CoA with the loss of one carbon atom as CO_2.
b. Pyruvate is reduced to a three-carbon molecule of lactate by lactate dehydrogenase and NADH.
c. Under anaerobic conditions, pyruvate is reduced to ethanol, a two-carbon molecule, and CO_2 by NADH.
d. Under aerobic conditions, pyruvate is oxidized to acetyl CoA and CO_2 by NAD^+.

STUDY CHECK 22.8

After strenuous exercise, lactate is oxidized back to pyruvate, with the accompanying reduction of NAD^+ in a reaction catalyzed by lactate dehydrogenase. Write an equation to show this reaction.

ANSWER

$$CH_3-CH(OH)-C(=O)-O^- + NAD^+ \xrightarrow{\text{Lactate dehydrogenase}} CH_3-C(=O)-C(=O)-O^- + NADH + H^+$$

TEST

Try Practice Problems 22.55 to 22.60

PRACTICE PROBLEMS

22.5 Pathways for Pyruvate

22.49 What condition is needed in the cell to convert pyruvate to acetyl CoA?

22.50 What coenzymes are needed for the oxidation of pyruvate to acetyl CoA?

22.51 Write the overall equation for the conversion of pyruvate to acetyl CoA.

22.52 Which compounds could be produced from pyruvate under anaerobic conditions?

22.53 In fermentation, a carbon atom is removed from pyruvate. What is the compound formed with that carbon atom?

22.54 In fermentation, what enzymes and coenzymes are used to produce ethanol?

22.55 Indicate which of the pathways for pyruvate, (1) acetyl CoA, (2) lactate, and/or (3) ethanol, occur under each of the following conditions:
a. The reaction is catalyzed by alcohol dehydrogenase.
b. NAD^+ is reduced.
c. A three-carbon compound is produced.
d. CO_2 is produced.

22.56 Indicate which of the pathways for pyruvate, (1) acetyl CoA, (2) lactate, and/or (3) ethanol, occur under each of the following conditions:
a. NAD^+ is produced.
b. HS—CoA is required.
c. NADH is required.
d. The reaction is catalyzed by pyruvate decarboxylase.

Clinical Applications

22.57 After a marathon, a runner has muscle pain and cramping. What might have occurred in the muscle cells to cause this?

22.58 Some students decided to make some wine by placing yeast and grape juice in a container with a tight lid. A few weeks later, the container exploded. What reaction could account for the explosion?

22.59 A defective *pyruvate dehydrogenase* causes a buildup of lactate in the body and lactic acidosis, which lead to neurological problems. In addition, the production of ATP is diminished. Write the chemical equation for the conversion of pyruvate to lactate.

22.60 Defective *pyruvate decarboxylase*, an enzyme in the pyruvate dehydrogenase complex, is a rare disorder that occurs in infancy and leads to an elevated level of pyruvate and lactate in the blood and urine. The inability to convert pyruvate to acetyl CoA and CO_2 leads to lack of energy, general weakness, and developmental delay. Write the chemical equation for the conversion of pyruvate to acetyl CoA.

22.6 Glycogen Synthesis and Degradation

LEARNING GOAL Describe the synthesis and breakdown of glycogen.

ENGAGE

Which organ in the body releases glucose from glycogen into the bloodstream?

Usually, your diet supplies you with all the glucose you need to produce pyruvate and ATP by glycolysis. Any excess glucose is used to replenish your energy reserves by synthesizing glycogen, which is stored in your muscles and liver. Glycogen is a polymer of glucose with $\alpha(1\rightarrow4)$-glycosidic bonds and multiple branches attached by $\alpha(1\rightarrow6)$-glycosidic bonds. When glycogen stores are full, any remaining glucose is converted to triacylglycerols and stored as body fat. When your diet does not supply sufficient glucose and your blood glucose is low, the glycogen in your muscles is converted to glucose for energy, and the glucose from glycogen in your liver is released into the bloodstream to raise your blood glucose levels. During conditions of fasting, glycogen breakdown supplies nearly all the glucose for the body.

CORE CHEMISTRY SKILL

Identifying the Compounds and Enzymes in Glycogenesis and Glycogenolysis

Glycogen Synthesis: Glycogenesis

Glycogenesis is the metabolic process of converting glucose molecules into glycogen. When glucose-6-phosphate, which is produced in reaction 1 of glycolysis, is not needed in the cell, it can be stored as glycogen.

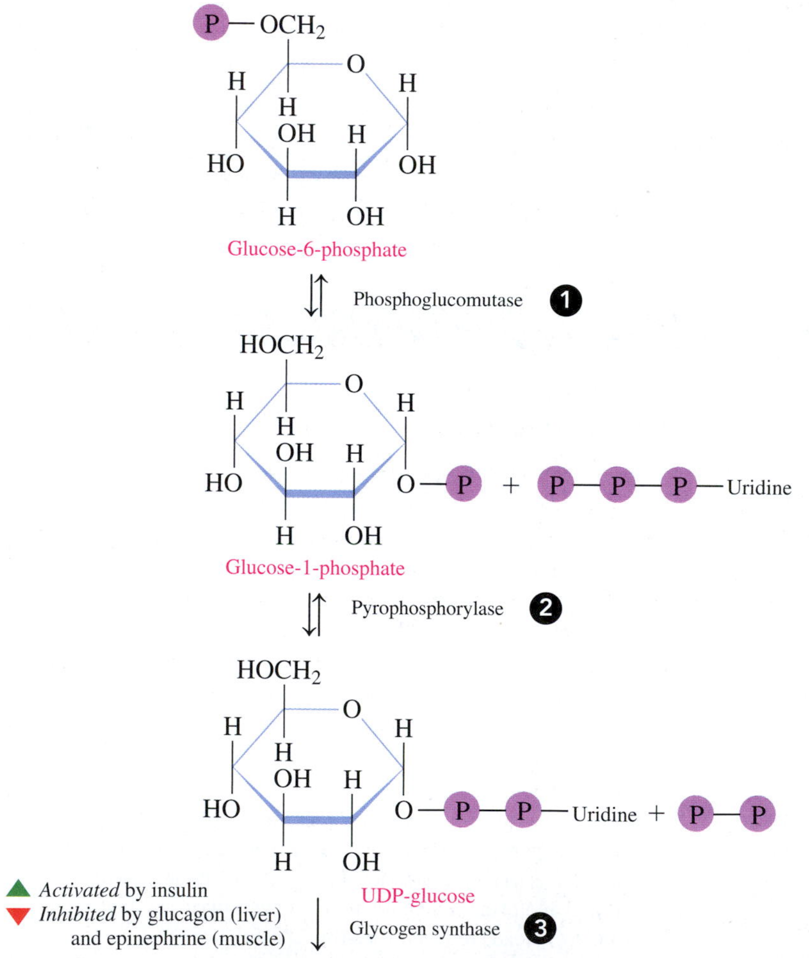
Glucose-6-phosphate
Phosphoglucomutase 1
Glucose-1-phosphate
Uridine
Pyrophosphorylase 2
UDP-glucose
Uridine
Activated by insulin
Inhibited by glucagon (liver) and epinephrine (muscle)
Glycogen synthase 3

Reaction 1 Isomerization

Glycogen synthesis begins when the enzyme *phosphoglucomutase* converts glucose-6-phosphate to the isomer glucose-1-phosphate. A mutase is a type of isomerase that shifts the location of the phosphate functional group (see **FIGURE 22.11**).

Reaction 2 Activation

Before glucose-1-phosphate can be added to the glycogen chain, it must be activated. Energy is released when the high-energy compound UTP (uridine triphosphate) transfers UMP to glucose-1-phosphate, giving UDP-glucose (uridine diphosphate glucose) and pyrophosphate, PP_i. This reaction is catalyzed by *pyrophosphorylase*.

Reaction 3 Glycogen Synthesis (Glucose Transfer to Glycogen)

Glycogen synthase catalyzes the breaking of the phosphate bond to glucose in UDP-glucose, releasing glucose, which forms an $\alpha(1\rightarrow4)$-glycosidic bond with the end of a glycogen chain.

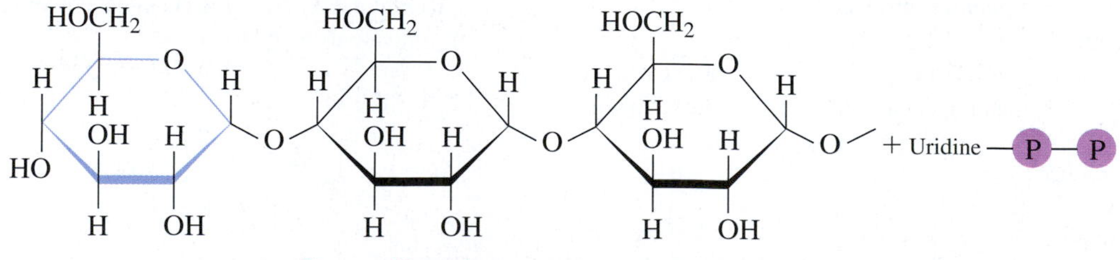
+ Uridine
Glucose adds to glycogen chain

FIGURE 22.11 ▶ In glycogenesis, glucose is used to synthesize glycogen.

Q What is the function of UTP in glycogen synthesis?

Glycogenolysis

Glycogen is a highly branched polysaccharide of glucose monomers with both $\alpha(1\rightarrow4)$- and $\alpha(1\rightarrow6)$-glycosidic bonds. Glucose is the primary energy source for muscle contractions, red blood cells, and the brain. When blood glucose is depleted, the glycogen stored in the muscle and liver is converted to glucose molecules in a process called **glycogenolysis** (see **FIGURE 22.12**).

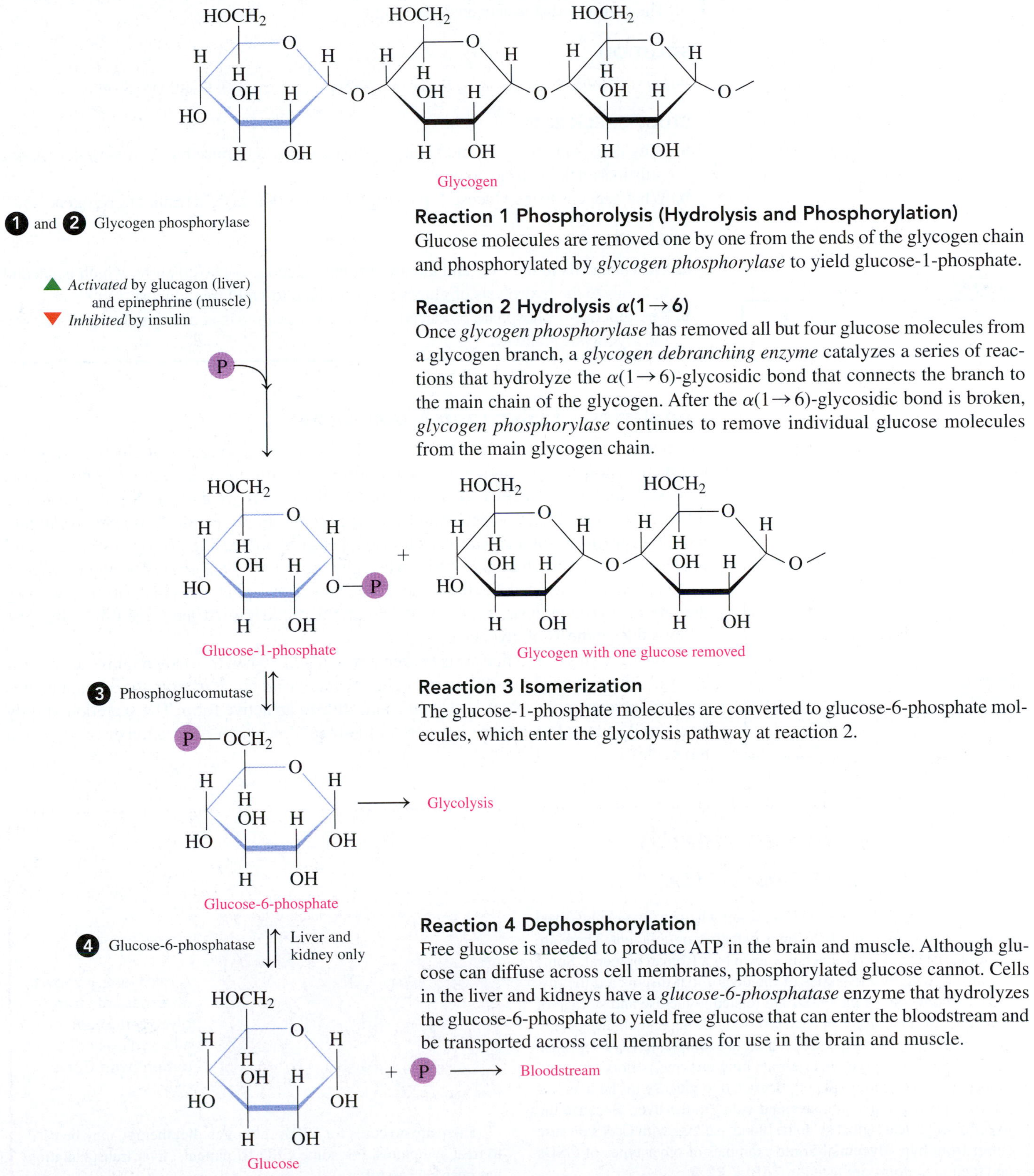
Glycogen
1 and 2 Glycogen phosphorylase
Activated by glucagon (liver) and epinephrine (muscle)
Inhibited by insulin
Reaction 1 Phosphorolysis (Hydrolysis and Phosphorylation)
Glucose molecules are removed one by one from the ends of the glycogen chain and phosphorylated by glycogen phosphorylase to yield glucose-1-phosphate.
Reaction 2 Hydrolysis α(1→6)
Once glycogen phosphorylase has removed all but four glucose molecules from a glycogen branch, a glycogen debranching enzyme catalyzes a series of reactions that hydrolyze the α(1→6)-glycosidic bond that connects the branch to the main chain of the glycogen. After the α(1→6)-glycosidic bond is broken, glycogen phosphorylase continues to remove individual glucose molecules from the main glycogen chain.
Glucose-1-phosphate
Glycogen with one glucose removed
3 Phosphoglucomutase
Reaction 3 Isomerization
The glucose-1-phosphate molecules are converted to glucose-6-phosphate molecules, which enter the glycolysis pathway at reaction 2.
Glycolysis
Glucose-6-phosphate
4 Glucose-6-phosphatase
Liver and kidney only
Reaction 4 Dephosphorylation
Free glucose is needed to produce ATP in the brain and muscle. Although glucose can diffuse across cell membranes, phosphorylated glucose cannot. Cells in the liver and kidneys have a glucose-6-phosphatase enzyme that hydrolyzes the glucose-6-phosphate to yield free glucose that can enter the bloodstream and be transported across cell membranes for use in the brain and muscle.
Bloodstream
Glucose

FIGURE 22.12 ▶ In glycogenolysis, glycogen is hydrolyzed to give glucose molecules.

Q What is the function of the debranching enzyme in glycogenolysis?

▶ SAMPLE PROBLEM 22.9 Glycogen Metabolism

TRY IT FIRST

Identify each of the following as part of the reaction pathways of glycogenolysis or glycogenesis:

a. Glucose-1-phosphate is converted to glucose-6-phosphate.
b. Glucose-1-phosphate forms UDP-glucose.
c. The pathway that is activated by insulin.

SOLUTION

a. glycogenolysis **b.** glycogenesis **c.** glycogenesis

STUDY CHECK 22.9

a. Why do cells in the liver and kidneys provide glucose to raise blood glucose levels, but cells in skeletal muscle do not?
b. What happens to the glucose-6-phosphate that is produced as a result of glycogenolysis?

ANSWER

a. Only the cells in the liver and kidneys provide glucose-6-phosphatase, which is needed to catalyze the hydrolysis of glucose-6-phosphate to yield free glucose.
b. The glucose-6-phosphate produced during glycogenolysis is converted to pyruvate via the glycolysis pathway.

TEST

Try Practice Problems 22.61 to 22.66

Regulation of Glycogen Metabolism

The brain, skeletal muscles, and red blood cells require large amounts of glucose every day to function properly. To protect the brain, hormones with opposing actions control blood glucose levels (see **FIGURES 22.11** and **22.12**). Soon after we eat and digest a meal, our blood glucose level rises, which stimulates the pancreas to secrete the hormone *insulin* into our bloodstream. Insulin signals cells to use glucose by increasing glycogen synthesis and oxidation reactions such as glycolysis. When glucose is low, *glucagon*, a hormone produced in the pancreas, is secreted into the bloodstream. Glucagon signals cells in the liver to increase the rate of glycogenolysis, which raises blood glucose levels. At the same time, glucagon inhibits the synthesis of glycogen.

Glycogen in skeletal muscle is broken down quickly when the body requires a "burst of energy," often referred to as "fight or flight." *Epinephrine* released from the adrenal glands converts glycogen phosphorylase from an inactive to an active form. The secretion of only a few molecules of epinephrine results in the breakdown of a huge number of glycogen molecules.

ENGAGE

Which hormone activates enzymes involved in glycogenesis?

TEST

Try Practice Problems 22.67 to 22.69

Chemistry Link to Health

Glycogen Storage Diseases (GSDs)

Glycogen storage diseases (GSDs) result when an enzyme involved in the use or storage of glycogen is defective or deficient. There are multiple types of GSDs. Each is known by a Roman numeral. Some are also known by the names of the doctor that discovered the disease. Although all of the GSDs are related to the use and storage of glycogen, the symptoms of each type of GSD depend on the specific enzyme that is defective or deficient. For example, in *Cori's disease* (GSD type III), a glycogen debranching enzyme is deficient, and glycogen cannot be completely degraded to glucose in the liver. As a result, glycogen accumulates and enlarges the liver. Because the liver doesn't release glucose in the blood, patients with Cori's disease suffer from hypoglycemia. Some examples of other types of GSDs and their symptoms are listed in **TABLE 22.3**.

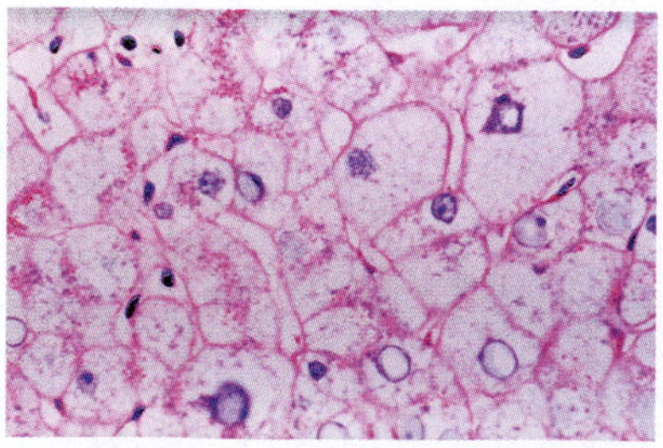

A micrograph shows an excess of stored glycogen (blue) in a liver biopsy of a patient with Cori's disease.

There are no cures for GSDs, although diet therapy may be used to treat symptoms. For some GSD III patients, liver transplantation has been successful.

(*continued*)

Chemistry Link to Health (*continued*)

TABLE 22.3 Examples of Glycogen Storage Diseases (GSDs)

Type	Organ	Defective Enzyme	Disease	Symptoms
0	Liver, muscle	Glycogen synthase	Glycogen synthase deficiency, cannot synthesize glycogen	Hypoglycemia, muscle weakness, exercise intolerance
I	Liver, kidneys	Glucose-6-phosphatase	*von Gierke's*, cannot convert glucose-6-phosphate to glucose	Severe hypoglycemia, enlarged liver, high concentrations of fats and lactic acid in the blood
II	Lysosomes of all organs	$\alpha(1\rightarrow4)$-Glucosidase	*Pompe's*, cannot degrade glycogen in the lysosomes	Enlarged heart, muscle weakness, heart failure, short life expectancy
III	Liver, muscle	Glycogen debranching enzyme	*Cori's*, cannot hydrolyze $\alpha(1\rightarrow6)$-glycosidic bonds to completely degrade glycogen to glucose	Mild hypoglycemia, enlarged liver, muscle weakness, high concentration of fats in the blood
IV	Liver	Glycogen branching enzyme	*Andersen's*, cannot form $\alpha(1\rightarrow6)$-glycosidic bonds in glycogen branches	Failure to thrive, short life expectancy due to heart or liver failure
V	Muscle	Glycogen phosphorylase	*McArdle's*, cannot degrade glycogen for use in muscle cells	Exercise intolerance, exercise-induced muscle cramps
VI	Liver	Glycogen phosphorylase	*Hers'*, cannot degrade glycogen in the liver	Clinical features are similar to those of Type I, but milder

TEST

Try Practice Problems 22.70 to 22.72

A summary of the processes of glycogenesis and glycogenolysis indicates that glucose-6-phosphate and glucose-1-phosphate are intermediates of both processes, but UDP-glucose is only involved in glycogenesis.

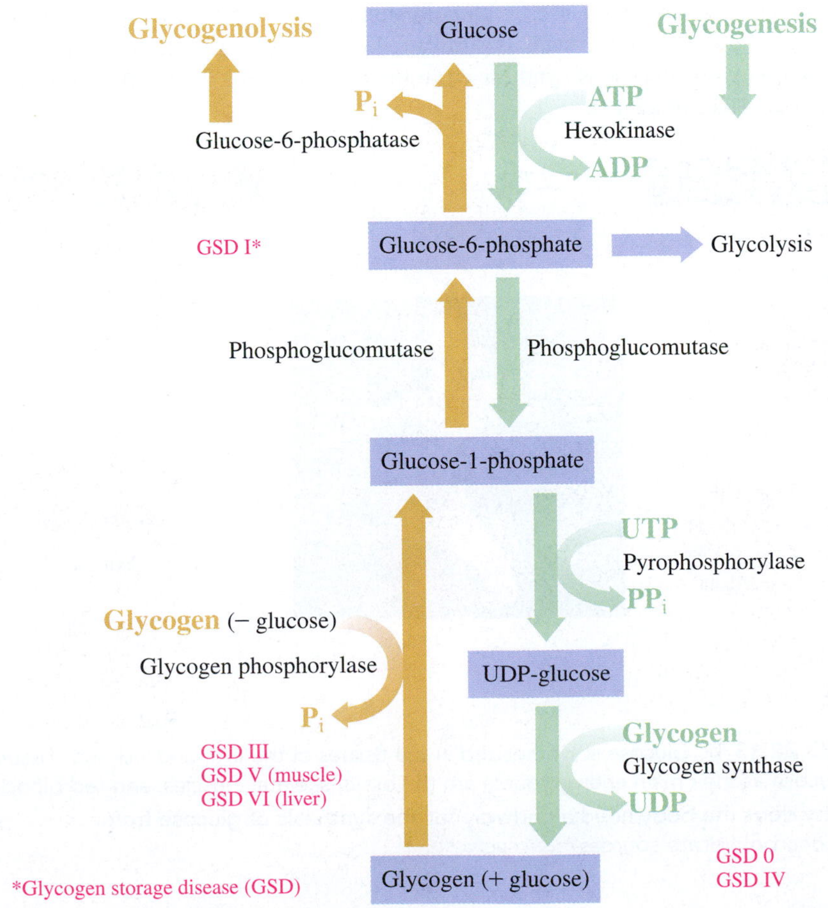

Glycogenolysis
Glucose
Glycogenesis
Pi
ATP
Hexokinase
ADP
Glucose-6-phosphatase
GSD I*
Glucose-6-phosphate
Glycolysis
Phosphoglucomutase
Phosphoglucomutase
Glucose-1-phosphate
UTP
Pyrophosphorylase
PPi
Glycogen (− glucose)
Glycogen phosphorylase
Pi
UDP-glucose
GSD III
GSD V (muscle)
GSD VI (liver)
Glycogen
Glycogen synthase
UDP
GSD 0
GSD IV
Glycogen (+ glucose)
*Glycogen storage disease (GSD)

Glucose-6-phosphate and glucose-1-phosphate are intermediates of both glycogenesis and glycogenolysis. Reactions affected in different glycogen storage diseases (GSDs) are indicated.

PRACTICE PROBLEMS

22.6 Glycogen Synthesis and Degradation

22.61 What is meant by the term *glycogenesis*?

22.62 What is meant by the term *glycogenolysis*?

22.63 How do muscle cells use glycogen to provide energy?

22.64 How does the liver raise blood glucose levels?

22.65 What is the function of glycogen phosphorylase?

22.66 Why is the enzyme phosphoglucomutase used in both glycogenolysis and glycogenesis?

22.67 Indicate whether each of the following hormones activates or inhibits glycogenesis:
- **a.** insulin
- **b.** glucagon
- **c.** epinephrine

22.68 Indicate whether each of the following hormones activates or inhibits glycogenolysis:
- **a.** insulin
- **b.** glucagon
- **c.** epinephrine

Clinical Applications

22.69 Diabetics often substitute fructose for glucose. Why is it acceptable for diabetics to include fructose in their diet?

22.70 Glycogen storage diseases are the result of a defective enzyme involved in pathways for glycogen storage or degradation. What is the effect of a defective debranching enzyme in Cori's disease (GSD III) (see Chemistry Link to Health: Glycogen Storage Diseases)?

22.71 Glycogen storage diseases are the result of a defective enzyme involved in pathways for glycogen storage or degradation. What is the effect of a defective liver glycogen phosphorylase in Hers' disease (GSD VI) (see Chemistry Link to Health: Glycogen Storage Diseases)?

22.72 Glycogen storage diseases are the result of a defective enzyme involved in pathways for glycogen storage or degradation. What is the effect of a defective glycogen synthase enzyme (GSD 0) (see Chemistry Link to Health: Glycogen Storage Diseases)?

22.7 Gluconeogenesis: Glucose Synthesis

LEARNING GOAL Describe how glucose is synthesized from molecules from noncarbohydrate sources.

Glycogen stored in our liver and muscles can supply us with about one day's requirement of glucose. However, glycogen stores are depleted in 10 to 12 h if we fast for more than one day or participate in heavy exercise. Glucose is the primary source of energy for the brain and the only source of energy for red blood cells. If our glycogen stores are depleted, our liver cells synthesize glucose by **gluconeogenesis** from carbon atoms we obtain from noncarbohydrate compounds such as pyruvate, lactate, glycerol from fats, fatty acids, and amino acids. Most glucose is synthesized in the cytosol of the liver cells and some in the kidneys (see **FIGURE 22.13**).

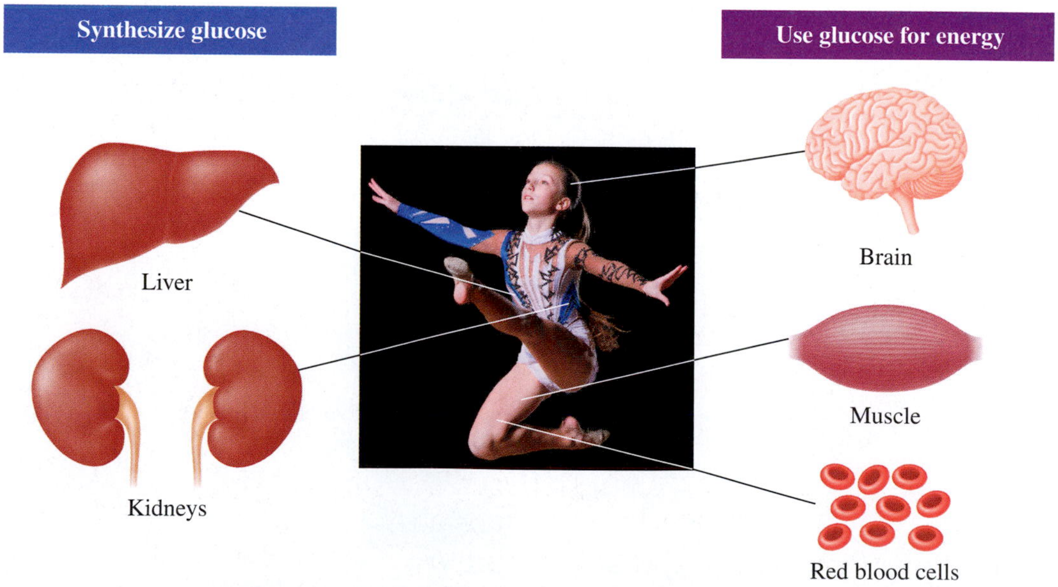

Synthesize glucose
Use glucose for energy
Liver
Kidneys
Brain
Muscle
Red blood cells

FIGURE 22.13 ▶ Glucose is synthesized in the tissues of the liver and kidneys. Tissues that use glucose as their main energy source are the brain, skeletal muscles, and red blood cells.

Q Why does the body need a pathway for the synthesis of glucose from noncarbohydrate sources?

Before gluconeogenesis can begin, the carbon atoms from noncarbohydrate food sources must be converted to pyruvate. From that point, many of the compounds and reactions in gluconeogenesis are the same in glycolysis and catalyzed by the same enzymes. There are 10 reactions in glycolysis and 11 reactions in gluconeogenesis. Three reactions (1, 3, and 10) in glycolysis, which are irreversible reactions, are catalyzed by different enzymes in gluconeogenesis.

A summary of the reactions and different enzymes in gluconeogenesis and glycolysis is shown in **FIGURE 22.14**. The overall equation for gluconeogenesis is written as follows:

$$2\text{ Pyruvate} + 4\text{ ATP} + 2\text{ GTP} + 2\text{ NADH} + 2H^+ + 6\text{ }H_2O \longrightarrow \text{glucose} + 4\text{ ADP} + 2\text{ GDP} + 6P_i + 2\text{ NAD}^+$$

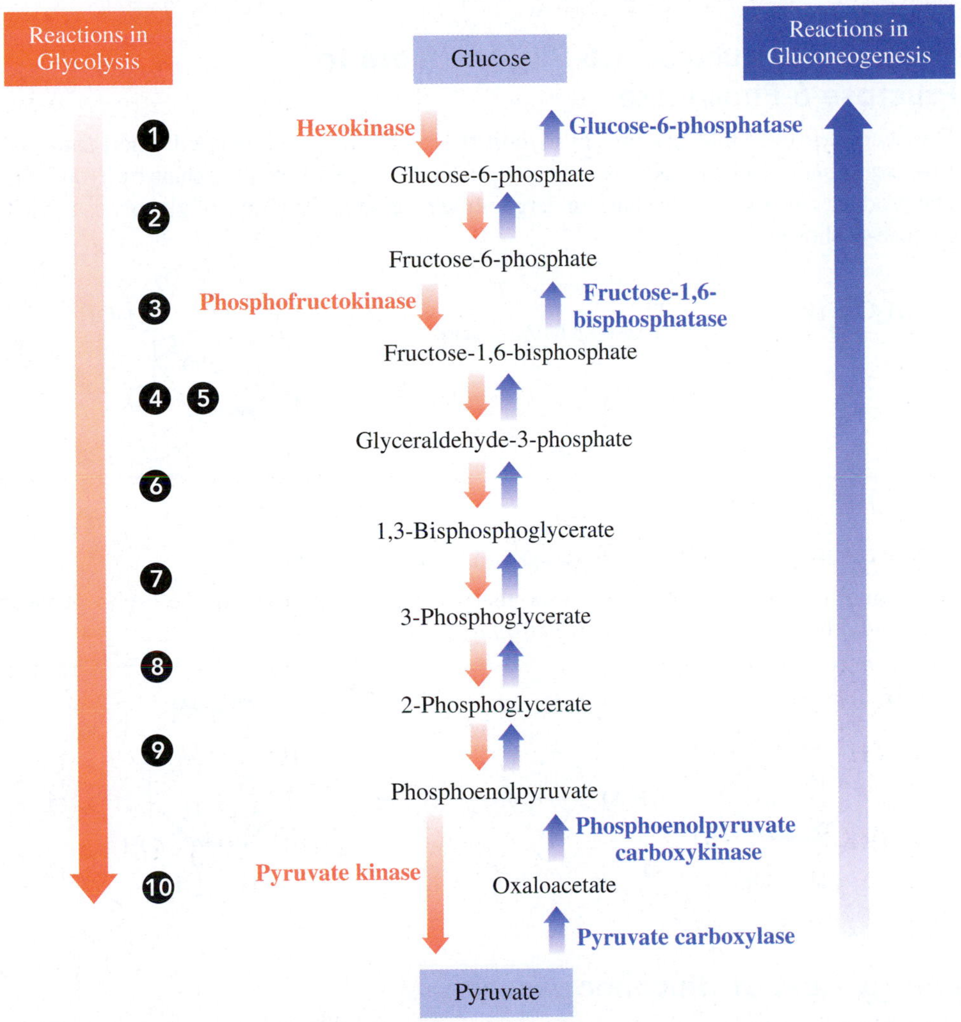

Reactions in Glycolysis
Glucose
Reactions in Gluconeogenesis
1
Hexokinase
Glucose-6-phosphatase
Glucose-6-phosphate
2
Fructose-6-phosphate
3
Phosphofructokinase
Fructose-1,6-bisphosphatase
Fructose-1,6-bisphosphate
4 5
Glyceraldehyde-3-phosphate
6
1,3-Bisphosphoglycerate
7
3-Phosphoglycerate
8
2-Phosphoglycerate
9
Phosphoenolpyruvate
Phosphoenolpyruvate carboxykinase
10
Pyruvate kinase
Oxaloacetate
Pyruvate carboxylase
Pyruvate

FIGURE 22.14 ▶ In gluconeogenesis, four different enzymes (in blue) are used to replace the enzymes that catalyze the three irreversible reactions of glycolysis (1, 3, and 10). All the other reactions are catalyzed in both directions by the same enzymes used for glycolysis.

Q Why are 11 enzymes required for gluconeogenesis and only 10 for glycolysis?

TEST

Try Practice Problems 22.73 to 22.75

In the sections below, we will describe the reactions that are unique to gluconeogenesis.

Converting Pyruvate to Phosphoenolpyruvate

To start the synthesis of glucose from pyruvate, two catalyzed reactions are needed to replace reaction 10 in glycolysis. In the first step, *pyruvate carboxylase* uses the energy of ATP hydrolysis to catalyze the addition of CO_2 to pyruvate to produce the four-carbon compound oxaloacetate. In the second step, *phosphoenolpyruvate carboxykinase* uses the energy of hydrolysis of the high-energy compound GTP to convert oxaloacetate to phosphoenolpyruvate. Enzymes from glycolysis then catalyze a series of reactions that convert phosphoenolpyruvate to fructose-1,6-bisphosphate.

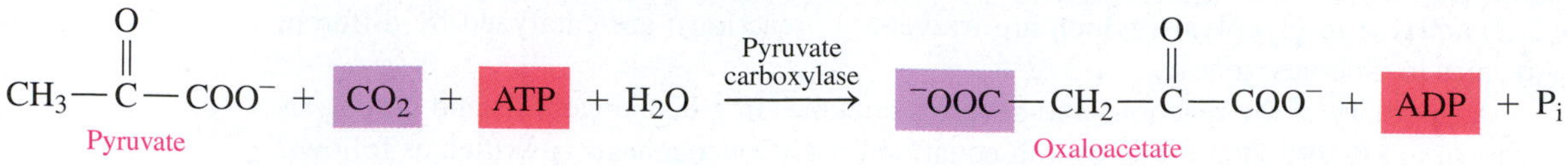

$$CH_3-\overset{\overset{O}{\|}}{C}-COO^- + CO_2 + ATP + H_2O \xrightarrow{\text{Pyruvate carboxylase}} {}^-OOC-CH_2-\overset{\overset{O}{\|}}{C}-COO^- + ADP + P_i$$

Pyruvate   Oxaloacetate

$${}^-OOC-CH_2-\overset{\overset{O}{\|}}{C}-COO^- + GTP \xrightarrow{\text{Phosphoenolpyruvate carboxykinase}} H_2C{=}\overset{\overset{O-P}{|}}{C}-COO^- + CO_2 + GDP$$

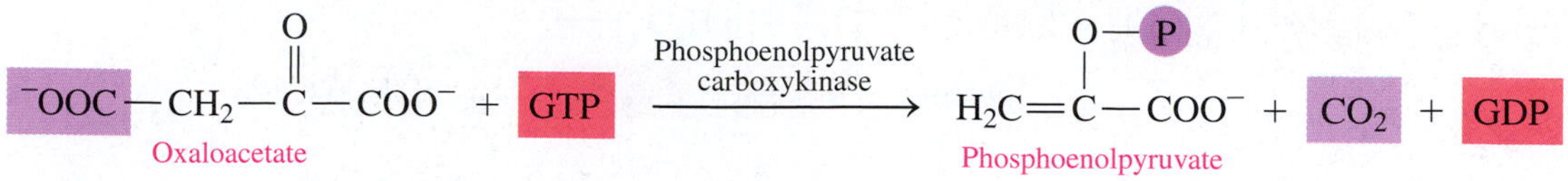

Oxaloacetate   Phosphoenolpyruvate

Converting Fructose-1,6-Bisphosphate to Fructose-6-Phosphate

The second irreversible reaction in glycolysis (reaction 3) is bypassed when *fructose-1,6-bisphosphatase* cleaves a phosphate group from fructose-1,6-bisphosphate by hydrolysis. The product, fructose-6-phosphate, undergoes the reversible reaction 2 of glycolysis to yield glucose-6-phosphate.

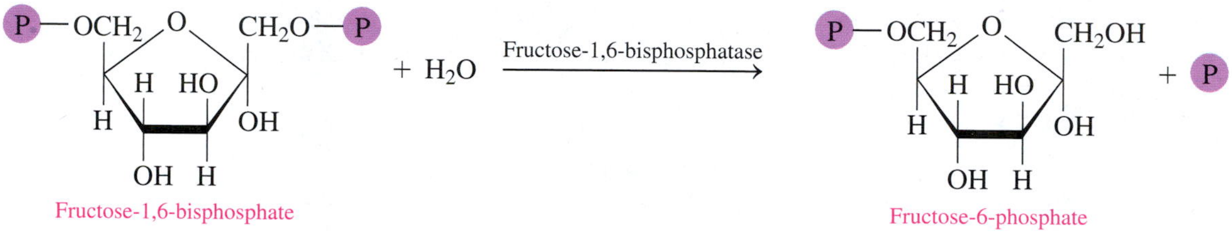

Fructose-1,6-bisphosphate   Fructose-1,6-bisphosphatase   Fructose-6-phosphate

Converting Glucose-6-Phosphate to Glucose

In the final reaction of gluconeogenesis, the phosphate group of glucose-6-phosphate is hydrolyzed by the enzyme *glucose-6-phosphatase* to form glucose.

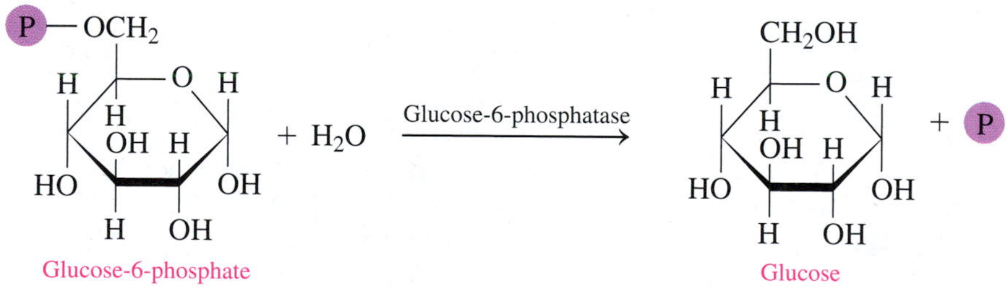

Glucose-6-phosphate   Glucose-6-phosphatase   Glucose

Energy Cost of Gluconeogenesis

The pathway of gluconeogenesis consists of seven reversible reactions of glycolysis and four new reactions that replace the three irreversible reactions of glycolysis. If all the reactions were simply the reverse of glycolysis, gluconeogenesis would not be energetically favorable. However, gluconeogenesis can take place because the formation of glucose from pyruvate is coupled with the hydrolysis of multiple high-energy compounds, processes that release energy.

Lactate and the Cori Cycle

When a person exercises vigorously, anaerobic conditions in the muscle reduce pyruvate to lactate. At the same time, NADH is converted to NAD$^+$ so that a small amount of ATP is formed when oxygen is lacking. Lactate is an important source of carbon atoms for gluconeogenesis. In the **Cori cycle**, lactate is transported to the liver where it is oxidized to pyruvate, which is used to synthesize glucose. Glucose enters the bloodstream and returns to the muscle to rebuild glycogen stores. The Cori cycle is very active when a person has just completed a period of vigorous exercise (see **FIGURE 22.15**).

TEST

Try Practice Problem 22.76

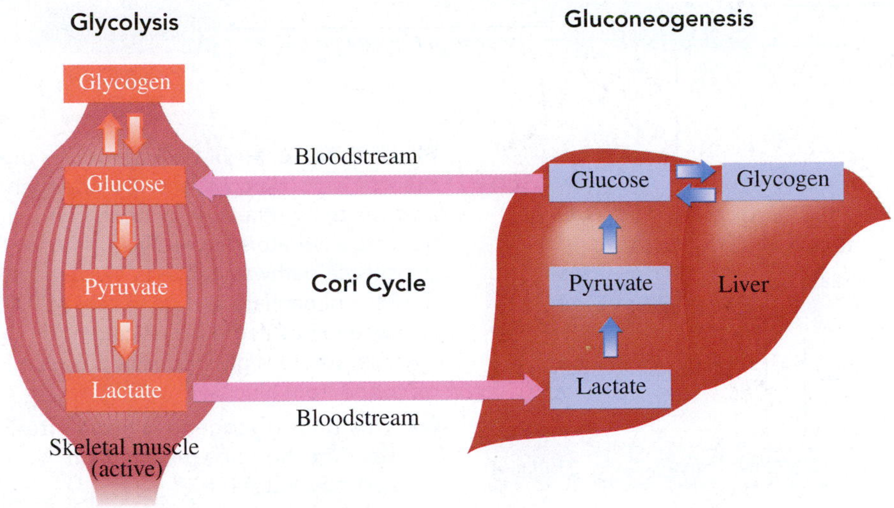
Glycolysis
Gluconeogenesis
Glycogen
Glucose
Pyruvate
Lactate
Skeletal muscle (active)
Bloodstream
Cori Cycle
Bloodstream
Glucose
Glycogen
Pyruvate
Liver
Lactate

FIGURE 22.15 ▶ In the Cori cycle, lactate from glycolysis in the muscles enters the bloodstream and travels to the liver where it is converted to glucose. The glucose travels through the bloodstream to muscles to produce ATP.

Q Why is lactate formed in the muscle converted to glucose in the liver?

Regulation of Gluconeogenesis

Gluconeogenesis is a pathway that protects the brain and nervous system from experiencing a loss of glucose, which causes impairment of function. It is also a pathway that is utilized when vigorous activity depletes blood glucose and glycogen stores. Thus, the level of carbohydrate available from the diet controls gluconeogenesis. When a diet is high in carbohydrate, the gluconeogenesis pathway is not utilized. However, when a diet is low in carbohydrate, the pathway is very active.

As long as conditions in a cell favor glycolysis, there is no synthesis of glucose. But when the cell requires the synthesis of glucose, glycolysis is turned off. **TABLE 22.4** lists compounds that activate or inhibit enzymes involved in the two processes.

TABLE 22.4 Regulation of Glycolysis and Gluconeogenesis

	Glycolysis	Gluconeogenesis
Enzyme	Hexokinase	Glucose-6-phosphatase
Activated by	High glucose levels, insulin	Low glucose levels, glucose-6-phosphate
Inhibited by	Glucose-6-phosphate, epinephrine	
Enzyme	Phosphofructokinase	Fructose-1,6-bisphosphatase
Activated by	AMP, insulin	Low glucose levels, glucagon
Inhibited by	ATP, glucagon	AMP, insulin
Enzyme	Pyruvate kinase	Pyruvate carboxylase
Activated by	Fructose-1,6-bisphosphate	Low glucose levels, glucagon
Inhibited by	ATP, acetyl CoA	Insulin

ENGAGE

Which compounds activate gluconeogenesis?

TEST

Try Practice Problems 22.77 and 22.78

In this chapter, we have discussed several metabolic pathways involving glucose. The relationship between these pathways are summarized in **FIGURE 22.16**.

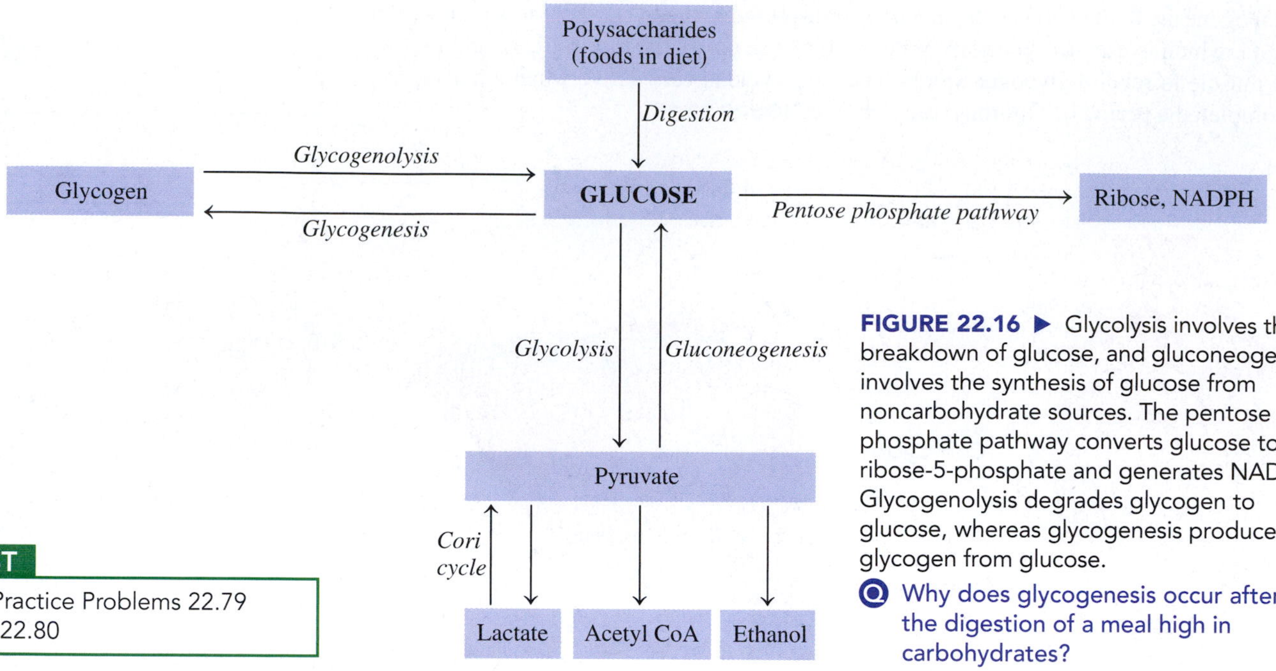
Polysaccharides (foods in diet)
Digestion
Glycogen
Glycogenolysis
Glycogenesis
GLUCOSE
Pentose phosphate pathway
Ribose, NADPH
Glycolysis
Gluconeogenesis
Pyruvate
Cori cycle
Lactate
Acetyl CoA
Ethanol

FIGURE 22.16 ▶ Glycolysis involves the breakdown of glucose, and gluconeogenesis involves the synthesis of glucose from noncarbohydrate sources. The pentose phosphate pathway converts glucose to ribose-5-phosphate and generates NADPH. Glycogenolysis degrades glycogen to glucose, whereas glycogenesis produces glycogen from glucose.

Q Why does glycogenesis occur after the digestion of a meal high in carbohydrates?

TEST

Try Practice Problems 22.79 and 22.80

Chemistry Link to Health

Glucocorticoids and Steroid-Induced Diabetes

Glucocorticoids are steroid hormones that have a number of beneficial medical applications. Examples include prednisone, hydrocortisone, dexamethasone, and prednisolone. Because of their anti-inflammatory effects, glucocorticoids are often used with patients that suffer from asthma, rheumatoid arthritis, or severe allergies. They are also used to suppress immune responses in patients that have received organ transplants or as anti-nausea medications for patients that are receiving chemotherapy.

Although the use of glucocorticoids has been lifesaving for many, glucocorticoids have some undesirable side effects on glucose metabolism. For example, glucocorticoid use results in (1) an increased level of gluconeogenesis in the liver and (2) a decrease in the amount of glucose that is taken up into muscle cells and adipose tissue. As a consequence, people who are being treated with glucocorticoids often display high levels of glucose in their blood (*hyperglycemia*). Because less glucose is being taken up into the cells, people suffering from hyperglycemia can experience fatigue or headache, among other symptoms.

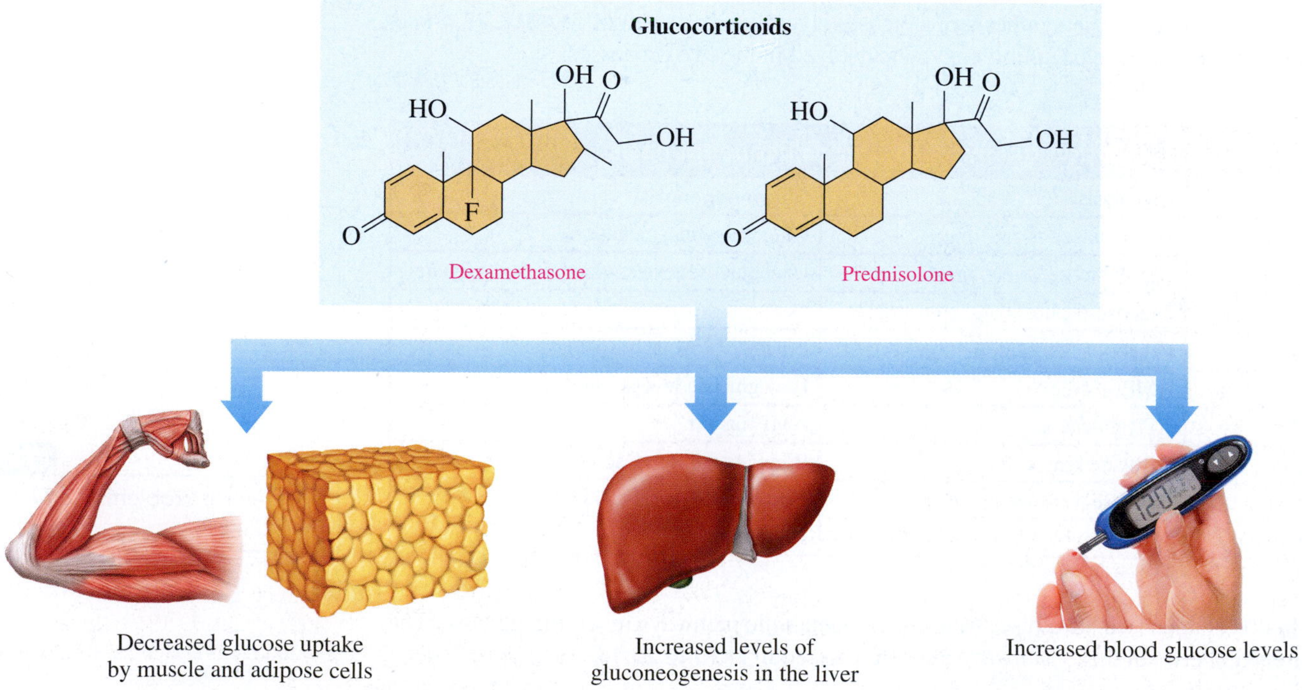
Glucocorticoids
HO
OH
O
OH
O
F
Dexamethasone
HO
OH
O
OH
O
Prednisolone
Decreased glucose uptake by muscle and adipose cells
Increased levels of gluconeogenesis in the liver
Increased blood glucose levels

(*continued*)

Chemistry Link to Health (*continued*)

In most cases, blood glucose levels return to normal after a person finishes taking glucocorticoids. However, when glucocorticoid doses have been particularly high or glucocorticoid treatment has continued for a prolonged amount of time, patients can develop *steroid-induced diabetes*, a condition that remains with them for the rest of their life, even if glucocorticoid use is discontinued. The development of steroid-induced diabetes has been observed more frequently with oral or intravenous use of glucocorticoids, as opposed to the use of inhaled glucocorticoids. People who have a genetic predisposition to develop diabetes are also more likely to develop steroid-induced diabetes when treated with glucocorticoids.

Because of their effect on blood glucose, it is very important that diabetics who are taking glucocorticoids monitor their blood sugar frequently and adjust their diets accordingly while taking these medications.

TEST

Try Practice Problems 22.81 and 22.82

PRACTICE PROBLEMS

22.7 Gluconeogenesis: Glucose Synthesis

22.73 What is the function of gluconeogenesis in the body?

22.74 What enzymes in glycolysis are not used in gluconeogenesis?

22.75 What enzymes in glycolysis are used in gluconeogenesis?

22.76 How is the lactate produced in skeletal muscle used for glucose synthesis?

22.77 Indicate whether each of the following activates or inhibits gluconeogenesis:
- **a.** low glucose levels
- **b.** glucagon
- **c.** insulin
- **d.** AMP

22.78 Indicate whether each of the following activates or inhibits glycolysis:
- **a.** low glucose levels
- **b.** insulin
- **c.** epinephrine
- **d.** ATP

Clinical Applications

22.79 Low carbohydrate diets ask dieters to lower or eliminate their intake of carbohydrates. Since we need glucose to fuel the brain, gluconeogenesis will produce glucose under these conditions. What carbohydrate store is lowered (and the source of weight loss) in people who do not eat carbohydrates?

22.80 When would gluconeogenesis be most active, after a meal, or when you wake up in the morning? Explain your reasoning.

22.81 What is the effect of glucocorticoids on gluconeogenesis in the liver (see Chemistry Link to Health: Glucocorticoids and Steroid-Induced Diabetes)?

22.82 What is the effect of glucocorticoids on glucose uptake by skeletal cells and adipose tissue (see Chemistry Link to Health: Glucocorticoids and Steroid-Induced Diabetes)?

Clinical Update Philip's Diet for von Gierke's Disease

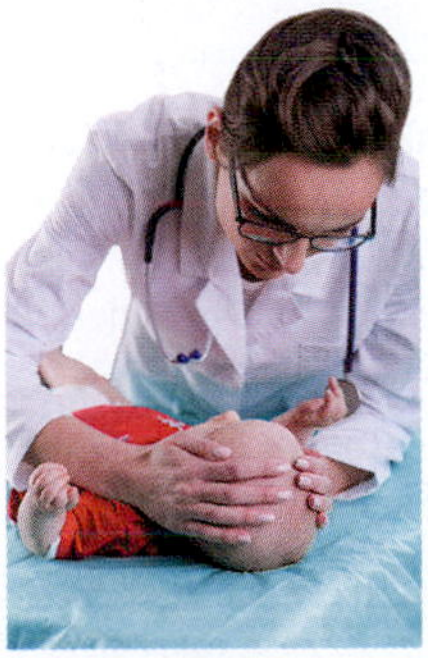

Much of the metabolism of carbohydrates depends on the storage and degradation of glycogen in the liver. If glucose-6-phosphatase is defective or deficient, the liver cannot properly break down glycogen. In *von Gierke's disease*, large amounts of glycogen accumulate, leading to an enlarged liver.

Because Philip cannot break down glycogen to maintain his blood glucose levels, he is hypoglycemic and weak. The hepatology nurse, Barbara, begins dietary management by giving Philip frequent feedings of glucose or starch by nasogastric tube to raise his blood glucose levels. A stomach tube (gastrostomy tube) is also used. Philip shows a rapid improvement in physical behavior in response to this treatment. In the days and months that follow, Philip is fed cornstarch, which is easily digested to provide small, but continuous, amounts of glucose in the blood. Foods containing fructose or galactose are avoided because they are converted to glucose-6-phosphate, which Philip cannot metabolize. Philip thrives as he follows a diet in which he eats frequently during the day, includes foods containing carbohydrates in each of his meals, and avoids foods containing fructose and galactose. As he grows older, he can be physically active, but should not participate in contact sports to avoid damage to his liver.

Clinical Applications

22.83 What are some symptoms of von Gierke's disease?

22.84 Write the chemical equation for the reaction catalyzed by glucose-6-phosphatase.

22.85 What is the impact of a defective glucose-6-phosphatase in von Gierke's disease (GSD I) on blood glucose and stored glycogen?

22.86 Explain why Philip's condition requires that he is fed cornstarch every few hours.

CONCEPT MAP

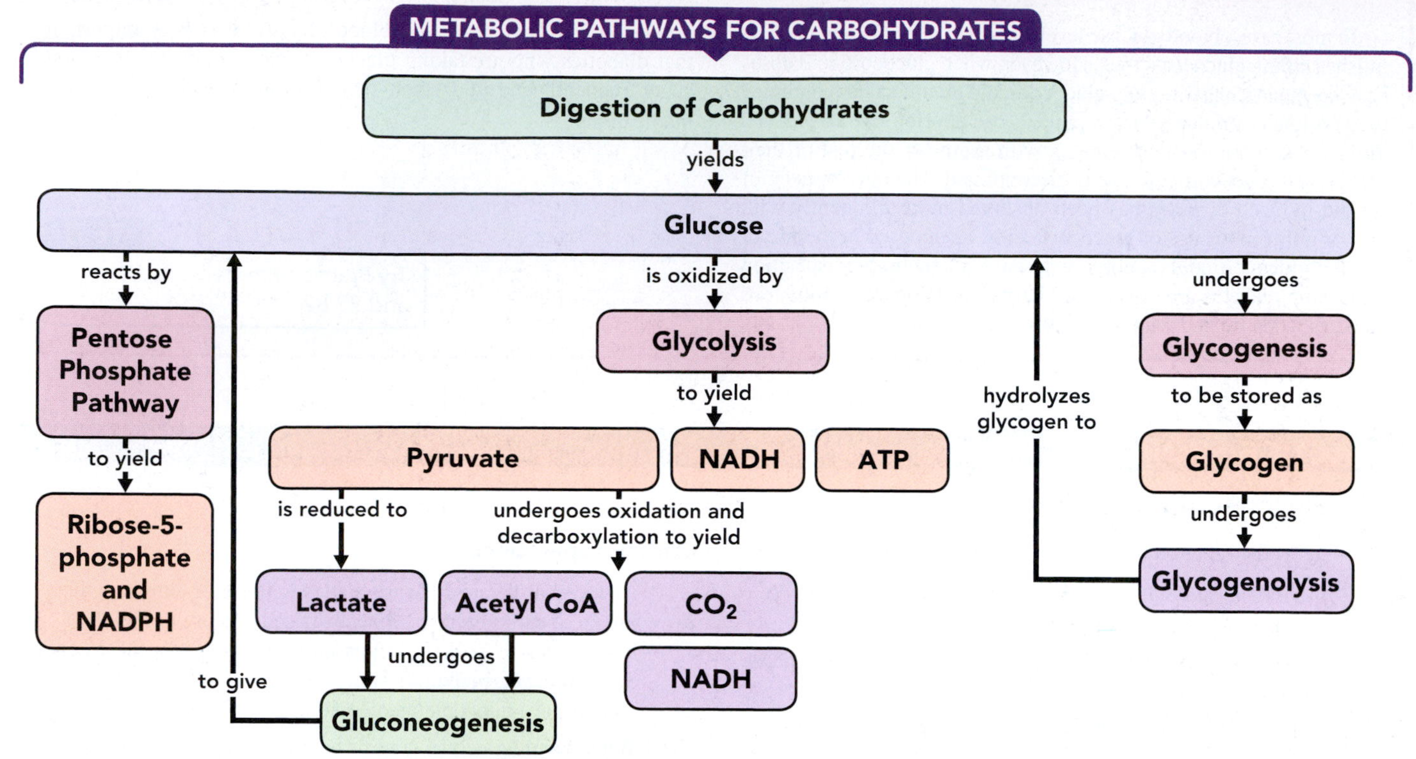

METABOLIC PATHWAYS FOR CARBOHYDRATES
Digestion of Carbohydrates
yields
Glucose
reacts by
Pentose Phosphate Pathway
to yield
Ribose-5-phosphate and NADPH
is oxidized by
Glycolysis
to yield
Pyruvate
NADH
ATP
is reduced to
Lactate
undergoes oxidation and decarboxylation to yield
Acetyl CoA
CO_2
NADH
undergoes
Gluconeogenesis
to give
hydrolyzes glycogen to
undergoes
Glycogenesis
to be stored as
Glycogen
undergoes
Glycogenolysis

CHAPTER REVIEW

22.1 Metabolism and Energy

LEARNING GOAL Describe the three stages of catabolism, the structure of ATP, and the role of ATP.

- Metabolism includes all the catabolic and anabolic reactions that occur in the cells.
- Catabolic reactions degrade large molecules into smaller ones with an accompanying release of energy.
- Anabolic reactions require energy to synthesize larger molecules from smaller ones.
- The three stages of catabolism are digestion of food, degradation of larger molecules into smaller groups such as the acetyl group, and the oxidation of the acetyl groups to release energy for ATP synthesis.
- Energy obtained from catabolic reactions is stored primarily in adenosine triphosphate (ATP), a high-energy compound.
- ATP is hydrolyzed when energy is required by anabolic reactions in the cells.

22.2 Important Coenzymes in Metabolic Pathways

LEARNING GOAL Describe the components and functions of the coenzymes NAD^+, $NADP^+$, FAD, and coenzyme A.

- NAD^+, $NADP^+$, and FAD are the oxidized forms of coenzymes that participate in oxidation–reduction reactions.

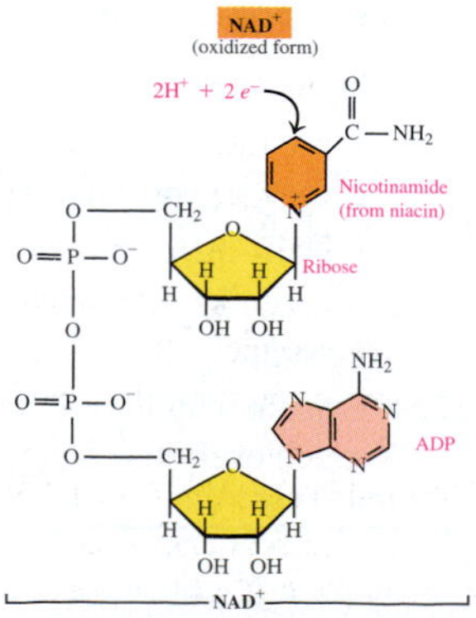

NAD^+
(oxidized form)
$2H^+ + 2e^-$
Nicotinamide (from niacin)
Ribose
ADP
NAD^+

- When NAD^+, $NADP^+$, and FAD pick up hydrogen ions and electrons, they are reduced to $NADH + H^+$, $NADPH + H^+$, and $FADH_2$.
- Coenzyme A contains a thiol group ($-SH$) that usually bonds with a two-carbon acetyl group (acetyl CoA).

22.3 Digestion of Carbohydrates

LEARNING GOAL Give the sites and products of the digestion of carbohydrates.

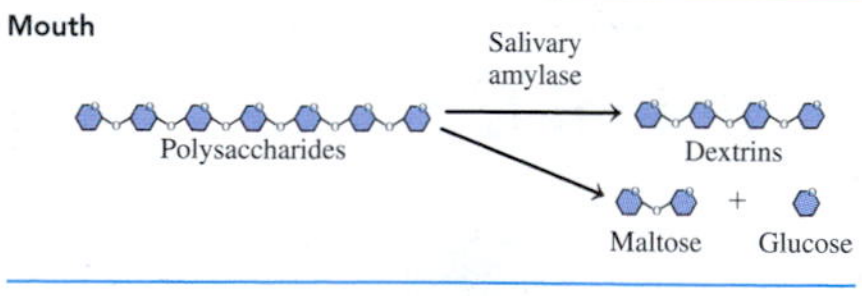

Mouth
Salivary amylase
Polysaccharides
Dextrins
Maltose
Glucose

- The digestion of carbohydrates is a series of reactions that breaks down polysaccharides into hexose monomers such as glucose, galactose, and fructose.
- The hexose monomers are absorbed through the intestinal wall into the bloodstream to be carried to cells where they provide energy and carbon atoms for the synthesis of new molecules.

22.4 Glycolysis: Oxidation of Glucose

LEARNING GOAL Describe the conversion of glucose to pyruvate in glycolysis.

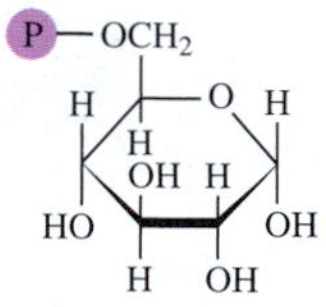
Glucose-6-phosphate

- Glycolysis, which occurs in the cytosol, consists of 10 reactions that degrade glucose (six carbons) to two pyruvate molecules (three carbons each).
- The overall series of reactions yields two molecules of the reduced coenzyme NADH and two ATP.
- The activity of three enzymes–hexokinase, phosphofructokinase, and pyruvate kinase–can be regulated to respond to cellular requirements for ATP and/or the end products of glycolysis.
- The pentose phosphate pathway utilizes glucose-6-phosphate to supply NADPH and pentose sugars for the biosynthesis of nucleotides, fatty acids, cholesterol, and amino acids.

22.5 Pathways for Pyruvate

LEARNING GOAL Give the conditions for the conversion of pyruvate to lactate, ethanol, and acetyl coenzyme A.

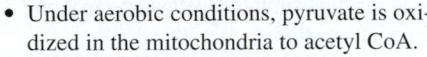

- Under aerobic conditions, pyruvate is oxidized in the mitochondria to acetyl CoA.
- In the absence of oxygen, pyruvate is reduced to lactate and NAD^+ is regenerated for the continuation of glycolysis, whereas microorganisms such as yeast reduce pyruvate to ethanol, a process known as fermentation.

22.6 Glycogen Synthesis and Degradation

LEARNING GOAL Describe the synthesis and breakdown of glycogen.

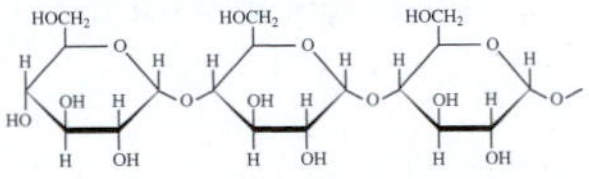

Glycogen

- When blood glucose levels are high, glycogenesis converts glucose to glycogen, which is stored in the liver.
- Glycogenolysis breaks down glycogen to glucose when glucose and ATP levels are low.

22.7 Gluconeogenesis: Glucose Synthesis

LEARNING GOAL Describe how glucose is synthesized from molecules from noncarbohydrate sources.

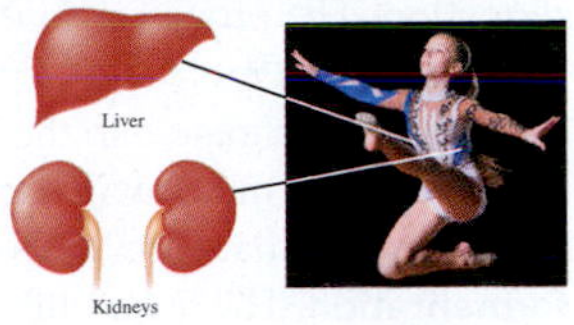
Liver
Kidneys

- When blood glucose levels are low and glycogen stores in the liver are depleted, gluconeogenesis occurs, synthesizing glucose from noncarbohydrate compounds such as pyruvate.
- Gluconeogenesis utilizes many of the same enzymes as glycolysis.

SUMMARY OF REACTIONS

The chapter Sections to review are shown after the name of each reaction.

Hydrolysis of ATP (22.1)

$$\text{ATP} + H_2O \longrightarrow \text{ADP} + P_i + 7.3\text{ kcal/mole (31 kJ/mole)}$$

Hydrolysis of ADP (22.1)

$$\text{ADP} + H_2O \longrightarrow \text{AMP} + P_i + 7.3\text{ kcal/mole (31 kJ/mole)}$$

Formation of ATP (22.1)

$$\text{ADP} + P_i + 7.3\text{ kcal/mole (31 kJ/mole)} \longrightarrow \text{ATP} + H_2O$$

Reduction of NAD$^+$, NADP$^+$, and FAD (22.2)

$$NAD^+ + 2H^+ + 2e^- \longrightarrow NADH + H^+$$
$$NADP^+ + 2H^+ + 2e^- \longrightarrow NADPH + H^+$$
$$FAD + 2H^+ + 2e^- \longrightarrow FADH_2$$

Glycolysis (22.4)

$$\underset{\text{Glucose}}{C_6H_{12}O_6} + 2NAD^+ + 2ADP + 2P_i \longrightarrow$$
$$2\underset{\text{Pyruvate}}{CH_3-\overset{\overset{\displaystyle O}{\|}}{C}-COO^-} + 2NADH + 2ATP + 4H^+ + 2H_2O$$

Pentose Phosphate Pathway (22.4)

$$\text{Glucose-6-P} + 2NADP^+ + H_2O \longrightarrow$$
$$\text{ribose-5-P} + 2NADPH + 2H^+ + CO_2$$

Oxidation of Pyruvate to Acetyl CoA (22.5)

$$\underset{\text{Pyruvate}}{CH_3-\overset{\overset{\displaystyle O}{\|}}{C}-COO^-} + HS-CoA + NAD^+ \xrightarrow{\text{Pyruvate dehydrogenase}}$$
$$\underset{\text{Acetyl CoA}}{CH_3-\overset{\overset{\displaystyle O}{\|}}{C}-S-CoA} + CO_2 + NADH$$

Reduction of Pyruvate to Lactate (22.5)

$$\underset{\text{Pyruvate}}{CH_3-\overset{\overset{\displaystyle O}{\|}}{C}-COO^-} + NADH + H^+ \longrightarrow$$
$$\underset{\text{Lactate}}{CH_3-\overset{\overset{\displaystyle OH}{|}}{CH}-COO^-} + NAD^+$$

Oxidation of Glucose to Lactate (22.5)

$$\text{Glucose} + 2ADP + 2P_i \longrightarrow 2\text{ lactate} + 2ATP$$

Reduction of Pyruvate to Ethanol (22.5)

$$\underset{\text{Pyruvate}}{CH_3-\overset{\overset{\displaystyle O}{\|}}{C}-COO^-} + NADH + H^+ \longrightarrow$$
$$\underset{\text{Ethanol}}{CH_3-CH_2-OH} + NAD^+ + CO_2$$

Glycogenesis (22.6)

$$\text{UDP-glucose} + \text{glycogen} \longrightarrow \text{glycogen–glucose} + \text{UDP}$$

Glycogenolysis (22.6)

$$\text{Glycogen–glucose} \longrightarrow \text{glycogen} + \text{glucose}$$

Gluconeogenesis (22.7)

$$2\text{ Pyruvate} + 4ATP + 2GTP + 2NADH + 2H^+ + 6H_2O \longrightarrow$$
$$\text{glucose} + 4ADP + 2GDP + 6P_i + 2NAD^+$$

KEY TERMS

acetyl CoA The compound that forms when a two-carbon acetyl unit bonds to coenzyme A.

ADP Adenosine diphosphate, formed by the hydrolysis of ATP; consists of adenine, a ribose sugar, and two phosphate groups.

aerobic An oxygen-containing environment in the cells.

anabolic reaction A metabolic reaction that requires energy to build large molecules from small molecules.

anaerobic A condition in cells when there is no oxygen.

ATP Adenosine triphosphate, a high-energy compound that stores energy in the cells; consists of adenine, a ribose sugar, and three phosphate groups.

catabolic reaction A metabolic reaction that releases energy for the cell by the degradation and oxidation of glucose and other molecules.

coenzyme A (CoA) A coenzyme that transports acyl and acetyl groups.

Cori cycle A cyclic process in which lactate produced in muscle is transferred to the liver to be converted to glucose, which can be used again by muscle.

cytosol The fluid of the cell, which is an aqueous solution of electrolytes and enzymes.

decarboxylation The loss of a carbon atom in the form of CO_2.

digestion The processes in the gastrointestinal tract that break down large food molecules to smaller ones that pass through the intestinal membrane into the bloodstream.

FAD A coenzyme (flavin adenine dinucleotide) for dehydrogenase enzymes that form carbon–carbon double bonds.

fermentation The anaerobic conversion of glucose by enzymes in yeast to yield ethanol and CO_2.

gluconeogenesis The synthesis of glucose from molecules from noncarbohydrate sources.

glycogenesis The synthesis of glycogen from glucose molecules.

glycogenolysis The breakdown of glycogen into glucose molecules.

glycolysis The 10 oxidation reactions of glucose that yield two pyruvate molecules.

metabolism All the chemical reactions in living cells that carry out molecular and energy transformations.

mitochondrion The organelle of cells where energy-releasing reactions take place.

NAD^+ The hydrogen acceptor used in oxidation reactions that form carbon–oxygen double bonds.

pentose phosphate pathway A metabolic pathway that oxidizes glucose to produce NADPH for anabolic reactions and five-carbon sugars required for the synthesis of nucleotides.

CORE CHEMISTRY SKILLS

The chapter Section containing each Core Chemistry Skill is shown in parentheses at the end of each heading.

Identifying Important Coenzymes in Metabolism (22.2)

- ATP is a nucleotide that transfers energy through metabolic reactions.
- NAD^+, $NADP^+$, and FAD are the oxidized forms of the coenzymes that move electrons and H^+ through metabolic reactions.
- CoA is a coenzyme that moves two-carbon acetyl groups through metabolic reactions.

Example: Provide the formulas of the reduced forms of the coenzymes that move electrons and H^+ through metabolism.

Answer: NADH, NADPH, and $FADH_2$

Identifying the Compounds in Glycolysis (22.4)

- Glycolysis, which occurs in the cytosol, consists of 10 reactions that degrade glucose (6C) to two pyruvate (3C).
- The overall series of reactions in glycolysis yields two NADH and two ATP.

Example: Identify the reaction(s) and compounds in glycolysis that have each of the following characteristics:

a. require ATP
b. convert a six-carbon compound to two three-carbon compounds
c. convert ADP to ATP

Answer: **a.** The phosphorylation of glucose in reaction 1 and fructose-6-phosphate in reaction 3 each require one ATP.
b. In reaction 4, the six-carbon fructose-1,6-bisphosphate is cleaved into two three-carbon compounds.
c. Phosphate transfer in reactions 7 and 10 converts ADP to ATP.

Identifying the Compounds and Enzymes in Glycogenesis and Glycogenolysis (22.6)

- Glucose undergoes glycogenesis when blood glucose levels are high and is stored as glycogen in the liver.
- Glycogenolysis breaks down glycogen to glucose when levels of glucose or ATP are low.

Example: Indicate which of the processes, glycogenesis and/or glycogenolysis, involve each of the following compounds:

a. glucose-1-phosphate
b. UDP-glucose
c. glycogen phosphorylase
d. UTP

Answer: **a.** glycogenesis, glycogenolysis
b. glycogenesis
c. glycogenolysis
d. glycogenesis

UNDERSTANDING THE CONCEPTS

The chapter Sections to review are given in parentheses at the end of each problem.

22.87 On a hike, you expend 350 kcal/h. How many moles of ATP will you use if you hike for 2.5 h? (22.1)

Vigorous hiking can expend 350 kcal per hour.

22.88 Identify each of the following as a six-carbon or a three-carbon compound and arrange them in the order in which they occur in glycolysis: (22.4)

a. 3-phosphoglycerate
b. pyruvate
c. glucose-6-phosphate
d. glucose
e. fructose-1,6-bisphosphate

22.89 Indicate whether each of the following enzymes is utilized in (1) glycogenesis but not glycogenolysis, (2) glycogenolysis but not glycogenesis, or (3) both glycogenesis and glycogenolysis: (22.6)

a. pyrophosphorylase
b. phosphoglucomutase
c. glycogen phosphorylase

22.90 Indicate whether each of the following enzymes is utilized in (1) glycogenesis but not glycogenolysis, (2) glycogenolysis but not glycogenesis, or (3) both glycogenesis and glycogenolysis: (22.6)

a. glycogen synthase
b. glucose-6-phosphatase
c. debranching enzyme

22.91 Indicate whether each of the following enzymes is utilized in (1) glycolysis but not gluconeogenesis, (2) gluconeogenesis but not glycolysis, or (3) both glycolysis and gluconeogenesis: (22.4, 22.7)

a. glucose-6-phosphatase **b.** aldolase
c. phosphofructokinase **d.** phosphoglycerate mutase

22.92 Indicate whether each of the following enzymes is utilized in (1) glycolysis but not gluconeogenesis, (2) gluconeogenesis but not glycolysis, or (3) both glycolysis and gluconeogenesis: (22.4, 22.7)

a. hexokinase **b.** pyruvate kinase
c. enolase **d.** pyruvate carboxylase

ADDITIONAL PRACTICE PROBLEMS

22.93 What is meant by the term metabolism? (22.1)

22.94 How do catabolic reactions differ from anabolic reactions? (22.1)

22.95 What stage of catabolism involves the digestion of large food polymers? (22.1)

22.96 What stage of catabolism degrades monomers such as glucose into smaller molecules? (22.1)

22.97 What type of cell has a nucleus? (22.1)

22.98 What is the function of each of the following cell components? (22.1)

a. cell membrane **b.** mitochondria **c.** cytosol

22.99 What is the full name of ATP? (22.1)

22.100 What is the full name of ADP? (22.1)

22.101 Write the abbreviated equation for the hydrolysis of ATP to ADP. (22.1)

22.102 Write the abbreviated equation for the hydrolysis of ADP to AMP. (22.1)

22.103 What are the reduced and oxidized forms of the coenzyme FAD? (22.2)

22.104 What type of reaction uses FAD as the coenzyme? (22.2)

22.105 What are the reduced and oxidized forms of the coenzyme NAD^+? (22.2)

22.106 What type of reaction uses NAD^+ as the coenzyme? (22.2)

22.107 Write the abbreviation for the reduced form of each of the following: (22.2)

a. FAD **b.** NAD^+ **c.** $NADP^+$

22.108 What is the name of the vitamin in the structure of each of the following? (22.2)

a. FAD **b.** NAD^+ **c.** coenzyme A

22.109 How and where does lactose undergo digestion in the body? What are the products? (22.3)

22.110 How and where does sucrose undergo digestion in the body? What are the products? (22.3)

22.111 What are the reactant and product of glycolysis? (22.4)

22.112 What is the coenzyme used in glycolysis? (22.4)

22.113 a. In glycolysis, which reactions involve phosphorylation? (22.4)
b. Which reactions involve a direct substrate phosphorylation to generate ATP?

22.114 How do ADP and ATP regulate the glycolysis pathway? (22.4)

22.115 What reaction and enzyme in glycolysis convert a hexose bisphosphate into two three-carbon intermediates? (22.4)

22.116 How does the investment and generation of ATP give a net gain of ATP for glycolysis? (22.4)

22.117 What compound is converted to fructose-6-phosphate by phosphoglucose isomerase? (22.4)

22.118 What product forms when glyceraldehyde-3-phosphate adds a phosphate group? (22.4)

22.119 When is pyruvate converted to lactate in the body? (22.5)

22.120 When pyruvate is used to form acetyl CoA in aerobic environments or ethanol in fermentation, the product has only two carbon atoms. What happened to the third carbon? (22.5)

22.121 How does phosphofructokinase regulate the rate of glycolysis? (22.5)

22.122 How does pyruvate kinase regulate the rate of glycolysis? (22.5)

22.123 When does the rate of glycogenolysis increase in the cells? (22.6)

22.124 If glucose-1-phosphate is the product from glycogen breakdown, how does it enter glycolysis? (22.6)

Clinical Applications

22.125 What is the end product of glycogenolysis in the liver? (22.6)

22.126 What is the end product of glycogenolysis in skeletal muscle? (22.6)

22.127 Indicate whether each of the following conditions would increase or decrease the rate of glycogenolysis in the liver: (22.6)

a. low blood glucose level **b.** secretion of insulin
c. secretion of glucagon **d.** high levels of ATP

22.128 Indicate whether each of the following conditions would increase or decrease the rate of glycogenesis in the liver: (22.6)

a. low blood glucose level **b.** secretion of insulin
c. secretion of glucagon **d.** high levels of ATP

22.129 Indicate whether each of the following conditions would increase or decrease the rate of gluconeogenesis: (22.7)

a. high blood glucose level **b.** secretion of insulin
c. secretion of glucagon **d.** high levels of ATP

22.130 Indicate whether each of the following conditions would increase or decrease the rate of glycolysis: (22.4)

a. high blood glucose level **b.** secretion of insulin
c. secretion of glucagon **d.** high levels of ATP

CHALLENGE PROBLEMS

The following problems are related to the topics in this chapter. However, they do not all follow the chapter order, and they require you to combine concepts and skills from several Sections. These problems will help you increase your critical thinking skills and prepare for your next exam.

22.131 Why is glucose provided by glycogenolysis in the liver but not in skeletal muscle? (22.6)

22.132 When does the rate of glycogenesis increase in the cells? (22.6)

22.133 How do the hormones insulin and glucagon affect the rates of glycogenesis, glycogenolysis, and glycolysis? (22.4, 22.6)

22.134 What is the function of gluconeogenesis? (22.7)

22.135 Where does the Cori cycle operate? (22.7)

22.136 Identify each of the following as part of glycolysis, glycogenolysis, glycogenesis, or gluconeogenesis: (22.4, 22.6, 22.7)
a. Glycogen is broken down to glucose in the liver.
b. Glucose is synthesized from noncarbohydrate sources.
c. Glucose is degraded to pyruvate.
d. Glycogen is synthesized from glucose.

22.137 The average daily diet of an adult woman is 2.0×10^3 kcal. The molar mass of ATP is 507 g/mole. (22.1)
a. If all the energy in her diet is converted to ATP, how many kilograms of ATP are synthesized in one day?
b. If the average amount of ATP in the body is 250 g, what is the lifetime, in minutes, of ATP?

22.138 A teaspoon containing 4.2 g of sucrose is added to a cup of tea. (22.1)
a. How much energy, in kilocalories, are in the sucrose if 1 g of sucrose provides 4 kcal?
b. How many grams of ATP (507 g/mole) are produced if all the energy in the sucrose is converted to ATP?

ANSWERS

22.1 The digestion of polysaccharides takes place in stage 1.

22.3 In metabolism, a catabolic reaction breaks apart large molecules, releasing energy.

22.5 a. anabolic **b.** anabolic **c.** catabolic **d.** catabolic

22.7 When a phosphate group is cleaved from ATP, sufficient energy is released for many energy-requiring processes in the cell.

22.9 a. PEP $\longrightarrow$ pyruvate + P_i + 14.8 kcal/mole
b. ADP + P_i + 7.3 kcal/mole $\longrightarrow$ ATP
c. PEP + ADP $\longrightarrow$ ATP + pyruvate + 7.5 kcal/mole

22.11 3.4×10^4 g

22.13 a. coenzyme A **b.** NAD^+, $NADP^+$ **c.** FAD

22.15 a. NADH **b.** FAD

22.17 FAD

22.19 Hydrolysis is the main reaction involved in the digestion of carbohydrates.

22.21 a. lactose **b.** glucose and fructose **c.** glucose

22.23 glucose

22.25 ATP is required in phosphorylation reactions.

22.27 glyceraldehyde-3-phosphate and dihydroxyacetone phosphate

22.29 ATP is produced in glycolysis by transferring phosphate groups from 1,3-bisphosphoglycerate and from phosphoenolpyruvate directly to ADP.

22.31 a. hexokinase; phosphofructokinase
b. phosphoglycerate kinase; pyruvate kinase

22.33 a. 1 ATP required **b.** 1 NADH produced
c. 2 ATP and 2 NADH produced

22.35 a. In reaction 1, a hexokinase uses ATP to phosphorylate glucose.
b. In reactions 7 and 10, phosphate groups are transferred from 1,3-bisphosphoglycerate and phosphoenolpyruvate directly to ADP to produce ATP.
c. In reaction 4, the six-carbon molecule fructose-1,6-bisphosphate is split into two three-carbon molecules, glyceraldehyde-3-phosphate and dihydroxyacetone phosphate.

22.37 In a series of enzymatic reactions, galactose is phosphorylated to yield glucose-1-phosphate, which is converted to glucose-6-phosphate, which enters glycolysis in reaction 2. In the muscles, fructose is phosphorylated to fructose-6-phosphate, which enters glycolysis in reaction 3, where its movement through glycolysis is regulated by the enzyme phosphofructokinase. In the liver, fructose is phosphorylated to fructose-1-phosphate, which is converted to glyceraldehyde-3-phosphate, which enters glycolysis at reaction 6.

22.39 a. activate **b.** inhibit

22.41 Glucose-6-phosphate is the initial substrate for the pentose phosphate pathway.

22.43 The pentose phosphate pathway produces NADPH required for anabolic reactions.

22.45 Sucrose contains the monosaccharides glucose and fructose. Once hydrolyzed during digestion, glucose directly enters glycolysis at step 1 and fructose enters glycolysis either at step 3 in the muscle and kidney or step 6 in the liver.

22.47 The disaccharide lactose can be hydrolyzed into glucose and galactose. Patients with galactosemia lack one of the enzymes needed to metabolize galactose, so galactose and its by-products can build up to toxic levels if products containing lactose are eaten.

22.49 Aerobic conditions are needed.

22.51 The oxidation of pyruvate converts NAD^+ to NADH and produces acetyl CoA and CO_2.

$$CH_3-\overset{\overset{\large O}{\|}}{C}-COO^- + NAD^+ + HS-CoA \xrightarrow{\text{Pyruvate dehydrogenase}} CH_3-\overset{\overset{\large O}{\|}}{C}-S-CoA + NADH + CO_2$$

22.53 carbon dioxide, CO_2

22.55 **a.** (3) ethanol **b.** (1) acetyl CoA
c. (2) lactate **d.** (1) acetyl CoA, (3) ethanol

22.57 During strenuous exercise, under anaerobic conditions, pyruvate is converted to lactate. The accumulation of lactate and the corresponding drop in pH causes the muscles to tire and become sore.

22.59 $$CH_3{-}\overset{\overset{\large O}{\|}}{C}{-}\overset{\overset{\large O}{\|}}{C}{-}O^- + NADH + H^+ \longrightarrow CH_3{-}\overset{\overset{\large OH}{|}}{CH}{-}\overset{\overset{\large O}{\|}}{C}{-}O^- + NAD^+$$

22.61 Glycogenesis is the synthesis of glycogen from glucose molecules.

22.63 Muscle cells break down glycogen to glucose-6-phosphate, which enters glycolysis.

22.65 Glycogen phosphorylase cleaves the glycosidic bonds at the ends of glycogen chains to remove glucose as glucose-1-phosphate.

22.67 **a.** activates **b.** inhibits **c.** inhibits

22.69 The movement of glucose out of the bloodstream and into the cells is regulated by insulin whereas fructose is not. Fructose can move into the cells and be degraded in glycolysis.

22.71 When the glycogen phosphorylase enzyme in the liver is defective, glycogen in the liver cannot be broken down to glucose.

22.73 When there are no glycogen stores remaining in the liver, gluconeogenesis synthesizes glucose from noncarbohydrate compounds such as pyruvate and lactate.

22.75 phosphoglucose isomerase, aldolase, triose phosphate isomerase, glyceraldehyde-3-phosphate dehydrogenase, phosphoglycerate kinase, phosphoglycerate mutase, and enolase

22.77 **a.** activates **b.** activates
c. inhibits **d.** inhibits

22.79 People on low carbohydrate diets do not store normal amounts of glycogen and therefore lose this weight and the water weight used to hydrolyze the glycogen.

22.81 Glucocorticoids increase gluconeogenesis levels in the liver.

22.83 Some of the symptoms of von Gierke's disease are: liver enlargement due to glycogen accumulation, and hypoglycemia, which can lead to a shortened life expectancy.

22.85 Glycogen cannot be degraded completely to glucose, causing low blood glucose. Glycogen accumulates as stored glycogen in the liver. Lower insulin levels inhibit glycolysis and activate gluconeogenesis.

22.87 120 moles of ATP

22.89 **a.** (1) glycogenesis but not glycogenolysis
b. (3) both glycogenesis and glycogenolysis
c. (2) glycogenolysis but not glycogenesis

22.91 **a.** (2) gluconeogenesis but not glycolysis
b. (3) both glycolysis and gluconeogenesis
c. (1) glycolysis but not gluconeogenesis
d. (3) both glycolysis and gluconeogenesis

22.93 Metabolism includes all the reactions in cells that provide energy and material for cell growth.

22.95 stage 1

22.97 eukaryotic cell

22.99 adenosine triphosphate

22.101 $ATP + H_2O \longrightarrow ADP + P_i + 7.3$ kcal/mole (31 kJ/mole)

22.103 The reduced form is $FADH_2$; the oxidized form is FAD.

22.105 The reduced form is NADH; the oxidized form is NAD^+.

22.107 **a.** $FADH_2$ **b.** $NADH + H^+$ **c.** $NADPH + H^+$

22.109 Lactose undergoes digestion in the small intestine to yield glucose and galactose.

22.111 Glucose is the reactant and pyruvate is the product of glycolysis.

22.113 **a.** Reactions 1 and 3 involve phosphorylation of hexoses with ATP.
b. Reactions 7 and 10 involve direct substrate phosphorylation that generates ATP.

22.115 Reaction 4, which converts fructose-1,6-bisphosphate into two three-carbon intermediates, is catalyzed by aldolase.

22.117 glucose-6-phosphate

22.119 Pyruvate is converted to lactate when oxygen is not present in the cell (anaerobic conditions) to regenerate NAD^+ for glycolysis.

22.121 Phosphofructokinase is an allosteric enzyme that is activated by high levels of AMP and ADP because the cell needs to produce more ATP. When ATP levels are high due to a decrease in energy needs, ATP inhibits phosphofructokinase, which reduces its catalysis of fructose-6-phosphate.

22.123 The rate of glycogenolysis increases when blood glucose levels are low and glucagon has been secreted, which accelerates the breakdown of glycogen.

22.125 glucose

22.127 **a.** increase **b.** decrease
c. increase **d.** decrease

22.129 **a.** decrease **b.** decrease
c. increase **d.** decrease

22.131 The cells in the liver, but not skeletal muscle, contain a phosphatase enzyme needed to convert glucose-6-phosphate to free glucose that can diffuse through cell membranes into the bloodstream. Glucose-6-phosphate, which is the end product of glycogenolysis in muscle cells, cannot diffuse easily across cell membranes.

22.133 Insulin increases the rate of glycogenesis and glycolysis and decreases the rate of glycogenolysis. Glucagon decreases the rate of glycogenesis and glycolysis and increases the rate of glycogenolysis.

22.135 The Cori cycle is a cyclic process that involves the transfer of lactate from muscle to the liver where glucose is synthesized, which can be used again by the muscle.

22.137 **a.** 140 kg **b.** 2.6 min

23

Metabolism and Energy Production

Brian, who is four years old, has difficulty with balance, walking, and coordination. Brian's parents take him to his pediatrician, who diagnoses Brian with mitochondrial myopathy, which is a genetic disorder that causes fatigue and weakness of muscles.

The mitochondria in the cells contain enzymes that catalyze reactions that release energy from carbohydrates and fats in our diet. When a mutation disrupts this system—as is the case with mitochondrial myopathy—the amount of energy released and, thus, the amount of ATP produced, is drastically reduced to levels that are insufficient for the proper functioning of the muscles of the body.

As part of his care, Brian works with Kate, a physical therapist who specializes in pediatric muscular therapy. At her first meeting with Brian and his parents, Kate assesses Brian's functional capacity. She tells Brian's parents that, before their next appointment, she will develop a program of physical therapy and dietary supplements that will reduce Brian's fatigue, increase his functional capacity, and increase his quality of life.

CAREER

Physical Therapist (PT)

A physical therapist is a health care professional who diagnoses and treats people who are restricted in their ability to move and perform daily activities. A physical therapist designs treatments that enable a patient to move more efficiently, reduce pain, and maintain daily activities, including sitting, eating, and walking. A physical therapist helps the patient to improve mobility and lessen pain using techniques that include exercise, massage, and specialized equipment. They also educate families about how to best manage the physical limitations of the patient.

CLINICAL UPDATE

Increasing Brian's Functional Capacity

In the **CLINICAL UPDATE Increasing Brian's Functional Capacity**, page 819, you can read more about Brian's exercise and supplement program and about the types of exercises that Kate develops to improve Brian's ability to participate in everyday activities and increase his quality of life.

LOOKING AHEAD

23.1 The Citric Acid Cycle

LEARNING GOAL Describe the oxidation of acetyl CoA in the citric acid cycle.

Although glycolysis produces a small amount of ATP, most of the ATP in the cells is produced through the citric acid cycle and electron transport. When oxygen is plentiful in the cell, the pyruvate obtained from the degradation of glucose during glycolysis is converted to acetyl CoA in a reaction catalyzed by the enzyme pyruvate dehydrogenase.

The citric acid cycle is a series of reactions that connects the intermediate acetyl CoA from the catabolic pathways in stage 2 with electron transport and the synthesis of ATP in stage 3. As a central pathway in metabolism, the **citric acid cycle** uses the two-carbon acetyl group of acetyl CoA to produce CO_2, and the reduced coenzymes NADH and $FADH_2$ (see **FIGURE 23.1**).

The citric acid cycle is named for the six-carbon citrate ion from citric acid ($C_6H_8O_7$), a tricarboxylic acid, which forms in the first reaction. The citric acid cycle is also known as the *tricarboxylic acid* (*TCA*) *cycle* or the *Krebs cycle*, named for H. A. Krebs, who received a Nobel Prize in 1953 for its discovery.

FIGURE 23.1 ▶ In the three stages of catabolism, large molecules from foods are digested and degraded to give smaller molecules that can be oxidized to produce ATP.

Q Where is most of the ATP produced in the cells?

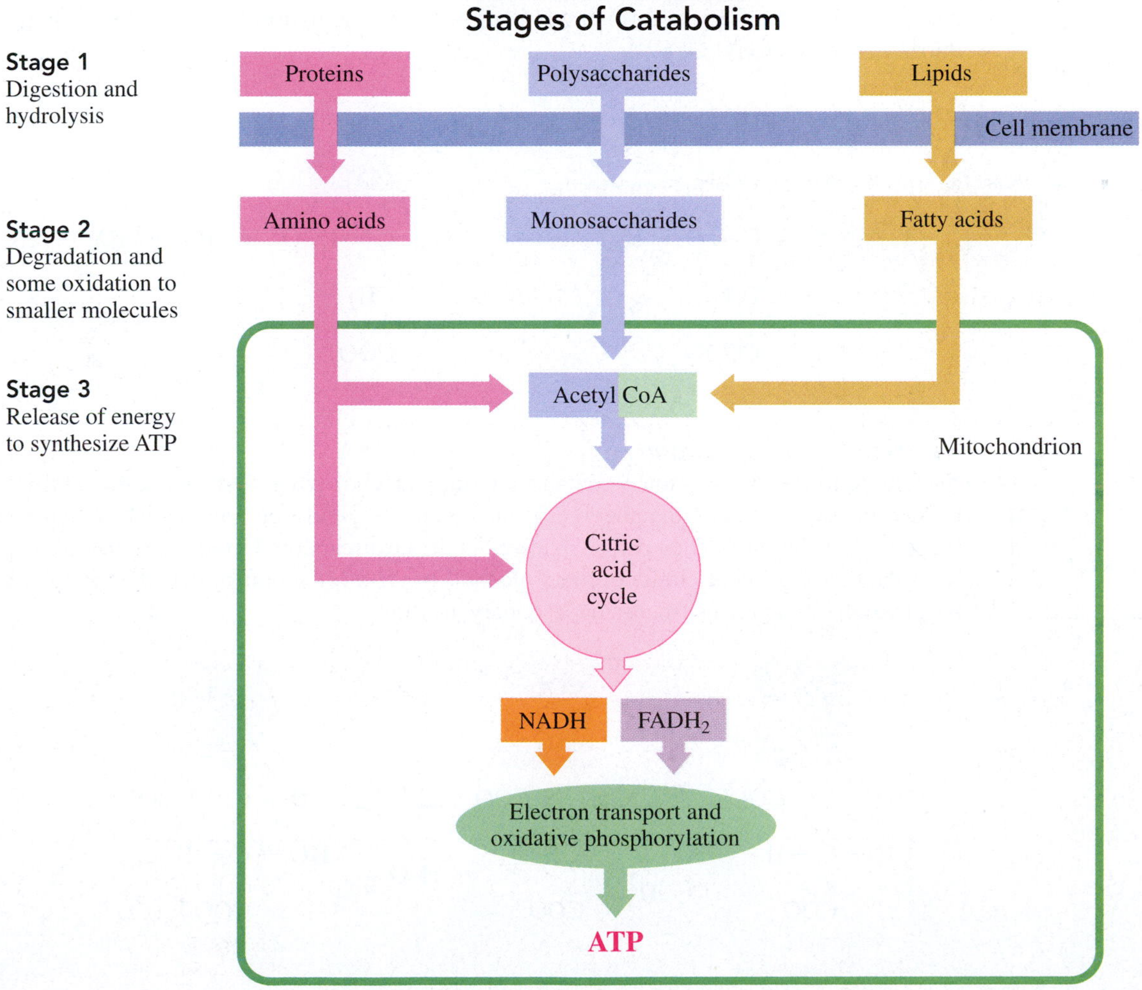

Stages of Catabolism
Stage 1
Digestion and hydrolysis
Stage 2
Degradation and some oxidation to smaller molecules
Stage 3
Release of energy to synthesize ATP
Proteins
Polysaccharides
Lipids
Cell membrane
Amino acids
Monosaccharides
Fatty acids
Acetyl CoA
Mitochondrion
Citric acid cycle
NADH
FADH2
Electron transport and oxidative phosphorylation
ATP

Overview of the Citric Acid Cycle

ENGAGE

Which reduced coenzymes are produced in each turn of the citric acid cycle?

Six carbons move through the eight reactions of the citric acid cycle, with each turn producing oxaloacetate (four carbons) and two CO_2 (see **FIGURE 23.2**). Each turn of the cycle includes four oxidation reactions, producing the reduced coenzymes NADH and $FADH_2$ from the energy released during the reactions. One GTP (converted to ATP in the cell) is also produced during the citric acid cycle. **FIGURE 23.3** shows the complete citric acid cycle.

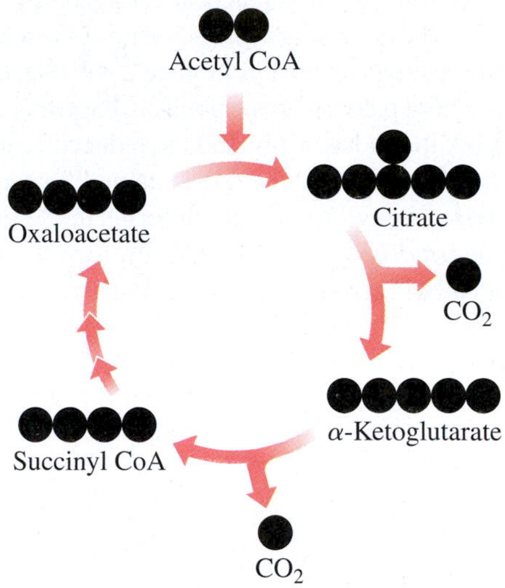
Acetyl CoA
Citrate
CO_2
α-Ketoglutarate
CO_2
Succinyl CoA
Oxaloacetate

FIGURE 23.2 ▶ In the citric acid cycle, two carbon atoms are removed as CO_2 from six-carbon citrate to give four-carbon succinyl CoA, which is converted to four-carbon oxaloacetate.

Q How many carbon atoms are removed in one turn of the citric acid cycle?

TEST

Try Practice Problems 23.1 to 23.4

CORE CHEMISTRY SKILL

Describing the Reactions in the Citric Acid Cycle

Reaction 1 Formation of Citrate

In the first reaction of the citric acid cycle, *citrate synthase* catalyzes the condensation of an acetyl group (2C) from acetyl CoA with oxaloacetate (4C) to yield citrate (6C) and coenzyme A. The energy to form citrate is provided by the hydrolysis of the high-energy thioester bond in acetyl CoA.

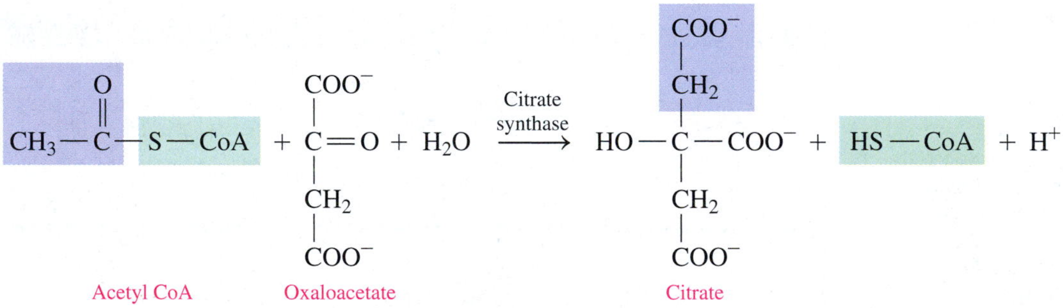
Citrate synthase
Acetyl CoA
Oxaloacetate
Citrate

ENGAGE

Why does citrate have to be isomerized to isocitrate before it can be oxidized?

Reaction 2 Isomerization

The citrate formed in reaction 1 contains a tertiary alcohol group that cannot be oxidized further. In reaction 2, citrate undergoes isomerization to yield isocitrate, which contains a secondary alcohol group that can be oxidized in the next reaction. Initially, *aconitase* catalyzes the dehydration of citrate (tertiary alcohol) to yield *cis*-aconitate, which is followed by a hydration that forms isocitrate (secondary alcohol).

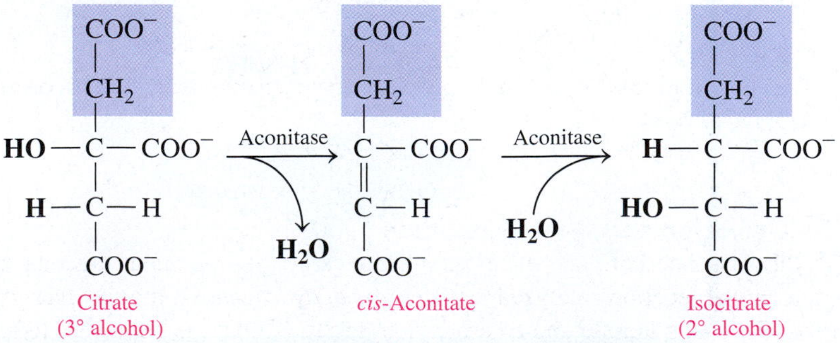
Aconitase
H_2O
Aconitase
H_2O
Citrate
(3° alcohol)
cis-Aconitate
Isocitrate
(2° alcohol)

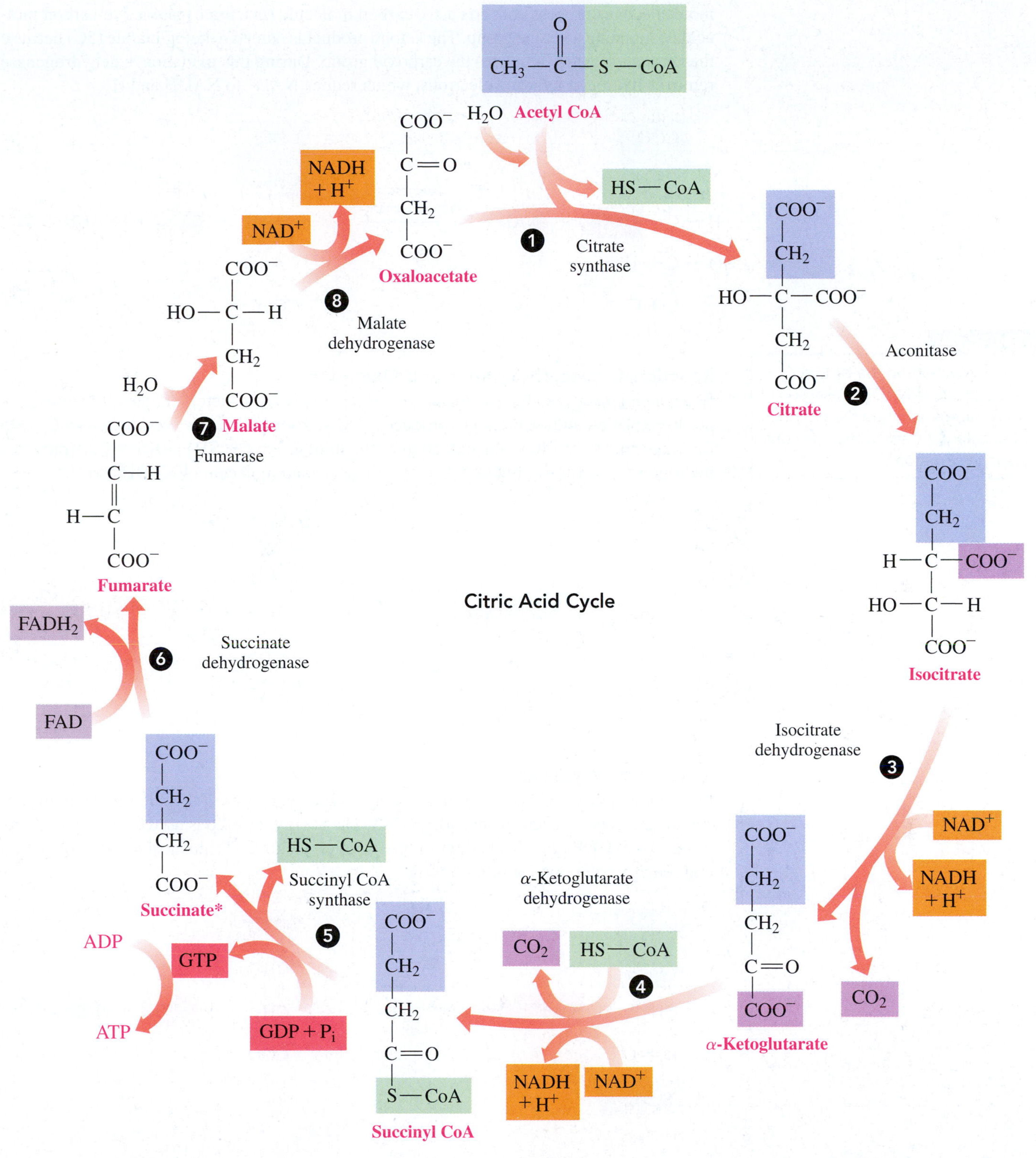

Acetyl CoA
H2O
HS—CoA
Oxaloacetate
1 Citrate synthase
Citrate
2 Aconitase
Isocitrate
3 Isocitrate dehydrogenase
NAD+
NADH + H+
CO2
α-Ketoglutarate
4 α-Ketoglutarate dehydrogenase
CO2
HS—CoA
NAD+
NADH + H+
Succinyl CoA
5 Succinyl CoA synthase
HS—CoA
GDP + Pi
GTP
ADP
ATP
Succinate*
6 Succinate dehydrogenase
FAD
FADH2
Fumarate
7 Fumarase
H2O
Malate
8 Malate dehydrogenase
NAD+
NADH + H+
Citric Acid Cycle
*Succinate is a symmetrical compound.

FIGURE 23.3 ▶ Each turn of the citric acid cycle regenerates oxaloacetate and produces 2 CO_2, 1 GTP, 3 NADH, and 1 $FADH_2$.

Q How many reactions in the citric acid cycle produce a reduced coenzyme?

Reaction 3 Oxidation and Decarboxylation

Reaction 3 is the first time both an oxidation and a decarboxylation occur together in the citric acid cycle. In the reaction catalyzed by *isocitrate dehydrogenase*, the secondary alcohol group is oxidized to a ketone and a carboxylate group (COO^-) is converted to a CO_2

molecule. This process converts a six-carbon molecule (isocitrate) into a five-carbon molecule containing a ketone group. The ketone product is named α-ketoglutarate (5C) because the ketone is next to (alpha to) the carboxyl group. During this oxidation, a dehydrogenase removes hydrogen ions and electrons, which reduce NAD^+ to NADH and H^+.

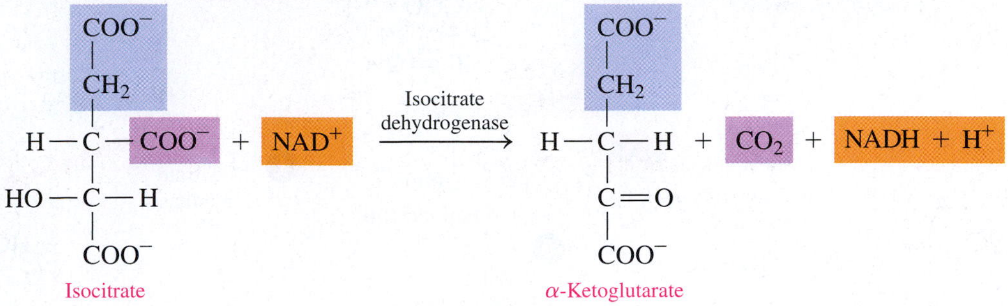
Isocitrate dehydrogenase
NAD+
CO2
NADH + H+
Isocitrate
α-Ketoglutarate

ENGAGE

Three reactions in the citric acid cycle produce NADH. Based on their names, what do the enzymes that catalyze these three reactions have in common?

Reaction 4 Decarboxylation and Oxidation

In a reaction catalyzed by *α-ketoglutarate dehydrogenase*, α-ketoglutarate (5C) undergoes decarboxylation and oxidation to produce a four-carbon group that combines with CoA to form succinyl CoA (4C). The oxidation of the thiol group (—SH) in HS—CoA provides hydrogen that is transferred to NAD^+ to form a second molecule of NADH and H^+.

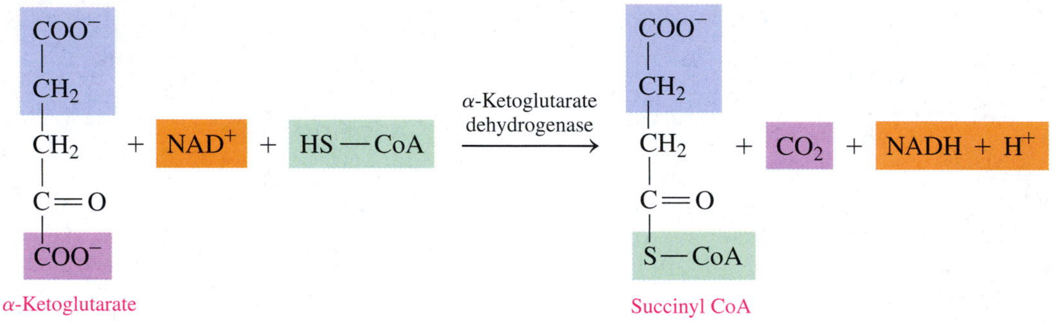
α-Ketoglutarate dehydrogenase
NAD+
HS—CoA
CO2
NADH + H+
α-Ketoglutarate
Succinyl CoA

Reaction 5 Hydrolysis

In reaction 5, *succinyl CoA synthetase* catalyzes the hydrolysis of the thioester bond in succinyl CoA to yield succinate and HS—CoA. The energy released is used to add a phosphate group (P_i) to GDP (guanosine diphosphate), which yields GTP (guanosine triphosphate), a high-energy compound similar to ATP.

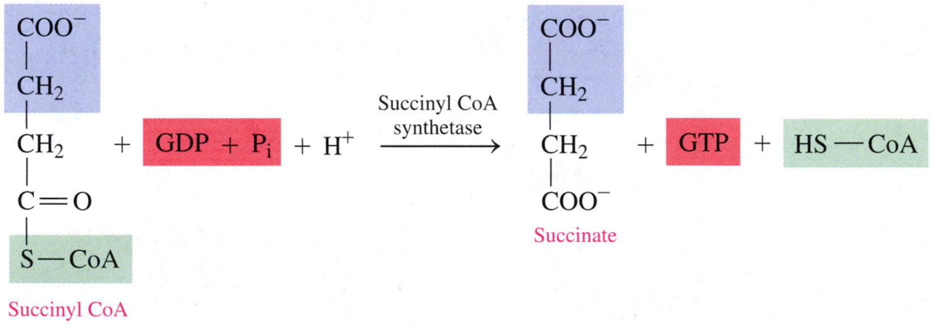
Succinyl CoA synthetase
GDP + Pi
GTP
HS—CoA
Succinyl CoA
Succinate

Eventually, the GTP undergoes hydrolysis with a release of energy that is used to add a phosphate group to ADP to form ATP. This reaction is the only time in the citric acid cycle that ATP is produced by a direct transfer of a phosphate group.

Reaction 6 Oxidation

In reaction 6, *succinate dehydrogenase* catalyzes the oxidation of succinate to yield fumarate, a compound with a trans carbon–carbon double bond (C=C). In this oxidation, the 2H lost from succinate are used to reduce the coenzyme FAD to $FADH_2$. This is the only reaction in the citric acid cycle in which FAD is reduced to $FADH_2$.

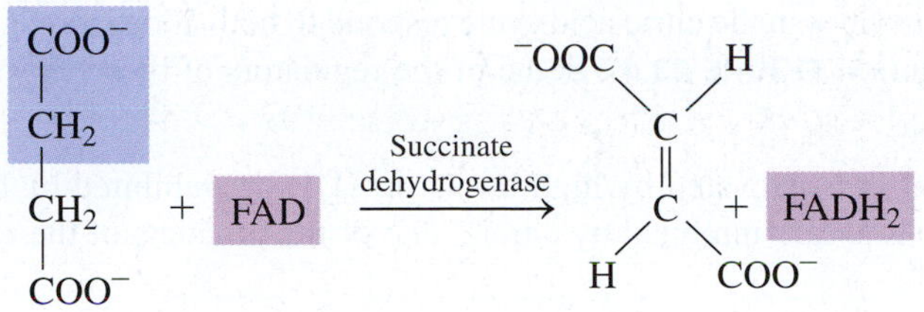
Succinate dehydrogenase

Succinate   Fumarate

Reaction 7 Hydration

In reaction 7, a hydration catalyzed by *fumarase* adds water to the double bond of fumarate to yield malate, which is a secondary alcohol.

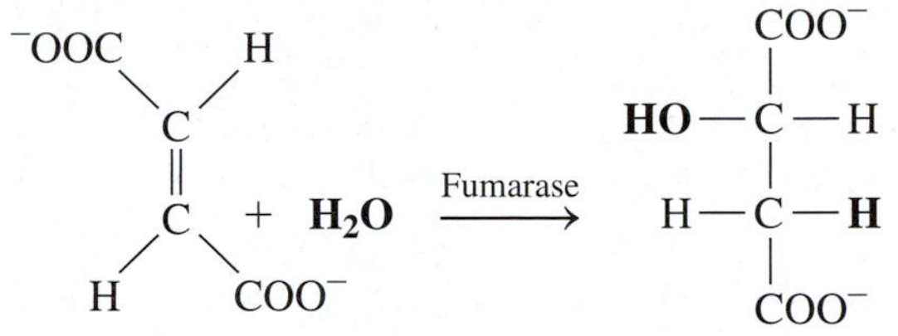
Fumarase

Fumarate   Malate

Reaction 8 Oxidation

In reaction 8, the last step of the citric acid cycle, *malate dehydrogenase* catalyzes the oxidation of the secondary alcohol group (—OH) in malate to a carbonyl group (C═O) yielding oxaloacetate. For the third time in the citric acid cycle, an oxidation by a dehydrogenase provides hydrogen ions and electrons for the reduction of NAD^+ to NADH and H^+.

ENGAGE

Which steps of the citric acid cycle involve an oxidation without a decarboxylation?

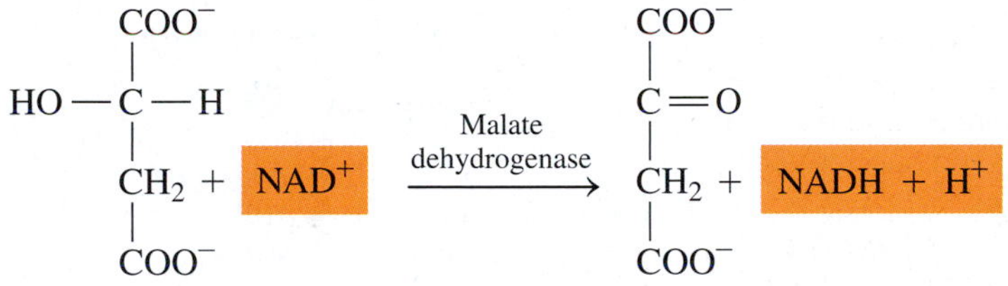
Malate dehydrogenase

Malate   Oxaloacetate

Summary of Products from the Citric Acid Cycle

We have seen that the citric acid cycle begins when a two-carbon acetyl group from acetyl CoA combines with a four-carbon oxaloacetate to form a six-carbon citrate. Through oxidation, reduction, and decarboxylation, two carbon atoms are removed from citrate to yield two CO_2 and a four-carbon compound that undergoes reactions to regenerate oxaloacetate.

In the four oxidation reactions for one turn of the citric acid cycle, three NAD^+ are reduced to three NADH and one FAD is reduced to one $FADH_2$. One GDP is converted to one GTP, which is used to convert one ADP and P_i to ATP. We can write an overall chemical equation for one complete turn of the citric acid cycle as follows:

$$\text{Acetyl CoA} + 3\text{NAD}^+ + \text{FAD} + \text{GDP} + \text{P}_i + 2\text{H}_2\text{O} \longrightarrow \text{HS—CoA} + 3\text{NADH} + 3\text{H}^+ + \text{FADH}_2 + \text{GTP} + 2\text{CO}_2$$

TEST

Try Practice Problems 23.5 to 23.12

Regulation of the Citric Acid Cycle

The primary function of the citric acid cycle is to produce high-energy compounds like NADH and $FADH_2$ for ATP synthesis. Therefore, it is important that the enzymes of the cycle are less active when the cell's energy needs are low and more active when the cell's energy needs are high. In general, high levels of ATP and NADH indicate that the cell's energy needs have been met and there is not a current need to synthesize a lot of ATP. On the other hand, high levels of ADP indicate that the cell's energy needs are high and that ATP should be synthesized.

The activities of three enzymes in the citric acid cycle respond to both allosteric regulation and feedback inhibition (see **FIGURE 23.4**). Some of the regulators of these enzymes are described below.

- Citrate synthase (reaction 1) is activated by high levels of ADP, but inhibited by high levels of ATP or NADH. It is also inhibited by citrate, one of the products of the reaction it catalyzes.
- Isocitrate dehydrogenase (reaction 3) is activated by high levels of ADP, but inhibited by high levels of ATP. It is also inhibited by NADH, one of the products of its reaction.
- α-Ketoglutarate dehydrogenase (reaction 4) is activated by high levels of ADP. However, it is inhibited by both of the products of the reaction it catalyzes: NADH and succinyl CoA.

TEST

Try Practice Problems 23.13 to 23.16

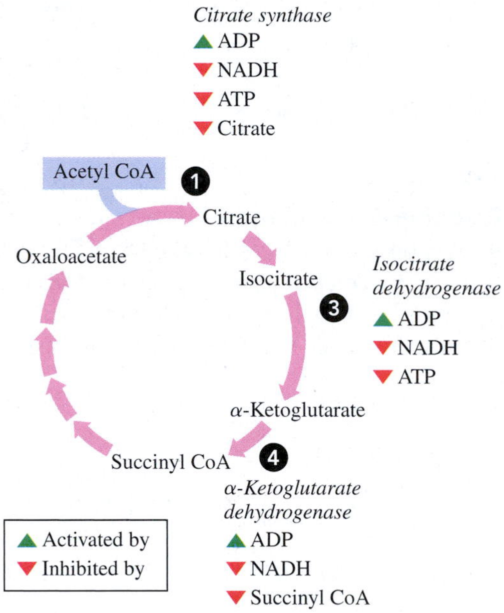
Citrate synthase
ADP
NADH
ATP
Citrate
Acetyl CoA
1
Citrate
Oxaloacetate
Isocitrate
Isocitrate dehydrogenase
3
ADP
NADH
ATP
α-Ketoglutarate
Succinyl CoA
4
α-Ketoglutarate dehydrogenase
ADP
NADH
Succinyl CoA
Activated by
Inhibited by

FIGURE 23.4 ▶ High levels of ADP activate the regulated enzymes in the citric acid cycle, whereas high levels of ATP, NADH, citrate, and succinyl CoA inhibit these enzymes.

Q How do high levels of ATP affect the rate of the citric acid cycle?

▶ SAMPLE PROBLEM 23.1 The Citric Acid Cycle

TRY IT FIRST

When one acetyl CoA moves through one turn of the citric acid cycle, how many of each of the following are produced?

a. NADH **b.** CO_2

SOLUTION

a. One turn of the citric acid cycle produces three molecules of NADH.

b. Two molecules of CO_2 are produced by the decarboxylation of isocitrate and α-ketoglutarate.

STUDY CHECK 23.1

When one acetyl CoA moves through one turn of the citric acid cycle, how many of each of the following are produced?

a. $FADH_2$ **b.** GTP

ANSWER

a. One turn of the citric acid cycle produces one molecule of $FADH_2$ (in reaction 6).

b. One molecule of GTP is produced in reaction 5 of the citric acid cycle.

Diseases of the Citric Acid Cycle

The citric acid cycle plays such a central role in aerobic metabolism that problems with any one of its enzymes could be fatal to a developing fetus. As such, diseases associated with defects or deficiencies in the enzymes of the citric acid cycle are extremely rare.

Although diseases of the citric acid cycle are rare, they share some common symptoms. Those with these diseases typically have neurological problems and excrete large amounts of compounds from the citric acid cycle in their urine. For example, people with *fumarase deficiency* often suffer from neurological impairments, seizures, weak muscle tone, developmental delays, and unusual facial features. Because their fumarase activity tends to be significantly less than normal, they cannot convert fumarate to malate effectively. As a consequence, fumarate builds up and is excreted in the urine. Most people with this disease do not survive into adulthood. Approximately 100 cases of fumarase deficiency have been reported across the world. Even fewer cases have been reported of deficiencies in the other enzymes in the citric acid cycle.

TEST

Try Practice Problems 23.17 to 23.20

PRACTICE PROBLEMS

23.1 The Citric Acid Cycle

23.1 What other names are used for the citric acid cycle?

23.2 What compounds are needed to start the citric acid cycle?

23.3 What are the products from one turn of the citric acid cycle?

23.4 What compound is regenerated in each turn of the citric acid cycle?

23.5 Identify the reaction(s) of the citric acid cycle that involve(s)
a. oxidation and decarboxylation
b. dehydration
c. reduction of NAD^+

23.6 Identify the reaction(s) of the citric acid cycle that involve(s)
a. reduction of FAD
b. direct phosphate transfer
c. hydration

23.7 Refer to the diagram of the citric acid cycle to answer each of the following:
a. What are the six-carbon compounds?
b. How is the number of carbon atoms decreased?
c. What is the five-carbon compound?
d. In which reactions are secondary alcohols oxidized?

23.8 Refer to the diagram of the citric acid cycle to answer each of the following:
a. What is the yield of CO_2 molecules?
b. What are the four-carbon compounds?
c. What is the yield of GTP molecules?
d. What are the decarboxylation reactions?

23.9 Indicate the name of the enzyme that catalyzes each of the following reactions in the citric acid cycle:
a. joins acetyl CoA to oxaloacetate
b. forms a carbon–carbon double bond
c. adds water to fumarate

23.10 Indicate the name of the enzyme that catalyzes each of the following reactions in the citric acid cycle:
a. isomerizes citrate
b. oxidizes and decarboxylates α-ketoglutarate
c. hydrolyzes succinyl CoA

23.11 Indicate the coenzyme that accepts hydrogen in each of the following:
a. isocitrate $\longrightarrow$ α-ketoglutarate
b. succinate $\longrightarrow$ fumarate

23.12 Indicate the coenzyme that accepts hydrogen in each of the following:
a. malate $\longrightarrow$ oxaloacetate
b. α-ketoglutarate $\longrightarrow$ succinyl CoA

23.13 What enzymes in the citric acid cycle are allosteric enzymes?

23.14 Why does the rate of the oxidation of pyruvate affect the rate of the citric acid cycle?

23.15 How do high levels of ADP affect the rate of the citric acid cycle?

23.16 How do high levels of NADH affect the rate of the citric acid cycle?

Clinical Applications

23.17 Diseases associated with defects or deficiencies of enzymes in the citric acid cycle are extremely rare. Why do you think this is, based on the role of the citric acid cycle in metabolism?

23.18 An enzyme deficiency in the citric acid cycle causes a buildup of fumarate. Symptoms of the enzyme deficiency include severe neurological problems, encephalopathy, and seizures, often leading to death in the early years. Which enzyme in the citric acid cycle is deficient?

23.19 An enzyme deficiency in the citric acid cycle causes a buildup of malate. Symptoms of the enzyme deficiency include polycythemia, severe neurological problems, and seizures. Which enzyme in the citric acid cycle is deficient?

23.20 An enzyme deficiency in the citric acid cycle produces a large amount of α-ketoglutarate in the blood and urine. Symptoms of the enzyme deficiency include neurological degeneration, rigid muscles, and encephalopathy. Which enzyme in the citric acid cycle is deficient?

23.2 Electron Transport and ATP

LEARNING GOAL Describe the transfer of hydrogen ions and electrons in electron transport and the process of oxidative phosphorylation in ATP synthesis.

For each glucose molecule that completes glycolysis, the oxidation of two pyruvate, and the citric acid cycle, four ATP, along with 10 NADH and two $FADH_2$, are produced.

From One Glucose	ATP	Reduced Coenzymes	
Glycolysis	2	2 NADH	
Oxidation of 2 Pyruvate		2 NADH	
Citric Acid Cycle with 2 Acetyl CoA	2	6 NADH	2 $FADH_2$
Total for One Glucose	**4**	**10 NADH**	**2 $FADH_2$**

In **electron transport**, hydrogen ions and electrons from NADH and $FADH_2$ are passed from one electron carrier to the next until they combine with oxygen to form H_2O. Each step in electron transport is an individual oxidation–reduction reaction. The energy released during electron transport is used to synthesize ATP from ADP and P_i, a process called *oxidative phosphorylation*. As long as oxygen is available for the mitochondria in the cell, electron transport and oxidative phosphorylation function to synthesize most of the ATP produced in the cell.

A mitochondrion consists of an outer membrane, an intermembrane space, and an inner membrane that surrounds the matrix. Along the highly folded inner membrane are the enzymes and electron carriers required for electron transport. Embedded within these membranes are four distinct protein complexes, labeled I, II, III, and IV. Three of the protein complexes (I, III, and IV) extend through the inner mitochondrial membrane, with one end of each complex in the matrix and the other end in the intermembrane space. Two electron carriers, *coenzyme Q* and *cytochrome c*, are not firmly attached to the membrane. They function as mobile carriers, shuttling electrons between the protein complexes that are bound to the inner membrane (see **FIGURE 23.5**).

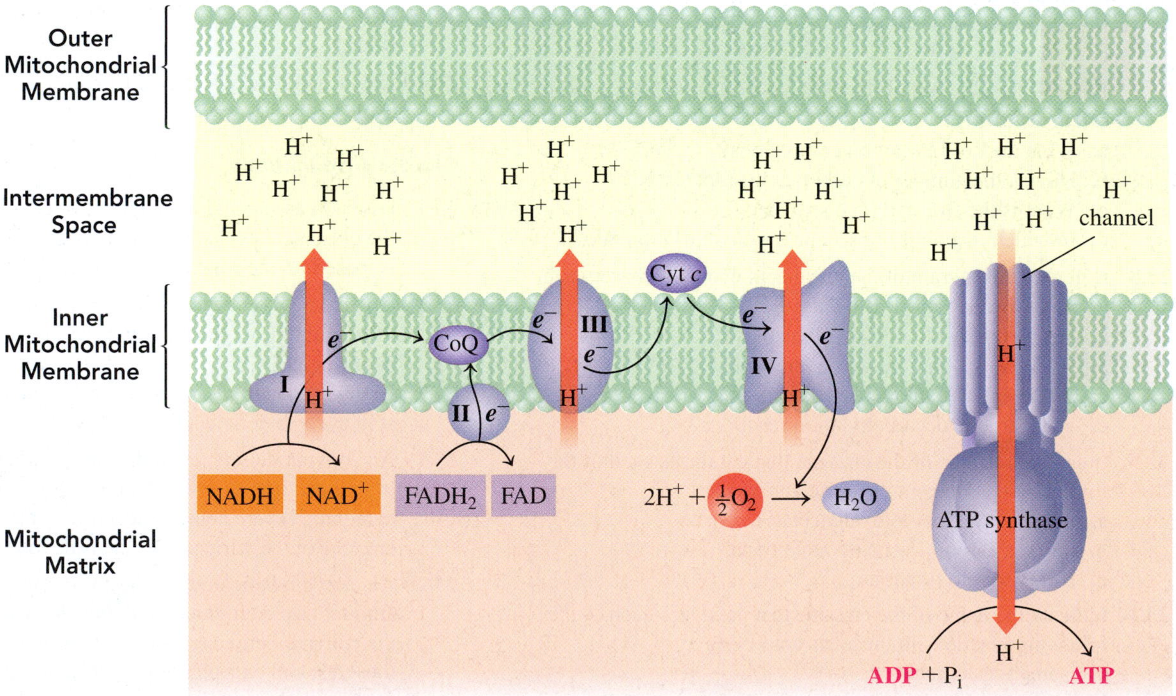
Outer Mitochondrial Membrane
Intermembrane Space
Inner Mitochondrial Membrane
Mitochondrial Matrix
H+
channel
Cyt c
CoQ
I
II
III
IV
e−
NADH
NAD+
FADH2
FAD
2H+ + ½O2 → H2O
ATP synthase
ADP + Pi
ATP

FIGURE 23.5 ▶ In electron transport, coenzymes NADH and $FADH_2$ are oxidized in enzyme complexes, providing electrons and hydrogen ions for ATP synthesis.

Q What pathway is the major source of NADH for electron transport?

Complex I

Electron transport begins when hydrogen ions and electrons are transferred from NADH to complex I. The loss of hydrogen from NADH regenerates NAD^+, which becomes available to oxidize more substrates in oxidative pathways such as the citric acid cycle. The hydrogen ions and electrons are transferred to the first mobile electron carrier **coenzyme Q (CoQ)**, which carries electrons to complex III (see **FIGURE 23.6**).

ENGAGE

Which electron transport complex accepts electrons from NADH?

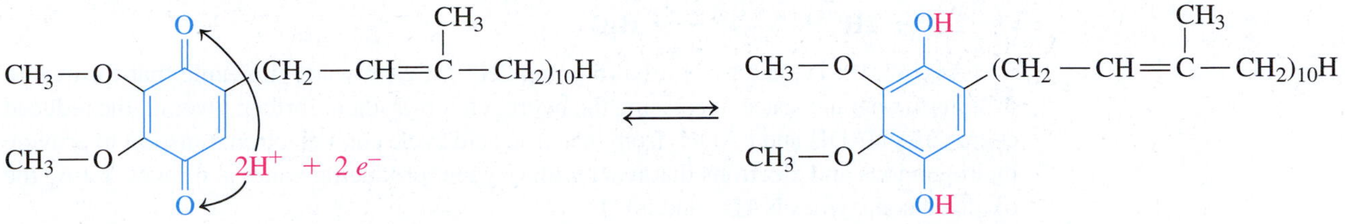

CH_3
$CH_3—O$
$CH_3—O$
O
O
$(CH_2—CH=C—CH_2)_{10}H$
$2H^+ + 2\,e^-$
OH
OH
CH_3
$CH_3—O$
$CH_3—O$
$(CH_2—CH=C—CH_2)_{10}H$

Oxidized coenzyme Q (CoQ) **Reduced coenzyme Q ($CoQH_2$)**

FIGURE 23.6 ▶ The electron carrier CoQ is reduced to $CoQH_2$ when it accepts $2H^+$ and $2\,e^-$ from NADH + H^+ or $FADH_2$.

Q How does reduced coenzyme $CoQH_2$ compare to the oxidized form?

During the electron transfer to CoQ, H^+ ions are pumped through complex I into the intermembrane space, producing a reservoir of H^+ called a **hydrogen ion gradient**. Each time two electrons pass from NADH to CoQ, a total of $4H^+$ are pumped into the intermembrane space. This movement of hydrogen ions produces a charge separation on opposite sides of the membrane, with the intermembrane space becoming more positively charged than the matrix. The overall reaction sequence in complex I is written as follows:

$$\text{NADH} + \text{H}^+ + \text{CoQ} \longrightarrow \text{NAD}^+ + \text{CoQH}_2$$

Complex II

Complex II includes the enzyme succinate dehydrogenase from the citric acid cycle. In complex II, CoQ obtains hydrogen ions and electrons directly from $FADH_2$ that was generated by the conversion of succinate to fumarate in the citric acid cycle. This produces $CoQH_2$ and regenerates the oxidized coenzyme FAD, which becomes available again to oxidize more substrates in oxidative pathways. $CoQH_2$ carries the electrons to complex III. Unlike complex I, complex II does not pump H^+ ions into the intermembrane space. Thus, it does not contribute to the hydrogen ion gradient.

The overall reaction sequence in complex II is written as follows:

$$\text{FADH}_2 + \text{CoQ} \longrightarrow \text{FAD} + \text{CoQH}_2$$

ENGAGE

Which electron transport complex accepts electrons from $FADH_2$?

Because complex II is at a lower energy than complex I, the electrons from $FADH_2$ enter electron transport at a lower energy level than those from NADH.

Complex III

In complex III, two electrons are transferred from the mobile carrier $CoQH_2$ to a series of iron-containing proteins called **cytochromes (cyt)** and eventually to cytochrome *c*, which is another mobile electron carrier. Cytochrome *c*, which contains Fe^{3+}/Fe^{2+}, is reduced to Fe^{2+} when it gains an electron, and oxidized to Fe^{3+} when it loses an electron. The energy released during this electron transfer is used to pump $4H^+$ into the intermembrane space, increasing the high-energy hydrogen ion gradient. As a mobile carrier, cytochrome *c* carries electrons to complex IV. The overall reaction in complex III is written as follows:

$$\text{CoQH}_2 + 2\text{cyt } c\,(\text{Fe}^{3+}) \longrightarrow \text{CoQ} + 2\text{H}^+ + 2\text{cyt } c\,(\text{Fe}^{2+})$$

(Oxidized) (Reduced)

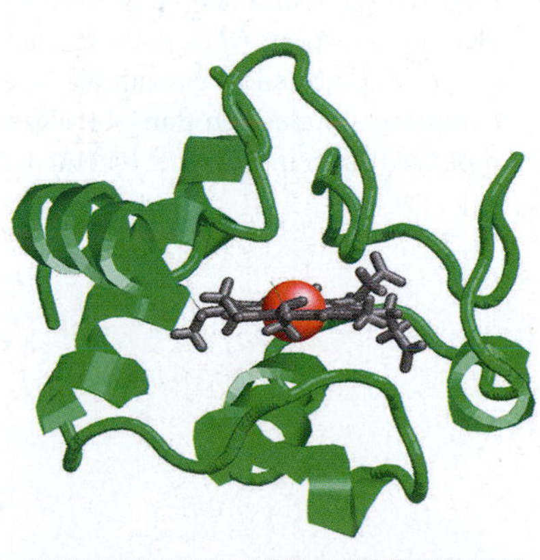

Cytochrome c carries one electron when Fe^{3+} is reduced to Fe^{2+} (orange sphere). The heme group (gray) holds the iron in place.

Complex IV

At complex IV, four electrons from four cytochrome *c* are passed to other electron carriers until the electrons combine with hydrogen ions and oxygen (O_2) to form two molecules of water. The overall reaction in complex IV is written as follows:

$$4\,e^- + \mathbf{4H^+} + O_2 \longrightarrow 2\mathbf{H_2}O$$

This equation may also be simplified:

$$2\,e^- + \mathbf{2H^+} + {}^1\!/_2\,O_2 \longrightarrow \mathbf{H_2}O$$

At complex IV, energy is used to pump $2H^+$ for each water molecule that forms into the intermembrane space, increasing the hydrogen ion gradient further. Overall, the reduced coenzymes NADH and $FADH_2$ from the citric acid cycle enter electron transport to provide hydrogen ions and electrons that react with oxygen, producing water and regenerating the oxidized coenzymes NAD^+ and FAD.

▶SAMPLE PROBLEM 23.2 Oxidation and Reduction

TRY IT FIRST

Identify each of the following steps in electron transport as oxidation or reduction:

a. $CoQH_2 \longrightarrow CoQ + 2H^+ + 2\,e^-$

b. cyt *c* (Fe^{3+}) + e^- $\longrightarrow$ cyt *c* (Fe^{2+})

SOLUTION

a. The loss of electrons is oxidation.

b. The gain of an electron is reduction.

STUDY CHECK 23.2

a. What is the final substance that accepts electrons in electron transport?

b. Which mobile electron carrier transfers electrons from complex II to complex III?

ANSWER

a. oxygen (O_2)

b. coenzyme Q (CoQ)

TEST

Try Practice Problems 23.21 to 23.34

Chemistry Link to Health

Toxins: Inhibitors of Electron Transport

When an inhibitor blocks a step in electron transport, the carriers preceding that step are unable to transfer electrons and remain in their reduced forms. All the carriers after the blocked step remain oxidized without a source of electrons. If an inhibitor shuts down electron transport, ATP synthesis cannot occur, and the cells die.

Several substances can inhibit the electron carriers in the different complexes of electron transport. Rotenone, a product from a plant root used as an insecticide, and the painkillers Amytal and Demerol block electron transport between complexes I and CoQ. Another inhibitor is the antibiotic antimycin A, which blocks the flow of electrons between complex III and cytochrome *c*. Another group of compounds, including cyanide (CN^-) and carbon monoxide, block the flow of electrons between cytochrome *c* and complex IV. The toxic nature of these compounds makes it clear that organisms rely heavily on the process of electron transport.

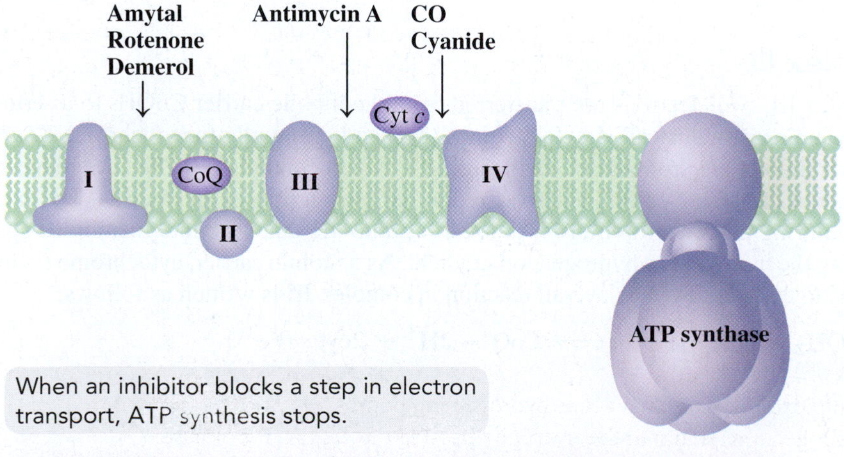
Amytal
Rotenone
Demerol
Antimycin A
CO
Cyanide
Cyt c
I
CoQ
II
III
IV
ATP synthase

When an inhibitor blocks a step in electron transport, ATP synthesis stops.

TEST

Try Practice Problems 23.41 to 23.44

Oxidative Phosphorylation

Energy is released as a result of the oxidation–reduction reactions that occur in electron transport. Now we will look at how that energy is coupled with the production of ATP in the process called **oxidative phosphorylation**.

In 1978, Peter Mitchell received the Nobel Prize in chemistry for his theory called the **chemiosmotic model**, which links the energy from electron transport to a hydrogen ion gradient that drives the synthesis of ATP. Three of the complexes (I, III, and IV) extend through the inner mitochondrial membrane with one end of each complex in the matrix and the other end in the intermembrane space. In the chemiosmotic model, these complexes act as hydrogen ion pumps by removing some H^+ ions from the matrix and releasing other H^+ ions into the intermembrane space. The increase in H^+ concentration lowers the pH in the intermembrane space and creates an H^+ gradient. Because H^+ ions are positively charged, the lower pH and the electrical charge of the H^+ gradient make it an *electrochemical gradient*.

To equalize the pH and the electrical charge between the intermembrane space and the matrix, H^+ ions must move to the matrix. However, H^+ cannot move through the inner membrane. Instead, H^+ must return to the matrix by passing through a protein complex in the inner membrane called **ATP synthase**. The flow of H^+ ions from the intermembrane space through the ATP synthase releases energy that is used to synthesize ATP from ADP and P_i. This process of oxidative phosphorylation couples the energy from electron transport to the synthesis of ATP.

TEST

Try Practice Problems 23.35 to 23.38

$$\text{ADP} + \text{P}_i + \text{energy} \xrightarrow{\text{ATP synthase}} \text{ATP}$$

Electron Transport and ATP Synthesis

When NADH enters electron transport at complex I, the energy transferred from its oxidation can be used to synthesize 2.5 ATP. When $FADH_2$ enters electron transport at complex II, which is at a lower energy level, its oxidation provides energy for the synthesis of 1.5 ATP. Older values of 3 ATP from NADH and 2 ATP from $FADH_2$ are still found in some chemistry texts and are often used in biology. However, current research indicates that the oxidation of one NADH yields 2.5 ATP and one $FADH_2$ yields 1.5 ATP.

ENGAGE

How many moles of ATP can be synthesized from the energy released by the oxidation of one mole of NADH?

Reduced Coenzyme		Oxidized Coenzyme		ATP Output
$NADH + H^+$	$\longrightarrow$	NAD^+	+	2.5 ATP
$FADH_2$	$\longrightarrow$	FAD	+	1.5 ATP

ENGAGE

How many moles of ATP can be synthesized from the energy released by the oxidation of one mole of $FADH_2$?

▶SAMPLE PROBLEM 23.3 ATP Synthesis

TRY IT FIRST

Why does the oxidation of NADH provide energy for the formation of more ATP molecules than does the oxidation of $FADH_2$?

SOLUTION

The more H^+ ions that are pumped into the intermembrane space during the oxidation of a given cofactor, the more ATP that can be produced. Electrons from the oxidation of NADH enter electron transport at complex I, passing through three complexes (I, III, and IV) that pump H^+ ions into the intermembrane space. Electrons from $FADH_2$ enter electron transport at complex II, passing through only two complexes (III and IV) that pump H^+ ions into the intermembrane space.

STUDY CHECK 23.3

a. Which complexes in electron transport act as hydrogen ion pumps?
b. How do hydrogen ions return to the matrix?

ANSWER

a. Complexes I, III, and IV act as hydrogen ion pumps.
b. Hydrogen ions return to the matrix by passing through ATP synthase.

Chemistry Link to Health

Uncouplers of ATP Synthase

Some types of compounds called *uncouplers* separate the electron transport system from ATP synthase. They do this by providing an alternate route for hydrogen ions to return to the matrix. The electrons are transported to O_2 in electron transport, but ATP is not formed by ATP synthase.

Some uncouplers transport H^+ through the inner mitochondrial membrane, which is normally impermeable to H^+. Compounds such as dicumarol and 2,4-dinitrophenol (DNP) are hydrophobic and bind with H^+ ions to carry them across the inner membrane. An antibiotic, oligomycin, blocks the channel through which H^+ ions normally flow through ATP synthase to return to the matrix. When protons do not flow through the ATP synthase channel, ATP synthesis does not occur.

When there is no mechanism for ATP synthesis, the energy of electron transport is released as heat. Certain animals that are adapted to cold climates have developed their own uncoupling system, which allows them to use electron transport energy for heat production. These animals have large amounts of a tissue called *brown fat*, which contains a high concentration of mitochondria. This tissue is brown because of the color of iron in the cytochromes of the mitochondria. In brown fat, a protein called thermogenin, embedded in the inner mitochondrial membrane of brown adipose tissues, allows the H^+ ions to bypass ATP synthase. The energy that would be used to synthesize ATP is released as heat.

In newborn babies, brown fat is used to generate heat because babies have a small mass but large surface area and need to produce more heat than adults. The brown fat deposits are located near major blood vessels, which carry the warmed blood to the body. Most adults have little or no brown fat, although someone who works outdoors in a cold climate will develop some brown fat deposits.

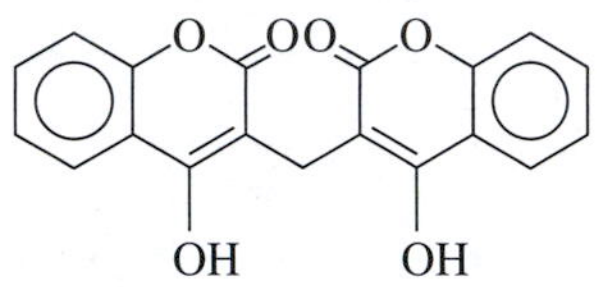
O
O
O
O
OH
OH
Dicumarol

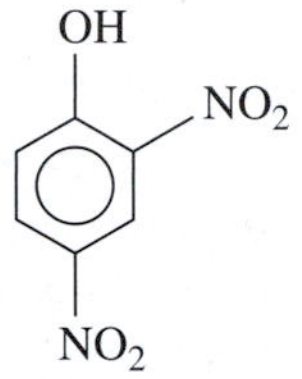
OH
NO2
NO2
2,4-Dinitrophenol (DNP)

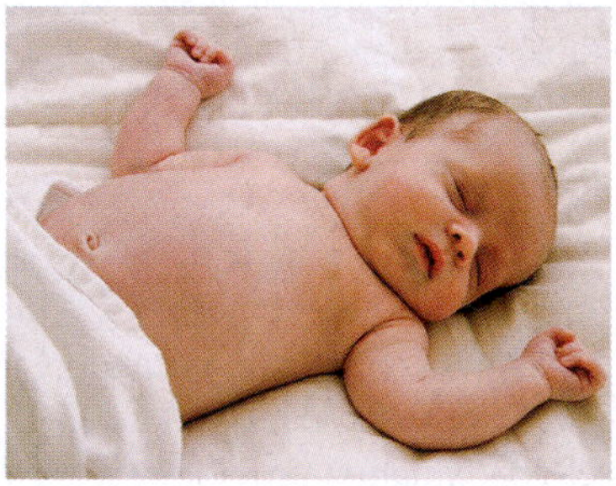
Brown fat helps babies to keep warm.

Regulation of Electron Transport and Oxidative Phosphorylation

When a cell is active and ATP is consumed rapidly, the elevated levels of ADP will activate the synthesis of ATP. Therefore, the activity of electron transport is strongly dependent on the availability of ADP for ATP synthesis. Electron transport is also regulated by the availability of P_i, oxygen (O_2), and NADH. Low levels of any of these compounds or of ADP will decrease the activity of electron transport and the rate of formation of ATP.

TEST

Try Practice Problems 23.39 and 23.40

PRACTICE PROBLEMS

23.2 Electron Transport and ATP

23.21 Is cyt *c* (Fe^{3+}) the abbreviation for the oxidized or reduced form of cytochrome *c*?

23.22 Is $FADH_2$ the abbreviation for the oxidized or reduced form of flavin adenine dinucleotide?

23.23 Identify each of the following as oxidation or reduction:
a. $NADH \longrightarrow NAD^+ + H^+ + 2\,e^-$
b. $CoQ + 2H^+ + 2\,e^- \longrightarrow CoQH_2$

23.24 Identify each of the following as oxidation or reduction:
a. cyt *c* (Fe^{2+}) $\longrightarrow$ cyt *c* (Fe^{3+}) $+ e^-$
b. $FAD + 2H^+ + 2\,e^- \longrightarrow FADH_2$

23.25 What reduced coenzyme provides hydrogen and electrons for electron transport at complex I?

23.26 What reduced coenzyme provides hydrogen and electrons for electron transport at complex II?

23.27 Arrange the following in the order in which they appear in electron transport: cytochrome *c* (Fe^{3+}), $FADH_2$, CoQ.

23.28 Arrange the following in the order in which they appear in electron transport: O_2, NAD^+, FAD.

23.29 How are electrons carried from complex I to complex III?

23.30 How are electrons carried from complex III to complex IV?

23.31 Where is NADH oxidized in electron transport, and what is its oxidized form?

23.32 Where is $FADH_2$ oxidized in electron transport, and what is its oxidized form?

23.33 Complete each of the following reactions in electron transport:
a. $NADH + H^+ +$ ________ $\longrightarrow$ ________ $+ CoQH_2$
b. $CoQH_2$ + 2cyt *c* (Fe^{3+}) $\longrightarrow$ CoQ + ________ + ________

23.34 Complete each of the following reactions in electron transport:
a. CoQ + ________ $\longrightarrow$ ________ + FAD
b. $4\,e^- + 4H^+ + O_2 \longrightarrow$ ________

23.35 What is meant by the term oxidative phosphorylation?

23.36 How is the hydrogen ion gradient established?

23.37 According to the chemiosmotic model, how does the hydrogen ion gradient provide energy to synthesize ATP?

23.38 Where does the phosphorylation of ADP occur?

23.39 How are glycolysis and the citric acid cycle linked to the production of ATP by electron transport?

23.40 What cellular conditions will result in a decrease in the activity of electron transport and in the rate of formation of ATP?

Clinical Applications

23.41 Potassium cyanide is a potent poison that kills by inhibiting electron transport when CN^- binds to the Fe^{3+} in the cytochrome heme groups. Why would this stop electron transport?

23.42 Coenzyme Q is sold as a dietary supplement called CoQ10, with recommended doses of 22 to 400 mg. It is said to boost energy and recovery from exercise. Based on its role in electron transport, explain how this might be possible.

23.43 Increasing thermogenesis has been thought to be a way to lose weight. In fact, 2,4-dinitrophenol is available as a supplement most often used by bodybuilders to burn fat. This dangerous drug is also marketed as a pesticide and kills insects in a similar manner. What might be a side effect of using this compound?

23.44 Oligomycin is an antibiotic that inhibits ATP synthase. If ATP synthase is not operable, what happens to the energy from electron transport?

23.3 ATP Energy from Glucose

LEARNING GOAL Account for the ATP produced by the complete oxidation of glucose.

The total ATP for the complete oxidation of glucose under aerobic conditions is calculated by combining the ATP produced from glycolysis, the oxidation of pyruvate, the citric acid cycle, and electron transport.

Malate–Aspartate Shuttle

The enzymes involved in electron transport are located in the inner mitochondrial membrane, but the NADH produced in the cytosol during glycolysis cannot pass through that membrane. However, the hydrogen ions and electrons from NADH in the cytosol can be transferred to compounds that can move in and out of the inner mitochondrial membrane. These compounds are called *transporters*. Although there is more than one way to transport the hydrogen ions and electrons from NADH across the inner mitochondrial membrane, we will focus on the *malate–aspartate shuttle*, which functions in the heart and in the liver.

In the cytosol, malate dehydrogenase catalyzes the reaction of oxaloacetate and NADH to yield malate and NAD^+. Then a transporter binds the malate and carries it across the inner mitochondrial membrane into the matrix, where a different malate dehydrogenase molecule oxidizes malate back to oxaloacetate. In the matrix, the oxidation provides hydrogen ions and electrons that are used to reduce NAD^+ to NADH, which can now enter electron transport to synthesize ATP (see **FIGURE 23.7**). Because the oxaloacetate produced in the matrix cannot cross the inner mitochondrial membrane, it is converted back to aspartate by aspartate transaminase so it can move out of the matrix back into the cytosol, where transamination converts it back to oxaloacetate.

ENGAGE
What is the purpose of the malate–aspartate shuttle?

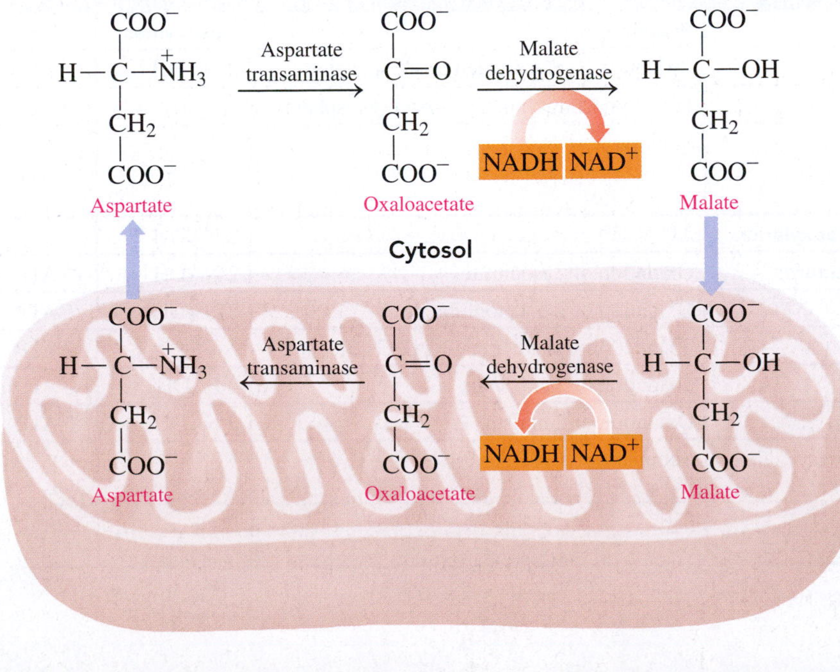
Aspartate transaminase
Malate dehydrogenase
NADH NAD+
Aspartate
Oxaloacetate
Malate
Cytosol
Aspartate transaminase
Malate dehydrogenase
NADH NAD+
Aspartate
Oxaloacetate
Malate

FIGURE 23.7 ▶ The malate–aspartate shuttle transfers the energy stored in NADH formed during glycolysis to transporters that move from the cytosol into the mitochondrial matrix where NADH is regenerated for use in electron transport.

Q Why does the NADH produced by glycolysis utilize the malate–aspartate shuttle?

ATP from Glycolysis

In glycolysis, the oxidation of glucose stores energy in two NADH molecules, as well as two ATP molecules from direct phosphate transfer. From glycolysis, glucose produces a total of seven ATP: five ATP from two NADH, and two ATP from direct phosphate transfer.

$$\text{Glucose} \longrightarrow 2\text{ pyruvate} + \boxed{2\text{ATP}} + \boxed{2\text{NADH}\ (5\text{ATP})}$$

ATP from the Oxidation of Two Pyruvate

Under aerobic conditions, pyruvate enters the mitochondria, where it is oxidized to give acetyl CoA, CO_2, and NADH. Because glucose yields two pyruvate, two NADH enter electron transport, where the oxidation of two pyruvate leads to the production of five ATP.

$$2\text{ Pyruvate} \longrightarrow 2\text{ acetyl CoA} + 2CO_2 + \boxed{2\text{NADH}\ (5\text{ATP})}$$

ATP from the Citric Acid Cycle

ENGAGE

How many ATP are produced during one turn of the citric acid cycle?

The two acetyl CoA produced from two pyruvate now enter the citric acid cycle. Each acetyl CoA makes one turn of the citric acid cycle and produces two CO_2, three NADH, one $FADH_2$, and one ATP (from GTP) by direct phosphate transfer. Thus, the two acetyl CoA initially from one glucose produce a total of six NADH, two $FADH_2$, and two ATP. In electron transport, six NADH produce 15 ATP, and two $FADH_2$ produce 3 ATP. In two turns of the citric acid cycle, a total of 20 ATP are produced.

$$\begin{aligned} 6\text{ NADH} \times 2.5\text{ ATP/NADH} &= 15\text{ ATP} \\ 2\text{ FADH}_2 \times 1.5\text{ ATP/FADH}_2 &= 3\text{ ATP} \\ 2\text{ GTP} \times 1\text{ ATP/GTP} &= 2\text{ ATP} \\ \hline \text{Total (two turns)} &= 20\text{ ATP} \end{aligned}$$

The overall equation for the reaction of two acetyl CoA is

$$2\text{ Acetyl CoA} \longrightarrow 4CO_2 + 20\text{ ATP (two turns of the citric acid cycle)}$$

ATP from the Complete Oxidation of Glucose

CORE CHEMISTRY SKILL

Calculating the ATP Produced from Glucose

The total ATP production for the complete oxidation of glucose is calculated by combining the ATP produced from glycolysis, the oxidation of pyruvate, and the citric acid cycle (see **FIGURE 23.8**). The ATP produced for these reactions is given in **TABLE 23.1**.

TABLE 23.1 ATP Produced by the Complete Oxidation of Glucose

Pathway	Reaction	Coenzymes	ATP Yield
Glycolysis	Oxidation of glyceraldehyde-3-phosphate	2 NADH	5 ATP
	Direct phosphorylation (2 triose phosphate)		2 ATP
	Summary: $C_6H_{12}O_6 \longrightarrow 2\text{ pyruvate} + 2H_2O$ (Glucose)		**7 ATP**
Oxidation and Decarboxylation	$2\text{ Pyruvate} \longrightarrow 2\text{ acetyl CoA} + 2CO_2$	2 NADH	**5 ATP**
Citric Acid Cycle (two turns)	Oxidation of 2 isocitrate	2 NADH	5 ATP
	Oxidation of 2 α-ketoglutarate	2 NADH	5 ATP
	2 Direct phosphate transfers (2 GTP)		2 ATP
	Oxidation of 2 succinate	2 $FADH_2$	3 ATP
	Oxidation of 2 malate	2 NADH	5 ATP
	Summary: $2\text{ Acetyl CoA} \longrightarrow 4CO_2 + 2H_2O$		**20 ATP**
Total Yield	$C_6H_{12}O_6 + 6O_2 \longrightarrow 6CO_2 + 6H_2O$ (Glucose)		**32 ATP**

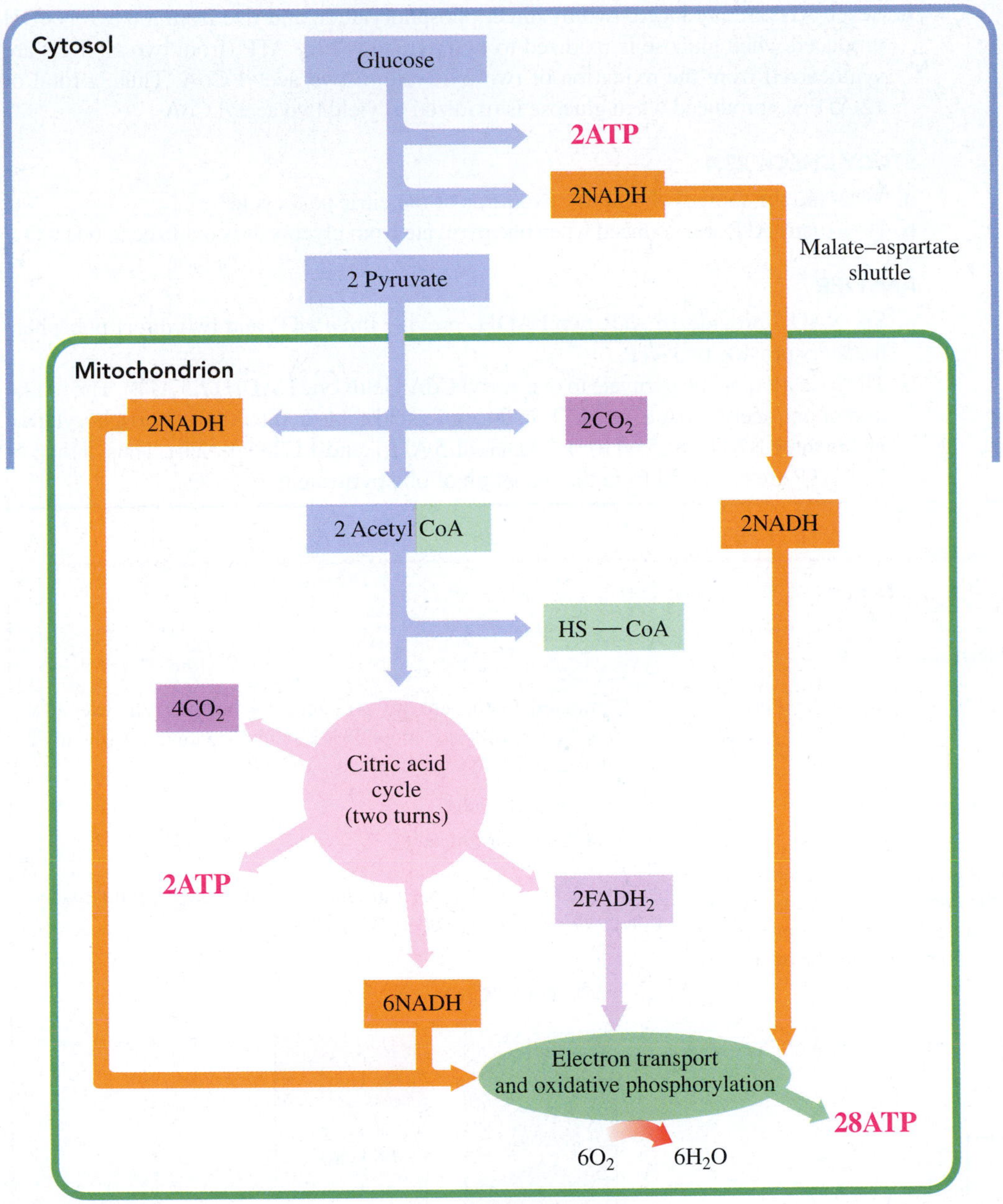

Cytosol
Glucose
2ATP
2NADH
Malate–aspartate shuttle
2 Pyruvate
Mitochondrion
2NADH
$2CO_2$
2NADH
2 Acetyl CoA
HS—CoA
$4CO_2$
Citric acid cycle (two turns)
2ATP
$2FADH_2$
6NADH
Electron transport and oxidative phosphorylation
28ATP
$6O_2$
$6H_2O$

FIGURE 23.8 ▶ The complete oxidation of glucose to CO_2 and H_2O yields a total of 32 ATP.

Q Which metabolic pathway produces most of the ATP from the oxidation of glucose?

▶ SAMPLE PROBLEM 23.4 ATP Production

TRY IT FIRST

How many ATP are produced during each of the following oxidations?

a. two pyruvate to two acetyl CoA **b.** one glucose to two acetyl CoA

SOLUTION

	Given	Need	Connect
ANALYZE THE PROBLEM	**a.** two pyruvate to two acetyl CoA **b.** one glucose to two acetyl CoA	number of ATP	ATP per NADH, ATP per $FADH_2$

a. The oxidation of two pyruvate to two acetyl CoA produces two NADH, which yields five ATP. We calculate this as:

$$2\ \text{NADH} \times 2.5\ \text{ATP/NADH} = 5\ \text{ATP}$$

b. Seven ATP are produced, two by direct phosphorylation and five from the two NADH produced when glucose is oxidized to two pyruvate. Five ATP (from two NADH) are synthesized from the oxidation of two pyruvate to two acetyl CoA. Thus, a total of 12 ATP are produced when glucose is oxidized to yield two acetyl CoA.

STUDY CHECK 23.4

a. What are the sources of ATP for two turns of the citric acid cycle?

b. How many ATP are produced when one pyruvate from glycolysis is oxidized to one CO_2?

ANSWER

a. Six NADH provide 15 ATP, two $FADH_2$ provide three ATP, and two direct phosphate transfers provide two ATP.

b. The oxidation of one pyruvate to one acetyl CoA yields one NADH (2.5 ATP). The oxidation of one acetyl CoA to one CO_2 in the citric acid cycle and oxidative phosphorylation yields three NADH (7.5 ATP), 1 $FADH_2$ (1.5 ATP), and 1 GTP (1 ATP). Thus, a total of 12.5 ATP are produced from the oxidation of one pyruvate to one CO_2.

TEST

Try Practice Problems 23.45 to 23.48

Chemistry Link to Health

Efficiency of ATP Production

In a laboratory, a calorimeter is used to measure the heat energy from the combustion of glucose. In a calorimeter, 1 mole of glucose produces 690 kcal.

$$C_6H_{12}O_6 + 6O_2 \longrightarrow 6CO_2 + 6H_2O + 690\text{ kcal}$$

We can compare the amount of energy produced from 1 mole of glucose in a calorimeter with the ATP energy produced in the mitochondria from glucose. We use the energy of the hydrolysis of ATP (7.3 kcal/mole of ATP). Because 1 mole of glucose generates energy for up to 32 moles of ATP, the total energy from the oxidation of 1 mole of glucose in the cells would be 230 kcal/mole.

$$\frac{32\ \cancel{\text{moles ATP}}}{1\text{ mole glucose}} \times \frac{7.3\text{ kcal}}{1\ \cancel{\text{mole ATP}}} = 230\text{ kcal/1 mole of glucose}$$

TEST

Try Practice Problems 23.49 to 23.50

Compared to the energy produced by burning glucose in a calorimeter, our cells are about 33% efficient in converting the total available chemical energy in glucose to ATP.

$$\frac{230\text{ kcal (cells)}}{690\text{ kcal (calorimeter)}} \times 100\% = 33\%$$

The rest of the energy from glucose produced during the oxidation of glucose in our cells is lost as heat.

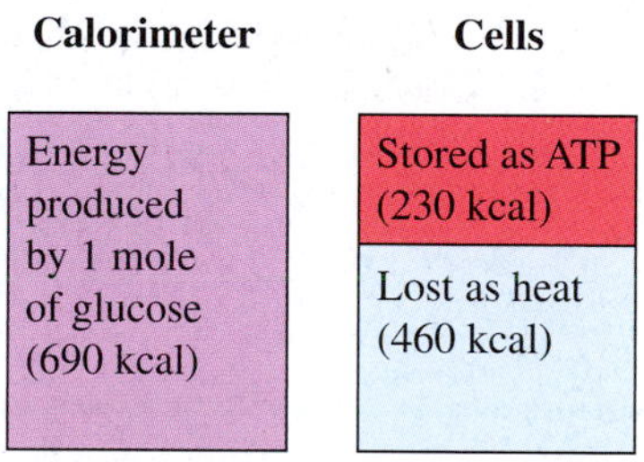
Calorimeter
Cells
Energy produced by 1 mole of glucose (690 kcal)
Stored as ATP (230 kcal)
Lost as heat (460 kcal)

PRACTICE PROBLEMS

23.3 ATP Energy from Glucose

23.45 How many turns of the citric acid cycle are required to oxidize one molecule of glucose?

23.46 Under aerobic conditions, what is the maximum number of ATP that can be produced from one glucose molecule?

23.47 What is the energy yield in ATP associated with each of the following?

a. NADH $\longrightarrow$ NAD^+ **b.** glucose $\longrightarrow$ 2 pyruvate

c. 2 pyruvate $\longrightarrow$ 2 acetyl CoA + $2CO_2$

23.48 What is the energy yield in ATP associated with each of the following?

a. $FADH_2 \longrightarrow FAD$

b. glucose + $6O_2 \longrightarrow 6CO_2 + 6H_2O$

c. acetyl CoA $\longrightarrow$ $2CO_2$

Clinical Applications

23.49 A person has a low O_2 blood level that gives only a 24% efficiency of ATP production from glucose. How many moles of ATP would be obtained from 25 g of glucose if glucose has a molar mass of 180.2 g/mole (see Chemistry Link to Health: Efficiency of ATP Production)?

23.50 An exercise program improves an O_2 blood level enough to give a 30.% efficiency of ATP production from glucose. How many moles of ATP are obtained from 25 g of glucose if glucose has a molar mass of 180.2 g/mole (see Chemistry Link to Health: Efficiency of ATP Production)?

Clinical Update Increasing Brian's Functional Capacity

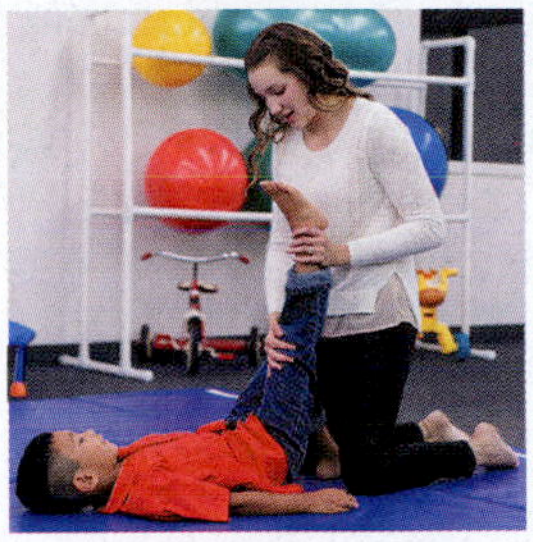

Patients with *mitochondrial myopathy* do not consume oxygen or produce ATP effectively. While exercise is difficult for Brian because he becomes fatigued quite easily, there is research that indicates that even a small amount of exercise increases the number of mitochondria in the body, as well as how efficiently those mitochondria use oxygen.

Therefore, Kate develops a program of physical activity for Brian that includes exercises focusing on stretching muscles, strengthening muscles, increasing balance, and building endurance. She recommends that Brian participate in physical activities every other day for a few minutes several times a day rather than once a day for a longer period of time. To help Brian increase his capacity to walk, several of the activities Kate recommends focus on building strength in Brian's leg muscles and increasing his ability to balance.

Kate also recommends that Brian's parents give him dietary supplements that other patients with mitochondrial myopathy have found beneficial in decreasing fatigue, such as NADH, coenzyme Q (CoQ), and vitamin B.

Each of these compounds participates in electron transport and, thus, the production of ATP. She suggests that Brian's parents start by giving him 100 mg of coenzyme Q daily. At Brian's next appointment with Kate, Brian's parents report that Brian's fatigue levels seem to be decreasing and that he appears to be able to participate in more physical activity than he did previously.

Clinical Applications

23.51 Many mitochondrial diseases are the result of genetic mutations that affect an enzyme involved in either electron transport or ATP production. For each mutation described below, identify the substance(s) that will accumulate, as well as the substance(s) that will not be produced in sufficient quantities:

a. produces an inactive enzyme for the oxidation of NADH

b. produces an inactive enzyme for the oxidation of $CoQH_2$

c. produces an inactive enzyme for the oxidation of cyt *c*

d. produces an inactive enzyme for the oxidation of ADP

23.52 Why might an oral supplement consisting of the following reduce the symptoms associated with mitochondrial myopathy?

a. NADH and CoQ

b. cyt *c* and ADP

CONCEPT MAP

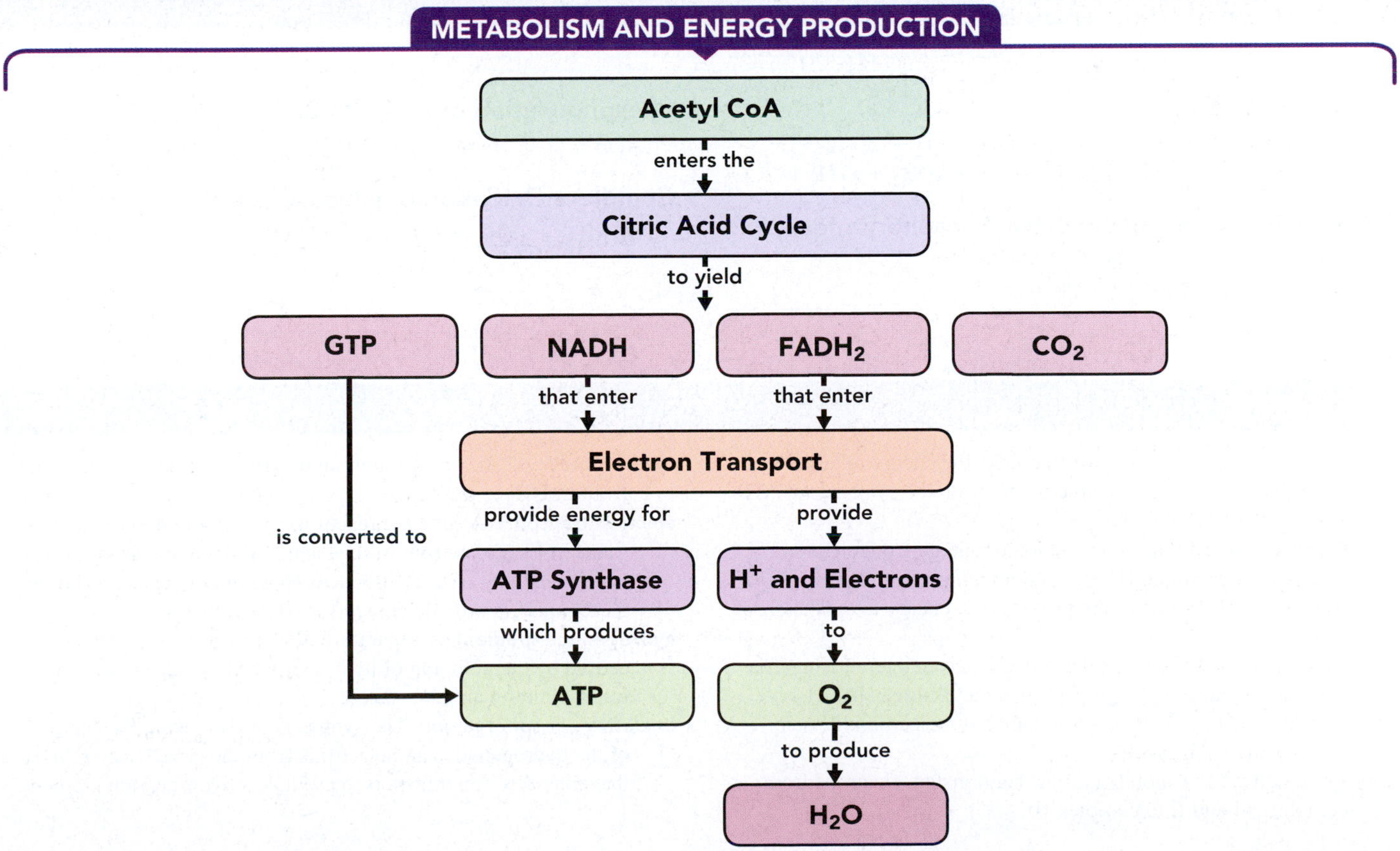
METABOLISM AND ENERGY PRODUCTION
Acetyl CoA
enters the
Citric Acid Cycle
to yield
GTP
NADH
FADH₂
CO₂
that enter
that enter
Electron Transport
is converted to
provide energy for
provide
ATP Synthase
H⁺ and Electrons
which produces
to
ATP
O₂
to produce
H₂O

CHAPTER REVIEW

23.1 The Citric Acid Cycle

LEARNING GOAL Describe the oxidation of acetyl CoA in the citric acid cycle.

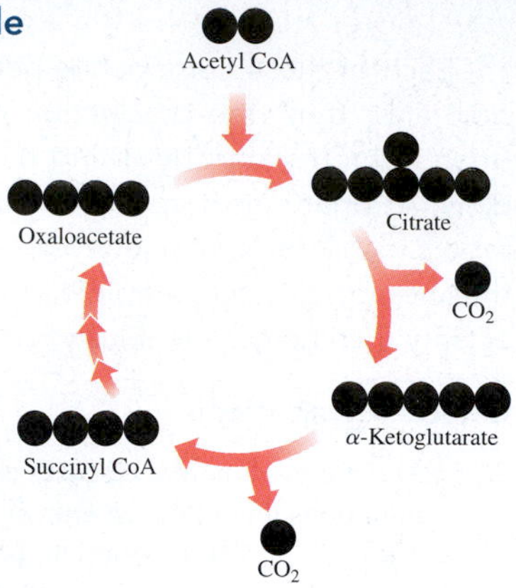

Acetyl CoA
Citrate
CO_2
α-Ketoglutarate
CO_2
Succinyl CoA
Oxaloacetate

- In a sequence of reactions called the citric acid cycle, an acetyl group is combined with oxaloacetate to yield citrate.
- Citrate undergoes oxidation and decarboxylation to yield two CO_2, one GTP, three NADH, and one $FADH_2$, and to regenerate oxaloacetate.
- The direct transfer of a phosphate group to ADP by GTP yields ATP.

23.2 Electron Transport and ATP

LEARNING GOAL Describe the transfer of hydrogen ions and electrons in electron transport and the process of oxidative phosphorylation in ATP synthesis.

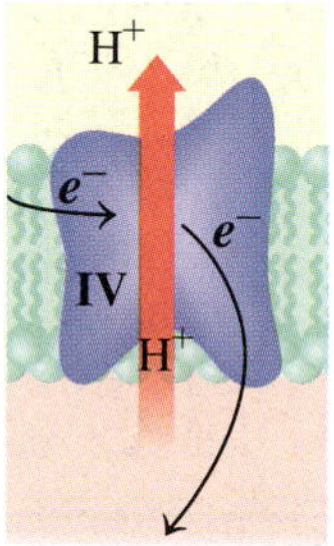

H^+
e^-
IV
e^-
H^+

- The reduced coenzymes NADH and $FADH_2$ from various metabolic pathways are oxidized to NAD^+ and FAD when their electrons are transferred through the electron transport system.
- Complexes I, III, and IV in electron transport act as hydrogen ion pumps producing an H^+ ion gradient across the inner mitochondrial membrane.
- As the hydrogen ions return to the matrix by way of ATP synthase, energy is released.
- This energy is used to drive the synthesis of ATP in a process known as oxidative phosphorylation.
- The availability of ADP, P_i, oxygen, and NADH controls the activity of electron transport.

23.3 ATP Energy from Glucose

LEARNING GOAL Account for the ATP produced by the complete oxidation of glucose.

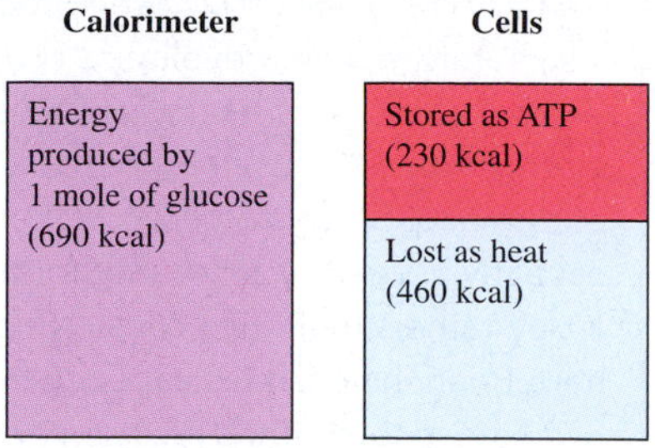

Calorimeter
Cells
Energy produced by 1 mole of glucose (690 kcal)
Stored as ATP (230 kcal)
Lost as heat (460 kcal)

- The oxidation of NADH yields 2.5 ATP, and $FADH_2$ yields 1.5 ATP.
- The energy from the NADH produced in the cytosol is transferred into the matrix via a shuttle. The malate–aspartate shuttle is used in the heart and liver.
- Under aerobic conditions, the complete oxidation of glucose yields a maximum of 32 ATP from direct phosphate transfer and from the oxidation of the reduced coenzymes NADH and $FADH_2$ by electron transport and oxidative phosphorylation.

SUMMARY OF REACTIONS

The chapter Sections to review are shown after the name of each reaction.

Citric Acid Cycle (23.1)

$$\text{Acetyl CoA} + 3NAD^+ + FAD + GDP + P_i + 2H_2O \longrightarrow HS{-}CoA + 3NADH + 3H^+ + FADH_2 + GTP + 2CO_2$$

Electron Transport and Oxidative Phosphorylation (23.2)

$$NADH + H^+ \longrightarrow NAD^+ + 2.5\ ATP$$

$$FADH_2 \longrightarrow FAD + 1.5\ ATP$$

Phosphorylation of ADP (23.2)

$$ADP + P_i + \text{energy} \longrightarrow ATP$$

Complete Oxidation of Glucose (23.3)

$$\underset{\text{Glucose}}{C_6H_{12}O_6} + 6O_2 \longrightarrow 6CO_2 + 6H_2O$$

KEY TERMS

ATP synthase A protein complex that links the energy released by hydrogen ions returning to the matrix with the synthesis of ATP from ADP and P_i.

chemiosmotic model The conservation of energy from electron transport by pumping H^+ ions into the intermembrane space to produce an H^+ gradient that provides the energy to synthesize ATP.

citric acid cycle A series of oxidation reactions in the mitochondria that converts acetyl CoA to CO_2 and yields three NADH, one $FADH_2$, and one GTP. It is also called the tricarboxylic acid cycle or the Krebs cycle.

coenzyme Q (CoQ) A mobile carrier that transfers electrons from complexes I and II to complex III.

cytochrome (cyt) An iron-containing protein that transfers electrons from $CoQH_2$ to oxygen.

electron transport A series of reactions in the mitochondria that transfers electrons from NADH and $FADH_2$ to electron carriers, and finally to O_2, which produces H_2O. Energy changes during these transfers provide energy for ATP synthesis.

hydrogen ion gradient A separation of charge across a membrane caused by the pumping of hydrogen ions from the matrix and into the intermembrane space.

oxidative phosphorylation The synthesis of ATP using the energy of the hydrogen ions and electrons from the NADH and $FADH_2$ that enter electron transport to produce a hydrogen ion gradient.

CORE CHEMISTRY SKILLS

The chapter Section containing each Core Chemistry Skill is shown in parentheses at the end of each heading.

Describing the Reactions in the Citric Acid Cycle (23.1)

- In the initial reaction of the citric acid cycle, an acetyl group combines with oxaloacetate to yield citrate.
- Citrate undergoes several reactions including oxidation, decarboxylation, and hydration to yield two CO_2, one GTP, three NADH, and one $FADH_2$, and also to regenerate oxaloacetate.

Example: Identify the compound(s) in the citric acid cycle that undergo each of the following changes:

a. loses a CO_2 molecule **b.** adds water

Answer: **a.** Isocitrate (6C) undergoes oxidation and decarboxylation to form α-ketoglutarate (5C), and α-ketoglutarate (5C) undergoes oxidation and decarboxylation to form succinyl CoA (4C).

b. Fumarate, which has a double bond, adds an H_2O molecule to form malate.

Calculating the ATP Produced from Glucose (23.3)

- The reduced coenzymes NADH and $FADH_2$ from various metabolic pathways are oxidized to NAD^+ and FAD when their H^+ ions and electrons are transferred to the electron transport system.
- The energy released is used to synthesize ATP from ADP and P_i.
- The final acceptor, O_2, combines with H^+ ions and electrons to yield H_2O.
- The protein complexes I, III, and IV in electron transport move hydrogen ions into the intermembrane space, which produces the H^+ gradient.
- As the H^+ ions return to the matrix by way of ATP synthase, ATP is produced in a process known as oxidative phosphorylation.
- The oxidation of one NADH yields 2.5 ATP, and the oxidation of one $FADH_2$ yields 1.5 ATP.
- Under aerobic conditions, the complete oxidation of glucose yields a maximum of 32 ATP.

Example: Calculate the ATP produced from each of the following:

a. one glucose to two pyruvate
b. one NADH to one NAD^+
c. complete oxidation of one glucose to 6 CO_2 and 6 H_2O

Answer: **a.** Glucose is converted to two pyruvate during glycolysis, which produces 2 ATP and 2 NADH for a total of 7 ATP.

b. One NADH is converted to one NAD^+ in the electron transport system, which produces 2.5 ATP.

c. Glucose is completely oxidized to 6 CO_2 and 6 H_2O in glycolysis, citric acid cycle, and electron transport, which produces a maximum of 32 ATP.

UNDERSTANDING THE CONCEPTS

The chapter Sections to review are given in parentheses at the end of each problem.

23.53 Identify each of the following as a substance that is part of the citric acid cycle, electron transport, or both: (23.1, 23.2)
a. succinate **b.** $CoQH_2$ **c.** FAD
d. cyt *c* (Fe^{2+}) **e.** citrate

23.54 Identify each of the following as a substance that is part of the citric acid cycle, electron transport, or both: (23.1, 23.2)
a. succinyl CoA **b.** acetyl CoA **c.** malate
d. NAD^+ **e.** α-ketoglutarate

23.55 Complete the names of the missing compounds in the citric acid cycle: (23.1)
a. citrate ⟶ __________
b. succinyl CoA ⟶ __________
c. malate ⟶ __________

23.56 Complete the names of the missing compounds in the citric acid cycle: (23.1)
a. oxaloacetate ⟶ __________
b. fumarate ⟶ __________
c. isocitrate ⟶ __________

23.57 Identify the reactant and product for each of the following enzymes in the citric acid cycle: (23.1)
a. aconitase **b.** succinate dehydrogenase
c. fumarase

23.58 Identify the reactant and product for each of the following enzymes in the citric acid cycle: (23.1)
a. isocitrate dehydrogenase **b.** succinyl CoA synthetase
c. malate dehydrogenase

23.59 For each of the given enzymes (**a** to **c**), indicate which of the following are needed: NAD^+, H_2O, FAD, GDP. (23.1)
a. aconitase
b. succinate dehydrogenase
c. isocitrate dehydrogenase

23.60 For each of the given enzymes (**a** to **c**), indicate which of the following are needed: NAD^+, H_2O, FAD, GDP. (23.1)
a. fumarase
b. succinyl CoA synthetase
c. malate dehydrogenase

23.61 Identify the type(s) of reaction(s)—(1) oxidation, (2) decarboxylation, (3) hydrolysis, (4) hydration—catalyzed by each of the following enzymes (**a** to **c**): (23.1)
a. aconitase
b. succinate dehydrogenase
c. isocitrate dehydrogenase

23.62 Identify the type(s) of reaction(s)—(1) oxidation, (2) decarboxylation, (3) hydrolysis, (4) hydration—catalyzed by each of the following enzymes (**a** to **c**): (23.1)
a. fumarase
b. α-ketoglutarate dehydrogenase
c. malate dehydrogenase

23.63 What is the maximum number of ATP produced by energy released when electrons flow from NADH to oxygen (O_2)? (23.3)

23.64 What is the maximum number of ATP produced by energy released when electrons flow from $FADH_2$ to oxygen (O_2)? (23.3)

ADDITIONAL PRACTICE PROBLEMS

23.65 What is the main function of the citric acid cycle in energy production? (23.1)

23.66 Most metabolic pathways are not considered cycles. Why is the citric acid cycle considered a metabolic cycle? (23.1)

23.67 If there are no reactions in the citric acid cycle that use oxygen, O_2, why does the cycle operate only in aerobic conditions? (23.1)

23.68 What products of the citric acid cycle are needed for electron transport? (23.1)

23.69 Identify the compounds in the citric acid cycle that have the following: (23.1)

a. six carbon atoms **b.** five carbon atoms
c. a keto group

23.70 Identify the compounds in the citric acid cycle that have the following: (23.1)

a. four carbon atoms **b.** a hydroxyl group
c. a carbon–carbon double bond

23.71 In which reaction of the citric acid cycle does each of the following occur? (23.1)

a. A five-carbon keto acid is decarboxylated.
b. A carbon–carbon double bond is hydrated.
c. NAD^+ is reduced.
d. A secondary hydroxyl group is oxidized.

23.72 In which reaction of the citric acid cycle does each of the following occur? (23.1)

a. FAD is reduced.
b. A six-carbon keto acid is decarboxylated.
c. A carbon–carbon double bond is formed.
d. GDP undergoes direct phosphate transfer.

23.73 Indicate the coenzyme(s) for each of the following reactions: (23.1)

a. isocitrate $\longrightarrow$ α-ketoglutarate
b. α-ketoglutarate $\longrightarrow$ succinyl CoA

23.74 Indicate the coenzyme(s) for each of the following reactions: (23.1)

a. succinate $\longrightarrow$ fumarate **b.** malate $\longrightarrow$ oxaloacetate

23.75 How does each of the following regulate the citric acid cycle? (23.1)

a. high levels of NADH **b.** high levels of ATP

23.76 How does each of the following regulate the citric acid cycle? (23.1)

a. high levels of ADP **b.** low levels of NADH

23.77 At which complexes in the electron transport system are hydrogen ions pumped into the intermembrane space? (23.2)

23.78 What is the effect of hydrogen ion accumulation in the intermembrane space? (23.2)

23.79 Which complex in electron transport is inhibited by each of the following? (23.2)

a. Amytal and rotenone **b.** antimycin A
c. cyanide and carbon monoxide

23.80 a. When an inhibitor blocks electron transport, how are the coenzymes that precede the blocked site affected? (23.2)
b. When an inhibitor blocks electron transport, how are the coenzymes that follow the blocked site affected? (23.2)

23.81 In the chemiosmotic model, how is energy provided to synthesize ATP? (23.2)

23.82 Where does the synthesis of ATP take place in electron transport? (23.2)

23.83 Why do hydrogen ions tend to leave the intermembrane space and return to the matrix within a mitochondrion? (23.2)

23.84 Why do the enzyme complexes that pump hydrogen ions extend across the mitochondrial membrane from the matrix to the intermembrane space? (23.2)

23.85 How is the energy from the hydrogen ion gradient utilized by ATP synthase? (23.2)

23.86 In electron transport, would the solution in the space between the outer and inner mitochondrial membrane be more or less acidic than the solution in the matrix? (23.2)

23.87 Where is ATP synthase for oxidative phosphorylation located in the cell? (23.2)

23.88 Why would an animal that lives in a warm climate have less brown fat than a similar animal that lives in a cold climate? (23.2)

23.89 Why would a bear that is hibernating have more brown fat than one that is active? (23.2)

23.90 How does the NADH generated in the cytosol during glycolysis get moved into the matrix? (23.3)

23.91 How many ATP are produced when glucose is oxidized to pyruvate compared to when glucose is oxidized to CO_2 and H_2O? (23.3)

23.92 Considering the efficiency of ATP synthesis, how many kilocalories of energy would be conserved from the complete oxidation of 4.0 moles of glucose? (23.3)

CHALLENGE PROBLEMS

The following problems are related to the topics in this chapter. However, they do not all follow the chapter order, and they require you to combine concepts and skills from several Sections. These problems will help you increase your critical thinking skills and prepare for your next exam.

23.93 Using the value 7.3 kcal/mole for ATP, how many kilocalories can be produced from the ATP provided by the reaction of 1 mole of glucose in each of the following? (23.1, 23.3)

a. glycolysis **b.** oxidation of pyruvate to acetyl CoA
c. citric acid cycle **d.** complete oxidation to CO_2 and H_2O

23.94 In a calorimeter, the combustion of 1 mole of glucose produces 690 kcal. What percentage of ATP energy is produced from 1 mole of glucose by each of the reactions in problem 23.93 **a** to **d**? (23.1, 23.3)

23.95 What does it mean to say that the cell is 33% efficient in storing the energy from the complete combustion of glucose? (23.3)

23.96 If acetyl CoA has a molar mass of 809 g/mole, how many moles of ATP are produced when 1.0 μg of acetyl CoA completes the citric acid cycle? (23.3)

Clinical Applications

23.97 A person's basal metabolic rate (BMR) gives the number of kilocalories utilized in one day to maintain weight. To lose weight, caloric intake needs to be lower or activity level can be increased. The resting BMR can be calculated in kilocalories by the following formula: (23.3)

Women: BMR kcal $= 655 + (4.35 \times \text{weight lb}) + (4.70 \times \text{height in.}) - (4.70 \times \text{age in years})$

a. If a woman weighs 115 lb, is 5 ft 4 in. tall, and is 21 years old, what is her basal metabolic rate, rounded to the nearest 10 kilocalories?

b. If the molar mass of ATP is 507 g/mole, how many kilograms of ATP are used in one day?

23.98 A person's basal metabolic rate (BMR) gives the number of kilocalories utilized in one day to maintain weight. To lose weight, caloric intake needs to be lower or activity level can be increased. The resting BMR can be calculated in kilocalories by the following formula: (23.3)

Men: BMR kcal $= 66 + (6.23 \times \text{weight lb}) + (12.7 \times \text{height in.}) - (6.80 \times \text{age in years})$

a. If a man weighs 184 lb, is 6 ft 1 in. tall, and is 32 years old, what is his basal metabolic rate, rounded to the nearest 10 kilocalories?

b. If the molar mass of ATP is 507 g/mole, how many kilograms of ATP are used in one day?

ANSWERS

23.1 Krebs cycle and tricarboxylic acid cycle

23.3 $2CO_2$, $3NADH + 3H^+$, $FADH_2$, GTP (ATP), and HS—CoA

23.5 **a.** Two reactions, reactions 3 and 4, involve oxidation and decarboxylation.
b. Reaction 6 involves dehydration.
c. Reactions 3, 4, and 8 involve reduction of NAD^+.

23.7 **a.** citrate and isocitrate
b. A carbon atom is lost as CO_2 in decarboxylation.
c. α-ketoglutarate
d. reactions 3 and 8

23.9 **a.** citrate synthase
b. succinate dehydrogenase and aconitase
c. fumarase

23.11 **a.** NAD^+ **b.** FAD

23.13 Isocitrate dehydrogenase and α-ketoglutarate dehydrogenase are allosteric enzymes.

23.15 High levels of ADP increase the rate of the citric acid cycle.

23.17 The citric acid cycle plays a central role in aerobic metabolism, and it is likely that defects or deficiencies in citric acid cycle enzymes would be fatal to a developing fetus.

23.19 malate dehydrogenase

23.21 oxidized

23.23 **a.** oxidation **b.** reduction

23.25 NADH

23.27 $FADH_2$, CoQ, cytochrome *c* (Fe^{3+})

23.29 The mobile carrier CoQ transfers electrons from complex I to complex III.

23.31 NADH transfers electrons to complex I to give NAD^+.

23.33 **a.** $NADH + H^+ + CoQ \longrightarrow NAD^+ + CoQH_2$
b. $CoQH_2 + 2\text{cyt } c\,(Fe^{3+}) \longrightarrow CoQ + 2\text{cyt } c\,(Fe^{2+}) + 2H^+$

23.35 In oxidative phosphorylation, the energy from the oxidation reactions in electron transport is used to synthesize ATP from ADP and P_i.

23.37 As hydrogen ions return to the lower energy environment in the matrix, they pass through ATP synthase, releasing energy to drive the synthesis of ATP.

23.39 Glycolysis and the citric acid cycle produce the reduced coenzymes NADH and $FADH_2$, which enter electron transport and are oxidized to provide energy for the synthesis of ATP.

23.41 If CN^- binds to the heme of a cytochrome, electrons cannot reduce Fe^{3+} to Fe^{2+} and will not be transported, stopping electron transport.

23.43 Because dinitrophenol induces thermogenesis and inhibits ATP synthase, it raises body temperature to unsafe levels and can cause constant sweating.

23.45 Two. In glycolysis, one glucose molecule forms two pyruvate, which oxidize to give 2 acetyl CoA.

23.47 **a.** 2.5 ATP **b.** 7 ATP **c.** 5 ATP

23.49 3.2 moles of ATP

23.51 **a.** accumulates: NADH; not produced in sufficient quantity: NAD^+
b. accumulates: $CoQH_2$; not produced in sufficient quantity: CoQ
c. accumulates: electrons, H^+, and O_2; not produced in sufficient quantity: H_2O
d. accumulates: ADP, P_i, and H^+; not produced in sufficient quantity: ATP

23.53 **a.** citric acid cycle **b.** electron transport
c. both **d.** electron transport
e. citric acid cycle

23.55 **a.** isocitrate **b.** succinate
c. oxaloacetate

23.57 **a.** citrate, isocitrate **b.** succinate, fumarate
c. fumarate, malate

23.59 **a.** Aconitase uses H_2O.
b. Succinate dehydrogenase uses FAD.
c. Isocitrate dehydrogenase uses NAD^+.

23.61 **a.** (4) hydration reaction **b.** (1) oxidation reaction
c. (1) oxidation and (2) decarboxylation reaction

23.63 2.5 ATP

23.65 The oxidation reactions of the citric acid cycle produce a source of reduced coenzymes for electron transport and ATP synthesis.

23.67 The oxidized coenzymes NAD^+ and FAD needed for the citric acid cycle are regenerated by electron transport, which requires oxygen.

23.69 **a.** citrate, isocitrate **b.** α-ketoglutarate
c. α-ketoglutarate, succinyl CoA, oxaloacetate

23.71 **a.** In reaction 4, α-ketoglutarate, a five-carbon keto acid, is decarboxylated.
b. In reactions 2 and 7, double bonds in aconitate and fumarate are hydrated.
c. NAD^+ is reduced in reactions 3, 4, and 8.
d. In reactions 3 and 8, a secondary hydroxyl group in isocitrate and malate is oxidized.

23.73 **a.** NAD^+ **b.** NAD^+ and CoA

23.75 **a.** High levels of NADH inhibit isocitrate dehydrogenase and α-ketoglutarate dehydrogenase to slow the rate of the citric acid cycle.
b. High levels of ATP inhibit isocitrate dehydrogenase to slow the rate of the citric acid cycle.

23.77 complexes I, III, and IV

23.79 **a.** electron flow from complex I to CoQ
b. electron flow from complex III to cyt *c*
c. electron flow from cyt *c* to complex IV

23.81 Energy is released as hydrogen ions flow through ATP synthase back to the matrix and is utilized for the synthesis of ATP.

23.83 Hydrogen ions flow through ATP synthase into the matrix where the H^+ concentration is lower.

23.85 As hydrogen ions from the hydrogen ion gradient move through the ATP synthase to return to the matrix, energy is released and used to synthesize ATP by ATP synthase.

23.87 ATP synthase is a protein complex which spans the inner mitochondrial membrane.

23.89 A hibernating bear has more brown fat because it can be used during the winter for heat rather than ATP energy.

23.91 The oxidation of glucose to pyruvate produces 7 ATP, whereas the oxidation of glucose to CO_2 and H_2O produces a maximum of 32 ATP.

23.93 **a.** 51 kcal **b.** 37 kcal
c. 150 kcal **d.** 230 kcal

23.95 If the combustion of glucose produces 690 kcal, but only 230 kcal (from 32 ATP) in cells, the efficiency of glucose use in the cells is 230 kcal/690 kcal or 33%.

23.97 **a.** 1360 kcal **b.** 94.5 kg

24

Metabolic Pathways for Lipids and Amino Acids

Luke is 48 years old and works as a paramedic. Recently, blood work from his annual physical examination indicated a plasma cholesterol level of 256 mg/dL. Clinically, cholesterol levels are considered elevated if the total plasma cholesterol level exceeds 200 mg/dL. Luke's doctor orders a liver profile that shows elevated liver enzymes: alanine transaminase (ALT) 282 Units/L (normal ALT 5 to 35 Units/L) and aspartate transaminase (AST) 226 Units/L (normal AST 5 to 50 Units/L). Luke's doctor orders a medication to lower Luke's plasma cholesterol.

During Luke's career as a paramedic, he was exposed several times to blood and was accidently stuck by a needle containing infected blood. Luke takes 8 to 10 ibuprofen tablets per month for pain and uses herbs, garlic, ginkgo, and antioxidants. Although herbs, antioxidants, and ibuprofen can cause liver inflammation, they would not usually cause the elevation of liver enzymes reported in Luke's blood tests. A hepatitis profile shows that Luke is positive for antibodies to both hepatitis B and C, so his doctor diagnoses Luke with chronic hepatitis C virus (HCV) infection. Hepatitis C is an infection caused by a virus that attacks the liver and leads to inflammation. Most people infected with the hepatitis C virus have no symptoms. Hepatitis C is usually passed by contact with contaminated blood or by needles shared during illegal drug use. As part of his treatment, Luke attends a class on living with hepatitis C given by Belinda, a public health nurse.

CAREER

Public Health Nurse (PHN)

Hepatitis C virus is a common cause of liver disease and a major health problem worldwide. Patients with HCV require lifelong monitoring and are usually cared for by specialist teams that include a public health nurse. A public health nurse works in public health departments, correctional facilities, occupational health facilities, schools, and organizations that aim to improve health at the community level. They often focus on high-risk populations such as the elderly, the homeless, teen mothers, and those at risk for a communicable disease such as hepatitis.

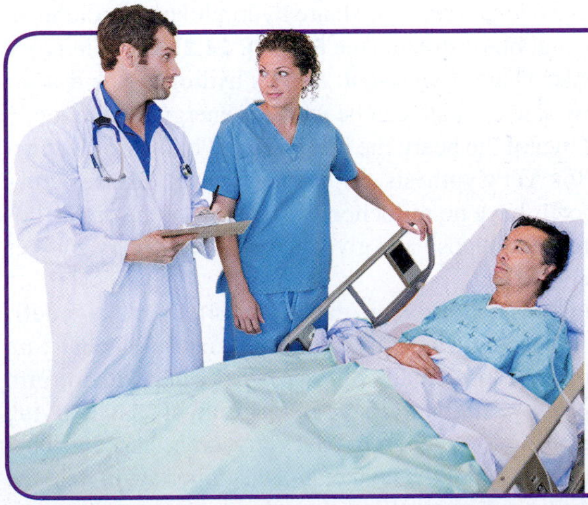

CLINICAL UPDATE

Treatment of Luke's Hepatitis C

When the levels of Luke's liver enzymes remain elevated, his doctor prescribes a therapy of antiviral agents that inhibit the replication of the hepatitis C virus. Read more about this treatment and how it affects the levels of Luke's liver enzymes in the **CLINICAL UPDATE** **Treatment of Luke's Hepatitis C**, page 855.

LOOKING AHEAD

ENGAGE

In which form is most of the fat in our bodies stored?

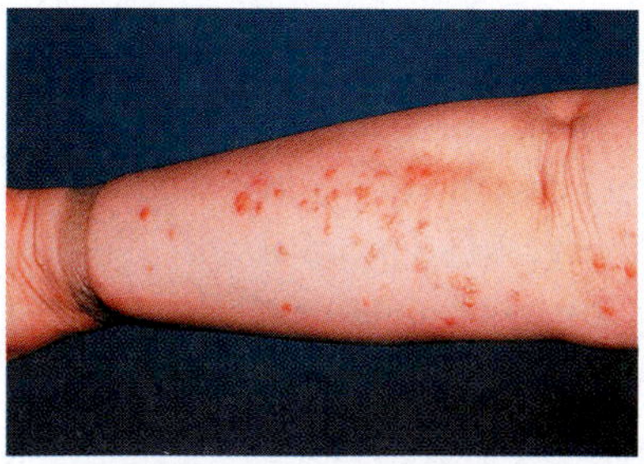

Eruptive xanthomas are often associated with elevated triacylglycerol levels in the bloodstream.

24.1 Digestion of Triacylglycerols

LEARNING GOAL Describe the process by which triacylglycerols are digested.

Our discussion in the last two chapters focused on the metabolic pathways of carbohydrates. However, lipids and proteins also play important roles in metabolism. In this chapter, our main focus is the processes by which lipids and proteins are digested to provide energy for ATP production.

Almost all of the stored energy in our bodies is in the form of triacylglycerols, which are stored in fat cells called *adipocytes* (see **FIGURE 24.1**). Let's compare the amount of energy stored in the fat cells to the energy from glucose, glycogen, and protein. A typical 70-kg (150-lb) person has about 140 000 kcal of energy stored as fat, 24 000 kcal as protein, 720 kcal as glycogen reserves, and 80 kcal as blood glucose. Therefore, the energy available from stored fats is about 85% of the total energy available in the body.

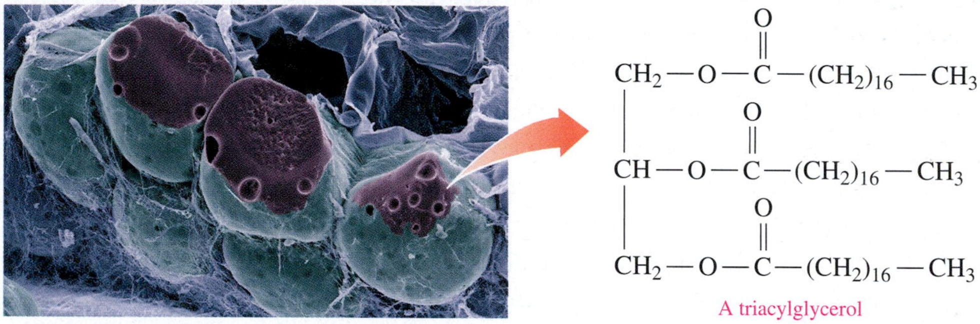

A triacylglycerol

FIGURE 24.1 ▶ The fat cells (adipocytes) that make up adipose tissue are capable of storing unlimited quantities of triacylglycerols.

Q What are some sources of fats in our diet?

Digestion of Dietary Fats

The initial digestion of dietary triacylglycerols (fats) starts in the small intestine and is catalyzed by enzymes called *lipases*. While fats are hydrophobic, lipases are soluble in hydrophilic aqueous environments and, thus, can only catalyze reactions at the surface of fat globules. Therefore, the first step in digestion is to increase the surface area of the fat. In a process called *emulsification*, molecules called *bile salts* break the fat globules into smaller droplets called *micelles*. Bile salts are secreted from the gallbladder into the small intestine. Because they contain hydrophobic and hydrophilic regions, they are able to interact with both the fat globules and the aqueous environment in which lipases are soluble.

Next, *pancreatic lipases* hydrolyze the triacylglycerols in the micelles to yield monoacylglycerols and fatty acids, both of which are small enough to be transported into the cells of the intestinal lining. Once there, the monoacylglycerols and fatty acids recombine to form triacylglycerols. These hydrophobic molecules are then coated with phospholipids and proteins to form lipoproteins called *chylomicrons*, which are hydrophilic and soluble in the aqueous environment of the lymph and bloodstream (see **FIGURE 24.2**). Chylomicrons transport the triacylglycerols to the cells. There, *lipoprotein lipases* hydrolyze the triacylglycerols to yield glycerol and free fatty acids, which can be used for energy production.

While fatty acids are the preferred fuel of the heart, the brain and red blood cells cannot oxidize fatty acids to provide energy for ATP synthesis. Fatty acids cannot diffuse across the blood–brain barrier, and red blood cells have no mitochondria, which are the organelles in which fatty acids are oxidized. Therefore, glucose and glycogen are the primary sources of energy for the brain and red blood cells.

Digestion of triacylglycerols depends strongly on the action of lipases. A deficiency in lipoprotein lipases results in elevated triacylglycerol levels in the bloodstream (*hypertriglyceridemia*), which can, in turn, cause small, yellowish, fatty deposits to form under the skin. These deposits, called *eruptive xanthomas*, are often a sign of an underlying abnormality in lipid metabolism. Because elevated levels of triacylglycerols can contribute to heart disease or *pancreatitis* (an inflammation of the pancreas), it is important to treat the underlying cause of the xanthomas. Most xanthomas clear up on their own once the cause of the high triacylglycerol levels is treated.

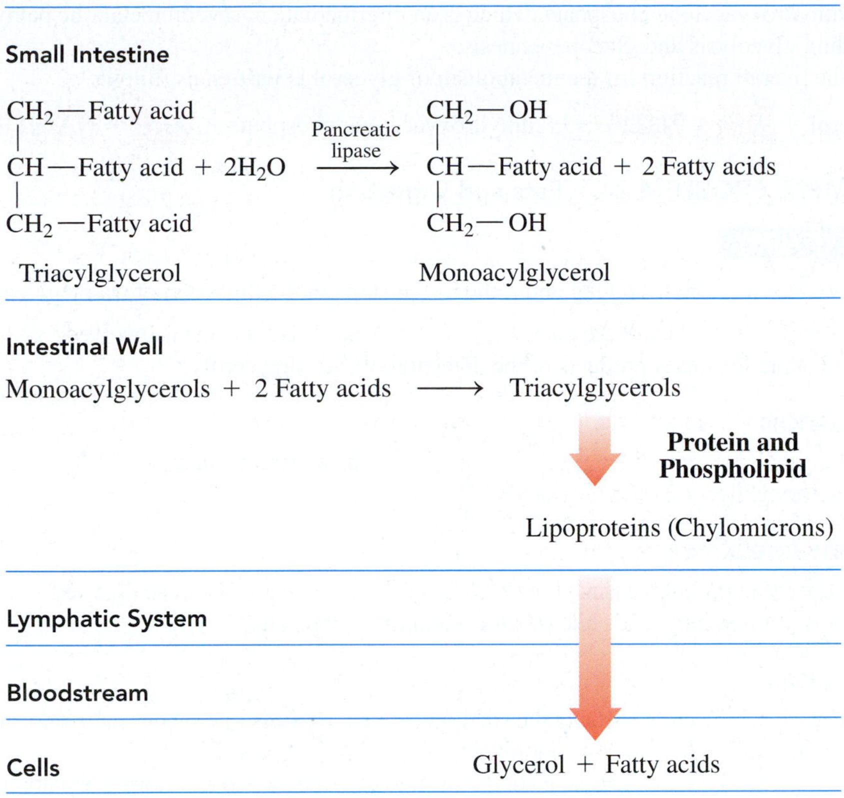
Small Intestine
CH2—Fatty acid
CH—Fatty acid + 2H2O
CH2—Fatty acid
Pancreatic lipase
CH2—OH
CH—Fatty acid + 2 Fatty acids
CH2—OH
Triacylglycerol
Monoacylglycerol
Intestinal Wall
Monoacylglycerols + 2 Fatty acids ⟶ Triacylglycerols
Protein and Phospholipid
Lipoproteins (Chylomicrons)
Lymphatic System
Bloodstream
Cells
Glycerol + Fatty acids

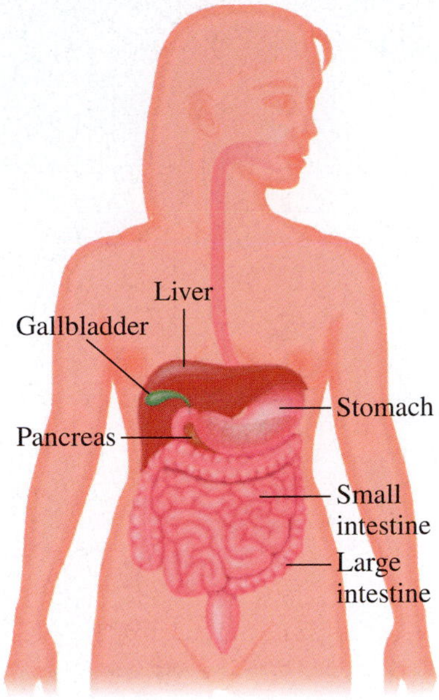
Liver
Gallbladder
Stomach
Pancreas
Small intestine
Large intestine

FIGURE 24.2 ▶ The digestion of fats begins in the small intestine when bile salts emulsify fats that undergo hydrolysis to monoacylglycerols and fatty acids.

Q What kinds of enzymes are secreted from the pancreas into the small intestine to hydrolyze triacylglycerols?

Utilization of Fat Stores

Our bodies use not only the fats we eat, but also the fats that are stored in our adipose tissue to provide energy for ATP synthesis. When blood glucose is depleted and glycogen stores are low, the process of stored fat utilization is stimulated as the hormones glucagon or epinephrine are secreted into the bloodstream, where they bind to receptors on the membrane of adipose tissue. A hormone-sensitive lipase within the fat cells catalyzes the hydrolysis of triacylglycerols to glycerol and free fatty acids, which then diffuse into the bloodstream and bind with plasma proteins (albumin) to be transported to the tissues, muscles, and fat cells.

TEST

Try Practice Problems 24.1 and 24.2

Metabolism of Glycerol

Most of the glycerol released when either dietary or stored triacylglycerols undergo hydrolysis is metabolized in the liver, where enzymes convert it to dihydroxyacetone phosphate in two steps. In the first step, ATP is used to phosphorylate glycerol, yielding glycerol-3-phosphate. In the second step, the secondary hydroxyl group is oxidized to

ENGAGE

In which organ is glycerol metabolized?

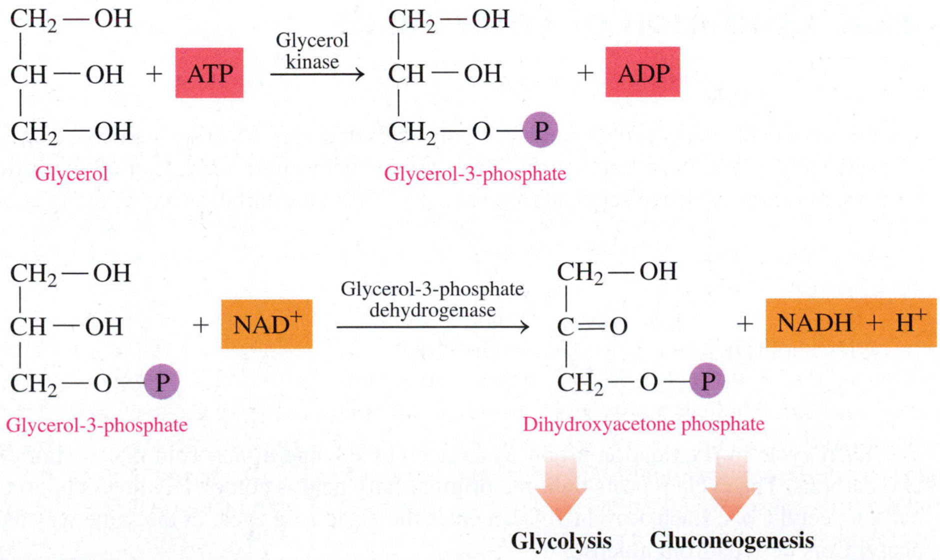
ATP
Glycerol kinase
ADP
Glycerol
Glycerol-3-phosphate
NAD+
Glycerol-3-phosphate dehydrogenase
NADH + H+
Glycerol-3-phosphate
Dihydroxyacetone phosphate
Glycolysis
Gluconeogenesis

yield dihydroxyacetone phosphate, which is an intermediate in several metabolic pathways, including glycolysis and gluconeogenesis.

The overall reaction for the metabolism of glycerol is written as follows:

$$\text{Glycerol} + \text{ATP} + \text{NAD}^+ \longrightarrow \text{dihydroxyacetone phosphate} + \text{ADP} + \text{NADH} + \text{H}^+$$

▶SAMPLE PROBLEM 24.1 Fats and Digestion

TRY IT FIRST

Answer each of the following about the initial steps in the digestion of triacylglycerols:

a. Where do these steps take place?
b. What enzyme is involved?
c. What are the initial products of the digestion of triacylglycerols?

SOLUTION

a. the small intestine
b. pancreatic lipase
c. monoacylglycerols and fatty acids

STUDY CHECK 24.1

a. How are hydrophobic triacylglycerols emulsified so that they can be digested?
b. How are free fatty acids released for oxidation in the cells?

ANSWER

a. The triacylglycerols entering the small intestine are emulsified by bile salts into smaller droplets of fat that can react more efficiently with the lipases.
b. The triacylglycerols that have been transported to the cells by chylomicrons are hydrolyzed by lipoprotein lipases to give glycerol and free fatty acids.

TEST

Try Practice Problems 24.3 to 24.8

PRACTICE PROBLEMS

24.1 Digestion of Triacylglycerols

24.1 What is the role of bile salts in lipid digestion?

24.2 How are insoluble triacylglycerols transported to the tissues and fat cells?

24.3 When are fats released from fat stores?

24.4 What happens to the glycerol produced from the hydrolysis of triacylglycerols in adipose tissues?

24.5 How is glycerol converted to an intermediate of glycolysis?

24.6 How can glycerol be used to synthesize glucose?

Clinical Applications

A person with *Duchenne muscular dystrophy* cannot convert glycerol to glycerol-3-phosphate.

24.7 What enzyme would be defective?

24.8 Why would a person with this enzyme deficiency have high blood glycerol levels?

REVIEW

Identifying Fatty Acids (17.2)

24.2 Oxidation of Fatty Acids

LEARNING GOAL Describe the metabolic pathway of β oxidation.

A large amount of energy is obtained when fatty acids undergo oxidation in the mitochondria to yield acetyl CoA. In fat catabolism, fatty acids undergo **beta oxidation (β oxidation),** which removes two-carbon segments, one at a time, from the carboxyl end of the fatty acid.

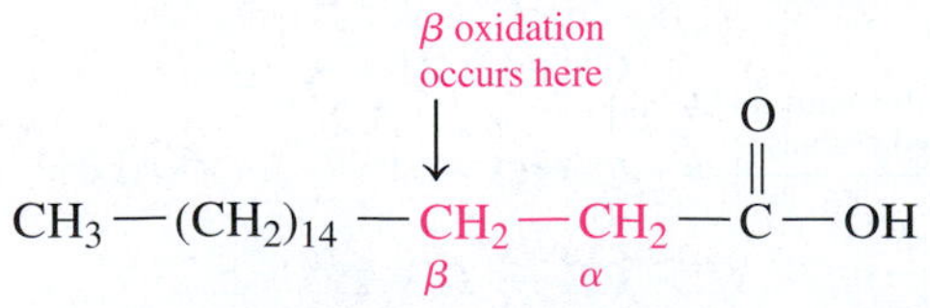
β oxidation occurs here
O
CH3—(CH2)14—CH2—CH2—C—OH
β
α

Stearic acid

Each cycle in β oxidation produces an acetyl CoA and a fatty acid that is shorter by two carbons. The cycle repeats until the original fatty acid is completely degraded to two-carbon acetyl CoA. Each acetyl CoA can enter the citric acid cycle in the same way as the acetyl CoA derived from glucose.

Fatty Acid Activation

Fatty acids, which are produced in the cytosol, must be transported through the inner mitochondrial membrane before they can undergo β oxidation in the matrix. In an *activation* process catalyzed by *acyl CoA synthetase*, a fatty acid is combined with CoA to yield a fatty acyl CoA, which can be transported across the membrane. The energy for the activation is obtained from the hydrolysis of ATP to give AMP and two phosphate (P_i) groups. This is equivalent to the energy released from the hydrolysis of two ATP to two ADP. In our discussion of fatty acid activation, we will use the terms *acyl CoA* and *acetyl CoA*. The term *acyl* refers to a long-chain fatty acid that is bonded to coenzyme A, whereas *acetyl* refers to a two-carbon acetyl group that is bonded to coenzyme A.

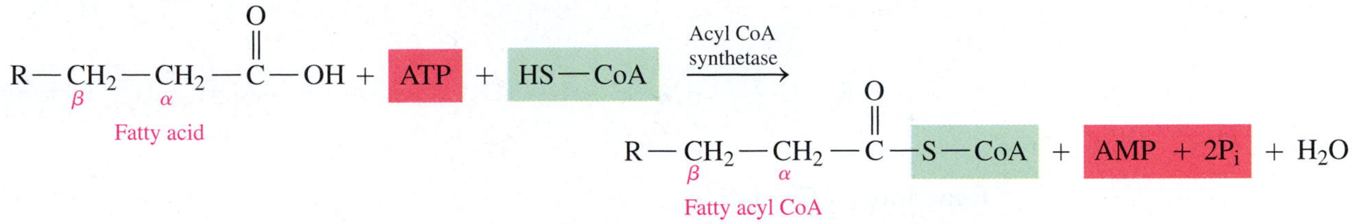
ATP
HS—CoA
Acyl CoA synthetase
Fatty acid
AMP + 2P_i
+ H_2O
Fatty acyl CoA

Transport of Fatty Acyl CoA into the Mitochondria

The acyl CoA molecules formed in the cytosol cannot pass directly through the inner mitochondrial membrane into the matrix where β oxidation of fatty acids takes place. A transport system called the *carnitine shuttle* is used to carry fatty acids into the mitochondria. In the cytosol, *carnitine acyltransferase* catalyzes the transfer of a fatty acyl group to the hydroxyl group of carnitine to produce fatty acyl carnitine, which can pass through the inner mitochondrial membrane into the matrix.

ENGAGE

In which part of the mitochondrion does β oxidation take place?

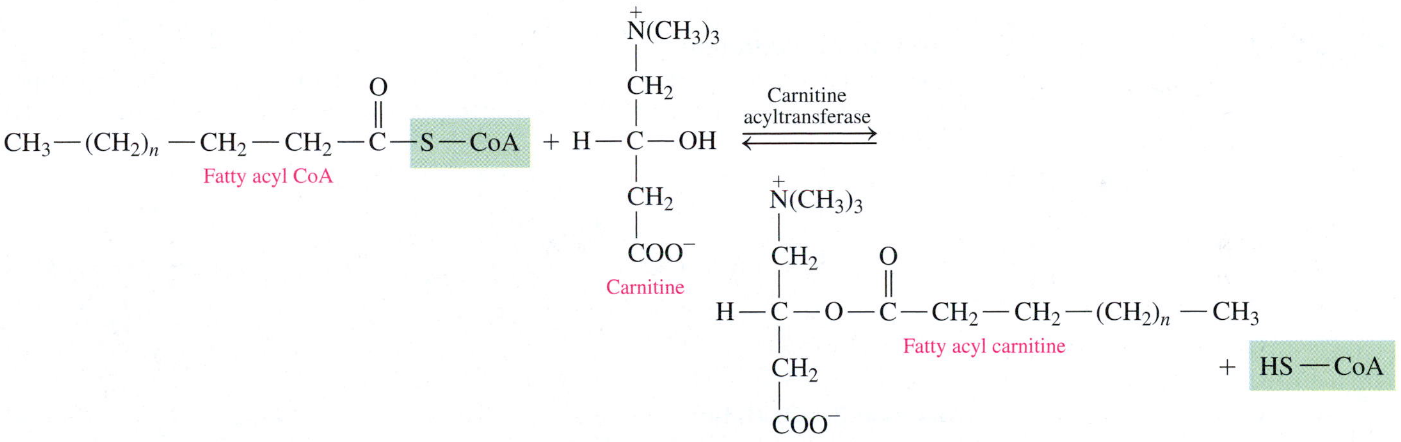
Fatty acyl CoA
Carnitine
Carnitine acyltransferase
Fatty acyl carnitine
+ HS—CoA

In the mitochondrial matrix, another *carnitine acyltransferase* catalyzes the reverse reaction, transferring the fatty acyl group to CoA to reform fatty acyl CoA. The carnitine is released and returns to the cytosol. Thus, the carnitine shuttle moves fatty acyl CoA from the cytosol into the mitochondrial matrix where the fatty acid can undergo β oxidation (see **FIGURE 24.3**).

TEST

Try Practice Problems 24.9 and 24.10

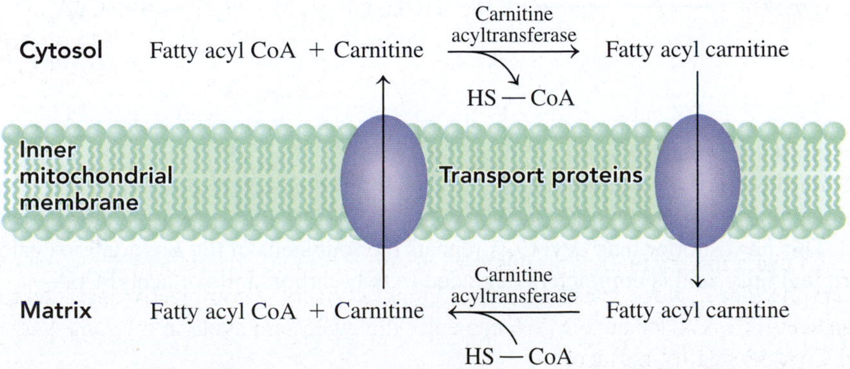
Cytosol
Fatty acyl CoA + Carnitine
Carnitine acyltransferase
Fatty acyl carnitine
HS—CoA
Inner mitochondrial membrane
Transport proteins
Matrix
Fatty acyl CoA + Carnitine
Carnitine acyltransferase
Fatty acyl carnitine
HS—CoA

FIGURE 24.3 ▶ In the carnitine shuttle system, fatty acids are activated and transported from the cytosol through the inner mitochondrial membrane into the matrix.

Q Why is carnitine needed to transport a fatty acid into the matrix?

Reactions of the β Oxidation Cycle

In the mitochondrial matrix, fatty acyl CoA molecules undergo β oxidation, which is a cycle of four reactions that convert the —CH_2— of the β-carbon (carbon 3) in a long-chain fatty acyl group into a keto group. Once the β-keto group is formed, a two-carbon acetyl CoA is split from the original fatty acyl chain, which gives a shortened fatty acyl chain. The reaction for one cycle of β oxidation is written as follows:

$$R—\underset{\beta}{CH}—\underset{\alpha}{CH_2}—\overset{\displaystyle O}{\overset{\|}{C}}—S—CoA + \boxed{NAD^+} + \boxed{FAD} + H_2O + \boxed{HS—CoA} \longrightarrow$$

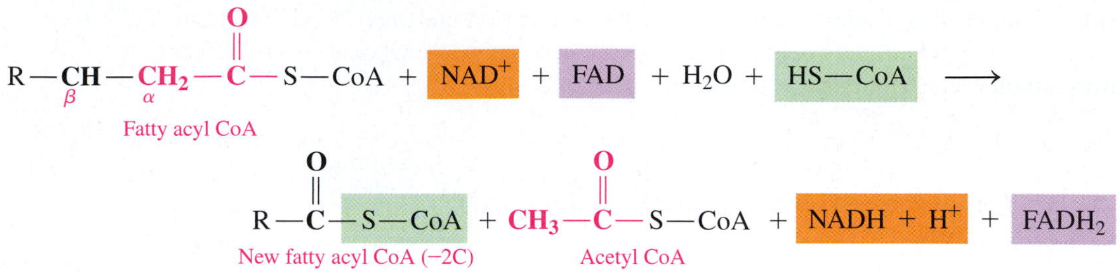

Fatty acyl CoA

$$R—\overset{\displaystyle O}{\overset{\|}{C}}—S—CoA + CH_3—\overset{\displaystyle O}{\overset{\|}{C}}—S—CoA + \boxed{NADH + H^+} + \boxed{FADH_2}$$

New fatty acyl CoA (−2C)   Acetyl CoA

Reaction 1   Oxidation

In the first step of β oxidation, *acyl CoA dehydrogenase* catalyzes the transfer of hydrogen atoms from the α- and β-carbons of the activated fatty acid to FAD, yielding a trans carbon–carbon double bond and $FADH_2$.

$$CH_3—(CH_2)_n—\underset{\beta}{CH_2}—\underset{\alpha}{CH_2}—\overset{\displaystyle O}{\overset{\|}{C}}—S—CoA + \boxed{FAD} \xrightarrow[\text{dehydrogenase}]{\text{Acyl CoA}} CH_3—(CH_2)_n—\underset{\underset{\displaystyle H}{|}}{\overset{\overset{\displaystyle H}{|}}{\underset{\beta}{C}}}=\underset{\alpha}{C}—\overset{\displaystyle O}{\overset{\|}{C}}—S—CoA + \boxed{FADH_2}$$

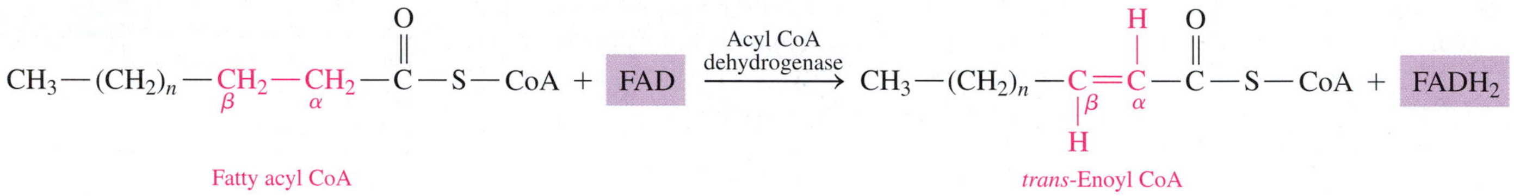

Fatty acyl CoA   *trans*-Enoyl CoA

ENGAGE

To which type of double bond are the components of water added in reaction 2 of β oxidation?

Reaction 2   Hydration

In the second step, a hydration reaction catalyzed by *enoyl CoA hydratase* adds the components of water to the trans double bond, which forms a secondary hydroxyl group (—OH) on the β-carbon.

$$CH_3—(CH_2)_n—\underset{\underset{\displaystyle H}{|}}{\overset{\overset{\displaystyle H}{|}}{\underset{\beta}{C}}}=\underset{\alpha}{C}—\overset{\displaystyle O}{\overset{\|}{C}}—S—CoA + H_2O \xrightarrow[\text{hydratase}]{\text{Enoyl CoA}} CH_3—(CH_2)_n—\underset{\underset{\displaystyle H}{|}}{\overset{\overset{\displaystyle OH}{|}}{\underset{\beta}{C}}}—\underset{\underset{\displaystyle H}{|}}{\overset{\overset{\displaystyle H}{|}}{\underset{\alpha}{C}}}—\overset{\displaystyle O}{\overset{\|}{C}}—S—CoA$$

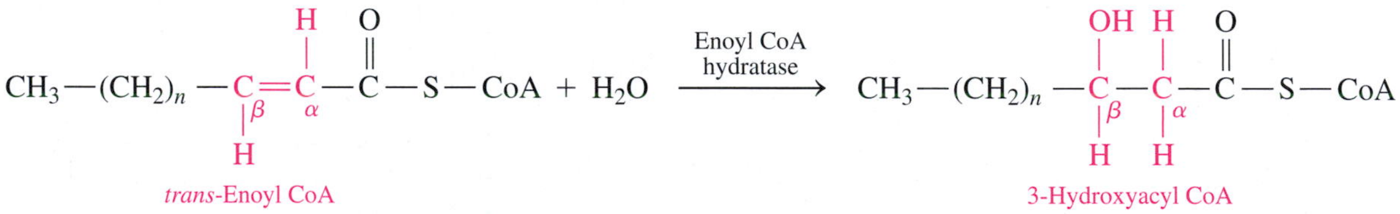

trans-Enoyl CoA   3-Hydroxyacyl CoA

Reaction 3   Oxidation

In the third step, the secondary hydroxyl group on the β-carbon, which is carbon 3, is oxidized by *3-hydroxyacyl CoA dehydrogenase* to yield a ketone. The hydrogen atoms removed in the oxidation are transferred to NAD^+ to yield a β-keto or 3-keto group, the reduced coenzyme NADH, and H^+.

$$CH_3—(CH_2)_n—\underset{\underset{\displaystyle H}{|}}{\overset{\overset{\displaystyle OH}{|}}{\underset{\beta}{C}}}—\underset{\underset{\displaystyle H}{|}}{\overset{\overset{\displaystyle H}{|}}{\underset{\alpha}{C}}}—\overset{\displaystyle O}{\overset{\|}{C}}—S—CoA + \boxed{NAD^+} \xrightarrow[\text{dehydrogenase}]{\text{3-Hydroxyacyl CoA}} CH_3—(CH_2)_n—\underset{\beta}{\overset{\displaystyle O}{\overset{\|}{C}}}—\underset{\underset{\displaystyle H}{|}}{\overset{\overset{\displaystyle H}{|}}{\underset{\alpha}{C}}}—\overset{\displaystyle O}{\overset{\|}{C}}—S—CoA + \boxed{NADH + H^+}$$

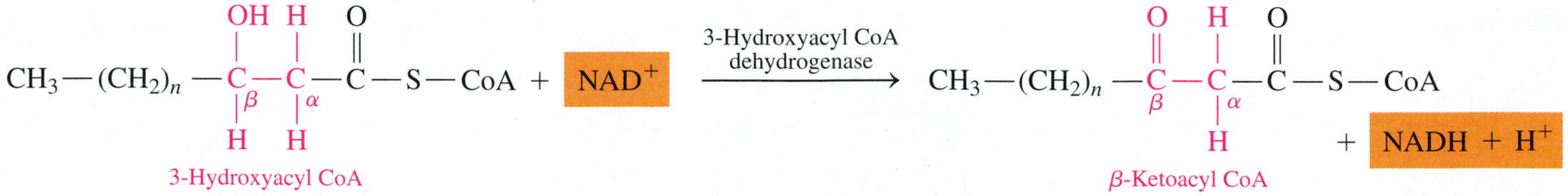

3-Hydroxyacyl CoA   β-Ketoacyl CoA

Reaction 4   Cleavage

In the fourth step of β oxidation, the C_α—C_β bond is cleaved by *β-ketoacyl CoA thiolase* to yield a two-carbon acetyl CoA and a new fatty acyl CoA that is shortened by two carbon atoms. This new, shorter fatty acyl CoA repeats the four steps of the β oxidation cycle until the original fatty acid is completely degraded to two-carbon units of acetyl CoA.

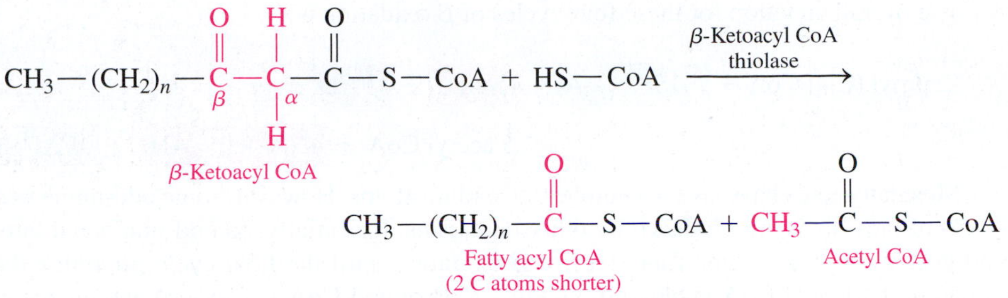
β-Ketoacyl CoA
thiolase
β-Ketoacyl CoA
Fatty acyl CoA
(2 C atoms shorter)
Acetyl CoA

TEST

Try Practice Problems 24.11 and 24.13 to 24.16

Fatty Acid Length Determines Cycle Repeats

The number of carbon atoms in a fatty acid determines the number of times the cycle repeats and the number of acetyl CoA units it produces. For example, the complete β oxidation of capric acid (C_{10}) produces five acetyl CoA, which is equal to one-half the number of carbon atoms in the fatty acid. Because the final turn of the cycle produces two acetyl CoA, the total number of times the cycle repeats is one less than the total number of acetyl groups it produces. Therefore, the C_{10} fatty acid goes through the cycle four times.

The four β oxidation cycles that capric acid goes through are shown in **FIGURE 24.4**.

ENGAGE

Why would caproic acid, a C_6 saturated fatty acid, go through two β oxidation cycles?

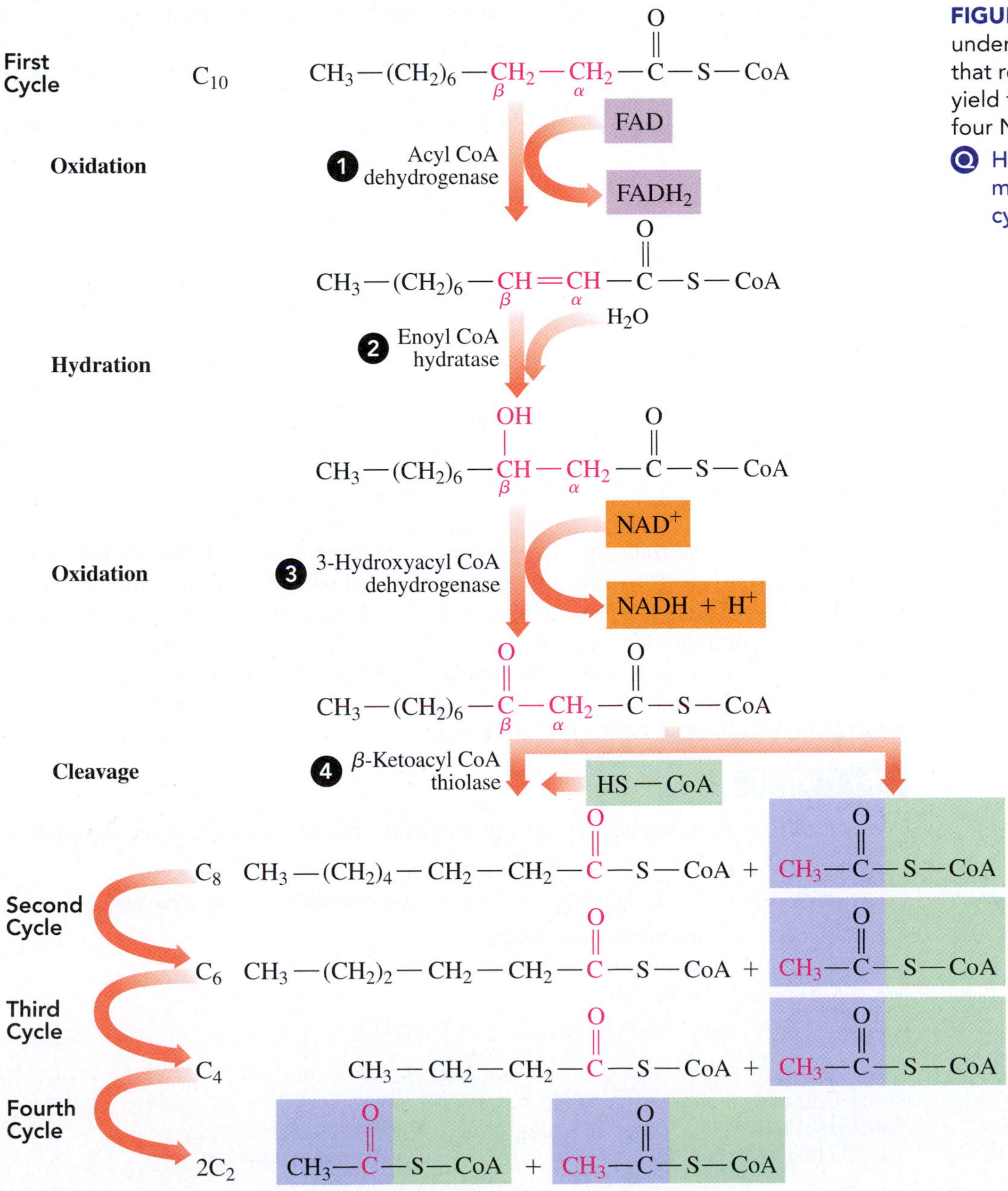
First Cycle
C_{10}
Oxidation
1 Acyl CoA dehydrogenase
FAD
$FADH_2$
Hydration
2 Enoyl CoA hydratase
H_2O
Oxidation
3 3-Hydroxyacyl CoA dehydrogenase
NAD^+
$NADH + H^+$
Cleavage
4 β-Ketoacyl CoA thiolase
HS—CoA
C_8
Second Cycle
C_6
Third Cycle
C_4
Fourth Cycle
$2C_2$

FIGURE 24.4 ▶ Capric acid (C_{10}) undergoes four oxidation cycles that repeat reactions 1 to 4 and yield five acetyl CoA molecules, four NADH, and four $FADH_2$.

Q How many NADH and $FADH_2$ molecules are produced in one cycle of β oxidation?

The overall equation for these four cycles of β oxidation is:

$$\text{Caproyl } (C_{10}) \text{ CoA} + 4\text{HS—CoA} + 4\text{NAD}^+ + 4\text{FAD} \longrightarrow$$
$$5 \text{ acetyl CoA} + 4\text{NADH} + 4\text{H}^+ + 4\text{FADH}_2$$

Most fatty acids have an even numbers of carbon atoms. However, some odd-numbered chains of fatty acids are produced by other organisms. Essentially, an odd-numbered fatty acid goes through the same four steps of β oxidation until the final cycle, in which the remaining fatty acyl CoA is cleaved, yielding a propionyl CoA (C_3 group) and an acetyl CoA.

Oxidation of Unsaturated Fatty Acids

Up to this point, we have discussed the β oxidation of saturated fatty acids. However, the fats in our diets, particularly the oils, contain unsaturated fatty acids, which have one or more cis double bonds. Reaction 2 of β oxidation is a hydration reaction in which the elements of water are added to a trans double bond. Before an unsaturated fatty acid can participate in this reaction, its cis double bond needs to be converted into a trans double bond. When the double bond in an unsaturated fatty acid is ready for hydration, an isomerase converts the cis double bond to a trans double bond between the α- and β-carbon atoms.

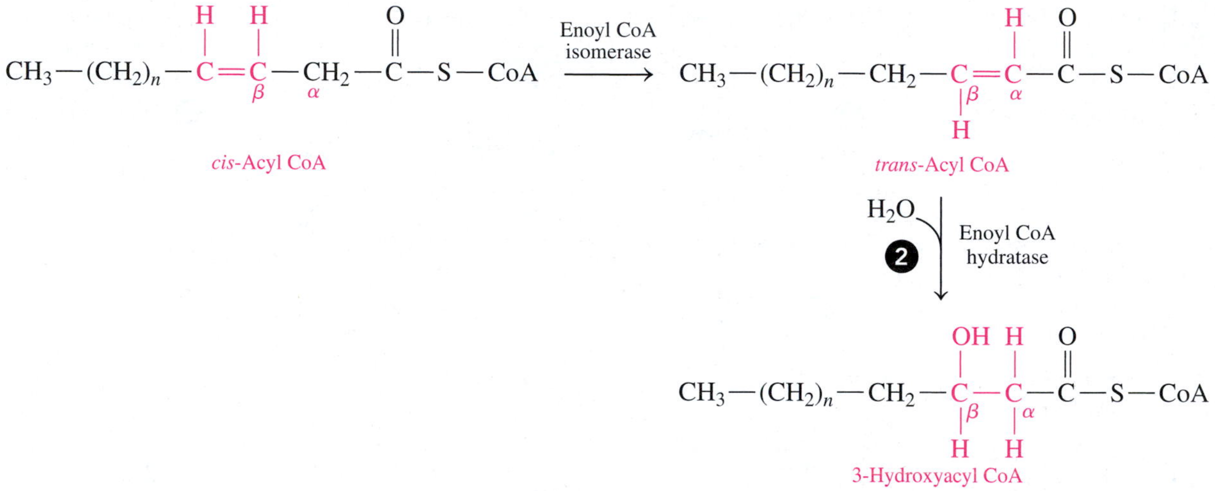
Enoyl CoA isomerase
cis-Acyl CoA
trans-Acyl CoA
H₂O
2
Enoyl CoA hydratase
3-Hydroxyacyl CoA

ENGAGE

Why is $FADH_2$ not produced during a β oxidation cycle involving the isomerization of a cis double bond?

Because the isomerization provides the trans double bond for the hydration in reaction 2, unsaturated fatty acids bypass the first reaction in β oxidation. Therefore, the energy released by the β oxidation of an unsaturated fatty acid is slightly less than that released by the β oxidation of a saturated fatty acid because no $FADH_2$ is produced in that cycle. For unsaturated fatty acids, we will decrease the total $FADH_2$ yield by one for each double bond.

▶ SAMPLE PROBLEM 24.2 β Oxidation

TRY IT FIRST

Match each of the following (**a** to **d**) with one of the reactions (**1** to **4**) in the β oxidation cycle:

1. first oxidation **2.** hydration **3.** second oxidation **4.** cleavage

a. Water is added to a trans double bond.
b. An acetyl CoA is removed.
c. FAD is reduced to $FADH_2$.
d. A 3-hydroxyacyl CoA is converted to 3-ketoacyl CoA.

SOLUTION

a. (2) hydration
b. (4) cleavage
c. (1) first oxidation
d. (3) second oxidation

STUDY CHECK 24.2

a. Which coenzyme is produced when a trans double bond is formed in a saturated fatty acid?

b. Which coenzyme is needed in reaction 3 when a β-hydroxyl group is converted to a β-keto group?

ANSWER

a. $FADH_2$ **b.** NAD^+

TEST

Try Practice Problems 24.12, 24.17, and 24.18

PRACTICE PROBLEMS

24.2 Oxidation of Fatty Acids

24.9 Where in the cell is a fatty acid activated?

24.10 What is the function of carnitine in the β oxidation of fatty acids?

24.11 What coenzymes are required for β oxidation?

24.12 When does an isomerization occur during the β oxidation of a fatty acid?

24.13 In each of the following acyl CoA molecules, identify the β-carbon:

a. $CH_3-CH_2-CH_2-CH_2-CH_2-CH_2-CH_2-\overset{O}{\overset{\|}{C}}-S-CoA$

b.

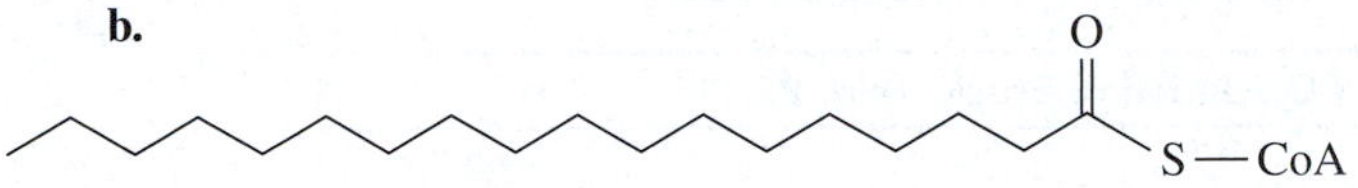
O
S—CoA

24.14 In each of the following acyl CoA molecules, identify the β-carbon:

a. $CH_3-(CH_2)_{14}-CH_2-CH_2-\overset{O}{\overset{\|}{C}}-S-CoA$

b.

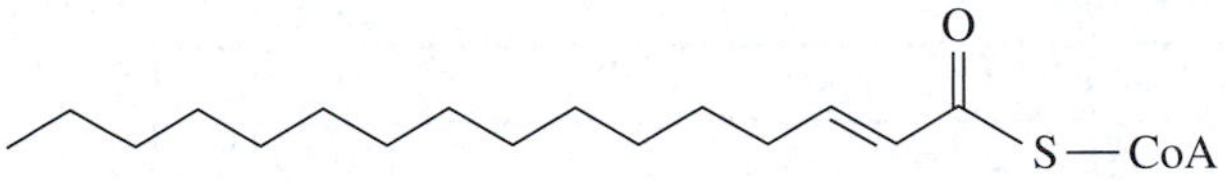
O
S—CoA

24.15 Draw the condensed structural or line-angle formula for the product(s) when each of the following undergoes the indicated reaction:

a.

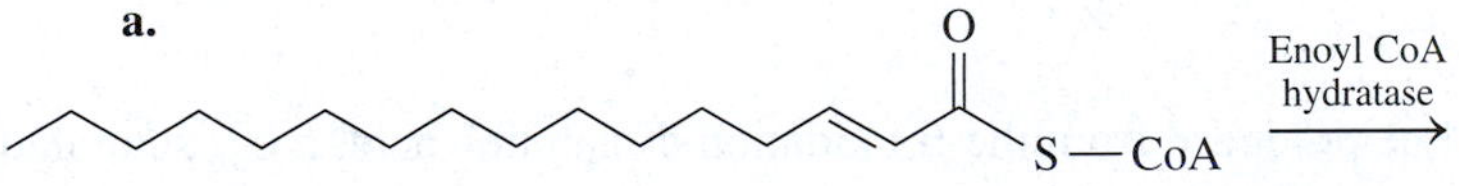
O
S—CoA
Enoyl CoA
hydratase

b. $CH_3-(CH_2)_6-CH_2-CH_2-\overset{O}{\overset{\|}{C}}-S-CoA \xrightarrow{\text{Acyl CoA dehydrogenase}}$

24.16 Draw the condensed structural or line-angle formula for the product(s) when each of the following undergoes the indicated reaction:

a.

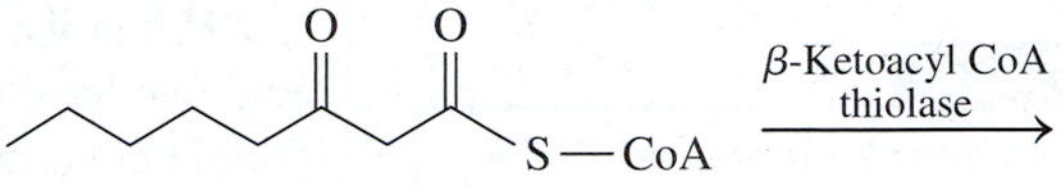
O
O
S—CoA
β-Ketoacyl CoA
thiolase

b. $CH_3-(CH_2)_8-\underset{H}{\overset{OH}{C}}-\underset{H}{\overset{H}{C}}-\overset{O}{\overset{\|}{C}}-S-CoA \xrightarrow{\text{3-Hydroxyacyl CoA dehydrogenase}}$

24.17 Caprylic acid (octanoic acid) is a saturated C_8 fatty acid found in coconut and palm oil as well as cow, goat, and human milk.

a. Draw the condensed structural formula for the activated form of caprylic acid.

b. Indicate the α- and β-carbon atoms in caprylic acid.

c. State the number of β oxidation cycles for the complete oxidation of caprylic acid.

d. State the number of acetyl CoA units obtained from the complete oxidation of caprylic acid.

24.18 Lignoceric acid is a saturated C_{24} fatty acid found in small amounts in peanut oil.

a. Draw the line-angle formula for the activated form of lignoceric acid.

b. Indicate the α- and β-carbon atoms in lignoceric acid.

c. State the number of β oxidation cycles for the complete oxidation of lignoceric acid.

d. State the number of acetyl CoA units obtained from the complete oxidation of lignoceric acid.

24.3 ATP and Fatty Acid Oxidation

LEARNING GOAL Calculate the total ATP produced by the complete oxidation of a fatty acid.

REVIEW

Identifying Important Coenzymes in Metabolism (22.2)

CORE CHEMISTRY SKILL

Calculating the ATP from Fatty Acid Oxidation (β Oxidation)

TEST

Try Practice Problems 24.19 and 24.20

The total energy yield from the oxidation of a particular fatty acid depends on the number of β oxidation cycles needed for the long-chain fatty acid. After the initial input of two ATP, each complete β oxidation cycle produces one NADH, one $FADH_2$, and one acetyl CoA. The oxidation of each NADH leads to the synthesis of 2.5 ATP, whereas the oxidation of each $FADH_2$ leads to the synthesis of 1.5 ATP. Thus, each β oxidation cycle produces coenzymes that can be oxidized through electron transport to release energy to synthesize a total of four ATP. However, the greatest amount of energy produced from a fatty acid is generated by the acetyl CoA molecules that enter the citric acid cycle. Each time an acetyl CoA is oxidized through the citric acid cycle, sufficient energy is released to synthesize 10 ATP.

We have seen that capric acid, C_{10}, goes through four turns of the β oxidation cycle, which produces five acetyl CoA units, four NADH, and four $FADH_2$. We also need to remember that activation of the capric acid requires the equivalent of two ATP. Therefore, we can set up the calculation of the amount of ATP produced from the β oxidation of capric acid as follows:

ATP Production from β Oxidation of Capric Acid (10:0)	
Activation	−2 ATP
5 Acetyl CoA (10 C atoms × 1 acetyl CoA/2 C atoms)	
5 acetyl CoA × 10 ATP/acetyl CoA	50 ATP
4 β Oxidation Cycles	
4 NADH × 2.5 ATP/NADH	10 ATP
4 $FADH_2$ × 1.5 ATP/$FADH_2$	6 ATP
Total	64 ATP

TEST

Try Practice Problems 24.21 and 24.22

ENGAGE

How must the number of $FADH_2$ produced during β oxidation be adjusted for the number of cis double bonds in an unsaturated fatty acid?

When we calculate the ATP generated from the β oxidation of an unsaturated fatty acid, we have to take into account the fact that one less $FADH_2$ is produced for every cis double bond that needs to be isomerized to a trans double bond. For example, we can calculate the ATP produced from the β oxidation of erucic acid, a C_{22} monounsaturated fatty acid found in wallflower seed, as shown below. Note that the number of $FADH_2$ produced from the β oxidation cycles is one less than the number of NADH produced because one cis double bond must be isomerized to a trans double bond.

ATP Production from β Oxidation of Erucic Acid (22:1)	
Activation	−2 ATP
11 Acetyl CoA (22 C atoms × 1 acetyl CoA/2 C atoms)	
11 acetyl CoA × 10 ATP/acetyl CoA	110 ATP
10 β Oxidation Cycles	
10 NADH × 2.5 ATP/NADH	25 ATP
9 $FADH_2$ × 1.5 ATP/$FADH_2$	13.5 ATP
Total	146.5 ATP

TEST

Try Practice Problems 24.23 and 24.24

▶ SAMPLE PROBLEM 24.3 ATP Production from β Oxidation

TRY IT FIRST

How many ATP will be produced from the β oxidation of palmitic acid, a C_{16} saturated fatty acid?

SOLUTION

ANALYZE THE PROBLEM	Given	Need	Connect
	β oxidation of palmitic acid (C_{16})	total ATP produced	NADH, $FADH_2$, acetyl CoA

Palmitic acid (C_{16}) requires seven β oxidation cycles, which produce seven NADH and seven $FADH_2$. The total number of acetyl CoA is eight. The activation of palmitic acid decreases the total produced by two ATP.

ATP Production from β Oxidation of Palmitic Acid (16:0)	
Activation	−2 ATP
8 Acetyl CoA (16 C atoms × 1 acetyl CoA/2 C atoms)	
8 acetyl CoA × 10 ATP/acetyl CoA	80 ATP
7 β Oxidation Cycles	
7 NADH × 2.5 ATP/NADH	17.5 ATP
7 $FADH_2$ × 1.5 ATP/$FADH_2$	10.5 ATP
Total	106 ATP

STUDY CHECK 24.3

How many ATP will be produced from the β oxidation of each of the following saturated fatty acids?

a. myristic acid (14:0), a fatty acid found in nutmeg
b. cerotic acid (26:0), a fatty acid found in beeswax and carnauba wax

ANSWER

a. 92 ATP **b.** 176 ATP

Chemistry Link to Health

Jamaican Vomiting Sickness

Almost all of the stored energy in our bodies is in the form of triacylglycerols. We store very little energy as glucose or glycogen. Therefore, any inhibitor of an enzyme in the β oxidation pathway can have significant effects on how we function.

Hypoglycin A is a molecule found in high concentrations in unripe ackee fruit, the national fruit of Jamaica. This molecule is a *protoxin*, meaning that it becomes a toxin when it is metabolized in the body. The metabolized version of hypoglycin A is an inhibitor of acyl CoA dehydrogenase, the first enzyme in the β oxidation pathway.

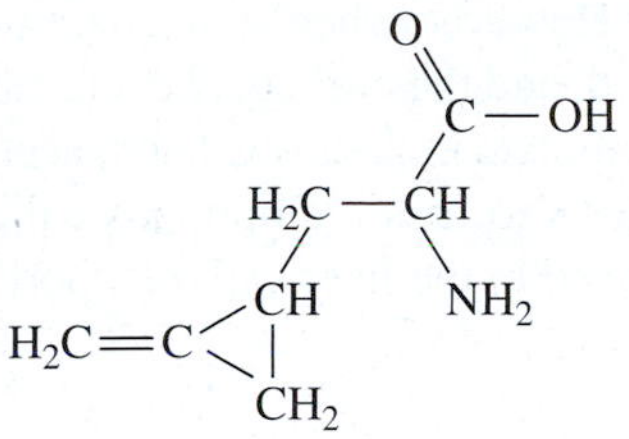

Hypoglycin A is a protoxin that is metabolized to become an inhibitor of acyl CoA dehydrogenase, the first enzyme in the β oxidation pathway.

People who eat unripe ackee fruit experience abdominal discomfort and begin to vomit violently. For this reason, the illness associated with ingesting hypoglycin A is known as *Jamaican vomiting sickness*. Because β oxidation is inhibited, people who suffer from this illness are forced to rely on their glucose and glycogen reserves to provide energy for ATP synthesis. Once these limited reserves are depleted, people with Jamaican vomiting sickness begin to experience symptoms of severe hypoglycemia, including sweating, headache, weakness, and increased heart rate. The effects of eating unripe ackee fruit can be so severe that, left untreated, they lead to death within 24 hours.

Treatment usually consists of maintaining fluid and glucose levels in the patient until the symptoms pass and the toxin has been cleared from the body. Fortunately, cases of this disease are rare because most people avoid eating the unripe fruit. Fruit that has been left to ripen on the tree to the point where it opens to reveal three large black seeds does not contain high levels of hypoglycin A. The ripened fruit is used to prepare the traditional Jamaican dish of ackee and saltfish.

Ripened ackee fruit does not contain high levels of hypoglycin A.

TEST

Try Practice Problems 24.25 and 24.26

PRACTICE PROBLEMS

24.3 ATP and Fatty Acid Oxidation

24.19 Why is the energy of fatty acid activation from ATP to AMP considered the same as the hydrolysis of 2 ATP $\longrightarrow$ 2 ADP?

24.20 What is the number of ATP molecules obtained from each acetyl CoA in the citric acid cycle?

Clinical Applications

24.21 Behenic acid is a saturated fatty acid found in peanut and canola oils. Answer each of the following (**a** to **c**) for behenic acid:

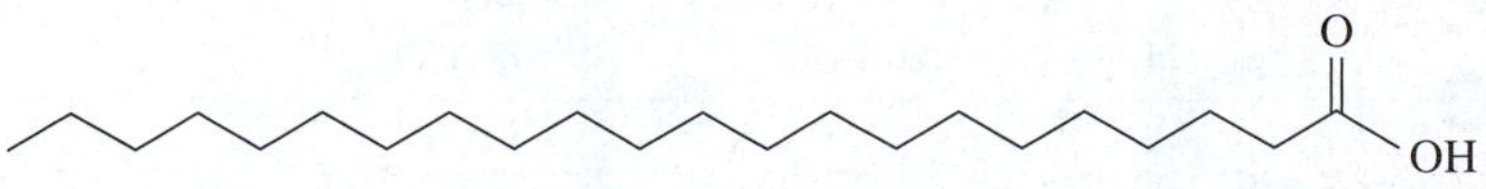

a. How many cycles of β oxidation are needed?
b. How many acetyl CoA units are produced?
c. How many ATP can be generated from the β oxidation of behenic acid?

24.22 Stearic acid is a saturated fatty acid found in animal products. Answer each of the following (**a** to **c**) for stearic acid:

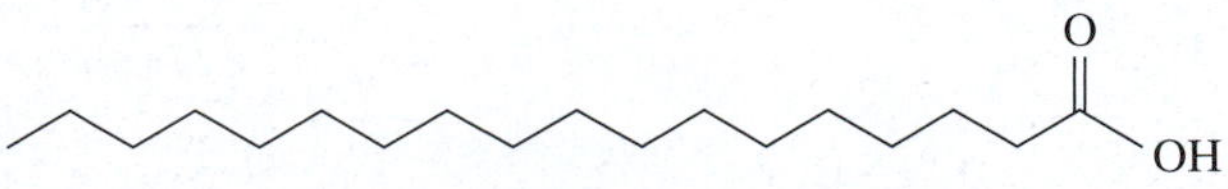

a. How many cycles of β oxidation are needed?
b. How many acetyl CoA units are produced?
c. How many ATP can be generated from the β oxidation of stearic acid?

24.23 Oleic acid is a monounsaturated fatty acid common in animal and vegetable oils. Answer each of the following (**a** to **c**) for oleic acid:

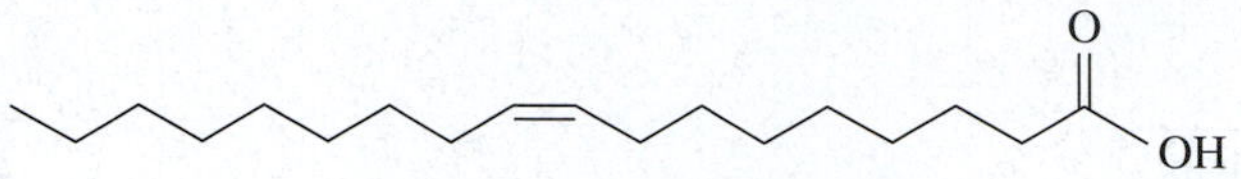
O
OH

a. How many cycles of β oxidation are needed?
b. How many acetyl CoA units are produced?
c. How many ATP can be generated from the β oxidation of oleic acid?

24.24 Palmitoleic acid is a monounsaturated fatty acid common in animal and vegetable oils. Answer each of the following (**a** to **c**) for palmitoleic acid:

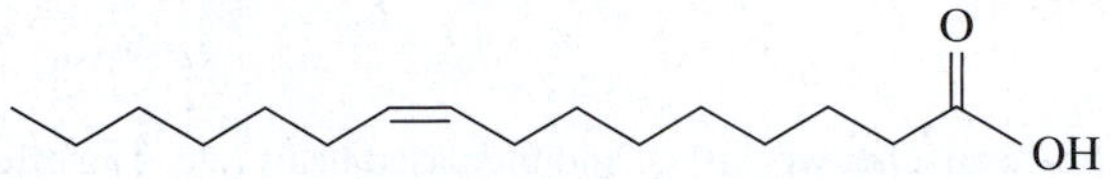
O
OH

a. How many cycles of β oxidation are needed?
b. How many acetyl CoA units are produced?
c. How many ATP can be generated from the β oxidation of palmitoleic acid?

24.25 Jamaican vomiting sickness is caused by eating the unripe ackee fruit, which contains the protoxin hypoglycin A. How does hypoglycin A inhibit β oxidation (see Chemistry Link to Health: Jamaican Vomiting Sickness)?

24.26 Based on the symptoms caused by eating the unripe ackee fruit, why do you think the protoxin that causes Jamaican vomiting sickness was named hypoglycin A (see Chemistry Link to Health: Jamaican Vomiting Sickness)?

24.4 Ketogenesis and Ketone Bodies

LEARNING GOAL Describe the pathway of ketogenesis.

CORE CHEMISTRY SKILL
Describing How Ketone Bodies are Formed

TEST
Try Practice Problems 24.27 and 24.28

When carbohydrates are not available to meet energy needs, the body breaks down fatty acids, which undergo β oxidation to acetyl CoA. Normally, acetyl CoA would enter the citric acid cycle for further oxidation and energy production. However, when large quantities of fatty acids are degraded, too much acetyl CoA is produced, and the citric acid cycle cannot oxidize it all. As a result, high levels of acetyl CoA accumulate in the liver. Then, acetyl CoA molecules combine in a pathway known as **ketogenesis** to form compounds called **ketone bodies** (see **FIGURE 24.5**). Ketone bodies are produced in the liver and transported for use in the cells of the heart, brain, and skeletal muscle.

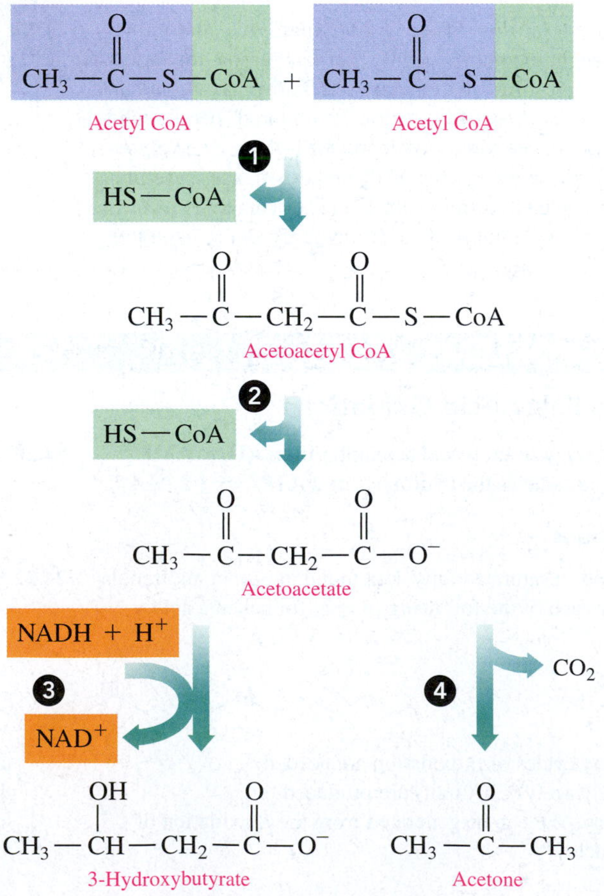
CH3—C(=O)—S—CoA + CH3—C(=O)—S—CoA
Acetyl CoA
Acetyl CoA
1
HS—CoA
CH3—C(=O)—CH2—C(=O)—S—CoA
Acetoacetyl CoA
2
HS—CoA
CH3—C(=O)—CH2—C(=O)—O−
Acetoacetate
NADH + H+
3
NAD+
4
CO2
CH3—CH(OH)—CH2—C(=O)—O−
3-Hydroxybutyrate
CH3—C(=O)—CH3
Acetone

FIGURE 24.5 ▶ In ketogenesis, molecules combine to produce ketone bodies: acetoacetate, 3-hydroxybutyrate, and acetone.

Q What condition in the body leads to the formation of ketone bodies?

Reaction 1 Condensation

In ketogenesis, two molecules of acetyl CoA combine to form acetoacetyl CoA and HS—CoA. This condensation is the reverse of the last step of β oxidation.

Reaction 2 Hydrolysis

The hydrolysis of acetoacetyl CoA forms acetoacetate, a ketone body, and HS—CoA. Once acetoacetate is formed, it can undergo either a hydrogenation or a decarboxylation (reactions 3 and 4 below) to produce other ketone bodies.

ENGAGE

Why is the coenzyme NADH required to convert acetoacetate to 3-hydroxybutyrate?

Reaction 3 Hydrogenation (Reduction)

Acetoacetate is reduced by 2H from NADH + H^+ to 3-hydroxybutyrate, which is considered a ketone body even though it does not contain a keto group.

Reaction 4 Decarboxylation

Acetoacetate can also undergo decarboxylation to yield acetone, a ketone body, and CO_2.

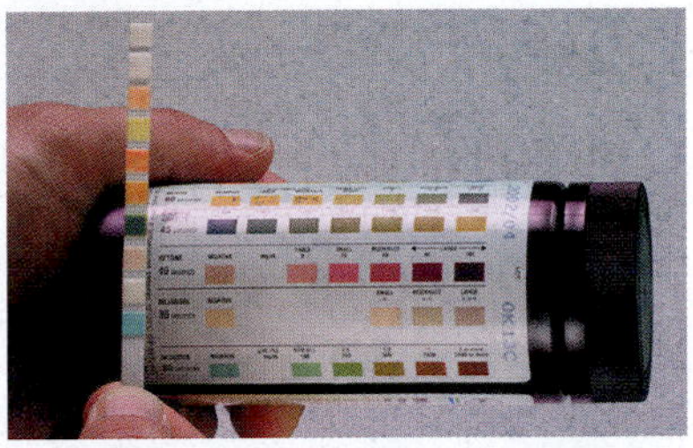

A ketone test strip indicates the level of ketone bodies in a urine sample.

Ketosis

Ketosis is the normal metabolic condition that occurs when the body uses ketone bodies for energy instead of glucose. This occurs when glucose stores are depleted and can be triggered by diets that are low in carbohydrates and high in fats, fasting, vigorous exercise, or diabetes. Both 3-hydroxybutyrate and acetoacetate can be converted back to acetyl CoA, which can then enter the citric acid cycle to provide energy for ATP synthesis.

$$\text{3-Hydroxybutyrate} \longrightarrow \text{acetoacetate} + \text{2CoA} \longrightarrow \text{2 acetyl CoA}$$

▶SAMPLE PROBLEM 24.4 Ketogenesis

TRY IT FIRST

The process called ketogenesis takes place in the liver.

a. What are some conditions that promote ketogenesis?
b. What are the names of the three compounds called ketone bodies?

SOLUTION

a. When excess acetyl CoA cannot be processed by the citric acid cycle, acetyl CoA molecules enter the ketogenesis pathway and form ketone bodies.
b. The ketone bodies are acetoacetate, 3-hydroxybutyrate, and acetone.

STUDY CHECK 24.4

Which ketone bodies contain each of the following functional groups?

a. ketone **b.** alcohol

ANSWER

a. Both acetoacetate and acetone contain ketone groups.
b. 3-Hydroxybutyrate contains an alcohol group.

TEST

Try Practice Problems 24.29 and 24.30

Chemistry Link to Health

Diabetes and Ketone Bodies

Blood glucose is elevated within 30 min following a meal containing carbohydrates. The elevated level of glucose stimulates the secretion of the hormone *insulin* from the pancreas, which increases the flow of glucose into muscle and adipose tissue for the synthesis of glycogen. As blood glucose levels drop, the secretion of insulin decreases. When blood glucose is low, another hormone, *glucagon*, is secreted by the pancreas, which stimulates the breakdown of glycogen in the liver to yield glucose.

(continued)

Chemistry Link to Health (*continued*)

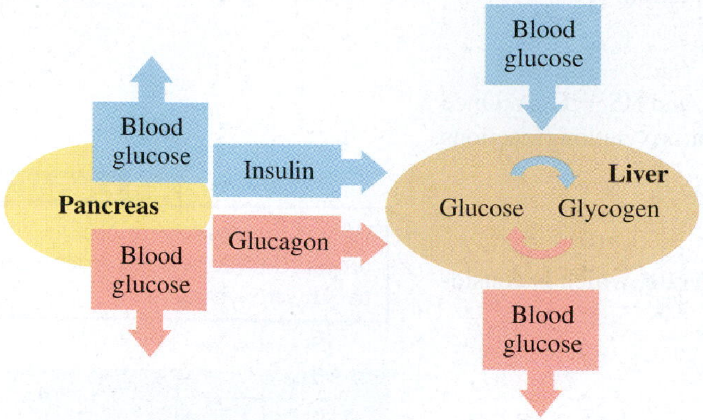
Blood glucose
Blood glucose
Insulin
Pancreas
Glucagon
Blood glucose
Liver
Glucose
Glycogen
Blood glucose

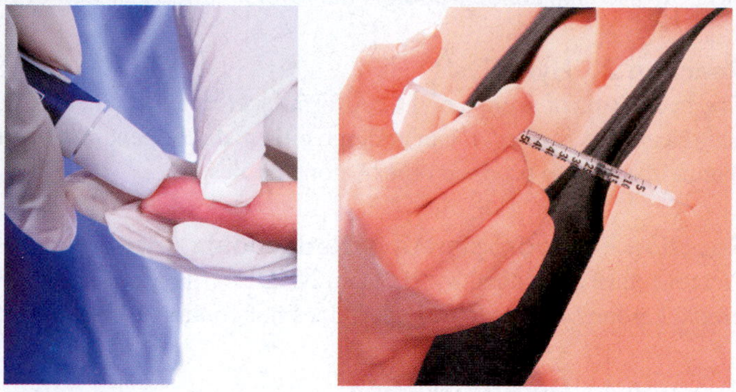

Type 1 diabetes can be treated with injections of insulin.

In *diabetes mellitus*, glucose cannot be utilized or stored as glycogen because insulin is not secreted or does not function properly. In type 1, *insulin-dependent diabetes*, which often begins in childhood, the pancreas produces inadequate levels of insulin. This type of diabetes can result from damage to the pancreas by viral infections or from genetic mutations. In type 2, *insulin-resistant diabetes*, which usually occurs in adults, insulin is produced, but insulin receptors are not responsive. Thus, a person with type 2 diabetes will not respond to insulin therapy. *Gestational diabetes* can occur during pregnancy, but blood glucose levels usually return to normal after the baby is born. Pregnant women with diabetes tend to gain weight and have large babies.

In all types of diabetes, insufficient amounts of glucose are available in the muscle, liver, and adipose tissue. As a result, liver cells synthesize glucose from noncarbohydrate sources via gluconeogenesis and use fatty acids for energy by breaking down fat, which elevates levels of acetyl CoA. Excess acetyl CoA undergoes ketogenesis, and ketone bodies accumulate in the blood.

When the levels of ketone bodies become extremely high, as can occur in the case of uncontrolled diabetes, a condition known as *ketoacidosis* can develop. Because two of the ketone bodies are carboxylic acids, they can lower the blood pH below 7.4 when they are present at high levels. Because changes in pH can affect the structure and function of biological molecules, ketoacidosis can have significant effects on the body, some of which can be life threatening. Symptoms of ketoacidosis include excessive thirst, frequent urination, vomiting, shortness of breath, fatigue, and confusion. Ketoacidosis occurs most often in type 1 diabetics, but it can also occur with type 2 diabetics. Although it occurs less often in nondiabetics, ketoacidosis can also develop in alcoholics or those who have undergone prolonged periods of starvation.

TEST

Try Practice Problems 24.31 and 24.32

PRACTICE PROBLEMS

24.4 Ketogenesis and Ketone Bodies

24.27 What is ketogenesis?

24.28 If a person were fasting, why would they have high levels of acetyl CoA?

24.29 What type of reaction converts acetoacetate to 3-hydroxybutyrate?

24.30 How is acetone formed from acetoacetate?

Clinical Applications

24.31 Diabetics often display signs of ketoacidosis if insulin levels are not maintained. What is ketoacidosis (see Chemistry Link to Health: Diabetes and Ketone Bodies)?

24.32 Why do diabetics produce high levels of ketone bodies (see Chemistry Link to Health: Diabetes and Ketone Bodies)?

24.5 Fatty Acid Synthesis

LEARNING GOAL Describe the synthesis of fatty acids from acetyl CoA.

When the body has met all its energy needs and the glycogen stores are full, acetyl CoA from the breakdown of carbohydrates and fatty acids is used to synthesize new fatty acids in the cytosol. In the pathway called **lipogenesis**, fatty acids are constructed through the successive addition of two-carbon acetyl units. Fatty acids up to 16 carbons in length can be constructed in this manner. Although the reactions appear much like the reverse of the reactions we discussed in fatty acid oxidation, the synthesis of fatty acids proceeds in a separate pathway with different enzymes. Fatty acid oxidation occurs in the mitochondria and uses FAD and NAD^+, whereas fatty acid synthesis occurs in the cytosol and uses the reduced coenzyme NADPH.

Acyl Carrier Protein (ACP)

Before a two-carbon acetyl unit can be added to a growing fatty acid chain, it must first be activated. In β oxidation, acetyl and acyl groups are activated through addition to coenzyme A (HS—CoA). In fatty acid synthesis, an acyl carrier protein (HS—ACP) activates the acetyl groups. HS—ACP contains the thiol and pantothenic acid (vitamin B_5) groups found in coenzyme A but these groups are attached to a protein instead of to an adenosine group.

$$HS-CH_2-CH_2-\underset{\underset{\text{Aminoethanethiol}}{}}{\overset{H}{N}}-\overset{O}{\overset{\|}{C}}-CH_2-CH_2-\underset{}{\overset{H}{N}}-\overset{O}{\overset{\|}{C}}-\underset{\underset{H}{\overset{OH}{|}}}{C}-\underset{\underset{CH_3}{\overset{CH_3}{|}}}{C}-CH_2-O-\underset{\underset{O^-}{\overset{O}{\|}}}{P}-O-CH_2-Protein$$

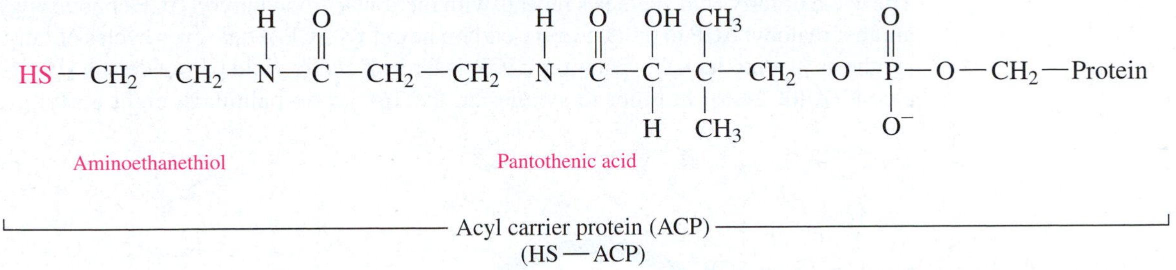

Aminoethanethiol   Pantothenic acid

Acyl carrier protein (ACP)
(HS—ACP)

Preparation of Activated Acyl and Acetyl Groups

In the first step of fatty acid synthesis, an ACP-activated acyl group (malonyl ACP) is combined with an ACP-activated acetyl group (acetyl ACP). These two molecules must be synthesized before the first step of fatty acid synthesis can occur.

The synthesis of the three-carbon malonyl ACP first requires the synthesis of malonyl CoA, which occurs when acetyl CoA combines with bicarbonate. The hydrolysis of ATP provides the energy for the reaction.

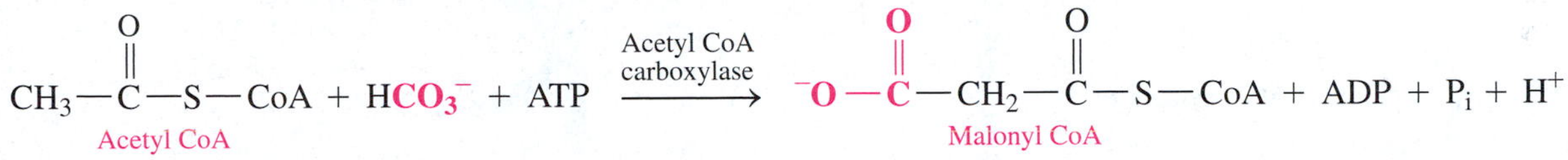

$$CH_3-\overset{O}{\overset{\|}{C}}-S-CoA + HCO_3^- + ATP \xrightarrow{\underset{\text{carboxylase}}{\text{Acetyl CoA}}} {}^-O-\overset{O}{\overset{\|}{C}}-CH_2-\overset{O}{\overset{\|}{C}}-S-CoA + ADP + P_i + H^+$$

Acetyl CoA   Malonyl CoA

Once malonyl CoA has been synthesized, it can be activated for fatty acid synthesis through addition to HS—ACP. An acetyl CoA group can be activated for fatty acid synthesis in a similar manner.

TEST

Try Practice Problems 24.33 to 24.36

$$^-O-\overset{O}{\overset{\|}{C}}-CH_2-\overset{O}{\overset{\|}{C}}-S-CoA + HS-ACP \xrightarrow{\underset{\text{transacylase}}{\text{Malonyl CoA}}} {}^-O-\overset{O}{\overset{\|}{C}}-CH_2-\overset{O}{\overset{\|}{C}}-S-ACP + HS-CoA$$

Malonyl CoA   Malonyl ACP

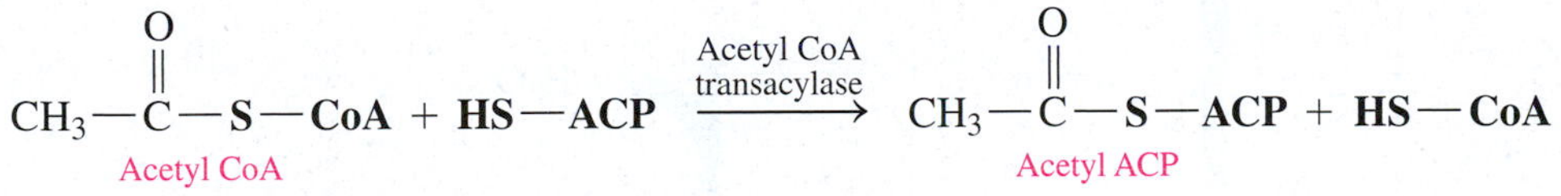

$$CH_3-\overset{O}{\overset{\|}{C}}-S-CoA + HS-ACP \xrightarrow{\underset{\text{transacylase}}{\text{Acetyl CoA}}} CH_3-\overset{O}{\overset{\|}{C}}-S-ACP + HS-CoA$$

Acetyl CoA   Acetyl ACP

Synthesis of Fatty Acids (Palmitate)

The synthesis of a fatty acid in the cytosol requires a cycle of four reactions that add two-carbon acetyl units to a growing carbon chain. Although these reactions can be used to construct fatty acids up to 16 carbons in length, we will focus specifically here on the reactions that produce palmitate (C_{16}).

Reaction 1   Condensation
In the first reaction, *3-ketoacyl ACP synthase* catalyzes the condensation of acetyl ACP and malonyl ACP to yield four-carbon acetoacetyl ACP and CO_2.

Reaction 2   Reduction
In the second reaction, *3-ketoacyl ACP reductase* reduces the 3-keto group using hydrogen from NADPH + H^+ to yield 3-hydroxyacyl ACP, an alcohol, and the oxidized coenzyme $NADP^+$.

Reaction 3   Dehydration
In the third reaction, *3-hydroxyacyl ACP dehydrase* catalyzes the dehydration of the hydroxyl group to form a trans double bond in *trans-2-enoyl ACP*.

ENGAGE

Which coenzyme is required for both reactions 2 and 4 in the fatty acid synthesis pathway?

Reaction 4 Reduction

In the fourth reaction, *enoyl ACP reductase* uses hydrogen from NADPH + H^+ to reduce the double bond to a single bond, which forms butyryl ACP, a saturated four-carbon compound, which is two carbon atoms longer that the original acetyl ACP.

Cycle of Fatty Acid Synthesis Repeats

The cycle of fatty acid synthesis repeats with the four-carbon butyryl ACP condensing with another malonyl ACP to produce a six-carbon hexanoyl ACP. After seven cycles of fatty acid synthesis, the product, C_{16} palmitoyl ACP, is hydrolyzed to yield palmitate and HS—ACP (see **FIGURE 24.6**). In order to synthesize the 16-carbon palmitate, eight acetyl groups

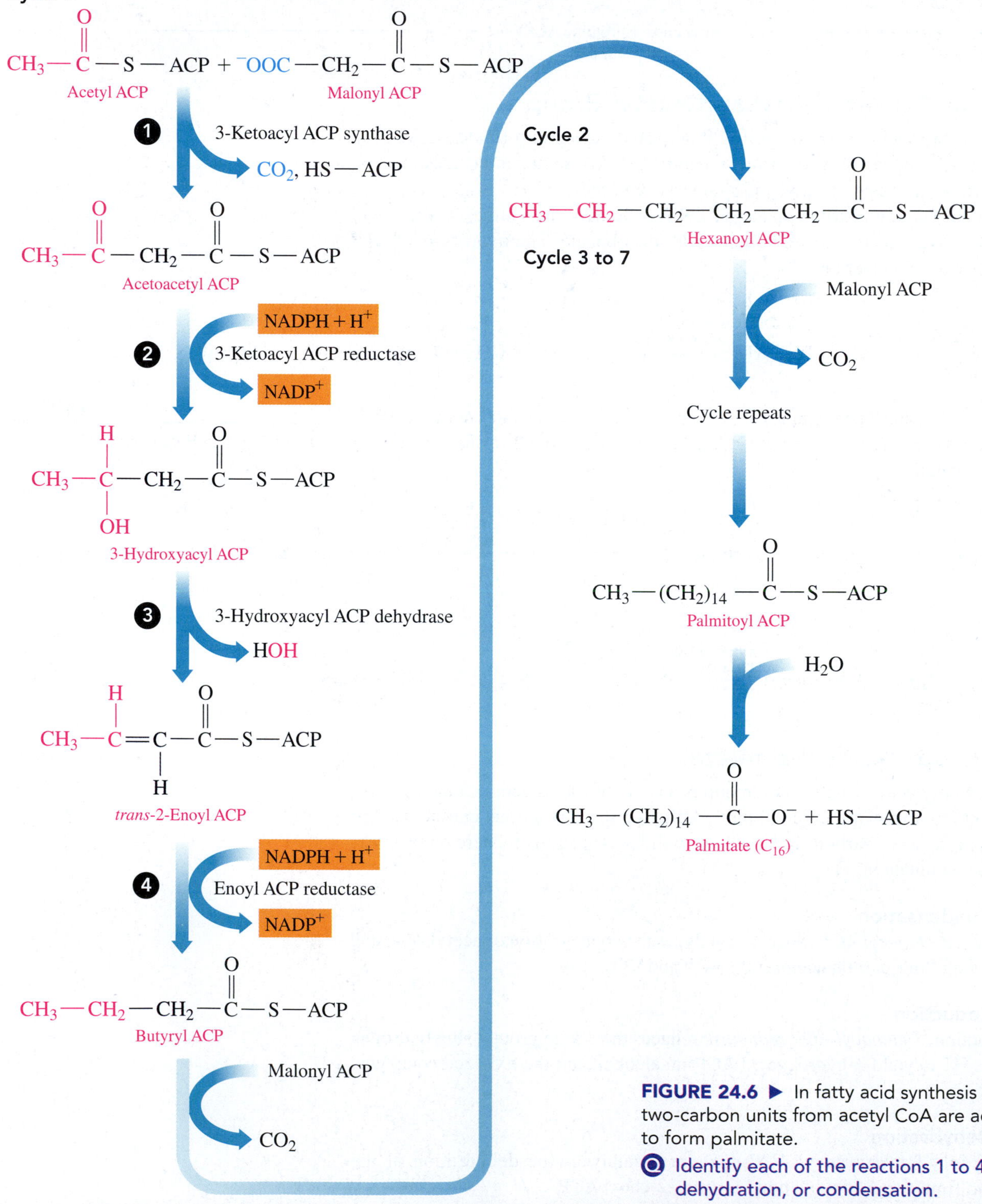
Cycle 1
Acetyl ACP
Malonyl ACP
1 3-Ketoacyl ACP synthase
CO2, HS—ACP
Acetoacetyl ACP
NADPH + H+
2 3-Ketoacyl ACP reductase
NADP+
3-Hydroxyacyl ACP
3 3-Hydroxyacyl ACP dehydrase
HOH
trans-2-Enoyl ACP
NADPH + H+
4 Enoyl ACP reductase
NADP+
Butyryl ACP
Malonyl ACP
CO2
Cycle 2
Hexanoyl ACP
Cycle 3 to 7
Malonyl ACP
CO2
Cycle repeats
Palmitoyl ACP
H2O
Palmitate (C16)

FIGURE 24.6 ▶ In fatty acid synthesis (lipogenesis), two-carbon units from acetyl CoA are added together to form palmitate.

Q Identify each of the reactions 1 to 4 as a reduction, dehydration, or condensation.

(one-half the number of carbons in the fatty acid) were combined through seven cycles of fatty acid synthesis (one less than the number of acetyl groups that were combined). The overall equation for the synthesis of palmitate from acetyl CoA is written as:

$$8\text{ Acetyl CoA} + 14\text{NADPH} + 14\text{H}^+ + 7\text{ATP} \longrightarrow$$
$$\text{palmitate} + 14\text{NADP}^+ + 8\text{HS—CoA} + 7\text{ADP} + 7\text{P}_i + 6\text{H}_2\text{O}$$

Longer and Shorter Fatty Acids

Palmitate is the longest fatty acid that can be produced in this manner. However, shorter and longer fatty acids are also produced in cells. Shorter fatty acids are released earlier in the fatty acid synthesis process before there are 16 carbon atoms in the chain. Longer fatty acids are produced with special enzymes that add two-carbon acetyl units to the carboxyl end of the fatty acid chain.

TEST

Try Practice Problems 24.37 to 24.42

Regulation of Fatty Acid Synthesis

Fatty acid synthesis takes place primarily in the adipose tissue, where triacylglycerols are formed and stored. The hormone *insulin* stimulates the formation of fatty acids. When blood glucose is high, insulin moves glucose into the cells. In the cell, insulin stimulates glycolysis and the oxidation of pyruvate, thereby producing acetyl CoA for fatty acid synthesis.

Comparison of β Oxidation and Fatty Acid Synthesis

We have seen that many of the steps in the synthesis of palmitate are similar to those that occur in the β oxidation of palmitate. Synthesis combines two-carbon acetyl units to make long carbon chains, whereas β oxidation removes two-carbon acetyl units from long carbon chains. Synthesis of fatty acids involves reduction and dehydration, whereas β oxidation of fatty acids involves oxidation and hydration. We can distinguish between the two pathways by comparing some of their features in **TABLE 24.1**.

ENGAGE

In which ways is β oxidation different from fatty acid synthesis?

TABLE 24.1 A Comparison of β Oxidation and Fatty Acid Synthesis

	β Oxidation	Fatty Acid Synthesis (Lipogenesis)
Site	Mitochondrial matrix	Cytosol
Activated by	Glucagon Low blood glucose	Insulin High blood glucose
Activator	Coenzyme A (HS—CoA)	Acyl carrier protein (HS—ACP)
Initial Substrate	Fatty acid	Acetyl CoA
Initial Coenzymes	FAD, NAD^+	NADPH + H^+
Types of Reactions	Oxidation Hydration Cleavage	Reduction Dehydration Condensation
Function	Cleaves two-carbon acetyl group	Adds two-carbon acetyl group
Final Product	Acetyl CoA	Palmitate (C_{16}) or other fatty acids
Final Coenzymes	$FADH_2$, NADH + H^+	$NADP^+$

▶SAMPLE PROBLEM 24.5 Fatty Acid Synthesis

TRY IT FIRST

Malonyl ACP is required for the elongation of fatty acid chains.

a. Complete the following equations for the formation of malonyl ACP from the starting material:

Acetyl CoA + HCO_3^- + ATP $\longrightarrow$

Malonyl CoA + HS—ACP $\longrightarrow$

b. What enzymes catalyze these reactions?

SOLUTION

a. Acetyl CoA combines with bicarbonate to form malonyl CoA, which reacts with ACP to form malonyl ACP.

$$\text{Acetyl CoA} + HCO_3^- + \text{ATP} \longrightarrow \text{malonyl CoA} + \text{ADP} + P_i + H^+$$
$$\text{Malonyl CoA} + \text{HS—ACP} \longrightarrow \text{malonyl ACP} + \text{HS—CoA}$$

b. The enzyme for the first reaction is acetyl CoA carboxylase. The enzyme for the second reaction is malonyl CoA transacylase.

STUDY CHECK 24.5

a. If malonyl ACP is a three-carbon acyl group, why are only two carbon atoms added each time malonyl ACP is combined with a fatty acid chain?

b. Which compound provides the energy for the synthesis of malonyl CoA from acetyl CoA?

ANSWER

a. In each cycle of fatty acid synthesis, a two-carbon acetyl group from the three-carbon group in malonyl ACP adds to the growing fatty acid chain and one carbon forms CO_2.

b. The hydrolysis of ATP provides the energy for this reaction.

PRACTICE PROBLEMS

24.5 Fatty Acid Synthesis

24.33 Where does fatty acid synthesis occur in the cell?

24.34 What compound is involved in the activation of acyl compounds in fatty acid synthesis?

24.35 What are the starting materials for fatty acid synthesis?

24.36 What is the function of malonyl ACP in fatty acid synthesis?

24.37 Identify the reaction (**a** to **c**) catalyzed by each of the following enzymes (**1** to **3**):

1. acetyl CoA carboxylase **2.** acetyl CoA transacylase
3. malonyl CoA transacylase

a. converts malonyl CoA to malonyl ACP
b. combines acetyl CoA with bicarbonate to give malonyl CoA
c. converts acetyl CoA to acetyl ACP

24.38 Identify the reaction (**a** to **d**) catalyzed by each of the following enzymes (**1** to **4**):

1. 3-ketoacyl ACP synthase
2. 3-ketoacyl ACP reductase
3. 3-hydroxyacyl ACP dehydrase
4. enoyl ACP reductase

a. catalyzes the dehydration of an alcohol
b. converts a carbon–carbon double bond to a carbon–carbon single bond
c. combines a two-carbon acetyl group with a three-carbon acyl group accompanied by the loss of CO_2
d. reduces a keto group to a hydroxyl group

24.39 Determine the number of each of the following components involved in the synthesis of one molecule of capric acid, a C_{10} fatty acid:

a. HCO_3^- **b.** ATP **c.** acetyl CoA
d. malonyl ACP **e.** NADPH **f.** CO_2 removed

24.40 Determine the number of each of the following components involved in the synthesis of one molecule of myristic acid, a C_{14} fatty acid:

a. HCO_3^- **b.** ATP **c.** acetyl CoA
d. malonyl ACP **e.** NADPH **f.** CO_2 removed

Clinical Applications

24.41 Triclosan, which is used in antiseptic hand washes and lotions to inhibit the growth of bacteria, is an inhibitor of enoyl ACP reductase.

a. What reaction in fatty acid synthesis is catalyzed by enoyl ACP reductase?
b. If fatty acid synthesis is required for the formation of bacterial membranes, why is triclosan used to control the growth of bacteria?

24.42 Researchers have synthesized a compound from biotin called CABI that inhibits the catalytic activity of acetyl CoA carboxylase in the cytosol. Mice without acetyl CoA carboxylase activity were found to lose weight even though they ate more food.

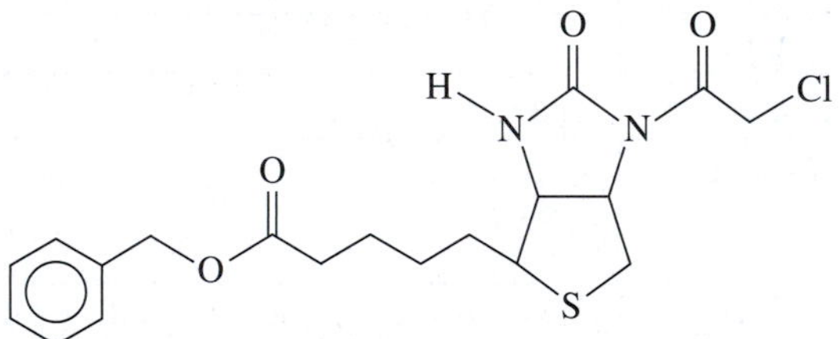
O
O
H
N
N
Cl
O
O
S

CABI

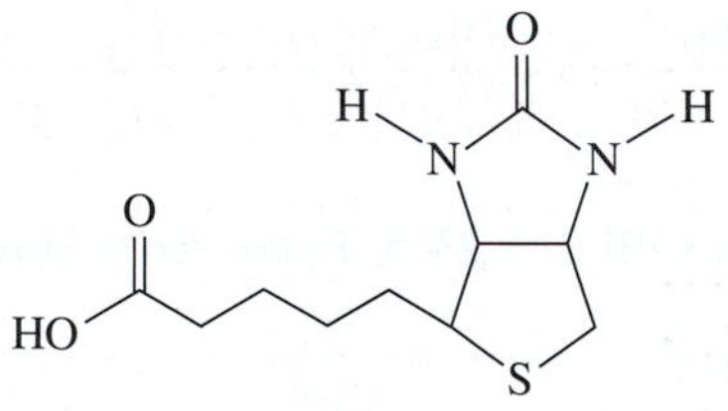
O
H
N
N
H
O
HO
S

Biotin

a. What reaction in fatty acid synthesis is catalyzed by acetyl CoA carboxylase?
b. How might CABI be useful in reducing obesity?

24.6 Degradation of Proteins and Amino Acids

LEARNING GOAL Describe the hydrolysis of dietary protein and the reactions of transamination and oxidative deamination in the degradation of amino acids.

The major role of protein in the diet is to provide amino acids for the synthesis of new proteins for the body and nitrogen atoms for the synthesis of compounds such as nucleotides. We have seen that carbohydrates and lipids are major sources of energy, but when they are not available, amino acids are degraded to substrates that enter energy-releasing pathways.

The digestion of proteins begins in the stomach, where hydrochloric acid (HCl) at pH 2 denatures the proteins and activates protease enzymes such as *pepsin* that begin to hydrolyze peptide bonds. Polypeptides move out of the stomach into the small intestine, where other proteases, such as *trypsin* and *chymotrypsin*, complete the hydrolysis of the peptides to amino acids. The amino acids are absorbed through the intestinal walls into the bloodstream for transport to the cells (see **FIGURE 24.7**).

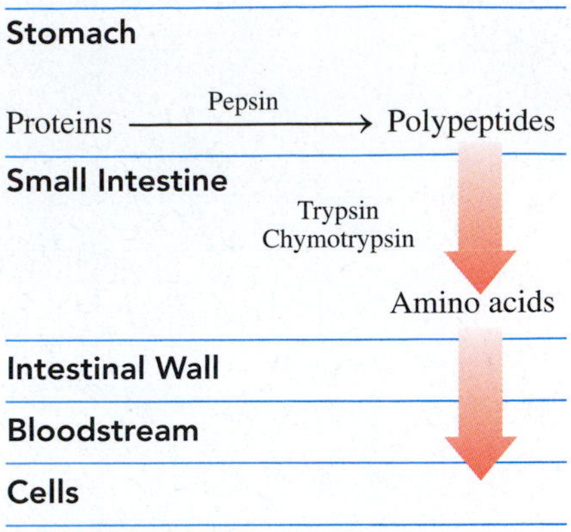

Stomach
Proteins
Pepsin
Polypeptides
Small Intestine
Trypsin
Chymotrypsin
Amino acids
Intestinal Wall
Bloodstream
Cells

FIGURE 24.7 ▶ Proteins are hydrolyzed to polypeptides in the stomach and to amino acids in the small intestine.

Q What enzyme, secreted into the small intestine, hydrolyzes peptides?

▶ SAMPLE PROBLEM 24.6 Digestion of Proteins

TRY IT FIRST

What are the sites and end products for the digestion of proteins?

SOLUTION

The digestion of proteins begins in the stomach and is completed in the small intestine to yield amino acids.

STUDY CHECK 24.6

a. What is the function of HCl in the stomach?
b. What is the function of proteases in the digestion of proteins?

ANSWER

a. HCl denatures proteins and activates enzymes such as pepsin.
b. Proteases hydrolyze peptide bonds to break proteins up into smaller polypeptide units.

Protein Turnover

Our bodies are constantly replacing old proteins with new ones. The process of breaking down proteins and synthesizing new proteins is called **protein turnover**. Many types of proteins, including enzymes, hormones, and hemoglobin, are synthesized in the cells and then degraded. For example, the hormone insulin has a half-life of 10 min, whereas the half-life of lactate dehydrogenase is about 2 days and that of hemoglobin is 120 days. Damaged and ineffective proteins are also degraded and replaced. Although most amino acids are used to build proteins, other compounds also require the nitrogen from amino acids for their synthesis, as seen in **TABLE 24.2** and **FIGURE 24.8**.

TABLE 24.2 Nitrogen-Containing Compounds

Type of Compound	Example
Amino alcohols	Choline, ethanolamine
Heme	Hemoglobin, cytochrome *c*
Hormones	Thyroxine, epinephrine, insulin
Neurotransmitters	Acetylcholine, dopamine, serotonin
Nonessential amino acids	Alanine, aspartate, cysteine, glycine
Nucleotides (nucleic acids)	Purines, pyrimidines
Proteins	Muscle protein, enzymes

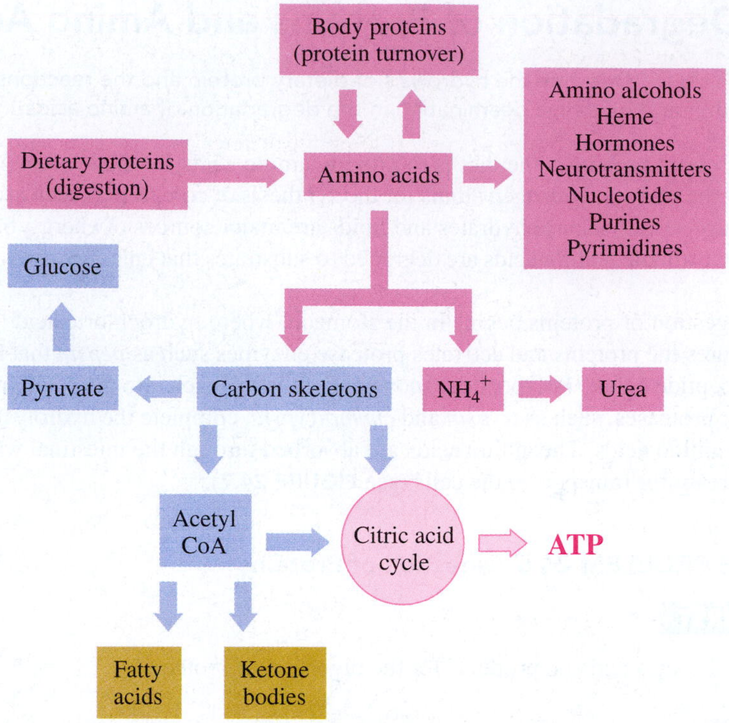
Body proteins
(protein turnover)
Amino alcohols
Heme
Hormones
Neurotransmitters
Nucleotides
Purines
Pyrimidines
Dietary proteins
(digestion)
Amino acids
Glucose
Pyruvate
Carbon skeletons
NH_4^+
Urea
Acetyl
CoA
Citric acid
cycle
ATP
Fatty
acids
Ketone
bodies

FIGURE 24.8 ▶ Proteins are used in the synthesis of nitrogen-containing compounds or degraded to urea and carbon skeletons that enter other metabolic pathways.

Q What are some compounds that require nitrogen for their synthesis?

Usually, we maintain a nitrogen balance in the cells so that the amount of protein we break down is equal to the amount that is reused. A diet that is high in protein, however, has a positive nitrogen balance because it supplies more nitrogen than we need. Because the body cannot store nitrogen, the excess is excreted as urea. Such high-protein diets put extra demands on the liver and kidneys. A diet that does not provide sufficient protein has a negative nitrogen balance, which is a condition that occurs during starvation and fasting.

Energy from Amino Acids

Normally, only a small amount (about 10%) of our energy needs is supplied by amino acids. However, more energy is extracted from amino acids in conditions such as fasting or starvation, when carbohydrate and fat stores are exhausted. If amino acids remain the only source of energy for a long period of time, the breakdown of body proteins eventually leads to a destruction of essential body tissues. In *anorexia*, the loss of protein decreases muscle mass and may severely weaken the heart muscle and impair heart function.

ENGAGE

Under what conditions are amino acids used for energy?

▶ SAMPLE PROBLEM 24.7 Nitrogen Balance

TRY IT FIRST

With a positive nitrogen balance, why are excess amino acids excreted?

SOLUTION

Because the body cannot store nitrogen, amino acids that are not needed for the synthesis of proteins are excreted.

STUDY CHECK 24.7

Under what condition does the body have a negative nitrogen balance?

ANSWER

In conditions such as fasting or starvation, a diet insufficient in protein leads to a negative nitrogen balance.

TEST
Try Practice Problems 24.43 to 24.46

Degradation of Amino Acids

When dietary protein exceeds the nitrogen needed for protein synthesis, the excess amino acids are degraded. The α-amino group is removed to yield an α-keto acid, which can be converted to an intermediate for other metabolic pathways. The carbon atoms from amino acids are used in the citric acid cycle as well as for the synthesis of fatty acids, ketone bodies, and glucose.

Transamination

The degradation of amino acids occurs primarily in the liver. In a **transamination** reaction, an α-amino group is transferred from an amino acid to an α-keto acid, usually α-ketoglutarate. A new amino acid and a new α-keto acid are produced. The enzymes that catalyze the transfer of amino groups are known as *transaminases* or *aminotransferases*.

We can write an equation to show the transfer of the amino group from alanine to α-ketoglutarate to yield glutamate, the new amino acid, and the α-keto acid pyruvate. Once formed, pyruvate can react to form acetyl CoA, which can enter the citric acid cycle.

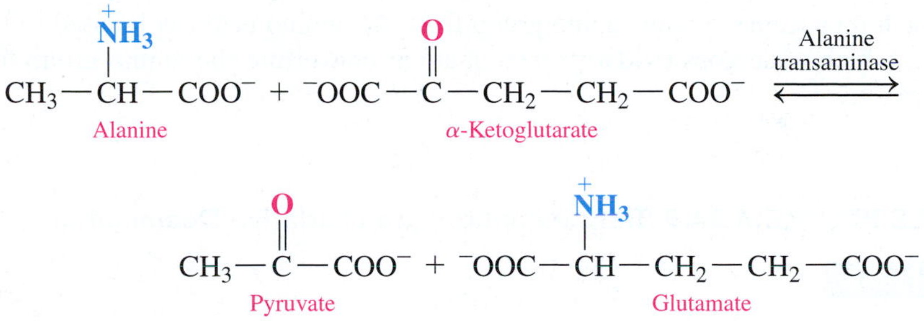

$$CH_3-\overset{\overset{+}{N}H_3}{\underset{|}{CH}}-COO^- + {}^-OOC-\overset{O}{\overset{||}{C}}-CH_2-CH_2-COO^- \underset{\text{Alanine transaminase}}{\rightleftharpoons}$$

Alanine / α-Ketoglutarate

$$CH_3-\overset{O}{\overset{||}{C}}-COO^- + {}^-OOC-\overset{\overset{+}{N}H_3}{\underset{|}{CH}}-CH_2-CH_2-COO^-$$

Pyruvate / Glutamate

▶ **SAMPLE PROBLEM 24.8 Transamination**

TRY IT FIRST

Write the equation for the transamination of glutamate and oxaloacetate by drawing the condensed structural formulas.

SOLUTION

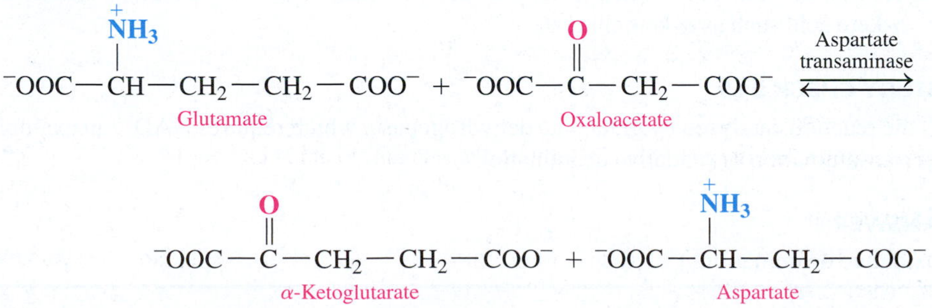

$$^-OOC-\overset{\overset{+}{N}H_3}{\underset{|}{CH}}-CH_2-CH_2-COO^- + {}^-OOC-\overset{O}{\overset{||}{C}}-CH_2-COO^- \underset{\text{Aspartate transaminase}}{\rightleftharpoons}$$

Glutamate / Oxaloacetate

$$^-OOC-\overset{O}{\overset{||}{C}}-CH_2-CH_2-COO^- + {}^-OOC-\overset{\overset{+}{N}H_3}{\underset{|}{CH}}-CH_2-COO^-$$

α-Ketoglutarate / Aspartate

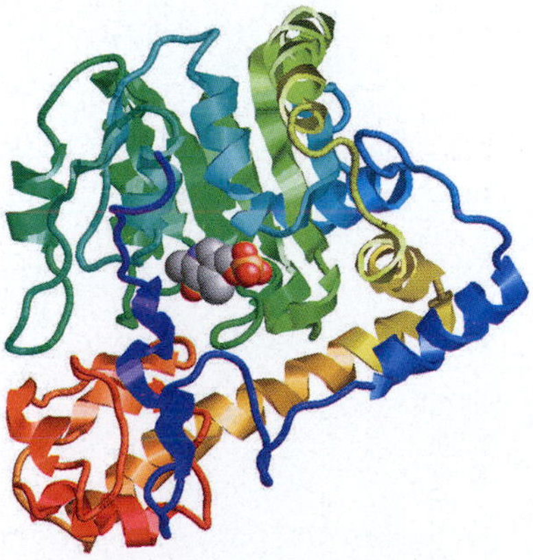

Aspartate transaminase (AST) catalyzes the reversible transfer of an amino group between glutamate and aspartate.

STUDY CHECK 24.8

What are some other possible names for the enzyme that catalyzes the reaction in Sample Problem 24.8?

ANSWER

glutamate aminotransferase, glutamate transaminase, aspartate aminotransferase

Oxidative Deamination

ENGAGE

What are the products of an oxidative deamination reaction?

In a process called **oxidative deamination**, the amino group ($-NH_3^+$) in glutamate is removed as an ammonium ion, NH_4^+. The deamination reaction produces α-ketoglutarate, which can participate in transamination with an amino acid. The reaction is catalyzed by *glutamate dehydrogenase*, which uses NAD^+ as a coenzyme.

$$\overset{\overset{\displaystyle +}{\underset{\displaystyle |}{NH_3}}}{^-OOC-CH-CH_2-CH_2-COO^-} + H_2O + \boxed{NAD^+} \underset{\xrightarrow{\hspace{1cm}}}{\overset{\text{Glutamate dehydrogenase}}{\xleftarrow{\hspace{1cm}}}}$$

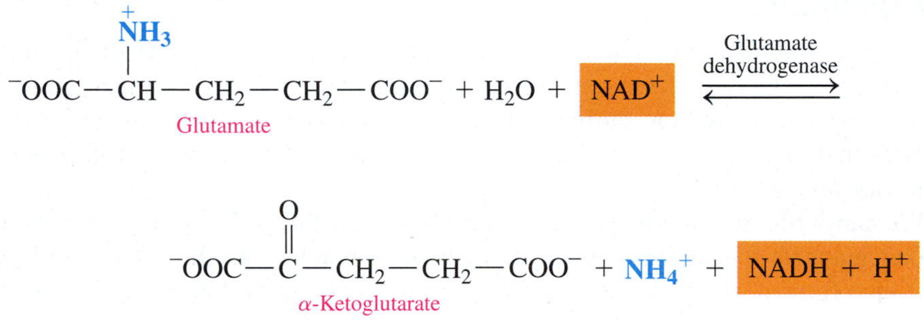

Glutamate

$$\overset{\overset{\displaystyle O}{\underset{\displaystyle ||}{}}}{^-OOC-C-CH_2-CH_2-COO^-} + NH_4^+ + \boxed{NADH + H^+}$$

α-Ketoglutarate

Through transamination, the amino group from any amino acid can be used to form glutamate, which undergoes oxidative deamination, converting the amino group to an ammonium ion.

▶ **SAMPLE PROBLEM 24.9** Transamination and Oxidative Deamination

TRY IT FIRST

Indicate whether each of the following represents a transamination or an oxidative deamination:

a. Glutamate is converted to α-ketoglutarate and NH_4^+.
b. Alanine and α-ketoglutarate react to form pyruvate and glutamate.

SOLUTION

a. Oxidative deamination occurs when the $-NH_3^+$ group in glutamate is removed as an ammonium ion, NH_4^+.
b. Transamination occurs when an amino group is transferred from an amino acid to an α-keto acid such as α-ketoglutarate.

STUDY CHECK 24.9

Is the reaction catalyzed by glutamate dehydrogenase, which requires NAD^+, an example of transamination or oxidative deamination?

TEST

Try Practice Problems 24.47 to 24.52

ANSWER

oxidative deamination

PRACTICE PROBLEMS

24.6 Degradation of Proteins and Amino Acids

24.43 Where do dietary proteins undergo digestion in the body?

24.44 What is meant by protein turnover?

24.45 What are some nitrogen-containing compounds that need amino acids for their synthesis?

24.46 What is the fate of the amino acids obtained from a high-protein diet?

24.47 What are the reactants and products in transamination reactions?

24.48 What types of enzymes catalyze transamination reactions?

24.49 Draw the condensed structural formula for the α-keto acid produced from each of the following in transamination:

a. $H-\underset{}{CH}(\overset{+}{N}H_3)-COO^-$ Glycine

b. $HS-CH_2-CH(\overset{+}{N}H_3)-COO^-$ Cysteine

c. $CH_3-CH(CH_3)-CH(\overset{+}{N}H_3)-COO^-$ Valine

24.50 Draw the condensed structural formula for the α-keto acid produced from each of the following in transamination:

a. $^-OOC-CH_2-CH(\overset{+}{N}H_3)-COO^-$ Aspartate

b. $CH_3-CH_2-CH(CH_3)-CH(\overset{+}{N}H_3)-COO^-$ Isoleucine

c. $HO-CH_2-CH(\overset{+}{N}H_3)-COO^-$ Serine

24.51 Write the equation, using condensed structural formulas, for the oxidative deamination of glutamate.

24.52 How do all 20 amino acids produce ammonium ions in oxidative deamination?

24.7 Urea Cycle

LEARNING GOAL Describe the formation of urea from an ammonium ion.

The ammonium ion, which is the end product of amino acid degradation, is toxic if it is allowed to accumulate. Thus, in the liver, a series of reactions called the **urea cycle** converts ammonium ions to urea, which is excreted in the urine.

$$2NH_4^+ + CO_2 \longrightarrow \underset{\text{Urea}}{H_2N-\overset{\overset{\displaystyle O}{\|}}{C}-NH_2} + 2H^+ + H_2O$$

In one day, a typical adult may excrete about 25 to 30 g of urea in the urine. This amount increases when a diet is high in protein. If urea is not properly excreted, as may be the case when the kidneys are not functioning properly, it builds up quickly to a toxic level. To detect renal disease, the *blood urea nitrogen* (BUN) level is measured. If the BUN is high, protein intake must be reduced, and hemodialysis may be needed to remove toxic nitrogen waste from the blood.

Urea Cycle

The urea cycle in the liver cells consists of reactions that take place in both the mitochondria and cytosol (see **FIGURE 24.9**). In preparation for the urea cycle, ammonium ions react with carbon dioxide and water, using the energy from two ATP to yield carbamoyl phosphate.

ENGAGE

Where do the reactions of the urea cycle take place in the cell?

$$NH_4^+ + CO_2 + 2ATP + H_2O \xrightarrow{\text{Carbamoyl phosphate synthetase}} \underset{\text{Carbamoyl phosphate}}{H_2N-\overset{\overset{\displaystyle O}{\|}}{C}-O-\overset{\overset{\displaystyle O}{\|}}{\underset{\underset{\displaystyle O^-}{|}}{P}}-O^-} + 2ADP + P_i + 3H^+$$

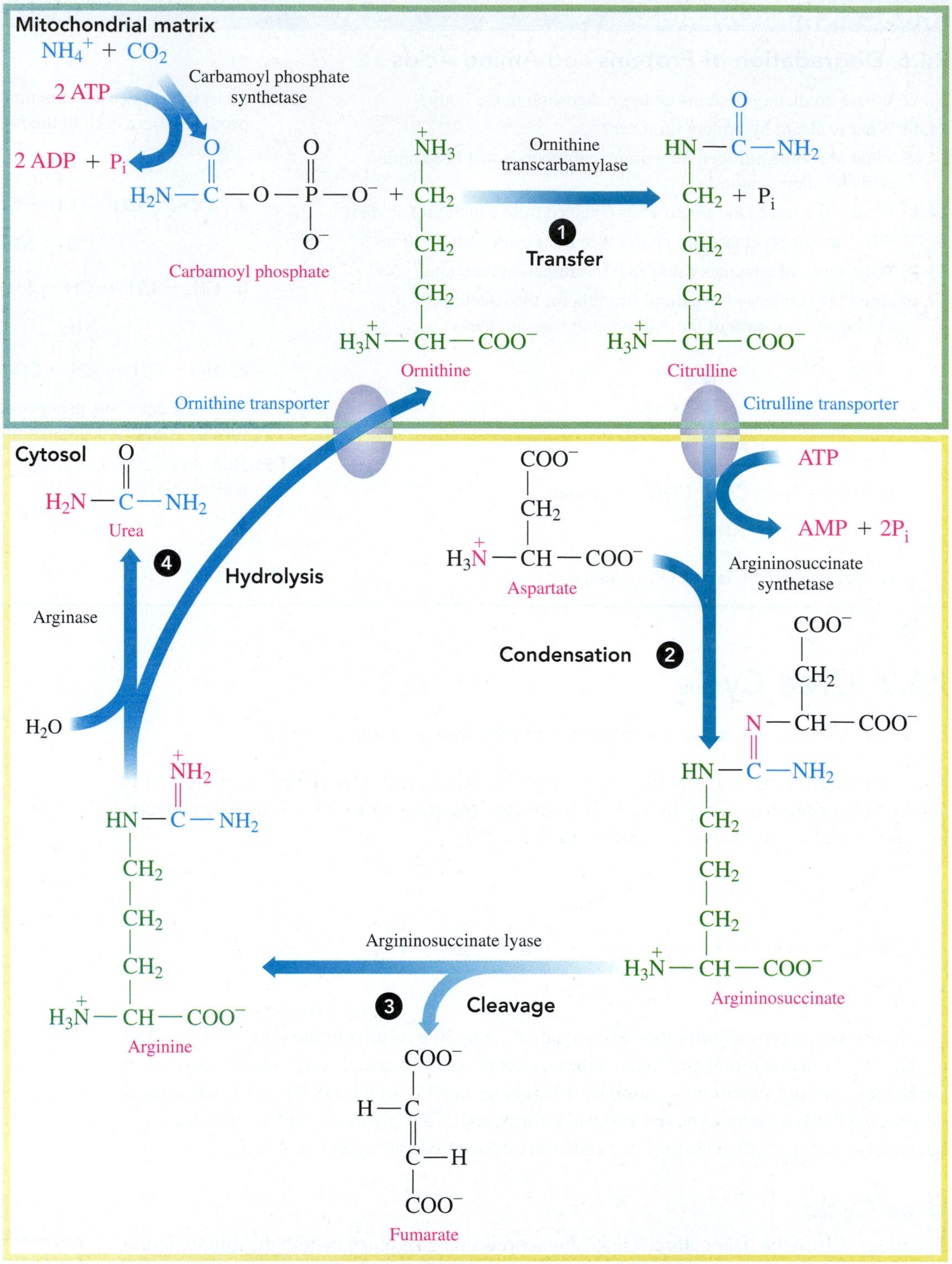
Mitochondrial matrix
$NH_4^+ + CO_2$
2 ATP
Carbamoyl phosphate synthetase
2 ADP + P_i
Carbamoyl phosphate
Ornithine transcarbamylase
1 Transfer
Ornithine
Citrulline
Ornithine transporter
Citrulline transporter
Cytosol
Urea
4 Hydrolysis
Arginase
H_2O
Aspartate
ATP
AMP + $2P_i$
Argininosuccinate synthetase
Condensation 2
Argininosuccinate
Argininosuccinate lyase
3 Cleavage
Arginine
Fumarate

FIGURE 24.9 ▶ In the urea cycle, urea is formed from a carbon and nitrogen (blue) from carbamoyl phosphate (initially an ammonium ion from oxidative deamination) and a nitrogen atom from aspartate (magenta).

Q Where in the cell is urea formed?

ENGAGE

What are the sources of the two nitrogen atoms in urea?

Reaction 1 Transfer of the Carbamoyl Group

In the mitochondrial matrix, reaction 1 utilizes *ornithine transcarbamylase* to catalyze the transfer of the carbamoyl group from carbamoyl phosphate to ornithine (an amino acid not found in proteins) to yield citrulline, which is transported across the mitochondrial membrane by the citrulline transporter protein to the cytosol. The hydrolysis of the high-energy phosphate bond in carbamoyl phosphate provides the energy to drive the reaction.

Reaction 2 Condensation of Citrulline with Aspartate

In the cytosol, reaction 2, a condensation of citrulline with aspartate catalyzed by *argininosuccinate synthetase*, produces argininosuccinate. The energy for the reaction is supplied by the hydrolysis of two phosphate bonds in ATP to yield AMP and two inorganic phosphates. The nitrogen atom in aspartate becomes one of the nitrogen atoms in the urea molecule that is produced in the final reaction.

Reaction 3 Cleavage of Argininosuccinate

Reaction 3 in the cytosol, catalyzed by *argininosuccinate lyase*, splits (cleaves) argininosuccinate to yield arginine, an amino acid, and fumarate, a citric acid cycle intermediate.

Reaction 4 Hydrolysis of Arginine to Form Urea

Reaction 4, catalyzed by *arginase*, is the hydrolysis of arginine to yield urea and ornithine. The ornithine returns to the mitochondrial matrix through the ornithine transporter protein to repeat the urea cycle.

▸SAMPLE PROBLEM 24.10 Urea Cycle

TRY IT FIRST

Indicate the reaction in the urea cycle where each of the following compounds is a reactant:

a. aspartate **b.** ornithine **c.** arginine

SOLUTION

a. Aspartate and citrulline undergo condensation in reaction 2.
b. A carbamoyl group is transferred to ornithine in reaction 1.
c. Arginine is cleaved in reaction 4.

STUDY CHECK 24.10

Name the products of each reaction in Sample Problem 24.10.

ANSWER

a. argininosuccinate **b.** citrulline **c.** urea and ornithine

TEST

Try Practice Problems 24.53 to 24.60

PRACTICE PROBLEMS

24.7 Urea Cycle

24.53 Draw the condensed structural formula for urea.

24.54 Draw the condensed structural formula for carbamoyl phosphate.

24.55 What is the source of carbon in urea?

24.56 How much ATP energy is required to drive one turn of the urea cycle?

Clinical Applications

24.57 Why does the body convert NH_4^+ to urea?

24.58 Why would a person on a high-protein diet be instructed to drink large quantities of water?

Use the following information for problems 24.59 and 24.60:

Research indicates that *sudden infant death syndrome* (SIDS) may be the result of a deficiency of any one of the enzymes in the urea cycle. The *urea cycle disorders* (UCD) are a result of a genetic mutation that leads to a deficient enzyme. When the urea cycle is disrupted, *hyperammonemia* (high ammonia levels) can cause irreversible changes in the brain.

24.59 In addition to ammonia, what intermediate of the urea cycle would be elevated if an infant has a deficiency of argininosuccinate synthetase?

24.60 In addition to ammonia, what intermediate of the urea cycle would be elevated if an infant has a deficiency of argininosuccinate lyase?

REVIEW

Identifying the Compounds in Glycolysis (22.4)

Identifying the Compounds and Enzymes in Glycogenesis and Glycogenolysis (22.6)

Describing the Reactions in the Citric Acid Cycle (23.1)

Calculating the ATP Produced from Glucose (23.3)

ENGAGE

Into which citric acid cycle intermediate are amino acids with four-carbon skeletons converted?

24.8 Fates of the Carbon Atoms from Amino Acids

LEARNING GOAL Describe where carbon atoms from amino acids enter the citric acid cycle or other pathways.

The carbon skeletons from the transamination of amino acids are used as intermediates in the citric acid cycle or other metabolic pathways. We can classify the amino acids according to the number of carbon atoms in those intermediates (see **FIGURE 24.10**). The amino acids that provide three-carbon compounds are converted to pyruvate. The amino acids with four carbon atoms are converted to oxaloacetate, and the five-carbon amino acids provide α-ketoglutarate. Some amino acids are listed twice because they can enter different pathways to form citric acid cycle intermediates.

A **glucogenic amino acid** has a carbon skeleton that can be used to produce pyruvate, α-ketoglutarate, succinyl CoA, fumarate, or oxaloacetate, each of which can be used to synthesize glucose by gluconeogenesis. A **ketogenic amino acid** has a carbon skeleton that can be used to produce acetoacetyl CoA or acetyl CoA, each of which can enter the ketogenesis pathway to form ketone bodies or the lipogenesis pathway to form fatty acids. Some amino acids can be both glucogenic and ketogenic.

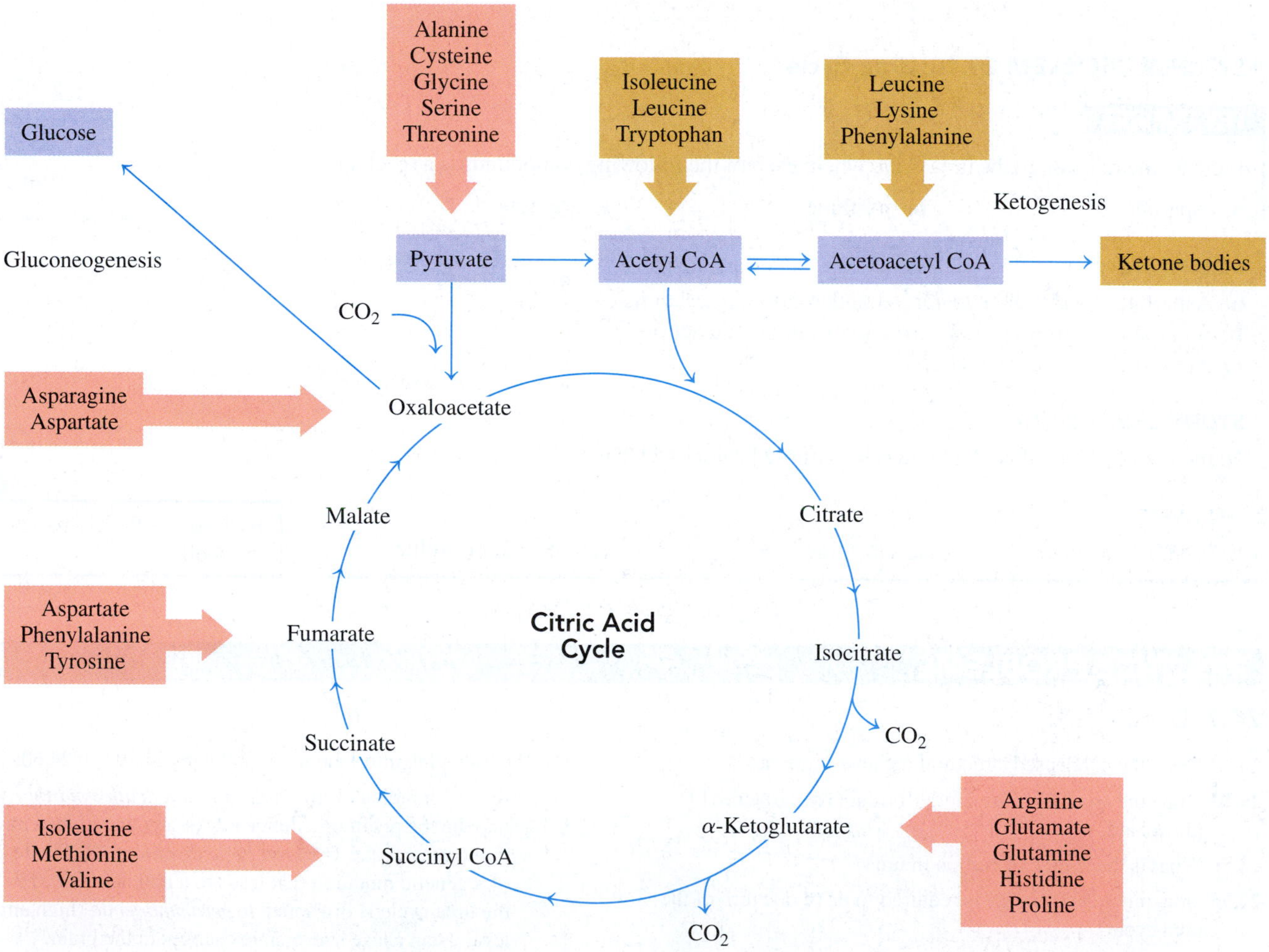

Alanine
Cysteine
Glycine
Serine
Threonine
Isoleucine
Leucine
Tryptophan
Leucine
Lysine
Phenylalanine
Glucose
Ketogenesis
Gluconeogenesis
Pyruvate
Acetyl CoA
Acetoacetyl CoA
Ketone bodies
CO_2
Asparagine
Aspartate
Oxaloacetate
Malate
Citrate
Aspartate
Phenylalanine
Tyrosine
Fumarate
Citric Acid Cycle
Isocitrate
CO_2
Succinate
Isoleucine
Methionine
Valine
Succinyl CoA
α-Ketoglutarate
Arginine
Glutamate
Glutamine
Histidine
Proline
CO_2

FIGURE 24.10 ▶ Carbon atoms from degraded amino acids are converted to the intermediates of the citric acid cycle or other pathways. Glucogenic amino acids (pink boxes) have carbon skeletons that can be used to form glucose, and ketogenic amino acids (gold boxes) have carbon skeletons that can be used to produce ketone bodies.

Q Why is aspartate glucogenic, but leucine ketogenic?

▶SAMPLE PROBLEM 24.11 Degradation of Amino Acids

TRY IT FIRST

Identify the citric acid cycle intermediate produced by each of the following amino acids. Determine the number of ATP that can be produced when that intermediate proceeds through the remaining steps of the citric acid cycle and the resulting coenzymes are oxidized in electron transport:

a. proline **b.** tyrosine **c.** tryptophan

SOLUTION

	Given	Need	Connect
ANALYZE THE PROBLEM	proline, tyrosine, tryptophan	ATP produced via the citric acid cycle and electron transport	NADH, $FADH_2$, GTP

a. Proline is converted to the citric acid cycle intermediate α-ketoglutarate. The reactions in the remaining part of the citric acid cycle from α-ketoglutarate to oxaloacetate produce two NADH, one GTP, and one $FADH_2$. The oxidation of two NADH provides energy to synthesize five ATP, one GTP provides one ATP, and the oxidation of one $FADH_2$ provides energy to synthesize 1.5 ATP, for a total of 7.5 ATP from proline.

b. Tyrosine is converted to the citric acid cycle intermediate fumarate. The remaining reactions of the citric acid cycle from fumarate to oxaloacetate produce one NADH, which can be oxidized via electron transport to provide the energy to synthesize 2.5 ATP.

c. Tryptophan is converted to the citric acid cycle intermediate acetyl CoA. The reactions in the citric acid cycle starting with acetyl CoA produce three NADH, one GTP, and one $FADH_2$. The oxidation of three NADH provides energy to synthesize 7.5 ATP, one GTP provides one ATP, and the oxidation of one $FADH_2$ provides the energy to synthesize 1.5 ATP, for a total of 10 ATP from tryptophan.

STUDY CHECK 24.11

a. Why does the citric acid cycle intermediate from leucine supply more ATP than the intermediate from phenylalanine?

b. Which amino acid produces the citric acid cycle intermediate that supplies more ATP, arginine or aspartate?

ANSWER

a. Leucine forms the citric acid cycle intermediate acetyl CoA, which enters the citric acid cycle to ultimately provide 10 ATP. Phenylalanine forms the citric acid cycle intermediate fumarate, producing one NADH, which is oxidized to provide the energy to synthesize 2.5 ATP.

b. Arginine, which forms the citric acid cycle intermediate α-ketoglutarate, supplies more ATP than aspartate, which enters the citric acid cycle later, as fumarate.

TEST

Try Practice Problems 24.61 to 24.64

PRACTICE PROBLEMS

24.8 Fates of the Carbon Atoms from Amino Acids

24.61 What is the function of a glucogenic amino acid?

24.62 What is the function of a ketogenic amino acid?

24.63 What component of the citric acid cycle can be produced from the carbon atoms of each of the following amino acids?

a. alanine **b.** asparagine
c. valine **d.** glutamine

24.64 What component of the citric acid cycle can be produced from the carbon atoms of each of the following amino acids?

a. leucine
b. threonine
c. cysteine
d. arginine

TABLE 24.3 Essential and Nonessential Amino Acids in Adults

Essential	Nonessential
Histidine	Alanine
Isoleucine	Arginine
Leucine	Asparagine
Lysine	Aspartate
Methionine	Cysteine
Phenylalanine	Glutamate
Threonine	Glutamine
Tryptophan	Glycine
Valine	Proline
	Serine
	Tyrosine

TEST

Try Practice Problems 24.65 and 24.66

ENGAGE

From which citric acid cycle intermediate can proline be synthesized?

24.9 Synthesis of Amino Acids

LEARNING GOAL Describe how some nonessential amino acids are synthesized from intermediates in the citric acid cycle and other metabolic pathways.

Plants and bacteria such as *E. coli* produce all of their amino acids using NH_4^+ and NO_3^-. However, humans can synthesize only 9 of the 20 amino acids found in their proteins. The *nonessential amino acids* are those that can be synthesized in the body, whereas the *essential amino acids* are those that must be obtained from the diet (see **TABLE 24.3**). The amino acids arginine, cysteine, and tyrosine are essential in diets for infants and children because of their rapid growth requirements, but they are not essential amino acids for adults.

Some Pathways for Amino Acid Synthesis

A variety of pathways are involved in the synthesis of nonessential amino acids. When the body synthesizes nonessential amino acids, the α-keto acid carbon skeletons are obtained from the citric acid cycle or glycolysis and converted to amino acids by transamination (see **FIGURE 24.11**).

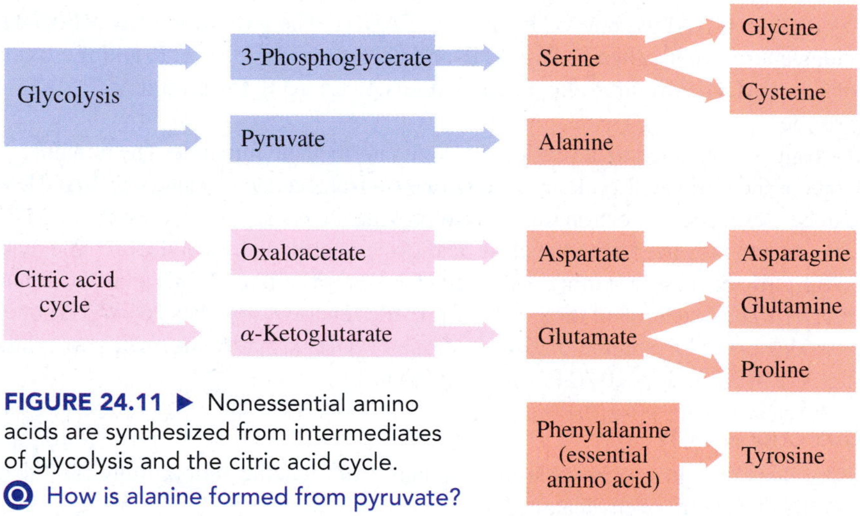
Glycolysis
3-Phosphoglycerate
Pyruvate
Serine
Glycine
Cysteine
Alanine
Citric acid cycle
Oxaloacetate
α-Ketoglutarate
Aspartate
Asparagine
Glutamate
Glutamine
Proline
Phenylalanine (essential amino acid)
Tyrosine

FIGURE 24.11 ▶ Nonessential amino acids are synthesized from intermediates of glycolysis and the citric acid cycle.

Q How is alanine formed from pyruvate?

Some of the amino acids are formed by reversing the transamination reaction that we saw in amino acid degradation. The transfer of an amino group from glutamate to pyruvate produces alanine in a reaction catalyzed by *alanine transaminase* (ALT).

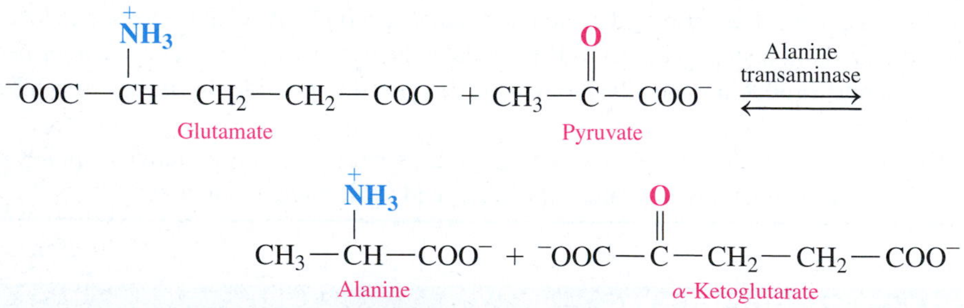
Alanine transaminase
Glutamate
Pyruvate
Alanine
α-Ketoglutarate

Oxaloacetate, a keto acid from the citric acid cycle, undergoes transamination with glutamate to produce aspartate, an amino acid, and α-ketoglutarate. The transamination is catalyzed by *aspartate transaminase* (AST).

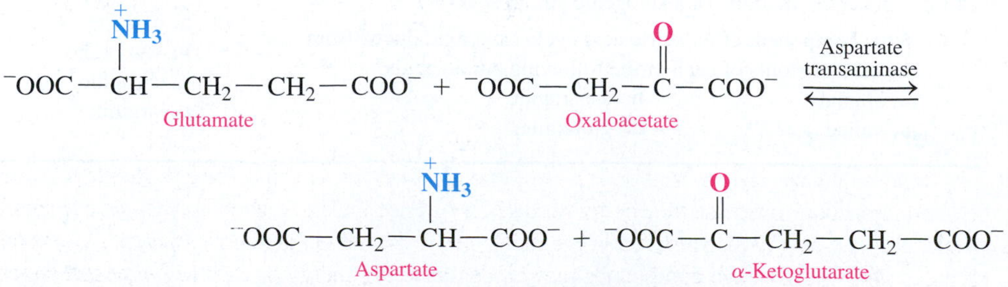
Aspartate transaminase
Glutamate
Oxaloacetate
Aspartate
α-Ketoglutarate

These two transaminases, ALT and AST, are abundant in the cells of the liver and heart, but they are present only in low levels in the bloodstream. When an injury or disease occurs, they are released from the damaged cells into the bloodstream. Elevated levels of serum ALT or AST provide a means to diagnose the extent of damage to the liver or the heart. Alanine transaminase is also known by the abbreviation SGPT (*serum glutamate pyruvate transaminase*) because it catalyzes the reversible transamination reaction between glutamate and pyruvate, and aspartate transaminase is also known by the abbreviation SGOT (*serum glutamate oxaloacetate transaminase*) because it catalyzes the reversible transamination reaction between glutamate and oxaloacetate.

The synthesis of the other nonessential amino acids may require several reactions in addition to transamination. For example, glutamine is synthesized when a second amino group is added to glutamate using the energy from the hydrolysis of ATP.

TEST

Try Practice Problems 24.67 and 24.68

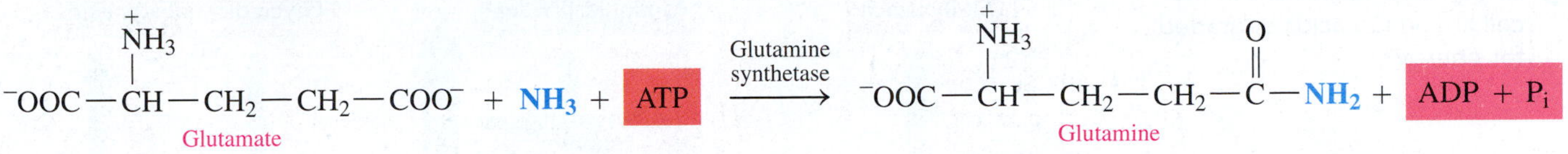
Glutamine synthetase
ATP
ADP + P_i
Glutamate
Glutamine

Tyrosine, an aromatic amino acid with a hydroxyl group, is formed from phenylalanine, an essential amino acid, and oxygen.

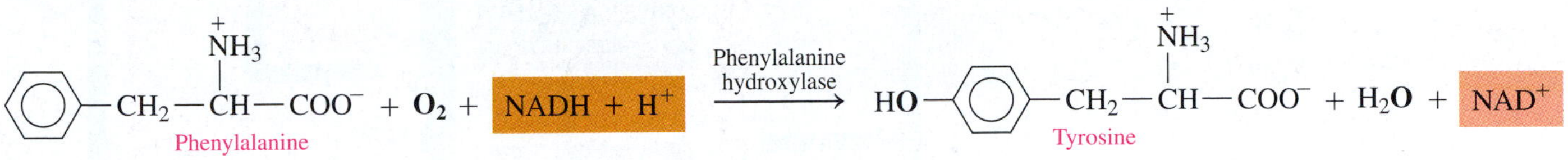
Phenylalanine hydroxylase
NADH + H^+
NAD^+
Phenylalanine
Tyrosine

Chemistry Link to Health

Phenylketonuria (PKU)

In the genetic disease *phenylketonuria* (PKU), a person cannot convert phenylalanine to tyrosine because the gene for the enzyme phenylalanine hydroxylase is defective. As a result, large amounts of phenylalanine accumulate. In this situation, a transaminase catalyzes the transfer of the $—NH_3^+$ from phenylalanine to pyruvate to form alanine and phenylpyruvate, which is then decarboxylated to phenylacetate. Large amounts of these compounds are excreted in the urine. Phenylacetate has a characteristic odor in the urine that can be used to recognize PKU in infants.

In infants, high levels of phenylpyruvate and phenylacetate cause severe mental retardation. However, the defect can be identified at birth, and all newborns are now tested for PKU. By detecting PKU early, retardation is avoided by using an infant diet with proteins that are low in phenylalanine and high in tyrosine. It is also important to avoid the use of sweeteners and soft drinks containing aspartame, which contains phenylalanine as one of the two amino acids in its structure. In adulthood, some people with PKU can eat a nearly normal diet as long as they are checked for phenylpyruvate periodically.

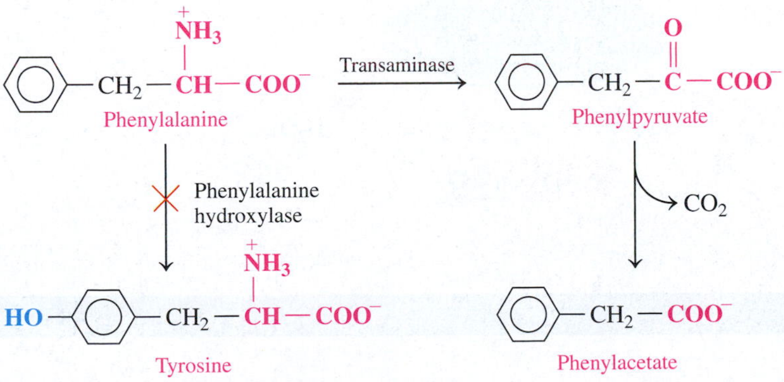
Phenylalanine
Transaminase
Phenylpyruvate
Phenylalanine hydroxylase
CO_2
Tyrosine
Phenylacetate

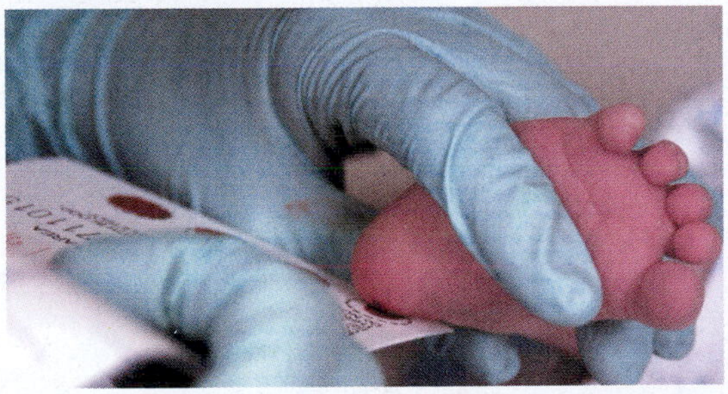

All newborns are tested for PKU.

TEST

Try Practice Problems 24.69 and 24.70

Overview of Metabolism

CORE CHEMISTRY SKILL

Distinguishing Anabolic and Catabolic Pathways

Catabolic pathways degrade large molecules to small molecules that enter the citric acid cycle and electron transport to release the energy needed for ATP synthesis. Anabolic pathways use small molecules and energy to synthesize larger molecules in the cell. In metabolism, there are several branch points from which compounds may be degraded for energy or used to synthesize larger molecules. For example, glucose can be degraded to pyruvate and then acetyl CoA for the citric acid cycle to ultimately produce ATP. Glucose may also be converted to glycogen for storage. When glycogen stores are depleted, fatty acids from fats are degraded for energy. Amino acids normally used to synthesize nitrogen-containing compounds in the cells can also be used for energy after they are degraded to intermediates of the citric acid cycle. In the synthesis of nonessential amino acids, α-keto acids of the citric acid cycle enter into a variety of reactions that convert them to amino acids through transamination by glutamate (see **FIGURE 24.12**).

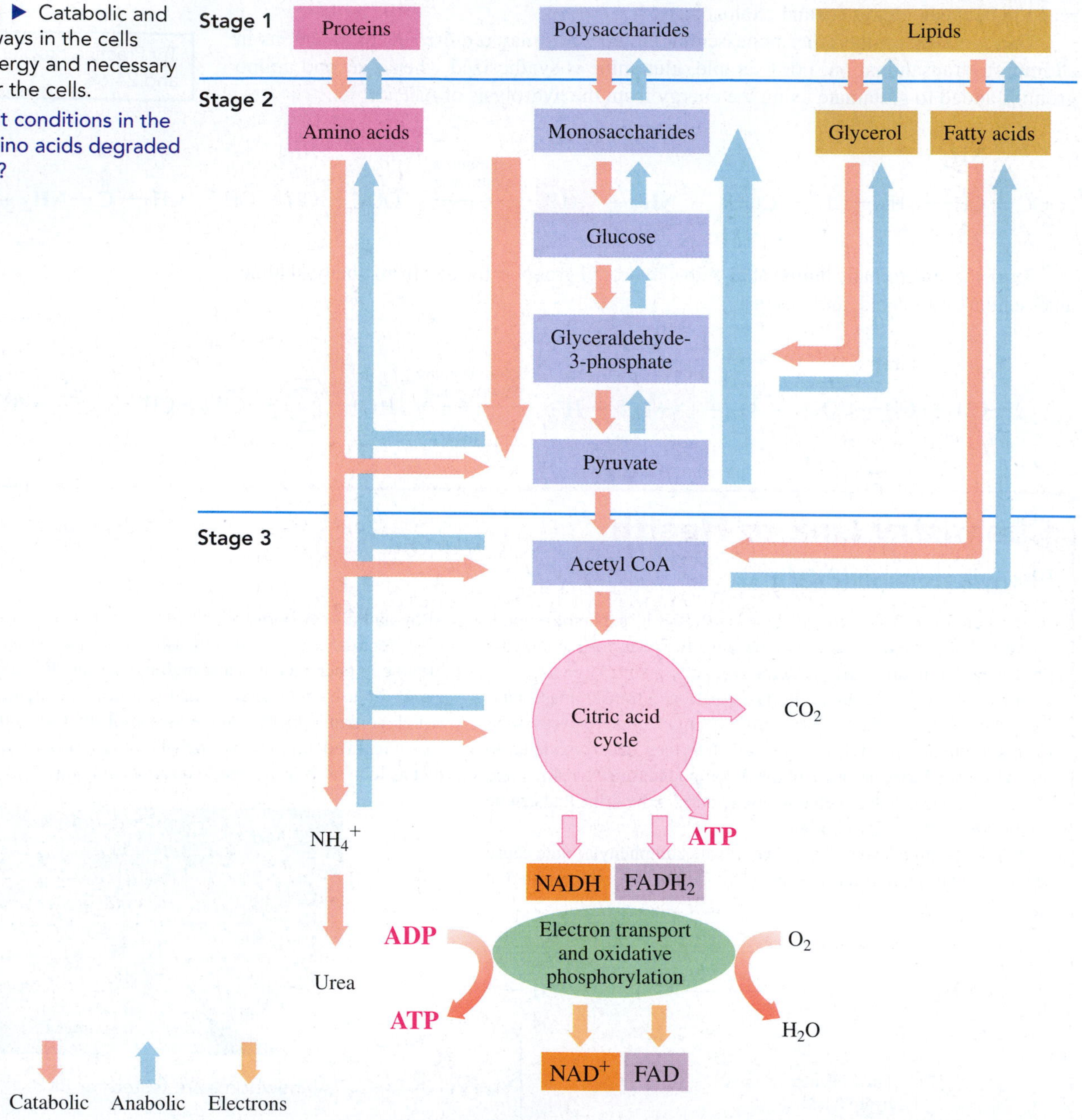
Stage 1
Proteins
Polysaccharides
Lipids
Stage 2
Amino acids
Monosaccharides
Glycerol
Fatty acids
Glucose
Glyceraldehyde-3-phosphate
Pyruvate
Stage 3
Acetyl CoA
Citric acid cycle
CO_2
ATP
NH_4^+
Urea
NADH
$FADH_2$
ADP
Electron transport and oxidative phosphorylation
O_2
ATP
H_2O
NAD^+
FAD
Catabolic
Anabolic
Electrons

FIGURE 24.12 ▶ Catabolic and anabolic pathways in the cells provide the energy and necessary compounds for the cells.

Q Under what conditions in the cell are amino acids degraded for energy?

PRACTICE PROBLEMS

24.9 Synthesis of Amino Acids

24.65 What do we call the amino acids that humans can synthesize?

24.66 How do humans obtain the amino acids that cannot be synthesized in the body?

24.67 How is glutamate converted to glutamine?

24.68 What amino acid can be converted into the amino acid tyrosine?

Clinical Applications

24.69 What does the abbreviation PKU represent (see Chemistry Link to Health: Phenylketonuria)?

24.70 How is PKU treated (see Chemistry Link to Health: Phenylketonuria)?

CLINICAL UPDATE Treatment of Luke's Hepatitis C

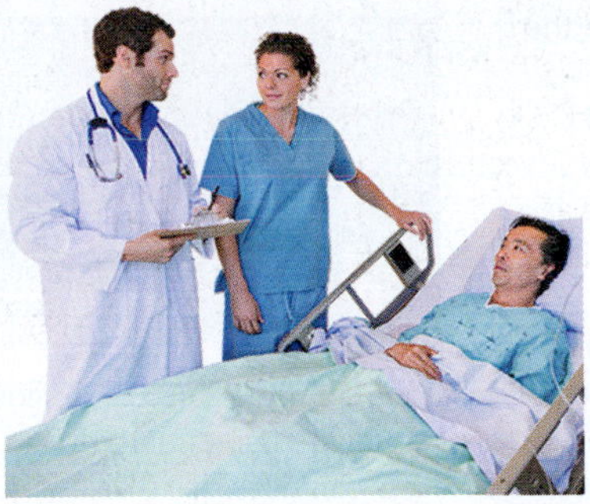

When Luke has a follow-up liver evaluation three months later, his liver enzymes are more elevated: ALT 356 Units/L and AST 418 Units/L. His doctor suggests that Luke begin interferon and ribavirin therapy for six months. Interferon and ribavirin work together as antiviral agents to inhibit the replication of the hepatitis C virus and strengthen the immune system. The interferon is injected subcutaneously or intramuscularly three times a week for nine months. The ribavirin is taken orally as 400-mg tablets twice a day.

After eight weeks of treatment, Luke's liver enzymes have lowered to ALT 85 Units/L and AST 115 Units/L. After four months of therapy, his ALT and AST enzyme levels are each within the normal range. Luke will continue to have a liver profile every three months for the next year, and then every six months. Monitoring of his liver enzymes will continue throughout his life.

Clinical Applications

24.71 Draw the condensed structural formulas for the products of the reaction of alanine and α-ketoglutarate, which is catalyzed by alanine transaminase (ALT).

24.72 Draw the condensed structural formulas for the products of the reaction of aspartate and α-ketoglutarate, which is catalyzed by aspartate transaminase (AST).

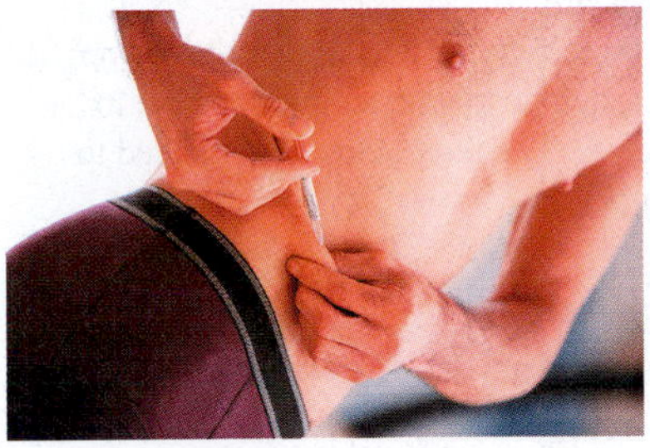

Interferon is one of the antiviral agents used to treat hepatitis C.

CONCEPT MAP

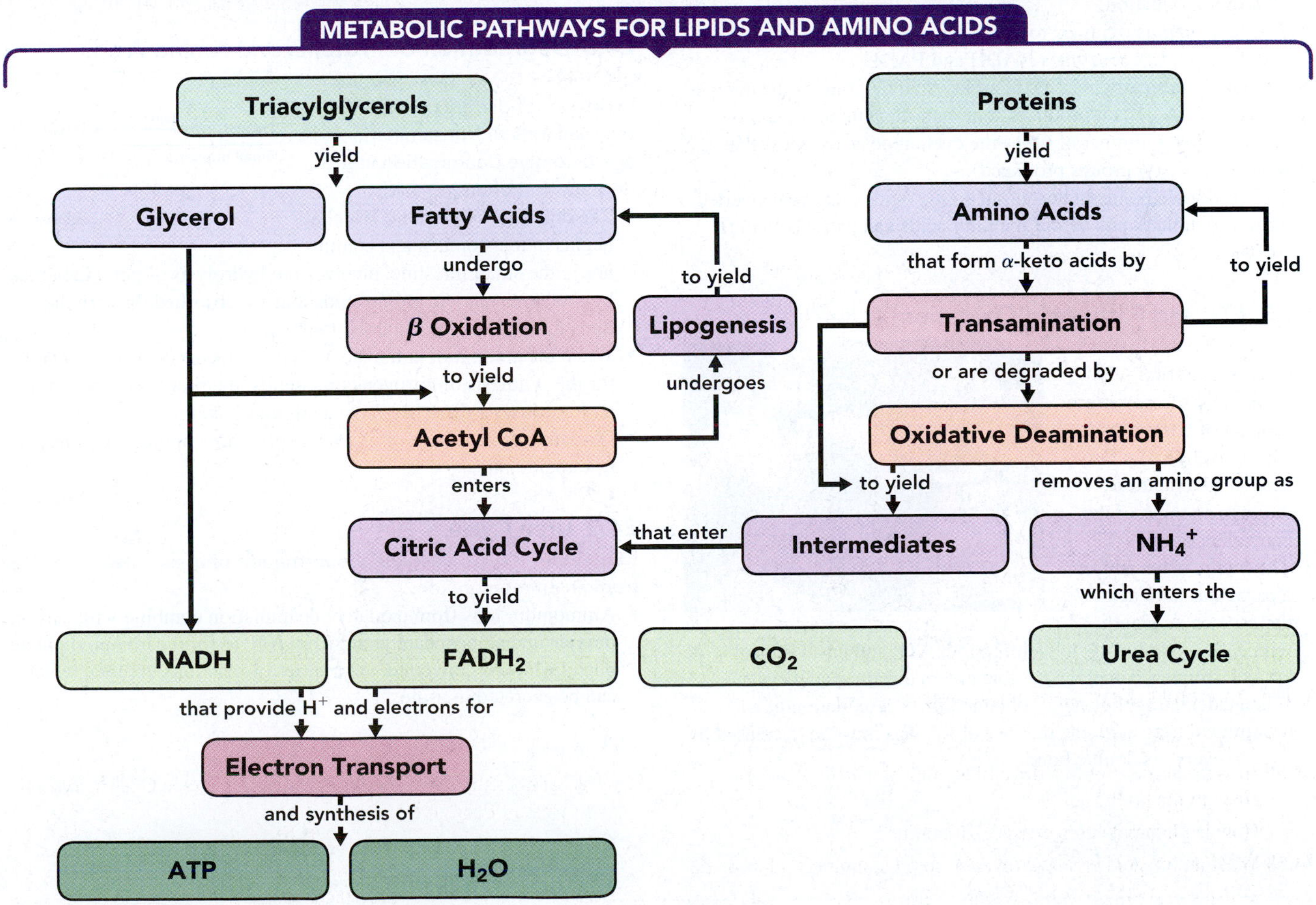
METABOLIC PATHWAYS FOR LIPIDS AND AMINO ACIDS
Triacylglycerols
yield
Glycerol
Fatty Acids
undergo
β Oxidation
to yield
Lipogenesis
undergoes
Acetyl CoA
enters
Citric Acid Cycle
to yield
NADH
FADH₂
CO₂
that provide H⁺ and electrons for
Electron Transport
and synthesis of
ATP
H₂O
Proteins
yield
Amino Acids
that form α-keto acids by
to yield
Transamination
or are degraded by
Oxidative Deamination
to yield
removes an amino group as
Intermediates
that enter
NH₄⁺
which enters the
Urea Cycle

CHAPTER REVIEW

24.1 Digestion of Triacylglycerols

LEARNING GOAL Describe the process by which triacylglycerols are digested.

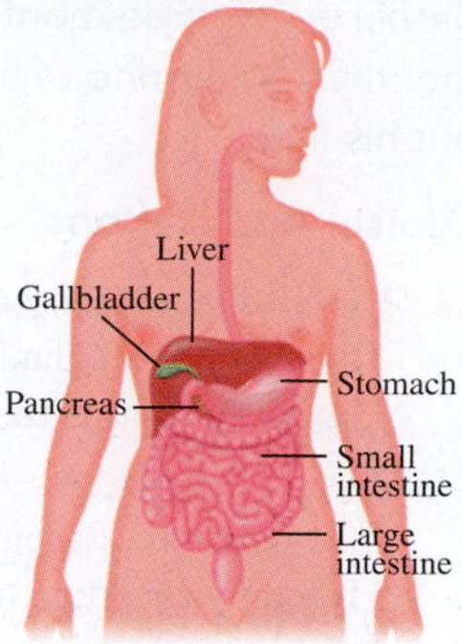

Liver
Gallbladder
Stomach
Pancreas
Small intestine
Large intestine

- Triacylglycerols are hydrolyzed in the small intestine to yield monoacylglycerols and fatty acids, which enter the intestinal wall and form new triacylglycerols.
- Triacylglycerols bind with proteins to form chylomicrons, which transport them through the lymphatic system and bloodstream to the tissues.
- In the tissues, triacylglycerols are hydrolyzed to glycerol and three fatty acids. Glycerol is converted in the liver to dihydroxyacetone phosphate, an intermediate in glycolysis and gluconeogenesis. The fatty acids are oxidized to provide energy for ATP synthesis.

24.2 Oxidation of Fatty Acids

LEARNING GOAL Describe the metabolic pathway of β oxidation.

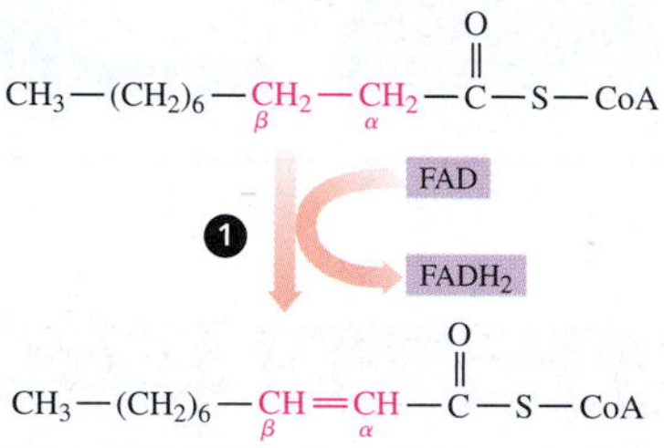

CH3—(CH2)6—CH2—CH2—C(=O)—S—CoA
β α
FAD
1
FADH2
CH3—(CH2)6—CH=CH—C(=O)—S—CoA
β α

- When used as an energy source, fatty acids link to coenzyme A to be transported into the mitochondria, where they undergo β oxidation.
- The acyl chain is oxidized to yield a shorter fatty acid, acetyl CoA, and the reduced coenzymes NADH and $FADH_2$.
- The number of carbon atoms in a fatty acid determines the number of acetyl CoA units it produces (one-half the number of carbon atoms) and the number of times the cycle repeats (one less than the number of acetyl groups produced).
- The cis double bonds in unsaturated fatty acids must be converted to trans double bonds before the fatty acids can participate in β oxidation.

24.3 ATP and Fatty Acid Oxidation

LEARNING GOAL Calculate the total ATP produced by the complete oxidation of a fatty acid.

- The activation of a fatty acid for β oxidation requires an input of the equivalent of two ATP.
- The energy obtained from a particular fatty acid depends on its length, with each oxidation cycle yielding four ATP and an additional 10 ATP from each acetyl CoA that enters the citric acid cycle.
- When calculating the energy obtained from β oxidation of an unsaturated fatty acid, the number of $FADH_2$ must be decreased by one for every cis double bond.

24.4 Ketogenesis and Ketone Bodies

LEARNING GOAL Describe the pathway of ketogenesis.

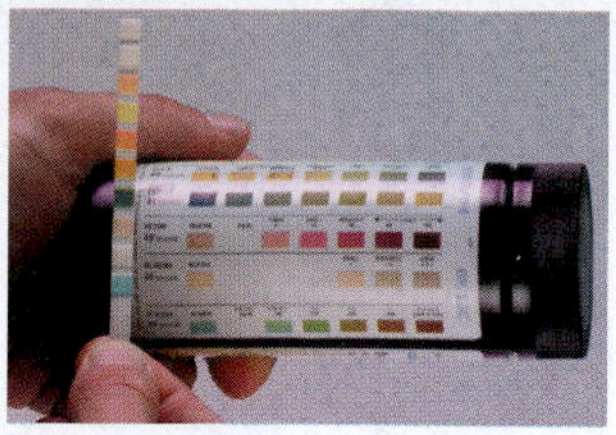

- When high levels of acetyl CoA are present in the cell, they enter the ketogenesis pathway, forming ketone bodies such as acetoacetate.
- Ketosis is the metabolic condition in which the body uses ketone bodies for energy instead of glucose.
- When ketone levels become excessively high, a potentially life-threatening condition called ketoacidosis can develop.

24.5 Fatty Acid Synthesis

LEARNING GOAL Describe the synthesis of fatty acids from acetyl CoA.

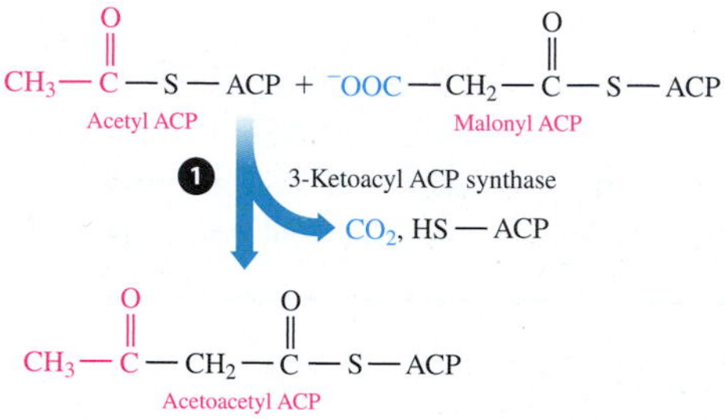

CH3—C(=O)—S—ACP + ⁻OOC—CH2—C(=O)—S—ACP
Acetyl ACP
Malonyl ACP
1
3-Ketoacyl ACP synthase
CO2, HS—ACP
CH3—C(=O)—CH2—C(=O)—S—ACP
Acetoacetyl ACP

- When there is an excess of acetyl CoA in the cell, the two-carbon acetyl CoA units link together to form fatty acids.

24.6 Degradation of Proteins and Amino Acids

LEARNING GOAL Describe the hydrolysis of dietary protein and the reactions of transamination and oxidative deamination in the degradation of amino acids.

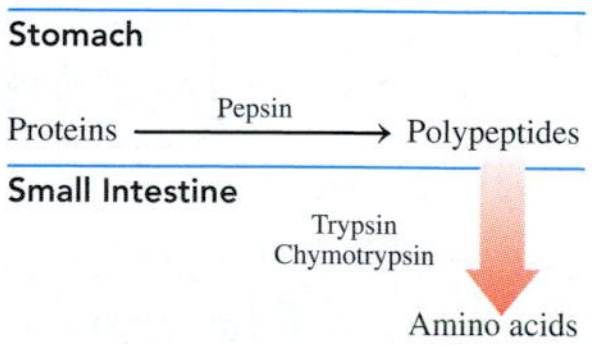

Stomach
Proteins
Pepsin
Polypeptides
Small Intestine
Trypsin
Chymotrypsin
Amino acids

- The digestion of proteins, which begins in the stomach and continues in the small intestine, involves the hydrolysis of peptide bonds by proteases to yield amino acids that are absorbed through the intestinal wall and transported to the cells.
- When the amount of amino acids in the cells exceeds that needed for the synthesis of nitrogen compounds, the process of transamination converts them to α-keto acids and glutamate.
- Oxidative deamination of glutamate produces ammonium ions and α-ketoglutarate.

24.7 Urea Cycle

LEARNING GOAL Describe the formation of urea from an ammonium ion.

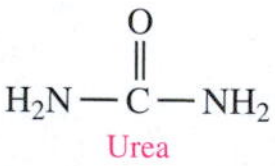

H2N—C(=O)—NH2
Urea

- Ammonium ions from oxidative deamination combine with carbon dioxide and a phosphate group from ATP to form carbamoyl phosphate, which is converted, via a series of reactions, to urea, which can be excreted in urine.

24.8 Fates of the Carbon Atoms from Amino Acids

LEARNING GOAL Describe where carbon atoms from amino acids enter the citric acid cycle or other pathways.

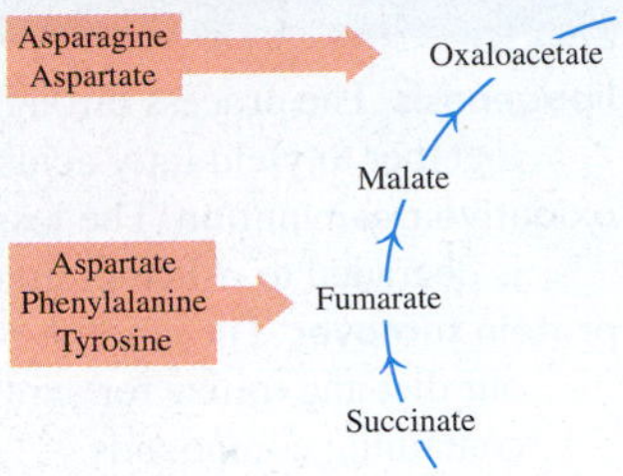
Asparagine
Aspartate
Oxaloacetate
Malate
Aspartate
Phenylalanine
Tyrosine
Fumarate
Succinate

- The carbon atoms from the degradation of glucogenic amino acids enter the citric acid cycle or gluconeogenesis.
- Ketogenic amino acids produce acetyl CoA or acetoacetyl CoA for ketogenesis.

24.9 Synthesis of Amino Acids

LEARNING GOAL Describe how some nonessential amino acids are synthesized from intermediates in the citric acid cycle and other metabolic pathways.

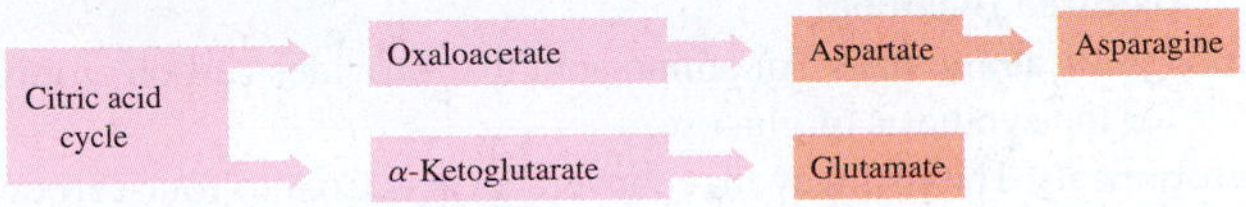
Citric acid cycle
Oxaloacetate
Aspartate
Asparagine
α-Ketoglutarate
Glutamate

- Nonessential amino acids are synthesized when amino groups from glutamate are transferred to an α-keto acid obtained from glycolysis or the citric acid cycle.

SUMMARY OF REACTIONS

The chapter Sections to review are shown after the name of each reaction.

Digestion of Triacylglycerols (24.1)

$$\text{Triacylglycerol} + 2H_2O \xrightarrow{\text{Pancreatic lipase}} \text{monoacylglycerol} + 2\text{ fatty acids}$$

Metabolism of Glycerol (24.1)

$$\text{Glycerol} + \text{ATP} + \text{NAD}^+ \longrightarrow \text{dihydroxyacetone phosphate} + \text{ADP} + \text{NADH} + H^+$$

Transamination (24.6)

$$\underset{\text{Alanine}}{CH_3-CH(\overset{+}{N}H_3)-COO^-} + \underset{\alpha\text{-Ketoglutarate}}{{}^-OOC-C(=O)-CH_2-CH_2-COO^-} \xrightleftharpoons{\text{Alanine transaminase}} \underset{\text{Pyruvate}}{CH_3-C(=O)-COO^-} + \underset{\text{Glutamate}}{{}^-OOC-CH(\overset{+}{N}H_3)-CH_2-CH_2-COO^-}$$

Oxidative Deamination (24.6)

$$\underset{\text{Glutamate}}{{}^-OOC-CH(\overset{+}{N}H_3)-CH_2-CH_2-COO^-} + H_2O + NAD^+ \xrightleftharpoons{\text{Glutamate dehydrogenase}} \underset{\alpha\text{-Ketoglutarate}}{{}^-OOC-C(=O)-CH_2-CH_2-COO^-} + NH_4^+ + \text{NADH} + H^+$$

Urea Cycle (24.7)

$$2NH_4^+ + CO_2 \longrightarrow \underset{\text{Urea}}{H_2N-C(=O)-NH_2} + 2H^+ + H_2O$$

KEY TERMS

beta oxidation (β oxidation) The degradation of fatty acids that removes two-carbon segments from a fatty acid chain at the oxidized β-carbon.

glucogenic amino acid An amino acid that provides carbon atoms for the synthesis of glucose.

ketogenesis The pathway that converts acetyl CoA to four-carbon acetoacetate and other ketone bodies.

ketogenic amino acid An amino acid that provides carbon atoms for the synthesis of fatty acids or ketone bodies.

ketone bodies The products of ketogenesis: acetoacetate, 3-hydroxybutyrate, and acetone.

lipogenesis The process through which two-carbon acetyl units link together to yield fatty acids.

oxidative deamination The loss of ammonium ion when glutamate is degraded to α-ketoglutarate.

protein turnover The amount of protein that we break down from our diet and utilize for synthesis of proteins and nitrogen-containing compounds.

transamination The transfer of an amino group from an amino acid to an α-keto acid.

urea cycle The process by which ammonium ions from the degradation of amino acids are converted to urea.

CORE CHEMISTRY SKILLS

The chapter Section containing each Core Chemistry Skill is shown in parentheses at the end of each heading.

Calculating the ATP from Fatty Acid Oxidation (β Oxidation) (24.3)

- When needed as an energy source, fatty acids are linked to coenzyme A and transported into the mitochondria where they undergo β oxidation.
- The fatty acyl CoA is oxidized to yield a shorter fatty acyl CoA, acetyl CoA, and reduced coenzymes NADH and $FADH_2$.
- Although the energy from a particular fatty acid depends on its length, each oxidation cycle yields four ATP with another 10 ATP from each acetyl CoA that enters the citric acid cycle.

Example: How many β oxidation cycles, acetyl CoA, and ATP are produced by the complete oxidation of caprylic acid, the C_8 saturated fatty acid?

Answer: An eight-carbon fatty acid will undergo three β oxidation cycles and produce four acetyl CoA. Each oxidation cycle produces four ATP (from one NADH and one $FADH_2$) for a total of 12 ATP. Each of the four acetyl CoA produces 10 ATP in the citric acid cycle or a total of 40 ATP. The 52 ATP minus two ATP for initial activation gives a total of 50 ATP from caprylic acid.

Describing How Ketone Bodies are Formed (24.4)

- Ketone bodies are formed when the body produces more acetyl CoA than it needs for energy.
- Excess acetyl CoA combines to form four-carbon and three-carbon ketone bodies.
- The ketone bodies are acetoacetate, 3-hydroxybutyrate, and acetone.

Example: Acetone is a three-carbon ketone body. How is it formed from the two-carbon acetyl CoA?

Answer: Two acetyl CoA combine to form the four-carbon molecule acetoacetate, which undergoes decarboxylation to form the three-carbon molecule acetone.

Distinguishing Anabolic and Catabolic Pathways (24.9)

- Pathways that degrade large molecules to small molecules to produce energy are catabolic.
- Those pathways that require energy for the biosynthesis of large molecules from small molecules are anabolic.

Example: Identify the following pathways as anabolic or catabolic:

a. β oxidation of fatty acids **b.** synthesis of fatty acids
c. degradation of amino acids **d.** synthesis of urea

Answer: **a.** catabolic **b.** anabolic
c. catabolic **d.** anabolic

UNDERSTANDING THE CONCEPTS

The chapter Sections to review are shown in parentheses at the end of each problem.

24.73 Lauric acid (12:0) is a fatty acid found in coconut oil. (24.2, 24.3)

Coconut oil is high in lauric acid.

a. Draw the line-angle formula for the activated form of lauric acid.
b. Indicate the α- and β-carbon atoms in the acyl molecule.
c. How many cycles of β oxidation are needed for the complete oxidation of lauric acid?
d. How many acetyl CoA units are produced from the complete oxidation of lauric acid?
e. Calculate the total ATP yield from the complete β oxidation of lauric acid by completing the following:

Activation	$\longrightarrow$	−2 ATP
_____ acetyl CoA	$\longrightarrow$	_____ ATP
_____ NADH	$\longrightarrow$	_____ ATP
_____ $FADH_2$	$\longrightarrow$	_____ ATP
Total		_____ **ATP**

24.74 Arachidic acid (20:0) is a fatty acid found in peanut and fish oils. (24.2, 24.3)

Peanuts contain arachidic acid.

a. Draw the line-angle formula for the activated form of arachidic acid.
b. Indicate the α- and β-carbon atoms in the acyl molecule.
c. How many cycles of β oxidation are needed for the complete oxidation of arachidic acid?
d. How many acetyl CoA units are produced from the complete oxidation of arachidic acid?
e. Calculate the total ATP yield from the complete β oxidation of arachidic acid by completing the following:

Activation	$\longrightarrow$	−2 ATP
______ acetyl CoA	$\longrightarrow$	______ ATP
______ NADH	$\longrightarrow$	______ ATP
______ $FADH_2$	$\longrightarrow$	______ ATP
Total		______ **ATP**

ADDITIONAL PRACTICE PROBLEMS

24.75 How are dietary triacylglycerols digested? (24.1)

24.76 What is a chylomicron? (24.1)

24.77 Why are the fats in the adipose tissues of the body considered the major form of stored energy? (24.1)

24.78 How are fatty acids obtained from stored fats? (24.1)

24.79 Why doesn't the brain utilize fatty acids for energy? (24.1)

24.80 Why don't red blood cells utilize fatty acids for energy? (24.1)

Clinical Applications

24.81 A triacylglycerol is hydrolyzed in the fat cells of adipose tissues, and the fatty acid is transported to the liver. (24.1, 24.2)
a. What happens to the glycerol?
b. Where in the liver cells is the fatty acid activated for β oxidation?
c. What is the energy cost for activation of the fatty acid?
d. What is the purpose of activating fatty acids?

24.82 Consider the β oxidation of a saturated fatty acid. (24.2, 24.3)
a. What is the activated form of the fatty acid?
b. Why is this oxidation called β oxidation?
c. Which reactions in the fatty acid cycle require coenzymes?
d. What is the yield in ATP for one cycle of β oxidation?

24.83 Identify each of the following as involved in β oxidation or in fatty acid synthesis: (24.2, 24.3, 24.5)
a. NAD^+
b. occurs in the mitochondrial matrix
c. malonyl ACP
d. cleavage of a two-carbon acetyl group
e. acyl carrier protein
f. carboxylase

24.84 Identify each of the following as involved in β oxidation or in fatty acid synthesis: (24.2, 24.3, 24.5)
a. NADPH **b.** takes place in the cytosol
c. FAD **d.** oxidation of a hydroxyl group
e. coenzyme A **f.** hydration of a double bond

24.85 The metabolism of triacylglycerols and carbohydrates is influenced by the hormones insulin and glucagon. Indicate the results of each of the following as stimulating fatty acid oxidation or fatty acid synthesis: (24.2, 24.3, 24.5)
a. high blood glucose **b.** secretion of glucagon

24.86 The metabolism of triacylglycerols and carbohydrates is influenced by the hormones insulin and glucagon. Indicate the results of each of the following as stimulating fatty acid oxidation or fatty acid synthesis: (24.2, 24.3, 24.5)
a. low blood glucose **b.** secretion of insulin

24.87 Why is ammonium ion that is produced in the liver converted immediately to urea? (24.7)

24.88 What compound is regenerated to repeat the urea cycle? (24.7)

24.89 Indicate the reactant in the urea cycle that reacts with each of the following compounds: (24.7)
a. aspartate **b.** ornithine

24.90 Indicate the products of the urea cycle that use each of the following compounds: (24.7)
a. arginine **b.** argininosuccinate

24.91 Which citric acid cycle intermediate can be produced from the carbon atoms of each of the following amino acids? (24.8)
a. serine **b.** lysine
c. methionine **d.** glutamate

24.92 Which citric acid cycle intermediate can be produced from the carbon atoms of each of the following amino acids? (24.8)
a. glycine **b.** isoleucine
c. histidine **d.** phenylalanine

24.93 How much ATP can be produced from the degradation of serine? (24.8)

24.94 Calculate the total ATP produced in the complete oxidation of caproic acid, $C_6H_{12}O_2$, and compare it with the total ATP produced from the oxidation of glucose, $C_6H_{12}O_6$. (24.3, 24.9)

CHALLENGE PROBLEMS

The following problems are related to the topics in this chapter. However, they do not all follow the chapter order, and they require you to combine concepts and skills from several Sections. These problems will help you increase your critical thinking skills and prepare for your next exam.

24.95 Identify each of the following reactions in the β oxidation of palmitic acid, a C_{16} fatty acid, as activation, transport into mitochondria, first oxidation, hydration, second oxidation, or cleavage: (24.2)

a. Palmitoyl CoA and FAD form α,β-unsaturated palmitoyl CoA and $FADH_2$.

b. 3-Ketopalmitoyl CoA forms myristoyl CoA and acetyl CoA.

c. Palmitic acid, HS—CoA, and ATP form palmitoyl CoA.

24.96 Identify each of the following reactions in the β oxidation of palmitic acid, a C_{16} fatty acid, as activation, transport into mitochondria, first oxidation, hydration, second oxidation, or cleavage: (24.2)

a. α,β-Unsaturated palmitoyl CoA and H_2O form 3-hydroxypalmitoyl CoA.

b. 3-Hydroxypalmitoyl CoA and NAD^+ form 3-ketopalmitoyl CoA and NADH + H^+.

c. Palmitoyl CoA condenses with carnitine to form palmitoyl carnitine.

24.97 Draw the condensed structural formula for and give the name of the amino acid formed when the following α-keto acids undergo transamination with glutamate: (24.9)

a. 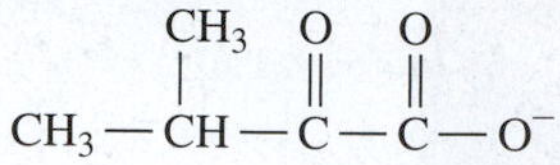

b. 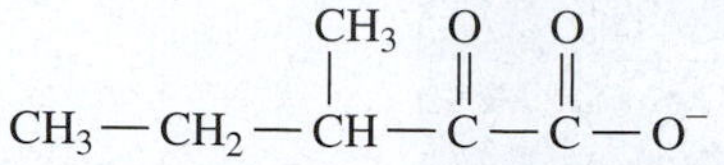

24.98 Draw the condensed structural formula for and give the name of the amino acid formed when the following α-keto acids undergo transamination with glutamate: (24.9)

a. 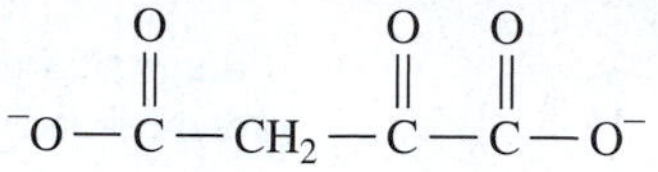

b. 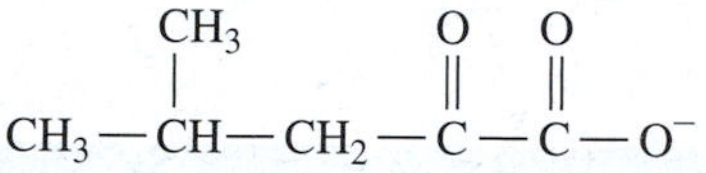

ANSWERS

24.1 The bile salts emulsify fat so that it forms small fat globules for hydrolysis by pancreatic lipase.

24.3 Fats are released from fat stores when blood glucose and glycogen stores are depleted.

24.5 Glycerol is converted to glycerol-3-phosphate, and then to dihydroxyacetone phosphate, an intermediate of glycolysis.

24.7 The enzyme glycerol kinase would be defective.

24.9 in the cytosol

24.11 FAD, NAD^+, and HS—CoA

24.13 a.

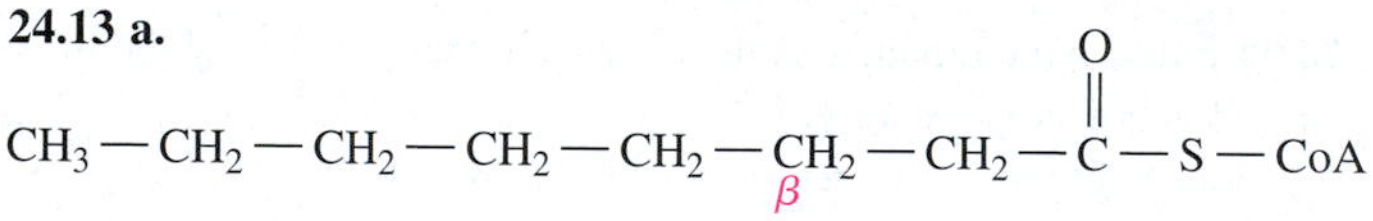

b.

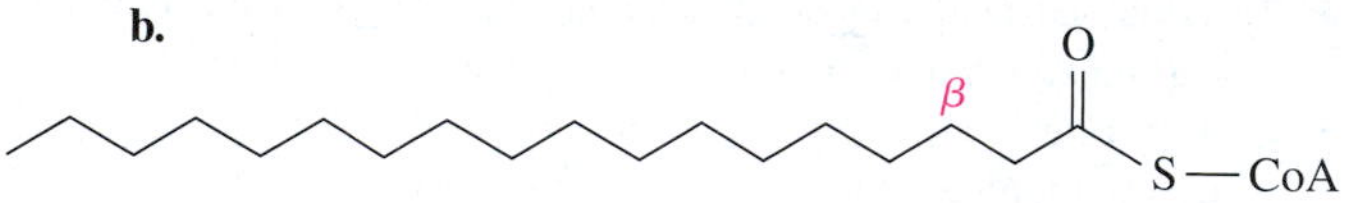

24.15 a.

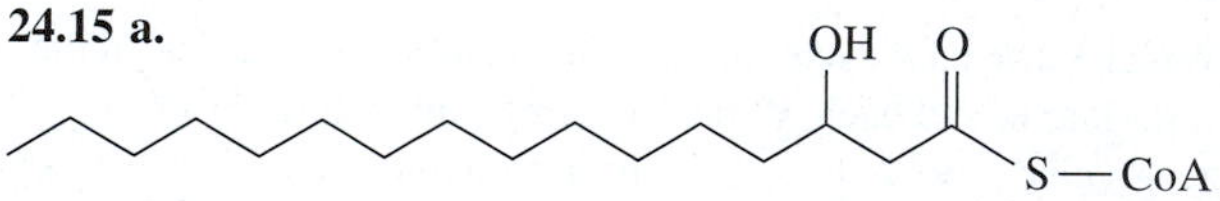

b. 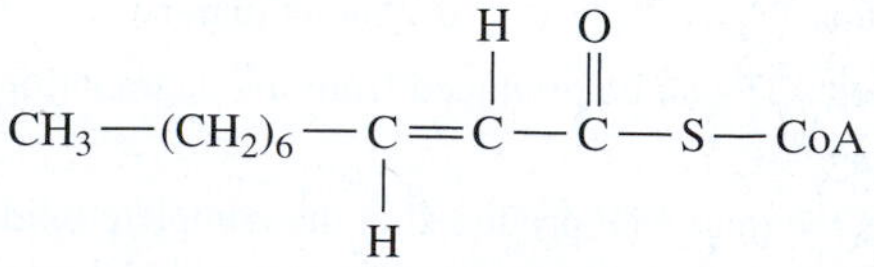

24.17 a. and **b.** 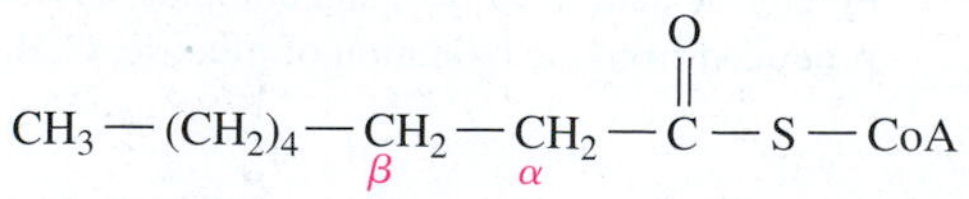

c. three β oxidation cycles

d. four acetyl CoA units

24.19 The hydrolysis of ATP to AMP involves the hydrolysis of ATP to ADP, and then ADP to AMP, which provides the same amount of energy as the hydrolysis of two ATP to two ADP.

24.21 a. 10 β oxidation cycles **b.** 11 acetyl CoA units

c. 110 ATP from 11 acetyl CoA (citric acid cycle) + 25 ATP from 10 NADH + 15 ATP from 10 $FADH_2$ − 2 ATP (activation) = 148 ATP (total)

24.23 a. 8 cycles of β oxidation **b.** 9 acetyl CoA units

c. 90 ATP from 9 acetyl CoA (citric acid cycle) + 20 ATP from 8 NADH + 10.5 ATP from 7 $FADH_2$ − 2 ATP (activation) = 118.5 ATP (total)

24.25 Hypoglycin A is metabolized in the body to produce a toxin that inhibits acyl CoA dehydrogenase, the first enzyme in the β oxidation pathway.

24.27 Ketogenesis is the synthesis of ketone bodies from excess acetyl CoA during fatty acid oxidation, which occurs when glucose is not available for energy, particularly in starvation, low-carbohydrate diets, fasting, alcoholism, and diabetes.

24.29 Acetoacetate undergoes reduction using NADH + H^+ to yield 3-hydroxybutyrate.

24.31 Ketoacidosis is a condition characterized by a drop in blood pH values, excessive urination, strong thirst, vomiting, shortness of breath, fatigue, and confusion.

24.33 Fatty acid synthesis primarily occurs in the cytosol of cells in adipose tissue.

24.35 acetyl CoA, HCO_3^-, and ATP

24.37 a. (3) malonyl CoA transacylase

b. (1) acetyl CoA carboxylase

c. (2) acetyl CoA transacylase

4

24.39 a. $4\ HCO_3^-$
b. 4 ATP
c. 5 acetyl CoA
d. 4 malonyl ACP
e. 8 NADPH
f. 4 CO_2 removed

24.41 a. Reaction 4, the reduction of the double bond to a single bond, is catalyzed by enoyl ACP reductase.
b. If fatty acid synthesis is inhibited, the bacteria cannot form cell membranes and will not thrive.

24.43 The digestion of proteins begins in the stomach and is completed in the small intestine.

24.45 Hormones, heme, purines and pyrimidines for nucleotides, proteins, nonessential amino acids, amino alcohols, and neurotransmitters require nitrogen obtained from amino acids.

24.47 The reactants are an amino acid and an α-keto acid, and the products are a new amino acid and a new α-keto acid.

24.49 a. $H - \overset{\overset{\displaystyle O}{\|}}{C} - COO^-$ 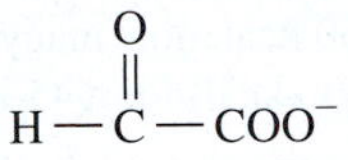

b. $HS - CH_2 - \overset{\overset{\displaystyle O}{\|}}{C} - COO^-$ 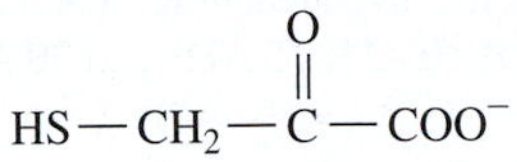

c. $CH_3 - \overset{\overset{\displaystyle CH_3}{|}}{CH} - \overset{\overset{\displaystyle O}{\|}}{C} - COO^-$ 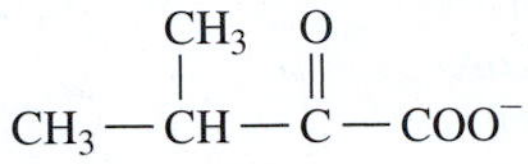

24.51

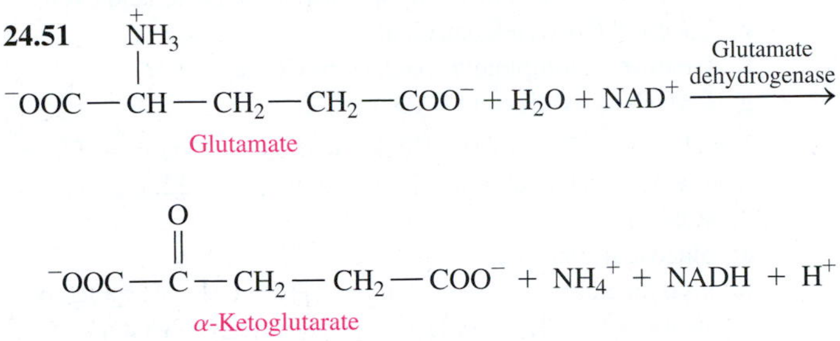

$\overset{\overset{\displaystyle +}{NH_3}}{|}$
$^-OOC - CH - CH_2 - CH_2 - COO^- + H_2O + NAD^+ \xrightarrow{\text{Glutamate dehydrogenase}}$
Glutamate

$^-OOC - \overset{\overset{\displaystyle O}{\|}}{C} - CH_2 - CH_2 - COO^- + NH_4^+ + NADH + H^+$
α-Ketoglutarate

24.53 $H_2N - \overset{\overset{\displaystyle O}{\|}}{C} - NH_2$ 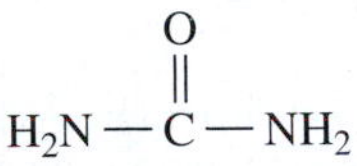

24.55 CO_2

24.57 NH_4^+ is toxic if allowed to accumulate.

24.59 If there is a deficiency of argininosuccinate synthetase, citrulline will accumulate.

24.61 Glucogenic amino acids are used to synthesize glucose.

24.63 a. oxaloacetate
b. oxaloacetate
c. succinyl CoA
d. α-ketoglutarate

24.65 nonessential amino acids

24.67 Glutamine synthetase catalyzes the addition of $-NH_3^+$ to glutamate using energy from the hydrolysis of ATP.

24.69 phenylketonuria

24.71 $CH_3 - \overset{\overset{\displaystyle O}{\|}}{C} - COO^- + {}^-OOC - \overset{\overset{\overset{\displaystyle +}{NH_3}}{|}}{CH} - CH_2 - CH_2 - COO^-$ 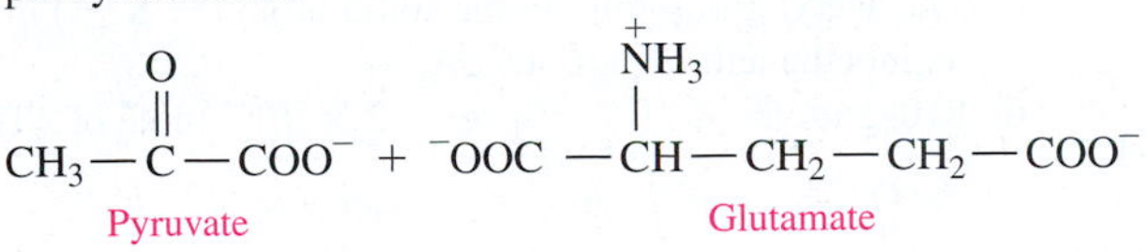
Pyruvate   Glutamate

24.73 a. and b. 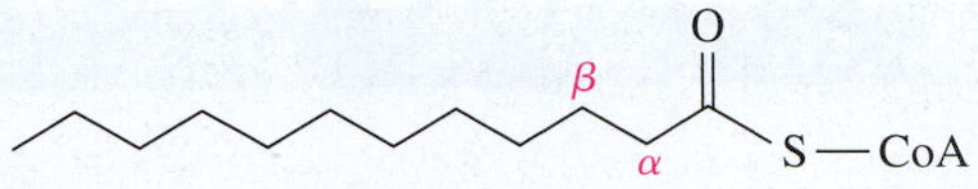

c. Five cycles of β oxidation are needed.
d. Six acetyl CoA units are produced.
e.

Activation	$\longrightarrow$	-2 ATP
6 acetyl CoA $\times$ 10 ATP/acetyl CoA	$\longrightarrow$	60 ATP
5 NADH $\times$ 2.5 ATP/NADH	$\longrightarrow$	12.5 ATP
5 $FADH_2$ $\times$ 1.5 ATP/$FADH_2$	$\longrightarrow$	7.5 ATP
Total		78 ATP

24.75 Triacylglycerols are hydrolyzed to monoacylglycerols and fatty acids in the small intestine, which reform as triacylglycerols in the intestinal lining for transport as lipoproteins to the tissues.

24.77 Fats can be stored in unlimited amounts in adipose tissue compared to the limited storage of carbohydrates as glycogen.

24.79 The fatty acids cannot diffuse across the blood–brain barrier.

24.81 a. Glycerol is converted to glycerol-3-phosphate and then to dihydroxyacetone phosphate, which can enter glycolysis or gluconeogenesis.
b. Activation of fatty acids occurs in the cytosol.
c. The energy cost is equal to two ATP.
d. Only acyl CoA can move into the intermembrane space for transport by carnitine into the matrix.

24.83 a. β oxidation
b. β oxidation
c. fatty acid synthesis
d. β oxidation
e. fatty acid synthesis
f. fatty acid synthesis

24.85 a. fatty acid synthesis
b. fatty acid oxidation

24.87 Ammonium ion is toxic if allowed to accumulate.

24.89 a. citrulline
b. carbamoyl phosphate

24.91 a. oxaloacetate
b. acetyl CoA
c. succinyl CoA
d. α-ketoglutarate

24.93 Serine is degraded to pyruvate, which is oxidized to acetyl CoA. The oxidation produces NADH + H^+, which can be oxidized to provide the energy to synthesize 2.5 ATP. From one turn of the citric acid cycle, the oxidation of the acetyl CoA eventually produces 10 ATP. Thus, serine can provide 12.5 ATP.

24.95 a. first oxidation
b. cleavage
c. activation

24.97 a. $CH_3 - \overset{\overset{\displaystyle CH_3}{|}}{CH} - \overset{\overset{\overset{\displaystyle +}{NH_3}}{|}}{CH} - \overset{\overset{\displaystyle O}{\|}}{C} - O^-$ Valine

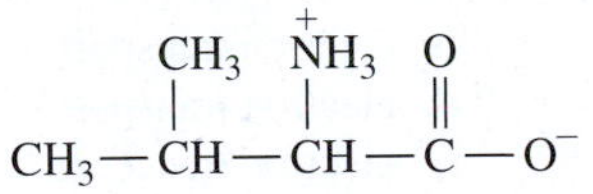

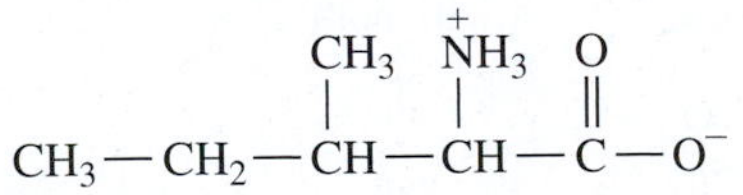

b. $CH_3 - CH_2 - \overset{\overset{\displaystyle CH_3}{|}}{CH} - \overset{\overset{\overset{\displaystyle +}{NH_3}}{|}}{CH} - \overset{\overset{\displaystyle O}{\|}}{C} - O^-$ Isoleucine

COMBINING IDEAS from Chapters 22 to 24

CI.47 Identify each of the following as a substance that is part of the citric acid cycle, electron transport, or both: (23.1, 23.2, 23.3)

a. GTP **b.** $CoQH_2$
c. $FADH_2$ **d.** cyt *c*
e. succinate dehydrogenase **f.** complex I
g. isocitrate **h.** NAD^+

CI.48 Use the value of 7.3 kcal per mole of ATP to determine the total kilocalories stored as ATP from each of the following: (23.4, 24.2, 24.3, 24.8)

a. the reactions of 1 mole of glucose in glycolysis
b. the oxidation of 2 moles of pyruvate to 2 moles of acetyl CoA
c. the complete oxidation of 1 mole of glucose to CO_2 and H_2O
d. the β oxidation of 1 mole of lauric acid, a saturated C_{12} fatty acid
e. the reaction of 1 mole of glutamate (from protein) in the citric acid cycle

CI.49 Acetyl CoA (HS—CoA) is the fuel for the citric acid cycle. It has the formula $C_{23}H_{38}N_7O_{17}P_3S$. (22.2, 23.1, 23.4)

a. What are the components of acetyl CoA?
b. What is the function of HS—CoA?
c. Where does the acetyl group attach to HS—CoA?
d. What is the molar mass (to three significant figures) of acetyl CoA?
e. How many moles of ATP are produced when 1.0 mg of acetyl CoA completes one turn of the citric acid cycle?

CI.50 State if each of the following produces or consumes ATP: (22.5, 22.6, 23.1, 23.4, 24.2)

a. citric acid cycle
b. glucose forms two pyruvate
c. pyruvate yields acetyl CoA
d. glucose forms glucose-6-phosphate
e. oxidation of α-ketoglutarate
f. transport of NADH across the mitochondrial membrane
g. activation of a fatty acid

CI.51 Butter is a fat that contains 80.% by mass triacylglycerols. Assume the triacylglycerol in butter is glyceryl tripalmitate. (17.4, 24.2, 24.3)

1 FIRST QUALITY 1
Unsalted
PLANT #49-34
ORGANIC
BUTTER
Sweet Cream
NET WT
4 OZ.
(113g)

Butter is high in triacylglycerols.

a. Write an equation for the hydrolysis of glyceryl tripalmitate.
b. What is the molar mass of glyceryl tripalmitate, $C_{51}H_{98}O_6$?
c. Calculate the ATP yield from the complete oxidation of 1 mole of palmitic acid, a saturated C_{16} fatty acid.
d. How many kilocalories are released from the palmitate in 0.50 oz of butter?
e. If running for exactly 1 h uses 750 kcal, how many ounces of butter would provide the energy (kcal) for a 45-min run?

CI.52 Match these ATP yields with the given reactions: 1 ATP, 1.5 ATP, 2 ATP, 2.5 ATP, 3 ATP, 5 ATP, 10 ATP, 15 ATP, and 30 ATP. (22.5, 22.6, 23.1, 23.4, 24.3)

a. Glucose yields two pyruvate.
b. Pyruvate yields acetyl CoA.
c. Glucose yields two acetyl CoA.
d. Acetyl CoA goes through one turn of the citric acid cycle.
e. Succinyl CoA yield succinate.
f. Glucose is completely oxidized to CO_2 and H_2O.
g. $FADH_2$ is oxidized to FAD.

CI.53 Which of the following molecules will produce more ATP per mole when each is completely oxidized? (22.4, 22.5, 23.1, 23.4, 24.3)

a. glucose or maltose
b. myristic acid, $CH_3—(CH_2)_{10}—CH_2—CH_2—COOH$, or stearic acid, $CH_3—(CH_2)_{14}—CH_2—CH_2—COOH$
c. glucose or two acetyl CoA

CI.54 Which of the following molecules will produce more ATP per mole when each is completely oxidized? (22.4, 22.5, 23.1, 23.4, 24.3)

a. glucose or caprylic acid C_8
b. citrate or succinate in one turn of the citric acid cycle
c. glutamate or tyrosine in one turn of the citric acid cycle

ANSWERS

CI.47 a. citric acid cycle **b.** electron transport
c. both **d.** electron transport
e. citric acid cycle **f.** electron transport
g. citric acid cycle **h.** both

CI.49 a. aminoethanethiol, pantothenic acid (vitamin B_5), and phosphorylated ADP
b. Coenzyme A carries an acetyl group to the citric acid cycle for oxidation.
c. The acetyl group links to the sulfur atom (—S—) in the aminoethanethiol part of CoA.
d. 810. g/mole **e.** 1.2×10^{-5} mole of ATP

CI.51 a.

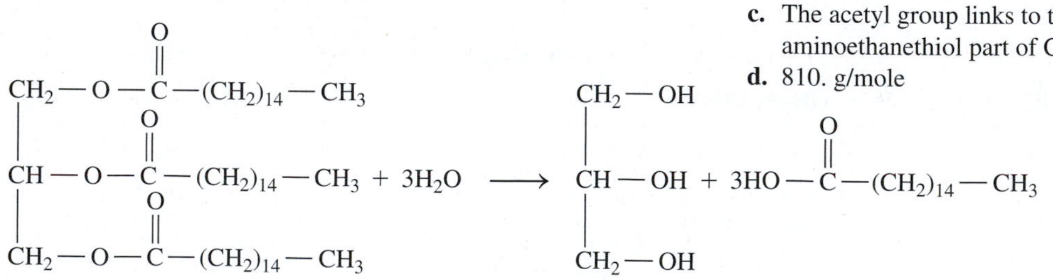

b. 807 g/mole **c.** 106 moles of ATP
d. 33 kcal **e.** 8.5 oz of butter

CI.53 a. maltose **b.** stearic acid **c.** glucose

Credits

Photo Credits

Chapter 1

p. 1 Anyaivanova/Shutterstock
p. 2 Maxx Studio/Shutterstock
p. 3 J.R. Eyerman/Time & Life Pictures/Getty Images
p. 4 *top:* Peter Van Evert/Alamy
p. 4 *center, left:* Slavko Sereda/Shutterstock
p. 4 *center, right:* Webphotographeer/E+/Getty Images
p. 4 *bottom:* Dmitriy Shironosov/Alamy
p. 5 Eugene Bochkarev/Getty Images
p. 8 *top:* Sergiy Kuzmin/123RF
p. 8 *bottom:* Track5/E+/Getty Images
p. 12 *top:* F9photos/Fotolia
p. 12 *bottom:* SpxChrome/E+/Getty Images
p. 13 Kevin Peschke/Alamy
p. 17 Jason Stitt/Shutterstock
p. 18 Heather Davies/Science Photo Library/Alamy

Chapter 2

p. 26 Avava/Shutterstock
p. 27 *top, right:* Rod Ferris/Shutterstock
p. 27 *bottom, left:* Andersen Ross/Blend Images/Getty Images
p. 28 *top:* Richard Megna/Fundamental Photographs, NYC
p. 31 Nick M. Do/iStock/Getty Images
p. 33 Radu Razvan/Shutterstock
p. 34 Emin kuliyev/Shutterstock
p. 37 *center, right:* Legolex/Fotolia
p. 37 *bottom, right:* Monkey Business Images/Shutterstock
p. 38 *top:* Chassenet/BSIP SA/Alamy
p. 42 *top, left:* Pongphan Ruengchai/Alamy
p. 42 *bottom, left:* Julie Woodhouse F/Alamy
p. 43 *top, right:* ExQuisine/Fotolia
p. 43 *center, right:* Paylessimages/123RF
p. 46 *center, left:* Stephen Mcsweeny/Shutterstock
p. 46 *bottom, left:* WavebreakMediaMicro/Fotolia
p. 47 Carlos Alvarez/Getty Images
p. 48 Gina Sanders/Fotolia
p. 51 *bottom, left:* Professor Pietro M. Motta/Science Source
p. 51 *bottom, right:* Voisin/Phanie/Science Source
p. 52 Syner-Comm/Alamy
p. 53 *top, right:* Jpc-Prod/Fotolia
p. 53 *bottom, left:* Dallas Events Inc/Shutterstock
p. 54 *right:* Helen Sessions/Alamy
p. 58 *bottom, left:* Romans14/Fotolia
p. 58 *top, right:* Bugtiger/Fotolia
p. 60 Cristian Ciureanu/Alamy

Chapter 3

p. 63 Network Photographer/Alamy
p. 64 *top, left:* Norman Chan/Fotolia
p. 64 *center, left:* VR Photos/Shutterstock
p. 64 *bottom, center:* Richard Megna/Fundamental Photographs, NYC
p. 64 *bottom, left:* Eric Schrader/Pearson Education, Inc.
p. 65 *top, left:* Rsooll/Fotolia
p. 65 *top, center:* Dvs71/Fotolia
p. 65 *center, right:* Jules Selmes/David Murray/Dorling Kindersley, Ltd.
p. 66 *top, left:* Richard Megna/Fundamental Photographs, NYC
p. 66 *top, right:* Eric Schrader/Pearson Education, Inc.
p. 66 *center:* Charles Stirling (Diving)/Alamy
p. 67 Siede Preis/Photodisc/Getty Images
p. 68 *top, left:* Justinb/Fotolia
p. 68 *top, right:* Lisay/E+/Getty Images
p. 68 *center, left:* Elena Elisseeva/Shutterstock
p. 69 *top:* Rafa Irusta/Shutterstock
p. 69 *center:* Westend61 GmbH/Alamy
p. 70 Jordache/Shutterstock
p. 72 Garo/Phanie/Alamy
p. 75 *top:* Jfergusonphotos/Fotolia
p. 75 *bottom:* Science Photo Library/Alamy
p. 79 Stockbyte/Getty Images
p. 80 Celso Pupo/Fotolia
p. 81 *top:* Alan Holden/Alamy
p. 81 *bottom:* Anna Moskvina/Shutterstock
p. 82 *bottom:* John A. Rizzo/Photodisc/Getty Images
p. 85 *top, right:* Wsphotos/E+/Getty Images
p. 87 Nikkytok/Fotolia
p. 92 *bottom, left:* Picsfive/Shutterstock
p. 92 *bottom, center:* Spencer Jones/Photodisc/Getty Images
p. 92 *bottom, right:* Teleginatania/Fotolia
p. 92 *top, right:* Franz Pfluegl/Getty Images
p. 93 Fuat Kose/E+/Getty Images
p. 94 Khoroshunova Olga/Shutterstock
p. 97 *top, left:* Impala/Fotolia
p. 97 *center, left:* Shalom Ormsby/Photodisc/Getty Images
p. 97 *center, right:* Kim Reinick/123RF
p. 98 *top, left:* Aleksandr Volkov/Fotolia
p. 98 *top, right:* Fuse/Getty Images

Chapter 4

p. 99 Martin Harvey/Alamy
p. 104 *bottom:* Polushkin Ivan/Shutterstock
p. 107 *top, right:* Sergiy Kuzmin/123RF
p. 107 *center, right:* Graham J. Hills/Science Source
p. 110 Eric Schrader/Fundamental Photographs, NYC
p. 112 *left:* Atiketta Sangasaeng/Shutterstock
p. 112 *right:* Gaby Jalbert/Getty Images
p. 114 Mindscape studio/Shutterstock
p. 118 *left:* Cindy Minear/Shutterstock
p. 118 *right:* Bruno Boissonnet/Science Source
p. 122 Olly/Fotolia
p. 135 *right:* Nigel Cattlin/Alamy
p. 140 Oktay Ortakcioglu/E+/Getty Images

Chapter 5

p. 145 Tyler Olson/Shutterstock
p. 147 Celig/Shutterstock
p. 148 Josh Sher/Science Source
p. 150 Uberphotos/E+/Getty Images
p. 151 Photo courtesy of Spruce Environmental Technologies, Inc.
p. 152 Laurent/B. HOP AME/BSIP SA/Alamy
p. 155 Australian Nuclear Science and Technology Organization
p. 156 *left:* Kyodo/Newscom
p. 156 *right:* Don Farrall/Stockbyte/Getty Images
p. 158 Jürgen Schulzki/Alamy
p. 161 Library of Congress (Photoduplication)
p. 162 Jihad Siqlawi/AFP/Getty Images
p. 164 *top, left* (a): Burger/Phanie/SuperStock
p. 164 *top, left* (b): Pasieka/Science Source
p. 164 *center, left:* Lawrence Berkeley/National Library/Photodisc/Getty Images
p. 164 *bottom, center:* Mehau Kulyk/Science Source
p. 164 *bottom, right:* GJLP/Science Source
p. 165 *center:* Karen C. Timberlake
p. 165 *right:* Cytyc Hologic Corporation
p. 168 *right:* CNRI/Science Source

Chapter 6

p. 174 Don Hammond/Design Pics Inc/Alamy
p. 179 *bottom, right:* J. Palys/Shutterstock
p. 180 *top, left:* Richard Megna/Fundamental Photographs, NYC
p. 182 Anthony Pleva/Alamy
p. 184 Gary Blakeley/Shutterstock
p. 185 Mark Huls/Fotolia
p. 186 *left:* Kameel4u/Shutterstock
p. 186 *right:* Singkham/Shutterstock
p. 188 *top:* Room27/Shutterstock
p. 188 *bottom:* Gerald Bernard/Shutterstock
p. 189 Studiomode/Alamy
p. 221 *top, right:* Augustin Ochsenreiter/Dapd/AP Images
p. 222 Photo courtesy of Spruce Environmental Technologies, Inc.

Chapter 7

p. 223 Javier Larrea/AGE Fotostock
p. 224 *bottom, right:* Sciencephotos/Alamy
p. 225 *top, left:* Lawrence Migdale/Science Source
p. 225 *top, right:* Paul Michael Hughes/Shutterstock
p. 229 Richard Megna/Fundamental Photographs, NYC
p. 233 *bottom, right:* Sergiy Zavgorodny/Shutterstock
p. 235 Mitch Hrdlicka/Getty Images
p. 236 *top, left:* Stinkyt/iStock/Getty Images
p. 239 Lissart/E+/Getty Images
p. 240 Reika/Shutterstock
p. 244 Norma Jean Gargasz/Alamy
p. 245 Ruzanna/Shutterstock
p. 246 Danny Smythe/Shutterstock
p. 247 Edward Westmacott/Shutterstock
p. 251 Thinkstock/Stockbyte/Getty Images
p. 255 Klaus Schräder/Westend61 GmbH/Alamy
p. 256 NASA
p. 261 Carolyn A McKeone/Getty Images
p. 262 *top, left:* Richard Megna/Fundamental Photographs, NYC
p. 262 *bottom, left:* Dario Secen/Lumi Images/Getty Images
p. 265 *top:* Gabriele Rohde/Fotolia
p. 265 *bottom:* Ljupco Smokovski/Fotolia

Chapter 8

p. 275 Adam Gault/Science Photo Library/Alamy
p. 277 NASA
p. 279 *top:* GybasDigiPhoto/Shutterstock
p. 279 *bottom:* Kenneth William Caleno/Shutterstock
p. 280 Koszivu/Fotolia
p. 281 Levent Konuk/Shutterstock
p. 284 Steve Bower/Shutterstock
p. 287 *top, right:* Prasit Rodphan/Shutterstock
p. 289 Andrey Nekrasov/Alamy
p. 292 Tomasz Wojnarowicz/Fotolia
p. 294 *top:* Yuri10/iStock/Getty Images
p. 294 *bottom:* Jaral Lertjamekorn/Shutterstock
p. 295 LoloStock/Shutterstock
p. 296 Mic Smith/Alamy
p. 297 Eric Schrader/Fundamental Photographs, NYC
p. 305 Library of Congress
p. 308 *bottom, left:* Flydragon/Shutterstock
p. 308 *top, right:* Mark Sykes/Science Photo Library/Alamy
p. 308 *center, right:* IS293/Alamy
p. 308 *bottom, right:* Oleksandr Kalinichenko/Shutterstock
p. 309 *left:* Patsy Michaud/Shutterstock
p. 309 *right:* Wrangler/Shutterstock

Chapter 9

p. 310 AJPhoto/Science Source
p. 313 *top, left:* Thinkstock Images/Getty Images
p. 318 Comstock Images/Getty Images
p. 320 *left:* Dr. P. Marazzi/Science Source
p. 320 *right:* Remik44992/Shutterstock
p. 321 Mega Pixel/Shutterstock
p. 323 CNRI/Science Source
p. 338 Ecologix Environmental Systems, LLC
p. 339 Jane/E+/Getty Images
p. 340 Photowind/Shutterstock
p. 341 Florida Images/Alamy
p. 343 *top, right:* Thinkstock Images/Getty Images
p. 345 Picsfive/Shutterstock
p. 350 Professor25/iStock/Getty Images
p. 351 Brand X Pictures/Getty Images

Chapter 10

p. 355 Fuse/Getty Images
p. 363 Gudellaphoto/Fotolia
p. 373 Incamerastock/Alamy
p. 375 *top:* Sirtravelalot/Shutterstock
p. 375 *bottom:* Photobac/Shutterstock

Chapter 11

p. 382 Lisa S./Shutterstock
p. 383 Magicinfoto/Shutterstock
p. 384 *center:* Kul Bhatia/Science Source
p. 384 *bottom:* Lukas Gojda/Shutterstock
p. 389 Magone/123RF
p. 390 *bottom, left:* Richard Megna/Fundamental Photographs, NYC
p. 391 Lana Langlois/Shutterstock
p. 399 *top, left:* George Tsartsianidis/123RF
p. 399 *top, center:* Fotofermer/Shutterstock
p. 399 *center, right:* PeterG/Shutterstock
p. 399 *bottom:* Luisa Leal/123RF
p. 400 *center:* Alexander Gospodinov/Fotolia
p. 400 *center, right:* Wayne Hutchinson/Alamy
p. 401 Eric Schrader/Pearson Education, Inc.
p. 403 Ana Bokan/Shutterstock
p. 404 Yenyu Shih/Shutterstock
p. 405 *top:* Richard Megna/Fundamental Photographs, NYC
p. 405 *bottom:* Eric Schrader/Pearson Education, Inc.
p. 407 *top:* Gerald Bernard/Shutterstock
p. 419 Geo Martinez/Getty Images
p. 421 Universal Images Group/SuperStock
p. 424 *center, left:* Rostislav Sedlacek/Fotolia
p. 424 *bottom, left:* Charles D. Winters/Science Source
p. 424 *center, right:* Masterchief_Productions/Shutterstock

Chapter 12

p. 426 Corepics/Fotolia
p. 428 *left:* lillisphotography/E+/Getty Images
p. 439 *bottom, left:* Natalie Jean/Shutterstock
p. 439 *bottom, right:* Nyvlt-art/Shutterstock
p. 441 *top:* Hans Chris/E+/Getty Images
p. 441 *center:* Roza/Fotolia
p. 442 Tim Hall/Photodisc/Getty Images
p. 446 *top, left:* James Trice/Getty Images
p. 447 Nigel Cattlin/Alamy
p. 452 *top, center:* Siede Preis/Photodisc/Getty Images
p. 452 *bottom, left:* Eduardo Luzzatti Buyé/Getty Images
p. 452 *bottom, center:* stocksnapper/123rf
p. 453 *left:* Alexandr Kanõkin/Shutterstock
p. 453 *center:* Vblinov/Shutterstock
p. 453 *right:* Dirkr Richter/E+/Getty Images
p. 456 *top, right:* Mediscan/Alamy
p. 457 Sumroeng/Fotolia
p. 461 *top, right:* Sean M. Carroll/Fotolia
p. 461 *center:* Eduardo Luzzatti Buyé/Getty Images
p. 464 *top:* Ingram Publishing/Getty Images
p. 464 *bottom:* Igor kisselev/Shutterstock

Chapter 13

p. 467 BSIP/UIG/Getty Images
p. 471 ASA studio/Shutterstock
p. 472 *right:* E+ /Getty Images
p. 473 *top, right:* Ultrashock/Shutterstock
p. 476 Renata Kazakova/Shutterstock
p. 479 *top:* Pictorial Press Ltd/Alamy
p. 479 *center:* Katharine Andriotis/Alamy
p. 479 *bottom:* Antagain/E+/Getty Images
p. 480 David Murray/Dorling Kindersley, Ltd.
p. 484 *top:* Chris Ryan/OJO Images Ltd/Alamy
p. 484 *bottom, left:* Schankz/Shutterstock
p. 484 *bottom, right:* Ekaterina Pokrovskaya/Shutterstock
p. 485 Marjan Laznik/E+/Getty Images
p. 487 *top, right:* Daniel Taeger/Shutterstock
p. 487 *top, left:* Carlos Romero/Shutterstock
p. 487 *center, right:* Elovich/Shutterstock
p. 487 *center, left:* WavebreakmediaMicro/Fotolia
p. 487 *bottom, left:* Tim UR/Fotolia
p. 493 Pixel 4 Images/Shutterstock

Chapter 14

p. 496 Viappy/Fotolia
p. 501 Robert Przybysz/Shutterstcok
p. 505 *bottom, right:* Richard Megna/Fundamental Photographs, NYC
p. 511 *top, right:* Bplanet/Shutterstock
p. 512 Wavebreakmedia/Shutterstock
p. 514 *left:* C Squared Studios/Photodisc/Getty Images
p. 514 *right:* Frank Greenaway/Dorling Kindersley Ltd.
p. 519 *top, left:* Gemphotography/Getty Images
p. 519 *bottom, left:* Allesalltag Bildagentur/AGE Fotostock
p. 519 *top, right:* GVictoria/Fotolia
p. 519 *bottom, right:* A-wrangler/Getty Images
p. 520 *left:* Marek Mierzejewski/Shutterstock
p. 520 *right:* Kakimage/Alamy

Chapter 15

p. 521 Rolf Bruderer/Blend Images/Alamy
p. 522 *right:* AZP Worldwide/Shutterstock
p. 523 *top, left:* Ljupco Smokovski/Shutterstock
p. 523 *center, left:* MOAimage/Shutterstock
p. 523 *bottom, left:* Nattika/Shutterstock
p. 523 *bottom, right:* Allesalltag Bildagentur/AGE Fotostock
p. 530 *left:* Q-Images/Alamy
p. 530 *center:* Steffen Hauser/Botanikfoto/Alamy
p. 530 *right:* Maksim Striganov/Shutterstock
p. 534 *top:* R McKown/Shutterstock
p. 534 *center:* Coprid/Shutterstock
p. 541 *top, left:* Ilya Andriyanov/Shutterstock
p. 541 *bottom, right:* Q-Images/Alamy
p. 548 Danny E Hooks/Shutterstock
p. 549 *top:* Used by permission of Harold E. Cross MD, PhD
p. 549 *bottom:* Sruilk/Shutterstock

Chapter 16

p. 560 Stefanolunardi/Shutterstock
p. 561 Skynet/Fotolia
p. 563 *bottom, left:* belushi/Shutterstock
p. 568 *left:* Jose Manuel Ribeiro/Reuters
p. 570 *top:* Egorxfi/Fotolia
p. 570 *bottom:* Mates/Fotolia
p. 571 *top:* Robophobic/Shutterstock
p. 572 Lynn Watson/Shutterstock
p. 573 *top, left:* Luisa Leal/Shutterstock
p. 573 *top, right:* travis manley/Shutterstock
p. 575 Podfoto/123RF
p. 576 HLPhoto/Shutterstock
p. 578 *top:* Alice S./BSIP/AGE Fotostock
p. 580 *left:* Denis Nata/Shutterstock
p. 580 *right:* Monika Wisniewska/Shutterstock
p. 582 *top:* Oksanka8306/Fotolia
p. 582 *bottom:* Malyshchyts Viktar/Fotolia

Chapter 17

p. 586 Michaeljung/Fotolia
p. 590 *left:* Chas53/Getty Images
p. 590 *right:* Dr3amer/Getty Images
p. 591 Jupiterimages/Getty Images
p. 594 PeJo/Shutterstock
p. 595 Tischenko Irina/Shutterstock
p. 596 Kuleczka/Shutterstock
p. 597 *right:* Dawn Wilson Photography/Moment/Getty Images
p. 606 Lotus_studio/Shutterstock
p. 608 National Heart, Lung, and Blood Institute
p. 609 Infinity/Alamy
p. 611 *center, left:* Science Stock Photography/Science Source
p. 611 *center, right:* Biophoto Associates/Science Source
p. 616 *left:* Robert Clare/Alamy
p. 616 *right:* Matt Rourke/AP Images
p. 619 *top:* Ekkachai/Shutterstock
p. 619 *bottom:* Sampsyseeds/Getty Images
p. 620 *center, left:* Egal/iStock/Getty Images
p. 620 *center, right:* Lauree Feldman/Photolibrary/Getty Images

Chapter 18

p. 626 *left:* Blickwinkel/Alamy
p. 626 *right:* Harvey Male/Imagestate Media Partners Limited - Impact Photos/Alamy
p. 632 *center:* Brasil2/Getty Images
p. 632 *bottom:* Darrin Jenkins/Alamy
p. 635 *top:* MikhailSh/Shutterstock
p. 635 *bottom:* De Meester Johan/Arterra Picture Library/Alamy
p. 638 National Institutes of Health
p. 640 David J. Green/Alamy
p. 641 *top, right:* Sherry Yates Young/Shutterstock
p. 650 *bottom, left:* Wayne Hutchinson/Alamy
p. 653 *bottom, left:* Denis Nata/Shutterstock
p. 653 *right:* Eric Schrader/Pearson Education, Inc.
p. 654 *bottom, left:* BartCo/E+/Getty Images
p. 655 Tereshchenko Dmitry/Shutterstock
p. 658 *top:* David Zaitz/Alamy
p. 658 *bottom:* Worldswildlifewonders/Shutterstock
p. 658 *center, right:* Sherry Yates Young/Shutterstock
p. 659 *left:* Analia Valeria Urani/Shutterstock
p. 659 *right:* Dustin Dennis/Shutterstock

Chapter 19

p. 660 Tetra Images/Alamy
p. 661 *top, right:* Fuse/Corbis/Getty Images
p. 661 *bottom, left:* Center for Disease Control (CDC) - PHIL
p. 665 Remik44992/Shutterstock
p. 673 *top:* Laurence Mouton/PhotoAlto sas/Alamy
p. 673 *bottom, right:* Ilknur Sen/Shutterstock
p. 684 *left:* Scott A. Frangos/Shutterstock
p. 684 *center:* Andy Crawford/Dorling Kindersley, Ltd.
p. 684 *right:* Dave Rudkin/Dorling Kindersley, Ltd.
p. 685 MaraZe/Shutterstock

Chapter 20

p. 688 Diedra Laird/The Charlotte Observer/MCT/Newscom
p. 690 Dr. Curt Samlaska, MD, FACP, FAAD, Asst. Professor, University of Nevada School of Medicine
p. 696 Natalia Pushchina/Shutterstock
p. 708 Sasimoto/Shutterstock
p. 714 Sheila Terry/Science Source
p. 717 *top, left:* ChiccoDodiFC/Shutterstock
p. 717 *bottom, left:* Arska N/Fotolia
p. 717 *top, right:* Jane/E+/Getty Images

Chapter 21

p. 721 Alex Traksel/Shutterstock
p. 745 Khunaspix/123RF
p. 746 *left:* Biophoto Associates/Science Source
p. 746 *right:* Mitskevich Uladzimir/Shutterstock
p. 749 Tek Image/Science Source
p. 753 Alfred Pasieka/Science Source
p. 762 Martin M. Rotker/Science Source

Chapter 22

p. 764 KatarzynaBialasiewicz/iStock/Getty Images
p. 765 Viafilms/iStock/Getty Images
p. 768 *top:* Rukxstockphoto/Fotolia
p. 768 *bottom:* Koji Aoki/Getty Images
p. 780 Stocksnapper/Shutterstock
p. 784 *top:* Andy Clark AC/TZ/Reuters
p. 784 *center:* Oleksii Sergieiev/Fotolia
p. 788 Biophoto Associates/Science Source
p. 790 Alexey Fursov/Shutterstock
p. 794 Dmitry Lobanov/Shutterstock
p. 798 Karen C. Timberlake

Chapter 23

p. 802 FatCamera/E+/Getty Images
p. 803 Viafilms/iStock/Getty Images
p. 814 Jill Lang/Shutterstock

Chapter 24

p. 825 Tyler Olson/Fotolia
p. 826 *top:* Steve Gschmeissner/Science Photo Library/Getty Images
p. 826 *bottom, left:* Clinical Photography/Central Manchester University Hospitals NHS Foundation Trust, UK/Science Source
p. 835 Findlay/Alamy
p. 838 *left:* Dewayne Flowers/Shutterstock
p. 838 *right:* Dmitry Lobanov/Shutterstock
p. 853 Staff Sgt Eric T. Sheler/U.S. Air Force
p. 855 *left:* Phanie/Alamy
p. 856 *bottom:* Findlay/Alamy
p. 858 Floortje/E+/Getty Images
p. 859 Matka Wariatka/Shutterstock
p. 862 Stargazer/Shutterstock

Text Credits

Chapter 1

p. 4 John Horgan, 1993. Profile: Linus C. Pauling, Stubbornly Ahead of His Time, *Scientific American,* March 1993, 36–37
p. 4 Mary Anne White, *Physical Properties of Materials,* Second Edition, CRC Press, Jun 28, 2011
p. 23 Sir Arthur Conan Doyle, *A Scandal in Bohemia,* June 25, 1891

Chapter 11

p. 415 Based on 2004 MedicineNet, Inc.

Chapter 20

p. 691 Lactase image by L. Frost of 1JYN 1D.H. Juers, T.D. Heightman, A. Vasella, J.D. McCarter, L. Mackenzie, S.G. Withers, B.W. Matthews 120012 Biochemistry 40: 14781-147942 created in Rasmol 2.7.5.2 1R. Sayle, E.J. White 119952 *Trends in Biochemical Sciences* (TIBS), 20192:374.2
p. 695 Alanine transaminase Image by L. Frost of 3IHJ 1M. Wisniewska, M.I. Siponen, C.H. Arrowsmith, H. Berglund, C. Bountra, R. Collins, A.M. Edwards, S. Flodin, A. Flores, S. Graslund, M. Hammarstrom, A. Johansson, I. Johansson, T. Karlberg, T. Kotenyova, A. Kotzsch, M. Moche, T.K. Nielsen, P. Nordlund, T. Nyman, C. Persson, A.K. Roos, P. Schutz, L. Svensson, A.G. Thorsell, L. Tresaugues, S. Van Den Berg, J. Weigelt, M. Welin, H. Schuler. Human alanine aminotransferase 2 in complex with PLP2 created in Rasmol 2.7.5.2 1R. Sayle, E.J. White 119952 *Trends in Biochemical Sciences* (TIBS), 20192:374.2
p. 701 Alpha-chymotrypsin image by L. Frost of 3CHA 1R.S. Blevins, A. Tulinsky 119852 J. Biol. Chem. 260: 4262–42752 created in Rasmol 2.7.5.2 1R. Sayle, E.J. White 119952 *Trends in Biochemical Sciences* (TIBS), 20192:374.2

Chapter 21

p. 754 Estrogen receptor image by L. Frost of 2ERD 1A.K. Shiau, D. Barstad, P.M. Loria, L. Chemg, P.J. Kushner, D.A. Agard, G.L. Greene 119982 Cell 1Cambridge, MA2 95:927–9372 created in Rasmol 2.7.5.2 1R. Sayle, E.J. White 119952 *Trends in Biochemical Sciences* (TIBS), 20192:374.2

Glossary/Index

Abbreviated electron configuration, 124, 125
Absolute zero, 71
Acceptor stem, 732
Acetal The product of the addition of two alcohols to an aldehyde or ketone, 508–511
Acetaminophen, 455, 647
Acetic acid, 561, 562, 563
Acetone, 500, 501
Acetyl ACP, 839
Acetylcholine, 638, 642
Acetyl CoA The compound that forms when a two-carbon acetyl unit bonds to coenzyme A, 773
Acetylsalicylic acid (aspirin), 2, 174, 401, 455, 571, 593
Acid A substance that dissolves in water and produces hydrogen ions (H^+), according to the Arrhenius theory. All acids are hydrogen ion donors, according to the Brønsted–Lowry theory, 383. *See also* Acids and bases
Acid–base titration, 407, 408
Acid dissociation constant, K_a The numerical value of the product of the ions from the dissociation of a weak acid divided by the concentration of the weak acid, 393
Acid dissociation expression, 393–395
Acid hydrolysis The splitting of an ester molecule in the presence of a strong acid to produce a carboxylic acid and an alcohol, 575–576, 648
Acidic amino acid An amino acid that has an R group with a carboxylate group ($-COO^-$), 662, 663
Acidosis, 413
Acid reflux, 414
Acid reflux disease, 382, 414–415
Acids and bases, 382–425
- acid dissociation expression, 393–395
- acids and carbonates or bicarbonates, 405
- acids and hydroxides (neutralization), 406
- acids and metals, 405
- Brønsted–Lowry theory, 385–388
- buffers, 409–414
- characteristics, 384, 394
- conjugate acid–base pair, 386–387, 417
- diprotic acids, 390
- direction of reaction, 392
- dissociation of water, 395–398
- dissociation of weak acids and bases, 393–395
- $[H_3O^+]$ and $[OH^-]$, 395–398, 418
- hydroxides, 384
- naming acids, 383–384
- naming bases, 384
- neutralization, 406
- pH. *See* pH
- reactions, 405–409, 418
- strong and weak acids, 388–389, 390, 394
- strong and weak bases, 391, 394
- titration, 407–408, 418
- water dissociation expression, 396

Aconitase, 804
ACP. *See* Acyl carrier protein (ACP)
Acquired immune deficiency syndrome (AIDS), 751–752
Actin, 661
Actinides, 102
Activation energy The energy that must be provided by a collision to break apart the bonds of the reacting molecules, 356, 357
Activator, 736
Active metals, 405
Active site A pocket in a part of the tertiary enzyme structure that binds to a substrate and catalyzes a reaction, 690–691
Active transport, 614, 615
Activity The rate at which an enzyme catalyzes the reaction that converts a substrate to a product, 696
Actron, 593
Actual yield The actual amount of product produced by a reaction, 256
Acyl carrier protein (ACP), 839
Acyl CoA dehydrogenase, 830
Acyl CoA synthetase, 829
Adderall, 639
Addition
- positive/negative numbers, 11
- significant figures (decimal places), 35

Addition A reaction in which atoms or groups of atoms bond to a carbon–carbon double bond. Addition reactions include the addition of hydrogen (hydrogenation), water (hydration), and monomers (polymerization), 447
Addition reactions, 447–453
- aldehydes and ketones, 507
- hydration, 448, 449–450, 459
- hydrogenation, 448, 449, 459
- polymerization, 448, 450–452, 459

Adenine (A), 722, 724
Adenine–thymine base pair, 728
Adenosine diphosphate. *See* ADP
Adenosine monophosphate (AMP), 767
Adenosine triphosphate. *See* ATP
Adipocytes, 826
ADP Adenosine diphosphate, formed by the hydrolysis of ATP; consists of adenine, a ribose sugar, and two phosphate groups, 723, 767
Adrenal corticosteroids, 612
Adrenaline, 640
Aerobic An oxygen-containing environment in the cells, 783

quantity and function in humans, 105
Phosphorylation, 701, 777, 778
Photosynthesis, 522
Phototherapy, 117, 118
pH paper, 399, 400
Physical change A change in which the physical properties of a substance change but its identity stays the same, 68, 69
Physical properties The properties that can be observed or measured without affecting the identity of a substance, 68, 69
Physical therapist (PT), 802
Physician assistant, 688
pico-, 37
Picocurie (pCi), 151
Piperidine, 634
PKU. *See* Phenylketonuria (PKU)
Place values, 10
Plaques, 672
Plastics, 572–573
Plum-pudding model, 108
Pneumocystis carinii, 751
Point mutation A mutation that replaces one base in a DNA with a different base, 742, 743
Polar amino acid An amino acid with a polar R group, 662, 663
Polar covalent bond A covalent bond in which the electrons are shared unequally between atoms, 200
Polar molecule A molecule containing bond dipoles that do not cancel, 206
Polar solvent, 312
Polarity A measure of the unequal sharing of electrons, indicated by the difference in electronegativities, 200
amino acids, 662
bonds, 199–202
molecules, 206–207
Polyatomic ion A group of covalently bonded nonmetal atoms that has an overall electrical charge, 186–190
common ions, listed, 187
equations, 228–230
formula, 187, 188–189
ionic compounds, 189
Lewis structure, 196–197
name, 187
Polycyclic aromatic hydrocarbons (PAHs), 456
Polydichloroethylene, 451, 452
Polyethylene, 450–452
Polyethyleneterephthalate (PETE), 573
Polymer A very large molecule that is composed of many small, repeating monomer units, 450, 451
Polymerase chain reaction (PCR) A procedure in which a strand of DNA is copied many times by mixing it with DNA polymerase and a mixture of deoxyribonucleotides, and subjecting it to repeated cycles of heating and cooling, 749
Polymerization, 448, 450–452, 459
Polypeptide, 665
Polypropylene, 451, 452
Polysaccharide A polymer of many monosaccharide units, usually glucose. Polysaccharides differ in the types of glycosidic bonds and the amount of branching in the polymer, 522, 546–548, 552
Polysome, 740
Polystyrene, 451, 452
Polytetrafluoroethylene (PTFE), 451, 452
Polyunsaturated fatty acid (PUFA) A fatty acid that contains two or more double bonds, 588, 589
Polyvinyl chloride (PVC), 451, 452
Pompe's disease, 789
p orbital, 120
Positive allosteric regulation, 699
Positive ion, 175–176
Positive numbers, 11
Positron A particle of radiation with no mass and a positive charge, symbol β^+ or ${}^{0}_{+1}e$, produced when a proton is transformed into a neutron and a positron, 147
Positron emission, 152–153, 154
Positron emission tomography (PET), 164
Potassium
atomic mass, 115
daily values (DV), 36
quantity and function in humans, 105
Potassium-40, 221
Potassium carbonate, 189
Potassium chloride, 221
Potassium-deficient potato plants, 135
Potassium ion, 318
Potential energy A type of energy related to position or composition of a substance, 74
Pounds per square inch (lb/in.^{2}, psi), 278, 279
ppb. *See* Parts per billion (ppb)
PPIs. *See* Proton pump inhibitors (PPIs)
ppm. *See* Parts per million (ppm)
Precipitate, 323
Prednisolone, 794
Prednisone, 612, 794
Prefix The part of the name of a metric unit that precedes the base unit and specifies the size of the measurement. All prefixes are related on a decimal scale, 36, 37
decreasing size of unit, 37
increasing size of unit, 37
Prefix in naming
naming alkanes, 430
naming molecular compounds, 191
Pre-mRNA, 735
Pressure The force exerted by gas particles that hit the walls of a container, 276
gas, 276–277, 278–280
ideal gas law, 296–297, 303
osmotic, 342
temperature, and (Gay-Lussac's law), 286–288
volume, and (Boyle's law), 281–283
Pressure–volume relationship (Boyle's law), 281–283, 289
Prilosec, 414

Metric and SI Units and Some Useful Conversion Factors

Length	SI Unit Meter (m)	Volume	SI Unit Cubic Meter (m^3)	Mass	SI Unit Kilogram (kg)
1 meter (m) = 100 centimeters (cm)		1 liter (L) = 1000 milliliters (mL)		1 kilogram (kg) = 1000 grams (g)	
1 meter (m) = 1000 millimeters (mm)		1 mL = 1 cm^3		1 g = 1000 milligrams (mg)	
1 cm = 10 mm		1 L = 1.06 quart (qt)		1 kg = 2.20 lb	
1 kilometer (km) = 0.621 mile (mi)		1 qt = 946 mL		1 lb = 454 g	
1 inch (in.) = 2.54 cm (exact)				1 mole = 6.02×10^{23} particles	
				Water density = 1.00 g/mL (at 4°C)	

Temperature	SI Unit Kelvin (K)	Pressure	SI Unit Pascal (Pa)	Energy	SI Unit Joule (J)
$T_F = 1.8(T_C) + 32$		1 atm = 760 mmHg		1 calorie (cal) = 4.184 J (exact)	
$T_C = \frac{T_F - 32}{1.8}$		1 atm = 101.325 kPa		1 kcal = 1000 cal	
$T_K = T_C + 273$		1 atm = 760 Torr			
		1 mole of gas = 22.4 L (STP) $R = 0.0821\ \text{L} \cdot \text{atm/mole} \cdot \text{K}$ $R = 62.4\ \text{mmHg} \cdot \text{atm/mole} \cdot \text{K}$		**Water** Heat of fusion = 334 J/g; 80. cal/g Heat of vaporization = 2260 J/g; 540 cal/g Specific heat (*SH*) = 4.184 J/g °C; 1.00 cal/g °C	

Metric and SI Prefixes

Prefix	Symbol	Scientific Notation
Prefixes That Increase the Size of the Unit		
peta	P	10^{15}
tera	T	10^{12}
giga	G	10^{9}
mega	M	10^{6}
kilo	k	10^{3}
Prefixes That Decrease the Size of the Unit		
deci	d	10^{-1}
centi	c	10^{-2}
milli	m	10^{-3}
micro	μ (mc)	10^{-6}
nano	n	10^{-9}
pico	p	10^{-12}
femto	f	10^{-15}

Formulas and Molar Masses of Some Typical Compounds

Name	Formula	Molar Mass (g/mole)
Ammonia	NH_3	17.03
Ammonium chloride	NH_4Cl	53.49
Ammonium sulfate	$(NH_4)_2SO_4$	132.15
Bromine	Br_2	159.80
Butane	C_4H_{10}	58.12
Calcium carbonate	$CaCO_3$	100.09
Calcium chloride	$CaCl_2$	110.98
Calcium hydroxide	$Ca(OH)_2$	74.10
Calcium oxide	CaO	56.08
Carbon dioxide	CO_2	44.01
Chlorine	Cl_2	70.90
Copper(II) sulfide	CuS	95.62
Hydrogen	H_2	2.016

Name	Formula	Molar Mass (g/mole)
Hydrogen chloride	HCl	36.46
Iron(III) oxide	Fe_2O_3	159.70
Magnesium oxide	MgO	40.31
Methane	CH_4	16.04
Nitrogen	N_2	28.02
Oxygen	O_2	32.00
Potassium carbonate	K_2CO_3	138.21
Potassium nitrate	KNO_3	101.11
Propane	C_3H_8	44.09
Sodium chloride	NaCl	58.44
Sodium hydroxide	NaOH	40.00
Sulfur trioxide	SO_3	80.07
Water	H_2O	18.02